Copyright © 2005 by William Andrew, Inc.
   No part of this book may be reproduced or utilized in any form or by any means, electronic or mechanical, including photocopying, recording, or by any information storage and retrieval system, without permission in writing from the Publisher.

Cover art © 2005 by Brent Beckley / William Andrew, Inc.

ISBN: 0-8155-1516-2 (William Andrew, Inc.)


Library or Congress Catalog Card Number: 2005008935
Library of Congress Cataloging-in-Publication Data

Sittig's handbook of pesticides and agricultural chemicals / edited by
Stanley A. Greene and Richard P. Pohanish.
    p. cm.
  Includes bibliographical references.
  ISBN 0-8155-1516-2 (alk. paper)
 1. Fertilizers--Handbooks, manuals, etc. 2. Fertilizers--Environmental aspects--Handbooks, manuals, etc. 3. Agricultural chemicals--Handbooks, manuals, etc. 4. Agricultural chemicals--Environmental aspects--Handbooks, manuals, etc. I. Title: Handbook of pesticides and agricultural chemicals.
II. Greene, Stanley A. III. Pohanish, Richard P.

 S633.S6265 2005
 632'.95--dc22
                              2005008935


Printed in the United States of America
This book is printed on acid-free paper.

10 9 8 7 6 5 4 3 2 1

Published by:
William Andrew Publishing
13 Eaton Avenue
Norwich, NY 13815
1-800-932-7045
www.williamandrew.com

# Sittig's Handbook of Pesticides and Agricultural Chemicals

Edited by

Stanley A. Greene

and

Richard P. Pohanish

William Andrew Publishing

Norwich, NY, U.S.A.

SOUTH UNIVERSITY
709 MALL BLVD.
SAVANNAH, GA 31406

# NOTICE

This reference is intended to provide data about chemical hazards and guidelines for those trained in the proper use and application of pesticides and agricultural chemicals and trained to respond to hazardous materials spills and accidents. It is not intended as a primary source of research information. **As with any reference, it cannot include all information or discuss all situations that might occur. It cannot be assumed that all necessary warnings and precautionary measures are contained in this work, and that other, or additional, information or assessments may not be required. Most of all, it cannot replace the training and experience of individual responders. Extreme care has been taken in the preparation of this work and, to the best knowledge of the publisher and the editors, the information presented is accurate and no warranty is expressed or implied. No warranty, express or implied, is made. Information may not be available for some chemicals; consequently, an absence of data does not necessarily mean that a substance is not hazardous. For major incidents it will be necessary to obtain additional detailed information from other resources as well as more expertise from those with more extensive training. Neither the publisher nor the editors assume any liability or responsibility for completeness or accuracy of the information presented or any damages of any kind alleged to result in connection with, or arising from, the use of this book. The publisher and the editors strongly encourage all readers, and users of chemicals, to follow the manufacturers' or suppliers' current instructions, technical bulletins, and material safety data sheets (MSDSs) for specific use, handling, and storage of all chemical materials.**

## About the Editors

Stanley A. Green and Richard P. Pohanish are the authors of numerous technical and scientific works, including the *Wiley Guide to Chemical Incompatibilities, Hazardous Materials Handbook: Storage, Handling, Transportation, and Emergency Management*, and the *Hazardous Substance Resource Guide, Second Edition*, which was selected by the American Library Association as a Notable Reference in 1994. Mr. Pohanish is also the editor of the best selling *Sittig's Handbook of Toxic and Hazardous Chemicals and Carcinogens, Fourth Edition, Volumes I and II*, also published by William Andrew Publishing.

# Contents

Preface ........................................................... ix
Introduction ...................................................... xi
   How to Use This Book ......................................... xi
Key to Abbreviations, Symbols, and Acronyms ...................... xxiii
Pesticide Records A to Z .......................................... 1
Bibliography ...................................................... 917
*Appendix A:* List of Companies Cited ............................. 921
*Appendix B:* Directory of Agrochemical Manufacturers ............. 928
*Appendix C:* Directory of Federal and International Regulatory Agencies
   for the Environment and Pesticides ........................... 971
*Appendix D:* Directory of State Regulatory Agencies
   for the Environment and Pesticides Management ................ 983
*Appendix E:* Directory of Industrial and Professional Agrochemical
   and Food-related Organizations ................................ 990
*Appendix F:* Directory of Useful Hotlines, Databases, and Web Sites .......... 1006
*Appendix G:* Agrochemical Web Sites: Sources of Information
   about Agrochemicals and Food Safety ........................... 1013
*Index 1:* Synonym and Trade Name-Cross Index ..................... 1017
*Index 2:* Index of EPA Product Codes ............................. 1179
*Index 3:* CAS Number-Cross Index ................................. 1184

# Preface

This companion to *Sittig's Handbook of Toxic and Hazardous Chemicals and Carcinogens* provides chemical, regulatory, health, and safety information on nearly 800 pesticides and other agricultural chemicals. These products are organized with common names, chemical synonyms, trade names, chemical formulae, US EPA (United States Environmental Protection Agency) pesticide codes, EEC (European Economic Community), EINECS (European Inventory of Existing Commercial Substances), RTECS (Registry of Toxic Effects of Chemical Substances), CAS (Chemical Abstract Service numbers), and other unique identifiers so that all who may have contact with, or interest in them can find needed information quickly.

For the most part, and in keeping with the broad changes initiated with the fourth edition of *Sittig's Handbook of Toxic and Hazardous Chemicals and Carcinogens,* this work is focused on "regulated chemicals." This implies recognition by some government agency, or rule-making body. For example, the "Regulatory Authority" section has been expanded, now containing U.S. federal listings as well as those for California, the largest agricultural state. Data are furnished, to the extent currently available, in a multi-section uniform format to make it easy for users who wish to find information quickly or to compare data within various records in any or all of these important categories:

- Chemical Name
- Use Type
- CAS Number
- Formula
- Alert
- Synonyms
- Trade Names
- Producers
- Chemical Class
- EPA/OPP PC Code
- California DPR Chemical Code
- ICSC Number
- RTECS Number
- EEC Number
- EINECS Number
- Uses
- Human toxicity (long-term)
- Fish toxicity (threshold)
- U.S. Maximum Allowable Residue Levels (with CFR citations)
- Carcinogen/Hazard Classifications
- U.S. EPA Carcinogens
- U.S. NTP
- California Prop. 65
- U.S. TRI
- IARC
- Label Signal Word
- WHO Acute Hazard
- Endocrine Disruptor
- Regulatory Authority
- Description (and physical properties)
- Incompatibilities
- Permissible Exposure Limits in Air
- Determination in Air
- Permissible Concentration in Water
- Determination in Water
- Routes of Entry
- Harmful Effects and Symptoms
- Short Term Exposure
- Long Term Exposure
- Points of Attack
- Medical Surveillance
- First Aid
- References

It should be noted that a "regulated chemical" need not indicate that it is a "registered product." Products are constantly being registered, canceled, or transferred in the U.S. by the EPA. Nevertheless they may be acceptable in other countries. When used on food products imported into the U.S., they may be subjected to inspection by the USDA (U.S. Department of Agriculture) at ports of entry.

Although every effort has been made to produce an accurate and highly useful handbook, the author appreciates the need for constant improvement. Any comments, corrections, or advice from users of this book are welcomed by the author who asks that all correspondence be submitted in writing and mailed to the publisher who maintains a file for reprints and future editions.

## ACKNOWLEDGMENTS

The editors would like to thank the following individuals and institutions, without whose encouragement and generous help, this work would not have been possible. Wendy Kramer, Administrative Librarian, and Judith Foster, Information Technology Assistant, both at the USDA Agriculture Research Service, Eastern Region Research Center in Wyndmoor, Pennsylvania; and Ken Pfeiffer, Pest Management Specialist/Agronomist, USDA/NRCS National Water and Climate Center. The library staff at the Delaware Valley College in Doylestown, Pennsylvania, was most helpful in describing the information needs of students and professionals in the agriculture industry. We are also extremely grateful to our publisher, Bill Woishnis, his excellent staff. We want to thank our editor, Millicent Treloar, for her many suggestions and encouragement. We appreciate the vigilant care of Valerie Haynes and Betty Leahy in the preparation of this work for publication.

# Introduction

Each year, over 350 billion pounds of toxic chemicals are manufactured worldwide. In the United States over 120 billion pounds of pesticides are used annually by industry, government and homeowners. In 2001, worldwide expenditures on pesticides totaled more than $32 billion, with herbicides accounting for 40% of the total, followed by insecticides, fungicides and other types of pesticides. One third of the world's overall expenditures for pesticides come from the U.S.[*]

The toxic chemicals problem in the United States, and indeed in all the world, is frightening and abundant news stories abound. In the night of December 2, 1984, in Bophal, India, 7,000 people died from a leak of methyl isocyanate gas, causing the world's worst industrial accident.

In the U.S. alone, nearly five million chemical poisonings occur annually, resulting in thousands of deaths. A good preponderance of those are from pesticides and other chemicals used in agriculture. Local news outlets frequently report incidents that do not reach national attention. Near Bakersfield, California, 22 farm workers were poisoned when a crop dusting plane applied a mixture of chlorpyrifos, fenpropathrin, and profenofos to a neighboring cotton field. In addition, as many as 225 farm workers in nearby grape fields were exposed and later released.[*]

The widespread use of pesticides and other chemicals on our food supply and other crops is a constant, potential threat to the health and economic livelihood of millions of farm workers. These workers face the highest rate of chemical-related illness of any occupational group in a workplace. Moreover, in many cases, collateral exposure in their homes and drinking water occurs. In June, 1999, the coalition Californians for Pesticide Reform (CPR) published *Fields of Poison: California Farmworkers and Pesticides* which presented data on the use and ramifications of the use of pesticides based on a report by the California Department of Pesticide Regulations (DPR). For five years beginning in 1991, DPR reported nearly 4,000 cases of pesticide poisoning in farmworkers. It can be assumed that not all cases are reported and that pesticides can have a long-term affect on workers and their families from run-off into their water supply, from ambient air, and from living in or near fields that have been treated with pesticides.

The effects of pesticide poisoning stretch far beyond state borders. U.S. sales for food and feed crops were halted in 1989 because of health considerations caused by the Alar scare on apples. An entire industry was halted for months. And today, the mere idea of intentionally poisoning our food supply with pesticides or microorganisms has given rise to a new threat– "agroterrorism."

The use of pesticides and agricultural chemicals has been the theme of numerous books and database materials. However, none of these resources publish a clear, quick, and concise reference containing hazard information with a focus on identification, regulation and food product usage. This is the goal of *Sittig's Handbook of Pesticides and Agricultural Chemicals*.

This book highlights critical data on nearly 800 important and/or regulated toxic and hazardous chemicals that are used as pesticides in agriculture or in residential and commercial applications. In addition to pesticides, hundreds of fertilizers, plant and insect growth regulators, and biocides are covered. Many of these chemicals are not allowed to be used in the United States but many have worldwide application. Others products are manufactured or formulated in one country (such as the United States) and shipped to other parts of the world to be used on another country's agricultural products.

Here is an important addition to the libraries of all persons engaged in agriculture, e.g., food processing, pesticide and agrochemical manufacturing/formulating personnel, agriculture extension personnel, agricultural and environmental management, pesticide applicators, and food safety scientists and toxicologists. In addition, industrial hygienists and industrial safety engineers, lawyers, physicians, legislators, enforcement officials, emergency response personnel, technical librarians, and waste disposal operators will also find essential information in this resource.

The use of this reference is not meant to be a substitute for environmental or workplace hazard communication programs required by regulatory bodies such as EPA, USDA, or OSHA, and/or any other U.S., foreign, or international government agencies. If data are required for legal purposes, the original source documents and appropriate agencies, which are referenced, should be consulted.

Following the Introduction is a key to the abbreviations and acronyms used in the handbook.

## How to Use This Book

Nearly 800 substances are profiled in this reference and the information is organized into sections described here. When a category is omitted, it indicates a lack of available information.

**Chemical Name:** Each record is arranged alphabetically by a chemical name used by regulatory and advisory bodies. In most cases, this is not a product name or trade name.

**Use Type:** The general agricultural use is given, e.g., herbicide, insecticide, fungicide, rodenticide, miticide, fertilizer, biocide, etc. The U.S. Department of Agriculture recognizes the following types of pesticides that are related because they are used against the same type of pests.

- Acaricides: Kill mites, ticks, and spiders that feed on plants and animals. Also called miticides.
- Algicides: Control algae in lakes, canals, swimming pools, water tanks, and other sites.

[*] http://www.pesticide.org\\index.html

- Antifouling agents: Kill or repel organisms that attach to underwater surfaces, such as boat bottoms.
- Antimicrobials: Kill microorganisms (such as bacteria and viruses).
- Attractants: Attract pests, e.g., to lure an insect or rodent to a trap. Food is not considered a pesticide when used as an attractant.
- Avicides: Kill birds
- Biopesticides: These are certain types of pesticides derived from such natural materials as animals, plants, bacteria, and certain minerals.
- Biocides: Kill microorganisms.
- Defoliants: Cause leaves or other foliage to drop from a tree or growing plant, usually to facilitate harvest. Various highly persistent types have been used by the military.
- Desiccants: Promote drying of living tissues, such as unwanted plant tops.
- Disinfectants and sanitizers: Kill or inactivate disease-producing microorganisms on inanimate objects.
- Fungicides: Kill fungi (including blights, mildews, molds, and rusts).
- Fumigants: Produce gas or vapor intended to destroy pests in buildings or soil.
- Herbicides: Kill weeds, grasses and other plants that grow where they are not wanted. May be organic or inorganic.
- Insect growth regulators: Disrupt the molting, maturity from pupal stage to adult, or other life processes of insects.
- Insecticides: Kill insects and arthropods.
- Miticides: Kill mites, ticks, and spiders that feed on plants and animals. Also called acaricides.
- Microbial pesticides: Microorganisms that kill, inhibit, or out compete pests, including insects or other microorganisms.
- Molluscicides: Kill snails and slugs.
- Nematicides: Kill nematodes (microscopic, worm-like organisms that feed on plant roots).
- Ovicides: Kill eggs of insects and mites.
- Pheromones: Biochemicals used to disrupt the mating behavior of insects.
- Piscicides: Kill fish.
- Plant growth regulators: Substances (excluding fertilizers or other plant nutrients) that alter the expected growth, flowering, or reproduction rate of plants.
- Predacides: Kill vertebrate predators.
- Repellents: Repel pests, including insects (such as mosquitoes) and birds.
- Rodenticides: Control mice and other rodents.
- Synergists: Improve the performance of another pesticide. Usually an inert ingredient.

**CAS Number:** The CAS number is a unique identifier assigned to each chemical registered with the Chemical Abstracts Service (CAS) of the American Chemical Society. This number is used to identify chemicals on the basis of their molecular structure. CAS numbers, in the format xxx-xx-x, can be used in conjunction with chemical names for positive identification. CAS numbers should always be used in conjunction with substance names to avoid confusion with like-sounding names, like benzene (71-43-2) and benzine (8032-32-4).

**Formula:** Generally, this has been limited to a commonly used one-line chemical formula. In the case of some organic compounds it has been possible to represent chemical structure.

**Alert:** This section serves as a notice of particular information that does not fit neatly into other categories in the chemical record, such as special regulatory actions and EPA designations.

**Synonyms:** This section contains scientific, product, trade, and other synonym names that are commonly used for each hazardous substance. Some of these names are registered trade names. Some are provided in other major languages other than English including Spanish, French, German, Dutch, Polish, and Italian. In some cases, "trivial" and important nicknames (such as TDE for tetrachlorodiphenylethane) have been included because they are frequently used in general communications, in the workplace, and in regulatory matters. This section is important because the various "regulatory" lists published by federal, state, international, and advisory bodies and agencies often use different names for the same pesticides and agricultural chemicals. Every attempt has been made to ensure the accuracy of the synonyms and trade names found in this volume, but errors are inevitable in compilations of this magnitude. Please note that this volume may not include the names of all products currently in commerce, particularly mixtures, that may contain regulated chemicals. The synonym index contains all synonym names listed in alphabetical order. It should be noted that organic chemical prefixes and interpolations such as $\alpha$, $\beta$, $\delta$, $\Delta$, or other Greek letters; $o$- (ortho-), $m$- (meta-); $p$- (para-); $as$- or $asym$- (asymmetric), $prim$- (primary), $sec$- (secondary), $trans$-, $cis$-, $n$- (normal-), and numerals are not treated as part of the chemical name for the purposes of alphabetization.

**Trade Names:** Most of the registered trade names included in this work are in current use or were registered at one time in the United States. In many cases, the trade names are also marketed in foreign markets; where available, trade names of pesticides made by foreign manufacturers and marketed primarily in foreign markets are also included.

**Producers:** The companies named here are major manufacturers of the agrochemicals or retain the registered trade names. Each company is identified with the country where their main headquarters are. Not included are companies that repackage the principal chemicals into their own products, such as formulators, wholesalers, distributors, farm co-ops, importers, and exporters. Also not included are those companies, many of whom are major manufacturers, who hold licenses to market products of other producers in various countries.

**Chemical Class:** This field describes pesticide family, i.e., carbamate, organophosphate, triazine, etc.; or the chemical family, i.e., aromatic amine, aldehyde, inorganic metals, halogenated organic compounds, etc.

**EPA/OPP PC Code:** This field contains the six-digit pesticide code assigned by the US EPA.

**California DPR Chemical Code:** This field contains the pesticide code assigned by the California Department of Pesticide Registration (DPR).

**ICSC Number:** International Chemical Safety Cards summarize essential information on chemical substances and are developed cooperatively by the International Programme on Chemical Safety (IPCS) and the Commission of the European Union (EC).

**RTECS Number:** The RTECS numbers (Registry of Toxic Effects of Chemical Substances) are assigned and published by NIOSH. The RTECS number in the format ABxxxxxxx may be useful for online searching for additional toxicologic information on specific substances. For example, it can be used to provide access to the MEDLARS® computerized literature retrieval services of the National Library of Medicine in Washington, DC.

**EEC Number:** The EEC and identification number is used by the European Economic Community.

**EINECS Number:** An identification number from *"European Inventory of Existing Commercial Chemical Substances,"* published by the European Community, Luxembourg, Brussels.

Use of these identification numbers for hazardous materials will (a) serve to verify descriptions of chemicals; (b) provide for rapid identification of materials when it might be inappropriate or confusing to require the display of lengthy chemical names on vehicles; (c) aid in speeding communication of information on materials from accident scenes and in the receipt of more accurate emergency response information; and (d) provide a means for quick access to immediate emergency response information in the *"North American Emergency Response Guidebook."*[31] In this latter volume, the various compounds have assigned "ID Numbers" or identification numbers which correspond closely (but not always precisely) to the UN listing.[20]

**Uses:** This is a brief summary of agricultural applications and uses found for the substance in other fields. For specific crops, target insects, weeds, varmints and fungi, the user is referred to the reference sources cited at the end of each chemical listing. In many cases, the pesticides are not registered for use in the U.S. but are used in other countries.

**U.S. Maximum Allowable Residue Levels:** These are the tolerance levels, in parts per million (ppm), for individual crops. The tolerance levels are established by the EPA and reported in 40 CFR 180. Readers are cautioned that these levels for country-wide usages and are reviewed and modified frequently to reflect regulatory changes, petitions by individual companies, or scientific developments; consequently, be encouraged to verify the tolerances in the EPA web site. State exceptions to the federal standards are not included here.

**Human toxicity (long-term):** This field contains 3 items: The hazard rating/long-term toxicity level (ppb)/toxicity type. This represents a relative long-term toxicity index for humans. The hazard rating (Extra High, High, Intermediate, Low, Very Low) are indicators of the relative risk to humans. The figure in ppb is calculated with respect for humans by the US EPA. The last item in this field is the toxicology "type" for the value expressed in ppb. Toxicities are based on availability in the priority order: MCL, HA, and CHCL. HA is used for Cancer Groups C, D, E, and Unclassified. CHCL is used for Cancer Groups A, B1 and B2 when MCL is unavailable.

- MCL (EPA's Maximum Contaminant Level). Maximum permissible long-term pesticide concentration allowed in a public water source.
- HA (Health Advisory) Determined by the US EPA Office of Water (OW). The concentration of a chemical in drinking water that is not expected to cause any adverse non-carcinogenic effects over the lifetime exposure with a margin of safety. In accordance with OW policy, Health Advisories are not calculated for chemicals that are known or probable carcinogens (EPA Cancer class A and B).
- CHCL* (Chronic Human Carcinogen Level, calculated). The concentration at which there is a 1 in 100,000 probability of contracting cancer. A CHCL provides a concentration comparable to an MCL.

**Fish toxicity (threshold):** This field contains three items: The hazard rating/toxicity threshold (ppb)/toxicity type. The hazard rating (Extra High, High, Intermediate, Low, Very Low) is based on Maximum Acceptable Toxicant Concentration (MATC), the soluble pesticide toxicity level for fish that is an indicator of the relative risk to the environment. The hazard rating is followed by the fish toxicity threshold for an active ingredient expressed in parts per billion (ppb) and determined empirically by performing long-term or life-stage toxicity tests. The final item in this field contains the type of toxicity for the toxicity threshold (MATC).

**Carcinogen/Hazard Classifications:**

**U.S. EPA Carcinogens:** The EPA has evaluated chemicals for their carcinogenic potential according to "Proposed Guidelines for Carcinogenic Risk Assessment," published April 23, 1996 in the Federal Register (FR: 17960-18011)

**NTP:** The National Toxicology Program, U.S. Department of Health and Human Services, *10th Report on Carcinogens*. Coordinates studies from several government agencies and classifies results in two categories:

- *Known to be a human carcinogen*: There is sufficient evidence of carcinogenicity from studies in humans which indicates a causal relationship between exposure to the agent, substance or mixture and human cancer.
- *Reasonably anticipated to be a human carcinogen*: There is limited evidence of carcinogenicity from studies in humans, which indicates that causal interpretation is credible, but that alternative explanations, such as chance, bias or confounding factors, could not adequately be excluded; or there is sufficient evidence of carcinogenicity from studies in experimental animals which indicates there is an increased incidence of

malignant and/or a combination of malignant and benign tumors: (1) in multiple species or at multiple tissue sites, or (2) by multiple routes of exposure, or (3) to an unusual degree with regard to incidence, site or type of tumor, or age at onset; or there is less than sufficient evidence of carcinogenicity in humans or laboratory animals; however, the agent, substance or mixture belongs to a well defined, structurally-related class of substances whose members are listed in a previous Report on Carcinogens as either a known to be human carcinogen or reasonably anticipated to be human carcinogen, or there is convincing relevant information that the agent acts through mechanisms indicating it would likely cause cancer in humans. See http://ehp.niehs.nih.gov/roc/toc10.html#toc

**California Prop. 65:** California's Proposition 65, The Safe Drinking Water and Toxic Enforcement Act of 1986, requires that warnings be given to individuals exposed to substances which cause cancer or reproductive toxicity. *Chemicals Known to the State to Cause Cancer or Reproductive Toxicity* is published each year and describes the type of toxicity (cancer; male and/or female developmental), the CAS number, and the date listed or delisted. For updated information, see http://www.oehha.ca.gov/prop65/prop65_list/files/P65single061104a.pdf

**U.S. TRI:** The Toxic Release Inventory (TRI) is an EPA database that contains information on toxic chemical releases and other waste management activities reported by industrial and federal facilities. If the agrichemical is reported by TRI as a known carcinogen, it is indicated here.

**IARC:** The International Agency for Research on Cancer (IARC) coordinates and conducts both epidemiological and laboratory research into the causes of cancer. See www.iarc.fr.

**Label Signal Word:** The EPA assigns a signal word that is a description of the short-term (acute) toxicity of a formulated product. It must be displayed on product labels to alert users to potential hazards. There are four categories of signal words; their levels are shown below. Formulated products contain both active and inert or other ingredients. Examples of inert ingredients are carriers, stickers, solvents, and adjuvants.

**WHO Acute Hazard:** The acute risk to health over a relatively short period of time, as established by the World Health Organization (WHO). See www.inchem.org/documents/pds/pdsother/class.pdf

| HAZARD INDICATORS | CATEGORY I | CATEGORY II | CATEGORY III | CATEGORY IV |
|---|---|---|---|---|
| Signal Words | Danger | Warning | Caution | Caution |
| Oral $LD_{50}$ | 0-50 mg/kg | 50-500 mg/kg | 500-5000 mg/kg | >5000 mg/kg |
| Dermal $LD_{50}$ | 0-200 mg/kg | 200-2000 mg/kg | 2000-20,000 mg/kg | >20,000 mg/kg |
| Inhalation $LC_{50}$ | 0-0.2 mg/l | 0.2-2.0 mg/l | 2.0-20 mg/l | >20 mg/l |
| Eye Effects | Corrosive corneal opacity not reversible within 7 days | Corrosive corneal opacity reversible within 7 days; irritation for 7 days | No corneal opacity; irritation reversible within 7 days | No irritation |
| Skin Effects | Corrosive | Severe irritation at 72 hours | Moderate irritation at 72 hours | Mild or slight irritation at 72 hours. |

*Source:* 40CFR156.64 (February 12, 2002)

The $LD_{50}$ value is a statistical estimate of the number of mg of toxicant per kg of body weight required to kill 50% of a large population of test animals.

| Class | | $LD_{50}$ for the rat (mg/kg body weight) | | | |
|---|---|---|---|---|---|
| | | Oral | | Dermal | |
| | | Solids* | Liquids* | Solids* | Liquids* |
| Ia | Extremely hazardous | 5 or less | 20 or less | 10 or less | 40 or less |
| Ib | Highly hazardous | 5 – 50 | 20 – 200 | 10–100 | 40 – 400 |
| II | Moderately hazardous | 50 – 500 | 200 - 2000 | 100 – 1000 | 400 – 4000 |
| III | Slightly hazardous | Over 500 | Over 2000 | Over 1000 | Over 4000b |

*Note:* The terms "solids" and "liquids" refer to the physical state of the active ingredient being classified. The $LD_{50}$ value is a statistical estimate of the number of mg of toxicant per kg of body weight required to kill 50% of a large population of test animals.

***Endocrine Disruptor:*** EPA defines endocrine disruptors as compounds that "interfere with the synthesis, secretion, transport, binding, action, or elimination of natural hormones in the body that are responsible for the maintenance of homeostasis (normal cell metabolism), reproduction, development, and/or behavior." Many endocrine disruptors are thought to mimic hormones, such as estrogen or testosterone. They have chemical properties similar to hormones that allow binding to hormone specific receptors on the cells of target organs. More information on endocrine effects can be found at the EPA Endocrine home page at http://www.epa.gov.endocrine.

**Regulatory Authority:** Contains a listing of major regulatory jurisdictions and authorities. Many law or regulatory references in this work have been abbreviated. For example, Title 40 of the Code of Federal Regulations, Part 261, subpart 32 has been abbreviated as 40CFR261.32. The symbol "§" may be used as well to designate a section or part. Under the title of each substance, there are designations indicating whether the substance is:

- A carcinogen (the agency making such a determination, the nature of the carcinogenicity — whether human or animal and whether positive or suspected, are given in each case). These are frequently cited by IARC (International Agency for Research on Cancer)[12], DFG (Deutche Forschungsgemeinschaft)[3], NIOSH (U.S. National Institute for Occupational Safety and Health),[58] or the NTP (U.S. National Toxicology Programs).[10] It should be noted that the DFG have designated some substances as carcinogens not so classified by other agencies.
- A banned or severely restricted product as designated by the United Nations[13] or by the U.S. EPA Office of Pesticide Programs under FIFRA (The Federal Insecticide, Fungicide and Rodenticide Act).[14]
- A substance cited by the World Bank[15]
- A substance with an air pollutant standard set or recommended by OSHA and/or NIOSH,[58] ACGIH,[1] DFG,[3] or HSE.[33] OSHA limits are the enforceable pre-1989 PELs. The transitional limits that were vacated by court order have not been included. The NIOSH and ACGIH airborne limits are recommendations that do not carry the force of law.
- A substance whose allowable concentrations in workplace air are adopted or proposed by the American Conference of Government Industrial Hygienists[1], Deutsche Forschungsgemeinschaft (German Research Society).[3] Substances whose allowable concentrations in air and other safety considerations have been considered by OSHA and NIOSH[2].Substances which have limits set in workplace air, in residential air, in water for domestic purposes or in water for fishery purposes as set forth by the former USSR-UNEP/IRPTC Project.[43]
- Substances that are specifically regulated by OSHA under 29CFR1910.1001 to 29CFR1910.1050
- Highly hazardous chemicals, toxics, and reactives regulated by OSHA's "Process Safety Management of Highly Hazardous Chemicals" under 29CFR1910.119, Appendix A.
- Substances that are Hazardous Air Pollutants (Title I, Part A, Section 112) as amended under 42USC7412. This list provided for regulating at least 189 specific substances using technology-based standards that employ Maximum Achievable Control Technology (MACT) standards; and, possibly health-based standards if required at a later time. Section 112 of the Clean Air Act (CAA) requires emission control by the EPA on a source-by-source basis. Therefore, the emission of substances on this list does not necessarily mean that a firm is subject to regulation
- Regulated Toxic Substances and Threshold Quantities for Accidental Release Prevention. These appear as Accidental Release Prevention/Flammable Substances, Clean Air Act (CAA) §112(r), Table 3, TQ (threshold quantity) in pounds and kilograms under 40 CFR68.130. The accidental release prevention regulations applies to stationary sources that have present more than a threshold quantity of a CAA Section 112(r) regulated substance.
- Clean Air Act (CAA)Public Law 101-549, Title VI, "Protection of Stratospheric Ozone," Subpart A, Appendix A-Class I and Appendix B, Class II, Controlled Substances, (CFCs) Ozone depleting substances under 40 CFR, Part 82.
- Clean Water Act (CWA) Priority toxic water pollutants defined by the U.S. Environmental Protection Agency for 65 pollutants and classes of pollutants which yielded 129 specific substances.[6]
- Chemicals designated by EPA as "Hazardous Substances"[4] under the Clean Water Act (CWA) 40 CFR§116.4, Table 116.4A.
- Clean Water Act (CWA) Section 311 Hazardous Materials Discharge Reportable Quantities (RQs). This regulation establishes reportable quantities for substances designated as hazardous (see §116.4, above) and sets forth requirements for notification in the event of discharges into navigable waters. Source: 40 CFR§117.3, amended at 60FR30937.
- Clean Water Act (CWA) Section 307 List of Toxic Pollutants. Source: 40 CFR§401.15.
- Clean Water Act (CWA) Section 307 Priority Pollutant List. This list was developed from the List of Toxic Pollutants classes discussed above and includes substances with known toxic effects on human and aquatic life, and those known to be, or suspected of being, carcinogens, mutagens, or teratogens. Source: 40CFR423, Appendix A.
- Clean water Act, Section 313 Water Priority Chemicals. Source: 57FR41331.
- RCRA Maximum Concentration of Contaminants for the Toxicity Characteristic with Regulatory levels in mg/L. Source: 40CFR§261.24.
- RCRA Hazardous Constituents. Source: 40CFR§261, Appendix VIII. Substances listed in this list have been shown, in scientific studies, to have carcinogenic, mutagenic, teratogenic or toxic effects on humans and other life forms. This list also contains RCRA waste codes. The words, "waste number not listed" appears when a RCRA number is NOT provided in Appendix VIII.

| Characteristic Hazardous Wastes | |
|---|---|
| Ignitability | •A nonaqueous solution containing less than 24% alcohol by volume and having a closed cup flashpoint below 60°C (140°F) using Pensky-Martens tester or equivalent.<br>•An ignitible compressed gas.<br>•A non-liquid capable of burning vigorously when ignited or causes fire by friction, moisture absorption, spontaneous chemical changes at standard pressure and temperature. •An oxidizer. See §261.21. |
| Corrosivity | Liquids with a pH equal to or less than 2 or equal to or more than 12.5 or which corrode steel at a rate greater than 6.35 mm (0.25 in) per year @ 55°C (130°F). See §261.22. |
| Reactivity | •Unstable substances that undergo violent changes without detonating. •Reacts violently with water or other substances to create toxic gases. •Forms potentially explosive mixtures with air. See §261.23. |
| Toxicity | A waste that leaches specified amounts of metals, pesticides, or organic chemicals using Toxicity Characteristic Leaching Procedure (TCLP). See §261, Appendix II, and §268, Appendix I. |
| **Listed Hazardous Wastes** | |
| "F" wastes | Hazardous wastes from nonspecific sources §261.31. |
| "K" Wastes | Hazardous wastes from specific sources §261.32. |
| "U" Wastes | Hazardous wastes from discarded commercial products, off-specification species, container residues §261.34. Covers some 455 compounds and their salts and some isomers of these compounds. |
| "P" Wastes | Acutely hazardous wastes from discarded commercial products, off-specification species, container residues §261.33. Covers some 203 compounds and their salts plus soluble cyanide salts. |

*Note*: If a waste is not found on any of these lists, it may be found on a state hazardous waste list.

RCRA Maximum Concentration of Contaminants for the Toxicity Characteristic. Source: 40CFR§261.24, Table 1. These are listed with regulatory level in mg/L and "D" waste numbers representing the broad waste classes of ignitability, corrosivity, and reactivity.
- EPA Hazardous Waste code(s), or RCRA number, appears in its own field. Acute hazardous wastes from commercial chemical products are identified with the prefix "P." Nonacutely hazardous wastes from commercial chemical products are identified with the prefix "U."
- RCRA Universal Treatment Standards. Lists hazardous wastes that are banned from land disposal unless treated to meet standards established by the regulations. Treatment standard levels for wastewater (reported in mg/L) and nonwastewater [reported in mg/kg or mg/L TCLP (Toxicity Characteristic Leachability Procedure)] have been provided. Source: 40CFR§268.48 and revision, 61FR15654.
- RCRA Ground Water Monitoring List. Sets standards for owners and operators of hazardous waste treatment, storage, and disposal facilities, and contains test methods suggested by the EPA (see Report SW-846) followed by the Practical Quantitation Limit (PQL) shown in parentheses. The regulation applies only to the listed chemical; and, although both the test methods and PQL are provided, they are *advisory only*. Source: 40CFR§264, Appendix IX.
- Safe Drinking Water Act (SDWA) Maximum Contaminant Level Goals (MCLG) for Organic Contaminants. Source: 40CFR§141 and §141.50, amended 57FR31776.
- Maximum Contaminant Levels (MCL) for Organic Contaminants. Source: 40CFR§141.61.
- Maximum Contaminant Level Goals (MCLG) for Inorganic Contaminants. Source: 40CFR§141.51.
- Maximum Contaminant Levels (MCL) for Inorganic Contaminants. Source: 40CFR§141.62.
- Maximum Contaminant Levels for Inorganic Chemicals. The maximum contaminant level for arsenic applies only to community water systems. Compliance with the MCL for arsenic is calculated pursuant to §141.23. Source: 40CFR§141.11.
- Secondary Maximum Contaminant Levels (SMCL). Federal advisory standards for the States concerning substances that affect physical characteristics (i.e., smell, taste, color, etc.) of public drinking water systems. Source: 40CFR§143.3.

- CERCLA Hazardous Substances ("RQ" Chemicals). From *Consolidated List of Chemicals Subject to the Emergency Planning and Community Right-to-Know Act (EPCRA) and Section 112(r) of the Clean Air Act, as Amended.* Source: EPA 550-B-98-017 "*Title III List of Lists.*"
- Releases of CERCLA hazardous substances in quantities equal to or greater than their reportable quantity (RQ), are subject to reporting to the National Response Center under CERCLA. Such releases are also subject to state and local reporting under §304 of SARA Title III (EPCRA). CERCLA hazardous substances, and their reportable quantities, are listed in 40CFR§302, Table 302.4. RQs are shown in pounds and kilograms for chemicals that are CERCLA hazardous substances. For metals listed under CERCLA (antimony, arsenic, beryllium, cadmium, chromium, copper, lead, nickel, selenium, silver, thallium, and zinc), no reporting of releases of the solid is required if the diameter of the pieces of solid metal released is 100 micrometers (0.004 inches) or greater. The RQs shown apply to smaller particles.
- EPCRA §302 Extremely Hazardous Substances (EHS). From *Consolidated List of Chemicals Subject to the Emergency Planning and Community Right-to-Know Act (EPCRA) and Section 112(r) of the Clean Air Act, as Amended.* Source: EPA document 550-B-98-017 "*Title III List of Lists*" The presence of Extremely Hazardous Substances in quantities in excess of the Threshold Planning Quantity (TPQ), requires certain emergency planning activities to be conducted. The Extremely Hazardous Substances and their TPQs are listed in 40CFR§355, Apendices A & B. For chemicals that are solids, there may be two TPQs given (e.g., 500/10,000). In these cases, the lower quantity applies for solids in powder form with particle size less than 100 microns; or, if the substance is in solution or in molten form. Otherwise, the higher quantity (10,000 pounds in the example) TPQ applies.
- EPCRA §304 Reportable Quantities (RQ). In the event of a release or spill exceeding the reportable quantity, facilities are required to notify State emergency response commissions (SERCs) and Local Emergency Planning Committees (LEPCs). From *Consolidated List of Chemicals Subject to the Emergency Planning and Community Right-to-Know Act (EPCRA) and Section 112(r) of the Clean Air Act, as Amended.* Source: EPA document 550-B-98-017, *Title III List of Lists*
- EPCRA Section 313 Toxic Chemicals. From *Consolidated List of Chemicals Subject to the Emergency Planning and Community Right-to-Know Act (EPCRA) and Section 112(r) of the Clean Air Act, as Amended.* Source: EPA document 550-B-98-017 Title III List of Lists." Chemicals on this list are reportable under §313 and §6607 of the Pollution Prevention Act. Some chemicals are reportable by category under §313. Category codes needed for reporting are provided for the EPCRA§313 categories. Information and Federal Register references have been provided where a chemical is subject to an administrative stay, and not reportable until further notice.
- From *Toxic Chemical Release Inventory Reporting Form R and Instructions, Revised March 1996*, EPA document 745-K-96-001 was used for *de minimis* concentrations, toxic chemical categories. Missing from the category listing was "chlorophenols" which has a reportable *de minimis* concentration of 1.0%.
- Chemicals which EPA has made the subject of Chemical Hazard Information Profiles or "CHIPS" review documents.
- Chemicals which NIOSH has made the subject of *Information Profile* review documents on *Current Intelligence Bulletins.*
- Carcinogens identified by the National Toxicology Program of the U.S. Department of Health and Human Services at Research Triangle Park, NC.[10]
- Chemicals that were covered in the periodical *Dangerous Properties of Industrial Materials Report* formerly edited by N. Irving Sax, Richard Lewis, and Jan C. Prager, published by Van Nostrand Reinhold, New York.
- Chemicals described in the 2-volume *Encyclopedia of Occupational Health and Safety* published by the International Labor Office.[30]
- Most of the chemicals covered in the Legal File published by International Register of Potentially Toxic Chemicals Program (IRPTC) of the United Nations.[35] The reader who is particularly concerned with legal standards (allowable concentration in air, water or in foods) is advised to check these most recent references because data may exist in this UN publication which has not been quoted *in toto* in this volume because of time and space limitations.
- A substance regulated by EPA[7] under the major environmental laws: Clean Air Act, Clean Water Act, Safe Drinking Water Act, RCRA, CERCLA, EPCRA, etc. A more detailed list appears above. And a substance with environmental standards set by some international bodies including Canada and the former USSR.[43]

If additional guidance or compliance assistance is needed, you are encouraged to use the information resources found in the *Appendix*. In particular, the *Directory of Industrial and Professional Agrichemical and Food-related Organizations*, the *Directory of Hotlines, Databases and Web Sites*, and the *Agrichemical Web Sites* directory provide gateways into a wealth of information on agrichemicals and food safety much too extensive to be included in this reference work. In addition, each state in the U.S. has its own statutes and regulations. Their agencies can be located in the *Directory of State Regulatory Agencies*.

**Description:** This section contains a summary of physical properties of the substance including state (solid, liquid or gas), color, odor description*, solubility or miscibility in water, molecular weight, boiling point, freezing/melting point, vapor pressure, flash point, autoignition temperature, explosion limits in air, Hazard Identification (based on NFPA-704 M Rating System) in the format, Health $\underline{x}$, Flammability $\underline{x}$, Reactivity $\underline{x}$ (see also below for a detailed explanation of the System and Fire Diamond), and

Octanol/Water coefficient. This section may also contain relevant comments about the substance.

*Boiling Point* at 1 atm: The value is the temperature of a liquid when its vapor pressure is 1 atm. For example, when water is heated to 100°C (212°F) its vapor pressure rises to 1 atm and the liquid boils. The boiling point at 1 atm indicates whether a liquid will boil and become a gas at any particular temperature and sea-level atmospheric pressure.

*Melting/Freezing Point*: The melting/freezing point is the temperature at which a solid changes to liquid or a liquid changes to a solid. For example, liquid water changes to solid ice at 0°C (32°F). Some liquids solidify very slowly even when cooled below their melting/freezing point. When liquids are not pure (for example, salt water) their melting/freezing points are lowered slightly.

*Vapor Pressure:* The pressure exerted by a vapor when it is in equilibrium with the liquid from which it is derived. Pesticides with vapor pressure values of more than $10^{-7}$ have a high potential for volatile loss. Pesticides with vapor pressure values of less than $10^{-7}$ have a low potential for volatile loss.

*Flash Point:* This is defined as the lowest temperature at which vapors above a volatile combustible substance will ignite in air when exposed to a flame. Depending on the test method used, the values given are either Tag Closed Cup (cc.) (ASTM D56) or Cleveland Open Cup (oc) (ASTM D93). The values, along with those in *Flammable Limits in Air* and *Autoignition Temperature* below, give an indication of the relative flammability of the chemical. In general, the open cup value is slightly higher (perhaps 10° to 15°F higher) than the closed cup value. The flash points of flammable gases are often far below 0° (F or C) and these values are of little practical value, so the term "flammable gas" is often used instead of the flash point value.

*Autoignition Temperature*: This is the minimum temperature at which the material will ignite without a spark or flame being present. Values given are only approximate and may change substantially with changes in geometry, gas, or vapor concentrations, presence of catalysts, or other factors.

*Flammable Limits in Air:* The percent concentration in air (by volume) is given for the LEL (lower explosive-flammable-limit in air, % by volume) and UEL (upper explosive-flammable-limit in air, % by volume), at room temperature, unless otherwise specified. The values, along with those in Flash Point and Autoignition Temperature give an indication of the relative flammability of the chemical.

*NFPA Hazard Classifications:* The NFPA 704 Hazard Ratings (Classifications) are reprinted with permission from "Fire Protection Guide to Hazardous Materials," 11th edition, National Fire Protection Association, Quincy, MA. ©1994. The classifications are defined in the Table below.

It should be noted that OSHA and DOT have differing definitions for the term "flammable liquid" and "combustible liquid." DOT defines a flammable liquid as one which, under specified procedures, has a flashpoint of 140°F (60°C) or less. A combustible liquid is defined as "having a flashpoint above 140°F (60°C) and below 200°F (93°C)." OSHA defines a combustible liquid as having a flash point above 100°F (37.7°C).

**National Fire Protection Association (NFPA) Fire Diamond**

| Health Hazard (Blue) | | Flammability (Red) | | Reactivity (Yellow) | |
|---|---|---|---|---|---|
| Signal | Type of Possible Injury | Signal | Susceptibility of Materials to Burn | Signal | Susceptibility to Release of Energy |
| 4 | Materials which on very short exposure could cause death or major residual injury. | 4 | Materials which will rapidly or completely vaporize at atmospheric pressure and normal ambient temperature, or which are readily dispersed in air and which will burn readily. | 4 | Materials that, in themselves, are readily capable of detonation, explosive decomposition or explosive reaction at normal temperatures and pressures. |
| 3 | Materials which on short exposure could cause serious temporary or residual injury. | 3 | Liquids and solids that can be easily ignited under almost all normal temperature conditions. | 3 | Materials that, in themselves, are capable of detonation or explosive decomposition but require a strong initiating source or which must be heated under confinement before initiation or which react explosively with water. |
| 2 | Materials that, on intense or continued (but not chronic) exposure, could cause temporary incapacitation or possible residual injury. | 2 | Materials that must be moderately heated or exposed to relatively high ambient temperatures before ignition can occur. | 2 | Materials that readily undergo violent chemical changes at elevated temperatures and pressure or which react violently with water or which may form explosive mixtures with water. |

| Signal | Type of Possible Injury | Signal | Susceptibility of Materials to Burn | Signal | Susceptibility to Release of Energy |
|---|---|---|---|---|---|
| 1 | Materials which on exposure would cause irritation but only minor residual injury. | 1 | Materials that must be preheated before ignition can occur. | 1 | Materials that in themselves are normally stable, but which become unstable at elevated temperatures and pressures. |
| 0 | Materials that, on exposure under fire conditions, would offer no hazard beyond that of ordinary combustible material. | 0 | Materials that will not burn. | 0 | Materials that in themselves are normally stable, even under fire exposure conditions, and which are not reactive with water. |

**Special Notice (White):** Water Reactive: Avoid the use of Water; OXY: Oxidizer; Radioactive

**Octanol/Water Coefficient (Log $K_{ow}$).** A statement of the ratio of the concentration in octanol to the concentration in water when a pesticide or agricultural chemical is dissolved in a mixture of these two liquids. This coefficient is used to indicate the pesticide's or agricultural chemical's relative potential for endangering the environment. The greater the value, the greater the chance of the pesticide or agricultural chemical accumulating in fats of living tissue, in particular at values greater than 3.0.

**Potential Exposure:** A brief indication is given of the nature of exposure to each compound in the industrial environment. Where pertinent, some indications are given of background concentration and occurrence from other than industrial discharges such as water purification plants. Obviously in a volume of this size, this coverage must be very brief. It is of course recognized that non-occupational exposures may be important as well.

**Incompatibilities:** Important, potentially hazardous incompatibilities of each substance are listed where available. Where a hazard with water exists, it is described. Reactivity with other materials are described including structural materials such as metal, wood, plastics, cement, and glass. The nature of the hazard, such as severe corrosion formation of a flammable gas, is described. This list is by no means complete or all inclusive. In some cases a very small quantity of material can act as a catalyst and produce violent reactions such as polymerization, disassociation and condensation. Some chemicals can undergo rapid polymerization to form sticky, resinous materials, with the liberation of much heat. The containers may explode. For these chemicals the conditions under which the reaction can occur are given.

**Permissible Exposure Limits in Air:** The permissible exposure limit (PEL), has been cited as the Federal Standard where one exists. Inasmuch as OSHA has made the decision to enforce only pre-1989 PELs, we decided to use these values rather than the transitional limits that were vacated by court order. Except where otherwise noted, the PELs are 8-hour work-shift time-weighted average (TWA) levels. Ceiling limits, Short Term Exposure Limits (STEL), and TWAs that are averaged over other than full work-shifts are noted.

The short-term exposure limit (STEL) values are derived from NIOSH[58], ACGIH[1] and HSE[33] publications. This value is the maximal concentration to which workers can be exposed for a period up to 15 minutes continuously without suffering from: irritation; chronic or irreversible tissue change; or narcosis of sufficient degree to increase accident proneness, impair self-rescue, or materially reduce work efficiency, provided that no more than four excursions per day are permitted, with at least 60 minutes between exposure periods, and provided that the daily TWA also is not exceeded.

The "Immediately Dangerous to Life or Health" (IDLH) concentration represents a maximum level from which one could escape within 30 minutes without any escape-impairing symptoms or any irreversible health effects. However, the 30-minute period is meant to represent a MARGIN OF SAFETY and is NOT meant to imply that any person should stay in the work environment any longer than necessary. In fact, every effort should be made to exit immediately. The concentrations are reported in either parts per million (ppm) or milligrams per cubic meter ($mg/m^3$).

Most U.S. specifications on permissible exposure limits in air have come from ACGIH[1] or NIOSH[2]. In the U.K. the Health and Safety Executive has set forth Occupational Exposure Limits.[33] In Germany the DFG has established Maximum Concentrations in the workplace[3] and the former USSR-UNEP/IRPTC project has set maximum allowable concentrations and tentative safe exposure levels of harmful substance in workplace air and residential air for many substances.[43] This section also contains numerical values for allowable limits of various materials in ambient air[60] as assembled by the US EPA.

Where available, this field contains legally enforceable airborne Permissible Exposure Limits (PELs) from OSHA. It also contains recommended airborne exposure limits from NIOSH, ACGIH, and international sources and special warnings when a chemical substance is a Special Health Hazard Substance. Each are described below.

- TLVs have not been developed as legal standards and the ACGIH does not advocate their use as such. The TLV is defined as the time weighted average (TWA) concentration for a normal 8-hour workday and a 40-hour workweek, to which nearly all workers may be repeatedly exposed, day after day, without adverse effects. A ceiling value (TLV-C) is the concentration that should not be exceeded during any part of the working exposure. If instantaneous monitoring is not feasible, then the TLV-C can be assessed by sampling over a 15-minute period except for those substances that may cause immediate irritation when exposures are short. As some people become ill after exposure to concentrations lower than the exposure limits, this value cannot be used to define exactly what is a "safe" or "dangerous" concentration. ACGIH threshold limit values (TLVs) are reprinted with permission of the American Conference of Governmental Industrial Hygienists, Inc., from the booklet entitled, *"Threshold Limit Values for Chemical Substances and Physical Agents and Biological Exposure Indices."* This booklet is revised on an annual basis. No entry appears when the chemical is a mixture; it is possible to calculate the TLV for a mixture only when the TLV for each component of the mixture is known and the composition of the mixture by weight is also known. According to ACGIH, *"Documentation of the Threshold Limit Values and Biological Exposure Indices"* is necessary to fully interpret and implement the TLVs.

- OSHA permissible exposure limits (PELs), are found in Tables Z-1, Z-2, and Z-3 of OSHA General Industry Air Contaminants Standard (29CFR1910.1000) that were effective on July 1, 2001 and which are currently enforced by OSHA. Unless otherwise noted, PELs are the time weighted average (TWA) concentrations that must not be exceeded during any 8-hour shift of a 40-hour workweek. An OSHA ceiling concentration must not be exceeded during any part of the workday; if instantaneous monitoring is not feasible, the ceiling must be assessed as a 15-minute TWA exposure. In addition there are a number of substances from Table Z-2 that have PEL ceiling values that must not be exceeded except for a maximum peak over a specified period (e.g., a 5-minute maximum peak in any 2 hours).

- NIOSH Recommended Exposure Limits (RELs) are time weighted average (TWA) concentrations for up to a 10-hour work day during a 40-hour work week. A ceiling REL should not be exceeded at any time. Exposure limits are usually expressed in units of parts per million (ppm) - i.e., the parts of vapor (gas) per million parts of contaminated air by volume at 25°C (77°F) and one atmosphere pressure. For a chemical that forms a fine mist or dust, the concentration is given in milligrams per cubic meter ($mg/m^3$).

- Short-Term Exposure Limits (15 minute TWA): This field contains Short Term Exposure Limits (STELs) from ACGIH, NIOSH and OSHA. The parts of vapor (gas per million parts of contaminated air by volume at 25°C (77°F) and one atmosphere pressure is given. The limits are given in milligrams per cubic meter ($mg/m^3$) for chemicals that can form a fine mist or dust. Unless otherwise specified, the STEL is a 15-minute TWA exposure that should not be exceeded at any time during the workday.

**Determination in Air:** The citations to analytical methods are drawn from various sources, such as the *NIOSH Manual of Analytical Methods*[18] In addition, methods have been cited in the latest US Department of Health and Human Services publications including the *NIOSH Pocket Guide to Chemical Hazards,* published June 1997.[2]

**Permissible Concentrations in Water:** The permissible concentrations in water are drawn from various sources also, including:

- The National Academy of Sciences/National Research Council publication, *Drinking Water and Health* published in 1980.[16]
- The priority toxic pollutant criteria published by U.S. EPA in draft form in 1979 and in final form in 1980.[6]
- The multimedia environmental goals for environmental assessment study conducted by EPA.[32] Values are cited from this source when not available from any other.

The U.S. EPA has come forth with a variety of allowable concentration levels:

- For allowable concentrations in "California List" wastes.[38] The California list consists of liquid hazardous wastes containing certain metals, free cyanides, polychlorinated biphenyls (PCBs), corrosives with a pH of less than or equal to 2.0, and liquid and nonliquid hazardous wastes containing halogenated organic compounds (HOCs).
- For regulatory levels in leachates from landfills.[37]
- For concentrations of various materials in effluents from the organic chemicals and plastics and synthetic fiber industries.[51]
- For contaminants in drinking water[36].
- For National Primary and Secondary Drinking Water Regulations[62].
- In the form of health advisories for 16 pesticides,[47] 25 organics,[48] and 7 inorganics.[49]
- For primary drinking water standards starting with a priority list of 8 Volatile Organic Chemicals.[40]
- State drinking water standards and guidelines[61] as assembled by the US EPA.

**Determination in Water:** The sources of information in this field have been primarily US EPA publications including the test procedures for priority pollutant analysis[25] and later modifications.[42]

**Routes of Entry:** The toxicologically important routes of entry of each substance are listed. These are primarily taken from the *NIOSH Pocket Guide*[2] but are drawn from other sources as well.

**Harmful Effects and Symptoms:** These are primarily drawn from NIOSH, EPA publications, and New Jersey and New York State fact sheets on individual chemicals, and are supplemented from information from the draft criteria documents for priority toxic pollutants[26] and from other sources. The other sources include:

- EPA Chemical Hazard Information Profiles (CHIPS) cited under individual entries.
- NIOSH Information Profiles cited under individual entries.
- EPA Health and Environmental Effect Profiles cited under individual entries.
- Particular attention has been paid to cancer as a "harmful

effect" and special effort has been expended to include the latest data on carcinogenicity.

**Short Term Exposure:** These are brief descriptions of the effects observed in humans when the vapor (gas) is inhaled, when the liquid or solid is ingested (swallowed), and when the liquid or solid comes in contact with the eyes or skin. The term $LD_{50}$ signifies that about 50% of the animals given the specified dose by mouth will die. Thus, for a Grade 4 chemical (below 50 mg/kg) the toxic dose for 50% of animals weighing 70 kg (150 lb) is 70 x 50 = 3500 mg = 3.5 g, or less than 1 teaspoonful; it might be as little as a few drops. For a Grade 1 chemical (5 to 15 g/k g), the $LD_{50}$ would be between a pint and a quart for a 150-lb man. All $LD_{50}$ values have been obtained using small laboratory animals such as rodents, cats, and dogs. The substantial risks taken in using these values for estimating human toxicity are the same as those taken when new drugs are administered to humans for the first time.

**Long Term Exposure:** Where there is evidence that the chemical can cause cancer, mutagenic effects, teratogenic effects, or a delayed injury to vital organs such as the liver or kidney, a description of the effect is given.

**Points of Attack:** This category is based in part on the "Target Organs" in the *NIOSH Pocket Guide*[2] but the title has been changed as many of the points of attack are not organs (blood and central nervous system for example).

**Medical Surveillance:** This information is often drawn from a NIOSH publication[27] but also from New Jersey State Fact Sheets on individual chemicals. Where additional information is desired in areas of diagnosis, treatment and medical control, the reader is referred to a private publication[28] which is adapted from the products of the NIOSH Standards Completion Program.

**First Aid:** Simple first aid procedures are listed for response to eye contact, skin contact, inhalation, and ingestion of the toxic substance as drawn to a large extent from the *NIOSH Pocket Guide*[2] but supplemented by information from recent commercially available volumes in the U.S.[29], in the U.K. and in Japan[24] as well as from state fact sheets. They deal with exposure to the vapor (gas), liquid, or solid and include inhalation, ingestion (swallowing) and contact with eyes or skin. The instruction "Do NOT induce vomiting" is given if an unusual hazard is associated with the chemical being sucked into the lungs (aspiration) while the patient is vomiting. "Seek medical attention" or "Call a doctor" is recommended in those cases where only competent medical personnel can treat the injury properly. In all cases of human exposure, seek medical assistance as soon as possible. In many cases, medical advice has been included for guidance only.

**References:** The general bibliography for this volume appears in the Appendix. It includes general reference sources and references dealing with analytical methods. The references at the end of individual product entries are generally restricted to references dealing only with that particular compound.

The following Internet web sites provided sources for much of the data in this reference and should be referred to for current and expanded information.

- U.S. Environmental Protection Agency, Office of Prevention, Pesticides and Toxic Substances, Washington, DC, for *Reregistration Eligibility Decisions (RED)*. These comprehensive documents present revised human health and ecological risk assessments and tolerance reassessments for products originally registered prior to November 1, 1984, as mandated by the Federal Insecticide, Fungicide, and Rodenticide Act (FIFRA) and the Food Quality Protection Act of 1996, which amended the Federal Food, Drug and Cosmetics Act ). See http://www.epa.gov/pesticides/reregistration
- EXTOXNET, Extension Toxicology Network, Oregon State University, Corvallis, OR , *Pesticide Information Profile*. This web page provides data about exposure to the most frequently encountered pesticides. EXTOXNET is a cooperative effort of University of California-Davis, Oregon State University, Michigan State University, Cornell University, and the University of Idaho. Primary files are maintained and archived at Oregon State University. See http://extoxnet.orst.edu/ghindex.html
- New Jersey Department of Health and Senior Services, Right-to-Know Project, *Hazardous Substance Fact Sheets*, Trenton, NJ (various dates from 1985-2004). See http://www.state.nj.us/health/eoh/rtkweb/rtkhsfs.htm
- Pesticides Action Network (PAN) Pesticides Database provides a compendium of current information on pesticides toxicity and regulatory information. It is maintained by Pesticides Action Network of North America, which, in turn, is affiliated with International Pesticide Action Network, the UK PAN and several other Pesticide Action Networks. See http://www.pesticideinfo.org/Index.html
- International Programme on Chemical Safety INCHEM web site consolidates current, internationally peer-reviewed chemical safety publications. Its *Pesticide Data Sheets* contain basic information on pesticides and are revised from time to time. See http://www.inchem.org/pages/pds.html

# Key to Abbreviations, Symbols, and Acronyms

| | | | |
|---|---|---|---|
| α- | Greek letter alpha; used as a prefix to denote the carbon atom in a straight chain compound to which the principal group is attached. | GUP | General Use Pesticide |
| | | h | hour(s) |
| | | HAPs | Hazardous Air Pollutants (CAA) |
| as- | Prefix for asymmetric | HCFC | hydrochlorofluorocarbons |
| ACGIH | American Conference of Governmental Industrial Hygienists | HCS | Hazard Communication Standard |
| | | HOC | Halogenated Organic Compounds |
| approx. | approximately | HSE | Health and Safety Executive (United Kingdom) |
| ASTDR | Agency for Toxic Substances and Disease Registry | | |
| asym- | Prefix for asymmetric | IARC | International Agency for Research on Cancer |
| @ | at | | |
| atm. | Atmosphere | ICSC | International Chemical Safety Cards |
| β | Greek letter beta | IDLH | Immediately Dangerous to Life or Health |
| BP | Boiling point | | |
| C | Centigrade | IPCS | International Programme on Chemical Safety |
| CAA | Clean Air Act | | |
| CAAA | Clean Air Act Amendments of 1990 | iso- | (Greek, equal, alike). Usually denoting an isomer of a compound |
| carc. | Carcinogen | | |
| CAS | Chemical Abstract Service | kg | Weight in kilograms (one thousand grams) |
| cc | Closed cup; cubic centimeter | L or l | Liter(s) |
| CEPA | Canadian Environmental Protection Act | lb | Weight in pound(s) |
| CERCLA | Comprehensive Environmental Response, Compensation, And Liability Act | LEL | Lower explosive (flammable) limit in air, % by volume at room temperature or other temperature as noted |
| CFR | Code of Federal Regulations | | |
| CHCL | Chronic Human Carcinogen Level | | |
| CHEMTREC | Chemical Manufacturers Association (CMA) Transportation Emergency Center | LEPC | Local Emergency Planning Committees |
| | | m- | an abbreviation for "meta-," a prefix used to distinguish between isomers or nearly related compounds |
| cis- | (Latin, on this side). Indicating one of two geometrical isomers in which certain atoms or groups are on the same side of a plane | | |
| | | $m^3$ | cubic meter |
| | | MACT | Maximum Achievable Control Technology (CAA) |
| comp. | Compound | MAK | airborne exposure limit used by the Deutsche Forschungsgemeinschaft (DFG) |
| CWA | Clean Water Act | | |
| cyclo- | (Greek, circle). Cyclic, ring structure; as cyclohexane | MATC | Maximum Acceptable Toxicant Concentration |
| Δ or δ | Greek letter delta | MCLs | Maximum Contaminant Levels (SDWA) |
| deriv. | Derivative | MCLGs | Maximum Contaminant Level Goals (SDWA) |
| DFG | Deutsche Forschungsgemeinschaft | | |
| DOT | U.S. Department of Transportation | $m^3$ | Cubic meter |
| DOT ID | U.S. Department of Transportation Identification Numbers | mg | Milligram |
| | | μ | Micro |
| Ed. | Editor(s) | μg | Microgram(s) |
| EEC | European Economic Community | min | Time in minute(s) |
| EHS | Extremely Hazardous Substances | mppcf | Million particles per cubic foot |
| EINECS | European Inventory of Existing Commercial Substances | MSDS | Material Safety Data Sheets |
| | | mbyp | Meat Byproducts |
| EPA | U.S. Environmental Protection Agency | n- | Abbreviation for "normal," referring to the arrangement of carbon atoms in a chemical molecule prefix for normal |
| EPCRA | Emergency Planning and Community Right-to-Know Act | | |
| | | N- | Symbol used in some chemical names, indicating that the next section of the name refers to a chemical group attached to a nitrogen atom. The bond to the nitrogen atom. |
| EXTOXNET | Extension Toxicology Network | | |
| F | Fahrenheit | | |
| FDA | U.S. Food and Drug Administration | | |
| FEMA | Federal Emergency Management Agency | | |
| FIFRA | Federal Insecticide, Fungicide, and Rodenticide Act | NCI | National Cancer Institute |
| | | NFPA | National Fire Protection Association |
| FR | Federal Register | | |

|  |  |
|---|---|
| NIOSH | National Institute for Safety and Occupational Health |
| NOAEL | No-Adverse-Effect-Level |
| NPRI | National Pollutant Release Inventory (Canada) |
| NTP | National Toxicology Program |
| *o-* | Abbreviation for *ortho-*, a prefix used to distinguish between isomers or nearly related compounds. |
| o.c. or oc | open cup |
| OPP | Office of Pesticides Programs of the EPA |
| OSHA | Occupational Safety and Health Administration |
| Oxy | Oxidizer or oxidizing agent |
| *p-* | Abbreviation for for *para-*, a prefix used to distinguish between isomers or nearly related compounds. |
| PCB | polychlorinated biphenyl |
| PEL | Permissible Exposure Limit (OSHA) |
| post-h | post harvest |
| pot. carc | potential carcinogen |
| POTW | Publicly Owned Treatments Works |
| PP | polypropylene |
| ppb | parts per billion |
| PPE | Personal Protective Equipment |
| ppm | parts per million |
| PQL | Practical Quantitation Limit (RCRA) |
| *prim-* | Prefix for primary |
| ® | Symbol for a registered trademark or proprietary product |
| RED | Reregistration Eligibility Decision |
| REL | Recommended Exposure Limits (NIOSH) |
| RQ | Reportable Quantity |
| RTECS | Registry of Toxic Effects of Chemical Substances |
| RTK | Right-to-Know |
| RUP | Regulated Use Pesticide |
| SARA | Superfund Amendments and Reauthorization Act |
| s. carc | Suspected Carcinogen |
| SCBA | Self-Contained Breathing Apparatus |
| SDWA | Safe Drinking Water Act |
| *sec-* | prefix for secondary |
| SERC | State emergency response commissions |
| SMCL | Secondary Maximum Contaminant Levels (SDWA) |
| soln. | Solution |
| STEL | Short-Term Exposure Limit |
| sus. carc. | Suspected Carcinogen |
| *sym-* | Abbreviation for "*symmetrical*," referring to a particular arrangement of elements within a chemical molecule |
| *t-* | prefix for tertiary |
| TRK | Technical Guiding Concentrations (DFG) for workplace control of carcinogens. |
| temp. | Temperature |
| TLV | Threshold Limit Value (ACGIH) |
| *tert-* | abbreviation for "*tertiary*," referring to a particular arrangement of elements within a chemical molecule |
| TQ | Threshold Quantity |
| *trans-* | (Latin, across). Indicating that one of two geometrical isomers in which certain atoms or groups are on opposite sides of a plane. |
| TRI | Toxics Release Inventory |
| TSCA | Toxic Substances Control Act |
| TWA | Time-Weighted Average |
| UEL | Upper explosive (flammable) limit in air, % by volume at room temperature or other temperature as noted |
| *unsym-* | Prefix for "*asymmetric*." |
| USDA | U.S. Department of Agriculture |
| USPHS | U.S. Public Health Service |
| VOCs | Volatile Organic Compounds |
| WHO | World Health Organization |
| > | Symbol for "greater than" |
| < | Symbol for "less than" |
| ° | Degrees of temperature |
| % | Percent |
| § | Symbol for "section." Used in regulatory matters. |

# A

## Abamectin (ANSI)

*Use Type:* Insecticide and miticide
*CAS Number:* 71751-41-2; 65195-55-3 ($B_{1a}$); 65195-56-4 ($B_{1b}$)
*Synonyms:* Avermectin; Avermectin $B_1$; Avermectin $B_{1a}$ + Avermectin $B_{1b}$ mixture
*Trade Names:* ABACIDE®, J. J. Mauget Co. (USA); AFFIRM®, Syngenta (Switzerland); AGRI-MEK®, Merck (Germany), canceled 7/21/1998; AVID®, Syngenta (Switzerland); AVOMEC®; INJECT-A-CIDE AV®, J. J. Mauget Co. (USA; MK 936®($B_{1a}$); L 676,863® ($B_{1a}$); MK 936®; VERTIMEC®, VIVID®; ZEPHEYR®, Syngenta (Switzerland)
*Producers:* J. J. Mauget Co. (USA); Ki-Hara Chemicals Ltd. (UK); Syngenta (Switzerland)
*Chemical Class:* Avermectins
*EPA/OPP PC Code:* 122804
*California DPR Chemical Code:* 2254
*Uses:* Used on citrus and nut crops and pears to control mite and insect pests, and also to control household and lawn insects.
*Human toxicity (long-term)*[77]: High–2.80 ppb, Health Advisory
*Fish toxicity (threshold)*[77]: Extra High–0.00570 ppb, MATC (Maximum Acceptable Toxicant Concentration)
*Carcinogen/Hazard Classifications*
**U.S. EPA Carcinogens:** Group E, unlikely carcinogen
**TRI Developmental Toxin:** Developmental toxin
**Label Signal Word:** WARNING, CAUTION, or DANGER, depending on the formulation
*Regulatory Authority:*
- EPCRA Section 313 Form R *de minimis* concentration reporting level: 1.0%
- Actively registered pesticide in California.

*Description:* Antibiotic. Vapor pressure = $1.5 \times 10^{-9}$ mmHg.
*Permissible Concentration in Water:* No criteria set. Runoff from spills or fire control may cause water pollution.
*Harmful Effects and Symptoms*
*Short Term Exposure:* Contact with eyes or skin may cause irritation or injury. Inhalation should be avoided; use NIOSH-approved air purifying respirators for pesticides. May be harmful if swallowed.
*First Aid:* If this chemical gets into the eyes, remove any contact lenses at once and irrigate immediately for at least 15 minutes, occasionally lifting upper and lower lids. Seek medical attention immediately. If this chemical contacts the skin, remove contaminated clothing and wash immediately with soap and water. Seek medical attention immediately. If this chemical has been inhaled, remove from exposure, begin rescue breathing (using universal precautions) if breathing has stopped and CPR if heart action has stopped. Transfer promptly to a medical facility. When this chemical has been swallowed, get medical attention. Give large quantities of water and induce vomiting. Do not make an unconscious person vomit.
*References:*
- Pesticide Management Education Program, "Avermectin (Agri-Mek, Affirm) EPA Pesticide Fact Sheet 9/89," http\\pmep.cce.cornell.edu/profiles/insect-mite/abamectin-bufencarb/avermectin/insect-prof-avermectin.html
- EXTOXNET, Extension Toxicology Network, "Pesticide Information Profile, Abamectin," Oregon State University, Corvallis, OR (June 1996). http://extoxnet.orst.edu/pips/abamecti.htm
- California Environmental Protection Agency Chemical List of Lists, Sacramento CA (February 1997)

## Acenaphthene

*Use Type:* Insecticide and fungicide intermediate
*CAS Number:* 83-32-9
*Formula:* $C_{12}H_{10}$
*Synonyms:* Acenafeno (Spanish); Acenaphthylene, 1,2-dihydro-; 1,2-Dihydroacenaphthylene; 1,8-Dihydroacenaphthalene; 1,8-Dihydroacenaphthylene; Ethylenenaphthalene; 1,8-Ethylenenaphthalene; Naphthyleneethylene; NSC 7657; *peri*-Ethylenenaphthalene
*Producers:* Deza AS (Czech Republic); Dow Chemical (USA); Fluorochem Ltd. (UK); Merck (Germany); Messer Group (Germany); Penta Manufacturing (USA); Sigma-Aldrich (USA); Sigma-Aldrich Laborchemikalien (Germany); TCI America (USA)
*EPA/OPP PC Code:* 442200
*RTECS Number:* AB1000000
*EINECS Number:* 201-469-6
*Uses:* Acenaphthene occurs in coal tar produced during the high-temperature carbonization or coking of coal, petroleum processing, shale oil processing. It is used as a dye intermediate, in the manufacture of some plastics as well as an insecticide and fungicide. It has been detected in cigarette smoke and gasoline exhaust condensates.
*Regulatory Authority:*
- OSHA, 29CFR1910 Specifically Regulated Chemicals (See CFR 1910.1002) as coal tar pitch volatiles
- List of priority pollutants (U.S. EPA)
- AB 1803-Well Monitoring Chemical (CAL)
- The "Director's List" (CAL/OSHA)
- Clean Water Act: 40CFR401.15 Section 307 Toxic Pollutants, 40CFR 413.02, Total Toxic Organics, 40CFR423, Priority Pollutants RCRA 40CFR258, Appendix 2
- RCRA 40CFR§268.48; 61FR15654, Universal Treatment Standards: Wastewater (mg/L), 0.059; Nonwastewater (mg/kg), 3.4

- RCRA, 40CFR264, Appendix 9, Ground Water Monitoring List, Suggested Testing Methods (PQL ug/L): 8100(200); 8270(10)
- Superfund/EPCRA 40CFR302.4, Appendix A, RQ: 100 lb (45.4 kg)
- Canada, WHMIS, Ingredients Disclosure List
- Mexico, Drinking Water, Criteria (Ecological): 0.02 mg/L; wastewater: organic toxic pollutant

*Description:* Acenaphthene is a combustible polyaromatic hydrocarbon (PAH). White or pale yellow crystalline solid. Insoluble in water. Boiling Point = 277°C. Melting/Freezing point = 95–97°C. Molecular weight = 154.23. Log $K_{ow}$ = 3.91–4.44. Values at or above 3.0 are likely to bioaccumulate in marine organisms.

*Incompatibilities:* Ozone and oxidizing agents such as perchlorates, peroxides, permanganates, chlorates, nitrates, chlorine, bromine and fluorine. It is also incompatible with ozone and chlorinating agents.

*Permissible Exposure Limits in Air:* No standards have been established for acenaphthene. The ACGIH[1] and OSHA[2] recommended TLV as coal tar pitch volatiles as benzene solubles is 0.2 mg/m³. NIOSH[2] considers coal tar products to be occupational carcinogens; the NIOSH[2] REL (10-hour TWA) for coal tar products is 0.1 mg/m³.

*Determination in Air:* See NIOSH Method 5506 (HPLC) and 5515 (GC)[18].

*Permissible Concentration in Water:* To protect freshwater aquatic life: 1,700 µg/L. To protect saltwater aquatic life–on an acute basis 970 µg/L and on a chronic basis 520 µg/L. To protect human health: 20.0 µg/L (based on organoleptic data)[6]. See also Regulatory Authority for U.S. and Mexico regulatory levels.

*Determination in Water:* Gas chromatography or high performance liquid chromatograph (EPA Method 610) or gas chromatography and mass spectrometry (EPA Method 625).

*Routes of Entry:* Ingestion, inhalation, eye and/or skin contact.

### Harmful Effects and Symptoms

*Short Term Exposure:* Acenaphthene is irritating to eyes, skin and respiratory tract causing coughing and wheezing. May cause vomiting if swallowed in large quantities.

*Long Term Exposure:* Although acenaphthene has not been identified as a carcinogen, it should be handled with care as several related polycyclic aromatic hydrocarbons (PAHs) are carcinogens. Repeated or high exposures may cause lung irritation, bronchitis with cough, phlegm, and/or shortness of breath. Acenaphthene may affect the liver and kidneys. The most thoroughly investigated effect of acenaphthene is its ability to produce nuclear and cytological changes in microbial and plant species. Most of these changes, such as an increase in cell size and DNA content, are associated with disruption of the spindle mechanism during mitosis and the biological impact of acenaphthene on mammalian cells, these effects are reported here because they are the only substantially investigated effects of acenaphthene. Reported to be a mutagen[11].

*Points of Attack:* Liver, kidneys and skin.

*Medical Surveillance:* Preplacement and regular physical examinations are indicated for workers having contact with acenaphthene in the workplace. Liver and kidney function tests recommended.

*First Aid:* If this chemical gets into the eyes, remove any contact lenses at once and irrigate immediately for at least 15 minutes, occasionally lifting upper and lower lids. If this chemical contacts the skin, remove contaminated clothing and wash with soap immediately. When this chemical has been swallowed, get medical attention. Give large quantities of water and induce vomiting. Do not make an unconscious person vomit. If this chemical has been inhaled, remove from exposure and transfer promptly to a medical facility.

*References:*
- U.S. Environmental Protection Agency, Acenaphthene: Ambient Water Criteria, Report PB 296-782, Washington, DC (1980).
- Lewis, Richard J, Ed., *Sax's Dangerous Properties of Industrial Materials*, 10th edition, New York, (1999).
- Sax, N.I., Ed., Dangerous Properties of Industrial Materials Report, 4, No. 1, 38-41 (1984).
- New Jersey Department of Health and Senior Services, "Hazardous Substance Fact Sheet Acenaphthene," Trenton NJ (November, 1998), http://www.state.nj.us/health/eoh/rtkweb/2958.pdf
- California Environmental Protection Agency Chemical List of Lists, Sacramento CA (February 1997).
- NTP Chemical Repository. http://ntp-server.niehs.nih.gov/htdocs/chem_h&s/ntp_Chem8/Radian83-32-9.html

# Acephate (ANSI)

*Use Type:* A contact and systemic insecticide.
*CAS Number:* 30560-19-1
*Formula:* $C_4H_{10}NO_3PS$
*Alert:* Effective October 31, 2002, homeowner use for lawns is discontinued except for treatment of fire ant mounds. Other indoor treatment has been discontinued. Human toxicity (long-term): High.

*Synonyms:* Acephat (German); Acetylphosphoramidothioic acid, *O,S*-dimethyl ester; ENT-27822; *O,S*-Dimethyl acetylphos-phoramidothioate; *O,S*-Dimethyl acetic phosphoramidothioate, *N*-[Methoxy(methylthio)phosphinoyl] acetamide; Phosphoramidothioic acid, *N*-acetyl-,*O,S*,-dimethyl ester

*Trade Names:* ACECAP SYSTEMIC INSECTICIDE IMPLANTS®, Pro-Outdoors, Inc. (USA); ACEFAL 75 PS®, Alcotan Laboratories (Spain); ACEHERO®, Sabero Organics Gujarat Ltd. (India); ACEPHATE 97 EG®, Micro Flo Co., LLC (USA); ACEPHATE 75SP®, Micro Flo Co., LLC (USA); ACEPHATE PCO SP INSECTICIDE®, Micro Flo Co., LLC (USA); ACESUL®, Sulphur Mills Ltd. (India); ACE-TOX®, Indiclay (India); ACHERO®, Sabero Organics (India); ACIFAT®, Crystal Chemical Inter-America (USA); ADDRESS®, Dow AgroSciences (USA); AIMTHENE®, Aimco Pesticides Ltd. (India); AMCOTHENE®, Sundat Pte. Ltd. (Singapore); ASATAF®, Rallis India Ltd. (India); ASIFY®, Nanjing Agrochemicals

Co., Ltd. (China); ATTACK®, Sudarshan India Pvt. Ltd. (India); CHEVRON RE 12420®, Chevron Phillips Chemical (USA); CLEAN CROP ACEPHATE 80 DF SEED PROTECTORANT®, United Agri Products (USA); DREXEL ACEPHATE 75 WSP®, Drexel Chemical Co. (USA); DREXEL ACEPHATE PCO SP INSECTICIDE®, Drexel Chemical Co. (USA); FATEL®, Sudarshan Chemical Industries (India); FORPHATE®, Forward (Beihai) Hepu Pesticide Co., Ltd. (China); GUSATAFSON ACEPHATE 90 SEED PROTECTORANT®, Gustafson LLC (USA), canceled; KITRON®, Saeryung Chemicals Co., Ltd. (Korea); KORANDA®, acephate + fenvelerate, Rallis India (India); LANCER®, United Phosphorus (India); ORCEPHATE®, Zagro Asia (Singapore); ORTHENE®, The Scotts Company (USA); ORTHENE 755®, Chevron Phillips Chemical (USA), The Scotts Company (USA); ORTHO 12420®, The Scotts Company (USA); ORTRAN®, Chevron Phillips Chemical (USA); ORTRIL®; PACE®, Nagarjuna Agrichem Ltd. (India); PAYLOAD®; PILARTHENE®, Pilarquim Corp. (Taiwan); PINPOINT®; POWER-X®, canceled; PRECISE ACEPHATE®, Pursell Technologies (USA); RACET®, Rotam Group, Agrochemical Div. (Hong Kong); RE 12420®; SAPHATE®, Saeryung Chemicals Co., Ltd. (Korea); 75 SP®, Chevron Phillips Chemical (USA); VALENT ORTHENE TECHNICAL®, Valent USAA Corp. (USA); VEGFRU TARGET®, PI Industries Ltd. (India)

**Producers:** Agrimor International Co. (USA); Agsin (Singapore); Aimco Pesticides Ltd. (India); Alcotan Laboratories (Spain); Ascot International Ltd. (UK); BEC Group (India); Bharat Pulverizing Mills (India); Bharat Rasayan (India); Biesterfeld Siemsgluess International. GmbH (Germany); Cangzhou Green Chemical Co. (China); Chevron Phillips Chemical (USA); China Chemical (China); Dow AgroSciences (USA); Drexel Chemical (USA); Forward (Beihai) Hepu Pesticide Co., Ltd. (China); Hokko Chemical Industry (Japan); ICI Group (UK); Indiclay (India); Jiangmen Pesticide Factory (China); Jingma Chemicals Ltd. (China); Ki-Hara Chemicals Ltd. (UK); Meghmani Organics (India); Micro Flo Co., LLC (USA); Nagarjuna Agrichem Ltd. (India); Nanjing Agrochemicals Co., Ltd. (China); Nufarm (AUSAtralia); Pazchem Ltd. (Israel); PI Industries Ltd. (India); Pilarquim Corp. (Taiwan); Pro-Outdoors, Inc. (USA); Pursell Technologies (USA); Rallis India Ltd. (India); Rhone-Poulenc Agro (France); Rotam Group, Agrochemical Division, (Hong Kong); Sabero Organics Gujarat Ltd. (India); Saeryung Chemicals Co., Ltd. (Korea); Sanonda Zhengzhou Pesticide Co., Ltd. (China); Scotts Company, The (USA); Shenzhen Guomeng Industry Co., Ltd. (China); Sinon Corporation (Taiwan); Sudarshan India Pvt. Ltd. (India); Sulphur Mills Ltd. (India); Sundat Pte. Ltd. (Singapore); Syngenta (Switzerland); Takeda Chemical IndusAtries (Japan); United Agri Products (USA); United Phosphorus (India); Vijayalakshmi Insecticides and Pesticides (India); Whitmire Micro-Gen (USA); Zagro Asia Ltd. (Singapore)

**Chemical Class:** Organophosphate
**Label Signal Word:** CAUTION
**EPA/OPP PC Code:** 103301
**California DPR Chemical Code:** 1685
**ICSC Number:** 0748
**RTECS Number:** TB4760000
**EEC Number:** 015-079-00-7
**EINECS Number:** 250-241-2

**Uses:** Acephate is a general use insecticide used on green- and lima beans, Brussels sprouts, cauliflower, celery, cotton, cottonseed, cranberries, head lettuce, macadamia nuts, peanuts, bell- and non-bell peppers, peppermint, spearmint, tobacco, and soybeans (Special Local Need Registration required in Mississippi and Texas only). Also used to control cockroach (spot treatment only) in residential and industrial buildings and insect control in forests, and on ornamental plants and to target armyworms, aphids, beetles, bollworms, borers, budworms, cankerworms, crickets, cutworms, fire ants, fleas, grasshoppers, leafhoppers, loopers, mealybugs, mites, moths, roaches, spiders, thirps, wasps, weevils, whiteflies, etc.

**Human toxicity (long-term)**[77]: High–2.80 ppb, Health Advisory
**Fish toxicity (threshold)**[77]: Very low–2725.54621 ppb, MATC (Maximum Acceptable Toxicant Concentration)

**U.S. Maximum Allowable Residue Levels for Acephate (40 CFR 180.108)**

| CROP | ppm |
| --- | --- |
| Bean, dry | 3.0 |
| Bean, succulent | 3.0 |
| Brussels sprouts | 3.0 |
| Cattle, fat | 0.1 |
| Cattle, meat | 0.1 |
| Cattle, mbyp | 0.1 |
| Cauliflower | 2.0 |
| Celery | 10.0 |
| Cotton, hulls | 4.0 |
| Cotton, meal | 8.0 |
| Cotton, undelinted seed | 2.0 |
| Cranberry | 0.5 |
| Egg | 0.1 |
| Goat, fat | 0.1 |
| Goat, meat | 0.1 |
| Goat, mbyp | 0.1 |
| Grass, hay | 15.0 |
| Grass, pasture & range | 15.0 |
| Hog, fat | 0.1 |
| Hog, meat | 0.1 |
| Hog, mbyp | 0.1 |
| Horse, fat | 0.1 |
| Horse, meat | 0.1 |
| Horse, mbyp | 0.1 |
| Lettuce, head | 10.0 |
| Milk | 0.1 |
| Mint, hay | 15.0 |
| Nut, macadamia | 0.05 |
| Peanut | 0.2 |
| Pepper | 4.0 |
| Poultry, fat | 0.1 |
| Poultry, meat | 0.1 |
| Poultry, mbyp | 0.1 |
| Processed food | 0.02 |

| | |
|---|---|
| Sheep, fat | 0.1 |
| Sheep, meat | 0.1 |
| Sheep, mbyp | 0.1 |
| Soybean | 1.0 |
| Soybean, meal | 4.0 |

***Carcinogen/Hazard Classifications***
**U.S. EPA Carcinogens:** Group C, possible carcinogen
**WHO Acute Hazard:** Class III, slightly hazardous
***Regulatory Authority:***
- AB 1803-Well Monitoring Chemical (CAL)
- Actively registered pesticide in California.
- FIFRA, 180.3(5); class of cholinesterase-inhibiting pesticide
- U.S. DOT Inhalation Hazard Chemicals as organophosphates

***Description:*** Colorless crystalline solid (80% or more pure) or white powder (technical). Has an odor. Readily soluble in water; solubility = $8.2 \times 10^{-5}$ ppm @ 25°C. Molecular weight = 183.19. Melting/Freezing point = 92–93°C; also listed at 82–89°C (technical grade 80 to 90% purity). Vapor pressure = $1.7 \times 10^{-6}$ mmHg @ 25°C. Log $K_{ow}$ = < 1.0. Unlikely to bioaccumulate in marine organisms.

***Incompatibilities:*** Strong oxidizers. Acephate emits toxic oxides of phosphorus, nitrogen, and sulfur when heated to decomposition.

***Determination in Air:*** OSHA versatile sampler-2; Toluene/Acetone; Gas chromatography/Flame ionization detection; NIOSH IV[18], Method #5600, Organophosphorus pesticides.

***Routes of Entry:*** Inhalation, ingestion, skin contact, passes through the skin.

*Harmful Effects and Symptoms*
***Short Term Exposure:*** Poisonous; may be fatal if inhaled, swallowed, or absorbed through skin. Because this material has a low vapor pressure, significant inhalation of vapors is unlikely at ordinary temperatures. Contact may cause burns to skin and eyes. Eye pupils appear small; blurred vision; eye watering; runny nose; cough; shortness of breath; salivation; dizziness; nausea, stomach cramps, diarrhea, and vomiting; increased blood pressure; profuse sweating; hypermotility, hallucinations; irritability; tingling of the skin; drowsiness; slow heartbeat; convulsions; fluid in lungs; loss of consciousness; incontinence; breathing stops; death. Orgnophosphates inhibit the action of acetylcholinesterase enzymes, and alter the way in which nervous impulses are transmitted. The effects can last for hours, days, or much longer. The action of the enzymes is reestablished after new enzymes are formed. Delayed pulmonary edema may occur after inhalation.

***Long Term Exposure:*** Cholinesterase inhibitor; cumulative effect is possible. Organophosphates may damage the nervous system with repeated exposure, resulting in convulsions, respiratory failure. May cause liver damage.

***Points of Attack:*** Respiratory system, central nervous system, cardiovascular system, blood cholinesterase.

***Medical Surveillance:*** Medical observation is recommended for 24 to 48 hours after breathing overexposure, as pulmonary edema may be delayed. As first aid for pulmonary edema, consider administering a corticosteroid spray. Cigarette smoking may exacerbate pulmonary injury and should be discouraged for at least 72 hours following exposure. Do not drink any alcoholic beverages before or during use; alcohol promotes absorption of organophosphates.

***First Aid:*** **Treatment for organophosphate poisoning consists of thorough decontamination, cardiorespiratory support, and administration of the antidotes atropine and pralidoxime. In cases of severe poisoning, diazepam, an anticonvulsant, should also be administered. Antidotes should be administered as prevention even if the diagnosis is in doubt.** Speed in removing material from eyes and skin is of extreme importance. *Eyes:* Eye contact can cause dangerous amounts of these chemicals to be quickly absorbed through the mucous membrane into the bloodstream. Immediately and gently flush eyes with plenty of warm or cold water (NO hot water) for at least 15 minutes, occasionally lifting the upper and lower eyelids. Get medical aid immediately. *Skin:* Get medical aid. Skin contact can cause dangerous amounts of these chemicals to be absorbed into the bloodstream. Wearing the appropriate PPE equipment and respirator for organophosphate pesticides, immediately flush skin with plenty of soap and water for at least 15 minutes while removing contaminated clothing and shoes. Shampoo hair promptly if contaminated. The removed, contaminated clothing and shoes should be double-bagged and left in Hot Zone for later disposal by hazardous materials experts. Skin may also be decontaminated with diluted hypochlorite solution. *Inhalation:* Get medical aid. Do not contaminate yourself. Wearing the appropriate PPE equipment and respirator for organophosphate pesticides, immediately remove the victim from the contaminated area to fresh air. If the victim is not breathing, administer artificial respiration. Do NOT use mouth-to-mouth resuscitation; use bag/mask apparatus. If breathing is difficult, administer oxygen through bag/mask apparatus until medical help arrives. Do not leave victim unattended. *Ingestion:* Call poison control. Loosen all clothing. Never give anything by mouth to an unconscious person. If victim is *unconscious or having convulsions,* do nothing except keep victim warm. Get medical aid. Transfer promptly to a medical facility. In cases of ingestion, **do not induce vomiting**. If the victim is alert and asymptomatic, administer a slurry of activated charcoal at a dose of 1 g/kg (infant, child, and adult dose). A soda can and straw may be of assistance when offering charcoal to a child. *In some cases you may be specifically instructed by poison control to induce vomiting by way of 2 tablespoons of syrup of ipecac (adult) washed down with a cup of water.* Do NOT give activated charcoal before or with ipecac syrup.

*Note to physician or authorized medical personnel:* Treat cases of respiratory compromise, coma, or excessive pulmonary secretions with respiratory support using protocols and techniques available and within the scope of training. Some cases may necessitate procedures such as endotracheal intubation or cricothyrotomy by properly trained and equipped personnel. When possible, atropine (see *Antidotes,* below) should be given under medical supervision. Patients who are comatose, hypotensive, or having seizures or cardiac arrhythmias should be treated according to advanced life support protocols. *Antidotes:* Two antidotes are administered to treat organophosphate poisoning. Atropine is a competitive antagonist of acetylcholine at muscarinic receptors and is used to control the excessive bronchial secretions which are often responsible for death. Pralidoxime relieves both the nicotinic and muscarine effects of organophosphate poisoning by regenerating acetylcholinesterase and can reduce both the bronchial secretions and the muscle weakness associated with poisoning. The initial intravenous dose of atropine in adults should be determined by the severity of symptoms: An initial adult dose of 1.0 to 2.0 mg or pediatric dose of 0.01 mg/kg (minimum 0.01 mg) should be administered intravenously. If intravenous access cannot be established, atropine may also be given intramuscularly, subcutaneously or via endotracheal tube. Doses should be repeated every 15 minutes until excessive secretions and sweating have been controlled. Once bronchial secretion has been controlled, atropine administration should be repeated whenever the secretions begin to recur. In seriously poisoned patients, very large doses may be required. Alterations of pulse rate and pupillary size should not be used as indicators of treatment adequacy. Pralidoxime should be administered as early in poisoning as possible as its efficacy may diminish when given more than 24 to 36 hours after exposure. Doses are as follows: adult 1.0 g; pediatric 25 to 50 mg/kg. The drug should be administered intravenously over 30 to 60 minutes, but in a life-threatening situation, one-half of the total dose can be given per minute for a total administration time of 2 minutes. Treatment should begin to take effect within 40 minutes with a reduction in symptoms and the amount of atropine necessary to control bronchial secretion. The initial dose can be repeated in 1 hour and then every 8 to 12 hours until the patient is clinically well and no longer requires atropine. If intravenous access cannot be established, pralidoxime may also be given intramuscularly. Early administration of diazepam in addition to the combined atropine and pralidoxime treatment may help prevent the onset of seizures and potential brain and cardiac morphologic damage following high-level organophosphate poisoning.

*References:*
- EPA, Office of Prevention, Pesticides and Toxic Substances, "Acephate Facts," September, 2001. http://www.epa.gov/REDs/factsheets/acephate_fs.pdf
- EPA, Office of Pesticide Programs, Pesticide Residue Limits, "Acephate," 40 CFR 180.108, www.epa.gov/cgi-bin/oppsrch
- International Chemical Safety Card, "Acephate," NIOSH, www.cdc.gov/niosh/ipcsneng/neng0748.html
- EPA, Office of Prevention, Pesticides and Toxic Substances, "Acetate Summary," February 2, 2000, www.epa.gov/pesticides/op/acephate/acephate_summ.htm
- California Environmental Protection Agency Chemical List of Lists, Sacramento CA (February 1997).
- Agency for Toxic Substances and Disease Registry, U.S. Department of Health and Human Services, Public Health Service, "Managing Hazardous Materials Incidents," Atlanta, GA (June 2003)
- EXTOXNET, Pesticide Information Profiles, "Acephate," University of Oregon, September, 1995, http://ace.orst.edu/cgi-bin/mfs/01/pips/acephate.htm?6#mfs

# Acetochlor (ANSI)

*Use Type:* Herbicide
*CAS Number:* 34256-82-1
*Formula:* $C_{14}H_{20}ClNO_2$
*Alert:* A Restricted Use Pesticide (RUP), depending on the formulation.
*Synonyms:* Acetamide, 2-chloro-*N*-(ethoxymethyl)-*N*-(2-ethyl-6-methylphenyl)-; *O*-Acetotoluidide, 2-chloro-*N*-(ethoxymethyl)-6'-ethyl-; Azetochlor; 2-chloro-*N*-(ethoxymethyl)-6'-ethyl-*O*-acetotoluidide; 2-Chloro-*N*-(ethoxymethyl)-*N*-(2-ethyl-6-methylphenyl)acetamide; 2'-Ethyl-6'-methyl-*N*-(ethoxymethyl)-2-chloroacetanilide
*Trade Names:* ACENIT®; CP 55097®; DEGREE®, Monsanto (USA); ERUNIT®, Nitrokemia 2000 (Hungary); FULTIME®, Dow AgroSciences (USA); GUARDIAN®; HARNESS®, Monsanto (USA); KEYSTONE LA®, Dow AgroSciences (USA); MG 02®; MON 097®, Monsanto (USA); MON 58420®, Monsanto (USA); NEVIREX®; RELAY®; SACEMID®; SURPASS®, Dow AgroSciences (USA); TOPHAND®; TOPNOTCH®, Dow AgroSciences (USA); TROPHY®; WINNER®
*Producers:* Agrimor International (USA); Agsin (Singapore); Dow AgroSciences (USA); Drexel Chemical (USA); Monsanto (USA); Nitrokemia 2000 (Hungary); Sanonda Ltd. (Australia); Syngenta (Switzerland)
*Chemical Class:* Organochlorine; Halo-organics
*EPA/OPP PC Code:* 121601
*California DPR Chemical Code:* 2349
*Uses:* A Restricted Use Pesticide (RUP). A pre-emergence herbicide for control of annual grasses and broadleaf weeds. It is used on cabbage, citrus cops, coffee, all types of corn, cotton, green peas, maize, onion, peanuts, potatoes, vineyards, sugar cane, and sugar beets, among others. It is compatible with most other pesticides.
*Human toxicity (long-term)*[77]: Intermediate–20.71006 ppb, CHCL (Chronic Human Carcinogen Level)
*Fish toxicity (threshold)*[77]: Low–187.34840 ppb, MATC (Maximum Acceptable Toxicant Concentration)
*U.S. Maximum Allowable Residue Levels for Acetochlor (40 CFR 180. 470):*

| CROP | ppm |
| --- | --- |
| Corn, field, forage | 1 |
| Corn, field, grain | 0.05 |
| Corn, field, stover | 1.5 |
| Sorghum, forage | 0.1 |
| Sorghum, grain, grain | 0.02 |
| Sorghum, grain, stover | 0.1 |
| Soybean | 0.1 |
| Soybean, forage | 0.7 |
| Soybean, hay | 1 |
| Wheat, forage | 0.5 |

Wheat, grain 0.02
Wheat, straw 0.1

*Carcinogen/Hazard Classifications*
**U.S. EPA Carcinogens:** Group B2, probable carcinogen
**California Prop. 65:** Carcinogen
**Label Signal Word:** CAUTION or WARNING
**WHO Acute Hazard:** Class III, slightly hazardous
**Endocrine Disruptor:** Suspected endocrine disruptor
*Regulatory Authority:*
- FIFRA, 40CFR180.470: Tolerances and exemptions from tolerances for pesticides in or on raw agricultural commodities
- AB 2588-Air Toxics "Hot Spots" Chemicals (CAL)
- Proposition 65 chemical (CAL)

*Description:* Pale yellow liquid. Soluble in water; solubility = 400 ppm @ 25°C. Molecular weight = 269.77.

*Incompatibilities:* Oxidizers (chlorates, nitrates, peroxides, permanganates, perchlorates, chlorine, bromine, fluorine, etc); strong acids, alkaline reagents. Slowly hydrolyzes in water, releasing ammonia and forming acetate salts. May attack some forms of plastic, rubber, and coatings.

*Permissible Concentration in Water:* Unknown, but runoff from spills or fire control may cause water pollution.

*Routes of Entry:* Inhalation, ingestion, absorbed through the intact skin

*Harmful Effects and Symptoms*

*Short Term Exposure:* Attacks central nervous system. Apprehension, anxiety, confusion, nervous excitation; dizziness; headache; numbness and weakness in limbs; muscle twitching, tremors; nausea and vomiting; slow, shallow respiration, bluish face; convulsions; loss of consciousness; breathing stops; death.

*Points of Attack:* CNS. May be fatal if inhaled, ingested, or absorbed through the skin.

*Medical Surveillance:* Medical observation is recommended for 24 to 48 hours after breathing overexposure, as pulmonary edema may be delayed. As first aid for pulmonary edema, consider administering a corticosteroid spray. Cigarette smoking may exacerbate pulmonary injury and should be discouraged for at least 72 hours following exposure.

*First Aid:* Speed in removing material from eyes and skin is of extreme importance. *Eyes:* Eye contact can cause dangerous amounts of these chemicals to be quickly absorbed through the mucous membrane into the bloodstream. Directly, irrigate with large amounts of plain, tepid water or saline for 20 minutes, occasionally lifting the lower and upper lids. During this time, remove contact lenses, if easily removable without additional trauma to the eye. Get medical aid immediately. Have physician check for possible delayed damage. *Skin:* Get medical aid. Skin contact can cause dangerous amounts of these chemicals to be absorbed into the bloodstream. Wearing the appropriate PPE equipment and respirator for organochlorine pesticides, immediately flush exposed skin, hair, and under nails with plain, running, tepid water for 20 minutes, then wash twice with mild soap. Shampoo hair promptly if contaminated; protect eyes. Do not scrub skin or hair, since this can increase absorption through the skin. Rinse thoroughly with water. Victims who are able and cooperative may assist with their own decontamination. Remove and double-bag contaminated clothing and personal belongings. Leather absorbs many organochlorines; therefore, items such as leather shoes, gloves, and belts should be discarded. If the skin is swollen or inflamed, cool affected areas with cold compresses. *Ingestion:* Call poison control. Loosen all clothing. Never give anything by mouth to an unconscious person. Get medical aid. Do NOT induce vomiting. The patient is at risk of CNS depression or seizures, which may lead to pulmonary aspiration during vomiting. If the victim is conscious and able to swallow, *administer an aqueous slurry of activated charcoal at 1 gm/kg (usual adult dose 60–90 g, child dose 25–50 g). A soda can and straw may be of assistance when offering charcoal to a child. The efficacy of activated charcoal for some organochlorine poisoning (such as chlordane) is uncertain. If victim is *unconscious or having convulsions,* do nothing except keep victim warm. *In some cases you may be specifically instructed by Poison Control to induce vomiting by way of 2 tablespoons of syrup of ipecac (adult) washed down with a cup of water.* Do NOT give activated charcoal *before or with* ipecac syrup. *Inhalation:* Get medical aid. Do not contaminate yourself. Wearing the appropriate PPE equipment and respirator for organochlorine pesticides, immediately remove the victim from the contaminated area to fresh air. For inhalation exposures, monitor for respiratory distress. If the victim is not breathing, administer artificial respiration. *Do NOT use mouth-to-mouth resuscitation; use bag/mask apparatus.* If cough or breathing difficulty develops, evaluate for respiratory tract irritation, bronchitis, or pneumonitis. If breathing is difficult, administer 100% humidified supplemental oxygen through bag/mask apparatus until medical help arrives. Do not leave victim unattended.

*References:*
- EXTOXNET, Extension Toxicology Network, "Pesticide Information Profile, Acetochlor," Oregon State University, Corvallis, OR. http://extoxnet.orst.edu/pips/acetochl.htm
- U.S. Environmental Protection Agency, Office of Pesticide Programs, Pesticide Residue Limits, "Acetochlor", 40 CFR 180.470. www.epa.gov/pesticides/food/viewtols.htm
- California Environmental Protection Agency Chemical List of Lists, Sacramento CA (February 1997).

# Acetylaminofluorene

*Use Type:* A carcinogenic pesticide that was never marketed in the United States. See Uses section, below.
*CAS Number:* 53-96-3
*Formula:* $C_{15}H_{13}NO$
*Alert:* This is a carcinogen and must be handled with extreme caution. It is not produced in the U.S.
*Synonyms:* AAF; 2-AAF; Acetamide,*N*-fluoren-2-yl-; Acetamide,*N*-9*H*-fluoren-2-yl; 2-Acetamidofluorene; 2-2-Acetylamidofluorene; 2-Acetylamino-fluoren (German); *N*-Acetyl-2-aminofluorene; 2-Acetylaminofluorene;

*N*-Acetyl-2-aminofluorene; 2-Acetylaminofluorene; Azetylaminofluoren; FAA; 2-FAA; 2-Fluorenylacetamide; *N*-2-Fluorenylacetamide; *N*-2-Fluoren-2-yl acetamide; *N*-2-Fluorenylacetamide

*Producers:* Sigma-Aldrich Fine Chemicals (USA)
*Chemical Class:* Aromatic amine
*RTECS Number:* AB9450000
*EINECS Number:* 200-188-6
*Uses:* 2-Acetylaminofluorene (AAF) was intended to be used as a pesticide, but it was never marketed because this chemical was found to be carcinogenic. AAF is used frequently by biochemists and technicians engaged in the study of liver enzymes and the carcinogenicity and mutagenicity of aromatic amines as a positive control.
*Carcinogen/Hazard Classifications*
**U.S. EPA Carcinogens:** Group B2, Probable Human Carcinogen
**California Prop. 65:** Carcinogen
*Regulatory Authority:*
- Carcinogen (OSHA, NTP, State of California)
- CAL/OSHA Carcinogen User Register Chemical
- AB 2588-Air Toxics "Hot Spots" Chemicals (CAL)
- CAL Air Resources Board/AB 1807 Toxic Air Contaminants
- Proposition 65 chemical (CAL)
- Permissible Exposure Limits for Chemical Contaminants (CAL/OSHA)
- The "Director's List" (CAL/OSHA)
- OSHA, 29CFR1910 Specifically Regulated Chemicals (See CFR 1910.1014)
- Clean Air Act 42USC7412; Title I, Part A,§112 hazardous pollutants
- RCRA 40CFR261, Appendix 8; 40CFR261.11 Hazardous Constituents
- RCRA 40CFR264, Appendix 9; Ground Water Monitoring List Suggested methods (PQL *ug*/L): 8270 (10)
- RCRA 40CFR§268.48; 61FR15654, Universal Treatment Standards: Wastewater (mg/L), 0.059; Nonwastewater (mg/kg), 140
- RCRA 40CFR266, Appendix 7, Basis for Listing Hazardous Waste
- EPA Hazardous Waste Number (RCRA No.): U005
- Superfund/EPCRA 40CFR302.4, Appendix A, RQ: 1lb (0.454 kg), SARA 313: Form R *de minimis* Concentration Reporting Level: 0.1%.
- Banned or Severely Restricted (Industrial Chemicals) (Belgium, Finland, Sweden) (UN)[13]
- Air Pollutant Standard Set (New York)[60]
- Canada WHMIS Ingredients Disclosure List. Concentration Reporting Level: 0.1%.

*Description:* 2-Acetylaminofluorene, $C_{15}H_{13}NO$, is a combustible, tan powder or crystalline solid. Insoluble in water. Molecular weight = 223.28. Melting/Freezing point = 194°C. Hazard Identification (based on NFPA-704 M Rating System): Health 1, Flammability 1, Reactivity 0.
*Incompatibilities:* Contact with strong oxidizers may cause fire and explosions. Not compatible with cyanides, acids, and acid anhydrides.

*Permissible Exposure Limits in Air:* NIOSH[2] recommends that exposure to occupational carcinogens be limited to the lowest feasible concentration. 0.03 $\mu g/m^3$ (New York)[60] for ambient air.
*Permissible Concentration in Water:* No criteria set, but runoff from spills or fire control may cause water pollution.
*Routes of Entry:* Ingestion, inhalation, mucous membrane, skin absorption, skin and/or eye contact.
*Harmful Effects and Symptoms*
*Short Term Exposure:* This chemical has limited use in industry, and contact is kept to a minimum to prevent cancer. Reduced function of liver, kidneys, bladder, pancreas; [Potential occupational carcinogen]. Contact with skin, eye, or respiratory tract may cause irritation.
*Long Term Exposure:* Incorporation of this compound in feed caused increased incidences of malignant tumors in a variety of organs in the rat. Long-term studies in which mice were given 2-acetylaminofluorene in their diet showed that this compound caused increased incidences of tumors and cancer of the liver, kidney, urinary bladder, lung, skin, and pancreas. There is limited evidence that this chemical is a teratogen in animals.
*Points of Attack:* Liver, bladder, kidney, pancreas, skin, lungs.
*Medical Surveillance:* Urine cytology for abnormal cells in the urine.
*First Aid:* If this chemical gets into the eyes, remove any contact lenses at once and irrigate immediately for at least 15 minutes, occasionally lifting upper and lower lids. Seek medical attention immediately. If this chemical contacts the skin, remove contaminated clothing and wash immediately with soap and water. Seek medical attention immediately. If this chemical has been inhaled, remove from exposure, begin rescue breathing (using universal precautions) if breathing has stopped, and CPR if heart action has stopped. Transfer promptly to a medical facility. When this chemical has been swallowed, get medical attention. Give large quantities of water and induce vomiting. Do not make an unconscious person vomit. Do not induce vomiting when formulations containing petroleum solvents are ingested.
*References:*
- New Jersey Department of Health and Senior Services, "Hazardous Substance Fact Sheet: 2-Acetylaminofluorene," Trenton, N.J. (June 1998). http://www.state.nj.us/health/eoh/rtkweb/0010.pdf
- US DOL OSHA, Reduced Immunologic Competence, Code of Federal Regulations. 29 CFR Part 1910, Air Contaminants. US GPO. July 1, 1996. US DHHS NIOSH and US DOL OSHA, Urine (chemical/metabolite) NIOSH/OSHA Occupational Health Guidelines for Chemical Hazards. DHHS (NIOSH) Pub Nos. 81-123.
- California Environmental Protection Agency "Chemical List of Lists," Sacramento CA (February 1997).

# Acifluorfen

*Use Type:* As sodium acifluorfen, a broad spectrum herbicide

# Acifluorfen

*CAS Number:* 50594-66-6; 62476-59-9 (sodium salt)
*Formula:* $C_{14}H_7ClF_3NO_5$; $F_3C-C_6H_3(Cl)-O-C_6H_3(NO_2)(COOH)$
*Alert:* A General Use Pesticide (GUP). Slated to be withdrawn by the European Commission in July 2003. Some exemptions may apply. Human toxicity (long-term): High
*Synonyms:* Acifluorfene; Benzoic acid, 5-(2-chloro-4-(trifluoromethyl)phenoxy)-2-nitro-; 5-(2-Chloro-4-(trifluoromethyl)phenoxy)-2-nitrobenzoic acid; 5-(2-Chloro-α-α-α-trifluoro-p-tolyloxy)-2-nitrobenzoic acid; Sodium acifluorfen
*Trade Names:* ASIF®, Sanonda Zhengzhou Pesticide (China); BLAZER®, BASF Canada (Canada); CARBOFUORFEN®; KLEENUP® Grass and Weed Killer, Bonide Products (USA); KLEERAWAY® Grass & Weed Killer, Solaris Group of Monsanto (USA); GALAXY®, I. Schneid Co. (canceled); RH-6201; TACKLE®, Rhone-Poulenc (France) (discontinued); SCEPTER O.T. HERBICIDE®, BASF Agriculture Products (Germany); STATUS®, BASF Agriculture Products (Germany); STORM® BASF Agriculture Products (Germany); ULTRA BLAZER®, BASF Agriculture Products (Germany)
*Producers:* Agrimor International (USA); BASF Agriculture Products (Germany); BASF Canada (Canada); Bonide Products, Inc. (USA); Monsanto Co. (USA); Sanonda Zhengzhou Pesticide (China)
*Chemical Class:* Diphenolic ether (diphenyl ether); an organofluorine pesticide
*EPA/OPP PC Code:* 114401; sodium salt: 114402, 209800
*California DPR Chemical Code:* 2218 as sodium salt
*RTECS Number:* DG5643200
*EINECS Number:* 256-634-5
*Uses:* Used to control pre-emergent and post-emergent broadleaf weeds and grasses in soybean, peanut, pea and rice crops. It is also registered for use by homeowners as a spot treatment on driveways, sidewalks, and patios. It should not be mixed with oils, surfactants, liquid fertilizers, and other pesticides.
*U.S. Maximum Allowable Residue Levels for Acifluorfen (40 CFR 180.383):*

| CROP | ppm |
|---|---|
| Cattle, kidney | 0.02 |
| Cattle, liver | 0.02 |
| Egg | 0.02 |
| Goat, kidney | 0.02 |
| Goat, liver | 0.02 |
| Hog, kidney | 0.02 |
| Hog, liver | 0.02 |
| Horse, kidney | 0.02 |
| Horse, liver | 0.02 |
| Milk | 0.02 |
| Peanut | 0.1 |
| Poultry, fat | 0.02 |
| Poultry, meat | 0.02 |
| Poultry, mbyp | 0.02 |
| Rice, grain | 0.1 |
| Rice, straw | 0.1 |
| Sheep, kidney | 0.02 |
| Sheep, liver | 0.02 |
| Soybean | 0.1 |
| Strawberry | 0.05 |

*Carcinogen/Hazard Classifications*
**U.S. EPA Carcinogens:** Group B2 as sodium salt; probable human carcinogen (possible human; animal positive)
**California Prop. 65:** Carcinogen
**Label Signal Word:** DANGER
**WHO Acute Hazard:** Class III, slightly hazardous
*Regulatory Authority:*
- FIFRA 40 CFR 180.383
- FIFRA 40 CFR 372
- AB 1803-Well Monitoring Chemical (CAL) [62476-59-9]
- AB 2588-Air Toxics "Hot Spots" Chemicals (CAL) [62476-59-9]
- Proposition 65 chemical (CAL) [62476-59-9]
- Actively registered pesticide in California. [62476-59-9]
- SARA 313: Form R *de minimis* Concentration Reporting Level: 1.0% (Sodium salt)
- US Code 40 CFR 455.65

*Description:* Acifluorfen is a combustible, off-white to light tan solid. The sodium salt is a white or brown crystalline powder. Molecular weight = 361.65. Melting/Freezing point = 152–157°C[23]; 124–126°C (sodium salt). The sodium salt is soluble in water.
*Incompatibilities:* Strong oxidizers. Avoid contact with all sources of ignition
*Permissible Exposure Limits in Air:* Not established.
*Permissible Concentration in Water:* A no-adverse-effect level (NOAEL) has been determined to be 20 mg/kg. Body weight/day based on fetotoxicity. However a NOAEL of 5.6 was determined based on increase in liver size of male rats; further a NOAEL of 1.25 mg/kg/day was determined in a 2-generation rat reproduction study. On this last basis, a long term health advisory of acifluorfen has been set at 0.44 mg/L for a 70-kg adult. A lifetime health advisory for that same adult of 0.009 mg/L. The U.S. EPA has also determined a reference dose (acceptable daily intake) of 0.013 mg/kg/day.
*Determination in Water:* Analysis of acifluorfen is by a gas chromatographic (GC) method applicable to the determination of certain chlorinated acid pesticides in water samples. In this method, approximately 1 liter of sample is acidified. The compounds are extracted with ethyl ether using a separatory funnel. The derivatives are hydrolyzed with potassium hydroxide, and extraneous organic material is removed by a solvent wash. After acidification, the acids are extracted and converted to their methyl esters using diazomethane as the derivatizing agent. Excess reagent is removed, and the esters are determined by electron capture GC. The method detection limit has not been determined for this compound, but it is estimated that the detection limits for analytes included in this method are in the range of 0.5 to 2 µg/L.
*Routes of Entry:* Ingestion
*Harmful Effects and Symptoms*
*Short Term Exposure:* Acifluorfen is a moderate dermal and eye irritant.
*Long Term Exposure:* A known animal carcinogen. Similar chlorinated diphenyl ethers have caused liver damage in

laboratory animals. The acute oral LD$_{50}$ for male rats is 2025 mg/kg; for female rats is 1370 mg/kg.
*Points of Attack:* Skin and liver.
*Medical Surveillance:* Liver function tests.
*First Aid:* If this chemical gets into the eyes, remove any contact lenses at once and irrigate immediately for at least 15 minutes, occasionally lifting upper and lower lids. Seek medical attention immediately. If this chemical contacts the skin, remove contaminated clothing and wash immediately with soap and water. Seek medical attention immediately. If this chemical has been inhaled, remove from exposure, begin rescue breathing (using universal precautions) if breathing has stopped and CPR if heart action has stopped. Transfer promptly to a medical facility. When this chemical has been swallowed, get medical attention. Give large quantities of water and induce vomiting. Do not make an unconscious person vomit. Do not induce vomiting when formulations containing petroleum solvents are ingested.
*References:*
- EPA, "Health Advisory: Acifluorfen," Washington DC, Office of Drinking Water (August 1987).
- "Sodium Acifluorfen Usage, Benefits, and Alternatives," EPA Office of Prevention, Pesticides and Toxic Substances, letter, December 6, 2001, www.epa.gov/oppsrrd1/reregistration/acifluorfen
- EPA, "Tolerance Reassessment & Registration", 40 CFR 180.383; FR April 12, 2002, www.epa.gov/oppsrrd1/reregistration/acifluorfe
- California Environmental Protection Agency Chemical List of Lists, Sacramento CA (February 1997)

# Acrolein

*Use Type:* Algaecide and rodenticide.
*CAS Number:* 107-02-8
*Formula:* $C_3H_4O$; $CH_2CHCHO$
*Alert:* A Restrictetd Use Pesticide (RUP). Exposure occurs mostly from breathing it in the air, from cigarette smoke, vehicle exhaust, and the burning of plants and trees.
*Synonyms:* Acrehyde; Acroleina (Italian); Acroleine (Dutch, French); Acrylehyd (German); Acrylehyde; Acrylic aldehyde; Acrylaldehyde; Acrylic aldehyde; Akrolein (Czech); Akroleina (Polish); Allylaldehyde; Aldehyde acrylique (French); Aldeide acrilica (Italian); Aqualine; Biocide; Ethylene aldehyde; NSC 8819; Propenal; 2-Propenal; Prop-2-en-1-al; 2-Propen-1-one; Propylene aldehyde; Slimicide
*Trade Names:* Aqualin®, Baker Petrolite (USA); Acquinite®; Crolean®; Magnacide® H, Baker Petrolite (USA)
*Producers:* Advanced Synthesis Technologies (USA); Air Products & Chemicals (USA); ATOFINA (France); Baker Petrolite (USA); BASF (Germany); Celanese (Germany); Ciba (Switzerland); Creanova (USA); Daicel Chemical Industries (Japan); Degussa (Germany); DSM (Netherlands); Goldschmidt (Germany); Nippon Kayaku Co. (Japan); Sumitomo Chemical (Japan); Union Carbide (USA)

*Chemical Class:* Aldehyde; Organics, non-halogenated
*EPA/OPP PC Code:* 000701
*California DPR Chemical Code:* 3
*ICSC Number:* 0090
*RTECS Number:* AS1050000
*EEC Number:* 605-008-00-3
*EINECS Number:* 203-453-4
*Uses:* Acrolein is produced by oxidation of propylene. Acrolein is principally used as a biocide to control plants, algae, molluscs, fungi, rodents, and microorganisms. Acrolein has also been used in the manufacture of other chemicals, plastics, and drugs; as a warning agent in gases, as a test gas for gas masks, in military poison gases, in the manufacture of colloidal metals, in leather tanning, and as a fixative in histology. Acrolein is primarily used as an intermediate in the production of glycerine and in the production of methionine analogs (poultry feed protein supplements). It is also used in chemical synthesis (1,3,6-hexametriol and glutaraldehyde), as a liquid fuel, antimicrobial agent, and as a slimicide in paper manufacture.
*Human toxicity (long-term)*[77]*:* Intermediate–14.00 ppb, Health Advisory
*Fish toxicity (threshold)*[77]*:* High–1.89088 ppb, MATC (Maximum Acceptable Toxicant Concentration)
*U.S. Maximum Allowable Residue Levels for Acrolein (40 CFR 180.1156)*

| CROP | ppm |
|---|---|
| Raw Agricultural Commodities | — |

*Carcinogen/Hazard Classifications*
**IARC:** Not classifiable as to human carcinogenicity
*Note:* The Department of Health and Human Services has determined that acrolein may possibly be a human carcinogen. Testing has not been completed by NIOSH to determine the carcinogenicity of acrolein and related low-molecular-weight-aldehydes. However, the limited studies to date indicate that these substances have chemical reactivity and mutagenicity similar to acetaldehyde and malonaldehyde. Therefore, NIOSH recommends that careful consideration should be given to reducing exposures to acrolein.
*Regulatory Authority:*
- Toxic Substance (World Bank)[15]
- Actively registered pesticide in California.
- AB 1803-Well Monitoring Chemical (CAL)
- AB 2588-Air Toxics "Hot Spots" Chemicals (CAL)
- CAL Air Resources Board/AB 1807 Toxic Air Contaminants
- Permissible Exposure Limits for Chemical Contaminants (CAL/OSHA)
- The "Director's List" (CAL/OSHA)
- U.S. DOT Inhalation Hazard Chemicals
- Air Pollutant Standard Set (ACGIH)[1] (DFG)[3] (HSE)[33] (Former USSR)[43] (OSHA)[58] (Various States)[60] (Various Canadian Provinces), Mexico, Israel, Australia
- OSHA 29CFR1910.119, Appendix A, Process Safety List of Highly Hazardous Chemicals, TQ = 150 lb
- Clean Air Act 42USC7412; Title I, Part A,§112 hazardous pollutants; Part A,§112(r), Accidental Release

Prevention/Flammable substances (Section 68.130) TQ = 5,000 lb (1,275 kg)
- Clean Water Act: 40CFR116.4 Hazardous Substances; RQ 40CFR117.3 (same as CERCLA); 40CFR423, Appendix A Priority Pollutants; 40CFR401.15 Toxic Pollutant
- EPA Hazardous Waste Number (RCRA No.): P003
- RCRA 40CFR261, Appendix 8; 40CFR261.11 Hazardous Constituents
- RCRA 40CFR§268.48; 61FR15654, Universal Treatment Standards: Wastewater (mg/L), 0.29; Nonwastewater, N/A.
- RCRA 40CFR264, Appendix 9; Ground Water Monitoring List Suggested methods (PQL $ug$/L): 8030(5); 8240(5)
- CERCLA/SARA Section 302, Extremely Hazardous Substances: TPQ = 500 lb (228 kg)
- Superfund/EPCRA 40CFR302.4, Appendix A, RQ: 1lb (0.454 kg), SARA 313: Form R *de minimis* Concentration Reporting Level: 1.0%.
- U.S. DOT Regulated Marine Pollutant (49CFR172.101, Appendix B)
- Canada, WHMIS, Ingredients Disclosure List
- Mexico: Drinking Water 0.3 mg/L (ecological criteria); Listed as an organic toxic pollutant in wasterwater.

***Description:*** At room temperature, acrolein is a clear, colorless to straw-colored liquid. It has a pungent, suffocating odor at 0.16 ppm and causes tears. Suffocating, pungent odor at 0.16 ppm. Soluble in water; solubility = 208 g/L @ 20 °C. Molecular weight = 56.06. Boiling point = 52.5°C. Melting/Freezing point = –87.7°C. Vapor pressure = 210 mmHg @ 20 °C. Flash point = –26°C (cc). Hazard Identification (based on NFPA-704 M Rating System): Health 4, Flammability 3, Reactivity 3. Explosive limits: LEL = 2.8%, UEL = 31.0%. Autoignition temperature (unstable) = 220°C. Odor threshold = 0.174 ppm. Soluble in water. The vapor is heavier than air and may travel along the ground. Distant ignition is possible. Log $K_{ow}$ = 0.89. Unlikely to bioaccumulate in marine organisms.

***Incompatibilities:*** Acrolein should be stored in a cool, dry, well-ventilated area in tightly sealed containers separated from alkaline materials such as caustics, ammonia, organic amines, or mineral acids, strong oxidizers, and oxygen. Forms explosive mixture with air. Elevated temperatures or sunlight may cause explosive polymerization. A strong reducing agent; reacts violently with oxidizers. Reacts with acids, alkalis, ammonia, organic amines, oxygen, peroxides, sulfur dioxide, thiourea and metal salts. Shock-sensitive peroxides or acids may be formed over time. Attacks zinc and cadmium.

***Permissible Exposure Limits in Air:*** The Federal OSHA[2] standard[58] the ACGIH[1] and NIOSH[2] recommendations for exposure to acrolein is 0.1 ppm (0.25 mg/m³)as an 8-hr TWA concentration. The ACGIH[1], NIOSH[2], and HSE[33] (U.K.) 15-minute TWA as STEL value is 0.3 ppm (0.8 mg/m³). The NIOSH[2] IDLH = 2 ppm. The Australian, Mexican, Israeli and Canadian Provincial (Alberta, British Columbia, Ontario Quebec) TWA and STEL values are the same as ACGIH[1] and NIOSH[2] and Israel has a Action Level of 0.05 ppm (0.115 mg/m³). The DFG (German)[3] TWA value is also the same, and Peak Limitation (5 min) is 2 times the normal MAK; do not exceed more than 8 times during a workshift. AIHA ERPG-2 (maximum airborne concentration below which it is believed that nearly all persons could be exposed for up to 1 hour without experiencing or developing irreversible or other serious health effects or symptoms that could impair their abilities to take protective action) = 0.5 ppm. The former USSR-UNEP/IRPTC project[43] has set a value in workplace air of 0.2 mg/m³ (1993) and of 0.03 mg/m³ for ambient air in residential areas on either a momentary or an average daily basis. In addition, a number of states have set guidelines or standards for acrolein in ambient air[60] ranging from 0.83 $\mu$g/m³ (N.Y.) to 1.25 $\mu$g/m³ (South Carolina) to 2.5 $\mu$g/m³ (Florida, North Dakota) to 4 $\mu$g/m³ (Virginia) to 5 $\mu$g/m³ (Connecticut) to 6.9 $\mu$g/m³ (Nevada) to 80.0 $\mu$g/m³ (North Carolina).

***Determination in Air:*** See NIOSH Method #2501[18] and OSHA Method #52[58]. See website http://www.osha-slc.gov/ (and applicable method, e.g., OSHA Method #21).[58]

***Permissible Concentration in Water:*** To protect freshwater aquatic life-on an acute basis 68 $\mu$g/L and on a chronic basis 21 $\mu$g/L. To protect saltwater aquatic life-55 $\mu$g/L on an acute toxicity basis. To protect human health-320 $\mu$g/L[6]. In addition, two states have set guidelines for acrolein in drinking water[61]. These are both 320 $\mu$g/L as set by Arizona and Kansas.

***Determination in Water:*** Gas chromatography (EPA Method #603) or gas chromatography and mass spectrometry (EPA Method #624).

***Routes of Entry:*** Acrolein is toxic by all exposure routes: Inhalation, ingestion, skin and/or eye contact. Absorbed through the skin. Systemic effects may occur after exposure by any route.

*Harmful Effects and Symptoms*

***Short Term Exposure:*** *Inhalation* Inhaled acrolein is highly toxic. Acrolein is irritating to the upper respiratory tract even at low concentrations. Its odor threshold of 0.16 ppm is similar to the OSHA permissible exposure limit (0.1 ppm); thus odor may provide an adequate warning of potentially hazardous concentrations. Acrolein vapor is heavier than air, but asphyxiation in enclosed, poorly ventilated, or low-lying areas is unlikely due to its strong odor. Children exposed to the same levels of acrolein vapor as adults may receive a larger dose because they have greater lung surface area:body weight ratios and higher minute volumes:weight ratios. In addition, they may be exposed to higher levels than adults in the same location because of their short stature and the higher levels of acrolein vapor found nearer to the ground. *Skin/Eye Contact* Direct contact with liquid acrolein causes rapid and severe eye and skin irritation or burns. Exposure to vapor produces inflammation of mucous membranes and it is a potent lacrimator. Because of their relatively larger surface area:body weight ratio, children are more vulnerable to toxicants affecting the skin. *Ingestion* Acrolein produces chemical burns of the lips, mouth, throat, esophagus, and stomach.

Nausea, vomiting, and diarrhea also occur.

Extremely toxic; probable oral human lethal dose is 5-50 mg/kg, between 7 drops and one teaspoon for a 70 kg (150 lb) person.

***Long Term Exposure:*** This chemical is a metabolite of cyclophosphamide, a well-recognized animal teratogen. Acrolein may cause mutations. Such chemicals have a cancer risk. Long-term exposure can cause drying and cracking of the skin. High or repeated lower exposure may cause permanent lung damage. NIOSH testing has not been completed to determine the carcinogenicity of acrolein. However, the limited studies to date indicate that this substance has chemical reactivity and mutagenicity similar to acetaldehyde and malonaldehyde. Therefore, NIOSH recommends that careful consideration should be given to reducing exposures to this related aldehyde. *Reproductive and Developmental Effects* No studies were located that address reproductive or developmental effects of acrolein in humans. Acrolein caused developmental effects when injected into rats, but did not cause developmental effects when ingested by rabbits. No information was found as to whether acrolein crosses the placenta, but it has been measured in breast milk.

***Points of Attack:*** Heart, lungs, eyes, skin and respiratory system.

***Medical Surveillance:*** Medical observation is recommended for 24 to 48 hours after breathing overexposure, as pulmonary edema may be delayed. Preplacement and periodic medical examinations should consider respiratory, skin, and eye disease. For those with frequent or potentially high exposure, lung function tests are recommended before beginning work and at regular times after that. If symptoms develop or overexposure is suspected, consider chest x-ray.

***First Aid:*** Victims exposed only to acrolein vapor do not pose contamination risks to rescuers. Victims whose clothing or skin is contaminated with liquid acrolein can secondarily contaminate response personnel by direct contact or by off-gassing vapor. If this chemical gets into the eyes, remove any contact lenses at once and irrigate immediately for at least 15 minutes, occasionally lifting upper and lower lids. If this chemical contacts the skin, remove contaminated clothing and wash immediately with soap and water. If this chemical has been inhaled, remove from exposure, begin rescue breathing (using universal precautions) if breathing has stopped and CPR if heart action has stopped. Transfer promptly to a medical facility. In cases of ingestion, **do not induce vomiting**. If the victim is alert, asymptomatic, and has a gag reflex, administer a slurry of activated charcoal at a dose of 1 g/kg (infant, child, and adult dose). A soda can and a straw may be of assistance when offering charcoal to a child. Victims who are conscious and able to swallow should be given 4 to 8 ounces of milk or water (not to exceed 15 mL/kg in a child). If the victim is symptomatic, delay decontamination until other emergency measures have been instituted. children at the exposure site. Provide reassurance to the child during decontamination, especially if separation from a parent occurs. Do not make an unconscious person vomit. Medical observation is recommended for 24 to 48 hours after breathing overexposure, as pulmonary edema may be delayed.

*Note to physician or authorized medical personnel.* In cases of respiratory compromise, secure airway and respiration via endotracheal intubation. If not possible, perform cricothyrotomy if equipped and trained to do so. Treat patients who have bronchospasm with an aerosolized bronchodilator such as albuterol. Consider that acrolein inhalation may cause hypertension and tachycardia, in which case the use of bronchodilators that are known cardiac sensitizing agents may pose enhanced risk. Administer corticosteroids as indicated to patients who have persistent wheezing or hypersensitivity pneumonitis. Consider racemic epinephrine aerosol for children who develop stridor. Dose 0.25–0.75 mL of 2.25% racemic epinephrine solution; repeat every 20 minutes as needed, cautioning for myocardial variability. Patients who are comatose, hypotensive, or having seizures or cardiac arrhythmias should be treated according to advanced life support protocols. If evidence of shock or hypotension is observed, begin fluid administration. For adults with systolic pressure less than 80 mmHg, bolus perfusion of 1000 mL/hour intravenous saline or lactated Ringer's solution may be appropriate. Higher adult systolic pressures may necessitate lower perfusion rates. For children with compromised perfusion, administer a 20 mL/kg bolus of normal saline over 10 to 20 minutes, then infuse at 2 to 3 mL/kg/hour.

***References:***
- U.S. Environmental Protection Agency, "Pesticide Residue Limits, Acrolein," 40 CFR 180.1156. www.epa.gov/cgi-bin/oppsrch
- U.S. Environmental Protection Agency, Chemical Hazard Information Profile: Acrolein, Washington DC (March 10, 1978).
- U.S. Environmental Protection Agency, Acrolein: Ambient Water Quality Criteria, Washington DC (1980).
- National Institute.for Occupational Safety and Health (NIOSH), Information Profiles on Potential Occupational Hazards-Single Chemicals: Acrolein, Report TR 79-607, Rockville, MD, pp 1-18 (December 1979).
- U.S. Environmental Protection Agency, "Acrolein, Health and Environmental Effects Profile No. 3," Washington, DC, Office of Solid Waste (April 30, 1980).
- Sax, N.I., Ed., Dangerous Properties of Industrial Materials Report, 1, No. 4, 28-31 (1981) and 3, No. 3, 36-41 (1983).
- Agency for Toxic Stances and Disease Registry, Center for Disease Control, USDHH, "Toxicology Profile for Acrolein," Atlanta GA (1990). http://www.atsdr.cdc.gov/tfacts124.html
- U.S. Environmental Protection Agency, "Chemical Profile: Acrolein," Washington, DC, Chemical Emergency Preparedness Program (Nov. 30, 1987).
- NIOSH, U.S. Department of Health and Human Services, Public Health Service, "NIOSH Current Intelligence Bulletin 55: Carcinogenicity of Acetaldehyde and Malonaldehyde, and Mutagenicity of Related Low-

Molecular-Weight Aldehydes, " NIOSH Publication No. 91-112, Cincinnati, OH (1991).
- ATSDR, U.S. Department of Health and Human Services, Public Health Service, "Managing Hazardous Materials Incidents," Atlanta, GA (June 2003).
- California Environmental Protection Agency Chemical List of Lists, Sacramento CA (February 1997).
- Agency for Toxic Substances and Disease Registry, U.S. Department of Health and Human Services, Public Health Service, "Managing Hazardous Materials Incidents," Atlanta, GA (June 2003)
- New Jersey Department of Health and Senior Services, "Hazardous Substance Fact Sheet: Acrolein," Trenton, NJ (May 1998).
http://www.state.nj.us/health/eoh/rtkweb/0021.pdf

## Acrolein Diacetate

*Use Type:* Herbicide
*CAS Number:* 869-29-4
*Formula:* $C_7H_{10}O_4$
*Synonyms:* Allylidene diacetate; Diacetoxypropene; 1,1-Diacetoxy-2-propene; 1,1-Diacetoxypropene-2; 3,3-Diacetoxypropene; 2-Propene-1,1-dioldiacetate
*Trade Names:* MAGNACIDE H®, Baker Petrolite (USA); SD-345®, Shell Chemicals (UK); SHELL 345®, Shell Chemicals (UK); SHELL SD 345®, Shell Chemicals (UK)
*Producers:* ABCR (Germany); Advanced Synthesis Technologies (USA); Air Products & Chemicals (USA); ATOFINA (France); Baker Petrolite (USA); BASF (Germany); Celanese (Germany); Ciba (Switzerland); Creanova (USA); Daicel Chemical Industries (Japan); Degussa (Germany); DSM (Netherlands); Goldschmidt (Germany); Lancaster Synthesis (UK); Nippon Kayaku Co. (Japan); Shell Chemicals (UK); Sumitomo Chemical (Japan); Union Carbide (USA)
*Chemical Class:* Organics, non-halogenated
*EPA/OPP PC Code:* 068402
*California DPR Chemical Code:* 196
*RTECS Number:* UC9625000
*EINECS Number:* 212-789-0
*Uses:* Not registered in the U.S.
*Carcinogen/Hazard Classifications*
*U.S. EPA Carcinogens:* Group C, possible carcinogen, as acrolein
*IARC:* Group 3, Unclassifiable, as acrolein
*WHO Acute Hazard:* Class 1B, highly hazardous, as acrolein
*Description:* Flammable liquid. Molecular weight = 158.17. Melting/Freezing point = –36.6°C. Boiling point = 107°C @ 50 mmHg, Flash point = 180°F (open cup). Density = 1.0749 @ 20°F/20°C.
*Incompatibilities:* Forms explosive mixture with air. A strong reducing agent; reacts violently with oxidizers. Reacts with acids, alkalis, ammonia, amines, oxygen, peroxides. Shock-sensitive peroxides or acids may be formed over time. Acetates are generally incompatible with nitrates. Moisture may cause hydrolysis or other forms of decomposition
*Permissible Concentration in Water:* The U.S. EPA recommends that levels in lakes and streams should be limited to 0.32 parts of acrolein per million parts of water (0.32 ppm) to prevent possible health effects from drinking water or eating fish contaminated with acrolein. Runoff from spills or fire control may cause water pollution.
*Routes of Entry:* Inhalation, ingestion, skin and/or eye contact. Absorbed through the skin.
*Harmful Effects and Symptoms*
*Short Term Exposure:* Poisonous. Severe irritant. This chemical can be absorbed through the skin, thereby increasing exposure. Eye and skin contact may cause intense tearing, irritation, blisters, and burns. Inhalation can irritate the lungs causing irritation, coughing, wheezing, and/or shortness of breath. Higher exposures can cause pulmonary edema, a medical emergency that can be delayed for several hours. This can cause death.
*Long Term Exposure:* Aldehydes such as acrolein may cause mutations. Such chemicals have a cancer risk. Long-term exposure can cause drying and cracking of the skin. High or repeated lower exposure may cause permanent lung damage.
*Points of Attack:* Heart, lungs, eyes, skin, respiratory system.
*Medical Surveillance:* Periodic medical examinations should consider respiratory, skin, and eye disease. For those with frequent or potentially high exposure, lung function tests are recommended. If symptoms develop or overexposure is suspected, consider chest x-ray.
*First Aid:* See also acrolein. If this chemical gets into the eyes, remove any contact lenses at once and irrigate immediately for at least 15 minutes, occasionally lifting upper and lower lids. Seek medical attention immediately. If this chemical contacts the skin, remove contaminated clothing and wash immediately with soap and water. Seek medical attention immediately. If this chemical has been inhaled, remove from exposure, begin rescue breathing (using universal precautions) if breathing has stopped and CPR if heart action has stopped. Transfer promptly to a medical facility. When this chemical has been swallowed, get medical attention. *Do not induce vomiting when formulations containing petroleum solvents are ingested.* Otherwise, give large quantities of water and induce vomiting. Do not make an unconscious person vomit.
*References:*
- California Environmental Protection Agency Chemical List of Lists, Sacramento CA (February 1997).
- U.S. Department of Health and Human Services, Public Health Service/Agency for Toxic Substances and Disease Registry / Division of Toxicology "ToxFAQs" Atlanta, GA (June 2003).

## Acrylamide

*Use Type:* Soil-conditioning agent
*CAS Number:* 79-06-1

# Acrylamide

*Formula:* $C_3H_5NO$; $CH_2CHCONH_2$

*Alert:* Acrylamide is a carcinogenic breakdown product from cooking. It should be handled with extreme caution. Although it is not used as an agricultural pesticide, its potential residual presence in prepared food is of importance to food chemists.

*Synonyms:* AAM; Acrilamida (Spanish); Acrylamide, 30%; Acrylamide, 50%; Acrylamide monomer; Acrylic acid amide (50%); Acrylic acid amide, (50%); Acrylic amide; Acrylic amide 30%; Acrylic amide 50%; Akrylamid (Czech); Ethylenecarboxamide; Ethylene monoclinic tablets carboxamide; Propenamide; 2-Propenamide Propenamide; Vinyl amide

*Trade Names:* ACRYLAGEL®; AMERESCO ACRYL-40®; OPTIMUM®

*Producers:* Cyanamid BV (Netherlands), now part of Degussa

*Chemical Class:* Organics, non-halogenated

*EPA/OPP PC Code:* 600008

*California DPR Chemical Code:* 02111 as polyacrylamide polymer

*ICSC Number:* 0091

*RTECS Number:* AS3325000

*EEC Number:* 616-003-00-0

*EINECS Number:* 201-173-7

*Uses:* Not actively registered as a pesticide in the U.S. The major application for monomeric acrylamide is in the production of polymers as polyacrylamides. Polyacrylamides are used for soil stabilization, gel chromatography, electrophoresis, papermaking strengtheners, clarifications and treatment of potable water, sewage treatment, and foods.

*Carcinogen/Hazard Classifications*

U.S. EPA Carcinogens: Group B2, probable carcinogen
U.S. NTP Carcinogen: Reasonably anticipated carcinogen
California Prop. 65: Carcinogen
IARC: Group 2A, probable carcinogen

*Regulatory Authority:*
- Carcinogen (Human Suspected) (ACGIH)[1] (IARC) (DRG)[3]
- Air Pollutant Standard Set (ACGIH)[1] (HSE)[33] (OSHA)[58] (Various States)[60] (Australia) (Various Canadian Provinces) (Israel) (Mexico)
- Water Pollution Standard Proposed (U.S. EPA)[48] (Minnesota)[61]
- Clean Air Act: Hazardous Air Pollutants, 42USC7412; Title I, Part A,§112.
- EPA/SARA 302 (EPCRA) Extremely hazardous substances
- AB 2588-Air Toxics "Hot Spots" Chemicals (CAL)
- CAL Air Resources Board/AB 1807 Toxic Air Contaminants
- Proposition 65 chemical (CAL)
- Permissible Exposure Limits for Chemical Contaminants (CAL/OSHA)
- The "Director's List" (CAL/OSHA)
- EPA Hazardous Waste Number (RCRA No.): U007
- RCRA 40CFR261, Appendix 8; 40CFR261.11 Hazardous Constituents
- RCRA 40CFR268.48; 61FR15654, Universal Treatment Standards: Wastewater (mg/L), 19; Nonwastewater (mgkg), 23
- Safe Drinking Water Act, MCL, treatment technique; MCLG, zero; Regulated Chemical (47FR 9352)
- EPCRA 40CFR302, Extremely Hazardous Substances: TPQ = 1000/10,000 lb (454/4,540 kg)
- EPCRA 40CFR302.4, Appendix A, RQ: 5,000 lb (2,270 kg)
- EPCRA Section 313 Form R *de minimis* Concentration Reporting Level: 0.1%
- TSCA: 716.120 (*a*), listed chemical
- U.S. DOT Regulated Marine Pollutant (49CFR172.101, Appendix B)
- Canada, WHMIS, Ingredients Disclosure List. Concentration Reporting Level: 0.1%

*Description:* Acrylamide in monomeric form is an odorless, flakelike crystals. Soluble in water. Molecular weight = 71.13. Melting/Freezing point = 84.5°C. Flash point = 138°C. Hazard Identification (based on NFPA-704 M Rating System): Health 2, Flammability 2, Reactivity 2. Autoignition temperature = 240°C. It is shipped as clear, colorless to pale-yellow 30% or 50% by weight aqueous solution, or white crystalline solid or pellets. Log $K_{ow}$ = –1.65 to –0.67. Unlikely to bioaccumulate in marine organisms.

*Incompatibilities:* Thermally unstable. Unless inhibited (with antioxidant like hydroquinine), ultraviolet light, oxidizers, peroxides, vinyl polymerization initiators or temperatures above melting point (85°C) can cause explosive polymerization. Reacts violently with reducing agents, peroxides, mineral acids, strong acids, oleum, ammonia and isocyanates. Finely divided particles form explosive mixture with air.

*Permissible Exposure Limits in Air:* The Federal OSHA[2] standard is 0.3 mg/m$^3$ as a time-weighted average (TWA) concentration for up to a 10-hour workshift. NIOSH[2] and ACGIH[1] has a TWA of 0.03 mg/m$^3$[58]. The notation "skin" indicated possibile cutaneous absorption. The NIOSH[2] IDLH, (potential occupational carcinogen)= 60 mg/m$^3$. HSE[33] (U.K.), Australia, Israel, and Mexico TWA is 0.3 mg/m$^3$ and Israel Action Limit is 0.015 mg/m$^3$ and Mexico STEL is 2.6 mg/m$^3$. Canadian Provincial TWAs are: Alberta and British Columbia: 0.3 mg/m$^3$ and STEL of 0.6 mg/m$^3$; Ontario and Quebec TWAEVs are 0.03 mg/m$^3$. California's PEL is 0.3 mg/m$^3$ TWA. In addition. Several states have set guidelines or standards for acrylamide ambient concentrations in air[60]: 0.3 $\mu$g/m$^3$ (South Carolina) to 1.0 $\mu$g/m$^3$ (New York) to 3.0 $\mu$g/m$^3$ (South Dakota) to 5.0 $\mu$g/m$^3$ (Virginia) to 6.0 $\mu$g/m$^3$ (Connecticut) to 7.0 $\mu$g/m$^3$ (Nevada).

*Determination in Air:* Filter/Si gel; Methanol; Gas chromatography/Nitrogen/phosphorus detection; OSHA (#21)

*Permissible Concentration in Water:* Health advisories have been developed by EPA[48] on a long term (7-year) basis as 0.02 mg/L for a 10 kg child and 0.07 mg/L for a 70 mg adult. A guideline for acrylamide in drinking water of 0.10 $\mu$g/L has been developed by the State of Minnesota[61].

*Determination in Water:* There is no standardized method for the determination of acrylamide in drinking water. An analytical procedure for the determination of acrylamide has been reported in the literature. This procedure consists of bromination, extraction of the brominated product from water with ethyl acetate and quantification using high performance liquid chromatography (HPLC) with an ultraviolet detector. The concentration of the ethyl acetate to dryness and dissolution in a small volume of distilled water prior to HPLC analysis allows the detection of acrylamide at concentrations of 0.2 $\mu g/L$[48].

*Routes of Entry:* Eyes, skin, central and peripheral nervous systems, reproductive system. Acrylamide can be absorbed through unbroken skin.

*Harmful Effects and Symptoms*

*Short Term Exposure:* Irritates the eyes, skin and respiratory tract. Symptoms of Exposure include complaints of drowsiness, fatigue, tingling of fingers, and a stumbling, propulsive type of walking with sense of unsteadiness have been reported. Motor and sensory impairment, numbness, tremor, abnormal feelings in the lower limbs accompanied by weakness, and speech disturbances were also reported. Classified as very toxic; probably oral lethal human dose is between 50 and 500 mg/kg or between 1 teaspoon and 1 ounce for a 150 lb. person. Polymerized acrylamide may not be toxic, but the monomer can cause peripheral nerve damage.

*Long Term Exposure:* There is evidence that acrylamide causes cancer in animals. It may cause skin and lung cancer in humans. There is limited evidence that this chemical damages the male testes. Can cause damage to the central nervous system, causing numbness, and weakness of the hands and feet. Acrylamide is a cumulative neurotoxin and repeated exposure to small amounts may cause serious injury to the nervous system. The neurological effects may be delayed. Polymer inhibitors or stabilizers added to the monomer may also produce toxicity. The symptoms of acrylamide toxicity are consistent with mid-brain lesions and blocked transport along both motor and sensory axons.

*Points of Attack:* Central nervous system, peripheral nervous system, skin and eyes.

*Medical Surveillance:* Since skin contact with the substance may result in localized or systemic effects, NIOSH recommends that medical surveillance be made available to all employees working in an area where acrylamide is stored, produced, processed, or otherwise used, except as an unintentional contaminant in other materials at a concentration of less than 1% by weight. For those with frequent or potentially high exposure, nerve condition tests should be considered The use of alcoholic beverages may enhance the harmful effects.

*First Aid:* If this chemical gets into the eyes, remove any contact lenses at once and irrigate immediately. If this chemical contacts the skin, flush with water immediately. If a person breathes in large amounts of this chemical, move the exposed person to fresh air at once and perform artificial respiration. When this chemical has been swallowed, get medical attention. Give large quantities of water and induce vomiting. Do not make an unconscious person vomit.

*References:*
- Cooperative Extension, Environmental Toxicology Newsletter, "Turning Up the Heat on Acrylamide," University of California at Davis, Davis, CA (January 2003). http://extoxnet.orst.edu/newsletters/ucd2003/nltrJan03.html
- New Jersey Department of Health and Senior Services, "Hazardous Substance Fact Sheet: Acrylamide," Trenton, NJ (April 1994, rev. December 1999). http://www.state.nj.us/health/eoh/rtkweb/0022.pdf
- FAO/WHO Acrylamide in Food Network, http://www.acrylamide-food.org/
- National Institute for Occupational Safety and Health, Criteria for a Recommended Standard: Occupational Exposure to Acrylamide, NIOSH Doc. No. 77-112, Washington, DC (1977).
- U.S. Environmental Protection Agency, Assessment of Testing Needs: Acrylamide, Report No. EPA-560/11-80-016, Washington, DC, Office of Toxic Substances (July 1980).
- Sax, N.I., Ed., Dangerous Properties of Industrial materials Report, 2, No. 4, 24-27 (1982).
- U.S. Environmental Protection Agency, "Chemical Profile: Acrylamide," Washington, DC, Chemical Emergency Preparedness Program (Nov. 30, 1987).
- New York State Department of Health, "Chemical Fact Sheet: Acrylamide," Albany, NY, Bureau of Toxic Substance Assessment (May 1986).
- LaDou, J., Nerve Conduction Studies, Occupational Medicine. Appleton and Lange. 1990
- California Environmental Protection Agency Chemical List of Lists, Sacramento CA (February 1997)

# Acrylonitrile

*Use Type:* Insecticide and fumigant
*CAS Number:* 107-13-1
*Formula:* $C_3H_3N_4$; $CH_2CHCN_4$
*Alert:* A Restricted Use Pesticide (RUP)
*Synonyms:* Acrilonitrilo (Spanish); Acrylnitril (Dutch, German); Acrylonitrile monomer; Akrylonitryl (Polish); AN; Carbacryl; Cianuro di vinile (Italian); Cyanoethylene; Cyanure de vinyle (French); ENT 54; NCI-C50215; Nitrile acrilico (Italian); Nitrile acrylique (French); Propenenitrile; 2-Propenenitrile; TL 314; VCN; Vinyl cyanide; Vinyl cyanide, propenenitrile

*Trade Names:* ACRITET® component of (with Carbon tetrachloride), canceled; ACRYLOFUME®, component of (with Carbon tetrachloride, Chloroform, and Chloropicrin); ACRYLON® component of (with Carbon tetrachloride); CARBACRYL®, component of (with Carbon tetrachloride); FUMIGRAIN®; MILLER'S FUMIGRAIN®; VENTOX® component of (with Carbon tetrachloride)

*Producers:* BASF (Germany); BP Chemicals (UK); China Petrochemical Development Corp. (Taiwan); Cytec (USA); DSM (Netherlands); DuPont (USA); EniChem (Italy); Huntsman (USA); Lukoil Oil Company (Russia); Mitsubishi

Chemical (Japan); Mitsui Chemicals (Japan); Petrobas Energia S.S. (Argentina); Showa Denko (Japan); Sigma-Aldrich Laborchemikalien (Germany); Sinopec Corporation (Singapore); Sterling Chemicals (USA); Sumitomo (Japan); Zeon (Japan)
*Chemical Class:* Organics, non-halogenated
*EPA/OPP PC Code:* 000601
*California DPR Chemical Code:* 3019
*ICSC Number:* 0092
*RTECS Number:* AT5250000
*EEC Number:* 608-00300-4
*EINECS Number:* 203-466-5
*Uses:* Acrylonitrile is used in the manufacture of synthetic fibers, polymers, acrylostyrene plastics, acrylonitrile-butadiene-styrene plastics, nitrile rubbers, chemicals, and adhesives. It is also used as a pesticide intermediate. In the past, this chemical was used as a room fumigant and pediculicide (an agent used to destroy lice).
*Carcinogen/Hazard Classifications*
**U.S. EPA Carcinogens:** Group B1, probable carcinogen
**U.S. NTP Carcinogen:** Reasonably anticipated
**California Prop. 65:** Carcinogen
**IARC:** Group 2B, possible carcinogen
**Label Signal Word:** DANGER
*Regulatory Authority:*
- Carcinogen (IARC)[12], (NTP)[9], (DFG)[3]
- CAL/OSHA Carcinogen User Register Chemical
- EPA/SARA 302 (EPCRA) Extremely hazardous substances
- AB 2588-Air Toxics "Hot Spots" Chemicals (CAL)
- CAL Air Resources Board/AB 1807 Toxic Air Contaminants
- Proposition 65 chemical (CAL)
- U.S. DOT Inhalation Hazard Chemicals
- Permissible Exposure Limits for Chemical Contaminants (CAL/OSHA)
- The "Director's List" (CAL/OSHA)
- OSHA, 29CFR1910 Specifically Regulated Chemicals (See CFR 1910.1045)
- Banned or Severely Restricted (Germany) (U.N.)[13]
- Toxic Substance (World Bank)[15]
- Air Pollutant Standard Set (ACGIH )[1] (HSE)[33] (UNEP)[43] (Several States and Canadian Provinces)[60] (Mexico)
- (Israel)(Australia)
- Clean Air Act, 42USC7412; Title I, Part A,§112 hazardous pollutants; Section 112[r], Accidental Release Prevention/Flammable Substances (Section 68.130), TQ = 20,000 lb (9,150 kg)
- Clean Water Act, 40CFR116.4 Hazardous Substances; RQ 40CFR117.3, (same as CERCLA)
- EPA Hazardous Waste Number (RCRA No.): U009
- RCRA 40CFR261, Appendix 8; 40CFR261.11 Hazardous Constituents
- RCRA Land Ban Waste Restrictions
- RCRA 40CFR§268.48; 61FR15654, Universal Treatment Standards: Wastewater (mg/L), 0.24; Nonwastewater, 84
- RCRA 40CFR264, Appendix 9; Ground Water
- Safe Drinking Water Act, 55FR1470 Priority List Monitoring List Suggested methods (PQL ug/L): 8030 (5); 8240 (5)
- CERCLA/SARA 40CFR302, Extremely Hazardous Substances: TPQ = 10,000 lb (4,540 kg).
- Superfund/EPCRA 40CFR302.4, Appendix A, RQ: 100 lb (45.5 kg), SARA 313: Form R *de minimis* Concentration Reporting Level: 0.1%.
- Canada, WHMIS, Ingredients Disclosure List. Concentration Reporting Level: 0.1%
- Mexico, Wastewater, organic pollutants

*Description:* Acrylonitrile, is a highly flammable, clear, colorless or light yellowish liquid. Inadequate; unpleasant onion or garlic odor at 17 ppm. Odor can only be detected above the PEL. Floats on water; solubility = 7% @ 20 °C. Molecular weight = 53.1. Boiling point = 77°C @ 760 mmHg. Melting/Freezing point = –82 °C. Specific gravity (water : 1) = 0.80. Gas density (air: 1) = 1.8. Vapor pressure = 83 mmHg @ 20 °C. Flash point = –1°C (cc). Hazard Identification (based on NFPA-704 M Rating System): Health 4, Flammability 3, Reactivity 2. Explosive limits: LEL= 3%, UEL = 17%. Log $K_{ow}$ = 0.28. Unlikely to bioaccumulate in marine organisms.

*Incompatibilities:* Forms explosive mixture with air. Reacts violently with strong acids, strong alkalies, bromine, and tetrahydrocarbazole. Copper, copper alloys, ammonia and amines may cause breakdown to poisonous products. Unless inhibited (usually with methylhydroquinone) acrylonitrile may polymerize spontaneously. It may also polymerize on contact with oxygen, heat, strong light, peroxides, and concentrated or heated alkalies. Reacts with oxidizers, acids, bromine, amines. Attacks copper and copper alloys. Attacks aluminum in high concentrations. Heat and flame may cause release of poisonous cyanide gas and nitrogen oxides.

*Permissible Exposure Limits in Air:* The Federal OSHA PEL is 2 ppm 8 hr. TWA and 10 ppm not to be exceeded during any 15 minute work period. This chemical can be absorbed through the skin, thereby increasing exposure. NIOSH[2] has a recommended airborne exposure limit of 1 ppm TWA on an 10-hour workshift and a 10 ppm ceiling not to be exceeded during any 15 minute work period. ACGIH's recommendation is a TLV of 2 ppm averaged over an 8-hour workshift, with the notation that acrylonitrile is a human carcinogen[1]. The odor threshold is about 10-fold greater than the OSHA PEL, so workers can be overexposed to acrylonitrile without being aware of its presence. The NIOSH[2] IDLH level = 85 ppm. This chemical is a probable carcinogen in humans; there may be no safe level and all contact should be reduced to lowest possible level. AIHA ERPG-2 (maximum airborne concentration below which it is believed that nearly all persons could be exposed for up to 1 hour without experiencing or developing irreversible or other serious health effects or symptoms that could impair their abilities to take protective action) = 35 ppm. The DFG[3] TRK is 3 ppm (7 mg/m$^3$), Animal Carcinogen, Suspected Human Carcinogen Australia's and Israel's limit is 2 ppm (4.3 mg/m$^3$) TWA. The (HSE)[33] (U.K) limit is 2 ppm (4 mg/m$^3$) TWA[33]. Mexico's limit is 2 ppm (5.4 mg/m$^3$) TWA. The former USSR-UNEP/IRPTC project[43] has set a MAC of

0.5 mg/m³ in workplace air and a limit of 0.03 mg/m³ in ambient air in residential areas on a daily, average basis. In addition, several states have set guidelines or standards for acrylinitrile in ambient air[60]: 0.0147 µg/m³ (Indiana) to 0.145 µg/m³ (North Carolina) to 0.15 µg/m³ (Massachusetts) to 11.3 µg/m³ (Pennsylvania) to 15.0 µg/m³ (New York) to 22.0 µg/m³ (Connecticut and South Dakota) to 22.5 µg/m³ (South Carolina) to 45.0 µg/m³ (Florida and Virginia).

*Determination in Air:* Charcoal adsorption followed by acetone extraction and gas chromatographic analysis. See NIOSH Method 1604[18], or OSHA Method 37[58]. See NIOSH Criteria Document 78-212 NITRILES

*Permissible Concentration in Water:* The substance is toxic to aquatic organisms. Acrylonitrile usually breaks down in about 1 or 2 weeks, but this can vary depending on conditions. For example, high concentrations (such as might occur following a spill) tend to be broken down more slowly. In one case, measurable amounts of acrylonitrile were found in nearby wells 1 year after a spill (ATSDR public Health Statement, December 1990). See RCRA and Clean Water Act under Regulatory Authority. The U.S. EPA has set a maximum contaminant level of cyanide in drinking water of 0.2 milligrams cyanide per liter of water (0.2 mg/L). To protect freshwater aquatic life–on an acute basis, 7,550 µg/L and on a chronic basis, 2,600 µg/L over 30 days. To protect saltwater aquatic life-insufficient data to yield a value. To protect human health-preferably zero. Water concentration should be below 0.58 µg/L to keep lifetime cancer risk below $10^{-5}$. The former USSR-UNEP/IRPTC project[43] has set a MAC of 2.0 mg/L for water bodies used for domestic purposes. The Mexico drinking water ecological criteria is 0.0006 mg/L, reduce human exposure to a minimum. In addition, several states have set guidelines for acrylonitrile in drinking water[61] ranging from 0.67 µg/L (Minnesota) to 3.8 µg/L (Kansas) to 10 µg/L (Arizona) to 35 µg/L (Connecticut).

*Determination in Water:* Charcoal tube; Acetone/$CS_2$; Gas chromatography/Flame ionization detection; NIOSH (IV) Method #1604.[18] Also, by gas chromatography (EPA Method #603) or gas chromatography plus mass spectrometry (EPA Method #624). Also, cyanide may be determined titrimetrically by EPA Methods 335.2 and 9010 which give total cyanide.

*Routes of Entry:* Persons whose clothing or skin are contaminated with liquid acrylonitrile can secondarily contaminate response personnel by direct contact or through off-gassing vapor. Inhalation and percutaneous absorption. It may be absorbed from contaminated rubber or leather. Routes include ingestion and eye and skin contact. Acrylonitrile vapor is absorbed readily from the lungs, and inhalation is an important route of exposure. This chemical's odor generally provides inadequate warning of hazardous concentrations and olfactory fatigue develops rapidly. The odor threshold is about 10-fold greater than the OSHAPEL, so response personnel and workers can be overexposed to acrylonitrile without being aware of its presence.

*Harmful Effects and Symptoms*
*Short Term Exposure:* Acrylonitrile is irritating to the skin, eyes, and respiratory tract. Toxic effects range from headache, fatigue, dyspnea, nausea and vomiting to asphyxiation, lactic acidosis and cardiovascular collapse. Toxic effects are due primarily to the bioreactivity of acrylonitrile with cellular proteins and to its epoxide intermediate that is mutagenic and genotoxic. Toxicity is also due to the release of cyanide during the metabolism of acrylonitrile. Persons whose clothing or skin are contaminated with liquid acrylonitrile can secondarily contaminate response personnel by direct contact or through off-gassing vapor. Splashes in the eye may result in corneal damage. Skin contact can cause severe irritation and blistering. Breathing acrylonitrile can irritate the lungs causing coughing and shortness of breath. Higher exposures can cause pulmonary edema, a medical emergency that can result in death. Skin contact contributes significantly in overall exposure and can lead to systemic toxicity. Acrylonitrile reaction causes redness, blisters and some systemic signs. Symptoms derive from tissue anoxia in order of onset: limb weakness, dyspnea (difficult breathing), burning sensation in throat, dizziness, impaired judgement, cyanosis (turning blue), nausea, collapse, irregular breathing, convulsions and death. In later stages collapse, irregular breathing or convulsions and cardiac arrest may occur without warning. Some patients appear hysterical or may even be violent. Acrylonitrile is classified as very toxic. Probable oral lethal dose for human is 50-500 mg/kg (between 1 teaspoon and 1 oz.) for a 70 kg (150 lb) person. Toxic concentrations have been reported at 16 ppm/20 min. Acute toxicity is similar to that due to cyanide poisoning and the level of cyanide ion in blood is related to the level of poisoning. Inhalation or ingestion can results in fatal systemic poisoning, collapse and death due to tissue anoxia (lack of oxygen) and cardiac arrest (heart failure). At higher concentrations there may be damage to red blood cells and the liver. Jaundice may develop 24 hours following exposure and persist for several days. Because of continued metabolic release of cyanide, symptoms of severe poisoning may recur and the patient may relapse.

*Long Term Exposure:* Chronic exposures to acrylonitrile have been associated with liver damage. Chronic exposure may be more serious for children because of their potential longer latency period. *Carcinogenicity* The Department of Health and Human Services has determined that acrylonitrile may reasonably be anticipated to be a carcinogen. IARC has determined that acrylonitrile is possibly carcinogenic to humans (Group 2B) based on sufficient evidence of carcinogenicity in experimental animals and inadequate evidence for carcinogenicity in humans. ACGIH classifies it as an a suspected human carcinogen. In animals, chronic exposure can cause tumors of the mammary gland, gastrointestinal tract, and CNS. Increased rates of lung and prostate cancer have been documented in some groups of chronically exposed workers, but not in others. *Reproductive and Developmental Effects:* According to *Shepard's Catalog of Teratogenic Agents*, when large doses of acrylonitrile were administered to experimental animals by oral, inhalation, or intraperitoneal routes, teratogenic effects were produced. In humans, there is no documented evidence that acrylonitrile is a reproductive or developmental toxicant. Acrylonitrile is not currently reviewed in the TERIS or Reprotext databases. Acrylonitrile is not included in *Reproductive and Developmental Toxicants*, a 1991 report published by the U.S. General Accounting Office that lists 30 chemicals of concern because of widely acknowledged reproductive and developmental consequences. There is no information regarding whether acrylonitrile can cross the placenta or whether it can accumulate in breast milk and be transferred to nursing infants.

*Points of Attack:* Eyes, skin, cardiovascular system, liver, kidneys, central nervous system. Cancer Site: brain tumors, lung and bowel cancer.

*Medical Surveillance:* Medical observation is recommended for 24 to 48 hours after breathing overexposure, as pulmonary edema may be delayed. For those with frequent or high exposure, consider urine thiocyanate levels, blood cyanide levels, liver function tests, fecal occult blood screening, pulmonary function tests. Consider chest x-ray following acute exposure. Consider the skin, respiratory tract, heart, central nervous system, renal and liver function in placement and periodic examinations. A history of fainting spells or convulsive disorders might present and added risk to persons working with toxic nitriles.

*First Aid:* Persons whose clothing or skin are contaminated with liquid acrylonitrile can secondarily contaminate response personnel by direct contact or through off-gassing vapor. If this chemical gets into the eyes, remove any contact lenses at once and irrigate immediately for at least 15 minutes, occasionally lifting upper and lower lids. Seek medical attention immediately. If this chemical contacts the skin, remove contaminated clothing and wash immediately with soap and water. Seek medical attention immediately. If this chemical has been inhaled, remove from exposure, begin rescue breathing (using universal precautions) if breathing has stopped and CPR if heart action has stopped. Transfer promptly to a medical facility. When this chemical has been swallowed, get medical attention. Give large quantities of water and induce vomiting. Do not make an unconscious person vomit. Medical observation is recommended for 24 to 48 hours after breathing overexposure, as pulmonary edema may be delayed.

*Note to physician or authorized medical personnel.* In cases of respiratory compromise secure airway and respiration via endotracheal intubation. If not possible, perform cricothyroidotomy if equipped and trained to do so. Administer 100% oxygen. Treat patients who have bronchospasm with aerosolized bronchodilators. The use of bronchial sensitizing agents in situations of multiple chemical exposures may pose additional risks. Also consider the health of the myocardium before choosing which type of bronchodilator should be administered. Cardiac sensitizing agents may be appropriate; however, the use of cardiac sensitizing agents after exposure to certain chemicals may pose enhanced risk of cardiac arrhythmias (especially in the elderly). Acrylonitrile poisoning is not known to pose additional risk during the use of bronchial or cardiac sensitizing agents. Consider racemic epinephrine aerosol for children who develop stridor. Dose 0.25–0.75 mL of 2.25% racemic epinephrine solution in water, repeat every 20 minutes as needed cautioning for myocardial variability. Patients who are comatose, hypotensive, or have seizures or cardiac dysrhythmias should be treated according to advanced life support protocols. These patients may be seriously acidotic; under medical control, consider giving them 1 ampule of sodium bicarbonate (pediatric dose: 1 mEq/kg may be appropriate). If massive exposure is suspected or if the patient is severely symptomatic with hypotension, infuse intravenous saline or lactated Ringer's solution. For adults, bolus 1000 mL/hour if blood pressure is under 80 mmHg; if systolic pressure is over 90 mmHg, an infusion rate of 150 to 200 mL/hour is sufficient. For children with compromised perfusion administer 20 mL/kg of normal saline or Ringer's lactate delivered over 10 to 20 minutes, then at a 2 to 3 mL/kg/hour infusion rate.

*Antidotes:* When possible, treatment with cyanide antidotes should be given under medical-base control to unconscious victims with known or strongly suspected acrylonitrile poisoning. Cyanide antidotes amyl nitrite perles and intravenous infusions of sodium nitrite and sodium thiosulfate are packaged in the cyanide antidote kit. Amyl nitrite perles (0.2 mL) should be broken onto a gauze pad and held under the nose, over the Ambu valve intake, or placed under the lip of the face mask. A new perle is crushed and inhaled for 30 seconds every minute until intravenous sodium nitrite is given. Infuse sodium nitrite intravenously as soon as possible. The usual adult dose is 10 to 20 mL of a 3% solution infused over no less than 5 minutes to produce a 20% methemoglobin level in adults. Children should receive 0.33 mL/kg of the 3% solution at an infusion rate of 2.5 mL/minute, up to a maximum of 10 mL. Administer sodium nitrite doses to children on the basis of body weight, since fatal methemoglobinemia has occurred in children dosed at adult rates. Monitor blood pressure during administration, and slow the rate of infusion if hypotension develops. Immediately after sodium nitrite infusion, administer sodium thiosulfate intravenously. The usual adult dose is 50 mL (12.5 g) of a 25% solution infused at a rate of 3 to 5 mL/minute; the average pediatric dose is 1.65 mL/kg (412.5 mg/kg) up to 50 mL. If symptoms reappear or persist within 1 hour, readminister sodium nitrite and sodium thiosulfate at 50% of the initial dose.

*References:*
- New Jersey Department of Health and Senior Services, Hazardous Substance Fact Sheet: "Acrylonitrile," Trenton, NJ (May, 1998).
  http://www.state.nj.us/health/eoh/rtkweb/0024.pdf
- ATSDR, U.S. Department of Health and Human Services, Public Health Service, "Managing Hazardous Materials Incidents," Atlanta, GA (June 2003)
- U.S. Department of Health and Human Services, Public Health Service, "Public Health Statements : Acrylonitrile," Atlanta, GA (December, 1990)
- National Institute for Occupational Safety and Health, Criteria for a Recommended Standard: Occupational

Exposure to Acrylonitrile, NIOSH Doc, No. 78-116, Washington DC (1978).
- Department of Labor, Economic Impact Assessment for Acrylontirile, Washington, DC, Occupational Safety and Health Administration (February 21, 1978).
- U.S. Environmental Protection Agency, Status Assessment of Toxic Chemicals: "Acrylonitrile," Report EPA-600/2-79-210A, Washington, DC (December 1979).
- U.S. Environmental Protection Agency, Acrylonitrile: Ambient Water Quality Criteria, Washington, DC (1980).
- U.S. Environmental Protection Agency, Investigation of Selected Potential Environmental Contaminants: Acrylonitrile, Report EPA-560/2-78-003, Washington, DC (May 1978).
- U.S. Environmental Protection Agency, Acrylonitrile, Health and Environmental Effects Profile No. 7, Washington, DC, Office of Solid Waste (April 30, 1980).
- Sax, N.I., Ed., Dangerous Properties of Industrial Materials Report, 1, No. 2, 25-27 (1980) and 3, No. 3, 41-46 (1988) and 5, No. 4, 31-33 (1985).
- U.S. Environmental Protection Agency, "Chemical Profile: Acrylonitrile," Washington, DC, Chemical Emergency Preparedness Program (November 30, 1987).
- New York State Department of Health, "Chemical Fact Sheet: Acrylonitrile," Albany, NY, Bureau of Toxic Substance Assessment (March 1986).
- Agency for Toxic Substances and Disease Registry, U.S. Department of Health and Human Services, Public Health Service, "Managing Hazardous Materials Incidents," Atlanta, GA (June 2003)
- California Environmental Protection Agency "Chemical List of Lists," Sacramento CA (February 1997).

# Alachlor

*Use Type:* A pre-emergence herbicide
*CAS Number:* 15972-60-8
*Formula:* $C_{14}H_{20}ClNO_2$
*Alert:* A Restricted Use Pesticide (RUP). Human toxicity (long-term): High.
*Synonyms:* Al3-51506; Acetamide, 2-chloro-*n*-(2,6-diethylphenyl)-*N*-(Methoxymethyl)-; Acetanilide, 2-chloro-2',6'-diethyl-*N*-methoxymethyl)-; Alachlore; α-Chloro-2',6'-diethyl-*N*-(methoxymethyl)acetanilide; 2-Chloro-*N*-(2,6-diethylphenyl)-*N*-(methoxymethyl) acetamide; Glyphosate isopropylamine salt; Metachlor; Methachlor; *N*-(Methoxymethyl)2,6-diethylchloro acetamide
*Trade Names:* AGIMIX®, Milenia Agro Ciencias (Brazil); ALAGAM®, Makhteshim-Agan Industries (Israel); Proficol (Colombia); ALAGAN®, Makhteshim-Agan Industries (Israel); Proficol (Colombia); ALANEX®, Makhteshim-Agan Industries (Israel); Proficol (Colombia); ALAPAZ®, Pazchem Ltd., (Israel), suspended; ALAZINE®, Makhteshim Agan (Israel); ALATOX 480®; ALCLOR 48 LE®, Alcotan Laboratories (Spain); BRONCO®, Monsanto (USA), canceled; CANNON HERBICIDE®, Monsanto (USA), canceled; CHIMICHLOR®; CROP STAR®, canceled; LARIAT®, Monsanto (USA); LASAGRIN®; LASSAGRIN®; LASSO®, Monsanto (USA); LASSO MICRO-TECH®, Monsanto (USA); LAZO®, Monsanto (USA), canceled; METACHLOR®, Syngenta (Switzerland); NIAGARA KOLO MALACHLOR DUST®, FMC Corp (USA) canceled; NIAGARA MALACHLOR LIVESTOCK SPRAY CODE 983®, FMC Corp (USA), canceled; PARTNER®, Monsanto (USA); PILLARZO®, Pilarquim Corp. (Taiwan); SANACHLOR®, Sanachem (Pty) Ltd. (South Africa); SHOLAY®, Rallis India (India), canceled
*Producers:* Agsin (Singapore); Alcotan Laboratories (Spain); ATOFINA Chemicals (France); Cerexagri Inc. (France); China Chemicals (China); Ehrenstorfer, Dr. (Germany); Makhteshim-Agan Industries (Israel); Milenia Agro Ciencias (Brazil); Monsanto (USA); Nissan Chemical Industries (Japan); Pazchem Ltd. (Israel); Pilarquim Corp (Taiwan); Proficol (Colombia); Shenzhen Guomeng Industry Co., Ltd. (China); Sigma-Aldrich Laborchemikalien (Germany); Sinon (Taiwan); Syngenta (Switzerland); Vijayalakshmi Insecticides and Pesticides (India); Zago Asia Ltd. (Singapore)
*Chemical Class:* Chloroacetanilide
*EPA/OPP PC Code:* 090501
*California DPR Chemical Code:* 678
*ICSC Number:* 0371
*RTECS Number:* AE1225000
*EEC Number:* 616-015-00-6
*EINECS Number:* 240-110-8
*Uses:* A pre-emergence herbicide for corn, soybeans, and peanuts, and other field crops. It is a selective systemic herbicide, absorbed by germinating shoots and by roots. It works by interfering with a plant's ability to produce protein and by interfering with root elongation. Alachlor is a Restricted Use Pesticide (RUP). This compound is one of the most highly used herbicides in the U.S. It is available as granules or emulsifiable concentrate. Alachlor is used in mixed formulations with atrazine, glyphosate, trifluralin, and imaquin.
*Human toxicity (long-term)[77]:* High–2.00 ppb, MCL (Maximum Contaminant Level)
*Fish toxicity (threshold)[77]:* Intermediate–26.19584 ppb, MATC (Maximum Acceptable Toxicant Concentration)
*U.S. Maximum Allowable Residue Levels for Alachlor (40 CFR 180.249):*

| CROP | ppm |
| --- | --- |
| Bean, dry | 0.1 |
| Bean, forage | 0.2 |
| Bean, hay | 0.2 |
| Bean, lima, green | 0.1 |
| Cattle, fat | 0.02 |
| Cattle, meat | 0.02 |
| Cattle, mbyp | 0.02 |
| Corn, fodder | 0.2 |
| Corn, forage | 0.2 |
| Corn, fresh (inc. sweet) | 0.05 |
| Corn, grain | 0.2 |
| Egg | 0.02 |
| Goat, fat | 0.02 |
| Goat, meat | 0.02 |

| | |
|---|---|
| Goat, mbyp | 0.02 |
| Hog, fat | 0.02 |
| Hog, meat | 0.02 |
| Hog, mbyp | 0.02 |
| Horse, fat | 0.02 |
| Horse, meat | 0.02 |
| Horse, mbyp | 0.02 |
| Milk | 0.02 |
| Peanut | 0.05 |
| Peanut, forage | 3.0 |
| Peanut, hay | 3.0 |
| Poultry, fat | 0.02 |
| Poultry, meat | 0.02 |
| Poultry, mbyp | 0.02 |
| Sheep, fat | 0.02 |
| Sheep, meat | 0.02 |
| Sheep, mbyp | 0.02 |
| Sorghum, forage | 2.0 |
| Sorghum, grain, milo | 0.1 |
| Sorghum, stover | 1.0 |
| Soybean | 0.2 |
| Soybean, forage | 0.75 |
| Soybean, hay | 0.2 |

**Carcinogen/Hazards Classifications:**
**U.S. EPA Carcinogens:** Likely for high doses; not likely for low doses
**California Prop. 65:** Carcinogen
**IARC:** Not evaluated
**Label Signal Word:** DANGER
**TRI Developmental Toxin:** Developmental toxin
**WHO Acute Hazard:** Class III, slightly hazardous
**Endocrine Disruptor:** Suspected
*Regulatory Authority:*
- AB 1803-Well Monitoring Chemical (CAL)
- AB 2588-Air Toxics "Hot Spots" Chemicals (CAL)
- Proposition 65 chemical (CAL)
- Actively registered pesticide in California.
- Banned or Severely Restricted (EPA-FIFRA)
- Safe Drinking Water Act, MCL, 0.002 mg/L; MGLC, zero; Regulated chemical (47 FR 9352)
- Superfund/EPCRA 40CFR302.4, Appendix A, RQ: 1lb (0.454 kg), SARA 313: Form R *de minimis* Concentration Reporting Level: 1.0%.
- Water Pollution Standard Proposed (Various States)[61]

*Description:* Alachlor, a poly-substituted single aromatic nucleus, is a cream-colored solid. Insoluble in water. Molecular weight = 269.79. Melting/Freezing point = 39.5–41.5°C. Flash point = 137°C (cc). Hazard Identification (based on NFPA-704 M Rating System): Health 2, Flammability 0, Reactivity 0. Log $K_{ow}$ = 3.48. Values at or above 3.0 are likely to bioaccumulate in marine organisms. Physical and toxicological properties may be affected by the carrier solvents used in commercial formulations.
*Incompatibilities:* Strong oxidizers. Corrosive to iron and steel.
*Permissible Exposure Limits in Air:* No standards set.
*Permissible Concentration in Water:* No adverse effect level in drinking water has been calculated by NSA/NRC[46] as 0.7 mg/L. Allowable daily intake (ADI) has been calculated at 0.1 mg/kg/day. More recently, the U.S. EPA[47] has reviewed alachlor and determined a ten-day health advisory value of 0.1 mg/L for a 10 kg child. An acceptable daily intake was calculated as 0.01 mg/kg/day in the study. A maximum level in drinking water of 0.002 mg/L has been proposed by EPA[62]. In addition, a number of states[61] have set guidelines for alachlor in drinking water ranging from 0.15 µg/L (Arizona) to 0.2 µg/L (Illinois) to 2.0 µg/L (Massachusetts) to 10.0 µg/L (Minnesota) to 15 µg/L (Kansas) to 200 µg/L (Maine).
*Determination in Water:* May be accomplished by liquid-liquid extraction gas chromatographic procedure[47].
*Routes of Entry:* Ingestion
*Harmful Effects and Symptoms*
*Short Term Exposure:* Toxic by skin contact, ingestion, and inhalation. Eye contact may cause severe irritation or injury. Skin contact may irritate and burn skin. No effects found in human studies[47]. Exhibits relatively low acute oral toxicity; the $LD_{50}$ value for rats is 0.93 g/kg. The technical product has only slight skin and eye irritation potential after an acute exposure[47]. However, alachlor feeding studies have demonstrated oncogenic effects including lung tumors in mice and stomach, thyroid and nasal turbinate tumors in rats.
*Long Term Exposure:* Human mutation data reported. Suspected carcinogen; some experimental data reported.
*First Aid:* If this chemical gets into the eyes, remove any contact lenses at once and irrigate immediately for at least 15 minutes, occasionally lifting upper and lower lids. If this chemical contacts the skin, remove contaminated clothing and wash immediately with soap and water. When this chemical has been swallowed, get medical attention. Give large quantities of water and induce vomiting. Do not make an unconscious person vomit. If this chemical has been inhaled, remove from exposure and transfer promptly to a medical facility.
*References:*
- U.S. Environmental Protection Agency, "Alachlor: Notice of Intent to Cancel Registrations; Conclusion of Special Review," Federal Register 52, No. 251, pp. 49480-49504 incl. (December 31, 1987).
- EPA, Office of Pesticide Programs, Pesticide Residue Limits, "Alachlor," 40 CFR 180.249. www.epa.gov/pesticides/food/viewtols.htm,
- EXTOXNET, Extension Toxicology Network, "Pesticide Information Profile, Alachlor," Oregon State University, Corvallis, OR (June 1996). http://ace.ace.orst.edu/info/extoxnet/pips/alachlor.htm
- Waxman, Michael F and Kammel, David W, A Guidebook for the Safe Use of Hazardous Agricultural Farm Chemicals and Pesticides, North Central Regional Publication 402, Madison WI, (1991).

- California Environmental Protection Agency Chemical List of Lists, Sacramento CA (February 1997)

## Aldicarb (ANSI)

*Use Type:* A systemic insecticide, nematicide and acaricide.
*CAS Number:* 116-06-3
*Formula:* $C_7H_{14}N_2O_2S$; $CH_3SC(CH_3)_2CH=NOCONHCH_3$; $C_7H_{14}N_2O_2S$
*Alert:* A Restricted Use Pesticide (RUP) for selected crops. Human toxicity (long-term): High.
*Synonyms:* A13-27093; Aldecarb; Aldecarbe (French); Carbamic acid, methyl-,*O*-[(2-methyl-2-(methylthio)propylidene)amino]derivative; Carbanolate; Caswell No. 011A; ENT 27093; NCI 08640; 2-Methyl-2-(methylthio)propanal, *O*-[(methylamino)carbonyl]oxime; 2-Methyl-2-(methylthio)propanaldehyde, *O*-(Methylcarbamoyl)oxime; 2-Methyl-2-methylthio-propionaldehyd-*O*-(*N*-methyl-carbamoyl)-oxim (German); 2-Metil-2-tiometil-propionaldeid-*O*-(*N*-metil-carbamoil)-ossima (Italian); Permethrin; Propionaldehyde,2-methyl-2-(methylthio)-,*O*-(methylcarbamoyl)oxime; Propanal,2-methyl-2-(methythio)-,*O*-[(methylamino)carbonyl]oxime; Sulfone aldoxycarb
*Trade Names:* TEMIC®; TEMIK®; Bayer CropScience (Germany); OMS 771®; TERNIC®; TEMIK 10 G®; TEMIK G 10®, Union Carbide (USA) canceled; UNION CARBIDE UC-21149®, Union Carbide (USA), canceled
*Producers:* Bayer CropScience (Germany); Ehrenstorfer, Dr.(Germany); Rhone-Poulenc (France); Shandong Huayang Pesticide Group (China); Shenzhen Guomeng Industry Co., Ltd. (China); Sigma-Aldrich Laborchemikalien (Germany); Union Carbide (Dow AgroScience) (USA)
*Chemical Class:* Carbamate (*N*-methyl)
*EPA/OPP PC Code:* 098301
*California DPR Chemical Code:* 00575
*ICSC Number:* 0094
*RTECS Number:* UE2275000
*EEC Number:* 006-017-00-X
*EINECS Number:* 204-123-2
*Uses:* A Restricted Use Pesticide (RUP) only on potatoes, citrus, cotton, ornamentals, peanuts, sorghum, soybeans, sugar beets, sugarcane, coffee, sweet potatoes, pecans, wheat and varius other crops. Aldicarb is effective against various insects (especially aphids, whiteflies, leaf miners, flea beetles and ground beetles), mites and nematodes. Use on bananas was revoked during special review process due to concerns about high residues. It is among the most toxic of all pesticides to birds, aquatic invertebrates, fish, and insect pollinators, such as bees.
*Human toxicity (long-term)*[77]*:* High–7.00 ppb, MCL (Maximum Contaminant Level)
*Fish toxicity (threshold)*[77]*:* High–3.75027 ppb, MATC (Maximum Acceptable Toxicant Concentration)
*U.S. Maximum Allowable Residue Levels for Aldicrab (40 CFR 180.269):*

| CROP | ppm |
|---|---|
| Bean, dry | 0.1 |
| Beet, sugar | 0.05 |
| Beet, sugar, tops | 1.0 |
| Cattle, fat | 0.01 |
| Cattle, meat | 0.01 |
| Cattle, mbyp | 0.01 |
| Citrus, dried pulp | 0.6 |
| Coffee, bean | 0.1 |
| Cotton, hulls | 0.3 |
| Cotton, undelinted seed | 0.1 |
| Goat, fat | 0.01 |
| Goat, meat | 0.01 |
| Goat, mbyp | 0.01 |
| Grapefruit | 0.3 |
| Hog, fat | 0.01 |
| Hog, meat | 0.01 |
| Hog, mbyp | 0.01 |
| Horse, fat | 0.01 |
| Horse, meat | 0.01 |
| Horse, mbyp | 0.01 |
| Lemon | 0.3 |
| Lime | 0.3 |
| Milk | 0.002 |
| Orange | 0.3 |
| Peanut | .05 |
| Pecan | 0.5 |
| Potato | 1.0 |
| Sheep, fat | 0.01 |
| Sheep, meat | 0.01 |
| Sheep, mbyp | 0.01 |
| Sorghum, bran | 0.5 |
| Sorghum, grain, grain | 0.2 |
| Sorghum, stover | 0.5 |
| Soybean | 0.02 |
| Sugarcane, cane | 0.02 |
| Sugarcane, fodder | 0.1 |
| Sugarcane, forage | 0.1 |
| Sweet potato | 0.1 |

*Carcinogen/Hazard Classifications*
**U.S. EPA Carcinogens:** Extremely hazardous substance
**IARC:** Group 3, unclassifiable
**Label Signal Word:** DANGER-POISON
**WHO Acute Hazard:** Class 1 a, extremely hazardous
**Endocrine Disruptor:** Suspected endocrine disruptor
*Regulatory Authority:*
- Classified by the U.S. EPA as Restricted Use Pesticide (RUP)

- Banned or Severely Restricted (Austria, Belgium, Germany, Israel, Norway, Philippines) (UN)[13]
- Very Toxic Substance (World Bank)[15]
- AB 1803-Well Monitoring Chemical (CAL)
- EPA/SARA 302 (EPCRA) Extremely hazardous substances
- The "Director's List" (CAL/OSHA)
- Actively registered pesticide in California.
- EPA Hazardous Waste Number (RCRA No.): P070
- RCRA 40CFR261, Appendix 8; 40CFR261.11 Hazardous Constituents
- RCRA Land Ban Waste Restrictions
- Safe Drinking Water Act, MCL, 0.003 mg/L; MCLG 0.001 mg/L; Regulated chemical (47 FR 9352)
- CERCLA/SARA 40CFR302, Extremely Hazardous Substances: TPQ = 100/10,000 lb (45.4/4,540 kg)
- Superfund/EPCRA 40CFR302.4, Appendix A, RQ: CERCLA 1lb (0.454 kg), SARA 313: Form R *de minimis* Concentration Reporting Level: 1.0%.
- U.S. DOT Regulated Marine Pollutant (49CFR172.101, Appendix B)
- Air Pollutant Standard Set (New York, South Carolina)[60] Water Pollution Standard Proposed (U.S. EPA)[47] (Several States)[61]

*Description:* Aldicarb is a broad spectrum carbamate pesticide; a fast-acting cholinesterase inhibiting agent with effective direct contact and stomach action; and, a plant systemic of extremely high acute toxicity to mammals. It is a noncombustible, white crystalline solid with a slight sulfurous odor. Slightly soluble in water. Melting/Freezing point = 98–100°C. Vapor pressure = 0.01 Pa @ 25°C. Log $K_{ow}$ = 1.35. Although this chemical is an environmental hazard, it is unlikely to bioaccumulate in marine organisms.

*Incompatibilities:* Strong alkalies.

*Permissible Exposure Limits in Air:* The American Industrial Hygiene Association (AIHA/WEEL) recommends a TWA level of 0.07 mg/m$^3$, and warns that aldicarb can be absorbed through the skin, thereby increasing exposure. Guidelines or standards have been set for aldicarb in ambient air[60] ranging from 2.0 $\mu\mu g/m^3$ (New York) to 6.0 $\mu\mu g/m^3$ (South Carolina).

*Permissible Concentration in Water:* EPA/Safe Drinking Water Act Levels for MCL, 0.003 mg/L; MCLG 0.001 mg/L. Canada Drinking Water Quality = 0.009 mg/L MAC.

*Determination in Water:* Aldicarb may be determined in water by gas-liquid chromatography with flame photometric detection after oxidation to the sulfone (aldoxycarb) by peracetic acid or 3-chloro-perbenzoic acid. Colorimetric methods have also been used based on hydrolysis to hydroxyl-amine which is oxidized to nitrous acid, the latter used to diazotize sulfanilic acid which is then coupled to give a dye[23].

*Routes of Entry:* Ingestion, skin contact.

*Harmful Effects and Symptoms*

*Short Term Exposure:* This chemical is one of the most highly toxic pesticides. It can be harmful or fatal if swallowed, inhaled or absorbed through the skin. Exposure can cause rapid severe poisoning with headache, blurred vision, sweating, nausea, abdominal pain, vomiting, diarrhea, loss of coordination, and death. In severe cases, unconsciousness and convulsions may occur. The probable oral lethal dose for humans is less than 5 mg/kg, or a taste (less than 7 drops) for a 150-lb person; it is extremely toxic by both oral and dermal routes. Carbamate insecticides inhibit the cholinesterase activity of enzymes, causing accumulation of acetylcholine at synapses and altering the way in which nervous impulses are transmitted. However, within several hours carbamates spontaneously detach from the enzymes.

*Long Term Exposure:* May affect the immune system. There is no evidence that aldicarb affects reproduction. Aldicarb is questionable carcinogen with no firm human evidence.

*Points of Attack:* Skin, lungs.

*Medical Surveillance:* Test plasma and red blood cell cholinesterase levels (for enzyme poisoned by this chemical). For this substance, these tests are accurate only if done within about two hours of exposure.

*First Aid:* Speed in removing material from eyes and skin is of extreme importance. *Eyes:* Eye contact can cause dangerous amounts of these chemicals to be quickly absorbed through the mucous membrane into the bloodstream. Immediately and gently flush eyes with plenty of warm or cold water (NO hot water) for at least 15 minutes, occasionally lifting the upper and lower eyelids. Get medical aid immediately. *Skin:* Get medical aid. Skin contact can cause dangerous amounts of these chemicals to be absorbed into the bloodstream. Wearing the appropriate PPE equipment and respirator for carbamate pesticides, immediately flush skin with plenty of soap and water for at least 15 minutes while removing contaminated clothing and shoes. Shampoo hair promptly if contaminated; protect eyes. *Ingestion:* Call poison control. Loosen all clothing. Never give anything by mouth to an unconscious person. Get medical aid. Do NOT induce vomiting.* If conscious, alert, and able to swallow, rinse mouth and have victim drink 4 to 8 ounces of water. Check to see if poison control instructs you to use ipecac syrup, otherwise administer slurry of activated charcoal (2 oz in 8 oz of water). If victim is UNCONSCIOUS OR HAVING CONVULSIONS, do nothing except keep victim warm. *\*In some cases you may be specifically instructed by poison control to induce vomiting by way of 2 tablespoons of syrup of ipecac (adult) washed down with a cup of water.* Do NOT give activated charcoal before or with ipecac syrup.

*Inhalation:* Get medical aid. Do not contaminate yourself. Wearing the appropriate PPE equipment and respirator for carbamate pesticides, immediately remove the victim from the contaminated area to fresh air. If the victim is not breathing, administer artificial respiration. Do NOT use mouth-to-mouth resuscitation; use bag/mask apparatus. If breathing is difficult, administer oxygen through bag/mask apparatus until medical help arrives. Do not leave victim unattended.

*Note to physician or authorized medical personnel.* Administer atropine, 2 mg (1/30 gr) intramuscularly or intravenously as soon as any local or systemic signs or symptoms of an intoxication are noted; repeat the administration of atropine every 3 to 8 minutes until signs of atropinization (mydriasis, dry mouth, rapid pulse, hot and dry skin) occur; initiate treatment in children with 0.05 mg mg/kg of atropine; repeat at 5 to 10 minute intervals. Watch respiration, and remove bronchial secretions if they appear to be obstructing the airway; intubate if necessary. *Medical note:* 2-PAMCI may be contraindicated in the case of some carbamate poisonings.

*References:*
- Sax, N.I., Ed., Dangerous Properties of Industrial Materials Report, 4, No. 2, 37-41 (1984).
- American Bird Conservancy, www.abcbirds.org/pesticides/Profiles/aldicarb.htm
- U.S. Environmental Protection Agency, "Chemical Profile: Aldicarb," Washington, DC, Chemical Emergency Preparedness Program (November 30, 1987).
- EPA, Office of Pesticide Programs, "Pesticide Residue Limits, Aldicarb," 40 CFR 180.269, http://www.epa.gov/pesticides/food/viewtols.htm
- New Jersey Department of Health and Senior Services, "Hazardous Substance Fact Sheet: Aldicarb," Trenton, NJ (January 2001), www.state.nj.us/health/eoh/rtkweb/rtkhsfs.htm
- California Environmental Protection Agency Chemical List of Lists, Sacramento CA (February 1997)

# Aldoxycarb (ANSI)

*Use Type:* Insecticide, acaricide and nematicide
*CAS Number:* 1646-88-4
*Formula:* $C_7H_{14}N_2O_4S$
*Synonyms:* Aldicarb sulfure (French); Carbamic acid, methyl-, *O*-[(2-methyl-2-(methylsulfonyl)propylidene)amino] derivative; ENT AI 3-29261; 2-Methyl-2-(methylsulfonyl)propanal-*O*-[(methylamino)carbonyl]oxime; 2-Methyl-2-(methylsulfonyl)propionaldehyde-*O*-(methylcarbamoyl)oxime; Propanal, 2-methyl-2-(methylsulfonyl)-, *O*-[(methylamino)carbonyl]oxime
*Trade Names:* ALDICARB SULFONE®; TEMIK SULFONE®; SULFOCARB®; STANDAK®, Bayer CropScience (Germany); UC-21865®, Union Carbide (USA)
*Producers:* Bayer CropScience (Germany); Union Carbide (USA)
*Chemical Class:* Carbamate
*EPA/OPP PC Code:* 110801
*California DPR Chemical Code:* 2265
*RTECS Number:* UE2080000
*Uses:* Used to control honey locust gall midge and other insects and mites on cotton, potatoes, sugar beets and ornamentals. Applied to the soil by soluble mixture or spikes.
*Human toxicity (long-term)*[77]: High–7.00 ppb, MCL (Maximum Contaminant Level)
*Fish toxicity (threshold)*[77]: Very low–7683.76758 ppb, MATC (Maximum Acceptable Toxicant Concentration)
*Description:* White, crystalline solid. Slight sulfur odor. Slightly soluble in water; solubility = 8,000 ppm @ 20°C. Molecular weight = 222.29. Melting/Freezing point = 141°C. Vapor pressure = 12 mPa @ 25°C; $1.0 \times 10^{-4}$ mmHg.
*Incompatibilities:* May react violently with bromine, ketones. Incompatible with strong acids, azo dyes, caustics, ammonia, amines, boranes, hydrazines, strong oxidizers. May form explosive materials with phosphorus pentachloride.
*Permissible Concentration in Water:* No criteria set. Runoff from spills or fire control may cause water pollution.

*Harmful Effects and Symptoms*
*Short Term Exposure:* Eye pupils are small; blurred vision; eye watering; runny nose; cough; shortness of breath; salivation; nausea, stomach cramps, diarrhea, and vomiting; increased blood pressure; profuse sweating; hypermotility, hallucinations; agitation; tingling of the skin; slow heartbeat; convulsions; fluid in lungs; loss of consciousness; incontinence; breathing stops; death. Carbamate insecticides inhibit the cholinesterase activity of enzymes, causing accumulation of acetylcholine at synapses and altering the way in which nervous impulses are transmitted. However, within several hours carbamates spontaneously detach from the enzymes.
*Long Term Exposure:* A potent cholinesterase inhibitor; cumulative effect is possible. This chemical may damage the nervous system with repeated exposure, resulting in convulsions, respiratory failure. May cause liver damage.
*Points of Attack:* Respiratory system, central nervous system, cardiovascular system, skin, eyes, plasma and red blood cell cholinesterase.
*Medical Surveillance:* Medical observation is recommended for 24 to 48 hours after breathing overexposure, as pulmonary edema may be delayed. As first aid for pulmonary edema, consider administering a

corticosteroid spray. Cigarette smoking may exacerbate pulmonary injury and should be discouraged for at least 72 hours following exposure. Before employment and at regular times after that, the following are recommended: Plasma and red blood cell cholinesterase levels (tests for the enzyme poisoned by this chemical). If exposure stops, plasma levels return to normal in 1-2 weeks while red blood cell levels may be reduced for 1-3 months. When acetylcholinesterase enzyme levels are reduced by 25% or more below preemployment levels, risk of poisoning is increased, even if results are in lower ranges of "normal." Reassignment to work not involving carbamate pesticides is recommended until enzyme levels recover. If symptoms develop or overexposure occurs, repeat the above tests as soon as possible and get an exam of the nervous system. Also consider complete blood count. Consider chest x-ray following acute overexposure.

*First Aid:* Speed in removing material from eyes and skin is of extreme importance. *Eyes:* Eye contact can cause dangerous amounts of these chemicals to be quickly absorbed through the mucous membrane into the bloodstream. Immediately and gently flush eyes with plenty of warm or cold water (NO hot water) for at least 15 minutes, occasionally lifting the upper and lower eyelids. Get medical aid immediately. *Skin:* Get medical aid. Skin contact can cause dangerous amounts of these chemicals to be absorbed into the bloodstream. Wearing the appropriate PPE equipment and respirator for carbamate pesticides, immediately flush skin with plenty of soap and water for at least 15 minutes while removing contaminated clothing and shoes. Shampoo hair promptly if contaminated; protect eyes. *Ingestion:* Call poison control. Loosen all clothing. Never give anything by mouth to an unconscious person. Get medical aid. Do NOT induce vomiting.* If conscious, alert, and able to swallow, rinse mouth and have victim drink 4 to 8 ounces of water. Check to see if poison control instructs you to use ipecac syrup, otherwise administer slurry of activated charcoal (2 oz in 8 oz of water). If victim is UNCONSCIOUS OR HAVING CONVULSIONS, do nothing except keep victim warm. **In some cases you may be specifically instructed by poison control to induce vomiting by way of 2 tablespoons of syrup of ipecac (adult) washed down with a cup of water.* Do NOT give activated charcoal before or with ipecac syrup. *Inhalation:* Get medical aid. Do not contaminate yourself. Wearing the appropriate PPE equipment and respirator for carbamate pesticides, immediately remove the victim from the contaminated area to fresh air. If the victim is not breathing, administer artificial respiration. Do NOT use mouth-to-mouth resuscitation; use bag/mask apparatus. If breathing is difficult, administer oxygen through bag/mask apparatus until medical help arrives. Do not leave victim unattended.

*Note to physician or authorized medical personnel.* Administer atropine, 2 mg (1/30 gr) intramuscularly or intravenously as soon as any local or systemic signs or symptoms of an intoxication are noted; repeat the administration of atropine every 3 to 8 minutes until signs of atropinization (mydriasis, dry mouth, rapid pulse, hot and dry skin) occur; initiate treatment in children with 0.05 mg mg/kg of atropine; repeat at 5 to 10 minute intervals. Watch respiration, and remove bronchial secretions if they appear to be obstructing the airway; intubate if necessary. *Medical note:* 2-PAMCI may be contraindicated in the case of some carbamate poisonings.
*References:*
- California Environmental Protection Agency Chemical List of Lists, Sacramento CA (February 1997).
- Pesticide Management Education Program, "Aldoxycarb (Standak) EPA Pesticide Fact Sheet 1/86," Cornell University, Ithaca, NY (January 1986). http://pmep.cce.cornell.edu/profiles/insect-mite/abamectin-bufencarb/aldoxycarb/insect-prof-aldoxycarb.html

# Aldrin

*Use Type:* Insecticide and termiticide
*CAS Number:* 309-00-2
*Formula:* $C_{12}H_8Cl_6$
*Alert:* There is no manufacture, sale, or commercial use permitted for aldrin in the U.S. EPA banned all uses of aldrin in 1974, except to control termites. In 1987, EPA banned all uses. Human toxicity (long-term): Very high.
*Synonyms:* Aldrine (French); Aldrina (Spanish); 1,4,5,8-Dimethanonaphthalene,1,2,3,4,10,10-hexachloro-1,4,4a,5,8,8a-hexahydro-(1α,4α,4β,5α,8α,8β)-; 1,4:5,8-Dimethanonaphthalene,1,2,3,4,10,10-hexachloro-1,4,4a,5,8,8a-hexahydro-, *endo-exo-*; ENT 15949; Hexachlorohexahydro-*endo-exo*-dimethanonaphthalene; 1,2,3,4,10,10-Hexachloro-1,4,4a,5,8,8a-hexahydro-1,4,5,8-dimethanonaphthalene; 1,2,3,4,10,10-Hexachloro-1,4,4a,5,8,8a-hexahydro-*exo*-1,4-*endo*-5,8-dimethanonaphthalene; 1,2,3,4,10,10-Hexachloro-1,4,4A,5,8,8A-Hexahydro-1,4-*endo-exo*-5,8-dimethanonaphthalene; HHDM; HHDN; HHPN; Octalene; 1,2,3,4,10-10-Hexachloro-1,4,4a,5,8,8a-hexahydro-1,4,5,8-*endo-exo*-dimethanonaphthalene
*Trade Names:* ALDOCIT®, canceled; ALDREC®, canceled; ALDRIN 37 EQUIVALENT SOLUTION, Zeneca (USA) (now Syngenta), canceled; ALDREX®, Shell Chemical (Netherlands), canceled; ALDREX-30®, Shell Chemical (Netherlands), canceled; ALDREX-40®, Shell Chemical (Netherlands), canceled; ALDRITE®, Shell Chemical (Netherlands), canceled; ALDRON®, canceled; ALDROSOL®, canceled; ALGRAN® canceled; ALTOX® All India Medical Corp. (India), canceled;

COMPOUND 118®, canceled; DAVCO®, W.R. Grace & Company (USA), canceled, 10/10/1989; DRINOX®, canceled; KORTOFIN®, canceled; MASTER BRAND®, canceled; OCTALENE®, canceled; OMS-194, canceled; PRENTOX®, Prentiss Inc. (USA), canceled; ROYAL BRAND®, canceled; ROYSTER®, canceled, 12/19/1988; SEEDRIN®; SD 2794, canceled; TATUZINHO®, canceled; TIPULA®, canceled; TOXADRIN®, canceled

*Producers:* Bharat Pulverizing Mills (India); Ehrenstorfer, Dr. (Germany); IMC Vigoro, see Vigoro (Canada);, Prentiss Inc. (USA); Rhone-Poulenc Agro France (France); Shell Chemical (Netherlands); Sigma-Aldrich Laborchemikalien (Germany); Zeneca Agro (UK) (now Syngenta) (Switzerland); Vigoro (Canada)

*Chemical Class:* Organochlorine, Halo-organics

*EPA/OPP PC Code:* 045101

*California DPR Chemical Code:* 9

*ICSC Number:* 0774

*RTECS Number:* IO2100000

*EEC Number:* 602-048-00-3

*EINECS Number:* 206-215-8.

*Uses:* There is no sale or commercial use permitted for aldrin. Formerly used as an insecticide against termites, grasshoppers and worms in the U.S. Some people may be exposed to aldrin (and dieldrin) in air, water and food because of its persistence in the environment. See dieldrin for more details.

*Human toxicity (long-term)*[77]*:* Extra high–0.02058 ppb, CHCL (Chronic Human Carcinogen Level)

*Fish toxicity (threshold)*[77]*:* Extra high–0.18049 ppb, MATC (Maximum Acceptable Toxicant Concentration)

*Carcinogen/Hazard Classifications*

**U.S. EPA Carcinogens:** Group B2, probable carcinogen

**California Prop. 65:** Carcinogen

**IARC:** Group 3, unclassifiable

**Label Signal Word:** DANGER

**Endocrine Disruptor:** Suspected endocrine disruptor

*Regulatory Authority:*

- Carcinogen (Probable Human) (USPHS) (See reference above), California, New Jersey, NIOSH[2] (potential occupational carcinogen)
- Banned or Severely Restricted (Many Countries) (UN)[13]
- Air Pollutant Standard Set (ACGIH)[1] (NIOSH)[2] (DFG)[3] (HSE)[33] (Former USSR)[43] (Several States)[60]
- (OSHA)[58] (Several Canadian Provinces) (Australia) (Israel)(Mexico)
- Water Pollution Standard Set (U.S. EPA)(Mexico)
- List of priority pollutants (U.S. EPA)
- AB 1803-Well Monitoring Chemical (CAL)
- EPA/SARA 302 (EPCRA) Extremely hazardous substances
- AB 2588-Air Toxics "Hot Spots" Chemicals (CAL)
- Proposition 65 chemical (CAL)
- Permissible Exposure Limits for Chemical Contaminants (CAL/OSHA)
- The "Director's List" (CAL/OSHA)
- Clean Water Act: 40CFR§116.4 Hazardous Substances; 40CFR§117.3 (same as CERCLA); 40CFR Part 423, Appendix A Priority Pollutants; 57FR41331 Priority Chemicals; 40CFR§401.15 Toxic Pollutant
- EPA Hazardous Waste Number (RCRA No.): P004
- RCRA 40CFR261, Appendix 8; 40CFR261.11 Hazardous Constituents
- RCRA 40CFR§268.48; 61FR15654, Universal Treatment Standards: Wastewater (mg/L), 0.021; Nonwastewater (mg/kg), 0.066
- RCRA 40CFR264, Appendix 9; Ground Water Monitoring List: Suggested methods (PQL ug/L): 8080(0.05); 8270(10)
- CERCLA/SARA 40CFR302, Extremely Hazardous Substances: Extremely Hazardous Substances: TPQ = 500/10,000 lb (227/4540 kg)
- Superfund/EPCRA 40CFR302.4, Appendix A, RQ: 1 lb (0.454 kg); SARA 313: Form R *de minimis* Concentration Reporting Level: 1.0%. Aldrin has been found in at least 207 of the 1,613 National Priorities List sites identified by the Environmental Protection Agency (EPA).
- U.S. DOT Regulated Marine Pollutant (49CFR172.101, Appendix B)

*Description:* Aldrin, is a colorless, crystalline solid. The technical grade is a tan to dark brown solid. It has a mild, chemical odor. Practically insoluble in water. Molecular weight = 364.89. Melting/Freezing point = 104°C.(pure); 49–60°C (technical grade). Vapor pressure = 6.6 x $10^{-6}$ mmHg. Although noncombustible, Aldrin may be dissolved in a flammable solvent that may change its physical and toxicological properties. Odor threshold = 0.3 mg/m$^3$. Hazard Identification (based on NFPA-704 M Rating System): Health 4, Flammability 0, Reactivity 0. Log $K_{ow}$ = 7.4. Values at or above 3.0 are likely to bioaccumulate in marine organisms.

*Incompatibilities:* Avoid concentrated mineral acids, acid catalysts, acid oxidizing agents, phenol, or active metals.

*Permissible Exposure Limits in Air:* The OSHA[2], ACGIH[1], and NIOSH[2] TWA is 0.25 mg/m$^3$ as is the DFG MAK[3] (total dust), Australian, (HSE)[33] (U.K.), Mexico and Israel value, with the notation "skin" indicating potential for cutaneous adsorption. (HSE)[33] and Mexico suggests a STEL of 0.75 mg/m$^3$. NIOSH[2] has recommends that Aldrin be held to the lowest feasible concentration (LFC). The NIOSH[2] IDLH level = 25 mg/m$^3$, with the notation [potential occupational carcinogen]. The former USSR-UNEP/IRPTC project[43] has set a MAC in workplace air of 0.01 mg/m$^3$. In addition, several states have set guidelines or standards for Aldrin in ambient air[60]: 0.035 $\mu$g/m$^3$ (Pennsylvania) to 0.595 $\mu$g/m$^3$ (Kansas) to 1.5 $\mu$g/m$^3$

(Connecticut) to 2.5 $\mu g/m^3$ (North Dakota) to 4.0 $\mu g/m^3$ (Virginia) to 6.0 $\mu g/m^3$ (Nevada).

*Determination in Air:* A filter plus bubbler containing isooctane followed by workup with isooctane and analysis by gas chromatography. See NIOSH Method #5502[18].

*Permissible Concentration in Water:* The U.S. EPA limits the amount of aldrin that may be present in drinking water to 0.001 and 0.002 milligrams per liter (mg/L) of water, respectively, for protection against health effects other than cancer. The U.S. EPA has determined that a concentration of aldrin of 0.0002 mg/L in drinking water limits the lifetime risk of developing cancer from exposure to each compound to 1 in 10,000. The former USSR-UNEP/IRPTC project[43] has set a MAC of 0.01 mg/L in water used for domestic purposes. Mexico limit in drinking water 0.00003 mg/L. In addition, several states have set guidelines and standards for aldrin in drinking water[61]. Illinois has set a standard of 0.1 $\mu g/L$. Guidelines in other states range from 0.013 $\mu g/L$ (Kansas) to 0.03 $\mu g/L$ (Minnesota) to 0.05 $\mu g/L$ (California). Aldrin is highly toxic to aquatic organisms and every care must be taken to avoid release to the environment. In the food chain important to humans, bioaccumulation takes place, specifically in aquatic organisms. It is strongly advised not to let the chemical enter into the environment because of its persistence.

*Determination in Water:* Gas chromatography (EPA Method 608) or gas chromatography plus mass spectrometry (EPA Method 625)

*Routes of Entry:* Inhalation, skin absorption, ingestion and eye and skin contact.

*Harmful Effects and Symptoms*

*Short Term Exposure:* High exposure can cause headache, dizziness, nausea and vomiting, muscle jerks, sever seizure and death. Eye and skin contact can cause irritation and burns. Can be fatal if swallowed or absorbed through the skin. Aldrin tends to produce convulsions before other, less serious signs of illness have appeared. Victims have reported headache, nausea, vomiting, dizziness, and mild clonic jerking. Some victims have convulsions without warning. Poisoning by Aldrin usually involves convulsions due to its effects on the central nervous system. Probable oral lethal dose for humans is between 7 drops and one ounce for a 150 pound adult human.

*Long Term Exposure:* Aldrin accumulates in the human body. It may cause tumors, cancer, mutations, and reproductive effects. Liver effects have also been reported. Drinking alcohol can increase the liver damage caused by aldrin. Aldrin may be a teratogen in humans since it has been shown to be a teratogen in animals. Aldrin may decrease fertility in males and females. It is classified as an extremely toxic chemical. Conflicting reports of carcinogenicity of this compound remain an area of controversy. Similar chemically and toxicologically to dieldrin.

*Points of Attack:* Central nervous system, liver, kidneys and skin.

*Medical Surveillance:* Consider the points of attack in preplacement and periodic physical examinations. Laboratory tests can measure aldrin and dieldrin in blood, urine, and body tissues. These tests are not routinely available because they require specialized equipment. Aldrin changes to dieldrin fairly quickly in the body, the test has to be done shortly after you are exposed to aldrin. Dieldrin can stay in the body for months, measurements of dieldrin can be made much longer after exposure to either aldrin or dieldrin. The tests cannot tell you whether harmful health effects will occur.

*First Aid:* If this chemical gets into the eyes, remove any contact lenses at once and irrigate immediately for at least 15 minutes, occasionally lifting upper and lower lids. Seek medical attention immediately. If this chemical contacts the skin, remove contaminated clothing and wash immediately with soap and water. Seek medical attention immediately. If this chemical has been inhaled, remove from exposure, begin rescue breathing (using universal precautions) if breathing has stopped and CPR if heart action has stopped. Transfer promptly to a medical facility. When this chemical has been swallowed, get medical attention. Give a slurry of activated charcoal in water to drink. Do NOT induce vomiting.

*References:*
- U.S. Environmental Protection Agency, Aldrin/Dieldrin: Ambient Water Quality Criteria, Washington, DC (1979).
- U.S. Environmental Protection Agency, Aldrin, Health and Environmental Effects Profile No. 8, Washington, DC, Office of Solid Waste (April 30, 1980).
- Sax, N.I., Ed., Dangerous Properties of Industrial Materials Report, 1, No. 5, 31-32 (1981) and 3, No. 5, 25-29 (1983), and 5, No. 2, 23-39 (1988).
- U.S. Environmental Protection Agency, "Chemical Profile: Aldrin," Washington, DC, Chemical Emergency Preparedness Program (November 30, 1987).
- New Jersey State Department of Health and Senior Services, "Hazardous Substance Fact Sheet: Aldrin," Trenton, NJ, (January 2001). http://www.state.nj.us/health/eoh/rtkweb/0033.pdf
- U.S. Public Health Service, "Toxicological Profile for Aldrin/Dieldrin," Atlanta, Georgia, Agency for Toxic Substances and Disease Registry (November 1987).
- California Environmental Protection Agency Chemical List of Lists, Sacramento CA (February 1997)

# Allethrins

*Use Type:* Non-systemic insecticide
*CAS Number:* 584-79-2 (I); 497-92-7 (II); 23031-36-9 (Prallethrin); 28434-00-6 (S-Bioallethrin); 28057-48-9 (Bioallethrin)

## Allethrins

*Formula:* $C_{19}H_{26}O_3$ (I); $C_{10}H_{26}O_5$ (II)

*Synonyms:* Allethrolone ester of chrysanthemum dicarboxylic acid monomethyl ester (II); (+)-Allelrethonyl; (+)-*cis,trans*-chrysanthemate; *d*-Allethrin; Allethrin I; Allyl cinerin; Allyl homolog of cinerin I; *d*, l-2-allyl-4-hydroxy-3-methyl-2-cyclopenten-1-one-*dl*-chrysanthemum monocarboxylate; 3-Allyl-4-keto-2-methylcyclopentenyl chrysanthemummonocarboxylate 3; 3-Allyl-2-methyl-4-*oxo*-2-cyclopenten-1-yl chrysanthemate; *dl*-3-allyl-2-methyl-4-oxocyclopent-2-enyl *dl-cis trans* chrysanthemate; Allylrethronyl *dl-cis*-trans-chrysanthemate; Bioaltrina; Cinerin I allyl homolog; *d*-Cisallethrin; (±)-*cis,trans*-Chrysanthemumic acid ester of (±) allethrolone; Depallethrin; EBT; ENT 17,510 (II); FDA 1446; 3-(3-Methoxy-2-methyl-3-*oxo*-1-propenyl)-2,2-dimethylcyclopropanecarboxylic acid 2 methyl-4-*oxo*-3-(2-propenyl)-2-cyclopenten-1-yl ester (II); Necarboxylic acid; OMS 468; Pallethrine (France); Synthetic pyrethrins

*Trade Names:* AGWAY FOOD PLANT FOGGING SPRAY®, Agway Inc. (USA), canceled; ALLEVIATE®, canceled 8/28/2002; BIOALLETHRIN®, Bayer CropScience (Germany); BIOALLETHRIN TECHNICAL®, Valent Biosciences Corp. (USA); *d*-CISALLETHRIN®; *d*-TRANS (TM) INTERMEDIATE 1828® Mclaughlin Gormley King Co. (USA), canceled; ESBIOTHRIN®; EXTHRIN®FMC 249®; MGK 264®, Mclaughlin Gormley King Co. (USA), canceled 7/1/1987; MGK INTERMEDIATE 10®, Mclaughlin Gormley King Co. (USA), canceled; NIA 249®; PYNAMIN®, Sumitomo Chemical Co., Ltd. (Japan); Mclaughlin Gormley King Co. (USA); PYNAMIN-FORTE®, Sumitomo Chemical Co., Ltd. (Japan); PYRESIN®; PYRESYN®; PYREXCEL®; PYROCIDE®; SBP 1382/BIOALLETHRIN CONCENTRATE®, Valent Biosciences Corp (USA); WHITMIRE PT 527 WITH ALLETHRIN®, Whitmire Micro-Gen Research Laboratories Inc. (USA), canceled 7/11/2001.

*Producers:* Agway Inc. (USA); Bayer CropScience (Germany); Changzhou Kangmei Chemical Industry Co., Ltd. (China); Hockley International Ltd. (UK); Ki-Hara Chemicals Ltd. (UK); Mclaughlin Gormley King Co. (USA); Navy Brand Manufacturing Co. (USA); Shenzhen Guomeng Industry Co., Ltd. (China); Shandong Huayang Pesticide Group (China); Shanghai Agricultural Chemical Industry Corp. (China); Sigma-Aldrich Laborchemikalien (Germany); Sumitomo Chemical Co., Ltd. (Japan); SuYan Agrochemical Group (China); Valent Biosciences Corporation (USA); Whitmire Micro-Gen Research Laboratories Inc. (USA); Zago Asia Ltd. (Singapore)

*Chemical Class:* Pyrethroids or synthetic pyrethrins

*EPA/OPP PC Code:* 004001 (Allethrin); 004002 (Allethrin Coil); 004003 [Bioallethrin (a mixture of *d-cis-trans*-Allethrin and *S*-Bioallethrin]; 004004 (*S*-Bioallethrin); 004005 (*d-cis-trans*-Allethrin); 128722 (Prallethrin)

*California DPR Chemical Code:* 12 (Allethrin); 90012 (Allethrin, other related)

*ICSC Number:* 212

*RTECS Number:* GZ1925000; GZ1472000 (*S*-Bio-allethrin).

*EEC Number:* 006-025-00-3

*EINECS Number:* 209-542-4

*Uses:* Allethrin is used almost exclusively to control flying and crawling insects in homes and industrial locations. Used extensively in pet animal shampoos, to treat lice in humans and in home and industrial sprays for flying insects, mosquitos, etc. It is available as mosquito coils, mats, oil formulations and as an aerosol spray. It may be hazardous to the environment; special attention should be given to fish and honey bees.

*US. Maximum Allowable Residue Levels for Allethrin (40 CFR 180.113 et sec.)*

| CROP | ppm |
|---|---|
| Apple | - |
| Apple, post-h | 4.0 |
| Blackberry, post-h | 4.0 |
| Boysenberry, post-h | 4.0 |
| Blueberry (huckleberry), post-h | 4.0 |
| Citrus | - |
| Currant, post-h | 4.0 |
| Dewberry, post-h | 4.0 |
| Loganberry, post-h | 4.0 |
| Raspberry, post-h | 4.0 |
| Gooseberry, post-h | 4.0 |
| Grape, post-h | 4.0 |
| Orange, post-h | 4.0 |
| Crabapples, post-h | 4.0 |
| Pear | - |
| Pear, post-h | 4.0 |
| Cherry, post-h | 4.0 |
| Peach | - |
| Peach, post-h | 4.0 |
| Plum (fresh prunes), post-h | 4.0 |
| Fig, post-h | 4.0 |
| Guava, post-h | 4.0 |
| Mango, post-h | 4.0 |
| Pineapple,, post-h | 4.0 |
| Horseradish | - |
| Muskmelon, post-h | 4.0 |
| Tomato | - |
| Tomato, post-h | 4.0 |
| Barley, grain,, post-h | 2.0 |
| Oat, grain, post-h | 2.0 |
| Rye, grain, post-h | 2.0 |
| Sorghum, grain, milo | - |
| Sorghum, grain, milo, post-h | 2.0 |
| Wheat, grain, post-h | 2.0 |
| Beet, sugar | - |

| | |
|---|---|
| Corn | - |
| Corn, grain, post-h | 2.0 |
| Pepper | - |
| Foods, processed | 1.0 |

*Carcinogen/Hazard Classifications*
**Label Signal Word:** CAUTION. Containers of technical grade *d*-trans-allethrin have the Signal Word WARNING.
**WHO Acute Hazard:** Group II, moderately hazardous
**Endocrine Disruptor:** Suspected endocrine disruptor
*Regulatory Authority:*
- AB 1803-Well Monitoring Chemical (CAL) as pyrethrins
- Permissible Exposure Limits for Chemical Contaminants (CAL/OSHA) as pyrethrum
- Clean water act: Section 311 Hazardous Substances/RQ (same as CERCLA) as pyrethrins
- Actively registered pesticide in California.
- EPCRA Section 304 RQ: CERCLA, 1 lb (0.454 kg) as pyrethrins

*Description:* Allethrins are synthetic analogs of naturally occurring insecticides. Clear, yellow to amber, oily liquids which is also available as wettable powder or granules. Sprays may be dissolved in xylene or kerosene. Slight aromatic odor. Practically insoluble in water (I & II). Molecular weight = 302.39 (*S*-bioallethrin). Boiling point = approx. 140–160°C. Flash point = approx 120–123°C. Hazard Identification (based on NFPA-704 M Rating System): Health 1, Flammability 1, Reactivity 0. Log $K_{ow}$ = 4.75–4.80. Values at or above 3.0 are likely to bioaccumulate in marine organisms. May be combined with organophosphates.

*Incompatibilities:* Strong alkalies and oxidizers. Unstable in light, UV, air, and alkaline conditions.
*Permissible Exposure Limits in Air:* Not established.
*Determination in Air:* Collection by impinger or fritted bubbler, analysis by gas liquid chromatography/ultraviolet. See NIOSH IV[18], Method #5008, Pyrethrum.
*Routes of Entry:* Skin, inhalation.
*Harmful Effects and Symptoms*
*Short Term Exposure:* Skin and eye contact causes irritation and burns. Inhalation can cause respiratory tract irritation with coughing and wheezing. High exposure may cause dizziness, shaking, irritability, seizures, and unconsciousness. Allethrin may cause effects on the nervous system.
*Long Term Exposure:* May cause skin allergy. If the allergy develops, very low future exposure can cause itching and skin rash. Allethrin may cause an asthma-like allergy. Future exposure can cause asthma attacks with shortness of breath, wheezing, cough and/or chest tightness. Allethrin can cause bronchitis to develop with cough, phlegm, and/or shortness of breath. This chemical may cause liver and kidney damage. There is no evidence that allethrin affects reproduction.[2] See also pyrethrins.
*Points of Attack:* Skin, lungs, liver, and kidneys.
*Medical Surveillance:* Liver and kidney function tests. Lung function test. (These could be normal if the person is not having an attack at the time of the test). Evaluation by a qualified allergist may help to diagnose skin allergy.
*First Aid:* If this chemical gets into the eyes, remove any contact lenses at once and irrigate immediately for at least 15 minutes, occasionally lifting upper and lower lids. Seek medical attention immediately. If this chemical contacts the skin, remove contaminated clothing and wash immediately with soap and water. Seek medical attention immediately. If this chemical has been inhaled, remove from exposure, begin rescue breathing (using universal precautions) if breathing has stopped and CPR if heart action has stopped. Transfer promptly to a medical facility. When this chemical has been swallowed, get medical attention. Give large quantities of water and induce vomiting. Do not make an unconscious person vomit.
*References:*
- New Jersey Department of Health and Senior Services, Hazardous Substance Fact Sheet, "Allethrin," Trenton NJ (December, 1998).
  http://www.state.nj.us/health/eoh/rtkweb/2102.pdf
- EPA, Office of Pesticide Programs, Pesticide Residue Limits, "Allethrin", 40 CFR 180.113 et sec;
  www.epa.gov/cgi-bin/oppsrch
- California Environmental Protection Agency Chemical List of Lists, Sacramento CA (February 1997)

# Allidochlor

*Use Type:* Herbicide
*CAS Number:* 93-71-0
*Formula:* $C_8H_{12}ClNO$
*Synonyms:* Acetamide, 2-chloro-*N*,*N*-di-2-propenyl-; Alidochlor; CDAA; CDAAT; 2-Chloro-*N*,*N*-diallylacetamide; α-Chloro-*N*,*N*-diallylacetamide; 2-Chloro-*N*,*N*-di-2-propenylacetamide; Diallylchloroacetamide; *N*,*N*-Diallylchloroacetamide; *N*,*N*-Diallyl-α-chloroacetamide; *N*,*N*-Diallyl-2-chloroacetamide; NCI-C04035
*Trade Names:* ACTOX®, canceled; CP 6343®; RADOX®; RANDOX®, Monsanto (USA), canceled; RANTOX T®; VEGA-RAND®, Helena Chemical (USA), canceled
*Producers:* Helena Chemical (USA); Monsanto (USA)
*Chemical Class:* Organochlorine; Halo-organics; acetamide herbicide
*EPA/OPP PC Code:* 019301
*California DPR Chemical Code:* 114
*RTECS Number:* AB5250000
*Uses:* There are no products registered with the U.S. EPA; all tolerances were revoked on July 21, 1999. It was primarily used to control weeds growing in onion crops. Used as a pre-emergence and post-emergence control for most annual grasses and broadleaf weeds on corn, sorghum,

lima beans, snap beans, soybeans, cabbage, peas for canning, celery, onions and some fruits and ornamentals.

*Fish toxicity (threshold)*[77]: Low–269.85657 ppb, MATC (Maximum Acceptable Toxicant Concentration)

***Carcinogen/Hazard Classifications***

**Label Signal Word:** WARNING

*Description:* An oily, amber liquid. Slightly irritating odor. Slightly soluble in water. Boiling point = 116°C; 74°C @ 0.3 mm. Molecular weight = 173.66. Vapor pressure = $1.0 \times 10^{-2}$ mmHg.

*Incompatibilities:* Oxidizers (chlorates, nitrates, peroxides, permanganates, perchlorates, chlorine, bromine, fluorine. etc); strong acids. Slowly hydrolyzes in water, releasing ammonia and forming acetate salts.

*Permissible Concentration in Water:* No criteria set. Runoff from spills or fire control may cause water pollution.

*Routes of Entry:* Inhalation, ingestion. Absorbed through the intact skin.

***Harmful Effects and Symptoms***

*Short Term Exposure:* Attacks the central nervous system. Apprehension, anxiety, confusion, nervous excitation; dizziness; headache; numbness and weakness in limbs; muscle twitching, tremors; nausea and vomiting; slow, shallow respiration, bluish face; convulsions; loss of consciousness; breathing stops; death.

*Long Term Exposure:* May cause kidney damage.

*Points of Attack:* May be fatal if inhaled, ingested, or absorbed through the skin

*Medical Surveillance:* Medical observation is recommended for 24 to 48 hours after breathing overexposure, as pulmonary edema may be delayed. As first aid for pulmonary edema, consider administering a corticosteroid spray. Cigarette smoking may exacerbate pulmonary injury and should be discouraged for at least 72 hours following exposure.

*First Aid:* Speed in removing material from eyes and skin is of extreme importance. *Eyes:* Contact can cause dangerous amounts of these chemicals to be quickly absorbed through the mucous membrane into the bloodstream. Directly, irrigate with large amounts of plain, tepid water or saline for 20 minutes, occasionally lifting the lower and upper lids. During this time, remove contact lenses, if easily removable without additional trauma to the eye. Get medical aid immediately. Have physician check for possible delayed damage. *Skin:* Get medical aid. Skin contact can cause dangerous amounts of these chemicals to be absorbed into the bloodstream. Wearing the appropriate PPE equipment and respirator for organochlorine pesticides, immediately flush exposed skin, hair, and under nails with plain, running, tepid water for 20 minutes, then wash twice with mild soap. Shampoo hair promptly if contaminated; protect eyes. Do not scrub skin or hair, since this can increase absorption through the skin. Rinse thoroughly with water. Victims who are able and cooperative may assist with their own decontamination. Remove and double-bag contaminated clothing and personal belongings. Leather absorbs many organochlorines; therefore, items such as leather shoes, gloves, and belts should be discarded. If the skin is swollen or inflamed, cool affected areas with cold compresses. *Ingestion:* Call poison control. Loosen all clothing. Never give anything by mouth to an unconscious person. Get medical aid. In cases of ingestion, Do not induce vomiting \*; the patient is at risk of CNS depression or seizures, which may lead to pulmonary aspiration during vomiting. If the victim is conscious and able to swallow, \*administer an aqueous slurry of activated charcoal at 1 gm/kg (usual adult dose 60–90 g, child dose 25–50 g). A soda can and straw may be of assistance when offering charcoal to a child. The efficacy of activated charcoal for some organochlorine poisoning (such as chlordane) is uncertain. If victim is UNCONSCIOUS OR HAVING CONVULSIONS, do nothing except keep victim warm. *\*In some cases you may be specifically instructed by Poison Control to induce vomiting by way of 2 tablespoons of syrup of ipecac (adult) washed down with a cup of water.* Do NOT give activated charcoal <u>before or with</u> ipecac syrup. *Inhalation:* Get medical aid. Do not contaminate yourself. Wearing the appropriate PPE equipment and respirator for organochlorine pesticides, immediately remove the victim from the contaminated area to fresh air. For inhalation exposures, monitor for respiratory distress. If the victim is not breathing, administer artificial respiration. Do NOT use mouth-to-mouth resuscitation; use bag/mask apparatus. If cough or breathing difficulty develops, evaluate for respiratory tract irritation, bronchitis, or pneumonitis. If breathing is difficult, administer 100% humidified supplemental oxygen through bag/mask apparatus until medical help arrives. Do not leave victim unattended.

*References:*
- Pesticide Management Education Program, "Allidochlor (Randox) Herbicide Profile 3/85," Cornell University, Ithaca, NY (March 1985). http://pmep.cce.cornell.edu/profiles/herb-growthreg/24-d-butylate/allidochlor/herb-prof-allidochlor.html
- California Environmental Protection Agency Chemical List of Lists, Sacramento CA (February 1997).

## Allyl Alcohol

*Use Type:* Fungicide and herbicide.
*CAS Number:* 107-18-6
*Formula:* $C_3H_6O$; $CH_2CHCH_2OH$
*Alert:* May be fatal if swallowed.
*Synonyms:* AA; Alcool allilco (Italian); Alcool allylique (French); Alilico alcohol (Spanish); Allyl Al; Allylalkohol (German); Allylic alcohol; 3-Hydroxypropene; Orvinylcarbinol; Propenol; 2-Propenol; 2-Propen-1-ol;

Propen-1-ol-3; 1-propen-3-ol; Propenyl alcohol; 2-Propenyl alcohol; Vinyl carbinol; Vinyl carbinol,2-propenol

*Trade Names:* SHELL UNDRAUTTED A®; WEED DRENCH®

*Producers:* BP Chemicals (UK); Lyondell Chemical (USA); Showa Denko (Japan); Sigma-Aldrich Fine Chemicals (USA)

*Chemical Class:* Alcohol/Ether; organics, non-halogenated

*EPA/OPP PC Code:* 068401

*California DPR Chemical Code:* 3023

*ICSC Number:* 0095

*RTECS Number:* BA5075000

*EEC Number:* 603-015-00-6

*EINECS Number:* 203-470-7

*Uses:* Allyl alcohol is used in the production of allyl esters. These compounds are used as monomers and prepolymers in the manufacture of resins and plastics. Allyl alcohol is also used in the preparation of pharmaceuticals, in organic syntheses of glycerol and acrolein and warfare gas.

*Carcinogen/Hazard Classifications*

**WHO Acute Hazard:** Class 1b, highly hazardous

*Regulatory Authority:*
- Toxic Substance (World Bank)[15]
- EPA/SARA 302 (EPCRA) Extremely hazardous substances
- AB 2588-Air Toxics "Hot Spots" Chemicals (CAL)
- U.S. DOT Inhalation Hazard Chemicals
- Permissible Exposure Limits for Chemical Contaminants (CAL/OSHA)
- The "Director's List" (CAL/OSHA)
- Air Pollutant Standard Set (ACGIH)[1] (DFG)[3] (HSE)[33] (OSHA)[58] (Several States)[60], (Various Canadian Provinces) (Australia) (Israel) (Mexico)
- Clean Air Act 42USC7412; Title I, Part A,§112(r), Accidental Release Prevention/Flammable substances (Section 68.130) TQ = 15,000 lb (5,825 kg)
- Clean Water Act,40CFR116.4 Hazardous Substances; RQ 40CFR117.3 (same as CERCLA)
- EPA Hazardous waste number (RCRA No.): P005
- RCRA Land Ban Waste Restrictions
- CERCLA/SARA 40CFR302, Extremely Hazardous Substances: TPQ = 1000 lb (454 kg)
- Superfund/EPCRA 40CFR302.4, Appendix A, RQ: 100 lb (45.5 kg), SARA 313: Form R *de minimis* Concentration Reporting Level: 1.0%.
- Canada, WHMIS, Ingredients Disclosure List; National pollutant Release Inventory (NPRI)

*Description:* Allyl alcohol is a flammable, colorless liquid. Pungent, mustard-like odor. Soluble in water. Odor threshold = 1.4 to 2.1 ppm. The odor and irritant properties of allyl alcohol should be sufficient warning to prevent serous injury. Boiling point = 97°C. Melting/Freezing point = –48°C. Flash point = 21°C (cc); 32°C (oc). Autoignition temperature = 378°C. Explosive limits : LEL = 2.5%, UEL = 18.0%. Hazard Identification (based on NFPA-704 M Rating System): Health 4, Flammability 3, Reactivity 1. Log $K_{ow}$ = 0.17. Unlikely to bioaccumulate in marine organisms.

*Incompatibilities:* Forms explosive mixture with air. Reacts explosively with carbon tetrachloride, strong bases. Also incompatible with strong acids, oxidizing agents, metal halides, oleum, diallyl phosphate, sodium, potassium, aluminum and magnesium. Contact with oxidizers may cause fire and explosions. Polymerization may be caused by heat, peroxides, or oxidizers.

*Permissible Exposure Limits in Air:* The Federal OSHA[2] standard (TWA), ACGIH[1], NIOSH[2], DFG MAK[3], Australian, Mexico, Israel value[3] is 2 ppm (5 mg/m$^3$). ACGIH[1] and NIOSH[2] add the notation "skin" indicating potential for cutaneous absorption. The NIOSH[2], ACGIH[1], HSE[33], Israel, and Mexico STEL is 4 ppm (10 mg/m$^3$). The DFG Peak Limitation (30 min.) is 2 times normal MAK; do not exceed more than 4 times during a workshift. The Canadian Provincial TWA limits are: 2 ppm (4.7-5.0 mg/m$^3$) TWA or TWAEV and STEL is 4 ppm (9.5-10 mg/m$^3$) [Alberta, British Columbia, Ontario, Quebec]. The NIOSH[2] IDLH value = 20 ppm. Guidelines or standards for allyl alcohol in ambient air have been set[60] by various states: 5 mg/m$^3$/STEL 10 mg/m$^3$ (California), 5 mg/m$^3$ (North Dakota); 8 mg/m$^3$ (Virginia); 10 mg/m$^3$ (Connecticut); 11.9 mg/m$^3$ (Nevada).

*Determination in Air:* Adsorption on charcoal, workup with $CS_2$ and gas chromatographic analysis. See NIOSH Method #1402[18].

*Permissible Concentration in Water:* No criteria set, but runoff from spills or fire control may cause water pollution.

*Routes of Entry:* Inhalation, ingestion, eye and/or skin contact. Absorbed through the skin.

*Harmful Effects and Symptoms*

*Short Term Exposure:* May be fatal if swallowed or inhaled. Allyl alcohol vapor can cause serious irritation and burns of eyes, nose and throat. Eye irritation may be accompanied by sensitivity to light, pain, blurred vision leading to permanent damage. The pain may not begin until 6 hours after exposure. Contact with the liquid may cause first and second degree burns of skin and blister formation. Areas of contact will become swollen and painful and local muscle spasms may occur. Allyl alcohol causes burns on contact, and may cause pulmonary edema, a medical emergency, if inhaled. It is poisonous in small quantities. The probable oral lethal dose is 50-500 mg/kg, or between 1 teaspoonful and 1 ounce for a 150 lb person.

*Long Term Exposure:* Allyl alcohol may cause mutations; such chemicals may have a cancer or reproductive risk. This chemical may cause liver and kidney damage. Repeated exposure may cause bronchitis with cough, phlegm, and/or shortness of breath.

*Points of Attack:* Eyes, skin, respiratory system.

*Medical Surveillance:* Preplacement and periodic

examinations should include lung function tests, liver and kidney function tests. Following acute exposure, chest x-ray should be considered.

*First Aid:* If this chemical gets into the eyes, remove any contact lenses at once and irrigate immediately for at least 15 minutes, occasionally lifting upper and lower lids. Seek medical attention immediately. If this chemical contacts the skin, remove contaminated clothing and wash immediately with soap and water. Seek medical attention immediately. If this chemical has been inhaled, remove from exposure, begin rescue breathing (using universal precautions) if breathing has stopped and CPR if heart action has stopped. Transfer promptly to a medical facility. When this chemical has been swallowed, get medical attention. Give large quantities of water and induce vomiting. Do not make an unconscious person vomit. Medical observation is recommended for 24 to 48 hours after breathing overexposure, as pulmonary edema may be delayed.

*References:*
- U.S. Environmental Protection Agency, Allyl Alcohol, Health and Environmental Effects Profile No. 9, Washington, DC, Office of Solid Waste (April 30, 1980).
- Sax, N.I., Ed., Dangerous Properties of Industrial Materials Report, 1, No. 7, 29-31 (1981).
- U.S. Environmental Protection Agency, "Chemical Profile: Allyl Alcohol," Washington, DC, Chemical Emergency Preparedness Program (November 30, 1987).
- California Environmental Protection Agency Chemical List of Lists, Sacramento CA (February 1997).
- New Jersey Department of Health and Senior Services, "Hazardous Substance Fact Sheet: Allyl Alcohol," Trenton, NJ (June 1998).
  http://www.state.nj.us/health/eoh/rtkweb/0036.pdf

# Allyl Bromide

*Use Type:* Insecticide
*CAS Number:* 106-95-6
*Formula:* $C_3H_5Br$
*Synonyms:* Bromuro de alilo (Spanish); Bromallylene; 1-Bromo, 2-propene; 3-Bromopropene; 3-Bromopropeno (Spanish); 3-Bromopropylene; 1-Propene, 3-bromo-; 3-Bromopropylene
*Producers:* Albemarle (USA); ATOFINA Chemicals (USA); Chemada Fine Chemicals (Israel); Ocean Chemicals Group (UK); Rhone-Poulenc (France); Shell Chemical (Netherlands)
*RTECS Number:* UC7090000
*EINECS Number:* 203-446-6
*Uses:* Also used as an intermediate in the manufacture of resins, pharmaceuticals, fragrances, and other chemicals.
*Regulatory Authority:*
- Air and Water Pollutant Standard Set: see Bromine
- U.S. DOT Regulated Marine Pollutant (49CFR172.101, Appendix B)
- Canada, WHMIS, Ingredients Disclosure List

*Description:* Allyl bromide is a highly flammable, colorless to light yellow liquid. Unpleasant, pungent odor. Slightly soluble in water. Boiling point = 71.3°C. Flash point = –2°C. Autoignition temperature = 295°C. Explosive limits: LEL = 4.4%, UEL = 7.3%. Hazard Identification (based on NFPA-704 M Rating System): Health 3, Flammability 3, Reactivity 1.

*Incompatibilities:* Forms explosive mixture with air. Contact with oxidizers may cause fire and explosions. Heat or light exposure may cause decomposition and corrosive vapors.

*Permissible Exposure Limits in Air:* The Federal OSHA[2] standard for *bromine* is 0.1 ppm (TWA, 8-hour workshift). NIOSH[2] recommends an airborne exposure limit of 0.1 ppm (TWA, 10-hour workshift), with an STEL of 0.3 ppm, not to be exceeded during any 15 minute work period. ACGIH[1] recommends 0.1 ppm (TWA, 8-hour workshift), and STEL of 0.2 ppm. *Note:* It should be recognized that allyl bromide can be absorbed through the skin, thereby increasing exposure. The NIOSH[2] IDLH for *bromine* = 3 ppm.

*Routes of Entry:* Skin, inhalation.
*Harmful Effects and Symptoms*
*Short Term Exposure:* Poisonous. This chemical can be absorbed through the skin, thereby increasing exposure. Irritates eyes, skin, and respiratory tract, and can cause burns and permanent damage. Inhalation can cause respiratory tract irritation with coughing and wheezing. Exposure can cause headache, dizziness, and severe digestive irritation with pain, nausea, vomiting and diarrhea. High exposure may cause pulmonary edema, a medical emergency, with severe shortness of breath. This can cause death.

*Long Term Exposure:* Allyl bromide can cause bronchitis to develop with cough, phlegm, and/or shortness of breath. This chemical may cause liver and kidney damage, and mutations. May cause skin disorders.

*Points of Attack:* Skin, lungs
*Medical Surveillance:* If symptoms develop or overexposure is suspected, chest x-ray should be considered.
*First Aid:* If this chemical gets into the eyes, remove any contact lenses at once and irrigate immediately for at least 15 minutes, occasionally lifting upper and lower lids. Seek medical attention immediately. If this chemical contacts the skin, remove contaminated clothing and wash immediately with soap and water. Seek medical attention immediately. If this chemical has been inhaled, remove from exposure, begin rescue breathing (using universal precautions) if breathing has stopped and CPR if heart action has stopped. Transfer promptly to a medical facility. When this chemical has been swallowed, get medical attention. Give large

quantities of water and induce vomiting. Do not make an unconscious person vomit. Medical observation is recommended for 24 to 48 hours after breathing overexposure, as pulmonary edema may be delayed. As first aid for pulmonary edema, a doctor or authorized paramedic may consider administering a corticosteroid spray.

*References:*
- Pohanish, Richard P., *Haz-Mat Data for first Response, Transportation, Storage, and Security*, John Wiley & Sons, New York, 2004.
- California Environmental Protection Agency Chemical List of Lists, Sacramento CA (February 1997)
- New Jersey Department of Health and Senior Services, "Hazardous Substance Fact Sheet, Allyl Bromide," Trenton NJ (November 1998). www.state.nj.us/health/eoh/rtkweb/0038.pdf

# Allyl Isothiocyanate

*Use Type:* Fumigant
*CAS Number:* 57-06-7
*Formula:* $C_4H_5NS$
*Synonyms:* AITC; Allyl isorhodanide; Allyl isosulfocyanate; Allyl isothiocyanate, stabilized; Allyl mustard oil; Allylsenfoel (German); Allyl sevenolum; Allyl thiocarbonimide; Artificial mustard oil; Isothiocyanic acid, allyl ester; Isothiocyanate d'allyle (French); 3-Isothiocyanato-1-propene; Mustard oil; NCI-C50464; Oil of mustard, artificial; Oleum sinapis volatile; 1-Propene, 3-isothiocyanato-; 2-Propenyl isothiocyanate; Senf oel (German); Synthetic mustard oil; Volatile oil of mustard
*Trade Names:* CARBOSPOL®; REDSKIN®
*Producers:* Fluorochem Ltd. (UK); Lancaster Synthesis (UK); Penta Manufacturing (USA); Sigma-Aldrich Laborchemikalien (Germany); Tokyo Kasei Kogyo (Japan)
*Chemical Class:* Thiocyanate; essential oil
*EPA/OPP PC Code:* 004901
*California DPR Chemical Code:* 1153 (Mustard oil); 1010 (Allyl isothiocyanate)
*ICSC Number:* 0372
*RTECS Number:* NX8225000
*EINECS Number:* 200-309-2
*Uses:* Used as an animal and insect repellant, food flavorings, ointments and mustard plasters, and in the manufacture of warfare gases.
*Carcinogen/Hazard Classifications*
**IARC:** Group 3, unclassifiable
**Label Signal Word:** CAUTION
*Regulatory Authority:*
- Carcinogen (animal evidence) (NTP)[9]
- Actively registered pesticide in California.
- Hazardous Substances List (The Director's List) (CAL/OSHA)
- U.S. DOT 49CFR172.101, Inhalation Hazardous Chemical

*Description:* Highly flammable, colorless to pale yellow, oily liquid. Pungent, irritating odor and acrid taste like mustard. Insoluble in water. Molecular weight : 99.18. Boiling point = 151°C. Melting/freezing point = −103°C. Density = 1.015 @ 15°F/4°C; 1.013-1.016 @ 25°F/25°C. Vapor pressure = 10 millimeter @ 38.3°C; Vapor density = 3.41. Flash point = 46°C. Hazard Identification (based on NFPA-704 M Rating System): Health 3, Flammability 2, Reactivity 0. Log $K_{ow}$ = 2.12. Unlikely to bioaccumulate in marine organisms.
*Incompatibilities:* Oxidizers (chlorates, nitrates, peroxides, permanganates, perchlorates, chlorine, bromine, fluorine, etc); strong acids may cause fire and explosion. Also incompatible with alcohols, strong bases, and amines.
*Permissible Concentration in Water:* No criteria set. Runoff from spills or fire control may cause water pollution.
*Routes of Entry:* Inhalation and skin contact
*Harmful Effects and Symptoms*
*Short Term Exposure:* Irritates the eyes, skin and respiratory tract. Eye and skin contact can cause skin irritation. Prolonged contact can cause burns and blisters. This chemical can be absorbed through the skin, thereby increasing exposure.
*Long Term Exposure:* There is limited evidence that this chemical causes cancer in animals; it may cause bladder cancer in male rats. It may damage the developing fetus. Exposure can cause an allergy-type reaction to develop with symptoms of asthma, watery eyes, sneezing, runny nose, couching, sneezing, chest tightness. Once allergy develops, small exposures can cause symptoms to develop.
*Developmental Effects:* No reproductive or developmental effects of this thiocyanate have been reported in experimental animals or humans. Increased levels of thiocyanate in the umbilical cords of fetuses whose mothers smoked compared to those whose mothers were non-smokers suggests that thiocyanate, and possibly also cyanide, can cross the placenta.
*Points of Attack:* Eyes, respiratory system.
*Medical Surveillance:* Pre-employment and regular lung function tests are recommended (these may be normal if the person is not having an attack at the time of the test). Urine thiocyanate levels. Blood cyanide levels. If symptoms develop or if overexposure is suspected, evaluation by a qualified allergist may help diagnose skin allergy.
*First Aid:* Treatment is as for aliphatic thiocyanates. If this chemical gets into the eyes, remove any contact lenses at once and irrigate immediately for at least 30 minutes, occasionally lifting upper and lower lids. Seek medical attention immediately. If this chemical contacts the skin, remove contaminated clothing and wash immediately with soap and water. Seek medical attention immediately. If this chemical has been inhaled, remove from exposure, begin

rescue breathing (using universal precautions) if breathing has stopped and CPR if heart action has stopped. Transfer promptly to a medical facility. Victims who are conscious and able to swallow should be given 4 to 8 ounces of water or milk. Gastric lavage with a small bore NG tube should be considered if it can be performed within 1 hour after ingestion. The effectiveness of activated charcoal administration is unknown, but it is suggested following lavage (administer activated charcoal at 1 gm/kg, usual adult dose 60–90 g, child dose 25–50 g). A soda can and straw may be of assistance when offering charcoal to a child. Medical observation is recommended for 24 to 48 hours after breathing overexposure, as pulmonary edema may be delayed. As first aid for pulmonary edema, a doctor or authorized paramedic may consider administering a corticosteroid spray. *Note:* Because cyanide is probably largely responsible for poisonings, antidotal measures against cyanide should be instituted promptly. Use amyl nitrate capsules if symptoms develop. All area employees should be trained regularly in emergency measures for cyanide poisoning and in CPR. A cyanide antidote kit should be kept in the immediate work area and must be rapidly available. Kit ingredients should be replaced every 1-2 years.

*References:*
- New Jersey Department of Health and Senior Services, "Hazardous Substance Fact Sheet, Allyl Isothiocyanate," Trenton NJ (June 1992, rev. June 1998). http://www.state.nj.us/health/eoh/rtkweb/0045.pdf
- California Environmental Protection Agency Chemical List of Lists, Sacramento CA (February 1997)

# Aluminum Phosphide

*Use Type:* Fumigant, fungicide, rodenticide, and insecticide
*CAS Number:* 20859-73-8
*Formula:* AlP
*Alert:* A Restricted Use Pesticide (RUP). Metallic phosphides on clothes, skin, or hair can react with water or moisture to generate phosphine gas. Vomitus containing phosphides can also off-gas phosphine.
*Synonyms:* AlP; Aluminum fosfide (Dutch); Aluminum monophosphide; Caswell No. 031; Fosfuri di alluminio (Italian); Fosfuro aluminico (Spanish); Phosphures d'aluminum (French)
*Trade Names:* AL-PHOS®; CELPHIDE®; CELPHOS®, Excel Industries (India); DELICIA®; DETIA®; DETIA-EX-B®; DETIA GAS EX®; DETIA-GAS-EX-B®; DELICIA GASTOXIN; FARMOZ®; FUMITOXIN®, Pestcon Systems (USA); PHOSTOXIN®, Degesch America (USA); PHOSTOXIN-A® Degesch America (USA); QUICKPHOS®, United Phosphorus (India); QUICK TOX®; RENTOKIL GASTION®, Rentokil (Australia)
*Producers:* Bhageria Dye-Chem (India); Biesterfeld Siemsgluess International. GmBH (Germany); China Chemical (China); Degesch America (USA); Excel Industries (India); Hunan Tianyu Pesticide Chemical Group (China); Pestcon Systems (USA); Quantum Chemicals (Australia); Rentokil (Australia); Shenzhen Guomeng Industry Co., Ltd. (China); United Phosphorus (India); Webcot (Australia)
*Chemical Class:* Inorganic phosphide
*EPA/OPP PC Code:* 066501
*California DPR Chemical Code:* 484
*ICSC Number:* 0472
*RTECS Number:* BD1400000
*EEC Number:* 015-004-00-8
*Uses:* Used as an insecticidal fumigant for grain, peanuts, processed food, animal feed, leaf tobacco, cottonseed, and as space fumigant for flour mills, warehouses and railcars. It is also used in baits for rodent and mole control in crops. Used as a source of phosphine; in semiconductor research. Zinc phosphide is often mixed with bait food such as cornmeal, which can be a danger to pets and children. When phosphides are ingested or exposed to moisture, they release phosphine gas.
*U.S. Maximum Allowable Residue Levels for Aluminum Phosphide (40 CFR 180.225):*
*Note:* The following residue limits are for phosphine compounds that produce phosphine gas.

| CROP | ppm |
|---|---|
| Almond | 0.1 |
| Animal feed | 0.1 |
| Avocado | 0.01 |
| Banana (incl. plantains) | 0.01 |
| Barley, grain | 0.1 |
| Brazil nuts | 0.1 |
| Cabbage, chinese, bok choy | 0.01 |
| Cacao bean, dried | 0.1 |
| Cashew | 0.1 |
| Citron, citrus | 0.01 |
| Coffee, bean | 0.1 |
| Corn, field, grain | 0.1 |
| Corn, pop, grain | 0.1 |
| Cotton, undelinted seed | 0.1 |
| Date, dried | 0.1 |
| Dill, seed | 0.01 |
| Eggplant | 0.01 |
| Endive | 0.01 |
| Filbert | 0.1 |
| Grapefruit | 0.01 |
| Kumquat | 0.01 |
| Lemon | 0.01 |
| Lettuce | 0.01 |
| Lime | 0.01 |
| Mango | 0.01 |
| Millet, grain | 0.1 |
| Mushroom | 0.01 |

| | |
|---|---|
| Oat, grain | 0.1 |
| Okra | 0.01 |
| Orange | 0.01 |
| Papaya | 0.01 |
| Peanut | 0.1 |
| Pecan | 0.1 |
| Pepper, black, post-h | 0.01 |
| Pepper, red, post-h | 0.01 |
| Pepper, white, post-h | 0.01 |
| Persimmon | 0.01 |
| Pimentos | 0.01 |
| Pistachio | 0.1 |
| Processed food | 0.01 |
| Raw agricultural commodities | 0.01 |
| Rice, grain | 0.1 |
| Rye, grain | 0.1 |
| Safflower, seed | 0.1 |
| Salsify, tops | 0.01 |
| Sesame, post-h | 0.1 |
| Sorghum, grain, grain | 0.1 |
| Soybean, seed | 0.1 |
| Sunflower, seed | 0.1 |
| Sweet potato | 0.01 |
| Tangelo | 0.01 |
| Tangerine | 0.01 |
| Tomato | 0.01 |
| Vegetable, legume (crop group 6), exc soybeans | 0.01 |
| Walnut | 0.1 |
| Wheat, grain | 0.1 |

*Carcinogen/Hazard Classifications*
**U.S. EPA Carcinogens:** Not listed; Group D, unclassifiable for phosphine, its parent chemical
**Label Signal Word:** DANGER
*Regulatory Authority:*
- EPA/SARA 302 (EPCRA) Extremely hazardous substances
- The "Director's List" (CAL/OSHA)
- Actively registered pesticide in California.
- U.S. DOT 49CFR172.101, Inhalation Hazardous Chemical
- Canada, WHMIS, Ingredients Disclosure List
- EPA Hazardous Waste Number (RCRA No.): P006[5]
- RCRA, 40CFR261, Appendix 8 Hazardous Constituents
- Superfund/EPCRA 40CFR355, Appendix B Extremely Hazardous Substances: TPQ = 500 lb (228 kg)
- Superfund/EPCRA 40CFR302.4 RQ: CERCLA, 100 lb (45.4 kg)
- EPCRA Section 313 Form R *de minimis* concentration reporting level: 1.0%
- Banned or Severely Restricted (Belgium)[13]

***Description:*** Aluminum phosphide is a pyrophoric, dark gray or dark yellow crystalline solid. Has an odor like decaying fish or garlic. Melting/Freezing point = > 1000°C. Decomposes in water forming highly poisonous and flammable phosphine gas. NFPA-704 Hazard Identification (based on NFPA-704 M Rating System): Health 4, Flammability 4, Reactivity 2, Water reactive; dangerous when wet.

***Incompatibilities:*** Reacts violently with water, carbon dioxide, and foam fire extinguishers. Contact with water and bases rapidly releases highly toxic and flammable phosphine gas. Contact with steam and acids may be violent. Can ignite spontaneously in moist air.

***Permissible Exposure Limits in Air:*** The NIOSH[2] recommended airborne exposure limit for soluble aluminum salts (measured as aluminum) is 2 mg/m$^3$ TWA for a 10-hour workshift. ACGIH recommends the same criterion for an 8-hour workshift[1]. *Note:* Metallic phosphides on clothes, skin, or hair can react with water or moisture to generate phosphine gas (colorless gas; odor of garlic or decaying fish). Vomitus containing phosphides can also off-gas phosphine. For *phosphine*: OSHA PEL = 0.3 ppm (averaged over an 8-hour workshift) NIOSH[2] IDLH (immediately dangerous to life or health) = 50 ppm ERPG-2 (Emergency Response Planning Guideline) (maximum airborne concentration below which it is believed nearly all individuals could be exposed for up to 1 hour without experiencing or developing irreversible or other serious adverse health effects or symptoms that could impair an individuals's ability to take protective action) = 0.5 ppm

***Routes of Entry:*** Inhalation, ingestion.

***Harmful Effects and Symptoms***

***Short Term Exposure:*** A severe health hazard. Irritates the eye, skin and respiratory tract. Inhalation can cause lung irritation with coughing, wheezing, and shortness of breath. Affects metabolism and the central nervous system; exposure can lead to death. Higher exposures can cause pulmonary edema, a medical emergency that can be delayed for several hours. This can cause death. Acute toxicity occurs primarily by the inhalation route when aluminum phosphide decomposes into the toxic gas, phosphine. The human median lethal dose for aluminum phosphide has been reported to be 20 mg/kg. Rated as super toxic: probable oral lethal dose is less than 5 mg/kg or less than 7 drops for a 70 kg (150 lb) person. Symptoms of phosphine gas poisoning include restlessness, headache, dizziness, fatigue, nausea, vomiting, coma, convulsions; lowered blood pressure, pulmonary edema, respiratory failure, and disorders of the kidney, liver, heart, and brain may be observed.

***Long Term Exposure:*** This chemical may cause lung, kidney, and liver damage. It may be able to cause skin rash or eczema

***Points of Attack:*** Central nervous system, liver, kidney, lungs.

***Medical Surveillance:*** Lung, liver, kidney, and nervous system function tests.

***First Aid:*** If this chemical gets into the eyes, remove any contact lenses at once and irrigate immediately for at least 15 minutes, occasionally lifting upper and lower lids. Seek medical attention immediately. If this chemical contacts the skin, remove contaminated clothing and wash immediately with soap and water. Seek medical attention immediately. If this chemical has been inhaled, remove from exposure, begin rescue breathing (using universal precautions) if breathing has stopped and CPR if heart action has stopped. Transfer promptly to a medical facility. When this chemical has been swallowed, get medical attention. Give large quantities of water and induce vomiting. Do not make an unconscious person vomit. Medical observation is recommended for 24 to 48 hours after breathing overexposure, as pulmonary edema may be delayed. As first aid for pulmonary edema, a doctor or authorized paramedic may consider administering a corticosteroid spray.

*Note to physician or authorized medical personnel (Advanced Treatment for phosphine exposure):* In cases of respiratory compromise secure airway and respiration via endotracheal intubation. If not possible, perform cricothyroidotomy if equipped and trained to do so. Treat patients who have bronchospasm with aerosolized bronchodilators. The use of bronchial sensitizing agents in situations of multiple chemical exposures may pose additional risks. Consider the health of the myocardium before choosing which type of bronchodilator should be administered. Cardiac sensitizing agents may be appropriate; however, the use of cardiac sensitizing agents after exposure to certain chemicals may pose enhanced risk of cardiac arrhythmias (especially in the elderly). Consider racemic epinephrine aerosol for children who develop stridor. Dose 0.25–0.75 mL of 2.25% racemic epinephrine solution in 2.5 cc water, repeat every 20 minutes as needed, cautioning for myocardial variability. Patients who are comatose, hypotensive, or having seizures or cardiac arrhythmias should be treated according to advanced life support protocols. If evidence of shock or hypotension is observed begin fluid administration. For adults, bolus 1000 mL/hour intravenous saline or lactated Ringer's solution if blood pressure is under 80 mmHg; if systolic pressure is over 90 mmHg, an infusion rate of 150 to 200 mL/hour is sufficient. For children with compromised perfusion administer a 20 mL/kg bolus of normal saline over 10 to 20 minutes, then infuse at 2 to 3 mL/kg/hour.

*References:*
- EPA, Office of Pesticide Programs, Pesticide Residue Limits, "Phosphine Compounds that Produce Phosphine Gas", 40 CFR 180.225.,
  www.epa.gov/pesticides/food/viewtols.htm
- U.S. Environmental Protection Agency, Reregistration Eligibility Decision (RED) Facts, Aluminum and Magnesium Phosphide, EPA-738-F-98-015, Washington, DC (December, 1998).
- U.S. Environmental Protection Agency, "Chemical Profile: Aluminum Phosphide," Washington, DC, Chemical Emergency Preparedness Program (November 30, 1987).
- Agency for Toxic Substances and Disease Registry, U.S. Department of Health and Human Services, Public Health Service, "Managing Hazardous Materials Incidents," Atlanta, GA (June 2003)
- California Environmental Protection Agency Chemical List of Lists, Sacramento CA (February 1997).
- New Jersey Department of Health and Senior Services, "Hazardous Substance Fact Sheet: Aluminum Phosphide," Trenton, NJ (April 1998). http://www.state.nj.us/health/eoh/rtkweb/0063.pdf

# Aluminum Sulfate

*Use Type:* Molluscicide and plant growth regulator
*CAS Number:* 10043-01-3
*Formula:* $Al_2S_3O_{12}$, $Al_2(SO_4)_3$
*Synonyms:* Alum; Aluminum alum; Aluminum sulphate; Aluminum trisulfate; Alunogenite mineral; Cake alum (octahydrate); Dialuminum sulfate; Diaaluminum trisulfate; Paper maker's alum; Patent alum (octahydrate); Sulfato aluminico (Spanish); Sulfuric acid, aluminum salt
*Producers:* Abaquim (Mexico); Adheswara Group of Companies (India); Alcan Chemicals (Canada); Central Glass (Japan); Chongqing Chuandong Chemical (Group) (China); Coogee Chemicals (Australia); Delta (USA); Dharamsi Morarji (India); General Alum (USA); GFS Chemicals (USA); Holland (USA); Kemira Chemicals (Finland); Marsulex (Canada); Rhodia Eco (France); Sumitomo Chemical (Japan); YiHua Group (China)
*Chemical Class:* Inorganic
*EPA/OPP PC Code:* 013906
*California DPR Chemical Code:* 1415
*ICSC Number:* 1191
*RTECS Number:* BD1700000
*Uses:* Widely used in tanning leather, sizing paper, mordant in dyeing, purifying water, fireproofing and waterproofing cloth, clarifying oils and fats, treating sewage, in antiperspirants, in manufacturing aluminum salts and others.
*Regulatory Authority:*
- Air Pollutant Standard Set (ACGIH)[1] (HSE)[33] (OSHA)[58] (North Dakota)[60]
- The "Director's List" (CAL/OSHA)
- Clean Water Act: Section 311 Hazardous Substances/RQ 40CFR117.3 (same as CERCLA, see below)
- Superfund/EPCRA 40CFR302.4 RQ: CERCLA, 5,000 lb (2270 kg)

*Description:* Aluminum sulfate is a white crystalline solid, powder, or granules; often used in water solution. Soluble in water. Molecular weight = 342.18. Melting/Freezing point = 770°C (decomposes). Noncombustible.
*Incompatibilities:* In aqueous solution aluminum sulfate

forms sulfuric acid; reacts with bases and many other substances.

***Permissible Exposure Limits in Air:*** The Federal OSHA[2] standard for soluble aluminum salt, including aluminum sulfate, is 2 mg/m$^3$ TWA averaged over an 8-hour workshift. This same standard is recommended by ACGIH[1], and HSE[33]. In addition, North Dakota[60] has set a guideline for in ambient air of 0.02 mg/m$^3$.

***Permissible Concentration in Water:*** An ambient water level of 73 µg/L for aluminum compounds has been suggested by EPA[32] based on health effects.

***Routes of Entry:*** Inhalation, ingestion.

***Harmful Effects and Symptoms***

***Short Term Exposure:*** Aluminum sulfate powder can irritate the eyes, skin and respiratory tract. It is capable of causing eye damage. Ingestion of large doses can cause stomach irritation, nausea and vomiting.

***Long Term Exposure:*** Aluminum sulfate may cause skin disorders, and may cause lung problems

***Points of Attack:*** Lungs and skin.

***Medical Surveillance:*** Lung function tests.

***First Aid:*** If this chemical gets into the eyes, remove any contact lenses at once and irrigate immediately for at least 15 minutes, occasionally lifting upper and lower lids. Seek medical attention immediately. If this chemical contacts the skin, remove contaminated clothing and wash immediately with soap and water. Seek medical attention immediately. If this chemical has been inhaled, remove from exposure, begin rescue breathing (using universal precautions) if breathing has stopped and CPR if heart action has stopped. Transfer promptly to a medical facility. When this chemical has been swallowed, get medical attention. If victim is *conscious*, administer water or milk. Do not induce vomiting.

***References:***

- New Jersey Department of Health and Senior Services, "Hazardous Substance Fact Sheet: Aluminum Sulfate," Trenton, NJ (November 1994, rev. January 2001). http://www.state.nj.us/health/eoh/rtkweb/0068.pdf
- California Environmental Protection Agency Chemical List of Lists, Sacramento CA (February 1997).
- New York State Department of Health, "Chemical Fact Sheet: Aluminum Sulfate," Albany, NY, Bureau of Toxic Substance Assessment (March 1986).

# Ametryn (ANSI)

***Use Type:*** A selective herbicide.
***CAS Number:*** 834-12-8
***Formula:*** $C_9H_{17}N_5S$
***Synonyms:*** A 1093; AI3-60365; Amyphyt; Cemerim; 2-Ethylamino-4-isopropylamino-6-methylmercarpo-*S*-triazine; Caswell No. 431; 2 Ethylamino-4-isopropylamino-6-methylthio-*S*-triazine; 2-Ethylamino-4-isopropylamino-6-methylthio-1,3,5-triazine; 2-(Ethylamino)-4-(isopropylamino)-6-(methylthio)-1,3,5-triazine; 2-(Ethylamino)-4-(isopropylamino)-6-(methylmercapto)-*S*-triazine; 2-(Ethylamino)-4-(isopropylamino)-6-(methylthio)-*S*-triazine; *N*-Ethyl-*N*-isopropyl-6-methylthio-1,3,5-triazine-2,4-diamine; $N^2$-Ethyl-$N^4$-isopropyl-6-methylthio-1,3,5-triazine-2,4-diamine; *N*-Ethyl-*N'*-isopropyl-6-methylthio-1,3,5-triazine-2,4-diyldiamine; *N*-Ethyl-*N'*-(1-methylethyl)-6-(methylthio)-1,3,5,-triazine-2,4-diamine; *N*-Ethyl-*N'*-(1-methylethyl)-6-(methylthio)-1,3,5-triazine-2,4-diamine; 2-Methylmercapto-4-ethylamino-6-isopropylamino-*S*-triazine; 2-Methylmercapto-4-isopropylamino-6-ethylamino-*S*-triazine; 2-Methylthio-4-ethylamino-6-isopropylamino-*S*-triazine; NSC 163044; 2-Triazine, 2-ethylamino-4-isopropylamino-6-methylthio-; 1,3,5-Triazine-2,4-diamine, *N*-ethyl-*N'*-(1-methylethyl)-6-(methylthio)-; *S*-Triazine, 2-(ethylamino)-4-(isopropylamino)-6-(methylthio)-; *S*-Triazine, 2-ethylamino-4-isopropylamino-6-methylthio-

***Trade Names:*** AMESIP®, OXON Italia S.p.A. (Italy); AMERTREX®, Makhteshim-Agan (Israel); AMETRON SC®, Milenia Agro Ciencias (Brazil); AMETRYNE TECHNICAL®, Syngenta Crop Protection (Switzerland); AMETRYNE 2E, Syngenta Crop Protection (Switzerland), canceled 10/10/1989; AMETRYNE 80W HERBICIDE®, Aceto Agriculture Chemicals (US), canceled 10/19/1988; AMIGAN®, Makhteshim-Agan (Israel); Proficol (Colombia); CRISATRINE®; CRISATRINA® Crystal Chemical Inter-America (US); DORUPLANT®; EVIK®, Syngenta (Switzerland); G-34162®, Ciba-Geigy (Switzerland); GESAPAX®, Syngenta (Switzerland); HERBIPAK®, Milenia Agro Ciencias (Brazil); KRISMAT®, Syngenta (Switzerland); OXON AMETRYN TECHNICAL®, Oxon Italia (Italy), canceled 6/9/1988; PRIMATOL Z 80®; PROKIL AMETRYNE 80W®, Gowan Co. (US), canceled 9/29/1988; SANCOPAX®, Sanachem (Pty) Ltd. (South Africa); TRINATOX-D®, Pyosa Agroquimicos (Mexico)

***Producers:*** Agsin (Singapore); Atanor S.A. (Argentina); Biesterfeld Siemsgluess International. GmbH (Germany); Fulon Chemical Industrial Co., Ltd. (Taiwan); Makhteshim-Agan (Israel); Milenia Agro Ciencias (Brazil); Nippon Kayaku (Japan); Nissan Chemical Industry (Japan); OXON Italia S.p.A. (Italy); Pharm-Chem Manufacturing Co., Ltd. (China); Proficol (Colombia); Pyosa Agroquimicos (Mexico); Sigma-Aldrich Laborchemikalien (Germany); Syngenta (Switzerland); Wuzhou International (China); Zhejiang Changxing Zhongshan Chemical Industry (China); Zagro Asia Ltd. (Singapore)

***Chemical Class:*** Triazine
***EPA/OPP PC Code:*** 080801
***California DPR Chemical Code:*** 18
***RTECS Number:*** XY91000000
***EINECS Number:*** 212-634-7

***Uses:*** Ametryn is a herbicide which inhibits photosynthesis and other enzymatic processes. Ametryn is an unrestricted or General Use Pesticide (GUP). It is used to control broadleaf weeds and annual grasses in pineapple, sugarcane and bananas. It is used on corn and potato crops for general weed control. It is also used as a vine desiccant on dry beans and potatoes. Ametryn is available as an emulsifiable concentrate, flowable wettable powder and a wettable powder. The U.S. EPA classifies ametryn as Toxicity Class III, slightly toxic. Used in premixes with atrazine, diuron, simazine, and terbutryn.

*Human toxicity (long-term)[77]:* Low–60.00 ppb, Health Advisory

*Fish toxicity (threshold)[77]:* Very low–989.94257 ppb, MATC (Maximum Acceptable Toxicant Concentration)

***U.S. Maximum Allowable Residue Levels for Ametryn: (40CFR 180.258):***

| CROP | ppm |
|---|---|
| Banana | 0.25 |
| Cassava, roots | 0.1 |
| Corn, fodder | 0.5 |
| Corn, forage | 0.5 |
| Corn, fresh (inc. sweet)(k+cwhr) | 0.25 |
| Corn, grain | 0.25 |
| Pineapple | 0.25 |
| Pineapple, fodder | 0.25 |
| Pineapple, forage | 0.25 |
| Sugarcane, cane | 0.25 |
| Sugarcane, fodder | 0.25 |
| Sugarcane, forage | 0.25 |
| Tanier | 0.25 |
| Yam, true, tuber | 0.25 |

*Carcinogen/Hazard Classifications*
**Label Signal Word:** CAUTION
**WHO Acute Hazard:** Class III, slightly hazardous
*Regulatory Authority:*
- Superfund/EPCRA 40CFR302.4, Appendix A, RQ: 100 lb (45.5 kg), 40CFR372.65: Form R *de minimis* Concentration Reporting Level: 1.0%.
- AB 1803-Well Monitoring Chemical (CAL)

*Description:* Ametryn is a colorless powder. Readily soluble in organic solvents. Melting/Freezing point = 84-86°C. Vapor pressure = $2.7 \times 10^{-6}$ mmHg.

*Incompatibilities:* Triazines are incompatible with nitric acid.

*Permissible Concentration in Water:* The No-Adverse-Effect-Level (NOAEL) has been found to be 100 mg/kg/day and on that basis a ten-day health advisory of 8.6 mg/L was determined for a 10-kg child. If, however, on assumes a NOAEL of 10 mg/kg/day one arrives at a long term health advisory of 0.86 mg/L for a 70-kg adult. The lifetime health advisory for an adult is 0.06 mg/L using a NOAEL of 10.

*Determination in Water:* Extraction with methylene chloride may be followed by gas chromatography using a nitrogen phosphorus detector. The detection limits are in the range of 0.1 to 2.0 µg/L.

*Routes of Entry:* Ingestion and skin.

***Harmful Effects and Symptoms***

*Short Term Exposure:* Ametryn is an eye and skin irritant. It is mildly toxic by skin contact. Poisonous if swallowed or inhaled.

*Long Term Exposure:* It apparently causes liver degeneration. The $LD_{50}$ value for male Charles River rats was 1207 mg/kg and 1543 mg/kg for female rats.

*Points of Attack:* Liver

*Medical Surveillance:* Liver function tests.

*First Aid:* If this chemical gets into the eyes, remove any contact lenses at once and irrigate immediately for at least 15 minutes, occasionally lifting upper and lower lids. Seek medical attention immediately. If this chemical contacts the skin, remove contaminated clothing and wash immediately with soap and water. Seek medical attention immediately. If this chemical has been inhaled, remove from exposure, begin rescue breathing (using universal precautions) if breathing has stopped and CPR if heart action has stopped. Transfer promptly to a medical facility. When this chemical has been swallowed, get medical attention. Give large quantities of water and induce vomiting. Do not make an unconscious person vomit.

*References:*
- U.S. Environmental Protection Agency, "Health Advisory: Ametryn," Washington, DC, Office of Drinking Water (August 1987).
- California Environmental Protection Agency Chemical List of Lists, Sacramento CA (February 1997)
- EXTOXNET, Pesticide Information Profiles, "Ametryn," Oregon State University (September, 1995), http://ace.orst.edu/cgi-bin/mfs/01/pips/ametryn.htm?33#mfs
- EPA, Office of Pesticide Programs, Pesticide Residue Limits, "Ametryn,", 40 CFR 180.258, www.epa.gov/pesticides/food/viewtols.htm
- "Registry of Toxic Effects of Chemical Substances, 2-Triazine, 2-ethylamino-4-isopropylamino-6-methylthio-," NIOSH, Sept., 1997.

# Aminoethoxyvinylglycine hydrochloride

*Use Type:* Plant growth regulator
*CAS Number:* 55720-26-8
*Formula:* $C_5ClF_2H_{10}MNO_4$
*Synonyms:* *trans*-L-2-Amino-4-(2-aminoethoxy)3-butenoic acid hydrochloride; AVG; L-α-(2-Aminoethoxyvinyl)glycine hydrochloride; (s)-*trans*-2-Amino-4-(2-aminoethyoxy)-3-butenoic acid hydrochloride; 3-Butenoic acid, 2-amino-4-(2-aminoethoxy)-,

monohydrochloride, [*s-(E)*]-
*Trade Names:* ABG-3097®, Valent BioSciences Corporation (USA); RETAIN®, Valent BioSciences Corporation (USA); X-11085®
*Producers:* Valent BioSciences Corporation (USA)
*EPA/OPP PC Code:* 129104
*California DPR Chemical Code:* 3907
*RTECS Number:* EM9080000
*Uses:* A plant growth regulator used on apples, pears and ornamentals. In apples, it delays fruit maturity and the resulting pre-harvest fruit drop, and in pears, it helps maintain fruit firmness. It is used as a spray solution.
*U.S. Maximum Allowable Residue Levels for Aminoethoxyvinylglycine Hydrochloride (40 CFR 180.502):*

| CROP | ppm |
|---|---|
| Apple | 0.08 |
| Fruit, stone, group 12, except cherry | 0.170 |
| Pear | 0.08 |

*Description:* Molecular weight = 183.61.
*Incompatibilities:* When heated to decomposition or on contact with acids or acid fumes, may produce highly toxic chloride fumes; deadly phosgene gas may be formed. It may cause pitting of some metals.
*Permissible Concentration in Water:* No criteria set. Runoff from spills or fire control may cause water pollution.
*Routes of Entry:* Inhalation
*Harmful Effects and Symptoms*
*Short Term Exposure:* Contact with eyes or skin may cause irritation or injury. Inhalation should be avoided; use NIOSH-approved air purifying respirators for pesticides. No significant risk from dietary exposure.
*First Aid:* If this chemical gets into the eyes, remove any contact lenses at once and irrigate immediately for at least 15 minutes, occasionally lifting upper and lower lids. Seek medical attention immediately. If this chemical contacts the skin, remove contaminated clothing and wash immediately with soap and water. Seek medical attention immediately. If this chemical has been inhaled, remove from exposure, begin rescue breathing (using universal precautions) if breathing has stopped and CPR if heart action has stopped. Transfer promptly to a medical facility. When this chemical has been swallowed, get medical attention. *Do not induce vomiting when formulations containing petroleum solvents are ingested.* Otherwise, give large quantities of water and induce vomiting. Do not make an unconscious person vomit.
*References:*
- U.S. Environmental Protection Agency, Office of Pesticide Programs, Pesticide Residue Limits, " Aminoethoxyvinylglycine hydrochloride," 40 CFR 180.502. http://www.epa.gov/fedrgstr/EPA-PEST/2004/February/Day-18/p3371.htm
- U.S. Environmental Protection Agency, Office of Pesticide Programs, "Aminoethoxyvinylglycine (AVG)(129104) Fact Sheet," Washington, DC (November 2001). http://www.epa.gov/pesticides/biopesticides/ingredients/factsheets/factsheet_129104.htm
- California Environmental Protection Agency Chemical List of Lists, Sacramento CA (February 1997)

# 4-Aminopyridine

*Use Type:* Avicide
*CAS Number:* 504-24-5 (*p*-isomer); 504-29-0 (*o*-isomer); 462-08-8 (*m*-isomer)
*Formula:* $C_5H_6N_2$; $C_5NH_4NH_2$
*Alert:* A Restricted Use Pesticide (RUP)
*Synonyms:* 4-Aminopiridina (Spanish); Amino-4-pyridine; γ-Aminopyridine; *p*-Aminopyridine; 4-Pyridinamine; Pyridine, 4-amino-
*Trade Names:* 4-AP®; AVITROL,® Avitrol Corporation (USA); AVITROL 200®, Avitrol Corporation (USA); COMPOUND 1861®; PRC-1237®; VMI 10-3®
*Producers:* Aldrich Chemical (USA); Avitrol Corporation (USA); Mitsubishi Corp. (Japan); Richman Chemical (USA); Seal Sands Chemicals (UK); Shanghai Chemical Reagent Co. (China); Sigma-Aldrich Laborchemikalien (Germany)
*Chemical Class:* Pyridine
*EPA/OPP PC Code:* 069201
*California DPR Chemical Code:* 50
*RTECS Number:* US1750000
*Uses:* A bird poison for control of crows, pigeons, grackles, gulls, blackbirds and other pests to crops, feed lots, grain processing plants and similar locations. Highly toxic to mammals. Also used as a chemical intermediate in pharmaceuticals and for treatment of certain nerve conditions.
*Regulatory Authority:*
- RCRA Section 261 Hazardous Constituents.
- EPA Hazardous Waste Number (RCRA No.): P008
- EPCRA Section 302, Extremely Hazardous Substances: TPQ = 500/10,000 lb (227/4,540 kg)
- EPCRA Section 304 RQ: CERCLA, 1000 lb (454 kg)
- The "Director's List" (CAL/OSHA)
- Actively registered pesticide in California.

*Description:* White to tan or brown powder. Odorless. Boiling point = 274°C. Melting/Freezing point = 159°C. Flash point = 164°C. Moderately soluble in water. Log $K_{ow}$ = < 1.0. Unlikely to bioaccumulate in marine organisms.
*Incompatibilities:* Sodium nitrite, strong oxidizers. Avoid contact with acid anhydrides, acid chlorides, and strong acids.
*Permissible Exposure Limits in Air:* ACGIH TLV 0.5 ppm (1.9 mg/m$^3$)[1]. NIOSH IDLH = 5 ppm (*o*-isomer).

*Permissible Concentration in Water:* No criteria set. Runoff from spills or fire control may cause water pollution.
*Routes of Entry:* Inhalation, ingestion, absorbed through the intact skin.
*Harmful Effects and Symptoms*
*Short Term Exposure:* Material may be fatal if inhaled, swallowed or absorbed through skin. Symptoms of exposure include rapid onset of disagreeable taste, immediate burning of throat, and abdominal discomfort; in addition, weakness, dizziness, disorientation, convulsions and seizures may occur. Delayed symptoms of oral ingestion include elevated liver enzymes, and respiratory arrest. Contact may cause burns to skin and eyes. Material attacks the nervous system and affects neural transmission. In sufficient concentration, material may cause metabolic acidosis, respiratory arrest, and cardiac arrhythmia. The fatal dose to a 70 kg. person is about 5 grams.
*Long Term Exposure:* High exposure or repeated exposure may cause liver damage.
*Points of Attack:* Central nervous system and liver.
*Medical Surveillance:* Pre-employment and regular physical examinations with emphasis on central nervous system. Liver function tests. Persons exposed to *strychnine* or other chemicals capable of causing seizures are probably at increased risk.
*First Aid:* If this chemical gets into the eyes, remove any contact lenses at once and irrigate immediately for at least 15 minutes, occasionally lifting upper and lower lids. Seek medical attention immediately. If this chemical contacts the skin, remove contaminated clothing and wash immediately with soap and water. Seek medical attention immediately. If this chemical has been inhaled, remove from exposure, begin rescue breathing (using universal precautions) if breathing has stopped and CPR if heart action has stopped. Transfer promptly to a medical facility. When this chemical has been swallowed, get medical attention. Give large quantities of water and induce vomiting. Do not make an unconscious person vomit.
*References:*
- EXTOXNET, Extension Toxicology Network, "Pesticide Information Profile, 4-Aminopyridine," Oregon State University, Corvallis, OR (June 1996). http://extoxnet.orst.edu/pips/4-aminop.htm
- Sax, N.I., Ed., Dangerous Properties of Industrial Materials Report 5, No. 5, 39-41 (1985).
- New Jersey Department of Health and Senior Services, "Hazardous Substance Fact Sheet: Avitrol," Trenton, NJ (April 1997). http://www.state.nj.us/health/eoh/rtkweb/0172.pdf
- U.S. Environmental Protection Agency, "Chemical Profile: 4-Aminopyridine," Washington, DC, Chemical Emergency Preparedness Program (November 30, 1987).
- California Environmental Protection Agency Chemical List of Lists, Sacramento CA (February 1997)

# Amiton

*Use Type:* Insecticide, miticide, and acaricide
*CAS Number:* 78-53-5
*Formula:* $C_{10}H_{24}NO_3PS$; $(C_2H_5O)_2POSCH_2CH_2N(C_2H_5)_2$
*Alert:* A Restricted Use Pesticide (RUP). Amiton is highly poisonous and has been suspected for being developed as a chemical weapon.
*Synonyms:* S-(2-Diethylamino) ethyl phosphorothioic acid-O,O-diethyl ester; Diethyl-S-2-diethylaminoethyl phosphorothioate; O,O-Diethyl-S-(2-diethylaminoethyl) thiophosphate; O,O-Diethyl-S-2-diethylaminoethyl phosphorothioate; O,O-Diethyl-S-(β-Diethylamino)ethyl phosphorothiolate; O,O-Diethyl-S-diethylamino ethyl phosphorothiolate; O,O-Diethyl-S-2-diethylaminoethyl phosphorothiolate; DSDP; ENT 24,980-X; R-5,158; Phosphorothioic acid, S-(2-(diethylamino)ethyl) O,O-diethyl ester
*Trade Names:* CHIPMAN 6200®, Chipman Chemicals (Canada); CITRAM®; INFERNO®; METRAMAC®; METRAMAK®; RHODIA-6200®, Rhodia Group (France); TETRAM®, ICI Group (UK), canceled
*Producers:* ICI Group (UK); Rhodia Group (France)
*Chemical Classes:* Organothiophosphate
*EPA/OPP PC Code:* 057302
*RTECS Number:* TF0525000
*Label Signal Word:* DANGER
*Regulatory Authority:*
- Very Toxic Substance (World Bank)[15]
- CERCLA/SARA 40CFR355 Extremely Hazardous Substances: TPQ = 500 lb (227 kgs)
- CERCLA/SARA Section 304 RQ: EHS, 1 lb (0.454 kg)
- Classified by EPA as a Restricted Use Pesticide (RUP)
- U.S. DOT Inhalation Hazard Chemicals as organophosphates
- Regulated under the Chemical Weapons Convention of 1994 as a toxic organophosphate nerve agent

*Description:* Amiton is a colorless liquid. Soluble in water. Molecular weight = 269.32. Boiling point = 110°C @ 0.2 mm pressure. Hazard Identification (based on NFPA-704 M Rating System): Health 4, Flammability 2, Reactivity 1. Emits highly toxic nitrogen oxides, phosphorus oxides, and sulfur oxides when heated to decomposition.
*Permissible Exposure Limits in Air:* No standards set.
*Determination in Air:* OSHA versatile sampler-2; Toluene/Acetone; Gas chromatography/Flame photometric detection for sulfur, nitrogen, or phosphorus; NIOSH Method IV Method #5600, Organophosphorus pesticides[18].
*Permissible Concentration in Water:* No criteria set, but runoff from spills or fire control may cause water pollution.
*Routes of Entry:* Inhalation, ingestion, skin and/or eye contact.

*Harmful Effects and Symptoms*
*Short Term Exposure:* Danger-poisonous; can be fatal if swallowed, inhaled, or absorbed through the skin or eyes. This material is highly toxic orally. It is a cholinesterase inhibitor. The $LD_{50}$ (oral, rat) = 3.3 mg/kg. The toxic effects are similar to parathion. Organic phosphorus insecticides are absorbed by the skin, as well as by the respiratory and gastrointestinal tracts. Symptoms of exposure include headache, giddiness, blurred vision, nervousness, weakness, nausea, cramps, diarrhea, and discomfort in the chest. Signs include sweating, tearing, salivation, vomiting, cyanosis, convulsions, coma, loss of reflexes and loss of sphincter control. Delayed pulmonary edema may occur after inhalation.

*Long Term Exposure:* Cholinesterase inhibitor; cumulative effect is possible. This chemical may damage the nervous system with repeated exposure, resulting in convulsions, respiratory failure. It may cause liver damage.

*Points of Attack:* Respiratory system, lungs, central nervous system, cardiovascular system, skin, eyes, plasma and red blood cell cholinesterase.

*Medical Surveillance:* Medical observation is recommended for 24 to 48 hours after breathing overexposure, as pulmonary edema may be delayed. Before employment and at regular times after that, the following are recommended: Plasma and red blood cell cholinesterase levels (tests for the enzyme poisoned by this chemical). If exposure stops, plasma levels return to normal in 1-2 weeks while red blood cell levels may be reduced for 1-3 months. When acetylcholinesterase enzyme levels are reduced by 25% or more below preemployment levels, risk of poisoning is increased, even if results are in lower ranges of "normal." Reassignment to work not involving organophosphate or carbamate pesticides is recommended until enzyme levels recover. If symptoms develop or overexposure occurs, repeat the above tests as soon as possible and get an exam of the nervous system. Also consider complete blood count. Consider chest x-ray following acute overexposure.

*First Aid:* **Treatment for organophosphate poisoning consists of thorough decontamination, cardiorespiratory support, and administration of the antidotes atropine and pralidoxime. In cases of severe poisoning, diazepam, an anticonvulsant, should also be administered. Antidotes should be administered as prevention even if the diagnosis is in doubt.** Speed in removing material from eyes and skin is of extreme importance. *Eyes:* Eye contact can cause dangerous amounts of these chemicals to be quickly absorbed through the mucous membrane into the bloodstream. Immediately and gently flush eyes with plenty of warm or cold water (NO hot water) for at least 15 minutes, occasionally lifting the upper and lower eyelids. Get medical aid immediately. *Skin:* Get medical aid. Skin contact can cause dangerous amounts of these chemicals to be absorbed into the bloodstream. Wearing the appropriate PPE equipment and respirator for organophosphate pesticides, immediately flush skin with plenty of soap and water for at least 15 minutes while removing contaminated clothing and shoes. Shampoo hair promptly if contaminated. The removed, contaminated clothing and shoes should be double-bagged and left in Hot Zone for later disposal by hazardous materials experts. Skin may also be decontaminated with diluted hypochlorite solution. *Ingestion:* Call poison control. Loosen all clothing. Never give anything by mouth to an unconscious person. Get medical aid. Do NOT induce vomiting.* If conscious, alert, and able to swallow, rinse mouth and have victim drink 4 to 8 ounces of water do NOT induce vomiting but immediately administer slurry of activated charcoal (2 oz in 8 oz of water). If victim is *unconscious or having convulsions,* do nothing except keep victim warm. *\*In some cases you may be specifically instructed by poison control to induce vomiting by way of 2 tablespoons of syrup of ipecac (adult) washed down with a cup of water.* Do NOT give activated charcoal before or with ipecac syrup. *Inhalation:* Get medical aid. Do not contaminate yourself. Wearing the appropriate PPE equipment and respirator for organophosphate pesticides, immediately remove the victim from the contaminated area to fresh air. If the victim is not breathing, administer artificial respiration. Do NOT use mouth-to-mouth resuscitation; use bag/mask apparatus. If breathing is difficult, administer oxygen through bag/mask apparatus until medical help arrives. Do not leave victim unattended.

*Note to physician or authorized medical personnel.* Administer atropine, 2 mg (1/30 gr) intramuscularly or intravenously as soon as any local or systemic signs or symptoms of an intoxication are noted; repeat the administration of atropine every 3 to 8 minutes until signs of atropinization (mydriasis, dry mouth, rapid pulse, hot and dry skin) occur; initiate treatment in children with 0.05 mg mg/kg of atropine; repeat at 5 to 10 minute intervals. Watch respiration, and remove bronchial secretions if they appear to be obstructing the airway; intubate if necessary. *Notes to physician or authorized medical personnel*: N-methylpyridinium-2-aldoxime (2-PAMCI) when used in conjunction with atropine reacts with the phosphorylated cholinesterase, thereby restoring normal activity to by removing the phosphorylating group. The combination of these two chemicals is synergistic and must be administered within minutes to a few hours following exposure (depending on the specific agent) to be effective. Give 2-PAMCI (Pralidoxime; Protopam), 2.5 gm in 100 ml of sterile water or in 5% dextrose and water, intravenously, slowly, in 15-30 minutes; if sufficient fluid is not available, give 1 gm of 2-PAMCI in 3 ml of distilled water by deep intramuscular injection; repeat this every half hour if respiration weakens or if muscle fasciculation or convulsions recur. Also, Diazepam, an anticonvulsant, or

1,1'-trimethylenebis(4-formylpyridinium bromide)dioxime (a.k.a TMB-4 Dibromide and TMV-4) have been used as an antidote for organophosphate poisoning.

*References:*
- New Jersey Department of Health and Senior Services, "Hazardous Substance Fact Sheet, Amiton," Trenton, NJ, (October, 2002), www.state.nj.us/health/eoh/rtkweb/2113.pdf
- U.S. Environmental Protection Agency, "Chemical Profile: Amiton," Washington, DC, Chemical Emergency Preparedness Program (November 30, 1987).
- California Environmental Protection Agency Chemical List of Lists, Sacramento CA (February 1997)

# Amiton Oxalate

*Use Type:* Insecticide and acaricide.
*CAS Number:* 3734-97-2
*Formula:* $C_{12}H_{26}NO_7PS$; $(C_2H_5O)_2POSCH_2CH_2N(C_2H_5)_2 \cdot HOOCCOOH$
*Alert:* A Restricted Use Pesticide (RUP)
*Synonyms:* Acid oxalate; [2-(2-Diethylamino)ethyl]-*O,O*-diethyl ester, oxalate (1:1); Amiton hydrogen oxalate; S-(2-Diethylaminoethyl)-*O,O*-diethylphosphorothioate hydrogen oxalate; *O,O*-Diethyl-*S*-(2-diethylamino)ethylphosphorothioate hydrogen oxalate; *O,O*-Diethyl-*S*-(β-diethylamino)ethylphosphorothioate hydrogen oxalate; *O,O*-Diethyl-*S*-(2-ethyl-*N,N*-diethylamino)ethylphosphorothioate hydrogen oxalate; Hydrogen oxalate of Amiton; Phosphorothioic acid, S-[2-(diethylamino)ethyl] *O,O*-diethyl ester, ethanedioate (1:1); Phosphorothioic acid, S-[(2-diethylamino)ethyl] *O,O*-diethyl ester, oxalate (1:1)
*Trade Names:* CHIPMAN® 6199, Nomix-Chipman Chemicals (UK); CHIPMAN® R-6, 199, Nomix-Chipman Chemicals (UK); CITRAM®; TETRAM® 75; TETRAM®, ACID OXALATE; TETRAM® MONOOXALATE S-
*Chemical Class:* Organophosphate
*EPA/OPP PC Code:* 057301
*RTECS Number:* TF1400000
*Regulatory Authority:*
- CERCLA/SARA 40CFR355 Extremely Hazardous Substances: TPQ = 100/10,000 lb (45.4/4,540 kg)
- CERCLA/SARA Section 304 RQ: EHS, 1 lb (0.454 kg)
- U.S. DOT Inhalation Hazard Chemicals as organophosphates
- Classified by EPA as a Restricted Use Pesticide (RUP)

*Description:* Amiton oxalate is a crystalline solid or powder. Molecular weight = 359.48. Melting/Freezing point = 98-99°C. Hazard Identification (based on NFPA-704 M Rating System): Health 4, Flammability 2, Reactivity 1. Soluble in water.
*Incompatibilities:* Avoid sources of heat including fire. Will emit very toxic fumes of nitrogen, phosphorus and sulfur oxides when heated to decomposition.
*Permissible Exposure Limits in Air:* No standards set.
*Determination in Air:* OSHA versatile sampler-2; Toluene/Acetone; Gas chromatography/Flame ionization detection; NIOSH IV[18], Method #5600, Organophosphorus pesticides.
*Permissible Concentration in Water:* No criteria set, but runoff from spills or fire control may cause water pollution.
*Routes of Entry:* Inhalation, ingestion, skin contact.
*Harmful Effects and Symptoms*
*Short Term Exposure:* Amiton oxalate is a cholinesterase inhibitor. Symptoms include headache, giddiness, nervousness, blurred vision, weakness, nausea, cramps, diarrhea, and discomfort in the chest. Signs include sweating, miosis, tearing, salivation and other excessive respiratory tract secretion, vomiting, cyanosis, uncontrollable muscle twitching followed by muscular weakness, convulsions, coma, loss of reflexes, and loss of muscular control. The $LD_{50}$ (oral, rat) = 3 mg/kg. Delayed pulmonary edema may occur after inhalation.
*Long Term Exposure:* Amiton oxalate may damage the nervous system causing numbness, tingling and/or weakness in the hands and feet. Repeated exposure may cause personality changes of depression, anxiety or irritablilty.
*Medical Surveillance:* Medical observation is recommended for 24 to 48 hours after breathing overexposure, as pulmonary edema may be delayed. See entry on parathion as referred to under amiton. Bear in mind that the oxalate is a solid whereas amiton is a high-boiling liquid.
*First Aid:* **Treatment for organophosphate poisoning consists of thorough decontamination, cardiorespiratory support, and administration of the antidotes atropine and pralidoxime. In cases of severe poisoning, diazepam, an anticonvulsant, should also be administered. Antidotes should be administered as prevention even if the diagnosis is in doubt.** Speed in removing material from eyes and skin is of extreme importance. *Eyes:* Eye contact can cause dangerous amounts of these chemicals to be quickly absorbed through the mucous membrane into the bloodstream. Immediately and gently flush eyes with plenty of warm or cold water (NO hot water) for at least 15 minutes, occasionally lifting the upper and lower eyelids. Get medical aid immediately. *Skin:* Get medical aid. Skin contact can cause dangerous amounts of these chemicals to be absorbed into the bloodstream. Wearing the appropriate PPE equipment and respirator for organophosphate pesticides, immediately flush skin with plenty of soap and water for at least 15 minutes while removing contaminated clothing and shoes. Shampoo hair promptly if contaminated. The removed, contaminated clothing and shoes should be double-bagged and left in Hot Zone for later disposal by hazardous materials experts. Skin may also be decontaminated with diluted hypochlorite solution.
*Inhalation:* Get medical aid. Do not contaminate yourself.

Wearing the appropriate PPE equipment and respirator for organophosphate pesticides, immediately remove the victim from the contaminated area to fresh air. If the victim is not breathing, administer artificial respiration. Do NOT use mouth-to-mouth resuscitation; use bag/mask apparatus. If breathing is difficult, administer oxygen through bag/mask apparatus until medical help arrives. Do not leave victim unattended. *Ingestion:* Call poison control. Loosen all clothing. Never give anything by mouth to an unconscious person. If victim is *unconscious or having convulsions,* do nothing except keep victim warm. Get medical aid. Transfer promptly to a medical facility. In cases of ingestion, **do not induce vomiting**. If the victim is alert and asymptomatic, administer a slurry of activated charcoal at a dose of 1 g/kg (infant, child, and adult dose). A soda can and straw may be of assistance when offering charcoal to a child. *In some cases you may be specifically instructed by poison control to induce vomiting by way of 2 tablespoons of syrup of ipecac (adult) washed down with a cup of water.* Do NOT give activated charcoal before or with ipecac syrup.

*Note to physician or authorized medical personnel:* Treat cases of respiratory compromise, coma, or excessive pulmonary secretions with respiratory support using protocols and techniques available and within the scope of training. Some cases may necessitate procedures such as endotracheal intubation or cricothyrotomy by properly trained and equipped personnel. When possible, atropine (see *Antidotes*, below) should be given under medical supervision. Patients who are comatose, hypotensive, or having seizures or cardiac arrhythmias should be treated according to advanced life support protocols. *Antidotes:* Two antidotes are administered to treat organophosphate poisoning. Atropine is a competitive antagonist of acetylcholine at muscarinic receptors and is used to control the excessive bronchial secretions which are often responsible for death. Pralidoxime relieves both the nicotinic and muscarine effects of organophosphate poisoning by regenerating acetylcholinesterase and can reduce both the bronchial secretions and the muscle weakness associated with poisoning. The initial intravenous dose of atropine in adults should be determined by the severity of symptoms: An initial adult dose of 1.0 to 2.0 mg or pediatric dose of 0.01 mg/kg (minimum 0.01 mg) should be administered intravenously. If intravenous access cannot be established, atropine may also be given intramuscularly, subcutaneously or via endotracheal tube. Doses should be repeated every 15 minutes until excessive secretions and sweating have been controlled. Once bronchial secretion has been controlled, atropine administration should be repeated whenever the secretions begin to recur. In seriously poisoned patients, very large doses may be required. Alterations of pulse rate and pupillary size should not be used as indicators of treatment adequacy. Pralidoxime should be administered as early in poisoning as possible as its efficacy may diminish when given more than 24 to 36 hours after exposure. Doses are as follows: adult 1.0 g; pediatric 25 to 50 mg/kg. The drug should be administered intravenously over 30 to 60 minutes, but in a life-threatening situation, one-half of the total dose can be given per minute for a total administration time of 2 minutes. Treatment should begin to take effect within 40 minutes with a reduction in symptoms and the amount of atropine necessary to control bronchial secretion. The initial dose can be repeated in 1 hour and then every 8 to 12 hours until the patient is clinically well and no longer requires atropine. If intravenous access cannot be established, pralidoxime may also be given intramuscularly. Early administration of diazepam in addition to the combined atropine and pralidoxime treatment may help prevent the onset of seizures and potential brain and cardiac morphologic damage following high-level organophosphate poisoning.

*References:*
- U.S. Environmental Protection Agency, "Chemical Profile: Amiton Oxalate," Washington, DC, Chemical Emergency Preparedness Program, November 30, 1987.
- California Environmental Protection Agency Chemical List of Lists, Sacramento CA (February 1997).
- Agency for Toxic Substances and Disease Registry, U.S. Department of Health and Human Services, Public Health Service, "Managing Hazardous Materials Incidents," Atlanta, GA (June 2003)
- New Jersey Department of Health and Senior Services, Hazardous Substance Fact Sheet, "Amiton Oxalate," Trenton, NJ (July 2000). www.state.nj.us/health/eoh/rtkweb/2114.pdf

# Amitraz (ANSI)

*Use Type:* Insecticide and acaricide
*CAS Number:* 33089-61-1
*Formula:* $C_{19}H_{23}N_3$
*Alert:* General Use Pesticide (GUP). Human toxicity (long-term): Very high.
*Synonyms:* A13-27967; Amitraze; Amitraz estrella; *N,N*-Bis(2,4-xylyliminomethyl)methylamine; 1,5-Di-(2,4-dimethylphenyl)-3-methyl-1,3,5-triazapenta-1,4-diene; *N'*-(2,4-Dimethylphenyl)-3-methyl-1,3,5-triazapenta-1,4-diene; *N'*-(2,4-Dimethylphenyl)-*N*-[((2,4-dimethylphenyl)imino)methyl]-*N*-methylmethanimidamide; *N,N*-Di-(2,4-xylyliminomethyl)methylamine; ENT 27967; Formamidine, *N*-methyl-*N'*-2,4-xylyl-*N*-(*N*-2,4-xylylformimidoyl)-; *N*-Methylbis(2,4-xylyliminomethyl)amine; 2-Methyl-1,3-di(2,4-xylylimino)-2-azapropane; *N,N'*-[(Methylimino)dimethylidyne]bis(2,4-xylidine); *N,N'*-[(Methylimino)dimethylidyne]d-2,4-xylidine; NSC 324552; OMS 1820; R.D. 27419; 2,4-Xylidine, *N,N'*-(methyliminodimethylidyne)bis-
*Trade Names:* AAZDIENO®; ACARAC®, Maag Agro

(Germany); ACADREX®, BASF CropScience (Germany); AMIPAZ®, Pazchem Ltd. (Israel), canceled; ARMY®, Wangs Ltd; AZODIENO®; BAAM®, NOR-AM Chemical Company (USA), canceled; BOOTS BTS 27419®; BTS 27,419®; BUMETRAN®, Bayer CropScience(Germany); COYOTE®, Makhteshim-Agan (Israel); CYTAC®, Schering AG (Germany), canceled; DANICUT®; ECTODEX® Hoechst/Roussel AG (Germany); EDRIZAN®; EDRIZAR®; GARIAL®; ISTAMBUL®; MITABAN®; MITAC®; OVASYN®, Bayer CropScience (Germany); OVIDREX®; PARSEC®, Makhteshim-Agan (Israel); ROTRAZ®, Rotam Agrochemical (HK) Co. (Hong Kong); SENDER®, Sanonda Zhengzhou Pesticide Co. (China); TAC-PLUS®, Makhteshim-Agan (Israel); TACTIK®; TRIATIX®, Schering-Plough Animal Health (USA); TRIATOX®, Schering-Plough Animal Health (USA); TUDY®, MFA Inc. (USA); VAPCOZIN TAKTIC®; UPJOHN U-36059®, Upjohn Inc. (USA)

*Producers:* Agrides (Spain); Agrimore International (USA); Agropharm Ltd. (UK); Agsin Pte. Ltd. (Singapore); BASF CropScience (Germany); Bayer CropScience (Germany); Biesterfeld Group (Germany); China Chemicals (China); Ehrenstorfer, Dr.(Germany); Fulon Chemical Industrial Co. (Taiwan); Hockley International Ltd. (UK); Jingma Chemicals Ltd. (China); Ki-Hara Chemicals Ltd. (UK); Kunshan Chemical Group (Industries) Corp. (China); Maag Agro (Germany), See Syngenta (Switzerland); Makhteshim-Agan (Israel); MFA Inc. (USA); Milenia Agro Ciencias S/A (Brazil); Nissan Chemical Industries (Japan); Pazchem Ltd. (Israel); Rotam Agrochemical (HK) Co., Ltd. (Hong Kong); Sanonda Zhengzhou Pesticide Co. (China); Schering AG (Germany); Schering-Plough Animal Health (USA); Shanghai Pesticide Research Institute (China); Shenzhen Guomeng Industry Co., Ltd. (China); Sigma-Aldrich Laborchemikalien (Germany); Upjohn Inc. (USA), see Pharmacia Animal Health (USA); Wangs Ltd. (China); Zago Asia Ltd. (Singapore)

*Chemical Class:* Formamidine
*EPA/OPP PC Code:* 106201
*California DPR Chemical Code:* 2016
*ICSC Number:* 0098
*RTECS Number:* ZF0480000
*EEC Number:* 612-086-00-2
*EINECS Number:* 251-375-4

*Uses:* An unrestricted or General Use Pesticide (GUP) registered for control of pear psylla on pears, whitefly and mites on pears and cotton; cattle, dogs, sheep, and hog dip to control ticks, mange mites, lice and other pests. Not permitted on apples. Used to control red spider mites, leaf miners, scale insects, and aphids. Also used on cotton to control bollworms, white fly, leaf worms, and tobacco budworms.

*Human toxicity (long-term)*[77]: Extra high–0.30973 ppb, CHCL (Chronic Human Carcinogen Level)

*Fish toxicity (threshold)*[77]: Very low–12422.37656 ppb, MATC (Maximum Acceptable Toxicant Concentration)

*U.S. Maximum Allowable Residue Levels for Amitraz (40 CFR 180.127):*

| CROP | ppm |
|---|---|
| Apple | - |
| Pear | 3.0 |
| Hop, dried cones | 60 |
| Honey | 1.0 |
| Honeycomb | 6.0 |
| Cotton, undelinted seed | 1.0 |
| Milk | 0.03 |
| Milk, fat | 0.3 |
| Cattle, mbyp | 0.3 |
| Cattle, fat | 0.1 |
| Cattle, meat | 0.05 |
| Goat, mbyp | 0.0 |
| Goat, fat | 0.0 |
| Goat, meat | 0.0 |
| Horse, mbyp | - |
| Horse, fat | - |
| Horse, meat | - |
| Sheep, mbyp | 0.0 |
| Sheep, fat | 0.0 |
| Sheep, meat | 0.0 |
| Hog, mbyp | 0.3 |
| Hog, fat | 0.1 |
| Hog, kidney | 0.2 |
| Hog, liver | 0.2 |
| Hog, meat | 0.05 |
| Poultry, mbyp | 0.05 |
| Poultry, fat | 0.01 |
| Poultry, meat | 0.01 |
| Egg | 0.01 |

*Carcinogen/Hazard Classifications*
**U.S. EPA Carcinogens:** Group C, possible carcinogen
**California Prop. 65:** Developmental toxin
**TRI Developmental Toxin:** Reproductive and developmental toxin
**Label Signal Word:** CAUTION. WARNING-Toxicity Class II for technical grade
**WHO Acute Hazard:** Class III, slightly hazardous

*Regulatory Authority:*
- Unrestricted or General Use Pesticide (GUP)
- Carcinogen (Animal Positive) (U.S. EPA)[13]
- Actively registered pesticide in California.
- Banned in Norway.
- CERCLA/SARA 40CFR372.65: Form R *de minimis* Concentration Reporting Level: 1.0%.

*Description:* Amitraz forms colorless needle-like crystals or needles. Insoluble in water. Molecular weight = 293.43. Melting/Freezing point = 86–87°C. Vapor pressure = 2.6 x $10^{-6}$ mmHg.

*Incompatibilities:* Decomposes on burning, producing toxic

fumes including nitrogen oxides. When stored for prolonged periods, slow decomposition occurs.

*Potential Exposure:* A rebuttable presumption against registration for amitraz was issued on April 6, 1977 by U.S. EPA on the basis of oncogenicity.

*Permissible Exposure Limits in Air:* No standards set.

*Permissible Concentration in Water:* No criteria set, but runoff from spills or fire control may cause water pollution.

*Routes of Entry:* Inhalation, ingestion, skin

*Harmful Effects and Symptoms*

*Short Term Exposure:* This chemical is poisonous if ingested or absorbed through the skin. Eye or skin contact can cause irritation. It may affect the central nervous system. May cause sedation with slow heart beat, low blood pressure, low body temperature.

Because Amitraz has a low vapor pressure, significant inhalation of vapors is unlikely at ordinary temperatures.

*Long Term Exposure:* May affect the central nervous system and liver. Amitraz metabolizes to 2,4-dimethylaniline which is a potential human carcinogen. A mouse oncogenic bioassay was conducted by Boots Chemical Company and reported by EPA; the results of that study have been disputed. Acute oral $LD_{50}$ for rats is 800 mg/kg; for mice is greater than 1600 mg/kg.

*Points of Attack:* Eyes, skin

*Medical Surveillance:* Liver function test.

*First Aid:* If this chemical gets into the eyes, remove any contact lenses at once and irrigate immediately for at least 15 minutes, occasionally lifting upper and lower lids. Seek medical attention immediately. If this chemical contacts the skin, remove contaminated clothing and wash immediately with soap and water. Seek medical attention immediately. If this chemical has been inhaled, remove from exposure, begin rescue breathing (using universal precautions) if breathing has stopped and CPR if heart action has stopped. Transfer promptly to a medical facility. When this chemical has been swallowed, get medical attention. Give large quantities of water and induce vomiting. Do not make an unconscious person vomit.

*References:*
- EPA, Office of Pesticide Programs, Pesticide Residue Limits, "Amitraz", 40 CFR 180.127, www.epa.gov/pesticides/food/viewtols.htm
- EXTOXNET, Extension Toxicology Network, Pesticide Information Profile "Amitraz," Oregon State University, Revised September, 1995, http://ace.orst.edu/cgi-bin/mfs/01/pips/amitraz.htm?6#mfs
- U.S. Environmental Protection Agency, "Reregistration Eligibility Decision (RED), Amitraz", Office of Prevention, Pesticides and Toxic Substances, Washington, DC http://www.epa.gov/REDs/0234red.pdf
- U.S. Environmental Protection Agency, "Rebuttable Presumption Against Registration (RPAR) of Pesticide Products Containing Amitraz," Washington, DC, April 6, 1977.
- California Environmental Protection Agency Chemical List of Lists, Sacramento CA (February 1997).

# Amitrole (ANSI)

*Use Type:* Herbicide and plant growth regulator
*CAS Number:* 61-82-5
*Formula:* $C_2H_4N_4$
*Alert:* Use limited to non-crop applications as a herbicide and plant growth regulator.
*Synonyms:* Aminotriazole; 2-Aminotriazole; 3-Aminotriazole; 3-Amino-S-triazole; 3-Amino-1,2,4-triazole; 2-Amino-1,3,4-triazole; 3-Amino-1H-1,2,4-triazole; ATA; ENT 25445; 1,2,4-Triazol-3-amine; Triazolamine; 1H-1,2,4-Triazol-3-amine; S-Triazole, 3-amino-; δ-2-1,2,2,4-Triazoline, 5-imino-; 1H-1,2,4-Triazol-3-ylamine
*Trade Names:* AMCHEM®, Bayer CropScience (Germany), canceled; AMEROL®, Syngenta (Switzerland), canceled 12/11/98; AMINOTRIAZOLE BAYER®, Bayer CropScience (Germany), canceled; AMINO TRIAZOLE WEEDKILLER 90®, Aceto Agriculture Chemicals (USA), canceled 7/1/87; AMITOL®, Syngenta (Switzerland), canceled 12/11/98; AMITROL 90®, BASF Corp. (Germany), canceled; AMITROL-T®, Bayer CropScience (Germany), canceled; AMITRIL®; AMIZOL®, Bayer CropScience (Germany), canceled; AMIZOL DP NAU®, Bayer CropScience (Germany), canceled; ATLAZIN®; ATLAZINE® FLOWABLE; AT®; 3-AT®; AT-90®; ATRAFLOW PLUS®; AZAPLANT®; AZAPLANT KOMBI®; AZOLAN®; AZOLE®; BOROFLOW® A/ATA; CAMPAPRIM® A 1544; CASWELL® No. 040; CDA SIMFLOW PLUS®; CHIPMAN® PATH WEEDKILLER; CLEARWAY®, Arborchem Products, canceled 7/19/95; CYTROL®, BASF (Germany), canceled 12/31/87; CYTROLE®; DIUROL®; DOMATOL®; ELMASIL®; EMISOL®; FARMCO®, Farmco Industries (USA); FENAMINE®, Bayer CropScience (Germany), canceled 12/24/86; FENAVAR®, Bayer CropScience (Germany), canceled 7/1/87; HERBAZIN PLUS SC®; HERBICIDE® TOTAL; HERBIZOLE®, Fair Products (USA), canceled 7/1/87; KLEER-LOT®, Bayer CropScience (Germany), canceled 12/24/86; MASCOT HIGHWAY®; MSS AMINOTRIAZOLE®; MSS SIMAZINE®; ORGA-414®; PRIMATOL®, Syngenta (Switzerland), canceled 12/11/98; RADOXONE® TL; RAMIZOL®; RASSAPRON®; SIMAZOL®; SIMFLOW PLUS®; SOLUTION CNCENTREE T271®; SYNCHEMICALS® TOTAL WEED KILLER; SYNTOX® TOTAL WEED KILLER, Crown Chemical Industries (USA), canceled 10/8/85; TORAPRON®; VOROX®; WEEDAR®; WEEDAZIN®; WEEDAZOL®, Bayer CropScience (Germany), canceled 10/25/90; WEEDEX®, Carroll Co. (USA), canceled 7/1/87;

WEEDOCLOR®; X-ALL® LIQUID, Bayer CropScience (Germany), canceled 8/9/85

**Producers:** Agan Chemical Manufacturers Ltd. (Israel); ATOFINA (France); BASF Agricultural Products (Germany); Bayer CropScience (Germany); Fairmount Chemical (USA); Kawaguchi Chemical Industry (Japan); Ki-Hara Chemicals Ltd. (UK); Merke (Germany); Mitsubishi Chemical (Japan); Nippon Carbide Industries (Japan); Nufarm (Australia); OxyChem (USA); Rhodia (France); Rhone-Poulenc Agro (France); Sigma-Aldrich Laborchemikalien (Germany); Zago Asia Ltd. (Singapore)

**Chemical Class:** Triazine
**EPA/OPP PC Code:** 004401
**California DPR Chemical Code:** 20
**ICSC Number:** 0631
**RTECS Number:** XZ3850000
**EINECS Number:** 200-521-5

**Uses:** All use of amitrole on food crops was canceled by the U.S. EPA in 1971 because it caused cancer in experimental animals.

**Carcinogen/Hazard Classifications**
**U.S. EPA Carcinogens:** Group B2, probable carcinogen
**U.S. NTP Carcinogen:** Reasonably anticipated carcinogen
**California Prop. 65:** Carcinogen
**IARC:** Group 3, unclassifiable
**Label Signal Word:** CAUTION, EPA Toxicity Class III, slightly toxic
**WHO Acute Hazard:** Class U, Unlikely to be hazardous
**Endocrine Disruptor:** Suspected

**Regulatory Authority:**
- Carcinogen (Animal Positive) (IARC) (suspected Carcinogen)(NTP)[9]
- AB 2588-Air Toxics "Hot Spots" Chemicals (CAL)
- Proposition 65 chemical (CAL)
- Permissible Exposure Limits for Chemical Contaminants (CAL/OSHA)
- The "Director's List" (CAL/OSHA)
- Air Pollutant Standard Set (ACGIH)[1] (DFG)[3] (OSHA)[58] (Several States)[60] (Several Canadian Provinces) (Australia) (Israel)
- EPA Hazardous Waste Number (RCRA No.): U011
- RCRA, 40CFR261, Appendix 8 Hazardous Constituents.
- RCRA Land Ban Waste Restrictions
- Superfund/EPCRA 40CFR302.4 RQ: CERCLA, 10 lb (4.54 kg)
- EPCRA Section 313 Form R *de minimus* concentration reporting level: 0.1%
- Canada, WHMIS, Ingredients Disclosure List
- Banned or Severely Restricted (UN) (Scandinavia)[13]
- U.S. DOT Regulated Marine Pollutant (49CFR172.101, Appendix B), severe pollutant

**Description:** Amitrol is a colorless to off white crystalline solid. Odorless when pure. Soluble in water; solubility = $2.8 \times 10^5$ ppm @ 25°C. Molecular weight = 84.08. Melting/Freezing point = 154–157°C. Vapor pressure = $4.4 \times 10^{-7}$ mmHg @ 20°C; $2.4 \times 10^{-4}$ mPa @ 60°C. Log $K_{ow}$ = –0.7. Unlikely to bioaccumulate in marine organisms.

**Incompatibilities:** Strong oxidizers, strong acids, and light (decomposes). Corrosive to iron, aluminum, and copper.

**Permissible Exposure Limits in Air:** NIOSH[2] recommends a limit of 0.2 mg/m³ for a 10-hour workshift. ACGIH has set a TLV of 0.2 mg/m³ TWA for an 8-hour workshift[1]. Australia and Israel use the same level. DFG set the MAK (total dust) at 0.2 mg/m³[3]. In addition, several states have set guidelines or standards for amitrole in ambient air[60]: 0.2 mg/m³ (California), 0.476 µg/m³ (Kansas), 1.8 µg/m³ (Pennsylvania), 2.0 µg/m³ (North Dakota), 3,000 µg/m³ (Virginia). Canadian province Levels for 0.2 mg/m³ TWA and 0.5 mg/m³ (Alberta), 0.2 mg/m³ TWA (Ontario, Quebec).

**Determination in Air:** NIOSH Method #0500[18]

**Permissible Concentration in Water:** No criteria set, but runoff from spills or fire control may cause water pollution.

**Harmful Effects and Symptoms**

**Short Term Exposure:** Amitrol can be absorbed through the skin, thereby increasing exposure. Because this material has a low vapor pressure, significant inhalation of vapors is unlikely at ordinary temperatures.

**Long Term Exposure:** Causes liver, thyroid, and pituitary cancer in animals. May damage the developing fetus. May cause liver, thyroid gland (possible goiter or underactive thyroid), and pituitary gland damage. Carcinogenicity is the primary observed effect. Amitrole is carcinogenic in mice and rats, producing thyroid and liver tumours following oral or subcutaneous administration. Railroad workers who were exposed to amitrole and other herbicides showed a slight (but statistically significant) excess of cancer when all sites were considered together. Because the workers were exposed to several different herbicides, however, no conclusions could be made regarding the carcinogenicity of amitrole alone.

**Points of Attack:** Liver, thyroid, and pituitary gland.

**Medical Surveillance:** Before beginning employment and at regular times after that, the following is recommended: Physical examination of the thyroid and thyroid function tests ($T_4$, TSH, and $T_3$). If symptoms develop or overexposure is suspected, the following may be useful: Liver function tests. Pituitary gland function tests.

**First Aid:** If this chemical gets into the eyes, remove any contact lenses at once and irrigate immediately for at least 15 minutes, occasionally lifting upper and lower lids. Seek medical attention immediately. If this chemical contacts the skin, remove contaminated clothing and wash immediately with soap and water. Seek medical attention immediately. If this chemical has been inhaled, remove from exposure, begin rescue breathing (using universal precautions) if breathing has stopped and CPR if heart action has stopped.

Transfer promptly to a medical facility. When this chemical has been swallowed, get medical attention. Give large quantities of water and induce vomiting. Do not make an unconscious person vomit.

*References:*
- EXTOXNET, Extension Toxicology Network, "Pesticide Information Profile, Amitrol," Oregon State University, Corvallis, OR (June 1996). http://ace.orst.edu/info/extoxnet/pips/amitrole.htm
- Sax, N.I., Ed., Dangerous Properties of Industrial materials Report, 1, No. 4, 34-35 (1981) and 4, No. 2, 41-43 (1984).
- California Environmental Protection Agency Chemical List of Lists, Sacramento CA (February 1997)
- New Jersey Department of Health and Senior Services, "Hazardous Substance Fact Sheet: Amitrol," Trenton, NJ (June 1998). http://www.state.nj.us/health/eoh/rtkweb/0083.pdf

# Ammonia

*Use Type:* Intermediate in fertilizer manufacturing; used as an insecticide, fungicide and deer repellent.
*CAS Number:* 7664-41-7
*Formula:* $NH_3$
*Synonyms:* AM-FOL; Ammonia gas; Ammonia, anhydrous; Amoniaco anhidro (Spanish); Ammoniac (French); Ammoniaca (Italian); Ammoniale (German); Ammonium amide; Ammonium hydroxide; Amoniaco (Spanish); Amoniaco anhidro (Spanish); Amoniak (Polish); Anhydrous ammonia; Aqua ammonia; Liquid ammonia; Spirit of Hartshorn
*Trade Names:* DAXAD-32s®; NITRO-SIL®; PRO 330 CLEAR THIN SPREAD®; R717®
*Producers:* Air Products & Chemicals (USA); Amomex (Mexico); ANWIL (Poland); BOC Gases (UK); Bombay Ammonia and Chemical Company (India); Caffaro (Italy); Carburos Metalicos (Spain); Cargill Crop Nutrition (USA); CF Industries (USA); Chimco (Bulgaria); DSM Agro (Netherlands); Grande Paroisse (France); Hoek Loos (Netherlands); Holox (USA); Hydro Agri Chemicals (Norway); Industria Quimica Loser (Mexico); Juhua Group Corp. (China); Linde Gas Group (Germany); Lonza (Switzerland); Messer Group (Germany); Nissan Chemical Industries (Japan); Occidental (USA); Petroquimica de Venezuela (Pequiven) (Venezuela); Praxair (USA); Qatar Fertiliser (Qatar); Rashtriya Chemicals & Fertilizers (India); Sasol Chemical (South Africa); Saudi Basic Industries Corp. (Saudi Arabia); Showa Denko Chemicals Group (Japan); Simplot (USA); Sumitomo Chemical Co., Ltd. (Japan); Terra Industries (USA); Ube Industries (Japan); Westfarmers CSBP (Australia)
*Chemical Class:* Inorganic

*EPA/OPP PC Code:* 005302 (anhydrous, gas); 005301 (monohydrate, aqua)
*California DPR Chemical Code:* 22
*ICSC Number:* 0414
*RTECS Number:* BO0875000
*EEC Number:* 007-001-00-5 (anhydrous)
*EINECS Number:* 231-635-3
*Uses:* Ammonia is used as an insecticide, deer repellant, and to control fungal growth during storage of citrus, e.g. oranges, grapefruit, and lemons. Ammonia is used as a nitrogen source for many nitrogen-containing compounds. It is used in the production of ammonium sulfate and ammonium nitrate for fertilizers and in the manufacture of nitric acid, soda, synthetic urea, synthetic fibers, dyes, and plastics. It is also utilized as a refrigerant and in the petroleum refining and chemical industries. It is used in the production of many drugs and pesticides. Other sources of occupational exposure include the silvering of mirrors, gluemaking, tanning of leather, and around nitriding furnaces. Ammonia is produced as a by-product in coal distillation and by the action of steam on calcium cyanamide, and from the decomposition of nitrogenous materials.

*U.S. Maximum Allowable Residue Levels for Ammonia (40 CFR 180.1003):*

| CROP | ppm |
|---|---|
| Grapefruit, post-h | none |
| Lemon, post-h | none |
| Orange, post-h | none |
| Corn, grain, post-h | none |

*Carcinogen/Hazard Classifications*
**U.S. EPA and IARC Carcinogens:** The Department of Health and Human Services, IARC, and the U.S. EPA have not classified ammonia for carcinogenicity.
*Regulatory Authority:*
- Toxic Substance (World Bank)[15]
- Air Pollutant Standard Set (NIOSH)[2] (ACGIH)[1] (DFG)[3] (HSE)[33] (Former USSR)[43] (OSHA)[58] (Several States)[60]
- (Several Canadian Provinces) (Australia) (Israel) (Mexico)
- Water Pollution Standard Set (Former USSR)[43]
- AB 1803-Well Monitoring Chemical (CAL)
- EPA/SARA 302 (EPCRA) Extremely hazardous substances
- AB 2588-Air Toxics "Hot Spots" Chemicals (CAL)
- Permissible Exposure Limits for Chemical Contaminants (CAL/OSHA)
- The "Director's List" (CAL/OSHA)
- OSHA 29CFR1910.119, Appendix A, Process Safety List of Highly Hazardous Chemicals, TQ = 10,000 lb (4,540 kg) (anhydrous); TQ = 15,000 lb (6,815 kg) (solution >44% $NH_3$)
- Clean Air Act 42USC7412; Title I, Part A,§112®),

Accidental Release Prevention/Flammable substances (Section 68.130); (anhydrous) TQ = 10,000 lb (4,540 kg) (anhydrous); (concentrations ≥ 20% NH$_3$) TQ =20,000 lb (9,150 kg)
- Clean Water Act: 40CFR116.4 Hazardous Substances; RQ 40CFR117.3 (same as CERCLA); Section 313 Water Priority Chemicals (57FR41331, 9/9/92).
- CERCLA/SARA 40CFR302, Extremely Hazardous Substances: TPQ = 500 lb (228 kg).
- Superfund/EPCRA 40CFR302.4, Appendix A, RQ: 100 lb (45.4 kg), 40CFR372.65: Form R *de minimis* Concentration Reporting Level: 1.0%; includes anhydrous ammonia and aqueous ammonia from water dissociable ammonium salts and other sources; 10% of total aqueous ammonia, and 100% of anhydrous forms of ammonia is reportable under this listing. If a facility manufactures, processes, or otherwise uses anhydrous ammonia or aqueous ammonia, they must report under the ammonia listing. Solutions containing aqueous ammonia at a concentration in excess of 1% of the 10% reportable under this listing should be factored into threshold and release determinations.
- U.S. DOT 49CFR172.10; Poisonous by inhalation substances (anhydrous UN1005)
- Canada, WHMIS, Ingredients Disclosure List; National Pollutant Release Inventory

***Description:*** Ammonia is a colorless, strongly alkaline, and extremely soluble gas. Pungent, suffocating odor at 5.75 ppm; eye irritation @ 20 ppm. Water solubility = 33.1% @20 °C Molecular weight = 17.0 daltons. Boiling point = –33°C @ 760 mmHg. Vapor pressure = >6,000 mmHg 20 °C. Gas density (air = 1) = 0.59. Flash point = (flammable gas). Autoignition temperature = 630°C. Hazard Identification (based on NFPA-704 M Rating System): Health 3, Flammability 1, Reactivity 0. Explosive limits: LEL = 15%; UEL = 28%. Anhydrous ammonia is a colorless, highly irritating gas at room temperature with a pungent, suffocating odor. Ammonia gas is lighter than air; hugs the ground when cool; and flammable at high concentrations and temperatures. Easily compressed, it forms a clear, colorless liquid under pressure. When released, the liquid under pressure floats and "boils" on water. The liquid dissolves in water and evaporates quickly, forming ammonium hydroxide, an alkaline, corrosive solution. A poisonous, visible vapor cloud is produced. The amount of ammonia produced by humans every year is almost equal to that produced by nature every year. Ammonia is produced naturally in soil by bacteria, decaying plants and animals, and animal wastes. Ammonia is essential for many biological processes. Ammonia does not build up in the food chain, but serves as a nutrient source for plants and bacteria. Plants and bacteria rapidly take ammonia from soil and water. Some ammonia in water and soil is changed to nitrate and nitrite by bacteria.

***Incompatibilities:*** Ammonia dissolves readily in water to form ammonium hydroxide, a corrosive, alkaline solution at high concentrations. Violent reaction with strong oxidizers and acids. Shock-sensitive compounds may be formed with gold, halogens, mercury, mercury oxide, and silver oxide. Fire and explosions may be caused by trimethylammonium amide, 1-chloro-2,4-dinitrobenzene, *o*-chloronitrobenzene, platinum, trioxygen difluoride, selenium difluoride dioxide, boron halides, mercury, chlorine, iodine, bromine, hypochlorites, chlorine bleach, amides, organic anhydrides, isocyanates, vinyl acetate, alkylene oxides, epichlorohydrin, and aldehydes. Attacks some coatings, plastics, and rubber, copper, brass, bronze, aluminum, steel, tin, zinc, and their alloys.

***Permissible Exposure Limits in Air:*** The Federal OSHA[2] standard is 50 ppm (35 mg/m$^3$)TWA, averaged over an 8-hour workshift. NIOSH[2] recommended limit is 25 ppm (17 mg/m$^3$) averaged over a 10-hour workshift and 35 ppm (27 mg/m$^3$) not to be exceeded during any 15 minute work period. Australia, HSE[33], Israel, and Mexico limits are similar to NIOSH. Israel has an action level of 12.5 ppm. ACGIH[1] recommends values of 25 ppm (18 mg/m$^3$) TWA and STEL 35 ppm (27 mg/m$^3$). 35 ppm. The NIOSH[2] IDLH value is 300 ppm. The DFG MAC is 20 ppm (14 mg/m$^3$) TWA[3]. In addition, several states have set airborne guidelines or standards for ammonia in ambient air[60]: 25 ppm (18 mg/m$^3$) TWA and STEL of 35 ppm (27 mg/m$^3$) (California); 0.024 mg/m$^3$ (Massachusetts), 0.042857 mg/m$^3$ (Kansas), 0.18 to 0.27 mg/m$^3$ (North Dakota), 0.25 mg/m$^3$ (Virginia), 0.36 mg/m$^3$ (Connecticut, Florida, New York, South Dakota), 0.429 mg/m$^3$ (Nevada, Wyoming), 2.7 mg/m$^3$ (North Carolina); Canadian Provinces of Alberta, British Columbia, Ontario, and Quebec have limits of 25 ppm TWA/TWAEV and STEL/STEV of 35 ppm. The former USSR-UNEP/IRPTC project[43] has set a MAC of 20 mg/m$^3$ in workplace air and a MAC of 0.2 mg/m$^3$ in ambient air in residential areas.

***Determination in Air:*** Sampling by absorption in sulfuric acid followed by measurement by ion chromatography, conductivity. See NIOSH Method #6015, #6016[18].

***Permissible Concentration in Water:*** The Former USSR-UNEP/IRPTC project[43] has set a MAC of 2.0 mg/ml in water bodies used for domestic purposes and 0.05 mg/ml in water bodies used for fishery purposes.

***Routes of Entry:*** Inhalation, ingestion, skin and eye contact.
***Harmful Effects and Symptoms***
***Short Term Exposure:*** Eye or skin contact with ammonia can cause irritation, burns, frostbite (anhydrous), and permanent damage. Ammonia is highly irritating to the eyes and respiratory tract. Swelling and narrowing of the throat and bronchi, coughing, and an accumulation of fluid in the lungs can occur. Ammonia causes rapid onset of a burning sensation in the eyes, nose, and throat, accompanied by lacrimation, rhinorrhea, and coughing. Upper airway

swelling and pulmonary edema may lead to airway obstruction. Prolonged skin contact (more than a few minutes) can cause pain and corrosive injury. Exposure can cause headache, loss of sense of smell, nausea, and vomiting. *Inhalation*: Nose and throat irritation have been reported at 72 ppm after 5 minutes exposure. Exposures of 500 ppm for 30 minutes have caused upper respiratory irritation, tearing, increased pulse rate and blood pressure. Death has been reported after an exposure to 10,000 ppm for an unknown duration. *Skin*: Solutions of 2% ammonia can cause burns and blisters after 15 minutes of exposure. These burns may be slow to heal. Anhydrous ammonia may cause skin to freeze. *Eyes*: Levels of 70 ppm (gas) have caused eye irritation. If not flushed with water immediately contact with eye may cause partial or complete blindness. *Ingestion*: ammonia will cause pain if swallowed and burning of the throat and stomach. May cause vomiting. One teaspoon of 28% aqua ammonia may cause death.

**Long Term Exposure:** Repeated exposure to ammonia may cause chronic irritation of the respiratory tract. Repeated lung irritation can result in bronchitis with coughing, shortness of breath, and phlegm. Levels of 170 ppm of ammonia vapor has caused mild changes in the spleens, kidneys and livers of guinea pigs. Chronic cough, asthma and lung fibrosis have been reported. Chronic irritation of the eye membranes and dermatitis have also been reported. *Carcinogenicity:* Ammonia has not been classified for carcinogenic effects. *Reproductive and Developmental Effects:* No data exist to evaluate the reproductive and developmental effects of ammonia in humans. Decreased egg production and conception rates have been observed in animals, and ammonia has been shown to cross the ovine placental barrier.

**Points of Attack:** Skin, respiratory system, eyes.

**Medical Surveillance:** Pre-employment physical examinations for workers in ammonia exposure areas should be directed toward significant changes in the skin, eyes, and respiratory system. Persons with corneal disease, and glaucoma, or chronic respiratory diseases may suffer increased risk. Periodic examinations should include evaluation of skin, eyes, and respiratory system, and pulmonary function test to compare with baselines established at pre-employment examination. Consider chest x-ray following acute exposure.

**First Aid:** Irrigate eyes that were exposed or that become irritated with plain water or saline for at least 15 minutes. Remove contact lenses, if easily removable without additional trauma to the eye. Seek medical attention immediately. Flush liquid-exposed skin and hair with water for at least 5 minutes. If feasible, wash exposed skin extremely thoroughly with soap and water. Use caution to avoid hypothermia when decontaminating of children or the elderly. Use blankets when appropriate. In cases of ingestion **do not induce vomiting**, perform gastric lavage, or attempt neutralization. **Do not administer activated charcoal.** Victims who are conscious and able to swallow should be given 4 to 8 ounces of water or milk. If this chemical has been inhaled, remove from exposure, begin rescue breathing (using universal precautions; assist ventilation with a bag-valve-mask device if necessary) if breathing has stopped and CPR if heart action has stopped. Transfer promptly to a medical facility. There is no antidote for ammonia poisoning. Treatment consists of supportive measures. These include administration of humidified oxygen and bronchodilators and airway management; treatment of skin and eyes with copious irrigation; and dilution of ingested ammonia with milk or water. Medical observation is recommended for 24 to 48 hours after breathing overexposure, as pulmonary edema may be delayed. As first aid for pulmonary edema, a doctor or authorized paramedic may consider administering a corticosteroid spray.

*Note to physician or authorized medical personnel:* In cases of respiratory compromise secure airway and respiration via endotracheal intubation. If not possible, perform cricothyroidotomy if equipped and trained to do so. Patients who are hypotensive or have seizures should be treated according to advanced life support protocols. Treat patients who have bronchospasm with aerosolized bronchodilators. The use of bronchial sensitizing agents in situations of multiple chemical exposures may pose additional risks. Also consider the health of the myocardium before choosing which type of bronchodilator should be administered. Cardiac sensitizing agents may be appropriate; however, the use of cardiac sensitizing agents after exposure to certain chemicals may pose enhanced risk of cardiac arrhythmias (especially in the elderly). Ammonia poisoning is not known to pose additional risk during the use of bronchial or cardiac sensitizing agents. Consider racemic epinephrine aerosol for children who develop stridor. Dose 0.25–0.75 mL of 2.25% racemic epinephrine solution in water, repeat every 20 minutes as needed cautioning for myocardial variability. Patients who are comatose, hypotensive, or are having seizures or have cardiac arrhythmias should be treated according to advanced life support protocols. Monitor fluid and electrolyte balance and restore if abnormal. Fluids should be administered cautiously to patients with pulmonary edema.

***References:***
- U.S. Environmental Protection Agency, Office of Pesticide Programs, Pesticide Residue Limits, "Ammonia", 40 CFR 180.1003, www.epa.gov/pesticides/food/viewtols.htm
- Agency for Toxic Substances and Disease Registry, "ToxFAQs™ for Ammonia," September, 2002, www.atsdr.cdc.gov/tfacts126.html
- Agency for Toxic Substances and Disease Registry. "Toxicological Profile for Ammonia" (Draft for Public Comment). Atlanta, GA, 2002. Department of Health and

Human Services, Public Health Service.
- National Institute for Occupational Safety and Health, Criteria for a Recommended Standard: Occupational Exposure to Ammonia, NIOSH Doc. No. 74-136, Washington, DC (1974).
- U.S. Environmental Protection Agency, "Toxic Pollutant List: Proposal to Add Ammonia," Federal Register, 45, No. 2, 803-806 (January 3, 1980) Rescinded by Federal Register, 45, No. 232, 79692-79693 (December 1, 1980).
- National Research Council, Committee on Medical and Biologic Effects of Environmental Pollutants, Ammonia, Baltimore, MD, University Park Press (1979).
- Sax, N. I., Ed., Dangerous Properties of Industrial materials Report 2, No. 1, 65-68 (1982) and 3, No. 3, 49-53, (1983).
- U.S. Environmental Properties Agency, "Chemical Profile: Ammonia," Washington, DC, Chemical Emergency Preparedness Program (November 30, 1987).
- New Jersey Department of Health and Senior Services, "Hazardous Substance Fact Sheet: Ammonia," Trenton, NJ (June 1998).
  http://www.state.nj.us/health/eoh/rtkweb/0084.pdf
- New York State Department of Health, "Chemical Fact Sheet: Ammonia," Albany, NY, Bureau of Toxic Substance Assessment (January 1986).
- Agency for Toxic Substances and Disease Registry, U.S. Department of Health and Human Services, Public Health Service, "Managing Hazardous Materials Incidents," Atlanta, GA (June 2003)
- California Environmental Protection Agency Chemical List of Lists, Sacramento CA (February 1997)

# Ammonium Carbamate

*Use Type:* A fertilizer and ammoniating agent.
*CAS Number:* 1111-78-0
*Formula:* $CH_6N_2O_2$; $NH_4COONH_2$
*Synonyms:* Ammonium aminoformate; Anhydride of ammonium carbonate; Carbamato amonico (Spanish); Carbamic acid, Monoammonium salt; Carbamic acid, Ammonium salt
*Producers:* BASF (Germany); Potash Corporation (Canada); SNPE Agro (France)
*Chemical Class:* Inorganic
*California DPR Chemical Code:* 3041
*RTECS Number:* EY8575000
*Uses:* A general fertilizer and used in combination with other fertilizing agents.
*Regulatory Authority:*
- Clean Water Act: 40CFR116.4 Hazardous Substances; RQ 40CFR117.3, (same as CERCLA)
- The "Director's List" (CAL/OSHA)
- Superfund/EPCRA 40CFR302.4, Appendix A, RQ: 5,000 lb (2270 kg); Section 313: Form R *de minimis* concentration reporting level: 1.0% (as ammonia); $NH_3$ Equivalent molecular weight: 21.81
- Canada, WHMIS, Ingredients Disclosure List

*Description:* Ammonium carbamate is a colorless crystalline powder or white powder with an ammonia odor. Melting/Freezing point = about 60°C (sublimes). The odor threshold is 5 ppm as $NH_3$ (detection) and 46.8 ppm as $NH_3$ (recognition). Highly soluble in water. Molecular weight = 99.10. Boiling point = 60°C.
*Incompatibilities:* Strong bases, strong oxidizers. Keep away from heat (forms urea), moisture, and direct sunlight.
*Permissible Exposure Limits in Air:* No standards set. Loses ammonia in air, changing to ammonia carbonate. *The following are Exposure Limits for ammonia:* The Federal OSHA[(2)] standard is 50 ppm (35 mg/m$^3$)TWA, averaged over an 8-hour workshift. NIOSH[(2)] recommended limit is 25 ppm (17 mg/m$^3$) averaged over a 10-hour workshift and 35 ppm (27 mg/m$^3$) not to be exceeded during any 15 minute work period. ACGIH[(1)] recommends values of 25 ppm (18 mg/m$^3$) TWA and STEL 35 ppm (27 mg/m$^3$). 35 ppm. The NIOSH[(2)] IDLH value is 300 ppm.
*Permissible Concentration in Water:* No criteria set, but runoff from spills or fire control may cause water pollution.
*Routes of Entry:* Inhalation, ingestion, eye or skin contact.
*Harmful Effects and Symptoms*
*Short Term Exposure:* Irritates skin, respiratory tract and mucous membranes on contact. Inhalation can irritate the nose and lungs with coughing, and/or shortness of breath.
*Long Term Exposure:* Repeated or prolonged exposure can cause lung irritation and the development of bronchitis.
*Points of Attack:* Respiratory system, eyes, skin.
*Medical Surveillance:* Lung function testing.
*First Aid:* If this chemical gets into the eyes, remove any contact lenses at once and irrigate immediately for at least 15 minutes, occasionally lifting upper and lower lids. Seek medical attention immediately. If this chemical contacts the skin, remove contaminated clothing and wash immediately with soap and water. Seek medical attention immediately. If this chemical has been inhaled, remove from exposure, begin rescue breathing (using universal precautions) if breathing has stopped and CPR if heart action has stopped. Transfer promptly to a medical facility. When this chemical has been swallowed, get medical attention. Give large quantities of water and induce vomiting. Do not make an unconscious person vomit.
*References:*
- Sax, N.I., "Dangerous Properties of Industrial Materials Report," 2, No. 3, 31-33 (1982).
- California Environmental Protection Agency Chemical List of Lists, Sacramento CA (February 1997).
- New Jersey Department of Health and Senior Services, "Hazardous Substance Fact Sheet: Ammonium Carbamate," Trenton, NJ (January, 1996, revised March 2002). www.state.nj.us/health/eoh/rtkweb/0091.pdf

# Ammonium Chromate

*Use Type:* Fungicide
*CAS Number:* 7788-98-9
*Formula:* $CrH_8N_2O_4$; $(NH_4)_2CrO_4$
*Synonyms:* Ammonium chromate(VI); Chromic acid, diammonium salt; Cromato amonico (Spanish); Diammonium chromate; Neutral ammonium chromate
*Producers:* Bayer Group (Germany); GFS Chemicals (USA)
*RTECS Number:* GB2880000
*EINECS Number:* 232-138-4
*Uses:* It is used to inhibit corrosion and in dyeing, photography and many chemical reactions. Used as a fungicide and fire retardant.
*Carcinogen/Hazard Classifications*
**U.S. EPA Carcinogens:** Group A, known carcinogen
**U.S. NTP Carcinogen:** Carcinogen
**California Prop. 65:** carcinogen, as chromium(VI).
**IARC:** Group 1,
**Label Signal Word:** DANGER,
*Regulatory Authority:*
- Actively registered pesticide in California.
- AB 2588-Air Toxics "Hot Spots" Chemicals (CAL)
- CAL Air Resources Board/AB 1807 Toxic Air Contaminants
- Permissible Exposure Limits for Chemical Contaminants (CAL/OSHA)
- The "Director's List" (CAL/OSHA)
- EPCRA Section 304 RQ: CERCLA, 10 lb (4.54 kg)
- Air Pollutant Standard Set (NIOSH)[2] (ACGIH)[1] (former USSR)[43] (Australia) (Various States), (Israel) (Various Canadian Provinces)
- Clean Air Act 42USC7412; Title I, Part A,§112 hazardous pollutants (as chromium compounds)
- Clean Water Act: 40CFR116.4 Hazardous Substances; RQ 40CFR117.3, (same as CERCLA); 40CFR423, Appendix A, Priority Pollutants
- Safe Drinking Water Act: MCL, 0.05 mg/L as chromium, hexavalent
- RCRA, 40CFR261, Appendix 8 Hazardous Constituents, waste number not listed (chromium compounds)
- Superfund/EPCRA 40CFR302.4, Appendix A, RQ: CERCLA, 10 lb (4.54 kg)
- EPCRA Section 313 Form R *de minimis* concentration reporting level: 1.0% (as ammonia) Molecular weight: 152.07; $NH_3$ Equivalent weight: 22.04. Also must be reported as a chromium compound: "Includes any unique chemical substances that contains chromium as part of that chemical's infrastructure." Form R *de minimus* concentration reporting level: Chromium (VI) compounds: 0.1%.
- Canada, WHMIS, Ingredients Disclosure List

*Description:* Yellow crystalline compound which can be used in solution. Ammonia odor. Soluble in water. Melting/Freezing point = 185°C (decomposes). Hazard Identification (based on NFPA-704 M Rating System): Health 2, Flammability 0, Reactivity 1.

*Incompatibilities:* A strong oxidizer. Contact with combustible, organic and other readily oxidizable substances may cause fire and explosions. Hydrazine, other reducing agents. Corrosive to metals.

*Permissible Exposure Limits in Air:* OSHA[2]: The legal airborne PEL is 0.1 mg/m$^3$ for chromic acid and chromates (as Cr), not to be exceeded at any time. NIOSH[2]: The recommended airborne exposure limit is 0.001 mg/m$^3$ averaged over a 10-hour workshift. ACGIH: The recommended airborne exposure limit is 0.05 mg/m$^3$ for Chromium compounds (as Cr) averaged over an 8-hour workshift[1]. These exposure limits are for air levels only. When skin contact also occurs, you may be overexposed, even though air levels are less than the limits listed. The NIOSH[2] IDLH for chromates is 15 mg/m$^3$ as Cr(VI) [Carcinogen]. The former USSR-UNEP/IRPTC project has set a MAK value of 0.01 mg/m$^3$ for chromates and bichromates in the workplace[43]. California Prop. 65 No significant risk level (inhalation) = 0.001 µg/day.

*Determination in Air:* Hexavalent chromium may be determined by filtration followed by visible absorption spectrophotometry according to NIOSH Method #7600[18]. Also, Filter (5.0-µm PVC membrane); Ion chromatography, conductivity detection; NIOSH Method #7604[18]

*Permissible Concentration in Water:* To protect human health, hexavalent chromium should be held below 0.05 mg/L according to EPA[6] in studies on priority toxic pollutants, This is also a WHO recommendation for total chromium in drinking water.

*Determination in Water:* Chromium(VI) may be determined by extraction and atomic absorption or colorimetry (using diphenylhydrazide).

*Routes of Entry:* Ingestion, skin and/or eye contact
*Harmful Effects and Symptoms*
*Short Term Exposure:* Eye contact can cause severe damage with possible loss of vision. Breathing Ammonium Chromate can cause a sore or hole through the inner nose (septum), sometimes with bleeding, discharge or crusting. Irritation of nose, throat and bronchial tubes can also occur, with cough and/or wheezing. Skin contact can cause deep ulcers or an allergic skin rash.

*Long Term Exposure:* Some water-soluble chromium [16] compounds are inferred non-carcinogens; the water-insoluble compounds are generally deemed to be carcinogens but the border line is not precise nor universally agreed to. Ammonium chromate is a hexavalent chromium compound which may be carcinogenic and should be handled with extreme caution. Breathing ammonium chromate can cause sores or hole in the septum dividing the inner nose, sometimes with bleeding, discharge, and/or formation of a crust. May cause skin allergy and kidney damage.

*Points of Attack:* Blood, respiratory system, liver kidneys, eyes, skin.

*Medical Surveillance:* Skin and nose examination, kidney function tests, evaluation by a qualified allergist.

*First Aid:* If this chemical gets into the eyes, remove any contact lenses at once and irrigate immediately for at least 30 minutes, occasionally lifting upper and lower lids. Seek medical attention immediately. If this chemical contacts the skin, remove contaminated clothing and wash immediately with soap and water. Seek medical attention immediately. If this chemical has been inhaled, remove from exposure, begin rescue breathing (using universal precautions) if breathing has stopped and CPR if heart action has stopped. Transfer promptly to a medical facility. When this chemical has been swallowed, get medical attention. Give large quantities of water and induce vomiting. Do not make an unconscious person vomit.

*References:*
- Sax, N.I., Ed., Dangerous Properties of Industrial Materials Report 2, No. 3, 36-38 (1982).
- New Jersey Department of Health and Senior Services, "Hazardous Substance Fact Sheet: Ammonium Chromate," Trenton, NJ (February 1998). http://www.state.nj.us/health/eoh/rtkweb/0095.pdf
- National Institute for Occupational Safety and Health, "Criteria for a Recommended Standard: Occupational Exposure to Chromium (VI), NIOSH Document No. 76-129 (1976).
- U.S. Environmental Protection Agency: "Chromium: Ambient Water Quality Criteria," Washington, DC (1980).
- Agency for Toxic Substances and Disease Registry, "Toxicological Profile for Chromium," Atlanta, Georgia (1988).

# Ammonium Hexafluorosilicate

*Use Type:* Insecticide, miticide, and preservative
*CAS Number:* 16919-19-0
*Formula:* $F_6H_8N_2Si$; $(NH_4)_2SiF_6$
*Alert:* There are no products currently registered with the U.S. EPA.
*Synonyms:* Ammonium fluorosilicate; Ammonium silicofluoride; Ammonium silicon fluoride; Crytophthalite; Diammonium fluosilicate; Diammonium silicon hexafluoride; Fluosilicate de ammonium (French); Picrato amonico (Spanish); Fluosilicato amonico (Spanish); Silicate(2-), hexafluoro-, diammonium; Ammonium hexafluorosilicate; Silicofluoruro amonico (Spanish)
*Trade Names:* ALL BUG®, canceled; BYE BUGS®, canceled; COMMON SENSE DRIONE 79700®, canceled; DEXOL EARWIG BAIT®, canceled; DRI-DYE®, canceled; XR-29®, canceled
*Producers:* Ozark Fluorine Specialties (USA)
*Chemical Class:* Inorganic
*EPA/OPP PC Code:* 075301
*California DPR Chemical Code:* 695
*RTECS Number:* GQ9450000
*EINECS Number:* 240-968-3
*Uses:* This material is also used as a wood preservative, soldering flux, as a sand inhibitor in magnesium light metal casting, and in the etching of glass.
*Regulatory Authority:*
- Air Pollutant Standard Set (NIOSH)[2] (ACGIH)[1] (HSE)[33] (DFG)[3] (Former USSR)[43]
- Water Pollution Standard Proposed (Former USSR)[43]
- Clean Water Act: 40CFR116.4 Hazardous Substances; RQ 40CFR117.3 (same as CERCLA); 40CFR423, Priority
- Pollutants (as inorganic fluorides)
- RCRA Universal Treatment Standards: Wastewater (mg/L), 0.059; Nonwastewater (mg/kg), 3.4 (as inorganic fluorides)
- RCRA, 40CFR264, Appendix 9, Ground Water Monitoring List, Suggested Testing Methods (PQL $ug$/L): 8100(200); 8270(10) (as inorganic fluorides)
- Superfund/EPCRA 40CFR302.4, Appendix A, RQ: CERCLA, 1000 lb (454 kg); Section 313: Form R *de minimis* concentration reporting level: 1.0% (as ammonia). $NH_3$ Equivalent weight: 19.12
- The "Director's List" (CAL/OSHA)
- Canada, WHMIS, Ingredients Disclosure List (as silicofluoride compounds)
- Mexico, Wastewater (inorganic fluorides)

*Description:* Ammonium hexafluorosilicate is a white crystalline powder. Odorless. Sinks and mixes with water. Molecular weight: 178.18. Boiling point = decomposes.

*Incompatibilities:* Liquid is corrosive. Contact with acids reacts to form hydrogen fluoride, which is a highly corrosive and toxic gas. Corrosive to aluminum. Keep away from strong oxidizers.

*Permissible Exposure Limits in Air:* OSHA[2]: The legal airborne PEL is 2.5 mg/m³ for fluorides (measured as fluorine) averaged over an 8-hour workshift. This is also the HSE[33] value (UK) and the MAC value in Germany[3]. NIOSH[2]: The recommended airborne exposure limit is 2.5 mg/m³ for fluorides, inorganic (measured as fluorine) averaged over a 10-hour workshift. *ACGIH:* The recommended airborne exposure limit is 2.5 mg/m³ for fluorides (measured as fluorine) averaged over an 8-hour workshift[1]. The NIOSH[2] IDLH is 250 mg/m³ as Fluorides (F). The former USSR-UNEP/IRPTC project[43] has not set a MAC in workplace air but readily soluble fluorides have assigned a MAC in residential air of 0.03 mg/m³ on a momentary basis and 0.01/mg/m³ on an average daily basis.

*Determination in Air:* Fluorides may be measured by collection on a filter and measurement by ion-specific electrode according to NIOSH Method 7902[18].

*Permissible Concentration in Water:* No criteria set for ammonium fluorosolicate as such. The former USSR-UNEP/IRPTC project[43] has set a limit of 1.5 mg/L for fluorine in water used for domestic purposes.

*Routes of Entry:* Inhalation, eyes and/or skin contact.

*Harmful Effects and Symptoms*

*Short Term Exposure:* Inhalation may cause difficult breathing and burning of the mouth, throat and nose which may result in bleeding. These may be felt at 7.5 mg/m$^3$. Nausea, vomiting, profuse sweating and excess thirst may occur at higher levels. May cause pulmonary edema, which can be delayed for several hours; there is a risk of death in serious cases. Skin contact may cause rash, itching and burning and ulceration of skin. Solutions of 1% strength may cause sores if not removed promptly. Eye contact may cause severe irritation. Most reported instances of fluoride toxicity are due to accidental ingestion and it is difficult to associate symptoms with dose. Five to 40 mg may cause nausea, diarrhea and vomiting. More severe symptoms of burning and painful abdomen, sores in mouth, throat and digestive tract, tremors, convulsions and shock will occur around a dose of 1 gm. Death may result by ingestion of 2 to 5 grams. Also reported as 1 teaspoon to 1 ounce.

*Long Term Exposure:* May cause chronic lung irritation and kidney and liver damage. Bronchitis may develop. Chronic exposure may cause weight loss, nausea, vomiting, weakness, shortness of breath. Fluoride may increase bone density, stimulate new bone growth or cause calcium deposits in ligaments. This may become a problem at levels of 20 to 50 mg/m$^3$ or higher. May cause mottling of teeth at this level.

*Points of Attack:* Lungs and eyes.

*Medical Surveillance:* Pre-employment and periodic examinations should consider possible effects on the skin, eyes, teeth, respiratory tract, and kidneys. Chest x-ray and pulmonary function should be followed. Kidney function should be evaluated. If exposures have been heavy and skeletal fluorosis is suspected, pelvic x-rays may be helpful. Intake of fluoride from natural sources in food or water should be known. In the case of exposure to fluoride dusts, periodic urinary fluoride excretion levels have been very useful in evaluating industrial exposures and environmental dietary sources.

*First Aid:* If this chemical gets into the eyes, remove any contact lenses at once and irrigate immediately for at least 15 minutes, occasionally lifting upper and lower lids. Seek medical attention immediately. If this chemical contacts the skin, remove contaminated clothing and wash immediately with soap and water. Seek medical attention immediately. If this chemical has been inhaled, remove from exposure, begin rescue breathing (using universal precautions) if breathing has stopped and CPR if heart action has stopped. Transfer promptly to a medical facility. When this chemical has been swallowed, get medical attention. Give large quantities of water and induce vomiting. Do not make an unconscious person vomit. Medical observation is recommended for 24 to 48 hours after breathing overexposure, as pulmonary edema may be delayed. As first aid for pulmonary edema, a physician or authorized paramedic may consider administering a corticosteroid spray.

*Note to physician or authorized medical personnel:* Ingestion: Give aluminum hydroxide gel, if conscious. Inject intravenously 10 ml of 10% calcium gluconate solution. Gastric lavage with lime water of 1% calcium chloride.

*References:*
- New Jersey Department of Health and Senior Services, "Hazardous Substance Fact Sheet, Ammonium Fluosilicate," Trenton NJ (May 2000). http://www.state.nj.us/health/eoh/rtkweb/0101.pdf
- National Institute for Occupational Safety and Health, "Criteria for a Recommended Standard: Occupational Exposure to Inorganic Fluorides," NIOSH Doc. No. 76-103 (1976).
- Sax, I.N., Ed., "Dangerous Properties of Industrial Materials Report," 4, No. 3, 36-38 (1984). (Al Ammonium Silicofluoride).
- California Environmental Protection Agency Chemical List of Lists, Sacramento CA (February 1997)
- New York State Department of Health, "Chemical Fact Sheet: Ammonium Hexafluorosilicate," Albany, NY, Bureau of Toxic Substance Assessment (March 1986).

# Ammonium Nitrate

*Use Type:* Used as a rodenticide, fertilizer and microbiocide.

*CAS Number:* 6484-52-2

*Formula:* $H_4N_2O_3$; $NH_4NO_3$

*Alert:* Relatively stable but when decomposition is accelerated or substance becomes contaminated, violent explosion may result. It is important that stored material be kept cool and well ventilated.

*Synonyms:* Ammoniumnitrat (German); Ammonium(I) nitrate(1:1); Ammonium Saltpeter; Ansax; Caswell No. 045; German Saltpeter; Nitram; Nitrate d'ammonium (French); Nitrato amonico (Spanish); Nitric acid, ammonium salt; Norway Saltpeter; Varioform I

*Trade Names:* AMTRATE®, Mississippi Chemical (USA); E-2®, El Dorado Chemical Company (USA); EASIGRAZE®, Terra Industries (USA); FIRST CUT No. 8®, Terra Industries (USA); HERCO PRILLS® (Hercules); KAYNITRO®, Terra Industries (USA); NITRAM®, Terra Industries (USA); OLD PLANTATION® (Columbia Nitrogen Corp.); SPRING-K®, Terra Industries (USA);

TURNOUT®, Terra Industries (USA)

**Producers:** Achema (Lithuania); Agrium (Canada); ANWIL (Poland); Apache Nitrogen Products (USA); Azot Association (Ukraine); Deepak Fertilizers and Petrochemicals (India); DSM Agro (Netherlands); Dynamit Nobel Group (Germany); El Dorado Chemical Company (USA); Fluorochem (UK); GFS Chemicals (USA); Grande Paroisse (France); Hydro Agri Chemicals (Norway); Incitec (Australia); Jilin Chemical (China); Kemira Agro (Finland); Kynoch (South Africa); LaRoche Industries (USA); Mallinckrodt Baker (USA); Mississippi Chemical (USA); Orica (Australia); Prodica (USA); Rashtriya Chemicals & Fertilizers (India); Sumitomo Chemical Co. (Japan); Terra Industries (USA); Ube Industries (Japan); Wesfarmers CSBP Ltd. (Australia)

**Chemical Class:** Inorganic
**EPA/OPP PC Code:** 076101
**California DPR Chemical Code:** 3052
**ICSC Number:** 0216
**RTECS Number:** BR9050000
**EINECS Number:** 299-347-8

**Uses:** Used as a nitrogen fertilizer, herbicide, insecticide, and as a desiccant for cotton. Widely used in the manufacture of liquid and solid fertilizers and has a broad application for all crops. California reports the top crops for ammonium nitrate usage in 2000 were cotton, oranges, white grapes, table and raisin grapes, figs and plumbs. Also used in explosives, propellants, matches, cosmetics, and antibiotics.

**Regulatory Authority:**
- EPA TSCA Section 8(b) Chemical Inventory
- AB 2588-Air Toxics "Hot Spots" Chemicals (CAL)
- Highly Reactive Substance and Explosive (World Bank)[15]
- Air Pollutant Standard Set (former USSR)[35]
- U.S. DOT Regulated Marine Pollutant (49CFR172.101, Appendix B) as nitrate compounds
- Clean Water Act: 40CFR116.4 Hazardous Substances; RQ 40CFR117.3 (same as CERCLA)
- CERCLA/SARA Section 313: Form R *de minimis* concentration reporting level: 1.0% (as ammonia) Molecular weight: 80.04; $NH_3$ Equivalent weight: 21.28 also reportable as a nitrate compound, water dissociable, (reportable only when in an aqueous solution), at the same reporting level (1.0%).
- Canada, National Pollutant Release Inventory (solution only)

**Description:** Ammonium nitrate is an odorless white to gray to brown, odorless beads, pellets or flakes. Melting/Freezing point = about 169°C with slow decomposition; the decomposition accelerates at about 210°C and may become explosive. Hazard Identification (based on NFPA-704 M Rating System): Health 0, Flammability 0, Reactivity 3, Oxidizer. Soluble in water.

**Incompatibilities:** A strong oxidizer. Reducing agents, combustible materials, organic materials, finely divided (powdered)metals may form explosive mixtures or cause fire and explosions. When contaminated with oil, charcoal or flammable liquids, can be considered an explosive which can be detonated by combustion or shock.

**Permissible Exposure Limits in Air:** A MAC value of 0.3 mg/m$^3$ on a daily average basis in ambient air has been set in the Former USSR[35].

**Permissible Concentration in Water:** A MAC value of 0.3 mg/m$^3$ on a daily average basis in ambient air has been set in the Former USSR[35].

**Routes of Entry:** Inhalation, skin and/or eye contact.

**Harmful Effects and Symptoms**

**Short Term Exposure:** The potential for ammonia poisoning in the course of $NH_4NO_3$ and fertilizer manufacture is the chief toxic effect associated with ammonium nitrate. Exposure may irritate the skin, eyes, nose, throat and lungs. Overexposure can cause nausea and vomiting, headaches, weakness, faintness and collapse. Severe overexposure may lower the ability of the blood to carry oxygen. This can result in a bluish color to skin and lips, headaches, dizziness, collapse and even death.

**Long Term Exposure:** Unknown at this time.

**Points of Attack:** Inhalation and skin.

**Medical Surveillance:** Consider the points of attack in preplacement and periodic physical examinations.

**First Aid:** If this chemical gets into the eyes, remove any contact lenses at once and irrigate immediately for at least 15 minutes, occasionally lifting upper and lower lids. Seek medical attention immediately. If this chemical contacts the skin, remove contaminated clothing and wash immediately with soap and water. Seek medical attention immediately. If this chemical has been inhaled, remove from exposure, begin rescue breathing (using universal precautions) if breathing has stopped and CPR if heart action has stopped. Transfer promptly to a medical facility. When this chemical has been swallowed, get medical attention. Give large quantities of water and induce vomiting. Do not make an unconscious person vomit.

**References:**
- National Institute for Occupational Safety and Health, Profiles on Occupational Hazard for Criteria Document Priorities: Ammonium Nitrate, pp 281-285, Report PB-274,073, Washington, DC (1977).
- Sax, N.I., Ed., Dangerous Properties of Industrial Materials Report, 2, No. 3, 44-46 (1982).
- Lewis, R. J. Sr., Ed., *Sax's Dangerous Properties of Industrial Materials, 9th Ed.*, John Wiley & Sons, NYC, 1998
- New Jersey Department of Health and Senior Services, "Hazardous Substance Fact Sheet: Ammonium Nitrate," Trenton, NJ (June 1998).
  http://www.state.nj.us/health/eoh/rtkweb/0106.pdf

- New York State Department of Health, "Chemical Fact Sheet: Ammonium Nitrate," Albany, NY, Bureau of Toxic Substance Assessment (January 1986).
- California Environmental Protection Agency Chemical List of Lists, Sacramento CA (February 1997)

## Ammonium Phosphate

*Use Type:* Fertilizer, insecticide, fungicide, herbicide, microbiocide, soil pH adjustment
*CAS Number:* 7783-28-0
*Formula:* $H_9N_2O_4P$; $(NH_4)_2HPO_4$
*Synonyms:* Ammonium phosphate, dibasic; Ammonium orthophosphate, monohydrogen; Ammonium orthophosphate, dibasic; Ammonium phosphate, hydrogen; Ammonium phosphate secondary; Diammonium hydrogen phosphate; Diammonium orthophosphate; Diammonium orthophosphate, hydrogen; Diammonium phosphate; Diammonium phosphate, hydrogen; Diammonium phosphate, monohydrogen; Dibasic ammonium phosphate; Secondary ammonium phosphate
*Trade Names:* 18-46-0 DI-AMMONIUM PHOSPHATE®, Simplot, J.R., Company (USA); DAP-DIAMMONIUM PHOSPHATE®, IMC Phosphates Company (USA)
*Producers:* Agrium (Canada); Albright & Wilson (UK); Astaris (USA); Cargill (USA); Central Glass (Japan); CF Industries (USA); Clariant Functional Chemicals (Germany); Coromandel Fertilisers Ltd. (India); EID Parry (India); Ercros (Spain); FMC (USA); GFS Chemicals (USA); Goldschmidt (Germany); IMC Global (USA); Jost Chemical (USA); Kemira Chemicals (Finland); Kynoch (South Africa); Potash Corporation (Canada); OxyChem (USA); Rhodia Specialty Phosphates (USA); Saudi Basic Industries Corp. (Saudi Arabia); Showa Denko Chemicals Group (Japan); Sichuan Chuanxi Xingda Chemical Plant (China); Simplot, J.R., Company (USA); Thermphos (Netherlands); WMC (Australia)
*Chemical Class:* Inorganic (Ammonium phosphates)
*ICSC Number:* 0217
*RTECS Number:* TB9375000
*EINECS Number:* 231-987-8
*Uses:* Used in fireproofing of textiles, wood, and paper; in soldering flux, as a fertilizer; a buffer; in baking powder and food additives.
*Regulatory Authority:*
- CERCLA/SARA Section 313: Form R *de minimis* concentration reporting level: 1.0% (as ammonia). $NH_3$ Equivalent weight: 25.79
- Actively registered pesticide in California.

*Description:* Ammonium phosphate is a white crystalline or powdery substance. Molecular weight: 132.06. Melting/Freezing point = 185°C (decomposes). Soluble in water. Odorless or weak ammonia odor. Not flammable or combustible.
*Incompatibilities:* Incompatible with strong oxidizers, strong bases. Contact with air causes this chemical to produce anhydrous ammonia fumes.
*Permissible Exposure Limits in Air:* No standards set.
*Permissible Concentration in Water:* No criteria set, but runoff from spills or fire control may cause water pollution.
*Routes of Entry:* Inhalation, ingestion, skin and/or eye contact.
*Harmful Effects and Symptoms*
*Short Term Exposure:* On short term exposure, may cause skin and eye irritation; ammonia fumes can cause eye irritation above 70 ppm. In closed spaces, inhalation of ammonia fumes may cause nose and throat irritation (70 ppm, 5 minutes). Levels of 500 ppm for 30 minutes may cause irritation to throat and lungs. High levels may result in accumulation of fluid in the lung and suffocation. Ammonia poisoning upon ingestion is characterized by sagging of facial muscles, tremors, anxiety, difficulty in controlling muscles, stupor and coma. There is only a slight chance of this happening from ingestion of ammonium phosphates, except in persons with impaired liver function. Large doses may cause calcium imbalance and an increased flow of urine.
*Points of Attack:* Liver, skin and eyes.
*Medical Surveillance:* Liver function tests.
*First Aid:* If this chemical gets into the eyes, remove any contact lenses at once and irrigate immediately for at least 15 minutes, occasionally lifting upper and lower lids. Seek medical attention immediately. If this chemical contacts the skin, remove contaminated clothing and wash immediately with soap and water. Seek medical attention immediately. If this chemical has been inhaled, remove from exposure, begin rescue breathing (using universal precautions) if breathing has stopped and CPR if heart action has stopped. Transfer promptly to a medical facility. When this chemical has been swallowed, get medical attention. Seek medical attention, if necessary. Give large quantities of water or milk Inhalation. Move to fresh air. Give oxygen or artificial respiration if required. Seek medical attention, if necessary.
*References:*
- New York State Department of Health, "Chemical Fact Sheet: Ammonium Phosphate," Albany, NY, Bureau of Toxic Substance Assessment (March 1986).
- California Environmental Protection Agency Chemical List of Lists, Sacramento CA (February 1997).

## Ammonium Sulfamate

*Use Type:* Herbicide and fertilizer
*CAS Number:* 7773-06-0
*Formula:* $H_6N_2O_3S$; $NH_2SO_3NH_4$
*Alert:* As with many herbicides used to clear woody growth, ammonium sulfate may be deadly to animals.
*Synonyms:* Ammonium aminosulfonate; Ammonium

amidosulphate; Ammonium salz der amidosulfonsaure (German); Ammonium sulphamate; Ammonium amidosulfonate; AMS; Monoammonium sulfamate; Monoammonium salt of sulfamic acid; Sulfamato amonico (Spanish); Sulfaminsaure (German); Sulfamic acid, monoammonium salt

*Trade Names:* AMCIDE®, Nufarm (Australia); AMICIDE®, Nufarm (Australia); AMIDOSULFATE®; AMMAT®; AMMATE®, DuPont Agricultural Products, USA, canceled; AMS® AMMONIUM SULFAMATE WEED & BRUSH KILLER, Clariant International, Switzerland, canceled, 3/11/83; BRUSH-OFF® AMMONIUM SULFAMATE BRUSH WEED KILLER, Crown Chemical, USA, canceled 10/8/85; FYRAN 206K®; IKURIN®, canceled; ROOT-OUT® (Dax Products, Ltd); SILVICIDE®; SOBIN® AMMONIUM SULFAMATE, IMC Chemicals, USA, canceled 3/11/83; SULFAMATE®, Southern Chemical Products, USA, canceled 1/22/91

*Producers:* Clariant International (Switzerland); GFS Chemicals (USA); IMC Chemicals (USA); Merck (Germany); Nufarm (Australia); Rhodia (France); Rutherford Chemicals (USA)

*Chemical Class:* Sulfamate

*EPA/OPP PC Code:* 005501

*RTECS Number:* WO6125000

*EINECS Number:* 231-871-7

*Uses:* Used as a contact herbicide and also as a fertilizer. Used to control herbaceous perennials and annual broadleaf weeds and grasses in fruit orchards. Also used for general weed and poison ivy control along rights-of-ways, commercial buildings and the home.

*Human toxicity (long-term)[77]:* Very low–2000.00 ppb, Health Advisory

*Fish toxicity (threshold)[77]:* Very low–43475.60557 ppb, MATC (Maximum Acceptable Toxicant Concentration)

*Carcinogen/Hazard Classifications*

**Label Signal Word:** CAUTION, Toxicity class III

*Regulatory Authority:*
- Air Pollutant Standard Set (NIOSH)[2] (DFG)[3] (ACGIH)[1] (Former USSR)[43] (HSE)[33] (OSHA)[58] (Several States)[60] (Several Canadian Provinces) (Australia) (Israel)
- Permissible Exposure Limits for Chemical Contaminants (CAL/OSHA)
- The "Director's List" (CAL/OSHA)
- Actively registered pesticide in California.
- RCRA 40CFR261, Appendix 8; 40CFR261.11 Hazardous Constituents
- Clean Water Act: 40CFR116.4 Hazardous Substances; 40CFR117.3, RQ (same as CERCLA)
- Superfund/EPCRA 40CFR302.4, Appendix A, RQ: CERCLA, 5,000 (2,270 kg). Section 313: Form R *de minimis* concentration reporting level: 1.0% (as ammonia) Molecular weight: 114.12; $NH_3$ Equivalent weight: 14.92
- Canada, WHMIS, Ingredients Disclosure List

*Description:* Ammonium sulfamate is a white to yellow crystalline solid although some products containing high levels of this chemical are brown. Highly soluble in water. Molecular weight = 114.15. Melting/Freezing point = 131°C (with decomposition). Boiling point = 160°C.

*Incompatibilities:* Strong oxidizers, potassium, potassium chlorate, sodium nitrite, metal chlorates, and hot acid solutions. Elevated temperatures cause a highly exothermic reaction with water.

*Permissible Exposure Limits in Air:* The OSHA[2] PEL is (total dust) 15 $mg/m^3$ TWA and (respirable fraction) 5 $mg/m^3$ TWA for an 8-hour workshift. The NIOSH[2] IDLH level = 1500 $mg/m^3$. NIOSH[2] REL is (total dust) 10 $mg/m^3$ and (respirable fraction) 5 $mg/m^3$. The MAK set by DFG[3] is 15 $mg/m^3$. The ACGIH TWA value is 10 $mg/m^{3(1)}$. Australia limit is 10 $mg/m^3$. Israel limit is 10 $mg/m^3$ with a 5 $mg/m^3$ Action Level. HSE[33] level is 10 $mg/m^3$ TWA and a STEL of 20 $mg/m^3$. The DFG MAK value for total dust is 15 $mg/m^{3(3)}$. In addition, several states have set guidelines or standards for ammonium sulfamate in ambient air[60]: (total dust) 15 $mg/m^3$ TWA and (respirable fraction). The California PEL is 5 $mg/m^3$ TWA. In addition, several states have set guidelines or standards for ammonium sulfamate in ambient air[60]: 0.1 $mg/m^3$ (North Dakota), 0.15 $mg/m^3$ (Virginia), 0.2 $mg/m^3$ (Connecticut), 0.238 $mg/m^3$ (Nevada). Canadian provincial guidelines follow: 10 $mg/m^3$ TWA (8-hour workshift) and STEL of 20 mg/m3 (Alberta, British Columbia), 10 $mg/m^3$ TWA (Ontario) 10 $mg/m^3$ TWAEV (Quebec).

*Determination in Air:* Collection on a filter followed by gravimetric analysis. See NIOSH Method # S348[18].

*Permissible Concentration in Water:* The No-Observed-Adverse-Effect-Level (NOAEL) is 250 mg/kg/day according to the U.S. EPA Health Advisory cited below. From this a health advisory of 21.4 mg/L of water was derived for a 10 kg child on a one-day, ten-day or longer term basis. An acceptable daily intake has been determined to be 0.214 mg/kg/day and a lifetime health advisory for a 70 kg adult is 1.5 mg/L.

*Determination in Water:* There is no standard method for determining ammonium sulfamate in water. There is, however, a method for detection in foods which is a colorimetric method based on liberation of $SO_4$, reduction to $H_2S$ which is measured after treatment with zinc, *p*-aminodimethylaniline and ferric chloride to give methylene blue.

*Routes of Entry:* Inhalation, ingestion, skin and/or eye contact.

*Harmful Effects and Symptoms*

*Short Term Exposure:* This material is moderately toxic by ingestion and may cause gastrointestinal disease. High levels may irritate the eyes, skin and respiratory tract, nausea and vomiting. The $LD_{50}$ (oral, rat) = 3900 mg/kg.

*Long Term Exposure:* Unknown at this time.
*Medical Surveillance:* Nothing special indicated.
*First Aid:* If this chemical gets into the eyes, remove any contact lenses at once and irrigate immediately for at least 15 minutes, occasionally lifting upper and lower lids. Seek medical attention immediately. If this chemical contacts the skin, remove contaminated clothing and wash immediately with soap and water. Seek medical attention immediately. If this chemical has been inhaled, remove from exposure, begin rescue breathing (using universal precautions) if breathing has stopped and CPR if heart action has stopped. Transfer promptly to a medical facility. When this chemical has been swallowed, get medical attention. Give large quantities of water and induce vomiting. Do not make an unconscious person vomit.

*References:*
- Sax, N.I., Ed., Dangerous Properties of Industrial materials Report, 2, No. 3, 52-54 (1982).
- Lewis, R.J., Sr., Ed., *Sax's Dangerous Properties of Industrial Materials*, 9$^{th}$ Ed., John Wiley & Sons, NYC, 1998.
- New Jersey Department of Health and Senior Services, "Hazardous Substance Fact Sheet: Ammonium Sulfamate," Trenton, NJ (December 1994, rev. January 2001). http://www.state.nj.us/health/eoh/rtkweb/0114.pdf
- U.S. Environmental Protection Agency, "Health Advisory: Ammonium Sulfamate," Washington, DC, Office of Drinking Water (August 1987).
- California Environmental Protection Agency Chemical List of Lists, Sacramento CA (February 1997).

# Ammonium Sulfite

*Use Type:* Preservative and treating agricultural grain.
*CAS Number:* 10196-04-0 (diammonium salt); 10192-30-0 (monoammonium salt)
*Formula:* $H_8N_2O_3S$; $(NH_4)_2SO_3$
*Synonyms:* *Monoammonium salt:* Ammonium acid sulfite; Ammonium hydrogen sulfite; Ammonium hydrosulfite; Ammonium monosulfite; Monosodium sulfite; Sulfito amonico (Spanish); Sulfurous acid, monoammonium salt; Ammonium sulfite, hydrogen
*Diammonium salt:* Diammonium sulfite; Sulfurous acid, diammonium salt; Sulfito amonico (Spanish)
*Producers:* Brotherton Specialty Products (UK); Rutherford Chemicals (USA); Shanghai Agricultural Chemical Industry (China); Showa Denko (Japan); William Blythe (UK)
*RTECS Number:* WT3505000 (diammonium); WT3595000 (monoammonium)
*Uses:* Ammonium sulfite is also used in medicines, metal lubricants, explosives, photography, hair wave solutions, and to make other chemicals.
*Regulatory Authority:*
- Clean Water Act: 40CFR116.4 Hazardous Substances; 40CFR117.3, RQ (same as CERCLA) (diammonium salt)
- Superfund/EPCRA 40CFR302.4, Appendix A, RQ: CERCLA, 5,000 lb (2,270 kg); Section 313: Form R *de minimis* concentration reporting level: 1.0% (as ammonia) Molecular weight: 99.10; $NH_3$ Equivalent weight: 17.18 (monoammonium salt)
- Superfund/EPCRA 40CFR302.4, Appendix A, RQ: CERCLA, 5,000 lb (2,270 kg); Section 313: Form R *de minimis* concentration reporting level: 1.0% (as ammonia) Molecular weight: 116.13; $NH_3$ Equivalent weight: 29.33
- The "Director's List" (CAL/OSHA)

*Description:* Ammonium sulfite is a colorless to yellow crystalline (sand-like or sugar-like) solid, normally sold or used in a solution. Melting/Freezing point = 150°C (sublimes). Soluble in water.
*Incompatibilities:* A strong reducing agent. Reacts violently with strong oxidizers, acids.
*Permissible Exposure Limits in Air:* No standards set
*Permissible Concentration in Water:* No criteria set, but runoff from spills or fire control may cause water pollution.
*Routes of Entry:* Inhalation of dust, ingestion
*Harmful Effects and Symptoms*
*Short Term Exposure:* Irritates the eyes, skin, and respiratory tract. Ammonium sulfite can affect you when breathed in; exposure can irritate the nose, throat, bronchial tubes, and lungs. Higher exposures can cause pulmonary edema, a medical emergency that can be delayed for several hours. This can cause death.
*Long Term Exposure:* Ammonium sulfite may cause an asthma-like allergy. Future exposures could then cause asthma attacks with cough, shortness of breath and wheezing. Very severe (anaphylactic) reactions could also occur, and could be fatal.
*Points of Attack:* Skin, eyes, respiratory system.
*Medical Surveillance:* Before beginning employment and at regular times after that, for those with frequent or potentially high exposure, the following are recommended: Lung function tests; Seek prompt medical attention if symptoms are suspected.
*First Aid:* If this chemical gets into the eyes, remove any contact lenses at once and irrigate immediately for at least 15 minutes, occasionally lifting upper and lower lids. Seek medical attention immediately. If this chemical contacts the skin, remove contaminated clothing and wash immediately with soap and water. Seek medical attention immediately. If this chemical has been inhaled, remove from exposure, begin rescue breathing (using universal precautions) if breathing has stopped and CPR if heart action has stopped. Transfer promptly to a medical facility. When this chemical has been swallowed, get medical attention. Give large quantities of water and induce vomiting. Do not make an unconscious person vomit. Medical observation is recommended for 24 to 48 hours after breathing overexposure, as pulmonary edema may be delayed. As first

aid for pulmonary edema, a doctor or authorized paramedic may consider administering a corticosteroid spray.

*References:*
- New Jersey Department of Health and Senior Services, "Hazardous Substance Fact Sheet: Ammonium Sulfite," Trenton, NJ (August 2000). www.state.nj.us/health/eoh/rtkweb/0116.pdf
- California Environmental Protection Agency Chemical List of Lists, Sacramento CA (February 1997).

## Ammonium Thiosulfate

*Use Type:* Insecticide, fungicide, and herbicide
*CAS Number:* 7783-18-8
*Formula:* $H_8N_2O_3S_2$
*Alert:* A notice was filed in the Federal Register, March 5, 2003, indicating the filing of a Pesticide Petition to establish a tolerance in or on food.
*Synonyms:* Ammonium hyposulfite; Amthio; Diammonium thiosulfate; Thiosulfuric acid, diammonium salt; Tiosulfato amonico (Spanish)
*Trade Names:* AMTHIO®
*Producers:* Agrium (Canada); Brotherton Specialty Products (UK); DuPont (USA); NorFalco (USA); Potash Corporation (Canada); Simplot, J.R., Company (USA); William Blythe (UK)
*Chemical Class:* Inorganic
*EPA/OPP PC Code:* 080103
*California DPR Chemical Code:* 892
*RTECS Number:* XN6465000
*Uses:* Used in poultry houses to reduce phosphorus runoffs into lakes and rivers. Also used as a metal lubricant, metal cleaner, and in photographic chemicals, making other chemicals, and as a laboratory reagent.
*Regulatory Authority:*
- CERCLA/SARA Section 313: Form R *de minimis* concentration reporting level: 1.0% (as ammonia) Molecular weight: 148.20; $NH_3$ Equivalent weight: 22.98
- Actively registered pesticide in California.

*Description:* Ammonium thiosulfate is a white crystalline solid with an ammonia odor. Melting/Freezing point = 150°C (decomposes below Melting/Freezing point. solution: <50°C, anhydrous crystals> 100°C). Hazard Identification (based on NFPA-704 M Rating System): Health 1, Flammability 0, Reactivity 0. Highly soluble in water.
*Incompatibilities:* Contact with sodium chlorate may cause a violent reaction. Corrodes brass, copper, and copper-based metals.
*Permissible Exposure Limits in Air:* None established. However, care must be taken in its use.
*Routes of Entry:* Skin contact
*Harmful Effects and Symptoms*
*Short Term Exposure:* Contact with eyes or skin may cause irritation or injury. Inhalation should be avoided; use NIOSH-approved air purifying respirators for pesticides. Harmful if swallowed.
*First Aid:* If this chemical gets into the eyes, remove any contact lenses at once and irrigate immediately for at least 15 minutes, occasionally lifting upper and lower lids. Seek medical attention immediately. If this chemical contacts the skin, remove contaminated clothing and wash immediately with soap and water. Seek medical attention immediately. If this chemical has been inhaled, remove from exposure, begin rescue breathing (using universal precautions) if breathing has stopped and CPR if heart action has stopped. Transfer promptly to a medical facility. When this chemical has been swallowed, get medical attention. Give large quantities of water and induce vomiting. Do not make an unconscious person vomit.
*References:*
- Lewis, Richard J., Sr., Ed., *Sax's Dangereous Proprties of Industrial Materials, 9th Ed.*, John Wiley & Sons, New York, 1998.
- California Environmental Protection Agency Chemical List of Lists, Sacramento CA (February 1997).

## Ampelomyces Quisqualis isolate M10

*Use Type:* Fungicide
*Trade Names:* AQ-10 Biofungicide®, Ecogen Inc. (USA); AQ-10 Technical Powder®, Ecogen Inc. (USA)
*Producers:* Ecogen Inc. (USA)
*Chemical Class:* Microbial
*EPA/OPP PC Code:* 021007
*Uses:* Used to treat powdery mildew on fruits, vegetables and ornamental crops. Can be used both outdoors and in controlled situations, e.g., greenhouses.
*Permissible Concentration in Water:* No criteria set. Runoff from spills or fire control may cause water pollution.
*Harmful Effects and Symptoms*
*Short Term Exposure:* Contact with eyes or skin may cause irritation or injury. Inhalation should be avoided; use NIOSH-approved air purifying respirators for pesticides. May be harmful if swallowed.
*First Aid:* If this chemical gets into the eyes, remove any contact lenses at once and irrigate immediately for at least 15 minutes, occasionally lifting upper and lower lids. Seek medical attention immediately. If this chemical contacts the skin, remove contaminated clothing and wash immediately with soap and water. Seek medical attention immediately. If this chemical has been inhaled, remove from exposure, begin rescue breathing (using universal precautions) if breathing has stopped and CPR if heart action has stopped. Transfer promptly to a medical facility. When this chemical has been swallowed, get medical attention. *Do not induce*

*vomiting when formulations containing petroleum solvents are ingested.* Otherwise, give large quantities of water and induce vomiting. Do not make an unconscious person vomit.

*References:*
- U.S. Environmental Protection Agency, Office of Pesticide Programs, Pesticide Fact Sheet, "Ampelomyces quisqualis isolate M-10," (April 13, 2004). http://www.epa.gov/pesticides/biopesticides/ingredients/factsheets/factsheet_021007.htm
- California Environmental Protection Agency Chemical List of Lists, Sacramento CA (February 1997)

## Anagrapha Falcifera

*Use Type:* Insecticide
*Synonyms:* Anagrapha falcifera multi-nuclear polyhedrosis virus (AfMNPV); Anagrapha falcifera multi-nuclear polyhedrosis virus polyhedral inclusion bodies in aqueous suspension; Anagrapha falcifera MNPV PIB's in aqueous suspension; Celery looper moth NPV
*Producers:* Certis USA (USA)
*Chemical Class:* Botanical insect virus
*EPA/OPP PC Code:* 127885
*California DPR Chemical Code:* 5089
*Uses:* This insect virus is used to kill various larval pests that feed on food crops and other plants. It targets various species of worms and moths including gypsy moths, codling moths and Indian meal moths and is used on vegetables, cotton, corn, peanuts, walnuts, apples, pears and ornamentals. It is sprayed on leaves during plant growth or on the crop after harvest. The virus interferes with the function of several larval organs, including food absorption in the gut. Larvae die after a few days. These viruses occur naturally and present no known risks to humans, other non-target organisms, or the environment.
*Permissible Concentration in Water:* No criteria set. Runoff from spills or fire control may cause water pollution.
*Harmful Effects and Symptoms*
*Short Term Exposure:* Contact with eyes or skin may cause irritation or injury. Inhalation should be avoided; use NIOSH-approved air purifying respirators for pesticides. May be harmful if swallowed.
*First Aid:* If this chemical gets into the eyes, remove any contact lenses at once and irrigate immediately for at least 15 minutes, occasionally lifting upper and lower lids. Seek medical attention immediately. If this chemical contacts the skin, remove contaminated clothing and wash immediately with soap and water. Seek medical attention immediately. If this chemical has been inhaled, remove from exposure, begin rescue breathing (using universal precautions) if breathing has stopped and CPR if heart action has stopped. Transfer promptly to a medical facility. When this chemical has been swallowed, get medical attention. *Do not induce vomiting when formulations containing petroleum solvents are ingested.* Otherwise, give large quantities of water and induce vomiting. Do not make an unconscious person vomit.

*References:*
- U.S. Environmental Protection Agency, Office of Pesticide Programs, "Biopesticide Active Ingredient Fact Sheet, Biopesticides" Washington, DC (October 22, 2002). http://www.epa.gov/oppbppd1/biopesticides/ingredients/factsheets/factsheet_107300.htm
- California Environmental Protection Agency Chemical List of Lists, Sacramento CA (February 1997)

## Ancymidol (ANSI)

*Use Type:* Plant growth regulator
*CAS Number:* 12771-68-5
*Formula:* $C_{15}H_{16}N_2O_2$
*Synonyms:* α-Cyclopropyl-α-(4-methoxyphenyl)-5-pyrimidinemethanol-; α-Cyclopropyl-4-methoxy-α-(pyrimidin-5-yl)benzyl alcohol; 5-Pyrimidinemethanol, α-cyclopropyl-α-(4-methoxyphenyl)
*Trade Names:* A-REST®, Sepro (USA); EL 531®; QUEL®; REDUCYMOL
*Producers:* Sepro (USA)
*Chemical Class:* Pyrimidine
*EPA/OPP PC Code:* 108601; (251200 old EPA code number)
*California DPR Chemical Code:* 1744
*Uses:* Ancymidol is a plant growth regulator registered for treating container-grown herbaceous plants, ornamental woody shrubs, and bedding plants grown in greenhouses and other plant bedding areas for primarily commercial production. Growth regulator effects produced by ancymidol are the result of inhibition of gibberellin biosynthesis. It produces a more compact growth form by suppressing internode elongation.
*Fish toxicity (threshold)[77]:* Very low–19952.6231500 ppb, MATC (Maximum Acceptable Toxicant Concentration)
*Carcinogen/Hazard Classifications*
**Label Signal Word:** CAUTION or WARNING
**WHO Acute Hazard:** Class U, unlikely to be hazardous
*Description:* White to buff crystalline granular. Slightly aromatic odor. Soluble in water; solubility = 645 ppm (approx). Melting/Freezing point = 112°C. Vapor pressure = $2.0 \times 10^{-7}$ mm Hg.
*Permissible Concentration in Water:* No criteria set. Runoff from spills or fire control may cause water pollution.
*First Aid:* If this chemical gets into the eyes, remove any contact lenses at once and irrigate immediately for at least 15 minutes, occasionally lifting upper and lower lids. Seek medical attention immediately. If this chemical contacts the skin, remove contaminated clothing and wash immediately with soap and water. Seek medical attention immediately. If

this chemical has been inhaled, remove from exposure, begin rescue breathing (using universal precautions) if breathing has stopped, and CPR if heart action has stopped. Transfer promptly to a medical facility. When this chemical has been swallowed, get medical attention. Give large quantities of water and induce vomiting. Do not make an unconscious person vomit.

*References:*
- U.S. Environmental Protection Agency, "Reregistration Eligibility Decision (RED), Ancymidol," Office of Prevention, Pesticides and Toxic Substances, Washington, DC (June 1995). http://www.epa.gov/REDs/3017.pdf
- California Environmental Protection Agency "Chemical List of Lists," Sacramento CA (February 1997)

# Anilazine

*Use Type:* Fungicide
*CAS Number:* 101-05-3
*Formula:* $C_9H_5Cl_3N_4$
*Alert:* Not registered in the U.S. Human toxicity (long-term): High.
*Synonyms:* 2-(2-Chloranilin)-4,6-dichlor-1,3,5-triazin (German); (o-Chloroanilino)dichlorotriazine; 2,4-Dichloro-6-(o-chloroanilino)-s-triazine; 2,4-Dichloro-6-o-chloranilino-s-triazine; 2,4-Dichloro-6-(2-chloroanilino)-1,3,5-triazine; 4,6-Dichloro-N-(2-chlorophenyl)-1,3,5-triazin-2-amine; ENT 26,058; NCI-C08684; 1,3,5-Triazine-2-amine, 4,6-dichloro-N-(2-chlorophenyl)-; s-Triazine, 2,4-dichloro-6-(o-chloroanilino)-
*Trade Names:* ANILAZIN®; B-622®, Bayer CropScience (Germany), canceled 7/1/1987; BORTRYSAN®; DIREZ®; DYRENE®, Bayer CropScience (Germany), canceled 11/27/1992; DYRENE 50W®, Bayer CropScience (Germany), canceled 8/31/1988; KEMATE®; NU-RENE 5 DUST®, J. R. Simplot (USA), canceled 7/1/1987; NUTRO®; PAX FUNGUS CONTROL®; TRIASYM®; TRIAZIN®; TRIAZINE®; ZINOCHLOR®
*Producers:* Bayer CropScience (Germany); J. R. Simplot (USA)
*Chemical Class:* Triazine
*EPA/OPP PC Code:* 080811
*California DPR Chemical Code:* 256
*EINECS Number:* 202-910-5
*Uses:* Broad spectrum fungicide used on fruits, vegetables, tobacco, cereals, potatoes, turf and ornamentals.
*Human toxicity (long-term)[77]:* High–2.80 ppb, Health Advisory
*Fish toxicity (threshold)[77]:* Intermediate–14.47908 ppb, MATC (Maximum Acceptable Toxicant Concentration)

*Carcinogen/Hazard Classifications*
**Label Signal Word:** CAUTION, WARNING or DANGER, depending upon formulation
*Regulatory Authority:*
- EPCRA Section 313 Form R *de minimis* concentration reporting level: 1.0%
- EPA 40 CFR 372.65, Specific Toxic Chemical Listings
- EPA 40 CFR 372.65, Specific Toxic Chemical Listings
- Actively registered pesticide in California.

*Description:* White to whitish-brown crystalline solid. Insoluble in water. Molecular weight = 275.53. Melting/Freezing point = 161°C. Vapor pressure = $6.2 \times 10^{-8}$ mmHg.
*Permissible Concentration in Water:* No criteria set. Runoff from spills or fire control may cause water pollution.
*Routes of Entry:* Inhalation, passing through the skin and ingestion.
*Harmful Effects and Symptoms*
*Short Term Exposure:* May cause skin and severe eye irritation. Moderately poisonous if ingested or inhaled. Exposure to a triazine (simazine) has caused acute and subacute dermatitis in the former USSR, characterized by erythema, slight edema, moderate pruritus, and burning lasting 4 to 5 days.
*Long Term Exposure:* May cause lung irritation and damage. May cause skin allergy. Contact with some triazine compounds (such as atrazine) may increase risks for tumors known to be associated with hormonal factors. These have been observed in both animals and human beings, and are consistent with the known effects on the hypothalamic pituitary gonadal axis. Repeated exposure may cause weight loss and reduced red blood cell count. May be mutagenic.
*Points of Attack:* Liver, lungs and skin.
*Medical Surveillance:* Before beginning employment and at regular times after that, for those with frequent or potentially high exposures, the following is recommended: Lung function tests. Consider chest x-ray following acute overexposure. Evaluation by a qualified allergist. Examination of the nervous system.
*First Aid:* If this chemical gets into the eyes, remove any contact lenses at once and irrigate immediately for at least 15 minutes, occasionally lifting upper and lower lids. Seek medical attention immediately. If this chemical contacts the skin, remove contaminated clothing and wash immediately with soap and water. Seek medical attention immediately. If this chemical has been inhaled, remove from exposure, begin rescue breathing (using universal precautions) if breathing has stopped, and CPR if heart action has stopped. Transfer promptly to a medical facility. When this chemical has been swallowed, get medical attention. *Do not induce vomiting when formulations containing petroleum solvents are ingested.* Otherwise, give large quantities of water and induce vomiting. Do not make an unconscious person vomit.

*References:*
- Pesticide Management Education Program, "Anilazine (Dyrene) Chemical Fact Sheet 12/83," Cornell University, Ithaca, NY (December 1983). http://pmep.cce.cornell.edu/profiles/fung-nemat/aceticacid-etridiazole/anilazine/fung-prof-anilazine.html
- California Environmental Protection Agency Chemical List of Lists, Sacramento CA (February 1997)

## Anisole

*Use Type:* A vermicide.
*CAS Number:* 100-66-3
*Formula:* $C_7H_8O$; $C_6H_5OCH_3$
*Synonyms:* Benzene, methoxy; Ether, methyl phenyl; Methoxybenzene; Methyl phenyl ether; Metil fenil eter (Spanish); Phenyl methyl ether
*Producers:* Atul (India); Bayer CropScience (Germany); Bhageria Dye-Chem (India); Cognis (Germany); Degussa (Germany); Great Lakes Chemical (USA); Penta Manufacturing (USA); Rhodia (France); Shell Chemical (UK); Syngenta (Switzerland)
*ICSC Number:* 1014
*RTECS Number:* BZ8050000
*EEC Number:* 202-876-1
*EINECS Number:* 202-876-1
*Uses:* Also used as a solvent, a flavoring, making perfumes, and in organic synthesis.
*Regulatory Authority:*
- Water pollution standard proposed (former USSR)[43]
- Canada, WHMIS, Ingredients Disclosure List

*Description:* Anisole is a colorless to yellowish liquid. Aromatic, spicy-sweet odor. Insoluble in water. Molecular weight = 108.13. Boiling point = 154°C. Melting/Freezing point = –37.3°C. Flash point = 51.6°C (oc). Autoignition temperature = 475°C. Hazard Identification (based on NFPA-704M Rating System): Health 1, Flammability 2, Reactivity 0.
*Incompatibilities:* Keep away from strong oxidizers; may cause violent reaction.
*Permissible Exposure Limits in Air:* No standards set.
*Permissible Concentration in Water:* The former USSR-UNEP/IRTC Project[43] has set a MAC of 0.05 mg/L in water bodies used for domestic purposes. Runoff from spills or fire control may cause water pollution.
*Routes of Entry:* Inhalation, absorbed through the skin, ingestion.
*Harmful Effects and Symptoms*
*Short Term Exposure:* A skin irritant since it degreases the skin; prolonged skin contact can cause drying and cracking. It irritates the eyes and respiratory tract if exposure occurs[57]. Exposure can cause dizziness, lightheadedness, and unconsciousness. It is moderately toxic by ingestion[44]. The $LD_{50}$ (oral, rat) = 3700 mg/kg; (oral, mouse) = 2800 mg/kg[9].
*Long Term Exposure:* Skin problems, dryness, cracking.
*Points of Attack:* Eyes, skin, respiratory system.
*First Aid:* If this chemical gets into the eyes, remove any contact lenses at once and irrigate immediately for at least 15 minutes, occasionally lifting upper and lower lids. Seek medical attention immediately. If this chemical contacts the skin, remove contaminated clothing and wash immediately with soap and water. Seek medical attention immediately. If this chemical has been inhaled, remove from exposure, begin rescue breathing (using universal precautions) if breathing has stopped and CPR if heart action has stopped. Transfer promptly to a medical facility. When this chemical has been swallowed, get medical attention. Give large quantities of water and induce vomiting. Do not make an unconscious person vomit.
*References:*
- Lewis, Richard J., Sr., Ed., *Sax's Dangerous Properties of Industrial Materials*, 9th Ed., John Wiley, NY, 1998.
- California Environmental Protection Agency Chemical List of Lists, Sacramento CA (February 1997).
- New Jersey Department of Health and Senior Services, Hazardous Substance Fact Sheet, Anisole, Trenton NJ (December, 1998). http://www.state.nj.us/health/eoh/rtkweb/0137.pdf

## Anthracene

*Use Type:* Insecticide, herbicide, and rodenticide
*CAS Number:* 120-12-7; 906-80-5 (anthracene oil)
*Formula:* $C_{14}H_{10}$
*Synonyms:* Anthracen (German); Anthracene oil; Anthracene polycyclic aromatic compound; Anthracin; Antraceno (Spanish); Carbolineun; Coal tar distillate (boiling beween 270-300° C); Green Oil; Paranaphthalene
*Trade Names:* STERILITE HOP DEFOLIANT®; TETRA OLIVE N2G®
*Producers:* ATOFINA (France); Bayer Group (Germany); Bilbaina de Alquitranes (Spain); Cindu (Netherlands); Crowley (USA); Deza (Czech Republic); Nippon Steel Chemical (Japan); Sigma-Aldrich Fine Chemicals (USA); Sigma-Aldrich Laborchemikalien (Germany); Wujin LinChuan Chemical Factory (China)
*Chemical Class:* A petroleum derivative
*EPA/OPP PC Code:* 006101 (Anthracene oil)
*California DPR Chemical Code:* 5042
*ICSC Number:* 0825
*RTECS Number:* CA9350000
*EINECS Number:* 204-371-1
*Uses:* It is also used in dye stuffs (alizarin), and wood preservatives, making synthetic fibers, anthraquinone and

other chemicals. May be present in coke oven emissions, diesel fuel, and coal tar pitch volatiles.

*Carcinogen/Hazard Classifications*
**IARC:** Group 3, unclassifiable
*Regulatory Authority:*
- Banned or Severely Restricted (UN) (in cosmetic products in the EEC)[35]
- Air Pollutant Standard Set (ACGIH)[1] (HSE)[33]
- List of priority pollutants (U.S. EPA)
- AB 1803-Well Monitoring Chemical (CAL)
- AB 2588-Air Toxics "Hot Spots" Chemicals (CAL)
- The "Director's List" (CAL/OSHA)
- Water Pollution Standard Set (U.S. EPA)[6] (Kansas)[61] (Mexico)
- Clean Water Act: 40CFR423, Appendix A, Priority Pollutants; Section 313 Water Priority Chemicals (57FR41331, 9/9/92)
- RCRA 40CFR§268.48; 61FR15654, Universal Treatment Standards: Wastewater (mg/L), 0.059; Nonwastewater (mg/kg), 3.4
- RCRA 40CFR264, Appendix 9; Ground Water Monitoring List: Suggested methods (PQL $ug/L$): 8100(200); 8270(10)
- Superfund/EPCRA 40CFR302.4, Appendix A, RQ: 5,000 lb (2,270 kg), 40CFR372.65: Form R *de minimis* Concentration Reporting Level: 1.0%.
- TSCA: 716.120 (*a*), listed chemical
- Canada, WHMIS, Ingredients Disclosure List; National Pollutant Release Inventory (NPRI)

*Description:* Anthracene, $C_{14}H_{10}$, is colorless, to pale yellow crystalline solid with a bluish fluorescence. Boiling point = 340°C. Melting/Freezing point = 216.5°C. Flash point = 121°C. Autoignition temperature = 540°C. NFPA 704 M Hazard Identification (based on NFPA-704M Rating System): Health 0, Flammability 1, Reactivity 0. Explosive limits: LEL = 0.6%[17]. NFPA 704 M Hazard Identification (based on NFPA-704M Rating System): Health 0, Flammability 1. Insoluble in water.

*Incompatibilities:* Dust or fine powder forms an explosive mixture with air. Contact with strong oxidizers, chromic acid, calcium hypochlorite, or fluorine may cause violent reactions.

*Permissible Exposure Limits in Air:* No occupational limits have been established for anthracene. However this chemical may be present as coke oven emissions and coal tar pitch volatiles.

*Determination in Air:* Use NIOSH Methods [for polynuclear aromatic hydrocarbons]: 5506 (HPLC), 5515 (GC), or OSHA Method #ID-58.[58]

*Permissible Concentration in Water:* Anthracene falls in the "polynuclear aromatic hydrocarbon" category of priority toxic pollutants as defined by EPA[6]. The U.S. EPA has considered setting criteria in the range from 0.097 to 9.7 nanograms/liter for the protection of human health from polynuclear aromatic hydrocarbons. In addition, Kansas has set forth a guideline for anthracene in drinking water[61] of 0.029 $\mu g/L$.

*Routes of Entry:* Inhalation, skin and/or eye contact.
*Harmful Effects and Symptoms*
*Short Term Exposure:* Anthracene can affect you when breathed in. Skin contact can cause irritation or a skin allergy which is greatly aggravated by sunlight on contaminated skin. Breathing irritates the nose, throat and bronchial tubes. Eye contact or "fume" exposure can cause irritation and burns.

*Long Term Exposure:* Repeated skin contact can cause thickening, pigment changes and growths. Anthacene may cause mutations. Handle with extreme caution. The carcinogenic status of anthracene is a bit confusing: Animal negative[9] compares with ACGIH[1] and DFG[3] categorization of coal tar volatiles as proven carcinogens. The Lewis/Sax reference below states that it is a mutagen and questionable carcinogen.

*Medical Surveillance:* Evaluation by a qualified allergist.
*First Aid:* If this chemical gets into the eyes, remove any contact lenses at once and irrigate immediately for at least 15 minutes, occasionally lifting upper and lower lids. Seek medical attention immediately. If this chemical contacts the skin, remove contaminated clothing and wash immediately with soap and water. Seek medical attention immediately. If this chemical has been inhaled, remove from exposure, begin rescue breathing (using universal precautions) if breathing has stopped and CPR if heart action has stopped. Transfer promptly to a medical facility. When this chemical has been swallowed, get medical attention. Give large quantities of water and induce vomiting. Do not make an unconscious person vomit.

*References:*
- Agency for Toxic Substances and Disease Registry, "ToxFAQs for Polycyclic Aromatic Hydrocarbons (PAHs)," Atlanta, GA (September 1996), http://www.atsdr.cdc.gov/tfacts69.html
- Sax, N.I., Ed., "Dangerous Properties of Industrial Materials Report" 4, No. 6, 18-43 (1984).
- Lewis, Richard J., Ed., *Dangerous Properties of Industrial Materials, 9th Ed.*, John Wiley & Sons, Inc. New York, 1998.
- California Environmental Protection Agency Chemical List of Lists, Sacramento CA (February 1997).
- New Jersey Department of Health and Senior Services, "Hazardous Substance Fact Sheet: Anthracene," Trenton, NJ (July 1996, Revised June 2002). www.state.nj.us/health/eoh/rtkweb/0139.pdf

# Anthraquinone

*Use Type:* Used as a bird repellent on seeds.
*CAS Number:* 84-65-1

*Formula:* $C_{14}H_8O_2$

*Synonyms:* 9,10-Anthracenedione; Anthradione; 9,10-Anthraquinone; Antraquinona (Spanish); 9,10-Dioxoanthracene; Anthradione

*Trade Names:* (p)ANTHRAPEL®; MORKIT®

*Producers:* BASF (Germany); Bayer Group (Germany); China Chemical (China); Clariant (Switzerland); Deza (Czech Republic); Kawasaki Kasei Chemicals (Japan); Orica (Australia); Quantum Chemicals (Australia); Rhone-Poulenc Agro (France); Sigma-Aldrich Laborchemikalien (Germany); Sumitomo Chemicals (Japan)

*EPA/OPP PC Code:* 122701

*RTECS Number:* CB4725000

*EINECS Number:* 201-549-0

*Uses:* Anthraquinone is an important starting material for vat dye manufacture. Also used in making organics.

*Regulatory Authority:*
- Air Pollutant Standard Set (former USSR)[43]
- TSCA 40CFR704.30; 40CFR716.120(a) List of substances; 40CFR712.30(m); 40CFR 799.500 Testing Requirements. Export notification required by §12(b).

*Description:* Colorless to yellow crystalline solid. Insoluble in water. Melting/Freezing point= 286°C. Boiling point = 380°C. Flash point = 185°C (cc). NFPA 704 M Hazard Identification (based on NFPA-704M Rating System): Health 0, Flammability 1, Reactivity unknown.

*Incompatibilities:* Contact with strong oxidizers may cause fire and explosions.

*Permissible Exposure Limits in Air:* The former USSR-UNEP/IRPTC project[43] has set a MAC value of 5 mg/m³ in workplace air.

*Permissible Concentration in Water:* No criteria set, but runoff from spills or fire control may cause water pollution.

*Routes of Entry:* Through the skin, inhalation.

*Harmful Effects and Symptoms*

*Short Term Exposure:* Can be absorbed through the skin, thereby increasing exposure. Eye or skin contact can cause irritation. An allergen, may cause skin irritation and sensitization. Severe poisoning may cause seizures and coma.

*Long Term Exposure:* May cause skin allergy, with itching and rash. It may be mutagenic.

*Points of Attack:* Skin and lungs.

*Medical Surveillance:* Evaluation by a qualified allergist.

*First Aid:* If this chemical gets into the eyes, remove any contact lenses at once and irrigate immediately for at least 15 minutes, occasionally lifting upper and lower lids. Seek medical attention immediately. If this chemical contacts the skin, remove contaminated clothing and wash immediately with soap and water. Seek medical attention immediately. If this chemical has been inhaled, remove from exposure, begin rescue breathing (using universal precautions) if breathing has stopped and CPR if heart action has stopped. Transfer promptly to a medical facility. When this chemical has been swallowed, get medical attention. Give large quantities of water and induce vomiting. Do not make an unconscious person vomit.

*References:*
- New Jersey Department of Health and Senior Services, "Hazardous Substance Fact Sheet, Anthraquinone," Trenton NJ (January 1999). http://www.state.nj.us/health/eoh/rtkweb/0140.pdf
- California Environmental Protection Agency Chemical List of Lists, Sacramento CA (February 1997).

## Antimony Potassium Tartrate

*Use Type:* Insecticide

*CAS Number:* 28300-74-5

*Formula:* $C_4H_4KO_7Sb$

*Synonyms:* Antimonate (2-), bis μ-2,3-dihydroxybutanedioata (4-)-01,02:03,04 di-, dipotassium, trihydrate, stereoisomer; Antimonate(2-), bis(μ-(2,3-dihydroxybutanedioato(4-)-01,02:03,04))di-, dipotassium, trihydrate, stereoisomer (9CI); Antimonyl potassium tartrate; Bis(μ-2,3-Dihydroxybutanedioato(4-)-0(2):0(3),0(4))diantimonate(2-), dipotassium, trihydrate, stereoisomer; Emetique (French); ENT 50,434; Potassium antimony tartrate; Potassium antimonyl tartrate; Potassium antimonyl-*d*-tartrate; Tartaric acid, antimony potassium salt; Tartar emetic; Tartarized antimony; Tartrated antimony; Tartrato de antimonio y potasio (Spanish)

*Trade Names:* TASTOX®, Zeneca Inc (USA) (now Syngenta), canceled 7/1/87

*Chemical Class:* Antimony

*EPA/OPP PC Code:* 006201

*California DPR Chemical Code:* 3057

*RTECS Number:* CC6825000

*EINECS Number:* 229-436-1

*Uses:* It is also used in medicine and textile and leather dyeing.

*U.S. Maximum Allowable Residue Levels for Antimony Potassium Tartrate (40 CFR 180.179):*

| CROP | ppm |
| --- | --- |
| Fruit, stone, group | 3.5 |
| Grape | 3.5 |
| Onion | 3.5 |

*Carcinogen/Hazard Classifications*

**Label Signal Word:** DANGER

**TRI Developmental Toxin:** Reproductive and developmental toxin as antimony compounds.

*Regulatory Authority:*
- Banned or Severely Restricted (New Zealand)[13] (Many countries, especially in food) (UN)[35]
- Air Pollutant Standard Set (ACGIH)[1] (OSHA)[2] (California) (HSE)[33] (Ontario, Quebec)
- AB 2588-Air Toxics "Hot Spots" Chemicals (CAL) as antimony compounds

- Permissible Exposure Limits for Chemical Contaminants (CAL/OSHA) as antimony compounds
- The "Director's List" (CAL/OSHA) as antimony compounds
- Clean Air Act, 42USC7412; Title I, Part A,§112 hazardous pollutants (as antimony compounds)
- Clean Water Act: 40CFR116.4 Hazardous Substances; 40CFR117.3, RQ (same as CERCLA); 40CFR423, Appendix A, Priority Pollutants; Section 313 Water Priority Chemicals (57FR41331, 9/9/92); 40CFR401.15 Section 307 Toxic
- Pollutants, as antimony compounds.
- RCRA, 40CFR261, Appendix 8 Hazardous Constituents, waste number not listed (as antimony compounds, n.o.s.)
- Safe Drinking Water Act, MCL, treatment technique; MCL, 0.006 mg/L; MCLG, 0.006 mg/L; Regulated Chemical (47FR9352)
- Superfund/EPCRA 40CFR302.4 RQ: CERCLA, 100 lb (45.4 kg)
- EPCRA Section 313: Includes any unique chemical substance that contains antimony as part of that chemical's infrastructure. Form R *de minimus* concentration reporting level: 0.1%
- Canada, WHMIS, Ingredients Disclosure List; National Pollutant Release Inventory as antimony compounds.

*Description:* Colorless, crystalline material or white powder. Odorless. Soluble in water; solution is slightly acidic. Molecular weight = 667.91. Hazard Identification (based on NFPA-704 M Rating System): Health 2, Flammability 0, Reactivity 0.

*Incompatibilities:* Solution will react with alkaline materials, mineral acids.

*Permissible Exposure Limits in Air:* The legal Federal OSHA[2] airborne PEL as Sb is 0.5 mg/m$^3$ averaged over an 8-hour workshift. The recommended NIOSH[2] airborne exposure limit (PEL) is 0.5 mg/m$^3$ averaged over a 10-hour workshift. and 2.5 m/gm$^3$ as F for an 8-hour workshift[9]. The recommended ACGIH[1] airborne exposure limit is 0.5 mg/m$^3$ averaged over a 8-hour workshift. The above exposure limits are for air levels only. When skin contact also occurs, you may be overexposed, even though air levels are lower than the limits listed above. The HSE[33] (U.K.), California, Ontario, and Quebec airborne exposure limits for antimony compounds are the same as the OSHA levels show above. The Former USSR-UNEP/IRPTC MAC value[43] is 0.3 mg/m$^3$. The NIOSH[2] IDLH value for antimony and compounds = 50 mg/m$^3$.

*Permissible Concentration in Water:* The U.S. EPA allows 0.006 ppm of antimony per million parts of drinking water.

*Routes of Entry:* Absorbed through the skin, inhalation, ingestion.

*Harmful Effects and Symptoms*

*Short Term Exposure:* Antimony Potassium Tartrate is poisonous if swallowed. It can affect you when breathed in and by passing through your skin. Eye and skin contact can cause irritation and skin rash. Exposure can cause poor appetite, abdominal pain, nausea, headaches, sore throat and irritation of air passages, with cough. Higher levels can cause pulmonary edema, a medical emergency that can be delayed for several hours. This can cause death. Exposure may make the heart beat irregularly or stop. High or repeated exposure may damage the liver or heart muscle.

*Long Term Exposure:* Prolonged or repeated contact can cause ulcers or sores in the nose, kidney, liver, and heart damage.

*Points of Attack:* Skin, eyes, respiratory system, cardiovascular system, kidneys and liver. Lung cancer has been observed in some studies of rats that breathed high levels of antimony. No human studies are available. It is unknown whether antimony will cause cancer in people.

*Medical Surveillance:* Antimony can be measured in the urine, feces, and blood for several days after exposure. EKG. Liver and kidney function tests. Consider chest x-ray following acute overexposure.

*First Aid:* If this chemical gets into the eyes, remove any contact lenses at once and irrigate immediately for at least 15 minutes, occasionally lifting upper and lower lids. Seek medical attention immediately. If this chemical contacts the skin, remove contaminated clothing and wash immediately with soap and water. Seek medical attention immediately. If this chemical has been inhaled, remove from exposure, begin rescue breathing (using universal precautions) if breathing has stopped and CPR if heart action has stopped. Transfer promptly to a medical facility. When this chemical has been swallowed, get medical attention. Give large quantities of water and induce vomiting. Do not make an unconscious person vomit. Medical observation is recommended for 24 to 48 hours after breathing overexposure, as pulmonary edema may be delayed. As first aid for pulmonary edema, a doctor or authorized paramedic may consider administering a corticosteroid spray.

*References:*
- U.S. Environmental Protection Agency, Office of Pesticide Programs, "Pesticide Residue Limits," 40 CFR180.179 www.epa.gov/cgi-bin/oppsrch
- Sax, N.I., Ed., "Dangerous Properties of Industrial Materials Report" 1, No. 8, 33-35 (1981).
- California Environmental Protection Agency Chemical List of Lists, Sacramento CA (February 1997).
- New Jersey Department of Health and Senior Services, Hazardous Substance Fact Sheet, "Antimony Potassium/Tartrate," Trenton, NJ (February 1998). www.state.nj.us/health/eoh/rtkweb/0145.pdf

# Antimycin A

*Use Type:* Specific uses for Antimycin A were not found;

however, Antimycin $A_1$, and Antimycin $A_3$ are reported to be antibiotic substances produced by Streptomyces for use as a fungicide, possible insecticide and miticide. Registered as a pesticide in the U.S.
*CAS Number:* 1397-94-0 ($A_1$-); 642-15-9 ($A_1$-) 11118-72-2 (antimycin)
*Note:* Both $A_1$ CAS numbers are found in RTECS, with the same chemical formula, although EPA regulates only 1397-94-0 as Antimycin A
*Formulas:* $C_{28}H_{40}N_2O_9$ (Antimycin $A_1$); $C_{26}H_{36}N_2O_9$ (Antimycin $A_3$); $C_{25}H_{34}N_2O_9$ (Antimycin $A_4$)
*Synonyms:* Antimicina A (Spanish); Antimycin A; Antipiricullin; Dihyrosamidin; Isovaleric acid-8-ester with 3-formamido-*N*-(7-hexyl-8-hydroxy-4,9-dimethyl-2,6-dioxo-1,5-dioxonan-3-yl)salicylamide isovaleric acid 8 ester
*Trade Names:* FINTROL®; VIROSIN®
*Producers:* BIMOL Research Laboratories (USA); Kyowa Hakko Kogyo (Japan)
*EPA/OPP PC Code:* 006314
*RTECS Number:* CD0350000
*Uses:* Fungicide, possible insecticide and miticide
*Regulatory Authority:*
- CERCLA/SARA 40CFR302 Extremely Hazardous Substances: TPQ = 1000/10,000 lb (454/4540 kg)[7]
- Superfund/EPCRA 40CFR302.4 RQ: CERCLA, 1 lb (0.454 kg)

*Description:* $C_{26}H_{36}N_2O_9$ (Antimycin $A_3$) and $C_{28}H_{40}N_2O_9$ (Antimycin $A_1$) are crystalline solids. Practically insoluble in water. Melting/Freezing point = 170–175°C; ($A_3$); 149–150°C ($A_1$). They are complex 9-membered (2 oxygens and 7 carbons) ring derivatives with complex side chains.
*Permissible Exposure Limits in Air:* No standards set.
*Permissible Concentration in Water:* No criteria set, but runoff from spills or fire control may cause water pollution.
*Routes of Entry:* Ingestion and intramuscular.
*Harmful Effects and Symptoms*
*Short Term Exposure:* Subcutaneous, intravenous, and intraperitoneal route poisons. Moderately toxic by ingestion and intramuscular routes. The $LD_{50}$ (oral rat) = 28 mg/kg[9].
*First Aid:* If this chemical gets into the eyes, remove any contact lenses at once and irrigate immediately for at least 15 minutes, occasionally lifting upper and lower lids. Seek medical attention immediately. If this chemical contacts the skin, remove contaminated clothing and wash immediately with soap and water. Seek medical attention immediately. If this chemical has been inhaled, remove from exposure, begin rescue breathing (using universal precautions) if breathing has stopped and CPR if heart action has stopped. Transfer promptly to a medical facility. When this chemical has been swallowed, get medical attention. Give large quantities of water and induce vomiting. Do not make an unconscious person vomit.
*References:*
- U.S. Environmental Protection Agency, "Chemical Profile: Antimycin A," Washington, DC, Chemical Emergency Preparedness Program (November 30, 1987).
- California Environmental Protection Agency Chemical List of Lists, Sacramento CA (February 1997).

# ANTU

*Use Type:* Rodenticide
*CAS Number:* 86-88-4
*Formula:* $C_{11}H_{10}N_2S$
*Synonyms:* α-Naphthyl thiourea; Alphanaphtyl thiouree (French); Chemical 109; 1-Naftil-tiourea (Italian); 1-Naftylthioureum (Dutch); α-Naphthothiourea; α-Naphthylthiocarbamide; 1-Naphthyl-thioharnstoff (German); 1-Naphthylthiourea; *N*-(1-Naphthyl)-2-thiourea; α-Naphthylthiourea; 1-(1-Naphthyl)-2-thiourea; 1-Naphthyl-thiouree (French); Thiourea, 1-naphthalenyl-; Urea,1-(1-naphthyl)-2-thio-
*Trade Names:* ALRATO®; ANTURAT®; BANTU®; DIRAX®; KILL KANTZ®; KRYSID®; KRYSID PI®; NAPHTOX®; RATTRACK®; RAT-TU®; SMEESANA®
*Producers:* Prentiss Inc. (USA); Sigma-Aldrich Laborchemikalien (Germany)
*Chemical Class:* Thiourea rodenticide
*EPA/OPP PC Code:* 004501
*California DPR Chemical Code:* 38
*ICSC Number:* 0973
*RTECS Number:* YT9275000
*EEC Number:* 006-008-00-0
*EINECS Number:* 201-706-3
*Uses:* Used specifically as a control for the adult Norway rat. It is less toxic to other rat species; relatively safe for domestic animals. It has been withdrawn from the market in some countries.
*Carcinogen/Hazard Classifications*
**IARC:** Group 3, unclassifiable
**Label Signal Word:** DANGER
*Regulatory Authority:*
- Air Pollutant Standard Set (ACGIH)[1] (DFG)[3] (OSHA)[58] (Several States)[60] (Australia) (Israel) (Mexico) (California) (Several Canadian Provinces)
- EPA Hazardous Waste Number (RCRA No.): P072
- Permissible Exposure Limits for Chemical Contaminants (CAL/OSHA)
- The "Director's List" (CAL/OSHA)
- EPA Hazardous Waste Number (RCRA No.): P072
- 40CFR261.11 Hazardous Constituents
- CERCLA/SARA 40CFR302 Extremely Hazardous Substances: TPQ = 500/10,000 lb (227/4540 kg)
- CERCLA/SARA Section 304 RQ: CERCLA, 100 lb (45.4 kg)

*Description:* White crystalline solid or gray powder. Odorless. Melting/Freezing point = 198°C. Hazard

Identification (based on NFPA-704 M Rating System): Health 4, Flammability 1, Reactivity 0. Slightly soluble in water. Noncombustible.

*Incompatibilities:* Strong oxidizers and silver nitrate.

*Permissible Exposure Limits in Air:* The Federal OSHA/NIOSH[2] standard[58] is 0.3 mg/m³ TWA. ACGIH TLV[1], DFG[3] MAK, Australia, Israel, Mexico have set the same value. The DFG peak limitation is 5 times the normal MAK (30 min.), do not exceed 2 times during a workshift[3]. Isarel has an Action Level of 0.15 mg/m³. The Mexico STEL is 0.9 mg/m³. The NIOSH[2] IDLH level = 100 mg/m³. Several states have set guidelines or standards for ANTU in ambient air[60] ranging from 3 $\mu$g/m³ (North Dakota) to 5 $\mu$g/m³ (Virginia) to 6 $\mu$g/m³ (Connecticut) to 7 $\mu$g/m³ (Nevada). Canadian Provincial level for Alberta, British Columbia, Ontario and Quebec are 0.2 mg/m³ TWA or TWAEV (Quebec). Alberta and British Columbia STEL is 0.9 mg/m³ (15 min.).

*Determination in Air:* Collection on a filter and analysis by gas-liquid chromatography. See NIOSH Method 5276[18].

*Permissible Concentration in Water:* No criteria set, but runoff from spills or fire control may cause water pollution.

*Routes of Entry:* Inhalation, ingestion and skin absorption.

*Harmful Effects and Symptoms*

*Short Term Exposure:* Poisonous. Symptoms include seizures, and dermal irritation. High exposures can cause pulmonary edema, a medical emergency that can be delayed for several hours. This can cause death. Ingestion may cause vomiting, shortness of breath, and bluish discoloration of the skin. ANTU is moderately toxic: probable oral lethal dose (human) 0.5-5 mg/kg, or between 1 ounce and 1 pint (or 1 lb) for 150 lb. person. The LD50 (oral, rat) = 6 mg/kg[9]. Chronic sublethal exposure may cause antithyroid activity. Can produce hyperglycemia of three times normal in three hours.

*Long Term Exposure:* May cause chronic dermatitis, increased production of white blood cells. A questionable carcinogen (IARC, Group 3, inadequate human evidence) and a possible mutagen.

*Points of Attack:* Respiratory system.

*Medical Surveillance:* Consider the points of attack in preplacement and periodic physical examinations. People with chronic respiratory disease or liver disease may be especially at risk. Lung function tests. Consider chest x-ray following acute overexposure. Evaluation by a dermatologist.

*First Aid:* If this chemical gets into the eyes, remove any contact lenses at once and irrigate immediately for at least 15 minutes, occasionally lifting upper and lower lids. Seek medical attention immediately. If this chemical contacts the skin, remove contaminated clothing and wash immediately with soap and water. Seek medical attention immediately. If this chemical has been inhaled, remove from exposure, begin rescue breathing (using universal precautions) if breathing has stopped and CPR if heart action has stopped. Transfer promptly to a medical facility. When this chemical has been swallowed, get medical attention. Give large quantities of water and induce vomiting. Do not make an unconscious person vomit. Medical observation is recommended for 24 to 48 hours after breathing overexposure, as pulmonary edema may be delayed. As first aid for pulmonary edema, a doctor or authorized paramedic may consider administering a corticosteroid spray.

*References:*

- New Jersey Department of Health and Senior Services, "Hazardous Substance Fact Sheet, alpha-Naphthyl Thiourea," Trenton NJ (December 1994, rev. February 2001). http://www.state.nj.us/health/eoh/rtkweb/0051.pdf
- Sax, N.I., Ed., "Dangerous Properties of Industrial Materials Report" 4, No. 2, 83-86 (1984).
- Lewis, Richard J. Sr., Ed., *Sax's Dangerous Properties of Industrial Materials*, 9$^{th}$ Ed., John Wiley & Sons, New York, 1998.
- California Environmental Protection Agency Chemical List of Lists, Sacramento CA (February 1997).
- U.S. Environmental Protection Agency, "Chemical Profile: ANTU," Washington, DC, Chemical Emergency Preparedness Program (November 30, 1987).

# Aramite®

*Use Type:* Insecticide and miticide

*CAS Number:* 140-57-8

*Formula:* $C_{15}H_{23}ClO_4S$; $(CH_3)_3C-C_6H_4-OCH_2CH(CH_3)OSO_2-(CH_2)_2Cl$

*Alert:* A carcinogen, handle with care. No products registered in the U.S.

*Synonyms:* Butylphenoxyisopropyl chloroethyl sulfite; 2-(*p*-Butylphenoxy)isopropyl 2-chloroethyl sulfite; 2-(4-*tert*-Butylphenoxy)isopropyl-2-chloroethyl sulfite; 2-(*p*-*tert*-Butylphenoxy)isopropyl 2'-chloroethyl sulphite; 2-(*p*-*tert*-Butylphenoxy)-1-methylethyl 2-chloroethyl ester of sulphurous acid; 2-(*p*-Butylphenoxy)-1-methylethyl 2-chloroethyl sulfite; 2-(*p*-*tert*-Butylphenoxy)-1-methylethyl-2-chloroethyl sulfite ester; 2-(*p*-*tert*-Butylphenoxy)-1-methylethyl 2'-chloroethyl sulphite; 2-(*p*-*tert*-Butylphenoxy)-1-methylethyl sulphite of 2-chloroethanol; 1-(*p*-*tert*-Butylphenoxy)-2-propanol-2-chloroethyl sulfite; CES; 2-Chloroethanol-2-(*p*-*tert*-butylphenoxy)-1-methylethyl sulfite; 2-Chloroethanol ester with 2-(*p*-*tert*-butylphenoxy)-1-methylethyl sulfite; $\beta$-Chloroethyl-$\beta$'-(*p*-*tert*-butylphenoxy)-$\alpha$'-methylethyl sulfite; $\beta$-Chloroethyl-$\beta$-(*p*-*tert*-butylphenoxy)-$\alpha$-methylethyl sulphite; 2-Chloroethyl 1-methyl-2-(*p*-*tert*-butylphenoxy)ethyl sulphite; 2-Chloroethyl sulfurous acid-2-[4-(1,1-dimethylethyl)phenoxy]-1-methylethyl ester; 2-Chloroethyl sulphite of 1-(*p*-*tert*-butylphenoxy)-2-propanol; ENT

16,519; Sulfurous acid 2-(*p-tert*-butylphenoxy)-1-methylethyl-2-chloroethyl ester
*Trade Names:* ACARACIDE®; ARACIDE®; ARARAMITE-15W®; ARATRON®; COMPOUND 88R®; NIAGARAMITE®, FMC Agricultural Products Group (USA), canceled 6/13/1977; ORTHO-MITE®, Scotts Company, The (USA), canceled; 88-R®
*Producers:* FMC Agricultural Products Group (USA)
*EPA/OPP PC Code:* 062501
*California DPR Chemical Code:* 39
*RTECS Number:* WT2975000
*Uses:* Aramite is a miticide and antimicrobial agent. Use of it was voluntarily canceled by the sole producer in 1975.
*Carcinogen/Hazard Classifications*
**U.S. EPA Carcinogens:** Group B2, probable carcinogen
**California Prop. 65:** Carcinogen
**IARC:** Group 2b, possible carcinogen
**WHO Acute Hazard:** Class III, slightly hazardous (as propargite)
*Regulatory Authority:*
- Carcinogen (Animal Positive) (IARC)[9]
- Air Pollutant Standard Set (Pennsylvania)[60]
- AB 2588-Air Toxics "Hot Spots" Chemicals (CAL)
- Proposition 65 chemical (CAL)
- The "Director's List" (CAL/OSHA)
- 40CFR261.11 Hazardous Constituents

*Description:* Aramite is a heavy liquid. Practically insoluble in water. Boiling point = 175°C @t 0.1 mmHg. Melting/Freezing point = –32°C.
*Incompatibilities:* Incompatible with alkaline material such as lime or Bordeaux mixture (slaked lime and copper sulfate solution.).
*Permissible Exposure Limits in Air:* A limit on aramite in ambient air has been set in Pennsylvania[60] at 18.07 $\mu g/m^3$.
*Permissible Concentration in Water:* No criteria set, but runoff from spills or fire control may cause water pollution.
*Routes of Entry:* Inhalation, ingestion, skin and/or eye contact.
*Harmful Effects and Symptoms*
*Short Term Exposure:* Contact may cause severe skin and eye irritation. This material is slightly toxic ($LD_{50}$ value for rats is 3,900 mg/kg) but it is carcinogenic to animals.
*Long Term Exposure:* Aramite is carcinogenic in rats and dogs (oral). It produced liver tumors in rats and carcinomas of the gall bladder and biliary ducts in dogs. Aramite was tested in two strains of mice by the oral route and produced a significant increase of hepatomas in males of one strain.
*Points of Attack:* Liver.
*First Aid:* If this chemical gets into the eyes, remove any contact lenses at once and irrigate immediately for at least 15 minutes, occasionally lifting upper and lower lids. Seek medical attention immediately. If this chemical contacts the skin, remove contaminated clothing and wash immediately with soap and water. Seek medical attention immediately. If this chemical has been inhaled, remove from exposure, begin rescue breathing (using universal precautions) if breathing has stopped and CPR if heart action has stopped. Transfer promptly to a medical facility. When this chemical has been swallowed, get medical attention. Give large quantities of water and induce vomiting. Do not make an unconscious person vomit.
*References:*
- New Jersey Department of Health and Senior Services, "Hazardous Substance Fact Sheet, Aramite," (May 2000). http://www.state.nj.us/health/eoh/rtkweb/0150.pdf
- Sax, N.I., Ed., "Dangerous Properties of Industrial Materials Report," 1, No. 3, 79-80 (1981).
- California Environmental Protection Agency Chemical List of Lists, Sacramento CA (February 1997).

# Arosurf® MSF

*Use Type:* Insecticide
*CAS Number:* 52292-17-8
*Synonyms:* POE isooctadecanol; Poly(oxy-1,2-ethanediyl), α-isooctadecyl-ω-hydroxy-; α-Isodecyl-ω-hydroxypoly(oxy-1,2-ethanediyl); Isosteareth-2
*Trade Names:* AGNIQUE MMF MOSQUITO LARVICIDE & PUPICIDE®, Goldschmidt (Germany); AROSURF® 66ES, Goldschmidt (Germany); AROSURF® 66E2, Goldschmidt (Germany); ISA-20E®
*Producers:* Cognis GmbH (Germany); Goldschmidt (Germany); Witco Corp (USA)
*Chemical Class:* Polyalkyloxy compound
*EPA/OPP PC Code:* 124601
*Uses:* Used to control mosquito larvae in fresh and salt water habitats such as irrigation and roadside ditches, reservoirs and other pollution sources. Also used as emulsifying agents for road paving and in cosmetics.
*Description:* Arosurf MSF is a clear to light amber liquid. Melting point = –3 to –7°C.
*Permissible Concentration in Water:* No criteria set. Runoff from spills or fire control may cause water pollution.
*Harmful Effects and Symptoms*
*Short Term Exposure:* Mildly toxic by ingestion and by skin contact.
*First Aid:* If this chemical gets into the eyes, remove any contact lenses at once and irrigate immediately for at least 15 minutes, occasionally lifting upper and lower lids. Seek medical attention immediately. If this chemical contacts the skin, remove contaminated clothing and wash immediately with soap and water. Seek medical attention immediately. If this chemical has been inhaled, remove from exposure, begin rescue breathing (using universal precautions) if breathing has stopped and CPR if heart action has stopped. Transfer promptly to a medical facility. When this chemical has been swallowed, get medical attention. *Do not induce vomiting when formulations containing petroleum solvents*

*are ingested.* Otherwise, give large quantities of water and induce vomiting. Do not make an unconscious person vomit.

*References:*
- Pesticide Management Education Program, "Arosurf MSF Chemical Fact Sheet 2/84", Cornell University, Ithaca, NY (February 1984).
http://pmep.cce.cornell.edu/profiles/insect-mite/abamectin-bufencarb/arosurf/insect-prof-arosurf.html
- California Environmental Protection Agency Chemical List of Lists, Sacramento CA (February 1997)

# ASPON®

*Use Type:* Insecticide
*CAS Number:* 3244-90-4
*Formula:* $C_{12}H_{28}O_5P_2S_2$
*Synonyms:* Bis-*O,O*-di-*n*-propylphosphorothionic anhydride; ENT 16,894; NPD; Propyl thiopyrophosphate; Tetra-*n*-propyl dithionopyrophosphate; Tetra-*n*-propyl dithiopyrophosphate; *O,O,O,O*-Tetrapropyl dithiopyrophosphate; Thiopyrophosphoric acid, tetrapropyl ester
*Trade Names:* A 42®; ASP 51®; STAUFFER ASP-51®
*Producers:* Stauffer Chemical (USA), Albright & Wilson (UK)
*Chemical Class:* Organophosphate
*EPA/OPP PC Code:* 079101
*California DPR Chemical Code:* 1693
*Uses:* Not registered in the U.S.
*Regulatory Authority:*
- DOT Inhalation Hazard Chemicals as organophosphates
- Actively registered pesticide in California.

*Description:* Amber liquid. Practically insoluble in water. Molecular weight = 378.46. Density = 1.12 @ 4°C. Boiling point = 104°C @ 0.1 mmHg; 148°C @ 2mmHg.
*Incompatibilities:* React violently with acetaldehyde. Incompatible with strong bases. May react violently with antimony(V)pentafluoride. Incompatible with lead diacetate, magnesium, silver nitrate.
*Determination in Air:* OSHA versatile sampler-2; Toluene/Acetone; Gas chromatography/Flame ionization detection; NIOSH IV[18], Method #5600, Organophosphorus Pesticides.
*Permissible Concentration in Water:* No criteria set. Runoff from spills or fire control may cause water pollution.
*Harmful Effects and Symptoms*
*Short Term Exposure:* Eye pupils are small; blurred vision; eye watering; runny nose; cough; shortness of breath; salivation; dizziness; nausea, stomach cramps, diarrhea, and vomiting; increased blood pressure; profuse sweating; hypermotility, hallucinations; irritability; tingling of the skin; drowsiness; slow heartbeat; convulsions; fluid in lungs; loss of consciousness; incontinence; breathing stops; death.

Orgnophosphates inhibit the action of acetylcholinesterase enzymes, and alter the way in which nervous impulses are transmitted. The effects can last for hours, days, or much longer. The action of the enzymes is reestablished after new enzymes are formed. Delayed pulmonary edema may occur after inhalation.
*Long Term Exposure:* Cholinesterase inhibitor; cumulative effect is possible. Organophosphates may damage the nervous system with repeated exposure, resulting in convulsions, respiratory failure. May cause liver damage.
*Points of Attack:* Respiratory system, central nervous system, cardiovascular system, blood cholinesterase.
*Medical Surveillance:* Medical observation is recommended for 24 to 48 hours after breathing overexposure, as pulmonary edema may be delayed. As first aid for pulmonary edema, consider administering a corticosteroid spray. Cigarette smoking may exacerbate pulmonary injury and should be discouraged for at least 72 hours following exposure. Do not drink any alcoholic beverages before or during use; alcohol promotes absorption of organophosphates.
*First Aid:* **Treatment for organophosphate poisoning consists of thorough decontamination, cardiorespiratory support, and administration of the antidotes atropine and pralidoxime. In cases of severe poisoning, diazepam, an anticonvulsant, should also be administered. Antidotes should be administered as prevention even if the diagnosis is in doubt.** Speed in removing material from eyes and skin is of extreme importance. *Eyes:* Eye contact can cause dangerous amounts of these chemicals to be quickly absorbed through the mucous membrane into the bloodstream. Immediately and gently flush eyes with plenty of warm or cold water (NO hot water) for at least 15 minutes, occasionally lifting the upper and lower eyelids. Get medical aid immediately. *Skin:* Get medical aid. Skin contact can cause dangerous amounts of these chemicals to be absorbed into the bloodstream. Wearing the appropriate PPE equipment and respirator for organophosphate pesticides, immediately flush skin with plenty of soap and water for at least 15 minutes while removing contaminated clothing and shoes. Shampoo hair promptly if contaminated. The removed, contaminated clothing and shoes should be double-bagged and left in Hot Zone for later disposal by hazardous materials experts. Skin may also be decontaminated with diluted hypochlorite solution. *Inhalation:* Get medical aid. Do not contaminate yourself. Wearing the appropriate PPE equipment and respirator for organophosphate pesticides, immediately remove the victim from the contaminated area to fresh air. If the victim is not breathing, administer artificial respiration. Do NOT use mouth-to-mouth resuscitation; use bag/mask apparatus. If breathing is difficult, administer oxygen through bag/mask apparatus until medical help arrives. Do not leave victim unattended. *Ingestion:* Call poison control. Loosen all

clothing. Never give anything by mouth to an unconscious person. If victim is *unconscious or having convulsions,* do nothing except keep victim warm. Get medical aid. Transfer promptly to a medical facility. In cases of ingestion, **do not induce vomiting**. If the victim is alert and asymptomatic, administer a slurry of activated charcoal at a dose of 1 g/kg (infant, child, and adult dose). A soda can and straw may be of assistance when offering charcoal to a child. *In some cases you may be specifically instructed by poison control to induce vomiting by way of 2 tablespoons of syrup of ipecac (adult) washed down with a cup of water.* Do NOT give activated charcoal before or with ipecac syrup.

*Note to physician or authorized medical personnel:* Treat cases of respiratory compromise, coma, or excessive pulmonary secretions with respiratory support using protocols and techniques available and within the scope of training. Some cases may necessitate procedures such as endotracheal intubation or cricothyrotomy by properly trained and equipped personnel. When possible, atropine (see *Antidotes*, below) should be given under medical supervision. Patients who are comatose, hypotensive, or having seizures or cardiac arrhythmias should be treated according to advanced life support protocols. *Antidotes:* Two antidotes are administered to treat organophosphate poisoning. Atropine is a competitive antagonist of acetylcholine at muscarinic receptors and is used to control the excessive bronchial secretions which are often responsible for death. Pralidoxime relieves both the nicotinic and muscarine effects of organophosphate poisoning by regenerating acetylcholinesterase and can reduce both the bronchial secretions and the muscle weakness associated with poisoning. The initial intravenous dose of atropine in adults should be determined by the severity of symptoms: An initial adult dose of 1.0 to 2.0 mg or pediatric dose of 0.01 mg/kg (minimum 0.01 mg) should be administered intravenously. If intravenous access cannot be established, atropine may also be given intramuscularly, subcutaneously or via endotracheal tube. Doses should be repeated every 15 minutes until excessive secretions and sweating have been controlled. Once bronchial secretion has been controlled, atropine administration should be repeated whenever the secretions begin to recur. In seriously poisoned patients, very large doses may be required. Alterations of pulse rate and pupillary size should not be used as indicators of treatment adequacy. Pralidoxime should be administered as early in poisoning as possible as its efficacy may diminish when given more than 24 to 36 hours after exposure. Doses are as follows: adult 1.0 g; pediatric 25 to 50 mg/kg. The drug should be administered intravenously over 30 to 60 minutes, but in a life-threatening situation, one-half of the total dose can be given per minute for a total administration time of 2 minutes. Treatment should begin to take effect within 40 minutes with a reduction in symptoms and the amount of atropine necessary to control bronchial secretion. The initial dose can be repeated in 1 hour and then every 8 to 12 hours until the patient is clinically well and no longer requires atropine. If intravenous access cannot be established, pralidoxime may also be given intramuscularly. Early administration of diazepam in addition to the combined atropine and pralidoxime treatment may help prevent the onset of seizures and potential brain and cardiac morphologic damage following high-level organophosphate poisoning.

*References:*
- California Environmental Protection Agency Chemical List of Lists, Sacramento CA (February 1997).
- Agency for Toxic Substances and Disease Registry, U.S. Department of Health and Human Services, Public Health Service, "Managing Hazardous Materials Incidents," Atlanta, GA (June 2003)

## Arsenic Acid

*Use Type:* Used as desiccant for cotton. Registered in California as herbicide, insecticide, and rodenticide.
*CAS Number:* 1327-52-2; 7778-39-4 (*o*-isomer) (These two CAS number are regulated by the US EPA, New Jersey, California, and others); 10102-53-1 (*m*-isomer)
*Formula:* $AsH_3O_4$ (*o*-isomer); $AsHO_3$ (*m*-isomer)
*Alert:* Arsenic compounds are generally regarded as carcinogens. Human toxicity (long-term): High.
*Synonyms:* Acido arsenico (Spanish); Arsenate; Arsenic pentoxide; *o*-Arsenic acid; Metaarsenic acid; Orthoarsenic acid (*o*-)
*Trade Names:* DESICCANT L-10®, Cerexagri (USA), canceled 5/6/1993; CRAB GRASS KILLER (*o*-isomer); DESICCANT L-10 (*o*-isomer); H-10®; HI-YIELD DESICCANT H-10® (*o*-isomer), Voluntary Purchasing Group, Inc. (USA), canceled 5/6/1993; INTRACEL-15®, Chemical Specialties (USA), canceled 4/1/1987; POLY B RAND DESICCANT®, Voluntary Purchasing Group, Inc. (USA), canceled 5/6/1993; SCORCH®, Cerexagri (USA), canceled 7/1/87; SYNERGIZED H-10®; ZOTOX® (*o*-isomer)
*Producers:* Drexel Chemical (USA); Great Western Inorganics (USA); Merck (Germany); Minerals Research & Development (USA); Rhodia (France); Rhone-Poulenc (France), See Aventis SA (France); William Blythe (UK)
*Chemical Class:* Inorganic arsenicals
*EPA/OPP PC Code:* 006801
*California DPR Chemical Code:* 40
*RTECS Number:* CG0700000
*EINECS Number:* 231-901-9
*Uses:* It is used as a cotton defoliant and soil sterilant, and on harvested cotton as a desiccant. Also used on seed crop okra in Arizona as a desiccant. It is also used as a wood treatment, drying agent, and to make other arsenates.

*Human toxicity (long-term)*[77]: High–5.00 ppb, MCL (Maximum Contaminant Level)
*Fish toxicity (threshold)*[77]: Very low–1357.56302 ppb, MATC (Maximum Acceptable Toxicant Concentration)
***Carcinogen/Hazard Classifications***
**U.S. NTP Carcinogen:** Known carcinogen
**California Prop. 65:** Carcinogen
**IARC:** Group 1, known carcinogen (human positive)
**Label Signal Word:** CAUTION
***Regulatory Authority:***
- Very Toxic Substance (World Bank)[15]
- OSHA, 29CFR1910 Specifically Regulated Chemicals (See CFR 1910.1018)
- Carcinogen (human positive) (DFG)[3]
- CAL/OSHA Carcinogen User Register Chemical
- AB 1803-Well Monitoring Chemical (CAL)
- AB 2588-Air Toxics "Hot Spots" Chemicals (CAL)
- CAL Air Resources Board/AB 1807 Toxic Air Contaminants
- Proposition 65 chemical (CAL)
- Permissible Exposure Limits for Chemical Contaminants (CAL/OSHA)
- The "Director's List" (CAL/OSHA)
- Actively registered pesticide in California.
- Banned or Severely Restricted (In Agricultural, Pharmaceutical and Industrial Chemicals) (Many Countries)[13, 35]
- Air Pollutant Standard Set (ACGIH)[1] (OSHA/NIOSH)[2] (HSE)[33] (former USSR)[43] (Several States)[60]
- Clean Air Act, 42USC7412; Title I, Part A,§112 hazardous pollutants
- Clean Water Act 40CFR401.15 Section 307 Toxic Pollutants; 40CFR423, Appendix A Priority Pollutants; §313 Priority Chemicals
- RCRA, 40CFR261, Appendix 8 Hazardous Constituents, waste number P010
- Superfund/EPCRA 40CFR302.4 RQ: CERCLA, 1 lb (0.454 kg)
- EPCRA Section 313: Form R *de minimis* concentration reporting level: 0.1%
- U.S. DOT Regulated Marine Pollutant (49CFR172.101, Appendix B)
- Canada, WHMIS, Ingredients Disclosure List. Concentration Reporting Level: 0.1%
- Canada: Priority Substance List & Restricted Substances/Ocean Dumping Forbidden (CEPA), National Pollutant Release Inventory (NPRI) (arsenic compounds)

***Description:*** White semi-transparent crystalline material or in a commercial grade that is a pale yellow syrup-like liquid. Odorless. Melting/Melting/Freezing point = 36°C. Hazard Identification (based on NFPA-704 M Rating System): Health 3, Flammability 0, Reactivity 0. It converts to $As_2O_5$ (arsenic pentoxide) when heated above 300°C. See also arsenic pentoxide. Noncombustible.

***Incompatibilities:*** Incompatible with sulfuric acid, caustics, ammonia, amines, isocyanates, alkylene oxides, oxidizers, epichlorohydrin, vinyl acetate, amides. Avoid contact with chemically active metals. Corrodes brass, mild steel and galvanized steel. Contact with acids or acid mists releases deadly arsine gas.

***Permissible Exposure Limits in Air:*** OSHA[2]: The legal airborne PEL is 0.01 mg/m$^3$ averaged over an 8-hour workshift for Arsenic and compounds as Arsenic, inorganic. NIOSH[2]: The recommended airborne exposure limit is 0.002 mg/m$^3$, which should not be exceeded during any 15 minute work period for Arsenic, inorganic. ACGIH: The recommended airborne exposure limit is 0.2 mg/m$^3$ average over an 8-hour workshift for Arsenic and soluble compounds[1]. The British HSE[33] has also adapted the ACGIH value of 0.2 mg/m$^3$ as an 8-hour TWA value. The DFG[3] has not set numerical limits for arsenic in air on the grounds that it is a proven human carcinogen. The former USSR-UNEP/IRPTC project[43] has set a MAC value for inorganic arsenic compounds (except arsine) of 0.003 mg/m$^3$ for ambient air in residential areas.

***Determination in Air:*** Collection on a filter and analysis by atomic absorption spectrometry. See NIOSH Methods 7900 and 73000, Elements[18]. See also OSHA Method ID 105.[58]

***Permissible Concentration in Water:*** EPA has set a limit of 0.05 parts per million (ppm) for arsenic in drinking water. The U.S. EPA arsenic drinking water standard of 0.01 ppm (10 ppb) is based on the U.S. EPA final rule for arsenic in drinking water published in. the January 22, 2001, *Federal Register*. However, the U.S. EPA is currently reviewing the science and cost estimate supporting this rule, and, in the interim, has reverted to the previous standard for arsenic. Thus, in the US, the current EPA arsenic drinking water standard remains at 0.05 ppm (50 ppb). To protect freshwater aquatic life-total recoverable trivalent inorganic arsenic never to exceed 440 µg/L. To protect saltwater aquatic life: 508 µg/L on an acute basis. To protect human health: preferably zero. The former USSR-UNEP/IRPTC project[43] has set MAC values for inorganic arsenic compounds in water for domestic purposes at 0.05 mg/L and in water bodies for fishery purposes of 0.5 mg/L also.

***Determination in Water:*** The atomic absorption graphite furnace technique is often used for measurement of total arsenic in water. It also has been standardized by EPA. Total arsenic may be determined by digestion followed by silver diethyldithiocarbamate; an alternative is atomic absorption; another is inductively coupled plasma optical emission spectrometry. See OSHA Method #ID-105 for arsenic[58].

***Routes of Entry:*** Inhalation, ingestion, and skin contact.
***Harmful Effects and Symptoms***
***Short Term Exposure:*** Skin contact can cause irritation, itching, burning sensation, and rash. Eye contact can cause irritation and burns. Inhalation can cause irritation of the

respiratory tract. High exposure can cause poor appetite, nausea, vomiting and muscle cramps. High exposure can cause nerve damage with numbness, "pins and needles" sensation, weakness of the arms and legs. Arsine, a very deadly gas is released in the presence of acid or acid mist. The $LD_{50}$ (oral, rat) = 48 mg/kg[9]. Ingestion of 130 mg of arsenic may be fatal to humans. Smaller doses may become fatal since arsenic accumulates in the body.

*Long Term Exposure:* Arsenic acid is a mutagen that may cause changes to genetic material and an animal teratogen. Can cause an ulcer of the septum dividing the inner nose. It can cause nerve damage, thickening of the skin with patch areas of darkening and loss of pigment, or the development of white lines in the nails.

*Points of Attack:* Several studies have shown that inorganic arsenic can increase the risk of lung cancer, skin cancer, bladder cancer, liver cancer, kidney cancer, and prostate cancer.

*Medical Surveillance:* Examination of the nose, skin, eyes, nails, and nervous system. Test for urine arsenic. At NIOSH recommended exposure limits, urine arsenic should not be greater than 50 to 100 micrograms per liter of urine. See also entry for Arsenic.

*First Aid:* If this chemical gets into the eyes, remove any contact lenses at once and irrigate immediately for at least 15 minutes, occasionally lifting upper and lower lids. Seek medical attention immediately. If this chemical contacts the skin, remove contaminated clothing and wash immediately with soap and water. Seek medical attention immediately. If this chemical has been inhaled, remove from exposure, begin rescue breathing (using universal precautions) if breathing has stopped and CPR if heart action has stopped. Transfer promptly to a medical facility. When this chemical has been swallowed, get medical attention. Give large quantities of water and induce vomiting. Do not make an unconscious person vomit.

*Antidotes and Special Procedures:* For severe poisoning BAL has been used. For milder poisoning *penicillamine (not penicillin)* has been used, both with mixed success. Side effects occur with such treatment and it is never a substitute for controlling exposure. It can only be done under strict medical care.

*References:*
- "Arsenic Acid Chemical Fact Sheet," Fact Sheet No. 91, Cornell University, Ithaca, NY. (September 1986).
- Sax, N.I., Ed., "Dangerous Properties of Industrial Materials Report" 2, No. 3, 59-61 (1982) and 8, No. 3, 45-55 (1988).
- California Environmental Protection Agency Chemical List of Lists, Sacramento CA (February 1997).
- Lewis, Richard J., Sr., Ed., *Sax's Dangerous Properties of Industrial Materials, 9th Ed.*, John Wiley & Sons, New York, 1998.
- New Jersey Department of Health and Senior Services, "Hazardous Substance Fact Sheet: Arsenic Acid," Trenton, NJ (April 1996, rev. April 2002). http://www.state.nj.us/health/eoh/rtkweb/0153.pdf

# Arsenic and inorganic arsenic compounds

*Use Type:* Arsenates and arsenites are used in agriculture as insecticides, herbicides, larvicides, and pesticides.

*CAS Number:* 7440-38-2 (metallic arsenic)

*Formula:* As

*Alert:* Arsenic compounds are generally considered carcinogens, handle with extreme caution.

*Synonyms:* Arsen (German, Polish); Arsenicals; Arsenic-75; Arsenic black; Arsenic, metallic; Arsenic, solid; Arsenico (Spanish); As-120; As-217; Colloidal arsenic; Grey arsenic; Metallic arsenic; Ruby arsenic; Realgar; Butter of Arsenic

*Note:* The above synonyms are for metallic arsenic. Other inorganic synonyms vary depending on the specific arsenic compound. The term "inorganic arsenic" does not include Arsine.

*Trade Names:* ACCUSPIN ASX-10 SPIN-ON DOPANT®

*Producers:* Air Products & Chemicals (USA); Aldrich Chemical (USA); ATOFINA N.A. (USA); ASARCO (USA); Cia. Universal de Industrias (Mexico); Degussa (Germany); Great Western Inorganics (USA); GFS Chemicals (USA); Mining & Chemical Products Ltd. (MCP) (UK); Mitsubishi Materials (Japan); Newmont Koch (UK); PPM Pure Metals (Germany); Union Miniere (Belgium)

*Chemical Class:* Inorganic arsenical

*EPA/OPP PC Code:* 006802 (arsenic pentoxide); 007001 (arsenic trioxide)

*California DPR Chemical Code:* 710 (arsenic); 11 (arsenic pentoxide); 1 (arsenic trioxide)

*ICSC Number:* 0013 (arsenic); 0377 (arsenic pentoxide)

*RTECS Number:* CG0525000 (arsenic)

*EEC Number:* 033-001-00-X (arsenic); 033-004-00-6 (arsenic pentoxide)

*EINECS Number:* 231-148-6 (arsenic)

*Uses:* When used as pesticides, organic compounds of arsenic are used primarily on cotton. Inorganic arsenic compounds are mainly used to preserve wood. Arsenic compounds have a variety of uses other than agriculture chemicals. Other arsenic compounds are used in pigment production, the manufacture of glass as a bronzing or decolorizing agent, the manufacture of opal glass and enamels, textile printing, tanning, taxidermy, and antifouling paints. They are also used to control sludge formation in lubricating oils. Metallic arsenic is used as an alloying agent for heavy metals, and in solders, medicines, herbicides.

*Carcinogen/Hazard Classifications*

**U.S. NTP Carcinogen:** Known carcinogen

**California Prop. 65:** Carcinogen. Developmental toxin for inorganic oxides
**IARC:** Group 1, known carcinogen (human positive)
**Label Signal Word:** CAUTION
**Endocrine Disruptor:** Suspected endocrine disruptor
*Regulatory Authority:*
- Carcinogen (human positive) (DFG)[3]
- CAL Air Resources Board/AB 1807 Toxic Air Contaminants
- Specific chemicals (EPA/NESHAP)
- Proposition 65 chemical (CAL)
- The "Director's List" (CAL/OSHA)
- AB 2588-Air Toxics "Hot Spots" Chemicals (CAL) as arsenic compounds
- Permissible Exposure Limits for Chemical Contaminants (CAL/OSHA) as arsenic compounds
- Carcinogen User Register Cemical (CAL/OSHA) as inorganic arsenic compound
- AB 1803-Well Monitoring Chemical (CAL) as inorganic arsenic compound
- CAL Air Resources Board/AB 1807 Toxic Air Contaminants as inorganic arsenic compound
- OSHA, 29CFR1910 Specifically Regulated Chemicals (See CFR 1910.1018) Inorganic compounds
- Banned or Severely Restricted (In Agricultural, Pharmaceutical and Industrial Chemicals) (Many Countries)[13, 35]
- Air Pollutant Standard Set (ACGIH)[1] (OSHA/NIOSH)[2] (HSE)[33] (former USSR)[43] (Several States)[60] (Australia) (Israel) (Mexico) (Several Canadian Provinces)
- Clean Air Act, 42USC7412; Title I, Part A,§112 hazardous pollutants
- Clean Water Act 40CFR401.15 Section 307Toxic Pollutants; 40CFR423, Appendix A Priority Pollutants; §313
- Priority Chemicals
- RCRA 40CFR§261.24 Toxicity Characteristics, Maximum Concentration of Contaminants, Regulatory level, 5.0 mg/L
- RCRA "D Series Waste" Number, D004, Chronic Toxicity Reference Level, 0.05 mg/L
- RCRA, 40CFR261, Appendix 8 Hazardous Constituents, waste number not listed
- RCRA 40CFR§268.48; 61FR15654, Universal Treatment Standards: Wastewater (mg/L), 1.4; Nonwastewater (mg/L), 5.0 TCLP
- RCRA 40CFR264, Appendix 9; TSD Facilities Ground Water Monitoring List Suggested methods (PQL $ug/L$): (total) 6010(500), 7060(10), 7061(20)
- Safe Drinking Water Act 47FR9352 Regulated chemical: MCL, 0.05 mg/L (Section 141.11) applies only to community water systems. *Note:* Effective January 2006 the new MCL will be 0.01mg/L 66CFR6976.
- Superfund/EPCRA 40CFR302.4 RQ: CERCLA, 1 lb (0.454 kg), no reporting required, if diameter of metal is equal to or exceeds 0.004 in.
- EPCRA Section 313: Form R *de minimis* concentration reporting level: 0.1%
- U.S. DOT Regulated Marine Pollutant (49CFR172.101, Appendix B)
- Canada, WHMIS, Ingredients Disclosure List. Concentration Reporting Level: 0.1%
- Canada: Priority Substance List & Restricted Substances/Ocean Dumping Forbidden (CEPA), National Pollutant Release Inventory (NPRI) (arsenic compounds)

*Description:* Elemental arsenic, As, occurs to a limited extent in nature as a steel-gray, amorphous metalloid. Boiling point = 612° (sublimes). Melting/Freezing point = 814° @ 36 atm; 817°C @ 28 atm. Hazard Identification (based on NFPA-704 M Rating System): Health 3, Flammability 1, Reactivity 0. Insoluble in water. Arsenic in this entry includes the element and any of its inorganic compounds <u>excluding</u> arsine. Arsenic trioxide ($As_2O_3$), the principal form in which the element is used, is frequently designated as arsenic, white arsenic, or arsenous oxide. Arsenic is present as an impurity in many other metal ores and is generally produced as arsenic trioxide as a by-product in the smelting of these ores, particularly copper. Most other arsenic compounds are produced from the trioxide.

*Incompatibilities:* Incompatible with strong acids, strong oxidizers, peroxides, bromine azide, bromine pentafluoride, bromine trifluoride, cesium acetylene carbide, chromium trioxide, nitrogen trichloride, silver nitrate. Can react vigorously with strong oxidizers (chlorine, dichromate, permanganate). Forms highly toxic fumes on contact with acids or active metals (iron, aluminum, zinc). Hydrogen gas can react with inorganic arsenic to form highly toxic arsine gas.

*Permissible Exposure Limits in Air:* The following exposure limits are for air levels only. When skin contact also occurs, overexposure is possible, even though air levels are less than the limits listed below. OSHA[2]: The legal airborne PEL is 0.010 mg/m$^3$ averaged over an 8-hour workshift. NIOSH[2]: The recommended airborne exposure limit is 0.002 mg/m$^3$ (ceiling), not to be exceeded during any 15 min. work period. ACGIH[1]: The recommended airborne exposure limit is 0.01 mg/m$^3$ averaged over an 8-hour workshift. The HSE[33] (U.K.) Maximum Exposure Limit as As is 0.1 mg/m$^3$TWA. California's workplace PEL is the same as ACGIH[1] and an Action Level of 0.005 mg/m$^3$. The Australia limit is 0.05 mg/m$^3$ TWA (confirmed carcinogen); Israel 0.01 mg/m$^3$ TWA and Action Level 0.005 mg/m$^3$. Mexico level 0.2 mg/m$^3$ TWA. Canada: Alberta level 0.2 mg/m$^3$ TWA and STEL of 0.6 mg/m$^3$ (15 min.); British Columbia level 0.5 mg/m$^3$ TWA; Ontario level 0.01 mg/m$^3$ TWAEV and STEV of 0.05; Quebec level 0.2 mg/m$^3$ TWAEV. The former USSR-UNEP/IRPTC project[43] has set a MAC of 0.003 mg/m$^3$ on an average daily basis for residential areas. In addition, several states have set guidelines or standards for arsenic in ambient air[60]: 0.06 mg/m³ (California Prop. 65), 0.0002 µg/m³ (Rhode Island), 0.00023 µg/m³ (North Carolina), 0.024 µg/m³ (Pennsylvania), 0.05 µg/m³ (Connecticut), 0.07 to 0.39 µg/m³ (Montana), 0.67 µg/m³ (New York), 1.0 µg/m³ (South Carolina), 2.0 µg/m³ (North Dakota), 3.3 µg/m³ (Virginia), 5 µg/m³ (Nevada).

***Determination in Air:*** Collection on a filter and analysis by atomic absorption spectrometry. See NIOSH Methods 7900 and 73000, Elements[18]. See also OSHA Method ID 105[58].

***Permissible Concentration in Water:*** EPA has set a limit of 0.05 ppm for arsenic in drinking water. The U.S. EPA arsenic drinking water standard of 0.01 ppm (10 ppb) is based on the U.S. EPA final rule for arsenic in drinking water published in. the January 22, 2001, *Federal Register*. However, the U.S. EPA is currently reviewing the science and cost estimate supporting this rule, and, in the interim, has reverted to the previous standard for arsenic. Thus, in the US, the current EPA arsenic drinking water standard remains at 0.05 ppm (50 ppb). To protect freshwater aquatic life-total recoverable trivalent inorganic arsenic never to exceed 440 µg/L. To protect saltwater aquatic life: 508 µg/L on an acute basis. To protect human health: preferably zero. The former USSR-UNEP/IRPTC project[43] has set MAC values for inorganic arsenic compounds in water for domestic purposes at 0.05 mg/L and in water bodies for fishery purposes of 0.5 mg/L also.

***Determination in Water:*** The atomic absorption graphite furnace technique is often used for measurement of total arsenic in water. It also has been standardized by EPA. Total arsenic may be determined by digestion followed by silver diethyldithiocarbamate; an alternative is atomic absorption; another is inductively coupled plasma optical emission spectrometry. See OSHA Method #ID-105 for arsenic[58].

***Routes of Entry:*** Inhalation, through the skin, and ingestion of dust and fumes.

### Harmful Effects and Symptoms

*Local:* Trivalent arsenic compounds are corrosive to the skin. Brief contact has no effect, but prolonged contact results in a local hyperemia and later vesicular or pustular eruption. The moist mucous membranes are most sensitive to the irritant action. Conjunctiva, moist and macerated areas of the skin, eyelids, the angles of the ears, nose, mouth, and respiratory mucosa are also vulnerable to the irritant effects. The wrists are common sites of dermatitis, as are the genitalia if personal hygiene is poor. Perforations of the nasal septum may occur. Arsenic trioxide and pentoxide are capable of producing skin sensitization and contact dermatitis. Arsenic is also capable of producing keratoses, especially of the palms and soles. Arsenic has been cited as a cause of skin cancer, but the incidence is low. *Systemic:* The acute toxic effects of arsenic are generally seen following ingestion of inorganic arsenical compounds. This rarely occurs in an industrial setting. Symptoms develop within ½ to 4 hours following ingestion and are usually characterized by constriction of the throat followed by dysphagia, epigastric pain, vomiting, and watery diarrhea. Blood may appear in vomitus and stools. If the amount ingested is sufficiently high, shock may develop due to sever fluid loss, and death may ensue in 24 hours. If the acute effects are survived, exfoliative dermatitis and peripheral neuritis may develop. Cases of acute arsenical poisoning due to inhalation are exceedingly rare in industry. When it does occur, respiratory tract symptoms-cough, chest pain, dyspnea-giddiness, headache, and extreme general weakness precede gastrointestinal symptoms. The acute toxic symptoms of trivalent arsenical poisoning are due to severe inflammation of the mucous membranes and greatly increased permeability of the blood capillaries. Chronic arsenical poisoning due to ingestion is rare and generally confined to patients taking prescribed medications. However, it can be a concomitant of inhaled inorganic arsenic from swallowed sputum and improper eating habits. Symptoms are weight loss, nausea and diarrhea alternating with constipation, pigmentation and eruption of the skin, loss of hair, and peripheral neuritis. Chronic hepatitis and cirrhosis have been described. Polyneuritis may be the salient feature, but more frequently there are numbness and paresthesias of "glove and sticking" distribution. The skin lesions are usually melanotic and keratotic and may occasionally take the form of an intradermal cancer of the squamous cell type, but without infiltrative properties. Horizontal white lines (striations) on the fingernails and toenails are commonly seen in chronic arsenical poisoning and are considered to be a diagnostic accompaniment of arsenical polyneuritis. Inhalation of inorganic arsenic compounds is the most common cause of chronic poisoning in the industrial situation. This condition is divided into three phases based on signs and symptoms. *First Phase*: The worker complains of weakness, loss of appetite, some nausea, occasional vomiting, a sense of heaviness in the stomach, and some diarrhea. *Second Phase:* The worker complains of conjunctivitis, and a catarrhal state of the mucous membranes of the nose, larynx, and respiratory passages. Coryza, hoarseness, and mild tracheobronchitis may occur. Perforation of the nasal septum is common, and is probably the most typical lesion of the upper respiratory tract in occupational exposure to arsenical dust. Skin lesions, eczematoid and allergic in type, are common. *Third Phase:* The worker complains of symptoms of peripheral neuritis, initially of hands and feet, which is essentially sensory. In more severe cases, motor paralyses occur; the first muscles affected are usually the toe extensors and the peronei. In only the most severe cases will paralysis of flexor muscles of the feet or of the extensor muscles of hands occur. Liver damage from chronic arsenical poisoning is still debated, and as yet the question is unanswered. In cases of chronic and acute arsenical poisoning, toxic effects to the myocardium have been reported based on EKG changes. These finding, however, are now largely discounted and the EKG changes are ascribed to electrolyte disturbances concomitant with arsenicalism. Inhalation of arsenic trioxide and other inorganic arsenical dusts does not give rise to radiological evidence of pneumoconiosis. Arsenic does have a depressant effect upon the bone marrow, with disturbances of both erythropoiesis and myelopoiesis. Evidence is now available incriminating arsenic compounds as a cause of lung cancer as well as skin cancer. Skin Cancer in humans is causally associated with exposure to inorganic arsenic compounds in drugs, drinking water and the occupational environment. The risk of lung cancer was increased 4 to 12 times in certain smelter workers who inhaled high levels of arsenic trioxide. However, the influence of other constituents of the working environment cannot be excluded in these studies. Case reports have suggested an association between exposure to arsenic compounds and blood dyscrasias and liver tumors.

***Short Term Exposure:*** Skin contact can cause irritation, itching, burning sensation, and rash. Eye contact can cause irritation and burns. Inhalation can cause irritation of the respiratory tract. High exposure can cause poor appetite, nausea, vomiting and muscle cramps. High exposure can cause nerve damage with numbness, "pins and needles" sensation, weakness of the arms and legs.

***Long Term Exposure:*** Arsenic is a carcinogen; causes skin, lung, bladder, liver, kidney, prostate, and lymphatic cancer, possible reproductive hazard (a teratogen in animals). Can cause an ulcer of the septum dividing the inner nose. It can cause hoarsness, sore eyes, nerve damage, thickening of the skin with patch areas of darkening and loss of pigment, liver damage and stomach problems. Small doses can accumulate in the body. Birth defects have been observed in animals exposed to inorganic arsenic. It is likely that health effects seen in children exposed to high amounts of arsenic will be similar to the effects seen in adults.

***Points of Attack:*** Several studies have shown that inorganic arsenic can increase the risk of lung cancer, skin cancer, bladder cancer, liver cancer, kidney cancer, and prostate cancer.

***Medical Surveillance:*** Before first exposure and every 6 to 12 months thereafter, OSHA 1910.1018 requires employers to provide (for persons exposed to 0.005 mg/m$^3$ of Arsenic) a medical history and exam which shall include : Chest x-ray, exam of the nose, skin, and nails, sputum cytology examination, test for urine Arsenic (may not be accurate within 2 days of eating shellfish or fish; most accurate at the end of a workday). Levels should not be greater than 100 micrograms per gram creatinine in the urine. Exam of the nervous system. After suspected overexposure, repeat these tests and consider complete blood count and liver function tests. Also examine skin periodically for abnormal growths. Skin cancer from arsenic can easily be cured when detected early. Employees have a legal right to testing information under OSHA 1910.20.

***First Aid:*** If this chemical gets into the eyes, remove any contact lenses at once and irrigate immediately for at least 15 minutes, occasionally lifting upper and lower lids. Seek medical attention immediately. If this chemical contacts the skin, remove contaminated clothing and wash immediately with soap and water. Seek medical attention immediately. If this chemical has been inhaled, remove from exposure, begin rescue breathing (using universal precautions) if breathing has stopped and CPR if heart action has stopped. Transfer promptly to a medical facility. When this chemical has been swallowed, get medical attention. Give large quantities of water and induce vomiting. Do not make an unconscious person vomit.

*Note:* For severe poisoning BAL has been used. For milder poisoning *penicillamine (not penicillin)* has been used, both with mixed success. Side effects occur with such treatment and it is never a substitute for controlling exposure. It can only be done under strict medical care.

***References:***
- U.S. Public Health Service, Agency for Toxic Substances and Disease Registry, "ToxFAQs for Arsenic,"
- Atlanta, GA, (December 2003). http://www.atsdr.cdc.gov/tfacts2.html
- New Jersey Department of Health and Senior Services, "Hazardous Substance Fact Sheet: Arsenic," Trenton, NJ (June 1998). http://www.state.nj.us/health/eoh/rtkweb/0152.pdf
- National Institute for Occupational Safety and Health, Criteria for a Recommended Standard: Occupational Exposure to Inorganic Arsenic, NIOSH Doc. No. 74-110, Washington, DC (1973).
- National Institute for Occupational Safety and Health, Criteria for a Recommended Standard: Occupational Exposure to Inorganic Arsenic (Revised), NIOSH Doc. No. 75-149, Washington, DC (1975).
- U.S. Environmental Protection Agency, Arsenic: Ambient Water Quality Criteria, Washington, DC (1979).
- U.S. Environmental Protection Agency, Status Assessment of Toxic Chemicals: Arsenic, Report No. EPA-600/2-79-210B, Washington, DC (December 1979).
- U.S. Environmental Protection Agency, Toxicology of Metals, Vol II: Arsenic, Report EPA-600/1-77-022, Research Triangle Park, NC, pp 30-70 (May 1977).
- National Academy of Sciences, Medical and Biological Effects of Environmental Pollutants: Arsenic, Washington, DC (1977).
- U.S. Environmental Protection Agency, Arsenic, Health and Environmental Effects Profile No. 11, Office of Solid Waste, Washington, DC (April 30, 1980).

- Sax, N.I., Ed., "Dangerous Properties of Industrial Materials Report" 1, No. 3, 32-34 (1981).
- Lederer, W.H. and Fensterheim, R.J., ARSENIC: Industrial, Biomedical and Environmental Perspectives, New York, Van Nostrand Reinhold Co. (1983).
- New York State Department of Health, "Chemical Fact Sheet: Arsenic," Albany, NY, Bureau of Toxic Substance Assessment (May 1986).
- California Environmental Protection Agency Chemical List of Lists, Sacramento CA (February 1997).

# Arsenic Pentoxide

*Use Type:* Herbicide, fungicide, rodenticide, insecticide and as a soil sterilant.
*CAS Number:* 1303-28-2
*Formula:* $As_2O_5$
*Alert:* Arsenic compounds are generally considered carcinogens.
*Synonyms:* Anhydride arsenique (French); Anidrino arsenioso (Italian); Arsenic acid anhydride; Arsenic anhydride; Arsenic oxide; Arsenic(V) oxide; Arsenic acid anhydride; Arsenic pentaoxide; Diarsenic pentoxide; Fotox; Peroxido de arsenico (Spanish)
*Producers:* GFS Chemicals (USA); Great Western Inorganics (USA); Merck (Germany); PPM Pure Metals (Germany)
*Chemical Class:* Inorganic arsenical
*EPA/OPP PC Code:* 006802
*California DPR Chemical Code:* 631
*ICSC Number:* 0377
*RTECS Number:* CG2275000
*EEC Number:* 033-004-00-6
*EINECS Number:* 215-116-9
*Uses:* This material is also used as a chemical intermediate and as an ingredient in wood preservatives and in glass.
*Carcinogen/Hazard Classifications*
**U.S. NTP Carcinogen:** Known carcinogen
**California Prop. 65:** Carcinogen
**IARC:** Group 1, known carcinogen (human positive)
**Label Signal Word:** CAUTION
**Endocrine Disruptor:** Suspected
*Regulatory Authority:*
*See also Arsenic and Inorganic Compounds*
- Banned or Severely Restricted (In Agricultural, Pharmaceutical and Industrial Chemicals) Many Countries[13, 35]
- OSHA, 29CFR1910 Specifically Regulated Chemicals (See CFR 1910.1018)
- Very Toxic Substance (World Bank)[15]
- Carcinogen (human positive) (DFG)[3]
- The "Director's List" (CAL/OSHA)
- Proposition 65 chemical (CAL) as inorganic arsenic compounds
- Actively registered pesticide in California.
- Carcinogen User Register Chemical (CAL/OSHA) as inorganic arsenic compound
- AB 1803-Well Monitoring Chemical (CAL) as inorganic arsenic compound
- CAL Air Resources Board/AB 1807 Toxic Air Contaminants as inorganic arsenic compound
- AB 2588-Air Toxics "Hot Spots" Chemicals (CAL) as arsenic compounds as inorganic arsenic compound
- Permissible Exposure Limits for Chemical Contaminants (CAL/OSHA) as arsenic compounds as inorganic arsenic compound
- Air Pollutant Standard Set (ACGIH)[1] (OSHA/NIOSH)[2] (HSE)[33] (former USSR)[43] (Several States)[60] (Australia) (Israel) (Mexico) (Several Canadian Provinces)
- Clean Air Act, 42USC7412; Title I, Part A,§112 hazardous pollutants
- Clean Water Act 40CFR401.15 Section 307Toxic Pollutants; 40CFR423, Appendix A Priority Pollutants; §313 Priority Chemicals
- RCRA 40CFR§261.24 Toxicity Characteristics, Maximum Concentration of Contaminants, Regulatory level, 5.0 mg/L
- RCRA, 40CFR261, Appendix 8 Hazardous Constituents, waste number P011
- CERCLA/SARA 40CFR302 Extremely Hazardous Substances: TPQ = 100/10,000 lb (454/4,540 kg)
- Superfund/EPCRA 40CFR302.4 RQ: CERCLA, 1 lb (0.454 kg)
- EPCRA Section 313: Form R *de minimis* concentration reporting level: 0.1%
- U.S. DOT Regulated Marine Pollutant (49CFR172.101, Appendix B)
- Canada, WHMIS, Ingredients Disclosure List
- Canada: Priority Substance List & Restricted Substances/Ocean Dumping Forbidden (CEPA), National Pollutant Release Inventory (NPRI)(arsenic compounds)

*Description:* Arsenic pentoxide, $As_2O_5$, is an odorless white lumpy solid powder and non-flammable. Melting/Freezing point = 315°C (decomposes). Hazard Identification (based on NFPA-704 M Rating System): Health 3, Flammability 0, Reactivity 0. Highly soluble in water.
*Incompatibilities:* Chemically active metals such as aluminum and zinc. Incompatible with acids, strong alkalis, halogens, rubidium carbide, zinc. Corrosive to metals in the presence of moisture. Contact with acids or acid mists releases deadly arsine gas.
*Permissible Exposure Limits in Air:* OSHA[2]: The legal airborne PEL is 0.01 mg/m$^3$ averaged over an 8-hour workshift for Arsenic and compounds as Arsenic, inorganic. NIOSH[2]: The recommended airborne exposure limit is 0.002 mg/m$^3$, which should not be exceeded during any 15

minute work period for Arsenic, inorganic. ACGIH: The recommended airborne exposure limit is 0.2 mg/m³ average over an 8-hour workshift for Arsenic and soluble compounds[1]. The British HSE[33] has also adapted the ACGIH value of 0.2 mg/m³ as an 8-hour TWA value. The DFG[3] has not set numerical limits for arsenic in air on the grounds that it is a proven human carcinogen. The former USSR-UNEP/IRPTC project[43] has set a MAC value for inorganic arsenic compounds (except arsine) of 0.003 mg/m³ for ambient air in residential areas. In addition, several states have set specific guidelines or standards for arsenic pentoxide in ambient air[60] ranging from zero (New York) to 0.0002 μg/m³ (North Carolina) to 1.0 μg/m³ (South Carolina).

*Determination in Air:* Collection on a filter and analysis by atomic absorption spectrometry. See NIOSH Methods 7900 and 73000, Elements[18]. See also OSHA Method ID 105.[58]

*Permissible Concentration in Water:* EPA has set a limit of 0.05 parts per million (ppm) for arsenic in drinking water. The U.S. EPA arsenic drinking water standard of 0.01 ppm (10 ppb) is based on the U.S. EPA final rule for arsenic in drinking water published in. the January 22, 2001, *Federal Register*. However, the U.S. EPA is currently reviewing the science and cost estimate supporting this rule, and, in the interim, has reverted to the previous standard for arsenic. Thus, in the US, the current EPA arsenic drinking water standard remains at 0.05 ppm (50 ppb). To protect freshwater aquatic life-total recoverable trivalent inorganic arsenic never to exceed 440 μg/L. To protect saltwater aquatic life: 508 μg/L on an acute basis. To protect human health: preferably zero. The former USSR-UNEP/IRPTC project[43] has set MAC values for inorganic arsenic compounds in water for domestic purposes at 0.05 mg/L and in water bodies for fishery purposes of 0.5 mg/L also.

*Determination in Water:* The atomic absorption graphite furnace technique is often used for measurement of total arsenic in water. It also has been standardized by EPA. Total arsenic may be determined by digestion followed by silver diethyldithiocarbamate; an alternative is atomic absorption; another is inductively coupled plasma optical emission spectrometry. See OSHA Method #ID-105 for arsenic[58].

*Routes of Entry:* Inhalation, ingestion, skin contact.

*Harmful Effects and Symptoms*

*Short Term Exposure:* It is irritating to eyes, nose, and respiratory system. This chemical can be absorbed through the skin, thereby increasing exposure. Skin contact can cause irritation, burning, itching, and a rash. Symptoms usually appear ½ to 1 hour after ingestion, but may be delayed. Symptoms include a sweetish, metallic taste and garlicky odor of breath; difficulty in swallowing; abdominal pain; vomiting and diarrhea; dehydration; feeble heart beat; dizziness and headache; and eventually coma, sometimes convulsions, general paralysis, and death. The $LD_{50}$ (oral, rat) = 8 mg/kg[9]. This material is extremely toxic; the probable oral lethal dose for humans is 5-50 mg/kg, or between 7 drops and 1 teaspoonful for a 150-lb. person.

*Long Term Exposure:* Arsenic pentoxide is a carcinogen in humans. It has been shown to cause skin cancer. May damage the male reproductive glands. Chronic exposure may cause nerve damage to the extremities, alter cellular composition of the blood, and cause structural changes in blood components. Repeated exposure can cause an ulcer in the septum dividing the inner nose. Long term skin contact can cause thickened skin and pigmentation changes. Some persons develop white lines in the finger nails. Birth defects have been observed in animals exposed to inorganic arsenic. It is likely that health effects seen in children exposed to high amounts of arsenic will be similar to the effects seen in adults.

*Points of Attack:* Several studies have shown that inorganic arsenic can increase the risk of lung cancer, skin cancer, bladder cancer, liver cancer, kidney cancer, and prostate cancer.

*Medical Surveillance:* See entry under Arsenic compounds.

*First Aid:* If this chemical gets into the eyes, remove any contact lenses at once and irrigate immediately for at least 15 minutes, occasionally lifting upper and lower lids. Seek medical attention immediately. If this chemical contacts the skin, remove contaminated clothing and wash immediately with soap and water. Seek medical attention immediately. If this chemical has been inhaled, remove from exposure, begin rescue breathing (using universal precautions) if breathing has stopped and CPR if heart action has stopped. Transfer promptly to a medical facility. When this chemical has been swallowed, get medical attention. Give large quantities of water and induce vomiting. Do not make an unconscious person vomit.

*References:*
- Sax, N.I., Ed., "Dangerous Properties of Industrial Materials Report" 2, No. 3, 59-61 (1982) and 8, No. 3, 45-55 (1988).
- Lewis, Richard J., Sr., Ed., *Sax's Dangerous Properties of Industrial Materials, 9th Ed.*, John Wiley & Sons, New York, 1998.
- U.S. Environmental Protection Agency, "Chemical Profile: Arsenic Pentoxide," Washington, DC, Chemical Emergency Preparedness Program (November 30, 1987).
- California Environmental Protection Agency Chemical List of Lists, Sacramento CA (February 1997).
- New Jersey Department of Health and Senior Services, "Hazardous Substance Fact Sheet: Arsenic Pentoxide," Trenton, NJ (January 1996, rev. April 2002). www.state.nj.us/health/eoh/rtkweb/0158.pdf

# Arsenous Oxide

*Use Type:* Herbicide, insecticide, and rodenticide

# Arsenous Oxide

*CAS Number:* 1327-53-3
*Formula:* $As_2O_3$
*Alert:* Highly toxic. Persons whose clothing or skin is contaminated with arsenic trioxide can secondarily contaminate rescuers by direct contact or through release of inhalable dust. May be fatal if swallowed. May cause allergic respiratory reaction. May act as a carcinogen; inorganic arsenic is a known cancer hazard.
*Synonyms:* Acide arsenieux (French); Anhydride arsenieux (French); Arsenic blanc (French); Arsenic(III) oxide; Arsenic sesquioxide; Arsenic trioxide, solid; Arsenicum album; Arsenigen saure (German); Arsenious acid; Arsenious oxide; Arsenious trioxide; Arsenite; Arsenolite; Arsenous acid; Arsenous acid anhydride; Arsenous anhydride; Arsenous oxide anhydride; Arsenic sesquioxide; Arsenic trioxide; Arsodent; Claudelite; Claudetite; Crude arsenic; Diarsenic trioxide; Spinrite arsenic; Trioxido de arsenico (Spanish); White arsenic
*Producers:* ESPI (USA); Great Western Inorganics (USA); Merck (Germany); Mitsubishi Chemicals (Japan); Mitsui Chemical (Japan); Noah Technologies (USA); PPM Pure Metals (Germany); Quantum Chemicals (USA); Sumitomo Chemical (Japan)
*Chemical Class:* Inorganic arsenic, heavy metal
*EPA/OPP PC Code:* 007001
*California DPR Chemical Code:* 42
*ICSC Number:* 0378
*RTECS Number:* CG3325000
*EEC Number:* 033-003-00-0
*EINECS Number:* 215-481-4
*Uses:* This is a primary raw material for all arsenic compounds. It is also an intermediate for insecticides, herbicides and fungicides. The material is used as a wood and tanning preservative, sheep dips, making enamels, and a decoloring and refining agent in glass manufacture. It is also used in pharmaceuticals and in the purification of synthetic gas.
*Carcinogen/Hazard Classifications*
**U.S. NTP Carcinogen:** Known carcinogen
**California Prop. 65:** Carcinogen
**IARC:** Group 1, carcinogen
**Label Signal Word:** DANGER
*Regulatory Authority:*
- Banned or Severely Restricted
- Carcinogen (human positive) (DFG)[3]
- OSHA, 29CFR1910 Specifically Regulated Chemicals (See CFR 1910.1018)
- Banned or Severely Restricted (In Agricultural, Pharmaceutical and Industrial Chemicals). Many Countries [13, 35]
- Air Pollutant Standard Set (ACGIH)[1] (OSHA/NIOSH)[2] (HSE)[33] (former USSR)[43] (Several States)[60] (Australia) (Israel) (Mexico) (Several Canadian Provinces)
- EPA/SARA 302 (EPCRA) Extremely hazardous substances
- The "Director's List" (CAL/OSHA)
- Actively registered pesticide in California.
- Carcinogen User Register Cemical (CAL/OSHA) as inorganic arsenic compound
- AB 1803-Well Monitoring Chemical (CAL) as inorganic arsenic compound
- CAL Air Resources Board/AB 1807 Toxic Air Contaminants as inorganic arsenic compound
- Proposition 65 chemical (CAL) as inorganic arsenic compound
- AB 2588-Air Toxics "Hot Spots" Chemicals (CAL) as arsenic compounds as inorganic arsenic compound
- Permissible Exposure Limits for Chemical Contaminants (CAL/OSHA) as arsenic compounds as inorganic arsenic compound
- Clean Air Act, 42USC7412; Title I, Part A,§112 hazardous pollutants
- Clean Water Act 40CFR401.15 Section 307 Toxic Pollutants; 40CFR423, Appendix A Priority Pollutants; §313 Priority Chemicals
- RCRA 40CFR§261.24 Toxicity Characteristics, Maximum Concentration of Contaminants (MCC), Regulatory level, 5.0 mg/L
- RCRA, 40CFR261, Appendix 8, Appendix 8, Hazardous Constituents, waste number P 012
- Safe Drinking Water Act 47FR9352 Regulated chemical: MCL, 0.05 mg/L (Section 141.11) applies only to community water systems
- CERCLA/SARA 40CFR302 Extremely Hazardous Substances: TPQ = 100/10,000 lb (45.4/4540 kg)
- Superfund/EPCRA 40CFR302.4 RQ: CERCLA, 1 lb (0.454 kg)
- EPCRA Section 313: Form R *de minimis* concentration reporting level: 0.1%
- U.S. DOT Regulated Marine Pollutant (49CFR172.101, Appendix B)
- Canada, WHMIS, Ingredients Disclosure List. Concentration Reporting Level: 0.1%
- Canada: Priority Substance List & Restricted Substances/Ocean Dumping Forbidden (CEPA), National Pollutant Release Inventory (NPRI) (arsenic compounds)]

*Description:* Arsenic trioxide is a white or transparent solid in the form of glassy, shapeless lumps or a crystalline powder that resembles sugar. It has no odor or taste. *Warning properties*: inadequate; odorless and tasteless; airborne arsenic trioxide may produce a burning sensation to the nose, mouth, and eyes and cause coughing, shortness of breath, headache, sore throat, and dizziness. Slightly soluble in water; solubility = 37 g/L @ 20 °C; 115 g/L @100°C. Molecular weight = 197.84 daltons. Density (solid-water = 1.00) = 3.74. Boiling point = 465 °C @ 760 mmHg. Sublimes at 193 °C. Melting/Freezing point = 312 °C.

Vapor pressure = 66.1 mmHg @ 312 °C. Hazard Identification (based on NFPA-704 M Rating System): Health 3, Flammability 0, Reactivity 0.

*Incompatibilities:* Sodium chlorate; sodium hydroxide, sulfuric acid, fluorine; chlorine trifluoride; chromic oxide; aluminum chloride; phosphorus pentoxide; hydrogen fluoride; oxygen difluoride; tannic acid; infusion cinchona and other vegetable astringent infusions and decoctions; iron in solution. Contact with acids or acid mists releases deadly arsine gas. Not flammable, but emits highly toxic arsine gas and oxides of arsenic fumes when burned.

*Permissible Exposure Limits in Air:* OSHA: The legal airborne PEL is 0.01 mg/m$^3$ averaged over an 8-hour workshift for Arsenic and compounds as Arsenic, inorganic. NIOSH[2]: The recommended airborne exposure limit is 0.002 mg/m$^3$, which should not be exceeded during any 15 minute work period for Arsenic, inorganic. ACGIH: The recommended airborne exposure limit is 0.2 mg/m$^3$ average over an 8-hour workshift for Arsenic and soluble compounds[1]. The British HSE[33] has also adapted the ACGIH value of 0.2 mg/m$^3$ as an 8-hour TWA value. The DFG[3] has not set numerical limits for arsenic in air on the grounds that it is a proven human carcinogen. The former USSR-UNEP/IRPTC project[43] has set a MAC value for inorganic arsenic compounds (except arsine) of 0.003 mg/m$^3$ for ambient air in residential areas. In addition, some states have set guidelines or standards for arsenic trioxide in ambient air[60]: zero (New York), 0.0002 µg/m$^3$ (North Carolina), 3.0 µg/m$^3$ (Virginia).

*Determination in Air:* Use NIOSH Method 7901, Arsenic trioxide[18].

*Permissible Concentration in Water:* EPA has set a limit of 0.05 parts per million (ppm) for arsenic in drinking water. The U.S. EPA arsenic drinking water standard of 0.01 ppm (10 ppb) is based on the U.S. EPA final rule for arsenic in drinking water published in. the January 22, 2001, *Federal Register*. However, the U.S. EPA is currently reviewing the science and cost estimate supporting this rule, and, in the interim, has reverted to the previous standard for arsenic. Thus, in the US, the current EPA arsenic drinking water standard remains at 0.05 ppm (50 ppb). To protect freshwater aquatic life-total recoverable trivalent inorganic arsenic never to exceed 440 µg/L. To protect saltwater aquatic life: 508 µg/L on an acute basis. To protect human health: preferably zero. The former USSR-UNEP/IRPTC project[43] has set MAC values for inorganic arsenic compounds in water for domestic purposes at 0.05 mg/L and in water bodies for fishery purposes of 0.5 mg/L also.

*Determination in Water: For arsenic:* The atomic absorption graphite furnace technique is often used for measurement of total arsenic in water. It also has been standardized by EPA. Total arsenic may be determined by digestion followed by silver diethyldithiocarbamate; an alternative is atomic absorption; another is inductively coupled plasma optical emission spectrometry. See OSHA Method #ID-105 for arsenic[58].

*Routes of Entry:* Inhalation, skin contact, ingestion.

*Harmful Effects and Symptoms*

*Short Term Exposure:* • Toxic effects of arsenic trioxide usually result from ingestion. Small amounts of arsenic trioxide can lead to multiple organ damage and death. Acute signs and symptoms include nausea, vomiting, diarrhea, gastrointestinal hemorrhage, cerebral edema, tachycardia, dysrhythmias, and hypovolemic shock. Symptoms are dose dependent and can be delayed. Dermal contact and inhalation of airborne arsenic trioxide can cause localized irritation and usually does not result in systemic effects. Skin contact can cause burning, itching, and rash. Inhalation can cause respiratory irritation. Eye contact can cause irritation and possible permanent damage. High exposures can cause an abnormal EKG. Symptoms of acute poisoning may take from ½ hour to several hours after ingestion to appear. They may include: sweetish metallic taste; garlicky odor of breath and feces; constriction in throat and difficulty in swallowing; burning and colicky pains in esophagus, stomach and bowel; vomiting and profuse painful diarrhea (stools are watery initially, later becoming bloody); dehydration with intense thirst and muscular cramps; bluing of skin; feeble pulse and cold extremities; vertigo, frontal headache, stupor, delirium and mania (these symptoms may occur without concurrent or preceding gastric symptoms); fainting, coma, convulsions, general paralysis and then death. This material is considered super toxic; probable oral lethal dose (human) is less than 5 mg/kg, i.e., a taste (less than 7 drops) for a 70 kg (150 lb) person. Material causes acute gastrointestinal and central nervous system symptoms.

*Long Term Exposure:* Chronic exposure is characterized by malaise, peripheral sensorimotor neuropathy, anemia, jaundice, gastrointestinal complaints, and characteristic skin lesions including hyperkeratosis (small corn-like elevations) and hyperpigmentation. Hyperkeratosis usually appears on the palms or soles. Pigmentation changes and hyperkeratosis can take 3 to 7 years to appear. Chronic inhalation can also lead to conjunctivitis, irritation of the throat and respiratory tract, and perforation of the nasal septum. Chronic exposure can cause allergic contact dermatitis. Chronic exposure may be more serious for children because of their potential longer latency period. *Carcinogenicity:* The Department of Health and Human Services, IARC, the U.S. EPA, and the NTP have classified arsenic as a human carcinogen based on sufficient evidence from human data. Arsenic trioxide causes skin and lung cancer, and may cause internal cancers such as liver, bladder, kidney, colon, and prostate cancers. Arsenic ions released from arsenic trioxide within the body can cross the placenta and affect the developing fetus; arsenic is also excreted in breast milk. Experimental animal studies support an association between high ingested arsenic dose and fetal toxicity. *Reproductive and Developmental*

*Toxicants*: Special consideration regarding the exposure of pregnant women may be warranted, since arsenic trioxide has been shown to be mutagenic and clastogenic, and is suspected of being teratogenic; thus, medical counseling is recommended for the acutely exposed pregnant woman. Birth defects have been observed in animals exposed to inorganic arsenic. It is likely that health effects seen in children exposed to high amounts of arsenic will be similar to the effects seen in adults.

*Points of Attack:* Several studies have shown that inorganic arsenic can increase the risk of lung cancer, skin cancer, bladder cancer, liver cancer, kidney cancer, and prostate cancer.

*Medical Surveillance:* Before first exposure and every 6 to 12 months thereafter, OSHA 1910.1018 requires employers to provide (for persons exposed to 0.005 mg/m$^3$ of Arsenic) a medical history and exam which shall include : Chest x-ray, Exam of the nose, skin, and nails, sputum cytology examination, test for urine arsenic (may not be accurate within 2 days of eating shellfish or fish; most accurate at the end of a workday). Levels should not be greater than 100 micrograms per gram creatinine in the urine. Exam of the nervous system. After suspected overexposure, repeat these tests and consider complete blood count and liver function tests. Also examine skin periodically for abnormal growths. Skin cancer from arsenic can easily be cured when detected early. Employees have a legal right to testing information under OSHA 1910.20.

*First Aid:* If this chemical gets into the eyes, remove any contact lenses at once and irrigate immediately for at least 30 minutes, occasionally lifting upper and lower lids. Seek medical attention immediately. If this chemical contacts the skin, remove contaminated clothing and wash immediately with large amounts of soap and water. Seek medical attention immediately. If this chemical has been inhaled, remove from exposure, begin rescue breathing (using universal precautions) if breathing has stopped and CPR if heart action has stopped. Transfer promptly to a medical facility. In cases of ingestion, **do not induce vomiting**. Aggressive decontamination with gastric lavage is recommended within 1 hour of ingestion of a life-threatening amount of poison. The effectiveness of activated charcoal in binding arsenic trioxide is questionable, but administration of a charcoal slurry is recommended pending further evaluation in cases of ingestion of unknown quantities (at 1 gm/kg, usual adult dose 60–90 g, child dose 25–50 g). A soda can and straw may be of assistance when offering charcoal to a child.

*Note to physician or authorized medical personnel:* There is no specific antidote for arsenic trioxide. In cases of respiratory compromise secure airway and respiration via endotracheal intubation. If not possible, perform cricothyroidotomy if equipped and trained to do so. Treat patients who have bronchospasm with aerosolized bronchodilators. The use of bronchial sensitizing agents in situations of multiple chemical exposures may pose additional risks. Also consider the health of the myocardium before choosing which type of bronchodilator should be administered. Cardiac sensitizing agents may be appropriate; however, the use of cardiac sensitizing agents after exposure to certain chemicals may pose enhanced risk of cardiac arrhythmias (especially in the elderly). Arsenic trioxide poisoning is not known to pose additional risk during the use of bronchial or cardiac sensitizing agents. Consider racemic epinephrine aerosol for children who develop stridor. Dose 0.25–0.75 mL of 2.25% racemic epinephrine solution in 2.5 cc water, repeat every 20 minutes as needed cautioning for myocardial variability. Patients who are comatose, hypotensive, or have seizures or cardiac dysrhythmias should be treated according to advanced life support protocols. If massive exposure is suspected or if the patient is hypotensive, infuse intravenous saline or lactated Ringer's solution. For adults, bolus 1000 mL/hour if blood pressure is under 80 mmHg; if systolic pressure is over 90 mmHg, an infusion rate of 150 to 200 mL/hour is sufficient. For children with compromised perfusion administer a 20 mL/kg bolus of normal saline over 10 to 20 minutes, then infuse at 2 to 3 mL/kg/hour. Chelation therapy is strongly recommended.

*References:*
- Sax, N.I., Ed., "Dangerous Properties of Industrial Materials Report" 3, No. 5, 50-58 (1983).
- U.S. Environmental Protection Agency, "Chemical Profile: Arsenous Oxide," Washington, DC, Chemical Emergency Preparedness Program (November 30, 1987).
- ATSDR, U.S. Department of Health and Human Services, Public Health Service, "Managing Hazardous Materials Incidents," Atlanta, GA (June 2003)
- California Environmental Protection Agency Chemical List of Lists, Sacramento CA (February 1997).
- New Jersey Department of Health and Senior Services, "Hazardous Substance Fact Sheet: Arsenic Trioxide," Trenton, NJ (December 1995, rev. January 2001). http://www.state.nj.us/health/eoh/rtkweb/0161.pdf

# Asulam (ANSI)

*Use Type:* Herbicide
*CAS Number:* 3337-71-1
*Formula:* $C_8H_{10}N_2O_4S$
*Synonyms:* 4-Amino-benzolsulfonyl-methylcarbamat (German); Methyl sulfanilylcarbamate; Carbamic acid, sulfanilyl-, methyl ester; 4-(Aminophenylsulfonyl) carbamate, methyl ester; Carbamic acid, [(4-aminophenyl)sulfonyl]-, methyl ester; Methyl-*N*-(4-amino benzenesulfonyl)carbamate; Methyl [(4-aminophenyl)sulfonyl]carbamate; Methyl sulfanilyl carbamate; Sulfanilylcarbamic acid, methyl ester

*Trade Names:* ASILAN®; ASULFOX F®; ASULOX®; ASULOX 40®; JONNIX®; MB 9057®
*Producers:* Agrimor International (USA); Rhone-Poulenc (France)
*Chemical Class:* Carbamate
*EPA/OPP PC Code:* 106901
*California DPR Chemical Code:* 5076
*EINECS Number:* 222-077-1
*Uses:* Not registered in the U.S. Use as a post-emergence herbicide to control broadleaf weeds, perennial grasses and nonflowering plants in sugarcane and reforestation areas, bananas, cocoa, coffee, coconuts and citrus crops.
*Human toxicity (long-term)[77]:* Very low–252.00 ppb, Health Advisory
*Fish toxicity (threshold)[77]:* Very low–841072.39665 ppb, MATC (Maximum Acceptable Toxicant Concentration)
*Carcinogen/Hazard Classifications*
**U.S. EPA Carcinogens:** Group C, possible carcinogen
**Label Signal Word:** CAUTION
**WHO Acute Hazard:** Class U, unlikely to be hazardous
*Description:* A colorless, crystalline solid. Odorless. Molecular weight = 230.25. Melting/Freezing point = 143-146°C. Moderately soluble in water. Solubility 5,000 mg/L.
*Incompatibilities:* May form explosive materials with phosphorus pentachloride.
*Determination in Air:* Filter; none; Gravimetric; NIOSH[2] IV [Particulates NOR; #0500 (total), #0600 (respirable)]
*Permissible Concentration in Water:* No criteria set. Runoff from spills or fire control may cause water pollution.
*Routes of Entry:* Skin absorption, ingestion, inhalation.
*Harmful Effects and Symptoms*
*Short Term Exposure:* Eye pupils are small; blurred vision; eye watering; runny nose; cough; shortness of breath; salivation; nausea, stomach cramps, diarrhea, and vomiting; increased blood pressure; profuse sweating; hypermotility, hallucinations; agitation; tingling of the skin; slow heartbeat; convulsions; fluid in lungs; loss of consciousness; incontinence; breathing stops; death. Carbamate insecticides inhibit the cholinesterase activity of enzymes, causing accumulation of acetylcholine at synapses and altering the way in which nervous impulses are transmitted. However, within several hours carbamates spontaneously detach from the enzymes.
*Long Term Exposure:* A potent cholinesterase inhibitor; cumulative effect is possible. This chemical may damage the nervous system with repeated exposure, resulting in convulsions, respiratory failure. May cause liver damage.
*Points of Attack:* Respiratory system, lungs, central nervous system, cardiovascular system, skin, eyes, plasma and red blood cell cholinesterase.
*Medical Surveillance:* Medical observation is recommended for 24 to 48 hours after breathing overexposure, as pulmonary edema may be delayed. As first aid for pulmonary edema, consider administering a corticosteroid spray. Cigarette smoking may exacerbate pulmonary injury and should be discouraged for at least 72 hours following exposure. Before employment and at regular times after that, the following are recommended: Plasma and red blood cell cholinesterase levels (tests for the enzyme poisoned by this chemical). If exposure stops, plasma levels return to normal in 1-2 weeks while red blood cell levels may be reduced for 1-3 months. When acetylcholinesterase enzyme levels are reduced by 25% or more below preemployment levels, risk of poisoning is increased, even if results are in lower ranges of "normal." Reassignment to work not involving carbamate pesticides is recommended until enzyme levels recover. If symptoms develop or overexposure occurs, repeat the above tests as soon as possible and get an exam of the nervous system. Also consider complete blood count. Consider chest x-ray following acute overexposure.
*First Aid:* Speed in removing material from eyes and skin is of extreme importance. *Eyes:* Eye contact can cause dangerous amounts of these chemicals to be quickly absorbed through the mucous membrane into the bloodstream. Immediately and gently flush eyes with plenty of warm or cold water (NO hot water) for at least 15 minutes, occasionally lifting the upper and lower eyelids. Get medical aid immediately. *Skin:* Get medical aid. Skin contact can cause dangerous amounts of these chemicals to be absorbed into the bloodstream. Wearing the appropriate PPE equipment and respirator for carbamate pesticides, immediately flush skin with plenty of soap and water for at least 15 minutes while removing contaminated clothing and shoes. Shampoo hair promptly if contaminated; protect eyes. *Ingestion:* Call poison control. Loosen all clothing. Never give anything by mouth to an unconscious person. Get medical aid. Do NOT induce vomiting.* If conscious, alert, and able to swallow, rinse mouth and have victim drink 4 to 8 ounces of water. Check to see if poison control instructs you to use ipecac syrup, otherwise administer slurry of activated charcoal (2 oz in 8 oz of water). If victim is *UNCONSCIOUS OR HAVING CONVULSIONS,* do nothing except keep victim warm. **In some cases you may be specifically instructed by poison control to induce vomiting by way of 2 tablespoons of syrup of ipecac (adult) washed down with a cup of water.* Do NOT give activated charcoal before or with ipecac syrup. *Inhalation:* Get medical aid. Do not contaminate yourself. Wearing the appropriate PPE equipment and respirator for carbamate pesticides, immediately remove the victim from the contaminated area to fresh air. If the victim is not breathing, administer artificial respiration. Do NOT use mouth-to-mouth resuscitation; use bag/mask apparatus. If breathing is difficult, administer oxygen through bag/mask apparatus until medical help arrives. Do not leave victim unattended. *Note to physician or authorized medical personnel.* Administer atropine, 2 mg (1/30 gr) intramuscularly or

intravenously as soon as any local or systemic signs or symptoms of an intoxication are noted; repeat the administration of atropine every 3 to 8 minutes until signs of atropinization (mydriasis, dry mouth, rapid pulse, hot and dry skin) occur; initiate treatment in children with 0.05 mg mg/kg of atropine; repeat at 5 to 10 minute intervals. Watch respiration, and remove bronchial secretions if they appear to be obstructing the airway; intubate if necessary.
*Medical note:* 2-PAMCI may be contraindicated in the case of some carbamate poisonings.
*References:*
- California Environmental Protection Agency Chemical List of Lists, Sacramento CA (February 1997)

## Atrazine (ANSI)

*Use Type:* Herbicide, plant growth regulator
*CAS Number:* 1912-24-9
*Formula:* $C_8H_{14}ClN_5$
*Alert:* A Restricted Use Product (RUP). Atrazine has been banned or restricted in Angola, South Africa, Austria, Denmark, Germany, Italy, the Netherlands, Norway, and Sweden. Human toxicity (long-term): High.
*Synonyms:* 2-Aethylamino-4-chlor-6-isopropylamino-1,3,5-triazin (German); Atrazin (German); Atrazina (Spanish); 2-Chloro-4-ethylamineisopropylamine-*s*-triazine; 1-Chloro-3-ethylamino-5-isopropylamino-*S*-triazine; 1-Chloro-3-ethylamino-5-isopropylamino-2,4,6-triazine; 2-Chloro-4-ethylamino-6-isopropylamino-*S*-triazine; 2-Chloro-4-ethylamino-6-isopropylamino-1,3,5-triazine; 6-Chloro-*N*-ethyl-*N*'-(1-methylethyl)-1,3,5-triazine-2,4-diamine; 2-Chloro-4-ethylamono-6-isopropylamino-; 6-Chloro-*N*-ethyl-*N*-isopropyl-1,3,5-triazinediyl-2,4-diamine; 2-Chloro-4-(2-propylamino)-6-ethylamino-*S*-triazine; 2-Ethylamino-4-isopropylamino-6-chloro-*S*-triazine; NSC 163046; Penatrol; *S*-Triazine, 2-chloro-4-(ethylamino)-6-(Isopropylamino)-; 1,3,5-Triazine-2,4-diamine, 6-chloro-*N*-ethyl-*N*-(1-methylethyl)-
*Trade Names:* A 361®, canceled; AI3-28244®; AATRAM®, Syngenta, (Switzerland); AATREX®, Syngenta, (Switzerland); ACTINITE PK®; ACTINIT A®; AGIMIX®, Milenia Agro Ciencias (Brazil); AKTIKON®, Chemol Trading Co., Ltd. (Hungary); AKTIKON PK®, Chemol Trading Co., Ltd. (Hungary); AKTINIT A®; ALAZINE®, Makhteshim Agan (Israel; ANELDAZIN® EMV North-Hungarian Chemicals Works Co. (Hungary) canceled; ARGEZIN®; ATAZINAX®; ATERBUTEX®, Makhteshim Agan (Israel); ATERBUTOX®, Pyosa Agroquimicos (Mexico); ATLAS ATRAZINE®, Whyte Agrochemicals (UK); ATLAZIN D-WEED®, Nomix-Chipman Ltd (UK); ATRANEX®, Makhteshim Agan (Israel); ATRASINE®; ATRATAF®, Rallis India Ltd. (India); ATRATOL®, Syngenta (Switzerland); ATRAZINEK®; ATRAZINE 90DF®, Drexel Chemical (USA); ATRED®, Agrimont S.p.A. (Italy) discontinued; ATREX®; AXIOM®, Bayer CropScience (Germany); AZINOTOX®, Pyosa Agroquimicos (Mexico); BICEP®, Syngenta (Switzerland); BLADEX/ATRAZINE (2:1) 80W®, E.I. DuPont (US) canceled 10/9/1992; BUCTRIL+ATRAZINE GEL®, Bayer CropScience (Germany) canceled 12/20/2000; CANDEX®; CEKUZINA-T®; CHROMOZIN®, Chromos Agro d.d. (Croatia); CO-OP ATRAZINE®, Land O'Lakes Farmland Feed LLC (USA), canceled 9/30/1991; CRISATRINA® Crystal Chemical (Mexico); CRISAZINE®; CYAZIN®; DOW ATRAZINE 80W HERBICIDE®, Dow Chemical Co. (US) canceled 11/3/1982; ERUNIT 500 FW®, Nitrokemia 2000, (Hungary); FARMCO® ATRAZINE, Farmco Australia (Australia); FENAMIN®; FENAMINE®, Adventis CropScience (France), canceled; FENATROL®, Adventis CropScience (France), canceled; FIELD MASTER®, Monsanto (USA); FLOWABLE ATRAZINE®, Nufarm (Australia); G 30027®, Syngenta (Switzerland); GEIGY 30,027®, Syngenta (Switzerland); GESAPRIM®, Syngenta (Switzerland); GESOPRIM®; GRIFFEX®, Griffin Pest Control Co. (USA); GRIFFIN ATRAZINE 90 DRY FLOWABLE HERBICIDE®, Griffin L.L.C. (USA), canceled 7/1/1987; HAVILAND ATRAZINE LINURON WEED KILLER®, Haviland Agricultural Inc. (USA), canceled 5/20/1983; HELENA ATRAZINE TECHNICAL®, Cedar Chemical Corp (USA), canceled 1/22/1991; GUARDSMAN® herbicide (mixture of Atrazine and Dimethenamid); HELENA BRAND ATRAZINE®, Helena Chemical Co. (USA), canceled 3/3/1983; HERBATOXOL®; HERBIMIX SC®, Milenia Agro Ciencias (Brazil); HERBITRIN 500 BR®, Milenia Agro Ciencias (Brazil); HUNGAZIN®; INAKOR®; LADDOK®, BASF Agricultural Products (Germany); LANCO ATRAZINE®, Landia Chemical Co. (USA), canceled 9/29/1988; LARIAT®, Monsanto (USA); LEADOFF®, DuPont Crop Protection (USA); MAGIC CARPET FERTILIZER WITH ATRAZINE®, canceled 10/01/1998; MALLET PM BROMOXYNIL, ATRAZINE BROADLEAF HERBICIDE®, Bayer CropScience (Germany) canceled 11/11/1988; MARKSMAN®, BASF Canada (Canada); MARZONE ATRAZINE®, Marzone Chemicals Ltd. (USA), canceled 10/10/1989; MITAC®, Aventis CropScience (Germany); Gowan Company (USA); NEW CHLOREA®, Nomix-Chipman Ltd (UK); NU-TRAZINE 900 DF®, Nufarm (Australia); NU-ZINOLE AA®, Nufarm (Australia); OLEOGESAPRIM®; PATRIOT®, BASF Canada (Canada); PITEZIN®; POSMIL®, Milenia Agro Ciencias (Brazil); PRIMATOL® Ciba (Switzerland) discontinued; PRIMATOP®, Syngenta (Switzerland); PRIMAZE®, Ciba-Geigy (Switzerland) discontinued; PRIMOLE®, Syngenta (Switzerland); PROKIL ATRAZINE 80W®, Gowan Company (USA), canceled 6/15/1987; RADAZIN®; RADIZINE®; READY MASTER®, Monsanto (USA);

RESIDOX®, Nomix-Chipman Ltd (UK); SHELL® ATRAZINE 80W HERBICIDE, DuPont Crop Protection (USA), canceled 11/22/1991; SIMAZAT®, Drexel Chemical (USA); STRAZINE® TRIAZINE A 1294; TRIPART® ATRAZINE 50 SC; VECTAL®; WEEDEX®, Syngenta, (Switzerland); WONUK®; ZEAPOS®, ICC-Chemol Trading and Distribution (Hungary) discontinued; ZEAZIN®; ZEAZINE®

***Producers:*** Agsin (Singapore); Alcotan Laboratories (Spain); Atanor S.A. (Argentina); BASF Agriculture Products Group (Germany); Bayer CropScience (Germany); Bhageria Dye-Chem (India); Bharat Pulverizing Mills (India); Biesterfeld Siemsgluess International. GmbH (Germany); Chromos Agro d.d. (Croatia); Drexel Chemical (USA); DuPont Crop Protection (USA); Ehrenstorfer, Dr.(Germany); Fulon Chemical Industrial Co., Ltd. (Taiwan); Gowan Company (USA); Helm AG (Germany); Kawaguchi Chemical Industry (Japan); Ki-Hara Chemicals Ltd. (UK); Makhteshim Agan (Israel); Meghmani Organics (India); Milenia Agro Ciencias (Brazil); Monsanto (USA); Nagarjuna Agrichem Ltd. (India); Nitrokemia 2000 (Hungary); Nomix-Chipman Ltd (UK); Nufarm (Australia); Oxon Italia S.p.A. (Italy); Proficol S.A (Colombia); Pyosa Agroquimicos (Mexico); Rallis India Ltd. (India); Sanonda Co., Ltd. (Australia); Shenzhen Guomeng Industry Co., Ltd. (China); Shenzhen Jiangshan Commerce & Industry (China); Sigma-Aldrich Laborchemikalien (Germany); Sipcam Agro USA (USA); Sulphur Mills Ltd. (India); Syngenta (Switzerland); United Agri Products (UAP) (Loveland Products) (USA); Vijayalakshmi Insecticides and Pesticides (India); Whyte Agrochemicals (UK); Zago Asia Ltd; Zhejiang Changxing Zhongshan Chemical Industry (China)

***Chemical Class:*** Triazine
***EPA/OPP PC Code:*** 080803
***California DPR Chemical Code:*** 00045
***ICSC Number:*** 0099
***RTECS Number:*** XY5600000
***EEC Number:*** 613-068-00-7
***EINECS Number:*** 217-617-8

***Uses:*** Atrazine is a selective pre- and post-emergence herbicide used for the control of broadleaf and grassy weeds in crops, such as corn (field and sweet), guava, hay, macadamia nuts, range grasses for the establishment of permanent grass cover on range lands and pastures in Oklahoma, Nebraska, Texas and Oregon, wheat, residential and recreational turf and sod farms, sorghum, sugarcane, pineapples, and Christmas trees and ornamentals. It is also used in forestry and, at higher application rates, for non-selective weed control in non-crop areas. It is the most widely used pesticide in the United States. Use data from 1900 to 1997 indicates that approximately 76.5 million pounds of atrazine active ingredient is used domestically each year. Certified herbicide workers may spread atrazine on crops or crop lands as a powder, liquid, or in a granular form. Atrazine is usually used in the spring and summer months. For it to be active, atrazine needs to dissolve in water and enter the plants through their roots. It then acts in the shoots and leaves of the weed to stop photosynthesis. Atrazine is taken up by all plants, but in plants not affected by atrazine, it is broken down before it can have an effect on photosynthesis. Atrazine degrades into hydroxy compounds and chlorotriazine degradates. The application of atrazine to crops as a herbicide accounts for almost all of the atrazine that enters the environment, but some may be released from manufacture, formulation, transport, and disposal. Atrazine does not tend to accumulate in living organisms such as algae, bacteria, clams, or fish, and, therefore, does not tend to build up in the food chain. Atrazine can be applied by ground boom sprayer, aircraft, tractor-drawn spreader, rights-of-way sprayer, hand-held sprayers, backpack sprayer, lawn handgun, push-type spreader, and bellygrinder.

***Human toxicity (long-term)[77]:*** High–3.00 ppb, MCL (Maximum Contaminant Level)
***Fish toxicity (threshold)[77]:*** Intermediate–88.31816 ppb, MATC (Maximum Acceptable Toxicant Concentration)
***U.S. Maximum Allowable Residue Levels for Atrazine (40 CFR 180.220):***

| CROP | ppm |
|---|---|
| Cattle, fat | 0.02 |
| Cattle, meat | 0.02 |
| Cattle, mbyp | 0.02 |
| Corn, field, forage | 15.0 |
| Corn, fodder, field | 15.0 |
| Corn, fodder, pop | 15.0 |
| Corn, fodder, sweet | 15.0 |
| Corn, field, stover | 15.0 |
| Corn, fresh (inc. sweet)(k=cwhr) | 0.25 |
| Corn, grain | 0.25 |
| Corn, pop, fodder | 15.0 |
| Corn, pop, forage | 15.0 |
| Corn, sweet, fodder | 15.0 |
| Corn, sweet, forage | 15.0 |
| Egg | 0.02 |
| Goat, fat | 0.02 |
| Goat, meat | 0.02 |
| Goat, mbyp | 0.02 |
| Grass, orchardgrass | - |
| Grass, range | 4.0 |
| Grass, rye, perennial | 15.0 |
| Guava | 0.05 |
| Hog, fat | 0.02 |
| Hog, meat | 0.02 |
| Hog, mbyp | 0.02 |
| Horse, fat | 0.02 |
| Horse, meat | 0.02 |
| Horse, mbyp | 0.02 |

| | |
|---|---|
| Milk | 0.02 |
| Nut, macadamia | 0.25 |
| Orchardgrass, hay | - |
| Poultry, fat | 0.02 |
| Poultry, meat | 0.02 |
| Poultry, mbyp | 0.02 |
| Sheep, fat | 0.02 |
| Sheep, meat | 0.02 |
| Sheep, mbyp | 0.02 |
| Sorghum, forage | 15.0 |
| Sorghum, grain, grain | 0.25 |
| Sorghum, stover | 15.0 |
| Sugarcane, cane | 0.25 |
| Sugarcane, fodder | 0.25 |
| Sugarcane, forage | 0.25 |
| Wheat, fodder | 5.0 |
| Wheat, grain | 0.25 |
| Wheat, straw | 5.0 |

*Note*: In Federal Register, July 17, 2002 (Volume 67, Number 137), EPA 40 CFR Part 180, **revokes** specific tolerances for residues of the insecticides phosphamidon and trimethacarb; the herbicides atrazine, *S-(O,O*-diisopropyl phosphorodithioate) ester of *N*-(2-mercaptoethyl) benzenesulfonamide, known as bensulide, *S*-propyl dipropylthiocarbamate, known as vernolate, and diphenamid; the fungicide imazalil; and the fungicide/insecticide 6-methyl-1,3-dithiolo(4,5-b)quinoxalin-2-one(oxythioquinox) because these pesticides are no longer registered on certain food uses in the United States. By law, EPA is required by August, 2002, to reassess 66% of the tolerances in existence on August 2, 1996, or about 6,400 tolerances.

*Carcinogen/Hazard Classifications*
**U.S. EPA Carcinogens:** Not a likely human carcinogen
**U.S. NTP Carcinogen**: Not listed
**California Prop. 65:** Not listed
**IARC:** Group 3. Not classifiable as to carcinogenicity to humans
**Label Signal Word:** CAUTION
**Endocrine Disruptor:** Suspected
*Regulatory Authority:*
- Air Pollutant Standard Set (ACGIH)[1] (DFG)[3] (HSE)[33] (former USSR)[43] (Several States)[60] (Australia) (Israel) (Mexico) (Various Canadian Provinces)
- AB 1803-Well Monitoring Chemical (CAL)
- MCL(Maximum Contaminants Levels) list of contaminants (CAL)
- Permissible Exposure Limits for Chemical Contaminants (CAL/OSHA)
- The "Director's List" (CAL/OSHA)
- Actively registered pesticide in California.
- Water Pollution Standard Set (U.S. EPA) (Canada)
- Safe Drinking Water Act: 40CFR141.61©)5, MCL, 0.003 mg/L; 40CFR141.50(*b*)7, MCGL 0.003 mg/L; 40CFR142.62, Variances and Exceptions from the MCLs; 40CFR9352 Regulated Chemical; 40CFR141.24, Requirements for Sampling and Analytical Testing; 40CFR141.32 Public Notification Requirements.
- FIFRA 40CFR180.220 tolerances on raw agricultural commodities.
- CERCLA/SARA 40CFR372.65: Form R *de minimis* Concentration Reporting Level: 0.1%.

*Description:* Atrazine is a colorless or white, crystalline solid or powder. Often mixed with a liquid. Soluble in water; solubility = 68 ppm @ 25°C. Odorless. Molecular weight = 215.71. Melting/Freezing point = 174–177°C. Hazard Identification (based on NFPA-704 M Rating System): Health 1, Flammability 0, Reactivity 0. Slightly soluble in water. Vapor pressure = $2.9 \times 10^{-7}$ mmHg @ 20°C. Log $K_{ow}$ = 2.24–2.78. Unlikely to bioaccumulate in marine organisms.

*Incompatibilities:* Strong oxidizers, acids.

*Permissible Exposure Limits in Air:* The NIOSH[2] and ACGIH[1] recommended airborne exposure limit is 5 mg/m$^3$ TWA for a 10-hour and 8-hour workshift, respectively[1]. The DFG[3] has set a MAK limit value of 2 mg/m$^3$. The Australian and Israel limit is 5 mg/m$^3$ and Israel's Action Limit is 2.5 mg/m$^3$ The HSE[33] (U.K.) and Mexico limit is 10 mg/m$^3$ TWA. Canadian Provincial Limts are: Alberta: 5 mg/m$^3$ TWA and 15 min. STEL of 10 mg/m$^3$; British Columbia: 10 mg/m$^3$; Ontario and Quebec: 5 mg/m$^3$ TWAEV. The former USSR-UNEP/IRPTC project[43] has set a MAC of 2 mg/m$^3$ in workplace air and of 0.02 mg/m$^3$ for both momentary and daily average exposure in ambient air in residential areas. In addition, several states have set guidelines or standards[60] for atrazine in ambient air ranging from 50 µg/m$^3$ (North Dakota) to 80 µg/m$^3$ (Virginia) to 100 µg/m$^3$ (Connecticut) to 119 µg/m$^3$ (Nevada).

*Determination in Air:* Use OSHA versatile sampler-2; Reagent; Gas chromatography/Electron capture detection; NIOSH #5602[18].

*Permissible Concentration in Water:* A maximum level (MCL and MCGL) in drinking water of 0.003 mg/L has been set by EPA[62]. The Canadian Drinking Water IMAC is 0.06 mg/L. A suggested no-adverse effect level in drinking water has been calculated by NAS/NRC as 0.15 mg/L. An acceptable daily intake (ADI) of 0.0215 mg/kg/day has been calculated for atrazine[46]. A limit of 0.5 mg/L of atrazine has been specified by the former USSR-UNEP/IRPTC program[43] in water bodies used for domestic purposes. Also, several states have set guidelines for atrazine in drinking water[61] ranging from 0.093 µg/L (Massachusetts) to 15 µg/L (California) to 25 µg/L (New York) to 43 µg/L (Maine) to 150 µg/L (Kansas).

*Determination in Water:* Analysis of atrazine is by a gas chromatographic (GC) method applicable to the determination of certain nitrogen-phosphorus containing pesticides in water samples. In this method, approximately

1 l of sample is extracted with methylene chloride. The extract is concentrated, and the compounds are separated using capillary column GC. Measurement is made using a nitrogen phosphorus detector. The method detection limit has not been determined for this compound, but it is estimated that the detection limits for the method analytes are in the range of 0.1 to 2 $\mu$g/L.

*Routes of Entry:* Inhalation, passing through the skin.

*Harmful Effects and Symptoms*

*Short Term Exposure:* Contact may cause congestion of heart, lungs and kidneys; low blood pressure; muscle spasms; weight loss; damage to adrenal glands, and skin and severe eye irritation. Because this material has a low vapor pressure, significant inhalation of vapors is unlikely at ordinary temperatures.

*Long Term Exposure:* Atrazine is a suspected human carcinogen when exposure is chronic. It causes many kinds of cancer, including cancer of the breast, ovaries, uterus, testicles, as well as leukemia and lymphoma. It is an endocrine disrupting chemical. It interrupts regular hormone function, causing birth defects, reproductive tumors, and weight loss in mothers and embryos. Atrazine may cause skin allergy. Atrazine has the potential to cause weight loss, cardiovascular damage, retinal and some muscle degeneration, and mammary tumors from a lifetime exposure at levels above the MCL. Atrazine is possibly carcinogenic to humans (IARC, 3). There is inadequate evidence to confirm carcinogenicity of atrazine in humans. However, there is the increased risks for tumors known to be associated with hormonal factors. These have been observed in both animals and human beings, and are consistent with the known effects of atrazine on the hypothalamic pituitary gonadal axis.

*Points of Attack:* Eyes, skin, respiratory system, central nervous system, liver.

*Medical Surveillance:* Evaluation by a qualified allergist. Examination of the nervous system.

*First Aid:* If this chemical gets into the eyes, remove any contact lenses at once and irrigate immediately for at least 15 minutes, occasionally lifting upper and lower lids. Seek medical attention immediately. If this chemical contacts the skin, remove contaminated clothing and wash immediately with soap and water. Seek medical attention immediately. If this chemical has been inhaled, remove from exposure, begin rescue breathing (using universal precautions) if breathing has stopped and CPR if heart action has stopped. Transfer promptly to a medical facility. When this chemical has been swallowed, get medical attention. Give large quantities of water or milk and induce vomiting. Do not make an unconscious person vomit.

*References:*
- EPA, Office of Pesticide Programs, 40 CFR 180.220, "Pesticide Residue Limits," www.epa.gov/pesticides/food/viewtols.htm
- Federal Register Environmental Documents, EPA, www.epa.gov/fedrgstr/EPA-PEST/2002/July/Day-17/p17870.htm
- FIFRA Scientific Advisory Panel, June 2000.
- California Environmental Protection Agency Chemical List of Lists, Sacramento CA (February 1997).
- ATSDR, U.S. Department of Health and Human Services, Public Health Service, "Managing Hazardous Materials Incidents," Atlanta, GA (June 2003)
- Agency for Toxic Substances and Disease Registry (ASTDR), "Public Health Statement for Atrazine," Sept., 2001, http://www.atsdr.cdc.gov/toxprofiles/phs153.htm
- Interim Reregistration Eligibility Decision for Atrazine, Case No. 0062, EPA, January 31, 2003. http://www.epa.gov/REDs/atrazine_ired.pdf
- "Summary of Atrazine Risk Assessment," EPA, May 2, 2002,
- www.epa.gov/oppsrrd1/reregistration/atrazine/index.htm
- New Jersey Department of Health and Senior Services, "Hazardous Substance Fact Sheet: Atrazine," Trenton, NJ (June 1998), www.state.nj.us/health/eoh/rtkweb/rtkhsfs.htm

# Auramine

*Use Type:* Fungicide
*CAS Number:* 492-80-8; 2465-27-2 (hydrochloride salt)
*Formula:* $C_{17}H_{21}N_3$; $(CH_3)_2N$-$C_6H_4CNH$-$C_6H_4$-$N(CH_3)_2$
*Alert:* Auramine should be handled as a carcinogen with extreme caution.
*Synonyms:* Apyonine auramarine base; Auramina (Spanish); Auramine base; Auramine $N$ base; Auramine OAF; Auramine O base; Auramine SS; Basic Yellow 2; Baso Yellow 124; Benzeneamine, 4,4'-cabonimidoylbis[$N$-dimethyl-; C. I. Solvent Yellow 34; Brilliant Oil Yellow; 4,4'-carbonimidoylbis($N,N$-dimethylbenzenamine); C.I. 41000B; C.I. Basic Yellow 2, free base; C.I. Solvent Yellow 34; 4,4'-Dimethylaminobenzophenonimide; Glauramine; 4,4-(Imidocarbonyl)bis($N,N$-dimethylaniline); Tetramethyldiaminodiphenylacetimine; Yellow pyoctanine
*Trade Names:* TOBAZ®, Mallinckrodt, Inc. (USA), canceled; WAXOLINE YELLOW O®
*Producers:* Aldrich Chemical (USA); Merck (Germany); Whyte Agrochemical (UK)
*Chemical Class:* Unclassified
*EPA/OPP PC Code:* 039501
*California DPR Chemical Code:* 702
*RTECS Number:* BY3500000; BY3675000 (hydrochloride salt)
*EINECS Number:* 219-567-2
*Uses:* Auramine is used industrially as a dye or dye intermediate for coloring textiles, paper, and leather. Also

used as an antiseptic (a powerful antiseptic in ear and nose surgery and in gonorrhea treatment).

***Carcinogen/Hazard Classifications***
**California Prop. 65:** Carcinogen
**IARC:** Group 2B; Carcinogenic, Sufficient animal data
**Label Signal Word:** CAUTION
***Regulatory Authority:***
- Carcinogenic (UN)[13] (Animal Positive) (DFG)[3] (IARC)
- AB 2588-Air Toxics "Hot Spots" Chemicals (CAL)
- Proposition 65 chemical (CAL)
- The "Director's List" (CAL/OSHA)
- Banned or Severely Restricted (Italy, Sweden) (UN)[13]
- Air Pollutant Standard Set (North Dakota, New York)[60]
- EPA Hazardous Waste Number (RCRA No.): U014 (as C.I. Solvent Yellow 34)
- RCRA, 40CFR261, Appendix 8 Hazardous Constituents
- Superfund/EPCRA 40CFR302.4, Appendix A, RQ: CERCLA, 100 lb (45.4 kg)
- CERCLA/SARA 40CFR372.65: Form R *de minimis* Concentration Reporting Level: 0.1%.

***Description:*** Auramine is a yellow, crystalline powder or flaky material. Melting/Freezing point = 136°C. Hazard Identification (based on NFPA-704 M Rating System): Health 2, Flammability 1, Reactivity 0. Soluble in water.
***Incompatibilities:*** Strong oxidizers. Emits nitrogen oxides and hydrogen chloride when heated to decomposition.
***Permissible Exposure Limits in Air:*** No occupational exposure limits have been established. However auramine may be a carcinogen; there may be no safe level of exposure. This chemical can be absorbed through the skin, thereby increasing the potential for exposure. Zero in New York, North Dakota[60] in ambient air.
***Permissible Concentration in Water:*** No criteria set, but runoff from spills or fire control may cause water pollution.
***Routes of Entry:*** Inhalation, ingestion, skin absorption. Low-level dermal exposure to the consumer may occur but would be limited to any migration of auramine from fabric, leather, or paper goods.
***Harmful Effects and Symptoms***
***Short Term Exposure:*** Contact can irritate the eyes, and may cause damage. Skin absorption may result in dermatitis and burns, nausea and vomiting.
***Long Term Exposure:*** Commercial auramine is carcinogenic in mice and rats after oral administration, producing liver tumors, and after subcutaneous injection in rats, producing local sarcomas. The manufacture of auramine (which also involves exposure to other chemicals) has been shown in one study to be causally associated with an increase in bladder cancer. The actual carcinogenic compound(s) has not been specified precisely.
***Points of Attack:*** Liver, bladder.
***Medical Surveillance:*** Monthly urinalysis. Physical exam every 6 months focused on bladder.
***First Aid:*** If this chemical gets into the eyes, remove any contact lenses at once and irrigate immediately for at least 15 minutes, occasionally lifting upper and lower lids. Seek medical attention immediately. If this chemical contacts the skin, remove contaminated clothing and wash immediately with soap and water. Seek medical attention immediately. If this chemical has been inhaled, remove from exposure, begin rescue breathing (using universal precautions) if breathing has stopped and CPR if heart action has stopped. Transfer promptly to a medical facility. When this chemical has been swallowed, get medical attention. Give large quantities of water and induce vomiting. Do not make an unconscious person vomit.
***References:***
- Sax, N.I., Ed., "Dangerous Properties of Industrial Materials Report" 1, No. 5, 37-38 (1981).
- California Environmental Protection Agency Chemical List of Lists, Sacramento CA (February 1997).
- New Jersey Department of Health and Senior Services, Hazardous Substance Fact Sheet, "Auramine," Trenton NJ (April 1997, rev. February 2004). http://www.state.nj.us/health/eoh/rtkweb/2894.pdf

## Azacosterol Dihydrochloride

***Use Type:*** A bird sterilant
***CAS Number:*** 1249-84-9
***Formula:*** $C_{25}H_{46}Cl_2N_2O$
***Synonyms:*** Androst-5-en-3-ol, 17-[((3-(dimethylamino)propyl)methyl)amino]-,dihydrochloride, (3$\beta$,17$\beta$)-; Azasterol HCL; Azacosterol hydrochloride; Diazacholesterol dihydrochloride; 20,25-Diazacosterol hydrochloride; 17-$\beta$-[(3-(Dimethylamino)-propyl)methylamino]androst-5-en-3-$\beta$-ol dihydrochloride
***Trade Names:*** AZASTEROL®; IMD-760®; ORNITROL®, Avitrol (USA), canceled; SC-12937®
***Producers:*** Avitrol (USA)
***Chemical Class:*** Organochlorine
***EPA/OPP PC Code:*** 098101
***California DPR Chemical Code:*** 2026
***Uses:*** Not registered in the U.S. Used as an avian sterilant.
***Carcinogen/Hazard Classifications***
**Label Signal Word:** CAUTION
***Description:*** Crystalline solid. Molecular weight = 388.71.
***Incompatibilities:*** When heated to decomposition or on contact with acids or acid fumes, may produce highly toxic chloride fumes; deadly phosgene gas may be formed. May cause pitting of some metals.
***Permissible Concentration in Water:*** No criteria set. Runoff from spills or fire control may cause water pollution.
***Routes of Entry:*** Inhalation, ingestion. Absorbed through the intact skin.
***Harmful Effects and Symptoms***
***Short Term Exposure:*** Irritating to eyes, skin, mucous

membranes. Large doses can cause central nervous system depression, dizziness, weakness, headache, nausea, vomiting, and difficult breathing. Apprehension, anxiety, confusion, nervous excitation; dizziness; headache; numbness and weakness in limbs; muscle twitching, tremors; nausea and vomiting; slow, shallow respiration, bluish face; convulsions; loss of consciousness; breathing stops; death. At fire temperatures can produce coughing, choking, difficult breathing, and cyanosis.

***Points of Attack:*** CNS. May be fatal if inhaled, ingested, or absorbed through the skin

***Medical Surveillance:*** Medical observation is recommended for 24 to 48 hours after breathing overexposure, as pulmonary edema may be delayed. As first aid for pulmonary edema, consider administering a corticosteroid spray. Cigarette smoking may exacerbate pulmonary injury and should be discouraged for at least 72 hours following exposure.

***First Aid:*** Speed in removing material from skin is of extreme importance. Eye contact can cause dangerous amounts of these chemicals to be quickly absorbed through the mucous membrane into the bloodstream. Directly, irrigate with large amounts of plain, tepid water or saline for 20 minutes, occasionally lifting the lower and upper lids. During this time, remove contact lenses, if easily removable without additional trauma to the eye. Get medical aid immediately. Have physician check for possible delayed damage. Skin: Get medical aid. Skin and/or eye contact can cause dangerous amounts of these chemicals to be absorbed into the bloodstream. Wearing the appropriate PPE equipment and respirator for organochlorine pesticides, immediately flush exposed skin, hair, and under nails with plain, running, tepid water for 20 minutes, then wash twice with mild soap. Shampoo hair promptly if contaminated; protect eyes. Do not scrub skin or hair, since this can increase absorption through the skin. Rinse thoroughly with water. Victims who are able and cooperative may assist with their own decontamination. Remove and double-bag contaminated clothing and personal belongings. Leather absorbs many organochlorines; therefore, items such as leather shoes, gloves, and belts should be discarded. If the skin is swollen or inflamed, cool affected areas with cold compresses. Ingestion: Call poison control. Loosen all clothing. Never give anything by mouth to an unconscious person. Get medical aid. Do NOT induce vomiting.* In cases of ingestion, Do not induce vomiting; the patient is at risk of CNS depression or seizures, which may lead to pulmonary aspiration during vomiting. If the victim is conscious and able to swallow, *administer an aqueous slurry of activated charcoal at 1 gm/kg (usual adult dose 60–90 g, child dose 25–50 g). A soda can and straw may be of assistance when offering charcoal to a child. The efficacy of activated charcoal for some organochlorine poisoning (such as chlordane) is uncertain. If victim is *UNCONSCIOUS OR HAVING CONVULSIONS,* do nothing except keep victim warm.

*\*In some cases you may be specifically instructed by Poison Control to induce vomiting by way of 2 tablespoons of syrup of ipecac (adult) washed down with a cup of water. Do NOT give activated charcoal before or with ipecac syrup.* Inhalation: Get medical aid. Do not contaminate yourself. Wearing the appropriate PPE equipment and respirator for organochlorine pesticides, immediately remove the victim from the contaminated area to fresh air. For inhalation exposures, monitor for respiratory distress. If the victim is not breathing, administer artificial respiration. Do NOT use mouth-to-mouth resuscitation; use bag/mask apparatus. If cough or breathing difficulty develops, evaluate for respiratory tract irritation, bronchitis, or pneumonitis. If breathing is difficult, administer 100% humidified supplemental oxygen through bag/mask apparatus until medical help arrives. Do not leave victim unattended.

***References:***
- California Environmental Protection Agency Chemical List of Lists, Sacramento CA (February 1997)

# Azadirachtin

***Use Type:*** Insecticide and nematicide
***CAS Number:*** 11141-17-6
***Formula:*** $C_{35}H_{44}O_{16}$
***Synonyms:*** Azadirachtin A
***Trade Names:*** ALIGN®; AZATIN EC®, Certis USA (USA); AZATIN®-XL PLUS, Certis USA (USA); AZATROL EC®, Pbi/Gordon (USA); AMAZIN®, AMVAC Chemical (USA); ECOZIN®, AMVAC Chemical (USA); EI-783® Pbi/Gordon (USA); MARGOSAN-O®, Certis USA (USA); NEEM®, Certis USA (USA); NEEMAZAL® Coromandel Fertilisers (India); ORNAZIN®, AMVAC Chemical (USA); SALANNIN®; SUPERNEEM®, Certis USA (USA); TURPLEX®
***Producers:*** AMVAC Chemical (USA); Certis USA (USA); Coromandel Fertilisers (India); Pbi/Gordon (USA); Pioneer Enterprise (India); United Agro Industries (India)
***Chemical Class:*** Tetranortriterpenoid
***EPA/OPP PC Code:*** 121701
***California DPR Chemical Code:*** 2328
***Uses:*** Azadirachtin is an extract of fruit from the Neem tree, which is largely grown in India. It is used as a commercial insect growth regulator that controls the metamorphosis process as the insect passes from the larva stage to the pupa stage. The Neem tree also yields extracts from its bark, leaves and wood that are used in medicine and cosmetics.
***Human toxicity (long-term)***[77]***:*** Very low–225.00 ppb, Health Advisory
***Fish toxicity (threshold)***[77]***:*** High–4.46000 ppb, MATC (Maximum Acceptable Toxicant Concentration)

*U.S. Maximum Allowable Residue Levels for* Azadirachtin (40 CFR 180.1119):

**CROP**                                 **ppm**
Raw agricultural commodities     -

*Carcinogen/Hazard Classifications*
**Label Signal Word:** CAUTION or WARNING
*Regulatory Authority:*
- Actively registered pesticide in California.

*Description:* A yellow to amber viscous liquid. Characteristic odor. Molecular weight = 720.69. Melting/Freezing point = 155°C.

*Harmful Effects and Symptoms*
*Short Term Exposure:* Contact with eyes or skin may cause irritation or injury. Inhalation should be avoided; use NIOSH-approved air purifying respirators for pesticides. May be harmful if swallowed.

*First Aid:* If this chemical gets into the eyes, remove any contact lenses at once and irrigate immediately for at least 15 minutes, occasionally lifting upper and lower lids. Seek medical attention immediately. If this chemical contacts the skin, remove contaminated clothing and wash immediately with soap and water. Seek medical attention immediately. If this chemical has been inhaled, remove from exposure, begin rescue breathing (using universal precautions) if breathing has stopped and CPR if heart action has stopped. Transfer promptly to a medical facility. When this chemical has been swallowed, get medical attention. *Do not induce vomiting when formulations containing petroleum solvents are ingested.* Otherwise, give large quantities of water and induce vomiting. Do not make an unconscious person vomit.

*References:*
- EXTOXNET, Extension Toxicology Network, "Pesticide Information Profile, Azadirachtin," Oregon State University, Corvallis, OR. http://extoxnet.orst.edu/pips/azadirac.htm
- California Environmental Protection Agency Chemical List of Lists, Sacramento CA (February 1997).
- U.S. Environmental Protection Agency, Office of Pesticide Programs, Pesticide Residue Limits, "Azadirachtin," 40 CFR 180.1119, http ://www.epa.gov/pesticides/food/viewtols.htm

# Azinphos-ethyl

*Use Type:* Insecticide and miticide
*CAS Number:* 2642-71-9
*Formula:* $C_{12}H_{16}N_3O_3PS_2$
*Alert:* Not registered for use in the United States.
*Synonyms:* Azinfos-ethyl (Dutch); Azinos; Azinphos-aethyl (German); Azinphos etile (Italian); Benzotriazine derivative of an ethyl dithiophosphate; Ethyl azinphos; *O,O*-Diethyl-*S*-(4-oxobezotriazin-3-methyl)-dithiophosphat (German); *O,O*-Diethyl-*S*-((4-*oxo*-3H-1,2,3-bezotriazin-3-yl)-methyl)-dithiophosphat (German); *O,O*-Diethyl-*S*-(4-oxo-3H-1,2,3-bezotriazine-3-yl)methyl)-dithiophosphate; *O,O*-Diethyl-*S*-(4-oxobezotriazino-3-methyl)-phosphorodithioate; *O,O*-Diethylphosphorodithioate-ester with 3-(mercaptomethyl)-1,2,3-benzotriazin-4(3H)-one; *O,O*-Diethyl-*S*-[(4-*oxo*-3H-1,2,3-benzotriazin-3yl)methyl]-dithio fosfaat (Dutch); *O,O*-Dietil-*S*-[(4-*oxo*-3H-1,2,3-bezotriazin-3il)metil]-ditiofosfato (Italian); 3,4-Dihydro-4-*oxo*-3-benzotriazinylmethyl *O,O*-diethyl phosphorodithioate; *S*-(3,4-Dihydro-4-oxo-1,2,3-benzotriazin-3-ylmethyl)*O,O*-diethyl phosphorodithioate; ENT 22,014; Ethyl guthion; Etil azinfos (Spanish); Etiltriazotion; Triazotion (Russian)

*Trade Names:* ATHYL-GUSATHION®; BAY 16225®, Bayer CropScience (Germany); BAYER 16259®, Bayer CropScience (Germany); COTNION-ETHYL®; CRYSTHION®; CRYSTHYON®; GUSATHION A®; GUSATHION A INSECTICIDE®; GUSATHION ETHYL®; GUTHION ETHYL®, Bayer CropScience (Germany), canceled; GUTHION INSECTICIDE®, Bayer CropScience (Germany), canceled; R 1513®

*Producers:* Bayer CropScience (Germany); Makhteshim Agan (Israel); Sigma-Aldrich Laborchemikalien (Germany)
*Chemical Class:* Organophosphate
*EPA/OPP PC Code:* 058002
*California DPR Chemical Code:* 4053
*RTECS Number:* TD8400000
*EINECS Number:* 220-147-6

*Uses:* It is a non-systemic organophosphate insecticide and miticide with good ovicidal properties and long persistence. It is not registered for use in the U.S. Among other crops, it is used on cotton, citrus, vegetables, potatoes, tobacco, rice and cereals to control caterpillars, beetles, aphids, spiders and many other insects.

*Carcinogen/Hazard Classifications*
**Label Signal Word:** DANGER
**WHO Acute Hazard:** Class 1b, highly hazardous
*Regulatory Authority:*
- Banned or Severely Restricted (UN)[35]
- Very Toxic Substance (World Bank)[15]
- AB 1803-Well Monitoring Chemical (CAL)
- U.S. DOT Inhalation Hazard Chemicals as organophosphates
- CERCLA/SARA 40CFR302, Extremely Hazardous Substances: TPQ = 100/10,000 lb (45.4/4,540 kg)
- Superfund/EPCRA 40CFR302.4, Appendix A, RQ: EHS, 1 lb (0.454 kg)
- U.S. DOT Regulated Marine Pollutant (49CFR172.101, Appendix B)

*Description:* Azinphos-ethyl is a colorless crystalline substance. Melting/Freezing point = 53°C. Boiling point = 111°C. Slightly soluble in water.
*Incompatibilities:* Strong oxidizers, strong acids.
*Permissible Exposure Limits in Air:* No standards set. However dusts or mists are poisonous.
*Determination in Air:* OSHA versatile sampler-2;

Toluene/Acetone; Gas chromatography/Flame photometric detection for sulfur, nitrogen, or phosphorus; NIOSH Method IV Method #5600, Organophosphorus pesticides[18].

***Permissible Concentration in Water:*** No criteria set, but runoff from spills or fire control may cause water pollution.

***Routes of Entry:*** Inhalation, ingestion, skin absorption.

***Harmful Effects and Symptoms***

***Short Term Exposure:*** The symptoms are similar to parathion. Nausea is often the first symptom followed by vomiting, abdominal cramps, diarrhea and excessive salivation. Also common in inhalation exposure are headache, giddiness, vertigo and weakness, nasal discharge and a sensation of tightness in the chest. Other symptoms include blurring or dimness of vision; tearing; eye muscle spasm and pain; pinpoint pupils; loss of muscle coordination; slurring of speech; muscle twitching (especially tongue and eyelids); difficulty in breathing; excessive secretions of mucous in mouth, nose, and respiratory tract; convulsions and coma. The systemic effects of this compound are similar to parathion. It is an extremely potent systemic toxicant via ingestion, inhalation and skin contact. It may cause death or permanent injury after very short exposure to small quantities. The $LD_{50}$ (oral, rat ) = 7 mg/kg[9]. A cholinesterase inhibitor. Like similar organic phosphorus poisons, guthion-ethyl may act as an irreversible inhibitor of the enzyme cholinesterase. This enzyme allows the accumulation of large amounts of acetylcholine. Death can be caused when a critical level of cholinesterase depletion is reached. Recovery may be complete when the poisoned victim has had time to recover and regenerate cholinesterase. Delayed pulmonary edema may occur after inhalation.

***Long Term Exposure:*** Cholinesterase inhibitor; cumulative effect is possible. This chemical may damage the nervous system with repeated exposure, resulting in convulsions, respiratory failure. May cause liver damage.

***Points of Attack:*** Respiratory system, lungs, central nervous system, cardiovascular system, skin, eyes, plasma and red blood cell cholinesterase.

***Medical Surveillance:*** Medical observation is recommended for 24 to 48 hours after breathing overexposure, as pulmonary edema may be delayed. Before employment and at regular times after that, the following are recommended: Plasma and red blood cell cholinesterase levels (tests for the enzyme poisoned by this chemical). If exposure stops, plasma levels return to normal in 1-2 weeks while red blood cell levels may be reduced for 1-3 months. When acetylcholinesterase enzyme levels are reduced by 25% or more below preemployment levels, risk of poisoning is increased, even if results are in lower ranges of "normal." Reassignment to work not involving organophosphate pesticides is recommended until enzyme levels recover. If symptoms develop or overexposure occurs, repeat the above tests as soon as possible and get an exam of the nervous system. Also consider complete blood count. Consider chest x-ray following acute overexposure. Do not drink any alcoholic beverages before or during use. Alcohol promotes absorption of organophosphates.

***First Aid:*** **Treatment for organophosphate poisoning consists of thorough decontamination, cardiorespiratory support, and administration of the antidotes atropine and pralidoxime. In cases of severe poisoning, diazepam, an anticonvulsant, should also be administered. Antidotes should be administered as prevention even if the diagnosis is in doubt.** Speed in removing material from eyes and skin is of extreme importance. *Eyes:* Eye contact can cause dangerous amounts of these chemicals to be quickly absorbed through the mucous membrane into the bloodstream. Immediately and gently flush eyes with plenty of warm or cold water (NO hot water) for at least 15 minutes, occasionally lifting the upper and lower eyelids. Get medical aid immediately. *Skin:* Get medical aid. Skin contact can cause dangerous amounts of these chemicals to be absorbed into the bloodstream. Wearing the appropriate PPE equipment and respirator for organophosphate/carbamate pesticides, immediately flush skin with plenty of soap and water for at least 15 minutes while removing contaminated clothing and shoes. Shampoo hair promptly if contaminated. The removed, contaminated clothing and shoes should be double-bagged and left in Hot Zone for later disposal by hazardous materials experts. Skin may also be decontaminated with diluted hypochlorite solution. *Inhalation:* Get medical aid. Do not contaminate yourself. Wearing the appropriate PPE equipment and respirator for organophosphate pesticides, immediately remove the victim from the contaminated area to fresh air. If the victim is not breathing, administer artificial respiration. Do NOT use mouth-to-mouth resuscitation; use bag/mask apparatus. If breathing is difficult, administer oxygen through bag/mask apparatus until medical help arrives. Do not leave victim unattended. *Ingestion:* Call poison control. Loosen all clothing. Never give anything by mouth to an unconscious person. If victim is *unconscious or having convulsions,* do nothing except keep victim warm. Get medical aid. Transfer promptly to a medical facility. In cases of ingestion, **do not induce vomiting**. If the victim is alert and asymptomatic, administer a slurry of activated charcoal at a dose of 1 g/kg (infant, child, and adult dose). A soda can and straw may be of assistance when offering charcoal to a child. *In some cases you may be specifically instructed by poison control to induce vomiting by way of 2 tablespoons of syrup of ipecac (adult) washed down with a cup of water.* Do NOT give activated charcoal before or with ipecac syrup.

*Note to physician or authorized medical personnel:* Treat cases of respiratory compromise, coma, or excessive pulmonary secretions with respiratory support using protocols and techniques available and within the scope of training. Some cases may necessitate procedures such as endotracheal intubation or cricothyrotomy by properly trained and equipped personnel. When possible, atropine (see *Antidotes*, below) should be given under medical supervision. Patients who are comatose, hypotensive, or having seizures or cardiac arrhythmias should be treated according to advanced life support protocols. *Antidotes:* Two antidotes are administered to treat organophosphate poisoning. Atropine is a competitive antagonist of acetylcholine at muscarinic receptors and is used to control the excessive bronchial secretions which are often responsible for death. Pralidoxime relieves both the nicotinic and muscarine effects of organophosphate poisoning by regenerating acetylcholinesterase and can reduce both the bronchial secretions and the muscle weakness associated with poisoning. The initial intravenous dose of atropine in adults should be determined by the severity of symptoms: An initial adult dose of 1.0 to 2.0 mg or pediatric dose of 0.01 mg/kg (minimum 0.01 mg) should be administered intravenously. If intravenous access cannot be established, atropine may also be given intramuscularly, subcutaneously or via endotracheal tube. Doses should be repeated every 15 minutes until excessive secretions and sweating have been controlled. Once bronchial secretion has been controlled, atropine administration should be repeated whenever the secretions begin to recur. In seriously poisoned patients, very large doses may be required. Alterations of pulse rate and pupillary size should not be used as indicators of treatment adequacy. Pralidoxime should be administered as early in poisoning as possible as its efficacy may diminish when given more than 24 to 36 hours after exposure. Doses are as follows: adult 1.0 g; pediatric 25 to 50 mg/kg. The drug should be administered intravenously over 30 to 60 minutes, but in a life-threatening situation, one-half of the total dose can be given per minute for a total administration time of 2 minutes. Treatment should begin to take effect within 40 minutes with a reduction in symptoms and the amount of atropine necessary to control bronchial secretion. The initial dose can be repeated in 1 hour and then every 8 to 12 hours until the patient is clinically well and no longer requires atropine. If intravenous access cannot be established, pralidoxime may also be given intramuscularly. Early administration of diazepam in addition to the combined atropine and pralidoxime treatment may help prevent the onset of seizures and potential brain and cardiac morphologic damage following high-level organophosphate poisoning.

*References:*
- New Jersey Department of Health and Senior Services, "Hazardous Substance Fact Sheet, Azinphos-ethyl," Trenton NJ (August 2002).
  http://www.state.nj.us/health/eoh/rtkweb/2140.pdf
- California Environmental Protection Agency Chemical List of Lists, Sacramento CA (February 1997).
- Agency for Toxic Substances and Disease Registry, U.S. Department of Health and Human Services, Public Health Service, "Managing Hazardous Materials Incidents," Atlanta, GA (June 2003).
- U.S. Environmental Protection Agency, "Chemical Profile: Azinphos-Ethyl," Washington, DC, Chemical Emergency Preparedness Program (November 30, 1987).

## Azinphos-methyl

*Use Type:* Insecticide and acaricide
*CAS Number:* 86-50-0
*Formula:* $C_{10}H_{12}N_3O_3PS_2$
*Alert:* Azinphos-methyl liquids with a concentration greater than 13.5% are classified as Restricted Use Pesticides (RUP).
*Synonyms:* Azinfos-methyl (Dutch); Azinphosmetile (Italian); Benzotriazine derivative of a methyl dithiophosphate; Benzotriazinedithiophosphoric acid dimethoxy ester; DBD; *S*-(3,4-Dihydro-4-oxobenzo[a][1,2,3]triazin-3-ylmethyl) *O,O*-dimethyl phosphorodithioate; *S*-(3,4-Dihydro-4-oxobenzol[d][1,2,3]triazin-3-ylmethyl) *O,O*-dimethyl phosphorodithioate; *S*-(3,4-Dihydro-4-oxo-1,2,3-benzotriazin-3-ylmethyl) *O,O*-dimethyl hosphorodithioate; Dimethoxy ester of (4-*oxo*-1,2,3-benzotriazin-3(4*H*)-yl)methyl ester of dithiophosphoric acid;*O,O*-Dimethyl-*S*-(1,2,3-bezotriazinyl-4-keto) methylphosphorodithioate; *O,O*-Dimethyl-*S*-(3,4-dihydro-4-keto-1,2,3-bezotriazinyl-3-methyl) dithiophosphate; *O,O*-Dimethyl-*S*-(4-oxo-1,2,3-bezotriazin-3(4H)-yl methyl)phosphorodithioate; Dimethyldithiophosphoric acid *N*-methylbenzazimide ester; *O,O*-Dimethyl-*S*-(4-oxo-3H-1,2,3-benzotriazine-3-methyl) phosphorodithioate; *O,O*-Dimethyl-*S*-(4-oxo-benzotriazino-3-methyl) phosphorodithioate; *O,O*-Dimethyl-*S*-(4-oxo-1,2,3-benzotriazino(3)-methyl) thiophosphorodithioate; *O,O*-Dimethyl-*S*-[(4-oxo-3H-1,2,3-benzotriazin-3-yl]-methyl)dithiophosphat (German); *O,O*-Dimethyl-*S*-oxo-1,2,3-benzotriazin-3-(4H)-yl-methyl) phosphodithioate; Ent 23,233; 3-(Mercaptomethyl)-1,2,3-benzotriazin-4(3H)-one-*O,O*-dimethyl phosphorodithioate; 3-(Mercaptomethyl)-1,2,3-benzotriazin-4(3H)-one-*O,O*-dimethyl phosphorodithioate-*S*-ester; Methyl azinphos; *N*-methylbenzazimide, dimethyldithiophosphoric acid ester; Metil azinfos (Spanish); Metiltriazotion (Russian); NCI-C00066
*Trade Names:* ACIFON®, General Quimica (Spain); AZINPHOS-METHYL GUTHION®, Bayer CropScience (Germany); BAY 9027®, Bayer CropScience (Germany); BAYER 17147®, Bayer CropScience (Germany); CARFENE®; COTNION-METHYL®, Bayer CropScience (Germany); CRYSTHION 2L®; CRYSTHYON®; DBD®; GOTHNION®, Bayer CropScience (Germany); GUSATHION®, Bayer CropScience (Germany);

GUSATHION M®, Bayer CropScience (Germany); GUTHION®, Bayer CropScience (Germany); METHYL GUTHION®, Bayer CropScience (Germany; R 1582®
*Producers:* Bayer CropScience (Germany); DuPont (USA); Ehrenstorfer, Dr.(Germany); General Quimica (Spain); Makhteshim Agan (Israel); Rhone-Poulenc Agro France (France); Sigma-Aldrich Laborchemikalien (Germany); Zago Asia Ltd. (Singapore)
*Chemical Class:* Organophosphate
*EPA/OPP PC Code:* 058001
*California DPR Chemical Code:* 314
*ICSC Number:* 0826
*RTECS Number:* TE1925000
*EEC Number:* 015-039-00-9
*EINECS Number:* 220-147-6
*Uses:* Also used as an intermediate in the manufacture, formulation and application of insecticides and acaricides
*Human toxicity (long-term)*[77]: Intermediate–10.50 ppb, Health Advisory
*Fish toxicity (threshold)*[77]: Extra high–0.28636 ppb, MATC (Maximum Acceptable Toxicant Concentration)
*U.S. Maximum Allowable Residue Levels for Azinphos-methyl (40 CFR 180.154):*

| CROP | ppm |
|---|---|
| Alfalfa | 2.0 |
| Alfalfa, hay | 5.0 |
| Almond | 0.2 |
| Almond, hulls | 5.0 |
| Apple | 1.5 |
| Bean, snap, succulent | 2.0 |
| Blackberry | 2.0 |
| Blueberry | 5.0 |
| Boysenberry | 2.0 |
| Broccoli | 2.0 |
| Brussels sprouts | 2.0 |
| Cabbage | 2.0 |
| Cauliflower | 2.0 |
| Celery | 2.0 |
| Cherry | 2.0 |
| Clover | 2.0 |
| Clover, hay | 5.0 |
| Cotton, undelinted seed | 0.5 |
| Crabapples | 1.5 |
| Cranberry | 0.5 |
| Cucumber | 2.0 |
| Eggplant | 0.3 |
| Filbert | 0.3 |
| Fruit, citrus, group 10 | 2.0 |
| Grape | 5.0 |
| Loganberry | 2.0 |
| Melon | 2.0 |
| Onion | 2.0 |
| Parsley, leaves | 5.0 |
| Parsley, roots | 2.0 |
| Peach | 2.0 |
| Pear | 1.5 |
| Pecan | 0.3 |
| Pepper | 0.3 |
| Pistachio | 0.3 |
| Plum, prune, fresh | 2.0 |
| Potato | 0.2 |
| Quince | 1.5 |
| Raspberry | 2.0 |
| Spinach | 2.0 |
| Strawberry | 2.0 |
| Sugarcane, cane | 0.3 |
| Tomato, post-h | 2.0 |
| Trefoil, birdsfoot | 2.0 |
| Trefoil, birdsfoot, hay | 5.0 |
| Walnut | 0.3 |

*Carcinogen/Hazard Classifications*
**U.S. EPA Carcinogens:** Not listed
**U.S. NTP Carcinogen:** Not listed
**California Prop. 65:** Not listed
**IARC:** Not listed
**Label Signal Word:** DANGER and POISON, depending on formulation
**WHO Acute Hazard:** Group 1B, highly hazardous
**Endocrine Disruptor:** Not listed
*Regulatory Authority:*
- Banned or Severely Restricted (various countries) (UN)[13][35]
- Very Toxic Substance (World Bank)[15]
- U.S. DOT Inhalation Hazard Chemicals as organophosphates
- Air Pollutant Standard Set (ACGIH)[1] (DFG)[3] (HSE)[33] (OSHA)[58] (Australia) (Israel) (Mexico) (California)
- Clean Water Act: 40CFR116.4 Hazardous Substances; 40CFR117.3, RQ (same as CERCLA) as guthion
- CERCLA/SARA 40CFR302, Extremely Hazardous Substances: TPQ = 100/10,000 lb (45.4/4,540 kg)
- Superfund/EPCRA 40CFR302.4, Appendix A, RQ: EHS, 1 lb (0.454 kg)
- U.S. DOT Regulated Marine Pollutant (49CFR172.101, Appendix B), severe pollutant
- Canada: Drinking Water MAC
- FIFRA, 40CFR152.175 (RUP)
- (RUP) Liquid formulations ≥13.5% are currently classified as restricted-use chemicals. (Purdue University, National Pesticide Information Retrieval System)
- Permissible Exposure Limits for Chemical Contaminants (CAL)
- The "Director's List" (CAL/OSHA)
- Actively registered pesticide in California.
- EPCRA Section 304 RQ: CERCLA, 1 lb (0.454 kg)
- EPCRA Section 302, Extremely Hazardous Substances: TPQ = 100/10,000 lb (45.4/4,540 kg)

*Description:* Colorless crystalline solid (pure) or brown waxy solid. Solubility in water; solubility = 29 mg/L @ 20°C. Molecular weight = 317.33. Density = 1.44 @ 20°C. Melting/Freezing point = 73°C. Vapor pressure = $2.0 \times 10^{-7}$ mmHg; 0.53 mPa @ 20°C. Log $K_{ow}$ = 2.78. Values of more than 3.0 are likely to bioaccumulate in marine organisms.

*Incompatibilities:* May be dissolved in flammable solvent; containers may explode at elevated temperatures. Keep away from oxidizers, sulfuric acid, caustics, ammonia, aliphatic amines, alkanolamines, isocyanates, alkylene oxides, epichlorohydrin.

*Permissible Exposure Limits in Air:* The Federal NIOSH/OSHA[2] value is 0.2 mg/m³ TWA with the notation "skin" indicating the potential for cutaneous absorption. The ACGIH value is the same, including "skin" notation[1]. The DFG[3] has set a MAK of 0.2 mg/m³ and Peak Limitation of 10 times normal MAK (30 min. average value; do not exceed during workshift). HSE[33] level is 0.2 mg/m³ and STEL value of 0.6 mg/m³. California, Australia, Mexico, and Israel values are 0.2 mg/m³ TWA. Israel set an Action Level of 0.1 mg/m³ TWA. Mexico's STEL is 0.6 mg/m³. Canadian Provincial: Alberta and British Columbia: 0.2 mg/m³ TWA and STEL of 0.6 mg/m³. Ontario and Quebec: 0.2 mg/m³ TWAEV. All regulatory authorities have warnings about skin absorption. The NIOSH[2] IDLH level = 10 mg/m³. Because no useful data on acute inhalation toxicity are available concerning the toxic effects produced by azinphos-methyl, the chosen IDLH has been based on an analogy with parathion, which has an IDLH of 20 mg/m3. (NIOSH[2])

*Determination in Air:* Collection by impinger or fritter bubbler, analysis by gas liquid chromatography. OSHA versatile sampler-2; Toluene/Acetone; Gas chromatography/Flame photometric detection for sulfur, nitrogen, or phosphorus; NIOSH Method (IV) #5600, Organophosphorus pesticides.[18]

*Permissible Concentration in Water:* For the protection of freshwater and marine aquatic life, a criterion of 0.01 µg/L has been suggested by EPA. For the protection of human health, a no-adverse effect level in drinking water has been calculated by NAS/NRC[46] as 0.088 mg/L. An allowable daily intake of 0.0125 mg/kg/day was calculated. Canada's maximum allowable concentration (MAC) in drinking water is 0.02 mg/L. The State of Maine[61] has set a guideline of 25 µg/L for Azinphos-methyl in drinking water.

*Determination in Water:* Pesticide residue methods which should be applicable involve hydrolysis with KOH in isopropanol to give anthranilic acid which is diazotized and coupled to give a measurable color.

*Routes of Entry:* Inhalation, skin absorption, ingestion, skin and/or eye contact.

*Harmful Effects and Symptoms*
Symptoms include nausea, vomiting, diarrhea, excessive salivation, blurring of vision and other signs of cholinesterase inhibition, loss of muscle coordination, twitching of muscles, confusion, difficulty breathing, convulsions, and death are observed with this organophosphate poison. The $LD_{50}$ (oral, rat) = 11 mg/kg. The acute toxicity rating is extremely toxic. Probable oral lethal dose in humans is 5-50 mg/kg. or between 7 drops and 1 teaspoon for a 70 kg (150 lb) person. This is a potent cholinesterase inhibitor which can cause death.

**Lethal concentration data:** Rat 69/1H*; Rat 79/1H**
*Newell & Dilley 1978; **Sanderson 1961.

*Short Term Exposure:* Exposure can cause rapid, fatal organophosphorus poisoning. Inhalation can irritate the lungs causing coughing and/or shortness of breath. Higher exposures can cause pulmonary edema, a medical emergency that can be delayed for several hours. This can cause death. Organic phosphorus insecticides are absorbed by the skin, as well as by the respiratory and gastrointestinal tracts. They are cholinesterase inhibitors. Symptoms of exposure include headache, giddiness, blurred vision, nervousness, weakness, nausea, cramps, diarrhea, and discomfort in the chest. Signs include sweating, tearing, salivation, vomiting, cyanosis, convulsions, coma, loss of reflexes and loss of sphincter control. Delayed pulmonary edema may occur after inhalation.

*Long Term Exposure:* Cholinesterase inhibitor; cumulative effect is possible. This chemical may damage the nervous system causing weakness, "pins and needles," and poor coordination in arms and legs, with repeated exposure, resulting in convulsions, respiratory failure. May cause liver damage. Repeated exposure may cause personality changes of depression, anxiety, or irritability.

*Points of Attack:* Respiratory system, lungs, central nervous system, cardiovascular system, blood cholinesterase.

*Medical Surveillance:* Medical observation is recommended for 24 to 48 hours after breathing overexposure, as pulmonary edema may be delayed. Before employment and at regular times after that, the following are recommended: Plasma and red blood cell cholinesterase levels (tests for the enzyme poisoned by this chemical). If exposure stops, plasma levels return to normal in 1-2 weeks while red blood cell levels may be reduced for 1-3 months. When acetylcholinesterase enzyme levels are reduced by 25% or more below preemployment levels, risk of poisoning is increased, even if results are in lower ranges of "normal." Reassignment to work not involving organophosphate or carbamate pesticides is recommended until enzyme levels recover. If symptoms develop or overexposure occurs, repeat the above tests as soon as possible and get an exam of the nervous system. Also consider complete blood count. Consider chest x-ray following acute overexposure. Do not drink any alcoholic beverages before or during use. Alcohol promotes absorption of organophosphates.

*First Aid:* **Treatment for organophosphate poisoning consists of thorough decontamination, cardiorespiratory** support, and administration of the antidotes atropine and pralidoxime. **In cases of severe poisoning, diazepam, an anticonvulsant, should also be administered. Antidotes should be administered as prevention even if the diagnosis is in doubt.** Speed in removing material from eyes and skin is of extreme importance. *Eyes:* Eye contact can cause dangerous amounts of these chemicals to be quickly absorbed through the mucous membrane into the bloodstream. Immediately and gently flush eyes with plenty of warm or cold water (NO hot water) for at least 15 minutes, occasionally lifting the upper and lower eyelids. Get medical aid immediately. *Skin:* Get medical aid. Skin contact can cause dangerous amounts of these chemicals to be absorbed into the bloodstream. Wearing the appropriate PPE equipment and respirator for organophosphate/carbamate pesticides, immediately flush skin with plenty of soap and water for at least 15 minutes while removing contaminated clothing and shoes. Shampoo hair promptly if contaminated. The removed, contaminated clothing and shoes should be double-bagged and left in Hot Zone for later disposal by hazardous materials experts. Skin may also be decontaminated with diluted hypochlorite solution. *Inhalation:* Get medical aid. Do not contaminate yourself. Wearing the appropriate PPE equipment and respirator for organophosphate pesticides, immediately remove the victim from the contaminated area to fresh air. If the victim is not breathing, administer artificial respiration. Do NOT use mouth-to-mouth resuscitation; use bag/mask apparatus. If breathing is difficult, administer oxygen through bag/mask apparatus until medical help arrives. Do not leave victim unattended. *Ingestion:* Call poison control. Loosen all clothing. Never give anything by mouth to an unconscious person. If victim is *unconscious or having convulsions,* do nothing except keep victim warm. Get medical aid. Transfer promptly to a medical facility. In cases of ingestion, **do not induce vomiting**. If the victim is alert and asymptomatic, administer a slurry of activated charcoal at a dose of 1 g/kg (infant, child, and adult dose). A soda can and straw may be of assistance when offering charcoal to a child. *In some cases you may be specifically instructed by poison control to induce vomiting by way of 2 tablespoons of syrup of ipecac (adult) washed down with a cup of water.* Do NOT give activated charcoal before or with ipecac syrup.

*Note to physician or authorized medical personnel:* Treat cases of respiratory compromise, coma, or excessive pulmonary secretions with respiratory support using protocols and techniques available and within the scope of training. Some cases may necessitate procedures such as endotracheal intubation or cricothyrotomy by properly trained and equipped personnel. When possible, atropine (see *Antidotes*, below) should be given under medical supervision. Patients who are comatose, hypotensive, or having seizures or cardiac arrhythmias should be treated according to advanced life support protocols. *Antidotes:* Two antidotes are administered to treat organophosphate poisoning. Atropine is a competitive antagonist of acetylcholine at muscarinic receptors and is used to control the excessive bronchial secretions which are often responsible for death. Pralidoxime relieves both the nicotinic and muscarine effects of organophosphate poisoning by regenerating acetylcholinesterase and can reduce both the bronchial secretions and the muscle weakness associated with poisoning. The initial intravenous dose of atropine in adults should be determined by the severity of symptoms: An initial adult dose of 1.0 to 2.0 mg or pediatric dose of 0.01 mg/kg (minimum 0.01 mg) should be administered intravenously. If intravenous access cannot be established, atropine may also be given intramuscularly, subcutaneously or via endotracheal tube. Doses should be repeated every 15 minutes until excessive secretions and sweating have been controlled. Once bronchial secretion has been controlled, atropine administration should be repeated whenever the secretions begin to recur. In seriously poisoned patients, very large doses may be required. Alterations of pulse rate and pupillary size should not be used as indicators of treatment adequacy. Pralidoxime should be administered as early in poisoning as possible as its efficacy may diminish when given more than 24 to 36 hours after exposure. Doses are as follows: adult 1.0 g; pediatric 25 to 50 mg/kg. The drug should be administered intravenously over 30 to 60 minutes, but in a life-threatening situation, one-half of the total dose can be given per minute for a total administration time of 2 minutes. Treatment should begin to take effect within 40 minutes with a reduction in symptoms and the amount of atropine necessary to control bronchial secretion. The initial dose can be repeated in 1 hour and then every 8 to 12 hours until the patient is clinically well and no longer requires atropine. If intravenous access cannot be established, pralidoxime may also be given intramuscularly. Early administration of diazepam in addition to the combined atropine and pralidoxime treatment may help prevent the onset of seizures and potential brain and cardiac morphologic damage following high-level organophosphate poisoning.

*References:*
- EPA, Office of Pesticide Programs, "Pesticide Residue Limits, Azinphos-methyl," 40 CFR 180.154, www.epa.gov/cgi-bin/oppsrch
- Sax, N.I., Ed., "Dangerous Properties of Industrial Materials Report" 3, No. 4, 60-65 (1983).
- U.S. Environmental Protection Agency, "Chemical Profile: Azinphos-Methyl," Washington, DC, Chemical
- Emergency Preparedness Program (November 30, 1987).
- New York State Department of Health, "Chemical Fact Sheet: Guthion," Albany, NY, Bureau of Toxic Substance Assessment (March 1, 1986).

- California Environmental Protection Agency Chemical List of Lists, Sacramento CA (February 1997).
- Agency for Toxic Substances and Disease Registry, U.S. Department of Health and Human Services, Public Health Service, "Managing Hazardous Materials Incidents," Atlanta, GA (June 2003).
- New Jersey Department of Health and Senior Services, "Hazardous Substance Fact Sheet: Guthion," Trenton, NJ (May 1999).
  http://www.state.nj.us/health/eoh/rtkweb/0966.pdf

## Azoxystrobin (BSI, ISO )

*Use Type:* Fungicide
*CAS Number:* 131860-33-8
*Synonyms:* Azoksystrobin; Azoxistrobin; Azoxystrolin
*Trade Names:* ABOUND®, Syngenta (Switzerland); AMISTAR®, Syngenta (Switzerland); BANKIT®; HERITAGE®, Syngenta (Switzerland); ICIA5504 80WG®, Syngenta (Switzerland); PROTEGE®, Syngenta (Switzerland); PROTEGE-ALLEGIANCE WP®, Gustafson (USA); PROTEGE-FL SEED APPLIED FUNGICIDE®, Gustafson (USA); QUADRIS OPTI®, Syngenta (Switzerland); QUILT®, Syngenta (Switzerland); SOYGARD WITH PROTEGE®, Gustafson (USA)
*Producers:* Gustafson (USA); Syngenta (Switzerland)
*Chemical Class:* Strobin
*EPA/OPP PC Code:* 128810
*California DPR Chemical Code:* 4037
*Uses:* Azoxystrobin is a systemic, broad-spectrum fungicide that was first introduced in 1998. It inhibits spore germination and is used on grape vines, cereals, potatoes, apples, bananas, citrus, tomatoes and other crops. Largest crop uses in California are on almonds, rice, pistachios, wine grapes, raisins and garlic. Among the diseases it controls are rusts, downey and powdery mildew, rice blast and apple scab.
*Human toxicity (long-term)*[77]*:* Very low–1260.00 ppb, Health Advisory
*Fish toxicity (threshold)*[77]*:* Intermediate–54.86655 ppb, MATC (Maximum Acceptable Toxicant Concentration)
*Carcinogen/Hazard Classifications*
**Label Signal Word:** CAUTION or WARNING
*U.S. Maximum Allowable Residue Levels for Azoxystrobin (40 CFR 180.507):*

| CROP | ppm |
|---|---|
| Acerola | 2 |
| Almond, hulls | 4 |
| Artichoke, globe | 4 |
| Asparagus | 0.04 |
| Atemoya | 2 |
| Avocado | 2 |
| Banana | 2 |
| Banana, pulp | 0.1 |
| Barley, bran | 0.2 |
| Barley, grain | 0.1 |
| Barley, hay | 15 |
| Barley, straw | 4 |
| Biriba | 2 |
| Brassica, head and stem, subgroup 5a | 3 |
| Brassica, leafy greens, subgroup 5b | 25 |
| Bushberry, subgroup 13b | 3 |
| Caneberry, subgroup 13a | 5 |
| Canistel | 2 |
| Canola, seed | 1 |
| Cattle, mbyp | 0.07 |
| Cherimoya | 2 |
| Chickpea, seed | 0.5 |
| Citrus, dried pulp | 2 |
| Citrus, oil | 4 |
| Coriander, leaves | 30 |
| Corn, field, forage | 12 |
| Corn, field, grain | 0.05 |
| Corn, field, refined oil | 0.3 |
| Corn, field, stover | 25 |
| Corn, pop, grain | 0.05 |
| Corn, pop, stover | 25 |
| Corn, sweet, forage | 12 |
| Corn, sweet, kernel plus cob with husks removed | 0.05 |
| Corn, sweet, stover | 25 |
| Cotton, gin byproducts | 0.02 |
| Cotton, undelinted seed | 0.02 |
| Cranberry | 0.5 |
| Custard apple | 2 |
| Eggplant | 2 |
| Feijoa | 2 |
| Fruit, citrus, group 10 | 1 |
| Fruit, stone, except cherry | 1.5 |
| Goat, mbyp | 0.07 |
| Grape | 1 |
| Grass, forage | 15 |
| Grass, hay | 20 |
| Guava | 2 |
| Herb subgroup 19a, dried, except chive | 260 |
| Herb subgroup 19a, fresh, Except chive | 50 |
| Hog, mbyp | 0.01 |
| Hop, dried cones | 20 |
| Horse, mbyp | 0.07 |
| Ilama | 2 |
| Jaboticaba | 2 |
| Jackfruit | 2 |
| Juneberry | 3 |
| Lingonberry | 3 |
| Longan | 2 |
| Loquat | 2 |

| | |
|---|---|
| Lychee | 2 |
| Mango | 2 |
| Nut, tree, group 14 | 0.02 |
| Okra | 2 |
| Onion, dry bulb | 1 |
| Onion, green | 7.5 |
| Papaya | 2 |
| Passionfruit | 2 |
| Pawpaw | 2 |
| Pea and bean, dried shelled, except soybean, subgroup 6c | 0.5 |
| Pea and bean, succulent shelled, subgroup 6b | 0.5 |
| Peanut | 0.2 |
| Peanut, hay | 15 |
| Peanut, refined oil | 0.6 |
| Pecan | 0.01 |
| Pepper | 2 |
| Peppermint, tops | 30 |
| Persimmon | 2 |
| Pistachio | 0.5 |
| Potato | 0.03 |
| Pulasan | 2 |
| Rambutan | 2 |
| Safflower, seed | 1 |
| Salal | 3 |
| Sapodilla | 2 |
| Sapote, black | 2 |
| Sapote, mamey | 2 |
| Sapote, white | 2 |
| Sheep, mbyp | 0.07 |
| Soursop | 2 |
| Soybean, forage | 25 |
| Soybean, hulls | 1 |
| Soybean, seed | 0.5 |
| Spanish lime | 2 |
| Spearmint, tops | 30 |
| Star apple | 2 |
| Starfruit | 2 |
| Sugar apple | 2 |
| Tamarind | 2 |
| Tomato | 0.2 |
| Tomato, paste | 0.6 |
| Turnip, greens | 25 |
| Vegetable, leafy, except brassica, group 4 | 30 |
| Vegetable, leaves of root and tuber, group 2 | 50 |
| Vegetable, legume, edible podded, subgroup 6a | 3 |
| Vegetable, root, subgroup 1a | 0.5 |
| Vegetable, tuberous and corm, subgroup 1c | 0.03 |
| Watercress | 3 |
| Wax jambu | 2 |
| Wheat, bran | 0.2 |
| Wheat, grain | 0.1 |
| Wheat, hay | 15 |
| Wheat, straw | 4 |

***Carcinogen/Hazard Classifications***
**Label Signal Word:** CAUTION or WARNING
**WHO Acute Hazard:** Class III, slightly hazardous
***Routes of Entry:*** Skin contact and inhalation
***Harmful Effects and Symptoms***
***Short Term Exposure:*** Contact with eyes or skin may cause irritation or injury. Inhalation should be avoided; use NIOSH-approved air purifying respirators for pesticides. May be harmful if swallowed. May cause skin sensitization.
***First Aid:*** If this chemical gets into the eyes, remove any contact lenses at once and irrigate immediately for at least 15 minutes, occasionally lifting upper and lower lids. Seek medical attention immediately. If this chemical contacts the skin, remove contaminated clothing and wash immediately with soap and water. Seek medical attention immediately. If this chemical has been inhaled, remove from exposure, begin rescue breathing (using universal precautions) if breathing has stopped and CPR if heart action has stopped. Transfer promptly to a medical facility. When this chemical has been swallowed, get medical attention. *Do not induce vomiting when formulations containing petroleum solvents are ingested.* Otherwise, give large quantities of water and induce vomiting. Do not make an unconscious person vomit.
***References:***
- Pesticide Management Education Program, "Azoxystrobin," Cornell University, Ithaca, NY (October 1997). http://pmep.cce.cornell.edu/profiles/extoxnet/24d-captan/azoxystrobin-ext.html
- California Environmental Protection Agency Chemical List of Lists, Sacramento CA (February 1997).
- U.S. Environmental Protection Agency, Office of Pesticide Programs, Pesticide Residue Limits, "Azoxystrobin," 40 CFR 180.507. http://www.epa.gov/pesticides/food/viewtols.htm

# B

## Barban (ANSI)

*Use Type:* Herbicide
*CAS Number:* 101-27-9
*Formula:* $C_{11}H_9Cl_2NO_2$
*Synonyms:* 2-Butynyl-4-chloro-*m*-chlorocarbanilate; Carbamic acid, 3-chlorophenyl-, 4-chloro-2-butynyl ester; Carbanilic acid, 3-chloro-, 4-chloro-2-butynyl ester; 4-Chloro-2-butynyl-(3-chlorophenyl)carbamate; CBN; (4-Chlor-but-2-in-yl)-*n*-(3-chlor-phenyl)-carbamat (German); Chlorinat; Chloro-2-butynyl-*m*-chlorocarbamate; 4-Chlorobut-2-ynyl-*m*-chlorocarbanilate; 4-Chloro-2-butynyl-*m*-chlorocarbanilate; 4-Chlorobut-2-ynyl-3-chlorophenylcarbamate; 4-Chloro-2-butynyl-*n*-(3-chlorophenyl)carbamate; *m*-Chloro carbanilic acid-4-chloro-2-butynyl ester; *n*-(3-Chlorophenyl)carbamate de 4-chloro 2-butynyle (French); (3-Chlorophenyl)carbamic acid 4-chloro-2-butynyl ester
*Trade Names:* A-980®; BARBAMATE®; BARBANE®; CARBIN®; CARBYNE®, Velsicol Chemical (USA) canceled 4/25/1986; CARYNE®; CS-847®; FISONS B25®; NEOBAN®; S-847®
*Producers:* Fisons, Ltd; Spencer Chemical (USA); Syngenta (Switzerland); Velsicol Chemical (USA)
*Chemical Class:* Carbamate
*EPA/OPP PC Code:* 017601
*California DPR Chemical Code:* 55
*EINECS Number:* 202-930-4
*Uses:* Not registered in the U.S. Used for post-emergence control of grasses and wild oats in barley, lentils, peas, lentils, sugar beets, wheat, sunflower and flax crops.
*Fish toxicity (threshold)*[77]: Low–168.01099 ppb, MATC (Maximum Acceptable Toxicant Concentration)
*Carcinogen/Hazard Classifications*
**Label Signal Word:** CAUTION
*Regulatory Authority:*
- The "Director's List" (CAL/OSHA)
- AB 1803 Well Monitoring Chemicals (CAL)
- RCRA Universal Treatment Standards: Wastewater (mg/L), 0.056; Nonwastewater (mg/kg), 1.4

*Description:* A colorless, crystalline solid. Practically insoluble in water. Molecular weight = 258.11. Melting/Freezing point = 75°C. Vapor pressure = $3.8 \times 10^{-7}$ mmHg.
*Incompatibilities:* May form explosive materials with phosphorus pentachloride. Contact with alkalis forms chlorine gas. Contact with acids forms β-chloroacrylic acid. When heated to decomposition, emits nitrogen oxides and chlorine fumes.
**Permissible Concentration in Water:** No criteria set. Runoff from spills or fire control may cause water pollution.
*Routes of Entry:* Skin absorption, ingestion and inhalation.
*Harmful Effects and Symptoms*
*Short Term Exposure:* Can cause contact dermatitis. Eye pupils are small; blurred vision; eye watering; runny nose; cough; shortness of breath; salivation; nausea, stomach cramps, diarrhea, and vomiting; increased blood pressure; profuse sweating; hypermotility, hallucinations; agitation; tingling of the skin; slow heartbeat; convulsions; fluid in lungs; loss of consciousness; incontinence; breathing stops; death. Carbamate insecticides inhibit the cholinesterase activity of enzymes, causing accumulation of acetylcholine at synapses and altering the way in which nervous impulses are transmitted. However, within several hours carbamates spontaneously detach from the enzymes.
*Long Term Exposure:* A potent cholinesterase inhibitor; cumulative effect is possible. This chemical may damage the nervous system with repeated exposure, resulting in convulsions, respiratory failure. May cause liver damage.
*Points of Attack:* Respiratory system, lungs, central nervous system, cardiovascular system, skin, eyes, plasma and red blood cell cholinesterase.
*Medical Surveillance:* Medical observation is recommended for 24 to 48 hours after breathing overexposure, as pulmonary edema may be delayed. As first aid for pulmonary edema, consider administering a corticosteroid spray. Cigarette smoking may exacerbate pulmonary injury and should be discouraged for at least 72 hours following exposure. Before employment and at regular times after that, the following are recommended: Plasma and red blood cell cholinesterase levels (tests for the enzyme poisoned by this chemical). If exposure stops, plasma levels return to normal in 1-2 weeks while red blood cell levels may be reduced for 1-3 months. When acetylcholinesterase enzyme levels are reduced by 25% or more below preemployment levels, risk of poisoning is increased, even if results are in lower ranges of "normal." Reassignment to work not involving carbamate pesticides is recommended until enzyme levels recover. If symptoms develop or overexposure occurs, repeat the above tests as soon as possible and get an exam of the nervous system. Also consider complete blood count. Consider chest x-ray following acute overexposure
*First Aid:* Speed in removing material from eyes and skin is of extreme importance. *Eyes:* Eye contact can cause dangerous amounts of these chemicals to be quickly absorbed through the mucous membrane into the bloodstream. Immediately and gently flush eyes with plenty of warm or cold water (NO hot water) for at least 15

minutes, occasionally lifting the upper and lower eyelids. Get medical aid immediately. *Skin:* Get medical aid. Skin contact can cause dangerous amounts of these chemicals to be absorbed into the bloodstream. Wearing the appropriate PPE equipment and respirator for carbamate pesticides, immediately flush skin with plenty of soap and water for at least 15 minutes while removing contaminated clothing and shoes. Shampoo hair promptly if contaminated; protect eyes. *Ingestion:* Call poison control. Loosen all clothing. Never give anything by mouth to an unconscious person. Get medical aid. Do NOT induce vomiting.* If conscious, alert, and able to swallow, rinse mouth and have victim drink 4 to 8 ounces of water. Check to see if poison control instructs you to use ipecac syrup, otherwise administer slurry of activated charcoal (2 oz in 8 oz of water). If victim is *UNCONSCIOUS OR HAVING CONVULSIONS*, do nothing except keep victim warm. **In some cases you may be specifically instructed by poison control to induce vomiting by way of 2 tablespoons of syrup of ipecac (adult) washed down with a cup of water.* Do NOT give activated charcoal before or with ipecac syrup. *Inhalation:* Get medical aid. Do not contaminate yourself. Wearing the appropriate PPE equipment and respirator for carbamate pesticides, immediately remove the victim from the contaminated area to fresh air. If the victim is not breathing, administer artificial respiration. Do NOT use mouth-to-mouth resuscitation; use bag/mask apparatus. If breathing is difficult, administer oxygen through bag/mask apparatus until medical help arrives. Do not leave victim unattended. *Note to physician or authorized medical personnel.* Administer atropine, 2 mg (1/30 gr) intramuscularly or intravenously as soon as any local or systemic signs or symptoms of an intoxication are noted; repeat the administration of atropine every 3 to 8 minutes until signs of atropinization (mydriasis, dry mouth, rapid pulse, hot and dry skin) occur; initiate treatment in children with 0.05 mg mg/kg of atropine; repeat at 5 to 10 minute intervals. Watch respiration, and remove bronchial secretions if they appear to be obstructing the airway; intubate if necessary. *Medical note:* 2-PAMCI may be contraindicated in the case of some carbamate poisonings.

*References:*
- California Environmental Protection Agency Chemical List of Lists, Sacramento CA (February 1997)

# Barium and Barium compounds

*Use Type:* Rodenticide and used in making other pesticides.
*CAS Number:* 7440-39-3
*Formula:* Ba
*Synonyms:* Bario (Spanish); Barium, elemental; Barium metal
*Producers:* Aisonschem (China); Ashland Specialty Chemical (USA); Alquimia Mexicana (Mexico); ATOFINA (France); Barium & Chemicals (USA); Bayer (Germany); Chemical Products (USA); Clariant (Switzerland); Degussa-Huls (Germany); Gayatri Minerals & Chemicals (India); GFS Chemicals (USA); Merck (Germany); Solvay Barium Strontium (Germany)
*Chemical Class:* Inorganic metal
*ICSC Number:* 1052
*RTECS Number:* CQ8370000
*EINECS Number:* 231-149-1
*Uses:* Metallic barium is used for removal of residual gas in vacuum tubes and in alloys with nickel, lead, calcium, magnesium, sodium, and lithium. Barium compounds are used in the manufacture of lithopone (a white pigment in paints), chlorine, sodium hydroxide, valves, and green flares; in synthetic rubber vulcanization, x-ray diagnostic work, glassmaking, papermaking, beet-sugar purification, animal and vegetable oil refining. They are used in the brick and tile, pyrotechnics, and electronics industries. They are found in lubricants, pesticides, glazes, textile dyes and finishes, pharmaceuticals, and in cements which will be exposed to saltwater; and barium is used as a rodenticide, a flux for magnesium alloys, a stabilizer and mold lubricant in the rubber and plastics industries, an extender in paints, a loader for paper, soap, rubber, and linoleum, and as a fire extinguisher for uranium or plutonium fires.

*Regulatory Authority:*
- Air Pollutant Standard Set (ACGIH)[1] (HSE)[33] (DFG)[3] (OSHA)[58] (Several States)[60] (Australia) (Israel) (Mexico)
- Water Pollution Standards Set (U.S. EPA)[49] (former USSR)[43] (Several States)[61] (Canada)(Mexico)
- AB 1803-Well Monitoring Chemical (CAL)
- MCL(Maximum Contaminants Levels) list of contaminants (CAL)
- AB 2588-Air Toxics "Hot Spots" Chemicals (CAL)
- Permissible Exposure Limits for Chemical Contaminants (CAL/OSHA)
- The "Director's List" (CAL/OSHA)
- EPA Hazardous Waste Number (RCRA No.): D005
- RCRA Toxicity Characteristic (Section 261.24), Maximum Concentration of Contaminants, regulatory level, 100.0 mg/L
- RCRA, 40CFR261, Appendix 8 Hazardous Constituents, waste number not listed.
- RCRA Maximum Concentration Limit for Ground Water Protection (40 CFR/264.94), 1.0 mg/L
- RCRA 40CFR268.48; 61FR15654, Universal Treatment Standards: Wastewater (mg/L), 1.2; Nonwastewater (mg/L), 7.6 TCLP
- RCRA 40CFR264, Appendix 9; TSD Facilities Ground Water Monitoring List, Suggested methods (PQL *ug*/L): 6010(20); 7080(1000)
- Safe Drinking Water Act: MCL, 1 mg/L; MCLG, 1 mg/L; Regulated chemical (47 FR 9352)

- EPCRA Section 313 Form R *de minimus* concentration reporting level: 1.0%.
- Canada MAC for drinking water quality: 1.0 mg/L
- Mexico, Drinking Water 1.0 mg/L

*Description:* Barium, Ba, a flammable, silver white or yellowish metal in various forms including powder. Barium may ignite spontaneously in air in the presence of moisture, evolving hydrogen. Density = 3.6. Melting/Freezing point = 725°C. Boiling point = 1640°C. Hazard Identification (based on NFPA-704 M Rating System): Health 1, Flammability 4, Reactivity 3. The primary sources are the minerals barite ($BaSO_4$) and witherite ($BaCO_3$).

*Incompatibilities:* Barium powder may spontaneously ignite on contact with air. It is a strong reducing agent and reacts violently with oxidizers and acids. Reacts violently with fire extinguishing agents including water, bicarbonate, powder, foam, and carbon dioxide. The reaction with water, forms combustible hydrogen gas and barium hydroxide. Reacts violently with halogenated hydrocarbon solvents, causing fire and explosion hazard.

*Permissible Exposure Limits in Air:* The Federal OSHA[2] standard and ACGIH[1] recommended airborne limit for soluble barium compounds is 0.5 mg/m³ TWA for an 8-hour workshift[1]. The NIOSH[2] level is the same for a 10-hour workshift. The DFG[3], HSE[33], Australia, Mexico, and Israel have adopted this same value and DFG has a Peak Limitation (5 min.) of 2 times normal MAK[3]; do not exceed more than 8 times during workshift. Israel's Action Level is one half the TWA. The NIOSH[2] IDLH level = 50 mg/m³ (Ba, soluble compounds). In addition, several states have set guidelines or standards for barium in ambient air[60] ranging from 0.67 μg/m³ (New York) to 5.0 μg/m³ (Florida and North Dakota) to 8.0 μg/m³ (Virginia) to 10.0 μg/m³ (Connecticut) to 12.0 μg/m³ (Nevada).

*Determination in Air:* Filter; Water; Flame atomic absorption spectrometry; NIOSH Methods (IV) #7056, Barium, soluble compounds. Collection on a cellulose membrane filter, workup with hot water, analysis by atomic absorption. See NIOSH Method #8310[18].

*Permissible Concentration in Water:* EPA allows 2 parts of barium per million parts of drinking water (2 ppm). See Regulatory Authority for Canadian and Mexican levels. The former USSR-UNEP/IRPTC project[43] has set a MAC of 4.0 mg/L in water bodies used for domestic purposes. Also, these states have set standards for barium in drinking water[61]: a standard of 100 μg/L in Massachusetts and guidelines of 1000 μg/L in Maine and 1500 μg/L in Minnesota.

*Determination in Water:* Conventional flame atomization does not have sufficient sensitivity to determine barium in most water samples; however, a barium detection limit of 10 μg/L can be achieved, if a nitrous oxide flame is used. A concentration procedure for barium uses thenoyltrifluoro-acetone-methylisobutylketone extraction at a pH of 6.8. With a tantalum liner insert, the barium detection limit of the flameless atomic absorption procedure can be improved to 0.1 μg/L according to NAS/NRC[46].

*Routes of Entry:* Ingestion or inhalation of dust or fume, skin and/or eye contact

*Harmful Effects and Symptoms*

*Short Term Exposure:* Alkaline barium compounds, such as the hydroxide and carbonate, may cause local irritation to the eyes, nose, throat, and skin.

*Long Term Exposure:* Barium poisoning is virtually unknown in industry, although the potential exists when the soluble forms are used. When ingested or given orally, the soluble, ionized barium compounds exert a profound effect on all muscles and especially smooth muscles, markedly increasing their contractility. The heart rate is slowed and may stop in systole. Other effects are increased intestinal peristalsis, vascular constriction, bladder contraction, and increased voluntary muscle tension. The inhalation of the dust of barium sulfate may lead to deposition in the lungs in sufficient quantities of produce "baritosis" (a benign pneumoconiosis). This produces a radiologic picture in the absence of symptoms and abnormal physical signs. x-rays, however, will show disseminated nodular opacities throughout the lung fields, which are discrete, but sometimes overlap.

*Points of Attack:* Heart, lungs, central nervous system, skin, respiratory system and eyes.

*Medical Surveillance:* Consideration should be given to the skin, eye, heart, and lung in any placement or periodic examination.

*First Aid:* If a soluble barium compound gets into the eyes, remove any contact lenses at once and irrigate immediately. If a soluble barium compound contacts the skin, flush with water immediately. If a person breathes in large amounts of a soluble barium compound, move the exposed person to fresh air at once and perform artificial respiration. When a soluble barium compound has been swallowed, get medical attention. Give large quantities of water and induce vomiting. Do not make an unconscious person vomit.

*References:*
- Agency for Toxic Substances and Disease Registry, "ToxFAQs for Barium and Barium Compounds," Atlanta, GA (September 1995). http://www.atsdr.cdc.gov/tfacts24.html
- U.S. Environmental Protection Agency, Toxicology of Metals, Vol. 2: Barium, pp 71-84, Report EPA-600/1-77-022, Research Triangle Park, NC (May 1977).
- U.S. Environmental Protection Agency, Barium, Health and Environmental Effects Profile No. 13, Washington, DC, Office of Solid Waste (April 30, 1980).
- Sax, N. I., Ed., "Dangerous Properties of Industrial Materials Report" 1, No. 7, 35-40 (1981) and 3, No. 4, 29-30 (1983).

- California Environmental Protection Agency Chemical List of Lists, Sacramento CA (February 1997).
- New Jersey Department of Health and Senior Services, "Hazardous Substance Fact Sheet: Barium," Trenton, NJ (January, 1996, rev. September, 2000). http://www.state.nj.us/health/eoh/rtkweb/0180.pdf

# Benazolin Ethyl

*Use Type:* Herbicide
*CAS Number:* 25059-80-7
*Synonyms:* Ethyl 4-chloro-2-*oxo*-3(2*H*)-benzothiazoleacetate; 4-Chloro-2-*oxo*-3(2*H*)-benzothiazoleacetic acid, ethyl ester; 3(2*H*)-Benzothiazoleacetic acid, 4-chloro-2-*oxo*-, ethyl ester
*Producers:* Anhui Huaxing Chemical Industry Co., Ltd. (China); SuYan Agrochemical Group (China)
*EPA/OPP PC Code:* 126801
*Uses:* Not registered in the U.S.
*Carcinogen/Hazard Classifications*
**WHO Acute Hazard:** Class U, unlikely to be hazardous (as benazolin)
*Incompatibilities:* Acetates are generally incompatible with nitrates. Moisture may cause hydrolysis or other forms of decomposition
*Permissible Concentration in Water:* No criteria set. Runoff from spills or fire control may cause water pollution.
*Harmful Effects and Symptoms*
*Short Term Exposure:* Contact with eyes or skin may cause irritation or injury. Inhalation should be avoided; use NIOSH-approved air purifying respirators for pesticides. May be harmful if swallowed.
*First Aid:* If this chemical gets into the eyes, remove any contact lenses at once and irrigate immediately for at least 15 minutes, occasionally lifting upper and lower lids. Seek medical attention immediately. If this chemical contacts the skin, remove contaminated clothing and wash immediately with soap and water. Seek medical attention immediately. If this chemical has been inhaled, remove from exposure, begin rescue breathing (using universal precautions) if breathing has stopped and CPR if heart action has stopped. Transfer promptly to a medical facility. When this chemical has been swallowed, get medical attention. *Do not induce vomiting when formulations containing petroleum solvents are ingested.* Otherwise, give large quantities of water and induce vomiting. Do not make an unconscious person vomit.

# Bendiocarb (ANSI)

*Use Type:* Insecticide
*CAS Number:* 22781-23-3
*Formula:* $C_{11}H_{13}NO_4$
*Alert:* The use of all bendiocarb products in the United States was canceled at the end of December, 2001, because of the potential of acute and subacute toxicity and cholinesterase inhibition.
*Synonyms:* AI3-27695; Bencarbate; Bendiocarbe; 1,3-Benzodioxole, 2,2-dimethyl-1,3-benzodioxol-4-ol methylcarbamate; 2,2-Dimethyl-4-(*N*-methylaminocarboxylato)-; 1,3-Benzodioxole, 2,2-dimethyl-4-(*N*-methylcarbamato)-; 1,3-Benzodioxol-4-ol, 2,2-dimethyl-,methylcrbamate; Bicam ULV; Carbamic acid, methyl-, 2,3-(isopropylidenedioxy)phenyl ester; Carbamic acid, methyl-, 2,3-(dimethylmethylenedioxy)phenyl ester; 2,2-Dimethyl-1,3-benzodioxol-4-yl-*N*-methylcarbamate; 2,2-Dimethylbenzo-1,3-benzodioxol-4-yl-*N*-methylcarbamate; 2,2-Dimethyl-4-(*N*-methylaminocarboxylato)-1,3-benxodioxole; 2,2-Dimethylbenzo-1,3-dioxol-4-yl methylcarbamate; 2,3-Isopropylidene-dioxyphenyl methylcarbamate; Methylcarbamic acid 2,3-(isopropylidenedioxy)phenyl ester
*Trade Names:* BENCARBATE®; DYCARB®, Scotts Company (USA), canceled; FICAM®, Bayer CropScience (USA), canceled 10/31/2000; FICAM 80W®; FICAM ULV®; FUAM®; GARVOX®, AgrEvo France; MC 6897®; MULTAMAT®; MULTIMET®; NC 6897®; NIOMIL®; OMS-1394®; ROTATE®; SEEDOX®; TATTOO®; TURCAM®
*Producers:* BASF (Germany); Bayer CropScience (Germany); NOR-AM Chemical (USA), see Bayer CropScience (Germany); Roussel Uclaf (France); Sigma-Aldrich Laborchemikalien (Germany); Zago Asia Ltd. (Singapore)
*Chemical Class:* *N*-Methyl Carbamate
*EPA/OPP PC Code:* 105201
*California DPR Chemical Code:* 1924
*RTECS Number:* FC1140000
*Uses:* Used against pests in the home, industrial sites and food storage sites; in agriculture as a seed treatment and as a foliar spray. It is used against insects in the soil, such as snails and slugs.
*Human toxicity (long-term)[77]:* Intermediate–35.00 ppb, Health Advisory
*Fish toxicity (threshold)[77]:* Low–379.47332 ppb, MATC (Maximum Acceptable Toxicant Concentration)
*U.S. Maximum Allowable Residue Levels for Bendiocarb (40 CFR 180.530):*

| CHEM. | CROP | ppm |
|---|---|---|
| Bendiocarb | ANIMAL FEED | – |
| Bendiocarb | PROCESSED FOOD | – |

*Carcinogen/Hazard Classifications*
**U.S. EPA Carcinogens:** Not likely carcinogenic.
**TRI Developmental Toxin:** Reproductive and developmental toxin.
**WHO Acute Hazard:** Class II, moderately hazardous
*Regulatory Authority:*
- EPA Hazardous Waste Number (RCRA No.): U278

- RCRA, 40CFR261, Appendix 8 Hazardous Constituents.
- Actively registered pesticide in California.
- RCRA 40CFR268.48; 61FR15654, UniversalTreatment Standards: Wastewater (mg/L), 0.056; Nonwastewater (mg/kg), 1.4
- EPCRA Section 313 Form R *de minimus* concentration reporting level: 1.0%.
- U.S. DOT 49CFR172.101, Appendix B, Regulated marine pollutant
- Canada: Drinking Water Quality Level Set

*Description:* Bendiocarb is a white odorless crystalline powder. Melting/Freezing point = 129–130°C. Slightly soluble in water.

*Incompatibilities:* Keep away from flammable materials and sources of heat and flame. Should not be mixed with alkaline preparations.

*Permissible Exposure Limits in Air:* No standards set.

*Permissible Concentration in Water:* Canada's Drinking Water Quality is 0.04 mg/L MAC.

*Routes of Entry:* Inhalation, skin, contact and ingestion.

*Harmful Effects and Symptoms*

*Short Term Exposure:* Bendiocarb is a toxic carbamate chemical. Bendiocarb can affect you when inhaled. Exposure can cause rapid poisoning, with headaches, sweating, nausea and vomiting, diarrhea, loss of concentration and death. Eye contact can cause irritation and blurred vision.

*Medical Surveillance:* Before starting work, at regular times after that, and if symptoms develop or over exposure occurs, the following is recommended: Serum and RBC cholinesterase levels (a test for the body substance affected by Bendiocarb). For this substance, these tests are accurate only if done within about two hours of exposure, and can return to normal before the person feels well.

*First Aid:* If this chemical gets into the eyes, remove any contact lenses at once and irrigate immediately for at least 15 minutes, occasionally lifting upper and lower lids. Seek medical attention immediately. If this chemical contacts the skin, remove contaminated clothing and wash immediately with soap and water. Shampoo hair. Seek medical attention immediately. If this chemical has been inhaled, remove from exposure, begin rescue breathing (using universal precautions) if breathing has stopped and CPR if heart action has stopped. Transfer promptly to a medical facility. When this chemical has been swallowed, get medical attention. Give large quantities of water and induce vomiting. Do not make an unconscious person vomit.

*References:*
- U.S. Environmental Protection Agency, Office of Pesticide Programs, Pesticide Residue Limits, "Bendiocarb," 40 CFR 180.530, www.epa.gov/pesticides/food/viewtols.htm
- New Jersey Department of Health and Senior Services, "Hazardous Substance Fact Sheet: Bendiocarb," Trenton, NJ (August 1987, rev. April, 1997). http://www.state.nj.us/health/eoh/rtkweb/0191.pdf
- California Environmental Protection Agency Chemical List of Lists, Sacramento CA (February 1997).
- EXTOXNET, Extension Toxicology Nerwork, "Pesticide Information Profiles, Bendiocarb," Oregon State University, Corvallis, OR (June 1996). http://ace.ace.orst.edu/info/extoxnet/pips/bendioca.htm

# Benefin

*Use Type:* Herbicide
*CAS Number:* 1861-40-1
*Formula:* $C_{13}H_{16}F_3N_3O_4$
*Synonyms:* Benfluralin; Benfluraline; Benzenamine, *N*-Butyl-*N*-ethyl-2,6-dinitro-4-(trifluoromethyl)-; *N*-Butyl-2,6-dinitro-*N*-ethyl-4-trifluoromethylaniline; *N*-Butyl-*N*-ethyl-2,6-dinitro-4-trifluoromethylbenzenamine; *N*-Butyl-*N*-ethyl-2,6-dinitro-4-(trifluromethyl)benzeneamine; *N*-Butyl-*N*-ethyl-2,6-dinitro-4-trifluoromethylaniline; *N*-Butyl-*N*-ethyl-$\alpha,\alpha,\alpha$-trifluoro-2,6-dinitro-*p*-toluidine; Caswell No. 130; *p*-Toluidine,*N*-butyl-*N*-ethyl-$\alpha,\alpha,\alpha$-trifluoro-2,6-dinitro-; $\alpha,\alpha,\alpha$-Trifluoro-2,6-dinitro-*N*,*N*-ethylbutyl-*p*-toluidine

*Trade Names:* BALAN®, Dow AgroSciences (USA); BALFIN®; BENEFEX®, Makhteshim-Agan Industries (Israel); BETHRODINE®; BHULAN®; BINNELL®; BONALAN®; CARPIDOR®; EL-110®; Dow AgroSciences (USA); EMBLEM®; FLUBALEX®; PEL-TECH®; QUILAN®; TEAM; XL 2G

*Producers:* Dow AgroSciences (USA); Makhteshim-Agan Industries (Israel)

*Chemical Class:* Organofluorine
*EPA/OPP PC Code:* 084301
*California DPR Chemical Code:* 53
*RTECS Number:* XU4550000
*EINECS Number:* 217-465-2

*Uses:* Selective pre-emergence herbicidal control of annual grasses and broad-leaf weeds. Used on alfalfa, red clover, seeded lettuce, trefoil; peanuts, certain tobaccos, vegetables such as endive, field and French beans and lentils.

*Human toxicity (long-term)[77]:* Very low–2100.00 ppb, Health Advisory

*Fish toxicity (threshold)[77]:* High–3.08219 ppb, MATC (Maximum Acceptable Toxicant Concentration)

*U.S. Maximum Allowable Residue Levels for Benfluralin (40 CFR 180.208):*

| CROP | ppm |
|---|---|
| Alfalfa, forage | 0.05 |
| Alfalfa, hay | 0.05 |
| Clover, forage | 0.05 |
| Clover, hay | 0.05 |
| Lettuce | 0.05 |
| Peanut | 0.05 |

| Trefoil, birdsfoot, forage | 0.05 |
| --- | --- |
| Trefoil, birdsfoot, hay | 0.05 |

*Carcinogen/Hazard Classifications*
**Label Signal Word:** WARNING or CAUTION
**TRI Developmental Toxin:** Reproductive toxin as dinitrobenzene isomer
**WHO Acute Hazard:** Class U, unlikely to be hazardous
*Regulatory Authority:*
- AB 1803-Well Monitoring Chemical (CAL)
- EPCRA Section 313 Form R *de minimis* concentration reporting level: 1.0%.

*Description:* Yellow-orange crystalline solid. Molecular weight = 335.31. Melting/Freezing point = 66°C. Vapor pressure = $1 \times 10^{-4}$ mmHg @ 20°C. Solubility = 70 mg/L @ 25°C; <1 mg/L @ 25°C. Log $K_{ow}$ = 4.67. Values above 3.0 are likely to bioaccumulate in aquatic organisms.
*Permissible Concentration in Water:* No criteria set. Runoff from spills or fire control may cause water pollution.
*Routes of Entry:* Absorbed through the skin, inhalation of the dust or vapor when these materials are heated.
*Harmful Effects and Symptoms*
*Short Term Exposure:* Severely irritates eyes, skin and respiratory tract, with burning sensation, pain, redness and swelling. Metabolic stimulant. If inhaled, causes coughing, dilated pupils, headache, profuse perpetration, intense thirst, extreme fatigue, rapid pulse, high fever, clammy, flushed skin, rapid breathing, nausea, vomiting, yellowish tint to skin and lips, anxiety and confusion, convulsions, risk of lung edema. If swallowed, face and lips turn bluish. Liver injury with associated jaundice, kidney failure, and cardiac arrhythmia is commonly noted. Because this material has a low vapor pressure, significant inhalation of vapors is unlikely at ordinary temperatures. Severe exposure can cause death from heart failure. $LD_{50}$ (rat oral) = 10,000 mg/kg. Dinitrobenzene materials are toxic to the blood; prevents hemoglobin from carrying oxygen. May be more toxic than aniline.
*Long Term Exposure:* May damage the liver, kidneys and blood cells. May stain yellow the skin, eyes, and fingernails. Repeated exposure can cause anxiety, fatigue, insomnia, excessive perspiration, unusual thirst, weight loss and cataracts in the eyes. May affect the thyroid gland.
*Points of Attack:* Skin, liver, kidneys, lungs, peripheral nervous system, eyes, thyroid gland, blood.
*Medical Surveillance:* Before beginning employment, at regular times after that and if symptoms develop or overexposure has occurred, the following may be useful: Exam of eyes for cataracts. Exam of skin and nails for staining. Liver and kidney function tests. Complete blood count. Blood methemoglobin levels.
*First Aid:* If this chemical gets into the eyes, remove any contact lenses at once and irrigate immediately for at least 15 minutes, occasionally lifting upper and lower lids. Seek medical attention immediately. If this chemical contacts the skin, remove contaminated clothing and wash immediately with soap and water. Seek medical attention immediately. If this chemical has been inhaled, remove from exposure, begin rescue breathing (using universal precautions) if breathing has stopped and CPR if heart action has stopped. Transfer promptly to a medical facility. When this chemical has been swallowed, get medical attention. *Do not induce vomiting when formulations containing petroleum solvents are ingested.* Otherwise, give large quantities of water and induce vomiting. Do not make an unconscious person vomit. *Note to Physician:* Treat for methemoglobinemia. Spectrophotometry may be required for precise determination of levels of methemoglobinemia in urine.
*References:*
- U.S. Environmental Protection Agency, Office of Pesticide Programs, Pesticide Residue Limits, "Benfluralin", 40 CFR 180.208. http://www.epa.gov/pesticides/food/viewtols.htm
- California Environmental Protection Agency Chemical List of Lists, Sacramento CA (February 1997)

# Benfuracarb

*Use Type:* Insecticide and nematicide
*CAS Number:* 82560-54-1
*Formula:* $C_{20}H_{30}N_2O_5S$
*Synonyms:* β-Alanine, N-[((((2,3-dihydro-2,2-dimethyl-7-benzofuranyl)oxy)carbonyl)methylamino)thio]-N-(1-methylethyl)-, ethyl ester; Aminofuracarb; Aminosulfulan; Benfuracarb; 2,3-Dihydro-2,2-dimethyl-7-benzofuranyl N-(N-2-(ethoxycarbonyl)ethyl-N-isopropylaminosulfenyl)-N-methylcarbamate; Ethyl N-(2,3-dihydro-2,2-dimethylbenzofuraN-7-yloxycarbonyl(methyl)aminothio)-N-isopropyl-β-alaninate; Ethyl [((((2,3-dihydro-2,2-dimethyl-7-benzofuranyl)oxy)carbonyl)methylamino)thio]-N-(1-methylethyl)-β-alanine
*Trade Names:* FURACON®; OC-11588®; OK 174®; ONCOL®, Otsuka Chemical (USA), Whyte Agrochemicals (UK); ONCOL 5G®, Otsuka Chemical (USA), Whyte Agrochemicals (UK)
*Producers:* Otsuka Chemical (USA); Whyte Agrochemicals (UK); Zhejiang Chem-tech Group Co., Ltd. (China)
*Chemical Class:* Carbamate
*EPA/OPP PC Code:* 123201
*Carcinogen/Hazard Classifications*
**WHO Acute Hazard:** Class II, moderately hazardous
*Description:* Molecular weight = 410.58
*Incompatibilities:* May form explosive materials with phosphorus pentachloride.
*Permissible Concentration in Water:* No criteria set. Runoff from spills or fire control may cause water pollution.
*Routes of Entry:* Inhalation, ingestion and skin absorption
*Harmful Effects and Symptoms*
*Short Term Exposure:* Eye pupils are small; blurred vision;

eye watering; runny nose; cough; shortness of breath; salivation; nausea, stomach cramps, diarrhea, and vomiting; increased blood pressure; profuse sweating; hypermotility, hallucinations; agitation; tingling of the skin; slow heartbeat; convulsions; fluid in lungs; loss of consciousness; incontinence; breathing stops; death. Carbamate insecticides inhibit the cholinesterase activity of enzymes, causing accumulation of acetylcholine at synapses and altering the way in which nervous impulses are transmitted. However, within several hours carbamates spontaneously detach from the enzymes.

*Long Term Exposure:* A potent cholinesterase inhibitor; cumulative effect is possible. This chemical may damage the nervous system with repeated exposure, resulting in convulsions, respiratory failure. May cause liver damage.

*Points of Attack:* Respiratory system, lungs, central nervous system, cardiovascular system, skin, eyes, plasma and red blood cell cholinesterase.

*Medical Surveillance:* Medical observation is recommended for 24 to 48 hours after breathing overexposure, as pulmonary edema may be delayed. As first aid for pulmonary edema, consider administering a corticosteroid spray. Cigarette smoking may exacerbate pulmonary injury and should be discouraged for at least 72 hours following exposure. Before employment and at regular times after that, the following are recommended: Plasma and red blood cell cholinesterase levels (tests for the enzyme poisoned by this chemical). If exposure stops, plasma levels return to normal in 1-2 weeks while red blood cell levels may be reduced for 1-3 months. When acetylcholinesterase enzyme levels are reduced by 25% or more below preemployment levels, risk of poisoning is increased, even if results are in lower ranges of "normal." Reassignment to work not involving carbamate pesticides is recommended until enzyme levels recover. If symptoms develop or overexposure occurs, repeat the above tests as soon as possible and get an exam of the nervous system. Also consider complete blood count. Consider chest x-ray following acute overexposure

*First Aid:* Speed in removing material from eyes and skin is of extreme importance. *Eyes:* Eye contact can cause dangerous amounts of these chemicals to be quickly absorbed through the mucous membrane into the bloodstream. Immediately and gently flush eyes with plenty of warm or cold water (NO hot water) for at least 15 minutes, occasionally lifting the upper and lower eyelids. Get medical aid immediately. *Skin:* Get medical aid. Skin contact can cause dangerous amounts of these chemicals to be absorbed into the bloodstream. Wearing the appropriate PPE equipment and respirator for carbamate pesticides, immediately flush skin with plenty of soap and water for at least 15 minutes while removing contaminated clothing and shoes. Shampoo hair promptly if contaminated; protect eyes. *Ingestion:* Call poison control. Loosen all clothing. Never give anything by mouth to an unconscious person. Get medical aid. Do NOT induce vomiting.* If conscious, alert, and able to swallow, rinse mouth and have victim drink 4 to 8 ounces of water. Check to see if poison control instructs you to use ipecac syrup, otherwise administer slurry of activated charcoal (2 oz in 8 oz of water). If victim is *UNCONSCIOUS OR HAVING CONVULSIONS,* do nothing except keep victim warm. *In some cases you may be specifically instructed by poison control to induce vomiting by way of 2 tablespoons of syrup of ipecac (adult) washed down with a cup of water. Do NOT give activated charcoal before or with ipecac syrup.*

*Inhalation:* Get medical aid. Do not contaminate yourself. Wearing the appropriate PPE equipment and respirator for carbamate pesticides, immediately remove the victim from the contaminated area to fresh air. If the victim is not breathing, administer artificial respiration. Do NOT use mouth-to-mouth resuscitation; use bag/mask apparatus. If breathing is difficult, administer oxygen through bag/mask apparatus until medical help arrives. Do not leave victim unattended. *Note to physician or authorized medical personnel.* Administer atropine, 2 mg (1/30 gr) intramuscularly or intravenously as soon as any local or systemic signs or symptoms of an intoxication are noted; repeat the administration of atropine every 3 to 8 minutes until signs of atropinization (mydriasis, dry mouth, rapid pulse, hot and dry skin) occur; initiate treatment in children with 0.05 mg mg/kg of atropine; repeat at 5 to 10 minute intervals. Watch respiration, and remove bronchial secretions if they appear to be obstructing the airway; intubate if necessary. *Medical note:* 2-PAMCI may be contraindicated in the case of some carbamate poisonings.

*References:*
- California Environmental Protection Agency Chemical List of Lists, Sacramento CA (February 1997)

# Benomyl (ANSI)

*Use Type:* Fungicide.
*CAS Number:* 17804-35-2
*Formula:* $C_{14}H_{18}N_4O_3$
*Alert:* All uses of benomyl products in the United States was phased out with a deadline of December 31, 2003.
*Synonyms:* BBC; Benomilo (Spanish); 2-Benzimidazolecarbamic acid, 1-(Butylcarbamoyl)-, methyl ester; BNM; 1-(Butylamino)carbonyl-1*H*-benzimidazol-2-yl-, methyl ester; 1-(Butylcarbamoyl)-2-benzimidazolecarbamic acid, methyl ester; 1-(*N*-Butylcarbamoyl)-2-(methoxy-carboxamido)-benzamidazol (German); 1-(*N*-Butylcarbamoyl)-2-(methoxy-carboxamido)-benzimidazol (German); Carbamic acid, 1-(butylamino)carbonyl-1*H*-benzimidazol-2yl, methyl ester; Carbamic acid, methyl-, 1-(butylcarbamoyl)-2-benzimidazole ester; Caswell No. 075A;

# Benomyl

MBC; Methyl 1-(butylcarbamoyl)-2-benzimidazolyl carbamate

***Trade Names:*** ABORTRINE®; AGROCITE®; ARILATE®; BBC 6597®; BENEX®; BENLAT®, DuPont Crop Protection (USA), canceled 12/31/2001; BENLATE®, DuPont Crop Protection (USA), withdrawn 5/7/01; BENLATE 50®; BENLATE 50 W®; BENLATE 50WP®; BENOMYL® 50W; BENOSAN®; D 1991®, DuPont Crop Protection (USA), canceled 12/31/2001; DuPont 1991®, DuPont Crop Protection (USA), canceled 12/31/2001; F 1991®; FUNDAZOL®; FUNGICIDE 1991®; FUNGACIDE D-1991®; FUNGOCHROM® (USA); TARSAN®; TERSAN®; TERSAN 1991®; UZGN®

***Producers:*** Agrimor International (USA); Agsin (Singapore); Bharat Pulverising Mills (India); Biesterfeld Siemsgluess International. GmbH (Germany); China Chemical (China); DuPont Crop Protection (USA); Ehrenstorfer, Dr.(Germany); Jingma Chemicals Ltd. (China); Ki-Hara Chemicals Ltd. (UK); Rhone-Poulenc Agro (France); Shenzhen Guomeng Industry Co., Ltd. (China); Sigma-Aldrich Laborchemikalien (Germany); Sinon (Taiwan); Sloss Industries (USA); Syngenta (Switzerland); Takeda Chemical Industries (Japan); Veterinary & Agricultural Products Manufacturing Co., Ltd. (VAPCO) (Jordan)

***Chemical Class:*** Benzinidizole
***EPA/OPP PC Code:*** 099101
***California DPR Chemical Code:*** 1552
***ICSC Number:*** 0382
***RTECS Number:*** DD6475000
***EEC Number:*** 613-049-00-3
***EINECS Number:*** 241-775-7

***Uses:*** Used as a pre-harvest systemic fungicide and as a post-harvest dip. Used on arable and vegetable crops, apples, soft fruit, nuts, mushrooms, lettuce, tomatoes and turf. In California, the top five crops for which benomyl is used are pistachios, table and raisin grapes, almonds, strawberries and wine grapes.

***Human toxicity (long-term)[77]:*** Intermediate–17.50 ppb, Health Advisory

***Fish toxicity (threshold)[77]:*** High–2.19089 ppb, MATC (Maximum Acceptable Toxicant Concentration)

***U.S. Maximum Allowable Residue Levels for Benomyl (40 CFR 180.294):***

| CROP | ppm |
|---|---|
| Almond, hulls | 1.0 |
| Apple (pre & post-h) | 7.0 |
| Apple, post-h | 7.0 |
| Apricot, post-h | 15 |
| Avocado | 3.0 |
| Banana, post-h | 1.0 |
| Banana, pulp | 0.2 |
| Barley, grain & straw | 0.20 |
| Bean | 2.0 |
| Bean, vines, forage | 50.0 |
| Beet, sugar, roots | 0.20 |
| Beet, sugar, tops | 15. |
| Blackberry | 7.0 |
| Blueberry | 7.0 |
| Boysenberry | 7.0 |
| Broccoli | 0.20 |
| Brussels sprouts | 15 |
| Cabbage | 0.20 |
| Cabbage, chinese, bok choy, napa | 10 |
| Carrot | 0.20 |
| Cattle, fat, meat, mbyp | 0.10 |
| Cauliflower | 0.20 |
| Celery | 3.0 |
| Cherry, post-h | 15 |
| Citrus, dried pulp | 50 |
| Collards | 0.20 |
| Corn, sweet, fodder & forage | 0.2 |
| Corn, sweet, kernels plus cob with husks removed | 0.20 |
| Corn, sweet, stover | 0.20 |
| Cucumber | 1.0 |
| Currant | 7.0 |
| Dandelion, leaves | 10 |
| Dewberry | 7.0 |
| Egg | 0.10 |
| Eggplant | 0.20 |
| Fruit, citrus, group 10 | 10 |
| Fruit, citrus, post-h | 10.0 |
| Garlic | 0.20 |
| Goat, fat, meat, mbyp | 0.10 |
| Grape | 10 |
| Grape, raisin | 50 |
| Hog, fat, meat, mbyp | 0.10 |
| Horse, fat, meat, mbyp | 0.10 |
| Kale | 0.20 |
| Kohlrabi | 0.20 |
| Loganberry | 7.0 |
| Mango | 3.0 |
| Melon | 1.0 |
| Milk | 0.10 |
| Mushroom & post-h | 10.0 |
| Mustard, greens | 0.20 |
| Nectarine, post-h | 15 |
| Nuts | 0.20 |
| Oat, grain & straw | 0.20 |
| Papaya | 3.0 |
| Peach, post-h | 15 |
| Peanut | 0.20 |
| Peanut, forage & hay | 15.0 |
| Pear, post-h | 7.0 |
| Pepper | 0.20 |
| Pineapple, post-h | 35 |

| | |
|---|---|
| Pistachio | 0.20 |
| Plum (fresh prunes) pre & post-h) | 15.0 |
| Plum, prune, fresh | 15 |
| Poultry, fat & meat | 0.10 |
| Poultry, liver | 0.20 |
| Poultry, meat by products, except liver | 0.10 |
| Pumpkin | 1.0 |
| Raspberry | 7.0 |
| Rice | 5.0 |
| Rice, hulls | 20 |
| Rice, straw | 15 |
| Rutabaga | 0.20 |
| Rye, grain & straw | 0.20 |
| Sheep, fat, meat, mbyp | 0.10 |
| Soybean | 0.20 |
| Spinach | 0.20 |
| Squash, summer & winter | 1.0 |
| Strawberry | 5.0 |
| Sweet potato | 0.20 |
| Tomato | 5.0 |
| Tomato, concentrated products | 50 |
| Turnip, greens | 6.0 |
| Turnip, roots | 0.20 |
| Watercress | 10 |
| Wheat, grain | 0.20 |
| Wheat, straw | 15 |

***Carcinogen/Hazards Classifications:***
**U.S. EPA Carcinogens:** Group C, possible carcinogen
**California Prop. 65:** Male developmental toxin
**TRI Developmental Toxin:** Developmental toxin
**Label Signal Word:** CAUTION
**Endocrine Disruptor:** Suspected endocrine disruptor

***Regulatory Authority:***
- Air Pollutant Standard Set (ACGIH)[1] (HSE)[33] (UNEP)[43] (former USSR)[35] (OSHA)[58] (Several States)[60] (Australia) (Israel) (Mexico) (Several Canadian Provinces)
- AB 1803-Well Monitoring Chemical (CAL)
- Proposition 65 chemical (CAL)
- Permissible Exposure Limits for Chemical Contaminants (CAL/OSHA)
- The "Director's List" (CAL/OSHA)
- Actively registered pesticide in California.
- EPA Hazardous Waste Number (RCRA No.): U271
- RCRA, 40CFR261, Appendix 8 Hazardous Constituents
- RCRA 40CFR268.48; 61FR15654, Universal Treatment Standards: Wastewater (mg/L), 0.056; Nonwastewater (mg/kg), 1.4
- EPCRA Section 313 Form R *de minimus* concentration reporting level: 1.0%
- Proposition 65 chemical (CAL)
- U.S. DOT 49CFR172.101, Appendix B, Regulated marine pollutant

***Description:*** White crystalline solid. Faint acrid odor. Slightly soluble in water; solubility = 1.8 ppm @ 25°C; also listed at 3.8 ppm (no temperature). Molecular weight = 290.42. Melting/Freezing point = >572°F (decomposes). Boiling point = decomposes. Vapor pressure = <1 x $10^{-10}$ mmHg. Log $K_{ow}$ = < 2.5. Unlikely to bioaccumulate in marine organisms.

***Incompatibilities:*** Strong bases [forms toxic oxides of nitrogen], strong acids, peroxides and oxidizers.

***Permissible Exposure Limits in Air:*** The Federal OSHA[2] standard is 10 mg/m$^3$ TWA for benomyl dust and 5 mg/m$^3$ TWA for the respirable fraction. California has set the same TWAs. ACGIH has set a TWA of 0.8 ppm (10 mg/m$^3$) but no STEL[1]. The HSE[33] has set the same TWA plus a STEL of 1.3 ppm (15 mg/m$^3$). The former USSR-UNEP/IRPTC project[43] has set a tentative safe exposure limit in the workplace of 0.01 mg/m$^3$. In addition, The former USSR has set[35] a limit in ambient air of 0.35 mg/m$^3$ on a once-a-day basis and 0.05 mg/m$^3$ on an average daily basis. Several states have set guidelines or standards for benomyl in ambient air[60] ranging from 100 µg/m$^3$ (North Dakota and Virginia) to 200 µg/m$^3$ (Connecticut) to 238 µg/m$^3$ (Nevada).

***Determination in Air:*** Filter; none; Gravimetric; NIOSH IV[18] [Particulates NOR; #0500 (total), #0600 (respirable)]

***Permissible Concentration in Water:*** The former USSR has set a MAC in surface water of 0.5 mg/L of benomyl[35].

***Routes of Entry:*** Inhalation

***Harmful Effects and Symptoms***

***Short Term Exposure:*** The substance irritates the skin eyes and upper respiratory system. Exposure could cause depression of the central nervous system and lack of muscular coordination.

***Long Term Exposure:*** Repeated or prolonged contact may cause skin sensitization and allergy. Human mutation data reported. Also experimental and reproductive effect. May damage the male reproductive system; cause heritable genetic damage in humans. Animal tests show that this substance possibly cause birth defects in human babies, miscarriage, or cancer. Benomyl is generally felt to have a low order of acute and chronic toxicity[53]. However, a rebuttable presumption against registration for benomyl was issued on December 6, 1978 by U.S. EPA on the basis of reduction in nontarget species, mutagenicity, teratogenicity, reproductive effects, and hazard to wildlife. The ADI for man is 0.02 mg/kg[23].

***Points of Attack:*** Eyes, skin, respiratory system, reproductive system

***Medical Surveillance:*** Evaluation by a qualified allergist.

***First Aid:*** If this chemical gets into the eyes, remove any contact lenses at once and irrigate immediately for at least 15 minutes, occasionally lifting upper and lower lids. Seek medical attention immediately. If this chemical contacts the skin, remove contaminated clothing and wash immediately with soap and water. Seek medical attention immediately. If this chemical has been inhaled, remove from exposure, begin rescue breathing (using universal precautions) if breathing has stopped and CPR if heart action has stopped. Transfer promptly to a medical facility. When this chemical has been swallowed, get medical attention. Give large quantities of water and induce vomiting. Do not make an unconscious person vomit.

*References:*
- U.S. Environmental Protection Agency, Office of Pesticide Programs, Pesticide Residue Limits, "Benomyl", 40 CFR 180.294, www.epa.gov/pesticides/food/viewtols.htm
- EXTOXNET, Extension Toxicology Network, "Pesticide Information Profile, Benomyl," Oregon State University, Corvallis, OR (June 1996). http://extoxnet.orst.edu/pips/benomyl.htm
- New Jersey Department of Health and Senior Services, "Hazardous Substance Fact Sheet: Benomyl," Trenton, NJ (February 1989, revised April 1997). http://www.state.nj.us/health/eoh/rtkweb/0192.pdf
- Sax, N.I., Ed., "Dangerous Properties of Industrial Materials Report" 4, No. 1, 20-21 (1984).
- California Environmental Protection Agency Chemical List of Lists, Sacramento CA (February 1997).

# Bensulfuron-methyl

*Use Type:* Herbicide
*CAS Number:* 83055-99-6
*Formula:* $C_{16}H_{18}N_4O_7S$
*Synonyms:* Benzoic acid, 2-[(((((4,6-dimethoxy-2-pyrimidinyl)amino)carbonyl)amino)sulfonyl)methyl]-, methyl ester; 2-[(((((4,6-Dimethoxy-2-pyrimidinyl)amino)carbonyl)amino)sulfonyl)methyl]benzoic acid, methyl ester; Methyl 2-[(((((4,6-dimethoxy-2-pyrimidinyl)amino)carbonyl)amino)sulfonyl)methyl]benzoate
*Trade Names:* DPX-F5384®, DuPont Crop Protection (USA); DUET®; F 5384®; LONDAX®, DuPont Crop Protection (USA); PRO-PACK®, Agriliance (USA)
*Producers:* Agrimor International (USA); Agsin (Singapore); Agriliance (USA); DuPont Crop Protection (USA); Shanghai Agricultural Chemical Industry (China)
*Chemical Class:* Sulfonylurea
*EPA/OPP PC Code:* 128820; (128880 old EPA code number)
*California DPR Chemical Code:* 2263
*Uses:* Used for weed control in rice crops.
*U.S. Maximum Allowable Residue Levels for Bensulfuron-methyl (40 CFR 180.445):*

| CROP | ppm |
|---|---|
| Crayfish | 0.05 |
| Rice | 0.02 |
| Rice, straw | 0.3 |

*Carcinogen/Hazard Classifications*
**Label Signal Word:** CAUTION or WARNING
**WHO Acute Hazard:** Class U, unlikely to be hazardous
*Description:* Yellowish to white powder. Odorless. Melting point = 185-188°C. Molecular weight = 410.40.
*Incompatibilities:* Esters are generally incompatible with nitrates. Moisture may cause hydrolysis or other forms of decomposition; will degrade under acid conditions.
*Permissible Concentration in Water:* No criteria set. Runoff from spills or fire control may cause water pollution.
*First Aid:* If this chemical gets into the eyes, remove any contact lenses at once and irrigate immediately for at least 15 minutes, occasionally lifting upper and lower lids. Seek medical attention immediately. If this chemical contacts the skin, remove contaminated clothing and wash immediately with soap and water. Seek medical attention immediately. If this chemical has been inhaled, remove from exposure, begin rescue breathing (using universal precautions) if breathing has stopped and CPR if heart action has stopped. Transfer promptly to a medical facility. When this chemical has been swallowed, get medical attention. Give large quantities of water and induce vomiting. Do not make an unconscious person vomit. Do not induce vomiting when formulations containing petroleum solvents are ingested. If this chemical gets into the eyes, remove any contact lenses at once and irrigate immediately for at least 15 minutes, occasionally lifting upper and lower lids. Seek medical attention immediately. If this chemical contacts the skin, remove contaminated clothing and wash immediately with soap and water. Seek medical attention immediately. If this chemical has been inhaled, remove from exposure, begin rescue breathing (using universal precautions) if breathing has stopped and CPR if heart action has stopped. Transfer promptly to a medical facility. When this chemical has been swallowed, get medical attention. Give large quantities of water and induce vomiting. Do not make an unconscious person vomit.

*References:*
- U.S. Environmental Protection Agency, Office of Pesticide Programs, Pesticide Residue Limits, "Bensulfuron-methyl," 40 CFR 180.445, www.epa.gov/pesticides/food/viewtols.htm
- California Environmental Protection Agency Chemical List of Lists, Sacramento CA (February 1997)

# Bensulide

*Use Type:* Herbicide
*CAS Number:* 741-58-2
*Formula:* $C_{14}H_{24}NO_4PS_3$
*Synonyms:* Benzulfide; *O,O*-Bis(1-methylethyl)-*S*-[2-

((phenylsulfonyl) amino)ethyl]pheosphorodithioate; *N*-[2-(*O,O*-Diisopropyldithiophosphoryl)ethyl] benzenesulfonamide; *N*-(*β*-*O,O*-Diisopropyl dithiophosphorylethyl)bezenesulfonamide; *S*-(*O,O*-Diisopropyl phosphorodithioate) ester of *N*-(2-mercaptoethyl)benzenesulfonamide; *N*-(2-Mercaptoethylbenzenesulfonamide)-*S*-(*O,O*-diisopropyl phosphorodithioate); Phosphorodithioic acid-*O,O*-bis(1-methylethyl)-*S*-[2-((phenylsulfonyl)amino)ethyl]ester

*Trade Names:* BENSUMEC®; BETAMEC®, Pbi/Gordon (USA); BETASAN®, Gowan Company (USA); BETASAN®-E, Gowan Company (USA); BETASAN®-G, Gowan Company (USA); DISAN®; EXPORSAN®; KAYAPHENONE®; PREFAR®, Gowan Company (USA); PREFAR®-E, Gowan Company (USA); PRE-SAN®, Pbi/Gordon (USA); PROTURF®, Scotts Company, The (USA); R-4461®, Gowan Company (USA); SAP (herbicide)

*Producers:* Gowan Company (USA); Pbi/Gordon (USA); Scotts Company, The (USA)

*Chemical Class:* Organophosphate

*EPA/OPP PC Code:* 009801

*California DPR Chemical Code:* 70

*ICSC Number:* 0383

*RTECS Number:* TE0250000

*EEC Number:* 015-083-00-9

*Uses:* A selective preemergence herbicide used to control bluegrass, crabgrass and other annual grasses and broadleaf weeds in agriculture crops, cotton and turf. It is widely used on golf courses and home lawns. Target weeds also include barnyardgrass, burning nettle and canarygrass.

*Human toxicity (long-term)[77]:* Intermediate–46.20 ppb, Health Advisory

*Fish toxicity (threshold)[77]:* Intermediate–43.42721 ppb, MATC (Maximum Acceptable Toxicant Concentration)

*U.S. Maximum Allowable Residue Levels for Bensulide (40 CFR 180.241):*

| CROP | ppm |
|---|---|
| Carrot, roots | 0.1 |
| Cucurbits | 0.1 |
| Onion, dry bulb | 0.1 |
| Vegetable, fruiting, group 8 | 0.1 |
| Vegetable, leafy | 0.1 |

*Carcinogen/Hazard Classifications*

**U.S. EPA Carcinogens:** Not likely a carcinogen

**Label Signal Word:** CAUTION or WARNING

**WHO Acute Hazard:** Class II, moderately hazardous

*Regulatory Authority:*
- Actively registered pesticide in California.
- U.S. Dot Inhalation Hazard Chemicals as organophosphates

*Description:* Amber, viscous liquid, or forms colorless or white crystalline solid below 34.4°C. Practically insoluble in water; solubility = 0.0025 g/ml @ 20°C. Molecular weight = 397.49. Insoluble and sinks in water. Density = 1.25 @ 22°C. Molecular weight = 397.54. Melting/Freezing point = 34°C. Vapor pressure = $8 \times 10^{-7}$ mmHg; 0.53 mPa @ 20°C. Log $K_{ow}$ = 4.2. Values above 3.0 are likely to bioaccumulate in aquatic organisms.

*Incompatibilities:* Corrosive to copper, aluminum, magnesium, zinc. Slowly hydrolyzes in water, releasing ammonia and forming acetate salts.

*Permissible Exposure Limits in Air:*

*Determination in Air:* OSHA versatile sampler-2; Toluene/Acetone; Gas chromatography/Flame ionization detection; NIOSH IV[18], Method #5600, Organophosphorus Pesticides.

*Permissible Concentration in Water:* No criteria set. Runoff from spills or fire control may cause water pollution.

*Routes of Entry:* Inhalation, ingestion, skin contact, passes through the skin.

*Harmful Effects and Symptoms*

*Short Term Exposure:* Eye pupils are small; blurred vision; eye watering; runny nose; cough; shortness of breath; salivation; dizziness; nausea, stomach cramps, diarrhea, and vomiting; increased blood pressure; profuse sweating; hypermotility, hallucinations; irritability; tingling of the skin; drowsiness; slow heartbeat; convulsions; fluid in lungs; loss of consciousness; incontinence; breathing stops; death. Orgnophosphates inhibit the action of acetylcholinesterase enzymes, and alter the way in which nervous impulses are transmitted. The effects can last for hours, days, or much longer. The action of the enzymes is reestablished after new enzymes are formed.

*Long Term Exposure:* Cholinesterase inhibitor; cumulative effect is possible. Organophosphates may damage the nervous system with repeated exposure, resulting in convulsions, respiratory failure. May cause liver damage.

*Points of Attack:* Respiratory system, central nervous system, cardiovascular system, blood cholinesterase.

*Medical Surveillance:* Medical observation is recommended for 24 to 48 hours after breathing overexposure, as pulmonary edema may be delayed. As first aid for pulmonary edema, consider administering a corticosteroid spray. Cigarette smoking may exacerbate pulmonary injury and should be discouraged for at least 72 hours following exposure. Do not drink any alcoholic beverages before or during use; alcohol promotes absorption of organophosphates.

*First Aid:* **Treatment for organophosphate poisoning consists of thorough decontamination, cardiorespiratory support, and administration of the antidotes atropine and pralidoxime. In cases of severe poisoning, diazepam, an anticonvulsant, should also be administered. Antidotes should be administered as prevention even if the diagnosis is in doubt.** Speed in removing material from eyes and skin is of extreme importance. *Eyes:* Eye contact can cause dangerous amounts of these chemicals to be quickly absorbed through the mucous membrane into the bloodstream. Immediately and gently flush eyes with plenty of warm or cold water (NO hot water) for at least 15 minutes, occasionally lifting the upper and lower eyelids. Get medical aid immediately. *Skin:* Get medical aid. Skin contact can cause dangerous amounts of these chemicals to be absorbed into the bloodstream. Wearing the appropriate PPE equipment and respirator for organophosphate/carbamate pesticides, immediately flush skin with plenty of soap and water for at least 15 minutes while removing contaminated clothing and shoes. Shampoo hair promptly if contaminated. The removed, contaminated clothing and shoes should be double-bagged and left in Hot Zone for later disposal by hazardous materials experts. Skin may also be decontaminated with diluted hypochlorite solution. *Inhalation:* Get medical aid. Do not contaminate yourself. Wearing the appropriate PPE equipment and respirator for organophosphate pesticides, immediately remove the victim from the contaminated area to fresh air. If the victim is not breathing, administer artificial respiration. Do NOT use mouth-to-mouth resuscitation; use bag/mask apparatus. If breathing is difficult, administer oxygen through bag/mask apparatus until medical help arrives. Do not leave victim unattended. *Ingestion:* Call poison control. Loosen all clothing. Never give anything by mouth to an unconscious person. If victim is *unconscious or having convulsions,* do nothing except keep victim warm. Get medical aid. Transfer promptly to a medical facility. In cases of ingestion, **do not induce vomiting**. If the victim is alert and asymptomatic, administer a slurry of activated charcoal at a dose of 1 g/kg (infant, child, and adult dose). A soda can and straw may be of assistance when offering charcoal to a child. *In some cases you may be specifically instructed by poison control to induce vomiting by way of 2 tablespoons of syrup of ipecac (adult) washed down with a cup of water.* Do NOT give activated charcoal before or with ipecac syrup.

*Note to physician or authorized medical personnel:* Treat cases of respiratory compromise, coma, or excessive pulmonary secretions with respiratory support using protocols and techniques available and within the scope of training. Some cases may necessitate procedures such as endotracheal intubation or cricothyrotomy by properly trained and equipped personnel. When possible, atropine (see *Antidotes*, below) should be given under medical supervision. Patients who are comatose, hypotensive, or having seizures or cardiac arrhythmias should be treated according to advanced life support protocols. *Antidotes:* Two antidotes are administered to treat organophosphate poisoning. Atropine is a competitive antagonist of acetylcholine at muscarinic receptors and is used to control the excessive bronchial secretions which are often responsible for death. Pralidoxime relieves both the nicotinic and muscarine effects of organophosphate poisoning by regenerating acetylcholinesterase and can reduce both the bronchial secretions and the muscle weakness associated with poisoning. The initial intravenous dose of atropine in adults should be determined by the severity of symptoms: An initial adult dose of 1.0 to 2.0 mg or pediatric dose of 0.01 mg/kg (minimum 0.01 mg) should be administered intravenously. If intravenous access cannot be established, atropine may also be given intramuscularly, subcutaneously or via endotracheal tube. Doses should be repeated every 15 minutes until excessive secretions and sweating have been controlled. Once bronchial secretion has been controlled, atropine administration should be repeated whenever the secretions begin to recur. In seriously poisoned patients, very large doses may be required. Alterations of pulse rate and pupillary size should not be used as indicators of treatment adequacy. Pralidoxime should be administered as early in poisoning as possible as its efficacy may diminish when given more than 24 to 36 hours after exposure. Doses are as follows: adult 1.0 g; pediatric 25 to 50 mg/kg. The drug should be administered intravenously over 30 to 60 minutes, but in a life-threatening situation, one-half of the total dose can be given per minute for a total administration time of 2 minutes. Treatment should begin to take effect within 40 minutes with a reduction in symptoms and the amount of atropine necessary to control bronchial secretion. The initial dose can be repeated in 1 hour and then every 8 to 12 hours until the patient is clinically well and no longer requires atropine. If intravenous access cannot be established, pralidoxime may also be given intramuscularly. Early administration of diazepam in addition to the combined atropine and pralidoxime treatment may help prevent the onset of seizures and potential brain and cardiac morphologic damage following high-level organophosphate poisoning.

***References:***
- EXTOXNET, Extension Toxicology Network, "Pesticide Information Profile, Bensulide," Oregon State University, Corvallis, OR. (June 1996). http://extoxnet.orst.edu/pips/bensulid.htm
- U.S. Environmental Protection Agency, "Interim Reregistration Eligibility Decision (IRED), Bensulide", Office of Prevention, Pesticides and Toxic Substances, Washington, DC (June 2000). http://www.epa.gov/REDs/2035ired.pdf
- California Environmental Protection Agency Chemical List of Lists, Sacramento CA (February 1997)
- U.S. Environmental Protection Agency, Office of Pesticide Programs, Pesticide Residue Limits, "Bensulide", 40 CFR 180.242. http://www.epa.gov/pesticides/food/viewtols.htm

# Bentazon (ANSI)

*Use Type:* Selective post-emergent herbicide.

*CAS Number:* 25057-89-0
*Formula:* $C_{10}H_{12}N_2O_3S$
*Alert:* A General Use Pesticide (GUP)
*Synonyms:* 1*H*-2,1,3-Benzothiadiazin-4(3*H*)-one, 3-(1-methylethyl)-, 2,2-dioxide (9CI); 1*H*-2,1,3-Benzothiadiazin-4(3*H*)-one, 3-isopropyl-, 2,2-dioxide (8CI); 3-Isopropyl-1H-2,1,3-benzothiadiazin-4(3H)-one-2,2-dioxide; 3-Isopropyl-1*H*-benzo-2,1,3-thiadiazin-4-one-2,2-dioxide; 3-Isopropyl-2,1,3-benzothiadiazininon-(4)-2,2-dioxid (German); 3-Isopropyl-1H-2,1,3-benzothiadiazin-4(3*H*)-one-2,2-dioxide; 3-(1-Methylethyl)-1*H*-2,1,3-benzothiazain-4(3*H*)-one-2,2-dioxide
*Trade Names:* ASAGIO®; BAS 351-H®; BASAGRAN®, BASF Agricultural Products Group (Germany); BENDIOXIDE®; BENTA®; BENTAZONE®; BLAST®, Veterinary & Agricultural Products Manufacturing Co., Ltd. (VAPCO) (Jordan); ENTRY®; LADDOK®, BASF Agricultural Products Group (Germany); LEADER®; PLEDGE®; STORM®
*Producers:* Agrimor International (USA); Agsin (Singapore); BASF Agricultural Products (Germany); Ehrenstorfer, Dr.(Germany); Kunshan Chemical Group (Industries) Corp. (China); Sevencontinent Agrichemical Co., Ltd. (China); Shenzhen Guomeng Industry Co., Ltd. (China); Sigma-Aldrich Laborchemikalien (Germany); Syngenta (Switzerland); Veterinary & Agricultural Products Manufacturing Co., Ltd. (VAPCO) (Jordan)
*Chemical Class:* Unclassified
*EPA/OPP PC Code:* 275200
*California DPR Chemical Code:* 2999 (Bentazon); 1944 (Bentazon, sodium salt)
*ICSC Number:* 0828
*RTECS Number:* DK9900000
*EEC Number:* 613-012-00-1

*Uses:* A post-emergence herbicide used to control broadleaf weeds in crops such as beans, corn, mint, soybeans, rice, and peanuts. All products marketed in the U.S. contain the sodium salt of bentazon as the active ingredient, referred to as sodium bentazon. Also used in selective post-emergent control of broadlelaf weeds and sedges in alfalfa, asparagus, cereals, clover, digitalis, dry peas, flax, garlic, grasses, green lima beans, mint, onions, potatoes, snap beans for seed, sorghum, soybeans and sugarcane.

*U.S. Maximum Allowable Residue Levels for Bentazon (40 CFR 180.355):*

| CROP | ppm |
|---|---|
| Bean, dry, seed | 0.05 |
| Bean, succulent | 0.5 |
| Cattle, meat, mbyp | 0.05 |
| Cattle, mbyp | 0.05 |
| Clover, forage | 1.0 |
| Clover, hay | 2.0 |
| Corn, field, forage | 3.0 |
| Corn, field, grain | 0.05 |
| Corn, field, stover | 3.0 |
| Corn, pop, grain | 0.05 |
| Corn, sweet, kernels plus cob with husks removed | 0.05 |
| Cowpeas, forage | 3.0 |
| Cowpeas, hay | 3.0 |
| Egg | 0.05 |
| Flax, seed | 1.0 |
| Goat, fat | 0.05 |
| Goat, meat, mbyp | 0.05 |
| Goat, mbyp | 0.05 |
| Hog, fat, meat, meat byproducts mbyp | 0.05 |
| Milk | 0.02 |
| Mint | 1.0 |
| Pea, dry | 0.05 |
| Pea, field, hay | 3.0 |
| Pea, field, vines | 0.3 |
| Pea, succulent | 3.0 |
| Peanut | 0.05 |
| Peanut, hay | 3.0 |
| Pepper, non-bell | 0.05 |
| Poultry, fat, meat, meat byproducts | 0.05 |
| Rice, grain | 0.05 |
| Rice, straw | 3.0 |
| Sheep, fat, meat, meat byproducts mbyp | 0.05 |
| Sorghum, forage | 0.2 |
| Sorghum, grain, stover | 0.05 |
| Soybean | 0.05 |
| Soybean, forage | 3.0 |
| Soybean, hay | 0.3 |

*Carcinogen/Hazard Classifications*
**U.S. EPA Carcinogens:** Group E, Unlikely
**Label Signal Word:** CAUTION
**WHO Acute Hazard:** Class III
*Regulatory Authority:*
• Safe Drinking Water Act, 55FR1470 Priority List
• AB 1803-Well Monitoring Chemical (CAL)
• MCL(Maximum Contaminants Levels) list of contaminants (CAL)
• The "Director's List" (CAL/OSHA)

*Description:* Bentazon is a colorless crystalline powder. Soluble in water; solubility = 490 ppm @ 20°C. Molecular weight = 240.28. Melting/Freezing point = 137-139°C. Boiling point = 200°C (decomposes). Vapor pressure = 0.53 mPa @ 20 °C. Hazard Identification (based on NFPA-704 M Rating System): Health 2, Flammability 2, Reactivity 0. Log $K_{ow}$ = –0.47. Unlikely to bioaccumulate in marine organisms.

*Incompatibilities:* Keep away from flammable materials, heat and flame. Risk of fire and explosion if formulations contain flammable/explosive solvents.

*Permissible Exposure Limits in Air:* No standards set.

*Permissible Concentration in Water:* A no-observed adverse effect level (NOAEL) of 2.5 mg/kg/day has been determined by EPA based on the absence of prostatic effects in dogs. This led to the determination of a longer-term health advisory of 0.875 mg/L for a 70-kg adult. It also led to the establishment of a lifetime health advisory of 0.0175 mg/L. In addition, California[61] has set a guideline in drinking water of 8.0 $\mu$g/L.

*Routes of Entry:* Ingestion., inhalation.

*Harmful Effects and Symptoms*

*Short Term Exposure:* The $LD_{50}$ (oral, rat) = 1100 mg/kg which puts in the "slightly toxic" category. Avoid eye contact; may cause severe irritation or injury. May cause skin burns.

*Long Term Exposure:* May be a reproductive hazard.

*First Aid:* If this chemical gets into the eyes, remove any contact lenses at once and irrigate immediately for at least 15 minutes, occasionally lifting upper and lower lids. Seek medical attention immediately. If this chemical contacts the skin, remove contaminated clothing and wash immediately with soap and water. Seek medical attention immediately. If this chemical has been inhaled, remove from exposure, begin rescue breathing (using universal precautions) if breathing has stopped and CPR if heart action has stopped. Transfer promptly to a medical facility. When this chemical has been swallowed, get medical attention. Give large quantities of water and induce vomiting. Do not make an unconscious person vomit.

*References:*
- U.S. Environmental Protection Agency, Office of Pesticide Programs, Pesticide Residue Limits, "Bentazon", 40 CFR 180.355, www.epa.gov/pesticides/food/viewtols.htm
- EXTOXNET, Extension Toxicolofy Network, "Pesticide Information Profile, Bentazon," Oregon State University, Corvallis, OR (September, 1993). http://pmep.cce.cornell.edu/profiles/extoxnet/24d-captan/bentazon-ext.html
- U.S. Environmental Protection Agency, "Health Advisory: Bentazon," Washington, DC, Office of Drinking Water (August 1987).
- California Environmental Protection Agency Chemical List of Lists, Sacramento CA (February 1997).

# 6-Benzaldenine

*Use Type:* Plant growth regulator
*CAS Number:* 1214-39-7
*Formula:* $C_{12}H_{11}N_5$
*Synonyms:* Adenine, *N*-benzyl-; 6-BAP; BAP; Benzyladenine; *N*-Benzyladenine; $N^6$-Benzyladenine; Benzylaminopurine; $N^6$-(Benzylamino)purine; 6-(Benzylamino)purine; 6-(*N*-Benzylamino)purine; *N*-(Phenylmethyl)-1*H*-purin-6-amine; 1*H*-Purin-6-amine, *N*-(phenylmethyl)-; Promalin, component of (with Gibberellin D); Verdan senescence inhibitor

*Trade Names:* ABG® 3034; ACCEL®, Valent BioSciences Corporation (USA); AGTROL®, Nufarm (Australia); 6-BA®, Nufarm (Australia); BA® (growth stimulant); CHRYSAL BVB®; EXILIS®, Fine Agrochemicals(UK); PERLAN®, Fine Agrochemicals (UK); PROMALIN®, Valent BioSciences Corporation (USA); SD® 4901; SQ® 4609

*Producers:* Fine Agrochemicals(UK); Fluorochem (UK); Lancaster Synthesis (UK); Nufarm (Australia); Valent BioSciences Corporation (USA)

*EPA/OPP PC Code:* 116901
*California DPR Chemical Code:* 2000 ($N^6$-Benzyladenine)
*EINECS Number:* 214-927-5

*Uses:* A plant growth regulator used to lengthen and enhance the shape of apples and to increase the fruit set in pears. It increases the yield of pistachios and tomatoes.

*Carcinogen/Hazard Classifications*

**Label Signal Word:** WARNING, CAUTION or DANGER, depending on formulation

*Regulatory Authority:*
- Actively registered pesticide in California.

*Description:* Molecular weight = 225.28

*Permissible Concentration in Water:* No criteria set. Runoff from spills or fire control may cause water pollution.

*Harmful Effects and Symptoms*

*Short Term Exposure:* Contact with eyes or skin may cause irritation or injury. Inhalation should be avoided; use NIOSH-approved air purifying respirators for pesticides. May be harmful if swallowed.

*First Aid:* If this chemical gets into the eyes, remove any contact lenses at once and irrigate immediately for at least 15 minutes, occasionally lifting upper and lower lids. Seek medical attention immediately. If this chemical contacts the skin, remove contaminated clothing and wash immediately with soap and water. Seek medical attention immediately. If this chemical has been inhaled, remove from exposure, begin rescue breathing (using universal precautions) if breathing has stopped and CPR if heart action has stopped. Transfer promptly to a medical facility. When this chemical has been swallowed, get medical attention. *Do not induce vomiting when formulations containing petroleum solvents are ingested.* Otherwise, give large quantities of water and induce vomiting. Do not make an unconscious person vomit.

*References:*
- California Environmental Protection Agency Chemical List of Lists, Sacramento CA (February 1997)

# Benzoic Acid

*Use Type:* Fungicide and insecticide
*CAS Number:* 65-85-0

*Formula:* $C_7H_6O_2$; $C_6H_5COOH$

*Synonyms:* Acide benzoique (French); Acido benzoico (Spanish); Benzenecarboxylic acid; Benzeneformic acid; Benzenemethanoic acid; Benzoate; Benzoesaeure (German); Carboxybenzene; Carboxylbenzene; Dracyclic acid; Oracylic acid; Phenyl carboxylic acid; Phenylformic acid

*Trade Names:* RETARDER BA®; TENN-PLAS®; RETARDEX®; SALVO LIQUID®; SALVO POWDER®

*Producers:* ATOFINA (France); Bayer Group (Germany); China Chemical (China); Cia. Quimica Universal de Industrias (Mexico); DSM Fine Chemicals (Netherlands); EniChem (Italy); Gayatri Minerals & Chemicals (India); GFS Chemicals; Gwalior Chemicals Industries (India); Merck (Germany); Penta Manufacturing (USA); Reilly Industries; TCI America (USA); Total Specialty Chemicals (USA), see ATOFINA (France); Varsal Instruments; Velsicol Chemical Corporation (USA)

*Chemical Class:* Benzoic Acid

*EPA/OPP PC Code:* 009101

*California DPR Chemical Code:* 1329

*ICSC Number:* 0103

*RTECS Number:* DG0875000

*EINECS Number:* 200-618-2

*Uses:* Used in the manufacture of benzoates; plasticizers, benzoyl chloride, alkyd resins, in the manufacture of food preservatives, in use as a dye binder in calico printing; in curing of tobacco, flavors, perfumes, dentifrices, standard in analytical chemistry.

*U.S. Maximum Allowable Residue Levels for Benzoic Acid (40 CFR 180.482):*

| CROP | ppm |
| --- | --- |
| Apple | 1.0 |
| Walnut | 0.1 |

*Carcinogen/Hazard Classifications*

**U.S. EPA Carcinogens:** Group D, unclassifiable, inadequate data

*Regulatory Authority:*
- Water Pollution Standard Proposed (former USSR)[35]
- The "Director's List" (CAL/OSHA)
- Actively registered pesticide in California.
- Clean Water Act: Section 311 Hazardous Substances/RQ 40CFR117.3 (same as CERCLA, see below).
- Superfund/EPCRA 40CFR302.4 RQ: CERCLA, 5,000 lb (2270 kg)
- Canada, WHMIS, Ingredients Disclosure List

*Description:* White crystalline solid, flakes, or powder. Faint, pleasant odor. Highly soluble in water; sinks and dissolves; solubility 3.1 x $10^3$ ppm. Molecular weight = 122.13. Melting/Freezing point = 122.4°C. Boiling point = 249°C. Flash point of 121°C. Hazard Identification (based on NFPA-704 M Rating System): Health 2, Flammability 1, Reactivity 0. Log $K_{ow}$ = < 1.90 @ 20°C. Unlikely to bioaccumulate in marine organisms.

*Incompatibilities:* Incompatible with strong oxidizers, caustics, ammonia, amines, isocyanates.

*Permissible Exposure Limits in Air:* No standards set.

*Permissible Concentration in Water:* The former USSR has proposed a MAC of 0.6 mg/Liter in surface water[35].

*Routes of Entry:* Inhalation and ingestion.

*Harmful Effects and Symptoms*

*Short Term Exposure:* Irritating to skin, eyes (possibly severe), and mucous membranes. Skin contact may cause irritation, skin rash, or burning feeling on contact. Ingestion causes nausea and G.I. troubles. For most people, ingestion of 1/10 to 2/10 ounce will have no effect although some sensitive people may experience allergic reactions. Larger amounts may cause stomach upset. Information from animal studies show that about 6 ounces may be lethal to a 150 pound person. $LD_{50}$ =1700 mg/kg (rat, oral).

*Long Term Exposure:* Repeated or prolonged contact may cause skin sensitization. Mutation data reported.

*Points of Attack:* Skin, eyes, and mucous membranes.

*First Aid:* If this chemical gets into the eyes, remove any contact lenses at once and irrigate immediately for at least 15 minutes, occasionally lifting upper and lower lids. Seek medical attention immediately. If this chemical contacts the skin, remove contaminated clothing and wash immediately with soap and water. Seek medical attention immediately. If this chemical has been inhaled, remove from exposure, begin rescue breathing (using universal precautions) if breathing has stopped and CPR if heart action has stopped. Transfer promptly to a medical facility. When this chemical has been swallowed, get medical attention. Give large quantities of water and induce vomiting. Do not make an unconscious person vomit.

*References:*
- EPA, Office of Pesticide Programs, Pesticide Residue Limits, "Benzoic Acid", 40 CFR 180.482, www.epa.gov/pesticides/food/viewtols.htm
- Sax, N.I., Ed., "Dangerous Properties of Industrial Materials Report" 1, No. 8, 38-40 (1981) and 3, No. 4, 37-40 (1983).
- New Jersey Department of Health and Senior Services, "Hazardous Substance Fact Sheet: Benzoic Acid," Trenton,
- NJ (May 1986, rev. October 2000). http://www.state.nj.us/health/eoh/rtkweb/0209.pdf
- New York State Department of Health, "Chemical Fact Sheet: Benzoic Acid," Albany, NY, Bureau of Toxic Substance Assessment (January 1986).
- California Environmental Protection Agency Chemical List of Lists, Sacramento CA (February 1997).

# Benzyl Alcohol

*Use Type:* Insecticide and fungicide
*CAS Number:* 100-51-6

*Formula:* $C_7H_8O$

*Synonyms:* Alcohol bencilico (Spanish); Benzenecarbinol; Benzenemethanol; $\alpha$-Hydroxytoluene; Methanol, phenyl-; NCI-C06111; Phenylcarbinol; Phenylmethanol; Phenylmethyl alcohol; $\alpha$-Toluenol

*Trade Names:* BENATOL®; BENTALOL®; EUXYL®-K-100; BENZYLICUM®

*Producers:* ATOFINA Chemicals (France); Bayer Chemicals (Germany); DSM Fine Chemicals (Netherlands); Gayatri Minerals & Chemicals (India); BP Chemicals (UK); Interstate Chemical (USA); Mallinckrodt Baker (USA); Rhone-Poulenc (France); Richman Chemical (USA); Tessenderlo (Belgium); Tokyo Ohka Kogyo (Japan); Velsicol Chemical Corporation (USA)

*Chemical Class:* Simple aromatic alcohol

*EPA/OPP PC Code:* 009502

*California DPR Chemical Code:* 915

*ICSC Number:* 0833

*EEC Number:* 603-057-00-5

*EINECS Number:* 202-859-9

*Uses:* Used as a solvent, degreaser, photographic developer, filament dying agent and chemical intermediate. It can also relieve itching.

*Regulatory Authority:*
- Actively registered pesticide in California.

*Description:* Combustible, clear, colorless liquid. Faint, pleasant odor. Sharp, burning taste. Moderately soluble in water; solubility = 1 gm/25 ml; 40,000 ppm @ 17°C. Molecular weight = 108.14. Density = 1.0454 @ 20°C. Boiling point @ 1 atm = 205°C. Melting/Freezing point = –15°C. Vapor pressure = 0.15 mmHg @ 25°C. Flash point = 104°C (oc); 101°C (cc); Hazard Identification (based on NFPA-704 M Rating System): Health Hazards (Blue): 2; Flammability (Red): 1; Reactivity (Yellow): 0. Log $K_{ow}$ = 1.11. Unlikely to bioaccumulate in aquatic organisms.

*Incompatibilities:* Strong reactions with oxidizers (chlorates, nitrates, peroxides, permanganates, perchlorates, chlorine, bromine, fluorine, etc); strong acids, bases, aliphatic amines, isocyanates.

*Permissible Exposure Limits in Air:*

*Determination in Air:* Collection by charcoal tube, 2-Butanol/$CS_2$; analysis by gas chromatography/flame ionization detection; NIOSH (IV). Method #1400, Alcohols I[18]

*Permissible Concentration in Water:* No criteria set. Toxic to aquatic organisms. Runoff from spills or fire control may cause water pollution.

*First Aid:* If this chemical gets into the eyes, remove any contact lenses at once and irrigate immediately for at least 15 minutes, occasionally lifting upper and lower lids. Seek medical attention immediately. If this chemical contacts the skin, remove contaminated clothing and wash immediately with soap and water. Seek medical attention immediately. If this chemical has been inhaled, remove from exposure, begin rescue breathing (using universal precautions) if breathing has stopped and CPR if heart action has stopped. Transfer promptly to a medical facility. When this chemical has been swallowed, get medical attention. Give large quantities of water and induce vomiting. Do not make an unconscious person vomit. Do not induce vomiting when formulations containing petroleum solvents are ingested.

*References:*
- California Environmental Protection Agency Chemical List of Lists, Sacramento CA (February 1997)

# Bifenox (ANSI)

*Use Type:* Herbicide

*CAS Number:* 42576-02-3

*Formula:* $C_{14}H_9Cl_2NO_5$

*Synonyms:* Benzoic acid, 5-(2,4-dichlorophenoxy)-2-nitro-, methyl ester; 5-(2,4-Dichlorophenoxy)-nitrobenzoic acid, methyl ester; 2,4-Dichlorophenyl 3-(methoxycarbonyl)-4-nitrophenyl ether; Methyl 5-(2,4-dichlorophenoxy)-2-nitrobenzoate

*Trade Names:* MODOWN®, Bayer CropScience (Germany), canceled; MC-4379®, Aventis (France) and Bayer CropScience (Germany)

*Producers:* Aventis (France); Bayer CropScience (Germany)

*Chemical Class:* Chlorophenoxy; diphenyl ether; nitrophenyl ether herbicide

*EPA/OPP PC Code:* 104301; (285200 old EPA code number)

*California DPR Chemical Code:* 1953

*Uses:* Not registered in the U.S. Used to control a variety of broadleaf weeds and grasses in legumes such as soybeans and peanuts, and post emergent weed control in wheat, barley and sugar beets. The Japan Food Chemical Research Foundation has established Maximum Residue Limits for bifenox of 0.1 ppm for rice, wheat, barley and other cereal grains; and 0.05 for potatoes.

*Carcinogen/Hazard Classifications*

**Label Signal Word:** CAUTION or DANGER

**WHO Acute Hazard:** Class U, unlikely to be hazardous

*Regulatory Authority:*
- AB 2588-Air Toxics "Hot Spots" Chemicals (CAL) as chlorophenoxy pesticides
- The "Director's List" (CAL/OSHA) as chlorophenoxy pesticides

*Description:* Yellow to beige crystalline powder. Practically insoluble in water. Molecular weight =342.22. Vapor pressure = Very low.

*Incompatibilities:* Strong oxidizers. When heated to decomposition, emits nitrogen oxides and chlorine gas.

*Permissible Concentration in Water:* No criteria set. Runoff from spills or fire control may cause water pollution.

*Routes of Entry:* Inhalation, ingestion, skin/eyes contact.
*Harmful Effects and Symptoms*
*Short Term Exposure:* Poisonous; may be fatal if inhaled, swallowed, or absorbed through skin. Severely irritates eyes, skin and respiratory tract, with burning sensation, pain, redness and swelling. Metabolic stimulant. If inhaled, causes coughing, dilated pupils, headache, profuse perpetration, intense thirst, extreme fatigue, rapid pulse, high fever, clammy, flushed skin, rapid breathing, nausea, vomiting, cyanosis (bluish tint to skin and lips), anxiety and confusion, convulsions, risk of lung edema. If swallowed, face and lips turn bluish. Liver injury with associated jaundice, kidney failure, and cardiac arrhythmias are commonly noted. Nerve damage, which may be delayed, may include swelling of legs and feet, muscle twitch and stupor. Severe exposure can cause death from heart failure. Dust or liquid left in contact with the skin for several hours may be absorbed. This may result in severe delayed symptoms as listed above. These symptoms may last for months or years.
*Long Term Exposure:* Workers exposed to chlorophenoxy compounds such as 2,4-D (in the manufacturing process) over a five to ten year period at levels above 10 mg/m$^3$ complained of weakness, rapid fatigue, headache and vertigo. Liver damage, low blood pressure and slowed heartbeat were also found. Based on animal tests, may affect human reproduction.
*Points of Attack:* Eyes, skin, respiratory system, central nervous system, cardiovascular system, liver, kidney
*Medical Surveillance:* If symptoms develop or overexposure is suspected, the following may be useful: Liver and kidney function tests. Exam of the nervous system.
*First Aid:* If this chemical gets into the eyes, remove any contact lenses at once and irrigate immediately for at least 15 minutes, occasionally lifting upper and lower lids. Seek medical attention immediately. If this chemical contacts the skin, remove contaminated clothing and wash immediately with soap and water. Seek medical attention immediately. If this chemical has been inhaled, remove from exposure, begin rescue breathing (using universal precautions) if breathing has stopped and CPR if heart action has stopped. Transfer promptly to a medical facility. When this chemical has been swallowed, get medical attention. *Do not induce vomiting when formulations containing petroleum solvents are ingested.* Otherwise, give large quantities of water and induce vomiting. Do not make an unconscious person vomit.
*References:*
- Food and Agricultural Organization of the United Nations (FAO), "FAO Specifications for Plant Protection Products, Bifenox, Rome, Italy (1994). www.fao.org/waicent/faoinfo/agricult/agp/agpp/pesticid/specs/docs/bife.doc
- California Environmental Protection Agency Chemical List of Lists, Sacramento CA (February 1997).

# Bifenthrin (ANSI)

*Use Type:* Insecticide and acaricide
*CAS Number:* 82657-04-3
*Formula:* $C_{23}H_{22}ClF_3O_2$
*Synonyms:* Biphenthrin; Caswell No. 463F; Cyclopropanecarboxylic acid,3-(2-chloro-3,3,3-trifluoro-1-propenyl)-2,2-dimethyl-,[2-methyl(1,1'-biphenyl)3-yl]methyl ester,(Z)-
*Trade Names:* BIFLEX®, FMC (USA); BISTAR®, FMC (USA); BRIGADE®, FMC (USA); CAPTURE®, FMC (USA); DISCIPLINE®, AMVAC Chemical (USA); DOUBLE THREAT®, FMC (USA); EMPOWER®, Helena Chemical (USA); FMC (USA); FMC® 54800, FMC (USA); FMC® 58000, FMC (USA); TALSTAR®, FMC (USA); TALSTAR LAWN & TREE®, FMC (USA) TORANT®; ZIPAK®
*Producers:* AMVAC Chemical (USA); Dow AgroSciences (USA); Helena Chemical (USA); FMC (USA); Ki-Hara Chemicals Ltd. (UK); Micro Flo (USA); Scotts Company, The (USA)
*Chemical Class:* Organofluorine
*EPA/OPP PC Code:* 128825
*California DPR Chemical Code:* 2300
*Uses:* Registered to control cone worms, seed bugs, seed worms and other insects and mites on rangeland, forests and right-of-ways. It is also used to control household and lawn pests.
*Human toxicity (long-term)$^{(77)}$:* Intermediate–10.50 ppb, Health Advisory
*Fish toxicity (threshold)$^{(77)}$:* Extra high–0.06197 ppb, MATC (Maximum Acceptable Toxicant Concentration)
*U.S. Maximum Allowable Residue Levels for Bifenthrin (40 CFR 180.442):*

| CROP | ppm |
|---|---|
| Almond, hulls | 2 |
| Animal feed | 0.05 |
| Artichoke, globe | 1 |
| Banana | 0.1 |
| Brassica, head and stem, subgroup 5a | 0.6 |
| Caneberry subgroup 13a | 1 |
| Cattle, fat | 1 |
| Cattle, meat | 0.5 |
| Cattle, mbyp | 0.1 |
| Corn, forage | 3 |
| Corn, grain | 0.05 |
| Corn, stover | 5 |
| Corn, sweet, kernel plus cob with husks removed | 0.05 |
| Cotton, undelinted seed | 0.5 |
| Egg | 0.05 |
| Eggplant | 0.05 |
| Fruit, citrus, group 10 | 0.05 |
| Goat, fat | 1 |

| | |
|---|---|
| Goat, meat | 0.5 |
| Goat, mbyp | 0.1 |
| Hog, fat | 1 |
| Hog, meat | 0.5 |
| Hog, mbyp | 0.1 |
| Hop, dried cones | 10 |
| Horse, fat | 1 |
| Horse, meat | 0.5 |
| Horse, mbyp | 0.1 |
| Lettuce, head | 3 |
| Milk | 0.1 |
| Milk, fat | 1 |
| Nut, tree, group 14 | 0.05 |
| Orchardgrass, forage | 0.05 |
| Orchardgrass, hay | 0.05 |
| Pea and bean, succulent shelled, subgroup 6b | 0.05 |
| Peanut | 0.05 |
| Pear | 0.5 |
| Pepper, bell | 0.5 |
| Pepper, nonbell | 0.5 |
| Poultry, fat | 0.05 |
| Poultry, meat | 0.05 |
| Poultry, mbyp | 0.05 |
| Processed food | 0.05 |
| Sheep, fat | 1 |
| Sheep, meat | 0.5 |
| Sheep, mbyp | 0.1 |
| Spinach | 0.2 |
| Strawberry | 3 |
| Sweet potato, roots | 0.05 |
| Tomato | 0.15 |
| Vegetable, cucurbit, group 9 | 0.4 |
| Vegetable, legume, edible podded, Subgroup 6a | 0.6 |

*Carcinogen/Hazard Classifications*
**U.S. EPA Carcinogens:** Group C, possible carcinogen
**Label Signal Word:** CAUTION or WARNING
**TRI Developmental Toxin:** Reproductive toxin
**WHO Acute Hazard:** Class II, moderately hazardous
**Endocrine Disruptor:** Suspected
*Regulatory Authority:*
1. EPCRA Section 313 Form R *de minimis* concentration reporting level: 1.0%
2. EPA 40 CFR 372.65, Specific Toxic Chemical Listings
3. Actively registered pesticide in California.
*Description:* Thick, dark yellow to light brown oily liquid. Practically insoluble in water; solubility = < 0.1 mg/L. Melting/freezing point = 57-63°C. Vapor pressure = 1.8 x $10^{-7}$ mmHg.
*Incompatibilities:* May react violently with strong oxidizers, bromine, 90% hydrogen peroxide, phosphorus trichloride, silver powders or dust. Incompatible with silver compounds. Mixture with some silver compounds forms explosive salts of silver oxalate.
*Permissible Concentration in Water:* No criteria set. Runoff from spills or fire control may cause water pollution.
*First Aid:* If this chemical gets into the eyes, remove any contact lenses at once and irrigate immediately for at least 15 minutes, occasionally lifting upper and lower lids. Seek medical attention immediately. If this chemical contacts the skin, remove contaminated clothing and wash immediately with soap and water. Seek medical attention immediately. If this chemical has been inhaled, remove from exposure, begin rescue breathing (using universal precautions) if breathing has stopped and CPR if heart action has stopped. Transfer promptly to a medical facility. When this chemical has been swallowed, get medical attention. *Do not induce vomiting when formulations containing petroleum solvents are ingested.* Otherwise, give large quantities of water and induce vomiting. Do not make an unconscious person vomit.
*References:*
- EXTOXNET, Extension Toxicology Network, "Pesticide Information Profile, Bifenthrin," Oregon State University, Corvallis, OR (September 1995). http://extoxnet.orst.edu/pips/bifenthr.htm
- California Environmental Protection Agency Chemical List of Lists, Sacramento CA (February 1997)
- U.S. Environmental Protection Agency, Office of Pesticide Programs, Pesticide Residue Limits, "Bifenthrin," 40 CFR 180.442. http//www.epa.gov/cgi-bin/oppsrch

## Binapacryl (ANSI)

*Use Type:* Fungicide, miticide
*CAS Number:* 485-31-4
*Formula:* $C_{15}H_{18}N_2O_6$
*Synonyms:* 2-*sec*-Butyl-4,6-dinitrophenyl-3,3-dimethylacrylate; 2-*sec*-Butyl-4,6-dinitrophenyl-3-methyl-2-butenoate; 2-*sec*-Butyl-4,6-dinitrophenyl-3-methylcrotonate; 2-*sec*-Butyl-4,5-dinitrophenyl senecioate; 3,3-Dimethyl-acrylate de 2,4-dinitro-6-(1-methylpropyle) phenyle (French); 3,3-Dimethylacrylic acid 2-*sec*-butyl-4,5-dinitrophenyl ester; 4,6-Dinitro-2-*sec*-butylphenyl β,β-dimethylacrylate; 2,4-Dinitro-6-*sec*-butylphenyl-2-methylcrotonate; 4,6-Dinitrophenyl-2-*sec*-butyl-3-methyl-2-butenonate; Dinoseb methacrylate; ENT 25,793; 3-Methylcrotonic acid 2-*sec*-butyl-4,6-dinitrophenyl ester; (6-(1-Methyl-propyl)-2,4-dinitro-phenyl)-3,3-dimethyl acrylat (German); 2-(1-Methylpropyl)-4,6-dinitrophenyl-β,β-dimethacrylate; 2-(1-Methylpropyl)-4,6-dinitrophenyl ester 3-methyl-2-butenoic acid
*Trade Names:* ACRICID®; AMBOX®; BP 855®; BP 736®; DAPACRYL®; DINAPACRYL®; ENDOSAN®; FMC 9044®, FMC (USA); HOE 2784®; MOROCIDE®;

MORROCID®; NIA 9044®, FMC (USA); NIAGARA® 9044, FMC (USA)
**Producers:** FMC (USA)
**Chemical Class:** Dinitrophenyl
**EPA/OPP PC Code:** 012201
**California DPR Chemical Code:** 73
**ICSC Number:** 0835
**RTECS Number:** GQ5600000
**EEC Number:** 609-024-00-1
**Uses:** Not registered in the U.S.
**Carcinogen/Hazard Classifications**
**U.S. EPA Carcinogens:** Group C, possible carcinogen as dinoseb, the parent chemical
**California Prop. 65:** Developmental toxin and male reproductive toxin, as dinoseb, the parent chemical.
**U.S. TRI:** Developmental and reproductive Toxin, as dinoseb, the parent chemical
**Regulatory Authority:**
- Clean Water Act: Section 311 Hazardous Substances/RQ (same as CERCLA); Section 307 Toxic Pollutants as nitrophenols.
- EPCRA Section 304 RQ: CERCLA, 100 lb (45.4 kg)as nitrophenols.
- U.S. DOT Regulated Marine Pollutant (49CFR172.101, Appendix B): Severe pollutant

**Description:** Binapacryl is the dimethyl acrylic ester of Dinoseb. Colorless, crystalline solid or powder. Crystals may turn pale-yellow to brownish on contact with air. Practically insoluble in water. Mild, ammonia odor. Molecular weight = 322. Density = 1.23 @ 20°C. Melting/Freezing = 66°C. Vapor pressure = $4.2 \times 10^{-7}$ mmHg @ 20°C. Log $K_{ow}$ = 4.72. Values above 3.0 are likely to bioaccumulate in marine organisms. See also Dinoseb.

**Incompatibilities:** Slight hydrolysis on long contact with water. Keep away from caustics and concentrated acids that can cause decomposition. Slowly decomposed by ultraviolet light

**Permissible Concentration in Water:** No criteria set. Runoff from spills or fire control may cause water pollution. To protect freshwater aquatic life–230 µg/L on an acute toxicity basis for nitrophenols as a class. To protect saltwater aquatic life 4,850 µg/L on an acute toxicity basis for nitrophenols as a class. To protect human health–70.0 µg/L[6].

**Determination in Water:** Filter/Bubbler; 2-Propanol; High-pressure liquid chromatography/Ultraviolet detection; NIOSH II[5] Method #S166[18]. Methylene chloride extraction followed by gas chromatography with flame ionization or electron capture detection. EPA Method 604, or gas chromatography plus mass spectrometry EPA Method 625.

**Routes of Entry:** Inhalation, skin contact, ingestion.

**Harmful Effects and Symptoms**
**Short Term Exposure:** Severely irritates eyes, skin and respiratory tract, with burning sensation, pain, redness and swelling. Metabolic stimulant. If inhaled, causes coughing, dilated pupils, headache, profuse perspiration, intense thirst, extreme fatigue, rapid pulse, high fever, clammy, flushed skin, rapid breathing, nausea, vomiting, yellowish tint to skin and lips, anxiety and confusion, convulsions, risk of lung edema. The effects may be delayed. Medical observation is recommended. If swallowed, face and lips turn bluish. Liver injury with associated jaundice, kidney failure, and cardiac arrhythmias are commonly noted. Severe exposure can cause death from heart failure.

**Long Term Exposure:** May damage the liver, kidneys and blood cells. May stain yellow the skin, eyes, and fingernails. Repeated exposure can cause anxiety, fatigue, insomnia, excessive perspiration, unusual thirst, weight loss and cataracts in the eyes.

**Points of Attack:** Skin, liver, kidneys, lungs, peripheral nervous system, eyes, thyroid gland, blood.

**Medical Surveillance:** Before beginning employment, at regular times after that and if symptoms develop or overexposure has occurred, the following may be useful: Exam of eyes for cataracts. Exam of skin and nails for staining. Blood tests for dinitro-*o*-cresol. Persons with blood levels over 10 ppm (10 mg/L) should be kept away from further exposure until levels return to normal. If symptoms develop or overexposure is suspected, the following may be useful: Liver and kidney function tests. Complete blood count.

**First Aid:** If this chemical gets into the eyes, remove any contact lenses at once and irrigate immediately for at least 15 minutes, occasionally lifting upper and lower lids. Seek medical attention immediately. If this chemical contacts the skin, remove contaminated clothing and wash immediately with soap and water. Seek medical attention immediately. If this chemical has been inhaled, remove from exposure, begin rescue breathing (using universal precautions) if breathing has stopped and CPR if heart action has stopped. Transfer promptly to a medical facility. When this chemical has been swallowed, get medical attention. *Do not induce vomiting when formulations containing petroleum solvents are ingested.* Otherwise, give large quantities of water and induce vomiting. Do not make an unconscious person vomit.
*Note to Physician:* Treat for methemoglobinemia. Spectrophotometry may be required for precise determination of levels of methemoglobinemia in urine.

**References:**
- U.S. Environmental Protection Agency, Nitrophenols: Ambient Water Quality Criteria, Washington, DC (1980).
- Sax, N.I., Ed., "Dangerous Properties of Industrial Materials Report," 2, No. 2, 25-27 (1982) and 3, No. 2, 38-44 (1983).
- New Jersey Department of Health, "Hazardous

Substance Fact Sheet: Dinitrophenol," Trenton, NJ (October 1996, rev. February 2003). http://www.state.nj.us/health/eoh/rtkweb/0780.pdf
- California Environmental Protection Agency Chemical List of Lists, Sacramento CA (February 1997)

# Biphenyl

*Use Type:* Fungicide
*CAS Number:* 92-52-4
*Formula:* $C_{12}H_{10}$; $C_6H_5C_6H_5$
*Synonyms:* Bibenzene; 1,1'-Biphenyl; Dibenzene; Diphenyl; 1,1'-Diphenyl; Phenylbenzene; PHPH; Xenene
*Trade Names:* LEMONENE®; PHENADOR-X®
*Producers:* Bayer (Germany); Koch Specialty Chemicals (USA; Rhodia (France); Sumika Fine Chemicals (Japan); Wuzhou International (China)
*Chemical Class:* Unclassified
*EPA/OPP PC Code:* 017002
*California DPR Chemical Code:* 227
*ICSC Number:* 0106
*RTECS Number:* DU8050000
*EEC Number:* 601-042-00-8
*EINECS Number:* 202-163-5
*Uses:* Biphenyl is used in household disinfectants and also as a heat transfer agent and as an intermediate in organic synthesis.
*Carcinogen/Hazard Classifications*
**U.S. EPA Carcinogens:** Group D, Unclassiable, insufficient data
**WHO Acute Hazard:** Class U, unlikely to be hazardous
*Regulatory Authority:*
- Air Pollutant Standard Set (ACGIH)[1] (DFG)[3] HSE[33] (OSHA)[58] (Several States)[60] Australia) (Israel) (Mexico) (Several Canadian Provinces)
- Clean Air Act: Hazardous Air Pollutants (Title I, Part A, Section 112)
- AB 2588-Air Toxics "Hot Spots" Chemicals (CAL)
- CAL Air Resources Board/AB 1807 Toxic Air Contaminants
- Permissible Exposure Limits for Chemical Contaminants (CAL/OSHA)
- The "Director's List" (CAL/OSHA)
- Superfund/EPCRA 40CFR302.4 RQ: CERCLA, 1 lb (0.454 kg)
- EPCRA Section 313 Form R *de minimus* concentration reporting level: 1.0%.
- Canada, WHMIS, Ingredients Disclosure List; National Pollutant Release Inventory (NPRI)

*Description:* White flakes or crystalline solid. Pleasant, characteristic odor. Practically insoluble in water; solubility = 8 ppm @ 25°C. Molecular weight 154.23. Melting/Freezing point = 69–70°C. Boiling point = 256°C. Flash point = 113°C. Autoignition temperature = 540°C. Explosive limits in air: LEL = 0.6%, UEL = 5.8%[17]. Hazard Identification (based on NFPA-704 M Rating System): Health 2, Flammability 1, Reactivity 0. Log $K_{ow}$ = 3.4. Values at or above 3.0 may bioaccumulate in marine organisms.

*Incompatibilities:* Mist forms explosive mixture with air. Strong oxidizers may cause fire and explosions.

*Permissible Exposure Limits in Air:* The Federal OSHA[2] PEL and NIOSH[2] recommended standard is 0.2 ppm (1 mg/m³)TWA. The ACGIH TLV is 0.2 ppm (1.3 mg/m³)[1]. The HSE[33] and DFG[3] use this same value as do several other countries[35] including Australia, Mexico, Israel. The Canadian Provinces of Alberta, British Columbia, Ontario and Quebect use the same TWA/TWAEV. The ACGIH has set no STEL value[1] but the HSE[33], Mexico, Alberta and British Columbia STEL value is 0.6 ppm (3.0 mg/m³). The NIOSH[2] IDLH level = 100 mg/m³. Several states have set guidelines or standards for biphenyl in ambient air[60] ranging from 0.086 µg/m³ (Massachusetts) to 0.4 µg/m³ (Rhode Island) to 5.0 µg/m³ (New York) to 15.0 µg/m³ (Florida and North Dakota) to 20.0 µg/m³ (Connecticut) to 25 µg/m³ (Virginia) to 36 µg/m³ (Nevada) to 40 µg/m³ (North Dakota).

*Determination in Air:* Tenax Gas chromatography; CCl4; Gas chromatography/Flame ionization detection; NIOSH(IV) Method #2530[18].

*Permissible Concentration in Water:* No criteria set but EPA has suggested an ambient limit of 13.8 µg/L based on health effects.

*Routes of Entry:* Inhalation of vapor or dust; percutaneous absorption, ingestion, eye and/or skin contact.

*Harmful Effects and Symptoms*
*Short Term Exposure:* Skin contact contributes significantly to overall exposure. Repeated exposure to dust may result in irritation of skin and respiratory tract. The vapor may cause moderate eye irritation. Repeated skin contact may produce a sensitization dermatitis. In acute exposure, biphenyl exerts a toxic action on the central nervous system, on the peripheral nervous system, and on the liver. Symptoms of poisoning are headache, diffuse, gastrointestinal pain, nausea, indigestion, numbness and aching of limbs, and general fatigue. The $LD_{50}$ (oral, rat) = 3280 mg/kg[9].

*Long Term Exposure:* Chronic exposure is characterized mostly be central nervous system symptoms, fatigue, headache, tremor, insomnia, sensory impairment, and mood changes. However, such symptoms may be rare. May cause lung irritation and bronchitis. liver and kidney damage. May cause skin allergy with itching and rash.

*Points of Attack:* Liver, skin, central nervous system, upper respiratory system, eyes.

*Medical Surveillance:* Consider skin, eye, liver function, and respiratory tract irritation in any preplacement or periodic examination. Examination by a qualified allergist. Examination of the nervous system.

*First Aid:* If this chemical gets into the eyes, remove any contact lenses at once and irrigate immediately for at least 15 minutes, occasionally lifting upper and lower lids. Seek medical attention immediately. If this chemical contacts the skin, remove contaminated clothing and wash immediately with soap and water. Seek medical attention immediately. If this chemical has been inhaled, remove from exposure, begin rescue breathing (using universal precautions) if breathing has stopped and CPR if heart action has stopped. Transfer promptly to a medical facility. When this chemical has been swallowed, get medical attention. Do not induce vomiting.

*References:*
- National Institute for Occupational Safety and Health, Profiles on Occupational Hazards for Criteria Document Priorities: Diphenyl, Report PB 274,073, Cincinnati, OH pp 274-276 (1977).
- Sax, N.I., Ed., "Dangerous Properties of Industrial Materials Report" 1, No. 5, 42-43 (1981).
- California Environmental Protection Agency Chemical List of Lists, Sacramento CA (February 1997).
- New Jersey Department of Health and Senior Services, "Hazardous Substance Fact Sheet: Diphenyl," Trenton, NJ (December 1998).
  http://www.state.nj.us/health/eoh/rtkweb/0795.pdf

# Bitertanol

*Use Type:* Fungicide
*CAS Number:* 55179-31-2
*Formula:* $C_{20}H_{23}N_3O_2$
*Synonyms:* β-[(1,1'-Biphenyl)-4-yloxy]-α-(1,1-dimethylethyl)-1$H$-1,2,4-triazole-1-ethanol; 1$H$-1,2,4-Triazole-1-ethanol, β-[(1,1'-biphenyl)-4-yloxy]-α-(1,1-dimethylethyl)-
*Trade Names:* BAYCOR®, Bayer CropScience (Germany); BAY KWG 0599®, Bayer CropScience (Germany); BAYMAT-SPRAY®, Bayer CropScience (Germany); BILOXAZOL®; BITERTANOL®; KWG 0599®; SIBUTOL®, component of (with Fuberidazole)
*Producers:* Bayer CropScience (Germany); Saeryung Chemicals (South Korea)
*Chemical Class:* Triazole
*EPA/OPP PC Code:* 117801
*Uses:* Not registered in the U.S. On July 23, 2004, the U.S. EPA announced the revocation of tolerances for residues of bitertanol. Registered in Australia for use on beans (all types) and various ornamentals. Also used for prunes when drying.
*Carcinogen/Hazard Classifications*
**WHO Acute Hazard:** Class U, unlikely to be hazardous
*Regulatory Authority:*

*Description:* Colorless crystalline solid. Aromatic odor. Slightly soluble in water; solubility = 5 ppm @ 20°C. Molecular weight 337.48. Melting/Freezing point = 126°C. Vapor pressure $10^{-5}$ mmHg @ 20°C. Log $K_{ow}$ = 4.11. Values greater than 3.0 may bioaccumulate in marine organisms.
*Permissible Concentration in Water:* No criteria set. Runoff from spills or fire control may cause water pollution.
*Routes of Entry:* Ingestion, skin contact
*Harmful Effects and Symptoms*
*Short Term Exposure:* Poisonous if swallowed. Contact may irritate skin and cause eye irritation and possible severe injury. Avoid inhalation.
*First Aid:* If this chemical gets into the eyes, remove any contact lenses at once and irrigate immediately for at least 15 minutes, occasionally lifting upper and lower lids. Seek medical attention immediately. If this chemical contacts the skin, remove contaminated clothing and wash immediately with soap and water. Seek medical attention immediately. If this chemical has been inhaled, remove from exposure, begin rescue breathing (using universal precautions) if breathing has stopped and CPR if heart action has stopped. Transfer promptly to a medical facility. When this chemical has been swallowed, get medical attention. Give large quantities of water and induce vomiting. Do not make an unconscious person vomit.

# Bithionol

*Use Type:* Fungicide (proposed)
*CAS Number:* 97-18-7
*Formula:* $C_{12}H_6Cl_4O_2S$
*Synonyms:* Bidiphenbis(2-hydroxy-3,5-dichlorophenyl) sulfide; CP 3438; 2,2'-Dihydroxy-3,3',5,5'-tetrachlorodiphenylsulfide; 2-Hydroxy-3,5-dichlorophenyl sulphide; NCI-C60628; TBP; 2,2'-Thiobis(4,6-dichlorophenol); USAF B-22
*Trade Names:* ACTAMER®; BIDIPHEN®; BITHIONOL®; BITHIN®; LOROTHIDOL®; NEOPELLIS®; NOBACTER®; PREVENOL®; VANCIDE BL®; VANCIDE BL® (sodium salt); XL 7®
*Producers:* Sigma-Aldrich Laborchemikalien (Germany); ISOCHEM (France)
*Chemical Class:* Chlorinated phenol
*EPA/OPP PC Code:* 064201
*RTECS Number:* SN0525000
*Uses:* It is used as a surfactant-formulated antimicrobial against bacteria, molds and yeast, and is an ingredient in household sprays for animal pets. It is proposed as an agricultural fungicide. Other uses include deodorant, germicide, fungistat and in the manufacture of pharmaceuticals. It is no longer allowed (banned by the FDA) to be used in cosmetics. A food additive in feed and

drinking water of animals. Also a food additive permitted in food for human consumption.

***Carcinogen/Hazard Classifications***
**IARC:** Group 2B, possible carcinogen
*Note:* In animal studies, one chlorophenol, 2,4,6-trichlorophenol, caused leukemia in rats and liver cancer in mice. The Department of Health and Human Services has determined that 2,4,6-trichlorophenol may reasonably be anticipated to be a carcinogen.

***Regulatory Authority:***
- Banned or Severely Restricted (USA, Japan) (UN)[13]
- TSCA 40CFR716.120(a)
- List of priority pollutants (U.S. EPA) as chlorophenols
- AB 2588-Air Toxics "Hot Spots" Chemicals (CAL) as chlorophenols
- The "Director's List" (CAL/OSHA) as chlorophenols
- Clean water act: Toxic Pollutant (Section 401.15) other than those listed elsewhere; includes trichloro phenols
- RCRA Section 261 Hazardous Constituents, waste number not listed as chlorophenols
- EPCRA Section 313 (as chlorophenols) Form R *de minimus* concentration reporting level: 1.0%.

***Description:*** White or grayish crystalline solid or powder. Slight phenolic odor. Practically insoluble in water. Molecular weight = 356.05. Melting/Freezing point = 188°C. Hazard Identification (based on NFPA-704 M Rating System): Health 3, Flammability 1, Reactivity 0.

***Incompatibilities:*** Strong oxidizers.

***Permissible Exposure Limits in Air:*** No standards set.

***Permissible Concentration in Water:*** No criteria set for this chemical; however, EPA recommends that drinking water contain no more than 0.04 mg/L of 2-chlorophenol for a lifetime exposure for an adult, and 0.05 mg/L for a 1-day, 10-day, or longer exposure for a child.

***Routes of Entry:*** Ingestion, skin and/or eye contact

***Harmful Effects and Symptoms***

**Short Term Exposure:** Contact with eyes or skin may cause irritation or injury. Inhalation should be avoided; use NIOSH-approved air purifying respirators for pesticides. May be harmful if swallowed. Probable oral lethal dose for humans is 5-15 g/kg for a 70 kg (150 lb) person. The toxicity of this compound is similar to that of phenol. Major hazard of phenol poisoning stems from its systemic effects which include central nervous system depression with coma, hypothermia, loss of vasoconstrictor tone, cardiac depression and respiratory arrest. Symptoms of exposure include burning pain in mouth and throat; white necrotic lesions in mouth, esophagus and stomach; abdominal pain; vomiting, bloody diarrhea; paleness; sweating; weakness; headache; dizziness; tinnitus; scanty, dark-colored urine; weak irregular pulse and shallow respiration.

**Long term Exposure:** Chlorophenols leave the body quickly, so they are not likely to accumulate in the mother's tissues or breast milk. There are no human studies on the effects of chlorophenols on developing fetuses. Studies in rats showed that chlorophenols can pass through the placenta and produce toxic effects to the developing fetuses. The most common problems in children are delayed hardening of the bones of the breastbone, spine, and skull.

***First Aid:*** If this chemical gets into the eyes, remove any contact lenses at once and irrigate immediately for at least 15 minutes, occasionally lifting upper and lower lids. Seek medical attention immediately. If this chemical contacts the skin, remove contaminated clothing and wash immediately with soap and water. Seek medical attention immediately. If this chemical has been inhaled, remove from exposure, begin rescue breathing (using universal precautions) if breathing has stopped and CPR if heart action has stopped. Transfer promptly to a medical facility. When this chemical has been swallowed, get medical attention. Give large quantities of water and induce vomiting. Do not make an unconscious person vomit.

***References:***
- New Jersey Department of Health and Senior Services, "Hazardous Substance Fact Sheet, 2,2'-Thiobis-4,6-dichlorophenol," Trenton NJ (June 2002). http://www.state.nj.us/health/eoh/rtkweb/2817.pdf
- U.S. Environmental Protection Agency, "Chemical Profile: 2,2'-Thiobis (4,6-Dichlorophenol)," Washington, DC, Chemical Emergency Preparedness Program (November 30, 1987).
- California Environmental Protection Agency Chemical List of Lists, Sacramento CA (February 1997).
- Agency for Toxic Substances and Disease Registry, "Toxicological profile for chlorophenols," 1999, Atlanta, GA.

# Borax and Boric Acid

***Use Type:*** Insecticide and herbicide
***CAS Number:*** 1303-96-4 (borax); 10043-35-3 (boric acid)
***Formula:*** $B_4H_2Na_2O_8$, $Na_2B_4O_7 \cdot H_2O$ (borax); $BH_3O_3$; $H_3BO_3$ (boric acid)
***Synonyms:*** *Borax*: Disodium tetraborate; Disodium tetraborate decahydrate; Sodium borate; Sodium borate decahydrate; Sodium tetraborate; Sodium tetraborate decahydrate ($Na2B_4O_7.10H_2O$). *Boric acid*: Boracic acid; Othroboric acid
***Trade Names:*** BORASCU®; BORICIN®; BOROFAX® (boric acid); GERSTLEY BORATE® (borax); KILL-OFF®; SOLUBOR® (borax)
***Producers:*** Agrium (Canada); American Borate (USA); Allipo Chemicals (India); Chemettal (Germany); Cia. Quimica Universal de Industrias (Mexico); Eagle-Picher Minerals (USA); Holvoet Chimie (Belgium); Humco (USA); IMC Global (USA); Kerr-McGee (USA); OxyChem (USA); Quantum Chemicals (Australia); Rio Tinto (UK);

Rohm & Haas (USA); SQM (Chile); Textron Systems (USA); SB Boron (USA); TCI America (USA); U.S. Borax (USA); Varsal Instruments (USA)
*Chemical Class:* Inorganic
*EPA/OPP PC Code:* 011102 (borax); 011001 (boric acid)
*California DPR Chemical Code:* 79 (borax); 769 (boric acid)
*ICSC Number:* 0567 (borax); 0991 (boric acid)
*RTECS Number:* VZ2275000 (borax); ED4550000 (boric acid)
*EINECS Number:* 233-139-2 (boric acid)
*Uses:* Borax and boric acid are commodity chemicals which are used by formulators, packagers and distributors under a variety of trade names for agricultural and domestic use. Boric acid is a fireproofing agent for wood, a preservative, and an antiseptic. It is used in the manufacture of glass, pottery, enamels, glazes, cosmetics, cements, porcelain, borates, leather, carpets, hats, soaps, and artificial gems, and in tanning, printing, dyeing, painting, and photography. It is a constituent in powders, ointments, nickeling baths, electric condensers and is used for impregnating wicks and hardening steel. Borax is used as a soldering flux, preservative against wood fungus, and as an antiseptic. It is used in the manufacture of enamels and glazes and in tanning, cleaning compounds, for fireproofing fabrics and wood, and in artificial aging of wood. Boron is used in metallurgy as a degasifying agent and is alloyed with aluminum, iron, and steel to increase hardness. It is also a neutron absorber in nuclear reactors.
*U.S. Maximum Allowable Residue Levels for Borax and Boric Acid (40 CFR 180.1121):*

| CHEMICAL | CROP | ppm |
|---|---|---|
| Borax $(B_4Na_2O_7 \cdot 10H_2O)$ | r.a.c | – |
| Boric acid | r.a.c. | – |
| Boric acid $(HBO_2)$, sodium salt | r.a.c. | – |

r.a.c. = raw agricultural commodities
*Carcinogen/Hazard Classifications*
**U.S. EPA Carcinogens:** Borax and Boric Acid: Group E, Unlikely to be Carcinogenic
*WHO Acute Hazard:* Borax: Class U, Unlikely to be hazardous; Boric acid: Not listed
*Regulatory Authority:*
• Banned or Severely Restricted (In many products) (UN)[13]
• Actively registered pesticide in California. as sodium tetraborate (borax) and boric acid
• Air Pollutant Standard Set (ACGIH)[1] (HSE)[33] (former USSR)[43] (OSHA)[58] (Several States)[60].
• Safe Drinking Water Act, 5FR1470 Priority List
• Canada Drinking Water Quality: 5.0 mg/L IMAC
• Mexico Drinking Water Criteria: 1.0 mg/L
*Description:* Boric acid is a white, amorphous powder or colorless, crystalline solid. Melting/Freezing point = 168° – 169°C (decomposes above 100°C). Saturated solutions: @ 0°C, 2.6% acid; @ 100°C, 28% acid. Boric acid is soluble in water (5 g/100 ml @ 20°C). Borax is a bluish-gray or green, odorless crystalline powder or granules. Boiling point = 320°C. Melting/Freezing point = 75°C (rapid heating). Borax is soluble in water (6g/100 ml @ 20°C). Boron, B is a yellow or brownish-black powder and may be either crystalline or amorphous. It does not occur free in nature and is found in the minerals borax, colemanite, boronatrocalcite, and boracite. Melting/Freezing point = 2190°C. Boiling point = 3660°C. Practically insoluble in water, although boron is reported to be slightly soluble under certain conditions.
*Incompatibilities:* Contact with strong oxidizers may cause explosions. Boron dust is explosive on exposure with air. Boron is incompatible with ammonia, bromine tetrafluoride, cesium carbide, chlorine, fluorine, interhalogens, iodic acid, lead dioxide, nitric acid, nitric oxide, nitrosyl fluoride, nitrous oxide, potassium nitrite, rubidium carbide, silver fluoride. Boric acid decomposes in heat above 100 °C forming boric anhydride and water. Boric acid aqueous solution is a weak acid; incompatible with alkali carbonates and hydroxides.
*Permissible Exposure Limits in Air:* The TWA set by ACGIH[1] and by HSE[33] for anhydrous sodium tetraborate (borax) is 1.0 mg/m$^3$; for sodium tetraborate decahydrate it is 5.0 mg/m$^3$; for the pentahydrate it is 1.0 mg/m$^3$. OSHA[2] has set a TWA of 10 mg/m$^3$ for all sodium tetraborates. A MAC value in workplace air has been set by the former USSR-UNEP/IRPTC project[43] at 10 mg/m$^3$ for boric acid. Several states have set guidelines or standards for sodium tetraborates in ambient air[60], ranging from 10 $\mu$g/m$^3$ (North Dakota) to 16 $\mu$g/m$^3$ (Virginia) to 20 $\mu$g/m$^3$ (Connecticut) to 24 $\mu$g/m$^3$ (Nevada) to 100 $\mu$g/m$^3$ (Connecticut).
*Permissible Concentration in Water:* EPA in July 1976 established a criterion for boron of 750 $\mu$g/L for long term irrigation on sensitive crops. More recently[32], EPA has suggested an ambient water limit of 43 $\mu$g/L based on health effects. The former USSR-UNEP/IRPTC project[43] has set MAC values in mg/L, in water used for fishery purposes of 0.1 for boric acid and 0.05 for sodium tetraborate. See Regulatory Authority for Canadian and Mexican levels.
*Routes of Entry:* Inhalation of dust, fumes, and aerosols; ingestion.
*Harmful Effects and Symptoms*
*Local:* These boron compounds may produce irritation of the nasal mucous membranes, the respiratory tract, and eyes.
*Systemic:* These effects vary greatly with the type of compound. Acute poisoning in man from boric acid or borax is usually the result of application of dressings, powders, or ointment to large areas of burned or abraded skin, or accidental ingestion. The sings are: nausea, abdominal pain, diarrhea and violent vomiting, sometimes

bloody, which may be accompanied by headache and weakness. There is a characteristic erythematous rash followed by peeling. In severe cases, shock with fall in arterial pressure, tachycardia, and cyanosis occur. Marked CNS irritation, oliguria, and anuria may be present. The oral lethal dose in adults is over 30 g. Little information is available on chronic oral poisoning, although it is reported to be characterized by mild Gl irritation, loss of appetite, disturbed digestion, nausea, possibly vomiting, and erythematous rash. The rash may be "hard" with a tendency to become purpuric. Dryness of skin and mucous membranes, reddening of tongue, cracking of lips, loss of hair, conjunctivitis, palpebral edema, gastro-intestinal disturbances, and kidney injury have also been observed. Workers manufacturing boric acid had some atrophic changes in respiratory mucous membranes, weakness, joint pains, and other vague symptoms. The biochemical mechanism of boron toxicity is not clear but seems to involve action on the nervous system, enzyme activity, carbohydrate metabolism, hormone function, and oxidation processes coupled with allergic effects. Borates are excreted principally by the kidneys. No toxic effects have been attributed to elemental boron. The $LD_{50}$ (oral mouse) for *boron* = 2000 mg/kg. The $LD_{50}$ (oral, rat) for *boric acid* = 2660 mg/kg (slightly toxic). The $LD_{50}$ (oral, rat) for *borax* = 2660 mg/kg (slightly toxic).

***Short Term Exposure:*** Boric acid irritates the eyes, skin, and the respiratory tract. High exposure may cause effects on the gastrointestinal tract, liver and kidneys. Borax and Boric acid may affect the nervous system. Serious overexposure can cause seizures, unconsciousness and death.

***Long Term Exposure:*** May cause brain, kidney and liver damage. Repeated or prolonged contact with skin may cause dermatitis. Animal tests show that this substance possibly causes toxic effects upon human reproduction. (WHO).

***Medical Surveillance:*** No specific considerations are needed for boric acid or borates except for general health and liver and kidney function. In the case of boron trifluoride, the skin, eyes, and respiratory tract should receive special attention. In the case of the boranes, central nervous system and lung function will also be of special concern.

***First Aid:*** If this chemical gets into the eyes, remove any contact lenses at once and irrigate immediately for at least 15 minutes, occasionally lifting upper and lower lids. Seek medical attention immediately. If this chemical contacts the skin, remove contaminated clothing and wash immediately with soap and water. Seek medical attention immediately. If this chemical has been inhaled, remove from exposure, begin rescue breathing (using universal precautions) if breathing has stopped and CPR if heart action has stopped. Transfer promptly to a medical facility. When this chemical has been swallowed, get medical attention. Give large quantities of water and induce vomiting. Do not make an unconscious person vomit.

***References:***
- EPA, Office of Pesticide Programs, Pesticide Residue Limits, "Borax" and "Boric Acid", 40 CFR 180.1121, http://www.epa.gov/pesticides/food/viewtols.htm
- Environmental Protection Agency, Preliminary Investigation of Effects on the Environment of Boron, Indium, Nickel, Selenium, Tin, Vanadium and Their Compounds, Volume 1: Boron, Report EPA-560/2-75-005A, Washington, DC, Office of Toxic Substances (August 1975).
- National Institute for Occupational Safety and Health, Information Profiles on Potential Occupational Hazards: Boron and Its Compounds, Report PB 276,678, Rockville, Maryland, pp 63-75 (October 1977).
- Sax, N.I., Ed., "Dangerous Properties of Industrial Materials Report" 1, No. 8, 42-45 (1981) (Boron and Boric Acid) and 3, No. 5, 65-67, New York, Van Nostrand Reinhold Co. (1983).
- Sax, N.I., Ed., "Dangerous Properties of Industrial Materials Report" 2, No. 6, 76-78 (1982) (Sodium Borate).
- California Environmental Protection Agency Chemical List of Lists, Sacramento CA (February 1997).
- New Jersey Department of Health and Senior Services, "Hazardous Substance Fact Sheet: Sodium Borates," Trenton, NJ (February 1989, rev. November 2001). http://www.state.nj.us/health/eoh/rtkweb/0241.pdf

# BPMC

***Use Type:*** Insecticide
***CAS Number:*** 3766-81-2
***Formula:*** $C_{12}H_{17}NO_2$
***Synonyms:*** o-sec-Butylphenyl methylcarbamate; 2-sec-Butylphenyl n-methylcarbamate; Fenobucarb; Methylcarbamic acid o-sec-butylphenyl ester; Phenol, 2-(1-methylpropyl)-, methylcarbamate; 2-(1-Methylpropyl)phenyl methylcarbamate
***Trade Names:*** BARIZON®; BASSA®; BAY-41637®, Bayer CropScience (Germany); BAY-41637-C®, Bayer CropScience (Germany); BAYCARB®, Bayer CropScience (Germany); BAYER-41367C®, Bayer CropScience (Germany); CARVIL®; FENOBCARB®; GEOCARB-50EC®; HOPCIN®; OSBAC®
***Producers:*** Bayer CropScience (Germany); Saeryung Chemicals (South Korea); Yellow River Enterprise (Taiwan)
***Chemical Class:*** N-Methyl carbamate
***Uses:*** Not registered in the U.S. BPMC, fenobucarb, is extensively used for control of plant hopper in rice, and for leafhoppers, rice stem-borers, and rice bugs on rice, and to control bollworms and aphids on cotton.

*Carcinogen/Hazard Classifications*
**WHO Acute Hazard:** Class II, moderately hazardous
*Description:* Yellowish liquid. Slight odor. Soluble in water; solubility = 610 ppm @ 20°C. Molecular weight = 207.32. Melting/freezing point = 31°C. Vapor pressure = 48 mPa @ 20°C. Density = 1.03; Log $K_{ow}$ = 3.2. Values greater than 3.0 may bioaccumulate in marine organisms.
*Incompatibilities:* May form explosive materials with phosphorus pentachloride.
*Permissible Concentration in Water:* No criteria set. Runoff from spills or fire control may cause water pollution.
*Routes of Entry:* Inhalation, skin contact, ingestion.
*Harmful Effects and Symptoms*
*Short Term Exposure:* Eye pupils are small; blurred vision; eye watering; runny nose; cough; shortness of breath; salivation; nausea, stomach cramps, diarrhea, and vomiting; increased blood pressure; profuse sweating; hypermotility, hallucinations; agitation; tingling of the skin; slow heartbeat; convulsions; fluid in lungs; loss of consciousness; incontinence; breathing stops; death. Carbamates inhibit the acetylcholinesterase enzymes and alter the way in which nervous impulses are transmitted. However, within several hours carbamates spontaneously detach from the enzymes.
*Long Term Exposure:* A potent cholinesterase inhibitor; cumulative effect is possible. This chemical may damage the nervous system with repeated exposure, resulting in convulsions, respiratory failure. May cause liver damage.
*Points of Attack:* Respiratory system, lungs, central nervous system, cardiovascular system, skin, eyes, plasma and red blood cell cholinesterase.
*Medical Surveillance:* Medical observation is recommended for 24 to 48 hours after breathing overexposure, as pulmonary edema may be delayed. As first aid for pulmonary edema, consider administering a corticosteroid spray. Cigarette smoking may exacerbate pulmonary injury and should be discouraged for at least 72 hours following exposure. Before employment and at regular times after that, the following are recommended: Plasma and red blood cell cholinesterase levels (tests for the enzyme poisoned by this chemical). If exposure stops, plasma levels return to normal in 1-2 weeks while red blood cell levels may be reduced for 1-3 months. When acetylcholinesterase enzyme levels are reduced by 25% or more below preemployment levels, risk of poisoning is increased, even if results are in lower ranges of "normal." Reassignment to work not involving carbamate pesticides is recommended until enzyme levels recover. If symptoms develop or overexposure occurs, repeat the above tests as soon as possible and get an exam of the nervous system. Also consider complete blood count. Consider chest x-ray following acute overexposure.
*First Aid:* Speed in removing material from skin is of extreme importance. Eye contact can cause dangerous amounts of these chemicals to be quickly absorbed through the mucous membrane into the bloodstream. Immediately and gently flush eyes with plenty of warm or cold water (NO hot water) for at least 15 minutes, occasionally lifting the upper and lower eyelids. Get medical aid immediately. Skin: Get medical aid. Skin and/or eye contact can cause dangerous amounts of these chemicals to be absorbed into the bloodstream. Wearing the appropriate PPE equipment and respirator for carbamate pesticides, immediately flush skin with plenty of soap and water for at least 15 minutes while removing contaminated clothing and shoes. Shampoo hair promptly if contaminated; protect eyes. *Ingestion:* Call poison control. Loosen all clothing. Never give anything by mouth to an unconscious person. Get medical aid. *Do NOT induce vomiting.*\* If conscious, alert, and able to swallow, rinse mouth and have victim drink 4 to 8 ounces of water. Check to see if poison control instructs you to use ipecac syrup, otherwise administer slurry of activated charcoal (2 oz in 8 oz of water). If victim is *unconscious or having convulsions,* do nothing except keep victim warm. \**In some cases you may be specifically instructed by poison control to induce vomiting by way of 2 tablespoons of syrup of ipecac (adult) washed down with a cup of water.* Do NOT give activated charcoal before or with ipecac syrup. Inhalation: Get medical aid. Do not contaminate yourself. Wearing the appropriate PPE equipment and respirator for carbamate pesticides, immediately remove the victim from the contaminated area to fresh air. If the victim is not breathing, administer artificial respiration. Do NOT use mouth-to-mouth resuscitation; use bag/mask apparatus. If breathing is difficult, administer oxygen through bag/mask apparatus until medical help arrives. Do not leave victim unattended. *Note to physician or authorized medical personnel.* Administer atropine, 2 mg (1/30 gr) intramuscularly or intravenously as soon as any local or systemic signs or symptoms of an intoxication are noted; repeat the administration of atropine every 3 to 8 minutes until signs of atropinization (mydriasis, dry mouth, rapid pulse, hot and dry skin) occur; initiate treatment in children with 0.05 mg mg/kg of atropine; repeat at 5 to 10 minute intervals. Watch respiration, and remove bronchial secretions if they appear to be obstructing the airway; intubate if necessary. *Medical note:* 2-PAMCI may be contraindicated in the case of some carbamate poisonings.

# Bromacil (ANSI)

*Use Type:* Herbicide
*CAS Number:* 314-40-9
*Formula:* $C_9H_{13}BrN_2O_2$
*Alert:* A General Use Pesticide (GUP)
*Synonyms:* Bromacil 1.5; Bromazil; 5-Bromo-3-*sec*-butyl-6-methyluracil; 5-Bromo-6-methyl-3-(1-methylpropyl)-2,4-(1H,3H)-pyrimidinedione; 5-Bromo-6-methyl-3-(1-methylpropyl)-2,4(1H,3H)-pyrimidinedione; 3-*sek*-Butyl-5-

brom-6-methyluracil (German); Cynogan; 2,4(1H,3H)-Pyrimidinedione, 5-bromo-6-methyl-3-(1-methylpropyl)-; Uracil, 5-Bromo-3-*sec*-butyl-6-methyl

*Trade Names:* BOREA®; BOROCIL EXTRA®; α-BROMACIL 80 WP®; BROMAX®; CROPTEX ONYX®; CYNOGEN®; DUPONT HERBICIDE 976®, DuPont Crop Protection (USA); EEREX®; FENOCIL®; HERBICIDE 976®; HIBOR; HYDON®; HYVAR®, DuPont Crop Protection (USA); HYVAR-X®, DuPont Crop Protection (USA); HYVAR X BROMACIL® DuPont Crop Protection (USA); HYVAR X-L® DuPont Crop Protection (USA); HYVAR X WEED KILLER® DuPont Crop Protection (USA); HYVAR X-WS® DuPont Crop Protection (USA); ISOCIL®; KROVAR®, DuPont Crop Protection (USA); NALKIL®; ROUT®; URAGAN®; URAGON®; UROX®; UROX B WATER SOLUBLE CONCENTRATE WEED KILLER®; UROX HX GRANULAR WEED KILLER®; WEED-BROOM® (mixture of DSMA, Bromacil & 2,4-D)

*Producers:* Agan Chemical Manufacturers Ltd. (Israel); DuPont Crop Protection (USA); Sigma-Aldrich Laborchemikalien (Germany)

*Chemical Class:* Uracil

*EPA/OPP PC Code:* 012301

*California DPR Chemical Code:* 83

*ICSC Number:* 1448

*RTECS Number:* YQ9100000

*EINECS Number:* 206-245-1

*Uses:* Bromacil is used primarily for the control of annual and perennial grasses and broadleaf weeds, both nonselectively on noncrop lands and selectively for weed-control in citrus and pineapple crops. The top five applications in California for which this is used are oranges, lemons, grapefruit, and right-of-ways and landscapes. A limit of 0.1 mg/kg of agricultural products is set in several countries[35].

*Human toxicity (long-term)[77]:* Low–90.00 ppb, Health Advisory

*Fish toxicity (threshold)[77]:* Very low–6485.34001 ppb, MATC (Maximum Acceptable Toxicant Concentration)

*U.S. Maximum Allowable Residue Levels for Bromacil (40 CFR 180.210):*

| CROP | ppm |
|---|---|
| Fruit, stone, group | 0.1 |
| Pineapple | 0.1 |
| Fruit, stone, group | 0.1 |
| Pineapple | 0.1 |

*Carcinogen/Hazard Classifications*

**U.S. EPA Carcinogens:** Group C, possible carcinogen

**Label Signal Word:** CAUTION (for dry formulations); WARNING (for liquid formulations)

**WHO Acute Hazard:** Class U, Not likely to be hazardous

*Regulatory Authority:*
- Air Pollutant Standard Set (ACGIH)[1] (NIOSH)[2] (HSE)[33] (Several States)[60] (Australia) (Israel) (Mexico) (Canada, Provincial)
- AB 1803-Well Monitoring Chemical (CAL)
- Permissible Exposure Limits for Chemical Contaminants (CAL/OSHA)
- The "Director's List" (CAL/OSHA)
- Actively registered pesticide in California.
- Safe Drinking Water Act: Priority List (55 FR 1470)
- EPCRA Section 313 Form R *de minimus* concentration reporting level: 1.0%.
- Canada, WHMIS, Ingredients Disclosure List

*Description:* White crystalline solid. May be dissolved in a flammable liquid. Odorless. Slightly soluble in water; solubility = 820 ppm @ 25°C. Boiling point = (sublimes). Melting/Freezing point = 158–159°C (sublimes). Vapor pressure = $3.1 \times 10^{-7}$ mmHg. Hazard Identification (based on NFPA-704 M Rating System): Health 1, Flammability 0, Reactivity 0

*Incompatibilities:* Incompatible with strong acids, oxidizers, heat. Decomposes slowly in strong acids. Emits toxic fumes of nitrogen oxides and bromine and when heated to decomposition.

*Permissible Exposure Limits in Air:* A TWA value of 1 ppm (10 mg/m$^3$) has been recommended by NIOSH[2] and ACGIH[1]. This same TWA has been set by Australia, Israel, Mexico, and HSE[33]. HSE[33] and Mexico set a STEL value at 2 ppm (20 mg/m$^3$). The Canadian provinces of Alberta, Ontario and Quebec have the same TWAs and Alberta's STEL is 2 ppm (21 mg/m$^3$). Four states have set guidelines or standards for bromacil in ambient air[60] ranging from 100 μg/m$^3$ (North Dakota) to 160 μg/m$^3$ (Virginia) to 200 μg/m$^3$ (Connecticut) to 238 μg/m$^3$ (Nevada).

*Determination in Air:* Filter; none; Gravimetri; NIOSH Methods (IV) #0500, Particulates NOR (total).[18]

*Permissible Concentration in Water:* A no-adverse effects level in drinking water has been calculated by NAS/NRC[46] as 0.086 mg/L. Some states have set guidelines for bromacil in drinking water[61], including Maine at 25 μg/L, and Kansas at 87.5 μg/L.

*Routes of Entry:* Inhalation, ingestion, skin and/or eye contact.

*Harmful Effects and Symptoms*

*Short Term Exposure:* Irritates the eyes, skin, upper respiratory system, lungs. Inhalation can cause irritation, coughing and wheezing.

*Long Term Exposure:* Has caused thyroid affects in animals.

*Points of Attack:* Eyes, skin, respiratory system, thyroid.

*Medical Surveillance:* Before beginning employment and at regular times after that, for those with frequent or potentially high exposures, the following is recommended: Lung function tests. Thyroid function tests. Consider x-ray following acute overexposure.

*First Aid:* If this chemical gets into the eyes, remove any contact lenses at once and irrigate immediately for at least 15 minutes, occasionally lifting upper and lower lids. Seek medical attention immediately. If this chemical contacts the skin, remove contaminated clothing and wash immediately with soap and water. Seek medical attention immediately. If this chemical has been inhaled, remove from exposure, begin rescue breathing (using universal precautions) if breathing has stopped and CPR if heart action has stopped. Transfer promptly to a medical facility. When this chemical has been swallowed, get medical attention. Give large quantities of water and induce vomiting. Do not make an unconscious person vomit.

*References:*
- U.S. Environmental Protection Agency, Office of Pesticide Programs, Pesticide Residue Limits, "Bromacil", 40 CFR 180.210, www.epa.gov/cgi-bin/oppsrch
- New Jersey Department of Health and Senior Services, "Hazardous Substance Fact Sheet: Bromacil," Trenton, NJ (August 1992, rev. July 1998). http://www.state.nj.us/health/eoh/rtkweb/0251.pdf
- EXTOXNET, Extension Toxicology Network, "Pesticide Information Profile, Bromacil," Oregon State University, Corvallis, OR (June 1996) http://ace.orst.edu/info/extoxnet/pips/bromacil
- California Environmental Protection Agency Chemical List of Lists, Sacramento CA (February 1997).

# Bromadiolone

*Use Type:* Rodenticide
*CAS Number:* 28772-56-7
*Formula:* $C_{30}H_{23}BrO_4$
*Synonyms:* 2H-1-Benzopyran-2-one, 3-(3-[4'-bromo(1,1'-biphenyl)-4-yl]-3-hydroxy-1-phenylpropyl-4-hydroxy-; Bromadialone; 3-[3-(4'-bromo(1,1'-biphenyl)-4-yl)3-hydroxy-1-phenylpropyl]-4-hydroxy-2H-1-benzopyran-2-one; 3-[3-(4'-Bromobiphenyl)-4-yl]3-hydroxy-1-phenylpropyl)-4-hydroxy-coumarin; 3-[α-(p-(p-Bromophenyl)-β-hydroxyphenethyl)benzyl]-4-hydroxy-coumarin; Coumarin, 3-[3-(4'-bromo-1,1'-biphenyl-4-yl)-3-hydroxy-1-phenylpropyl]-4-hydroxy-; Coumarin, 3-[α-(p-(p-bromophenyl)-β-hydroxyphenethyl)benzyl]-4-hydroxy-; (Hydroxy-4-coumarinyl 3)-3 phenyl-3(bromo-4 biphenyl-4)-1 propanol-1 (French)
*Trade Names:* BOLDO®; BOOT HILL®; BROMONE®; CANADIEN 2000®; CONTRAC®, Bell Laboratories (USA); HAWK®; LM-637®; MAKI®, Liphatech, Inc. (USA); RAT ARREST®; RAT FREE®; RATIMUS®; RENTOKIL DEADLINE®, Rentokil Environmental Service (Netherlands); SLAYMOR®; SUPER-CAID®, Liphatech, Inc. (USA); SUPER-ROZOL®, Liphatech, Inc. (USA); SUP'ORATS®; TEMUS®
*Producers:* Agrimor International (USA); Agro-care Chemical Industry Group (China); Bell Laboratories (USA); BP Chemicals (UK); I.N.D.I.A. Industrie Chimiche (Italy); Liphatech, Inc (USA); Sigma-Aldrich Laborchemikalien (Germany); Sumitomo (Japan); Wuhan kernel Bio-pesticide (China)
*Chemical Class:* Coumarin
*EPA/OPP PC Code:* 112001; (208500 & 214600 use code No. 112001)
*California DPR Chemical Code:* 2135
*RTECS Number:* GN4934700
*EINECS Number:* 249-205-9
*Uses:* Bromadiolone is used as bait for rodent control against house mice, roof rats and warfarin-resistant Norway rats. It is also authorized by USDA for use in official establishments operating under the Federal meat, poultry, shell egg grading and egg products inspection program.
*Carcinogen/Hazard Classifications*
**Label Signal Word:** CAUTION
**WHO Acute Hazard:** Class 1 a, extremely hazardous
*Regulatory Authority:*
- Superfund/EPCRA 40CFR355, Appendix B Extremely Hazardous Substances: TPQ = 100/10,000 lb (45.4/4,540 kg)
- Superfund/EPCRA 40CFR302.4 RQ: EHS, 100 lb (45.4 kg)
- Actively registered pesticide in California.

*Description:* Bromadiolone is a white to yellowish powder. Practically insoluble in water. Melting/Freezing point = 200-210°C. Hazard Identification (based on NFPA-704 M Rating System): Health 4, Flammability 1, Reactivity 0.
*Permissible Exposure Limits in Air:* No standards set.
*Permissible Concentration in Water:* No criteria set, but runoff from spills or fire control may cause water pollution.
*Routes of Entry:* Ingestion, inhalation, skin and/or eye contact
*Harmful Effects and Symptoms*
*Short Term Exposure:* Contact with eyes or skin may cause irritation or injury. Inhalation should be avoided; use NIOSH-approved air purifying respirators for pesticides. May be harmful if swallowed. The compound is toxic by oral exposure, the $LD_{50}$ (rabbit, oral) = 1.0 mg/kg and the $LD_{50}$-(oral, rat) = 1.125 mg/kg. (Extremely toxic)[9].
*Long Term Exposure:* May cause skin and eye irritation.
*First Aid:* If this chemical gets into the eyes, remove any contact lenses at once and irrigate immediately for at least 15 minutes, occasionally lifting upper and lower lids. Seek medical attention immediately. If this chemical contacts the skin, remove contaminated clothing and wash immediately with soap and water. Seek medical attention immediately. If this chemical has been inhaled, remove from exposure, begin rescue breathing (using universal precautions) if breathing has stopped and CPR if heart action has stopped. Transfer promptly to a medical facility. When this chemical has been swallowed, get medical attention. Give large

quantities of water and induce vomiting. Do not make an unconscious person vomit. Keep victim quiet and maintain normal body temperature. Effects may be delayed; keep victim under observation.

*References:*
- New Jersey Department of Health and Senior Services, "Hazardous Substance Fact Sheet, Bromadiolone, Trenton NJ (July 2000). http://www.state.nj.us/health/eoh/rtkweb/2179.pdf
- California Environmental Protection Agency Chemical List of Lists, Sacramento CA (February 1997).
- U.S. Environmental Protection Agency, "Chemical Profile: Bromadiolone," Washington, DC, Chemical Emergency Preparedness Program (November 30, 1987).

# Bromophos

*Use Type:* Insecticide and acaricide
*CAS Number:* 2104-96-3
*Formula:* $C_8H_8BrCl_2O_3PS$
*Synonyms:* O-(4-Brom-2,5-dichlor-phenyl)-O,O-dimethyl-monothiophosphat (German); 4-Bromo-2,5-dichlorophenyl dimethyl phosphorothionate; O-(4-bromo-2,5-dichlorophenyl) O,O-dimethylphosphorothioic acid; Bromofos; Bromofos methyl; Bruomophos (Russian); O,O-Dimethyl-O-(4-bromo-2,5-dichlorophenyl) phosphorothioate; O,O-Dimethyl-O-(2,5-dichlor-4-bromphenyl)-thionophosphat (German); O,O-dimethyl-O-(2,5-dichloro-4-bromophenyl)phosphorothioate; O,O-Dimethyl-O-(2,5-dichloro-4-bromophenyl) thiophosphate; ENT 27,162; Methyl bromofos; Mtehyl bromophos; Phosphorothioic acid, O-(4-bromo-2,5-dichlorophenyl) O,O-dimethyl ester; Thiophosphate de O,O-dimethyle et de O-4-bromo-2,5-dichlorophenyle (French)
*Trade Names:* BROFENE®, canceled; BROPHENE®; CELA S-1942®; EL 400®; MONSANTO®-CP 51969; NETAL®; NEXION®; NEXION®, canceled; NEXION®-40, canceled; OMEXAN®; OMS-658®; S 1942®
*Producers:* Epochem (China); Monsanto (USA)
*Chemical Class:* Organophosphate
*EPA/OPP PC Code:* 008706
*RTECS Number:* TE7175000
*Uses:* Not registered in the U.S. Bromophos is a broad spectrum, non-cumulative non-systemic organophosphorus insecticide. It is used on field crops, vegetable and fruit crops, on ornamentals, and on grain storage. It is also used as a sheep dip.
*Regulatory Authority:*
- DOT Inhalation Hazard Chemicals as organophosphates.

*Description:* White to slightly yellow crystalline solid. Very slightly soluble in water. Molecular weight = 366.00. Melting point: 53°C.
*Incompatibilities:* Keep away from oxidizers, sulfuric acid, caustics, ammonia, aliphatic amines, alkanolamines, isocyanates, alkylene oxides, epichlorohydrin. May react violently with antimony(V) pentafluoride. Incompatible with lead diacetate, magnesium, silver nitrate.
*Determination in Air:* OSHA versatile sampler-2; Toluene/Acetone; Gas chromatography/Flame ionization detection; NIOSH IV[18], Method #5600, Organophosphorus Pesticides.
*Permissible Concentration in Water:* No criteria set. Runoff from spills or fire control may cause water pollution.
*Routes of Entry:* Inhalation, skin absorption, ingestion, skin and/or eye contact.
*Harmful Effects and Symptoms*
*Short Term Exposure:* Eye pupils are small; blurred vision; eye watering; runny nose; cough; shortness of breath; salivation; dizziness; nausea, stomach cramps, diarrhea, and vomiting; increased blood pressure; profuse sweating; hypermotility, hallucinations; irritability; tingling of the skin; drowsiness; slow heartbeat; convulsions; fluid in lungs; loss of consciousness; incontinence; breathing stops; death. Orgnophosphates inhibit the action of acetylcholinesterase enzymes, and alter the way in which nervous impulses are transmitted. The effects can last for hours, days, or much longer. The action of the enzymes is reestablished after new enzymes are formed.
*Long Term Exposure:* Cholinesterase inhibitor; cumulative effect is possible. Organophosphates may damage the nervous system with repeated exposure, resulting in convulsions, respiratory failure. May cause liver damage.
*Points of Attack:* Respiratory system, central nervous system, cardiovascular system, blood cholinesterase.
*Medical Surveillance:* Medical observation is recommended for 24 to 48 hours after breathing overexposure, as pulmonary edema may be delayed. As first aid for pulmonary edema, consider administering a corticosteroid spray. Cigarette smoking may exacerbate pulmonary injury and should be discouraged for at least 72 hours following exposure.
*First Aid:* Speed in removing material from skin is of extreme importance. Eye contact can cause dangerous amounts of these chemicals to be quickly absorbed through the mucous membrane into the bloodstream. Immediately and gently flush eyes with plenty of warm or cold water (NO hot water) for at least 15 minutes, occasionally lifting the upper and lower eyelids. Get medical aid immediately. Skin: Get medical aid. Skin and/or eye contact can cause dangerous amounts of these chemicals to be absorbed into the bloodstream. Wearing the appropriate PPE equipment and respirator for organophosphate/carbamate pesticides, immediately flush skin with plenty of soap and water for at least 15 minutes while removing contaminated clothing and shoes. Shampoo hair promptly if contaminated. *Ingestion:* Call poison control. Loosen all clothing. Never give anything by mouth to an unconscious person. Get medical aid. Do NOT induce vomiting.* If conscious, alert, and able

to swallow, rinse mouth and have victim drink 4 to 8 ounces of water do NOT induce vomiting but immediately administer slurry of activated charcoal (2 oz in 8 oz of water). If victim is *unconscious or having convulsions,* do nothing except keep victim warm. *In some cases you may be specifically instructed by poison control to induce vomiting by way of 2 tablespoons of syrup of ipecac (adult) washed down with a cup of water.* Do NOT give activated charcoal before or with ipecac syrup. Inhalation: Get medical aid. Do not contaminate yourself. Wearing the appropriate PPE equipment and respirator for organophosphate pesticides, immediately remove the victim from the contaminated area to fresh air. If the victim is not breathing, administer artificial respiration. Do NOT use mouth-to-mouth resuscitation; use bag/mask apparatus. If breathing is difficult, administer oxygen through bag/mask apparatus until medical help arrives. Do not leave victim unattended. *Note to physician or authorized medical personnel.* Administer atropine, 2 mg (1/30 gr) intramuscularly or intravenously as soon as any local or systemic signs or symptoms of an intoxication are noted; repeat the administration of atropine every 3 to 8 minutes until signs of atropinization (mydriasis, dry mouth, rapid pulse, hot and dry skin) occur; initiate treatment in children with 0.05 mg mg/kg of atropine; repeat at 5 to 10 minute intervals. Watch respiration, and remove bronchial secretions if they appear to be obstructing the airway; intubate if necessary. *Notes to physician or authorized medical personnel*: $N$-methylpyridinium-2-aldoxime (2-PAMCI) when used in conjunction with atropine reacts with the phosphorylated cholinesterase, thereby restoring normal activity to by removing the phosphorylating group. The combination of these two chemicals is synergistic and must be administered within minutes to a few hours following exposure (depending on the specific agent) to be effective. Give 2-PAMCI (Pralidoxime; Protopam), 2.5 gm in 100 ml of sterile water or in 5% dextrose and water, intravenously, slowly, in 15-30 minutes; if sufficient fluid is not available, give 1 gm of 2-PAMCI in 3 ml of distilled water by deep intramuscular injection; repeat this every half hour if respiration weakens or if muscle fasciculation or convulsions recur. Also Diazepam, an anticonvulsant, might be considered.

*References:*
- International Programme on Chemical Safety (IPCS), "Data Sheet on Pesticides No. 76, Bromophos," Geneva, Switzerland (1994).
  http://www.inchem.org/documents/pds/pds/pest76_e.htm
- International Programme on Chemical Safety (IPCS), "Pesticide Residues in Food–1982, Bromophos," Geneva, Switzerland (1982).
  http://www.inchem.org/documents/jmpr/jmpmono/v82pr06.htm
- California Environmental Protection Agency Chemical List of Lists, Sacramento CA (February 1997)

# Bromophos-ethyl

*Use Type:* Insecticide
*CAS Number:* 4824-78-6
*Formula:* $C_{10}H_{12}BrCl_2O_3PS$
*Synonyms:* 4-Bromo-2,5-dichlorophenol-$O$-ester with $O,O$-diethyl phosphorothioate; $O$-(4-Bromo-2,5-dichlorophenyl) $O,O$-diethyl phosphorothioate; Bromophos-ethyl; $O,O$-Diaethyl-$O$-(4-brom-2,5-dichlor)-phenyl-monothiophosphat (German); $O,O$-Diaethyl-$O$-(2,5-dichlor-4-bromphenyl)-thionophosphat (German); $O,O$-Diethyl $O$-2,5-dichloro-4-bromophenyl-phosphorothioate; $O,O$-Diethyl $O$-(2,5-dichloro-4-bromophenyl)thiophosphate; ENT 27,258; Ethyl bromophos; Phosphorothioic acid, $O$-(4-bromo-2,5-dichlorophenyl) $O,O$-diethyl ester; Thiophosphate de $O,O$-diethyle et de $O$-(2,5-dichloro-4-bromo) phenyle (French)
*Trade Names:* CELA S-2225®; FILARIOL®; NEXAGAN®; OMS-659®; S 2225®
*Chemical Class:* Organophosphate
*EPA/OPP PC Code:* 214500
*Uses:* Not registered in the U.S.
*Regulatory Authority:*
- DOT Inhalation Hazard Chemicals as organophosphates
- List of priority pollutants (EPA) as chlorophenols
- AB 2588-Air Toxics "Hot Spots" Chemicals (CAL) as chlorophenols
- The "Director's List" (CAL/OSHA) as chlorophenols
- Clean Water Act: Toxic Pollutant (Section 401.15) other than those listed elsewhere; includes trichloro phenols
- RCRA Section 261 Hazardous Constituents, waste number not listed as chlorophenols
- EPCRA Section 313 (as chlorophenols) Form R *de minimis* concentration reporting level: 1.0%.

*Description:* Pale-yellow liquid. Molecular weight = 394.05. Density = 1.53 @ 20°C. Boiling point = 123°C @ $4 \times 10^{-3}$ mmHg.
*Incompatibilities:* May react violently with antimony(V) pentafluoride. Incompatible with lead diacetate, magnesium, silver nitrate.
*Permissible Exposure Limits in Air:*
*Determination in Air:* OSHA versatile sampler-2; Toluene/Acetone; Gas chromatography/Flame ionization detection; NIOSH IV, Method #5600, Organophosphorus Pesticides.[18]
*Permissible Concentration in Water:* No criteria set. Runoff from spills or fire control may cause water pollution.
*Routes of Entry:* Inhalation, skin absorption, ingestion, skin and/or eye contact.
*Harmful Effects and Symptoms*
*Short Term Exposure:* Eye pupils are small; blurred vision; eye watering; runny nose; cough; shortness of breath;

salivation; dizziness; nausea, stomach cramps, diarrhea, and vomiting; increased blood pressure; profuse sweating; hypermotility, hallucinations; irritability; tingling of the skin; drowsiness; slow heartbeat; convulsions; fluid in lungs; loss of consciousness; incontinence; breathing stops; death. Orgnophosphates inhibit the action of acetylcholinesterase enzymes, and alter the way in which nervous impulses are transmitted. The effects can last for hours, days, or much longer. The action of the enzymes is reestablished after new enzymes are formed.

*Long Term Exposure:* Cholinesterase inhibitor; cumulative effect is possible. Organophosphates may damage the nervous system with repeated exposure, resulting in convulsions, respiratory failure. May cause liver damage.

*Points of Attack:* Respiratory system, central nervous system, cardiovascular system, blood cholinesterase.

*Medical Surveillance:* Medical observation is recommended for 24 to 48 hours after breathing overexposure, as pulmonary edema may be delayed. As first aid for pulmonary edema, consider administering a corticosteroid spray. Cigarette smoking may exacerbate pulmonary injury and should be discouraged for at least 72 hours following exposure.

*First Aid:* Speed in removing material from skin is of extreme importance. Eye contact can cause dangerous amounts of these chemicals to be quickly absorbed through the mucous membrane into the bloodstream. Immediately and gently flush eyes with plenty of warm or cold water (NO hot water) for at least 15 minutes, occasionally lifting the upper and lower eyelids. Get medical aid immediately. Skin: Get medical aid. Skin and/or eye contact can cause dangerous amounts of these chemicals to be absorbed into the bloodstream. Wearing the appropriate PPE equipment and respirator for organophosphate/carbamate pesticides, immediately flush skin with plenty of soap and water for at least 15 minutes while removing contaminated clothing and shoes. Shampoo hair promptly if contaminated. *Ingestion:* Call poison control. Loosen all clothing. Never give anything by mouth to an unconscious person. Get medical aid. Do NOT induce vomiting.* If conscious, alert, and able to swallow, rinse mouth and have victim drink 4 to 8 ounces of water do NOT induce vomiting but immediately administer slurry of activated charcoal (2 oz in 8 oz of water). If victim is *unconscious or having convulsions,* do nothing except keep victim warm. **In some cases you may be specifically instructed by poison control to induce vomiting by way of 2 tablespoons of syrup of ipecac (adult) washed down with a cup of water.* Do NOT give activated charcoal before or with ipecac syrup. Inhalation: Get medical aid. Do not contaminate yourself. Wearing the appropriate PPE equipment and respirator for organophosphate pesticides, immediately remove the victim from the contaminated area to fresh air. If the victim is not breathing, administer artificial respiration. Do NOT use mouth-to-mouth resuscitation; use bag/mask apparatus. If breathing is difficult, administer oxygen through bag/mask apparatus until medical help arrives. Do not leave victim unattended. *Note to physician or authorized medical personnel.* Administer atropine, 2 mg (1/30 gr) intramuscularly or intravenously as soon as any local or systemic signs or symptoms of an intoxication are noted; repeat the administration of atropine every 3 to 8 minutes until signs of atropinization (mydriasis, dry mouth, rapid pulse, hot and dry skin) occur; initiate treatment in children with 0.05 mg mg/kg of atropine; repeat at 5 to 10 minute intervals. Watch respiration, and remove bronchial secretions if they appear to be obstructing the airway; intubate if necessary. *Notes to physician or authorized medical personnel*: N-methylpyridinium-2-aldoxime (2-PAMCl) when used in conjunction with atropine reacts with the phosphorylated cholinesterase, thereby restoring normal activity to by removing the phosphorylating group. The combination of these two chemicals is synergistic and must be administered within minutes to a few hours following exposure (depending on the specific agent) to be effective. Give 2-PAMCl (Pralidoxime; Protopam), 2.5 gm in 100 ml of sterile water or in 5% dextrose and water, intravenously, slowly, in 15-30 minutes; if sufficient fluid is not available, give 1 gm of 2-PAMCl in 3 ml of distilled water by deep intramuscular injection; repeat this every half hour if respiration weakens or if muscle fasciculation or convulsions recur. Also Diazepam, an anticonvulsant, might be considered.

*References:*
- California Environmental Protection Agency Chemical List of Lists, Sacramento CA (February 1997).
- Agency for Toxic Substances and Disease Registry "Toxicological profile for chlorophenols," 1999, Atlanta, GA.

# Bromoxynil (ANSI)

*Use Type:* Herbicide
*CAS Number:* 1689-84-5; 1689-99-2 (octanoate)
*Formula:* $C_7H_3Br_2NO$; $C_{15}H_{17}Br_2NO$ (octanoate)
*Alert:* A Restricted Use Pesticide (RUP).
*Synonyms:* Bensonitrile, 3,5-dibromo-4-hydroxy-; Benzonitrile, 3,5-dibromo-4-hydroxy-; Butilchlorofos; Caswell No. 119; 2,6-Dibromo-4-cyanophenol; 2,6-Dibromo-4-hydroxybenzonitrile; 2,6-Dibromo-4-phenylcyanide; 3,5-Dibromo-4-hydroxybenzonitrile; 3,5-Dibromo-4-hydroxyphenyl cyanide; ENT 20,852; 4-Hydroxy-3,5-dibromobenzonitrile
*octanoate*: 3,5-Dibromo-4-octanoyloxybenzonitrile
*Trade Names:* BRIOTRIL®, Makhteshim-Agan Industries (Israel); BRUCIL®; BRITTOX®; BROMINAL®; BROMINEX®; BROMINAL®; BROMINAL ME-4®;

BROMINIL®; Bromox 2E, Micro-Flo (USA); BROMOTRIL®; BROMOXYNIL NITRILE HERBICIDE®; BRONATE®, Bayer CropScience (Germany); BROXYNIL®; BUCTRIL®, Bayer CropScience (Germany); BUCTRIL® GEL HERBICIDE (octanoate), Bayer CropScience (Germany); BUCTRIL® 4EC GEL (mixture of bromoxynil octanoate and bromoxynil heptanoate), Bayer CropScience (Germany); BUCTRIL INDUSTRIAL®, Bayer CropScience (Germany); CHIPCO BUCTRIL®, Bayer CropScience (Germany); CHIPCO CRAB-KLEEN®; FLAGON®, 400 EC, Makhteshim-Agan Industries (Israel); HOBANE®; LABUCTRIL®; LITAROL®; M&B 10064®; MB 10064®; MB 10731® (octanoate); M&B 10731®; ME4 BROMINAL®; MERIT®; MEXTROL-BIOX®, Whyte Agrochemicals (UK); MOXY 2E®, Agriliance (USA); NCR CE EE DOV7® (octanoate); NU-LAWN WEEDER®; OXYTRIL M®; PARDNER®; SABRE®; TORCH®

*Producers:* Agriliance (USA); Agsin (Singapore); Bayer CropScience (Germany); Makhteshim-Agan Industries (Israel); Micro-Flo (USA); Sanonda Ltd. (Australia); Whyte Agrochemicals (UK

*Chemical Class:* Nitriles (organic cyanides)
*EPA/OPP PC Code:* 035301; 035302 (octanoate)
*California DPR Chemical Code:* 2429
*RTECS Number:* DI3150000
*EINECS Number:* 216-882-7
*Uses:* For post-emergent control of broadleaf weeds. Used on alfalfa, garlic, corn, sorghum, flax, cereals, turf and on pasture and rangelands.
*Human toxicity (long-term)*[77]*:* Intermediate–10.50 ppb, Health Advisory; (octanoate) Very low–140.00 ppb, Health Advisory
*Fish toxicity (threshold)*[77]*:* Low–283.24412 ppb, MATC (Maximum Acceptable Toxicant Concentration); (octanoate) High–4.40230 ppb, MATC (Maximum Acceptable Toxicant Concentration)
*U.S. Maximum Allowable Residue Levels for Bromoxynil [40 CFR 180.324 (B) and (C)]:*

| CROP | ppm |
|---|---|
| Alfalfa | 0.1 |
| Barley, grain & straw | 0.1 |
| Canarygrass, annual, seed | 0.1 |
| Cattle, fat | 0.1 |
| Cattle, meat | 0.5 |
| Cattle, mbyp | 3.5 |
| Corn, field, forage | 0.1 |
| Corn, field, grain | 0.1 |
| Corn, field, stover | 0.1 |
| Corn, forage | 0.1 |
| Corn, grain | 0.1 |
| Corn, stover | 0.1 |
| Cotton, gin byproducts | 7 |
| Cotton, hulls | 5 |
| Cotton, undelinted seed | 1.5 |
| Egg | 0.05 |
| Flax, seed | 0.1 |
| Flax, straw | 0.1 |
| Garlic | 0.1 |
| Goat, fat | 1.0 |
| Goat, meat | 0.5 |
| Goat, mbyp | 3.5 |
| Hog, fat | 1.0 |
| Hog, meat | 0.5 |
| Hog, mbyp | 3.5 |
| Horse, fat | 1.0 |
| Horse, meat | 0.5 |
| Horse, mbyp | 3.5 |
| Milk | 0.1 |
| Mint, hay | 0.1 |
| Oat, forage and hay | 0.1 |
| Oat, grain | 0.1 |
| Oat, straw | 0.1 |
| Onion, dry bulb | 0.1 |
| Poultry, fat | 0.05 |
| Poultry, meat | 0.05 |
| Poultry, mbyp | 0.3 |
| Rye, forage | 0.1 |
| Rye, grain | 0.1 |
| Rye, straw | 0.1 |
| Sheep, fat | 1.0 |
| Sheep, meat | 0.5 |
| Sheep, mbyp | 3.5 |
| Sorghum, forage | 0.1 |
| Sorghum, grain, grain | 0.1 |
| Sorghum, grain, stover | 0.1 |
| Wheat, forage | 0.1 |
| Wheat, grain | 0.1 |
| Wheat, straw | 0.1 |

*Carcinogen/Hazard Classifications*
**U.S. EPA Carcinogens:** Group C, possible carcinogen
**California Prop. 65:** Reproductive toxin
**Label Signal Word:** WARNING
**TRI Developmental Toxin:** Developmental toxin
**WHO Acute Hazard:** Class II, moderately hazardous
*Regulatory Authority:*
- Actively registered pesticide in California. (octanoate)
- EPA 40 CFR 372.65, Specific Toxic Chemical Listings
- EPCRA Section 313 Form R *de minimis* concentration reporting level: 1.0%. (also, octanoate)
- AB 2588-Air Toxics "Hot Spots" Chemicals (CAL)
- The "Director's List" (CAL/OSHA)
- U.S. DOT Regulated Marine Pollutant (49CFR172.101, Appendix B)
- Proposition 65 chemical (CAL): Reproductive toxin

*Description:* Colorless to white crystalline solid; needles. Odorless. Slightly soluble in water; solubility = 130 ppm @ 20–25°C. Molecular weight 276.93; 403.15 (octanoate). Melting/Freezing point = 187°–194°C; 45-46°C (octanoate);

360°C (sodium or potassium salt). Boiling point sublimes @ 135°C @ 0.2 mmHg. Vapor pressure = <$10^{-5}$ mmHg @ 20°C; 4.8 x $10^{-6}$ mmHg (octanoate). Log $K_{ow}$ = Very low. Unlikely to bioaccumulate in aquatic organisms.

*Incompatibilities:* React with boranes, alkalies, aliphatic amines, amides, nitric acid, sulfuric acid. Keep away from oxidizers (chlorates, nitrates, peroxides, permanganates, perchlorates, chlorine, bromine, fluorine, etc) and strong acids.

*Determination in Air:* Bromoxynil/bromoxynil octanoate, NIOSH: #5010. See NIOSH Criteria Document 78-212 NITRILES[18]

*Permissible Concentration in Water:* No criteria set. Runoff from spills or fire control may cause water pollution.

*Routes of Entry:* Inhalation, skin contact.

*Harmful Effects and Symptoms*

*Short Term Exposure:* Contact may cause burns to skin and eyes. Because this material has a low vapor pressure, significant inhalation of vapors is unlikely at ordinary temperatures. May affect the iron metabolism, causing asphyxia. It is highly toxic. Forms cyanide in the body. Exposure results in headache, dizziness, rapid pulse, deep-rapid breathing, nausea, vomiting, unconsciousness, convulsions and sometimes death. May cause cyanosis (blue coloration of skin and lips caused by lack of oxygen). Some nitriles have a probable oral lethal dose in humans of less than 5 mg/kg or a taste (less than 7 drops) for a 70 kg (150 lb) person.

*Long Term Exposure:* Chronic exposure over long periods may cause fatigue and weakness. Can cause same general symptoms as hydrogen cyanide but onset of symptoms is likely to be slower. May cause liver and kidney damage.

*Points of Attack:* In animals: liver, kidney damage.

*Medical Surveillance:* Liver and kidney function tests.

*First Aid:* If this chemical gets into the eyes, remove any contact lenses at once and irrigate immediately for at least 15 minutes, occasionally lifting upper and lower lids. Seek medical attention immediately. If this chemical contacts the skin, remove contaminated clothing and wash immediately with soap and water. Seek medical attention immediately. If this chemical has been inhaled, remove from exposure, begin rescue breathing (using universal precautions) if breathing has stopped and CPR if heart action has stopped. Transfer promptly to a medical facility. When this chemical has been swallowed, get medical attention. *Do not induce vomiting when formulations containing petroleum solvents are ingested.* Otherwise, give large quantities of water and induce vomiting. Do not make an unconscious person vomit. Use amyl nitrate capsules if symptoms of cyanide poisoning develop. All area employees should be trained regularly in emergency measures for cyanide poisoning and in CPR. A cyanide antidote kit should be kept in the immediate work area and must be rapidly available. Kit ingredients should be replaced every 1-2 years to ensure freshness. Persons trained in the use of this kit, oxygen use, and CPR must be quickly available.

*References:*
- EXTOXNET, Extension Toxicology Network, "Pesticide Information Profile, Bromoxynil," Oregon State University, Corvallis, OR (June 1996). http://extoxnet.orst.edu/pips/bromoxyn.htm
- U.S. Environmental Protection Agency, Office of Pesticide Programs, Pesticide Residue Limits, "Bromoxynil," 40 CFR 180.324, http://www.epa.gov/pesticides/food/viewtols.htm
- California Environmental Protection Agency Chemical List of Lists, Sacramento CA (February 1997)

# Brucine

*Use Type:* Rodenticide
*CAS Number:* 357-57-3
*Formula:* $C_{23}H_{26}O_4 \cdot 4H_2O$
*Synonyms:* Brucina (Italian, Spanish); (-)Brucine; (-)Brucine dihydrate; Brucine hydrate; Dimethoxy strychnine; 2,3-Dimethoxystrichnidin-10-one; 2,3-Dimethoxystrychnine; 10,11-Dimethoxystrychnine; 10,11-Dimethylstrychnine; Strychnidin-10-one, 2,3-dimethoxy-(9CI); Strychnine, 2,3-dimethoxy-
*Trade Names:* DOLCO MOUSE CEREAL®; PIED PIPER MOUSE SEED®
*Producers:* Kothari Phytochemicals International (India); Merck (Germany)
*ICSC Number:* 1017
*RTECS Number:* EH8925000
*EEC Number:* 614-006-00-1
*EINECS Number:* 206-614-7
*Uses:* Used in the manufacture of other chemicals, in perfumes, as a medication for animals, and as a poison for rodents.

*Regulatory Authority:*
- EPA Hazardous Waste Number (RCRA No.): P018
- RCRA, 40CFR261, Appendix 8 Hazardous Constituents.
- Superfund/EPCRA 40CFR302.4 RQ: CERCLA, 100 lb (45.4 kg)
- EPCRA Section 313 Form R *de minimus* concentration reporting level: 1.0%.
- Canada, WHMIS, Ingredients Disclosure List

*Description:* Brucine is colorless to white, crystalline solid. Odorless with a very bitter taste. Slightly soluble in water. Molecular weight = 394.5. Freezing/ Melting point = 178°C. Hazard Identification (based on NFPA-704 M Rating System): Health 2, Flammability 1, Reactivity 0. Log $K_{ow}$ = < 1.0. Unlikely to bioaccumulate in marine organisms.

*Incompatibilities:* Reacts with strong oxidizers. Finely dispersed material in air can cause dust explosions.

*Permissible Exposure Limits in Air:* None established.

*Routes of Entry:* Inhalation, ingestion, eye and/or skin contact. Absorbed through the skin.

*Harmful Effects and Symptoms*

*Short Term Exposure:* Irritates the eyes and respiratory tract. Exposure can cause headache, nausea, vomiting, ringing in the ears, disturbed vision, restlessness, excitement, twitching and convulsions, seizures, breathing difficulties. An alkaloid. Severe poisoning can cause paralysis, unconsciousness, and death.

*First Aid:* If this chemical gets into the eyes, remove any contact lenses at once and irrigate immediately for at least 15 minutes, occasionally lifting upper and lower lids. Seek medical attention immediately. If this chemical contacts the skin, remove contaminated clothing and wash immediately with soap and water. Seek medical attention immediately. If this chemical has been inhaled, remove from exposure, begin rescue breathing (using universal precautions) if breathing has stopped and CPR if heart action has stopped. Transfer promptly to a medical facility. When this chemical has been swallowed, get medical attention. Give large quantities of water and induce vomiting. Do not make an unconscious person vomit.

*References:*
- EXTOXNET, Extension Toxicology Network, "Pesticide Information Profile, Brucine," Oregon State University, Corvallis, OR (January 1999). http://www.state.nj.us/health/eoh/rtkweb/0270.pdf
- California Environmental Protection Agency Chemical List of Lists, Sacramento CA (February 1997).
- New Jersey Department of Health and Senior Services, "Hazardous Substance Fact Sheet: "Brucine," Trenton, NJ (January 1999). http://www.state.nj.us/health/eoh/rtkweb/0270.pdf

# Bufencarb (ANSI)

*Use Type:* Insecticide
*CAS Number:* 8065-36-9
*Formula:* $C_{13}H_{19}NO_2$
*Alert:* In April of 1986, all registered products containing bufencarb in the U.S. were cancelled, although existing stocks were allowed to be used until exhausted.
*Synonyms:* 3-(1-Methylbutyl)phenyl methylcarbamate and 3-(1-ethylpropyl)phenyl methylcarbamate, mixed esters (3:1); Carbamic acid, methyl-, mixed (1-methylbutyl)phenyl and (1-ethylpropyl)phenyl esters; Methylcarbamic acid-*m*-[(1-methyl)butyl]phenyl ester mixed with carbamic acid, methyl-*m*-(1-ethylpropyl)phenyl ester (3:1)
*Trade Names:* METALKAMATE®; BUX®; BUX-TEN®; ORTHO 5353®, Chevron (Ortho) (USA)
*Producers:* Chevron (Ortho) (USA)
*Chemical Class:* Carbamate
*EPA/OPP PC Code:* 059303
*California DPR Chemical Code:* 91

*Uses:* Bufencarb was used in the U.S. on corn fodder and forage, fresh corn (includes sweet corn kernels plus cob with husk removed), corn grain, rice grain, and rice straw.

*Regulatory Authority:*
- Actively registered pesticide in California.

*Description:* Yellow-amber solid. Practically insoluble in water. Molecular weight = 221.30. Melting/Freezing point = 26.4-39°C. Density = 1.024. Boiling point = 125°C.

*Incompatibilities:* Strong oxidizers. When heated to decomposition, emits nitrogen oxides and carbon dioxide.

*Determination in Air:* Filter; none; Gravimetric; NIOSH IV [Particulates NOR; #0500 (total), #0600 (respirable)].[18]

*Permissible Concentration in Water:* No criteria set. Runoff from spills or fire control may cause water pollution.

*Routes of Entry:* Inhalation, ingestion, absorbed through the skin.

*Harmful Effects and Symptoms*

*Short Term Exposure:* Eye pupils are small; blurred vision; eye watering; runny nose; cough; shortness of breath; salivation; nausea, stomach cramps, diarrhea, and vomiting; increased blood pressure; profuse sweating; hypermotility, hallucinations; agitation; tingling of the skin; slow heartbeat; convulsions; fluid in lungs; loss of consciousness; incontinence; breathing stops; death. Carbamate insecticides inhibit the cholinesterase activity of enzymes, causing accumulation of acetylcholine at synapses and altering the way in which nervous impulses are transmitted. However, within several hours carbamates spontaneously detach from the enzymes.

*Long Term Exposure:* A potent cholinesterase inhibitor; cumulative effect is possible. This chemical may damage the nervous system with repeated exposure, resulting in convulsions, respiratory failure. May cause liver damage.

*Points of Attack:* Respiratory system, lungs, central nervous system, cardiovascular system, skin, eyes, plasma and red blood cell cholinesterase.

*Medical Surveillance:* Medical observation is recommended for 24 to 48 hours after breathing overexposure, as pulmonary edema may be delayed. As first aid for pulmonary edema, consider administering a corticosteroid spray. Cigarette smoking may exacerbate pulmonary injury and should be discouraged for at least 72 hours following exposure. Before employment and at regular times after that, the following are recommended: Plasma and red blood cell cholinesterase levels (tests for the enzyme poisoned by this chemical). If exposure stops, plasma levels return to normal in 1-2 weeks while red blood cell levels may be reduced for 1-3 months. When acetylcholinesterase enzyme levels are reduced by 25% or more below preemployment levels, risk of poisoning is increased, even if results are in lower ranges of "normal." Reassignment to work not involving carbamate pesticides is recommended until enzyme levels recover. If symptoms develop or overexposure occurs, repeat the above tests as soon as possible and get an exam of the nervous system. Also consider complete blood count. Consider chest x-ray following acute overexposure.

*First Aid:* Speed in removing material from eyes and skin is of extreme importance. *Eyes:* Eye contact can cause dangerous amounts of these chemicals to be quickly absorbed through the mucous membrane into the bloodstream. Immediately and gently flush eyes with plenty of warm or cold water (NO hot water) for at least 15 minutes, occasionally lifting the upper and lower eyelids. Get medical aid immediately. *Skin:* Get medical aid. Skin contact can cause dangerous amounts of these chemicals to be absorbed into the bloodstream. Wearing the appropriate PPE equipment and respirator for carbamate pesticides, immediately flush skin with plenty of soap and water for at least 15 minutes while removing contaminated clothing and shoes. Shampoo hair promptly if contaminated; protect eyes. *Ingestion:* Call poison control. Loosen all clothing. Never give anything by mouth to an unconscious person. Get medical aid. Do NOT induce vomiting.* If conscious, alert, and able to swallow, rinse mouth and have victim drink 4 to 8 ounces of water. Check to see if poison control instructs you to use ipecac syrup, otherwise administer slurry of activated charcoal (2 oz in 8 oz of water). If victim is UNCONSCIOUS OR HAVING CONVULSIONS, do nothing except keep victim warm. *In some cases you may be specifically instructed by poison control to induce vomiting by way of 2 tablespoons of syrup of ipecac (adult) washed down with a cup of water.* Do NOT give activated charcoal before or with ipecac syrup. *Inhalation:* Get medical aid. Do not contaminate yourself. Wearing the appropriate PPE equipment and respirator for carbamate pesticides, immediately remove the victim from the contaminated area to fresh air. If the victim is not breathing, administer artificial respiration. Do NOT use mouth-to-mouth resuscitation; use bag/mask apparatus. If breathing is difficult, administer oxygen through bag/mask apparatus until medical help arrives. Do not leave victim unattended. *Note to physician or authorized medical personnel.* Administer atropine, 2 mg (1/30 gr) intramuscularly or intravenously as soon as any local or systemic signs or symptoms of an intoxication are noted; repeat the administration of atropine every 3 to 8 minutes until signs of atropinization (mydriasis, dry mouth, rapid pulse, hot and dry skin) occur; initiate treatment in children with 0.05 mg mg/kg of atropine; repeat at 5 to 10 minute intervals. Watch respiration, and remove bronchial secretions if they appear to be obstructing the airway; intubate if necessary. *Medical note:* 2-PAMCI may be contraindicated in the case of some carbamate poisonings.

*References:*
- California Environmental Protection Agency Chemical List of Lists, Sacramento CA (February 1997).

# Buprofezin

*Use Type:* Insect growth regulator
*CAS Number:* 69327-76-0
*Synonyms:* 2-*tert*-Butylimino-3-isopropyl-5-phenylperhydro-1,3,5-thidiazin-4-one; 4*H*-1,3,5-Thiadiazin-4-one, 2-[(1,1-dimethylethyl)imino]tetrahydro-3-(1-methylethyl)-5-phenyl-
*Trade Names:* APPLAUD®, Nihon Nohyaku Co., Ltd. (Japan)
*Producers:* Agsin (Singapore); Ki-Hara Chemicals Ltd. (UK); Nihon Nohyaku Co., Ltd. (Japan); SuYan Agrochemical Group (China)
*EPA/OPP PC Code:* 275100
*California DPR Chemical Code:* 3947
*Uses:* For insect control in food crops and greenhouse ornamentals.
*Carcinogen/Hazard Classifications*
**U.S. EPA Carcinogens:** Possible Carcinogen
*Regulatory Authority:*
- Actively registered pesticide in California.

**WHO Acute Hazard:** Class U, unlikely to be hazardous
*Permissible Concentration in Water:* No criteria set. Runoff from spills or fire control may cause water pollution.
*Harmful Effects and Symptoms*
*Short Term Exposure:* Contact with eyes or skin may cause irritation or injury. Inhalation should be avoided; use NIOSH-approved air purifying respirators for pesticides. May be harmful if swallowed.
*First Aid:* If this chemical gets into the eyes, remove any contact lenses at once and irrigate immediately for at least 15 minutes, occasionally lifting upper and lower lids. Seek medical attention immediately. If this chemical contacts the skin, remove contaminated clothing and wash immediately with soap and water. Seek medical attention immediately. If this chemical has been inhaled, remove from exposure, begin rescue breathing (using universal precautions) if breathing has stopped and CPR if heart action has stopped. Transfer promptly to a medical facility. When this chemical has been swallowed, get medical attention. Give large quantities of water and induce vomiting. Do not make an unconscious person vomit.
*References:*
- California Environmental Protection Agency Chemical List of Lists, Sacramento CA (February 1997)

# Busulfan

*Use Type:* An insect sterilant.
*CAS Number:* 55-98-1
*Formula:* $C_6H_{14}O_6S_2$; $CH_3SO_2O(CH_2)_4OSO_2CH_3$
*Alert:* Busulfan is a carcinogen and teratogen and should be handled with extreme caution.

*Synonyms:* 1,4-Bis(methanesulfonoxy)butane; [1,4-Bis(methanesulfonyloxy)butane]; Bisulfan; Bisulphane; 1,4-Butanediol dimethyl sulfonate; 1,4-Butanediol dimethanesulphonate; Buzulfan; C.B. 2041; Citosulfan; 1,4-Dimesyloxybutane; 1,4-Dimethanesulfonoxbutane; 1,4-Di(methanesulfonyloxy)buane; 1,4-Dimethanesulphonyloxy butan; 1,4-Dimethylsulfonoxybutane; Methanesulfonic acid tetramethylene ester; NCI-C01592; NSC-750; Sulphabutin; Tetramethylene bis(methanesulfonate); Tetramethylene dimethane sulfonate

*Trade Names:* BUSULFEX®, Orphan Drug Company; GT41®; GT 2041®; LEUCOSULFAN®; MABLIN®; MIELUCIN®; MISULBAN®; MITOSTAN®; MYELOLEUKON®; MYLERAN®, GlaxoSmithKline (UK) (a drug for treating chronic myelogenous leukemia); X 149®

*Producers:* GlaxoSmithKline (UK)

*Chemical Class:* A bifunctional alkylating agent

*RTECS Number:* EK1750000

*Uses:* This compound has applications as an insect sterilant and as a chemotherapeutic agent taken orally or by injection to treat chronic myelogenouos leukemia.

*Carcinogen/Hazard Classifications*

**California Prop. 65:** Carcinogen as 1,4-BUTANEDIOL DIMETHANESULFONATE

IARC: Carcinogen (human carcinogen)

*Regulatory Authority:*
- Carcinogen (human carcinogen) (IARC) (NTP)[10]
- AB 2588-Air Toxics "Hot Spots" Chemicals (CAL)
- Proposition 65 chemical (CAL)
- The "Director's List" (CAL/OSHA)

*Description:* Busulfan is a white crystalline powder. Practically insoluble in water; hydrolyzes slowly. Molecular weight = 246.33. Melting/Freezing point = 117°C.

*Incompatibilities:* Oxidizers, moist air and water.

*Permissible Concentration in Water:* No criteria set, but runoff from spills or fire control may cause water pollution.

*Routes of Entry:* Inhalation

*Harmful Effects and Symptoms*

*Short Term Exposure:* Irritates the skin causing rash. Exposure can cause nausea, vomiting, diarrhea, and seizures.

*Long Term Exposure:* A carcinogen in humans, causes leukemia and kidney and uterine cancer. A probable teratogen in humans. May damage the developing fetus. May cause testes damage in males (decrease sperm count, cause impotence), and decrease fertility in females. Long term exposure may cause cataracts, lung irritation, permanent lung scarring, bone marrow damage, liver damage. Symptoms of exposure include bleeding tendencies, decreased leucoyte count or depressed bone marrow activity[52].

*Points of Attack:* See above.

*Medical Surveillance:* Completed blood count, chest x-ray, lung function tests, liver function tests.

*First Aid:* If this chemical gets into the eyes, remove any contact lenses at once and irrigate immediately with water or normal saline for at least 20-30 minutes, occasionally lifting upper and lower lids. Seek medical attention immediately. If this chemical contacts the skin, remove contaminated clothing and wash immediately with soap and water. Seek medical attention immediately. If this chemical has been inhaled, remove from exposure, begin rescue breathing (using universal precautions) if breathing has stopped and CPR if heart action has stopped. Transfer promptly to a medical facility. When this chemical has been swallowed, get medical attention. Give large quantities of water and induce vomiting. Do not make an unconscious person vomit.

*References:*
- New Jersey Department of Health and Senior Services,"Hazardous Substance Fact Sheet, Busulfan," Trenton NJ (December, 1998). http://www.state.nj.us/health/eoh/rtkweb/0271.pdf
- California Environmental Protection Agency Chemical List of Lists, Sacramento CA (February 1997).

## Butachlor (ANSI)

*Use Type:* Herbicide

*CAS Number:* 23184-66-9

*Formula:* $C_{17}H_{26}ClNO_2$

*Alert:* Not registered in the U.S.

*Synonyms:* Butoxymethyl; N-(Butoxymethyl)-2-chloro-N-(2,6-diethylphenyl)acetamide; 2-Chloro-2',6'-diethyl-N-(butoxymethyl)acetanalide; N-(Butoxymethyl)-2-chloro-2',6'-diethylacetanilide; Acetanilide, 2-chloro-2',6'-diethyl-N-(butoxymethyl)-; Acetamide, N-(butoxymethyl)-2-chloro-N-(2,6-diethylphenyl)-

*Trade Names:* BUTANEX®; BUTANOX®; CP 53619®; HILTACHLOR®; LAMBAST® Monsanto (USA); MACHETE® Monsanto (USA); MACHETTE® Monsanto (USA); PILLARSET®; RASAYANCHLOR®

*Producers:* Agrimor International (USA); Agsin (Singapore); BEC Group Ltd. (India); Hindustan Insecticides (India); Monsanto (USA); Nagarjuna Agrichem (India); Vijayalakshmi Insecticides and Pesticides (India)

*Chemical Class:* Chloroacetanilida

*EPA/OPP PC Code:* 112301

*California DPR Chemical Code:* 4056

*EINECS Number:* 245-477-8

*Uses:* Used for preemergence control of annual grasses, sedges and broadleaf weeds in rice crops. Used primarily in Asia, South America, Europe and Africa.

*Human toxicity (long-term)[77]:* Very low–259.00 ppb, Health Advisory

*Fish toxicity (threshold)[77]:* Intermediate–14.47908 ppb, MATC (Maximum Acceptable Toxicant Concentration)

*Carcinogen/Hazard Classifications*
**U.S. EPA Carcinogens:** Likely Carcinogen
**Label Signal Word:** CAUTION
**WHO Acute Hazard:** Class U, unlikely to be hazardous
*Regulatory Authority:*
• Actively registered pesticide in California.
*Description:* Light yellow or amber oily liquid. Slightly soluble in water; solubility = 20 mg/L @ 20°C; Molecular weight = 311.88; D: 1.070 @ 30°F/4°C. Melting/Freezing point −5°C. Boiling point = 156°C @ 0.66 mmHg (decomposes @ 165°C); 196°C @ 0.5 mm. Vapor pressure = $4.5 \times 10^{-6}$ mmHg.
*Incompatibilities:* Slowly hydrolyzes in water, releasing ammonia and forming acetate salts.
*Permissible Concentration in Water:* No criteria set. Runoff from spills or fire control may cause water pollution.
*Harmful Effects and Symptoms*
*Short Term Exposure:* Apprehension, anxiety, confusion, nervous excitation; dizziness; headache; numbness and weakness in limbs; muscle twitching, tremors; nausea and vomiting; slow, shallow respiration, bluish face; convulsions; loss of consciousness; breathing stops; death.
*Points of Attack:* May be fatal if inhaled, ingested, or absorbed through the skin
*Medical Surveillance:* Medical observation is recommended for 24 to 48 hours after breathing overexposure, as pulmonary edema may be delayed. As first aid for pulmonary edema, consider administering a corticosteroid spray. Cigarette smoking may exacerbate pulmonary injury and should be discouraged for at least 72 hours following exposure.
*First Aid:* Speed in removing material from eyes and skin is of extreme importance. *Eyes:* Eye contact can cause dangerous amounts of these chemicals to be quickly absorbed through the mucous membrane into the bloodstream. Directly, irrigate with large amounts of plain, tepid water or saline for 20 minutes, occasionally lifting the lower and upper lids. During this time, remove contact lenses, if easily removable without additional trauma to the eye. Get medical aid immediately. Have physician check for possible delayed damage. *Skin:* Get medical aid. Skin contact can cause dangerous amounts of these chemicals to be absorbed into the bloodstream. Wearing the appropriate PPE equipment and respirator for organochlorine pesticides, immediately flush exposed skin, hair, and under nails with plain, running, tepid water for 20 minutes, then wash twice with mild soap. Shampoo hair promptly if contaminated; protect eyes. Do not scrub skin or hair, since this can increase absorption through the skin. Rinse thoroughly with water. Victims who are able and cooperative may assist with their own decontamination. Remove and double-bag contaminated clothing and personal belongings. Leather absorbs many organochlorines; therefore, items such as leather shoes, gloves, and belts should be discarded. If the skin is swollen or inflamed, cool affected areas with cold compresses. *Ingestion:* Call poison control. Loosen all clothing. Never give anything by mouth to an unconscious person. Get medical aid. Do NOT induce vomiting. The patient is at risk of CNS depression or seizures, which may lead to pulmonary aspiration during vomiting. If the victim is conscious and able to swallow, *administer an aqueous slurry of activated charcoal at 1 gm/kg (usual adult dose 60–90 g, child dose 25–50 g). A soda can and straw may be of assistance when offering charcoal to a child. The efficacy of activated charcoal for some organochlorine poisoning (such as chlordane) is uncertain. If victim is *unconscious or having convulsions,* do nothing except keep victim warm. *Inhalation:* Get medical aid. Do not contaminate yourself. Wearing the appropriate PPE equipment and respirator for organochlorine pesticides, immediately remove the victim from the contaminated area to fresh air. For inhalation exposures, monitor for respiratory distress. If the victim is not breathing, administer artificial respiration. Do NOT use mouth-to-mouth resuscitation; use bag/mask apparatus. If cough or breathing difficulty develops, evaluate for respiratory tract irritation, bronchitis, or pneumonitis. If breathing is difficult, administer 100% humidified supplemental oxygen through bag/mask apparatus until medical help arrives. Do not leave victim unattended.
*\*Note:* In some cases you may be specifically instructed by Poison Control to induce vomiting by way of 2 tablespoons of syrup of ipecac (adult) washed down with a cup of water. Do NOT give activated charcoal <u>before or with</u> ipecac syrup.
*References:*
• California Environmental Protection Agency Chemical List of Lists, Sacramento CA (February 1997)

# Butralin

*Use Type:* Herbicide
*CAS Number:* 33629-47-9
*Formula:* $C_{14}H_{21}N_3O_4$
*Synonyms:* Aniline, *N-sec*-butyl-4-*tert*-butyl-2,6-dinitro-; Benzenamine, 4-(1,1-dimethylethyl)-*N*-(1-methylpropyl)-2,6-dinitro-; Butalin; Butraline; *N-sec*-Butyl-4-*tert*-butyl-2,6-dinitroaniline; Dibutalin; 4-(1,1-Dimethylethyl)-*N*-(1-methylpropyl)-2,6-dinitrobenzenamine; Dimethyl-ethyl-*n*-(1-methylpropyl)-2,6-dinitrobenzeneamine[4-(1,1-)]; Rutralin
*Trade Names:* A-820®, Amchem (USA), canceled; 72-A34®; AMCHEM 70-25®, Amchem (USA), canceled; AMCHEM A-280®, Amchem (USA), canceled; AMEX®, Cfpi Agro (France), canceled; AMEX 820®, Cfpi Agro (France), canceled; AMEXINE®, canceled; NO CRAB®, Cfpi Agro (France), canceled; SECTOR®; STIFLE®,

Crompton Corporation (USA); TAMEX®, Cfpi Agro (France)
***Producers:*** Amchem (USA); Crompton Corporation (USA); Cfpi Agro (France), Sub. of Nufarm (Australia)
***Chemical Class:*** Dinitroaniline
***EPA/OPP PC Code:*** 106501
***California DPR Chemical Code:*** 1756
***Uses:*** Butralin is used as a plant grow regulator on tobacco after the tobacco is topped. All tobacco is topped to stimulate desirable chemical and physical characteristics but also stimulates the growth of suckers. Butralin is a contact-local systemic type of plant growth regulator that inhibits sucker growth. All uses on food crops were canceled in January, 1991, and all uses on turf and ornamental grasses were voluntarily canceled in March, 1997. Butralin is been found effective in destroying opium poppy crops.
***Human toxicity (long-term):*** Low–70.00 ppb, Health Advisory
***Fish toxicity (threshold):*** Intermediate–42.17178 ppb, MATC (Maximum Acceptable Toxicant Concentration)
***Carcinogen/Hazard Classifications***
**Label Signal Word:** CAUTION or DANGER
**WHO Acute Hazard:** Class U, unlikely to be hazardous
***Description:*** Yellow-orange crystalline solid. Slightly soluble in water; solubility = <1.0 ppm. Molecular weight = 295.37. Flash point = 35.8°C.
***Incompatibilities:*** Oxidizers (chlorates, nitrates, peroxides, permanganates, perchlorates, chlorine, bromine, fluorine, etc); strong acids.
***Permissible Concentration in Water:*** No criteria set. Runoff from spills or fire control may cause water pollution.
***Routes of Entry:*** Inhalation, skin contact, ingestion.
***Harmful Effects and Symptoms***
***Short Term Exposure:*** May cause irritation to the eyes, skin, or respiratory tract. May be toxic if ingested or upon skin contact.
***First Aid:*** If this chemical gets into the eyes, remove any contact lenses at once and irrigate immediately for at least 15 minutes, occasionally lifting upper and lower lids. Seek medical attention immediately. If this chemical contacts the skin, remove contaminated clothing and wash immediately with soap and water. Seek medical attention immediately. If this chemical has been inhaled, remove from exposure, begin rescue breathing (using universal precautions) if breathing has stopped and CPR if heart action has stopped. Transfer promptly to a medical facility. When this chemical has been swallowed, get medical attention. Give large quantities of water and induce vomiting. Do not make an unconscious person vomit.
***References:***
- U.S. Environmental Protection Agency, "Reregistration Eligibility Decision (RED), Butralin," Office of Prevention, Pesticides and Toxic Substances, Washington, DC (May 1998).

http://www.epa.gov/REDs/2075red.pdf
- California Environmental Protection Agency Chemical List of Lists, Sacramento CA (February 1997).

# Butylate

***Use Type:*** Herbicide
***CAS Number:*** 2008-41-5
***Formula:*** $C_{11}H_{23}NOS$
***Alert:*** Classified as a General Use Pesticide (GUP) with applications limited to corn fields.
***Synonyms:*** Bis(2-methylpropyl)carbamothioic acid-S-ethyl ester; Butilate; Diisobutylthiocarbamic acid-S-ethyl ester; Diisocarb; S-Ethyl N,N-diisobutylthiocarbamate; S-Ethyl bis(2-methylpropyl)carbamothioate; Ethyl-N,N-diisobutylthiocarbamate; S-Ethyldiisobutyl thiocarbamate; Ethyl-N,N-diisobutyl thiolcarbamate
***Trade Names:*** ANELDA PLUS®; ANELDAZIN®; ANELIROX®; ATRA-BUTE®, Zeneca Ag Products (USA) (now Syngenta), canceled Nov. 1992; BUTILATE®; GENATE®, Zeneca Ag Products (USA) (now Syngenta), canceled August 1994; R-1910®, Zeneca Ag Products (USA) (now Syngenta), canceled Dec. 1987; STAUFFER R-1910®, Zeneca Ag Products (USA) (now Syngenta), canceled Dec. 1987; SUTAN®, Zeneca Ag Products (USA) (now Syngenta), canceled Sept. 1994; SUTAZINE®, Zeneca Ag Products (USA) (now Syngenta), canceled Dec. 1987; TOMAHAWK®
***Producers:*** Sigma-Aldrich Laborchemikalien (Germany)
***Chemical Class:*** Thiocarbamate
**Label Signal Word:** CAUTION
***EPA/OPP PC Code:*** 041405
***California DPR Chemical Code:*** 565
***RTECS Number:*** EZ7525000
***EINECS Number:*** 217-916-3
***Uses:*** A selective herbicide for use on field corn, sweet corn, and popcorn to control grassy and broadleaf weeds and seeds in the soil prior to sowing a crop. Often applied in combination with atrazine and/or cyanazine.
***Human toxicity (long-term)[77]:*** Very low–400.00 ppb, Health Advisory
***Fish toxicity (threshold)[77]:*** Intermediate–22.61732 ppb, MATC (Maximum Acceptable Toxicant Concentration)
***U.S. Maximum Allowable Residue Levels for Butylate, Federal Register (67FR49689-49691, July 31, 2002):*** "By itself, butylate poses no risk concerns within the limits of the existing tolerances, which will remain in effect at 0.1 part per million (ppm) for all registered commodities; however, the Agency intends to revise the commodity definitions in accordance with current Agency administrative practice."
***Carcinogen/Hazard Classifications***
**U.S. EPA Carcinogens:** Group E, Unlikely to be Carcinogen

**WHO Acute Hazard:** Class U, unlikely to be hazardous

*Regulatory Authority:*
- Water Pollution Standard Proposed (U.S. EPA) (See reference below) (Wisconsin)[61]
- CAL-DHS/DHS Drinking Water Action Levels
- Actively registered pesticide in California.
- RCRA 40CFR268.48; 61FR15654, Universal Treatment Standards: Wastewater (mg/L), 0.003; Nonwastewater (mg/kg), 1.4

*Description:* Clear liquid with an aromatic odor. Slightly soluble in water; solubility = 4ppm @ 25°C. Boiling point = 130°C @ 10 mmHg. Vapor pressure = $1 \times 10^{-3}$ mmHg.

*Permissible Exposure Limits in Air:* No standards set.

*Permissible Concentration in Water:* A No-Adverse Effects Level (NOAEL) in the range of 24-40 mg/kg body weight/day/has been determined by USEPA. This leads to derivation of a 10-day health advisory for Butylate of 2.4 mg/L and a lifetime health advisory of 0.05 mg/L for a 70-kg man. Wisconsin has set a guideline for butylate in drinking water of 200 μg/L[61].

*Determination in Water:* Analysis of butylate is by a gas chromatographic (GC) method applicable to the determination of certain nitrogen- and phosphorus-containing pesticides in water samples. In this method, approximately 1 L of sample is extracted with methylene chloride. The extract is concentrated and the compounds are separated using capillary column GC. Measurement is made using a nitrogen-phosphorus detector. The method detection limit has not been determined for butylate, but it is estimated that the detection limits for analytes included in this method are in the range of 0.1 to 2 μg/L.

*Harmful Effects and Symptoms*
The $LD_{50}$-(oral, rat) = 4,000 mg/kg (slightly toxic). Applying the criteria described in EPA's guidelines for assessment of carcinogenic risk, butylate may be placed in Group C: a possible human carcinogen. This category is for substances that show limited evidence of carcinogeniciy in animals and inadequate evidence in humans.

*First Aid:* If this chemical gets into the eyes, remove any contact lenses at once and irrigate immediately for at least 15 minutes, occasionally lifting upper and lower lids. Seek medical attention immediately. If this chemical contacts the skin, remove contaminated clothing and wash immediately with soap and water. Seek medical attention immediately. If this chemical has been inhaled, remove from exposure, begin rescue breathing (using universal precautions) if breathing has stopped and CPR if heart action has stopped. Transfer promptly to a medical facility. When this chemical has been swallowed, get medical attention. Give large quantities of water and induce vomiting. Do not make an unconscious person vomit.

*References:*
- U.S. Environmental Protection Agency, Office of Preventions, Pesticides and Toxic Substances, Reregistration Statement, "Butylate Facts," Washington, DC, (September 2001).
- EXTOXNET, Extension Toxicology Network, "Pesticide Information Profile, Butylate," Oregon State University, Corvallis, OR (June 1996). http://pmep.cce.cornell.edu/profiles/extoxnet/24d-captan/butylate-ext.htm
- California Environmental Protection Agency Chemical List of Lists, Sacramento CA (February 1997).
- U.S. Environmental Protection Agency, "Health Advisory: Butylate," Washington, DC, Office of Drinking Water, August 1987.

# Butylphenols

*Use Type:* Insecticide and fumigant.

*CAS Number:* 3180-09-4 (*o*-isomer); 89-72-5 (*o-sec*-isomer); 99-71-8 (*p-sec*-isomer); 4074-43-5 (*m*-isomer); 88-18-6 (*o-tert*-isomer); 98-54-4 (*p-tert*-isomer); 1638-22-8 (*p*-isomer); 28805-86-9 (mixed isomers)

*Formula:* $C_{10}H_{14}O$, $C_4H_9C_6H_4OH$

*Synonyms:* *(o-n-):* 2-*n*-Butylphenol; *(o-sec-):* 2-*sec*-Butylphenol; *o-sec*-Butylphenol; *(o-tert-):* 2-*tert*-Butylphenol; Phenol, *o-(tert*-butyl)-; *(p-sec-):* 4-*sec*-Butylphenol; *p-sec*-Butylphenol; *(p-tert-):* 4-*tert*-Butylphenol; 4-*tert*-Butylphenol; 4-(1,1-Demethylethyl)phenol; Butylphen; 1-Hydroxy-4-*tert*-Butylbenzene; UCAR butylphenol 4-*tert*

*Producers:* Albemarle (USA); CONDEA (Germany); Dainippon Ink & Chemicals (Japan); Degussa (Germany); Fluorochem Ltd. (UK); Great Lakes Chemical (USA); Honshu Chemical Industry (Japan); Mitsubishi Chemical (Japan); Schenectady International (USA); Shell Chemical (Netherlands); Tokyo Chemical Industry (Japan)

*Chemical Class:* Phenols

*EPA/OPP PC Code:* 064113 (*p-tert*-)

*California DPR Chemical Code:* 1000 (*p-tert*-)

*ICSC Number:* 1472 (o-sec-); 0637 (*p-tert*-)

*RTECS Number:* SJ8850000 (*o-n*-); SJ8920000 (*o-sec*-); SJ8810000 (*m*-); SJ8924000 (*p-sec*-); SJ8925000 (*p-tert*-); SJ8922500 (*p-n*-)

*EINECS Number:* 201-933-8 (*o-sec*-); 202-679-0 (*p-tert*-)

*Uses:* Butylphenols may be used as intermediates in manufacturing varnish and lacquer resins; as a germicidal agent in detergent disinfectants; as a pour point depressant, motor-oil additive, de-emulsifier for oil, soap-antioxidant, plasticizer, fumigant and insecticide.

*Regulatory Authority:*
- Actively registered pesticide in California.
- (*o-sec*-isomer)Air Pollutant Standard Set (ACGIH)[1] (OSHA)[58] (DFG)[3] (HSE)[33] (Several States)[60](Australia)(Israel)(Several Canadian Provinces)
- U.S. DOT 49CFR172.101, Appendix B, Regulated marine pollutant (as butyl phenol)

- Canada, WHMIS, Ingredients Disclosure List (*o-sec-*; *p-tert-*; *p-*; *o-*; *m-*)
- Permissible Exposure Limits for Chemical Contaminants (CAL/OSHA) (*o-sec-*)
- The "Director's List" (CAL/OSHA) (*o-sec-*)

*Description:* The butylphenols, $C_4H_9C_6H_4OH$ include a number of isomers, the two most highly regulated are *o-sec-*butylphenol and *p-tert-*butylphenol. Their properties are as follows: (*o-sec-*): Colorless liquid or solid (below 61°F). Boiling point = 108°C. Flash point = 108°C. Insoluble in water. (*p-tert-*): White crystalline solid. Melting/Freezing point = 97°C. Hazard Identification (based on NFPA-704 M Rating System): Health 1, Flammability 1, Reactivity 0. Insoluble in water.

*Incompatibilities:* Incompatible with strong acids, caustics, aliphatic amines, amides, oxidizers.

*Permissible Exposure Limits in Air:* There is no OSHA[2] PEL for the *o-sec-*isomer. The NIOSH[2] recommended REL is 5 ppm (30 mg/m³) TWA with the notation that skin absorption is of concern. The ACGIH[1], Australia, HSE, and Israel TWA is the same as NIOSH[2]. In Canada Alberta, Ontario, and Quebec TWA is 5 ppm and Alberta's STEL is 10 ppm. *p-tert-isomer:* The DFG[3] has set a MAK of 0.08 ppm (0.5 mg/m³) and Peak Limitation of 5 times the normal MAK (30 min.), not to be exceeded 2 times during a workshift. Mexico's TWA is 10 ppm and STEL of 20 ppm. Several states have set guidelines or standards for the *o-sec-isomer* in ambient air[60] ranging from 300 µg/m³ (North Dakota) to 500 µg/m³ (Virginia) to 600 µg/m³ (Connecticut) to 714 µg/m³ (Nevada).

*Permissible Concentration in Water:* No criteria set, but runoff from spills or fire control may cause water pollution.

*Routes of Entry:* Inhalation, skin absorption, ingestion, skin and/or eye contact.

*Harmful Effects and Symptoms*

*Short Term Exposure:* Inhalation may cause irritation to nose, throat and lungs. Sensitization may occur. Skin contact studies with animals suggest that severe irritation at concentrations above 10% may occur. May cause rash, redness and irritation, especially when skin is wet. Absorption is significant and contact may lead to allergic reaction. Eye studies with animals suggest that severe irritation may occur. Ingestion studies on animals suggest that 8 oz. May by lethal to a 150 lb. person.

*Long Term Exposure:* May cause skin color changes by contact or inhalation of levels between 10 and 100 ppm. Allergy may develop after repeated exposure. Liver damage may also occur. There is limited evidence that butylphenol causes skin cancer in animals. Repeated or prolonged skin contact can cause skin ulcers and lead to permanent loss of skin pigment in affected areas.

*Points of Attack:* Eyes, skin, respiratory system.

*Medical Surveillance:* Lung function tests. Liver function tests.

*First Aid:* If this chemical gets into the eyes, remove any contact lenses at once and irrigate immediately for at least 15 minutes, occasionally lifting upper and lower lids. Seek medical attention immediately. If this chemical contacts the skin, remove contaminated clothing and wash immediately with soap and water. Seek medical attention immediately. If this chemical has been inhaled, remove from exposure, begin rescue breathing (using universal precautions) if breathing has stopped and CPR if heart action has stopped. Transfer promptly to a medical facility. When this chemical has been swallowed, get medical attention. Give large quantities of water and induce vomiting. Do not make an unconscious person vomit.

*References:*
- New Jersey Department of Health and Senior Services, "Hazardous Substance Fact Sheet: *o-sec-*Butylphenol," Trenton, NJ (April 1986, rev. August 2001). http://www.state.nj.us/health/eoh/rtkweb/1440.pdf
- New York State Department of Health, "Chemical Fact Sheet: *p-tert-*Butylphenol," Albany, NY, Bureau of Toxic Substance Assessment (March 1986).
- California Environmental Protection Agency Chemical List of Lists, Sacramento CA (February 1997).

# C

## Cacodylic Acid

*Use Type:* Herbicide and defoliant
*CAS Number:* 75-60-5
*Formula:* $C_2H_7AsO_2$
*Synonyms:* Acide cacodylique (French); Acide dimethylarsinique (French); Acido cacodilico (Spanish); Agent Blue; Arsinic acid, dimethyl-; Dimethylarsenic acid; Dimethylarsinic arsinic acid; DMAA; Hydroxydimethylarsine oxide; Dimethylarsinic acid
*Trade Names:* ANSAR®; ARSAN®; BOLLS-EYE®, Vertac Chemical Corp. (USA), canceled 3/20/1987; BROADSIDE®, Vertac Chemical Corp. (USA), canceled 3/20/1987; CHECK-MATE®; CHEM-AX®, Monterey Chemical (USA), canceled 3/01/1990; CLEAN-UP®, Sylorr Plant Corp. (USA); COTTON AIDE HC®; DILIC®, Inter-Ag Corp. (USA), canceled 11/10/1989; EZY-PICKIN' COTTON DEFOLIANT®, Drexel Chemical (USA); ERASE®, Scott Company (USA), canceled 10/10/1989; HERB-ALL®, Luxembourg-Pamol Inc. (USA); KACK®, Drexel Chemical (USA); MONOCIDE®, Monterey Chemical (USA); MONTAR®, Monterey Chemical (USA); PHYTAR® (with Sodium cacodylate), APC Holdings Corp. (USA), canceled; PHYTAR 138®, APC Holdings Corp. (USA), canceled 9/30/1991; PHYTAR 560® (with Sodium cacodylate), APC Holdings Corp. (USA), canceled 11/30/1992; PHYTAR 600®, APC Holdings Corp. (USA), canceled; RAD-E-CATE 25®, Vineland Chemical Co. (USA), canceled 8/12/1987; SALVO®; SILVISAR 510®, TSI Company (USA), canceled 10/10/1989; SYLVICOR®
*Producers:* Drexel Chemical (USA); Research Organics (USA); Scott Company (USA); Sigma-Aldrich Laborchemikalien (Germany)
*Chemical Class:* Organoarsenic
*EPA/OPP PC Code:* 012501
*California DPR Chemical Code:* 32
*RTECS Number:* CH7525000
*Uses:* Cacodylic acid is an arsenical non-selective contact herbicide which will defoliate or desiccate a wide variety of plant species. It is used as a cotton defoliant and for lawn renovation, for weed control in non-crop areas such as around buildings, near perennial ornamentals, along fence rows, and in forest management. It is used as a soil sterilant and in timber thinning. *Agent Blue* (0.37 kg/L cacodylic acid) was used to control vegetation during the Vietnam war. The top five crop usages in California are on cotton, rights of way, landscaping, uncultivated agricultural areas, and in outdoor container nurseries.

*Human toxicity (long-term)*[77]: High–5.61798 ppb, CHCL (Chronic Human Carcinogen Level)
*Fish toxicity (threshold)*[77]: Very low–2841.14769 ppb, MATC (Maximum Acceptable Toxicant Concentration)
*U.S. Maximum Allowable Residue Levels for Cacodylic Acid (40 CFR 180.311):*

| CROP | ppm |
|---|---|
| Cotton, undelinted seed | 2.8 |
| Cattle, mbyp, except kidney and liver | 0.7 |
| Cattle, fat | 0.7 |
| Cattle, kidney | 1.4 |
| Cattle, meat | 0.7 |

*Carcinogen/Hazard Classifications*
**U.S. EPA Carcinogens:** Group B2, probable carcinogen
**California Prop. 65:** Carcinogen
**IARC:** Group 1, known carcinogen
**Label Signal Word:** CAUTION
**WHO Acute Hazard:** Class III, slightly hazardous
*Regulatory Authority:*
- Superfund/EPCRA 40CFR302.4 RQ: CERCLA, 1 lb (0.454 kg)
- EPA Hazardous Waste Number (RCRA No.): U136
- RCRA, 40CFR261, Appendix 8 Hazardous Constituents.
- EPCRA Section 313 Form R (as organic arsenic compound) *de minimus* concentration reporting level: 1.0%
- U.S. DOT Regulated Marine Pollutant (49CFR172.101, Appendix B) as dimethylarsinic acid.
- Actively registered pesticide in California.
- Canada, WHMIS, Ingredients Disclosure List
- Canada: Priority Substance List & Restricted Substances/Ocean Dumping Forbidden (CEPA), National Pollutant Release Inventory (NPRI) (arsenic compounds)

*Description:* Cacodylic acid is a colorless, odorless, crystalline solid arsenic compound. Melting/Freezing point = 192°C. Hazard Identification (based on NFPA-704 M Rating System): Health 1, Flammability 0, Reactivity 0. Highly soluble in water.
*Incompatibilities:* Aqueous solution react violently with chemically active metals releasing toxic arsenic fumes. Incompatible with oxidizers, sulfuric acid, caustics (strong bases), reducing agents, ammonia, amines, isocyanates, alkylene oxides, epichlorohydrin.
*Permissible Exposure Limits in Air:* The following exposure limits are for air levels only. When skin contact also occurs, overexposure is possible, even though air levels are less than the limits listed below. The OSHA[2] PEL 8-hour TWA is 0.5 mg(As)/m$^3$. ACGIH recommends a TWA of 0.2 mg(As)/m$^3$.[1]

*Determination in Air:* Filter; Reagent; Ion chromatography/Hydride generation atomic absorption spectrometry; NIOSH IV Method #5022, Arsenic, Organo-[18].

*Permissible Concentration in Water:* EPA has set a limit of 0.05 parts per million (ppm) for arsenic in drinking water. The U.S. EPA arsenic drinking water standard of 0.01 ppm (10 ppb) is based on the U.S. EPA final rule for arsenic in drinking water published in. the January 22, 2001, *Federal Register*. However, the U.S. EPA is currently reviewing the science and cost estimate supporting this rule, and, in the interim, has reverted to the previous standard for arsenic. Thus, in the US, the current EPA arsenic drinking water standard remains at 0.05 ppm (50 ppb). To protect freshwater aquatic life-total recoverable trivalent inorganic arsenic never to exceed 440 µg/L. To protect saltwater aquatic life: 508 µg/L on an acute basis. To protect human health: preferably zero. The former USSR-UNEP/IRPTC project[43] has set MAC values for inorganic arsenic compounds in water for domestic purposes at 0.05 mg/L and in water bodies for fishery purposes of 0.5 mg/L also.

*Determination in Water:* For arsenic: The atomic absorption graphite furnace technique is often used for measurement of total arsenic in water. It also has been standardized by EPA. Total arsenic may be determined by digestion followed by silver diethyldithiocarbamate; an alternative is atomic absorption; another is inductively coupled plasma optical emission spectrometry. See OSHA Method #ID-105 for arsenic[58].

*Routes of Entry:* Inhalation, ingestion, skin and/or eye contact This chemical can be absorbed through the skin, thereby increasing exposure.

*Harmful Effects and Symptoms*

*Short Term Exposure:* Irritates the eyes, skin, and respiratory tract. Skin contact can also cause burning, itching, thickening and color changes.

*Long Term Exposure:* Certain other arsenic compounds have been identified as carcinogen. Although this chemical has not been identified as a carcinogen it should be handled with extreme caution. May cause an ulcer of the septum dividing the inner nose. It can cause nerve damage, thickening of the skin with patch areas of darkening and loss of pigment, or the development of white lines in the nails. May cause liver and kidney damage. Repeated exposure can lead to a metallic or garlic taste in the mouth, loss of appetite, nausea, vomiting, difficulty swallowing, stomach pain, diarrhea, and death. Birth defects have been observed in animals exposed to inorganic arsenic. It is likely that health effects seen in children exposed to high amounts of arsenic will be similar to the effects seen in adults.

*Points of Attack:* Several studies have shown that inorganic arsenic can increase the risk of lung cancer, skin cancer, bladder cancer, liver cancer, kidney cancer, and prostate cancer.

*Medical Surveillance:* Test for urine arsenic. Levels should not be greater than 100 micrograms per gram of creatinin in the urine. Examine the skin for abnormal growths. Liver and kidney function tests.

*First Aid:* If this chemical gets into the eyes, remove any contact lenses at once and irrigate immediately for at least 15 minutes, occasionally lifting upper and lower lids. Seek medical attention immediately. If this chemical contacts the skin, remove contaminated clothing and wash immediately with soap and water. Seek medical attention immediately. If this chemical has been inhaled, remove from exposure, begin rescue breathing (using universal precautions) if breathing has stopped and CPR if heart action has stopped. Transfer promptly to a medical facility. When this chemical has been swallowed, get medical attention. Give large quantities of water and induce vomiting. Do not make an unconscious person vomit.

*References:*
- EPA, Office of Pesticide Programs, Pesticide Residue Limits, "Cacodylic Acid", 40 CFR 180.311, http://www.epa.gov/pesticides/food/viewtols.htm
- EXTOXNET, Extension Toxicology Network, "Pesticide Information Profile, Cacodylic Acid," Oregon State University, Corvallis, OR (September 1993). http://extoxnet.orst.edu/pips/cacodyli.html
- New Jersey Department of Health and Senior Services, Hazardous Substance Fact Sheet, "Cacodylic Acid," Trenton NJ (January 1999). http://www.state.nj.us/health/eoh/rtkweb/0304.pdf
- California Environmental Protection Agency Chemical List of Lists, Sacramento CA (February 1997).

# Cadmium

*Use Type:* Compounds used for fungicides, insecticides, and nematocides.
*CAS Number:* 7440-43-9 (metal)
*Formula:* Cd
*Synonyms:* Cadmio (Spanish); C.I. 77180; Colloidal cadmium; Elemental cadmium; Kadmium (German); Kadmu (Polish)
*Producers:* ASARCO (USA); Boliden (Sweden); Degussa (Germany); Dowa Mining (Japan); Goldschmidt (Germany); Great Western Inorganics (USA); Hall (USA); Industrias Penoles (Mexico); Metaleurop (France); Noranda (Canada); PPM Pure Metals (Germany); Sigma-Aldrich Fine Chemicals; Teck Cominco (Canada); UMICORE (Belgium); Zinc Corp. (USA)
*Chemical Class:* Inorganic-cadmium, heavy metal
*ICSC Number:* 0020
*RTECS Number:* EU9800000 (metal)
*EINECS Number:* 231-152-8 (cadmium)
*Uses:* EPA doesn't allow cadmium in pesticides. Food products account for most of the human exposure to

cadmium, except in the vicinity of cadmium-emitting industries. Human cadmium toxicity caused by contaminated rice plants was first reported in Japan in the 1950s. These studies showed that subsistence rice farmers had been sickened by ingesting cadmium that had passed from municipal sewage sludge used as fertilizer through the rice crop. Cadmium is taken up through the roots of plants to edible leaves, fruits and seeds. It will also build up in animal milk and fatty tissues. Therefore, people are exposed to cadmium upon consumption of cadmium containing plants or animals. Seafood, such as mollusks and crustaceans, can be a source of cadmium, as well. Cadmium is highly corrosion resistant and is used as a protective coating for iron, steel, and copper; it is generally applied by electroplating, but hot dipping and spraying are possible. Cadmium may be alloyed with copper, nickel, gold, silver, bismuth, and aluminum to form easily fusible compounds. These alloys may be used as coatings for other materials, welding electrodes, solders, etc. It is also utilized in electrodes of alkaline storage batteries, as a neutron absorber in nuclear reactors, a stabilizer for polyvinyl chloride plastics, a deoxidizer in nickel plating, an amalgam in dentistry, in the manufacture of fluorescent lamps, semiconductors, photocells, and jewelry, in process engraving, in the automobile and aircraft industries, and to charge Jones reductors. Various cadmium compounds find use as fungicides, insecticides, nematocides, polymerization catalysts, pigments, paints, and glass; they are used in the photographic industry and in glazes. Cadmium is also a contaminant of superphosphate fertilizers. Human exposure to cadmium and certain cadmium compounds occurs through inhalation and ingestion. The entire population is exposed to low levels of cadmium in the diet because of the entry of cadmium into the food chain as a result of its natural occurrence. Tobacco smokers are exposed to an estimated 17 μg/cigarette. Cadmium is present in relatively low amounts in the earth's crust; as a component of zinc ores, cadmium may be released into the environment around smelters.

***U.S. Maximum Allowable Levels for Cadmium***:
The Food and Drug Administration (FDA) limits the amount of cadmium in food colors to 15 parts per million (15 ppm).

***Carcinogen/Hazard Classifications***
**U.S. EPA Carcinogens:** Group B1, probable carcinogen
**U.S. NTP Carcinogen:** Known carcinogen
**California Prop. 65:** Listed; Developmental toxicity, male; carcinogen
**TRI Developmental Toxin:** Reproductive and developmental toxin
**Endocrine Disruptor:** Probable ED

***Regulatory Authority:***
- Carcinogen (Animal Positive) (IARC) (DFG)[9] (Suspected) (ACGIH)[1] (NTP)(DFG)[3]
- Banned or Severely Restricted (Many Countries) (UN) (13, 35)
- Air Pollutant Standard Set (OSHA) (ACGIH)[1] (Australia) (Israel) (Mexico) (Several States) (Several Canadian Provinces)
- AB 1803-Well Monitoring Chemical (CAL)
- MCL(Maximum Contaminants Levels) list of contaminants (CAL)
- AB 2588-Air Toxics "Hot Spots" Chemicals (CAL)
- CAL Air Resources Board/AB 1807 Toxic Air Contaminants
- Permissible Exposure Limits for Chemical Contaminants (CAL/OSHA)
- Clean Air Act: Hazardous Air Pollutants (Title I, Part A, Section 112). Includes any unique chemical substance that contains cadmium as part of that chemical's infrastructure.
- Clean Water Act: 40CFR423, Appendix A, Priority Pollutants; Section 313 Water Priority Chemicals
- (57FR41331, 9/9/92); Toxic Pollutant (Section 401.15)
- EPA Hazardous Waste Number (RCRA No.): D006
- RCRA, 40CFR261, Appendix 8 Hazardous Constituents, waste number not listed.
- RCRA Toxicity Characteristic (Section 261.24), Maximum Concentration of Contaminants, regulatory level, 1.0 mg/L
- RCRA 40CFR268.48; 61FR15654, Universal Treatment Standards: Wastewater (mg/L), 0.69; Nonwastewater (mg/L), 0.19 TCLP
- RCRA 40CFR264, Appendix 9; TSD Facilities Ground Water Monitoring List. Suggested test method(s) (PQL $ug$/L): 6010(40); 7130(50); 7131[1]
- Safe Drinking Water Act: MCL, 0.005 mg/L; MCLG, 0.005 mg/L; Regulated chemical (47 FR 9352); Priority List (55 FR 1470)
- Superfund/EPCRA 40CFR302.4 RQ: CERCLA, 10 lb (4.54 kg). *Note*: No release report required if diameter of pieces is equal to or exceeds 0.004 in.
- EPCRA Section 313 Form R *de minimus* concentration reporting level: 0.1%.
- U.S. DOT Regulated Marine Pollutant (49CFR172.101, Appendix B): Severe pollutant, as cadmium compounds.
- Canada, WHMIS, Ingredients Disclosure List; National Pollutant Release Inventory (NPRI); CEPA Priority Substance List, Ocean dumping prohibited; Drinking Water Quality: 0.005 mg/L MAC.
- Mexico Drinking Water Criteria: 0.01 mg/L

***Description:*** Cadmium is a bluish-white metal. Boiling point = 765°C. Melting/Freezing point = 321°C. Cadmium metal dust has an Autoignition temperature = 250°C. Hazard Identification (based on NFPA-704 M Rating System): (powder) Health 2, Flammability 2, Reactivity 0. The only cadmium mineral, greenockite (CdS), is rare; however, small amounts of cadmium are found in zinc, copper, and lead ores. It is generally produced as a by-product of these metals, particularly zinc. Cadmium is insoluble in water but is soluble in acids. "Cadmium dust" includes dust of various cadmium compounds such as $CdCl_2$. "Cadmium fume" has the composition $Cd/CdO$.

*Incompatibilities:* Air exposure with cadmium powder may cause self ignition. Moist air slowly oxidizes cadmium forming cadmium oxide. Cadmium dust is incompatible with strong oxidizers, ammonium nitrate, elemental sulfur, hydrazoic acid, selenium, zinc, tellurium. Contact with acids cause a violent reaction producing flammable hydrogen gas.

*Permissible Exposure Limits in Air:* The Federal standard[58] for cadmium fume or dust is 0.005mg/m3 (as Cd) as an 8-hour TWA. ACGIH has recommended a TWA value of 0.01 mg/m$^3$ for Cd dusts/salts and 0.002 mg/m3 for respirable dust[1]. NIOSH[2] recommends that exposure to cadmium be at the lowest feasible level. The NIOSH[2] IDLH: Carcinogen [9 mg/m3 (as Cd)]. Several states have set guidelines or standards for cadmium in ambient air[60] ranging from zero (North Dakota) to 0.0006 µg/m3 (Rhode Island) to 0.0055 µg/m3 (North Carolina) to 0.0056 µg/m3 (Massachusetts) to 0.07 (Montana) to 0.12 µg/m3 (Pennsylvania) to 0.25 µg/m3 (South Carolina) to 0.4 µg/m3 (Connecticut and South Dakota) to 0.8 µg/m3 (Virginia) to 1.0 µg/m3 (Nevada) to 2.0 µg/m3 (New York).

*Determination in Air:* Collection of particles on a filter, workup with acid and measurement by atomic absorption has been specified by NIOSH. See NIOSH Methods #7048 (Cd) and #7300 (Elements), #7200 (Welding and Brazing Fume)[18] or OSHA: #ID-125G.[58]

*Permissible Concentration in Water:* The U.S. EPA has set a limit of 5 ppb. Canada set a limit of 0.005 mg/L in drinking water. Mexico has set a limit of 0.01 mg/L in drinking water. Effluent standards for cadmium in water have been set by Argentina, 0.1 mg/L; Japan, 0.1 mg/L. Drinking water standards have been set[35] by Czechoslovakia, 0.010 mg/L; EEC, 5.0 µg/L (0.005 mg/L); Japan, <0.01 mg/L; USSR/UNEP, 0.01 mg/L; WHO, 0.005 mg/L. Further, guidelines for cadmium in drinking water have been set[61] ranging from 5 µg/L (Kansas and Minnesota) to 10 µg/L (Maine).

*Determination in Water:* Total cadmium may be determined by digestion followed by atomic absorption of colorimetric (Dithizone) analysis or by Inductively Coupled Plasma Optical Emission Spectrometry. Dissolved cadmium is determined by 0.45 µ filtration followed by the previously cited methods. See also reverence[49].

*Routes of Entry:* Inhalation or ingestion of fumes or dust.

*Harmful Effects and Symptoms*

*Short Term Exposure: Inhalation:* Dust may cause irritation of the nose and throat. Non-fatal lung inflammation has been reported from concentrations of 0.5 to 2.5 mg/m$^3$. In 4 to 10 hours after exposure severe chest pain, with persistent cough and difficult breathing, headache, chills, muscle aches, nausea, vomiting, diarrhea. Fluid in the lungs (pulmonary edema) and dark-purple coloration of the skin may occur. Pulmonary edema is a medical emergency that can be delayed for several hours. This can cause death. Breathing becomes more difficult and is accompanied by wheezing or coughing of blood. Other symptoms which may occur 12 to 36 hours after exposure in addition to those above include dizziness, irritability, gastrointestinal disturbances, shortness of breath, fever, profuse sweating, exhaustion and inflammation of the lungs. Death may result within 7 to 10 days after exposure. The average concentrations of fume responsible for fatalities have been 40 to 50 mg/m$^3$ for 1 hour, 9 mg/m$^3$ for 5 hours, or 5 mg/m$^3$ for 8 hours. *Skin:* Absorption is negligible. *Eyes:* Cadmium compound dust may cause irritation. *Ingestion:* A dose of 15-30 mg (1/1000 oz.) of metal or soluble compounds may cause increased salivation, choking, vomiting, abdominal pain, anemia, kidney malfunction, diarrhea, and persistent desire to empty the bladder. Symptoms may occur within 15-30 minutes after ingestion. May cause heart and lung failure.

*Long Term Exposure:* Continued exposure to low levels of cadmium in air may cause irreversible lung injury, abnormal lung function and kidney disease. Other consequences of cadmium exposures are inflammation of the nose and throat, open sores in the nose, soreness, bleeding and reduced nose size, loss of sense of smell, damage to the olfactory nerve, yellow cadmium stains on teeth, sleeplessness, nausea, lack of appetite, weight loss, anemia, lung distention with scar formation, and liver damage. May cause bone disease characterized by softening, bending and reduction in bone size. Difficulty walking, pain in back and extremities, and spontaneous fractures may result. Inhalation of 0.06 mg/m$^3$ to 0.68 mg/m$^3$ for 4 to 8 years may cause throat irritation, cough, chest pain, upset stomach and fatigue. Exposure to levels of 3.0 to 15.0 mg/m$^3$ of fumes or dust over a period of 20 years has caused lung distention, anemia, protein in urine and kidney dysfunction. Studies indicate that there is an increased incidence of prostatic cancer and possible kidney and respiratory cancer in cadmium workers. Cadmium causes birth defects in rats, mice and hamsters. Cadmium does not readily go from a pregnant woman's body into the developing child, but some portion can cross the placenta. It can also be found in breast milk. Animals given cadmium in food or water had high blood pressure, iron-poor blood, liver disease, and nerve or brain damage. We don't know if humans get any of these diseases from eating or drinking cadmium. Skin contact with cadmium is not known to cause health effects in humans or animals. The babies of animals exposed to high levels of cadmium during pregnancy had changes in behavior and learning ability. Cadmium may also affect birth weight and the skeleton in developing animals.

*Points of Attack:* Respiratory system, lungs, kidneys, prostate, blood.

*Medical Surveillance:* In preemployment physical examinations emphasis should be given to a history of, or actual presence of, significant kidney disease, smoking history, and respiratory disease. A chest x-ray and baseline pulmonary function study is recommended. Periodic examinations should emphasize the respiratory system, including pulmonary function tests, kidneys and blood.

A low molecular weight proteinuria may be the earliest indication of renal toxicity. The trichloroacetic acid test may pick this up, but more specific quantitative studies would be preferable. If renal disease due to cadmium is present, there may also be increased excretion of calcium, amino acids, glucose, and phosphates.

*First Aid:* If this chemical gets into the eyes, remove any contact lenses at once and irrigate immediately for at least 15 minutes, occasionally lifting upper and lower lids. Seek medical attention immediately. If this chemical contacts the skin, remove contaminated clothing and wash immediately with soap and water. Seek medical attention immediately. If this chemical has been inhaled, remove from exposure, begin rescue breathing (using universal precautions) if breathing has stopped and CPR if heart action has stopped. Transfer promptly to a medical facility. When this chemical has been swallowed, get medical attention. Give large quantities of water and induce vomiting. Do not make an unconscious person vomit. Medical observation is recommended for 24 to 48 hours after breathing overexposure, as pulmonary edema may be delayed. As first aid for pulmonary edema, a doctor or authorized paramedic may consider administering a corticosteroid spray.

*References:*
- EXTOXNET FAQs, Extension Toxicology Network, "Cadmium Contamination of Food," Oregon State University, Corvallis, OR. (December 1997). *Note:* This site has links for more information and studies. http://ace.orst.edu/info/extoxnet/faqs/foodcon/cadmium.htm
- Agency for Toxic Substances and Disease Registry, "ToxFAQs for Cadmium," Atlanta, GA (June 1999). http://www.atsdr.cdc.gov/tfacts5.html
- National Institute for Occupational Safety and Health, Criteria for a Recommended Standard: Occupational Exposure to Cadmium, NIOSH Doc. No. 76-192, Washington, DC (1976).
- U.S. Environmental Protection Agency, Cadmium: Ambient Water Quality Criteria, Washington, DC (1980).
- U.S. Environmental Protection Agency, Status Assessment of Toxic Chemicals: Cadmium, Report EPA-600/2-9-210F, Washington, DC (December 1979).
- U.S. Environmental Protection Agency, Toxicology of Metals, Vol II: Cadmium, Report EPA-600/1-77-022, Research Triangle Park, NC, pp 124-163 (May 1977).
- U.S. Environmental Protection Agency, Reviews of the Environmental Effects of Pollutants, IY: Cadmium,
- Report EPA-600/1-78-026, Health Effects Research Laboratory, Cincinnati, OH (1978).
- U.S. Environmental Protection Agency, Health Assessment Document for Cadmium, Report EPA-600/8-79-003, Research Triangle Park, NC (1979).
- U.S. Environmental Protection Agency, Cadmium, Health and Environmental Effects Profile No. 31, Office of Solid Wastes, Washington, DC (April 30, 1980).
- Sax, N.I., Ed., "Dangerous Properties of Industrial Materials Report" 1, No. 1, 36-38 (1980) and 3, No. 5, 72-76 (1983).
- New Jersey Department of Health and Senior Services, "Hazardous Substance Fact Sheet: Cadmium," Trenton, NJ (April 1994, rev. December 1999). http://www.state.nj.us/health/eoh/rtkweb/0305.pdf
- New York State Department of Health, "Chemical Fact Sheet: Cadmium Compounds," Albany, NY, Bureau of Toxic Substance Assessment (January 1986).
- California Environmental Protection Agency Chemical List of Lists, Sacramento CA (February 1997).
- U.S. Environmental Protection Agency, "Health Advisory: Cadmium," Washington, DC, Office of Drinking Water (March 31, 1987).

## Cadmium Chloride

*Use Type:* Fungicide
*CAS Number:* 10108-64-2
*Formula:* $CdCl_2$
*Synonyms:* Cadmium dichloride; Cloruro de cadmio (Spanish); Dichlorocadmium; Kadmiumchlorid (Germany)
*Trade Names:* CADDY®, Cleary Chemical Corp. (USA), canceled 8/13/1990; VI-CAD®, Vineland Chemical (USA), canceled 12/31/1987
*Producers:* Advance Research Chemicals, Inc. (USA); Degussa (Germany); Great Western Inorganics (USA); Nihon Kagaku Sangyo (Japan); Honeywell Performance P & C (USA); Rhone-Poulenc (France)
*Chemical Class:* Inorganic, heavy metal
*EPA/OPP PC Code:* 012902
*California DPR Chemical Code:* 94
*ICSC Number:* 0116
*RTECS Number:* EV0175000
*EEC Number:* 048-008-00-3
*EINECS Number:* 233-296-7
*Uses:* Cadmium chloride is also used in dyeing and printing of fabrics; in electronic component manufacture; in photography.
*U.S. Maximum Allowable Levels for Cadmium:*
The Food and Drug Administration (FDA) limits the amount of cadmium in food colors to 15 parts per million (15 ppm).
*Carcinogen/Hazard Classifications*
**U.S. NTP Carcinogen:** Known carcinogen
**California Prop. 65:** Carcinogen
**IARC:** Group 1, known carcinogen
**Label Signal Word:** CAUTION

**Endocrine Disruptor:** Not listed as cadmium chloride; Probable ED as cadmium.

*Regulatory Authority:*
- Carcinogen (UK) (DFG)[3] (suspected) (NTP)
- Air Pollutant Standard Set (see Cadmium)
- Banned or Severely Restricted (In pesticides in UK)[13] (other) (UN)[35]
- Air Pollutant Standard Set (ACGIH)[1] (Several States)
- AB 2588-Air Toxics "Hot Spots" Chemicals (CAL)
- CAL Air Resources Board/AB 1807 Toxic Air Contaminants
- Proposition 65 chemical (CAL)
- The "Director's List" (CAL/OSHA)
- Clean Air Act: Hazardous Air Pollutants (Title I, Part A, Section 112).
- Clean Water Act: Toxic Pollutant (Section 401.15); Section 311 Hazardous Substances/RQ 40CFR117.3 (same as CERCLA, see below); Section 313 Water Priority Chemicals (57FR41331, 9/9/92).
- Superfund/EPCRA 40CFR302.4 RQ: CERCLA, 10 lb (4.54 kg).
- EPCRA Section 313 Form R *de minimus* concentration reporting level: 0.1%.
- U.S. DOT Regulated Marine Pollutant (49CFR172.101, Appendix B): Severe pollutant, as cadmium compounds.
- Canada, WHMIS, Ingredients Disclosure List; National Pollutant Release Inventory (NPRI); CEPA Priority Substance List, Ocean dumping prohibited, as cadmium compounds.

*Description:* Colorless, crystalline solid or powder. Odorless. Boiling point = 960°C. Melting/Freezing point = 568°C. Hazard Identification (based on NFPA-704 M Rating System): Health 2, Flammability 0, Reactivity 0. Soluble in water.

*Incompatibilities:* Sulfur, selenium, potassium, and strong oxidizers.

*Permissible Exposure Limits in Air:* The Federal standard[58] for cadmium compounds is 0.005mg/m$^3$ (as Cd) as an 8-hour TWA. ACGIH[1] has recommended a TWA value of 0.01 mg/m$^3$ for Cd dusts/salts and 0.002 mg/m$^3$ for respirable dust. NIOSH[2] recommends that exposure to cadmium be at the lowest feasible level. See also this section in the entry on Cadmium. Guidelines for cadmium chloride in ambient air have been set[60] ranging from 1.67 µg/m$^3$ (New York) to 5.0µg/m$^3$ (Florida).

*Determination in Air:* Collection of particles on a filter, workup with acid and measurement by atomic absorption has been specified by NIOSH. See NIOSH Methods #7048 (Cd).[18]

*Permissible Concentration in Water:* The U.S. EPA has set a limit of 5 ppb. Canada set a limit of 0.005 mg/L in drinking water. Mexico has set a limit of 0.01 mg/L in drinking water. Effluent standards for cadmium in water have been set by Argentina, 0.1 mg/L; Japan, 0.1 mg/L. Drinking water standards have been set[35] by Czechoslovakia, 0.010 mg/L; EEC, 5.0 µg/L (0.005 mg/L); Japan, <0.01 mg/L; USSR/UNEP, 0.01 mg/L; WHO, 0.005 mg/L. Further, guidelines for cadmium in drinking water have been set[61] ranging from 5 µg/L (Kansas and Minnesota) to 10 µg/L (Maine).

*Determination in Water:* Total cadmium may be determined by digestion followed by atomic absorption of colorimetric (Dithizone) analysis or by Inductively Coupled Plasma Optical Emission Spectrometry. Dissolved cadmium is determined by 0.45 µ filtration followed by the previously cited methods. See also reverence[49].

*Routes of Entry:* Inhalation of dust, ingestion

*Harmful Effects and Symptoms*

*Short Term Exposure:* Eye contact can cause irritation. Cadmium chloride can cause severe irritation of the gastrointestinal tract and the respiratory tract. Inhalation can cause nose, throat, and lung irritation. Fumes can cause flu-like illness with chills, headache, aching muscles and/or fever. Higher exposures can cause nausea, salivation, vomiting, cramps, diarrhea, and pulmonary edema, a medical emergency that can be delayed for several hours. This can cause death. Cadmium chloride is highly toxic. As little as 14.5 mg of Cd orally causes nausea and vomiting; 8.9 grams has caused death. Cadmium salts cause cramps, nausea, vomiting and diarrhea. Acute poisoning causes lung damage. Delayed pulmonary edema may occur after inhalation.

*Long Term Exposure:* This chemical is a probable carcinogen in humans, with some evidence that it causes prostate and kidney cancer in humans and it has been shown to cause lung and testes cancer in animals. It may also be a reproductive hazard in humans. Repeated low exposures may cause permanent kidney and liver damage, anemia, and/or loss of the sensed of smell. Chronic poisoning damages kidneys, lungs, bones and causes blood changes (anemia). Cadmium does not readily go from a pregnant woman's body into the developing child, but some portion can cross the placenta. It can also be found in breast milk. The babies of animals exposed to high levels of cadmium during pregnancy had changes in behavior and learning ability. Cadmium may also affect birth weight and the skeleton in developing animals.

*Points of Attack:* Respiratory system, lungs, kidneys, prostate, blood.

*Medical Surveillance:* Medical observation is recommended for 24 to 48 hours after breathing overexposure, as pulmonary edema may be delayed. Urine test for cadmium (levels should be less than 10 µgms/L of urine). Urine test for "low molecular weight proteins" to detect kidney damage. Urinalysis. Complete Blood Count. Lung function tests. Consider chest x-ray following acute overexposure.

**138    Cadmium Sulfate**

*First Aid:* If this chemical gets into the eyes, remove any contact lenses at once and irrigate immediately for at least 15 minutes, occasionally lifting upper and lower lids. Seek medical attention immediately. If this chemical contacts the skin, remove contaminated clothing and wash immediately with soap and water. Seek medical attention immediately. If this chemical has been inhaled, remove from exposure, begin rescue breathing (using universal precautions) if breathing has stopped and CPR if heart action has stopped. Transfer promptly to a medical facility. When this chemical has been swallowed, get medical attention. Give large quantities of water and induce vomiting. Do not make an unconscious person vomit. Medical observation is recommended for 24 to 48 hours after breathing overexposure, as pulmonary edema may be delayed. As first aid for pulmonary edema, a doctor or authorized paramedic may consider administering a corticosteroid spray.

*References:*
- New Jersey Department of Health and Senior Services and Senior Services, Hazardous Substance Fact Sheet, " Cadmium Chloride," Trenton NJ (January, 1996, rev. April, 2002).
  http://www.state.nj.us/health/eoh/rtkweb/0308.pdf
- California Environmental Protection Agency Chemical List of Lists, Sacramento CA (February 1997).
- Sax, N.I., Ed., "Dangerous Properties of Industrial Materials Report" 2, No. 3, 73-76 (1982)

# Cadmium Sulfate

*Use Type:* Fungicide
*CAS Number:* 10124-36-4
*Formula:* $O_4S \cdot Cd$
*Synonyms:* Cadmium monosulfate; Cadmium sulphate; Sulphuric acid, cadmium salt; Sulfuric acid, cadmium(2+) salt; Sulfuric acid, cadmium(II) salt
*Producers:* Degussa (Germany); Fluorochem Ltd. (UK); Goldschmidt (Germany); Merke (Germany); Rhodia Group (France)
*Chemical Class:* Inorganic heavy metal
*EPA/OPP PC Code:* 012905
*ICSC Number:* 1318
*EEC:* 048-009-00-9
*EINECS Number:* 233-331-6
*Uses:* It is used in pigments, electroplating., and in synthetic and analytical chemistry. Also used in fluorescent screens, as an electrolyte. See also "Cadmium."
*U.S. Maximum Allowable Levels for Cadmium:*
The Food and Drug Administration (FDA) limits the amount of cadmium in food colors to 15 parts per million (15 ppm).
*Carcinogen/Hazard Classifications*
**U.S. EPA Carcinogens:** Group B1, Probably Carcinogen (as cadmiuim)
**U.S. NTP Carcinogen:** Known carcinogen
**California Prop. 65:** Carcinogen
**IARC:** Group 1, known carcinogen
**Endocrine Disruptor:** Probable ED (as cadmium)
*Regulatory Authority:*
- Carcinogen (suspect) (NTP) (IARC) (animal positive) (DFG)
- AB 2588-Air Toxics "Hot Spots" Chemicals (CAL)
- CAL Air Resources Board/AB 1807 Toxic Air Contaminants
- Proposition 65 chemical (CAL)
- The "Director's List" (CAL/OSHA)
- Clean Air Act: Hazardous Air Pollutants (Title I, Part A, Section 112). Includes any unique chemical substance that contains cadmium as part of that chemical's infrastructure.
- Clean Water Act: 40CFR401.15 Section 307 Toxic Pollutants
- RCRA, 40CFR261, Appendix 8 Hazardous Constituents, waste number not listed.
- EPCRA (Section 313): Form R *de minimus* concentration reporting level: (inorganic compounds: 0.1%; organic compounds: 1.0%).
- Canada, WHMIS, Ingredients Disclosure List; National Pollutant Release Inventory (NPRI); CEPA Priority Substance List, Ocean dumping prohibited.
- U.S. DOT Regulated Marine Pollutant (49CFR172.101, Appendix B).: Severe pollutant, as cadmium compounds.
- Canada, WHMIS, Ingredients Disclosure List

*Description:* Cadmium sulfate is a white or colorless, crystalline solid. Odorless. Molecular weight = 208.48. Melting/Freezing point = 1000°C. Hazard Identification (based on NFPA-704 M Rating System): Health 2, Flammability 0, Reactivity 0. Soluble in water.

*Incompatibilities:* Incompatible with strong oxidizers, sulfur, selenium, tellurium, zinc. May ignite combustible materials.

*Permissible Exposure Limits in Air:* The Federal standard[58] for cadmium compounds is 0.005mg/m$^3$ (as Cd) as an 8-hour TWA. ACGIH[1] has recommended a TWA value of 0.01 mg/m$^3$ for Cd dusts/salts and 0.002 mg/m$^3$ for respirable dust[1]. NIOSH[2] recommends that exposure to cadmium be at the lowest feasible level.

*Determination in Air:* Collection of particles on a filter, workup with acid and measurement by atomic absorption has been specified by NIOSH. See NIOSH Methods #7048 (Cd).[18]

*Permissible Concentration in Water:* The U.S. EPA has set a limit of 5 ppb. Canada set a limit of 0.005 mg/L in drinking water. Mexico has set a limit of 0.01 mg/L in drinking water. Effluent standards for cadmium in water have been set by Argentina, 0.1 mg/L; Japan, 0.1 mg/L. Drinking water standards have been set[35] by Czechoslovakia, 0.010 mg/L; EEC, 5.0 µg/L (0.005 mg/L); Japan, <0.01 mg/L; USSR/UNEP, 0.01 mg/L; WHO, 0.005 mg/L. Further, guidelines for cadmium in drinking water have been set[61] ranging from 5 µg/L (Kansas and Minnesota) to 10 µg/L (Maine).

***Determination in Water:*** Total cadmium may be determined by digestion followed by atomic absorption of colorimetric (Dithizone) analysis or by Inductively Coupled Plasma Optical Emission Spectrometry. Dissolved cadmium is determined by 0.45 µ filtration followed by the previously cited methods. See also reverence[49].

***Routes of Entry:*** Inhalation, ingestion

***Harmful Effects and Symptoms***

***Short Term Exposure:*** Irritates the eyes on contact. Inhalation irritates the nose, throat and lungs with coughing, and/or shortness of breath. Higher exposures can cause pulmonary edema, a medical emergency that can be delayed for several hours. This can cause death. Cadmium sulfate can cause nausea, salivation, vomiting, cramps and diarrhea, metal fume fever with flu-like symptoms, chills, headache, weakness, metallic tasted in the mouth. IDLH level = 50 mg/m$^3$ dust; 9 mg/m$^3$ fume. Delayed pulmonary edema may occur after inhalation.

***Long Term Exposure:*** Repeated exposure can cause anemia, brittle and painful bones, diminished or loss of the sense of smell, fatigue, and/or yellow staining of the teeth. May cause lung and prostate cancer, kidney damage with kidney stones, liver damage, lung damage with bronchitis, cough, phlegm, and/or shortness of breath. There is some evidence that cadmium sulfate is a teratogen in humans. It is unknown if cadmium causes birth defects in people. Cadmium does not readily go from a pregnant woman's body into the developing child, but some portion can cross the placenta. It can also be found in breast milk. The babies of animals exposed to high levels of cadmium during pregnancy had changes in behavior and learning ability. Cadmium may also affect birth weight and the skeleton in developing animals.

***Points of Attack:*** Lungs, liver, kidneys, blood.

***Medical Surveillance:*** Medical observation is recommended for 24 to 48 hours after breathing overexposure, as pulmonary edema may be delayed. Urine test for cadmium (levels should be less than 10 µgms/L of urine). Urine test for low molecular weight proteins ($\beta$-2-microglobulin) to detect kidney damage. Complete blood Count. Lung function tests. Liver function tests. Consider chest x-ray following acute overexposure.

***First Aid:*** If this chemical gets into the eyes, remove any contact lenses at once and irrigate immediately for at least 15 minutes, occasionally lifting upper and lower lids. Seek medical attention immediately. If this chemical contacts the skin, remove contaminated clothing and wash immediately with soap and water. Seek medical attention immediately. If this chemical has been inhaled, remove from exposure, begin rescue breathing (using universal precautions) if breathing has stopped and CPR if heart action has stopped. Transfer promptly to a medical facility. When this chemical has been swallowed, get medical attention. Give large quantities of water and induce vomiting. Do not make an unconscious person vomit. Medical observation is recommended for 24 to 48 hours after breathing overexposure, as pulmonary edema may be delayed. As first aid for pulmonary edema, a doctor or authorized paramedic may consider administering a corticosteroid spray.

*Note to physician or authorized medical personnel:* In case of fume inhalation, treat pulmonary edema. Give prednisone or other corticosteroid orally to reduce tissue response to fume. Positive pressure ventilation may be necessary. Treat metal fume fever with bed rest, analgesics and antipyretics. The symptoms of metal fume fever may be delayed for 4-12 hours following exposure: it may last less than 36 hours

***References:*** (See also references for "Cadmium.")
- New Jersey Department of Health and Senior Services, Hazardous Substance Fact Sheet, "Cadmium Sulfate," Trenton NJ (September 1998). http://www.state.nj.us/health/eoh/rtkweb/3073.pdf
- California Environmental Protection Agency Chemical List of Lists, Sacramento CA (February 1997).

# Cadusafos

***Use Type:*** Insecticide and nematicide
***CAS Number:*** 95465-99-9
***Formula:*** $C_{10}H_{23}O_2PS_2$
***Synonyms:*** *O*-Ethyl *S,S*-[di(sec-butyl)]phosphorodithioate; *O*-Ethyl *S,S*-[bis(1-methylpropyl)]phosphorodithioate; Phosphorodithioic acid, *O*-ethyl *S,S*-bis(1-methylpropyl) ester
***Trade Names:*** APACHE®; EBUFOS®; FMC® 67825, FMC (USA); RUGBY®, FMC (USA); TAREDAN®
***Producers:*** FMC (USA)
***Chemical Class:*** Organophosphate
***EPA/OPP PC Code:*** 128864
***RTECS Number:***
***EEC Number:***
***EINECS Number:***
***Uses:*** Not registered in the U.S. It is used to control nematodes and soil insects on bananas in Mexico, Guatemala, Honduras, Costa Rica, and Ecuador. Although it is not registered for use in the U.S., an import tolerance was established for cadusafos in or on bananas in 1987.

***U.S. Maximum Allowable Residue Levels for Cadusafos (40 CFR 180.461):***

| CROP | ppm |
|---|---|
| Banana | 0.01 |

***Carcinogen/Hazard Classifications***
**U.S. EPA Carcinogens:** Group E, unlikely carcinogen
**WHO Acute Hazard:** Class 1b, highly hazardous
***Regulatory Authority:***
- DOT Inhalation Hazard Chemicals as organophosphates

***Description:*** Molecular weight = 270.41

***Determination in Air:*** OSHA versatile sampler-2; Toluene/Acetone; Gas chromatography/Flame ionization detection; NIOSH IV, Method #5600, Organophosphorus Pesticides.[18]

***Permissible Concentration in Water:*** No criteria set. Runoff from spills or fire control may cause water pollution.

***Routes of Entry:*** Inhalation, skin absorption, ingestion, skin and/or eye contact.

*Harmful Effects and Symptoms*

***Short Term Exposure:*** Eye pupils are small; blurred vision; eye watering; runny nose; cough; shortness of breath; salivation; dizziness; nausea, stomach cramps, diarrhea, and vomiting; increased blood pressure; profuse sweating; hypermotility, hallucinations; irritability; tingling of the skin; drowsiness; slow heartbeat; convulsions; fluid in lungs; loss of consciousness; incontinence; breathing stops; death. Orgnophosphates inhibit the action of acetylcholinesterase enzymes, and alter the way in which nervous impulses are transmitted. The effects can last for hours, days, or much longer. The action of the enzymes is reestablished after new enzymes are formed.

***Long Term Exposure:*** Cholinesterase inhibitor; cumulative effect is possible. Organophosphates may damage the nervous system with repeated exposure, resulting in convulsions, respiratory failure. May cause liver damage.

***Points of Attack:*** Respiratory system, central nervous system, cardiovascular system, blood cholinesterase.

***Medical Surveillance:*** Medical observation is recommended for 24 to 48 hours after breathing overexposure, as pulmonary edema may be delayed. As first aid for pulmonary edema, consider administering a corticosteroid spray. Cigarette smoking may exacerbate pulmonary injury and should be discouraged for at least 72 hours following exposure. Do not drink any alcoholic beverages before or during use; alcohol promotes absorption of organophosphates.

***First Aid:*** **Treatment for organophosphate poisoning consists of thorough decontamination, cardiorespiratory support, and administration of the antidotes atropine and pralidoxime. In cases of severe poisoning, diazepam, an anticonvulsant, should also be administered. Antidotes should be administered as prevention even if the diagnosis is in doubt.** Speed in removing material from eyes and skin is of extreme importance. *Eyes:* Eye contact can cause dangerous amounts of these chemicals to be quickly absorbed through the mucous membrane into the bloodstream. Immediately and gently flush eyes with plenty of warm or cold water (NO hot water) for at least 15 minutes, occasionally lifting the upper and lower eyelids. Get medical aid immediately. *Skin:* Get medical aid. Skin contact can cause dangerous amounts of these chemicals to be absorbed into the bloodstream. Wearing the appropriate PPE equipment and respirator for organophosphate/carbamate pesticides, immediately flush skin with plenty of soap and water for at least 15 minutes while removing contaminated clothing and shoes. Shampoo hair promptly if contaminated. The removed, contaminated clothing and shoes should be double-bagged and left in Hot Zone for later disposal by hazardous materials experts. Skin may also be decontaminated with diluted hypochlorite solution. *Inhalation:* Get medical aid. Do not contaminate yourself. Wearing the appropriate PPE equipment and respirator for organophosphate pesticides, immediately remove the victim from the contaminated area to fresh air. If the victim is not breathing, administer artificial respiration. Do NOT use mouth-to-mouth resuscitation; use bag/mask apparatus. If breathing is difficult, administer oxygen through bag/mask apparatus until medical help arrives. Do not leave victim unattended. *Ingestion:* Call poison control. Loosen all clothing. Never give anything by mouth to an unconscious person. If victim is *unconscious or having convulsions,* do nothing except keep victim warm. Get medical aid. Transfer promptly to a medical facility. In cases of ingestion, **do not induce vomiting**. If the victim is alert and asymptomatic, administer a slurry of activated charcoal at a dose of 1 g/kg (infant, child, and adult dose). A soda can and straw may be of assistance when offering charcoal to a child. *In some cases you may be specifically instructed by poison control to induce vomiting by way of 2 tablespoons of syrup of ipecac (adult) washed down with a cup of water.* Do NOT give activated charcoal before or with ipecac syrup.

*Note to physician or authorized medical personnel:* Treat cases of respiratory compromise, coma, or excessive pulmonary secretions with respiratory support using protocols and techniques available and within the scope of training. Some cases may necessitate procedures such as endotracheal intubation or cricothyrotomy by properly trained and equipped personnel. When possible, atropine (see *Antidotes*, below) should be given under medical supervision. Patients who are comatose, hypotensive, or having seizures or cardiac arrhythmias should be treated according to advanced life support protocols. *Antidotes:* Two antidotes are administered to treat organophosphate poisoning. Atropine is a competitive antagonist of acetylcholine at muscarinic receptors and is used to control the excessive bronchial secretions which are often responsible for death. Pralidoxime relieves both the nicotinic and muscarine effects of organophosphate poisoning by regenerating acetylcholinesterase and can reduce both the bronchial secretions and the muscle weakness associated with poisoning. The initial intravenous dose of atropine in adults should be determined by the severity of symptoms: An initial adult dose of 1.0 to 2.0 mg or pediatric dose of 0.01 mg/kg (minimum 0.01 mg) should be administered intravenously. If intravenous access cannot be established, atropine may also be given intramuscularly, subcutaneously or via endotracheal tube. Doses should be repeated every 15 minutes until excessive secretions and sweating have been controlled. Once bronchial secretion has been controlled, atropine administration should be repeated whenever the secretions begin to recur. In seriously poisoned patients, very large doses may be required. Alterations of pulse rate and pupillary size should not be used as indicators of treatment adequacy. Pralidoxime should be administered as early in poisoning as possible as its efficacy may diminish when given more than 24 to 36 hours after exposure. Doses are as follows: adult 1.0 g; pediatric 25 to 50 mg/kg. The drug should be administered intravenously over 30 to 60 minutes, but in a life-threatening situation, one-half of the total dose can be given per minute for a total administration time of 2 minutes. Treatment should begin to take effect within 40 minutes with a reduction in symptoms and the amount of atropine necessary to control bronchial secretion. The initial dose can be repeated in 1 hour and then every 8 to 12 hours until the patient is clinically well and no longer requires atropine. If intravenous access cannot be established, pralidoxime may also be given intramuscularly. Early administration of diazepam in addition to the combined atropine and pralidoxime treatment may help prevent the onset of seizures and potential brain and cardiac morphologic damage following high-level organophosphate poisoning.

*References:*
- California Environmental Protection Agency Chemical List of Lists, Sacramento CA (February 1997)
- U.S. Environmental Protection Agency, Office of Pesticide Programs, Pesticide Residue Limits, "Cadusafos", 40 CFR 180.461, www.epa.gov/pesticides/food/viewtols.htm
- U.S. Environmental Protection Agency, Office of Pesticide Programs, "Cadusafos Summary," (June 30, 1999). http://www.epa.gov/pesticide/op/cadusafos/cadsum.htm

# Calcium Arsenate

*Use Type:* Insecticide, herbicide, rodenticide, and molluscicide
*CAS Number:* 7778-44-1
*Formula:* $As_2Ca_3O_8$; $Ca_3(AsO_4)_2$
*Synonyms:* Arseniato calcico (Spanish); Arsenic acid, calcium salt (2:3); Arsenate de calcium (French); Calciumarsenat (German); Calcium orthoarsenate; Kalziumarseniat (German); Tricalcium arsenate; Tricalciumarsenat (German); Tricalcium orthoarsenate
*Trade Names:* CAL-META®, Zeneca Inc. (USA) (now Syngenta), canceled 7/01/1987; CHIP-CAL®, Rhone-Poulenc Agrochemical (France) canceled 9/27/1985; CUCUMBER DUST®; FENCAL®; FLAC®; KALO®; KILMAG®; METAG®, Ortho Business Group (USA), canceled 10/04/1985; NIAGARA®, FMC (USA), canceled 5/01/1987; PENCAL®; PROTARS®; SECURITY® (trade name for lead arsenate also), Mallinckrodt Inc. (USA), canceled 10/10/1989; SPRA-CAL®; TURF-CAL®
*Producers:* Great Western Inorganics (USA); Helena Chemical Company (USA)
*Chemical Class:* Inorganic arsenical
*EPA/OPP PC Code:* 013501
*California DPR Chemical Code:* 96
*ICSC Number:* 0765
*RTECS Number:* CG0830000
*RCRA Number:* 231-904-5
*EEC Number:* 033-005-00-1
*EINECS Number:* 231-904-5
*Uses:* In 1988, EPA revoked the registration for the pesticide calcium arsenate, which in years past was used to control insects on a variety of raw agricultural commodities. Calcium arsenate was registered for use on asparagus, beans, blackberries, blueberries (huckleberries), boysenberries, broccoli, Brussels sprouts, cabbage, carrots, cauliflower, celery, collards, corn, cucumbers, dewberries, eggplant, kale, kohlrabi, loganberries, melons, peppers, pumpkins, raspberries, rutabagas, spinach, squash and youngberries.
*Carcinogen/Hazard Classifications*
**U.S. NTP Carcinogen:** Known carcinogen
**California Prop. 65:** Listed; carcinogen
**IARC:** Group 1, known carcinogen
**Label Signal Word:** WARNING
**WHO Acute Hazard:** Class 1b, highly hazardous
**Endocrine Disruptor:** Suspected endocrine disruptor
*Regulatory Authority:*
- Air Pollutant Standard Set (former USSR/UNEP),[43)(60)] (HSE),[33] and (Several Canadian Provinces) as arsenic
- Carcinogen (Human) (IARC)[9] (DFG)[3] (NTP) (OSHA)
- Banned or Severely Restricted (UK, India) (UN)[13] (Various) (UN)[35] as arsenic compounds
- AB 2588-Air Toxics "Hot Spots" Chemicals (CAL)
- CAL Air Resources Board/AB 1807 Toxic Air Contaminants
- Proposition 65 chemical (CAL)
- The "Director's List" (CAL/OSHA)
- Carcinogen User Register Chemical (CAL/OSHA) as inorganic arsenic compound
- AB 1803-Well Monitoring Chemical (CAL) as inorganic arsenic compound
- Permissible Exposure Limits for Chemical Contaminants (CAL/OSHA) as inorganic arsenic compound
- Clean Air Act: List of high risk pollutants (Section 63.74) as arsenic compounds.
- Clean Water Act: Section 311 Hazardous Substances/RQ 40CFR117.3 (same as CERCLA, see below); Section 313 Water Priority Chemicals (57FR41331, 9/9/92); Toxic Pollutant (Section 401.15) as arsenic and compounds.
- RCRA, 40CFR261, Appendix 8 Hazardous Constituents, waste number not listed.

- Superfund/EPCRA 40CFR302.4 RQ: CERCLA, 1 lb (0.454 kg)
- EPCRA Section 313 Form R *de minimus* concentration reporting level: 0.1%.
- Superfund/EPCRA 40CFR355, Appendix B Extremely Hazardous Substances: TPQ = 500/10,000 lb (227/4,540 kg).
- U.S. DOT Regulated Marine Pollutant (49CFR172.101, Appendix B), listed by name; also listed as calcium arsenate and calcium arsenite, mixtures, solid.
- Canada, WHMIS, Ingredients Disclosure List; National Pollutant Release Inventory (NPRI); CEPA Priority Substance List, Ocean dumping prohibited.

***Description:*** $Ca_3(AsO_4)_2$, a white flocculent powder. Molecular weight = 398.10. Slightly soluble in water. Boiling point = (decomposes). Hazard Identification (based on NFPA-704 M Rating System): Health 4, Flammability 0, Reactivity 0.

***Incompatibilities:*** None reported, according to NIOSH. When heated produces As fumes.

***Permissible Exposure Limits in Air:*** The Federal limit is 0.010 mg/m$^3$ (2). NIOSH[2] has recommended a limit of 0.002 g/m$^3$ with a 15-minute ceiling. The NIOSH[2] IDLH level = 5 mg/m$^3$ (as As). The German DFG[3] has set no MAK value because of the notation that this is a proven human carcinogen. The former USSR-UNEP/IPRTC project[43] has set a MAC for calcium arsenate in ambient air of residential areas as 0.009 mg/m$^3$ on a momentary basis and 0.004 mg/m$^3$ on an average daily basis. North Carolina has set a limit in ambient air[60] of 0.0002 μg/m$^3$.

***Determination in Air:*** Filter collection followed by atomic absorption analysis. NIOSH 7900; OSHA ID105.

***Permissible Concentration in Water:*** EPA has set a limit of 0.05 parts per million (ppm) for arsenic in drinking water. The U.S. EPA arsenic drinking water standard of 0.01 ppm (10 ppb) is based on the U.S. EPA final rule for arsenic in drinking water published in. the January 22, 2001, *Federal Register*. However, the U.S. EPA is currently reviewing the science and cost estimate supporting this rule, and, in the interim, has reverted to the previous standard for arsenic. Thus, in the US, the current EPA arsenic drinking water standard remains at 0.05 ppm (50 ppb). To protect freshwater aquatic life-total recoverable trivalent inorganic arsenic never to exceed 440 μg/L. To protect saltwater aquatic life: 508 μg/L on an acute basis. To protect human health: preferably zero. The former USSR-UNEP/IRPTC project[43] has set MAC values for inorganic arsenic compounds in water for domestic purposes at 0.05 mg/L and in water bodies for fishery purposes of 0.5 mg/L also.

***Determination in Water:*** *For arsenic:* The atomic absorption graphite furnace technique is often used for measurement of total arsenic in water. It also has been standardized by EPA. Total arsenic may be determined by digestion followed by silver diethyldithiocarbamate; an alternative is atomic absorption; another is inductively coupled plasma optical emission spectrometry. See OSHA Method #ID-105 for arsenic[58].

***Routes of Entry:*** Inhalation, skin absorption, ingestion, eye and skin contact.

***Harmful Effects and Symptoms***

***Short Term Exposure:*** Irritates eyes, skin and respiratory tract. Can cause poor appetite, a metallic or garlic taste, nausea, vomiting, stomach pain, diarrhea, abnormal heart rhythm, seizures, pain in extremities and muscles, weakness, flushing of skin, numbness and tingling in extremities, intense thirst, and muscular cramps, delerium and even death. Kidney failure may occur. Jaundice may appear within an hour. In severe poisoning, death can occur within an hour, but the usual interval is 24 hours. This material is extremely toxic; the probable oral lethal dose for humans is 5-50 mg/kg; or between 7 drops and 1 teaspoonful for a 150 lb. person. It is an irritant to eyes, respiratory tract, mouth and stomach. IDLH level = 100 mg/m$^3$.

***Long Term Exposure:*** Calcium arsenate is a carcinogen in humans and may be a reproductive hazard. Damage to kidneys, liver and the nervous system have been reported. Chronic exposure can cause bone marrow damage, often leading to aplastic anemia. Long term exposure can cause an ulcer in the septum dividing the inner nose. High or repeated exposure may damage the nerves causing weakness, and poor coordination in the limbs. Repeated skin contact can cause thickened skin and/or patch areas of darkening and loss of pigment. Some may develop white lines on the nails. There is epidemiological evidence that chronic ingestion or arsenic compounds causes a predisposition to skin cancers. A rebuttable presumption against pesticide registration was issued on October 18, 1978 by U.S. EPA on the basis of oncogenicity, teratogenicity and mutagenicity. Birth defects have been observed in animals exposed to inorganic arsenic. It is likely that health effects seen in children exposed to high amounts of arsenic will be similar to the effects seen in adults.

***Points of Attack:*** Several studies have shown that inorganic arsenic can increase the risk of lung cancer, skin cancer, bladder cancer, liver cancer, kidney cancer, and prostate cancer.

***Medical Surveillance:*** Consider the points of attack in preplacement and periodic physical examinations. Examination of the nose, eyes, nails, and nervous system. Liver function tests. Tests for Urine arsenic. NIOSH[2] recommends urine arsenic should not exceed 100 micrograms per gram of creatine in the urine.

***First Aid:*** If this chemical gets into the eyes, remove any contact lenses at once and irrigate immediately for at least 15 minutes, occasionally lifting upper and lower lids. Seek medical attention immediately. If this chemical contacts the skin, remove contaminated clothing and wash immediately with soap and water. Seek medical attention immediately. If this chemical has been inhaled, remove from exposure, begin rescue breathing (using universal precautions) if breathing has stopped and CPR if heart action has stopped. Transfer promptly to a medical facility. When this chemical has been swallowed, get medical attention. Give large quantities of water and induce vomiting. Do not make an unconscious person vomit. Medical observation is recommended for 24 to 48 hours after breathing overexposure.

*References:*
- National Institute for Occupational Safety and Health, Criteria for a Recommended Standard: Occupational Exposure to Inorganic Arsenic, NIOSH Doc. No. 74-110, Washington, DC (1974).
- Sax, N.I., Ed., "Dangerous Properties of Industrial Materials Report" 2, No. 1, 89-91 (1982).
- U.S. Environmental Protection Agency, "Chemical Profile: Calcium Arsenate," Washington, DC, Chemical Emergency Preparedness Program (November 30, 1987).
- California Environmental Protection Agency Chemical List of Lists, Sacramento CA (February 1997).
- New Jersey Department of Health and Senior Services, "Hazardous Substance Fact Sheet, Calcium Arsenate," Trenton, NJ (January 1999). www.state.nj.us/health/eoh/rtkweb/0310.pdf

# Calcium Arsenite

*Use Type:* Herbicide, insecticide and rodenticide
*CAS Number:* 52740-16-6
*Formula:* $CaAsO_3H$; $AsH_2O_3 \cdot Ca$
*Synonyms:* Arsenito calcico (Spanish); Arsenous acid, calcium salt; Arsonic acid, calcium salt (1:1); Calcium meta-arsenate; Monocalcium arsenite
*Chemical Class:* Arsenical
*EPA/OPP PC Code:* 013602
*RTECS Number:* CG3380000
*Uses:* Not registered in the U.S. Used in dormant and delayed dormant application to apricots, cherries, and peaches.
*U.S. Maximum Allowable Residue Levels:*
Tolerances for residue limits of combined arsenic (calculated as As) in food are established in edible tissues and in eggs of chickens and turkeys and in edible tissues of swine. (total arsenic). See 21 CFR 556.60 (4/1/90) for U.S. Residue Limits for Veterinary Drugs and Unavoidable Contaminants in Meat, Poultry, and Egg Products.
*Carcinogen/Hazard Classifications*
**U.S. NTP Carcinogen:** Known carcinogen
**California Prop. 65:** Carcinogen (as arsenic) and developmental toxin
**IARC:** Group 1, known carcinogen
**Endocrine Disruptor:** Suspected endocrine disruptor as an arsenic compound

*Regulatory Authority:*
- Carcinogen User Register Chemical (CAL/OSHA)
- AB 1803-Well Monitoring Chemical (CAL)
- EPCRA Section 313 Form R *de minimis* concentration reporting level: 0.1%.
- AB 2588-Air Toxics "Hot Spots" Chemicals (CAL)
- CAL Air Resources Board/AB 1807 Toxic Air Contaminants
- Specific chemicals (EPA/NESHAP)
- Proposition 65 chemical (CAL)
- The "Director's List" (CAL/OSHA)
- Permissible Exposure Limits for Chemical Contaminants (CAL/OSHA) as arsenic compounds
- Clean air act: List of high risk pollutants (Section 63.74) as arsenic compounds.
- Clean Water Act: Section 311 Hazardous Substances/RQ (same as CERCLA); Section 313 Priority Chemicals; Toxic Pollutant (Section 401.15) as arsenic and compounds.
- RCRA Section 261 Hazardous Constituents, waste number not listed.
- EPCRA Section 304 RQ: CERCLA, 1 lbs(0.454 kgs)
- Marine pollutant (49CFR, Subchapter 172.101, Appendix B).

*Description:* White, granular powder. Odorless. Slightly soluble in water. Molecular weight =164. Hazard Identification (based on NFPA-704 M Rating System): Health Hazards (Blue): 4; Flammability (Red): 0; Reactivity (Yellow): 0
*Incompatibilities:* Strong oxidizers.
*Permissible Exposure Limits in Air:* ACGIH[1] TLV 0.01 mg/m$^3$; OSHA PEL[2] [1910.1080] 0.010 mg/m$^3$; NIOSH[2] REL ceiling 0.002 mg/m$^3$/15 min, as arsenic; potential human carcinogen; reduce exposure to lowest feasible level
*Determination in Air:* Filter collection followed by atomic absorption analysis. NIOSH 7900; OSHA ID105.
*Permissible Concentration in Water:* No criteria set. Runoff from spills or fire control may cause water pollution. EPA[6] recommends a zero concentration of arsenic for human health reasons but has set a guideline of 50 μg/L[61] for drinking water.
*Determination in Water:* See OSHA Method ID-105 for arsenic.
*Routes of Entry:* Inhalation, absorbed through the skin, eyes, ingestion.
*Harmful Effects and Symptoms*
*Short Term Exposure:* IDLH: 5 mg/m$^3$ as inorganic arsenic. *Inhalation* : Dust is readily absorbed from the lungs, but inhaled quantities are usually insufficient to cause acute systemic toxicity. Can cause cough with foamy sputum and rales. Inhalation Limit: 10 mg/m$^3$ *Skin/Eye Contact:* Dust can cause localized skin irritation, but systemic absorption through the skin is negligible. Skin contact is unlikely to cause systemic effects unless the dermal barrier is

compromised. Eye exposure may produce burns and ulcerations. *Ingestion*: The most important route of acute exposure. Arsenicals are quickly absorbed and can be extremely hazardous. Significant tissue and organ damage and death may result. *Symptoms of exposure include*: headache (lethargy, delirium, hallucinations, seizures, or coma can occur); dizziness; acute nausea and occasional vomiting; labored breathing; restlessness; cyanosis (pale or bluish lips, face and fingernails); perspiration; difficult breathing; abdominal pain with diarrhea; trembling and feeling of "pins and needles" or electrical shock like pains in the lower extremities.; convulsions; unconsciousness; possible pulmonary edema, a medical emergency. Very acute poisoning: extreme headache; muscular paralysis, and liver and kidney dysfunction, loss of consciousness; death.

*Long Term Exposure:* Chronic exposure is characterized by malaise, peripheral sensorimotor neuropathy, anemia, jaundice, gastrointestinal complaints, and characteristic skin lesions including hyperkeratosis (small corn-like elevations) and hyperpigmentation. Hyperkeratosis usually appears on the palms or soles. Pigmentation changes and hyperkeratosis can take 3 to 7 years to appear. Chronic inhalation can also lead to conjunctivitis, irritation of the throat and respiratory tract, and perforation of the nasal septum. Chronic exposure can cause allergic contact dermatitis. Chronic exposure may be more serious for children because of their potential longer latency period. *Carcinogenicity:* The Department of Health and Human Services, IARC, EPA, and NTP have classified arsenic as a human carcinogen based on sufficient evidence from human data. Arsenic trioxide causes skin and lung cancer, and may cause internal cancers such as liver, bladder, kidney, colon, and prostate cancers. Arsenic ions released from arsenic trioxide within the body can cross the placenta and affect the developing fetus; arsenic is also excreted in breast milk. Experimental animal studies support an association between high ingested arsenic dose and fetal toxicity.

*Points of Attack:* Muscles, liver, and kidney.

*Medical Surveillance:* In acutely ill patients, the agent most frequently recommended is dimercaprol, also known as BAL (British Anti Lewisite). The standard dosage regimen is 3 to 5 mg/kg IM every 4–6 hours until the 24-hour urinary arsenic level falls below 50 µg/L, unless an orally administered chelating agent (e.g., DMSA, see below) is substituted. This regimen may be adjusted depending upon the severity of the exposure and the symptoms. Do not chelate asymptomatic patients without the guidance of a 24-hour urinary arsenic level. Contraindications to BAL include preexisting renal disease, pregnancy (except in life-threatening circumstances) and concurrent use of medicinal iron (BAL and iron together form a complex that is very toxic).

*First Aid:* Do not contaminate yourself. Positive-pressure, self-contained breathing apparatus (SCBA) is recommended in response situations that involve exposure to potentially unsafe levels of arsenicals or combustion products which may include arsine and arsenic trioxide fumes. Eyes: Immediately and gently flush eyes with plenty of warm or cold water (NO hot water) for at least 15 minutes, occasionally lifting the upper and lower eyelids. Get medical aid immediately. Skin: Although it is poorly absorbed dermally, skin contact should be avoided because arsenicals may irritate the skin. Wearing the appropriate PPE equipment and respirator for arsenicals. Immediately flush skin with plenty of soap and water for at least 15 minutes while removing contaminated clothing and shoes. Ingestion: Call poison control. Loosen all clothing. Never give anything by mouth to an unconscious person. *Get medical aid. Do NOT induce vomiting.* If conscious, alert, and able to swallow, aggressive decontamination with gastric lavage is recommended within 1 hour of ingestion of a life-threatening amount of poison. The effectiveness of activated charcoal in binding arsenic trioxide is questionable, but administration of a charcoal slurry is recommended pending further evaluation in cases of ingestion of unknown quantities (at 1 gm/kg, usual adult dose 60–90 g, child dose 25–50 g). A soda can and straw may be of assistance when offering charcoal to a child. If victim is *UNCONSCIOUS OR HAVING CONVULSIONS,* do nothing except keep victim warm. Inhalation: Get medical aid. Wearing the appropriate PPE equipment and respirator for arsenicals, immediately remove the victim from the contaminated area to fresh air. If the victim is not breathing, administer artificial respiration. Do NOT use mouth-to-mouth resuscitation; use bag/mask apparatus. If breathing is difficult, administer oxygen through bag/mask apparatus until medical help arrives. Do not leave victim unattended.

*References:*
- New Jersey Department of Health and Senior Services, "Hazardous Substance Fact Sheet, Calcium Arsenite," Trenton NJ (December 2002). http://www.state.nj.us/health/eoh/rtkweb/0311.pdf
- U.S. Department of Agriculture, U.S. Residue Limits for Veterinary Drugs and Unavoidable Contaminants in Meat, Poultry, and Egg Products. 21 CFR 556.60. http://www.fsis.usda.gov/OPHS/red2000/appendix3.pdf
- California Environmental Protection Agency Chemical List of Lists, Sacramento CA (February 1997).

# Calcium Carbonate

*Use Type:* Fungicide, herbicide, microbiocide, and used as a pH soil adjustment
*CAS Number:* 471-34-1 (monocarbonate); 1317-65-3
*Formula:* $CCaO_3$
*Synonyms:* Agricultural limestone; Aragonite (mineral); Calcite (mineral); Calcium(II)carbonate (1:1); Carbonic acid, calcium salt (1:1); Chalk; Domolite; Limestone;

Lithographic stone; Marble; Parcipitated chalk; Portland stone; Slaker rejects; Sohnhofen stone; Vaterite (mineral); Whiting

*Trade Names:* AGLIME®; AGSTONE®; ATOMIT®; BELL MINE PULVERIZED LIMESTONE®; FRANKLIN®

*Producers:* Advance Research Chemicals, Inc; Ashland Chemical (USA); BASF (Germany); Bombay Mineral Supply (India); Chemettal (Germany); Clariant (Switzerland); C.P. Hall (USA); Dankalk (Denmark); Franklin Industrial Minerals (USA); General Chemical (USA); Humco (USA); ICI Group (UK); Infinity Industries (USA); Lhoist Group (France); Lime Chemicals (India); Martin Marietta (USA); Microfill K. Zafranas (Greece); Minerals Technologies (USA); Mississippi Lime (USA); Ocean Chemicals Group (UK); Omya (Switzerland); Pfizer (USA); Quimica Industrial Agricola (Mexico); Solvay Group (Belgium); Specialty Minerals (USA)

*Chemical Class:* Inorganic
*EPA/OPP PC Code:* 073502
*California DPR Chemical Code:* 1007
*ICSC Number:* 1193
*RTECS Number:* EV9580000; FF9335000 (mono-carbonate)
*EINECS Number:* 207-439-9

*Uses:* Calcium carbonate is used as a source of lime; neutralizing agent; manufacturing or rubber, plastics, paint, paper, dentifrices, ceramics, putty, polishes, insecticides, inks and cosmetics; whitewash; Portland cement; antacid; analytical chemistry and others.

*U.S. Maximum Allowable Residue Levels for Calcium Carbonate [40 CFR 180.1001© and 180.1001(e)]:*

| CROP | ppm |
|---|---|
| Raw Agricultural Commodities | – |
| Animal Products | – |

*Regulatory Authority:*
- Air Pollutant Standard Set (ACGIH)[1] (HSE)[33] (OSHA)[58] (Australia) (Israel) (Mexico) (Several Canadian Provinces)
- Permissible Exposure Limits for Chemical Contaminants (CAL/OSHA)
- Actively registered pesticide in California.

*Description:* Calcium carbonate, $CaCO_3$ (CAS 471-34-1) is a white, odorless powder or crystalline solid. Very slightly soluble in water. Molecular weight = 100.05. Melting/Freezing point = 1517-2442°F (decomposes). Boiling point = 825°C (decomposes)

*Incompatibilities:* Calcium carbonate decomposes in high temperature forming carbon dioxide and corrosive materials. Reacts with acids producing carbon dioxide gas release. Incompatible with acids, ammonium salts, fluorine. NIOSH[2] also lists alum as incompatible, but this is questionable.

*Permissible Exposure Limits in Air:* ACGIH[1] has set a TWA of 10 mg/m³ (for dust containing no asbestos and <1% free silica). The HSE[33] has set a TWA of 10 mg/m³ for total inhalable dust and 5 mg/m³ for respirable dust. OSHA[2] has set a TWA of 15 mg/m³ on a total dust basis and 5 mg/m³ on a respirable fraction basis. NIOSH[2] has set a TWA of 10 mg/m³ on a total dust basis and 5 mg/m³ on a respirable fraction basis.

*Determination in Air:* Filter; Acid; Flame atomic absorption spectrometry; IV NIOSH Method #7020 (Calcium)[18]. Calcium Carbonate may be measured by gravimetric means after filter collection. See NIOSH analytical methods #0500[18] for nuisance dust, total and 0600 for nuisance dust, respirable.

*Permissible Concentration in Water:* No criteria set, but runoff from spills or fire control may cause water pollution.

*Routes of Entry:* Inhalation of dust, ingestion.

*Harmful Effects and Symptoms*

*Short Term Exposure:* Inhalation can cause irritation to nose. *Eyes:* Contact can cause irritation. *Ingestion:* Large amounts can cause irritability, nausea, dehydration and constipation. Estimated lethal dose is over 2 lb.

*Long Term Exposure:* Ingestion of more than 8 grams (1/3 ounce) a day can cause blood and kidney disorders.

*Points of Attack:* Eyes, respiratory system, digestive system.

*First Aid:* If this chemical gets into the eyes, remove any contact lenses at once and irrigate immediately for at least 15 minutes, occasionally lifting upper and lower lids. Seek medical attention immediately. If this chemical contacts the skin, remove contaminated clothing and wash immediately with soap and water. Seek medical attention immediately. If this chemical has been inhaled, remove from exposure, begin rescue breathing (using universal precautions) if breathing has stopped and CPR if heart action has stopped. Transfer promptly to a medical facility. When this chemical has been swallowed, get medical attention. Give large quantities of water and induce vomiting. Do not make an unconscious person vomit.

*References:*
- U.S. Environmental Protection Agency, Office of Pesticide Programs, Pesticide Residue Limits, "Calcium Carbonate," 40 CFR 180.1001(c) and 180.1001(e). http://www.epa.gov/pesticides/food/viewtols.htm
- New York State Department of Health, "Chemical Fact Sheet: Calcium Carbonate," Albany, NY, Bureau of Toxic Substance Assessment (March 1986).
- California Environmental Protection Agency Chemical List of Lists, Sacramento CA (February 1997).

# Calcium Cyanamide

*Use Type:* Herbicide, insecticide, fertilizer and defoliant
*CAS Number:* 156-62-7
*Formula:* $CCaN_2$; $CaCN_2$

# Calcium Cyanamide

*Synonyms:* Calcium carbimide; Calcium cyanamid; CCC; Cianamida calcica (Spanish); Cy-L 500; Cyanamid; Cyanamide; Cyanamide, calcium salt (1:1); Cyanamide calcique (French); Cyanamid granular; Cyanamid special grade; Lime nitrogen; NCI-C02937; Nitrogen lime; Nitrolime; USAF Cy-2

*Trade Names:* AERO-CYANAMID®; AERO-CYANAMID®, SPECIAL GRADE; ALZODEF®

*Producers:* Degussa (Germany); Fluorochem Ltd. (UK); ICI Group (UK); Denki Kagaku Kogyo (Japan); Hunan Darong Chemical & Pesticide Company (China); Jilin Chemical Industrial (China); Juhua Group Corp. (China); Nippon Carbide Industries (Japan); Orica (Australia); Quantum Chemicals (Australia)

*Chemical Class:* Inorganic

*EPA/OPP PC Code:* 014001

*RTECS Number:* GS6000000

*EINECS Number:* 205-861-8

*Uses:* Calcium cyanamide is used a defoliant for cotton plants in addition to its other agriculture uses. It is also used in the manufacture of dicyandiamide and calcium cyanide as a desulfurizer in the iron and steel industry, and in steel hardening.

*Carcinogen/Hazard Classifications*

U.S. EPA Carcinogens: Not listed. Group C, possible carcinogen (as cyanide)

*Regulatory Authority:*
- Air Pollutant Standard Set (ACGIH)[1] (HSE)[33] (UNEP)[43] (OSHA)[58] (Several States)[60] (Australia) (DFG) (Israel)
- (Mexico) (Several Canadian Provinces)
- AB 2588-Air Toxics "Hot Spots" Chemicals (CAL)
- CAL Air Resources Board/AB 1807 Toxic Air Contaminants
- Permissible Exposure Limits for Chemical Contaminants (CAL/OSHA)
- The "Director's List" (CAL/OSHA)
- Clean Air Act: Hazardous Air Pollutants (Title I, Part A, Section 112)
- Clean water act: Section 307 Priority Pollutants as cyanide, total; Toxic Pollutant (Section 401.15)
- EPA Hazardous Waste Number (RCRA No.): P030
- RCRA Section 261 Hazardous Constituents.
- RCRA Universal Treatment Standards: Wastewater (mg/L), 1.2 (total); 0.86 (amenable); Nonwastewater (mg/kg), 590 (total); 30 (amenable)
- RCRA Ground Water Monitoring List. Suggested test method(s) (PQL *ug*/L): 9010(40)
- Safe Drinking Water Act: MCL, 0.1 mg/L; MCLG, 0.1 mg/L; Regulated chemical (47 FR 9352)
- U.S. DOT Regulated Marine Pollutant (49CFR, Subchapter 172.101, Appendix B) as cyanides, mixtures or solutions
- Superfund/EPCRA 40CFR302.4 RQ: CERCLA, 1 lb (0.454 kg).
- EPCRA Section 313 Form R de minimus concentration reporting level: 1.0%.
- U.S. DOT Regulated Marine Pollutant (49CFR172.101, Appendix B) as cyanide mixtures, cyanide solutions or cyanides, inorganic, n.o.s.
- Canada, WHMIS, Ingredients Disclosure List; National Pollutant Release Inventory (NPRI)

*Description:* Calcium cyanamide is a blackish-grey, shiny crystalline material or powder. Insoluble in water. Molecular weight = 80.09. Melting/Freezing point = ~1342°C (sublimes >1200°C).

*Incompatibilities:* Commercial grades of calcium cyanamide may contain calcium carbide; contact with any form of moisture solutions may cause decomposition, liberating explosive acetylene gas and ammonia. May polymerize in water or alkaline solutions to dicyanamide. Contact with all solvents tested also cause decomposition.

*Permissible Exposure Limits in Air:* NIOSH[2] and ACGIH[1] TLV recommends 0.5 mg/m$^3$. The MAK set by DFG[3] is 1.0 mg/m$^3$. The HSE[33] TLV is 0.5 mg/m$^3$ and they have set a STEL of 1.0 mg/m$^3$. The former USSR/UNEP joint project[43] has set a MAC in ambient residential air of 0.02 mg/m$^3$ either on a momentary or a daily average basis. Several states have set guidelines or standards for calcium cyanamide in ambient air[60] ranging from 5.0 $\mu$g/m$^3$ (North Dakota) to 8.0 $\mu$g/m$^3$ (Virginia) to 10.0 $\mu$g/m$^3$ (Connecticut) to 12.0 $\mu$g/m$^3$ (Nevada).

*Determination in Air:* Filter; none; Gravimetric; IV NIOSH Method #0500, Particulates NOR (total)[18]

*Permissible Concentration in Water:* The U.S. EPA has set a maximum contaminant level of cyanide in drinking water of 0.2 milligrams cyanide per liter of water (0.2 mg/L). The former USSR/UNEP joint project[43] has set a MAC of 1.0 mg/L in water for domestic purposes.

*Determination in Water:* Cyanide may be determined titrimetrically by EPA Methods 335.2 and 9010 which give total cyanide.

*Routes of Entry:* Inhalation of dust, ingestion

*Harmful Effects and Symptoms*

*Short Term Exposure:* Calcium cyanamide can cause nausea, headache, dizziness, and flushing of the skin. It is a primary irritant of the mucous membranes of the respiratory tract, eyes, and skin. Drinking alcohol shortly before or within 1-2 days after exposure can cause a severe reaction. Inhalation may result in rhinitis, pharyngitis, laryngitis, and bronchitis. Conjunctivitis, keratitis, and corneal ulceration may occur. An itchy erythematous dermatitis has been reported and continued skin contact leads to the formation of slowly healing ulcerations on the palms and between the fingers. Sensitization occasionally develops. Chronic rhinitis and perforation on the nasal septum have been reported after long exposures. All local effects appear to be due to the caustic nature of cyanamide.

*Long Term Exposure:* Calcium cyanamide may damage the developing fetus. This chemical may damage the nervous system, causing numbness, and weakness in the hands and feet. Prolonged contact can cause skin ulcers. It causes a characteristic vasomotor reaction. There is erythema of the upper portions of the body, face and arms accompanied by nausea, fatigue, headache, dyspnea, vomiting, oppression in the chest and shivering. Circulatory collapse may follow in the more serious cases. The vasomotor response may be triggered or intensified by alcohol ingestion. Pneumonia or lung edema may develop. Cyanide ion is not released in the body, and the mechanism of toxic action is unknown.

*Points of Attack:* Eyes, skin, respiratory system, vasomotor system

*Medical Surveillance:* Examination of the nervous system. Evaluation by a qualified allergist. Evaluate skin, respiratory tract and history of alcohol intake in placement or periodic examinations.

*First Aid:* If this chemical gets into the eyes, remove any contact lenses at once and irrigate immediately for at least 15 minutes, occasionally lifting upper and lower lids. Seek medical attention immediately. If this chemical contacts the skin, remove contaminated clothing and wash immediately with soap and water. Seek medical attention immediately. If this chemical has been inhaled, remove from exposure, begin rescue breathing (using universal precautions) if breathing has stopped and CPR if heart action has stopped. Transfer promptly to a medical facility. When this chemical has been swallowed, get medical attention. Give large quantities of water and induce vomiting. Do not make an unconscious person vomit.

*References:*
- Sax, N.I., Ed., "Dangerous Properties of Industrial Materials Report" 2, No. 6, 38-41 (1982).
- California Environmental Protection Agency Chemical List of Lists, Sacramento CA (February 1997).
- New Jersey Department of Health and Senior Services, "Hazardous Substance Fact Sheet: Calcium Cyanamide," Trenton, NJ (April 1998). www.state.nj.us/health/eoh/rtkweb/0316.pdf

# Calcium Cyanide

*Use Type:* Rodenticide and fumigant
*CAS Number:* 592-01-8
*Formula:* $C_2CaN_2$, $Ca(CN)_2$
*Synonyms:* Calcid; Calcyan; Calcyanide; Cianuro calcico (Spanish); Cyanogas; Cyanure de calcium (French)
*Trade Names:* AERO®, Sasol Ltd. (South Africa)
*Producers:* Sasol Ltd. (South Africa)
*Chemical Class:* Inorganic
*EPA/OPP PC Code:* 074001
*California DPR Chemical Code:* 176
*ICSC Number:* 0407
*RTECS Number:* EW0700000
*EEC Number:* 020-002-00-5

*Uses:* Calcium cyanide is used as a fumigant, as a rodenticide, and for killing honey bees in the hives. Also used in leaching precious metal ores; in the manufacture of stainless steel and as a stabilizer for cement.

*Carcinogen/Hazard Classifications*
WHO Acute Hazard: Class 1 a, extremely hazardous
*Regulatory Authority:*
- Air Pollutant Standard Set (ACGIH)[1] (DFG)[3] (HSE)[33] (UNEP)[43] (OSHA)[58]
- The "Director's List" (CAL/OSHA)
- AB 2588-Air Toxics "Hot Spots" Chemicals (CAL)
- CAL Air Resources Board/AB 1807 Toxic Air Contaminants
- Clean Air Act: Hazardous Air Pollutants (Title I, Part A, Section 112)
- Clean Water Act: Section 311 Hazardous Substances/RQ 40CFR117.3 (same as CERCLA, see below); Section 313 Water Priority Chemicals (57FR41331, 9/9/92)
- EPA Hazardous Waste Number (RCRA No.): P021
- RCRA, 40CFR261, Appendix 8 Hazardous Constituents.
- RCRA Universal Treatment Standards: Wastewater (mg/L), 1.2 (total); 0.86 (amenable); Nonwastewater (mg/kg), 590 (total); 30 (amenable)
- RCRA Ground Water Monitoring List. Suggested test method(s) (PQL *ug*/L): 9010(40)
- Safe Drinking Water Act: MCL, 0.1 mg/L; MCLG, 0.1 mg/L; Regulated chemical (47 FR 9352)
- EPCRA Section 304 RQ: CERCLA, 10 lb (4.54 kg)
- EPCRA Section 313 Form R *de minimus* concentration reporting level: 1.0%
- U.S. DOT Regulated Marine Pollutant (49CFR, Subchapter 172.101, Appendix B) as cyanides, mixtures or solutions
- Canada, WHMIS, Ingredients Disclosure List (cyanide compounds, inorganic, n.o.s)

*Description:* Calcium cyanide is a white crystalline solid or powder. Almond odor. Melting/Freezing point = >350°C (dangerous decomposition below m.p.). Hazard Identification (based on NFPA-704 M Rating System): Health 3, Flammability 0, Reactivity 1. Soluble in water; violent reaction.

*Incompatibilities:* Contact with water, acids, acidic salts, moist air, or carbon dioxide, forms highly toxic and flammable hydrogen cyanide. Incompatible with fluorine, magnesium. Reacts violently when heated with nitrites, nitrates, chlorates and perchlorates. Calcium cyanide decomposes in high heat forming hydrogen cyanide and nitrous oxides fumes.

*Permissible Exposure Limits in Air:* OSHA[2] and ACGIH[1]: 5 mg/m³ TWA; NIOSH[2]: Ceiling limit, 4.7 ppm;

5 mg/m³ per 10 minutes as cyanides. All have notations that skin contact contributes significantly in overall exposure. IDLH level = 25 mg/m³ as CN. The former USSR/UNEP Project[43] has set a MAC of 0.3 mg/m³.

*Determination in Air:* Filter/Bubbler; Potassium hydroxide; Ion-specific electrode; IV NIOSH Method #7904, Cyanides[18].

*Permissible Concentration in Water:* The U.S. EPA has set a maximum contaminant level of cyanide in drinking water of 0.2 milligrams cyanide per liter of water (0.2 mg/L).[6] as part of the priority toxic pollutant program.

*Determination in Water:* Cyanide may be determined titrimetrically by EPA Methods 335.2 and 9010 which give total cyanide.

*Routes of Entry:* Can be absorbed through the skin, inhalation, ingestion.

*Harmful Effects and Symptoms*

*Short Term Exposure:* The substance is corrosive to the eyes, skin, and respiratory tract. Higher exposures can cause pulmonary edema, a medical emergency that can be delayed for several hours. This can cause death. My affect the central nervous system, blood, heart and respiratory tract. Delayed pulmonary edema may occur after inhalation. Calcium cyanide is highly toxic. The lethal human dose is 18 mg/kg. The hazard is that of hydrogen cyanide. The dust is irritating to the eyes, nose and throat[41]. Inhalation or ingestion causes headache, nausea, vomiting and weakness; high concentrations are rapidly fatal.

*Long Term Exposure:* Repeated or prolonged contact with skin may cause dermatitis. May be a reproductive toxin in humans.

*Points of Attack:* Skin, lungs.

*Medical Surveillance:* Medical observation is recommended for 24 to 48 hours after breathing overexposure, as pulmonary edema may be delayed. Lung function tests. Examination by a qualified allergist. Consider chest x-ray following acute overexposure.

*First Aid:* If this chemical gets into the eyes, remove any contact lenses at once and irrigate immediately for at least 15 minutes, occasionally lifting upper and lower lids. Do not allow water to enter nose or mouth. Seek medical attention immediately. If this chemical contacts the skin, remove contaminated clothing and wash immediately with soap and water. Seek medical attention immediately. If this chemical has been inhaled, remove from exposure, begin rescue breathing (using universal precautions) if breathing has stopped and CPR if heart action has stopped. Transfer promptly to a medical facility. When this chemical has been swallowed, get medical attention. Give large quantities of water and induce vomiting. Do not make an unconscious person vomit. Medical observation is recommended for 24 to 48 hours after breathing overexposure, as pulmonary edema may be delayed. As first aid for pulmonary edema, a doctor or authorized paramedic may consider administering a corticosteroid spray. Use amyl nitrate capsules if symptoms of cyanide poisoning develop. All area employees should be trained regularly in emergency measures for cyanide poisoning and in CPR. A cyanide antidote kit should be kept in the immediate work area and must be rapidly available. Kit ingredients should be replaced every 1-2 years to ensure freshness. Persons trained in the use of this kit, oxygen use, and CPR must be quickly available.

*References:*
- New Jersey Department of Health and Senior Services, "Hazardous Substance Fact Sheet, Calcium Cyanide," Trenton NJ (November 2000). http://www.state.nj.us/health/eoh/rtkweb/0317.pdf
- Agency for Toxic Substances and Disease Registry, "ToxFAQs for Cyanide," Atlanta, GA (September 1997). http://www.atsdr.cdc.gov/tfacts8.html
- Sax, N.I., Ed., "Dangerous Properties of Industrial Materials Report" 2, No. 1, 95-96 (1982).

# Calcium Hydroxide

*Use Type:* Soil pH adjustment, fungicide, herbicide
*CAS Number:* 1305-62-0
*Formula:* $CaH_2O_2$; $Ca(OH)_2$
*Synonyms:* Bell mine; Calcium hydrate; Carboxide; Hydrated kemikal; Hydrated lime; Slaked lime; Lime water
*Producers:* Chemical Lime (USA); Continental Lime (USA); C.P. Hall (USA); Dankalk (Denmark); ICI Group (UK); Lhoist (France); Merck (Germany); Omya (Switzerland); Orica (Australia); Showa Denko (Japan)
*Chemical Class:* Inorganic
*EPA/OPP PC Code:* 075601
*California DPR Chemical Code:* 99, 3669
*RTECS Number:* 1305-62-0
*EINECS Number:* 215-137-3
*Uses:* Calcium hydroxide is used in agriculture as fertilizer and in fertilizer manufacturing; it is used in the formulation of mortar, plasters and cements and as a scrubbing and neutralizing agent in the chemical industry.

*Regulatory Authority:*
- Air Pollutant Standard Set (ACGIH)[1] (HSE)[33] (OSHA)[58] (Several States)[60] (Australia) (Israel) (Mexico) (Several Canadian Provinces)
- Permissible Exposure Limits for Chemical Contaminants (CAL/OSHA)
- The "Director's List" (CAL/OSHA)
- Actively registered pesticide in California.
- Canada, WHMIS, Ingredients Disclosure List

*Description:* Calcium hydroxide is a soft white crystalline, odorless powder with an alkaline, bitter taste. Melting/Freezing point (decomposes; dehydrates to calcium oxide) = 580°C. Hazard Identification (based on NFPA-704 M Rating System): Health 1, Flammability 0, Reactivity 0.

Insoluble in water. *Note*: Readily absorbs $CO_2$ from the air to form calcium carbonate.
*Incompatibilities:* May react violently with acids, maleic anhydride, nitromethane, nitroethane, nitropropane, nitroparaffins and phosphorus.
*Permissible Exposure Limits in Air:* NIOSH[2] REL: TWA 5 mg/m3. OSHA[2] PEL: TWA 15 mg/m³ (total); 5 mg/m³ (respirable fraction). ACGIH recommends a TWA value of 5 mg/m³[1]. Several states have set guidelines or standards for calcium hydroxide in ambient air[60] ranging from 50 µg/m³ (North Dakota) to 80 µg/m³ (Virginia) to 100 µg/m³ (Connecticut) to 119 µg/m³ (Nevada).
*Determination in Air:* Filter collection followed by atomic absorption analysis. See NIOSH Method #7020 for Calcium[18]. See also OSHA Method ID 121.[58]
*Permissible Concentration in Water:* No criteria set, but runoff from spills or fire control may cause water pollution.
*Routes of Entry:* Inhalation of dust, ingestion.
*Harmful Effects and Symptoms*
*Short Term Exposure: Inhalation:* May cause severe irritation to mouth, throat and lungs if dust is inhaled. Higher exposures can cause pulmonary edema, a medical emergency that can be delayed for several hours. This can cause death. *Skin:* May cause painful irritation and chemical burns on contact with open cuts or sores or on prolonged contact with intact skin. *Eyes:* Powders and slurries may cause severe chemical burns. Blindness can result. *Ingestion:* Powders, crystals or slurries may give rise to irritation, soreness and chemical burns. The estimated lethal dose is about one pound.
*Long Term Exposure:* Repeated or prolonged contact with skin may cause dermatitis. Lungs may be affected by repeated or prolonged exposure to dust particles.
*Points of Attack:* Eyes, skin, respiratory system.
*Medical Surveillance:* Lung function tests. Consider chest x-ray following acute overexposure.
*First Aid:* If this chemical gets into the eyes, remove any contact lenses at once and irrigate immediately for at least 30 minutes, occasionally lifting upper and lower lids. Seek medical attention immediately. If this chemical contacts the skin, remove contaminated clothing and wash immediately with soap and water. Seek medical attention immediately. If this chemical has been inhaled, remove from exposure, begin rescue breathing (using universal precautions) if breathing has stopped and CPR if heart action has stopped. Transfer promptly to a medical facility. When this chemical has been swallowed, get medical attention. Give large quantities of water and induce vomiting. Do not make an unconscious person vomit. Medical observation is recommended for 24 to 48 hours after breathing overexposure, as pulmonary edema may be delayed. As first aid for pulmonary edema, a doctor or authorized paramedic may consider administering a corticosteroid spray.

*References:*
- Sax, N.I., Ed., "Dangerous Properties of Industrial Materials Report" 1, No. 8, 48-50 (1981).
- New York State Department of Health, "Chemical Fact Sheet: Calcium Hydroxide," Albany, NY, Bureau of Toxic Substance Assessment (January 1986).
- California Environmental Protection Agency Chemical List of Lists, Sacramento CA (February 1997).
- New Jersey Department of Health and Senior Services and Senior Services, "Hazardous Substance Fact Sheet: Calcium Hydroxide," Trenton, NJ (February 1989, rev. April 1998).
http://www.state.nj.us/health/eoh/rtkweb/0322.pdf

# Calcium Methanearsonate

*Use Type:* Herbicide, defoliant
*CAS Number:* 5902-95-4
*Formula:* $C_2H_8As_2O_6 \cdot Ca$
*Synonyms:* Arsonic acid, methyl-, calcium salt (2:1); Calcium acid methanearsonate; Calcium hydrogen methanearsonate; Calcium methanearsonate; Methanearsonic acid, calcium salt (2:1); Methylarsonic acid, calcium salt (2:1)
*Trade Names:* CALAR®, Drexel Chemical (USA); CAMA®, canceled; SUPER CRAB-E-RAD-CALAR®; SUPER DAL-E-RAD®, Drexel Chemical (USA); SUPER DAL-E-RAD-CALAR®, Drexel Chemical (USA); WEED-B-GON®; The Scotts Company (USA)
*Producers:* Drexel Chemical (USA); Ortho Business Group (USA); The Scotts Company (USA)
*Chemical Class:* Organoarsenical
*EPA/OPP PC Code:* 013806
*California DPR Chemical Code:* 101
*Uses:* Used to control a wide variety of broadleaf weeds and annual grasses, primarily in industrial and residential environments. Tolerances for residue limits of combined arsenic (calculated as As) in food are established in edible tissues and in eggs of chickens and turkeys and in edible tissues of swine. (total arsenic). See 21 CFR 556.60 (4/1/90) for U.S. Residue Limits for Veterinary Drugs and Unavoidable Contaminants in Meat, Poultry, and Egg Products.
*Carcinogen/Hazard Classifications*
**U.S. EPA Carcinogens:** Not likely a carcinogen; Group B2, Probable carcinogen as cacodylic acid, its parent chemical
**California Prop. 65:** Carcinogen as cacodylic acid, its parent chemical.
**IARC:** Group 1, known carcinogen
**Label Signal Word:** CAUTION
**WHO Acute Hazard:** Class III, slightly hazardous as cacodylic acid, the parent chemical

***Regulatory Authority:***
- CLEAN AIR ACT: Hazardous Air Pollutants (Title I, Part A, Section 112) as arsenic compounds.
- Clean Water Act: Toxic Pollutant (Section 401.15) as arsenic and compounds.
- RCRA Section 261 Hazardous Constituents, waste number not listed.
- EPCRA Section 304 RQ: CERCLA, 1 lb (0.454 kg).
- EPCRA (Section 313): Includes any unique chemical substance that contains arsenic as part of that chemical's infrastructure. Form R *de minimis* concentration reporting level: (inorganics) 0.1%.; organics 1.0%.
- Marine pollutant (49CFR, Subchapter 172.101, Appendix B) as arsenates, liquid, n.o.s.; arsenates, solid, n.o.s.; arsenical pesticides liquid, toxic, flammable, n.o.s.
- AB 2588-Air Toxics "Hot Spots" Chemicals (CAL) as arsenic compounds
- Permissible Exposure Limits for Chemical Contaminants (CAL/OSHA) as arsenic compounds
- The "Director's List" (CAL/OSHA) as arsenic compounds

***Description:*** Colorless. Odorless. Soluble in water.
***Incompatibilities:*** Strong oxidizers.
***Permissible Exposure Limits in Air:*** The OSHA limit for organic arsenic compounds is 0.5 mg/m$^3$ TWA. For an 8-hour workshift. There is no NIOSH recommendation. The ACGIH TLV is 0.2 mg/m$^3$ TWA.
***Determination in Air:*** Filter; Reagent; Ion chromatography/Hydride generation atomic absorption spectrometry; NIOSH IV[18], Method #5022, Arsenic, organo-
***Permissible Concentration in Water:*** No criteria set. Runoff from spills or fire control may cause water pollution. EPA[6] recommends a zero concentration of arsenic for human health reasons but has set a guideline of 50 μg/L[61] for drinking water.
***Determination in Water:*** See OSHA Method ID-105 for arsenic.
***Routes of Entry:*** Inhalation, ingestion, skin and/or eye contact.

***Harmful Effects and Symptoms***
***Short Term Exposure:*** Irritating to skin and eyes. If swallowed, will cause nausea, vomiting, or loss of consciousness. Poisonous. Grade 2; LD$_{50}$ = 0.5 to 5 g/kg (rat).
***Long Term Exposure:*** Repeated contact may cause skin sensitivity. Chronic exposure to arsenic compounds can cause dermatitis and digestive disorders. Renal damage may develop. In animals: kidney damage; muscle tremor, seizure; possible gastrointestinal tract, reproductive effects; possible liver damage
***Points of Attack:*** Skin, kidneys, liver.
***Medical Surveillance:*** Kidney function tests. Examination by a qualified allergist. Kidney, liver, lung function tests. Consider chest x-ray following acute overexposure.
***First Aid:*** If artificial respiration is administered, *avoid mouth-to-mouth resuscitation; use bag/mask apparatus.* Remove contaminated clothing and shoes. Flush affected areas with plenty of water. If in eyes, hold eyelids open and flush with plenty of water. If swallowed and victim is conscious and able to swallow, have victim drink 4 to 8 ounces of water and have victim induce vomiting. If swallowed and victim is unconscious or having convulsions, do nothing except keep victim warm.
***References:***
- U.S. Department of Agriculture, U.S. Residue Limits for Veterinary Drugs and Unavoidable Contaminants in Meat, Poultry, and Egg Products. 21 CFR 556.60. http://www.fsis.usda.gov/OPHS/red2000/appendix3.pdf
- California Environmental Protection Agency Chemical List of Lists, Sacramento CA (February 1997)

# Calcium Nitrate

***Use Type:*** Fertilizer
***CAS Number:*** 10124-37-5
***Formula:*** CaN$_2$O$_6$; Ca(NO$_3$)$_2$
***Synonyms:*** Calcium (II) Nitrate (1:2); Nitric acid, calcium salt; Lime saltpeter; Nitric acid, calcium salt; Lime saltpeter; Norwegian saltpeter; Nitrocalcite
***Producers:*** GFS Chemicals (USA); Mineral Research & Development (USA); Hydro Agri Chemicals (Norway); Kynoch (South Africa); Omnia Group (South Africa); Pechiney (France); Rallis India (India); Timminco (Canada); Varsal Instruments (USA)
***Chemical Class:*** Inorganic
***EPA/OPP PC Code:*** 070582
***California DPR Chemical Code:*** 5809
***ICSC Number:*** 1037
***RTECS Number:*** EW2985000
***EINECS Number:*** 233-332-1
***Uses:*** It is used to make explosives, fertilizers, matches, fireworks and other industrial products.
***Regulatory Authority:***
- Actively registered pesticide in California.

***Description:*** Calcium Nitrate is a colorless, moisture absorbing crystalline material. Melting/Freezing point = 561°C. Hazard Identification (based on NFPA-704 M Rating System): Health 2, Flammability 0, Reactivity 1. Oxidizer. Soluble in water.
***Incompatibilities:*** A strong oxidizer. Incompatible with combustible materials, reducing agents, organics and other oxidizable materials, chemically active metals, aluminum nitrate, ammonium nitrate.
***Permissible Exposure Limits in Air:*** No standards set.
***Permissible Concentration in Water:*** No criteria set, but runoff from spills or fire control may cause water pollution.
***Routes of Entry:*** Eyes, skin, respiratory system.

*Harmful Effects and Symptoms*
*Short Term Exposure:* Calcium nitrate can the skin, eyes, nose, throat, bronchial tubes and lungs. Overexposure may cause nausea and vomiting, headaches, flushing, weakness, faintness and collapse. Severe overexposure can cause nausea, vomiting, flushing of the head and neck, headache, weakness faintness and collapse; the lips and fingernails may become bluish. There may be shortness of breath. Coma, convulsions and death are possible.
*Points of Attack:* Eyes, skin, respiratory system.
*Medical Surveillance:* For those with frequent or potentially high exposure, the following are recommended before beginning work and at regular times after that: Lung function tests. If overexposure is suspected, also consider: Complete blood count and test for methemoglobin.
*First Aid:* If this chemical gets into the eyes, remove any contact lenses at once and irrigate immediately for at least 15 minutes, occasionally lifting upper and lower lids. Seek medical attention immediately. If this chemical contacts the skin, remove contaminated clothing and wash immediately with soap and water. Seek medical attention immediately. If this chemical has been inhaled, remove from exposure, begin rescue breathing (using universal precautions) if breathing has stopped and CPR if heart action has stopped. Transfer promptly to a medical facility. When this chemical has been swallowed, get medical attention. Give large quantities of water and induce vomiting. Do not make an unconscious person vomit.
*Note to physician or authorized medical personnel:* Treat for methemoglobinemia. Spectrophotometry may be required for precise determination of levels of methemoglobinemia in urine.
*References:*
- New Jersey Department of Health and Senior Services and Senior Services, "Hazardous Substance Fact Sheet: Calcium Nitrate," Trenton, NJ (September 1986, rev. October 2000).
  http://www.state.nj.us/health/eoh/rtkweb/0324.pdf
- Sax, N.I., Ed., "Dangerous Properties of Industrial Materials Report" 2, No. 1, 96-98 (1982) (Calcium Nitrate Tetrahydrate)
- California Environmental Protection Agency Chemical List of Lists, Sacramento CA (February 1997).

# Calcium Oxide

*Use Type:* Fungicides, herbicides, and in soil treatment
*CAS Number:* 1305-78-8
*Formula:* CaO
*Synonyms:* Burnt lime; Calcia; Calx; Fluxing lime; Lime; Lime, burned; Lime, unslaked; Oxyde de calcium (French); Quicklime; Pebble lime
*Producers:* Chemical Lime (USA); Continental Lime (USA); GFS Chemicals (USA); C.P. Hall (USA); Dankalk (Denmark); Holvoet Chimie (Belgium); Lhoist Group (France); Lime Industries (Australia); Mississippi Lime (USA); Ube Industries (Japan)
*Chemical Class:* Inorganic
*EPA/OPP PC Code:* 075604
*California DPR Chemical Code:* 977
*ICSC Number:* 0409
*RTECS Number:* EW3100000
*EINECS Number:* 215-138-9
*Uses:* Calcium oxide is used as a refractory material, a binding agent in bricks, plaster, mortar, stucco and other building materials, a dehydrating agent, a flux in steel manufacturing, and a laboratory agent to absorb $CO_2$; in the manufacture of aluminum, magnesium, glass, pulp and paper, sodium carbonate, calcium hydroxide, chlorinated lime, calcium salts, and other chemicals; in the flotation of nonferrous ores, water and sewage treatment, dehairing hides, the clarification of cane and beet sugar juice, and in drilling fluids and lubricants.
*U.S. Maximum Allowable Residue Levels for Calcium Oxide [40 CFR 180.2 and 180.1001 (c)]:*

| CROP | ppm |
|---|---|
| Animal products | - |
| Raw agricultural commodities | - |

*Regulatory Authority:*
- Air Pollutant Standard Set (ACGIH)[1] (DFG)[3] (HSE)[33] (OSHA)[58] (Several States)[60] (Australia) (Israel) (Mexico) (Several Canadian Provinces)
- Permissible Exposure Limits for Chemical Contaminants (CAL/OSHA)
- The "Director's List" (CAL/OSHA)

*Description:* Calcium oxide, CaO, occurs as white or grayish-white lumps or granular powder. The presence of iron gives it a yellowish or brownish tint. Soluble in water (reactive). Molecular weight = 56.09. Melting/Freezing point = 2575°C. Boiling point = 2858°C. Hazard Identification (based on NFPA-704 M Rating System): Health 1, Flammability 0, Reactivity 0.
*Incompatibilities:* The water solution is a medium strong base. Reacts with water generating calcium hydroxide and sufficient heat to ignite nearby combustible materials. Reacts violently with acids, halogens, metals.
*Permissible Exposure Limits in Air:* The OSHA[2] TWA calcium oxide is 5 mg/m³. NIOSH[2], ACGIH[1], and HSE[33] recommend a TWA value of 2 mg/m³. The NIOSH[2] IDLH level = 25 mg/m³. There is no DFG MAK at present. Other countries regulations[35] include: 10 mg/m³ (Argentina:), 0.3 mg/m³ (former USSR). In addition, several states have set guidelines or standards for calcium oxide in ambient air[60] ranging from 20 µg/m³ (North Dakota) to 35 µg/m³ (Virginia) to 40 µg/m³ (Connecticut) to 48 µg/m³ (Nevada).

***Determination in Air:*** Filtration, workup with acid and analysis by atomic absorption are specified in NIOSH Method 7020 for calcium. See also OSHA Method ID121.[58]

***Permissible Concentration in Water:*** No criteria set, but runoff from spills or fire control may cause water pollution.

***Routes of Entry:*** Inhalation of dust.

***Harmful Effects and Symptoms***

***Short Term Exposure:*** The corrosive action of calcium oxide is due primarily to its alkalinity and exothermic reaction with water. It is irritating and may be caustic to the skin, conjunctiva, cornea and mucous membranes of upper respiratory tract; may produce burns or dermatitis with desquamation and vesicular rash, lacrimation, spasmodic blinking, ulceration, and ocular perforation, ulceration and inflammation of the respiratory passages, ulceration of nasal and buccal mucosa, and perforation of nasal septum. Bronchitis and pneumonia have been reported from inhalation of dust. Higher exposures can cause pulmonary edema, a medical emergency that can be delayed for several hours. This can cause death. The lower respiratory tract may not be affected because irritation of upper respiratory passages is so severe that workers may be forced to leave the area.

***Long Term Exposure:*** Repeated or prolonged contact with skin may cause brittle nails and thickening and cracking of the skin. Repeated or prolonged exposure to dust particles may cause lung problems. Calcium oxide may cause ulceration and perforation of the cartilage separating the nose (septum).

***Points of Attack:*** Respiratory system, skin and eyes.

***Medical Surveillance:*** Preemployment physical examinations should be directed to significant problems of the eyes, skin and the upper respiratory tract. Periodic examinations should evaluate the skin, changes in the eyes, especially the cornea and conjunctiva, mucosal ulcerations of the nose, mouth and nasal septum, and any pulmonary symptoms. Smoking history should be known.

***First Aid:*** If this chemical gets into the eyes, remove any contact lenses at once and irrigate immediately for at least 15 minutes, occasionally lifting upper and lower lids. Seek medical attention immediately. If this chemical contacts the skin, remove contaminated clothing and wash immediately with soap and water. Seek medical attention immediately. If this chemical has been inhaled, remove from exposure, begin rescue breathing (using universal precautions) if breathing has stopped and CPR if heart action has stopped. Transfer promptly to a medical facility. When this chemical has been swallowed, get medical attention. If victim is conscious, administer water or milk. Do not induce vomiting. Medical observation is recommended for 24 to 48 hours after breathing overexposure, as pulmonary edema may be delayed. As first aid for pulmonary edema, a doctor or authorized paramedic may consider administering a corticosteroid spray.

***References:***
- U.S. Environmental Protection Agency, Office of Pesticide Programs, Pesticide Residue Limits, "Calcium Oxide", 40 CFR 180.2 and 180.1001(c), http://www.epa.gov/pesticides/food/viewtols.htm
- Sax, N.I., Ed., "Dangerous Properties of Industrial Materials Report" 2, No. 1, 98-99 (1982).
- New Jersey Department of Health and Senior Services and Senior Services, "Hazardous Substance Fact Sheet: Calcium Oxide," Trenton, NJ (October 1985, rev. April 2003). http://www.state.nj.us/health/eoh/rtkweb/0325.pdf
- New York State Department of Health, "Chemical Fact Sheet: Calcium Oxide," Albany, NY, Bureau of Toxic Substance Assessment (January 1996).
- California Environmental Protection Agency Chemical List of Lists, Sacramento CA (February 1997).

## Calcium Phosphide

***Use Type:*** Rodenticide
***CAS Number:*** 1305-99-3
***Formula:*** $Ca_3P_2$
***Synonyms:*** Calcium phosphid; Phosphure de calcium (French); Photophor; Tricalcium diphosphide
***Trade Names:*** POLYTANOL®
***EPA/OPP PC Code:*** 066503
***ICSC Number:*** 1126
***RTECS Number:*** EW3860000
***EEC Number:*** 015-003-00-2
***EINECS Number:*** 215-142-0

***Uses:*** Calcium phosphide is used to kill rodents and in explosives and fireworks.

***Description:*** Calcium phosphide is a grey granular solid or reddish-brown crystalline solid with a musty odor, somewhat like acetylene. Water reactive. Melting/Freezing point = about 1600°C. Hazard Identification (based on NFPA-704 M Rating System): Health 4, Flammability 0, Reactivity 2.

***Incompatibilities:*** A strong reducing agent. Forms spontaneously flammable phosphine gas in moist air. Contact with water or acids release spontaneously flammable phosphine gas, and can cause explosions. Incompatible with oxidizers, acids, chlorine, chlorine monoxide, halogens, halogen acids, oxygen, sulfur.

***Permissible Exposure Limits in Air:*** No standards set.

***Permissible Concentration in Water:*** No criteria set, but runoff from spills or fire control may cause water pollution.

***Routes of Entry:*** Inhalation of dust, Eye or skin contact.

***Harmful Effects and Symptoms***

***Short Term Exposure:*** Calcium phosphide can affect you when breathed. Phosphine gas is a highly toxic gas released when calcium phosphide is wet or has contacted moisture. Consult the entry on phosphine. As noted by Sax (reference below), phosphine is an acute local irritant, is toxic upon inhalation causing restlessness, tremors, fatigue, gastric pain, diarrhea, coma and convulsions. Also, calcium phosphide is a dangerous fire and explosion hazard.

***Long Term Exposure:*** Long term or high exposure can cause pulmonary edema, a buildup of fluid in the lungs.

***First Aid:*** If this chemical gets into the eyes, remove any contact lenses at once and irrigate immediately for at least 15 minutes, occasionally lifting upper and lower lids. Seek medical attention immediately. If this chemical contacts the skin, remove contaminated clothing and wash immediately with soap and water. Seek medical attention immediately. If this chemical has been inhaled, remove from exposure, begin rescue breathing (using universal precautions) if breathing has stopped and CPR if heart action has stopped. Transfer promptly to a medical facility. When this chemical has been swallowed, get medical attention. Give large quantities of water and induce vomiting. Do not make an unconscious person vomit.

***References:***
- New Jersey Department of Health and Senior Services and Senior Services, "Hazardous Substance Fact Sheet: Calcium Phosphide," Trenton, NJ (February 1987, rev. October 2000).
http://www.state.nj.us/health/eoh/rtkweb/0329.pdf
- Sax, N.I., Ed., "Dangerous Properties of Industrial Materials Report" 2, No. 1, 102-103 (1982).
- California Environmental Protection Agency Chemical List of Lists, Sacramento CA (February 1997).

# Capsaicin

***Use Type:*** Bird, animal and insect repellent
***CAS Number:*** 404-86-4
***Formula:*** $C_{18}H_{27}NO_3$
***Synonyms:*** Capsaicin (in oleoresin of capsicum); Capsaicine; 8-Methyl-*N*-vanillyl-6-nonenamide,(*E*)- (8CI); *N*-[(4-Hydroxy-3-methoxyphenyl)methyl]-8-methyl-6-nonenamide; *trans-N*-[(4-Hydroxy-3-methoxyphenyl)methyl]-8-methyl-6-noneamide; *trans*-8-Methyl-*N*-vanillyl-6-noneamide; Major Capsaicinoids (Capsaicin+Dihydrocapsaicin+Nordihydrocapsacin); 6-Nonenamide, *N*-[(4-hydroxy-3-methoxyphenyl)methyl]-8-methyl-, (*E*)- (9CI); 6-Nonenamide, 8-methyl-*N*-vanillyl-, (*E*)- (8CI)
***Trade Names:*** AGRIGARD®; CAPSYN® (Capsaicin+Dihydrocapsaicin+Nordihydrocapsacin); DEER-OFF®; FRONTIERSMAN®; HALT®; KEEPOUT®; STYPTYSAT®
***Chemical Class:*** Botanical
***EPA/OPP PC Code:*** 070701
***California DPR Chemical Code:*** 470
***Uses:*** Capsaicin, which is made from the Capsicum red chili pepper, is used as a bird, animal and insect repellent. Specifically, it is used to repel birds, voles, deer, rabbits, squirrels, insects and attacking dogs. Capsaicin repellents are used indoors to protect carpets and upholstered furniture, and outdoors to protect fruit and vegetable crops, flowers, ornamental plants, shrubbery, trees, and lawns. It is also used in pepper sprays such as MACE, and as an analgesic in creams, lotions and solid sticks to reduce arthritic, post-operative and neuopathic pain, such as shingles. Capsaicin is obtained by grinding dried, ripe Capsicum frutescens L. chili peppers into a fine powder. The oleoresin is derived by distilling the powder in a solvent and evaporating the solvent. The resulting highly concentrated liquid has little odor but has an extremely pungent taste.

***U.S. Maximum Allowable Residue Levels for Capsaicin (40 CFR 180.1165):***
Capsaicin is exempt from the requirement of a tolerance in or on all food commodities when used in accordance with approved label rates and good agricultural practice.

***Carcinogen/Hazard Classifications***
**Label Signal Word:** CAUTION or DANGER
***Regulatory Authority:***
- Actively registered pesticide in California.

***Description:*** Crystalline solid, rectangular plates, or scales. Pungent odor and burning taste. Practically insoluble in cold water. Molecular weight = 305.40. Boiling point = 210 –218°C. Melting/Freezing point = 65°C.

***Incompatibilities:*** Slowly hydrolyzes in water, releasing ammonia and forming acetate salts.

***Routes of Entry:*** Eyes and skin
***Harmful Effects and Symptoms***
***Short Term Exposure:*** Irritates skin, eyes and mucous membranes. Poisonous if ingested; with severe gastritis, stomach pain, and diarrhea.

***Long Term Exposure:*** Mutagen.
***Points of Attack:*** Nerve endings and eyes.
***First Aid:*** If this chemical gets into the eyes, remove any contact lenses at once and irrigate immediately for at least 15 minutes, occasionally lifting upper and lower lids. Seek medical attention immediately. If this chemical contacts the skin, remove contaminated clothing and wash immediately with soap and water. Seek medical attention immediately. If this chemical has been inhaled, remove from exposure, begin rescue breathing (using universal precautions) if breathing has stopped and CPR if heart action has stopped. Transfer promptly to a medical facility. When this chemical has been swallowed, get medical attention. *Do not induce vomiting when formulations containing petroleum solvents are ingested.* Otherwise, give large quantities of water and induce vomiting. Do not make an unconscious person vomit.

***References:***
- U.S. Environmental Protection Agency, "Reregistration Eligibility Decision (RED) Facts, Capsaicin," Office of Prevention, Pesticides and Toxic Substances,

Washington, DC (June 1992).
http://www.epa.gov/REDs/factsheets/4018fact.pdf
- U.S. Environmental Protection Agency, Office of Pesticide Programs, Pesticide Residue Limits, "Capsaicin," 40 CFR 180.1165, www.epa.gov/pesticides/food/viewtols.htm
- California Environmental Protection Agency Chemical List of Lists, Sacramento CA (February 1997).

# Captafol

*Use Type:* Fungicide
*CAS Number:* 2939-80-2; 2425-06-1 (this CAS used in the California "Lists of Lists" and other reference sources)
*Formula:* $C_{10}H_9Cl_4NO_2S$
*Alert:* Captafol is no longer sold in the United States. Human toxicity (long-term): High.
*Synonyms:* 4-Cyclohexene-1,2-dicarboximide, *N*-(1,1,2,2-Tetrachloroethyl)thiol-1*H*-Isoindole-1,3(2*H*)-dione,3A,4,7,7A-tetrahydro-2-(1,1,2,2-tetrachloroethyl)thio-; *N*-[(1,1,2,2-Tetrachloroethyl)thio]-4-cyclohexene-1,2-dicarboximide *N*-[(1,1,2,2-Tetrachloroethyl)thio]-4-cyclohexene-1,2-dicarboximide *N*-[(1,1,2,2-Tetrachloroehtyl)-thio]4-cyclohexene-1,2-dicarboximide; *N*-(1,1,2,2-Tetrachloraethylthio)-cyclohex-4-en-1,4-diacarboximid (German); *N*-1,1,2,2-Tetrachloroethylmercapto-4-cyclohexene-1,2-carboximide; *N*-[(1,1,2,2-Tetrachloroethyl)sulfenyl]-*cis*-4-cyclohexene-1,2-dicarboximide; *N*-(1,1,2,2-Tetrachloroethylthio)-4-cyclohexene-1,2-dicarboximide
*Trade Names:* CAPTAFOL®, The Scotts Company (USA), canceled 4/23/1986; CAPTATOL®; CAPTOFOL®; CRISFOLATAN®; DIFOLATAN®, The Scotts Company (USA), canceled 5/1/1987; DIFOCAP®, The Scotts Company (USA), canceled 4/23/1986; DIFOSAN®; FOLCID®; HAIPEN®; KENOFOL®; MERPAFOL®; ORTHO® 5865, The Scotts Company (USA), canceled 5/1/1987; PILLARTAN®; SANSEAL®; SANSPOR®; SANTAR-SM®; SULFONIMIDE®; SULPHEIMIDE®
*Producers:* Agsin (Singapore); Chevron Phillips Chemical (USA); Rallis India (India); Rhone-Poulenc (France); Sigma-Aldrich Laborchemikalien (Germany)
*Chemical Class:* Carboximide
*EPA/OPP PC Code:* 081701, 081702 (unspecified isomer)
*California DPR Chemical Code:* 292
*ICSC Number:* 0119
*RTECS Number:* GW4900000
*EEC Number:* 613-046-00-7
*EINECS Number:* 219-363-3
*Uses:* Captafol was a General Use Pesticide and used for the control of practically all forms of fungal diseases except powdery mildew, but is not permitted for use in the United States. Outside of the US, it is used to control fruit disease on apples, citrus, tomato, cranberry, coffee, pineapple, potato, onion, stone fruit, cucumber, watermelon, and much more. It is also used as a seed protectorant on cotton, rice and peanut crops.
*Human toxicity (long-term)[77]:* Not listed.
*Fish toxicity (threshold)[77]:* High–1.75895 ppb, MATC (Maximum Acceptable Toxicant Concentration)
***U.S. Maximum Allowable Residue Levels for Captafol (40 CFR 180.267):***

| Chemical name | Crop | ppm |
|---|---|---|
| Captafol | onion | 0.1 |
| Captafol | potato | 0.5 |
| Captafol | tomato | 15 |
| Captafol (cis-*N*-[(1,1,2,2-tetrachloroethyl)thio]-4-cyclohex | onion | 0.1 |
| Captafol (cis-*N*-[(1,1,2,2-tetrachloroethyl)thio]-4-cyclohex | potato | 0.5 |
| Captafol (cis-*N*-[(1,1,2,2-tetrachloroethyl)thio]-4-cyclohex | tomato | 15 |

*Carcinogen/Hazard Classifications*
**U.S. EPA Carcinogens:** Group B2, probable carcinogen
**California Prop. 65:** Carcinogen
**IARC:** Group 2A, probable carcinogen
**Label Signal Word:** WARNING
**WHO Acute Hazard:** Class 1 a, extremely hazardous
*Regulatory Authority:*
- Banned or Severely Restricted (E. Germany, Norway) (UN)[13]
- AB 1803-Well Monitoring Chemical (CAL)
- AB 2588-Air Toxics "Hot Spots" Chemicals (CAL)
- Proposition 65 chemical (CAL)
- Permissible Exposure Limits for Chemical Contaminants (CAL/OSHA)
- The "Director's List" (CAL/OSHA)
- Air Pollutant Standard Set (ACGIH)[1] (HSE)[33] (OSHA)[58] (Several States)[60] (Australia) (Israel) (Mexico) (Several
- Canadian Provinces)
- Canada, WHMIS, Ingredients Disclosure List

*Description:* Captafol is a colorless or white to pale yellow crystalline solid. Weak, mercaptan odor. Practically insoluble in water; solubility = 1.6 ppm @ 20°C. Molecular weight = 349.09. Melting/Freezing point = 160–161°C. Vapor pressure $1.22 \times 10^{-8}$ mmHg @ 20°C.

*Incompatibilities:* Reacts violently with bases causing fire and explosion hazard. Not compatible with strong acids or acid vapor, oxidizers. Strong alkaline conditions contribute to instability. Attacks some metals.

*Permissible Exposure Limits in Air:* NIOSH[2] and ACGIH[1] recommended airborne exposure limit is 0.1 mg/m$^3$ TWA with the notation "skin" indicating the possibility of cutaneous absorption. HSE[33], Mexico, Israel, and Australia have set this same value as has NIOSH. In addition, several states have set guidelines or standards for

captafol in air[60] ranging from 1.0 $\mu g/m^3$ (North Dakota) to 1.5 $\mu g/m^3$ (Virginia) to 2.0 $\mu g/m^3$ (Connecticut and Nevada).
***Permissible Concentration in Water:*** No criteria set, but runoff from spills or fire control may cause water pollution.
***Routes of Entry:*** Inhalation, ingestion, skin.
***Harmful Effects and Symptoms***
***Short Term Exposure:*** Irritates eyes, skin and respiratory tract. Captafol can affect you when breathed in and by passing through your skin. Because this material has a low vapor pressure, significant inhalation of vapors is unlikely at ordinary temperatures. Captafol may cause an asthma-like allergy. Future exposures can cause asthma attacks with shortness of breath, wheezing, cough, and/or chest tightness. Exposure can irritate the skin. It can also cause a skin allergy to develop. Exposure to the sun (or other ultraviolet light) after exposure to captafol may cause severe rash with itching, swelling, and blistering.
***Long Term Exposure:*** Repeated or prolonged contact cause skin sensitization, dermatitis, allergic conjunctivitis. Repeated or prolonged inhalation exposure may cause asthma. The substance may have damaging effects on the liver and kidneys. Captafol is a probable carcinogen in humans. There is some evidence that it causes liver cancer in humans and it has caused kidney cancer in animals. Captafol may cause mutations. Handle with extreme caution.
***Points of Attack:*** Skin, respiratory system, liver and kidneys.
***Medical Surveillance:*** If symptoms develop or overexposure is suspected, the following may be useful: Liver and kidney function test; lung function test. Skin testing with dilute Captafol may help diagnose allergy, if done by a qualified allergist.
***First Aid:*** If this chemical gets into the eyes, remove any contact lenses at once and irrigate immediately for at least 15 minutes, occasionally lifting upper and lower lids. Seek medical attention immediately. If this chemical contacts the skin, remove contaminated clothing and wash immediately with soap and water. Seek medical attention immediately. If this chemical has been inhaled, remove from exposure, begin rescue breathing (using universal precautions) if breathing has stopped and CPR if heart action has stopped. Transfer promptly to a medical facility. When this chemical has been swallowed, get medical attention. Give large quantities of water and induce vomiting. Do not make an unconscious person vomit.
***References:***
- EPA, Office of Pesticide Programs, Pesticide Residue Limits, "Captafol ", 40 CFR 180.267. http://www.epa.gov/pesticides/food/viewtols.htm
- EXTOXNET, Extension Toxicology Network, "Pesticide Information Profile, Captafol," Oregon State University, Corvallis, OR. (September 1995). ace.orst.edu/info/extoxnet/pips/captafol.htm

- New Jersey Department of Health and Senior Services and Senior Services, "Hazardous Substance Fact Sheet: Captafol," Trenton, NJ (April 1998). http://www.state.nj.us/health/eoh/rtkweb/0338.pdf
- California Environmental Protection Agency Chemical List of Lists, Sacramento CA (February 1997).

# Captan (ANSI)

***Use Type:*** Fungicide
***CAS Number:*** 133-06-2
***Formula:*** $C_9H_8Cl_3NO_2S$
***Alert:*** Most uses of captan on food crops in the United States have been canceled since 1989. It is still used in apple production, almonds and strawberries.
***Synonyms:*** 4-Cyclohexene-1,2-dicarboximide,*N*-[(trichloromethyl)mercapto; ENT 26538; 1H-Isoindole-1,3(2H)-dione,3a,4,7,7a-tetrahydro-2-[(trichloromethyl)thiol]-; Isopto carbachol; le Captane (French); NCI-0077; *N*-Trichloromethylmercapto-4-cyclohexene-1,2-dicarboximide; *N*-(Trichloromethylmercapto)-$\delta^4$-tetrahydrophthalimide; *N*-Trichloromethylthiocyclohex-4-ene-1,2-dicarboximide; *N*-Trichloromethylthio-*cis*-$\delta^4$-cyclohexene-1,2-dicarboximide; *N*-trichloromethylthiocyclohex-4-ene-1,2-dicarboximide; *N*-[(trichloromethyl)thio]-4-cyclohexene-1,2-dicarboximide; *N*-[(trichloromethyl)thio]tetrahydrophthalimide; *N*-[(trichloromethyl)thio]-$\delta^4$-tetrahydrophthalimide; *N*-trichloromethylthio-3a,4,7,7a-tetrahydrophthalimide;
***Trade Names:*** AACAPTAN®; AGROSOL S®, Agrocliance, LLC (UK), canceled 7/11/2002; AGROX® 2-WAY and 3-WAY, Chipman (Canada), Zeneca Ag Products (USA) (now Syngenta), canceled; AMERCIDE®; APRON®, Syngenta (Switzerland), canceled 5/31/1996; BEISTERGARD®, Biesterfeld Siemsgluess International. GmbH (Germany); BANGTON®; BEAN SEED PROTECTANT®; CAPTANCAPTENEET® 26,538; CAPTAF®, Rallis India (India); CAPTAF®, Rallis India (India); CAPTAN® 50W, Milenia Agro Ciencias (Brazil); CAPTAN SC®, Milenia Agro Ciencias (Brazil); CAPTEX®; CRIPTAN®, Veterinary & Agricultural Products Manufacturing Co., Ltd. (VAPCO) (Jordan); ESSO® FUNGICIDE 406, Exxon (USA); FLIT® 406; FUNGUS BAN® TYPE II; FUNGICIDE 406®; GLYODEX® 37-22; GRANOX PFM®; GUSTAFSON CAPTAN 30-DD; HEXACAP®; ISOTOX SEED TREATER® "D" and "F"; KAPTAN®; MALIPUR®; MERPAN®, Makhteshim Agan (Israel); MICRO-CHECK® 12; MIOSTAT®; NERACID®; ORTHOCIDE®, The Scotts Company, (USA); OSOCIDE®; POTATO SEED PIECE PROTECTANT®; SR 406®; STAUFFER CAPTAN®; TRIMEGOL®; VANICIDE®; VANICIDE® P-75; VANICIDE® 89; VANICIDE® 89RE; VANGARD® K; VANGUARD® K; VONDCAPTAN®

*Producers:* AMVAC Chemical (USA); Bharat Pulverising Mills (India); Biesterfeld Siemsgluess International. GmbH (Germany); Calliope (France); Chevron Phillips Chemical (USA); Cyanamid Agricultural Products (Germany); Dow AgroScience (USA); Drexel Chemical (USA); DuPont (USA); Hokko Chemical Industry (Japan); Indiclay (India); Makhteshim Agan (Israel); Milenia Agro Ciencias (Brazil); Nippon Carbide Industries (Japan); Nippon Soda (Japan); Nissan Chemical Industries (Japan); Nufarm (Australia); Occidental Chemical (USA); Rallis India (India); Rhone-Poulenc Agro (France); Sankei Chemical (Japan); Sigma-Aldrich Laborchemikalien (Germany); Syngenta (Switzerland); Veterinary & Agricultural Products Manufacturing Co., Ltd. (VAPCO) (Jordan); Zhejiang Heben Pesticide & Chemicals Co., Ltd. (China)
*Chemical Class:* Thiophthalimide
*EPA/OPP PC Code:* 081301
*California DPR Chemical Code:* 104
*ICSC Number:* 0120
*RTECS Number:* GW5075000
*EEC Number:* 613-044-00-6
*EINECS Number:* 205-087-0
*Uses:.* Most uses of captan on food crops in the United States have been canceled since 1989. It is often applied to packing and shipping boxes for fruits and vegetables. Captan is rapidly degraded in natural soil by chemical as well as biologic means (estimated half-life, days to weeks). It is also used as a preservative for awnings, draperies and leather, as a root dip and seed treatment. It is also added to paints, wallpaper paste, plastic and leather goods. There are over 320 federally registered pesticide products that contain captan.
*Human toxicity (long-term)*[77]: Very low–145.83333 ppb, CHCL (Chronic Human Carcinogen Level)
*Fish toxicity (threshold)*[77]: Intermediate–25.36705 ppb, MATC (Maximum Acceptable Toxicant Concentration)
*U.S. Maximum Allowable Residue Levels for Captan (40 CFR 180.103):*

| CROP | ppm |
|---|---|
| Almond | 2.0 |
| Almond hulls | 100.0 |
| Apple | 25.0 |
| Apricot | 50.0 |
| Bean dry | 25.0 |
| Bean, succulent | 25.0 |
| Beet, greens | 100.0 |
| Beet roots | 2.0 |
| Blackberry | 25.0 |
| Blueberry (huckleberry) | 25.0 |
| Broccoli | 2.0 |
| Brussels sprouts | 2.0 |
| Cabbage | 2.0 |
| Cantaloupe | 25.0 |
| Carrot | 2.0 |
| Cattle fat | 0.05 |
| Cattle meat | 0.05 |
| Cattle, mbyp | 0.05 |
| Cauliflower | 2.0 |
| Celery | 50.0 |
| Cherry | 100.0 |
| Collards | 2.0 |
| Corn, sweet, kernels & cob with husks (removed) | 2.0 |
| Cotton undelinted seed | 2.0 |
| Cucumber | 25.0 |
| Dewberry | 25.0 |
| Eggplant | 25.0 |
| Grape | 50.0 |
| Hog fat | 0.05 |
| Hog meat | 0.05 |
| Hog mbyp | 0.05 |
| Honeydew | 25.0 |
| Kale | 2.0 |
| Lettuce | 100.0 |
| Mango | 50.0 |
| Muskmelon | 25.0 |
| Mustard greens | 2.0 |
| Nectarine | 50.0 |
| Onion dry bulb | 25.0 |
| Onion green | 50.0 |
| Pea, dry | 2.0 |
| Pea, succulent | 2.0 |
| Peach | 50.0 |
| Pear | 25.0 |
| Pepper | 25.0 |
| Plum, prune, fresh | 100.0 |
| Potato | 25.0 |
| Pumpkin | 25.0 |
| Raspberry | 25.0 |
| Rutabaga roots | 2.0 |
| Soybean, dry | 2.0 |
| Soybean, succulent | 2.0 |
| Spinach | 100.0 |
| Squash, summer | 25.0 |
| Squash, winter | 25.0 |
| Strawberry | 25.0 |
| Tomato | 25.0 |
| Turnip greens | 2.0 |
| Turnip roots | 2.0 |
| Watermelon | 25.0 |

*Carcinogen/Hazard Classifications*
**U.S. EPA Carcinogens:** Group B2, probable carcinogen
**California Prop. 65:** Carcinogen
**IARC:** Group 3, unclassifiable
**Label Signal Word:** DANGER or CAUTION if packaged in concentrated form.
**WHO Acute Hazard:** Class U, unlikely to be hazardous

***Regulatory Authority:***
- Carcinogen (Animal positive-mice) (NTP)
- Banned or Severely Restricted (Finland, Sweden) (UN)[13]
- Air Pollutant Standard Set (ACGIH)[1] (Australia) (HSE)[33] (Israel) (Mexico) (OSHA)[58] (Several States)[60] (Several
- Canadian Provinces)
- AB 1803-Well Monitoring Chemical (CAL)
- AB 2588-Air Toxics "Hot Spots" Chemicals (CAL)
- CAL-DHS/DHS Drinking Water Action Levels
- CAL Air Resources Board/AB 1807 Toxic Air Contaminants
- Proposition 65 chemical (CAL)
- Permissible Exposure Limits for Chemical Contaminants (CAL/OSHA)
- The "Director's List" (CAL/OSHA)
- Actively registered pesticide in California.
- Clean Air Act: Hazardous Air Pollutants (Title I, Part A, Section 112)
- Clean Water Act: Section 311 Hazardous Substances/RQ 40CFR117.3 (same as CERCLA, see below); Section 313 Water Priority Chemicals (57FR41331, 9/9/92).
- Superfund/EPCRA 40CFR302.4 RQ: CERCLA, 10 lb (4.54 kg)
- EPCRA Section 313 Form R *de minimus* concentration reporting level: 1.0%.

***Description:*** Captan, when pure, is a colorless crystalline solid. The technical grade is a cream to yellow powder with a strong odor. It is commonly dissolved in a "carrier" which may be combustible or flammable. Very slightly soluble in water; solubility = 3.3 ppm. Melting/Freezing point = 178°C (decomposes). Vapor pressure $8 \times 10^{-8}$ mmHg @ 20°C. Hazard Identification (based on NFPA-704 M Rating System): Health 3, Flammability 2, Reactivity 0. Log $K_{ow}$ = 2.34. Unlikely to bioaccumulate in marine organisms. Physical and toxicological properties may be affected by the carrier solvents used in commercial formulations.

***Incompatibilities:*** Incompatible with tetraethyl pyrophosphate, parathion. Keep away from strong alkaline materials (e.g., hydrated lime) as captan may become unstable. May react with water releasing hydrogen chloride gas. Corrosive to metals in the presence of moisture.

***Permissible Exposure Limits in Air:*** ACGIH[1] and NIOSH[2] have recommended a TWA for captan of 5 mg/m³. In addition, several states have set guidelines or standards for captan in ambient air[60] ranging from 11.9050 µg/m³ (Kansas) to 35 µg/m³ (Pennsylvania) to 50 µg/m³ (North Dakota) to 100 µg/m³ (Connecticut) to 119 µg/m³ (Nevada).

***Determination in Air:*** OSHA versatile sampler-2; Reagent; High-pressure liquid chromatography/Ultraviolet detection; IV NIOSH Method #5601[18].

***Permissible Concentration in Water:*** A no-adverse-effect level of drinking water has been calculated by NAS/NRC as 0.35 mg/L. The former USSR/UPEN joint project[43] has set a MAC of 2.0 mg/L in water bodies used for domestic purposes. Guidelines have been set in two states for Captan in drinking water ranging from 100 µg/L in Maine to 350 µg/L in California.

***Routes of Entry:*** Skin contact, inhalation of dust, ingestion.

***Harmful Effects and Symptoms***

***Short Term Exposure:*** The substance irritates the eyes and the skin. The acute $LD_{50}$ (oral, rat) = 9,000 mg/kg (insignificantly toxic). Most of the chronic-oral-toxicity data on captan suggest that the no-adverse-effect or toxicologically safe dosage is about 1000ppm (50 mg/kg/day). However, on the basis of fetal mortality observed in monkeys exposed to captan (12.5 mg/kg/day), the acceptable daily intake of captan has been established at 0.1 mg/kg of body weight by the FAO/WHO. Based of long-term feeding studies results in rats and dogs, ADIs were calculated at 0.05 mg/kg/day for captan. A rebuttal presumption against registration for captan was issued on August 19, 1980 by EPA on the basis of possible oncogenicity, mutagenicity, and teratogenicity.

***Long Term Exposure:*** Repeated or prolonged contact with skin may cause skin allergy to develop. Once this occurs, even very small future exposures can cause itching and a skin rash. Exposure may cause mutations or damage the developing fetus; however, this needs further study. Animal studies have found the development of cancer in animals. Whether captan is a human cancer hazard may require further study.

***Points of Attack:*** Eyes, skin, respiratory system, gastrointestinal tract, liver, kidneys. Cancer Site in animals: duodenal tumors.

***Medical Surveillance:*** If symptoms develop or overexposure is suspected, the following may be useful: Skin testing with dilute captan may help diagnose allergy, if done by a qualified allergist.

***First Aid:*** If this chemical gets into the eyes, remove any contact lenses at once and irrigate immediately for at least 15 minutes, occasionally lifting upper and lower lids. Seek medical attention immediately. If this chemical contacts the skin, remove contaminated clothing and wash immediately with soap and water. Seek medical attention immediately. If this chemical has been inhaled, remove from exposure, begin rescue breathing (using universal precautions) if breathing has stopped and CPR if heart action has stopped. Transfer promptly to a medical facility. When this chemical has been swallowed, get medical attention. Give large quantities of water and induce vomiting. Do not make an unconscious person vomit.

***References:***
- EPA, Office of Pesticide Programs, Pesticide Residue Limits, "Captan", 40 CFR 180.103. www.epa.gov/cgi-bin/oppsrc
- EXTOXNET, Extension Toxicology Network, "Pesticide

Information Profile, Captan," Oregon State University, Corvallis, OR (June, 1996).
http://ace.orst.edu/info/extoxnet/pips/captan.htm
- U.S. Environmental Protection Agency, "Reregistration Eligibility Decision (RED) Facts, Captan," Office of Prevention, Pesticides and Toxic Substances, Washington, DC (September 1999).
http://www.epa.gov/REDs/factsheets/0120fact.pdf
- U.S. Environmental Protection Agency, "Rebuttable Presumption Against Registration (RPAR) and Continued Registration of Pesticide Products Containing Captan," Federal Register, 45, No. 161, 54938-54986 (August 18, 1980).
- National Cancer Institute, Bioassay of Captan for Possible Carcinogenicity, Technical Report Series No. 15, Bethesda, MD (1977).
- Sax, N.I., Ed., "Dangerous Properties of Industrial Materials Report," 1, No. 4, 93-94 (1981); and 3, No. 5, 80-84 (1983).
- California Environmental Protection Agency Chemical List of Lists, Sacramento CA (February 1997).
- New Jersey Department of Health and Senior Services and Senior Services, "Hazardous Substance Fact Sheet: Captan," Trenton, NJ (August 1998).
http://www.state.nj.us/health/eoh/rtkweb/0339.pdf

# Carbaryl

*Use Type:* Insecticide, nematicide and plant growth regulator
*CAS Number:* 63-25-2
*Formula:* $C_{12}H_{11}NO_2$; $C_{10}H_7OOCNHCH_3$
*Alert:* A General Use Pesticide (GUP), but various formulations vary in toxicity.
*Synonyms:* Carbamic acid, methyl-, 1-naphthyl ester; Carbaril (Italian); Carbaryl, NAC; ENT 23969; *N*-Methylcarbamate de 1-naphtyle (French); Methylcarbamate 1-naphthalenol; Methylcarbamic acid, 1-naphthyl ester; *N*-Methyl-1-naphthyl-carbamat (German); *N*-Methyl-α-naphthylcarbamate; *N*-Methyl-1-naphthyl carbamate; *N*-Methyl-α-naphthylurethan; *N*-Metil-1-naftil-carbammato (Italian); α-Naphthyl *N*-methylcarbamate; 1-Naphthylmethylcarbamate; 1-Naphthol; 1-Naphthyl *N*-Methylcarbamate; 1-Naphthyl *N*-methyl-carbamate
*Trade Names:* ADIOS®, Micro-Flo Company (USA); ARILAT®; ARILATE®; ARYLAM®; BERCEMA NMC50®; BUGMASTER®, Southern National Manufacturing Co. (USA), canceled 7/1/1987; CAPROLIN®; CARBAMEC®; CARBAMINE®; CARBATOX®; CARBAVUR®; CARBOMATE®; CARPOLIN®; COMPOUND 7744®, Union Carbide (USA); CARYLDERM®; CRAG SEVIN®; CRUNCH®; DENAPON®; DICARBAM®; DYNA-CARBYL®; EXPERIMENTAL INSECTICIDE 7744®, Union Carbide (USA); GAMONIL®; GERMAIN'S®; HEXAVIN®; KARBASPRAY®; KARBATOX®; KARBOSEP®; MENAPHAM®; MICROCARB®; MUGAN®; MURVIN®; NAC®; NMC® 50; OMS-29®; OMS 629®; OLTITOX®; PANAM®; POMEX®; PROSEVOR® 85; RAVYON®; SAVIT®, Griffin L.L.C. (USA), canceled 1/22/1991; SEPTENE®; SEFFEIN®; SEVIMOL®; SEVIN®, Aventis CropScience (France); Bayer CropScience (Germany); Rhone-Poulenc Agro (France) plus several more; SEWIN®; SOK®; TERCYL®; THINSEC®; TORNADO®; TRICAR®; UNION CARBIDE 7,744®, Union Carbide (USA); VIOXAN®
*Producers:* Agrimor International (USA); Agsin (Singapore); Atul (India); BASF (Germany); Bayer CropScience (Germany); Bharat Pulverising Mills (India); China Chemical (China); Coromandel Fertilisers (India); Diachem (Italy); Drexel Chemical (USA); Ehrenstorfer, Dr. (Germany); Griffin L.L.C. (USA); Hokko Chemical Industry (Japan); Jingma Chemicals Ltd. (China); Nagarjuna Agrichem (India); Nihon Nohyaku (Japan); Nissan Chemical Industries (Japan); Nufarm (Australia); Rhone-Poulenc Argo (France); Saeryung Chemicals (South Korea); Sankei Chemical (Japan); Shenzhen Guomeng Industry Co., Ltd. (China); Sigma-Aldrich Laborchemikalien (Germany)
*Chemical Class:* N-Methyl Carbamate
*EPA/OPP PC Code:* 056801
*California DPR Chemical Code:* 105
*ICSC Number:* 0121
*RTECS Number:* FC5950000
*EEC Number:* 006-011-00-7
*EINECS Number:* 200-555-0
*Uses:* Carbaryl is one of the most widely used insecticides in agriculture, professional turf management and ornamental production, as well as in residential pet, lawn, and garden markets. It controls over 100 species of insects that infect citrus, cotton, nuts, and forest and ornaments trees, as well as poultry and livestock. Carbaryl also is used as a mosquito adulticide. It is available in a variety of formulations –bait, dust, wettable powders, granules, dispersions and suspensions. Washington State, for example, has a Special Local Needs registration to control burrowing shrimp in oyster beds.
*Human toxicity (long-term)*[77]*:* Low–70.00 ppb, Health Advisory
*Fish toxicity (threshold)*[77]*:* Intermediate–27.39896 ppb, MATC (Maximum Acceptable Toxicant Concentration)
*U.S. Maximum Allowable Residue Levels for Carbaryl (40 CFR 180.169):*

| CROP | ppm |
| --- | --- |
| Alfalfa hay | 100.0 |
| Almond | 1.0 |
| Almond, hulls | 40.0 |
| Apricot | 10.0 |
| Asparagus | 10.0 |
| Avocado | – |

# Carbaryl

| | | | |
|---|---|---|---|
| Banana | 10.0 | Horse, meat | 0.1 |
| Barley, fodder, green | – | Horse, mbyp | 0.1 |
| Barley, grain | – | Horseradish | 5.0 |
| Barley, straw | – | Kale | 12.0 |
| Bean | 10.0 | Kohlrabi | 10.0 |
| Bean, forage | 100.0 | Lentil | 10.0 |
| Bean, hay | 100.0 | Lettuce | 10.0 |
| Beet, garden, roots | 5.0 | Loganberry | 12.0 |
| Beet, garden, tops | 12.0 | Maple sap | – |
| Beet, sugar, tops | 100.0 | Melon | 10.0 |
| Blackberry | 12.0 | Milk | 0.3 |
| Blueberry | 10.0 | Millet, proso, grain | 3.0 |
| Boysenberry | 12.0 | Millet, proso, straw | 100.0 |
| Broccoli | 10.0 | Mustard, greens | 12.0 |
| Brussels sprouts | 10.0 | Nectarine | 10.0 |
| Cabbage | 10.0 | Oat, fodder, green | – |
| Cabbage, chinese, bok choy | 10.0 | Oat, grain & straw | – |
| Carrot | 10.0 | Okra | 10.0 |
| Cattle, fat | 0.1 | Olive | 10.0 |
| Cattle, kidney & liver | 1.0 | Oyster | 0.25 |
| Cattle, meat & byproducts | 0.1 | Parsley | 12.0 |
| Cauliflower | 10.0 | Parsnip | 5.0 |
| Celery | 10.0 | Pea, cowpea | 5.0 |
| Cherry | 10.0 | Pea, cowpea, forage | 100.0 |
| Chestnut | 1.0 | Pea, cowpea, hay | 100.0 |
| Clover & clover hay | 100.0 | Pea, vines | 100.0 |
| Collards | 12.0 | Pea, with pods | 10.0 |
| Corn (inc. sweet)(k+cwhr) | 5.0 | Peach | 10.0 |
| Corn, fodder & forage | 100.0 | Peanut | 5.0 |
| Cotton, forage | – | Peanut, hay | 100.0 |
| Cotton, undelinted seed | 5.0 | Pecan | 1.0 |
| Cranberry | 10.0 | Pepper | 10.0 |
| Cucumber | 10.0 | Pineapple | 2.0 |
| Dandelion, leaves | 12.0 | Pineapple, bran | 20 |
| Dewberry | 12.0 | Pistachio | 1.0 |
| Dill | 0.2 | Plum, prune, fresh | 10.0 |
| Egg | 0.5 | Potato | 0.2 |
| Eggplant | 10.0 | Poultry, meat & fat | 5.0 |
| Endive (escarole) | 10.0 | Prickly pear cactus, fruit | 12.0 |
| Filbert (hazelnuts) | 1.0 | Prickly pear cactus, pads | 12.0 |
| Flax, seed | 5.0 | Pumpkin | 10.0 |
| Flax, straw | 100.0 | Radish | 5.0 |
| Fruit, pome | 10.0 | Raspberry | 12.0 |
| Fruit, stone, group | 10.0 | Rice | 5.0 |
| Goat, fat | 0.1 | Rice, straw | 100.0 |
| Goat, kidney | 1.0 | Rutabaga | 5.0 |
| Goat, liver | 1.0 | Rye, fodder, green | – |
| Goat, meat & byproducts | 0.1 | Rye, grain | – |
| Grape | 10.0 | Rye, straw | – |
| Grass & grass hay | 100.0 | Salsify, roots | 5.0 |
| Hog, fat | 0.1 | Salsify, tops | 10.0 |
| Hog, kidney & liver | 1.0 | Sheep, fat | 0.1 |
| Hog, meat & byproducts | 0.1 | Sheep, kidney | 1.0 |
| Horse, fat | 0.1 | Sheep, meat & liver | 0.1 |
| Horse, kidney & liver | 1.0 | Sheep, mbyp | 0.1 |

| | |
|---|---|
| Sorghum, forage | 100.0 |
| Sorghum, grain, grain | 10.0 |
| Soybean | 5.0 |
| Soybean, forage & hay | 100.0 |
| Spinach | 12.0 |
| Squash, summer & winter | 10.0 |
| Strawberry | 10.0 |
| Sunflower, seed | 1.0 |
| Sweet potato | 0.2 |
| Swiss chard | 12.0 |
| Tomato | 10.0 |
| Trefoil, birdsfoot, forage | 100.0 |
| Trefoil, birdsfoot, hay | 100.0 |
| Turnip, greens | 12.0 |
| Turnip, roots | 5.0 |
| Walnut | 1.0 |
| Wheat, fodder, green | 100.0 |
| Wheat, grain | 3.0 |
| Wheat, straw | 100.0 |

*Carcinogen/Hazard Classifications*
**U.S. EPA Carcinogens:** Group C, possible carcinogen
**IARC:** Group 3, unclassifiable
**Label Signal Word:** DANGER, WARNING, or CAUTION, depending on formulation
**WHO Acute Hazard:** Class II, moderately hazardous
**Endocrine Disruptor:** Suspected endocrine disruptor
*Regulatory Authority:*
- Air Pollutant Standard Set (ACGIH)[1] (Australia) (DFG)[3] (HSE)[33] (Israel) (Mexico) (former USSR)[43] (OSA)[58] (Several States)[60] (Several Canadian Provinces)Clean Air Act: Hazardous Air Pollutants (Title I, Part A, Section 112)
- Clean Water Act: Section 311 Hazardous Substances/RQ 40CFR117.3 (same as CERCLA, see below); Section 313 Water Priority Chemicals (57FR41331, 9/9/92).
- RCRA 40CFR268.48; 61FR15654, Universal Treatment Standards: Wastewater (mg/L), 0.006; Nonwastewater (mg/kg), 0.14
- Superfund/EPCRA 40CFR302.4 RQ: CERCLA, 100 lb (45.4 kg).
- EPCRA Section 313 Form R *de minimus* concentration reporting level: 1.0%.
- AB 1803-Well Monitoring Chemical (CAL)
- AB 2588-Air Toxics "Hot Spots" Chemicals (CAL)
- CAL-DHS/DHS Drinking Water Action Levels
- CAL Air Resources Board/AB 1807 Toxic Air Contaminants
- Permissible Exposure Limits for Chemical Contaminants (CAL/OSHA)
- The "Director's List" (CAL/OSHA)
- Actively registered pesticide in California.
- U.S. DOT Regulated Marine Pollutant (49CFR172.101, Appendix B)
- Canada: Drinking water MAC = 0.09 mg/L

*Description:* Carbaryl is a white or grayish, odorless, crystalline solid, or various other forms including liquid and paste. Soluble in water; solubility = 115 ppm @ 25°C. Molecular weight = 201.19. Boiling point = (decomposes below BP). Melting/Freezing point = 142°C. Vapor pressure 1.2 x $10^{-6}$ mmHg @ 20°C. Flash point = 203°C. Hazard Identification (based on NFPA-704 M Rating System): Health 2, Flammability 0, Reactivity 0. Log $K_{ow}$ = 2.34–2.38. Unlikely to bioaccumulate in marine organisms.
*Incompatibilities:* Contact with strong oxidizers can cause fire and explosions.
*Permissible Exposure Limits in Air:* The OSHA[2] PEL is 5 mg/m³ TWA for an 8-hour workshift. NIOSH[2] and ACGIH[1] recommended limit is also 5 mg/m³. Australia, Israel, Mexico, DFG[3] and HSE[33] have all set this same value. The STEL set by HSE[33] is 10 mg/m³. The NIOSH[2] IDLH level = 100 mg/m³. The former USSR/UNEP joint project[43] sets a MAC in workplace air of 1 mg/m³ and limits in the ambient air in residential areas of 0.02 mg/m³ on a momentary basis and 0.01 mg/m³ on an average daily basis. In addition, several states have set guidelines or standards for carbaryl in ambient air[60] ranging from 3.5 μg/m³ (Pennsylvania) to 11.9050 μg/m³ (Kansas) to 50 μg/m³ (North Dakota) to 80 μg/m³ (Virginia) to 100 μg/m³ (Connecticut) to 119 μg/m³ (Nevada).
*Determination in Air:* OSHA versatile sampler-2; Reagent; High-pressure liquid chromatography/Ultraviolet detection; IV NIOSH Method #5601; also NIOSH Method #5006[18].
*Permissible Concentration in Water:* A no-adverse effect level in drinking water has been calculated as 0.574 mg/L by NAS/NRC. The UNEP/USSR joint project[43] has set a MAC of 0.1 mg/L in water used for domestic purposes and 0.0005 mg/L in water bodies used for fishery purposes. Further, several states have set guidelines for carbaryl in drinking water[61] ranging from 10 μg/L (Wisconsin) to 60 μg/L (California) to 164 μg/L (Maine) to 574 μg/L (Kansas). See Regulatory section for Canada drinking water level.
*Routes of Entry:* Inhalation, skin contact or eye contact, skin absorption.
*Harmful Effects and Symptoms*
*Short Term Exposure:* Carbaryl is a acetylcholinesterase inhibitor. Carbaryl irritates the eyes, skin, and respiratory tract. The hot liquid may cause severe skin burns. The substance may affect the nervous system, resulting in convulsions and respiratory failure. The effects may be delayed. The $LD_{50}$ (oral, rat) = 250 mg/kg (moderately toxic). Single doses of up to about 140 mg (0.005 oz.) have been reported to cause no effect. However, a single dose of about 200 mg has caused stomach pain and excessive sweating. Individual responses may vary. Several milliliters (0.1 fluid oz) of an 80% solution of carbaryl has caused nausea, salivation, headache, tremors, and excessive tearing. 500 ml (1 pint) of a 80% solution has resulted in death.

*Long Term Exposure:* The major health problem associated with occupational exposure to carbaryl is related to its inhibition of the enzyme cholinesterase in the central, autonomic and peripheral nervous systems. The inhibition of cholinesterase allows acetylcholine to accumulate at these sites and thereby leads to over stimulation of innervated organs. The signs and symptoms observed as a consequence of exposure to carbaryl in the workplace environment are manifestations of excessive cholinergic stimulation, e.g., nausea, vomiting, mild abdominal cramping, dimness of vision, dizziness, headache, difficulty in breathing, and weakness. Carbaryl may affect the kidneys and nervous system. It may cause mutations and be a teratogen in humans. There is limited evidence that it reduces fertility in both males and females.

*Points of Attack:* Respiratory system, skin, central nervous system, cardiovascular system.

*Medical Surveillance:* NIOSH recommends that workers subject to carbaryl exposure have comprehensive preplacement medical examinations, with subsequent annual medical surveillance. If symptoms develop or overexposure has occurred, the following may be useful: Kidney function tests. Exam of the nervous system. If done within 2-3 hours after exposure, serum and RBC cholinesterase levels may be helpful. Levels can return to normal before the exposed person feels well.

*First Aid:* If this chemical gets into the eyes, remove any contact lenses at once and irrigate immediately for at least 15 minutes, occasionally lifting upper and lower lids. Seek medical attention immediately. If this chemical contacts the skin, remove contaminated clothing and wash immediately with soap and water. Seek medical attention immediately. If this chemical has been inhaled, remove from exposure, begin rescue breathing (using universal precautions) if breathing has stopped and CPR if heart action has stopped. Transfer promptly to a medical facility. When this chemical has been swallowed, get medical attention. Give large quantities of water and induce vomiting. Do not make an unconscious person vomit

*References:*
- U.S. Environmental Protection Agency, Office of Pesticide Programs, Pesticide Residue Limits, "Carbaryl", 40 CFR 180.169, www.epa.gov/cgi-bin/oppsrch
- EXTOXNET, Extension Toxicology Network, "Pesticide Information Profile, Carbaryl," Oregon State University, Corvallis, OR (June 1996). http://ace.orst.edu/info/extoxnet/pips/carbaryl.htm
- National Institute for Occupational Safety and Health, Criteria for a Recommended Standard: Occupational Exposure to Carbaryl, NIOSH Document No. 77-107 (1977).
- Sax, N.I., Ed., "Dangerous Properties of Industrial Materials Report," 1, No. 5, 45-46 (1981) and 3, No. 6, 42-48 (1983).
- U.S. Environmental Protection Agency, "Health Advisory: Carbaryl," Washington, DC, Office of Drinking Water (August 1987).
- New Jersey Department of Health and Senior Services and Senior Services, "Hazardous Substance Fact Sheet: Carbaryl," Trenton, NJ (September 1996, rev. February 2002). http://www.state.nj.us/health/eoh/rtkweb/0340.pdf
- New York Department of Health and Senior Services and Senior Services, "Chemical Fact Sheet: Carbaryl," Albany, NY, Bureau of Toxic Substance Assessment (March 1986).
- California Environmental Protection Agency Chemical List of Lists, Sacramento CA (February 1997).

# Carbendazim

*Use Type:* Fungicide
*CAS Number:* 10605-21-7
*Formula:* $C_9H_9N_3O_2$
*Synonyms:* BCM; Benzimidazole-2-carbamic acid, methyl ester; *N*-2-(Benzimidazolyl) carbamate; 2-Benzimidazolecarbamic acid, methyl ester; 1*H*-benzimidazol-2-ylcarbamic acid methyl ester; BMC; Carbamic acid, 1*H*-benzimidazole-2-yl-, Methyl ester; MBC; 2-(Methoxy-carbonylamino)-benzimidazol; 2-(Methoxycarbonylamino)-benzimidazole; Methyl 1*H*-benzemedazol-2-yl carbamate; Methyl 2-benzimidazolecarbamate; Methyl benzimidazole-2-yl carbamate

*Trade Names:* ABACOL®, J. J. Mauget (USA); AIMCOZIM; BAS® 3460; BAS® 67054; BATTAL®,; BAVISTIN®; BENDAZIM®,; CARBATE®,; CARBENDAZIME®,; CARBENDAZOL®; CARBENDAZOLE®; CARBENDAZYM®; CARBENDOR; CEKUDAZIM®; CORBEL; CTR® 6669; CUSTOS®; DEFENSOR®; DELSENE®; DEROSAL®; E-965®; DERROPRENE®; EQUITDAZIN®; FUNGISOL®, J. J. Mauget (USA); HOE 17411®; LIGNASAN®; IMISOL®, J. J. Mauget (USA); KEMDAZIN®; MERGAL®; PILLARSTIN®; POLYPHASE®; RODAZIM®, Rotam Agrochemical (Hong Kong); STEMPOR®; SUPERCARB, TRITICOL®; TRITICOL®; U-32.104®

*Producers:* Agrimor International (USA); Agsin (Singapore); Gharda Chemicals (India); Hebei Huafeng Chemical Group (China); Hindustan Insecticides (India); J. J. Mauget (USA); Ki-Hara Chemicals (UK); Nagarjuna Agrichem (India); Sanonda Zhengzhou Pesticide Co., Ltd. (China); Shandong Huayang Pesticide Group (China); Rotam Agrochemical (Hong Kong); Vijayalakshmi Insecticides and Pesticides (India); Whyte Agrochemicals (UK); Yellow River Enterprise Co., Ltd. (Taiwan)

*Chemical Class:* Carbamate; Benzimidazole
*EPA/OPP PC Code:* 128872
*California DPR Chemical Code:* 2176

## Carbendazim

*ICSC Number:* 1277
*RTECS Number:* DD6500000
*EEC Number:* 613-048-00-8
*EINECS Number:* 234-232-0

*Uses:* Carbendazim is a systemic fungicide used to control a broad range of fungi in cereals, vegetables, oilseed rape, sugar beets, grapes, tomatoes, pome fruit and stone fruit. It is also used in post-harvest storage and as treatment in seed pre-planting, frequently in combination with other fungicides. In some areas, it has been used to combat Dutch Elm disease.

*Human toxicity (long-term)*[77]*:* High–7.00 ppb, Health Advisory

*Fish toxicity (threshold)*[77]*:* High–1.44530 ppb, MATC (Maximum Acceptable Toxicant Concentration)

*Carcinogen/Hazard Classifications*

**U.S. EPA Carcinogens:** Group C, possible carcinogen
**Label Signal Word:** WARNING, CAUTION or DANGER, depending on formulation
**WHO Acute Hazard:** Class U, unlikely to be hazardous
**Endocrine Disruptor:** Suspected endocrine disruptor

*Regulatory Authority:*
- AB 1803-Well Monitoring Chemical (CAL)
- U.S. DOT Regulated Marine Pollutant (49CFR, Subchapter 172.101, Appendix B)

*Description:* Colorless crystalline solid or light-gray powder. Slightly soluble in water; solubility = 5.8 ppm @ 24°C; 4.8 ppm @ pH 7. Molecular weight = 191.20. Melting/Freezing point = (decomposes) 300–305°C. Vapor pressure $4.9 \times 10^{-10}$ mmHg @ 20°C. Log $K_{ow}$ = 1.51. Unlikely to bioaccumulate in marine organisms.

*Incompatibilities:* Oxidizers (chlorates, nitrates, peroxides, permanganates, perchlorates, chlorine, bromine, fluorine, etc); strong acids. May form explosive materials with phosphorus pentachloride.

*Permissible Concentration in Water:* No criteria set. Runoff from spills or fire control may cause water pollution.

*Routes of Entry:* Skin absorption, ingestion, inhalation.

*Harmful Effects and Symptoms*

*Short Term Exposure:* Eye pupils are small; blurred vision; eye watering; runny nose; cough; shortness of breath; salivation; nausea, stomach cramps, diarrhea, and vomiting; increased blood pressure; profuse sweating; hypermotility, hallucinations; agitation; tingling of the skin; slow heartbeat; convulsions; fluid in lungs; loss of consciousness; incontinence; breathing stops; death. Carbamate insecticides inhibit the cholinesterase activity of enzymes, causing accumulation of acetylcholine at synapses and altering the way in which nervous impulses are transmitted. However, within several hours carbamates spontaneously detach from the enzymes.

*Long Term Exposure:* A potent cholinesterase inhibitor; cumulative effect is possible. This chemical may damage the nervous system with repeated exposure, resulting in convulsions, respiratory failure. May cause liver damage.

*Points of Attack:* Respiratory system, lungs, central nervous system, cardiovascular system, skin, eyes, plasma and red blood cell cholinesterase.

*Medical Surveillance:* Medical observation is recommended for 24 to 48 hours after breathing overexposure, as pulmonary edema may be delayed. As first aid for pulmonary edema, consider administering a corticosteroid spray. Cigarette smoking may exacerbate pulmonary injury and should be discouraged for at least 72 hours following exposure. Before employment and at regular times after that, the following are recommended: Plasma and red blood cell cholinesterase levels (tests for the enzyme poisoned by this chemical). If exposure stops, plasma levels return to normal in 1-2 weeks while red blood cell levels may be reduced for 1-3 months. When acetylcholinesterase enzyme levels are reduced by 25% or more below preemployment levels, risk of poisoning is increased, even if results are in lower ranges of "normal." Reassignment to work not involving carbamate pesticides is recommended until enzyme levels recover. If symptoms develop or overexposure occurs, repeat the above tests as soon as possible and get an exam of the nervous system. Also consider complete blood count. Consider chest x-ray following acute overexposure

*First Aid:* Speed in removing material from eyes and skin is of extreme importance. *Eyes:* Eye contact can cause dangerous amounts of these chemicals to be quickly absorbed through the mucous membrane into the bloodstream. Immediately and gently flush eyes with plenty of warm or cold water (NO hot water) for at least 15 minutes, occasionally lifting the upper and lower eyelids. Get medical aid immediately. *Skin:* Get medical aid. Skin contact can cause dangerous amounts of these chemicals to be absorbed into the bloodstream. Wearing the appropriate PPE equipment and respirator for carbamate pesticides, immediately flush skin with plenty of soap and water for at least 15 minutes while removing contaminated clothing and shoes. Shampoo hair promptly if contaminated; protect eyes. *Ingestion:* Call poison control. Loosen all clothing. Never give anything by mouth to an unconscious person. Get medical aid. Do NOT induce vomiting.* If conscious, alert, and able to swallow, rinse mouth and have victim drink 4 to 8 ounces of water. Check to see if poison control instructs you to use ipecac syrup, otherwise administer slurry of activated charcoal (2 oz in 8 oz of water). If victim is *UNCONSCIOUS OR HAVING CONVULSIONS,* do nothing except keep victim warm. **In some cases you may be specifically instructed by poison control to induce vomiting by way of 2 tablespoons of syrup of ipecac (adult) washed down with a cup of water.* Do NOT give activated charcoal before or with ipecac syrup. *Inhalation:* Get medical aid. Do not contaminate yourself. Wearing the appropriate PPE equipment and respirator for carbamate pesticides, immediately remove the victim from the contaminated area to fresh air. If the victim is not breathing, administer artificial respiration. Do NOT use mouth-to-mouth resuscitation; use bag/mask apparatus. If breathing is difficult, administer oxygen through bag/mask apparatus until medical help arrives. Do not leave victim unattended. *Note to physician or authorized medical personnel.* Administer atropine, 2 mg (1/30 gr) intramuscularly or intravenously as soon as any local or systemic signs or symptoms of an intoxication are noted; repeat the administration of atropine every 3 to 8 minutes until signs of atropinization (mydriasis, dry mouth, rapid pulse, hot and dry skin) occur; initiate treatment in children with 0.05 mg mg/kg of atropine; repeat at 5 to 10 minute intervals. Watch respiration, and remove bronchial secretions if they appear to be obstructing the airway; intubate if necessary. *Medical note:* 2-PAMCI may be contraindicated in the case of some carbamate poisonings.

*References:*
- California Environmental Protection Agency Chemical List of Lists, Sacramento CA (February 1997)

# Carbetamide (ANSI)

*Use Type:* Herbicide
*CAS Number:* 16118-49-3
*Formula:* $C_{12}H_{16}N_2O_3$
*Synonyms:* d-(-)-Carbanilic acid(1-ethylcarbamoyl)ethyl ester; Carbetamid (German); N-Ethyl-2-[((phenylamino)carbonyl)oxy]propanamide, (d) isomer; d-N-Ethylacetamide carbanilate; d-N-Ethyllactamide carbanilate (ester); (R)-N-Ethyl-2-[((phenylamino)carbonyl)oxy]propanamide; d-(-)-1-(Ethylcarbamoyl)ethylphenyl carbamate; 2-Phenyl-carbamoyloxy-N-aethyl-propionamid (German); (Phenylcarbamoyloxy)-2-N-ethylpropionamide; N-Phenyl-1-(ethylcarbamoyl-1)-ethylcarbamate, d-isomer; 11,561 RP
*Trade Names:* CARBETAMEX®, Makhteshim-Agan Industries (Israel); HELMSMAN®, Bayer CropScience (Germany); LEGURAME®
*Producers:* Bayer CropScience (Germany); Makhteshim-Agan Industries (Israel)
*Chemical Class:* Carbamate
*EPA/OPP PC Code:* 259200
*EINECS Number:* 240-286-6
*Uses:* Not registered in the U.S.
*Carcinogen/Hazard Classifications*
**WHO Acute Hazard:** Class U, unlikely to be hazardous
*Description:* Soluble in water; solubility = 3,500 ppm @ 20°C. Molecular weight 236.27. Melting/Freezing point =110°C. Boiling point = decomposes. Vapor pressure = <1 x $10^{-5}$ mmHg @ 20°C.

*Incompatibilities:* Slowly hydrolyzes in water, releasing ammonia and forming acetate salts. May form explosive materials with phosphorus pentachloride.
*Permissible Concentration in Water:* No criteria set. Runoff from spills or fire control may cause water pollution.
*Routes of Entry:* Skin absorption, ingestion, inhalation.
*Harmful Effects and Symptoms*
*Short Term Exposure:* Eye pupils are small; blurred vision; eye watering; runny nose; cough; shortness of breath; salivation; nausea, stomach cramps, diarrhea, and vomiting; increased blood pressure; profuse sweating; hypermotility, hallucinations; agitation; tingling of the skin; slow heartbeat; convulsions; fluid in lungs; loss of consciousness; incontinence; breathing stops; death. Carbamate insecticides inhibit the cholinesterase activity of enzymes, causing accumulation of acetylcholine at synapses and altering the way in which nervous impulses are transmitted. However, within several hours carbamates spontaneously detach from the enzymes.
*Long Term Exposure:* A potent cholinesterase inhibitor; cumulative effect is possible. This chemical may damage the nervous system with repeated exposure, resulting in convulsions, respiratory failure. May cause liver damage.
*Points of Attack:* Respiratory system, lungs, central nervous system, cardiovascular system, skin, eyes, plasma and red blood cell cholinesterase.
*Medical Surveillance:* Medical observation is recommended for 24 to 48 hours after breathing overexposure, as pulmonary edema may be delayed. As first aid for pulmonary edema, consider administering a corticosteroid spray. Cigarette smoking may exacerbate pulmonary injury and should be discouraged for at least 72 hours following exposure. Before employment and at regular times after that, the following are recommended: Plasma and red blood cell cholinesterase levels (tests for the enzyme poisoned by this chemical). If exposure stops, plasma levels return to normal in 1-2 weeks while red blood cell levels may be reduced for 1-3 months. When acetylcholinesterase enzyme levels are reduced by 25% or more below preemployment levels, risk of poisoning is increased, even if results are in lower ranges of "normal." Reassignment to work not involving carbamate pesticides is recommended until enzyme levels recover. If symptoms develop or overexposure occurs, repeat the above tests as soon as possible and get an exam of the nervous system. Also consider complete blood count. Consider chest x-ray following acute overexposure
*First Aid:* Speed in removing material from eyes and skin is of extreme importance. *Eyes:* Eye contact can cause dangerous amounts of these chemicals to be quickly absorbed through the mucous membrane into the bloodstream. Immediately and gently flush eyes with plenty of warm or cold water (NO hot water) for at least 15 minutes, occasionally lifting the upper and lower eyelids.

Get medical aid immediately. *Skin:* Get medical aid. Skin contact can cause dangerous amounts of these chemicals to be absorbed into the bloodstream. Wearing the appropriate PPE equipment and respirator for carbamate pesticides, immediately flush skin with plenty of soap and water for at least 15 minutes while removing contaminated clothing and shoes. Shampoo hair promptly if contaminated; protect eyes. *Ingestion:* Call poison control. Loosen all clothing. Never give anything by mouth to an unconscious person. Get medical aid. Do NOT induce vomiting.* If conscious, alert, and able to swallow, rinse mouth and have victim drink 4 to 8 ounces of water. Check to see if poison control instructs you to use ipecac syrup, otherwise administer slurry of activated charcoal (2 oz in 8 oz of water). If victim is *UNCONSCIOUS OR HAVING CONVULSIONS,* do nothing except keep victim warm. **In some cases you may be specifically instructed by poison control to induce vomiting by way of 2 tablespoons of syrup of ipecac (adult) washed down with a cup of water.* Do NOT give activated charcoal before or with ipecac syrup. *Inhalation:* Get medical aid. Do not contaminate yourself. Wearing the appropriate PPE equipment and respirator for carbamate pesticides, immediately remove the victim from the contaminated area to fresh air. If the victim is not breathing, administer artificial respiration. Do NOT use mouth-to-mouth resuscitation; use bag/mask apparatus. If breathing is difficult, administer oxygen through bag/mask apparatus until medical help arrives. Do not leave victim unattended. *Note to physician or authorized medical personnel.* Administer atropine, 2 mg (1/30 gr) intramuscularly or intravenously as soon as any local or systemic signs or symptoms of an intoxication are noted; repeat the administration of atropine every 3 to 8 minutes until signs of atropinization (mydriasis, dry mouth, rapid pulse, hot and dry skin) occur; initiate treatment in children with 0.05 mg mg/kg of atropine; repeat at 5 to 10 minute intervals. Watch respiration, and remove bronchial secretions if they appear to be obstructing the airway; intubate if necessary. *Medical note:* 2-PAMCl may be contraindicated in the case of some carbamate poisonings.

*References:*
- International Programme on Chemical Safety (IPCS), "Carbamate Pesticides: A General Introduction," World Health Organization, Geneva, Switzerland (1986). http://www.inchem.org/documents/ehc/ehc/ehc64.htm
- California Environmental Protection Agency Chemical List of Lists, Sacramento CA (February 1997)

## Carbofuran (ANSI)

*Use Type:* Insecticide, acaricide and nematicide.
*CAS Number:* 1563-66-2
*Formula:* $C_{12}H_{15}NO_3$

*Alert:* A Restricted Use Pesticide (RUP). According to the Ecological Incident Investigation System, carbofuran has been responsible for more avian deaths than any other pesticide. Toxicity (human): Intermediate.
*Synonyms:* 7-Benzofuranol, 2,3-dihydro-2,2-dimethyl-, methylcarbamate; Carbamic acid, methyl-, 2,2-dimethyl-2,3-dihydrobenzofuran-7-yl ester; Carbofurano (Spanish); 2,3-Dihydro-2,2-dimethylbenzofuranyl-7-*N*-methylcarbamate; 2,3-Dihydro-2,2-dimethyl-7-benzofuranolmethylcarbamate; 2,3-Dihydro-2,2-dimethyl-7-benzofuranol-*N*-methylcarbamate; 2,3-Dihydro-2,2-dimethylbenzofuranyl-7-*N*-methylcarbamate; 2,3-Dihydro-2,2-dimethylbenzofuran-7-yl methylcarbamate; 2,2-Dimethyl-7-coumaranyl *N*-methylcarbamate; 2,2-Dimethyl-2,2-dihydrobenzofuranyl-7 *N*-methylcarbamate; 2,2-Dimethyl-2,3-dihydro-7-benzofuranyl-*N*-methylcarbamate; ENT 27,164; Methyl carbamic acid 2,3-dihydro-2,2-dimethyl-7-benzofuranyl ester; NSC 167822
*Trade Names:* A13-27164®; AU'ULTRAMICIN®; BAY 704143®, Bayer CropScience (Germany); BAY 78537®, Bayer CropScience (Germany); BRIFUR®; CARBODAN®, Agan Chemical Manufacturers Ltd. (Israel); Makhteshim Agan (Israel); CARBOSIP 5G®; CRISFURAN®; CURETERR®, Bayer CropScience (Germany); CHINUFUR®; D 1221®; DIAFURAN®, Calliope (France); FMC 10242®, FMC Agricultural Products Group (USA); FURACARB®; FURADAN®, FMC Agricultural Products Group (USA); Rallis India (India); FURAN®, United Phosphorus (India); FURODAN®; KENFURAN®, Kenso Corp. (Malaysia); KENOFURAN®; NEX®; NIA-10242, FMC Agricultural Products Group (USA); NIAGARA 10242, FMC Agricultural Products Group (USA); NIAGARA NIA-10242, FMC Agricultural Products Group (USA); PILLARFURAN®; RAMPART®; NIAGARA 10242®; YALTOX®
*Producers:* Agan Chemical Manufacturers Ltd. (Israel); Agrimor International (USA); Agro-care Chemical Industry Group (China); Agsin (Singapore); Alcotan Laboratories (Spain); Bayer CropScience (Germany); Biesterfeld Siemsgluess International. GmbH (Germany); Borregaard (Norway); Calliope (France); China Chemical (China); Ehrenstorfer, Dr. (Germany); Chromos Agro (Croatia); EniChem (Italy); FMC Agricultural Products Group (USA); Kenso Corp. (Malaysia); Luxembourg Industries (PAMOL) (Israel); Makhteshim Agan (Israel); Nagarjuna Agrichem (India); Rallis India (India); Saeryung Chemicals (South Korea); Sanonda Zhengzhou Pesticide Co., Ltd. (China); Shandong Huayang Pesticide Group (China); Shenzhen Guomeng Industry Co., Ltd. (China); Sigma-Aldrich Laborchemikalien (Germany); Sinon (Taiwan); United Phosphorus (India); Vijayalakshmi Insecticides Pesticides (India); Zago Asia Ltd. (Singapore)
*Chemical Class:* *N*-Methyl Carbamate
*EPA/OPP PC Code:* 090601

*California DPR Chemical Code:* 106
*ICSC Number:* 0122
*RTECS Number:* FB9450000
*EEC Number:* 006-026-00-9
*EINECS Number:* 216-353-0

*Uses:* Carbofuran is a broad spectrum carbamate pesticide that kills insects, mites, and nematodes on contact or after ingestion. It is used against soil and foliar pests of field, fruit, vegetable, and forest crops. Carbofuran is available in liquid and granular formulations, but the granule form is banned in the U.S.

*Human toxicity (long-term)[77]:* Intermediate–40.00 ppb, MCL (Maximum Contaminant Level)

*Fish toxicity (threshold)[77]:* Intermediate–17.59989 ppb, MATC (Maximum Acceptable Toxicant Concentration)

**U.S. Maximum Allowable Residue Levels for Carbofuran (40 CFR 180.254):**

| CROP | ppm |
|---|---|
| Alfalfa, fresh | 10.0 |
| Alfalfa, hay | 40.0 |
| Artichoke | 0.4 |
| Banana | 0.1 |
| Barley, grain | 0.2 |
| Barley, straw | 5.0 |
| Beet, sugar | 0.1 |
| Beet, sugar, tops | 2.0 |
| Cattle, fat | 0.05 |
| Cattle, meat | 0.05 |
| Cattle, mbyp | 0.05 |
| Coffee, bean | 0.1 |
| Corn, fodder | 25.0 |
| Corn, forage | 25.0 |
| Corn, fresh (Inc. sweet)(k+cwhr) | 1.0 |
| Corn, grain (inc. pop) | 0.2 |
| Cotton, undelinted seed | 1.0 |
| Cranberry | 0.5 |
| Cucumber | 0.4 |
| Goat, fat | 0.05 |
| Goat, meat | 0.05 |
| Goat, mbyp | 0.05 |
| Grape | 0.4 |
| Grape, raisin | 2.0 |
| Hog, fat | 0.05 |
| Hog, meat | 0.05 |
| Hog, mbyp | 0.05 |
| Horse, fat | 0.05 |
| Horse, meat | 0.05 |
| Horse, mbyp | 0.05 |
| Melon | 0.4 |
| Milk | 0.1 |
| Oat, grain | 0.2 |
| Oat, straw | 5.0 |
| Pepper | 1.0 |
| Potato | 2.0 |
| Pumpkin | 0.8 |
| Raisin, waste | 6.0 |
| Rice | – |
| Rice, straw- | – |
| Sheep, fat | 0.05 |
| Sheep, meat | 0.05 |
| Sheep, mbyp | 0.05 |
| Sorghum, forage | 3.0 |
| Sorghum, grain, | 0.1 |
| Sorghum, stover | 3.0 |
| Soybean | 1.0 |
| Soybean, forage | 35.0 |
| Soybean, hay | 35.0 |
| Squash | 0.8 |
| Strawberry | 0.5 |
| Sugarcane, cane | 0.1 |
| Sunflower, seed | 1.0 |
| Wheat, grain | 0.2 |
| Wheat, straw | 5.0 |

*Carcinogen/Hazard Classifications*

**Label Signal Word:** DANGER or WARNING, depending upon the formulation

**WHO Acute Hazard:** Class 1b, highly hazardous

*Regulatory Authority:*

- Very Toxic Substance (World Bank)[15]
- Air Pollutant Standard Set (ACGIH)[1] (HSE)[33] (OSHA)[58] (Several States)[60]
- Clean Water Act: Section 311 Hazardous Substances/RQ 40CFR117.3 (same as CERCLA, see below).
- AB 1803-Well Monitoring Chemical (CAL)
- EPA/SARA 302 (EPCRA) Extremely hazardous substances
- MCL (Maximum Contaminants Levels) list of contaminants (CAL)
- Permissible Exposure Limits for Chemical Contaminants (CAL/OSHA)
- The "Director's List" (CAL/OSHA)
- Actively registered pesticide in California.
- RCRA 40CFR268.48; 61FR15654, Universal Treatment Standards: Wastewater (mg/L), 0.006; Nonwastewater (mg/kg), 0.14
- Safe Drinking Water Act: MCL, 0.04 mg/L; MCLG, 0.04 mg/L; Regulated chemical (47 FR 9352)
- Superfund/EPCRA 40CFR355, Appendix B Extremely Hazardous Substances: TPQ = 10/10,000 lb (4.54/4,540 kg).
- Superfund/EPCRA 40CFR302.4 RQ: CERCLA, 10 lb (4.54 kg).
- EPCRA Section 313 Form R *de minimus* concentration reporting level: 1.0%
- U.S. DOT Regulated Marine Pollutant (49CFR172.101, Appendix B)
- Canada: Drinking water quality, 0.09 mg/L MAC

*Description:* Carbofuran is a white, crystalline solid. Odorless. Slightly soluble in water; solubility = about 500-750 ppm @ 25°C; 325 ppm @ 20°C. Molecular weight = 221.29. Melting/Freezing point = 150-152°C. Vapor pressure $6 \times 10^{-6}$ mmHg @ 20°C; $1.8 \times 10^{-5}$ mmHg @ 30°C; $2.7 \times 10^{-3}$ Pa @ 33°C. Log $K_{ow}$ = 2.33. Unlikely to bioaccumulate in marine organisms. Physical and toxicological properties may be affected by the carrier solvents used in commercial formulations.

*Incompatibilities:* Alkaline substances, acid, strong oxidizers such as perchlorates, peroxides, chlorates, nitrates, permanganates.

*Permissible Exposure Limits in Air:* NIOSH[2] and ACGIH[1] recommend an airborne exposure limit of 0.1 mg/m³ TWA. This is also the British HSE, Israeli, Australian, Mexican, and Canadian Provincial (Alberta, British Columbia, Ontario, and Quebec) limit, and Alberta's STEL 0.3 mg/m³. Several States have set guidelines or standards for carbofuran in ambient air[60] ranging from 1.0 μg/m³ (North Dakota) to 1.6 μg/m³ (Virginia) to 2.0 μg/m³ (Connecticut and Nevada).

*Permissible Concentration in Water:* EPA[47] has determined one-day, ten-day and longer-term health advisories for a 10 kg. child of 50 μg/L of carbofuran. The longer term (1 year) value for a 70 kg. adult is 0.18 mg/L or 180 μg/L. A lifetime health advisory for a 70 kg. adult has been determined to be 36 μg/L of carbofuran. Most recently, EPA has proposed a limit of 40 μg/L in drinking water[62]. Further, several states have set guidelines for carbofuran in drinking water[61] ranging from 10 μg/L (Massachusetts) to 15 μg/L (New York) to 36 μg/L (Arizona and Minnesota) to 50 μg/L (Kansas and Wisconsin).

*Determination in Water:* Analysis of carbofuran is by a high performance liquid chromatographic procedure used for the determination of N-methyl carbamoyloximes and N-methylcarbamates in drinking water (U.S. EPA 1984). In this method, the water sample is filtered and a 400 μl aliquot is injected into a reverse phase HPLC column. Separation of compounds is achieved using gradient elution chromatography. After elution from the HPLC column, the compounds are hydrolyzed with sodium hydroxide. The methylamine formed during hydrolysis is reacted with o-phthalaldehyde (OPA) to form a fluorescent derivative which is detected using a fluorescence detector. The method detection limit has been estimated to be approximately 0.9 μg/L for carbofuran.

*Routes of Entry:* Inhalation, ingestion, skin contact.

*Harmful Effects and Symptoms*

*Short Term Exposure:* Carbofuran may affect the nervous system, resulting in convulsions and respiratory failure. Cholinesterase inhibitor. Exposure may result in death. The effects may be delayed and exposed personnel should be kept under medical observation. Symptoms include headache, giddiness, blurred vision, weakness; nausea, cramps, diarrhea, chest discomfort, sweating, contraction of pupils, tearing; salivation, blue lips, lungs and abdomen fill with fluid, convulsions, coma, loss of reflexes and sphincter control. This material is extremely poisonous. The $LD_{50}$ (oral, rat) = 5.3 mg/kg. May be fatal if swallowed, inhaled, or absorbed through skin. Contact may burn skin or eyes. Probable lethal oral dose to humans 5 to 50 mg/kg or 7 drops to 1 teaspoon for 150 lb. person.

*Long Term Exposure:* The major health problem associated with occupational exposure to carbaryl is related to its inhibition of the enzyme cholinesterase in the central, autonomic and peripheral nervous systems. The inhibition of cholinesterase allows acetylcholine to accumulate at these sites and thereby leads to over stimulation of innervated organs. The signs and symptoms observed as a consequence of exposure to carbaryl in the workplace environment are manifestations of excessive cholinergic stimulation, e.g., nausea, vomiting, mild abdominal cramping, dimness of vision, dizziness, headache, difficulty in breathing, and weakness. Carbofuran may affect the immune system.

*Points of Attack:* Central nervous system, peripheral nervous system, blood cholinesterase

*Medical Surveillance:* Before starting work, at regular times after that, and if any symptoms develop, or overexposure occurs, the following is recommended: Serum and red blood cell cholinesterase levels (a special test for the substance in the body that carbofuran affects). For this substance these tests are accurate only if done within about two hours of exposure.

*First Aid:* If this chemical gets into the eyes, remove any contact lenses at once and irrigate immediately for at least 15 minutes, occasionally lifting upper and lower lids. Seek medical attention immediately. If this chemical contacts the skin, remove contaminated clothing and wash immediately with soap and water. Seek medical attention immediately. If this chemical has been inhaled, remove from exposure, begin rescue breathing (using universal precautions) if breathing has stopped and CPR if heart action has stopped. Transfer promptly to a medical facility. When this chemical has been swallowed, get medical attention. Give large quantities of water and induce vomiting. Do not make an unconscious person vomit. Effects may be delayed; keep victim under observation.

*References:*

- EPA, Office of Pesticide Programs, Pesticide Residue Limits, "Carbofuran", 40 CFR 180.254. www.epa.gov/cgi-bin/oppsrch
- EXTOXNET, Extension Toxicology Network, "Pesticide Information Profile, Carbofuran," Oregon State University, Corvallis, OR (June 1996), ace.ace.orst.edu/info/extoxnet/pips/carbofur
- U.S. Environmental Protection Agency, "Chemical Profile: Carbofuran," Washington, DC, Chemical Emergency Preparedness Program (November 30, 1987).

- California Environmental Protection Agency Chemical List of Lists, Sacramento CA (February 1997).
- New Jersey Department of Health and Senior Services and Senior Services, "Hazardous Substance Fact Sheet: Carbofuran," Trenton, NJ (April 1998). http://www.state.nj.us/health/eoh/rtkweb/0341.pdf
- U.S. Environmental Protection Agency, "Preliminary Determination to Cancel Registrations of Carbofuran Products," Federal Register 54, No. 15, 3744-3754 (January 25, 1989).

# Carbon Disulfide

*Use Type:* Fumigant, insecticide, miticide, and as a pesticide intermediate
*CAS Number:* 75-15-0
*Formula:* $CS_2$
*Alert:* Persons whose skin or clothing is contaminated with liquid carbon disulfide can secondarily contaminate rescuers by direct contact or through off-gassing of vapor. Carbon disulfide is readily absorbed through the upper respiratory tract. Carbon disulfide can also be readily absorbed through the digestive tract or skin.
*Synonyms:* Carbon bisulfide; Carbon bisulphide; Carbon disulphide; Carbone (sufure de) (French); Carbonio (solfuro di) (Italian); Carbon sulfide; Dithiocarbonic anhydride; Kohlendisulfid (schwefelkohlenstoff) (German); NCI-C04591; Schwefelkohlenstoff (German); Sulphocarbonic anhydride
*Trade Names:* DE-PESTER FUMIGANT®, Crompton (USA), canceled 11/22/1985; DOWFUME®, Dow Chemical (USA), canceled 11/10/1983; SERFUME®, Dow Chemical (USA), canceled 11/10/1983; WEEVILTOX®
*Producers:* Akzo Nobel Chemicals (Netherlands); ATOFINA Chemicals (France); Chimco (Bulgaria); GFS Chemicals (USA); Industrial Quimica de Mexico (Mexico); Merck (German); Rhodia Eco (France); Shanxi Friends Union Chemicals (China)
*Chemical Class:* Inorganic
*EPA/OPP PC Code:* 016401
*California DPR Chemical Code:* 108
*RTECS Number:* FF6650000
*RCRA Number:* P022
*EEC Number:* 006-003-00-3
*EINECS Number:* 200-843-6
*Uses:* Carbon disulfide is widely used as a pesticide intermediate. It is also used in the manufacture of viscose rayon, ammonium salts, carbon tetrachloride, carbanilide, xanthogenates, flotation agents, soil disinfectants, dyes, electronic vacuum tubes, optical glass, paints, enamels, paint removers, varnishes, varnish removers, tallow, textiles, explosives, rocket fuel, putty, preservatives, and rubber cement; as a solvent for phosphorus, sulfur, selenium, bromine, iodine, alkali cellulose, fats, waxes, lacquers, camphor, resins, and cold vulcanized rubber. It is also used in degreasing, chemical analysis, electroplating, grain fumigation, oil extraction, and dry-cleaning.
*Human toxicity (long-term)[77]:* Very low–700.00 ppb, Health Advisory
*Fish toxicity (threshold)[77]:* Very low–27756.65644 ppb, MATC (Maximum Acceptable Toxicant Concentration)
*U.S. Maximum Allowable Residue Levels for Carbon Disulfide (40 CFR 180.467):*

| CROP | ppm |
|---|---|
| Almond | 0.1 |
| Almond, hulls | 0.1 |
| Grape | 0.1 |
| Grapefruit | 0.1 |
| Lemon | 0.1 |
| Orange | 0.1 |
| Peach | 0.1 |
| Plum, prune, fresh | 0.1 |

*Carcinogen/Hazard Classifications*
*California Prop. 65:* Listed; Developmental, female
*Regulatory Authority:*
- Banned or Severely Restricted (In Agriculture) (Several Countries) (UN)[13]
- Toxic Substance (World Bank)[15]
- Air Pollutant Standard Set (ACGIH)[1] (DFG)[3] (HSE)[33] (OSHA)[58] (Several States)[60] (Australia) (Israel) (Mexico) (Several Canadian Provinces)
- AB 1803-Well Monitoring Chemical (CAL)
- AB 2588-Air Toxics "Hot Spots" Chemicals (CAL)
- CAL Air Resources Board/AB 1807 Toxic Air Contaminants
- Proposition 65 chemical (CAL)
- Permissible Exposure Limits for Chemical Contaminants (CAL/OSHA)
- The "Director's List" (CAL/OSHA)
- Clean Air Act: Hazardous Air Pollutants (Title I, Part A, Section 112); Accidental Release Prevention/Flammable substances, (Section 112[r], Table 3), TQ = 20,000 lb (9080kg).
- Clean Water Act: Section 311 Hazardous Substances/RQ 40CFR117.3 (same as CERCLA, see below); Section 313 Water Priority Chemicals (57FR41331, 9/9/92)
- EPA Hazardous Waste Number (RCRA No.): P022
- RCRA, 40CFR261, Appendix 8 Hazardous Constituents.
- RCRA 40CFR268.48; 61FR15654, Universal Treatment Standards: Wastewater (mg/L), 3.8; Non-wastewater (mg/L), 4.8 TCLP
- RCRA 40CFR264, Appendix 9; TSD Facilities Ground Water Monitoring List. Suggested test method(s) (PQL ug/L): 8240(5)
- Superfund/EPCRA 40CFR355, Appendix B Extremely Hazardous Substances: TPQ = 10,000 lb (4,540 kg).
- Superfund/EPCRA 40CFR302.4 RQ: CERCLA, 100 lb (45.4 kg).

- EPCRA Section 313 Form R *de minimus* concentration reporting level: 1.0%.
- U.S. DOT Regulated Marine Pollutant (49CFR172.101, Appendix B) as carbon bisulphide.
- Canada, WHMIS, Ingredients Disclosure List; National Pollutant Release Inventory (NPRI)

*Description:* Gaseous carbon disulfide is more than twice as heavy as air. Pure carbon disulfide liquid is colorless with a pleasant odor. Most industrially-used carbon disulfide liquid is yellowish in color and has an unpleasant foul smelling odor, characteristic of hydrogen sulfide (a contaminant of technical grade carbon disulfide). *Warning properties:* Sweet odor of pure carbon disulfide and foul odor of commercial and technical grade carbon disulfide are usually adequate to warn of acute exposure. Most people can detect carbon disulfide at levels of 0.02 to 0.1 ppm (1 ppm is equivalent to 3.1 mg/m$^3$). However, the sense of smell fatigues rapidly and, therefore, odor does not serve as a good warning property. Slightly soluble in water; solubility = 0.23 g/100 mL @ 22 °C. Molecular weight = 76.14 daltons. Boiling point = 46.3 °C. Melting/Freezing point = –111.6 °C. Vapor density (air = 1.00) = 2.67 g/mL. Vapor pressure = 352.6 mmHg @ 25 °C. At room temperature, carbon disulfide is a very flammable liquid that readily evaporates when exposed to air. It is handled and transported as a very flammable and explosive liquid. Flash point = –30°C. Autoignition temperature = 90°C. The explosive limits are LEL = 1.3%; UEL = 50%. Hazard Identification (based on NFPA-704 M Rating System): Health 3, Flammability 4, Reactivity 0. Log $K_{ow}$ = 1.85. Unlikely to bioaccumulate in marine organisms.

*Incompatibilities:* Carbon disulfide is incompatible with air, alkali metals, aluminum, azides, many oxidants, and phenyl copper-triphenylphosphine complexes. Such incompatible mixtures may result in violent, and possibly explosive, reactions. Chemically active metals (such as sodium, potassium, zinc), organic amines, halogens. May explosively decompose on shock, friction, or concussion. May explode on heating. The substance may spontaneously ignite on contact with air and on contact with hot surfaces producing toxic fumes of sulphur dioxide. Reacts violently with oxidants to produce oxides of sulfur and carbon monoxide and causing fire and explosion hazard. Attacks some forms of plastic, rubber, and coatings.

*Permissible Exposure Limits in Air:* The OSHA[2] legal PEL is 20 ppm TWA for an 8-hour work shift and 30 ppm as an acceptable ceiling, and 100 ppm as a maximum peak above the acceptance ceiling concentration not to be exceeded during any 30-minute work period. NIOSH[2] recommended exposure limit is 1 ppm TWA for a 10-hour work shift, and 10 ppm not to be exceeded during any 15-minute work period. The ACGIH recommended TWA value is 10 ppm (30 mg/m$^3$) for an 8-hour work shift[1]. They add the notation "skin" indicating the possibility of cutaneous absorption. AIHA ERPG-2 (maximum airborne concentration below which it is believed that nearly all persons could be exposed for up to 1 hour without experiencing or developing irreversible or other serious health effects or symptoms that could impair their abilities to take protective action) = 50 ppm. The NIOSH[2] IDLH level = 500 ppm. The odor threshold is approximately 200 to 1000 times lower than the OSHA PEL-TWA (20 ppm). Australia, DFG, HSE, Israel, and Mexico airborne limit is 10 ppm. The former USSR-UNEP/IRPTC project[43] has set a MAC of 1.0 mg/m$^3$ in workplace air, of 0.03 mg/m$^3$ in ambient residential air on a momentary basis, and 0.003 mg/m$^3$ in residential ambient air on a daily average basis. Further, several states have set guidelines or standards for carbon disulfide in ambient air[60] ranging from 60 $\mu$g/m$^3$ (Connecticut) to 100 $\mu$g/m$^3$ (New York) to 150 $\mu$g/m$^3$ (South Carolina) to 186 $\mu$g/m$^3$ (North Carolina) to 300 $\mu$g/m$^3$ (Florida and North Dakota) to 714 $\mu$g/m$^3$ (Nevada). The WHO[35] has recommended a TWA of 10 mg/m$^3$ for male workers but a TWA of 3 mg/m$^3$ for women of fertile age.

*Determination in Air:* Adsorption on charcoal, workup with benzene, gas chromatographic analysis per NIOSH Method #1600[18].

*Permissible Concentration in Water:* In view of the relative paucity of data on the mutagenicity, carcinogenicity, and long-term oral toxicity of carbon disulfide, it was stated that estimates of the effects of chronic oral exposure at low levels cannot be made with any confidence. It was recommended by NAS/NRC that studies to produce such information be conducted before limits in drinking water are established. Now, however, EPA[32] has suggested a permissible ambient goal of 830 $\mu$g/L. The former USSR-UNEP/IRPTC project[43] has suggested that limits in drinking water be set on an organoleptic basis and that a MAC of 1 mg/L be set in water bodies used for fishery purposes. Arizona has set a guideline for $CS_2$ in drinking water of 830 $\mu$g/L[61].

*Routes of Entry:* Carbon disulfide is readily absorbed through the upper respiratory tract. Carbon disulfide can also be readily absorbed through the digestive tract or skin.

*Harmful Effects and Symptoms*

*Short Term Exposure:* Carbon disulfide is irritating to the eyes, mucous membranes, and skin. Acute neurological effects may result from all routes of exposure and may include headache, confusion, psychosis, and coma. Acute exposure to extremely high levels of carbon disulfide may result in death. The neurotoxic effects caused by carbon disulfide may be due, in part, to its metabolic conversion to dithiocarbamates. Individuals especially susceptible to the toxic effects of carbon disulfide include those with existing disorders of the nervous system, respiratory system, cardiovascular system, or eyes. Swallowing the liquid may cause aspiration into the lungs with the risk of chemical pneumonitis. This chemical may affect the central nervous system. In acute poisoning, early excitation of the central nervous system occurs, followed by depression with stupor, restlessness, and unconsciousness. If recovery occurs, the patient usually passes through the after-stage of narcosis, with nausea, vomiting, headache, etc. Also possible are motor disturbances of the bowel, anemia, disturbances of cardiac rhythm, loss of weight, polyuria and menstrual disorders. Severe chronic poisoning may also result in liver degeneration and jaundice. Exposure can cause a loss of consciousness. Exposure far above the PEL may result in death. The probable oral lethal dose for a human is between 0.5 and 5 g/kg or between 1 ounce and 1 pint (or 1 pound) for a 70 kg (150 lb) person. In chronic exposures, the central nervous system is damaged and results in the disturbance of vision and sensory changes as the most common early symptoms. Lowest lethal dose for humans has been reported at 14 mg/kg or 0.98 grams for a 70 kg person. Alcoholics and those suffering from neuropsychic trouble are at special risk.

*Long Term Exposure:* Chronic exposure to carbon disulfide can result in neurological effects similar to those experienced during acute exposure, but at much lower exposure levels. In addition, chronic exposure may cause effects such as permanent central and peripheral nervous system damage, atherosclerotic tendencies, ECG abnormalities, gastrointestinal disturbances, fatty degeneration of the liver, kidney damage, sexual dysfunction, hearing loss, visual disturbances, retinal microaneurism, and blood dyscrasia. Chronic exposure may be more serious for children because of their potential for a longer latency period. *Carcinogenicity:* A carcinogenicity classification for carbon disulfide has not been established by the Department of Health and Human Services, the International Association for Research on Cancer, or the U.S. EPA. *Reproductive and Developmental Effects:* Carbon disulfide is included in the list of *"Reproductive and Developmental Toxicants,"* a 1991 report published by the U.S. General Accounting Office. Carbon disulfide-induced reproductive effects include alterations of the menstrual cycle in women and altered libido and abnormalities in spermatogenesis in men. There is no conclusive evidence that carbon disulfide is a genotoxin in humans.

*Points of Attack:* The material affects the central nervous system, cardiovascular system, eyes, kidneys, liver, and skin.

*Medical Surveillance:* Preplacement and periodic medical examinations should be concerned especially with skin, eyes, central and peripheral nervous system, cardiovascular disease, as well as liver and kidney function. Routine laboratory studies include chest radiography, electrocardiogram, blood chemistry, and arterial blood gases. $CS_2$, can be determined in expired air, blood, and urine. The iodine-azide test detects carbon disulfide metabolites in the urine, and it may indicate other sulfur compounds. Examination of the nervous system, including mental status. If a work-related incident has occurred, you might be legally required to file a report; contact your state or local health department. Cigarette smoking can exacerbate pulmonary injury and should be discouraged for 72 hours after exposure.

*First Aid:* Persons exposed only to carbon disulfide vapor pose no risk of secondary contamination. Persons whose skin or clothing is contaminated with liquid carbon disulfide can secondarily contaminate rescuers by direct contact or through off-gassing of vapor. There is no antidote for carbon disulfide. Treatment consists of removal of the victim from the contaminated area, support of respiratory and cardiovascular functions, and irrigation of contaminated eyes or skin. If this chemical gets into the eyes, remove any contact lenses at once and irrigate immediately for at least 15 minutes, occasionally lifting upper and lower lids. Seek medical attention immediately. If this chemical contacts the skin, remove contaminated clothing and wash immediately with soap and water. Seek medical attention immediately. If this chemical has been inhaled, remove from exposure, begin rescue breathing (using universal precautions) if breathing has stopped and CPR if heart action has stopped. Transfer promptly to a medical facility. In cases of ingestion, **do not induce vomiting**. Rinse the mouth and administer water for dilution if the patient can swallow, has a strong gag reflex, and does not drool. If the victim is not symptomatic, administer activated charcoal at a dose of 1 g/kg (infant, child, and adult dose). A soda can and straw may be of assistance when offering charcoal to a child. occurs Do not make an unconscious person vomit.

*References:*
- U.S. Environmental Protection Agency, Office of Pesticide Programs, Pesticide Residue Limits, "Carbon Disulfide," 40 CFR 180.467. http://www.epa.gov/pesticides/food/viewtols.htm
- U.S. Department of Health and Human Services, Agency for Toxic Substances and Disease Registry, "ToxFAQs for Carbon Disulfide," Atlanta, GA (September, 1997). http://www.atsdr.cdc.gov/tfacts82.html
- National Institute for Occupational Safety and Health, Criteria for a Recommended Standard: Occupational Exposure to Carbon Disulfide, NIOSH Doc. No. 77-156 (1977).
- ATSDR, U.S. Department of Health and Human Services, Public Health Service, "Managing Hazardous Materials Incidents," Atlanta, GA (June 2003)
- World Health Organization, Carbon Disulfide, Environmental Health Criteria No. 10, Geneva (1979).
- U.S. Environmental Protection Agency, Carbon Disulfide, Health and Environmental Effects Profile N0. 32, Office of Solid Waste, Washington, DC (April 30, 1980).
- Sax, N.I., Ed., "Dangerous Properties of Industrial Materials Report," 1, N0. 2, 28-30 (1980); and 3, No. 5, 84-87, New York, Van Nostrand Reinhold Co. (1983).

- U.S. Environmental Protection Agency, "Chemical Profile: Carbon Disulfide," Washington, DC, Chemical Emergency Preparedness Program (October 31, 1985).
- California Environmental Protection Agency Chemical List of Lists, Sacramento CA (February 1997).
- New Jersey Department of Health and Senior Services and Senior Services, "Hazardous Substance Fact Sheet: Carbon Disulfide," Trenton, NJ (November 1986, rev. August 2001).
- http://www.state.nj.us/health/eoh/rtkweb/0344.pdf
- New York State Department of Health, "Chemical Fact Sheet: Carbon Disulfide," Albany, NY, Bureau of Toxic Substance Assessment (May 1986).

# Carbon Tetrachloride

*Use Type:* Fumigant and solvent
*CAS Number:* 56-23-5
*Formula:* $CCl_4$
*Synonyms:* Benzinoform; Carbona; Carbon chloride; Carbon tet; Czterochlorek wegla (Polish); ENT 4705; Fasciolin; Flukoids; Freon 10; Halon 104; Katharin; Methane Tetrachloride; Methane, tetrachloro-; Necatorina; Necatorine; Perchloromethane; R 10; Tetrachloorkoolstof (Dutch); Tetrachloormetan; Tetrachlorkohlenstoff, tetra (German); Tetrachlormethan (German); Tetrachlorocarbon; Tetrachloromethane; Tetrachlorure de carbone (French); Tetraclorometano (Italian); Tetracloruro di carbonio (Italian); Tetracloruro de carbono (Spanish); UN 1846
*Trade Names:* ACRITET ® (mixture of Acrylonitrile and Carbon tetrachloride); DOWFUME®, Dow Chemical (USA), canceled 11/22/1985; F.C.® FORMULAS, Vulcan Chemicals (USA), canceled 11/10/1983; F.i.a. GRAIN FUMIGANT®, Zeneca (UK) (now Syngenta), canceled 11/22/1985; TETRAFINOL®; TETRAFORM®; TETRASOL®; TWAWPIT®; UNIVERM; Ventox (mixture of Acrylonitrile and Carbon tetrachloride); VERMOESTRICID®
*Producers:* Akzo Nobel Chemicals (Netherlands); Asahi Glass (Japan); Alquimia Mexicana (Mexico); ATOFINA (France); Degussa (Germany); Dow Chemical (USA); EniChem (Italy); GFS Chemicals (USA); ICC Chemical Industries (USA); Juhua Group Corp. (China); Merke (Germany); OxyChem (USA); Richman Chemical (USA); Samsung Fine Chemicals (South Korea); Scott Specialty Gases (USA); Shanghai Tian Yuan (Group) Corp. (China); Shin-Etsu Chemical (Japan); Solvay Group (Belgium); Vulcan Chemicals (USA)
*Chemical Class:* Halogenated organic compound
*EPA/OPP PC Code:* 016501; (Use two codes: 000601 & 016501 for mixture of Acrylonitrile and Carbon tetrachloride)
*California DPR Chemical Code:* 109
*RTECS Number:* FG4900000
*ICSC Number:* 0024
*EEC Number:* 602-008-00-5
*EINECS Number:* 200-262-8
*Uses:* Wide use as a grain fumigant and wood preserver. Used as a solvent and carrier for pesticides. Carbon tetrachloride is used as a solvent for oils, fats, lacquers, varnishes, rubber, waxes, and resins. Fluorocarbons are chemically synthesized from it. It is also used as an azeotropic drying agent for spark plugs, a dry-cleaning agent, a fire extinguishing agent, a fumigant, and an anthelmintic agent. The use of this solvent is widespread, and substitution of less toxic solvents when technically possible is recommended.
*Carcinogen/Hazard Classifications*
**U.S. EPA Carcinogens:** Group 2B, probable carcinogen
**U.S. NTP Carcinogen:** Reasonably anticipated carcinogen
**California Prop. 65:** Carcinogen
**IARC:** Group 2B, possible carcinogen
*Regulatory Authority:*
- Carcinogen (Suspected Human) (ACGIH)[1] (Animal Positive)
- Banned or Severely Restricted (Several Countries) (UN)[13]
- Air Pollutant Standard Set (ACGIH)[1] (DFG)[3] (HSE)[33] (UNEP)[43] (OSHA)[58] (Several States)[60] (Australia) (Israel) (Mexico) (Several Canadian Provinces)
- List of priority pollutants (U.S. EPA)
- AB 1803-Well Monitoring Chemical (CAL)
- MCL (Maximum Contaminants Levels) list of contaminants (CAL)
- AB 2588-Air Toxics "Hot Spots" Chemicals (CAL)
- CAL Air Resources Board/AB 1807 Toxic Air Contaminants
- Proposition 65 chemical (CAL)
- Permissible Exposure Limits for Chemical Contaminants (CAL/OSHA)
- The "Director's List" (CAL/OSHA)
- Clean Air Act: Hazardous Air Pollutants (Title I, Part A, Section 112); Stratospheric ozone protection (Title VI, Subpart A, Appendix A), Class I, Ozone Depletion Potential = 1.1
- Clean Water Act: Section 311 Hazardous Substances/RQ 40CFR117.3 (same as CERCLA, see below); 40CFR423, Appendix A, Priority Pollutants; Section 313 Water Priority Chemicals (57FR41331, 9/9/92); Toxic Pollutant (Section 401.15)
- EPA Hazardous Waste Number (RCRA No.): U211, D019
- RCRA Toxicity Characteristic (Section 261.24), Maximum
- Concentration of Contaminants, regulatory level, 0.5 mg/L
- RCRA, 40CFR261, Appendix 8 Hazardous Constituents.

- RCRA 40CFR268.48; 61FR15654, Universal Treatment Standards: Wastewater (mg/L), 0.057; Nonwastewater (mg/kg), 6.0
- RCRA Maximum Concentration Limit for Ground Water Protection (Section 264.94): 8010[1]; 8240(5)
- Safe Drinking Water Act: MCL, 0.005 mg/L; MCLG, zero; Regulated chemical (47 FR 9352)
- Superfund/EPCRA 40CFR302.4 RQ: CERCLA, 10 lb (4.54 kg).
- EPCRA Section 313 Form R *de minimus* concentration reporting level: 0.1%.
- Proposition 65 chemical (CAL)
- U.S. DOT Regulated Marine Pollutant (49CFR172.101, Appendix B)
- Canada, WHMIS, Ingredients Disclosure List; National Pollutant Release Inventory (NPRI); CEPA Toxic Substance List
- Mexico, Drinking water criteria, 0.004 mg/L

*Description:* Colorless, nonflammable liquid with a characteristic ether-like odor. The odor threshold is 0.52 mg/L in water and 140 to 548 ppm in air. Boiling point = 76.5°C. Melting/Freezing point = –23°C. Very slightly soluble in water. Log $K_{ow}$ = 2.6. Unlikely to bioaccumulate in marine organisms.

*Incompatibilities:* Oxidative decomposition on contact with hot surfaces, flames or welding arcs. Decomposition forms toxic phosgene fumes and hydrogen chloride. Violent decomposition, producing heat, on contact with chemically active metals such as aluminium, barium, magnesium, potassium, sodium; with fluorine gas, allyl alcohol, and other substances, causing fire and explosion hazard. Attacks copper, lead and zinc. Attacks some coatings, plastics and rubber. Becomes corrosive when in contact with water; corrosive to metals in the presence of moisture.

*Permissible Exposure Limits in Air:* The OSHA[2] PEL is 10 ppm TWA for an 8-hour workshift; 25 ppm not to exceeded during any 15-minute workperiod, and 200 ppm as an acceptable maximum peak level for 5-minutes in any 4-hours. NIOSH[2] recommends an airborne exposure limit of 2 ppm which should not be exceeded during any 1-hour period. ACGIH recommends a limit of 5 ppm TWA for an 8-hour workshift and STEL of 10ppm[1]. These limits contain notations that $CCL_4$ is a substance suspect of carcinogenic potential for man and with the further notation "skin" indicating the possibility of cutaneous absorption. The NIOSH[2] IDLH level = 200 ppm with the notation "occupational carcinogen." HSE[33] has set a TWA of 10 pm (65 mg/m$^3$) and a STEL of 20 ppm (130 mg/m$^3$). DFG[3] has set a TWA of 10 ppm (65 mg/m$^3$). Australia's, Israel's and Mexico's TWA is 5 ppm (30 mg/m$^3$) and Mexico's STEL is 20 ppm (120 mg/m$^3$). The former-USSR/UNEP joint project[43] has set a MAC in workplace air of 20 mg/m$^3$ and a MAC in ambient air of residential areas of 4 mg/m$^3$ on a momentary basis and 2 mg/m$^3$ on an average daily basis. In addition, several states have set guidelines or standards for $CCl_4$ in ambient air[60] ranging from zero (North Dakota) to 0.03 µg/m$^3$ (Rhode Island) to 0.0667 µg/m$^3$ (Indiana) to 0.67 µg/m$^3$ (Massachusetts and North Carolina) to 72 µg/m$^3$ (Pennsylvania) to 100 µg/m$^3$ (New York) to 150 µg/m$^3$ (South Carolina) to 300 µg/m$^3$ (Connecticut and Virginia) to 714 µg/m$^3$ (Nevada).

*Determination in Air:* Charcoal adsorption followed by workup with $CS_2$ and analysis by gas chromatography; see NIOSH Method #1003 for Hydrocarbons, Chlorinated[18].

*Permissible Concentration in Water:* To protect freshwater aquatic life: 35,2000 µg/L on an acute toxicity basis. To protect saltwater aquatic life: 50,000 µg/L on an acute toxicity basis. To protect human health: preferably zero. An additional lifetime cancer risk of 1 in 100,000 is presented by a concentration of 4.0 µg/L. Mexico's drinking water criteria is 0.004 mg/L The former USSR/UNEP joint project[43] has set a MAC in water bodies used for domestic purposes of 0.3 mg/L. The USEPA[48] has set a lifetime health advisory of 0.0007 mg/kg/day and a drinking water equivalent level of 25 µg/L.

In addition, several states have set standards and guidelines for $CCl_4$ in drinking water[61] ran ranging from standards of 2 µg/L (New Jersey) to 3 µg/L (Florida) to 10 µg/L (New Mexico) and guidelines of 2.7 µg/L (Minnesota) to 5 µg/L (California, Kansas and Maine, Colorado).

*Determination in Water:* Gas chromatography (EPA Method 601) or gas chromatography plus mass spectrometry (EPA Method 624).

*Routes of Entry:* Inhalation of vapor, percutaneous absorption, ingestion and skin and eye contact.

*Harmful Effects and Symptoms*

*Short Term Exposure:* Carbon tetrachloride irritates the eyes, causing redness. *Inhalation:* Levels of 20 ppm may cause dizziness, headache, vomiting, visual disturbances, extreme fatigue and nose and throat irritation. Other symptoms may include restlessness, loss of balance, twitching and tremors. Severe exposure can lead to liver, kidney, eye and nerve damage that may be delayed after exposure; can cause breathing stoppage, coma and death. 1000ppm for an unspecified time has caused death. *Skin:* May cause irritation and redness; carbon tetrachloride is readily absorbed through the skin. Symptoms as listed above may occur through skin absorption even when vapor concentrations are below OSHA standards. *Ingestion:* May cause severe abdominal pain with diarrhea, followed by symptoms described under inhalation. Death may occur by ingestion of as little as ½ teaspoon. Between 45 and 100 ppm, carbon tetrachloride may cause headache, drowsiness, fatigue, nausea and vomiting. 100 to 300 ppm may cause additional effects of mental confusion, weight loss and sluggishness. Liver, kidney, eye and nerve damage can result from more severe exposures. Coma and death may occur.

*Long Term Exposure:* Repeated or prolonged skin contact may cause dermatitis. Carbon tetrachloride is a possible human carcinogen. Cancer site in animals: liver cancer.

*Points of Attack:* Central nervous system, eyes, lungs, liver, kidneys and skin.

*Medical Surveillance:* Preplacement and periodic examinations should include an evaluation of alcohol intake and appropriate tests for liver and kidney functions. Special attention should be given to the central and peripheral nervous system, the skin, and blood. Expired air and blood levels may be useful as indicators of exposure.

*First Aid:* If this chemical gets into the eyes, remove any contact lenses at once and irrigate immediately for at least 15 minutes, occasionally lifting upper and lower lids. Seek medical attention immediately. If this chemical contacts the skin, remove contaminated clothing and wash immediately with soap and water. Seek medical attention immediately. If this chemical has been inhaled, remove from exposure, begin rescue breathing (using universal precautions) if breathing has stopped and CPR if heart action has stopped. Transfer promptly to a medical facility. When this chemical has been swallowed, get medical attention. Give large quantities of water and induce vomiting. Do not make an unconscious person vomit.

*References:*

- National Institute for Occupational Safety and Health, Criteria for a Recommended Standard: Occupational Exposure to Carbon Tetrachloride, NIOSH Doc. No. 76-133 (1976).
- U.S. Environmental Protection Agency, Carbon Tetrachloride: Ambient Water Quality Criteria, Washington, DC
- (1980).
- National Academy of Sciences, Chloroform, Carbon Tetrachloride and Other Halomethanes: An Environmental Assessment, Washington, DC (1978).
- Sax, N.I., Ed., "Dangerous Properties of Industrial Materials Report," 1, No. 2, 30-32 (1980) and 3, No. 5, 88-94 (1983).
- Agency for Toxic Substances and Disease Registry, "ToxFAQs for Carbon Tetrachloride," Atlanta, GA (September 2003). http://www.atsdr.cdc.gov/tfacts30.html
- New Jersey Department of Health and Senior Services and Senior Services, "Hazardous Substance Fact Sheet: Carbon Tetrachloride," Trenton, NJ (August 1998). www.state.nj.us/health/eoh/rtkweb/0347.pdf
- New York State Department of Health, "Chemical Fact Sheet: Carbon Tetrachloride," Albany, NY, Bureau of Toxic Substance Assessment (January 1986, Version 2).
- California Environmental Protection Agency Chemical List of Lists, Sacramento CA (February 1997).

# Carbophenothion (ANSI)

*Use Type:* Insecticide and acaricide
*CAS Number:* 786-19-6
*Formula:* $C_{11}H_{16}ClO_2PS_3$
*Alert:* A Restricted Use Pesticide (RUP). Human toxicity (long-term): Very high.

*Synonyms:* Akarithion; Carbofenothion (Dutch); $S$-[($p$-Chlorophenylthio)methyl]-$O,O$-diethyl Phosphorodithioate; $S$-(4-chlorophenylthiomethyl)diethyl Phosphorothiolothionate; $O,O$-Diethy-$S$-$p$-chlorophenylthiomethyl dithiophosphate; $O,O$-Diaethy-$S$-[(4-chlor-phenyl-thio)-methyl]dithiophosphat (German); $O,O$-Diethy-$S$-[(4-chloor-fenyl-thio)-methyl]dithiofosfaat (Dutch); $O,O$-diethy-$S$-($p$-chlorophenylthiomethyl)phosphorodithioate; $O,O$-Diethyl-$p$-chlorophenylmercaptomethyl dithiophosphate; $O,O$-Diethyl-4-chlorophenylmercaptomethyl dithiophosphate; $O,O$-Diethy-dithiophosphoric acid, $p$-chlorophenylthiomethyl ester; $O,O$-Dietil-$S$-[(4-clorofenil-tio)-metile]-ditiofosfato (Italian); $O,O$-Dietil-$S$-[($p$-clorofenil-tio)-metile]-ditiofosfato (Italian); Dithiophosphate de $O,O$-Diethyle et de (4-chlorophenyl) thiomethyle (French); ENT 23,708; Oleoakarithion

*Trade Names:* ACARITHION®; DAGADIP®; ENDYL®; GARRATHION®; HEXATHION®; LETHOX®; NEPHOCARP®; R-1303®, Zeneca (USA) (now Syngenta) canceled 7/1/1987; STAUFFER R-1,303®, Zeneca (USA) (now Syngenta) canceled 7/1/1987; TRITHION® MITICIDE

*Producers:* Calliope (France); ICI Americas (USA); Sigma-Aldrich Laborchemikalien (Germany)
*Chemical Class:* Organophosphate
*EPA/OPP PC Code:* 058102
*California DPR Chemical Code:* 110
*ICSC Number:* 0410
*RTECS Number:* ID5250000
*EEC Number:* 015-044-00-6
*EINECS Number:* 212-324-1

*Uses:* Primarily used to control mites, suckers, and aphids on citrus crops, deciduous fruits and nuts, vegetables, sorghum and maize. Also used to enhance the toxicity of malathion and other similar compounds.

*Human toxicity (long-term)*[77]: Extra high–0.91 ppb, Health Advisory

*Fish toxicity (threshold)*[77]: High–1.06008 ppb, MATC (Maximum Acceptable Toxicant Concentration)

*Carcinogen/Hazard Classifications*
**Label Signal Word:** DANGER
*Regulatory Authority:*

- Banned or Severely Restricted Use (India) (UN)[13]
- Very Toxic Substance (World Bank)[15]
- AB 1803-Well Monitoring Chemical (CAL)
- EPA/SARA 302 (EPCRA) Extremely hazardous substances
- CAL-DHS/DHS Drinking Water Action Levels

- The "Director's List" (CAL/OSHA)
- U.S. DOT Inhalation Hazard Chemicals as organophosphates
- Superfund/EPCRA 40CFR355, Appendix B Extremely Hazardous Substances: TPQ = 500 lb (227 kg).
- Superfund/EPCRA 40CFR302.4 RQ: EHS, 1 lb (0.454 kg).
- U.S. DOT Regulated Marine Pollutant (49CFR172.101, Appendix B), severe pollutant

*Description:* Carbophenothion is a colorless to light amber liquid. Mercaptan odor. Practically insoluble in water; solubility = < 35 ppm. Boiling point = 82°C @ 0.01 mmHg. Vapor pressure = $8.3 \times 10^{-6}$ mmHg. Log $K_{ow}$ = 5.1. Values at or above 3.0 are likely to bioaccumulate in marine organisms.

*Incompatibilities:* The substance decomposes on heating or on burning producing toxic fumes including phosphorus oxides, sulfur oxides, hydrogen chloride.

*Permissible Exposure Limits in Air:* No criteria set, but runoff from spills or fire control may cause water pollution.

*Determination in Air:* OSHA versatile sampler-2; Toluene/Acetone; Gas chromatography/Flame ionization detection; NIOSH IV, Method #5600, Organophosphorus pesticides[18].

*Permissible Concentration in Water:* No criteria set, but runoff from spills or fire control may cause water pollution.

*Routes of Entry:* Inhalation, ingestion and skin contact.

*Harmful Effects and Symptoms*

*Short Term Exposure:* The substance may affect the nervous system, resulting in convulsions and respiratory failure. Cholinesterase inhibitor. Exposure may result in death. Produces headaches, nausea, weakness and dizziness. Symptoms may include nausea, vomiting, abdominal cramps, diarrhea, excessive salivation, headache, giddiness, weakness, muscle twitching, difficult breathing, blurring or dimness of vision, and loss of muscle coordination. Death may occur from failure of the respiratory center, paralysis of the respiratory muscles, intense bronchoconstriction, or all three. This material is highly toxic; the estimated fatal oral dose is 0.6 g for a 150 lb. (70 kg) person. $LD_{50}$ (oral, rat) = 6.8 mg/kg. Delayed pulmonary edema may occur after inhalation.

*Long Term Exposure:* Cholinesterase inhibitor; cumulative effect is possible: see acute hazards/symptoms. The state of Massachusetts lists this chemical as a neurotoxin.

*Points of Attack:* Respiratory system, lungs, central nervous system, cardiovascular system, skin, eyes, plasma and red blood cell cholinesterase.

*Medical Surveillance:* Medical observation is recommended for 24 to 48 hours after breathing overexposure, as pulmonary edema may be delayed. Before employment and at regular times after that, the following are recommended: Plasma and red blood cell cholinesterase levels (tests for the enzyme poisoned by this chemical). If exposure stops, plasma levels return to normal in 1-2 weeks while red blood cell levels may be reduced for 1-3 months. When acetylcholinesterase enzyme levels are reduced by 25% or more below preemployment levels, risk of poisoning is increased, even if results are in lower ranges of "normal." Reassignment to work not involving organophosphate or carbamate pesticides is recommended until enzyme levels recover. If symptoms develop or overexposure occurs, repeat the above tests as soon as possible and get an exam of the nervous system. Also consider complete blood count. Consider chest x-ray following acute overexposure. Do not drink any alcoholic beverages before or during use. Alcohol promotes absorption of organophosphates.

*First Aid:* **Treatment for organophosphate poisoning consists of thorough decontamination, cardiorespiratory support, and administration of the antidotes atropine and pralidoxime. In cases of severe poisoning, diazepam, an anticonvulsant, should also be administered. Antidotes should be administered as prevention even if the diagnosis is in doubt.** Speed in removing material from eyes and skin is of extreme importance. *Eyes:* Eye contact can cause dangerous amounts of these chemicals to be quickly absorbed through the mucous membrane into the bloodstream. Immediately and gently flush eyes with plenty of warm or cold water (NO hot water) for at least 15 minutes, occasionally lifting the upper and lower eyelids. Get medical aid immediately. *Skin:* Get medical aid. Skin contact can cause dangerous amounts of these chemicals to be absorbed into the bloodstream. Wearing the appropriate PPE equipment and respirator for organophosphate/carbamate pesticides, immediately flush skin with plenty of soap and water for at least 15 minutes while removing contaminated clothing and shoes. Shampoo hair promptly if contaminated. The removed, contaminated clothing and shoes should be double-bagged and left in Hot Zone for later disposal by hazardous materials experts. Skin may also be decontaminated with diluted hypochlorite solution. *Inhalation:* Get medical aid. Do not contaminate yourself. Wearing the appropriate PPE equipment and respirator for organophosphate pesticides, immediately remove the victim from the contaminated area to fresh air. If the victim is not breathing, administer artificial respiration. Do NOT use mouth-to-mouth resuscitation; use bag/mask apparatus. If breathing is difficult, administer oxygen through bag/mask apparatus until medical help arrives. Do not leave victim unattended. *Ingestion:* Call poison control. Loosen all clothing. Never give anything by mouth to an unconscious person. If victim is *unconscious or having convulsions,* do nothing except keep victim warm. Get medical aid. Transfer promptly to a medical facility. In cases of ingestion, **do not induce vomiting**. If the victim is alert and asymptomatic, administer a slurry of activated charcoal at a dose of 1 g/kg (infant, child, and adult dose). A soda can and straw may be of assistance when offering charcoal to a child. *In some cases you may be specifically* *instructed by poison control to induce vomiting by way of 2 tablespoons of syrup of ipecac (adult) washed down with a cup of water.* Do NOT give activated charcoal before or with ipecac syrup.

*Note to physician or authorized medical personnel:* Treat cases of respiratory compromise, coma, or excessive pulmonary secretions with respiratory support using protocols and techniques available and within the scope of training. Some cases may necessitate procedures such as endotracheal intubation or cricothyrotomy by properly trained and equipped personnel. When possible, atropine (see *Antidotes*, below) should be given under medical supervision. Patients who are comatose, hypotensive, or having seizures or cardiac arrhythmias should be treated according to advanced life support protocols. *Antidotes:* Two antidotes are administered to treat organophosphate poisoning. Atropine is a competitive antagonist of acetylcholine at muscarinic receptors and is used to control the excessive bronchial secretions which are often responsible for death. Pralidoxime relieves both the nicotinic and muscarine effects of organophosphate poisoning by regenerating acetylcholinesterase and can reduce both the bronchial secretions and the muscle weakness associated with poisoning. The initial intravenous dose of atropine in adults should be determined by the severity of symptoms: An initial adult dose of 1.0 to 2.0 mg or pediatric dose of 0.01 mg/kg (minimum 0.01 mg) should be administered intravenously. If intravenous access cannot be established, atropine may also be given intramuscularly, subcutaneously or via endotracheal tube. Doses should be repeated every 15 minutes until excessive secretions and sweating have been controlled. Once bronchial secretion has been controlled, atropine administration should be repeated whenever the secretions begin to recur. In seriously poisoned patients, very large doses may be required. Alterations of pulse rate and pupillary size should not be used as indicators of treatment adequacy. Pralidoxime should be administered as early in poisoning as possible as its efficacy may diminish when given more than 24 to 36 hours after exposure. Doses are as follows: adult 1.0 g; pediatric 25 to 50 mg/kg. The drug should be administered intravenously over 30 to 60 minutes, but in a life-threatening situation, one-half of the total dose can be given per minute for a total administration time of 2 minutes. Treatment should begin to take effect within 40 minutes with a reduction in symptoms and the amount of atropine necessary to control bronchial secretion. The initial dose can be repeated in 1 hour and then every 8 to 12 hours until the patient is clinically well and no longer requires atropine. If intravenous access cannot be established, pralidoxime may also be given intramuscularly. Early administration of diazepam in addition to the combined atropine and pralidoxime treatment may help prevent the onset of seizures and potential brain and cardiac morphologic damage following high-level organophosphate poisoning.

*References:*
- EXTOXNET, Extension Toxicology Network, "Pesticide Information Profiles, Carbophenothion," Oregon State University, Corvallis, OR (September 1995). ace.orst.edu/cgi-webglimpse/mfs
- Sax, N.I., Ed., "Dangerous Properties of Industrial Materials Report," 2, No. 4, 55-59, New York, Van Nostrand Reinhold Co. (1982).
- U.S. Environmental Protection Agency, "Chemical Profile: Carbophenothion," Washington, DC, Chemical Emergency Preparedness Program (October 31, 1985).
- Agency for Toxic Substances and Disease Registry, U.S. Department of Health and Human Services, Public Health Service, "Managing Hazardous Materials Incidents," Atlanta, GA (June 2003).
- California Environmental Protection Agency Chemical List of Lists, Sacramento CA (February 1997).

## Carbosulfan (ANSI)

*Use Type:* A broad spectrum insecticide
*CAS Number:* 55285-14-8
*Formula:* $C_{20}H_{32}N_2O_3S$
*Synonyms:* Carbamic acid,[(dibutylamino)thio]methyl-, 2,2-dimethyl-2,3-dihydro-7-benzofuranyl ester; [(Dibutylamino)thio]methylcarbamic acid, 2,2-dimethyl-2,3-dihydro-7-benzofuranyl ester; 2,2-Dihydro-2,2-dimethyl-7-benzofuranyl[(dibutylamino)thio]methylcarbamate; 2,3-Dihydro-2,2-dimethyl-7-benzofuranyl(di-*N*-butylaminosulfenyl)methylcarbamate; Carbamic acid, [(dibutylamino)thio]methyl-, 2,3-dihydro-2,2-dimethyl-7-benzofuranyl ester
*Trade Names:* ADVANTAGE®; FMC 35001®, FMC (USA); MARSHAL®, Rallis India (India); MARSHALL®, Rallis India (India); POSSE®
*Producers:* Agsin (Singapore); FMC (USA); Rallis India (India); Yellow River Enterprise (Taiwan)
*Chemical Class:* Carbamate
*EPA/OPP PC Code:* 090602
*California DPR Chemical Code:* 2182
*EINECS Number:* 259-565-9
*Uses:* Carbosulfan is a low toxic derivative from cabofuran. It is a broad spectrum insecticide, nematicide, miticide, effective against pests and mites. It is used to protect alfalfa, apple, citrus, corn, deciduous fruit, potato, rice, sorghum, soybean, sugar beets, sugarcane, and other vegetable, field, tree and orchard crops. It is used for seed treatments.
*Carcinogen/Hazard Classifications*
**WHO Acute Hazard:** Class II, moderately hazardous
*Description:* Molecular weight = 380.59
*Incompatibilities:* May form explosive materials with phosphorus pentachloride.
*Permissible Concentration in Water:* No criteria set. Runoff from spills or fire control may cause water pollution.

*Routes of Entry:* Skin absorption, ingestion, inhalation.
*Harmful Effects and Symptoms*
*Short Term Exposure:* Eye pupils are small; blurred vision; eye watering; runny nose; cough; shortness of breath; salivation; nausea, stomach cramps, diarrhea, and vomiting; increased blood pressure; profuse sweating; hypermotility, hallucinations; agitation; tingling of the skin; slow heartbeat; convulsions; fluid in lungs; loss of consciousness; incontinence; breathing stops; death. Carbamate insecticides inhibit the cholinesterase activity of enzymes, causing accumulation of acetylcholine at synapses and altering the way in which nervous impulses are transmitted. However, within several hours carbamates spontaneously detach from the enzymes.
*Long Term Exposure:* A potent cholinesterase inhibitor; cumulative effect is possible. This chemical may damage the nervous system with repeated exposure, resulting in convulsions, respiratory failure. May cause liver damage.
*Points of Attack:* Respiratory system, lungs, central nervous system, cardiovascular system, skin, eyes, plasma and red blood cell cholinesterase.
*Medical Surveillance:* Medical observation is recommended for 24 to 48 hours after breathing overexposure, as pulmonary edema may be delayed. As first aid for pulmonary edema, consider administering a corticosteroid spray. Cigarette smoking may exacerbate pulmonary injury and should be discouraged for at least 72 hours following exposure. Before employment and at regular times after that, the following are recommended: Plasma and red blood cell cholinesterase levels (tests for the enzyme poisoned by this chemical). If exposure stops, plasma levels return to normal in 1-2 weeks while red blood cell levels may be reduced for 1-3 months. When acetylcholinesterase enzyme levels are reduced by 25% or more below preemployment levels, risk of poisoning is increased, even if results are in lower ranges of "normal." Reassignment to work not involving carbamate pesticides is recommended until enzyme levels recover. If symptoms develop or overexposure occurs, repeat the above tests as soon as possible and get an exam of the nervous system. Also consider complete blood count. Consider chest x-ray following acute overexposure
*First Aid:* Speed in removing material from eyes and skin is of extreme importance. *Eyes:* Eye contact can cause dangerous amounts of these chemicals to be quickly absorbed through the mucous membrane into the bloodstream. Immediately and gently flush eyes with plenty of warm or cold water (NO hot water) for at least 15 minutes, occasionally lifting the upper and lower eyelids. Get medical aid immediately. *Skin:* Get medical aid. Skin contact can cause dangerous amounts of these chemicals to be absorbed into the bloodstream. Wearing the appropriate PPE equipment and respirator for carbamate pesticides, immediately flush skin with plenty of soap and water for at least 15 minutes while removing contaminated clothing and shoes. Shampoo hair promptly if contaminated; protect eyes. *Ingestion:* Call poison control. Loosen all clothing. Never give anything by mouth to an unconscious person. Get medical aid. Do NOT induce vomiting.* If conscious, alert, and able to swallow, rinse mouth and have victim drink 4 to 8 ounces of water. Check to see if poison control instructs you to use ipecac syrup, otherwise administer slurry of activated charcoal (2 oz in 8 oz of water). If victim is UNCONSCIOUS OR HAVING CONVULSIONS, do nothing except keep victim warm. *In some cases you may be specifically instructed by poison control to induce vomiting by way of 2 tablespoons of syrup of ipecac (adult) washed down with a cup of water.* Do NOT give activated charcoal before or with ipecac syrup. *Inhalation:* Get medical aid. Do not contaminate yourself. Wearing the appropriate PPE equipment and respirator for carbamate pesticides, immediately remove the victim from the contaminated area to fresh air. If the victim is not breathing, administer artificial respiration. Do NOT use mouth-to-mouth resuscitation; use bag/mask apparatus. If breathing is difficult, administer oxygen through bag/mask apparatus until medical help arrives. Do not leave victim unattended. *Note to physician or authorized medical personnel.* Administer atropine, 2 mg (1/30 gr) intramuscularly or intravenously as soon as any local or systemic signs or symptoms of an intoxication are noted; repeat the administration of atropine every 3 to 8 minutes until signs of atropinization (mydriasis, dry mouth, rapid pulse, hot and dry skin) occur; initiate treatment in children with 0.05 mg mg/kg of atropine; repeat at 5 to 10 minute intervals. Watch respiration, and remove bronchial secretions if they appear to be obstructing the airway; intubate if necessary. *Medical note:* 2-PAMCI may be contraindicated in the case of some carbamate poisonings.
*References:*
- California Environmental Protection Agency Chemical List of Lists, Sacramento CA (February 1997)

# Carboxin (ANSI)

*Use Type:* Fungicide
*CAS Number:* 5234-68-4
*Formula:* $C_{12}H_{13}O_2NS$; $C_6H_2NHCO-C_5OSH_7$
*Alert:* General Use Pesticide (GUP). Human toxicity (long-term): Very high.
*Synonyms:* Carbathiin; 5-Carboxanilido-2,3-dihydro-6-methyl-1,4-oxathiin; Carboxine; DCMO; 2,3-Dihydro-5-carboxanilido-6-methyl-1,4-oxathiin; 2,3-Dihydro-6-methyl-1,4-oxathiin-5-carboxanilide; 5,6-Dihydro-2-methyl-1,4-oxathiin-3-carboxanilide; 5,6-Dihydro-2-methyl-*N*-phenyl-1,4-oxathiin-3-carboxamide; 2,3-Dihydro-6-methyl-5-phenylcarbamoyl-1,4-oxathiin; DMOC; NSC 263492; 1,4-Oxathiin-3-carboxamide,5,6-dihydro-2-methyl-n-phenyl; 1,4-Oxathiin-3-carboxanilide,5,6-dihydro-2-methyl; 1,4-

# Carboxin

Oxathiin-3-carboxanilide,5,6-dihydro-2-methyl-; 1,4-Oxathiin-2,3-dihydro-5-carboxanilido-6-methyl-

*Trade Names:* CADAN®; CARBOXIN OXATHION PESTICIDE®; CASWELL No. 165 A®; D-735®; F-735®; FLO PRO V SEED PROTECTANT®, Cargill, Inc. (USA) canceled 12/12/1983; KEMIKER®; KISVAX®; OXALIN®; PADAN®; SANVEX®; THIOBEL®; VEGETOX®; VITAFLO®; VITAVAX®, Gustafson LLC (USA); V 4X®

*Producers:* Agro-care Chemical Industry Group (China); Bharat Pulverising Mills (India); Crompton Corp. (USA); Epochem Co., (China); Gustafson LLC (USA); Hindustan Insecticides (India); Nufarm (Australia); Sigma-Aldrich Laborchemikalien (Germany); Uniroyal Crop Protection Div (USA)

*Chemical Class:* Carboxamide or anilide

*EPA/OPP PC Code:* 090201

*California DPR Chemical Code:* 1755

*RTECS Number:* RP4550000

*EINECS Number:* 226-031-1

*Uses:* Carboxin is a General Use Pesticide (GUP) and is used as a seed protectant and wood preservative. It is often used in combination with other fungicides such as thiram or captan. Carboxin is a systemic anilide fungicide. It is used as a seed treatment for control of smut, rot, and blight on barley, oats, rice, cotton, vegetables, corn and wheat. It is also used to control fairy rings on turf grass. Carboxin may be used to prevent the formation of these diseases or may be used to cure existing plant diseases.

*Human toxicity (long-term)*[77]: Extra high–0.91 ppb, Health Advisory

*Fish toxicity (threshold)*[77]: High–1.06008 ppb, MATC (Maximum Acceptable Toxicant Concentration)

*U.S. Maximum Allowable Residue Levels for Carboxin (40 CFR 180.301):*

| CROP | ppm |
|---|---|
| Barley, grain | 0.2 |
| Barley, straw | 0.2 |
| Bean, dry | 0.2 |
| Bean, forage | 0.5 |
| Bean, hay | 0.5 |
| Bean, straw | 0.5 |
| Bean, succulent | 0.2 |
| Canola, seed | 0.03 |
| Cattle, fat | 0.1 |
| Cattle, meat | 0.1 |
| Cattle, mbyp | 0.1 |
| Corn, fodder | 0.2 |
| Corn, forage | 0.2 |
| Corn, fresh, inc.sweet | 0.2 |
| Corn, grain | 0.2 |
| Cotton, undelinted seed | 0.2 |
| Egg | 0.01 |
| Goat, fat | 0.1 |
| Goat, meat | 0.1 |
| Goat, mbyp | 0.1 |
| Hog, fat | 0.1 |
| Hog, meat | 0.1 |
| Hog, mbyp | 0.1 |
| Horse, fat | 0.1 |
| Horse, meat | 0.1 |
| Horse, mbyp | 0.1 |
| Milk | 0.01 |
| Oat, forage | 0.5 |
| Oat, seed | 0.2 |
| Oat, straw | 0.2 |
| Onion, dry bulb | 0.20 |
| Onion, dry bulb (40 CFR 180.130b) | 0.20 |
| Peanut | 0.2 |
| Peanut, hay | 0.2 |
| Poultry, fat | 0.1 |
| Poultry, meat | 0.1 |
| Poultry, mbyp | 0.1 |
| Rice | 0.2 |
| Rice, straw | 0.2 |
| Safflower, seed | 0.2 |
| Sheep, fat | 0.1 |
| Sheep, meat | 0.1 |
| Sheep, mbyp | 0.1 |
| Sorghum, forage | 0.2 |
| Sorghum, grain | 0.2 |
| Sorghum, stover | 0.2 |
| Soybean | 0.2 |
| Wheat, forage | 0.5 |
| Wheat, grain | 0.2 |
| Wheat, straw | 0.2 |

*Carcinogen/Hazard Classifications*

**Label Signal Word:** CAUTION

**WHO Acute Hazard:** Group U, unlikely to be hazardous

*Regulatory Authority:*
- EPCRA Section 313 Form R *de minimus* concentration reporting level: 1.0%
- Cited in U.S. State Regulations: Actively registered pesticide in California.

*Description:* Carboxin is a crystalline solid. Soluble in water; solubility = 170 ppm @ 25°C. Molecular weight 235.31. Melting/Freezing point = 93-101°C. Vapor pressure = $1.78 \times 10^{-7}$ mmHg.

*Permissible Concentration in Water:* The no-observed-adverse-effect level has been determined by EPA to be 10 mg/kg body weight/day. This results in a long-term health advisory of 3.5 mg/L and a lifetime health advisory of 0.7 mg/L.

*Determination in Water:* Analysis of carboxin is by a gas chromatographic (GC) method applicable to the determination of certain nitrogen-phosphorus containing pesticides in water samples. In this method, approximately

1 liter of sample is extracted with Methylene chloride. The extract is concentrated and the compounds are separated using capillary column GC. Measurement is made using a nitrogen phosphorus detector. The method detection limit has not been determined for carboxin but it is estimated that detection limits for analyses included in this method are in the range of 0.1 to 2 µg/L.

*Harmful Effects and Symptoms*

*Short Term Exposure:* Contact with eyes or skin may cause irritation or injury. Inhalation should be avoided; use NIOSH-approved air purifying respirators for pesticides. May be harmful if swallowed. The $LD_{50}$-oral for mice has been reported to be 3550 mg/kg (slightly toxic). A value for $LD_{50}$-rat of 430 mg/kg puts carboxin in the moderately toxic category. Because this material has a low vapor pressure, significant inhalation of vapors is unlikely at ordinary temperatures.

*First Aid:* If this chemical gets into the eyes, remove any contact lenses at once and irrigate immediately for at least 15 minutes, occasionally lifting upper and lower lids. Seek medical attention immediately. If this chemical contacts the skin, remove contaminated clothing and wash immediately with soap and water. Seek medical attention immediately. If this chemical has been inhaled, remove from exposure, begin rescue breathing (using universal precautions) if breathing has stopped and CPR if heart action has stopped. Transfer promptly to a medical facility. When this chemical has been swallowed, get medical attention. Give large quantities of water and induce vomiting. Do not make an unconscious person vomit.

*References:*
- U.S. Environmental Protection Agency, Office of Pesticide Programs, Pesticide Residue Limits, " Carboxin", (40 CFR 180.301), www.epa.gov/cgi-bin/oppsrch
- EXTOXNET, Extension Toxicology Network, "Pesticide Information Profiles, Carboxin," Oregon State University Corvallis, OR (June 1996), ace.orst.edu/info/extoxnet/pips/carboxin
- U.S. Environmental Protection Agency, "Health Advisory: Carboxin," Washington, DC, Office of Drinking Water (August 1987).
- California Environmental Protection Agency Chemical List of Lists, Sacramento CA (February 1997).

# Cartap Hydrochloride

*Use Type:* Insecticide
*CAS Number:* 15263-52-2
*Formula:* $C_7H_{15}N_3O_2N_2S_2Cl$
*Synonyms:* 1,3-Bis(carbamoylthio)-2-(N,N-dimethylamino)propane hydrochloride; Carbamothioic acid-$S,S'$-[2-(dimethylamino)-1,3-propanediyl]ester, monohydrochloride; Cartap monohydrochloride; $S,S$-[2-(Dimethylamino)trimethylene]bis(thiocarbamate)hydrochloride; Thiocarbamic acid-$S,S$-[2-(Dimethylamino)trimethylene]ester hydrochloride

*Trade Names:* CADAN®, Sumika Takeda Agrochemical (Japan); CALDAN®; KRITAP®;NTD 2®; PADAN®; PATAP®, Sumika Takeda Agrochemical (Japan); SANVEX®; THIOBEL®, Sumika Takeda Agrochemical (Japan); TI-1258®; VEGETOX®, Sumika Takeda Agrochemical (Japan)

*Producers:* Saeryung Chemicals (South Korea); Sumika Takeda Agrochemical (Japan)

*Chemical Class:* Nereistoxin insecticide

*EPA/OPP PC Code:* 041415

*Uses:* Not registered in the U.S. Cartap hydrochloride is used to control chewing and sucking insects on many crops, including rice, potatoes, cabbage and other vegetables, soy beans, peanuts, sunflowers, maize, sugar beet, wheat, pearl barley, pome fruit, stone fruit, citrus fruit, vines, chestnuts, ginger, tea, cotton, and sugar cane.

*Regulatory Authority:*
U.S. DOT Regulated Marine Pollutant (49CFR, Subchapter 172.101, Appendix B)

*Description:* Molecular weight = 273.81

*Incompatibilities:* When heated to decomposition or on contact with acids or acid fumes, may produce highly toxic chloride fumes; deadly phosgene gas may be formed. May cause pitting of some metals.

*Permissible Concentration in Water:* No criteria set. Runoff from spills or fire control may cause water pollution.

*Routes of Entry:* Inhalation, skin, ingestion.

*Harmful Effects and Symptoms*

*Short Term Exposure:* Eye pupils are small; blurred vision; eye watering; runny nose; cough; shortness of breath; salivation; nausea, stomach cramps, diarrhea, and vomiting; increased blood pressure; profuse sweating; hypermotility, hallucinations; agitation; tingling of the skin; slow heartbeat; convulsions; fluid in lungs; loss of consciousness; incontinence; breathing stops; death. Carbamate insecticides inhibit the cholinesterase activity of enzymes, causing accumulation of acetylcholine at synapses and altering the way in which nervous impulses are transmitted. However, within several hours carbamates spontaneously detach from the enzymes.

*Long Term Exposure:* A potent cholinesterase inhibitor; cumulative effect is possible. This chemical may damage the nervous system with repeated exposure, resulting in convulsions, respiratory failure. May cause liver damage.

*Points of Attack:* Respiratory system, central nervous system, cardiovascular system, skin, eyes.

*Medical Surveillance:* Medical observation is recommended for 24 to 48 hours after breathing overexposure, as pulmonary edema may be delayed. As first aid for pulmonary edema, consider administering a corticosteroid spray. Cigarette smoking may exacerbate pulmonary injury and should be discouraged for at least 72 hours following exposure. Before employment and at regular times after that, the following are recommended: Plasma and red blood cell cholinesterase levels (tests for the enzyme poisoned by this chemical). If exposure stops, plasma levels return to normal in 1-2 weeks while red blood cell levels may be reduced for 1-3 months. When acetylcholinesterase enzyme levels are reduced by 25% or more below preemployment levels, risk of poisoning is increased, even if results are in lower ranges of "normal." Reassignment to work not involving carbamate pesticides is recommended until enzyme levels recover. If symptoms develop or overexposure occurs, repeat the above tests as soon as possible and get an exam of the nervous system. Also consider complete blood count. Consider chest x-ray following acute overexposure.

*First Aid:* Speed in removing material from eyes and skin is of extreme importance. *Eyes:* Eye contact can cause dangerous amounts of these chemicals to be quickly absorbed through the mucous membrane into the bloodstream. Immediately and gently flush eyes with plenty of warm or cold water (NO hot water) for at least 15 minutes, occasionally lifting the upper and lower eyelids. Get medical aid immediately. *Skin:* Get medical aid. Skin contact can cause dangerous amounts of these chemicals to be absorbed into the bloodstream. Wearing the appropriate PPE equipment and respirator for carbamate pesticides, immediately flush skin with plenty of soap and water for at least 15 minutes while removing contaminated clothing and shoes. Shampoo hair promptly if contaminated; protect eyes. *Ingestion:* Call poison control. Loosen all clothing. Never give anything by mouth to an unconscious person. Get medical aid. *Do NOT induce vomiting.*\* If conscious, alert, and able to swallow, rinse mouth and have victim drink 4 to 8 ounces of water. Check to see if poison control instructs you to use ipecac syrup, otherwise administer slurry of activated charcoal (2 oz in 8 oz of water). If victim is *unconscious or having convulsions,* do nothing except keep victim warm. *\*In some cases you may be specifically instructed by poison control to induce vomiting by way of 2 tablespoons of syrup of ipecac (adult) washed down with a cup of water.* Do NOT give activated charcoal before or with ipecac syrup. Inhalation: Get medical aid. Do not contaminate yourself. Wearing the appropriate PPE equipment and respirator for carbamate pesticides, immediately remove the victim from the contaminated area to fresh air. If the victim is not breathing, administer artificial respiration. Do NOT use mouth-to-mouth resuscitation; use bag/mask apparatus. If breathing is difficult, administer oxygen through bag/mask apparatus until medical help arrives. Do not leave victim unattended. *Note to physician or authorized medical personnel.* Administer atropine, 2 mg (1/30 gr) intramuscularly or intravenously as soon as any local or systemic signs or symptoms of an intoxication are noted; repeat the administration of atropine every 3 to 8 minutes until signs of atropinization (mydriasis, dry mouth, rapid pulse, hot and dry skin) occur; initiate treatment in children with 0.05 mg mg/kg of atropine; repeat at 5 to 10 minute intervals. Watch respiration, and remove bronchial secretions if they appear to be obstructing the airway; intubate if necessary. *Medical Note:* 2-PAMCI may be contraindicated in the case of some carbamate poisonings.

*References:*
- California Environmental Protection Agency Chemical List of Lists, Sacramento CA (February 1997)

# Chloramben (ANSI)

*Use Type:* Herbicide
*CAS Number:* 133-90-4
*Formula:* $C_7H_5Cl_2NO_2$
*Alert:* A General Use Pesticide (GUP) that is no longer produced or sold in the United States.
*Synonyms:* 3-Amino-2,5-dichlorobenzoic acid; Benzoic acid, 3-amino-2,5-dichloro-; Chlorambed; Chlorambene; Chloramben, Aromatic carboxylic acid; Chloramben benzoic acid herbicide; Benzoic acid, 3-amino-2,5-dichloro-; 3-Amino-2,6-dichlorobenzoic acid; 2,5-Dichloro-3-aminobenzoic acid; NCI-C00055
*Trade Names:* ACP-M-728®; AMBEN®, Aventis CropScience (France); Bayer CropScience (Germany); Rhone-Poulenc (France); AMBIBEN®; AMIBEN®, Bayer CropScience (Germany) canceled 1/22/1991; AMIBIN®; AMOBEN®; ORNAMENTAL WEEDER®, Bayer CropScience (Germany) canceled 10/10/1989; VEGABEN®; VEGIBEN®, Bayer CropScience (Germany) canceled 4/9/1982; WEEDONE® GARDEN WEEDER, Bayer CropScience (Germany)
*Producers:* Aventis CropScience (France); Bayer CropScience (Germany); Flourochem (UK); Nufarm (Australia); Rhone-Poulenc (France); Sigma-Aldrich Laborchemikalien (Germany)
*Chemical Class:* Benzoic acid
*EPA/OPP PC Code:* 029901
*California DPR Chemical Code:* 2184
*RTECS Number:* DG1925000
*Uses:* A herbicide for grasses, and broadleaf weeds. Mostly used on soybeans, and also on corn, beans, asparagus, pumpkins, peanuts, sunflowers, peppers, cotton, sweet potatoes, squash, melons, hardwood trees, and some conifers, but is no longer produced or sold in the U.S.
*Human toxicity (long-term)[77]:* Very low–100.00 ppb, Health Advisory
*Fish toxicity (threshold)[77]:* Very low–1584.89319 ppb, MATC (Maximum Acceptable Toxicant Concentration)
*Carcinogen/Hazard Classifications*
*U.S. EPA Carcinogens:* Deferred
**Label Signal Word:** CAUTION

**WHO Acute Hazard:** Class U, unlikely to be hazardous
***Regulatory Authority:***
- Carcinogen, (animal positive) (NTP)
- Air Pollutant Standard Set (former USSR)[35] (Pennsylvania)[60]
- AB 2588-Air Toxics "Hot Spots" Chemicals (CAL)
- CAL Air Resources Board/AB 1807 Toxic Air Contaminants
- The "Director's List" (CAL/OSHA)Clean Air Act: Hazardous Air Pollutants (Title I, Part A, Section 112)
- Superfund/EPCRA 40CFR302.4 RQ: CERCLA, 1 lb (0.454 kg).
- EPCRA Section 313 Form R *de minimus* concentration reporting level: 1.0%.

***Description:*** Chloramben is a colorless, crystalline solid. Odorless. Molecular weight = 206.03. Melting/Freezing point = 200° to 201°C. Soluble in water; solubility = 690 ppm @ 25°C. It is practically nontoxic. Vapor pressure = 6.8 x $10^{-3}$ mmHg @ 100°C.

***Incompatibilities:*** Rapidly decomposed by light. Strong acids and acid fumes.

***Permissible Exposure Limits in Air:*** Although no occupational exposure limits have been established, this chemical can be absorbed through the skin. The former USSR[35] has set a MAC in ambient air in residential areas of 0.01 mg/m³ on a momentary basis and 0.006 mg/m³ on an average daily basis. Pennsylvania[60] has set a guideline for chloramben in ambient air of 1.3333 mg/m³.

***Permissible Concentration in Water:*** The USR[35] has set a MAC of 0.5 mg/L in surface water. A lifetime health advisory of 0.105 mg/L has been determined by EPA (see reference below).

***Routes of Entry:*** Inhalation, passes through the skin.

***Harmful Effects and Symptoms***

***Short Term Exposure:*** Contact with eyes or skin may cause irritation or injury. Inhalation should be avoided; use NIOSH-approved air purifying respirators for pesticides. May be harmful if swallowed. The available data on Chloramben are very sparse. Much additional information is needed regarding its chronic toxicity, teratogenicity, and carcinogenicity before limits can be confidently set. No-observed-adverse-effect doses for chloramben were 15 mg/kg/day. Based on these data an ADI was calculated at 0.015 mg/kg/day

***Long Term Exposure:*** There is evidence that this chemical causes cancer in animals. It may cause cancer of the liver. Repeated exosure may cause skin rash with itching.

***Points of Attack:*** Skin, liver.

***Medical Surveillance:*** Liver function tests. Examination by a qualified allergist.

***First Aid:*** If this chemical gets into the eyes, remove any contact lenses at once and irrigate immediately for at least 15 minutes, occasionally lifting upper and lower lids. Seek medical attention immediately. If this chemical contacts the skin, remove contaminated clothing and wash immediately with soap and water. Seek medical attention immediately. If this chemical has been inhaled, remove from exposure, begin rescue breathing (using universal precautions) if breathing has stopped and CPR if heart action has stopped. Transfer promptly to a medical facility. When this chemical has been swallowed, get medical attention. Give large quantities of water and induce vomiting. Do not make an unconscious person vomit.

***References:***
- National Cancer Institute, Bioassay of Chloramben for Possible Carcinogenicity, Technical Report Series No. 25, Bethesda, Maryland (1977).
- EXTOXNET, Extension Toxicology Network, "Pesticide Information Profiles, Chloramben," Oregon State University Corvallis, OR (June 1996). ace.orst.edu/info/extoxnet/pips/chloramb
- Sax, N.I., Ed., "Dangerous Properties of Industrial Materials Report," 1, No. 3, 28-29, New York, Van Nostrand Reinhold Co. (1981). (As 3-Amino-2,5-Dichlorobenzoic Acid).
- U.S. Environmental Protection Agency, "Health Advisory: Chloramben." Washington, DC, Office of Drinking Water (August 1987).
- California Environmental Protection Agency Chemical List of Lists, Sacramento CA (February 1997).
- New Jersey Department of Health and Senior Services, Hazardous Substance Fact Sheet, Chloramben. Trenton NJ (September, 1998). http://www.state.nj.us/health/eoh/rtkweb/0357.pdf

# Chlordane

***Use Type:*** Insecticide
***CAS Number:*** 57-74-9; 12789-03-6 (technical grade chemical)
***Formula:*** $C_{10}H_6Cl_8$
***Alert:*** Since 1988, the use and commercial production of chlordane (except for export) has been prohibited in the United States and many other countries. However, chlordane residue is still present from prior use in many homes and other structures, as well as the surrounding soil. Old supplies of chlordane may still exist in locations such as warehouses, garages, and landfills. The only commercial use still permitted is for fire ant control in power transformers. Human toxicity (long-term): High.

***Synonyms:*** Chloordaan (Dutch); Chlordan; Clordan (Italian); Clordano (Spanish); Cortilan-Neu; Dichlorochlordene; ENT 9,932; ENT 25,552-X; 4,7-Methanoindan, 1,2,3,4,5,6,7,8,8-octachloro-2,3,3a,4,7,7a-hexahydro-; 4,7-Methano-1H-indene,1,2,4,5,6,7,8,8-octachloro-2,3,3a,4,7,7a-hexahydro-; 4,7-Methanoindan, 1,2,4,5,6,8,8-octachloro 3a,4,7,7a-tetrahydro; NCI-C00099; Octachlor; Octachlorodihydrodicyclopentadiene;

1,2,4,5,6,7,8,8-Octachloro-2,3,3a,4,7,7a-hexahydro-4,7-methanoindene; 1,2,4,5,6,7,8,8-Octachloro-2,3,3a,4,7,7a-hexahydro-4,7-methano-1H-indene; 1,2,4,5,6,7,8,8-Octachloro-3a,4,7,7a-hexahydro-4,7-methylene indane; Octachloro-4,7-methanohydroindane; Octachloro-4,7-methanotetrahydroindane; 1,2,4,5,6,7,8,8-octachloro-4,7-methano-3a,4,7,7a-tetrahydroindane; 1,2,4,5,6,7,8,8-Octachloor-3a,4,7,7a-tetrahydro-4,7-*endo*-methano-indaan (Dutch); 1,2,4,5,6,7,8,8-Octachloro-3a,4,7,7a-tetrahydro-4,7-methanoindan; 1,2,4,5,6,7,8,8-Octachloro-3a,4,7,7a-tetrahydro-4,7-methanoindane; 1,2,4,5,6,7,10,10-Octachloro-4,7,8,9-tetrahydro-4,7-methyleneindane; 1,2,4,5,6,7,8,8-Octachlor-3a,4,7,7a-tetrahydro-4,7-*endo*-methano-indan (German); 1,2,4,5,6,7,8,8-Ottochloro-3a,4,7,7a-tetraidro-4,7-*endo*-metano-indano (Italian)

*Trade Names:* ASPON-CHLORDANE®; BELT®; CD 68®; CHLORINDAN®; CHLOR KIL®; CHLORODANE®; CORODANE®; CHLORTOX®; DOWCHLOR®, Dow AgroSciences (USA), canceled; DOW-KLOR®, Dow AgroSciences (USA), canceled; GOLD CREST®, Velsicol Chemical Corporation (USA) canceled 8/11/1987; KILEX LINDANE®; HCS 3260®; KYPCHLOR®; M 140®; M 410®; NIRAN®; OCTACHLOR®; OKTATERR®; OMS 1437®; ORTHO-KLOR®, Ortho Business Group (USA), canceled 5/1/1987; SD 5532®; SHELL SD-5532®, Shell Chemical (UK), canceled; SYNKLOR®; TAT®; TAT CHLOR® 4; TERMEX®; TOPICHLOR® 20; TOPICLOR®; TOXICHLOR®; VELSICOL® 1068, Velsicol Chemical Corporation (USA), canceled

*Producers:* Bharat Pulverizing Mills (India); EID Parry (India); Ehrenstorfer, Dr. (Germany); Rallis India (India); Sigma-Aldrich Laborchemikalien (Germany); Velsicol Chemical Corporation (USA)

*Chemical Class:* Organochlorine; Halo-organics

*EPA/OPP PC Code:* 058201, 058202 (technical grade chemical)

*California DPR Chemical Code:* 130

*ICSC Number:* 0740

*RTECS Number:* PB9800000

*RCRA Number:* U036

*EEC Number:* 602-047-00-8

*Uses:* Chlordane is a broad spectrum insecticide of the group of polycyclic chlorinated hydrocarbons called cyclodiene insecticides. Chlordane has been used extensively since the 1950's for termite control, as an insecticide for homes and gardens, and as a control for soil insects during the production of crops such as corn. Both the uses and the production volume of chlordane have decreased extensively since the issuance of a registration suspension notice for all food crops and home and garden uses of chlordane by the U.S. Environmental Protection Agency, However, significant commercial use of chlordane for termite control continues. Special groups at risk include children as a result of milk consumed; fishermen and their families because of the high consumption of fish and shellfish, especially freshwater fish; persons living downwind from treated fields; and persons living in houses treated with chlordane pesticide control agents. Formulations include dusts, emulsifiable concentrates, granules, oil solutions, and wettable powders.

*Human toxicity (long-term)*[77]: High–2.00 ppb, MCL (Maximum Contaminant Level)

*Fish toxicity (threshold)*[77]: Extra high–0.04936 ppb, MATC (Maximum Acceptable Toxicant Concentration)

*Carcinogen/Hazard Classifications*

**U.S. EPA Carcinogens:** Group B2, Probably Carcinogen

**California Prop. 65:** Carcinogen

**IARC:** Group 2B, Possibly Carcinogenic to Humans

**WHO Acute Hazard:** Class II, moderately hazardous

**Endocrine Disruptor:** Known ED

*Regulatory Authority:*

- Carcinogen (Animal Positive) (IARC) (NCI)[9] (DFG)[3]
- Banned or Severely Restricted (In Agriculture) (Many Countries) (UN)[13][35]
- Air Pollutant Standard Set (ACGIH)[1] (DFG)[3] (HSE)[33] (UNEP)[43] (OSHA)[58] (Several States)[60] (Australia) (Israel) (Mexico) (Several Canadian Provinces)
- List of priority pollutants (U.S. EPA)
- AB 1803-Well Monitoring Chemical (CAL)
- EPA/SARA 302 (EPCRA) Extremely hazardous substances
- MCL(Maximum Contaminants Levels) list of contaminants (CAL)
- AB 2588-Air Toxics "Hot Spots" Chemicals (CAL)
- CAL-DHS/DHS Drinking Water Action Levels
- CAL Air Resources Board/AB 1807 Toxic Air Contaminants
- Proposition 65 chemical (CAL)
- Permissible Exposure Limits for Chemical Contaminants (CAL/OSHA)
- The "Director's List" (CAL/OSHA)
- Clean Air Act: Hazardous Air Pollutants (Title I, Part A, Section 112)
- Clean Water Act: Section 311 Hazardous Substances/RQ 40CFR117.3 (same as CERCLA, see below);
- 40CFR423, Appendix A, Priority Pollutants; Section 313 Water Priority Chemicals (57FR41331, 9/9/92); Toxic Pollutant (Section 401.15) as technical mixture and metabolites.
- EPA Hazardous Waste Number (RCRA No.): U036
- RCRA Toxicity Characteristic (Section 261.24), Maximum Concentration of Contaminants, regulatory level, 0.03 mg/L
- RCRA, 40CFR261, Appendix 8 Hazardous Constituents
- RCRA 40CFR268.48; 61FR15654, UniversalTreatment Standards: Wastewater (mg/L), (*alpha-* and *gamma-*isomers)
- 0.0033; Nonwastewater (mg/kg), 0.26

- RCRA 40CFR264, Appendix 9; TSD Facilities Ground Water Monitoring List. Suggested test method(s) (PQL ug/L): 8080(0.1); 8250(10)
- Safe Drinking Water Act: MCL, 0.002 mg/L; MCLG, zero; Regulated chemical (47 FR 9352)
- Superfund/EPCRA 40CFR355, Appendix B Extremely Hazardous Substances: TPQ = 1000 lb (454 kg). Chlordane has been found in at least 171 of 1,416 National Priorities List sites identified by the Environmental Protection Agency.
- Superfund/EPCRA 40CFR302.4 RQ: CERCLA, 1 lb (0.454 kg).
- EPCRA Section 313 Form R *de minimus* concentration reporting level: 1.0%.
- U.S. DOT Regulated Marine Pollutant (49CFR172.101, Appendix B), severe pollutant
- Canada, Drinking water quality: 0.007 mg/L MAC
- Mexico, Drinking water quality: 0.003 mg/L

*Description:* Technical-grade chlordane is a mixture of chlordane isomers and more than 140 related reaction products. Depending on the composition, the mixture may be an amber-to-brown, viscous liquid or a white powder. As a commercial pesticide, chlordane is usually dissolved in hydrocarbons and used as a spray. At room temperature, chlordane is almost odorless or may have a slight chlorine-like odor, but the odor is inadequate as a warning of exposure. Practically insoluble in water. Molecular weight = 409.8 daltons. Boiling point = 175°C @ 2 mmHg. Melting/Freezing point = (*cis* isomer) 107–109°C; (*trans* isomer) 103–105°C). Vapor pressure = $1 \times 10^{-5}$ mmHg @ 25°C). Noncombustible liquid (but the commercial product may be dissolved in various flammable solvents). Hazard Identification (based on NFPA-704 M Rating System): (in a flammable solution)Health 3, Flammability 3 [may vary depending on carrier], Reactivity 0.

*Incompatibilities:* Contact with strong oxidizers may cause fire and explosions. High heat and contact with alkaline solutions cause decomposition with the production of toxic fumes including chlorine, phosgene, hydrogen chloride. Attacks iron, zinc, plastics, rubber and coatings.

*Permissible Exposure Limits in Air:* The Federal (OSHA PEL) limit[58] is 0.5 mg/m$^3$ TWA for an 8-hour workshift. NIOSH[2] and ACGIH[1] recommend the same limit. Australia, DFG (MAK)[3], Israel and Mexico TWA value is 0.5 mg/m$^3$ and Mexico's STEL value is 2.0 mg/m$^3$. Many of these regulatory and advisories add the notation "skin," indicating the possibility of cutaneous absorption. The NIOSH[2] IDLH level = 100 mg/m$^3$. The Canadian provinces of Alberta, British Columbia, Ontario, and Quebec set a limit of 0.5 mg/m$^3$ TWA and STEL value of 2.0 mg/m$^3$. The former USSR-UNEP/IRPTC project has set a MAC of 0.01 mg/m$^3$ in workplace air[43]. Several states have set guidelines or standards for chlordane in ambient air[60] ranging from 0.068 µg/m$^3$ (Massachusetts) to 0.36 µg/m$^3$ (Pennsylvania) to 1.19 µg/m$^3$ (Kansas) to 1.7 µg/m$^3$ (New York) to 2.5 µg/m$^3$ (Connecticut and South Carolina) to 5.0 µg/m$^3$ (Florida) to 5-20 µg/m$^3$ (North Dakota) to 8.0 µg/m$^3$ (Virginia) to 12.0 µg/m$^3$ (Nevada).

*Determination in Air:* Filter/Chromosorb tube-102; Toluene; Gas chromatography/Electrochemical detection; NIOSH IV, Method #5510[18].

*Permissible Concentration in Water:* EPA has set a limit in drinking water of 2 ppb. The U.S. EPA recommends that a child should not drink water with more than 60 parts of chlordane per billion parts of drinking water (60 ppb) for longer than 1 day. To protect freshwater aquatic life: 0.0043 µg/L as a 24-hour average, not to exceed 2.4 µg/L at any time. To protect saltwater aquatic life: 0.0040 µg/L as a 24-hour average, never to exceed 0.09 µg/L. To protect human health: preferably zero. An additional lifetime cancer risk of 1 in 100,000 is presented by a concentration of 0.0046 µg/L[6]. The U.S. EPA[47] has found a lowest-observed-adverse-effect-level (LOAEL) of 0.045 mg/kg body weight/day which results in a lifetime health advisory of 2 µg/L. Several states have set standards and guidelines for chlordane in drinking water[61]. Standards range from 0.5 µg/L (New Jersey) to 3.0 µg/L (Illinois) and guidelines range from 0.055 µg/L (California) to 0.22 µg/L (Kansas and Minnesota) to 0.50 µg/L (Arizona) to 0.55 µ/l (Maine). See values listed under Regulatory Authority for Canada and Mexico. It is strongly advised not to let the chemical enter into the environment because it persists in the environment. The substance may cause long-term effects in the aquatic environment.

*Determination in Water:* Filter/Chromosorb tube-102; Toluene; Gas chromatography/Electrochemical detection; NIOSH IV, Method #5510[18]. Gas chromatography (EPA Method 608) or gas chromatography plus mass spectrometry (EPA Method 625).

*Routes of Entry:* Chlordane is absorbed well by the lungs and gastrointestinal tract and through intact skin. Exposure by any route can cause systemic effects.

*Harmful Effects and Symptoms*

*Short Term Exposure:* Significant chlordane exposure by any route disrupts the transmission of nerve impulses, resulting in CNS excitation, convulsions, and respiratory depression. Chlordane is absorbed well through intact skin. Early formulations contained impurities that were skin and mucous-membrane irritants. Common symptoms of chlordane poisoning include headache, nausea, excitability, confusion, and muscle tremors that may precede convulsions. Chlordane can irritate the eyes and skin and can cause burns on contact. Skin rash or acne may develop. Exposure can cause blurred vision, nausea, headache, abdominal pain and vomiting, Exposure at high levels may result in disorientation, tremors, convulsions, respiratory failure and death. Medical observation is indicated. Symptoms include increased sensitivity to stimuli, tremors, muscular incoordination, and convulsions with or without coma. Fatal oral dose to adult humans is between 6 and 60 g with onset of symptoms within 45 minutes to several hours after ingestion, although symptoms have occurred following very small doses either orally or by skin exposure. Some reports of delayed development of liver disease, blood disorders and upset stomach. Chlordane is considered to be borderline between a moderately and highly toxic substance. The $LD_{50}$ (oral, rat) = 283 mg/kg.

*Long Term Exposure:* Chronic chlordane exposure can cause permanent alterations of nervous system function, including problems with memory, learning, thinking, sleeping, personality changes, depression, numbness in the extremities, headache, and sensory and perceptual changes. It has been suggested that chronic exposure can cause blood disorders, but these disorders were not shown to have an increased incidence in heavily exposed groups of workers. Besides blood disorders, jaundice has been reported in persons living in homes treated with chlordane for termite control, but liver-function tests were normal in workers who manufactured chlordane. Chronic exposures may be more serious for children because of their potential longer latency period. *Carcinogenicity:* IARC has determined that chlordane is possibly carcinogenic to humans (group 2B). The U.S. EPA has determined that technical grade chlordane is a likely human carcinogen. Chlordane is structurally similar to rodent carcinogens and chronic chlordane exposure can cause hepatocellular carcinoma in several strains of mice. The evidence for carcinogenicity in humans is weak: a few case reports and mixed or equivocal case-control study results associating exposure to chlordane with leukemia and non-Hodgkin's lymphoma. Epidemiological evidence for an association is supported by limited evidence of mutagenicity in human and rodent lymphocytes tested *in vitro. Reproductive and Developmental Effects:* The TERIS database states that no epidemiologic studies have reported birth defects among infants born to mothers exposed to chlordane during pregnancy. Chlordane is excreted in breast milk. No teratogenic effects from acute exposure have been reported. Prenatal exposure to chlordane has been reported for a few cases of neuroblastoma, blood dyscrasias, and depressed cell-mediated immunity, but no direct link with the chemical was established. Chlordane induces liver enzymes and enhances metabolism of steroid hormones, including oral contraceptives; sterility has been reported in animals.

*Points of Attack:* Central nervous system, eyes, lungs, liver, kidneys, skin.

*Medical Surveillance:* Consider the points of attack in preplacement and periodic physical examinations. Liver and kidney function tests. Examination by a qualified allergist. Complete blood count.

*First Aid:* Persons exposed only to chlordane vapor do not pose risks of secondary contamination to others. Persons whose skin or clothing is contaminated with liquid or powdered chlordane can cause secondary contamination by direct contact. Inhalation: Monitor for respiratory distress. If cough or breathing difficulty develops, evaluate for respiratory tract irritation, bronchitis, or pneumonitis. Administer 100% humidified supplemental oxygen if breathing is difficult. *Basic Decontamination:* Victims who are able and cooperative may assist with their own decontamination. Remove and double-bag contaminated clothing and personal belongings. Leather absorbs chlordane; therefore, items such as leather shoes, gloves, and belts should be discarded. If there has been direct contact with liquid or powdered chlordane, flush exposed skin, hair, and under nails with plain, running, tepid water for 20 minutes, then wash twice with mild soap. **Do not scrub**, since this can increase absorption through the skin. Rinse thoroughly with water. *Eyes:* irrigate with large amounts of plain, tepid water or saline for 20 minutes, occasionally lifting the lower and upper lids. During this time, remove contact lenses, if easily removable without additional trauma to the eye. Keep victims (adults or children) warm and quiet to avoid triggering seizures and the complication of hypothermia. *Ingestion:* **Do not induce vomiting**. If the victim is conscious and able to swallow, administer an aqueous slurry of activated charcoal at 1 gm/kg (usual adult dose 60–90 g, child dose 25–50 g). A soda can and straw may be of assistance when offering charcoal to a child. (The efficacy of activated charcoal for chlordane poisoning is uncertain).

*Note to physician or authorized medical personnel:* In cases of respiratory compromise secure airway and respiration via endotracheal intubation. If not possible, perform cricothyroidotomy if equipped and trained to do so. To control metabolic acidosis, treat with sodium bicarbonate under medical base control (adult dose = 1 ampule; pediatric dose = 1 mEq/kg). Patients who are comatose, hypotensive, or have seizures or cardiac dysrhythmias should be treated according to advanced life support protocols. High concentrations of chlordane can increase cardiac irritability, use caution with cardiac or bronchial sensitizing agents.

*References:*
- EXTOXNET, Extension Toxicology Network, "Pesticide Information Profile, Chlordane," Oregon State University, Corvallis, OR (June 1996). ace.orst.edu/info/extoxnet/pips/chlordan
- Agency for Toxic Substances and Disease Registry, "ToxFAQs for Chlordane," Atlanta, GA (September 1995). http://www.atsdr.cdc.gov/tfacts31.html
- U.S. Environmental Protection Agency, Chlordane: Ambient Water Quality Criteria, Washington, DC (1980).
- U.S. Environmental Protection Agency, Chlordane, Health and Environmental Effects Profile No. 35, Office of Solid Waste, Washington, DC (April 30, 1980).

- Agency for Toxic Substances and Disease Registry, U.S. Department of Health and Human Services, Public Health Service, "Managing Hazardous Materials Incidents," Atlanta, GA (June 2003).
- Sax, N.I., Ed., "Dangerous Properties of Industrial Materials Report," 1, No. 2, 33-35 (1980) and 3, No. 5, 94-99 (1983) and 7, No. 6, 46-55 (1987).
- U.S. Environmental Protection Agency, "Chemical Profile: Chlordane," Washington, DC, Chemical Emergency Preparedness Program (October 31, 1985).
- New Jersey Department of Health and Senior Services and Senior Services, "Hazardous Substance Fact Sheet: Chlordane," Trenton, NJ (April 1998). http://www.state.nj.us/health/eoh/rtkweb/0361.pdf
- New York State Department of Health, "Chemical Fact Sheet: Chlordane," Albany, NY, Bureau of Toxic Substance Assessment (March 1986).
- California Environmental Protection Agency Chemical List of Lists, Sacramento CA (February 1997).
- Agency for Toxic Substances and Disease Registry, U.S. Department of Health and Human Services, Public Health Service, "Managing Hazardous Materials Incidents," Atlanta, GA (June 2003)

# Chlordecone (Kepone®)

*Use Type:* Insecticide
*CAS Number:* 143-50-0
*Formula:* $C_{10}Cl_{10}O$
*Alert:* Chlordecone is no longer used in the U.S. Introduced in 1958, it was used until 1978 as an insecticide for leaf-eating insects, ants, and cockroaches, and as a larvicide for flies.
*Synonyms:* 1,2,3,5,6,7,8,9,10,10-Decachloro($5.2.2.0^{2.6}.0^{3.9}.0^{5.8}$)decano-4-one; Decachloroketone; Decachlorooctahydro-1,3,4-metheno-2*H*-cyclobuta[c,d]-pentalen-2-one; Decachlorooctahydrokepone-2-one; 1,1a,3,3a,4,5,5,5a,5b,6-Decachloro-octahydro-1,3,4-metheno-2*H*-cyclobuta[c,d]pentalen-2-one; Decachlorooctahydro-1,3,4-metheno-2*H*-cyclobuta[c,d]pentalen-2-one; 1,1a, 3, 3a, 4,5,5,5a, 5b, 6-Decachlorooctahydro-1,3,4-metheno-2H-cyclobuta[c,d]pentalen-2-one; Decachlorotetracyclodecanone; Decachlorotetrahydro-4,7-methanoindeneone; ENT 16,391; 1,3,4-Metheno-2*H*-cyclobuta[c,d]pentalen-2-one,1,1a,3,3a,4,5,5a,5b,6-decachloro-octahydro-Kepone®; NCI-C00191
*Trade Names:* CIBA® 8514, Ciba-Geigy (Switzerland); COMPOUND 1189®, General Chemicals (USA); GC-1189®, General Chemicals (USA); GENERAL CHEMICALS 1189®, General Chemicals (USA); KEPONE®, Allied Signal Inc. (USA) canceled 12/31/1987; MEREX®
*Producers:* Ehrenstorfer, Dr. (Germany); Sigma-Aldrich Laborchemikalien (Germany)

*Chemical Class:* Cyclodiene; Halo-organics
*EPA/OPP PC Code:* 027701
*California DPR Chemical Code:* 347
*ICSC Number:* 1432
*RTECS Number:* PC8575000
*RCRA Number:* U142
*EEC Number:* 606-019-00-6
*Uses:* Kepone was registered for the control or root borers on bananas with a residue tolerance of 0.01 ppm. This constituted the only food or feed use of Kepone. Non-food uses included wireworm control in tobacco fields and bait to control ants and other insects in indoor and outdoor areas. A rebuttable presumption against registration of chlordecone was issued by the U.S. EPA on March 25, 1976 on the basis of oncogenicity. The trademarked Kepone and products of six formulations were the subject of voluntary cancellation according to a U.S. EPA notice dated July 27, 1977. In a series of decisions, the first of which was issued on June 17, 1976, the U.S. EPA effectively canceled all registered products containing kepone as of May 1, 1978.
*Carcinogen/Hazard Classifications*
**IARC:** Group 2B, suspected human carcinogen
**California Prop. 65:** Carcinogen; developmental toxin
*Regulatory Authority:*
- Carcinogen (Animal Positive, Human Suspected) (IARC)[9] (NPT)[10] (DFG)[3]
- Banned or Severely Restricted (Many Countries) (UN)[13][35]
- Air Pollutant Standard Set (NIOSH) (See Reference below) (Several States)[60]
- AB 1803-Well Monitoring Chemical (CAL)
- AB 2588-Air Toxics "Hot Spots" Chemicals (CAL)
- Proposition 65 chemical (CAL); Reproductive toxin
- The "Director's List" (CAL/OSHA)
- Clean Water Act: Section 311 Hazardous Substances/RQ 40CFR117.3 (same as CERCLA, see below)
- EPA Hazardous Waste Number (RCRA No.): U142
- RCRA, 40CFR261, Appendix 8 Hazardous Constituents.
- RCRA 40CFR268.48; 61FR15654, Universal Treatment Standards: Wastewater (mg/L), 0.0011; Non-wastewater (mg/kg), 0.13
- RCRA 40CFR264, Appendix 9; TSD Facilities Ground Water Monitoring List. Suggested test method(s) (PQL $ug$/L): 8270(10).
- Superfund/EPCRA 40CFR302.4 RQ: CERCLA, 1 lb (0.454 kg)

*Description:* Kepone is a tan to white, crystalline solid. Odorless. Very slightly soluble in water; solubility = 8.1 ppm @ 25°C; about 0.4% @ 100°C. Molecular weight = 490.58. Melting/Freezing point = 360°C (sublimes without melting). Vapor pressure = $< 3.1 \times 10^{-7}$ mmHg @ 25°C. Hazard Identification (based on NFPA-704 M Rating System): Health 3, Flammability 1, Reactivity 0.

*Incompatibilities:* Acids, acid fumes. Technical grade kepone contains small amounts of hexachlorocyclopentadiene.

*Permissible Exposure Limits in Air:* NIOSH[2] recommends that the workplace environmental level be limited to 0.001 g/m$^3$ TWA for up to a 10-hour workday, 40-hour workweek, as an emergency standard. Guidelines or standards for Kepone in ambient air have been set[60] ranging from zero (South Carolina) to 0.03 $\mu$g/m$^3$ (New York) to 0.88 $\mu$g/L (Pennsylvania).

*Determination in Air:* Collection by membrane filter and backup impinger containing NaOH solution, workup with benzene, analysis by gas chromatography with electron capture detector. See NIOSH Method 5508.[18]

*Permissible Concentration in Water:* No criteria set, but runoff from spills or fire control may cause water pollution.

*Routes of Entry:* Inhalation of dust, ingestion, skin absorption.

*Harmful Effects and Symptoms*

In July 1975, a private physician submitted a blood sample to the Center for Disease Control (CDC) to be analyzed for Kepone, a chlorinated hydrocarbon pesticide. The sample had been obtained from a Kepone production worker who suffered from weight loss, nystagmus, and tremors. CDC notified the State epidemiologist that high levels of Kepone were present in the blood sample, and he initiated an epidemiologic investigation which revealed other employees suffering with similar symptoms. It was evident to the State official after visiting the plant that the employees had been exposed to Kepone at extremely high concentrations through inhalation, ingestion, and skin absorption. He recommended that the plant be closed, and company management complied. Of the 113 current and former employees of this Kepone-manufacturing plant examined, more than half exhibited clinical symptoms of Kepone poisoning. Medical histories of tremors (called "Kepone shakes" by employees), visual disturbances, loss of weight, nervousness, insomnia, pain in the chest and abdomen and, in some cases, infertility and loss of libido were reported. The employees also complained of vertigo and lack of muscular coordination. The intervals between exposure and onset of the signs and symptoms varied between patients but appeared to be dose related. NIOSH has received a report on a carcinogenesis bioassay of technical grade Kepone which was conducted by the National Cancer Institute using Osborne-Mendel rats and mice, Kepone was administered in the diet at two tolerated dosages. In addition to the clinical signs of toxicity, which were seen in both species, a significant increase (P<0.05) of hepatocellular carcinoma in rats given large dosages of Kepone and in mice at both dosages was found. Rats and mice also had extensive hyperplasia of the liver. In view of these findings, NIOSH must assume that Kepone is a potential human carcinogen. The LD$_{50}$ (oral, rat) = 95 mg/kg (moderately toxic).

*Short Term Exposure:* May be poisonous if absorbed through the skin. Eye or skin contact may cause irritation and rash. Poisonous if swallowed. Because this material has a low vapor pressure, significant inhalation of vapors is unlikely at ordinary temperatures. Exposure can cause headache, nervousness, tremor; liver, kidney damage; visual disturbance; ataxia, chest pain, skin erythema (skin redness). The Food and Drug Administration (FDA) suggests that eating fish and other foods with concentrations of chlordecone below 400 ppt, will not cause harmful health effects in people.

*Long Term Exposure:* Has been shown to cause liver cancer in animals; potential human carcinogen. May cause testicular atrophy, low sperm count, damage to the developing fetus, reproductive damage; sterility, breast enlargement; skin changes; liver and kidney damage; brain and nervous system damage with hyperactivity, hyperexcitability, muscle spasms, tremors.

*Points of Attack:* Eyes, skin, respiratory system, central nervous system, liver, kidneys, reproductive system

*Medical Surveillance:* Employers shall make medical surveillance available to all workers occupationally exposed to Kepone, including personnel periodically exposed during routine maintenance or emergency operations. Periodic examinations shall be made available at least on an annual basis.

*First Aid:* Speed in removing material from eyes and skin is of extreme importance. If this chemical gets into the eyes, remove any contact lenses at once and irrigate immediately for at least 15 minutes, occasionally lifting upper and lower lids. Seek medical attention immediately. If this chemical contacts the skin, remove contaminated clothing and wash immediately with soap and water. Shampoo hair promptly if contaminated. Seek medical attention immediately. If this chemical has been inhaled, remove from exposure, begin rescue breathing (using universal precautions) if breathing has stopped and CPR if heart action has stopped. Transfer promptly to a medical facility. When this chemical has been swallowed, get medical attention. Give large quantities of water and induce vomiting. Do not make an unconscious person vomit. Qualified medical personnel may consider the administration of cholestyramine resin (QUESTRAN®). Medical personnel should wear neoprene gloves as protection against contamination. (Dreisbach)

*References:*

- Agency for Toxic Substances and Disease Registry, "ToxFAQS for Mirex and Chlordecone," Atlanta, GA (September 1996). http://www.atsdr.cdc.gov/tfacts66.html
- National Institute for Occupational Safety and Health, Recommended Standard for Occupational Exposure to Kepone, Washington, DC (January 27, 1976).
- U.S. Environmental Protection Agency, Reviews of the Environmental Effects of Pollutants: I. Mirex and

Kepone, Report EPA 600/1-78-013, Cincinnati, OH (1978).
- National Academy of Sciences, Kepone, Mirex, Hexachlorocyclopentadiene: An Environmental Assessment, Washington, DC (1978).
- Sax, N.I., Ed., "Dangerous Properties of Industrial Materials Report," 1, No. 4, 77-79 (1981) and 4, No. 4, 10-44 (1984).
- California Environmental Protection Agency Chemical List of Lists, Sacramento CA (February 1997).

# Chlordimeform (ANSI)

*Use Type:* Acaricide and insecticide
*CAS Number:* 6164-98-3
*Formula:* $C_{10}H_{13}ClN_2$
*Synonyms:* CDM; Chlorfenamidine; *N'*-(4-chloro-2-methylphenyl)-*N,N*-dimethylmethanimidamide; Chlorphenamidine; *N'*-(4-Chloro-*o*-tolyl)-*N,N*-dimethylformamidine; Dimethylformamidine *N'*-(4-chlor-*o*-tolyl)-*N,N*-dimethylformamidin (German); *N,N*-Dimethyl-*N'*-(2-methyl-4-chlorophenyl)formamidine; *N,N*-Dimethyl-*N'*-(2-Methyl-4-chlorphenyl)-formadin (German); ENT 27,335; ENT 27,567; EP-333; Formamidine, *N'*-(4-chloro-*o*-tolyl)-*N,N*-dimethyl-; Methanimidamide, *N'*-(4-chloro-2-methylphenyl)-*N,N*-dimethyl-; *N'*-(2-Methyl-4-chlorophenyl)-*N,N*-Dimethylformamidine; *N'*-(2-Methyl-4-chlorphenyl)-formamidin-hydrochlorid (German); NSC-190935
*Trade Names:* ACARON®; BERMAT®; C 8514®; CARZOL®; CIBA® 8514; FUNDAL®; FUNDAL® 500; FUNDEX®; GALECRON®; RS 141®; SCHERING® 36,268, Schering (Germany); SN® 36268; SPANON®; SPANONE®
*Producers:* Ciba-Geigy (Switzerland); Schering (Germany); Syngenta (Switzerland)
*Chemical Class:* Formamidine; Halo-organics
*EPA/OPP PC Code:* 059701
*California DPR Chemical Code:* 300
*ICSC Number:* 0124
*RTECS Number:* LQ4375000
*EEC Number:* 650-007-00-3
*EINECS Number:* 228-200-5
*Uses:* Not registered in the U.S. Was a Restricted Use Pesticide (RUP). Used for control of bollworm and tobacco budworm in cotton, and to control mites and their eggs. Banned for use on cotton in 1989.
*Human toxicity (long-term)[77]:* Extra high–0.26923 ppb, CHCL (Chronic Human Carcinogen Level)
*Fish toxicity (threshold)[77]:* Very low–2150.95498 ppb, MATC (Maximum Acceptable Toxicant Concentration)
*Carcinogen/Hazard Classifications*
**U.S. EPA Carcinogens:** Group B2, probable carcinogen
**California Prop. 65:** Carcinogen
**IARC:** Group 3, unclassifiable

**Label Signal Word:** CAUTION or WARNING
*Regulatory Authority:*
- AB 1803-Well Monitoring Chemical (CAL)
- AB 2588-Air Toxics "Hot Spots" Chemicals (CAL)
- The "Director's List" (CAL/OSHA)
- Proposition 65 chemical (CAL)
- U.S. DOT Regulated Marine Pollutant (49CFR172.101, Appendix B)

*Description:* Colorless, crystalline solid. Slightly soluble in water; solubility = 250 ppm @ 20°C. Molecular weight 196.68. Melting/Freezing point = 35°C. Boiling point = 156°C (decomposes below BP). Vapor pressure = $3.5 \times 10^{-4}$ mmHg @ 20°C. Log $K_{ow}$ = 0.11. Unlikely to bioaccumulate in marine organisms.

*Incompatibilities:* May form explosive materials with phosphorus pentachloride. Oxidizers (chlorates, nitrates, peroxides, permanganates, perchlorates, chlorine, bromine, fluorine, etc); strong acids; alkaline reagents. May attack some forms of plastic, rubber, and coatings. Slowly hydrolyzes in water, releasing ammonia and forming acetate salts.

*Permissible Concentration in Water:* No criteria set. Runoff from spills or fire control may cause water pollution.

*Routes of Entry:* Skin absorption, ingestion and inhalation.

*Harmful Effects and Symptoms*

*Short Term Exposure:* Apprehension, anxiety, confusion, nervous excitation; dizziness; headache; numbness and weakness in limbs; muscle twitching, tremors; nausea and vomiting; slow, shallow respiration; bluish lips, fingernails or face; convulsions; loss of consciousness; breathing stops; death. May have blood in the urine or a sweet taste in the mouth. Because this material has a low vapor pressure, significant inhalation of vapors is unlikely at ordinary temperatures.

*Points of Attack:* May be fatal if inhaled, ingested, or absorbed through the skin

*Medical Surveillance:* Medical observation is recommended for 24 to 48 hours after breathing overexposure, as pulmonary edema may be delayed. As first aid for pulmonary edema, consider administering a corticosteroid spray. Cigarette smoking may exacerbate pulmonary injury and should be discouraged for at least 72 hours following exposure.

*First Aid:* If this chemical gets into the eyes, remove any contact lenses at once and irrigate immediately for at least 15 minutes, occasionally lifting upper and lower lids. Seek medical attention immediately. If this chemical contacts the skin, remove contaminated clothing and wash immediately with soap and water. Seek medical attention immediately. If this chemical has been inhaled, remove from exposure, begin rescue breathing (using universal precautions) if breathing has stopped and CPR if heart action has stopped. Transfer promptly to a medical facility. When this chemical has been swallowed, get medical attention. Give large

quantities of water and induce vomiting. Do not make an unconscious person vomit. Do not induce vomiting when formulations containing petroleum solvents are ingested.

*References:*
- Pesticide Management Education Program, "Chlordimeform (Galecron, Fundal) Chemical Profile 5/85," Cornell University, Ithaca, NY (May 1985). http://pmep.cce.cornell.edu/profiles/insect-mite/cadusafos-cyromazine/chlordimeform/insect-prof-chlordim.html
- California Environmental Protection Agency Chemical List of Lists, Sacramento CA (February 1997)

## Chlorethoxyfos (ANSI)

*Use Type:* Insecticide
*CAS Number:* 54593-83-8
*Formula:* $C_6H_{11}Cl_4O_3PS$
*Alert:* A Restricted Use Pesticide (RUP). Human toxicity (long-term): High.
*Synonyms:* Chlorethoxyphos; *O,O*-Diethyl *O*-(1,2,2,2-tetrachloroethyl) phosphorothioate; *O,O*-Diethyl *O*-(1,2,2,2-tetrachloroethyl) thionophosphate; Phosphorothioic acid, *O,O*-diethyl *O*-(1,2,2,2-tetrachloroethyl)ester
*Trade Names:* DPX-43898®; FORTRESS®, Amvac Chemicals Corp (USA); SD 208304®
*Producers:* Amvac Chemicals Corp (USA)
*Chemical Class:* Organophosphate; Halo-organics
*EPA/OPP PC Code:* 129006
*California DPR Chemical Code:* 5106
*Uses:* Chlorethoxyfos is a restricted use organophosphate insecticide registered for use on field corn, seed corn, sweet corn, and popcorn for the control of corn rootworms, wireworms, cutworms, seed corn maggots, white grubs, and symphylans.
*Human toxicity (long-term)[77]:* High–4.20 ppb, Health Advisory
*Fish toxicity (threshold):* Extra high–0.40299 ppb, MATC (Maximum Acceptable Toxicant Concentration)
*U.S. Maximum Allowable Residue Levels for Chlorethoxyfos (40 CFR 180.486):*

| CROP | ppm |
|---|---|
| Corn, field, forage | 0.01 |
| Corn, field, grain | 0.01 |
| Corn, field, stover | 0.01 |
| Corn, pop, grain | 0.01 |
| Corn, pop, stover | 0.01 |
| Corn, sweet, forage | 0.01 |
| Corn, sweet, kernel plus cob with husks removed | 0.01 |
| Corn, sweet, stover | 0.01 |

*Carcinogen/Hazard Classifications*
**Label Signal Word:** WARNING or DANGER
**WHO Acute Hazard:** Class 1 a, extremely hazardous
*Regulatory Authority:*
- U.S. Dot Inhalation Hazard Chemicals as organophosphates

*Description:* Molecular weight = 335.98.
*Incompatibilities:* May react violently with antimony(V) pentafluoride. Incompatible with lead diacetate, magnesium, silver nitrate.
*Determination in Air:* OSHA versatile sampler-2; Toluene/Acetone; Gas chromatography/Flame ionization detection; NIOSH IV, Method #5600, Organophosphorus Pesticides.[18]
*Permissible Concentration in Water:* No criteria set. Runoff from spills or fire control may cause water pollution.
*Routes of Entry:* Inhalation, skin absorption, ingestion, skin and/or eye contact.
*Harmful Effects and Symptoms*
*Short Term Exposure:* Eye pupils are small; blurred vision; eye watering; runny nose; cough; shortness of breath; salivation; dizziness; nausea, stomach cramps, diarrhea, and vomiting; increased blood pressure; profuse sweating; hypermotility, hallucinations; irritability; tingling of the skin; drowsiness; slow heartbeat; convulsions; fluid in lungs; loss of consciousness; incontinence; breathing stops; death. Organophosphates inhibit the action of acetylcholinesterase enzymes, and alter the way in which nervous impulses are transmitted. The effects can last for hours, days, or much longer. The action of the enzymes is reestablished after new enzymes are formed.
*Long Term Exposure:* Cholinesterase inhibitor; cumulative effect is possible. Organophosphates may damage the nervous system with repeated exposure, resulting in convulsions, respiratory failure. May cause liver damage.
*Points of Attack:* Respiratory system, central nervous system, cardiovascular system, blood cholinesterase.
*Medical Surveillance:* Medical observation is recommended for 24 to 48 hours after breathing overexposure, as pulmonary edema may be delayed. As first aid for pulmonary edema, consider administering a corticosteroid spray. Cigarette smoking may exacerbate pulmonary injury and should be discouraged for at least 72 hours following exposure.
*First Aid:* **Treatment for organophosphate poisoning consists of thorough decontamination, cardiorespiratory support, and administration of the antidotes atropine and pralidoxime. In cases of severe poisoning, diazepam, an anticonvulsant, should also be administered. Antidotes should be administered as prevention even if the diagnosis is in doubt.** Speed in removing material from eyes and skin is of extreme importance. *Eyes:* Eye contact can cause dangerous amounts of these chemicals to be quickly absorbed through the mucous membrane into the bloodstream. Immediately and gently flush eyes with plenty of warm or cold water (NO hot water) for at least 15 minutes, occasionally lifting the upper and lower eyelids. Get medical aid immediately. *Skin:* Get medical aid. Skin

contact can cause dangerous amounts of these chemicals to be absorbed into the bloodstream. Wearing the appropriate PPE equipment and respirator for organophosphate/carbamate pesticides, immediately flush skin with plenty of soap and water for at least 15 minutes while removing contaminated clothing and shoes. Shampoo hair promptly if contaminated. The removed, contaminated clothing and shoes should be double-bagged and left in Hot Zone for later disposal by hazardous materials experts. Skin may also be decontaminated with diluted hypochlorite solution. *Inhalation:* Get medical aid. Do not contaminate yourself. Wearing the appropriate PPE equipment and respirator for organophosphate pesticides, immediately remove the victim from the contaminated area to fresh air. If the victim is not breathing, administer artificial respiration. Do NOT use mouth-to-mouth resuscitation; use bag/mask apparatus. If breathing is difficult, administer oxygen through bag/mask apparatus until medical help arrives. Do not leave victim unattended. *Ingestion:* Call poison control. Loosen all clothing. Never give anything by mouth to an unconscious person. If victim is *unconscious or having convulsions,* do nothing except keep victim warm. Get medical aid. Transfer promptly to a medical facility. In cases of ingestion, **do not induce vomiting**. If the victim is alert and asymptomatic, administer a slurry of activated charcoal at a dose of 1 g/kg (infant, child, and adult dose). A soda can and straw may be of assistance when offering charcoal to a child. *In some cases you may be specifically instructed by poison control to induce vomiting by way of 2 tablespoons of syrup of ipecac (adult) washed down with a cup of water.* Do NOT give activated charcoal before or with ipecac syrup. *Note to physician or authorized medical personnel:* Treat cases of respiratory compromise, coma, or excessive pulmonary secretions with respiratory support using protocols and techniques available and within the scope of training. Some cases may necessitate procedures such as endotracheal intubation or cricothyrotomy by properly trained and equipped personnel. When possible, atropine (see *Antidotes,* below) should be given under medical supervision. Patients who are comatose, hypotensive, or having seizures or cardiac arrhythmias should be treated according to advanced life support protocols. *Antidotes:* Two antidotes are administered to treat organophosphate poisoning. Atropine is a competitive antagonist of acetylcholine at muscarinic receptors and is used to control the excessive bronchial secretions which are often responsible for death. Pralidoxime relieves both the nicotinic and muscarine effects of organophosphate poisoning by regenerating acetylcholinesterase and can reduce both the bronchial secretions and the muscle weakness associated with poisoning. The initial intravenous dose of atropine in adults should be determined by the severity of symptoms: An initial adult dose of 1.0 to 2.0 mg or pediatric dose of 0.01 mg/kg (minimum 0.01 mg) should be administered intravenously. If intravenous access cannot be established, atropine may also be given intramuscularly, subcutaneously or via endotracheal tube. Doses should be repeated every 15 minutes until excessive secretions and sweating have been controlled. Once bronchial secretion has been controlled, atropine administration should be repeated whenever the secretions begin to recur. In seriously poisoned patients, very large doses may be required. Alterations of pulse rate and pupillary size should not be used as indicators of treatment adequacy. Pralidoxime should be administered as early in poisoning as possible as its efficacy may diminish when given more than 24 to 36 hours after exposure. Doses are as follows: adult 1.0 g; pediatric 25 to 50 mg/kg. The drug should be administered intravenously over 30 to 60 minutes, but in a life-threatening situation, one-half of the total dose can be given per minute for a total administration time of 2 minutes. Treatment should begin to take effect within 40 minutes with a reduction in symptoms and the amount of atropine necessary to control bronchial secretion. The initial dose can be repeated in 1 hour and then every 8 to 12 hours until the patient is clinically well and no longer requires atropine. If intravenous access cannot be established, pralidoxime may also be given intramuscularly. Early administration of diazepam in addition to the combined atropine and pralidoxime treatment may help prevent the onset of seizures and potential brain and cardiac morphologic damage following high-level organophosphate poisoning.

*References:*
- U.S. Environmental Protection Agency, Office of Pesticide Programs, " Chlorethoxyfos Facts," (June 2000). http://www.epa.gov/REDs/factsheets/chlorethfact.pdf
- California Environmental Protection Agency Chemical List of Lists, Sacramento CA (February 1997).
- U.S. Environmental Protection Agency, Office of Pesticide Programs, Pesticide Residue Limits, "Chlorethoxyfos", 40 CFR 180.486. www.epa.gov/pesticides/food/viewtols.htm

## Chlorfenapyr

*Use Type:* Insecticide and miticide
*CAS Number:* 122453-73-0
*Formula:* $C_{15}H_{11}BrClF_3N_2O$
*Alert:* Chlorfenapyr is a new pesticide that, in 2000, American Cyanamid applied for registration with the U.S. EPA for food crop use. Objections arose because it could greatly harm the bird population. The application was withdrawn and reapplied for, and accepted for non food use in January, 2001.
*Synonyms:* 1*H*-Pyrrole-3-carbonitrile, 4-bromo-2-(4-chlorophenyl)-1-(ethoxymethyl)-5-(trifluoromethyl)-; 4-Bromo-2-(4-chlorophenyl)-1-(ethoxymethyl)-5-(trifluoromethyl)pyrrole-3-carbonitrile (IUPAC)

*Trade Names:* AC 303630®; CL 303630®; PHANTOM®, BASF Agricultural Products Group (Germany); PIRATE®; PYLON®, BASF Agricultural Products Group (Germany); SUMILARV®

*Producers:* Agsin (Singapore); BASF Agricultural Products Group (Germany)

*Chemical Class:* Pyrazole; Pyrroles (new); Halo-organics

*EPA/OPP PC Code:* 129093

*California DPR Chemical Code:* 3938

*Uses:* Used on ornamental crops in commercial greenhouses to control mites, caterpillar pests, thirps and fungus gnats. Not for food use. Registered in the U.S. January, 2001. Chlorfenapyr has not been previously registered in the United States. However, it has been used on cotton (under the trade name Pirate) under Sec. 18 of the Federal Insecticide Fungicide and Rodenticide Act (FIFRA).

*Carcinogen/Hazard Classifications*

**U.S. EPA Carcinogens:** Unclassifiable

**Label Signal Word:** CAUTION or WARNING

**WHO Acute Hazard:** Class II, moderately hazardous

*Harmful Effects and Symptoms*

*Short Term Exposure:* Contact with eyes or skin may cause irritation or injury. Inhalation should be avoided; use NIOSH-approved air purifying respirators for pesticides. May be harmful if swallowed.

*First Aid:* If this chemical gets into the eyes, remove any contact lenses at once and irrigate immediately for at least 15 minutes, occasionally lifting upper and lower lids. Seek medical attention immediately. If this chemical contacts the skin, remove contaminated clothing and wash immediately with soap and water. Seek medical attention immediately. If this chemical has been inhaled, remove from exposure, begin rescue breathing (using universal precautions) if breathing has stopped and CPR if heart action has stopped. Transfer promptly to a medical facility. When this chemical has been swallowed, get medical attention. Give large quantities of water and induce vomiting. Do not make an unconscious person vomit.

*References:*
- U.S. Environmental Protection Agency, Office of Pesticide Programs, Pesticide Fact Sheet, "Chlorfenapry," Washington, DC (January 2001). http://www.epa.gov/opprd001/factsheets/chlorfenapyr.pdf
- California Environmental Protection Agency Chemical List of Lists, Sacramento CA (February 1997).

# Chlorfenvinphos

*Use Type:* Insecticide
*CAS Number:* 470-90-6
*Formula:* $C_{12}H_{14}Cl_3O_4P$
*Alert:* Banned in the U.S. since 1991 and in Australia since December, 2000.

*Synonyms:* Benzyl alcohol,2,4-dichloro-α-(chloromethylene)-, Diethyl phosphate; Clorfenvinfos (Spanish); O-2-Chloor-1-(2,4-dichloor-fenyl)-vinyl-$O,O$-diethylfosfaat (Dutch); O-2-Chlor-1-(2,4-dichlor-phenyl)-vinyl-$O,O$-diaethylphosphat (German); 2-Chloro-1-(2,4-dichlorophenyl)vinyl diethyl phosphate; β-2-Chloro-1-(2',4'-dichlorophenyl) vinyl dethylphosphate; Chlorofenvinphos; Chlorphenvinfos; Chlorphenvinphos; Clofenvinfos; O-2-Cloro-1-(2,4-dicloro-fenil)-vinil-$O,O$-di etilfosfato (Italian); CVP; $O,O$-Diethylo-(2-chloro-1-(2',4'-dichlorophenyl)vinyl) phosphate; Diethyl1-(2,4-dichlorophenyl)-2-chlorovinyl phosphate; ENT 24969; Phosphate de $O,O$-diethyle etdeo-2-chloro-1-(2,4-dichlorophenyl) vinyle (French); Phosphoric acid, 2-chloro-1-(2,4-dichlorophenyl)ethenyldiethyl ester

*Trade Names:* APACHLOR®; BIRLANE®; BIRLANE LIQUID®; C8949®; C-10015®; CFV®; CGA 26351®; COMPOUND 4072®, Shell Chemical Co. (UK) canceled 10/10/1989; DERMATON®, Coopers Animal Health (USA) canceled 7/1/1987; GC 4072®, General Chemicals (USA); OMS 1328®; SAPECRON®; SAPECRON® 240; SAPECRON® 10FGEC; SD 4072®; SD 7859®; SHELL 4072®; STELADONE®; SUPONA®; SUPONE®; UNITOX®; VINYLPHATE®

*Producers:* Cyanamid Agriculture Products (USA); Ehrenstorfer, Dr. (Germany); Shell Chemical (UK); Sigma-Aldrich Laborchemikalien (Germany)

*Chemical Class:* Organophosphate; Halo-organics

*EPA/OPP PC Code:* 084101

*California DPR Chemical Code:* 306

*ICSC Number:* 1305

*RTECS Number:* TB8750000

*EEC Number:* 015-071-00-3

*EINECS Number:* 207-432-0

*Uses:* Since 1991, all products containing chlorfenvinphos have been banned in the U.S. It was widely used to control pests such as flies, fleas and mice. It is used in veterinary applications to control ticks and buffalo flies on cattle and blowflies on sheep. It is also used to control insects in mushroom, potato, lucern, dairy barns and pasture crops.

*U.S. Maximum Allowable Residue Levels*

The Food and Drug Administration (FDA) has set tolerances for chlorfenvinphos for agricultural products ranging from 0.005 to 0.2 parts chlorfenvinphos per million parts of food (0.005-0.2 ppm).

*Carcinogen/Hazard Classifications*

**WHO Acute Hazard:** Class 1b, highly hazardous

*Regulatory Authority:*
- Very Toxic Substance (World Bank)[15]
- The "Director's List" (CAL/OSHA)
- Superfund/EPCRA 40CFR355, Appendix B Extremely Hazardous Substances: TPQ = 500 lb (227 kg)
- Superfund/EPCRA 40CFR302.4 RQ: EHS, 1 lb (0.454 kg)

- U.S. DOT Inhalation Hazard Chemicals as organophosphates
- U.S. DOT Regulated Marine Pollutant (49CFR172.101, Appendix B)

*Description:* Chlorfenvinphos is a nonflammable, yellow or amber liquid with a mild odor. Slightly soluble in water; solubility = 145 ppm @ 23°C. Boiling point = 110°C @ 0.001 mmHg; 168–170°C @ 0.5 mmHg. Melting/Freezing point = –16 to –23°C. Vapor pressure = $1.9 \times 10^{-7}$ mmHg @ 25 °C. Hazard Identification (based on NFPA-704 M Rating System): Health 4, Flammability 1, Reactivity 0. Log $K_{ow}$ = 3.82. Values at or above 3.0 are likely to bioaccumulate in marine organisms.

*Incompatibilities:* Strong oxidizers, strong bases, moisture. May be corrosive to metals in the presence of moisture.

*Permissible Exposure Limits in Air:* No standards set. However, it should be recognized that this chemical can be absorbed through the skin, thereby increasing exposure.

*Determination in Air:* OSHA versatile sampler-2; Toluene/Acetone; Gas chromatography/Flame photometric detection for sulfur, nitrogen, or phosphorus; NIOSH Method IV #5600, Organophosphorus pesticides.[18]

*Permissible Concentration in Water:* 0.1 mg/L in drinking water is a recommended drinking water limit.

*Routes of Entry:* Inhalation, passes through the unbroken skin.

*Harmful Effects and Symptoms*

*Short Term Exposure:* Highly toxic ($LD_{50}$ rats = 10 mg/kg). Symptoms exhibited on chlorfenvinphos exposure are typical of cholinesterase poisoning. Nausea is often first symptom, with vomiting, abdominal cramps, diarrhea, and excessive salivation. Headache, giddiness, weakness, tightness in chest, blurring of vision, pinpoint pupils, loss of muscle coordination, and difficulty breathing. Convulsions and coma precede death. Higher exposures can cause pulmonary edema, a medical emergency that can be delayed for several hours. This can cause death. Chlorfenvinphos can cause the heart to beat slower (bradycardia) or irregularly (arrhythmia). Delayed pulmonary edema may occur after inhalation.

*Long Term Exposure:* There is no evidence that long-term exposure to small amounts of chlorfenvinphos causes any harmful health effects in people. It is not known whether chlorfenvinphos can affect reproduction or cause birth defects in people. One animal study reported decreased fertility in rats given chlorfenvinphos in their food, and another study reported that chlorfenvinphos interfered with the development of rats when the pregnant animals were fed chlorfenvinphos. May cause liver damage.

*Points of Attack:* Respiratory system, lungs, central nervous system, cardiovascular system, skin, eyes, plasma and red blood cell cholinesterase.

*Medical Surveillance:* Medical observation is recommended for 24 to 48 hours after breathing overexposure, as pulmonary edema may be delayed. Before employment and at regular times after that, the following are recommended: Plasma and red blood cell cholinesterase levels (tests for the enzyme poisoned by this chemical). If exposure stops, plasma levels return to normal in 1-2 weeks while red blood cell levels may be reduced for 1-3 months. When acetylcholinesterase enzyme levels are reduced by 25% or more below preemployment levels, risk of poisoning is increased, even if results are in lower ranges of "normal." Reassignment to work not involving organophosphate or carbamate pesticides is recommended until enzyme levels recover. If symptoms develop or overexposure occurs, repeat the above tests as soon as possible and get an exam of the nervous system. Also consider complete blood count. Consider chest x-ray following acute overexposure. Do not drink any alcoholic beverages before or during use. Alcohol promotes absorption of organophosphates.

*First Aid:* **Treatment for organophosphate poisoning consists of thorough decontamination, cardiorespiratory support, and administration of the antidotes atropine and pralidoxime. In cases of severe poisoning, diazepam, an anticonvulsant, should also be administered. Antidotes should be administered as prevention even if the diagnosis is in doubt.** Speed in removing material from eyes and skin is of extreme importance. *Eyes:* Eye contact can cause dangerous amounts of these chemicals to be quickly absorbed through the mucous membrane into the bloodstream. Immediately and gently flush eyes with plenty of warm or cold water (NO hot water) for at least 15 minutes, occasionally lifting the upper and lower eyelids. Get medical aid immediately. *Skin:* Get medical aid. Skin contact can cause dangerous amounts of these chemicals to be absorbed into the bloodstream. Wearing the appropriate PPE equipment and respirator for organophosphate/carbamate pesticides, immediately flush skin with plenty of soap and water for at least 15 minutes while removing contaminated clothing and shoes. Shampoo hair promptly if contaminated. The removed, contaminated clothing and shoes should be double-bagged and left in Hot Zone for later disposal by hazardous materials experts. Skin may also be decontaminated with diluted hypochlorite solution. *Inhalation:* Get medical aid. Do not contaminate yourself. Wearing the appropriate PPE equipment and respirator for organophosphate pesticides, immediately remove the victim from the contaminated area to fresh air. If the victim is not breathing, administer artificial respiration. Do NOT use mouth-to-mouth resuscitation; use bag/mask apparatus. If breathing is difficult, administer oxygen through bag/mask apparatus until medical help arrives. Do not leave victim unattended. *Ingestion:* Call poison control. Loosen all clothing. Never give anything by mouth to an unconscious person. If victim is *unconscious or having convulsions,* do nothing except keep victim warm. Get medical aid. Transfer promptly to a medical facility. In cases of ingestion, **do not induce vomiting.** If the victim is alert and asymptomatic, administer a slurry of activated charcoal at a dose of 1 g/kg (infant, child, and adult dose). A soda can and straw may be of assistance when offering charcoal to a child. *In some cases you may be specifically instructed by poison control to induce vomiting by way of 2 tablespoons of syrup of ipecac (adult) washed down with a cup of water.* Do NOT give activated charcoal before or with ipecac syrup.

*Note to physician or authorized medical personnel:* Treat cases of respiratory compromise, coma, or excessive pulmonary secretions with respiratory support using protocols and techniques available and within the scope of training. Some cases may necessitate procedures such as endotracheal intubation or cricothyrotomy by properly trained and equipped personnel. When possible, atropine (see *Antidotes*, below) should be given under medical supervision. Patients who are comatose, hypotensive, or having seizures or cardiac arrhythmias should be treated according to advanced life support protocols. *Antidotes:* Two antidotes are administered to treat organophosphate poisoning. Atropine is a competitive antagonist of acetylcholine at muscarinic receptors and is used to control the excessive bronchial secretions which are often responsible for death. Pralidoxime relieves both the nicotinic and muscarine effects of organophosphate poisoning by regenerating acetylcholinesterase and can reduce both the bronchial secretions and the muscle weakness associated with poisoning. The initial intravenous dose of atropine in adults should be determined by the severity of symptoms: An initial adult dose of 1.0 to 2.0 mg or pediatric dose of 0.01 mg/kg (minimum 0.01 mg) should be administered intravenously. If intravenous access cannot be established, atropine may also be given intramuscularly, subcutaneously or via endotracheal tube. Doses should be repeated every 15 minutes until excessive secretions and sweating have been controlled. Once bronchial secretion has been controlled, atropine administration should be repeated whenever the secretions begin to recur. In seriously poisoned patients, very large doses may be required. Alterations of pulse rate and pupillary size should not be used as indicators of treatment adequacy. Pralidoxime should be administered as early in poisoning as possible as its efficacy may diminish when given more than 24 to 36 hours after exposure. Doses are as follows: adult 1.0 g; pediatric 25 to 50 mg/kg. The drug should be administered intravenously over 30 to 60 minutes, but in a life-threatening situation, one-half of the total dose can be given per minute for a total administration time of 2 minutes. Treatment should begin to take effect within 40 minutes with a reduction in symptoms and the amount of atropine necessary to control bronchial secretion. The initial dose can be repeated in 1 hour and then every 8 to 12 hours until the patient is clinically well and no longer requires atropine. If intravenous access cannot be established, pralidoxime may also be given intramuscularly. Early administration of diazepam in addition to the combined atropine and pralidoxime treatment may help prevent the onset of seizures and potential brain and cardiac morphologic damage following high-level organophosphate poisoning.

*References:*
- Agency for Toxic Substances and Disease Registry, "ToxFAQs for Chlorfenvinphos," Atlanta, GA (September 1997). http://www.atsdr.cdc.gov/tfacts83.html
- New Jersey Department of Health and Senior Services, "Hazardous Substance Fact Sheet, Chlorfenvinphos," Trenton NJ (March 1989, rev. September 1998). http://www.state.nj.us/health/eoh/rtkweb/0364.pdf
- Sax, N.I., Ed., "Dangerous Properties of Industrial Materials Report," 2, No. 4, 63-67, New York, Van Nostrand Reinhold Co. (1982).
- U.S. Environmental Protection Agency, "Chemical Profile: Chlorfenvinphos," Washington, DC, Chemical Emergency Preparedness Program (October 31, 1985).
- Agency for Toxic Substances and Disease Registry, U.S. Department of Health and Human Services, Public Health Service, "Managing Hazardous Materials Incidents," Atlanta, GA (June 2003).
- California Environmental Protection Agency Chemical List of Lists, Sacramento CA (February 1997).
- Toxic Substances and Disease Registry, "ToxFAQs for Chlorfenvinphos," 1997, Atlanta, GA.

# Chlorfluazuron

*Use Type:* Insecticide
*CAS Number:* 71422-67-8
*Formula:* $C_{20}H_9Cl_3F_3N_3O_3$
*Alert:* Withdrawn, as one of 320 pesticides, by the European Commission July 25, 2003. Also withdrawn in Australia where contamination of livestock has resulted in a multitude of law suits.
*Synonyms:* Benzamide, *N*-[((3,5-dichloro-4-{{3-chloro-5-(trifluoromethyl)-2-pyridinyl}oxy}phenyl)amino)carbonyl]-2,6-difluoro- (9CI); *N*-[4-(3-Chloro-5-trifluoromethyl-2-pyridinyl-oxy)-3,5-dichloro-phenyl-aminocarbonyl]-2,6-difluorobenzamide; CFZ
*Trade Names:* ATABRON®; CGA-112913®; EXTERRA REQUIEM TERMITE BAIT®, HELIX®, Crop Care Australasia (Australia), canceled; IKI-7899®; PP-145®; UC-62644®
*Producers:* Crop Care Australasia (Australia); ICI Group (UK); Union Carbide (USA)
*Chemical Class:* Pyridine
*EPA/OPP PC Code:* 108202
*Uses:* Not registered in the U.S. Used as an insect development inhibitor insecticide for control of termites in the soil, moths and other insects, primarily on cotton.

*Carcinogen/Hazard Classifications*
**WHO Acute Hazard:** Class U, unlikely to be hazardous

*Synonyms:* Benzamide, *N*-[((3,5-dichloro-4-{{3-chloro-5-(trifluoromethyl)-2-pyridinyl}oxy}phenyl)amino)carbonyl]-2,6-difluoro- (9CI); *N*-[4-(3-Chloro-5-trifluoromethyl-2-pyridinyl-oxy)-3,5-dichloro-phenyl-aminocarbonyl]-2,6-difluorobenzamide; CFZ
*Trade Names:* ATABRON®; CGA-112913®; EXTERRA REQUIEM TERMITE BAIT®, HELIX®, Crop Care Australasia (Australia), canceled; IKI-7899®; PP-145®; UC-62644®
*Producers:* Crop Care Australasia (Australia); ICI Group (UK); Union Carbide (USA)
*Chemical Class:* Pyridine
*EPA/OPP PC Code:* 108202
*Uses:* Not registered in the U.S. Used as an insect development inhibitor insecticide for control of termites in the soil, moths and other insects, primarily on cotton.
*Carcinogen/Hazard Classifications*
**WHO Acute Hazard:** Class U, unlikely to be hazardous
*Description:* White powder. Odorless. Disperses in water.
*Permissible Concentration in Water:* No criteria set. Runoff from spills or fire control may cause water pollution.
*Routes of Entry:* Eyes and inhalation.
*Harmful Effects and Symptoms*
*Short Term Exposure:* Mechanical action of the powder may cause mild eye irritation and coughing.
*Points of Attack:* Eyes and throat
*Medical Surveillance:*
*First Aid:* If this chemical gets into the eyes, remove any contact lenses at once and irrigate immediately for at least 15 minutes, occasionally lifting upper and lower lids. Seek medical attention immediately. If this chemical contacts the skin, remove contaminated clothing and wash immediately with soap and water. Seek medical attention immediately. If this chemical has been inhaled, remove from exposure, begin rescue breathing (using universal precautions) if breathing has stopped and CPR if heart action has stopped. Transfer promptly to a medical facility. When this chemical has been swallowed, get medical attention. Give large quantities of water and induce vomiting. Do not make an unconscious person vomit.

# Chlorimuron-ethyl

*Use Type:* Herbicide
*CAS Number:* 90982-32-4
*Formula:* $C_{13}H_{11}ClN_4O_6S$
*Synonyms:* Benzoic acid,2-[(((((4-chloro-6-methoxy-2-pyrimidinyl)amino)carbonyl)amino)sulfonyl]-,ethyl ester; Caswell No. 193B; Chlorimuronethyl ester; Chlorimuronethyl [Ethyl-2-[(((4-chloro-6-methoxyprimidin-2-yl)-carbonyl)-amino)sulfonyl)benzoate]; 2-[((((4-Chloro-6-methoxy-2-pyrimidinyl)amino)carbonyl)amino)sulfonyl]benzoic acid, ethyl ester; Ethyl-2-[(((4-chloro-6-methoxyprimidin-2-yl)-carbonyl)-amino)sulfonyl]benzoate; Ethyl 2-[(((4-chloro-6-methoxypyrimidine-2-yl)aminocarbonyl)aminosulfonyl]benzoate; Ethyl 2-[(((((4-chloro-6-methoxy-2-pyrimidinyl)amino)carbonyl)amino)sulfonyl]benzoate
*Trade Names:* AUTHORITY®, FMC Agricultural Products (USA); CANOPY® (chlorimuron-ethyl + Metribuzin), DuPont Crop Protection (USA); CLASSIC®, DuPont (USA); CONCERT®, DuPont Crop Protection (USA); DPX-F6025®, DuPont (USA); GEMINI®, DuPont Crop Protection (USA), canceled 10/11/1994; LOROX®, DuPont Crop Protection (USA), canceled 1/17/1996; PREVIEW®, DuPont Crop Protection (USA), canceled 3/9/2000; RELIANCE®, DuPont Crop Protection (USA); SKIRMISH®, FMC Agricultural Products (USA); SYNCHRONCY®, DuPont Crop Protection (USA); SYNCHRONY®, DuPont Crop Protection (USA), canceled 1/17/1996
*Producers:* Agrimor International (USA); DuPont Crop Protection (USA); FMC Agricultural Products (USA)
*Chemical Class:* Sulfonylurea
*EPA/OPP PC Code:* 128901
*California DPR Chemical Code:* 2458
*Uses:* Used to control broadleaf weeds and annual morning glory in soybeans and peanuts.
*Human toxicity (long-term):* Very low–140.00 ppb, Health Advisory
*Fish toxicity (threshold):* Very low–1308.29960 ppb, MATC (Maximum Acceptable Toxicant Concentration)
*U.S. Maximum Allowable Residue Levels for Chlorimuron-ethyl (40 CFR 180.429):*

| CROP | ppm |
|---|---|
| Peanuts | 0.02 |
| Soybeans | 0.05 |

*Carcinogen/Hazard Classifications*
**Label Signal Word:** CAUTION or WARNING
*Regulatory Authority:*
- EPCRA Section 313 Form R *de minimis* concentration reporting level: 1.0%.
- EPA 40 CFR 372.65, Toxic Chemical Release, Specific Toxic Chemical Listings

*Description:* Off white to pale yellow crystals or powder. Odorless. Soluble in water. Melting point = 181°C.
*Incompatibilities:* Nitrates.
*Permissible Concentration in Water:* No criteria set. Runoff from spills or fire control may cause water pollution.
*Routes of Entry:* Skin and eye contact.
*Harmful Effects and Symptoms*
*Short Term Exposure:* Contact with eyes or skin may cause irritation or burns. Inhalation of dust should be avoided; use NIOSH-approved air purifying respirators for pesticides. Because this material has a low vapor pressure, significant inhalation of vapors is unlikely at ordinary temperatures. May be harmful if swallowed. Skin contact may cause allergic reaction.

*Long Term Exposure:* Repeated or prolonged contact may cause skin and lung sensitization, resulting in allergies.
*Points of Attack:* Skin.
*Medical Surveillance:* Evaluation by a qualified allergist, including careful exposure history and special testing, may help diagnose skin allergy.
*First Aid:* If this chemical gets into the eyes, remove any contact lenses at once and irrigate immediately for at least 15 minutes, occasionally lifting upper and lower lids. Seek medical attention immediately. If this chemical contacts the skin, remove contaminated clothing and wash immediately with soap and water. Seek medical attention immediately. If this chemical has been inhaled, remove from exposure, begin rescue breathing (using universal precautions) if breathing has stopped and CPR if heart action has stopped. Transfer promptly to a medical facility. When this chemical has been swallowed, get medical attention. Give large quantities of water and induce vomiting. Do not make an unconscious person vomit.
*References:*
- Pesticide Management Education Program, "Chlorimuron ethyl Herbicide Profile 4/86," Cornell University, Ithaca, NY (April 4, 1986). http://pmep.cce.cornell.edu/profiles/herb-growthreg/cacodylic-cymoxanil/chlorimuron-ethyl/herb-prof-chlorim-ethyl.html
- U.S. Environmental Protection Agency, Office of Pesticide Programs, Pesticide Residue Limits, "Chlorimuron ethyl," 40 CFR 180.429. http://www.setonresourcecenter.com/40CFR/Docs/wcd0004d/wcd04d68.asp
- California Environmental Protection Agency Chemical List of Lists, Sacramento CA (February 1997).

# Chlormephos

*Use Type:* A soil insecticide
*CAS Number:* 24934-91-6
*Formula:* $C_5H_{12}ClO_2PS_2$
*Alert:* No longer sold in the United States.
*Synonyms:* S-(Chloromethyl)-O,O-diethyl phosphorodithioate; S-Chloromethyl-O,O-diethyl phosphorodithioate; S-Chloromethyl-O,O-diethyl phosphorodithiothiolothionate; S-(Chloromethyl)-O,O-diethyl phosphorodithioic acid; S-(Chloromethyl) O,O-diethyl ester phosphorodithioic acid; Phosphorodithioic acid, S-(chloromethyl) O,O-diethyl ester
*Trade Names:* DOTAN®; MC2188®; SHERMAN®, Calliope (France)
*Producers:* Calliope (France); Feinchemie Schwebda (Germany); Sigma-Aldrich Laborchemikalien (Germany)
*Chemical Class:* Organophosphate; halo-organics
*EPA/OPP PC Code:* 295300
*RTECS Number:* TD5170000

*Uses:* Not registered as a pesticide in the U.S.
*Carcinogen/Hazard Classifications*
**WHO Acute Hazard:** Class 1 a, extremely hazardous
*Regulatory Authority:*
- Superfund/EPCRA 40CFR355, Appendix B Extremely Hazardous Substances: TPQ = 500 lb (227 kg)
- Superfund/EPCRA 40CFR302.4 RQ: EHS, 1 lb (0.454 kg)
- U.S. DOT Inhalation Hazard Chemicals as organophosphates
- U.S. DOT Regulated Marine Pollutant (49CFR172.101, Appendix B)

*Description:* Chlormephos is an organophosphate, colorless to pale yellow liquid. Slightly soluble in water; solubility = 60 ppm @ 20°C. Molecular weight = 234.72. Boiling point = 81-85°C @ 0.1 mmHg. Vapor pressure $5 \times 10^{-3}$ mmHg @ 20°C. Hazard Identification (based on NFPA-704 M Rating System): Health 3, Flammability 1, Reactivity 0.
*Incompatibilities:* Contact with strong oxidizers may cause fire and explosion. May corrode metals.
*Permissible Exposure Limits in Air:* No standards set. Passes through the skin.
*Determination in Air:* OSHA versatile sampler-2; Toluene/Acetone; Gas chromatography/Flame photometric detection for sulfur, nitrogen, or phosphorus; NIOSH Method IV #5600, Organophosphorus pesticides.[18]
*Routes of Entry:* Inhalation, ingestion, skin contact, passes through the skin.
*Harmful Effects and Symptoms*
*Short Term Exposure:* This material is poisonous; it may be fatal if inhaled, swallowed, or absorbed through the skin. The acute $LD_{50}$ (oral, rat) = 7 mg/kg (highly toxic). Symptoms exhibited on chlormephos exposure are typical of cholinesterase poisoning. Nausea is often first symptom, with vomiting, abdominal cramps, diarrhea, and excessive salivation. Headache, giddiness, weakness, tightness in chest, blurring of vision, pinpoint pupils, loss of muscle coordination, and difficulty breathing. Death may occur from failure of the respiratory center, paralysis of the respiratory muscles, intense bronchoconstriction, or all three. Convulsions and coma precede death. Higher exposures can cause pulmonary edema, a medical emergency that can be delayed for several hours. This can cause death. Chlormephos can cause the heart to beat slower (bradycardia) or irregularly (arrhythmia). Delayed pulmonary edema may occur after inhalation.
*Long Term Exposure:* There is limited evidence that this chemical may damage the developing fetus. Can affect the nervous system and cause impaired memory, depression, anxiety, or irritability. Repeated exposure may damage the nerves, causing weakness, prickly feeling, and affect coordination in the arms and legs. Symptoms resembling influenza with headache, nausea, weakness have been reported.

*Points of Attack:* Respiratory system, lungs, central nervous system, cardiovascular system, skin, eyes, plasma and red blood cell cholinesterase.

*Medical Surveillance:* Medical observation is recommended for 24 to 48 hours after breathing overexposure, as pulmonary edema may be delayed. Before employment and at regular times after that, the following are recommended: Plasma and red blood cell cholinesterase levels (tests for the enzyme poisoned by this chemical). If exposure stops, plasma levels return to normal in 1-2 weeks while red blood cell levels may be reduced for 1-3 months. When acetylcholinesterase enzyme levels are reduced by 25% or more below preemployment levels, risk of poisoning is increased, even if results are in lower ranges of "normal." Reassignment to work not involving organophosphate or carbamate pesticides is recommended until enzyme levels recover. If symptoms develop or overexposure occurs, repeat the above tests as soon as possible and get an exam of the nervous system. Also consider complete blood count. Consider chest x-ray following acute overexposure. Do not drink any alcoholic beverages before or during use. Alcohol promotes absorption of organophosphates.

*First Aid:* **Treatment for organophosphate poisoning consists of thorough decontamination, cardiorespiratory support, and administration of the antidotes atropine and pralidoxime. In cases of severe poisoning, diazepam, an anticonvulsant, should also be administered. Antidotes should be administered as prevention even if the diagnosis is in doubt.** Speed in removing material from eyes and skin is of extreme importance. *Eyes:* Eye contact can cause dangerous amounts of these chemicals to be quickly absorbed through the mucous membrane into the bloodstream. Immediately and gently flush eyes with plenty of warm or cold water (NO hot water) for at least 15 minutes, occasionally lifting the upper and lower eyelids. Get medical aid immediately. *Skin:* Get medical aid. Skin contact can cause dangerous amounts of these chemicals to be absorbed into the bloodstream. Wearing the appropriate PPE equipment and respirator for organophosphate pesticides, immediately flush skin with plenty of soap and water for at least 15 minutes while removing contaminated clothing and shoes. Shampoo hair promptly if contaminated. The removed, contaminated clothing and shoes should be double-bagged and left in Hot Zone for later disposal by hazardous materials experts. Skin may also be decontaminated with diluted hypochlorite solution. *Inhalation:* Get medical aid. Do not contaminate yourself. Wearing the appropriate PPE equipment and respirator for organophosphate pesticides, immediately remove the victim from the contaminated area to fresh air. If the victim is not breathing, administer artificial respiration. Do NOT use mouth-to-mouth resuscitation; use bag/mask apparatus. If breathing is difficult, administer oxygen through bag/mask apparatus until medical help arrives. Do not leave victim unattended. *Ingestion:* Call poison control. Loosen all clothing. Never give anything by mouth to an unconscious person. If victim is *unconscious or having convulsions,* do nothing except keep victim warm. Get medical aid. Transfer promptly to a medical facility. In cases of ingestion, **do not induce vomiting**. If the victim is alert and asymptomatic, administer a slurry of activated charcoal at a dose of 1 g/kg (infant, child, and adult dose). A soda can and straw may be of assistance when offering charcoal to a child. *In some cases you may be specifically instructed by poison control to induce vomiting by way of 2 tablespoons of syrup of ipecac (adult) washed down with a cup of water.* Do NOT give activated charcoal before or with ipecac syrup.

*Note to physician or authorized medical personnel:* Treat cases of respiratory compromise, coma, or excessive pulmonary secretions with respiratory support using protocols and techniques available and within the scope of training. Some cases may necessitate procedures such as endotracheal intubation or cricothyrotomy by properly trained and equipped personnel. When possible, atropine (see *Antidotes*, below) should be given under medical supervision. Patients who are comatose, hypotensive, or having seizures or cardiac arrhythmias should be treated according to advanced life support protocols. *Antidotes:* Two antidotes are administered to treat organophosphate poisoning. Atropine is a competitive antagonist of acetylcholine at muscarinic receptors and is used to control the excessive bronchial secretions which are often responsible for death. Pralidoxime relieves both the nicotinic and muscarine effects of organophosphate poisoning by regenerating acetylcholinesterase and can reduce both the bronchial secretions and the muscle weakness associated with poisoning. The initial intravenous dose of atropine in adults should be determined by the severity of symptoms: An initial adult dose of 1.0 to 2.0 mg or pediatric dose of 0.01 mg/kg (minimum 0.01 mg) should be administered intravenously. If intravenous access cannot be established, atropine may also be given intramuscularly, subcutaneously or via endotracheal tube. Doses should be repeated every 15 minutes until excessive secretions and sweating have been controlled. Once bronchial secretion has been controlled, atropine administration should be repeated whenever the secretions begin to recur. In seriously poisoned patients, very large doses may be required. Alterations of pulse rate and pupillary size should not be used as indicators of treatment adequacy. Pralidoxime should be administered as early in poisoning as possible as its efficacy may diminish when given more than 24 to 36 hours after exposure. Doses are as follows: adult 1.0 g; pediatric 25 to 50 mg/kg. The drug should be administered intravenously over 30 to 60 minutes, but in a life-threatening situation, one-half of the total dose can be given per minute for a total administration time of 2 minutes. Treatment should begin to take effect within 40 minutes with a reduction in symptoms and the amount of atropine necessary to control bronchial secretion. The initial dose can be repeated in 1 hour and then every 8 to 12 hours until the patient is clinically well and no longer requires atropine. If intravenous access cannot be established, pralidoxime may also be given intramuscularly. Early administration of diazepam in addition to the combined atropine and pralidoxime treatment may help prevent the onset of seizures and potential brain and cardiac morphologic damage following high-level organophosphate poisoning.

*References:*
- New Jersey Department of Health and Senior Services, "Hazardous Substances Fact Sheets, Chlormephos," Trenton, NJ (August 2000). www.state.nj.us/health/eoh/rtkweb/2235.pdf
- U.S. Environmental Protection Agency, "Chemical Profile: Chlormephos," Washington, DC, Chemical Emergency Preparedness Program (October 31, 1985).
- Agency for Toxic Substances and Disease Registry, U.S. Department of Health and Human Services, Public Health Service, "Managing Hazardous Materials Incidents," Atlanta, GA (June 2003).
- California Environmental Protection Agency "Chemical List of Lists," Sacramento CA (February 1997).

# Chlormequat Chloride

*Use Type:* Plant growth regulator
*CAS Number:* 999-81-5
*Formula:* $C_5H_{13}Cl_2N$; $ClCH_2CH_2N(CH_3)_3Cl$
*Synonyms:* Ammonium, (2-chloroethyl)trimethyl-, Chloride 2-chloro-$N,N,N$-trimethylethanaminium chloride; Choline dichloride; 2-Chloraethyl-trimethylammoniumchlorid (German); 2-Chloroethyl trimethylammonium chloride; Chlorcholinchlorid; Chlorcholine chloride; Chlormequat; Chlorocholine chloride; ($\beta$-Chloroethyl)trimethylammonium chloride; 2-Chloro-$N,N,N$-trimethylammonium chloride; (2-Chloroethyl)trimethylammonium chloride; 2-Chloro-$N,N,N$-ethyl)trimethylethanaminium chloride; Clormecuato de cloroacetilo (Spanish); Ethanaminium, 2-Chloro-$N,N,N$-trimethyl-, Chloride (9CI); NCI-C02960; Trimethyl-$\beta$-chlorethylammoniumchlorid; Trimethyl-$\beta$-chloroethyl ammonium chloride
*Trade Names:* AC 38555®; ADJUST®, Mandops (UK); ATLAS CHLORMEQUAT®, Whyte Agrochemicals (UK); CCC PLANT GROWTH REGULANT®; 60-CS-16®; CYCLOCEL®; CYCOCEL®, BASF (Germany); CYCOCEL-EXTRA®, BASF (Germany); CYCOGAN®; CYCOGAN EXTRA®; CYOCEL®; EI 38,555®; HICO CCC®; HORMOCEL-2CCC®; INCRECEL®; LIHOCIN®; RETACEL®; RETACIL®; STABILAN®; TUR®; WR 62®
*Producers:* Agan Chemical Manufacturers Ltd. (Israel); Agsin (Singapore); BASF (Germany); Fluorochem (UK); Rhodia (France); Rhone-Poulenc Agro (France); Sigma-Aldrich Laborchemikalien (Germany); TCI America (USA); Whyte Agrochemicals (UK)
*Chemical Class:* Quaternary ammonium compound; Haloorganics
*EPA/OPP PC Code:* 018101
*California DPR Chemical Code:* 1512
*ICSC Number:* 0781
*RTECS Number:* BP5250000
*EEC Number:* 007-003-00-6
*EINECS Number:* 213-666-4
*Uses:* Said to be effective for cereal grains, tomatoes, and peppers. In California, major uses are in greenhouse plants and flowers and in nurseries.

*Carcinogen/Hazard Classifications*
WHO Acute Hazard: Class III, slightly hazardous
*Regulatory Authority:*
- Air Pollutant Standard Set (former USSR)[35]
- Actively registered pesticide in California.
- Superfund/EPCRA 40CFR355, Appendix B Extremely Hazardous Substances: TPQ = 100/10,000 lb (45.4/4,540 kg)
- Superfund/EPCRA 40CFR302.4 RQ: EHS, 1 lb (0.454 kg)

*Description:* Chlormequat chloride is a white to yellowish crystalline solid. Fishy odor. Soluble in water; solubility = 75 g/100 ml @ 20°C. Molecular weight = 158.13. Melting/Freezing point = 245°C (decomposes). Vapor pressure = very low. Hazard Identification (based on NFPA-704 M Rating System): Health 3, Flammability 0, Reactivity 0. Highly soluble in water. Carrier solvents used in commercial products may alter physical and toxicological properties

*Incompatibilities:* Chlormequat chloride decomposes on heating or in fire forming nitrogen oxides, carbon monoxide, and hydrogen chloride fumes. This chemical decomposes on heating with strong aqueous alkali solutions forming trimethylamine and other gaseous products. Contact with strong oxidizers may cause fire and explosions. Attacks many metals in presence of water.

*Permissible Exposure Limits in Air:* There is no U.S. airborne exposure limit. This chemical can be absorbed through the skin, thereby increasing exposure. The former USSR[35] has set a ceiling TLV in workplace air of 0.3 mg/m$^3$.

*Permissible Concentration in Water:* The former USSR[35] has set a MAC in surface water of 0.2 mg/L.

*Routes of Entry:* Inhalation, absorbed through the skin.

*Harmful Effects and Symptoms*
*Short Term Exposure:* The $LD_{low}$ (oral, human) = 10 mg/kg. It is an irritant and can be absorbed through the skin. Irritates the eyes and the respiratory tract. Higher exposures can cause pulmonary edema, a medical emergency that can be delayed for several hours. This can cause death. Exposure can cause nausea and vomiting. Higher levels can

cause slow or irregular heartbeat, tremors, seizures, and coma. This can be fatal. Chlormequat chloride may affect the nervous system.
*Long Term Exposure:* May cause liver damage.
*Points of Attack:* Lungs, liver, nervous system.
*Medical Surveillance:* Liver function tests. Consider chest x-ray following acute overexposure. EKG examination of the nervous system.
*First Aid:* If this chemical gets into the eyes, remove any contact lenses at once and irrigate immediately for at least 15 minutes, occasionally lifting upper and lower lids. Seek medical attention immediately. If this chemical contacts the skin, remove contaminated clothing and wash immediately with soap and water. Seek medical attention immediately. If this chemical has been inhaled, remove from exposure, begin rescue breathing (using universal precautions) if breathing has stopped, and CPR if heart action has stopped. Transfer promptly to a medical facility. When this chemical has been swallowed, get medical attention. Give large quantities of water and induce vomiting. Do not make an unconscious person vomit. Medical observation is recommended for 24 to 48 hours after breathing overexposure, as pulmonary edema may be delayed. As first aid for pulmonary edema, a doctor or authorized paramedic may consider administering a corticosteroid spray.
*References:*
- New Jersey Department of Health and Senior Services, Hazardous Substance Fact Sheet, "Chlormequat Chloride," Trenton NJ (April 1999). http://www.state.nj.us/health/eoh/rtkweb/2236.pdf
- U.S. Environmental Protection Agency, "Chemical Profile: Chlormequat Chloride," Washington, DC, Chemical Emergency Preparedness Program (October 31, 1985).
- California Environmental Protection Agency "Chemical List of Lists," Sacramento CA (February 1997).

# Chloroacetaldehyde

*Use Type:* Fungicide
*CAS Number:* 107-20-0
*Formula:* $C_2H_3ClO$; $ClCH_2CHO$
*Synonyms:* Acetaldehyde, chloro-; 2-Chloroacetaldehyde; Chloroacetaldehyde monomer; 2-Chloroethanal; 2-Chloro-1-ethanal; Cloroacetaldehido (Spanish); Monochloroacetaldehyde
*Producers:* Cllariant (Switzerland); Daicel Chemical Industries (Japan); Fluorochem (UK); Junsei Chemical Co., Ltd. (Japan); Showa Denko (Japan); Sumitomo Seika Chemicals (Japan)
*ICSC Number:* 0706
*RTECS Number:* AB2450000
*RCRA Number:* P023
*EINECS Number:* 203-472-8

*Uses:* Chloroacetaldehyde is used as a fungicide, as an intermediate in 2-aminothiazole manufacture. It is used to control algae, bacteria and fungi in water. It is also used in dentistry and in bark removal from tree trunks.
*Regulatory Authority:*
- Air Pollutant Standard Set (ACGIH)[1] (Australia) (DFG)[3] (HSE)[33] (Israel) (Mexico) (Several States)[60] (Several Canadian Provinces)
- U.S. DOT Inhalation Hazard Chemicals
- Permissible Exposure Limits for Chemical Contaminants (CAL/OSHA)
- The "Director's List" (CAL/OSHA)
- EPA Hazardous Waste Number (RCRA No.): P023
- RCRA, 40CFR261, Appendix 8 Hazardous Constituents.
- Superfund/EPCRA 40CFR302.4 RQ: CERCLA, 1000 lb (454 kg)
- U.S. DOT 49CFR172.101, Inhalation Hazardous Chemical
- Canada, WHMIS, Ingredients Disclosure List

*Description:* Chloroacetaldehyde is a combustible, colorless liquid. Very sharp, irritating odor. Soluble in water; solubility = $4.42 \times 10^5$ ppm @ 20°C. Molecular weight = 78.51. Boiling point = 85–90°C (decomposes) @ 760 mmHg. Melting/Freezing point = −16.3; also 43–48°C (Merck/12). Soluble in water. Flash point = 88°C (40% solutions). Autoignition = 88 °C. Vapor pressure = 139 hPa @ 25°C; 95 torr @ 20 °C. Log $K_{ow}$ = < 0.5. Unlikely to bioaccumulate in marine organisms.
*Incompatibilities:* Reacts with oxidizers, acids. On heating, chloroacetaldehyde forms chlorine fumes
*Permissible Exposure Limits in Air:* The Federal (OSHA PEL) ceiling limit is 1 ppm (3 mg/m³), not to be exceeded at any time. NIOSH[2] and ACGIH[1] also recommend the same ceiling limit. The British (HSE)[33], Australian, Israli, Mexican and Canadian Provinces of Alberta, British Columbia, Ontario, and Quebec have set a limit of 1 ppm (3 mg/m³) as a ceiling value and the British HSE's[33] STEL is 1 ppm (3 mg/m³). The German DFG MAK is also 1 ppm and Peak Limitation is 2 times normal MAK (5 minute) not to be exceeded 8 times during a workshift[3]. The NIOSH[2] IDLH level = 45 ppm. The odor threshold = 0.92 ppm. In addition, several states have set guidelines or standards for chloroacetaldehyde in ambient air[60] ranging from 1.0 µg/m³ (New York) to 25 µg/m³ (Virginia) to 30 µg/m³ (North Dakota) to 60 µg/m³ (Connecticut) to 71 µg/m³ (Nevada).
*Determination in Air:* Si gel; Methanol; Gas chromatography/Electrochemical detection; NIOSH IV Method #2015.[18]
*Permissible Concentration in Water:* No criteria set, but runoff from spills or fire control may cause water pollution.
*Routes of Entry:* Inhalation, ingestion, eye and skin contact.
*Harmful Effects and Symptoms*
*Short Term Exposure:* Corrosive to the eyes, skin, and respiratory tract. Contact can cause burns and permanent

damage. Higher exposures can cause pulmonary edema, a medical emergency that can be delayed for several hours. This can cause death. Can cause sensitization of skin and respiratory system. The $LD_{50}$ (oral, mouse) = 21 mg/kg.

*Long Term Exposure:* This chemical may cause mutations. It can cause skin allergy and an asthma-like lung allergy.

*Points of Attack:* Eyes, skin, respiratory system, lungs.

*Medical Surveillance:* Consider the points of attack in preplacement and periodic physical examinations. Lung function tests., evaluation by a qualified allergist. Consider chest x-ray following acute overexposure.

*First Aid:* If this chemical gets into the eyes, remove any contact lenses at once and irrigate immediately for at least 30 minutes, occasionally lifting upper and lower lids. Seek medical attention immediately. If this chemical contacts the skin, remove contaminated clothing and wash immediately with soap and water. Seek medical attention immediately. If this chemical has been inhaled, remove from exposure, begin rescue breathing (using universal precautions) if breathing has stopped, and CPR if heart action has stopped. Transfer promptly to a medical facility. When this chemical has been swallowed, get medical attention. Give large quantities of water and induce vomiting. Do not make an unconscious person vomit. Medical observation is recommended for 24 to 48 hours after breathing overexposure, as pulmonary edema may be delayed. As first aid for pulmonary edema, a doctor or authorized paramedic may consider administering a corticosteroid spray.

*References:*
- U.S. Environmental Protection Agency, Chloroacetaldehyde, Health and Environmental Effects Profile No. 40, Office of Solid Waste, Washington, DC (April 30, 1980).
- Sax, N.I., Ed., "Dangerous Properties of Industrial Materials Report," 2, No. 4, 70-72, New York, Van Nostrand Reinhold Co. (1982).
- California Environmental Protection Agency "Chemical List of Lists," Sacramento CA (February 1997).
- U.S. Environmental Protection Agency, "Chemical Profile: Chloroacetaldehyde," Washington, DC, Chemical Emergency Preparedness Program (October 31, 1985).
- New Jersey Department of Health and Senior Services and Senior Services, "Hazardous Substance Fact Sheet: Chloroacetaldehyde," Trenton, NJ (April 1998). www.state.nj.us/health/eoh/rtkweb/0372.pdf

# Chloroacetic Acid

*Use Type:* Herbicide and defoliant
*CAS Number:* 79-11-8
*Formula:* $C_2H_3ClO_2$; $ClCH_2COOH$
*Synonyms:* Acide chloracetique (French); Acide monochloracetique (French); Acido cloroacetico (Spanish); Acidomonocloroacetico (Italian); Acetic acid, chloro-; Chloracetic acid; Chloroethanoic acid; MCA; Monochloracetic acid; Monochloorazijnzuur (Dutch); Monochloressigsaeure (German); Monochloroacetic acid; Monochloroethanoic acid; NCI-C60231

*Producers:* Acetex (Canada); Akzo Nobel (Netherlands); ATOFINA Chemicals (USA); Clariant (Switzerland); Daicel Chemical Industries (Japan); Dow Chemical (USA); GFS Chemicals (USA); ICC (USA); Merck (Germany); Mitsui Chemicals (Japan); Niacet (USA); Sigma-Aldrich Laborchemikalien (Germany); United Phosphorus (India); Whyte (UK)

*Chemical Class:* Halo-organics
*EPA/OPP PC Code:* 279400
*California DPR Chemical Code:* 3102
*ICSC Number:* 0235
*RTECS Number:* AF8575000
*EEC Number:* 607-003-00-1
*EINECS Number:* 201-178-4

*Uses:* Chloroacetic acid major use is a defoliant. It is also used as a chemical intermediate in the synthesis of sodium carboxymethyl cellulose (used for drilling muds, detergents, food, and pharmaceuticals), and such other diverse substances as ethyl chloroacetate, glycine, synthetic caffeine, sarcosine, thioglycolic acid, and various dyes, and synthetic caffeine.

*Carcinogen/Hazard Classifications*

**WHO Acute Hazard:** Class III, slightly hazardous

*Regulatory Authority:*
- Air Pollutant Standard Set (former USSR)[43] (HSE)[33]
- EPA/SARA 302 (EPCRA) Extremely hazardous substances
- AB 2588-Air Toxics "Hot Spots" Chemicals (CAL)
- CAL Air Resources Board/AB 1807 Toxic Air Contaminants
- U.S. DOT Inhalation Hazard Chemicals
- Clean Air Act: Hazardous Air Pollutants (Title I, Part A, Section 112)
- Superfund/EPCRA 40CFR355, Appendix B Extremely Hazardous Substances: TPQ = 100/10,000 lb (45.4/4,540 kg).
- Superfund/EPCRA 40CFR302.4 RQ: CERCLA, 1 lb (0.454 kg).
- EPCRA Section 313 Form R *de minimus* concentration reporting level: 1.0%
- Canada, WHMIS, Ingredients Disclosure List; National Pollutant Release Inventory (NPRI)

*Description:* Colorless to white crystalline solid. It has a strong vinegar-like odor and an odor threshold of 0.15 mg/m³. Very soluble in water; solubility = 85 g/100 ml. Molecular weight = 94.48. Boiling point = 188°C ($\alpha, \beta, \gamma$ forms). Melting/Freezing point = 63°C ($\alpha$); 5-56°C ($\beta$); 50°C ($\gamma$). Vapor pressure = 1 mmHg @ 43. Flash point = 126°C. Autoignition temperature = >500°C. Hazard Identification (based on NFPA-704 M Rating System):

Health 3, Flammability 1, Reactivity 0. Log $K_{ow}$ = 0.24. Unlikely to bioaccumulate in marine organisms.

***Incompatibilities:*** The solution in water is a strong acid. Contact with strong oxidizers, strong bases, and strong reducing agents can cause violent reactions. Chloroacetic acid decomposes on heating producing toxic and corrosive hydrogen chloride, phosgene gases. Attacks metals in the presence of moisture.

***Permissible Exposure Limits in Air:*** No U.S. exposure limits have been set by OSHA[2], NIOSH[2], or ACGIH[1], however, the HSE[33] (UK) and the American Industrial Hygiene Association recommends an airborne exposure limit of 0.3 ppm (1 mg/m$^3$) TWA and the AIHA recommends a STEL of 1 ppm (4 mg/m$^3$). Skin absorbtion is indicated. This chemical can be absorbed through the skin, thereby increasing exposure. The former USSR-UNEP/IRPTC project[43] has set a MAC in workplace air of 1.0 mg/m$^3$.

***Permissible Concentration in Water:*** No criteria set, but runoff from spills or fire control may cause water pollution.

***Routes of Entry:*** Inhalation, ingestion and skin contact. This chemical can be absorbed through the skin, thereby increasing exposure.

***Harmful Effects and Symptoms***

***Short Term Exposure:*** Corrosive to the eyes, skin, and respiratory tract. Contact can cause severe irritation and burns. Inhaling thid chemical can cause pulmonary edema, a medical emergency that can be delayed for several hours. This can cause death. Chloroacetic acid can cause a feeling of anxiety, restlessness, blurred vision, a feeling of "pins and needles" in the limbs, muscle twitching, and/or hallucinations, may affect the cardiovascular system, central nervous system and kidneys, resulting in heart problems, convulsions, and kidney damage. These effects may be delayed. Symptoms of exposure include irritation and pain in skin. If chloroacetic acid is inhaled the patient may exhibit difficulty in breathing. Vomiting may occur if the material is ingested. It can burn the skin, cornea and respiratory tract. This material is very toxic. The probable lethal oral dose is 50-500 mg/kg of body weight, between one teaspoon and one ounce, for a 150 lb. person. Chloroacetic acid is irritating to the skin, cornea, and respiratory tract and causes burns. It may severely damage skin and mucous membranes. Ingestion may interfere with essential enzyme systems and cause perforation and peritonitis. Burns to skin result in marked fluid and electrolyte loss. Death may follow if more than 3% of the skin is exposed to this material. Other health hazards include central nervous system depression, and respiratory system depression.

***Long Term Exposure:*** Repeated exposure may cause kidney damage and affect the lungs.

***Points of Attack:*** Lungs, kidneys, central nervous system.

***Medical Surveillance:*** Lung function tests, kidney function tests. Examination of the nervous system.

***First Aid:*** If this chemical gets into the eyes, remove any contact lenses at once and irrigate immediately for at least 15 minutes, occasionally lifting upper and lower lids. Seek medical attention immediately. If this chemical contacts the skin, remove contaminated clothing and wash immediately with soap and water. Seek medical attention immediately. If this chemical has been inhaled, remove from exposure, begin rescue breathing (using universal precautions) if breathing has stopped, and CPR if heart action has stopped. Transfer promptly to a medical facility. When this chemical has been swallowed, get medical attention. If victim is conscious, administer water or milk. Do not induce vomiting. Medical observation is recommended for 24 to 48 hours after breathing overexposure, as pulmonary edema may be delayed. As first aid for pulmonary edema, a doctor or authorized paramedic may consider administering a corticosteroid spray.

***References:***

- New Jersey Department of Health and Senior Services and Senior Services, "Hazardous Substance Fact Sheet: Chloroacetic Acid," Trenton, NJ (July 1996, rev. July 2002). www.state.nj.us/health/eoh/rtkweb/0373.pdf
- National Institute for Occupational Safety and Health, Profiles on Occupational Hazards for Criteria Document Priorities: Monochloroacetic Acid, pp 309-11, Report PB-274,073, Rockville, MD (1977).
- Sax, N.I., Ed., "Dangerous Properties of Industrial Materials Report," 3, No. 5, 99-101, New York, Van Nostrand Reinhold Co. (1983).
- U.S. Environmental Protection Agency, "Chemical Profile: Chloroacetic Acid," Washington, DC, Chemical Emergency Preparedness Program (October 31, 1985).
- California Environmental Protection Agency "Chemical List of Lists," Sacramento CA (February 1997).
- New York State Department of Health, "Chemical Fact Sheet: Chloroacetic Acid," Albany, NY, Bureau of Toxic Substance Assessment (January 1986).

# 1-Chloro-1-nitropropane

***Use Type:*** Fungicide
***CAS Number:*** 600-25-9
***Formula:*** $C_3H_6ClNO_2$; $C_2H_5CHClNO_2$
***Synonyms:*** 1-Chloro-1-nitropropano (Spanish); Chloronitropropane; 1,1-Chloronitropropane; Propane, 1-chloro-1-nitro-
***Trade Names:*** KORAX®; KORAX 6®; LANSTAN®
***Chemical Class:*** Halo-organics
***ICSC Number:*** 1423
***RTECS Number:*** TX5075000
***EEC Number:*** 610-007-00-6

*Uses:* This compound is also used in the synthetic rubber industry and as a component in rubber cements.

*Regulatory Authority:*
- Air Pollutant Standard Set (ACGIH)[1] (Australia) (DFG)[3] (Israel) (Mexico) (OSHA)[58] (Several Canadian Provinces) (Several States)[60]
- Permissible Exposure Limits for Chemical Contaminants (CAL/OSHA)
- The "Director's List" (CAL/OSHA)
- Canada, WHMIS, Ingredients Disclosure List

*Description:* Chloronitroproane is a flammable, colorless liquid with an unpleasant odor that causes tears (lachrymator). Slightly soluble in water; solubility = 6ppm @ 20°C. Molecular weight = 123.53. Boiling point = 141°C. Vapor pressure 6 mmHg @ 25°C. Flash point = 62°C. Hazard Identification (based on NFPA-704 M Rating System): Health U (unknown), Flammability 3, Reactivity 2.

*Incompatibilities:* Forms explosive mixture with air. Strong oxidizers may cause a fire and explosion hazard. May explode when exposed to heat.

*Permissible Exposure Limits in Air:* The Federal limit (OSHA PEL)[58] is 20 ppm (100 mg/m$^3$) as is the Mexico and DFG MAK value[3]. The TWA recommended by NIOSH[2] and ACGIH[1] is 2 ppm (10 mg/m$^3$) as is the Australian and Israeli value. The NIOSH[2] IDLH level = 100 ppm. Canadian provincial limts are: British Columbia TWA 20 ppm and STEL of 30 ppm; Alberta, Ontario and Quebec TWA is 2 ppm and Alberta's STEL is 4 ppm. In addition, several states have set guidelines or standards for chloronitropropane in ambient air[60] ranging from 10 $\mu$g/m$^3$ (North Dakota) to 150 $\mu$g/m$^3$ (Virginia) to 200 $\mu$g/m$^3$ (Connecticut) to 238 $\mu$g/m$^3$ (Nevada).

*Determination in Air:* Chromosorb tube-108; Ethyl acetate; Gas chromatography/Flame ionization detection; See NIOSH II(5) Method #S-211[18].

*Permissible Concentration in Water:* No criteria set, but runoff from spills or fire control may cause water pollution.

*Routes of Entry:* Inhalation, ingestion, skin and eye contact.

*Harmful Effects and Symptoms*

*Short Term Exposure:* 1-Chloro-1-Nitropropane can affect you when breathed in. Exposure can irritate and burn the eyes, skin and cause respiratory tract irritation with coughing and/or shortness of breath. Higher exposures can cause pulmonary edema, a medical emergency that can be delayed for several hours. This can cause death.

*Long Term Exposure:* May cause damage to the heart, liver, and kidneys. Similar irritating substances can cause lung injury and bronchitis.

*Points of Attack:* In animals: respiratory system, lungs, liver, kidneys, cardiovascular system.

*Medical Surveillance:* Consider the points of attack in preplacement and periodic physical examinations. EKG. Lung function tests. Liver and kidney function tests. Consider chest x-ray following acute overexposure.

*First Aid:* If this chemical gets into the eyes, remove any contact lenses at once and irrigate immediately for at least 15 minutes, occasionally lifting upper and lower lids. Seek medical attention immediately. If this chemical contacts the skin, remove contaminated clothing and wash immediately with soap and water. Seek medical attention immediately. If this chemical has been inhaled, remove from exposure, begin rescue breathing (using universal precautions) if breathing has stopped, and CPR if heart action has stopped. Transfer promptly to a medical facility. When this chemical has been swallowed, get medical attention. Give large quantities of water and induce vomiting. Do not make an unconscious person vomit. Medical observation is recommended for 24 to 48 hours after breathing overexposure, as pulmonary edema may be delayed. As first aid for pulmonary edema, a doctor or authorized paramedic may consider administering a corticosteroid spray.

*References:*
- New Jersey Department of Health and Senior Services and Senior Services, "Hazardous Substance Fact Sheet: 1-Chloro-1-Nitropropane," Trenton, NJ (March 2000). http://www.state.nj.us/health/eoh/rtkweb/0395.pdf
- California Environmental Protection Agency "Chemical List of Lists," Sacramento CA (February 1997).

# *p*-Chloro-*m*-cresol

*Use Type:* Fungicide and microbiocide
*CAS Number:* 59-50-7
*Formula:* $C_7H_7ClO$; $C_6H_3OHCH_3Cl$
*Synonyms:* Chlorocresol; *p*-chlorocresol; 4-Chloro-*m*-cresol; 6-Chloro-*m*-cresol; 2-Chloro-hydroxytoluene; 6-Chloro-3-hydroxytoluene; 4-Chloro-3-methylphenol; 4-Cloro-3-metilfenol (Spanish); 3-Methyl-4-chlorophenol; Parachlorometacresol; PCMC
*Trade Names:* APTAL®; BAKTOL®; BAKTOLAN®; CANDASEPTIC®; CMK®; NIPACIDE PCMC®; Clariant (Switzerland); OTTAFACT®; PARMETOL®; PARSOL®; PREVENTOL®, Bayer CropScience (Germany); RASCHIT®; RASCHIT K®; RASEN-ANICON®
*Producers:* Bayer CropScience (Germany); Clariant (Switzerland); Lancaster Synthesis (UK); Molekula Fine Chemicals (UK)
*Chemical Class:* Chlorophenols; Halo-organics
*EPA/OPP PC Code:* 064206
*California DPR Chemical Code:* 1813
*ICSC Number:* 0131
*RTECS Number:* GO7100000
*RCRA Number:* U039
*EEC Number:* 604-014-00-3
*EINECS Number:* 200-431-6

*Uses:* Primarily registered for non-food use. Used to control bacteria, yeasts and fungi in industrial environments, oil recovery operations and industrial preservatives, e.g., for paints, inks, glues, gums, textiles and leather goods.

***Carcinogen/Hazard Classifications***

**U.S. EPA Carcinogens:** Group D, unclassifiable, inadequate data

***California Prop. 65:*** Listed

***Regulatory Authority:***
- Proposition 65 chemical (CAL)
- Clean Water Act: Section 307 Priority Pollutants; Section 313 Priority Chemicals.
- RCRA Section 261 Hazardous Constituents.
- RCRA Universal Treatment Standards: Wastewater (mg/L), 0.018; Nonwastewater (mg/kg), 14
- RCRA Ground Water Monitoring List. Suggested test method(s) (PQL $ug/L$): 8040(5); 8270(20)
- EPCRA Section 304 RQ: CERCLA, 5,000 lb (2270 kg)
- AB 2588-Air Toxics "Hot Spots" Chemicals (CAL)
- The "Director's List" (CAL/OSHA) as chlorophenols
- List of priority pollutants (U.S. EPA)
- EPCRA Section 313 (as chlorophenols) Form R *de minimis* concentration reporting level: 1.0%

*Description:* Crystalline solid when pure. Odorless. Partially soluble in water. Molecular weight=142.57. Melting/Freezing point = 56°C. Boiling point =235°C. Log $K_{ow}$ = 3.12. Values above 3.0 are likely to bioaccumulate in aquatic organisms.

*Incompatibilities:* React with boranes, alkalies, aliphatic amines, amides, nitric acid, sulfuric acid. Contact with oxidizing agents may cause a fire and explosion hazard. Heat produces hydrogen chloride and chlorine. Corrosive to aluminum, copper, tin, and other chemically active metals.

*Permissible Concentration in Water:* No criteria set for this chemical; however, EPA recommends that drinking water contain no more than 0.04 mg/L of 2-chlorophenol for a lifetime exposure for an adult, and 0.05 mg/L for a 1-day, 10-day, or longer exposure for a child.

*Determination in Water:* Gas chromatography (EPA Method 604) or gas chromatography plus mass spectrometry (EPA Method 625).

*Routes of Entry:* Skin absorption, inhalation, ingestion, skin and/or eye contact.

***Harmful Effects and Symptoms***

***Short Term Exposure:*** Inhalation can cause severe irritation, burns to the nose and throat, headache, dizziness, vomiting, lung damage, muscle twitchings, spasms, tremors, weakness, staggering and collapse. Skin contact can cause severe irritation and burns. Can be absorbed through the skin to cause or increase the severity of symptoms listed above. Eye contact causes severe irritation. May cause burns. Ingestion can cause irritation, burns to the mouth and throat, low blood pressure, profuse sweating, intense thirst, nausea, abdominal pain, stupor, vomiting, red blood cell damage and accumulation of fluid in the lungs followed by pneumonia. May also cause restlessness and increased breathing rate followed by rapidly developing muscle weakness. Tremors, convulsions and coma can promptly set in and will continue until death. Based on animal studies, the estimated lethal dose is between one teaspoon and one ounce for a 150 pound adult. *p*-isomer: The substance irritates the eyes, the skin and the respiratory tract. The substance may cause effects on the central nervous system and bladder.

***Long Term Exposure:*** Skin sensitivity may develop. May have effects on the blood, heart, liver, lung, kidney. The state of New Jersey lists the 2-chloro-isomer as probable carcinogen in humans, and that it causes leukemia and soft tissue cancers in humans. Chlorophenols leave the body quickly, so they are not likely to accumulate in the mother's tissues or breast milk. There are no human studies on the effects of chlorophenols on developing fetuses. Studies in rats showed that chlorophenols can pass through the placenta and produce toxic effects to the developing fetuses. The most common problems in children are delayed hardening of the bones of the breastbone, spine, and skull.

***Points of Attack:*** Liver, kidneys. An allergen, may cause skin irritation and sensitization.

***Medical Surveillance:*** If symptoms develop or overexposure is suspected, the following may be useful: Liver function tests. Kidney function tests. Examination by a qualified allergist. EKG.

***First Aid:*** If this chemical gets into the eyes, remove any contact lenses at once and irrigate immediately for at least 15 minutes, occasionally lifting upper and lower lids. Seek medical attention immediately. If this chemical contacts the skin, remove contaminated clothing and wash immediately with soap and water. Seek medical attention immediately. If this chemical has been inhaled, remove from exposure, begin rescue breathing (using universal precautions) if breathing has stopped, and CPR if heart action has stopped. Transfer promptly to a medical facility. When this chemical has been swallowed, get medical attention. *Do not induce vomiting when formulations containing petroleum solvents are ingested.* Otherwise, give large quantities of water and induce vomiting. Do not make an unconscious person vomit.

***References:***
- U.S. Environmental Protection Agency, "Reregistration Eligibility Decision (RED), *p*-Chloro-*m*-cresol", Office of Prevention, Pesticides and Toxic Substances, Washington, DC (January 1997). http://www.epa.gov/oppsrrd1/REDs/3046red.pdf
- California Environmental Protection Agency "Chemical List of Lists," Sacramento CA (February 1997)
- Agency for Toxic Substances and Disease Registry "Toxicological profile for chlorophenols," 1999, Atlanta, GA.

# Chlorobenzilate

*Use Type:* Miticide and insecticide
*CAS Number:* 510-15-6
*Formula:* $C_{16}H_{14}Cl_2O_3$
*Synonyms:* Benzeneacetic acid, 4-chloro-α-(4-chlorophenyl)-α-hydroxy-,ethyl ester; Benzilan; benzilic acid, 4,4'-dichloro-,ethyl ester; Benzilic acid 4,4'-dichloro, ethyl ester; Benz-*o*-chlor; Chlorbenzilat; Chlorbenzalate; 4-Chloro-α-(4-chlorophenyl)-α-hydroxybenzeneacetic acid ethyl ether; 4,4'-Cichlorbenzilsaeureaethylester (German); 4,4'-Dichlorobenzilic acid ethyl ester; 4,4'-dichlorobenzilate; ECB; ENT 18,596; Ethyl 4-chloro-α-(4-chlorophenyl)-α-hydroxybenzene acetate; Ethyl-*p,p*'-dichlorobenzilate; Ethyl 4,4'-dichlorobenzilate; Ethyl-4,4'-dichlorodiphenyl glycollate; Ethyl-4,4'-dichlorophenyl glycollate; Ethyl ester of 4,4'-dichlorobenzilic acid; Ethyl 2-hydroxy-2,2-bis(4-chlorophenyl)acetate; NCI-C00408; NCI-C60413
*Trade Names:* ACAR®; ACARABEN®, Syngenta (Switzerland), canceled; ACARABEN 4E®, Syngenta (Switzerland), canceled 12/23/1988; AKAR 338®; BENZILAN®, Makhteshim-Agan Industries (Israel), canceled 7/01/1987; BENZ-O-CHLOR®, Tower Chemical (USA), canceled 7/01/1987; COMPOUND 338®; FOLBEX®; FOLBEX SMOKE STRIPS®; G 338®, Syngenta (Switzerland); G 23992®, Syngenta (Switzerland); GEIGY 338®, Syngenta (Switzerland); KOP MITE®; MAL-O-CHLOR®, Tower Chemical (USA), canceled 7/01/1987; M ITROL®, American Refining (USA), canceled 1/06/1983; RID-A-MITE®, Tex-Ag Company (USA), canceled 1.06/1983
*Producers:* Helena Chemicals (USA); Makhteshim-Agan Industries (Israel); Syngenta (Switzerland)
*Chemical Class:* Organochlorine
*EPA/OPP PC Code:* 028801
*California DPR Chemical Code:* 132
*ICSC Number:* 0749
*RTECS Number:* DD2275000
*EEC Number:* 607-159-00-0
*EINECS Number:* 208-110-2
*Uses:* It is used to kill mites, ticks, and other insects, and as a synergist for DDT.
*Human toxicity (long-term)*[77]: Very low–140.00 ppb, Health Advisory
*Fish toxicity (threshold)*[77]: Intermediate–81.03680 ppb, MATC (Maximum Acceptable Toxicant Concentration)
*Carcinogen/Hazard Classifications*
**U.S. EPA Carcinogens:** Group B2, Probable carcinogen
**IARC:** Group 3, unclassifiable
**Label Signal Word:** CAUTION or WARNING
*Regulatory Authority:*

- Clean Air Act: Hazardous Air Pollutants (Title I, Part A, Section 112)
- AB 1803-Well Monitoring Chemical (CAL)
- AB 2588-Air Toxics "Hot Spots" Chemicals (CAL)
- CAL Air Resources Board/AB 1807 Toxic Air Contaminants
- Proposition 65 chemical (CAL)
- Permissible Exposure Limits for Chemical Contaminants (CAL/OSHA)
- The "Director's List" (CAL/OSHA)
- EPA Hazardous Waste Number (RCRA No.): U038
- RCRA, 40CFR261, Appendix 8 Hazardous Constituents.
- RCRA 40CFR268.48; 61FR15654, Universal Treatment Standards: Wastewater (mg/L), 0.10; Nonwastewater (mg/kg), N/A.
- RCRA 40CFR264, Appendix 9; TSD Facilities Ground Water Monitoring List. Suggested test method(s) (PQL $ug$/L): 8270(10)
- Superfund/EPCRA 40CFR302.4 RQ: CERCLA, 10 lb (4.54 kg)
- EPCRA Section 313 Form R *de minimus* concentration reporting level: 1.0%
- U.S. DOT Regulated Marine Pollutant (49CFR172.101, Appendix B).

*Description:* Ethyl 4,4'-dichlorobenzilate is a colorless solid when pure. The technical product is a yellow or brown liquid. Slightly soluble in water; solubility = 13 ppm @ 20°C. Molecular weight = 325.20. Boiling point = 146–148°C. Melting/Freezing point = 36–37°C. Vapor pressure = $6.8 \times 10^{-6}$ mmHg. @ 20°C. This chemical is not expected to bioconcentrate in aquatic organisms (SRC). However its metabolite, dichlorobenzophenone, has a Log $K_{ow}$ = 4.8. Values at or above 3.0 are likely to bioaccumulate in marine organisms.
*Incompatibilities:* Keep away from oxidizers, sulfuric acid, caustics, ammonia, aliphatic amines, alkanolamines, isocyanates, alkylene oxides, epichlorohydrin. Moisture may cause hydrolysis or other forms of decomposition.
*Permissible Exposure Limits in Air:* No OELs have been established.
*Permissible Concentration in Water:* No criteria set. Runoff from spills or fire control may cause water pollution.
*Routes of Entry:* Inhalation, absorbed through the skin, eye contact, ingestion
*Harmful Effects and Symptoms*
*Short Term Exposure:* Symptoms include apprehension, anxiety, confusion, nervous excitation; dizziness; headache; numbness and weakness in limbs; muscle twitching, tremors; nausea and vomiting; slow, shallow respiration, bluish face; convulsions; loss of consciousness; breathing stops; death.
*Long Term Exposure:* This chemical may damage the male reproductive glands. May cause liver and kidney damage. Repeated or prolonged exposure may affect the nervous system, causing a loss of coordination, muscle weakness, tremors, convulsions, dizziness, and possible coma. This chemical may be a human carcinogen.

*Points of Attack:* Liver, kidneys, nervous system.

*Medical Surveillance:* Examination of the nervous system. Liver and kidney function tests. More than light alcohol consumption may exacerbate liver damage. Medical observation is recommended for 24 to 48 hours after breathing overexposure, as pulmonary edema may be delayed.

*First Aid:* Speed in removing material from skin is of extreme importance. Eye contact can cause dangerous amounts of these chemicals to be quickly absorbed through the mucous membrane into the bloodstream. Directly, irrigate with large amounts of plain, tepid water or saline for 20 minutes, occasionally lifting the lower and upper lids. During this time, remove contact lenses, if easily removable without additional trauma to the eye. Get medical aid immediately. Have physician check for possible delayed damage. *Skin:* Get medical aid. Skin and/or eye contact can cause dangerous amounts of these chemicals to be absorbed into the bloodstream. Wearing the appropriate PPE equipment and respirator for organochlorine pesticides, immediately flush exposed skin, hair, and under nails with plain, running, tepid water for 20 minutes, then wash twice with mild soap. Shampoo hair promptly if contaminated; protect eyes. Do not scrub skin or hair, since this can increase absorption through the skin. Rinse thoroughly with water. Victims who are able and cooperative may assist with their own decontamination. Remove and double-bag contaminated clothing and personal belongings. Leather absorbs many organochlorines; therefore, items such as leather shoes, gloves, and belts should be discarded. If the skin is swollen or inflamed, cool affected areas with cold compresses. Ingestion: Call poison control. Loosen all clothing. Never give anything by mouth to an unconscious person. Get medical aid. *Do not induce vomiting.\** In cases of ingestion, the patient is at risk of CNS depression or seizures, which may lead to pulmonary aspiration during vomiting. If the victim is conscious and able to swallow, \*administer an aqueous slurry of activated charcoal at 1 gm/kg (usual adult dose 60–90 g, child dose 25–50 g). A soda can and straw may be of assistance when offering charcoal to a child. The efficacy of activated charcoal for some organochlorine poisoning (such as chlordane) is uncertain. If victim is *unconscious or having convulsions,* do nothing except keep victim warm. *\*In some cases you may be specifically instructed by Poison Control to induce vomiting by way of 2 tablespoons of syrup of ipecac (adult) washed down with a cup of water. Do not give activated charcoal before or with ipecac syrup. Inhalation:* Get medical aid. Do not contaminate yourself. Wearing the appropriate PPE equipment and respirator for organochlorine pesticides, immediately remove the victim from the contaminated area to fresh air. For inhalation exposures, monitor for respiratory distress. If the victim is not breathing, administer artificial respiration. *Do not use mouth-to-mouth resuscitation; use bag/mask apparatus.* If cough or breathing difficulty develops, evaluate for respiratory tract irritation, bronchitis, or pneumonitis. If breathing is difficult, administer 100% humidified supplemental oxygen through bag/mask apparatus until medical help arrives. Do not leave victim unattended.

*References:*
- New Jersey Department of Health and Senior Services, "Hazardous Substance Fact Sheet, Ethyl-4,4'-dichlorobenzilate," Trenton NJ (May 1999). http://www.state.nj.us/health/eoh/rtkweb/0205.pdf
- EXTOXNET, Extension Toxicology Network, "Pesticide Information Profile, Chlorobenzilate," Oregon State University, Corvallis, OR (June 1996). http://extoxnet.orst.edu/pips/chlorobe.htm
- California Environmental Protection Agency "Chemical List of Lists," Sacramento CA (February 1997).

## Chloroneb (ANSI)

*Use Type:* Fungicide
*CAS Number:* 2675-77-6
*Formula:* $C_8H_8Cl_2O_2$
*Synonyms:* Chloronebe (French); 1,4-Dichloro-2,5-dimethoxybenzene
*Trade Names:* CHLORAXYL® SEED TREATER, Micro-Flo (USA); DELTA-COAT® II, Agriliance (USA); DEMOSAN®; SOIL FUNGICIDE®-1823; TERSAN® SP; TERRANEB® SP
*Producers:* Agriliance (USA); Gustafson (USA); Micro-Flo (USA); Pbi/Gordon (USA)
*Chemical Class:* Organochlorine; Halo-organics
*EPA/OPP PC Code:* 027301
*California DPR Chemical Code:* 135
*Uses:* A systemic fungicide used to control snow mold on turf grass; used on cotton, sugar beets and bean seeds to control seedling disease.
*Human toxicity (long-term)*[77]: Low–91.00 ppb, Health Advisory
*Fish toxicity (threshold)*[77]: Very low–530.91122 ppb, MATC (Maximum Acceptable Toxicant Concentration)
*U.S. Maximum Allowable Residue Levels for Chloroneb (40 CFR 180.257):*

| CROP | ppm |
|---|---|
| Bean | 0.1 |
| Bean, forage | 2.0 |
| Beet, sugar, roots | 0.1 |
| Beet, sugar, tops | 0.1 |
| Cattle, fat | 0.2 |
| Cattle, meat | 0.2 |
| Cattle, mbyp | 0.2 |
| Cotton, forage | 2.0 |
| Cotton, undelinted seed | 0.1 |
| Goat, fat | 0.2 |

| | |
|---|---|
| Goat, meat | 0.2 |
| Goat, mbyp | 0.2 |
| Hog, fat | 0.2 |
| Hog, meat | 0.2 |
| Hog, mbyp | 0.2 |
| Horse, fat | 0.2 |
| Horse, meat | 0.2 |
| Horse, mbyp | 0.2 |
| Milk | 0.05 |
| Sheep, fat | 0.2 |
| Sheep, meat | 0.2 |
| Sheep, mbyp | 0.2 |
| Soybean | 0.1 |
| Soybean, forage | 2.0 |

***Carcinogen/Hazard Classifications***
**Label Signal Word:** WARNING, CAUTION or DANGER
***Regulatory Authority:***
- Actively registered pesticide in California.
- AB 1803-Well Monitoring Chemical (CAL)

***Description:*** A solid. Molecular weight = 207.06. Melting/Freezing point = 134°C. Boiling point = 262° @ 744 mmHg. Vapor pressure = $3 \times 10^{-3}$ mmHg. @ 20°C.

***Determination in Air:*** Filter; none; Gravimetric; NIOSH IV [Particulates NOR; #0500 (total), #0600 (respirable)].[18]

***Permissible Concentration in Water:*** No criteria set. Runoff from spills or fire control may cause water pollution.

***Harmful Effects and Symptoms***

***Short Term Exposure:*** Symptoms include apprehension, anxiety, confusion, nervous excitation; dizziness; headache; numbness and weakness in limbs; muscle twitching, tremors; nausea and vomiting; slow, shallow respiration, bluish face; convulsions; loss of consciousness; breathing stops; death.

***Points of Attack:*** May be fatal if inhaled, ingested, or absorbed through the skin

***Medical Surveillance:*** Medical observation is recommended for 24 to 48 hours after breathing overexposure, as pulmonary edema may be delayed. As first aid for pulmonary edema, consider administering a corticosteroid spray. Cigarette smoking may exacerbate pulmonary injury and should be discouraged for at least 72 hours following exposure.

***First Aid:*** Speed in removing material from eyes and skin is of extreme importance. *Eyes:* Eye contact can cause dangerous amounts of these chemicals to be quickly absorbed through the mucous membrane into the bloodstream. Directly, irrigate with large amounts of plain, tepid water or saline for 20 minutes, occasionally lifting the lower and upper lids. During this time, remove contact lenses, if easily removable without additional trauma to the eye. Get medical aid immediately. Have physician check for possible delayed damage. *Skin:* Get medical aid. Skin contact can cause dangerous amounts of these chemicals to be absorbed into the bloodstream. Wearing the appropriate PPE equipment and respirator for organochlorine pesticides, immediately flush exposed skin, hair, and under nails with plain, running, tepid water for 20 minutes, then wash twice with mild soap. Shampoo hair promptly if contaminated; protect eyes. **Do not scrub skin or hair**, since this can increase absorption through the skin. Rinse thoroughly with water. Victims who are able and cooperative may assist with their own decontamination. Remove and double-bag contaminated clothing and personal belongings. Leather absorbs many organochlorines; therefore, items such as leather shoes, gloves, and belts should be discarded. If the skin is swollen or inflamed, cool affected areas with cold compresses. *Ingestion:* Call poison control. Loosen all clothing. Never give anything by mouth to an unconscious person. Get medical aid. *Do not induce vomiting.** In cases of ingestion, the patient is at risk of CNS depression or seizures, which may lead to pulmonary aspiration during vomiting. If the victim is conscious and able to swallow, *administer an aqueous slurry of activated charcoal at 1 gm/kg (usual adult dose 60–90 g, child dose 25–50 g). A soda can and straw may be of assistance when offering charcoal to a child. The efficacy of activated charcoal for some organochlorine poisoning (such as chlordane) is uncertain. If victim is *unconscious or having convulsions, do nothing except keep victim warm. *In some cases you may be specifically instructed by Poison Control to induce vomiting by way of 2 tablespoons of syrup of ipecac (adult) washed down with a cup of water. Do not give activated charcoal before or with ipecac syrup. Inhalation:* Get medical aid. Do not contaminate yourself. Wearing the appropriate PPE equipment and respirator for organochlorine pesticides, immediately remove the victim from the contaminated area to fresh air. For inhalation exposures, monitor for respiratory distress. If the victim is not breathing, administer artificial respiration. *Do not use mouth-to-mouth resuscitation; use bag/mask apparatus.* If cough or breathing difficulty develops, evaluate for respiratory tract irritation, bronchitis, or pneumonitis. If breathing is difficult, administer 100% humidified supplemental oxygen through bag/mask apparatus until medical help arrives. Do not leave victim unattended.

***References:***
- Pesticide Management Education Program, "Chloroneb (Demosan, Tersan-SP) Chemical Profile 1/85," Cornell University, Ithaca, NY (January 1985). http://pmep.cce.cornell.edu/profiles/fung-nemat/aceticacid-etridiazole/chloroneb/fung-prof-chloroneb.html
- California Environmental Protection Agency "Chemical List of Lists," Sacramento CA (February 1997).
- U.S. Environmental Protection Agency, Office of Pesticide Programs, Pesticide Residue Limits, "Chloroneb", 40 CFR 180.257, www.epa.gov/pesticides/food/viewtols.htm

# Chlorophacinone

**Use Type:** Rodenticide
**CAS Number:** 3691-35-8
**Formula:** $C_{23}H_{15}ClO_3$
**Alert:** Chlorophacinone could be hazardous to other small mammals and birds if used indiscriminately. Persons with bleeding problems and children should not come in contact. Human toxicity (long-term): Very high.
**Synonyms:** Chlorfacinon (German); 2-(α-p-chlorophenylacetyl)indane-1,3-dione; 2-[(p-Chlorophenyl)phenylacetyl]-1,3-indandione; 2[2-(4-Chlorophenyl)-2-phenylacetyl]indan-1,3-dione; 2[(4-Chlorophenyl)phenylacetyl]-1H-indene-1,3(2H)-dione; Clorphacinon (Italian); [(4-Chlorophenyl)-1-phenyl]-acetyl-1,3-indandion (German); 1-(4-Chlorphenyl)-1-phenyl-acetyl indan-1,3-dion (German); Clorofacinona (Spanish); Chlorphacinon (Italian); 1H-Indene-1,3(2H)-dione, 2-[(4-chlorophenyl)phenylacetyl]-; 1,3-Indandione, 2-[(p-chlorophenyl)phenylacetyl]-; 2-[2-Phenyl-2-(4-chlorophenyl)acetyl]-1,3-indandione
**Trade Names:** AFNOR®; CAID®; DELTA®; DRAT®; LIPHADIONE®, LiphaTech (USA); LM 91®; MICROZUL®; MURIOL®; PARTOX®, LiphaTech (USA) canceled 7/1/1987; QUICK®; RAMUCIDE®; RANAC®; RATOMET®; RAVIAC®; ROZOL®, LiphaTech (USA) canceled 7/1/1987; TOPITOX®
**Producers:** Agrimor International (USA); BASF Agricultural Products Group (Germany); I.N.D.I.A. Industrie Chemiche (Italy); Jingma Chemicals Ltd. (China); LiphaTech (USA); Rhone-Poulenc Jardin (France)
**Chemical Class:** 1,3-Indandione; Halo-organics
**EPA/OPP PC Code:** 067707
**California DPR Chemical Code:** 1625
**RTECS Number:** NK5335000
**EINECS Number:** 223-003-0
**Uses:** This material is an anticoagulant rodenticide used around livestock and also on crops such as artichokes. Some brands are labeled for use indoors and outdoors for the control of mice, rats, moles, muskrats, voles and vampire bats.
**Human toxicity (long-term)[77]:** Extra high–0.03500 ppb, Health Advisory
**Fish toxicity (threshold)[77]:** Intermediate–27.64016 ppb, MATC (Maximum Acceptable Toxicant Concentration)
**Carcinogen/Hazard Classifications**
**WHO Acute Hazard:** Class 1 a, extremely hazardous
**Regulatory Authority:**

- Superfund/EPCRA 40CFR355, Appendix B Extremely Hazardous Substances: TPQ = 100/10,000 lb (45.4/4,540 kg).
- Superfund/EPCRA 40CFR302.4 RQ: EHS, 1 lb (0.454 kg).
- EPA/SARA 302 (EPCRA) Extremely hazardous substances
- The "Director's List" (CAL/OSHA)
- Actively registered pesticide in California.

**Description:** Chlorophacinone is a pale yellow crystalline solid. Slightly soluble in water. Melting/Freezing point = 140°C.
**Permissible Exposure Limits in Air:** No standards set.
**Permissible Concentration in Water:** No criteria set, but runoff from spills or fire control may cause water pollution.
**Routes of Entry:** Ingestion and skin contact.
**Harmful Effects and Symptoms**
**Short Term Exposure:** Contact with eyes or skin may cause irritation or injury. Contact may cause burns to skin and eyes. Symptoms of exposure are similar to those of warfarin. Inhalation should be avoided; use NIOSH-approved air purifying respirators for pesticides. May be harmful if swallowed. Symptoms develop after a few days or a few weeks or repeated ingestion and include nosebleed and bleeding gums; pallor and sometimes a rash; massive bruises, especially of the elbow, knees, and buttocks; blood in urine and feces; occasionally paralysis from cerebral hemorrhage; and hemorrhagic shock and death. Chlorophacinone is highly toxic orally and by skin adsorption. The probable oral lethal dose for humans is less than 5 mg/kg to 50 mg/kg, or between a taste (less than 7 drops) and 1 teaspoonful for a 150 lb. (70 kg) person. The $LD_{50}$ (oral, mouse) = 1.06 mg/kg.
**Long Term Exposure:** See above
**Points of Attack:** Blood, cardiovascular system
**First Aid:** If this chemical gets into the eyes, remove any contact lenses at once and irrigate immediately for at least 15 minutes, occasionally lifting upper and lower lids. Seek medical attention immediately. If this chemical contacts the skin, remove contaminated clothing and wash immediately with soap and water. Seek medical attention immediately. If this chemical has been inhaled, remove from exposure, begin rescue breathing (using universal precautions) if breathing has stopped, and CPR if heart action has stopped. Transfer promptly to a medical facility. When this chemical has been swallowed, get medical attention. Give large quantities of water and induce vomiting. Do not make an unconscious person vomit.
**References:**

- New Jersey Department of Health and Senior Services, "Hazardous Substance Fact Sheet, Chlorophacinone," Trenton NJ (July 2000).
  http://www.state.nj.us/health/eoh/rtkweb/0400.pdf
- Pesticide Management Education Program, Chemical Profile 1/85 on Chlorophacionone (Rozol), Cornell University, Ithaca, NY (Modified March 1, 2002).
  http://pmep.cce.cornell.edu/profiles/rodent/chlorophacinone/rod-prof-chlorophacinone.html

- California Environmental Protection Agency "Chemical List of Lists," Sacramento CA (February 1997).
- U.S. Environmental Protection Agency, "Chemical Profile: Chlorophacinone," Washington, DC, Chemical Emergency Preparedness Program (October 31, 1985).

# Chloropicrin

*Use Type:* A soil fumigant and nematicide
*CAS Number:* 76-06-2
*Formula:* $CCl_3NO_2$
*Alert:* A Restricted Use Pesticide (RUP) which is undergoing EPA reregistration under FIFRA.
*Synonyms:* AI3-00027; Caswell No. 214; Chloroform, nitro-; Chloorpikrine (Dutch); Chloropicrine (French); Chlorpikrin (German); Clorpicrina (Italian, Spanish); Methane, trichloronitro-; Mycrolysin; NCI-C00533; Nitrotrichloromethane; Nitrochloroform; Trichloronitromethane; Trichlor
*Trade Names:* ACQUINITE®; BROM-O-GAS®, Great Lakes Chemical (USA) canceled 7/1/1987; BROZONE®, Dow AgroSciences (USA) canceled 3/11/1983; CHLOR-O-PIC®, Great Lakes Chemical (USA); DOWFUME®, Dow AgroSciences (USA); Albemarle Corp. (USA); FUM-A-CIDE® 15, AMVAC Chemicals (USA) canceled 6/19/1984; KLOP®; LARVACIDE®, Soweco (USA) canceled 11/10/1983; LARVACIDE 100®; METAPICRIN®, Dead Sea Bromine Group (Israel); NAMFUME®, Univar USA Inc. (USA) canceled 9/30/1991; NEMAX®; OG-25®; PESTMASTER® FUMIGANT 1, Velsicol Chemical Corp. (USA) canceled 5/12/1986; PICFUME®; PIC-CHLOR® 16, Soil Chemicals Corp. (USA); PICRIDE®; PROFUME A®; PS®; TELONE®, Dow AgroSciences (USA); TELONE® C, Dow AgroSciences (USA) canceled 3/18/1987; TERR-O-CIDE® 15, Great Lakes Chemical (USA) canceled 11/10/1983; TERR-O-GAS®, Great Lakes Chemical (USA); TIMBERFUME II®, Osmose Railroad Services, Inc. (USA); TRI-CLOR®, Trical Inc. (USA) canceled 5/26/1987; TRI-CON®; TRI-FORM®; TRIFUME®, Univar USA Inc., (USA) canceled 9/30/1991
*Producers:* Albemarle Corp. (USA); AMVAC Chemical Corp. (USA); Arvesta Corporation (USA); Dead Sea Bromine Group (Israel); Dow AgroSciences (USA); Great Lakes Chemical (USA); ICI Group (UK); ISOCHEM (France); Mitsui Chemicals (Japan); Niklor Chemical (USA); Nippon Kayaku (Japan); Orica (Australia); Sigma-Aldrich Laborchemikalien (Germany)
*Chemical Class:* Halo-organics
*EPA/OPP PC Code:* 081501
*California DPR Chemical Code:* 136
*ICSC Number:* 0750
*RTECS Number:* PB6300000
*EINECS Number:* 200-930-9

*Uses:* Chloropicrin is used in the manufacture of the dyestuff methyl violet and in other organic syntheses. It has been used as a chemical warfare gas. It is used as a preplant soil fumigant in seed beds and transplant nurseries for control of verticillium wilt, nematodes, weed seeds and insects. In grain elevators, it is used to control insects and rodents. The top five uses in California are on strawberries, tomatoes, bell peppers, and outdoor nursery plants.
*Human toxicity (long-term)[77]:* Low–56.00 ppb, Health Advisory
*Fish toxicity (threshold)[77]:* High–1.37795 ppb, MATC (Maximum Acceptable Toxicant Concentration)
*U.S. Maximum Allowable Residue Levels for Chloropicrin (40 CFR 180.1008):* (*Note:* No ppm values appear in this residue table issued by the USEPA)

| CROP | ppm |
|---|---|
| Barley, grain, post-h | – |
| Buckwheat, grain, post-h | – |
| Corn, field, grain, post-h | – |
| Corn, pop, grain, post-h | – |
| Oat, grain, post-h | – |
| Rice, grain, post-h | – |
| Rye, grain, post-h | – |
| Sorghum, grain, post-h | – |
| Wheat, grain, post-h | – |

*Carcinogen/Hazard Classifications*
**Label Signal Word: DANGER**
*Regulatory Authority:*
- Banned or Severely Restricted (Various Countries) (UN)[13, 35]
- AB 1803-Well Monitoring Chemical (CAL)
- AB 2588-Air Toxics "Hot Spots" Chemicals (CAL)
- CAL-DHS/DHS Drinking Water Action Levels
- DOT Inhalation Hazard Chemicals
- Permissible Exposure Limits for Chemical Contaminants (CAL/OSHA)
- The "Director's List" (CAL/OSHA)
- Actively registered pesticide in California
- OSHA 29CFR1910.119, Appendix A. Process Safety List of Highly Hazardous Chemicals, TQ = 500 lb (227 kg)
- Air Pollutant Standard Set (ACGIH)[1] (Australia) (DFG)[3] (HSE)[33] (Israel) (Mexico) (USSR, Japan)[35] (Several States)[60] (Several Canadian Provinces)
- Safe Drinking Water Act: Priority List (55 FR 1470)
- EPCRA Section 313 Form R *de minimus* concentration reporting level: 1.0%
- Canada, WHMIS, Ingredients Disclosure List

*Description:* Chloropicrin is a highly reactive, colorless, oily liquid. Sharp, penetrating odor that causes tearing. Odor threshold = 1.1 ppm. Molecular weight = 164.38. Boiling point = 112°C. Melting/Freezing point = –64°. Slightly soluble in water; solubility = 1950 ppm; 0.2272 g/100 ml @ 0°C, 0.1621 g/100 ml at 25°C. Vapor pressure = 16.9 mmHg @ 20°C.

*Incompatibilities:* Can be self reactive. Chloropicrin may explode when heated under confinement. Quickly elevated temperatures, shock, contact with alkali metals or alkaline earth may cause explosions. A strong oxidizer; violent reaction with reducing agents, aniline (especially in presence of heat), alcoholic sodium hydroxide, combustible substances, sodium methoxide, propargyl bromide, metallic powders. Liquid attacks some plastics, rubber and coatings.

*Permissible Exposure Limits in Air:* The Federal limit (OSHA PEL)[58], the DFG[3], the HSE[33], the Japanese[35], Australian, Israeli, Mexican, and the ACGIH[1] TWA value is 0.1 ppm (0.7 mg/m$^3$) and the ACGIH[1], Mexican, and HSE[33] STEL value is 0.3 ppm. The NIOSH[2] IDLH level = 2 ppm. The former USSR[35] has set a MAC for ambient air in residential areas of 0.01 mg/m$^3$ on a momentary basis and 0.007 mg/m$^3$ on a daily average basis. Several states have set guidelines or standards for chloropicrin in ambient air[60] ranging from 7-20 $\mu$g/m$^3$ (North Dakota) to 11.7 $\mu$g/m$^3$ (Virginia) to 14 $\mu$g/m$^3$ (Connecticut) to 17 $\mu$g/m$^3$ (Nevada).

*Determination in Air:* No method available.

*Permissible Concentration in Water:* California[61] has set guidelines for chloropicrin in drinking water of 50 $\mu$g/L on a taste basis and 37 $\mu$g/L on an odor basis.

*Routes of Entry:* Inhalation, ingestion, eye and skin contact.

*Harmful Effects and Symptoms*

*Short Term Exposure:* Chlorpicrin was used as poison gas in WW I. Exposure causes intense tearing of the eyes, headache, nausea and vomiting, diarrhea, and cough. Contact can severely irritate the skin causing rash or burning sensation. Higher exposures can irritate and burn the lungs, causing a build-up of fluid (pulmonary edema), a medical emergency that can be delayed for several hours. This can cause death. The LD$_{50}$ (oral, rat) = 250 mg/kg (moderately toxic). IDLH level = 2 ppm.[58]

*Long Term Exposure:* Repeated exposure can damage the lungs, causing bronchitis. It may also damage the liver and kidneys.

*Points of Attack:* Eyes, skin, respiratory system, liver, kidneys.

*Medical Surveillance:* Before beginning employment and at regular times after that, the following is recommended: Lung function test. If symptoms develop or overexposure has occurred, the following may be useful: Liver and kidney function tests. Consider chest X-ray following acute overexposure.

*First Aid:* If this chemical gets into the eyes, remove any contact lenses at once and irrigate immediately for at least 15 minutes, occasionally lifting upper and lower lids. Seek medical attention immediately. If this chemical contacts the skin, remove contaminated clothing and wash immediately with soap and water. Seek medical attention immediately. If this chemical has been inhaled, remove from exposure, begin rescue breathing (using universal precautions) if breathing has stopped, and CPR if heart action has stopped. Transfer promptly to a medical facility. When this chemical has been swallowed, get medical attention. Give large quantities of water and induce vomiting. Do not make an unconscious person vomit. Medical observation is recommended for 24 to 48 hours after breathing overexposure, as pulmonary edema may be delayed. As first aid for pulmonary edema, a doctor or authorized paramedic may consider administering a corticosteroid spray.

*References:*
- EXTOXNET, Extension Toxicology Network, "Pesticide Information Profile, Chloropicrin," Oregon State University, Corvallis, OR. http://ace.ace.orst.edu/info/extoxnet/pips/chloropi.htm
- EPA, Office of Pesticide Programs, Pesticide Residue Limits, " Chloropicrin", 40 CFR 180.1008, http://www.epa.gov/pesticides/food/viewtols.htm
- Sax, N.I., Ed., "Dangerous Properties of Industrial Materials Report," 2, No. 2, 17-19, New York, Van Nostrand Reinhold Co. (1982).
- California Environmental Protection Agency "Chemical List of Lists," Sacramento CA (February 1997).
- New Jersey Department of Health and Senior Services and Senior Services, "Hazardous Substance Fact Sheet: Chloropicrin," Trenton, NJ (April 1998). http://www.state.nj.us/health/eoh/rtkweb/0405.pdf

## Chlorothalonil (ANSI)

*Use Type:* Fungicide
*CAS Number:* 1897-45-6
*Formula:* $C_8Cl_4N_2$; $C_6Cl_4(CN)_2$
*Alert:* A General Use Pesticide (RUP). In is the second most used fungicide in the U.S. It can be found in formulations with many other pesticides.

*Synonyms:* 1,3-Benzenedicarbonitrile,2,4,6,6-tetrachloro-; Chlorthalonil (German); 1,3-Dicyanotetrachlorobenzene; Isophthalonitrile, tetrachloro; Metatetrachloro phthalodinitrile; NCI-C00102; 2,4,5,6-Tetrachloro-1,3-benzenedicarbonitrile; 2,4,5,6-Tetrachloro-1,3-dicyanobenzene; Tetrachloroisophthalonitrile; *meta*-tetrachlorophthalodinitrile; Tetrachlorophthalodinitrile, *meta*-; Thalonil

*Trade Names:* ATLAS CROPGARD®, Whyte Agrochemicals (UK); BANOL C®, Bayer CropScience (Germany); BB CHLOROTHALONIL®; BOMmHgDIER®; BRAVO®, Syngenta (Switzerland); Zeneca Ag Products (USA) (now Syngenta); BRAVO® 6F, Syngenta (Switzerland); Zeneca Ag Products (USA) (now Syngenta); BRAVO® 500, Syngenta (Switzerland); Zeneca Ag Products (USA) (now Syngenta); BRAVO® 6F, Syngenta (Switzerland); Zeneca Ag Products (USA) (now Syngenta); BRAVO ULTREX®, Syngenta (Switzerland); Zeneca Ag Products (USA) (now Syngenta); BRAVO-W-75®, Syngenta

(Switzerland); Zeneca Ag Products (USA) (now Syngenta); CHILTERN OLE®; CHLORONIL®, Veterinary & Agricultural Products Manufacturing Co., Ltd. (VAPCO) (Jordan); CONTACT® 75; DAC® 2787; DACONIL®, Syngenta (Switzerland); DACONIL® 2787 FUNGICIDE, Syngenta (Switzerland); DACONIL® 2787 W, Syngenta (Switzerland); DACONIL® F, Syngenta (Switzerland); DACONIL® M, Syngenta (Switzerland); DACONIL® TURF, Syngenta (Switzerland); DACOSOIL®, Syngenta (Switzerland); DIVA FUNGICIDE®, Bayer CropScience (Germany) canceled 9/13/2001; ECHO®, Sipcam Agro USA; EXOTHERM®; EXOTHERM TERMIL®, Griffin L.L.C. (USA); FORTURF®; FUNGINIL®, Milenia Agro Ciencias (Brazil); IMPACT EXCEL®; JUPITAL®; NUOCIDE®; OLE®; PILLARICH®; POWER CHLOROTHALONIL® 50; REPULSE®; RIDOMIL GOLD/BRAVO®, Milenia Agro Ciencias (Brazil); SICLOR®; SIPCAM® UK ROVER 5000, Sipcam Agro USA; SWEEP®; TER-MIL®; TPN®; TPN (PESTICIDE)®; TRIPART FABER®, Makhteshim-Agan Industries (Israel); TRIPART ULTRAFABER®, Makhteshim-Agan Industries (Israel); TUFFCIDE®

*Producers:* Agrimor International (USA); Agsin (Singapore); Bayer CropScience (Germany); Biesterfeld Siemsgluess International. GmbH (Germany); China Chemical (China); Ehrenstorfer, Dr. (Germany); Fluorochem (UK); Gharda Chemicals (India); Hebei Huafeng Chemical Group (China); Ishihara Sangyo Kaisha (Japan); Jiangmen Pesticide Factory (China); Jingma Chemicals Ltd. (China); Ki-Hara Chemicals Ltd. (UK); KMG Chemicals (USA); Luxembourg Industries (PAMOL) (Israel); Makhteshim-Agan Industries (Israel); Milenia Agro Ciencias (Brazil); Shenzhen Guomeng Industry Co., Ltd. (China); Sigma-Aldrich Laborchemikalien (Germany); Sipcam Agro USA; Syngenta (Switzerland); Veterinary & Agricultural Products Manufacturing Co., Ltd. (VAPCO) (Jordan); Whyte Agrochemicals (UK); Zeneca Ag Products (USA) (now Syngenta)

*Chemical Class:* Chloronitrile; Halo-organics
*EPA/OPP PC Code:* 081901
*California DPR Chemical Code:* 677
*ICSC Number:* 0134
*RTECS Number:* NT2600000
*EEC Number:* 608-014-00-4
*EINECS Number:* 217-588-1

*Uses:* Chlorothalonil is a broad spectrum fungicide. It is used on vegetables, peanuts, potatoes, small fruits, trees, turf, roses, ornamentals, and other crops. In California, the top crops are tomatoes, onions, celery, and landscaping. It targets fungal blights, needlecasts, and cankers on conifer trees. It is also used as a fungicide additive in paints and grouts.

*Human toxicity (long-term)[77]:* Intermediate–45.69190 ppb, CHCL (Chronic Human Carcinogen Level)

*Fish toxicity (threshold)[77]:* High–4.41591 ppb, MATC (Maximum Acceptable Toxicant Concentration)

*U.S. Maximum Allowable Residue Levels for Chlorothalonil (40 CFR 180.275):*

| CROP | ppm |
|---|---|
| Almond | 0.05 |
| Almond, hulls | 1.0 |
| Apricot | 0.5 |
| Asparagus | 0.1 |
| Banana | 0.5 |
| Banana, pulp | 0.05 |
| Bean, dry | 0.1 |
| Bean, snap | 5.0 |
| Blueberry | 1.0 |
| Broccoli | 5.0 |
| Brussels sprouts | 5.0 |
| Cabbage | 5.0 |
| Carrot | 1.0 |
| Cattle, fat | 0.1 |
| Cattle, kidney | 0.5 |
| Cattle, meat | 0.03 |
| Cattle, mbyp, except kidney | 0.05 |
| Cauliflower | 5.0 |
| Celery | 15.0 |
| Cherry, sweet | 0.50 |
| Cherry, tart | 0.50 |
| Cocoa beans | 0.05 |
| Coffee, bean | 0.2 |
| Corn sweet kernels plus cob with husks removed | 1.0 |
| Cranberry | 5.0 |
| Cucumber | 5.0 |
| Filbert | 0.1 |
| Fruit, passion | 3.0 |
| Ginseng, roots | 0.10 |
| Goat, fat | 0.1 |
| Goat, kidney | 0.5 |
| Goat, meat | 0.03 |
| Goat, mbyp, except kidney | 0.05 |
| Hog, fat | 0.1 |
| Hog, kidney | 0.5 |
| Hog, meat | 0.03 |
| Hog, mbyp, except kidney | 0.05 |
| Horse, fat | 0.1 |
| Horse, kidney | 0.5 |
| Horse, meat | 0.03 |
| Horse, mbyp, except kidney | 0.05 |
| Mango | 1.0 |
| Melon | 5.0 |
| Milk | 0.1 |
| Mint, hay | 2.0 |

| | |
|---|---|
| Mushroom | 1.0 |
| Nectarine | 0.5 |
| Onion, dry bulb | 0.5 |
| Onion, green | 5.0 |
| Papaya | 15.0 |
| Parsnip, roots | 1.0 |
| Peach | 0.5 |
| Peanut | 0.3 |
| Pepper, non-bell | 5 |
| Pistachio | 0.2 |
| Plum | 0.2 |
| Plum, prune | 0.2 |
| Potato | 0.1 |
| Pumpkin | 5.0 |
| Sheep, fat | 0.1 |
| Sheep, kidney | 0.5 |
| Sheep, meat | 0.03 |
| Sheep, mbyp, except kidney | 0.05 |
| Soybean | 0.2 |
| Squash, summer | 5.0 |
| Squash, winter | 5.0 |
| Tomato | 5.0 |

*Carcinogen/Hazard Classifications*
**U.S. EPA Carcinogens:** Likely carcinogen
**California Prop. 65:** Carcinogen
**IARC:** Group 2B, possible carcinogen
**Label Signal Word:** Chlorothalonil is often used in combination with other compounds and, depending upon the formulations, requires different levels of Signal Words–WARNING, CAUTION or DANGER.
**WHO Acute Hazard:** Class U, unlikely to be hazardous

*Regulatory Authority:*
- Carcinogen (Animal Positive) (NCI) (NTP) (California)
- EPCRA Section 313 Form R *de minimus* concentration reporting level: 1.0%
- AB 1803-Well Monitoring Chemical (CAL)
- AB 2588-Air Toxics "Hot Spots" Chemicals (CAL)
- Proposition 65 chemical (CAL)
- The "Director's List" (CAL/OSHA)
- Actively registered pesticide in California.
- CAL Air Resources Board/AB 1807 Toxic Air Contaminants as cyanide compound
- Clean Air Act: Hazardous Air Pollutants (Title I, Part A, Section 112) as cyanide compound
- Clean water act: Section 307 Priority Pollutants as cyanide, total; Toxic Pollutant (Section 401.15) as cyanide compound
- EPA Hazardous Waste Number (RCRA No.): P030 as cyanide compound
- RCRA Section 261 Hazardous Constituents. as cyanide compound
- RCRA Universal Treatment Standards: Wastewater (mg/L), 1.2 (total); 0.86 (amenable); Nonwastewater (mg/kg), 590 (total); 30 (amenable) as cyanide compound
- RCRA Ground Water Monitoring List. Suggested test method(s) (PQL $ug/L$): 9010(40) as cyanide compound
- Safe Drinking Water Act: MCL, 0.2 mg/L; MCLG, 0.2 mg/L;Regulated chemical (47 FR 9352) as cyanide compoundEPCRA Section 304 RQ: CERCLA, 10 lb (4.54 kg) as cyanide compound
- EPCRA Section 313 Form R *de minimus* concentration reporting level: 1.0% as cyanide compound
- U.S. DOT Regulated Marine Pollutant (49CFR, Subchapter 172.101, Appendix B) as cyanides, mixtures or solutions

***Description:*** White, crystalline solid. Odorless. Soluble in water; solubility = 0.6 ppm @ 25°C. Molecular weight 265.91.
Boiling point = 350°C. Melting/Freezing point = 260°C. Vapor pressure = $1 \times 10^{-3}$ mmHg. @ 20°C. Based on the NFPA rating system. (FEMA) Hazard Identification: Health 3, Flammability 1, Reactivity 0. BCF = –820. Bioconcentration in aquatic organisms may occur.

***Incompatibilities:*** Contact with strong oxidizers may cause a fire and explosion hazard. Thermal decomposition may include fumes of hydrogen cyanide.

***Permissible Exposure Limits in Air:*** No standards set. However, inasmuch as it is a cyanide compound, the exposure limits are listed here: OSHA[2] and ACGIH: 5 mg/m$^3$ TWA; NIOSH[2]: Ceiling limit, 4.7 ppm; 5 mg/m$^3$ per 10 minutes as cyanides[1]. All have notations that skin contact contributes significantly in overall exposure. IDLH level = 25 mg/m$^3$ as CN

***Determination in Air:*** NIOSH REL: (Nitriles) TWA 22 mg/m$^3$; NIOSH REL: (Nitriles) CL 4 mg/m$^3$/15minute. See also NIOSH Criteria Document 78-212 NITRILES.[18]

***Permissible Concentration in Water:*** The U.S. EPA has set a maximum contaminant level of cyanide in drinking water of 0.2 ppm (0.2 mg/L). A ten-day health advisory for a 10 kg child has been calculated by EPA to be 0.25 mg/L. A longer term health advisory for a 10-kg child was calculated to be 0.15 mg/L and for a 70-kg adult was calculated to be 0.525 mg/L. The estimated excess cancer risk associated with lifetime exposure to drinking water containing 0.525 mg/L of Chlorothalonil is $3.5 \times 10^{-4}$.

***Determination in Water:*** Analysis of chlorothalonil is by a gas chromatographic (GC) method applicable to the determination of certain chlorinated pesticides in water samples. In this method, approximately 1 liter of sample is extracted with Methylene chloride. The extract is concentrated and the compounds are separated using capillary column GC. Measurement is made using an electron capture detector. The method detection limit has not been determined for chlorothalonil, but it is estimated that the detection limits for analytes included in this method are in the range of 0.01 to 0.1 µg/L.

***Routes of Entry:*** Inhalation, skin contact.

***Harmful Effects and Symptoms***

*Johnson et al. (1983)* reported that chlorothalonil exposure resulted in contact dermatitis in 14 of 20 workers involved in woodenware preservation. The wood preservative used by the workers consisted mainly of "white spirit," with 0.5% chlorothalonil as a fungicide. Workers exhibited erythema and edema of the eyelids, especially the upper eyelids, and eruptions on the wrist and forearms. Results of patch test conducted with 0.1% chlorothalonil in acetone were positive in 7 of 14 subjects. Reactions ranged from a few erythematous papules to marked popular erythema with a brownish hue without infiltration. *Wilson et al. (1985)* gave chlorothalonil (98.1% pure with less than 0.03% hexachlorobenzene) to Fischer 344 rats (60/sex/dose) in their diet at dose levels of 0, 40, 80 or 175 mg/kg/day. Males were treated for 116 weeks, while females received the chemical for 129 weeks. Survival among the various groups was comparable. In both sexes, at the high dose level, there were significant decreases in body weights. In addition, there were also significant increases in blood urea nitrogen and creatinine, while there were decreases in serum glucose and albumin levels. In both sexes, there were dose-dependent increases in kidney carcinomas and adenomas at doses above 40 mg/kg/day. In the high-dose females, there was also a significant increase in stomach papillomas. The data show that, in the Fischer 344 rat, chlorothalonil is a carcinogen. The $LD_{50}$ (oral, rat) = 10 g/kg (insignificantly toxic).

***Short Term Exposure:*** Irritates the eyes, skin, respiratory tract. Inhalation can cause coughing, phlegm, and/or tightness in the chest.

***Long Term Exposure:*** Repeated or prolonged contact with skin may cause nose bleeding, skin sensitization and dermatitis with skin rash. May affect the kidneys and gastrointestinal tract. This chemical causes cancer of the kidneys in animals.

***Points of Attack:*** Skin, lungs and kidneys.

***Medical Surveillance:*** Complete blood count. Lung function tests. Kidney function tests.

***First Aid:*** If this chemical gets into the eyes, remove any contact lenses at once and irrigate immediately for at least 20-30 minutes, occasionally lifting upper and lower lids. Seek medical attention immediately. If this chemical contacts the skin, remove contaminated clothing and wash immediately with soap and water. Seek medical attention immediately. If this chemical has been inhaled, remove from exposure, begin rescue breathing (using universal precautions) if breathing has stopped, and CPR if heart action has stopped. Transfer promptly to a medical facility. When this chemical has been swallowed, get medical attention. Give large quantities of water and induce vomiting. Do not make an unconscious person vomit.

***References:***

- U.S. Environmental Protection Agency, Office of Pesticide Programs, Pesticide Residue Limits, "Chlorothalonil," 40 CFR 180.275, http://www.epa.gov/pesticides/food/viewtols.htm
- EXTOXNET, Extension Toxicology Network, "Pesticide Information Profile, Chlorothalonil," Oregon State University, Corvallis, OR. (June 1996). http://ace.ace.orst.edu/info/extoxnet/pips/chloroth.htm
- New Jersey Department of Health and Senior Services, Hazardous Substance Fact Sheet, Chlorothalonil, Trenton NJ (April 1998). http://www.state.nj.us/health/eoh/rtkweb/0415.pdf
- U.S. Environmental Protection Agency, "Chemical Profile: Chlorothalonil," Washington, DC, Office of Drinking Water (August 1987).
- California Environmental Protection Agency "Chemical List of Lists," Sacramento CA (February 1997).
- "Handbook of Chemical Property Estimation Methods." Amer Chemical Society, Washington DC, 1990.

## o-Chlorotoluene

***Use Type:*** Insecticide and bactericide

***CAS Number:*** 95-49-8

***Formula:*** $C_7H_7Cl$; $CH_3C_6H_4Cl$

***Synonyms:*** Benzene, 1-chloro-2-methyl-; 2-Chloro-1-methylbenzene; 2-Chlorotoluene; 1-Methyl-2-chlorobenzene; 2-Methylchlorobenzene; Toluene, o-chloro-; o-Tolylchloride

***Producers:*** ATOFINA (France); Bayer Chemicals (Germany); China Chemical (China); EniChem (Italy); Kawaguchi Chemical Industry (Japan); OxyChem (USA); Sigma-Aldrich Laborchemikalien (Germany); Tessenderlo (Belgium)

***Chemical Class:*** Halo-organics

***EPA/OPP PC Code:*** 216300

***California DPR Chemical Code:*** 2467

***ICSC Number:*** 1458

***RTECS Number:*** XS9000000

***EEC Number:*** 602-040-00-X

***EINECS Number:*** 202-424-3

***Uses:*** o-Chlorotoluene is also widely used as a solvent and intermediate in the synthesis of dyes, synthetic rubber, pharmaceuticals, and other organic chemicals.

***Regulatory Authority:***

- Air Pollutant Standard Set (ACGIH)[1] (Australia) (HSE)[33] (Israel) (OSHA)[58] (Several States)[60] (Several Canadian Provinces)
- EPA Toxic Substance Control Act (TSCA) Sec. 8(b) Chemical Inventory
- AB 1803-Well Monitoring Chemical (CAL)

- Permissible Exposure Limits for Chemical Contaminants (CAL/OSHA)
- The "Director's List" (CAL/OSHA)
- Canada, WHMIS, Ingredients Disclosure List

*Description:* Colorless liquid with an aromatic odor. The odor threshold = 0.32 ppm. Slightly soluble in water. Molecular weight = 126.59. Boiling point = 159°C. Flash point = 52°C. Hazard Identification (based on NFPA-704 M Rating System): Health 2, Flammability 2, Reactivity 0.

*Incompatibilities:* Incompatible with acids, alkalis, oxidizers, reducing materials, water.

*Permissible Exposure Limits in Air:* There is no OSHA[2] PEL. NIOSH[2] recommends an airborne exposure limit of 50 ppm (250 mg/m3) TWA for a 10-hour workshift and STEL of 75 ppm (375 mg/m$^3$) not to be exceeded during any 15 minute work period., with the notation "skin" indicating the possibility of cutaneous absorption. The ACGIH recommends the same TWA averaged over an 8-hour workshift and does not have an STEL value[1]. This chemical can be absorbed through the skin, thereby increasing exposure. There is no NIOSH[2] IDLH. The UK HSE, Australia, Israel, and Canadian provinces of Alberta, BC, Ontario, and Quebec have adopted the same TWA as NIOSH[2] and Alberta, BC, Ontario use the same STEL value. Several states have set guidelines or standards for chlorotoluene in ambient air[60] ranging from 2.5–3.75 mg/m$^3$ (North Dakota) to 4.0 mg/m$^3$ (Virginia) to 5.0 mg/m$^3$ (Connecticut) to 5.95 mg/m$^3$ (Nevada).

*Permissible Concentration in Water:* No criteria set, but EPA[32] has suggested a permissible ambient goal of 3,450 µg/L based on health effects.

*Routes of Entry:* Inhalation, skin absorption, ingestion, skin and/or eye contact.

*Harmful Effects and Symptoms*

*Short Term Exposure:* Contact can irritate and burn the eyes and skin. Inhalation can irritate the respiratory tract, causing coughing, and/or shortness of breath. High exposure can cause dizziness, loss of coordination, convulsions and coma. Vasodilatation, labored respiration, and narcosis have been observed in test animals.

*Long Term Exposure:* May affect the liver and kidneys.

*Points of Attack:* Eyes, skin, respiratory system, central nervous system, liver, kidneys. Prolonged or repeated contact may cause dermatitis.

*Medical Surveillance:* If symptoms develop or overexposure is suspected, the following may be useful: liver function tests. Kidney function tests. Examination by a dermatologist.

*First Aid:* If this chemical gets into the eyes, remove any contact lenses at once and irrigate immediately for at least 15 minutes, occasionally lifting upper and lower lids. Seek medical attention immediately. If this chemical contacts the skin, remove contaminated clothing and wash immediately with soap and water. Seek medical attention immediately. If this chemical has been inhaled, remove from exposure, begin rescue breathing (using universal precautions) if breathing has stopped, and CPR if heart action has stopped. Transfer promptly to a medical facility. When this chemical has been swallowed, get medical attention. Give large quantities of water and induce vomiting. Do not make an unconscious person vomit.

*References:*
- New Jersey Department of Health and Senior Services and Senior Services, "Hazardous Substance Fact Sheet: o-Chlorotoluene," Trenton, NJ (November 1998). http://www.state.nj.us/health/eoh/rtkweb/1425.pdf
- California Environmental Protection Agency "Chemical List of Lists," Sacramento CA (February 1997).

## Chloroxuron (ANSI)

*Use Type:* Herbicide
*CAS Number:* 1982-47-4
*Formula:* $C_{15}H_{15}ClN_2O_2$; $(CH_3)_2NCONHC_6H_4OC_6H_4Cl$
*Synonyms:* N'-[4-(4-chlorophenoxy)phenyl]-N,N-dimethylurea; 3-[4-(4-Chlorophenoxy)phenyl]-1,1-dimethylurea; 3-[p-(p-chlorophenoxy)phenyl-1,1]-dimethylurea; Chloroxifenidum; Cloroxuron (Spanish); Urea, N'-[4-(4-Chlorophenoxy)phenyl]-N,N-dimethyl-; Urea, 3-[p-(p-chlorophenoxy)phenyl]-1,1-dimethyl-
*Trade Names:* C 1983®, Ciba Specialty Chemicals (Switzerland)U.S. DOT Regulated Marine PollutantCIBA 1983®, Ciba Specialty Chemicals (Switzerland); NOREX®; TENORAN®, Syngenta (Switzerland), canceled 7/2/1985
*Producers:* Ciba Specialty Chemicals (Switzerland); Sigma-Aldrich Laborchemikalien (Germany); Syngenta (Switzerland)
*Chemical Class:* Urea; Halo-organics
*EPA/OPP PC Code:* 025501
*California DPR Chemical Code:* 576
*RTECS Number:* YS6125000
*EINECS Number:* 217-843-7
*Uses:* A selective pre- and early post-emergency herbicide in soybeans, strawberries, onions, celery and ornamentals. It is a root- and foliage-absorbed herbicide selective in leek, celery, onion, carrot and strawberry.
*Human toxicity (long-term)*[77]: Intermediate–35.00 ppb, Health Advisory
*Fish toxicity (threshold)*[77]: Intermediate–49.75254 ppb, MATC (Maximum Acceptable Toxicant Concentration)
*Regulatory Authority:*
- AB 2588-Air Toxics "Hot Spots" Chemicals (CAL) as chlorophenoxy pesticides
- The "Director's List" (CAL/OSHA) as chlorophenoxy pesticides
- Superfund/EPCRA 40CFR355, Appendix B Extremely Hazardous Substances: TPQ = 500/10,000 lb (227/4,540 kg)

- Superfund/EPCRA 40CFR302.4 RQ: EHS, 1 lb (0.454 kg)

*Description:* White crystalline solid. Practically insoluble in water; solubility = 3 ppm. Molecular weight = 290.77. Melting/Freezing point = 151-152°C. Vapor pressure = 3.9 x $10^{-9}$ mmHg. @ 20°C. Log $K_{ow}$ = 3.9. Values at or above 3.0 are likely to bioaccumulate in marine organisms.

*Incompatibilities:* Contact with strong oxidizers may cause fire and explosion hazard. When heated to decomposition this chemical may emit toxic nitrogen oxides and chlorine gas.

*Permissible Exposure Limits in Air:* No standards set.

*Permissible Concentration in Water:* No criteria set, but runoff from spills or fire control may cause water pollution.

*Routes of Entry:* Ingestion

*Harmful Effects and Symptoms*

*Short Term Exposure:* Slightly irritating to eyes and skin. The LD50-oral (dog) is 10 mg/kg (highly toxic). The $LD_{50}$ (oral, rat) = 3700 mg/kg. Chloroxuron is stated to be highly toxic to humans by ingestion and under certain conditions, it can form carcinogenic dimethylnitrosamine. A regulated, marked area should be established where this chemical is handled, used, or stored in compliance with OSHA standard 1910.1045.

*First Aid:* If this chemical gets into the eyes, remove any contact lenses at once and irrigate immediately for at least 15 minutes, occasionally lifting upper and lower lids. Seek medical attention immediately. If this chemical contacts the skin, remove contaminated clothing and wash immediately with soap and water. Seek medical attention immediately. If this chemical has been inhaled, remove from exposure, begin rescue breathing (using universal precautions) if breathing has stopped, and CPR if heart action has stopped. Transfer promptly to a medical facility. When this chemical has been swallowed, get medical attention. Give large quantities of water and induce vomiting. Do not make an unconscious person vomit.

*References:*
- New Jersey Department of Health and Senior Services, "Hazardous Substance Fact Sheet, Chloroxuron," Trenton NJ (August 2000). http://www.state.nj.us/health/eoh/rtkweb/2246.pdf
- Pesticide Management Education Program, Cornell University, "Chloroxuron Herbicide Profile 2/85," Ithica, NY (February 1985). http://pmep.cce.cornell.edu/profiles/herb-growthreg/cacodylic-cymoxanil/chloroxuron/herb-prof-chloroxuron.html
- U.S. Environmental Protection Agency, "Chemical Profile: Chloroxuron," Washington, DC, Chemical Emergency Preparedness Program (October 31, 1985).
- California Environmental Protection Agency "Chemical List of Lists," Sacramento CA (February 1997).

# 4-Chloro-3,5-xylenol

*Use Type:* Microbiocide
*CAS Number:* 88-04-0
*Formula:* $C_8H_9ClO$
*Synonyms:* Benzytol; 4-Chloro-3,5-dimethylphenol; Chloroxylenol; Chloro-*m*-xylenol; 3,5-Dimethyl-4-chlorophenol; Parachlorometaxylenol; *p*-Chloro-*m*-xylenol
*Trade Names:* BAR-FUNGAL PLUS®, canceled; BIO SLIME®, canceled; DESSON®; DETTOL®; ECOTRU®; ESPADOL®; HUSEPT®; HUSEPT EXTRA®; NIPACIDE®, Clariant International (Switzerland); OTTASEPT®, Clariant International (Switzerland), canceled; OTTASEPT® EXTRA, Clariant International (Switzerland), canceled; PCMX; RBA 777®
*Producers:* Clariant International (Switzerland)
*Chemical Class:* Chlorophenol
*EPA/OPP PC Code:* 086801
*California DPR Chemical Code:* 925
*Uses:* Chloroxylenol was first registered in the U.S. for use as a fungicide and is now applied as an antimicrobial for control of bacteria, algae and fungi in adhesives, emulsions, paints and wash tanks. It is registered for use in household and domestic dwellings, laundry equipment, bathrooms, diaper pails and in industrial water and aqueous systems. It is also used to sanitize industrial and hospital premises and pet living quarters, and in hand and body lotions.

*Carcinogen/Hazard Classifications*

**Label Signal Word:** CAUTION or DANGER *Note:* In animal studies, one chlorophenol, 2,4,6-trichlorophenol, caused leukemia in rats and liver cancer in mice. The Department of Health and Human Services has determined that 2,4,6-trichlorophenol may reasonably be anticipated to be a carcinogen.

*Regulatory Authority:*
- Actively registered pesticide in California.
- List of priority pollutants (EPA) as chlorophenols
- AB 2588-Air Toxics "Hot Spots" Chemicals (CAL) as chlorophenols
- The "Director's List" (CAL/OSHA) as chlorophenols
- Clean Water Act: Toxic Pollutant (Section 401.15) other than those listed elsewhere; includes trichloro phenols.
- RCRA Section 261 Hazardous Constituents, waste number not listed as chlorophenols
- EPCRA Section 313 (as chlorophenols) Form R *de minimis* concentration reporting level: 1.0%

*Description:* A white to buff crystalline powder. Phenolic odor. Slight soluble in water. Boiling point = 246°C. Melting/Freezing point = 116°C. Molecular weight = 156.65.

*Incompatibilities:* React with boranes, alkalies, aliphatic amines, amides, nitric acid, sulfuric acid. Keep away from oxidizers (chlorates, nitrates, peroxides, permanganates, perchlorates, chlorine, bromine, fluorine, etc).

*Permissible Concentration in Water:* No criteria set for this chemical; however, EPA recommends that drinking water contain no more than 0.04 mg/L of 2-chlorophenol for a lifetime exposure for an adult, and 0.05 mg/L for a 1-day, 10-day, or longer exposure for a child.

*Routes of Entry:* Inhalation, absorbed by skin, ingestion.

*Harmful Effects and Symptoms*

*Short Term Exposure:* Inhalation can cause severe irritation, burns to the nose and throat, headache, dizziness, vomiting, lung damage, muscle twitching, spasms, tremors, weakness, staggering and collapse. Skin contact can cause severe irritation and burns. May be absorbed through the skin to cause or increase the severity of symptoms listed above. Eye contact may cause severe irritation. May cause burns. Ingestion may cause irritation, burns to the mouth and throat, low blood pressure, profuse sweating, intense thirst, nausea, abdominal pain, stupor, vomiting, red blood cell damage and accumulation of fluid in the lungs followed by pneumonia. May also cause restlessness and increased breathing rate followed by rapidly developing muscle weakness. Tremors, convulsions and coma can promptly set in and will continue until death.

*Long Term Exposure:* Skin sensitivity may develop. May have effects on the blood, heart, liver, lung, kidney. The state of New Jersey lists at least one chlorophenol as a probable carcinogen in humans; may cause leukemia and soft tissue cancers in humans. The substance may cause effects on the central nervous system and bladder. Chlorophenols leave the body quickly, so they are not likely to accumulate in the mother's tissues or breast milk. There are no human studies on the effects of chlorophenols on developing fetuses. Studies in rats showed that chlorophenols can pass through the placenta and produce toxic effects to the developing fetuses. The most common problems in children are delayed hardening of the bones of the breastbone, spine, and skull.

*Points of Attack:* Eyes, skin, respiratory system.

*Medical Surveillance:* If symptoms develop or overexposure is suspected, the following may be useful: Liver function tests. Kidney function tests. Examination by a qualified allergist. EKG.

*First Aid:* If this chemical gets into the eyes, remove any contact lenses at once and irrigate immediately for at least 15 minutes, occasionally lifting upper and lower lids. Seek medical attention immediately. If this chemical contacts the skin, remove contaminated clothing and wash immediately with soap and water. Seek medical attention immediately. If this chemical has been inhaled, remove from exposure, begin rescue breathing (using universal precautions) if breathing has stopped, and CPR if heart action has stopped. Transfer promptly to a medical facility. When this chemical has been swallowed, get medical attention. If victim is conscious and able to swallow, have victim drink 4 to 8 ounces of water. Do not induce vomiting. *Note to physician or authorized medical personnel*: Medical observation is recommended for 24 to 48 hours after breathing overexposure, as pulmonary edema may be delayed. As first aid for pulmonary edema, consider administering a corticosteroid spray. Cigarette smoking may exacerbate pulmonary injury and should be discouraged for at least 72 hours following exposure.

*References:*
- U.S. Environmental Protection Agency, "Reregistration Eligibility Decision (RED), Chloroxylenol," Office of Prevention, Pesticides and Toxic Substances, Washington, DC (September 1994). http://www.epa.gov/REDs/3045.pdf
- California Environmental Protection Agency "Chemical List of Lists," Sacramento CA (February 1997)
- Agency for Toxic Substances and Disease Registry "Toxicological profile for chlorophenols," 1999, Atlanta, GA.

# Chlorpropham

*Use Type:* Herbicide (pre-emergence) and plant growth regulator

*CAS Number:* 101-21-3

*Formula:* $C_{10}H_{12}ClNO_2$

*Synonyms:* Carbamic acid, (3-chlorophenyl)-, 1-methylethyl ester; Chlor-IFC; Chlor-IPC; *m*-Chlorocarbanilic acid, isopropyl ester; 3-Chlorocarbanilic acid, isopropyl ester; Chloro IPC; *N*-(3-Chloro phenyl)carbamate disopropyle (French); *N*-3-Chlorophenylisopropylcarbamate; *N*-(3-Chlorophenyl)carbamic acid, isopropyl ester; (3-Chlorophenyl)carbamic acid, 1-methylethyl ester; Chloropropham; *N*-(3-chlor-phenyl)-isopropyl-carbamat (German); Chlorpropname (French); CICP; CI-IPC; CIPC; ENT 18,060; Isopropyl-*m*-chlorocarbanilate; Isopropyl-3-chlorophenylcarbamate; Isopropyl-*N*-(3-chlorophenyl)carbamate; *o*-Isopropyl-*N*-(3-chlorophenyl)carbamate; Isopropyl-*N*-(3-chlorphenyl)-carbamat (German)

*Trade Names:* ATLAS® CIPC 40; BEET-KLEEN® (with Fenuron® and isopropyl carbanilate); BUD-NIP®; CAMPBELL'S® CIPC 40%; CHLORO IPC®; ELBANIL®; FASCO® WY-HOE; FURLOE®; FURLOE® 4EC; JACK WILSON® CHLORO 51 (OIL); LIRO METOXON®; MIRVALE®; MORCRAN® (with *n*-1-naphthylphthalamic acid); MSS® CICP; NEXOVAL®; PREVENOL® 56, Bayer CropScience (Germany); PREVENTOL®, Bayer CropScience (Germany); PREVENTOL® 56, Bayer CropScience (Germany); PREWEED®; RESIDUREN®; RESIDUREN® EXTRA; SPROUT NIP®; SPROUT-NIP® EC; SPUD-NIC®; SPUD-NIE®; STOPGERME®-S; TATERPEX®; TRIPEC®, (with carbamic acid, phenyl-, 1-methylethyl ester); TRIHERBICIDE® CIPC; UNICROP® CIPC; WAREFOG®; Y3®

# Chlorpropham

***Producers:*** Cerexagri (France); Epochem Co., (China); Ki-Hara Chemicals Ltd. (UK); Whyte Agrochemicals (UK
***Chemical Class:*** Carbamate; Halo-organics
***EPA/OPP PC Code:*** 018301
***California DPR Chemical Code:*** 141
***EINECS Number:*** 202-925-7
***Uses:*** Chlorpropham is a plant growth regulator that is used primarily in the U.S. to inhibit post-harvest potato sprouting. Other uses include pre-emergence control of grass weeds in alfalfa, beans, blueberries, cane berries, carrots, cranberries, ladino clover, garlic, seed grass, onions, spinach, sugar beets, tomatoes, safflower, soybeans, gladioli and woody nursery stock. It is used to control suckers in tobacco.
***Human toxicity (long-term)***[77]***:*** Very low–350.00 ppb, Health Advisory
***Fish toxicity (threshold)***[77]***:*** Low–424.62703 ppb, MATC (Maximum Acceptable Toxicant Concentration)
***Carcinogen/Hazard Classifications***
**U.S. EPA Carcinogens:** Group E, unlikely carcinogen
**IARC:** Group 3, unclassifiable
**Label Signal Word:** CAUTION or DANGER
**WHO Acute Hazard:** Class U, unlikely to be hazardous
***Regulatory Authority:***
- AB 1803 Well monitoring chemical (CAL)
- DHS drinking water action levels (CAL)
- Actively registered pesticides in California

***Description:*** Light-tan powder or commercial liquid. Slightly soluble in water. Solubility 89 mg/L @ 25°C. Molecular weight 213.67. Density = 1.18 @ 30°C. Melting/Freezing point = 41°C. Boiling point = 149°C. Vapor pressure = $8 \times 10^{-6}$ mmHg; 2 mmHg @ 149°C.
***Incompatibilities:*** May form explosive materials with phosphorus pentachloride.
***Permissible Concentration in Water:*** No criteria set. Runoff from spills or fire control may cause water pollution.
***Routes of Entry:*** Skin absorption, ingestion, inhalation.
***Harmful Effects and Symptoms***
***Short Term Exposure:*** Eye pupils are small; blurred vision; eye watering; runny nose; cough; shortness of breath; salivation; nausea, stomach cramps, diarrhea, and vomiting; increased blood pressure; profuse sweating; hypermotility, hallucinations; agitation; tingling of the skin; slow heartbeat; convulsions; fluid in lungs; loss of consciousness; incontinence; breathing stops; death. Carbamate insecticides inhibit the cholinesterase activity of enzymes, causing accumulation of acetylcholine at synapses and altering the way in which nervous impulses are transmitted. However, within several hours carbamates spontaneously detach from the enzymes. Because this material has a low vapor pressure, significant inhalation of vapors is unlikely at ordinary temperatures.
***Long Term Exposure:*** A potent cholinesterase inhibitor; cumulative effect is possible. This chemical may damage the nervous system with repeated exposure, resulting in convulsions, respiratory failure. May cause liver damage.
***Points of Attack:*** Respiratory system, lungs, central nervous system, cardiovascular system, skin, eyes, plasma and red blood cell cholinesterase.
***Medical Surveillance:*** Medical observation is recommended for 24 to 48 hours after breathing overexposure, as pulmonary edema may be delayed. As first aid for pulmonary edema, consider administering a corticosteroid spray. Cigarette smoking may exacerbate pulmonary injury and should be discouraged for at least 72 hours following exposure. Before employment and at regular times after that, the following are recommended: Plasma and red blood cell cholinesterase levels (tests for the enzyme poisoned by this chemical). If exposure stops, plasma levels return to normal in 1-2 weeks while red blood cell levels may be reduced for 1-3 months. When acetylcholinesterase enzyme levels are reduced by 25% or more below preemployment levels, risk of poisoning is increased, even if results are in lower ranges of "normal." Reassignment to work not involving carbamate pesticides is recommended until enzyme levels recover. If symptoms develop or overexposure occurs, repeat the above tests as soon as possible and get an exam of the nervous system. Also consider complete blood count. Consider chest x-ray following acute overexposure
***First Aid:*** Speed in removing material from eyes and skin is of extreme importance. *Eyes:* Eye contact can cause dangerous amounts of these chemicals to be quickly absorbed through the mucous membrane into the bloodstream. Immediately and gently flush eyes with plenty of warm or cold water (NO hot water) for at least 15 minutes, occasionally lifting the upper and lower eyelids. Get medical aid immediately. *Skin:* Get medical aid. Skin contact can cause dangerous amounts of these chemicals to be absorbed into the bloodstream. Wearing the appropriate PPE equipment and respirator for carbamate pesticides, immediately flush skin with plenty of soap and water for at least 15 minutes while removing contaminated clothing and shoes. Shampoo hair promptly if contaminated; protect eyes. *Ingestion:* Call poison control. Loosen all clothing. Never give anything by mouth to an unconscious person. Get medical aid. Do NOT induce vomiting.* If conscious, alert, and able to swallow, rinse mouth and have victim drink 4 to 8 ounces of water. Check to see if poison control instructs you to use ipecac syrup, otherwise administer slurry of activated charcoal (2 oz in 8 oz of water). If victim is *UNCONSCIOUS OR HAVING CONVULSIONS,* do nothing except keep victim warm. **In some cases you may be specifically instructed by poison control to induce vomiting by way of 2 tablespoons of syrup of ipecac (adult) washed down with a cup of water.* Do NOT give activated charcoal before or with ipecac syrup. *Inhalation:* Get medical aid. Do not contaminate yourself. Wearing the appropriate PPE

equipment and respirator for carbamate pesticides, immediately remove the victim from the contaminated area to fresh air. If the victim is not breathing, administer artificial respiration. Do NOT use mouth-to-mouth resuscitation; use bag/mask apparatus. If breathing is difficult, administer oxygen through bag/mask apparatus until medical help arrives. Do not leave victim unattended. *Note to physician or authorized medical personnel.* Administer atropine, 2 mg (1/30 gr) intramuscularly or intravenously as soon as any local or systemic signs or symptoms of an intoxication are noted; repeat the administration of atropine every 3 to 8 minutes until signs of atropinization (mydriasis, dry mouth, rapid pulse, hot and dry skin) occur; initiate treatment in children with 0.05 mg mg/kg of atropine; repeat at 5 to 10 minute intervals. Watch respiration, and remove bronchial secretions if they appear to be obstructing the airway; intubate if necessary. *Medical note:* 2-PAMCI may be contraindicated in the case of some carbamate poisonings.

*References:*
- U.S. Environmental Protection Agency, "Reregistration Eligibility Decision (RED), Chlorpropham", Office of Prevention, Pesticides and Toxic Substances, Washington, DC (October 1996). http://www.epa.gov/REDs/0271red.pdf
- EXTOXNET, Extension Toxicology Network, "Pesticide Information Profile, Chlorpropham," Oregon State University, Corvallis, OR (June 1996). http://extoxnet.orst.edu/pips/choropro.htm
- California Environmental Protection Agency "Chemical List of Lists," Sacramento CA (February 1997)

# Chlorpyrifos (ANSI)

*Use Type:* Insecticide and nematicide
*CAS Number:* 2921-88-2
*Formula:* $C_9H_{11}Cl_3NO_3PS$
*Alert:* A General Use Pesticide (GUP). Under a Revised Risk Assessment, EPA and the registrants have agreed to the following modifications: (The complete Agreement can be found at the U.S. EPA page shown in the footnote below).
*Reducing Food Risks:* The agreement will expeditiously address food uses posing the greatest risks to children. It decreases the use of chlorpyrifos on apples, terminating or canceling all post-bloom applications, and cancels the use on tomatoes. EPA will also propose to (1) lower the tolerance or maximum residue limit on apples to reflect this change in use, (2) revoke the tolerance on tomatoes, and (3) lower the tolerance on grapes to a level that will allow for dormant applications (the only use allowed domestically), but not foliar applications typically made in foreign countries on grapes that are imported into the U.S. These actions will reduce acute dietary risk by 75 percent, effectively eliminating dietary risk concerns for children and others.
*Reducing Residential Risks:* About 50 percent of chlorpyrifos is used in and around the home. The agreement will cancel and phase out nearly all indoor and outdoor residential uses. It effectively eliminates the use of chlorpyrifos by homeowners, limiting use to certified, professional, or agricultural applicators. Those uses that pose the most immediate risks to children, including home lawn, indoor crack and crevice treatments, and whole house "postconstruction" termiticide treatments, will be canceled first. Spot and local post-construction and pre-construction termiticide uses will be phased out over the next several years.
*Reducing Drinking Water Risks:* The actions on residential uses also will reduce exposure to chlorpyrifos through drinking water, since residential applications are potentially a major source of drinking water contamination.
*Reducing Non-Residential Risks:* Chlorpyrifos use in schools, parks, and other settings where children may be exposed will be canceled. Only use in some limited commercial settings, like warehouses, ship holds and railroad boxcars, may continue.
*Synonyms:* α-Chlorpyrifos 48EC (α); Chlorpyrifos-ethyl; Clorpirifos (Spanish); O,O-DIethyl; ENT 27311; O-3,5,6-Trichloro-2-pyridylphosphorothioate; O,O-Diaethyl-O-3,5,6-trichlor-2-pyridylmonothiophosphat (German); O,O-Dimethyl O-(3,5,6-trichloro-2-pyridinyl)phosphorothioate; Phosphorothioic acid, O.O-diethyl O-(3,5,6-trichloro-2-pyridinyl)ester; 2-Pyridinol, 3,5,6-trichloro-,O-ester with O,O-Diethylphosphorothioate
*Trade Names:* (Note: EPA Office of Pesticide Programs lists 2135 products, both active and past-registered) ALUDOR®, Calliope (France); BAR 500 EC®, Sanonda (Australia); BRODAN®; CHLORBAN®, United Phosphorus (India); CHLORPIRIFOS 480 CE MILENIA®, Milenia Agro Ciencias (Brazil); CHOIR®, Whyte Agrochemicals (UK); COROBAN®; CURIGNA®, Milenia Agro Ciencias (Brazil); CYREN®, Cheminova (Denmark); DETMOL U.A.®; DORSAN®, Luxembourg Industries (PAMOL) (Israel); DORSAN®-C, Luxembourg Industries (PAMOL) (Israel); DOWCO® 179, Dow AgroSciences (USA); DURSBAN®, Dow AgroSciences (USA); EF 121®; EMPIRE®; ERADEX®; GLOBAL CRAWLING INSECT BAIT®; KENSBAN®, Kenso Corp. (Malaysia); LORSBAN®, Dow AgroSciences (USA), Gowan Dompany (USA); MURPHY SUPER ROOT GUARD®; PAQEANT®; PILOT®, Drexel Company (USA); PYRINEX®, Makhteshim Agan (Israel); SCOUT®; SPANNIT®; STIPEND; TALON®; TAFABAN®, Rallis India (India); TERIAL®; TWINSPAN®
*Producers:* Agan Chemical Manufacturers Ltd. (Israel); Agrimor International (USA); Agsin (Singapore); Aimco Pesticides Ltd. (India); Alcotan Laboratories (Spain);

## Chlorpyrifos

Atanor S.A. (Argentina); BEC Group (India); Bharat Rasayan (India); Biesterfeld Siemsgluess International. GmbH (Germany); Calliope (France); Cangzhou Green Chemical Co. (China); Cheminova (Denmark); China Chemical (China); Chromos Agro (Croatia); Drexel Chemical (USA); Dow AgroSciences (USA); Ehrenstorfer, Dr. (Germany); Excel Industries (India); Gharda Chemicals (India); Godavari Fertilisers and Chemicals (India); Gowan Company (USA); Hindustan Insecticides (India); Hockley International (UK); Indiclay (India); Indo Gulf (India); Kenso Corp. (Malaysia); Ki-Hara Chemicals Ltd. (UK); Luxembourg Industries (PAMOL) (Israel); Makhteshim Agan (Israel); Milenia Agro Ciencias (Brazil); Nagarjuna Agrichem (India); Rallis India (India); Saeryung Chemicals (South Korea); Shandong Huayang Pesticide Group (China); Sanonda (Australia); Shenzhen Guomeng Industry Co., Ltd. (China); United Phosphorus (India); Vijayalakshmi Insecticides and Pesticides (India); YiHua Group (China); Whyte Agrochemicals (UK); Zago Asia Ltd. (Singapore)

**Chemical Class:** Organophosphate
**EPA/OPP PC Code:** 059101
**California DPR Chemical Code:** 253
**ICSC Number:** 0851
**RTECS Number:** TF6300000
**EEC Number:** 015-084-00-4
**EINECS Number:** 220-864-4

**Uses:** Chlorpyrifos is one of the most widely-used insecticide in the U.S., both around the home and in agriculture. A broad-spectrum insecticide, originally used primarily to kill mosquitoes but no longer registered for that use. Chlorpyrifos is effective in controlling cutworms, corn rootworms, cockroaches, grubs, flea beetles, flies, termites, fire ants, and lice. It is used as an insecticide on grain, cotton, field, fruit, nut and vegetable crops, and well as on lawns and ornamental plants. It is also registered for direct use on sheep and turkeys, for horse site treatment, dog kennels, domestic dwellings, farm buildings, storage bins, and commercial establishments. Chlorpyrifos acts on pests primarily as a contact poison, with some action as a stomach poison. It is available as granules, wettable powder, dustable powder and emulsifiable concentrate. Top crop uses in California include cotton, alfalfa, almonds, and oranges.

**Human toxicity (long-term)[77]:** Intermediate–20.00 ppb, Health Advisory
**Fish toxicity (threshold)[77]:** Extra high–0.36660 ppb, MATC (Maximum Acceptable Toxicant Concentration)

**U.S. Maximum Allowable Residue Levels for Chlorpyrifos (40 CFR 180.342):**

| CROP | ppm |
|---|---|
| Alfalfa, green, forage | 3.0 |
| Alfalfa, hay | 13.0 |
| Almond | 0.2 |
| Almond, hulls | 12.0 |
| Apple | 1.5 |
| Asparagus | 5.0 |
| Banana | 0.1 |
| Banana, pulp (no peel) | 0.01 |
| Bean, forage | – |
| Bean, lima, forage | – |
| Bean, lima, seed | – |
| Bean, snap | 0.05 |
| Bean, snap, forage | – |
| Beet, sugar, dried pulp | 5.0 |
| Beet, sugar, molasses | 15.0 |
| Beet, sugar, roots | 1.0 |
| Beet, sugar, tops | 8.0 |
| Blueberry | 2.0 |
| Broccoli | 1 |
| Brussels sprouts | 1 |
| Cabbage | 1 |
| Cabbage, chinese, bok choy | 1 |
| Caneberry, subgroup 13a | – |
| Cattle, fat | 0.3 |
| Cattle, meat | 0.05 |
| Cattle, mbyp | 0.05 |
| Cauliflower | 1 |
| Cherimoya | 0.05 |
| Cherry | 1.0 |
| Citrus, dried pulp | 5.0 |
| Citrus, oil | 25.0 |
| Corn, field, grain | 0.05 |
| Corn, fodder | 8.0 |
| Corn, forage | 8.0 |
| Corn, fresh (inc. sweet)(k+cwhr) | 0.1 |
| Corn, oil | 3.0 |
| Cotton, undelinted seed | 0.2 |
| Cranberry | 1.0 |
| Cucumber | 0.05 |
| Date, dried fruit | – |
| Egg | 0.01 |
| Feijoa | 0.05 |
| Fig | 0.01 |
| Fruit, stone, group | 1.0 |
| Goat, fat | 0.2 |
| Goat, meat | 0.05 |
| Goat, mbyp | 0.05 |
| Grape | 0.50 |
| Hog, fat | 0.2 |
| Hog, meat | 0.05 |
| Hog, mbyp | 0.05 |
| Horse, fat | 0.25 |
| Horse, meat | 0.25 |
| Horse, mbyp | 0.25 |
| Kiwifruit | 2.0 |
| Leek | 0.50 |
| Legume vegetables,succulent or dried (ex soybeans) | 0.05 |
| Milk, fat | 0.25 |

| | |
|---|---|
| Milk, whole | 0.01 |
| Mint, hay | 0.8 |
| Mint, oil | 8.0 |
| Mushroom | – |
| Nectarine | 0.05 |
| Nut, tree, group 14 | 0.2 |
| Onion, dry bulb | 0.5 |
| Pea, forage | – |
| Peach | 0.05 |
| Peanut | 0.2 |
| Peanut, oil | 0.4 |
| Pear | 0.05 |
| Pepper | 1.0 |
| Plum | 0.05 |
| Poultry, fat (inc. turkeys) | 0.1 |
| Poultry, mbyp (inc. turkeys) | 0.1 |
| Poultry, meat (inc. turkeys) | 0.1 |
| Processed food | 0.1 |
| Pumpkin | 0.05 |
| Radish | 2.0 |
| Rutabaga | 0.5 |
| Sapote, white | 0.05 |
| Sheep, fat | 0.2 |
| Sheep, meat | 0.05 |
| Sheep, mbyp | 0.05 |
| Sorghum, forage | 1.5 |
| Sorghum, grain, grain | 0.75 |
| Sorghum, milled fractions | – |
| Sorghum, stover | 6.0 |
| Soybean | 0.3 |
| Soybean, forage | 0.7 |
| Soybean, straw | 15.0 |
| Strawberry | 0.2 |
| Sugarcane, cane | – |
| Sunflower, seed | 0.25 |
| Sweet potato | 0.05 |
| Tomato | 0.5 |
| Turnip, greens | 0.3 |
| Turnip, roots | 1.0 |
| Vegetable, seed & pod | – |
| Vegetables, brassica, leafy, group 5 | 2.0 |
| Walnut | 0.2 |
| Wheat, forage | 3.0 |
| Wheat, grain | 0.5 |
| Wheat, milled fractions, except flour | 1.5 |
| Wheat, straw | 6.0 |

### Carcinogen/Hazard Classifications

**U.S. EPA Carcinogens:** Group E, unlikely carcinogen
**Label Signal Word:** WARNING or CAUTION, depending on the formulation.
**WHO Acute Hazard:** Class II, moderately hazardous
**Endocrine Disruptor:** Suspected endocrine disruptor

### Regulatory Authority:
- Air Pollutant Standard Set (ACGIH)[1] (Australia) (HSE)[33] (Israel) (Mexico) (OSHA)[58] (former USSR)[35] (Several States)[60] (Several Canadian Provinces)
- AB 1803-Well Monitoring Chemical (CAL)
- Permissible Exposure Limits for Chemical Contaminants (CAL/OSHA)
- The "Director's List" (CAL/OSHA)
- Actively registered pesticide in California.
- Clean Water Act: Section 311 Hazardous Substances/RQ 40CFR117.3 (same as CERCLA, see below)
- Superfund/EPCRA 40CFR302.4 RQ: CERCLA, 1 lb (0.454 kg)
- U.S. DOT Inhalation Hazard Chemicals as organophosphates
- U.S. DOT Regulated Marine Pollutant (49CFR172.101, Appendix B), severe pollutant
- Canada, Drinking Water Quality MAC = 0.09 mg/L

**Description:** White crystalline compound. Mild mercaptan odor. The odor is also described as like natural gas. Practically insoluble in water; solubility = 2.2 ppm; 0.4 ppm @ 23°C. Molecular weight = 350.58. Melting/Freezing point = 41–43°C. Vapor pressure = $1.89 \times 10^{-5}$ mmHg @ 20°C. Log $K_{ow}$ = 4.98. Values at or above 3.0 are likely to bioaccumulate in marine organisms.

**Incompatibilities:** Above 130°C this chemical may undergo violent exothermic decomposition. The substance decomposes on heating at approximately 160°C and on burning producing toxic and corrosive fumes including hydrogen chloride, nitrogen oxides, phosphorous oxides, sulfur oxides. Reacts with strong acids, strong bases, causing hydrolysis. Attacks copper and brass.

**Permissible Exposure Limits in Air:** There is no OSHA[2] PEL. NIOSH[2] recommends a REL of 0.2 mg/m$^3$ TWA for a 10-hour workshift. and STEL of 0.6 mg/m3 [skin]. ACGIH, has recommended the same TWA (for an 8-hour workshift)[1]. Both limits bear the notation "skin" indicating the cutaneous absorption should be prevented so the threshold limit value is not invalidated. UK's HSE[33], Australia, Mexico, and the Canadian provinces of Alberta, BC, Ontario, and Quebec have set these same TWA values as ACGIH. The former Soviet Union[35] has set a MAC in workplace air of 0.3 mg/m$^3$. Several states have set guidelines or standards for Chlorpyrifos in ambient air[60] ranging from 2-6 $\mu$g/m$^3$ (North Dakota) to 3.0 $\mu$g/m$^3$ (Virginia) to 4.0 $\mu$g/m$^3$ (Connecticut) to 5 $\mu$g/m$^3$ (Nevada).

**Determination in Air:** OSHA versatile sampler-2; Toluene/Acetone; Gas chromatography/Flame photometric detection for sulfur, nitrogen, or phosphorus; NIOSH IV Method #5600, Organophosphorus pesticides.[18]

**Permissible Concentration in Water:** Mexico[35] has set a limit of 3.0 $\mu$g/L in coastal waters and 0.03 mg/L in estuaries. The former USSR[35] has set a MAC in water bodies used for fishery purposes of 5.0 $\mu$g/L.

# Chlorpyrifos

***Routes of Entry:*** Skin absorption, inhalation of dust, ingestion.

***Short Term Exposure:*** May cause eye and skin irritation. Cholinesterase inhibitor. Exposure at high levels may result in death. The effects may be delayed. The $LD_{50}$ (rat) = 82 mg/kg (moderately toxic). Chlorpyrifos can affect you when breathed in and quickly enters the body by passing through the skin. Severe poisoning can occur from skin contact. It is a moderately toxic organophosphate chemical. Exposure can cause rapid severe poisoning with headache, sweating, nausea and vomiting, diarrhea, loss of coordination, and death. Delayed pulmonary edema may occur after inhalation.

***Long Term Exposure:*** Cholinesterase inhibitor; cumulative effect is possible. Chlorpyrifos may damage the nervous system with repeated exposure, resulting in convulsions, respiratory failure. May cause liver damage.

***Points of Attack:*** Respiratory system, central nervous system, peripheral nervous system, plasma cholinesterase

***Medical Surveillance:*** Medical observation is recommended for 24 to 48 hours after breathing overexposure, as pulmonary edema may be delayed. Before employment and at regular times after that, the following are recommended: Plasma and red blood cell cholinesterase levels (tests for the enzyme poisoned by this chemical). If exposure stops, plasma levels return to normal in 1-2 weeks while red blood cell levels may be reduced for 1-3 months. When acetylcholinesterase enzyme levels are reduced by 25% or more below preemployment levels, risk of poisoning is increased, even if results are in lower ranges of "normal." Reassignment to work not involving organophosphate or carbamate pesticides is recommended until enzyme levels recover. If symptoms develop or overexposure occurs, repeat the above tests as soon as possible and get an exam of the nervous system. Also consider complete blood count. Consider chest x-ray following acute overexposure. Do not drink any alcoholic beverages before or during use. Alcohol promotes absorption of organophosphates.

***First Aid:*** **Treatment for organophosphate poisoning consists of thorough decontamination, cardiorespiratory support, and administration of the antidotes atropine and pralidoxime. In cases of severe poisoning, diazepam, an anticonvulsant, should also be administered. Antidotes should be administered as prevention even if the diagnosis is in doubt.** Speed in removing material from eyes and skin is of extreme importance. *Eyes:* Eye contact can cause dangerous amounts of these chemicals to be quickly absorbed through the mucous membrane into the bloodstream. Immediately and gently flush eyes with plenty of warm or cold water (NO hot water) for at least 15 minutes, occasionally lifting the upper and lower eyelids. Get medical aid immediately. *Skin:* Get medical aid. Skin contact can cause dangerous amounts of these chemicals to be absorbed into the bloodstream. Wearing the appropriate PPE equipment and respirator for organophosphate/carbamate pesticides, immediately flush skin with plenty of soap and water for at least 15 minutes while removing contaminated clothing and shoes. Shampoo hair promptly if contaminated. The removed, contaminated clothing and shoes should be double-bagged and left in Hot Zone for later disposal by hazardous materials experts. Skin may also be decontaminated with diluted hypochlorite solution. *Inhalation:* Get medical aid. Do not contaminate yourself. Wearing the appropriate PPE equipment and respirator for organophosphate pesticides, immediately remove the victim from the contaminated area to fresh air. If the victim is not breathing, administer artificial respiration. Do NOT use mouth-to-mouth resuscitation; use bag/mask apparatus. If breathing is difficult, administer oxygen through bag/mask apparatus until medical help arrives. Do not leave victim unattended. *Ingestion:* Call poison control. Loosen all clothing. Never give anything by mouth to an unconscious person. If victim is *unconscious or having convulsions,* do nothing except keep victim warm. Get medical aid. Transfer promptly to a medical facility. In cases of ingestion, **do not induce vomiting**. If the victim is alert and asymptomatic, administer a slurry of activated charcoal at a dose of 1 g/kg (infant, child, and adult dose). A soda can and straw may be of assistance when offering charcoal to a child. *In some cases you may be specifically instructed by poison control to induce vomiting by way of 2 tablespoons of syrup of ipecac (adult) washed down with a cup of water.* Do NOT give activated charcoal before or with ipecac syrup.

*Note to physician or authorized medical personnel:* Treat cases of respiratory compromise, coma, or excessive pulmonary secretions with respiratory support using protocols and techniques available and within the scope of training. Some cases may necessitate procedures such as endotracheal intubation or cricothyrotomy by properly trained and equipped personnel. When possible, atropine (see *Antidotes,* below) should be given under medical supervision. Patients who are comatose, hypotensive, or having seizures or cardiac arrhythmias should be treated according to advanced life support protocols. *Antidotes:* Two antidotes are administered to treat organophosphate poisoning. Atropine is a competitive antagonist of acetylcholine at muscarinic receptors and is used to control the excessive bronchial secretions which are often responsible for death. Pralidoxime relieves both the nicotinic and muscarine effects of organophosphate poisoning by regenerating acetylcholinesterase and can reduce both the bronchial secretions and the muscle weakness associated with poisoning. The initial intravenous dose of atropine in adults should be determined by the severity of symptoms: An initial adult dose of 1.0 to 2.0 mg or pediatric dose of 0.01 mg/kg (minimum 0.01 mg) should be administered intravenously. If intravenous access cannot be established, atropine may also be given intramuscularly, subcutaneously or via endotracheal tube. Doses should be repeated every 15 minutes until excessive secretions and sweating have been controlled. Once bronchial secretion has been controlled, atropine administration should be repeated whenever the secretions begin to recur. In seriously poisoned patients, very large doses may be required. Alterations of pulse rate and pupillary size should not be used as indicators of treatment adequacy. Pralidoxime should be administered as early in poisoning as possible as its efficacy may diminish when given more than 24 to 36 hours after exposure. Doses are as follows: adult 1.0 g; pediatric 25 to 50 mg/kg. The drug should be administered intravenously over 30 to 60 minutes, but in a life-threatening situation, one-half of the total dose can be given per minute for a total administration time of 2 minutes. Treatment should begin to take effect within 40 minutes with a reduction in symptoms and the amount of atropine necessary to control bronchial secretion. The initial dose can be repeated in 1 hour and then every 8 to 12 hours until the patient is clinically well and no longer requires atropine. If intravenous access cannot be established, pralidoxime may also be given intramuscularly. Early administration of diazepam in addition to the combined atropine and pralidoxime treatment may help prevent the onset of seizures and potential brain and cardiac morphologic damage following high-level organophosphate poisoning.

*References:*
- EXTOXNET, Extension Toxicology Network, "Pesticide Information Profile, Chlorpyrifos," Oregon State University, Corvallis, OR (June 1996). ace.orst.edu/info/extoxnet/pips/chlorpyr
- EPA, Office of Pesticide Programs, Pesticide Residue Limits, " Chlorpyrifos", 40 CFR 180.342, www.epa.gov/cgi-bin/oppsrch
- EPA, Office of Prevention, Pesticides and Toxic Substances, "Chlorpyrifos Revised Risk Assessment and Agreement with Registrants," (June 2000). www.epa.gov/pesticides/op/chlorpyrifos/agreement.pdf
- New Jersey Department of Health and Senior Services and Senior Services, "Hazardous Substance Fact Sheet: Chlorpyrifos," Trenton, NJ (July 1996, rev. July 2002). http://www.state.nj.us/health/eoh/rtkweb/0426.pdf
- California Environmental Protection Agency "Chemical List of Lists," Sacramento CA (February 1997).
- Agency for Toxic Substances and Disease Registry, U.S. Department of Health and Human Services, Public Health Service, "Managing Hazardous Materials Incidents," Atlanta, GA (June 2003).

# Chlorpyrifos-methyl (ANSI)

*Use Type:* Insecticide
*CAS Number:* 5598-13-0
*Formula:* $C_7H_7Cl_3NO_3PS$

*Alert:* In 2000, the registrants of Chlorpyrifos-methyl requested voluntary cancellation of their products rather than committing to develop additional data for reregistration.

*Synonyms:* Clorpirifos metil (Spanish); Chlormethylfos; O,O-Dimethyl O-(3,5,6-trichloro-2-pyridyl)phosophorothioate; Dursban methyl; ENT 27,520; Methyl chlorpyrifos; Methyl dursban; NSC-60380; OMS 1155; Phosphorothioic acid, O,O-dimethyl O-(3,5,6-trichloro-2-pyridyl)ester; Trichlormethylfos

*Trade Names:* DOWCO-217®, Dow Chemical (USA); NOLTRAN®; RELDAN®. Gustafson (USA); STORCIDE®. Gustafson (USA); ZERTELL®

*Producers:* Dow Chemical (USA); Gustafson (USA)

*Chemical Class:* Organophosphate; Halo-organics

*EPA/OPP PC Code:* 059102

*California DPR Chemical Code:* 2468

*RTECS Number:* TG0700000

*EINECS Number:* 227-011-5

*Uses:* Chlorpyrifos-methyl is a general use organophosphate insecticide for use on stored grain (for protection of stored food, feed oil, and seed grains against injury from stored grain weevils, moths, borers, beetles and mealworms including granary weevil, rice weevil, red flour beetle, confused flour beetle, saw-toothed grain beetle, Indian meal moth, and Angoumois grain moth, lessor grain borers), seed treatment, grain bin and warehouse.

*Human toxicity (long-term)[77]:* Low–70.00 ppb, Health Advisory

*Fish toxicity (threshold)[77]:* High–1.02425 ppb, MATC (Maximum Acceptable Toxicant Concentration)

*Carcinogen/Hazard Classifications*

**Label Signal Word:** DANGER

**WHO Acute Hazard:** Class U, unlikely to be hazardous; Class II, moderately hazardous as chlorpyrifos

*Regulatory Authority:*
- U.S. Dot Inhalation Hazard Chemicals as organophosphates
- EPA 40 CFR 372.65, Specific Toxic Chemical Listings
- EPCRA Section 313 Form R *de minimis* concentration reporting level: 1.0%
- FIFRA, 180.3(4); class of chlorinated organic pesticide
- FIFRA, 40CFR186: tolerances for pesticides in animal feeds

*Description:* Amber chrystalline solid; Slightly soluble in water; solubility = 4 mg/L. Molecular weight = 322.5. Melting/Freezing point = 46°C. Vapor pressure = $4.2 \times 10^{-5}$ mmHg @ 25°C. Density = 1.39 @ 50°C.

*Determination in Air:* OSHA versatile sampler-2; Toluene/Acetone; Gas chromatography/Flame ionization detection; NIOSH IV, Method #5600, Organophosphorus Pesticides.[18]

*Permissible Concentration in Water:* No criteria set. Runoff from spills or fire control may cause water pollution.

## Chlorpyrifos-methyl

*Harmful Effects and Symptoms*

**Short Term Exposure:** Eye pupils are small; blurred vision; eye watering; runny nose; cough; shortness of breath; salivation; dizziness; nausea, stomach cramps, diarrhea, and vomiting; increased blood pressure; profuse sweating; hypermotility, hallucinations; irritability; tingling of the skin; drowsiness; slow heartbeat; convulsions; fluid in lungs; loss of consciousness; incontinence; breathing stops; death. Organophosphates inhibit the action of acetylcholinesterase enzymes, and alter the way in which nervous impulses are transmitted. The effects can last for hours, days, or much longer. The action of the enzymes is reestablished after new enzymes are formed.

**Long Term Exposure:** Cholinesterase inhibitor; cumulative effect is possible. Organophosphates may damage the nervous system with repeated exposure, resulting in convulsions, respiratory failure. May cause liver damage.

**Points of Attack:** Respiratory system, central nervous system, cardiovascular system, blood cholinesterase.

**Medical Surveillance:** Medical observation is recommended for 24 to 48 hours after breathing overexposure, as pulmonary edema may be delayed. As first aid for pulmonary edema, consider administering a corticosteroid spray. Cigarette smoking may exacerbate pulmonary injury and should be discouraged for at least 72 hours following exposure.

***First Aid:*** **Treatment for organophosphate poisoning consists of thorough decontamination, cardiorespiratory support, and administration of the antidotes atropine and pralidoxime. In cases of severe poisoning, diazepam, an anticonvulsant, should also be administered. Antidotes should be administered as prevention even if the diagnosis is in doubt.** Speed in removing material from eyes and skin is of extreme importance. *Eyes:* Eye contact can cause dangerous amounts of these chemicals to be quickly absorbed through the mucous membrane into the bloodstream. Immediately and gently flush eyes with plenty of warm or cold water (NO hot water) for at least 15 minutes, occasionally lifting the upper and lower eyelids. Get medical aid immediately. *Skin:* Get medical aid. Skin contact can cause dangerous amounts of these chemicals to be absorbed into the bloodstream. Wearing the appropriate PPE equipment and respirator for organophosphate/carbamate pesticides, immediately flush skin with plenty of soap and water for at least 15 minutes while removing contaminated clothing and shoes. Shampoo hair promptly if contaminated. The removed, contaminated clothing and shoes should be double-bagged and left in Hot Zone for later disposal by hazardous materials experts. Skin may also be decontaminated with diluted hypochlorite solution. *Inhalation:* Get medical aid. Do not contaminate yourself. Wearing the appropriate PPE equipment and respirator for organophosphate pesticides, immediately remove the victim from the contaminated area to fresh air. If the victim is not breathing, administer artificial respiration. Do NOT use mouth-to-mouth resuscitation; use bag/mask apparatus. If breathing is difficult, administer oxygen through bag/mask apparatus until medical help arrives. Do not leave victim unattended. *Ingestion:* Call poison control. Loosen all clothing. Never give anything by mouth to an unconscious person. If victim is *unconscious or having convulsions,* do nothing except keep victim warm. Get medical aid. Transfer promptly to a medical facility. In cases of ingestion, **do not induce vomiting**. If the victim is alert and asymptomatic, administer a slurry of activated charcoal at a dose of 1 g/kg (infant, child, and adult dose). A soda can and straw may be of assistance when offering charcoal to a child. *In some cases you may be specifically instructed by poison control to induce vomiting by way of 2 tablespoons of syrup of ipecac (adult) washed down with a cup of water.* Do NOT give activated charcoal before or with ipecac syrup.

*Note to physician or authorized medical personnel:* Treat cases of respiratory compromise, coma, or excessive pulmonary secretions with respiratory support using protocols and techniques available and within the scope of training. Some cases may necessitate procedures such as endotracheal intubation or cricothyrotomy by properly trained and equipped personnel. When possible, atropine (see *Antidotes*, below) should be given under medical supervision. Patients who are comatose, hypotensive, or having seizures or cardiac arrhythmias should be treated according to advanced life support protocols. *Antidotes:* Two antidotes are administered to treat organophosphate poisoning. Atropine is a competitive antagonist of acetylcholine at muscarinic receptors and is used to control the excessive bronchial secretions which are often responsible for death. Pralidoxime relieves both the nicotinic and muscarine effects of organophosphate poisoning by regenerating acetylcholinesterase and can reduce both the bronchial secretions and the muscle weakness associated with poisoning. The initial intravenous dose of atropine in adults should be determined by the severity of symptoms: An initial adult dose of 1.0 to 2.0 mg or pediatric dose of 0.01 mg/kg (minimum 0.01 mg) should be administered intravenously. If intravenous access cannot be established, atropine may also be given intramuscularly, subcutaneously or via endotracheal tube. Doses should be repeated every 15 minutes until excessive secretions and sweating have been controlled. Once bronchial secretion has been controlled, atropine administration should be repeated whenever the secretions begin to recur. In seriously poisoned patients, very large doses may be required. Alterations of pulse rate and pupillary size should not be used as indicators of treatment adequacy. Pralidoxime should be administered as early in poisoning as possible as its efficacy may diminish when given more than 24 to 36 hours after exposure. Doses are as follows: adult 1.0 g; pediatric 25 to 50 mg/kg. The drug should be administered intravenously over 30 to 60 minutes, but in a life-threatening situation, one-half of the total dose can be given per minute for a total administration time of 2 minutes. Treatment should begin to take effect within 40 minutes with a reduction in symptoms and the amount of atropine necessary to control bronchial secretion. The initial dose can be repeated in 1 hour and then every 8 to 12 hours until the patient is clinically well and no longer requires atropine. If intravenous access cannot be established, pralidoxime may also be given intramuscularly. Early administration of diazepam in addition to the combined atropine and pralidoxime treatment may help prevent the onset of seizures and potential brain and cardiac morphologic damage following high-level organophosphate poisoning.

*References:*
- U.S. Environmental Protection Agency, "Reregistration Chlorpyrifos-methyl Facts," Office of Prevention, Pesticides and Toxic Substances, Washington, DC (October 2000).
  http://www.epa.gov/REDs/factsheets/cpm_fs.htm
- California Environmental Protection Agency "Chemical List of Lists," Sacramento CA (February 1997)

# Chlorsulfuron (ANSI)

*Use Type:* Herbicide
*CAS Number:* 64902-72-3
*Formula:* $C_{12}H_{12}ClN_5O_4S$
*Synonyms:* 2-Chloro-*N*-[((4-methoxy-6-methyl-1,3,5-triazin-2-yl)amino)carbonyl]benzenesulfonamide; 2-chloro-*N*-[(4-methoxy-6-methyl-1,3,5-triazin-2-yl)aminocarbonyl]-benzenesulfonamide; 1-[(O-chlorophenyl)sulfonyl)-3-(4-methoxy-6-methyl-*S*-triazin-2-yl)urea; Benzenesulfonamide, 2-chloro-*N*-[((4-methoxy-6-methyl-1,3, 5-triazin-2-yl)amino)carbonyl]-; Urea, 1-[(Ochlorophenyl)sulfonyl]-3-(4-methoxy-6-methyl-*S*-triazin-2-yl)-
*Trade Names:* DPX 4189®; FINESSE®, DuPont Crop Protection (USA); GLEAN®, DuPont Crop Protection (USA); GLEAN 20DF®, DuPont Crop Protection (USA); LANDMARK® MP, DuPont Crop Protection (USA); LASHER®, Sanonda (Australia); RIVERDALE CORSAIR®, Nufarm Ltd. (Australia); TELAR® DF, DuPont Crop Protection (USA)
*Producers:* Agsin (Singapore); DuPont Crop Protection (USA); Ki-Hara Chemicals Ltd. (UK); Nufarm Ltd. (Australia); Sanonda (Australia); Yellow River Enterprise (China)
*Chemical Class:* Sulfonylurea; Halo-organics
*EPA/OPP PC Code:* 118601
*California DPR Chemical Code:* 2143
*EINECS Number:* 265-268-5
*Uses:* A selective systemic herbicide used to control most broadleaf weeds and some annual grasses in wheat, barley, oats, durum, rye, triticale and flax. Applied to non-crop sites such as rights-of-way, fence rows and roadsides.

*Human toxicity (long-term)*[77]: Very low–350.00 ppb, Health Advisory
*Fish toxicity (threshold)*[77]: Very low–45107.62902 ppb, MATC (Maximum Acceptable Toxicant Concentration)
*U.S. Maximum Allowable Residue Levels for Chlorsulfuron (40 CFR 180.405):*

| CROP | ppm |
|---|---|
| Barley, grain | 0.1 |
| Barley, straw | 0.5 |
| Cattle, fat | 0.3 |
| Cattle, meat | 0.3 |
| Cattle, mbyp | 0.3 |
| Goat, fat | 0.3 |
| Goat, meat | 0.3 |
| Goat, mbyp | 0.3 |
| Grass, forage | 11 |
| Grass, hay | 19 |
| Hog, fat | 0.3 |
| Hog, meat | 0.3 |
| Hog, mbyp | 0.3 |
| Horse, fat | 0.3 |
| Horse, meat | 0.3 |
| Horse, mbyp | 0.3 |
| Milk | 0.1 |
| Oat, forage | 20 |
| Oat, grain | 0.1 |
| Oat, straw | 0.5 |
| Sheep, fat | 0.3 |
| Sheep, meat | 0.3 |
| Sheep, mbyp | 0.3 |
| Wheat, forage | 20 |
| Wheat, grain | 0.1 |
| Wheat, straw | 0.5 |

*Carcinogen/Hazard Classifications*
**California Prop. 65:** Listed: Developmental, female, male
**Label Signal Word:** CAUTION
**TRI Developmental Toxin:** Reproductive and developmental toxin
**WHO Acute Hazard:** Class U, unlikely to be hazardous
*Regulatory Authority:*
- Actively registered pesticide in California.
- AB 1803-Well Monitoring Chemical (CAL)
- Proposition 65 chemical (CAL)
- EPA 40 CFR 372.65, Specific Toxic Chemical Listings
- EPCRA Section 313 Form R *de minimis* concentration reporting level: 1.0%

*Description:* Colorless, odorless crystals. Soluble in water; solubility = 113 mg/L @ 25°C. Molecular weight = 357.79. Melting/Freezing point = 176°C; decomposes @ 192°C. Vapor pressure = $4.6 \times 10^{-6}$ mmHg @ 20°C.
*Incompatibilities:* Slowly hydrolyzes in water, releasing ammonia and forming acetate salts. When heated to decomposition, emits oxides of sulfur and nitrogen and chlorine fumes.

*Permissible Concentration in Water:* No criteria set. Runoff from spills or fire control may cause water pollution.
*Routes of Entry:* Inhalation, passing through the skin, ingestion.

*Harmful Effects and Symptoms*

*Short Term Exposure:* Contact with eyes or skin may cause irritation or burns. Inhalation of dust should be avoided; use NIOSH-approved air purifying respirators for pesticides. Because this material has a low vapor pressure, significant inhalation of vapors is unlikely at ordinary temperatures. May be harmful if swallowed. Skin contact may cause allergic reaction.

*Long Term Exposure:* Repeated or prolonged contact may cause skin and lung sensitization, resulting in allergies.

*Points of Attack:* Skin.

*Medical Surveillance:* Evaluation by a qualified allergist, including careful exposure history and special testing, may help diagnose skin allergy.

*First Aid:* If this chemical gets into the eyes, remove any contact lenses at once and irrigate immediately for at least 15 minutes, occasionally lifting upper and lower lids. Seek medical attention immediately. If this chemical contacts the skin, remove contaminated clothing and wash immediately with soap and water. Seek medical attention immediately. If this chemical has been inhaled, remove from exposure, begin rescue breathing (using universal precautions) if breathing has stopped, and CPR if heart action has stopped. Transfer promptly to a medical facility. When this chemical has been swallowed, get medical attention. Give large quantities of water and induce vomiting. Do not make an unconscious person vomit.

*References:*
- California Environmental Protection Agency "Chemical List of Lists," Sacramento CA (February 1997).
- EPA, Office of Pesticide Programs, Pesticide Residue Limits, "Chlorsulfuron", 40 CFR 180.405. www.epa.gov/pesticides/food/viewtols.htm

# Chlorthiophos

*Use Type:* Insecticide and acaricide
*CAS Number:* 60238-56-4; 21923-23-9 (Chlorthiophos I isomer )
*Formula:* $C_{11}H_{15}Cl_2O_3PS_2$
*Alert:* The registered use of chlorthiophos was canceled by the U.S. EPA in 1984. Residue tolerances were revoked in 1996.
*Synonyms:* O-[Dichloro(methylthio)phenyl] O,O-diethyl phosphorothioate (3 isomers); O,O-[Diethyl-O-2,4,5-dichloro(methylthio)phenyl]thionophosphate; ENT 27,635; NSC 195164
*Trade Names:* CELAMERCK S-2957®, canceled; CELA S-2957®, canceled; CELATHION®, canceled; CM S® 2957, canceled; OMS® 1342, canceled
*Producers:* Ehrenstorfer, Dr. (Germany); Sigma-Aldrich Laborchemikalien (Germany)
*Chemical Class:* Organophosphate; Halo-organics
*EPA/OPP PC Code:* 111811
*California DPR Chemical Code:* 2469
*RTECS Number:* TF1590000
*Uses:* Registration and residue tolerances for this chemical in the U.S. have been canceled by the U.S. Environmental Protection Agency.

*Carcinogen/Hazard Classifications*

**WHO Acute Hazard:** Class 1b, Extremely Hazardous

*Regulatory Authority:*
- Banned or Severely Restricted (In Agriculture) (Germany, Malaysia) (UN)[13]
- U.S. DOT Inhalation Hazard Chemicals as organophosphates
- Superfund/EPCRA 40CFR355, Appendix B Extremely Hazardous Substances: TPQ = 500 lb (227 kg)
- U.S. DOT Regulated Marine Pollutant (49CFR172.101, Appendix B)

*Description:* Chlorthiophos is a yellowish-brown liquid. Boiling point = 153-158°C @ 13 mmHg pressure and crystallizes at less than 25°C. Hazard Identification (based on NFPA-704 M Rating System): Health 3, Flammability 1, Reactivity 0.

*Incompatibilities:* Strong acids, strong bases, strong oxidizers

*Permissible Exposure Limits in Air:* No standards set. However, this chemical can be absorbed through the skin, thereby increasing exposure.

*Determination in Air:* OSHA versatile sampler-2; Toluene/Acetone; Gas chromatography/Flame photometric detection for sulfur, nitrogen, or phosphorus; NIOSH Method IV Method #5600, Organophosphorus pesticides[18].

*Permissible Concentration in Water:* No criteria set, but runoff from spills or fire control may cause water pollution.

*Routes of Entry:* Inhalation, ingestion, skin contact. This chemical can be absorbed through the skin, thereby increasing exposure.

*Harmful Effects and Symptoms*

*Short Term Exposure:* The LD50-oral (rabbit) is 20 mg/kg which is in the highly toxic class. Symptoms of chlorthiophos exposure include headache, giddiness, blurred vision, nervousness, weakness, nausea, cramps, diarrhea, and discomfort in the chest. Signs include sweating, tearing, salivation, vomiting, cyanosis, convulsions, coma, loss of reflexes and loss of sphincter control. Organic phosphorus insecticides are absorbed by the skin, as well as by the respiratory and gastrointestinal tracts. They are cholinesterase inhibitors. Delayed pulmonary edema may occur after inhalation.

*Long Term Exposure:* Cholinesterase inhibitor; cumulative effect is possible. Chlorthiophos may damage the nervous system with repeated exposure, resulting in convulsions, respiratory failure. May cause liver damage.

***Points of Attack:*** Respiratory system, central nervous system, peripheral nervous system, plasma cholinesterase

***Medical Surveillance:*** Medical observation is recommended for 24 to 48 hours after breathing overexposure, as pulmonary edema may be delayed. Before employment and at regular times after that, the following are recommended: Plasma and red blood cell cholinesterase levels (tests for the enzyme poisoned by this chemical). If exposure stops, plasma levels return to normal in 1-2 weeks while red blood cell levels may be reduced for 1-3 months. When acetylcholinesterase enzyme levels are reduced by 25% or more below preemployment levels, risk of poisoning is increased, even if results are in lower ranges of "normal." Reassignment to work not involving organophosphate or carbamate pesticides is recommended until enzyme levels recover. If symptoms develop or overexposure occurs, repeat the above tests as soon as possible and get an exam of the nervous system. Also consider complete blood count. Consider chest x-ray following acute overexposure Do not drink any alcoholic beverages before or during use. Alcohol promotes absorption of organophosphates.

***First Aid:*** **Treatment for organophosphate poisoning consists of thorough decontamination, cardiorespiratory support, and administration of the antidotes atropine and pralidoxime. In cases of severe poisoning, diazepam, an anticonvulsant, should also be administered. Antidotes should be administered as prevention even if the diagnosis is in doubt.** Speed in removing material from eyes and skin is of extreme importance. *Eyes:* Eye contact can cause dangerous amounts of these chemicals to be quickly absorbed through the mucous membrane into the bloodstream. Immediately and gently flush eyes with plenty of warm or cold water (NO hot water) for at least 15 minutes, occasionally lifting the upper and lower eyelids. Get medical aid immediately. *Skin:* Get medical aid. Skin contact can cause dangerous amounts of these chemicals to be absorbed into the bloodstream. Wearing the appropriate PPE equipment and respirator for organophosphate/carbamate pesticides, immediately flush skin with plenty of soap and water for at least 15 minutes while removing contaminated clothing and shoes. Shampoo hair promptly if contaminated. The removed, contaminated clothing and shoes should be double-bagged and left in Hot Zone for later disposal by hazardous materials experts. Skin may also be decontaminated with diluted hypochlorite solution. *Inhalation:* Get medical aid. Do not contaminate yourself. Wearing the appropriate PPE equipment and respirator for organophosphate pesticides, immediately remove the victim from the contaminated area to fresh air. If the victim is not breathing, administer artificial respiration. Do NOT use mouth-to-mouth resuscitation; use bag/mask apparatus. If breathing is difficult, administer oxygen through bag/mask apparatus until medical help arrives. Do not leave victim unattended. *Ingestion:* Call poison control. Loosen all clothing. Never give anything by mouth to an unconscious person. If victim is *unconscious or having convulsions,* do nothing except keep victim warm. Get medical aid. Transfer promptly to a medical facility. In cases of ingestion, **do not induce vomiting**. If the victim is alert and asymptomatic, administer a slurry of activated charcoal at a dose of 1 g/kg (infant, child, and adult dose). A soda can and straw may be of assistance when offering charcoal to a child. *In some cases you may be specifically instructed by poison control to induce vomiting by way of 2 tablespoons of syrup of ipecac (adult) washed down with a cup of water.* Do NOT give activated charcoal before or with ipecac syrup.

*Note to physician or authorized medical personnel:* Treat cases of respiratory compromise, coma, or excessive pulmonary secretions with respiratory support using protocols and techniques available and within the scope of training. Some cases may necessitate procedures such as endotracheal intubation or cricothyrotomy by properly trained and equipped personnel. When possible, atropine (see *Antidotes*, below) should be given under medical supervision. Patients who are comatose, hypotensive, or having seizures or cardiac arrhythmias should be treated according to advanced life support protocols. *Antidotes:* Two antidotes are administered to treat organophosphate poisoning. Atropine is a competitive antagonist of acetylcholine at muscarinic receptors and is used to control the excessive bronchial secretions which are often responsible for death. Pralidoxime relieves both the nicotinic and muscarine effects of organophosphate poisoning by regenerating acetylcholinesterase and can reduce both the bronchial secretions and the muscle weakness associated with poisoning. The initial intravenous dose of atropine in adults should be determined by the severity of symptoms: An initial adult dose of 1.0 to 2.0 mg or pediatric dose of 0.01 mg/kg (minimum 0.01 mg) should be administered intravenously. If intravenous access cannot be established, atropine may also be given intramuscularly, subcutaneously or via endotracheal tube. Doses should be repeated every 15 minutes until excessive secretions and sweating have been controlled. Once bronchial secretion has been controlled, atropine administration should be repeated whenever the secretions begin to recur. In seriously poisoned patients, very large doses may be required. Alterations of pulse rate and pupillary size should not be used as indicators of treatment adequacy. Pralidoxime should be administered as early in poisoning as possible as its efficacy may diminish when given more than 24 to 36 hours after exposure. Doses are as follows: adult 1.0 g; pediatric 25 to 50 mg/kg. The drug should be administered intravenously over 30 to 60 minutes, but in a life-threatening situation, one-half of the total dose can be given per minute for a total administration time of 2 minutes. Treatment should begin to take effect within 40 minutes with a reduction in symptoms and the amount of atropine necessary to control bronchial secretion. The initial dose can be repeated in 1 hour and then every 8 to 12 hours until the patient is clinically well and no longer requires atropine. If intravenous access cannot be established, pralidoxime may also be given intramuscularly. Early administration of diazepam in addition to the combined atropine and pralidoxime treatment may help prevent the onset of seizures and potential brain and cardiac morphologic damage following high-level organophosphate poisoning.

*References:*
- U.S. Environmental Protection Agency, "Chemical Profile: Chlorthiophos," Washington, DC, Chemical Emergency Preparedness Program (November 30, 1987).
- California Environmental Protection Agency "Chemical List of Lists," Sacramento CA (February 1997).
- Agency for Toxic Substances and Disease Registry, U.S. Department of Health and Human Services, Public Health Service, "Managing Hazardous Materials Incidents," Atlanta, GA (June 2003).

# Cholecalciferol

*Use Type:* Rodenticide
*CAS Number:* 67-97-0
*Formula:* $C_{27}H_{44}O$
*Synonyms:* Colecalciferol; 7-Dehydrochloesterol; Oleovitamin D3; 9,10-Secocholesta-5,7,10(19)-trien-3-$\beta$-ol; 9,10-Secocholesta-5,7,10(19)-trien-3-ol, (3.$\beta$,5Z,7E)-; Vitamin $D_3$
*Trade Names:* DELSTEROL®; DEPARAL®; D3-VIGANTOL®; QUINTOX®, Bell Laboratories (USA); RAMPAGE®; RICKETON®; TRIVITAN®; VIGORSAN®; VITINC DAN-DEE-3®
*Producers:* Bell Laboratories (USA)
*Chemical Class:* Unclassified
*EPA/OPP PC Code:* 202901; (208700 old EPA code number)
*California DPR Chemical Code:* 2232
*Uses:* Used in bait for vermin control and in vitamin supplements. Vitamin D is a steroid hormone that has an important role in regulating body levels of calcium and phosphorus, and in mineralization of bone.
*Carcinogen/Hazard Classifications*
**Label Signal Word:** CAUTION
*Regulatory Authority:*
- Actively registered pesticide in California.

*Description:* White crystalline solid. Odorless. Insoluble in water. Melting/freezing point = 87°C.
*First Aid:* If this chemical gets into the eyes, remove any contact lenses at once and irrigate immediately for at least 15 minutes, occasionally lifting upper and lower lids. Seek medical attention immediately. If this chemical contacts the skin, remove contaminated clothing and wash immediately with soap and water. Seek medical attention immediately. If this chemical has been inhaled, remove from exposure, begin rescue breathing (using universal precautions) if breathing has stopped, and CPR if heart action has stopped. Transfer promptly to a medical facility. When this chemical has been swallowed, get medical attention. *Do not induce vomiting when formulations containing petroleum solvents are ingested.* Otherwise, give large quantities of water and induce vomiting. Do not make an unconscious person vomit.

*References:*
- U.S. Environmental Protection Agency, "Reregistration Eligibility Decision (RED), Rodenticide Cluster," Office of Prevention, Pesticides and Toxic Substances, Washington, DC (July 1998).
http://www.epa.gov/REDs/2100red.pdf
- California Environmental Protection Agency "Chemical List of Lists," Sacramento CA (February 1997)

# C. I. Basic Green 1

*Use Type:* Fungicide
*CAS Number:* 633-03-4
*Formula:* $C_{27}H_{34}N_2O_4S$; $C_{27}H_{33}N_2 \cdot HSO_4$
*Alert:* Not registered as a pesticide in the U.S.
*Synonyms:* Brilliant green; Ethyl green; Emerald green; Malachite green G
*Producers:* Aldrich Chemical (USA); China Chemicals (China); Deepak Group (India); Flourochem (UK); Merck (Germany)
*RTECS Number:* BP6825000
*EINECS Number:* 211-190-1
*Uses:* C. I. Basic Green 1 is used in dyeing silk, wool, leather, jute and cotton yellowish-green; manufacturing green ink; as staining constituent of bacteriological media; indicator, an intestinal anthelmintic; a wound antiseptic; treatment of mycotic infections; agricultural fungicide.
*Description:* C.I. Basic Green 1 is a metallic green, odorless crystal or powder. Soluble in water.
*Incompatibilities:* Oxidizing agents, reducing agents, anionics, and aqueous solutions of bentonite. Keep away from moisture. When heated to decomposition, forms carbon dioxide and carbon monoxide.
*Permissible Exposure Limits in Air:* No occupational exposure limits have been established
*Permissible Concentration in Water:* No criteria set, but runoff from spills or fire control may cause water pollution.
*Routes of Entry:* Ingestion
*Harmful Effects and Symptoms*
*Short Term Exposure:* C.I. Basic Green can irritate and burn the skin and eyes. Ingestion causes nausea, vomitine, diarrhea and abdominal pain. It is classified as very toxic; probable lethal dose is 50-500 mg/kg in humans (between 1 teaspoon and 1 ounce for a 150-lb. person). This substance

is classified as very toxic; probable lethal dose is 50-500 mg/kg in humans (between 1 teaspoon and 1 ounce for a 150-lb. person). It is a skin irritant. Ingestion causes diarrhea and abdominal pain.
*Long Term Exposure:* Skin contact can cause drying and cracking.
*Points of Attack:* Skin
*Medical Surveillance:* There is no special test for this substance.
*First Aid:* If this chemical gets into the eyes, remove any contact lenses at once and irrigate immediately for at least 15 minutes, occasionally lifting upper and lower lids. Seek medical attention immediately. If this chemical contacts the skin, remove contaminated clothing and wash immediately with soap and water. Seek medical attention immediately. If this chemical has been inhaled, remove from exposure, begin rescue breathing (using universal precautions) if breathing has stopped, and CPR if heart action has stopped. Transfer promptly to a medical facility. When this chemical has been swallowed, get medical attention. Give large quantities of water and induce vomiting. Do not make an unconscious person vomit.
*References:*
- U.S. Environmental Protection Agency, "Chemical Profile: C. I. Basic Green 1," Washington, DC, Chemical Emergency Preparedness Program (October 31, 1985).
- New Jersey Department of Health and Senior Services, "Hazardous Substance Fact Sheet, C. I. Basic Green 1," Trenton, NJ (January, 1999).
  www.state.nj.us/health/eoh/rtkweb/2249.pdf

# Cimectacarb

*Use Type:* Plant growth regulator
*CAS Number:* 95266-40-3
*Formula:* $C_{13}H_{16}O_5$
*Alert:* A Restricted Use Pesticide (RUP)
*Synonyms:* Cyclohexanecarboxylic acid, 4-(cyclopropylhydroxymethylene)-3,5-dioxo-, ethyl ester; 4-(Cyclo-α-hydroxymethylene)-3,5-dioxocyclohexane carboxylic acid ethyl ester; Ethyl 4-(cyclopropylhydroxy methylene)-3,5-dioxocyclohexanecarboxylate; Trinexapac-ethyl; Ethyl 4-(cyclopropylhydroxymethylene)-3,5-dioxocyclohexanecarboxylate
*Trade Names:* CGA 163935®; VISION®; PRIMO®, Syngenta Crop Protection (Switzerland); PRIMO® WSB, Syngenta Crop Protection (Switzerland)
*Producers:* Syngenta Crop Protection (Switzerland)
*Chemical Class:* Carbamate (cyclopropyl derivative of cyclohexenone plant growth Inhibitor)
*EPA/OPP PC Code:* 112602
*California DPR Chemical Code:* 2345
*ICSC Number:* 1268
*RTECS Number:* GU8473500

*Uses:* A Restricted Use Pesticide (RUP) used on commercial and residential turf. Provides chemical alternatives to mechanical and hand removal of aggressive stoloniferous grasses such as Bermudagrass, zoysiagrass, St. Augustinegrass and kikuyigrass.
*Carcinogen/Hazard Classifications*
**Label Signal Word:** WARNING or CAUTION
*Description:* White powder. Soluble in water; solubility = 0.28-2.11 g/100 ml @ 20°C (depends on pH). Molecular weight = 252.31. Melting Point=36°C. Flash point = 132°C. Autoignition temperature = 355°C. Vapor pressure = $2.9 \times 10^{-3}$ Pa @ 20°C. Log $K_{ow}$ = 2.43. Unlikely to bioaccumulate in marine organisms.
*Incompatibilities:* May react violently with strong oxidizers, bromine, 90% hydrogen peroxide, phosphorus trichloride, silver powders or dust. Incompatible with silver compounds. Mixture with some silver compounds forms explosive salts of silver oxalate. May form explosive materials with phosphorus pentachloride.
*Permissible Concentration in Water:* No criteria set. Runoff from spills or fire control may cause water pollution.
*Routes of Entry:* Skin adsorption, ingestion, inhalation.
*Harmful Effects and Symptoms*
*Short Term Exposure:* Eye pupils are small; blurred vision; eye watering; runny nose; cough; shortness of breath; salivation; nausea, stomach cramps, diarrhea, and vomiting; increased blood pressure; profuse sweating; hypermotility, hallucinations; agitation; tingling of the skin; slow heartbeat; convulsions; fluid in lungs; loss of consciousness; incontinence; breathing stops; death. Carbamate insecticides inhibit the cholinesterase activity of enzymes, causing accumulation of acetylcholine at synapses and altering the way in which nervous impulses are transmitted. However, within several hours carbamates spontaneously detach from the enzymes.
*Long Term Exposure:* A potent cholinesterase inhibitor; cumulative effect is possible. This chemical may damage the nervous system with repeated exposure, resulting in convulsions, respiratory failure. May cause liver damage.
*Points of Attack:* Respiratory system, lungs, central nervous system, cardiovascular system, skin, eyes, plasma and red blood cell cholinesterase.
*Medical Surveillance:* Medical observation is recommended for 24 to 48 hours after breathing overexposure, as pulmonary edema may be delayed. As first aid for pulmonary edema, consider administering a corticosteroid spray. Cigarette smoking may exacerbate pulmonary injury and should be discouraged for at least 72 hours following exposure. Before employment and at regular times after that, the following are recommended: Plasma and red blood cell cholinesterase levels (tests for the enzyme poisoned by this chemical). If exposure stops, plasma levels return to normal in 1-2 weeks while red blood cell levels may be reduced for 1-3 months. When

acetylcholinesterase enzyme levels are reduced by 25% or more below preemployment levels, risk of poisoning is increased, even if results are in lower ranges of "normal." Reassignment to work not involving carbamate pesticides is recommended until enzyme levels recover. If symptoms develop or overexposure occurs, repeat the above tests as soon as possible and get an exam of the nervous system. Also consider complete blood count. Consider chest x-ray following acute overexposure

*First Aid:* Speed in removing material from eyes and skin is of extreme importance. *Eyes:* Eye contact can cause dangerous amounts of these chemicals to be quickly absorbed through the mucous membrane into the bloodstream. Immediately and gently flush eyes with plenty of warm or cold water (NO hot water) for at least 15 minutes, occasionally lifting the upper and lower eyelids. Get medical aid immediately. *Skin:* Get medical aid. Skin contact can cause dangerous amounts of these chemicals to be absorbed into the bloodstream. Wearing the appropriate PPE equipment and respirator for carbamate pesticides, immediately flush skin with plenty of soap and water for at least 15 minutes while removing contaminated clothing and shoes. Shampoo hair promptly if contaminated; protect eyes. *Ingestion:* Call poison control. Loosen all clothing. Never give anything by mouth to an unconscious person. Get medical aid. Do NOT induce vomiting.* If conscious, alert, and able to swallow, rinse mouth and have victim drink 4 to 8 ounces of water. Check to see if poison control instructs you to use ipecac syrup, otherwise administer slurry of activated charcoal (2 oz in 8 oz of water). If victim is *UNCONSCIOUS OR HAVING CONVULSIONS,* do nothing except keep victim warm. *\*In some cases you may be specifically instructed by poison control to induce vomiting by way of 2 tablespoons of syrup of ipecac (adult) washed down with a cup of water.* Do NOT give activated charcoal before or with ipecac syrup. *Inhalation:* Get medical aid. Do not contaminate yourself. Wearing the appropriate PPE equipment and respirator for carbamate pesticides, immediately remove the victim from the contaminated area to fresh air. If the victim is not breathing, administer artificial respiration. Do NOT use mouth-to-mouth resuscitation; use bag/mask apparatus. If breathing is difficult, administer oxygen through bag/mask apparatus until medical help arrives. Do not leave victim unattended. *Note to physician or authorized medical personnel.* Administer atropine, 2 mg (1/30 gr) intramuscularly or intravenously as soon as any local or systemic signs or symptoms of an intoxication are noted; repeat the administration of atropine every 3 to 8 minutes until signs of atropinization (mydriasis, dry mouth, rapid pulse, hot and dry skin) occur; initiate treatment in children with 0.05 mg mg/kg of atropine; repeat at 5 to 10 minute intervals. Watch respiration, and remove bronchial secretions if they appear to be obstructing the airway; intubate if necessary. *Medical note:* 2-PAMCI may be contraindicated in the case of some carbamate poisonings.

*References:*
- California Environmental Protection Agency "Chemical List of Lists," Sacramento CA (February 1997).

# Cinnamaldehyde

*Use Type:* Funicide, insecticide and dog and cat repellent
*CAS Number:* 104-55-2; 14371-10-9 (*trans*-isomer)
*Formula:* $C_9H_8O$
*Synonyms:* Benzylideneacetaldehyde; Cassia aldehyde; Cinnamal; Cinnamyl aldehyde; Cinnimic aldehyde; NCI-C56111; Phenylacrolein; β-phenylacrolein; 3-Phenylacrolein; 3-Phenylpropenal; 3-Phenyl-2-propenal; Mixture of cinnamaldehyde, cinnamyl alcohol, 4-methoxycinnamaldehyde, 3-phenyl-1-propanol, 4-methoxyphenethyl alcohol, indole, and 1,2,4-trimethoxybenzene. *trans-isomer:* (*E*)-Cinnamaldehyde; *trans*-Cinnamaldehyde; *trans*-Cinnamic aldehyde; *trans*-Cinnamylaldehyde; (*E*)-3-Phenylpropenal; (*E*)-3-Phenyl-2-propenal; 2-Propenal, 3-phenyl-, (*E*)-
*Trade Names:* ADIOS® Micro-flo (USA); ZIMTALDEHYDE®; ZIMTALDEHYDE® LIGHT
*Producers:* DSM Agro (Netherlands); Micro-flo (USA)
*Chemical Class:* Botanical
*EPA/OPP PC Code:* 040506; 040516 (*trans*-isomer)
*California DPR Chemical Code:* 2277
*EINECS Number:* 203-213-9
*Uses:* Used as an antifungal agent, corn rootworm attractant, and dog and cat repellent. Can be used on soil casing for mushrooms, row crops, turf and all food commodities.
*U.S. Maximum Allowable Residue Levels for Cinnamaldehyde (40 CFR 180.1127 and 1156):*

| Chemical | Crop | ppm | CFR |
|---|---|---|---|
| Cinnamaldehyde | r.a.c* | – | 180.1156 |
| Mixture of cinnamaldehyde, cinnamyl alcohol, 4-methoxy cinnamaldehyde, 3-phenyl-1-propanol, 4-methoxyphenethyl alcohol, indole, and 1,2,4-trimethoxy-benzene | r.a.c* | – | 180.1127 |

* r.a.c.= raw agricultural commodities

*Carcinogen/Hazard Classifications*
**Label Signal Word:** CAUTION
*Description:* Combustible, yellowish, oily liquid (thickens on exposure to air); strong cinnamon odor. Very slightly soluble in water. Molecular weight = 132.16. Density = 1.048-1.052 @ 25°C; 1.1 @ 4°C. Melting/Freezing point = –7.5°C. Boiling point = 251°C. Flash point = 120°C. Vapor

pressure = 1 mmHg @ 76.1°C, 40 mmHg @ 152°C. Log $K_{ow}$ = 1.9. Unlikely to bioaccumulate in aquatic organisms.

***Incompatibilities:*** May react violently with bromine, ketones. Incompatible with strong acids, azo dyes, caustics, ammonia, amines, boranes, hydrazines, strong oxidizers.

***Harmful Effects and Symptoms***

***Short Term Exposure:*** Contact with eyes or skin may cause severe irritation or injury. Inhalation should be avoided; use NIOSH-approved air purifying respirators for pesticides. May be harmful if swallowed. Has anesthetic properties.

***First Aid:*** If this chemical gets into the eyes, remove any contact lenses at once and irrigate immediately for at least 15 minutes, occasionally lifting upper and lower lids. Seek medical attention immediately. If this chemical contacts the skin, remove contaminated clothing and wash immediately with soap and water. Seek medical attention immediately. If this chemical has been inhaled, remove from exposure, begin rescue breathing (using universal precautions) if breathing has stopped, and CPR if heart action has stopped. Transfer promptly to a medical facility. When this chemical has been swallowed, get medical attention. Give large quantities of water and induce vomiting. Do not make an unconscious person vomit.

***References:***

- U.S. Environmental Protection Agency, Office of Pesticide Programs, "Cinnamaldehyde Fact Sheet (040506), Washington, DC (December 2000). http://www.epa.gov/oppbppd1/biopesticides/ingredients/factsheets/factsheet_040506.htm
- California Environmental Protection Agency "Chemical List of Lists," Sacramento CA (February 1997).
- U.S. Environmental Protection Agency, Office of Pesticide Programs, Pesticide Residue Limits, "Cinnamaldehyde", 40 CFR 180.1127, 1156. www.epa.gov/pesticides/food/viewtols.htm

# Clethodim (ANSI)

***Use Type:*** Herbicide
***CAS Number:*** 99129-21-2
***Synonyms:*** 2-Cyclohexen-1-one, 2-[1-(((3-chloro-2-propenyl)oxy)imino)propyl]-5-[2-(ethylthio)propyl]-3-hydroxy-; (E)-2-[1-(((3-Chloro-2-propenyl)oxy)imino)propyl]-5-[2-(ethylthio)propyl]-3-hydroxy-2-cyclohexen-1-one
***Trade Names:*** CLETODIME®; PRISM®, Valent BioSciences (USA); RE 45601; SELECT®, Valent BioSciences (USA)
***Producers:*** Agan Chemical Manufacturers (Israel); Arvesta (USA); Makhteshim-Agan Industries (Israel); Micro Flo (USA); Valent BioSciences (USA)
***Chemical Class:*** Cyclohexanone derivative
***EPA/OPP PC Code:*** 121011
***California DPR Chemical Code:*** 3566

***Uses:*** A selective post-emergence herbicide used to control annual and perennial grasses to a range of crops including cotton, flax, peanuts, soybeans, sugarbeets, potatoes, alfalfa, sunflowers and most vegetables.

***Human toxicity (long-term)[77]:*** Low–70.00 ppb, Health Advisory

***Fish toxicity (threshold)[77]:*** Very low–3210.91612 ppb, MATC (Maximum Acceptable Toxicant Concentration)

***U.S. Maximum Allowable Residue Levels for Clethodim (40 CFR 180.458):***

| CROP | ppm |
|---|---|
| Alfalfa, forage | 6 |
| Alfalfa, hay | 10 |
| Bean, dry | 2.5 |
| Beet, sugar, molasses | 1 |
| Beet, sugar, roots | 0.2 |
| Beet, sugar, tops | 1 |
| Brassica, head and stem, subgroup 5a | 3 |
| Brassica, leafy greens, subgroup 5b | 3 |
| Canola, meal | 1 |
| Canola, seed | 0.5 |
| Cattle, fat | 0.2 |
| Cattle, meat | 0.2 |
| Cattle, mbyp | 0.2 |
| Clover, forage | 10 |
| Clover, hay | 20 |
| Cotton, meal | 2 |
| Cotton, undelinted seed | 1 |
| Cranberry | 0.5 |
| Egg | 0.2 |
| Fescue, forage | 10 |
| Fescue, hay | 20 |
| Flax, meal | 1 |
| Flax, seed | 0.5 |
| Goat, fat | 0.2 |
| Goat, meat | 0.2 |
| Goat, mbyp | 0.2 |
| Hog, fat | 0.2 |
| Hog, meat | 0.2 |
| Hog, mbyp | 0.2 |
| Horse, fat | 0.2 |
| Horse, meat | 0.2 |
| Horse, mbyp | 0.2 |
| Leaf petioles subgroup 4b | 0.6 |
| Lettuce, leaf | 2 |
| Melon subgroup 9a | 2 |
| Milk | 0.05 |
| Mustard, seed | 0.5 |
| Onion | 0.2 |
| Onion, dry bulb | 0.2 |
| Onion, green | 2 |
| Peanut | 3 |
| Peanut, hay | 3 |
| Peanut, meal | 5 |
| Peppermint, tops | 5 |

| | |
|---|---|
| Potato | 0.5 |
| Potato, granules/flakes | 2 |
| Poultry, fat | 0.2 |
| Poultry, meat | 0.2 |
| Poultry, mbyp | 0.2 |
| Radish, tops | 0.7 |
| Sheep, fat | 0.2 |
| Sheep, meat | 0.2 |
| Sheep, mbyp | 0.2 |
| Soybean | 10 |
| Spearmint, tops | 5 |
| Spinach | 2 |
| Squash/cucumber subgroup 9b | 0.5 |
| Strawberry | 3 |
| Sunflower, meal | 10 |
| Sunflower, seed | 5 |
| Turnip, greens | 3 |
| Vegetable, fruiting, group 8 | 1 |
| Vegetable, root, except sugarbeet, subgroup 1b | 1 |
| Vegetable, tuberous and corm, Subgroup 1c | 1 |

**Carcinogen/Hazard Classifications**
**Label Signal Word:** WARNING or CAUTION
*Description:*
*Harmful Effects and Symptoms*
*Short Term Exposure:* Contact with eyes or skin may cause irritation or injury. Inhalation should be avoided; use NIOSH-approved air purifying respirators for pesticides. May be harmful if swallowed.
*First Aid:* If this chemical gets into the eyes, remove any contact lenses at once and irrigate immediately for at least 15 minutes, occasionally lifting upper and lower lids. Seek medical attention immediately. If this chemical contacts the skin, remove contaminated clothing and wash immediately with soap and water. Seek medical attention immediately. If this chemical has been inhaled, remove from exposure, begin rescue breathing (using universal precautions) if breathing has stopped, and CPR if heart action has stopped. Transfer promptly to a medical facility. When this chemical has been swallowed, get medical attention. Give large quantities of water and induce vomiting. Do not make an unconscious person vomit.
*References:*
- EXTOXNET, Extension Toxicology Network, "Pesticide Information Profile, Clethodim," Oregon State University, Corvallis, OR (July 1995). http://ace.orst.edu/cgi-bin/mfs/01/pips/clethodi.htm
- California Environmental Protection Agency "Chemical List of Lists," Sacramento CA (February 1997).
- EPA, Office of Pesticide Programs, Pesticide Residue Limits, "Clethodim", 40 CFR 180.458. www.epa.gov/pesticides/food/viewtols.htm

# Clofentezine (ANSI)

*Use Type:* Miticide
*CAS Number:* 74115-24-5
*Formula:* $C_{14}H_8Cl_2N_4$
*Alert:* A Restricted Use Pesticide (RUP). Human toxicity (long-term): High.
*Synonyms:* 3,6-Bis(2-chlorophenyl)-1,2,4,5-tetrazine; Bisclofentezin; NC-21314; 1,2,4,5-Tetrazine, 3,6-bis(2-chlorophenyl)-; NC
*Trade Names:* APOLLO®, Makhteshim-Agan Industries (Israel); OVATION®
*Producers:* Makhteshim-Agan Industries (Israel)
*EPA/OPP PC Code:* 125501
*California DPR Chemical Code:* 2249
*EINECS Number:* 277-728-2
*Uses:* Used on a wide variety of crops–fruit, cereals, rangeland, ornamentals, etc.
*Human toxicity (long-term)[77]:* High–9.10 ppb, Health Advisory
*Fish toxicity (threshold)[77]:* High–7.00003 ppb, MATC (Maximum Acceptable Toxicant Concentration)
*U.S. Maximum Allowable Residue Levels Clofentezine (40 CFR 180.446):*

| CROP | ppm |
|---|---|
| Almond | 0.5 |
| Almond, hulls | 5 |
| Apple | 0.5 |
| Apple, pomace, wet & dry | 3 |
| Apricot | 1 |
| Cattle, fat | 0.05 |
| Cattle, liver | 0.4 |
| Cattle, meat | 0.05 |
| Cattle, mbyp, except liver | 0.05 |
| Cherry | 1 |
| Goat, fat | 0.05 |
| Goat, liver | 0.4 |
| Goat, meat | 0.05 |
| Goat, mbyp, except liver | 0.05 |
| Hog, fat | 0.05 |
| Hog, liver | 0.4 |
| Hog, meat | 0.05 |
| Hog, mbyp, except liver | 0.05 |
| Horse, fat | 0.05 |
| Horse, liver | 0.4 |
| Horse, meat | 0.05 |
| Horse, mbyp, except liver | 0.05 |
| Milk | 0.01 |
| Nectarine | 1 |
| Peach | 1 |
| Pear | 0.5 |
| Sheep, fat | 0.05 |
| Sheep, liver | 0.4 |
| Sheep, meat | 0.05 |

| | |
|---|---|
| Sheep, mbyp, except liver | 0.05 |
| Walnut | 0.02 |

*Carcinogen/Hazard Classifications*
**U.S. EPA Carcinogens:** Group C, possible carcinogen
**Label Signal Word:** CAUTION
**WHO Acute Hazard:** Class U, unlikely to be hazardous
**Endocrine Disruptor:** Suspected endocrine disruptor
*Regulatory Authority:*
• Actively registered pesticide in California.
*Description:* Magenta crystalline solid. Molecular weight = 303.16. Melting/Freezing point = 180°C. Vapor pressure = $1 \times 10^{-9}$ mmHg @ 20°C.
*Incompatibilities:* Oxidizers (chlorates, nitrates, peroxides, permanganates, perchlorates, chlorine, bromine, fluorine, etc); strong acids.
*Harmful Effects and Symptoms*
*Short Term Exposure:* Contact with eyes or skin may cause irritation or injury. Inhalation should be avoided; use NIOSH-approved air purifying respirators for pesticides. May be harmful if swallowed.
*First Aid:* If this chemical gets into the eyes, remove any contact lenses at once and irrigate immediately for at least 15 minutes, occasionally lifting upper and lower lids. Seek medical attention immediately. If this chemical contacts the skin, remove contaminated clothing and wash immediately with soap and water. Seek medical attention immediately. If this chemical has been inhaled, remove from exposure, begin rescue breathing (using universal precautions) if breathing has stopped, and CPR if heart action has stopped. Transfer promptly to a medical facility. When this chemical has been swallowed, get medical attention. Give large quantities of water and induce vomiting. Do not make an unconscious person vomit.
*References:*
• California Environmental Protection Agency "Chemical List of Lists," Sacramento CA (February 1997).
• U.S. Environmental Protection Agency, Office of Pesticide Programs, Pesticide Residue Limits, "Clofentezine", 40 CFR 180.446. http://www.epa.gov/pesticides/food/viewtols.htm

# Clomazone (ANSI)

*Use Type:* Herbicide
*CAS Number:* 81777-89-1
*Formula:* $C_{12}H_{14}ClNO_2$
*Synonyms:* 2-[(2-Chlorophenyl)methyl]-4,4-dimethyl-3-isoxazolidinone; 2-(2-Chlorobenzyl)-4,4-dimethyl-1,2-oxazolidin-3-one; Dimethazone
*Trade Names:* CERANO®, FMC Agricultural Products (USA); COLZOR TRIO®, Syngenta (Switzerland); COMMAND® FMC Agricultural Products (USA) and Rallis India (India); COMMENCE®, DIBEL®, FMC Agricultural Products (USA); FMC® 57020, FMC Agricultural Products (USA); GAMBIT®; MAGISTER®; MERIT®; STRATEGY®
*Producers:* Agsin (Singapore); Epochem Co., (China); FMC Agricultural Products (USA); Rallis India (India); Syngenta (Switzerland)
*Chemical Class:* Oxazolidione
*EPA/OPP PC Code:* 125401
*California DPR Chemical Code:* 3537
*Uses:* Clomazone is a broad-spectrum herbicide used on rice, peas, pumpkins, soybeans, sweet potatoes, winter squash, cotton, tobacco and fallow wheat fields to control annual grasses and broadleaf weeds.
*Human toxicity (long-term)[77]:* Very low–301.00 ppb, Health Advisory
*Fish toxicity (threshold)[77]:* Very low–3156.16718 ppb, MATC (Maximum Acceptable Toxicant Concentration)
*U.S. Maximum Allowable Residue Levels for Clomazone (40 CFR 180.425):*

| CROP | ppm |
|---|---|
| Bean, snap, succulent | 0.05 |
| Cabbage | 0.1 |
| Cotton, undelinted seed | 0.05 |
| Cucumber | 0.1 |
| Pea, succulent | 0.05 |
| Pepper | 0.05 |
| Peppermint, tops | 0.05 |
| Pumpkin | 0.1 |
| Rice, grain | 0.02 |
| Rice, straw | 0.02 |
| Soybean | 0.05 |
| Spearmint, tops | 0.05 |
| Squash, summer | 0.1 |
| Squash, winter | 0.1 |
| Sugarcane, cane | 0.05 |
| Sweet potato | 0.05 |
| Vegetable, cucurbit, group 9 | 0.05 |
| Vegetable, tuberous and corm, except potato, subgroup 1(d) | 0.05 |
| Watermelon | 0.1 |

*Carcinogen/Hazard Classifications*
**U.S. EPA Carcinogens:** Not Likely a Carcinogen
**Label Signal Word:** WARNING or CAUTION
**WHO Acute Hazard:** Class II, moderately hazardous
*Description:* Vapor pressure = $1 \times 10^{-4}$ mmHg @ 20°C.
*Harmful Effects and Symptoms*
*Short Term Exposure:* Contact with eyes or skin may cause irritation or injury. Inhalation should be avoided; use NIOSH-approved air purifying respirators for pesticides. May be harmful if swallowed.
*First Aid:* If this chemical gets into the eyes, remove any contact lenses at once and irrigate immediately for at least 15 minutes, occasionally lifting upper and lower lids. Seek medical attention immediately. If this chemical contacts the skin, remove contaminated clothing and wash immediately

with soap and water. Seek medical attention immediately. If this chemical has been inhaled, remove from exposure, begin rescue breathing (using universal precautions) if breathing has stopped, and CPR if heart action has stopped. Transfer promptly to a medical facility. When this chemical has been swallowed, get medical attention. Give large quantities of water and induce vomiting. Do not make an unconscious person vomit.

*References:*
- EXTOXNET, Extension Toxicology Network, "Pesticide Information Profile, Clomazone," Oregon State University, Corvallis, OR (June 1996). http://extoxnet.orst.edu/pips/clomazon.htm
- California Environmental Protection Agency "Chemical List of Lists," Sacramento CA (February 1997).
- U.S. Environmental Protection Agency, Office of Pesticide Programs, Pesticide Residue Limits, "Clomazone", 40 CFR 180.425. www.epa.gov/pesticides/food/viewtols.htm

# Clonitralid

*Use Type:* Molluscicide
*CAS Number:* 1420-04-8
*Formula:* $C_{15}H_{15}Cl_2N_3O_5$
*Synonyms:* 2-Aminoethanol salt of 2',5-dichloro-4'-nitrosalicylanilide; 5-Chloro-*N*-(2-chloro-4-nitrophenyl)-2-hydroxybenzamide with 2-aminoethanol (1:1); Clonitarlid; 5-Chloro-*N*-(2-chloro-4-nitrophenyl)-2-hydroxybenzamide, 2-aminoethanol salt; 2',5-Dichloro-4'-nitrosalicylanilide, 2-aminoethanol salt; 5,2'-Dichloro-4'-nitrosalicylanilide ethanolamine salt; 5,2-Dichloro-4-nitrosalicylicanilide-2-aminoethanol salt; 2',5-Dichloro-4'-nitrosalicyloylanilide ethanolamine salt; Ethanolamine salt of 5,2'-dichloro-4'-nitrosalicyclicanilide; NCI-C00431; Niclosamide; 2-Aminoethanol salt of 5-chloro-N-(2-chloro-4-nitrophenyl)-2-hydroxybenzamide
*Trade Names:* BAY 73®, Bayer CropScience (Germany) and U.S. Fish and Wildlife Service; BAY 6076®, Bayer CropScience (Germany); BAYER 73®, Bayer CropScience (Germany); BAYER 25648®, Bayer CropScience (Germany); BAYLUSCID®, Bayer CropScience (Germany), canceled; BAYLUSCIDE®, Bayer CropScience (Germany) and U.S. Fish and Wildlife Service; M 73®; MOLLUSCICIDE BAYER 73®, Bayer CropScience (Germany); NICLOSAMIDE®; SR 73®
*Producers:* Bayer CropScience (Germany); U.S. Fish and Wildlife Service (USA)
*Chemical Class:* Chloronitrophenol derivative; chlorinated aromatic amide
*EPA/OPP PC Code:* 077401
*California DPR Chemical Code:* 2472
*Uses:* Niclosamide, the parent chemical, is a relatively selective, non-cumulative chlorinated aromatic amide pesticide. Clonitralid (ethanolamine salt of Niclosamide) is used principally against molluscs, especially fresh water snails and to control sea lamprey larvae and also as an antiparasitic drug in human, pets and livestock. Niclosamide is toxic to all fish species at 0.5 mg/L (48 hours).
*Carcinogen/Hazard Classifications*
**Label Signal Word:** CAUTION or DANGER
**WHO Acute Hazard:** Class U, unlikely to be hazardous as niclosamide, the parent chemical.
*Regulatory Authority:*
- FIFRA, 40CFR152.175 (RUP)
- The "Director's List" (CAL/OSHA)

*Description:* A yellow or brownish-yellow crystaooline solid. Odorless. Soluble in water. Molecular weight = 388.23. Melting point = 191°C.
*Incompatibilities:* Slowly hydrolyzes in water, releasing ammonia and forming acetate salts.
***Permissible Concentration in Water:*** No criteria set. Runoff from spills or fire control may cause water pollution.
*Routes of Entry:* Ingestion, inhalation or skin contact.
*Harmful Effects and Symptoms*
***Short Term Exposure:*** May result in nausea and abdominal pain.
*Points of Attack:* Gastrointestinal system
*Medical Surveillance:*
*First Aid:* If this chemical gets into the eyes, remove any contact lenses at once and irrigate immediately for at least 15 minutes, occasionally lifting upper and lower lids. Seek medical attention immediately. If this chemical contacts the skin, remove contaminated clothing and wash immediately with soap and water. Seek medical attention immediately. If this chemical has been inhaled, remove from exposure, begin rescue breathing (using universal precautions) if breathing has stopped, and CPR if heart action has stopped. Transfer promptly to a medical facility. When this chemical has been swallowed, get medical attention. Give large quantities of water and induce vomiting. Do not make an unconscious person vomit. Do not induce vomiting when formulations containing petroleum solvents are ingested.
*References:*
- International Programme on Chemical Safety (IPCS), "Data Sheet on Pesticides No. 673, Niclosamide," Geneva, Switzerland (March 1988). http://www.inchem.org/documents/pds/pds/pest63_e.htm
- California Environmental Protection Agency "Chemical List of Lists," Sacramento CA (February 1997).

# Clopyralid (ANSI)

*Use Type:* Herbicide
*CAS Number:* 1702-17-6
*Formula:* $C_7H_3Cl_2NO_2$
*Synonyms:* 3,6-Dichloropicolinic acid; 3,6-Dichloro-2-pyridinecarboxylic acid; 2-Pyridinecarboxylic acid, 3,6-

dichloro-; 3,6-Dichloro-2-picolinic acid; 3,6-DCP; Picolinic acid, 3,6-dichloro-

*Trade Names:* ACCENT®, DuPont Crop Protection (USA); CONFRONT®, Dow AgroSciences (USA); CURTAIL®, Dow AgroSciences (USA); CURTAIL M®, Dow AgroSciences (USA); DOWCO®-290, Dow AgroSciences (USA); HORNET®, Dow AgroSciences (USA); LONTREL®, Dow AgroSciences (USA); LONTREL® 3, Dow AgroSciences (USA); LONTRIL® F, Dow AgroSciences (USA); LONTRIL® T, Dow AgroSciences (USA); MATRIGON®; MILLENNIUM®, Nufarm (Australia); NAF®-280, Dow AgroSciences (USA); PARADIGM®, Dow AgroSciences (USA); RECLAIM®; REDEEM®, Dow AgroSciences (USA); RIVERDALE®, Nufarm (Australia); SCORPION®; STINGER®, Dow AgroSciences (USA); TRANSLINE®; WIDEMATCH®, Dow AgroSciences (USA); XRM-3972®

*Producers:* Dow AgroSciences (USA); DuPont Crop Protection (USA); Nufarm (Australia)

*Chemical Class:* Chlorophenoxy

*EPA/OPP PC Code:* 117403

*California DPR Chemical Code:* 5135

*ICSC Number:* 0443

*RTECS Number:* TJ550700

*EEC Number:* 607-231-00-1

*Uses:* Clopyralid is used to control annual and perennial broadleaf weeds on rangeland, pastures, turf and lawns, rights-of-way and a few agricultural products such as sugarbeets, oats, barley, mint and wheat.

*Human toxicity (long-term)$^{(77)}$:* Very low–3500.00 ppb, Health Advisory

*Fish toxicity (threshold)$^{(77)}$:* Very low–20832.27357 ppb, MATC (Maximum Acceptable Toxicant Concentration)

*Carcinogen/Hazard Classifications*

**U.S. EPA Carcinogens:** Not likely a carcinogen

**Label Signal Word:** WARNING, CAUTION or WARNING, depending on formulation

*Regulatory Authority:*
- AB 2588-Air Toxics "Hot Spots" Chemicals (CAL) as chlorophenoxy pesticides
- The "Director's List" (CAL/OSHA) as chlorophenoxy pesticides

*Description:* Odorless and colorless crystalline solid. Combustible. Poor solubility in water. Molecular weight = 192.00. Melting/Freezing point = 152°C. Vapor pressure = $2 \times 10^{-3}$ Pa @ 20°C.

*Incompatibilities:* May react violently with strong oxidizers, bromine, 90% hydrogen peroxide, phosphorus trichloride, silver powders or dust. Incompatible with silver compounds. Mixture with some silver compounds forms explosive salts of silver oxalate. Keep away from oxidizers, sulfuric acid, caustics, ammonia, aliphatic amines, alkanolamines, isocyanates, alkylene oxides, epichlorohydrin. Solutions are corrosive to aluminum, iron and tin. When heated to decomposition, this chemical may form nitrogen oxides and chlorine gas.

*Harmful Effects and Symptoms*

*Short Term Exposure:* Poisonous; may be fatal if inhaled, swallowed, or absorbed through skin. Severely irritates eyes, skin and respiratory tract, with burning sensation, pain, redness and swelling. Metabolic stimulant. If inhaled, causes coughing, dilated pupils, headache, profuse persperation, intense thirst, extreme fatigue, rapid pulse, high fever, clammy, flushed skin, rapid breathing, nausea, vomiting, cyanosis (bluish tint to skin and lips), anxiety and confusion, convulsions, risk of lung edema. If swallowed, face and lips turn bluish. Liver injury with associated jaundice, kidney failure, and cardiac arrhythmias are commonly noted. Nerve damage, which may be delayed, may include swelling of legs and feet, muscle twitch and stupor. Severe exposure can cause death from heart failure. Dust or liquid left in contact with the skin for several hours may be absorbed. This may result in severe delayed symptoms as listed above. These symptoms may last for months or years. Can cause permanent impairment of vision or irreversible damage.

*Long Term Exposure:* Workers exposed to chlorophenoxy compounds such as 2,4-D (in the manufacturing process) over a five to ten year period at levels above 10 mg/m³ complained of weakness, rapid fatigue, headache and vertigo. Liver damage, low blood pressure and slowed heartbeat were also found. Based on animal tests, may affects human reproduction.

*Points of Attack:* Eyes, skin, respiratory system, central nervous system, cardiovascular system, liver and kidney

*Medical Surveillance:* If symptoms develop or overexposure is suspected, the following may be useful: Liver and kidney function tests. Exam of the nervous system.

*First Aid:* If this chemical gets into the eyes, remove any contact lenses at once and irrigate immediately for at least 15 minutes, occasionally lifting upper and lower lids. Seek medical attention immediately. If this chemical contacts the skin, remove contaminated clothing and wash immediately with soap and water. Seek medical attention immediately. If this chemical has been inhaled, remove from exposure, begin rescue breathing (using universal precautions) if breathing has stopped, and CPR if heart action has stopped. Transfer promptly to a medical facility. When this chemical has been swallowed, get medical attention. *Do not induce vomiting when formulations containing petroleum solvents are ingested.* Otherwise, give large quantities of water and induce vomiting. Do not make an unconscious person vomit.

*References:*
- "Clopyralid Herbicide Fact Sheet," Caroline Cox, *Journal of Pesticide Reform,* Winter 1998, Northwest Coalition for Alternatives to Pesticides, Eugene, OR.
- California Environmental Protection Agency "Chemical List of Lists," Sacramento CA (February 1997).

## Cloransulam-methyl

*Use Type:* Herbicide
*CAS Number:* 147150-35-4
*Synonyms:* N-(2-carboxymethyl-6-chlorophenyl)-5-ethoxy-7-fluoro(1,2,4)triazolo-(1,5-c)pyrimidine-2-sulfonamide; Benzoic acid, 3-chloro-2-[((5-ethoxy-7-fluoro(1,2,4)triazolo(1,5-c)pyrimidin-2-yl)sulfonyl)amino]-, methyl ester
*Trade Names:* AMPLIFY™, Monsanto (USA); FIRSTRATE®, Dow AgroSciences (USA); FRONTROW, Dow AgroSciences (USA); GAUNTLET®, FMC (USA); XDE-565
*Producers:* Dow AgroSciences (USA); FMC (USA); Monsanto (USA); Nufarm (Australia)
*Chemical Class:* Triazolopyrimidine sulfonanilide
*EPA/OPP PC Code:* 129116
*California DPR Chemical Code:* 5781
*Uses:* Used as a pre-emergence or post-emergence control of broadleaf weeds in soybeans.
*Carcinogen/Hazard Classifications*
**U.S. EPA Carcinogens:** Not likely to be a carcinogen
**Label Signal Word:** CAUTION
**WHO Acute Hazard:** Class U, unlikely to be hazardous
*Description:* Off-white powder. Slight mint odor. Melting Point= 216-218°C.
*Incompatibilities:* Slowly hydrolyzes in water, releasing ammonia and forming acetate salts.
*Harmful Effects and Symptoms*
*Short Term Exposure:* Contact with eyes or skin may cause irritation or injury. Inhalation should be avoided; use NIOSH-approved air purifying respirators for pesticides. May be harmful if swallowed.
*First Aid:* If this chemical gets into the eyes, remove any contact lenses at once and irrigate immediately for at least 15 minutes, occasionally lifting upper and lower lids. Seek medical attention immediately. If this chemical contacts the skin, remove contaminated clothing and wash immediately with soap and water. Seek medical attention immediately. If this chemical has been inhaled, remove from exposure, begin rescue breathing (using universal precautions) if breathing has stopped, and CPR if heart action has stopped. Transfer promptly to a medical facility. When this chemical has been swallowed, get medical attention. Give large quantities of water and induce vomiting. Do not make an unconscious person vomit.
*References:*
- U.S. Environmental Protection Agency, Office of Pesticide Programs, "Pesticide Fact Sheet, Cloransulam-methyl," Washington, DC (October 29, 1997). http://www.epa.gov/opprd001/factsheets/cloransulam.pdf
- California Environmental Protection Agency "Chemical List of Lists," Sacramento CA (February 1997)

## Copper Ammonium Carbonate

*Use Type:* Fungicide
*CAS Number:* 33113-08-5
*Synonyms:* Carbonic acid, ammonium copper salt; Copper Count N
*Trade Names:* ACQ®, Chemical Specialties (USA); COPSOL®, Chemical Specialties (USA), canceled
*Producers:* Chemical Specialties (USA)
*Chemical Class:* Inorganic copper compound
*EPA/OPP PC Code:* 022703
*California DPR Chemical Code:* 1762
*Uses:* Used to treat wood and burlap. All food uses were canceled by Chemical Specialties, Inc. in March, 2001. Canceled were uses on mangos, beans, carrots, celery, citrus, grapes, cucurbits, peanuts, potatoes, peppers, strawberries, tomatoes and sugarbeets.
*Carcinogen/Hazard Classifications*
**Label Signal Word:** DANGER or CAUTION
*Regulatory Authority:*
- AB 2588-Air Toxics "Hot Spots" Chemicals (CAL)
- The "Director's List" (CAL/OSHA)
- Clean Water Act: Toxic Pollutant (Section 401.15) as copper and compounds.
- RCRA Ground Water Monitoring List. Suggested test method(s) (PQL ug/L): 6010(60); 7210(200) Note: All species in the ground water that contain copper are included
- EPCRA (Section 313): Includes any unique chemical substance that contains copper as part of that chemical's infrastructure. Form R *de minimus* concentration reporting level: 1.0%

*Description:* White solid.
*Incompatibilities:* Incompatible with germanium, lead diacetate, magnesium, mercurous chloride, silicon, silver nitrate, titanium.
*Harmful Effects and Symptoms*
*Short Term Exposure:* Contact with eyes or skin may cause irritation or injury. Inhalation should be avoided; use NIOSH-approved air purifying respirators for pesticides. May be harmful if swallowed.
*First Aid:* If this chemical gets into the eyes, remove any contact lenses at once and irrigate immediately for at least 15 minutes, occasionally lifting upper and lower lids. Seek medical attention immediately. If this chemical contacts the skin, remove contaminated clothing and wash immediately with soap and water. Seek medical attention immediately. If this chemical has been inhaled, remove from exposure, begin rescue breathing (using universal precautions) if breathing has stopped, and CPR if heart action has stopped. Transfer promptly to a medical facility. When this chemical has been swallowed, get medical attention. Give large quantities of water and induce vomiting. Do not make an unconscious person vomit.

*References:*
- California Environmental Protection Agency "Chemical List of Lists," Sacramento CA (February 1997).

# Copper and Copper compounds

*Use Type:* Copper is used as a fungicide. Copper compounds are used as insecticides, algicides, molluscicide, and plant fungicides.
*CAS Number:* 7440-50-8; 1317-38-0 (CuO, copper fume)
*Formula:* Cu
*Synonyms:* Bronze powder; C.I. 77400; C.I. Pigment metal 2; Cobre (Spanish); Copper bronze; Elemental copper; 1721 Gold; Gold bronze; Kafar copper; M2 Copper; M1 (Copper); OFHC Cu; Raney copper
*Trade Names:* ALLBRI NATURAL COPPER®; ANAC 110®; ARWOOD COPPER®; CDA 101®; CDA 110®; CDA 122®; CDA 102®
*Producers:* Adrian Resources (Canada); Aldrich Chemical (USA); Anglo American (UK); Antofagasta (UK); ASARCO (USA); BHP Billiton (Australia & UK); Boliden (Sweden); Codelco (Chile); Dowa Mining (Japan); First Quantum Minerals (Canada); Goldschmidt (Germany); Great Western Inorganics (USA); Grupo Mexico (Mexico); Indo Gulf (India); Ingenieria Industrial (Mexico); Metorex (South Africa); Minera Escondida (Chile); Minerals Research & Development (USA); Mitsubishi Materials (Japan); Mitsui Mining (Japan); Nippon Mining & Metals (Japan); Noranda (Canada); PCF Chimie (France); Phelps Dodge (USA); Philbro-Tech (USA); Rio Tinto (UK); Shyam (India); Southern Peru (Peru); Spiess-Urania Chemicals (Germany); Teck Cominco (Canada); UMICORE (Belgium); William Blythe (UK); WMC (Australia)
*Chemical Class:* Inorganic; Metals
*EPA/OPP PC Code:* 022501
*California DPR Chemical Code:* 714
*ICSC Number:* 0240
*RTECS Number:* GL5325000; GL7900000 (CuO, copper fume)
*EINECS Number:* 231-159-6 (copper)
*Uses:* Copper is used as a fungicide on table and raisin grapes, wine grapes, oranges and pears. Copper compounds are used as copper supplements for pasture lands. Metallic copper is an excellent conductor of electricity and is widely used in the electrical industry in all gauges of wire for circuitry, coil, and armature windings, high conductivity tubes, commutator bars, etc. It is made into castings, sheets, rods, tubing and wire, and is used in water and gas piping, roofing materials, cooking utensils, chemical and pharmaceutical equipment and coinage. Copper forms many important alloys: Be-Cu alloy, brass, bronze, gunmetal, bell metal, German silver, aluminum bronze, silicon bronze, phosphor bronze, and manganese bronze. Copper compounds are used as mordants, pigments, catalysts, and in the manufacture of powdered bronze paint and percussion caps. They are also utilized in analytical reagents, in paints for ships' bottoms, in electroplating, and in the solvent for cellulose in rayon manufacture.

*U.S. Maximum Allowable Residue Levels for Copper and Copper Compounds (40 CFR 180.1021; 180.136; 180.538):*

| CHEMICAL | CROP | ppm | CFR |
| --- | --- | --- | --- |
| Basic copper carbonate | pear(post-h) | 3.0 | 180.136 |
| Basic copper carbonate | r.a.c | — | 180.1021 |
| Basic copper sulfate | r.a.c | — | 180.1021 |
| Copper (II), bis(1,2-ethane-diamine-*N,N*)- | r.a.c. | — | 180.1021 |
| Copper(II) oxide | r.a.c. | — | 180.1021 |
| Copper (metallic) water, potable | | 1 | 180.538 |
| Copper oxide | crops, irrigated | — | 180.1021 |
| Copper oxide | egg | — | 180.1021 |
| Copper oxide | fish | — | 180.1021 |
| Copper I oxide | meat | — | 180.1021 |
| Copper I oxide | milk | — | 180.1021 |
| Copper I oxide | poultry | — | 180.1021 |
| Copper I oxide | r.a.c. | — | 180.1021 |
| Copper I oxide | shellfish | — | 180.1021 |
| Copper(II) oxide | r.a.c. | — | 180.1021 |
| Copper carbonate post-h | pear | 3.0 | 180.136 |
| Copper carbonate | r.a.c. | — | 180.1021 |
| Copper carbonate | r.a.c. | — | 180.1021 |
| Copper hydroxide | r.a.c | — | 180.1021 |
| Copper linoleate | r.a.c. | — | 180.1021 |
| Copper octanoate | r.a.c. | — | 180.1021 |
| Copper oleate | r.a.c. | — | 180.1021 |
| Copper oxychloride | r.a.c. | — | 180.1021 |
| Copper powder water, potable | | 1 | 180.538 |
| Copper sulfate, | r.a.c. | | |

| | | | |
|---|---|---|---|
| pre- and post-h | — | | 180.1021 |
| Copper sulfate | r.a.c. | — | 180.1021 |
| Copper sulfate | r.a.c. | — | 180.1021 |
| Copper sulfate pentahydrate, pre- & post-h | r.a.c. | — | 180.1021 |
| Copper sulfate pentahydrate | r.a.c. | — | 180.1021 |
| Copper-ethylene diamine complex | r.a.c. | — | 180.1021 |
| Octanoic acid, copper salt | r.a.c. | — | 180.1021 |

"r.a.c." = raw agricultural commodities

### Carcinogen/Hazard Classifications

**U.S. EPA Carcinogens:** Group D, unclassifiable, inadequate data

### Regulatory Authority:

- Air Pollutant Standard Set (ACGIH)[1] (Australia) (DFG)[3] (HSE)[33] (Israel) (Mexico) (OSHA) (former USSR)[43] (Several States)[60] (Several Canadian Provinces)
- AB 1803-Well Monitoring Chemical (CAL)
- MCL (Maximum Contaminants Levels) list of contaminants (CAL)
- AB 2588-Air Toxics "Hot Spots" Chemicals (CAL)
- Permissible Exposure Limits for Chemical Contaminants (CAL/OSHA)
- The "Director's List" (CAL/OSHA)
- Actively registered pesticide in California.
- Clean Water Act: 40CFR423, Appendix A, Priority Pollutants; Section 313 Water Priority Chemicals (57FR41331, 9/9/92); Toxic Pollutant (Section 401.15).
- RCRA 40CFR264, Appendix 9; TSD Facilities Ground Water Monitoring List. Suggested test method(s) (PQL $\mu$g/L): (total) 6010(60); 7210(200)
- Safe Drinking Water Act: MCL, 1.0 mg/L; MCLG, 1 mg/L; SMLC, 1.0 mg/L; Regulated chemical (47 FR 9352)
- Superfund/EPCRA 40CFR302.4 RQ: CERCLA, 5,000 lb (2270 kg) (no reporting of releases of this hazardous substance is required if the diameter of the pieces of solid metal released is equal to 0.004 in)
- EPCRA Section 313 Form R *de minimus* concentration reporting level: 1.0%
- Canada, WHMIS, Ingredients Disclosure List; National Pollutant Release Inventory (NPRI); CEPA Priority Substance List, Ocean dumping prohibited; Drinking Water Quality $\leq$ 1.0 mg/L.
- Mexico, Drinking Water = 1.0 mg/L

**Description:** Copper is a reddish-brown metal which occurs free or in ores such as malachite, cuprite, and chalcopyrite. Copper fume is finely divided black particulate dispersed in air. Copper dusts and mists have been assigned the formula $CuSO_4 \cdot 5H_2O/CuCl$ by NIOSH. Copper fume has been designated as $Cu/CuO/Cu_2O$ by NIOSH. Boiling point = 2595°C. Melting/Freezing point = 1083°C It may form both mono- and divalent-compounds. Copper is insoluble in water, but soluble in nitric acid and hot sulfuric acid.

**Incompatibilities:** Copper dust, fume, and mists form shock-sensitive compounds with acetylene gas, acetylenic compounds, azides, and ethylene oxides. Incompatible with acids, chemically active metals such as potassium, sodium, magnesium, and zinc, zirconium, strong bases. Violent reaction, possibly explosive, if finely-divided material come in contact with strong oxidizers.

**Permissible Exposure Limits in Air:** NIOSH[2] recommends the same level for a 10-hour workshift. ACGIH[1] recommends a TWA of 0.2 mg/m$^3$ for copper fume and 1 mg/m$^3$ for dusts and mists, as has HSE[33], Australia, Israel, and the Canadian provinces of Alberta, BC, Ontario, and Quebec and HSE, Alberta, BC set a STEL for dusts and mists of 2.0 mg/m$^3$. The NIOSH[2] IDLH is 100 mg/m$^3$ (as Cu). The DFG MAK for toatal dust is 1 mg/m$^3$; 0.1 mg/m$^3$ for fine dust and Peak Limitation is 2 times MAK (30 min), not to be exceeded 4 times during a workshift[3]. Mexico set a limit of 0.2 mg/m$^3$ TWA and STEL of 2 mg/m$^3$. The former USSR-UNEP/IRPTC project[43] has set a MAC value for copper in workplace air of 1.0 mg/m$^3$ and 0.5 mg/m$^3$ on an average value per workshift basis. Several states have set guidelines or standards for copper in ambient air[60] ranging from 0.26-1.57 $\mu$g/m$^3$ (Montana) to 2.0 $\mu$g/m$^3$ (North Dakota) to 2.0-20.0 $\mu$g/m$^3$ (Connecticut) to 4.0-20.0 $\mu$g/m$^3$ (Florida) to 5.0 $\mu$g/m$^3$ (Nevada) to 16.0 $\mu$g/m$^3$ (Virginia) to 20.0 $\mu$g/m$^3$ (New York).

**Determination in Air:** Copper dusts and mists are collected on a filter, worked up with acid, measured by atomic absorption. See NIOSH Method #7029 for copper[18]. For copper fume: filter collection, acid digestion, measurement by atomic absorption. See NIOSH Method #7200 for welding and brazing fume[18].

**Permissible Concentration in Water:** To protect freshwater aquatic life: 5.6 $\mu$g/L as a 24-hour average, never to exceed $e^{[0.94 \ln (\text{hardness}) -1.23]}\mu$g/L. To protect human health: 1000$\mu$g/L[6]. Canada: Drinking Water Quality (AO) <= 1.0 mg/L. Mexico, Drinking Water = 1.0 mg/L. Czechoslovakia[35] has set a MAC in surface water of 0.1 mg/L and in drinking water of 0.05 mg/L. The former USSR[35] and The former USSR-UNEP/IRPTC joint project have set a MAC in water used for domestic purposes of 1.0 mg/L and 0.001 mg/L in fresh water and 0.005 mg/L in seawater used for fishery purposes. Two states have set guidelines for copper in drinking water[61]; they are Kansas at 1000 $\mu$g/L and Minnesota at 1300 $\mu$g/L.

**Determination in Water:** Total copper may be determined by digestion followed by atomic absorption or by colorimetry (using neocuproine) or by Inductively Coupled Plasma Optical Emission Spectrometry. Dissolved Copper may be determined by 0.45 $\mu$ filtration followed by the preceding methods.

*Routes of Entry:* Inhalation of dust or fume, ingestion, or skin or eye contact.

*Harmful Effects and Symptoms*

*Short Term Exposure:* Copper salts act as irritants to the intact skin causing itching, erythema, and dermatitis. In the eyes, copper salts may cause conjunctivitis and even ulceration and turbidity of the cornea. Metallic copper may cause keratinization of the hands and soles of the feet, but it is not commonly associated with industrial dermatitis. The fumes and dust cause irritation of the upper respiratory tract, metallic taste in the mouth, nausea, metal fume fever. Inhalation of dusts, fumes, and mists of copper salts may cause congestion of the nasal mucous membranes. If the salts reach the gastrointestinal tract, they act as irritants producing salivation, nausea, vomiting, gastric pain, hemorrhagic gastritis, and diarrhea. It is unlikely that poisoning by ingestion in industry would progress to a serious point as small amounts induce vomiting, emptying the stomach of copper salts. Chronic human intoxication occurs rarely and then only in individuals with Wilson's disease (hepatolenticular degeneration). This is a genetic condition caused by the pairing of abnormal autosomal recessive genes in which there is abnormally high absorption, retention, and storage of copper by the body. The disease is progressive and fatal if untreated.

*Long Term Exposure:* Copper may decrease fertility in both males and females. Repeated or prolonged contact may cause skin sensitization and allergy, thickening of the skin, and greenish color to the skin, teeth, and hair. Repeated exposure can cause chronic irritation of the nose and cause ulcers and hole in the septum dividing the inner nose. Repeated high exposure to copper can cause liver damage. There is evidence that workers in copper smelting plants have an increased risk of lung cancer, but this is thought to be due to arsenic trioxide and not copper.

*Points of Attack:* For copper dusts and mists: respiratory system, lungs, skin, liver, including risk with Wilson's disease, kidneys. For copper fume: respiratory system, skin, eyes, and risk with Wilson's disease.

*Medical Surveillance:* Serum and urine copper levels. Evaluation by a qualified allergist. Liver function tests. Copper often contains arsenic as an impurity. Wilson's disease is a rare hereditary condition which interferes with the body's ability to get rid of copper. If you have this condition, consult your doctor about copper exposure.

*First Aid:* If copper dust or powder gets into the eyes, remove any contact lenses at once and irrigate immediately for at least 15 minutes, occasionally lifting upper and lower lids. Seek medical attention immediately. If copper dust or powder contacts the skin, remove contaminated clothing and wash immediately with soap and water. Seek medical attention immediately. If this chemical has been inhaled, remove from exposure, begin rescue breathing (using universal precautions) if breathing has stopped, and CPR if heart action has stopped. Transfer promptly to a medical facility. When this chemical has been swallowed, get medical attention. Give large quantities of water and induce vomiting. Do not make an unconscious person vomit.

*Note to physician or authorized medical personnel:* In case of fume inhalation, treat pulmonary edema. Give prednisone or other corticosteroid orally to reduce tissue response to fume. Positive pressure ventilation may be necessary. Treat metal fume fever with bed rest, analgesics and antipyretics. The symptoms of metal fume fever may be delayed for 4-12 hours following exposure: it may last less than 36 hours.

*References:*

- U.S. Environmental Protection Agency, Office of Pesticide Programs, Pesticide Residue Limits, "Copper", 40 CFR 180.1021; 180.136; 180.538). www.epa.gov/cgi-bin/oppsrch
- U.S. Environmental Protection Agency, Toxicology of Metals, Vol. II: Copper, Report EPA-600/1-77-022, pp 206-221, Research Triangle Park, NC (May 1977).
- U.S. Environmental Protection Agency, Copper: Ambient Water Quality Criteria, Washington, DC (1980).
- National Academy of Sciences, Medical and Biologic Effects of Environmental Pollutants: Copper, Washington, DC (1977).
- Sax, N.I., Ed., "Dangerous Properties of Industrial Materials Report," 1, No. 5, 48-49, New York, Van Nostrand Reinhold Co. (1981). (Copper).
- New Jersey Department of Health and Senior Services and Senior Services, "Hazardous Substance Fact Sheet: Copper," Trenton, NJ (January 1999). http://www.state.nj.us/health/eoh/rtkweb/0528.pdf
- New York State Department of Health, "Chemical Fact Sheet: Copper," Albany, NY, Bureau of Toxic Substance Assessment (January 1986 and Version 3).
- California Environmental Protection Agency "Chemical List of Lists," Sacramento CA (February 1997).

# Copper Carbonate, Basic

*Use Type:* Algaecide, insecticide and fungicide
*CAS Number:* 12069-69-1
*Formula:* $Cu_2(OH)_2CO_3$
*Synonyms:* Basic cupric carbonate; (Carbonato)dihydroxydicopper; Chestnut compound; Copper carbonate hydroxide; Cupric carbonate; Dicopper dihydroxycarbonate; Kupfercarbonat (German); Malachite
*Trade Names:* KOP KARB®; NOAH GOLD®, Osmose (USA); NW 200®, Osmose (USA); SPIN OUT 400®, Griffin (USA)
*Producers:* Griffin (USA); Nihon Kagaku Sangyo (Japan); Nufarm (Australia); Osmose (USA); Philbro-Tech (USA); William Blythe (UK)
*Chemical Class:* Inorganic copper compounds
*EPA/OPP PC Code:* 022901

*California DPR Chemical Code:* 60
*EINECS Number:* 235-113-6
*Uses:* A fungicide used against a broad spectrum of fungi. Also used as an insecticide.
*U.S. Maximum Allowable Residue Levels for Copper Carbonate (40 CFR 180.136; 180.1001 (b); 180.1021):*

| CROP | ppm | CFR |
| --- | --- | --- |
| Pear, post-h | 3 | 180.136 |
| Raw agricultural commodities | — | 180.1001 (b) |
| Raw agricultural commodities | — | 180.1021 |

*Carcinogen/Hazard Classifications*
**Label Signal Word:** CAUTION, WARNING, DANGER
*Regulatory Authority:*
- AB 2588-Air Toxics "Hot Spots" Chemicals (CAL) (as copper compounds)
- The "Director's List" (CAL/OSHA) (as copper compounds)
- Clean Water Act: Toxic Pollutant (Section 401.15) as copper and compounds.
- RCRA Ground Water Monitoring List. Suggested test method(s) (PQL µg/L): 6010(60); 7210(200) Note: All species in the ground water that contain copper are included.
- FIFRA, 40CFR185: tolerances for pesticides in food, as copper compounds, total.
- EPCRA (Section 313): Includes any unique chemical substance that contains copper as part of that chemical's infrastructure. Form R *de minimis* concentration reporting level: 1.0%

*Description:* Green powder. Insoluble in water. Molecular weight = 221.10. Density 3.9. Melting/Freezing point = decomposes @ 197°C.
*Incompatibilities:* Incompatible with germanium, lead diacetate, magnesium, mercurous chloride, silicon, silver nitrate, titanium. Incompatible with calcium (metal hydroxides), nitroethane, nitromethane, 1-nitropropane, zirconium.
*Permissible Exposure Limits in Air:* As copper: ACGIH TLV 0.2 mg/m$^3$ (fume); 1 mg/m$^3$ (dusts and mists); NIOSH/OSHA 0.1 mg/m$^3$ (fume); 1 mg/m$^3$ as (dusts and mists).
*Determination in Air:* Copper dusts and mists are collected on a filter, worked up with acid, measured by atomic absorption. See NIOSH Method #7029 for copper. For copper fume: filter collection, acid digestion, measurement by atomic absorption.
*Permissible Concentration in Water:* No criteria set. Runoff from spills or fire control may cause water pollution. The U.S. EPA (6) has set a maximum of 1.0 mg/L in water to protect human health.
*Determination in Water:* Total copper may be determined by digestion followed by atomic absorption or by colorimetry (using neocuproine) or by Inductively Coupled Plasma Optical Emission Spectrometry. Dissolved Copper may be determined by 0.45 $\mu$ filtration followed by the preceding methods.

*Harmful Effects and Symptoms*
*Short Term Exposure:* Poisonous if ingested. Irritates eyes, skin and mucous membranes. Possible skin discoloration.
*Long Term Exposure:* May cause mutations in humans. May damage the testes and decrease fertility in both males and females. May cause skin allergy and thickening of the skin; copper deposits can cause discoloration in the skin and hair, leaving a green color. Repeated exposure can cause shrinking of the lining of the inner nose with watery discharge, liver damage. Individuals with Wilson's disease absorb, retain, and store copper excessively.
*Points of Attack:* Skin, reproductive system., liver
*Medical Surveillance:* Serum and urine copper level. Liver and kidney tests. Examination by a qualified allergist. More than light alcohol consumption may exacerbate the liver damage caused by copper sulfate.
*First Aid:* If this chemical gets into the eyes, remove any contact lenses at once and irrigate immediately for at least 15 minutes, occasionally lifting upper and lower lids. Seek medical attention immediately. If this chemical contacts the skin, remove contaminated clothing and wash immediately with soap and water. Seek medical attention immediately. If this chemical has been inhaled, remove from exposure, begin rescue breathing (using universal precautions) if breathing has stopped, and CPR if heart action has stopped. Transfer promptly to a medical facility. When this chemical has been swallowed, get medical attention. Give large quantities of water and induce vomiting. Do not make an unconscious person vomit.
*References:*
- Environmental Protection Agency, Office of Pesticide Programs, Pesticide Residue Limits, "Copper Carbonate, Basic," 40 CFR 40 CFR 180.136; 180.1001 (b); 180.1021, www.epa.gov/pesticides/food/viewtols.htm
- California Environmental Protection Agency "Chemical List of Lists," Sacramento CA (February 1997)
- U.S. Environmental Protection Agency, Toxicology of Metals, Vol. II: Copper, Report EPA-600/1-77-022, pp 206-221, Research Triangle Park, NC (May 1977).
- U.S. Environmental Protection Agency, Copper: ambient Water Quality Criteria, Washington, DC (1980).U.S.

# Copper Acetoarsenite

*Use Type:* Insecticide and fungicide
*CAS Number:* 12002-03-8
*Formula:* $C_4H_6As_6Cu_4O_{16}$
*Synonyms:* Acetoarsenito de cobre (Spanish); Acetato(trimetaarsenito)dicopper; Acetoarsenite de cuivre (French); Basle green; C.I. Pigment Green 21; Copper aceto-arsenite; Cupric acetoarsenite; Emerald green; French green; Imperial green; King's green; Meadow green; Mitis

green; Moss green; Mountain green; Paris green; Parrot green; Patent green; Schweinfurt green; Swedish green; Vienna green

***Trade Names:*** FASCO PARIS GREEN®, Kerr-McGee Chemical (USA); LACCO PARIS GREEN®, Los Angeles Chemical Co. (USA); ORTHO P-G BAIT®

***Producers:*** Bayer CropScience (Germany); Kerr-McGee Chemical (USA)

***Chemical Class:*** Inorganic, arsenite, copper

***EPA/OPP PC Code:*** 022601

***California DPR Chemical Code:*** 2485 (cupric acetoarsenite) and 460 (Paris green)

***ICSC Number:*** 0013

***RTECS Number:*** GL6475000

***Uses:*** It is also used as a wood preservative and in paints for marine vessels.

***Carcinogen/Hazard Classifications***

**U.S. NTP:** Carcinogen

**California Prop. 65:** Listed

**IARC:** Group 1, known carcinogen

**WHO Acute Hazard:** Group 1b, highly hazardous

***Regulatory Authority:***

- OSHA, 29CFR1910 Specifically Regulated Chemicals (See CFR 1910.1018)
- The "Director's List" (CAL/OSHA)

*Arsenic compounds:*

- Carcinogen (arsenic compounds, n.o.s.) (IARC, Group I, carcinogenic to humans)(OSHA, select carcinogens)
- Clean Air Act, 42USC7412; Title I, Part A,§112 hazardous pollutants (arsenic compounds)
- Clean Water Act 40CFR401.15 Section 307,Toxic Pollutants, as arsenic and compounds
- RCRA, 40CFR261, Appendix 8 Hazardous Constituents, waste number not listed (arsenic compounds)
- Superfund/EPCRA 40CFR302.4 RQ: CERCLA, 1 lb (0.454 kg) (arsenic compounds)
- EPCRA Section 313: Form R *de minimis* concentration reporting level: 0.1%(inorganic arsenic)
- OSHA: Subpart Z-Toxic and Hazardous Substance 29 CFR 1910, Specifically Regulated Chemicals, regulates arsenic compounds.
- U.S. DOT 49CFR172.101, Appendix B, Regulated marine pollutant (arsenic compounds)
- Canada: Priority Substance List & Restricted Substances/Ocean Dumping Forbidden (CEPA), National Pollutant Release Inventory (NPRI)(arsenic compounds)

*Copper compounds:*

- Clean Water Act: Section 311 Hazardous Substances/RQ 40CFR117.3 (same as CERCLA, see below); Section 313
- Water Priority Chemicals (57FR41331, 9/9/92); 40CFR401.15 Section 307 Toxic Pollutants, as copper and compounds.
- RCRA 40CFR264, Appendix 9; TSD Facilities Ground Water Monitoring List. Suggested test method(s) (PQL µg/L): 6010(60); 7210(200) Note: All species in the ground water that contain copper are included.
- Superfund/EPCRA 40CFR302.4 RQ: CERCLA, 100 lb (45.4 kg)
- EPCRA Section 313 Form R *de minimus* concentration reporting level: 1.0%
- Canada, WHMIS, Ingredients Disclosure List; National Pollutant Release Inventory (NPRI); CEPA Priority Substance List, Ocean dumping prohibited.

***Description:*** Copper acetoarsenite is an emerald-green, crystalline powder. Hazard Identification (based on NFPA-704 M Rating System): Health 3, Flammability 0, Reactivity 0. Insoluble in water.

***Incompatibilities:*** Decomposes in water with prolonged heating. Incompatible with strong bases, strong acids.

***Permissible Exposure Limits in Air:***

*Copper compounds:* The Federal standard (OSHA PEL 8-hour TWA)[58] for copper fume is 0.1 mg/m$^3$, and 1 mg/m$^3$ for copper dusts and mists. NIOSH[2] recommends the same level for a 10-hour workshift. ACGIH[1] recommends a TWA of 0.2 mg/m$^3$ for copper fume and 1 mg/m$^3$ for dusts and mists. The NIOSH[2] IDLH level = 100 mg/m$^3$ (as copper)

*Arsenic compounds:* The following exposure limits are for air levels only. When skin contact also occurs, overexposure is possible, even thought air levels are less than the limits listed below. OSHA[2]: The legal airborne PEL is 0.010 mg/m$^3$ averaged over an 8-hour workshift. NIOSH[2]: The recommended airborne exposure limit is 0.002 mg/m$^3$ (ceiling), not to be exceeded during any 15 min. work period. ACGIH[1]: The recommended airborne exposure limit is 0.01 mg/m$^3$ averaged over an 8-hour workshift. The HSE[33] (U.K.) Maximum Exposure Limit (as As) is 0.1 mg/m$^3$ TWA. California's workplace PEL is the same as ACGIH[1] and an Action Level of 0.005 mg/m$^3$. The Australia limit is 0.05 mg/m$^3$ TWA (confirmed carcinogen); Israel 0.01 mg/m$^3$ TWA and Action Level 0.005 mg/m$^3$. Mexico level 0.2 mg/m$^3$ TWA. Canada: Alberta level 0.2 mg/m$^3$ TWA and STEL of 0.6 mg/m$^3$ (15 min.); British Columbia level 0.5 mg/m$^3$ TWA; Ontario level 0.01 mg/m$^3$ TWAEV and STEV of 0.05; Quebec level 0.2 mg/m$^3$ TWAEV. The former USSR-UNEP/IRPTC project[43] has set a MAC of 0.003 mg/m$^3$ on an average daily basis for residential areas. In addition, several states have set guidelines or standards for arsenic in ambient air[60]: 0.06 mg/m$^3$ (California Prop. 65), 0.0002 µg/m$^3$ (Rhode Island), 0.00023 µg/m$^3$ (North Carolina), 0.024 µg/m$^3$ (Pennsylvania), 0.05 µg/m$^3$ (Connecticut), 0.07 to 0.39 µg/m$^3$ (Montana), 0.67 µg/m$^3$ (New York), 1.0 µg/m$^3$ (South Carolina), 2.0 µg/m$^3$ (North Dakota), 3.3 µg/m$^3$ (Virginia), 5 µg/m$^3$ (Nevada).

***Determination in Air:*** Copper dusts and mists are collected on a filter, worked up with acid, measured by atomic absorption. See NIOSH Method #7029 for copper[18]. For copper fume: filter collection, acid digestion, measurement by atomic absorption.

***Permissible Concentration in Water:*** EPA has set a limit of 0.05 parts per million (ppm) for arsenic in drinking water. The U.S. EPA arsenic drinking water standard of 0.01 ppm (10 ppb) is based on the U.S. EPA final rule for arsenic in drinking water published in the January 22, 2001, *Federal Register*. However, the U.S. EPA is currently reviewing the science and cost estimate supporting this rule, and, in the interim, has reverted to the previous standard for arsenic. Thus, in the US, the current EPA arsenic drinking water standard remains at 0.05 ppm (50 ppb). To protect freshwater aquatic life-total recoverable trivalent inorganic arsenic never to exceed 440 $\mu$g/L. To protect saltwater aquatic life: 508 $\mu$g/L on an acute basis. To protect human health: preferably zero. The former USSR-UNEP/IRPTC project[43] has set MAC values for inorganic arsenic compounds in water for domestic purposes at 0.05 mg/L and in water bodies for fishery purposes of 0.5 mg/L also.

***Determination in Water:*** *For arsenic:* The atomic absorption graphite furnace technique is often used for measurement of total arsenic in water. It also has been standardized by EPA. Total arsenic may be determined by digestion followed by silver diethyldithiocarbamate; an alternative is atomic absorption; another is inductively coupled plasma optical emission spectrometry. See OSHA Method #ID-105 for arsenic[58].

***Routes of Entry:*** Inhalation, absorbed through the skin.

***Harmful Effects and Symptoms***

***Short Term Exposure:*** Eye contact can cause severe irritation and burns. Skin contact can cause irritation, burning sensation, itching, thickening and color changes. This chemical can be absorbed through the skin, thereby increasing exposure.

***Long Term Exposure:*** Repeated exposure can cause copper to deposit in the liver, kidneys, and other body organs, causing damage, atrophy of the inner lining of the nose and possible hole in the nasal septum, with a watery or bloody discharge. Metallic or garlic taste may also occur. Repeated skin exposure can cause skin allergy and possibly a green discoloration of the skin and hair. Some copper and arsenic compounds, but not this one, have been identified as carcinogens and certain arsenic compounds have been determined to be reproductive hazards. Therefore this chemical should be handled with extreme caution. May damage the nervous system. Birth defects have been observed in animals exposed to inorganic arsenic. It is likely that health effects seen in children exposed to high amounts of arsenic will be similar to the effects seen in adults.

***Points of Attack:*** Several studies have shown that inorganic arsenic can increase the risk of lung cancer, skin cancer, bladder cancer, liver cancer, kidney cancer, and prostate cancer.

***Medical Surveillance:*** Urine arsenic test. Examine skin for abnormal growths. Examination of the nose, skin, nails, and nervous system. Liver function tests, Kidney function tests.

***First Aid:*** If this chemical gets into the eyes, remove any contact lenses at once and irrigate immediately for at least 15 minutes, occasionally lifting upper and lower lids. Seek medical attention immediately. If this chemical contacts the skin, remove contaminated clothing and wash immediately with soap and water. Seek medical attention immediately. If this chemical has been inhaled, remove from exposure, begin rescue breathing (using universal precautions) if breathing has stopped, and CPR if heart action has stopped. Transfer promptly to a medical facility. When this chemical has been swallowed, get medical attention. Give large quantities of water and induce vomiting. Do not make an unconscious person vomit.

***References:***
- New Jersey Department of Health and Senior Services, Hazardous Substance Fact Sheet, "Copper Acetoarsenite,"Trenton NJ (January 1999). www.state.nj.us/health/eoh/rtkweb/0529.pdf
- California Environmental Protection Agency "Chemical List of Lists," Sacramento CA (February 1997).

# Copper Arsenite

***Use Type:*** Insecticide, herbicide and rodenticide
***CAS Number:*** 10290-12-7
***Formula:*** $CuHAsO_3$
***Alert:*** Not registered in the U.S. A teratogen.
***Synonyms:*** Acid orthoarsenite; Arsenito de cobre (Spanish); Arsonic acid, copper(2+) salt (1:1); Arsenious acid, copper(2+) salt (1:1); Cupric arsenite; Cupric green; Copper orthoarsenite; Scheele's green; Scheele's mineral; Swedish green
***Trade Names:*** KOCIDE® 2000, Griffin (USA)
***Producers:*** Griffin (USA)
***Chemical Class:*** Inorganic arsenic; Metals
***EPA/OPP PC Code:*** 022401
***ICSC Number:*** 1211
***RTECS Number:*** CG3385000
***EEC Number:*** 033-002-00-5
***EINECS Number:*** 233-644-8
***Uses:*** No longer registered in the U.S. Formerly used in agriculture as and insecticide, fungicide, rodenticide and fungicide. Also used in pigments and animal medications.
***Carcinogen/Hazard Classifications***
***U.S. NTP Carcinogen:*** Known carcinogen
***California Prop. 65:*** Carcinogen
***IARC:*** Group 1, known carcinogen
***Regulatory Authority:***
- AB 2588-Air Toxics "Hot Spots" Chemicals (CAL)
- The "Director's List" (CAL/OSHA)

- Clean Water Act: Toxic Pollutant (Section 401.15) as copper and compounds
- RCRA Ground Water Monitoring List. Suggested test method(s) (PQL ug/L): 6010(60); 7210(200) Note: All species in the ground water that contain copper are included
- Carcinogen User Register Chemical (CAL/OSHA)
- AB 1803-Well Monitoring Chemical (CAL)
- EPCRA Section 313 Form R de minimis concentration reporting level: 0.1%
- CAL Air Resources Board/AB 1807 Toxic Air Contaminants
- Specific chemicals (EPA/NESHAP)
- Proposition 65 chemical (CAL)
- Permissible Exposure Limits for Chemical Contaminants (CAL/OSHA) as arsenic compounds
- Clean Air Act: List of high risk pollutants (Section 63.74) as arsenic compounds
- Clean Water Act: Section 311 Hazardous Substances/RQ (same as CERCLA); Section 313 Priority Chemicals; Toxic Pollutant (Section 401.15) as arsenic and compounds
- RCRA Section 261 Hazardous Constituents, waste number not listed
- EPCRA Section 304 RQ: CERCLA, 1 lbs(0.454 kgs)
- Marine pollutant (49CFR, Subchapter 172.101, Appendix B). as arsenates, liquid, n.o.s; arsenates, solid, n.o.s; arsenical pesticides liquid, toxic, flammable, n.o.s.

***Description:*** An odorless, yellow-green powder that gives off irritating and/or toxic fumes in a fire.

***Permissible Exposure Limits in Air:*** OSHA[(2)]: 0.01 mg/m$^3$; NIOSH[(2)]: Ceiling 0.002 mg/m$^3$/15 min; carcinogen; ACGIH[(1)]: 0.01 mg/m$^3$; carcinogen

***Determination in Air:*** Collection on a filter and analysis by atomic absorption spectrometry. See NIOSH Methods #7900 and #73000, Elements[(18)]. See also OSHA Method ID 105.[(58)]

***Permissible Concentration in Water:*** No criteria set. Runoff from spills or fire control may cause water pollution

***Routes of Entry:*** Inhalation, skin, eyes and ingestion

***Harmful Effects and Symptoms***

***Short Term Exposure:*** *Inhalation:* Dust is readily absorbed from the lungs, but inhaled quantities are usually insufficient to cause acute systemic toxicity. Can cause cough with foamy sputum and rales. *Skin/Eye Contact:* Dust can cause localized skin irritation, but systemic absorption through the skin is negligible. Skin contact is unlikely to cause systemic effects unless the dermal barrier is compromised. Eye exposure may produce burns and ulcerations. *Ingestion:* The most important route of acute exposure. Arsenicals are quickly absorbed and can be extremely hazardous. Significant tissue and organ damage and death may result. *Symptoms of exposure include*: headache (lethargy, delirium, hallucinations, seizures, or coma can occur); dizziness; acute nausea and occasional vomiting; labored breathing; restlessness; cyanosis (pale or bluish lips, face and fingernails); perspiration; difficult breathing; abdominal pain with diarrhea; trembling and feeling of "pins and needles" or electrical shock like pains in the lower extremities; convulsions; unconsciousness; possible pulmonary edema, a medical emergency. Very acute poisoning: extreme headache; muscular paralysis, and liver and kidney dysfunction, loss of consciousness; death.

***Long Term Exposure:*** Chronic exposure is characterized by malaise, peripheral sensorimotor neuropathy, anemia, jaundice, gastrointestinal complaints, and characteristic skin lesions including hyperkeratosis (small corn-like elevations) and hyperpigmentation. Hyperkeratosis usually appears on the palms or soles. Pigmentation changes and hyperkeratosis can take 3 to 7 years to appear. Chronic inhalation can also lead to conjunctivitis, irritation of the throat and respiratory tract, and perforation of the nasal septum. Chronic exposure can cause allergic contact dermatitis. Chronic exposure may be more serious for children because of their potential longer latency period. *Carcinogenicity:* The Department of Health and Human Services, IARC, U.S. EPA, and NTP have classified arsenic as a human carcinogen based on sufficient evidence from human data. Arsenic trioxide causes skin and lung cancer, and may cause internal cancers such as liver, bladder, kidney, colon, and prostate cancers. Arsenic ions released from arsenic trioxide within the body can cross the placenta and affect the developing fetus; arsenic is also excreted in breast milk. Experimental animal studies support an association between high ingested arsenic dose and fetal toxicity.

***Points of Attack:*** Muscles, liver, and kidney.

***Medical Surveillance:*** In acutely ill patients, the agent most frequently recommended is dimercaprol, also known as BAL (British Anti Lewisite). The standard dosage regimen is 3 to 5 mg/kg IM every 4–6 hours until the 24-hour urinary arsenic level falls below 50 µg/L, unless an orally administered chelating agent (e.g., DMSA, see below) is substituted. This regimen may be adjusted depending upon the severity of the exposure and the symptoms. Do not chelate asymptomatic patients without the guidance of a 24-hour urinary arsenic level. Contraindications to BAL include preexisting renal disease, pregnancy (except in life-threatening circumstances) and concurrent use of medicinal iron (BAL and iron together form a complex that is very toxic).

***First Aid:*** Do not contaminate yourself. Positive-pressure, self-contained breathing apparatus (SCBA) is recommended in response situations that involve exposure to potentially unsafe levels of arsenicals or combustion products which may include arsine and arsenic trioxide fumes. *Eyes:* Immediately and gently flush eyes with plenty of warm or cold water (NO hot water) for at least 15 minutes, occasionally lifting the upper and lower eyelids. Get medical aid immediately. *Skin:* Although it is poorly absorbed dermally, skin contact should be avoided because arsenicals may irritate the skin. Wearing the appropriate PPE equipment and respirator for arsenicals. Immediately flush skin with plenty of soap and water for at least 15 minutes while removing contaminated clothing and shoes. *Ingestion:* Call poison control. Loosen all clothing. Never give anything by mouth to an unconscious person. Get medical aid. Do NOT induce vomiting. If conscious, alert, and able to swallow, aggressive decontamination with gastric lavage is recommended within 1 hour of ingestion of a life-threatening amount of poison. The effectiveness of activated charcoal in binding arsenic trioxide is questionable, but administration of a charcoal slurry is recommended pending further evaluation in cases of ingestion of unknown quantities (at 1 gm/kg, usual adult dose 60–90 g, child dose 25–50 g). A soda can and straw may be of assistance when offering charcoal to a child. If victim is *UNCONSCIOUS OR HAVING CONVULSIONS*, do nothing except keep victim warm. *Inhalation:* Get medical aid. Wearing the appropriate PPE equipment and respirator for arsenicals, immediately remove the victim from the contaminated area to fresh air. If the victim is not breathing, administer artificial respiration. Do NOT use mouth-to-mouth resuscitation; use bag/mask apparatus. If breathing is difficult, administer oxygen through bag/mask apparatus until medical help arrives. Do not leave victim unattended.

*References:*
- New Jersey Department of Health and Senior Services, "Hazardous Substance Fact Sheet, Copper Arsenite," Trenton NJ (December 2002). http://www.state.nj.us/health/eoh/rtkweb/0530.pdf
- California Environmental Protection Agency "Chemical List of Lists," Sacramento CA (February 1997)

# Copper Cyanide

*Use Type:* Insecticide
*CAS Number:* 544-92-3 (I); 14763-77-0 (II)
*Formula:* CCuN, $C_2CuN_2$; CuCN, $Cu(CN)_2$
*Synonyms: cuprous cyanide:* Cianuro de cobre (Spanish); Copper(1+) cyanide; Copper(I) cyanide; Cupricin
*cupric cyanide:* Copper(II) cyanide; Copper cynanamide; Cyanure de cuivre (French)
*Producers:* Degussa (Germany); DSM (Netherlands); DuPont (USA); Fluorochem (UK); ICI Group (UK); Philbro-Tech (USA); Univertical (USA)
*Chemical Class:* Inorganic; Metals
*RTECS Number:* GL7150000 (I); GL7175000 (II)
*EINECS Number:* 208-883-6
*Uses:* Copper Cyanide is also used in electroplating copper on iron, and as a catalyst.

*Regulatory Authority:*
- Air Pollutant Standard Set (ACGIH)[1] (DFG)[3] (HSE)[33] (former USSR)[43] (OSHA)[58]
- AB 2588-Air Toxics "Hot Spots" Chemicals (CAL) as cyanide compounds
- CAL Air Resources Board/AB 1807 Toxic Air Contaminants as cyanide compounds

*as copper(I) cyanide:*
- Clean Air Act: Hazardous Air Pollutants (Title I, Part A, Section 112) as cyanide compounds
- Clean Water Act: Toxic Pollutant (Section 401.15)
- EPA Hazardous Waste Number (RCRA No.): P029
- RCRA, 40CFR261, Appendix 8 Hazardous Constituents.
- Superfund/EPCRA 40CFR302.4 RQ: CERCLA, 10 lb (4.54 kg)
- RCRA 40CFR264, Appendix 9; TSD Facilities Ground Water Monitoring List. Suggested test method(s) (PQL ug/L): 6010(60); 7210(200) *Note:* All species in the ground water that contain copper are included.
- EPCRA Section 313 Form R *de minimus* concentration reporting level: 1.0% (copper)
- EPCRA Section 313 Form R *de minimus* concentration reporting level: 1.0% (cyanide)
- U.S. DOT Regulated Marine Pollutant (49CFR172.101, Appendix B)
- Canada, WHMIS, Ingredients Disclosure List; National Pollutant Release Inventory (NPRI); CEPA Priority Substance List, Ocean dumping prohibited, as copper compounds, n.o.s; Drinking Water Quality = 0.2 mg (CN)/L MAC as cyanide compounds.

*Description:* Copper(I) cyanide is a white crystalline substance. Cupric cyanide, $Cu(CN)_2$ is a yellowish-green powder which decomposes on heating. Melting/Freezing point = 473°C (in nitrogen). Hazard Identification (based on NFPA-704 M Rating System): Health 4, Flammability 0, Reactivity 0. Insoluble in water.

*Incompatibilities:* Contact with heat forms deadly hydrogen cyanide gas. May form hydrogen cyanide with water. Keep away from acetylene gas and chemically active metals such as potassium., sodium, magnesium, and zinc, strong oxidizers (chlorine, fluorine, peroxides, etc).

*Permissible Exposure Limits in Air:* The legal airborne permissible exposure limit (OSHA) PEL[58] and the ACGIH[1] recommended TLV is 1 mg/m$^3$ as copper dust and mist. The NIOSH[2] IDLH is 100 mg (Cu)/m$^3$. The limit set by the former USSR-UNEP/IRPTC project[43] is 0.3 mg/m$^3$ as a MAC in workplace air and 0.009 mg/m$^3$ as a momentary value in ambient air of residential areas; the daily average MAC allowable in residential areas is 0.004 mg/m$^3$.

*Determination in Air:* Collection by a filter and bubbler followed by measurement with an ion-specific electrode. See NIOSH Method #7904[18].

***Permissible Concentration in Water:*** The permissible concentration for copper set by USEPA to protect human health is 1 mg/Liter. The U.S. EPA has set a maximum contaminant level of cyanide in drinking water of 0.2 milligrams cyanide per liter of water (0.2 mg/L)[6]. The Canadian MAC is the same. The former USSR-UNEP/IRPTC project[43] has set a MAC of 0.1 mg/Liter in water bodies used for domestic purposes and 0.05 mg/Liter in water bodies used for fishery purposes.

***Determination in Water:*** Cyanide may be determined titrimetrically by EPA Methods 335.2 and 9010 which give total cyanide.

***Routes of Entry:*** Inhalation, ingestion.

***Harmful Effects and Symptoms***

***Short Term Exposure:*** Copper Cyanide can affect you when breathed in. Eye contact can cause severe burns with loss of vision. Skin contact can cause irritation or burns. Breathing Copper Cyanide causes irritation of respiratory tract, and may cause nose bleeds or sores in the nose. Delayed pulmonary edema may occur after inhalation.

***Long Term Exposure:*** Repeated exposure can cause copper to deposit in the liver and other body organs, causing damage, atrophy of the inner lining of the nose, with a watery discharge. Metallic taste may also occur. Repeated skin exposure can cause skin allergy and possibly a green discoloration of the skin and hair. May be able to affect the lungs.

***Points of Attack:*** Skin, lungs, possibly other body organs.

***Medical Surveillance:*** Medical observation is recommended for 24 to 48 hours after breathing overexposure, as pulmonary edema may be delayed. For those with frequent or potentially high exposure (half the TLV or greater), the following are recommended before beginning work and at regular times after that: Lung function tests. If symptoms develop or overexposure is suspected, the following may be useful. Urine copper test.

***First Aid:*** If this chemical gets into the eyes, remove any contact lenses at once and irrigate immediately for at least 15 minutes, occasionally lifting upper and lower lids. Seek medical attention immediately. If this chemical contacts the skin, remove contaminated clothing and wash immediately with soap and water. Seek medical attention immediately. If this chemical has been inhaled, remove from exposure, begin rescue breathing (using universal precautions) if breathing has stopped, and CPR if heart action has stopped. Transfer promptly to a medical facility. When this chemical has been swallowed, get medical attention. Give large quantities of water and induce vomiting. Do not make an unconscious person vomit.

***References:***
- New Jersey Department of Health and Senior Services and Senior Services, "Hazardous Substance Fact Sheet, Cupric Cyanide," Trenton, NJ (October 1994, rev. February 2001). http://www.state.nj.us/health/eoh/rtkweb/0533.pdf
- California Environmental Protection Agency "Chemical List of Lists," Sacramento CA (February 1997).

## Copper Hydroxide

***Use Type:*** Fungicide, nematicide and microbiocide
***CAS Number:*** 20427-59-2
***Formula:*** $Cu(OH)_2$
***Synonyms:*** Copper dihydroxide; Copper hydrate; Copper(II) hydroxide; Copper oxide hydrated; Cupravit blau (German); Cupravit blue; Cupric hydroxide; Kuprablau (German)
***Trade Names:*** CHAMPION®, Nufarm (Australia); CHILTERN KOCIDE® 101, Griffin (USA); COMAC PARASOL®; CRISCOBRE®; KOCIDE® 101, Griffin (USA); KOZINC®, Micro-Flo (USA); MEFENOXAM/COPPER®, Syngenta Crop Protection (Switzerland); NU-COP®, Micro-Flo (USA); PARASOL®; SPIN-OUT®, Griffin (USA)
***Producers:*** American Chemet; Cuproquim (Mexico); Drexel Chemical (USA); Feasy & Besthoff (USA); Gowan (USA); Griffin (USA); Hebei Huafeng Chemical Group (China); Micro-Flo (USA); Nufarm (Australia); Philbro-Tech (USA); Syngenta Crop Protection (Switzerland); William Blythe (UK)
***Chemical Class:*** Inorganic copper
***EPA/OPP PC Code:*** 023401
***California DPR Chemical Code:*** 151
***EINECS Number:*** 243-815-9
***U.S. Maximum Allowable Residue Levels for Copper Hydroxide [40 CFR 180.1001(b); 180.102]:***

| CROP | ppm |
|---|---|
| Raw agricultural commodities | – |

***Carcinogen/Hazard Classifications***
**Label Signal Word:** WARNING, CAUTION or DANGER
**WHO Acute Hazard:** Class III, slightly hazardous
***Regulatory Authority:***
- Actively registered pesticide in California.
- AB 2588-Air Toxics "Hot Spots" Chemicals (CAL)
- The "Director's List" (CAL/OSHA)
- Clean Water Act: Toxic Pollutant (Section 401.15) as copper and compounds
- RCRA Ground Water Monitoring List. Suggested test method(s) (PQL ug/L): 6010(60); 7210(200) Note: All species in the ground water that contain copper are included.
- EPCRA (Section 313): Includes any unique chemical substance that contains copper as part of that chemical's infrastructure.
- Form R *de minimis* concentration reporting level: 1.0%

*Description:* Blue, gelatinous or amorphous powder. Insoluble in water. Molecular weight = 97.56. Density 3.368. Melting/Freezing point (decomposes).

*Incompatibilities:* Incompatible with calcium (metal hydroxides), nitroethane, nitromethane, 1-nitropropane, zirconium.

*Harmful Effects and Symptoms*

*Short Term Exposure:* Contact with eyes or skin may cause irritation or injury. Inhalation should be avoided; use NIOSH-approved air purifying respirators for pesticides. May be harmful if swallowed.

*First Aid:* If this chemical gets into the eyes, remove any contact lenses at once and irrigate immediately for at least 15 minutes, occasionally lifting upper and lower lids. Seek medical attention immediately. If this chemical contacts the skin, remove contaminated clothing and wash immediately with soap and water. Seek medical attention immediately. If this chemical has been inhaled, remove from exposure, begin rescue breathing (using universal precautions) if breathing has stopped, and CPR if heart action has stopped. Transfer promptly to a medical facility. When this chemical has been swallowed, get medical attention. If victim is conscious and able to swallow, have victim drink 4 to 8 ounces of water. Do not induce vomiting. *Note to physician or authorized medical personnel*: Medical observation is recommended for 24 to 48 hours after breathing overexposure, as pulmonary edema may be delayed. As first aid for pulmonary edema, consider administering a corticosteroid spray. Cigarette smoking may exacerbate pulmonary injury and should be discouraged for at least 72 hours following exposure.

*References:*
- California Environmental Protection Agency "Chemical List of Lists," Sacramento CA (February 1997)
- U.S. Environmental Protection Agency, Office of Pesticide Programs, Pesticide Residue Limits, " ", (40 CFR 180.1001(b); 180.102).
www.epa.gov/pesticides/food/viewtols.htm

# Copper(II)-8-hydroxyquinoline

*Use Type:* Fungicide and microbiocide
*CAS Number:* 10380-28-6
*Formula:* $C_{18}H_{12}CuN_2O_2$
*Synonyms:* Bis(8-quinolinato)copper; Bis(8-quinolinolato)copper; Bis(8-quinolinolato-N1,O8)-copper; Bis(8-oxyquinoline)copper; Copper-8; Copper hydroxyquinolate; Copper-8-hydroxyquinolate; Copper-8-hydroxyquinolinate; Copper-8-hydroxyquinoline; Copper oxinate; Copper(II) oxinate; Copper oxine; Copper oxyquinolate; Copper oxyquinoline; Copper quinolate; Copper-8-quinolate; Copper-8-quinolinol; Copper quinolinolate; Copper-8-quinolinolate; Cupric-8-hydroxyquinolate; Cupric-8-quinolinolate; 8-Hydroxyquinoline copper complex; Oxime-copper; Oxine-copper; Oxine-Cu; Oxine cuivre (French); Oxyquinolinoleate de cuivre (French)

*Trade Names:* BIOQUIN®; BIOQUIN®-1; BLUE CONTROL®; CELLU-QUIN®; CHAMPMAN PQ-8; CUNILATE®, Rohm and Haas (USA); CUNILATE®-2472, Rohm and Haas (USA); DOKIRIN®; FRUITDO®; MILMER®; NYTEK®, Rohm and Haas (USA); QUINONDO®

*Producers:* Rohm and Haas (USA)
*Chemical Class:* Inorganic copper compound
*EPA/OPP PC Code:* 024002
*California DPR Chemical Code:* 159
*ICSC Number:* 0756
*RTECS Number:* VC5250000
*Carcinogen/Hazard Classifications*
**IARC:** Group 3, unclassifiable
**Label Signal Word:** CAUTION, WARNING or DANGER
**WHO Acute Hazard:** Class U, Unlikely to be hazardous
*Regulatory Authority:*
- AB 2588-Air Toxics "Hot Spots" Chemicals (CAL) as copper compounds
- The "Director's List" (CAL/OSHA) as copper compounds
- Clean Water Act: Toxic Pollutant (Section 401.15) as copper and compounds.
- RCRA Ground Water Monitoring List. Suggested test method(s) (PQL ug/L): 6010(60); 7210(200) Note: All species in the ground water that contain copper are included.
- EPCRA (Section 313): Includes any unique chemical substance that contains copper as part of that chemical's infrastructure.
- Form R *de minimis* concentration reporting level: 1.0%

*Description:* Yellow-green crystalline solid or powder. Odorless. Insoluble in water. Molecular weight = 351.85.

*Incompatibilities:* May form highly unstable acetylides. Decomposes on burning producing toxic and corrosive fumes including copper and nitrogen oxides.

*Permissible Exposure Limits in Air:* As copper: ACGIH TLV 0.2 mg/m$^3$ (fume); 1 mg/m$^3$ (dusts and mists); NIOSH/OSHA 0.1 mg/m$^3$ (fume); 1 mg/m$^3$ as (dusts and mists).

*Determination in Air:* Copper dusts and mists are collected on a filter, worked up with acid, measured by atomic absorption. See NIOSH Method #7029 for copper. For copper fume: filter collection, acid digestion, measurement by atomic absorption.

*Permissible Concentration in Water:* No criteria set. Runoff from spills or fire control may cause water pollution. The U.S. EPA (6) has set a maximum of 1.0 mg/L in water to protect human health.

*Determination in Water:* Total copper may be determined by digestion followed by atomic absorption or by colorimetry (using neocuproine) or by Inductively Coupled

Plasma Optical Emission Spectrometry. Dissolved Copper may be determined by 0.45 $\mu$ filtration followed by the preceding methods.

*Harmful Effects and Symptoms*

*Long Term Exposure:* Repeated or prolonged inhalation exposure may cause asthma. May cause skin allergy and thickening of the skin; copper deposits can cause discoloration in the skin and hair, leaving a green color. Repeated exposure can cause shrinking of the lining of the inner nose with watery discharge, liver damage. Individuals with Wilson's disease absorb, retain, and store copper excessively. May cause mutations in humans. May damage the testes and decrease fertility in both males and females.

*Points of Attack:* Skin, reproductive system., liver

*Medical Surveillance:* Serum and urine copper level. Liver and kidney tests. Examination by a qualified allergist. More than light alcohol consumption may exacerbate the liver damage caused by copper sulfate.

*First Aid:* If this chemical gets into the eyes, remove any contact lenses at once and irrigate immediately for at least 15 minutes, occasionally lifting upper and lower lids. Seek medical attention immediately. If this chemical contacts the skin, remove contaminated clothing and wash immediately with soap and water. Seek medical attention immediately. If this chemical has been inhaled, remove from exposure, begin rescue breathing (using universal precautions) if breathing has stopped, and CPR if heart action has stopped. Transfer promptly to a medical facility. When this chemical has been swallowed, get medical attention. Give large quantities of water and induce vomiting. Do not make an unconscious person vomit.

*References:*
- California Environmental Protection Agency "Chemical List of Lists," Sacramento CA (February 1997)
- U.S. Environmental Protection Agency, Toxicology of Metals, Vol. II: Copper, Report EPA-600/1-77-022, pp 206-221, Research Triangle Park, NC (May 1977).
- U.S. Environmental Protection Agency, Copper: Ambient Water Quality Criteria, Washington, DC (1980).U.S.

# Copper Napthenate

*Use Type:* Fungicide, insecticide, dog and cat repellant and wood preservative

*CAS Number:* 1338-02-9

*Formula:* Mixture

*Synonyms:* CNC; Copper uversol; Naphthenic acids, copper salts

*Trade Names:* CHAPCO® Cu-NAP; CUNAPSOL®; CUPRINOL®; TROYSAN® COPPER 8%; TROYSAN® COPPER 11.5%; WILTZ®-65; WITTOX®-C

*Producers:* Akzo Nobel (Netherlands)

*Chemical Class:* Inorganic copper compound

*EPA/OPP PC Code:* 023102; (006000 and 006300 are old EPA code numbers)

*California DPR Chemical Code:* 153

*ICSC Number:* 0303

*RTECS Number:* QK9100000

*EEC Number:* 029-003-00-5

*Carcinogen/Hazard Classifications*

**Label Signal Word:** CAUTION or WARNING

*Regulatory Authority:*
- Actively registered pesticide in California.
- AB 2588-Air Toxics "Hot Spots" Chemicals (CAL) (as copper compounds)
- The "Director's List" (CAL/OSHA) (as copper compounds)
- Clean Water Act: Toxic Pollutant (Section 401.15) as copper and compounds.
- RCRA Ground Water Monitoring List. Suggested test method(s) (PQL ug/L): 6010(60); 7210(200) Note: All species in the ground water that contain copper are included.
- EPCRA (Section 313): Includes any unique chemical substance that contains copper as part of that chemical's infrastructure. This category does not include copper phthalocyanide compounds that are substituted with only hydrogen, and/or chlorine, and/or bromine.
- Form R *de minimis* concentration reporting level: 1.0%

*Description:* Dark green, thick liquid or blue-green solid. Generally used only as a solution, usually in oils or mineral spirits. Gasoline-like odor (liquid). Insoluble in water. Density = 1.055. Boiling point = 154-202°C. Flash point (typical) = 38°C (cc). Hazard Identification (based on NFPA-704 M Rating System): Health Hazards (Blue): 0; Flammability (Red): 2; Reactivity (Yellow): 0

*Incompatibilities:* Reaction with strong oxidizers may be violent.

*Permissible Exposure Limits in Air:* As copper: ACGIH TLV 0.2 mg/m$^3$ (fume); 1 mg/m$^3$ (dusts and mists); NIOSH/OSHA 0.1 mg/m$^3$ (fume); 1 mg/m$^3$ as (dusts and mists).

*Determination in Air:* Copper dusts and mists are collected on a filter, worked up with acid, measured by atomic absorption. See NIOSH Method #7029 for copper. For copper fume: filter collection, acid digestion, measurement by atomic absorption.

*Permissible Concentration in Water:* No criteria set. Runoff from spills or fire control may cause water pollution. The U.S. EPA [6] has set a maximum of 1.0 mg/L in water to protect human health.

*Determination in Water:* Total copper may be determined by digestion followed by atomic absorption or by colorimetry (using neocuproine) or by Inductively Coupled Plasma Optical Emission Spectrometry. Dissolved Copper may be determined by 0.45 $\mu$ filtration followed by the preceding methods.

### Harmful Effects and Symptoms

**Short Term Exposure:** Grade 1; oral rat $LD_{50}$ = 4-6 g/kg.

**Long Term Exposure:** May cause mutations in humans. May damage the testes and decrease fertility in both males and females. May cause skin allergy and thickening of the skin; copper deposits can cause discoloration in the skin and hair, leaving a green color. Repeated exposure can cause shrinking of the lining of the inner nose with watery discharge, liver damage. Individuals with Wilson's disease absorb, retain, and store copper excessively.

**Points of Attack:** Skin, reproductive system., liver

**Medical Surveillance:** Serum and urine copper level. Liver and kidney tests. Examination by a qualified allergist. More than light alcohol consumption may exacerbate the liver damage caused by copper sulfate.

**First Aid:** If this chemical gets into the eyes, remove any contact lenses at once and irrigate immediately for at least 15 minutes, occasionally lifting upper and lower lids. Seek medical attention immediately. If this chemical contacts the skin, remove contaminated clothing and wash immediately with soap and water. Seek medical attention immediately. If this chemical has been inhaled, remove from exposure, begin rescue breathing (using universal precautions) if breathing has stopped, and CPR if heart action has stopped. Transfer promptly to a medical facility. When this chemical has been swallowed, get medical attention. Give large quantities of water and induce vomiting. Do not make an unconscious person vomit. Do not induce vomiting when formulations containing petroleum solvents are ingested.

**References:**
- California Environmental Protection Agency "Chemical List of Lists," Sacramento CA (February 1997)
- U.S. Environmental Protection Agency, Toxicology of Metals, Vol. II: Copper, Report EPA-600/1-77-022, pp 206-221, Research Triangle Park, NC (May 1977).
- U.S. Environmental Protection Agency, Copper: ambient Water Quality Criteria, Washington, DC (1980).

## Copper Octanoate

**Use Type:** Fungicide
**CAS Number:** 20543-04-8
**Synonyms:** Octanoic acid, copper salt
**Trade Names:** CONCERN®; NEU® 1140F, W. Neudorff GmbH (Germany)
**Producers:** W. Neudorff GmbH (Germany)
**Chemical Class:** Inorganic copper
**EPA/OPP PC Code:** 023306
**California DPR Chemical Code:** 5225
**Uses:** A cooper soap fungicide that controls many common diseases on vegetables, fruits and ornamentals, both indoors and outdoors. Treats mildew, powdery mildew, fruit rots, white rust, blue mold, downy mildew and anthracnose. Registered on beans, peas, beets, broccoli, Brussels sprouts, cantaloupes, cucumbers, pumpkins, squash, carrot, celeriac, celery chicory, chive, corn, currant, gooseberry, eggplant, pepper, tomato, endive, lettuce, garlic, leek, onion, shallots, ginseng, grape, hop, kale, kohlrabi, potato, quince, spinach, chard, strawberry, sunflowers, and turnip

**Carcinogen/Hazard Classifications**
**Label Signal Word:** CAUTION
**Regulatory Authority:**
- AB 2588-Air Toxics "Hot Spots" Chemicals (CAL)
- The "Director's List" (CAL/OSHA)
- Clean Water Act: Toxic Pollutant (Section 401.15) as copper and compounds.
- RCRA Ground Water Monitoring List. Suggested test method(s) (PQL $ug/L$): 6010(60); 7210(200) Note: All species in the ground water that contain copper are included.
- EPCRA (Section 313): Includes any unique chemical substance that contains copper as part of that chemical's infrastructure.
- Form R *de minimis* concentration reporting level: 1.0%

### Harmful Effects and Symptoms
**Short Term Exposure:** Contact with eyes or skin may cause irritation or injury. Inhalation should be avoided; use NIOSH-approved air purifying respirators for pesticides. May be harmful if swallowed.

**First Aid:** If this chemical gets into the eyes, remove any contact lenses at once and irrigate immediately for at least 15 minutes, occasionally lifting upper and lower lids. Seek medical attention immediately. If this chemical contacts the skin, remove contaminated clothing and wash immediately with soap and water. Seek medical attention immediately. If this chemical has been inhaled, remove from exposure, begin rescue breathing (using universal precautions) if breathing has stopped, and CPR if heart action has stopped. Transfer promptly to a medical facility. When this chemical has been swallowed, get medical attention. Give large quantities of water and induce vomiting. Do not make an unconscious person vomit.

**References:**
- Pesticide Management Education Program, "Copper Octanoate Pesticide Tolerance Petition Filing 1/97," Cornell University, Ithaca, NY (January 1997). http://pmep.cce.cornell.edu/profiles/fung-nemat/aceticacid-etridiazole/copper-octanoate/copper-octanoate-tol.html
- California Environmental Protection Agency "Chemical List of Lists," Sacramento CA (February 1997).

## Copper Oxychloride

**Use Type:** Fungicide
**CAS Number:** 1332-40-7; 1332-65-6
**Formula:** $Cl_2Cu_4H_6O_6$

*Synonyms:* Basic copper chloride; Basic cupric chloride; Blue copper; Copper chloride, basic; Copper chloride, mixed with copper oxide, hydrate; Copper chloride oxide; Copper chloride oxide, hydrate; Copper chloroxide; Copper OC fungicide; Copper oxychloride; Cupric oxide chloride; Kupferoxychlorid (German); Oxychlorue de cuivre (French)
*Trade Names:* AGRIZAN®; BASF® GRUNKUPFER, BASF Group (Germany); BLITOX®; BLITOX® 50; BLUE COPPER-50®; CHEMOCIN®; CHEMPAR®; COBOX®; COBOX BLUE®; COLLOIDOX®; COPPERSAN®; COPPESAN®; COPPESAN BLUE®; COPRANTOL®; COPREX®; COPROSAN BLUE®; COP-TOX®; COXYSAN®; CU-56®; CUPRAL 45®; CUPRAMAR®; CUPRAMER®; CUPRANTOL®; CUPRAVET®; CUPRAVIT®; CUPRAVIT® FORTE; CUPRAVIT GREEN®; CUPRICOL®; CUPRITOX®; CUPROKYLT®; CUPROL®; CUPROSAN®; CUPROSANA®; CUPROSAN BLUE®; CUPROVINOL; CUPROX®; CUPROXOL®; DEVICOPPER®; FALIGRUEN®; FYCOL 8®; FYTOLAN®; KAURITIL®; KILEX®; KT 35®; KUPRICOL®; KUPRIKOL®; MICROCOP®; MIEDZIAN®; MIEDZIAN 50®; NEORAM BLU®; OXICOB®; OXIVOR®; OXYCLOR®; OXYCUR®; PARRYCOP®; PEPROSAN; RECOP®; RHODIACUIVRE®; SUTOX®, Sudarshan Chemical Industries (India); TAMRAGHOL®; TRICOP 50®; VIRICUIVRE®; VITIGRAN®; VITIGRAN BLUE®
*Producers:* Agrimor International (USA); BASF Group (Germany); Drexel Chemical (USA); Griffin (USA); Hindustan Insecticides (India); Micro-Flo (USA); Nufarm (Australia); Philbro-Tech (USA); Sudarshan Chemical Industries (India)
*Chemical Class:* Inorganic copper
*EPA/OPP PC Code:* 008001
*California DPR Chemical Code:* 156
*Uses:* Used to control fungi on beets, fruit crops, grapes, olive trees, potatoes, vegetables, tomatoes, ornamental plants and many more. Used as a bird repellant.
*U.S. Maximum Allowable Residue Levels for Copper Oxychloride ($Cu_2Cl(OH)_3$) [40 CFR 180.1001 (b), 180.1021]:*

| CROP | ppm | CFR |
|---|---|---|
| Raw agricultural commodities | — | 180.1001(b) |
| Raw agricultural commodities | — | 180.1021 |

*Carcinogen/Hazard Classifications*
**Label Signal Word:** WARNING, CAUTION or DANGER
**WHO Acute Hazard:** Class III, slightly hazardous
*Regulatory Authority:*
- AB 2588-Air Toxics "Hot Spots" Chemicals (CAL)
- The "Director's List" (CAL/OSHA)
- Clean Water Act: Toxic Pollutant (Section 401.15) as copper and compounds.
- RCRA Ground Water Monitoring List. Suggested test method(s) (PQL $ug/L$): 6010(60); 7210(200) Note: All species in the ground water that contain copper are included.
- EPCRA (Section 313): Includes any unique chemical substance that contains copper as part of that chemical's infrastructure.
- Form R *de minimus* concentration reporting level: 1.0%

*Description:* Bluish-green, odorless powder. Soluble in acids, ammonia; insoluble in water. Molecular weight = 427.12.
*Incompatibilities:* When heated to decomposition or on contact with acids or acid fumes, may produce highly toxic chloride fumes; deadly phosgene gas may be formed. May cause pitting of some metals.
*Permissible Concentration in Water:* No criteria set. Runoff from spills or fire control may cause water pollution.
*Harmful Effects and Symptoms*
*Short Term Exposure:* Contact with eyes or skin may cause irritation or injury. Inhalation should be avoided; use NIOSH-approved air purifying respirators for pesticides. May be harmful if swallowed.
*First Aid:* If this chemical gets into the eyes, remove any contact lenses at once and irrigate immediately for at least 15 minutes, occasionally lifting upper and lower lids. Seek medical attention immediately. If this chemical contacts the skin, remove contaminated clothing and wash immediately with soap and water. Seek medical attention immediately. If this chemical has been inhaled, remove from exposure, begin rescue breathing (using universal precautions) if breathing has stopped, and CPR if heart action has stopped. Transfer promptly to a medical facility. When this chemical has been swallowed, get medical attention. Give large quantities of water and induce vomiting. Do not make an unconscious person vomit.
*References:*
- California Environmental Protection Agency "Chemical List of Lists," Sacramento CA (February 1997)
- U.S. Environmental Protection Agency, Office of Pesticide Programs, Pesticide Residue Limits, "Copper Oxychloride," 40 CFR 180.1001 (b), 180.1021. www.epa.gov/pesticides/food/viewtols.htm

# Copper Sulfate

*Use Type:* Fungicide, algaecide, and molluscicide
*CAS Number:* 7758-98-7; 7758-99-8 (pentahydrate)
*Formula:* $CuO_4S$, $CuSO_4$
*Synonyms:* Blue copper; Blue stone; Blue vitriol; Copper monosulfate; Copper sulfate (1:1); Copper(II) sulfate; Copper(2+) sulfate; Copper(2+) sulfate (1:1); Copper sulfate pentahydrate; Cu basic sulfate; Cupric sulfate anhydrous; Cupric sulphate; Griffin super Cu; Kupfersulfat (German); Phyto-bordeaux; Roman vitriol; Sulfate de cuivre (French); Sulfuric acid, copper(2+) Salt (1:1); Sulfate of copper; Sulfato de cobre (Spanish); Tri-basic copper sulfate

# Copper Sulfate

*Trade Names:* AGRITOX®; BASICOP®, Griffin (USA); BCS COPPER FUNGICIDE®, Pesticide Service Consultants (USA); BSC FLOWABLE®, Newfarm (Australia) canceled 8/25/2000; COPSIN®, Newfarm (Australia); CP BASIC SULFATE®, Phibro-Tech (USA); CUPROFIX®, Cerexagi (France); FUNGI-SPERSE II, Micro Flo (USA) canceled 11/30/1992; SULTRACOB®, Ingenieria Industrial (Mexico); TNCS® 53; TRIANGLE®

*Producers:* Adheswara Group of Companies (India); Ashland Chemical (USA); Bhageria Dye-Chem (India); Boliden (Sweden); Celtic Chemicals (UK); Cerexagri (France); Coogee Chemicals (Australia); Drexel Chemical (USA); Great Western Inorganics (USA); Griffin (USA); Harcros Chemicals (USA); Helena Chemical (USA); Ingenieria Industrial (Mexico); Micro Flo (USA); Newfarm (Australia); Noranda (Canada; Philbro-Tech (USA); Prince Manufacturing (USA); Teck Cominco (Canada); Univertical (USA); World Metal (USA)

*Chemical Class:* Inorganic, a sulfate

*EPA/OPP PC Code:* 024408; 024401 (pentahydrate)

*California DPR Chemical Code:* 1778

*ICSC Number:* 0751

*RTECS Number:* GL8800000

*EINECS Number:* 231-847-6 (Copper II Sulfate)

*Uses:* Copper sulfate is a fungicide used to control bacterial and fungal diseases of fruit, vegetable, nut, and field crops. These diseases include mildew, leaf spots, blights, and apple scab. It is used as a protective fungicide (Bordeaux mixture) for leaf application and seed treatment. It is also used as an algacide and herbicide, and to kill slugs and snails in irrigation and municipal water treatment systems. It has been used to control Dutch elm disease. It is available as a dust, wettable powder, or liquid concentrate. Used as a fungicide and algicide, in veterinary medicine and others. Copper sulfate is also used to detect and to remove trace amounts of water from alcohols and organic compounds.

*Fish toxicity (threshold)[77]:* High–2.85539 ppb, MATC (Maximum Acceptable Toxicant Concentration)

*Carcinogen/Hazard Classifications*

**Label Signal Word:** DANGER–POISON (Toxicity Class I–Highly toxic)

*Regulatory Authority:*
- Air Pollutant Standard Set (Czechoslovakia)[35] (former USSR)[43]
- The "Director's List" (CAL/OSHA)
- Actively registered pesticide in California.
- AB 2588-Air Toxics "Hot Spots" Chemicals (CAL) (as copper compounds)
- Clean water act: Toxic Pollutant (Section 401.15) as copper and compounds.
- RCRA 40CFR264, Appendix 9; TSD Facilities Ground Water Monitoring List. Suggested test method(s) (PQL $ug/L$): 6010(60); 7210(200) Note: All species in the ground water that contain copper are included.
- EPCRA (Section 313): Includes any unique chemical substance that contains copper as part of that chemical's infrastructure. Form R *de minimus* concentration reporting level: 1.0%
- Canada, WHMIS, Ingredients Disclosure List; National Pollutant Release Inventory (NPRI); CEPA Priority Substance List, Ocean dumping prohibited.

*Description:* Copper sulfate is a greenish-white crystalline solid. Boiling point = 650°C (decomposes to CuO). Melting/Freezing point = decomposes slightly above 200°C. Hazard Identification (based on NFPA-704 M Rating System): Health 2, Flammability 0, Reactivity 0. Highly soluble in water; forms bright blue solution.

*Incompatibilities:* Aqueous solution is an acid. May form explosive materials on contact with acetylene and nitromethane. Incompatible with strong bases, hydroxylamine, magnesium; zirconium., sodium hpobromite, hydrazine.

*Permissible Exposure Limits in Air:* The Federal standard (OSHA PEL 8-hour TWA)[58] for copper fume is 0.1 mg/m$^3$, and 1 mg/m$^3$ for copper dusts and mists. NIOSH recommends the same level for a 10-hour workshift. ACGIH[1] recommends a TWA of 0.2 mg/m$^3$ for copper fume and 1 mg/m$^3$ for dusts and mists. The former USSR-UNEP/IRPTC project[43] has set limits in the ambient air of residential areas of 0.009 mg/m$^3$ on a momentary basis and 0.004 mg/m$^3$ on a daily average basis. The NIOSH[2] IDLH is 100 mg (Cu)/m$^3$. Czechoslovakia[35] has set a MAC of 0.1 mg/m$^3$ on a daily average basis; a MAC of 0.3 mg/m$^3$ on a 30-minute basis.

*Determination in Air:* Copper dusts and mists are collected on a filter, worked up with acid, measured by atomic absorption. See NIOSH Method #7029 for copper[18]. For copper fume: filter collection, acid digestion, measurement by atomic absorption.

*Permissible Concentration in Water:* The former USSR-UNEP/IRPTC joint project[43] has set a MAC in water used for fishery purposes of 0.004 mg/L (0.001 as Cu). The U.S. EPA[6] has set a maximum of 1.0 mg/L in water to protect human health.

*Routes of Entry:* Inhalation, ingestion.

*Harmful Effects and Symptoms*

*Short Term Exposure:* Inhalation: May cause irritation to nose, throat and lungs causing coughing and wheezing. Skin : May cause irritation of skin, localized coloration, itching and burns. *Eyes:* May cause severe irritation, inflammation, burns, excessive tissue fluid and a cloudy cornea; possible permanent damage. *Ingestion:* Poisonous if swallowed. May cause burning and metallic taste in mouth, blue skin coloration, intense inflammation of the stomach and intestines, abdominal pain, vomiting, diarrhea, blood in feces, headache, cold sweat, weak pulse, salivation, nausea, dehydration, low blood pressure, jaundice, and kidney failure. Death may result from a dose of a little as 1 teaspoon for a 150 pound person.

***Long Term Exposure:*** May cause mutations in humans. May damage the testes and decrease fertility in both males and females. May cause skin allergy and thickening of the skin; copper deposits can cause discoloration in the skin and hair, leaving a green color. Repeated exposure can cause shrinking of the lining of the inner nose with watery discharge, liver damage. Individuals with Wilson's disease absorb, retain, and store copper excessively.

***Points of Attack:*** Skin, reproductive system., liver

***Medical Surveillance:*** Serum and urine copper level. Liver and kidney tests. Examination by a qualified allergist. More than light alcohol consumption may exacerbate the liver damage caused by copper sulfate.

***First Aid:*** If this chemical gets into the eyes, remove any contact lenses at once and irrigate immediately for at least 15 minutes, occasionally lifting upper and lower lids. Seek medical attention immediately. If this chemical contacts the skin, remove contaminated clothing and wash immediately with soap and water. Seek medical attention immediately. If this chemical has been inhaled, remove from exposure, begin rescue breathing (using universal precautions) if breathing has stopped, and CPR if heart action has stopped. Transfer promptly to a medical facility. When this chemical has been swallowed, get medical attention. Give large quantities of water and induce vomiting. Do not make an unconscious person vomit.

*Note to physician or authorized medical personnel:* Empty stomach by lavage with 0.1% solution of potassium ferrocyanide or milk. Liver or kidney function tests may be indicated. May result in methaemoglobinemia.

***References:***
- EXTOXNET, Extension Toxicology Network, "Pesticide Information Profile,Copper Sulfate," Oregon State University, Corvallis, OR (June 1996) ace.orst.edu/info/extoxnet/pips/coppersu
- New York State Department of Health, "Chemical Fact Sheet: Copper Sulfate," Albany, NY, Bureau of Toxic Substance Assessment (January 1986 and Version 3).
- California Environmental Protection Agency "Chemical List of Lists," Sacramento CA (February 1997).
- New Jersey Department of Health and Senior Services and Senior Services, "Hazardous Substance Fact Sheet: Cupric Sulfate," Trenton, NJ (January 1999). www.state.nj.us/health/eoh/rtkweb/0549.pdf

# Coumafuryl

***Use Type:*** Rodenticide
***CAS Number:*** 117-52-2
***Formula:*** $C_{17}H_{14}O_5$

***Alert:*** As with many rodenticides used in domestic situations, this substance may severely harm pets, squirrels and otherwise harmless species.

***Synonyms:*** 3-($\alpha$-Acetonylfurfuryl)-4-hydroxycoumarin; Cumafuryl (German); 3-($\alpha$-Furyl-$\beta$-acetylaethyl)-4-hydroxycumarin (German); 3-(1-Furyl-3-acetylethyl)-4-hydroxycoumarin

***Trade Names:*** FOUMARIN®; FUMARIN®, Bayer CropScience (Germany) canceled 4/1/1987; KRUMKIL®; RATAFIN®, HGP Inc. (USA) canceled 4/26/1988; RAT-A-WAY®; TOMARIN®

***Producers:*** Bayer CropScience (Germany); Sigma-Aldrich Laborchemikalien (Germany)

***Chemical Class:*** Coumarin

***EPA/OPP PC Code:*** 086001

***California DPR Chemical Code:*** 298

***RTECS Number:*** GN4850000

***Uses:*** This material is an anticoagulant rodenticide. It is used for rodent control in landscapes, right-of-ways and structures. It is also used in almond groves.

***Carcinogen/Hazard Classifications***
Label Signal Word: DANGER

***Regulatory Authority:***
- Extremely Hazardous Substance (EPA-SARA) (TPQ = 10,000)[7] (Dropped from listing in 1988)

***Description:*** Coumafuryl is a colorless, white, crystalline solid. Melting/Freezing point = 124°C. Insoluble in water.

***Incompatibilities:*** Strong oxidizers may cause a fire and explosive hazard.

***Permissible Exposure Limits in Air:*** No standards set

***Permissible Concentration in Water:*** No criteria set, but runoff from spills or fire control may cause water pollution.

***Routes of Entry:*** Ingestion, skin contact.

***Harmful Effects and Symptoms***

***Short Term Exposure:*** Coumafuryl is very similar to warfarin, a hemorrhagic agent. Inhalation may cause symptoms described in long term exposure. With a single large ingested dose may cause hemorrhagic shock. The LD50-oral (mouse) is 14.7 mg/kg (highly toxic). This chemical can be absorbed through the skin, thereby increasing exposure or hemorrhagic effect. High exposure can cause death.

***Long Term Exposure:*** Chronic exposure may cause death by hemorrhagic shock. Absorption by the lungs or after a few days or few weeks of repeated ingestion, may cause inhibition of prothrombin synthesis, nose bleeds and bleeding gums, hematoma, small reddish spots like a rash, bruises of the elbows, knees and buttocks, blood in urine and stools, anemia, occasional paralysis due to a stroke.

***First Aid:*** Speed in removing material from eyes and skin is of extreme importance. If this chemical gets into the eyes, remove any contact lenses at once and irrigate immediately for at least 30 minutes, occasionally lifting upper and lower lids. Seek medical attention immediately. If this chemical contacts the skin, remove contaminated clothing and wash immediately with soap and water. Seek medical attention immediately. If this chemical has been inhaled, remove from exposure, begin rescue breathing (using universal precautions) if breathing has stopped, and CPR if heart action has stopped. Transfer promptly to a medical facility. When this chemical has been swallowed, get medical attention. Give large quantities of water and induce vomiting. Do not make an unconscious person vomit. Remove and isolate contaminated clothing and shoes at the site. Keep victim quiet and maintain normal body temperature. Effects may be delayed. Keep victim under observation.

*References:*
- U.S. Environmental Protection Agency, "Chemical Profile: Coumafuryl," Washington, DC, Chemical Emergency Preparedness Program (October 31, 1985).
- California Environmental Protection Agency "Chemical List of Lists," Sacramento CA (February 1997).

# Coumaphos

*Use Type:* Used as a feed additive to control larvae of flies that fecal breed around cattle and poultry. Also used as a dip or dust for control of mange, horn flies, face flies, lice, cattle grubs and parasites around cattle, swine, poultry, horses and other farm animals.

*CAS Number:* 56-72-4

*Formula:* $C_{14}H_{16}ClO_5PS$

*Alert:* The U.S. Environmental Protection Agency (U.S. EPA) classifies most formulations of coumaphos as General Use Pesticides (GUPs). The formulations 11.6% EC and 42% flowable concentrate end-use products have been classified as Restricted Use Pesticides (RUPs) because they pose a hazard of acute poisoning from ingestion. Human toxicity (long-term): High.

*Synonyms:* 3-Chloro-7-hydroxy-4-methyl-coumarin $O,O$-diethyl phosphorothioate; 3-Chloro-7-hydroxy-4-methyl-coumarin-$O,O$-diethylphosphorothionate; 3-Chloro-7-hydroxy-4-methyl-coumarin $O$-ester with $O,O$-diethylphosphorothioate; 3-Chloro-4-methyl-7-coumarinyldiethyl phosphorothioate; $O$-3-Chloro-4-methyl-7-coumarinyl $O,O$-Diethyl phosphorothioate; 3-Chloro-4-methyl-7-hydroxycoumarindiethyl thiophosphoric acid ester; 3-Chloro-4-methylumbelliferoneo-ester with $O,O$-diethyl phosphorothioate; Coumafos; Cumafos (Dutch, Spanish); $O,O$-Diaethyl-$O$-(3-chlor-4-methyl-cu marin-7-yl)-monothiophosphat (German); $O,O$-Diethyl-$O$-(3-chloor-4-methyl-cumarin-7-yl)monothiofosfaat (Dutch); $O,O$-Diethylo-(3-chloro-4-methyl-7-coumarinyl) phosphorothioate; $O,O$-Diethylo-(3-chloro-4-methylcoumarinyl-7)thiophosphate; $O,O$-Diethylo-(3-chloro-4-methyl-2-oxo-2H-benzopyran-7-yl)phosphorothioate; $O,O$-Diethyl3-chloro-4-methyl-7-umbelliferone thiophosphate; $O,O$-Diethylo-(3-chloro-4-methylumbelliferyl)phosphorothioate; Diethyl3-chloro-4-methylumbelliferyl thionophosphate; Diethylthiophosphoric acid ester of 3-chloro-4-methyl-7-hydroxycoumarin; $O,O$-Dietil-$O$-(3-cloro-4-metil-cumarin-7-il-monotiofosfato) (Italian); ENT 17,957; NCI-C08662; Phosphorothioic acid, $O$-(3-chloro-4-methyl-2-oxo-2H-1-benzopyran-7-yl) $O,O$-diethyl ester; Phosphorothioic acid, $O,O$-diethyl ester, $O$-ester with 3-chloro-7-hydroxy-4-methylcoumarin; Thiophosphate de $O,O$-diethyle et de $O$-(3-chloro-4-methyl-7-coumarinyle) (French)

*Trade Names:* AGRIDIP®; ASUNTOL®; AZUNTHOL®; BAY® 21/199, Bayer CropScience (Germany); BAYER® 21/199, Bayer CropScience (Germany); BAYMIX®, Bayer CropScience (Germany); BAYMIX® 50, Bayer CropScience (Germany); CHECKMITE®; CO-RAL®, Bayer Healthcare (USA) canceled 3/10/2000; DELICE®; MELDANE®; MELDONE®; MUSCATOX®; NEGASHUNT®; DIOLICE®; RESITOX®; SUNTOL®; UMBETHION®

*Producers:* Bayer CropScience (Germany); Miles (USA); Sigma-Aldrich Laborchemikalien (Germany)

*Chemical Class:* Organophosphate

*EPA/OPP PC Code:* 036501

*California DPR Chemical Code:* 165

*ICSC Number:* 0422

*RTECS Number:* GN6300000

*EEC Number:* 015-038-00-3

*Uses:* Coumaphos is an insecticide/acaricide used to control a wide variety of liver stock insects including cattle grubs, fleeceworms, lice scabies, flies, and ticks; the common ectoparasites of beef cattle, dairy cows, sheep, goats, horse, swine, and poultry as well as for screw worms in all these animals. The USDA uses coumaphos in dip vats along the U.S./Texas border to control ticks that carry Texas Cattle Fever. It is rated Category I, II, or III depending upon the routes of exposure. It is added to cattle and poultry feed to control the development of fly larvae that breed in manure. It has applications in beekeeping.

*Human toxicity (long-term)[77]:* High–1.75 ppb, Health Advisory

*Fish toxicity (threshold)[77]:* Intermediate–16.96524 ppb, MATC (Maximum Acceptable Toxicant Concentration)

*U.S. Maximum Allowable Residue Levels for Coumaphos (40 CFR 180.189):*

| CROP | ppm |
| --- | --- |
| Cattle, fat | 1.0 |
| Cattle, meat | 1.0 |
| Cattle, mbyp | 1.0 |
| Goat, fat | 1.0 |
| Goat, meat | 1.0 |
| Goat, mbyp | 1.0 |
| Hog, fat | 1.0 |
| Hog, meat | 1.0 |
| Hog, mbyp | 1.0 |

| | |
|---|---|
| Honey | 0.10 |
| Honeycomb | 100 |
| Horse, fat | 1.0 |
| Horse, meat | 1.0 |
| Horse, mbyp | 1.0 |
| Milk, fat (= n in whole milk) | 0.5 |
| Sheep, fat | 1.0 |
| Sheep, meat | 1.0 |
| Sheep, mbyp | 1.0 |

***Carcinogen/Hazard Classifications***
**U.S. EPA Carcinogens:** Group E, noncarcinogenic to humans
**Label Signal Word:** WARNING
**WHO Acute Hazard:** Group 1b, highly hazardous
***Regulatory Authority:***
- Clean Water Act: Section 311 Hazardous Substances/RQ 40CFR117.3 (same as CERCLA, see below).
- Superfund/EPCRA 40CFR355, Appendix B Extremely Hazardous Substances: TPQ = 100/10,000 lb (455/4,540 kg)
- Superfund/EPCRA 40CFR302.4 RQ: CERCLA, 10 lb (4.54 kg)
- U.S. DOT Regulated Marine Pollutant (49CFR172.101, Appendix B), severe pollutant
- U.S. DOT Inhalation Hazard Chemicals as organophosphates
- EPA/SARA 302 (EPCRA) Extremely hazardous substances
- The "Director's List" (CAL/OSHA)
- Actively registered pesticide in California.
- Water Pollution Standard Proposed (Mexico)[35]

***Description:*** White to tan crystalline solid. Slight, sulfurous odor. Practically insoluble in water. Melting/Freezing point = 91°C. Odor threshold = 0.02 ppm. Molecular weight = 362.79. Vapor pressure = $1 \times 10^{-7}$ mmHg @ 20°C. Hazard Identification (based on NFPA-704 M Rating System): Health 3, Flammability 1, Reactivity 0. Insoluble in water. Log $K_{ow}$ = 4.12. Values at or above 3.0 are likely to bioaccumulate in marine organisms.

***Incompatibilities:*** Reacts with piperonyl butoxide, oxidizers, strong bases. Slowly reacts with caustics to be hydrolyzed. Keep away from water and heat. When heated to decomposition this chemical may emit toxic nitrogen oxides.

***Permissible Exposure Limits in Air:*** No standards set.

***Determination in Air:*** OSHA versatile sampler-2; Toluene/Acetone; Gas chromatography/Flame photometric detection for sulfur, nitrogen, or phosphorus; NIOSH Method IV Method #5600, Organophosphorus pesticides[18].

***Permissible Concentration in Water:*** Mexico has set a maximum permissible concentration in estuaries of 0.02 mg/Liter.

***Routes of Entry:*** Inhalation, ingestion; this chemical can be absorbed through the skin, thereby increasing exposure.

***Harmful Effects and Symptoms***
***Short Term Exposure:*** Contact may cause burns to skin and eyes. Fatal skin absorption can occur even if there is no feeling of irritation after contact. Cholinesterase inhibitor. Exposure can cause rapid, fatal organophosphate poisoning: with headache, sweating, nausea and vomiting, diarrhea, salivation, abdominal cramps, difficult breathing, stiffness of legs, blurring of vision, followed by loss of muscle coordination, muscle twitching, convulsions, coma, and death. The $LD_{50}$ (oral, rat) = 13 mg/kg (highly toxic). The probable oral lethal dose is 50-500 mg/kg; or between 1 teaspoonful and 1 oz. For a 70 kg (150 lb) person. May be fatal if inhaled, swallowed, or absorbed through skin. The effects may be delayed. Delayed pulmonary edema may occur after inhalation.

***Long Term Exposure:*** High or repeated exposure may cause nerve damage causing weakness, poor coordination in the arms and legs. May cause personality changes, depression, memory loss, or irritability. Cholinesterase inhibitor; cumulative effect is possible. This chemical may damage the nervous system with repeated exposure, resulting in convulsions, respiratory failure. May cause liver damage.

***Points of Attack:*** Respiratory system, central nervous system, peripheral nervous system, plasma cholinesterase

***Medical Surveillance:*** Medical observation is recommended for 24 to 48 hours after breathing overexposure, as pulmonary edema may be delayed. Before employment and at regular times after that, the following are recommended: Plasma and red blood cell cholinesterase levels (tests for the enzyme poisoned by this chemical). If exposure stops, plasma levels return to normal in 1-2 weeks while red blood cell levels may be reduced for 1-3 months. When acetylcholinesterase enzyme levels are reduced by 25% or more below preemployment levels, risk of poisoning is increased, even if results are in lower ranges of "normal." Reassignment to work not involving organophosphate or carbamate pesticides is recommended until enzyme levels recover. If symptoms develop or overexposure occurs, repeat the above tests as soon as possible and get an exam of the nervous system. Also consider complete blood count. Consider chest x-ray following acute overexposure. Do not drink any alcoholic beverages before or during use. Alcohol promotes absorption of organophosphates.

***First Aid:*** **Treatment for organophosphate poisoning consists of thorough decontamination, cardiorespiratory support, and administration of the antidotes atropine and pralidoxime. In cases of severe poisoning, diazepam, an anticonvulsant, should also be administered. Antidotes should be administered as prevention even if the diagnosis is in doubt.** Speed in removing material from eyes and skin is of extreme importance. *Eyes:* Eye contact can cause dangerous amounts of these chemicals to be quickly absorbed through the mucous membrane into the bloodstream. Immediately and gently flush eyes with plenty of warm or cold water (NO hot water) for at least 15 minutes, occasionally lifting the upper and lower eyelids. Get medical aid immediately. *Skin:* Get medical aid. Skin contact can cause dangerous amounts of these chemicals to be absorbed into the bloodstream. Wearing the appropriate PPE equipment and respirator for organophosphate pesticides, immediately flush skin with plenty of soap and water for at least 15 minutes while removing contaminated clothing and shoes. Shampoo hair promptly if contaminated. The removed, contaminated clothing and shoes should be double-bagged and left in Hot Zone for later disposal by hazardous materials experts. Skin may also be decontaminated with diluted hypochlorite solution. *Inhalation:* Get medical aid. Do not contaminate yourself. Wearing the appropriate PPE equipment and respirator for organophosphate pesticides, immediately remove the victim from the contaminated area to fresh air. If the victim is not breathing, administer artificial respiration. Do NOT use mouth-to-mouth resuscitation; use bag/mask apparatus. If breathing is difficult, administer oxygen through bag/mask apparatus until medical help arrives. Do not leave victim unattended. *Ingestion:* Call poison control. Loosen all clothing. Never give anything by mouth to an unconscious person. If victim is *unconscious or having convulsions,* do nothing except keep victim warm. Get medical aid. Transfer promptly to a medical facility. In cases of ingestion, **do not induce vomiting**. If the victim is alert and asymptomatic, administer a slurry of activated charcoal at a dose of 1 g/kg (infant, child, and adult dose). A soda can and straw may be of assistance when offering charcoal to a child. *In some cases you may be specifically instructed by poison control to induce vomiting by way of 2 tablespoons of syrup of ipecac (adult) washed down with a cup of water.* Do NOT give activated charcoal before or with ipecac syrup.

*Note to physician or authorized medical personnel:* Treat cases of respiratory compromise, coma, or excessive pulmonary secretions with respiratory support using protocols and techniques available and within the scope of training. Some cases may necessitate procedures such as endotracheal intubation or cricothyrotomy by properly trained and equipped personnel. When possible, atropine (see *Antidotes*, below) should be given under medical supervision. Patients who are comatose, hypotensive, or having seizures or cardiac arrhythmias should be treated according to advanced life support protocols. *Antidotes:* Two antidotes are administered to treat organophosphate poisoning. Atropine is a competitive antagonist of acetylcholine at muscarinic receptors and is used to control the excessive bronchial secretions which are often responsible for death. Pralidoxime relieves both the nicotinic and muscarine effects of organophosphate poisoning by regenerating acetylcholinesterase and can reduce both the bronchial secretions and the muscle weakness associated with poisoning. The initial intravenous dose of atropine in adults should be determined by the severity of symptoms: An initial adult dose of 1.0 to 2.0 mg or pediatric dose of 0.01 mg/kg (minimum 0.01 mg) should be administered intravenously. If intravenous access cannot be established, atropine may also be given intramuscularly, subcutaneously or via endotracheal tube. Doses should be repeated every 15 minutes until excessive secretions and sweating have been controlled. Once bronchial secretion has been controlled, atropine administration should be repeated whenever the secretions begin to recur. In seriously poisoned patients, very large doses may be required. Alterations of pulse rate and pupillary size should not be used as indicators of treatment adequacy. Pralidoxime should be administered as early in poisoning as possible as its efficacy may diminish when given more than 24 to 36 hours after exposure. Doses are as follows: adult 1.0 g; pediatric 25 to 50 mg/kg. The drug should be administered intravenously over 30 to 60 minutes, but in a life-threatening situation, one-half of the total dose can be given per minute for a total administration time of 2 minutes. Treatment should begin to take effect within 40 minutes with a reduction in symptoms and the amount of atropine necessary to control bronchial secretion. The initial dose can be repeated in 1 hour and then every 8 to 12 hours until the patient is clinically well and no longer requires atropine. If intravenous access cannot be established, pralidoxime may also be given intramuscularly. Early administration of diazepam in addition to the combined atropine and pralidoxime treatment may help prevent the onset of seizures and potential brain and cardiac morphologic damage following high-level organophosphate poisoning.

*References:*
- EXTOXNET, Extension Toxicology Network, "Pesticide Information Profile, Coumaphos," Oregon State University, Corvallis, OR, June 1996. http://ace.orst.edu/info/extoxnet/pips/coumapho.htm
- U.S. Environmental Protection Agency, Office of Pesticide Programs, Pesticide Residue Limits, "Coumaphos", 40 CFR 180.189. www.epa.gov/cgi-bin/oppsrch
- U.S. Environmental Protection Agency, Office of Prevention, Pesticides and Toxic Substances, "Reregistration Eligibility Decision Facts, Coumaphos," EPA-738-F-96-014, August 1996. www.epa.gov/oppsrrd1/REDs/factsheets/0018fact.pdf
- U.S. Environmental Protection Agency, "Chemical Profile: Coumaphos," Washington, DC, Chemical Emergency Preparedness Program (October 31, 1985).
- Agency for Toxic Substances and Disease Registry, U.S. Department of Health and Human Services, Public Health Service, "Managing Hazardous Materials Incidents," Atlanta, GA (June 2003).
- California Environmental Protection Agency "Chemical List of Lists," Sacramento CA (February 1997).

- New Jersey Department of Health and Senior Services and Senior Services, "Hazardous Substance Fact Sheet: Coumaphos," Trenton, NJ (December 1998). http://www.state.nj.us/health/eoh/rtkweb/0536.pdf
- Sax, N.I., Ed., "Dangerous Properties of Industrial Materials Report," 4, No. 1, 53-56 (1984) and 9, No. 1, 19-29 (1989).

## Coumatetralyl

*Use Type:* Rodenticide
*CAS Number:* 5836-29-3
*Formula:* $C_{19}H_{16}O_3$
*Synonyms:* 2H-1-Benzopyran-2-one, 4-hydroxy-3-(1,2,3,4-tetrahydro-1-naphthalenyl)-; Cumatetralyl (German, Dutch); Coumarin, 4-hydroxy-3-(1,2,3,4-tetrahydro-1-naphthyl)-; ENE 11183; 4-Hydroxy-3-(1,2,3,4-tetrahydro-1-naftyl)-4-cumarine (Dutch); 4-Hydroxy-3-(1,2,3,4-tetrahydro-1-napthalenyl)-2H-1-benzopyran-2-one; 4-Hydroxy-3-(1,2,3,4-tetrahydro-1-napthyl)cumarin; 3-($\alpha$-Tetral)-4-oxycoumarin; 3-($\alpha$-Tetrayl)-4-hydroxycoumarin; 3-(D-Tetrayl)-4-hydroxycoumarin
*Trade Names:* BAY® 25634, Bayer CropScience (Germany); BAY ENE® 11183B, Bayer CropScience (Germany); BAYER® 25,634, Bayer CropScience (Germany); ENDOX®; ENDROCID®; ENDROCIDE®; RACUMIN®; RAUCUMIN® 57; RODENTIN®
*Producers:* Agrimor International (USA); BASF Agricultural Products Group (Germany); Bayer CropScience (Germany)
*Chemical Class:* Coumarin
*EPA/OPP PC Code:* 496100
*RTECS Number:* GN7630000
*EINECS Number:* 227-424-0
*Uses:* Coumatetralyl is used for rat control and functions as an anticoagulant, of the warfarin type, that does not induce bait-shyness.
*Carcinogen/Hazard Classifications*
**Label Signal Word:** DANGER
**WHO Acute Hazard:** Group 1b, highly hazardous
*Regulatory Authority:*
- Superfund/EPCRA 40CFR355, Appendix B Extremely Hazardous Substances: TPQ = 500/10,000 lb (227/4,540 kg)
- Superfund/EPCRA 40CFR302.4 RQ: EHS, 1 lb (0.454 kg)

*Description:* Coumatetralyl is a yellowish-white, crystalline solid. Boiling point = 290°C. Melting/Freezing point = 69–70°C. Hazard Identification (based on NFPA-704 M Rating System): Health 4, Flammability 1, Reactivity 0. Soluble in hot water.
*Incompatibilities:* Oxidizers may cause fire and explosion hazard. Keep away from metals.
*Permissible Exposure Limits in Air:* No standards set.
*Permissible Concentration in Water:* No criteria set, but runoff from spills or fire control may cause water pollution.
*Routes of Entry:* Inhalation, ingestion, skin contact.
*Harmful Effects and Symptoms*
*Short Term Exposure:* Inhalation may cause symptoms described in long term exposure. With a single large ingested dose may cause hemorrhagic shock. This chemical can be absorbed through the skin, thereby increasing exposure or hemorrhagic effect. High exposure can cause death. The $LD_{50}$-(oral, rat) = 16.5 mg/kg (highly toxic).
*Long Term Exposure:* Chronic exposure may cause death by hemorrhagic shock. Absorption by the lungs or after a few days or few weeks of repeated ingestion, may cause inhibition of prothrombin synthesis, nose bleeds and bleeding gums, hematoma, small reddish spots like a rash, bruises of the elbows, knees and buttocks, blood in urine and stools, anemia, occasional paralysis due to a stroke. Pre-existing blood clotting disease or liver disease are aggravated by Coumatetralyl exposure.
*First Aid:* Speed in removing material from eyes and skin is of extreme importance. If this chemical gets into the eyes, remove any contact lenses at once and irrigate immediately for at least 15 minutes, occasionally lifting upper and lower lids. Seek medical attention immediately. If this chemical contacts the skin, remove contaminated clothing and wash immediately with soap and water. Shampoo hair promptly if contaminated. Seek medical attention immediately. If this chemical has been inhaled, remove from exposure, begin rescue breathing (using universal precautions) if breathing has stopped, and CPR if heart action has stopped. Transfer promptly to a medical facility. When this chemical has been swallowed, get medical attention. Give large quantities of water and induce vomiting. Do not make an unconscious person vomit.Medical observation is recommended for 24 to 36 hours following overexposure, as effects may be delayed.
*References:*
- U.S. Environmental Protection Agency, "Chemical Profile: Coumatetralyl," Washington, DC, Chemical Emergency Preparedness Program (November 30, 1987).
- California Environmental Protection Agency "Chemical List of Lists," Sacramento CA (February 1997).

## Crimidine

*Use Type:* Rodenticide
*CAS Number:* 535-89-7
*Formula:* $C_7H_{10}ClN_3$; $C_7H_{10}N_3Cl$
*Alert:* Not registered in the United States.
*Synonyms:* 2-Chloor-4-dimethylamino-6-methyl-pyrimidine (Dutch); 2-Chloro-4-methyl-6-dimethylaminopyrimidine; Crimidin (German); Crimidina (Italian); Pyrimidine, 2-Chloro-4-(dimethylamino)-6-methyl-
*Trade Names:* CASTRIX®; CRIMITOX®; W 491®

*Producers:* Sigma-Aldrich Laborchemikalien (Germany)
*EPA/OPP PC Code:* 288200
*RTECS Number:* UV8050000
*EINECS Number:* 208-622-6
*Uses:* Crimidine is used as a rodenticide but is not registered in the U.S. as a pesticide.
*Regulatory Authority:*
- Banned or Severely Restricted (In Agriculture) (Germany) (UN)[13]
- Very Toxic Substance (World Bank)[15]
- Superfund/EPCRA 40CFR355, Appendix B Extremely Hazardous Substances: TPQ = 100/10,000 lb (4.54/4,540 kg)
- Superfund/EPCRA 40CFR302.4 RQ: EHS, 1 lb (0.454 kg)

*Description:* Brown, waxy solid. Practically insoluble in water. Boiling point = 144°C @ 4 mmHg. Melting/Freezing point = 87°C. Hazard Identification (based on NFPA-704 M Rating System): Health 4, Flammability 1, Reactivity 0.
*Incompatibilities:* Acids and acid fumes, strong bases.
*Permissible Exposure Limits in Air:* No standards set. This chemical can be absorbed through the skin, thereby increasing exposure.
*Permissible Concentration in Water:* No criteria set, but runoff from spills or fire control may cause water pollution.
*Routes of Entry:* Ingestion, absorbed through the skin.
*Harmful Effects and Symptoms*
*Short Term Exposure:* Contact can cause eye and skin irritation and burns. Inhalation can irritate the nose and throat. Exposure may result in serious central nervous system damage with anxiety, restlessness, muscle stiffness, light sensitivity, noise sensitivity, touch sensitivity, cold sweat, and leading to convulsions that may be fatal. If patient survives 5 to 6 hours there may not be serious problems. Extremely toxic; the $LD_{50}$ (oral, rat) = 1.25 mg/kg. The probable oral lethal dose in humans is less than 5 mg/kg or less than 7 drops for a 70 kg (150 lb) person.
*Long Term Exposure:* May cause central nervous system damage.
*Points of Attack:* Central nervous system.
*Medical Surveillance:* There is no special test for this chemical. However, if illness occurs or overexposure is suspected, medical attention is recommended.
*First Aid: Eye Contact:* Immediately remove any contact lenses and flush with large amounts of water for at least 15 minutes, occasionally lifting upper and lower lids. Seek medical attention immediately. *Skin Contact:* Quickly remove contaminated clothing. Immediately wash area with large amounts of water. Seek medical attention immediately. *Inhalation:* Remove the person from exposure, trying to avoid rapid, jerky motions or noise. Begin rescue breathing if breathing has stopped, and CPR if heart action has stopped. If seizures occur, begin seizure first aid measures. Call for immediate medical attention to visit the patient prior to transfer if possible. Any facility using this chemical should have 24 hour rapid access to medical personnel with training and equipment for emergency treatment. All area employees should be trained in first aid measures for Castrix, including seizure management and CPR.
*References:*
- U.S. Environmental Protection Agency, "Chemical Profile: Crimidine," Washington, DC, Chemical Emergency Preparedness Program (November 30, 1987).
- California Environmental Protection Agency "Chemical List of Lists," Sacramento CA (February 1997).
- New Jersey Department of Health and Senior Services and Senior Services, "Hazardous Substance Fact Sheet: Castrix," Trenton, NJ (September 1999). www.state.nj.us/health/eoh/rtkweb/0351.pdf

# Crotoxyphos

*Use Type:* Insecticide and acaricide
*CAS Number:* 7700-17-6
*Formula:* $C_{14}H_{19}O_6P$
*Synonyms:* Crotonic acid, 3-hydroxy-, α-methylbenzyl ester, dimethylphosphate, (E)-; (E)-3-; [(dimethoxyphosphinyl)oxy]-2-butenoic acid 1-phenylethyl ester; O,O-Dimethyl-O-(1-methyl-2-carboxy-α-phenylethyl)vinyl phosphate; dimethyl-cis-1-methyl-2-(1-phenylethoxycarbonyl)vinyl phosphate; Dimethyl phosphate ester of α-methylbenzyl-3-hydroxy-cis-crotonate; ENT 24,717; (E)-3-Hydroxy-crotonic acid, α-methylbenzyl ester, dimethyl phosphate; 1-Methylbenzyl-3-(dimethoxyphosphinyloxo) isocrotonate; α-Methylbenzyl 3-(dimethoxyphosphinoxy)-cis-crotonate; α-Methylbenzyl-3-hydroxy-crotonate dimethyl phosphate; cis-2-(1-Phenylethoxy)carbonyl-1-methylvinyl dimethylphosphate
*Trade Names:* CIODRIN®, Pbi/Gordon (USA), canceled, Boehringer Ingelheim (Germany), canceled; CIODRIN VINYL PHOSPHATE; CIOVAP®; COOP RTU® CATTLE SPRAY; CYODRIN®; CYPONA E.C.®; DECROTOX®; DUO-KILL®; PANTOZOL-1®; SD 4294®; SHELL® SD 4294, Shell Chemical (UK); VAPORIN® DAIRY SPRAY; VOLFAZOL®
*Producers:* Boehringer Ingelheim (Germany); Pbi/Gordon (USA); Shell Chemical (UK)
*Chemical Class:* Organophosphate
*EPA/OPP PC Code:* 058801
*California DPR Chemical Code:* 140
*Uses:* A contact and stomach poison; controls flies, lice, and ticks on lactating dairy and beef cattle; may also be used on swine, goats, horses, and sheep.
*Carcinogen/Hazard Classifications*
**Label Signal Word:** WARNING, CAUTION
*Regulatory Authority:*
- U.S. DOT Inhalation Hazard Chemicals

***Description:*** Yellowish liquid. Soluble in water; solubility = 1,200 ppm. Molecular weight = 314.28. Density = 1.19; 1.2 @ 15°C. Boiling point = 135°C @ 0.04 mmHg. Vapor pressure = $1.8 \times 10^{-5}$ mmHg @ 20°C.

***Incompatibilities:*** May react violently with antimony(V) pentafluoride. Incompatible with lead diacetate, magnesium, silver nitrate.

***Determination in Air:*** OSHA versatile sampler-2; Toluene/Acetone; Gas chromatography/Flame ionization detection; NIOSH IV, Method #5600, Organophosphorus Pesticides.[18]

***Permissible Concentration in Water:*** No criteria set. Runoff from spills or fire control may cause water pollution.

### Harmful Effects and Symptoms

**Short Term Exposure:** Eye pupils are small; blurred vision; eye watering; runny nose; cough; shortness of breath; salivation; dizziness; nausea, stomach cramps, diarrhea, and vomiting; increased blood pressure; profuse sweating; hypermotility, hallucinations; irritability; tingling of the skin; drowsiness; slow heartbeat; convulsions; fluid in lungs; loss of consciousness; incontinence; breathing stops; death. Organophosphates inhibit the action of acetylcholinesterase enzymes, and alter the way in which nervous impulses are transmitted. The effects can last for hours, days, or much longer. The action of the enzymes is reestablished after new enzymes are formed.

***Long Term Exposure:*** Cholinesterase inhibitor; cumulative effect is possible. Organophosphates may damage the nervous system with repeated exposure, resulting in convulsions, respiratory failure. May cause liver damage.

***Points of Attack:*** Respiratory system, central nervous system, cardiovascular system, blood cholinesterase.

***Medical Surveillance:*** Medical observation is recommended for 24 to 48 hours after breathing overexposure, as pulmonary edema may be delayed. As first aid for pulmonary edema, consider administering a corticosteroid spray. Cigarette smoking may exacerbate pulmonary injury and should be discouraged for at least 72 hours following exposure.

***First Aid:*** **Treatment for organophosphate poisoning consists of thorough decontamination, cardiorespiratory support, and administration of the antidotes atropine and pralidoxime. In cases of severe poisoning, diazepam, an anticonvulsant, should also be administered. Antidotes should be administered as prevention even if the diagnosis is in doubt.** Speed in removing material from eyes and skin is of extreme importance. *Eyes:* Eye contact can cause dangerous amounts of these chemicals to be quickly absorbed through the mucous membrane into the bloodstream. Immediately and gently flush eyes with plenty of warm or cold water (NO hot water) for at least 15 minutes, occasionally lifting the upper and lower eyelids. Get medical aid immediately. *Skin:* Get medical aid. Skin contact can cause dangerous amounts of these chemicals to be absorbed into the bloodstream. Wearing the appropriate PPE equipment and respirator for organophosphate pesticides, immediately flush skin with plenty of soap and water for at least 15 minutes while removing contaminated clothing and shoes. Shampoo hair promptly if contaminated. The removed, contaminated clothing and shoes should be double-bagged and left in Hot Zone for later disposal by hazardous materials experts. Skin may also be decontaminated with diluted hypochlorite solution. *Inhalation:* Get medical aid. Do not contaminate yourself. Wearing the appropriate PPE equipment and respirator for organophosphate pesticides, immediately remove the victim from the contaminated area to fresh air. If the victim is not breathing, administer artificial respiration. Do NOT use mouth-to-mouth resuscitation; use bag/mask apparatus. If breathing is difficult, administer oxygen through bag/mask apparatus until medical help arrives. Do not leave victim unattended. *Ingestion:* Call poison control. Loosen all clothing. Never give anything by mouth to an unconscious person. If victim is *unconscious or having convulsions,* do nothing except keep victim warm. Get medical aid. Transfer promptly to a medical facility. In cases of ingestion, **do not induce vomiting.** If the victim is alert and asymptomatic, administer a slurry of activated charcoal at a dose of 1 g/kg (infant, child, and adult dose). A soda can and straw may be of assistance when offering charcoal to a child. *In some cases you may be specifically instructed by poison control to induce vomiting by way of 2 tablespoons of syrup of ipecac (adult) washed down with a cup of water.* Do NOT give activated charcoal before or with ipecac syrup.

*Note to physician or authorized medical personnel:* Treat cases of respiratory compromise, coma, or excessive pulmonary secretions with respiratory support using protocols and techniques available and within the scope of training. Some cases may necessitate procedures such as endotracheal intubation or cricothyrotomy by properly trained and equipped personnel. When possible, atropine (see *Antidotes*, below) should be given under medical supervision. Patients who are comatose, hypotensive, or having seizures or cardiac arrhythmias should be treated according to advanced life support protocols. *Antidotes:* Two antidotes are administered to treat organophosphate poisoning. Atropine is a competitive antagonist of acetylcholine at muscarinic receptors and is used to control the excessive bronchial secretions which are often responsible for death. Pralidoxime relieves both the nicotinic and muscarine effects of organophosphate poisoning by regenerating acetylcholinesterase and can reduce both the bronchial secretions and the muscle weakness associated with poisoning. The initial intravenous dose of atropine in adults should be determined by the severity of symptoms: An initial adult dose of 1.0 to 2.0 mg or pediatric dose of 0.01 mg/kg (minimum 0.01 mg) should be administered intravenously. If intravenous access cannot be established, atropine may also be given intramuscularly, subcutaneously or via endotracheal tube. Doses should be repeated every 15 minutes until excessive secretions and sweating have been controlled. Once bronchial secretion has been controlled, atropine administration should be repeated whenever the secretions begin to recur. In seriously poisoned patients, very large doses may be required. Alterations of pulse rate and pupillary size should not be used as indicators of treatment adequacy. Pralidoxime should be administered as early in poisoning as possible as its efficacy may diminish when given more than 24 to 36 hours after exposure. Doses are as follows: adult 1.0 g; pediatric 25 to 50 mg/kg. The drug should be administered intravenously over 30 to 60 minutes, but in a life-threatening situation, one-half of the total dose can be given per minute for a total administration time of 2 minutes. Treatment should begin to take effect within 40 minutes with a reduction in symptoms and the amount of atropine necessary to control bronchial secretion. The initial dose can be repeated in 1 hour and then every 8 to 12 hours until the patient is clinically well and no longer requires atropine. If intravenous access cannot be established, pralidoxime may also be given intramuscularly. Early administration of diazepam in addition to the combined atropine and pralidoxime treatment may help prevent the onset of seizures and potential brain and cardiac morphologic damage following high-level organophosphate poisoning.

*References:*
- Pesticide Management Education Program, "Crotoxyphos (Ciodrin, Ciovap) Chemical Profile 4/85," Cornell University, Ithaca, NY (April 1985). http://pmep.cce.cornell.edu/profiles/insect-mite/cadusafos-cyromazine/crotoxyphos/insect-prof-crotoxyphos.html
- California Environmental Protection Agency "Chemical List of Lists," Sacramento CA (February 1997)

## Crufomate (ANSI)

*Use Type:* Insecticide and anthelmintic for cattle
*CAS Number:* 299-86-5
*Formula:* $C_{12}H_{19}ClNO_3P$
*Alert:* All residue limits for crufomate, as listed in 40 CFR 180.295, were revoked by the U.S. EPA on June 9, 1993. All registered uses of crufomate in or on fat, meat, and mbyp of cattle, goats, and sheep were canceled.
*Synonyms:* 4-*t*-Butyl-2-chlorophenyl methyl methylphosphoramidate; O-(4-*tert*-Butyl-2-chlor-phenyl)-O-methyl-phosphorsaeure-*N*-methylamid (German); 4-*tert* Butyl 2-chlorophenyl methylphosphoramidate de methyle (French); ENT 25,602-X; O-Methyl O-2-chloro-4-*tert*-butylphenyl *N*-methylamidophosphate; Methylphosphoramidic acid,4-*t*-butyl-2-chlorophenyl methyl ester; Phenol,4-*t*-butyl-2-chloro-, ester with methyl methylphosphoramidate; Phosphoramidic acid, 4-*tert*-butyl-2-chlorophenylphosphor amidate; Phosphoramidic acid, methyl-,2-chloro-4-(1,1-dimethylethyl)phenyl methyl ester; Phosphoramidic acid, methyl-,4-*tert*-butyl-2-chlorophenyl; O-(4-Tertbutyl-2-chloor-fenyl)-O-methyl-fosforzuur-*N*-methyl-amide (Dutch)
*Trade Names:* DOWCO® 132, Dow Chemical Co. (USA) canceled; MONTREL®, canceled; RUELENE®, Dow Chemical Co. (USA) canceled 10/1/1988; DRENCH®, canceled; RULENE®, canceled
*Producers:* Dow Chemical Co. (USA)
*Chemical Class:* Organophosphate
*EPA/OPP PC Code:* 012101
*California DPR Chemical Code:* 519
*ICSC Number:* 1143
*RTECS Number:* TB3850000
*EEC Number:* 015-074-00-X
*Uses:* It is applied as a spray on plants to kill insects and worms.
*Regulatory Authority:*
- Air Pollutant Standard Set (ACGIH)[1] (Australia) (Israel) (Mexico) (OSHA)[58] (Several States)[60] (Several Canadian Provinces)
- U.S. DOT Inhalation Hazard Chemicals as organophosphates
- Permissible Exposure Limits for Chemical Contaminants (CAL/OSHA)
- The "Director's List" (CAL/OSHA)

*Description:* Colorless, crystalline compound. Commercial product is a yellow oil. Very slightly soluble in water. Molecular weight = 291.47. Melting/Freezing point = 60–63°C (decomposes). Vapor pressure = 0.107 Pa @ 25°C.
*Incompatibilities:* The substance decomposes on heating forming corrosive and toxic fumes of hydrogen chloride, nitrogen oxides and phosphorous oxides. Incompatible with strongly alkaline and strongly acidic media. Unstable over long periods in aqueous preparations or above 140°F.
*Permissible Exposure Limits in Air:* There is no OSHA[2] PEL. The NIOSH[2] REL 10-hour TWA and ACGIH[1] recommended 8-hour TWA is 5 mg/m$^3$ and the NIOSH[2] STEL is 20 mg/m$^3$, not to be exceeded during any 15 minute work period. The Australian, Mexican, and Israeli TWA is the same as NIOSH and Mexico's STEL is 20 mg/m$^3$. The Canadian provinces of Alberta, BC, Ontario, and Quebec have adopted a TWA of 5 mg/m$^3$ and except for Quebec use the STEL of 20 mg/m$^3$. Several states have set guidelines or standards for crufomate in ambient air[60] ranging from 50-200 μg/m$^3$ (North Dakota) to 80 μg/m$^3$ (Virginia) to 100 μg/m$^3$ (Connecticut) to 119 μg/m$^3$ (Nevada).
*Determination in Air:* Filter; none; Gravimetric; NIOSH IV Method #0500, Particulates NOR (total)[18]. OSHA versatile sampler-2; Toluene/Acetone; Gas chromatography/Flame photometric detection for sulfur, nitrogen, or phosphorus;

NIOSH Method IV Method #5600, Organophosphorus pesticides[18].

***Permissible Concentration in Water:*** No criteria set, but runoff from spills or fire control may cause water pollution.

***Routes of Entry:*** Skin absorption, inhalation of dust, ingestion.

***Harmful Effects and Symptoms***

***Short Term Exposure:*** Cholinesterase inhibitor. Crufomate irritates the eyes, skin, and respiratory tract. Crufomate can affect you when breathed in and quickly enters the body by passing through the skin. Severe poisoning can occur from skin contact. It is a moderately toxic organophosphate chemical. Exposure can cause effects on the nervous system, rapid severe poisoning with headache, sweating, nausea and vomiting, diarrhea, loss of coordination, convulsions, respiratory failure, and death. The $LD_{50}$ (oral, rat) = 460 mg/kg (slightly toxic). The health effects may be delayed. Delayed pulmonary edema may occur after inhalation.

***Long Term Exposure:*** Exposure may affect the developing fetus. Crufomate may damage the testes. High or repeated exposure may cause nerve damage and poor coordination in arms and legs. Repeated exposure may cause personality changes of depression, anxiety, or irritability.

***Points of Attack:*** Respiratory system, central nervous system, peripheral nervous system, plasma cholinesterase

***Medical Surveillance:*** Medical observation is recommended for 24 to 48 hours after breathing overexposure, as pulmonary edema may be delayed. Before employment and at regular times after that, the following are recommended: Plasma and red blood cell cholinesterase levels (tests for the enzyme poisoned by this chemical). If exposure stops, plasma levels return to normal in 1-2 weeks while red blood cell levels may be reduced for 1-3 months. When acetylcholinesterase enzyme levels are reduced by 25% or more below preemployment levels, risk of poisoning is increased, even if results are in lower ranges of "normal." Reassignment to work not involving organophosphate or carbamate pesticides is recommended until enzyme levels recover. If symptoms develop or overexposure occurs, repeat the above tests as soon as possible and get an exam before or during use. Alcohol promotes absorption of organophosphates.

***First Aid:*** **Treatment for organophosphate poisoning consists of thorough decontamination, cardiorespiratory support, and administration of the antidotes atropine and pralidoxime. In cases of severe poisoning, diazepam, an anticonvulsant, should also be administered. Antidotes should be administered as prevention even if the diagnosis is in doubt.** Speed in removing material from eyes and skin is of extreme importance. *Eyes:* Eye contact can cause dangerous amounts of these chemicals to be quickly absorbed through the mucous membrane into the bloodstream. Immediately and gently flush eyes with plenty of warm or cold water (NO hot water) for at least 15 minutes, occasionally lifting the upper and lower eyelids. Get medical aid immediately. *Skin:* Get medical aid. Skin contact can cause dangerous amounts of these chemicals to be absorbed into the bloodstream. Wearing the appropriate PPE equipment and respirator for organophosphate pesticides, immediately flush skin with plenty of soap and water for at least 15 minutes while removing contaminated clothing and shoes. Shampoo hair promptly if contaminated. The removed, contaminated clothing and shoes should be double-bagged and left in Hot Zone for later disposal by hazardous materials experts. Skin may also be decontaminated with diluted hypochlorite solution. *Inhalation:* Get medical aid. Do not contaminate yourself. Wearing the appropriate PPE equipment and respirator for organophosphate pesticides, immediately remove the victim from the contaminated area to fresh air. If the victim is not breathing, administer artificial respiration. Do NOT use mouth-to-mouth resuscitation; use bag/mask apparatus. If breathing is difficult, administer oxygen through bag/mask apparatus until medical help arrives. Do not leave victim unattended. *Ingestion:* Call poison control. Loosen all clothing. Never give anything by mouth to an unconscious person. If victim is *unconscious or having convulsions,* do nothing except keep victim warm. Get medical aid. Transfer promptly to a medical facility. In cases of ingestion, **do not induce vomiting**. If the victim is alert and asymptomatic, administer a slurry of activated charcoal at a dose of 1 g/kg (infant, child, and adult dose). A soda can and straw may be of assistance when offering charcoal to a child. *In some cases you may be specifically instructed by poison control to induce vomiting by way of 2 tablespoons of syrup of ipecac (adult) washed down with a cup of water.* Do NOT give activated charcoal before or with ipecac syrup.

*Note to physician or authorized medical personnel:* Treat cases of respiratory compromise, coma, or excessive pulmonary secretions with respiratory support using protocols and techniques available and within the scope of training. Some cases may necessitate procedures such as endotracheal intubation or cricothyrotomy by properly trained and equipped personnel. When possible, atropine (see *Antidotes*, below) should be given under medical supervision. Patients who are comatose, hypotensive, or having seizures or cardiac arrhythmias should be treated according to advanced life support protocols. *Antidotes:* Two antidotes are administered to treat organophosphate poisoning. Atropine is a competitive antagonist of acetylcholine at muscarinic receptors and is used to control the excessive bronchial secretions which are often responsible for death. Pralidoxime relieves both the nicotinic and muscarine effects of organophosphate poisoning by regenerating acetylcholinesterase and can reduce both the bronchial secretions and the muscle weakness associated with poisoning. The initial intravenous dose of atropine in adults should be determined by the severity of symptoms: An initial adult dose of 1.0 to 2.0 mg or pediatric dose of 0.01 mg/kg (minimum 0.01 mg) should be administered intravenously. If intravenous access cannot be established, atropine may also be given intramuscularly, subcutaneously or via endotracheal tube. Doses should be repeated every 15 minutes until excessive secretions and sweating have been controlled. Once bronchial secretion has been controlled, atropine administration should be repeated whenever the secretions begin to recur. In seriously poisoned patients, very large doses may be required. Alterations of pulse rate and pupillary size should not be used as indicators of treatment adequacy. Pralidoxime should be administered as early in poisoning as possible as its efficacy may diminish when given more than 24 to 36 hours after exposure. Doses are as follows: adult 1.0 g; pediatric 25 to 50 mg/kg. The drug should be administered intravenously over 30 to 60 minutes, but in a life-threatening situation, one-half of the total dose can be given per minute for a total administration time of 2 minutes. Treatment should begin to take effect within 40 minutes with a reduction in symptoms and the amount of atropine necessary to control bronchial secretion. The initial dose can be repeated in 1 hour and then every 8 to 12 hours until the patient is clinically well and no longer requires atropine. If intravenous access cannot be established, pralidoxime may also be given intramuscularly. Early administration of diazepam in addition to the combined atropine and pralidoxime treatment may help prevent the onset of seizures and potential brain and cardiac morphologic damage following high-level organophosphate poisoning.

*References:*
- New Jersey Department of Health and Senior Services and Senior Services, "Hazardous Substance Fact Sheet: Crufomate," Trenton, NJ (May 1998). www.state.nj.us/health/eoh/rtkweb/0541.pdf
- California Environmental Protection Agency "Chemical List of Lists," Sacramento CA (February 1997).
- Agency for Toxic Substances and Disease Registry, U.S. Department of Health and Human Services, Public Health Service, "Managing Hazardous Materials Incidents," Atlanta, GA (June 2003).

# Cupric Acetate

*Use Type:* Fungicide
*CAS Number:* 142-71-2
*Formula:* $C_4H_6CuO_4$; $Cu(OOCCH_3)_2$
*Alert:* Not registered with the U.S. EPA.
*Synonyms:* Acetate de cuivre (French); Acetato de cobre (Spanish); Acetic acid, copper(2+) salt; Acetic acid, copper(II) salt; Acetic acid, cupric salt; Copper(2+) acetate; Copper(II) acetate; Copper acetate; Copper diacetate; Copper(2+) diacetate; Copper(II) diacetate; Crystallized verdigris; Crystals of Venus; Cupric diacetate; Neutral verdigris
*Producers:* Celtic Chemicals (UK); GFS Chemicals (USA); Goldschmidt (Germany); Nihon Kagaku Sangyo (Japan); Omya (Switzerland); Philipp Brothers Chemicals (USA); Shyam Chemicals (India)
*Chemical Class:* Inorganic
*California DPR Chemical Code:* 3013
*RTECS Number:* AG3480000
*EINECS Number:* 205-553-3
*Uses:* Cupric acetate is also used as a catalyst for organic reactions, in textile dyeing and as a pigment for ceramics.
*Regulatory Authority:*
- Water Pollution Standard Proposed (U.S. EPA)[6] (former USSR)[35]
- The "Director's List" (CAL/OSHA)
- AB 2588-Air Toxics "Hot Spots" Chemicals (CAL) as copper compounds
- Clean Water Act: Section 311 Hazardous Substances/RQ 40CFR117.3 (same as CERCLA, see below); Section 313 Water Priority Chemicals (57FR41331, 9/9/92); Toxic Pollutant (Section 401.15), as copper and compounds.
- RCRA 40CFR264, Appendix 9; TSD Facilities Ground Water Monitoring List. Suggested test method(s) (PQL $ug/L$): 6010(60); 7210(200) Note: All species in the ground water that contain copper are included.
- Superfund/EPCRA 40CFR302.4 RQ: CERCLA, 100 lb (45.4 kg).
- EPCRA Section 313 Form R *de minimus* concentration reporting level: 1.0%
- Canada, WHMIS, Ingredients Disclosure List; National Pollutant Release Inventory (NPRI); CEPA Priority Substance List, Ocean dumping prohibited.

*Description:* Greenish blue powder or small crystals. Soluble in water; solubility = 7.2 g/100 ml. Molecular weight = 181.65. Boiling point = 240°C (decomposes). Melting/Freezing point = 115°C.
*Incompatibilities:* Forms explosive materials with acetylene gas, ammonia, caustic solutions, sodium hypobromite, notromethane. Keep away from chemically active metals, strong acids, nitrates. Decomposes above 240°C forming acetic acid fumes.
*Permissible Exposure Limits in Air:* The Federal standard (OSHA PEL 8-hour TWA)[58] for copper fume is 0.1 mg/m³, and 1 mg/m³ for copper dusts and mists. NIOSH[2] recommends the same level for a 10-hour workshift. ACGIH[1] recommends a TWA of 0.2 mg/m³ for copper fume and 1 mg/m³ for dusts and mists. The NIOSH[2] IDLH is 100 mg/m³.
*Determination in Air:* Copper dusts and mists are collected on a filter, worked up with acid, measured by atomic absorption. See NIOSH Method #7029 for copper[18]. For copper fume: filter collection, acid digestion, measurement by atomic absorption

*Permissible Concentration in Water:* A limit of 1.0 µg/L in drinking water has been set for copper by EPA[6] and by The former USSR[35] as noted in the entry on copper.

*Routes of Entry:* Inhalation, ingestion.

*Harmful Effects and Symptoms*

*Short Term Exposure:* Inhaling cupric acetate dust and vapors can irritate the respiratory tract causing coughing and wheezing. High levels may cause fluid to build up in the lungs (pulmonary edema), This can cause death. Corrosive: contact can irritate and may burn the skin and eyes. The $LD_{50}$ (oral,rat) = 595 mg/kg (slightly toxic).

*Long Term Exposure:* Repeated exposure can cause skin allergy, thickening of the skin, and/or a green discoloration of the skin and hair. Repeated exposure can cause shrinking (atrophy) of the inner lining of the nose an may cause sores in the nose. Can cause liver and kidney damage.

*Points of Attack:* Skin, lung, liver, and kidney damage.

*Medical Surveillance:* For those with frequent or potentially high exposure (half the TLV or greater), the following are recommended before beginning work and at regular times after that: Lung function tests. If symptoms develop or overexposure is suspected, the following may be useful: Consider chest x-ray after acute overexposure. Serum and urine tests for copper can measure recent exposure. Liver and kidney function tests. Evaluation by a qualified allergist.

*First Aid:* If this chemical gets into the eyes, remove any contact lenses at once and irrigate immediately for at least 15 minutes, occasionally lifting upper and lower lids. Seek medical attention immediately. If this chemical contacts the skin, remove contaminated clothing and wash immediately with soap and water. Seek medical attention immediately. If this chemical has been inhaled, remove from exposure, begin rescue breathing (using universal precautions) if breathing has stopped, and CPR if heart action has stopped. Transfer promptly to a medical facility. When this chemical has been swallowed, get medical attention. Give large quantities of water and induce vomiting. Do not make an unconscious person vomit. Medical observation is recommended for 24 to 48 hours after breathing overexposure, as pulmonary edema may be delayed. As first aid for pulmonary edema, a doctor or authorized paramedic may consider administering a corticosteroid spray.

*References:*
- New Jersey Department of Health and Senior Services and Senior Services, "Hazardous Substance Fact Sheet: Cupric Acetate," Trenton, NJ (February 1999). www.state.nj.us/health/eoh/rtkweb/0546.pdf
- California Environmental Protection Agency "Chemical List of Lists," Sacramento CA (February 1997).

# Cupric Nitrate

*Use Type:* Insecticide
*CAS Number:* 3251-23-8
*Formula:* $CuN_2O_6$; $Cu(NO_3)_2$

*Synonyms:* Copper dinitrate; Copper(2+) nitrate; Copper(II) nitrate; Nitrato de cobre (Spanish); Nitric acid, copper(2+) salt; Nitric acid, copper(II) salt; Cupric dinitrate

*Producers:* Celtic Chemicals (UK); Goldschmidt (Germany); Merck (Germany); PCF Chimie (France); Pechiney (France); Philbro-Tech (USA); Rhodia (France); Rhone-Poulenc (France); Shyam Chemicals (India); William Blythe (UK)

*Chemical Class:* Inorganic, metal salt

*EPA/OPP PC Code:* 076102

*California DPR Chemical Code:* 3118

*RTECS Number:* WU7400000

*EINECS Number:* 221-838-5

*Uses:* Cupric nitrate is used as an insecticide, in paint, varnish, enamel and in wood preservatives. Metal compounds are often used in "hot" operations in the workplace. These may include, but are not limited to, welding, brazing, soldering, plating, cutting, and metalizing. At the high temperatures reached in these operations, metals often form metal fumes which have different health effects and exposure standards than the original metal compound and require specialized controls.

*Regulatory Authority:*
- Air Pollutant Standard Set (ACGIH)[1] (DFG)[3] (HSE)[33] (OSHA)[58]
- The "Director's List" (CAL/OSHA)
- Clean Water Act: Section 311 Hazardous Substances/RQ 40CFR117.3 (same as CERCLA, see below); Section 313 Water Priority Chemicals (57FR41331, 9/9/92); Toxic Pollutant (Section 401.15), as copper and compounds.
- RCRA 40CFR264, Appendix 9; TSD Facilities Ground Water Monitoring List. Suggested test method(s) (PQL µg/L): 6010(60); 7210(200) Note: All species in the ground water that contain copper are included.
- Superfund/EPCRA 40CFR302.4 RQ: CERCLA, 100 lb (45.4 kg).
- EPCRA Section 313 Form R *de minimus* concentration reporting level: 1.0%
- Canada, WHMIS, Ingredients Disclosure List; National Pollutant Release Inventory (NPRI); CEPA Priority Substance List, Ocean dumping prohibited.

*Description:* Blue crystalline solid. Soluble in water. Boiling point = 172°C (decomposes below this point). Melting/Freezing point = 255°C.

*Incompatibilities:* A strong oxidizer. Aqueous solution is acidic; incompatible with bases. Violent reaction with potassium hexacyanoferrate; ammonia and potassium amide mixtures; acetic anhydrides, cyanides, ethers. Forms explosive materials with nitromethanes, sodium hypobromite, acetylene, chemically active metals such as potassium, sodium, etc. May ignite on contact with aluminum foil or tin. Risk of spontaneous combustion with combustibles (wood, cloth, etc.) organics, or reducing agents and readily oxidizable materials. Attacks metals in the presence of moisture.

***Permissible Exposure Limits in Air:*** The Federal standard (OSHA PEL 8-hour TWA)[58] for copper fume is 0.1 mg/m$^3$, and 1 mg/m$^3$ for copper dusts and mists. NIOSH[2] recommends the same level for a 10-hour workshift. ACGIH[1] recommends a TWA of 0.2 mg/m$^3$ for copper fume and 1 mg/m$^3$ for dusts and mists. The NIOSH[2] IDLH is 100 mg/m$^3$.

***Determination in Air:*** Copper dusts and mists are collected on a filter, worked up with acid, measured by atomic absorption. See NIOSH Method #7029 for copper[18]. For copper fume: filter collection, acid digestion, measurement by atomic absorption

***Permissible Concentration in Water:*** A limit of 1.0 μg/L in drinking water has been set for copper by EPA[6] and by The former USSR[35] as noted in the entry on copper.

***Routes of Entry:*** Inhalation, ingestion or skin contact

***Harmful Effects and Symptoms***

***Short Term Exposure:*** Skin and eye contact can cause irritation and burns. Inhalation can irritate the nose and throat, causing coughing and wheezing. Cupric nitrate may produce fumes that can cause "metal fume fever." Ingestion cause salivation, nausea, vomiting, stomach pain. May cause blood effects if swallowed. High exposure can cause unconsciousness. The $LD_{50}$ (oral, rat) = 940 mg/kg (slightly toxic).

***Long Term Exposure:*** Repeated exposure can cause copper to deposit in various parts of the body. Large deposits can make the skin and hair a green color. Repeated exposure can cause shrinking of the inner lining of the nose and may cause runny nose and sores. Excess deposits in the liver can cause liver damage. Metallic taste may also occur. Skin allergy with rash sometimes occurs. If allergy develops, even small future exposures can trigger rash. Repeated exposures can also cause thickening of the skin not caused by allergy.

***Points of Attack:*** Skin, liver.

***Medical Surveillance:*** For those with frequent or potentially high exposure (half the TLV or greater), the following are recommended before beginning work and at regular times after that: Lung function tests. If symptoms develop or overexposure is suspected, the following may be useful: Urine test for copper can measure recent exposure. Evaluation by a qualified allergist, including careful exposure history and special testing, may help diagnose skin allergy. Liver function tests.

***First Aid:*** If this chemical gets into the eyes, remove any contact lenses at once and irrigate immediately for at least 15 minutes, occasionally lifting upper and lower lids. Seek medical attention immediately. If this chemical contacts the skin, remove contaminated clothing and wash immediately with soap and water. Seek medical attention immediately. If this chemical has been inhaled, remove from exposure, begin rescue breathing (using universal precautions) if breathing has stopped, and CPR if heart action has stopped. Transfer promptly to a medical facility. When this chemical has been swallowed, get medical attention. Give large quantities of water and induce vomiting. Do not make an unconscious person vomit.

***References:***

- New Jersey Department of Health and Senior Services and Senior Services, "Hazardous Substance Fact Sheet: Cupric Nitrate," Trenton, NJ (February 1999). www.state.nj.us/health/eoh/rtkweb/0547.pdf
- Sax, N.I., Ed., "Dangerous Properties of Industrial Materials Report," 2, No. 5, 35-38 (1982) and 5, No. 6, 45-49 (1985).
- California Environmental Protection Agency "Chemical List of Lists," Sacramento CA (February 1997).

## Cuprous Chloride

***Use Type:*** Fungicide, plant growth regulator
***CAS Number:*** 7758-89-6
***Formula:*** ClCu
***Synonyms:*** Copper(I) chloride; Copper monochloride; Cuprous dichloride; Dicopper dichloride
***Trade Names:*** CUPROID®
***Producers:*** Celtic (UK); GFS Chemicals (USA); Goldschmidt (Germany); Kawaguchi Chemical Industry (Japan); Merck (Germany); Mineral Research (USA); Shyam (India); World Metal (USA)
***Chemical Class:*** Inorganic copper compound
***EPA/OPP PC Code:*** 108303
***California DPR Chemical Code:*** 5597
***EINECS Number:*** 231-842-9
***Uses:*** Used to control root growth of plants grown in nursery pots. Not widely used in agricultural applicaions.
***Carcinogen/Hazard Classifications***
**Label Signal Word:** CAUTION, DANGER
***Regulatory Authority:***

- AB 2588-Air Toxics "Hot Spots" Chemicals (CAL) (as copper compounds)
- The "Director's List" (CAL/OSHA) (as copper compounds)
- Clean Water Act: Toxic Pollutant (Section 401.15) as copper and compounds.
- RCRA Ground Water Monitoring List. Suggested test method(s) (PQL ug/L): 6010(60); 7210(200) Note: All species in the ground water that contain copper are included.
- EPCRA (Section 313): Includes any unique chemical substance that contains copper as part of that chemical's infrastructure. Form R *de minimus* concentration reporting level: 1.0%

***Description:*** White, odorless crystalline solid. Stable in dry air; becomes green on exposure to moist air and brown on exposure to light. Slightly soluble in water. Molecular weight = 98.99. Melting/Freezing point = 430°C. Undergoes a phase transition @ 4°C. Boiling point = 1490°C. Density = 3.58. Vapor pressure = 1 mmHg @ 546°C.

*Incompatibilities:* When heated to decomposition or on contact with acids or acid fumes, may produce highly toxic chloride fumes; deadly phosgene gas may be formed. May cause pitting of some metals.

*Permissible Exposure Limits in Air:* As copper: ACGIH[1] TLV 0.2 mg/m$^3$ (fume); 1 mg/m$^3$ (dusts and mists); NIOSH/OSHA 0.1 mg/m$^3$ (fume); 1 mg/m$^3$ as (dusts and mists).

*Determination in Air:* Copper dusts and mists are collected on a filter, worked up with acid, measured by atomic absorption. See NIOSH Method #7029 for copper. For copper fume: filter collection, acid digestion, measurement by atomic absorption.[18]

*Permissible Concentration in Water:* No criteria set. Runoff from spills or fire control may cause water pollution. The U.S. EPA[6] has set a maximum of 1.0 mg/L in water to protect human health.

*Determination in Water:* Total copper may be determined by digestion followed by atomic absorption or by colorimetry (using neocuproine) or by Inductively Coupled Plasma Optical Emission Spectrometry. Dissolved Copper may be determined by 0.45 $\mu$ filtration followed by the preceding methods.

*Routes of Entry:* Ingestion, inhalation.

*Harmful Effects and Symptoms*

*Long Term Exposure:* May cause mutations in humans. May damage the testes and decrease fertility in both males and females. May cause skin allergy and thickening of the skin; copper deposits can cause discoloration in the skin and hair, leaving a green color. Repeated exposure can cause shrinking of the lining of the inner nose with watery discharge, liver damage. Individuals with Wilson's disease absorb, retain, and store copper excessively.

*Points of Attack:* Skin, reproductive system., liver

*Medical Surveillance:* Serum and urine copper level. Liver and kidney tests. Examination by a qualified allergist. More than light alcohol consumption may exacerbate the liver damage caused by copper sulfate.

*First Aid:* If this chemical gets into the eyes, remove any contact lenses at once and irrigate immediately for at least 15 minutes, occasionally lifting upper and lower lids. Seek medical attention immediately. If this chemical contacts the skin, remove contaminated clothing and wash immediately with soap and water. Seek medical attention immediately. If this chemical has been inhaled, remove from exposure, begin rescue breathing (using universal precautions) if breathing has stopped, and CPR if heart action has stopped. Transfer promptly to a medical facility. When this chemical has been swallowed, get medical attention. If victim is conscious and able to swallow, have victim drink 4 to 8 ounces of water. Do not induce vomiting. *Note to physician or authorized medical personnel*: Medical observation is recommended for 24 to 48 hours after breathing overexposure, as pulmonary edema may be delayed. As first aid for pulmonary edema, consider administering a corticosteroid spray. Cigarette smoking may exacerbate pulmonary injury and should be discouraged for at least 72 hours following exposure.

*References:*
- U.S. Environmental Protection Agency, Office of Pesticide Programs, Pesticide Fact Sheet, Cuprous Chloride," Washington, DC. (September 1998). http://www.epa.gov/opprd001/factsheets/cuprous.pdf
- California Environmental Protection Agency "Chemical List of Lists," Sacramento CA (February 1997).
- U.S. Environmental Protection Agency, Toxicology of Metals, Vol. II: Copper, Report EPA-600/1-77-022, pp 206-221, Research Triangle Park, NC (May 1977).
- U.S. Environmental Protection Agency, Copper: ambient Water Quality Criteria, Washington, DC (1980).

## Cuprous Oxide

*Use Type:* Fungicide and insecticide
*CAS Number:* 1317-39-1
*Formula:* $CuO_2$
*Synonyms:* Brown copper oxide; C.I. 77402; Copper(I) oxide; Copper oxide ($Cu_2O$); Copper suboxide; Dicopper monoxide; Kupferoxydul (German); Oleocuivre(French); Red copper oxide; Yellow cuprocide
*Trade Names:* COPOX®; FUNGI-RHAP®; COPPER NORDOX®; COPPER SARDEX®; Cu-75®; FUNGIMAR®; OLEO NORDOX®, PERECOT®, PERENOX®
*Chemical Class:* Inorganic copper compound
*EPA/OPP PC Code:* 025601
*California DPR Chemical Code:* 175
*Uses:* In December, 1991, the U.S. EPA announced its intention to cancel the registration of products containing cuprous oxide as a fungicide used on fruits and vegetables. Used primarily as a wood preservative and produced mainly by paint companies.
*U.S. Maximum Allowable Residue Levels for Cuprous Oxide (40 CFR 180.1021):*

| CROP | ppm |
|---|---|
| Egg | – |
| Fish | – |
| Meat | – |
| Milk | – |
| Poultry | – |
| Raw agricultural commodities | – |
| Shellfish | – |

*Regulatory Authority:*
- Actively registered pesticide in California.

- AB 2588-Air Toxics "Hot Spots" Chemicals (CAL) (as copper compounds)
- The "Director's List" (CAL/OSHA) (as copper compounds)
- Clean Water Act: Toxic Pollutant (Section 401.15) as copper and compounds.
- RCRA Ground Water Monitoring List. Suggested test method(s) (PQL $ug/L$): 6010(60); 7210(200) Note: All species in the ground water that contain copper are included.
- EPCRA (Section 313): Includes any unique chemical substance that contains copper as part of that chemical's infrastructure. Form R *de minimis* concentration reporting level: 1.0%

*Description:* Octahedral, cubic, red or yellow crystals or powder. Practically insoluble in water. Molecular weight = 143.08.

Density = 6.05. Melting/Freezing point = 1233°C. Boiling point = (decomposes @ 1026°C); loses $O_2$ @ 1800°C.

*Permissible Exposure Limits in Air:* As copper: ACGIH[1] TLV 0.2 mg/m$^3$ (fume); 1 mg/m$^3$ (dusts and mists); NIOSH[2]/OSHA 0.1 mg/m$^3$ (fume); 1 mg/m$^3$ as (dusts and mists).

*Determination in Air:* Copper dusts and mists are collected on a filter, worked up with acid, measured by atomic absorption. See NIOSH Method #7029 for copper. For copper fume: filter collection, acid digestion, measurement by atomic absorption.[18]

*Permissible Concentration in Water:* No criteria set. Runoff from spills or fire control may cause water pollution. The U.S. EPA[6] has set a maximum of 1.0 mg/L in water to protect human health.

*Determination in Water:* Total copper may be determined by digestion followed by atomic absorption or by colorimetry (using neocuproine) or by Inductively Coupled Plasma Optical Emission Spectrometry. Dissolved Copper may be determined by 0.45 $\mu$ filtration followed by the preceding methods.

*Harmful Effects and Symptoms*

*Long Term Exposure:* May cause mutations in humans. May damage the testes and decrease fertility in both males and females. May cause skin allergy and thickening of the skin; copper deposits can cause discoloration in the skin and hair, leaving a green color. Repeated exposure can cause shrinking of the lining of the inner nose with watery discharge, liver damage. Individuals with Wilson's disease absorb, retain, and store copper excessively.

*Points of Attack:* Skin, reproductive system., liver

*Medical Surveillance:* Serum and urine copper level. Liver and kidney tests. Examination by a qualified allergist. More than light alcohol consumption may exacerbate the liver damage caused by copper sulfate.

*First Aid:* If this chemical gets into the eyes, remove any contact lenses at once and irrigate immediately for at least 15 minutes, occasionally lifting upper and lower lids. Seek medical attention immediately. If this chemical contacts the skin, remove contaminated clothing and wash immediately with soap and water. Seek medical attention immediately. If this chemical has been inhaled, remove from exposure, begin rescue breathing (using universal precautions) if breathing has stopped, and CPR if heart action has stopped. Transfer promptly to a medical facility. When this chemical has been swallowed, get medical attention. Give large quantities of water and induce vomiting. Do not make an unconscious person vomit.

*References:*
- California Environmental Protection Agency "Chemical List of Lists," Sacramento CA (February 1997)
- U.S. Environmental Protection Agency, Toxicology of Metals, Vol. II: Copper, Report EPA-600/1-77-022, pp 206-221, Research Triangle Park, NC (May 1977).
- U.S. Environmental Protection Agency, Copper: ambient Water Quality Criteria, Washington, DC (1980).
- U.S. Environmental Protection Agency, Office of Pesticide Programs, Pesticide Residue Limits, "Cuprous Oxide," 40 CFR 180.1021. www.epa.gov/pesticides/food/viewtols.htm

# Cuprous Thiocyanate

*Use Type:* Microbiocide, algaecide

*CAS Number:* 1111-67-7

*Formula:* CuSCN

*Alert:* Some formulations may be a Restricted Use Pesticide (RUP)

*Synonyms:* Copper(I) thiocyanate; Cuprous sulfocyanate; Cuprous sulfocyanide

*Trade Names:* BARDIKE®; MICRON®; OSP 3506-35®; SELF POLISHING COPOLYMER®; SUPERYACHT®; TRI-LUX®

*Chemical Class:* Inorganic cyanate

*EPA/OPP PC Code:* 025602

*California DPR Chemical Code:* 2108

*RTECS Number:* GS7175000 (cyanide ion)

*RCRA Number:* P030 as cyanides

*Uses:* Used in anti-fouling paints

*Carcinogen/Hazard Classifications*

**Label Signal Word:** CAUTION, WARNING or DANGER

**WHO Acute Hazard:** Class III, slightly hazardous as elemental copper

*Regulatory Authority:*
- Actively registered pesticide in California.
- AB 2588-Air Toxics "Hot Spots" Chemicals (CAL) (as copper compounds)
- The "Director's List" (CAL/OSHA) (as copper compounds)

- RCRA Ground Water Monitoring List. Suggested test method(s) (PQL ug/L): 6010(60); 7210(200) Note: All species in the ground water that contain copper are included.
- AB 2588-Air Toxics "Hot Spots" Chemicals (CAL)
- CAL Air Resources Board/AB 1807 Toxic Air Contaminants
- Clean Air Act: Hazardous Air Pollutants (Title I, Part A, Section 112);
- Clean Water Act: Toxic Pollutant (Section 401.15) as copper and compounds.; Section 307 Priority Pollutants as cyanide, total; Toxic Pollutant (Section 401.15)
- RCRA Section 261 Hazardous Constituents.
- RCRA Universal Treatment Standards: Wastewater (mg/L), 1.2 (total); 0.86 (amenable); Nonwastewater (mg/kg), 590 (total); 30 (amenable).
- RCRA Ground Water Monitoring List. Suggested test method(s) (PQL ug/L): 9010(40)
- Safe Drinking Water Act: MCL, 0.01 mg/L; MCLG, 0.01 mg/L; Regulated chemical (47 FR 9352)
- EPCRA Section 304 RQ: CERCLA, 10 lb (4.54 kg)
- EPCRA Section 313 Form R *de minimis* concentration reporting level: 1.0%
- Marine pollutant (49CFR, Subchapter 172.101, Appendix B) as cyanides, mixtures or solutions
- Canada, WHMIS, Ingredients Disclosure List (cyanide compounds, inorganic, n.o.s)

*Description:* White to yellowish powder. Practically insoluble in water. Molecular weight = 121.65.

*Incompatibilities:* Violent reaction with chlorates, nitrates including nitric acid, oxidizers, especially peroxides.

*Permissible Exposure Limits in Air:* For cyanides, the OSHA PEL 8-hour TWA is 5 mg/m$^3$ (4.7 ppm). NIOSH recommended ceiling (10 minute) is also 5 mg/m$^3$ (4.7 ppm). ACGIH recommends 5 mg/m$^3$ as a TWA. The NIOSH[2] IDLH level = 25 mg/m$^3$. ACGIH adds the notation "skin" indicating the possibility off cutaneous absorption.

*Permissible Concentration in Water:* No criteria set. Runoff from spills or fire control may cause water pollution. As of 1980, the U.S. EPA criteria are: To protect freshwater aquatic life: 3.5 $\mu$g/L as a 24-hour average, never to exceed 52.0 $\mu$g/L. To protect saltwater aquatic life: 30.0 $\mu$g/L on an acute toxicity basis; 2.0 $\mu$g/L on a chronic toxicity basis. To protect human health: 200 $\mu$g/L. The allowable daily intake for man is 8.4 mg/day (6).

*Determination in Water:* Distillation followed by silver nitrate titration or colorimetric analysis using pyridine pyrazolone (or barbituric acid).

*Routes of Entry:* Inhalation, skin absorption, ingestion.

*First Aid:* If this chemical gets into the eyes, remove any contact lenses at once and irrigate immediately for at least 15 minutes, occasionally lifting upper and lower lids. Seek medical attention immediately. If this chemical contacts the skin, remove contaminated clothing and wash immediately with soap and water. Seek medical attention immediately. If this chemical has been inhaled, remove from exposure, begin rescue breathing (using universal precautions) if breathing has stopped, and CPR if heart action has stopped. Transfer promptly to a medical facility. When this chemical has been swallowed, get medical attention. Give large quantities of water and induce vomiting. Do not make an unconscious person vomit. Do not induce vomiting when formulations containing petroleum solvents are ingested.

*References:*
- U.S. Environmental Protection Agency, Cyanides: Ambient Water Quality Criteria, Washington, DC (1980).
- U.S. Environmental Protection Agency, Reviews of the Environmental Effects of Pollutants; V: Cyanide, Report No. EPA-600/1-78-027, Washington, DC (1978).
- U.S. Environmental Protection Agency, Cyanides, Health and Environmental Effects Profile No. 56, Office of Solid Waste, Washington, DC (April 30, 1980).
- U.S. Public Health Service, "ToxFAQs for Cyanide," Atlanta, Georgia, Agency for Toxic Substances and Disease Registry (September 1997). http://www.atsdr.cdc.gov/tfacts8.html
- California Environmental Protection Agency "Chemical List of Lists," Sacramento CA (February 1997)

# Cyanamide

*Use Type:* Herbicide and plant growth regulator
*CAS Number:* 420-04-2
*Formula:* $CH_2N_2$; $H_2NCN$
*Alert:* A Restricted Use Pesticide (RUP)
*Synonyms:* Amidocyanogen; Carbamonitrile; Carbimide; Carbodiimide; Cyanogen nitride; Cyanogenamide; Hydrogen cyanamide
*Trade Names:* DORMEX®, Degussa (Germany); SKW 83010®, SKW (Germany)
*Producers:* Degussa (Germany); Denki Kagaku Kogyo (Japan); Nippon Carbide Industries (Japan); Odda (Norway); Richman Chemicals (USA); SKW (Germany); Union Carbide (USA)
*Chemical Class:* Inorganic
*EPA/OPP PC Code:* 014002; (011139 old EPA code number)
*California DPR Chemical Code:* 2238
*ICSC Number:* 0424
*RTECS Number:* GS5950000
*EEC Number:* 615-013-00-2
*Uses:* Used primarily as a plant growth regulator. Cyanamide may be melted to give a dimer, dicyandiamide or cyanoguanidine. At higher temperatures it gives the trimer, melamine, a raw material for melamine-formaldehyde resins.

*Carcinogen/Hazard Classifications*
**U.S. EPA Carcinogens:** Group C, possible carcinogen

# Cyanazine

**Label Signal Word:** DANGER

*Regulatory Authority:*

- Permissible Exposure Limits for Chemical Contaminants (CAL/OSHA)
- The "Director's List" (CAL/OSHA)
- Actively registered pesticide in California.
- AB 2588-Air Toxics "Hot Spots" Chemicals (CAL)
- CAL Air Resources Board/AB 1807 Toxic Air Contaminants
- Clean Air Act: Hazardous Air Pollutants (Title I, Part A, Section 112);
- Section 307 Priority Pollutants as cyanide, total; Toxic Pollutant (Section 401.15)
- RCRA Section 261 Hazardous Constituents.
- RCRA Universal Treatment Standards: Wastewater (mg/L), 1.2 (total); 0.86 (amenable); Nonwastewater (mg/kg), 590 (total); 30 (amenable).
- RCRA Ground Water Monitoring List. Suggested test method(s) (PQL ug/L): 9010(40)
- Safe Drinking Water Act: MCL, 0.01 mg/L; MCLG, 0.01 mg/L; Regulated chemical (47 FR 9352)
- EPCRA Section 304 RQ: CERCLA, 10 lb (4.54 kg)
- EPCRA Section 313 Form R *de minimis* concentration reporting level: 1.0%
- EPA Hazardous Waste Number (RCRA No.): P030 as cyanides
- Marine pollutant (49CFR, Subchapter 172.101, Appendix B) as cyanides, mixtures or solutions

**Description:** Cyanamide is a combustible crystalline solid, but it is usually found as a 25% liquid solution. Boiling point = (decomposes) 260°C. Melting/Freezing point = 45°C. Flash point = 141°C. Molecular weight = 42.00. Hazard Identification (based on NFPA-704 M Rating System): Health 4, Flammability 1, Reactivity 3. Soluble in water.

**Incompatibilities:** Slowly hydrolyzes in water, releasing ammonia and forming acetate salts. Cyanamide may polymerize at temperatures above 122°C, or on evaporation of aqueous solutions. Reacts with acids, strong oxidants, strong reducing agents and water causing explosion and toxic hazard. Attacks various metals. Decomposes when heated above 49°C, on contact with acids, bases, 1,2-phenylene diamine salts, and moisture producing toxic fumes including nitrogen oxides and cyanides.

**Permissible Exposure Limits in Air:** There is no OSHA[2] PEL. NIOSH[2] and ACGIH[1] recommend a TWA of 2 mg/m$^3$ as does HSE[33], Australia, Israel, and Mexico. The Canadian provinces of Alberta, BC, Ontario, and Quebec use the same TWA as well, and Alberta's STEL is 4 mg/m$^3$. The former USSR-UNEP/IRPTC project[43] has set an MAC of 0.5 mg/m$^3$ in workplace air and limits for ambient air in residential areas of 0.01 mg/m$^3$ on an average daily basis. Several states have set guidelines or standards for Cyanamide in ambient air (60) ranging from 6.7 µg/m$^3$ (New York) to 20.0 µg/m$^3$ (Florida and North Dakota) to 35 µg/m$^3$ (Virginia) to 40 µg/m$^3$ (Connecticut) to 50 µg/m$^3$ (South Carolina).

**Determination in Air:** Filter; none; Gravimetric; NIOSH Methods (IV) #0500, Particulates NOR (total)

**Permissible Concentration in Water:** No criteria set. Runoff from spills or fire control may cause water pollution.

**Routes of Entry:** Inhalation, skin absorption, ingestion, skin and/or eye contact.

*Harmful Effects and Symptoms*

**Short Term Exposure:** Cyanamide is caustic and severely irritates the eyes, skin, and respiratory tract, and may affect the liver. Ingestion or inhalation may cause transitory intense redness of the face, headache, vertigo, increased respiration, tachycardia and hypotensions. The adverse effects off Cyanamide are potentiated by the ingestion of alcohol (beer, wine or liquor) within 1-2 days before or after exposure. Cyanamide is a highly reactive chemical and is a dangerous explosion hazard.

**Long Term Exposure:** Repeated or prolonged contact may cause skin sensitization and allergy. Exposure may cause liver and nervous system damage.

**Points of Attack:** Liver and skin

**Medical Surveillance:** If overexposure occurs or if illness is suspected, the following are recommended: Liver function tests. Exam of the nervous system. Examination by a qualified allergist.

**First Aid:** If this chemical gets into the eyes, remove any contact lenses at once and irrigate immediately for at least 15 minutes, occasionally lifting upper and lower lids. Seek medical attention immediately. If this chemical contacts the skin, remove contaminated clothing and wash immediately with soap and water. Seek medical attention immediately. If this chemical has been inhaled, remove from exposure, begin rescue breathing (using universal precautions) if breathing has stopped, and CPR if heart action has stopped. Transfer promptly to a medical facility. When this chemical has been swallowed, get medical attention. Give large quantities of water and induce vomiting. Do not make an unconscious person vomit. Do not induce vomiting when formulations containing petroleum solvents are ingested.

*References:*

- New Jersey Department of Health and Senior Services and Senior Services, "Hazardous Substance Fact Sheet: Cyanamide," Trenton, NJ (November 1994, rev. February 2001). http://www.state.nj.us/health/eoh/rtkweb/0552.pdf
- Sax, N.I., Ed., "Dangerous Properties of Industrial Materials Report," 8, No. 5, 65-68 (1988).
- California Environmental Protection Agency "Chemical List of Lists," Sacramento CA (February 1997)

# Cyanazine

*Use Type:* Herbicide

# Cyanazine

*CAS Number:* 21725-46-2
*Formula:* $C_9H_{13}ClN_6$
*Alert:*. On August 2, 1995, the U.S. EPA announced a voluntary phase-out of the use of cyanazine. On January 6, 2000, the U.S. EPA announced the cancellation of all cyanazine registrations. It had been a Restricted Use Pesticide (RUP) because of its teratogenicity and because it has been found in groundwater. Human toxicity (long-term): High.

*Synonyms:* 2-Chloro-4-((1-cyano-1-methylethyl)amino)-6-(ethylamino)-*s*-triazine; 2-Chloro-4-(1-cyano-1-methylethylamino)-6-ethylamino-1,3,5-triazine; 2-Chloro-4-ethylamino-6-(1-cyano-1-methyl)ethylamino-*s*-triazine; 2-(4-Chloro-6-ethylamino-1,3,5-triazin-2-ylamino)-2-methylpropionitrile; 2-[(4-Chloro-6-(Ethylamino)-S-triazin-2-yl)amino]-2-methylpropionitrile; 2-[(4-Chloro-6-(ethylamino)-1,3,5-triazin-2-yl)amino]-2-methylpropanenitrile; 2-((4-Chloro-6-(ethylamino)-*s*-triazin-2-yl)amino)-2-methylpropanenitrile; Cyanazine triazine pesticide; Propanenitrile, 2-((4-Chloro-6-(ethylamino)-1,3,5-triazin-2-yl)amino)-2-methyl-; Propanenitrile, 2-((4-Chloro-6-(ethylamino)-S-triazin-2-yl)amino)-2-methyl-; *S*-Triazine, 2-chloro-4-ethylamino-6-(1-cyano-1-methyl)ethylamino-

*Trade Names:* BLADEX®, DuPont Crop Protection (USA) canceled; BLADEX® 80WP, DuPont Crop Protection (USA) canceled; BULLET®, Makhteshim-Agan Industries (Israel); CYCLE®, Syngenta (Switzerland); CY-PRO®, Griffin Corp. (USA) canceled; CYNEX® 41, Griffin Corp. (USA) canceled; DW 3418®; EXTRAZINE®, DuPont Crop Protection (USA) canceled; FORTROL®, Makhteshim-Agan Industries (Israel); MATCH®; PAYZE®; SD 15418®; WL 19805®

*Producers:*; BASF Agricultural Products Group (Germany); DuPont Crop Protection (USA); Ehrenstorfer, Dr. (Germany); Griffin Corp. (USA); Makhteshim-Agan Industries (Israel); Sanonda Ltd. (Australia); Sanonda Zhengzhou Pesticide Co., Ltd. (China); Shenzhen Guomeng Industry Co., Ltd. (China); Sigma-Aldrich Laborchemikalien (Germany); Syngenta (Switzerland)

*Chemical Class:* Triazine
*EPA/OPP PC Code:* 100101
*California DPR Chemical Code:* 1640
*ICSC Number:* 0391
*RTECS Number:* UG1490000
*EEC Number:* 613-013-00-7
*EINECS Number:* 244-544-9

*Uses:* Cyanazine is used as a pre- and post-emergent herbicide to control annual grasses and broadleaf weeds. It is used mostly on corn, some on cotton, and less than 1% on grain sorghum and wheat fallow. The compound is formulated as a wettable powder, a flowable suspension, or as granules. In California, major usages are on cotton, corn and corn for forage, almonds, and uncultivated agricultural areas. In 1995, cyanazine was the fourth most widely used synthetic pesticide in the U.S. Cyanazine, atrazine, and simazine are collectively referred to as the triazines and may be alternatively used for each other on corn and in some other situations.

*Human toxicity (long-term)*[77]: High–1.00 ppb, Health Advisory
*Fish toxicity (threshold)*[77]: Very low–1411.45410 ppb, MATC (Maximum Acceptable Toxicant Concentration)

*Carcinogen/Hazard Classifications*
**U.S. EPA Carcinogens:** Group C, possible carcinogen
**California Prop. 65:** Developmental toxin
**Label Signal Word:** WARNING–Toxicity Class II, Moderately Toxic
**TRI Developmental Toxin:** Developmental toxin
**WHO Acute Hazard:** Class II, moderately hazardous
**Endocrine Disruptor:** Suspected

*Regulatory Authority:*
- Safe Drinking Water Act: Priority List (55 FR 1470)
- EPCRA Section 313 Form R *de minimus* concentration reporting level: 1.0%
- Clean Air Act: Hazardous Air Pollutants (Title I, Part A, Section 112)
- Clean Water Act: 40CFR423, Appendix A, Priority Pollutants as cyanide, total.
- EPA Hazardous Waste Number (RCRA No.): P030 as cyanides soluble salts and complexes, n.o.s.
- RCRA, 40CFR261, Appendix 8 Hazardous Constituents, as cyanides, soluble salts and complexes, n.o.s.
- EPCRA (Section 313): X+CN- where X = H+ or any other group where a formal dissociation may occur. For example, KCN or $Ca(CN)_2$. Form R *de minimus* concentration reporting level: 1.0%
- AB 1803-Well Monitoring Chemical (CAL)
- AB 2588-Air Toxics "Hot Spots" Chemicals (CAL)
- Proposition 65 chemical (CAL)
- Actively registered pesticide in California.
- U.S. DOT Regulated Marine Pollutant (49CFR172.101, Appendix B) as cyanide mixtures, cyanide solutions or cyanides, inorganic, n.o.s.
- Canada, Drinking Water Quality, IMAC = 0.01 mg/L

*Description:* Off-white to tan crystalline solid. Soluble in water; solubility = 170 ppm @ 20°C; 0.02 g/100 mL @ 25°C. Molecular weight = 240.69. Melting/Freezing point = 167–169°C. Hazard Identification (based on NFPA-704 M Rating System): Health 2, Flammability 1, Reactivity 0. Vapor pressure = $1.6 \times 10^{-9}$ mmHg @ 20°C; $2.2 \times 10^{-7}$ Pa @ 20°C. Physical properties may be altered by carrier solvents used in commercial formulations. Log $K_{ow}$ = 2.2. Unlikely to bioaccumulate in marine organisms.

*Incompatibilities:* Cyanazine decomposes in heat producing very toxic fumes and gases of hydrogen cyanide, hydrogen chloride, ethyl chloride, ammonia, acetone, and ethylene. Attacks metals in the presence of heat and moisture.

*Permissible Exposure Limits in Air:* No standards set. This chemical can be absorbed through the skin, thereby increasing exposure.

*Determination in Air:* NIOSH[2] REL: (Nitriles) TWA 22 mg/m$^3$; NIOSH[2] REL: (Nitriles); ceiling limit 4 mg/m$^3$/15minute. See also NIOSH Criteria Document 78-212 NITRILES.

*Permissible Concentration in Water:* The U.S. EPA has set a maximum contaminant level of cyanide in drinking water of 0.2 milligrams cyanide per liter of water (0.2 mg/L). Kansas[61] has set a guideline for cyanazine in drinking water of 42 $\mu$g/L. Canada's IMAC for drinking water is 0.01 mg?L.

*Determination in Water:* High-performance liquid chromatography is applicable to the determination of cyanazine in water according to EPA.

*Routes of Entry:* Inhalation, passing through the unbroken skin.

*Harmful Effects and Symptoms*

*Short Term Exposure:* This chemical can be absorbed through the skin, thereby increasing exposure. Inhalation of dust can irritate the nose, throat and bronchial tubes. Because this material has a low vapor pressure, significant inhalation of vapors is unlikely at ordinary temperatures. Contact can irritate the skin or eyes. Overexposure can cause weakness, nausea and difficulty breathing. The LD$_{50}$ (oral,rat) = 149 mg/kg (moderately toxic). Toxicological properties may be altered by carrier solvents used in commercial formulations.

*Long Term Exposure:* Long-term effects are unknown. Related chemicals in the triazin chemical groups can cause liver damage, reduce thyroid function and/or cause skin allergy. May cause reproductive toxicity in humans.

*Medical Surveillance:* Liver function tests. Thyroid function tests. Evaluation by a qualified allergist.

*First Aid:* If this chemical gets into the eyes, remove any contact lenses at once and irrigate immediately for at least 15 minutes, occasionally lifting upper and lower lids. Seek medical attention immediately. If this chemical contacts the skin, remove contaminated clothing and wash immediately with soap and water. Seek medical attention immediately. If this chemical has been inhaled, remove from exposure, begin rescue breathing (using universal precautions) if breathing has stopped, and CPR if heart action has stopped. Transfer promptly to a medical facility. When this chemical has been swallowed, get medical attention. Give large quantities of water and induce vomiting. Do not make an unconscious person vomit.

*References:*
- EXTOXNET, Extension Toxicology Network, "Pesticide Information Profile, Cyanazine," Oregon State University, Corvallis, OR. June 1996. ace.orst.edu/info/extoxnet/pips/cyanazin.htm
- Sax, N.I., Ed., "Dangerous Properties of Industrial Materials Report," 3, No. 1, 47-50, New York, Van Nostrand Reinhold Co. (1983).
- U.S. Environmental Protection Agency, "Chemical Profile: Cyanazine," Washington, DC, Office of Drinking Water (August 1987).
- California Environmental Protection Agency "Chemical List of Lists," Sacramento CA (February 1997).
- New Jersey Department of Health and Senior Services and Senior Services, "Hazardous Substance Fact Sheet: Bladex," Trenton, NJ (November, 1986, Rev. October, 2000). www.state.nj.us/health/eoh/rtkweb/0240.pdf
- Environment Working Group, "Background Information on Cyanazine," Washington, DC, August 2, 1995. www.ewg.org/pub/home/reports/Cyanazine/Cyanazine.html

# Cyanogen Bromide

*Use Type:* Fumigant and pesticide
*Alert:* Not registered with the U.S. EPA
*CAS Number:* 506-68-3
*Formula:* BrCN
*Synonyms:* Bromine cyanide; Bromocyan; Bromocyanide; Bromocyanogen; Bromure de cyanogen (French); Bromuro de cianogeno (Spanish); Campilit; Cyanobromide; Cyanogen monobromide
*Producers:* Eastman Chemical (USA)
*Chemical Class:* Inorganic
*ICSC Number:* 0136
*RTECS Number:* GT2100000
*EINECS Number:* 208-051-2
*Uses:* Used in organic synthesis or as a fumigant, in textile treatment, in gold cyaniding or as a military poison gas.
*Regulatory Authority:*
- EPA Hazardous Waste Number (RCRA No.): U246
- RCRA, 40CFR261, Appendix 8 Hazardous Constituents.
- Superfund/EPCRA 40CFR355, Appendix B Extremely Hazardous Substances: TPQ = 500/10,000 lb (227/4,540 kg).
- Superfund/EPCRA 40CFR302.4 RQ: CERCLA, 1000 lb (454 kg).
- Clean Air Act: Hazardous Air Pollutants (Title I, Part A, Section 112) as cyanide compounds
- Clean water act: Section 307 Priority Pollutants as cyanide, total; Toxic Pollutant (Section 401.15) as cyanide compounds
- RCRA Universal Treatment Standards: Wastewater (mg/L), 1.2 (total); 0.86 (amenable); Nonwastewater (mg/kg), 590 (total); 30 (amenable) as cyanide compounds
- RCRA Ground Water Monitoring List. Suggested test method(s) (PQL $\mu$g/L): 9010(40) as cyanide compounds

# Cyanogen Bromide

- Safe Drinking Water Act: MCL, 0.01 mg/L; MCLG, 0.01 mg/L; Regulated chemical (47 FR 9352) as cyanide compounds
- EPCRA Section 304 RQ: CERCLA, 10 lb (4.54 kg) as cyanide compounds
- EPCRA Section 313 Form R de minimus concentration reporting level: 1.0% as cyanide compounds
- U.S. DOT Inhalation Hazard Chemicals
- U.S. DOT Regulated Marine Pollutant (49CFR, Subchapter 172.10)
- Canada, WHMIS, Ingredients Disclosure List; National Pollutant Release Inventory (NPRI); CEPA Priority Substance
- List, Canada, WHMIS, Ingredients Disclosure List; National Pollutant Release Inventory (NPRI); CEPA Priority Substance List, Ocean dumping prohibited.

***Description:*** Colorless or white, crystalline solid. Penetrating odor. Soluble in water; 10%. Molecular weight = 105.93.
Boiling point = 61-62°C @ 760 mmHg. Melting/Freezing point = 49–51°C; decomposes slowly forming hydrogen cyanide and hydrogen bromide. Vapor pressure =92 mmHg @ 20°C. Hazard Identification (based on NFPA-704 M Rating System): Health 3 (severe, poison), Flammability 0, Reactivity 3 (severe, water reactive).

***Incompatibilities:*** May be unstable unless dry and pure. Impure material may explode. Violent reaction with acids, ammonia, amines. The substance decomposes on heating or on contact with water, acids, or acid vapors producing highly toxic and flammable hydrogen cyanide and corrosive hydrogen bromide.

***Permissible Exposure Limits in Air:*** No standards set.

***Determination in Air:*** No criteria set. However, inasmuch as it is a cyanide compound, the exposure limits are listed here: OSHA[(2)] and ACGIH[(1)] : 5 mg/m$^3$ TWA; NIOSH[(2)]: Ceiling limit, 4.7 ppm; 5 mg/m$^3$ per 10 minutes as cyanides. All have notations that skin contact contributes significantly in overall exposure. IDLH level = 25 mg/m$^3$ as CN

***Permissible Concentration in Water:*** The U.S. EPA has set a maximum contaminant level of cyanide in drinking water of 0.2 milligrams cyanide per liter of water (0.2 mg/L). Runoff from spills or fire control may cause water pollution.

***Routes of Entry:*** Inhalation, ingestion, skin absorption.

***Harmful Effects and Symptoms***

***Short Term Exposure:*** Cyanogen bromide's toxic action resembles that of hydrocyanic acid. Cyanogen bromide is corrosive to the eyes, skin, and respiratory tract. Higher exposures can cause pulmonary edema, a medical emergency that can be delayed for several hours. This can cause death. Exposure may result in death. Super toxic; probable oral lethal dose in humans in less than 5 mg/kg or a taste (less than 7 drops) for a 70 kg (150 lb) person. Vapors are highly irritant and very poisonous. High concentrations produce excessive respiration (causing increased uptake of cyanide), then labored breathing, paralysis, unconsciousness, convulsions and respiratory arrest. Headache, dizziness, nausea, and vomiting may occur with lesser concentrations. Patients may experience confusion, anxiety, an initial rise in blood pressure with a decreased heart heat followed by an increased heart beat; cyanosis is not a consistent finding, in fact, the patient may be reddish. An odor of bitter almonds on the patient's breath may be present. Individuals with chronic diseases of the kidneys, respiratory tract, skin, or thyroid are at greater risk of developing toxic cyanide effects. Delayed pulmonary edema may occur after inhalation.

***Long Term Exposure:*** Repeated or prolonged exposure to cyanogen bromide may cause thyroid gland enlargement. Chronic exposure may cause fatigue and weakness.

***Points of Attack:*** Eyes, respiratory system, cardiovascular system, central nervous system and thyroid gland.

***Medical Surveillance:*** Medical observation is recommended for 24 to 48 hours after breathing overexposure, as pulmonary edema may be delayed. Thyroid gland examination.

***First Aid:*** If this chemical gets into the eyes, remove any contact lenses at once and irrigate immediately for at least 15 minutes, occasionally lifting upper and lower lids. Seek medical attention immediately. If this chemical contacts the skin, remove contaminated clothing and wash immediately with soap and water. Seek medical attention immediately. Do not perform direct mouth to mouth resuscitation; use bag/mask apparatus. If this chemical has been inhaled, remove from exposure, begin rescue breathing (using universal precautions) if breathing has stopped, and CPR if heart action has stopped. Transfer promptly to a medical facility. When this chemical has been swallowed, get medical attention. If victim is conscious, administer water or milk. Do not induce vomiting. Medical observation is recommended for 24 to 48 hours after breathing overexposure, as pulmonary edema may be delayed. As first aid for pulmonary edema, a doctor or authorized paramedic may consider administering a corticosteroid spray. Use amyl nitrate capsules if symptoms of cyanide poisoning develop. All area employees should be trained regularly in emergency measures for cyanide poisoning and in CPR. A cyanide antidote kit should be kept in the immediate work area and must be rapidly available. Kit ingredients should be replaced every 1-2 years to ensure freshness. Persons trained in the use of this kit, oxygen use, and CPR must be quickly available.

***References:***
- Sax, N.I., Ed., "Dangerous Properties of Industrial Materials Report," 1, No. 8, 60-62, New York, Van Nostrand Reinhold Co. (1981).
- U.S. Environmental Protection Agency, "Chemical Profile: Cyanogen Bromide," Washington, DC, Chemical Emergency Preparedness Program (November 30, 1987).

- California Environmental Protection Agency "Chemical List of Lists," Sacramento CA (February 1997).

# Cyanogen Chloride

*Use Type:* Fumigant
*CAS Number:* 506-77-4
*Formula:* CClN; CNCl
*Synonyms:* Chlorcyan; Chlorine cyanide; Chlorocyan; Chlorocyanide; Chlorocyanogen; Chlorure de cyanogene (French); Cloruro de cianogeno (Spanish); Cyanogen chloride [(CN)Cl]; Cyanogen chloride, containing less than 0.9% Water
*Producers:* Air Liquid (France); American Gas Group (USA); Degussa (Germany); Praxair (USA)
*Chemical Class:* Organic cyano halide
*EPA/OPP PC Code:* 025801
*ICSC Number:* 1053
*RTECS Number:* GT2275000
*EINECS Number:* 208-052-8
*Uses:* Cyanogen chloride is used as a fumigant, metal cleaner, in ore refining, production of synthetic rubber and in chemical synthesis. Cyanogen chloride can be used in the military as a poison gas.
*Carcinogen/Hazard Classifications*
**WHO Acute Hazard:** Not listed as cyanogen chloride. Class 1b, highly hazardous as sodium cyanide.
*Regulatory Authority:*
- Air Pollutant Standard Set (ACGIH)[1] (Australia) (HSE)[33] (Israel) (OSHA)[58] (Several States)[60] (Several Canadian Provinces)
- U.S. DOT Inhalation Hazard Chemicals
- Permissible Exposure Limits for Chemical Contaminants (CAL/OSHA)
- The "Director's List" (CAL/OSHA)
- RCRA Universal Treatment Standards: Wastewater (mg/L), 1.2 (total); 0.86 (amenable); Nonwastewater (mg/kg), 590 (total); 30 (amenable) as cyanide compounds
- RCRA Ground Water Monitoring List. Suggested test method(s) (PQL ug/L): 9010(40) as cyanide compounds
- EPCRA Section 304 RQ: CERCLA, 10 lb (4.54 kg) as cyanide compounds
- EPCRA Section 313 Form R *de minimus* concentration reporting level: 1.0% as cyanide compounds
- OSHA Process Safety Management of Highly Hazardous Chemicals (29CFR, Part 1910.119, Appendix A): Threshold Quantity: 500 pounds.
- Clean Air Act: Accidental Release Prevention/Flammable substances, (Section 112[r], Table 3), TQ = 10,000 lb (4540 kg)
- Clean Air Act: Hazardous Air Pollutants (Title I, Part A, Section 112) as cyanide compounds
- Clean Water Act: Section 311 Hazardous Substances/RQ 40CFR117.3 (same as CERCLA, see below); Section 313
- Water Priority Chemicals (57FR41331, 9/9/92)
- Clean Water Act: Section 307 Priority Pollutants as cyanide, total; Toxic Pollutant (Section 401.15) as cyanide compounds
- EPA Hazardous Waste Number (RCRA No.): P033
- RCRA, 40CFR261, Appendix 8 Hazardous Constituents.
- Safe Drinking Water Act: Priority List (55 FR 1470)
- Safe Drinking Water Act: MCL, 0.01 mg/L; MCLG, 0.01 mg/L; Regulated chemical (47 FR 9352) as cyanide compounds
- Superfund/EPCRA 40CFR302.4 RQ: CERCLA, 10 lb (4.54 kg).
- U.S. DOT Regulated Marine Pollutant (49CFR172.101, Appendix B)
- U.S. DOT 49CFR172.101, Inhalation Hazardous Chemical
- Canada, WHMIS, Ingredients Disclosure List; National Pollutant Release Inventory (NPRI); CEPA Priority Substance List, Ocean dumping prohibited.

*Description:* Cyanogen chloride is a colorless gas or liquid (below 55°F/13°C) with a pungent, irritating odor. Shipped as a liquefied gas. A solid below 20°F/–7°C. Soluble in water (slowly decomposes). Boiling point = 14°C. Melting/Freezing point = –6°C. Vapor pressure = 1987 kPa @ 21°C. Flash point = 51°C.

*Incompatibilities:* May be stabilized to prevent polymerization. Cyanogen chloride may polymerize violently if contaminated with chlorine. In crude form chemical trimerizes violently if catalyzed by traces of hydrogen chloride or ammonium chloride. Contact with alcohols, acids, acid salts, amines, strong alkalis, olefins, strong oxidizers may cause fire and explosion. Heat causes decomposition producing toxic and corrosive fumes of hydrogen cyanide, hydrochloric acid, nitrogen oxides. Reacts slowly with water or water vapor to form hydrogen chloride. Attacks copper, brass, and bronze in the presence of moisture.

*Permissible Exposure Limits in Air:* No PEL has been set. NIOSH[2] and ACGIH[1] recommend a ceiling value of 0.3 ppm (0.6 mg/m$^3$). HSE[33] and Israel set a STEL 0.3 ppm (0.6 mg/m$^3$). The Canadian provinces of Alberta, British Columbia, Ontario and Quebec set the same STEL value. Some states have set guidelines and standards for cyanogen chloride in ambient air[60] ranging from 5.0 $\mu$g/m$^3$ (Virginia) to 6.0 $\mu$g/m$^3$ (North Dakota) to 14.0 $\mu$g/m$^3$ (Nevada).

*Determination in Air:* No method available.

*Permissible Concentration in Water:* The U.S. EPA has set a maximum contaminant level of cyanide in drinking water of 0.2 milligrams cyanide per liter of water (0.2 mg/L). Runoff from spills or fire control may cause water pollution.

*Routes of Entry:* Inhalation, skin absorption (liquid), ingestion (liquid), skin and/or eye contact (liquid)

## Harmful Effects and Symptoms

**Short Term Exposure:** Cyanogen chloride is converted to cyanide in the body. A lacrimator. Cyanogen chloride severely irritates the eyes, skin, and respiratory tract. Inhalation can cause weakness, headache, giddiness, dizziness, confusion, nausea, vomiting; irregular/irregularities heartbeat, and pulmonary edema, a medical emergency that can be delayed for several hours. This can cause death. Skin contact with the liquid may may cause frostbite and irritaion. The toxicity of cyanogen chloride resides very largely on its pharmacokinetic property of yielding readily to hydrocyanic acid in vivo. Inhaling small amounts of cyanogen chloride causes dizziness, weakness, congestion of the lungs, hoarseness, conjunctivitis, loss of appetite, weight loss, and mental deterioration. These effects are similar to those found from inhalation of cyanide. Ingestion or inhalation of a lethal dose of cyanogen chloride ($LD_{50}$ = 13 mg/kg), as for cyanide or other cyanogenic compounds, causes dizziness, rapid respiration, vomiting, flushing, headache, drowsiness, drop in blood pressure, rapid pulse, unconsciousness, convulsions, with death occurring within 4 hours. Delayed pulmonary edema may occur after inhalation.

**Points of Attack:** Eyes, skin, respiratory system, central nervous system, cardiovascular system

**Medical Surveillance:** Medical observation is recommended for 24 to 48 hours after breathing overexposure, as pulmonary edema may be delayed. Lung function tests. EKG.

**First Aid:** If this chemical gets into the eyes, remove any contact lenses at once and irrigate immediately for at least 15 minutes, occasionally lifting upper and lower lids. Seek medical attention immediately. If this chemical contacts the skin, remove contaminated clothing and wash immediately with soap and water. Seek medical attention immediately. If this chemical has been inhaled, remove from exposure, begin rescue breathing (using universal precautions) if breathing has stopped, and CPR if heart action has stopped. Transfer promptly to a medical facility. When this chemical has been swallowed, get medical attention. Give large quantities of water and induce vomiting. Do not make an unconscious person vomit. Medical observation is recommended for 24 to 48 hours after breathing overexposure, as pulmonary edema may be delayed. As first aid for pulmonary edema, a doctor or authorized paramedic may consider administering a corticosteroid spray. If frostbite has occurred, seek medical attention immediately; do *NOT* rub the affected areas or flush them with water. In order to prevent further tissue damage, do *NOT* attempt to remove frozen clothing from frostbitten areas. If frostbite has *NOT* occurred, immediately and thoroughly wash contaminated skin with soap and water. Use amyl nitrate capsules if symptoms develop. All area employees should be trained regularly in emergency measures for cyanide poisoning and in CPR. A cyanide antidote kit should be kept in the immediate work area and must be rapidly available. Kit ingredients should be replaced every 1-2 years to ensure freshness. Persons trained in the use of this kit, oxygen use, and CPR must be quickly available.

**References:**
- U.S. Environmental Protection Agency, Cyanogen Chloride, Health and Environmental Effects Profile No. 57, Office of Solid Waste, Washington, DC (April 30, 1980).
- Sax, N.I., Ed., "Dangerous Properties of Industrial Materials Report," 1, No. 8, 62-63 (1981) and 6, No. 1, 46-49 (1986).
- California Environmental Protection Agency "Chemical List of Lists," Sacramento CA (February 1997).

# Cyanophos

**Use Type:** Insecticide and avicide
**CAS Number:** 2636-26-2
**Formula:** $C_8H_{16}N_5O_6P_2S_2$
**Alert:** Not registered as a pesticide in the U.S.
**Synonyms:** Ciafos; Cianofos (Spanish); *O-p*-Cyanophenyl *O,O*-dimethyl phosphorothioate; *O*-(4-Cyanophenyl) *O,O*-dimethyl Phosphorothioate; Cyanophos organophosphate compound; *O,O*-Dimethyl-*O*-(4-cyano-phenyl)-monothiophosphat (German); *O,O*-dimethyl-*O-p*-cyanophenyl-phosphorothioate; *O,O*-Dimethyl-*O*-4-cyanophenyl-phosphorothioate; *O,O*-Dimethyl-*O*-4-cyanophenyl-phosphorothioate; ENT 25,675; Phosphorothioic acid, *O*-(4-cyanophenyl)-9,9-dimethyl ester; Phosphorothioic acid, *O*-(4-cyanophenyl)-*O,O*-dimethyl ester; Phosphorothioic acid, *O,O*-dimethyl ester, *O*-ester with *p*-hydroxybenzonitrile
**Trade Names:** BAY® 34727, Bayer CropScience (Germany); BAYER 34727, Bayer CropScience (Germany); CYANOX®; CYAP®; S 4084®, Sumitomo Chemical (Japan); SUMITOMO S® 4084, Sumitomo Chemical (Japan)
**Producers:** Bayer CropScience (Germany); Samsung Fine Chemicals (South Korea); Sigma-Aldrich Laborchemikalien (Germany); Sumitomo Chemical (Japan)
**Chemical Class:** Organophosphate
**EPA/OPP PC Code:** 268200
**RTECS Number:** TF7600000
**Uses:** This insecticide is used against rice stem borers and house flies. It is not registered as a pesticide in the U.S.
**Carcinogen/Hazard Classifications**
**WHO Acute Hazard:** Group II, moderately hazardous
**Regulatory Authority:**
- Superfund/EPCRA 40CFR355, Appendix B Extremely Hazardous Substances: TPQ = 1000 lb (454 kg)
- Superfund/EPCRA 40CFR302.4 RQ: EHS, 1 lb (0.454 kg)

- U.S. DOT Inhalation Hazard Chemicals as organophosphates
- U.S. DOT Regulated Marine Pollutant (49CFR172.101, Appendix B)
- Canada, WHMIS, Ingredients Disclosure List; National Pollutant Release Inventory (NPRI); CEPA Priority Substance List, Ocean dumping prohibited.

***Description:*** Cyanophos is a yellow to reddish-yellow or amber liquid. Slightly soluble in water. Boiling point = 119°C (decomposes). Melting/Freezing point = 14–15°C.

***Incompatibilities:*** Alkaline materials and exposure to light can cause rapid decomposition.

***Permissible Exposure Limits in Air:*** No standards set. However, inasmuch as it is a cyanide compound, the exposure limits are listed here: OSHA[2] and ACGIH[1] : 5 mg/m$^3$ TWA; NIOSH[2]: Ceiling limit, 4.7 ppm; 5 mg/m$^3$ per 10 minutes as cyanides. All have notations that skin contact contributes significantly in overall exposure. IDLH level = 25 mg/m$^3$ as CN

***Determination in Air:*** Filter/Bubbler; Potassium hydroxide; Ion-specific electrode; NIOSH IV Method #7904, Cyanides[18]. OSHA versatile sampler-2; Toluene/Acetone; Gas chromatography/Flame photometric detection for sulfur, nitrogen, or phosphorus; NIOSH Method IV Method #5600[18], Organophosphorus pesticides.

***Permissible Concentration in Water:*** The U.S. EPA has set a maximum contaminant level of cyanide in drinking water of 0.2 milligrams cyanide per liter of water (0.2 mg/L). Runoff from spills or fire control may cause water pollution.

***Routes of Entry:*** Inhalation, ingestion, skin contact. Absorbed through the skin

***Harmful Effects and Symptoms***

***Short Term Exposure:*** Cyanophos is an organophosphorus insecticide. It is a cholinesterase inhibitor. Death may occur after a large oral dose; with smaller accidental doses, onset of illness may be delayed. The $LD_{50}$ (oral, rat) = 25 mg/kg (highly toxic). Symptoms of organophosphorus pesticide poisoning include: headache, giddiness, nervousness, blurred vision, weakness, nausea, cramps, diarrhea, and discomfort in the chest. Signs include: sweating, pinpoint pupils, tearing, salivation and other excessive respiratory tract secretion, vomiting, cyanosis, papilledema, uncontrollable muscle twitches followed by muscular weakness, convulsions, coma, loss of sphincter control. Delayed pulmonary edema may occur after inhalation.

***Long Term Exposure:*** Cholinesterase inhibitor; possible cumulative effect. Cyanophos may damage the nervous system, resulting in convulsions, respiratory failure. May cause liver damage.

***Points of Attack:*** Respiratory system, central nervous system, peripheral nervous system, plasma cholinesterase

***Medical Surveillance:*** Medical observation is recommended for 24 to 48 hours after breathing overexposure, as pulmonary edema may be delayed. Before employment and at regular times after that, the following are recommended: Plasma and red blood cell cholinesterase levels (tests for the enzyme poisoned by this chemical). If exposure stops, plasma levels return to normal in 1-2 weeks while red blood cell levels may be reduced for 1-3 months. When acetylcholinesterase enzyme levels are reduced by 25% or more below preemployment levels, risk of poisoning is increased, even if results are in lower ranges of "normal." Reassignment to work not involving organophosphate or carbamate pesticides is recommended until enzyme levels recover. If symptoms develop or overexposure occurs, repeat the above tests as soon as possible and get an exam of the nervous system. Also consider complete blood count. Consider chest x-ray following acute overexposure. Do not drink any alcoholic beverages before or during use. Alcohol promotes absorption of organophosphates.

***First Aid:*** **Treatment for organophosphate poisoning consists of thorough decontamination, cardiorespiratory support, and administration of the antidotes atropine and pralidoxime. In cases of severe poisoning, diazepam, an anticonvulsant, should also be administered. Antidotes should be administered as prevention even if the diagnosis is in doubt.** Speed in removing material from eyes and skin is of extreme importance. *Eyes:* Eye contact can cause dangerous amounts of these chemicals to be quickly absorbed through the mucous membrane into the bloodstream. Immediately and gently flush eyes with plenty of warm or cold water (NO hot water) for at least 15 minutes, occasionally lifting the upper and lower eyelids. Get medical aid immediately. *Skin:* Get medical aid. Skin contact can cause dangerous amounts of these chemicals to be absorbed into the bloodstream. Wearing the appropriate PPE equipment and respirator for organophosphate pesticides, immediately flush skin with plenty of soap and water for at least 15 minutes while removing contaminated clothing and shoes. Shampoo hair promptly if contaminated. The removed, contaminated clothing and shoes should be double-bagged and left in Hot Zone for later disposal by hazardous materials experts. Skin may also be decontaminated with diluted hypochlorite solution. *Inhalation:* Get medical aid. Do not contaminate yourself. Wearing the appropriate PPE equipment and respirator for organophosphate pesticides, immediately remove the victim from the contaminated area to fresh air. If the victim is not breathing, administer artificial respiration. Do NOT use mouth-to-mouth resuscitation; use bag/mask apparatus. If breathing is difficult, administer oxygen through bag/mask apparatus until medical help arrives. Do not leave victim unattended. *Ingestion:* Call poison control. Loosen all clothing. Never give anything by mouth to an unconscious person. If victim is *unconscious or having convulsions,* do nothing except keep victim warm. Get medical aid. Transfer promptly to a medical facility. In cases of ingestion, **do not induce vomiting**. If the victim is alert and asymptomatic, administer a slurry of activated charcoal at a dose of 1 g/kg (infant, child, and adult dose). A soda can and straw may be of assistance when offering charcoal to a child. *In some cases you may be specifically instructed by poison control to induce vomiting by way of 2 tablespoons of syrup of ipecac (adult) washed down with a cup of water.* Do NOT give activated charcoal before or with ipecac syrup.

*Note to physician or authorized medical personnel:* Treat cases of respiratory compromise, coma, or excessive pulmonary secretions with respiratory support using protocols and techniques available and within the scope of training. Some cases may necessitate procedures such as endotracheal intubation or cricothyrotomy by properly trained and equipped personnel. When possible, atropine (see *Antidotes*, below) should be given under medical supervision. Patients who are comatose, hypotensive, or having seizures or cardiac arrhythmias should be treated according to advanced life support protocols. *Antidotes:* Two antidotes are administered to treat organophosphate poisoning. Atropine is a competitive antagonist of acetylcholine at muscarinic receptors and is used to control the excessive bronchial secretions which are often responsible for death. Pralidoxime relieves both the nicotinic and muscarine effects of organophosphate poisoning by regenerating acetylcholinesterase and can reduce both the bronchial secretions and the muscle weakness associated with poisoning. The initial intravenous dose of atropine in adults should be determined by the severity of symptoms: An initial adult dose of 1.0 to 2.0 mg or pediatric dose of 0.01 mg/kg (minimum 0.01 mg) should be administered intravenously. If intravenous access cannot be established, atropine may also be given intramuscularly, subcutaneously or via endotracheal tube. Doses should be repeated every 15 minutes until excessive secretions and sweating have been controlled. Once bronchial secretion has been controlled, atropine administration should be repeated whenever the secretions begin to recur. In seriously poisoned patients, very large doses may be required. Alterations of pulse rate and pupillary size should not be used as indicators of treatment adequacy. Pralidoxime should be administered as early in poisoning as possible as its efficacy may diminish when given more than 24 to 36 hours after exposure. Doses are as follows: adult 1.0 g; pediatric 25 to 50 mg/kg. The drug should be administered intravenously over 30 to 60 minutes, but in a life-threatening situation, one-half of the total dose can be given per minute for a total administration time of 2 minutes. Treatment should begin to take effect within 40 minutes with a reduction in symptoms and the amount of atropine necessary to control bronchial secretion. The initial dose can be repeated in 1 hour and then every 8 to 12 hours until the patient is clinically well and no longer requires atropine. If intravenous access cannot be established, pralidoxime may also be given intramuscularly. Early administration of diazepam in addition to the combined atropine and pralidoxime treatment may help prevent the onset of seizures and potential brain and cardiac morphologic damage following high-level organophosphate poisoning.

**References:**
- U.S. Environmental Protection Agency, "Chemical Profile: Cyanophos," Washington, DC, Chemical Emergency Preparedness Program (November 30, 1987).
- California Environmental Protection Agency "Chemical List of Lists," Sacramento CA (February 1997).
- Agency for Toxic Substances and Disease Registry, U.S. Department of Health and Human Services, Public Health Service, "Managing Hazardous Materials Incidents," Atlanta, GA (June 2003).

## Cyclanilide

*Use Type:* Plant growth regulator
*CAS Number:* 113136-77-9
*Formula:* $C_{11}H_9Cl_2NO_3$
*Synonyms:* 1-(2,4-Dichlorophenylaminocarbonyl) cyclopropanecarboxylic acid; Cyclopropanecarboxylic acid, 1-[((2,4-dichlorophenyl)amino)carbonyl]-; Cyclopropanecarboxamide, 1-carboxy-, *N*-(2,4-dichlorophenyl)-
*Trade Names:* FINISH®, Bayer CropScience (Germany); RPA 90946®; EXP 31039B®
*Producers:* Bayer CropScience (Germany)
*Chemical Class:* Malonanilate
*EPA/OPP PC Code:* 026201
*California DPR Chemical Code:* 4030
*Uses:* Cyclanilide is a plant growth regulator used as a cotton harvest aid.

*U.S. Maximum Allowable Residue Levels for Cyclanilide (40 CFR 180.506 ):*

| CROP | ppm |
|---|---|
| Cattle, fat | 0.10 |
| Cattle, meat | 0.02 |
| Cattle, mbyp (except kidney) | 0.2 |
| Cattle, kidney | 2.0 |
| Cottonseed | 0.60 |
| Cotton gin byproducts | 25.0 |
| Goat, fat | 0.10 |
| Goat, meat | 0.02 |
| Goat, mbyp (except kidney) | 0.20 |
| Goat, kidney | 2.0 |
| Horse, fat. | 0.10 |
| Horse, meat | 0.02 |
| Horse, mbyp (except kidney) | 0.20 |
| Horse, kidney | 2.0 |
| Hog, fat | 0.10 |
| Hog, meat | 0.02 |
| Hog, mbyp (except kidney) | 0.20 |
| Hog, kidney | 2.0 |
| Milk | 0.04 |
| Sheep, fat | 0.10 |

Sheep, meat. 0.20
Sheep, mbyp (except kidney). 0.20
Sheep, kidney. 2.0

*Carcinogen/Hazard Classifications*
**U.S. EPA Carcinogens:** Not likely a carcinogen
**Label Signal Word:** DANGER
*Description:* A white powdery solid. No characteristic odor. Melting point=195.5°C.
*Incompatibilities:* May react violently with strong oxidizers, bromine, 90% hydrogen peroxide, phosphorus trichloride, silver powders or dust. Incompatible with silver compounds. Mixture with some silver compounds forms explosive salts of silver oxalate. Keep away from oxidizers, sulfuric acid, caustics, ammonia, aliphatic amines, alkanolamines, isocyanates, alkylene oxides, epichlorohydrin.

*Harmful Effects and Symptoms*
*Short Term Exposure:* Contact with eyes or skin may cause irritation or injury. Inhalation should be avoided; use NIOSH-approved air purifying respirators for pesticides. May be harmful if swallowed.
*First Aid:* If this chemical gets into the eyes, remove any contact lenses at once and irrigate immediately for at least 15 minutes, occasionally lifting upper and lower lids. Seek medical attention immediately. If this chemical contacts the skin, remove contaminated clothing and wash immediately with soap and water. Seek medical attention immediately. If this chemical has been inhaled, remove from exposure, begin rescue breathing (using universal precautions) if breathing has stopped, and CPR if heart action has stopped. Transfer promptly to a medical facility. When this chemical has been swallowed, get medical attention. Give large quantities of water and induce vomiting. Do not make an unconscious person vomit.

*References:*
- U.S. Environmental Protection Agency, Office of Pesticide Programs, Pesticide Residue Limits, "Cyclanilide," 40 CFR 180.506. http://frwebgate.access.gpo.gov/cgi-bin/get-cfr.cgi
- U.S. Environmental Protection Agency, Office of Pesticide Programs, "Pesticide Fact Sheet, Cyclanilide," Washington, DC (May 19, 1997). http://www.epa.gov/opprd001/factsheets/cyclanilide.pdf
- California Environmental Protection Agency "Chemical List of Lists," Sacramento CA (February 1997)

# Cycloate

*Use Type:* Herbicide
*CAS Number:* 1134-23-2
*Formula:* $C_{11}H_{21}NOS$
*Synonyms:* Cyclohexylethylcarbamothioic acid-S-ethyl ester; Cyclohexylethylthiocarbamic acid-S-ethyl ester; S-Ethyl cyclohexylethylthiocarbamate; S-Ethyl-N-ethyl-N-cyclohexylthiolcarbamate; Hexylthiocarbam

*Trade Names:* ETSAN®; EUREX®; R-2063®; RO-NEET®; RO-NEET®-6E; RO-NEET® 10G; RONIT®; SABET®
*Chemical Class:* Thiocarbamate
*EPA/OPP PC Code:* 041301
*California DPR Chemical Code:* 516
*EINECS Number:* 214-482-7
*Uses:* Used to control broadleaf weeds, annual and perennial grasses and nutgrass in spinach, beets, and sugar beets.
*Human toxicity (long-term)[77]:* Intermediate–35.00 ppb, Health Advisory
*Fish toxicity (threshold)[77]:* Very low–658.46663 ppb, MATC (Maximum Acceptable Toxicant Concentration)
*U.S. Maximum Allowable Residue Levels for Cycloate (40 CFR 180.212):*

| CROP | ppm |
| --- | --- |
| Beet, garden, roots | 0.05 |
| Beet, garden, tops | 0.05 |
| Beet, sugar, roots | 0.05 |
| Beet, sugar, tops | 0.05 |
| Spinach | 0.05 |

*Carcinogen/Hazard Classifications*
**California Prop. 65:** Developmental toxin
**Label Signal Word:** WARNING, CAUTION
**TRI Developmental Toxin:** Reproductive and developmental toxin
**WHO Acute Hazard:** Class III, slightly hazardous
*Regulatory Authority:*
- RCRA Universal Treatment Standards: Wastewater (mg/L), 0.003; Nonwastewater (mg/kg), 1.4
- EPA 40 CFR 372.65, Specific Toxic Chemical Listings
- EPCRA Section 313 Form R *de minimis* concentration reporting level: 1.0%

*Description:* An oily, clear amber to yellow liquid. Aromatic odor. Slightly soluble in water; solubility = 100 mg/L. Molecular weight 215.36. Density = 1.0 @ 20°C. Boiling point = 145–146°C @10 mmHg. Vapor pressure = $1.6 \times 10^{-3}$ mmHg @ 20°C.
*Incompatibilities:* Reacts violently with powerful oxidizers such as calcium hypochlorite.
*Permissible Concentration in Water:* No criteria set. Runoff from spills or fire control may cause water pollution.

*Harmful Effects and Symptoms*
*Short Term Exposure:* Low levels of toxicity. Concentrated solutions are slightly corrosive to eyes and mucous membranes. Dust inhalation can cause irritation of the respiratory system with sneezing. Eye contact can cause irritation, watering, pain, and inflammation of the eyelids. Skin contact can cause irritation and minor ulceration. Ingestion can cause nausea, vomiting, fever, muscle twitching, seizure, rapid respiration, slow heart beat. Severe exposure may result in death.
*Long Term Exposure:* High or repeated exposures may cause nerve damage.

*Points of Attack:* Respiratory system, central nervous system, cardiovascular system, skin, eyes.

*Medical Surveillance:* Medical observation is recommended for 24 to 48 hours after breathing overexposure, as pulmonary edema may be delayed. As first aid for pulmonary edema, consider administering a corticosteroid spray. Cigarette smoking may exacerbate pulmonary injury and should be discouraged for at least 72 hours following exposure. Before employment and at regular times after that, the following are recommended: If symptoms develop or overexposure occurs, repeat the above tests as soon as possible and get an exam of the nervous system. Also consider complete blood count. Consider chest x-ray following acute overexposure.

*First Aid:* Eyes: If this chemical gets into the eyes, remove any contact lenses at once and irrigate immediately for at least 15 minutes, occasionally lifting upper and lower lids. Seek medical attention immediately. If this chemical contacts the skin, remove contaminated clothing and wash immediately with soap and water. Seek medical attention immediately. If this chemical has been inhaled, remove from exposure, begin rescue breathing (using universal precautions) if breathing has stopped, and CPR if heart action has stopped. Transfer promptly to a medical facility. When this chemical has been swallowed, get medical attention. If victim is conscious and able to swallow, have victim drink 4 to 8 ounces of water. Do not induce vomiting.

*Note to physician or authorized medical personnel*: Medical observation is recommended for 24 to 48 hours after breathing overexposure, as pulmonary edema may be delayed. As first aid for pulmonary edema, consider administering a corticosteroid spray. Cigarette smoking may exacerbate pulmonary injury and should be discouraged for at least 72 hours following exposure.

*References:*
- California Environmental Protection Agency "Chemical List of Lists," Sacramento CA (February 1997).
- U.S. Environmental. Protection Agency, Office of Pesticide Programs, Pesticide Residue Limits, "Cycloate," 40 CFR 180.212.
  www.epa.gov/pesticides/food/viewtols.htm

# Cycloheximide

*Use Type:* Fungicide; plant growth regulator
*CAS Number:* 66-81-9
*Formula:* $C_{15}H_{23}NO_4$
*Alert:* Restricted Use Pesticide (RUP)
*Synonyms:* 3[2-(3,5-Dimethyl-2-oxocyclohexyl)-2-hydroxyethyl]glutarimide; NSC-185; 2,6-Piperidinedione, 4-(2-3,5-dimethyl-2-oxocyclohexyl)-2-hydroxyethyl-, (IS)-($1\alpha(S^*),3\alpha,5\beta$)-
*Trade Names:* ACTI-AID®, Pharmacia & Upjohn (USA) canceled 4/22/1985; ACTIDIONE®, Pharmacia & Upjohn (USA) canceled 4/22/1985; ACTIDIONE® TGF, Pharmacia & Upjohn (USA) canceled 4/22/1985; ACTIDONE®, Pharmacia & Upjohn (USA) canceled 4/22/1985; ACTISPRAY; HIZAROCIN®; KAKEN®; NARAMYCIN®; NARAMYCIN A®; NEOCYCLOHEXIMIDE®; U-4527

*Producers:* Aldrich Chemical (USA); Merck (Germany); Pharmacia & Upjohn (USA)
*EPA/OPP PC Code:* 043401
*California DPR Chemical Code:* 5
*ICSC Number:* 0244
*RTECS Number:* MA4375000
*EINECS Number:* 200-636-0

*Uses:* Used as an antibiotic, plant growth regulator, and protein synthesis inhibitor. Inhibits growth of many plant pathogemic fungi. Effective for control of powdery mildew on roses and many other ornamentals, rusts and leaf spots on lawn grasses, and azalea petal blight. Also used as a repellent for rodents and other animal pests, and in cancer therapy.

*Human toxicity (long-term):* Very low–140.00 ppb, Health Advisory
*Fish toxicity (threshold):* Very low–1308.29960 ppb, MATC (Maximum Acceptable Toxicant Concentration)

*Carcinogen/Hazard Classifications*
**California Prop. 65:** Developmental toxin
*Regulatory Authority:*
- Banned or Severely Restricted (In Agriculture) (Malaysia) (UN)[13]
- Very Toxic Substance (World Bank)[15]
- EPA/SARA 302 (EPCRA) Extremely hazardous substances
- AB 2588-Air Toxics "Hot Spots" Chemicals (CAL)
- Proposition 65 chemical (CAL)
- The "Director's List" (CAL/OSHA)
- Superfund/EPCRA 40CFR302.4 RQ: EHS, 1 lb (0.454 kg).
- Superfund/EPCRA 40CFR355, Appendix B Extremely Hazardous Substances: TPQ = 100/10,000 lb (45.4/4,540 kg)
- Canada, WHMIS, Ingredients Disclosure List

*Description:* Cycloheximide is a colorless crystalline solid. Slightly soluble in water; solubility = 2.08 g/100 ml @ 2°C. Melting/Freezing point = 119–121°C. Log $K_{ow}$ = < 0.75. Unlikely to bioaccumulate in marine organisms.

*Incompatibilities:* Incompatible with oxidizers, acid anhydrides, strong bases.
*Permissible Exposure Limits in Air:* No OEL has been established
*Permissible Concentration in Water:* No criteria set, but runoff from spills or fire control may cause water pollution.
*Routes of Entry:* Inhalation, ingestion and skin
*Harmful Effects and Symptoms*
*Short Term Exposure:* Contact can cause eye and skin irritation. Exposure can cause excessive salivation, nausea,

vomiting, diarrhea, and elevated blood urea nitrogen (BUN). High exposures can also cause imbalance, tremors, seizures and coma. Extremely toxic [$LD_{50}$ (oral, rat) = 3.7 mg/kg]. The probable oral lethal dose in humans is 5-50 mg/kg, or 7 drops to 1 teaspoonful for a 150-lb. person. Signs of skin irritation may appear as much as 6 to 24 hours after exposure.

*Long Term Exposure:* May cause mutations and damage the developing fetus. May cause liver and kidney damage.

*Points of Attack:* Reproductive system. Liver and kidneys.

*Medical Surveillance:* Liver and kidney function tests.

*First Aid:* If this chemical gets into the eyes, remove any contact lenses at once and irrigate immediately for at least 15 minutes, occasionally lifting upper and lower lids. Seek medical attention immediately. If this chemical contacts the skin, remove contaminated clothing and wash immediately with soap and water. Seek medical attention immediately. If this chemical has been inhaled, remove from exposure, begin rescue breathing (using universal precautions) if breathing has stopped, and CPR if heart action has stopped. Transfer promptly to a medical facility. When this chemical has been swallowed, get medical attention. Give large quantities of water and induce vomiting. Do not make an unconscious person vomit. Medical observation is recommended for 24 to 48 hours following skin contact.

*References:*
- Sax, N.I., Ed., "Dangerous Properties of Industrial Materials Report," 2, No. 5, 41-43 (1982) and 9, No. 1, 55-64 (1989).
- U.S. Environmental Protection Agency, "Chemical Profile: Cycloheximide," Washington, DC, Chemical Emergency Preparedness Program (November 30, 1987).
- New Jersey Department of Health and Senior Services, Hazardous Substance Fact Sheet, "Cycloheximide," Trenton NJ (January 1999). www.state.nj.us/health/eoh/rtkweb/0574.pdf
- Pesticide Management Education Program (PMEP), Cornell University, "Cycloheximide (Acti-dione) Chemical Profile 2/85," Ithaca, NY. http://pmep.cce.cornell.edu/profiles/fung-nemat/aceticacid-etridiazole/cycloheximide/fung-prof-cycloheximide.html
- California Environmental Protection Agency "Chemical List of Lists," Sacramento CA (February 1997).

# Cyfluthrin

*Use Type:* Insecticide

*CAS Number:* 68359-37-5

*Formula:* $C_{22}H_{18}Cl_2FNO_3$

*Alert:* Cyfluthrin can be found in both Restricted Use Pesticides (RUP) and General Use Pesticides (GUP).

*Synonyms:* Cyano(4-fluoro-3-phenoxyphenyl)methyl 3-(2,2-dichloroethenyl)-2,2-dimethylcyclopropanecarboxylate; Cyclopropanecarboxylic acid, 3-(2,2-dichloroethenyl)-2,2-dimethyl-, cyano(4-fluoro-3-phenoxyphenyl)methyl ester; Cyclopropanecarboxylic acid, 2-(2,2-dichlorovinyl)-3,3-dimethyl-, ester with (4-fluoro-3-phenoxyphenyl)hydroxyacetonitrile; Cyfluthin; Cyfluthrine; Cyfoxylate

*Trade Names:* AZTEC®, Bayer CropScience (Germany); ATTATOX®; BAY FCR 1272®, Bayer CropScience (Germany); BAYTHROID® Bayer CropScience (Germany); BAYTHROID® H, Bayer CropScience (Germany); BAYTHROID® TECHNICAL, Bayer CropScience (Germany); BUG-B-GON®, Scotts (USA); CONTUR®; CYLATHRIN®, Bayer CropScience (Germany); EULAN SP®; FCR 1272®; INTUDER® Bayer CropScience (Germany); INTUDER HPX®, Bayer CropScience (Germany); LASER®, Bayer CropScience (Germany); RENOUNCE®, Bayer CropScience (Germany); RESPONSAR®; SOLFAC®; TEMPO®; TEMPO® H; TEMPO® 20WP

*Producers:* Agsin (Singapore); Bayer CropScience (Germany); Changzhou Kangmei Chemical Industry Co., Ltd. (China); Scotts Company, The (USA)

*Chemical Class:* Organofluorine

*EPA/OPP PC Code:* 128831

*California DPR Chemical Code:* 2223

*EINECS Number:* 269-855-7

*Uses:* Cyfluthrin is a non-systemic insecticide used to control a variety of chewing and sucking insects on cotton, hops, cereals, corn, peanuts, fruit, potatoes and other crops and vegetables. It is also used o control structural pests such as termites.

*Human toxicity (long-term)*[77]*:* Very low–175.00 ppb, Health Advisory

*Fish toxicity (threshold)*[77]*:* Extra high–0.01330 ppb, MATC (Maximum Acceptable Toxicant Concentration)

*U.S. Maximum Allowable Residue Levels for Cyfluthrin (40 CFR 180.436):*

| CROP | ppm |
|---|---|
| Alfalfa | 5 |
| Alfalfa, forage | 5 |
| Alfalfa, hay | 10 |
| Animal feed | 0.05 |
| Barley, bran | 5 |
| Barley, grain | 2 |
| Brassica, head and stem, subgroup 5a | 2.5 |
| Carrot, roots | 0.2 |
| Cattle, fat | 10 |
| Cattle, meat | 0.4 |
| Cattle, mbyp | 0.4 |
| Citrus, dried pulp | 0.3 |
| Citrus, oil | 0.3 |
| Corn, field, forage | 3 |
| Corn, field, grain | 0.01 |

| | |
|---|---|
| Corn, field, milled byproducts | 7 |
| Corn, field, refined oil | 30 |
| Corn, field, stover | 6 |
| Corn, pop, forage | 0.01 |
| Corn, pop, grain | 0.01 |
| Corn, pop, stover | 0.01 |
| Corn, sweet, forage | 15.0 |
| Corn, sweet, kernel plus cob with husks removed | 0.05 |
| Corn, sweet, stover | 15.0 |
| Cotton, hulls | 2 |
| Cotton, refined oil | 2 |
| Cotton, undelinted seed | 1 |
| Egg | 0.01 |
| Fruit, citrus, group 10 | 0.2 |
| Goat, fat | 10 |
| Goat, meat | 0.4 |
| Goat, mbyp | 0.4 |
| Grain, aspirated fractions | 300 |
| Grain, cereal, group 15 | 4 |
| Grape | 1 |
| Grape, raisin | 1.5 |
| Hog, fat | 6 |
| Hog, meat | 0.4 |
| Hog, mbyp | 0.4 |
| Hop, dried cones | 20 |
| Hop, fresh | 4 |
| Horse, fat | 6 |
| Horse, meat | 0.4 |
| Horse, mbyp | 0.4 |
| Lettuce, head | 2 |
| Lettuce, leaf | 3 |
| Milk | 1 |
| Milk, fat | 15 |
| Mustard greens | 7 |
| Oat, bran | 5 |
| Oat, grain | 2 |
| Pea, dry | 0.15 |
| Pea, southern, succulent | 0.25 |
| Pepper | 0.5 |
| Potato | 0.01 |
| Poultry, fat | 0.01 |
| Poultry, meat | 0.01 |
| Poultry, mbyp | 0.01 |
| Processed food | 0.05 |
| Radish | 1 |
| Radish, roots | 1 |
| Rice, bran | 6 |
| Rice, hulls | 18 |
| Rye, bran | 5 |
| Sheep, fat | 6 |
| Sheep, meat | 0.4 |
| Sheep, mbyp | 0.4 |
| Sorghum, forage | 2 |
| Sorghum, grain, forage | 2 |
| Sorghum, grain, grain | 4 |
| Sorghum, grain, stover | 5 |
| Soybean, forage | 8 |
| Soybean, hay | 4 |
| Soybean, seed | 0.03 |
| Sugarcane, cane | 0.05 |
| Sugarcane, molasses | 0.2 |
| Sunflower, forage | 5 |
| Sunflower, seed | 0.02 |
| Tomato | 0.2 |
| Tomato, concentrated products | 0.5 |
| Tomato, dried pomace | 5 |
| Tomato, paste | 0.5 |
| Tomato, wet pomace | 5 |
| Wheat, grain | 2 |
| Wheat, milled fractions, except flour | 5 |

*Carcinogen/Hazard Classifications*
**U.S. EPA Carcinogens:** Not likely a carcinogen
**Label Signal Word:** WARNING, CAUTION, DANGER
**WHO Acute Hazard:** Class II, moderately hazardous
**Endocrine Disruptor:** Possible ED
*Regulatory Authority:*
- Actively registered pesticide in California.
- EPA 40 CFR 372.65, Specific Toxic Chemical Listings
- FIFRA, 40CFR185: tolerances for pesticides in food
- FIFRA, 40CFR186: tolerances for pesticides in animal feeds
- EPCRA Section 313 Form R *de minimis* concentration reporting level: 1.0%

*Description:* Molecular weight = 434.33. Vapor pressure - $1.6 \times 10^{-8}$ mmHg @ 20°C.

*Incompatibilities:* May react violently with strong oxidizers, bromine, 90% hydrogen peroxide, phosphorus trichloride, silver powders or dust. Incompatible with silver compounds. Mixture with some silver compounds forms explosive salts of silver oxalate.

*Determination in Air:* See NIOSH Criteria Document 78-212 NITRILES

*Permissible Concentration in Water:* No criteria set. Runoff from spills or fire control may cause water pollution.

*Harmful Effects and Symptoms*
*Short Term Exposure:* Contact with eyes or skin may cause irritation or injury. Inhalation should be avoided; use NIOSH-approved air purifying respirators for pesticides. May be harmful if swallowed.

*First Aid:* If this chemical gets into the eyes, remove any contact lenses at once and irrigate immediately for at least 15 minutes, occasionally lifting upper and lower lids. Seek medical attention immediately. If this chemical contacts the skin, remove contaminated clothing and wash immediately with soap and water. Seek medical attention immediately. If this chemical has been inhaled, remove from exposure, begin rescue breathing (using universal precautions) if breathing has stopped, and CPR if heart action has stopped. Transfer promptly to a medical facility. When this chemical has been swallowed, get medical attention. Give large quantities of water and induce vomiting. Do not make an unconscious person vomit.

*References:*
- EXTOXNET, Extension Toxicology Network, "Pesticide Information Profile, Cyfluthrin," Oregon State University, Corvallis, OR (March 2001). http://pmep.cce.cornell.edu/profiles/extoxnet/carbaryl-dicrotophos/cyfluthrin-ext.html
- California Environmental Protection Agency "Chemical List of Lists," Sacramento CA (February 1997).
- U.S. Environmental Protection Agency, Office of Pesticide Programs, Pesticide Residue Limits, "Cyfluthrin," 40 CFR 180.436. http://www.epa.gov/pesticides/food/viewtols.htm

# lambda-Cyhalothrin

*Use Type:* Insecticide
*CAS Number:* 91465-08-6
*Formula:* $C_{23}H_{19}ClF_3NO_3$
*Alert:* A Restricted Use Pesticide (RUP)
*Synonyms:* (R+S)-α-Cyano-3-phenoxybenzyl (1S+1R)-cis-3-(Z-2-Chloro-3,3,3-trifluoroprop-1-enyl)-2,2-dimethylcyclopropanecarboxylate; (RS)-α-Cyano-3-phenoxybenzyl(Z)-(1RS,3RS)-(2-chloro-3,3,3-trifluoropropenyl)-2,2-dimethylcyclopropanecarboxylate; Cyclopropanecarboxylic acid, 3-(2-chloro-3,3,3-trifluoro-1-propenyl)-2,2-dimethyl-, cyano(3-phenoxyphenyl)methyl ester, [1α(S*),3α(Z)]-(+)-; λ-Cyhalothrin; Cyhalothrin-K
*Trade Names:* CHARGE®; COMMODORE® Syngenta Crop Protection (Switzerland); DEMAND® Syngenta Crop Protection (Switzerland); DEMAND CS® Syngenta Crop Protection (Switzerland); DOUBLE BARREL®, Schering-Plough Animal Health (USA); EXCALIBER®; GRENADE®; HALLMARK®; ICON®; IMPASSE® Syngenta Crop Protection (Switzerland); KARATE® Syngenta Crop Protection (Switzerland); MATADOR®; NINJA®; PP-321® Syngenta Crop Protection (Switzerland); RATE®; SABER®, Schering-Plough Animal Health (USA); SAMURAI®; SCIMITAR® Syngenta Crop Protection (Switzerland); SENTINEL®; WARRIOR® Syngenta Crop Protection (Switzerland)
*Producers:* Schering-Plough Animal Health (USA); Syngenta Crop Protection (Switzerland)
*Chemical Class:* Pyrethroid
*EPA/OPP PC Code:* 128897
*California DPR Chemical Code:* 2297
*ICSC Number:* 0859
*RTECS Number:* GZ1227780

*Uses:* Used to control a variety of pests in many crops. Also used in structural pest situations. A Restricted Use Pesticide (RUP)
*Fish toxicity (threshold)[77]:* Extra high–0.04384 ppb, MATC (Maximum Acceptable Toxicant Concentration)

*U.S. Maximum Allowable Residue Levels for lambda-Cyhalothrin (40 CFR 180.438)*

| CROP | ppm |
|---|---|
| Alfalfa, forage | 5 |
| Alfalfa, hay | 6 |
| Almond, hulls | 1.5 |
| Apple, wet pomace | 2.5 |
| Avocado | 0.2 |
| Barley, bran | 0.2 |
| Barley, grain | 0.05 |
| Barley, hay | 2 |
| Barley, straw | 2 |
| Brassica, head and stem, subgroup 5a | 0.4 |
| Canola, refined oil | 2 |
| Canola, seed | 0.15 |
| Cattle, fat | 3 |
| Cattle, meat | 0.2 |
| Cattle, mbyp | 0.2 |
| Clover, forage | 5 |
| Clover, hay | 6 |
| Corn, field, flour | 0.15 |
| Corn, forage | 6 |
| Corn, grain | 0.05 |
| Corn, stover | 1 |
| Corn, sweet, kernel plus cob With husks removed | 0.05 |
| Cotton, undelinted seed | 0.05 |
| Egg | 0.01 |
| Fruit, pome, group 11 | 0.3 |
| Fruit, stone, group 12 | 0.5 |
| Garlic | 0.1 |
| Goat, fat | 3 |
| Goat, meat | 0.2 |
| Goat, mbyp | 0.2 |
| Grain, aspirated fractions | 2 |
| Grass, forage | 5 |
| Grass, hay | 6 |
| Hog, fat | 3 |
| Hog, meat | 0.2 |
| Hog, mbyp | 0.2 |
| Hop, dried cones | 10 |
| Horse, fat | 3 |
| Horse, meat | 0.2 |
| Horse, mbyp | 0.2 |
| Lettuce, head | 2 |
| Lettuce, leaf | 2 |
| Milk | 0.2 |
| Milk, fat | 5 |
| Nut, tree, group 14 | 0.05 |
| Onion, dry bulb | 0.1 |

| | |
|---|---|
| Pea and bean, dried shelled, Except soybean, subgroup 6c | 0.1 |
| Pea and bean, succulent shelled, Subgroup 6b | 0.01 |
| Peanut | 0.05 |
| Peanut, hay | 3 |
| Poultry, fat | 0.03 |
| Poultry, meat | 0.01 |
| Poultry, mbyp | 0.01 |
| Processed food | 0.01 |
| Rice, grain | 1 |
| Rice, hulls | 5 |
| Rice, straw | 1.8 |
| Rice, wild | 1 |
| Sheep, fat | 3 |
| Sheep, meat | 0.2 |
| Sheep, mbyp | 0.2 |
| Sorghum, grain, forage | 0.3 |
| Sorghum, grain, grain | 0.2 |
| Sorghum, grain, stover | 0.5 |
| Soybean | 0.01 |
| Sugarcane, cane | 0.05 |
| Sunflower, forage | 0.2 |
| Sunflower, hulls | 0.5 |
| Sunflower, refined oil | 0.3 |
| Sunflower, seed | 0.2 |
| Tomato | 0.1 |
| Tomato, dried pomace | 6 |
| Tomato, wet pomace | 6 |
| Vegetable, fruiting, group 8 | 0.2 |
| Vegetable, legume, edible podded, Subgroup 6a | 0.2 |
| Wheat, bran | 0.2 |
| Wheat, forage | 2 |
| Wheat, grain | 0.05 |
| Wheat, hay | 2 |
| Wheat, straw | 2 |

*Carcinogen/Hazard Classifications*
**Label Signal Word:** CAUTION, DANGER
**WHO Acute Hazard:** Class II, moderately hazardous
**Endocrine Disruptor:** Suspected endocrine disruptor
**Description:** Colorless to beige powder. Mild odor. Molecular weight 449.9; solubility = <1 ppm. Hydrolyzed by water (slowly @ pH 7–9, rapidly @ pH >9). Liquid formulations containing organic solvents may be flammable. Vapor pressure = $1.5 \times 10^{-9}$ mmHg @ 20°C.
**Incompatibilities:** May react violently with strong oxidizers, bromine, 90% hydrogen peroxide, phosphorus trichloride, silver powders or dust. Incompatible with silver compounds. Mixture with some silver compounds forms explosive salts of silver oxalate.
**Permissible Concentration in Water:** No criteria set. Runoff from spills or fire control may cause water pollution.
**Routes of Entry:** Inhalation, skin contact, eyes and ingestion

*Harmful Effects and Symptoms*
**Short Term Exposure:** Contact with eyes or skin may cause irritation or injury. Inhalation should be avoided; use NIOSH-approved air purifying respirators for pesticides. May be harmful if swallowed. *Inhalation:* Convulsions, cough, trouble breathing, sore throat. *Skin:* Pain and redness. *Eyes:* pain and redness. *Ingestion:* Abdominal pain and coughing.
*First Aid:* If this chemical gets into the eyes, remove any contact lenses at once and irrigate immediately for at least 15 minutes, occasionally lifting upper and lower lids. Seek medical attention immediately. If this chemical contacts the skin, remove contaminated clothing and wash immediately with soap and water. Seek medical attention immediately. If this chemical has been inhaled, remove from exposure, begin rescue breathing (using universal precautions) if breathing has stopped, and CPR if heart action has stopped. Transfer promptly to a medical facility. When this chemical has been swallowed, get medical attention. Give large quantities of water and induce vomiting. Do not make an unconscious person vomit.
*References:*
- National Pesticide Telecommunications Network (Now NPIC), "Lambda-cyhalothrin General Fact Sheet," Corvallis, OR (January 2001). http://www.npic.orst.edu/factsheets/l_cyhalogen.pdf
- EXTOXNET, Extension Toxicology Network, "Pesticide Information Profile, lambda Cyhalothrin," Oregon State University, Corvallis, OR. http://extoxnet.orst.edu/pips/lambdacy.htm
- California Environmental Protection Agency "Chemical List of Lists," Sacramento CA (February 1997).
- U.S. Environmental Protection Agency, Office of Pesticide Programs, Pesticide Residue Limits, "lambda-Cyhalothrin," 40 CFR 180.438. www.epa.gov/pesticides/food/viewtols.htm

# Cyhexatin (ANSI)

*Use Type:* Insecticide and acaricide (miticide)
*CAS Number:* 13121-70-5
*Formula:* $C_{18}H_{34}O_4Sn$; $(C_6H_{11}O)_3SnOH$
*Synonyms:* Cihexatin; Cyhexatin; TCTH; Tricyclohexyltin hydroxide; Tricyclohexylhydroxystannane; ENT 27395-x
*Trade Names:* ACARSTIN®; DOWCO® 213, Dow Chemical Co. (USA); PLICTRAN®
*Producers:* Agrimor International (USA); Dow Chemical Co. (USA); Jingma Chemicals (Japan); Oxon Italia S.p.A. (Italy); Rhone-Poulenc Agro France (France)
*Chemical Class:* Organotin, heavy metal
*EPA/OPP PC Code:* 101601
*California DPR Chemical Code:* 1638
*RTECS Number:* WH8750000
*EINECS Number:* 236-049-1

# Cyhexatin

*Uses:* Used to control plant-feeding mites, that are resistant to other acaricides, on almonds, apples, citrus fruit, peaches, nectarines, walnuts, hops, some fruit, and ornamental plants. Used as a spray in aerial and ground applications. Carried as a wettable powder.

*Human toxicity (long-term)*[77]: High–5.25 ppb, Health Advisory

*Fish toxicity (threshold)*[77]: Extra high–0.08348 ppb, MATC (Maximum Acceptable Toxicant Concentration)

*U.S. Maximum Allowable Residue Levels for Cyhexatin (40 CFR 180.144):*

| CROP | ppm |
|---|---|
| Almond | 0.5 |
| Almond, hulls | 60 |
| Apple | 2 |
| Cattle, fat | 0.2 |
| Cattle, kidney | 0.5 |
| Cattle, liver | 0.5 |
| Cattle, mbyp, except kidney and liver | 0.2 |
| Cattle, meat | 0.2 |
| Citrus, dried pulp | 8 |
| Fruit, citrus | 2 |
| Goat, fat | 0.2 |
| Goat, kidney | 0.5 |
| Goat, liver | 0.5 |
| Goat, mbyp, except kidney and liver | 0.2 |
| Goat, meat | 0.2 |
| Hog, fat | 0.2 |
| Hog, kidney | 0.5 |
| Hog, liver | 0.5 |
| Hog, mbyp, except kidney and liver | 0.2 |
| Hog, meat | 0.2 |
| Hop | 30 |
| Hop, dried cone | 90 |
| Horse, fat | 0.2 |
| Horse, kidney | 0.5 |
| Horse, liver | 0.5 |
| Horse, mbyp, Except kidney and liver | 0.2 |
| Horse, meat | 0.2 |
| Milk, fat (=n in whole milk) | 0.05 |
| Nectarine | 4 |
| Nut, macadamia | 0.5 |
| Peach | 4 |
| Pear | 2 |
| Plum, prune, dried | 4 |
| Plum, prune, fresh | 1 |
| Sheep, fat | 0.2 |
| Sheep, kidney | 0.5 |
| Sheep, liver | 0.5 |
| Sheep, mbyp, except kidney and liver | 0.2 |
| Sheep, meat | 0.2 |
| Strawberry | 3 |
| Walnut | 0.5 |

*Carcinogen/Hazard Classifications*

**California Prop. 65:** Developmental toxin

**Label Signal Word:** WARNING, Toxicity Class II

**WHO Acute Hazard:** Class III, slightly hazardous

*Regulatory Authority:*

- Air Pollutant Standard Set (ACGIH)[1] (Australia) (HSE)[33] (Israel) (Mexico) (OSHA)[58] (Several States)[60]
- EPCRA Section 313: Form R *de minimus* concentration reporting level: 1.0%
- U.S. DOT Regulated Marine Pollutant (49CFR, Subchapter 172.101, Appendix B), severe pollutant, as organotin pesticide compounds. The "Director's List" (CAL/OSHA)
- Canada, CEPA Prohibited Export Substance List

*Description:* Cyhexatin is a colorless to white, nearly odorless, crystalline powder. Boiling point = 227°C (decomposes). Melting/Freezing point = 195–198°C. Practicall insoluble in water. An organotin compound.

*Incompatibilities:* Strong oxidizers, ultraviolet light

*Permissible Exposure Limits in Air:* The OSHA[2] PEL for organotin compounds is 0.32 mg/m$^3$ [0.1 mg/m$^3$ (as Sn)]. The ACGIH[1] recommended TWA value is 5 mg/m$^3$. Australia, Israel, Mexico, and the HSE[33] have set that same TWA as ACGIH[1] and HSE's STEL value is 10 mg/m$^3$. In Canada the provinces of Alberta, BC, Ontario, and Quebec have set aTWA of 5 mg/m$^3$ and the STEL for Alberta and BC is 10 mg/m$^3$. Several states have set guidelines or standards for cyhexatin in ambient air[60] ranging from 50 $\mu$g/m$^3$ (North Dakota) to 80 $\mu$g/m$^3$ (Virginia) to 100 $\mu$g/m$^3$ (Connecticut) to 119 $\mu$g/m$^3$ (Nevada). The NIOSH[2] IDLH is 80 mg/m$^3$ [25 mg/m$^3$ (as Sn)].

*Determination in Air:* Filter/XAD-2® (tube); Acetic acid/$CH_3CN$; High-pressure liquid chromatography/Graphite furnace atomic absorption spectrometry; NIOSH IV, Method #5504, Organotin compounds.[18]

*Permissible Concentration in Water:* No criteria set, but EPA[32] has suggested a permissible ambient goal of 1.4 $\mu$g/L based on health effects (organotin)

*Routes of Entry:* Inhalation, skin absorption, ingestion, skin and/or eye contact

*Harmful Effects and Symptoms*

*Short Term Exposure:* These chemicals are strong poisons that can cause neurologic emergencies. Irritates eyes, skin, and respiratory system. Symptoms of exposure include headache, vertigo (an illusion of movement); sore throat, cough; abdominal pain, vomiting; skin burns, pruritus. Cyhexatin is moderate in acute oral toxicity to animals. This is in contrast to alkyl tin compounds with smaller (methyl and ethyl) radicals which are highly toxic. A diet including 6 mg/kg of body weight of cyhexatin for two years showed no effect in rats. The $LD_{50}$ (oral, rat) = 180 mg/kg. Breathing, swallowing, or skin contact with some organotins, such as trimethyltin and triethyltin compounds, can interfere with the way the brain and nervous system work. In severe cases, they can cause death.

*Long Term Exposure:* This chemical has been shown to cause liver and kidney damage in animals. Some organotin compounds, such as dibutyltins and tributyltins, have been shown to affect the immune system in animals, but this has not been examined in people. Studies in animals also have shown that some organotins, such as dibutyltins, tributyltins, and triphenyltins, can affect the reproductive system. This, also, has not been examined in people.

*Points of Attack:* Eyes, skin, respiratory system, liver, kidneys

*Medical Surveillance:* Liver function tests. Kidney function tests.

*First Aid:* Speed in removing material from eyes and skin is of extreme importance. If this chemical gets into the eyes, remove any contact lenses at once and irrigate immediately for at least 15 minutes, occasionally lifting upper and lower lids. Seek medical attention immediately. If this chemical contacts the skin, remove contaminated clothing and wash immediately with soap and water. Shampoo hair promptly if contaminated. Seek medical attention immediately. If this chemical has been inhaled, remove from exposure, begin rescue breathing (using universal precautions) if breathing has stopped, and CPR if heart action has stopped. Transfer promptly to a medical facility. When this chemical has been swallowed, get medical attention. Give large quantities of water and induce vomiting. Do not make an unconscious person vomit.

*References:*
- U.S. Environmental Protection Agency, Office of Pesticide Programs, Pesticide Residue Limits, "Cyhexatin", 40 CFR 180.144, http://www.epa.gov/pesticides/food/viewtols.htm
- National Institute for Occupational Safety and Health, Criteria for a Recommended Standard: Occupational Exposure to Organotin Compounds, NOSH Document No. 77-115 (1977).
- California Environmental Protection Agency "Chemical List of Lists," Sacramento CA (February 1997).
- Pesticide Management Education Program, Cornell University, "Cyhexatin (Plictran) Chemical Fact Sheet 6/85,", Ithaca, NY. (June 30, 1985). http://pmep.cce.cornell.edu/profiles/insect-mite/cadusafos-cyromazine/cyhexatin/insect-prof-cyhexatin.html

# Cymoxanil (ANSI)

*Use Type:* Fungicide
*CAS Number:* 57966-95-7
*Formula:* $C_7H_{10}N_4O_3$
*Synonyms:* Acetamide, 2-cyano-N-[(ethylamino)carbonyl]-2-(methoxyimino)-; 2-Cyano-N-[(ethylamino)carbonyl]-2-(methoxyimino)acetamide; Acetamide, 2-cyano-N-[(ethylamino)carbonyl]-2-(methoxyimino)-; 2-Cyano-N-ethylcarbamoyl-2-methoxyiminoacetamide
*Trade Names:* CURZATE®, DuPont Crop Protection (USA); DPX 3217®; DPX 3217M®; DPX-T3217®; EVOLVE®, Gustafson (USA); MZ-CURZATE®, Gustafson (USA); TANOS®, DuPont Crop Protection (USA)
*Producers:* DuPont Crop Protection (USA); Gustafson (USA); Ki-Hara Chemicals Ltd. (UK); Limin Chemical (China); OXON Italia (Italy)
*Chemical Class:* Aliphatic nitrogen fungicide
*EPA/OPP PC Code:* 129106
*California DPR Chemical Code:* 4002
*EINECS Number:* 261-043-0
*Uses:* Cymoxanil is applied as a seed treatment to cut potato seed pieces or as a foliar to control late blight.
*Human toxicity (long-term)*[77]*:* Low–91.00 ppb, Health Advisory
*Fish toxicity (threshold)*[77]*:* High–1.53362 ppb, MATC (Maximum Acceptable Toxicant Concentration)
*U.S. Maximum Allowable Residue Levels for Cymoxanil (40 CFR 180.503):*

| CROP | ppm |
| --- | --- |
| Grape | 0.1 |
| Hop, dried cones | 1 |
| Lettuce, head | 4 |
| Lychee | 1 |
| Potato | 0.05 |
| Vegetable, cucurbit, group 9 | 0.05 |
| Vegetable, fruiting, group 8 | 0.2 |

*Carcinogen/Hazard Classifications*
**U.S. EPA Carcinogens:** Not likely a carcinogen
**Label Signal Word:** WARNING, CAUTION
**WHO Acute Hazard:** Class III, slightly hazardous
*Regulatory Authority:*
- Clean Air Act: Hazardous Air Pollutants (Title I, Part A, Section 112)
- Clean water act: Section 307 Priority Pollutants as cyanide, total; Toxic Pollutant (Section 401.15)
- RCRA Section 261 Hazardous Constituents.
- RCRA Universal Treatment Standards: Wastewater (mg/L), 1.2 (total); 0.86 (amenable); Nonwastewater (mg/kg), 590 (total); 30 (amenable).
- RCRA Ground Water Monitoring List. Suggested test method(s) (PQL *ug*/L): 9010(40)
- Safe Drinking Water Act: MCL, 0.2 mg/L; MCLG, 0.2 mg/L; Regulated chemical (47 FR 9352)
- EPCRA Section 304 RQ: CERCLA, 10 lb (4.54 kg)
- EPCRA Section 313 Form R de minimis concentration reporting level: 1.0%

- Marine pollutant (49CFR, Subchapter 172.101, Appendix B) as cyanides, mixtures or solutions

*Description:* A peach color powder. Molecular weight = 198.21. Melting point=159-160°C.

*Incompatibilities:* Slowly hydrolyzes in water, releasing ammonia and forming acetate salts.

*Harmful Effects and Symptoms*

*Short Term Exposure:* Contact with eyes or skin may cause irritation or injury. Inhalation should be avoided; use NIOSH-approved air purifying respirators for pesticides. May be harmful if swallowed.

*First Aid:* If this chemical gets into the eyes, remove any contact lenses at once and irrigate immediately for at least 15 minutes, occasionally lifting upper and lower lids. Seek medical attention immediately. If this chemical contacts the skin, remove contaminated clothing and wash immediately with soap and water. Seek medical attention immediately. If this chemical has been inhaled, remove from exposure, begin rescue breathing (using universal precautions) if breathing has stopped, and CPR if heart action has stopped. Transfer promptly to a medical facility. When this chemical has been swallowed, get medical attention. Give large quantities of water and induce vomiting. Do not make an unconscious person vomit.

*References:*
- U.S. Environmental Protection Agency, Office of Pesticide Programs, "Pesticide Fact Sheet, Cymoxanil," Washington DC (April 21, 1998). http://www.epa.gov/opprd001/factsheets/cymozanil.pdf
- California Environmental Protection Agency "Chemical List of Lists," Sacramento CA (February 1997)
- U.S. Environmental Protection Agency, Office of Pesticide Programs, Pesticide Residue Limits, "Cymoxanil" 40 CFR 180.503. www.epa.gov/pesticides/food/viewtols.htm

# Cypermethrin (ANSI)

*Use Type:* Insecticide

*CAS Number:* 52315-07-8 (former numbers: 69865-47-0; 86752-99-0; 86753-92-6; 88161-75-5; 97955-44-7)

*Formula:* $C_{22}H_{19}Cl_2NO_3$

*Alert:* A Restricted Use Pesticide (RUP). Human toxicity (long-term): High.

*Synonyms:* (RS)-α-Cyano-3-phenoxybenzyl (1RS)-*cis, trans*-3-(2,2-dichlorovinyl)-2,2-dimethylcyclopropanecarboxylate; Cyano(3-phenoxyphenyl)methyl 3-(2,2-dichloroethenyl)-2,2-dimethylcyclopropanecarboxylate; Cyano (3-phenoxyphenyl]methyl 3-(2,2-dichlorovinyl-2,2-dimethylcyclopropanecarboxylate; Cyclopropanecarboxylic acid, 3-(2,2-dichloroethenyl)-2,2-dimethyl-, cyano(3-phenoxyphenyl)methyl ester; (±)-α-cyano-3-phenoxybenzyl 2,2-dimethyl-3-(2,2-dichlorovinyl)cyclopropanecarboxylate; Cyano(3-phenoxyphenyl)methyl 3-(2,2-dichloroethenyl)-2,2-dimethylcyclopropanecarboxylate

*Trade Names:* AMMO®; FMC (USA); AGROTHRIN®; ARDAP®; ARRIVO®; AVICADE®; BARRICADE®; CCN52®; CNN 52®; CYMBUSH® 2E; CYMBUSH® 3E; CYMPERATOR®; CYNOFF®; FMC (USA); CYPERCARE®; CYPERSECT®; CYPERKILL®; CYRUX®; DEMON® Syngenta (Switzerland); DORSAN-C®; Luxembourg Industries (PAMOL) (Israel); DYSECT®; FASTAC®; FLECTRON®; FMC® 30980; FMC 45497; FMC® 45806; FOLCORD®; IMPERATOR®; JF 5705F®; KAFIL® SUPER; KENCIS®, Kenso Corp. (Malaysia); NAGATA®, Rallis India (India); NRDC 149®; NRDC 160®; NRDC 166®; NURELLE; POLYTRIN®; PERMASECT C®,Whyte Agrochemicals (UK); PP383®; PREVAIL®, FMC (USA); RALO 10®, Rallis India (India); RIPCORD®, BASF Agricultural Products Group (Germany); ROCYPER®; Rotam Agrochemical (Hong Kong); RYCOPEL®; SHERPA®; SIPERIN®; STOCKADE®; SUPERSECT®, Calliope (France); TOPCLIP-PARASOL®; USTAAD®; WL 43467®; WRDC149®

*Producers:* Agrimor International (USA); Agsin (Singapore); Alcotan Laboratories (Spain); Ascot Fine Chemicals (UK); Atanor S.A. (Argentina); BASF Agricultural Products Group (Germany); Bayer CropScience (Germany); BEC Group (India); Bhageria Dye-Chem (India); Bharat Pulverizing Mills (India); Bonide Products (USA); Calliope (France); Cangzhou Green Chemical Co. (China); Changzhou Kangmei Chemical Industry Co., Ltd. (China); China Chemical (China); Epochem Co., (China); FMC (USA); Gharda Chemicals (India); Godavari Fertilisers and Chemicals (India); Hebei Huafeng Chemical Group (China); Hindustan Insecticides (India); Hockley International (UK); ICI Group (UK); I.N.D.I.A. Industrie Chimiche (Italy); Indiclay (India); Jiangmen Pesticide Factory (China); Kenso Corp. (Malaysia); Ki-Hara Chemicals Ltd. (UK); Luxembourg Industries (PAMOL) (Israel); Meghmani Organics (India); Nagarjuna Agrichem (India); Rallis India (India); Rotam Agrochemical (Hong Kong); Sanonda (Australia); Sanonda Zhengzhou Pesticide Co., Ltd. (China); Shandong Huayang Pesticide Group (China); Shell Chemical (Netherlands); Shenzhen Guomeng Industry Co., Ltd. (China); Sheyang Pesticides and Chemical Industry Co.(China); Sinon (Taiwan); Sudarshan Chemical Industries (India); Sumitomo Chemical (Japan); SuYan Agrochemical Group (China); Syngenta (Switzerland); United Phosphorus (India); Valent BioSciences (USA); Whyte Agrochemicals (UK); Zago Asia Ltd. (Singapore)

*Chemical Class:* Pyrethroid; Botanical

*EPA/OPP PC Code:* 109702

*California DPR Chemical Code:* 2171

*ICSC Number:* 0246

*RTECS Number:* GZ1250000

*EINECS Number:* 257-842-9

*Uses:* A Restricted Use Pesticide (RUP) used to control a variety of insects on cotton, fruit and vegetable crops. Also used in commercial and residential settings, ships, laboratories and food processing plants.

*Human toxicity (long-term)*[77]: High–7.00 ppb, Health Advisory

*Fish toxicity (threshold)*[77]: Extra high–0.21494 ppb, MATC (Maximum Acceptable Toxicant Concentration)

*Carcinogen/Hazard Classifications*

**U.S. EPA Carcinogens:** Group C, possible carcinogen

**Label Signal Word:** WARNING, CAUTION, DANGER

**Endocrine Disruptor:** Suspected endocrine disruptor

*Regulatory Authority:*
- AB 1803-Well Monitoring Chemical (CAL) as pyrethrins
- Permissible Exposure Limits for Chemical Contaminants (CAL/OSHA) as pyrethrum
- Actively registered pesticide in California, as pyrethrins
- Clean Water Act: Section 311 Hazardous Substances/RQ (same as CERCLA) as pyrethrins
- EPCRA Section 304 RQ: CERCLA, 1 lb (0.454 kg) as pyrethrins

*Description:* Thick, yellow-brown liquid or semisolid mass (technical product). Very sparingly soluble in water; solubility = 0.041 ppm @ room temp; about 0.01 ppm @ 20°C. Molecular weight = 416.31. Density = 1.12 @ 22°C. Melting/Freezing point = 60-80°C. Vapor pressure = $5.1 \times 10^{-8}$ mmHg @ 70°C. Log $K_{ow}$ = 4.47. Values above 3.0 are likely to bioaccumulate in aquatic organisms.

*Incompatibilities:* May react violently with strong oxidizers, bromine, 90% hydrogen peroxide, phosphorus trichloride, silver powders or dust. Incompatible with silver compounds. Mixture with some silver compounds forms explosive salts of silver oxalate.

*Determination in Air:* Collection by impinger or fritted bubbler, analysis by gas liquid chromatography/ultraviolet. See NIOSH IV, Method #5008[18]. (pyrethrum)

*Permissible Concentration in Water:* No criteria set. Runoff from spills or fire control may cause water pollution.

*Determination in Water:* Collection by impinger or fritted bubbler, analysis by gas liquid chromatography/ultraviolet. See NIOSH IV, Method #5008[18].

*Harmful Effects and Symptoms*

*Short Term Exposure:* Pyrethrins can affect you when breathed in and by passing through your skin. Irritates the eyes and respiratory tract. High exposure can affect the nervous system causing headache, nausea, vomiting, fatigue, and restlessness, rhinorrhea (discharge of thin nasal mucous).

*Long Term Exposure:* High or repeated exposure can cause lung allergy (with cough, wheezing and/or shortness of breath) or hay fever symptoms (sneezing, runny or stuffy nose). Allergic "pneumonia" can also occur with cough, chest pain, breathing difficulty and abnormal chest x-ray. Repeated attacks may lead to permanent scarring. Skin allergy may also develop with rash and itching, even with lower exposures. Skin contact can cause rash with redness, blisters and intense itching. A severe generalized allergy can occur with weakness and collapse.

*Points of Attack:* Respiratory system, skin, central nervous system.

*Medical Surveillance:* Before beginning employment and at regular times after that, the following are recommended: Lung function tests. These may be normal if the person is not having an attack at the time of the test. Consider chest x-ray if lung symptoms are present. Evaluation by a qualified allergist, including careful exposure history and special testing, may help diagnose skin allergy.

*First Aid:* If this chemical gets into the eyes, remove any contact lenses at once and irrigate immediately for at least 15 minutes, occasionally lifting upper and lower lids. Seek medical attention immediately. If this chemical contacts the skin, remove contaminated clothing and wash immediately with soap and water. Seek medical attention immediately. If this chemical has been inhaled, remove from exposure, begin rescue breathing (using universal precautions) if breathing has stopped, and CPR if heart action has stopped. Transfer promptly to a medical facility. When this chemical has been swallowed, get medical attention. *Do not induce vomiting when formulations containing petroleum solvents are ingested.* Otherwise, give large quantities of water and induce vomiting. Do not make an unconscious person vomit.

*References:*
- EXTOXNET, Extension Toxicology Network, "Pesticide Information Profile, Cypermethrin," Oregon State University, Corvallis, OR (September 1993). http://pmep.cce.cornell.edu/profiles/extoxnet/carbaryl-dicrotophos/cypermet-ext.html
- California Environmental Protection Agency "Chemical List of Lists," Sacramento CA (February 1997)

# *alpha*-Cypermethrin

*Use Type:* Insecticide

*CAS Number:* 67375-30-8; 66841-24-5 (d-trans-$\beta$-Cypermethrin)

*Formula:* $C_{22}H_{19}Cl_2NO_3$

*Synonyms:* Alphacypermethrin; (+)-Alphamethrin; $\alpha$-Cyano-3-phenoxybenzyl 3-(2,2-dichlorovinyl)-2,2-dimethylcyclopropanecarboxylate, (±)-*cis* isomer; Cyano(3-phenoxyphenyl)methyl 3-(2,2-dichlorovinyl)-2,2-dimethylcyclopropanecarboxylate, (±)-*cis* isomer; Cyclopropanecarboxylic acid, 3-(2,2-dichloroethenyl)-2,2-dimethyl-, cyano(3-phenoxyphenyl)methyl ester, [1$\alpha$(S*), 3$\alpha$]-(+)-

*Cypermethrin-s:* s-Cyano(3-phenoxyphenyl)methyl (+)-*cis/trans*-3-(2,2-dichloethenyl)-2,2-dimethylcyclopropane arboxylate; Cyclopropanecarboxylic acid, 3-(2,2-

dichloroethenyl)-2,2-dimethyl-, cyano(3-phenoxyphenyl)methyl ester, (S)-; Cypermethrin-minus; zeta-Cypermethrin

***Trade Names:*** BESTOX®; CONCORD®; DOMINEX®; FASTAC®; FENDONA®; FMC 45497®; FMC (USA), canceled; NRDC 160®; RENEGADE®; WL-85871®
*Cypermethrin-s:* FURY® (s-isomer); FMC 56701® (s-isomer)

***Producers:*** FMC (USA)

***Chemical Class:*** Pyrethroid; botanical

***EPA/OPP PC Code:*** 209600; 129064; (109702 use code 129064) cypermethrin-*s*

***California DPR Chemical Code:*** 3866

***Uses:*** Not registered in the U.S. Formerly used for the control of a wide range of chewing and sucking insects (particularly Lepidotera, Coleoptera, and Hemiptera) in fruit (including citrus), vegetables, vines, cereals, maize, beet, oilseed rape, potatoes, cotton, rice, soya beans, forestry, and other crops. Control of cockroaches, mosquitoes, flies, and other insect pests in public health; and flies in animal houses. Also used as an animal ectoparasiticide. (Hartley, D., and Kidd, H. (eds.), *"The Agrochemicals Handbook,"* 2nd Ed, The Royal Society of Chemistry, p. A649, Lechworth, Herts, England, August, 1987). Control of cotton leaf perforator, cotton semi-looper, false codling moth, red bollworm, cotton stainers, spiny bollworm, American bollworm, cotton bollworm, cotton spotted bollworm, native budworm, pink bollworm, tobacco budworm, cotton leafworm, boll weevil (cotton). For major pests in coffee, maize, sweet corn, sorghum, flax, soybean, mung bean, navy bean, sunflower, tobacco, rice, bush, and trellis tomato, cruciferous crops, field peas, lupines, and pasture. Cutworm control in all row crops (Farm Chemicals Handbook 1993. Willoughby, OH: Meister Publishing Co., 1993)

***Carcinogen/Hazard Classifications***

**WHO Acute Hazard:** Class II, moderately hazardous

**Endocrine Disruptor:** Suspected endocrine disruptor

***Regulatory Authority:***
- AB 1803-Well Monitoring Chemical (CAL) as pyrethrins
- Permissible Exposure Limits for Chemical Contaminants (CAL/OSHA) as pyrethrum
- Actively registered pesticide in California, as pyrethrins
- Clean Water Act: Section 311 Hazardous Substances/RQ (same as CERCLA) as pyrethrins
- EPCRA Section 304 RQ: CERCLA, 1 lb (0.454 kg) as pyrethrins
- U.S. DOT Regulated Marine Pollutant (49CFR, Subchapter 172.101, Appendix B), severe pollutant

***Description:*** Viscous yellowish-brown semisolid mass; colorless crystalline solid; mild aromatic odor. Very low solubility in water; solubility = 0.2 ppm @ 20°C. Boiling point = 200°C @ 0.07 mmHg. Melting/Freezing point = 80.5°C. Vapor pressure = 170 nPa @ 20°C. Log $K_{ow}$ = 6.29. Values at or above 3.0 are likely to bioaccumulate in marine organisms.

***Incompatibilities:*** May react violently with strong oxidizers, bromine, 90% hydrogen peroxide, phosphorus trichloride, silver powders or dust. Incompatible with silver compounds. Mixture with some silver compounds forms explosive salts of silver oxalate. When heated to decomposition this chemical emits toxic nitrogen oxides and chlorine gas.

***Permissible Concentration in Water:*** No criteria set. Runoff from spills or fire control may cause water pollution.

***Determination in Water:*** Collection by impinger or fritted bubbler, analysis by gas liquid chromatography/ultraviolet. See NIOSH IV, Method #5008[18].

***Harmful Effects and Symptoms***

***Short Term Exposure:*** Pyrethrins can affect you when breathed in and by passing through your skin. Irritates the eyes and respiratory tract. High exposure can affect the nervous system causing headache, nausea, vomiting, fatigue, and restlessness, rhinorrhea (discharge of thin nasal mucous).

***Long Term Exposure:*** High or repeated exposure can cause lung allergy (with cough, wheezing and/or shortness of breath) or hay fever symptoms (sneezing, runny or stuffy nose). Allergic "pneumonia" can also occur with cough, chest pain, breathing difficulty and abnormal chest x-ray. Repeated attacks may lead to permanent scarring. Skin allergy may also develop with rash and itching, even with lower exposures. Skin contact can cause rash with redness, blisters and intense itching. A severe generalized allergy can occur with weakness and collapse.

***Points of Attack:*** Respiratory system, skin, central nervous system.

***Medical Surveillance:*** Before beginning employment and at regular times after that, the following are recommended: Lung function tests. These may be normal if the person is not having an attack at the time of the test. Consider chest x-ray if lung symptoms are present. Evaluation by a qualified allergist, including careful exposure history and special testing, may help diagnose skin allergy.

***First Aid:*** If this chemical gets into the eyes, remove any contact lenses at once and irrigate immediately for at least 15 minutes, occasionally lifting upper and lower lids. Seek medical attention immediately. If this chemical contacts the skin, remove contaminated clothing and wash immediately with soap and water. Seek medical attention immediately. If this chemical has been inhaled, remove from exposure, begin rescue breathing (using universal precautions) if breathing has stopped, and CPR if heart action has stopped. Transfer promptly to a medical facility. When this chemical has been swallowed, get medical attention. *Do not induce vomiting when formulations containing petroleum solvents are ingested.* Otherwise, give large quantities of water and induce vomiting. Do not make an unconscious person vomit.

*References:*
- International Programme on Chemical Safety (IPCS), "Environmental Health Criteria, alpha-Cypermethrin," Geneva, Switzerland (1992). http://www.inchem.org/documents/ehc/ehc/ehc142.htm
- California Environmental Protection Agency "Chemical List of Lists," Sacramento CA (February 1997)

# Cyphenothrin

*Use Type:* Insecticide
*CAS Number:* 39515-40-7
*Formula:* $C_{24}H_{25}NO_3$
*Synonyms:* α-Cyano-3-phenoxybenzyl 2,2-dimethyl-3-(2-methylpropenyl)cyclopropanecarboxylate; Cyclopropanecarboxylic acid, 2,2-dimethyl-3-(2-methyl-1-propenyl)-, cyano(3-phenoxyphenyl)methyl ester; [(R,S-α-Cyano-3-phenoxybenzyl (1R)-cis,trans-crysanthemate; Cyphenothrin, (35% cis-; 65% trans-)
*Trade Names:* GOKILAHT®, Sumitomo Chemical (Japan); MULTICIDE®, Mclaughlin Gormley King (USA); S-2703®; S-2703 FORTE®
*Producers:* Ascot International (UK); Mclaughlin Gormley King (USA); Sumitomo Chemical (Japan)
*Chemical Class:* Pyrethroid (synthetic)
*EPA/OPP PC Code:* 129013
*California DPR Chemical Code:* 3885
*Carcinogen/Hazard Classifications*
**Label Signal Word:** WARNING, CAUTION
**WHO Acute Hazard:** Class II, moderately hazardous
**Endocrine Disruptor:** Suspected endocrine disruptor
*Regulatory Authority:*
- AB 1803-Well Monitoring Chemical (CAL) as pyrethrins
- Permissible Exposure Limits for Chemical Contaminants (CAL/OSHA) as pyrethrum
- Actively registered pesticide in California, as pyrethrins
- Clean Water Act: Section 311 Hazardous Substances/RQ (same as CERCLA) as pyrethrins
- EPCRA Section 304 RQ: CERCLA, 1 lb (0.454 kg) as pyrethrins

*Description:* Thick, yellow liquid. Molecular weight = 375.49. Vapor pressure = $3.13 \times 10^{-6}$ mmHg.
*Incompatibilities:* May react violently with strong oxidizers, bromine, 90% hydrogen peroxide, phosphorus trichloride, silver powders or dust. Incompatible with silver compounds. Mixture with some silver compounds forms explosive salts of silver oxalate.
*Permissible Concentration in Water:* No criteria set. Runoff from spills or fire control may cause water pollution.
*Determination in Water:* Collection by impinger or fritted bubbler, analysis by gas liquid chromatography/ultraviolet. See NIOSH IV, Method #5008[18].
*Harmful Effects and Symptoms*

*Short Term Exposure:* Pyrethrins can affect you when breathed in and by passing through your skin. Irritates the eyes and respiratory tract. High exposure can affect the nervous system causing headache, nausea, vomiting, fatigue, and restlessness, rhinorrhea (discharge of thin nasal mucous).
*Long Term Exposure:* High or repeated exposure can cause lung allergy (with cough, wheezing and/or shortness of breath) or hay fever symptoms (sneezing, runny or stuffy nose). Allergic "pneumonia" can also occur with cough, chest pain, breathing difficulty and abnormal chest x-ray. Repeated attacks may lead to permanent scarring. Skin allergy may also develop with rash and itching, even with lower exposures. Skin contact can cause rash with redness, blisters and intense itching. A severe generalized allergy can occur with weakness and collapse.
*Points of Attack:* Respiratory system, skin, central nervous system.
*Medical Surveillance:* Before beginning employment and at regular times after that, the following are recommended: Lung function tests. These may be normal if the person is not having an attack at the time of the test. Consider chest x-ray if lung symptoms are present. Evaluation by a qualified allergist, including careful exposure history and special testing, may help diagnose skin allergy.
*First Aid:* If this chemical gets into the eyes, remove any contact lenses at once and irrigate immediately for at least 15 minutes, occasionally lifting upper and lower lids. Seek medical attention immediately. If this chemical contacts the skin, remove contaminated clothing and wash immediately with soap and water. Seek medical attention immediately. If this chemical has been inhaled, remove from exposure, begin rescue breathing (using universal precautions) if breathing has stopped, and CPR if heart action has stopped. Transfer promptly to a medical facility. When this chemical has been swallowed, get medical attention. *Do not induce vomiting when formulations containing petroleum solvents are ingested.* Otherwise, give large quantities of water and induce vomiting. Do not make an unconscious person vomit.
*References:*
- California Environmental Protection Agency "Chemical List of Lists," Sacramento CA (February 1997)

# Cyproconazole

*Use Type:* Fungicide
*CAS Number:* 94361-06-5
*Formula:* $C_{15}H_{18}ClN_3O$
*Synonyms:* α-(4-Chlorophenyl)-α-(1-cyclopropylethyl)-1H-1,2,4-triazole-1-ethanol; (2RS,3RS)-2-(4-Chlorophenyl)-3-cyclopropyl-1-(1H-1,2,4-triazol-1-yl)butan-2-ol; 1H-1,2,4-Triazole-1-ethanol, α-(4-chlorophenyl)-α-(1-cyclopropylethyl)-

*Trade Names:* ALTO®, Syngenta (Switzerland); ATEMI®, Syngenta (Switzerland); ATEMI-50-SL®, Syngenta (Switzerland); EVIPOL®, Osmose (USA); FLINT®, Bayer CropScience (Germany); NOAH GOLD®, Osmose (USA); SAN-619F®; SENTINEL, Syngenta (Switzerland); SN 108266®
*Producers:* Bayer CropScience (Germany); Osmose (USA); Syngenta (Switzerland)
*Chemical Class:* Triazole
*EPA/OPP PC Code:* 128993
*California DPR Chemical Code:* 5105
*Uses:* Used to control fungus on cereals, coffee beans; anthracnose and other diseases on turfgrass; used against rust and leaf spot disease.
*Human toxicity (long-term)[77]:* High–1.15894 ppb, CHCL (Chronic Human Carcinogen Level)
*Fish toxicity (threshold)[77]:* Very low–3210.91612 ppb, MATC (Maximum Acceptable Toxicant Concentration)
*U.S. Maximum Allowable Residue Levels for Cyproconazole (40 CFR 180.485):*

| CROP | ppm |
|---|---|
| Coffee, bean | 0.1 |

*Carcinogen/Hazard Classifications*
**U.S. EPA Carcinogens:** Group 2B, probable carcinogen
**Label Signal Word:** CAUTION, DANGER
**WHO Acute Hazard:** Class III, slightly hazardous
*Description:* Colorless, crystalline solid. Practically insoluble in water; solubility = <0.015 @25°C. Melting/Freezing point = 104°C. Vapor pressure = $2.5 \times 10^{-7}$ mPa @ 20°C.
*Routes of Entry:* Ingestion, skin contact
*Harmful Effects and Symptoms*
**Short Term Exposure:** Contact may irritate skin and cause eye irritation and possible severe injury. Inhalation should be avoided; use NIOSH-approved air purifying respirators for pesticides. Poisonous if swallowed.
*Medical Surveillance:* Contact physician if poisoning is suspected or if redness, itching, burning of the skin or eyes develop.
*First Aid:* If this chemical gets into the eyes, remove any contact lenses at once and irrigate immediately for at least 15 minutes, occasionally lifting upper and lower lids. Seek medical attention immediately. If this chemical contacts the skin, remove contaminated clothing and wash immediately with soap and water. If this chemical has been inhaled, remove from exposure, begin rescue breathing (using universal precautions) if breathing has stopped, and CPR if heart action has stopped. Transfer promptly to a medical facility. When this chemical has been swallowed, get medical attention. Give large quantities of water and induce vomiting. Do not make an unconscious person vomit
*References:*
- California Environmental Protection Agency "Chemical List of Lists," Sacramento CA (February 1997)

# Cyprodinil

*Use Type:* Fungicide
*CAS Number:* 121552-61-2
*Formula:* $C_{14}H_{15}N_3$
*Synonyms:* N-(4-Cyclopropyl-6-methyl-pyrimidin-2-yl)-; 4-Cyclopropyl-6-methyl-N-phenyl-2-pyrimidinamine; 2-Pyrimidinamine, 4-cyclopropyl-6-methyl-N-phenyl-; aniline
*Trade Names:* CGA 219417® technical; CHORUS®, Syngenta (Switzerland); SWITCH®, Syngenta (Switzerland); UNIX®; VANGUARD®, Syngenta (Switzerland)
*Producers:* Syngenta (Switzerland)
*Chemical Class:* Anilino-pyrimidine
*EPA/OPP PC Code:* 288202
*California DPR Chemical Code:* 4000
*Uses:* Cyprodinil is applied to the foliage of almonds, grapes, stone fruit crops, and pome fruit crops to control plant diseases. Target fungi for cyprodinil include scab and brown rot blossom.
*Human toxicity (long-term)[77]:* Very low–262.50 ppb, Health Advisory
*Fish toxicity (threshold)[77]:* Low–160.91666 ppb, MATC (Maximum Acceptable Toxicant Concentration)
*Carcinogen/Hazard Classifications*
**U.S. EPA Carcinogens:** Not likely a carcinogen
**Label Signal Word:** CAUTION
*Description:* White crystalline solid. Soluble in water; solubility = 14 ppm. Molecular weight = 225.30. Melting/Freezing point = 80°C. Vapor pressure = $5.0 \times 10^{-4}$. Log $K_{ow}$ = 4.0. Values of more than 3.0 are likely to bioaccumulate in marine organisms.
*Harmful Effects and Symptoms*
**Short Term Exposure:** Contact with eyes or skin may cause irritation or injury. Inhalation should be avoided; use NIOSH-approved air purifying respirators for pesticides. May be harmful if swallowed.
*First Aid:* If this chemical gets into the eyes, remove any contact lenses at once and irrigate immediately for at least 15 minutes, occasionally lifting upper and lower lids. Seek medical attention immediately. If this chemical contacts the skin, remove contaminated clothing and wash immediately with soap and water. Seek medical attention immediately. If this chemical has been inhaled, remove from exposure, begin rescue breathing (using universal precautions) if breathing has stopped, and CPR if heart action has stopped. Transfer promptly to a medical facility. When this chemical has been swallowed, get medical attention. Give large quantities of water and induce vomiting. Do not make an unconscious person vomit. Do not induce vomiting when formulations containing petroleum solvents are ingested.
*References:*
- U.S. Environmental Protection Agency, Office of Pesticide Programs, "Pesticide Fact Sheet, Cyprodinil,"

## Cyromazine (ANSI)

*Use Type:* Insecticide (insect growth regulator)
*CAS Number:* 66215-27-8
*Formula:* $C_6H_{10}N_6$
*Synonyms:* N-Cyclopropyl-1,3,5-triazine-2,4,6-triamine; 2-Cyclopropylamino-4,6-diamino-s-triazine; Cyclopropylmelamine; 2,4-Diamino-6-(cyclopropylamino)-s-triazine; 1,3,5-Triazine-2,4,6-triamine, N-cyclopropyl-
*Trade Names:* ARMOR®, Syngenta (Switzerland); CITATION®, Syngenta (Switzerland); CGA-72662®; LARVADEX®, Syngenta (Switzerland); PATRON®, Syngenta (Switzerland); TRIGARD®, Syngenta (Switzerland); VETRAZIN®
*Producers:* AJE (Switzerland); Ki-Hara Chemicals Ltd. (UK); SuYan Agrochemical Group (China); Syngenta (Switzerland)
*Chemical Class:* Triazine
*EPA/OPP PC Code:* 121301
*California DPR Chemical Code:* 2286
*Uses:* As an insect growth regulator, cyromazine is feed to caged poultry and is passed through the chicken, leaving a residue in the manure. The chemical controls the growth of the fly larvae developing in the manure. Used as a foliar spray to control leaf miners in vegetables, mushrooms, potatoes and ornamentals and to control flies on animals.
*Human toxicity (long-term)*[77]: Low–52.50 ppb, Health Advisory
*Fish toxicity (threshold)*[77]: Very low–22450.18892 ppb, MATC (Maximum Acceptable Toxicant Concentration)
*U.S. Maximum Allowable Residue Levels for Cyromazine (40 CFR 180.414):*

| CROP | ppm |
|---|---|
| Bean, dry, seed, except cowpea | 3 |
| Bean, lima, seed | 1 |
| Broccoli | 1 |
| Cabbage, abyssinian | 10 |
| Cabbage, chinese, bok choy | 3 |
| Cabbage, seakale | 10 |
| Cattle, fat | 0.05 |
| Cattle, kidney | 0.2 |
| Cattle, meat | 0.05 |
| Cattle, mbyp | 0.05 |
| Cattle, mbyp, except kidney | 0.05 |
| Corn, sweet, forage | 0.5 |
| Corn, sweet, kernel plus cob with husks removed | 0.5 |
| Corn, sweet, stover | 0.5 |
| Cotton, undelinted seed | 0.1 |
| Egg | 0.25 |
| Garlic, bulb | 0.2 |
| Garlic, great headed, bulb | 0.2 |
| Goat, fat | 0.05 |
| Goat, kidney | 0.2 |
| Goat, meat | 0.05 |
| Goat, mbyp | 0.05 |
| Goat, mbyp, except kidney | 0.05 |
| Hanover salad, leaves | 10 |
| Hog, fat | 0.05 |
| Hog, kidney | 0.2 |
| Hog, meat | 0.05 |
| Hog, mbyp | 0.05 |
| Hog, mbyp, except kidney | 0.05 |
| Horse, fat | 0.05 |
| Horse, kidney | 0.2 |
| Horse, meat | 0.05 |
| Horse, mbyp | 0.05 |
| Horse, mbyp, except kidney | 0.05 |
| Leek | 3 |
| Mango | 0.3 |
| Milk | 0.05 |
| Mushroom | 1 |
| Mustard greens | 3 |
| Onion, dry bulb | 0.2 |
| Onion, green | 3 |
| Onion, potato, bulb | 3 |
| Onion, tree, tops | 3 |
| Onion, welsh, tops | 3 |
| Pepper | 1 |
| Potato | 0.8 |
| Poultry feed, for chicken layers and breeder hens | 5 |
| Poultry, fat | 0.05 |
| Poultry, meat | 0.05 |
| Poultry, mbyp | 0.05 |
| Radish, roots | 0.5 |
| Radish, tops | 0.5 |
| Rakkyo, bulb | 0.2 |
| Shallot, bulb | 0.2 |
| Shallot, fresh leaves | 3 |
| Sheep, fat | 0.05 |
| Sheep, kidney | 0.2 |
| Sheep, meat | 0.05 |
| Sheep, mbyp | 0.05 |
| Sheep, mbyp, Except kidney | 0.05 |
| Tomato | 0.5 |
| Turnip, greens | 10 |
| Vegetable, brassica, leafy, group 5, except broccoli | 10 |
| Vegetable, cucurbit, group 9 | 1 |
| Vegetable, leafy, except brassica, Group 4 | 7 |

(preceding references:)
- Washington, DC (April 6, 1998). http://www.epa.gov/opprd001/factsheets/cyprodinil.pdf
- California Environmental Protection Agency "Chemical List of Lists," Sacramento CA (February 1997).

# Cyromazine

*Carcinogen/Hazard Classifications*
**U.S. EPA Carcinogens:** Group E, unlikely carcinogen
**Label Signal Word:** CAUTION
**WHO Acute Hazard:** Class U, unlikely to be hazardous
*Regulatory Authority:*
- Actively registered pesticide in California.
- FIFRA, 40CFR186: tolerances for pesticides in animal feeds.

*Description:* White, odorless rystalline solid. Soluble in water. Molecular weight = 166.20. Vapor pressure = $3.36 \times 10^{-9}$ mmHg @ 20°C.

*Incompatibilities:* Strong oxidizers.

*Permissible Concentration in Water:* No criteria set. Runoff from spills or fire control may cause water pollution.

*Routes of Entry:* Inhalation, passing through the skin, ingestion.

*Harmful Effects and Symptoms*

*Short Term Exposure:* May cause skin and severe eye irritation. Moderately poisonous if ingested or inhaled. Exposure to a triazine (simazine) has caused acute and subacute dermatitis in the former USSR, characterized by erythema, slight edema, moderate pruritus, and burning lasting 4 to 5 days.

*Long Term Exposure:* May cause lung irritation and damage. May cause skin allergy. Contact with some triazine compounds (such as atrazine) may increase risks for tumors known to be associated with hormonal factors. These have been observed in both animals and human beings, and are consistent with the known effects on the hypothalamic pituitary gonadal axis. Repeated exposure may cause weight loss and reduced red blood cell count. May be mutagenic.

*Points of Attack:* Liver, lungs and skin.

*Medical Surveillance:* Before beginning employment and at regular times after that, for those with frequent or potentially high exposures, the following is recommended: Lung function tests. Consider chest x-ray following acute overexposure. Evaluation by a qualified allergist. Examination of the nervous system.

*First Aid:* If this chemical gets into the eyes, remove any contact lenses at once and irrigate immediately for at least 15 minutes, occasionally lifting upper and lower lids. Seek medical attention immediately. If this chemical contacts the skin, remove contaminated clothing and wash immediately with soap and water. Seek medical attention immediately. If this chemical has been inhaled, remove from exposure, begin rescue breathing (using universal precautions) if breathing has stopped, and CPR if heart action has stopped. Transfer promptly to a medical facility. When this chemical has been swallowed, get medical attention. Give large quantities of water or milk and induce vomiting. Do not make an unconscious person vomit.

*References:*
- Pesticide Management Education Program, "Cyromazine (Larvadex, Trigard) Chemical Fact Sheet 12/86," Cornell University, Ithaca, NY (December 1986). http://pmep.cce.cornell.edu/profiles/insect-mite/cadusafos-cyromazine/cyromazine/insect-prof-cyromazine.html
- California Environmental Protection Agency "Chemical List of Lists," Sacramento CA (February 1997)
- U.S. Environmental Protection Agency, Office of Pesticide Programs, Pesticide Residue Limits, "Cyromazine," 40 CFR 180.414. www.epa.gov/pesticides/food/viewtols.htm

# D

## 2,4-D

*Use Type:* Herbicide and plant growth regulator
*CAS Number:* 94-75-7
*Formula:* $C_8H_6Cl_2O_3$; $Cl_2C_6H_3OCH_2COOH$
*Alert:* A General Use Pesticide (GUP)
*Synonyms:* Acetic acid (2,4-dichlorophenoxy)-; Acide 2,4-dichloro phenoxyacetique (French); Acido (2,4-diclorofenossi)-acetico (Italian); Acido 2,4-diclorofenoxiacetico (Spanish); 2,4-D acid; (2,4-Dichloor-fenoxy)-azijnzuur (Dutch); Dichlorophenoxyacetic acid; 2,4-Dichlorphenoxyacetic acid; 2,4-Dichlorophenoxyacetic acid, salts and esters; (2,4-Dichlor-phenoxy)-essigsaeure (German); 2,4-D, salts and esters; 2,4-Dwuchlorofenoksyoctowy kwas (Polish); ENT 8,538; Kwas 2,4-dwuchlorofenoksyoctowy; Kwasu 2,4-dwuchlorofenoksoctowego; Kyselina 2,4-dichlorfenoxyoctova; NSC 423; 2,4-PA (in Japan)
*Trade Names:* ACME®, Pbi/Gordon (USA); AGROCER COMPLEX®, Alcotan Laboratories (Spain); AGROTECT®; AMIDOX®; AMINOL 806®, Milenia Agro Ciencias (Brazil); AMINOZ®, Sanonda (Australia); AMOXONE®; AQUA-KLEEN®, Newfarm (Australia); BARRAGE®, Helena Chemical (USA); BH 2,4-D®; BRUSH-RHAP®, Helena Chemical Corp. (USA), active; Vertac Chemical Corp. (USA), canceled; BUSH KILLER®; B-SELEKTONON®; CAMPAIGN®, Monsanto (USA); CHIPCO TURF HERBICIDE "D"®; CHLOROXONE®; CITRUS FIX®, AMVAC Chemical (USA); CROP RIDER®, Occidental Chemical (USA), canceled; CROTILIN®; CURTAIL®, Dow AgroSciences (USA); D 50®; 2,4-D PHENOXY PESTICIDE®; DACAMID®, CONDEA (Germany); DACAMINE®, Occidental Chemical Corp. (USA) canceled 10/28/1983; DECAMINE®; DEBROUSSAILLANT 600®; DED-WEED®, several manufacturers, canceled; DEHERBAN®; DESORMONE®; DICOPUR®; DICOTOX®; DIKAMIN®, Nitrokemia 2000 (Hungary); DIKONIRT®, Nitrokemia 2000 (Hungary); DINOXOL®, Bayer CropScience (Germany), canceled 11/28/1984; DMA-4®, Dow AgroSciences (USA); DORMONE®; DYMEC®, Pbi/Gordon Corp. (USA); EMULSAMINE BK®, Bayer CropScience (Germany), canceled 11/28/1984; EMULSAMINE E-3®, Bayer CropScience (Germany), canceled 11/28/1984; ENVERT®, Bayer CropScience (Germany), canceled 1/11/1980; ESTERON®, Dow AgroSciences (USA); ESTERON BRUSH KILLER®, Dow AgroSciences (USA); ESTERON 99 CONCENTRATE®, Dow AgroSciences (USA); ESTERONE FOUR®, ESTERON 44 WEED KILLER®, Dow AgroSciences (USA); ESTONE®; FARMCO®, Farmco Industries (USA), canceled; FERNESTA®; FERNIMINE®; FERNOXONE®; FERXONE®; FOREDEX 75®; FORMULA 40®; HEDONAL®; HERBANIL®, Milenia Agro Ciencias (Brazil); HERBI D-480®, Milenia Agro Ciencias (Brazil); HERBIDAL®; HERBOXONE®, A H Marks (UK); HI-DEP®, Pbi/Gordon Corp. (USA); HORMOTOX®; IPANER®; KROTILINE®; LAND MASTER®, Monsanto (USA); LAWN-KEEP®; MACRONDRAY®; MALERBANE®; MATON®, A H Marks (UK); MIRACLE®, Dragon Chemical Corp. (USA), canceled; MONOSAN®; MOTA MASKROS®; MOXONE®; NETAGRONE®; PENNAMINE®, Cerexagri Inc., North America (USA), canceled 10/10/1989; PHENOX®; PIELIK®; PLANOTOX®; PLANTGARD®; RHODIA®, Rhodia Group (France); SALVO®, Arch Chemicals (USA), canceled; SAVAGE®, United Agri Products (UAP) (USA); SPRITZ-HORMIN/2,4-D®; SPRITZ-HORMIT/2,4-D®; SUPER D WEEDONE®, Bayer CropScience (Germany); SUPERORMONE CONCENTRE®; TRANSAMINE®, United Agri Products (UAP) (USA), canceled; Vertac Chemical (USA), canceled; TRIBUTON®; TRINOXOL®, Bayer CropScience (Germany), canceled 11/28/1984; U 46®; U-5043®; VERGEMASTER®; VERTON®, Dow Chemical Company (USA), canceled 10/7/1983; VIDON 638®; VISKO®, Hercules Inc. (USA), canceled; Rhone-Poulenc Agrochemical Div. (France), canceled; VISKO-RHAP®, Hercules Inc. (USA), canceled; Rhone-Poulenc Agrochemical Div. (France), canceled; WEED-AG-BAR®; WEEDAR®, Bayer CropScience (Germany); WEED-B-GON®, The Scotts Company (USA) (active in several forms); WEEDEZ WONDER BAR®; WEEDONE®, Bayer CropScience (Germany); WEED-RHAP®, Transvaal Inc. (USA), canceled; WEED TOX®; WEEDTROL®
*Producers:* Aero Agro Chemical Industries (India); Agrimor International (USA); A H Marks (UK); Akzo Nobel Chemicals (Netherlands); Alcotan Laboratories (Spain); AMVAC Chemical (USA); Atanor S.A. (Argentina); Atul (India); BASF (Germany); Bayer CropScience (Germany); Bhageria Dye-Chem (India); Cerexagri Inc., North America (USA); CONDEA (Germany); China Chemical (China); Cyanamid Agriculture Products; Dow AgroSciences (USA); Ehrenstorfer, Dr. (Germany); Fluorochem (UK); Helena Chemical (USA); Ishihara Sangyo Kaisha (Japan); Kenso Corp. (Malayasia); Ki-Hara Chemicals Ltd. (UK); Milenia Agro Ciencias (Brazil); Monsanto (USA); Newfarm (Australia); Nissan Chemical Industries (Japan); Nitrokemia

2000 (Hungary); Occidental Chemical Corp. (USA); Pbi/Gordon Corp. (USA); Proficol (Colombia); Rhodia Group (France); Sanonda (Australia); The Scotts Company (USA); Shenzhen Guomeng Industry Co., Ltd. (China); Sigma-Aldrich Laborchemikalien (Germany); United Agri Products (UAP) (USA); Whyte Agrochemicals (UK); Zago Asia Ltd. (Singapore)

*Chemical Class:* Chlorophenoxy acid or ester; Halo-organics; Chloro-organics

*EPA/OPP PC Code:* 030001

*California DPR Chemical Code:* 636

*ICSC Number:* 0033

*RTECS Number:* AG6825000

*EEC Number:* 607-039-00-8

*EINECS Number:* 202-361-1

*Uses:* 2,4-Dichlorophenoxyacetic acid was introduced as a plant growth-regulator in 1942. It is registered in the United States as an herbicide for control of broadleaf plants and as a plant growth-regulator. There are many forms or derivatives of 2,4-D including esters, amines, and salts. It is used in cultivated agriculture, in pasture and rangeland applications, forest management, home, garden, and to control aquatic vegetation. It may be found in emulsion form, in aqueous solutions (salts), and as a dry compound. The product *Agent Orange*, made by Monsanto Chemical and used extensively throughout Vietnam, was about 50% 2,4-D. However, the controversies associated with the use of Agent Orange were associated with a contaminant (dioxin) in the 2,4,5-T component of the defoliant. In 1964 *Agent Orange* replaced *Agent Purple* a mixture of the *n*-butyl esters of 2,4-D and 2,4,5-T plus the isobutyl ester of 2,4,5-T.

*Human toxicity (long-term)*[77]: Low–70.00 ppb, MCL (Maximum Contaminant Level)

*Fish toxicity (threshold)*[77]: Very low–4247.00420 ppb, MATC (Maximum Acceptable Toxicant Concentration)

*U.S. Maximum Allowable Residue Levels for 2,4-D:*

*Note:* There are 1,847 tolerances listed by EPA for 2,4-D and its salts, amines and esters. These can be found at the U.S. EPA site

http://www.epa.gov/pesticides/food/viewtols.htm, along with their CFR citations.

*Carcinogen/Hazard Classifications*

**U.S. EPA Carcinogens:** Group D, unclassifiable, ambiguous data

**IARC:** Group 2B, possible carcinogen

**WHO Acute Hazard:** Class II, moderately hazardous

**Endocrine Disruptor:** Suspected

*Regulatory Authority:*

- Carcinogen (Human Suspected) (IARC)[9]
- Air Pollutant Standard Set (ACGIH)[1] (Australia) (DFG)[3] (HSE)[33] (Israel) (Mexico) (former USSR)[35] (OSHA)[58] (Several States)[60] (Several Canadian Provinces)
- AB 1803-Well Monitoring Chemical (CAL)
- MCL (Maximum Contaminants Levels) list of contaminants (CAL)
- AB 2588-Air Toxics "Hot Spots" Chemicals (CAL)
- Permissible Exposure Limits for Chemical Contaminants (CAL/OSHA)
- The "Director's List" (CAL/OSHA)
- Actively registered pesticide in California.
- Water Pollution Standard Proposed (EPA, Mexico)[35] (Maine, Minnesota)[61]
- Clean Air Act: Hazardous Air Pollutants (Title I, Part A, Section 112)
- Clean Water Act: Section 311 Hazardous Substances/RQ 40CFR117.3 (same as CERCLA, see below); Section 313 Water Priority Chemicals (57FR41331, 9/9/92)
- RQ: CERCLA, 100 lb (45.5 kgs)
- EPA Hazardous Waste Number (RCRA No.): U240, D016
- RCRA, 40CFR261, Appendix 8 Hazardous Constituents
- RCRA Toxicity Characteristic (Section 261.24), Maximum
- Concentration of Contaminants, regulatory level, 10.0 mg/L
- RCRA 40CFR268.48; 61FR15654, Universal Treatment Standards: Wastewater (mg/L), 0.72; Nonwastewater (mg/kg), 10
- RCRA 40CFR264, Appendix 9; TSD Facilities Ground Water Monitoring List. Suggested test method(s) (PQL, $ug/L$): 8150(10).
- Safe Drinking Water Act: MCL, 0.1 mg/L; MCGL, 0.07 mg/L; Regulated chemical (47 FR 9352) as 2,4-D.
- CERCLA/SARA 313: Form R *de minimus* concentration reporting level: 1.0%
- U.S. DOT Regulated Marine Pollutant (49CFR172.101, Appendix B)
- Canada, Drinking Water Quality, 0.1 mg/L IMAC.
- Mexico, Drinking Water Criteria, 0.1 mg/L

*Description:* White to yellow crystalline powder. Slight phenolic odor. Very slightly soluble in water; solubility = 900 ppm @ 25°C. The taste and odor threshold in water is 3.0 ppm. Molecular weight = 221.04. Boiling point = (decomposes). Melting/Freezing point = 138°C. Vapor pressure = $8 \times 10^{-6}$ mmHg @ 20°C. Log $K_{ow}$ = 2.79. Values at or above 3.0 are likely to bioaccumulate in marine organisms.

*Incompatibilities:* A weak acid, incompatible with bases. Decomposes in sunlight or heat, forming hydrogen chloride and phosgene. Contact with strong oxidizers may cause fire and explosions.

*Permissible Exposure Limits in Air:* The legal limit (OSHA PEL), and ACGIH[1] recommended TWA is 10 mg/m³ as is the value for DFG[3], HSE[33] value, Israel, Mexico, and Australia. The HSE[33] and Mexico adds an STEL of 20 mg/m³. The NIOSH[2] IDLH value is 100 mg/m³. The former

USSR[35] has set an MAC in workplace air of 1.0 mg/m³. The Canadian provincial limit for Alberta, BC, Ontario, and Quebec is the same as OSHA and Alberta and BC set a STEL of 20 mg/m³. The former USSR-UNEP/IRPTC project[43] has set an MAC in ambient air in residential areas of 0.02 mg/m³ on a momentary basis and 0.01 mg/m³ on a daily average basis for the sodium salt of 2,4-D. In addition, several states have set guidelines or standards for 2,4-D in ambient air[60] ranging from 100 μg/m³ (North Dakota) to 105 μg/m³ (Pennsylvania) to 160 μg/m³ (Virginia) to 200 μg/m³ (Connecticut) to 238 μg/m³ (Nevada).

*Determination in Air:* Collection on a glass fiber filter and analysis by HPLC with UV detection. See NIOSH Method #5001[18].

*Permissible Concentration in Water:* The U.S.[35] has set an MPC in bottled water intended for human consumption of 0.1 mg/L. Mexico[35] has set levels in ambient water of 0.1 mg/L in estuaries and 0.01 mg/L in coastal waters. The former USSR-UNEP/IRPTC project[43] has set an MAV of 1.0 mg/L in water bodies used for drinking purposes for the sodium salt and 0.62 mg/L in water bodies used for fishery purposes. A no-observed-adverse-effect-level (NOAEL) of 1 mg/kg/day has been determined[47] which results in the calculation of a lifetime health advisory of 0.070 mg/L. This level has been proposed by EPA[62] as a maximum level in drinking water. Drinking water levels for Canada and Mexico are 0.1 mg/L. States which have set guidelines for 2,4-D in drinking water[61] include Minnesota at 70 μg/L and Maine at 100 μg/L.

*Determination in Water:* Filter; Methanol; High-pressure liquid chromatography/Ultraviolet detection; NIOSH IV Method #5001[18]

*Routes of Entry:* Inhalation, skin absorption, ingestion, skin and eye contact.

*Harmful Effects and Symptoms*

*Short Term Exposure: Inhalation:* May cause irritation of the mouth, nose and throat, headache, nausea, vomiting, and diarrhea at levels above 10 mg/m³. Nerve damage, which may be delayed, may include swelling of legs and feet, muscle twitch and stupor. Severe exposures may result in death. *Skin:* Dust or liquid left in contact with the skin for several hours may be absorbed. This may result in severe delayed symptoms as listed above. These symptoms may last for months or years. *Eyes:* Irritation may occur. *Ingestion:* The oral dose (human) required to produce symptoms is about 1/12 ounce (1/2 teaspoon). Increasing amounts may result in increasingly severe symptoms as listed above. Death has resulted from as little as 1/5 ounce. Survival for more than 48 hours is usually followed by complete recovery although symptoms may last for several months.

*Long Term Exposure:* Workers exposed to 2,4-D in the manufacturing process over a five to ten year period at levels above 10 mg/m³ complained of weakness, rapid fatigue, headache and vertigo. Liver damage, low blood pressure and slowed heartbeat were also found. Based on animal tests, 2,4-D may affects human reproduction

*Points of Attack:* Skin, central nervous system, liver, kidneys.

*Medical Surveillance:* If symptoms develop or overexposure is suspected, the following may be useful: Liver and kidney function tests. Exam of the nervous system.

*First Aid:* If this chemical gets into the eyes, remove any contact lenses at once and irrigate immediately for at least 15 minutes, occasionally lifting upper and lower lids. Seek medical attention immediately. If this chemical contacts the skin, remove contaminated clothing and wash immediately with soap and water. Seek medical attention immediately. If this chemical has been inhaled, remove from exposure, begin rescue breathing (using universal precautions) if breathing has stopped, and CPR if heart action has stopped. Transfer promptly to a medical facility. When this chemical has been swallowed, get medical attention. Give large quantities of water and induce vomiting. Do not make an unconscious person vomit.

*References:*
- EXTOXNET, Extension Toxicology Network, "Pesticide Information Profile, 2,4-D," Oregon State University, Corvallis, OR. (June, 1996) http:\\ace.ace.orst.edu/info/extoxnet/pips/24-D.htm
- EPA, Office of Pesticide Programs, Pesticide Residue Limits, "2,4-D", 40 CFR 180.xxxx., www.epa.gov/cgi-bin/oppsrch
- U.S. Environmental Protection Agency, 2,4-Dichlorophenoxy Acetic Acid, Health and Environmental Effects Profile No. 77, Washington, DC, Office of Solid Waste (April 30, 1980).
- New Jersey Department of Health, "Hazardous Substance Fact Sheet: 2,4-Dichlorophenoxyacetic Acid Ester," Trenton, NJ (March 1989, rev. January 1999). www.state.nj.us/health/eoh/rtkweb/0593.pdf
- New York State Department of Health, "Chemical Fact Sheet: 2,4-D," Albany, NY, Bureau of Toxic Substance Assessment (March 1986 and Version 2).
- California Environmental Protection Agency "Chemical List of Lists," Sacramento CA (February 1997).
- Sax, N.I., Ed., "Dangerous Properties of Industrial Materials Report," 1, No. 6, 49-52 (1981) and 7, No. 6, 11-46 (1987).

## 2,4-D, butoxyethyl ester

*Use Type:* Herbicide
*CAS Number:* 1929-73-3
*Formula:* $C_{14}H_{18}Cl_2O_4$
*Synonyms:* Acetic acid, (2,4-dichlorophenoxy)-, 2-butoxyethyl ester; Butoxy-D3; Butoxyethanol ester of 2,4-

dichlorophenoxyacetic acid; Butoxyethyl 2,4-dichlorophenoxyacetate; 2-Butoxyethyl 2,4-dichlorophenoxyacetate; 2,4-D 2-Butoxyethyl ester; 2,4-D, Butoxyethyl ester; 2,4-D, Butoxyethanol ester; 2,4-D (Butoxyethyl); 2,4-D, 2-Butoxyethyl ester; Caswell No. 315AI; 2,4-Dichlorophenoxyacetic acid butoxyethyl ester; 2,4-Dichlorophenoxyacetic acid butoxyethanol ester; (2,4-Dichlorophenoxy)acetic acid butoxyethyl ester; 2,4-Dichlorophenoxyacetic acid 2-butoxyethyl ester; 2,4-Dichlorophenoxyacetic acid ethylene glycol butyl ether ester

*Trade Names:* AQUA-KLEEN®, Nufarm (Australia); 2,4-D-BEE®, Dow AgroSciences (USA); BRUS KILLER 64®; CROSSBOW® (with Triclopyr, butoxyethyl ester), Dow AgroSciences (USA); EASTERON® 99 CONCENTRATE, Dow AgroSciences (USA); E-99®, Dow AgroSciences (USA); LO-ESTASOL®; SILVAPROP® 1;TURFLON®, Dow AgroSciences (USA); WEEDONE® 100 EMULSIFIABLE, Bayer CropScience (Germany), canceled; WEEDONE® 638, Nufarm (Australia); WEEDONE® LV 4, Bayer CropScience (Germany), canceled; WEEDONE® LV-6, Bayer CropScience (Germany), canceled; WEED-RHAP® LV

*Producers (Makers of 2,4-D):* Aero Agro Chemical Industries (India); Agrimor International (USA); A H Marks (UK); Akzo Nobel Chemicals (Netherlands); Alcotan Laboratories (Spain); Atanor S.A. (Argentina); Atul (India); BASF (Germany); Bayer CropScience (Germany); Bhageria Dye-Chem (India); China Chemical (China); Cyanamid Agriculture Products; Dow AgroSciences; Ishihara Sangyo Kaisha (Japan); Kenso Corp. (Malayasia); Ki-Hara Chemicals Ltd. (UK); Milenia Agro Ciencias (Brazil); Monsanto (USA); Nissan Chemical Industries (Japan); Nitrokemia 2000 (Hungary); Nufarm (Australia); Proficol (Colombia); Sanonda (Australia); Shenzhen Guomeng Industry Co., Ltd. (China); Sigma-Aldrich Laborchemikalien (Germany); Whyte Agrochemicals (UK); Zago Asia Ltd. (Singapore)

*Chemical Class:* Chlorophenoxy acid or ester

*EPA/OPP PC Code:* 030053; (030061 old EPA code number)

*California DPR Chemical Code:* 802

*Uses:* Used to control weeds and annual grasses on rights of way, rangeland and pastureland, wheat and barley and other crops.

*Human toxicity (long-term)[77]:* Low–70.00 ppb, MCL (Maximum Contaminant Level) as 2,4-D

*Fish toxicity (threshold)[77]:* Very low–4247.00420 ppb, MATC (Maximum Acceptable Toxicant Concentration) as 2,4-D

*U.S. Maximum Allowable Residue Levels for 2,4-D and its butoxyethyl ester (40 CFR 180.142):*

*Note:* Residues on all the of the following may result from application of 2,4-D in acid form, or in the form of its butoxyethyl ester salt.

| CROP | ppm |
|---|---|
| Barley, grain | 0.5 |
| Blueberrie | 0.1 |
| Corn, fodder | 20 |
| Corn, forage | 20 |
| Corn, fresh, sweet | 0.5 |
| Corn, grain | 0.5 |
| Cranberries | 0.5 |
| Grapes | 0.5 |
| Grass hay | 300 |
| Grasses, pasture | 1000 |
| Grasses, rangeland | 1000 |
| Millet, forage | 20 |
| Millet, grain | 0.5 |
| Millet, straw | 20 |
| Nuts | 0.2 |
| Oats, forage | 20 |
| Oats, grain | 0.5 |
| Pistachios | 0.2 |
| Rice | 0.1 |
| Rice, straw | 20 |
| Rye, forage | 20 |
| Rye, grain | 0.5 |
| Sorghum, fodder | 20 |
| Sorghum, forage | 20 |
| Sorghum, grain | 0.5 |
| Stone fruits | 0.2 |
| Sugarcane | 2 |
| Sugarcane, forage | 20 |
| Wheat, forage | 20 |
| Wheat, grain | 0.5 |

*Carcinogen/Hazard Classifications*

**U.S. EPA Carcinogens:** Group D, unclassifiable as 2,4-D, the parent chemical

**IARC:** Group 2B, possible carcinogen

**Label Signal Word:** CAUTION or WARNING

**WHO Acute Hazard:** Class II, moderately hazardous as 2,4-D, the parent chemical.

**Endocrine Disruptor:** Probable ED as 2,4-D, the parent chemical

*Regulatory Authority:*

- AB 2588-Air Toxics "Hot Spots" Chemicals (CAL) as chlorophenoxy pesticides
- The "Director's List" (CAL/OSHA) as chlorophenoxy pesticides
- Actively registered pesticide in California.
- Clean Water Act: Section 311 Hazardous Substances/RQ (same as CERCLA)
- RCRA Section 261 Hazardous Constituents
- Safe Drinking Water Act: MCL, 0.1 mg/L; MCGL, 0.07 mg/L; Regulated chemical (47 FR 9352) as 2,4-D.
- EPA Hazardous Waste Number (RCRA No.): U240
- EPCRA Section 304 RQ: CERCLA, 100 lb (45.4 kg)

- EPCRA Section 313 Form R *de minimis* concentration reporting level: 0.1%
- U.S. DOT Regulated Marine Pollutant (49CFR172.101, Appendix B) as 2,4-D

*Description:* Colorless to amber viscous liquid. Odorless (pure). Very slightly soluble in water; solubility = 12 ppm Molecular weight = 321.21. Density = 1.23 @ 20°C. Boiling point = 190°C @ 7 mmHg. Decomposes above 200°C. Vapor Pressure = $3.8 \times 10^{-6}$ @ 25°C.

*Incompatibilities:* May attack some forms of plastics. Keep material dissolved in organic solvents (such as hexane, benzene, acetone, and alcohols) away from oxidizers. Keep dry material away from oxidizers, sulfuric acid, caustics, ammonia, aliphatic amines, alkanolamines, isocyanates, alkylene oxides, epichlorohydrin. Moisture may cause hydrolysis or other forms of decomposition.

*Permissible Exposure Limits in Air:* The legal limit (OSHA PEL), and ACGIH[1] recommended TWA is 10 mg/m$^3$ as is the value for DFG[3], HSE[33] value, Israel, Mexico, and Australia. The HSE[33] and Mexico adds an STEL of 20 mg/m$^3$. The NIOSH[2] IDLH value is 100 mg/m$^3$. The former USSR[35] has set an MAC in workplace air of 1.0 mg/m$^3$. The Canadian provincial limit for Alberta, BC, Ontario, and Quebec is the same as OSHA and Alberta and BC set a STEL of 20 mg/m$^3$. The former USSR-UNEP/IRPTC project[43] has set an MAC in ambient air in residential areas of 0.02 mg/m$^3$ on a momentary basis and 0.01 mg/m$^3$ on a daily average basis for the sodium salt of 2,4-D. In addition, several states have set guidelines or standards for 2,4-D in ambient air[60] ranging from 100 $\mu$g/m$^3$ (North Dakota) to 105 $\mu$g/m$^3$ (Pennsylvania) to 160 $\mu$g/m$^3$ (Virginia) to 200 $\mu$g/m$^3$ (Connecticut) to 238 $\mu$g/m$^3$ (Nevada).

*Determination in Air:* Collection on a glass fiber filter and analysis by HPLC with UV detection. See NIOSH Method #5001[18], for 2,4-D.[18]

*Permissible Concentration in Water:* No criteria set. Runoff from spills or fire control may cause water pollution. For 2,4-D, the U.S. EPA[35] has set an MPC in bottled water intended for human consumption of 0.1 mg/L.

*Determination in Water:* Filter; Methanol; High-pressure liquid chromatography/Ultraviolet detection; NIOSH IV Method #5001[18] as 2,4-D.

*Routes of Entry:* Inhalation, skin absorption, ingestion, skin and eye contact.

*Harmful Effects and Symptoms*

*Short Term Exposure:* Poisonous; may be fatal if inhaled, swallowed, or absorbed through skin. Severely irritates eyes, skin and respiratory tract, with burning sensation, pain, redness and swelling. Metabolic stimulant. If inhaled, causes coughing, dilated pupils, headache, profuse perspiration, intense thirst, extreme fatigue, rapid pulse, high fever, clammy, flushed skin, rapid breathing, nausea, vomiting, cyanosis (bluish tint to skin and lips), anxiety and confusion, convulsions, risk of lung edema. If swallowed, face and lips turn bluish. Liver injury with associated jaundice, kidney failure, and cardiac arrhythmias are commonly noted. Nerve damage, which may be delayed, may include swelling of legs and feet, muscle twitch and stupor. Severe exposure can cause death from heart failure. Dust or liquid left in contact with the skin for several hours may be absorbed. This may result in severe delayed symptoms as listed above. These symptoms may last for months or years.

*Long Term Exposure:* Workers exposed to chlorophenoxy compounds such as 2,4-D (in the manufacturing process) over a five to ten year period at levels above 10 mg/m$^3$ complained of weakness, rapid fatigue, headache and vertigo. Liver damage, low blood pressure and slowed heartbeat were also found. Based on animal tests, may affects human reproduction

*Points of Attack:* Eyes, skin, respiratory system, central nervous system, cardiovascular system, liver, kidney

*Medical Surveillance:* If symptoms develop or overexposure is suspected, liver or kidney function tests may be useful. Liver function tests.

*First Aid:* If this chemical gets into the eyes, remove any contact lenses at once and irrigate immediately for at least 15 minutes, occasionally lifting upper and lower lids. Seek medical attention immediately. If this chemical contacts the skin, remove contaminated clothing and wash immediately with soap and water. Seek medical attention immediately. If this chemical has been inhaled, remove from exposure, begin rescue breathing (using universal precautions) if breathing has stopped, and CPR if heart action has stopped. Transfer promptly to a medical facility. When this chemical has been swallowed, get medical attention. *Do not induce vomiting when formulations containing petroleum solvents are ingested.* Otherwise, give large quantities of water and induce vomiting. Do not make an unconscious person vomit. *Note to Physician:* If ingested, remove by lavage or vomiting. Use general supportive measures for CNS depression. Consider the use of quinidine for myotonia.

*References:*
- U.S. Environmental Protection Agency, Office of Pesticide Programs, Pesticide Residue Limits, "2,4-D and its butoxyethyl ester," 40 CFR 180.142, www.epa.gov/pesticides/food/viewtols.htm
- California Environmental Protection Agency "Chemical List of Lists," Sacramento CA (February 1997).

# 2,4-D, isooctyl ester

*Use Type:* Herbicide
*CAS Number:* 25168-26-7
*Formula:* $C_{16}H_{22}Cl_2O_3$
*Alert:* Herbicides containing 2,4-D use the amine salt or ester form. The amine and ester form of 2,4-D may differ in health-related activity and environmental fate and effects from the parent 2,4-D.

# 2,4-D, isooctyl ester

*Synonyms:* Acetic acid(2,4-dichlorophenoxy)-,isooctyl ester; 2,4-Dichlorophenoxyacetic acid isooctyl ester; 2,4-D Esters; 2,4-D (IOE); Isooctyl alcohol (2,4-dichlorophenoxy)acetate; Isooctyl 2,4-dichlorophenoxyacetate; Isooctyl ester of dichloro 2,4-chloroacetic acid

*Trade Names:* BANVEL®-520, canceled; CLEAN KILL®, canceled; 2,4-D L.V. 4 ESTER®, Nufarm (Australia); REED® LV 2,4-D; REED® LV 400 2,4-D; REED® LV 600 2,4-D; WEEDTRINE®-II

*Producers:* Nufarm (Australia)

*Chemical Class:* Chlorophenoxy acid or ester

*EPA/OPP PC Code:* 030064

*California DPR Chemical Code:* 809

*RTECS Number:* AG8575000

*Uses:* Not registered in the U.S. 2,4-D and its salts and esters are used to control broadleaf weeds, grasses, woody plants, aquatic weeds and non-flowering plants on rangeland, non-crop areas, rights-of-way and in forestry management. As a plant growth regulator it is used to cause all tomato crops to ripen at the same time and other fruit crop control. Formulations are used to control weeds in cereal, rice and other crops and to increase the output of latex from rubber trees.

*Human toxicity (long-term)$^{(77)}$:* Low–70.00 ppb, MCL (Maximum Contaminant Level) as 2,4-D

*Fish toxicity (threshold)$^{(77)}$:* Very low–4247.00420 ppb, MATC (Maximum Acceptable Toxicant Concentration) as 2,4

*U.S. Maximum Allowable Residue Levels:* For allowable levels for the parent chemical, see 2,4-D.

*Carcinogen/Hazard Classifications*

**U.S. EPA Carcinogens:** Group D, unclassifiable because of ambiguous data, as 2,4-D, the parent chemical

**IARC:** Group 2B, possible carcinogen

**Label Signal Word:** WARNING or DANGER

**WHO Acute Hazard:** Class II, moderately hazardous as 2,4-D, the parent chemical

**Endocrine Disruptor:** Probable ED

*Regulatory Authority:*
- AB 2588-Air Toxics "Hot Spots" Chemicals (CAL) as chlorophenoxy pesticides
- The "Director's List" (CAL/OSHA) as chlorophenoxy pesticides
- Actively registered pesticide in California.
- Clean Water Act: Section 311 Hazardous Substances/RQ (same as CERCLA).
- RCRA Section 261 Hazardous Constituents
- Safe Drinking Water Act: MCL, 0.1 mg/L; MCGL, 0.07 mg/L; Regulated chemical (47 FR 9352) as 2,4-D.
- EPA Hazardous Waste Number (RCRA No.): U240
- EPCRA Section 304 RQ: CERCLA, 100 lbs. (45.4 kg)
- U.S. DOT Regulated Marine Pollutant (49CFR172.101, Appendix B) as 2,4-D

*Description:* Dark viscous liquid. Odorless (pure); may develop a gasoline like odor. Soluble in water; solubility = 10 ppm. Molecular weight =333.28. Density = 1.16 Boiling point = 315°C.

*Incompatibilities:* May attack some forms of plastics. Keep material dissolved in organic solvents (such as hexane, benzene, acetone, and alcohols) away from oxidizers. May not be compatible with nitrates. Moisture may cause hydrolysis or other forms of decomposition.

*Permissible Exposure Limits in Air:* See 2,4-D.

*Determination in Air:* Collection on a glass fiber filter and analysis by HPLC with UV detection. See NIOSH Method #5001$^{(18)}$, for 2,4-D$^{(18)}$.

*Permissible Concentration in Water:* No criteria set. Runoff from spills or fire control may cause water pollution. For 2,4-D, the U.S. EPA$^{(35)}$ has set an MPC in bottled water intended for human consumption of 0.1 mg/L.

*Determination in Water:* Filter; Methanol; High-pressure liquid chromatography/Ultraviolet detection; NIOSH IV Method #5001$^{(18)}$

*Routes of Entry:* Inhalation, ingestion, absorbed through skin.

*Harmful Effects and Symptoms*

*Short Term Exposure:* Poisonous; may be fatal if inhaled, swallowed, or absorbed through skin. Severely irritates eyes, skin and respiratory tract, with burning sensation, pain, redness and swelling. Metabolic stimulant. If inhaled, causes coughing, dilated pupils, headache, profuse persperation, intense thirst, extreme fatigue, rapid pulse, high fever, clammy, flushed skin, rapid breathing, nausea, vomiting, cyanosis (bluish tint to skin and lips), anxiety and confusion, convulsions, risk of lung edema. If swallowed, face and lips turn bluish. Liver injury with associated jaundice, kidney failure, and cardiac arrhythmias are commonly noted. Nerve damage, which may be delayed, may include swelling of legs and feet, muscle twitch and stupor. Severe exposure can cause death from heart failure. Dust or liquid left in contact with the skin for several hours may be absorbed. This may result in severe delayed symptoms as listed above. These symptoms may last for months or years.

*Long Term Exposure:* Workers exposed to chlorophenoxy compounds such as 2,4-D (in the manufacturing process) over a five to ten year period at levels above 10 mg/m³ complained of weakness, rapid fatigue, headache and vertigo. Liver damage, low blood pressure and slowed heartbeat were also found. Based on animal tests, may affects human reproduction

*Points of Attack:* Eyes, skin, respiratory system, central nervous system, cardiovascular system, liver, kidney

*Medical Surveillance:* If symptoms develop or overexposure is suspected, liver or kidney function tests may be useful. Liver function tests.

*First Aid:* If this chemical gets into the eyes, remove any contact lenses at once and irrigate immediately for at least

15 minutes, occasionally lifting upper and lower lids. Seek medical attention immediately. If this chemical contacts the skin, remove contaminated clothing and wash immediately with soap and water. Seek medical attention immediately. If this chemical has been inhaled, remove from exposure, begin rescue breathing (using universal precautions) if breathing has stopped, and CPR if heart action has stopped. Transfer promptly to a medical facility. When this chemical has been swallowed, get medical attention. *Do not induce vomiting when formulations containing petroleum solvents are ingested.* Otherwise, give large quantities of water and induce vomiting. Do not make an unconscious person vomit. *Note to Physician:* If ingested, remove by lavage or vomiting. Use general supportive measures for CNS depression. Consider the use of quinidine for myotonia.

*References:*
- International Programme on Chemical Safety (IPCS), "Environmental Health Criteria, 2,4-Dichlorophenoxyacetic Acid (2,4-D)," Geneva, Switzerland (1984). http://www.inchem.org/documents/ehc/ehc/ehc29.htm
- California Environmental Protection Agency "Chemical List of Lists," Sacramento CA (February 1997).

## 2,4-D, isopropyl ester

*Use Type:* Herbicide
*CAS Number:* 94-11-1
*Formula:* $C_{11}H_{12}Cl_2O_3$
*Alert:* Herbicides containing 2,4-D use the amine salt or ester form. The amine and ester form of 2,4-D may differ in health-related activity and environmental fate and effects from the parent 2,4-D.
*Synonyms:* Acetic acid, (2,4-dichlorophenoxy)-, isopropyl ester; Acetic acid, (2,4-dichlorophenoxy)-,1-methylethyl ester; AI3-16667; Caswell No. 315AV; 2,4-D esters; 2,4-Dichlorophenoxyacetic acid isopropyl ester; (2,4-Dichlorophenoxy)acetic acid isopropyl ester; 2,4-D-isopropyl; isopropyl 2,4-D ester; isopropyl (2,4-dichlorophenoxy)acetate; Isopropyl 2,4-dichlorophenoxyacetate
*Trade Names:* ALCO CITRUS FIX®, Amvac Chemical (USA); ALPHASET IPE, Nufarm (Australia); AMCHEM® WEED KILLER 650, Amvac Chemical (USA); BARBER'S® WEED KILLER (ESTER FORMULATION); BRIDGEPORT® SPOT WEED KILLER; CHEMICAL INSECTICIDE'S® ISOPROPYL ESTER OF 2,4-D LIQUID CONCENTRATE; CROP RIDER® 3.34D; CROP RIDER® 3-34D-2; ESTERON® 44; HIVOL 44, Amvac Chemical (USA); MONSANTO® 2,4-D ISOPROPYL ESTER, Monsanto (USA); NIA® ESTASOL, FMC Agricultural Products (USA), canceled; NIAGARA® ESTASOL, FMC Agricultural Products (USA), canceled; NSC® 521749; PARSONS® 2,4-D WEED KILLER ISOPROPYL ESTER; SWIFT'S® GOLD BEAR 44 ESTER; WEEDONE® 128
*Producers:* Amvac Chemical (USA); Bayer CropScience (Germany); FMC Agricultural Products (USA); Monsanto (USA); Nufarm (Australia); Syngenta (Switzerland)
*Chemical Class:* Chlorophenoxy acid or ester
*EPA/OPP PC Code:* 030066
*California DPR Chemical Code:* 810
*RTECS Number:* AG8750000
*Uses:* 2,4-D and its salts and esters are used to control broadleaf weeds, grasses, woody plants, aquatic weeds and non-flowering plants on rangeland, non-crop areas, rights-of-way and in forestry management. As a plant growth regulator it is used to cause all tomato crops to ripen at the same time and other fruit crop control. Formulations are used to control weeds in cereal, rice and other crops and to increase the output of latex from rubber trees.
*Human toxicity (long-term)[77]:* Low–70.00 ppb. MCL (Maximum Contaminant Level) as 2,4-D
*Fish toxicity (threshold)[77]:* Very low–4247.00420 ppb. MATC (Maximum Acceptable Toxicant Concentration) as 2,4-D
*U.S. Maximum Allowable Residue Levels:* For allowable levels for the parent chemical, see 2,4-D.
*Carcinogen/Hazard Classifications*
**U.S. EPA Carcinogens:** Group D, unclassifiable because of ambiguous date as 2,4-D, its parent chemical
**IARC:** Group 2B, possible carcinogen
**Label Signal Word:** WARNING or CAUTION
**WHO Acute Hazard:** Class II, moderately hazardous as 2,4-D, its parent chemical
**Endocrine Disruptor:** Probable ED
*Regulatory Authority:*
- AB 2588-Air Toxics "Hot Spots" Chemicals (CAL) as chlorophenoxy pesticides
- CAL Air Resources Board/AB 1807 Toxic Air Contaminants
- The "Director's List" (CAL/OSHA)
- Actively registered pesticide in California.
- Clean Water Act: Section 311 Hazardous Substances/RQ (same as CERCLA).
- RCRA Section 261 Hazardous Constituents
- RCRA Land Ban Waste.
- Safe Drinking Water Act: Regulated chemical (47FR9352) as 2,4-D.
- EPA Hazardous Waste Number (RCRA No.): U240
- EPCRA Section 304 RQ: CERCLA, 100 lb (45.4 kg).
- EPCRA Section 313 Form R *de minimis* concentration reporting level: 0.1%
- U.S. DOT Regulated Marine Pollutant (49CFR172.101, Appendix B) as 2,4-D

*Description:* Colorless solid (pure). Colorless to yellowish brown liquid. Technical, 99%; 64% in petroleum oil. Fuel oil-like odor. Sinks in water; practically insoluble.

Molecular weight = 234–291, also listed at 263.11. Density: 1.255-1.270 @ 25. Boiling point: 140°C @ 1 mmHg. Melting/Freezing point = 25°C. Flash point = > 79°C (oc).

*Incompatibilities:* May attack some forms of plastics. Keep material dissolved in organic solvents (such as hexane, benzene, acetone, and alcohols) away from oxidizers. May not be compatible with nitrates. Moisture may cause hydrolysis or other forms of decomposition.

*Permissible Exposure Limits in Air:* The legal limit (OSHA PEL), and ACGIH recommended TWA is 10 mg/m$^3$ The NIOSH IDLH value is 100 mg/m$^3$.

*Determination in Air:* Collection on a glass fiber filter and analysis by HPLC with UV detection. See NIOSH Method #5001[18], for 2,4-D.[18]

*Permissible Concentration in Water:* No criteria set. Runoff from spills or fire control may cause water pollution. For 2,4-D, the U.S. EPA[35] has set an MPC in bottled water intended for human consumption of 0.1 mg/L. Aquatic toxicity: 350 ppm/24 hr/bass, bluegill/50% kill/fresh water; 1.0–5.0 ppm/96 hr/oyster/39% shell growth disease/salt water. Waterfowl toxicity: $LD_{50}$ = 2025.0 mg/kg.

*Determination in Water:* Filter; Methanol; High-pressure liquid chromatography/Ultraviolet detection; NIOSH IV Method #5001[18]

*Routes of Entry:* Inhalation, ingestion, absorbed through skin.

*Harmful Effects and Symptoms*

*Short Term Exposure:* Poisonous; may be fatal if inhaled, swallowed, or absorbed through skin. Severely irritates eyes, skin and respiratory tract, with burning sensation, pain, redness and swelling. Metabolic stimulant. If inhaled, causes coughing, dilated pupils, headache, profuse persperation, intense thirst, extreme fatigue, rapid pulse, high fever, clammy, flushed skin, rapid breathing, nausea, vomiting, cyanosis (bluish tint to skin and lips), anxiety and confusion, convulsions, risk of lung edema. If swallowed, face and lips turn bluish. Liver injury with associated jaundice, kidney failure, and cardiac arrhythmias are commonly noted. Nerve damage, which may be delayed, may include swelling of legs and feet, muscle twitch and stupor. Severe exposure can cause death from heart failure. Dust or liquid left in contact with the skin for several hours may be absorbed. This may result in severe delayed symptoms as listed above. These symptoms may last for months or years.

*Long Term Exposure:* Workers exposed to chlorophenoxy compounds such as 2,4-D (in the manufacturing process) over a five to ten year period at levels above 10 mg/m$^3$ complained of weakness, rapid fatigue, headache and vertigo. Liver damage, low blood pressure and slowed heartbeat were also found. Based on animal tests, may affects human reproduction

*Points of Attack:* Eyes, skin, respiratory system, central nervous system, cardiovascular system, liver, kidney

*Medical Surveillance:* If symptoms develop or overexposure is suspected, liver or kidney function tests may be useful. Liver function tests.

*First Aid:* If this chemical gets into the eyes, remove any contact lenses at once and irrigate immediately for at least 15 minutes, occasionally lifting upper and lower lids. Seek medical attention immediately. If this chemical contacts the skin, remove contaminated clothing and wash immediately with soap and water. Seek medical attention immediately. If this chemical has been inhaled, remove from exposure, begin rescue breathing (using universal precautions) if breathing has stopped, and CPR if heart action has stopped. Transfer promptly to a medical facility. When this chemical has been swallowed, get medical attention. *Do not induce vomiting when formulations containing petroleum solvents are ingested.* Otherwise, give large quantities of water and induce vomiting. Do not make an unconscious person vomit. *Note to Physician:* If ingested, remove by lavage or vomiting. Use general supportive measures for CNS depression. Consider the use of quinidine for myotonia.

*References:*
- International Programme on Chemical Safety (IPCS), "Environmental Health Criteria, 2,4-Dichlorophenoxy acetic Acid (2,4-D)," Geneva, Switzerland (1984). http://www.inchem.org/documents/ehc/ehc/ehc29.htm
- California Environmental Protection Agency "Chemical List of Lists," Sacramento CA (February 1997)

# Dalapon (ANSI)

*Use Type:* Herbicide and plant growth regulator
*CAS Number:* 75-99-0
*Formula:* $C_3H_4Cl_2O_2$; $CH_3CCl_2COOH$
*Synonyms:* Acido 2,2-dicloropropionico (Spanish); Dalapon aliphatic acid herbicide; 2,2-Dichloropropionic acid; $\alpha$-Dichloropropionic acid; $\alpha,\alpha$-Dichloropropionic acid; Propanoic acid, 2,2-dichloro-; Proprop (South Africa)
*Trade Names:* ATLAS LIGNUM (FORMULATION)®; ALATEX®; BASFAPON®, BASF Agricultural Products Group (Germany); BASFAPON® B, BASF Agricultural Products Group (Germany); BASFAPON® N, BASF Agricultural Products Group (Germany); BASINEX®, BASF Agricultural Products Group (Germany); BH® DALAPON; BH RASINOX R (FORMULATION)®; BH TOTAL (FORMULATION)®; CRISAPON®; DALAPON® 85; DED-WEED®; DESTRAL®; DEVIPON®; DOWPON®, Dow AgroSciences (USA); DOWPON® M, Dow AgroSciences (USA); DOWPON®-RAE.; FYDULAN (FORMULATION)®; GRAMEVIN®; KENAPON®; LIROPON®; PROPROP®; RADAPON®; REVENGE®; S95®; S 1315®; SYNCHEMICALS COUCH AND GRASS KILLER®; UNIPON®; VOLUNTEERED®
*Producers:* BASF Agricultural Products Group (Germany); Bharat Pulverizing Mills (India); Daicel Chemical Industries

(Japan); Dow AgroSciences (USA); Ki-Hara Chemicals Ltd. (UK); R.S.A. Corp. (USA); Sigma-Aldrich (USA); Sigma-Aldrich Laborchemikalien (Germany)
*Chemical Class:* Organochlorine; Halo-organics
*EPA/OPP PC Code:* 028901
*California DPR Chemical Code:* 180
*ICSC Number:* 1509
*RTECS Number:* UF0690000
*EEC Number:* 607-162-00-7
*EINECS Number:* 200-923-0
*Uses:* Not registered in the U.S. Used on sugar cane and sugar beets to control annual and perennial grasses, e.g., Bermuda grass, Johnson grass and quackgrass, and rushes and cattails. Also used on fruits, potatoes, asparagus, alfalfa, flax and carrots. Commercial products consist of the sodium salt or of mixed sodium and magnesium salts of dalapon. Also used on non-crop areas such as ditches, railways and fence lines and as a defoliant.
*Carcinogen/Hazard Classifications*
**Label Signal Word:** CAUTION
**WHO Acute Hazard:** Class U, unlikely to be hazardous
*Regulatory Authority:*
- AB 1803-Well Monitoring Chemical (CAL)
- MCL (Maximum Contaminants Levels) list of contaminants (CAL)
- Permissible Exposure Limits for Chemical Contaminants (CAL/OSHA)
- The "Director's List" (CAL/OSHA)
- Clean Water Act: Section 311 Hazardous Substances/RQ 40CFR117.3 (same as CERCLA, see below).
- Safe Drinking Water Act: MCL, 0.2 mg/L; MCLG, 0.2 mg/L
- Superfund/EPCRA 40CFR302.4 RQ: CERCLA, 5,000 lb (2270 kg)

*Description:* Colorless liquid or a white to tan powder below 8°C. The sodium salt, a white powder, is often used. Acrid odor. Highly soluble in water (slowly reactive). Water solubility = $5 \times 10^5$ ppm; 8% @ 25°C. Molecular weight = 142.99. Boiling point = 185–190°C. Melting/Freezing point = 8°C; also reported @ 20°C. Hazard Identification (based on NFPA-704 M Rating System): Health 1, Flammability 0, Reactivity 0. Log $K_{ow}$ = 0.8. Unlikely to bioaccumulate in marine organisms.
*Incompatibilities:* Keep away from oxidizers, sulfuric acid, caustics, ammonia, aliphatic amines, alkanolamines, isocyanates, alkylene oxides, epichlorohydrin. Reacts slowly in water to form hydrochloric and pyruvic acids
Very corrosive to iron and to aluminum and copper alloys. Flammable and explosive hydrogen gas may form in enclosed spaces.
*Permissible Exposure Limits in Air:* There is no OSHA PEL[2]. NIOSH[2] and ACGIH[1] recommend a TWA value of 1 ppm (6 mg/m³). Israel, Mexico, DFG[3], and the Canadian provinces of Alberta, Ontario, and Quebec set the same value as ACGIH[1] and Alberta's STEL is 2 ppm. The former USSR[35, 43] has set an MAC in workplace air of 10 mg/m³ and a value for ambient air in residential areas of 0.03 mg/m³[35].
*Permissible Concentration in Water:* No criteria set. Runoff from spills or fire control may cause water pollution.
*Routes of Entry:* Inhalation, ingestion. Absorbed through the intact skin.
*Harmful Effects and Symptoms*
**Short Term Exposure:** Corrosive to eyes and skin. *Ingestion:* Grade 2; $LD_{50}$ (oral, mouse) = 3.65 g/kg; (oral, rat) = 7.57 g/kg. 2,2-Dichloropropionic acid is a corrosive chemical and can cause irritation and burn the skin and eyes, causing permanent damage. Symptoms include apprehension, anxiety, confusion, nervous excitation; dizziness; headache; numbness and weakness in limbs; muscle twitching, tremors; nausea and vomiting; slow, shallow respiration, bluish face; convulsions; loss of consciousness; breathing stops; death. Exposure can irritate the upper respiratory tract. 2,2-dichloropropionic acid may cause a skin allergy. If allergy develops, very low future exposures can cause itching and a skin rash.
*Long Term Exposure:* May cause liver and kidney damage. May cause skin allergy.
*Points of Attack:* Eyes, skin, respiratory system, gastrointestinal tract, central nervous system
*Medical Surveillance:* If overexposure or illness is suspected, consider: Lung function tests. Kidney and liver function tests.
Medical observation is recommended for 24 to 48 hours after breathing overexposure, as pulmonary edema may be delayed. As first aid for pulmonary edema, consider administering a corticosteroid spray. Cigarette smoking may exacerbate pulmonary injury and should be discouraged for at least 72 hours following exposure.
*First Aid: Eyes:* Speed in removing material from eyes and skin is of extreme importance. Eye contact can cause dangerous amounts of these chemicals to be quickly absorbed through the mucous membrane into the bloodstream. Directly, irrigate with large amounts of plain, tepid water or saline for 20 minutes, occasionally lifting the lower and upper lids. During this time, remove contact lenses, if easily removable without additional trauma to the eye. Get medical aid immediately. Have physician check for possible delayed damage. *Skin:* Get medical aid. Skin contact can cause dangerous amounts of these chemicals to be absorbed into the bloodstream. Wearing the appropriate PPE equipment and respirator for organochlorine pesticides, immediately flush exposed skin, hair, and under nails with plain, running, tepid water for 20 minutes, then wash twice with mild soap. Shampoo hair promptly if contaminated; protect eyes. **Do not scrub skin or hair**, since this can increase absorption through the skin. Rinse thoroughly with water. Victims who are able and cooperative may assist with

their own decontamination. Remove and double-bag contaminated clothing and personal belongings. Leather absorbs many organochlorines; therefore, items such as leather shoes, gloves, and belts should be discarded. If the skin is swollen or inflamed, cool affected areas with cold compresses. *Ingestion:* Call poison control. Loosen all clothing. Never give anything by mouth to an unconscious person. Get medical aid. *Do not induce vomiting.** In cases of ingestion, the patient is at risk of CNS depression or seizures, which may lead to pulmonary aspiration during vomiting. If the victim is conscious and able to swallow, *administer an aqueous slurry of activated charcoal at 1 gm/kg (usual adult dose 60–90 g, child dose 25–50 g). A soda can and straw may be of assistance when offering charcoal to a child. The efficacy of activated charcoal for some organochlorine poisoning (such as chlordane) is uncertain. If victim is *unconscious or having convulsions, do nothing except keep victim warm. *In some cases you may be specifically instructed by Poison Control to induce vomiting by way of 2 tablespoons of syrup of ipecac (adult) washed down with a cup of water. Do not give activated charcoal before or with ipecac syrup. Inhalation:* Get medical aid. Do not contaminate yourself. Wearing the appropriate PPE equipment and respirator for organochlorine pesticides, immediately remove the victim from the contaminated area to fresh air. For inhalation exposures, monitor for respiratory distress. If the victim is not breathing, administer artificial respiration. *Do not use mouth-to-mouth resuscitation; use bag/mask apparatus.* If cough or breathing difficulty develops, evaluate for respiratory tract irritation, bronchitis, or pneumonitis. If breathing is difficult, administer 100% humidified supplemental oxygen through bag/mask apparatus until medical help arrives. Do not leave victim unattended.

*References:*
- EXTOXNET, Extension Toxicology Network, "Pesticide Information Profile, Dalapon," Oregon State University, Corvallis, OR (June 1996). http://extoxnet.orst.edu/pips/dalapon.htm
- California Environmental Protection Agency "Chemical List of Lists," Sacramento CA (February 1997).
- Sax, N.I., Ed., "Dangerous Properties of Industrial Materials Report," 3, No. 2, 74-77 (1983).
- U.S. Environmental Protection Agency, "Health Advisory: Dalapon," Washington, DC, Office of Drinking Water (August 1987).
- New Jersey Department of Health, "Hazardous Substance Fact Sheet: 2,2-Dichloropropionic Acid," Trenton, NJ (February 1989, revised June 2001). http://www.state.nj.us/health/eoh/rtkweb/0668.pdf

# Daminozide (ANSI)

*Use Type:* Plant growth regulator

*CAS Number:* 1596-84-5
*Formula:* $C_6H_{12}N_2O_3$
*Synonyms:* Bernsteinsaeure-2,2-dimethylhydrazid (German); Butanedioic acid mono(2,2-dimethylhydrazide); DIMAS; N-Dimethyl amino-β-carbamyl propionic acid; N-(Dimethylamino)succinamic acid; N-dimethylamino-succinamidsaeure (German) DMASA; DMSA; NCI-C03827; Succinic acid, 2,2-dimethylhydrazide; Succinic-1,1-dimethyl hydrazide
*Trade Names:* ALAR®, Crompton (USA); ALAR-85®, Crompton (USA); AMINOZID®; AMINOZIDE®; B-9®; B-995®; B-NINE®, Crompton (USA); DAZIDE®, Fine Agrochemicals(UK); DAZIDE® ENHANCE, Fine Agrochemicals(UK); DIMAS®; KYLAR®; SADH®
*Producers:* Crompton (USA); Fine Agrochemicals(UK); Rhone-Poulenc Agro France (France); Sigma-Aldrich (USA)
*EPA/OPP PC Code:* 035101
*California DPR Chemical Code:* 7
*EINECS Number:* 216-485-9
*Uses:* Daminozide is a systemic growth regulator registered for use on ornamentals, including potted chrysanthemums and poinsettias, and bedding plants in enclosed structures. U.S. sales for food and feed crops were halted in 1989 because of health considerations, i.e., the Alar scare on apples.
*Human toxicity (long-term)[77]:* Intermediate–40.22989 ppb, CHCL (Chronic Human Carcinogen Level)
*Fish toxicity (threshold)[77]:* Very low–31007.43129 ppb, MATC (Maximum Acceptable Toxicant Concentration)
*Carcinogen/Hazard Classifications*
**U.S. EPA Carcinogens:** Group B2, probable carcinogen
**U.S. NTP Carcinogen:** Carcinogen (Animal Positive) (NCI) (NTP)[9]
*California Prop. 65:* Carcinogen
**Label Signal Word:** CAUTION, DANGER
**WHO Acute Hazard:** Class U, unlikely to be hazardous
*Regulatory Authority:*
- AB 2588-Air toxics "hot spots" chemicals (CAL)
- Prop 65 chemicals (CAL)
- Actively registered pesticides in California

*Description:* White crystalline solid. Soluble in water; solubility = 100 g/kg @ 25°C. Molecular weight = 160.20. Melting/Freezing point =154°C. Vapor pressure = 1 x 10⁻⁸ mmHg @ 20°C. Log $K_{ow}$ = very low. Unlikely to bioaccumulate in marine organisms.
*Incompatibilities:* Keep away from oxidizers, sulfuric acid, caustics, ammonia, aliphatic amines, alkanolamines, isocyanates, alkylene oxides, epichlorohydrin.
*Determination in Air:* Filter; none; Gravimetric; NIOSH IV [Particulates NOR; #0500 (total), #0600 (respirable)].[18]
*Harmful Effects and Symptoms*
*Short Term Exposure:* The acute $LD_{50}$ (oral, rat) = 8400 mg/kg (insignificantly toxic). However, Daminozide

metabolizes to diamethylhydrazine which is a proven carcinogen in animal tests. In has a low dermal irritation potential and it is neither teratogenic nor mutagenic[55]. It is not an acute toxicant to fish or wildlife.

*First Aid:* **Skin:** Flood all areas of body that have contacted the substance with water. Don't wait to remove contaminated clothing; do it under the water stream. Use soap to help assure removal. Isolate contaminated clothing when removed to prevent contact by others. **Eyes:** Remove any contact lenses at once. Flush eyes well with copious quantities of water or normal saline for at least 20-30 minutes. Seek medical attention. **Inhalation:** Leave contaminated area immediately; breathe fresh air. Proper respiratory protection must be supplied to any rescuers. If coughing, difficult breathing or any other symptoms develop, seek medical attention at once, even if symptoms develop many hours after exposure. *Ingestion:* If convulsions are not present, give a glass or two of water or milk to dilute the substance. Assure that the person's airway is unobstructed and contact a hospital or poison center immediately for advice on whether or not to induce vomiting

*References:*
- EXTOXNET, Extension Toxicology Network, "Pesticide Information Profile, Daminozide," Oregon State University, Corvallis, OR (September 1996). http://ace.orst.edu/cgi-bin/mfs/01/pips/daminozi.htm?8
- U.S. Environmental Protection Agency, "Reregistration Eligibility Decision (RED), Daminozide," Office of Prevention, Pesticides and Toxic Substances, Washington, DC (September 1993). http://www.epa.gov/REDs/factsheets/0032fact.pdf
- California Environmental Protection Agency "Chemical List of Lists," Sacramento CA (February 1997)

# Dazomet

*Use Type:* Fumigant, fungicide and nematicide
*CAS Number:* 533-74-4
*Formula:* $C_5H_{10}N_2S_2$
*Synonyms:* Caswell No. 840; Dimethylformocarbothialdine; 3,5-Dimethylperhydro-1,3,5-thiadiazin-2-thion (Czech, German); 3,5-Dimethyl-1,2,3,5-tetrahydro-1,3,5-thiadiazinethione-2; 3,5-Dimethyltetrahydro-1,3,5-2H-thiadiazine-2-thione; 3,5-Dimethyltetrahydro-1,3,5-thiadiazine-2-thione; 3,5-Dimethyl-1,3,5-thiadiazinane-2-thione; 3,5-Dimethyl-2-thionotetrahydro-1,3,5-thiadiazine; DMTT; Tetrahydro-2H-3,5-dimethyl-1,3,5-thiadiazine-2-thione; Tetrahydro-3,5-dimethyl-1,3,5-thiadiazine-2-thione; Tetrahydro-3,5-dimethyl-2H-1,3,5-thiadiazine-2-thione; 2H-1,3,5-Thiadiazine-2-thione, tetrahydro-3,5-dimethyl-; 2-Thio-3,5-dimethyltetrahydro-1,3,5-thiadiazine
*Trade Names:* AMA-20®, Kemira Chemicals (Finland); BASAMID®, BASF Agricultural Products (Germany); BASAMID® G, BASF Agricultural Products (Germany); BASAMID®-GRANULAR, BASF Agricultural Products (Germany); BASAMID® P, BASF Agricultural Products (Germany); BASAMID-PUDER®, BASF Agricultural Products (Germany); CARBOTHIALDIN®; CARBOTHIALDINE®; CRAG®; CRAG® FUNGICIDE 974; CRAG® NEMACIDE; DAZOMET®-POWDER BASF, BASF Agricultural Products (Germany); FENNOSAN® B 100; MICO-FUME®; MYLON®; MYLONE®; MYLONE® 85; N 521®; NALCON 243®; PREZERVIT®; STAUFFER N® 521; THIADIAZIN® (PESTICIDE); TIAZON®; TROYSAN® 142; UCC 974®

*Producers:* BASF Agricultural Products (Germany), Kemira Chemicals (Finland); Union Carbide (USA)
*Chemical Class:* Unclassified
*EPA/OPP PC Code:* 035602
*California DPR Chemical Code:* 233
*ICSC Number:* 0786
*RTECS Number:* XI2800000
*EEC Number:* 613-008-00-X
*Uses:* Dazomet is a soil fumigant used against germinating weed seeds, soil insects, nematodes, and soil-borne diseases in forest nursery seed beds, tobacco crops, greenhouse crops, and substrates for potted plants, turf, and ornamentals. It is also used as an antimicrobial agent for slimicide preparations and for adhesives, paper-mill slimicide, paint, and cooling water slimicides.
*Human toxicity (long-term)[77]:* Intermediate 24.500 ppb. Health Advisory
*Fish toxicity (threshold)[77]:* Intermediate 16.769960 ppb. MATC (Maximum Acceptable Toxicant Concentration)
*Carcinogen/Hazard Classifications*
**U.S. EPA Carcinogens:** Group D, unclassifiable, ambiguous data
**Label Signal Word:** CAUTION
**WHO Acute Hazard:** Class III, slightly hazardous
*Regulatory Authority:*
- Actively registered pesticide in California.
- RCRA Section 261 Hazardous Constituents
- EPA Hazardous Waste Number (RCRA No.): U366
- EPCRA Section 313 Form R *de minimis* concentration reporting level: 1.0%

*Description:* White crystalline solid. Nearly odorless. Slightly soluble in water; solubility = 1,300 mg/l. @ 25°C. Molecular weight = 162.27. Melting/Freezing point = 102-105°C (decomposes). Vapor pressure = 3.2 x 10$^{-6}$ mm @ 20°C. Flash point = 93°C. Combustible in dust form. Log $K_{ow}$ = 1.38. Unlikely to bioaccumulate in marine organisms.
*Incompatibilities:* Sensitive to moisture and heat. Decomposes on heating above 102°C, producing toxic fumes including nitrogen oxides and sulfur oxides. Decomposes on contact with acids producing carbon disulfide. Decomposes on contact with water or moisture producing toxic gases.

*Permissible Concentration in Water:* No criteria set. Runoff from spills or fire control may cause water pollution.
*Routes of Entry:* Absorbed through the skin, inhalation, ingestion.
*Harmful Effects and Symptoms*
*Short Term Exposure:* Irritates skin and a severe eye irritant. Poisonous if swallowed.
*Long Term Exposure:* Skin sensitizer.
*Points of Attack:* Skin.
*First Aid:* If this chemical gets into the eyes, remove any contact lenses at once and irrigate immediately for at least 15 minutes, occasionally lifting upper and lower lids. Seek medical attention immediately. If this chemical contacts the skin, remove contaminated clothing and wash immediately with soap and water. Seek medical attention immediately. If this chemical has been inhaled, remove from exposure, begin rescue breathing (using universal precautions) if breathing has stopped, and CPR if heart action has stopped. Transfer promptly to a medical facility. When this chemical has been swallowed, get medical attention. Give large quantities of water and induce vomiting. Do not make an unconscious person vomit. Do not induce vomiting when formulations containing petroleum solvents are ingested.
*References:*
- California Environmental Protection Agency "Chemical List of Lists," Sacramento CA (February 1997)

# 2,4-DB

*Use Type:* Herbicide
*CAS Number:* 94-82-6
*Formula:* $C_{10}H_{10}Cl_2O_3$
*Synonyms:* Acido 2,4-diclorofenoxibutirico (Spanish); 2,4-D butyric; Butanoic acid, 4-(2,4-dichlorophenoxy)-; Butyric acid, 4-(2,4-dichlorophenoxy)-; Caswell No. 316; 4(2,4-DB); 4-(2,4-Dichlorophenoxy)butyric acid; γ-(2,4-Dichlorophenoxy)butyric acid; 2,4-DM
*Trade Names:* BUTIREX®; BUTORMONE®; BUTOXON®; BUTOXONE®; BUTOXONE® ESTER; BUTOXONE® AMINE; BUTYRAC®; BUTYRAC® 118; BUTYRAC® 200; BUTYRAC® ESTER; CAMPBELL'S® DB STRAIGHT; CAMPBELL'S® REDLEGOR; DESORMONE®; EMBUTOX KLEAN-UP®; EMBUTONE®; EMBUTOX®; LEGUMEX® D
*Producers:* A H Marks (UK); Lancaster Synthesis (UK); Makhteshim-Agan (Israel); United Phosphorus (India)
*Chemical Class:* Chlorophenoxy compound; Halo-organics
*EPA/OPP PC Code:* 030801
*California DPR Chemical Code:* 5020
*RTECS Number:* ES9100000
*EINECS Number:* 202-366-9
*Uses:* 2,4-DB is a selective systemic herbicide used to control annual and perennial broadleaf weeds in many field crops such as alfalfa, peanuts, cereals and soybeans; used as a defoliant. In the plant, the compound changes to 2,4-D and inhibits growth at the tips of stems and roots.
*Human toxicity (long-term)[77]:* Low–70.00 ppb, Health Advisory
*Fish toxicity (threshold)[77]:* Low–269.85657 ppb, MATC (Maximum Acceptable Toxicant Concentration)
*Carcinogen/Hazard Classifications*
*Label Signal Word:* CAUTION
*TRI Developmental Toxin:* Reproductive and developmental toxin
*Regulatory Authority:*
- EPCRA Section 313 Form R *de minimis* concentration reporting level: 1.0%
- EPA 40 CFR 372.65, Specific Toxic Chemical Listings
- AB 2588-Air Toxics "Hot Spots" Chemicals (CAL) as chlorophenoxy pesticides
- The "Director's List" (CAL/OSHA) as chlorophenoxy pesticides

*Description:* White to light-brown crystalline solid. Slight phenolic odor. Soluble in water; solubility = 46 mg/L @ 25°C. Molecular weight = 249.09. Melting/Freezing point =119°C. Vapor pressure = $3.5 \times 10^{-6}$ mmHg @ 25°C(est). Log $K_{ow}$ = >3.5. Values above 3.0 are likely to bioaccumulate in marine organisms.
*Incompatibilities:* Keep away from oxidizers, sulfuric acid, caustics, ammonia, aliphatic amines, alkanolamines, isocyanates, alkylene oxides, epichlorohydrin; slightly corrosive to iron, aluminum, zinc.
*Permissible Exposure Limits in Air:*
*Determination in Air:* Filter: none; Gravimetric; NIOSH IV [Particulates NOR; #0500 (total), #0600 (respirable)].[18]
*Permissible Concentration in Water:* No criteria set. Runoff from spills or fire control may cause water pollution.
*Routes of Entry:* Inhalation, ingestion, absorbed through the skin.
*Harmful Effects and Symptoms*
*Short Term Exposure:* Poisonous; may be fatal if inhaled, swallowed, or absorbed through skin. Severely irritates eyes, skin and respiratory tract, with burning sensation, pain, redness and swelling. Metabolic stimulant. If inhaled, causes coughing, dilated pupils, headache, profuse persperation, intense thirst, extreme fatigue, rapid pulse, high fever, clammy, flushed skin, rapid breathing, nausea, vomiting, cyanosis (bluish tint to skin and lips), anxiety and confusion, convulsions, risk of lung edema. If swallowed, face and lips turn bluish. Liver injury with associated jaundice, kidney failure, and cardiac arrhythmias are commonly noted. Nerve damage, which may be delayed, may include swelling of legs and feet, muscle twitch and stupor. Severe exposure can cause death from heart failure. Dust or liquid left in contact with the skin for several hours may be absorbed. This may result in severe delayed symptoms as listed above. These symptoms may last for months or years.

*Long Term Exposure:* Workers exposed to chlorophenoxy compounds such as 2,4-D (in the manufacturing process) over a five to ten year period at levels above 10 mg/m³ complained of weakness, rapid fatigue, headache and vertigo. Liver damage, low blood pressure and slowed heartbeat were also found. Based on animal tests, may affects human reproduction

*Points of Attack:* Eyes, skin, respiratory system, central nervous system, cardiovascular system, liver, kidney

*Medical Surveillance:* See a physician if poisoning is suspected for if redness, itching, or burning of the skin or eyes develop.

*First Aid:* If this chemical gets into the eyes, remove any contact lenses at once and irrigate immediately for at least 15 minutes, occasionally lifting upper and lower lids. Seek medical attention immediately. If this chemical contacts the skin, remove contaminated clothing and wash immediately with soap and water. Seek medical attention immediately. If this chemical has been inhaled, remove from exposure, begin rescue breathing (using universal precautions) if breathing has stopped, and CPR if heart action has stopped. Transfer promptly to a medical facility. When this chemical has been swallowed, get medical attention. *Do not induce vomiting when formulations containing petroleum solvents are ingested.* Otherwise, give large quantities of water and induce vomiting. Do not make an unconscious person vomit.

*References:*
- EXTOXNET, Extension Toxicology Network, "Pesticide Information Profile, 2,4-DB," Oregon State University, Corvallis, OR (June 1996). http://extoxnet.orst.edu/pips/24-DB.htm
- California Environmental Protection Agency "Chemical List of Lists," Sacramento CA (February 1997).

# DCPA

*Use Type:* Herbicide
*CAS Number:* 1861-32-1
*Formula:* $C_{10}H_6Cl_4O_4$
*Alert:* A General Use Pesticide (GUP)
*Synonyms:* Chlorthal-dimethyl; Chlorthal-methyl; Dimethyl tetrachloroterephthalate; Dimethyl 2,3,5,6-tetrachloroterephthalate; Chlorthal dimethyl; 2,3,5,6-Tetrachloro-1,4-benzenedicarboxylic acid, dimethyl ester; Tetrachloroterephthalic acid, dimethyl ester; 2,3,5,6-Tetrachlorphthalsaure-dimethylester (German)
*Note:* "DCPA" is also a synonym for Propanil (CAS 709-98-8)
*Trade Names:* ACME®, Pbi/Gordon (USA); DAC 893®, Amvac Chemical (USA); DACTHAL®, Amvac Chemical (USA); DACTHAL® W-75; DECIMATE® (with Propachlor); FATAL®, GREEN WEEDER®, Pbi/Gordon (USA)

*Producers:* Amvac Chemical (USA); Epochem (China); ISK Biosciences (USA); Pbi/Gordon (USA); Nufarm (Australia)
*Chemical Class:* Organochlorine; alkyl phthalate; chlorinated benzoic acid
*EPA/OPP PC Code:* 078701
*California DPR Chemical Code:* 179
*Uses:* This preemergent herbicide is used on annual broadleaf weeds and grasses in a wide spectrum of vegetable crops.
*U.S. Maximum Allowable Residue Levels for Chlorthal-dimethyl (40 CFR 180.185):*

| CROP | ppm |
|---|---|
| Basil, dried leaves | 5 |
| Basil, fresh leaves | 20 |
| Bean, mung, seed | 2 |
| Bean, snap, succulent | 2 |
| Cantaloupe | 1 |
| Celeriac, roots | 2 |
| Chicory, tops | 5 |
| Chive, leaves | 5 |
| Coriander, leaves | 5 |
| Corn, field, forage | 0.4 |
| Corn, field, stover | 0.4 |
| Corn, grain | 0.05 |
| Corn, pop, forage | 0.4 |
| Corn, pop, stover | 0.4 |
| Corn, sweet, forage | 0.4 |
| Corn, sweet, kernel plus cob with husks removed | 0.05 |
| Corn, sweet, stover | 0.4 |
| Cotton, undelinted seed | 0.2 |
| Cowpea, seed | 2 |
| Cress, upland | 5 |
| Cucumber | 1 |
| Dill, seed | 5 |
| Eggplant | 1 |
| Garlic | 1 |
| Ginseng, roots | 2 |
| Honeydew | 1 |
| Horseradish | 2 |
| Lettuce | 2 |
| Marjoram, tops | 5 |
| Onion | 1 |
| Parsley, dried leaves | 20 |
| Parsley, leaves | 5 |
| Pea, blackeyed | 2 |
| Pepper | 2 |
| Pepper, pimento | 2 |
| Potato | 2 |
| Radicchio | 5 |
| Radish, oriental, roots | 2 |
| Radish, roots | 2 |
| Radish, tops | 15 |

| | |
|---|---|
| Rutabaga | 2 |
| Soybean | 2 |
| Squash, summer | 1 |
| Squash, winter | 1 |
| Strawberry | 2 |
| Sweet potato | 2 |
| Tomato | 1 |
| Turnip | 2 |
| Turnip, greens | 5 |
| Vegetable, brassica, leafy, group 5 | 5 |
| Watermelon | 1 |

*Carcinogen/Hazard Classifications*
**U.S. EPA Carcinogens:** Group C, possible carcinogen
**Label Signal Word:** CAUTION
**WHO Acute Hazard:** Class U, unlikely to be hazardous
*Regulatory Authority:*
- Actively registered pesticide in California.
- Washington state requires supplemental labeling as DACTHAL® W-75

*Description:* White crystalline solid. Essentially odorless. Practically insoluble in water; solubility = 0.5 ppm @ 25°C. Molecular weight 331.97. Melting/Freezing point = 155°C. Vapor pressure <0.4 mmHg @ 30°C.

*Incompatibilities:* May react violently with strong oxidizers, bromine, 90% hydrogen peroxide, phosphorus trichloride, silver powders or dust. Incompatible with silver compounds. Mixture with some silver compounds forms explosive salts of silver oxalate.

*Permissible Concentration in Water:* No criteria set. Runoff from spills or fire control may cause water pollution.

*Determination in Water:* EPA Methods 8081, 608.2, 515.2.

*Routes of Entry:* Inhalation, ingestion. Absorbed through the intact skin.

*Harmful Effects and Symptoms*
*Short Term Exposure:* Symptoms include apprehension, anxiety, confusion, nervous excitation; dizziness; headache; numbness and weakness in limbs; muscle twitching, tremors; nausea and vomiting; slow, shallow respiration, bluish face; convulsions; loss of consciousness; breathing stops; death.

*Points of Attack:* CNS. May be fatal if inhaled, ingested, or absorbed through the skin

*Medical Surveillance:* Medical observation is recommended for 24 to 48 hours after breathing overexposure, as pulmonary edema may be delayed. As first aid for pulmonary edema, consider administering a corticosteroid spray. Cigarette smoking may exacerbate pulmonary injury and should be discouraged for at least 72 hours following exposure.

*First Aid:* Eyes: Speed in removing material from skin is of extreme importance. Eye contact can cause dangerous amounts of these chemicals to be quickly absorbed through the mucous membrane into the bloodstream. Directly, irrigate with large amounts of plain, tepid water or saline for 20 minutes, occasionally lifting the lower and upper lids. During this time, remove contact lenses, if easily removable without additional trauma to the eye. Get medical aid immediately. Have physician check for possible delayed damage. *Skin:* Get medical aid. Skin and/or eye contact can cause dangerous amounts of these chemicals to be absorbed into the bloodstream. Wearing the appropriate PPE equipment and respirator for organochlorine pesticides, immediately flush exposed skin, hair, and under nails with plain, running, tepid water for 20 minutes, then wash twice with mild soap. Shampoo hair promptly if contaminated; protect eyes. Do not scrub skin or hair, since this can increase absorption through the skin. Rinse thoroughly with water. Victims who are able and cooperative may assist with their own decontamination. Remove and double-bag contaminated clothing and personal belongings. Leather absorbs many organochlorines; therefore, items such as leather shoes, gloves, and belts should be discarded. If the skin is swollen or inflamed, cool affected areas with cold compresses. *Ingestion:* Call poison control. Loosen all clothing. Never give anything by mouth to an unconscious person. Get medical aid. *Do not induce vomiting.*  * In cases of ingestion, the patient is at risk of CNS depression or seizures, which may lead to pulmonary aspiration during vomiting. If the victim is conscious and able to swallow, *administer an aqueous slurry of activated charcoal at 1 gm/kg (usual adult dose 60–90 g, child dose 25–50 g). A soda can and straw may be of assistance when offering charcoal to a child. The efficacy of activated charcoal for some organochlorine poisoning (such as chlordane) is uncertain. If victim is *unconscious or having convulsions,* do nothing except keep victim warm. *In some cases you may be specifically instructed by Poison Control to induce vomiting by way of 2 tablespoons of syrup of ipecac (adult) washed down with a cup of water. Do not give activated charcoal before or with ipecac syrup. Inhalation:* Get medical aid. Do not contaminate yourself. Wearing the appropriate PPE equipment and respirator for organochlorine pesticides, immediately remove the victim from the contaminated area to fresh air. For inhalation exposures, monitor for respiratory distress. If the victim is not breathing, administer artificial respiration. *Do not use mouth-to-mouth resuscitation; use bag/mask apparatus.* If cough or breathing difficulty develops, evaluate for respiratory tract irritation, bronchitis, or pneumonitis. If breathing is difficult, administer 100% humidified supplemental oxygen through bag/mask apparatus until medical help arrives. Do not leave victim unattended.

*References:*
- U.S. Environmental Protection Agency, Office of Pesticide Programs, Pesticide Residue Limits, "Chlorthal-dimethly," 40 CFR 180.185, www.epa.gov/pesticides/food/viewtols.htm

- U.S. Environmental Protection Agency, "Reregistration Eligibility Decision (RED), DCPA," Office of Prevention, Pesticides and Toxic Substances, Washington, DC (November 1998). http://www.epa.gov/REDs/0270red.pdf
- U.S. Environmental Protection Agency, "Reregistration Eligibility Decision (RED) Facts, DCPA," Office of Prevention, Pesticides and Toxic Substances, Washington, DC (November 1998). http://www.epa.gov/REDs/factsheets/0270fact.pdf
- EXTOXNET, Extension Toxicology Network, "Pesticide Information Profile, DCPA, Chlorthal, Clorthal-dimethyl," Oregon State University, Corvallis, OR (June 1996). http://extoxnet.orst.edu/pips/DCPA.htm
- California Environmental Protection Agency "Chemical List of Lists," Sacramento CA (February 1997).

# Decanoic Acid

*Use Type:* Microbiocide
*CAS Number:* 334-48-5
*Formula:* $C_{10}H_{20}O_2$
*Synonyms:* Acido decanoico (Spanish); Capric acid; n-Capric acid; Caprinic acid; Caprynic acid; n-Decanoic acid; n-Decoic acid; Decylic acid; n-Decylic acid; 1-Nonanecarboxylic acid
*Trade Names:* ECONOSAN®, West Agro (USA); HEXACID®-1095; NEO-FAT 10®
*Producers:* West Agro (USA)
*Chemical Class:* Unclassified
*EPA/OPP PC Code:* 128955
*California DPR Chemical Code:* 2315
*Uses:* Used in cleaning, sanitizing and disinfecting applications for food processors and dairy farmers.
*Carcinogen/Hazard Classifications*
**Label Signal Word:** DANGER
*Regulatory Authority:*
- Actively registered pesticide in California.

*Description:* White crystalline solid or needles. Unpleasant, rancid odor. Insoluble in water. Molecular weight = 172.26. Density = 0.8858 @ 40°C. Boiling point = 268–270°C @ 30 mmHg. Melting/Freezing point = 31–32°C. Vapor pressure = 1 mm @ 125°C. Flash point = >110°C (cc). Hazard Identification (based on NFPA-704 M Rating System): Health Hazards (Blue): 0; Flammability (Red): 1; Reactivity (Yellow): 0. Log $K_{ow}$ = 1.86. Unlikely to bioaccumulate in marine organisms.
*Incompatibilities:* An organic acid. Keep away from oxidizers, sulfuric acid, caustics, ammonia, aliphatic amines, alkanolamines, isocyanates, alkylene oxides, epichlorohydrin. Corrosive solution; attacks most common metals. React violently with strong oxidizers, bromine, 90% hydrogen peroxide, phosphorus trichloride, silver powders or dust. Mixture with some silver compounds forms explosive salts of silver oxalate. Incompatible with silver compounds.
*Routes of Entry:* Inhalation, skin and eye contact, ingestion.
*Harmful Effects and Symptoms*
*Short Term Exposure:* Vapors cannot be tolerated even at low concentrations; can cause severe irritation of eyes and throat; may cause eye and lung injury. If inhaled, causes coughing or difficult breathing. Contact with liquid or solid causes second and third degree burns in a short time; very injurious to the eyes. If swallowed, causes nausea and vomiting. Toxicity by ingestion: Grade 3: $LD_{50}$ = 129 mg/kg mouse, intravenous.
*Long Term Exposure:* May cause lung damage.
*Medical Surveillance:* Lung x-ray.
*First Aid:* If this chemical gets into the eyes, remove any contact lenses at once and irrigate immediately for at least 15 minutes, occasionally lifting upper and lower lids. Seek medical attention immediately. If this chemical contacts the skin, remove contaminated clothing and wash immediately with soap and water. Seek medical attention immediately. If this chemical has been inhaled, remove from exposure, begin rescue breathing (using universal precautions) if breathing has stopped, and CPR if heart action has stopped. Transfer promptly to a medical facility. When this chemical has been swallowed, get medical attention. If victim is conscious and able to swallow, have victim drink 4 to 8 ounces of water. Do not induce vomiting. *Note to physician or authorized medical personnel*: Medical observation is recommended for 24 to 48 hours after breathing overexposure, as pulmonary edema may be delayed. As first aid for pulmonary edema, consider administering a corticosteroid spray. Cigarette smoking may exacerbate pulmonary injury and should be discouraged for at least 72 hours following exposure.
*References:*
- California Environmental Protection Agency "Chemical List of Lists," Sacramento CA (February 1997)

# D-D mixture

*Use Type:* Fumigant and nematicide
*CAS Number:* 8003-19-8
*Formula:* $C_3H_6Cl_2 \cdot C_3H_4Cl_2$
*Synonyms:* 1,3-Dichloro-1-propene, mixture with 1,2-dichloropropane; 1,3-Dichloropropene and 1,2-Dichloropropane mixture; Dichlorpropan-Dichlorpropengemisch (German); ENT 8,420; Mezcla de dicloropropeno y dicloropropano (Spanish); Mixture of 1,3-dichloropropane, 1,3-dichloropropene, and related $C_3$ compounds
*Trade Names:* DOWFUME®-N, Dow AgroSciences (USA); NEMAFENE®; TELONE®; NEMAFENE®; VIDDEN®-D
*Producers:* Dow AgroSciences (USA)

*Chemical Class:* Substituted allyls; Chlorinated hydrocarbon, aliphatic
*EPA/OPP PC Code:* 029003
*California DPR Chemical Code:*
*RTECS Number:* TX9800000
*EINECS Number:* 208-826-5
*Uses:* Controls plant parasites (nematodes, symphylids, and wireworms) on a broad range of crops; pre-planting treatment to cankers on peach trees and other fruit; controls quackgrass in white potatoes crops.
*Regulatory Authority:*
- The "Director's List" (CAL/OSHA)
- Clean Water Act: Section 311 Hazardous Substances/RQ (same as CERCLA); Toxic Pollutant (Section 401.15)
- RCRA Section 261 Hazardous Constituents, waste number not listed
- EPCRA Section 304 RQ: CERCLA, 100 lb (45.4 kg)
- Marine pollutant (49CFR, Subchapter 172.101, Appendix B)

*Description:* Straw to amber liquid. Pungent, garlic-like odor. Sinks and slowly mixes with water. Molecular weight = 223.96. Density = 1.4 @ 4°C (approx.). Boiling point = 103–171°C. Melting/Freezing point = (estimated) –61°C; Specific gravity = 1.2. Flash point = 19°C (cc); LEL: 5.3%; UEL: 14.5%; Hazard Identification (based on NFPA-704 M Rating System): Health Hazards (Blue): 3; Flammability (Red): 3; Reactivity (Yellow): 0. Vapor pressure = 4.6 kPa @ 20°C. Log $K_{ow}$ = 2.0. Unlikely to bioaccumulate in marine organisms.
*Incompatibilities:* Inorganic bases; concentrated acids; halogens; metal salts. Corrosive to iron, aluminum, magnesium, zinc, and other active metals and alloys.
*Permissible Concentration in Water:* No criteria set. Runoff from spills or fire control may cause water pollution.
*Routes of Entry:* Inhalation, ingestion, absorbed by skin.
*Harmful Effects and Symptoms*
*Short Term Exposure:* Liquid: May be fatal if swallowed or absorbed through skin. Will burn exposed tissues. Vapors: Cause eye and upper respiratory irritation and drowsiness. Corrosive to skin or eyes; severe irritation and burns. Ingestion will cause acute gastrointestinal distress, with congestion. Inhalation will cause gasping, refusal to breathe, respiratory distress; pulmonary edema may develop.
*Long Term Exposure:* Repeated exposure may cause central nervous system depression, and liver and kidney damage.
*Points of Attack:* Liver, kidneys, CNS.
*Medical Surveillance:* Liver and kidney function tests.
*First Aid:* Move victim to fresh air. If breathing has stopped, give artificial respiration. If breathing is difficult, administer oxygen. *Eyes:* hold eyelids open and flush with water for 15 minutes. *Skin:* flush with water for 15 minutes; wash with soap and water. Remove and double bag contaminated clothing and shoes at the site. *If ingested* and victim is *conscious and able to swallow*, have victim drink water and induce vomiting. *If ingested* and victim is *unconscious or having convulsions*, do nothing except keep victim warm.
*References:*
- Pesticide Management Education Program, "D-D Mixture (Nemafene) Chemical Profile 6/84," Cornell University, Ithaca, NY (June 1984). http://pmep.cce.cornell.edu/profiles/fumigant/d-dmixture/fumi-prof-dd.html
- California Environmental Protection Agency "Chemical List of Lists," Sacramento CA (February 1997)

# DDT

*Use Type:* Insecticide
*CAS Number:* 50-29-3 9 (*p,p'*-DDT); 789-02-6 (*o,p'*-DDT)
*Formula:* $C_{14}H_9Cl_5$
*Alert:* DDT was banned form use in the U.S. on December 31, 1972, but it is still used in other countries, primarily in the subtropics to control malaria. It is in the U.S. EPA Toxicity Class II, moderately toxic. Human toxicity (long-term): High.
*Synonyms:* Benzene, 1,1'-(2,2,2-trichloroethylidene)bis(4-chloro); α,α-Bis(*p*-chlorophenyl)-β,β,β-trichlorethane; 1,1-Bis-(*p*-chlorophenyl)-2,2,2-trichloroethane; 2,2-Bis(*p*-chlorophenyl)-1,1-trichloroethane; Chlorophenothan; Chlorophenothane; α.-Chlorophenothane; Chlorophenotoxum; Dichlorodiphenyl trichloroethane 2,2-Bis(*p*-chlorophenyl)-1,1,1-trichloroethane; *p,p'*-DDT; 4,4' DDT; Dichlorodiphenyltrichloroethane; *p,p'*-Dichlorodiphenyltrichloroethane; 4,4'-Dichlorodiphenyltrichloroethane; Diclorodifeniltricloroetano (Spanish); Diphenyltrichloroethane; ENT 1,506; Ethane, 1,1,1-trichloro-2,2-bis(*p*-chlorophenyl)-; NA 2761; NCI-C00464; OMS 16; Parachlorocidum; PEB1; Pentachlorin; 1,1,1-Trichloor-2,2-bis(4-chloorfenyl)-ethaan (Dutch); 1,1,1-Trichlor-2,2-bis(4-chlor-phenyl)-aethan (German); Trichlorobis(4-chlorophenyl)ethane; 1,1,1-Trichloro-2,2-bis(*p*-chlorophenyl)ethane; 1,1,1-Trichloro-2,2-di(4-chlorophenyl)-ethane; 1,1,1-Tricloro-2,2-bis(4-cloro-fenil)-etano (Italian)
*Trade Names:* ARKOTINE®; AGRITAN®; ANOFEX®; AZOTOX®; BOSAN SUPRA®; BOVIDERMOL®; CESAREX®; CITOX®; CLOFENOTANE®; DEDELO®; DICOPHANE®; DIDIGAM®; DIDIMAC®; DINOCIDE®; DODAT®; DYKOL®; ESTONATE®; GENITOX®; GESAFID®; GESAPON®; GESAREX®; GESAROL®; GENITOX®; GEXAREX®; GUESAPON®; GUESAROL®; GYRON®; HAVERO-EXTRA®; HILDIT®; IVORAN®; IXODEX®; KOPSOL®; MUTOXIN; NEOCID®; NIAGARA ZINEB®, FMC Agricultural Products Group

(USA) canceled 2/21/1986; PENTECH®; PZEIDAN®; R 50®; RUKSEAM®; SANTOBANE®; ZEIDANE®; ZERDANE®

***Producers:*** Fluorochem (UK); Hindustan Insecticides (India); Shenzhen Guomeng Industry Co., Ltd. (China)

***Chemical Class:*** Organochlorine; Halo-organics

***EPA/OPP PC Code:*** 029201

***California DPR Chemical Code:*** 186

***ICSC Number:*** 0034

***RTECS Number:*** KJ3325000

***EEC Number:*** 602-045-00-7

***Uses:*** DDT is an organochlorine insecticide used mainly to control mosquito-borne malaria. It was extensively used during the Second World War among Allied troops and certain civilian populations to control insect typhus and malaria vectors, bubonic plague and body lice and was then extensively used as an agricultural insecticide after 1945. By 1970, the Bald Eagle and the Osprey population in the U.S. were near extinction. DDT was banned for use in Sweden in 1970 and in the United States in 1972. DDT is a relatively low-cost broad-spectrum insecticide. However, following an extensive review of health and environmental hazards of the use of DDT, U.S. EPA decided to ban further use of DDT in December 1972. This decision was based on several properties of DDT that had been well evidenced: (1) DDT and its metabolites are toxicants with long-term persistence in soil and water; (2) it is widely dispersed by erosion, runoff and volatization; and (3) the low-water solubility and high lipophilicity of DDT result in concentrated accumulation of DDT in the fat of wildlife and humans which may be hazardous. Nevertheless, DDT is used to control malaria is subtropical countries, in Africa and some South American locales, where malaria is of epidemic proportions.

***Human toxicity (long-term)***[77]***:*** High–1.02941 ppb, CHCL (Chronic Human Carcinogen Level)

***Fish toxicity (threshold)***[77]***:*** Extra high–0.09855 ppb, MATC (Maximum Acceptable Toxicant Concentration)

***Carcinogen/Hazard Classifications***

**U.S. EPA Carcinogens:** Group B2, Probably Carcinogen

**U.S. NTP Carcinogen:** Reasonably anticipated carcinogen

**California Prop. 65:** Listed; Developmental, female, male.

**IARC:** Group 2B, possible carcinogen

**WHO Acute Hazard:** Class II, moderately hazardous

**Endocrine Disruptor:** Known ED

***Regulatory Authority:***
- Carcinogen (Animal Positive) (IARC)[9]
- Banned or Severely Restricted (Many Countries) (UN)[13]
- Air Pollutant Standard Set (ACGIH)[1] (Australia) (DFG)[3] (HSE)[33] (Israel) (Mexico) (former USSR)[43] (OSHA)[58]
- (Several States)[60] (Several Canadian Provinces)
- AB 1803-Well Monitoring Chemical (CAL)
- AB 2588-Air Toxics "Hot Spots" Chemicals (CAL)
- Proposition 65 chemical (CAL)
- Permissible Exposure Limits for Chemical Contaminants (CAL/OSHA)
- The "Director's List" (CAL/OSHA)
- Clean Water Act: Section 311 Hazardous Substances/RQ 40CFR117.3 (same as CERCLA, see below);
- 40CFR423, Appendix A, Priority Pollutants; Section 313 Water Priority Chemicals (57FR41331, 9/9/92); Toxic Pollutant (Section 401.15)
- EPA Hazardous Waste Number (RCRA No.): U061
- RCRA, 40CFR261, Appendix 8 Hazardous Constituents
- RCRA 40CFR268.48; 61FR15654, Universal Treatment Standards: Wastewater (mg/L), 0.0039; Nonwastewater (mg/kg), 0.087
- RCRA 40CFR264, Appendix 9; TSD Facilities Ground Water Monitoring List. Suggested test method(s) (PQL, ug/L): 8080(0.1); 8270(10)
- Superfund/EPCRA 40CFR302.4 RQ: CERCLA, 1 lb (0.454 kg)
- U.S. DOT Regulated Marine Pollutant (49CFR172.101, Appendix B), severe pollutant.
- Canada, Drinking Water Quality, 0.03 mg/L MAC
- Mexico, Drinking Water Criteria, 0.001 mg/L

***Description:*** DDT is a waxy solid, colorless crystalline solid, or slightly off-white powder. Weak, chemical odor. Practically insoluble in water; solubility = $3.2 \times 10^{-4}$ ppm @ 25°C. Molecular weight = 354.49. Melting/Freezing point = 107–109°C. Flash point = 72–75°C. Vapor pressure = $2.1 \times 10^{-7}$ mmHg @ 20°C. Hazard Identification (based on NFPA-704 M Rating System): Health 2, Flammability 2, Reactivity 0. Log Kow = 6.15 @ 20°C. Values at or above 3.0 are likely to bioaccumulate in marine organisms.

***Incompatibilities:*** Contact with strong oxidizers may cause fire and explosion hazard. Incompatible with salts of iron or aluminum, and bases. Do not store in iron containers.

***Permissible Exposure Limits in Air:*** The Federal limit (OSHA PEL)[58] and the ACGIH[1] recommended TWA value is 1 mg/m$^3$ as is the Australian, Israeli, Mexican, DFG[3], HSE[33], and Canadian provincial (Alberta, BC, Ontario, Quebec) value. The NIOSH IDLH = 500 mg/m$^3$. The HSE, Mexican and Alberta, and BC STEL value is 3 mg/m$^3$. NIOSH[2] recommended REL is 0.5 mg/m$^3$. The former USSR-UNEP/IRPTC project[43] has set an MAC in workplace air of 0.1 mg/m$^3$ and values for ambient air in residential areas of 0.005 mg/m$^3$ on a momentary basis and 0.001 mg/m$^3$ on a daily average basis. Several states have set guidelines or standards for DDT in ambient air[60] ranging from 1.8 $\mu$g/m$^3$ (Pennsylvania) to 2.38 $\mu$g/m$^3$ (Kansas) to 5.0 $\mu$g/m$^3$ (Connecticut) to 10.0 $\mu$g/m$^3$ (North Dakota) to 16.0 $\mu$g/m$^3$ (Virginia) to 24 $\mu$g/m$^3$ (Nevada).

***Determination in Air:*** Collection on a filter, workup with isooctane, analysis by gas chromatography. See NIOSH Method #S-274[18]

a 24 hr average; never to exceed 0.13 µg/L. To protect human health: preferably zero. An additional lifetime cancer risk of 1 in 100,000 is imposed by a level of 0.24 ng/l (0.00024µg/L).

Various states have set guidelines and standards for DDT in drinking water[61] ranging from guidelines of 0.42 µg/L (Kansas) to 0.83 µg/L (Maine) to 1.0 µg/L (Minnesota) and a standard of 50 µg/L (Illinois). The former USSR has set an MAC of 0.1 mg/L in water used for domestic purposes and zero in surface water for fishing[35]. Canada has set a water quality MAC of 0.03 mg/L. Mexico[35] has set an MPC of 0.001 mg/L in drinking water supply; of 0.006 mg/L in estuaries and 0.6 µg/L in coastal waters.

*Determination in Water:* Gas chromatography (EPA Method 608) or gas chromatography plus mass spectrometry (EPA Method 625).

*Routes of Entry:* Inhalation, skin absorption, ingestion, eye and skin contact.

*Harmful Effects and Symptoms*

*Short Term Exposure: Inhalation:* Can cause irritation. 500-4200 mg/m$^3$ has produced dizziness. *Skin:* Can cause irritation in very high concentrations. DDT can be absorbed through the skin if dissolved in vegetable oils or other solvents. *Eyes:* Can cause irritation. *Ingestion:* 1/30- to 1/4-ounce has caused nausea, vomiting, headache and convulsions. Other symptoms include weakness, restlessness, dizziness, incoordination, numbness of face and extremities, abdominal pain, diarrhea, tremors, and death. Symptoms may be delayed from ½ to 3 hours. Estimated lethal dose is between 1 teaspoon and 1 ounce. Can cause a prickling or tingling sensation in the mouth, tongue, lower face, nausea, vomiting, confusion, a sense of apprehension, weakness, loss of muscle control, and tremors, paresthesia tongue, lips, face; tremor; dizziness, confusion, malaise (vague feeling of discomfort), headache, fatigue; convulsions; paresis hands. High exposures can cause convulsions and death.

*Long Term Exposure:* DDT may cause liver and kidney damage. Prolonged or repeated exposure can cause irritation of the eyes, skin, and throat. Occupational exposure to DDT has been associated with changes in genetic material. DDT levels build up and stay in the body for long periods of time. Exposure to DDT and Aldrin may increase retention of DDT in the body. DDT causes cancer in laboratory animals. The U.S. EPA determined that DDT, DDE, and DDD are probable human carcinogens. There is no evidence that DDT, DDE, or DDD cause birth defects in people. Studies in rats have shown that DDT and DDE can mimic the action of natural hormones and in this way affect the development of the reproductive and nervous systems. Puberty was delayed in male rats given high amounts of DDE as juveniles. This could possibly happen in humans. A study in mice showed that exposure to DDT during the first weeks of life may cause neurobehavioral problems later in life.

*Points of Attack:* Eyes, skin, central nervous system, kidneys, liver, peripheral nervous system. Cancer Site in animals: liver, lung, and lymphatic tumors.

*Medical Surveillance:* Serum DDT level. Urine *dichlorodiphenyl acetic acid* level. Liver and kidney function tests.

*First Aid:* Speed in removing material from eyes and skin is of extreme importance. If this chemical gets into the eyes, remove any contact lenses at once and irrigate immediately for at least 15 minutes, occasionally lifting upper and lower lids. Seek medical attention immediately. If this chemical contacts the skin, remove contaminated clothing and wash immediately with soap and water. Shampoo hair promptly if contaminated. Seek medical attention immediately. If this chemical has been inhaled, remove from exposure, begin rescue breathing (using universal precautions) if breathing has stopped, and CPR if heart action has stopped. Transfer promptly to a medical facility. When this chemical has been swallowed, get medical attention. Give large quantities of water and induce vomiting. Do not make an unconscious person vomit.

*References:*

- EXTOXNET, Extension Toxicology Network, "Pesticide Information Profile, DDT," Oregon State University, Corvallis, OR (June, 1996). ace.orst.edu/info/extoxnet/pips/ddt.htm
- National Pesticide Information Center, "General Fact Sheet, DDT," Oregon State University, Corvallis, OR. (December, 1999). npic.orst.edu/factsheets/ddtgen.pdf
- New Jersey Department of Health and Senior Services, Hazardous Substance Fact Sheet, "DDT," Trenton NJ (September 1996, Rev. July, 2002). www.state.nj.us/health/eoh/rtkweb/0596.pdf
- U.S. Environmental Protection Agency, DDT: ambient Water Quality Criteria, Washington, DC (1980).
- U.S. Environmental Protection Agency, DDT Health and Environmental Effects Profile No. 60, Washington, DC, Office of Solid Waste (April 30, 1980).
- California Environmental Protection Agency "Chemical List of Lists," Sacramento CA (February 1997).
- Sax, N.I., Ed., "Dangerous Properties of Industrial Materials Report," 1, No. 3, 51-54 (1981) and 5, No. 1, 12-20 (1985).
- New York State Department of Health, "Chemical Fact Sheet: DDT." Albany, NY, Bureau of Toxic Substance Assessment (Mar. 1986 and Version 2).

# Deltamethrin

*Use Type:* Insecticide
*CAS Number:* 52918-63-5
*Formula:* $C_{22}H_{19}Br_2NO_3$
*Alert:* Some formulations are Restricted Use Pesticides (RUP)

# Deltamethrin

*Synonyms:* (*S*)-α-Cyano-*m*-phenoxybenzyl (1R,3R)-3-(2,2-dibromovinyl)-2,2-dimethylcyclopropanecarboxylate; 1R-[1-α(S*),3-α)]-cyano(3-phenoxyphenyl)methyl-3-(2,2-dibromovinyl)-2,2-dimethylcyclopropanecarboxylate; Cyclopropanecarboxylic acid, 3-(2,2-dibromoethenyl)-2,2-dimethyl-, cyano(3-phenoxyphenyl)methyl ester, [1R-(1α(S*),3α)]-; Decamethrin; (1R,3R)-3-(2,2-Dibromovinyl)-2,2-dimethylcyclopropane carboxylic acid, (*S*)-α-cyano-3-phenoxybenzyl ester

*Trade Names:* BUTOFLIN®; BUTOSS®; BUTOX®; CISLIN®, Bayer CropScience (Germany); CRACKDOWN®; DECIS®, Bayer CropScience (Germany); DELTA®, Bayer CropScience (Germany); DELTAGUARD®, Bayer CropScience (Germany); ESBECYTHRIN®; FMC 45498®, FMC (USA); JMC 45498®; K-OTHRINE® dust; NRDC 161®; RU 22974®; STRIKER® IEC insecticide (mixture of deltamethrin and tralomethrin)

*Producers:* Agrimor International (USA); Agsin (Singapore); Bayer CropScience (Germany); Bonide Products (USA); Chromos Agro (Croatia); FMC (USA); Hockley International (UK); Ki-Hara Chemicals Ltd. (UK); Nagarjuna Agrichem (India); Valent BioSciences (USA)

*Chemical Class:* Pyrethroid; botanical

*EPA/OPP PC Code:* 097805

*California DPR Chemical Code:* 3010

*ICSC Number:* 0247

*RTECS Number:* GZ1233000

*EINECS Number:* 258-256-6

*Uses:* Deltamethrin is a synthetic pyrethroid insecticide that kills insects on contact and through digestion. It is used to control a variety of chewing and sucking insects that infest fruit, vegetables and field crops, including apples, pears and plums; peas, glasshouse cucumbers, tomatoes, peppers, potted plants, and ornamentals; hops, oats, cotton and other field crops. Deltamethrin is also used to control residential and commercial insect pests.

*Human toxicity (long-term)[77]:* Low–70.00 ppb, Health Advisory

*Fish toxicity (threshold):* Extra high–0.02258 ppb, MATC (Maximum Acceptable Toxicant Concentration)

*U.S. Maximum Allowable Residue Levels for Deltamethrin (40 CFR 180.435):*

| CROP | ppm |
|---|---|
| Animal feed | 0.05 |
| Cotton, refined oil | 0.2 |
| Cotton, undelinted seed | 0.04 |
| Processed food | 0.05 |
| Tomato | 0.2 |
| Tomato, concentrated products | 1 |

*Carcinogen/Hazard Classifications*

**IARC:** Group 3, unclassifiable

**Label Signal Word:** CAUTION

**WHO Acute Hazard:** Class II, moderately hazardous

**Endocrine Disruptor:** Suspected

*Regulatory Authority:*
- AB 1803-Well Monitoring Chemical (CAL) as pyrethrins
- FIFRA, 40CFR185: tolerances for pesticides in food
- Permissible Exposure Limits for Chemical Contaminants (CAL/OSHA) as pyrethrum
- Actively registered pesticide in California. as pyrethrins
- Clean Water Act: Section 311 Hazardous Substances/RQ (same as CERCLA) as pyrethrins
- EPCRA Section 304 RQ: CERCLA, 1 lb (0.454 kg) as pyrethrins

*Description:* Crystalline solid or powder. Odorless. Combustible. Practically insoluble in water; solubility = ≤0.002 mg/L. Molecular weight = 505.22. Melting/Freezing point = 98-100°C. Boiling point decomposes on distillation. Vapor pressure = 2 x $10^{-8}$ mmHg @ 25°C; 0.002 mPa @ 25°C. Density = 1.22 @ 25°C. Log $K_{ow}$ = 5.41. Values above 3.0 are likely to bioaccumulate in marine organisms.

*Incompatibilities:* May react violently with strong oxidizers, bromine, 90% hydrogen peroxide, phosphorus trichloride, silver powders or dust. Incompatible with silver compounds. Mixture with some silver compounds forms explosive salts of silver oxalate.

*Determination in Air:* Collection by impinger or fritted bubbler, analysis by gas liquid chromatography/ultraviolet. See NIOSH IV, Method #5008[18]. (pyrethrum)

*Permissible Concentration in Water:* No criteria set. Runoff from spills or fire control may cause water pollution.

*Determination in Water:* Collection by impinger or fritted bubbler, analysis by gas liquid chromatography/ultraviolet. See NIOSH IV, Method #5008[18].

*Routes of Entry:* Inhalation, ingestion, absorbed through the skin.

*Harmful Effects and Symptoms*

*Short Term Exposure:* Poisonous. May be fatal if inhaled, swallowed, or absorbed through skin. Contact may cause severe irritation and burns to skin and eyes. Pyrethrins can affect you when breathed in and by passing through your skin. Irritates the eyes and respiratory tract. High exposure can affect the nervous system causing headache, nausea, vomiting, fatigue, and restlessness, rhinorrhea (discharge of thin nasal mucous).

*Long Term Exposure:* High or repeated exposure can cause lung allergy (with cough, wheezing and/or shortness of breath) or hay fever symptoms (sneezing, runny or stuffy nose). Allergic "pneumonia" can also occur with cough, chest pain, breathing difficulty and abnormal chest x-ray. Repeated attacks may lead to permanent scarring. Skin allergy may also develop with rash and itching, even with lower exposures. Skin contact can cause rash with redness, blisters and intense itching. A severe generalized allergy can occur with weakness and collapse.

*Points of Attack:* Respiratory system, skin, central nervous system.

*Medical Surveillance:* Before beginning employment and at regular times after that, the following are recommended:

Lung function tests. These may be normal if the person is not having an attack at the time of the test. Consider chest x-ray if lung symptoms are present. Evaluation by a qualified allergist, including careful exposure history and special testing, may help diagnose skin allergy.

*First Aid:* If this chemical gets into the eyes, remove any contact lenses at once and irrigate immediately for at least 15 minutes, occasionally lifting upper and lower lids. Seek medical attention immediately. If this chemical contacts the skin, remove contaminated clothing and wash immediately with soap and water. Seek medical attention immediately. If this chemical has been inhaled, remove from exposure, begin rescue breathing (using universal precautions) if breathing has stopped, and CPR if heart action has stopped. Transfer promptly to a medical facility. When this chemical has been swallowed, get medical attention. *Do not induce vomiting when formulations containing petroleum solvents are ingested.* Otherwise, give large quantities of water and induce vomiting. Do not make an unconscious person vomit.

*References:*
- EXTOXNET, Extension Toxicology Network, "Pesticide Information Profile, Deltamethrin," Oregon State University, Corvallis, OR (September 1995). http://extoxnet.orst.edu/pips/deltamet.htm
- California Environmental Protection Agency "Chemical List of Lists," Sacramento CA (February 1997).
- U.S. Environmental Protection Agency, Office of Pesticide Programs, Pesticide Residue Limits, "Deltamethrin", 40 CFR 180.435. www.epa.gov/pesticides/food/viewtols.htm

# Demeton

*Use Type:* Insecticide and acaricide
*CAS Number:* 8065-48-3 (mixture of demeton-*O* + demeton-*S*); 298-03-3 (demeton-*O*); 126-75-0 (demeton-*S*)
*Formula:* $C_8H_{19}O_3PS_2$
*Alert:* Demeton is a mixture of Demeton-*O* (I) and Demeton-*S* (II). Human toxicity (long-term): Very high.
*Synonyms: Demeton:* Demetona (Spanish); Demeton-*O* + demeton-*S* mixture; Diethoxy thiophosphoric acid ester of 2-ethylmercaptoethanol; *O,O*-Diethyl *S*-2-(ethylthio)ethyl phosphorothioate mixed with phosphorothioic acid,*O,O*-diethyl *O*-2-(ethylthio)ethyl ester; *O,O*-Diethyl-2-ethylmercaptoethyl thiophosphate, diethoxythiophosphoric acid; ENT 17295; Mercaptofos (in former USSR); Phosphorothioic acid,*O,O*-diethyl *O*-2-(ethylthio)ethyl ester, mixed with *O,O*-diethyl *S*-2-(ethylthio)ethyl phosphorothioate
*Demeton-O:* *O,O*-Diethyl *O*-(2-(ethylthio)ethyl) phosphorothioate; *O*-(2-(Ethylthio)ethyl) *O,O*-diethyl thiophosphate; Phosphorothioic acid, *O,O*-diethyl *O*-(2-ethylthio)ethyl ester; Thionodemeton
*Demeton-S:* *O,O*-Diaethyl-*S*(2-aethyltio-aethyl)monothiophosphat (Russia); *O,O*-Diethyl-*S*-ethyl-2-ethylmercaptophosphorothiolate;*O,O*-Diethyl-*S*-(2-ethylthio-ethyl)-monothiofosfaat; Diethyl-*S*-(2-(ethylthio)ethyl)phosphorothiolate; *O,O*-Diethyl *S*-(2-(ethylthio)ethyl) phosphorothioate; *O,O*-Dietil-*S*-(2-etiltio-etil) monotiofosfato; *O,O*-Diethyl *S*-(2-ethioethyl)phosphorothioate; Ethanethiol, 2-(ethylthio)-, *S*-ester with *O*, salt *O*-diethylphosphorothioate; *S*-2-(Ethylthio)ethyl *O,O*-diethyl thiophosphate; *S*-2-(Ethylthio)ethyl phosphoric acid, *O,O*-diethyl ester; Isodemeton; Mercaptofos teolery; Phosphorothioic acid, *O,O*-diethyl *S*-(2-ethylthio)ethyl ester; Thiolodemeton; Thiophosphate de *O,O*-diethyle et de *S*-(2-ethylthioethyle)
*Trade Names:* BAYER 8169®; BAYER 10756®; BAYER 25/154; BAY 18436®; DEMOX®, Aceto Agricultural Chemicals (USA) canceled 7/1/1987; DEMOX®; DENOX®; DURATOX®; E-1059®; ISOMETASYSTOX®; ISOMETHYLSYSTOX®; ISOSYSTOX®; METAISOSEPTOX®; METAISOSYSTOX®; METASYSTOX FORTE®; METHYL ISOSYSTOX®; MITOL®, American Refining & Manufacturing (USA) canceled 1/6/1983; PO-SYSTOX®; PS-SYSTOX®; SYSTEMOX®; SYSTOX®, Bayer CropScience (Germany) canceled 4/23/1986; SYSTOX THIOL®; THIOL-SYSTOX®; UL®
*Demeton-O:* BAYER® 8169; DI-SEPTON®; E-1059; DISYSTON®; MERCAPTOFOS® (Russian); THIODEMETON®; THIONODEMETON®; THIOLMECAPTOPHOS®
*Producers:* Bayer CropScience (Germany)
*Chemical Class:* Organophosphate
*EPA/OPP PC Code:* 057601 (mixture of demeton-*O* + demeton-*S*); 057602 (demeton-*O*); 057603 (demeton-*S*)
*California DPR Chemical Code:* 566 (Dementon)
*ICSC Number:* 0864 (dementon-*S*)
*RTECS Number:* TF3150000 (mixture of demeton-*O* + demeton-*S*); TF3125000 (demeton-*O*); FT3130000 (demeton-*S*)
*EEC Number:* 015-031-00-5 (demeton-*S*); 015-030-00-X (demeton-*O*)
*Uses:* Demeton mixture is a plant systemic and extremely toxic to bees, fish and wildlife.
*Human toxicity (long-term)[77]:* Extra high–0.28 ppb, Health Advisory
*Fish toxicity (threshold)[77]:* High–3.85101 ppb, MATC (Maximum Acceptable Toxicant Concentration)
*Carcinogen/Hazard Classification*
**Label Signal Word:** DANGER; WARNING, DANGER (*demeton-s*)
**WHO Acute Hazard:** Class 1b, highly hazardous (*demeton-s*)
*Regulatory Authority:*
- Banned or Severely Restricted (in agriculture) (Germany and Russia)

- Very Toxic Substance (World Bank)
- AB 1803-Well Monitoring Chemical (CAL)
- EPA/SARA 302 (EPCRA) Extremely hazardous substances
- Permissible Exposure Limits for Chemical Contaminants (CAL/OSHA)
- The "Director's List" (CAL/OSHA)
- Air Pollutant Standard Set (ACGIH)[1] (Australia) (DFG)[3] (Israel) (Mexico) (OSHA)[58] (Several States)[60] (Several Canadian Provinces)
- U.S. DOT Inhalation Hazard Chemicals as organophosphates

For demeton and demeton-S:
- Superfund/EPCRA 40CFR355, Appendix B Extremely Hazardous Substances: TPQ = 500 lb (227kg)
- Superfund/EPCRA 40CFR302.4 RQ: EHS, 1 lb (0.454 kg)

*Description:* Clear, light yellow to light brown, oily liquids. Colorless to pale yellow, oily liquid (demeton-*O*). Odor of sulfur compounds. Slightly soluble in water; solubility = 25 mg/L (demeton-*O*). Molecular weight = 258.38 (demeton-*O*). Melting/Freezing point = < −13°C. Boiling point = 92–93° @ 0.15 mmHg (demeton-*O*); 123°C @ 1 mmHg; 134°C @ 2 mmHg. Vapor pressure = 1.79 to $3.4 \times 10^{-3}$ mmHg @ 20°C. Flash point = 45°C. Explosive limits: LEL = 1%; UEL = 5.3%. Hazard Identification (based on NFPA-704 M Rating System): Health 3, Flammability 2, Reactivity 0.

*Incompatibilities:* Forms explosive mixture with air. Strong oxidizers, strong bases, soluble mercury, other pesticides, and water.

*Permissible Exposure Limits in Air:* The Federal limit (OSHA[2] PEL) is 0.1 mg/m³. The ACGIH[1] recommended TWA value is 0.01 ppm (0.11 mg/m³. ). The DFG, Australian, Israeli, Mexican, and Canadian provincial (Alberta, BC, Ontario, Quebec) TWA value[3] is the same as OSHA and Mexico, Alberta, and BC set a STEL of 0.03 ppm (0.3 mg/m³). OSHA[2], NIOSH[2] and ACGIH[1] adds the notation "skin" indicating the possibility of cutaneous absorption. The NIOSH[2] IDLH level = 10 mg/m³. The MAC in workplace air set in the former USSR is 0.02 mg/m³[35, 43]. Brazil has set an MAC in workplace air of 0.08 mg/m³[35]. States which have set guidelines or standards for Demeton in ambient air[60] include North Dakota at 1.0 μg/m³ and Connecticut and Nevada at 2.0 μg/m³.

*Determination in Air:* Filter/XAD-2® (tube); Toluene; Gas chromatography/Flame photometric detection for sulfur, nitrogen, or phosphorus; NIOSH IV Method #5514.[18]

*Permissible Concentration in Water:* An MAC of 0.01 mg/L in water bodies used for domestic purposes has been set in The former USSR[35, 43].

*Routes of Entry:* Inhalation, skin absorption, ingestion, eye and skin contact.

## Harmful Effects and Symptoms

*Short Term Exposure:* Demeton can be absorbed through the skin, thereby increasing exposure. Demeton may cause effects on the nervous system by cholinesterase inhibiting effect, causing convulsions, respiratory failure and possible death. High exposure (above OEL) may result in unconsciousness and death. Acute exposure to Demeton may produce the following symptoms of exposure: pinpoint pupils, blurred vision, headache, dizziness, muscle spasms, and profound weakness. Vomiting, diarrhea, abdominal pain, seizures, and coma may also occur. The heart rate may decrease following oral exposure or increase following dermal exposure. Chest pain may be noted. Hypotension (low blood pressure) may occur, although hypertension (high blood pressure) is not uncommon. Respiratory symptoms include dyspnea (shortness of breath), respiratory depression, and respiratory paralysis. Psychosis may occur. This material is a cholinesterase inhibitor. It is readily absorbed through the skin and is extremely toxic. Probable human lethal oral dose is 5-50 mg/kg or 7 drops to 1 teaspoonful for 150 lb. person. Acute dose is believed to be 12 to 20 mg by oral route. The effects may be delayed. Medical observation is indicated.

*Long Term Exposure:* May cause mutations. May damage the developing fetus. May damage the nervous system, causing sensation of "pins and needles" in the hands and feet. May cause depression, irritability and personality changes. Cumulative effect is possible. Demeton may affect cholinesterase, causing significant depression of blood cholinesterase.

*Points of Attack:* Respiratory system, lungs, central nervous system, cardiovascular system, skin, eyes, plasma and red blood cell cholinesterase.

*Medical Surveillance:*.Before employment and at regular times after that, the following are recommended: Plasma and red blood cell cholinesterase levels (tests for the enzyme poisoned by this chemical). If exposure stops, plasma levels return to normal in 1-2 weeks while red blood cell levels may be reduced for 1-3 months. When acetylcholinesterase enzyme levels are reduced by 25% or more below preemployment levels, risk of poisoning is increased, even if results are in lower ranges of "normal." Reassignment to work not involving organophosphate or carbamate pesticides is recommended until enzyme levels recover. If symptoms develop or overexposure occurs, repeat the above tests as soon as possible and get an exam of the nervous system. Also consider complete blood count. Consider chest x-ray following acute overexposure. Do not drink any alcoholic beverages before or during use. Alcohol promotes absorption of organophosphates.

*First Aid:* Speed in removing material from eyes and skin is of extreme importance. If this chemical gets into the eyes, remove any contact lenses at once and irrigate immediately for at least 15 minutes, occasionally lifting upper and lower lids. Seek medical attention immediately. If this chemical contacts the skin, remove contaminated clothing and wash immediately with soap and water. Shampoo hair promptly if contaminated. Seek medical attention immediately. If this chemical has been inhaled, remove from exposure, begin rescue breathing (using universal precautions) if breathing has stopped, and CPR if heart action has stopped. Transfer promptly to a medical facility. When this chemical has been swallowed, get medical attention. Give large quantities of water and induce vomiting. Do not make an unconscious person vomit. Effects may be delayed; medical observation is recommended.

*References:*
- U.S. Environmental Protection Agency, "Chemical Profile: Demeton," Washington, DC, Chemical Emergency Preparedness Program (November 30, 1987).
- California Environmental Protection Agency "Chemical List of Lists," Sacramento CA (February 1997).
- New Jersey Department of Health, "Hazardous Substance Fact Sheet: Demeton," Trenton, NJ (April 1999). http://www.state.nj.us/health/eoh/rtkweb/0604.pdf
- New York State Department of Health, "Chemical Fact Sheet: Demeton," Albany, NY, Bureau of Toxic Substance Assessment (April 1986).

# Demeton-methyl

*Use Type:* Insecticide and miticide
*CAS Number:* 8022-00-2; 301-12-2 (oxydemeton-methyl); 919-86-8 (demeton-S-methyl); 867-27-6 (demeton-O-methyl)
*Formula:* $C_6H_{15}O_3PS_2$
*Alert:* Demeton-S-methyl is no longer registered for use in the U.S. Human toxicity (long-term): (oxydemeton-methyl) High High
*Synonyms:* Demethon-methyl; ENT 18,862; S (and O)-2-(Ethylthio)ethyl-O,O-dimethyl phosphorothioate; Methyl demeton; Methyl mercaptophos; Phosphorothioic acid, O-2-(ethylthio)ethyl O,O-dimethyl ester mixed with S-2-(ethylthio)ethyl O,O-dimethyl phosphorothioate
*Demeton-O-methyl:* O-(2-(Ethylthio)ethyl) O,O-dimethyl phosphorothioate; Phosphorothioic acid, O-(2-(ethylthio)ethyl) O,O-dimethyl ester; Methyl demeton-O; Methyl-O-demeton
*Oxydemeton-methyl:* AI3-24964; Caswell No. 455; O,O-Dimethyl S-[2-(ethylsulfinyl)ethyl]phosphorothioate; O,O-Dimethyl S-[2-(ethylsulfinyl)ethyl]thiophosphate; S-[2-(Ethylsulfinyl)ethyl]O,O-dimethyl phosphorothioate; Isomethylsystox sulfoxide; Demeton-O-methyl sulfoxide; Demeton-S-methyl sulfoxide; Demeton-methyl sulphoxide; O,O-Dimethyl S-(2-eththionylethyl) phosphorothioate; Dimethyl S-(2-eththionylethyl) thiophosphate; O,O-Dimethyl S-[2-(ethylsulfinyl)ethyl] monothiophosphate; O,O-Dimethyl S-2-(ethyl sulfinylethyl)phosphorothioate; O,O-Dimethyl S-[2-(ethylsulfinyl)ethyl] phosphorothioate; O,O-Dimethyl S-(2-ethylsulfinyl)ethyl thiophosphate; O,O-Dimethyl S-ethylsulphinylethyl phosphorothiolate; ENT 24,964; Ethanethiol, 2-(ethylsulfinyl)-, S-ester with O,O-dimethylphosphorothioate; S-[2-(Ethylsulfinyl)ethyl] O,O-dimethyl ester phosphorothioic acid; S-(2-(Ethylsulfinyl)ethyl) O,O-dimethyl phosphorothioate; S-2-Ethylsulfinylethyl O,O-dimethyl phosphorothioate; S-2-Ethylsulphinylethyl O,O-dimethyl phosphorothioate; Isomethylsystox sulfoxide; Metaisosystox sulfoxide; Metasystemox; Metasystemox R; Metasystox R; Methyl demeton-O-sulfoxide; Methyl oxydemeton S; NSC 370785; Oxydemetonmethyl; Oxydemeton-methyl; Oxydemeton methyl [S-2-(Ethylsulfinyl)ethyl]O,O-dimethyl ester phosphorothioic acid; Phosphorothioic acid, O,O-dimethyl S-[2-(ethylsulfinyl)ethyl] ester; Phosphorothioic acid, S-[2-(ethylsulfinyl)ethyl] O,O-dimethyl ester
*Demeton-s-methyl:* O,O-Dimethyl-S-(2-aethtyl-thio-aethyl)-monothiophosphat (German); O,O-Dimethyl-S-(2-eththioethyl)phosphorothioate; Dimethyl-S-(2-eththioethyl)thiophosphate; O,O-Dimethyl-S-ethylmercaptoethyl thiophosphate; O,O-Dimethyl-S-ethylmercaptoethyl thiophosphate, thiolo isomer; O,O-Dimethyl-S-(2-(eththio)ethyl)phosphorthioate; O,O-Dimethyl-S-(3-thia-pentyl)-monothiophosphat (German); S-(2-(ethylthio)ethyl)-O,O-dimethylphosphorothioate; S-(2-(Ethylthio)ethyl)dimethyl phosphorothiolate; S-(2-(Ethylthio)ethyl)-O,O-dimethyl thiophosphate; Methyl demeton thioester; Methyl isosystox; Thiophosphate de O,O-dimethyle et de S-2-ethylthioethyle (French)
*Trade Names:* AIMCO SYSTOX® (oxydemeton-methyl); BAY 15203®, Bayer CropScience (Germany); BAY 21097® (oxydemeton-methyl); BAYER 21097® (oxydemeton-methyl); BAYER 21/116®, Bayer CropScience (Germany); DURATOX®; MATA-SYSTOX®, Bayer CropScience (Germany); Gowan Company (USA) canceled 3/18/1987; METASYSTOX-R®, Bayer CropScience (Germany); METHYL SYSTOX®, Bayer CropScience (Germany), canceled 3/18/1987; ODM®; MSR®; R 2170®
*Demeton-S-methyl:* BAY 18436®, Bayer CropScience (Germany); BAYER 25-154®, Bayer CropScience (Germany); DSM®; DURATOX®; ISOMETASYSTOX®, Bayer CropScience (Germany), canceled; ISOMETHYLSYSTOX®; METAISOSEPTOX®; METAISOSYSTOX®, Gowan (USA); METASYSTOX® FORTE, Gowan (USA); MIFATOX®; PERSYST®
*Producers:* Bayer CropScience (Germany); Bharat Pulverizing Mills (India); Ehrenstorfer, Dr. (Germany); Gowan Company (USA); Sigma-Aldrich Laborchemikalien (Germany); *Demeton-S-methyl:* Bayer CropScience (Germany); Gowan Company (USA)
*Chemical Class:* Organophosphate
*EPA/OPP PC Code:* 058701; 058702 (oxydemeton-methyl); 058703 (demeton-O-methyl); 057603 (demeton-S-methyl)
*California DPR Chemical Code:* 4063

*ICSC Number:* 0862
*RTECS Number:* TG1760000; TG1750000 (demeton-*S*-methyl)
*EEC Number:* 015-031-00-5
*Carcinogen/Hazard Classifications*
**WHO Acute Hazard:** Class 1b, highly hazardous
**California Prop. 65:** Listed; Developmental toxin, female, male for the oxymeton-methyl form.
**TRI Developmental Toxin:** Reproductive and developmental toxin
**Label Signal Word:** WARNING, DANGER (Demeton-*S*-methyl)
**WHO Acute Hazard:** Class 1b, highly hazardous (Demeton-*S*-methyl)
*Uses:* Used to control leafminers, thirps, whiteflies, aphids, leafhoppers and mealybugs, primarily on flowering ornamental plants. Demeton-*S*-methyl not registered for use in the U.S. Demeton-*S*-methyl is a systemic and contact insecticide that is used to kill sucking insects such as aphids, sawflies and spider mites in fruits, vegetables, potato, cereals, ornamentals and forestry management.
*Human toxicity (long-term):* (oxydemeton-methyl) High–3.50 ppb, Health Advisory
*Fish toxicity (threshold)[77]:* Low–159.50116 ppb, MATC (Maximum Acceptable Toxicant Concentration); (oxydemeton-methyl) Intermediate–89.05435 ppb, MATC (Maximum Acceptable Toxicant Concentration)
*Regulatory Authority:*
- Banned or Severely Restricted (Restricted In Many Countries) (UN)[35]
- Proposition 65 chemical (CAL) (oxydemeton-methyl)
- Air Pollutant Standard Set (DFG)[3] (ACGIH)[1]
- The "Director's List" (CAL/OSHA) (Demeton-*S*-methyl)
- EPCRA Section 302 Extremely Hazardous Substances: TPQ = 500 lb (227kg) (Demeton-*S*-methyl)
- EPCRA Section 304 RQ: EHS, 1 lb (0.454 kg) (Demeton-*S*-methyl)

*Description:* Demeton-methyl is a colorless to pale yellow oily liquid with an unpleasant odor. Slightly soluble in water. Molecular weight : 230.31. Hazard Identification (based on NFPA-704 M Rating System): Health 2, Flammability 2, Reactivity 0. Vapor pressure = very low. Log $K_{ow}$ = 1.33 (Demeton-*S*-methyl). Unlikely to bioaccumulate in marine organisms. Physical and toxicological properties may be affected by the carrier solvents used in commercial formulations.
*Incompatibilities:* Strong oxidizers such as chlorine, bromine, fluorine. Demeton-*S*-methyl may react violently with antimony(V) pentafluoride. Incompatible with lead diacetate, magnesium, silver nitrate. May be dissolved in hydrocarbon solvent such as xylene which forms an explosive mixture with air. Oxidizers (chlorates, nitrates, peroxides, permanganates, perchlorates, chlorine, bromine, fluorine, etc); strong acids, strong bases, soluble mercury, other pesticides, and water.

*Permissible Exposure Limits in Air:* ACGIH[1] recommended TLV is 0.5 mg/m³ TWA, with skin notation. The DFG MAK is 0.5 ppm (4.8 mg/m³)[3]. This chemical can be absorbed through the skin, thereby increasing exposure. *Demeton-S-methyl*: The OSHA PEL[58] is 0.1 mg/m³. The ACGIH TWA value is 0.01 ppm (0.11 mg/m³). OSHA, NIOSH and ACGIH adds the notation "skin" indicating the possibility of cutaneous absorption. The NIOSH IDLH level for demeton is 10 mg/m³.
*Determination in Air:* OSHA versatile sampler-2; Toluene/Acetone; Gas chromatography/Flame photometric detection for sulfur, nitrogen, or phosphorus; NIOSH Method IV Method #5600, Organophosphorus pesticides[18]. For *Demeton-S-methyl*:
Filter/XAD-2® (tube); Toluene; Gas chromatography/Flame photometric detection for sulfur, nitrogen, or phosphorus; NIOSH IV Method #5514[18].
*Permissible Concentration in Water:* For *Demeton-S-methyl*: A MAC of 0.01 mg/L in water bodies used for domestic purposes has been set in the former USSR[35, 43].
*Routes of Entry:* Inhalation and through the skin
*Harmful Effects and Symptoms*
*Short Term Exposure:* Methyl demeton can be fatal by skin contact even if there is no feeling of irritation. Exposure can cause rapid, fatal organophosphate poisoning. Acute exposure to this chemical may produce the following signs and Symptoms: pinpoint pupils, blurred vision, headache, dizziness, muscle spasms, and profound weakness, vomiting, diarrhea, abdominal pain, loss of coordination, seizures, coma and death. The heart rate may decrease following oral exposure or increase following dermal exposure. Hypotension (low blood pressure) is not uncommon. Respiratory symptoms include dyspnea (shortness of breath), respiratory depression, and respiratory paralysis. Psychosis may occur. Eye contact may cause irritation. Delayed pulmonary edema may occur after inhalation. See also Demeton entry.
*Long Term Exposure:* May cause mutations. In animal studies this chemical causes a decrease in fertility and is toxic to the animal fetus. See also Demeton entry.
*Points of Attack:* Respiratory system, lungs, central nervous system, cardiovascular system, skin, eyes, plasma and red blood cell cholinesterase.
*Medical Surveillance:* Medical observation is recommended for 24 to 48 hours after breathing overexposure, as pulmonary edema may be delayed. Before employment and at regular times after that, the following are recommended: Plasma and red blood cell cholinesterase levels (tests for the enzyme poisoned by this chemical). If exposure stops, plasma levels return to normal in 1-2 weeks while red blood cell levels may be reduced for 1-3 months. When acetylcholinesterase enzyme levels are reduced by 25% or more below preemployment levels, risk of poisoning is increased, even if results are in lower ranges of "normal." Reassignment to work not involving organophosphate or

carbamate pesticides is recommended until enzyme levels recover. If symptoms develop or overexposure occurs, repeat the above tests as soon as possible and get an exam of the nervous system. Also consider complete blood count. Consider chest x-ray following acute overexposure. Do not drink any alcoholic beverages before or during use. Alcohol promotes absorption of organophosphates.

*First Aid:* **Treatment for organophosphate poisoning consists of thorough decontamination, cardiorespiratory support, and administration of the antidotes atropine and pralidoxime. In cases of severe poisoning, diazepam, an anticonvulsant, should also be administered. Antidotes should be administered as prevention even if the diagnosis is in doubt.** Speed in removing material from eyes and skin is of extreme importance. *Eyes:* Eye contact can cause dangerous amounts of these chemicals to be quickly absorbed through the mucous membrane into the bloodstream. Immediately and gently flush eyes with plenty of warm or cold water (NO hot water) for at least 15 minutes, occasionally lifting the upper and lower eyelids. Get medical aid immediately. *Skin:* Get medical aid. Skin contact can cause dangerous amounts of these chemicals to be absorbed into the bloodstream. Wearing the appropriate PPE equipment and respirator for organophosphate/carbamate pesticides, immediately flush skin with plenty of soap and water for at least 15 minutes while removing contaminated clothing and shoes. Shampoo hair promptly if contaminated. The removed, contaminated clothing and shoes should be double-bagged and left in Hot Zone for later disposal by hazardous materials experts. Skin may also be decontaminated with diluted hypochlorite solution. *Inhalation:* Get medical aid. Do not contaminate yourself. Wearing the appropriate PPE equipment and respirator for organophosphate pesticides, immediately remove the victim from the contaminated area to fresh air. If the victim is not breathing, administer artificial respiration. Do NOT use mouth-to-mouth resuscitation; use bag/mask apparatus. If breathing is difficult, administer oxygen through bag/mask apparatus until medical help arrives. Do not leave victim unattended. *Ingestion:* Call poison control. Loosen all clothing. Never give anything by mouth to an unconscious person. If victim is *unconscious or having convulsions,* do nothing except keep victim warm. Get medical aid. Transfer promptly to a medical facility. In cases of ingestion, **do not induce vomiting**. If the victim is alert and asymptomatic, administer a slurry of activated charcoal at a dose of 1 g/kg (infant, child, and adult dose). A soda can and straw may be of assistance when offering charcoal to a child. *In some cases you may be specifically instructed by poison control to induce vomiting by way of 2 tablespoons of syrup of ipecac (adult) washed down with a cup of water.* Do NOT give activated charcoal before or with ipecac syrup.

*Note to physician or authorized medical personnel:* Treat cases of respiratory compromise, coma, or excessive pulmonary secretions with respiratory support using protocols and techniques available and within the scope of training. Some cases may necessitate procedures such as endotracheal intubation or cricothyrotomy by properly trained and equipped personnel. When possible, atropine (see *Antidotes*, below) should be given under medical supervision. Patients who are comatose, hypotensive, or having seizures or cardiac arrhythmias should be treated according to advanced life support protocols. *Antidotes:* Two antidotes are administered to treat organophosphate poisoning. Atropine is a competitive antagonist of acetylcholine at muscarinic receptors and is used to control the excessive bronchial secretions which are often responsible for death. Pralidoxime relieves both the nicotinic and muscarine effects of organophosphate poisoning by regenerating acetylcholinesterase and can reduce both the bronchial secretions and the muscle weakness associated with poisoning. The initial intravenous dose of atropine in adults should be determined by the severity of symptoms: An initial adult dose of 1.0 to 2.0 mg or pediatric dose of 0.01 mg/kg (minimum 0.01 mg) should be administered intravenously. If intravenous access cannot be established, atropine may also be given intramuscularly, subcutaneously or via endotracheal tube. Doses should be repeated every 15 minutes until excessive secretions and sweating have been controlled. Once bronchial secretion has been controlled, atropine administration should be repeated whenever the secretions begin to recur. In seriously poisoned patients, very large doses may be required. Alterations of pulse rate and pupillary size should not be used as indicators of treatment adequacy. Pralidoxime should be administered as early in poisoning as possible as its efficacy may diminish when given more than 24 to 36 hours after exposure. Doses are as follows: adult 1.0 g; pediatric 25 to 50 mg/kg. The drug should be administered intravenously over 30 to 60 minutes, but in a life-threatening situation, one-half of the total dose can be given per minute for a total administration time of 2 minutes. Treatment should begin to take effect within 40 minutes with a reduction in symptoms and the amount of atropine necessary to control bronchial secretion. The initial dose can be repeated in 1 hour and then every 8 to 12 hours until the patient is clinically well and no longer requires atropine. If intravenous access cannot be established, pralidoxime may also be given intramuscularly. Early administration of diazepam in addition to the combined atropine and pralidoxime treatment may help prevent the onset of seizures and potential brain and cardiac morphologic damage following high-level organophosphate poisoning.

*References:*
- EXTOXNET, Extension Toxicology Network, "Pesticide Information Profile, Dementon-S-Methyl," Oregon State University, Corvallis, OR (September 1995). http://extoxnet.orst.edu/pips/demetons.htm

- Sax, N.I., Ed., "Dangerous Properties of Industrial Materials Report," 1, No. 68-69 (1981). (As Meta-Systox).
- U.S. Environmental Protection Agency, "Chemical Profile: Demeton-S-Methyl," Washington, DC, Chemical Emergency Preparedness Program (November 30, 1987).
- California Environmental Protection Agency "Chemical List of Lists," Sacramento CA (February 1997).
- Agency for Toxic Substances and Disease Registry, U.S. Department of Health and Human Services, Public Health Service, "Managing Hazardous Materials Incidents," Atlanta, GA (June 2003).
- New Jersey Department of Health and Senior Services, Hazardous Substance Fact Sheet, "Methyl Demeton," Trenton NJ (March 1989, rev. September 2001). http://www.state.nj.us/health/eoh/rtkweb/1246.pdf

# 2,4-DES-sodium

*Use Type:* Herbicide
*CAS Number:* 136-78-7
*Formula:* $C_8H_7Cl_2NaO_5$; $Cl_2C_6H_3O(CH_2)_2OSO_3Na$
*Alert:* Tolerances for Sesone were revoked by the U.S. EPA March, 1999.
*Synonyms:* 2,4-DES-Na; 2,4-DES-natrium (German); 2-(2,4-Dichlorophenoxy)ethanol hydrogen sulfate sodium salt; 2,4-Dichlorophenoxyethyl sulfate, sodium salt; Disul; Disul-Na; Disul-sodium; Natrium-2,4-dichlorphenoxyathylsulfat (German); SES; Sesone; Sodium-2-(2,4-dichlorophenoxy)ethyl sulfate; Sodium-2,4-dichlorophenoxyethyl sulphate; Sodium-2,4-dichlorophenyl cellosolve sulfate
*Trade Names:* CRAG HERBICIDE 1®; CRAG SESONE®; SCATHE PEANUT HERBICIDE®, Landia Chemical Co. (USA), canceled 11/13/1986
*Chemical Class:* Chlorophenoxy acid or ester
*EPA/OPP PC Code:* 030602
*ICSC Number:* 1142
*RTECS Number:* KK4900000
*U.S. Maximum Allowable Residue Levels for Sesone (40 CFR 180.102):*

| CROP | ppm |
|---|---|
| Asparagus | 2.0 |
| Peanut hay | 6.0 |
| Peanuts | 6.0 |
| Potatoes | 6.0 |
| Strawberries | 2.0 |

*Carcinogen/Hazard Classifications*
**IARC:** Group 2B, possible carcinogen
**Label Signal Word:** CAUTION
*Regulatory Authority:*
- Air Pollutant Standard Set (ACGIH)[1] (OSHA)[58] (Several States)[60]
- Permissible Exposure Limits for Chemical Contaminants (CAL/OSHA)
- AB 2588-Air Toxics "Hot Spots" Chemicals (CAL) as chlorophenoxy pesticides
- The "Director's List" (CAL/OSHA)

*Description:* Colorless, crystalline solid. Odorless. Very soluble in water; solubility = 25% @ 25°C. Molecular weight = 309.10. Melting/Freezing point = 245°C (decomposes). Vapor pressure = 0.1 mmHg @ 20°C. May be used in a carrier solvent which may change its physical properties.
*Incompatibilities:* Strong oxidizers, acids.
*Permissible Exposure Limits in Air:* The Federal Limit (OSHA PEL) is 10 mg/m$^3$, respirable fraction is 5 mg/m$^{3(58)}$. ACGIH[1] as of has set a TWA of 10 mg/m$^3$. The NIOSH[2] IDLH level = 500 mg/m$^3$. Several states have set guidelines or standards for sesone in ambient air[60] ranging from 100 $\mu$g/m$^3$ (North Dakota) to 160 $\mu$g/m$^3$ (Virginia) to 200 $\mu$g/m$^3$ (Connecticut) to 238 $\mu$g/m$^3$ (Nevada).
*Determination in Air:* See NIOSH Method # S-356 (II)5[18]
*Permissible Concentration in Water:* No criteria set, but runoff from spills or fire control may cause water pollution.
*Routes of Entry:* Inhalation, ingestion skin and eye contact.
*Harmful Effects and Symptoms*
*Short Term Exposure:* Irritates eyes, skin, and respiratory tract. High levels of exposure may cause central nervous system effects, convulsions. May affect the kidneys and liver.
*Long Term Exposure:* May cause liver and kidney damage.
*Points of Attack:* Eyes, skin, central nervous system, liver, kidneys.
*Medical Surveillance:* Liver function. Kidney function. Tests of nervous system.
*First Aid:* If this chemical gets into the eyes, remove any contact lenses at once and irrigate immediately for at least 15 minutes, occasionally lifting upper and lower lids. Seek medical attention immediately. If this chemical contacts the skin, remove contaminated clothing and wash immediately with soap and water. Seek medical attention immediately. If this chemical has been inhaled, remove from exposure, begin rescue breathing (using universal precautions) if breathing has stopped, and CPR if heart action has stopped. Transfer promptly to a medical facility. When this chemical has been swallowed, get medical attention. Give large quantities of water and induce vomiting. Do not make an unconscious person vomit.
*References:*
- U.S. Environmental Protection Agency, Office of Pesticide Programs, Pesticide Residue Limits, "Sesone", 40 CFR 180.102.,http://frwebgate.access.gpo.gov/cgi-bin/get-cfr.cgi
- New Jersey Department of Health and Senior Services, Hazardous Substance Fact Sheet," Sesone," Trenton, NJ, (October 2001) http://www.state.nj.us/health/eoh/rtkweb/1654.pdf

# Desmedipham (ANSI)

*Use Type:* Herbicide
*CAS Number:* 13684-56-5
*Formula:* $C_{16}H_{16}N_2O_4$
*Synonyms:* 3-(Aethoxycarbonylaminophenyl)-N-phenyl-carbamat (German); Carbamic acid, N-phenyl-, 3-[(ethoxycarbonyl)amino]phenyl ester; m-Carbaniloyloxycarbanilic acid ethyl ester; 3-[(Ethoxycarbonyl)amino]phenyl N-phenylcarbamate; Ethyl m-hydroxycarbanilate carbanilate; Ethyl phenylcarbamoyloxyphenylcarbamate; Phenylcarbamoyloxyphenylcarbamate
*Trade Names:* BETANAL® AM; BETANAL®-475; BETANEX®, Bayer CropScience (Germany); BETAMIX® 70 WP, Bayer CropScience (Germany); BETANEX® 70 WP, Bayer CropScience (Germany); EP 475®; PROGRESS®, Bayer CropScience (Germany); SCHERING® 38107; SN-475®; SN-38107®
*Producers:* Bayer CropScience (Germany)
*Chemical Class:* Carbamate
*EPA/OPP PC Code:* 104801
*California DPR Chemical Code:* 1748
*EINECS Number:* 237-198-5
*Uses:* Post-emergence sugarbeet herbicide for control of annual weeds such as pigweed, wild mustard, lamb's quarters, nightshade, chickweed, buckwheat, goosefoot, ragweed, fiddleneck, and kochia.
*Human toxicity (long-term)[77]:* Very low–280.00 ppb, Health Advisory
*Fish toxicity (threshold)[77]:* Low–225.68038 ppb, MATC (Maximum Acceptable Toxicant Concentration)
*U.S. Maximum Allowable Residue Levels for Desmedipham (40 CFR 180.353):*

| CROP | ppm |
| --- | --- |
| Beet, garden, roots | 0.2 |
| Beet, garden, tops | 15 |
| Beet, sugar, roots | 0.2 |
| Beet, sugar, tops | 0.2 |

*Carcinogen/Hazard Classifications*
**U.S. EPA Carcinogens:** Group E, unlikely carcinogen
**Label Signal Word:** WARNING, CAUTION, DANGER
**WHO Acute Hazard:** Class U, unlikely to be hazardous
*Regulatory Authority:*
- Actively registered pesticide in California.
- EPA 40 CFR 372.65, Specific Toxic Chemical Listings
- EPCRA Section 313 Form R *de minimis* concentration reporting level: 1.0%

*Description:* Colorless, crystalline solid. Molecular weight = 300.34. Melting/Freezing point = 120°C. Vapor pressure = $2.9 \times 10^{-9}$ mmHg @ 20°C.
*Incompatibilities:* May form explosive materials with phosphorus pentachloride. Forms Carbon monoxide and toxic nitrogen oxides when heated to decomposition.
*Determination in Air:* Filter; none; Gravimetric; NIOSH IV [Particulates NOR; #0500 (total), #0600 (respirable)].[18]
*Permissible Concentration in Water:* No criteria set. Runoff from spills or fire control may cause water pollution.
*Routes of Entry:* Skin absorption, ingestion, inhalation.
*Harmful Effects and Symptoms*
*Short Term Exposure:* Eye pupils are small; blurred vision; eye watering; runny nose; cough; shortness of breath; salivation; nausea, stomach cramps, diarrhea, and vomiting; increased blood pressure; profuse sweating; hypermotility, hallucinations; agitation; tingling of the skin; slow heartbeat; convulsions; fluid in lungs; loss of consciousness; incontinence; breathing stops; death. Carbamate insecticides inhibit the cholinesterase activity of enzymes, causing accumulation of acetylcholine at synapses and altering the way in which nervous impulses are transmitted. However, within several hours carbamates spontaneously detach from the enzymes.
*Long Term Exposure:* A potent cholinesterase inhibitor; cumulative effect is possible. This chemical may damage the nervous system with repeated exposure, resulting in convulsions, respiratory failure. May cause liver damage.
*Points of Attack:* Respiratory system, lungs, central nervous system, cardiovascular system, skin, eyes, plasma and red blood cell cholinesterase.
*Medical Surveillance:* Medical observation is recommended for 24 to 48 hours after breathing overexposure, as pulmonary edema may be delayed. As first aid for pulmonary edema, consider administering a corticosteroid spray. Cigarette smoking may exacerbate pulmonary injury and should be discouraged for at least 72 hours following exposure. Before employment and at regular times after that, the following are recommended: Plasma and red blood cell cholinesterase levels (tests for the enzyme poisoned by this chemical). If exposure stops, plasma levels return to normal in 1-2 weeks while red blood cell levels may be reduced for 1-3 months. When acetylcholinesterase enzyme levels are reduced by 25% or more below preemployment levels, risk of poisoning is increased, even if results are in lower ranges of "normal." Reassignment to work not involving carbamate pesticides is recommended until enzyme levels recover. If symptoms develop or overexposure occurs, repeat the above tests as soon as possible and get an exam of the nervous system. Also consider complete blood count. Consider chest x-ray following acute overexposure
*First Aid:* Speed in removing material from eyes and skin is of extreme importance. *Eyes:* Eye contact can cause dangerous amounts of these chemicals to be quickly absorbed through the mucous membrane into the bloodstream. Immediately and gently flush eyes with plenty of warm or cold water (NO hot water) for at least 15 minutes, occasionally lifting the upper and lower eyelids. Get medical aid immediately. *Skin:* Get medical aid. Skin

contact can cause dangerous amounts of these chemicals to be absorbed into the bloodstream. Wearing the appropriate PPE equipment and respirator for carbamate pesticides, immediately flush skin with plenty of soap and water for at least 15 minutes while removing contaminated clothing and shoes. Shampoo hair promptly if contaminated; protect eyes. *Ingestion:* Call poison control. Loosen all clothing. Never give anything by mouth to an unconscious person. Get medical aid. Do NOT induce vomiting.* If conscious, alert, and able to swallow, rinse mouth and have victim drink 4 to 8 ounces of water. Check to see if poison control instructs you to use ipecac syrup, otherwise administer slurry of activated charcoal (2 oz in 8 oz of water). If victim is *UNCONSCIOUS OR HAVING CONVULSIONS,* do nothing except keep victim warm. **In some cases you may be specifically instructed by poison control to induce vomiting by way of 2 tablespoons of syrup of ipecac (adult) washed down with a cup of water.* Do NOT give activated charcoal before or with ipecac syrup. *Inhalation:* Get medical aid. Do not contaminate yourself. Wearing the appropriate PPE equipment and respirator for carbamate pesticides, immediately remove the victim from the contaminated area to fresh air. If the victim is not breathing, administer artificial respiration. Do NOT use mouth-to-mouth resuscitation; use bag/mask apparatus. If breathing is difficult, administer oxygen through bag/mask apparatus until medical help arrives. Do not leave victim unattended. *Note to physician or authorized medical personnel.* Administer atropine, 2 mg (1/30 gr) intramuscularly or intravenously as soon as any local or systemic signs or symptoms of an intoxication are noted; repeat the administration of atropine every 3 to 8 minutes until signs of atropinization (mydriasis, dry mouth, rapid pulse, hot and dry skin) occur; initiate treatment in children with 0.05 mg mg/kg of atropine; repeat at 5 to 10 minute intervals. Watch respiration, and remove bronchial secretions if they appear to be obstructing the airway; intubate if necessary. *Medical note:* 2-PAMCI may be contraindicated in the case of some carbamate poisonings.

*References:*
- Pesticide Management Education Program, "Desmedipham (Betanex) Herbicide Profile 3/85," Cornell University, Ithaca, NY (March 1985). http://pmep.cce.cornell.edu/profiles/herb-growthreg/dalapon-ethephon/desmedipham/herb-prof-desmedipham
- California Environmental Protection Agency "Chemical List of Lists," Sacramento CA (February 1997).
- U.S. Environmental Protection Agency, Office of Pesticide Programs, Pesticide Residue Limits, "Desmedipham", 40 CFR 180.353. www.epa.gov/pesticides/food/viewtols.htm

# Dialifor (ANSI)

*Use Type:* Insecticide
*CAS Number:* 10311-84-9
*Formula:* $C_{14}H_{17}ClNO_4PS_2$
*Alert:* Tolerances were revoked in 1998.
*Synonyms:* N-(2-Chloro-1-(diethoxyphosphinpthioylthio) ethyl)phthalimide; S-(2-Chloro-1-(1,3-dihydro-1,3-dioxo-2H-isoindol-2-yl)ethyl)-O,O-diethyl phosphorodithioate; S-(2-Chloro-1-phthalimidoethyl)-O,O-diethylphosphorodithioate; Dialifos; O,O-DIethyl-S-(2-chloro-1-phthalimidoethyl)phosphorodithioate; O,O-Diethyl phosphorodithioate S-ester with N-(2-Chloro-1-mercaptoethyl)phthalimide; ENT 27,320; Phosphorodithioic acid S-[2-chloro-1-(1,3-dihydro-1,3-dioxo-2H-isoindol-2-yl)ethyl]O,O-diethyl ester; Phosphorodithioic acid-S-(2-chloro-1-phthalimidoethyl)-O,O-diethyl ester
*Trade Names:* HERCULES 14503®, Hercules (USA) canceled 7/6/1979; TORAK®, Hercules (USA) canceled 7/6/1979
*Producers:* Hercules (USA); Sigma-Aldrich Laborchemikalien (Germany)
*Chemical Class:* Organophosphate
*EPA/OPP PC Code:* 102501
*California DPR Chemical Code:* 1799
*RTECS Number:* TD5165000
*Regulatory Authority:*
- Banned or Severely Restricted (Malaysia, DDR)[13]
- Very Toxic Substance (World Bank)[15]
- The "Director's List" (CAL/OSHA)
- U.S. DOT Inhalation Hazard Chemicals as organophosphates
- Superfund/EPCRA 40CFR355, Appendix B Extremely Hazardous Substances: TPQ = 100/10,000 lb (45.4/4,540 kg)
- U.S. DOT Regulated Marine Pollutant (49CFR172.101, Appendix B)

*Description:* White crystalline solid or oily liquid. Also reported as an oil. Insoluble in water. Molecular weight = 393.85. Melting/Freezing point (solid) = 67–69°C.
*Incompatibilities:* Strong bases.
*Permissible Exposure Limits in Air:* No standards set.
*Determination in Air:* OSHA versatile sampler-2; Toluene/Acetone; Gas chromatography/Flame photometric detection for sulfur, nitrogen, or phosphorus; NIOSH Method IV Method #5600, Organophosphorus pesticides[18].
*Permissible Concentration in Water:* No criteria set, but runoff from spills or fire control may cause water pollution.
*Routes of Entry:* Inhalation, passing through the skin, ingestion.

## Harmful Effects and Symptoms

*Short Term Exposure:* This material is highly toxic (the $LD_{50}$ for rats is 5 mg/kg) This material can cause serious symptoms and in extreme cases death by respiratory arrest.

Organic phosphorus insecticides are absorbed by the skin, as well as by the respiratory and gastrointestinal tracts. They are cholinesterase inhibitors. Symptoms of exposure include headache, giddiness, blurred vision, nervousness, weakness, nausea, cramps, diarrhea, and discomfort in the chest. Signs include sweating, tearing, salivation, vomiting, cyanosis, convulsions, coma, loss of reflexes and loss of sphincter control. Delayed pulmonary edema may occur after inhalation.

*Long Term Exposure:* Cholinesterase inhibitor; cumulative effect is possible. This chemical may damage the nervous system with repeated exposure, resulting in convulsions, respiratory failure. May cause liver damage.

*Points of Attack:* Respiratory system, lungs, central nervous system, cardiovascular system, skin, eyes, plasma and red blood cell cholinesterase.

*Medical Surveillance:* Medical observation is recommended for 24 to 48 hours after breathing overexposure, as pulmonary edema may be delayed. Before employment and at regular times after that, the following are recommended: Plasma and red blood cell cholinesterase levels (tests for the enzyme poisoned by this chemical). If exposure stops, plasma levels return to normal in 1-2 weeks while red blood cell levels may be reduced for 1-3 months. When acetylcholinesterase enzyme levels are reduced by 25% or more below preemployment levels, risk of poisoning is increased, even if results are in lower ranges of "normal." Reassignment to work not involving organophosphate or carbamate pesticides is recommended until enzyme levels recover. If symptoms develop or overexposure occurs, repeat the above tests as soon as possible and get an exam of the nervous system. Also consider complete blood count. Consider chest x-ray following acute overexposure. Do not drink any alcoholic beverages before or during use. Alcohol promotes absorption of organophosphates.

*First Aid:* **Treatment for organophosphate poisoning consists of thorough decontamination, cardiorespiratory support, and administration of the antidotes atropine and pralidoxime. In cases of severe poisoning, diazepam, an anticonvulsant, should also be administered. Antidotes should be administered as prevention even if the diagnosis is in doubt.** Speed in removing material from eyes and skin is of extreme importance. *Eyes:* Eye contact can cause dangerous amounts of these chemicals to be quickly absorbed through the mucous membrane into the bloodstream. Immediately and gently flush eyes with plenty of warm or cold water (NO hot water) for at least 15 minutes, occasionally lifting the upper and lower eyelids. Get medical aid immediately. *Skin:* Get medical aid. Skin contact can cause dangerous amounts of these chemicals to be absorbed into the bloodstream. Wearing the appropriate PPE equipment and respirator for organophosphate/carbamate pesticides, immediately flush skin with plenty of soap and water for at least 15 minutes while removing contaminated clothing and shoes. Shampoo hair promptly if contaminated. The removed, contaminated clothing and shoes should be double-bagged and left in Hot Zone for later disposal by hazardous materials experts. Skin may also be decontaminated with diluted hypochlorite solution. *Inhalation:* Get medical aid. Do not contaminate yourself. Wearing the appropriate PPE equipment and respirator for organophosphate pesticides, immediately remove the victim from the contaminated area to fresh air. If the victim is not breathing, administer artificial respiration. Do NOT use mouth-to-mouth resuscitation; use bag/mask apparatus. If breathing is difficult, administer oxygen through bag/mask apparatus until medical help arrives. Do not leave victim unattended. *Ingestion:* Call poison control. Loosen all clothing. Never give anything by mouth to an unconscious person. If victim is *unconscious or having convulsions,* do nothing except keep victim warm. Get medical aid. Transfer promptly to a medical facility. In cases of ingestion, **do not induce vomiting**. If the victim is alert and asymptomatic, administer a slurry of activated charcoal at a dose of 1 g/kg (infant, child, and adult dose). A soda can and straw may be of assistance when offering charcoal to a child. *In some cases you may be specifically instructed by poison control to induce vomiting by way of 2 tablespoons of syrup of ipecac (adult) washed down with a cup of water.* Do NOT give activated charcoal before or with ipecac syrup.

*Note to physician or authorized medical personnel:* Treat cases of respiratory compromise, coma, or excessive pulmonary secretions with respiratory support using protocols and techniques available and within the scope of training. Some cases may necessitate procedures such as endotracheal intubation or cricothyrotomy by properly trained and equipped personnel. When possible, atropine (see *Antidotes*, below) should be given under medical supervision. Patients who are comatose, hypotensive, or having seizures or cardiac arrhythmias should be treated according to advanced life support protocols. *Antidotes:* Two antidotes are administered to treat organophosphate poisoning. Atropine is a competitive antagonist of acetylcholine at muscarinic receptors and is used to control the excessive bronchial secretions which are often responsible for death. Pralidoxime relieves both the nicotinic and muscarine effects of organophosphate poisoning by regenerating acetylcholinesterase and can reduce both the bronchial secretions and the muscle weakness associated with poisoning. The initial intravenous dose of atropine in adults should be determined by the severity of symptoms: An initial adult dose of 1.0 to 2.0 mg or pediatric dose of 0.01 mg/kg (minimum 0.01 mg) should be administered intravenously. If intravenous access cannot be established, atropine may also be given intramuscularly, subcutaneously or via endotracheal tube. Doses should be repeated every 15 minutes until excessive secretions and sweating have been controlled. Once bronchial secretion has been controlled, atropine administration should be repeated whenever the secretions begin to recur. In seriously poisoned patients, very large doses may be required. Alterations of pulse rate and pupillary size should not be used as indicators of treatment adequacy. Pralidoxime should be administered as early in poisoning as possible as its efficacy may diminish when given more than 24 to 36 hours after exposure. Doses are as follows: adult 1.0 g; pediatric 25 to 50 mg/kg. The drug should be administered intravenously over 30 to 60 minutes, but in a life-threatening situation, one-half of the total dose can be given per minute for a total administration time of 2 minutes. Treatment should begin to take effect within 40 minutes with a reduction in symptoms and the amount of atropine necessary to control bronchial secretion. The initial dose can be repeated in 1 hour and then every 8 to 12 hours until the patient is clinically well and no longer requires atropine. If intravenous access cannot be established, pralidoxime may also be given intramuscularly. Early administration of diazepam in addition to the combined atropine and pralidoxime treatment may help prevent the onset of seizures and potential brain and cardiac morphologic damage following high-level organophosphate poisoning.

*References:*
- Sax, N.I., Ed., "Dangerous Properties of Industrial Materials Report," 2, No. 5, 43-45 (1982).
- U.S. Environmental Protection Agency, "Chemical Profile: Dialifor," Washington, DC, Chemical Emergency Preparedness Program (November 30, 1987).
- Agency for Toxic Substances and Disease Registry, U.S. Department of Health and Human Services, Public Health Service, "Managing Hazardous Materials Incidents," Atlanta, GA (June 2003).
- California Environmental Protection Agency "Chemical List of Lists," Sacramento CA (February 1997).

# Diallate

*Use Type:* Herbicide.
*CAS Number:* 2303-16-4
*Formula:* $C_{10}H_{17}Cl_2NOS$; $[(CH_3)_2CH]_2NCOSCH_2CCl=CHCl$
*Alert:* The U.S. Environmental Protection Agency revoked all tolerances for pesticide residue of diallate on August 30, 1996. Affected crops were alfalfa, barley, clover, field corn, flaxseed, lentils, peas, potatoes, safflower, soybeans, and sugar beets.
*Synonyms:* Bis(1-methylethyl) carbamothioic acid, S-(2,3-dichloro-2-propenyl)ester; Carbamothioic acid, Bis(1-methylethyl)-S-(2,3-dichloro-2-propenyl) ester; CP 15,336; DATC; 2,3-DCDT; Diallaat (Dutch); Diallat (German); Di-allate; Diallate carbamate herbicide; Dichloroallyldiisopropylthiocarbamate; S-2,3-Dichloroallyldiisopropylthiocarbamate; 2,3-Dichloroallyl N,N-diisopropylthiolcarbamate; 2,3-Dichloro-2-propene-1-thiol, Iisopropylcarbamate; Diisopropylthiocarbamic acid,-(2,3-dichloroallyl) ester; Di-isopropylthiolocarbamate des-(2,3-dichloro allyle) (French); 2-Propene-1-thiol, 2,3-dichloro-,diisopropylcarbamate; S-2,3-Dichloroallyl diisopropylthiocarbamate; S-2,3-Dichloroallyl di-isopropyl(thiocarbamate); S-(2,3-Dichloroallyl) diisopropylthiocarbamate); S-(2,3-Dichloro-2-propenyl)bis(1-methylethtl)carbamothioate
*Trade Names:* AVADEX®, Monsanto (USA), canceled 10/10/1989
*Producers:* Monsanto (USA)
*Chemical Class:* Thiocarbamate
*EPA/OPP PC Code:* 078801
*California DPR Chemical Code:* 48
*RTECS Number:* EX8225000
*EINECS Number:* 218-961-1
*Uses:* Used as a before or after planting treatment depending on the crop for control of wild oats. For use on alfalfa, alsike clover, barley, corn, flax, soybeans, lentils, peas, potatoes, red clover, sugar beets and sweet clover.
*Human toxicity (long-term)[77]:* Intermediate–35.00 ppb, Health Advisory
*Fish toxicity (threshold)[77]:* Low–344.93242 ppb, MATC (Maximum Acceptable Toxicant Concentration)
*Carcinogen/Hazard Classifications*
**IARC:** Group 3, unclassifiable
**Label Signal Word:** WARNING
*Regulatory Authority:*
- Carcinogen (Animal Positive) (IARC)[9]
- AB 2588-Air Toxics "Hot Spots" Chemicals (CAL)
- The "Director's List" (CAL/OSHA)
- EPA Hazardous Waste Number (RCRA No.): U062
- RCRA, 40CFR261, Appendix 8 Hazardous Constituents
- RCRA 40CFR264, Appendix 9; TSD Facilities Ground Water Monitoring List. Suggested test method(s) (PQL $ug/L$): 8270(10)
- Superfund/EPCRA 40CFR302.4 RQ: CERCLA, 100 lb (45.4 kg)
- EPCRA Section 313 Form R *de minimus* concentration reporting level: 1.0%
- U.S. DOT Regulated Marine Pollutant (49CFR172.101, Appendix B)

*Description:* Diallate is a brown liquid. Slightly soluble in water; solubility =14 ppm @ 25°C. Molecular weight = 270.22. Boiling point = 150°C @ 9 mmHg. Melting/Freezing point = 25–30°C. Vapor pressure = 1.9 x $10^{-4}$ mmHg @ 25°C Hazard Identification (based on NFPA-704 M Rating System): Health 2, Flammability 0, Reactivity 0. Log $K_{ow}$ = >3.1. Values at or above 3.0 are likely to bioaccumulate in marine organisms.
*Incompatibilities:* Alkalis.
*Permissible Exposure Limits in Air:* No standards set.
*Permissible Concentration in Water:* The former USSR has set an MAC of 0.03 mg/L in water used for domestic purposes[35, 43].

*Routes of Entry:* Inhalation, passing through the skin, ingestion.

### Harmful Effects and Symptoms

*Short Term Exposure:* Eye contact can irritate and possibly cause burns. Inhalation caused irritation of the respiratory tract with chest tightness and/or difficulty breathing. Higher levels can affect the nervous system. With nausea, vomiting, diarrhea, abdominal pain, reduced muscle coordination, blurred vision, muscle twitching, convulsions, coma and possible death.

*Long Term Exposure:* High or repeated exposures can cause liver and kidney damage. There is limited evidence that diallate causes liver cancer in animals.

*Points of Attack:* Skin, eyes, nervous system.

*Medical Surveillance:* Lung function tests. Kidney and liver function tests. Examination of the nervous system. Interview exposed person for brain effects, including memory, mood, concentration, headaches, malaise, and altered sleep patterns.

*First Aid:* If this chemical gets into the eyes, remove any contact lenses at once and irrigate immediately for at least 15 minutes, occasionally lifting upper and lower lids. Seek medical attention immediately. If this chemical contacts the skin, remove contaminated clothing and wash immediately with soap and water. Seek medical attention immediately. If this chemical has been inhaled, remove from exposure, begin rescue breathing (using universal precautions) if breathing has stopped, and CPR if heart action has stopped. Transfer promptly to a medical facility. When this chemical has been swallowed, get medical attention. Give large quantities of water and induce vomiting. Do not make an unconscious person vomit.

*References:*
- New Jersey Department of Health and Senior Services, Hazardous Substance Fact Sheet, "Diallate," Trenton NJ (April 1997). www.state.nj.us/health/eoh/rtkweb/0608.pdf
- Pesticide Management Education Program (PEMP), "Diallate (Avadex) Herbicide Profile 3/85," Cornell University, Ithaca, NY. (March 11, 1985, modified January 16, 2003). http://pmep.cce.cornell.edu/profiles/herb-growthreg/dalapon-ethephon/diallate/herb-prof-diallate.html
- California Environmental Protection Agency "Chemical List of Lists," Sacramento CA (February 1997).
- Sax, N.I., Ed., "Dangerous Properties of Industrial Materials Report," 3, No. 1, 50-53 (1983).

# Diatomaceous Earth

*Use Type:* Diatomaceous earth is used an insecticide, molluscicide, and as a filler in pesticides.

*CAS Number:* 61790-53-2; 7631-86-9 [Silica, amorphous-diatomaceous earth (uncalcined)]

*Formula:* $SiO_2$

*Synonyms:* Amorphous Silica; Diatomaceous silica; Diatomite, uncalcined; Kieselguhr (German); Silica, amorphous diatomaceous earth; Precipitated amorphous silica; Silicon dioxide (amorphous)

*Trade Names:* CELITE®, Celite (USA); CROP GUARD®, Eagle-Picher Minerals (USA); DIACTIV®, Celite (USA); DIAFIL®, Celite (USA); DI-ATOMATE®, White Mountain Natural Products (USA); DIE-SECTICIDE®, White Mountain Natural Products (USA); KENITE®, Celite (USA); PRIMISIL®, Celite (USA)

*Producers:* ATOFINA (France); Bombay Mineral Supply (India); CECA (France); Celite (USA); Eagle-Picher Minerals (USA); Harcros Chemicals (USA); Hellenic Corundum (Greece); Merck (Germany); Omya (Switzerland); White Mountain Natural Products (USA); World Minerals (USA)

*Chemical Class:* Inorganic

*EPA/OPP PC Code:* 072605

*California DPR Chemical Code:* 195

*ICSC Number:* 0428

*RTECS Number:* HL8600000

*Uses:* Diatomaceous earth kills and repels insects by dehydrating them, rather than chemically poisoning them, and has been used since early times. It is also used as a filtering agent and as a filler in construction materials, pesticides, paints, and varnishes. The calcined version (which has been heat treated) is the most dangerous and contains crystallized silica, and should be handled as silica.

*Carcinogen/Hazard Classifications*

**IARC:** Group 3, unclassifiable

**Label Signal Word:** CAUTION, Group 3

*Regulatory Authority:*
- Air Pollutant Standard Set (ACGIH)[1] (former USSR)[43] (OSHA)[58]
- Permissible Exposure Limits for Chemical Contaminants (CAL/OSHA)

*Description:* Diatomaceous earth is a transparent to gray, amorphous powder. Odorless. Insoluble in water. Molecular weight =60.09. Boiling point = 2230°C. Melting/Freezing point = 1610°C. Vapor Pressure = 10 mmHg @ 1732°C.

*Incompatibilities:* High temperatures cause the formation of crystalline silica. Incompatible with oxygen difluoride, chlorine difluoride.

*Permissible Exposure Limits in Air:* The OSHA[2] PEL TWA 20 mppcf [80 mg/m$^3$/%SiO$_2$]. NIOSH[2] recommends 6 mg/m$^3$ as the limit. The ACGIH[1] recommended airborne exposure limit for respirable (small particles) diatomaceous earth is 1.0 mg/m$^3$ of total dust (large and small particles containing no asbestos and less than 1% free silica) averaged over an 8-hour workshift. The NIOSH[2] IDLH is 3,000 mg/m$^3$ The former USSR-UNEP/IRPTC project has set an MAC of 2 mg/m$^3$ in workplace air[43].

*Determination in Air:* Filter; Low-temperature ashing; x-ray diffraction spectrometry; NIOSH IV, Method #7501[18]; or, Gravimetric plus OSHA Method ID/42[58].

*Permissible Concentration in Water:* No criteria set, but runoff from spills or fire control may cause water pollution.

*Routes of Entry:* Inhalation

*Harmful Effects and Symptoms*

*Short Term Exposure:* Unknown at this time

*Long Term Exposure:* Exposure can cause permanent scarring of the lungs, especially if diatomaceous earth has been calcined (heat treated). Symptoms include shortness of breath and cough. This can begin anywhere from months to years after exposure. The name of this disease is silicosis. With heavy exposure, individuals may become respiratory cripples. This can be fatal.

*Points of Attack:* Lungs.

*Medical Surveillance:* Before first exposure to calcined diatomaceous earth and at regular times after, the following are recommended: Medical exam of the lungs. Lung function tests. Chest x-ray (every two to five years).

*First Aid:* If this chemical gets into the eyes, remove any contact lenses at once and irrigate immediately for at least 15 minutes, occasionally lifting upper and lower lids. Seek medical attention immediately.If this chemical has been inhaled, remove from exposure, begin rescue breathing (using universal precautions) if breathing has stopped, and CPR if heart action has stopped. Transfer promptly to a medical facility.

*References:*
- New Jersey Department of Health, "Hazardous Substance Fact Sheet: Diatomaceous Earth," Trenton, NJ (August 1985, revised May 1999). www.state.nj.us/health/eoh/rtkweb/0616.pdf
- California Environmental Protection Agency "Chemical List of Lists," Sacramento CA (February 1997).

# Diazinon (ANSI)

*Use Type:* Non-systemic insecticide and acaricide

*CAS Number:* 333-41-5

*Formula:* $C_{12}H_{21}O_3N_2SP$

*Alert:* A Restricted Use Pesticide (RUP). The U.S. EPA initiated a program to phase out all non-agricultural uses of diazinon commencing in March, 2001. Many commercial outdoor uses of diazinon have been canceled or restricted to license pesticide applicators because of its know toxicity to birds and aquatic life. Diazinon is highly toxic to bees and very highly toxic to birds, fish and aquatic invertebrates. Diazinon was canceled for use on golf courses and sod farms in 1988 because of its high risk to birds.

*Synonyms:* O,O-Diaethyl-O-(2-isopropyl-4-methyl-pyrimidin-6-yl)-monothiophosphat (German); O,O-Diaethyl-O-(2-isopropyl-4-methyl-6-pyrimidyl)-thionophosphat (German); O,O-Diethyl-O-(2-isopropyl-4-methyl-pyrimidin-6-yl)-monothiofospaat (Dutch); Diethyl 4-(2-isopropyl-6-methylpyrimidinl)phosphorothionate; Diethyl 2-isopropyl-4-methyl-6-pyrimidinl phosphorothionate; O,O-Diethyl O-2-isopropyl-6-methylpyrimidin-4-ylphosphorothionate; O,O-Diethyl-O-(2-isopropyl-4-methyl-6-pyrimidyl)phosphorothionate; O,O-Diethyl 2-isopropyl-4-methylpyrimidyl-6-thiophosphate; O,O-Diethyl O-(2-isopropyl-4-methyl-6-pyrimidyl)thionophosphate; Diethyl 2-isopropyl-4-methyl-6-pyrimidylthionophosphate; O,O-Diethyl O-6-methyl-2-isopropyl-4-pyrimidinyl phosphorthioate; O,O-Diethyl O-(6-methyl-2-(1-methylethyl)-4-pyrimidinyl) phosphorthioate; ENT 19,507; Isopropylmethylpyrimidyl diethyl thiophosphate; O-2-Isopropyl-4-methylpyrimyl-O,O-diethyl phosphorothioate; NA2783 (DOT); NCI-C08673; Phosphoric acid, O,O-diethyl O-6-methyl-2-(1-methylethyl)-4-pyrimidinyl ester; Phosphorothioate, O,O-diethyl O-6-(2-isopropyl-4-methylpyrimidyl; Phosphorothioic acid, O,O-diethyl O-(2-isopropyl-6-methyl-4-pyrimidinl) ester; Phosphorothioic acid, O,O-diethyl O-(isopropylmethylpyrimidyl) ester; Phosphorothioic acid, O,O-diethyl O-(6-methyl-2-(1-methylethyl)-4-pyrimidinyl) ester; 4-Pyrimidinol, 2-isopropyl-6-methyl-, O-ester with O,O-diethylphosphorothioate; Thiophosphate de O,O-diethyle et de O-2-isopropyl-4-methyl 6-pyrimidyle (French); Thiophosphoric acid 2-isopropyl-4-methyl-6-pyrimidyl diethyl ester

*Trade Names:* (Pesticide Action Network [PAN] lists 2155 active and canceled/transferred diazinon products) AG-500®; AI3-19507®; ALFA-TOX®, Syngenta (Switzerland) canceled 5/4/1987; ANTIGAL®; ANTLAK®; BASUDIN®; BAZUDEN®; CASWELL No. 342®; DACUTOX®; DASSITOX®; DAZZEL®; DIAGRAN®; DIANON®; DIATERR-FOS®; DIAZAJET®; DIAZATOL®; DIAZIDE®; DIAZINON AG 500 WBC®, Syngenta (Switzerland); DIAZINONE®; DIAZITOL®; DIAZOL®, Makhteshim Agan (Israel) some non-agricultural products canceled; DICID®; DIMPYLATE®; DIPOFENE®; DIZIKTOL®; DIZINON®, Pbi/Gordon (USA) canceled 1/22/1991; DRAWIZON®; DYMET®; DYZOL®, Makhteshim Agan (Israel); D.Z.N.®, Syngenta (Switzerland); EXODIN®; FEZUDIN®; FLYTROL®; G 301®; G-24480®; GALESAN®; GARDENTOX®; GEIGY 24480®; KAYAZINON®; KAYAZOL®; NEOCIDOL® (OIL); NEOCIDOL®; NIPSAN®; NUCIDOL®; OLEODIAZINON®; ROOT GUARD; SAROLEX®, Syngenta (Switzerland) canceled 1/22/1991; SPECTRACIDE®, Syngenta (Switzerland); SROLEX®; SUZON®, Sudarshan Chemical Industries Ltd. (India)

*Producers:* Agro-care Chemical Industry Group (China); Agsin (Singapore); Alcotan Laboratories (Spain); Bayer CropScience (Germany); BASF (Germany); Ciba Specialty Chemicals (Switzerland); Cognis (Germany); Drexel Chemical (USA); Ehrenstorfer, Dr. (Germany); Gowan Company (USA); Hockley International (UK); Hokko

Chemical Industry (Japan); Ki-Hara Chemicals Ltd. (UK); Makhteshim Agan (Israel); Nippon Kayaku (Japan); Nissan Chemical Industries (Japan); Nufarm (Australia); Proficol (Colombia); Rhone-Poulenc Agro (France); Roussel Uclaf (France); Shenzhen Guomeng Industry Co., Ltd. (China); Sigma-Aldrich Laborchemikalien (Germany); Sudarshan Chemical Industries Ltd. (India); Syngenta (Switzerland); Zago Asia Ltd. (Singapore); Zhejiang Heben Pesticide & Chemicals Co., Ltd. (China)

*Chemical Class:* Organophosphate
*EPA/OPP PC Code:* 057801
*California DPR Chemical Code:* 198
*ICSC Number:* 0137
*RTECS Number:* TF3325000
*EEC Number:* 015-040-00-4
*EINECS Number:* 206-373-8

*Uses:* Diazinon is the most widely used pesticide by homeowners on lawns, and is one of the most widely used pesticide ingredients for application around the home and in gardens. It is used to control insects and grub worms. It is a nonsystemic organophosphate insecticide used to control cockroaches, silverfish, ants, and fleas in residential, non-food buildings. Bait is used to control scavenger yellow jackets in the western U.S. It is used on home gardens and farms to control a wide variety of sucking and leaf eating insects. It is used on rice, fruit trees, sugarcane, corn, tobacco, potatoes and on horticultural plants, and is also an ingredient in pest strips. Diazinon has veterinary uses against fleas and ticks. It is available in dust, granules, seed dressings, wettable powder, and emulsifiable solution formulations. In 1988, there were 500 different products containing diazinon on the market, and used in such products as agricultural sprays and granules, animal ear tags, household sprays and dust and veterinary products.

*Human toxicity (long-term)*[77]: Extra high–0.60 ppb, Health Advisory

*Fish toxicity (threshold)*[77]: Extra high–0.09200 ppb, MATC (Maximum Acceptable Toxicant Concentration)

*U.S. Maximum Allowable Residue Levels for Diazinon (40 CFR 180.153):*

| CROP | ppm |
|---|---|
| Alfalfa, hay | 10 |
| Alfalfa, forage | 40 |
| Almond, hulls | 3 |
| Almond | 0.5 |
| Animal feed | – |
| Apple | 0.5 |
| Apricot | 0.5 |
| Banana, pulp | 0.1 |
| Banana | 0.2 |
| Bean, lima, seed | 0.5 |
| Bean, snap, succulent | 0.5 |
| Beet, garden, tops | 0.7 |
| Beet, sugar, tops | 10 |
| Beet, garden, roots | 0.75 |
| Beet, sugar, roots | 0.5 |
| Blackberry | 0.5 |
| Blueberry | 0.5 |
| Carrot, roots | 0.75 |
| Cattle, mbyp | 0.7 |
| Cattle, fat | 0.7 |
| Cattle, meat | 0.7 |
| Celery | 0.7 |
| Cherry | 0.75 |
| Clover, forage | 40 |
| Clover, hay | 10 |
| Coffee, bean | 0.2 |
| Corn, forage | 40 |
| Corn | 0.7 |
| Cotton, undelinted seed | 0.2 |
| Cowpea, seed | 0.1 |
| Cowpea, forage | 0.1 |
| Cranberry | 0.5 |
| Cucumber | 0.75 |
| Dandelion, leaves | 0.7 |
| Endive | 0.7 |
| Fig | 0.5 |
| Filbert | 0.5 |
| Fruit, stone, group 12 | 0.7 |
| Ginseng, roots | 0.75 |
| Grape | 0.75 |
| Guar | 0.1 |
| Hop | 0.75 |
| Kiwifruit | 0.75 |
| Lespedeza | 1 |
| Lettuce | 0.7 |
| Loganberry | 0.75 |
| Melon | 0.75 |
| Mushroom | 0.75 |
| Nectarine | 0.5 |
| Olive | 1 |
| Onion | 0.75 |
| Parsley | 0.75 |
| Parsnip | 0.5 |
| Pea, shelled | 0.5 |
| Pea, field, vines | 25 |
| Pea, field, hay | 10 |
| Peach | 0.7 |
| Pear | 0.5 |
| Pepper | 0.5 |
| Pineapple | 0.5 |
| Plum, prune, fresh | 0.5 |
| Potato | 0.1 |
| Processed food | – |
| Radicchio | 0.7 |
| Radish, oriental, tops | 0.1 |
| Radish, oriental, roots | 0.1 |
| Radish | 0.5 |
| Raspberry | 0.5 |
| Rutabaga | 0.75 |

| | |
|---|---|
| Sheep, fat | 0.7 |
| Sheep, mbyp | 0.7 |
| Sheep, meat | 0.7 |
| Sorghum, forage | 10 |
| Sorghum, grain, grain | 0.75 |
| Spinach | 0.7 |
| Squash, summer | 0.5 |
| Squash, winter | 0.75 |
| Strawberry | 0.5 |
| Sweet potato | 0.1 |
| Swiss chard | 0.7 |
| Tomato | 0.75 |
| Turnip, greens | 0.75 |
| Turnip, roots | 0.5 |
| Vegetable, brassica, leafy, group 5 | 0.7 |
| Walnut | 0.5 |
| Watercress | 0.7 |

*Carcinogen/Hazard Classifications*
**Label Signal Word:** CAUTION, Moderately Toxic
**WHO Acute Hazard:** Class II, moderately hazardous
**TRI Developmental Toxin:** Developmental toxin
*Regulatory Authority:*
- Air Pollutant Standard Set (ACGIH)[1] (Argentina) (Australia) (DFG)[3] (HSE)[33] (Israel) (Mexico) (former USSR)[35] (Several States)[60] (Several Canadian Provinces)
- AB 1803-Well Monitoring Chemical (CAL)
- CAL-DHS/DHS Drinking Water Action Levels
- Permissible Exposure Limits for Chemical Contaminants (CAL/OSHA)
- The "Director's List" (CAL/OSHA)
- Actively registered pesticide in California.
- Clean Water Act: Section 311 Hazardous Substances/RQ 40CFR117.3 (same as CERCLA, see below).
- Superfund/EPCRA 40CFR302.4 RQ: CERCLA, 1 lb (0.454 kg)
- EPCRA Section 313 Form R *de minimus* concentration reporting level: 1.0%
- U.S. DOT Regulated Marine Pollutant (49CFR172.101, Appendix B), severe pollutant
- Canada, Drinking Water Quality, 0.02 mg/L MAC

**Description:** Colorless, oily liquid. Technical grade is pale to dark brown. Faint amine odor. Slightly soluble in water; solubility = 40 ppm. Molecular weight = 304.39. Boiling point = (decomposes)120°C; 83-84°C @ 0.002 mmHg. Flash point = 82°C. Vapor Pressure = $1.7 \times 10^{-4}$ mmHg @ 20. Hazard Identification (based on NFPA-704 M Rating System): Health 3, Flammability 1, Reactivity 0. Log $K_{ow}$ = 3.09. Values at or above 3.0 are likely to bioaccumulate in marine organisms.

*Incompatibilities:* Hydrolyzes slowly in water and dilute acid. Reacts with strong acids and alkalis with possible formation of highly toxic tetraethyl thiopyrophosphates. Incompatible with copper-containing compounds.

*Permissible Exposure Limits in Air:* There is no OSHA[2] PEL. NIOSH[2] and ACGIH[1] recommend an airborne exposure limit of 0.1 mg/m³ TWA with the notation "skin" indicating the possibility of cutaneous absorption. Argentina, Australia, Israel, Mexico, HSE[33], and the Canadian provinces of Alberta, BC, Ontario, and Quebec use this same value but HSE, Mexico, Alberta, and BC add the STEL of 0.3 mg/m³ as does Argentina[35]. The DFG, however, has set an MAK of 1.0 mg/m³ and an STEL of 10 mg/m³ [3]. The former USSR[35] has set an MAC of 0.2 mg/m³ in workplace air and an MAC for ambient air in residential areas of 0.01 mg/m³ on either a momentary or a daily average basis. In addition, several states have set guidelines or standards for Diazinon in ambient air[60] ranging from 1.0 μg/m³ (North Dakota) to 1.6 μg/m³ (Virginia) to 2.0 μg/m³ (Connecticut and Nevada).

*Determination in Air:* OSHA versatile sampler-2; Toluene/Acetone; Gas chromatography/Flame photometric detection for sulfur, nitrogen, or phosphorus; NIOSH IV Method #5600[18], Organophosphorus pesticides. See also OSHA Method #62[58].

*Permissible Concentration in Water:* The U.S. EPA has developed one- and 10-day health advisories (maximum recommended drinking water concentrations) for adults and children of 20 micrograms per liter of water (20 *m*g/L). The USEPA has determined a NOAEL of 0.05 mg/kg/day which gives a long-term health advisory of 0.0175 mg/L and a lifetime health advisory of 0.00063 mg/L. Canada set a MAC of 0.02 mg/L. The former USSR[35] has set an MAC in water bodies for domestic purposes of 0.3 mg/L. Several states have set guidelines for Diazinon in drinking water[61] ranging from 4 μg/L (Maine) to 14 μg/L (California and Kansas).

*Determination in Water:* By Methylene chloride extraction followed by gas chromatography.

*Routes of Entry:* Inhalation, skin absorption, ingestion, skin and/or eye contact

*Harmful Effects and Symptoms*
*Short Term Exposure:* Diazinon can affect you when breathed in and quickly enters the body by passing through the skin. May cause skin and eye irritation. Exposure can cause organophosphate poisoning with headache, sweating, nausea, and vomiting, diarrhea, muscle twitching and possible death. It is a moderately toxic organophosphate chemical. The $LD_{50}$ (oral, rat) = 66 mg/kg (moderately toxic). Delayed pulmonary edema may occur after inhalation.

*Long Term Exposure:* Damage to the pancreas has developed in some people and in laboratory animals exposed to large amounts of diazinon. In animal studies, high doses of diazinon produced effects on the nervous system similar to those seen in people. There is no evidence that long-term exposure to *low* levels of diazinon causes any harmful health effects in people.

*Points of Attack:* Eyes, respiratory system, central nervous system, cardiovascular system, blood cholinesterase.

*Medical Surveillance:* Medical observation is recommended for 24 to 48 hours after breathing overexposure, as pulmonary edema may be delayed. Before employment and at regular times after that, the following are recommended: Plasma and red blood cell acetylcholinesterase levels (tests for the enzyme poisoned by this chemical). If exposure stops, plasma levels return to normal in 1-2 weeks while red blood cell levels may be reduced for 1-3 months. When acetylcholinesterase enzyme levels are reduced by 25% or more below preemployment levels, risk of poisoning is increased, even if results are in lower ranges of "normal." Reassignment to work not involving organophosphate or carbamate pesticides is recommended until enzyme levels recover. If symptoms develop or overexposure occurs, repeat the above tests as soon as possible and get an exam of the nervous system. Also consider complete blood count. Consider chest x-ray following acute overexposure. Do not drink any alcoholic beverages before or during use. Alcohol promotes absorption of organophosphates. Liver function tests. Exam of the nervous system. Complete blood count.

*First Aid:* **Treatment for organophosphate poisoning consists of thorough decontamination, cardiorespiratory support, and administration of the antidotes atropine and pralidoxime. In cases of severe poisoning, diazepam, an anticonvulsant, should also be administered. Antidotes should be administered as prevention even if the diagnosis is in doubt.** Speed in removing material from eyes and skin is of extreme importance. *Eyes:* Eye contact can cause dangerous amounts of these chemicals to be quickly absorbed through the mucous membrane into the bloodstream. Immediately and gently flush eyes with plenty of warm or cold water (NO hot water) for at least 15 minutes, occasionally lifting the upper and lower eyelids. Get medical aid immediately. *Skin:* Get medical aid. Skin contact can cause dangerous amounts of these chemicals to be absorbed into the bloodstream. Wearing the appropriate PPE equipment and respirator for organophosphate/carbamate pesticides, immediately flush skin with plenty of soap and water for at least 15 minutes while removing contaminated clothing and shoes. Shampoo hair promptly if contaminated. The removed, contaminated clothing and shoes should be double-bagged and left in Hot Zone for later disposal by hazardous materials experts. Skin may also be decontaminated with diluted hypochlorite solution. *Inhalation:* Get medical aid. Do not contaminate yourself. Wearing the appropriate PPE equipment and respirator for organophosphate pesticides, immediately remove the victim from the contaminated area to fresh air. If the victim is not breathing, administer artificial respiration. Do NOT use mouth-to-mouth resuscitation; use bag/mask apparatus. If breathing is difficult, administer oxygen through bag/mask apparatus until medical help arrives. Do not leave victim unattended. *Ingestion:* Call poison control. Loosen all clothing. Never give anything by mouth to an unconscious person. If victim is *unconscious or having convulsions,* do nothing except keep victim warm. Get medical aid. Transfer promptly to a medical facility. In cases of ingestion, **do not induce vomiting**. If the victim is alert and asymptomatic, administer a slurry of activated charcoal at a dose of 1 g/kg (infant, child, and adult dose). A soda can and straw may be of assistance when offering charcoal to a child. *In some cases you may be specifically instructed by poison control to induce vomiting by way of 2 tablespoons of syrup of ipecac (adult) washed down with a cup of water.* Do NOT give activated charcoal before or with ipecac syrup.

*Note to physician or authorized medical personnel:* Treat cases of respiratory compromise, coma, or excessive pulmonary secretions with respiratory support using protocols and techniques available and within the scope of training. Some cases may necessitate procedures such as endotracheal intubation or cricothyrotomy by properly trained and equipped personnel. When possible, atropine (see *Antidotes,* below) should be given under medical supervision. Patients who are comatose, hypotensive, or having seizures or cardiac arrhythmias should be treated according to advanced life support protocols. *Antidotes:* Two antidotes are administered to treat organophosphate poisoning. Atropine is a competitive antagonist of acetylcholine at muscarinic receptors and is used to control the excessive bronchial secretions which are often responsible for death. Pralidoxime relieves both the nicotinic and muscarine effects of organophosphate poisoning by regenerating acetylcholinesterase and can reduce both the bronchial secretions and the muscle weakness associated with poisoning. The initial intravenous dose of atropine in adults should be determined by the severity of symptoms: An initial adult dose of 1.0 to 2.0 mg or pediatric dose of 0.01 mg/kg (minimum 0.01 mg) should be administered intravenously. If intravenous access cannot be established, atropine may also be given intramuscularly, subcutaneously or via endotracheal tube. Doses should be repeated every 15 minutes until excessive secretions and sweating have been controlled. Once bronchial secretion has been controlled, atropine administration should be repeated whenever the secretions begin to recur. In seriously poisoned patients, very large doses may be required. Alterations of pulse rate and pupillary size should not be used as indicators of treatment adequacy. Pralidoxime should be administered as early in poisoning as possible as its efficacy may diminish when given more than 24 to 36 hours after exposure. Doses are as follows: adult 1.0 g; pediatric 25 to 50 mg/kg. The drug should be administered intravenously over 30 to 60 minutes, but in a life-threatening situation, one-half of the total dose can be given per minute for a total administration time of 2 minutes. Treatment should begin to take effect within 40 minutes with a reduction in symptoms and the amount of atropine necessary to control bronchial secretion. The initial dose can be repeated in 1 hour and then every 8 to 12 hours until the patient is clinically well and no longer requires atropine. If intravenous access cannot be established, pralidoxime may also be given intramuscularly. Early administration of diazepam in addition to the combined atropine and pralidoxime treatment may help prevent the onset of seizures and potential brain and cardiac morphologic damage following high-level organophosphate poisoning.

*References:*
- EXTOXNET, Extension Toxicology Network, "Pesticide Information Profile, Diazinon," Oregon State University, Corvallis, OR (June, 1996). ace.orst.edu/info/extoxnet/pips/diazinon.htm
- EPA, Office of Pesticide Programs, Pesticide Residue Limits, "Diazinon," 40 CFR 180.153, www.epa.gov/cgi-bin/oppsrch
- National Pesticide Information Center (NPIC), Fact Sheet "Diazinon," EPA and Oregon State University, Corvallis, OR (October 1988). ace.orst.edu/info/npic/factsheets/diazinon.pdf
- New Jersey Department of Health, "Hazardous Substance Fact Sheet: Diazinon," Trenton, NJ (March 1998). www.state.nj.us/health/eoh/rtkweb/0618.pdf
- U.S. Environmental Protection Agency, "Health Advisory: Diazinon," Washington, DC, Office of Drinking Water (August 1987).
- Sax, N.I., Ed., "Dangerous Properties of Industrial Materials Report," 7, No. 5, 36-43 (1987).
- Agency for Toxic Substances and Disease Registry, U.S. Department of Health and Human Services, Public Health Service, "Managing Hazardous Materials Incidents," Atlanta, GA (June 2003).
- California Environmental Protection Agency "Chemical List of Lists," Sacramento CA (February 1997).

# Dibenzofuran

*Use Type:* Insecticide
*CAS Number:* 132-64-9
*Formula:* $C_{12}H_8O$
*Synonyms:* [1,1'-Biphenyl]-2,2'-diyl oxide; 2,2'-Biphenylene oxide; 2,2'-Biphenylyleme oxide; Dibenzo[b,d]furan; Dibenzofurano (Spanish); Diphenylene oxide
*Producers:* Fluorochem (UK); Nippon Steel (Japan); Sigma-Aldrich Laborchemikalien (Germany)
*RTECS Number:* HP4450000
*EINECS Number:* 205-071-3
*Uses:* This material is used as an insecticide and in organic synthesis to make other chemicals. It is derived from coal tar creosote.

*Regulatory Authority:*
- AB 2588-Air Toxics "Hot Spots" Chemicals (CAL)
- CAL Air Resources Board/AB 1807 Toxic Air Contaminants
- RCRA 40CFR264, Appendix 9; TSD Facilities Ground Water Monitoring List. Suggested test method(s) (PQL ug/L): 8270(10)
- Superfund/EPCRA 40CFR302.4 RQ: CERCLA, 1 lb (0.454 kg)
- EPCRA Section 313 Form R *de minimus* concentration reporting level: 1.0%
- Canada, CEPA Toxic Substances

*Description:* Dibenzofuran, $C_{12}H_8O$ is a white crystalline powder. Very slightly soluble in water. Molecular weight = 168.21. Boiling point = 285–288°C. Melting/Freezing point = 85–87°C. Hazard Identification (based on NFPA-704 M Rating System): Health 1, Flammability 1, Reactivity 0. Log $K_{ow}$ = 4.1. Values at or above 3.0 are likely to bioaccumulate in marine organisms.

*Incompatibilities:* Strong oxidizers.
*Permissible Exposure Limits in Air:* No standards set.
*Permissible Concentration in Water:* No criteria set, but runoff from spills or fire control may cause water pollution.
*Routes of Entry:* Inhalation, passing through the skin.

*Harmful Effects and Symptoms*

*Short Term Exposure:* Dibenzofuran can be absorbed through the skin, thereby increasing exposure. Exposure irritates the eyes, skin and respiratory tract. Poisonous if ingested.

*Long Term Exposure:* Repeated contact may cause skin growths, rashes, and changes in skin color. Exposure to sunlight may make rash worse.

*Points of Attack:* Skin

*Medical Surveillance:* Evaluation by a qualified allergist

*First Aid: Skin Contact:* Flood all areas of body that have contacted the substance with water. Don't wait to remove contaminated clothing; do it under the water stream. Use soap to help assure removal. Isolate contaminated clothing when removed to prevent contact by others. *Eye Contact:* Remove any contact lenses at once. Flush eyes well with copious quantities of water or normal saline for at least 20-30 minutes. Seek medical attention. *Inhalation*: Leave contaminated area immediately; breathe fresh air. Proper respiratory protection must be supplied to any rescuers. If coughing, difficult breathing or any other symptoms develop, seek medical attention at once, even if symptoms develop many hours after exposure. *Ingestion:* If convulsions are not present, give a glass or two of water or milk to dilute the substance. Assure that the person's airway is unobstructed and contact a hospital or poison center immediately for advice on whether or not to induce vomiting.

*References:*
- New Jersey Department of Health and Senior Services, Hazardous Substance Fact Sheet, "Dibenzofuran,"

Trenton NJ (March 1992, rev. May, 1998). http://www.state.nj.us/health/eoh/rtkweb/2230.pdf
- California Environmental Protection Agency "Chemical List of Lists," Sacramento CA (February 1997).

# Dibromochloropropane (DBCP)

*Use Type:* nematicide and fumigant
*CAS Number:* 96-12-8
*Formula:* $C_3H_5Br_2Cl$; $CH_2BrCHBrCH_2Cl$
*Alert:* Withdrawn by the U.S. EPA on November 3, 1977. No longer available in the United States. Human toxicity (long-term): Very high.
*Synonyms:* 1,2-Dibromo-3-cloropropano (Spanish); 1-Chloro-2,3-dibromopropane; 3-Chloro-1,2-dibromopropane; DBCP; Dibromchlorpropan (German); 1,2-Dibrom-3-chlorpropan (German); 1,2-Dibromo-3-cloro-propano (Italian); 1,2-Dibromo-3-cloropropano (Spanish); 1,2-Dibroom-3-chloorpropaan (Dutch); NCI-C00500; Propane, 1,2-dibromo-3-chloro-
*Trade Names:* BBC 12®; FUMAGONE®; FUMAZONE®, Dow Chemical Co. (USA), canceled 12/9/1979; MEMATOCIDE®; NEMABROM®; NEMAFUM®; NEMAGON®, Chem-nut Inc. (USA), canceled; Helena Chemical (USA), canceled, plus several other companies; NEMAGON SOIL FUMIGANT®, Chem-nut Inc. (USA), canceled; NEMANAX®; NEMAPAZ®; NEMASET®; NEMATOCIDE®, Amvac Chemical Co. (USA), canceled 7/1/1987; NEMATOX®; NEMAZON®; OS 1897®; OXY BCP®, Occidental Chemical Corp. (USA), canceled 12/9/1979; SD 1897®
*Producers:* Amvac Chemical Co. (USA); Dow Chemical Co. (USA); Helena Chemical ((USA); Occidental Chemical Corp. (USA); Simplot, J.R., Company (USA)
*Chemical Class:* Halogenated organic compound
*EPA/OPP PC Code:* 011301
*California DPR Chemical Code:* 183
*ICSC Number:* 0002
*RTECS Number:* TX8750000
*EEC Number:* 602-021-00-6
*EINECS Number:* 202-479-3
*Uses:* DBCP has been used in agriculture as a nematicide since 1955, being supplied for such use in the forms of liquid concentrate, emulsifiable concentrate, powder, granules, and solid material. A rebuttable presumption against registration for pesticide uses was issued by U.S. EPA on September 22, 1977, on the basis of oncogenicity and reproductive effects. Then, as of November 3, 1977, EPA in a further action suspended all registrations of end use products, subject to various specific restrictions.
*Human toxicity (long-term)*[77]*:* Extra high–0.20 ppb, MCL (Maximum Contaminant Level)
*Carcinogen/Hazard Classifications*
**U.S. EPA Carcinogens:** Group B2, probable carcinogen
**U.S. NTP Carcinogen:** Reasonably anticipated carcinogen
**California Prop. 65:** Carcinogen and male reproductive toxin
**IARC:** Group 2B, possible carcinogen
**Endocrine Disruptor:** Known ED
*Regulatory Authority:*
- Carcinogen (Animal Positive) (IARC) (NCI) (NTP)[9] (DFG)[3]
- Banned or Severely Restricted (Several Countries) (UN)[35]
- OSHA, 29CFR1910 Specifically Regulated Chemicals (See CFR 1910.1044)
- Air Pollutant Standard Set (US, Argentina)[35] (former USSR)[43] (Several States)[60]
- CAL/OSHA Carcinogen User Register Chemical
- AB 1803-Well Monitoring Chemical (CAL)
- MCL (Maximum Contaminants Levels) list of contaminants (CAL)
- AB 2588-Air Toxics "Hot Spots" Chemicals (CAL)
- CAL-DHS/DHS Drinking Water Action Levels
- CAL Air Resources Board/AB 1807 Toxic Air Contaminants
- Proposition 65 chemical (CAL)
- Permissible Exposure Limits for Chemical Contaminants (CAL/OSHA)
- The "Director's List" (CAL/OSHA)
- Clean Air Act: Hazardous Air Pollutants (Title I, Part A, Section 112)
- EPA Hazardous Waste Number (RCRA No.): U066
- RCRA, 40CFR261, Appendix 8 Hazardous Constituents
- RCRA 40CFR268.48; 61FR15654, Universal Treatment Standards: Wastewater (mg/L), 0.11; Nonwastewater (mg/kg), 15
- RCRA 40CFR264, Appendix 9; TSD Facilities Ground Water Monitoring List. Suggested test method(s) (PQL $ug/L$): 8010(100); 8240(5); 8270(10)
- Safe Drinking Water Act: MCL, 0.0002 mg/L; MCLG, zero; Regulated chemical (47 FR 9352)
- Superfund/EPCRA 40CFR302.4 RQ: CERCLA, 1 lb (0.454 kg)
- EPCRA Section 313 Form R *de minimus* concentration reporting level: 0.1%
- Canada, WHMIS, Ingredients Disclosure List

*Description:* Amber to brown liquid (a solid below 6°C). Strong, pungent odor. It has an odor and taste threshold at 0.01
mg/L in water. Slightly soluble in water; solubility = 1000ppm. Molecular weight = 236.35. Boiling point = 196°C.
(decomposes). Melting/Freezing point = 6.7°C. Vapor pressure = 0.9 mmHg @ 20°C. Flash point = 77°C. Hazard Identification (based on NFPA-704 M Rating System): Health 2, Flammability 2, Reactivity 0.
*Incompatibilities:* Reacts with oxidizers and chemically active metals (i.e., aluminum, magnesium and tin alloys).

Attacks some rubber materials and coatings. Corrosive to metals.

*Permissible Exposure Limits in Air:* The OSHA[2] PEL is 1 ppb (0.01 mg/m$^3$) TWA and that no exposure to eyes or skin should occur[35]. Argentina has set a TWA of 0.25 mg/m$^3$ with an STEL of 0.75 mg/m$^3$. Sweden and Germany have set no limits; simply stated that DBCP is a carcinogenic substance and should be avoided. The former USSR-UNEP/IRPTC project[43] has set an MAC in ambient air in residential areas of 0.0004 mg/m$^3$ on a momentary basis and 0.00003 mg/m$^3$ on a daily average basis. In addition, several states have set guidelines or standards for DBCP in ambient air[60] ranging from zero (North Dakota) to 0.05 $\mu$g/m$^3$ (Connecticut) to 1.0 $\mu$g/m$^3$ (Pennsylvania).

*Permissible Concentration in Water:* The former USSR-UNEP/IRPTC project[43] has set an MAC of 0.01 mg/L in water bodies used for domestic purposes. The USEPA[47] has set a one-day health advisory of 0.2 mg/L and a ten-day health advisory of 0.02 mg/L both for a 10 kg child. Longer term health advisories could not be calculated because of the carcinogenicity of DBCP. EPA has recently proposed[62] a maximum drinking water level of 0.0002 mg/L (0.2 $\mu$g/L). Several states have set guidelines for DBCP in dirking water[61] ranging from 0.025 $\mu$g/L (Arizona) to 0.25 $\mu$g/L (Minnesota) to 0.5 $\mu$g/L (Wisconsin) to 1.0 $\mu$g/L (California).

*Determination in Water:* By purge-and-trap gas chromatography[47].

*Routes of Entry:* Inhalation, skin absorption, ingestion, skin and/or eye contact

*Harmful Effects and Symptoms*

*Short Term Exposure:* Symptoms include severe local irritation to eyes, skin and mucous membranes. Nausea and vomiting may occur after ingestion. Exposure to DBCP can cause headache, nausea, vomiting, weakness, lightheadedness, unconsciousness, and possible death. Higher exposures can cause pulmonary edema, a medical emergency that can be delayed for several hours. This can cause death. Narcotic at high levels of concentration.

*Long Term Exposure:* The possible effects on the health of employees chronically exposed to repeated or lower expsures of DBCP may include sterility, diminished renal function, and degeneration and cirrhosis of the liver. DBCP is a probable carcinogen in humans. It has been shown to cause stomach, breast, tongue, and nasal cavity cancer in animals. May damage the testes and decrease fertility in males and females. Repeated exposure can damage the eyes causing clouding of lens or cornea, opens sores on the skin, liver and kidney damage.

*Points of Attack:* Eyes, skin, respiratory system, central nervous system, liver, kidneys, spleen, reproductive system, digestive system

*Medical Surveillance:* Medical surveillance shall be made available to employees as outlined below:

Comprehensive preplacement or initial medical and work histories with emphasis on reproductive experience and menstrual history. Comprehensive physical examination with emphasis on the genito-urinary tract including testicle size and consistency in males. Semen analysis to include sperm count, motility and morphology. Other tests, such as serum testosterone, serum follicle stimulating hormone (FSH), and serum lutenizing hormone (LH) may be carried out if, in the opinion of the responsible physician, they are indicated. In addition, screening tests of the renal and hepatic systems may be considered.

A judgment of the worker's ability to use positive pressure respirators. Employees shall be counseled by the physician to ensure that each employee is aware that DBCP has been implicated in the production of effects on the reproductive system including sterility in male workers. In addition, they should be made aware that cancer was produced in some animals. While the relevancy of these findings is not yet clearly defined, they do indicate that both employees and employers should do everything possible to minimize exposure to DBCP. Periodic examinations containing the elements of the preplacement or initial examination shall be made available on at least an annual basis. Examinations of current employees shall be made available as soon as practicable after the promulgation of a standard for DBCP. Medical surveillance shall be made available to any worker suspected of having been exposed to DBCP. Pertinent medical records shall be maintained for all employees subject to exposure to DBCP in the workplace. Such records shall be maintained for 30 years and shall be available to medical representatives of the U.S. Government, the employer and the employee.

*First Aid:* If this chemical gets into the eyes, remove any contact lenses at once and irrigate immediately for at least 20-30 minutes, occasionally lifting upper and lower lids. Seek medical attention immediately. If this chemical contacts the skin, remove contaminated clothing and wash immediately with soap and water. Seek medical attention immediately. If this chemical has been inhaled, remove from exposure, begin rescue breathing (using universal precautions) if breathing has stopped, and CPR if heart action has stopped. Transfer promptly to a medical facility. When this chemical has been swallowed, get medical attention. Contact local poison control center for advice about inducing vomiting. Medical observation is recommended for 24 to 48 hours after breathing overexposure, as pulmonary edema may be delayed. As first aid for pulmonary edema, a doctor or authorized paramedic may consider administering a corticosteroid spray.

*References:*
- National Institute for Occupational Safety and Health, Criteria for a Recommended Standard: Occupational Exposure to Dibromochloropropane, NIOSH Doc. No. 78-115 (1978).

- Sax, N.I., Ed., "Dangerous Properties of Industrial Materials Report," 1, No. 3, 55-57 (1981).
- California Environmental Protection Agency "Chemical List of Lists," Sacramento CA (February 1997).
- New Jersey Department of Health and Senior Services, Hazardous Substance Fact Sheet, "DBCP." Trenton NJ (June 1998). http://www.state.nj.us/health/eoh/rtkweb/0595.pdf

# 1,2-Dibromo-2,4-dicyanobutane

*Use Type:* Microbiocide
*CAS Number:* 35691-65-7
*Formula:* $C_6H_6Br_2N_2$
*Synonyms:* 2-Bromo-2-(bromomethyl)glutaronitrile; 1-Bromo-1-(bromomethyl)-1,3-propanedicarbonitrile; Dibromodicyanobutane; Glutaronitrile, 2-bromo-2-(bromomethyl)-; Pentanedinitrile, 2-bromo-2-(bromomethyl)-
*Trade Names:* BIOCHEK®, Bayer Group (Germany); BIOCLEAR®; MERCK® 48051, Merck (Germany); METACIDE® 38; METASOL, Bayer Group (Germany); TEKTAMER, Bayer Group (Germany)
*Producers:* Bayer Group (Germany); Merck (Germany)
*Chemical Class:* Nitrile
*EPA/OPP PC Code:* 111001; (217700 and 218500 are old EPA code numbers)
*California DPR Chemical Code:* 2313
*Uses:* Tolerances have been established by the U.S. Food and Drug Administration when this substances is used as a preservative in food grade adhesives and as a slimicide in the manufacture of food grade paper and paperboard. Used to control slime-forming bacteria and fungi in recirculating water cooling system; oil recovery drilling mud systems; paper mill and pulp mill water systems and similar industrial processing and chemical systems.
*Carcinogen/Hazard Classifications*
**Label Signal Word:** DANGER
*Regulatory Authority:*
- Actively registered pesticide in California.

*Description:* Crystalline solid. Insoluble in water. Mild, acrid, bitter odor. Molecular weight = 266.02.
*Incompatibilities:* Strong oxidizers and reducing agents, strong acids and bases. Reacts with acids, steam, warm water producing toxic and flammable hydrogen cyanide fumes. Hydrogen cyanide is produced when propionitrile is heated to decomposition.
*Permissible Concentration in Water:* No criteria set. Runoff from spills or fire control may cause water pollution.
*Routes of Entry:* Inhalation, skin absorption, ingestion, skin and/or eye contact.
*Harmful Effects and Symptoms*
*Short Term Exposure:* Contact may cause burns to skin and eyes. May affect the iron metabolism, causing asphyxia. It is highly toxic. Forms cyanide in the body. Exposure results in headache, dizziness, rapid pulse, deep-rapid breathing, nausea, vomiting, unconsciousness, convulsions and sometimes death. May cause cyanosis (blue coloration of skin and lips caused by lack of oxygen).
*Long Term Exposure:* Chronic exposure over long periods may cause fatigue and weakness. Can cause same general symptoms as hydrogen cyanide but onset of symptoms is likely to be slower. May cause liver and kidney damage.
*Points of Attack:* In animals: liver, kidney damage
*Medical Surveillance:* Liver and kidney function tests.
*First Aid:* If this chemical gets into the eyes, remove any contact lenses at once and irrigate immediately for at least 15 minutes, occasionally lifting upper and lower lids. Seek medical attention immediately. If this chemical contacts the skin, remove contaminated clothing and wash immediately with soap and water. Seek medical attention immediately. If this chemical has been inhaled, remove from exposure, begin rescue breathing (using universal precautions) if breathing has stopped, and CPR if heart action has stopped. Transfer promptly to a medical facility. When this chemical has been swallowed, get medical attention. Give large quantities of water and induce vomiting. Do not make an unconscious person vomit. Use amyl nitrate capsules if symptoms develop. All area employees should be trained regularly in emergency measures for cyanide poisoning and in CPR. A cyanide antidote kit should be kept in the immediate work area and must be rapidly available. Kit ingredients should be replaced every 1-2 years to ensure freshness. Persons trained in the use of this kit, oxygen use, and CPR must be quickly available.
*References:*
- U.S. Environmental Protection Agency, "Reregistration Eligibility Decision (RED), Dibromodicyanobutane," Office of Prevention, Pesticides and Toxic Substances, Washington, DC (June 1996). http://www.epa.gov/REDs/factsheets/2780fact.pdf
- California Environmental Protection Agency "Chemical List of Lists," Sacramento CA (February 1997).

# Dicamba (ANSI)

*Use Type:* Post-emergence herbicide
*CAS Number:* 1918-00-9
*Formula:* $C_8H_6Cl_2O_3$
*Alert:* General Use Pesticide (GUP)
*Synonyms:* Acido (3,6-dichloro-2-metossi)-benzoico (Italian); AI3-27556; o-Anisic acid, 3,6-dichloro-; Benzoic acid, 3,6-dichloro-2-methoxy-; Dianat (Russian); Dicamba benzoic acid herbicide; 3,6-Dichloor-2-methoxy-benzoeizuur (Dutch); 3,6-Dichlor-3-methoxy-benzoesaeure (German); 3,6-Dichloro-o-anisic acid; 2,5-Dichloro-6-methoxybenzoic acid; 3,6-Dichloro-2-methoxybenzoic acid;

3,6-Dichloro-2-methoxybenzoic acid; MDBA; 2-Methoxy-3,6-dichlorobenzoic acid

*Trade Names:* BANEX®; BANLEN®; BANVEL®, BASF Agricultural Products Group (Germany); Syngenta (Switzerland); BANVEL 4S®, Syngenta (Switzerland); BANVEL 4WS®, Syngenta (Switzerland); BANVEL CST®, Syngenta (Switzerland); BANVEL HERBICIDE®, Syngenta (Switzerland); BANVEL II HERBICIDE®, Syngenta (Switzerland); BRUSH BUSTER®; BUSHWHACKER®, Albaugh Inc. (USA); CADENCE®, Syngenta (Switzerland); CASWELL No. 295®; CLARITY®, BASF Agricultural Products Group (Germany); COMPOUND B DICAMBA®; DIANATE®; DISTINCT®, BASF Agricultural Products (Germany); DYVEL®, BASF Agricultural Products (Germany); FALLOWMASTER®, Monsanto (USA); FLOWMASTER®, Monsanto (USA); GORDON'S TRIGUARD®, Pbi/Gordon Corporation (USA); GORDON'S TRI-MEC®, Pbi/Gordon Corporation (USA); MARKSMAN®, BASF (Canada); MEDIBEN®; NORTHSTAR®, Syngenta (Switzerland); SUMMIT®, Syngenta (Switzerland); TARGET®, Syngenta (Switzerland); TRACKER®; TROOPER®; VANQUISH®, Syngenta (Switzerland); VELSICOL 58-CS-11®, Velsicol Chemical Corp. (USA); VELSICOL COMPOUND R®, Velsicol Chemical Corp. (USA); WEEDMASTER®, BASF Agricultural Products (Germany); YUKON®, Monsanto (USA)

*Producers:* Agro-care Chemical Industry Group (China); BASF Agricultural Products (Germany); Gharda Chemicals (India); Monsanto (USA); Pbi/Gordon Corporation (USA); Rhone-Poulenc (France); Sanex (Canada); Sigma-Aldrich Laborchemikalien (Germany); Syngenta (Switzerland); Velsicol Chemical Corp. (USA); Vigoro (Canada); Zago Asia Ltd. (Singapore)

*Chemical Class:* Benzoic acid
*EPA/OPP PC Code:* 029801
*California DPR Chemical Code:* 200
*ICSC Number:* 0139
*RTECS Number:* DG7525000
*EEC Number:* 607-043-00-X
*EINECS Number:* 217-635-6

*Uses:* Used to control annual and perennial broad leaf weeds in corn, sorghum, small grains, pastures, hay, rangeland, sugarcane, asparagus, turf, grass-seed crops, and non-croplands. It can be applied to the leaves or to the soil. Dicamba controls annual and perennial broadleaf weeds in grain crops and grasslands, and it is used to control brush and bracken in pastures. It will kill broadleaf weeds before and after they sprout. Legumes will be killed by dicamba. In combination with a phenoxyalkanoic acid or other herbicide, dicamba is used in pastures, range land, and non-crop areas such as fence-rows and roadways to control weeds.

*Human toxicity (long-term)*[77]: Very low–200.00 ppb, Health Advisory

*Fish toxicity (threshold)*[77]: Very low–4918.96611 ppb, MATC (Maximum Acceptable Toxicant Concentration)

*U.S. Maximum Allowable Residue Levels for Dicamba with its metabolites (40 CFR 180.227):*

| CROP | ppm |
| --- | --- |
| Asparagus | 4 |
| Barley, grain | 6 |
| Barley, straw | 15 |
| Barley, hay | 2 |
| Cattle, mbyp | 0.2 |
| Cattle, fat | 0.2 |
| Cattle, kidney | 1.5 |
| Cattle, liver | 1.5 |
| Cattle, meat | 0.2 |
| Corn, field, forage | 3 |
| Corn, field, stover | 3 |
| Corn, forage | 0.5 |
| Corn, grain | 0.5 |
| Corn, pop, stover | 3 |
| Corn, stover | 0.5 |
| Cotton, meal | 5 |
| Cotton, undelinted seed | 5 |
| Goat, mbyp | 0.2 |
| Goat, fat | 0.2 |
| Goat, kidney | 1.5 |
| Goat, liver | 1.5 |
| Goat, meat | 0.2 |
| Grain, aspirated fractions | 5100 |
| Grass, forage | 125 |
| Grass, hay | 200 |
| Hog, fat | 0.2 |
| Hog, kidney | 1.5 |
| Hog, liver | 1.5 |
| Hog, mbyp | 0.2 |
| Hog, meat | 0.2 |
| Horse, fat | 0.2 |
| Horse, kidney | 1.5 |
| Horse, liver | 1.5 |
| Horse, mbyp | 0.2 |
| Horse, meat | 0.2 |
| Milk | 0.3 |
| Millet, proso, grain | 0.5 |
| Millet, proso, straw | 0.5 |
| Oat, forage | 80 |
| Oat, grain | 0.5 |
| Oat, hay | 20 |
| Oat, straw | 0.5 |
| Sheep, fat | 0.2 |
| Sheep, kidney | 1.5 |
| Sheep, liver | 1.5 |
| Sheep, mbyp | 0.2 |
| Sheep, meat | 0.2 |
| Sorghum, forage | 3 |
| Sorghum, grain, grain | 3 |
| Sorghum, grain, stover | 3 |

| | |
|---|---|
| Soybean, forage | 0.01 |
| Soybean, hay | 0.01 |
| Soybean, hulls | 13 |
| Soybean, seed | 10 |
| Sugarcane, cane | 0.1 |
| Sugarcane, fodder | 0.1 |
| Sugarcane, forage | 0.1 |
| Sugarcane, molasses | 2 |
| Wheat, grain | 2 |
| Wheat, forage | 80 |
| Wheat, hay | 20 |
| Wheat, straw | 30 |

*Carcinogen/Hazard Classifications*

**U.S. EPA Carcinogens:** Group D, unclassifiable, inadequate data

**Label Signal Word:** Signal words for products containing dicamba range from CAUTION to DANGER, reflecting the combined toxicity of dicamba and other ingredients in each product.

**TRI Developmental Toxin:** Developmental toxin

**WHO Acute Hazard:** Class III, slightly hazardous

*Regulatory Authority:*
- Air Pollutant Standard Set (former USSR)[35]
- AB 1803-Well Monitoring Chemical (CAL)
- The "Director's List" (CAL/OSHA)
- Actively registered pesticide in California.
- Clean Water Act: Section 311 Hazardous Substances/RQ 40CFR117.3 (same as CERCLA, see below)
- Safe Drinking Water Act: Priority List (55 FR 1470)
- Superfund/EPCRA 40CFR302.4 RQ: CERCLA, 1000 lb (454 kg)
- EPCRA Section 313 Form R *de minimus* concentration reporting level: 1.0%
- Canada, Drinking Water Quality, 0.12 mg/L MAC

*Description:* White or brown crystalline solid. Slightly soluble in water; solubility = $8.0 \times 10^3$ ppm @ 25°C; also reported at $5.6 \times 10^3$ ppm. Molecular weight = 221.03. Boiling point = 200°C (decomposes below B.P.). Melting/Freezing point = 115°C. Hazard Identification (based on NFPA-704 M Rating System): Health 1, Flammability 0, Reactivity 0. Log $K_{ow}$ = 2.19. Uunlikely to bioaccumulate in marine organisms.

*Incompatibilities:* Incompatible with sulfuric acid, bases, ammonia, aliphatic amines, alkanolamines, isocyanates, alkylene oxides, epichlorohydrin. Dicamba decomposes in heat producing toxic and corrosive fumes including hydrogen chloride.

*Permissible Exposure Limits in Air:* The former USSR[35] has set an MAC in workplace air of 1.0 mg/m³. Although no U.S. exposure limits have been established, this chemical can be absorbed through the skin, thereby increasing exposure.

*Permissible Concentration in Water:* A no-adverse effect level in drinking water has been calculated by NSA/NRC[46] at 0.009 mg/L. States which have set guidelines for dicamba in drinking water[61] include Maine at 9.0 µg/L and Wisconsin at 12.5 µg/L. Canada's MAC in drinking water is 0.12 mg/L.

*Routes of Entry:* Ingestion, inhalation, and through the skin.

*Harmful Effects and Symptoms*

*Short Term Exposure:* Dicamba irritates the eyes, skin, and respiratory tract. Exposure can cause nausea, vomiting, loss of appetite and weight, muscle weakness, and exhaustion. The acute toxicity of dicamba is relatively low. Dicamba produced no adverse effect when fed to rats at up to 19.3 mg/kg/day and 25 mg/kg/day in subchronic and chronic studies. The no-adverse-effect dose in dogs was 1.25 mg/kg/day in a 2-year feeding study. Based on these data, an ADI was calculated at 0.0012 mg/kg/day. The $LD_{50}$ (oral, rat) = 1037 mg/kg (slightly toxic).

*Long Term Exposure:* May affect the liver.

*Points of Attack:* Liver

*Medical Surveillance:* Liver function tests.

*First Aid:* If this chemical gets into the eyes, remove any contact lenses at once and irrigate immediately for at least 15 minutes, occasionally lifting upper and lower lids. Seek medical attention immediately. If this chemical contacts the skin, remove contaminated clothing and wash immediately with soap and water. Seek medical attention immediately. If this chemical has been inhaled, remove from exposure, begin rescue breathing (using universal precautions) if breathing has stopped, and CPR if heart action has stopped. Transfer promptly to a medical facility. When this chemical has been swallowed, get medical attention. Give large quantities of water and induce vomiting. Do not make an unconscious person vomit.

*References:*
- EXTOXNET, Extension Toxicology Network, "Pesticide Information Profile, Dicamba," Oregon State University, Corvallis, OR (June 1996). http://ace.orst.edu/info/extoxnet/pips/dicamba.htm
- U.S. Environmental Protection Agency, Office of Pesticide Programs, Pesticide Residue Limits, "Dicamba", 40 CFR 180.227, http://www.epa.gov/pesticides/food/viewtols.htm
- U.S. Environmental Protection Agency, National Pesticide Information Center, "NPIC Technical Fact Shhet, Dicamba," Washington, DC. (January 2002). http://npic.orst.edu/factsheets/dicamba_tech.pdf
- U.S. Environmental Protection Agency, "Health Advisory: Dicamba," Washington, DC, Office of Drinking Water (August 1987).
- California Environmental Protection Agency "Chemical List of Lists," Sacramento CA (February 1997).
- New Jersey Department of Health and Senior Services, Hazardous Substance Fact Sheet, "Dicamba," Trenton NJ (January 1999). http://www.state.nj.us/health/eoh/rtkweb/0634.pdf

# Dichlobenil (ANSI)

*Use Type:* Herbicide
*CAS Number:* 1194-65-6
*Formula:* $C_7H_3Cl_2N$
*Synonyms:* Benzonitrile, 2,6-Dichloro-; DCBN; DBN; 2,6-DBN; DCB; Decabane; 2,6-Dichlorobenzonitrile; 2,6-Dichlorocyanobenzene
*Trade Names:* BARRIER®, Pbi/Gordon (USA); BH Prefix D®; CARSORON®, Crompton Corporation (USA); CASORON® 133, Crompton Corporation (USA); CARSORON® G, Crompton Corporation (USA); CARSORON® G4, Crompton Corporation (USA); CARSORON® G20-SR, Crompton Corporation (USA); CODE H 133®; DECABANE®; DU-SPREX®; DYCLOMEC®; FYDULAN; FYDUMAS; FYDUSIT; H 133®; H 1313®; NIA 5996®; NIAGARA® 5006; NIAGARA® 5,996; NOROSAC®, Pbi/Gordon (USA); PREFIX D®
*Producers:* Crompton Corporation (USA); Fluorochem (UK); Pbi/Gordon (USA)
*Chemical Class:* Benzonitrile
*EPA/OPP PC Code:* 027401
*California DPR Chemical Code:* 112
*ICSC Number:* 0867
*RTECS Number:* DI3500000
*EEC Number:* 608-015-00-X
*EINECS Number:* 214-787-5
*Uses:* Dichlobenil is a herbicide used on cranberry bogs, dichondra, ornamentals, blackberry, raspberry, and blueberry fields, apple, pear, filbert and cherry orchards, vineyards, hybrid poplar-cottonwood plantations, and rights-of-way to control weeds; and sewers to remove roots. It acts on dandelion, prickly oxtongue (pre-emergence), and tree roots.
*Human toxicity (long-term)[77]:* High–9.10 ppb, Health Advisory
*Fish toxicity (threshold)[77]:* Very low–973.64357 ppb, MATC (Maximum Acceptable Toxicant Concentration)
*U.S. Maximum Allowable Residue Levels for Dichlobenil (40 CFR 180.231):*

| CROP | ppm |
| --- | --- |
| Apple | 0.5 |
| Blackberry | 0.1 |
| Blueberry | 0.15 |
| Cranberry | 0.1 |
| Filbert | 0.1 |
| Fruit, stone, except cherry | 0.15 |
| Grape | 0.15 |
| Pear | 0.5 |
| Raspberry | 0.1 |

*Carcinogen/Hazard Classifications*
**U.S. EPA Carcinogens:** Group C, possible carcinogen
**Label Signal Word:** CAUTION
**WHO Acute Hazard:** Class U, unlikely to be hazardous
*Regulatory Authority:*
- The "Director's List" (CAL/OSHA)
- Actively registered pesticide in California.
- Clean Water Act: Section 311 Hazardous Substances/RQ (same as CERCLA)
- EPCRA Section 304 RQ: CERCLA, 100 lb (45.4 kg)

*Description:* White solid. Gives off irritating or toxic fumes or gases in a fire. Practically insoluble in water; solubility = 18 ppm @ 20°C. Molecular weight = 172.01. Melting/Freezing point = 143–146°C. Boiling point = 269°C. Vapor pressure = $1 \times 10^{-3}$ mmHg @ 25. Log $K_{ow}$ = 2.63. Unlikely to bioaccumulate in marine organisms.
*Determination in Air:* Filter; none; Gravimetric; NIOSH IV [Particulates NOR; #0500 (total), #0600 (respirable)]. Determination in Air.[18]
*Permissible Concentration in Water:* No criteria set. Runoff from spills or fire control may cause water pollution.
*Routes of Entry:* Absorbed by dry skin and inhalation
*Harmful Effects and Symptoms*
*Short Term Exposure:* Contact can irritate the skin and eyes. Inhalation may irritate the nose and throat. High exposure can cause headache, dizziness, a drop in blood pressure, rapid pulse, loss of appetite, seizures, coma and death. Repeated high exposure may result in a loss of smell, acne-like rash.
*Points of Attack:* Eyes, skin, liver and kidneys.
*First Aid:* If this chemical gets into the eyes, remove any contact lenses at once and irrigate immediately for at least 15 minutes, occasionally lifting upper and lower lids. Seek medical attention immediately. If this chemical contacts the skin, remove contaminated clothing and wash immediately with soap and water. Seek medical attention immediately. If this chemical has been inhaled, remove from exposure, begin rescue breathing (using universal precautions) if breathing has stopped, and CPR if heart action has stopped. Transfer promptly to a medical facility. When this chemical has been swallowed, get medical attention. Give large quantities of water and induce vomiting. Do not make an unconscious person vomit. Do not induce vomiting when formulations containing petroleum solvents are ingested.
*References:*
- New Jersey Department of Health and Senior Services, "Hazardous Substance Fact Sheet, Dichlobenil," Trenton NJ (October 2000).
  http://www.state.nj.us/health/eoh/rtkweb/0636.pdf
- U.S. Environmental Protection Agency, "Reregistration Eligibility Decision (RED), Dichlobenil" Office of Prevention, Pesticides and Toxic Substances, Washington, DC (October 1998).
  http://www.epa.gov/REDs/0263red.pdf
- California Environmental Protection Agency "Chemical List of Lists," Sacramento CA (February 1997)
- U.S. Environmental Protection Agency, Office of Pesticide Programs, Pesticide Residue Limits,

"Dichlobenil", 40 CFR 180.231.
www.epa.gov/pesticides/food/viewtols.htm

# Dichlone

*Use Type:* Fungicide
*CAS Number:* 117-80-6
*Formula:* $C_{10}H_4Cl_2O_2$
*Synonyms:* 2,3-Dichlor-1,4-naphthochinon (German); 2,3-Dichloro-1,4-naphthalenedione; 2,3-Dichloro-1,4-naphthoquinone; Dichloronaphthoquinone; 2,3-Dichloronaphthoquinone; 2.3-Dichloro-α-naphthoquinone; 2,3-Dichloronaphthoquinone-1,4; Diclona (spanish); Diclone; ENT 3,776; 1,4-Napthalenedione, 2,3-dichloro-
*Trade Names:* ALGISTAT®; COMPOUND 604®; PHYGON®; PHYGON® PASTE; PHYGON® SEED PROTECTANT; PHYGON® XL; QUINTAR®; QUINTAR® 540F; SANQUINON®; UNIROYAL® 604; USR® 604; U.S. RUBBER® 604
*Producers:* Whyte Agrochemical (UK)
*Chemical Class:* Organochlorine; Halo-organics; napthoquinone fungicide
*EPA/OPP PC Code:* 029601
*California DPR Chemical Code:* 202
*RTECS Number:* QL7525000
*EINECS Number:* 204-210-5
*Uses:* Not currently registered in the U.S. Dichlone is used as a fungicide for foliage and to control blue algae in ponds, swimming pools and lakes. As a substitute for copper and sulfur to control of rot on fruit trees, vegetables, field crops, ornamentals, resident and commercial outdoor areas.
*Human toxicity (long-term)[77]:* Very low–560.00 ppb, Health Advisory
*Fish toxicity (threshold)[77]:* High–4.56263 ppb, MATC (Maximum Acceptable Toxicant Concentration)
*Regulatory Authority:*
- The "Director's List" (CAL/OSHA)
- Clean Water Act: Section 311 Hazardous Substances/RQ (same as CERCLA)
- EPCRA Section 304 RQ: CERCLA, 1 lb (0.454 kg)

*Description:* Golden-yellow crystals, leaflets, or needles from alcohol. Practically insoluble in water; solubility = <0.1 mg/L @ 25°C. Molecular weight = 227.05. Melting/Freezing point = 193°C. Boiling point = 275°C @ 2 mmHg;. Vapor Density = 7.8. Vapor pressure = 8.2 mmHg @ 20°C. Toxicity (human): Low. Log $K_{ow}$ = 3.15. Values above 3.0 are likely to bioaccumulate in marine organisms.
*Routes of Entry:* Dangerous through all routes of contact. Absorbed through intact skin.

*Harmful Effects and Symptoms*
*Short Term Exposure:* Central nervous system depressant. Irritating to eyes, skin, mucous membranes. Large doses can cause central nervous system depression, dizziness, weakness, headache, nausea, vomiting, and difficult breathing. Apprehension, anxiety, confusion, nervous excitation; dizziness; headache; numbness and weakness in limbs; muscle twitching, tremors; nausea and vomiting; slow, shallow respiration, bluish face; convulsions; loss of consciousness; breathing stops; death. At fire temperatures can produce coughing, choking, difficult breathing, and cyanosis.
*Long Term Exposure:* Irritating to the skin; may cause dermatitis.
*Points of Attack:* May be fatal if inhaled, ingested, or absorbed through the skin
*Medical Surveillance:* Medical observation is recommended for 24 to 48 hours after breathing overexposure, as pulmonary edema may be delayed. As first aid for pulmonary edema, consider administering a corticosteroid spray. Cigarette smoking may exacerbate pulmonary injury and should be discouraged for at least 72 hours following exposure.
*First Aid:* Speed in removing material from eyes and skin is of extreme importance. *Eyes:* Eye contact can cause dangerous amounts of these chemicals to be quickly absorbed through the mucous membrane into the bloodstream. Directly, irrigate with large amounts of plain, tepid water or saline for 20 minutes, occasionally lifting the lower and upper lids. During this time, remove contact lenses, if easily removable without additional trauma to the eye. Get medical aid immediately. Have physician check for possible delayed damage. *Skin:* Get medical aid. Skin contact can cause dangerous amounts of these chemicals to be absorbed into the bloodstream. Wearing the appropriate PPE equipment and respirator for organochlorine pesticides, immediately flush exposed skin, hair, and under nails with plain, running, tepid water for 20 minutes, then wash twice with mild soap. Shampoo hair promptly if contaminated; protect eyes. Do not scrub skin or hair, since this can increase absorption through the skin. Rinse thoroughly with water. Victims who are able and cooperative may assist with their own decontamination. Remove and double-bag contaminated clothing and personal belongings. Leather absorbs many organochlorines; therefore, items such as leather shoes, gloves, and belts should be discarded. If the skin is swollen or inflamed, cool affected areas with cold compresses. *Ingestion:* Call poison control. Loosen all clothing. Never give anything by mouth to an unconscious person. Get medical aid. Do NOT induce vomiting. The patient is at risk of CNS depression or seizures, which may lead to pulmonary aspiration during vomiting. If the victim is conscious and able to swallow, *administer an aqueous slurry of activated charcoal at 1 gm/kg (usual adult dose 60–90 g, child dose 25–50 g). A soda can and straw may be of assistance when offering charcoal to a child. The efficacy of activated charcoal for some organochlorine poisoning (such as chlordane) is uncertain. If victim is *UNCONSCIOUS OR HAVING CONVULSIONS,* do nothing except keep victim warm.

*In some cases you may be specifically instructed by Poison Control to induce vomiting by way of 2 tablespoons of syrup of ipecac (adult) washed down with a cup of water. Do NOT give activated charcoal <u>before or with</u> ipecac syrup.
*Inhalation:* Get medical aid. Do not contaminate yourself. Wearing the appropriate PPE equipment and respirator for organochlorine pesticides, immediately remove the victim from the contaminated area to fresh air. For inhalation exposures, monitor for respiratory distress. If the victim is not breathing, administer artificial respiration. Do NOT use mouth-to-mouth resuscitation; use bag/mask apparatus. If cough or breathing difficulty develops, evaluate for respiratory tract irritation, bronchitis, or pneumonitis. If breathing is difficult, administer 100% humidified supplemental oxygen through bag/mask apparatus until medical help arrives. Do not leave victim unattended.

*References:*
- Pesticide Management Education Program, "Dichlone (Phygon, Quintar) Chemical Profile 2/85," Cornell University, Ithaca, NY (February 1985). http://pmep.cce.cornell.edu/profiles/fung-nemat/aceticacid-etridiazole/dichlone/fung-prof-dichlone.html
- New Jersey Department of Health and Senior Services, "Hazardous Substance Fact Sheet, Dichlone," Trenton NJ (March 2001). http://www.state.nj.us/health/eoh/rtkweb/0637.pdf
- California Environmental Protection Agency "Chemical List of Lists," Sacramento CA (February 1997)

# Dichloran

*Use Type:* Fungicide
*CAS Number:* 99-30-9
*Formula:* $C_6H_4Cl_2N_2O_2$
*Synonyms:* AI3-08870; AI-50; Aniline, 2,6-dichloro-4-nitro-; Benzenamine, 2,6-Dichloro-4-nitro-; Caswell No. 311; CDNA; CNA; DCNA; 2,6-Dichloro-4-nitroaniline; 2,6-Dichloro-4-nitrobenzenamine; Dicloran; 4-Nitroaniline, 2,6-dichloro-; 4-Nitro-2,6-dichloroaniline
*Trade Names:* AL-50®; ALLISAN®, Gowan Company (USA); BORTRAN® Gowan Company (USA); BOTRAN®, Gowan Company (USA); DITRANIL®; FUMITE DICLORAN SMOKE ACARICIDE®; RESISAN®; RD-6584®; U-2069®
*Producers:* Gowan Company (USA); Whyte Agrochemical (UK)
*Chemical Class:* Substituted benzene
*EPA/OPP PC Code:* 031301
*California DPR Chemical Code:* 81
*ICSC Number:* 0871
*RTECS Number:* BX2975000
*EINECS Number:* 202-746-4

*Uses:* Used to control fungi on a variety of crops. The top crop usages in California are on celery, head lettuce, and grapes (table, wine and raisin).
*U.S. Maximum Allowable Residue Levels for Dicloran (40 CFR 180.200):*

| CROP | ppm |
|---|---|
| Apricot, post-h | 20 |
| Bean, snap, succulent | 20 |
| Carrot, roots, post-h | 10 |
| Celery | 15 |
| Cherry, sweet, post-h | 20 |
| Cucumber | 5 |
| Endive | 10 |
| Garlic | 5 |
| Grape | 10 |
| Lettuce | 10 |
| Nectarine, post-h | 20 |
| Onion | 10 |
| Peach, post-h | 20 |
| Plum, fresh prune, post-h | 15 |
| Potato | 0.25 |
| Rhubarb | 10 |
| Sweet potato, roots, post-h | 10 |
| Tomato | 5 |

*Carcinogen/Hazard Classifications*
**Label Signal Word:** WARNING, CAUTION
**WHO Acute Hazard:** Class U, unlikely to be hazardous
*Regulatory Authority:*
- FIFRA, 180.3(4); class of chlorinated organic pesticide
- EPA 40 CFR 372.65, Specific Toxic Chemical Listings
- EPCRA Section 313 Form R *de minimus* concentration reporting level: 1.0%
- Actively registered pesticide in California.

*Description:* Yellow needles from ethyl alcohol. Molecular weight = 207.02. Melting/Freezing point = 190°C.
*Harmful Effects and Symptoms*
*Short Term Exposure:* Contact with eyes or skin may cause irritation or injury. Inhalation should be avoided; use NIOSH-approved air purifying respirators for pesticides. May be harmful if swallowed.
*Routes of Entry:* Inhalation, skin contact, eyes
*First Aid:* If this chemical gets into the eyes, remove any contact lenses at once and irrigate immediately for at least 15 minutes, occasionally lifting upper and lower lids. Seek medical attention immediately. If this chemical contacts the skin, remove contaminated clothing and wash immediately with soap and water. Seek medical attention immediately. If this chemical has been inhaled, remove from exposure, begin rescue breathing (using universal precautions) if breathing has stopped, and CPR if heart action has stopped. Transfer promptly to a medical facility. When this chemical has been swallowed, get medical attention. Give large quantities of water and induce vomiting. Do not make an unconscious person vomit. Do not induce vomiting when formulations containing petroleum solvents are ingested.

*References:*
- U.S. Environmental Protection Agency, Office of Pesticide Programs, Pesticide Residue Limits, "Dicloran", 40 CFR 180.200. www.epa.gov/pesticides/food/viewtols.htm
- California Environmental Protection Agency "Chemical List of Lists," Sacramento CA (February 1997)

## Dichloroacetic Acid

*Use Type:* Fungicide
*CAS Number:* 79-43-6
*Formula:* $C_2H_2C_{l2}O_2$
*Synonyms:* Acetic acid; Bichloro-; Acetic acid, dichloro-; Bichloroacetic acid; Dichlorethanoic acid; DCA
*Producers:* Clariant (Switzerland); Fluorochem (UK); Merck Eurolab, Ltd. (UK); Sigma-Aldrich Laborchemikalien (Germany); TCI America (USA)
*Chemical Class:* Unclassified
*California DPR Chemical Code:* 3133
*ICSC Number:* 0868
*RTECS Number:* AG6125000
*EEC Number:* 607-066-00-5
*EINECS Number:* 201-207-0
*Uses:* Used as a fungicide, a medication, and a chemical intermediate in pharmaceuticals.
*Carcinogen/Hazard Classifications*
**California Prop. 65:** Carcinogen
**IARC:** Group 3, unclassifiable
*Regulatory Authority:*
- Air Pollutant Standard Set (former USSR)

*Description:* Colorless liquid. Pungent odor. Soluble in water; solubility = $8.5 \times 10^4$ ppm. Molecular weight = 128.95. Boiling point = 193–194°C. Melting/Freezing point = 9.7°C. Flash point = 110°C. Vapor pressure = 1 mmHg @ 44°C; 20 Pa @ 20°C. Hazard Identification (based on NFPA-704 M Rating System): Health 3, Flammability 1, Reactivity 0. Log $K_{ow}$ = –1.1 to 0.9. Ulikely to bioaccumulate in marine organisms.

*Incompatibilities:* DCA is a medium strong acid; incompatible with non-oxidizing mineral acids, organic acids, bases, acrylates, aldehydes, alcohols, alkylene oxides, ammonia, aliphatic amines, alkanolamines, aromatic amines, amides, glycols, isocyanates, ketones. Attacks metals generating flammable hydrogen gas. Attacks some plastics, rubber and coatings.
*Permissible Exposure Limits in Air:* The former USSR set an airborne exposure limit of 4 mg/m$^3$.
*Routes of Entry:* Inhalation, ingestion.
*Harmful Effects and Symptoms*
*Short Term Exposure:* Corrosive to the eyes, skin, and respiratory tract; causes severe irritation and burns. Eye contact may cause permanent damage. Higher exposures can cause pulmonary edema, a medical emergency that can be delayed for several hours. This can cause death. Corrosive if swallowed.
*Long Term Exposure:* May cause damage to the developing fetus. May affect the liver and kidneys. May damage the nervous system causing numbness, "pins and needles," and/or weakness in the hands and feet. Repeated exposure may cause lung irritation, bronchitis. There is limited evidence that DCA causes liver cancer in animals.
*Points of Attack:* Lungs, liver, kidneys, nervous system.
*Medical Surveillance:* Liver and kidney function tests. Lung function tests. Examination of the nervous system. Consider chest x-ray following acute overexposure.
*First Aid:* If this chemical gets into the eyes, remove any contact lenses at once and irrigate immediately for at least 15 minutes, occasionally lifting upper and lower lids. Seek medical attention immediately. If this chemical contacts the skin, remove contaminated clothing and wash immediately with soap and water. Seek medical attention immediately. If this chemical has been inhaled, remove from exposure, begin rescue breathing (using universal precautions) if breathing has stopped, and CPR if heart action has stopped. Transfer promptly to a medical facility. When this chemical has been swallowed, get medical attention. If victim is conscious, administer water or milk. Do not induce vomiting. Medical observation is recommended for 24 to 48 hours after breathing overexposure, as pulmonary edema may be delayed. As first aid for pulmonary edema, a doctor or authorized paramedic may consider administering a corticosteroid spray.
*References:*
- New Jersey Department of Health and Senior Services, "Hazardous Substance Fact Sheet, Dichloroacetic Acid," Trenton NJ (February 1999). http://www.state.nj.us/health/eoh/rtkweb/0638.pdf
- California Environmental Protection Agency "Chemical List of Lists," Sacramento CA (February 1997).

## *para*-Dichlorobenzene

*Use Type:* Insecticide, rodenticide, and fungicide
*CAS Number:* 106-46-7
*Formula:* $C_6H_4Cl_2$
*Alert:* The *para* isomer of dichlorobenzene is the isomer most prominently used in agriculture.
*Synonyms:* *p-DCB:* 1,4-DCB: Benzene, 1,4-dichloro-; Benzene, *p*-dichloro-; *p*-Chlorophenyl chloride; PDCB; Dichloricide; 1,4-Diclorobenceno (Spanish); *p*-Diclorobenceno (Spanish); *p*-Dichlorobenzene; Paradichlorobenzene
*Mixed Isomers:* Amisia-mottenschutz; Benzene, dichloro-; Diclorobenceno (Spanish); DCB; Diclorobenceno (Spanish); Dichlorobenzene (Mixed Isomers)
*Trade Names:* DowTHERM®, Dow Chemical Co. (USA); EVOLA; PARACIDE®; PARA CRYSTALS®; PARADI®;

PARADOW®; PARAMOTH®; PARANUGGETS®; PARAZENE®; PERSIA-PERAZOL®; SANTOCHLOR®, Solutia (USA); *Mixed isomers:* DILATIN DBI®; MOTTENSCHUTZMITTEL EVAU P®; MOTT-EX®; TOTAMOTT®

*Producers of Chlorobenzenes:* ATOFINA (France); Bayer Chemicals (Germany); Changshan Chemical (China); Clariant (Switzerland); Clorobencenos (Mexico); Degussa (Germany); Dow Chemical (USA); EniChem (Italy); ICI Group (UK); Ishihara Sangyo Kaisha (Japan); Kureha (Japan); Mitsubishi Gas Chemical (Japan); Monsanto (USA); Nippon Kayaku (Japan); Piedmont (USA); PPG Industries (USA); Sigma-Aldrich Laborchemikalien (Germany); Solution (USA); Sumitomo Chemical (Japan); Tekchem (Mexico); Toagosei (Japan); Toray Industries (Japan)

*Chemical Class:* Halogenated orgnic compound
*EPA/OPP PC Code:* 061501
*California DPR Chemical Code:* 455
*ICSC Number:* 0037
*RTECS Number:* CZ4550000 (*p*-DCB); CZ4430000 (mixed isomers)
*EEC Number:* 602-035-00-2
*EINECS Number:* 202-425-9

*Uses:* *p*-Dichlorbenzene is used primarily as an air deodorant, as moth balls, and in insecticides, which accounts for 90% of the total production of this isomer. Information is not available concerning the production and use of *m*-DCB. However, it may occur as a contaminant of *o*- or *p*-DCB formulations. Both *o*- and *p*-isomers are produced almost entirely as by-products during the production of monochlorobenzene. The major uses of *o*-DCB are as a process solvent in the manufacturing of toluene diisocynate and as an intermediate in the synthesis of dyestuffs, herbicides, and degreasers.

*Carcinogen/Hazard Classifications*
**U.S. EPA Carcinogens:** Group C, possible carcinogen
**U.S. NTP Carcinogen:** Reasonably anticipated Carcinogen
**California Prop. 65:** Carcinogen
**IARC:** Group 2B, possible carcinogen
**WHO Acute Hazard:** Class III, slightly hazardous

*Regulatory Authority:*
- Air Pollutant Standard Set (ACGIH)[1] (Australia) (DFG)[3] (HSE)[33] (Israel) (Mexico) (OSHA)[58] (former USSR)[35, 43] (Several States)[60] (Several Canadian Provinces)
- List of priority pollutants (U.S. EPA)
- AB 1803-Well Monitoring Chemical (CAL)
- MCL (Maximum Contaminants Levels) list of contaminants (CAL)
- AB 2588-Air Toxics "Hot Spots" Chemicals (CAL)
- CAL Air Resources Board/AB 1807 Toxic Air Contaminants
- Proposition 65 chemical (CAL)
- Permissible Exposure Limits for Chemical Contaminants (CAL/OSHA)
- Actively registered pesticide in California.

*1,4-DCB:*
- Carcinogen, (animal positive) (IARC), (NTP), (ACGIH)[1]
- Clean Air Act: Hazardous Air Pollutants (Title I, Part A, Section 112)
- Clean Water Act: Section 311 Hazardous Substances/RQ 40CFR117.3 (same as CERCLA, see below);
- 40CFR423, Appendix A, Priority Pollutants; Section 313 Water Priority Chemicals (57FR41331, 9/9/92); Toxic Pollutant (Section 401.15).
- EPA Hazardous Waste Number (RCRA No.): U072; D027
- RCRA, 40CFR261, Appendix 8 Hazardous Constituents
- RCRA Toxicity Characteristic (Section 261.24), Maximum Concentration of Contaminants, regulatory level, 7.5 mg/L
- RCRA 40CFR268.48; 61FR15654, Universal Treatment Standards: Wastewater (mg/L), 0.090; Nonwastewater (mg/kg), 6.0
- RCRA 40CFR264, Appendix 9; TSD Facilities Ground Water Monitoring List. Suggested test method(s) (PQL *ug*/L): 8010(2); 8020(5); 8120(15); 8270(10)
- Safe Drinking Water Act: MCL, 0.075 mg/L; MCLG, 0.075 mg/L; Regulated chemical (47 FR 9352).
- Superfund/EPCRA 40CFR302.4 RQ: CERCLA, 100 lb (45.4 kg).
- EPCRA Section 313 Form R *de minimus* concentration reporting level: 0.1%.
- U.S. DOT Regulated Marine Pollutant (49CFR172.101, Appendix B)
- Canada, WHMIS, Ingredients Disclosure List, CEPA Priority Substance List, National Pollutant Release Inventory
- (NPRI); Drinking Water Quality 0.005 mg/L MAC and <=0.001 mg/L AO

*Mixed isomers:*
- Clean Water Act: Section 311 Hazardous Substances/RQ 40CFR117.3 (same as CERCLA, see below);
- 40CFR423, Appendix A, Priority Pollutants; Section 313 Water Priority Chemicals (57FR41331, 9/9/92); Toxic Pollutant (Section 401.15).
- RCRA, 40CFR261, Appendix 8 Hazardous Constituents, waste number not listed
- Safe Drinking Water Act: Regulated chemical, MCL, 0.075 mg/L, *p*-isomer (47 FR 9352)
- Superfund/EPCRA 40CFR302.4 RQ: CERCLA, 100 lb (45.4 kg)
- EPCRA Section 313 Form R *de minimus* concentration reporting level: 0.1%.
- U.S. DOT Regulated Marine Pollutant (49CFR172.101, Appendix B)

- Canada, WHMIS, Ingredients Disclosure List, CEPA Priority Substance List, National Pollutant Release Inventory (NPRI)
- Mexico, Drinking Water, 0.4 mg/L

***Description:*** There are three isomeric forms of dichlorobenzene, $C_6H_4Cl_2$: o-DCB is a colorless to pale yellow liquid with a pleasant, aromatic odor. Odor threshold = 0.30 ppm. Boiling point = 117°. Melting/Freezing point = –17°C. Flash point = 66°C (cc). Autoignition temperature: 648°C. Explosive limits: LEL = 2.2%; UEL = 9.2%. Hazard Identification (based on NFPA-704 M Rating System): Health 2, Flammability 2, Reactivity 0. Insoluble in water. m-DCB is a liquid. Boiling point =172°C. p-DCB is a colorless or white solid with a mothball-like odor. Odor threshold = 0.18 ppm. Boiling point = 174°C. Melting/Freezing point = 53°C. Flash point = 66°C. Insoluble in water. Explosive limits: LEL = 2.5%; UEL = unknown. Hazard Identification (based on NFPA-704 M Rating System): Health 2, Flammability 2, Reactivity 0. Insoluble in water. Log $K_{ow}$ = 3.36. Values at or above 3.0 are likely to bioaccumulate in marine organisms (p-DCB).

***Incompatibilities:*** o-DCB and m-DCB: acid fumes, chlorides, strong oxidizers, hot aluminum or aluminum alloys. p-DCB: Strong oxidizers; although, incompatibilities for this chemical may also include other materials listed for o-DCB.

***Permissible Exposure Limits in Air:*** p-DCB: OSHA[2] PEL, 75 ppm (450 mg/m³) TWA; ACGIH[1] TLV, 10 ppm TWA (60 mg/m³) TWA; NIOSH[2] IDLH, 150 ppm [Carcinogen]. Foreign Regulations (ppm) TWA (except as noted) p-DCB for Australia, Germany (DFG), United Kingdom (HSE), Mexico and the Canadian provinces of Alberta, British Columbia, Ontario, and Quebec.

| Australia | DFG | HSE | Israel | Mexico | Canada |
|---|---|---|---|---|---|
| 75 | 50 | 75 | 75 | 75 | 75 |
| 110* | 100* | 110* | 110* | 110* | 110* |

*As a ceiling or peak limitation value, not to be exceeded at any time, except for DFG Peak Limitation (30 minute) which is not to be exceeded 4 times during a normal workshift[3]. The former USSR seta a MAC for ambient air in residential areas of 0.035 mg/m³ on a once daily basis for p-DCB[35].

***Determination in Air:*** Charcoal adsorption followed by $CS_2$ workup and gas chromatographic analysis. See NIOSH Method 1003 for halogenated hydrocarbons.[18]

***Permissible Concentration in Water:*** To protect freshwater aquatic life: 1,120 µg/L on an acute toxicity basis and 763 µg/L on a chronic basis. To protect saltwater aquatic life: 1,970 µg/L on an acute toxicity basis. To protect human health: 400 µg/L for all isomers[6]. The former USSR established a MAC of 0.002 mg/L for water bodies used for domestic purposes[35,43]. The USEPA[48] has derived lifetime health advisories for o- and m-DCB as 0.62 mg/L (620 µg/L) and for p-DCB of 0.075 mg/L (75 µg/L). EPA[62] has recently proposed a maximum level for o-DCB of 0.6 mg/L in drinking water. See also Regulatory section for drinking water criteria for Mexico and Canada.

***Determination in Water:*** Gas chromatography (EPA Methods 601, 602, 612) or gas chromatography plus mass spectrometry (EPA Method 625). Gas-chromatographic methods have been developed for PDB with a sensitivity of 380 pg/cm peak high, and PDB concentrations as low as 1.0 ppb in water have been analyzed according to NAS/NRC.

***Routes of Entry:*** For p-DCB: Inhalation, ingestion, eye and skin contact

***Harmful Effects and Symptoms***

Human exposure to dichlorobenzene is reported to cause hemolytic anemia and liver necrosis, and 1,4-dichlorobenzene has been found in human adipose tissue. In addition, the dichlorobenzenes are toxic to nonhuman mammals, birds, and aquatic organisms and impart an offensive taste and odor to water. The dichlorobenzenes are metabolized by mammals, including humans, to various dichlorophenols, some of which are as toxic as the dichlorobenzenes. Exposure can damage blood cells. Contact can cause irritation of the skin and eyes. Prolonged contact can cause severe burns. It may damage the liver, kidneys and lungs. Exposure can cause headache, dizziness, swelling of the eyes, hands and feet, and nausea. Higher levels can cause severe liver damage and death. Persons with preexisting pathology (hepatic, renal, central nervous system, blood) or metabolic disorders, who are taking certain drugs (hormones or otherwise metabolically active), or who are otherwise exposed to DCBs or related (chemically or biologically) chemicals by such means as occupation, or domestic use or abuse (e.g., pica or "sniffing") of DCB products, might well be considered at increased risk from exposure to DCBs.

***Short Term Exposure:*** For p-DCB: Can be absorbed through the skin, thereby increasing exposure. Exposure can cause headache, dizziness, nausea, swelling of the hands and feet. Contact with the dust can irritate and burn the eyes and skin. Skin allergy may develop.

***Long Term Exposure:*** For p-DCB: May be carcinogenic to humans; it causes kidney and liver cancer in animals. There is a suggested association between this chemical and leukemia. There is evidence that p-DCB can damage the developing animal fetus. Repeated exposure can damage the nervous system, skin allergy and damage the lungs, liver, and kidneys. p-May affect the blood and cause hemolytic anemia.

***Points of Attack:*** For p-DCB-liver, respiratory system, eyes, kidneys, skin.

***Medical Surveillance:*** For those with frequent or potentially high exposure (half the TLV or greater, or significant skin contact) the following are recommended before beginning work and at regular times after that: Liver, kidney and lung function tests. Complete blood count. If symptoms develop or overexposure is suspected, the following may be useful: Evaluation by a qualified allergist, including careful exposure history and special testing, may help diagnose skin allergy.

*First Aid:* If this chemical gets into the eyes, remove any contact lenses at once and irrigate immediately for at least 15 minutes, occasionally lifting upper and lower lids. Seek medical attention immediately. If this chemical contacts the skin, remove contaminated clothing and wash immediately with soap and water. Seek medical attention immediately. If this chemical has been inhaled, remove from exposure, begin rescue breathing (using universal precautions) if breathing has stopped, and CPR if heart action has stopped. Transfer promptly to a medical facility. When this chemical has been swallowed, get medical attention. Give large quantities of water and induce vomiting. Do not make an unconscious person vomit.

*References:*
- U.S. Environmental Protection Agency, Dichlorobenzene: Ambient Water Quality Criteria, Washington, DC (1980).
- U.S. Environmental Protection Agency, 1,4-Dichlorobenzene, Health and Environmental Effects Profile No. 66,
- Washington, DC, Office of Solid Waste (April 30, 1980).
- U.S. Environmental Protection Agency, Dichlorobenzenes, Health and Environmental Effects Profile No. 67,
- Washington, DC, Office of Solid Waste (April 30, 1980).
- Sax, N.I., Ed., "Dangerous Properties of Industrial Materials Report," 4, No. 2, 45-48 (1984) (1,3-Dichlorobenzene); 4,
- No. 2, 49-52, and 6, No. 2, 50-57 (1986) (Mixed isomers).
- U.S. Public Health Service, "Toxicological Profile for 1,4-Dichlorobenzene," Atlanta, Georgia, Agency for Toxic
- Substances & Disease Registry (December 1987).
- New Jersey Department of Health, "Hazardous Substance Fact Sheet: 1,4-Dichlorobenzene," Trenton, NJ (June 1998).
- http://www.state.nj.us/health/eoh/rtkweb/0643.pdf
- New York State Department of Health, "Chemical Fact Sheet: para-Dichlorobenzene," Albany, NY, Bureau of Toxic Substance Assessment (April 1986).
- California Environmental Protection Agency "Chemical List of Lists," Sacramento CA (February 1997).

# Dichloroethyl Ether

*Use Type:* Soil fumigant
*CAS Number:* 111-44-4
*Formula:* $C_4H_8Cl_2O$; $ClCH_2CH_2OCH_2CH_2Cl$
*Alert:* This should be handled as a carcinogen, with extreme caution.
*Synonyms:* BCEE; Bis(β-chloroethyl) ether; Bis(2-chloroethyl) ether; Bis(2-cloroetil)eter (Spanish); Chlorex; 1-Chloro-2-(ß-chloroethoxy)ethane; Chloroethyl ether (DOT); Clorex; DCEE; Dichloroethyl ether; 2,2'-Dichloorethylether (Dutch); 2,2'-Dichloro-diethylether; 2,2'-Dichlor-diaethylaether (German); Dichloroethyl ether; 2,2'-Dichlorethyl Ether (DOT); ß,β'-Dichlorodiethyl ether; Dichloroether; Di(β-chloroethyl)ether; Di(2-chloroethyl) ether; ß,β'-Dichloroethyl ether; *sym*-Dichloroethyl ether; 2,2'-Dichloroethyl ether; Dichloroethyl oxide; 2,2'-Dicloroetiletere (Italian); ENT 4,504; Ethane, 1,1'-oxybis 2-chloro-; Ether dichlore (French); 1,1'-Oxybis(2-chloro)ethane; Oxyde de chlorethyle (French)
*Producers:* Richman Chemical (USA)
*EPA/OPP PC Code:* 029501
*California DPR Chemical Code:* 203
*ICSC Number:* 0417
*RTECS Number:* KN0875000
*Uses:* Dichloroethyl ether is used as an acaricide and in an oil solution to control earworms on corn silk. It is also used in the manufacture of paint, varnish, lacquer, soap, and finish remover. It is also used as a solvent for cellulose esters, naphthalenes, oils, fats, greases, pectin, tar, and gum; in dry cleaning and in textile scouring.
*Carcinogen/Hazard Classifications*
**U.S. EPA Carcinogens:** Group B2, probable carcinogen
**California Prop. 65:** Carcinogen
**IARC:** Group 3, unclassifiable
**Label Signal Word:** CAUTION
*Regulatory Authority:*
- Carcinogen (Animal Positive) (IARC)[9]
- Banned or Severely Restricted (Finland, Sweden) (UN)[13]
- Air Pollutant Standard Set (ACGIH)[1] (Australia) (DFG) (Israel) (former USSR)[43] (OSHA)[58] (Several States)[60]
- List of priority pollutants (U.S. EPA)
- AB 1803-Well Monitoring Chemical (CAL)
- AB 2588-Air Toxics "Hot Spots" Chemicals (CAL)
- CAL Air Resources Board/AB 1807 Toxic Air Contaminants
- Proposition 65 chemical (CAL)
- U.S. DOT Inhalation Hazard Chemicals
- Permissible Exposure Limits for Chemical Contaminants (CAL/OSHA)
- The "Director's List" (CAL/OSHA)
- Clean Air Act: Hazardous Air Pollutants (Title I, Part A, Section 112)
- Clean Water Act: 40CFR423, Appendix A, Priority Pollutants; Section 313 Water Priority Chemicals (57FR41331, 9/9/92)
- EPA Hazardous Waste Number (RCRA No.): U025
- RCRA, 40CFR261, Appendix 8 Hazardous Constituents
- RCRA 40CFR268.48; 61FR15654, Universal Treatment Standards: Wastewater (mg/L), 0.033; Nonwastewater (mg/kg), 6.0
- RCRA 40CFR264, Appendix 9; TSD Facilities Ground Water Monitoring List. Suggested test method(s) (PQL ug/L): 8270(10)

## Dichloroethyl Ether

- Superfund/EPCRA 40CFR302.4 RQ: CERCLA, 10 lb (4.54 kg)
- Superfund/EPCRA 40CFR355, Appendix B Extremely Hazardous Substances: TPQ = 10,000 lb (4,550 kg)
- EPCRA Section 313 Form R *de minimus* concentration reporting level: 0.1%
- Canada, WHMIS, Ingredients Disclosure List
- Mexico, drinking water criteria = 0.0003 mg/L.

*Description:* Dichloroethyl ether is a clear, colorless liquid with a pungent, fruity odor. It is also described as having a chlorinated solvent-like odor. Soluble in water; solubility = $1.3 \times 10^4$ ppm. Molecular weight = 143.02. Boiling point = 176–178°C. Vapor pressure = 0.73 mmHg @ 20. Flash point = 55°C[17]. Autoignition temperature = 369°C. Explosive limits: LEL = 2.7%; UEL = unknown. Hazard Identification (based on NFPA-704 M Rating System): Health 3, Flammability 2, Reactivity 1. Insoluble in water. Log $K_{ow}$ = 1.12. Unlikely to bioaccumulate in marine organisms.

*Incompatibilities:* Contact with moisture caused decomposition producing hydrochloric acid. Can form peroxides. Forms explosive mixture with air. Contact with strong oxidizers may cause fire and explosion hazard. Attacks some plastics, rubber and coatings. Attacks metals in the presence of moisture.

*Permissible Exposure Limits in Air:* The Federal Limit (OSHA PEL) and the ACGIH[1] recommended TLV is 5 ppm (30 mg/m$^3$) and the STEL value in both cases is 10 ppm (60 mg/m$^3$). They both add the notation "skin" indicating the possibility of cutaneous absorption. Australia, Israel and the Canadian provinces of Alberta, BC, Ontario, and Quebec set the same TWA and STEL as OSHA. The DFG MAK value is 10 ppm[3]. NIOSH[2] IDLH level = [Ca]100 ppm. The former USSR-UNEP/IRPTC project[43] has set an MAC in workplace air of 2 mg/m$^3$. Several states have set guidelines or standards for BCEE in ambient air[60] ranging from 0.0714 mg/m$^3$ (Kansas) to 0.3-0.6 mg/m$^3$ (North Dakota) to 0.5 mg/m$^3$ (Virginia) to 0.6 mg/m$^3$ (Connecticut) to 0.714 mg/m$^3$ (Nevada) to 0.72 mg/m$^3$ (Pennsylvania).

*Determination in Air:* Charcoal tube; $CS_2$; Gas chromatography/Flame ionization detection; NIOSH IV Method #1004[18]

*Permissible Concentration in Water:* To protect freshwater aquatic life: 238,000 µg/L, for chloroalkyl ethers in general. No criteria developed for protection of saltwater aquatic life due to insufficient data. For the protection of human health: preferably zero. An additional lifetime cancer risk of 1 in 100,000 is posed by a concentration of 0.3 µg/L[6]. Some states have set guidelines for BCEE in drinking water[61] ranging from 0.31 µg/L (Minnesota) to 4.2 µg/L (Kansas) to 8.3 µg/L (Maine) to 10.0 µg/L (Arizona).

*Determination in Water:* $CH_2Cl_2$ extraction followed be gas chromatography with halogen; specific detector (EPA Method 611) or gas chromatography plus mass spectrometry (EPA Method 625).

*Routes of Entry:* Inhalation, skin absorption, ingestion, skin and/or eye contact

*Harmful Effects and Symptoms*

*Short Term Exposure:* BCEE can be absorbed through the skin, thereby increasing exposure. Exposure irritates the eyes, skin, and respiratory tract. Skin and eye contact may cause burns. Higher exposures can cause pulmonary edema, a medical emergency that can be delayed for several hours. This can cause death. At concentrations above 500 ppm, coughing, retching, and vomiting may occur, as well as profuse tearing. There can be irritation at lower concentrations. This material is very toxic; the probable oral lethal dose is 50-500 mg/kg, or between 1 teaspoon and 1 ounce for a 150 pound person. It can be a central nervous system depressant in high concentrations. It is extremely irritating to the eyes, nose, and respiratory passages. It can penetrate the skin to cause serious and even fatal poisoning. Poisonous; may be fatal if inhaled, swallowed or absorbed through skin.

*Long Term Exposure:* BCEE may damage the liver and kidneys. Can irritate the lungs; repeated exposures may cause bronchitis.

*Points of Attack:* Respiratory system, skin, eyes, liver and kidneys. This chemical causes liver cancer in animals and may be a potential human carcinogen.

*Medical Surveillance:* Before beginning employment and at regular times after that, the following are recommended: Lung function tests. If symptoms develop or overexposure is suspected, the following may also be useful: Liver, kidney, and lung function tests. Consider chest x-ray following acute overexposure.

*First Aid:* If this chemical gets into the eyes, remove any contact lenses at once and irrigate immediately for at least 15 minutes, occasionally lifting upper and lower lids. Seek medical attention immediately. If this chemical contacts the skin, remove contaminated clothing and wash immediately with soap and water. Seek medical attention immediately. If this chemical has been inhaled, remove from exposure, begin rescue breathing (using universal precautions) if breathing has stopped, and CPR if heart action has stopped. Transfer promptly to a medical facility. When this chemical has been swallowed, get medical attention. Give large quantities of water and induce vomiting. Do not make an unconscious person vomit. Medical observation is recommended for 24 to 48 hours after breathing overexposure, as pulmonary edema may be delayed. As first aid for pulmonary edema, a doctor or authorized paramedic may consider administering a corticosteroid spray.

*References:*

- U.S. Environmental Protection Agency, Chloroalkyl Ethers: Ambient Water Quality Criteria, Washington, DC (1980).

- U.S. Environmental Protection Agency, Bis (2-Chloroethyl) Ether, Health and Environmental Effects Profile No. 24, Washington, DC, Office of Solid Waste (April 30, 1980).
- International Agency for Research on Cancer, IARC Monographs on the Carcinogenic Risks of Chemicals to Humans, Lyon, France 9, 117 (1975).
- U.S. Environmental Protection Agency, "Chemical Profile: Dichloroethyl Ether," Washington, DC, Chemical Emergency Preparedness Program (November 30, 1987).
- California Environmental Protection Agency "Chemical List of Lists," Sacramento CA (February 1997).
- New Jersey Department of Health, "Hazardous Substance Fact Sheet: Bis (2-Chloroethyl) Ether," Trenton, NJ (January 1986, rev. September 1996). http://www.state.nj.us/health/eoh/rtkweb/0232.pdf
- Sax, N.I., Ed., "Dangerous Properties of Industrial Materials Report," 1, No. 4, 62-76 (1987).

# Dichloroisocyanuric Acid

*Use Type:* Biocide and water treatment chemical
*CAS Number:* 2782-57-2
*Formula:* $C_3H_2Cl_2N_3O_3$
*Synonyms:* Dichloroisocyanurate; Dichloroisocyanuric acid; Dichloroisocyanuric acid, dry or dichloroisocyanuric acid salts; Dichloro-*s*-triazinetrione; 1,3-Dichloro-*s*-triazine-2,4,6-(1*H*,3*H*,5*H*)-trione; Isocyanuric acid, dichloro-; Isocyanuric dichloride
*Trade Names:* ACL 70®; CDB 60®; FI CLOR 71®; HILITE 60®; ORCED®; TROCLOSENE®
*Chemical Class:* Triazinetrione
*EPA/OPP PC Code:* 081401
*California DPR Chemical Code:* 204
*Uses:* Used in disinfectants and cleaning solution in domestic products and in food processing plants.
*Regulatory Authority:*
- Actively registered pesticide in California.
- AB 2588-Air Toxics "Hot Spots" Chemicals (CAL)
- CAL Air Resources Board/AB 1807 Toxic Air Contaminants
- Clean Air Act: Hazardous Air Pollutants (Title I, Part A, Section 112);
- Clean Water Act: Section 307 Priority Pollutants as cyanide, total; Toxic Pollutant (Section 401.15)
- RCRA Section 261 Hazardous Constituents
- RCRA Universal Treatment Standards: Wastewater (mg/L), 1.2 (total); 0.86 (amenable); Nonwastewater (mg/kg), 590 (total); 30 (amenable).
- RCRA Ground Water Monitoring List. Suggested test method(s) (PQL $\mu$g/L): 9010(40)
- Safe Drinking Water Act: MCL, 0.2 mg/L; MCLG, 0.2 mg/L; Regulated chemical (47 FR 9352)
- EPA Hazardous Waste Number (RCRA No.): P030 as cyanide compound
- EPCRA Section 304 RQ: CERCLA, 10 lb (4.54 kg)
- EPCRA Section 313 Form R *de minimis* concentration reporting level: 1.0%
- U.S. DOT Regulated Marine Pollutant (49CFR172.101, Appendix B) as cyanides, mixtures or solutions

*Description:* White, crystalline solid or powder, or granules. Chlorine odor. Soluble in water. Melting/Freezing point = 225°C (decomposes). Molecular weight = 197.97.
*Incompatibilities:* A strong oxidizer. Violent reaction with organic and flammable materials.
*Permissible Exposure Limits in Air:* No standards set. However, inasmuch as it is a cyanide compound, the exposure limits are listed here: OSHA and ACGIH: 5 mg/m$^3$ TWA; NIOSH: Ceiling limit, 4.7 ppm; 5 mg/m$^3$ per 10 minutes as cyanides. All have notations that skin contact contributes significantly in overall exposure. IDLH = 25 mg/m$^3$ as CN
*Determination in Air:* Filter/Bubbler; Potassium hydroxide; Ion-specific electrode; NIOSH IV Method #7904, Cyanides. See also Method #6010, Hydrogen Cyanide.[18]
*Permissible Concentration in Water:* No criteria set. Runoff from spills or fire control may cause water pollution. The U.S.EPA[49] has determined a no-observed-adverse-effect-level (NOAEL) of 10.8 mg/kg/day which yields a lifetime health advisory of 154 $\mu$g/L. States which have set guidelines for cyanides in drinking water[61]
*Determination in Water:* Distillation followed by silver nitrate titration or colorimetric analysis using pyridine pyrazolone (or barbituric acid).
*Routes of Entry:* Inhalation, passing through the skin, ingestion.
*Harmful Effects and Symptoms*
*Short Term Exposure:* May cause skin and severe eye irritation. Moderately poisonous if ingested or inhaled. Exposure to a triazine (simazine) has caused acute and subacute dermatitis in the former USSR, characterized by erythema, slight edema, moderate pruritus, and burning lasting 4 to 5 days.
*Long Term Exposure:* May cause lung irritation and damage. May cause skin allergy. Contact with some triazine compounds (such as atrazine) may increase risks for tumors known to be associated with hormonal factors. These have been observed in both animals and human beings, and are consistent with the known effects on the hypothalamic pituitary gonadal axis. Repeated exposure may cause weight loss and reduced red blood cell count. A mutagen.
*Points of Attack:* Liver, lungs, skin.
*Medical Surveillance:* Before beginning employment and at regular times after that, for those with frequent or potentially high exposures, the following is recommended: Lung function tests. Consider chest x-ray following acute overexposure. Evaluation by a qualified allergist. Examination of the nervous system.

*First Aid:* If this chemical gets into the eyes, remove any contact lenses at once and irrigate immediately for at least 15 minutes, occasionally lifting upper and lower lids. Seek medical attention immediately. If this chemical contacts the skin, remove contaminated clothing and wash immediately with soap and water. Seek medical attention immediately. If this chemical has been inhaled, remove from exposure, begin rescue breathing (using universal precautions) if breathing has stopped, and CPR if heart action has stopped. Transfer promptly to a medical facility. When this chemical has been swallowed, get medical attention. Give large quantities of water or milk and induce vomiting. Do not make an unconscious person vomit.

*References:*
- California Environmental Protection Agency "Chemical List of Lists," Sacramento CA (February 1997).

# Dichlorophene

*Use Type:* Fungicide, herbicide and bactericide
*CAS Number:* 97-23-4
*Formula:* $C_{13}H_{10}Cl_2O_2$
*Synonyms:* AI3-02370; Bis(5-chloro-2-hydroxyphenyl)methane; Bis(chlorohydroxyphenyl) methane; Bis(2-hydroxy-5-chlorophenyl)methane; Caswell No. 563; DDDM; DDM; 5,5'-Dichloro-2,2'-dihydroxydiphenylmethane; Dichlorofen; Di(5-chloro-2-hydroxyphenyl)methane; 4,4'-Dichloro-2,2'-methylenediphenol; diclorofeno (Spanish); Dihydroxy dichlorodiphenylmethane; 2,2'-Dihydroxy-5,5'-dichloro diphenylmethane; [(Dihydroxydichloro)diphenyl] methane; GH; 2,2'-Methylenebis(4-chlorophenol); 2,2'-Methylenebis(4-chlorophenol; NSC 38642; Phenol, 2,2'-methylenebis(4-chloro-
*Trade Names:* ANTHIPHEN®; DIPHENTANE 70®; DICHLOROPHEN®; DICHLOROPHEN B®; DICHLOROPHENE 10®; DICHLORPHEN®; DIDROXANE®; DIPHENTHANE 70®; FUNGICIDE F®; FUNGICIDE GM®; FUNGICIDE M®; G 4®; GEFIR®; HYOSAN; KORIUM®; PLATH-LYSE®; PREVENTAL®, Bayer CropScience (Germany); PREVENTOL®, Bayer CropScience (Germany); PREVENTOL GD®, Bayer CropScience (Germany); PREVENTOL GDC®, Bayer CropScience (Germany); SUPER MOSSTOX®; TAENIATOL®; TENIATOL®; TENIATHANE®; TRIVEX®; VERMITHANA®; WESPURIL®
*Producers:* GlaxoSmithKline Animal Health (UK)
*Chemical Class:* Chlorophenol
*EPA/OPP PC Code:* 055001
*California DPR Chemical Code:* 206
*RTECS Number:* SM0175000
*Uses:* Not registered in the U.S. Dichlorophene is a wide spectrum, non-oxidizing biocide used against all types of algae and bacteria. Widely used to treat fungi, fleas and worm conditions in pet animals and livestock. See U.S. Food and Drug Administration 20 CFR 520.580 and 20 CFR 520.581.

*Carcinogen/Hazard Classifications*
*California Prop. 65:* Developmental toxin
*U.S. TRI:* Developmental toxin
*IARC:* Group 2B, possible carcinogen
*Label Signal Word:* CAUTION or WARNING
*WHO Acute Hazard:* Class III, slightly hazardous
*Regulatory Authority:*
- Actively registered pesticide in California.
- EPCRA Section 313 Form R *de minimis* concentration reporting level: 1.0%

*Description:* White crystalline solid or tan, sand-like powder. Odorless when pure; weak phenol odor from contact with air. Practically insoluble in water; solubility = 29.5 ppm @ 25°C. Molecular weight = 269.12. Melting/Freezing point = 177°C. Vapor pressure = $1 \times 10^{-10}$ mmHg @ 25°C.

*Incompatibilities:* React with boranes, alkalies, aliphatic amines, amides, nitric acid, sulfuric acid.
*Routes of Entry:* Skin contact
*Harmful Effects and Symptoms*
*Short Term Exposure:* Skin irritation. May damage liver, kidneys, spleen and central nervous system.
*Points of Attack:* Liver, kidneys, spleen and CNS.
*First Aid:* If this chemical gets into the eyes, remove any contact lenses at once and irrigate immediately for at least 15 minutes, occasionally lifting upper and lower lids. Seek medical attention immediately. If this chemical contacts the skin, remove contaminated clothing and wash immediately with soap and water. Seek medical attention immediately. If this chemical has been inhaled, remove from exposure, begin rescue breathing (using universal precautions) if breathing has stopped, and CPR if heart action has stopped. Transfer promptly to a medical facility. When this chemical has been swallowed, get medical attention. Give large quantities of water and induce vomiting. Do not make an unconscious person vomit.

*References:*
- California Environmental Protection Agency "Chemical List of Lists," Sacramento CA (February 1997)

# 1,2-Dichloropropane

*Use Type:* Fumigant and nematicide
*CAS Number:* 78-87-5
*Formula:* $C_3H_6Cl_2$; $ClCH_2CHClCH_3$
*Alert:* 1,2-Dichloropropane production in the United States has declined over the past 20 years. It was used in the past as a soil fumigant, chemical intermediate, and industrial solvent and was found in paint strippers, varnishes, and furniture finish removers. Most of these uses were discontinued. Today, almost all of the 1,2-dichloropropane

is used as a chemical intermediate to make perchloroethylene and several other related chlorinated chemicals. Human toxicity (long-term): High.

*Synonyms:* Bichlorure de propylene (French); α,β-Dichloropropane; 1,2-Dicloroprpano (Spanish); Dwuchloropropan (Polish); ENT 15,406; NCI-C55141; Propane, 1,2-dichloro-; Propylene chloride; α,β-Propylene dichloride; Propylene dichloride

*Trade Names:* D-D SOIL FUMIGANT®, Shell Chemical (Netherlands), canceled 9/29/1988; VIDDEN D®, Dow Chemical Co. (USA), canceled 7/1/1987

*Producers:* ATOFINA (France); BASF (Germany); Bayer Chemicals (Germany); Cyanamid Agro Products; Dow Chemical Co. (USA); Mitsui Chemicals (Japan); Shell Chemical (Netherlands); Showa Denko (Japan); Sigma-Aldrich Laborchemikalien (Germany)

*Chemical Class:* Halogenated organic compound

*EPA/OPP PC Code:* 029002; 600030 (as impurity only; no longer cleared as inert)

*California DPR Chemical Code:* 2501

*ICSC Number:* 0441

*RTECS Number:* TX9625000

*EEC Number:* 602-020-00-0

*EINECS Number:* 201-152-2

*Uses:* Dichloropropane is used as a soil fumigant, alone and in combination with dichloropropane, for nematodes on stored grain, and as an insecticide for livestock. It is used as a chemical intermediate in perchloroethylene and carbon tetrachloride synthesis and as a lead scavenger for antiknock fluids. It is also used as a solvent for fats, oils, waxes, gums and resins, and in solvent mixtures for cellulose esters and ethers. Other applications include scouring compounds and metal degreasing agents.

*Human toxicity (long-term)[77]:* High–5.00 ppb, MCL (Maximum Contaminant Level)

*Fish toxicity (threshold)[77]:* Very low–25952.79223 ppb, MATC (Maximum Acceptable Toxicant Concentration)

*Carcinogen/Hazard Classifications*

**California Prop. 65:** Carcinogen

**IARC:** Group 3, unclassifiable

*Regulatory Authority:*
- Carcinogen (Animal Suspected) (IARC)[9]
- Air Pollutant Standard Set (ACGIH)[1] (Australia) (DFG)[3] (Israel) (Mexico) (former USSR)[35, 43] (OSHA)[58] (Several
- States)[60] (Several Canadian Provinces)
- List of priority pollutants (U.S. EPA)
- AB 1803-Well Monitoring Chemical (CAL)
- MCL (Maximum Contaminants Levels) list of contaminants (CAL)
- AB 2588-Air Toxics "Hot Spots" Chemicals (CAL)
- CAL Air Resources Board/AB 1807 Toxic Air Contaminants
- Proposition 65 chemical (CAL)
- Permissible Exposure Limits for Chemical Contaminants (CAL/OSHA)
- The "Director's List" (CAL/OSHA)
- Clean Air Act: Hazardous Air Pollutants (Title I, Part A, Section 112)
- Clean Water Act: Section 311 Hazardous Substances/RQ 40CFR117.3 (same as CERCLA, see below); 40CFR423, Appendix A, Priority Pollutants; Section 313 Water Priority Chemicals (57FR41331, 9/9/92)
- EPA Hazardous Waste Number (RCRA No.): U083
- RCRA, 40CFR261, Appendix 8 Hazardous Constituents
- RCRA 40CFR268.48; 61FR15654, UniversalTreatment Standards: Wastewater (mg/L), 0.85; Nonwastewater (mg/kg), 18
- RCRA 40CFR264, Appendix 9; TSD Facilities Ground Water Monitoring List. Suggested test method(s) (PQL ug/L): 8010(0.5); 8240(5)
- Safe Drinking Water Act: MCL, 0.005 mg/L; MCLG, zero; Regulated chemical (47 FR 9352); Priority List (55 FR 1470)
- Superfund/EPCRA 40CFR302.4 RQ: CERCLA, 1000 lb (454 kg)
- EPCRA Section 313 Form R *de minimus* concentration reporting level: 1.0%
- U.S. DOT Regulated Marine Pollutant (49CFR172.101, Appendix B)
- Canada, WHMIS, Ingredients Disclosure List, National Pollution Release Inventory (NPRI)

*Description:* Colorless stable liquid with an odor similar to chloroform. The odor threshold in air is 0.25 ppm. Somewhat soluble in water; solubility =0.26 g/100mL @ 20°C. Boiling point = 96°C. Melting/Freezing point = –100°C. Flash point = 16°C (cc). Autoignition temperature = 557°C. Explosive limits: LEL= 3.4%; UEL = 14.5%. Vapor pressure = 30 kPa @ 20°C; 50 mmHg @ 20°C. Hazard Identification (based on NFPA-704 M Rating System): Health 2, Flammability 3, Reactivity 0. Log $K_{ow}$ = 2.02. Unlikely to bioaccumulate in marine organisms.

*Incompatibilities:* Forms explosive mixture with air. May accumulate static electrical charges, and may cause ignition of its vapors. contact with strong oxidizers, powdered aluminum may cause fire and explosion hazard. Strong acids can cause decomposition and the formation of hydrogen chloride vapors. Reacts with strong bases, *o*-dichlorobenzene, 1,2-dichloroethane. Corrosive to aluminum and its alloys. Attacks some plastics, rubber and coatings.

*Permissible Exposure Limits in Air:* The Federal Limit (OSHA PEL)[58], DFG MAK[3] and the ACGIH[1] TWA recommended value is 75 ppm (350 mg/m³) TWA and the ACGIH[1] STEL is 110 ppm (510 mg/m³). Australia, Israel, Mexico, and the Canadian provinces of Alberta, BC, Ontario, and Quebec use the same values as ACGIH[1]. The NIOSH[2] IDLH level = 400 ppm [Carcinogen]. Brazil[35] has set a workplace limit of 59 ppm (275 mg/m³). Russia[35, 43] has set an MAC in workplace air of 10 mg/m$^3$ and a limit in ambient air in residential areas of 0.18 mg/m$^3$ on a daily average basis. Several states have set guidelines or standards for propylene dichloride in ambient air[60] ranging from 5.1 µg/m$^3$ (Massachusetts) to 13.89 µg/m$^3$ (Kansas) to 3,500-5,100 µg/m$^3$ (North Dakota) to 5,800 µg/m$^3$ (Virginia) to 7,000 µg/m$^3$ (Connecticut) to 8,330 µg/m$^3$ (Nevada).

*Determination in Air:* Charcoal tube (petroleum-based); Acetone/Cyclohexane; Gas chromatography /Electrochemical detection; NIOSH IV Method #1013, 1,2-Dichloropropane[18].

*Permissible Concentration in Water:* To protect freshwater aquatic life: 23,000 µg/L on an acute toxicity basis and 5,700 µg/L on a chronic basis. To protect saltwater aquatic life: 10,300 µg/L on an acute toxicity basis and 3,040 µg/L on a chronic basis. To protect human health: No value set because of insufficient data[6]. EPA[62] has recently proposed a limit in drinking water of 0.005 mg/L (5 µg/L). Several states have set guidelines for propylene dichloride in drinking water[61] ranging from 1 µg/L (Arizona, Massachusetts) to 6 µg/L (Kansas and Minnesota) to 10 µg/L (California and Connecticut). The former USSR[35, 43] has set an MAC in water bodies used for domestic purposes of 0.4 mg/L. The USEPA[47] has derived a no-observed adverse effects level (NOAEL) of 8.8 mg/kg/day which gives a 10-day health advisory for a 10-kg child of 0.09 mg/L.

*Determination in Water:* Inert gas purge followed by gas chromatography with halide specific detection (EPA Method 601) or gas chromatography plus mass spectrometry (EPA Method 624).

*Routes of Entry:* Inhalation of vapor, ingestion, eye and skin contact.

*Harmful Effects and Symptoms*

*Short Term Exposure:* 1,2-Dichloropropane irritates the eyes, skin, and respiratory tract. It may affects on the nervous system. High concentration exposure can result in lightheadedness, dizziness, and unconsciousness. Propylene dichloride may cause dermatitis be defatting the skin. More severe irritation may occur if it is confined against the skin by clothing. Undiluted, it is moderately irritating to the eyes, but does not cause permanent injury. The vapor can irritate the nose, throat, eyes and air passages. Repeated or prolonged skin contact can cause rash.

*Long Term Exposure:* Repeated exposure can cause skin drying and dermatitis. There is limited evidence that this chemical causes cancer in animals. It may cause liver cancer. In animal experiments, acute exposure to propylene dichloride produced central nervous system narcosis, fatty degeneration of the liver and kidneys. High or repeated exposure can damage the liver, kidneys and brain. Early symptoms include headaches, nausea, personality changes. Based on animal tests this chemical may affect reproduction and may cause malformations in the human fetus.

*Points of Attack:* Eyes, skin, respiratory system, liver, kidneys, central nervous system. *Cancer Site:* (in animals) liver & mammary gland tumors.

*Medical Surveillance:* Liver and kidney function tests. Examination of the nervous system. Evaluate the skin condition.

*First Aid:* If this chemical gets into the eyes, remove any contact lenses at once and irrigate immediately for at least 15 minutes, occasionally lifting upper and lower lids. Seek medical attention immediately. If this chemical contacts the skin, remove contaminated clothing and wash immediately with soap and water. Seek medical attention immediately. If this chemical has been inhaled, remove from exposure, begin rescue breathing (using universal precautions) if breathing has stopped, and CPR if heart action has stopped. Transfer promptly to a medical facility. When this chemical has been swallowed, get medical attention. Give large quantities of water and induce vomiting. Do not make an unconscious person vomit.

*References:*

- Agency for Toxic Substances and Disease Registry (ASTDR), "ToxFAQs, 1,2-Dichloropropane," Atlanta, GA. (July 1999). http://www.atsdr.cdc.gov/tfacts134.pdf
- U.S. Environmental Protection Agency, Dichloropropanes/Dichloropopenes: Ambient Water Quality Criteria, Washington, DC (1980).
- National Institute for Occupational Safety and Health, Profiles on Occupational Hazards for Criteria Document Priorities: Dichloropropane, pp 292-294, Report PB-274–73, Cincinnati, OH (1977).
- U.S. Environmental Protection Agency, 1,2-Dichloropopane, Health and Environmental Effects Profile No. 78, Washington, DC, Office of Solid Waste (April 30, 1980).
- U.S. Environmental Protection Agency, Dichloropropanes/Dichloropropenes, Health and Environmental Effects Profile No. 79, Washington, DC, Office of Solid Waste (April, 30, 1980).
- California Environmental Protection Agency "Chemical List of Lists," Sacramento CA (February 1997).
- New Jersey Department of Health, "Hazardous Substance Fact Sheet: 1,2-Dichloropropane," Trenton, NJ (May 1986, rev. July 2002). http://www.state.nj.us/health/eoh/rtkweb/0664.pdf

# 1,3-Dichloropropene

*Use Type:* A soil fumigant and nematicide.
*CAS Number:* 542-75-6; 10061-01-5 (*cis*-); 10061-02-6 (*trans*-)
*Formula:* $C_3H_4Cl_2$; CHCl=CHCH$_2$Cl
*Alert:* 1,3-Dichloropropene should be handled with extreme caution, as a Carcinogen.

*Synonyms:* β-Chloroallyl chloride; 3-Chloroallyl chloride; 3-Chloropropenyl chloride; 1,3-D; 1,3-Dichloro-1-propene; 1,3-Dichloro-2-propene; 1,3-Dicloropropeno (Spanish); α, β-Dichloropropylene; 1,3-Dichloropropylene; 1-Propene, 1,3-dichloro-; Propene, 1,3-dichloro-

*Trade Names:* DURHAM NEMATOCIDE®, AMVAC chemical Corp. (USA), canceled 7/1/1987; FUMAZONE®, Dow chemical Co. (USA), canceled 11/10/1983; NEMAGON®, Shell Chemical Co. (UK), canceled 12/9/1979; NEMEX®; PRO-KILL NEMATOCIDE®, Gowan Co. (USA), canceled 7/31/1979; TELONE®; TELONE II®; VIDDEN D®; VORLEX®

*Producers:* Asahi Glass (Japan); Dow AgroSciences (USA); Solvay Group (Belgium)

*Chemical Class:* Chlorinated hydrocarbon

*EPA/OPP PC Code:* 029001

*California DPR Chemical Code:* 573

*RTECS Number:* UC8310000

*EINECS Number:* 201-153-8

*Uses:* It is also used in combinations with dichloropropanes as a soil fumigant to kill nematodes, insects and fungus on cotton, potatoes, tobacco, sugar beets, vegetables, grain, citrus planting sites, deciduous fruit and nut-tree planting cites, and ornamental trees and floral sites. Top five applications in California are on sweet potatoes, carrots, wine grapes and outdoor propagation nurseries. It is used on a wide variety of crops.

*Carcinogen/Hazard Classifications*

**U.S. EPA Carcinogens:** Group B2, probable carcinogen

**U.S. NTP Carcinogen:** Reasonably anticipated carcinogen

**California Prop. 65:** Carcinogen

**IARC:** Group 2B, possible carcinogen

**Label Signal Word:** DANGER

*Regulatory Authority:*
- Carcinogen (Animal Positive) (IARC)[9] (Australia) (NTP)[10] (Israel) (DFG, *cis*- and *trans*-isomers)[3] (Several Canadian Provinces)
- Air Pollutant Standard Set (ACGIH)[1] (HSE)[33] (former USSR)[35, 43] (Several States)[60]
- List of priority pollutants (U.S. EPA)
- AB 1803-Well Monitoring Chemical (CAL)
- MCL (Maximum Contaminants Levels) list of contaminants (CAL)
- AB 2588-Air Toxics "Hot Spots" Chemicals (CAL)
- CAL Air Resources Board/AB 1807 Toxic Air Contaminants
- Proposition 65 chemical (CAL)
- Permissible Exposure Limits for Chemical Contaminants (CAL/OSHA)
- The "Director's List" (CAL/OSHA)
- Actively registered pesticide in California.
- Clean Air Act: Hazardous Air Pollutants (Title I, Part A, Section 112)
- Clean Water Act: Section 311 Hazardous Substances/RQ 40CFR117.3 (same as CERCLA, see below); 40CFR423, Appendix A, Priority Pollutants; Section 313 Water Priority Chemicals (57FR41331, 9/9/92)
- EPA Hazardous Waste Number (RCRA No.): U084
- RCRA, 40CFR261, Appendix 8 Hazardous Constituents
- Superfund/EPCRA 40CFR302.4 RQ: CERCLA, 100 lb (45.4 kg)
- EPCRA Section 313 Form R de minimus concentration reporting level: 0.1%.
- U.S. DOT Regulated Marine Pollutant (49CFR172.101, Appendix B)
- Canada, WHMIS, Ingredients Disclosure List
- Mexico, drinking water, 0.09 mg/L

*trans-isomer:*
- RCRA 40CFR268.48; 61FR15654, UniversalTreatment Standards: Wastewater (mg/L), 0.036; Nonwastewater (mg/kg), 18
- RCRA 40CFR264, Appendix 9; TSD Facilities Ground Water Monitoring List. Suggested test method(s) (PQL *ug*/L): 8010(5); 8240(5)
- EPCRA Section 313 Form R de minimus concentration reporting level: 0.1%.

*Description:* Colorless to straw-colored liquid with a sharp, sweet, irritating, chloroform-like odor. Boiling point = 106°C; 103–110°C (mixed *cis*- and *trans*-isomers). Melting/Freezing point = –84°C. Flash point = 35°C. Vapor pressure = 29 mmHg @ 20°C. The explosive limits are: LEL = 5.3%; UEL = 14.5%. Hazard Identification (based on NFPA-704 M Rating System): Health 2, Flammability 3, Reactivity 0. Practically insoluble in water.

*Incompatibilities:* Forms explosive mixture with air. Violent reaction with strong oxidizers. May accumulate static electrical charges, and may cause ignition of its vapors. Incompatible with strong acids, oxidizers, aluminum or magnesium compounds, aliphatic amines, alkanolamines, alkaline matrials, halogens, or corrosives. *Note*: Epichlorohydrin may be added as a stabilizer.

*Permissible Exposure Limits in Air:* There is no OSHA[2] PEL. NIOSH[2] and ACGIH[1] recommend a TWA of 1.0 ppm (5 mg/m$^3$). HSE, Australia, Israel, and the Canadian provinces of Alberta, Ontario, and Quebec set the same TWA as ACGIH[1] and HSE[33], and Alberta add an STEL of 10 ppm (50 mg/m$^3$). DFG has no numerical limit, but notes carcinogenic effect in animals[3]. The additional notation "skin" indicates the possibility of cutaneous absorption. The former USSR[35, 43] has set an MAC in workplace air of 5.0 mg/m$^3$ and MAC values for ambient air in residential areas of 0.1 mg/m$^3$ on a momentary basis and 0.01 mg/m$^3$ on a daily average basis. Several states have set guidelines or standards for 1,3-dichloropropene in ambient air[60] ranging from 50 mg/m$^3$ (North Dakota) to 80 µg/m$^3$ (Virginia) to 100 mg/m$^3$ (Connecticut) to 119 mg/m$^3$ (Nevada).

*Determination in Air:* No method listed.

*Permissible Concentration in Water:* To protect freshwater aquatic life: 6,060 µg/L on an acute toxicity basis and 244 µg/L on a chronic basis. To protect saltwater aquatic life:

790 µg/L on an acute toxicity basis. To protect human health: 87.0 µg/L[6]. The former USSR[35, 43] has set an MAC in water bodies used for domestic purposes of 0.4 mg/L. Mexico's drinking water criteria is 0.09 mg/L. A no-observed-adverse-effects level (NOAEL) of 3.0 mg/kg/day has been determined by the U.S. EPA (see Health Advisory reference below). This results in a drinking water level on a lifetime basis of 0.011 mg/L (11.0 µg/L). Several states have set guidelines for 1,3-dichloropopene in drinking water[61] ranging from 10 µg/L (Connecticut) to 87 µg/L (Arizona and Kansas) to 89 µg/L (Vermont).

*Determination in Water:* Inert gas purge followed by gas chromatography with halide specific detection (EPA Method 601) or gas chromatography plus mass spectrometry (EPA Method 624).

*Routes of Entry:* Inhalation, skin absorption, ingestion, skin and/or eye contact.

*Harmful Effects and Symptoms*

*Short Term Exposure:* This chemical can be absorbed through the skin, thereby increasing exposure. Exposure can cause headaches, chest pain, and dizziness. High levels can cause you to pass out. Contact can severely burn the eyes and skin, with permanent damage. High exposures can damage the kidneys, liver and lungs

*Long Term Exposure:* Ther is evidence that 1,3-dichloropropene causes cancer in animals and humans. May damage the kidneys, liver and lungs. May cause choronic headache, and personality changes.

*Points of Attack:* Eyes, skin, respiratory system, central nervous system, liver, kidneys. Cancer Site [in animals: cancer of the bladder, liver, lung & stomach].

*Medical Surveillance:* Before beginning employment and at regular times after that, the following are recommended: Liver function tests. Lung function tests. Kidney function tests.

*First Aid:* If this chemical gets into the eyes, remove any contact lenses at once and irrigate immediately for at least 15 minutes, occasionally lifting upper and lower lids. Seek medical attention immediately. If this chemical contacts the skin, remove contaminated clothing and wash immediately with soap and water. Seek medical attention immediately. If this chemical has been inhaled, remove from exposure, begin rescue breathing (using universal precautions) if breathing has stopped, and CPR if heart action has stopped. Transfer promptly to a medical facility. When this chemical has been swallowed, get medical attention. Give large quantities of water and induce vomiting. Do not make an unconscious person vomit.

*References:*
- Agency for Toxic Substances and Disease Registry, "ToxFAQs for 1,3-Dichloropropene," Atlanta, GA
- (September 1995). http://www.atsdr.cdc.gov/tfacts40.html
- Pesticide Management Education Program, "Chemical Fact Sheet 9/96: 1,3-Dichloropropene (Telon II)," Cornell University, Ithaca, NY (September 1986) http://pmep.cce.cornell.edu/profiles/fumigant/dichloropropene/fumi-prof-dichloropropene.html
- New Jersey Department of Health, "Hazardous Substance Fact Sheet: 1,3-Dichloropropene," Trenton, NJ (April 1986, rev. December 1999). http://www.state.nj.us/health/eoh/rtkweb/0666.pdf
- U.S. Environmental Protection Agency, Dichloropropanes/Dichloropropenes: Ambient Water Quality Criteria, Washington, DC (1980)/U.S. Environmental Protection Agency, 1,3-Dichloropropene, Health and Environmental Effects Profile No. 81, Washington, DC, Office of Solid Waste (April 30, 1980).
- U.S. Environmental Protection Agency, Dichloropropanes/Dichloropropenes: Health and Environmental Effects Profile No. 79, Washington, DC, Office of Solid Waste (April 30, 1980).
- U.S. Environmental Protection Agency, "Health Advisory: 1,3-Dichloropropene," Washington, DC, Office of Drinking Water (August 1987).
- California Environmental Protection Agency "Chemical List of Lists," Sacramento CA (February 1997).
- Sax, N.I., Ed., "Dangerous Properties of Industrial Materials Report," 6, No. 5, 88-93 (1986).

# Dichlorprop

*Use Type:* Herbicide and plant growth regulator
*CAS Number:* 120-36-5
*Formula:* $C_9H_8C_{12}O_3$
*Synonyms:* Acide-2-(2,4-dichloro-phenoxy)propionique (French); Acido-2-(2,4-dicloro-fenossi)propionico (Italian); Acido 2-(2,4-diclorofenoxi)propionico (Spanish); BH 2,4-DP; Caswell No. 320; 2(2,4-Dichloor-fenoxy)propionzuur (Dutch); 2,4-Dichlorophenoxy-α-propionic acid; α-(2,4-Dichlorophenoxy)propionic acid; (±)-2-(2,4-Dichlorophenoxy)propionic acid; 2,4-Dichlorophenoxy propionic acid; 2-(2,4-Dichlorophenoxy) propionic acid; Dichloroprop; 2-(2,4-Dichlor-phenoxy)-propionsaeure (German); 2,4-DP (U.S. EPA); 2-(2,4-DP); NSC 39624; Propanoic acid, 2-(2,4-dichlorophenoxy)-; Propionic acid, 2-(2,4-dichlorophenoxy)-

*Trade Names:* BH 2,4-DP®; CAMPBELL'S® REDIPON; CELATOX-DP®; CORNOX RD®; DESORMONE®; EMBUTOX®; GRAMINON-PLUS®; HEDONAL®; HERBIZID DP®; HORMATOX®; KILDIP®; POLYCLENE®; POLYMONE®; POLYTOX®; RD 406®; SERITOX 50®; U46®; DP-FLUID®; VISKO-RHAP®; WEEDONE 170®; WEEDONE DP®

*Producers:* A H Marks & Company (UK); Bayer CropScience (Germany); Dow AgroSciences (USA)
*Chemical Class:* Chlorophenoxy acid or ester
*EPA/OPP PC Code:* 031401
*California DPR Chemical Code:* 2503
*ICSC Number:* 0038

**RTECS Number:** UF1050000
**EEC Number:** 607-045-00-0
**EINECS Number:** 204-390-5
*Uses:* A phenoxy herbicide. Dichlorprop is a translocated post-emergence herbicide effective in the control of some Polygonum persicaria and other weeds in small grain cereals. It may be used either alone or in combination with other herbicides such as MCPA, benazolin, and dicamba. The latter mixture is an extremely effective broad spectrum herbicide controlling such weeds as Anthemis spp., Matricaria spp., Stellaria media, Galium aparine, as well as Polygonum spp. Dichlorprop is also used for brush control in nonagricultural land, for weed control in aquatic environments, and to prevent premature fall of apples.
*Human toxicity (long-term)[77]:* Intermediate–35.00 ppb, Health Advisory
*Fish toxicity (threshold)[77]:* Intermediate–58.73095 ppb, MATC (Maximum Acceptable Toxicant Concentration)
*Carcinogen/Hazard Classifications*
**IARC:** Group 2B, possible carcinogen
**TRI Developmental Toxin:** Developmental toxin
**WHO Acute Hazard:** Class III, Slightly Hazardou
*Regulatory Authority:*
- AB 2588-Air Toxics "Hot Spots" Chemicals (CAL) as chlorophenoxy pesticides
- The "Director's List" (CAL/OSHA) as chlorophenoxy pesticides
- EPCRA Section 313 Form R *de minimus* concentration reporting level: 0.1%
- U.S. DOT Regulated Marine Pollutant (49CFR, Subchapter 172.101, Appendix B) as phenoxy herbicide

*Description:* White, crystalline powder. Faint phenolic odor. May be applied as a liquid containing a flammable carrier. Slightly soluble in water; solubility = $2.3 \times 10^3$ ppm @ 20°C. Molecular weight = 235.07. Boiling point = 215°C. Melting/Freezing point = 114 (technical grade) –118°C. Vapor pressure = $2.9 \times 10^{-6}$ mmHg @ 20°C. Log $K_{ow}$ = 5.89. Values at or above 3.0 are likely to bioaccumulate in marine organisms.
*Incompatibilities:* Keep away from oxidizers, sulfuric acid, caustics, ammonia, aliphatic amines, alkanolamines, isocyanates, alkylene oxides, epichlorohydrin. The aqueous solution is a weak acid. Attacks many metals in presence of moisture.
*Permissible Exposure Limits in Air:* No occupational exposure limits have been established
*Permissible Concentration in Water:* No criteria set. Runoff from spills or fire control may cause water pollution.
*Routes of Entry:* Inhalation, ingestion, absorbed through the skin.
*Harmful Effects and Symptoms*
*Short Term Exposure:* Poisonous; may be fatal if inhaled, swallowed, or absorbed through skin. Severely irritates eyes, skin and respiratory tract, with burning sensation, pain, redness and swelling. Metabolic stimulant. If inhaled, causes coughing, dilated pupils, headache, profuse perspiration, intense thirst, extreme fatigue, rapid pulse, high fever, clammy, flushed skin, rapid breathing, nausea, vomiting, cyanosis (bluish tint to skin and lips), anxiety and confusion, convulsions, risk of lung edema. If swallowed, face and lips turn bluish. Liver injury with associated jaundice, kidney failure, and cardiac arrhythmias are commonly noted. Nerve damage, which may be delayed, may include swelling of legs and feet, muscle twitch and stupor. Severe exposure can cause death from heart failure. Dust or liquid left in contact with the skin for several hours may be absorbed. This may result in severe delayed symptoms as listed above. These symptoms may last for months or years.
*Long Term Exposure:* Workers exposed to chlorophenoxy compounds such as 2,4-D (in the manufacturing process) over a five to ten year period at levels above 10 mg/m³ complained of weakness, rapid fatigue, headache and vertigo. Liver damage, low blood pressure and slowed heartbeat were also found. Based on animal tests, may affects human reproduction
*Points of Attack:* Eyes, skin, respiratory system, central nervous system, cardiovascular system, liver, kidney
*Medical Surveillance:* If symptoms develop or overexposure is suspected, the following may be useful: Liver and kidney function tests. Exam of the nervous system.
*First Aid:* If this chemical gets into the eyes, remove any contact lenses at once and irrigate immediately for at least 15 minutes, occasionally lifting upper and lower lids. Seek medical attention immediately. If this chemical contacts the skin, remove contaminated clothing and wash immediately with soap and water. Seek medical attention immediately. If this chemical has been inhaled, remove from exposure, begin rescue breathing (using universal precautions) if breathing has stopped, and CPR if heart action has stopped. Transfer promptly to a medical facility. When this chemical has been swallowed, get medical attention. *Do not induce vomiting when formulations containing petroleum solvents are ingested.* Otherwise, give large quantities of water and induce vomiting. Do not make an unconscious person vomit.
*References:*
- Pesticide Management Education Program, "Pesticide Information Profile, Dichlorprop Herbicide Profile 9/88," Cornell University, Ithaca, NY (September 1988). http://pmep.cce.cornell.edu/profiles/herb-growthreg/24-d-butylate/24-dp/herb-prof-24dp.html
- New Jersey Department of Health and Senior Services, "Hazardous Substance Fact Sheet, 2-(2,4-Dichlorophenoxy) Propionic Acid," Trenton NJ (February 1999). http://www.state.nj.us/health/eoh/rtkweb/3076.pdf
- California Environmental Protection Agency "Chemical List of Lists," Sacramento CA (February 1997).

# Diclofop-methyl

*Use Type:* Herbicide
*CAS Number:* 51338-27-3
*Formula:* $C_{16}H_{14}Cl_2O_4$
*Alert:* Some uses are classified Restricted Use Pesticide (RUP). Human toxicity (long-term): High.
*Synonyms:* Caswell No. 319A; Diclofop methyl ester; Dichlordiphenprop; dichlorfop-methyl; 2-[4-(2,4-Dichlorophenoxy)phenoxy]-methyl-propionate; 2-[4-(2,4-Dichlorophenoxy)phenoxy]propanoic acid methyl ester; Methyl 2-[2-(2,4-dichlorophenoxy)phenoxy]propanoate; Methyl ester of 2-[4-(2,4-dichlorophenoxy)phenoxy] propanoic acid; Propanoic acid, 2-[4-(2,4-dichlorophenoxy) phenoxy]-, methyl ester
*Trade Names:* DICHLORDIPHENPROP®; HOELON®, Bayer CropScience (Germany); HOELON® 3EC, Bayer CropScience (Germany); HOE-GRASS®, Bayer CropScience (Germany); HOEGRASS®, Bayer CropScience (Germany); HOE® 23408, Bayer CropScience (Germany); ILOXAN®, Bayer CropScience (Germany); ILLOXAN®, Bayer CropScience (Germany); ONE SHOT®, Clariant (Switzerland), canceled
*Producers:* Bayer CropScience (Germany); China Shenghua Group Agrochemical (China)
*Chemical Class:* Chlorophenoxy
*EPA/OPP PC Code:* 110902
*California DPR Chemical Code:* 2034
*RTECS Number:* UF1180000
*EINECS Number:* 257-141-8
*Uses:* Some uses are classified Restricted Use Pesticide (RUP). Diclofop-methyl is a selective post-emergence herbicide used to control wild oats and annual grassy weeds in grain and vegetable crops: alfalfa, carrots, celery, box, field and french beans, barley, wheat, brassicas, parsnips, peas, potatoes, rapeseed (canola), soy beans, oilseed rape, onions, sugar beets and lettuce.
*Human toxicity (long-term)[77]:* High–1.40 ppb, Health Advisory
*Fish toxicity (threshold)[77]:* Intermediate–10.60669 ppb, MATC (Maximum Acceptable Toxicant Concentration)
*U.S. Maximum Allowable Residue Levels for Diclofop-methyl (40 CFR 180.385):*

| CROP | ppm |
|---|---|
| Barley, grain | 0.1 |
| Barley, straw | 0.1 |
| Lentil | 0.1 |
| Pea, dry | 0.1 |
| Wheat, grain | 0.1 |
| Wheat, straw | 0.1 |

*Carcinogen/Hazard Classifications*
**U.S. EPA Carcinogens:** Likely carcinogen
**Label Signal Word:** WARNING or DANGER
**TRI Developmental Toxin:** Developmental toxin
**California Prop. 65:** Listed; Developmental toxin
*Regulatory Authority:*
- Actively registered pesticide in California.
- EPA 40 CFR 372.65, Specific Toxic Chemical Listings
- AB 2588-Air Toxics "Hot Spots" Chemicals (CAL) as chlorophenoxy pesticides
- The "Director's List" (CAL/OSHA) as chlorophenoxy pesticides
- Proposition 65 chemical (CAL)
- EPCRA Section 313 Form R *de minimis* concentration reporting level: 1.0%

*Description:* Slightly soluble in water; solubility = 0.03 ppm. Molecular weight = 341.20. Melting/Freezing point = 40°C. $3.5 \times 10^{-6}$ mmHg @ 20°C.
*Incompatibilities:* Keep away from oxidizers, sulfuric acid, caustics, ammonia, aliphatic amines, alkanolamines, isocyanates, alkylene oxides, epichlorohydrin.
*Permissible Concentration in Water:* No criteria set. Runoff from spills or fire control may cause water pollution.
*Routes of Entry:* Inhalation, ingestion, absorbed through the skin.
*Harmful Effects and Symptoms*
*Short Term Exposure:* Poisonous; may be fatal if inhaled, swallowed, or absorbed through skin. Severely irritates eyes, skin and respiratory tract, with burning sensation, pain, redness and swelling. Metabolic stimulant. If inhaled, causes coughing, dilated pupils, headache, profuse persperation, intense thirst, extreme fatigue, rapid pulse, high fever, clammy, flushed skin, rapid breathing, nausea, vomiting, cyanosis (bluish tint to skin and lips), anxiety and confusion, convulsions, risk of lung edema. If swallowed, face and lips turn bluish. Liver injury with associated jaundice, kidney failure, and cardiac arrhythmias are commonly noted. Nerve damage, which may be delayed, may include swelling of legs and feet, muscle twitch and stupor. Severe exposure can cause death from heart failure. Dust or liquid left in contact with the skin for several hours may be absorbed. This may result in severe delayed symptoms as listed above. These symptoms may last for months or years.
*Long Term Exposure:* Workers exposed to chlorophenoxy compounds such as 2,4-D (in the manufacturing process) over a five to ten year period at levels above 10 mg/m³ complained of weakness, rapid fatigue, headache and vertigo. Liver damage, low blood pressure and slowed heartbeat were also found. Based on animal tests, may affects human reproduction
*Points of Attack:* Eyes, skin, respiratory system, central nervous system, cardiovascular system, liver, kidney
*Medical Surveillance:* If symptoms develop or overexposure is suspected, the following may be useful: Liver and kidney function tests. Exam of the nervous system.
*First Aid:* If this chemical gets into the eyes, remove any contact lenses at once and irrigate immediately for at least 15 minutes, occasionally lifting upper and lower lids. Seek

medical attention immediately. If this chemical contacts the skin, remove contaminated clothing and wash immediately with soap and water. Seek medical attention immediately. If this chemical has been inhaled, remove from exposure, begin rescue breathing (using universal precautions) if breathing has stopped, and CPR if heart action has stopped. Transfer promptly to a medical facility. When this chemical has been swallowed, get medical attention. *Do not induce vomiting when formulations containing petroleum solvents are ingested.* Otherwise, give large quantities of water and induce vomiting. Do not make an unconscious person vomit.

*References:*
- EXTOXNET, Extension Toxicology Network, "Pesticide Information Profile, Diclofop-methyl," Oregon State University, Corvallis, OR (September 1995). http://extoxnet.orst.edu/pips/diclofop.htm
- California Environmental Protection Agency "Chemical List of Lists," Sacramento CA (February 1997)
- U.S. Environmental Protection Agency, Office of Pesticide Programs, Pesticide Residue Limits, "Diclifop-methyl", 40 CFR 180.385. www.epa.gov/pesticides/food/viewtols.htm

# Dichlorvos

*Use Type:* Insecticide, acaricide and nematicide.
*CAS Number:* 62-73-7
*Formula:* $C_4H_7Cl_2O_4P$
*Alert:* A Restricted Use Pesticide (RUP). Dichlorvos also exists as a breakdown product. Dichlorvos should be handled as a carcinogen, with extreme caution. Human toxicity (long-term): Very high.
*Synonyms:* Chlorvinphos; Cyanophos; DDVF; DDVP (Insecticide); Dichlofos; (2,2-Dichloor-vinyl)-dimethyl-fosfaat (Dutch); Dichoorvo (Dutch); Dichlorfos (Polish); Dichlorman; 2,2-Dichloroethenol dimethyl phosphate; 2,2-Dichloroethenyl dimethyl phosphate; 2,2-Dichlorovinyl dimethyl phosphate; Dichlorovos; Dimethyl 2,2-dichloroethenyl phosphate; *O,O*-Dimethyl 2,2-dichlorovinyl phosphate; Dimethyl 2,2-dichlorovinyl phosphate; (2,2-Dichlorvinyl)-dimethyl-phosphat (German); *O*-(2,2-Dichlorvinyl)-*O,O*-dimethylphosphat (German); (2,2-Dicloro-vinil)dimetilfosfato (Italian); Dimethyl dichlorovinyl phosphate; ENT 20,738; Ethenol, 2,2-dichloro-, dimethyl phosphate; NCI-C00113; NSC-6738; Phosphate de dimethyle et de 2,2-dichlorovinyle (French); Phosphoric acid, 2,2-dichloroethenyl dimethyl ester; Phosphoric acid, 2,2-dichlorovinyl dimethyl ester
*Trade Names:* AGWAY®, Agway Inc. (USA), canceled; ALCO®, Amvac Chemical Corp. (USA); APAVAP®; ASTROBOT®; ATGARD®; BAY 19149®, Bayer CropScience (Germany); BAYER 19149®, Bayer CropScience (Germany); BENFOS®; BIBESOL®; BREVINYL®; BREVINYL E 50®; CANOGARD®; CEKUSAN®; CYPONA®; DEDEVAP®; DERRIBAN®; DERRIBANTE®; DES®; DEVIKOL®; DICLORCAL 50®, Calliope (France); DIDIVANE; DIVIPAN®; DOOM®, United Phosphorus (India); DQUIGARD®; DUO-KILL®; DURAVOS®; ELASTREL®; EQUIGARD®, Shell Chemical (UK), canceled; EQUIGEL®, Shell Chemical (UK), canceled; ESTROSEL®; ESTROSOL®; FECAMA®; FEKAMA®; FLY-DIE®; FLY FIGHTER®; HERKOL®; INSECTIGAS D®; KRECALVIN®; LINDAN®; MAFU®; MARVEX®; MOPARI®; NEFRAFOS®; NERKOL®; NOGOS®; NO-PEST®; NOVOTOX®; NUVA®; NUVAN®; OKO®; OMS 14®; PANAPLATE®; PHOSVIT®; PRENTOX®, Prentiss Inc. (USA); SD 1750®; SUCHLOR®, Sudarshan Chemical Industries (India); SZKLARNIAK®; TAP 9VP®; TASK®; TENAC®; TETRAVOS®; UNIFOS (PESTICIDE)®; UNITOX®, Platte Chemical (USA); VAPONA®, Shell Chemical (UK) et al, canceled; VAPONITE®; VERDICAN®; VERDIPOR®; VERDISOL®; VINYLOFOS®; VINYLOPHOS®; WINYLOPHOS®

*Producers:* Agrimor International (USA); Agsin (Singapore); Amvac Chemical Corp. (USA); Bayer CropScience (Germany); BEC Group (India); Bhageria Dye-Chem (India); Bharat Pulverizing Mills (India); Bharat Rasayan (India); Biesterfeld Siemsgluess International. GmbH (Germany); Calliope (France); China Chemical (China); Dainippon Ink and Chemicals (Japan); Drexel Chemical (USA); Hebei Huafeng Chemical Group (China); Hindustan Insecticides (India); Hockley International (UK); Hokko Chemical Industry (Japan); Hunan Tianyu Pesticide Chemical Group (China); I.N.D.I.A. Industrie Chimiche (Italy); Ki-Hara Chemicals Ltd. (UK); Nihon Kagaku Sangyo (Japan); Nippon Chemical Industrial Co. (Japan); Nippon Soda (Japan); Nissan Chemical Industries (Japan);, Prentiss Inc. (USA); Saeryung Chemicals (South Korea); Sankyo Organic Chemicals (Japan); Shenzhen Guomeng Industry Co., Ltd. (China); Sigma-Aldrich Laborchemikalien (Germany); Sudarshan Chemical Industries (India); United Phosphorus (India); YiHua Group (China)
*Chemical Class:* Organophosphate; Halo-organics
*EPA/OPP PC Code:* 084001 and 600020
*California DPR Chemical Code:* 187
*ICSC Number:* 0690
*RTECS Number:* TC0350000
*EEC Number:* 015-019-00-X
*EINECS Number:* 200-547-7
*Uses:* Dichlorvos is used for insect control in food storage areas, green houses, and barns, and control of insects on livestock. It is not generally used on outdoor crops. Dichlorvos is sometimes used for insect control in workplaces and in the home. Dichlorvos used in pest control is diluted with other chemicals and used as a spray. It can also be incorporated into plastic that slowly releases the chemical. Veterinarians use it to control parasites on pets. Dichlorvos is effective against mushroom flies, aphids,

spider mites, caterpillars, thrips, and white flies in greenhouse, outdoor fruit, and vegetable crops. It is used to treat a variety of parasitic worm infections in dogs, livestock, and humans. Dichlorvos can be fed to livestock to control botfly larvae in the manure. It acts against insects as both a contact and a stomach poison. It is used as a fumigant and has been used to make pet collars and pest strips. It is available as an aerosol and soluble concentrate.

*Human toxicity (long-term)*[77]: Extra high–0.35 ppb, Health Advisory

*Fish toxicity (threshold)*[77]: High–7.24703 ppb, MATC (Maximum Acceptable Toxicant Concentration)

*Carcinogen/Hazard Classifications*

**U.S. EPA Carcinogens:** Group C, possible carcinogen

**California Prop. 65:** Listed

**IARC:** Group 2B, possible carcinogen

**Label Signal Word:** DANGER-POISON

**WHO Acute Hazard:** Class 1B, highly hazardous

*Regulatory Authority:*

- Air Pollutant Standard Set (ACGIH)[1] (Australia) (DFG)[3] (HSE)[33] (Israel) (Mexico) (Argentina) (OSHA) (former USSR)[35] (Several Canadian Provinces)
- AB 1803-Well Monitoring Chemical (CAL)
- EPA/SARA 302 (EPCRA) Extremely hazardous substances
- AB 2588-Air Toxics "Hot Spots" Chemicals (CAL)
- CAL Air Resources Board/AB 1807 Toxic Air Contaminants
- Proposition 65 chemical (CAL)
- Permissible Exposure Limits for Chemical Contaminants (CAL/OSHA)
- The "Director's List" (CAL/OSHA)
- Actively registered pesticide in California.
- Clean Air Act: Hazardous Air Pollutants (Title I, Part A, Section 112)
- Clean Water Act: Section 311 Hazardous Substances/RQ 40CFR117.3 (same as CERCLA, see below); Section
- 313 Water Priority Chemicals (57FR41331, 9/9/92)
- Superfund/EPCRA 40CFR355, Appendix B Extremely Hazardous Substances: TPQ = 1000 lb (454 kg)
- Superfund/EPCRA 40CFR302.4 RQ: CERCLA, 10 lb (4.54 kg)
- EPCRA Section 313 Form R de minimus concentration reporting level: 1.0%
- U.S. DOT Regulated Marine Pollutant (49CFR172.101, Appendix B)

*Description:* Colorless to amber liquid. Mild aromatic odor. Moderate solubility in water; solubility = $1.0 \times 10^4$ ppm. Molecular weight = 220.98. Boiling point = 140°C @ 20 mmHg. Vapor pressure = 0.01 mmHg @ 20°C. Flash point = > 79°C. Hazard Identification (based on NFPA-704 M Rating System): Health 3, Flammability 1, Reactivity 0. Log $K_{ow}$ = <1.5. Unlikely to bioaccumulate in marine organisms. Physical and toxicological properties may be affected by the organic carrier solvents used in commercial formulations.

*Incompatibilities:* Strong acids, strong alkalis. Corrosive to iron, mild steel, and some forms of plastics, rubber, and coatings.

*Permissible Exposure Limits in Air:* The Federal limit (OSHA PEL), Australian, Israeli, DFG MAK[3], HSE, Canadian provincial (Alberta, BC, Ontario, Quebec), and ACGIH[1] value is 0.1 ppm (1 mg/m$^3$) TWA. and HSE[33], Alberta, BC add STEL value of 0.3 mg/m$^3$. The Mexican TWA is 0.16 ppm. The NIOSH[2] IDLH level = 100 mg/m$^3$. The notation "skin" is added indicating the possibility of cutaneous absorption. Argentina[35] uses the same TWA limits set forth above. The former USSR[35] has set an MAC in workplace air of 0.2 mg/m$^3$ and an MAC for ambient air in residential areas of 0.07 mg/m$^3$ on a once-a-day basis and 0.002 mg/m$^3$ on a daily average basis.

*Determination in Air:* XAD-2® (tube); Toluene; Gas chromatography/Flame photometric detection for sulfur, nitrogen, or phosphorus; NIOSH II(5), P&CAM Method #295[18].

*Permissible Concentration in Water:* The former USSR[35] has set an MAC in water bodies used for domestic purposes of 1.0 mg/L and in water for fishing of zero.

*Routes of Entry:* Inhalation, skin absorption, ingestion, eye and/or skin contact.

*Harmful Effects and Symptoms*

*Short Term Exposure:* Dichlorvos irritates the eyes and skin. Symptoms include miosis, aching eyes; rhinorrhea (discharge of thin nasal mucous); headache; chest tightness, wheezing, laryngeal spasm, salivation; cyanosis; anorexia, nausea, vomiting, diarrhea; sweating; muscle fasiculation, paralysis, giddiness, ataxia; convulsions; low blood pressure, cardiac irregular/irregularities. The substance may cause effects on the central nervous system. Cholinesterase inhibitor. High levels of exposure may result in death. Delayed pulmonary edema may occur after inhalation. Symptoms of exposure include sweating, twitching, contracted pupils, respiratory distress (tightness in the chest and wheezing), salivation (drooling), lacrimation (tearing), nausea, vomiting, abdominal cramps, diarrhea, involuntary defecation and urination, slurred speech, coma, apnea (cessation of breathing), and death. Dichlorvos is a very toxic compound with a probable lethal oral dose in humans between 50 and 500 mg/kg, or between 1 teaspoon and 1 oz. for a 70 kg (150 lb) person. However, brief exposure (30-60 minutes) to vapor concentrations as high as 6.9 mg/Liter did not result in clinical signs or depressed serum cholinesterase levels. Toxic changes are typical of organophosphate insecticide poisoning with progression to respiratory distress, respiratory paralysis, and death if there is no clinical intervention.

*Long Term Exposure:* Repeated or prolonged contact with skin may cause skin sensitization and dermatitis. Cholinesterase inhibitor; cumulative effect is possible: see short tem exposure. A study in rats and mice reported that rats had an increase in cancer of the pancreas and in leukemia, and female mice had an increase in stomach cancer after they were fed dichlorvos for 2 years. Animal studies have also shown effects on the nervous system when animals drank water or ate food containing dichlorvos. It is not known whether dichlorvos can affect reproduction or cause birth defects in people. Animal studies have not reported effects on reproduction or birth defects when animals were exposed to dichlorvos.

*Points of Attack:* Eyes, skin, respiratory system, cardiovascular system, central nervous system, blood acetylcholinesterase

*Medical Surveillance:* Medical observation is recommended for 24 to 48 hours after breathing overexposure, as pulmonary edema may be delayed. Before employment and at regular times after that, the following are recommended: Plasma and red blood cell acetylcholinesterase levels (tests for the enzyme poisoned by this chemical). If exposure stops, plasma levels return to normal in 1-2 weeks while red blood cell levels may be reduced for 1-3 months.

When acetylcholinesterase enzyme levels are reduced by 25% or more below preemployment levels, risk of poisoning is increased, even if results are in lower ranges of "normal." Reassignment to work not involving organophosphate or carbamate pesticides is recommended until enzyme levels recover. If symptoms develop or overexposure occurs, repeat the above tests as soon as possible and get an exam of the nervous system. Also consider complete blood count. Consider chest x-ray following acute overexposure. Do not drink any alcoholic beverages before or during use. Alcohol promotes absorption of organophosphates.

*First Aid:* **Treatment for organophosphate poisoning consists of thorough decontamination, cardiorespiratory support, and administration of the antidotes atropine and pralidoxime. In cases of severe poisoning, diazepam, an anticonvulsant, should also be administered. Antidotes should be administered as prevention even if the diagnosis is in doubt.** Speed in removing material from eyes and skin is of extreme importance. *Eyes:* Eye contact can cause dangerous amounts of these chemicals to be quickly absorbed through the mucous membrane into the bloodstream. Immediately and gently flush eyes with plenty of warm or cold water (NO hot water) for at least 15 minutes, occasionally lifting the upper and lower eyelids. Get medical aid immediately. *Skin:* Get medical aid. Skin contact can cause dangerous amounts of these chemicals to be absorbed into the bloodstream. Wearing the appropriate PPE equipment and respirator for organophosphate/carbamate pesticides, immediately flush skin with plenty of soap and water for at least 15 minutes while removing contaminated clothing and shoes. Shampoo hair promptly if contaminated. The removed, contaminated clothing and shoes should be double-bagged and left in Hot Zone for later disposal by hazardous materials experts. Skin may also be decontaminated with diluted hypochlorite solution. *Inhalation:* Get medical aid. Do not contaminate yourself. Wearing the appropriate PPE equipment and respirator for organophosphate pesticides, immediately remove the victim from the contaminated area to fresh air. If the victim is not breathing, administer artificial respiration. Do NOT use mouth-to-mouth resuscitation; use bag/mask apparatus. If breathing is difficult, administer oxygen through bag/mask apparatus until medical help arrives. Do not leave victim unattended. *Ingestion:* Call poison control. Loosen all clothing. Never give anything by mouth to an unconscious person. If victim is *unconscious or having convulsions,* do nothing except keep victim warm. Get medical aid. Transfer promptly to a medical facility. In cases of ingestion, **do not induce vomiting**. If the victim is alert and asymptomatic, administer a slurry of activated charcoal at a dose of 1 g/kg (infant, child, and adult dose). A soda can and straw may be of assistance when offering charcoal to a child. *In some cases you may be specifically instructed by poison control to induce vomiting by way of 2 tablespoons of syrup of ipecac (adult) washed down with a cup of water.* Do NOT give activated charcoal before or with ipecac syrup. *Note to physician or authorized medical personnel:* Treat cases of respiratory compromise, coma, or excessive pulmonary secretions with respiratory support using protocols and techniques available and within the scope of training. Some cases may necessitate procedures such as endotracheal intubation or cricothyrotomy by properly trained and equipped personnel. When possible, atropine (see *Antidotes,* below) should be given under medical supervision. Patients who are comatose, hypotensive, or having seizures or cardiac arrhythmias should be treated according to advanced life support protocols. *Antidotes:* Two antidotes are administered to treat organophosphate poisoning. Atropine is a competitive antagonist of acetylcholine at muscarinic receptors and is used to control the excessive bronchial secretions which are often responsible for death. Pralidoxime relieves both the nicotinic and muscarine effects of organophosphate poisoning by regenerating acetylcholinesterase and can reduce both the bronchial secretions and the muscle weakness associated with poisoning. The initial intravenous dose of atropine in adults should be determined by the severity of symptoms: An initial adult dose of 1.0 to 2.0 mg or pediatric dose of 0.01 mg/kg (minimum 0.01 mg) should be administered intravenously. If intravenous access cannot be established, atropine may also be given intramuscularly, subcutaneously or via endotracheal tube. Doses should be repeated every 15 minutes until excessive secretions and sweating have been controlled. Once bronchial secretion has been controlled, atropine administration should be repeated whenever the secretions begin to recur. In seriously poisoned patients, very large doses may be required. Alterations of pulse rate and pupillary size should not be used as indicators of treatment adequacy. Pralidoxime should be administered as early in poisoning as possible as its efficacy may diminish when given more than 24 to 36 hours after exposure. Doses are as follows: adult 1.0 g; pediatric 25 to 50 mg/kg. The drug should be administered intravenously over 30 to 60 minutes, but in a life-threatening situation, one-half of the total dose can be given per minute for a total administration time of 2 minutes. Treatment should begin to take effect within 40 minutes with a reduction in symptoms and the amount of atropine necessary to control bronchial secretion. The initial dose can be repeated in 1 hour and then every 8 to 12 hours until the patient is clinically well and no longer requires atropine. If intravenous access cannot be established, pralidoxime may also be given intramuscularly. Early administration of diazepam in addition to the combined atropine and pralidoxime treatment may help prevent the onset of seizures and potential brain and cardiac morphologic damage following high-level organophosphate poisoning.

*References:*
- EXTOXNET, Extension Toxicology Network, "Pesticide Information Profile, Dichlorvos," Oregon State University, Corvallis, OR. (June 1996). http://ace.ace.orst.edu/info/extoxnet/pips/dichlorv.htm
- U.S. Environmental Protection Agency, Investigation of Selected Potential Environmental Contaminants: Haloalkyl Phosphates, Report EPA-560/2076-007, Washington, DC (August 1976).
- Sax, N.I., Ed., "Dangerous Properties of Industrial Materials Report," 1, No. 3, 57-59 (1981).
- U.S. Environmental Protection Agency, "Chemical Profile: Dichlorvos," Washington, DC, Chemical Emergency Preparedness Program (November 30, 1987).
- California Environmental Protection Agency "Chemical List of Lists," Sacramento CA (February 1997).
- Agency for Toxic Substances and Disease Registry, U.S. Department of Health and Human Services, Public Health Service, "Managing Hazardous Materials Incidents," Atlanta, GA (June 2003).
- New Jersey Department of Health, "Hazardous Substance Fact Sheet: Dichlorvos," Trenton, NJ (October 1996). http://www.state.nj.us/health/eoh/rtkweb/0674.pdf

# Dicofol

*Use Type:* Insecticide
*CAS Number:* 115-32-2
*Formula:* $C_{14}H_9Cl_5$
*Alert:* A General Use Pesticide (GUP). Human toxicity (long-term): Very high.
*Synonyms:* Benzenemethanol, 4-chloro-α-(-4-chlorophenyl)-α.-(trichloromethyl)-; Benzhydrol, 4,4'-dichloro-.alpha.-(trichloromethyl)-; 1,1-Bis(p-chlorophenyl)-2,2,2-trichloro ethanol; 1,1-Bis(4-chlorophenyl)-2,2,2-trichloroethanol; 4-Chloro-α-(4-chlorophenyl)-α-(trichloromethyl)benzene methanol; CPCA; 4,4'-Dichloro-α-(trichloromethyl) benzhydrol; Dichlorokelthane; Di(p-chlorophenyl) trichloromethyl carbinol; DTMC; ENT 23,648; Ethanol, 2,2,2-trichloro-1,1-bis(4-chlorophenyl)-; NCI-C00486; 2,2,2-Trichloro-1,1-bis(4-chlorophenyl) ethanol; 2,2,2-Trichloro-1,1-bis(p-chlorophenyl)ethanoL; 2,2,2-Trichloro-1,1-di(4-chlorophenyl)ethanol

*Trade Names:* ACARIN®, Makhteshim-Agan Industries (Israel), canceled 1/30/1984; CALLIFOL®, Calliope (France); CARBAX®; CEKUDIFOL®; DECOFOL®; DICOMITE®; DIFOL®; FUMITE DICOFOL®; FW 293®; HIFOL®; KELTANE®; KELTHANE®; P,P'-KELTHANE®; KELTHANETHANOL®; MILBOL®; MITIGAN®; TIKTOK®, United Phosphorus (India); VAPCOTHION®, dicofol, Veterinary & Agricultural Products Manufacturing Co., Ltd. (VAPCO) (Jordan)

*Producers:* Agan Chemical Manufacturers Ltd. (Israel); Agrimor International (USA); Alcotan Laboratories (Spain); BASF Agricultural Products Group (Germany); Calliope (France); Dow AgroSciences (USA); Drexel Chemical (USA); DuPont (USA); Eli Lilly (USA); Gowan Company (USA); Hindustan Insecticides (India); Hockley International (UK); Jingma Chemicals Ltd. (China); Ki-Hara Chemicals Ltd. (UK); Makhteshim-Agan Industries (Israel); Nagarjuna Agrichem (India); Nissan Chemical Industries (Japan); Rhone-Poulenc Agro (France); Shenzhen Guomeng Industry Co., Ltd. (China); Sigma-Aldrich Laborchemikalien (Germany); United Phosphorus (India); Veterinary & Agricultural Products Manufacturing Co., Ltd. (VAPCO) (Jordan)

*Chemical Class:* Organochlorine; Halo-organics
*EPA/OPP PC Code:* 010501
*California DPR Chemical Code:* 346
*ICSC Number:* 0752
*RTECS Number:* DC8400000
*EEC Number:* 603-044-00-4
*EINECS Number:* 204-082-0

*Uses:* Dicofol is an organochlorine miticide/pesticide used for foliar applications, mostly on cotton, apples, and citrus crops. Other crops include: strawberries, mint, beans, peppers, tomatoes, pecans, walnuts, stonefruit, cucurbits, and non-residential lawns/ornamentals. Formulations registered for use on food/feed crops include emulsifiable concentrates, and wettable powder formulations. These formulations may be applied as concentrated or dilute sprays using aircraft, duster, groundboom, and sprayer. Dicofol is manufactured from DDT. In 1986, use of dicofol was temporarily canceled by the U.S. EPA because of concerns raised by high levels of DDT contamination. However, it was reinstated when it was shown that modern manufacturing processes can produce technical grade dicofol which contains less than 0.1% DDT.

*Human toxicity (long-term)*[77]: Extra high–0.84 ppb, Health Advisory
*Fish toxicity (threshold)*[77]: High–5.89576 ppb, MATC (Maximum Acceptable Toxicant Concentration)

*U.S. Maximum Allowable Residue Levels for Dicofol (40 CFR 180.163):*[3]

| CROP | ppm |
|---|---|
| Apple | 5 |
| Apricot | 10 |
| Bean, lima, succulent | 5 |
| Bean, snap, succulent | 5 |
| Bean, dry | 5 |
| Blackberry | 5 |
| Boysenberry | 5 |
| Butternut | 5 |
| Cantaloupe | 5 |
| Cherry | 5 |
| Chestnut | 5 |
| Cotton, undelinted seed | 0.1 |
| Crabapple | 5 |
| Cucumber | 5 |
| Dewberry | 5 |
| Eggplant | 5 |
| Fig | 5 |
| Filbert | 5 |
| Grape | 5 |
| Grapefruit | 10 |
| Hop | 30 |
| Kumquat | 10 |
| Lemon | 10 |
| Lime | 10 |
| Loganberry | 5 |
| Melon | 5 |
| Muskmelon | 5 |
| Nectarine | 10 |
| Nut, hickory | 5 |
| Nut, macadamia | 5 |
| Orange | 10 |
| Peach | 10 |
| Pear | 5 |
| Pecan | 5 |
| Pepper, pimento | 5 |
| Pepper | 5 |
| Peppermint, tops | 25 |
| Plum, prune, fresh | 5 |
| Pumpkin | 5 |
| Quince | 5 |
| Raspberry | 5 |
| Spearmint, tops | 25 |
| Squash, summer | 5 |
| Squash, winter | 5 |
| Strawberry | 5 |
| Tangerine | 10 |
| Tea, dried | 45 |
| Tomato | 5 |
| Walnut | 5 |
| Watermelon | 5 |

*Carcinogen/Hazard Classifications*
**U.S. EPA Carcinogens:** Group C, possible carcinogen
**IARC:** Group 3, unclassifiable
**Label Signal Word:** WARNING or CAUTION, depending on the formulation.
**WHO Acute Hazard:** Class III, slightly hazardous
**Endocrine Disruptor:** Suspected endocrine disruptor
*Regulatory Authority:*
- Carcinogen (animal positive) NTP
- AB 1803-Well Monitoring Chemical (CAL)
- AB 2588-Air Toxics "Hot Spots" Chemicals (CAL)
- The "Director's List" (CAL/OSHA)
- Actively registered pesticide in California.
- Clean Water Act: Section 311 Hazardous Substances/RQ 40CFR117.3 (same as CERCLA, see below); Section
- 313 Water Priority Chemicals (57FR41331, 9/9/92)
- Superfund/EPCRA 40CFR302.4 RQ: CERCLA, 10 lb (4.54 kg)
- EPCRA Section 313 Form R *de minimus* concentration reporting level: 1.0%

*Description:* Dicofol is a white or brown waxy solid. Practically insoluble in water; solubility = 1.5 ppm @ 20°C. Molecular weight = 370.47. Boiling point = 180°C @ 0.13 mmHg. Melting/Freezing point = 79°C. Vapor pressure = 4 x $10^{-7}$ mmHg. Flash point = 120°C. Hazard Identification (based on NFPA-704 M Rating System): Health 2, Flammability 1, Reactivity 0. Log $K_{ow}$ = 3.55. Values at or above 3.0 are likely to bioaccumulate in marine organisms. Physical and toxicological properties may be affected by the carrier solvents used in commercial formulations.

*Incompatibilities:* Incompatible with alkaline pesticides, strong acids, acid fumes, aliphatic amines, isocyanates, and steel.

*Permissible Exposure Limits in Air:* No occupational exposure limits have been established.

*Routes of Entry:* Inhalation, passing through the skin, ingestion.

*Harmful Effects and Symptoms*

*Short Term Exposure:* Dicofol can be absorbed through the skin, thereby increasing exposure. It irritates the skin and the respiratory tract. Exposure can cause headache, nausea, vomiting and poor appetite. Dicofol may affect the central nervous system causing numbness and weakness in the hands and feet., muscle twitching, seizures, unconsciousness and death.

*Long Term Exposure:* May affect the liver and kidneys. May cause personality changes with depression, anxiety, and irritability. May decrease fertility in females. Prolonged or repeated skin contact may cause dermatitis. There is limited evidence that dicofol causes liver cancer in animals.

*Points of Attack:* Skin, nervous system. liver, kidneys.

*Medical Surveillance:* Liver and kidney function tests. Examination of the nervous system. Dermatological examination.

*First Aid:* If this chemical gets into the eyes, remove any contact lenses at once and irrigate immediately for at least 15 minutes, occasionally lifting upper and lower lids. Seek

medical attention immediately. If this chemical contacts the skin, remove contaminated clothing and wash immediately with soap and water. Seek medical attention immediately. If this chemical has been inhaled, remove from exposure, begin rescue breathing (using universal precautions) if breathing has stopped, and CPR if heart action has stopped. Transfer promptly to a medical facility. When this chemical has been swallowed, get medical attention. Give large quantities of water and induce vomiting. Do not make an unconscious person vomit.

*References:*
- EPA, Office of Pesticide Programs, Reregistration Eligibility Decision (RED), "Dicofol," Washington, DC (November 1998).
  http://www.epa.gov/oppsrrd1/REDs/factsheets/0021fact.pdf
- EXTOXNET, Extension Toxicology Network, "Pesticide Information Profile, Dicofol," Oregon State University, Corvallis, OR (June 1996).
  http://ace.orst.edu/info/extoxnet/pips/dicofol.htm
- EPA, Office of Pesticide Programs, Pesticide Residue Limits, "Dicofol", 40 CFR 180.163, http://www.epa.gov/pesticides/food/viewtols.htm
- New Jersey Department of Health and Senior Services, "Hazardous Substance Fact Sheet, Dicofol," Trenton NJ (October 1998).
  http://www.state.nj.us/health/eoh/rtkweb/0675.pdf
- California Environmental Protection Agency "Chemical List of Lists," Sacramento CA (February 1997).

# Dicrotophos

*Use Type:* Insecticide and acaricide.
*CAS Number:* 141-66-2
*Formula:* $C_8H_{16}NO_5P$
*Alert:* A Restricted Use Pesticide (RUP). One of the major degradates of dicrotophos is monocrotophos. All uses of monocrotophos have been voluntarily cancelled in the United States due to its extreme toxicity to humans and wildlife. Human toxicity (long-term): Very high.
*Synonyms:* Crotonamide, 3-hydroxy-*N-N*,-dimethyl-, *cis*-, Dimethyl phosphate; Crotonamide, 3-hydroxy-*N-N*-dimethyl-, dimethylphosphate, *cis*-; Crotonamide, 3-hydroxy-*N-N*-dimethyl-, dimethylphosphate, *(E)*-; Dicrotofos (Dutch); Dicroptophos; 3-(Dimethoxyphosphinyloxy)-*N, N*-dimethyl-*(E)*-crotonamide; 3-(Dimethoxyphosphinyloxy)-*N, N*-dimethyl-*cis*-crotonamide; 3-(Dimethoxyphosphinyloxy)-*N, N*-dimethylisocrotonamide; 3-(Dimethylamino)-1-methyl-3-oxo-1-propenyl dimethyl phosphate; *cis*-2-Dimethylcarbamoyl-1-methylvinyl dimethylphosphate; *(E)*-2-Dimethylcarbamoyl-1-methylvinyl dimethylphosphate; *O,O*-Dimethyl-*O*-(2-dimethyl-carbamoyl-1-methyl-vinyl)phosphat (German); *O,O*-Dimethylo-(*N,N*-dimethylcarbamoyl-1-methylvinyl) phosphate; *O,O*-Dimethyl-*O*-(1,4-dimethyl-3-oxo-4-aza-pent-1-enyl)fosfaat (Dutch); *O,O*-Dimethyl-*O*-(1,4-dimethyl-3-oxo-4-aza-pent-1-enyl)phosphate; Dimethyl phosphate of 3-hydroxy-*N,N*-dimethyl-*cis*-crotonamide; Dimethyl phosphate ester with 3-hydroxy-*N,N*-dimethyl-*cis*-crotonamide; *O,O*-Dimetil-*O*-(1,4-dimetil-3-oxo-4-aza-pent-1-enil)-fosfato (Italian); ENT 24,482; 3-Hydroxydimethyl crotonamide dimethyl phosphate; 3-Hydroxy-*N,N*-dimethyl-*cis*-crotonamide dimethyl phosphate; 3-Hydroxy-*N,N*-dimethyl-*(E)*-crotonamide dimethyl phosphate; Phosphatede dimethyle et de 2-dimethylcarbamoyl 1-methyl vinyle (French); Phosphoric acid, 3-(dimethylamino)-1-methyl-3-oxo-1-propenyl dimethyl ester, *(E)*-; Phosphoric acid, dimethyl ester with *cis*-3-hydroxy-*N,N*-dimethylcrotonamide; Phosphoric acid, dimethyl ester with *(E)*-3-hydroxy-*N,N*-dimethylcrotonamide

*Trade Names:* BIDIRL®; BIDRIN®, Amvac Chemical Corp. (USA); DuPont (USA) et al, active and canceled; BIDRIN-R®, Gowan (USA), canceled; BIDRIN (SHELL)®, Shell Chemical (UK), canceled 8/15/1986; C-709®, Ciba Specialty Chemicals (Switzerland); C-709 (CIBA-GEIGY)®, Ciba Specialty Chemicals (Switzerland); CARBICRIN®; CARBICRON®; CARBOMICRON®; CIBA 709®, Ciba Specialty Chemicals (Switzerland); DIAPADRIN®; DICRON®; DIDRIN®; EKTAFOS®; EKTOFOS®; KARBICRON®; SD 3562.®, Shell Chemical (UK); SHELL SD-3562®, Shell Chemical (UK)
*Producers:* Amvac Chemical Corp. (USA); Ciba Specialty Chemicals (Switzerland); Shell Chemical (UK); Sigma-Aldrich Laborchemikalien (Germany)
*Chemical Class:* Organophosphate
*EPA/OPP PC Code:* 035201
*California DPR Chemical Code:* 72
*ICSC Number:* 0872
*RTECS Number:* TC3850000
*EEC Number:* 015-073-00-4
*Uses:* Dicrotophos was introduced in 1956 as a contact systemic pesticide with a wide range of applications. Today, dicrotophos is currently used mainly as an insecticide for apples and other fruit crops, and for cotton pests, mostly in the Mississippi Valley. It is acutely toxic to birds, especially those that follow their migratory corridors and feed in the farmlands that have been treated with this pesticide. Internationally, dicrotophos is used on rice, coffee and citrus.
*Human toxicity (long-term)*[77]*:* Extra high–0.70 ppb, Health Advisory
*Fish toxicity (threshold)*[77]*:* Very low–953.39880 ppb, MATC (Maximum Acceptable Toxicant Concentration)
*U.S. Maximum Allowable Residue Levels for Dicrotophos (40 CFR 180.299):*

| CROP | ppm |
|---|---|
| Cotton, undelinted seed | .05 |

*Carcinogen/Hazard Classifications*

**U.S. EPA Carcinogens:** A suggestive carcinogen
**Label Signal Word:** DANGER
**WHO Acute Hazard:** Class 1B, highly hazardous
*Regulatory Authority:*
- Banned or Severely Restricted (DDR and Malaysia) (UN)[13]
- Air Pollutant Standard Set (ACGIH)[1] (Australia) (Israel) (Mexico) (Several States)[60] (Several Canadian Provinces)
- AB 1803-Well Monitoring Chemical (CAL)
- EPA/SARA 302 (EPCRA) Extremely hazardous substances
- Permissible Exposure Limits for Chemical Contaminants (CAL/OSHA)
- The "Director's List" (CAL/OSHA)
- Superfund/EPCRA 40CFR355, Appendix B Extremely Hazardous Substances: TPQ = 100 lb (45.4 kg)
- Superfund/EPCRA 40CFR302.4 RQ: EHS, 1 lb (0.454 kg)
- U.S. DOT Inhalation Hazard Chemicals as organophosphate
- U.S. DOT Regulated Marine Pollutant (49CFR172.101, Appendix B)

*Description:* Amber liquid. Mild ester odor. Soluble in water. Boiling point = –400°C. Decomposes below boiling point @ 75°C after storage for 31 days. Flash point = 93°C (cc). Vapor pressure = $2 \times 10^{-4}$ mmHg; 1.29 mPa @ 20°C. Hazard Identification (based on NFPA-704 M Rating System): Health 3, Flammability 1, Reactivity 0. Log $K_{ow}$ = –0.5. Unlikely to bioaccumulate in marine organisms.

*Incompatibilities:* Attacks some metals: Corrosive to cast iron, mild steel, brass, and stainless steel304. Decomposes after prolonged storage, but is stable when stored in glass or polyethylene containers with temperatures to 40°C. Forms highly toxic fumes of phosphorous and nitrogen oxides when heated to decomposition.

*Permissible Exposure Limits in Air:* There is no OSHA[2] PEL. The ACGIH[1] recommended airborne exposure limit 0.25 mg/m³ TWA. NIOSH[2] recommends the same exposure limit. Australia, Israel, Mexico, and the Canadian provinces of Alberta, BC, Ontario, and Quebec set the same limit as ACGIH[1] and Alberta's STEL is 0.75 mg/m³. The TWA bears the notation "skin" indicating the possibility of cutaneous absorption. Several states have set guidelines or standards for dicrotophos in ambient air[60] ranging from 2.5 µ/m³ (North Dakota) to 4.0 µg/m³ (Virginia) to 5.0 µg/m³ (Connecticut) to 6.0 µg/m³ (Nevada).

*Determination in Air:* OSHA versatile sampler-2; Toluene/Acetone; Gas chromatography/Flame photometric detection for sulfur, nitrogen, or phosphorus; NIOSH IV, Method #5600, Organophosphorus pesticides[18].

*Permissible Concentration in Water:* No criteria set, but runoff from spills or fire control may cause water pollution.

*Routes of Entry:* Inhalation, skin absorption, ingestion, skin and/or eye contact.

*Harmful Effects and Symptoms*
*Short Term Exposure:* Dicrotophos may affects the nervous system, causing convulsions, respiratory failure. Dicrotophos is a cholinesterase inhibitor which can penetrate the skin. effects may be cumulative. It is extremely toxic. Probable human oral lethal dose is 5 to 50 mg/kg, 7 drops to one teaspoonful for a 70 kg (150 lb) person. Closely related in toxicity to azodrin. Acute exposure to dicrotophos may produce the following signs and symptoms; pinpoint pupils, blurred vision, headache, dizziness, muscle spasms, and profound weakness. Vomiting, diarrhea, abdominal pain, seizures, and coma may also occur. The heart rate may decrease following oral exposure or increase following dermal exposure. Hypotension (low blood pressure) is not uncommon. Respiratory symptoms include dyspnea (shortness of breath), respiratory depression, and respiratory paralysis. Psychosis may occur. The effects may be delayed. Delayed pulmonary edema may occur after inhalation.

*Long Term Exposure:* Dicrotophos is a cholinesterase inhibitor; cumulative effect is possible. May damage the nervous system causing numbness, "pins and needles," sensation and/or weakness of the hands and feet. Repeated exposure may cause personality changes of depression, anxiety or irritability.

*Points of Attack:* Respiratory system, lungs, central nervous system, cardiovascular system, skin, eyes, plasma and red blood cell cholinesterase.

*Medical Surveillance:* Medical observation is recommended for 24 to 48 hours after breathing overexposure, as pulmonary edema may be delayed. Before employment and at regular times after that, the following are recommended: Plasma and red blood cell cholinesterase levels (tests for the enzyme poisoned by this chemical). If exposure stops, plasma levels return to normal in 1-2 weeks while red blood cell levels may be reduced for 1-3 months. When acetylcholinesterase enzyme levels are reduced by 25% or more below preemployment levels, risk of poisoning is increased, even if results are in lower ranges of "normal." Reassignment to work not involving organophosphate or carbamate pesticides is recommended until enzyme levels recover. If symptoms develop or overexposure occurs, repeat the above tests as soon as possible and get an exam of the nervous system. Also consider complete blood count. Consider chest x-ray following acute overexposure. Do not drink any alcoholic beverages before or during use. Alcohol promotes absorption of organophosphates.

*First Aid:* **Treatment for organophosphate poisoning consists of thorough decontamination, cardiorespiratory support, and administration of the antidotes atropine and pralidoxime. In cases of severe poisoning, diazepam, an anticonvulsant, should also be administered. Antidotes should be administered as prevention even if the diagnosis is in doubt.** *Eyes:* Speed in removing material from eyes and skin is of extreme importance. Eye contact can cause dangerous amounts of these chemicals to be quickly absorbed through the mucous membrane into the bloodstream. Immediately and gently flush eyes with plenty of warm or cold water (NO hot water) for at least 15 minutes, occasionally lifting the upper and lower eyelids. Get medical aid immediately. *Skin:* Get medical aid. Skin contact can cause dangerous amounts of these chemicals to be absorbed into the bloodstream. Wearing the appropriate PPE equipment and respirator for organophosphate/carbamate pesticides, immediately flush skin with plenty of soap and water for at least 15 minutes while removing contaminated clothing and shoes. Shampoo hair promptly if contaminated. The removed, contaminated clothing and shoes should be double-bagged and left in Hot Zone for later disposal by hazardous materials experts. Skin may also be decontaminated with diluted hypochlorite solution. *Inhalation:* Get medical aid. Do not contaminate yourself. Wearing the appropriate PPE equipment and respirator for organophosphate pesticides, immediately remove the victim from the contaminated area to fresh air. If the victim is not breathing, administer artificial respiration. Do NOT use mouth-to-mouth resuscitation; use bag/mask apparatus. If breathing is difficult, administer oxygen through bag/mask apparatus until medical help arrives. Do not leave victim unattended. *Ingestion:* Call poison control. Loosen all clothing. Never give anything by mouth to an unconscious person. If victim is *unconscious or having convulsions,* do nothing except keep victim warm. Get medical aid. Transfer promptly to a medical facility. In cases of ingestion, **do not induce vomiting**. If the victim is alert and asymptomatic, administer a slurry of activated charcoal at a dose of 1 g/kg (infant, child, and adult dose). A soda can and straw may be of assistance when offering charcoal to a child. *In some cases you may be specifically instructed by poison control to induce vomiting by way of 2 tablespoons of syrup of ipecac (adult) washed down with a cup of water.* Do NOT give activated charcoal before or with ipecac syrup.

*Note to physician or authorized medical personnel:* Treat cases of respiratory compromise, coma, or excessive pulmonary secretions with respiratory support using protocols and techniques available and within the scope of training. Some cases may necessitate procedures such as endotracheal intubation or cricothyrotomy by properly trained and equipped personnel. When possible, atropine (see *Antidotes*, below) should be given under medical supervision. Patients who are comatose, hypotensive, or having seizures or cardiac arrhythmias should be treated according to advanced life support protocols. *Antidotes:* Two antidotes are administered to treat organophosphate poisoning. Atropine is a competitive antagonist of acetylcholine at muscarinic receptors and is used to control the excessive bronchial secretions which are often responsible for death. Pralidoxime relieves both the nicotinic and muscarine effects of organophosphate poisoning by regenerating acetylcholinesterase and can reduce both the bronchial secretions and the muscle weakness associated with poisoning. The initial intravenous dose of atropine in adults should be determined by the severity of symptoms: An initial adult dose of 1.0 to 2.0 mg or pediatric dose of 0.01 mg/kg (minimum 0.01 mg) should be administered intravenously. If intravenous access cannot be established, atropine may also be given intramuscularly, subcutaneously or via endotracheal tube. Doses should be repeated every 15 minutes until excessive secretions and sweating have been controlled. Once bronchial secretion has been controlled, atropine administration should be repeated whenever the secretions begin to recur. In seriously poisoned patients, very large doses may be required. Alterations of pulse rate and pupillary size should not be used as indicators of treatment adequacy. Pralidoxime should be administered as early in poisoning as possible as its efficacy may diminish when given more than 24 to 36 hours after exposure. Doses are as follows: adult 1.0 g; pediatric 25 to 50 mg/kg. The drug should be administered intravenously over 30 to 60 minutes, but in a life-threatening situation, one-half of the total dose can be given per minute for a total administration time of 2 minutes. Treatment should begin to take effect within 40 minutes with a reduction in symptoms and the amount of atropine necessary to control bronchial secretion. The initial dose can be repeated in 1 hour and then every 8 to 12 hours until the patient is clinically well and no longer requires atropine. If intravenous access cannot be established, pralidoxime may also be given intramuscularly. Early administration of diazepam in addition to the combined atropine and pralidoxime treatment may help prevent the onset of seizures and potential brain and cardiac morphologic damage following high-level organophosphate poisoning.

***References:***
- EXTOXNET, Extension Toxicology Network, "Pesticide Information Profile, Dicrotophos," Oregon State University, Corvallis, OR. (September 1995). http://ace.ace.orst.edu/info/extoxnet/pips/dicrotop.htm
- EPA, Office of Pesticide Programs, Pesticide Residue Limits, "Dicrotophos," 40 CFR 180.299, http://www.epa.gov/pesticides/food/viewtols.htm
- New Jersey Department of Health and Senior Services, "Hazardous Substance Fact Sheet, Dicrotophos," Trenton NJ (March 1989, revised October 1998). http://www.state.nj.us/health/eoh/rtkweb/0676.pdf
- Sax, N.I., Ed., "Dangerous Properties of Industrial Materials Report," 2, No. 5, 49-54 (1982).
- California Environmental Protection Agency "Chemical List of Lists," Sacramento CA (February 1997).
- Agency for Toxic Substances and Disease Registry, U.S. Department of Health and Human Services, Public Health Service, "Managing Hazardous Materials Incidents," Atlanta, GA (June 2003).

- U.S. Environmental Protection Agency, "Chemical Profile: Dicrotophos," Washington, DC, Chemical Emergency Preparedness Program (October 1998).

# Dieldrin

*Use Type:* Insecticide and also a breakdown product
*CAS Number:* 60-57-1
*Formula:* $C_{12}H_8Cl_6O$
*Alert:* Manufacture in the United States prohibited since 1974. In 1987, EPA banned all uses. Human toxicity (long-term): Very high.
*Synonyms:* Dieldrina (Spanish); Dieldrine (French); 2,7:3,6-Dimethanonaphtha[2,3B]oxirene,3,4,5,6,9,9-hexachloro-1a,2,2a,3,6,6a,7,7a-octahydro-(1a α,2.β,2A.α,3β,6.β,6Aα,7β,7Aα); 1,2,3,4,10,10-Hexachloro-6,7-epoxy-1,4,4a,5,6,7,8,8a-octahydro-1,4-*endo,exo*-5,8-dimethanonaphthaleNE; 3,4,5,6,9,9-Hexachloro-1a,2,2a,3,6,6a,7,7a-octahydro-2,7:3,6-dimethano; ENT 16,225; HEOD; Hexachloroepoxyoctahydro-*endo,exo*-dimethanonaphthalene; 3,4,5,6,9,9-Hexachloro-1a, 2, 2a, 3, 6, 6a, 7, 7a-octahydro-2,7:3,6-dimethanonaphth(2,3-b)oxirene; NCI-C00124; (1R,4S,4AS,5R,6R,7S,8S,8AR) 1,2,3,4,10,10-Hexachloro-1,4,4a,5,6,7,8,8a-octahydro-6,7-epoxy-1,4:5,8-dimethanonaphthalene
*Trade Names:* ALVIT®; BELCO®, Bell Laboratories (USA), canceled; COMPOUND 497®; D-31®; DIELDREX®; DIELDRITE®; ILLOXOL®; KILLGERM DETHLAC INSECTICIDAL LAQUER®; OCTALOX®; OXRALOX®; PANORAM®; PANORAM D-31®; PRENTOX®, Prentiss (USA), canceled; QUINTOX®; ROYAL BRAND®, Amvac Chemical Corp. (USA), canceled; SD 3417®
*Producers:* Aldrich Chemical (USA); Amvac Chemical Corp. (USA); Bell Laboratories (USA); Ehrenstorfer, Dr. (Germany); Prentiss Inc. (USA); Shell Chemical (UK); Sigma-Aldrich Laborchemikalien (Germany)
*Chemical Class:* Organochlorine; Halo-organics
*EPA/OPP PC Code:* 045001
*California DPR Chemical Code:* 210
*ICSC Number:* 0787
*RTECS Number:* IO1750000
*EEC Number:* 602-049-00-9
*Uses:* Aldrin and dieldrin are manmade compounds belonging to the group of cyclodiene insecticides. They are a subgroup of the chlorinated cyclic hydrocarbon insecticides which include DDT, BHC, etc. From the 1950s until 1970, aldrin and dieldrin were widely used pesticides for crops like corn and cotton, and also on citrus. Because of concerns about damage to the environment and potentially to human health, EPA banned all uses of aldrin and dieldrin in 1974, except to control termites. In 1987, EPA banned all uses. Dieldrin's persistence in the environment is due to its extremely low volatility (i.e., a vapor pressure of $1.78 \times 10^{-7}$ mm mercury @ 20°C), and low solubility in water (186 $\mu$g/L at 25° to 29°C). In addition, dieldrin is extremely apolar, resulting in a high affinity for fat which accounts for its retention in animal fats, plant waxes, and other such organic matter in the environment. The fat solubility of dieldrin results in the progressive accumulation in the food chain which may result in a concentration in an organism which would exceed the lethal limit for a consumer.
*Human toxicity (long-term)[77]:* Extra high–0.02187 ppb, CHCL (Chronic Human Carcinogen Level)
*Fish toxicity (threshold)[77]:* Extra high–0.07710 ppb, MATC (Maximum Acceptable Toxicant Concentration)
*Carcinogen/Hazard Classifications*
**U.S. EPA Carcinogens:** Group 3B, probable carcinogen
**U.S. NTP Carcinogen:** Not listed
**California Prop. 65:** Carcinogen
**IARC:** Group 3, unclassifiable
**Label Signal Word:** WARNING
**Endocrine Disruptor:** Suspected
*Regulatory Authority:*
- Carcinogen (Animal Positive) (IARC) (NCI)[9]
- Banned or Severely Restricted (Many Countries) (UN)[13]
- Air Pollutant Standard Set (ACGIH)[1] (Australia) (DFG)[3] (HSE)[33] (Israel) (Mexico) (former USSR)[35, 43] (OSHA)[58] (Several States)[60]
- List of priority pollutants (U.S. EPA)
- AB 1803-Well Monitoring Chemical (CAL)
- AB 2588-Air Toxics "Hot Spots" Chemicals (CAL)
- CAL-DHS/DHS Drinking Water Action Levels
- Proposition 65 chemical (CAL)
- Permissible Exposure Limits for Chemical Contaminants (CAL/OSHA)
- The "Director's List" (CAL/OSHA)
- Clean Water Act: Section 311 Hazardous Substances/RQ 40CFR117.3 (same as CERCLA, see below); 40CFR423, Appendix A, Priority Pollutants; Section 313 Water Priority Chemicals (57FR41331, 9/9/92); Toxic Pollutant (Section 401.15).
- EPA Hazardous Waste Number (RCRA No.): P037
- RCRA, 40CFR261, Appendix 8 Hazardous Constituents
- RCRA 40CFR268.48; 61FR15654, UniversalTreatment Standards: Wastewater (mg/L), 0.017; Nonwastewater (mg/kg), 0.13
- RCRA 40CFR264, Appendix 9; TSD Facilities Ground Water Monitoring List. Suggested test method(s) (PQL ug/L): 8080(0.05); 8270(10)
- Superfund/EPCRA 40CFR302.4 RQ: CERCLA, 1 lb (0.454 kg). Dieldrin has been found in at least 287 of the 1,613 National Priorities List sites identified by the Environmental Protection Agency (EPA).
- U.S. DOT Regulated Marine Pollutant (49CFR172.101, Appendix B)
- Mexico, Drinking Water Criteria, $0.7 \times 10^{-7}$ mg/L

# Dieldrin

***Description:*** Dieldrin is a colorless to light tan solid with a mild chemical odor. The odor threshold in water is 0.04 mg/L. Practically insoluble in water; solubility = 0.1 ppm. Boiling point = (decomposes). Molecular weight = 380. 91. Melting/Freezing point = 175–176°C. Vapor pressure = $3 \times 10^{-6}$ mmHg @ 20°C. Hazard Identification (based on NFPA-704 M Rating System): Health 3, Flammability 0, Reactivity 0. Log $K_{ow}$ = 6.24. Values at or above 3.0 are likely to bioaccumulate in marine organisms. Physical and toxicological properties may be affected by the carrier solvents used in commercial formulations.

***Incompatibilities:*** Incompatible with strong acids: concentrated mineral acids, acid catalysts, phenols, strong oxidizers, phenols, active metals, like sodium, potassium, magnesium, and zinc. Keep away from copper, iron, and their salts.

***Permissible Exposure Limits in Air:*** The Federal Limit ( OSHA[2] PEL), as well as the recommended ACGIH[1] TWA value is 0.25 mg/m$^3$. The DFG MAK[3], HSE, Australian, Israeli, Mexican and Canadian provincial (Alberta, BC, Ontario, and Quebec) TWA is the same as OSHA and the STEL value set by Mexico, Alberta, British columbia, and HSE[33] is 0.75 mg/m$^3$. The ceiling value in Germany[35] is 2.5 mg/m$^3$. The notation "skin" indicates the possibility of cutaneous absorption. The NIOSH[2] IDLH level = [Ca] 50 mg/m$^3$. The former USSR[35, 43] has set an MAC in workplace air of 0.01 mg/m$^3$. Several states have set guidelines or standards for dieldrin in ambient air[60] ranging from 0.035 $\mu$g/m$^3$ (Pennsylvania) to 0.595 $\mu$g/m$^3$ (Kansas) to 2.5 $\mu$g/m$^3$ (North Dakota) to 4.0 $\mu$g/m$^3$ (Virginia) to 5.0 $\mu$g/m$^3$ (Connecticut) to 6.0 $\mu$g/m$^3$ (Nevada).

***Determination in Air:*** Filter; Isooctane; Gas chromatography/Electrochemical detection; NIOSH II(3), Method #S283[18].

***Permissible Concentration in Water:*** The U.S. EPA limits the amount of dieldrin that may be present in drinking water to 0.001 and 0.002 milligrams per liter (mg/L) of water, respectively, for protection against health effects other than cancer. The U.S. EPA has determined that a concentration of dieldrin of 0.0002 mg/L in drinking water limits the lifetime risk of developing cancer from exposure to each compound to 1 in 10,000.[6]. Mexico[35] has set MAC values for dieldrin of 0.0000007 mg/L in water used for drinking water supply; of 0.003 mg/L in estuaries and 0.03 $\mu$g/L in estuaries. WHO[35] has set a limit of 0.03 $\mu$g/L in drinking water. A NOAEL (no observed adverse effects level) of 0.005 mg/kg/day has been calculated by EPA which results in the calculation of a drinking water equivalent of 1.75 $\mu$g/L. No lifetime health advisory could be calculated in view of the cancer risk. Several states have set standards or guidelines for dieldrin in drinking water[61] ranging from 0.01 $\mu$g/L (Minnesota) to 0.019 $\mu$g/L (Kansas) to 0.05 $\mu$g/L (California) to 1.0 $\mu$g/L (Illinois).

***Determination in Water:*** Methylene chloride extraction followed by gas chromatography with electron capture or halogen specific detection (EPA Method 608) or gas chromatography plus mass spectrometry (EPA Method 625).

***Routes of Entry:*** Inhalation, skin absorption, ingestion, eye and/or skin contact.

***Harmful Effects and Symptoms***

During the past decade, considerable information has been generated concerning the toxicity and potential carcinogenicity of the two organochlorine pesticides, aldrin and dieldrin. These two pesticides are usually considered together since aldrin is readily epoxidized to dieldrin in the environment. Both are acutely toxic to most forms of life including arthropods, mollusks, invertebrates, amphibians, reptiles, fish, birds and mammals. Dieldrin is extremely persistent in the environment. By means of bioaccumulation it is concentrated many times as it moves up the food chain.

***Short Term Exposure:*** *Inhalation:* May cause nausea, drowsiness, loss of appetite, visual disturbances and insomnia. Sprays of 1 to 21/2% have caused giddiness, headache, muscle twitching, convulsions and loss of consciousness. *Skin*: Can be absorbed to cause or increase the severity of symptoms as listed under ingestion. Contact may cause skin rash. *Eyes:* May cause irritation, redness, and affect vision. *Ingestion*: Can cause headache, nausea, irritability, insomnia, high blood pressure, vision problems, loss of coordination, profuse sweating, dizziness, frothing at the mouth, convulsions and loss of consciousness. Death may occur from as little as 1/20 ounce (1.4 gram). Some symptoms may be delayed up to 12 hours. Exposure to dieldrin may affects the central nervous system, resulting in convulsions.

***Long Term Exposure:*** May cause liver damage. Dieldrin accumulates in the human body. Dieldrin has caused cancer in laboratory animals. It is considered a suspect occupational carcinogen. may damage the developing fetus. May reduce fertility in males and females. Dieldrin concentrates in breast milk, and therefore, may be transferred to breast feeding infants. Repeated higher exposure can cause tremors. muscle twitching and seizures (convulsions) and may lead to coma and death. Convulsions are somewhat delayed and may occur weeks or months following exposure. Repeated exposure may cause personality changes of depression, anxiety or irritability.

***Points of Attack:*** Central nervous system, liver, kidneys, skin. Cancer Site in animals: lung, liver, thyroid and adrenal gland tumors.

***Medical Surveillance:*** Before employment and at regular times after that, the following are recommended: Plasma and red blood cell cholinesterase levels (tests for the enzyme poisoned by this chemical). If exposure stops, plasma levels return to normal in 1-2 weeks while red blood cell levels may be reduced for 1-3 months. When acetylcholinesterase enzyme levels are reduced by 25% or more below preemployment levels, risk of poisoning is increased, even if results are in lower ranges of "normal." Reassignment to work not involving organophosphate or carbamate pesticides is recommended until enzyme levels recover. If symptoms develop or overexposure occurs, repeat the above tests as soon as possible and get an exam of the nervous system. Also consider complete blood count. Consider chest x-ray following acute overexposure. Do not drink any alcoholic beverages before or during use. Alcohol promotes absorption of organophosphates. Blood dieldrin level. Examination of the nervous system. If symptoms develop or overexposure is suspected, the following may be useful: Liver function tests. EEG. Blood dieldrin levels (Normal = less than 1 mg/100 ml; level should not exceed 15 mg/100 ml. Examination of the nervous system.

*First Aid:* If this chemical gets into the eyes, remove any contact lenses at once and irrigate immediately for at least 15 minutes, occasionally lifting upper and lower lids. Seek medical attention immediately. If this chemical contacts the skin, remove contaminated clothing and wash immediately with soap and water. Seek medical attention immediately. If this chemical has been inhaled, remove from exposure, begin rescue breathing (using universal precautions) if breathing has stopped, and CPR if heart action has stopped. Transfer promptly to a medical facility. When this chemical has been swallowed, get medical attention. Give large quantities of water and induce vomiting. Do not make an unconscious person vomit. Medical observation is recommended for 12 hours after overexposure.

*References:*
- U.S. Public Health Service, "ToxFAQs for Aldrin/Dieldrin," Atlanta, Georgia, Agency for Toxic Substances & Disease Registry (September 2002). http://www.atsdr.cdc.gov/tfacts1.html
- U.S. Environmental Protection Agency, Aldrin/Dieldrin: Ambient Water Quality Criteria, Washington, DC (1980).
- U.S. Environmental Protection Agency, Dieldrin, Health and Environmental Effects Profile No. 82, Washington, DC, Office of Solid Waste (April 30, 1980).
- Sax, N.I., Ed., "Dangerous Properties of Industrial Materials Report," 1, No. 4, 52-55 (1981) an 6, No. 1, 9-16 (1986).
- Adema, D.M.M., and G.J. Vink. 1981. "A comparative study of the toxicity of 1,1,2- trichloroethane, dieldrin, pentacholorophenol, and 3,4 dichloroaniline for marine and fresh water," *Chemosphere,* Pergamon Press, Elmsford NY.
- U.S. Environmental Protection Agency, "Health Advisory: Dieldrin," Washington, DC, Office of Drinking Water (August 1987).
- New Jersey Department of Health, "Hazardous Substance Fact Sheet: Dieldrin," Trenton, NJ (November 1998). http://www.state.nj.us/health/eoh/rtkweb/0683.pdf
- New York State Department of Health, "Chemical Fact Sheet: Dieldrin," Albany, NY, Bureau of Toxic Substance Assessment (January 1986 & Version 2).

## Dienochlor

*Use Type:* Acaricide, miticide
*CAS Number:* 2227-17-0
*Formula:* $C_{10}Cl_{10}$
*Synonyms:* Bis(pentachlor-2,4-cyclopentadien-1-yl); Bis(pentachlorocyclopentadienyl); Bis(pentachloro-2,4-cyclopentadien-1-yl); Decachlor; ENT 25,718; 1,1',2,2',3,3',4,4',5,5'-Decachloro-bis(2,4-cyclopentadien-1-yl); Decachlorobis(2,4-cyclopentadiene-1-yl); Dienochlor; Bis(pentachloro-2,4-cyclopentadien-1-yl)
*Trade Names:* HOOKER® HRS-16; HOOKER® HRS 1654; HRS-16®; HRS 16A®; HRS 1654®; PENTAC® Syngenta (Switzerland), canceled; PENTAC® Aquaflow, Syngenta (Switzerland), canceled; PENTAC® WP, Syngenta (Switzerland), canceled; SATHON®, Sandoz (Switzerland), canceled; ZOECON®, Sandoz (Switzerland), canceled
*Producers:* Dow Elanco (USA); Occidental (USA), Syngenta (Switzerland)
*Chemical Class:* Chlorinated hydrocarbons (aliphatic); Organochlorine; Halo-organics
*EPA/OPP PC Code:* 027501
*California DPR Chemical Code:* 468
*EINECS Number:* 218-763-5
*Uses:* No products currently registered in the U.S. Used in formulations in combination with many other pesticides for the control of mites on ornamental trees and shrubs, and other non-food crops.
*Fish toxicity (threshold)*[77]: High–4.66516 ppb, MATC (Maximum Acceptable Toxicant Concentration)
*Carcinogen/Hazard Classifications*
**Label Signal Word:** WARNING or CAUTION
**WHO Acute Hazard:** Class III, slightly hazardous
*Regulatory Authority:*
- Actively registered pesticide in California.

*Description:* Tan crystalline solid or yellow prisms. Molecular weight = 474.5. Melting/Freezing point = 122°C.
*Harmful Effects and Symptoms*
*Short Term Exposure:* Symptoms include apprehension, anxiety, confusion, nervous excitation; dizziness; headache; numbness and weakness in limbs; muscle twitching, tremors; nausea and vomiting; slow, shallow respiration, bluish face; convulsions; loss of consciousness; breathing stops; death.
*Long Term Exposure:*
*Points of Attack:* May be fatal if inhaled, ingested, or absorbed through the skin
*Medical Surveillance:* Medical observation is recommended for 24 to 48 hours after breathing overexposure, as pulmonary edema may be delayed. As first

aid for pulmonary edema, consider administering a corticosteroid spray. Cigarette smoking may exacerbate pulmonary injury and should be discouraged for at least 72 hours following exposure.

*First Aid:* Speed in removing material from eyes and skin is of extreme importance. *Eyes:* Eye contact can cause dangerous amounts of these chemicals to be quickly absorbed through the mucous membrane into the bloodstream. Directly, irrigate with large amounts of plain, tepid water or saline for 20 minutes, occasionally lifting the lower and upper lids. During this time, remove contact lenses, if easily removable without additional trauma to the eye. Get medical aid immediately. Have physician check for possible delayed damage. *Skin:* Get medical aid. Skin contact can cause dangerous amounts of these chemicals to be absorbed into the bloodstream. Wearing the appropriate PPE equipment and respirator for organochlorine pesticides, immediately flush exposed skin, hair, and under nails with plain, running, tepid water for 20 minutes, then wash twice with mild soap. Shampoo hair promptly if contaminated; protect eyes. **Do not scrub skin or hair**, since this can increase absorption through the skin. Rinse thoroughly with water. Victims who are able and cooperative may assist with their own decontamination. Remove and double-bag contaminated clothing and personal belongings. Leather absorbs many organochlorines; therefore, items such as leather shoes, gloves, and belts should be discarded. If the skin is swollen or inflamed, cool affected areas with cold compresses. *Ingestion:* Call poison control. Loosen all clothing. Never give anything by mouth to an unconscious person. Get medical aid. *Do not induce vomiting.*\* In cases of ingestion, the patient is at risk of CNS depression or seizures, which may lead to pulmonary aspiration during vomiting. If the victim is conscious and able to swallow, \*administer an aqueous slurry of activated charcoal at 1 gm/kg (usual adult dose 60–90 g, child dose 25–50 g). A soda can and straw may be of assistance when offering charcoal to a child. The efficacy of activated charcoal for some organochlorine poisoning (such as chlordane) is uncertain. If victim is *unconscious or having convulsions,* do nothing except keep victim warm. \**In some cases you may be specifically instructed by Poison Control to induce vomiting by way of 2 tablespoons of syrup of ipecac (adult) washed down with a cup of water. Do not give activated charcoal before or with ipecac syrup. Inhalation:* Get medical aid. Do not contaminate yourself. Wearing the appropriate PPE equipment and respirator for organochlorine pesticides, immediately remove the victim from the contaminated area to fresh air. For inhalation exposures, monitor for respiratory distress. If the victim is not breathing, administer artificial respiration. *Do not use mouth-to-mouth resuscitation; use bag/mask apparatus.* If cough or breathing difficulty develops, evaluate for respiratory tract irritation, bronchitis, or pneumonitis. If breathing is difficult, administer 100% humidified supplemental oxygen through bag/mask apparatus until medical help arrives. Do not leave victim unattended.

*References:*
- EXTOXNET, Extension Toxicology Network, "Pesticide Information Profile, Dienochlor," Oregon State University, Corvallis, OR (June 1996). http://extoxnet.orst.edu/pips/dienochl.htm
- California Environmental Protection Agency "Chemical List of Lists," Sacramento CA (February 1997)

# Diethatyl-ethyl

*Use Type:* Herbicide
*CAS Number:* 38727-55-8
*Formula:* $C_{14}H_{18}ClNO_3$
*Synonyms:* Caswell No. 179; *N*-Chloroacetyl-*N*-(2,6-diethylphenyl)glycine, ethyl ester; Ethyl *N*-(chloroacetyl)-*N*-(2,6-diethylphenyl)glycinate; Glycine, *N*-(chloroacetyl)-*N*-(2,6-diethylphenyl)-, ethyl ester
*Trade Names:* AC 22,234®; ANTOR®, Bayer CropScience (Germany), canceled; BAY NTN 6867®, Bayer CropScience (Germany); H 22234®; HERCULES® 22234
*Producers:* Bayer CropScience (Germany)
*Chemical Class:* Chloracetanilide
*EPA/OPP PC Code:* 279500
*California DPR Chemical Code:* 1995
*RTECS Number:* MB9200000
*Uses:* No products currently registered in the U.S. Diethatyl-ethyl is a selective pre-emergence herbicide used to control many annual grasses and broadleaf weeds in alfalfa, carrots, dry beans, flax, lima beans, peanuts, peas, southern peas, potatoes, red beets, soybeans, spinach, sugar beets, tomatoes, cotton and others.
*Human toxicity (long-term)*[77]: High–1.75 ppb, Health Advisory
*Fish toxicity (threshold)*[77]: Low–243.26438 ppb, MATC (Maximum Acceptable Toxicant Concentration)
*Carcinogen/Hazard Classifications*
**Label Signal Word:** WARNING or CAUTION
*Regulatory Authority:*
- EPCRA Section 313 Form R *de minimis* concentration reporting level: 1.0%
- EPA 40 CFR 372.65, Specific Toxic Chemical Listings

*Description:* White crystalline solid. Soluble in water. Vapor pressure = $3.2 \times 10^{-6}$ mmHg @ 20°C.
*Harmful Effects and Symptoms*
*Short Term Exposure:* Toxic if swallowed. Contact with skin or eyes may cause severe irritation, burns, and severe eye injury. Inhalation should be avoided; use NIOSH-approved air purifying respirators for pesticides.
*Medical Surveillance:* Consult physician if the solvent has been inhaled or if poisoning is suspected or of burning or itching skin develops.

*First Aid:* If this chemical gets into the eyes, remove any contact lenses at once and irrigate immediately for at least 15 minutes, occasionally lifting upper and lower lids. Seek medical attention immediately. If this chemical contacts the skin, remove contaminated clothing and wash immediately with soap and water. Seek medical attention immediately. If this chemical has been inhaled, remove from exposure, begin rescue breathing (using universal precautions) if breathing has stopped, and CPR if heart action has stopped. Transfer promptly to a medical facility. When this chemical has been swallowed, get medical attention. Give large quantities of water. Do not induce vomiting when formulations containing petroleum solvents are ingested. Do not make an unconscious person vomit.

*References:*
- Pesticide Management Education Program, "Diethatyl ethyl (Antor) Herbicide Profile 3/85," Cornell University, Ithaca, NY (March 1985). http://pmep.cce.cornell.edu/profiles/herb-growthreg/dalapon-ethephon/diethatyl-ethyl/herb-prof-diethatyl-ethyl.html
- California Environmental Protection Agency "Chemical List of Lists," Sacramento CA (February 1997)

# Difenacoum

*Use Type:* Rodenticide
*CAS Number:* 56073-07-5
*Formula:* $C_{31}H_{24}O_3$
*Synonyms:* 3-(3,1,1'-Biphenyl-4-yl-1,2,3,4-tetrahydro-1-napthalenyl)-4-hydroxy-1(2*H*)-benzopyran-2-one
*Trade Names:* COMPO®, MATRAK®, NEOSOREXA, Sorex (UK); NEOSOREXA PP580®, Sorex (UK), RASTOP®, RATAK®, RATRICK®, SILO; SOREXA, Sorex (UK); STORM, Sorex (UK)
*Producers:* I.N.D.I.A. Industrie Chimiche (Italy); Sorex (UK)
*Chemical Class:* Coumarin
*EPA/OPP PC Code:* 119901
*California DPR Chemical Code:*
*RTECS Number:* GN4934500
*EINECS Number:* 259-978-4
*Uses:* Not registered in the U.S. Difenacoum is an anticoagulant that is effective against rats and mice, including warfarin-resistant strains. It is used in agriculture and urban rodent control as ready-to-use baits.
*Carcinogen/Hazard Classifications*
**WHO Acute Hazard:** Class 1 a, extremely hazardous
*Description:* Off-white powder. Low solubility in water. A weak acid. Melting point = 215-219°C.
*Points of Attack:* Liver
*Medical Surveillance:* Gastric lavage and repeated administration of charcoal. Blood sample to measure hemoglobin level, prothrombin time, blood grouping and cross-matching. (See Pied Piper site in reference.)

*First Aid:* If this chemical gets into the eyes, remove any contact lenses at once and irrigate immediately for at least 15 minutes, occasionally lifting upper and lower lids. Seek medical attention immediately. If this chemical contacts the skin, remove contaminated clothing and wash immediately with soap and water. Seek medical attention immediately. If this chemical has been inhaled, remove from exposure, begin rescue breathing (using universal precautions) if breathing has stopped, and CPR if heart action has stopped. Transfer promptly to a medical facility. When this chemical has been swallowed, get medical attention. Give large quantities of water and induce vomiting. Do not make an unconscious person vomit. Do not induce vomiting when formulations containing petroleum solvents are ingested.

*References:*
- International Programme on Chemical Safety (IPCS), "Health and Safety Guide, Difenacoum," Geneva, Switzerland(1995). http://www.inchem.org/documents/hsg/hsg/hsg095.htm
- The PiedPiper Northern Ltd, Stuart M. Bennett, http://www. The-piedpiper.co.uk/th15(c).htm

# Difenoconazole

*Use Type:* Fungicide
*CAS Number:* 119446-68-3
*Formula:* $C_{19}H_{17}Cl_2N_3O_3$
*Synonyms:* CGA 169374; 1*H*-1,2,4-Triazole, 1-[[2-(2-chloro-4-(4-chlorophenoxy)phenyl)-4-methyl-1,3-dioxolan-2-yl)methyl]- (9CI)
*Trade Names:* CGA 169374®, Syngenta (Switzerland); DIVIDEND®, Syngenta (Switzerland); DIVIDEND® EXTREME FUNGICIDE, Syngenta (Switzerland); HELIX®, Syngenta (Switzerland); SCORE®, Syngenta (Switzerland); TECHNICAL CGA-169374®, Syngenta (Switzerland)
*Producers:* Syngenta (Switzerland)
*Chemical Class:* Chlorophenoxy; azole
*EPA/OPP PC Code:* 128847
*California DPR Chemical Code:* 5024
*Uses:* For suppression of fungi diseases in crops and seeds.
*U.S. Maximum Allowable Residue Levels for Difenoconazole (40 CFR 180.475):*

| CROP | ppm |
|---|---|
| Banana | 0.2 |
| Barley, grain | 0.1 |
| Canola, seed | 0.01 |
| Cattle, fat | 0.05 |
| Cattle, meat | 0.05 |
| Cattle, mbyp | 0.05 |
| Corn, sweet, forage | 0.1 |

# Difenzoquat

| | |
|---|---|
| Corn, sweet, kernel plus cob with husks removed | 0.1 |
| Corn, sweet, stover | 0.1 |
| Egg | 0.05 |
| Goat, fat | 0.05 |
| Goat, meat | 0.05 |
| Goat, mbyp | 0.05 |
| Hog, fat | 0.05 |
| Hog, meat | 0.05 |
| Hog, mbyp | 0.05 |
| Horse, fat | 0.05 |
| Horse, meat | 0.05 |
| Horse, mbyp | 0.05 |
| Milk | 0.01 |
| Poultry, fat | 0.05 |
| Poultry, meat | 0.05 |
| Poultry, mbyp | 0.05 |
| Rye, grain | 0.1 |
| Sheep, fat | 0.05 |
| Sheep, meat | 0.05 |
| Sheep, mbyp | 0.05 |
| Wheat, forage | 0.1 |
| Wheat, grain | 0.1 |
| Wheat, straw | 0.1 |

*Carcinogen/Hazard Classifications*
**U.S. EPA Carcinogens:** Group C, possible carcinogen
**Label Signal Word:** CAUTION
**WHO Acute Hazard:** Class III, slightly hazardous
*Regulatory Authority:*
- AB 2588-Air Toxics "Hot Spots" Chemicals (CAL) as chlorophenoxy pesticides
- The "Director's List" (CAL/OSHA) as chlorophenoxy pesticides

*Description:* Molecular weight = 406.30
*Routes of Entry:* Ingestion, skin contact
*Harmful Effects and Symptoms*
*Short Term Exposure:* Poisonous; may be fatal if inhaled, swallowed, or absorbed through skin. Severely irritates eyes, skin and respiratory tract, with burning sensation, pain, redness and swelling. Metabolic stimulant. If inhaled, causes coughing, dilated pupils, headache, profuse persperation, intense thirst, extreme fatigue, rapid pulse, high fever, clammy, flushed skin, rapid breathing, nausea, vomiting, cyanosis (bluish tint to skin and lips), anxiety and confusion, convulsions, risk of lung edema. If swallowed, face and lips turn bluish. Liver injury with associated jaundice, kidney failure, and cardiac arrhythmias are commonly noted. Nerve damage, which may be delayed, may include swelling of legs and feet, muscle twitch and stupor. Severe exposure can cause death from heart failure. Dust or liquid left in contact with the skin for several hours may be absorbed. This may result in severe delayed symptoms as listed above. These symptoms may last for months or years.
*Long Term Exposure:* Workers exposed to chlorophenoxy compounds such as 2,4-D (in the manufacturing process) over a five to ten year period at levels above 10 mg/m$^3$ complained of weakness, rapid fatigue, headache and vertigo. Liver damage, low blood pressure and slowed heartbeat were also found. Based on animal tests, may affects human reproduction
*Points of Attack:* Eyes, skin, respiratory system, central nervous system, cardiovascular system, liver, kidney
*Medical Surveillance:* If symptoms develop or overexposure is suspected, liver or kidney function tests may be useful. Liver function tests.
*First Aid:* If this chemical gets into the eyes, remove any contact lenses at once and irrigate immediately for at least 15 minutes, occasionally lifting upper and lower lids. Seek medical attention immediately. If this chemical contacts the skin, remove contaminated clothing and wash immediately with soap and water. Seek medical attention immediately. If this chemical has been inhaled, remove from exposure, begin rescue breathing (using universal precautions) if breathing has stopped, and CPR if heart action has stopped. Transfer promptly to a medical facility. When this chemical has been swallowed, get medical attention. Give large quantities of water and induce vomiting. Do not make an unconscious person drink or vomit. *Note to Physician:* If ingested, remove by lavage or vomiting. Use general supportive measures for CNS depression. Consider the use of quinidine for myotonia.
*References:*
- U.S. Environmental Protection Agency, Office of Pesticide Programs, Pesticide Residue Limits, "Difenoconazole", 40 CFR 180.475. www.epa.gov/pesticides/food/viewtols.htm
- California Environmental Protection Agency "Chemical List of Lists," Sacramento CA (February 1997)

# Difenzoquat (ANSI)

*Use Type:* Herbicide
*CAS Number:* 43222-48-6 (salt)
*Formula:* $C_{17}H_{20}N_2O_4S$
*Synonyms:* 1,2-Dimethyl-3,5-diphenyl-1$H$-pyrazolium methyl sulfate; Difenzoquat methyl sulfate; 1$H$-Pyrazolium, 1,2-dimethyl-3,5-diphenyl-, methyl sulfate
*Trade Names:* AVENGE®, American Cyanamid (USA); AC 84777®; FINAVEN®; MATAVEN®; PYRAZOLIUM®; EH-YAN-KU®
*Producers:* American Cyanamid Agricultural Products Group (USA)
*Chemical Class:* Pyrazolium
*EPA/OPP PC Code:* 106401 (salt)
*California DPR Chemical Code:* 1930 (Difenzoquat methyl sulfate)
*EINECS Number:* 256-152-5
*Uses:* Registered to control wild oats in alfalfa, wheat and barley crops.

*U.S. Maximum Allowable Residue Levels for Difenzoquat (40 CFR 180.369):*

| CROP | ppm |
|---|---|
| Barley, grain | 0.2 |
| Barley, straw | 20 |
| Cattle, fat | 0.05 |
| Cattle, meat | 0.05 |
| Cattle, mbyp | 0.05 |
| Goat, fat | 0.05 |
| Goat, meat | 0.05 |
| Goat, mbyp | 0.05 |
| Hog, fat | 0.05 |
| Hog, meat | 0.05 |
| Hog, mbyp | 0.05 |
| Horse, fat | 0.05 |
| Horse, meat | 0.05 |
| Horse, mbyp | 0.05 |
| Poultry, fat | 0.05 |
| Poultry, meat | 0.05 |
| Poultry, mbyp | 0.05 |
| Sheep, fat | 0.05 |
| Sheep, meat | 0.05 |
| Sheep, mbyp | 0.05 |
| Wheat, grain | 0.05 |
| Wheat, straw | 20 |

*Carcinogen/Hazard Classifications*
**WHO Acute Hazard:** Class II, moderately hazardous
*Description:* White to off-white solid. Odorless. Highly soluble in water; solubility = 765,000 mg/L. Molecular weight = 348.58. Melting/Freezing point = 156°C.
*Incompatibilities:* May react violently with carbon dust, finely divided aluminum, magnesium, potassium
*Permissible Exposure Limits in Air:*
*Determination in Air:* Filter; none; Gravimetric; NIOSH IV[18] [Particulates NOR; #0500 (total), #0600 (respirable)]
*Permissible Concentration in Water:* No criteria set. Runoff from spills or fire control may cause water pollution.
*Routes of Entry:* Inhalation and skin contact
*First Aid:* If this chemical gets into the eyes, remove any contact lenses at once and irrigate immediately for at least 15 minutes, occasionally lifting upper and lower lids. Seek medical attention immediately. If this chemical contacts the skin, remove contaminated clothing and wash immediately with soap and water. Seek medical attention immediately. If this chemical has been inhaled, remove from exposure, begin rescue breathing (using universal precautions) if breathing has stopped, and CPR if heart action has stopped. Transfer promptly to a medical facility. When this chemical has been swallowed, get medical attention. Give large quantities of water and induce vomiting. Do not make an unconscious person vomit. Do not induce vomiting when formulations containing petroleum solvents are ingested.
*References:*
- Pesticide Management Education Program, "Difenzoquat (Avenge) Herbicide Profile 12/88," Cornell University, Ithaca, NY (December 1988). http://pmep.cce.cornell.edu/profiles/herb-growthreg/dalapon-ethephon/difenzoquat/herb-prof-difenzoquat.html
- U.S. Environmental Protection Agency, "Reregistration Eligibility Decision (RED), Difenzoquat" Office of Prevention, Pesticides and Toxic Substances, Washington, DC (September 1994). http://www.epa.gov/REDs/0223.pdf
- U.S. Environmental Protection Agency, Office of Pesticide Programs, Pesticide Residue Limits, "Difenzoquat", 40 CFR 180.369. www.epa.gov/pesticides/food/viewtols.htm
- California Environmental Protection Agency "Chemical List of Lists," Sacramento CA (February 1997)

## Diflubenzuron (ANSI)

*Use Type:* Insecticide, larvicide
*CAS Number:* 35367-38-5
*Formula:* $C_{14}H_9ClF_2N_2O_2$
*Synonyms:* AI 329054; Benzamide, *N*-[((4-chlorophenyl)amino)carbonyl]-2,6-difluoro; *N*-[((4-Chlorophenyl)amino)carbonyl]-2,6-difluorobenzamide; 1-(4-Chlorophenyl)-3-(2,6-difluorobenzoyl)urea; Diflubenzuron (Spanish); ENT 29,054; Urea, 1-(*p*-chlorophenyl)-3-(2,6-difluorobenzoyl)-
*Trade Names:* ADEPT®, Crompton Corporation (USA); ASTONEX®; DIMILIN®, Crompton Corporation (USA); DU-112307®; DUPHAR® PH 60-40; ODC-45®; DIFLURON®; DU 112307®; LARGON®; LARVAKIL®; MICROMITE®, Crompton Corporation (USA); OMS 1804®; PDD 60401®; PH 60-40®; PHILIPS-DUPHAR® PH 60-40; TH 60-40®; THOMPSON-HAYWARD® 6040; VIGILANTE®, Crompton Corporation (USA)
*Producers:* Agrimor International (USA); Crompton Corporation (USA); Hebei Huafeng Chemical Group (China)
*Chemical Class:* Organofluorine
*EPA/OPP PC Code:* 108201
*California DPR Chemical Code:* 1992
*RTECS Number:* YS6200000
*EINECS Number:* 252-529-3
*Uses:* Diflubenzuron is used primarily on citrus, cattle feed, cotton, forestry, mushrooms, ornamentals, pastures, soybeans, standing water, sewage systems, and wide-area general outdoor treatment sites. The insecticide behaves as a chitin inhibitor to inhibit the growth of many leaf-eating larvae, mosquito larvae, aquatic midges, rust mite, bollweevil, and flies. Diflubenzuron was first registered in the United States in 1979 for use as an insecticide.
*Human toxicity (long-term)[77]:* Very low–140.00 ppb, Health Advisory

## Diflufenican

*Fish toxicity (threshold)*[77]: Intermediate–20.00 ppb, MATC (Maximum Acceptable Toxicant Concentration)

*Carcinogen/Hazard Classifications*

**U.S. EPA Carcinogens:** Group E, Unlikely a Carcinogen
**Label Signal Word:** CAUTION
**WHO Acute Hazard:** Class U, unlikely to be hazardous

*Regulatory Authority:*
- 40CFR186: tolerances for pesticides in animal foods
- EPA 40 CFR 372.65, Specific Toxic Chemical Listings
- Actively registered pesticide in California.
- EPCRA Section 313 Form R *de minimis* concentration reporting level: 1.0%

*Description:* Colorless or white crystalline solid. Practically insoluble in water; solubility = $2 \times 10^{-5}$ % @ 20°C. Molecular weight = 310.70. Melting/Freezing point = 220°C. Boiling point = decomposes. Vapor pressure = $9 \times 10^{-10}$ mmHg @ 20°C. Log $K_{ow}$ = 3.10. Values above 3.0 are likely to bioaccumulate in marine organisms.

*Incompatibilities:* Slowly hydrolyzes in water, releasing ammonia and forming acetate salts.

*Permissible Exposure Limits in Air:*

*Determination in Air:* Filter; none; Gravimetric; NIOSH IV[18] [Particulates NOR; #0500 (total), #0600 (respirable)]

*Permissible Concentration in Water:* No criteria set. Runoff from spills or fire control may cause water pollution.

*Harmful Effects and Symptoms*

*Short Term Exposure:* Contact with eyes or skin may cause irritation or burns. Inhalation should be avoided; use NIOSH-approved air purifying respirators for pesticides. May be harmful if swallowed. Skin contact may cause allergic reaction.

*Long Term Exposure:* Repeated or prolonged contact may cause skin and lung sensitization, resulting in allergies.

*Points of Attack:* Skin.

*Medical Surveillance:* Evaluation by a qualified allergist, including careful exposure history and special testing, may help diagnose skin allergy.

*First Aid:* If this chemical gets into the eyes, remove any contact lenses at once and irrigate immediately for at least 15 minutes, occasionally lifting upper and lower lids. Seek medical attention immediately. If this chemical contacts the skin, remove contaminated clothing and wash immediately with soap and water. Seek medical attention immediately. If this chemical has been inhaled, remove from exposure, begin rescue breathing (using universal precautions) if breathing has stopped, and CPR if heart action has stopped. Transfer promptly to a medical facility. When this chemical has been swallowed, get medical attention. Give large quantities of water and induce vomiting. Do not make an unconscious person vomit.

*References:*
- EXTOXNET, Extension Toxicology Network, "Pesticide Information Profile, Diflubenzuron," Oregon State University, Corvallis, OR (June 1996). http://extoxnet.orst.edu/pips/difluben.htm
- U.S. Environmental Protection Agency, "Reregistration Eligibility Decision (RED), Duflubenzuron" Office of Prevention, Pesticides and Toxic Substances, Washington, DC (August 1997). http://www.epa.gov/REDs/0144red.pdf
- California Environmental Protection Agency "Chemical List of Lists," Sacramento CA (February 1997)

# Diflufenican

*Use Type:* Herbicide
*CAS Number:* 83164-33-4
*Formula:* $C_{19}H_{11}F_5N_2O_2$
*Synonyms:* N-(2,4-Difluorophenyl)-2-[3-(trifluoromethyl)phenoxy]-3-pyridinecarboxamide; 2',4'-Difluoro-2-(α-α-α-trifluoro-*m*-tolyloxy)nicotinanilide; DFF; Diflufenicanil (French); 3-Pyridinecarboxamide, N-(2,4-difluorophenyl)-2-[3-(trifluoromethyl)phenoxy]-

*Trade Names:* ARDENT®; BACARA®; CAPTURE®; COUGAR®; CUB®; GRENADIER®; IONIZ®; JAVELIN®; KWARC®; MB-38183®; M&B 38544®; PANTHER®; QUARTZ®; SPEARHEAD®

*Producers:* Bayer CropScience (Germany); Ki-Hara Chemicals Ltd. (UK)

*Chemical Class:* Anilide

*Uses:* Not registered in the U.S. Used on crops such as apples, apricot, barley, beans (dry), broad beans, cherry, grapefruit, lemon, lime, loquat, natsudaidai (whole), nectarine, orange, peach, pear, peas, persimon, plum, quince, rye, soybeans, unshu orange and wheat.

*Carcinogen/Hazard Classifications*

**WHO Acute Hazard:** Class U, unlikely to be hazardous

*Description:* Light brown oily suspension. Miscible in water.

*Incompatibilities:* Slowly hydrolyzes in water, releasing ammonia and forming acetate salts.

*Routes of Entry:* Ingestion, eye, skin contact

*Harmful Effects and Symptoms*

*Short Term Exposure:* May irritate mucous membranes in the mouth and nose. Mild eye irritation.

*First Aid:* If this chemical gets into the eyes, remove any contact lenses at once and irrigate immediately for at least 15 minutes, occasionally lifting upper and lower lids. Seek medical attention immediately. If this chemical contacts the skin, remove contaminated clothing and wash immediately with soap and water. Seek medical attention immediately. If this chemical has been inhaled, remove from exposure, begin rescue breathing (using universal precautions) if breathing has stopped, and CPR if heart action has stopped. Transfer promptly to a medical facility. When this chemical has been swallowed, get medical attention. Give large quantities of water and induce vomiting. Do not make an unconscious person vomit. Do not induce vomiting when formulations containing petroleum solvents are ingested.

# Dihydroazadirachtin

*Use Type:* Insecticide and nematicide
*CAS Number:* 108189-58-8
*Synonyms:* 22,23-Dihydroazadirchtin
*Trade Names:* DAZA®, Certis USA (USA), canceled 7/25/2000
*Producers:* Certis USA (USA)
*Chemical Class:* Biochemical
*EPA/OPP PC Code:* 121702
*California DPR Chemical Code:* 3994
*Uses:* Exempted from the requirement of a tolerance for residues. Used to control insects, centipedes, millipedes, mites, nematodes, sowbugs and more, for horticulture and ornamental plants, trees, shrubs and agricultural crops.
*Regulatory Authority:*
- CFR 180.1169, Dihydroazadirachtin: exemption from the requirement of a tolerance

*Description:* Dihydroazadirachtin is a reduced (hydrogenated) form of the naturally occurring azadirachtin (AZA) obtained from the seed kernels of the neem tree, *Azadirachta indica A. Juss.*

### Harmful Effects and Symptoms

*Short Term Exposure:* Contact with eyes or skin may cause irritation or injury. Inhalation should be avoided; use NIOSH-approved air purifying respirators for pesticides. May be harmful if swallowed.

*First Aid:* If this chemical gets into the eyes, remove any contact lenses at once and irrigate immediately for at least 15 minutes, occasionally lifting upper and lower lids. Seek medical attention immediately. If this chemical contacts the skin, remove contaminated clothing and wash immediately with soap and water. Seek medical attention immediately. If this chemical has been inhaled, remove from exposure, begin rescue breathing (using universal precautions) if breathing has stopped, and CPR if heart action has stopped. Transfer promptly to a medical facility. When this chemical has been swallowed, get medical attention. Give large quantities of water and induce vomiting. Do not make an unconscious person vomit. Do not induce vomiting when formulations containing petroleum solvents are ingested.

*References:*
- U.S. Environmental Protection Agency, "Dihydroazadirachtin (121702) Technical Fact Sheet," Washington, DC (October 1998). http://www.epa.gov/pesticides/biopesticides/ingredients/tech_docs/tech_121702.htm

# Dimefox

*Use Type:* Insecticide
*CAS Number:* 115-26-4
*Formula:* $C_4H_{12}FN_2OP$
*Synonyms:* BFPO; Bis(dimethylamido)fluorophosphate; Bis(dimethylamido)fluorophosphine oxide; Bis(dimethylamido)phosphoryl fluoride; Bis(dimethylamino)fluorophosphate; Bisdimethylaminofluorophosphine oxide; BPF; CR 409; DIFO; DMF; ENT 19,109; Fluophosphoric acid di(dimethylamide); Fluorure de *N,N,N',N'*-tetramethyle phosphoro-diamide (French); *N,N,N',N'*-Tetramethyl-diamido-fosforzuur-fluoride (Dutch); Tetramethyldiamidophosphoric fluoride; *N,N,N',N'*-Tetramethyl-diamido-phosphorsaeure-fluorid (German); Tetramethylphosphorodiamidic fluoride; *N,N,N,N*-Tetramethylphosphorodiamidic fluoride; *N,N,N',N'*-Tetrametil-fosforodiammido-fluoruro (Italian)
*Trade Names:* HANANE®; PESTOX IV®; PESTOX XIV®; PESTOX 14®; T-2002®; TERRA-SYSTAM®; TERRA-SYTAM®; TERRASYTUM®; TL 792®; WACKER 14/10®
*Producers:* Sigma-Aldrich Laborchemikalien (Germany)
*Chemical Class:* Organophosphate
*EPA/OPP PC Code:* 443100
*RTECS Number:* TD4025000
*Uses:* A selective systemic insecticide which is absorbed into the plant sap and remains active for long periods of time. Selective systemic organophosphate insecticides are toxic to plant pests but not to their predators.
*Regulatory Authority:*
- Very Toxic Substance (World Bank)[15]
- EPA/SARA 302 (EPCRA) Extremely hazardous substances
- U.S. DOT Inhalation Hazard Chemicals as organophosphates
- Superfund/EPCRA 40CFR355, Appendix B Extremely Hazardous Substances: TPQ = 500 lb (227kg)
- Superfund/EPCRA 40CFR302.4 RQ: EHS, 1 lb (0.454 kg)

*Description:* Dimefox is a clear liquid. Fishy odor. Soluble in water. Molecular weight = 154.17. Boiling point = 86°C @15 mmHg. Hazard Identification (based on NFPA-704 M Rating System): Health 4, Flammability 1, Reactivity 1.
*Incompatibilities:* Strong oxidants, strong acids and halogens.
*Permissible Exposure Limits in Air:* No standards set.
*Determination in Air:* OSHA versatile sampler-2; Toluene/Acetone; Gas chromatography/Flame photometric detection for sulfur, nitrogen, or phosphorus; NIOSH Method IV Method #5600, Organophosphorus pesticides[18].
*Permissible Concentration in Water:* No criteria set, but runoff from spills or fire control may cause water pollution.

### Harmful Effects and Symptoms

*Short Term Exposure:* Organic phosphorus insecticides are absorbed by the skin, as well as by the respiratory and gastrointestinal tracts. They are cholinesterase inhibitors. Symptoms of exposure include headache, giddiness, blurred vision, nervousness, weakness, nausea, cramps, diarrhea, and discomfort in the chest. Signs include sweating, tearing, salivation, vomiting, cyanosis, convulsions, coma, loss of reflexes and loss of sphincter control. This material is extremely toxic; the probable oral lethal dose (human) is 5-50 mg/kg, or 7 drops to 1 teaspoonful for a 150-lb. person. Death may occur from respiratory arrest. Hazards of vapor toxicity are high. Delayed pulmonary edema may occur after inhalation.

*Long Term Exposure:* Cholinesterase inhibitor; cumulative effect is possible. Dimefox may damage the nervous system with repeated exposure, resulting in convulsions, respiratory failure. May cause liver damage.

*Points of Attack:* Respiratory system, lungs, central nervous system, cardiovascular system, skin, eyes, plasma and red blood cell cholinesterase.

*Medical Surveillance:* Medical observation is recommended for 24 to 48 hours after breathing overexposure, as pulmonary edema may be delayed. Before employment and at regular times after that, the following are recommended: Plasma and red blood cell cholinesterase levels (tests for the enzyme poisoned by this chemical). If exposure stops, plasma levels return to normal in 1-2 weeks while red blood cell levels may be reduced for 1-3 months. When acetylcholinesterase enzyme levels are reduced by 25% or more below preemployment levels, risk of poisoning is increased, even if results are in lower ranges of "normal." Reassignment to work not involving organophosphate or carbamate pesticides is recommended until enzyme levels recover. If symptoms develop or overexposure occurs, repeat the above tests as soon as possible and get an exam of the nervous system. Also consider complete blood count. Consider chest x-ray following acute overexposure. Do not drink any alcoholic beverages before or during use. Alcohol promotes absorption of organophosphates.

*First Aid:* **Treatment for organophosphate poisoning consists of thorough decontamination, cardiorespiratory support, and administration of the antidotes atropine and pralidoxime. In cases of severe poisoning, diazepam, an anticonvulsant, should also be administered. Antidotes should be administered as prevention even if the diagnosis is in doubt.** Speed in removing material from eyes and skin is of extreme importance. *Eyes:* Eye contact can cause dangerous amounts of these chemicals to be quickly absorbed through the mucous membrane into the bloodstream. Immediately and gently flush eyes with plenty of warm or cold water (NO hot water) for at least 15 minutes, occasionally lifting the upper and lower eyelids. Get medical aid immediately. *Skin:* Get medical aid. Skin contact can cause dangerous amounts of these chemicals to be absorbed into the bloodstream. Wearing the appropriate PPE equipment and respirator for organophosphate/carbamate pesticides, immediately flush skin with plenty of soap and water for at least 15 minutes while removing contaminated clothing and shoes. Shampoo hair promptly if contaminated. The removed, contaminated clothing and shoes should be double-bagged and left in Hot Zone for later disposal by hazardous materials experts. Skin may also be decontaminated with diluted hypochlorite solution. *Inhalation:* Get medical aid. Do not contaminate yourself. Wearing the appropriate PPE equipment and respirator for organophosphate pesticides, immediately remove the victim from the contaminated area to fresh air. If the victim is not breathing, administer artificial respiration. Do NOT use mouth-to-mouth resuscitation; use bag/mask apparatus. If breathing is difficult, administer oxygen through bag/mask apparatus until medical help arrives. Do not leave victim unattended. *Ingestion:* Call poison control. Loosen all clothing. Never give anything by mouth to an unconscious person. If victim is *unconscious or having convulsions,* do nothing except keep victim warm. Get medical aid. Transfer promptly to a medical facility. In cases of ingestion, **do not induce vomiting**. If the victim is alert and asymptomatic, administer a slurry of activated charcoal at a dose of 1 g/kg (infant, child, and adult dose). A soda can and straw may be of assistance when offering charcoal to a child. *In some cases you may be specifically instructed by poison control to induce vomiting by way of 2 tablespoons of syrup of ipecac (adult) washed down with a cup of water.* Do NOT give activated charcoal before or with ipecac syrup.

*Note to physician or authorized medical personnel:* Treat cases of respiratory compromise, coma, or excessive pulmonary secretions with respiratory support using protocols and techniques available and within the scope of training. Some cases may necessitate procedures such as endotracheal intubation or cricothyrotomy by properly trained and equipped personnel. When possible, atropine (see *Antidotes*, below) should be given under medical supervision. Patients who are comatose, hypotensive, or having seizures or cardiac arrhythmias should be treated according to advanced life support protocols. *Antidotes:* Two antidotes are administered to treat organophosphate poisoning. Atropine is a competitive antagonist of acetylcholine at muscarinic receptors and is used to control the excessive bronchial secretions which are often responsible for death. Pralidoxime relieves both the nicotinic and muscarine effects of organophosphate poisoning by regenerating acetylcholinesterase and can reduce both the bronchial secretions and the muscle weakness associated with poisoning. The initial intravenous dose of atropine in adults should be determined by the severity of symptoms: An initial adult dose of 1.0 to 2.0 mg or pediatric dose of 0.01 mg/kg (minimum 0.01 mg) should be administered intravenously. If intravenous access cannot be established, atropine may also be given intramuscularly, subcutaneously or via endotracheal tube. Doses should be repeated every 15 minutes until excessive secretions and sweating have been controlled. Once bronchial secretion has been controlled, atropine administration should be repeated whenever the secretions begin to recur. In seriously poisoned patients, very large doses may be required. Alterations of pulse rate and pupillary size should not be used as indicators of treatment adequacy. Pralidoxime should be administered as early in poisoning as possible as its efficacy may diminish when given more than 24 to 36 hours after exposure. Doses are as follows: adult 1.0 g; pediatric 25 to 50 mg/kg. The drug should be administered intravenously over 30 to 60 minutes, but in a life-threatening situation, one-half of the total dose can be given per minute for a total administration time of 2 minutes. Treatment should begin to take effect within 40 minutes with a reduction in symptoms and the amount of atropine necessary to control bronchial secretion. The initial dose can be repeated in 1 hour and then every 8 to 12 hours until the patient is clinically well and no longer requires atropine. If intravenous access cannot be established, pralidoxime may also be given intramuscularly. Early administration of diazepam in addition to the combined atropine and pralidoxime treatment may help prevent the onset of seizures and potential brain and cardiac morphologic damage following high-level organophosphate poisoning.

*References:*
- U.S. Environmental Protection Agency, "Chemical Profile: Dimefox," Washington, DC, Chemical Emergency Preparedness Program (November 30, 1987).
- California Environmental Protection Agency "Chemical List of Lists," Sacramento CA (February 1997).
- Agency for Toxic Substances and Disease Registry, U.S. Department of Health and Human Services, Public Health Service, "Managing Hazardous Materials Incidents," Atlanta, GA (June 2003).
- New Jersey Department of Health and Senior Services, Hazardous Substance Fact Sheet, "Dimefox." Trenton NJ (February 1999).
http://www.state.nj.us/health/eoh/rtkweb/2342.pdf

# Dimethenamid

*Use Type:* Herbicide
*CAS Number:* 87674-68-8
*Formula:* $C_{12}H_{18}ClNO_2S$
*Synonyms:* Acetamide, 2-chloro-*N*-(2,4-dimethyl-3-thienyl)-*N*-(2-methoxy-1-methylethyl)-; 2-Chloro-*N*-(2,4-dimethyl-3-thienyl)-*N*-(2-methoxy-1-methylethyl)acetamide; 2-Chloro-*N*-[(1-methyl-2-methoxy)-ethyl]-*N*-(2,4-dimethyl-thien-3-yl)acetamide
*Trade Names:* DETAIL®, BASF Agricultural Products Group (Germany); DPX-PM082® (formulation containing Dimethenamid, Chlorimuron-ethyl, and Sodium sulforicinol); FRONTIER®, BASF Agricultural Products Group (Germany); GUARDSMAN®, BASF Agricultural Products Group (Germany); LEADOFF®, (atrazine + dimethenamid), DuPont Crop Protection; PURSUIT®, BASF Agricultural Products Group (Germany); OPTILL®, BASF Agricultural Products Group (Germany); SAN-582H®

*Producers:* BASF Agricultural Products Group (Germany); DuPont Crop Protection (USA)
*Chemical Class:* Amide
*EPA/OPP PC Code:* 129051
*California DPR Chemical Code:* 5112
*Human toxicity (long-term)*[77]*:* Intermediate–35.00 ppb, Health Advisory
*Fish toxicity (threshold)*[77]*:* Low–169.70710 ppb, MATC (Maximum Acceptable Toxicant Concentration)
*U.S. Maximum Allowable Residue Levels for Dimethenamid (40 CFR 180.464):*

| CROP | ppm |
|---|---|
| Bean, dry | 0.01 |
| Corn, fodder | 0.01 |
| Corn, forage | 0.01 |
| Corn, grain | 0.01 |
| Corn, stover | 0.01 |
| Corn, sweet, foder (stover) | 0.01 |
| Corn, sweet, (kernals plus cobs with husks removed) | 0.01 |
| Peanut, hay | 0.01 |
| Peanut, nutmeat | 0.01 |
| Sorghum, grain, fodder | 0.01 |
| Sorghum, grain, forage | 0.01 |
| Sorghum, grain | 0.01 |
| Soybeans | 0.01 |

*Carcinogen/Hazard Classifications*
**U.S. EPA Carcinogens:** Group C, possible carcinogen
**Label Signal Word:** WARNING, DANGER
*Description:* Thick, dark yellow-brown liquid. Soluble in water; solubility = 1165 ppm. Molecular weight = 275.78. Boiling point = 130°C.
*Incompatibilities:* Slowly hydrolyzes in water, releasing ammonia and forming acetate salts.
*Permissible Concentration in Water:* No criteria set. Runoff from spills or fire control may cause water pollution.
*Harmful Effects and Symptoms*
*Short Term Exposure:* Contact with eyes or skin may cause irritation or injury. Inhalation should be avoided; use NIOSH-approved air purifying respirators for pesticides. May be harmful if swallowed.
*First Aid:* If this chemical gets into the eyes, remove any contact lenses at once and irrigate immediately for at least 15 minutes, occasionally lifting upper and lower lids. Seek medical attention immediately. If this chemical contacts the skin, remove contaminated clothing and wash immediately with soap and water. Seek medical attention immediately. If this chemical has been inhaled, remove from exposure, begin rescue breathing (using universal precautions) if breathing has stopped, and CPR if heart action has stopped. Transfer promptly to a medical facility. When this chemical has been swallowed, get medical attention. Give large quantities of water and induce vomiting. Do not make an unconscious person vomit.

*References:*
- California Environmental Protection Agency "Chemical List of Lists," Sacramento CA (February 1997).
- U.S. Environmental Protection Agency, Office of Pesticide Programs, Pesticide Residue Limits, "Dimethenamid", 40 CFR 180.464. www.epa.gov/pesticides/food/viewtols.htm

# Dimethipin

*Use Type:* Plant growth regulator; defoliant
*CAS Number:* 55290-64-7
*Formula:* $C_6H_{10}O_4S_2$
*Synonyms:* Caswell No. 472AA; 2,3-Dehydro-2,3-dimethyl-tetroxide; 2,3-dihydro-5,6-dimethyl-1,4-dithiin 1,1,4,4-tetraoxide; 2,3,-Dihydro-5,6-dimethyl-1,4-dithiin 1,1,4,4-tetraoxide; 2,3-Dihydro-5,6-dimethyl-1,4-dithiin 1,1,4,4-tetraoxide; *p*-Dithiane, dimethipin [2,3,-Dihydro-5,6-dimethyl-1,4-dithiin-1,1,4,4-tetraoxide]; 2,3-*p*-Dithiane, 2,3-dehydro-2,3-dimethyl-, tetroxide; 1,4-Dithiin, 2,3-dihydro-5,6-dimethyl-,1,1,4,4-tetraoxide; Oxidimethiin; Oxydimethiin
*Trade Names:* HARVADE®, Crompton (USA); HARVADE-5F®, Crompton (USA); LEAFLESS®, Crompton (USA); N 252®; TETRATHIIN®; UBI-N 252®
*Producers:* Crompton (USA)
*Chemical Class:* Unclassified
*EPA/OPP PC Code:* 118901; (210600 old EPA code number)
*California DPR Chemical Code:* 2159
*Uses:* Used principally on cotton crops.
*Feed Additives Permitted in Animal Feed for dimethipin in animal feed (40 CFR 186.2050):*

| CROP | ppm |
|---|---|
| Cottonseed hulls | 0.7 |

*Carcinogen/Hazard Classifications*
**U.S. EPA Carcinogens:** Group C, possible carcinogen
**Label Signal Word:** CAUTION or DANGER
**WHO Acute Hazard:** Class III, slightly hazardous
*Regulatory Authority:*
- EPCRA Section 313 Form R *de minimis* concentration reporting level: 1.0%
- FIFRA, 40 CFR 186.2050: tolerances for pesticides in animal feeds.
- Actively registered pesticide in California.

*Description:* White crystalline solid or needles. Molecular weight = 210.30.
*Permissible Concentration in Water:* No criteria set. Runoff from spills or fire control may cause water pollution.
*First Aid:* If this chemical gets into the eyes, remove any contact lenses at once and irrigate immediately for at least 15 minutes, occasionally lifting upper and lower lids. Seek medical attention immediately. If this chemical contacts the skin, remove contaminated clothing and wash immediately with soap and water. Seek medical attention immediately. If this chemical has been inhaled, remove from exposure, begin rescue breathing (using universal precautions) if breathing has stopped, and CPR if heart action has stopped. Transfer promptly to a medical facility. When this chemical has been swallowed, get medical attention. Give large quantities of water and induce vomiting. Do not make an unconscious person vomit.
*References:*
- U.S. Environmental Protection Agency, Office of Pesticide Programs, Feed Additives Limits, "Dimethipin," 40 CFR 186.2050, http://frwebgate.access.gpo.gov/cgi-bin/get-cfr.cgi
- California Environmental Protection Agency "Chemical List of Lists," Sacramento CA (February 1997).

# Dimethoate (ANSI)

*Use Type:* Insecticide and miticide
*CAS Number:* 60-51-5
*Formula:* $C_5H_{12}NO_3PS_2$; $H_3COP(S)(OCH_3)SCH_2CONHCH_3$
*Alert:* A General Use Pesticide (GUP). Human toxicity (long-term): Very high.
*Synonyms:* Acetic acid, *O,O*-dimethyldithiophosphoryl-, *N*-monomethylamide Salt; *O,O*-Dimethyl *S*-(*N*-methylcarbamoylmethyl) dithiophosphate; Phosphorodithioic acid, *O,O*-dimethyl *S*-[2-(Methylamino)-2-oxoethyl] ester; Phosphamide
*Trade Names:* REBELATE®; CEKUTHOATE®; CHIMIGOR 40®; CYGON 400®, Amvac Chemicals (USA), BASF Agricultural Products Group (Germany), canceled; DEFEND®; DAPHENE®; DANADIM®, Cheminova (Denmark); DE-FEND®; DEMOS NF®; DEVIGON®; DICAP®; DIMATE 267®; DIMET®; DIMETHOPGAN®; FERKETHION®; FOSTION MM®; KENLOGO®, Kenso Corp. (Malaysia); NUGOR®, United Phosphorus (India); PERFEKTION®; ROGODAN®; ROGODIAL®; ROGOR®, Rallis India (India); Whyte Agrochemicals (UK); ROXION®; SEVIGOR®
*Producers:* Agrimor International (USA); Agro-care Chemical Industry Group (China); Agsin (Singapore); Alcotan Laboratories (Spain); Amvac Chemicals (USA); Atanor S.A. (Argentina); BASF Agricultural Products Group (Germany); Bayer Chemicals (Germany); Bharat Pulverizing Mills (India); Bharat Rasayan (India); Biesterfeld Siemsgluess International. GmbH (Germany); Bonide Products (USA); Calliope (France); Cheminova (Denmark); China Chemical (China); Chromos Agro (Croatia); Drexel Chemical (USA); Ehrenstorfer, Dr. (Germany); Gowan Company (USA); Hockley International (UK); Indiclay (India); Godavari Fertilisers and Chemicals (India); Gowan Company); Hockley International (UK); Jiangmen Pesticide Factory (China); Jingma Chemicals Ltd. (China); Kenso Corp. (Malaysia); Microflo Company

(USA); Montedison (Italy); Nagarjuna Agrichem (India); Nissan Chemical Industries (Japan); Rallis India (India); Rhone-Poulenc Agro (France); Shenzhen Guomeng Industry Co., Ltd. (China); Sigma-Aldrich Laborchemikalien (Germany); Sinon Corporation (Taiwan); United Phosphorus (India); Whyte Agrochemicals (UK) Zago Asia Ltd. (Singapore)

***Chemical Class:*** Organophosphate
***EPA/OPP PC Code:*** 035001
***California DPR Chemical Code:*** 216
***ICSC Number:*** 0741
***RTECS Number:*** TE1750000
***EEC Number:*** 015-051-00-4
***EINECS Number:*** 200-480-3

***Uses:*** Dimethoate is used to kill mites and insects systemically and on contact. It is used against a wide range of insects, including aphids, thrips, planthoppers, and whiteflies on ornamental plants, alfalfa, apples, corn, cotton, grapefruit, grapes, lemons, melons, oranges, pears, pecans, safflower, sorghum, soybeans, tangerines, tobacco, tomatoes, watermelons, wheat, and other vegetables. It is also used as a residual wall spray in farm buildings for house flies. Dimethoate has been administered to livestock for control of botflies. Dimethoate is available in aerosol spray, dust, emulsifiable concentrate, and ULV concentrate formulations. It has not been produced in the U.S. Since 1982.

***Human toxicity (long-term)***[77]: Extra high–0.35 ppb, Health Advisory

***Fish toxicity (threshold)***[77]: Very low–600.99381 ppb, MATC (Maximum Acceptable Toxicant Concentration)

***U.S. Maximum Allowable Residue Levels for Dimethoate (40 CFR 180.204):***

| CROP | ppm |
|---|---|
| Alfalfa | 2 |
| Apple | 2 |
| Asparagus | 0.15 |
| Bean, dry | 2 |
| Bean, snap | 2 |
| Bean, lima | 2 |
| Broccoli | 2 |
| Brussels sprouts | 5 |
| Cabbage | 2 |
| Cattle, fat | 0.02 |
| Cattle, meat | 0.02 |
| Cattle, mbyp | 0.02 |
| Cauliflower | 2 |
| Celery | 2 |
| Cherries | 2 |
| Citrus, dried pulp | 5 |
| Collards | 2 |
| Corn, fodder | 1 |
| Corn, forage | 1 |
| Corn, grain | 0.1 |
| Cotton, undelinted seed | 0.1 |
| Eggs | 0.02 |
| Endive (escarole) | 2 |
| Goat, fat | 0.02 |
| Goat, meat | 0.02 |
| Goat, mbyp | 0.02 |
| Grape | 1 |
| Grapefruit | 2 |
| Hog, fat | 0.02 |
| Hog, mbyp | 0.02 |
| Hog, meat | 0.02 |
| Horse, fat | 0.02 |
| Horse, meat | 0.02 |
| Horse, mbyp | 0.02 |
| Kale | 2 |
| Lemon | 2 |
| Lentil | 2 |
| Lettuce | 2 |
| Melon | 1 |
| Milk | 0.002 |
| Mustard greens | 2 |
| Orange | 2 |
| Pears | 2 |
| Peas | 2 |
| Pecan | 0.1 |
| Peppers | 2 |
| Potatoes | 0.2 |
| Poultry, fat | 0.02 |
| Poultry, mbyp | 0.02 |
| Poultry, meat | 0.02 |
| Safflower, seed | 0.1 |
| Sheep, fat | 0.02 |
| Sheep, mbyp | 0.02 |
| Sheep, meat | 0.02 |
| Sorghum, forage | 0.2 |
| Sorghum, grain, grain | 0.1 |
| Soybeans | 0.05 |
| Soybean, forage | 2 |
| Soybean, hay | 2 |
| Spinach | 2 |
| Swiss chard | 2 |
| Tangerines | 2 |
| Tomatoes | 2 |
| Turnip, greens | 2 |
| Turnip, roots | 2 |
| Wheat, grain | 0.04 |
| Wheat, hay | 2 |
| Wheat, straw | 2 |

***Carcinogen/Hazard Classifications***
**U.S. EPA Carcinogens:** Group C, possible carcinogen
**Label Signal Word:** WARNING, Toxicity Class II, or CAUTION, depending on the formulation
**TRI Developmental Toxin:** Developmental toxin
**WHO Acute Hazard:** Class II, moderately hazardous
***Regulatory Authority:***
• Banned or Severely Restricted (US EPA) (UN)[13]

- Air Pollutant Standard Set (former USSR)[35, 43]
- AB 1803-Well Monitoring Chemical (CAL)
- CAL-DHS/DHS Drinking Water Action Levels
- Actively registered pesticide in California.
- EPA Hazardous Waste Number (RCRA No.): P044
- RCRA, 40CFR261, Appendix 8 Hazardous Constituents
- RCRA 40CFR264, Appendix 9; TSD Facilities Ground Water Monitoring List. Suggested test method(s) (PQL $ug$/L): 8270(10)
- Superfund/EPCRA 40CFR355, Appendix B Extremely Hazardous Substances: TPQ = 500/10,000 lb (227/4,540 kg)
- Superfund/EPCRA 40CFR302.4 RQ: CERCLA, 10 lb (4.54 kg)
- EPCRA Section 313 Form R *de minimus* concentration reporting level: 1.0%
- U.S. DOT Regulated Marine Pollutant (49CFR172.101, Appendix B), severe pollutant
- Canada Drinking Water Quality, 0.02 mg/L MAC

*Description:* Colorless crystalline solid. Camphor-like odor. Slightly soluble in water; solubility = > $5.0 \times 10^3$ ppm @ 20°C. Boiling point = 107°C @ 0.05 mmHg. Melting/Freezing point = 52°C. Vapor Pressure = 1.1 mPa @ 25°C. Hazard Identification (based on NFPA-704 M Rating System): Health 3, Flammability 2, Reactivity 0. Log $K_{ow}$ = 0.5 to 0.8. Unlikely to bioaccumulate in marine organisms.

*Incompatibilities:* Strong bases (alkalis). Do not store solid above 77–86°F (25–30°C). Liquid solutions must be stored above 45°F/7°C.

*Permissible Exposure Limits in Air:* The former USSR[35, 43] has set air MAC in workplace air of 0.5 mg/m3 and limits in the ambient air of residential areas of 0.003 mg/m3 on either a momentary or a daily average basis.

*Determination in Air:* OSHA versatile sampler-2; Toluene/Acetone; Gas chromatography/Flame photometric detection for sulfur, nitrogen, or phosphorus; NIOSH Method IV Method #5600, Organophosphorus pesticides[18].

*Permissible Concentration in Water:*

*Determination in Water:* An MAC in water bodies used for domestic purposes of 0.03 mg/L has been set by the former USSR[35,43]. States which have set guidelines for Dimethoate in drinking water include California at 140 $\mu$g/L and Wisconsin at 10 $\mu$g/L.

*Routes of Entry:* Inhalation, skin absorption, eyes and ingestion.

*Harmful Effects and Symptoms*

*Short Term Exposure:* Acute exposure to Dimethoate may produce the following signs and symptoms: pinpoint pupils, blurred vision, headache, dizziness, muscle spasms, and profound weakness. Vomiting, diarrhea, abdominal pain, seizures, and coma may also occur. The heart rate may decrease following oral exposure or increase following dermal exposure. Hypotension (low blood pressure) and chest pain may be noted. Hypertension (high blood pressure) is not uncommon. Respiratory effects may include dyspnea (shortness of breath), respiratory depression, and respiratory paralysis. Psychosis may occur. Dimethoate is very toxic; the probably oral lethal dose in humans is between 50-500 mg/kg, or between 1 teaspoon and 1 ounce for a 70 kg (150 lb) person. Dimethoate is a cholinesterase inhibitor, meaning it affects the central nervous system. Death is due to respiratory arrest arising from failure of respiratory center, paralysis of respiratory muscles, intense bronchoconstriction or all three, Dimethoate is a mutagen. Mutagens may have a cancer risk. All contact with this chemical should be reduced to the lowest possible level. Delayed pulmonary edema may occur after inhalation.

*Long Term Exposure:* Repeated or prolonged contact with skin may cause dermatitis. Cholinesterase inhibitor; cumulative effect is possible. This chemical may damage the nervous system with repeated exposure, resulting in convulsions, respiratory failure. May cause liver damage. Animal tests indicate that this chemical possibly causes toxic effects upon human reproduction.

*Points of Attack:* Respiratory system, lungs, central nervous system, cardiovascular system, skin, eyes, plasma and red blood cell cholinesterase.

*Medical Surveillance:* Medical observation is recommended for 24 to 48 hours after breathing overexposure, as pulmonary edema may be delayed. Before employment and at regular times after that, the following are recommended: Plasma and red blood cell cholinesterase levels (tests for the enzyme poisoned by this chemical). If exposure stops, plasma levels return to normal in 1-2 weeks while red blood cell levels may be reduced for 1-3 months. When acetylcholinesterase enzyme levels are reduced by 25% or more below preemployment levels, risk of poisoning is increased, even if results are in lower ranges or "normal." Reassignment to work not involving organophosphate or carbamate pesticides is recommended until enzyme levels recover. If symptoms develop or overexposure occurs, repeat the above tests as soon as possible and get an exam of the nervous system. Do not drink any alcoholic beverages before or during use. Alcohol promotes absorption of organophosphates.

*First Aid:* **Treatment for organophosphate poisoning consists of thorough decontamination, cardiorespiratory support, and administration of the antidotes atropine and pralidoxime. In cases of severe poisoning, diazepam, an anticonvulsant, should also be administered. Antidotes should be administered as prevention even if the diagnosis is in doubt.** Speed in removing material from eyes and skin is of extreme importance. *Eyes:* Eye contact can cause dangerous amounts of these chemicals to be quickly absorbed through the mucous membrane into the bloodstream. Immediately and gently flush eyes with plenty of warm or cold water (NO hot water) for at least 15 minutes, occasionally lifting the upper and lower eyelids. Get medical aid immediately. *Skin:* Get medical aid. Skin contact can cause dangerous amounts of these chemicals to be absorbed into the bloodstream. Wearing the appropriate PPE equipment and respirator for organophosphate/carbamate pesticides, immediately flush skin with plenty of soap and water for at least 15 minutes while removing contaminated clothing and shoes. Shampoo hair promptly if contaminated. The removed, contaminated clothing and shoes should be double-bagged and left in Hot Zone for later disposal by hazardous materials experts. Skin may also be decontaminated with diluted hypochlorite solution. *Inhalation:* Get medical aid. Do not contaminate yourself. Wearing the appropriate PPE equipment and respirator for organophosphate pesticides, immediately remove the victim from the contaminated area to fresh air. If the victim is not breathing, administer artificial respiration. Do NOT use mouth-to-mouth resuscitation; use bag/mask apparatus. If breathing is difficult, administer oxygen through bag/mask apparatus until medical help arrives. Do not leave victim unattended. *Ingestion:* Call poison control. Loosen all clothing. Never give anything by mouth to an unconscious person. If victim is *unconscious or having convulsions,* do nothing except keep victim warm. Get medical aid. Transfer promptly to a medical facility. In cases of ingestion, **do not induce vomiting**. If the victim is alert and asymptomatic, administer a slurry of activated charcoal at a dose of 1 g/kg (infant, child, and adult dose). A soda can and straw may be of assistance when offering charcoal to a child. *In some cases you may be specifically instructed by poison control to induce vomiting by way of 2 tablespoons of syrup of ipecac (adult) washed down with a cup of water.* Do NOT give activated charcoal before or with ipecac syrup.

*Note to physician or authorized medical personnel:* Treat cases of respiratory compromise, coma, or excessive pulmonary secretions with respiratory support using protocols and techniques available and within the scope of training. Some cases may necessitate procedures such as endotracheal intubation or cricothyrotomy by properly trained and equipped personnel. When possible, atropine (see *Antidotes,* below) should be given under medical supervision. Patients who are comatose, hypotensive, or having seizures or cardiac arrhythmias should be treated according to advanced life support protocols. *Antidotes:* Two antidotes are administered to treat organophosphate poisoning. Atropine is a competitive antagonist of acetylcholine at muscarinic receptors and is used to control the excessive bronchial secretions which are often responsible for death. Pralidoxime relieves both the nicotinic and muscarine effects of organophosphate poisoning by regenerating acetylcholinesterase and can reduce both the bronchial secretions and the muscle weakness associated with poisoning. The initial intravenous dose of atropine in adults should be determined by the severity of symptoms: An initial adult dose of 1.0 to 2.0 mg or pediatric dose of 0.01 mg/kg (minimum 0.01 mg) should be administered intravenously. If intravenous access cannot be established, atropine may also be given intramuscularly, subcutaneously or via endotracheal tube. Doses should be repeated every 15 minutes until excessive secretions and sweating have been controlled. Once bronchial secretion has been controlled, atropine administration should be repeated whenever the secretions begin to recur. In seriously poisoned patients, very large doses may be required. Alterations of pulse rate and pupillary size should not be used as indicators of treatment adequacy. Pralidoxime should be administered as early in poisoning as possible as its efficacy may diminish when given more than 24 to 36 hours after exposure. Doses are as follows: adult 1.0 g; pediatric 25 to 50 mg/kg. The drug should be administered intravenously over 30 to 60 minutes, but in a life-threatening situation, one-half of the total dose can be given per minute for a total administration time of 2 minutes. Treatment should begin to take effect within 40 minutes with a reduction in symptoms and the amount of atropine necessary to control bronchial secretion. The initial dose can be repeated in 1 hour and then every 8 to 12 hours until the patient is clinically well and no longer requires atropine. If intravenous access cannot be established, pralidoxime may also be given intramuscularly. Early administration of diazepam in addition to the combined atropine and pralidoxime treatment may help prevent the onset of seizures and potential brain and cardiac morphologic damage following high-level organophosphate poisoning.

*References:*
- EXTOXNET, Extension Toxicology Network, "Pesticide Information Profile, Dimethoate," Oregon State University, Corvallis, OR. (June 1996). http://ace.orst.edu/info/extoxnet/pips/dimethoa.htm
- EPA, Office of Pesticide Programs, Pesticide Residue Limits, "Dimethoate", 40 CFR 180.204, http://www.epa.gov/pesticides/food/viewtols.htm
- U.S. Environmental Protection Agency, "Chemical Profile: Dimethoate," Washington, DC, Chemical Emergency Preparedness Program (November 30, 1987).
- California Environmental Protection Agency "Chemical List of Lists," Sacramento CA (February 1997).
- Agency for Toxic Substances and Disease Registry, U.S. Department of Health and Human Services, Public Health Service, "Managing Hazardous Materials Incidents," Atlanta, GA (June 2003).
- New Jersey Department of Health, "Hazardous Substance Fact Sheet: Dimethoate," Trenton, NJ (February 1999). http://www.state.nj.us/health/eoh/rtkweb/0733.pdf

# Dimethomorph

*Use Type:* Fungicide
*CAS Number:* 110488-70-5
*Formula:* $C_{21}H_{22}ClNO_4$

**Synonyms:** (*E,Z*)-4-[3-(4-Chlorophenyl)-3-(3,4-dimethoxyphenyl)acryloyl]morpholine; 4-[3-(4-Chlorophenyl)-3-(3,4-dimethoxyphenyl)-1-*oxo*-2-propenyl]morpholine; 3-(4-Chlorophenyl)-3-(3,4-dimethyphenyl)acrylic acid morpholide; Morpholine, 3-[3-(4-chlorophenyl)-3-(3,4-dimethoxyphenyl)-1-oxo-2-propenyl]-

**Trade Names:** ACROBAT® WP; FORUM DC®, [mancozeb + dimethomorph], BASF Canada (Canada), BASF Agricultural Products Group (Germany); BASF Canada (Canada); CME 151®; STATURE®, BASF Agricultural Products Group (Germany)

**Producers:** BASF Agricultural Products Group (Germany); BASF Canada (Canada); Epochem Co., (China); Shenyang Harvest Agrochemical (China); Wuzhou International (China)

**Chemical Class:** Morpholine

**EPA/OPP PC Code:** 268800

**California DPR Chemical Code:** 4003

**Uses:** A systemic fungicide that protects crops from mold. It also kills mold and prevents their spread; controls late blight on tomatoes; and is used as a wood preservative to control downey mildew.

**Human toxicity (long-term)**[77]: Very low–700.00 ppb, Health Advisory

**Fish toxicity (threshold)**[77]: Intermediate–34.00 ppb, MATC (Maximum Acceptable Toxicant Concentration)

**U.S. Maximum Allowable Residue Levels for Dimethomorph (40 CFR 180.493):**

| CROP | ppm |
|---|---|
| Brassica, leafy greens, subgroup 5B. | 20.0 |
| Grape | 3.5 |
| Grape, raisin | 6.0 |
| Hop, dried cones | 60 |
| Lettuce, head | 10 |
| Lettuce, leaf | 10 |
| Potato, wet peel | 0.15 |
| Taro, corm | 0.5 |
| Taro, leaves | 6.0 |
| Vegetable, bulb, group 3 | 2.0 |
| Vegetable, cucurbit, group 9 | 0.5 |
| Vegetable, fruiting, group 8 | 1.5 |

* There are no U.S. registrations as of August 25, 2000, for the use of dimethomorph on the growing crops, grape, hop, and raisins.

**Carcinogen/Hazard Classifications**

**U.S. EPA Carcinogens:** Not likely a carcinogen

**Label Signal Word:** CAUTION

**WHO Acute Hazard:** Class U, unlikely to be hazardous

**Description:** Colorless or gray crystalline solid. Odorless. Commercial product is a mixture of *cis*- and *trans*-isomers. Practically insoluble in water; solubility = <0.005 ppm.

**Harmful Effects and Symptoms**

**Short Term Exposure:** Contact with eyes or skin may cause irritation or injury. Inhalation should be avoided; use NIOSH-approved air purifying respirators for pesticides. May be harmful if swallowed.

**First Aid:** If this chemical gets into the eyes, remove any contact lenses at once and irrigate immediately for at least 15 minutes, occasionally lifting upper and lower lids. Seek medical attention immediately. If this chemical contacts the skin, remove contaminated clothing and wash immediately with soap and water. Seek medical attention immediately. If this chemical has been inhaled, remove from exposure, begin rescue breathing (using universal precautions) if breathing has stopped, and CPR if heart action has stopped. Transfer promptly to a medical facility. When this chemical has been swallowed, get medical attention. Give large quantities of water and induce vomiting. Do not make an unconscious person vomit.

**References:**
- U.S. Environmental Protection Agency, Office of Pesticide Programs, Pesticide Residue Limits, "Dimethomorph," 40 CFR 180.493. http://ecfr.gpoaccess.gov/cgi/t/text/text-idx?c=ecfr&sid=8c864e7c447828dfa4ba332edba93ad6&rgn=div8&view=text&node=40:21.0.1.1.27.3.27.264&idno=40
- EXTOXNET, Extension Toxicology Network, "Pesticide Information Profile, Dimethomorph," Oregon State University, Corvallis, OR (September 1995). http://extoxnet.orst.edu/pips/dimetomo.htm

# Dimetilan

**Use Type:** Insecticide

**CAS Number:** 644-64-4

**Formula:** $C_{10}H_{16}N_4O_3$

**Alert:** No longer produced in the U.S.

**Synonyms:** Carbamic acid, dimethyl-, 1-[(dimethylamino)carbonyl]-5-methyl-1*H*-pyrazol-2-yl ester; Carbamic acid, dimethyl-, ester with 3-hydroxy-*N,N*-5-trimethylpyrazole-1-carboxamide; Dimethyl carbamate ester of 3-hydroxy-*N,N*-5-trimethylpyrazole-1-carboxamide; Dimethylcarbamic acid-1-[(dimethylamino)carbonyl]-5-methyl-1*H*-pyrazol-3-yl ester; Dimethylcarbamic acid ester with 3-hydroxy-*N,N*,5-trimethylpyrazole-1-carboxamide; Dimethylcarbamic acid-5-methyl-1*H*-carboxamine; Dimethylcarbamic acid-5-methyl-1*H*-pyrazol-3-yl ester; 2-Dimethylcarbamoyl-3-methylpyrazolyl-(5)-*N,N*-dimethylcarbamat; Dimethylcarbamoyl-3-methyl-5-pyrazolyldimethylcarbamate; 1-Dimethylcarbamoyl-5-methylpyrazol-3-yl dimethylcarbamate; Dimetilane; ENT 25,595-X; ENT 25,922; 3-Hydroxy-*N,N*,5-trimethylpyrazole-1-carboxamidedimethylcarbamate (ester); 5-Methyl-1*H*-pyrazol-3-yl dimethylcarbamate

**Trade Names:** FLYBANDS®, canceled; GEIGY 22870®, Syngenta Crop Protection (Switzerland), canceled; GEIGY GS-13332®, Syngenta Crop Protection (Switzerland),

canceled; SNIP®, Syngenta Crop Protection (Switzerland), canceled; SNIP FLY®, Syngenta Crop Protection (Switzerland), canceled
**Producers:** Sigma-Aldrich Laborchemikalien (Germany); Syngenta Crop Protection (Switzerland)
**Chemical Class:** Carbamate
**EPA/OPP PC Code:** 090101
**RTECS Number:** EZ9084000
**Uses:** Formerly an insecticide for insect control on livestock, especially housefly control. It is no longer produced commercially in the U.S.
**Regulatory Authority:**
- Banned or Severely Restricted (Portugal) (UN)[13]
- RCRA, 40CFR261, Appendix 8 Hazardous Constituents
- RCRA 40CFR268.48; 61FR15654, Universal Treatment Standards: Wastewater (mg/L), 0.056; Nonwastewater (mg/kg), 1.4
- Superfund/EPCRA 40CFR355, Appendix B Extremely Hazardous Substances: TPQ = 500/10,000 lb (227/4,540 kg)
- Superfund/EPCRA 40CFR302.4 RQ: EHS, 1 lb (0.454 kg)

**Description:** Dimetilan, Yellow to reddish-brown solid. Highly soluble in water. Molecular weight = 240.30. Melting/Freezing point = 68-71°C (the technical grade at 55-65°C). Hazard Identification (based on NFPA-704 M Rating System): Health 3, Flammability 1, Reactivity 0.
**Incompatibilities:** Hydrolyzed by acids and alkalis.
**Permissible Exposure Limits in Air:** No standards set.
**Permissible Concentration in Water:** No criteria set, but runoff from spills or fire control may cause water pollution.
**Routes of Entry:** Skin contact, inhalation.
**Harmful Effects and Symptoms**
**Short Term Exposure:** Very toxic; probable oral lethal dose for humans is 50-500 mg/kg or between 1 teaspoon and 1 oz for a 70 kg (150 lb) person. Dimetilan is highly toxic by ingestion and moderately toxic by contact with the skin. Death is primarily due to respiratory arrest of central origin, paralysis off the respiratory muscles, intense bronchoconstriction, or all three. This compound is a cholinesterase inhibitor. Symptoms are similar to carbaryl poisoning: nausea, vomiting, abdominal cramps, diarrhea, pinpoint pupils, excessive salivation, and sweating are common symptoms. Running nose and tightness in chest are common in inhalation exposures. Difficulty in breathing, raspy breathing, and loss of muscle coordination may also be seen. Exposure may also result in random jerky movements, incontinence, convulsions, and coma and death.
**Long Term Exposure:** Many carbamates affect the central nervous system.
**Points of Attack:**
**Medical Surveillance:** Before employment and at regular times after that, the following are recommended: Plasma and red blood cell cholinesterase levels (tests for the enzyme poisoned by this chemical). If exposure stops, plasma levels return to normal in 1-2 weeks while red blood cell levels may be reduced for 1-3 months. When acetylcholinesterase enzyme levels are reduced by 25% or more below preemployment levels, risk of poisoning is increased, even if results are in lower ranges of "normal." Reassignment to work not involving organophosphate or carbamate pesticides is recommended until enzyme levels recover. If symptoms develop or overexposure occurs, repeat the above tests as soon as possible and get an exam of the nervous system. Also consider complete blood count. Consider chest x-ray following acute overexposure. Do not drink any alcoholic beverages before or during use. Alcohol promotes absorption of organophosphates. Carbamate insecticides inhibit the cholinesterase activity of enzymes, causing accumulation of acetylcholine at synapses and altering the way in which nervous impulses are transmitted. However, within several hours carbamates spontaneously detach from the enzymes.
**First Aid:** Speed in removing material from eyes and skin is of extreme importance. *Eyes:* Eye contact can cause dangerous amounts of these chemicals to be quickly absorbed through the mucous membrane into the bloodstream. Immediately and gently flush eyes with plenty of warm or cold water (NO hot water) for at least 15 minutes, occasionally lifting the upper and lower eyelids. Get medical aid immediately. *Skin:* Get medical aid. Skin contact can cause dangerous amounts of these chemicals to be absorbed into the bloodstream. Wearing the appropriate PPE equipment and respirator for carbamate pesticides, immediately flush skin with plenty of soap and water for at least 15 minutes while removing contaminated clothing and shoes. Shampoo hair promptly if contaminated; protect eyes. *Ingestion:* Call poison control. Loosen all clothing. Never give anything by mouth to an unconscious person. Get medical aid. Do NOT induce vomiting.* If conscious, alert, and able to swallow, rinse mouth and have victim drink 4 to 8 ounces of water. Check to see if poison control instructs you to use ipecac syrup, otherwise administer slurry of activated charcoal (2 oz in 8 oz of water). If victim is *UNCONSCIOUS OR HAVING CONVULSIONS,* do nothing except keep victim warm. *In some cases you may be specifically instructed by poison control to induce vomiting by way of 2 tablespoons of syrup of ipecac (adult) washed down with a cup of water.* Do NOT give activated charcoal before or with ipecac syrup. *Inhalation:* Get medical aid. Do not contaminate yourself. Wearing the appropriate PPE equipment and respirator for carbamate pesticides, immediately remove the victim from the contaminated area to fresh air. If the victim is not breathing, administer artificial respiration. Do NOT use mouth-to-mouth resuscitation; use bag/mask apparatus. If breathing is difficult, administer oxygen through bag/mask apparatus until medical help arrives. Do not leave victim unattended. *Note to physician or authorized medical personnel.* Administer atropine, 2 mg (1/30 gr) intramuscularly or

intravenously as soon as any local or systemic signs or symptoms of an intoxication are noted; repeat the administration of atropine every 3 to 8 minutes until signs of atropinization (mydriasis, dry mouth, rapid pulse, hot and dry skin) occur; initiate treatment in children with 0.05 mg mg/kg of atropine; repeat at 5 to 10 minute intervals. Watch respiration, and remove bronchial secretions if they appear to be obstructing the airway; intubate if necessary.

*Medical note:* 2-PAMCI may be contraindicated in the case of some carbamate poisonings.

*References:*
- U.S. Environmental Protection Agency, "Chemical Profile: Dimetilan," Washington, DC, Chemical Emergency Preparedness Program (November 30, 1987).
- California Environmental Protection Agency "Chemical List of Lists," Sacramento CA (February 1997).

# Dinex

*Use Type:* Pesticide, acaricide
*CAS Number:* 131-89-5
*Formula:* $C_{12}H_{14}N_2O_5$
*Synonyms:* 6-Cyclohexyl-2,4-dinitrophenol; 2-Cyclohexyl-4,6-dinitrophenol; Dinitrocyclophenol; 4,6-Dinitro-*o*-cyclohexylphenol; 2-Cyclohexyl-4,6-dinitrophenol; Dinitrocyclohexylphenol; Dinitro-*o*-cyclohexylphenol; 2,4-Dinitro-6-cyclohexylphenol; 4,6-Dinitro-*o*-cyclohexylphenol; DNOCHP; ENT 157; Phenol, 2-cyclohexyl-4,6-dinitro-
*Trade Names:* DN®; DN-111®; DN dry mix; MIX No. 1®; DN dust No. 12®; DOW SPRAY®-17, Dow AgroSciences (USA); DRY MIX No. 1®; PEDINEX®; SN 46®
*Producers:* Dow AgroSciences (USA)
*Chemical Class:* Dinitrophenol
*EPA/OPP PC Code:* 037501
*RTECS Number:* SK6650000
*RCRA Number:* P034
*Uses:* Not registered in the U.S. Used to control mites on citrus fruits.
*Regulatory Authority:*
- Clean Water Act: Section 311 Hazardous Substances/RQ (same as CERCLA); Section 307 Toxic Pollutants as nitrophenols.
- EPCRA Section 304 RQ: CERCLA, 100 lb (45.4 kg)as nitrophenols.
- Marine pollutant (49CFR, Subchapter 172.101, Appendix B) as nitrophenols.

*Description:* Yellow, crystalline solid. Very slightly soluble in water. Molecular weight = 266.25. Melting/Freezing point = 107°C; (NFPA) Health Hazards (Blue): 2; Flammability (Red): 2; Reactivity (Yellow): 2; Relative vapor Density = 9.2. Log $K_{ow}$ = 4.63. Values above 3.0 are likely to bioaccumulate in marine organisms.

*Incompatibilities:* Reacts with oxidizing materials and combustibles; risk of fire and explosion. Incompatible with sulfuric acid, nitric acid, caustics, aliphatic amines, isocyanates. React with boranes, alkalies, aliphatic amines, amides, nitric acid, sulfuric acid. May detonate when heated under confinement.

*Determination in Air:* Filter; none; Gravimetric; NIOSH IV [Particulates NOR; #0500 (total), #0600 (respirable)][18]

*Permissible Concentration in Water:* No criteria set. Runoff from spills or fire control may cause water pollution. To protect freshwater aquatic life: 230 µg/L on an acute toxicity basis for nitrophenols as a class. To protect saltwater aquatic life: 4,850 µg/L on an acute toxicity basis for nitrophenols as a class. To protect human health: 70.0 µg/L[6].

*Determination in Water:* Filter/Bubbler; 2-Propanol; High-pressure liquid chromatography/Ultraviolet detection; NIOSH II[5] Method #S166. Methylene chloride extraction followed by gas chromatography with flame ionization or electron capture detection. EPA Method 604, or gas chromatography plus mass spectrometry EPA Method 625.

*Routes of Entry:* Skin contact, inhalation and ingestion.

*Harmful Effects and Symptoms*

*Short Term Exposure:* Solid is poisonous if swallowed. Grade 4: $LD_{50}$ = 50 mg/kg (mouse). Severely irritates eyes, skin and respiratory tract, with burning sensation, pain, redness and swelling. Metabolic stimulant. If inhaled, causes coughing, dilated pupils, headache, profuse persperation, intense thirst, extreme fatigue, rapid pulse, high fever, clammy, flushed skin, rapid breathing, yellowish face and lips, anxiety, convulsions, risk of lung edema. If swallowed, face and lips turn bluish. Severe exposure can cause death from heart failure.

*Long Term Exposure:* May damage the liver, kidneys and blood cells. May stain yellow the skin, eyes, and fingernails. Repeated exposure can cause anxiety, fatigue, insomnia, excessive perspiration, unusual thirst, weight loss and cataracts in the eyes.

*Points of Attack:* Skin, liver, kidneys, lungs, peripheral nervous system, eyes, thyroid gland and blood.

*Medical Surveillance:* Before beginning employment, at regular times after that and if symptoms develop or overexposure has occurred, the following may be useful: Exam of eyes for cataracts. Exam of skin and nails for staining. Blood tests for dinitro-*o*-cresol. Persons with blood levels over 10 ppm (10 mg/L) should be kept away from further exposure until levels return to normal. If symptoms develop or overexposure is suspected, the following may be useful: Liver and kidney function tests. Complete blood count.

*First Aid:* If this chemical gets into the eyes, remove any contact lenses at once and irrigate immediately for at least 15 minutes, occasionally lifting upper and lower lids. Seek medical attention immediately. If this chemical contacts the skin, remove contaminated clothing and wash immediately

with soap and water. Seek medical attention immediately. If this chemical has been inhaled, remove from exposure, begin rescue breathing (using universal precautions) if breathing has stopped, and CPR if heart action has stopped. Transfer promptly to a medical facility. When this chemical has been swallowed, get medical attention. *Do not induce vomiting when formulations containing petroleum solvents are ingested.* Otherwise, give large quantities of water and induce vomiting. Do not make an unconscious person vomit. *Note to Physician:* Treat for methemoglobinemia. Spectrophotometry may be required for precise determination of levels of methemoglobinemia in urine.

*References:*
- New Jersey Department of Health and Senior Services, "Hazardous Substance Fact Sheet, Dinex," Trenton NJ (May 2002). http://www.state.nj.us/health/eoh/rtkweb/0774.pdf
- U.S. Environmental Protection Agency, Nitrophenols: Ambient Water Quality Criteria, Washington, DC (1980).
- U.S. Environmental Protection Agency, 2,4-Dinitrophenol, Health and Environmental Effects Profile No. 91, Washington, DC, Office of Solid Waste (April 30, 1980).
- Sax, N.I., Ed., "Dangerous Properties of Industrial Materials Report," 2, No. 2, 25-27 (1982) and 3, No. 2, 38-44 (1983).

# Diniconazole

*Use Type:* Fungicide
*CAS Number:* 83657-18-5
*Formula:* $C_{15}H_{17}Cl_2N_3O$
*Synonyms:* [(R-(E)]-1-(2,4-Dichlorophenyl)-4,4-dimethyl-2-(1H-1,2,4-triazol-1-yl)pent-1-en-3-ol; Diniconazole M; 1H-1,2,4-Triazole-1-ethanol, β-[(2,4-dichlorophenyl)methylene]-α-(1,1,-dimethylethyl)-, [R-(E)]-
*Trade Names:* EMBASSADOR®, Sanonda (Australia); (R)-S 3308®; S-3308 L®; SPOTLESS (XE-779L)W®; SUMI-EIGHT® 12.5 WP
*Producers:* Ascot International UK); Sanonda (Australia); SuYan Agrochemical Group (China)
*Chemical Class:* Azole
*EPA/OPP PC Code:* 128932
*California DPR Chemical Code:* 2500
*Uses:* Not registered in the U.S. It is a broad spectrum fungicide that is used on corn, wheat, peanut, apple, pear etc. to control covered smut, powdery mildew, etc. It is effective on powdery mildew, rust of coffee, flowers and vegetable.
*Carcinogen/Hazard Classifications*
**WHO Acute Hazard:** Class III, slightly hazardous
*Description:* White, odorless crystals. Molecular weight = 326.24.
*Incompatibilities:* Strong oxidizers.

*Permissible Concentration in Water:* No criteria set. Runoff from spills or fire control may cause water pollution.
*Routes of Entry:* Ingestion, skin contact
*Harmful Effects and Symptoms*
*Short Term Exposure:* Poisonous if swallowed. Contact may irritate skin and cause eye irritation and possible severe injury. Avoid inhalation.
*First Aid:* If this chemical gets into the eyes, remove any contact lenses at once and irrigate immediately for at least 15 minutes, occasionally lifting upper and lower lids. Seek medical attention immediately. If this chemical contacts the skin, remove contaminated clothing and wash immediately with soap and water. Seek medical attention immediately. If this chemical has been inhaled, remove from exposure, begin rescue breathing (using universal precautions) if breathing has stopped, and CPR if heart action has stopped. Transfer promptly to a medical facility. When this chemical has been swallowed, get medical attention. Give large quantities of water and induce vomiting. Do not make an unconscious person vomit. Do not induce vomiting when formulations containing petroleum solvents are ingested.

# Dinitro-o-cresol (DNOC)

*Use Type:* Herbicide, fungicide and pesticide
*CAS Number:* 534-52-1
*Formula:* $C_7H_6N_2O_5$
*Alert:* DNOC's registration in the U.S. as a pesticide was canceled in 1991.
*Synonyms:* o-Cresol, 4,6-dinitro-; Dinitro; Dinitrocresol; Dinitro-*ortho*-cresol; Dinitrodendtroxal; 3,5-Dinitro-2-hydroxytoluene; Dinitrol; 4,6-Dinitro-2-methylphenol; 2,4-Dinitro-6-methylphenol; Dinitrocresol; 3,5-Dinitro-o-cresol; 4,6-Dinitro-o-cresol and salts; 2-Methyl-4,6-dinitrophenol; 6-Methyl-2,4-dinitrophenol; Phenol, 2-methyl-4,6-dinitro-
*Trade Names:* ANTINONIN®; ANTINONNIN®; ARBOROL®; DEGRASSAN®; DEKRYSIL®; DETAL®; DILLEX®; DINOC®; DINURANIA®; DITROSOL®; DNOC®, Gowan Co. (USA), canceled; EFFUSAN®; EFFUSAN 3436®; ELGETOL®; ELGETOL 30®; ELIPOL®; EXTRAR®; FLAVIN-SANDOZ®; HEDOLIT®; HEDOLITE®; K III®; K IV®; KREOZAN®; KREZOTOL 50®; LIPAN®; NEUDORFF DN 50®; NITROFAN®; PROKARBOL®; RAFEX®; RAFEX 35®; RAPHATOX®; SANDOLIN®; SANDOLIN A®; SELINON®; SINOX®; WINTERWASH®
*Producers:* A H Marks (UK); ATOFINA (France); BASF Agricultural Products Group (Germany); Bayer Chemicals (Germany); Calliope (France); Cyanamid Agro Products Group (USA); Eli Lilly (USA); Gowan Co. (USA); Rhone-Poulenc Agro (France); Shell Chemical (UK); SNPE Chemie (France)
*Chemical Class:* Derivative of dinitrophenol
*EPA/OPP PC Code:* 037507 and 600023

# Dinitro-o-cresol (DNOC)

*California DPR Chemical Code:* 3170
*RTECS Number:* GO9625000
*Uses:* DNOC is widely used in agriculture as a herbicide and pesticide; it is also used in the dyestuff industry. Although 4,6-dinitro-*o*-cresol (DNOC) is no longer manufactured in the United States (since 1991), it was used as a blossom-thinning agent on fruit trees and as a fungicide, insecticide, and miticides on fruit trees during the dormant season. It is used in mushroom houses to control foreign fungi; to kill locusts and other insects; and as a pre-harvest desiccant of potatoes and leguminous seed crops.
*Fish toxicity (threshold)*[77]: High–6.33138 ppb, MATC (Maximum Acceptable Toxicant Concentration)
*Carcinogen/Hazard Classifications*
**Label Signal Word:** CAUTION
**WHO Acute Hazard:** Group 1B, highly hazardous
*Regulatory Authority:*
- Banned or Severely Restricted (Sweden) (UN)[13]
- Air Pollutant Standard Set (ACGIH)[1] (DFG)[3] (HSE)[33] ()SHA)[58] (former USSR)[43] (Several States)[60]
- List of priority pollutants (U.S. EPA)
- AB 1803-Well Monitoring Chemical (CAL)
- EPA/SARA 302 (EPCRA) Extremely hazardous substances
- AB 2588-Air Toxics "Hot Spots" Chemicals (CAL)
- CAL Air Resources Board/AB 1807 Toxic Air Contaminants
- Permissible Exposure Limits for Chemical Contaminants (CAL/OSHA)
- The "Director's List" (CAL/OSHA)
- Clean Air Act: Hazardous Air Pollutants (Title I, Part A, Section 112)
- Clean Water Act: Section 313 Water Priority Chemicals (57FR41331, 9/9/92)
- EPA Hazardous Waste Number (RCRA No.): P047
- RCRA, 40CFR261, Appendix 8 Hazardous Constituents
- RCRA 40CFR268.48; 61FR15654, Universal Treatment Standards: Wastewater (mg/L), 0.28; Nonwastewater (mg/kg), 160
- RCRA 40CFR264, Appendix 9; TSD Facilities Ground Water Monitoring List. Suggested test method(s) (PQL *ug*/L): 8040(150); 8270(50)
- Superfund/EPCRA 40CFR355, Appendix B Extremely Hazardous Substances: TPQ = 10/10,000 lb (4.54/4,540 kg)
- Superfund/EPCRA 40CFR302.4 RQ: CERCLA, 10 lb (4.54 kg)
- EPCRA Section 313 Form R *de minimus* concentration reporting level: 1.0%
- U.S. DOT Regulated Marine Pollutant (49CFR172.101, Appendix B)
- Canada, WHMIS, Ingredients Disclosure List

*Description:* This chemical exists in nine isomeric forms of which 4,6-dinitro-*o*-cresol is the most important commercially, and the most heavily regulated. It is a noncombustible, yellow crystalline solid. Practically insoluble in water; solubility = 130 ppm @ 15°C. Molecular weight = 198.13. Boiling point = 312°C. Melting/Freezing point = 85–87°C. Vapor Pressure = $1.1 \times 10^{-4}$ mmHg @ 25°C. Log $K_{ow}$ = 2.6. Unlikely to bioaccumulate in marine organisms.
*Incompatibilities:* Dust can form an explosive mixture with air. Strong oxidizers, oxidizers, strong bases. Protect from heat and shock.
*Incompatibilities:* Strong oxidizers.
*Permissible Exposure Limits in Air:* The Federal Limit (OSHA PEL)[58], the DFG[3] and HSE[33] values and recommended ACGIH[1] TWA value for all isomers of DNOC is 0.2 mg/m³. The notation "skin" indicates the possibility of cutaneous absorption. The HSE[33] STEL value is 0.6 mg/m³. The NIOSH[2] IDLH level = 5.0 mg/m³. The MAC in workplace air set by The former USSR-UNEP/IRPTC project[43] is 0.05 mg/m³. They have also set MAC values for the ambient air in residential areas of 0.003 mg/m³ on a momentary basis and 0.0008 mg/m³ on a daily average basis. Several states have set guidelines or standards for DNOC in ambient air[60] ranging from 2.0 µg/m³ (North Dakota) to 3.0 µg/m³ (Virginia) to 4.0 µg/m³ (Connecticut) to 5.0 µg/m³ (Nevada).
*Determination in Air:* Collection by charcoal tube, analysis by gas liquid chromatography.
*Permissible Concentration in Water:* To protect human health, 13.4 µg/L[6]. The former USSR[35] has set an MAC in water bodies used for domestic purposes of 0.05 mg/L and in water bodies used for fishery purposes of 0.002 mg/L[43].
*Determination in Water:* Filter/Bubbler; 2-Propanol; High-pressure liquid chromatography/Ultraviolet detection; NIOSH II(5) Method #S166[18]. Methylene chloride extraction followed by gas chromatography with flame ionization or electron capture detection. EPA Method 604, or gas chromatography plus mass spectrometry EPA Method 625.
*Routes of Entry:* Inhalation, percutaneous absorption, ingestion, eye and/or skin contact.
*Harmful Effects and Symptoms*
*Short Term Exposure:* Early manifestations of acute dinitrocresol exposure include fever, sweating, headache, and confusion. Blood pressure, pulse, and respiratory rate are often elevated. Severe exposure may result in restlessness, seizures, and coma. Other signs and symptoms include dyspnea (shortness of breath), cyanosis (blue tint to skin and mucous membranes), pulmonary edema, nausea, vomiting, and abdominal pain. Liver injury with associated jaundice, kidney failure, and cardiac arrhythmias are commonly noted. Dermal exposure results in yellow staining of the skin and may produce burns. Dinitrocresol may irritate and burn the eyes and mucous membranes. DNOC is an extremely toxic material; probable oral lethal dose is 5-50 mg/kg in humans or between 7 drops and 1 teaspoon for

a 70 kg (150 lb) person. Delayed pulmonary edema may occur after inhalation.

***Long Term Exposure:*** May damage the liver, kidneys and blood cells. May stain yellow the skin, eyes, and fingernails. Repeated exposure can cause anxiety, fatigue, insomnia, excessive perperation, unusual thirst, weight loss and cataracts in the eyes.

***Points of Attack:*** Cardiovascular system, endocrine system.

***Medical Surveillance:*** Medical observation is recommended for 24 to 48 hours after breathing overexposure, as pulmonary edema may be delayed. Before beginning employment, at regular times after that and if symptoms develop or overexposure has occurred, the following may be useful: Exam of eyes for cataracts. Exam of skin and nails for staining. Blood tests for 4,6-dinitro-*o*-cresol. Persons with blood levels over 10 ppm (10 mg/L) should be kept away from further exposure until levels return to normal. If symptoms develop or overexposure is suspected, the following may be useful: Liver and kidney function tests. Complete blood count.

***First Aid:*** If this chemical gets into the eyes, remove any contact lenses at once and irrigate immediately for at least 15 minutes, occasionally lifting upper and lower lids. Seek medical attention immediately. If this chemical contacts the skin, remove contaminated clothing and wash immediately with soap and water. Seek medical attention immediately. If this chemical has been inhaled, remove from exposure, begin rescue breathing (using universal precautions) if breathing has stopped, and CPR if heart action has stopped. Transfer promptly to a medical facility. When this chemical has been swallowed, get medical attention. Give large quantities of water and induce vomiting. Do not make an unconscious person vomit. If high fever is present, drench victim's clothes in cool water, or immerse person in cool bath before transfer.

***References:***
- EXTOXNET, Extension Toxicology Network, "Pesticide Information Profile," Oregon State University, Corvallis, OR
- U.S. Department of Health and Human Services, Agency for Toxic Substances and Disease Registry, "ToxFAQs for Dinitrocresols," Atlanta, GA (September 1996). http://www.atsdr.cdc.gov/tfacts63.html
- National Institute for Occupational Safety and Health, "Criteria for a Recommended Standards: Occupational Exposure to Dinitro-ortho-Cresol," NIOSH Publication No. 78-131, Washington, DC (1978)/U.S. Environmental Protection Agency, Nitrophenols: Ambient Water Quality Criteria, Washington, DC (1980).
- U.S. Environmental Protection Agency, "4,6-Dinitro-*o*-Cresol, Health and Environmental Effects," Profile No. 90, Washington, DC, Office of Solid Waste (April 30, 1980).
- Sax, N.I., Ed., "Dangerous Properties of Industrial Materials Report," 2, No. 5, 54-59 (1982) and 4, No. 1, 62-66 (1984).
- U.S. Environmental Protection Agency, "Chemical Profile: Dinitrocresol," Washington, DC, Chemical Emergency Preparedness Program (November 30, 1987).
- California Environmental Protection Agency "Chemical List of Lists," Sacramento CA (February 1997).
- New Jersey Department of Health, "Hazardous Substance Fact Sheet: 4,6-Dinitro-*o*-Cresol," Trenton, NJ (June 1998). http://www.state.nj.us/health/eoh/rtkweb/0779.pdf

# Dinocap

***Use Type:*** Fungicide and acaricide
***CAS Number:*** 39300-45-3
***Formula:*** $C_{18}H_{24}N_2O_6$
***Alert:*** Highly explosive and flammable. Highly reactive when dry.
***Synonyms:*** 2-Butenoic acid, 2-isooctyl-4,6-dinitrophenyl ester; 2-Butenoic acid, 4-isooctyl-2,6-dinitrophenyl ester; 2-Butenoic acid 2-(1-methylheptyl)-4,6-dinitrophenyl ester; Capryldinitrophenyl crotonate; 2-Capryl-4,6-dinitrophenyl crotonate; Caswell No. 391D; CPC; Crotonic acid 2,4-dinitro-6-(1-methylheptyl)phenyl ester; Crotonic acid 2,4-dinitro-6-(2-octyl)phenyl ester; Crotonic acid, 2-(1-methylheptyl)-4,6-dinitrophenyl ester; Crotonic acid, 4-(1-methylheptyl)-2,6-dinitrophenyl ester; Crotonic acid 2-(1-methylheptyl)-4,6-dinitrophenyl ester; DCPC; 4,6-Dinitro-2-(2-capryl)phenyl crotonate; 4,6-Dinitro-2-caprylphenyl crotonate; Dinitrocaprylphenyl crotonate; Dinitro(1-methylheptyl)phenyl crotonate; 2,4-Dinitro-6-(1-methylheptyl)phenyl crotonate; 4,6-Dinitro-2-(1-methylheptyl)phenyl crotonate; 2,4-Dinitro-6-octyl* phenyl crotonate, 2,6-Dinitro-4-octyl* phenylcrotonate, and nitrooctylphenols (principally dinitro); Dinitromethylheptyphenyl crotonate; Dinitro(1-methylheptyl)phenyl crotonate; 2,4-Dinitro-6-(2-octyl)phenyl crotonate; 2,4-Dinitro-6-octyl-phenyl crotonate; 2,6-Dinitro-4-octyl-phenyl crotonate; DNOCP; DNOPC; DPC; ENT 24727; [6-(1-Methyl-heptyl)-2,3-dinitro-phenyl]-crotonat (German); 2-(1-Methylheptyl)-4,6-dinitrophenylcrotonate; Phenol, 2-(1-methylheptyl)-4,6-dinitro-, crotonate (ester)
*(A mixture of 1-methylheptyl, 1-ethylhexyl and 1-propylpentyl derivatives, chiefly 2-(1-methyl)-4,6-dinitrophenol)
***Trade Names:*** ACTUAL DINOCAP®; ARATHANE®; CAPRANE®; CR 1639®; CEKUCAP® 25 WP; CROTOTHANE®; DICAP®; DIKAR®, EZENOAN®; ISCOTHANE®; ISOCOTHANE®; KARATHANE®, canceled; KARATHANE® WD, canceled; KARATHENE®, canceled; MILDANE®; MILDEX®
***Chemical Class:*** Dinitrophenyl

# Dinocap

*EPA/OPP PC Code:* 036001
*California DPR Chemical Code:* 344
*ICSC Number:* 0881(as isomer mixture)
*RTECS Number:* GQ5775000
*EEC Number:* 609-023-00-6
*EINECS Number:* 254-408-0

*Uses:* Not registered in the U.S. Dinocap is a foliar fungicide/miticide used to control powdery mildew. It is applied to limit mites in apples and grapes crops outside of the U.S., mainly in Europe, the Middle East and northern Africa. It is also used to control powdery mildew on fruit, vegetables and ornamentals. There are currently no registered dinocap products in the U.S. DAS, the registrant of dinocap, intends to support tolerances for dinocap residues in/on apples and grapes to permit legal importation of these commodities into the U.S.

*Human toxicity (long-term)*[77]: Intermediate–28.00 ppb, Health Advisory

*Fish toxicity (threshold)*[77]: High–1.06008 ppb, MATC (Maximum Acceptable Toxicant Concentration)

*U.S. Maximum Allowable Residue Levels for Dinocap (40 CFR 180.341):*

| CROP | ppm |
|---|---|
| Apple | 0.1 |
| Grape | 0.1 |

*Carcinogen/Hazard Classifications*
**U.S. EPA Carcinogens:** Group E, Unlikely a carcinogen
**California Prop. 65:** Male developmental toxin
**TRI Developmental Toxin:** Developmental toxin
**Label Signal Word:** CAUTION or DANGER
**WHO Acute Hazard:** Class III, slightly hazardous

*Regulatory Authority:*
- EPA 40 CFR 372.65, Specific Toxic Chemical Listings
- EPCRA Section 313 Form R *de minimus* concentration reporting level: 1.0%
- AB 2588-Air Toxics "Hot Spots" Chemicals (CAL)
- Proposition 65 chemical (CAL)
- Clean Water Act: Section 311 Hazardous Substances/RQ (same as CERCLA); Section 307 Toxic Pollutants as nitrophenols.
- EPCRA Section 304 RQ: CERCLA, 100 lb (45.4 kg)as nitrophenols.
- U.S. DOT Marine pollutant (49CFR, Subchapter 172.101, Appendix B) as nitrophenols.

*Description:* Brown liquid. Insoluble in water. Combustible. Physical and toxicological properties may be affected by the carrier solvents used in commercial formulations. Molecular weight = 364.43. Boiling point = 139°C @ 0.07 mmHg. Vapor pressure = $4 \times 10^{-8}$ mmHg @ 20°C.

*Permissible Exposure Limits in Air:* The former USSR-UNEP/IRPTC joint project[43] has set an MAC in workplace air of 0.05 mg/m$^3$.

*Permissible Concentration in Water:* No criteria set. Runoff from spills or fire control may cause water pollution. To protect freshwater aquatic life: 230 μg/L on an acute toxicity basis for nitrophenols as a class. To protect saltwater aquatic life: 4,850 μg/L on an acute toxicity basis for nitrophenols as a class. To protect human health: 70.0 μg/L[6].

*Determination in Water:* Filter/Bubbler; 2-Propanol; High-pressure liquid chromatography/Ultraviolet detection; NIOSH II[5] Method #S166. Methylene chloride extraction followed by gas chromatography with flame ionization or electron capture detection. EPA Method 604, or gas chromatography plus mass spectrometry EPA Method 625.

*Routes of Entry:* Skin contact, inhalation and ingestion.

*Harmful Effects and Symptoms*

*Short Term Exposure:* Severely irritates eyes, skin and respiratory tract, with burning sensation, pain, redness and swelling. Metabolic stimulant. If inhaled, causes coughing, dilated pupils, headache, profuse persperation, intense thirst, extreme fatigue, rapid pulse, high fever, clammy, flushed skin, rapid breathing, nausea, vomiting, yellowish tint to skin and lips, anxiety and confusion, convulsions, risk of lung edema. If swallowed, face and lips turn bluish. Liver injury with associated jaundice, kidney failure, and cardiac arrhythmias are commonly noted. Severe exposure can cause death from heart failure.

*Long Term Exposure:* May damage the liver, kidneys and blood cells. May stain yellow the skin, eyes, and fingernails. Repeated exposure can cause anxiety, fatigue, insomnia, excessive persperation, unusual thirst, weight loss and cataracts in the eyes.

*Points of Attack:* Skin, liver, kidneys, lungs, peripheral nervous system, eyes, thyroid gland, blood.

*Medical Surveillance:* Before beginning employment, at regular times after that and if symptoms develop or overexposure has occurred, the following may be useful: Exam of eyes for cataracts. Exam of skin and nails for staining. Blood tests for dinitro-*o*-cresol. Persons with blood levels over 10 ppm (10 mg/L) should be kept away from further exposure until levels return to normal. If symptoms develop or overexposure is suspected, the following may be useful: Liver and kidney function tests. Complete blood count.

*First Aid:* If this chemical gets into the eyes, remove any contact lenses at once and irrigate immediately for at least 15 minutes, occasionally lifting upper and lower lids. Seek medical attention immediately. If this chemical contacts the skin, remove contaminated clothing and wash immediately with soap and water. Seek medical attention immediately. If this chemical has been inhaled, remove from exposure, begin rescue breathing (using universal precautions) if breathing has stopped, and CPR if heart action has stopped. Transfer promptly to a medical facility. When this chemical has been swallowed, get medical attention. *Do not induce vomiting when formulations containing petroleum solvents are ingested.* Otherwise, give large quantities of water and induce vomiting. Do not make an unconscious person vomit. *Note to physician or authorized medical personnel:* Treat for methemoglobinemia. Spectrophotometry may be required for precise determination of levels of methemoglobinemia in urine.

*References:*
- U.S. Environmental Protection Agency, "Reregistration Eligibility Decision (RED), Dinocap" Office of Prevention, Pesticides and Toxic Substances, Washington, DC (May 29, 2003). http://www.epa.gov/REDs/dinocap_red.pdf
- U.S. Environmental Protection Agency, Office of Pesticide Programs, Pesticide Residue Limits, "Dinocap," 40 CFR 180.341, http://www.epa.gov/pesticides/food/viewtols.htm
- EXTOXNET, Extension Toxicology Network, "Pesticide Information Profile, Dinocap," Oregon State University, Corvallis, OR (June 1996). http://extoxnet.orst.edu/pips/dinocap.htm
- California Environmental Protection Agency "Chemical List of Lists," Sacramento CA (February 1997)
- U.S. Environmental Protection Agency, Nitrophenols: Ambient Water Quality Criteria, Washington, DC (1980).
- U.S. Environmental Protection Agency, 2,4-Dinitrophenol, Health and Environmental Effects Profile No. 91, Washington, DC, Office of Solid Waste (April 30, 1980).
- Sax, N.I., Ed., "Dangerous Properties of Industrial Materials Report," 2, No. 2, 25-27 (1982) and 3, No. 2, 38-44 (1983).
- New Jersey Department of Health, "Hazardous Substance Fact Sheet: Dinitrophenol," Trenton, NJ (October 1996, rev. February 2003). http://www.state.nj.us/health/eoh/rtkweb/0780.pdf

# Dinoseb (ANSI)

*Use Type:* A plant growth regulator and herbicide.
*CAS Number:* 88-85-7
*Formula:* $C_{10}H_{12}N_2O_5$; $C_6H_2(NO_2)_2(C_4H_9)OH$
*Alert:* It was a Restricted Use Pesticide (RUP). The use of dinoseb was canceled in the U.S. in 1986. This action was based on the potential risk of birth defects and other adverse health effects for applicators and other persons with substantial dinoseb exposure. This pesticide is not commercially available in the U.S. Human toxicity (long-term): Very high.
*Synonyms:* 2-*sec*-Butyl-4,6-dinitrophenol; DBNF; Dinitrall; Dinitrobutylphenol; 4,6-Dinitro-2-*sec*-butylfenol (Czech); 2,4-Dinitro-6-*sec*-butylphenol; 4,6-Dinitro-*o-sec*-butylphenol; 4,6-Dinitro-2-*sec*-butylphenol; 4,6-Dinitro-2-(1-methyl-*N*-propyl)phenol; 2,4-Dinitro-6-(1-methylpropyl)phenol; 4,6-Dinitro-2-(1-methylpropyl)phenol; Dinitro-*ortho-sec*-butylphenol; 4,6-Dinitro-*o-sec*-butylphenol; 2,4-Dinitro-6-*sec*-butylphenol; 4,6-Dinitro-2-*sec*-butylphenol; DN 289; DNBP; DNOSBP; DNPB; DNSBP; ENT 1,122; 6-(1-Methyl-propyl)-2,4-dinitrofenol (Dutch); 2-(1-Methylpropyl)-4,6-dinitrophenol; 6-(1-Metil-propil)-2,4-dinitrnolo (Italian); NSC 202753; Phenol, 2-*sec*-butyl-4,6-dinitro-; Phenol, 2-(1-methylpropyl)-4,6-dinitro-

*Trade Names:* AATOX®; AI3-01122®; ARETIT®; BASANITE®; BNP 20®; BNP 30®; BUTAPHENE®; CALDON®; CASWELL No. 392DD®; CHEMOX®, Mid America Chemical Co. (USA), canceled; CHEMOX GENERAL®, Mid America Chemical Co. (USA), canceled; CHEMOX P.E.®, Mid America Chemical Co. (USA), canceled; CHEMSECT DNBP®; DESICOIL®; DIBUTOX®; DINITRALL®; DINITRO®; DN 289®; DOW GENERAL®, Dow Chemical (USA, canceled; DOW GENERAL WEED KILLER®, Dow Chemical (USA), canceled; DOW SELECTIVE WEED KILLER®, Dow Chemical (USA, canceled; DYNAMYTE®, Drexel Chemical (USA), canceled 10/10/1989; DYTOP®; ELGETOL 318®; FANICIDE®; GEBUTOX®; HEL-FIRE®, Helena Chemical Co. (USA), canceled 3/22/1984; HIVERTOX®; HOE 26150®; IVOSIT®; KILOSEB®; KNOWX-WEED®; KNOX-WEED®; LADOB®; LASEB®; LIRO DNBP®; NITROPONE C®; PERSEVTOX®; PHENOTAN®; PREMERGE®; PROKIL®, Gowan Co. (USA), canceled; SINOX GENERAL®, FMC Agricultural Products Group (USA), canceled 10/14/1986; SPARIC®; SPURGE®; SUBITEX®; UNICROP DNBP®; VERTAC DINITRO WEED KILLER®, canceled; VERTAC GENERAL WEED KILLER®, canceled; VERTAC SELECTIVE WEED KILLER®, canceled

*Producers:* A H Marks (UK); Clariant Corp. (Switzerland); Crompton Corp. (USA); Dow Chemical (USA); Drexel Chemical (USA); FMC Agricultural Products Group (USA); Gowan Co. (USA); Helena Chemical Co. (USA); SNPE Agro (France); Rhone-Poulenc Agro (France); Sigma-Aldrich Laborchemikalien (Germany)
*Chemical Class:* Derivitive of dinitrophenol.
*EPA/OPP PC Code:* 037505
*California DPR Chemical Code:* 238
*ICSC Number:* 1049
*RTECS Number:* SJ9800000
*EEC Number:* 609-025-00-7
EINECS: 201-861-7
*Uses:* Dinoseb is a phenolic herbicide used in soybeans, vegetables, fruits and nuts, citrus, and other field crops for the selective control of grass and broadleaf weeds (e.g., in corn). It is also used as an insecticide in grapes, and as a seed crop drying agent. It is produced in emuslifiable concentrates or as water soluble ammonium or amine salts. It is no longer available in the U.S. Formerly widely used in the UK for the fumigation of potatoes, however dinoseb acetate and dinoseb amine were banned from use in 1988.
*Human toxicity (long-term)*[77]*:* Extra high–0.70 ppb, Health Advisory

*Fish toxicity (threshold)*[77]: High–2.85539 ppb, MATC (Maximum Acceptable Toxicant Concentration)

***Carcinogen/Hazard Classifications***

**U.S. EPA Carcinogens:** Group C, possible carcinogen
**Label Signal Word:** DANGER
***Regulatory Authority:***

- Banned or Severely Restricted (Several Countries) (UN)[13]
- Air Pollutant Standard Set (former USSR)[43]
- AB 1803-Well Monitoring Chemical (CAL)
- EPA/SARA 302 (EPCRA) Extremely hazardous substances
- AB 2588-Air Toxics "Hot Spots" Chemicals (CAL)
- Proposition 65 chemical (CAL)
- The "Director's List" (CAL/OSHA)
- EPA Hazardous Waste Number (RCRA No.): P020
- RCRA, 40CFR261, Appendix 8 Hazardous Constituents
- RCRA 40CFR268.48; 61FR15654, Universal Treatment Standards: Wastewater (mg/L), 0.066; Nonwastewater (mg/kg), 2.515
- RCRA 40CFR264, Appendix 9; TSD Facilities Ground Water Monitoring List. Suggested test method(s) (PQL $ug/L$): 8150[1]; 8270(10)
- Safe Drinking Water Act: MCL, 0.007 mg/L; MCLG, 0.007 mg/L; Regulated chemical (47 FR 9352)
- Superfund/EPCRA 40CFR355, Appendix B Extremely Hazardous Substances: TPQ = 100/10,000 lb (45.4/4,540 kg)
- Superfund/EPCRA 40CFR302.4 RQ: CERCLA, 1000 lb (454 kg)
- EPCRA Section 313 Form R *de minimus* concentration reporting level: 1.0%
- U.S. DOT Regulated Marine Pollutant (49CFR172.101, Appendix B)

***Description:*** Orange-brown viscous liquid or an orange crystalline solid. Pungent odor. Very slightly soluble in water; solubility = 50 ppm. Molecular weight = 240.25. Melting/Freezing point = 38–42°C. Vapor pressure = 1 mmHg @ 151°C; 7 x 10–3 Pa @ 20°C. Flash point = 16–29°C (for 3 commercial products). Physical and toxicological properties may be affected by the carrier solvents used in commercial formulations.

***Incompatibilities:*** The solution in water is a weak acid. Attacks many metals in presence of water.

***Permissible Exposure Limits in Air:*** The former USSR-UNEP/IRPTC project[43] has set an MAC in workplace air of 0.05 mg/m3.

***Permissible Concentration in Water:*** A health advisory of 3,5 μg/L has been developed by EPA based on possible teratogenic action of dinoseb as described in the U.S. EPA document referred to below. In addition, several states have set guidelines for dinoseb in drinking water[61] ranging from 2.0 μg/L (Maine) to 5.0 μg/L (Massachusetts) to 13.0 μg/L (Wisconsin) to 39.0 μg/L (Kansas). The former USSR[35] has set an MAC in surface water of 0.1 mg/L.

***Determination in Water:*** Extraction with ether, conversion to methyl ester and determination by electron capture gas chromatography.

***Routes of Entry:*** Inhalation, skin and eye contact, and ingestion.

***Harmful Effects and Symptoms***

***Short Term Exposure:*** Dinoseb causes eye irritation. May affect the gastrointestinal tract and central nervous system. Early manifestations of dinoseb exposure include fever, sweating, headache, and confusion. Elevations of blood pressure, pulse, and respiratory rate are common. Severe exposure may result in restlessness, seizures, and coma. Other signs and symptoms include dyspnea (shortness of breath), nausea, vomiting, and abdominal pain. Liver injury with associated jaundice, kidney failure, and cardiac arrhythmias may be noted. Inhalation of the aerosol may cause pulmonary edema, a medical emergency that can be delayed for several hours. This can cause death. Muscle weakness may be pronounced. Dermal exposure results in yellow staining of the skin and may produce burns.

***Warning:*** Exposure to dinoseb fumes or aerosol in hot environment may cause death. Effects may be delayed from several hours to 2 days. Caution is advised. Toxicity of dinoseb is enhanced by high ambient temperature and physical activity. Dinoseb is extremely toxic: Probable oral lethal dose is 5-50 mg/kg; between 7 drops and 1 teaspoonful for 70 kg person (150 lb). Delayed pulmonary edema may occur after inhalation.

***Long Term Exposure:*** Dinoseb may affect the kidneys, liver, blood, immune system, and eyes; may cause cataracts. May cause reproductive toxicity in humans.

***Points of Attack:*** Liver, kidneys, blood, cardiovascular system, immune system, eyes.

***Medical Surveillance:*** Medical observation is recommended for 24 to 48 hours after breathing overexposure, as pulmonary edema may be delayed. Liver and kidney function tests. Complete blood count. Eye examination. EKG.

***First Aid:*** If this chemical gets into the eyes, remove any contact lenses at once and irrigate immediately for at least 15 minutes, occasionally lifting upper and lower lids. Seek medical attention immediately. If this chemical contacts the skin, remove contaminated clothing and wash immediately with soap and water. Seek medical attention immediately. If this chemical has been inhaled, remove from exposure, begin rescue breathing (using universal precautions) if breathing has stopped, and CPR if heart action has stopped. Transfer promptly to a medical facility. When this chemical has been swallowed, get medical attention. Give large quantities of water and induce vomiting. Do not make an unconscious person vomit. Consult poison center on use of antidotes.

***References:***

- EXTOXNET, Extension Toxicology Network, "Pesticide Information Profile, Dinoseb," Oregon State University,

University, Corvallis, OR (June 1996). http://ace.orst.edu/info/extoxnet/pips/dinoseb.htm
- U.S. Environmental Protection Agency, "Chemical Profile: Dinoseb," Washington, DC, Chemical Emergency Preparedness Program (November 30, 1987).
- California Environmental Protection Agency "Chemical List of Lists," Sacramento CA (February 1997).
- U.S. Environmental Protection Agency, "Health Advisory: Dinoseb," Washington, DC, Office of Drinking Water (August 1987).

# Dinoterb

*Use Type:* Herbicide and rodenticide.
*CAS Number:* 1420-07-1
*Formula:* $C_{10}H_{12}N_2O_5$; $C_6H_2(NO_2)_2(C_4H_9)(OH)$
*Alert:* Banned from use in the United Kingdom and Thailand
*Synonyms:* o-*tert*-Butyl-4,6-dinitrophenol; 2-(1,1-Dimethylethyl)-4,6-dinitrophenol; 2,4-Dinitro-6-*tert*-butylphenol; Dinitroterb; DNTBP; Phenol, 2-(1,1-dimethylethyl)4,6-dinitro-; Phenol-2-*tert*-butyl-4,6-dinitro-
*Trade Names:* HERBOGIL®
*Producers:* Rhone-Poulenc Agro France (France); Sigma-Aldrich Laborchemikalien (Germany)
*Chemical Class:* Derivative of dinitrophenol
*EPA/OPP PC Code:* 228400
*RTECS Number:* SK0160000
*EINECS Number:* 215-813-8
*Carcinogen/Hazard Classifications*
**WHO Acute Hazard:** Class 1B, highly hazardous
*Regulatory Authority:*
- Superfund/EPCRA 40CFR355, Appendix B Extremely Hazardous Substances: TPQ = 500/10,000 lb (227/4,540 kg)
- Superfund/EPCRA 40CFR302.4 RQ: EHS, 1 lb (0.454 kg) (TPQ = 500)[7]

*Description:* Dinoterb is a yellow solid. Melting/Freezing point = 126°C. Hazard Identification (based on NFPA-704 M Rating System): Health 3, Flammability 2, Reactivity 3.
*Incompatibilities:* Strong caustics. Heat may cause material to explode.
*Permissible Exposure Limits in Air:* No standards set.
*Permissible Concentration in Water:* No criteria set, but runoff from spills or fire control may cause water pollution.
*Routes of Entry:* Inhalation and ingestion
*Harmful Effects and Symptoms*
*Short Term Exposure:* Symptoms of poisoning are similar to other dinitrophenols and may include nausea, gastric distress, restlessness, sensation of heat, flushed skin, sweating, thirst, deep and rapid breathing, rapid heart rate, fever, and lack of oxygen to tissues (blueness of skin). This compound is toxic by all routes of exposure. The dangerous single oral dose of dinitro-*o*-cresol, a structurally similar compound, is estimated to be about 29 mg/kg. Delayed pulmonary edema may occur after inhalation.
*Medical Surveillance:* Medical observation is recommended for 24 to 48 hours after breathing overexposure, as pulmonary edema may be delayed.
*First Aid:* If this chemical gets into the eyes, remove any contact lenses at once and irrigate immediately for at least 15 minutes, occasionally lifting upper and lower lids. Seek medical attention immediately. If this chemical contacts the skin, remove contaminated clothing and wash immediately with soap and water. Seek medical attention immediately. If this chemical has been inhaled, remove from exposure, begin rescue breathing (using universal precautions) if breathing has stopped, and CPR if heart action has stopped. Transfer promptly to a medical facility. When this chemical has been swallowed, get medical attention. Give large quantities of water and induce vomiting. Do not make an unconscious person vomit.
*References:*
- U.S. Department of Health and Human Services, Agency for Toxic Substances and Disease Registry, "ToxFAQs for Nitrophenols," Atlanta, GA (September 1995). http://www.atsdr.cdc.gov/tfacts50.html
- New Jersey Department of Health, "Hazardous Substance Fact Sheet: Dinitrophenol," Trenton, NJ (October 1996). http://www.state.nj.us/health/eoh/rtkweb/0780.pdf
- U.S. Environmental Protection Agency, "Chemical Profile: Dinoterb," Washington, DC, Chemical Emergency Preparedness Program (November 30, 1987).
- California Environmental Protection Agency "Chemical List of Lists," Sacramento CA (February 1997).

# Dioxathion

*Use Type:* Insecticide and acaricide
*CAS Number:* 78-34-2
*Formula:* $C_{12}H_{26}O_6P_2S_4$
*Synonyms:* 2,3-Dioxanedithiol S,S-bis(O,O-diethylphosphorodithioate); 1,4-Dioxan-2,3-diyl S,S-di(O,O-diethyl phosphorodithioate); S,S'-1,4-Dioxane-2,3-diyl bis(O,O-diethyl phosphorodithioate); S,S'-(1,4-Dioxane-2,3-diyl) O,O,O',O'-tetraethylbis phosphorodithioate); S,S'-*para*-Dioxane-2,3-diyl bis(O,O-diethylphosphorodithioate); ENT 22879; NCI-C00395; Phosphorodithioic acid, S,S'-1,4-dioxane-2,3-diyl-O,O,O',O'-tetraethyl ester; Phosphorodithioic acid, O,O-diethyl ester, S,S-diester with *p*-dioxane-2,3-dithiol; Phosphorodithioic acid-S,S'-1,4-dioxane-2,3-diyl, O,O,O',O'-tetraethyl ester; Phosphorodithioic acid-S,S'-*para*-dioxane-2,3-diyl, O,O,O',O'-tetraethyl Ester; Bis(dithiophospate de O,O-diethyle) de S,S'-(1,4-dioxanne-2,3-diyle) (French)
*Trade Names:* AC 528®; ALCOV®, Amvac Chemical (USA), canceled; DELNAV®, Agrevo USA Co., Helena

Chemical (USA); Scott Company (USA), canceled; DELNATEX®; HERCULES AC528®; KAVADEL®; NAVADEL®

**Producers:** Amvac Chemical (USA); Helena Chemical (USA); Hercules (USA)
**Chemical Class:** Organophosphate
**EPA/OPP PC Code:** 037801
**California DPR Chemical Code:** 192
**ICSC Number:** 0883
**RTECS Number:** TE335000
**EEC Number:** 015-063-00-X
**EINECS Number:** 201-107-7

**Uses:** Dioxathion is registered and used in many countries principally for treating livestock, citrus fruit, pome fruit, grapes and stone fruit. Dioxathion products are used on livestock principally cattle in the U.S.A., Australia, East and South Africa, South and Central America, France and Italy. Countries using these products on citrus fruit include the U.S.A., Africa, Japan, Turkey and Italy. There is significant use on pome fruit in the U.S.A., France, United Kingdom, South America and Italy. Use on grapes extends to the U.S.A., Germany, France and Italy. In some countries use on cattle precludes application to dairy cattle. However, in countries where ticks are a serious problem and frequent dipping is needed, dairy cows must be treated as well as beef cattle.

**Regulatory Authority:**
- Air Pollutant Standard Set (ACGIH)[1] (HSE)[33] (OSHA)[58] (Several States)[60]
- AB 1803-Well Monitoring Chemical (CAL)
- EPA/SARA 302 (EPCRA) Extremely hazardous substances
- Permissible Exposure Limits for Chemical Contaminants (CAL/OSHA)
- The "Director's List" (CAL/OSHA)
- U.S. DOT Inhalation Hazard Chemicals as organophosphates
- Superfund/EPCRA 40CFR355, Appendix B Extremely Hazardous Substances: TPQ = 500 lb (227kg)
- Superfund/EPCRA 40CFR302.4 RQ: EHS, 1 lb (0.454 kg)
- U.S. DOT Regulated Marine Pollutant (49CFR172.101, Appendix B)

**Description:** Reddish-brown, thick liquid or powder. Garlic-like odor. Molecular weight = 456.58. Boiling point = decomposes @ > 135°C. Melting/Freezing point = –20°C. Hazard Identification (based on NFPA-704 M Rating System): Health 3, Flammability 0, Reactivity 0. Insoluble in water. Log $K_{ow}$ = 2.9 to 3.1. Values at or above 3.0 are likely to bioaccumulate in marine organisms.

**Incompatibilities:** Incompatible with strong acids. Dioxathion is hydrolyzed by strong bases attacks iron and tin surfaces.

**Permissible Exposure Limits in Air:** There is no OSHA[2] PEL. NIOSH[2], ACGIH[1] TLV and HSE[33] TWA value is 0.2 mg/m3 with the notation "skin" indicating the possibility of cutaneous absorption. There is no NIOSH[2] ILDH value. Several states have set guidelines or standards for Dioxathion in ambient air[60] ranging from zero (New York) to 2.0 $\mu$g/m3 (North Dakota) to 3.0 $\mu$g/m3 (Virginia) to 4.0 $\mu$g/m3 (Connecticut) to 5.0 $\mu$g/m3 (Nevada).

**Determination in Air:** Although NIOSH lists "none available," the organophosphate method is listed: OSHA versatile sampler-2; Toluene/Acetone. Gas chromatography/Flame photometric detection for sulfur, nitrogen, or phosphorus. NIOSH Method IV Method #5600, Organophosphorus pesticides[18].

**Permissible Concentration in Water:** No criteria set, but runoff from spills or fire control may cause water pollution.

**Routes of Entry:** Inhalation, skin absorption, ingestion, skin and/or eye contact

**Harmful Effects and Symptoms**

**Short Term Exposure:** Dioxathion is a cholinesterase inhibitor. It may affect the nervous system, resulting in convulsions, respiratory failure. Exposure to high level may cause death. Acute exposure to Dioxathion may produce the following signs and symptoms: pinpoint pupils, blurred vision, headache, dizziness, muscle spasms, and profound weakness. Vomiting, diarrhea, abdominal pain, seizures, and coma may also occur. The heart rate may decrease following oral exposure or increase following dermal exposure. Hypotension (low blood pressure) may occur although hypertension (high blood pressure) is not uncommon. Chest pain may be noted. Respiratory symptoms include dyspnea (shortness of breath), respiratory depression, and respiratory paralysis. Psychosis may occur. Dioxathion is very toxic. $LD_{50}$ female rat, skin = 63 mg/kg. Probable oral lethal dose for humans is 50-500 mg/kg or between 1 teaspoonful and 1 oz for a 70 kg (150 lb) person. It is a cholinesterase inhibitor. Death is primarily due to respiratory arrest arising from failure of the respiratory center, paralysis of respiratory muscles, intense bronchoconstriction, or all three. Delayed pulmonary edema may occur after inhalation.

**Long Term Exposure:** Cholinesterase inhibitor; cumulative effect is possible. This chemical may damage the nervous system with repeated exposure, resulting in convulsions, respiratory failure. May cause liver damage.

**Points of Attack:** Respiratory system, lungs, central nervous system, cardiovascular system, skin, eyes, plasma and red blood cell cholinesterase.

**Medical Surveillance:** Medical observation is recommended for 24 to 48 hours after breathing overexposure, as pulmonary edema may be delayed. Before employment and at regular times after that, the following are recommended: Plasma and red blood cell cholinesterase levels (tests for the enzyme poisoned by this chemical). If exposure stops, plasma levels return to normal in 1-2 weeks while red blood cell levels may be reduced for 1-3 months. When acetylcholinesterase enzyme levels are reduced by 25% or more below preemployment levels, risk of poisoning is increased, even if results are in lower ranges of "normal." Reassignment to work not involving organophosphate or carbamate pesticides is recommended until enzyme levels recover. If symptoms develop or overexposure occurs, repeat the above tests as soon as possible and get an exam of the nervous system. Also consider complete blood count. Consider chest x-ray following acute overexposure Do not drink any alcoholic beverages before or during use; alcohol promotes absorption of organophosphates.

*First Aid:* **Treatment for organophosphate poisoning consists of thorough decontamination, cardiorespiratory support, and administration of the antidotes atropine and pralidoxime. In cases of severe poisoning, diazepam, an anticonvulsant, should also be administered. Antidotes should be administered as prevention even if the diagnosis is in doubt.** Speed in removing material from eyes and skin is of extreme importance. *Eyes:* Eye contact can cause dangerous amounts of these chemicals to be quickly absorbed through the mucous membrane into the bloodstream. Immediately and gently flush eyes with plenty of warm or cold water (NO hot water) for at least 15 minutes, occasionally lifting the upper and lower eyelids. Get medical aid immediately. *Skin:* Get medical aid. Skin contact can cause dangerous amounts of these chemicals to be absorbed into the bloodstream. Wearing the appropriate PPE equipment and respirator for organophosphate/carbamate pesticides, immediately flush skin with plenty of soap and water for at least 15 minutes while removing contaminated clothing and shoes. Shampoo hair promptly if contaminated. The removed, contaminated clothing and shoes should be double-bagged and left in Hot Zone for later disposal by hazardous materials experts. Skin may also be decontaminated with diluted hypochlorite solution. *Inhalation:* Get medical aid. Do not contaminate yourself. Wearing the appropriate PPE equipment and respirator for organophosphate pesticides, immediately remove the victim from the contaminated area to fresh air. If the victim is not breathing, administer artificial respiration. Do NOT use mouth-to-mouth resuscitation; use bag/mask apparatus. If breathing is difficult, administer oxygen through bag/mask apparatus until medical help arrives. Do not leave victim unattended. *Ingestion:* Call poison control. Loosen all clothing. Never give anything by mouth to an unconscious person. If victim is *unconscious or having convulsions,* do nothing except keep victim warm. Get medical aid. Transfer promptly to a medical facility. In cases of ingestion, **do not induce vomiting**. If the victim is alert and asymptomatic, administer a slurry of activated charcoal at a dose of 1 g/kg (infant, child, and adult dose). A soda can and straw may be of assistance when offering charcoal to a child. *In some cases you may be specifically instructed by poison control to induce vomiting by way of 2 tablespoons of syrup of ipecac (adult) washed down with a cup of water.* Do NOT give activated charcoal before or with ipecac syrup.

*Note to physician or authorized medical personnel:* Treat cases of respiratory compromise, coma, or excessive pulmonary secretions with respiratory support using protocols and techniques available and within the scope of training. Some cases may necessitate procedures such as endotracheal intubation or cricothyrotomy by properly trained and equipped personnel. When possible, atropine (see *Antidotes*, below) should be given under medical supervision. Patients who are comatose, hypotensive, or having seizures or cardiac arrhythmias should be treated according to advanced life support protocols. *Antidotes:* Two antidotes are administered to treat organophosphate poisoning. Atropine is a competitive antagonist of acetylcholine at muscarinic receptors and is used to control the excessive bronchial secretions which are often responsible for death. Pralidoxime relieves both the nicotinic and muscarine effects of organophosphate poisoning by regenerating acetylcholinesterase and can reduce both the bronchial secretions and the muscle weakness associated with poisoning. The initial intravenous dose of atropine in adults should be determined by the severity of symptoms: An initial adult dose of 1.0 to 2.0 mg or pediatric dose of 0.01 mg/kg (minimum 0.01 mg) should be administered intravenously. If intravenous access cannot be established, atropine may also be given intramuscularly, subcutaneously or via endotracheal tube. Doses should be repeated every 15 minutes until excessive secretions and sweating have been controlled. Once bronchial secretion has been controlled, atropine administration should be repeated whenever the secretions begin to recur. In seriously poisoned patients, very large doses may be required. Alterations of pulse rate and pupillary size should not be used as indicators of treatment adequacy. Pralidoxime should be administered as early in poisoning as possible as its efficacy may diminish when given more than 24 to 36 hours after exposure. Doses are as follows: adult 1.0 g; pediatric 25 to 50 mg/kg. The drug should be administered intravenously over 30 to 60 minutes, but in a life-threatening situation, one-half of the total dose can be given per minute for a total administration time of 2 minutes. Treatment should begin to take effect within 40 minutes with a reduction in symptoms and the amount of atropine necessary to control bronchial secretion. The initial dose can be repeated in 1 hour and then every 8 to 12 hours until the patient is clinically well and no longer requires atropine. If intravenous access cannot be established, pralidoxime may also be given intramuscularly. Early administration of diazepam in addition to the combined atropine and pralidoxime treatment may help prevent the onset of seizures and potential brain and cardiac morphologic damage following high-level organophosphate poisoning.

*References:*
- International Programme for Chemical Safety (IPCS), "INCHEM Report, Dioxathion," Geneva, Switzerland (1972).

- http://www.inchem.org/documents/jmpr/jmpmono/v072pr13.htm
- New Jersey Department of Health, "Hazardous Substance Fact Sheet: Dioxathion," Trenton, NJ (December 1998). http://www.state.nj.us/health/eoh/rtkweb/0790.pdf
- Sax, N.I., Ed., "Dangerous Properties of Industrial Materials Report," 2, No. 5, 60-63 (1982) New York, Van Nostrand Reinhold Co. (1982).
- U.S. Environmental Protection Agency, "Chemical Profile: Dioxathion," Washington, DC, Chemical Emergency Preparedness Program (November 30, 1987).
- Agency for Toxic Substances and Disease Registry, U.S. Department of Health and Human Services, Public Health Service, "Managing Hazardous Materials Incidents," Atlanta, GA (June 2003).
- California Environmental Protection Agency "Chemical List of Lists," Sacramento CA (February 1997).

# Diphacinone

*Use Type:* Rodenticide
*CAS Number:* 82-66-6
*Formula:* $C_{23}H_{16}O_3$
*Alert:* A Restricted Use Pesticide (RUP) if the formulation contains 3% or more of diphacinone.
*Synonyms:* Dipazin; Diphacin (Italy and Turkey); Diphacinon; Diphenacin; Diphenadion; Diphenadione; 2-Diphenylacetyl-1,3-diketohydrindene; 2-(Diphenylacetyl)indan-1,3-indandione; 2-(Diphenylacetyl)-1H-indene-1,3(2h)-dione; Ratindan (Russia)
*Trade Names:* DE-PESTER®, Harcros Chemicals (USA), canceled 7/1/1987; DIDANDIN®; DIPAXIN®; DITRAC®, Bell Laboratories (USA); GOLD CREST®; KILL-RO RAT KILLER®; LIQUA-TOX®, diphacinone sodium salt, Bell Laboratories (USA); ORAGULANT®; P.C.Q.®; PID®; PROMAR®; RAMIK®; RAT KILLER®; RODENT CAKE®, Bell Laboratories (USA), canceled 10/10/1989; SOLVAN®; TOMCAT®; U 1363®
*Producers:* Agrimor International (USA); Bell Laboratories (USA); Haco (USA); Harcros Chemicals (USA)
*Chemical Class:* 1,3-Indandione
*EPA/OPP PC Code:* 067701
*California DPR Chemical Code:* 225
*RTECS Number:* NK5600000
*Uses:* Diphacinone is an anti-coagulant rodenticide bait used for control of rats, mice, voles and other rodents. It is available in meal, pellet, wax block, and liquid bait formulations, as well as in tracking powder and concentrate formulations. It is used in general agriculture and in food processing areas. The top five uses for diphacinone in California are on landscapes, general vertebrate pest control, around structures and right of ways, and on oranges. This material is also used as an anticoagulant medication.

*Carcinogen/Hazard Classifications*
**Label Signal Word:** DANGER for technical material; WARNING for concentrate formulations; CAUTION for bait formulations.
**WHO Acute Hazard:** Class 1b, Extremely Hazardous
*Regulatory Authority:*
- Very Toxic Substance (World Bank)[15]
- Actively registered pesticide in California.
- Superfund/EPCRA 40CFR355, Appendix B Extremely Hazardous Substances: TPQ = 10/10,000 lb (4.54/4,540 kg)
- Superfund/EPCRA 40CFR302.4 RQ: EHS, 1 lb (0.454 kg)

*Description:* Pale yellow, crystalline solid. Odorless. Very slightly soluble in water. Melting/Freezing point = 146-147°C. Hazard Identification (based on NFPA-704 M Rating System): Health 3, Flammability 1, Reactivity 0.
*Permissible Exposure Limits in Air:* No standards set.
*Permissible Concentration in Water:* No criteria set, but runoff from spills or fire control may cause water pollution.
*Harmful Effects and Symptoms*
*Short Term Exposure:* This material is extremely toxic; probable oral lethal dose in humans is 5-50 mg/kg, or between 7 drops and 1 teaspoonful for a 150 lb persons. Diphacinone is an anticoagulant (inhibits blood clotting). Hemorrhage is the most common effect and may be manifested by nose bleeding, gum bleeding, bloody stools and urine, ecchymoses (extravasations of blood into skin), and hemoptysis (coughing up blood). Bruising is heightened. Abdominal and flank pain are also common. Other signs and symptoms include flushing, dizziness, hypotension (low blood pressure), dyspnea (shortness of breath), cyanosis (blue tint to the skin and mucous membranes), fever, and diarrhea.
*Long Term Exposure:* May affect the liver and kidneys. Repeated exposure may cause low white blood cell count and affect the brain.
*Points of Attack:* Blood, liver, kidneys.
*Medical Surveillance:* Blood test for clotting time (PT, INR, or PTT). Stool and urine tests for blood. Liver and kidney function tests. Complete blood count. EEG.
*First Aid:* If this chemical gets into the eyes, remove any contact lenses at once and irrigate immediately for at least 15 minutes, occasionally lifting upper and lower lids. Seek medical attention immediately. If this chemical contacts the skin, remove contaminated clothing and wash immediately with soap and water. Seek medical attention immediately. If this chemical has been inhaled, remove from exposure, begin rescue breathing (using universal precautions) if breathing has stopped, and CPR if heart action has stopped. Transfer promptly to a medical facility. When this chemical has been swallowed, get medical attention. Give large quantities of water and induce vomiting. Do not make an unconscious person vomit. Obtain authorization and/or further instructions from the local hospital for administration

of an antidote or performance of other invasive procedures. Rush to a health care facility. Acute exposure to Diphacinone may require decontamination and life support for the victims. Emergency personnel should wear protective clothing appropriate to the type and degree of contamination. Air-purifying or supplied-air respiratory equipment should also be worn, as necessary. Rescue vehicles should carry supplies such as plastic sheeting and disposable plastic bags to assist in preventing spread of contamination.

*References:*
- EXTOXNET, Extension Toxicology Network, "Pesticide Information Profile, Diphacinone," Oregon State University, Corvallis, OR. (June 1996). http://ace.orst.edu/info/extoxnet/pips/diphacin.htm
- New Jersey Department of Health and Senior Services, "Hazardous Substance Fact Sheet, Diphacinone," Trenton NJ (May 1999). http://www.state.nj.us/health/eoh/rtkweb/0794.pdf
- U.S. Environmental Protection Agency, "Chemical Profile: Diphacinone," Washington, DC, Chemical Emergency Preparedness Program (November 30, 1987).
- California Environmental Protection Agency "Chemical List of Lists," Sacramento CA (February 1997).

# Diphenamid (ANSI)

*Use Type:* Herbicide
*CAS Number:* 957-51-7
*Formula:* $C_{16}H_{17}NO$; $(CH_3)_2NCOCH(C_6H_5)_2$
*Synonyms:* Acetamide, N,N-dimethyl-2,2-diphenyl-; Benzeneacetamide, N,N-dimethyl-α-phenyl-; Difenamid (Spanish); N,N-Dimethyl-α,α-diphenylacetamide; N,N-Dimethyl-α-phenylbenzeneacetamide; N,N-Dimethyldiphenylacetamide; N,N-Dimethyl-2,2-diphenylacetamide; N,N-Dimethyl-α-phenylbenzeneacetamide; Dimid; Diphenamide; Diphenylamide; 2,2-Diphenyl-N,N-dimethylacetamide
*Trade Names:* DIF 4®; DYMID®, Dow Chemical (USA), canceled 9/27/1985; ENIDE®, Pharmacia & Upjohn (USA), canceled 4/22/1985; FDN®; FENAM®; L 34314®, Eli Lilly (USA); LILLY 34314®, Eli Lilly (USA); RIDEON®; U 4513®; ZARUR®
*Producers:* Dow Chemical (USA); Scott Company (USA); Sigma-Aldrich Laborchemikalien (Germany)
*Chemical Class:* Amide
*EPA/OPP PC Code:* 036601
*California DPR Chemical Code:* 226
*ICSC Number:* 0763
*RTECS Number:* AB8050000
*EINECS Number:* 213-482-4
*Uses:* This material is used as a pre-emergent and selective herbicide for tomatoes, peanuts, alfalfa, soybeans, cotton and other crops.

*Human toxicity (long-term)*[77]: Very low–200.00 ppb, Health Advisory
*Fish toxicity (threshold)*[77]: High–5.00495 ppb, MATC (Maximum Acceptable Toxicant Concentration)
*U.S. Maximum Allowable Residue Levels for Diphenamid (40 CFR 180.230):*
Note: Tolerances are established for residues of the herbicide diphenamid (N,N,-Dimethyl-2,2,-diphenylacetamide) including its desmethyl metabolite (N-Methyl-2,2-diphenylacetamide) in or on raw agricultural commodities as follows:

| CROP | ppm |
|---|---|
| Apples | 0.1 |
| Cattles meat, fat & byproducts | 0.05 |
| Cotton forage | 0.5 |
| Cottonseed | 0.1 |
| Fruiting vegetables | 0.1 |
| Goat meat, fat & byproducts | 0.05 |
| Hog meat, fat & byproducts | 0.05 |
| Horse meat, fat & byproducts | 0.05 |
| Milk | 0.01 |
| Okra | 0.1 |
| Peaches | 0.1 |
| Peanut hay and forage | 2.0 |
| Peanuts | 0.1 |
| Potatoes | 1.0 |
| Raspberries | 1.0 |
| Sheep meat, fat & byproducts | 0.05 |
| Soybeans | 0.1 |
| Strawberries | 1.0 |
| Sweet potatoes | 0.1 |

Note: Cited in 37 FR 738, Jan. 18, 1972, as amended at 46 FR 18315, Mar. 24, 1981; 55 FR 26440, June 28, 1990.
*Carcinogen/Hazard Classifications*
**WHO Acute Hazard:** Class III, slightly hazardous
*Regulatory Authority:*
- Air Pollutant Standard Set (former USSR)[35]
- AB 1803-Well Monitoring Chemical (CAL)
- MCL (Maximum Contaminants Levels) list of contaminants (CAL)
- CAL-DHS/DHS Drinking Water Action Levels
- EPCRA Section 313 Form R *de minimus* concentration reporting level: 1.0%

*Description:* White crystalline solid in various forms. Very slightly soluble in water; solubility = 255 ppm @ 27°C. Molecular weight = 239.30. Melting/Freezing point = 135°C. Vapor pressure = 2.8 x $10^{-8}$ mmHg @ 20°C. Hazard Identification (based on NFPA-704 M Rating System): Health 1, Flammability 0, Reactivity 0. Log $K_{ow}$ = 2.19. Unlikely to bioaccumulate in marine organisms. Physical and toxicological properties may be affected by the carrier solvents used in commercial formulations.
*Incompatibilities:* Reacts with strong oxidants, strong acids and alkalies.

*Permissible Exposure Limits in Air:* The former USSR[35] has set an MAC in workplace air of 5.0 mg/m³.

*Permissible Concentration in Water:* The former USSR[35] has set an MAC in water bodies used for domestic purposes of 1.2 mg/L. A lifetime health advisory of 0.2 mg/L has been calculated by EPA. California[61] has set a guideline for diphenamid in drinking water of 40 μg/L.

*Routes of Entry:* Ingestion

*Harmful Effects and Symptoms*

*Short Term Exposure:* The $LD_{50}$ (oral, rat) = 685 mg/kg (slightly toxic).

*Long Term Exposure:* A slight increase in liver weights was observed in long-term animal feeding studies. Mutation data reported.

*First Aid:* If this chemical gets into the eyes, remove any contact lenses at once and irrigate immediately for at least 15 minutes, occasionally lifting upper and lower lids. Seek medical attention immediately. If this chemical contacts the skin, remove contaminated clothing and wash immediately with soap and water. Seek medical attention immediately. If this chemical has been inhaled, remove from exposure, begin rescue breathing (using universal precautions) if breathing has stopped, and CPR if heart action has stopped. Transfer promptly to a medical facility. When this chemical has been swallowed, get medical attention. Give large quantities of water and induce vomiting. Do not make an unconscious person vomit.

*References:*
- 40 CFR 180.230, Diphenamid. http://www.setonresourcecenter.com/40CFR/Docs/wcd0004c/wcd04cdd.asp
- U.S. Environmental Protection Agency, "Chemical Profile: Diphenamid," Washington, DC, Office of Drinking Water (August 1987).
- California Environmental Protection Agency "Chemical List of Lists," Sacramento CA (February 1997).

# Dipropetryn (ANSI)

*Use Type:* Herbicide
*CAS Number:* 4147-51-7
*Formula:* $C_{11}H_{21}N_5S$
*Synonyms:* 2,4-Bis(isopropylamino)-6-ethylthio-*S*-triazine; Dipropetryne; Dipropetryn [2-(ethylthio)-4,6-bis(isopropylamino)-*S*-triazine]; 2-(Ethylthio)-4,6-bis(isopropylamino)-*S*-triazine; 6-(Ethylthio)*N,N'*-bis(1-methylethyl)-1,3,5-triazine-2,4-diamine
*Trade Names:* COTOFOR®; GS 16068®; SANCAP®, Syngenta (Switzerland), canceled
*Producers:* Syngenta (Switzerland)
*Chemical Class:* Triazine
*EPA/OPP PC Code:* 104401
*California DPR Chemical Code:* 2532

*Uses:* Not registered in the U.S. Dipropetryn was used for pre-emergence control of pigweed and Russian thistle on cotton. In addition, dipropetryn was registered for use only on cotton grown on the sandy soils in Oklahoma, Texas, Arizona, and New Mexico.

*Fish toxicity (threshold)[77]:* Low–211.12127 ppb, MATC (Maximum Acceptable Toxicant Concentration)

*Carcinogen/Hazard Classifications*

**Label Signal Word:** CAUTION

*Description:* A solid. Molecular weight = 255.43. Melting/Freezing point = 105°C. Vapor pressure = 7.4 x $10^{-7}$ mmHg @ 20°C.

*Determination in Air:* Filter; none; Gravimetric; NIOSH IV [Particulates NOR; #0500 (total), #0600 (respirable)].[18]

*Permissible Concentration in Water:* No criteria set. Runoff from spills or fire control may cause water pollution.

*Routes of Entry:* Inhalation, passing through the skin, ingestion.

*Harmful Effects and Symptoms*

*Short Term Exposure:* May cause skin and severe eye irritation. Moderately poisonous if ingested or inhaled. Exposure to a triazine (simazine) has caused acute and subacute dermatitis in the former USSR, characterized by erythema, slight edema, moderate pruritus, and burning lasting 4 to 5 days.

*Long Term Exposure:* May cause lung irritation and damage. May cause skin allergy. Contact with some triazine compounds (such as atrazine) may increase risks for tumors known to be associated with hormonal factors. These have been observed in both animals and human beings, and are consistent with the known effects on the hypothalamic pituitary gonadal axis. Repeated exposure may cause weight loss and reduced red blood cell count. May be mutagenic.

*Points of Attack:* Liver, lungs and skin.

*Medical Surveillance:* Before beginning employment and at regular times after that, for those with frequent or potentially high exposures, the following is recommended: Lung function tests. Consider chest x-ray following acute overexposure. Evaluation by a qualified allergist. Examination of the nervous system.

*First Aid:* If this chemical gets into the eyes, remove any contact lenses at once and irrigate immediately for at least 15 minutes, occasionally lifting upper and lower lids. Seek medical attention immediately. If this chemical contacts the skin, remove contaminated clothing and wash immediately with soap and water. Seek medical attention immediately. If this chemical has been inhaled, remove from exposure, begin rescue breathing (using universal precautions) if breathing has stopped, and CPR if heart action has stopped. Transfer promptly to a medical facility. When this chemical has been swallowed, get medical attention. Give large quantities of water or milk and induce vomiting. Do not make an unconscious person vomit.

*References:*
- Pesticide Management Education Program, "Dipropetryn (Sancap) Herbicide Profile 6/85," Cornell University, Ithaca, NY (June 1985). http://pmep.cce.cornell.edu/profiles/herb-growthreg/dalapon-ethephon/dipropetryn/herb-prof-dipropetryn.html

## Dipropyl Isocinchomeronate

*Use Type:* Insect repellent
*CAS Number:* 136-45-8; 3737-22-2
*Formula:* $C_{13}H_{17}NO_4$
*Synonyms:* AI3-17591; Caswell No. 400; Dipropyl isocincnomeronate; Di-*N*-propyl isocinchomeronate; Dipropylisocinchomeronate; Dipropyl 2,5-pyridinedicarboxylate; Di-*N*-propyl 2,5-pyridinedicarboxylate; Dipropyl pyridine-2,5-dicarboxylate; Dipropyl 2,5-pyridinedicarboxylate; Di-*N*-propyl 2,5-pyridinedicarboxylate; ENT 17591; Isocinchomeronic acid, dipropyl ester; Isocinchomeronyl dipropylester; NSC 22364; Pyridin-2,5-dicarbonsaeure-di-*N*-propylester (German); 2,5-Pyridinedicarboxylic acid, dipropyl ester
*Trade Names:* ADAMS®; AERO-FLYING INSECT SPRAY®; ALCO®, Amvac Chemical (USA); BUG BAN PLUS®; 856 INSECT REPELLENT®; ENT-17591®; MGK REPELLENT-326®; MGK-326®; REPELLENT-333®; R-326®; REPPER-333®
*Producers:* Amvac Chemical (USA)
*Chemical Class:* Unclassified
*EPA/OPP PC Code:* 047201
*California DPR Chemical Code:* 681
*RTECS Number:* US8000000
*Uses:* Used as insect repellent around livestock and in dips.
*U.S. Maximum Allowable Residue Levels for Dipropyl Isocinchomeronate (40 CFR 180.143):*

| CROP | ppm |
| --- | --- |
| Cattle, meat, fat, mbyp | 0.1 |
| Goat, meat, fat, mbyp | 0.1 |
| Hog, meat, fat, mbyp | 0.1 |
| Horse, meat, fat, mbyp | 0.1 |
| Milk | 0.004 |
| Sheep, meat, fat, mbyp | 0.1 |

*Carcinogen/Hazard Classifications*
**U.S. EPA Carcinogens:** Group C, possible carcinogen
**California Prop. 65:** Carcinogen
**U.S. TRI:** Carcinogen
**Label Signal Word:** CAUTION, WARNING, DANGER
**WHO Acute Hazard:** Class U, unlikely to be hazardous
*Regulatory Authority:*
- Actively registered pesticide in California.
- EPCRA Section 313 Form R *de minimis* concentration reporting level: 1.0%

*Incompatibilities:* May react violently with strong oxidizers, bromine, 90% hydrogen peroxide, phosphorus trichloride, silver powders or dust. Incompatible with silver compounds. Mixture with some silver compounds forms explosive salts of silver oxalate.
*Permissible Concentration in Water:* No criteria set. Runoff from spills or fire control may cause water pollution.
*First Aid:* If this chemical gets into the eyes, remove any contact lenses at once and irrigate immediately for at least 15 minutes, occasionally lifting upper and lower lids. Seek medical attention immediately. If this chemical contacts the skin, remove contaminated clothing and wash immediately with soap and water. Seek medical attention immediately. If this chemical has been inhaled, remove from exposure, begin rescue breathing (using universal precautions) if breathing has stopped, and CPR if heart action has stopped. Transfer promptly to a medical facility. When this chemical has been swallowed, get medical attention. Give large quantities of water and induce vomiting. Do not make an unconscious person vomit.
*References:*
- U.S. Environmental Protection Agency, Office of Pesticide Programs, Pesticide Residue Limits, "Dipropyl Isocinchomeronate," 40 CFR 180.143, www.epa.gov/pesticides/food/viewtols.htm
- California Environmental Protection Agency "Chemical List of Lists," Sacramento CA (February 1997).

## Diquat

*Use Type:* Herbicide and desiccant
*CAS Number:* 85-00-7
*Formula:* $C_{12}H_{12}N_2Br_2$
*Alert:* a General Use Pesticide (GUP).
*Synonyms:* 9,10-Dihydro-8a,10,-diazoniaphenanthrene dibromide; 9,10-Dihydro-8a,10a-diazoniaphenanthrene(1,1'-ethylene-2,2'-bipyridylium)dibromide; 5,6-Dihydrodipyrido(1,2a,2,1c)pyrazinium dibromide; 5,6-Dihydrodipyrido(1,2-a:2,1'-c)pyrazinium dibromide; 6,7-Dihydropyrido(1,2-a:2',1'-c)pyrazinedium dibromide; 6,7-Dihydropyridol(1,2-a:2',1'-c)pyrazinedium dibromide; Dipyrido(1,2-a:2',1'-c)pyrazinediium, 6,7-dihydro-, dibromide; *o*-Diquat; Diquat dibromide; 1,1'-Ethylene-2,2'-bipyridyliumdibromide; Ethylene dipyridylium dibromide; 1,1-Ethylene 2,2-dipyridylium dibromide; 1,1'-Ethylene-2,2'-dipyridylium dibromide
*Trade Names:* AQUACIDE®; AQUA-CLEAR®, I. Schneid Co. (USA), canceled 1/9/2002; AQUAKILL®; CLEANSWEEP®; DEIQUAT®; DEXTRONE®; FARMON PDQ®; FB/2®; FEGLOX®; GROUNDHOG SOLTAIR®; ORTHO DIQUAT®, The Scott Co. (USA); PATHCLEAR®; PREEGLONE®; REGLON®; REGLONE®, Syngenta (Switzerland); Zeneca Ag Products; REGLOX®; REWARD®, Syngenta (Switzerland); TAG®; TORPEDO®;

VEGATROLE®; WEEDOL (ICI)®, ICI Group (Imperial Chemical Industries) (UK); WEEDTRINE-D®

*Producers:* ICI Group (Imperial Chemical Industries) (UK); Nihon Nohyaku (Japan); Scotts Company, The (USA); Sigma-Aldrich Laborchemikalien (Germany); Syngenta (Switzerland); Zeneca Ag Products (USA) (now Syngenta)

*Chemical Class:* Bipyridylium

*EPA/OPP PC Code:* 032201

*California DPR Chemical Code:* 229

*ICSC Number:* 1363

*RTECS Number:* JM5690000

*EEC Number:* 613-089-00-1

*Uses:* Diquat or diquat dibromide is a nonselective, quick-acting herbicide and plant growth regulator, causing injury only to the parts of the plant to which it is applied. Diquat is referred to as a desiccant because it causes a leaf or an entire plant to dry out quickly. It is used to desiccate potato vines and seed crops, to control flowering of sugarcane, and for industrial and aquatic weed control in environments such as catfish farms. It is not residual; it does not leave any trace of herbicide on or in plants, soil, or water.

*Human toxicity (long-term)[77]:* Intermediate–35.00 ppb, Health Advisory

*Fish toxicity (threshold)[77]:* Very low–932.73871 ppb, MATC (Maximum Acceptable Toxicant Concentration)

*Carcinogen/Hazard Classifications*

**Label Signal Word:** WARNING, Class II, Moderately toxic

*Regulatory Authority:*

- Air Pollutant Standard Set (ACGIH)[1] (HSE)[33] (former USSR)[35] (OSHA)[58] (Several States)[60]
- AB 1803-Well Monitoring Chemical (CAL)
- MCL (Maximum Contaminants Levels) list of contaminants (CAL)
- Permissible Exposure Limits for Chemical Contaminants (CAL/OSHA)
- The "Director's List" (CAL/OSHA)
- Actively registered pesticide in California.
- Clean Water Act: Section 311 Hazardous Substances/RQ 40CFR117.3 (same as CERCLA, see below).
- Safe Drinking Water Act: MCL, 0.02 mg/L; MCLG, 0.02 mg/L; Regulated chemical (47 FR 9352)
- Superfund/EPCRA 40CFR302.4 RQ: CERCLA, 1000 lb (454 kg)

*Description:* yellow crystalline solid. The commercial product may be found in a liquid concentrate or a solution. Soluble in water; solubility = $7.0 \times 10^5$ ppm @ 20°C. Molecular weight = 344.08. Boiling point = decomposes. Melting/Freezing point = 335–340°C. Vapor pressure = <12 mPa @ 20. Hazard Identification (based on NFPA-704 M Rating System): Health 2, Flammability 0, Reactivity 0.

*Incompatibilities:* Incompatible with alkalis, UV light, basic solutions. The active ingredient is corrosive to metals. Concentrated solutions attack aluminum. Diquat cylinder may explode in heat of fire.

*Permissible Exposure Limits in Air:* There is no OSHA[2] PEL. NIOSH[2] (10-hr.workshift) and ACGIH[1] (8-hr. workshift) recommend a TWA value of 0.5 mg/m³. The HSE[33] uses the same TWA and has set an STEL of 1.0 mg/m³. The former USSR[35] has set an MAC in workplace air of 0.2 mg/m³. Several states have set guidelines or standards for diquat in ambient air[60] ranging from 5 μg/m³ (North Dakota) to 8.0 μg/m³ (Virginia) to 10.0 μg/m³ (Connecticut) to 12.0 μg/m³ (Nevada).

*Determination in Air:* No method available.

*Permissible Concentration in Water:* No criteria set

*Routes of Entry:* Inhalation, skin absorption, ingestion, skin and/or eye contact.

*Harmful Effects and Symptoms*

*Short Term Exposure:* Diquat can affect you when breathed in and by passing through your skin. Skin contact can cause burns. High exposure can cause nausea, diarrhea, lung, liver, and kidney damage. Higher exposures can cause pulmonary edema, a medical emergency that can be delayed for several hours.

*Long Term Exposure:* Long-term or repeated exposure may cause cataracts. Repeated contact causes dry, cracked skin and nail damage. Exposure can cause nosebleeds. Diquat may cause mutations. Handle with extreme caution. Diquat may damage the developing fetus. Lung damage may occur.

*Points of Attack:* Eyes, skin, respiratory system, kidneys, liver, central nervous system.

*Medical Surveillance:* If symptoms develop or overexposure has occurred, the following may be useful: Lung function tests. Examination of the eyes. Kidney function tests.

*First Aid:* If this chemical gets into the eyes, remove any contact lenses at once and irrigate immediately for at least 15 minutes, occasionally lifting upper and lower lids. Seek medical attention immediately. If this chemical contacts the skin, remove contaminated clothing and wash immediately with soap and water. Seek medical attention immediately. If this chemical has been inhaled, remove from exposure, begin rescue breathing (using universal precautions) if breathing has stopped, and CPR if heart action has stopped. Transfer promptly to a medical facility. When this chemical has been swallowed, get medical attention. Give large quantities of water and induce vomiting. Do not make an unconscious person vomit.

*References:*

- EXTOXNET, Extension Toxicology Network, "Pesticide Information Profile, Diquat Dibromide," Oregon State University, Corvallis, OR (June 1996). http://ace.orst.edu/info/extoxnet/pips/diquatdi.htm
- New Jersey Department of Health, "Hazardous Substance Fact Sheet: Diquat," Trenton, NJ (June 1986, rev. January 2001). http://www.state.nj.us/health/eoh/rtkweb/0808.pdf
- California Environmental Protection Agency "Chemical List of Lists," Sacramento CA (February 1997).

# Diquat Dibromide

*Use Type:* Herbicide and desiccant
*CAS Number:* 85-00-7
*Formula:* $C_{12}H_{12}Br_2N_2$
*Synonyms:* 1,1'-Aethylen-2,2'-bipyridinium-dibromid (German); DDB; 9,10-Dihydro-8A,10A,-diazoniaphenanthrene dibromide; 9,10-Dihydro-8A,10A-diazoniaphenanthrene(1,1'-ethylene-2,2'-bipyridylium)dibromide; 5,6-Dihydro-dipyrido(1,2A,2,1C) pyrazinium dibromide; 9,10-Dihydro-8A,10A-diazoniaphenanthrene(1,1'-ethylene-2,2'-bipyridylium) dibromide; 5,6-Dihydro-dipyrido(1,2A. 2,1C)pyrazinium dibromide; 5,6-Dihydro-dipyrido(1,2-A:2,1'-C)pyrazinium dibromide; 6,7-Dihydrodipyrido(1,2-A:2',1'-C)pyrazinediium dibromide; 6,7-Dihydropyridol(1,2-A:2',1'-C)pyrazinedium dibromide; Dipyrido(1,2-A:2',1'-C)pyrazinediium, 6,7-dihydro-, dibromide; 1,1'-Ethylene-2,2'-bipyridylium dibromide; Ethylene dipyridylium dibromide; 1,1'-ethylene-2,2'-dipyridylium dibromide; 1,1'-Ethylene-2,2'-dipyridylium dibromide; Diquat bromide
*Trade Names:* AQUACIDE®; AQUAKILL®; CLEANSWEEP®; CONKILL®; DEIQUAT®; DEXTRONE®; DIQUAT WEED KILLER® Syngenta (Switzerland); ENFORCER®; FARMON PDQ®; FB/2®; FEGLOX®; GROUNDHOG SOLTAIR®; KLEENUP®; MON 78567®; Monsanto (USA); PP 100®; PREEGLONE®; RAPID KILL #1®, Syngenta (Switzerland); RAZOROOTER II®; REAL-KILL®; REGLON®, Syngenta (Switzerland); REGLONE®, Syngenta (Switzerland); REGLOX®; REWARD®, Syngenta (Switzerland); SUPER K-GRO®; TAG®; TORPEDO®; TOUCHDOWN® Syngenta (Switzerland); VEGETROLE®; WEEDKILLER Conc. D®; WEEDOL®; WEEDTRINE-D®
*Producers:* ICI Corp (UK); Monsanto (USA); Sigma-Aldrich Laborchemikalien (Germany); Syngenta (Switzerland); Zeneca Ag Products (USA) (now Syngenta)
*Chemical Class:* Bipyridilium
*EPA/OPP PC Code:* 032201
*California DPR Chemical Code:* 229
*ICSC Number:* 1363
*RTECS Number:* JM5690000
*EEC Number:* 613-089-00-1
*Uses:* Diquat dibromide is a non-selective contact herbicide, desiccant and plant growth regulator for use as a general herbicide for control of broadleaf and grassy weeds in terrestrial non-crop and aquatic areas; as a desiccant in seed crops and potatoes; and for tassel control and spot weed control in sugarcane. It is used in agricultural drainage systems, ornamental ponds and reservoirs; in greenhouse food crops; indoor storage areas; and on feed crops and food crops. As a desiccant, diquat dibromide causes a leaf or plant to dry out rapidly. It is used to desiccate potato vines and seed crops, and to control the flowering of sugarcane. It does not leave any trace of herbicide on or in plants, soil, or water

*Carcinogen/Hazard Classifications*
**U.S. EPA Carcinogens:** Group E, Unlikely carcinogen
**Label Signal Word:** CAUTION or WARNING
*Regulatory Authority:*
- Clean Water Act: Section 311 Hazardous Substances/RQ (same as CERCLA).
- Safe Drinking Water Act: MCL, 0.02 mg/L; MCLG, 0.02 mg/L; Regulated chemical (47 FR 9352)
- EPCRA Section 304 RQ: CERCLA, 1000 lb (454 kg).
- AB 1803-Well Monitoring Chemical (CAL)
- MCL (Maximum Contaminants Levels) list of contaminants (CAL)
- Permissible Exposure Limits for Chemical Contaminants (CAL/OSHA)
- The "Director's List" (CAL/OSHA)
- Actively registered pesticide in California.

*Description:* Colorless to yellow crystalline solid. Soluble in water; solubility = 70 g/100 ml @ 20°C. Molecular weight = 344.0. Density = 1.24 – 1.25 @ 20 °C. Melting/Freezing point = < 320°C; 335–340°C. Vapor pressure = <13 mPa @ 20°C;<$10^{-5}$ mmHg @ 20°C. Hazard Identification (based on NFPA-704 M Rating System): Health 2, Flammability 0, Reactivity 0. Log $K_{ow}$ = < –3.00. Unlikely to bioaccumulate in marine organisms.

*Incompatibilities:* The dry material is UV sensitive. Keep away from alkalis, basic solutions. Concentrated diquat solutions attacks aluminum. May explode at elevated temperatures.

*Permissible Exposure Limits in Air:* There is no OSHA PEL[58]. NIOSH (10-hr.workshift) and ACGIH (8-hr. workshift) recommend a TWA value of 0.5 mg/m³.

*Permissible Concentration in Water:* No criteria set. Runoff from spills or fire control may cause water pollution.

*Routes of Entry:* Inhalation, ingestion, skin and/or eye contact. Slowly absorbed through the skin.

*Harmful Effects and Symptoms*
*Short Term Exposure:* Diquat can affect you when breathed in and by passing through your skin. Skin contact can cause burns. High exposure can cause nausea, diarrhea, lung, liver, and kidney damage. Higher exposures can cause pulmonary edema, a medical emergency that can be delayed for several hours.

*Long Term Exposure:* Long-term or repeated exposure may cause cataracts. Repeated contact causes dry, cracked skin and nail damage. Exposure can cause nosebleeds. Handle with extreme caution. Diquat may damage the developing fetus. Diquat may cause mutations. Lung, liver, and kidney damage may occur.

*Points of Attack:* Eyes, skin, respiratory system, kidneys, liver, central nervous system.

*Medical Surveillance:* If symptoms develop or overexposure has occurred, the following may be useful:

Lung function tests. Examination of the eyes. Kidney and liver function tests.

*First Aid:* If this chemical gets into the eyes, remove any contact lenses at once and irrigate immediately for at least 15 minutes, occasionally lifting upper and lower lids. Seek medical attention immediately. If this chemical contacts the skin, remove contaminated clothing and wash immediately with soap and water. Seek medical attention immediately. If this chemical has been inhaled, remove from exposure, begin rescue breathing (using universal precautions) if breathing has stopped, and CPR if heart action has stopped. Transfer promptly to a medical facility. When this chemical has been swallowed, get medical attention. *Do not induce vomiting when formulations containing petroleum solvents are ingested.* Otherwise, give large quantities of water and induce vomiting. Do not make an unconscious person vomit.

*References:*
- U.S. Environmental Protection Agency, "Reregistration Eligibility Decision (RED), Diquat dibromide," Office of Prevention, Pesticides and Toxic Substances, Washington, DC (July 1995). http://www.epa.gov/REDs/0288.pdf
- EXTOXNET, Extension Toxicology Network, "Pesticide Information Profile, Diquat dibromide," Oregon State University, Corvallis, OR (June 1996). http://extoxnet.orst.edu/pips/diquatdi.htm
- New Jersey Department of Health, "Hazardous Substance Fact Sheet: Diquat," Trenton, NJ (June 1986, rev. November 1994, rev. January 2001). http://www.state.nj.us/health/eoh/rtkweb/0808.pdf
- U.S. Environmental Protection Agency, Office of Pesticide Programs, Pesticide Residue Limits, " ", 40 CFR 180., www.epa.gov/pesticides/food/viewtols.htm
- California Environmental Protection Agency "Chemical List of Lists," Sacramento CA (February 1997).

# Disulfiram

*Use Type:* A seed disinfectant and fungicide.
*CAS Number:* 97-77-8
*Formula:* $C_{10}H_{20}N_2S_4$
*Synonyms:* (Bis(diethylamino)thioxomethyl) disulphide; Bis(diethylthiocarbamoyl) disulfide; Bis($N,N$-diethylthiocarbamoyl) disulfide; Bis($N,N$-diethylthiocarbamoyl) disulphide; Disulfan; Disulfuram; Disulphuram; 1,1'-Dithiobis($N,N$-diethylthioformamide); Esperal (France); Etabus; Ethyldithiourame; Ethyldithiurame; Ethyl thiram; Ethyl thiudad; Ethyl thiurad; Ethyl tuads; Ethyl tuex; NCI-C02959; Refusal (Netherlands); TETD; Tetraethylthioperoxydicarbonic diamide; Tetraethylthiram disulphide; Tetraethylthiuram; Tetraethylthiuram disulfide: Tetraethylthiuram disulphide; $N,N,N',N'$-tetraethylthiuram disulphide; Tetraetil (Spanish); TTD; TTS

*Trade Names:* ABSTENSIL®; ABSTINYL®; ALCOPHOBIN®; ALK-AUBS®; ANTADIX®; ANTAENYL®; ANTAETHAN®; ANTAETHYL®; ANTAETIL®; ANTALCOL®; ANTETAN®; ANTETHYL®; ANTETIL®; ANTEYL®; ANTIAETHAN®; ANTIETANOL®; ANTI-ETHYL®; ANTIETIL®; ANTIKOL®; ANTIVITIUM®; AVERSAN®; AVERZAN®; BONIBAL®; CONTRALIN®; CONTRAPOT®; CRONETAL®; DICUPRAL®; DISETIL®; EKAGOM TEDS®; EPHORRAN®; ESPENAL®; EXHORAN®; EXHORRAN®; HOCA®; KROTENAL®; NOCBIN®; NOXAL®; RO-SULFIRAM®; STOPAETHYL®; STOPETHYL®; STOPETYL®; TATD®; TENURID®; TENUTEX®; TETIDIS®; TETRADIN®; TETRADINE®; TETURAM®; TETURAMIN®; THIOSAN®; THIOSCABIN®; THIRERANIDE®; THIURAM E®; THIURANIDE®; TILLRAM®; TIURAM®

*Producers:* Abbott Laboratories (USA); Bayer Chemicals (Germany); Kawaguchi Chemical Industry (Japan); Sumitomo Chemical (Japan)
*Chemical Class:* Dithiocarbamate
*ICSC Number:* 1438
*RTECS Number:* JO1225000
*EEC Number:* 006-079-00-8
*EINECS Number:* 202-607-8
*Uses:* Disulfiram is also used as a rubber accelerator and vulcanizer, and in adhesives. It is widely used in the treatment of chronic alcoholism or alcohol therapy because of its effectiveness in maintaining abstinence. Its medical trade name is Antabuse®.

*Regulatory Authority:*
- Air Pollutant Standard Set (ACGIH)[1] (DFG)[3]
- Permissible Exposure Limits for Chemical Contaminants (CAL/OSHA)
- The "Director's List" (CAL/OSHA)
- Canada, WHMIS, Ingredients Disclosure List

*Description:* Disulfiram is a white to off-white or light gray powder with a slight odor. Very slightly soluble in water. Molecular weight = 296.54. Boiling point = 117°C @ 17 mmHg. Melting/Freezing point = 70°C. Log $K_{ow}$ = 3.89. Values at or above 3.0 are likely to bioaccumulate in marine organisms.

*Incompatibilities:* Oxidizers. Forms toxic sulfur and nitrogen oxides when heated to decomposition.

*Permissible Exposure Limits in Air:* There is no OSHA[2] PEL. NIOSH[2] recommended REL, DFG MAK[3], and the ACGIH[1] recommended TLV is 2 mg/m³ TWA. NIOSH[2] warns that precautions should be taken to avoid concurrent exposure to ethylene dibromide.

*Determination in Air:* No method available.
*Permissible Concentration in Water:* No method available.
*Routes of Entry:* Inhalation, ingestion, skin and/or eye contact.

*Harmful Effects and Symptoms*
**Short Term Exposure:** Irritates the eyes, skin, and respiratory system. Eye contact can lead to damage. Disulfiram can affect you when breathed in and by passing through your skin. Exposure to Disulfiram and alcohol or alcohol-containing products such as medications, cold remedies, and cough syrups within 1-2 days to each other can cause a reaction. Usually within ½-hour following ingestion of alcohol the following symptoms may occur: flushing of the face and neck, severe headache, burning eyes, rapid heart beat, nausea, and vomiting. Tightness in the chest may be mistaken for a cardiac problem. Shock may occur and this reaction could be fatal. If working with Disulfiram you should never be exposed to ethylene dibromide because of possible severe reaction. Symptoms of exposure include lassitude (weakness, exhaustion), fatigue, tremor, restlessness, headache, dizziness; metallic taste; vomiting, peripheral neuropathy
**Long Term Exposure:** May cause liver and kidney damage. It may damage the developing fetus. Damage to vision, nervous system with numbness, "pins and needles," weakness and poor coordination can result from repeated exposure. May cause personality changes of depression, anxiety or irritability. Can cause skin sensitization dermatitis. Enlarged thyroid and skin rash can also occur.
**Points of Attack:** Eyes, skin, respiratory system, central nervous system, peripheral nervous system, liver
**Medical Surveillance:** If symptoms develop or overexposure is suspected, the following may be useful: Liver, kidney, and thyroid function tests. Skin testing with dilute Disulfiram may help diagnose allergy, if done by a qualified allergist. Exam of the nervous system, eyes, and vision. Evaluate for brain effects. Alcohol use may increase liver damage.
**First Aid:** If this chemical gets into the eyes, remove any contact lenses at once and irrigate immediately for at least 15 minutes, occasionally lifting upper and lower lids. Seek medical attention immediately. If this chemical contacts the skin, remove contaminated clothing and wash immediately with soap and water. Seek medical attention immediately. If this chemical has been inhaled, remove from exposure, begin rescue breathing (using universal precautions) if breathing has stopped, and CPR if heart action has stopped. Transfer promptly to a medical facility. When this chemical has been swallowed, get medical attention. Give large quantities of water and induce vomiting. Do not make an unconscious person vomit.
*Note:* For alcohol–disulfiram or ethylene dibromide–disulfiram reaction, remove the person from exposure. Begin rescue breathing if breathing has stopped, and CPR if heart action has stopped. Treat for shock and transfer to a medical facility. For alcohol-related reaction, physician or authorized medical personnel may administer antihistaminics, either orally or IV.

*References:*
- Sax, N.I., Ed., "Dangerous Properties of Industrial Materials Report," 1, No. 5, 40 (1981).
- California Environmental Protection Agency "Chemical List of Lists," Sacramento CA (February 1997).
- New Jersey Department of Health, "Hazardous Substance Fact Sheet: Disulfiram," Trenton, NJ (March 1999). http://www.state.nj.us/health/eoh/rtkweb/0811.pdf

# Disulfoton

*Use Type:* Insecticide and acaricide
*CAS Number:* 298-04-4
*Formula:* $C_8H_{19}O_2PS_3$
*Alert:* All products formulated at greater than 2% disulfoton are classified as Restricted Use Pesticides (RUP).
*Synonyms:* O,O-Diaethyl-S-(3-thia-pentyl)-dithiophosphat (German); O,O-Diaethyl-S-(2-aethylthio-aethyl)-dithiophosphat (German); O,O-diethyl S-(2-eththioethyl) phosphorodithioate; O,O-diethyl S-(2-eththioethyl) thiothionophosphate; O,O-Diethyl S-(2-ethylmercaptoethyl) dithiophosphate; O,O-Diethyl-S-(2-ethylthio-ethyl)-dithiofosfaat (Dutch); O,O-Diethyl 2-ethylthioethylphosphorodithioate; O,O-Diethyls-2-(ethylthio)ethylphosphorodithioate; O,O-Dietil-S-(2-etiltio-etil)-ditiofosfato (Italian); Dithiophosphate de O,O-diethyle etde S-(2-ethylthio-ethyle) (French); ENT 23,437; O,O-Ethyl S-2(ethylthio)ethylphosphorodithioate; S-2-(Ethylthio)ethyl O,O-diethylester of phosphorodithioic acid; Ethylthiodemeton; Phosphorodithionic acid, S-(2-(ethylthio)ethyl-O,O-diethylester; Phosphorodithionic acid, O,O-diethyl S-2-(ethylthio)ethyl ester
*Trade names:* BAY 19639®, Bayer CropScience (Germany); BAYER 19639®, Bayer CropScience (Germany); DIMAZ®; DISULFATON®; DI-SYSTON®, Bayer CropScience (Germany), canceled; DISYSTON®, Bayer CropScience (Germany), canceled; DISYSTOX®; DITHIODEMETON®; DITHIOSYSTOX®; EKATIN TD®; FRUMIN-AL®; FRUMIN G®; GLEBOFOS®; M-74®; S 276®; SOLVIREX®; THIODEMETON®; THIODEMETRON®
*Producers:* BASF Agricultural Products Group (Germany); Bayer CropScience (Germany); Bonide Products (USA); Ehrenstorfer, Dr. (Germany); Sigma-Aldrich Laborchemikalien (Germany); Simplot, J.R, Company (USA)
*Chemical Class:* Organophosphate
*EPA/OPP PC Code:* 032501
*California DPR Chemical Code:* 230
*ICSC Number:* 1408
*RTECS Number:* TD9275000
*EEC Number:* 015-060-00-3

# Disulfoton

*EINECS Number:* 206-054-3

*Uses:* Disulfoton is a selective, systemic organophosphate insecticide and acaricide that is especially effective against sucking insects. It is used to control aphids, leafhoppers, thrips, beet flies, spider mites, and coffeeleaf miners. Disulfoton products are used on fruit and nut crops, small grains, sorghum, cotton, tobacco, sugar beets, cole crops, corn, peanuts, wheat, ornamentals and potatoes.

*Human toxicity (long-term):* Extra high–0.30 ppb, Health Advisory

*Fish toxicity (threshold):* High–2.90001 ppb, MATC (Maximum Acceptable Toxicant Concentration)

***U.S. Maximum Allowable Residue Levels for Disulfoton (40 CFR 180.183):***

| CROP | ppm |
|---|---|
| Asparagus | 0.1 |
| Barley, grain | 0.75 |
| Barley, straw | 5.0 |
| Beans, dry | 0.75 |
| Beans, lima | 0.75 |
| Beans, snap | 0.75 |
| Beans, vines | 5.0 |
| Beets, sugar, roots | 0.5 |
| Beets, sugar, tops | 2.0 |
| Broccoli | 0.75 |
| Brussels sprouts | 0.75 |
| Cabbage | 0.75 |
| Cauliflower | 0.75 |
| Coffee, bean | 0.3 |
| Corn, field, forage | 5.0 |
| Corn, field, kernel plus cob with husks removed | 0.3 |
| Corn, field, fodder | 5.0 |
| Corn, field, forage | 5.0 |
| Corn, grain | 0.3 |
| Corn, pop | 0.3 |
| Corn, pop, fodder | 5.0 |
| Corn, pop, forage | 5.0 |
| Corn, sweet, fodder | 5.0 |
| Corn, sweet, forage | 5.0 |
| Corn, sweet, kernel plus cob with husks removed | 0.3 |
| Cottonseed | 0.75 |
| Hop, dried cones | 0.5 |
| Lettuce | 0.75 |
| Oats, fodder, green | 5.0 |
| Oat, grain | 0.75 |
| Oat, straw | 5.0 |
| Peas | 0.75 |
| Pea, field, vines | 5.0 |
| Peanuts | 0.75 |
| Pecans | 0.75 |
| Peppers | 0.1 |
| Pineapples | 0.75 |
| Potatoes | 0.75 |
| Rice | 0.75 |
| Rice, straw | 5.0 |
| Sorghum, fodder | 5.0 |
| Sorghum, forage | 5.0 |
| Sorghum, grain | 0.75 |
| Soybeans | 0.1 |
| Soybean, forage | 0.25 |
| Soybean, hay | 0.25 |
| Spinach | 0.75 |
| Sugarcane | 0.3 |
| Tomatoes | 0.75 |
| Wheat, fodder, green | 5.0 |
| Wheat, grain | 0.3 |
| Wheat, straw | 5.0 |

***Carcinogen/Hazard Classifications***

**U.S. EPA Carcinogens:** Group E, unlikely carcinogen

**Label Signal Word:** WARNING or DANGER, Toxicity class I-highly toxic, according to formulation

**WHO Acute Hazard:** Class 1 a, extremely hazardous

***Regulatory Authority:***

- Banned or Severely Restricted (Various Countries) (UN)[13]
- Very Toxic Substance (World Bank)[15]
- Air Pollutant Standard Set (ACGIH)[1] (HSE)[33] (Several States)[60]
- CLEAN WATER ACT: Section 311 Hazardous Substances/RQ 40CFR117.3 (same as CERCLA, see below).
- EPA HAZARDOUS WASTE NUMBER (RCRA No.): P039
- RCRA, 40CFR261, Appendix 8 Hazardous Constituents
- RCRA 40CFR268.48; 61FR15654, Universal Treatment Standards: Wastewater (mg/L), 0.017; Nonwastewater (mg/kg), 6.2
- RCRA 40CFR264, Appendix 9; TSD Facilities Ground Water Monitoring List. Suggested test method(s) (PQL ug/L): 8140(2)
- SUPERFUND/EPCRA 40CFR355, Appendix B Extremely Hazardous Substances: TPQ = 500 lb (227kg)
- SUPERFUND/EPCRA 40CFR302.4 RQ: CERCLA, 1 lb (0.454 kg)
- U.S. DOT Regulated Marine Pollutant (49CFR172.101, Appendix B)
- Canada, WHMIS, Ingredients Disclosure List

*Description:* Disulfoton is a combustible, colorless to yellowish oil with a characteristic odor. Technical product is a brown liquid. Boiling point = 132-133°C @ 1.5 mm pressure. Freezing/Melting point = > – 25°C. Flash point = >82°C. Hazard Identification (based on NFPA-704 M Rating System): Health 4, Flammability 1, Reactivity 0. Practically insoluble in water.

*Incompatibilities:* Alkalies, strong oxidizers. Forms toxic nitrogen and sulfur oxides when heated to decomposition.

*Permissible Exposure Limits in Air:* There is no OSHA PEL[2]. HSE[33] and the recommended ACGIH[1] TWA value is 0.1 mg/m³ and the HSE[33] has set an STEL of 0.3 mg/m³. Several states have set guidelines or standards for disulfoton in ambient air[60] ranging from 1.0 μg/m³ (North Dakota) to 1.6 μg/m³ (Virginia) to 2.0 μg/m³ (Connecticut and Nevada).

*Determination in Air:* OSHA versatile sampler-2; Toluene/Acetone; Gas chromatography/Flame photometric detection for sulfur, nitrogen, or phosphorus; NIOSH Method IV, Method #5600, organophosphorus pesticides.

*Permissible Concentration in Water:* The EPA recommends that no more than 10 parts of disulfoton per billion parts of water (10 ppb) be present in water that children drink for periods of up to 10 days. They also recommend that disulfoton should not exceed 3 ppb for children or 9 ppb for adults if they drink water for longer periods of time, and it should not exceed 0.3 ppb for adults who drink the water for a lifetime.

*Determination in Water:* Extraction with methylene chloride followed by measurement by gas chromatography using a nitrogen-phosphorus detector.

*Routes of Entry:* Inhalation, skin absorption, ingestion, skin and/or eye contact.

*Harmful Effects and Symptoms*

*Short Term Exposure:* Contact may cause burns to skin and eyes. Symptoms include pinpoint pupils, blurred vision, headache, dizziness, muscle spasms, and profound weakness. Vomiting, diarrhea, abdominal pain, seizures, and coma may also occur. The heart rate may decrease following oral exposure or increase following dermal exposure. Hypotension (low blood pressure) and chest pain may be noted. Hypertension (high blood pressure) is not uncommon. Respiratory symptoms include dyspnea (shortness of breath), respiratory depression, and respiratory paralysis. Psychosis may occur. It is classified as super toxic. Probable oral lethal dose in humans is less that 5 mg/kg or a taste (less than 7 drops) for a 70 kg (150 lb) person. It is poisonous and may be fatal if inhaled, swallowed, or absorbed through the skin.

*Long Term Exposure:* Cholinesterase inhibitor; cumulative effect is possible. This chemical may damage the nervous system with repeated exposure, resulting in convulsions, respiratory failure. May cause liver damage.

*Points of Attack:* Respiratory system, lungs, central nervous system, cardiovascular system, skin, eyes, plasma and red blood cell cholinesterase.

*Medical Surveillance:* Before employment and at regular times after that, the following are recommended: Plasma and red blood cell cholinesterase levels (tests for the enzyme poisoned by this chemical). If exposure stops, plasma levels return to normal in 1-2 weeks while red blood cell levels may be reduced for 1-3 months. When acetylcholinesterase enzyme levels are reduced by 25% or more below preemployment levels, risk of poisoning is increased, even if results are in lower ranges of "normal." Reassignment to work not involving organophosphate or carbamate pesticides is recommended until enzyme levels recover. If symptoms develop or overexposure occurs, repeat the above tests as soon as possible and get an exam of the nervous system. Also consider complete blood count. Consider chest x-ray following acute overexposure. Do not drink any alcoholic beverages before or during use. Alcohol promotes absorption of organic phosphates.

*First Aid:* **Treatment for organophosphate poisoning consists of thorough decontamination, cardiorespiratory support, and administration of the antidotes atropine and pralidoxime. In cases of severe poisoning, diazepam, an anticonvulsant, should also be administered. Antidotes should be administered as prevention even if the diagnosis is in doubt.** Speed in removing material from eyes and skin is of extreme importance. *Eyes:* Eye contact can cause dangerous amounts of these chemicals to be quickly absorbed through the mucous membrane into the bloodstream. Immediately and gently flush eyes with plenty of warm or cold water (NO hot water) for at least 15 minutes, occasionally lifting the upper and lower eyelids. Get medical aid immediately. *Skin:* Get medical aid. Skin contact can cause dangerous amounts of these chemicals to be absorbed into the bloodstream. Wearing the appropriate PPE equipment and respirator for organophosphate/carbamate pesticides, immediately flush skin with plenty of soap and water for at least 15 minutes while removing contaminated clothing and shoes. Shampoo hair promptly if contaminated. The removed, contaminated clothing and shoes should be double-bagged and left in Hot Zone for later disposal by hazardous materials experts. Skin may also be decontaminated with diluted hypochlorite solution. *Inhalation:* Get medical aid. Do not contaminate yourself. Wearing the appropriate PPE equipment and respirator for organophosphate pesticides, immediately remove the victim from the contaminated area to fresh air. If the victim is not breathing, administer artificial respiration. Do NOT use mouth-to-mouth resuscitation; use bag/mask apparatus. If breathing is difficult, administer oxygen through bag/mask apparatus until medical help arrives. Do not leave victim unattended. *Ingestion:* Call poison control. Loosen all clothing. Never give anything by mouth to an unconscious person. If victim is *unconscious or having convulsions,* do nothing except keep victim warm. Get medical aid. Transfer promptly to a medical facility. In cases of ingestion, **do not induce vomiting**. If the victim is alert and asymptomatic, administer a slurry of activated charcoal at a dose of 1 g/kg (infant, child, and adult dose). A soda can and straw may be of assistance when offering charcoal to a child. *In some cases you may be specifically instructed by poison control to induce vomiting by way of 2 tablespoons of syrup of ipecac (adult) washed down with a*

with ipecac syrup. *Note to physician or authorized medical personnel:* Treat cases of respiratory compromise, coma, or excessive pulmonary secretions with respiratory support using protocols and techniques available and within the scope of training. Some cases may necessitate procedures such as endotracheal intubation or cricothyrotomy by properly trained and equipped personnel. When possible, atropine (see *Antidotes*, below) should be given under medical supervision. Patients who are comatose, hypotensive, or having seizures or cardiac arrhythmias should be treated according to advanced life support protocols. *Antidotes:* Two antidotes are administered to treat organophosphate poisoning. Atropine is a competitive antagonist of acetylcholine at muscarinic receptors and is used to control the excessive bronchial secretions which are often responsible for death. Pralidoxime relieves both the nicotinic and muscarine effects of organophosphate poisoning by regenerating acetylcholinesterase and can reduce both the bronchial secretions and the muscle weakness associated with poisoning. The initial intravenous dose of atropine in adults should be determined by the severity of symptoms: An initial adult dose of 1.0 to 2.0 mg or pediatric dose of 0.01 mg/kg (minimum 0.01 mg) should be administered intravenously. If intravenous access cannot be established, atropine may also be given intramuscularly, subcutaneously or via endotracheal tube. Doses should be repeated every 15 minutes until excessive secretions and sweating have been controlled. Once bronchial secretion has been controlled, atropine administration should be repeated whenever the secretions begin to recur. In seriously poisoned patients, very large doses may be required. Alterations of pulse rate and pupillary size should not be used as indicators of treatment adequacy. Pralidoxime should be administered as early in poisoning as possible as its efficacy may diminish when given more than 24 to 36 hours after exposure. Doses are as follows: adult 1.0 g; pediatric 25 to 50 mg/kg. The drug should be administered intravenously over 30 to 60 minutes, but in a life-threatening situation, one-half of the total dose can be given per minute for a total administration time of 2 minutes. Treatment should begin to take effect within 40 minutes with a reduction in symptoms and the amount of atropine necessary to control bronchial secretion. The initial dose can be repeated in 1 hour and then every 8 to 12 hours until the patient is clinically well and no longer requires atropine. If intravenous access cannot be established, pralidoxime may also be given intramuscularly. Early administration of diazepam in addition to the combined atropine and pralidoxime treatment may help prevent the onset of seizures and potential brain and cardiac morphologic damage following high-level organophosphate poisoning.

*References:*
- EXTOXNET, Extension Toxicology Network, "Pesticide Information Profile, Disulfoton," Oregon State University, Corvallis, OR (June 1996). http://ace.orst.edu/info/extoxnet/pips/disulfot.htm
- U.S. Environmental Protection Agency, Office of Pesticide Programs, Pesticide Residue Limits, "Disulfoton", 40 CFR 180.183, http://www.epa.gov/pesticides/food/viewtols.htm
- Agency for Toxic Substances and Disease Registry, "ToxFAQs for Disulfoton," Atlanta, GA (September 1996). http://www.atsdr.cdc.gov/tfacts65.html
- New Jersey Department of Health, "Hazardous Substance Fact Sheet: Disulfoton," Trenton, NJ (January 1999). http://www.state.nj.us/health/eoh/rtkweb/0812.pdf
- International Programme on Chemical Safety (IPCS), "Data Sheets on Pesticides No. 68, Disulfoton," Geneva, Switzerland (March, 1988). http://www.inchem.org/documents/pds/pds/pest68_e.htm
- U.S. Environmental Protection Agency, Disulfoton, Health and Environmental Effects Profile No. 97, Washington, DC, Office of Solid Waste (April 30, 1980).
- U.S. Environmental Protection Agency, "Chemical Profile: Disulfoton," Washington, DC, Chemical Emergency Preparedness Program (November 30, 1987).
- U.S. Environmental Protection Agency, "Health Advisory: Disulfoton," Washington, DC, Office of Drinking Water (August 1987).
- Sax, N.I., Ed., "Dangerous Properties of Industrial Materials Report," 8, No. 5, 74-85 (1988).

## Dithiazanine Iodide

*Use Type:* An insecticide for treating animals
*CAS Number:* 514-73-8
*Formula:* $C_{23}H_{23}IN_2S_2$
*Alert:* Not registered in the United States
*Synonyms:* Diethylthiadicarbocyanine iodide; 3,3'-Diethylthiadicarbocyanine iodide; Dilombrin; Dithiazinine; Dithiazanine iodide; Dithiazanin iodide; 3-Ethyl-2-(5-(3-ethyl-2-benzothiazolinylidene)-1,3-pentadienyl)benzothiazolium iodide
*Trade Names:* ABMINTHIC®; ANELMID®; ANGUIFUGAN®; COMPOUND 01748®; DEJO®; DELVEX®; EASTMAN 7663®; L-01748®; NETOCYD®; NK 136®; OMNIPASSIN®; PARTEL®; TELMICID®; TELMID®; TELMIDE®; VERCIDON®
*Producers:* Molekula Fine Chemicals (UK)
*RTECS Number:* DL7060000
*Uses:* This material is used as a veterinary anthelmintic for treating dogs for large roundworms, hookworms, whipworms, and strongyloides. It is also a sensitizer for photographic emulsions and for insecticides. Not registered as a pesticide in the U.S.
*Regulatory Authority:*
- Banned or Severely Restricted (Several Countries) (UN)[13]

- Superfund/EPCRA 40CFR355, Appendix B Extremely Hazardous Substances: TPQ = 500/10,000 lb (227/4,540 kg)
- Superfund/EPCRA 40CFR302.4 RQ: EHS, 1 lb (0.454 kg)

*Description:* Green, needle-like crystalline solid. Practically insoluble in water. Molecular weight = 518.69. Melting/Freezing point = (decomposes) 248°C.

*Harmful Effects and Symptoms*

*Short Term Exposure:* Contact with eyes or skin may cause irritation or injury. Inhalation should be avoided; use NIOSH-approved air purifying respirators for pesticides. Poisonous if swallowed, or if dust is inhaled. $LD_{50}$ (oral, mouse) = 20 mg/kg (highly toxic).

*First Aid:* If this chemical gets into the eyes, remove any contact lenses at once and irrigate immediately for at least 15 minutes, occasionally lifting upper and lower lids. Seek medical attention immediately. If this chemical contacts the skin, remove contaminated clothing and wash immediately with soap and water. Seek medical attention immediately. If this chemical has been inhaled, remove from exposure, begin rescue breathing (using universal precautions) if breathing has stopped, and CPR if heart action has stopped. Transfer promptly to a medical facility. When this chemical has been swallowed, get medical attention. Give large quantities of water and induce vomiting. Do not make an unconscious person vomit.

*References:*
- U.S. Environmental Protection Agency, "Chemical Profile: Dithiazanine Iodine," Washington, DC, Chemical Emergency Preparedness Program (November 30, 1987).
- California Environmental Protection Agency "Chemical List of Lists," Sacramento CA (February 1997).

# Dithiopyr (ANSI)

*Use Type:* Herbicide
*CAS Number:* 97886-45-8
*Formula:* $C_{15}H_{16}F_5NO_2S_2$
*Synonyms:* 3,5-Pyridinedicarbothioic acid, 2-(difluoromethyl)-4-(2-methylpropyl)-6-(trifluoromethyl)-, *S,S*-dimethyl ester; *S,S*-Dimethyl 2-(difluoromethyl)-4-(2-methylpropyl)-6-(trifluoromethyl)-3,5-pyridinedicarbothioate
*Trade Names:* DIMENSION®, Dow AgroSciences (USA); EH-1400®, Pbi/Gordon (USA); MON 7200®, Monsanto (USA); MON-15100®, Monsanto (USA)
*Producers:* Dow AgroSciences (USA); Monsanto (USA); Pbi/Gordon (USA); J.R. Simplot (USA)
*Chemical Class:* Thiocarbamate
*EPA/OPP PC Code:* 128994
*California DPR Chemical Code:* 2308
*Uses:* Used primarily for pre-emergence control of crabgrass and goosgrass on turf and lawns.

*Human toxicity (long-term)[77]:* Intermediate–25.20 ppb, Health Advisory
*Fish toxicity (threshold)[77]:* Intermediate–27.92865 ppb, MATC (Maximum Acceptable Toxicant Concentration)

*Carcinogen/Hazard Classifications*

**U.S. EPA Carcinogens:** Group E, Unlikely a carcinogen
**Label Signal Word:** CAUTION
**WHO Acute Hazard:** Class U, unlikely to be hazardous
*Description:* Colorless, crystalline solid. Soluble in water; solubility = 1.41 ppm. Molecular weight = 401.45.
*Permissible Concentration in Water:* No criteria set. Runoff from spills or fire control may cause water pollution.

*Harmful Effects and Symptoms*

*Short Term Exposure:* Low levels of toxicity. Concentrated solutions are slightly corrosive to eyes and mucous membranes. Dust inhalation can cause irritation of the respiratory system with sneezing. Eye contact can cause irritation, watering, pain, and inflammation of the eyelids. Skin contact can cause irritation and minor ulceration. Ingestion can cause nausea, vomiting, fever, muscle twitching, seizure, rapid respiration, slow heart beat. Severe exposure may result in death.

*Points of Attack:* Respiratory system, central nervous system, cardiovascular system, skin, eyes.

*Medical Surveillance:* Medical observation is recommended for 24 to 48 hours after breathing overexposure, as pulmonary edema may be delayed. As first aid for pulmonary edema, consider administering a corticosteroid spray. Cigarette smoking may exacerbate pulmonary injury and should be discouraged for at least 72 hours following exposure. Before employment and at regular times after that, the following are recommended: If symptoms develop or overexposure occurs, repeat the above tests as soon as possible and get an exam of the nervous system. Also consider complete blood count. Consider chest x-ray following acute overexposure.

*First Aid:* If this chemical gets into the eyes, remove any contact lenses at once and irrigate immediately for at least 15 minutes, occasionally lifting upper and lower lids. Seek medical attention immediately. If this chemical contacts the skin, remove contaminated clothing and wash immediately with soap and water. Seek medical attention immediately. If this chemical has been inhaled, remove from exposure, begin rescue breathing (using universal precautions) if breathing has stopped, and CPR if heart action has stopped. Transfer promptly to a medical facility. When this chemical has been swallowed, get medical attention. If victim is conscious and able to swallow, have victim drink 4 to 8 ounces of water. Do not induce vomiting. *Note to physician or authorized medical personnel*: Medical observation is recommended for 24 to 48 hours after breathing overexposure, as pulmonary edema may be delayed. As first aid for pulmonary edema, consider administering a corticosteroid spray. Cigarette smoking may exacerbate pulmonary injury and should be discouraged for at least 72 hours following exposure.

## Diuron (ANSI)

*Use Type:* Herbicide
*CAS Number:* 330-54-1
*Formula:* $C_9H_{10}Cl_2N_2O$; $Cl_2C_6H_3NHCON(CH_3)_2$
*Alert:* A General Use Pesticide (GUP).
*Synonyms:* DCMU (In Japan); 3-(3,4-Dichloor-fenyl)-1,1-dimethylureum (Dutch); Dichlorfenidim; 3-(3,4-Dichlorophenol)-1,1-dimethylurea; 3-(3,4-Dichlorophenyl)-1,1-demethylurea; $N'$-(3,4-dichlorophenyl)-$N,N$-dimethylurea; $N$-(3,4-Dichlorophenyl)-$N',N$-dimethylurea; 1-(3,4-Dichlorophenyl)-3,3-dimethylurea; 1,1-Dimethyl-3-(3,4-dichlorophenyl)urea; 1(3,4-Dichlorophenyl)-3,3-dimethyluree (French); 3-(3,4-Dichlor-phenyl)-1,1-dimethylharnstoff (German); 3-(3,4-Diclorofenil)-1,1-dimetilurea (Spanish); 3-(3,4-Dicloro-fenyl)-1,1-dimetil-urea (Italian); Urea, $N'$-(3,4-dichlorophenyl)-$N,N$-dimethyl-; Urea, 3-(3,4-dichlorophenyl)-1,1-dimethyl-
*Trade Names:* 330541®; AF 101®; AI3-61438®; AMETRON SC®, Milenia Agro Ciencias (Brazil); BOUNDRY®, Syngenta (Switzerland), canceled 7/21/1998; CASWELL No. 410®; CHEMIURON®, Agrimont S.p.A. (Italy), canceled 2/21/1986; CEKIURON®; CRISURON®; DAILON®; DIATER®; DI-ON®; DIREX®, Griffin (USA); DITOX®, Pyosa Agroquimicos (Mexico); DIUMATE®, Drexel Chemical (USA); DIUREX®, Makhteshim-Agan Industries (Israel), canceled; DIUROL®; DIURON 4L®; DMU®; DREXEL DIURON 4L®, Drexel Chemical (USA); DROPP ULTRA®, Bayer CropScience (Germany); DURAN®; DYNEX®, Cedar Chemical Corporation (USA), canceled 2/3/2003; FARMCO DIURON®; FORTEX SC®, Milenia Agro Ciencias (Brazil); FREEFLO®, Kenso Corp. (Malaysia); GINSTAR®, Bayer CropScience (Germany); HERBURON 500 BR®, Milenia Agro Ciencias (Brazil); HW 920®; KARMEX®, DuPont Crop Protection (USA), canceled 5/2/1996; K-4®, DuPont Crop Protection (USA); KARMEX DIURON HERBICIDE®; KARMEX DW®; KROVAR IDF®, DuPont Crop Protection (USA), canceled 5/2/1996; MARMER®; STRIKER®, Sanonda (Australia); SUP'R FLO®; TELVAR®; TIGREX®; TREVISSIMO®, Calliope (France); UNIDRON®; UROX D®, Haco (USA), canceled 5/1/1987; VONDURON®
*Producers:* Agan Chemical Manufacturers Ltd. (Israel); Air Products & Chemicals (USA); Atul (India); Bayer CropScience (Germany); Bharat Pulverizing Mills (India); Biesterfeld Siemsgluess International. GmbH (Germany); Cedar Chemical Corporation (USA); Degussa (Germany); Drexel Chemical (USA); Dow AgroSciences (USA); DuPont Crop Protection (USA); Ehrenstorfer, Dr. (Germany); Griffin (USA); Harcros Chemicals (USA); Helena Chemical (USA); Hockley International (UK); ICI Group (UK); Kenso Corp. (Malaysia); Makhteshim-Agan Industries (Israel); Milenia Agro Ciencias (Brazil); Montedison (Italy); Pyosa Agroquimicos (Mexico); Rhone-Poulenc Agro (France); Sanonda (Australia); Shenzhen Guomeng Industry Co., Ltd. (China); Sigma-Aldrich Laborchemikalien (Germany); Syngenta (Switzerland); United Phosphorus (India); Zago Asia Ltd. (Singapore)
*Chemical Class:* Urea
*EPA/OPP PC Code:* 035505
*California DPR Chemical Code:* 231
*RTECS Number:* YS8925000
*EINECS Number:* 206-354-4
*Uses:* Diuron is a substituted urea herbicide used to control a wide variety of annual and perennial broadleaf and grassy weeds, as well as mosses. It is used on non-crop areas and many agricultural crops such as fruit, cotton, sugar cane, alfalfa, and wheat. Diuron works by inhibiting photosynthesis. It may be found in formulations as wettable powders and suspension concentrates.
*Human toxicity (long-term)[77]:* Intermediate–18.32460 ppb, CHCL (Chronic Human Carcinogen Level)
*Fish toxicity (threshold)[77]:* Intermediate–40.39243 ppb, MATC (Maximum Acceptable Toxicant Concentration)
*U.S. Maximum Allowable Residue Levels for Diuron (40 CFR 180.106):*

| CROP | ppm |
| --- | --- |
| Alfalfa | 2 |
| Apple | 1 |
| Artichoke, globe | 1 |
| Asparagus | 7 |
| Banana | 0.1 |
| Barley, grain | 1 |
| Barley, hay | 2 |
| Blackberry | 1 |
| Blueberry | 1 |
| Boysenberry | 1 |
| Catfish | 2 |
| Cattle, fat | 1 |
| Cattle, mbyp | 1 |
| Cattle, meat | 1 |
| Citrus, dried pulp | 4 |
| Clover, forage | 2 |
| Clover, hay | 2 |
| Corn, sweet, kernel plus cob with husks removed | 1 |
| Corn, pop, grain | 1 |
| Corn, sweet, grain | 1 |
| Corn, field, ear | 1 |
| Corn, field, grain | 1 |
| Corn, sweet, stover | 2 |
| Corn, sweet, forage | 2 |
| Cotton, undelinted seed | 1 |
| Currant | 1 |
| Dewberry | 1 |
| Fruit, stone, group 12 | 1 |

| | |
|---|---|
| Goat, fat | 1 |
| Goat, meat | 1 |
| Goat, mbyp | 1 |
| Gooseberry | 1 |
| Grape | 1 |
| Grass, except bermuda grass | 2 |
| Grass, hay, except bermudagrass | 2 |
| Hog, fat | 1 |
| Hog, mbyp | 1 |
| Hog, meat | 1 |
| Horse, fat | 1 |
| Horse, meat | 1 |
| Horse, mbyp | 1 |
| Loganberry | 1 |
| Nuts | 0.1 |
| Oat, forage | 2 |
| Oat, grain | 1 |
| Oat, hay | 2 |
| Oat, straw | 2 |
| Olive | 1 |
| Papaya | 0.5 |
| Pea, field, vines | 2 |
| Pea, field, hay | 2 |
| Pea | 1 |
| Peach | 0.1 |
| Pear | 1 |
| Peppermint, tops | 2 |
| Pineapple | 1 |
| Potato | 1 |
| Raspberry | 1 |
| Rye, forage | 2 |
| Rye, hay | 2 |
| Rye, grain | 1 |
| Rye, straw | 2 |
| Sheep, fat | 1 |
| Sheep, mbyp | 1 |
| Sheep, meat | 1 |
| Sorghum, forage | 2 |
| Sorghum, grain, stover | 2 |
| Sorghum, grain, grain | 1 |
| Sugarcane, cane | 1 |
| Trefoil, birdsfoot, forage | 2 |
| Trefoil, birdsfoot, hay | 2 |
| Vetch, seed | 1 |
| Vetch, forage | 2 |
| Vetch, hay | 2 |
| Wheat, hay | 2 |
| Wheat, forage | 2 |
| Wheat, grain | 1 |
| Wheat, straw | 2 |

***Carcinogen/Hazard Classifications***

**U.S. EPA Carcinogens:** Known/likely to be carcinogen

**California Prop. 65:** Carcinogen

**Label Signal Word:** WARNING, Toxiciity Class III–Slightly Toxic

**WHO Acute Hazard:** Class U, Unlikely to be hazardous

***Regulatory Authority:***

- Air Pollutant Standard Set (ACGIH)[1] (Argentina) (HSE)[33] (former USSR)[35, 43] (Several States)[60]
- Clean Water Act: Section 311 Hazardous Substances/RQ 40CFR117.3 (same as CERCLA, see below)Superfund/EPCRA 40CFR302.4 RQ: CERCLA, 100 lb (45.4 kg)
- AB 1803-Well Monitoring Chemical (CAL)
- Proposition 65 chemical (CAL)
- Permissible Exposure Limits for Chemical Contaminants (CAL/OSHA)
- The "Director's List" (CAL/OSHA)
- Actively registered pesticide in California
- EPCRA Section 313 Form R *de minimus* concentration reporting level: 1.0%
- Canada, WHMIS, Ingredients Disclosure List

***Description:*** Diuron is a white, odorless crystalline solid. Very slightly soluble in water; solubility = 75 ppm @ 25°C. Molecular weight = 233.1. Boiling point = (decomposes) 180°C. Melting/Freezing point = 158–159°C. Vapor pressure = $6.9 \times 10^{-8}$ mmHg @ 20°C; $3.1 \times 10^{-6}$ mmHg @ 50°C. Hazard Identification (based on NFPA-704 M Rating System): Health 1, Flammability 0, Reactivity 0. Log $K_{ow}$ = 2.85. Values at or above 3.0 are likely to bioaccumulate in marine organisms. Harmful to aquatic life in very low concentrations.

***Incompatibilities:*** Strong acids.

***Permissible Exposure Limits in Air:*** There is no OSHA[2] PEL. The HSE[33] and recommended ACGIH[1] TLVis 10 mg/m³ TWA. Argentina[35] has adopted 10 mg/m³ as a TWA and set 20 mg/m³ as a STEL. The former USSR has set an MAC in ambient air of residential areas of 0.5 mg/m³ either on a momentary or a daily average basis[35, 43]. Several states have set guidelines or standards for diuron in ambient air[60] ranging from 100 μg/m3 (North Dakota) to 160 μg/m3 (Virginia) to 200 μg/m3 (Connecticut) to 238 μg/m3 (Nevada).

***Determination in Air:*** OSHA versatile sampler-2; Reagent; High-pressure liquid chromatography/Ultraviolet detection; NIOSH IV, Method #5601[18].

***Permissible Concentration in Water:*** The former USSR[35, 43] has set an MAC in water bodies used for domestic purposes of 1.0 mg/L and an MAC in water bodies used for fishery purposes of 1.5 μg/L. A long-term health advisory of 0.875 mg/L has been calculated by EPA and a lifetime health advisory of 0.014 mg/L (14 μg/L).

***Determination in Water:*** High performance liquid chromatography may be used after extraction with Methylene chloride. Measurement is made using an ultraviolet detector.

***Routes of Entry:*** Inhalation, ingestion, skin and/or eye contact.

*Harmful Effects and Symptoms*
*Short Term Exposure:* Exposure may irritate the skin, eyes, and throat.
*Long Term Exposure:* May damage the developing fetus. In animals: anemia, methemoglobinemia
*Points of Attack:* Eyes, skin, respiratory system, blood
*Medical Surveillance:* If symptoms develop or overexposure is suspected, the following may be useful: Complete blood count.
*First Aid:* If this chemical gets into the eyes, remove any contact lenses at once and irrigate immediately for at least 15 minutes, occasionally lifting upper and lower lids. Seek medical attention immediately. If this chemical contacts the skin, remove contaminated clothing and wash immediately with soap and water. Seek medical attention immediately. If this chemical has been inhaled, remove from exposure, begin rescue breathing (using universal precautions) if breathing has stopped, and CPR if heart action has stopped. Transfer promptly to a medical facility. When this chemical has been swallowed, get medical attention. Give large quantities of water and induce vomiting. Do not make an unconscious person vomit.
*Note to physician or authorized medical personnel:* Treat for methemoglobinemia. Spectrophotometry may be required for precise determination of levels of methemoglobinemia in urine.
*References:*
- EXTOXNET, Extension Toxicology Network, "Pesticide Information Profile, Diuron," Oregon State University, Corvallis, OR (June 1996). http://ace.orst.edu/info/extoxnet/pips/diuron.htm
- EPA, Office of Pesticide Programs, Pesticide Residue Limits, "Diuron", 40 CFR 180.106, http://www.epa.gov/pesticides/food/viewtols.htm
- New Jersey Department of Health, "Hazardous Substance Fact Sheet: Diuron," Trenton, NJ (April 1997). http://www.state.nj.us/health/eoh/rtkweb/0819.pdf
- Sax, N.I., Ed., "Dangerous Properties of Industrial Materials Report," 7, No. 5, 49-55 (1987).
- California Environmental Protection Agency "Chemical List of Lists," Sacramento CA (February 1997).
- U.S. Environmental Protection Agency, "Health Advisory: Diuron," Washington, DC, Office of Drinking Water (August 1987).

# Dodecylbenzenesulfonic Acid

*Use Type:* Insecticide, adjuvant, and biocide
*CAS Number:* 27176-87-0
*Formula:* $C_{18}H_{30}O_3S$
*Synonyms:* Acido dodecilbencenosulfonico (Spanish); Benzenesulfonic acid, dodecyl-; Benzenesulphonic acid, dodecyl-; Benzene sulfonic acid, dodecyl ester; Benzene sulphonic acid, dodecyl ester; DDBSA; Dodanic acid 83; Dodecyl benzenesulfonate; Dodecyl benzenesulphonate; *N*-Dodecyl benzenesulfonic acid; *N*-Dodecyl benzenesulphonic acid; Dodecylbenzenesulphonic acid; Laurylbenzenesulfonate; Laurylbenzenesulphonate; Laurylbenzenesulfonic acid; Laurylbenzenesulphonic acid; NANSA SSA; Pentine acid 5431; Richonic acid; Sulframin acid 1298
*Trade Names:* ACCOMPLISH®, Cognis (Germany), canceled; ACIDET®; ACIDISOL®; CALSOFT LAS 99®; E 7256®; ELFAN WA sulphonic acid®; NACCONOL 98 SA®; RHODACAL ABSA®
*Producers:* Cognis (Germany)
*Chemical Class:* Soap
*EPA/OPP PC Code:* 098002
*California DPR Chemical Code:* 941
*ICSC Number:* 1470
*RTECS Number:* DB6600000
*EINECS Number:* 248-289-4
*Uses:* Dodecylbenzenesulfonic acid and its salts are used as degreasers in meat cutting rooms. Also used as a laboratory chemical, to make detergents, and in electronically cleaning and pickling baths.
*Carcinogen/Hazard Classifications*
**Label Signal Word:** DANGER
*Regulatory Authority:*
- Permissible Exposure Limits for Chemical Contaminants (CAL/OSHA)
- The "Director's List" (CAL/OSHA)
- Actively registered pesticide in California.
- Rating System): Section 311 Hazardous Substances/RQ (same as CERCLA)
- EPCRA Section 304 RQ: CERCLA, 1000 lb (454 kg)

*Description:* Dodecylbenzene sulfonic acid is a light yellow to brown liquid with a slight $SO_2$ odor. Molecular weight = 326.50. Boiling point = 315°C. Melting/Freezing point = 10°C. Hazard Identification (based on NFPA-704 M Rating System): Health 0, Flammability 1, Reactivity 1. Soluble in water.
*Incompatibilities:* Keep away from oxidizers, sulfuric acid, caustics, ammonia, aliphatic amines, alkanolamines, isocyanates, alkylene oxides, epichlorohydrin. May attack metals, forming flammable hydrogen gas.
*Permissible Concentration in Water:* No criteria set. Runoff from spills or fire control may cause water pollution.
*Routes of Entry:* Inhalation
*Harmful Effects and Symptoms*
*Short Term Exposure:* A corrosive. Contact with the eyes and skin can cause severe irritation and burns. Inhalation can irritate the respiratory tract.
*Long Term Exposure:* Repeated skin contact may cause dermatitis. Corrosive materials may affect the lungs or cause bronchitis with coughing, phlegm and/or shortness of breath.
*Points of Attack:* Lungs
*Medical Surveillance:* Lung function tests. Consider chest x-ray following acute overexposure.

*First Aid:* If this chemical gets into the eyes, remove any contact lenses at once and irrigate immediately for at least 15 minutes, occasionally lifting upper and lower lids. Seek medical attention immediately. If this chemical contacts the skin, remove contaminated clothing and wash immediately with soap and water. Seek medical attention immediately. If this chemical has been inhaled, remove from exposure, begin rescue breathing (using universal precautions) if breathing has stopped, and CPR if heart action has stopped. Transfer promptly to a medical facility. When this chemical has been swallowed, get medical attention. Give large quantities of water and induce vomiting. Do not make an unconscious person vomit. Do not induce vomiting when formulations containing petroleum solvents are ingested.

*References:*
- New Jersey Department of Health and Senior Services, "Hazardous Substance Fact Sheet, Dodecylbenzene sulfonic Acid," Trenton NJ (October 1996, rev. May 2003).
  http://www.state.nj.us/health/eoh/rtkweb/0822.pdf
- California Environmental Protection Agency "Chemical List of Lists," Sacramento CA (February 1997).

## Dodemorph Acetate

*Use Type:* Fungicide
*CAS Number:* 31717-87-0
*Formula:* $C_{20}H_{40}NO_3$; $C_{18}H_{36}NO \cdot C_2H_4O_2$
*Synonyms:* Cyclododecyl-2,6-dimethylmorpholine acetate; 4-Cyclododecyl-2,6-dimethylmorpholine acetate; Cyclododecyl(4)-2,6-dimethylmorpholine acetate; N-Cyclododecyl-2,6-dimethylmorpholinium acetate; Dodemorfe (French); Morpholine, N-cyclododecyl-2,6-dimethyl-, acetate
*Trade Names:* CYCLOMORPH®; MELTATOX® BASF Canada; MILBAN®
*Producers:* BASF Canada
*Chemical Class:* Morpholine
*EPA/OPP PC Code:* 110401 (213600 and 268601 are old PC codes)
*California DPR Chemical Code:* 2120
*Uses:* Not registered in the U.S. A system fungicide, one of the most recently developed, that can be applied as soil treatments. They are slowly absorbed through the roots. Dodemorph acetate is used to inhibit rusts and mildews.
*Description:* Yellowish liquid, aromatic odor. Molecular weight = 342.65. Melting point = 0°C. Emulsifies in water.
*Incompatibilities:* Acetates are generally incompatible with nitrates. Moisture may cause hydrolysis or other forms of decomposition.
*Determination in Air:* Ionization. NIOSH IV Method#1450, Esters.[18]
*Permissible Concentration in Water:* No criteria set. Runoff from spills or fire control may cause water pollution.

*Routes of Entry:* Eye, skin, ingestion and inhalation
*Harmful Effects and Symptoms*
*Short Term Exposure:* Irritation of eyes, skin, digestive tract and respiratory tract. Inhalation should be avoided; use NIOSH-approved air purifying respirators for pesticides. May be harmful if swallowed.
*First Aid:* If this chemical gets into the eyes, remove any contact lenses at once and irrigate immediately for at least 15 minutes, occasionally lifting upper and lower lids. Seek medical attention immediately. If this chemical contacts the skin, remove contaminated clothing and wash immediately with soap and water. Seek medical attention immediately. If this chemical has been inhaled, remove from exposure, begin rescue breathing (using universal precautions) if breathing has stopped, and CPR if heart action has stopped. Transfer promptly to a medical facility. When this chemical has been swallowed, get medical attention. *Do not induce vomiting when formulations containing petroleum solvents are ingested.* Otherwise, give large quantities of water and induce vomiting. Do not make an unconscious person vomit.

## Dodine (ANSI)

*Use Type:* Fungicide and microbiocide
*CAS Number:* 2439-10-3
*Formula:* $C_{15}H_{33}N_3O_2$
*Synonyms:* Aceto de N-dodecilguanidina (Spanish); Caswell No. 419; N-Dodecylguanidineacetat (German); Dodecylguanidine acetate; N-Dodecylguanidine acetate; Dodecylguanidine monoacetate; 1-Dodecylguanidinium acetate; Dodguadine; Dodin; Dodine acetate; Dodine, mixture with glyodin; Doguadine; ENT 16,436; Guanidine, dodecyl-, acetate; Guanidine, dodecyl-, monoacetate; Laurylguanidine acetate
*Trade Names:* AC 5223®; AMERICAN CYANAMID® 5223; APADODINE®; CARPENE®; CURITAN®; CYPREX®, BASF Agricultural Products (Germany); CYPREX® 65W, BASF Agricultural Products (Germany); CYTOX® 2160; DOQUADINE®; EFUZIN®; KARPEN®; MELPREX®; MELPREX® 65; MILPREX®; QUESTURAN®; SULGEN®; SYLLIT® Chimac-Agriphar (Belgium); SYLLIT® 65, Chimac-Agriphar (Belgium); TEBULAN®; TSITREX®; VANDODINE®; VENTUROL®; VONDODINE®
*Producers:* BASF Agricultural Products (Germany); Chimac-Agriphar (Belgium)
*Chemical Class:* Guanidine
*EPA/OPP PC Code:* 044301
*California DPR Chemical Code:* 245
*RTECS Number:* MF1750000
*EINECS Number:* 219-459-5
*Uses:* Used to control black spot on apples, pears and pecans; brown rot and foliar diseases on peaches and nectarines, cherries, strawberries, black walnuts and

sycamore trees. In industry, used as a biocide and preservative.

***Human toxicity (long-term)***[77]***:*** Intermediate–28.00 ppb, Health Advisory

***Fish toxicity (threshold)***[77]***:*** Low–140.71163 ppb, MATC (Maximum Acceptable Toxicant Concentration)

***U.S. Maximum Allowable Residue Levels for Dodine (40 CFR 180.172):***

| CROP | ppm |
|---|---|
| Apple | 5 |
| Cherry, sweet | 5 |
| Cherry, tart | 5 |
| Meat | 0 |
| Milk | 0 |
| Peach | 5 |
| Pear | 5 |
| Pecan | 0.3 |
| Spinach | 12 |
| Strawberry | 5 |
| Walnut, black | 0.3 |

***Carcinogen/Hazard Classifications***

**Label Signal Word:** CAUTION or DANGER

**WHO Acute Hazard:** Class III, slightly hazardous

***Regulatory Authority:***
- EPA 40 CFR 372.65, Specific Toxic Chemical Listings
- EPCRA Section 313 Form R *de minimis* concentration reporting level: 1.0%

***Description:*** Colorless crystalline solid. Soluble in hot water; solubility = 630 mg/L @ 25°C. Molecular weight = 287.44. Melting/Freezing point = 136°C. Vapor pressure = $1 \times 10^{-7}$ mmHg @ 20°C.

***Incompatibilities:*** May not be compatible with nitrates. Moisture may cause hydrolysis or other forms of decomposition

***Determination in Air:*** Ionization. NIOSH IV Method#1450, Esters.[18]

***Permissible Concentration in Water:*** No criteria set. Runoff from spills or fire control may cause water pollution.

***Routes of Entry:*** Inhalation, ingestion, skin contact.

***Harmful Effects and Symptoms***

***Short Term Exposure:*** Contact with eyes or skin may cause irritation, burns, or pemanent injury. Inhalation should be avoided; use NIOSH-approved air purifying respirators for pesticides. Poisonous if inhaled or ingested.

***First Aid:*** If this chemical gets into the eyes, remove any contact lenses at once and irrigate immediately for at least 15 minutes, occasionally lifting upper and lower lids. Seek medical attention immediately. If this chemical contacts the skin, remove contaminated clothing and wash immediately with soap and water. Seek medical attention immediately. If this chemical has been inhaled, remove from exposure, begin rescue breathing (using universal precautions) if breathing has stopped, and CPR if heart action has stopped. Transfer promptly to a medical facility. When this chemical has been swallowed, get medical attention. *Do not induce vomiting when formulations containing petroleum solvents are ingested.* Otherwise, give large quantities of water and induce vomiting. Do not make an unconscious person vomit.

***References:***
- EXTOXNET, Extension Toxicology Network, "Pesticide Information Profile, Dodine," Oregon State University, Corvallis, OR (June 1996). http://extoxnet.orst.edu/pips/dodine.htm
- U.S. Environmental Protection Agency, Office of Pesticide Programs, Pesticide Residue Limits, "Dodine", 40 CFR 180.172, www.epa.gov/pesticides/food/viewtols.htm

# E

## Emamectin Benzoate

*Use Type:* Insecticide
*CAS Number:* 137512-74-4; 155569-91-8
*Formula:* $C_{49}H_{75}NO_{13}$ (emanectin $B_{1a}$) + $C_{48}H_{73}NO_{13}$ (emamectin $B_{1b}$)
*Synonyms:* Avermectin $B_1$, 4"-deoxy-4"-(methylamino)-, (4"R)-,benzoate (salt); 4"-Epimethylamino-4"-deoxyavermectin $B_{1a}$ and $B_{1b}$ benzoates; 4"-Epimethylamino-4"-deoxyavermectin $B_1$ benzoate (a mixture of minimum of 80% 4"-epimethylamino-4"-deoxyavermectin $B_{1a}$ (25-*sec*-butyl) and a maximum of 20% 4"-epimethylamino-4"-deoxyavermectin $B_{1b}$ (25-*iso*-propyl)benzoate
*Trade Names:* AFFIRM®, Syngenta (Switzerland); DENIM®, Syngenta (Switzerland); MK 244®; MK-0244®; PROCLAIM® Syngenta (Switzerland)
*Producers:* Syngenta (Switzerland)
*Chemical Class:* Botanical
*EPA/OPP PC Code:* 122806
*California DPR Chemical Code:* 4020
*U.S. Maximum Allowable Residue Levels for Emamactin Benzoate (40 CFR 180.505):*

| CROP | ppm |
|---|---|
| Brassica, head and stem, subgroup 5a | 0.025 |
| Celery | 0.025 |
| Lettuce, head | 0.025 |

*Carcinogen/Hazard Classifications*
**U.S. EPA Carcinogens:** Not likely a carcinogen
**Label Signal Word:** CAUTION or DANGER
*Description:* White to off-white crystalline solid or powder. Melting/Freezing point = 143°C. Log $K_{ow}$ = 4.95 @ pH 7.0. Values above 3.0 are likely to bioaccumulate in marine organisms.
*Permissible Concentration in Water:* No criteria set. Runoff from spills or fire control may cause water pollution.
*Harmful Effects and Symptoms*
*Short Term Exposure:* Contact with eyes or skin may cause irritation or injury. Inhalation should be avoided; use NIOSH-approved air purifying respirators for pesticides. May be harmful if swallowed.
*First Aid:* If this chemical gets into the eyes, remove any contact lenses at once and irrigate immediately for at least 15 minutes, occasionally lifting upper and lower lids. Seek medical attention immediately. If this chemical contacts the skin, remove contaminated clothing and wash immediately with soap and water. Seek medical attention immediately. If this chemical has been inhaled, remove from exposure, begin rescue breathing (using universal precautions) if breathing has stopped, and CPR if heart action has stopped. Transfer promptly to a medical facility. When this chemical has been swallowed, get medical attention. Give large quantities of water and induce vomiting. Do not make an unconscious person vomit.
*References:*
- U.S. Environmental Protection Agency, Office of Pesticide Programs, Pesticide Residue Limits, "Emamectin benzoate", 40 CFR 180.505, www.epa.gov/pesticides/food/viewtols.htm

## Endosulfan (ANSI)

*Use Type:* Insecticide and acaricide
*CAS Number:* 115-29-7; 959-98-8 (*alpha*-isomer); 33213-65-9 (*beta*-isomer)
*Formula:* $C_9H_6Cl_6O_3S$
*Alert:* A Restricted Use Pesticide (RUP).
*Synonyms:* Benzoepin (in Japan); Endosulphan; Ensodulfan (Spanish); ENT 23,979; A,β-1,2,3,4,7,7-Hexachlorobiclo(2,2,1)hepten-5,6-bioxymethylenesulfite; 1,2,3,4,7,7-Hexachlorobiclo(2,2,1)hepten-5,6-bioxymethylenesulfite; Hexachlorohexahydromethano 2,4,3-benzodioxathiepin-3-oxide; 6,7,8,9,10,10-Hexachloro-1,5,5a,6,9,9a-hexahydro-6,9-methano-2,4,3-benzodioxathiepin-3-oxide; 1,4,5,6,7,7-Hexachloro-5-norborene-2,3-dimethanol cyclic sulfite; C,c'-(1,4,5,6,7,7-hexachloro-8,9,10-trinorborn-5-en-2,3-ylene)(dimethylsulphite)6,7,8,9,10,10-hexachloro-1,5,5a,6,9,9a-hexahydro-6,9-methano-2,4,3-benzodioxathiepin 3-oxide; 6,9-Methano-2,4,3-benzodioxathiepin,6,7,8,9,10,10-hexachloro-1,5,5a,6,9,9a-hexahydro-, 3-oxide; NCI-C00566; OMS570; Rasayansulfan; Sulfurous acid cyclic ester with 1,4,5,6,7,7-hexachloro-5-norborene-2,3-dimethanol

*alpha*-isomer: α-Benzoepin; Endosulfan-1; α-Endosulfan; Endosulfan-A; Endosulfan-α; 1,4,5,6,7,7-Hexachloro-5-norbornene-2,3-dimethanol, cyclic sulfite, *endo*-; 6,9-Methano-2,4,3-benzodioxathiepin, 6,7,8,9,10,10-hexachloro-1,5,5α,6,9,9α-hexahydro-, 3-oxide, (3, 5aβ,6,9,9a.β)-; 5-Norbornene-2,3-dimethanol, 1,4,5,6,7,7-hexachloro-, cyclic sulfite, *endo*-

*beta*-isomer: β-Benzoepin; Endosulfan-2; Endosulfan, *beta*; β-Endosulfan; β-Ensodulfan (Spanish);1,4,5,6,7,7-Hexachloro-5-norbornene-2,3-dimethanol, cyclic sulfite, *exo*-; 6,9-Methano-2,4,3-benzodioxathiepin,6,7,8,9,10,10-hexachloro-1,5,5A,6,9,9A-hexahydro-, 3-oxide, (3α,5Aα,6β,9β,9Aα)-

## Endosulfan

*Trade Names:* AFIDEN®; BEOSIT®; BIO 5,462®; CHLORTHIEPIN®; CLEAN-CROP®; CRISUFAN®; CYCLODAN®; DE-PESTER®; DESTROY®, Pyosa Agroquimicos (Mexico); DEVISULPHAN®; DISSULFAN CE®, Milenia Agro Ciencias (Brazil); ENDOCEL®, Excel Industries (India); ENDOCIDE®; ENDOSOL®; END-O-SULFAN®, Helena Chemical Co. (USA); ENDOTAF®, Rallis India (India); ENDOX®; ENSURE®; E-Z FLO®; FMC 5462®; HEXASULFAN®; HILDAN®; HOE 2671®; INSECTO®; INSECTOPHENE®; KENDAN®, Kenso Corp. (Malaysia); KERNTOX®; KOP-THIODAN®; MALIX®; MAUX®; MOS-570; METHOFAN®, Makhteshim Agan (Israel); NIA 5462®, FMC Agricultural Products Group (USA), canceled 7/1/1987; NIAGARA 5,462®, FMC Agricultural Products Group (USA), canceled 7/1/1987; PHASER®, Bayer CropScience (Germany); ROCKY®, Calliope (France); THIFOR®; THIDAN®; THIMUL®; α-THIODAN®; β-THIODAN®; α-THIONEX®; β-THIONEX®; THIOKILL®, United Phosphorus (India); THIOFOR®; THIONEX®, Makhteshim Agan (Israel); THIOSULFAN®; THIOSULFAN THIONEL®; TIOVEL®

*Producers:* Agsin (Singapore); Aimco Pesticides Ltd. (India); Alcotan Laboratories (Spain); Bayer CropScience (Germany); BEC Group (India); Bharat Pulverizing Mills (India); Bharat Rasayan (India); Biesterfeld Siemsgluess International. GmbH (Germany); Calliope (France); Chromos Agro (Croatia); Drexel Chemical (USA); Ehrenstorfer, Dr. (Germany); Excel Industries (India); FMC Agricultural Products Group (USA); Godavari Fertilisers and Chemicals (India); Gowan (USA); Helena Chemical Co. (USA); Hindustan Insecticides (India); Indiclay (India); Indo Gulf (India); Kenso Corp. (Malaysia); Ki-Hara Chemicals Ltd. (UK); Makhteshim Agan (Israel); Milenia Agro Ciencias (Brazil); Nagarjuna Agrichem (India); Pyosa Agroquimicos (Mexico); Rallis India (India); Shenzhen Guomeng Industry Co., Ltd. (China); Sigma-Aldrich Laborchemikalien (Germany); Sinon (Taiwan); United Phosphorus (India; Vijayalakshmi Insecticides and Pesticides (India); Zago Asia Ltd. (Singapore)

*Chemical Class:* A chlorinated cyclodiene

*EPA/OPP PC Code:* 079401; 079402 (*alpha*-isomer); 079403 (*beta*-isomer)

*California DPR Chemical Code:* 259

*ICSC Number:* 0742

*RTECS Number:* RB9275000 (*alpha*-isomer); RB9875200 (*beta*-isomer)

*EEC Number:* 602-052-00-5

*EINECS Number:* 204-079-4

*Uses:* Endosulfan is a chlorinated hydrocarbon insecticide and acaricide of the cyclodiene subgroup which acts as a poison to a wide variety of insects and mites on contact. Although it may also be used as a wood preservative, it is used primarily on a wide variety of food crops including tea, coffee, fruits, and vegetables, as well as on rice, cereals, maize, sorghum, or other grains. Formulations of endosulfan include emsulsifiable concentrate, wettable powder, ultra-low volume (ULV) liquid, and smoke tablets. It is compatible with many other pesticides and may be found in formulations with dimethoate, malathion, methomyl, monocrotophos, pirimicarb, triazophos, fenoprop, parathion, petroleum oils, and oxine-copper. It is not compatible with alkaline materials. Technical endosulfan is made up of a mixture of two molecular forms (isomers) of endosulfan, the *alpha*- and *beta*-isomers.

*Human toxicity (long-term)*[77]: Intermediate–42.00 ppb, Health Advisory

*Fish toxicity (threshold)*[77]: Extra high–0.00446 ppb, MATC (Maximum Acceptable Toxicant Concentration)

*U.S. Maximum Allowable Residue Levels for Endosulfan (40 CFR 180.182):*

| CROP | ppm |
|---|---|
| Alfalfa, hay | 1 |
| Alfalfa, forage | 0.3 |
| Almond | 0.2 |
| Almond, hulls | 1 |
| Apple | 2 |
| Apricot | 2 |
| Artichoke, globe | 2 |
| Barley, straw | 0.2 |
| Barley, grain | 0.1 |
| Bean | 2 |
| Beet, sugar, roots | 0.1 |
| Blueberry | 0.1 |
| Broccoli | 2 |
| Brussels sprouts | 2 |
| Cabbage | 2 |
| Carrot, roots | 0.2 |
| Cattle, meat | 0.2 |
| Cattle, fat | 0.2 |
| Cattle, mbyp | 0.2 |
| Cauliflower | 2 |
| Celery | 2 |
| Cherry | 2 |
| Collards | 2 |
| Corn, sweet, kernel plus cob with husks removed | 0.2 |
| Cotton, undelinted seed | 1 |
| Cucumber | 2 |
| Eggplant | 2 |
| Filbert | 0.2 |
| Goat, meat | 0.2 |
| Goat, fat | 0.2 |
| Goat, mbyp | 0.2 |
| Grape | 2 |
| Hog, mbyp | 0.2 |
| Hog, fat | 0.2 |
| Hog, meat | 0.2 |
| Horse, mbyp | 0.2 |
| Horse, meat | 0.2 |
| Horse, fat | 0.2 |

| | |
|---|---|
| Kale | 2 |
| Lettuce | 2 |
| Melon | 2 |
| Milk, fat | 0.5 |
| Mustard greens | 2 |
| Mustard, seed | 0.2 |
| Nectarine | 2 |
| Nut, macadamia | 0.2 |
| Oat, grain | 0.1 |
| Oat, straw | 0.2 |
| Pea, succulent | 2 |
| Peach | 2 |
| Pear | 2 |
| Pecan | 0.2 |
| Pepper | 2 |
| Pineapple | 2 |
| Plum, prune | 2 |
| Plum | 2 |
| Potato | 0.2 |
| Pumpkin | 2 |
| Rapeseed, seed | 0.2 |
| Raspberry | 0.1 |
| Rye, straw | 0.2 |
| Rye, grain | 0.1 |
| Safflower, seed | 0.2 |
| Sheep, mbyp | 0.2 |
| Sheep, meat | 0.2 |
| Sheep, fat | 0.2 |
| Spinach | 2 |
| Squash, winter | 2 |
| Squash, summer | 2 |
| Strawberry | 2 |
| Sugarcane, cane | 0.5 |
| Sunflower, seed | 2 |
| Sweet potato | 0.2 |
| Tea, dried | 24 |
| Tomato | 2 |
| Turnip, greens | 2 |
| Walnut | 0.2 |
| Watercress | 2 |
| Wheat, grain | 0.1 |
| Wheat, straw | 0.2 |

*Carcinogen/Hazard Classifications*
**Label Signal Word:** DANGER–POISON
**WHO Acute Hazard:** Class II, moderately hazardous
**Endocrine Disruptor:** Known ED
*Regulatory Authority:*

- Banned or Severely Restricted (Many Countries) (UN)[13]
- Air Pollutant Standard Set (ACGIH)[1] (HSE)[33] (Argentina)[35] (former USSR)[35, 43] (OSHA)[58] (Several States)[60]
- AB 1803-Well Monitoring Chemical (CAL)
- EPA/SARA 302 (EPCRA) Extremely hazardous substances
- Permissible Exposure Limits for Chemical Contaminants (CAL/OSHA)
- The "Director's List" (CAL/OSHA)
- Actively registered pesticide in California.
- Clean Water Act: Section 311 Hazardous Substances/RQ 40CFR117.3 (same as CERCLA, see below); Toxic Pollutant (Section 401.15)
- EPA Hazardous Waste Number (RCRA No.): P050
- RCRA, 40CFR261, Appendix 8 Hazardous Constituents
- Superfund/EPCRA 40CFR355, Appendix B Extremely Hazardous Substances: TPQ = 10/10,000 lb (4.54/4,540 kg)
- Superfund/EPCRA 40CFR302.4 RQ: CERCLA, 1 lb (0.454 kg)
- U.S. DOT Regulated Marine Pollutant (49CFR172.101, Appendix B). Severe pollutant

*alpha*-isomer:

- Clean Water Act: 40CFR423, Appendix A, Priority Pollutants; Toxic Pollutant (Section 401.15)
- RCRA 40CFR268.48; 61FR15654, Universal Treatment Standards: Wastewater (mg/L), 0.023; Nonwastewater (mg/kg), 0.066
- RCRA 40CFR264, Appendix 9; TSD Facilities Ground Water Monitoring List. Suggested test method(s) (PQL $ug/L$): 8080(0.1); 8250(10)
- Superfund/EPCRA 40CFR302.4 RQ: CERCLA, 1 lb (0.454 kg)
- U.S. DOT Regulated Marine Pollutant (49CFR172.101, Appendix B). Severe pollutant; as endosulfan

*beta*-isomer:

- Clean Water Act: 40CFR423, Appendix A, Priority Pollutants; Toxic Pollutant (Section 401.15)
- RCRA 40CFR268.48; 61FR15654, Universal Treatment Standards: Wastewater (mg/L), 0.029; Nonwastewater (mg/kg), 0.13
- RCRA 40CFR264, Appendix 9; TSD Facilities Ground Water Monitoring List. Suggested test method(s) (PQL $ug/L$): 8080(0.05)
- Superfund/EPCRA 40CFR302.4 RQ: CERCLA, 1 lb (0.454 kg)
- U.S. DOT Regulated Marine Pollutant (49CFR172.101, Appendix B). Severe pollutant; as endosulfan

*U.S. Maximum Allowable Residue Levels for Endosulfan:*

| CROP | ppm |
|---|---|
| Dried tea | 24 |
| Raw agricultural products | 0.1 to 2 |

*Description:* The pure product is a colorless crystalline solid. The technical product is a light to dark brown waxy solid. "Technical grade endosulfan is composed of two stereochemical isomers: *alpha*-endosulfan and *beta*-endosulfan, in concentrations of approximately 70% and 30%, respectively." [RED (EPA)] It has a rotten egg or sulfur odor. Insoluble in water. Melting/Freezing point = 108–110°C ($\alpha$-isomer); 208–212°C ($\beta$-isomer); 80°C (technical). Vapor pressure = $1.7 \times 10^{-7}$ mmHg @ 20°C.

Hazard Identification (based on NFPA-704 M Rating System): Health 4, Flammability 1, Reactivity 0. Log $K_{ow}$ = 3.58. Values at or above 3.0 are likely to bioaccumulate in marine organisms.

*Incompatibilities:* Strong acids, strong bases. Hydrolysed by acids. Contact with alkalis forms toxic sulfur dioxide fumes. Corrosive to iron in the presence of moisture.

*Permissible Exposure Limits in Air:* There is no OSHA[2] PEL. The ACGIH[1], NIOSH[2], and HSE[33] has set a TWA of 0.1 mg/m³. HSE[33] adds an STEL of 0.3 mg/m³. The notation "skin" is added to indicate the possibility of cutaneous absorption. Argentina[35] has the same limits. The former USSR[35, 43] has set an MAC of 0.1 mg/m³ in workplace air as well as an MAC in ambient air of residential areas of 0.005 mg/m³ on a momentary basis and 0.001 mg/m³ on a daily average basis. Several states have set guidelines or standards for endosulfan in ambient air[60] ranging from 0.238 µg/m³ (Kansas) to 1.0 µg/m³ (North Dakota) to 1.6 µg/m³ (Virginia) to 2.0 µg/m³ (Connecticut and Nevada) to 2.4 µg/m³ (Pennsylvania).

*Determination in Air:* No test available.

*Permissible Concentration in Water:* The U.S. EPA recommends that the amount of endosulfan in rivers, lakes, and streams should not be more than 74 parts per billion (74 ppb). To protect freshwater aquatic life: 0.056 µg/L as a 24 hr average, never to exceed 0.22 µg/L. To protect saltwater aquatic life: 0.0087 µg/L as a 24 hr average, never to exceed 0.034 µg/L. To protect human health: 74.0 µg/L[6]. Mexico[35] has set an MAC of 2 µg/L in estuaries and 0.2 µg/L in coastal waters. Kansas[61] has set a guideline of 74.0 µg/L for endosulfan in drinking water.

*Determination in Water:* Methylene chloride extraction followed by gas chromatography with electron capture or halogen specific detection (EPA Method 608) or gas chromatography plus mass spectrometry (EPA Method 625).

*Routes of Entry:* Inhalation, ingestion, eye and/or skin contact. Can be absorbed through the skin.

*Harmful Effects and Symptoms*

*Short Term Exposure:* Endosulfan may affect the central nervous system, blood, resulting in irritability, convulsions and renal failure. High level exposure at may result in death. The effects may be delayed. Ingestion of endosulfan may result in nausea, vomiting, and diarrhea. Dizziness, agitation, nervousness, tremor, incoordination, and convulsions may also occur. Central nervous system depression may terminate in respiratory failure. Contact with endosulfan may irritate or burn the skin, eyes, and mucous membranes. The probable oral lethal dose is 50 to 500 mg/kg, or 1 teaspoonful to 1 ounce for a 150 lb person. The $LD_{50}$ (oral, rat) = 18 mg/kg (highly toxic). Death has occurred within 2 hours of heavy dust exposure during bagging operations.

*Long Term Exposure:* Repeated exposure may cause brain damage, causing convulsions, loss of coordination, and memory loss. May cause liver and kidney damage. Studies of the effects of endosulfan on animals suggest that long-term exposure to endosulfan can also damage the kidneys, testes, and liver and may possibly affect the body's ability to fight infection. However, it is not known if these effects also occur in humans. The U.S. EPA does not know if children are more sensitive to endosulfan than adults. The U.S. EPA does not know if endosulfan can affect the ability of people to have children or if it causes birth defects. Large amounts of endosulfan damaged the testes of animals, but it is not known if this damaged their ability to reproduce. Some birth defects have been seen in the offspring of animals ingesting endosulfan during pregnancy.

*Points of Attack:* Respiratory system, lungs, central nervous system, cardiovascular system, skin, eyes, plasma and red blood cell cholinesterase.

*Medical Surveillance:* Before employment and at regular times after that, the following are recommended: Plasma and red blood cell cholinesterase levels (tests for the enzyme poisoned by this chemical). If exposure stops, plasma levels return to normal in 1-2 weeks while red blood cell levels may be reduced for 1-3 months. When acetylcholinesterase enzyme levels are reduced by 25% or more below preemployment levels, risk of poisoning is increased, even if results are in lower ranges of "normal." Reassignment to work not involving organophosphate or carbamate pesticides is recommended until enzyme levels recover. If symptoms develop or overexposure occurs, repeat the above tests as soon as possible and get an exam of the nervous system. Also consider complete blood count. Consider chest x-ray following acute overexposure. Do not drink any alcoholic beverages before or during use. Alcohol promotes absorption of organophosphates. Liver and kidney function tests. Examination of the nervous system. EEG.

*First Aid:* Speed in removing material from eyes and skin is of extreme importance. If this chemical gets into the eyes, remove any contact lenses at once and irrigate immediately for at least 15 minutes, occasionally lifting upper and lower lids. Seek medical attention immediately. If this chemical contacts the skin, remove contaminated clothing and wash immediately with soap and water. Shampoo hair promptly if contaminated. Seek medical attention immediately. If this chemical has been inhaled, remove from exposure, begin rescue breathing (using universal precautions) if breathing has stopped, and CPR if heart action has stopped. Transfer promptly to a medical facility. When this chemical has been swallowed, get medical attention. Consult hospital or poison control center on use of antidotes. Transport to health care facility.

*References:*
- EXTOXNET, Extension Toxicology Network, "Pesticide Information Profile, Endosulfan," Oregon State University, Corvallis, OR (June 1996). http://ace.orst.edu/info/extoxnet/pips/endosulf.htm

- U.S. Environmental Protection Agency, Office of Pesticide Programs, Pesticide Residue Limits, "Endosulfan", 40 CFR 180.182, http://www.epa.gov/pesticides/food/viewtols.htm
- New Jersey Department of Health and Senior Services, "Hazardous Substance Fact Sheet, Endosulfan,"Trenton NJ (May 1999), http://www.state.nj.us/health/eoh/rtkweb/0824.pdf
- U.S. Environmental Protection Agency, Office of Prevention, Pesticides and Toxic Substances, "Reregistration Eligibility Decision for Endosulfan," Washington, DC. (November 2002). http://www.epa.gov/REDs/endosulfan_red.pdf
- U.S. Department of Health and Human Services, Agency for Toxic Substances and Disease Control (ATSDR), "ToxFAQs for Endosulfan," Atlanta, GA. (February 2001). http://www.atsdr.cdc.gov/tfacts41.html
- U.S. Environmental Protection Agency, Endosulfan: Ambient Water Quality Criteria, Washington, DC (1980).
- U.S. Environmental Protection Agency, Endosulfan, Health and Environmental Effects Profile No. 98, Washington, DC, Office of Solid Waste (April 30, 1980).
- U.S. Environmental Protection Agency, "Chemical Profile: Endosulfan," Washington, DC, Chemical Emergency Preparedness Program (November 30, 1987).
- New York State Department of Health, "Chemical Fact Sheet: Endosulfan," Albany, NY, Bureau of Toxic Substance Assessment (April 1986).
- California Environmental Protection Agency "Chemical List of Lists," Sacramento CA (February 1997).

# Endothall (ANSI)

*Use Type:* Herbicide
*CAS Number:* 145-73-3; 129-67-9 (disodium salt)
*Formula:* $C_8H_{10}NO_5PS$
*Alert:* A General Use Pesticide (GUP)
*Synonyms:* 1,2-Dicarboxy 3,6-endoxocyclohexane; 3,6-Endooxohexahydrophthalic acid; Endothal (Great Britian); Endothal chlorophenoxy herbicide; Endothall technical; 3,6-Endoxohexahydrophthalic acid; 3,6-Epoxycyclohexane-1,2-dicarboxylic acid; 3,6-Endo-epoxy-1,2-cyclohexanedicarboxylic acid; Hexahydro-3,6-*endo*-oxyphthalic acid; 7-Oxabicyclo(2.2.1)heptane-2,3-dicarboxylic acid
*Trade Names:* ACCELERATE®, Cerexagri North America (USA); AQUATHOL®; DES-I-CATE®, Cerexagri North America (USA); HYDOUT®; HYDROTHAL-47®; HYDROTHOL®; NIAGARATHOL®; RIPENTHOL®; TRI-ENDOTHAL®
*Producers:* ATOFINA Chemicals (France); Cerexagri North America (USA); Ehrenstorfer, Dr. (Germany); Sigma-Aldrich Laborchemikalien (Germany)
*Chemical Class:* Dicarboxylic acid

*EPA/OPP PC Code:* 038901
*California DPR Chemical Code:* 5813
*RTECS Number:* RN7875000; RN8225000 (disodium salt)
*Uses:* Endothall is used as a cotton defoliant and as a selective contact herbicide on both terrestrial and aquatic weeds. The potassium and amine salts of endothall are used as aquatic herbicides to control a variety of plants including plankton, pondweed, niad, coontail, milfoil, elodea, and algaes in water bodies and rice fields. Endothall is also used to control annual grass and broadleaf weeds in sugar beets, spinach and turf. It reduces sucker branch growth in hops. Endothall is a desiccant to aid the harvest of alfalfa, potatoes, clover, and cotton.
*Human toxicity (long-term)*[77]: Very low–100.00 ppb, MCL (Maximum Contaminant Level)
*Fish toxicity (threshold)*[77]: Low–240.32545 ppb, MATC (Maximum Acceptable Toxicant Concentration)
*U.S. Maximum Allowable Residue Levels for Endothall (40 CFR 180.293 and 180.319):*

| CROP | ppm |
|---|---|
| Beet, sugar | 0.2 |
| Hop | 0.1 |
| Potato | 0.1 |
| Rice, straw | 0.05 |
| Rice, grain | 0.05 |
| Cotton, undelinted seed | 0.1 |
| Water, potable | 0.2 |

*Carcinogen/Hazard Classifications*
**Label Signal Word:** WARNING–EPA Toxicity Class II, moderately toxic
**WHO Acute Hazard:** Group II, moderately hazardous for the disodium salt
*Regulatory Authority:*
- Banned or Severely Restricted (Several Countries) (UN)[13]
- AB 1803-Well Monitoring Chemical (CAL)
- AB 2588-Air Toxics "Hot Spots" Chemicals (CAL) as chlorophenoxy pesticides
- The "Director's List" (CAL/OSHA) as chlorophenoxy pesticides
- MCL (Maximum Contaminants Levels) list of contaminants (CAL)
- EPA Hazardous Waste Number (RCRA No.): P088
- RCRA, 40CFR261, Appendix 8 Hazardous Constituents
- Safe Drinking Water Act: MCL, 0.1 mg/L; MCLG, 0.1 mg/L; Regulated chemical (47 FR 9352)
- Superfund/EPCRA 40CFR302.4 RQ: CERCLA, 1000 lb (454 kg)

*Description:* When pure, a white crystalline solid. The technical grade is a light brown liquid. Soluble in water; solubility = 100 g/l @ 20°C. Molecular weight = 186.18. Melting/Freezing point = 144°C with conversion to the anhydride.
*Permissible Exposure Limits in Air:* No standard set.

*Permissible Concentration in Water:* A no-observed-adverse effects-level (NOAEL) of 2 mg/kg/day has been determined by EPA. This gives a reference dose (or acceptable daily intake) of 0.02 mg/kg/day on the basis of which a lifetime health advisory of 0.14 mg/L (140 $\mu$g/L) was calculated.

*Routes of Entry:* Inhalation, ingestion, eye and/or skin contact.

*Harmful Effects and Symptoms*

*Short Term Exposure:* Irritates eyes, skin and respiratory tract. Poisonous: approximate lethal dose (human) is about 2.5 teaspoonful. Little information was found in the available literature on the health effects of endothall in humans except for one case history of a young male suicide victim who ingested an estimated 7 to 8 g of disodium Endothall in solution (approximately 100 mg Endothall ion/kg) Repeated vomiting was evident. Autopsy revealed focal hemorrhages and edema in the lungs and gross hemorrhage of the gastrointestinal (GI) tract.

*Long Term Exposure:* May be mutagenic.

*First Aid:* If this chemical gets into the eyes, remove any contact lenses at once and irrigate immediately for at least 15 minutes, occasionally lifting upper and lower lids. Seek medical attention immediately. If this chemical contacts the skin, remove contaminated clothing and wash immediately with soap and water. Seek medical attention immediately. If this chemical has been inhaled, remove from exposure, begin rescue breathing (using universal precautions) if breathing has stopped, and CPR if heart action has stopped. Transfer promptly to a medical facility. When this chemical has been swallowed, get medical attention. Give large quantities of water and induce vomiting. Do not make an unconscious person vomit.

*References:*
- EXTOXNET, Extension Toxicology Network, "Pesticide Information Profile, Endothall," Oregon State University, Corvallis, OR (September 1995). http://ace.orst.edu/info/extoxnet/pips/endothal.htm
- U.S. Environmental Protection Agency, Office of Pesticide Programs, Pesticide Residue Limits, "Endothall", 40 CFR 180.293 and 180.319, http://www.epa.gov/pesticides/food/viewtols.htm
- Sax, N.I., Ed., "Dangerous Properties of Industrial Materials Report," 8, No. 6, 51-56 (1988).
- U.S. Environmental Protection Agency, "Health Advisory: Endothall," Washington, DC, Office of Drinking Water (August 1987).

# Endothion

*Use Type:* A systemic insecticide
*CAS Number:* 2778-04-3
*Formula:* $C_9H_{13}O_6PS$
*Alert:* Endothion is not sold in the U.S. or Canada.

*Synonyms:* $O,O$-Dimethyl-$S$-(5-methoxy-4-oxo-4H-pyran-2-yl)phosphorothioate; $O,O$-Dimethyl-$S$-((5-methoxy-pyron-2-yl)-methyl)-thiolphosphat (German); Phosphorothioate; $O,O$-Dimethyl-$S$-(5-methoxypyronyl-2-methyl) thiolphosphate; Endotiona (Spanish); ENT 24,653; 5-Methoxy-2-(Dimethoxyphosphinylthiomethyl)pyrone-4; $S$-5-Methoxy-4-oxopyran-2-ylmethyl dimethyl phosphorothioate; $S$-((5-Methoxy-4H-pyron-2-yl)-methyl)-$O,O$-dimethyl-monothiofosfaat (Dutch); $S$-((5-Methoxy-4H-pyron-2-yl)-methyl)-$O,O$-dimethyl-monothiophosphat (German); $S$-(5-Methoxy-4-pyron-2-ylmethyl)dimethyl phosphorothiolate; Thiophosphate de $O,O$-dimethyle et de $S$-((5-methoxy-4-pyronyl)-methyle) (French)

*Trade Names:* AC-18,737®; ENDOCID®; ENDOCIDE®; EXOTHION®; NIA-5767®, FMC Agricultural Products Group (USA); NIAGARA 5767®, FMC Agricultural Products Group (USA); PHOSPHATE 10®; PHOSPHOPYRON®; PHOSPHOPYRONE®

*Producers:* FMC Agricultural Products Group (USA)

*Chemical Class:* Organophosphate

*EPA/OPP PC Code:* 422100

*RTECS Number:* IF8225000

*Uses:* This material is a systemic insecticide. It is not sold in the U.S. or Canada.

*Regulatory Authority:*
- Superfund/EPCRA 40CFR355, Appendix B Extremely Hazardous Substances: TPQ = 500/10,000 lb (227/4,540 kg)
- Superfund/EPCRA 40CFR302.4 RQ: EHS, 1 lb (0.454 kg)
- AB 1803-Well Monitoring Chemical (CAL)

*Description:* Endothion is a white crystalline solid. Slight odor. Highly soluble in water. Molecular weight = 280.27. Melting/Freezing point = 96°C. Hazard Identification (based on NFPA-704 M Rating System): Health 4, Flammability 1, Reactivity 0.

*Incompatibilities:* Strong oxidizers.

*Permissible Exposure Limits in Air:* No standards set.

*Permissible Concentration in Water:* No criteria set, but runoff from spills or fire control may cause water pollution.

*Routes of Entry:* Inhalation, ingestion, skin contact.

*Harmful Effects and Symptoms*

*Short Term Exposure:* Organic phosphorus insecticides are absorbed by the skin, as well as by the respiratory and gastrointestinal tracts. They are cholinesterase inhibitors. Symptoms of exposure include headache, giddiness, blurred vision, nervousness, weakness, nausea, cramps, diarrhea, and discomfort in the chest. Signs include sweating, tearing, salivation, vomiting, cyanosis, convulsions, coma, loss of reflexes and loss of sphincter control. Exposure may cause psychotic behavior, loss of coordination, unconsciousness, and rarely, convulsions. This material is poisonous to humans. Its toxic effects are most likely related to action on the nervous system, The $LD_{50}$ (oral, rat) = 23 mg/kg (highly toxic). Delayed pulmonary edema may occur after inhalation.

*Long Term Exposure:* Cholinesterase inhibitor; cumulative effect is possible. Endothion may damage the nervous system with repeated exposure, resulting in convulsions, respiratory failure. May cause liver damage.

*Points of Attack:* Cholinesterase inhibitor; cumulative effect is possible. Endothion may damage the nervous system with repeated exposure, resulting in convulsions, respiratory failure. May cause liver damage.

*Medical Surveillance:* Medical observation is recommended for 24 to 48 hours after breathing overexposure, as pulmonary edema may be delayed. Before employment and at regular times after that, the following are recommended: Plasma and red blood cell cholinesterase levels (tests for the enzyme poisoned by this chemical). If exposure stops, plasma levels return to normal in 1-2 weeks while red blood cell levels may be reduced for 1-3 months. When acetylcholinesterase enzyme levels are reduced by 25% or more below preemployment levels, risk of poisoning is increased, even if results are in lower ranges of "normal." Reassignment to work not involving organophosphate or carbamate pesticides is recommended until enzyme levels recover. If symptoms develop or overexposure occurs, repeat the above tests as soon as possible and get an exam of the nervous system. Also consider complete blood count. Consider chest x-ray following acute overexposure. Do not drink any alcoholic beverages before or during use. Alcohol promotes absorption of organophosphates.

*First Aid:* **Treatment for organophosphate poisoning consists of thorough decontamination, cardiorespiratory support, and administration of the antidotes atropine and pralidoxime. In cases of severe poisoning, diazepam, an anticonvulsant, should also be administered. Antidotes should be administered as prevention even if the diagnosis is in doubt.** Speed in removing material from eyes and skin is of extreme importance. *Eyes:* Eye contact can cause dangerous amounts of these chemicals to be quickly absorbed through the mucous membrane into the bloodstream. Immediately and gently flush eyes with plenty of warm or cold water (NO hot water) for at least 15 minutes, occasionally lifting the upper and lower eyelids. Get medical aid immediately. *Skin:* Get medical aid. Skin contact can cause dangerous amounts of these chemicals to be absorbed into the bloodstream. Wearing the appropriate PPE equipment and respirator for organophosphate/carbamate pesticides, immediately flush skin with plenty of soap and water for at least 15 minutes while removing contaminated clothing and shoes. Shampoo hair promptly if contaminated. The removed, contaminated clothing and shoes should be double-bagged and left in Hot Zone for later disposal by hazardous materials experts. Skin may also be decontaminated with diluted hypochlorite solution. *Inhalation:* Get medical aid. Do not contaminate yourself. Wearing the appropriate PPE equipment and respirator for organophosphate pesticides, immediately remove the victim from the contaminated area to fresh air. If the victim is not breathing, administer artificial respiration. Do NOT use mouth-to-mouth resuscitation; use bag/mask apparatus. If breathing is difficult, administer oxygen through bag/mask apparatus until medical help arrives. Do not leave victim unattended. *Ingestion:* Call poison control. Loosen all clothing. Never give anything by mouth to an unconscious person. If victim is *unconscious or having convulsions,* do nothing except keep victim warm. Get medical aid. Transfer promptly to a medical facility. In cases of ingestion, **do not induce vomiting**. If the victim is alert and asymptomatic, administer a slurry of activated charcoal at a dose of 1 g/kg (infant, child, and adult dose). A soda can and straw may be of assistance when offering charcoal to a child. *In some cases you may be specifically instructed by poison control to induce vomiting by way of 2 tablespoons of syrup of ipecac (adult) washed down with a cup of water.* Do NOT give activated charcoal before or with ipecac syrup.

*Note to physician or authorized medical personnel:* Treat cases of respiratory compromise, coma, or excessive pulmonary secretions with respiratory support using protocols and techniques available and within the scope of training. Some cases may necessitate procedures such as endotracheal intubation or cricothyrotomy by properly trained and equipped personnel. When possible, atropine (see *Antidotes*, below) should be given under medical supervision. Patients who are comatose, hypotensive, or having seizures or cardiac arrhythmias should be treated according to advanced life support protocols. *Antidotes:* Two antidotes are administered to treat organophosphate poisoning. Atropine is a competitive antagonist of acetylcholine at muscarinic receptors and is used to control the excessive bronchial secretions which are often responsible for death. Pralidoxime relieves both the nicotinic and muscarine effects of organophosphate poisoning by regenerating acetylcholinesterase and can reduce both the bronchial secretions and the muscle weakness associated with poisoning. The initial intravenous dose of atropine in adults should be determined by the severity of symptoms: An initial adult dose of 1.0 to 2.0 mg or pediatric dose of 0.01 mg/kg (minimum 0.01 mg) should be administered intravenously. If intravenous access cannot be established, atropine may also be given intramuscularly, subcutaneously or via endotracheal tube. Doses should be repeated every 15 minutes until excessive secretions and sweating have been controlled. Once bronchial secretion has been controlled, atropine administration should be repeated whenever the secretions begin to recur. In seriously poisoned patients, very large doses may be required. Alterations of pulse rate and pupillary size should not be used as indicators of treatment adequacy. Pralidoxime should be administered as early in poisoning as possible as its efficacy may diminish when given more than 24 to 36 hours after exposure. Doses are as follows: adult 1.0 g; pediatric 25 to 50 mg/kg. The drug should be administered intravenously over 30 to 60 minutes, but in a life-threatening situation, one-half of the total dose can be given per minute for a total administration time of 2 minutes. Treatment should begin to take effect within 40 minutes with a reduction in symptoms and the amount of atropine necessary to control bronchial secretion. The initial dose can be repeated in 1 hour and then every 8 to 12 hours until the patient is clinically well and no longer requires atropine. If intravenous access cannot be established, pralidoxime may also be given intramuscularly. Early administration of diazepam in addition to the combined atropine and pralidoxime treatment may help prevent the onset of seizures and potential brain and cardiac morphologic damage following high-level organophosphate poisoning.

*References:*
- U.S. Environmental Protection Agency, "Chemical Profile: Endothion," Washington, DC, Chemical Emergency Preparedness Program (November 30, 1987).
- Agency for Toxic Substances and Disease Registry, U.S. Department of Health and Human Services, Public Health Service, "Managing Hazardous Materials Incidents," Atlanta, GA (June 2003).
- California Environmental Protection Agency "Chemical List of Lists," Sacramento CA (February 1997).

# Endrin

*Use Type:* Insecticide and avicide
*CAS Number:* 72-20-8
*Formula:* $C_{12}H_8Cl_6O$
*Alert:* A Restricted Use Pesticide (RUP). Most uses were canceled in 1980 and it has not been produced or sold for general use in the United States since 1986. It is a persistent and acutely toxic insecticide. Human toxicity (long-term): High.
*Synonyms:* 2,7:3,6-Dimethanonaphth(2, 3-b)oxirene, 3,4,5,6,9,9-hexachloro-1a,2,2a,3,6,6a,7,7a-octahydro-,(aα,2.β,2aβ,2aβ,3α,6α,6aβ,7β,7aα)-; Endrina (Spanish); Endrine (French); ENT 17,251; (1r, 4s, 4as, ss, 7r, 8r, 8ar)-1,2,3,4,10-Hexachloro-1,4,4a,5,6,7,8,8a-octahydro-6,7-epoxy-1,4:5,8-dimethano naphthalene; Hexachloroepoxyoctahydro-endo,endo-dimethano napthalene; 1,2,3,4,10,10-Hexachloro-6,7-epoxy-1,4,4a,5,6,7,8,8a-octahydro-1,4-endo-endo-1,4,5,8-dimethanonaphthalene; NCI-C00157
*Trade Names:* COMPOUND 269®; EN 57®; ENDREX®; ENDRICOL®; ENDRIN CHLORINATED HYDROCARBON INSECTICIDE®; HEXADRIN®; MENDRIN®; NENDRIN®; OKTANEX®
*Producers:* Sigma-Aldrich Laborchemikalien (Germany)
*Chemical Class:* Organochlorine; Halo-organics
*EPA/OPP PC Code:* 041601
*California DPR Chemical Code:* 262
*ICSC Number:* 1023
*RTECS Number:* IO1575000
*EEC Number:* 602-051-00-X

*Uses:* Endrin is an insecticide which has been used to control insects, rodents, and birds, mainly on field crops such as cotton, maize, sugarcane, rice, cereals, ornamentals, and other crops. It has also been used for grasshoppers in non-cropland and to control voles and mice in orchards. Once widely used in the US, most uses were canceled in 1980 and it has not been produced nor sold in the U.S. since 1986. It is not easily dissolved in water and can remain in the soil for more than 14 years.

*Human toxicity (long-term)*[77]: High–2.00 ppb, MCL (Maximum Contaminant Level)
*Fish toxicity (threshold)*[77]: Extra high–0.00440 ppb, MATC (Maximum Acceptable Toxicant Concentration)

*Carcinogen/Hazard Classifications*
**U.S. EPA Carcinogens:** Group D, unclassifiable
**California Prop 65:** Listed; Developmental toxin
**IARC:** Group 3, unclassifiable
**Label Signal Word:** DANGER
**Endocrine Disruptor:** Suspected endocrine disruptor

*Regulatory Authority:*
- Banned or Severely Restricted (Many Countries) (UN)[13]
- Air Pollutant Standard Set (ACGIH)[1] (Australia) (HSE)[33] (DFG)[3] (Israel) (Mexico) (OSHA)[58] (Several States)[60] (Several Canadian Provinces)
- AB 1803-Well Monitoring Chemical (CAL)
- EPA/SARA 302 (EPCRA) Extremely hazardous substances
- MCL (Maximum Contaminants Levels) list of contaminants (CAL)
- Permissible Exposure Limits for Chemical Contaminants (CAL/OSHA)
- The "Director's List" (CAL/OSHA)
- Clean Water Act: Section 311 Hazardous Substances/RQ 40CFR117.3 (same as CERCLA, see below); 40CFR423, Appendix A, Priority Pollutants.
- EPA Hazardous Waste Number (RCRA No.): P051; D012
- RCRA Toxicity Characteristic (Section 261.24), Maximum Concentration of Contaminants, regulatory level, 0.02 mg/L
- RCRA, 40CFR261, Appendix 8 Hazardous Constituents
- RCRA 40CFR268.48; 61FR15654, Universal Treatment Standards: Wastewater (mg/L), 0.0028; Nonwastewater (mg/kg), 0.13
- RCRA 40CFR264, Appendix 9; TSD Facilities Ground Water Monitoring List. Suggested test method(s) (PQL $ug/L$): 8080(0.1); 8250(10)
- Safe Drinking Water Act: MCL, 0.0002 mg/L; MCLG, 0.002 mg/L; Regulated chemical (47 FR 9352)

- Superfund/EPCRA 40CFR355, Appendix B Extremely Hazardous Substances: TPQ = 500/10,000 lb (227/4,540 kg)
- Superfund/EPCRA 40CFR302.4 RQ: CERCLA, 1 lb (0.454 kg)
- U.S. DOT Regulated Marine Pollutant (49CFR172.101, Appendix B), severe pollutant
- Mexico, Drinking Water Criteria, 0.07 mg/L.

*Description:* Endrin is the common name of one member of the cyclodiene group of pesticides. It is a cyclic hydrocarbon having a chlorine-substituted, methano-bridge structure. Endrin is a white, crystalline solid. Almost odorless. Insoluble in water. Molecular weight = 380.88. Melting/Freezing point = 230°C (decomposes). Vapor pressure = $1.9 \times 10^{-7}$ mmHg @ 25°C; Mixture in xylene: Flash point = 27°C. Explosive limits: LEL = 1.1%; UEL = 7.0%. Hazard Identification (based on NFPA-704 M Rating System): Health 4, Flammability 1, Reactivity 0. Log $K_{ow}$ = 5.6. Values at or above 3.0 are likely to bioaccumulate in marine organisms.

*Incompatibilities:* Parathion, strong acids (forms explosive vapors), strong oxidizers. Slightly corrosive to metal.

*Permissible Exposure Limits in Air:* The OSHA[2] PEL, the HSE[33] TWA and the DFG[3] and ACGIH[1] value is 0.1 mg/m³ TWA. The notation "skin" indicates the possibility of cutaneous absorption. Australia, Israel, Mexico, and the Canadian provinces of Alberta, BC, Ontario, and Quebec use the same TWA as OSHA and Alberta, BC, HSE[33] and Mexico set a STEL of 0.3 mg/m³ and that set in Germany is 1.0 mg/m³[35]. The NIOSH[2] IDLH level = 2 mg/m³. Several states have set guidelines or standards for Endrin in ambient air[60] ranging from 0.07 μg/m³ (Pennsylvania) to 0.238 μg/m³ (Kansas) to 1.0 μg/L (North Dakota) to 1.6 μg/L (Virginia) to 2.0 μg/L (Connecticut and Nevada).

*Determination in Air:* Collection by filter/chromosorb tube-102; workup with toluene; analysis by gas chromatography/electrochemical detection; NIOSH IV, Method #5519[18].

*Permissible Concentration in Water:* To protect freshwater aquatic life: 0.0023 μg/L as a 24 hr average, never to exceed 0.18 μg/L. To protect saltwater aquatic life: 0.0023 μg/L as a 24 hr average, never to exceed 0.037 μg/L. To protect human health: 1.0 μg/L[6]. According to a UN publication[35], the limit on Endrin in drinking water delivered at the tap is 0.2 μg/L. Mexico has imposed limits on Endrin in water[35] as follows: 2 μg/L in estuaries; 1 μg/L in receiving waters used for drinking water supply and 0.2 μg/L in coastal waters. The U.S. EPA has derived a no-observed-adverse effects-level (NOAE) of 0.045 mg/kg/day on the basis of which they have arrived at a long-term health advisory of 16 μg/L and a lifetime health advisory of 0.32 μg/L. Massachusetts has set a standards of 0.2 μg/L and Maine a guideline of 0.2 μg/L for Endrin in drinking water[61]. Mexico's drinking water criteria is 0.07 mg/L.

*Determination in Water:* Methylene chloride extraction followed by gas chromatography with electron capture or halogen specific detection (EPA Method 608) or gas chromatography plus mass spectrometry (EPA Method 625).

*Routes of Entry:* Inhalation, skin absorption, ingestion, skin and/or eye contact. Quickly passes through the skin.

*Harmful Effects and Symptoms*

*Short Term Exposure:* Contact can irritate the skin and eyes and may affect vision. Inhalation can cause irritation of the respiratory tract. Exposure can cause headache, nausea, vomiting, diarrhea, loss of appetite, sweating and weakness, lightheadedness, dizziness, convulsions and unconciousness. Lower exposure can affect concentration, memory and muscle coordination. Endrin can cause death by respiratory arrest. Symptoms include headache, nausea, vomiting, dizziness, tremors, convulsions, loss of consciousness, rise in blood pressure, fever, frothing of the mouth, deafness, coma, and death. This material is extremely toxic. It is rapidly absorbed through the skin. Symptoms appear between 20 minutes and 12 hours after exposure. Doses of 1 mg/kg can cause symptoms. Also, it is a central nervous system depressant and hepatotoxin. Inhalation may cause irritation to nose and throat, and sudden convulsions, which may occur from 30 minutes to 10 hours after exposure. Recovery is usually rapid, but headache, dizziness, lethargy, weakness, and weight loss may persist to 2 to 4 weeks. Prolonged breathing or ingestion can result in an onset of symptoms in 3 hours at a dose of 1 mg per kg of body weight. Ingestion of 12 grams has caused death. Pregnant women are considered to be at special risk. A rebuttable presumption notice against pesticide registration was issued on July 27, 1976 by EPA on the basis of oncogenicity, teratogenicity, and reductions in endangered species and non-target species.

*Long Term Exposure:* May damage the developing fetus. May damage the nervous system causing numbness and weakness in the extremities. Repeated exposure may cause personality changes of depression, anxiety and/or irritability. May cause anorexia. High or repeated exposure may cause liver damage. Studies in animals confirm that endrin's main target is the nervous system. Birth defects, especially abnormal bone formation, have been seen in some animal studies.

*Points of Attack:* Respiratory system, lungs, central nervous system, cardiovascular system, skin, eyes, plasma and red blood cell cholinesterase.

*Medical Surveillance:* Before employment and at regular times after that, the following are recommended: Plasma and red blood cell cholinesterase levels (tests for the enzyme poisoned by this chemical). If exposure stops, plasma levels return to normal in 1-2 weeks while red blood cell levels may be reduced for 1-3 months.

When acetylcholinesterase enzyme levels are reduced by 25% or more below preemployment levels, risk of poisoning

is increased, even if results are in lower ranges of "normal." Reassignment to work not involving organophosphate or carbamate pesticides is recommended until enzyme levels recover. If symptoms develop or overexposure occurs, repeat the above tests as soon as possible and get an exam of the nervous system. Also consider complete blood count. Consider chest x-ray following acute overexposure
Do not drink any alcoholic beverages before or during use. Alcohol promotes absorption of organophosphates. Examination of the nervous system. Electroencephalogram (a test for abnormal seizure activity). Blood Endrin level. Liver and kidney function tests.

*First Aid:* Speed in removing material from eyes and skin is of extreme importance. If this chemical gets into the eyes, remove any contact lenses at once and irrigate immediately for at least 15 minutes, occasionally lifting upper and lower lids. Seek medical attention immediately. If this chemical contacts the skin, remove contaminated clothing and wash immediately with soap and water. Shampoo hair promptly if contaminated. Seek medical attention immediately. If this chemical has been inhaled, remove from exposure, begin rescue breathing (using universal precautions) if breathing has stopped, and CPR if heart action has stopped. Transfer promptly to a medical facility. When this chemical has been swallowed, get medical attention. Give large quantities of water and induce vomiting. Do not make an unconscious person vomit.

*References:*
- U.S. Department of Health and Human Services; Agency for Toxic Substances and Disease Registry, "ToxFAQs for Endrin," Atlanta, GA (September 1997). http://www.atsdr.cdc.gov/tfacts89.html
- New Jersey Department of Health and Senior Services, "Hazardous Substance Fact Sheet: Endrin," Trenton, NJ (December, 1998). http://www.state.nj.us/health/eoh/rtkweb/0825.pdf
- U.S. Environmental Protection Agency, Endrin: Ambient Water Quality Criteria, Washington, DC (1980).
- U.S. Environmental Protection Agency, Reviews of the Environmental Effects of Pollutants: XIII, Endrin, Report EPA-600/1-79-005, Cincinnati, OH (1979).
- U.S. Environmental Protection Agency, Endrin, Health and Environmental Effects Profile No. 99, Washington, DC, Office of Solid Waste (April 30, 1980).
- Sax, N.I., Ed., "Dangerous Properties of Industrial Materials Report," 1, No. 5, 55-57 (1981).
- U.S. Environmental Protection Agency, "Chemical Profile: Endrin," Washington, DC, Chemical Emergency Preparedness Program (November 30, 1987).
- New York State Department of Health, "Chemical Fact Sheet: Endrin," Albany, NY, Bureau of Toxic Substance Assessment (August 1987).
- California Environmental Protection Agency "Chemical List of Lists," Sacramento CA (February 1997).

# EPN

*Use Type:* Insecticide
*CAS Number:* 2104-64-5
*Formula:* $C_{14}H_{14}NO_4PS$
*Alert:* All registered uses of EPN in the U.S. were canceled by the U.S. EPA on August 31, 1988. Effective June 9, 1993, the U.S. EPA announced the revocation of all tolerances for residues of the insecticide EPN in or on various agricultural commodities. These tolerances are listed in 40 CFR 180.119. Human toxicity (long-term): Very high.
*Synonyms:* O-Aethyl-O-N(4-nitrophenyl)-phenylmonothiophosphonat (German); ENT 17,798; O-Ester-p-nitrophenol with O-ethylphenyl phosphonothioate; Ethoxy-4-nitrophenoxyphenylphosphine sulfide; O-Ethyl-O-[(4-nitrofenyl)-fenyl]monothiofosfonaat (Dutch); O-Ethyl-O-(4-nitrophenyl)-benzenethionophosphonate; Ethyl-p-nitrophenylbenzenethionophosphonate; Ethyl-p-nitrophenyl benzenethionophosphate; Ethyl-p-nitrophenyl benzenethiophosphonate; Ethyl-p-nitrophenyl phenylphosphonothioate; O-Ethyl-O-p-nitrophenyl phenylphosphonothioate; O-Ethyl-O-(4-nitrophenyl)phenylphosphonothioate; O-Ethyl-O-p-nitrophenyl phenylphosphonothioate; Ethyl-p-nitrophenyl thionobenzenephosphate; O-Ethyl phenyl-p-nitrophenylthiophosphonate; Thionobenzenephosphonic acid Ethyl-p-nitrophenyl ester
*Trade Names:* NIAGARA®, FMC Agricultural Products Group (USA), 10/23/1983; PIN®; SANTOX®, TRIPLE KILL T®, Helena Chemical (USA), canceled 3/3/1983; VETO®, Drexel Chemical (USA), canceled 7/1/1987
*Producers:* Drexel Chemical (USA); DuPont (USA); FMC Agricultural Products Group (USA); Helena Chemical (USA); Hokko Chemical Industry (Japan); Nissan Chemical Industries (Japan); Sankei Chemical (Japan); Sigma-Aldrich Laborchemikalien (Germany); Sumitomo Chemical (Japan)
*Chemical Class:* Organophosphate
*EPA/OPP PC Code:* 041801
*California DPR Chemical Code:* 263
*ICSC Number:* 0753
*RTECS Number:* TB1925000
*EEC Number:* 015-036-00
*Human toxicity (long-term)[77]:* Extra high–0.07 ppb, Health Advisory
*Fish toxicity (threshold)[77]:* High–1.51635 ppb, MATC (Maximum Acceptable Toxicant Concentration)
*Carcinogen/Hazard Classifications*
**Label Signal Word:** DANGER
**WHO Acute Hazard:** Class 1 a, extremely hazardous
*Regulatory Authority:*
- Banned or Severely Restricted (Several Countries) (UN)[13]
- Very Toxic Substance (World Bank)[15]
- AB 1803-Well Monitoring Chemical (CAL)

- Permissible Exposure Limits for Chemical Contaminants (CAL/OSHA)
- The "Director's List" (CAL/OSHA)
- U.S. DOT Inhalation Hazard Chemicals as organophosphates
- Air Pollutant Standard Set (ACGIH)[1] (DFG)[3] (Argentina)[35] (Several States)[60]
- Water Pollution Standard Proposed (Japan)[35]
- Superfund/EPCRA 40CFR355, Appendix B Extremely Hazardous Substances: TPQ = 100/10,000 lb (45.4/4,540 kg)
- Superfund/EPCRA 40CFR302.4 RQ: EHS, 1 lb (0.454 kg)

*Description:* EPN is a light yellow crystalline solid with an aromatic odor [pesticide]or a n amber-brown liquid above 36°C (technical grade). Molecular weight = 323.29. Melting/Freezing point = 36°C. Vapor pressure = $3.4 \times 10^{-7}$ mmHg @ 20°C; $2.9 \times 10^{-4}$ mmHg @ 100°C. Hazard Identification (based on NFPA-704 M Rating System): Health 4, Flammability 1, Reactivity 0. Insoluble in water. Log $K_{ow}$ = 4.97. Also reported at 3.89. Values at or above 3.0 are likely to bioaccumulate in marine organisms. Physical and toxicological properties may be affected by the carrier solvents used in commercial formulations.

*Incompatibilities:* Reaction with oxidizers. Contact with alkalies causes decomposition (hydrolysis) producing *p*-nitrophenol.

*Permissible Exposure Limits in Air:* The OSHA[2] PEL, the DFG[3] an Argentina[35] TWA value[1] is 0.5 mg/m³. The ACGIH[1] recommends a TWA of 0.1 mg/m³. The notation "skin" indicates the possibility of cutaneous absorption. The NIOSH[2] IDLH level = 5 mg/m³. Various STEL values have been set in various countries: 1.5 mg/m³ in Argentina, 2.0 mg/m³ in the US and 5.0 in Germany[35]. Some states have set guidelines or standards for EPN in ambient air[60] ranging from 5.0 μg/m³ (North Dakota) to 10.0 μg/m³ (Connecticut) to 12.0 μg/m³ (Nevada).

*Determination in Air:* OSHA versatile sampler-2; Toluene/Acetone; Gas chromatography/Flame photometric detection for sulfur, nitrogen, or phosphorus; NIOSH IV Method #5600[18], Organophosphorus pesticides. Collection on a filter, workup with isooctane, Gas chromatography/Flame photometric detection for sulfur, nitrogen, or phosphorus; NIOSH IV, Method #5012[18].

*Permissible Concentration in Water:* Japan[35] has set an effluent maximum of 1 mg/L and an environmental water quality standard of zero.

*Routes of Entry:* Inhalation, ingestion, skin absorption, skin and/or eye contact. Passes through the skin.

*Harmful Effects and Symptoms*

*Short Term Exposure:* A cholinesterase inhibitor. EPN can affect the nervous system, causing convulsions and possible respiratory failure. Exposure may result in unconsciousness or death. The effects may be delayed. Medical observation is indicated. This material may be fatal is swallowed. It is poisonous if inhaled and extremely hazardous by skin contact. Repeated exposure may, without symptoms, be increasingly hazardous. The estimated fatal oral dose is 0.3 grams for a 150 lb (70 kg) person. Acute exposure to EPN may produce the following signs and symptoms: pinpoint pupils, blurred vision, headache, dizziness, muscle spasms, and profound weakness. Vomiting, diarrhea, abdominal pain, seizures, and coma may also occur. The heart rate may decrease following oral exposure or increase following dermal exposure. Hypertension (high blood pressure) is not uncommon. Respiratory symptoms include dyspnea (shortness of breath), respiratory depression and respiratory paralysis. Giddiness, slurred speech, confusion, and psychosis may also be observed. A rebuttable presumption against pesticide registration was issued for EPN on September 19, 1979 by EPA on the basis of neurotoxicity. Delayed pulmonary edema may occur after inhalation.

*Long Term Exposure:* Cholinesterase inhibitor; cumulative effect is possible. EPN may damage the nervous system with repeated exposure, resulting in convulsions, respiratory failure. May cause liver damage.

*Points of Attack:* Respiratory system, lungs, cardiovascular system, central nervous system, eyes, skin, blood cholinesterase.

*Medical Surveillance:* Medical observation is recommended for 24 to 48 hours after breathing overexposure, as pulmonary edema may be delayed. Before employment and at regular times after that, the following are recommended: Plasma and red blood cell cholinesterase levels (tests for the enzyme poisoned by this chemical). If exposure stops, plasma levels return to normal in 1-2 weeks while red blood cell levels may be reduced for 1-3 months. When acetylcholinesterase enzyme levels are reduced by 25% or more below preemployment levels, risk of poisoning is increased, even if results are in lower ranges of "normal." Reassignment to work not involving organophosphate or carbamate pesticides is recommended until enzyme levels recover. If symptoms develop or overexposure occurs, repeat the above tests as soon as possible and get an exam of the nervous system. Also consider complete blood count. Consider chest x-ray following acute overexposure

Do not drink any alcoholic beverages before or during use. Alcohol promotes absorption of organophosphates.

*First Aid:* **Treatment for organophosphate poisoning consists of thorough decontamination, cardiorespiratory support, and administration of the antidotes atropine and pralidoxime. In cases of severe poisoning, diazepam, an anticonvulsant, should also be administered. Antidotes should be administered as prevention even if the diagnosis is in doubt.** Speed in removing material from eyes and skin is of extreme importance. *Eyes:* Eye contact can cause dangerous amounts of these chemicals to be quickly absorbed through the mucous membrane into the bloodstream. Immediately and gently flush eyes with plenty of warm or cold water (NO hot water) for at least 15 minutes, occasionally lifting the upper and lower eyelids. Get medical aid immediately. *Skin:* Get medical aid. Skin contact can cause dangerous amounts of these chemicals to be absorbed into the bloodstream. Wearing the appropriate PPE equipment and respirator for organophosphate pesticides, immediately flush skin with plenty of soap and water for at least 15 minutes while removing contaminated clothing and shoes. Shampoo hair promptly if contaminated. The removed, contaminated clothing and shoes should be double-bagged and left in Hot Zone for later disposal by hazardous materials experts. Skin may also be decontaminated with diluted hypochlorite solution. *Inhalation:* Get medical aid. Do not contaminate yourself. Wearing the appropriate PPE equipment and respirator for organophosphate pesticides, immediately remove the victim from the contaminated area to fresh air. If the victim is not breathing, administer artificial respiration. Do NOT use mouth-to-mouth resuscitation; use bag/mask apparatus. If breathing is difficult, administer oxygen through bag/mask apparatus until medical help arrives. Do not leave victim unattended. *Ingestion:* Call poison control. Loosen all clothing. Never give anything by mouth to an unconscious person. If victim is *unconscious or having convulsions,* do nothing except keep victim warm. Get medical aid. Transfer promptly to a medical facility. In cases of ingestion, **do not induce vomiting**. If the victim is alert and asymptomatic, administer a slurry of activated charcoal at a dose of 1 g/kg (infant, child, and adult dose). A soda can and straw may be of assistance when offering charcoal to a child. *In some cases you may be specifically instructed by poison control to induce vomiting by way of 2 tablespoons of syrup of ipecac (adult) washed down with a cup of water.* Do NOT give activated charcoal before or with ipecac syrup.

*Note to physician or authorized medical personnel:* Treat cases of respiratory compromise, coma, or excessive pulmonary secretions with respiratory support using protocols and techniques available and within the scope of training. Some cases may necessitate procedures such as endotracheal intubation or cricothyrotomy by properly trained and equipped personnel. When possible, atropine (see *Antidotes*, below) should be given under medical supervision. Patients who are comatose, hypotensive, or having seizures or cardiac arrhythmias should be treated according to advanced life support protocols. *Antidotes:* Two antidotes are administered to treat organophosphate poisoning. Atropine is a competitive antagonist of acetylcholine at muscarinic receptors and is used to control the excessive bronchial secretions which are often responsible for death. Pralidoxime relieves both the nicotinic and muscarine effects of organophosphate poisoning by regenerating acetylcholinesterase and can reduce both the bronchial secretions and the muscle weakness associated with poisoning. The initial intravenous dose of atropine in adults should be determined by the severity of symptoms: An initial adult dose of 1.0 to 2.0 mg or pediatric dose of 0.01 mg/kg (minimum 0.01 mg) should be administered intravenously. If intravenous access cannot be established, atropine may also be given intramuscularly, subcutaneously or via endotracheal tube. Doses should be repeated every 15 minutes until excessive secretions and sweating have been controlled. Once bronchial secretion has been controlled, atropine administration should be repeated whenever the secretions begin to recur. In seriously poisoned patients, very large doses may be required. Alterations of pulse rate and pupillary size should not be used as indicators of treatment adequacy. Pralidoxime should be administered as early in poisoning as possible as its efficacy may diminish when given more than 24 to 36 hours after exposure. Doses are as follows: adult 1.0 g; pediatric 25 to 50 mg/kg. The drug should be administered intravenously over 30 to 60 minutes, but in a life-threatening situation, one-half of the total dose can be given per minute for a total administration time of 2 minutes. Treatment should begin to take effect within 40 minutes with a reduction in symptoms and the amount of atropine necessary to control bronchial secretion. The initial dose can be repeated in 1 hour and then every 8 to 12 hours until the patient is clinically well and no longer requires atropine. If intravenous access cannot be established, pralidoxime may also be given intramuscularly. Early administration of diazepam in addition to the combined atropine and pralidoxime treatment may help prevent the onset of seizures and potential brain and cardiac morphologic damage following high-level organophosphate poisoning.

*References:*
- Pesticide Management Education Program, ""EPN Tolerance Revocation 5/93," Cornell University, Ithaca, NY. (May 1993)
  http://pmep.cce.cornell.edu/profiles/insect-mite/ddt-famphur/epn/epn-rev-tol.html
- U.S. Environmental Protection Agency, "Chemical Profile: EPN," Washington, DC, Chemical Emergency Preparedness Program (November 30, 1987).
- Agency for Toxic Substances and Disease Registry, U.S. Department of Health and Human Services, Public Health Service, "Managing Hazardous Materials Incidents," Atlanta, GA (June 2003).
- California Environmental Protection Agency "Chemical List of Lists," Sacramento CA (February 1997).
- New Jersey Department of Health and Senior Services, "Hazardous Substance Fact Sheet, EPN," Trenton NJ (May 1999).
  http://www.state.nj.us/health/eoh/rtkweb/0829.pdf

# EPTC

*Use Type:* Herbicide
*CAS Number:* 759-94-4
*Formula:* $C_9H_{19}NOS$

*Alert:* Some formulations are Restricted Group Pesticides (RUP)

*Synonyms:* S-Aethyl-*N,N*-dipropylthiocarbamat (German); Carbamic acid, dipropylthio-, S-ethyl ester; Carbamothioic acid, dipropyl-, S-ethyl ester; Caswell No. 435; Dipropylcarbamothioic acid S-ethyl ester; *N,N*-Dipropylthiocarbamic acid S-ethyl ester; Ethyl di-*N*-Propylthiolcarbamate; S-Ethyl dipropylcarbamothioate; Ethyl *N,N*-dipropylthiocarbamate; Ethyl Dipropylthiocarbamate; S-Ethyl dipropylthiocarbamate; S-Ethyldipropylthiocarbamate; S-Ethyl-*N,N*-Di-*N*-propylthiocarbamate; Ethyl *N,N*-Di-*N*-propylthiolcarbamate; Ethyl *N,N*-dipropylthiolcarbamate; FDA 1541; NSC 40486

*Trade Names:* ALIROX®; EPTAM®; Makhteshim-Agan (Israel); EPTAM® 6E, Makhteshim-Agan (Israel); EPTAM 2.3G (granular, 2.3% by weight); EPTAM 10G (granular, 10% by weight); ERADICANE®, Makhteshim-Agan (Israel) and Epochem (China); EPTC, Makhteshim-Agan (Israel); GENEP® EPTC; R-1608®; SHORTSTOP®; STAUFFER® R 1608; TORBIN®

*Producers:* Drexel Chemical (USA); Epochem (China); Makhteshim-Agan (Israel); Syngenta (Switzerland)

*Chemical Class:* Thiocarbamate

*EPA/OPP PC Code:* 041401

*California DPR Chemical Code:* 264

*ICSC Number:* 0469

*RTECS Number:* FA4550000

*EEC Number:* 006-030-00-0

*EINECS Number:* 212-073-8

*Uses:* EPTC is a pre-emergence and early post-emergence herbicide used to control the growth of germinating annual weeds, including broadleaves, grasses, and sedges. It is used in every region of the United States in the production of a wide variety of food crops. The heaviest usage is in the Corn Belt, Northeastern and Mid-Atlantic states, Coastal and Northern Great Plains and in the Pacific Northwest on corn, potatoes, sweet potatoes, dry beans, peas, alfalfa, and snap beans. EPTC is also used on home-grown vegetables and ornamentals.

*Human toxicity (long-term)*[77]: Very low–175.00 ppb, Health Advisory

*Fish toxicity (threshold)*[77]: Very low–1848.27939 ppb, MATC (Maximum Acceptable Toxicant Concentration)

*Carcinogen/Hazard Classifications*

**U.S. EPA Carcinogens:** Not likely a carcinogen

**California Prop. 65:** Developmental toxin

**TRI Developmental Toxin:** Reproductive and developmental toxin

**Label Signal Word:** WARNING or CAUTION

**WHO Acute Hazard:** Class II, moderately hazardous

*Regulatory Authority:*
- RCRA Section 261 Hazardous Constituents
- EPA 40 CFR 372.65, Specific Toxic Chemical Listings
- EPA Hazardous Waste Number (RCRA No.): U390
- RCRA Universal Treatment Standards: Wastewater (mg/L), 0.003; Nonwastewater (mg/kg), 1.4
- EPCRA Section 313 Form R *de minimus* concentration reporting level: 1.0%
- Actively registered pesticide in California.

*Description:* Liquid or granular material. Slightly soluble in water; solubility = 370 ppm. Molecular weight = 189.34. Vapor pressure = 0.035 mmHg @25°C. Boiling point = 231°C. Flash point = 115°C. Log $K_{ow}$ = 3.22. Values at or above 3.0 are likely to bioaccumulate in marine organisms. Physical and toxicological properties may be affected by the carrier solvents used in commercial formulations.

*Incompatibilities:* Reacts violently with powerful oxidizers such as calcium hypochlorite.

*Permissible Concentration in Water:* No criteria set. Runoff from spills or fire control may cause water pollution.

*Routes of Entry:* Inhalation, ingestion, skin contact.

*Harmful Effects and Symptoms*

*Short Term Exposure:* Low levels of toxicity. Concentrated solutions are slightly corrosive to eyes and mucous membranes. Dust inhalation can cause irritation of the respiratory system with sneezing. Eye contact can cause irritation, watering, pain, and inflammation of the eyelids. Skin contact can cause irritation and minor ulceration. Ingestion can cause nausea, vomiting, fever, muscle twitching, seizure, rapid respiration, slow heart beat. Severe exposure may result in death.

*Long Term Exposure:* A cholinesterase inhibitor.

*Points of Attack:* Respiratory system, central nervous system, cardiovascular system, skin, eyes.

*Medical Surveillance:* Medical observation is recommended for 24 to 48 hours after breathing overexposure, as pulmonary edema may be delayed. As first aid for pulmonary edema, consider administering a corticosteroid spray. Cigarette smoking may exacerbate pulmonary injury and should be discouraged for at least 72 hours following exposure. Before employment and at regular times after that, the following are recommended: If symptoms develop or overexposure occurs, repeat the above tests as soon as possible and get an exam of the nervous system. Also consider complete blood count. Consider chest x-ray following acute overexposure.

*First Aid:* If this chemical gets into the eyes, remove any contact lenses at once and irrigate immediately for at least 15 minutes, occasionally lifting upper and lower lids. Seek medical attention immediately. If this chemical contacts the skin, remove contaminated clothing and wash immediately with soap and water. Seek medical attention immediately. If this chemical has been inhaled, remove from exposure, begin rescue breathing (using universal precautions) if breathing has stopped, and CPR if heart action has stopped. Transfer promptly to a medical facility. When this chemical has been swallowed, get medical attention. If victim is conscious and able to swallow, have victim drink 4 to 8 ounces of water. Do not induce vomiting. *Note to physician or authorized medical personnel*: Medical observation is recommended for 24 to 48 hours after breathing overexposure, as pulmonary edema may be delayed. As first aid for pulmonary edema, consider administering a corticosteroid spray. Cigarette smoking may exacerbate pulmonary injury and should be discouraged for at least 72 hours following exposure.

*References:*
- EXTOXNET, Extension Toxicology Network, "Pesticide Information Profile, EPTC," Oregon State University, Corvallis, OR. (June 1996). http://extoxnet.orst.edu/pips/eptc.htm
- U.S. Environmental Protection Agency, "Reregistration Eligibility Decision (RED), Facts EPTC" Office of Prevention, Pesticides and Toxic Substances, Washington, DC (September 1999). http://www.epa.gov/REDs/factsheets/0064fact.pdf
- California Environmental Protection Agency "Chemical List of Lists," Sacramento CA (February 1997)

# Esfenvalerate

*Use Type:* Insecticide
*CAS Number:* 66230-04-4
*Formula:* $C_{25}H_{22}ClNO_3$
*Synonyms:* Benzeneacetic acid, 4-chloro-α-(1-methylethyl)-, cyano (3-phenoxyphenyl)methyl ester, [s-(R*,R*)]-; [s-(R*,R*)]-4-Chloro-α-(1-methylethyl)benzeneacetic acid, cyano(3-phenoxyphenyl)methyl ester; s-(R*,R*)-Cyano (3-phenoxyphenyl) methyl 4-chloro-2-(1-methylethyl) benzene-; Fenvalerate A-α; (S)-α-Cyano-3-phenoxybenzyl (S)-2-(4-chlorophenyl)isovalerate; s-Fenvalerate (S)-α-Cyano-3-phenoxybenzyl (S)-2-(4-chlorophenyl)-3-methylbutyrate
*Trade Names:* AMERICARE®, Boehringer Ingelheim (Germany); ASANA®, DuPont Crop Protection (USA); ASANA-XL®, DuPont Crop Protection (USA); ASANA® DPX-YB656-84, DuPont Crop Protection (USA); ENFORCER®; EVERCIDE®, Mclaughlin Gormley King (USA); HALMARK®; OMS-3023®; S-1844®; S-5602 ALPHA®; SUMI-ALFA®; SUMI-ALPHA®; SS-PYDRIN®; SUMICIDIN A-ALPHA®
*Producers:* Agsin (Singapore); Boehringer Ingelheim (Germany); Bonide Products (USA); DuPont Crop Protection (USA); Mclaughlin Gormley King (USA); Shandong Huayang Pesticide Group (China)
*Chemical Class:* Pyrethroid; botanical
*EPA/OPP PC Code:* 109303
*California DPR Chemical Code:* 2321
*Uses:* Esfenvalerate is a synthetic insecticide used to control wide range of pests such as moths, flies, beetles, and other insects. It is used on vegetable crops (soya beans, sugar cane), tree fruit, cotton, maize, sorghum and nut crops, and non-crop lands. It also is used on a wide variety of household pests. It is usually mixed with a wide variety of other types of pesticides such as carbamate compounds or organophosphates and has the naturally occurring compound fenvalerate for use in the U.S. Esfenvalerate is almost identical to fenvalerate. Much of the data for fenvalerate is applicable to the pesticide esfenvalerate because the two compounds contain the same components. The only differences in the two products are the relative proportions of the four separate constituents (isomers). Esfenvalerate has become the preferred compound because it requires lower applications rates than fenvalerate, is less chronically toxic, and is a more powerful insecticide.

*Human toxicity (long-term)*[77]: Very low–140.00 ppb, Health Advisory
*Fish toxicity (threshold)*[77]: Extra high–0.00338 ppb, MATC (Maximum Acceptable Toxicant Concentration)
*U.S. Maximum Allowable Residue Levels for Esfenvalerate (40 CFR 180.533):*

| CROP | ppm |
|---|---|
| Artichoke, globe | 1 |
| Beet, sugar, dried pulp | 2.5 |
| Beet, sugar, roots | 0.5 |
| Beet, sugar, tops | 5 |
| Egg | 0.03 |
| Kiwifruit | 0.5 |
| Kohlrabi | 2 |
| Lettuce, head | 5 |
| Mustard greens | 5 |
| Poultry, fat | 0.3 |
| Poultry, liver | 0.03 |
| Poultry, meat | 0.03 |
| Poultry, mbyp, except liver | 0.3 |
| Sorghum, forage | 10 |
| Sorghum, grain, grain | 5 |
| Sorghum, grain, stover | 10 |

*Carcinogen/Hazard Classifications*
**Label Signal Word:** WARNING or CAUTION
*Regulatory Authority:*
- AB 1803-Well Monitoring Chemical (CAL) as pyrethrins
- Permissible Exposure Limits for Chemical Contaminants (CAL/OSHA) as pyrethrum
- Actively registered pesticide in California. as pyrethrins
- Clean Water Act: Section 311 Hazardous Substances/RQ (same as CERCLA) as pyrethrins
- EPCRA Section 304 RQ: CERCLA, 1 lb (0.454 kg) as pyrethrins

*Description:* Viscous yellow or amber liquid or white crystalline solid. Soluble in water; solubility = <1 ppm @ 20°C. Molecular weight = 419.9. Density = 1.175 @ 25°C. Boiling point = 151–167°C. Melting/Freezing point = 59.7°C. Vapor pressure = $1.1 \times 10^{-8}$ mmHg @ 20°C; 0.037 mPa @ 25°C.

*Incompatibilities:* Oxidizers (chlorates, nitrates, peroxides, permanganates, perchlorates, chlorine, bromine, fluorine, etc); strong acids. Moisture may cause hydrolysis/decomposition.

*Determination in Air:* Filter; none; Gravimetric; NIOSH IV[18] [Particulates NOR; #0500 (total), #0600 (respirable)]. Collection by impinger or fritted bubbler, analysis by gas liquid chromatography/ultraviolet. See NIOSH IV, Method #5008[18]. (pyrethrum)

*Permissible Concentration in Water:* No criteria set. Runoff from spills or fire control may cause water pollution.

*Determination in Water:* Collection by impinger or fritted bubbler, analysis by gas liquid chromatography/ultraviolet. See NIOSH IV, Method #5008[18].

*Routes of Entry:* Inhalation, absorbed through the skin.

*Harmful Effects and Symptoms*

*Short Term Exposure:* Pyrethroids can affect you when breathed in and by passing through your skin. Irritates the eyes and respiratory tract. High exposure can affect the nervous system causing headache, nausea, vomiting, fatigue, and restlessness, rhinorrhea (discharge of thin nasal mucous).

*Long Term Exposure:* High or repeated exposure can cause lung allergy (with cough, wheezing and/or shortness of breath) or hay fever symptoms (sneezing, runny or stuffy nose). Allergic "pneumonia" can also occur with cough, chest pain, breathing difficulty and abnormal chest x-ray. Repeated attacks may lead to permanent scarring. Skin allergy may also develop with rash and itching, even with lower exposures. Skin contact can cause rash with redness, blisters and intense itching. A severe generalized allergy can occur with weakness and collapse.

*Points of Attack:* Respiratory system, skin, central nervous system.

*Medical Surveillance:* Before beginning employment and at regular times after that, the following are recommended: Lung function tests. These may be normal if the person is not having an attack at the time of the test. Consider chest x-ray if lung symptoms are present. Evaluation by a qualified allergist, including careful exposure history and special testing, may help diagnose skin allergy.

*First Aid:* If this chemical gets into the eyes, remove any contact lenses at once and irrigate immediately for at least 15 minutes, occasionally lifting upper and lower lids. Seek medical attention immediately. If this chemical contacts the skin, remove contaminated clothing and wash immediately with soap and water. Seek medical attention immediately. If this chemical has been inhaled, remove from exposure, begin rescue breathing (using universal precautions) if breathing has stopped, and CPR if heart action has stopped. Transfer promptly to a medical facility. When this chemical has been swallowed, get medical attention. *Do not induce vomiting when formulations containing petroleum solvents are ingested.* Otherwise, give large quantities of water and induce vomiting. Do not make an unconscious person vomit.

*References:*
- EXTOXNET, Extension Toxicology Network, "Pesticide Information Profile, Esfenverate," Oregon State University, Corvallis, OR (June 1996). http://extoxnet.orst.edu/pips/esfenval.htm
- U.S. Environmental Protection Agency, Office of Pesticide Programs, Pesticide Residue Limits, "Esfenvalerate", 40 CFR 180.533, www.epa.gov/pesticides/food/viewtols.htm
- California Environmental Protection Agency "Chemical List of Lists," Sacramento CA (February 1997)

# Ethalfluralin (ANSI)

*Use Type:* Herbicide
*CAS Number:* 5523-68-6
*Formula:* $C_{13}H_{14}F_3N_3O_4$
*Synonyms:* Benzenamine, *N*-ethyl-*N*-(2-methyl-2-propenyl)-2,6-dinitro-4-(trifluoromethyl)-;
*N*-Ethyl-*N*-(2-methyl-2-propenyl)-2,6-dinitro-4-(trifluoromethyl)benzenamine

*Trade Names:* COMPOUND 94961®; COBEX®; EL 161®; ETHALFLURLIN®; ETHALFLURALIN®; SOMILAN®, Dow AgroSciences (USA); SONALAN®, Dow AgroSciences (USA); SONALEN®, Dow AgroSciences (USA)

*Producers:* Dow AgroSciences (USA)
*Chemical Class:* Fluorodinitrotoluidine; dinitroanaline
*EPA/OPP PC Code:* 113101
*California DPR Chemical Code:* 2166
*EINECS Number:* 259-564-3

*Uses:* Ethalfluralin is a selective herbicide used for the pre-emergence control of annual grasses and broadleaf weeds in certain food and feed crops. Ethalfluralin may be used in growing a variety of grain, seed, and cucurbit crops. The greatest amounts of ethalfluralin are used in growing soybeans, dry beans, and sunflower seeds. Ethalfluralin is used only outdoors, in agriculture; no residential uses were registered. The U.S. EPA revoked, in July, 2002, tolerances for residue of ethalfluralin in or on goat fat, goat meat, and goat meat by products (mbyp).

*Carcinogen/Hazard Classifications*
*U.S. EPA Carcinogens:* Group C, possible carcinogen
*Label Signal Word:* WARNING, CAUTION or DANGER
*WHO Acute Hazard:* Class U, unlikely to be hazardous

*Regulatory Authority:*
- Actively registered pesticide in California.

*Description:* Yellow-orange crystalline solid. Very slightly soluble in water; solubility = 0.35 ppm. @ 25°C. Molecular weight = 333.30. Melting/Freezing point = 55-58°C.

*Determination in Air:* Filter; none; Gravimetric; NIOSH IV [Particulates NOR; #0500 (total), #0600 (respirable)].[18]

*Permissible Concentration in Water:* No criteria set. Runoff from spills or fire control may cause water pollution.

*Routes of Entry:* Absorbed through the skin. Inhalation, ingestion.

*Harmful Effects and Symptoms*
*Short Term Exposure:* May be absorbed through the skin. Irritates the skin. Eye contact may cause severe irritation and serious damage. Inhalation should be avoided; use NIOSH-approved air purifying respirators for pesticides. May be harmful if swallowed.
*Medical Surveillance:* Contact physician if poisoning is suspected or if redness, itching, or a burning sensation develops in the eyes or skin.
*First Aid:* If this chemical gets into the eyes, remove any contact lenses at once and irrigate immediately for at least 15 minutes, occasionally lifting upper and lower lids. Seek medical attention immediately. If this chemical contacts the skin, remove contaminated clothing and wash immediately with soap and water. Seek medical attention immediately. If this chemical has been inhaled, remove from exposure, begin rescue breathing (using universal precautions) if breathing has stopped, and CPR if heart action has stopped. Transfer promptly to a medical facility. When this chemical has been swallowed, get medical attention. Give large quantities of water and induce vomiting. Do not make an unconscious person vomit.
*References:*
- U.S. Environmental Protection Agency, "Reregistration Eligibility Decision (RED), Ethalfluralin" Office of Prevention, Pesticides and Toxic Substances, Washington, DC (March 1995). http://www.epa.gov/REDs/2260.pdf
- California Environmental Protection Agency "Chemical List of Lists," Sacramento CA (February 1997).

# Ethametsulfuron-methyl

*Use Type:* Herbicide
*CAS Number:* 97780-06-8
*Formula:* $C_{14}H_{16}N_6O_6S$
*Synonyms:* Benzoic acid, 2-[(((((4-ethoxy-6-(methylamino)-1,3,5-triazin-2-yl)amino)carbonyl)amino)sulfonyl]-, methyl ester; Methyl 2-[(((((4-ethoxy-6-(methylamino)-1,3,5-triazin-2-yl)amino)carbonyl)amino)sulfonyl]benzoate; Ethametsulfuron-methyl; Methyl 2-[(4-ethoxy-6-methylamino-1,3,5-triazin-2-yl)carbamoylsulfamoyl]benzoate
*Trade Names:* A-7881®, DuPont Crop Protection (USA); DPX-A 7881® (application withdrawn); MUSTER®, DuPont Crop Protection (USA)
*Producers:* DuPont Crop Protection (USA); Wuzhou International (China)
*Chemical Class:* Sulfonylurea
*EPA/OPP PC Code:* 129091
*Carcinogen/Hazard Classifications*
**U.S. EPA Carcinogens:** Unclassifiable
**Label Signal Word:** WARNING or CAUTION

*Routes of Entry:* Inhalation, passing through the skin, ingestion.
*Harmful Effects and Symptoms*
*Short Term Exposure:* May cause skin and severe eye irritation. Moderately poisonous if ingested or inhaled. Exposure to a triazine (simazine) has caused acute and subacute dermatitis in the former USSR, characterized by erythema, slight edema, moderate pruritus, and burning lasting 4 to 5 days.
*Long Term Exposure:* May cause lung irritation and damage. May cause skin allergy. Contact with some triazine compounds (such as atrazine) may increase risks for tumors known to be associated with hormonal factors. These have been observed in both animals and human beings, and are consistent with the known effects on the hypothalamic pituitary gonadal axis. Repeated exposure may cause weight loss and reduced red blood cell count. May be mutagenic.
*Points of Attack:* Liver, lungs and skin.
*Medical Surveillance:* Before beginning employment and at regular times after that, for those with frequent or potentially high exposures, the following is recommended: Lung function tests. Consider chest x-ray following acute overexposure. Evaluation by a qualified allergist. Examination of the nervous system.
*First Aid:* If this chemical gets into the eyes, remove any contact lenses at once and irrigate immediately for at least 15 minutes, occasionally lifting upper and lower lids. Seek medical attention immediately. If this chemical contacts the skin, remove contaminated clothing and wash immediately with soap and water. Seek medical attention immediately. If this chemical has been inhaled, remove from exposure, begin rescue breathing (using universal precautions) if breathing has stopped, and CPR if heart action has stopped. Transfer promptly to a medical facility. When this chemical has been swallowed, get medical attention. Give large quantities of water or milk and induce vomiting. Do not make an unconscious person vomit.

# Ethephon (ANSI)

*Use Type:* Plant growth regulator
*CAS Number:* 16672-87-0
*Formula:* $C_2H_6ClO_3P$
*Synonyms:* CEP; 2-CEPA; (2-Chloroethyl)phosphonic acid; 2-Chloraethyl phosphonsaeure (German); Chlorethephon; 2-Chlorethylphosphonic acid; 2-Chloroethanephosphonic acid; Ethefon; Ethephon; Phosphonic acid, (2-chloroethyl)-
*Trade Names:* AMCHEM® 68-250; ARVEST®; BASE® 250, Bayer CropScience (Germany); BOLL'D®, Agriliance (USA); BROMEFLOR®; BROMOFLOR®; CAMPOSAN®; CEPHA®; CEPHA® 10LS; CERONE®, Bayer CropScience (Germany); CHIPCO® FLOREL PRO; ETHEPON®, Micro-Flo (USA); ETHEL®; ETHEVERSE®; ETHREL®, Bayer CropScience (Germany); FINISH®, Bayer CropScience

(Germany); FLORDIMEX®; FLOREL®, Bayer CropScience (Germany); G-996®; KAMPOSAN®; PREP®, Bayer CropScience (Germany); ROLL-FRUCT®; TERPAL® (with mepiquat chloride); T-EXTRA®; TOMATHREL®

*Producers:* Agriliance (USA); Agsin (Singapore); Bayer CropScience (Germany); Griffin (USA); Ki-Hara Chemicals Ltd. (UK); Lancaster Synthesis (UK); Micro-Flo (USA)

*Chemical Class:* Organophosphate

*EPA/OPP PC Code:* 099801

*California DPR Chemical Code:* 1626

*EINECS Number:* 240-718-3

*Uses:* Ethephon is a plant growth regulator used to promote fruit ripening, abscission, flower induction, and other responses. It is registered for use on a number of food, feed and nonfood crops (rubber plants, flax), greenhouse nursery stock, and outdoor residential ornamental plants, but is used primarily on cotton. Ethephon is applied to plant foliage by either ground or aerial equipment. It also may be applied by hand sprayer to certain home garden vegetables and ornamentals. Use practice limitations include prohibitions against applying ethephon through any type of irrigation system; feeding or grazing livestock in treated areas; and treating within 2 to 60 days of harvest, depending on the crop.

*Human toxicity (long-term)[77]:* Very low–126.00 ppb, Health Advisory

*Fish toxicity (threshold)[77]:* Very low–26627.94738 ppb, MATC (Maximum Acceptable Toxicant Concentration)

*U.S. Maximum Allowable Residue Levels for Ethephon (40 CFR 180.300):*

| CROP | ppm |
|---|---|
| Apple | 5 |
| Barley, bran | 5 |
| Barley, grain | 2 |
| Barley, pearled barley | 5 |
| Barley, straw | 10 |
| Blackberry | 30 |
| Blueberry | 20 |
| Cantaloupe | 2 |
| Cattle, fat | 0.1 |
| Cattle, meat | 0.1 |
| Cattle, mbyp | 0.1 |
| Cherry | 10 |
| Coffee, bean | 0.1 |
| Cotton, undelinted seed | 2 |
| Cranberry | 5 |
| Cucumber | 0.1 |
| Fig | 5 |
| Goat, fat | 0.1 |
| Goat, meat | 0.1 |
| Goat, mbyp | 0.1 |
| Grape | 2 |
| Grape, raisin | 12 |
| Hog, fat | 0.1 |
| Hog, meat | 0.1 |
| Hog, mbyp | 0.1 |
| Horse, fat | 0.1 |
| Horse, meat | 0.1 |
| Horse, mbyp | 0.1 |
| Milk | 0.1 |
| Nut, macadamia | 0.5 |
| Pepper | 30 |
| Pineapple | 2 |
| Pumpkin | 0.1 |
| Sheep, fat | 0.1 |
| Sheep, meat | 0.1 |
| Sheep, mbyp | 0.1 |
| Sugarcane, cane | 0.1 |
| Sugarcane, molasses | 1.5 |
| Tomato | 2 |
| Walnut | 0.5 |
| Wheat, bran | 5 |
| Wheat, grain | 2 |
| Wheat, milled fractions, except flour | 5 |
| Wheat, straw | 10 |

*Carcinogen/Hazard Classifications*

**U.S. EPA Carcinogens:** Group D, unclassifiable, inadequate data

**Label Signal Word:** CAUTION or DANGER

**WHO Acute Hazard:** Class U, unlikely to be hazardous

*Regulatory Authority:*

FIFRA, 180.3(5); class of cholinesterase-inhibiting pesticide
- FIFRA, 40CFR185: tolerances for pesticides in food
- FIFRA, 40CFR186: tolerances for pesticides in animal feeds
- U.S. DOT Inhalation Hazard Chemicals
- Actively registered pesticide in California.

*Description:* Hygroscopic needles from benzene. Commercial product is a white, waxy solid. Readily soluble in water; solubility = <1 kg/L. Molecular weight = 144.50. Melting/Freezing point = 74°C. Vapor pressure = $1.1 \times 10^{-7}$ mmHg @ 20°C; <0.01 mPa @ 20°C.

*Incompatibilities:* Contact with flammable material may cause fire and explosions. Contact with combustible or oxidizable materials may form heat-, shock-, and friction-sensitive explosive mixtures. Static electricity may also cause explosions. Keep away from all acids, especially dibasic organic acids, ammonium compounds, antimony sulfide, arsenic trioxide, metal sulfides, powdered metals, calcium aluminum hydride, cyanides, manganese dioxide, phosphorus, selenium, sulfur, thiocyanates, zinc.

*Determination in Air:* OSHA versatile sampler-2; Toluene/Acetone; Gas chromatography/Flame ionization detection; NIOSH IV, Method #5600, Organophosphorus Pesticides.[18]

*Permissible Concentration in Water:* No criteria set. Runoff from spills or fire control may cause water pollution.

*Short Term Exposure:* Because this material has a low vapor pressure, significant inhalation of vapors is unlikely at ordinary temperatures. Delayed pulmonary edema may occur after inhalation.

*Medical Surveillance:* Medical observation is recommended for 24 to 48 hours after breathing overexposure, as pulmonary edema may be delayed. Do not drink any alcoholic beverages before or during use; alcohol promotes absorption of organophosphates.

*First Aid:* **Treatment for organophosphate poisoning consists of thorough decontamination, cardiorespiratory support, and administration of the antidotes atropine and pralidoxime. In cases of severe poisoning, diazepam, an anticonvulsant, should also be administered. Antidotes should be administered as prevention even if the diagnosis is in doubt.** Speed in removing material from eyes and skin is of extreme importance. *Eyes:* Eye contact can cause dangerous amounts of these chemicals to be quickly absorbed through the mucous membrane into the bloodstream. Immediately and gently flush eyes with plenty of warm or cold water (NO hot water) for at least 15 minutes, occasionally lifting the upper and lower eyelids. Get medical aid immediately. *Skin:* Get medical aid. Skin contact can cause dangerous amounts of these chemicals to be absorbed into the bloodstream. Wearing the appropriate PPE equipment and respirator for organophosphate/carbamate pesticides, immediately flush skin with plenty of soap and water for at least 15 minutes while removing contaminated clothing and shoes. Shampoo hair promptly if contaminated. The removed, contaminated clothing and shoes should be double-bagged and left in Hot Zone for later disposal by hazardous materials experts. Skin may also be decontaminated with diluted hypochlorite solution. *Inhalation:* Get medical aid. Do not contaminate yourself. Wearing the appropriate PPE equipment and respirator for organophosphate pesticides, immediately remove the victim from the contaminated area to fresh air. If the victim is not breathing, administer artificial respiration. Do NOT use mouth-to-mouth resuscitation; use bag/mask apparatus. If breathing is difficult, administer oxygen through bag/mask apparatus until medical help arrives. Do not leave victim unattended. *Ingestion:* Call poison control. Loosen all clothing. Never give anything by mouth to an unconscious person. If victim is *unconscious or having convulsions,* do nothing except keep victim warm. Get medical aid. Transfer promptly to a medical facility. In cases of ingestion, **do not induce vomiting**. If the victim is alert and asymptomatic, administer a slurry of activated charcoal at a dose of 1 g/kg (infant, child, and adult dose). A soda can and straw may be of assistance when offering charcoal to a child. *In some cases you may be specifically instructed by poison control to induce vomiting by way of 2 tablespoons of syrup of ipecac (adult) washed down with a cup of water.* Do NOT give activated charcoal before or with ipecac syrup.

*Note to physician or authorized medical personnel:* Treat cases of respiratory compromise, coma, or excessive pulmonary secretions with respiratory support using protocols and techniques available and within the scope of training. Some cases may necessitate procedures such as endotracheal intubation or cricothyrotomy by properly trained and equipped personnel. When possible, atropine (see *Antidotes*, below) should be given under medical supervision. Patients who are comatose, hypotensive, or having seizures or cardiac arrhythmias should be treated according to advanced life support protocols. *Antidotes:* Two antidotes are administered to treat organophosphate poisoning. Atropine is a competitive antagonist of acetylcholine at muscarinic receptors and is used to control the excessive bronchial secretions which are often responsible for death. Pralidoxime relieves both the nicotinic and muscarine effects of organophosphate poisoning by regenerating acetylcholinesterase and can reduce both the bronchial secretions and the muscle weakness associated with poisoning. The initial intravenous dose of atropine in adults should be determined by the severity of symptoms: An initial adult dose of 1.0 to 2.0 mg or pediatric dose of 0.01 mg/kg (minimum 0.01 mg) should be administered intravenously. If intravenous access cannot be established, atropine may also be given intramuscularly, subcutaneously or via endotracheal tube. Doses should be repeated every 15 minutes until excessive secretions and sweating have been controlled. Once bronchial secretion has been controlled, atropine administration should be repeated whenever the secretions begin to recur. In seriously poisoned patients, very large doses may be required. Alterations of pulse rate and pupillary size should not be used as indicators of treatment adequacy. Pralidoxime should be administered as early in poisoning as possible as its efficacy may diminish when given more than 24 to 36 hours after exposure. Doses are as follows: adult 1.0 g; pediatric 25 to 50 mg/kg. The drug should be administered intravenously over 30 to 60 minutes, but in a life-threatening situation, one-half of the total dose can be given per minute for a total administration time of 2 minutes. Treatment should begin to take effect within 40 minutes with a reduction in symptoms and the amount of atropine necessary to control bronchial secretion. The initial dose can be repeated in 1 hour and then every 8 to 12 hours until the patient is clinically well and no longer requires atropine. If intravenous access cannot be established, pralidoxime may also be given intramuscularly. Early administration of diazepam in addition to the combined atropine and pralidoxime treatment may help prevent the onset of seizures and potential brain and cardiac morphologic damage following high-level organophosphate poisoning.

*References:*
- EXTOXNET, Extension Toxicology Network, "Pesticide Information Profile, Ethephon," Oregon State University, Corvallis, OR (September 1995). http://extoxnet.orst.edu/pips/ethephon.htm

- U.S. Environmental Protection Agency, "Reregistration Eligibility Decision (RED), Ethephon" Office of Prevention, Pesticides and Toxic Substances, Washington, DC (April 1995).
  http://www.epa.gov/REDs/0382.pdf
- U.S. Environmental Protection Agency, Office of Pesticide Programs, Pesticide Residue Limits, "Ethephon", 40 CFR 180.300,
  www.epa.gov/pesticides/food/viewtols.htm
- California Environmental Protection Agency "Chemical List of Lists," Sacramento CA (February 1997)

# Ethion (ANSI)

*Use Type:* Insecticide and acaricide
*CAS Number:* 563-12-2
*Formula:* $C_9H_{22}O_4P_2S_4$
*Alert:* A General Use Pesticide (GUP). Human toxicity (long-term): High.
*Synonyms:* Diethion (France); Bis(S-(Diethoxyphosphinothioyl)mercapto)methane; Bis (dithiophosphatede O,O-diethyle) de S,S'-methylene (French); ENT 24,105; Ethyl methylene phosphorodithioate; Etion (Spanish); Methanedithiol, S,S-diester with O,O-diethyl phosphorodithioate acid; Methyleen-S,S'-bis(O,O-diethyl-dith iofosfaat) (Dutch); Methylene-S,S'-bis(O,O-diaethyl-dithiophosphat) (German); S,S'-Methylene O,O,O',O'-tetraethyl phosphorodithioate; Metilen-S,S'-bis(O,O-dietil-ditiofosfato) (Italian); Phosphorodithioic acid, O,O-diethyl ester, S,S-diester with methanedithiol; STCC 4921565; O,O,O',O'-Tetraaethyl-bis(dithiophosphat) (German); O,O,O',O'-Tetraethyl S,S'-methylenebis(dithiophosphate); O,O,O',O'-Tetraethyl S,S'-methylenebisphosphordithioate; Tetraethyl S,S'-methylene bis(phosphorothiolthionate); O,O,O',O'-Tetraethyl S,S'-methylene di(phosphorodithioate); a,S'-Methylene O,O,O',O'-tetraethyl ester phosphorodithioic acid
*Trade Names:* AC 3422®; EACITHION®; EAQUA ETHION®; EBLADAN®; ECOMMANDO INSECTICIDE CATTLE EAR TAG®, Boehringer Ingelheim (Germany); EDRASTIC®, Sudarshan Chemical Industries (India); EEMBATHION®; EETHANOX®; EETHIOL®; EETHODAN®; EETHOPAZ®; EFMC-1240®; EFOSFATOXE®; EFOSFONO 50®; EHYLEMOX®; EITOPAZ®; EKWIT®, Scott Company (USA), canceled 10/10/1989; EMITKILL®, United Phosphorus (India); ENAGATA®, Rallis India (India); ENIA 1240®, FMC Agricultural Products Group (USA), canceled; ENIAGARA 1240®, FMC Agricultural Products Group (USA), canceled; ENIALATE®; EPHOSPHOTOX E®; EPROKIL®, Gowan (USA), canceled 8/22/1986; ERHODIACIDE®; ERHODOCIDE®; ERODOCID®; ERP-THION®; ESENTRY®, Boehringer Ingelheim (Germany), canceled 5/12/1997; ESOPRATHION®; ETAFETHION®, Rallis India (India); EVEGFRUFOSMITE®; EVEGFRU FOSMITE®
*Producers:* Aimco Pesticides Ltd. (India); EAlcotan Laboratories (Spain); EAmvac (USA); EBayer Chemicals (Germany); EBEC Group (India); EBharat Pulverizing Mills (India); EBharat Rasayan (India); EBoehringer Ingelheim (Germany); EEhrenstorfer, Dr. (Germany); EFMC Agricultural Products Group (USA); EGowan (USA); EHelena Chemical (USA); EHindustan Insecticides (India); EPI Industries (India); ERallis India (India); ERhone-Poulenc Agro (France); EScott Company (USA); EShenzhen Guomeng Industry Co., Ltd. (China); ESigma-Aldrich Laborchemikalien (Germany); ESudarshan Chemical Industries (India); ESyngenta (Switzerland); EUnited Phosphorus (India)
*Chemical Class:* Organophosphate
*EPA/OPP PC Code:* 058401
*California DPR Chemical Code:* 268
*ICSC Number:* 0888
*RTECS Number:* TE4550000
*EEC Number:* 015-047-00-2
*EINECS Number:* 209-242-3
*Uses:* Ethion is an organophosphate pesticide used to kill aphids, mites, scales, thrips, leafhoppers, maggots, and foliar feeding larvae. It may be used on a wide variety of food, fiber, and ornamental crops, including greenhouse crops, lawns, and turf. Ethion is often used on citrus, apples, nuts and cotton. It is mixed with oil and sprayed on dormant trees to kill eggs and scales. It is also used as a cattle dip for ticks and as a treatment for buffalo flies. It is available in dust, emulsifiable concentrate, emulsifiable solution, granular, and wettable powder formulations.
*Human toxicity (long-term)[77]:* High–3.50 ppb, Health Advisory
*Fish toxicity (threshold)[77]:* Intermediate–18.38485 ppb, MATC (Maximum Acceptable Toxicant Concentration)
*U.S. Maximum Allowable Residue Levels for Ethion (40 CFR 180.173):*

| CROP | ppm |
| --- | --- |
| Cattle, mbyp | 1.0 |
| Cattle, fat | 2.5 |
| Cattle, meat (fat basis) | 0.5 |
| Citrus fruits | 2.0 |
| Citrus pulp, dehydrated | 10.0 |
| Goats, fat | 0.2 |
| Goats, mbyp | 0.2 |
| Goat, meat | 0.2 |
| Hogs, fat | 0.2 |
| Hogs, mbyp | 0.2 |
| Hogs, meat | 0.2 |
| Horses, fat | 0.2 |
| Horses, mbyb | 0.2 |
| Horses, meat | 0.2 |
| Milk fat (residues in milk) | 0.5 |
| Raisins | 4.0 |

| | |
|---|---|
| Sheep, fat | 0.2 |
| Sheep, mbyb | 0.2 |
| Sheep, meat | 0.2 |
| Tea, dried | 10.0 |

*Carcinogen/Hazard Classifications*
**U.S. EPA Carcinogens:** Group E, unlikely carcinogen
**Label Signal Word:** WARNING
**WHO Acute Hazard:** Class II, moderately hazardous
*Regulatory Authority:*
- Very Toxic Substance (World Bank)[15]
- Air Pollutant Standard Set (ACGIH)[1] (Several States)[60]
- AB 1803-Well Monitoring Chemical (CAL)
- EPA/SARA 302 (EPCRA) Extremely hazardous substances
- MCL (Maximum Contaminants Levels) list of contaminants (CAL)
- CAL-DHS/DHS Drinking Water Action Levels
- Permissible Exposure Limits for Chemical Contaminants (CAL/OSHA)
- The "Director's List" (CAL/OSHA)
- U.S. DOT Inhalation Hazard Chemicals as organophosphates
- Clean Water Act: Section 311 Hazardous Substances/RQ 40CFR117.3 (same as CERCLA, see below)
- Superfund/EPCRA 40CFR355, Appendix B Extremely Hazardous Substances: TPQ = 1000 lb (454 kg)
- Superfund/EPCRA 40CFR302.4 RQ: CERCLA, 10 lb (4.54 kg)
- Canada, WHMIS, Ingredients Disclosure List

*Description:* Ethion is a colorless to amber-colored, odorless liquid. The technical product has a very disagreeable odor. Soluble in water; solubility = $10^{-4}$ g/100 mL @ 20°C. Molecular weight = 384.48. Melting/Freezing point = –12 to –13°C. Boiling point = 164°C. Vapor pressure = $1.5 \times 10^{-6}$ mmHg @ 25°C. Flash point = 176°C. Hazard Identification (based on NFPA-704 M Rating System): Health 3, Flammability 1, Reactivity 0. Log $K_{ow}$ = 5.1. Values at or above 3.0 are likely to bioaccumulate in marine organisms.

*Incompatibilities:* Incompatible with alkaline formulations and strong acids. Decomposes violently when heated above 150°C. Mixtures with magnesium may be explosive.

*Permissible Exposure Limits in Air:* There is no OSHA[2] PEL. NIOSH[2] and ACGIH[1] recommend a TWA value of 0.4 mg/m³. The notation "skin" is added to the TWA indicating the possibility of cutaneous absorption. Argentina[35] has set a TWA of 0.4 mg/m³ also and has added an STEL at the same level. Several states have set guidelines or standards for Ethion in ambient air[60] ranging from 4.0 µg/m³ (North Dakota) to 6.0 µg/m³ (Connecticut) to 9.0 µg/m³ (Nevada).

*Determination in Air:* OSHA versatile sampler-2; Toluene/Acetone; Gas chromatography/Flame photometric detection for sulfur, nitrogen, or phosphorus; NIOSH IV, Method #5600, Organophosphorus Pesticide.[18]

*Permissible Concentration in Water:* California[61] has set a guideline for Ethion in drinking water of 35 µg/L.
*Routes of Entry:* Inhalation, skin contact, and ingestion.
*Harmful Effects and Symptoms*
*Short Term Exposure:* Organic phosphorus insecticides are absorbed by the skin, as well as by the respiratory and gastrointestinal tracts. They are cholinesterase inhibitors. Symptoms may include nausea, vomiting, abdominal cramps, diarrhea, excessive salivation, headache, giddiness, weakness, muscle twitching, difficult breathing, blurring or dimness of vision, and loss of muscle coordination. Death may occur from failure of the respiratory center, paralysis of the respiratory muscles, intense bronchoconstriction, or all three. This material is very toxic; the probable oral lethal dose for humans is 50-500 mg/kg, which is between one teaspoonful and one ounce for a 150 lb person. Delayed pulmonary edema may occur after inhalation.

*Long Term Exposure:* Cholinesterase inhibitor; cumulative effect is possible. Ethion may damage the nervous system with repeated exposure, resulting in convulsions, respiratory failure. May cause liver damage.

*Points of Attack:* Respiratory system, lungs, central nervous system, cardiovascular system, skin, eyes, plasma and red blood cell cholinesterase.

*Medical Surveillance:* Medical observation is recommended for 24 to 48 hours after breathing overexposure, as pulmonary edema may be delayed. Before employment and at regular times after that, the following are recommended: Plasma and red blood cell cholinesterase levels (tests for the enzyme poisoned by this chemical). If exposure stops, plasma levels return to normal in 1-2 weeks while red blood cell levels may be reduced for 1-3 months. When acetylcholinesterase enzyme levels are reduced by 25% or more below preemployment levels, risk of poisoning is increased, even if results are in lower ranges of "normal." Reassignment to work not involving organophosphate or carbamate pesticides is recommended until enzyme levels recover. If symptoms develop or overexposure occurs, repeat the above tests as soon as possible and get an exam of the nervous system. Also consider complete blood count. Consider chest x-ray following acute overexposure. Do not drink any alcoholic beverages before or during use. Alcohol promotes absorption of organophosphates.

*First Aid:* **Treatment for organophosphate poisoning consists of thorough decontamination, cardiorespiratory support, and administration of the antidotes atropine and pralidoxime. In cases of severe poisoning, diazepam, an anticonvulsant, should also be administered. Antidotes should be administered as prevention even if the diagnosis is in doubt.** Speed in removing material from eyes and skin is of extreme importance. *Eyes:* Eye contact can cause dangerous amounts of these chemicals to be quickly absorbed through the mucous membrane into the bloodstream. Immediately and gently flush eyes with plenty of warm or cold water (NO hot water) for at least 15 minutes, occasionally lifting the upper and lower eyelids. Get medical aid immediately. *Skin:* Get medical aid. Skin contact can cause dangerous amounts of these chemicals to be absorbed into the bloodstream. Wearing the appropriate PPE equipment and respirator for organophosphate pesticides, immediately flush skin with plenty of soap and water for at least 15 minutes while removing contaminated clothing and shoes. Shampoo hair promptly if contaminated. The removed, contaminated clothing and shoes should be double-bagged and left in Hot Zone for later disposal by hazardous materials experts. Skin may also be decontaminated with diluted hypochlorite solution. *Inhalation:* Get medical aid. Do not contaminate yourself. Wearing the appropriate PPE equipment and respirator for organophosphate pesticides, immediately remove the victim from the contaminated area to fresh air. If the victim is not breathing, administer artificial respiration. Do NOT use mouth-to-mouth resuscitation; use bag/mask apparatus. If breathing is difficult, administer oxygen through bag/mask apparatus until medical help arrives. Do not leave victim unattended. *Ingestion:* Call poison control. Loosen all clothing. Never give anything by mouth to an unconscious person. If victim is *unconscious or having convulsions,* do nothing except keep victim warm. Get medical aid. Transfer promptly to a medical facility. In cases of ingestion, **do not induce vomiting**. If the victim is alert and asymptomatic, administer a slurry of activated charcoal at a dose of 1 g/kg (infant, child, and adult dose). A soda can and straw may be of assistance when offering charcoal to a child. *In some cases you may be specifically instructed by poison control to induce vomiting by way of 2 tablespoons of syrup of ipecac (adult) washed down with a cup of water.* Do NOT give activated charcoal before or with ipecac syrup.

*Note to physician or authorized medical personnel:* Treat cases of respiratory compromise, coma, or excessive pulmonary secretions with respiratory support using protocols and techniques available and within the scope of training. Some cases may necessitate procedures such as endotracheal intubation or cricothyrotomy by properly trained and equipped personnel. When possible, atropine (see *Antidotes*, below) should be given under medical supervision. Patients who are comatose, hypotensive, or having seizures or cardiac arrhythmias should be treated according to advanced life support protocols. *Antidotes:* Two antidotes are administered to treat organophosphate poisoning. Atropine is a competitive antagonist of acetylcholine at muscarinic receptors and is used to control the excessive bronchial secretions which are often responsible for death. Pralidoxime relieves both the nicotinic and muscarine effects of organophosphate poisoning by regenerating acetylcholinesterase and can reduce both the bronchial secretions and the muscle weakness associated with poisoning. The initial intravenous dose of atropine in adults should be determined by the severity of symptoms: An initial adult dose of 1.0 to 2.0 mg or pediatric dose of 0.01 mg/kg (minimum 0.01 mg) should be administered intravenously. If intravenous access cannot be established, atropine may also be given intramuscularly, subcutaneously or via endotracheal tube. Doses should be repeated every 15 minutes until excessive secretions and sweating have been controlled. Once bronchial secretion has been controlled, atropine administration should be repeated whenever the secretions begin to recur. In seriously poisoned patients, very large doses may be required. Alterations of pulse rate and pupillary size should not be used as indicators of treatment adequacy. Pralidoxime should be administered as early in poisoning as possible as its efficacy may diminish when given more than 24 to 36 hours after exposure. Doses are as follows: adult 1.0 g; pediatric 25 to 50 mg/kg. The drug should be administered intravenously over 30 to 60 minutes, but in a life-threatening situation, one-half of the total dose can be given per minute for a total administration time of 2 minutes. Treatment should begin to take effect within 40 minutes with a reduction in symptoms and the amount of atropine necessary to control bronchial secretion. The initial dose can be repeated in 1 hour and then every 8 to 12 hours until the patient is clinically well and no longer requires atropine. If intravenous access cannot be established, pralidoxime may also be given intramuscularly. Early administration of diazepam in addition to the combined atropine and pralidoxime treatment may help prevent the onset of seizures and potential brain and cardiac morphologic damage following high-level organophosphate poisoning.

*References:*
- EXTOXNET, Extension Toxicology Network, "Pesticide Information Profile, Ethion," Oregon State University, Corvallis, OR (June 1996). http://ace.orst.edu/info/extoxnet/pips/ethion.htm
- EPA, Office of Pesticide Programs, Pesticide Residue Limits, "Ethion", 40 CFR 180.173, http://www.epa.gov/pesticides/food/viewtols.htm
- Department of Health and Human Resources, Agency for Toxic substances and Disease Registry, "ToxFAQs for Ethion," Atlanta, GA (February 2001). http://www.atsdr.cdc.gov/tfacts152.html
- New Jersey Department of Health and Senior Services, "Hazardous Substance Fact Sheet: Ethion," Trenton, NJ (December 1998). http://www.state.nj.us/health/eoh/rtkweb/0837.pdf
- U.S. Environmental Protection Agency, S,S'-Methylene-O,O,O',O'-Tetraethyl Phosphorodithioate, Health and Environmental Effects Profile No. 127, Washington, DC, Office of Solid Waste (April 30, 1980).
- Agency for Toxic Substances and Disease Registry, U.S. Department of Health and Human Services, Public Health Service, "Managing Hazardous Materials Incidents," Atlanta, GA (June 2003).

- Sax, N.I., Ed., "Dangerous Properties of Industrial Materials Report," 4, No. 1, 69-74 (1984) and 7, No. 1, 9-37 (1987).
- U.S. Environmental Protection Agency, "Chemical Profile: Ethion," Washington, DC, Chemical Emergency Preparedness Program (November 30, 1987).

## Ethofenprox

*Use Type:* Insecticide
*CAS Number:* 80844-07-1
*Formula:* $C_{25}H_{28}O_3$
*Synonyms:* Benzene, 1-[(2-(4-ethoxyphenyl)-2-methylpropoxy)methyl]-3-phenoxy-; 2-(4-Ethoxyphenyl)-2-methylpropyl 3-phenoxybenzyl ether; Etofenprox
*Trade Names:* MTI 500®; PUNKASO®, Rallis India (India); TREBON®, Kenso Corp. (Malaysia); ZOECON® RF-316
*Producers:* Kenso Corp. (Malaysia); Mitsui Chemicals (Japan); Rallis India (India
*EPA/OPP PC Code:* 128965
*California DPR Chemical Code:* 2292
*Carcinogen/Hazard Classifications*
*U.S. EPA Carcinogens:* Group C, possible carcinogen
*Label Signal Word:* CAUTION, DANGER
*WHO Acute Hazard:* Class U, unlikely to be hazardous
*Harmful Effects and Symptoms*
*Short Term Exposure:* Contact with eyes or skin may cause irritation or injury. Inhalation should be avoided; use NIOSH-approved air purifying respirators for pesticides. May be harmful if swallowed.
*First Aid:* If this chemical gets into the eyes, remove any contact lenses at once and irrigate immediately for at least 15 minutes, occasionally lifting upper and lower lids. Seek medical attention immediately. If this chemical contacts the skin, remove contaminated clothing and wash immediately with soap and water. Seek medical attention immediately. If this chemical has been inhaled, remove from exposure, begin rescue breathing (using universal precautions) if breathing has stopped, and CPR if heart action has stopped. Transfer promptly to a medical facility. When this chemical has been swallowed, get medical attention. Give large quantities of water and induce vomiting. Do not make an unconscious person vomit. Do not induce vomiting when formulations containing petroleum solvents are ingested.

## Ethofumesate (ANSI)

*Use Type:* Herbicide
*CAS Number:* 26225-79-6
*Formula:* $C_{13}H_{18}O_5S$
*Synonyms:* 5-Benzofuranol, 2-ethoxy-2,3-dihydro-3,3-dimethyl-, methanesulfonate (+)-; 2-Ethoxy-2,3-dihydro-3,3-dimethyl-5-benzofuranyl methanesulfonate,(+)-; 2-ethoxy-2,3-dihydro-3,3-dimethylbenzofuran-5-yl methanesulfonate
*Trade Names:* BATAMIX PROGRESS®, Bayer CropScience (Germany); BETANAL®; ETHOSAT® 500, Makhteshim-Agan Industries (Israel); KEMIRON®; NORTRON®, Bayer CropScience (Germany); NC-8438®; POWERTWIN®, Makhteshim-Agan Industries (Israel); PROGRESS®, Bayer CropScience (Germany); TANDEM®; TORERO®, Makhteshim-Agan Industries (Israel)
*Producers:* Bayer CropScience (Germany); Calliope (France); Ki-Hara Chemicals Ltd. (UK); Makhteshim-Agan Industries (Israel); Rhodia Group (France)
*Chemical Class:* Miscellaneous hydrocarbon
*EPA/OPP PC Code:* 110601
*California DPR Chemical Code:* 1900
*EINECS Number:* 247-525-3
*Uses:* A major use is to control weeds in sugar beet crops.
*Human toxicity (long-term)[77]:* Very low–2800.00 ppb, Health Advisory
*Fish toxicity (threshold)[77]:* Intermediate–58.73095 ppb, MATC (Maximum Acceptable Toxicant Concentration)
*Carcinogen/Hazard Classifications*
*U.S. EPA Carcinogens:* Group D, unclassifiable, inadequate data
*Label Signal Word:* WARNING, CAUTION or DANGER
*WHO Acute Hazard:* Class U, unlikely to be hazardous
*Regulatory Authority:*
- Actively registered pesticide in California.

*Description:* Colorless to whitish crystalline solid. Odorless. Soluble in water; solubility = 110 ppm @ 25°C. Molecular weight 286.3. Melting/Freezing point = 71°C. Vapor pressure = $4.9 \times 10^{-6}$ mmHg @ 20°C; $8.6 \times 10^{-7}$ mmHg @ 25°C.
*Determination in Air:* Filter; none; Gravimetric; NIOSH IV [Particulates NOR; #0500 (total), #0600 (respirable)].[18]
*Permissible Concentration in Water:* No criteria set. Runoff from spills or fire control may cause water pollution.
*Harmful Effects and Symptoms*
*Short Term Exposure:* May irritate eyes, skin, and respiratory tract. Inhalation should be avoided; use NIOSH-approved air purifying respirators for pesticides. May be harmful if swallowed.
*First Aid:* If this chemical gets into the eyes, remove any contact lenses at once and irrigate immediately for at least 15 minutes, occasionally lifting upper and lower lids. Seek medical attention immediately. If this chemical contacts the skin, remove contaminated clothing and wash immediately with soap and water. Seek medical attention immediately. If this chemical has been inhaled, remove from exposure, begin rescue breathing (using universal precautions) if breathing has stopped, and CPR if heart action has stopped. Transfer promptly to a medical facility. When this chemical has been swallowed, get medical attention. Give large quantities of water and induce vomiting. Do not make an unconscious person vomit.

*References:*
- California Environmental Protection Agency "Chemical List of Lists," Sacramento CA (February 1997).

# Ethoprop (ANSI)

*Use Type:* A nematicide and soil insecticide
*CAS Number:* 13194-48-4
*Formula:* $C_8H_{19}O_2PS_2$
*Synonyms:* AI3-27318; Caswell No. 434C; ENT 27,318; Ethoprophos; *O*-ethyl *S,S*-dipropyl dithiophosphate; *O*-Ethyl *S,S*-dipropyl phosphorodithioate; Phosethoprop; Phosphorodithioic acid, *O*-ethyl *S,S*-dipropyl ester
*Trade Names:* AI3-27318®; CASWELL No. 434C®; JOLT®, Velsicol Chemical Corporation (USA), canceled 2/21/1986; MENAP®; MOBIL V-C 9-104®, Mobil Chemical (USA); MOCAP®, Mobile Chemical (USA); Bayer CropScience (Germany), canceled 2/9/2000; MOCAP 10G®, Mobile Chemical (USA); Bayer CropScience (Germany), canceled 2/9/2000; PHOSETHOPROP®; ROVOKIL®; V-C 9-104®, Virginia-Carolina Corp. USA), canceled; V-C CHEMICAL V-C 9-104®, Virginia-Carolina Corp. USA), canceled; VIRGINIA-CAROLINA VC 9-104®, Virginia-Carolina Corp. USA), canceled
*Producers:* Aventis CropScience (France); Bayer CropScience (Germany; Hokko Chemical Industry (Japan); Shenzhen Guomeng Industry Co., Ltd. (China); Sigma-Aldrich Laborchemikalien (Germany); Velsicol Chemical Corporation (USA)
*Chemical Class:* Organophosphate
*EPA/OPP PC Code:* 041101
*California DPR Chemical Code:* 404
*RTECS Number:* TE4025000
*EINECS Number:* 236-152-1
*Uses:* Ethoprop is used as a pre-plant, soil application to control wireworms and nematodes in potatoes, sugar cane, sweet potatoes, and tobacco, with lesser usage on corn (field and sweet), beans (lima and snap), cucumbers, and cabbage. In addition, it is used to treat pineapples, bananas, and plantains, as well as field-grown ornamentals and non-bearing citrus trees, and commercial turf. Roughly 60% of ethoprop is applied to potatoes.
*Human toxicity (long-term)[77]:* Intermediate–12.45550 ppb, CHCL (Chronic Human Carcinogen Level)
*Fish toxicity (threshold)[77]:* Extra high–0.47783 ppb, MATC (Maximum Acceptable Toxicant Concentration)
*U.S. Maximum Allowable Residue Levels for Ethoprop (40 CFR 180.262):*

| CROP | ppm |
|---|---|
| Banana | 0.02 |
| Bean, lima, seed | 0.02 |
| Bean, snap, succulent | 0.02 |
| Cabbage | 0.02 |
| Corn, forage | 0.02 |
| Corn, grain | 0.02 |
| Corn, stover | 0.02 |
| Corn, sweet, kernel plus cob with husks removed | 0.02 |
| Cucumber | 0.02 |
| Peanut | 0.02 |
| Peanut, hay | 0.02 |
| Pineapple | 0.02 |
| Potato | 0.02 |
| Sugarcane, cane | 0.02 |
| Sweet potato | 0.02 |

*Carcinogen/Hazard Classifications*
*U.S. EPA Carcinogens:* Likely carcinogen
*California Prop. 65:* Carcinogen
*WHO Acute Hazard:* Class I, Extremely Hazardous
*Regulatory Authority:*
- Banned or Severely Restricted (E. Germany, Malaysia, Philippines) (UN)[13]
- U.S. DOT Inhalation Hazard Chemicals as organophosphates
- Permissible Exposure Limits for Chemical Contaminants (CAL/OSHA)
- The "Director's List" (CAL/OSHA)
- Proposition 65 chemical (CAL)
- Superfund/EPCRA 40CFR355, Appendix B Extremely Hazardous Substances: TPQ = 1000 lb (454 kg)
- Superfund/EPCRA 40CFR302.4 RQ: EHS, 1 lb (0.454 kg)
- EPCRA Section 313 Form R *de minimus* concentration reporting level: 1.0%
- U.S. DOT Regulated Marine Pollutant (49CFR172.101, Appendix B)

*Description:* Ethoprophos is a pale yellow liquid. Molecular weight = 242.35. Boiling point = 86–91°C @ 0.2 mmHg. Vapor pressure = $4.2 \times 10^{-4}$ mmHg @ 26°C Hazard Identification (based on NFPA-704 M Rating System): Health 3, Flammability 1, Reactivity 0. Slightly soluble in water; solubility = 675 ppm @ 20°C. Log $K_{ow}$ = < 2.5. Unlikely to bioaccumulate in marine organisms.
*Permissible Exposure Limits in Air:* No standards set.
*Determination in Air:* OSHA versatile sampler-2; Toluene/Acetone; Gas chromatography/Flame photometric detection for sulfur, nitrogen, or phosphorus; NIOSH Method IV Method #5600, Organophosphorus pesticides[18].
*Permissible Concentration in Water:* No criteria set, but runoff from spills or fire control may cause water pollution.
*Routes of Entry:* Inhalation, ingestion, eye and/or contact.
*Harmful Effects and Symptoms*
*Short Term Exposure:* Symptoms are similar to parathion and may include nausea, vomiting, abdominal cramps, diarrhea, excessive salivation, headache, giddiness, weakness, muscle twitching, difficult breathing, blurring or dimness of vision, and loss of muscle coordination. Death may occur from failure of the respiratory center, paralysis of the respiratory muscles, intense bronchoconstriction, or all

three. This material is extremely toxic; the probable oral lethal dose for humans is 5-50 mg/kg, or between 7 drops and 1 teaspoonful for a 150 lb person. It is a cholinesterase inhibitor which affects the nervous system. Delayed pulmonary edema may occur after inhalation.

*Long Term Exposure:* Cholinesterase inhibitor; cumulative effect is possible. This chemical may damage the nervous system with repeated exposure, resulting in convulsions, respiratory failure. May cause liver damage.

*Points of Attack:* Respiratory system, lungs, central nervous system, cardiovascular system, skin, eyes, plasma and red blood cell cholinesterase.

*Medical Surveillance:* Medical observation is recommended for 24 to 48 hours after breathing overexposure, as pulmonary edema may be delayed. Before employment and at regular times after that, the following are recommended: Plasma and red blood cell cholinesterase levels (tests for the enzyme poisoned by this chemical). If exposure stops, plasma levels return to normal in 1-2 weeks while red blood cell levels may be reduced for 1-3 months. When acetylcholinesterase enzyme levels are reduced by 25% or more below preemployment levels, risk of poisoning is increased, even if results are in lower ranges of "normal." Reassignment to work not involving organophosphate or carbamate pesticides is recommended until enzyme levels recover. If symptoms develop or overexposure occurs, repeat the above tests as soon as possible and get an exam of the nervous system. Also consider complete blood count. Consider chest x-ray following acute overexposure. Do not drink any alcoholic beverages before or during use. Alcohol promotes absorption of organophosphates.

*First Aid:* **Treatment for organophosphate poisoning consists of thorough decontamination, cardiorespiratory support, and administration of the antidotes atropine and pralidoxime. In cases of severe poisoning, diazepam, an anticonvulsant, should also be administered. Antidotes should be administered as prevention even if the diagnosis is in doubt.** Speed in removing material from eyes and skin is of extreme importance. *Eyes:* Eye contact can cause dangerous amounts of these chemicals to be quickly absorbed through the mucous membrane into the bloodstream. Immediately and gently flush eyes with plenty of warm or cold water (NO hot water) for at least 15 minutes, occasionally lifting the upper and lower eyelids. Get medical aid immediately. *Skin:* Get medical aid. Skin contact can cause dangerous amounts of these chemicals to be absorbed into the bloodstream. Wearing the appropriate PPE equipment and respirator for organophosphate pesticides, immediately flush skin with plenty of soap and water for at least 15 minutes while removing contaminated clothing and shoes. Shampoo hair promptly if contaminated. The removed, contaminated clothing and shoes should be double-bagged and left in Hot Zone for later disposal by hazardous materials experts. Skin may also be decontaminated with diluted hypochlorite solution. *Inhalation:* Get medical aid. Do not contaminate yourself. Wearing the appropriate PPE equipment and respirator for organophosphate pesticides, immediately remove the victim from the contaminated area to fresh air. If the victim is not breathing, administer artificial respiration. Do NOT use mouth-to-mouth resuscitation; use bag/mask apparatus. If breathing is difficult, administer oxygen through bag/mask apparatus until medical help arrives. Do not leave victim unattended. *Ingestion:* Call poison control. Loosen all clothing. Never give anything by mouth to an unconscious person. If victim is *unconscious or having convulsions,* do nothing except keep victim warm. Get medical aid. Transfer promptly to a medical facility. In cases of ingestion, **do not induce vomiting**. If the victim is alert and asymptomatic, administer a slurry of activated charcoal at a dose of 1 g/kg (infant, child, and adult dose). A soda can and straw may be of assistance when offering charcoal to a child. *In some cases you may be specifically instructed by poison control to induce vomiting by way of 2 tablespoons of syrup of ipecac (adult) washed down with a cup of water.* Do NOT give activated charcoal before or with ipecac syrup.

*Note to physician or authorized medical personnel:* Treat cases of respiratory compromise, coma, or excessive pulmonary secretions with respiratory support using protocols and techniques available and within the scope of training. Some cases may necessitate procedures such as endotracheal intubation or cricothyrotomy by properly trained and equipped personnel. When possible, atropine (see *Antidotes*, below) should be given under medical supervision. Patients who are comatose, hypotensive, or having seizures or cardiac arrhythmias should be treated according to advanced life support protocols. *Antidotes:* Two antidotes are administered to treat organophosphate poisoning. Atropine is a competitive antagonist of acetylcholine at muscarinic receptors and is used to control the excessive bronchial secretions which are often responsible for death. Pralidoxime relieves both the nicotinic and muscarine effects of organophosphate poisoning by regenerating acetylcholinesterase and can reduce both the bronchial secretions and the muscle weakness associated with poisoning. The initial intravenous dose of atropine in adults should be determined by the severity of symptoms: An initial adult dose of 1.0 to 2.0 mg or pediatric dose of 0.01 mg/kg (minimum 0.01 mg) should be administered intravenously. If intravenous access cannot be established, atropine may also be given intramuscularly, subcutaneously or via endotracheal tube. Doses should be repeated every 15 minutes until excessive secretions and sweating have been controlled. Once bronchial secretion has been controlled, atropine administration should be repeated whenever the secretions begin to recur. In seriously poisoned patients, very large doses may be required. Alterations of pulse rate and pupillary size should not be used as indicators of treatment adequacy. Pralidoxime should be administered as early in poisoning as possible as its efficacy may diminish when given more than 24 to 36 hours after exposure. Doses are as follows: adult 1.0 g; pediatric 25 to 50 mg/kg. The drug should be administered intravenously over 30 to 60 minutes, but in a life-threatening situation, one-half of the total dose can be given per minute for a total administration time of 2 minutes. Treatment should begin to take effect within 40 minutes with a reduction in symptoms and the amount of atropine necessary to control bronchial secretion. The initial dose can be repeated in 1 hour and then every 8 to 12 hours until the patient is clinically well and no longer requires atropine. If intravenous access cannot be established, pralidoxime may also be given intramuscularly. Early administration of diazepam in addition to the combined atropine and pralidoxime treatment may help prevent the onset of seizures and potential brain and cardiac morphologic damage following high-level organophosphate poisoning.

*References:*
- EXTOXNET, Extension Toxicology Network, "Pesticide Information Profile," Oregon State University, Corvallis, OR
- N. J. Department of Health and Senior Services, "Hazardous Substance Fact Sheet, Ethoprophos," Trenton, NJ (April 2002). http://www.state.nj.us/health/eoh/rtkweb/2395.pdf
- Sax, N.I., Ed., "Dangerous Properties of Industrial Materials Report," 2, No. 4, 85-88 (1982).
- U.S. Environmental Protection Agency, "Chemical Profile: Ethoprophos," Washington, DC, Chemical Emergency Preparedness Program (November 30, 1987).
- Agency for Toxic Substances and Disease Registry, U.S. Department of Health and Human Services, Public Health Service, "Managing Hazardous Materials Incidents," Atlanta, GA (June 2003).
- California Environmental Protection Agency "Chemical List of Lists," Sacramento CA (February 1997).

# Ethoxyquin

*Use Type:* Insecticide, fungicide and plant growth regulator
*CAS Number:* 91-53-2
*Formula:* $C_{14}H_{19}NO$
*Synonyms:* 1,2-Dihydro-6-ethoxy-2,2,4-trimethylquinoline; 1,2-Dihydro-2,2,4-trimethyl-6-ethoxyquinoline; 6-Ethoxy-1,2-dihydro-2,2,4-trimethyl quinoline; Ethoxyquine; 6-Ethoxy-2,2,4-trimethyl-1,2-dihydroquinoline; 2,2,4-Trimethyl-6-ethoxy-1,2-dihydroquinoline
*Trade Names:* CHEMLEY®, Cerexagri (France), canceled; DECCOQUIN 305®, Cerexagri (France); EMQ®; EQ®; NIFLEX®; NIX-SCALD®, canceled; SANTOFLEX A®; SANTOFLEX AW®; SANTOQUIN®; SANTOQUINE®; STOP-SCALD®
*Producers:* Cerexagri (France)
*Chemical Class:* Unclassified
*EPA/OPP PC Code:* 055501
*California DPR Chemical Code:* 269
*Uses:* Used for preharvest or post-h preservation of color in apples and pears. It is used as an anti-oxidant to preserve color in paprika and ground and powdered chili. Ethoxyquin also is a chemical preservative used in animal feed to prevent ingredients from reacting with oxygen and becoming rancid. It has been known to cause birth defects in pet birds and dogs.
*U.S. Maximum Allowable Residue Levels for ethoxyquin (40 CFR 180.178):*

| CROP | ppm |
|---|---|
| Pear | 3.0 |

*Carcinogen/Hazard Classifications*
**Label Signal Word:** CAUTION
*Description:* Transparent, yellow liquid. Molecular weight = 217.34. Boiling point 125C (2 millimeter mercury). Melting/freezing point = < 0°C. Density = 1.029 @ 25. Combustible if exposed to heat or flame.
*Incompatibilities:* Oxidizers (chlorates, nitrates, peroxides, permanganates, perchlorates, chlorine, bromine, fluorine, etc); strong acids.
*Permissible Concentration in Water:* No criteria set. Runoff from spills or fire control may cause water pollution.
*First Aid:* If this chemical gets into the eyes, remove any contact lenses at once and irrigate immediately for at least 15 minutes, occasionally lifting upper and lower lids. Seek medical attention immediately. If this chemical contacts the skin, remove contaminated clothing and wash immediately with soap and water. Seek medical attention immediately. If this chemical has been inhaled, remove from exposure, begin rescue breathing (using universal precautions) if breathing has stopped, and CPR if heart action has stopped. Transfer promptly to a medical facility. When this chemical has been swallowed, get medical attention. Give large quantities of water and induce vomiting. Do not make an unconscious person vomit.

*References:*
- U.S. Environmental Protection Agency, Office of Pesticide Programs, Pesticide Residue Limits, "Ethoxyquin," 40 CFR 180.178, www.epa.gov/pesticides/food/viewtols.htm
- California Environmental Protection Agency "Chemical List of Lists," Sacramento CA (February 1997).

# Ethylan

*Use Type:* Insecticide
*CAS Number:* 72-56-0
*Formula:* $C_{18}H_{20}Cl_2$
*Synonyms:* Benzene, 1,1'-(2,2-dichloroethylidene)bis(4-ethyl-; 1,1-Bis(*p*-ethylphenyl)-2,2-dichloroethane; 1,1-Dichloro-2,2-bis(*p*-ethylphenyl)ethane and related compounds; α,α-Dichloro-2,2-bis(*p*-ethylphenyl)ethane; 1,1-

Dichloro-2,2-bis(*p*-ethylphenyl)ethane; 1,1-Dichloro-2,2-bis(4-ethylphenyl)ethane; 2,2-Dichloro-1,1-bis(*p*-ethylphenyl)ethane; Diethyldiphenyl; DI(*p*-ethylphenyl)dichloroethane; Ethane, 1,1-dichloro-2,2-bis(*p*-ethylphenyl)-; *p,p*-Ethyl DDD; *p,p'*-Ethyl-DDD; NCI-C02868
*Trade Names:* PERTHANE®, canceled; Q-137®
*Chemical Class:* Chlorinated hydrocarbons (aliphatic)
*EPA/OPP PC Code:* 032101
*California DPR Chemical Code:* 472
*Uses:* Not registered in the U.S. Originally registered for limited use to control insects on fruit and vegetables. Also used for moth and carpet beetles in dry cleaning. Product was discontinued by Rohm and Haas.
*Regulatory Authority:*
- FIFRA, 180.3(4); class of chlorinated organic pesticide.
- AB 1803-Well Monitoring Chemical (CAL)
- The "Director's List" (CAL/OSHA)

*Description:* Crystals from ethanol; when pure, Cream to tan, waxy, semi-solid (technical product). formulated as emulsifiable concentrate, wettable powder, or dust. Wettable forms will dissolve in water; solubility = 100 μg/L @24°C. Pure material will sink. Molecular weight = 307.28. Melting/Freezing point = decomposes above 52°C. Log $K_{ow}$ = 6.67. Values above 3.0 are likely to bio-accumulate in marine organisms.
*Incompatibilities:* Slightly corrosive to iron, copper, zinc, aluminum and their alloys.
*Permissible Concentration in Water:* No criteria set. Runoff from spills or fire control may cause water pollution.
*Harmful Effects and Symptoms*
*Short Term Exposure:* Contact with eyes or skin may cause irritation or injury. Inhalation should be avoided; use NIOSH-approved air purifying respirators for pesticides. May be harmful if swallowed.
*First Aid:* If this chemical gets into the eyes, remove any contact lenses at once and irrigate immediately for at least 15 minutes, occasionally lifting upper and lower lids. Seek medical attention immediately. If this chemical contacts the skin, remove contaminated clothing and wash immediately with soap and water. Seek medical attention immediately. If this chemical has been inhaled, remove from exposure, begin rescue breathing (using universal precautions) if breathing has stopped, and CPR if heart action has stopped. Transfer promptly to a medical facility. When this chemical has been swallowed, get medical attention. Give large quantities of water and induce vomiting. Do not make an unconscious person vomit. Do not induce vomiting when formulations containing petroleum solvents are ingested.
*References:*
- Pesticide Management Education Program, "Ethylan (Perthane) Chemical Profile 4/85," Cornell University, Ithaca, NY (April 1985).
  http://pmep.cce.cornell.edu/profiles/insect-mite/ddt-famphur/ethylan/insect-prof-ethylan.html
- California Environmental Protection Agency "Chemical List of Lists," Sacramento CA (February 1997).

# Ethyl Formate

*Use Type:* Fumigant and insecticide
*CAS Number:* 109-94-4
*Formula:* $C_3H_6O_2$; $HCOOC_2H_5$
*Synonyms:* Aethylformiat (German); Areginal; Ethyle (Formiate D') (French); Ethylformiaat (Dutch); Ethyl formic ester; Ethyl methanoate; Etile (Formiato di) (Italian); Formic acid, ethyl ester; Formic ether; Mrowczan etylu (Polish)
*Chemical Class:* Unclassified
*EPA/OPP PC Code:* 043102
*California DPR Chemical Code:* 278
*ICSC Number:* 0623
*RTECS Number:* LQ8400000
*EEC Number:* 607-015-00-7
*Uses:* Not registered in the U.S. Used as a fumigant and also as a solvent for cellulose nitrate and acetate and in the production of synthetic flavors. It is also a raw material in pharmaceutical manufacture.
*Carcinogen/Hazard Classifications*
**Label Signal Word:** DANGER
*Regulatory Authority:*
- FIFRA, 40CFR185: tolerances for pesticides in food.
- Permissible Exposure Limits for Chemical Contaminants (CAL/OSHA)
- The "Director's List" (CAL/OSHA)

*Description:* Ethyl formate is a colorless liquid with a fruity odor. Slightly soluble in water; solubility = 11 g/100 ml @ 20°C. Molecular weight = 74.10. Boiling point = 52–55°C. Melting/Freezing point = –80°C. Vapor pressure = 25.6 kPa @ 20°C. Flash point = –20°C. Autoignition temperature = 440°C. Explosive limits: LEL = 2.8%; UEL = 16.0%. Hazard Identification (based on NFPA-704 M Rating System): Health 2, Flammability 3, Reactivity 0. Log $K_{ow}$ = 0.23. Unlikely to bioaccumulate in marine organisms.
*Incompatibilities:* Forms explosive mixture with air. Reacts violently with nitrates, strong oxidizers, strong alkalis, and strong acids. Decomposes slowly in water to form ethyl alcohol and formic acid. May accumulate static electrical charges, and may cause ignition of its vapors.
*Permissible Exposure Limits in Air:* The Federal limit [OSHA PEL][58], the DFG MAK[3], the ACGIH TWA[1] and the HSE[33] TWA is 100 ppm (300 mg/m$^3$). The STEL set by HSE[33] is 150 ppm (450 mg/m$^3$). The NIOSH IDLH level is 1500 ppm. Several states have set guidelines or standards for ethyl formate in ambient air[60] ranging from 3.0-4.5 mg/m$^3$ (North Dakota) to 5.0 mg/m$^3$ (Virginia) to 6.0 mg/m$^3$ (Connecticut) to 7.143 mg/m$^3$ (Nevada).

*Determination in Air:* Charcoal (tube) adsorption, workup with $CS_2$; analysis by gas chromatography/flame ionization detection; NIOSH IV, Method #1452.

*Permissible Concentration in Water:* No criteria set. Runoff from spills or fire control may cause water pollution.

*Routes of Entry:* Inhalation, ingestion, eye and/or skin contact.

*Harmful Effects and Symptoms*

*Short Term Exposure:* Either contact or the vapor can cause skin and eye irritation. Inhalation irritates the respiratory tract. Higher exposures can cause pulmonary edema, a medical emergency that can be delayed for several hours. This can cause death. Ethyl formate may affect the central nervous system. Exposure can cause headache, nausea, and vomiting.

*Long Term Exposure:* Prolonged or repeated contact can cause skin dryness and cracking. May affect the nervous system.

*Points of Attack:* Eyes, respiratory system, central nervous system.

*Medical Surveillance:* Consider the points of attack in preplacement and periodic physical examinations. Consider chest x-ray following acute overexposure. Nervous system tests.

*First Aid:* If this chemical gets into the eyes, remove any contact lenses at once and irrigate immediately for at least 15 minutes, occasionally lifting upper and lower lids. Seek medical attention immediately. If this chemical contacts the skin, remove contaminated clothing and wash immediately with soap and water. Seek medical attention immediately. If this chemical has been inhaled, remove from exposure, begin rescue breathing (using universal precautions) if breathing has stopped, and CPR if heart action has stopped. Transfer promptly to a medical facility. When this chemical has been swallowed, get medical attention. Give large quantities of water and induce vomiting. Do not make an unconscious person vomit. Do not induce vomiting when formulations containing petroleum solvents are ingested.

*References:*
- New Jersey Department of Health and Senior Services, "Hazardous Substance Fact Sheet, Ethyl Formate," Trenton NJ (March 1999). http://www.state.nj.us/health/eoh/rtkweb/0885.pdf
- California Environmental Protection Agency "Chemical List of Lists," Sacramento CA (February 1997).

# Ethyl Mercuric Chloride

*Use Type:* Fungicide and insecticide
*CAS Number:* 107-27-7
*Formula:* $C_2H_5ClHg$; $C_2H_5HgCl$
*Alert:* This chemical has been banned in most countries and is no longer permitted in the U.S.

*Synonyms:* Ceresan; EMC; Chloroethyl mercury; Ethylmercuric chloride; Ethylmercury chloride

*Trade Names:* CERESAN®; GRANOZAN®, banned; GRANOSAN®, banned; HEXASAN®

*Producers:* Alfa Aesar (USA)

*RTECS Number:* OV9800000

*Uses:* It is used as an organic fungicide for seed treatment.

*Regulatory Authority:*
- Air Pollutant Standard Set (ACGIH)[1] (former USSR)[43] (OSHA)[58]
- Clean Air Act: Hazardous Air Pollutants (Title I, Part A, Section 112)
- Clean Water Act: Section 307 Toxic Pollutants as mercury and compounds
- Safe Drinking Water Act: MCL, 0.002 mg/L; MCLG, 0.002 mg/L
- RCRA Section 261 Hazardous Constituents, waste number not listed, as mercury compounds, n.o.s.
- EPCRA Section 313: Form R *de minimus* concentration reporting level: 1.0%
- U.S. DOT Regulated Marine Pollutant (49CFR172.101, Appendix B), severe pollutant as mercury based pesticides
- Proposition 65 chemical (CAL)
- Water Pollution Standard Proposed (former USSR)[43]
- Canada, WHMIS, Ingredients Disclosure List

*Description:* Ethyl mercuric chloride is silvery white, forming leaflike crystals. Insoluble in water. Molecular weight = 265.12. Melting/Freezing point = 192°C.

*Incompatibilities:* Oxidizers. Do not use in water.

*Permissible Exposure Limits in Air:* The OSHA[2] PEL is 0.01 mg/m³ TWA, averaged over an 8-hour workshift and 0.04 mg/m³ as a ceiling, not to be exceeded at any time. The NIOSH[2] IDLH level = 2 mg/m³ (as Hg) The ACGIH[1] recommended TWA is the same as OSHA and a STEL of 0.04 mg/m³. The former USSR-UNEP/IRPTC project[43] has set an MAC in workplace air of 0.005 mg/m³ and an MAC for ambient air in residential areas of 0.0009 mg/m³ on a momentary basis and 0.0001 mg/m³ on an average daily basis.

*Determination in Air:* No test available.

*Permissible Concentration in Water:* An MAC in water bodies used for domestic purposes has been set by the former USSR-UNEP/IRPTC project[43] at 0.0001 mg/L.

*Routes of Entry:* Inhalation, skin absorption, ingestion, skin and/or eye contact.

*Harmful Effects and Symptoms*

*Short Term Exposure:* Ethyl mercuric chloride is an extremely toxic chemical that can cause permanent brain damage weeks after exposure with little or no warning during exposure. Severe poisoning can cause death. It enters the body through the lungs, skin and contaminated hands. Poisoning causes a "pins and needles" feeling, becoming clumsy and weak, hearing loss, abnormal walking, tremors, personality changes and other brain damage. Eye contact may cause severe irritation.

*Long Term Exposure:* Ethyl mercuric chloride should be handled as a teratogen-with extreme caution. It may cause mutations. Handle with extreme caution. Mercury accumulates in the body. May cause kidney damage. It can take months or years for the body to get rid of excess mercury.

*Points of Attack:* Eyes, skin. Central nervous system, peripheral nervous system, kidneys.

*Medical Surveillance:* Before first exposure and every 6 to 12 months after, a complete medical history and exam is strongly recommended, with: Exam of the nervous system including handwriting. Visual exam, including "visual field" exam. Hearing tests. Test for mercury in hair and blood. After suspected illness or overexposure, repeat these tests promptly, again in 4 to 6 weeks and then as recommended by your doctor. Kidney function tests.

*First Aid:* If this chemical gets into the eyes, remove any contact lenses at once and irrigate immediately for at least 15 minutes, occasionally lifting upper and lower lids. Seek medical attention immediately. If this chemical contacts the skin, remove contaminated clothing and wash immediately with soap and water. Seek medical attention immediately. If this chemical has been inhaled, remove from exposure, begin rescue breathing (using universal precautions) if breathing has stopped, and CPR if heart action has stopped. Transfer promptly to a medical facility. When this chemical has been swallowed, get medical attention. Give large quantities of water and induce vomiting. Do not make an unconscious person vomit.

*References:*
- New Jersey Department of Health and Senior Services, "Hazardous Substance Fact Sheet: Ethyl Mercuric Chloride," Trenton, NJ (January 11, 1988, rev. January 2000). http://www.state.nj.us/health/eoh/rtkweb/0895.pdf
- For a related compound which is not regulated but which was on the USEPA Extremely Hazardous Chemical List in 1985 (dropped in 1988), see the following: U.S. Environmental Protection Agency, "Chemical Profile: Ethylmercuric Phosphate," Washington, DC, Chemical Emergency Preparedness Program (October 31, 1985).
- California Environmental Protection Agency "Chemical List of Lists," Sacramento CA (February 1997).

# Ethylene

*Use Type:* Plant growth regulator
*CAS Number:* 74-85-1
*Formula:* $C_2H_4$
*Synonyms:* Acetene; Athylen (German); Bicarburretted hydrogen; Dicarburetted hydrogen; Elayl; Eteno (Spanish); Ethene; Etherin; heavy carburetted hydrogen; Olefiant gas
*Producers:* Air Liquide Group (France); Air Products & Chemicals (USA); ATOFINA (France); BASF (Germany); Bayer Chemicals (Germany); BOC Gases (UK); Borealis (Denmark); BP P.L.C. (UK); Chevron Phillips Chemical (USA); Cia. Petroquimica do Sul (COPESUL) (Brazil); CONDEA (Germany); Copene (Brazil); Donghae Gas Industrial Co. (South Korea); Dow Chemical (USA); DSM (Netherlands); DuPont (USA); EniChem (Italy); Equistar (USA); Exxon Mobil Chemical (USA); Formosa Plastics (Taiwan); Hoek Loos (Netherlands); Holox (USA); Huntsman (USA); Jilin Chemical (China); Lyondell (USA); Matheson Tri-Gas (USA); Messer Group (Germany); MG Industries (USA); Mitsubishi Chemical (Japan); Mitsui Chemicals (Japan); BASF Agricultural Products Group (Germany); Occidental (USA); Petromont (Canada); Petroquimica de Venezuela (Pequiven); Petroquimica Uniao S.A. (PQU) (Brazil); Polifin (South Africa); Praxair (USA); Qatar General Petroleum (Qatar); Reliance (India); Samsung General Chemicals (South Korea); Saudi Basic Industries Corp. (Saudi Arabia); Scott Specialty Gases (USA); Sinopec Corporation (China); Shell Chemical (USA); Showa Denko (Japan); Sunoco Chemicals (USA); Tosoh Corp. (Japan); Total (France); and most petroleum refineries

*Chemical Class:* Petroleum derivative
*EPA/OPP PC Code:* 041901
*California DPR Chemical Code:* 270
*ICSC Number:* 0475
*RTECS Number:* KU5340000
*EEC Number:* 601-010-00-3
*EINECS Number:* 200-815-1

*Uses:* Ethylene gas is used to accelerate ripening of bananas, tomatoes, tobacco, avocados and other fruits. Ethylene gas is a ripening agent which occurs naturally in nature and causes fruits to ripen and decay, vegetables and floral to wilt. Methods to control ethylene gas after picking extends the life cycle of the produce, extending their storage and shipping life. Ethylene gas is also used in ripening rooms to color up the fruit before it is moved to a regular cold storage room. It is used to manufacture ethylene oxide, polyethylene for plastics, alcohol, mustard gas and other organics. It is also used as an anesthetic and for oxyethylene welding and cutting of metals.

*U.S. Maximum Allowable Residue Levels for Ethylene (40 CFR 180.1016):*

Ethylene is exempted from the requirement of a tolerance for residues under the following conditions: (a) For all food commodities, it is used as a plant regulator on plants, seeds, or cuttings and on all commodities after harvest when applied in accordance with good agricultural practices. (b) Injected into the soil to cause premature germination of witchhweed in bean (lima and string), cabbage, cantaloupe, collard, corn, cotton, cucumber, eggplant, okra, onion, pasture grass, pea (field and sweet), peanut, pepper, potato, sweet potato, sorghum, soybean, squash, tomato, turnip, and watermelon fields as part of the U.S. Department of Agriculture witch weed control program.

*Carcinogen/Hazard Classifications*
**IARC:** Group 3, unclassifiable
**TRI Developmental Toxin:** Developmental toxin
*Regulatory Authority:*
- Air Pollutant Standard Set (former USSR)[43] (Virginia)[60]
- AB 2588-Air Toxics "Hot Spots" Chemicals (CAL)
- Permissible Exposure Limits for Chemical Contaminants (CAL/OSHA)
- Actively registered pesticide in California.
- Clean Air Act: Accidental Release Prevention/Flammable substances, (Section 112[r], Table 3), TQ = 10,000 lb (4540 kg)
- EPCRA Section 313 Form R *de minimus* concentration reporting level: 1.0%

*Description:* Colorless gas (at room temperature). Sweet odor. The minimum detectable by odor is 260 ppm. Soluble in water; solubility = 256 $cm^3$/L @ 0°C; 1:4 @ 0°C. Molecular weight = 28.06. Boiling point = –104°C. Melting/Freezing point = –169°C. Autoignition temperature = 450°C. Explosive limits: LEL = 2.7%; UEL = 36.0%[17]. Hazard Identification (based on NFPA-704 M Rating System): Health 1, Flammability 4, Reactivity 2. Log $K_{ow}$ = < 1.5. Unlikely to bioaccumulate in marine organisms.

*Incompatibilities:* A highly flammable gas at room temperature. Contact with oxidizers may cause explosive polymerization and fire. May be spontaneously explosive in sunlight or ultraviolet light when mixed with chlorine. Reacts violently with mixtures of carbon tetrachloride and benzoyl peroxide, bromotrichloromethane, aluminum chloride and ozone. Incompatible with acids, halogens, nitrogen oxides, hydrogen bromide, aluminum chloride, chlorine dioxide, nitrogen dioxide. May accumulate static electrical charges, and may cause ignition of its vapors.

*Permissible Exposure Limits in Air:* ACGIH[1] and HSE[33] classify ethylene as an asphyxiant and specify no TWA values. The former USSR-UNEP/IRPTC project cites an MAC for ambient air of residential areas of 3 mg/$m^3$ on either a momentary or a daily average basis. Virginia[60] has set a guideline for ethylene in ambient air of 3.0 μg/m3.

*Permissible Concentration in Water:* The former USSR-UNEP/IRPTC project[43] gives an MAC in water bodies used for domestic purposes of 0.5 mg/L.

*Routes of Entry:* Inhalation

*Harmful Effects and Symptoms*

**Short Term Exposure:** *Inhalation:* Increasingly severe exposures may cause faintness, incoordination, excitement, stupor, unconsciousness, convulsions, stopped breathing, paralysis, and heart, liver and kidney damage. 20-25% (200,000-250,000 ppm) gas has caused loss of sense of pain. 80-90% (800,000-900,000 ppm) has caused anesthesia. *Skin:* Contact with liquid can cause a "freezing burn." *Eyes:* Same as skin. *Ingestion:* No information available.

*Long Term Exposure:* Inhalation may cause loss of appetite and weight, irritability, insomnia, increase in red blood cell count, and inflammation of the kidneys.
*Points of Attack:* Kidneys
*Medical Surveillance:* Kidney function tests.
*First Aid:* If this chemical gets into the eyes, remove any contact lenses at once and irrigate immediately for at least 15 minutes, occasionally lifting upper and lower lids. Seek medical attention immediately. If this chemical contacts the skin, remove contaminated clothing and wash immediately with soap and water. Seek medical attention immediately. If this chemical has been inhaled, remove from exposure, begin rescue breathing (using universal precautions) if breathing has stopped, and CPR if heart action has stopped. Transfer promptly to a medical facility. When this chemical has been swallowed, get medical attention. Give large quantities of water and induce vomiting. Do not make an unconscious person vomit. If frostbite has occurred, seek medical attention immediately; do *NOT* rub the affected areas or flush them with water. In order to prevent further tissue damage, do *NOT* attempt to remove frozen clothing from frostbitten areas. If frostbite has *NOT* occurred, immediately and thoroughly wash contaminated skin with soap and water.

*References:*
- EPA, Office of Pesticide Programs, Pesticide Residue Limits, "Ethylene", 40 CFR 180.1016, http://www.epa.gov/pesticides/food/viewtols.htm
- Ethylene Control Products, Inc., Selma, CA.
- Sax, N.I., Ed., "Dangerous Properties of Industrial Materials Report," 4, No. 1, 79-81 (1984).
- New Jersey Department of Health and Senior Services, "Hazardous Substance Fact Sheet: Ethylene," Trenton, NJ (June 1996, rev. August 2002). http://www.state.nj.us/health/eoh/rtkweb/0873.pdf
- New York State Department of Health, "Chemical Fact Sheet: Ethylene," Albany, NY, Bureau of Toxic Substance Assessment (March 1986).
- California Environmental Protection Agency "Chemical List of Lists," Sacramento CA (February 1997).

# Ethylene Dibromide

*Use Type:* Fumigant and nematicide
*CAS Number:* 106-93-4
*Formula:* $C_2H_4Br_2$; $BrCH_2CH_2Br$
*Alert:* Persons whose clothing or skin is contaminated with liquid ethylene dibromide (above 50 °F) can secondarily contaminate others by direct contact or through offgassing vapor. Human toxicity (long-term): Very high.
*Synonyms:* Aethylenbromid (German); Bromuro di etile (Italian); Celmide; DBE; 1,2-Dibromaethan (German); 1,2-Dibromoetano (Italian, Spanish); Dibromoethane; α,β-Dibromoethane; *sym*-Dibromoethane; 1,2-Dibromoethane;

# Ethylene Dibromide

Dibromure d'ethylene (French); 1,2-Dibroomethaan (Dutch); Dibromuro de etileno (Spanish); Dwubromoetan (Polish); EDB; EDB-85; ENT 15,349; Ethane, 1,2-dibromo-; Ethylene bromide; 1,2-Ethylene dibromide; Glycol bromide; Glycol dibromide; NCI-C00522

*Trade Names:* AADIBROOM®; BROMOFUME®, Kerr-McGee Chemical (USA), canceled 11/10/1983; DOWFUME®, Dow Chemical (USA), canceled 11/10/1983; E-D-BEE®, Great Lakes Chemical (USA), canceled 2/01/1985; FUM-A-CIDE®, Amvac Chemical (USA), canceled 6/19/1984; FUMO-GAS®; ISCOBROME D®; KOPFUME®; NEFIS®; NEPHIS®; NIAGARA SOILFUME 85®, FMC Chemical (USA), canceled 11/10/1983; PESTMASTER®, Velsicol Chemical Corp. (USA), canceled 11/10/1983; SANHYUUM®; SOILBROM®, Great Lakes Chemical (USA), canceled 11/10/1983; SOILFUME®; TERR-O-CIDE®, Great Lakes Chemical (USA), canceled 11/10/1983; UNIFUME®

*Producers:* Albemarle (USA); Amvac Chemical (USA); ATOFINA (France); Dead Sea Bromine Group (Israel); Dow Chemical (USA); FMC Chemical (USA); Great Lakes Chemical (USA); Kerr-McGee Chemical (USA); Ocean Chemicals Group (UK); PPG Industries (USA); Sigma-Aldrich Laborchemikalien (Germany); Tosoh Corp. (Japan); United Phosphorus Ltd (India); Velsicol Chemical Corp. (USA)

*Chemical Class:* Halogenated organic compound
*EPA/OPP PC Code:* 042002
*California DPR Chemical Code:* 271
*ICSC Number:* 0045
*RTECS Number:* KH9275000
*EEC Number:* 602-010-00-6
*EINECS Number:* 203-444-5

*Uses:* Ethylene dibromide was used extensively as a pesticide and an ingredient of soil, vegetable, fruit, and grain fumigant formulations. However, these uses have almost disappeared in the United States. It is primarily used in fire extinguishers, gauge fluids, and waterproofing preparations; and it is used as a solvent for celluloid, fats, oils, and waxes. It is used to some extent as a chemical intermediate.

*Human toxicity (long-term)[77]:* Extra high–0.05 ppb, MCL (Maximum Contaminant Level)

*Fish toxicity (threshold)[77]:* Very low–4841.72356 ppb, MATC (Maximum Acceptable Toxicant Concentration)

*U.S. Maximum Allowable Residue Levels for Fumigants Including Ethylene Dibromide (40 CFR 180.521, 180.522, 193.225):*

| CROP | ppm |
|---|---|
| Grain, cereal, milled fractions | 125 |
| Corn, field, grits | 125 |
| Rice, cracked | 125 |

*Carcinogen/Hazard Classifications*
**U.S. EPA Carcinogens:** Group 2B, probable carcinogen
**U.S. NTP Carcinogen:** Reasonably anticipated carcinogen
**California Prop. 65:** Carcinogen and developmental toxin.
**IARC:** Group 2A, probable carcinogen
**Label Signal Word:** DANGER, POISON
**Endocrine Disruptor:** Suspected endocrine disruptor
*Regulatory Authority:*
- Carcinogen (Animal Positive) (IARC) (NCI) (NTP)[9] (ACGIH)[1] (DFG)[3] (Several Canadian Provinces)
- Banned or Severely Restricted (Many Countries) (UN)[13][35]
- Toxic Substance (World Bank)[15]
- Air Pollutant Standard Set (ACGIH)[1] (Other Countries)[35] (OSHA)[58] (Several States)[60]
- CAL/OSHA Carcinogen User Register Chemical
- AB 1803-Well Monitoring Chemical (CAL)
- MCL (Maximum Contaminants Levels) list of contaminants (CAL)
- AB 2588-Air Toxics "Hot Spots" Chemicals (CAL)
- CAL Air Resources Board/AB 1807 Toxic Air Contaminants
- Proposition 65 chemical (CAL)
- U.S. DOT Inhalation Hazard Chemicals
- Permissible Exposure Limits for Chemical Contaminants (CAL/OSHA)
- The "Director's List" (CAL/OSHA)
- Clean Air Act: Hazardous Air Pollutants (Title I, Part A, Section 112); Accidental Release Prevention/Flammable substances, (Section 112[r], Table 3), TQ = 20,000 lb (9080kg)
- Clean Water Act: Section 311 Hazardous Substances/RQ 40CFR117.3 (same as CERCLA, see below); Section 313 Water Priority Chemicals (57FR41331, 9/9/92)
- EPA Hazardous Waste Number (RCRA No.): U067
- RCRA, 40CFR261, Appendix 8 Hazardous Constituents
- RCRA 40CFR268.48; 61FR15654, Universal Treatment Standards: Wastewater (mg/L), 0.028; Nonwastewater (mg/kg), 15
- RCRA 40CFR264, Appendix 9; TSD Facilities Ground Water Monitoring List. Suggested test method(s) (PQL $ug$/L): 8010(10); 8240(5)
- Safe Drinking Water Act: MCL, 0.00005 mg/L; MCGL, zero
- Superfund/EPCRA 40CFR355, Appendix B Extremely Hazardous Substances: TPQ = 10,000 lb (4,540 kg)
- Superfund/EPCRA 40CFR302.4 RQ: CERCLA, 1 lb (0.454 kg)
- EPCRA Section 313 Form R *de minimus* concentration reporting level: 0.1%
- Canada, WHMIS, Ingredients Disclosure List
- IMO Marine Pollutant

*Description:* Ethylene dibromide is a colorless liquid or solid (below 10°C). Sweet, chloroform-like odor. *Warning properties*: Inadequate for exposure to vapors. The minimum concentration detectable by odor is 10 ppm. Soluble in water; solubility = 0.43% @ 30°C. Molecular weight = 187.9. Molecular weight = 187.88. Boiling point

= 131°C @ 760 mmHg. Melting/Freezing point = 10°C. Vapor pressure = 11 mmHg @ 25 °C. Liquid specific gravity = 2.172 @ 25 °C. Gas density (air: 1) = 6.48. Nonflammable. Log $K_{ow}$ = 1.94. Unlikely to bioaccumulate in marine organisms.

*Incompatibilities:* Reacts vigorously with chemically active metals, alkali metals, liquid ammonia, strong bases, strong oxidizers, causing fire and explosion hazard. Light, heat, and moisture can cause slow decomposition, forming hydrogen bromide. Attacks fats, rubber, some plastics and coatings.

*Permissible Exposure Limits in Air:* ACGIH[1], DFG[3] and Sweden have set no numerical limits, stating only that the substance is a carcinogen and that exposure should be avoided. OSHA[2] has set a TWA of 20 ppm, a ceiling of 30 ppm and a maximum peak above the ceiling of 50 ppm for 5 minutes. NIOSH[2] IDLH (immediately dangerous to life or health) = 100 ppm. Argentina[35] has set a TWA of 20 ppm (140 mg/m$^3$) and an STEL of 30 ppm (220 mg/m$^3$). Czechoslovakia[35] has set a TWA of 10 mg/m$^3$ with a ceiling value of 20 mg/m$^3$. Several states have set guidelines or standards for ethylene dibromide in ambient air[60] ranging from zero (North Dakota and New York) to 0.045 µg/m$^3$ (North Carolina) to 2.47 µg/m$^3$ (Pennsylvania) to 720 µg/m$^3$ (Indiana) to 770 µg/m$^3$ (South Carolina) to 1500 µg/m$^3$ (Virginia) to 1550 µg/m$^3$ (Connecticut).

*Determination in Air:* Charcoal tube absorption; workup with Benzene/Methanol; Gas chromatography/ Electrochemical detection; NIOSH IV, Method #1008[18].

*Permissible Concentration in Water:* Several states have set guidelines or standards for ethylene dibromide in drinking water[61] ranging from 0.005 µg/L (Kansas) to 0.008 µg/L (Minnesota) to 0.01 µg/L (Arizona) to 0.02 µg/L (California and Washington) to 0.04 µg/L (Massachusetts) to 0.10 µg/L (Connecticut and New Mexico) to 0.50 µg/L (Wisconsin) to 1.0 µg/L (Maine). EPA[62] has proposed a maximum concentration level in drinking water of 0.05 µg/L.

*Routes of Entry:* Absorption can occur by the inhalation, oral, and dermal routes. It is toxic by these three routes of exposure. Toxicity is thought to be due to metabolic products of ethylene dibromide.

*Harmful Effects and Symptoms*

*Short Term Exposure:* Contact can cause severe skin and eye burns, with permanent eye damage. Exposure to the vapor may also damage the eyes. Inhalation may irritate and damage the lungs. Higher exposures can cause pulmonary edema, a medical emergency that can be delayed for several hours. This can cause death. High exposure can cause dizziness, drowsiness, vomiting, unconsciousness, and death. High exposure can damage the liver or kidneys enough to cause death. The systemic effects of ethylene dibromide are in part due to metabolic conversion to the cell toxicant 2-bromoacetaldehyde. Persons with pre-existing skin disorders or eye problems, or impaired liver, kidney, or respiratory tract function may be more susceptible to the effects of ethylene dibromide. *Inhalation:* Levels of 75 ppm may cause irritation of the nose, throat and lungs. 100-200 ppm for 1 hour may cause diarrhea, abdominal pain and vomiting. Other symptoms may include headache, loss of appetite, swollen glands, pale skin coloring, insomnia, dizziness and depression. Accidental high exposure has caused symptoms as listed above, internal bleeding and death. *Skin:* Contact with as little as 1 gram (1/28 ounce) may cause itching, swelling, redness, burning and blistering. May be absorbed through the skin and cause symptoms as listed under inhalation. *Eyes:* May cause irritation of eyes and eyelids. *Ingestion:* May cause vomiting, diarrhea, abdominal pain, nausea and damage to the liver and kidneys. As little as 4.5 ml (about 1 teaspoon) has caused death.

*Long Term Exposure:* No reliable reports exist of adverse health effects in humans exposed chronically to ethylene dibromide. Chronic exposure may be more serious for children because of their potential for a longer latency period. *Carcinogenicity:* The Department of Health and Human Services has determined that ethylene dibromide can reasonably be anticipated to be a human carcinogen, based on ethylene dibromide-induced tumors in multiple sites and by various routes of exposure in animals. Results from epidemiological studies have been inconclusive. *Reproductive and Developmental Effects:* There is inconclusive but suggestive evidence that ethylene dibromide may reduce fertility in men. Antispermatogenic effects have been demonstrated in various animal species. Ethylene dibromide is included in *Reproductive and Developmental Toxicants,* a 1991 report published by the U.S. General Accounting Office. Special consideration regarding the exposure of pregnant women is warranted, since ethylene dibromide has been shown to be a genotoxin; thus, medical counseling is recommended for pregnant women.

*Points of Attack:* Eyes, skin, respiratory system, liver, kidneys, reproductive system. Cancer Site in animals: skin and lung tumors.

*Medical Surveillance:* Preemployment and periodic examinations should evaluate the skin and eyes, respiratory tract, and liver and kidney functions. Medical observation is recommended for 24 to 48 hours after breathing overexposure, as pulmonary edema may be delayed. Serum bromide levels can be used to document that exposure did occur. However, bromide levels do not accurately predict the clinical course. Routine laboratory studies include CBC, glucose, and electrolyte determinations. Additional studies for patients exposed to ethylene dibromide include liver function tests and renal function tests. In cases of inhalation exposure, chest radiography and arterial blood gas measurements may be helpful.

*First Aid:* Persons whose clothing or skin is contaminated with liquid ethylene dibromide (above 50°F) can secondarily contaminate others by direct contact or through offgassing vapor. If this chemical gets into the eyes, remove any contact lenses at once and irrigate immediately for at least 15 minutes, occasionally lifting upper and lower lids. Seek medical attention immediately. If this chemical contacts the skin, remove contaminated clothing and wash immediately with soap and water. Seek medical attention immediately. If this chemical has been inhaled, remove from exposure, begin rescue breathing (using universal precautions) if breathing has stopped, and CPR if heart action has stopped. Transfer promptly to a medical facility. If ingestion of liquid ethylene dibromide occurs, **do not induce vomiting**. If the victim is alert and able to swallow, administer a slurry of activated charcoal at a dose of 1 g/kg (infant, child, and adult dose). A soda can and straw may be of assistance when offering charcoal to a child. As first aid for pulmonary edema, a doctor or authorized paramedic may consider administering a corticosteroid spray.

*Note to physician or authorized medical personnel:* In cases of respiratory compromise secure airway and respiration via endotracheal intubation. If not possible, perform cricothyrotomy if equipped and trained to do so. Treat patients who have bronchospasm with an aerosolized bronchodilator such as albuterol. Consider racemic epinephrine aerosol for children who develop stridor. Dose 0.25–0.75 mL of 2.25% racemic epinephrine solution, repeat every 20 minutes as needed, cautioning for myocardial variability. Patients who are comatose, hypotensive, or are having seizures or cardiac arrhythmias should be treated according to advanced life support protocols. If evidence of shock or hypotension is observed, begin fluid administration. For adults with systolic pressure less than 80 mm Hg, bolus perfusion of 1000 mL/hour intravenous saline or lactated Ringer's solution may be appropriate. Higher adult systolic pressures may necessitate lower perfusion rates. For children with compromised perfusion administer a 20 mL/kg bolus of normal saline over 10 to 20 minutes, then infuse at 2 to 3 mL/kg/hour.

*References:*
- EPA, Office of Pesticide Programs, Pesticide Residue Limits, "Fumigants Including Ethylene Dibromide", 40 CFR 180.521, 40CFR 180.522, 40CFR 193.225, http://www.epa.gov/pesticides/food/viewtols.htm
- Environmental Protection Agency, Sampling and Analysis of Selected Toxic Substances, Task II-Ethylene Dibromide, Final Report. Office of Toxic Substances, EPA, Washington, DC (September 1975).
- Occupational Health and Safety Administration, Criteria for a Recommended Standard: Occupational Exposure to ethylene Dibromide, NIOSH Doc. No. 77-221 (1977).
- National Institute for Occupational Safety and Health, Current Intelligence Bulletin No. 3: Ethylene Dibromide, Rockville, MD (July 7, 1975), and Current Intelligence Bulletin No. 37, Ethylene Dibromide, Cincinnati, Ohio (October 26, 1981).
- Agency for Toxic Substances and Disease Registry, U.S. Department of Health and Human Services, Public Health Service, "Managing Hazardous Materials Incidents," Atlanta, GA (June 2003).
- Sax, N.I., Ed., "Dangerous Properties of Industrial Materials Report," 1, No. 5, 58-60 (1981).
- New York State Department of Health, "Chemical Fact Sheet: Ethylene Dibromide," Albany, NY, Bureau of Toxic Substance Assessment (March 1986, Version 2).
- California Environmental Protection Agency "Chemical List of Lists," Sacramento CA (February 1997).
- New Jersey Department of Health and Senior Services, "Hazardous Substance Fact Sheet, Ethylene Dibromide," Trenton, NJ (October 1995, rev. August 2001). http://www.state.nj.us/health/eoh/rtkweb/0877.pdf

## Ethylene Dichloride

*Use Type:* Fumigant and insecticide
*CAS Number:* 107-06-2
*Formula:* $C_2H_4Cl_2$; $ClCH_2CH_2Cl$
*Synonyms:* Aethylenchlorid (German); 1,2-Bichloroethane; Bichlorure d'ethylene (French); Chlorure d'ethylene (French); Cloruro di ethene (Italian); 1,2-Dichloorethaan (Dutch); 1,2-Dichlor-aethan (German); Dichloremulsion; Di-chlor-mulsion; Dichloro-1,2-ethane (French); $\alpha,\beta$-Dichloroethane; *sym*-Dichloroethane; 1,2-Dichloroethane; Dichloroethylene; 1,2-Dicloroetano (Italian, Spanish); EDC; ENT 1,656; Ethane, 1,2-dichloro-; Ethane dichloride; Ethyleendichloride (Dutch); Ethylene chloride; 1,2-Ethylene dichloride; Freon 150; Glycol dichloride; NCI-C00511
*Trade Names:* BORER SOL®; BROCIDE®; DESTRUXOL BORER-SOL®; DOWFUME®, Dow Chemical (USA), canceled 11/22/1985; DUTCH LIQUID®; DUTCH OIL®
*Producers:* Albright & Wilson Pty. (UK); Atofina (France); BF Goodrich Performance Materials (USA); Borden Chemical (USA); Carbocloro (Brazil); Central Glass (Japan); Chlor-Chemicals (UK); Dead Sea Bromine Group (Israel); Dow Chemical (USA); Degussa (Germany); EniChem (Italy); Formosa Plastics (Formosa); Georgia Gulf (USA); GFS Chemicals (USA); Merck (Germany); OxyChem (USA); Petroquimica de Venezuela (Pequiven); PPG Industries (USA); Reliance (India); Saudi Basic Industries Corp. (Saudi Arabia); Sigma-Aldrich Laborchemikalien (Germany); Tokuyama Group (Japan); Tosoh Corp. (Japan); Union Carbide (USA); Vista Chemical (USA); Vulcan Chemicals (USA)
*Chemical Class:* Chlorinated hydrocarbon
*EPA/OPP PC Code:* 042003
*California DPR Chemical Code:* 274
*ICSC Number:* 0250
*RTECS Number:* KI0525000
*EEC Number:* 602-012-00-7
*EINECS Number:* 203-458-1

*Uses:* When mixed with carbon tetrachloride, ethylene dichloride is used as a grain fumigant for bulk storage in bags, sealed containers, bins or on floors. In recent years, 1,2-dichloroethane has found wide use in the manufacture to ethylene glycol, diaminoethylene, polyvinyl chloride, nylon, viscose rayon, styrenebutadiene rubber, and various plastics. It is a solvent for resins, asphalt, bitumen, rubber, cellulose acetate, cellulose ester, and paint; a degreaser in the engineering, textile and petroleum industries; and an extracting agent for soybean oil and caffeine. It is also used as an antiknock agent in gasoline, a pickling agent and a dry-cleaning agent. It has found use in photography, xerography, water softening, and also in the production of adhesives, cosmetics, pharmaceuticals, and varnishes.

**U.S. Maximum Allowable Residue Levels for Fumigants, including ethylene dichloride (40 CFR 193.225, 180.521, 180.522):**

| CROP | ppm |
|---|---|
| Grain, cereal, milled fractions | 125 |
| Corn, field, grits | 125 |
| Rice, cracked | 125 |

*Carcinogen/Hazard Classifications*
**U.S. EPA Carcinogens:** Group B2, probable carcinogen
**U.S. NTP Carcinogen:** Reasonably anticipated carcinogen
**California Prop. 65:** Carcinogen
**IARC:** Group 2B, possible carcinogen

*Regulatory Authority:*
- Carcinogen (Animal Positive) (IARC) (NCI)[9] (ACGIH)[1]
- Banned or Severely Restricted (E. Germany and Saudi Arabia) (UN)[13]
- Air Pollutant Standard Set (ACGIH)[1] (DFG)[3] (HSE)[33] (OSHA)[58] (Other Countries)[35] (Several States)[60]
- List of priority pollutants (U.S. EPA)
- AB 1803-Well Monitoring Chemical (CAL)
- MCL (Maximum Contaminants Levels) list of contaminants (CAL)
- AB 2588-Air Toxics "Hot Spots" Chemicals (CAL)
- CAL Air Resources Board/AB 1807 Toxic Air Contaminants
- Proposition 65 chemical (CAL)
- U.S. DOT Inhalation Hazard Chemicals
- Permissible Exposure Limits for Chemical Contaminants (CAL/OSHA)
- The "Director's List" (CAL/OSHA)
- Clean Air Act: Hazardous Air Pollutants (Title I, Part A, Section 112)
- Clean Water Act: Section 311 Hazardous Substances/RQ 40CFR117.3 (same as CERCLA, see below); 40CFR423, Appendix A, Priority Pollutants; Section 313 Water Priority Chemicals (57FR41331, 9/9/92); Toxic Pollutant (Section 401.15)
- EPA Hazardous Waste Number (RCRA No.): U077, D028
- RCRA, 40CFR261, Appendix 8 Hazardous Constituents
- RCRA Toxicity Characteristic (Section 261.24), Maximum Concentration of Contaminants, regulatory level, 0.5 mg/L
- RCRA 40CFR268.48; 61FR15654, Universal Treatment Standards: Wastewater (mg/L), 0.21; Nonwastewater (mg/kg), 6.0
- RCRA 40CFR264, Appendix 9; TSD Facilities Ground Water Monitoring List. Suggested test method(s) (PQL $ug$/L): 8010(0.5); 8240(5)
- Safe Drinking Water Act: MCL, 0.005 mg/L; MCLG, zero; Regulated chemical (47 FR 9352)
- Superfund/EPCRA 40CFR302.4 RQ: CERCLA, 100 lb (45.4 kg)
- U.S. DOT Regulated Marine Pollutant (49CFR172.101, Appendix B)
- Canada, WHMIS, Ingredients Disclosure List

*Description:* 1,2-Dichloroethane is a colorless liquid. Pleasant, chloroform-like odor. Sweetish taste. Decomposes slowly: turns dark and acidic on contact with air, moisture, and light. The odor threshold is 100 ppm. Boiling point = 84°C. Melting point = −35.4°C. Flash point = 13°C. Autoignition temperature = 413°C. Explosive limits: LEL = 6.2%; UEL = 16.0%. Hazard Identification (based on NFPA-704 M Rating System): Health 2, Flammability 3, Reactivity 0. Insoluble in water; solubility = 8.7 x $10^3$ ppm @ 20°C.

*Incompatibilities:* Forms explosive mixture with air. Reacts violently with strong oxidizers and caustics; chemically-active metals such as magnesium or aluminum powder, sodium and potassium, alkali metals, alkali amides; liquid ammonia Decomposes to vinyl chloride and HCl above 1112°F. Attacks plastics, rubber, coatings. Attacks many metals in presence of water.

*Permissible Exposure Limits in Air:* The OSHA[2] PEL is 50 ppm TWA and ceiling of 100 ppm, 200 ppm 5-minute maximum peak in any 3 hours. ACGIH[1] recommends a limit of 10 ppm (40 mg/m$^3$). The NIOSH[2] IDLH level = potential occupational carcinogen [50 ppm]. The former USSR-UNEP/IRPTC project[43] has set MAC values in ambient air in residential areas of 3 mg/m$^3$ on a momentary basis and 1 mg/m$^3$ on a daily average basis. Several states have set guidelines or standards for ethylene dichloride in ambient air[60] ranging from 0-400 $\mu$g/m$^3$ (North Dakota) to 0.038 $\mu$g/m$^3$ (North Carolina) to 0.04 $\mu$g/m$^3$ (Rhode Island) to 0.2 $\mu$g/m$^3$ (New York) to 0.39 $\mu$g/m$^3$ (Massachusetts) to 20.0 $\mu$g/m$^3$ (Connecticut) to 148.0 $\mu$g/m$^3$ (Pennsylvania) to 200.0 $\mu$g/m$^3$ (South Carolina) to 650.0 $\mu$g/m$^3$ (Nevada) to 1000.0 $\mu$g/m$^3$ (Indiana).

*Determination in Air:* Charcoal adsorption, workup with $CS_2$, analysis by gas chromatography/flame ionization. See NIOSH IV, Method #1003 for halogenated hydrocarbons[18].

*Permissible Concentration in Water:* To protect freshwater aquatic life: 118,000 $\mu$g/L on an acute toxicity basis and 20,000 $\mu$g/L on a chronic basis. To protect saltwater aquatic life: 113,000 $\mu$g/L on an acute toxicity basis. To protect human health: preferably zero. An additional lifetime cancer risk of 1 in 100,000 occurs at a concentration of 9.4 µg/L[6]. The WHO has set a limit for drinking water of 0.01 mg/L (10 µg/L)[35]. The former USSR has set a limit in water bodies used for domestic purposes variously quoted at 2.0 mg/L[43] and 0.02 mg/L[35]. EPA[48] has set a longer term health advisory of 2.6 mg/L for an adult. Several states have set guidelines or standards for ethylene dichloride in drinking water[61]. The standards set range from 2.0 µg/L (New Jersey) to 3.0 µg/L (Florida) to 10.0 µg/L (New Mexico). The guidelines range from 0.38 µg/L (New Hampshire) to 1.0 µg/L (California and Connecticut) to 3.8 (Minnesota) to 5.0 µg/L (Maine).

*Determination in Water:* Inert gas purge followed by chromatography with halide specific detection (EPA Method 601) or gas chromatography plus mass spectrometry (EPA Method 624).

*Routes of Entry:* Inhalation of vapor, skin absorption of liquid, ingestion, skin and/or eye contact.

*Harmful Effects and Symptoms*

*Short Term Exposure:* Irritates the eyes, skin, and respiratory tract. Inhalation of the vapors can cause pulmonary edema, a medical emergency that can be delayed for several hours. This can cause death. Exposure can cause nausea, vomiting, headaches, drowsisness and loss of consciousness. Overexposure to ethylene dichloride may damage the central nervous system, kidneys, liver. *Inhalation:* Levels of 10 to 30 ppm may cause dizziness, nausea, and vomiting. Levels up to 50 pm may cause weakness, trembling, headaches, abdominal cramps, liver and kidney damage, and fluid build up in lungs. May cause coma and death at high levels. *Skin:* Contact may cause irritation and skin rash, and irritates the eyes. *Eyes:* May cause redness, pain, and blurred vision. Vapor can damage the cornea. *Ingestion:* Ingestion of 2 ounces has resulted in nausea, vomiting, faintness, drowsiness, difficulty breathing, pale skin, internal bleeding, kidney damage, and death due to respiratory failure. Other possible symptoms may include abdominal spasms, severe headache, lethargy, lowered blood pressure, diarrhea, shock, physical collapse, and coma.

*Long Term Exposure:* Repeated or prolonged contact can chronically irritate the skin causing dryness, redness and a rash. Prolonged or repeated exposure may cause eye, nose and throat irritation, nerve damage, liver and kidney damage. This substance has been determined to cause cancer of the lung, stomach, breast and other sites in laboratory animals, and may be a human carcinogen. Can irritate the lungs and bronchitis may develop. Repeated or prolonged exposure can cause loss of appetite, nausea and vomiting, trembling and low blood sugar.

*Points of Attack:* Eyes, skin, kidneys, liver, respiratory system, central nervous system, cardiovascular system. Cancer Site in animals: forestomach, mammary gland and circulatory system cancer.

*Medical Surveillance:* Before beginning employment and at regular times after that, the following are recommended: Lung function tests. Liver and kidney function tests. If symptoms develop or overexposure is suspected, the following may be useful: Consider chest x-ray after acute overexposure.

*First Aid:* If this chemical gets into the eyes, remove any contact lenses at once and irrigate immediately for at least 15 minutes, occasionally lifting upper and lower lids. Seek medical attention immediately. If this chemical contacts the skin, remove contaminated clothing and wash immediately with soap and water. Seek medical attention immediately. If this chemical has been inhaled, remove from exposure, begin rescue breathing (using universal precautions) if breathing has stopped, and CPR if heart action has stopped. Transfer promptly to a medical facility. When this chemical has been swallowed, get medical attention. Give large quantities of water and induce vomiting. Do not make an unconscious person vomit. Medical observation is recommended for 24 to 48 hours after breathing overexposure, as pulmonary edema may be delayed. As first aid for pulmonary edema, a doctor or authorized paramedic may consider administering a corticosteroid spray.

*References:*

- U.S. Environmental Protection Agency, Office of Pesticide Programs, Pesticide Residue Limits, "Fumigants, including Ethylene Dichloride", 40 CFR 193.225, 180.521, 180.522. http://www.epa.gov/pesticides/food/viewtols.htm
- New Jersey Department of Health and Senior Services, "Hazardous Substance Fact Sheet: 1,2-Dichloroethane," Trenton, NJ (July 1986, rev. February 2001). http://www.state.nj.us/health/eoh/rtkweb/0652.pdf
- National Institute for Occupational Safety and Health, Criteria for a Recommended Standard: Occupational Exposure to Ethylene Dichloride, NIOSH Doc. No. 76-139 (1976).
- National Institute for Occupational Safety and Health, Ethylene Dichloride, NIOSH Current Intelligence Bulletin No. 25, Washington, DC (April 19, 1978).
- National Institute for Occupational Safety and Health, Chloroethanes: Review of Toxicity, Current Intelligence Bulletin No. 27, Washington, DC (August 21, 1978).
- U.S. Environmental Protection Agency, "Chemical Hazard Information Profile: 1,2-Dichloroethane," Washington, DC (September 1, 1977).
- U.S. Environmental Protection Agency, Chlorinated Ethanes: Ambient Water Quality Criteria, Washington, DC (1980).
- U.S. Environmental Protection Agency, 1,2-Dichloroethane, Health and Environmental Effects Profile No. 70, Washington, DC, Office of Solid Waste (April 31, 1980).
- Sax, N.I., Ed., "Dangerous Properties of Industrial Materials Report," 1, No. 4, 50-52 (1981).

- U.S. Environmental Protection Agency, "Health Advisory: 1,2-Dichloroethane," Washington, DC, Office of Drinking Water (March 31, 1987).
- U.S. Public Health Service, "Toxicological Profile for 1,2-Dichloroethane," Atlanta, Georgia, Agency for Toxic Substance & Disease Registry (December 1988).
- California Environmental Protection Agency "Chemical List of Lists," Sacramento CA (February 1997).
- New York State Department of Health, "Chemical Fact Sheet: 1,2-Dichloroethane," Albany, NY, Bureau of Toxic Substance Assessment (Version 2).

## Ethylene Fluorohydrin

*Use Type:* Rodenticide, insecticide, and acaricide
*CAS Number:* 371-62-0
*Formula:* $C_2H_5FO$
*Alert:* Not registered as a pesticide in the U.S. It is an acutely toxic chemical.
*Synonyms:* 2-Fluoetanol (Spanish); 2-Fluroethanol; β-Fluoroethanol; TL 741
*RTECS Number:* KL1575000
*Uses:* Ethylene fluorohydrin is not registered as a pesticide in the U.S.
*Regulatory Authority:*
- OSHA 29CFR1910.119, Appendix A, Process Safety List of Highly Hazardous Chemicals, TQ = 100 lb (45 kg)
- Superfund/EPCRA 40CFR355, Appendix B Extremely Hazardous Substances: TPQ = 10 lb (4.54 kg)
- Superfund/EPCRA 40CFR302.4 RQ: EHS, 1 lb (0.454 kg)

*Description:* Ethylene fluorohydrin is a colorless liquid. Soluble in water. Molecular weight = 64.05. Boiling point = 103.5°C. Melting/Freezing point = –26.45°C. Flash point = 31°C.
*Incompatibilities:* Strong oxidizers.
*Permissible Exposure Limits in Air:* No standards set.
*Permissible Concentration in Water:* No criteria set, but runoff from spills or fire control may cause water pollution.
*Routes of Entry:* Eye or skin contact.
*Harmful Effects and Symptoms*
*Short Term Exposure:* Symptoms include tremors, severe muscular weakness, nausea, headache, and slight swelling of the liver. Delayed convulsant. Toxicity rating is the same as for fluoroacetate, super toxic. The probable oral lethal dose in humans is a taste (less than 7 drops) for a 70 kg (150 lb) person. The chemical is highly toxic when inhaled or absorbed through the skin. Toxicity depends on its oxidation to fluoroacetate by tissue alcohol dehydrogenase.
*Points of Attack:* Heart (irregular heart beat), liver, kidneys, central nervous system ('pins and needles,' weakness in hands and feet).
*Medical Surveillance:* If symptoms develop or overexposure is suspected, the following are recommended: liver and kidney function tests; exam of the nervous system, and EKG. Any evaluation should include a careful history of past and present symptoms with an exam. Medical tests that look for damage already done are not a substitute for controlling exposure.
*First Aid:* Acute poisoning should be treated like poisoning by fluoroacetate. Ethylene fluorohydrin (2-fluoroethanol) is listed among the organic fluorine derivatives of fluoroacetic acid. The emergency procedures for fluoroacetic acid are: move victim to fresh air; call emergency medical care. If not breathing, give artificial respiration. If breathing is difficult, give oxygen. In case of contact with material, immediately flush skin or eyes with running water for at least 15 minutes. Remove and isolate contaminated clothing and shoes at the site. Keep victim quiet and maintain normal body temperature. Effects may be delayed; keep victim under observation.
*References:*
- New Jersey Department of Health and Senior Services, "Hazardouos Substance Fact Sheet, Ethylene Fluorohydrin," Trenton, NJ (September 2000). http://www.state.nj.us/health/eoh/rtkweb/2400.pdf
- U.S. Environmental Protection Agency, "Chemical Profile: Ethylene Fluorohydrin," Washington, DC, Chemical Emergency Preparedness Program (November 30, 1987).
- California Environmental Protection Agency "Chemical List of Lists," Sacramento CA (February 1997).

## Ethylene Oxide

*Use Type:* Fungicide and fumigant
*CAS Number:* 75-21-8
*Formula:* $C_2H_4O$
*Alert:* Although ethylene oxide is not currently registered for use specifically against anthrax spores, EPA has determined that emergency conditions exist which necessitate its limited sale, use, and distribution for this purpose. This exemption was granted after considering available data and includes requirements which ensure protection of health and the environment. Therefore, based upon sampling results and a review of cleanup options, EPA has issued the following two crisis exemptions: EPA has allowed ethylene oxide to be used in fumigating items retrieved from congressional offices that were potentially contaminated with anthrax. EPA also allowed ethylene oxide to be used by the U.S. Department of Justice (DOJ) in order for the department to test the fumigation process for mail received by DOJ that may be potentially contaminated with anthrax.

Persons whose clothing or skin is contaminated with ethylene oxide liquid or solution can secondarily contaminate personnel by direct contact or through off-gassing vapor.

## Ethylene Oxide

*Synonyms:* Aethylenoxid (German); Dihydrooxirene; Dimethylene oxide; ENT 26,263; E.O; 1,2-Epoxyaethan (German); Epoxyethane; 1,2-Epoxyethane; Ethene oxide; Ethyleenoxide (Dutch); Ethylene (oxyde d') (French); Etilene(ossido di) (Italian); ETO; Etylenu Tlenek (Polish); NCI-C50088; Oxacyclopropane; Oxane; A,β-Oxidoethane; Oxiraan (Dutch); Oxirane; Oxirene, Dihydro-; UN 1040

*Trade Names:* AMPROLENE®; ANPROLENE®; ANPROLINE®; BIODAC®, CONDEA (Germany); MERPOL®; OXYFUME®; OXYFUME 12®; T-GAS®; STERILIZING GAS ETHYLENE OXIDE 100%®

*Producers:* Air Products & Chemicals (USA); Albright & Wilson Pty. (UK); ARC Specialty Products (USA); BASF (Germany); Bayer Chemicals (Germany); BOC Gases (UK); BP Chemicals (UK); Celanese (Germany); CONDEA (Germany); Dow Chemical (USA); Degussa (Germany); Eastman Chemical (USA); EniChem (Italy); Equistar (USA); Formosa Plastics (Taiwan); Huntsman (USA); Messer Group (Germany); MG Industries (USA); Mitsubishi Chemical (Japan); Mitsui Chemicals (Japan); Nippon Shokubai (Japan); Olin (USA); OxyChem (USA); Petroquimica de Venezuela (Pequiven); Praxair (USA); Reliance (India); Rhodia (France); Samsung General Chemicals (South Korea); Sasol (South Africa); Scott Specialty Gases (USA); Shell Chemicals (USA); Sunoco Chemicals (USA); Union Carbide (USA)

*Chemical Class:* Alcohol/ether
*EPA/OPP PC Code:* 042301
*California DPR Chemical Code:* 277
*ICSC Number:* 0155
*RTECS Number:* KX2450000
*EEC Number:* 603-023-00-X
*EINECS Number:* 200-849-9

*Uses:* Ethylene oxide is used as a fumigant for spices, seasonings, and foodstuffs and as an agricultural fungicide. When used directly in the gaseous form or in nonexplosive gaseous mixtures with nitrogen or carbon dioxide, ethylene oxide can act as a disinfectant, fumigant, sterilizing agent, and insecticide. It is a man-made chemical used as an intermediate in organic synthesis for ethylene glycol, polyglycols, glycol ethers, esters, ethanolamines, acrylonitrile, plastics, and surface-active agents. It is also used as a fumigant for textiles and for sterilization, especially for surgical instruments. It is used in drug synthesis and as a pesticide intermediate.

*U.S. Maximum Allowable Residue Levels for Ethylene Oxide (40 CFR 180.151, 185.2850):*

| CROP | ppm |
| --- | --- |
| Walnut, black, post-h | 50 |
| Coconut, copra, post-h | 50 |
| Spice, post-h | 50 |
| Seasonings, processed natural, including spices, except salt mixture | 50 |
| Spices, processed | 50 |

*Carcinogen/Hazard Classifications*
**U.S. EPA Carcinogens:** Group B1, probable carcinogen
**U.S. NTP Carcinogen:** Known carcinogen
**California Prop. 65:** Carcinogen and female developmental toxin
**IARC:** Group 1, known carcinogen
**Label Signal Word:** HIGHLY TOXIC

*Regulatory Authority:*
- Carcinogen (Animal Positive) (IARC)[9] (DFG)[3] (ACGIH)[1]
- Banned or Severely Restricted ((In Agriculture) (Germany)[13]
- Highly Reactive Substance and Explosive (World Bank)[15]
- OSHA, 29CFR1910 Specifically Regulated Chemicals (See CFR 1910.1047)
- Air Pollutant Standard Set (ACGIH)[1] (HSE)[33] (OSHA)[63] (Other Countries)[35] (Several States)[60]
- CAL/OSHA Carcinogen User Register Chemical
- EPA/SARA 302 (EPCRA) Extremely hazardous substances
- AB 2588-Air Toxics "Hot Spots" Chemicals (CAL)
- CAL Air Resources Board/AB 1807 Toxic Air Contaminants
- Proposition 65 chemical (CAL)
- U.S. DOT Inhalation Hazard Chemicals
- Permissible Exposure Limits for Chemical Contaminants (CAL/OSHA)
- The "Director's List" (CAL/OSHA)
- Actively registered pesticide in California.
- OSHA 29CFR1910.119, Appendix A. Process Safety List of Highly Hazardous Chemicals, TQ = 5,000 lb (2,270 kg)
- Clean Air Act: Hazardous Air Pollutants (Title I, Part A, Section 112); Accidental Release Prevention/Flammable substances, (Section 112[r], Table 3), TQ = 10,000 lb (4540 kg)
- EPA Hazardous Waste Number (RCRA No.): U115
- RCRA, 40CFR261, Appendix 8 Hazardous Constituents
- RCRA 40CFR268.48; 61FR15654, Universal Treatment Standards: Wastewater (mg/L), 0.12; Nonwastewater (mg/kg), N/A
- Superfund/EPCRA 40CFR355, Appendix B Extremely Hazardous Substances: TPQ = 1000 lb (454 kg)
- Superfund/EPCRA 40CFR302.4 RQ: CERCLA, 10 lb (4.54 kg)
- EPCRA Section 313 Form R *de minimus* concentration reporting level: 0.1%
- FDA tolerance limit, 50 ppm in ground spices
- Canada, WHMIS, Ingredients Disclosure List

*Description:* A colorless, compressed, liquefied gas or liquid (below 11°C). Sweet odor. The odor threshold is 50 pm. Molecular weight = 44.1 daltons. Boiling point = 10.7°C. Melting/Freezing point = –113°C. Vapor pressure = $1.1 \times 10^3$ mmHg @ 20°C. Flash point = –20°C.

Autoignition temperature = 1058°C. Explosive limits: LEL = 3.0%; UEL = 100%. Hazard Identification (based on NFPA-704 M Rating System): Health 3, Flammability 4, Reactivity 3. Easily dissolved in water. Log $K_{ow}$ = Negative (< –0.5). Unlikely to bioaccumulate in marine organisms.

*Incompatibilities:* Forms explosive mixture with air. Dangerously reactive; may rearrange chemically and/or polymerize violently with evolution of heat, when in contact with highly active catalytic surfaces such as anhydrous chlorides of iron, tin and aluminum, pure oxides of iron and aluminum, and alkali metal hydroxides. Even small amounts of strong acids, alkalis, oxidizers can cause a reaction. Avoid contact with copper. Protect container from physical damage, sun and heat. Attacks some plastics, rubber or coatings.

*Permissible Exposure Limits in Air:* The OSHA[2] PEL and the recommended ACGIH[1] TLV is 1 ppm (2 mg/m$^3$) TWA. The HSE[33] and Sweden[35] have set a TWA of 5 ppm (10 mg/m$^3$); Sweden adds an STEL of 10 ppm (18 mg/m$^3$), Japan[35] has set a TWA of 50 ppm (90 mg/m$^3$). At the other extreme, the former USSR[35, 43] has set an MAC in workplace air of 1 mg/m$^3$; Czechoslovakia has adopted this level also. NIOSH[2] IDLH level = Ca [800 ppm]. The former USSR[35, 43] ha also set an MAC for ambient air in residential areas of 0.3 mg/m$^3$ on a momentary basis and 0.03 mg/m$^3$ on a daily average basis. The NIOSH[2] IDLH level = 800 ppm. Several states have set guidelines or standards for ethylene oxide in ambient air[60] ranging from zero (North Dakota) to 0.01 mg/m$^3$ (Rhode Island) to 0.1 $\mu$g/m$^3$ (North Carolina) to 4.87 $\mu$g/m$^3$ (Pennsylvania) to 6.67 $\mu$g/m$^3$ (New York) to 10.0 $\mu$g/m$^3$ (South Carolina) to 20.0 $\mu$g/m$^3$ (Connecticut, South Dakota and Virginia) to 48.0 $\mu$g/m$^3$ (Nevada) to 450.0 $\mu$g/m$^3$ (Indiana).

*Determination in Air:* Collection by charcoal tube(petroleum-based); DMF Any dust and mist respirator with a full facepiece; analysis by gas chromatography/electrochemical detection; NIOSH IV, Method #1614[18].

*Permissible Concentration in Water:* No criteria set, but runoff from spills or fire control may cause water pollution.

*Routes of Entry:* Ethylene oxide is rapidly absorbed after inhalation, and solutions of ethylene oxide can penetrate human skin.

## Harmful Effects and Symptoms

*Short Term Exposure:* Ethylene oxide gas may produce immediate local irritation of the skin, eyes, and upper respiratory tract. At high concentrations, it may cause an immediate or delayed accumulation of fluid in the lungs. Inhalation of ethylene oxide can produce CNS depression, and in extreme cases, respiratory distress and coma. The onset of symptoms may be delayed for up to 72 hours. In some persons, ethylene oxide exposure may result in allergic sensitization, and future exposure may cause hives or a life-threatening allergic reaction. Signs and symptoms of acute exposure to ethylene oxide may be severe, and include dyspnea (shortness of breath), cough, pulmonary edema, pneumonia, and respiratory failure. Lethargy, headache, dizziness, twitching, convulsions, paralysis, and coma may be observed. Cardiac arrhythmias and cardiovascular collapse may also occur. Gastrointestinal effects of acute exposure may include nausea, vomiting, and abdominal pain. Ethylene oxide irritates the eyes, skin, and respiratory tract. Very high exposures can cause pulmonary edema, a medical emergency that can be delayed for several hours. *Inhalation:* Exposure to 500-700 ppm for 2 to 3 minutes, resulted in nausea, vomiting, headache, disorientation, fluid in the lungs, followed by seizures. Human volunteers breathing a concentration of about 2500 ppm experienced slight irritation of the respiratory tract; breathing in 12,500 ppm showed definite respiratory tract irritation within 10 seconds. Symptoms may not occur for hours after exposure. Other symptoms reported at unknown concentrations include headache, nausea, coughing, vomiting, difficult breathing, respiratory tract irritation, weakness, incoordination, seizures and fluid in the lungs. *Skin:* The pure liquid may cause frostbite. A 1% water solution can cause irritation and redness. A 40-80% water solution may cause extensive blister formation. Ethylene oxide may severely irritate or burn mucous membranes and moist skin. *Eyes:* May cause irritation and severe burns. May affect the eyes, causing delayed development of cataract. Eye contact may result in conjunctivitis (red, inflamed eyes) and erosion of the cornea. *Ingestion:* May cause gastric irritation and liver injury.

*Long Term Exposure:* Chronic ethylene oxide exposure may cause delayed peripheral nerve damage (neuropathy), especially in the lower extremities. Although the results are inconclusive, some data suggest that chronic ethylene oxide exposure impairs cognitive function. Ethylene oxide may also damage the liver and kidneys. Skin allergy can occur, and some persons may become sensitized to the chemical. Cataracts and corneal burns have been reported from occupational exposure. Chronic exposure may be more serious for children because of their potential longer latency period. *Carcinogenicity* The DHHS has determined that ethylene oxide may reasonably be anticipated to be a human carcinogen (NTP 2000). In animals, chronic exposure causes leukemia and intra-abdominal cancer, and there is some evidence that it increases the risk of leukemia in human workers. The International Agency for Research on Cancer has determined that ethylene oxide is carcinogenic to humans. *Reproductive and Developmental Effects* Shepard's Catalog of Teratogenic Agents describes one study in which the spontaneous abortion frequency in hospital workers exposed to ethylene oxide during pregnancy was average for the general population (6.7%); however, the frequency for appropriate hospital controls was below average for the general population (5.6%). Ethylene oxide is included in *Reproductive and Developmental Toxicants*, a 1991 report published by the U.S. General

Accounting Office. Special consideration regarding the exposure of pregnant women is warranted, since ethylene oxide has been shown to be a teratogen and genotoxin; thus, medical counseling is recommended for the acutely exposed pregnant woman. ETO has caused cancer in several species of laboratory animals. It has also caused changes in genetic material and reproductive problems in laboratory animals. may damage the developing fetus. It may cause inheritable genetic damage in humans. There is an increased incidence of gynecological disorders and spontaneous abortions among workers in ethylene oxide production. Its role in this increase is unclear at this time. (NJ) Increased incidence of leukemia and stomach cancer have been reported; however the evidence is not considered conclusive. Leukemia, brain tumors, lung tumors, and other cancers have been observed in laboratory animals.

*Points of Attack:* Eyes, skin, respiratory system, liver, central nervous system, blood, kidneys, reproductive system. Cancer Site: peritoneal cancer, leukemia.

*Medical Surveillance:* For those with frequent or potentially high exposure (half the TLV or greater), the following are recommended before beginning work and at regular times after that: Lung function tests. If symptoms develop or overexposure is suspected, the following may be useful: Consider chest x-ray after acute overexposure. Evaluation by a qualified allergist, including careful exposure history and special testing, may help diagnose skin allergy. Liver and kidney function tests. Medical observation is recommended for 24 to 48 hours after breathing overexposure, as pulmonary edema may be delayed.

*First Aid:* If this chemical gets into the eyes, remove any contact lenses at once and irrigate immediately for at least 15 minutes, occasionally lifting upper and lower lids. Seek medical attention immediately. If this chemical contacts the skin, remove contaminated clothing and wash immediately with soap and water. Seek medical attention immediately. If this chemical has been inhaled, remove from exposure, begin rescue breathing (using universal precautions) if breathing has stopped, and CPR if heart action has stopped. Transfer promptly to a medical facility. When this chemical has been swallowed, get medical attention. Give large quantities of water and induce vomiting. Do not make an unconscious person vomit. If frostbite has occurred, seek medical attention immediately; do *NOT* rub the affected areas or flush them with water. In order to prevent further tissue damage, do *NOT* attempt to remove frozen clothing from frostbitten areas. If frostbite has *NOT* occurred, immediately and thoroughly wash contaminated skin with soap and water. As first aid for pulmonary edema, a doctor or authorized paramedic may consider administering a corticosteroid spray.

*Note to physician or authorized medical personnel:* In cases of respiratory compromise secure airway and respiration via endotracheal intubation. If not possible, perform cricothyroidotomy if equipped and trained to do so. Treat patients who have bronchospasm with aerosolized bronchodilators. The use of bronchial sensitizing agents in situations of multiple chemical exposures may pose additional risks. Consider the health of the myocardium before choosing which type of bronchodilator should be administered. Cardiac sensitizing agents may be appropriate; however, the use of cardiac sensitizing agents after exposure to certain chemicals may pose enhanced risk of cardiac arrhythmias (especially in the elderly). Ethylene oxide poisoning is not known to pose additional risk during the use of bronchial or cardiac sensitizing agents. Consider racemic epinephrine aerosol for children who develop stridor. Dose 0.25–0.75 mL of 2.25% racemic epinephrine solution in 2.5 cc water, repeat every 20 minutes as needed, cautioning for myocardial variability. Patients who are comatose, hypotensive, or are having seizures or cardiac arrhythmias should be treated according to advanced life support protocols.

*References:*
- U.S. Environmental Protection Agency, Office of Pesticide Programs, Topical & Chemical Fact Sheets, "Ethylene Oxide," http://www.epa.gov/pesticides/factsheets/chemicals/etofactsheet.htm#bkmrk1
- U.S. Environmental Protection Agency, Office of Pesticide Programs, Pesticide Residue Limits, "Ethylene Oxide", 40 CFR 180.151, 185.2850, http://www.epa.gov/pesticides/food/viewtols.htm
- New Jersey Department of Health and Senior Services, "Hazardous Substance Fact Sheet: Ethylene Oxide," Trenton, NJ (December 1994, rev. May 2001). http://www.state.nj.us/health/eoh/rtkweb/0882.pdf
- U.S. Department of Health and Human Services; Agency for Toxic Substances and Disease Registry, "ToxFAQs, Ethylene Oxide," Atlanta GA (July 1999). http://www.atsdr.cdc.gov/tfacts137.html
- Bogyo, D.A., Lande, S.S., Meylan, W.M., Howard, P.H. and Santodonate, J., (Syracuse Research Corp. Center for Chemical Hazard Assessment), Investigation of Selected Potential Environmental Contaminants: Epoxides, Report 560/11-80-005, Washington, DC, U.S. Environmental Protection Agency, (March 1980).
- Agency for Toxic Substances and Disease Registry, U.S. Department of Health and Human Services, Public Health Service, "Managing Hazardous Materials Incidents," Atlanta, GA (June 2003).
- National Institute for Occupational Safety and Health, Ethylene Oxide, Current Intelligence Bulletin N. 35, DHHS (NIOSH) Publication No. 81-130, Cincinnati, Ohio (May 22, 1981).
- Sax, N.I., Ed., "Dangerous Properties of Industrial Materials Report," 4, No. 2, 70-73 (1984).
- U.S. Environmental Protection Agency, "Chemical Profile: Ethylene Oxide," Washington, DC, Chemical Emergency Preparedness Program (November 30, 1987).

- California Environmental Protection Agency "Chemical List of Lists," Sacramento CA (February 1997).

# Ethylene Thiourea

*Use Type:* A fungicide contaminant and break-down substance
*CAS Number:* 96-45-7
*Formula:* $C_3H_6N_2S$
*Synonyms:* 4,5-Dihydro-2-mercaptoimidazole; 4,5-Dihydroimidazole-2(3*H*)-thione; 1,3-Ethylenethiourea; *N,N'*-Ethylenethiourea; Etilentiourea (Spanish); ETU; Imidazolidinethione; 2-Imidazolidinethione; 2-Imidazoline-2-thiol; Imidazoline-2-thiol; Imidazoline-2(3*H*)-thione; Mercaptoimidazoline; 2-Mercaptoimidazoline; 2-Mercapto-2-imidazoline; 2-Merkaptoimidazolin (Czech); NCI-C03372; Rhenogran ETU; Rhodanin S-62 (Czech); Tetrahydro-2H-imidazole-2-thione; 2-Thioimidazolidine; 2-Thionoimidazolidine; 2-Thiol-dihydroglyoxaline; Thiourea, *N,N'*-(1,2-ethanediyl)-
*Trade Names:* ACCEL 22®; AKROCHEM ETU-22®; MERCAZIN I®; NA 22®; NOCCELER 22®; PENNAC CRA®; SOXINOL 22®; VULKACIT NPV/C2®; WARECURE C®
*Producers:* Akzo Nobel Chemicals (Germany); ATOFINA (France); Bayer Chemicals (Germany); Degussa (Germany); DuPont (USA); EniChem (Italy); Sumitomo Chemical (Japan)
*EPA/OPP PC Code:* 600016
*California DPR Chemical Code:* 2559
*ICSC Number:* 1148
*RTECS Number:* NI9625000
*EEC Number:* 613-039-00-9
*EINECS Number:* 202-506-9
*Uses:* Exposure to ethylene thiourea results from the very widely used ethylene bisdithiocarbamate fungicides. Ethylene thiourea may be present as a contaminant in the ethylene bisdithiocarbamate fungicides and can also be formed when food containing the fungicides is cooked. Ethylene thiourea is also used extensively as an accelerator in the curing of polychloroprene (Neoprene) and other elastomers.
*Carcinogen/Hazard Classifications*
**U.S. EPA Carcinogens:** Group B2, probable carcinogen
**U.S. NTP Carcinogen:** Reasonably anticipated carcinogen
**California Prop. 65:** Carcinogen and developmental toxin
**IARC:** Group 3, unclassifiable
**Endocrine Disruptor:** Probable ED, especially from its parent chemical, Menab.
*Regulatory Authority:*
- Carcinogen (Animal Positive) (IARC)[9]
- Banned or Severely Restricted (Sweden)[13]
- Air Pollutant Standard Set (North Dakota, Pennsylvania)[6]
- AB 1803-Well Monitoring Chemical (CAL)
- AB 2588-Air Toxics "Hot Spots" Chemicals (CAL)
- CAL Air Resources Board/AB 1807 Toxic Air Contaminants
- Proposition 65 chemical (CAL)
- The "Director's List" (CAL/OSHA)
- Clean Air Act: Hazardous Air Pollutants (Title I, Part A, Section 112)
- EPA Hazardous Waste Number (RCRA No.): U116
- RCRA, 40CFR261, Appendix 8 Hazardous Constituents
- Safe Drinking Water Act: Priority List (55 FR 1470)
- Superfund/EPCRA 40CFR302.4 RQ: CERCLA, 10 lb (4.54 kg)
- EPCRA Section 313 Form R *de minimus* concentration reporting level: 0.1%

*Description:* Ethylene thiourea is a white to light green, needle-like crystalline solid. Faint amine odor. Boiling point = 230–313°C. Melting/Freezing point = 203–204°C. Molecular weight = 102.15. Flash point = 252°C. Hazard Identification (based on NFPA-704 M Rating System): Health 2, Flammability 1, Reactivity 0. Slightly soluble in cold water; highly soluble in hot. Commercial ethylene thiourea is available as a solid powder, as a dispersion in oil (which retards the formation of fine dust dispersions in workplace air), and "encapsulated" in a matrix of compatible elastomers. In this latter form, ethylene thiourea may be least likely to escape into the work-place air. Log $K_{ow}$ = Negative (< –0.7). Unlikely to bioaccumulate in marine organisms.
*Incompatibilities:* Strong oxidizers, acids, acid anhydrides, acrolein
*Permissible Exposure Limits in Air:* There is no current OSHA[2] or NIOSH exposure standard for ethylene thiourea. The ACGIH[1] has set no limits either. NIOSH[2] IDLH level = not determined but contains potential occupational carcinogen notation. States which have set guidelines or standards for ethylene thiourea in ambient air[60] include North Dakota (zero level) and Pennsylvania (0.7 µg/m3).
*Determination in Air:* Filter collection, extraction with water, complexation with pentacyanoamineferrate and spectrophotometric measurement. See NIOSH IV, Method #5011[18].
*Permissible Concentration in Water:* Maine[61] has set a guideline for drinking water of 4.4 µg/L. The USEPA in a health advisory (see reference below) has developed a no-observed-adverse-effect level (NOAEL) of 1.25 mg/kg/day based on absence of thyroid effects in male rats exposed to ETU in the diet for up to 12 months. This results in a longer term health advisory for an adult of 0.44 mg/L.
*Routes of Entry:* Inhalation, ingestion, skin and/or eye contact.
*Harmful Effects and Symptoms*
*Short Term Exposure:* Inhalation can cause irritation of the respiratory tract with soreness, hoarseness, cough and phlegm. High exposure can cause sweating, thirst, nausea,

an increase in the heart rate and blood pressure that can last for hours or days. Higher exposures can cause pulmonary edema, a medical emergency that can be delayed for several hours. This can cause death. Contact can cause irritation of the skin and eyes, and may cause eye burns. A related chemical, ziram, can cause brain swelling and hemorrhage with muscle weakness and liver and kidney effects.

*Long Term Exposure:* Ethylene thiourea has been shown to be carcinogenic and teratogenic (causing malformation in offspring) in laboratory animals. In addition, ethylene thiourea can cause myxedema (the drying and thickening of skin, together with a slowing down of physical and mental activity), goiter, and other effects related to decreased output of thyroid hormone. Maneb, a related fungicide, can cause nerve damage.

*Points of Attack:* Eyes, skin, thyroid, reproductive system. Cancer Site in animals: liver, thyroid and lymphatic system tumors.

*Medical Surveillance:* Initial and routine employee exposure surveys should be made by competent industrial hygiene and engineering personnel. These surveys are necessary to determine the extent of employee exposure and to ensure that controls are effective. The NIOSH Occupational Exposure Sampling Strategy Manual, NIOSH Publication #77-173, may be helpful in developing efficient programs to monitor employee exposures to ethylene thiourea. The manual discusses determination of the need for exposure measurements, selection of appropriate employees for exposure evaluation, and selection of sampling times. Employee exposure measurements should consist of 8-hour TWA (time-weighted average) exposure estimates calculated from personal or breathing zone samples (air that would most nearly represent that inhaled by the employees). Area and source measurements may be useful to determine problem areas, processes, and operations. Thyroid function tests. Examination of the nervous system. Consider chest x-ray following acute overexposure.

*First Aid:* If this chemical gets into the eyes, remove any contact lenses at once and irrigate immediately for at least 15 minutes, occasionally lifting upper and lower lids. Seek medical attention immediately. If this chemical contacts the skin, remove contaminated clothing and wash immediately with soap and water. Seek medical attention immediately. If this chemical has been inhaled, remove from exposure, begin rescue breathing (using universal precautions) if breathing has stopped, and CPR if heart action has stopped. Transfer promptly to a medical facility. When this chemical has been swallowed, get medical attention. Give large quantities of water and induce vomiting. Do not make an unconscious person vomit. Medical observation is recommended for 24 to 48 hours after breathing overexposure, as pulmonary edema may be delayed. As first aid for pulmonary edema, a doctor or authorized paramedic may consider administering a corticosteroid spray.

*References:*
- National Institute for Occupational Safety and Health, Ethylene Thiourea, Current Intelligence Bulletin 22, Washington, DC (April 11, 1978).
- Sax, N.I., Ed., "Dangerous Properties of Industrial Materials Report," 1, No. 2, 38-39 (1980); 7 No. 3, 106-111 (1987).
- U.S. Environmental Protection Agency, "Health Advisory: Ethylene Thiourea," Washington, DC, Office of Drinking Water (August 1987).
- California Environmental Protection Agency "Chemical List of Lists," Sacramento CA (February 1997).

# Ethylthiocyanate

*Use Type:* Insecticide
*CAS Number:* 542-90-5
*Formula:* $C_3H_5NS$; $C_2H_5SCN$
*Synonyms:* Aethylrhodanid (German); Ethane, thiocyanato- (Italian); Ethyl rhodanate; Ethyl sulfocyanate; Thiocyanatoethane; Thiocyanic acid, ethyl ester
*Producers:* Carbolabs Div., Sigma-Aldrich (USA); Sigma-Aldrich Fine Chemicals (USA)
*Chemical Class:* Thiocyanate (aliphatic)
*RTECS Number:* XK99000000
*Regulatory Authority:*
- Superfund/EPCRA 40CFR355, Appendix B Extremely Hazardous Substances: TPQ = 10,000 lb (4,540 kg)
- Superfund/EPCRA 40CFR302.4 RQ: EHS, 1 lb (0.454 kg)
- AB 2588-Air Toxics "Hot Spots" Chemicals (CAL)
- CAL Air Resources Board/AB 1807 Toxic Air Contaminants
- Clean Air Act: Hazardous Air Pollutants (Title I, Part A, Section 112) as cyanide compounds
- lean water act: Section 307 Priority Pollutants as cyanide, total; Toxic Pollutant (Section 401.15) as cyanide compounds
- EPA Hazardous Waste Number (RCRA No.): P030 as cyanide compounds
- RCRA Section 261 Hazardous Constituents as cyanide compounds
- RCRA Universal Treatment Standards: Wastewater (mg/L), 1.2 (total); 0.86 (amenable); Nonwastewater (mg/kg), 590 (total); 30 (amenable) as cyanide compounds
- RCRA Ground Water Monitoring List. Suggested test method(s) (PQL $ug/L$): 9010(40) as cyanide compounds
- Safe Drinking Water Act: MCL, 0.01 mg/L; MCLG, 0.01 mg/L;Regulated chemical (47 FR 9352) as cyanide compounds
- U.S. DOT Regulated Marine Pollutant (49CFR, Subchapter 172.101, Appendix B) as cyanides, mixtures or solutions

*Description:* Ethyl thiocyanate is a liquid. Insoluble in water. Boiling point = 146°C. Melting/Freezing point = –86°C.

*Incompatibilities:* Contact with chlorates, nitrates, nitric acid, organic peroxides, peroxides may cause a violent reaction.

*Permissible Exposure Limits in Air:* No standards set

*Permissible Concentration in Water:* The U.S. EPA has set a maximum contaminant level of cyanide in drinking water of 0.2 milligrams cyanide per liter of water (0.2 mg/L). Runoff from spills or fire control may cause water pollution.

*Routes of Entry:* Inhalation, skin contact.

*Harmful Effects and Symptoms*

*Short Term Exposure:* The ingestion of a concentrated solution may lead to vomiting. The principal systemic reaction is probably one of central nervous depression, interrupted by periods of restlessness, abnormally fast and deep respiratory movements and convulsions. Death is usually due to respiratory arrest from paralysis of the medullary centers. In nonfatal cases injures to the liver and kidneys may appear.

*Long Term Exposure:* Prolonged absorption may produce various skin eruptions, runny nose, and occasionally dizziness, cramps, nausea, vomiting and mild or severe disturbances of the nervous system. *Developmental Effects:* No reproductive or developmental effects of this thiocyanate have been reported in experimental animals or humans. Increased levels of thiocyanate in the umbilical cords of fetuses whose mothers smoked compared to those whose mothers were non-smokers suggests that thiocyanate, and possibly also cyanide, can cross the placenta.

*Points of Attack:* Nervous system. liver, kidneys.

*Medical Surveillance:* Urine thiocyanate levels. Blood cyanide levels. Liver function tests. Kidney function tests. Examination of the nervous system.

*First Aid:* Treatment is as for aliphatic thiocyanates. If this chemical gets into the eyes, remove any contact lenses at once and irrigate immediately for at least 30 minutes, occasionally lifting upper and lower lids. Seek medical attention immediately. If this chemical contacts the skin, remove contaminated clothing and wash immediately with soap and water. Seek medical attention immediately. If this chemical has been inhaled, remove from exposure, begin rescue breathing (using universal precautions) if breathing has stopped, and CPR if heart action has stopped. Transfer promptly to a medical facility. Victims who are conscious and able to swallow should be given 4 to 8 ounces of water or milk. Gastric lavage with a small bore NG tube should be considered if it can be performed within 1 hour after ingestion. The effectiveness of activated charcoal administration is unknown, but it is suggested following lavage (administer activated charcoal at 1 gm/kg, usual adult dose 60–90 g, child dose 25–50 g). A soda can and straw may be of assistance when offering charcoal to a child.

Medical observation is recommended for 24 to 48 hours after breathing overexposure, as pulmonary edema may be delayed. As first aid for pulmonary edema, a doctor or authorized paramedic may consider administering a corticosteroid spray.

*Note:* Because cyanide is probably largely responsible for poisonings, antidotal measures against cyanide should be instituted promptly. Use amyl nitrate capsules if symptoms develop. All area employees should be trained regularly in emergency measures for cyanide poisoning and in CPR. A cyanide antidote kit should be kept in the immediate work area and must be rapidly available. Kit ingredients should be replaced every 1-2 years.

*References:*
- U.S. Environmental Protection Agency, "Chemical Profile: Ethylthiocyanate," Washington, DC, Chemical Emergency Preparedness Program (November 30, 1987).
- California Environmental Protection Agency "Chemical List of Lists," Sacramento CA (February 1997).

# Etridiazole

*Use Type:* Fungicide

*CAS Number:* 2593-15-9

*Formula:* $C_5H_5Cl_3N_2OS$

*Synonyms:* 5-Aethoxy-3-trichlormethyl-1,2,4-thiadiazol (German); Echlomezole (Japan); Ethazole; 5-Ethoxy-3-(trichloromethyl)-1,2,4-thiadiazole; ETMT; Etridiazole; 3-(Trichloromethyl)-5-ethoxy-1,2,4-thiadiazole; 1,2,4-Thiadiazole, 5-ethoxy-3-(trichloromethyl)-

*Trade Names:* AATERRA®; BANROT® (With Thiophanate-methyl); DWELL®, Crompton Corporation, (USA), canceled; ETHAZOLE®; 4-WAY®; KOBAN®; MF-344®; OLIN MATHIESON® 2,424; OM® 2424; PANSOIL®; TERRACHLOR-SUPER X®, Crompton Corporation (USA) (with Pentachloronitrobenzene); TERRACLOR SUPER X® (with Pentachloronitrobenzene), Crompton Corporation (USA); TERRACOAT®, Crompton Corporation (USA); TERRAFLO®, Crompton Corporation (USA); TERRAMASTER®, Crompton Corporation, (USA); TERRAZOLE®, Crompton Corporation (USA); TRUBAN®, Scotts-Sierra (USA)

*Producers:* Crompton Corporation (USA); Gustafson (USA)

*Chemical Class:* Thiazole

*EPA/OPP PC Code:* 084701

*California DPR Chemical Code:* 580

*Uses:* Etridiazole is a fungicide registered for use as a seed treatment on barley, beans, corn, cotton, peanuts, peas, sorghum, soybeans, safflower, and wheat. It is also registered for use on cotton for in-furrow application at planting, on ornamental plants and shrubs by horticultural nurseries, on non-bearing citrus and non-bearing coffee, and for golf course fairways, tees and greens. Some states hold Special Local Need registrations for use on tobacco transplants.

***Human toxicity (long-term)***[77]***:*** Intermediate–10.51051 ppb, CHCL (Chronic Human Carcinogen Level)

***Fish toxicity (threshold)***[77]***:*** Low–369.29192 ppb, MATC (Maximum Acceptable Toxicant Concentration)

***U.S. Maximum Allowable Residue Levels for Etridiazole (40 CFR 180.370):***

| CROP | ppm |
|---|---|
| Cattle, fat | 0.1 |
| Cattle, meat | 0.1 |
| Cattle, mbyp | 0.1 |
| Corn, field, grain | 0.05 |
| Corn, forage | 0.1 |
| Corn, stover | 0.1 |
| Cotton, undelinted seed | 0.2 |
| Egg | 0.05 |
| Goat, fat | 0.1 |
| Goat, meat | 0.1 |
| Goat, mbyp | 0.1 |
| Hog, fat | 0.1 |
| Hog, meat | 0.1 |
| Hog, mbyp | 0.1 |
| Horse, fat | 0.1 |
| Horse, meat | 0.1 |
| Horse, mbyp | 0.1 |
| Milk | 0.05 |
| Poultry, fat | 0.1 |
| Poultry, meat | 0.1 |
| Poultry, mbyp | 0.1 |
| Sheep, fat | 0.1 |
| Sheep, meat | 0.1 |
| Sheep, mbyp | 0.1 |
| Strawberry | 0.2 |
| Tomato | 0.15 |
| Wheat, forage | 0.1 |
| Wheat, grain | 0.05 |
| Wheat, straw | 0.1 |

***Carcinogen/Hazard Classifications***

**U.S. EPA Carcinogens:** Group B2, probable carcinogen

**California Prop. 65:** Carcinogen

**WHO Acute Hazard:** Class III, slightly hazardous

***Regulatory Authority:***
- Proposition 65 Chemical (CAL).

***Description:*** Pale-yellow (pure) or reddish-brown (technical) liquid. Mild odor. Practically insoluble in water; solubility = 50 mg/L @ 25°C. Molecular weight = 247.53. Melting/Freezing point = 20°. Boiling point = 95°C@1 mmHg. Vapor pressure =1.3 x$10^{-4}$ mmHg @ 20°C; 9.75 x$10^{-5}$ mmHg @ 25°C.

***Incompatibilities:*** Alkali materials can cause material to become hydrolyzed. Toxic fumes of nitrogen and sulfur oxides and chlorine gas are formed when heated to decomposition.

***Permissible Concentration in Water:*** No criteria set. Runoff from spills or fire control may cause water pollution.

***Harmful Effects and Symptoms***

***Short Term Exposure:*** Contact with eyes or skin may cause irritation or injury. Inhalation should be avoided; use NIOSH-approved air purifying respirators for pesticides. May be harmful if swallowed.

***First Aid:*** If this chemical gets into the eyes, remove any contact lenses at once and irrigate immediately for at least 15 minutes, occasionally lifting upper and lower lids. Seek medical attention immediately. If this chemical contacts the skin, remove contaminated clothing and wash immediately with soap and water. Seek medical attention immediately. If this chemical has been inhaled, remove from exposure, begin rescue breathing (using universal precautions) if breathing has stopped, and CPR if heart action has stopped. Transfer promptly to a medical facility. When this chemical has been swallowed, get medical attention. Give large quantities of water and induce vomiting. Do not make an unconscious person vomit. Do not induce vomiting when formulations containing petroleum solvents are ingested.

***References:***
- U.S. Environmental Protection Agency, "Reregistration Eligibility Decision (RED), Etridiazole" Office of Prevention, Pesticides and Toxic Substances, Washington, DC (September 2000). http://www.epa.gov/REDs/0009red.pdf
- U.S. Environmental Protection Agency, Office of Pesticide Programs, Pesticide Residue Limits, "Etridiazole," 40 CFR 180.370, http://www.epa.gov/pesticides/food/viewtols.htm
- California Environmental Protection Agency "Chemical List of Lists," Sacramento CA (February 1997)

# F

## Famphur

*Use Type:* Insecticide (grubcide)
*CAS Number:* 52-85-7
*Formula:* $C_{10}H_{16}NO_5PS$
*Synonyms:* AI3-25644; Benzenesulfonamide, *p*-hydroxy-*N,N*-dimethyl-, *O*-ester with *O,O*-dimethyl phosphorothioate; Caswell No. 456D; CL 38023; *O*-[4-((Dimethylamino)sulfonyl)phenyl]*O,O*-dimethyl phosphorothioate; *O*-[4-((dimethylamino)sulphonyl)phenyl]*O,O*-dimethyl thiophosphate; *O,O*-Dimethyl *O*-[*p*-(dimethylsulfamoyl)phenyl]phosphorothioate; *O,O*-Dimethyl *O*-[*p*-(*N,N*-dimethylsulfamoyl)phenyl]phosphorothioate; *O,O*-Dimethyl phosphorothioate *O*-ester with *p*-hydroxy-N,N-dimethylbenzenesulfonamide; *O*-4-Dimethylsulfamoylphenyl *O,O*-dimethyl phosphorothioate; *O*-4-Dimethylsulphamoylphenyl *O,O*-dimethylphosphorothioate; ENT 25,644; Famfur (Spanish); *p*-Hydroxy-*N,N*-dime thylbenzenesulfonamide ester with phosphorothioic acid *O,O*-dimethyl ester; Phosphorothioic acid, O-[4-((dimethylamino)sulfonyl)phenyl] *O,O*-dimethyl ester; Phosphorothioic acid, *O,O*-dimethyl ester, *O*-ester with *p*-hydroxy-*N,N*-dimethylbenzenesulfonamide
*Trade Names:* AC 38023®; AMERICAN CYANAMID® 38023, BASF Agricultural Products (Germany), canceled; BO-ANA®; CYFLEE®; DOVIP®; FAMFOS®; FAMIX®, BASF Agricultural Products (Germany), canceled; FAMOPHOS®; FAMOPHOS WARBEX®; FAMPHOS®; FANFOS®; NEMACUR®; VARBEX®; WARBEX®, BASF Agricultural Products (Germany), canceled
*Producers:* BASF Agricultural Products (Germany)
*Chemical Class:* Organophosphate
*EPA/OPP PC Code:* 059901
*California DPR Chemical Code:* 282
*RTECS Number:* TF7640000
*Uses:* Not registered in the U.S. For use as a veterinary medication against lice and cattle grub of cattle and other livestock. Administered to livestock in their feed; travels out and kills larvae.
*Carcinogen/Hazard Classifications*
**Label Signal Word:** WARNING or CAUTION
**WHO Acute Hazard:** Class 1b, highly hazardous
*Regulatory Authority:*
- RCRA Section 261 Hazardous Constituents
- EPA 40 CFR 372.65, Specific Toxic Chemical Listings
- EPA Hazardous Waste Number (RCRA No.): P097
- RCRA Universal Treatment Standards: Wastewater (mg/L), 0.017; Nonwastewater (mg/kg), 15
- RCRA Ground Water Monitoring List. Suggested test method(s) (PQL ug/L): 8270(10)
- EPCRA Section 313 Form R *de minimis* concentration reporting level: 1.0%
- U.S. DOT Inhalation Hazard Chemicals as organophosphates

*Description:* Colorless sand-like powder. Molecular weight 325.349
*Incompatibilities:* May react violently with antimony(V) pentafluoride. Incompatible with lead diacetate, magnesium, silver nitrate. Slowly hydrolyzes in water, releasing ammonia and forming acetate salts.
*Determination in Air:* OSHA versatile sampler-2; Toluene/Acetone; Gas chromatography/Flame ionization detection; NIOSH IV, Method #5600, Organophosphorus Pesticides.[18]
*Permissible Concentration in Water:* No criteria set. Runoff from spills or fire control may cause water pollution.
*Harmful Effects and Symptoms*
*Short Term Exposure:* Eye pupils are small; blurred vision; eye watering; runny nose; cough; shortness of breath; salivation; dizziness; nausea, stomach cramps, diarrhea, and vomiting; increased blood pressure; profuse sweating; hypermotility, hallucinations; irritability; tingling of the skin; drowsiness; slow heartbeat; convulsions; fluid in lungs; loss of consciousness; incontinence; breathing stops; death. Organophosphates inhibit the action of acetylcholinesterase enzymes, and alter the way in which nervous impulses are transmitted. The effects can last for hours, days, or much longer. The action of the enzymes is reestablished after new enzymes are formed.
*Long Term Exposure:* Cholinesterase inhibitor; cumulative effect is possible. Organophosphates may damage the nervous system with repeated exposure, resulting in convulsions, respiratory failure. May cause liver damage.
*Points of Attack:* Respiratory system, central nervous system, cardiovascular system, blood cholinesterase.
*Medical Surveillance:* Medical observation is recommended for 24 to 48 hours after breathing overexposure, as pulmonary edema may be delayed. As first aid for pulmonary edema, consider administering a corticosteroid spray. Cigarette smoking may exacerbate pulmonary injury and should be discouraged for at least 72 hours following exposure. Do not drink any alcoholic beverages before or during use; alcohol promotes absorption of organophosphates.
*First Aid:* **Treatment for organophosphate poisoning consists of thorough decontamination, cardiorespiratory support, and administration of the antidotes atropine and pralidoxime. In cases of severe poisoning, diazepam, an anticonvulsant, should also be administered.**

**Antidotes should be administered as prevention even if the diagnosis is in doubt.** *Eyes:* Speed in removing material from eyes and skin is of extreme importance. Eye contact can cause dangerous amounts of these chemicals to be quickly absorbed through the mucous membrane into the bloodstream. Immediately and gently flush eyes with plenty of warm or cold water (NO hot water) for at least 15 minutes, occasionally lifting the upper and lower eyelids. Get medical aid immediately. *Skin:* Get medical aid. Skin contact can cause dangerous amounts of these chemicals to be absorbed into the bloodstream. Wearing the appropriate PPE equipment and respirator for organophosphate/carbamate pesticides, immediately flush skin with plenty of soap and water for at least 15 minutes while removing contaminated clothing and shoes. Shampoo hair promptly if contaminated. The removed, contaminated clothing and shoes should be double-bagged and left in Hot Zone for later disposal by hazardous materials experts. Skin may also be decontaminated with diluted hypochlorite solution. *Inhalation:* Get medical aid. Do not contaminate yourself. Wearing the appropriate PPE equipment and respirator for organophosphate pesticides, immediately remove the victim from the contaminated area to fresh air. If the victim is not breathing, administer artificial respiration. Do NOT use mouth-to-mouth resuscitation; use bag/mask apparatus. If breathing is difficult, administer oxygen through bag/mask apparatus until medical help arrives. Do not leave victim unattended. *Ingestion:* Call poison control. Loosen all clothing. Never give anything by mouth to an unconscious person. If victim is *unconscious or having convulsions,* do nothing except keep victim warm. Get medical aid. Transfer promptly to a medical facility. In cases of ingestion, **do not induce vomiting**. If the victim is alert and asymptomatic, administer a slurry of activated charcoal at a dose of 1 g/kg (infant, child, and adult dose). A soda can and straw may be of assistance when offering charcoal to a child. *In some cases you may be specifically instructed by poison control to induce vomiting by way of 2 tablespoons of syrup of ipecac (adult) washed down with a cup of water.* Do NOT give activated charcoal before or with ipecac syrup.

*Note to physician or authorized medical personnel:* Treat cases of respiratory compromise, coma, or excessive pulmonary secretions with respiratory support using protocols and techniques available and within the scope of training. Some cases may necessitate procedures such as endotracheal intubation or cricothyrotomy by properly trained and equipped personnel. When possible, atropine (see *Antidotes*, below) should be given under medical supervision. Patients who are comatose, hypotensive, or having seizures or cardiac arrhythmias should be treated according to advanced life support protocols. *Antidotes:* Two antidotes are administered to treat organophosphate poisoning. Atropine is a competitive antagonist of acetylcholine at muscarinic receptors and is used to control the excessive bronchial secretions which are often responsible for death. Pralidoxime relieves both the nicotinic and muscarine effects of organophosphate poisoning by regenerating acetylcholinesterase and can reduce both the bronchial secretions and the muscle weakness associated with poisoning. The initial intravenous dose of atropine in adults should be determined by the severity of symptoms: An initial adult dose of 1.0 to 2.0 mg or pediatric dose of 0.01 mg/kg (minimum 0.01 mg) should be administered intravenously. If intravenous access cannot be established, atropine may also be given intramuscularly, subcutaneously or via endotracheal tube. Doses should be repeated every 15 minutes until excessive secretions and sweating have been controlled. Once bronchial secretion has been controlled, atropine administration should be repeated whenever the secretions begin to recur. In seriously poisoned patients, very large doses may be required. Alterations of pulse rate and pupillary size should not be used as indicators of treatment adequacy. Pralidoxime should be administered as early in poisoning as possible as its efficacy may diminish when given more than 24 to 36 hours after exposure. Doses are as follows: adult 1.0 g; pediatric 25 to 50 mg/kg. The drug should be administered intravenously over 30 to 60 minutes, but in a life-threatening situation, one-half of the total dose can be given per minute for a total administration time of 2 minutes. Treatment should begin to take effect within 40 minutes with a reduction in symptoms and the amount of atropine necessary to control bronchial secretion. The initial dose can be repeated in 1 hour and then every 8 to 12 hours until the patient is clinically well and no longer requires atropine. If intravenous access cannot be established, pralidoxime may also be given intramuscularly. Early administration of diazepam in addition to the combined atropine and pralidoxime treatment may help prevent the onset of seizures and potential brain and cardiac morphologic damage following high-level organophosphate poisoning.

*References:*
- New Jersey Department of Health and Senior Services, "Hazardous Substance Fact Sheet, Famphur," Trenton NJ (June 2002).
  http://www.state.nj.us/health/eoh/rtkweb/2915.pdf
- Pesticide Management Education Program, "Famphur (Bo-Ana, Warbex) Chemical Profile 4/85," Cornell University, Ithaca, NY (April 1985).
  http://pmep.cce.cornell.edu/profiles/insect-mite/ddt-famphur/famphur/insect-prof-famphur.html
- California Environmental Protection Agency "Chemical List of Lists," Sacramento CA (February 1997).

# Fenac

*Use Type:* Herbicide
*CAS Number:* 85-34-7

*Formula:* $C_8H_5Cl_3O_2$

*Synonyms:* Benzeneacetic acid, 2,3,6-trichloro-; 2,3,6-Trichlorobenzeneacetic acid; 2,3,6-Trichlorophenylacetic acid; 2,3,6-Trichlorphenylessigsaeure (German)

*Trade Names:* CHLORFENAC®; FENAB®; FENATROL®, Union Carbide (USA), canceled; KANEPAR®; TCPA®; TRI-FEN®; TRIFENE®

*Producers:* Union Carbide (USA)

*Chemical Class:* Organochlorine; Halo-organics

*EPA/OPP PC Code:* 082601

*EINECS Number:* 201-599-3

*Uses:* Not registered in the U.S. For pre-emergence season-long control of weeds, particularly johnsongrass in sugarcane; also used on many annual grasses and perennial broadleaf weeds in noncrop areas such as railroad rights-of-way, around buildings, and under highway guard rails.

*Human toxicity (long-term)[77]:* Very low–140000.00 ppb, Health Advisory

*Fish toxicity (threshold)[77]:* Very low–2294.77886 ppb, MATC (Maximum Acceptable Toxicant Concentration)

*Description:* Colorless and odorless crystalline solid or white powder. Sparingly soluble in water; solubility = 200 ppm @ 28°C. Molecular weight 239.48. Melting/Freezing point = 156-159°C. Vapor pressure = $8.5 \times 10^{-3}$ mmHg @ 100°C.

*Incompatibilities:* Keep away from oxidizers, sulfuric acid, caustics, ammonia, aliphatic amines, alkanolamines, isocyanates, alkylene oxides, epichlorohydrin.

*Permissible Concentration in Water:* No criteria set. Runoff from spills or fire control may cause water pollution.

*Harmful Effects and Symptoms*

*Short Term Exposure:* Contact with eyes or skin may cause irritation or injury. Inhalation should be avoided; use NIOSH-approved air purifying respirators for pesticides. May be harmful if swallowed.

*First Aid:* If this chemical gets into the eyes, remove any contact lenses at once and irrigate immediately for at least 15 minutes, occasionally lifting upper and lower lids. Seek medical attention immediately. If this chemical contacts the skin, remove contaminated clothing and wash immediately with soap and water. Seek medical attention immediately. If this chemical has been inhaled, remove from exposure, begin rescue breathing (using universal precautions) if breathing has stopped, and CPR if heart action has stopped. Transfer promptly to a medical facility. When this chemical has been swallowed, get medical attention. Give large quantities of water and induce vomiting. Do not make an unconscious person vomit.

*References:*
- Pesticide Management Education Program, "Fenac (Fenatrol) Herbicide Profile 2/85," Cornell University, Ithaca, NY (February 1985).
  http://pmep.cce.cornell.edu/profiles/herb-growthreg/fatty-alcohol-monuron/fenac/herb-prof-fenac.html

# Fenaminosulf

*Use Type:* Fungicide

*CAS Number:* 140-56-7

*Formula:* $C_8H_{10}N_3NaO_3S$

*Alert:* May cause mutations. Handle with extreme care.

*Synonyms:* p-Dimethylaminobenzenediazosodium sulphonate; p-Dimethylaminobenzene diazo sodium sulfonate; p-(Dimethylamino)benzenediazo sodium sulfonate; p-(Dimethylamino)benzenediazosulfonate; p-Dimethylaminobenzenediazosulfonic acid, sodium salt; 4-Dimethylaminobenzenediazosulfonic acid, sodium salt; p-(Dimethylamino)benzenediazosulphonate; p-(Dimethylamino)benzenediazosulphonic acid, sodium salt; 4-Dimethylaminobenzenediazosulphonic acid, sodium salt; p-Dimethylaminobenzoldiazosulfonat (Natriumsalz) (German);[4-(Dimethylamino)phenyl]diazenesulfonic acid, sodium salt; 4-[(Dimethylamino)phenyl]diazenesulfonic acid, sodium salt; p-(Dimethylamino)-phenyldiazo-natriumsulfonat (German); N,N-Dimethyl-p-anilinediazosulfonic acid sodium salt; NCI-C03010; Sodium-p-(dimethylamino)benzenediazosulfonate sodium; Sodium-4-(dimethylamino)benzenediazosulfonate; Sodium-p-(dimethylamino)benzenediazosulphonate; Sodium-4-(dimethylamino)benzenediazosulphonate; Sodium-[4-(dimethylamino)phenyl]diazenesulfonate

*Trade Names:* BAY-22555®, Bayer CropScience (Germany); BAYER-5072®, Bayer CropScience (Germany); BRAVO D®; DAPA®; DAS®; DEKSONAL®; DEXON®, Bayer CropScience (Germany), canceled; DIAZOBEN®; FENAMINOSULF®; GOLD ORANGE MP®; LESAN®, Bayer CropScience (Germany), canceled; TROPAEOLIN D®

*Producers:* Bayer CropScience (Germany)

*Chemical Class:* Sulfonated phenyl diazo

*EPA/OPP PC Code:* 034201

*California DPR Chemical Code:* 194

*Uses:* Not registered in the U.S. Used on vegetables, ornamentals, lawns, and turf.

*Fish toxicity (threshold)[77]:* Very low–3772.81013 ppb, MATC (Maximum Acceptable Toxicant Concentration)

*Carcinogen/Hazard Classifications*

**IARC:** Group 3, unclassifiable

**Label Signal Word:** WARNING, CAUTION, DANGER

*Description:* Odorless yellow-brown crystalline solid or powder. Moderately soluble in water; solubility = 2-3% @ 25°C. Molecular weight 251.25. Melting/Freezing point = >200°C (decomposes).

*Permissible Concentration in Water:* No criteria set. Runoff from spills or fire control may cause water pollution.

*Routes of Entry:* Inhalation and skin contact

*Harmful Effects and Symptoms*

*Short Term Exposure:* High exposure may cause weakness, tremors, convulsions and death.

*Points of Attack:* Liver and kidneys

*First Aid:* If this chemical gets into the eyes, remove any contact lenses at once and irrigate immediately for at least 15 minutes, occasionally lifting upper and lower lids. Seek medical attention immediately. If this chemical contacts the skin, remove contaminated clothing and wash immediately with soap and water. Seek medical attention immediately. If this chemical has been inhaled, remove from exposure, begin rescue breathing (using universal precautions) if breathing has stopped, and CPR if heart action has stopped. Transfer promptly to a medical facility. When this chemical has been swallowed, get medical attention. Give large quantities of water and induce vomiting. Do not make an unconscious person vomit.

*References:*
- Pesticide Management Education Program, "Fenaminosulf (Dexon) Chemical Fact Sheet 10/83," Cornell University, Ithaca, NY (October 1983). http://pmep.cce.cornell.edu/profiles/fung-nemat/febuconazole-sulfur/fenaminosulf/fung-prof-fenaminosulf.html
- New Jersey Department of Health and Senior Services, "Hazardous Substance Fact Sheet, Fenaminosulf," Trenton NJ (April 2001). http://www.state.nj.us/health/eoh/rtkweb/0913.pdf

# Fenamiphos

*Use Type:* nematicide and insecticide
*CAS Number:* 22224-92-6
*Formula:* $C_{13}H_{22}NO_3PS$
*Alert:* A Restricted Use Pesticide (RUP). EPA Toxicity Class 1, a highly toxic compound. Bayer Corporation announced on March 15, 2002, due to the escalating costs of defending fenamiphos relative to its limited use, Bayer requests voluntary cancellation of all uses of fenamiphos effective May 31, 2005. In addition, Bayer agrees that the FIFRA 6(f) provisions for a 180-day comment period can be waived. Human toxicity (long-term): High.

*Synonyms:* *O*-Aethyl-*O*-(3-methyl-4-methylthiophenyl)-isopropylamido-phosphorsaeure ester (German); ENT 27,572; Ethyl 3-methyl-4-(methylthio)phenyl(1-methylethyl)phosphoramidate; Ethyl 4-(methylthio)-*m*-tolylisopropylphosphoramidate; Isopropylamino-*O*-ethyl-(4-methylmer capto-3-methylphenyl)phosphate; (1-Methylethyl) phosphoramidic acid ethyl 3-methyl-4-(methylthio)phenyl ester; Isopropylphosphoramidic acid ethyl 4-(methylthio)-*m*-toyl ester; 1-(Methylethyl)-ethyl 3-methyl-4-(methylthio)phenylphosphoramidate; NSC-195106; Phenamiphos; Phosphoramidic acid, (1-methylethyl)-, ethyl(3-methyl-4-(methylthio)phenyl)ester; Phosphoramidic acid, (1-methylethyl)-, ethyl 3-methyl-4-(methylthio)phenyl ester; Phosphoramidic acid,isopropyl-, ethyl 4-(methylthio)-*m*-tolyl ethyl ester

*Trade Names:* BAY 68138®, Bayer CropScience (Germany); Bayer 68138®, Bayer CropScience (Germany); NEMACUR®, Bayer CropScience (Germany); NEMACURP®

*Producers:* Bayer CropScience (Germany); Ehrenstorfer, Dr. (Germany); Miles (USA); Shenzhen Guomeng Industry Co., Ltd. (China); Sigma-Aldrich Laborchemikalien (Germany); Yashima Chemical Industry (Japan)

*Chemical Class:* Organophosphate
*EPA/OPP PC Code:* 100601
*California DPR Chemical Code:* 1857
*ICSC Number:* 0483
*RTECS Number:* TB3675000
*EEC Number:* 015-123-00-5
*EINECS Number:* 244-848-1

*Uses:* Fenamiphos is an organophosphate nematicide used to control a wide variety of nematode (roundworm) pests. Nematodes can live as parasites on the outside or the inside of a plant. They may be free living or associated with cyst and root-knot formations in plants. Fenamiphos is used on a variety of plants including tobacco, turf, bananas, pineapples, citrus and other fruit vines, some vegetables, and grains. The compound is absorbed by roots and is then distributed throughout the plant. Fenamiphos, as is typical of other organophosphates, blocks the enzyme acetylcholinesterase in the target pest. The pesticide also has secondary activity against other invertebrates such as sucking insects and spider mites. It is available in emulsifiable concentrate, granular, or emulsion formulations.

*Human toxicity (long-term)*[77]: High–2.00 ppb, Health Advisory

*Fish toxicity (threshold)*[77]: Extra high–0.33001 ppb, MATC (Maximum Acceptable Toxicant Concentration)

*U.S. Maximum Allowable Residue Levels for Fenamiphos (40 CFR 180.349 ): Note:* There are currently no pesticide products containing fenamiphos approved for use in the UK. See UK Maximum Residue Levels, 1 August 2004, http://www.pesticides.gov.uk/applicant/aahip/aahl0404.htm

| CROP | ppm |
|---|---|
| Apple | 0.25 |
| Asparagus | 0.02 |
| Banana | 0.1 |
| Beet, garden, roots | 1.5 |
| Beet, garden, tops | 1 |
| Brussels sprouts | 0.1 |
| Cabbage | 0.1 |
| Cabbage, chinese, bok choy | 0.5 |
| Cattle, fat | 0.05 |
| Cattle, meat | 0.05 |
| Cattle, mbyp | 0.05 |
| Cherry | 0.25 |
| Citrus, dried pulp | 2.5 |
| Citrus, oil | 25 |
| Cotton, undelinted seed | 0.05 |

| | |
|---|---|
| Eggplant | 0.1 |
| Garlic | 0.5 |
| Goat, fat | 0.05 |
| Goat, meat | 0.05 |
| Goat, mbyp | 0.05 |
| Grape | 0.1 |
| Grape, raisin | 0.3 |
| Grapefruit | 0.6 |
| Hog, fat | 0.05 |
| Hog, meat | 0.05 |
| Hog, mbyp | 0.05 |
| Horse, fat | 0.05 |
| Horse, meat | 0.05 |
| Horse, mbyp | 0.05 |
| Kiwifruit | 0.1 |
| Lemon | 0.6 |
| Lime | 0.6 |
| Milk | 0.01 |
| Okra | 0.3 |
| Orange | 0.6 |
| Peach | 0.25 |
| Peanut | 0.02 |
| Pepper, nonbell | 0.6 |
| Pineapple | 0.3 |
| Pineapple, bran | 10 |
| Raspberry | 0.1 |
| Sheep, fat | 0.05 |
| Sheep, meat | 0.05 |
| Sheep, mbyp | 0.05 |
| Strawberry | 0.6 |
| Tangerine | 0.6 |

*Carcinogen/Hazard Classifications*
**U.S. EPA Carcinogens:** Group E, unlikely carcinogen
**Label Signal Word:** DANGER
**WHO Acute Hazard:** Class 1b, highly hazardous
*Regulatory Authority:*
- Air Pollutant Standard Set (ACGIH)[1] (Several States)[60] (Several Canadian Provinces)
- AB 1803-Well Monitoring Chemical (CAL)
- Permissible Exposure Limits for Chemical Contaminants (CAL/OSHA)
- The "Director's List" (CAL/OSHA)
- Actively registered pesticide in California.
- Superfund/EPCRA 40CFR355, Appendix B Extremely Hazardous Substances: TPQ = 10/10,000 lb (4.54/4,540 kg)
- Superfund/EPCRA 40CFR302.4 RQ: EHS, 1 lb (0.454 kg)
- MARINE POLLUTANT (49CFR, Subchapter 172.101, Appendix B
- U.S. DOT Inhalation Hazard Chemicals as organophosphates
- Canada, WHMIS, Ingredients Disclosure List

*Description:* Fenamiphos is an off-white to tan, waxy solid. Found commercially as a granular ingredient (5-15%) or in an emulsifiable concentrate (400 g/l). Molecular weight = 303.38. Melting/Freezing point = 40°C (technical grade) and 49°C (pure compound). Vapor pressure = $1 \times 10^{-6}$ mmHg @ 20°C. Hazard Identification (based on NFPA-704 M Rating System): Health 3, Flammability 1, Reactivity 0. Slightly soluble in water. Log $K_{ow}$ = 3.33. Values at or above 3.0 are likely to bioaccumulate in marine organisms.
*Incompatibilities:* May hydrolyze under alkaline conditions. Keep away from moisture.
*Permissible Exposure Limits in Air:* There is no OSHA[2] PEL. ACGIH[1] and NIOSH[2] recommend a TWA of 0.1 mg/m$^3$, with the notation that skin absorption is possible. Several states have set guidelines or standards for fenamiphos in ambient air[60] ranging from 1.0 $\mu$g/m$^3$ (North Dakota) to 1.6 $\mu$g/m$^3$ (Virginia) to 2.0 $\mu$g/m$^3$ (Connecticut).
*Determination in Air:* OSHA versatile sampler-2; Toluene/Acetone; Gas chromatography/Flame ionization detection; NIOSH IV, Method #5600, Organophosphorus pesticides[18].
*Permissible Concentration in Water:* A long term health advisory set by EPA is 18 $\mu$g/L and a lifetime health advisory is 9 $\mu$g/L.
*Routes of Entry:* Inhalation, skin absorption, ingestion, skin and/or eye contact
*Harmful Effects and Symptoms*
*Short Term Exposure:* This is a highly toxic chemical (the LD-50 for rats is 8 mg/kg) It is a cholinesterase inhibitor with effects typical of such compounds. Acute exposure to fenamiphos may produce the following sings and symptoms: pinpoint pupils, blurred vision, headaches, dizziness, muscle spasm, and profound weakness. Vomiting, diarrhea, abdominal pain, seizures, and coma may also occur. The heart rate may increase following oral exposure or decrease following dermal exposure. Hypotension (low blood pressure) may occur although hypertension (high blood pressure) is not uncommon. Chest pain may be noted. Respiratory symptoms include dyspnea (shortness of breath), respiratory depression, and respiratory paralysis. Psychosis may occur. This material is highly toxic orally, by inhalation, and by absorption through the skin. Death may occur from respiratory failure. Delayed pulmonary edema may occur after inhalation.
*Long Term Exposure:* Cholinesterase inhibitor; cumulative effect is possible. Organophosphates may damage the nervous system with repeated exposure, resulting in convulsions, respiratory failure. May cause liver damage.
*Points of Attack:* Respiratory system, central nervous system, cardiovascular system, blood cholinesterase
*Medical Surveillance:* Medical observation is recommended for 24 to 48 hours after breathing overexposure, as pulmonary edema may be delayed. Before employment and at regular times after that, the following are recommended: Plasma and red blood cell cholinesterase levels (tests for the enzyme poisoned by this chemical). If exposure stops, plasma levels return to normal in 1-2 weeks

while red blood cell levels may be reduced for 1-3 months. When acetylcholinesterase enzyme levels are reduced by 25% or more below preemployment levels, risk of poisoning is increased, even if results are in lower ranges of "normal." Reassignment to work not involving organophosphate or carbamate pesticides is recommended until enzyme levels recover. If symptoms develop or overexposure occurs, repeat the above tests as soon as possible and get an exam of the nervous system. Also consider complete blood count. Consider chest x-ray following acute overexposure. Do not drink any alcoholic beverages before or during use. Alcohol promotes absorption of organophosphates.

*First Aid:* **Treatment for organophosphate poisoning consists of thorough decontamination, cardiorespiratory support, and administration of the antidotes atropine and pralidoxime. In cases of severe poisoning, diazepam, an anticonvulsant, should also be administered. Antidotes should be administered as prevention even if the diagnosis is in doubt.** Speed in removing material from eyes and skin is of extreme importance. *Eyes:* Eye contact can cause dangerous amounts of these chemicals to be quickly absorbed through the mucous membrane into the bloodstream. Immediately and gently flush eyes with plenty of warm or cold water (NO hot water) for at least 15 minutes, occasionally lifting the upper and lower eyelids. Get medical aid immediately. *Skin:* Get medical aid. Skin contact can cause dangerous amounts of these chemicals to be absorbed into the bloodstream. Wearing the appropriate PPE equipment and respirator for organophosphate/carbamate pesticides, immediately flush skin with plenty of soap and water for at least 15 minutes while removing contaminated clothing and shoes. Shampoo hair promptly if contaminated. The removed, contaminated clothing and shoes should be double-bagged and left in Hot Zone for later disposal by hazardous materials experts. Skin may also be decontaminated with diluted hypochlorite solution. *Inhalation:* Get medical aid. Do not contaminate yourself. Wearing the appropriate PPE equipment and respirator for organophosphate pesticides, immediately remove the victim from the contaminated area to fresh air. If the victim is not breathing, administer artificial respiration. Do NOT use mouth-to-mouth resuscitation; use bag/mask apparatus. If breathing is difficult, administer oxygen through bag/mask apparatus until medical help arrives. Do not leave victim unattended. *Ingestion:* Call poison control. Loosen all clothing. Never give anything by mouth to an unconscious person. If victim is *unconscious or having convulsions,* do nothing except keep victim warm. Get medical aid. Transfer promptly to a medical facility. In cases of ingestion, **do not induce vomiting**. If the victim is alert and asymptomatic, administer a slurry of activated charcoal at a dose of 1 g/kg (infant, child, and adult dose). A soda can and straw may be of assistance when offering charcoal to a child. *In some cases you may be specifically instructed by poison control to induce vomiting by way of 2 tablespoons of syrup of ipecac (adult) washed down with a cup of water.* Do NOT give activated charcoal before or with ipecac syrup. *Note to physician or authorized medical personnel:* Treat cases of respiratory compromise, coma, or excessive pulmonary secretions with respiratory support using protocols and techniques available and within the scope of training. Some cases may necessitate procedures such as endotracheal intubation or cricothyrotomy by properly trained and equipped personnel. When possible, atropine (see *Antidotes*, below) should be given under medical supervision. Patients who are comatose, hypotensive, or having seizures or cardiac arrhythmias should be treated according to advanced life support protocols. *Antidotes:* Two antidotes are administered to treat organophosphate poisoning. Atropine is a competitive antagonist of acetylcholine at muscarinic receptors and is used to control the excessive bronchial secretions which are often responsible for death. Pralidoxime relieves both the nicotinic and muscarine effects of organophosphate poisoning by regenerating acetylcholinesterase and can reduce both the bronchial secretions and the muscle weakness associated with poisoning. The initial intravenous dose of atropine in adults should be determined by the severity of symptoms: An initial adult dose of 1.0 to 2.0 mg or pediatric dose of 0.01 mg/kg (minimum 0.01 mg) should be administered intravenously. If intravenous access cannot be established, atropine may also be given intramuscularly, subcutaneously or via endotracheal tube. Doses should be repeated every 15 minutes until excessive secretions and sweating have been controlled. Once bronchial secretion has been controlled, atropine administration should be repeated whenever the secretions begin to recur. In seriously poisoned patients, very large doses may be required. Alterations of pulse rate and pupillary size should not be used as indicators of treatment adequacy. Pralidoxime should be administered as early in poisoning as possible as its efficacy may diminish when given more than 24 to 36 hours after exposure. Doses are as follows: adult 1.0 g; pediatric 25 to 50 mg/kg. The drug should be administered intravenously over 30 to 60 minutes, but in a life-threatening situation, one-half of the total dose can be given per minute for a total administration time of 2 minutes. Treatment should begin to take effect within 40 minutes with a reduction in symptoms and the amount of atropine necessary to control bronchial secretion. The initial dose can be repeated in 1 hour and then every 8 to 12 hours until the patient is clinically well and no longer requires atropine. If intravenous access cannot be established, pralidoxime may also be given intramuscularly. Early administration of diazepam in addition to the combined atropine and pralidoxime treatment may help prevent the onset of seizures and potential brain and cardiac morphologic damage following high-level organophosphate poisoning.

*References:*
- EXTOXNET, Extension Toxicology Network, "Pesticide Information Profile, Fenamiphos," Oregon State University, Corvallis, OR (June 1996). http://extoxnet.orst.edu/pips/fenamiph.htm
- EPA, Office of Pesticide Programs, Pesticide Residue Limits, "Fenamiphos", 40 CFR 180.349, http://www.epa.gov/pesticides/food/viewtols.htm
- New Jersey Department of Health and Senior Services, "Hazardous Substance Fact Sheet: Fenamiphos," Trenton, NJ (February 1999). http://www.state.nj.us/health/eoh/rtkweb/0914.pdf
- Sax, N.I., Ed., "Dangerous Properties of Industrial Materials Report," 3, No. 1, 52-56, New York, Van Nostrand Reinhold Co. (1983).
- Agency for Toxic Substances and Disease Registry, U.S. Department of Health and Human Services, Public Health Service, "Managing Hazardous Materials Incidents," Atlanta, GA (June 2003).
- U.S. Environmental Protection Agency, "Chemical Profile: Fenamiphos," Washington, DC, Chemical Emergency Preparedness Program (November 30, 1987).
- U.S. Environmental Protection Agency, "Health Advisory: Fenamiphos," Washington, DC, Office of Drinking Water (August 1987).

# Fenarimol (ANSI)

*Use Type:* Fungicide
*CAS Number:* 60168-88-9
*Formula:* $C_{17}H_{12}Cl_2N_2O$
*Synonyms:* Caswell No. 207AA; (2-Chlorophenyl)-α-(4-chlorophenyl)-5-pyrimidinemethanol; α-(2-Chlorophenyl)-α-(4-chlorophenyl)-5-pyridinemethanol; α-(2-Chlorophenyl)-α-(4-chlorophenyl)-5-pyrimidinemethanol; (±)-2,4'-Dichloro-α-(pyrimidin-5-yl)benzhydryl alcohol; 2,4'-Dichloro-α-(pyrimidin-5-yl)benzhydryl alcohol; 5-Pyrimidinemethanol, α-(2-chlorophenyl)-α-(4-chlorophenyl)-
*Trade Names:* BLOC®; EL 222®; COMPOUND 56722®; RIMIDIN®; RUBIGAN®, Gowan Company (USA)
*Producers:* Gowan Company (USA); Nufarm (Australia)
*Chemical Class:* Pyrimidine
*EPA/OPP PC Code:* 206600;
*California DPR Chemical Code:* 1980
*RTECS Number:* UV9279400;
*EINECS Number:* 262-095-7
*Uses:* Used to manage fungi diseases on turf, golf courses, field and greenhouse grown roses and ornamentals.
*Human toxicity (long-term)[77]:* Very low–455.00 ppb, Health Advisory
*Fish toxicity (threshold)[77]:* Low–430.00153 ppb, MATC (Maximum Acceptable Toxicant Concentration)
*Carcinogen/Hazard Classifications*
**U.S. EPA Carcinogens:** Not likely a carcinogen
**Label Signal Word:** CAUTION, DANGER
**WHO Acute Hazard:** Class U, unlikely to be hazardous
**Endocrine Disruptor:** Suspected endocrine disruptor
*Regulatory Authority:*
- EPCRA Section 313 Form R *de minimus* concentration reporting level: 1.0%
- Actively registered pesticide in California.

*Description:* White to off-white crystalline solid. Odorless. Practically insoluble in water. Melting/Freezing point = 118°C. Vapor pressure = $2.2 \times 10^{-7}$ mmHg @ 20°C.
*Determination in Air:* Filter; none; Gravimetric; NIOSH IV [Particulates NOR; #0500 (total), #0600 (respirable)].[18]
*Permissible Concentration in Water:* No criteria set. Runoff from spills or fire control may cause water pollution.
*Harmful Effects and Symptoms*
*Short Term Exposure:* Contact with eyes or skin may cause irritation or injury. Inhalation should be avoided; use NIOSH-approved air purifying respirators for pesticides. May be harmful if swallowed.
*First Aid:* If this chemical gets into the eyes, remove any contact lenses at once and irrigate immediately for at least 15 minutes, occasionally lifting upper and lower lids. Seek medical attention immediately. If this chemical contacts the skin, remove contaminated clothing and wash immediately with soap and water. Seek medical attention immediately. If this chemical has been inhaled, remove from exposure, begin rescue breathing (using universal precautions) if breathing has stopped, and CPR if heart action has stopped. Transfer promptly to a medical facility. When this chemical has been swallowed, get medical attention. Give large quantities of water and induce vomiting. Do not make an unconscious person vomit. Do not induce vomiting when formulations containing petroleum solvents are ingested.
*References:*
- Pesticide Management Education Program, "Fenarimol (Rubigan) Chemical Fact Sheet 2/85," Cornell University, Ithaca, NY (February 1985). http://pmep.cce.cornell.edu/profiles/fung-nemat/febuconazole-sulfur/fenarimol/fung-prof-fenarimol.html
- California Environmental Protection Agency "Chemical List of Lists," Sacramento CA (February 1997)

# Fenbuconazole (ANSI)

*Use Type:* Fungicide
*CAS Number:* 114369-43-6
*Formula:* $C_{19}H_{17}ClN_4$
*Synonyms:* 2-Cyano-2-phenyl-2-(β-p-chlorophenethyl)ethyl-1H-1,2,4-triazole; α-[2-(4-Chlorophenyl)ethyl]-α-phenyl-3-(1H-1,2,4-triazole)-1-propanenitrile; α-[2-(4-Chlorophenyl)ethyl]-α-phenyl-1H-1,2,4-triazole-1-propane nitrile; 4-(4-Chlorophenyl)-2-phenyl-2-(1H-1,2,4-triazol-1-

ylmethyl)butyronitrile; 1H-1,2,4-Triazole-1-propanenitrile, α-[2-(4-chlorophenyl)ethyl]-α-phenyl-

*Trade Names:* ENABLE®, Dow AgroSciences (USA) and Nufarm (Australia); FENETHANIL®; INDAR®, Dow AgroSciences (USA); RH-7592®, Dow AgroSciences (USA); RH-7592-2F®, Dow AgroSciences (USA); RH-7592® Technical Fungicide, Dow AgroSciences (USA); RH-7592-2F® Experimental Fungicide, Dow AgroSciences (USA)

*Producers:* Dow AgroSciences (USA); Nufarm (Australia)

*Chemical Class:* Triazole

*EPA/OPP PC Code:* 129011

*California DPR Chemical Code:* 3905

*Uses:* Fenbuconazole is a triazole fungicide used as an agricultural and horticultural fungicide spray for the control of leaf spot, yellow and brown rust, powdery mildew and net blotch on wheat and barley and apple scab, pear scab and apple powdery mildew on cherries, apples and pears.

*Human toxicity (long-term)*[77]*:* Intermediate–21.00 ppb, Health Advisory

*Fish toxicity (threshold)*[77]*:* Intermediate–34.85700 ppb, MATC (Maximum Acceptable Toxicant Concentration)

*U.S. Maximum Allowable Residue Levels for Fenbuconazole (40 CFR 180.480):*

| CROP | ppm |
| --- | --- |
| Banana | 0.3 |
| Blueberry | 1 |
| Cattle, fat | 0.01 |
| Cattle, meat | 0.01 |
| Cattle, mbyp | 0.01 |
| Fruit, stone, except fresh prune plum | 2 |
| Goat, fat | 0.01 |
| Goat, meat | 0.01 |
| Goat, mbyp | 0.01 |
| Grapefruit | 0.5 |
| Grapefruit, dried pulp | 4 |
| Grapefruit, oil | 35 |
| Hog, fat | 0.01 |
| Hog, meat | 0.01 |
| Hog, mbyp | 0.01 |
| Horse, fat | 0.01 |
| Horse, meat | 0.01 |
| Horse, mbyp | 0.01 |
| Pecan | 0.1 |
| Sheep, fat | 0.01 |
| Sheep, meat | 0.01 |
| Sheep, mbyp | 0.01 |

*Carcinogen/Hazard Classifications*

**U.S. EPA Carcinogens:** Group C, possible carcinogen

**Label Signal Word:** CAUTION or DANGER

**Endocrine Disruptor:** Suspected endocrine disruptor

*Description:* Off white to white powder. Faint sulfur-like odor. Melting point = 126–127°C. Molecular weight = 336.85

*Permissible Concentration in Water:* No criteria set. Runoff from spills or fire control may cause water pollution.

*Short Term Exposure:* Poisonous if swallowed. Contact may irritate skin and cause eye irritation and possible severe injury. Avoid inhalation.

*Medical Surveillance:* If this chemical gets into the eyes, remove any contact lenses at once and irrigate immediately for at least 15 minutes, occasionally lifting upper and lower lids. Seek medical attention immediately. If this chemical contacts the skin, remove contaminated clothing and wash immediately with soap and water. If this chemical has been inhaled, remove from exposure, begin rescue breathing (using universal precautions) if breathing has stopped, and CPR if heart action has stopped. Transfer promptly to a medical facility. When this chemical has been swallowed, get medical attention. Give large quantities of water and induce vomiting. Do not make an unconscious person vomit

*References:*
- U.S. Environmental Protection Agency, Office of Pesticide Programs, Pesticide Residue Limits, "Fenbuconazole," 40 CFR 180.480, www.epa.gov/pesticides/food/viewtols.htm
- California Environmental Protection Agency "Chemical List of Lists," Sacramento CA (February 1997).

# Fenbutatin Oxide

*Use Type:* Insecticide, miticide

*CAS Number:* 13356-08-6

*Formula:* $C_{60}H_{78}OSn_2$

*Synonyms:* A13-27738; Bis(trineophyltin) oxide; Bis[tris($\beta,\beta$-dimethylphenethyl)tin]oxide; Bis[tris(2-methyl-2-phenylpropyl)tin]oxide; Caswell No. 481DD; Distannoxane, hexakis($\beta,\beta$-dimethylphenethyl)-; Distannoxane, hexakis(2-methyl-2-phenylpropyl)-; Di[tri(2,2-dimethyl-2-phenylethyl)tin]oxide; ENT 27738; Fenbutatin-oxyde; Hexakis($\beta,\beta$-dimethylphenethyl)distannoxane; Hexakis; Hexakis(2-methyl-2-phenylpropyl)distannoxane; 2-(Methyl-2-phenylpropyl)distannoxane

*Trade Names:* BENDEX®; NEOSTANOX®; OSDARAN®; SD-14114®; SHELL SD-14114®; TORQUE®; VENDEX®, DuPont Crop Protection (USA)

*Producers:* Agrimor International (USA); DuPont Crop Protection (USA); Griffin (USA); Ortho Business Group (USA)

*Chemical Class:* Organotin, heavy metal

*EPA/OPP PC Code:* 104601; 596300 (obsolete)

*California DPR Chemical Code:* 1876

*RTECS Number:* JN8770000

*Uses:* A selective miticide for deciduous pome and stone fruits, citrus fruits, grapes, vegetables, berry fruit, nut crops (selected), ornamentals and greenhouse crops.

*U.S. Maximum Allowable Residue Levels for Fenbutatin Oxide (40 CFR 180.362):*

| CROP | ppm |
|---|---|
| Almond | 0.5 |
| Almond, hulls | 80 |
| Apple | 15 |
| Cattle, fat | 0.5 |
| Cattle, meat | 0.5 |
| Cattle, mbyp | 0.5 |
| Cherry, sweet | 6 |
| Cherry, tart | 6 |
| Citrus, dried pulp | 100 |
| Citrus, oil | 140 |
| Cucumber | 4 |
| Egg | 0.1 |
| Eggplant | 6 |
| Fruit, stone, group 12 | 20 |
| Goat, fat | 0.5 |
| Goat, meat | 0.5 |
| Goat, mbyp | 0.5 |
| Grape | 5 |
| Grape, raisin | 20 |
| Hog, fat | 0.5 |
| Hog, meat | 0.5 |
| Hog, mbyp | 0.5 |
| Horse, fat | 0.5 |
| Horse, meat | 0.5 |
| Horse, mbyp | 0.5 |
| Milk, fat | 0.1 |
| Papaya | 2 |
| Peach | 10 |
| Pear | 15 |
| Pecan | 0.5 |
| Plum | 4 |
| Plum, prune | 4 |
| Plum, prune, dried | 20 |
| Poultry, fat | 0.1 |
| Poultry, meat | 0.1 |
| Poultry, mbyp | 0.1 |
| Raspberry | 10 |
| Sheep, fat | 0.5 |
| Sheep, meat | 0.5 |
| Sheep, mbyp | 0.5 |
| Strawberry | 10 |
| Walnut | 0.5 |

*Carcinogen/Hazard Classifications*
**U.S. EPA Carcinogens:** Group E, Unlikely a carcinogen
**TRI Developmental Toxin:** Reproductive and developmental toxin
**Label Signal Word:** DANGER
**WHO Acute Hazard:** Class U, unlikely to be hazardous
*Regulatory Authority:*
- EPCRA Section 313 Form R *de minimus* concentration reporting level: 1.0%
- FIFRA, 40CFR186: tolerances for pesticides in animal feeds.
- Actively registered pesticide in California.
- U.S. DOT Regulated Marine Pollutant (49CFR, Subchapter 172.101, Appendix B), severe pollutant, as organotin pesticide compounds.

*Description:* White crystalline solid or powder. Mild odor. Practically insoluble in water; solubility = 0.005 mg/L @ 23°C. Molecular weight = 1052.66. Boiling point = 238°C @ 0.05 mm. Melting/Freezing point = 138°C. Vapor pressure = $1.8 \times 10^{-11}$ mmHg @ 20°C.

*Permissible Exposure Limits in Air:* NIOSH[2]/OSHA: 0.1 mg/m$^3$; ACGIH STEL®: 0.2 mg/m$^3$. The notation "skin" is added indicating the possibility of cutaneous absorption. as organic tin compounds

*Determination in Air:* Filter/XAD-2® (tube); $CH_3COOH/CH_3CN$; High-pressure liquid chromatography/Graphite furnace atomic absorption spectrometry; NIOSH IV Method #5504, as organotin compounds.[18]

*Permissible Concentration in Water:* No criteria set. Runoff from spills or fire control may cause water pollution. A MAC in water bodies used for domestic purposes of 0.2 µg/L has been set by the former USSR-UNEP/IRPTC joint project[43]. Log $K_{ow}$ = 4.09. Highly likely to bioaccumulate in marine organisms.

*Determination in Water:*

*Routes of Entry:* Inhalation, skin and/or eye contact. Absorbed through the skin.

*Harmful Effects and Symptoms*

*Short Term Exposure:* Irritates the eyes, skin, and respiratory tract. Contact may cause skin burns. Inhalation can cause coughing, wheezing and/or shortness of breath. Toxic hazard rating is high for oral, intravenous, intraperitoneal administration. This material causes swelling of the brain and spinal cord. Exposure may result in muscular weakness and paralysis, leading to respiratory failure; convulsive movements; closure of eyelids and sensitivity to light; headaches, and EEG changes, headache, dizziness, psychological and neurological disturbances, vertigo (an illusion of movement), sore throat, cough, abdominal pain, nausea, vomiting, diarrhea; urine retention; paresis, focal anesthesia; pruritus. Higher levels can cause unconsciousness, collapse and death.

*Long Term Exposure:* Repeated or prolonged contact can cause dermatitis; dry and cracked skin. May cause brain damage, hepatic necrosis; kidney damage. Some organotin compounds, such as dibutyltins and tributyltins, have been shown to affect the immune system in animals, but this has not been examined in people. Studies in animals also have shown that some organotins, such as dibutyltins, tributyltins, and triphenyltins, can affect the reproductive system. This, also, has not been examined in people.

*Points of Attack:* Skin, brain, kidneys.

*Medical Surveillance:* Kidney function tests. Psychological testing. Examination of the nervous system. EEG

*First Aid:* If this chemical gets into the eyes, remove any contact lenses at once and irrigate immediately for at least 15 minutes, occasionally lifting upper and lower lids. Seek medical attention immediately. If this chemical contacts the skin, remove contaminated clothing and wash immediately with soap and water. Seek medical attention immediately. If this chemical has been inhaled, remove from exposure, begin rescue breathing (using universal precautions) if breathing has stopped, and CPR if heart action has stopped. Transfer promptly to a medical facility. When this chemical has been swallowed, get medical attention. *Do not induce vomiting when formulations containing petroleum solvents are ingested.* Otherwise, give large quantities of water and induce vomiting. Do not make an unconscious person vomit.

*References:*
- U.S. Environmental Protection Agency, Office of Pesticide Programs, Pesticide Residue Limits, "Fenbutatin Oxide," 40 CFR 180.362, http://www.epa.gov/pesticides/food/viewtols.htm
- International Programme on Chemical Safety (IPCS), "Health and Safety Guide, Fenbutatin Oxide," Geneva, Switzerland (1977). http://www.inchem.org/documents/jmpr/jmpmono/v077pr28.htm
- California Environmental Protection Agency "Chemical List of Lists," Sacramento CA (February 1997)

# Fenfluthrin

*Use Type:* Insecticide
*CAS Number:* 75867-00-4
*Formula:* $C_{15}H_{11}Cl_2F_5O_2$
*Synonyms:* Cyclopropanecarboxylic acid, 3-(2,2-dichloroethenyl)-2,2-dimethyl-, (pentafluorophenyl)methyl ester, (1R-*trans*)-; (1R)-*trans*-(2,2-Dichlorovinyl)-2,2-dimethylcyclopropanecarboxylate; Pentafluorobenzyl; (+)-(Pentafluorophenyl)methyl (1R-*trans*)-3-(2,2-dichlorovinyl)-2,2-dimethylcyclopropanecarboxylate
*Trade Names:* BAYNAC®, Bayer CropScience (Germany); BAY NAK 1654®, Bayer CropScience (Germany);
*Producers:* Ascot International UK); Bayer CropScience (Germany); Changzhou Kangmei Chemical Industry Co., Ltd. (China)
*Chemical Class:* Pyrethroid
*EPA/OPP PC Code:* 109705
*Uses:* Not registered in the U.S.
*Carcinogen/Hazard Classifications*
**Endocrine Disruptor:** Suspected endocrine disruptor
*Incompatibilities:* May react violently with strong oxidizers, bromine, 90% hydrogen peroxide, phosphorus trichloride, silver powders or dust. Incompatible with silver compounds. Mixture with some silver compounds forms explosive salts of silver oxalate.

*Harmful Effects and Symptoms*
**Short Term Exposure:** Contact with eyes or skin may cause irritation or injury. Inhalation should be avoided; use NIOSH-approved air purifying respirators for pesticides. Harmful if swallowed. Skin contact is toxic and may cause severe irritation or burns.

*First Aid:* If this chemical gets into the eyes, remove any contact lenses at once and irrigate immediately for at least 15 minutes, occasionally lifting upper and lower lids. Seek medical attention immediately. If this chemical contacts the skin, remove contaminated clothing and wash immediately with soap and water. Seek medical attention immediately. If this chemical has been inhaled, remove from exposure, begin rescue breathing (using universal precautions) if breathing has stopped, and CPR if heart action has stopped. Transfer promptly to a medical facility. When this chemical has been swallowed, get medical attention. Give large quantities of water and induce vomiting. Do not make an unconscious person vomit. Do not induce vomiting when formulations containing petroleum solvents are ingested.

# Fenitrothion

*Use Type:* Insecticide
*CAS Number:* 122-14-5
*Formula:* $C_9H_{12}NO_5PS$; $(CH_3O)_2PSO-C_6H_3(NO_2)(CH_3)$
*Alert:* A General Use Pesticide (GUP). Human toxicity (long-term): High.
*Synonyms:* O,O-Dimethyl-O-(3-methyl-4-nitrofenyl)-monothiofosfaat (Dutch); O,O-Dimethyl-O-(3-methyl-4-nitrophenyl)-monothiophosphat (German); O,O-Dimethyl-O-(3-methyl-4-nitrophenyl)-phosphorothioate; O,O-Dimethyl-O-(3-methyl-4-nitrophenyl)-thiophosphate; O,O-Dimethyl-O-(3-methyl)phosphorothioate; O,O-Dimethyl-O-(4-nitro-3-methylphenyl)thiophosphate; O,O-Dimethyl-O-4-nitro-*m*-toylphosphorothioate; ENT 25,715; Fenitrotion (Hungarian); Methylnitrophos (used in Eastern Europe); Phosphorothioic acid, O,O-dimethyl O-(3-methyl-4-nitrophenyl)ester; Phosphorothioic acid, O,O-dimethyl O-(4-nitro-*m*-tolyl)ester; Phenitrothion; Thiophosphate de O,O-dimethyle et de O-(3-methyl-4-nitrophenyle)(french)
*Trade Names:* ACCOTHION®; ACEOTHION®; AGRIA 1050®; AGRIYA 1050®; AGROTHION®; AMERICAN CYANAMID CL-47,300®; ARBOGAL®; BAY 41831®, Bayer Chemicals (Germany); BAYER 41831®, Bayer Chemicals (Germany); BAYER S 5660®, Bayer Chemicals (Germany); CEKUTROTHION®; CL 47300®; CP47114®; CYFEN®; CYTEL®; CYTEN®; DICATHION®; DICOFEN®; DYBAR®; EI 47300®; FALITHION®; FENITEX®; FENITOX®; FENSTAN®; FOLETHION®; FOLITHION®; H-35-F 87 (BVM)®; 8057HC®; KALEIT®; KEEN SUPERKILL ANT AND ROACH

EXTERMINATOR®; KILLGERM TETRACIDE INSECTICIDAL SPRAY®; KOTION®; MEP (PESTICIDE)®; METATHION®; METATHIONE®; METATION®; MICROMITE®; MONSANTO CP 47114®, Dow Chemical (USA); NITROPHOS®; NOVATHION®; NUVAND®; NUVANOL®; OLEOSUMIFENE®; OMS 43®; OVADOFOS®; PENNWALT C-4852®, Atofina (France); PESTROY®; S 112A®; S 5660®; SMT®; SUMITHION®, Sumitomo Chemical (Japan), canceled 10/05/1995; TURBAIR GRAIN STORAGE INSECTICIDE®; VERTHION®

***Producers:*** Agrimor International (USA); Agsin (Singapore); Alcotan Laboratories (Spain); Atofina (France); Bayer Chemicals (Germany); Bharat Pulverizing Mills (India); Biesterfeld Siemsgluess International. GmbH (Germany); Cyanamid (USA); Dow Chemical (USA); Eli Lilly (USA); Hockley International (UK); Hokko Chemical Industry (Japan); Ki-Hara Chemicals Ltd. (UK); Jingma Chemicals Ltd. (China); Luxembourg Industries (PAMOL) (Israel); Nissan Chemical Industries (Japan); Rallis India (India); Saeryung Chemicals (South Korea); Shenzhen Guomeng Industry Co., Ltd. (China); Sigma-Aldrich Laborchemikalien (Germany); Sinon (Taiwan); Sumitomo Chemical (Japan); Takeda Chemical Industries (Japan)

***Chemical Class:*** Organophosphate
***EPA/OPP PC Code:*** 105901
***California DPR Chemical Code:*** 2520
***ICSC Number:*** 0622
***RTECS Number:*** TG0350000
***EEC Number:*** 015-054-00-0
***EINECS Number:*** 204-524-2

***Uses:*** This is a selective acaricide and a contact and stomach insecticide. Fenitrothion is a contact insecticide and selective acaricide of low ovicidal properties. It is considered a cholinesterase inhibitor. Fenitrothion is effective against a wide range of pests, i.e. penetrating, chewing and sucking insect pests (coffee leafminers, locusts, rice stem borers, wheat bugs, flour beetles, grain beetles, grain weevils) on cereals, cotton, orchard fruits, rice, vegetables, and forests. It may also be used as a fly, mosquito, and cockroach residual contact spray for farms and public health programs. Fenitrothion is also effective against household insects and all of the nuisance insects listed by the World Health Organization. Its effectiveness as a vector control agent for malaria is confirmed by the World Health Organization. Fenitrothion is non-systemic, and non-persistent. Fenitrothion was introduced in 1959 by both Sumitomo Chemical Company and Bayer Leverkusen and later by American Cyanamid Company. ). Fenitrothion is far less toxic than parathion with a range of insecticidal activity that is very similar and is similar enough in structure to be produced in the same factories. The difference in precursor chemicals might make it somewhat more expensive, but it is heavily used in other countries, including Japan, where parathion has been banned. Fenitrothion comes in dust, emulsifiable concentrate, flowable, fogging concentrate, granules, ULV, oil-based liquid spray, and wettable powder formultaions. It is compatible with other neutral insecticides.

***Human toxicity (long-term)[77]:*** High–9.10 ppb, Health Advisory

***Fish toxicity (threshold)[77]:*** Intermediate–63.62386 ppb, MATC (Maximum Acceptable Toxicant Concentration)

***U.S. Maximum Allowable Residue Levels for Fenitrothion (40 CFR 180.540):***

| CROP | ppm |
|---|---|
| Wheat, gluten | 15 |

***Carcinogen/Hazard Classifications***
**U.S. EPA Carcinogens:** Group E, unlikely carcinogen
**Label Signal Word:** CAUTION
**WHO Acute Hazard:** Group II, moderately hazardous
**Endocrine Disruptor:** Suspected endocrine disruptor
***Regulatory Authority:***
- Air Pollutant Standard Set (former USSR)[43] (Japan)[35]
- U.S. DOT Inhalation Hazard Chemicals as organophosphates
- Superfund/EPCRA 40CFR355, Appendix B Extremely Hazardous Substances: TPQ = 500 lb (227kg)
- Superfund/EPCRA 40CFR302.4 RQ: EHS, 1 lb (0.454 kg)
- U.S. DOT Regulated Marine Pollutant (49CFR, Subchapter 172.101, Appendix B), severe pollutant

***Description:*** Fenitrothion is a volatile brownish-yellow oil. Practically insoluble in water; solubility = 30 ppm. Molecular weight = 277.2. Boiling point = 118°C @ 0.05 mmHg. (also found in the literature @140 –145°C, decomposes). Decomposes below the BP. Melting/Freezing point = 0.3°C. Vapor pressure = $5.3 \times 10^{-5}$ mmHg @ 20. Flash point = 157°C. Hazard Identification (based on NFPA-704 M Rating System): Health 3, Flammability 1, Reactivity 0. Log $K_{ow}$ = 3.4 @ 20°C. Values at or above 3.0 are likely to bioaccumulate in marine organisms.

***Incompatibilities:*** Strong oxidizers, strong bases.

***Permissible Exposure Limits in Air:*** The former USSR-UNEP/IRPTC project[43] has set an MAC in workplace air of 0.1 mg/m³ and an MAC in ambient air in residential areas of 0.008 mg/m³ on a momentary basis and 0.001 mg/m³ on a daily average basis. Japan[35] has set an MAC in workplace air of 1.0 mg/m³ in workplace air.

***Determination in Air:*** OSHA versatile sampler-2; Toluene/Acetone; Gas chromatography/Flame photometric detection for sulfur, nitrogen, or phosphorus; NIOSH Method IV Method #5600, Organophosphorus pesticides[18].

***Permissible Concentration in Water:*** The former USSR-UNEP/IRPTC project[43] has set an MAC in water bodies for domestic purposes of 0.25 mg/L; in water bodies used for fishery purposes of zero.

***Routes of Entry:*** Inhalation, through the skin, ingestion.

***Harmful Effects and Symptoms***

***Short Term Exposure:*** Irritates the eyes and skin. Nausea is often the first symptom, followed by vomiting; abdominal cramps; diarrhea; excessive salivation; headache; giddiness; dizziness; weakness; tightness in the chest; loss of muscle coordination; slurring of speech, muscle twitching (particularly the tongue and eyelid); respiratory difficulty; blurring or dimness of vision; pinpoint pupils; profound weakness; mental confusion; disorientation and drowsiness. This compound is an organophosphate insecticide. It is a highly toxic cholinesterase inhibitor that acts on the nervous system. Does not cause delayed neurotoxicity and contact produces little irritation. The effects may be delayed. Delayed pulmonary edema may occur after inhalation. Keep exposed victim under medical observation.

*Long Term Exposure:* Cholinesterase inhibitor; cumulative effect is possible. This chemical may damage the nervous system with repeated exposure, resulting in convulsions, respiratory failure. May cause liver damage.

*Points of Attack:* Respiratory system, kidneys, lungs, central nervous system, cardiovascular system, skin, eyes, plasma and red blood cell cholinesterase; personality change, depression, anxiety or irritability.

*Medical Surveillance:* Medical observation is recommended for 24 to 48 hours after breathing overexposure, as pulmonary edema may be delayed. Before employment and at regular times after that, the following are recommended: Plasma and red blood cell cholinesterase levels (tests for the enzyme poisoned by this chemical). If exposure stops, plasma levels return to normal in 1-2 weeks while red blood cell levels may be reduced for 1-3 months. When acetylcholinesterase enzyme levels are reduced by 25% or more below preemployment levels, risk of poisoning is increased, even if results are in lower ranges of "normal." Reassignment to work not involving organophosphate or carbamate pesticides is recommended until enzyme levels recover. If symptoms develop or overexposure occurs, repeat the above tests as soon as possible and get an exam of the nervous system. Also consider complete blood count. Consider chest x-ray following acute overexposure. Do not drink any alcoholic beverages before or during use. Alcohol promotes absorption of organophosphates. Also consider complete blood count. Consider chest x-ray following acute overexposure.

*First Aid:* **Treatment for organophosphate poisoning consists of thorough decontamination, cardiorespiratory support, and administration of the antidotes atropine and pralidoxime. In cases of severe poisoning, diazepam, an anticonvulsant, should also be administered. Antidotes should be administered as prevention even if the diagnosis is in doubt.** Speed in removing material from eyes and skin is of extreme importance. *Eyes:* Eye contact can cause dangerous amounts of these chemicals to be quickly absorbed through the mucous membrane into the bloodstream. Immediately and gently flush eyes with plenty of warm or cold water (NO hot water) for at least 15 minutes, occasionally lifting the upper and lower eyelids. Get medical aid immediately. *Skin:* Get medical aid. Skin contact can cause dangerous amounts of these chemicals to be absorbed into the bloodstream. Wearing the appropriate PPE equipment and respirator for organophosphate/carbamate pesticides, immediately flush skin with plenty of soap and water for at least 15 minutes while removing contaminated clothing and shoes. Shampoo hair promptly if contaminated. The removed, contaminated clothing and shoes should be double-bagged and left in Hot Zone for later disposal by hazardous materials experts. Skin may also be decontaminated with diluted hypochlorite solution. *Inhalation:* Get medical aid. Do not contaminate yourself. Wearing the appropriate PPE equipment and respirator for organophosphate pesticides, immediately remove the victim from the contaminated area to fresh air. If the victim is not breathing, administer artificial respiration. Do NOT use mouth-to-mouth resuscitation; use bag/mask apparatus. If breathing is difficult, administer oxygen through bag/mask apparatus until medical help arrives. Do not leave victim unattended. *Ingestion:* Call poison control. Loosen all clothing. Never give anything by mouth to an unconscious person. If victim is *unconscious or having convulsions,* do nothing except keep victim warm. Get medical aid. Transfer promptly to a medical facility. In cases of ingestion, **do not induce vomiting**. If the victim is alert and asymptomatic, administer a slurry of activated charcoal at a dose of 1 g/kg (infant, child, and adult dose). A soda can and straw may be of assistance when offering charcoal to a child. *In some cases you may be specifically instructed by poison control to induce vomiting by way of 2 tablespoons of syrup of ipecac (adult) washed down with a cup of water.* Do NOT give activated charcoal before or with ipecac syrup.

*Note to physician or authorized medical personnel:* Treat cases of respiratory compromise, coma, or excessive pulmonary secretions with respiratory support using protocols and techniques available and within the scope of training. Some cases may necessitate procedures such as endotracheal intubation or cricothyrotomy by properly trained and equipped personnel. When possible, atropine (see *Antidotes*, below) should be given under medical supervision. Patients who are comatose, hypotensive, or having seizures or cardiac arrhythmias should be treated according to advanced life support protocols. *Antidotes:* Two antidotes are administered to treat organophosphate poisoning. Atropine is a competitive antagonist of acetylcholine at muscarinic receptors and is used to control the excessive bronchial secretions which are often responsible for death. Pralidoxime relieves both the nicotinic and muscarine effects of organophosphate poisoning by regenerating acetylcholinesterase and can reduce both the bronchial secretions and the muscle weakness associated with poisoning. The initial intravenous dose of atropine in adults should be determined by the severity of symptoms: An initial adult dose of 1.0 to 2.0 mg or pediatric dose of 0.01 mg/kg (minimum 0.01 mg) should be administered intravenously. If intravenous access cannot be established, atropine may also be given intramuscularly, subcutaneously or via endotracheal tube. Doses should be repeated every 15 minutes until excessive secretions and sweating have been controlled. Once bronchial secretion has been controlled, atropine administration should be repeated whenever the secretions begin to recur. In seriously poisoned patients, very large doses may be required. Alterations of pulse rate and pupillary size should not be used as indicators of treatment adequacy. Pralidoxime should be administered as early in poisoning as possible as its efficacy may diminish when given more than 24 to 36 hours after exposure. Doses are as follows: adult 1.0 g; pediatric 25 to 50 mg/kg. The drug should be administered intravenously over 30 to 60 minutes, but in a life-threatening situation, one-half of the total dose can be given per minute for a total administration time of 2 minutes. Treatment should begin to take effect within 40 minutes with a reduction in symptoms and the amount of atropine necessary to control bronchial secretion. The initial dose can be repeated in 1 hour and then every 8 to 12 hours until the patient is clinically well and no longer requires atropine. If intravenous access cannot be established, pralidoxime may also be given intramuscularly. Early administration of diazepam in addition to the combined atropine and pralidoxime treatment may help prevent the onset of seizures and potential brain and cardiac morphologic damage following high-level organophosphate poisoning.

*References:*
- EXTOXNET, Extension Toxicology Network, "Pesticide Information Profile, Fenitrothion," Oregon State University, Corvallis, OR. (September 1995). http://extoxnet.orst.edu/pips/fenitrot.htm
- EPA, Office of Pesticide Programs, Pesticide Residue Limits, "Fenitrothion", 40 CFR 180.540, http://www.epa.gov/pesticides/food/viewtols.htm
- New Jersey Department of Health and Senior Services, "Hazardous substance Fact Sheet, Fenitrothion," Trenton, NJ (August 2000). http://www.state.nj.us/health/eoh/rtkweb/2410.pdf
- Sax, N.I., Ed., "Dangerous Properties of Industrial Materials Report," 2, No. 4, 88-92 (1982).
- California Environmental Protection Agency "Chemical List of Lists," Sacramento CA (February 1997).
- Lee, C.C., "Environmental Law Index to Chemicals," Government Institutes, Inc., Rockville, MD (1996).
- Agency for Toxic Substances and Disease Registry, U.S. Department of Health and Human Services, Public Health Service, "Managing Hazardous Materials Incidents," Atlanta, GA (June 2003).
- U.S. Environmental Protection Agency, "Chemical Profile: Fenitrothion," Washington, DC, Chemical Emergency Preparedness Program (November 30, 1987).

# Fenoxaprop-ethyl

*Use Type:* Herbicide
*CAS Number:* 66441-23-4
*Formula:* $C_{18}H_{16}ClNO_5$
*Synonyms:* Caswell No. 431C; 2-[4-((6-Chloro-2-benzoxazolyl)oxy)phenoxy]propionic acid, ethyl ester, (±)-; (±)-Ethyl 2-[4-((6-chloro-2-benzoxazolyl)oxy)phenoxy]propanoate; Ethyl-2-[(4-(6-chloro-2-benzoxazolyloxy))-phenoxy]propionate; Ethyl-2-[4-((6-chlorobenzoxazol-2-yl)oxy)phenoxy]propionate; Ethyl (D+)-2-[4-(6-chlor-2-benzoxazolyloxy)phenoxy]propanoate; (±)Ethyl-2-[-((6-chloro-2-benzoxazolyl)oxy)phenoxy]ropionate; Propionic acid, 2-[4-((6-chloro-2-benzoxazolyl)oxy)phenoxy]ethylester, (±)-
*Trade Names:* ACCLAIM®, Bayer CropScience (Germany), canceled; DEPON®; EXCEL®; FENOXYPROP®; FURORE®; HOE 033171®; HOE-A 25-01®; OPTION®; PUMA®; WHIP®, Bayer CropScience (Germany)
*Producers:* Agrimor International (USA); Bayer CropScience (Germany)
*Chemical Class:* Chlorophenoxy
*EPA/OPP PC Code:* 128701
*California DPR Chemical Code:* 2311
*EINECS Number:* 266-362-9
*Uses:* Used to control annual and perennial grassy weeds in potatoes, soy beans, beans, beets, vegetables, flax, ground nuts, rape and cotton.
*Human toxicity (long-term)*[77]: Intermediate–17.50 ppb, Health Advisory
*Fish toxicity (threshold)*[77]: Intermediate–34.71346 ppb, MATC (Maximum Acceptable Toxicant Concentration)
*U.S. Maximum Allowable Residue Levels for Fenoxaprop-Ethyl (40 CFR 180.430):*

| CROP | ppm |
|---|---|
| Barley, grain | 0.05 |
| Barley, straw | 0.1 |

*Carcinogen/Hazard Classifications*
**U.S. EPA Carcinogens:** Deferred decision
**California Prop. 65:** Developmental toxin
**TRI Developmental Toxin:** Reproductive and developmental toxin
**Label Signal Word:** WARNING, CAUTION or DANGER
*Regulatory Authority:*
- AB 2588-Air Toxics "Hot Spots" Chemicals (CAL) as chlorophenoxy pesticides
- The "Director's List" (CAL/OSHA) as chlorophenoxy pesticides
- Proposition 65 chemical (CAL)
- EPCRA Section 313 Form R *de minimis* concentration reporting level: 1.0%

*Description:* Colorless solid. Melting point = 84–85°C. Vapor pressure = $3.2 \times 10^{-8}$ mmHg @ 20°C.

*Incompatibilities:* Decomposed by acids and alkalis.
*Permissible Concentration in Water:* No criteria set. Runoff from spills or fire control may cause water pollution.
*Routes of Entry:* Inhalation, ingestion, absorbed through the skin.

*Harmful Effects and Symptoms*

*Short Term Exposure:* Irritates the skin, eyes, and respiratory tract. Eye contact may cause irritation, burning sensation, and damage. Harmful if ingested, inhaled or absorbed through the skin. Inhalation should be avoided; use NIOSH-approved air purifying respirators for pesticides. May be harmful if swallowed.

*Medical Surveillance:* Consult a physician if poisoning is suspected or if redness, itching, or burning of the eyes or skin develop.

*First Aid:* If this chemical gets into the eyes, remove any contact lenses at once and irrigate immediately for at least 15 minutes, occasionally lifting upper and lower lids. Seek medical attention immediately. If this chemical contacts the skin, remove contaminated clothing and wash immediately with soap and water. Seek medical attention immediately. If this chemical has been inhaled, remove from exposure, begin rescue breathing (using universal precautions) if breathing has stopped, and CPR if heart action has stopped. Transfer promptly to a medical facility. When this chemical has been swallowed, get medical attention. Give large quantities of water and induce vomiting. Do not make an unconscious person vomit. Do not induce vomiting when formulations containing petroleum solvents are ingested.

*References:*
- California Environmental Protection Agency "Chemical List of Lists," Sacramento CA (February 1997)
- U.S. Environmental Protection Agency, Office of Pesticide Programs, Pesticide Residue Limits, "Fenoxaprop-Ethyl," 40 CFR 180.430, http://pmep.cce.cornell.edu/profiles/herb-growthreg/fatty-alcohol-monuron/fenoxaprop-ethyl/Fenoxaprop-ethyl_tol_498.html

## Fenoxycarb (ANSI)

*Use Type:* Insecticide
*CAS Number:* 79127-80-3
*Formula:* $C_{17}H_{19}NO_4$
*Synonyms:* AI3-29460; Carbamic acid, [2-(4-phenoxyphenoxy)ethyl]-, ethyl ester; Carbamic acid, [2-4(-phenoxyphenoxy)ethyl]-, ethyl ester; Caswell No. 652C; Ethyl [2-(4-phenoxyphenoxy)ethyl]carbamate; Ethyl 2-(*p*-phenoxyphenoxy)ethyl carbamate; Ethyl[2-(*p*-phenoxyphenoxy)ethyl]carbamate; 2-(4-Phenoxyphenoxy)ethylcarbamic acid ethyl ester; [2-(4-Phenoxyphenoxy)ethyl]carbamic acid ethyl ester; *N*-[2-(*p*-Phenoxyphenoxy)ethyl]carbamic acid; [2-(4-Phenoxyphenoxy)-ethyl]carbamic acid ethyl ester

*Trade Names:* AWARD; ABG 6215; ACR®-2984F; ACR® 2913; BASUS®, Syngenta (Switzerland), canceled; COMPLY®, Syngenta (Switzerland); ECTOGARD®, Boehringer Ingelheim (Germany), canceled; ELIMINATOR®, Boehringer Ingelheim (Germany), canceled; INSEGAR®, Syngenta (Switzerland). LogIC®, Syngenta (Switzerland); LUFOX®, Syngenta (Switzerland); PICTYL®; PRECISION®, Syngenta (Switzerland); TORUS®, Syngenta (Switzerland), canceled; RO 13-5223®; VARIKILL®

*Producers:* AJE (Switzerland); Ascot International UK); Sanonda Ltd. (Australia); Syngenta (Switzerland)

*Chemical Class:* Carbamate
*EPA/OPP PC Code:* 125301; (128801 old code no.)
*California DPR Chemical Code:* 2283

*Uses:* Fenoxycarb is an insect growth regulator used as bait to control a broad spectrum of insects on olives, vines, fruit, cotton and stored products. It is used to control fire ants, cockroaches, mosquitos, beetles, moths, scale and sucking insects.

*Carcinogen/Hazard Classifications*
**U.S. EPA Carcinogens:** Likely carcinogen
**TRI Developmental Toxin:** Developmental toxin
**Label Signal Word:** CAUTION
**WHO Acute Hazard:** Class U, unlikely to be hazardous
*Regulatory Authority:*
- EPCRA Section 313 Form R *de minimis* concentration reporting level: 1.0%

*Description:* Light tan to white solid. Soluble in water; solubility = 5.95 ppm @ 25°C. Melting point = 49-55°C. Molecular weight = 301.35

*Incompatibilities:* May form explosive materials with phosphorus pentachloride. Keep away from oxidizers, sulfuric acid, caustics, ammonia, aliphatic amines, alkanolamines, isocyanates, alkylene oxides, epichlorohydrin.

*Permissible Concentration in Water:* No criteria set. Runoff from spills or fire control may cause water pollution.
*Routes of Entry:* Skin absorption, ingestion, inhalation.

*Harmful Effects and Symptoms*

*Short Term Exposure:* Eye pupils are small; blurred vision; eye watering; runny nose; cough; shortness of breath; salivation; nausea, stomach cramps, diarrhea, and vomiting; increased blood pressure; profuse sweating; hypermotility, hallucinations; agitation; tingling of the skin; slow heartbeat; convulsions; fluid in lungs; loss of consciousness; incontinence; breathing stops; death. Carbamates inhibit the acetylcholinesterase enzymes and alter the way in which nervous impulses are transmitted. However, within several hours carbamates spontaneously detach from the enzymes.

*Long Term Exposure:* A potent cholinesterase inhibitor; cumulative effect is possible. This chemical may damage the nervous system with repeated exposure, resulting in convulsions, respiratory failure. May cause liver damage.

*Points of Attack:* Respiratory system, lungs, central nervous system, cardiovascular system, skin, eyes, plasma and red blood cell cholinesterase.

*Medical Surveillance:* Medical observation is recommended for 24 to 48 hours after breathing overexposure, as pulmonary edema may be delayed. As first aid for pulmonary edema, consider administering a corticosteroid spray. Cigarette smoking may exacerbate pulmonary injury and should be discouraged for at least 72 hours following exposure. Before employment and at regular times after that, the following are recommended: Plasma and red blood cell cholinesterase levels (tests for the enzyme poisoned by this chemical). If exposure stops, plasma levels return to normal in 1-2 weeks while red blood cell levels may be reduced for 1-3 months. When acetylcholinesterase enzyme levels are reduced by 25% or more below preemployment levels, risk of poisoning is increased, even if results are in lower ranges of "normal." Reassignment to work not involving carbamate pesticides is recommended until enzyme levels recover. If symptoms develop or overexposure occurs, repeat the above tests as soon as possible and get an exam of the nervous system. Also consider complete blood count. Consider chest x-ray following acute overexposure

*First Aid:* Speed in removing material from eyes and skin is of extreme importance. *Eyes:* Eye contact can cause dangerous amounts of these chemicals to be quickly absorbed through the mucous membrane into the bloodstream. Immediately and gently flush eyes with plenty of warm or cold water (NO hot water) for at least 15 minutes, occasionally lifting the upper and lower eyelids. Get medical aid immediately. *Skin:* Get medical aid. Skin contact can cause dangerous amounts of these chemicals to be absorbed into the bloodstream. Wearing the appropriate PPE equipment and respirator for carbamate pesticides, immediately flush skin with plenty of soap and water for at least 15 minutes while removing contaminated clothing and shoes. Shampoo hair promptly if contaminated; protect eyes. *Ingestion:* Call poison control. Loosen all clothing. Never give anything by mouth to an unconscious person. Get medical aid. Do NOT induce vomiting.* If conscious, alert, and able to swallow, rinse mouth and have victim drink 4 to 8 ounces of water. Check to see if poison control instructs you to use ipecac syrup, otherwise administer slurry of activated charcoal (2 oz in 8 oz of water). If victim is *UNCONSCIOUS OR HAVING CONVULSIONS,* do nothing except keep victim warm. **In some cases you may be specifically instructed by poison control to induce vomiting by way of 2 tablespoons of syrup of ipecac (adult) washed down with a cup of water.* Do NOT give activated charcoal before or with ipecac syrup. *Inhalation:* Get medical aid. Do not contaminate yourself. Wearing the appropriate PPE equipment and respirator for carbamate pesticides, immediately remove the victim from the contaminated area to fresh air. If the victim is not breathing, administer artificial respiration. Do NOT use mouth-to-mouth resuscitation; use bag/mask apparatus. If breathing is difficult, administer oxygen through bag/mask apparatus until medical help arrives. Do not leave victim unattended. *Note to physician or authorized medical personnel.* Administer atropine, 2 mg (1/30 gr) intramuscularly or intravenously as soon as any local or systemic signs or symptoms of an intoxication are noted; repeat the administration of atropine every 3 to 8 minutes until signs of atropinization (mydriasis, dry mouth, rapid pulse, hot and dry skin) occur; initiate treatment in children with 0.05 mg mg/kg of atropine; repeat at 5 to 10 minute intervals. Watch respiration, and remove bronchial secretions if they appear to be obstructing the airway; intubate if necessary. *Medical note:* 2-PAMCI may be contraindicated in the case of some carbamate poisonings.

*References:*
- EXTOXNET, Extension Toxicology Network, "Pesticide Information Profile, Fenoxycarb," Oregon State University, Corvallis, OR (June 1996). http://extoxnet.orst.edu/pips/fenoxyca.htm
- California Environmental Protection Agency "Chemical List of Lists," Sacramento CA (February 1997)

# Fenpropathrin (ANSI)

*Use Type:* Insecticide
*CAS Number:* 39515-41-8
*Formula:* $C_{22}H_{23}NO_3$
*Alert:* Some formulations may be Restricted Use Pesticides (RUP)
*Synonyms:* AI3-29234; Caswell No. 273H; Cyano-3-phenoxybenzyl-2,2,3,3-tetramethylcyclopropane carboxylate; α-Cyano-3-phenoxybenzyl 2,2,3,3-tetramethylcyclopropanecarboxylate; α-Cyano-3-phenoxybenzyl 2,2,3,3-tetramethyl-1-cyclopropane carboxylate; Cyclopropanecarboxylic acid, 2,2,3,3-tetramethyl-,cyano(3-phenoxyphenyl)methyl ester; 2,2,3,3-tetramethylcyclopropane carboxylic acid, cyano(3-phenoxy phenyl)methyl ester
*Trade Names:* FENPROPANATE®; DANITOL®,Rallis India (India); and Valent BioSciences (USA); GENPROPATHRIN®; HERALD®; RODY®; S 3206®; SD 41706®; TAME®, Valent BioSciences (USA); WL 417-06®; XE-938®
*Producers:* Agsin (Singapore); Bharat Rasayan (India); Cangzhou Green Chemical Co. (China); Hockley International (UK); Rallis India (India); Valent BioSciences (USA); Whitmire Micro-Gen Research Laboratories (USA)
*Chemical Class:* Pyrethroid
*EPA/OPP PC Code:* 127901
*California DPR Chemical Code:* 2234

# Fenpropathrin

*Uses:* Originally registered for commercial greenhouses to control mites, aphids, armyworms, mealybugs and other pests.

*U.S. Maximum Allowable Residue Levels for Fenpropathrin (40 CFR 180.466):*

| CROP | ppm |
|---|---|
| Brassica, head and stem, subgroup 5a | 3 |
| Cattle, fat | 1 |
| Cattle, meat | 0.1 |
| Cattle, mbyp | 0.1 |
| Citrus, dried pulp | 4 |
| Citrus, oil | 75 |
| Cotton, refined oil | 3 |
| Cotton, undelinted seed | 1 |
| Currant | 15 |
| Egg | 0.05 |
| Fruit, citrus, group 10 | 2 |
| Fruit, pome, group 11 | 5 |
| Goat, fat | 1 |
| Goat, meat | 0.1 |
| Goat, mbyp | 0.1 |
| Grape | 5 |
| Grape, raisin | 10 |
| Hog, fat | 1 |
| Hog, meat | 0.1 |
| Hog, mbyp | 0.1 |
| Horse, fat | 1 |
| Horse, meat | 0.1 |
| Horse, mbyp | 0.1 |
| Melon subgroup 9a | 0.5 |
| Milk, fat | 2 |
| Peanut | 0.01 |
| Peanut, hay | 20 |
| Poultry, fat | 0.05 |
| Poultry, meat | 0.05 |
| Poultry, mbyp | 0.05 |
| Sheep, fat | 1 |
| Sheep, meat | 0.1 |
| Sheep, mbyp | 0.1 |
| Squash/cucumber subgroup 9b | 0.5 |
| Strawberry | 2 |
| Tomato | 0.6 |

*Carcinogen/Hazard Classifications*
**U.S. EPA Carcinogens:** Group E, Unlikely a carcinogen
**Label Signal Word:** WARNING or DANGER
**WHO Acute Hazard:** Class II, moderately hazardous
**Endocrine Disruptor:** Possible ED
*Regulatory Authority:*
- AB 1803-Well Monitoring Chemical (CAL) as pyrethrins
- FIFRA, 40CFR185: tolerances for pesticides in food
- FIFRA, 40CFR186: tolerances for pesticides in animal feeds.
- Permissible Exposure Limits for Chemical Contaminants (CAL/OSHA) as pyrethrum
- Actively registered pesticide in California. as pyrethrins
- Clean Water Act: Section 311 Hazardous Substances/RQ (same as CERCLA) as pyrethrins
- EPCRA Section 304 RQ: CERCLA, 1 lb (0.454 kg) as pyrethrins
- EPCRA Section 313 Form R *de minimus* concentration reporting level: 1.0%
- U.S. DOT Marine pollutant (49CFR, Subchapter 172.101, Appendix B), severe pollutant

*Description:* Pale yellow oil or solid. Faint odor. Melting point = 25-50°C. Molecular weight = 349.44. Vapor pressure = $5.5 \times 10^{-6}$ mmHg @ 20°C.

*Incompatibilities:* May react violently with strong oxidizers, bromine, 90% hydrogen peroxide, phosphorus trichloride, silver powders or dust. Incompatible with silver compounds. Mixture with some silver compounds forms explosive salts of silver oxalate.

*Determination in Air:* Collection by impinger or fritted bubbler, analysis by gas liquid chromatography/ultraviolet. See NIOSH IV, Method #5008[18]. (pyrethrum).

*Permissible Concentration in Water:* No criteria set. Runoff from spills or fire control may cause water pollution.

*Determination in Water:* Collection by impinger or fritted bubbler, analysis by gas liquid chromatography/ultraviolet. See NIOSH IV, Method #5008[18].

*Harmful Effects and Symptoms*

*Short Term Exposure:* Pyrethrins can affect you when breathed in and by passing through your skin. Irritates the eyes and respiratory tract. High exposure can affect the nervous system causing headache, nausea, vomiting, fatigue, and restlessness, rhinorrhea (discharge of thin nasal mucous).

*Long Term Exposure:* High or repeated exposure can cause lung allergy (with cough, wheezing and/or shortness of breath) or hay fever symptoms (sneezing, runny or stuffy nose). Allergic "pneumonia" can also occur with cough, chest pain, breathing difficulty and abnormal chest x-ray. Repeated attacks may lead to permanent scarring. Skin allergy may also develop with rash and itching, even with lower exposures. Skin contact can cause rash with redness, blisters and intense itching. A severe generalized allergy can occur with weakness and collapse.

*Points of Attack:* Respiratory system, skin, central nervous system.

*Medical Surveillance:* Before beginning employment and at regular times after that, the following are recommended: Lung function tests. These may be normal if the person is not having an attack at the time of the test. Consider chest x-ray if lung symptoms are present. Evaluation by a qualified allergist, including careful exposure history and special testing, may help diagnose skin allergy.

*First Aid:* If this chemical gets into the eyes, remove any contact lenses at once and irrigate immediately for at least

15 minutes, occasionally lifting upper and lower lids. Seek medical attention immediately. If this chemical contacts the skin, remove contaminated clothing and wash immediately with soap and water. Seek medical attention immediately. If this chemical has been inhaled, remove from exposure, begin rescue breathing (using universal precautions) if breathing has stopped, and CPR if heart action has stopped. Transfer promptly to a medical facility. When this chemical has been swallowed, get medical attention. *Do not induce vomiting when formulations containing petroleum solvents are ingested.* Otherwise, give large quantities of water and induce vomiting. Do not make an unconscious person vomit.

*References:*
- Pesticide Management Education Program, "Fenpropathrin (Danitol) chemical Fact Sheet 12/89," Cornell University, Ithaca, NY (December 1989). http://pmep.cce.cornell.edu/profiles/insect-mite/fenitrothion-methylpara/fenpropathrin/insect-prof-fenpropathrin.html
- U.S. Environmental Protection Agency, Office of Pesticide Programs, Pesticide Residue Limits, "Fenpropathrin," 40 CFR 180.466, http://www.epa.gov/pesticides/food/viewtols.htm
- California Environmental Protection Agency "Chemical List of Lists," Sacramento CA (February 1997)

# Fenpyroximate

*Use Type:* Acaricide
*CAS Number:* 134098-61-6
*Formula:* $C_{24}H_{27}N_3O_4$
*Synonyms:* *tert*-Butyl (*E*)-4-[((((1,3-dimethyl-5-phenoxy-1*H*-pyrazol-4-yl)methylene)amino)oxy)methyl]benzoate; *tert*-Butyl (*E*)-α-(1,3-dimethyl-5-phenoxypyrazol-4-methyleneaminooxy)-*p*-toluate; Benzoic acid, 4-[((((1,3-dimethyl-5-phenoxy-1*H*-pyrazol-4-yl)methylene)amino)oxy)methyl]-, 1,1-dimethylethyl ester, (*E*)-; 4-[((((1,3-Dimethyl-5-phenoxy-1*H*-pyrazol-4-yl)methylene)amino)oxy)methyl]benzoic acid
*Trade Names:* AKARI®; HOE® 555-02A; NNI®-850; SEQUEL®
*Chemical Class:* Phenoxypyrazole
*EPA/OPP PC Code:* 129131
*California DPR Chemical Code:* 5784
*Uses:* Used to control spider mites greenhouses.
*Description:* White crystalline solid or powder. Practically insoluble in water; solubility = 0.000015 ppm. Molecular weight = 421.48. Melting/Freezing point = 101°C. Log $K_{ow}$ = 4.99. Values above 3.0 are likely to bio-accumulate in marine organisms.
*Permissible Concentration in Water:* No criteria set. Runoff from spills or fire control may cause water pollution.

*Harmful Effects and Symptoms*
*Short Term Exposure:* Contact with eyes or skin may cause irritation or injury. Inhalation should be avoided; use NIOSH-approved air purifying respirators for pesticides. May be harmful if swallowed.
*First Aid:* If this chemical gets into the eyes, remove any contact lenses at once and irrigate immediately for at least 15 minutes, occasionally lifting upper and lower lids. Seek medical attention immediately. If this chemical contacts the skin, remove contaminated clothing and wash immediately with soap and water. Seek medical attention immediately. If this chemical has been inhaled, remove from exposure, begin rescue breathing (using universal precautions) if breathing has stopped, and CPR if heart action has stopped. Transfer promptly to a medical facility. When this chemical has been swallowed, get medical attention. Give large quantities of water and induce vomiting. Do not make an unconscious person vomit.

# Fentin Hydroxide

*Use Type:* Fungicide, herbicide and molluscicide
*CAS Number:* 76-87-9
*Formula:* $C_{18}H_{16}OSn$
*Alert:* Restricted Use Pesticide (RUP). Human toxicity (long-term): Very high.
*Synonyms:* AI3-28009; Caswell No. 896E; ENT 28,009; Fentin; Fintine hydroxyde (French); Fintin hydroxid (German); Hydroxyde de triphenyl-etain (French); Hydroxytriphenylstannane; Hydroxytriphenyltin; NCI-C00260; NSC 113243; Stannane, Hydroxytriphenyl-; Stannol, triphenyl-; Tin, hydroxytriphenyl-; TNIV; TPTH; TPTH technical; Triphenylstannanol; Triphenylstannium hydroxide; Triphenyltin(IV) hydroxide; Triphenyltin hydroxide organotin fungicide; Triphenyltin oxide; Triphenyl-zinnhydroxid (German)
*Trade Names:* AGRI-TIN®, Nufarm (Australia); AGTROL®, Nufarm (Australia); BRESTAN® H; DOWCO® 186, Dow AgroSciences (USA); DUTER®; DU-TER®, Solvay (Belgium), canceled; ENABLE®, Dow AgroSciences (USA) and Nufarm (Australia); ERITHANE®; FENOLOVO®; FLO TIN 4L®; HAITIN®; HAITIN WP 60®; IDA FLO-TIN 4L®, Nufarm (Australia), canceled 10/10/1989; K19®; MSS FLOTIN®, Whyte Agrochemicals (UK); ORBIT®, Griffin (USA); PHENOSTAT®-H; PHOTON®; PRO-TEX®, Griffin (USA), canceled 4/13/1992; SUPER TIN 4L®, Griffin (USA); SUZU H®; TRIPLE-TIN®; TRIPLE TIN 4L®; TUBOTIN®; VANCIDE KS®, canceled 10/12/1983; VITO SPOT FUNGICIDE®, canceled; WESLEY®, Griffin (USA), canceled 9/1/1992
*Producers:* Bayer CropScience (Germany); Dow AgroSciences (USA); Fluorochem (UK); Griffin (USA); Nufarm (Australia); Whyte Agrochemicals (UK)

# Fentin Hydroxide

*Chemical Class:* Organotin, heavy metal
*EPA/OPP PC Code:* 083601
*California DPR Chemical Code:* 599
*ICSC Number:* 1283
*RTECS Number:* WH8575000
*EEC Number:* 050-004-00-1
*EINECS Number:* 200-990-6

*Uses:* Triphenyltin compounds are used as fungicides on potatoes, sugar beets, pecans, coffee, cocoa and bananas. There are no residential, public health or other non-food uses. It targets potato blight; Colorado potato beetles; sugar beet leaf spot; and scab, leaf spots and other diseases on pecans.

*Human toxicity (long-term)[77]:* Extra high–0.19125 ppb, CHCL (Chronic Human Carcinogen Level)

*Fish toxicity (threshold)[77]:* Extra high–0.00650 ppb, MATC (Maximum Acceptable Toxicant Concentration)

*U.S. Maximum Allowable Residue Levels for Fentin Hydroxide (40 CFR 180.236):*

| CROP | ppm |
| --- | --- |
| Beet, sugar, roots | 0.1 |
| Cattle, kidney | 0.05 |
| Cattle, liver | 0.05 |
| Goat, kidney | 0.05 |
| Goat, liver | 0.05 |
| Hog, kidney | 0.05 |
| Hog, liver | 0.05 |
| Horse, kidney | 0.05 |
| Horse, liver | 0.05 |
| Pecan | 0.05 |
| Potato | 0.05 |
| Sheep, kidney | 0.05 |
| Sheep, liver | 0.05 |

*Carcinogen/Hazard Classifications*
**U.S. EPA Carcinogens:** Group B2, probable carcinogen
**California Prop. 65:** Carcinogen and developmental toxin
**Label Signal Word:** DANGER
**WHO Acute Hazard:** Class II, moderately hazardous
**Endocrine Disruptor:** Suspected endocrine disruptor

*Regulatory Authority:*
- Restricted Use Pesticide (RUP)
- Proposition 65 chemical (CAL)
- EPCRA Section 313: Form R *de minimus* concentration reporting level: 1.0%
- U.S. DOT Regulated Marine Pollutant (49CFR, Subchapter 172.101, Appendix B), severe pollutant, as organotin pesticide compounds.

*Description:* White granular powder. Odorless. Very slightly soluble in water; solubility = 1.2 mg/L @ 20°C. Molecular weight = 367.02. Melting/Freezing point = 118°C (decomposition). Vapor pressure = $3.5 \times 10^{-7}$ mmHg @ 20°C; $4.7 \times 10^{-7}$ mmHg @ 50°C. Combustible.

*Incompatibilities:* Incompatible with calcium (metal hydroxides), nitroethane, nitromethane, 1-nitropropane, zirconium.

*Permissible Exposure Limits in Air:* OSHA/NIOSH[2], and ACGIH[1] TWA is 0.1 mg/m³ (for tin organic compounds) and STEL is 0.2 mg/m³. The notation "skin" is added indicating the possibility of cutaneous absorption. The NIOSH[2] IDLH level = 25 mg/m³ (as Sn).

*Determination in Air:* Filter/XAD-2® (tube); $CH_3COOH/CH_3CN$; High-pressure liquid chromatography/Graphite furnace atomic absorption spectrometry; NIOSH IV Method #5504, as organotin compounds.[18]

*Permissible Concentration in Water:* No criteria set. Runoff from spills or fire control may cause water pollution. No criteria set, but EPA[32] has suggested a permissible ambient goal of 1.4 µg/L based on health effects as organotin.

*Routes of Entry:* Inhalation, skin and/or eye contact. Absorbed through the skin.

*Harmful Effects and Symptoms*

*Short Term Exposure:* Irritates the eyes, skin, and respiratory tract. Contact may cause skin burns. Inhalation can cause coughing, wheezing and/or shortness of breath. Toxic hazard rating is high for oral, intravenous, intraperitoneal administration. This material causes swelling of the brain and spinal cord. Exposure may result in muscular weakness and paralysis, leading to respiratory failure; convulsive movements; closure of eyelids and sensitivity to light; headaches, and EEG changes, headache, dizziness, psychological and neurological disturbances, vertigo (an illusion of movement), sore throat, cough, abdominal pain, nausea, vomiting, diarrhea; urine retention; paresis, focal anesthesia; pruritus. Higher levels can cause unconsciousness, collapse and death.

*Long Term Exposure:* Repeated or prolonged contact can cause dermatitis; dry and cracked skin. May cause brain damage, hepatic necrosis; kidney damage. Some organotin compounds, such as dibutyltins and tributyltins, have been shown to affect the immune system in animals, but this has not been examined in people. Studies in animals also have shown that some organotins, such as dibutyltins, tributyltins, and triphenyltins, can affect the reproductive system. This, also, has not been examined in people.

*Points of Attack:* Skin, brain and kidneys.

*Medical Surveillance:* Kidney function tests. Psychological testing. Examination of the nervous system. EEG

*First Aid:* If this chemical gets into the eyes, remove any contact lenses at once and irrigate immediately for at least 15 minutes, occasionally lifting upper and lower lids. Seek medical attention immediately. If this chemical contacts the skin, remove contaminated clothing and wash immediately with soap and water. Seek medical attention immediately. If this chemical has been inhaled, remove from exposure, begin rescue breathing (using universal precautions) if breathing has stopped, and CPR if heart action has stopped. Transfer promptly to a medical facility. When this chemical has been swallowed, get medical attention. *Do not induce vomiting when formulations containing petroleum solvents are ingested.* Otherwise, give large quantities of water and induce vomiting. Do not make an unconscious person vomit.

*References:*
- U.S. Environmental Protection Agency, "Reregistration Eligibility Decision (RED), Triphenyltin Hydroxide (TPTH)," Office of Prevention, Pesticides and Toxic Substances, Washington, DC (September 1999). http://www.epa.gov/REDs/0099red.pdf
- U.S. Environmental Protection Agency, Office of Pesticide Programs, Pesticide Residue Limits, "Fentin Hydroxide," 40 CFR 180.236, http://www.epa.gov/pesticides/food/viewtols.htm
- California Environmental Protection Agency "Chemical List of Lists," Sacramento CA (February 1997)

# Fenvalerate

*Use Type:* Insecticide
*CAS Number:* 51630-58-1
*Formula:* $C_{25}H_{22}ClNO_3$
*Synonyms:* A13-29235; Benzeneacetic acid, 4-chloro-α-(1-methylethyl)-, cyano(3-phenoxyphenyl)methyl ester; Caswell No. 077A; 4-Chloro-α-(1-methylethyl)benzeneacetic acid, cyano(3-phenoxyphenyl)methyl ester; α-Cyano-3-phenoxybenzyl-2-(4-chlorophenyl)-3-methybutyrate; Cyano(3-phenoxyphenyl)methyl ester of 4-chloro-α-(1-methylethyl)benzeneacetic acid; Cyano(3-phenoxyphenyl)methyl 4-chloro-α-(1-methylethyl)benzeneacetate; (IRS)-α-Cyano-3-phenoxybenzyl (RS)-2-(4-chlorophenyl)-3-methylbutyrate; α-Cyano-*m*-phenoxybenzyl 2-(*p*-chlorophenyl)-3-methylbutyrate; α-Cyano-3-phenoxybenzyl 2-(4-chlorophenyl)isovalerate; Cyano-(3-phenoxybenzyl)methyl 2-(4-chlorophenyl)-3-methylbutyrate; Cyano-(3-phenoxyphenyl)methyl 4-chloro-α-(1-methylethyl)benzeneacetate; 4-Chloro-α-(1-methylethyl)benzeneacetic acid cyano(3-phenoxyphenyl)methyl ester; Fenvaleriato (Spanish); Phenvalerate
*Trade Names:* BELMARK®; ECTIN®, Boehringer Ingelheim (Germany); FENKILL®; EVERCIDE®, Mclaughlin Gormley King (USA); KORANDA®, Rallis India (India); PYDRIN®; S 5602®; SANMARTON®; SD 43775®; SUMICIDE®; SUMICIDIN®; SUMICIDINE®; SUMIFLEECE®; SUMIFLY®; SUMIPOWER®; SUMITICK®; TIRADE®; WL 43775®
*Producers:* Agrimor International (USA); Agsin (Singapore); Bharat Rasayan (India); Boehringer Ingelheim (Germany); Cangzhou Green Chemical Co. (China); Chongqing Chuandong Chemical (Group) (China); Gharda Chemicals (India); Hebei Huafeng Chemical Group (China); Hindustan Insecticides (India); Hockley International (UK); Mclaughlin Gormley King (USA); Nagarjuna Agrichem (India); Rallis India (India); Sudarshan Chemical Industries (India)
*Chemical Class:* Pyrethroid; botanical
*EPA/OPP PC Code:* 109301 (295700 or 296700 old code Nos.)
*California DPR Chemical Code:* 1963
*ICSC Number:* 0273
*RTECS Number:* CY1576300
*EINECS Number:* 257-326-3
*Uses:* Fenvalerate is one of the most versatile synthetic pyrethroid insecticides. It is mostly used in agriculture and on cattle, but also in homes and gardens. It acts as a stomach poison against a wide variety of leaf and fruit eating such as bollworm fruit and shoot borers and aphids. Crops on which it is used include cotton, cauliflower, okra, vines and fruits. It is also used in public health and animal husbandry. It is effective against pests whose strains are resistant to organochlorine, organophosphorus, and carbamate insecticides.
*Human toxicity (long-term)*[77]: Very low–175.00 ppb, Health Advisory
*Fish toxicity (threshold)*[77]: Extra high–0.01433 ppb, MATC (Maximum Acceptable Toxicant Concentration)
*U.S. Maximum Allowable Residue Levels for Fenvalerate (40 CFR 180.533): Note:* At the present time, the U.S. EPA lists 100 tolerances for fenvalerate. They can be found at the web site in the Reference section of this compound shown below.
*Carcinogen/Hazard Classifications*
**U.S. EPA Carcinogens:** Group E, Unlikely a carcinogen
**IARC:** Group 3, unclassifiable
**Label Signal Word:** WARNING or CAUTION
**WHO Acute Hazard:** Class II, moderately hazardous
**Endocrine Disruptor:** Suspected endocrine disruptor
*Regulatory Authority:*
- EPA 40 CFR 372.65, Specific Toxic Chemical Listings
- AB 1803-Well Monitoring Chemical (CAL) as pyrethrins
- Permissible Exposure Limits for Chemical Contaminants (CAL/OSHA) as pyrethrum
- Actively registered pesticide in California. as pyrethrins
- Clean Water Act: Section 311 Hazardous Substances/RQ (same as CERCLA) as pyrethrins
- EPCRA Section 304 RQ: CERCLA, 1 lb (0.454 kg) as pyrethrins
- U.S. DOT Marine pollutant (49CFR, Subchapter 172.101, Appendix B), severe pollutant

*Description:* Yellowish to brown viscous liquid. Technical grade is a brown viscous liquid. Faint chemical odor. Practically insoluble in water; solubility = 0.085 ppm @ 20°C. Boiling point = decomposes. Molecular weight 419.93. Density =1.17 @ 23°C. Vapor pressure = $1.1 \times 10^{-8}$ mmHg @ 25°C; $3.7 \times 10^{-7}$ mmHg @ 25°C. Log $K_{ow}$ = 4.40. Values above 3.0 are likely to bioaccumulate in marine organisms.

*Incompatibilities:* Keep away from oxidizers, sulfuric acid, caustics, ammonia, aliphatic amines, alkanolamines, isocyanates, alkylene oxides, epichlorohydrin. Moisture may cause hydrolysis or other forms of decomposition.

*Determination in Air:* Collection by impinger or fritted bubbler, analysis by gas liquid chromatography/ultraviolet. See NIOSH IV, Method #5008[18]. (pyrethrum).

*Permissible Concentration in Water:* No criteria set. Runoff from spills or fire control may cause water pollution.

*Determination in Water:* Collection by impinger or fritted bubbler, analysis by gas liquid chromatography/ultraviolet. See NIOSH IV, Method #5008[18].

*Routes of Entry:* Inhalation, absorbed through the skin.

*Harmful Effects and Symptoms*

*Short Term Exposure:* Pyrethrins can affect you when breathed in and by passing through your skin. Irritates the eyes and respiratory tract. High exposure can affect the nervous system causing headache, nausea, vomiting, fatigue, and restlessness, rhinorrhea (discharge of thin nasal mucous).

*Long Term Exposure:* High or repeated exposure can cause lung allergy (with cough, wheezing and/or shortness of breath) or hay fever symptoms (sneezing, runny or stuffy nose). Allergic "pneumonia" can also occur with cough, chest pain, breathing difficulty and abnormal chest x-ray. Repeated attacks may lead to permanent scarring. Skin allergy may also develop with rash and itching, even with lower exposures. Skin contact can cause rash with redness, blisters and intense itching. A severe generalized allergy can occur with weakness and collapse.

*Points of Attack:* Respiratory system, skin, central nervous system.

*Medical Surveillance:* Before beginning employment and at regular times after that, the following are recommended: Lung function tests. These may be normal if the person is not having an attack at the time of the test. Consider chest x-ray if lung symptoms are present. Evaluation by a qualified allergist, including careful exposure history and special testing, may help diagnose skin allergy.

*First Aid:* If this chemical gets into the eyes, remove any contact lenses at once and irrigate immediately for at least 15 minutes, occasionally lifting upper and lower lids. Seek medical attention immediately. If this chemical contacts the skin, remove contaminated clothing and wash immediately with soap and water. Seek medical attention immediately. If this chemical has been inhaled, remove from exposure, begin rescue breathing (using universal precautions) if breathing has stopped, and CPR if heart action has stopped. Transfer promptly to a medical facility. When this chemical has been swallowed, get medical attention. *Do not induce vomiting when formulations containing petroleum solvents are ingested.* Otherwise, give large quantities of water and induce vomiting. Do not make an unconscious person vomit.

*References:*

- U.S. Environmental Protection Agency, Office of Pesticide Programs, Pesticide Residue Limits, "Fenvalerate," 40 CFR 180.533, http://www.epa.gov/pesticides/food/viewtols.htm
- International Programme on Chemical Safety (IPCS), "Health and Safety Guide, Fenvalerate," Geneva, Switzerland (1990). http://www.inchem.org/documents/ehc/ehc/ehc95.htm
- California Environmental Protection Agency "Chemical List of Lists," Sacramento CA (February 1997)

# Ferbam

*Use Type:* Fungicide

*CAS Number:* 14484-64-1

*Formula:* $C_9H_{18}FeN_3S_6$; $Fe[(CH_3)_2NCS_2]_3Fe$

*Synonyms:* Carbamic acid, aimethyldithio-, iron salt; Dimethylcarbamo dithioic acid, iron complex; Dimethylcarbamodithioic acid, iron(3+) salt; Dimethyldithiocarbamic acid, iron salt; Dimethyldithiocarbamic acid, iron(3+) salt; Eisendimethyldithiocarbamat (German); Eisen(III)-tris(N,N-dimethyldithiocarbamat) (German); ENT 14,689; Ferbam, iron salt; Ferric dimethyl dithiocarbamate; Iron dimethyldithiocarbamate; Iron(III) dimethyldithio carbamate; Iron, tris(dimethylcarbamodithioato-$S,S'$)-, (OC-6-11)-; Iron, tris(dimethylcarbamodithioato-$S,S'$-)-; Iron, tris(dimethyldithiocarbamato)-; Iron tris(dimethyldithiocarbamate); Tris(dimethylcarbamo dithioato-$S,S'$)iron; (OC-6-11)-tris(Dimethylcarbamo dithioato-$S,S'$)iron; Tris(Dimethyldithiocarbamato)iron; Tris(N,N-dimethyldithiocarbamato)iron(III)

*Trade Names:* AI3-14689®; AAFERTIS®; APPLE DUST No. 1®, Amvac Chemicals (USA), canceled 7/19/1995; BERCEMA FERTAM 50®; CASWELL No. 458®; FERBAM 50®; FERBECK®; FERMATE FERBAM FUNGICIDE®, Daly-Herring (USA), canceled 10/10/1989; FERMOCIDE®; FERRADOUR®; FERRADOW®; FUKLASIN ULTRA®; HEXAFERB®; HOKMATE®; KARBAM BLACK®; KARBAM CARBAMATE®; KNOCKMATE®; NIACIDE®; STAUFFER FERBAM®; SUP'R-FLO FERBAM FLOWABLE®; TRICARBAMIX®, Cerexagri Inc, North America (USA), canceled 7/01/1987; TRIFUNGOL®; VANCIDE FE95®, R. T. Vanderbilt Co. (USA), canceled 4/23/1986

*Producers:* Amvac Chemicals (USA); Bayer CropScience (Germany); Cerexagri Inc, North America (USA); Ehrenstorfer, Dr. (Germany); FMC Agricultural Products Group (USA); Fluorochem (UK); Gowan Company (USA); Sigma-Aldrich Laborchemikalien (Germany); Syngenta (Switzerland); UCB Group (Belgium)

*Chemical Class:* Dithiocarbamate

*EPA/OPP PC Code:* 034801
*California DPR Chemical Code:* 288
*ICSC Number:* 0792
*RTECS Number:* NO8750000
*EEC Number:* 006-051-00-5
*EINECS Number:* 238-484-2

*Uses:* A dithiocarbamate fungicide. It is widely used, together with other fungicides, to control Postbloom Fruit Drop (PFD) on citrus crops, and as a foliar protectant against scab, rust, mold and many fungus disease on fruits, vegetables, melons and ornamentals. It is registered in several states for use on currents and gooseberries to control leaf spot disease, and on apple, crabapple, hawthorn and quince to control cedar-apple rust disease. It is used to control rust disease on shrubs and ornamentals.

*Human toxicity (long-term)[77]:* Very low–140.00 ppb, Health Advisory

*Fish toxicity (threshold)[77]:* Low–437.01653 ppb, MATC (Maximum Acceptable Toxicant Concentration)

*U.S. Maximum Allowable Residue Levels for Ferbam (40 CFR 180.114):*

| CROP | ppm |
| --- | --- |
| Apple | 7 |
| Apricot | 7 |
| Asparagus | 7 |
| Bean | 7 |
| Blackberry | 7 |
| Blueberry | 7 |
| Boysenberry | 7 |
| Cabbage | 7 |
| Cherry | 7 |
| Cranberry | 7 |
| Cucumber | 7 |
| Dewberry | 7 |
| Fruit, stone, group 12 | 7 |
| Grape | 7 |
| Guava | 7 |
| Lettuce | 7 |
| Loganberry | 7 |
| Mango | 7 |
| Nectarine | 7 |
| Papaya | 7 |
| Pea | 7 |
| Peach | 7 |
| Pear | 7 |
| Raspberry | 7 |
| Squash | 7 |
| Tomato | 7 |
| Youngberry | 7 |

*Carcinogen/Hazard Classifications*
**IARC:** Group 3, unclassifiable
**Label Signal Word:** CAUTION
**WHO Acute Hazard:** Class U, unlikely to be hazardous
*Regulatory Authority:*

- Air Pollutant Standard Set (ACGIH)[1] (DFG)[3] (HSE)[33] (Argentina)[35] (OSHA)[58] (Several States)[60] (Several Canadian Provinces)
- AB 1803-Well Monitoring Chemical (CAL)
- Permissible Exposure Limits for Chemical Contaminants (CAL/OSHA)
- The "Director's List" (CAL/OSHA)
- Actively registered pesticide in California.
- EPA Hazardous Waste Number (RCRA No.): U396
- RCRA, 40CFR261, Appendix 8 Hazardous Constituents
- EPCRA Section 313 Form R *de minimus* concentration reporting level: 1.0%
- Canada, WHMIS, Ingredients Disclosure List

*Description:* Ferbam is an combustible, odorless dark brown to black powder or granular solid. Practically insoluble in water; solubility = 120 ppm @ 20°C. Molecular weight = 416.50. Melting/Freezing point = 180°C (decomposes). Vapor pressure = <$10^{-5}$ mmHg @ 20°C.

*Incompatibilities:* Strong oxidizers, strong bases. Heat alkalies (lime), moisture can cause decomposition. Decomposes on prolonged storage. Degradation produces ethylene thiourea.

*Permissible Exposure Limits in Air:* The Federal standard [OSHA PEL] is 15 ppm (total dust) TWA(58). NIOSH, the HSE[33], Argentina[35], and ACGIH[1] have adopted a TWA value of 10 mg/m$^3$. The DFG[3] has no MAK at present. The HSE[33] has set an STEL value of 20 mg/m$^3$. The NIOSH[2] IDLH level = 800 mg/m$^3$. Several states have set guidelines or standards for ferbam in ambient air[60] ranging from 100 μg/m$^3$ (North Dakota) to 200 μg/m$^3$ (Connecticut) to 238 μg/m$^3$ (Nevada).

*Determination in Air:* Collection by filter; Gravimetric; NIOSH IV, Method #0500, Particulates NOR (total)[18]

*Permissible Concentration in Water:* No criteria set, but runoff from spills or fire control may cause water pollution.

*Routes of Entry:* Inhalation, ingestion, eye and/or skin contact.

*Harmful Effects and Symptoms*

*Short Term Exposure:* Ferbam can affect you when breathed in. Breathing ferbam can irritate the nose and throat. Ferbam can cause skin and eye irritation. High exposure to ferbam may affect the nervous system and thyroid; dizziness, confusion, loss of coordination, seizures, paralysis, and coma. Unlike carbamates the dithiocarbamates are not cholinesterase inhibitors, but some of them may react with recently ingested alcohol or alcohol-containing products including wine, medications, and cold remedies such as cough-syrups.

*Long Term Exposure:* Repeated or prolonged contact with skin may cause allergy with skin rash and itching. Exposure to ferbam may damage the kidneys and liver. Ferbam may damage the developing fetus.

*Points of Attack:* Eyes, skin, respiratory system, gastrointestinal tract.

*Medical Surveillance:* For those with frequent or potentially high exposure (half the TLV or greater), the following are recommended before beginning work and at regular times after that: Kidney function tests. Liver function tests. Evaluation by a qualified allergist, including careful exposure history and special testing, may help diagnose skin allergy.

*First Aid:* If this chemical gets into the eyes, remove any contact lenses at once and irrigate immediately for at least 15 minutes, occasionally lifting upper and lower lids. Seek medical attention immediately. If this chemical contacts the skin, remove contaminated clothing and wash immediately with soap and water. Seek medical attention immediately. If this chemical has been inhaled, remove from exposure, begin rescue breathing (using universal precautions) if breathing has stopped, and CPR if heart action has stopped. Transfer promptly to a medical facility. When this chemical has been swallowed, get medical attention. Give large quantities of water and induce vomiting. Do not make an unconscious person vomit.

*References:*
- Pesticide Education Management Program, "Ferbam Chemical Profile 2/85," Cornell University, Ithaca, NY. (February 1985). http://pmep.cce.cornell.edu/profiles/fung-nemat/febuconazole-sulfur/ferbam/fung-prof-ferbam.html
- California Environmental Protection Agency "Chemical List of Lists," Sacramento CA (February 1997).
- EPA, Office of Pesticide Programs, Pesticide Residue Limits, " Ferbam," 40 CFR 180.114., http://www.epa.gov/pesticides/food/viewtols.htm
- New Jersey Department of Health and Senior Services, "Hazardous Substance Fact Sheet: Ferbam," Trenton, NJ (April 1999). http://www.state.nj.us/health/eoh/rtkweb/0917.pdf
- Sax, N.I., Ed., "Dangerous Properties of Industrial Materials Report," 1, No. 6, 56-58 (1981) and 8, No. 6, 57-63 (1988).

# Ferric Sulfate

*Use Type:* Herbicide
*CAS Number:* 10028-22-5
*Formula:* $Fe_2O_{12}S_3$
*Synonyms:* Diiron trisulfate; Iron persulfate; Iron sesquisulfate; Iron(III) sulfate; Iron sulfate (2:3); Iron (3+) sulfate; Iron tersulfate; Sulfato ferrico (Spanish); Sulfuric acid, iron(III) salt (3:2); Sulfuric acid, iron(3+) salt (3:2)
*Trade Names:* GREENMASTER AUTUMN®; MAXICROP MOSS KILLER®; VITAX MICRO GRAN®; VITAX TURF TONIC®
*Producers:* Ashland (USA); Celtic Chemicals (UK); Dankalk (Denmark); GFS Chemicals (USA); Honeywell Performance P & C (USA); Huntsman (USA); ICI Group (UK); Mallinckrodt Baker (USA); Merck (Germany)
*Chemical Class:* Inorganic
*EPA/OPP PC Code:* 034902
*California DPR Chemical Code:* 1811
*RTECS Number:* NO8505000
*EINECS Number:* 233-072-9
*Uses:* Ferric sulfate is used on forage alfalfa, almonds, nurseries and structural pest control. This material is also used in pigments, textile dyeing, water treatment, and metal pickling.

*Regulatory Authority:*
- Air Pollutant Standard Set (ACGIH)[1] (HSE)[33]
- The "Director's List" (CAL/OSHA)
- Actively registered pesticide in California.
- Clean Water Act: Section 311 Hazardous Substances/RQ 40CFR117.3 (same as CERCLA, see below)
- Superfund/EPCRA 40CFR302.4 RQ: CERCLA, 1000 lb (454 kg)
- Canada, WHMIS, Ingredients Disclosure List

*Description:* Ferric Sulfate is a grayish-white powder or yellow lumpy crystals. Slightly soluble in water. Molecular weight = 399.89. Melting/Freezing point = 480°C (decomposes). Hazard Identification (based on NFPA-704 M Rating System): Health 1, Flammability 0, Reactivity 0.

*Incompatibilities:* Hydrolyzed slowly in aqueous solution. Incompatible with magnesium, aluminum. Corrosive to copper and its alloys, mild and galvanized steel. Light sensitive.

*Permissible Exposure Limits in Air:* The recommended airborne exposure limit is 1 mg/m³ (as iron) averaged over an 8-hour workshift. (This exposure limit is recommended for all soluble iron salts.) This is the value set by ACGIH[1], NIOSH, and by HSE[33]. HSE[33] adds an STEL of 2.0 mg/m³. Other countries including Australia, Belgium, Finland, Italy, Netherlands, and Switzerland set the same TWA

*Determination in Air:* Filter; Acid; Inductively coupled plasma; NIOSH IV, Method #7300, Elements[18].

*Permissible Concentration in Water:* The former USSR-UNEP/IRPTC project[43] has set an MAC for iron in water bodies used for domestic purposes of 0.5 mg/L and an MAC in sea water bodies used for fishery purposes of 0.05 mg/L.

*Routes of Entry:* Inhalation, ingestion, skin and/or eye contact.

*Harmful Effects and Symptoms*
*Short Term Exposure: Inhalation:* May cause irritation of nose and throat, coughing and difficulty in breathing. 0.075 mg/m³ for 2 hours did not cause any change in breathing functions. Inhaling iron oxide fumes may cause a pneumoconiosis in the lungs. Iron oxide fumes can cause "metal fume fever." irritation eyes, skin, mucous membrane; abdominal pain, diarrhea, vomiting. *Skin:* Contact causes irritation Remove promptly. *Eyes:* Contact causes irritation.

*Ingestion:* May cause irritation of mouth and stomach, nausea, vomiting, diarrhea, drowsiness, liver damage, coma and death. The estimated lethal dose is 30 kg (one ounce).

*Long Term Exposure:* Excessive intake of iron compounds may result in increased accumulations of iron in body, especially the liver, spleen and lymphatic system. May cause nausea, vomiting, stomach pain, constipation, and black bowel movements. Inhalation of iron dusts may cause mottling of the lung. Prolonged or repeated high exposure may cause liver damage Prolonged eye contact can cause a brownish discoloration of the eye. Repeated overexposure may cause kidney stones.

*Points of Attack:* Eyes, skin, respiratory system, liver, lungs, gastrointestinal tract

*Medical Surveillance:* If symptoms develop or overexposure is suspected, the following may be useful: Blood test for iron level (serum iron). Liver function tests. Lung function tests. For those exposed to this chemical, taking dietary supplements or vitamins containing iron is not recommended without medical advice.

*First Aid:* If this chemical gets into the eyes, remove any contact lenses at once and irrigate immediately for at least 15 minutes, occasionally lifting upper and lower lids. Seek medical attention immediately. If this chemical contacts the skin, remove contaminated clothing and wash immediately with soap and water. Seek medical attention immediately. If this chemical has been inhaled, remove from exposure, begin rescue breathing (using universal precautions) if breathing has stopped, and CPR if heart action has stopped. Transfer promptly to a medical facility. When this chemical has been swallowed, get medical attention. Give large quantities of water and induce vomiting. Do not make an unconscious person vomit. The symptoms of metal fume fever may be delayed for 4-12 hours following exposure: it may last less than 36 hours.

*References:*
- New Jersey Department of Health and Senior Services, "Hazardous Substance Fact Sheet: Ferric Sulfate," Trenton, NJ (March 1999). http://www.state.nj.us/health/eoh/rtkweb/0925.pdf
- New York State Department of Health, "Chemical Fact Sheet: Iron (III) Sulfate," Albany, NY, Bureau of Toxic Substance Assessment (March 1986).
- California Environmental Protection Agency "Chemical List of Lists," Sacramento CA (February 1997).
- Sax, N.I., Ed., "Dangerous Properties of Industrial Materials Report," 3, No. 4, 45-47 (1983) and 7, No. 2, 75-79 (1987).

# Ferrous Sulfate

*Use Type:* Herbicide and molluscicide and as a fertilizer
*CAS Number:* 7720-78-7
*Formula:* $FeO_4S$; $FeSO_4$

*Synonyms:* Exsiccated ferrous sulfate; Exsiccated ferrous sulphate; Ferrosulfat (German); Ferrosulfate; Ferrosulphate; Ferro-theron; Ferrous sulphate (1:1); Fersolate; Green vitriol iron monosulfate; Iron protosulfate; Iron sulfate (1:1); Iron(II) sulfate; Iron(2+) sulfate; Iron(2+) sulfate (1:1); Iron vitriol; Sulfato ferroso (Spanish); Sulferrous; sulfuric acid iron salt (1:1); Sulfuric acid, iron(2+) salt (1:1); Sulfuric acid, iron(II) salt (1:1)

*Trade Names:* COPPERAS®; DURETTER®; DUROFERON®; FEOSOL®; FEOSPAN®; FER-IN-SOL®; FERRO-GRADUMET®; FERRALYN®; IROSPAN®; IROSUL®; SLOW-FE®

*Producers:* Agrium (Canada); Alfa Rio Quimica Ltda. (Brazil); Ashland (USA); Celtic Chemicals (UK); Coogee Chemicals (Australia); Crown Technology (USA); Dankalk (Denmark); GFS Chemicals (USA); Hindustan Basic Drugs (India); Honeywell Performance P & C (USA); Prince Manufacturing (USA); Salvi Chemical Industries (India); Synergy Production Group (USA)

*Chemical Class:* Inorganic
*California DPR Chemical Code:* 289
*RTECS Number:* NO8500000
*EINECS:* 231-753-5

*Uses:* It is used as a fertilizer, food or feed additive, and in herbicides, process engraving, dyeing, and water treatment. It is also used as a dietary supplement.

*U.S. Maximum Allowable Residue Levels for Ferrous sulfate heptahydrate (40 CFR 180.2):*

| CROP | ppm |
|---|---|
| Animal products | – |
| Raw agricultural commodities | – |

*Regulatory Authority:*
- Air Pollutant Standard Set (ACGIH)[1] (HSE)[33] (OSHA)[58]
- The "Director's List" (CAL/OSHA)
- Actively registered pesticide in California.
- Clean Water Act: Section 311 Hazardous Substances/RQ 40CFR117.3 (same as CERCLA, see below)
- Superfund/EPCRA 40CFR302.4 RQ: CERCLA, 1000 lb (454 kg)
- Canada, WHMIS, Ingredients Disclosure List

*Description:* Ferrous sulfate is a greenish or yellowish, solid in fine or lumpy crystals. Hazard Identification (based on NFPA-704 M Rating System): Health 2, Flammability 0, Reactivity 0. Slowly soluble in water.

*Incompatibilities:* Aqueous solution is acidic. Contact with alkalies form iron. Keep away from alkalies, soluble carbonates, gold and silver salts, lead acetate, lime water, potassium iodide, potassium and sodium tartrate, sodium borate, tannin.

*Permissible Exposure Limits in Air:* The recommended airborne exposure limit is 1 mg/m³ (as iron) averaged over an 8-hour workshift. (This exposure limit is recommended for all soluble iron salts.) This is the value set by ACGIH[1], NIOSH, and by HSE[33]. HSE[33] adds an STEL of 2.0

mg/m³. Other countries including Australia, Belgium, Finland, Italy, Netherlands, and Switzerland set the same TWA

***Determination in Air:*** Filter; Acid; Inductively coupled plasma; NIOSH IV, Method #7300, Elements[18].

***Permissible Concentration in Water:*** The former USSR-UNEP/IRPTC project[43] has set an MAC for iron in water bodies used for domestic purposes of 0.5 mg/L and an MAC in sea water bodies used for fishery purposes of 0.05 mg/L.

***Routes of Entry:*** Inhalation, ingestion, skin and/or eye contact.

***Harmful Effects and Symptoms***

***Short Term Exposure:*** Ferrous sulfate can affect you when breathed in. Irritates the eyes, skin and respiratory tract. *Ingestion:* Less than 5 grams (1/6 ounce) can cause drowsiness, irritability, weakness, abdominal pain, nausea, vomiting and black, bloody stools. Delayed symptoms include fluid in the lungs, liver abnormalities, shock, coma, intestinal blockage and breakdown of the stomach and intestinal lining. Death has resulted from ingestion off less than an ounce.

***Long Term Exposure:*** Excessive intake of iron compounds may result in increased accumulation of iron in body, especially the liver, spleen and lymphatic system. Inhalation of iron dusts may cause mottling of the lung. Prolonged eye contact can cause a brownish discoloration of the eye. Repeated overexposure may cause kidney stones.

***Points of Attack:*** Eyes, skin, respiratory system, liver, gastrointestinal tract.

***Medical Surveillance:*** If symptoms develop or overexposure is suspected, the following may be useful: Serum iron test. Liver function test. Kidney function tests. For those exposed to this chemical, taking dietary supplements or vitamins containing iron is not recommended without medical advice.

***First Aid:*** If this chemical gets into the eyes, remove any contact lenses at once and irrigate immediately for at least 15 minutes, occasionally lifting upper and lower lids. Seek medical attention immediately. If this chemical contacts the skin, remove contaminated clothing and wash immediately with soap and water. Seek medical attention immediately. If this chemical has been inhaled, remove from exposure, begin rescue breathing (using universal precautions) if breathing has stopped, and CPR if heart action has stopped. Transfer promptly to a medical facility. When this chemical has been swallowed, get medical attention. Give large quantities of water and induce vomiting. Do not make an unconscious person vomit. *Note to physician or authorized medical personnel:* Gastric lavage with large amounts of 5% sodium phosphate or water. Follow this with a large amount of 1% sodium bicarbonate over a 3 hour period.

***References:***
- EPA, Office of Pesticide Programs, Pesticide Residue Limits, "Ferrous sulfate heptahydrate," 40 CFR 180.2., http://www.epa.gov/pesticides/food/viewtols.htm
- Sax, N.I., Ed., "Dangerous Properties of Industrial Materials Report," 3, No. 4, 45-47 (1983) and 7, No. 1, 55-60 (1987).
- California Environmental Protection Agency "Chemical List of Lists," Sacramento CA (February 1997).New Jersey Department of Health and Senior Services, "Hazardous Substance Fact Sheet, Ferrous Sulfate," Trenton NJ (February 1999). http://www.state.nj.us/health/eoh/rtkweb/0931.pdf

# Fipronil

***Use Type:*** Insecticide
***CAS Number:*** 120068-37-3
***Formula:*** $C_{12}H_4Cl_2F_6N_4OS$
***Synonyms:*** 5-Amino-1-(2,6-dichloro-4-(trifluoromethyl)phenyl)-4-(1,*R*,*S*)-(trifluoromethyl)sulfinyl-1*H*-pyrazole-3-carbonitrile; 5-Amino-3-cyano-1-(2,6-dichloro-4-trifluoromethylphenyl)-4-trifluoromethylsulfinylpyrazole; (±)-5-Amino-1-(2,6-dichloro-α,α,α-trifluoro-*p*-tolyl)-4-trifluoromethylsulfinylpyrazole-3-carbonitrile; 1*H*-Pyrazole-3-carbonitrile, 5-amino-1-(2,6-dichloro-4-(trifluoromethyl)phenyl)-4-[(trifluoromethyl)sulfinyl]-

***Trade Names:*** BES® 602, BASF Agricultural Products (Germany); CEASEFIRE®, Bayer CropScience (Germany); CHIPCO®, Bayer CropScience (Germany); COMBAT®; FRONTLINE; MB-46030®; H&G®, BASF Agricultural Products (Germany); ICON®, Bayer CropScience (Germany); MAXFORCE® ANT STATION, Bayer CropScience (Germany); MAXFORCE® ROACH STATION, Bayer CropScience (Germany); REGENT®, BASF Agricultural Products (Germany); TERMIDOR®, BASF Agricultural Products (Germany)

***Producers:*** BASF Agricultural Products (Germany); Bayer CropScience (Germany)

***Chemical Class:*** Organofluorine

***EPA/OPP PC Code:*** 129121

***California DPR Chemical Code:*** 3995

***Uses:*** Fipronil was introduced into the U.S. in 1996 for use in animal health and indoor pest control. It is the constituent of many products for controlling a wide spectrum of domestic animal and residential pests.

***Human toxicity (long-term)***[77]***:*** Extra high–0.14 ppb, Health Advisory

***Fish toxicity (threshold)***[77]***:*** Extra high–0.31368 ppb, MATC (Maximum Acceptable Toxicant Concentration)

***U.S. Maximum Allowable Residue Levels for Fipronil (40 CFR 180.517):***

| CROP | ppm |
|---|---|
| Cattle, fat | 0.4 |
| Cattle, liver | 0.1 |
| Cattle, meat | 0.04 |

| | |
|---|---|
| Cattle, mbyp, except liver | 0.04 |
| Corn, field, forage | 0.15 |
| Corn, field, grain | 0.02 |
| Corn, field, stover | 0.3 |
| Egg | 0.03 |
| Goat, fat | 0.4 |
| Goat, liver | 0.1 |
| Goat, meat | 0.04 |
| Goat, mbyp, except liver | 0.04 |
| Hog, fat | 0.04 |
| Hog, liver | 0.02 |
| Hog, meat | 0.01 |
| Hog, mbyp, except liver | 0.01 |
| Horse, fat | 0.4 |
| Horse, liver | 0.1 |
| Horse, meat | 0.04 |
| Horse, mbyp, except liver | 0.04 |
| Milk, fat | 1.5 |
| Poultry, fat | 0.05 |
| Poultry, meat | 0.02 |
| Poultry, mbyp | 0.02 |
| Rice, grain | 0.04 |
| Rice, straw | 0.1 |
| Sheep, fat | 0.4 |
| Sheep, liver | 0.1 |
| Sheep, meat | 0.04 |
| Sheep, mbyp, except liver | 0.04 |

*Carcinogen/Hazard Classifications*
**U.S. EPA Carcinogens:** Group C, possible carcinogen
**Label Signal Word:** CAUTION
**WHO Acute Hazard:** Class II, moderately hazardous
**Endocrine Disruptor:** Suspected endocrine disruptor
*Description:* White crystalline solid. Soluble in water; solubility = 1.9 ppm. Molecular weight = 437.18. Melting/Freezing point = 201°C. Log $K_{ow}$ = 4.10. Values above 3.0 are likely to bioaccumulate in marine organisms.
*Permissible Concentration in Water:* No criteria set. Runoff from spills or fire control may cause water pollution.
*Harmful Effects and Symptoms*
*Short Term Exposure:* Contact with eyes or skin may cause irritation or injury. Inhalation should be avoided; use NIOSH-approved air purifying respirators for pesticides. May be harmful if swallowed.
*First Aid:* If this chemical gets into the eyes, remove any contact lenses at once and irrigate immediately for at least 15 minutes, occasionally lifting upper and lower lids. Seek medical attention immediately. If this chemical contacts the skin, remove contaminated clothing and wash immediately with soap and water. Seek medical attention immediately. If this chemical has been inhaled, remove from exposure, begin rescue breathing (using universal precautions) if breathing has stopped, and CPR if heart action has stopped. Transfer promptly to a medical facility. When this chemical has been swallowed, get medical attention. Give large quantities of water and induce vomiting. Do not make an unconscious person vomit.

*References:*
- U.S. Environmental Protection Agency, Office of Pesticide Programs, Pesticide Residue Limits, "Fipronil," 40 CFR 180.517, http://www.epa.gov/pesticides/food/viewtols.htm
- National Pesticide Information Center (NPIC), "Fipronil Fact Sheet," Corvallis, OR

## Fluazifop-butyl

*Use Type:* Herbicide
*CAS Number:* 69806-50-4; 79241-46-6 (*p*-butyl isomer)
*Formula:* $C_{19}H_{20}F_3NO_4$
*Synonyms:* Butyl(*RS*)-2-[4-((5-(trifluoromethyl)-2-pyridinyl)oxy)phenoxy]propanoate; (±)-Butyl-2-[4-(((5-trifluoro-methyl)-2-pyridinyl)oxy)phenoxy]propanoate; Butyl 2-[4-((5-(trifluoromethyl)-2-pyridyl)oxy)phenoxy]propionate]; Caswell No. 460C; Propanoic acid, 2-[4-((5-(trifluoromethyl)-2-pyridinyl)oxy)phenoxy]-,butyl ester; Propionic acid, 2-[*p*-((5-(trifluoromethyl)-2-pyridyl)oxy)phenoxy]-, butylester; (*RS*)-2-[4-(5-Trifluoromethyl-2-pyridyloxy)]-phenoxy]propanoic acid, butyl ester; 2-[4-((5-(Trifluoromethyl)-2-pyridinyl)oxy)-phenoxy]propanoic acid, butyl ester
*Trade Names:* FUSILADE®, Syngenta (Switzerland), canceled; FUSION®, Syngenta (Switzerland); GRASS-B-GONE®, The Scotts Company (USA); HACHE UNO SUPER®; HORIZON®; ICI-A0009®; ONESIDE®; ONESIDE EC®; ORNAMEC®; PP 009®; SL-236®; TF 1169®; TS-7236®; TORNADO®
*Producers:* Pbi/Gordon (USA); The Scotts Company (USA); Syngenta (Switzerland); Zenica Ag Products (USA).
*Chemical Class:* Organofluorine
*EPA/OPP PC Code:* 122805 (Fluazifop-butyl); 122809 (Fluazifop-*p*-butyl)
*California DPR Chemical Code:* 2186 (Fluazifop-butyl); 5815 (Fluazifop-*p*-butyl)
*RTECS Number:* UA3000000
*Uses:* Fluazifop-butyl is a selective post-emergence herbicide. Its principal uses In California is on rights-of-way, landscapes, almonds, cotton, and outdoor container nurseries.
*Human toxicity (long-term)*[77]*:* Low–70.00 ppb, Health Advisory
*Fish toxicity (threshold)*[77]*:* Intermediate–62.61862 ppb, MATC (Maximum Acceptable Toxicant Concentration)

## Fluazifop-butyl

*U.S. Maximum Allowable Residue Levels for Fluazifop-butyl (40 CFR 180.411):*

*Note:* For tolerances for residues of the resolved isomer of fluazifop, see also 40 CFR 180.411

| CROP | ppm |
|---|---|
| Cattle, fat. | 0.05 |
| Cattle, meat | 0.05 |
| Cattle, mbyp | 0.05 |
| Cotton, undelinted seed | 0.1 |
| Cotton, oil | 0.2 |
| Egg | 0.05 |
| Goat, fat | 0.05 |
| Goat, meat | 0.05 |
| Goat, mbyp | 0.05 |
| Hog, fat | 0.05 |
| Hog, meat | 0.05 |
| Hog, mbyp | 0.05 |
| Horse, fat | 0.05 |
| Horse, meat | 0.05 |
| Horse, mbyp | 0.05 |
| Milk | 0.05 |
| Poultry, fat | 0.05 |
| Poultry, meat | 0.05 |
| Poultry, mbyp | 0.05 |
| Sheep, fat | 0.05 |
| Sheep, meat | 0.05 |
| Sheep, mbyp | 0.05 |
| Soybean. | 1.0 |
| Soybean, meal. | 2.0 |
| Soybean, refined oil | 2.0 |

*Carcinogen/Hazard Classifications*
**California Prop. 65:** Listed; Developmental toxin
**TRI Developmental Toxin:** Reproductive and developmental toxin
**Label Signal Word:** CAUTION
*Regulatory Authority:*
- Actively registered pesticide in California. as Fluazifop-butyl
- FIFRA, 40CFR185: tolerances for pesticides in food
- FIFRA, 40CFR186: tolerances for pesticides in animal feeds
- AB 2588-Air Toxics "Hot Spots" Chemicals (CAL) as chlorophenoxy pesticides
- Proposition 65 chemical (CAL)
- The "Director's List" (CAL/OSHA) as chlorophenoxy pesticides
- EPCRA Section 313 Form R *de minimus* concentration reporting level: 1.0% as Fluazifop-butyl

**Description:** Pale yellow liquid. Slightly soluble in water; solubility = 1 mg/L @ 25°C. Molecular weight = 383.39. Melting/freezing point = 4.8°C. Vapor pressure = $4.1 \times 10^{-7}$ mmHg @ 20°C; 0.055 mPa @ 20°C. Log $K_{ow}$ = 4.49. Values above 3.0 are likely to bioaccumulate in marine organisms. According to Fluoride Action Network Pesticde Project, "the potential for bioconcentration in aquatic organisms is very high. Highly toxic to zooplankton."

***Permissible Concentration in Water:*** No criteria set. Runoff from spills or fire control may cause water pollution.
***Routes of Entry:*** Inhalation, ingestion, absorbed through the skin.

*Harmful Effects and Symptoms*

***Short Term Exposure:*** Poisonous; may be fatal if inhaled, swallowed, or absorbed through skin. Severely irritates eyes, skin and respiratory tract, with burning sensation, pain, redness and swelling. Metabolic stimulant. If inhaled, causes coughing, dilated pupils, headache, profuse persperation, intense thirst, extreme fatigue, rapid pulse, high fever, clammy, flushed skin, rapid breathing, nausea, vomiting, cyanosis (bluish tint to skin and lips), anxiety and confusion, convulsions, risk of lung edema. If swallowed, face and lips turn bluish. Liver injury with associated jaundice, kidney failure, and cardiac arrhythmias are commonly noted. Nerve damage, which may be delayed, may include swelling of legs and feet, muscle twitch and stupor. Severe exposure can cause death from heart failure. Dust or liquid left in contact with the skin for several hours may be absorbed. This may result in severe delayed symptoms as listed above. These symptoms may last for months or years.

***Points of Attack:*** Eyes, skin, respiratory system, central nervous system, cardiovascular system, liver, kidney

***Medical Surveillance:*** If symptoms develop or overexposure is suspected, the following may be useful: Liver and kidney function tests. Exam of the nervous system.

***First Aid:*** If this chemical gets into the eyes, remove any contact lenses at once and irrigate immediately for at least 15 minutes, occasionally lifting upper and lower lids. Seek medical attention immediately. If this chemical contacts the skin, remove contaminated clothing and wash immediately with soap and water. Seek medical attention immediately. If this chemical has been inhaled, remove from exposure, begin rescue breathing (using universal precautions) if breathing has stopped, and CPR if heart action has stopped. Transfer promptly to a medical facility. When this chemical has been swallowed, get medical attention. *Do not induce vomiting when formulations containing petroleum solvents are ingested.* Otherwise, give large quantities of water and induce vomiting. Do not make an unconscious person vomit.

*References:*
- U.S. Environmental Protection Agency, Office of Pesticide Programs, Pesticide Residue Limits, "Fluazifop Butyl," 40 CFR 180.411, http://www.ehso.com/ehso.php
- EXTOXNET, Extension Toxicology Network, "Pesticide Information Profile, Fluazifop-p-butyl," Oregon State University, Corvallis, OR (June 1996). http://extoxnet.orst.edu/pips/fluazifo.htm
- California Environmental Protection Agency "Chemical List of Lists," Sacramento CA (February 1997)

# Fluazinam®

*Use Type:* Fungicide
*CAS Number:* 79622-59-6
*Formula:* $C_{13}H_4Cl_2F_6N_4O_4$
*Synonyms:* 3-Chloro-*N*-[3-chloro-2,6-dinitro-4-(trifluoromethyl)phenyl]-5-(trifluoromethyl)-2-pyridinamine; 2-*N*-(3-Chloro-5-trifluoromethyl-2-pyridyl)-2,6-dinitro-3-chloro-4-trifluoromethylaniline; 3-Chloro-*N*-(3-chloro-5-trifluoromethyl-2-pyridinyl)-α,α,α-trifluoro-2,6-dinitro-*p*-toluidine; Pyridinamine, 3-chloro-*N*-[3-chloro-2,6-dinitro-4-(trifluoromethyl)phenyl]-5-(trifluoromethyl)-
*Trade Names:* FLUAZINAM 50 WP®; FROWNCIDE®; IKF-1216®; ICIA-192®; OMEGA; PP-192®; SHIRLAN®
*Producers:* Ishihara Sangyo Kaisha (Japan); ISK Biosciences (USA) [now Syngenta]
*Chemical Class:* Phenyl-pyridinamine
*EPA/OPP PC Code:* 129098
*California DPR Chemical Code:* 3898
*Uses:* Used to control Sclerotinia blight on peanuts and late blight and white mold on potatoes.
*U.S. Maximum Allowable Residue Levels for Fluazinam (40 CFR 180.575):*

| CROP | ppm |
| --- | --- |
| Peanuts | 0.02 |
| Potatoes | 0.02 |

*Carcinogen/Hazard Classifications*
**U.S. EPA Carcinogens:** Deferred decision
**Label Signal Word:** WARNING or DANGER
*Description:* Yellow liquid suspension. Pungent odor. Molecular weight = 464.99. Melting/Freezing point = 101°C.
*Permissible Concentration in Water:* No criteria set. Runoff from spills or fire control may cause water pollution.
*Harmful Effects and Symptoms*
*Short Term Exposure:* Contact with eyes or skin may cause irritation or injury. Inhalation should be avoided; use NIOSH-approved air purifying respirators for pesticides. May be harmful if swallowed.
*First Aid:* If this chemical gets into the eyes, remove any contact lenses at once and irrigate immediately for at least 15 minutes, occasionally lifting upper and lower lids. Seek medical attention immediately. If this chemical contacts the skin, remove contaminated clothing and wash immediately with soap and water. Seek medical attention immediately. If this chemical has been inhaled, remove from exposure, begin rescue breathing (using universal precautions) if breathing has stopped, and CPR if heart action has stopped. Transfer promptly to a medical facility. When this chemical has been swallowed, get medical attention. Give large quantities of water and induce vomiting. Do not make an unconscious person vomit.
*References:*
- U.S. Environmental Protection Agency, Office of Pesticide Programs, "Pesticide Fact Sheet, Fluazinam," (August 10, 2001). http://www.epa.gov/opprd001/factsheets/fluazinam.pdf
- U.S. Environmental Protection Agency, Office of Pesticide Programs, Pesticide Residue Limits, "Fluazinam," 40 CFR 180.574, http://www.epa.gov/fedrgstr/EPA-PEST/2002/April/Day-18/p9497.htm

# Fluchloralin (ANSI)

*Use Type:* Herbicide
*CAS Number:* 33245-39-5
*Formula:* $C_{12}H_{13}ClF_3N_3O_4$
*Synonyms:* Benzenamine, *N*-(2-chloroethyl)-2,6-dinitro-*N*-propyl-4-(trifluoromethyl)-; *N*-(2-Chloroethyl)-2,6-dinitro-*n*-propyl-4-(trifluoromethyl)aniline; *N*-(2-Chloroethyl)-2,6-dinitro-*n*-propyl-4-(trifluoromethyl)benzenamide; *N*-(2-Chloroethyl)-α,α,α-trifluoro-2,6-dinitro-*N*-propyl-*p*-toluidine; *N*-(2-Chloroethyl)-2,6-dinitro-*N*-propyl-4-(trifluoromethyl)benzenamine; *N*-Propyl-*N*-(2-chloroethyl)-2,6-dinitro-4-trifluoromethylaniline; *N*-Propyl-*N*-(2-chloroethyl)-α,α,α-trifluoro-2,6-dinitro-*p*-toluidine
*Trade Names:* BAS 392-H® BASF Agricultural Products (Germany); BASALIN®, BASF Agricultural Products (Germany), canceled
*Producers:* BASF Agricultural Products (Germany)
*Chemical Class:* Dinitroaniline; chloroaniline
*EPA/OPP PC Code:* 108701; (460200 old code)
*California DPR Chemical Code:* 1848
*Uses:* Not registered in the U.S. U.S. tolerances were revoked in 1999. A pre-emergence herbicide used on dry and succulent peas and beans, okra, peanuts, soybeans, sunflowers and cotton.
*Human toxicity (long-term)[77]:* Very low–46200.00 ppb, Health Advisory
*Fish toxicity (threshold)[77]:* High–7.09937 ppb, MATC (Maximum Acceptable Toxicant Concentration)
*Carcinogen/Hazard Classifications*
**Label Signal Word:** WARNING
**WHO Acute Hazard:** Class III, slightly hazardous
*Regulatory Authority:*
- AB 1803-Well Monitoring Chemical (CAL)

*Description:* Orange-yellow crystalline solid. Faint and unusual odor. Solubility in water: 10 ppm;<1 ppm @ 20°C. Molecular weight = 355.71. Melting/Freezing point = 42°C. Vapor pressure = $4 \times 10^{-5}$ mmHg @ 20°C.
*Incompatibilities:* Slowly hydrolyzes in water, releasing ammonia and forming acetate salts. Forms toxic fumes of nitrogen oxides, chlorine, and fluorine when heated to decomposition.
*Determination in Air:* Filter; none; Gravimetric; NIOSH IV[18] [Particulates NOR; #0500 (total), #0600 (respirable)]

*Permissible Concentration in Water:* No criteria set. Runoff from spills or fire control may cause water pollution.

*Harmful Effects and Symptoms*

*Short Term Exposure:* Contact with eyes or skin may cause irritation or injury. Inhalation should be avoided; use NIOSH-approved air purifying respirators for pesticides. May be harmful if swallowed.

*First Aid:* If this chemical gets into the eyes, remove any contact lenses at once and irrigate immediately for at least 15 minutes, occasionally lifting upper and lower lids. Seek medical attention immediately. If this chemical contacts the skin, remove contaminated clothing and wash immediately with soap and water. Seek medical attention immediately. If this chemical has been inhaled, remove from exposure, begin rescue breathing (using universal precautions) if breathing has stopped, and CPR if heart action has stopped. Transfer promptly to a medical facility. When this chemical has been swallowed, get medical attention. Give large quantities of water and induce vomiting. Do not make an unconscious person vomit. Do not induce vomiting when formulations containing petroleum solvents are ingested.

*References:*
- Pesticide Management Education Program, "Fluchloralin (Basalin) Herbicide Profile 6/85," Cornell University, Ithaca, NY (June 30, 1985). http://pmep.cce.cornell.edu/profiles/herb-growthreg/fatty-alcohol-monuron/fluchloralin/herb-prof-fluchloralin.html
- California Environmental Protection Agency "Chemical List of Lists," Sacramento CA (February 1997)

# Flucythrinate (ANSI)

*Use Type:* Insecticide, acaricide
*CAS Number:* 70124-77-5
*Formula:* $C_{26}H_{23}F_2NO_4$
*Alert:* Some formulations were Restricted Use Pesticides (RUP)
*Synonyms:* AC 222705; Benzeneacetic acid, 4-(difluoromethoxy)-α-(1-methylethyl)-, cyano(3-phenoxyphenyl)methyl ester; (±)-Cyano(3-phenoxyphenyl)methyl(+)-4-(difluoromethoxy)-α(1-methylethyl)benzeneacetate; (+)-Cyano-(3-phenoxyphenyl)methyl (+)-4-(difluoromethoxy)-α-(1-methylethyl) benzene acetate; (RS)-Cyano-(3-phenoxyphenyl)methyl (S)-4-(difluoromethoxy)-α-(1-methylethyl)-benzeneacetate; (+)-α-Cyano-m-phenoxybenzyl alcohol ester of (+)-2-(p-difluoromethoxy)phenyl-3-methylbutyric acid; (RS)-α-cyano-3-phenoxybenzyl(S)-2-(4-difluoromethoxyphenyl)-3-methylbutyrate; Fluorocythrin; OMS 2007 (WHO)
*Trade Names:* AASTAR, BASF Agricultural Products (Germany), canceled; CYBOLT®; CYTHRIN®; FUCHING JUJR®; GUARDIAN®, BASF Agricultural Products (Germany), canceled; PAYOFF®, DuPont Crop Protection (USA), canceled; STOCK GUARD®; TOMAHAWK®
*Producers:* BASF Agricultural Products (Germany); DuPont Crop Protection (USA)
*Chemical Class:* Pyrethroids; botanical
*EPA/OPP PC Code:* 118301
*California DPR Chemical Code:* 2168
*Uses:* Not registered in the U.S. It is a synthetic pyrethroid used to control pests in apples, cabbage, head lettuce, pears, corn and cotton, but it was used primarily on cotton.
*Human toxicity (long-term)[77]:* Very low–140.00 ppb, Health Advisory
*Fish toxicity (threshold)[77]:* Extra high–0.00707 ppb, MATC (Maximum Acceptable Toxicant Concentration)
*Carcinogen/Hazard Classifications*
**Label Signal Word:** CAUTION or DANGER
**WHO Acute Hazard:** Class 1b, highly hazardous
**Endocrine Disruptor:** Possible ED
*Regulatory Authority:*
- AB 1803-Well Monitoring Chemical (CAL) as pyrethrins
- Permissible Exposure Limits for Chemical Contaminants (CAL/OSHA) as pyrethrum
- Actively registered pesticide in California. as pyrethrins
- Clean Water Act: Section 311 Hazardous Substances/RQ (same as CERCLA) as pyrethrins
- EPCRA Section 304 RQ: CERCLA, 1 lb (0.454 kg) as pyrethrins

*Description:* Dark amber, viscous liquid. Faint odor, like esters. Practically insoluble in water; solubility = 0.5 ppm @ 21°C. Molecular weight = 451.47. Density = 1.188 @ 22°C. Boiling point = 108°C @ 0.35 mmHg. Vapor pressure = $8.7 \times 10^{-9}$ mmHg @ 20°C; $1.16 \times 10^{-8}$ mmHg @ 25°C; 0.0012 mPa @ 25°C. Log $K_{ow}$ = 6.19. Values above 3.0 are likely to bioaccumulate in marine organisms.

*Incompatibilities:* Keep away from oxidizers, sulfuric acid, caustics, ammonia, aliphatic amines, alkanolamines, isocyanates, alkylene oxides, epichlorohydrin. Moisture may cause hydrolysis or other forms of decomposition.

*Determination in Air:* Collection by impinger or fritted bubbler, analysis by gas liquid chromatography/ultraviolet. See NIOSH IV, Method #5008[18]. (pyrethrum)

*Permissible Concentration in Water:* No criteria set. Runoff from spills or fire control may cause water pollution.

*Determination in Water:* Collection by impinger or fritted bubbler, analysis by gas liquid chromatography/ultraviolet. See NIOSH IV, Method #5008[18].

*Routes of Entry:* Inhalation, absorbed through the skin.

*Harmful Effects and Symptoms*

*Short Term Exposure:* Flucythrinate can cause extreme eye irritation. Pyrethrins can affect you when breathed in and by passing through your skin. Irritates the eyes and respiratory tract. High exposure can affect the nervous system causing headache, nausea, vomiting, fatigue, and restlessness, rhinorrhea (discharge of thin nasal mucous).

*Long Term Exposure:* High or repeated exposure can cause lung allergy (with cough, wheezing and/or shortness of breath) or hay fever symptoms (sneezing, runny or stuffy nose). Allergic "pneumonia" can also occur with cough, chest pain, breathing difficulty and abnormal chest x-ray. Repeated attacks may lead to permanent scarring. Skin allergy may also develop with rash and itching, even with lower exposures. Skin contact can cause rash with redness, blisters and intense itching. A severe generalized allergy can occur with weakness and collapse.

*Points of Attack:* Respiratory system, skin, eyes and central nervous system.

*Medical Surveillance:* Before beginning employment and at regular times after that, the following are recommended: Lung function tests. These may be normal if the person is not having an attack at the time of the test. Consider chest x-ray if lung symptoms are present. Evaluation by a qualified allergist, including careful exposure history and special testing, may help diagnose skin allergy.

*First Aid:* If this chemical gets into the eyes, remove any contact lenses at once and irrigate immediately for at least 15 minutes, occasionally lifting upper and lower lids. Seek medical attention immediately. If this chemical contacts the skin, remove contaminated clothing and wash immediately with soap and water. Seek medical attention immediately. If this chemical has been inhaled, remove from exposure, begin rescue breathing (using universal precautions) if breathing has stopped, and CPR if heart action has stopped. Transfer promptly to a medical facility. When this chemical has been swallowed, get medical attention. *Do not induce vomiting when formulations containing petroleum solvents are ingested.* Otherwise, give large quantities of water and induce vomiting. Do not make an unconscious person vomit.

*References:*
- EXTOXNET, Extension Toxicology Network, "Pesticide Information Profile, Flucythrinate," Oregon State University, Corvallis, OR (June 1996). http://extoxnet.orst.edu/pips/flucythr.htm
- International Programme on Chemical Safety (IPCS), "Health and Safety Guide, Flucythrinate," Geneva, Switzerland. http://www.inchem.org/documents/jmpr/jmpmono/v85pr09.htm
- California Environmental Protection Agency "Chemical List of Lists," Sacramento CA (February 1997)

# Fluenetil

*Use Type:* Acaricide and insecticide
*CAS Number:* 4301-50-2
*Formula:* $C_{16}H_{15}FO_2$
*Alert:* This chemical is believed to be obsolete or discontinued for use as a pesticide.

*Synonyms:* Acetic acid, diphenyl-, 2-fluoroethyl ester; β-Fluoroethyl-4-biphenylacetate; 4-Biphenylacetic acid,2-fluoroethyl ester; (1,1'-Biphenyl)-4-acetic acid, 2-fluoroethyl ester; Fluenethyl; Fluenyl; Fluoroethylic ester of xenylacetic acid

*Trade Names:* LAMBROL®; M 2060®; MYTROL®; TH 3671®; TH 367-1®; UC 20299®

*EPA/OPP PC Code:* 462200

*RTECS Number:* DV8335000

*Uses:* Fluenetil's main use was as a dormant spray for orchard fruit. It is no longer made. It is not registered as a pesticide in the U.S.

*Regulatory Authority:*
- Very Toxic Substance (World Bank)[15]
- Superfund/EPCRA 40CFR355, Appendix B Extremely Hazardous Substances: TPQ = 100/10,000 lb (45.4/4,540 kg)
- Superfund/EPCRA 40CFR302.4 RQ: EHS, 1 lb (0.454 kg)

*Description:* $C_{16}H_{15}FO_2$ is a crystalline solid. Molecular weight = 258.32. Hazard Identification (based on NFPA-704 M Rating System): Health 3, Flammability 1, Reactivity 0.

*Incompatibilities:* Nitrates. Moisture may cause material to hydrolyze.

*Permissible Exposure Limits in Air:* No standards set.

*Permissible Concentration in Water:* No criteria set, but runoff from spills or fire control may cause water pollution.

*Harmful Effects and Symptoms*

*Short Term Exposure:* Contact with eyes or skin may cause irritation or injury. Inhalation should be avoided; use NIOSH-approved air purifying respirators for pesticides. May be harmful if swallowed. Fluenetil is highly toxic. The LD-50 (oral, rat) = 6 mg/kg.

*First Aid:* Speed in removing material from eyes and skin is of extreme importance. If this chemical gets into the eyes, remove any contact lenses at once and irrigate immediately for at least 15 minutes, occasionally lifting upper and lower lids. Seek medical attention immediately. If this chemical contacts the skin, remove contaminated clothing and wash immediately with soap and water. Seek medical attention immediately. If this chemical has been inhaled, remove from exposure, begin rescue breathing (using universal precautions) if breathing has stopped, and CPR if heart action has stopped. Transfer promptly to a medical facility. When this chemical has been swallowed, get medical attention. Give large quantities of water and induce vomiting. Do not make an unconscious person vomit. Effects may be delayed; keep victim under observation.

*References:*
- Fluoride Action Network, http://www.fluorideaction.org/pesticides/fluenetil-page.htm

- U.S. Environmental Protection Agency, "Chemical Profile: Fluenetil," Washington, DC, Chemical Emergency Preparedness Program (November 30, 1987).
- California Environmental Protection Agency "Chemical List of Lists," Sacramento CA (February 1997).

# Flumetsulam (ANSI)

*Use Type:* Herbicide
*CAS Number:* 98967-40-9
*Formula:* $C_{12}H_9F_2N_5O_2S$
*Synonyms:* (1,2,4)Triazolo(1,5-α)pyrimidine-2-sulfonamide, N-(2,6-difluorophenyl)-5-methyl-; N-(2,6-Difluorophenyl)-5-methyl-(1,2,4)triazolo-(1,5-α)pyrimidine-2-sulfonamide
*Trade Names:* ACCENT®, DuPont Crop Protection (USA); BROADSTRIKE® (flumetsulam plus metolachlor); DE-498®; FRONTROW®, (cloransulam-methyl + flumetsulam), Dow AgroSciences (USA); HORNET®, (flumetsulam + clopyralid), Dow AgroSciences (USA); NAF-9® (flumetsulam + metolachlor); NAF-2® (flumetsulam + metolachlor), Dow AgroSciences (USA); PYTHON®, Dow AgroSciences (USA); SCORPION®, Dow AgroSciences (USA); XRD-498®; XRM-5313® (flumetsulam plus trifluralin); XRM-5019® Herbicide
*Producers:* Dow AgroSciences (USA); DuPont Crop Protection (USA)
*Chemical Class:* Triazolopyrimidine; sulfonanilide
*EPA/OPP PC Code:* 129016
*California DPR Chemical Code:* 3927
*U.S. Maximum Allowable Residue Levels for Flumetsulam (40 CFR 180.468):*

| CROP | ppm |
| --- | --- |
| Corn, field, forage | 0.05 |
| Corn, field, grain | 0.05 |
| Corn, field, stover | 0.05 |
| Soybean | 0.05 |

*Carcinogen/Hazard Classifications*
**U.S. EPA Carcinogens:** Group E, Unlikely a carcinogen
**Label Signal Word:** WARNING, CAUTION or DANGER
**WHO Acute Hazard:** Class U, unlikely to be hazardous
*Description:* White, crystalline powder. Soluble in water. Molecular weight = 325.29. Log $K_{ow}$ = < 1.0. Unlikely to bioaccumulate in marine organisms.
*Permissible Concentration in Water:* No criteria set. Runoff from spills or fire control may cause water pollution.
*Harmful Effects and Symptoms*
*Short Term Exposure:* Contact with eyes or skin may cause irritation or injury. Inhalation should be avoided; use NIOSH-approved air purifying respirators for pesticides. May be harmful if swallowed.
*First Aid:* If this chemical gets into the eyes, remove any contact lenses at once and irrigate immediately for at least 15 minutes, occasionally lifting upper and lower lids. Seek medical attention immediately. If this chemical contacts the skin, remove contaminated clothing and wash immediately with soap and water. Seek medical attention immediately. If this chemical has been inhaled, remove from exposure, begin rescue breathing (using universal precautions) if breathing has stopped, and CPR if heart action has stopped. Transfer promptly to a medical facility. When this chemical has been swallowed, get medical attention. Give large quantities of water and induce vomiting. Do not make an unconscious person vomit. Do not induce vomiting when formulations containing petroleum solvents are ingested.
*References:*
- U.S. Environmental Protection Agency, Office of Pesticide Programs, Pesticide Residue Limits, "Flumetsulam," 40 CFR 180.468, http://www.epa.gov/pesticides/food/viewtols.htm

# Fluometuron (ANSI)

*Use Type:* Herbicide
*CAS Number:* 2164-17-2
*Formula:* $C_{10}H_{11}F_3N_2O$
*Synonyms:* N,N-Dimethyl-N'-[3-(trifluoromethyl)phenyl]urea; 1,1-Dimethyl-3-(3-trifluoromethylphenyl)urea; Meturone; NCI-C08695; 3-(5-Trifluormethylphenyl)-, dimethylharnstoff (German); N-(m-Trifluoromethylphenyl)-N',N'-dimethylurea; N-(3-Trifluoromethylphenyl)-N',N'-dimethylurea; 3-(3-Trifluoromethylphenyl)-1,1-dimethylurea; 3-(m-Trifluoromethylphenyl)-1,1-dimethylurea; Urea, N,N-dimethyl-N'-[3-(trifluoromethyl)phenyl]-; Urea, 1,1-dimethyl-3-(α,α,α-trifluoro-m-tolyl)-
*Trade Names:* C 2059®, Syngenta (Switzerland); CIBA 2059®, Syngenta (Switzerland); COTORAN®, Griffin LLC (USA); Makhteshim-Agan Industries (Israel); COTORAN MULTI 50WP®, Griffin LLC (USA); COTOREX®; COTTONEX®; DREXEL CROAK®, Drexel Chemical (USA); FLO-MET®, Micro Flo (USA); HERBICIDE C-2059®; HIGALCOTON®; LANEX®, Agrevo USA Co., canceled 7/01/1987; METURON 80 DF®, Griffin LLC (USA); PAKHTARAN®; SETRE FLUOMETURON 80 WP®, Helena Chemical (USA)
*Producers:* Drexel Chemical (USA); Epochem Co., (China); Griffin L.L.C. (USA); Helena Chemical (USA); Makhteshim-Agan Industries (Israel); Micro Flo (USA); Sigma-Aldrich Laborchemikalien (Germany); Syngenta (Switzerland)
*Chemical Class:* Urea
*EPA/OPP PC Code:* 035503
*California DPR Chemical Code:* 166
*RTECS Number:* YT1575000
*EINECS Number:* 218-500-4
*Uses:* A General Use Pesticide (GUP). Fluometuron is a selective herbicide which acts on susceptible plants by

inhibiting photosynthesis. Fluometuron is registered by EPA exclusively for use on cotton and sugarcane. It can be applied pre-emergence, for weed control before planting, or post-emergence, after target crops and weeds come up, and may have residual activity for several months. Fluometuron is available in liquid, dry flowable, and wettable powder formulations.

*U.S. Maximum Allowable Residue Levels for Fluometuron (40 CFR 180.229):*

| CROP | ppm |
|---|---|
| Cotton, undelinted seed | 0.1 |

*Carcinogen/Hazard Classifications*
**U.S. EPA Carcinogens:** Group C, possible carcinogen
**IARC:** Group 3, unclassifiable
**Label Signal Word:** WARNING. EPA Toxicity Class II.
**WHO Acute Hazard:** Class U, unlikely to be hazardous
*Regulatory Authority:*
- EPCRA Section 313 Form R *de minimus* concentration reporting level: 1.0%
- Permissible Exposure Limits for Chemical Contaminants (CAL/OSHA)
- The "Director's List" (CAL/OSHA)

*Description:* Colorless, crystalline solid often used in liquid "carrier" that may be flammable. Slightly soluble in water; solubility = 103 ppm @ 20. Molecular weight = 232.24 Melting/Freezing point = 163–165°C. Vapor pressure = $6.5 \times 10^{-7}$ mmHg @ 20°C. Hazard Identification (based on NFPA-704 M Rating System): Health 1, Flammability 1, Reactivity 0. Log $K_{ow}$ = 2.3. Unlikely to bioaccumulate in marine organisms.

*Incompatibilities:* Liquid solutions are incompatible with oxidizers.
*Permissible Exposure Limits in Air:* No standards set.
*Permissible Concentration in Water:* A no-observed-adverse effect level (NOAEL) of 0.0125 mg/kg/day has been calculated by EPA. On this basis a long-term health advisory of 5.3 mg/L and a lifetime health advisory of 0.09 mg/L have been calculated.
*Routes of Entry:* Inhalation, through the skin.
*Harmful Effects and Symptoms*
*Short Term Exposure:* Contact can cause eye and skin irritation. Inhalation can irritate the respiratory tract. Symptoms of exposure include increased leukocyte content in circulation blood. The material is a mild cholinesterase inhibitor. The $LD_{50}$ (oral, rat) = 6400 mg/kg (insignificantly toxic).
*Long Term Exposure:* May cause skin allergy. Mild cholinesterase inhibitor; cumulative effect is possible. Repeated exposure may cause in the red blood cell count. May cause liver damage
*Points of Attack:* Respiratory system, lungs, central nervous system, skin, eyes, plasma and red blood cell cholinesterase.
*Medical Surveillance:* Liver function tests. Complete blood count. Evaluation by a qualified allergist.

*First Aid: Skin Contact:* Flood all areas of body that have contacted the substance with water. Don't wait to remove contaminated clothing; do it under the water stream. Use soap to help assure removal. Isolate contaminated clothing when removed to prevent contact by others. *Eye Contact:* Remove any contact lenses at once. Flush eyes well with copious quantities of water or normal saline for at least 20-30 minutes. Seek medical attention. *Inhalation:* Leave contaminated area immediately; breathe fresh air. Proper respiratory protection must be supplied to any rescuers. If coughing, difficult breathing or any other symptoms develop, seek medical attention at once, even if symptoms develop many hours after exposure. *Ingestion:* Consult a physician, hospital or poison center at once. If the victim is unconscious or convulsing, do not induce vomiting or give anything by mouth. Assure that the airway is open, lay on side and keep head lower than body and transport immediately to medical facility. If conscious and not convulsing, give a glass of water to dilute the substance. Do not induce vomiting without a physician's advice.

*References:*
- EXTOXNET, Extension Toxicology Network, "Pesticide Information Profile, Fluometuron," Oregon State University, Corvallis, OR. (June 1996). http://extoxnet.orst.edu/pips/fluometu.htm
- EPA, Office of Pesticide Programs, Pesticide Residue Limits, "Fluometuron," 40 CFR 180.229, http://www.epa.gov/pesticides/food/viewtols.htm
- U.S. Environmental Protection Agency, "Health Advisory: Fluometuron," Washington, DC, Office of Drinking Water (August 1987).
- California Environmental Protection Agency "Chemical List of Lists," Sacramento CA (February 1997).
- New Jersey Department of Health and Senior Services, "Hazardous Substance Fact Sheet, FLUMETURON," Trenton NJ (April 1999). http://www.state.nj.us/health/eoh/rtkweb/0935.pdf

# Fluoroacetamide

*Use Type:* Rodenticide and insecticide
*CAS Number:* 640-19-7
*Formula:* $C_2H_4FNO$; $CH_2FCONH_2$
*Alert:* A Restricted Use Pesticide (RUP)
*Synonyms:* FAA; 2-Fluoroacetamide; Fluoroacetic acid amide; Monofluoroacetamide
*Trade Names:* AFL 1081®; COMPOUND 1081®; FLUORAKIL 100®; FUSSOL®; MEGATOX®; NAVRON®; RODEX®; YANOCK®
*Producers:* Halocarbon Products (USA); Molekula Fine Chemicals (UK)
*Chemical Class:* Unclassified
*EPA/OPP PC Code:* 075002
*ICSC Number:* 1434

*RTECS Number:* AC1225000
*EEC Number:* 616-002-00-5
*Uses:* This rodenticide is a Restricted Use Pesticide; insecticide proposed mainly for use on fruits to combat scale insects, aphids, and mites.
*Carcinogen/Hazard Classifications*
**Label Signal Word:** DANGER
**WHO Acute Hazard:** Class 1b, highly hazardous
*Regulatory Authority:*
* Banned or Severely Restricted (In Agriculture) (Several Countries) (UN)[13]
* The "Director's List" (CAL/OSHA)
* EPA Hazardous Waste Number (RCRA No.): P057
* RCRA, 40CFR261, Appendix 8 Hazardous Constituents
* Superfund/EPCRA 40CFR355, Appendix B Extremely Hazardous Substances: TPQ = 100/10,000 lb (45.4/4,540 kg)
* Superfund/EPCRA 40CFR302.4 RQ: CERCLA, 100 lb (45.4 kg)

*Description:* Fluoroacetamide is a colorless crystalline solid. Soluble in water. Melting/Freezing point = 107–109°C. Hazard Identification (based on NFPA-704 M Rating System): Health 4, Flammability 1, Reactivity 0.
*Incompatibilities:* Strong oxidizers.
*Permissible Exposure Limits in Air:* No standards set.
*Permissible Concentration in Water:* No criteria set, but runoff from spills or fire control may cause water pollution.
*Routes of Entry:* Inahalation, skin and/or eyes.
*Harmful Effects and Symptoms*
*Short Term Exposure:* Signs and symptoms may be extremely severe and range from nausea, vomiting, and diarrhea to convulsions, coma, and heart failure. Other symptoms include hyperactivity, respiratory depression or arrest, cyanosis (blue tint to the skin and mucous membranes), and ventricular fibrillation. This material is super toxic; probable oral lethal dose in humans is less than 5 mg/kg, or a taste (less than 7 drops) for a 150-lb person. Chemically inhibits oxygen metabolism by cells with critical damage occurring to the heart, brain, and lungs resulting in heart failure, respiratory arrest, convulsions, and death.
*Warning:* Effects usually appear within 30 minutes of exposure but may be delayed as long as 20 hours. Caution is advised. Vital signs should be monitored closely.
*Long Term Exposure:* Repeated high exposures may affect kidneys. Repeated high exposures can cause deposits of fluorides in the bones (fluorosis) that may cause pain, disability and mottling of the teeth. Repeated exposure can cause nausea, vomiting, loss of appetite, diarrhea or constipation. Nosebleeds and sinus problems can also occur.
*First Aid:* Acute exposure to fluoroacetamide may require decontamination and life support for the victim. Emergency personnel should wear protective clothing appropriate to the type and degree of contamination. Air-purifying or supplied-air respiratory equipment should also be worn, as necessary. Rescue vehicles should carry supplies such as plastic sheeting and disposable plastic bags to assist in preventing spread of contamination. *Inhalation:* Move victim to fresh air. Evaluate vital signs. If no pulse is detected, provide CPR. If not breathing, provide artificial respiration. If breathing is labored, administer oxygen. Rush to a health care facility. *Eye Exposure:* Remove any contact lenses at once and flush eyes with lukewarm water for 15 minutes. *Skin Exposure:* Follow steps under inhalation above. Wash exposed skin areas 3 times with soap and water. Rush to health care facility. *Ingestion:* Evaluate vital signs including pulse and respiratory rate, and note any trauma. If no pulse is detected, provide CPR. If not breathing, provide artificial respiration. If breathing is labored, administer oxygen or other respiratory support. Rush to health care facility. Obtain authorization and/or further instructions from the local hospital for performance of other invasive procedures.
*Warning:* Effects usually appear within 30 minutes of exposure but may be delayed as long as 20 hours. Caution is advised. Vital signs should be monitored closely.
*References:*
* U.S. Environmental Protection Agency, "Chemical Profile: Fluoroacetamide," Washington, DC, Chemical Emergency Preparedness Program (November 30, 1987).
* California Environmental Protection Agency "Chemical List of Lists," Sacramento CA (February 1997).

# Fluoroacetic Acid

*Use Type:* Rodenticide
*CAS Number:* 144-49-0
*Formula:* $C_2H_3FO_2$; $FCH_2COOH$
*Synonyms:* Acide monofluoracetique (French); Acido fluoroacetico (Spanish); Acido monofluoroacetio (Italian); Cymonic acid; FAA; Fluoroacetate; Fluoroethanoic acid; 2-Fluoroacetic acid; Gifblaar Poison; HFA; MFA; Monofluorazijnzuur (Dutch); Monofluoressigsaeure (German); Monofluoroacetate; Monofluoroacetic acid
*Trade Names:* COMPOUND 1809®; RATBANE 1080®
*Producers:* Halocarbon Products (USA)
*Chemical Class:* Unclassified
*EPA/OPP PC Code:* 075001
*ICSC Number:* 0274
*RTECS Number:* AH5950000
*EEC Number:* 607-081-00-7
*Regulatory Authority:*
* Very Toxic Substance (World Bank)[13]
* Superfund/EPCRA 40CFR355, Appendix B Extremely Hazardous Substances: TPQ = 10/10,000 lb (4.54/4,540 kg)
* Superfund/EPCRA 40CFR302.4 RQ: EHS, 1 lb (0.454 kg)

*Description:* Fluoroacetic acid is a colorless crystalline solid. Highly soluble in water. Molecular weight = 78.04. Boiling point = 165°C. Melting/Freezing point = 35°C.

Vapor pressure = 5.34 hPa @ 20°C. Log $K_{ow}$ = Negative. Unlikely to bioaccumulate in marine organisms. Liquid formulations containing organic solvents may be flammable.
*Incompatibilities:* Strong oxidizers, strong bases. The aqueous solution is a weak acid.
*Permissible Exposure Limits in Air:* The sodium salt has a TWA of 0.05 mg/m³ as set by ACGIH[(1)], DFG[(3)] and HSE[(33)] as well as OSHA[(2)], and a STEL of 0.15 mg/m³ with the notation "skin" indication the possibility of cutaneous absorption.
*Permissible Concentration in Water:* No standards set for the acid but see the entry on "Sodium Fluoroacetate."
*Routes of Entry:* Inhalation, ingestion, eye and skin exposure.
*Harmful Effects and Symptoms*
*Short Term Exposure:* Corrosive to the eyes, skin, and respiratory tract. The major symptoms of fluoroacetic acid poisoning include severe epileptiform convulsions alternating with coma and depression; death may result from asphyxiation during convulsion or from respiratory failure. Cardiac irregularities, such as ventricular fibrillation and sudden cardiac arrest, nausea, vomiting, excessive salivation, numbness, tingling sensations, epigastric pain, mental apprehension, muscular twitching, low blood pressure, and blurred vision may also occur. This material is very toxic; the $LD_{50}$ (oral, rat) = 4.7 mg/kg (extremely toxic), and may affect the cardiovascular system, central nervous system, and kidneys and may cause cardiac and renal failure. This may cause death.
*Long Term Exposure:* See information for short-term exposure.
*Points of Attack:* Central nervous system, heart, kidneys, lungs.
*Medical Surveillance:* Kidney function tests. EKG. Lung function tests. Examination of the nervous system. Consider chest x-ray following acute exposure.
*First Aid:* If this chemical gets into the eyes, remove any contact lenses at once and irrigate immediately for at least 15 minutes, occasionally lifting upper and lower lids. Seek medical attention immediately. If this chemical contacts the skin, remove contaminated clothing and wash immediately with soap and water. Seek medical attention immediately. If this chemical has been inhaled, remove from exposure, begin rescue breathing (using universal precautions) if breathing has stopped, and CPR if heart action has stopped. Transfer promptly to a medical facility. When this chemical has been swallowed, get medical attention. Give large quantities of water and induce vomiting. Do not make an unconscious person vomit. The symptoms of central nervous system, cardiac, and renal failure do not become manifest until a few hours have passed. Specific treatment is necessary in case of poisoning with this substance; the appropriate means with instructions must be available.

*References:*
- U.S. Environmental Protection Agency, "Chemical Profile: Fluoroactic Acid," Washington, DC, Chemical Emergency Preparedness Program (November 30, 1987).
- California Environmental Protection Agency "Chemical List of Lists," Sacramento CA (February 1997).

# Fluorobenzene

*Use Type:* Insecticide
*CAS Number:* 462-06-6
*Formula:* $C_6H_5F$
*Synonyms:* Benzene, fluoro-; Phenyl fluoride; Monofluorobenzene; Benzene fluoride; MFB
*Producers:* Aldrich Chemical (USA); Archimica (UK); Avecia (UK); Clariant (Switzerland); DuPontl (USA); EniChem (Italy); Honeywell Performance P & C (USA); Indofine Chemicall (USA); Jiangsu Wujin Zhenhua Chemical Plant (China); Miteni (Italy); Oakwood Products (USA); Rhodia (France); Rhone-Poulenc (France)
*RTECS Number:* DA0800000
*EINECS Number:* 207-321-7
*Uses:* Fluorobenzene is also used as a reagent for plastic or resin polymers. It is used to control the carbon content in steel manufacturing and as an intermediate for pharmaceuticals, pesticides and other organic compounds.
*Description:* Fluorobenzene is a colorless liquid. Boiling point = 85°C. Flash point = –15°C. Insoluble in water. Highly flammable.
*Incompatibilities:* Oxidizers
*Permissible Exposure Limits in Air:* No standards set.
*Permissible Concentration in Water:* No criteria set, but runoff from spills or fire control may cause water pollution.
*Routes of Entry:* Inhalation, passing through the skin, eye contact.
*Harmful Effects and Symptoms*
*Short Term Exposure:* Fluorobenzene can irritate the eyes, nose, throat and lungs. Higher exposures can cause pulmonary edema, a medical emergency that can be delayed for several hours. This can cause death. A closely related chemical, chlorobenzene, can damage the liver and kidneys with high or repeated exposure. It is unknown if fluorobenzene causes these effects. Overexposure could cause headache, nausea and make you dizzy.
*Long Term Exposure:* May cause liver and kidney damage. Repeated exposure may damage the lungs and affect the nervous system.
*Points of Attack:* Lungs, liver, kidney, nervous system.
*Medical Surveillance:* Before beginning employment and at regular times after that, for those with frequent or potentially high exposures, the following are recommended: Periodic lung function tests. If symptoms develop or overexposure is suspected, the following may also be useful:

Tests for kidney and liver function. Examination of the nervous system. Consider chest x-ray after acute overexposure.

*First Aid:* If this chemical gets into the eyes, remove any contact lenses at once and irrigate immediately for at least 15 minutes, occasionally lifting upper and lower lids. Seek medical attention immediately. If this chemical contacts the skin, remove contaminated clothing and wash immediately with soap and water. Seek medical attention immediately. If this chemical has been inhaled, remove from exposure, begin rescue breathing (using universal precautions) if breathing has stopped, and CPR if heart action has stopped. Transfer promptly to a medical facility. When this chemical has been swallowed, get medical attention. Give large quantities of water and induce vomiting. Do not make an unconscious person vomit. Medical observation is recommended for 24 to 48 hours after breathing overexposure, as pulmonary edema may be delayed.

*References:*
- New Jersey Department of Health and Senior Services, "Hazardous Substance Fact Sheet: Fluorobenzene," Trenton, NJ (March 1999). http://www.state.nj.us/health/eoh/rtkweb/0939.pdf

# Fluridone (ANSI)

*Use Type:* Herbicide
*CAS Number:* 59756-60-4
*Formula:* $C_{19}H_{14}F_3NO$
*Synonyms:* 1-Methyl-3-phenyl-5-[3-(trifluoromethyl)phenyl]-4(1$H$)-pyridinone; 4(1$H$)-Pyridinone, 1-methyl-3-phenyl-5-[3-(trifluoromethyl)phenyl]-
*Trade Names:* AVAST®, Griffin (USA); EL 171®; PRIDE®; SONAR®, Sepro (USA)
*Producers:* Griffin (USA)
*EPA/OPP PC Code:* 112900; (215900 use code No.112900)
*Uses:* Used is horticulture, ornamental, aquatic and greenhouse environments. It is used to treat large areas of water (lakes, ponds, reservoirs, etc.) for Eurasian watermilfoil. Not used on food crops.
*Human toxicity (long-term)$^{(77)}$:* Very low–560.00 ppb, Health Advisory
*Fish toxicity (threshold)$^{(77)}$:* Very low–678.82839 ppb, MATC (Maximum Acceptable Toxicant Concentration)
*Carcinogen/Hazard Classifications*
**U.S. EPA Carcinogens:** Group E, Unlikely a carcinogen
**Label Signal Word:** WARNING or CAUTION
**WHO Acute Hazard:** Class U, unlikely to be hazardous
*Description:* Vapor pressure = $1 \times 10^{-7}$ mmHg @ 20°C.
*Harmful Effects and Symptoms*
*Short Term Exposure:* Contact with eyes or skin may cause irritation or injury. Inhalation should be avoided; use NIOSH-approved air purifying respirators for pesticides. May be harmful if swallowed.

*First Aid:* If this chemical gets into the eyes, remove any contact lenses at once and irrigate immediately for at least 15 minutes, occasionally lifting upper and lower lids. Seek medical attention immediately. If this chemical contacts the skin, remove contaminated clothing and wash immediately with soap and water. Seek medical attention immediately. If this chemical has been inhaled, remove from exposure, begin rescue breathing (using universal precautions) if breathing has stopped, and CPR if heart action has stopped. Transfer promptly to a medical facility. When this chemical has been swallowed, get medical attention. Give large quantities of water and induce vomiting. Do not make an unconscious person vomit.

# Fluroxypyr 1-methylheptyl Ester

*Use Type:* Herbicide
*CAS Number:* 81406-37-3
*Formula:* $C_7H_5Cl_2FN_2O_3$
*Synonyms:* 1-Methylheptyl [(4-amino-3,5-dichloro-6-fluoro-2-pyridinyl)oxy]acetate; Acetic acid, [(4-amino-3,5-dichloro-6-fluoro-2-pyridinyl)oxy]-,1-methylheptyl ester; [(4-Amino-3,5-dichloro-6-fluoro-2-pyridinyl)oxy]acetic acid, 1-methylheptyl ester
*Trade Names:* DOWCO® 433 MHE, Dow AgroSciences (USA); PARADIGM®, Dow AgroSciences (USA); PASTUREGARD®, Dow AgroSciences (USA); STARANE®, Dow AgroSciences (USA); TOMAHAWK®, Makhteshim-Agan Industries (Israel); VISTA®, AgroSciences (USA); WIDEMATCH®, (fluroxypyr + clopyralid), Dow AgroSciences (USA); XRM-5084®
*Producers:* Dow AgroSciences (USA); Makhteshim-Agan Industries (Israel)
*EPA/OPP PC Code:* 128968
*U.S. Maximum Allowable Residue Levels for Fluroxypyr 1-methylheptyl Ester (40 CFR 180.535):*

| CROP | ppm |
|---|---|
| Barley, grain | 0.5 |
| Barley, forage | 12.0 |
| Barley, hay | 20.0 |
| Barley, straw | 12.0 |
| Cattle, fat | 0.1 |
| Cattle, kidney | 1.5 |
| Cattle, meat | 0.1 |
| Cattle, mbyp | 0.1 |
| Corn, field, forage | 1.0 |
| Corn, field, grain | 0.02 |
| Corn, field, stover | 0.5 |
| Corn, sweet, forage | 1.0 |
| Corn, sweet, kernel plus cob with husks removed | 0.02 |
| Corn, sweet, stover | 2.0 |

| | |
|---|---|
| Goat, fat | 0.1 |
| Goat, kidney | 1.5 |
| Goat, meat | 0.1 |
| Goat, mbyp | 0.1 |
| Grain, aspirated fractions | 0.6 |
| Grass, forage | 120 |
| Grass, hay | 160 |
| Hog, fat | 0.1 |
| Hog, kidney | 1.5 |
| Hog, meat | 0.1 |
| Hog, mbyp | 0.1 |
| Horse, fat | 0.1 |
| Horse, kidney | 1.5 |
| Horse, meat | 0.1 |
| Horse, mbyp | 0.1 |
| Milk | 0.3 |
| Oat, forage | 12.0 |
| Oat, grain | 0.5 |
| Oat, hay | 20.0 |
| Oat, straw | 12.0 |
| Sheep, fat | 0.1 |
| Sheep, kidney | 1.5 |
| heep, meat | 0.1 |
| Sheep, mbyp | 0.1 |
| Sorghum, grain, forage | 2.0 |
| Sorghum, grain, grain | 0.02 |
| Sorghum, grain, stover | 4.0 |
| Wheat, forage | 12.0 |
| Wheat, grain | 0.5 |
| Wheat, hay | 20.0 |
| Wheat, straw | 12.0 |

*Carcinogen/Hazard Classifications*
**Label Signal Word:** WARNING, CAUTION or DANGER
*Description:* Crystalline solid.
*Incompatibilities:* May not be compatible with nitrates. Moisture may cause hydrolysis or other forms of decomposition.
*Permissible Concentration in Water:* No criteria set. Runoff from spills or fire control may cause water pollution.
*Harmful Effects and Symptoms*
*Short Term Exposure:* Contact with eyes or skin may cause irritation or injury. Inhalation should be avoided; use NIOSH-approved air purifying respirators for pesticides. May be harmful if swallowed.
*Points of Attack:* Adrenal and testicular glands
*First Aid:* If this chemical gets into the eyes, remove any contact lenses at once and irrigate immediately for at least 15 minutes, occasionally lifting upper and lower lids. Seek medical attention immediately. If this chemical contacts the skin, remove contaminated clothing and wash immediately with soap and water. Seek medical attention immediately. If this chemical has been inhaled, remove from exposure, begin rescue breathing (using universal precautions) if breathing has stopped, and CPR if heart action has stopped. Transfer promptly to a medical facility. When this chemical has been swallowed, get medical attention. Give large quantities of water and induce vomiting. Do not make an unconscious person vomit.
*References:*
- U.S. Environmental Protection Agency, Office of Pesticide Programs, Pesticide Residue Limits, "Fluroxypyr 1-methylheptyl Ester", 40 CFR 180.535, www.epa.gov/pesticides/food/viewtols.htm

# Flurprimidol

*Use Type:* Plant growth regulator
*CAS Number:* 56425-91-3
*Formula:* $C_{15}H_{15}F_3N_2O_2$
*Synonyms:* α-Isopropyl-α-[*p*-(trifluoromethoxy)phenyl]-5-pyrimidinemethanol; α-(1-Methylethyl)-α-[4-(trifluoromethoxy)phenyl]-5-pyrimidinemethanol; 5-Pyrimidinemethanol, α-(1-methylethyl)-α-[4-(trifluoromethoxy)phenyl]-
*Trade Names:* EL-500®; COMPOUND-72500®; CUTLESS®, SEPRO (US)
*Producers:* SePRO (US)
*Chemical Class:* Pyrimidine
*EPA/OPP PC Code:* 125701
*California DPR Chemical Code:* 2320
*Uses:* Used on turf and ornamental trees which reduces internode and leaf elongation in cool and warm seasons.
*Carcinogen/Hazard Classifications*
**Label Signal Word:** CAUTION
**WHO Acute Hazard:** Class III, slightly hazardous
*Regulatory Authority:*
- Actively registered pesticide in California.

*Description:* White crystalline solid. Molecular weight = 312.29.
*Permissible Concentration in Water:* No criteria set. Runoff from spills or fire control may cause water pollution.
*Harmful Effects and Symptoms*
*Short Term Exposure:* Contact with eyes or skin may cause irritation or injury. Inhalation should be avoided; use NIOSH-approved air purifying respirators for pesticides. May be harmful if swallowed.
*Points of Attack:* May affect the adrenal and ovary glands.
*First Aid:* If this chemical gets into the eyes, remove any contact lenses at once and irrigate immediately for at least 15 minutes, occasionally lifting upper and lower lids. Seek medical attention immediately. If this chemical contacts the skin, remove contaminated clothing and wash immediately with soap and water. Seek medical attention immediately. If this chemical has been inhaled, remove from exposure, begin rescue breathing (using universal precautions) if breathing has stopped, and CPR if heart action has stopped. Transfer promptly to a medical facility. When this chemical has been swallowed, get medical attention. Give large quantities of water and induce vomiting. Do not make an

unconscious person vomit. Do not induce vomiting when formulations containing petroleum solvents are ingested.

*References:*
- Pesticide Management Education Program, "Flurprimidol (Cutless) EPA Pesticide Fact Sheet 2/89," Cornell University, Ithaca, NY (February 1989). http://pmep.cce.cornell.edu/profiles/herb-growthreg/fatty-alcohol-monuron/flurprimidol/herb-prof-flurprimidol.html
- California Environmental Protection Agency "Chemical List of Lists," Sacramento CA (February 1997)

# Flutolanil

*Use Type:* Fungicide
*CAS Number:* 66332-96-5
*Formula:* $C_{17}H_{16}F_3NO_2$
*Synonyms:* Benzamide, N-[3-(1-methylethoxy)phenyl]-2-(trifluoromethyl)-; 3'-Isopropoxy-2-trifluoromethylbenzanilide; α,α,α-Trifluoro-3'-isopropoxy-o-toluanalide; N-[3-(1-Methylethoxy)phenyl]-2-(trifluoromethyl)benzamide
*Trade Names:* ARTESIAN®, Nihon Nohyaku (Japan); FOLISTAR®, Bayer CropScience (Germany), canceled; MONCUT®, Gowan Company (USA) and Nihon Nohyaku (Japan); NNF-136®; PROSTAR-50 WP®, Bayer CropScience (Germany)
*Producers:* Bayer CropScience (Germany); Gowan Company (USA); Nihon Nohyaku (Japan)
*Chemical Class:* Anilide
*EPA/OPP PC Code:* 128975
*California DPR Chemical Code:* 2305
*ICSC Number:* 1265
*RTECS Number:* CV5581320
*Uses:* Used on a number of crops. For potatoes, it is used for the suppression of seed borne Fusarium dry rot, Black scurf, Silver scurf, and Rhizoctonia stem canker. Flutolanil is also used on golf courses and other turf applications.
*Human toxicity (long-term)[77]:* Very low–4200.00 ppb, Health Advisory
*Fish toxicity (threshold)[77]:* Low–336.51157 ppb, MATC (Maximum Acceptable Toxicant Concentration)
*U.S. Maximum Allowable Residue Levels for Flutolanil (40 CFR 180.484):*

| CROP | ppm |
|---|---|
| Cattle, fat | 0.1 |
| Cattle, kidney | 1 |
| Cattle, liver | 2 |
| Cattle, meat | 0.05 |
| Cattle, mbyp | 0.05 |
| Egg | 0.05 |
| Goat, fat | 0.1 |
| Goat, kidney | 1 |
| Goat, liver | 2 |
| Goat, meat | 0.05 |
| Goat, mbyp | 0.05 |
| Hog, fat | 0.1 |
| Hog, kidney | 1 |
| Hog, liver | 2 |
| Hog, meat | 0.05 |
| Hog, mbyp | 0.05 |
| Horse, fat | 0.1 |
| Horse, kidney | 1 |
| Horse, liver | 2 |
| Horse, meat | 0.05 |
| Horse, mbyp | 0.05 |
| Milk | 0.05 |
| Peanut | 0.5 |
| Peanut, hay | 15 |
| Peanut, meal | 1 |
| Potato | 0.2 |
| Potato, wet peel | 0.3 |
| Poultry, fat | 0.05 |
| Poultry, meat | 0.05 |
| Poultry, mbyp | 0.05 |
| Rice, bran | 10 |
| Rice, grain | 7 |
| Rice, hulls | 25 |
| Rice, straw | 10 |
| Sheep, fat | 0.1 |
| Sheep, kidney | 1 |
| Sheep, liver | 2 |
| Sheep, meat | 0.05 |
| Sheep, mbyp | 0.05 |

*Carcinogen/Hazard Classifications*
**U.S. EPA Carcinogens:** Group E, Unlikely a carcinogen
**Label Signal Word:** CAUTION
**WHO Acute Hazard:** Class U, unlikely to be hazardous
*Description:* Colorless crystalline solid. Odorless. Practically insoluble in water; solubility = < 0.001 g/100 ml @ 20°C. Molecular weight = 323.29. Melting/Freezing point = 102–103°C. Vapor pressure = $18 \times 10^{-3}$ Pa @ 20°C. Log $K_{ow}$ = 3.7. Values above 3.0 are likely to bioaccumulate in marine organisms.
*Permissible Concentration in Water:* No criteria set. Runoff from spills or fire control may cause water pollution.
*Harmful Effects and Symptoms*
*Short Term Exposure:* Contact with eyes or skin may cause irritation or injury. Inhalation should be avoided; use NIOSH-approved air purifying respirators for pesticides. May be harmful if swallowed.
*First Aid:* If this chemical gets into the eyes, remove any contact lenses at once and irrigate immediately for at least 15 minutes, occasionally lifting upper and lower lids. Seek medical attention immediately. If this chemical contacts the skin, remove contaminated clothing and wash immediately with soap and water. Seek medical attention immediately. If this chemical has been inhaled, remove from exposure, begin rescue breathing (using universal precautions) if breathing has stopped, and CPR if heart action has stopped.

Transfer promptly to a medical facility. When this chemical has been swallowed, get medical attention. Give large quantities of water and induce vomiting. Do not make an unconscious person vomit. Do not induce vomiting when formulations containing petroleum solvents are ingested.

*References:*
- U.S. Environmental Protection Agency, Office of Pesticide Programs, Pesticide Residue Limits, "Flutolanil," 40 CFR 180.484, http://www.epa.gov/pesticides/food/viewtols.htm

# Fluvalinate (ANSI)

*Use Type:* Insecticide
*CAS Number:* 69409-94-5; 102851-06-9
*Formula:* $C_{26}H_{22}ClF_3N_2O_3$
*Alert:* Some applications may be classified as a Restricted Use Pesticide (RUP)
*Synonyms:* AI3-29426; Caswell No. 934; N-2-Chloro-α,α,α-(trifluoro-p-tolyl)-dl-valinealpha-cyano-phenoxybenzyl ester; N-(2-Chloro-4-(trifluoromethyl)phenyl-dl-valinecyano(3-phenoxylphenyl)methyl ester; N-[2-chloro-4-(trifluoromethyl)phenyl]-dl-valine(±)-cyano(3-phenoxylphenyl)methyl ester; (RS)-α-(Cyano-3-phenoxybenzyl n-(2-chloro-α,α,α-trifluoro-p-tolyl)-d-valinate; (RS)-α-Cyano-3-phenoxybenzyl (R)-2-[2-chloro-4-(trifluoromethyl)anilino]-3-methylbutanoate; Cyano(3-phenoxyphenyl)methyl N-[((2-chloro-4-trifluoromethyl)phenyl)]-d-valinate; dl-Valine,n-[2-chloro-4-(trifluoromethyl)phenyl]-cyano(3-phenoxylphenyl)methyl ester; D-Valine, N-(2-chloro-4-(trifluoromethyl)phenyl)-, cyano(3-phenoxyphenyl)methyl ester
*Trade Names:* APISTAN®, Wellmark International (USA); KARTAN®; KLARTAN®; MAVRIK®, Wellmark International (USA); MAVRIK AQUAFLOW®, Wellmark International (USA); SPUR®, Sandoz Agro (USA), canceled; TAUFLUALINATE®, Wellmark International (USA); YARDER®; ZEOCON®, Wellmark International (USA); ZR 3210®
*Producers:* Wellmark International (USA)
*Chemical Class:* Organofluorine
*EPA/OPP PC Code:* 109302
*California DPR Chemical Code:* 2195
*RTECS Number:* YV9397100
*Uses:* Used as a broad spectrum insecticide to control moths, beetles and other pests on cereals, potatoes, fruit trees, vegetables, fleas, cotton, turf and ornamentals. It is used in Apistan® to control varroa mites in honey bees.
*Human toxicity (long-term)[77]:* Low–70.00 ppb, Health Advisory
*Fish toxicity (threshold)[77]:* Extra high–0.09863 ppb, MATC (Maximum Acceptable Toxicant Concentration)

*U.S. Maximum Allowable Residue Levels for Fluvalinate (40 CFR 180.427):*
**CROP   ppm**
Honey   0.05

*Carcinogen/Hazard Classifications*
**California Prop. 65:** Developmental toxin
**TRI Developmental Toxin:** Reproductive and developmental toxin
**Label Signal Word:** WARNING, CAUTION or DANGER
**WHO Acute Hazard:** Class U, unlikely to be hazardous
**Endocrine Disruptor:** Possible ED
*Regulatory Authority:*
- EPCRA Section 313 Form R *de minimis* concentration reporting level: 1.0%
- Actively registered pesticide in California.
- Proposition 65 chemical (CAL)
- AB 1803-Well Monitoring Chemical (CAL) as pyrethrins
- Permissible Exposure Limits for Chemical Contaminants (CAL/OSHA) as pyrethrum
- Clean Water Act: Section 311 Hazardous Substances/RQ (same as CERCLA) as pyrethrins
- EPCRA Section 304 RQ: CERCLA, 1 lb (0.454 kg) as pyrethrins

*Description:* Yellow to yellow-amber, viscous liquid. Practically insoluble in water; solubility = 0.005 ppm @ 20-25°C. Molecular weight = 502.95. Density = 1.29 @ 25°C. Boiling point = > 450°C. Vapor pressure = $1 \times 10^{-7}$ mmHg @ 25°C. Log $K_{ow}$ = >3.3. Values above 3.0 are likely to bioaccumulate in marine organisms.
*Incompatibilities:* May be corrosive to some metals.
*Determination in Air:*; Collection by impinger or fritted bubbler, analysis by gas liquid chromatography/ultraviolet. See NIOSH IV, Method #5008[18]. (pyrethrum)
*Permissible Concentration in Water:* No criteria set. Runoff from spills or fire control may cause water pollution.
*Determination in Water:* Collection by impinger or fritted bubbler, analysis by gas liquid chromatography/ultraviolet. See NIOSH IV, Method #5008[18].
*Routes of Entry:* Inhalation, absorbed through the skin.
*Harmful Effects and Symptoms*
*Short Term Exposure:* Pyrethrins can affect you when breathed in and by passing through your skin. Irritates the eyes and respiratory tract. High exposure can affect the nervous system causing headache, nausea, vomiting, fatigue, and restlessness, rhinorrhea (discharge of thin nasal mucous).
*Long Term Exposure:* High or repeated exposure can cause lung allergy (with cough, wheezing and/or shortness of breath) or hay fever symptoms (sneezing, runny or stuffy nose). Allergic "pneumonia" can also occur with cough, chest pain, breathing difficulty and abnormal chest x-ray. Repeated attacks may lead to permanent scarring. Skin allergy may also develop with rash and itching, even with lower exposures. Skin contact can cause rash with redness,

blisters and intense itching. A severe generalized allergy can occur with weakness and collapse.

*Points of Attack:* Respiratory system, skin, central nervous system.

*Medical Surveillance:* Before beginning employment and at regular times after that, the following are recommended: Lung function tests. These may be normal if the person is not having an attack at the time of the test. Consider chest x-ray if lung symptoms are present. Evaluation by a qualified allergist, including careful exposure history and special testing, may help diagnose skin allergy.

*First Aid:* If this chemical gets into the eyes, remove any contact lenses at once and irrigate immediately for at least 15 minutes, occasionally lifting upper and lower lids. Seek medical attention immediately. If this chemical contacts the skin, remove contaminated clothing and wash immediately with soap and water. Seek medical attention immediately. If this chemical has been inhaled, remove from exposure, begin rescue breathing (using universal precautions) if breathing has stopped, and CPR if heart action has stopped. Transfer promptly to a medical facility. When this chemical has been swallowed, get medical attention. *Do not induce vomiting when formulations containing petroleum solvents are ingested.* Otherwise, give large quantities of water and induce vomiting. Do not make an unconscious person vomit.

*References:*
- EXTOXNET, Extension Toxicology Network, "Pesticide Information Profile, Fluvalinate," Oregon State University, Corvallis, OR (June 1996). http://extoxnet.orst.edu/pips/fluvalin.htm
- U.S. Environmental Protection Agency, Office of Pesticide Programs, Pesticide Residue Limits, "Fluvalinate," 40 CFR 180.427, http://www.epa.gov/pesticides/food/viewtols.htm
- California Environmental Protection Agency "Chemical List of Lists," Sacramento CA (February 1997)

# Fomesafen

*Use Type:* Herbicide
*AS Number:* 72178-02-0; 108731-70-0 (sodium salt)
*Formula:* $C_{15}H_{10}ClF_3N_2O_6S$
*Synonyms:* Benzamide, 5-[2-chloro-4-(trifluoromethyl)phenoxy]-N-(methylsulfonyl)-2-nitro-; 5-[2-Chloro-4-(trifluoromethyl)phenoxy]-N-methylsulfonyl)-2-nitrobenzamide; 5-(2-Chloro-α,α α-trifluoro-p-tolyloxy)-N-methylsulfonyl-2-nitrobenzamide; 5-[2-Chloro-4-(trifluoromethyl)phenoxy]-N-(methylsulphonyl)-2-nitrobenzamide

*Trade Names:* BAS 530 04®, BASF Agricultural Products (Germany); FASTER®, BASF Agricultural Products (Germany); FLEX®; FLEXSTAR, Syngenta (Switzerland); FOMESAFEN® SODIUM, Syngenta (Switzerland); PP 021®; REFLEX®, Syngenta (Switzerland); REFLEX® 2LC Herbicide (sodium salt), Syngenta (Switzerland); TORNADO®, Syngenta (Switzerland), canceled; TWISTE®, Syngenta (Switzerland), canceled; TYPHOON®, Syngenta (Switzerland)

*Producers:* BASF Agricultural Products (Germany); Syngenta (Switzerland); Wuzhou International (China)

*Chemical Class:* Chlorophenoxy; Diphenyl ether
*EPA/OPP PC Code:* 123802 (sodium salt)
*California DPR Chemical Code:* 5083 (sodium salt)
*Uses:* After July 25, 2003, flumesafen was not permitted in many countries to be used as an active ingredient on crops except to control weeds in soybean crops, and white, kidney and snap beans.

*Human toxicity (long-term)[77]:* (sodium salt) High–1.75 ppb, Health Advisory

*Fish toxicity (threshold)[77]:* (sodium salt) Very low–69325.01943 ppb, MATC (Maximum Acceptable Toxicant Concentration)

*U.S. Maximum Allowable Residue Levels for Fomesafen sodium salt (40 CFR 180.433):*

| CROP | ppm |
|---|---|
| Soybean | 0.05 |

*Carcinogen/Hazard Classifications*
*U.S. EPA Carcinogens:* Group C, possible carcinogen
*Label Signal Word:* WARNING or DANGER
*WHO Acute Hazard:* Class III, slightly hazardous
*Regulatory Authority:*
- AB 2588-Air Toxics "Hot Spots" Chemicals (CAL) as chlorophenoxy pesticides
- The "Director's List" (CAL/OSHA) as chlorophenoxy pesticides
- Safe Drinking Water Act: Priority List (55 FR 1470)
- EPCRA Section 313 Form R *de minimis* concentration reporting level: 1.0%

*Incompatibilities:* Slowly hydrolyzes in water, releasing ammonia and forming acetate salts.

*Permissible Concentration in Water:* No criteria set. Runoff from spills or fire control may cause water pollution.

*Routes of Entry:* Inhalation, ingestion, absorbed through the skin.

*Harmful Effects and Symptoms*

*Short Term Exposure:* Poisonous; may be fatal if inhaled, swallowed, or absorbed through skin. Severely irritates eyes, skin and respiratory tract, with burning sensation, pain, redness and swelling. Metabolic stimulant. If inhaled, causes coughing, dilated pupils, headache, profuse persperation, intense thirst, extreme fatigue, rapid pulse, high fever, clammy, flushed skin, rapid breathing, nausea, vomiting, cyanosis (bluish tint to skin and lips), anxiety and confusion, convulsions, risk of lung edema. If swallowed, face and lips turn bluish. Liver injury with associated jaundice, kidney failure, and cardiac arrhythmias are commonly noted. Nerve damage, which may be delayed, may include swelling of legs and feet, muscle twitch and stupor. Severe exposure can cause death from heart failure. Dust or liquid left in contact with the skin for several hours may be absorbed. This may result in severe delayed symptoms as listed above. These symptoms may last for months or years.

*Long Term Exposure:* Workers exposed to chlorophenoxy compounds such as 2,4-D (in the manufacturing process) over a five to ten year period at levels above 10 mg/m$^3$ complained of weakness, rapid fatigue, headache and vertigo. Liver damage, low blood pressure and slowed heartbeat were also found. Based on animal tests, may affects human reproduction

*Points of Attack:* Eyes, skin, respiratory system, central nervous system, cardiovascular system, liver, kidney

*Medical Surveillance:* If symptoms develop or overexposure is suspected, the following may be useful: Liver and kidney function tests. Exam of the nervous system.

*First Aid:* If this chemical gets into the eyes, remove any contact lenses at once and irrigate immediately for at least 15 minutes, occasionally lifting upper and lower lids. Seek medical attention immediately. If this chemical contacts the skin, remove contaminated clothing and wash immediately with soap and water. Seek medical attention immediately. If this chemical has been inhaled, remove from exposure, begin rescue breathing (using universal precautions) if breathing has stopped, and CPR if heart action has stopped. Transfer promptly to a medical facility. When this chemical has been swallowed, get medical attention. *Do not induce vomiting when formulations containing petroleum solvents are ingested.* Otherwise, give large quantities of water and induce vomiting. Do not make an unconscious person vomit.

*References:*
- U.S. Environmental Protection Agency, Office of Pesticide Programs, Pesticide Residue Limits, "Fomesafen sodium salt", 40 CFR 180.433, www.epa.gov/pesticides/food/viewtols.htm
- California Environmental Protection Agency "Chemical List of Lists," Sacramento CA (February 1997)

# Fonofos

*Use Type:* A soil insecticide
*CAS Number:* 944-22-9
*Formula:* $C_{10}H_{15}OPS_2$; $C_6H_5SPS(OC_2H_5)C_2H_5$
*Alert:* A Restricted Use Pesticide (RUP). The use of fonophos was discontinued in Canada in 2000.
*Synonyms:* O-Aethyl-S-phenyl-aethyl-dithiophosphonat (German); ENT 25,796; O-Ethyl-S-phenyl ethylphosphono dithioate; O-Ethyl-S-phenyl(RS)-ethylphosphonodithioate; O-Ethyl-S-phenyl ethyldithio phosphonate; Phosphonodithioic acid, ethyl-O-ethyl, S-phenyl ester
*Trade Names:* CAPFOS®; CUDGEL®; DIFONATE®; DYFONATE®, Zeneca Ag Products (USA) (now Syngenta), canceled 4/26/1988; DYPHONATE®; DOUBLE DOWN®; STAUFFER-2790

*Producers:* Sigma-Aldrich Laborchemikalien (Germany); Zeneca Ag Products (USA) (now Syngenta); Syngenta (Switzerland)
*Chemical Class:* Organophosphate
*EPA/OPP PC Code:* 041701
*California DPR Chemical Code:* 254
*ICSC Number:* 0708
*RTECS Number:* TA5950000
*EEC Number:* 015-091-00-2
*EINECS Number:* 213-408-0
*Uses:* Fonofos is a soil organophosphate insecticide primarily used on corn. It is also used on maize, cereals, sorghum, fruit, olives, potatoes, sugar cane, peanuts, tobacco, turf, and some vegetable crops. It controls aphids, corn borer, corn rootworm, corn wireworm, cutworms, white grubs, and some maggots. It is available in granular, microgranular, emusifiable concentrate, suspension concentrate, microcapsule suspension, and seed treatment.
*Human toxicity (long-term)*[77]: Intermediate–10.00 ppb, Health Advisory
*Fish toxicity (threshold)*[77]: High–3.50002 ppb, MATC (Maximum Acceptable Toxicant Concentration)
*Carcinogen/Hazard Classifications*
**U.S. EPA Carcinogens:** Group E, unlikely carcinogen
**Label Signal Word:** DANGER–POISON, EPA Toxicity Class I
**WHO Acute Hazard:** Class 1 a, extremely hazardous
*Regulatory Authority:*
- Banned or Severely Restricted (In Agriculture) (Malaysia) (UN)[13]
- Air Pollutant Standard Set (ACGIH)[1] (Several States)[60]
- U.S. DOT Inhalation Hazard Chemicals as organophosphates
- Permissible Exposure Limits for Chemical Contaminants (CAL/OSHA)
- The "Director's List" (CAL/OSHA)
- Actively registered pesticide in California.
- Superfund/EPCRA 40CFR355, Appendix B Extremely Hazardous Substances: TPQ = 500 lb (227kg)
- Superfund/EPCRA 40CFR302.4 RQ: EHS, 1 lb (0.454 kg)
- U.S. DOT Regulated Marine Pollutant (49CFR172.101, Appendix B), severe pollutant

*Description:* Fonofos is a pale yellow liquid. Pungent, mercaptan-like odor. Practically insoluble in water; solubility = 12 ppm @ 20°C. Molecular weight = 246.36. Boiling point = 130°C @ 0.1 mm. Melting/Freezing point = 30°C. Flash point = 94°C (cc). Vapor pressure = $2.68 \times 10^{-4}$ mmHg @ 25°C. Hazard Identification (based on NFPA-704 M Rating System): Health 4, Flammability 1, Reactivity 0. Log $K_{ow}$ = 3.92. Values at or above 3.0 are likely to bioaccumulate in marine organisms.

*Incompatibilities:* Contact with strong acids or alkalies causes chemical to be hydrolyzed.

*Permissible Exposure Limits in Air:* There is no OSHA[2] PEL. NIOSH[2] as well as the ACGIH[1] recommends a TWA value of 0.1 mg/m$^3$ with the notation "skin" indicating the possibility of skin absorption. Several states have developed guidelines or standards for fonofos in ambient air[60] ranging from 1.0 $\mu$g/m$^3$ (North Dakota) to 2.0 $\mu$g/m$^3$ (Connecticut, Nevada and Virginia).

*Determination in Air:* OSHA versatile sampler-2; Toluene/Acetone; Gas chromatography/Flame photometric detection for sulfur, nitrogen, or phosphorus; NIOSH IV Method #5600, Organophosphorus pesticides[18].

*Permissible Concentration in Water:* The U.S. EPA has developed health advisories for fonofos as follows; long term health advisory is 70 $\mu$g/L and lifetime health advisory is 14 $\mu$g/L (See Reference Below).

*Routes of Entry:* Inhalation, skin absorption, ingestion, skin and/or eye contact.

*Harmful Effects and Symptoms*

*Short Term Exposure:* Symptoms include nausea, vomiting, abdominal cramps, diarrhea, excessive salivation, headache, giddiness, vertigo, sensation of tightness in chest, blurring of vision, ocular pain, loss of muscle coordination, slurring in speech, muscle twitching, drowsiness, excessive secretion of respiratory tract mucous, and convulsions. This material is cholinesterase inhibitor. It can cause severe symptoms and death from respiratory arrest. The LD-50 (oral, rat) = 3 mg/kg (extremely toxic). Exposure above the airborne exposure limit may result in death. The effects may be delayed. Delayed pulmonary edema may occur after inhalation. Medical observation is recommended.

*Long Term Exposure:* Cholinesterase inhibitor; cumulative effect is possible. Fonofos may damage the nervous system with repeated exposure, resulting in convulsions, respiratory failure. May cause liver damage.

*Points of Attack:* Respiratory system, central nervous system, cardiovascular system, blood cholinesterase

*Medical Surveillance:* Medical observation is recommended for 24 to 48 hours after breathing overexposure, as pulmonary edema may be delayed. Before employment and at regular times after that, the following are recommended: Plasma and red blood cell cholinesterase levels (tests for the enzyme poisoned by this chemical). If exposure stops, plasma levels return to normal in 1-2 weeks while red blood cell levels may be reduced for 1-3 months. When acetylcholinesterase enzyme levels are reduced by 25% or more below preemployment levels, risk of poisoning is increased, even if results are in lower ranges of "normal." Reassignment to work not involving organophosphate or carbamate pesticides is recommended until enzyme levels recover. If symptoms develop or overexposure occurs, repeat the above tests as soon as possible and get an exam of the nervous system. Also consider complete blood count. Consider chest x-ray following acute overexposure. Do not drink any alcoholic beverages before or during use. Alcohol promotes absorption of organophosphates.

*First Aid:* **Treatment for organophosphate poisoning consists of thorough decontamination, cardiorespiratory support, and administration of the antidotes atropine and pralidoxime. In cases of severe poisoning, diazepam, an anticonvulsant, should also be administered. Antidotes should be administered as prevention even if the diagnosis is in doubt.** Speed in removing material from eyes and skin is of extreme importance. *Eyes:* Eye contact can cause dangerous amounts of these chemicals to be quickly absorbed through the mucous membrane into the bloodstream. Immediately and gently flush eyes with plenty of warm or cold water (NO hot water) for at least 15 minutes, occasionally lifting the upper and lower eyelids. Get medical aid immediately. *Skin:* Get medical aid. Skin contact can cause dangerous amounts of these chemicals to be absorbed into the bloodstream. Wearing the appropriate PPE equipment and respirator for organophosphate pesticides, immediately flush skin with plenty of soap and water for at least 15 minutes while removing contaminated clothing and shoes. Shampoo hair promptly if contaminated. The removed, contaminated clothing and shoes should be double-bagged and left in Hot Zone for later disposal by hazardous materials experts. Skin may also be decontaminated with diluted hypochlorite solution. *Inhalation:* Get medical aid. Do not contaminate yourself. Wearing the appropriate PPE equipment and respirator for organophosphate pesticides, immediately remove the victim from the contaminated area to fresh air. If the victim is not breathing, administer artificial respiration. Do NOT use mouth-to-mouth resuscitation; use bag/mask apparatus. If breathing is difficult, administer oxygen through bag/mask apparatus until medical help arrives. Do not leave victim unattended. *Ingestion:* Call poison control. Loosen all clothing. Never give anything by mouth to an unconscious person. If victim is *unconscious or having convulsions,* do nothing except keep victim warm. Get medical aid. Transfer promptly to a medical facility. In cases of ingestion, **do not induce vomiting**. If the victim is alert and asymptomatic, administer a slurry of activated charcoal at a dose of 1 g/kg (infant, child, and adult dose). A soda can and straw may be of assistance when offering charcoal to a child. *In some cases you may be specifically instructed by poison control to induce vomiting by way of 2 tablespoons of syrup of ipecac (adult) washed down with a cup of water.* Do NOT give activated charcoal before or with ipecac syrup.

*Note to physician or authorized medical personnel:* Treat cases of respiratory compromise, coma, or excessive pulmonary secretions with respiratory support using protocols and techniques available and within the scope of training. Some cases may necessitate procedures such as endotracheal intubation or cricothyrotomy by properly trained and equipped personnel. When possible, atropine (see *Antidotes*, below) should be given under medical supervision. Patients who are comatose, hypotensive, or having seizures or cardiac arrhythmias should be treated according to advanced life support protocols. *Antidotes:* Two antidotes are administered to treat organophosphate poisoning. Atropine is a competitive antagonist of acetylcholine at muscarinic receptors and is used to control the excessive bronchial secretions which are often responsible for death. Pralidoxime relieves both the nicotinic and muscarine effects of organophosphate poisoning by regenerating acetylcholinesterase and can reduce both the bronchial secretions and the muscle weakness associated with poisoning. The initial intravenous dose of atropine in adults should be determined by the severity of symptoms: An initial adult dose of 1.0 to 2.0 mg or pediatric dose of 0.01 mg/kg (minimum 0.01 mg) should be administered intravenously. If intravenous access cannot be established, atropine may also be given intramuscularly, subcutaneously or via endotracheal tube. Doses should be repeated every 15 minutes until excessive secretions and sweating have been controlled. Once bronchial secretion has been controlled, atropine administration should be repeated whenever the secretions begin to recur. In seriously poisoned patients, very large doses may be required. Alterations of pulse rate and pupillary size should not be used as indicators of treatment adequacy. Pralidoxime should be administered as early in poisoning as possible as its efficacy may diminish when given more than 24 to 36 hours after exposure. Doses are as follows: adult 1.0 g; pediatric 25 to 50 mg/kg. The drug should be administered intravenously over 30 to 60 minutes, but in a life-threatening situation, one-half of the total dose can be given per minute for a total administration time of 2 minutes. Treatment should begin to take effect within 40 minutes with a reduction in symptoms and the amount of atropine necessary to control bronchial secretion. The initial dose can be repeated in 1 hour and then every 8 to 12 hours until the patient is clinically well and no longer requires atropine. If intravenous access cannot be established, pralidoxime may also be given intramuscularly. Early administration of diazepam in addition to the combined atropine and pralidoxime treatment may help prevent the onset of seizures and potential brain and cardiac morphologic damage following high-level organophosphate poisoning.

*References:*
- EXTOXNET, Extension Toxicology Network, "Pesticide Information Profile, Fonofos," Oregon State University, Corvallis, OR (June 1996). http://extoxnet.orst.edu/pips/fonofos.htm
- New Jersey Department of Health and Senior Services, "Hazardous Substance Fact Sheet: Fonofos," Trenton, NJ (August 1985, rev. April 1999). http://www.state.nj.us/health/eoh/rtkweb/0945.pdf
- U.S. Environmental Protection Agency, "Chemical Profile: Fonofos," Washington, DC, Chemical Emergency Preparedness Program (November 30, 1987).
- Agency for Toxic Substances and Disease Registry, U.S. Department of Health and Human Services, Public Health Service, "Managing Hazardous Materials Incidents," Atlanta, GA (June 2003).
- California Environmental Protection Agency "Chemical List of Lists," Sacramento CA (February 1997).
- U.S. Environmental Protection Agency, "Health Advisory: Fonofos," Washington, DC, Office of Drinking Water (August 1987).

# Forchlorfenuron

*Use Type:* Plant growth regulator
*CAS Number:* 68157-60-8
*Formula:* $C_{12}H_{10}ClN_3O$
*Synonyms:* $N$-(2-Chloro-4-pyridinyl)-$N'$-phenylurea; 1-(2-Chloro-4-pyridyl)-3-phenylurea; CPPU; $N$-Phenyl-$N'$-(2-chloro-4-pyridyl)urea; Urea, $N$-(2-chloro-4-pyridinyl)-$N'$-phenyl-
*Trade Names:* KT-30®; CN-11-3183; SKW 20010
*Producers:* Agrochem (USA)
*Chemical Class:* Substituted urea
*EPA/OPP PC Code:* 128819
*California DPR Chemical Code:* 5557
*Uses:* Forchlorfenuron is as a plant growth regulator widely used in agriculture on fruits to increase their size, to promote cell division, and to improve the quality and the yield of fruits. In some parts of California, forchlorfenuron is said to double the size of Thompson Seedless berries and delay crop maturity up to a couple of weeks.

*U.S. Maximum Allowable Residue Levels for Forchlorfenuron (40 CFR 180.569):*

| CROP | ppm |
| --- | --- |
| Almond | 0.01 |
| Apple | 0.01 |
| Blueberry | 0.01 |
| Cranberry | 0.01 |
| Fig | 0.01 |
| Grape | 0.01 |
| Kiwi fruit | 0.01 |
| Olive | 0.01 |
| Pear | 0.01 |
| Plum | 0.01 |

*Description:* White crystalline powder. Melting point = 171–172°C.
*Short Term Exposure:* Contact with eyes or skin may cause irritation or injury. Inhalation should be avoided; use NIOSH-approved air purifying respirators for pesticides. May be harmful if swallowed. Skin contact may cause severe irritation or burns.
*Points of Attack:* Skin.
*First Aid:* If this chemical gets into the eyes, remove any contact lenses at once and irrigate immediately for at least 15 minutes, occasionally lifting upper and lower lids. Seek medical attention immediately. If this chemical contacts the skin, remove contaminated clothing and wash immediately

with soap and water. Seek medical attention immediately. If this chemical has been inhaled, remove from exposure, begin rescue breathing (using universal precautions) if breathing has stopped, and CPR if heart action has stopped. Transfer promptly to a medical facility. When this chemical has been swallowed, get medical attention. Give large quantities of water. Do not induce vomiting when formulations containing petroleum solvents are ingested. Do not make an unconscious person vomit.

*References:*
- U.S. Environmental Protection Agency, Office of Pesticide Programs, Pesticide Residue Limits, "Forchlorfenuron ", 40 CFR 180.569, www.epa.gov/pesticides/food/viewtols.htm
- California Environmental Protection Agency "Chemical List of Lists," Sacramento CA (February 1997)

# Formaldehyde

*Use Type:* Microbiocide and fungicide
*CAS Number:* 50-00-0
*Formula:* $CH_2O$; HCHO
*Synonyms:* Aldehyde formique (French); Aldeide formica (Italian); BFV; FA; Formaldehido (Spanish); Formaldehyd (Czech); Formaldehyd (Polish); Formalin; Formalin 40; Formalina (Italian, Spanish); Formaline (German); Formalin-loesungen (German); Formic aldehyde; Methanal; Methyl aldehyde; Methylene glycol; Methylene oxide; NCI-C02799; Oplossingen (Dutch); Oxomethane; Oxymethylene; Polyoxymethylene glycols; Tetraoxymethylene; Trioxane
*Trade Names:* DYNOFORM®; FANNOFORM®; FORMALITH®; FORMOL®; FYDE®; HERCULES 37M6-8®, Hercules (USA); HOCH®; IVALON®; KARSAN®; LYSOFORM®; MAGNIFLOC 156C FLOCCULANT®; MORBICID®; STERIFORM®; SUPERLYSOFORM®
*Producers:* Albright & Wilson Pty. (UK); Ashland (USA); ATOFINA Chemicals (France); Atul (India); BASF (Germany); Bayer Chemicals (Germany); Borden Chemical (USA); Caldic Group (Netherlands); Celanese Acetyl Products (Germany); Degussa (Germany); DuPont (USA); Georgia-Pacific (USA); GFS Chemicals (USA); Hercules (USA); Juhua Group Corp. (China); KCIL Group (India); Neste (Finland); Orica (Australia); Sumitomo (Japan); Total Specialty Chemicals (USA); YiHua Group (China)
*Chemical Class:* Unclassified
*EPA/OPP PC Code:* 043001
*California DPR Chemical Code:* 295
*ICSC Number:* 0275
*RTECS Number:* LP8925000
*EINECS Number:* 200-001-8
*Uses:* Formaldehyde has found wide industrial usage as a fungicide, germicide, and in disinfectants and embalming fluids. It is used most often in an aqueous solution stabilized with methanol (formalin). It is also used in the manufacture of artificial silk and textiles, latex, phenol, urea, thiourea and melamine resins, dyes, and inks, cellulose esters and other organic molecules, mirrors, and explosives. It is also used in the paper, photographic, and furniture industries. It is an intermediate in drug manufacture and is also a pesticide intermediate. It has been identified as a major indoor air pollutant.
*U.S. Maximum Allowable Residue Levels for Formaldehyde (40 CFR 180.180.1001 (d) and (e), 180.1024):*
Formaldehyde is exempted from the requirement of a tolerance with good agricultural practice as inert (or occasionally active) ingredients in pesticide formulations applied to growing crops only. The insecticide Paraformaldehyde is exempted from the requirement of a tolerance for residues in or on sugar beets (roots and tops) when applied to the soil not later than planting.
*Carcinogen/Hazard Classifications*
**U.S. EPA Carcinogens:** Group B1, probable carcinogen
**U.S. NTP Carcinogen:** Reasonably anticipated carcinogen
**California Prop. 65:** Carcinogen
**IARC:** Group 2A, probable carcinogen
*Regulatory Authority:*
- Carcinogen (Animal Positive) (IARC)[9] (ACGIH)[1] (DFG)[3]
- Banned or Severely Restricted (Several Countries) (UN)[13]
- Toxic Substance (World Bank)[15]
- OSHA, 29CFR1910 Specifically Regulated Chemicals (See CFR 1910.1048)
- Air Pollutant Standard Set (ACGIH)[1] (OSHA)[58] (DFG)[3] (HSE)[33] (Other Countries)[35] (Several States)[60] (Several Canadian Provinces)
- CAL/OSHA Carcinogen User Register Chemical
- AB 1803-Well Monitoring Chemical (CAL)
- MCL (Maximum Contaminants Levels) list of contaminants (CAL)
- AB 2588-Air Toxics "Hot Spots" Chemicals (CAL)
- CAL-DHS/DHS Drinking Water Action Levels
- CAL Air Resources Board/AB 1807 Toxic Air Contaminants
- Proposition 65 chemical (CAL)
- Permissible Exposure Limits for Chemical Contaminants (CAL/OSHA)
- The "Director's List" (CAL/OSHA)
- Actively registered pesticide in California.
- Clean Air Act: Hazardous Air Pollutants (Title I, Part A, Section 112); Accidental Release Prevention/Flammable substances, (Section 112[r], Table 3), TQ = 15,000 lb (6810 kg)
- OSHA 29CFR1910.119, Appendix A. Process Safety List of Highly Hazardous Chemicals, TQ = 1000 lb (450 kg)
- Clean Water Act: Section 311 Hazardous Substances/RQ 40CFR117.3 (same as CERCLA, see below); Section 313 Water Priority Chemicals (57FR41331, 9/9/92)

- EPA Hazardous Waste Number (RCRA No.): U122
- RCRA, 40CFR261, Appendix 8 Hazardous Constituents
- Superfund/EPCRA 40CFR355, Appendix B Extremely Hazardous Substances: TPQ = 500 lb (227kg)
- Superfund/EPCRA 40CFR302.4 RQ: CERCLA, 100 lb (45.4 kg)
- EPCRA Section 313 Form R *de minimus* concentration reporting level: 0.1%
- Canada, WHMIS, Ingredients Disclosure List

***Description:*** Formaldehyde is a colorless, pungent gas. The odor threshold is 0.8 ppm[41]. Formalin (as flormaldehyde) is sold as an aqueous solution containing 30 to 50% formaldehyde and 6 to 15% methanol, which is added to prevent polymerization. *Warning properties*: Odor is detectable at less than 1 ppm, but many sensitive persons experience symptoms below the odor threshold. Water solubility = 55% @ 20°C. Molecular weight = 30.0 daltons. Boiling point (gas) = –21°C. The 37% commercial solution BP = 101°C. Vapor pressure = 3883 mmHg at @ 25°C. Gas density (air:1) = 1.07. Flash point = 50°C (commercial 37% solution, 15% methanol). Explosive limits (gas): LEL = 7.0%; UEL = 73.0%. NFPA 704 M Hazard Identification (gas): Health 3, Flammability 4, Reactivity 0. NFPA 704 M Hazard Identification (37% solution, 15% methanol): Health 3, Flammability20, Reactivity 0. Log $K_{ow}$ = < 1.0. Unlikely to bioaccumulate in marine organisms.

***Incompatibilities:*** Pure formaldehyde may polymerize unless properly inhibited (usually with methanol). Forms explosive mixture with air Incompatible with strong acids, amines, strong oxidizers, alkaline materials, nitrogen dioxide, performic acid, phenols,urea. Reaction with hydrochloric acid forms bis-chloromethyl ether, a carcinogen. Formalin is incompatible with strong oxidizers, alkalis, acids, phenols, urea, oxides, isocyanates, caustics, anhydrides.

***Permissible Exposure Limits in Air:*** The OSHA[2] TWA is 3 ppm determined as a TWA. The ceiling concentration is 5 ppm which shall not be exceeded at any time, and 10 ppm which shall not be exceeded in any 30 minute period. ACGIH[1] has recommended a TWA of 1.0 ppm (1.5 mg/m$^3$) and STEL of 2.0 ppm (3 mg/m$^3$) with the notation that formaldehyde is "an industrial substance suspect of carcinogenic potential for man." AIHA ERPG-2 (emergency response planning guideline) (the maximum airborne concentration below which it is believed that nearly all individuals could be exposed for up to 1 hour without experiencing or developing irreversible or other serious health effects or symptoms which could impair an individual's ability to take protective action) = 10 ppm. The NIOSH[2] IDLH level = Ca [20 ppm]. *For sensitized persons, odor is not an adequate indicator of formaldehyde's presence and may not provide reliable warning of hazardous concentrations.* There are a number of values set in other countries for TWA values (some using different conversion factors from ppm to mg/m$^3$).

Some off these are as follows[35]:

| Country | TWA ppm | TWA mg/m$^3$ |
|---|---|---|
| former USSR | – | 0.5 |
| Czechoslovakia | – | 0.5 |
| W. Germany | 0.5 | 0.6 |
| Sweden | 0.8 | 1.0 |
| Brazil | 1.6 | 2.3 |
| Japan | 2.0 | 2.5 |
| U.K.[33] | 2.0 | 3.0 |

The former USSR has also set MAC values for ambient air in residential areas of 0.035 mg/m$^3$ on a momentary basis and 0.012 mg/m$^3$ on a daily average basis[43], also cited as 0.003 mg/m$^{3[35]}$. A number of states have set guidelines or standards for formaldehyde in ambient air[60] ranging from zero (North Carolina and North Dakota) to 0.77 μg/m$^3$ (Massachusetts) to 5.0 μg/m$^3$ (New York) to 7.2 μg/m$^3$ (Pennsylvania) to 7.5 μg/m$^3$ (South Carolina) to 12.0 μg/m$^3$ (Connecticut, South Dakota and Virginia) to 18.0 μg/m$^3$ (Indiana) to 71.0 μg/m$^3$ (Nevada) to 75.0 μg/m$^3$ (Washington).

***Determination in Air:*** Collection with Si gel cartridge coated with DNPH; Workup with acetonitrile; analysis with high-pressure liquid chromatography/ultraviolet detection; NIOSH IV, Method #2016[18]. See also Method(s) #2541, #3500. See NIOSH Criteria Document 78-212 NITRILES.[18]

***Permissible Concentration in Water:*** EPA[32] has suggested a permissible ambient goal of 41.4 μg/L based on health effects. The former USSR has set an MAC in water bodies used for domestic purposes of 0.01 mg/L[43], also quoted as 0.05 mg/L[35]. Further, they have set an MAC in water bodies used for fishery purposes of 0.25 mg/L. Several states have set guidelines for formaldehyde in drinking water[61] ranging from 0.7 μg/L (New Jersey)[59] to 10.0 μg/L (Maryland) to 30.0 μg/L (California and Maine).

***Routes of Entry:*** Inhalation, ingestion, skin and/or eye contact.

***Harmful Effects and Symptoms***

***Short Term Exposure:*** Formaldehyde is a highly toxic systemic poison that is absorbed well by inhalation. The vapor is a severe respiratory tract and skin irritant and may cause dizziness or suffocation. Contact with formaldehyde solution may cause severe burns to the eyes and skin. Inhalation of vapors can produce narrowing of the bronchi and an accumulation of fluid in the lungs. Children may be more susceptible than adults to the respiratory effects of formaldehyde. Formaldehyde solution (formalin) causes corrosive injury to the gastrointestinal tract, especially the pharynx, epiglottis, esophagus, and stomach. The systemic effects of formaldehyde are due primarily to its metabolic conversion to formate, and may include metabolic acidosis, circulatory shock, respiratory insufficiency, and acute renal failure. Formaldehyde is a potent sensitizer and a probable human carcinogen. Acute exposure to formaldehyde may result in burns to the skin, eyes, and mucous membranes; lacrimation (tearing); nausea; vomiting (may be bloody); abdominal pain; and diarrhea. Difficulty in breathing, cough, pneumonia, and pulmonary edema may occur. Sensitized people may experience asthmatic reactions, even when exposed briefly. Hypotension (low blood pressure) and hypothermia (reduced body temperature) may precede cardiovascular collapse. Lethargy, dizziness, convulsions, and coma may be noted. Nephritis (inflammation of the kidneys), hematuria (bloody urine), and liver toxicity have been reported. Exposure at concentrations well above the PEL may cause death. The effects may be delayed

*Note:* There is considerable individual variation in sensitivity to formaldehyde. *Inhalation:* Irritation of the nose and throat can occur after an exposure of 0.25 ppm to 0.45 ppm. Levels between 0.4 ppm and 0.8 ppm can give rise to coughing and wheezing, tightness of the chest and shortness of breath. Sudden exposures to concentrations of 4 ppm may lead to irritation of lung and throat severe enough to give rise to bronchitis and laryngitis. Breathing may be impaired at levels above 100 ppm and serious lung damage may occur at 50 ppm. *Skin:* Direct contact with the liquid can lead to irritation, itching, burning and drying. It is also possible to develop an allergic reaction to the compound following exposure by any route. *Eyes:* Exposure to airborne levels of formaldehyde of 0.4 ppm have brought on tearing and irritation. Small amounts of liquid splashed in the eye can cause damage to the cornea. Eye irritation was reported at levels between 0.05-2.0 ppm. *Ingestion:* As little as 1 liquid ounce has resulted in death to humans. Smaller amounts can damage the throat, stomach and intestine resulting in nausea, vomiting, abdominal pain and diarrhea. Accidental exposure may also cause loss of consciousness, lowered blood pressure, kidney damage and, if the person is pregnant, the possibility of the fetus being aborted.

**Long Term Exposure:** The major concerns of repeated formaldehyde exposure are sensitization and cancer. In sensitized persons, formaldehyde can cause asthma and contact dermatitis. In persons who are not sensitized, prolonged inhalation of formaldehyde at low levels is unlikely to result in chronic pulmonary injury. Adverse effects on the central nervous system such as increased prevalence of headache, depression, mood changes, insomnia, irritability, attention deficit, and impairment of dexterity, memory, and equilibrium have been reported to result from long-term exposure. Inhalation can result in respiratory congestion with associated coughing and shortness of breath. Repeated skin contact can lead to drying and scaling. Some individuals may experience allergic reactions after initial contact with the chemical. Subsequent contact may cause skin rashes and asthma and reactions may become more severe if exposure persists. Chronic exposure may be more serious for children because of their potential longer latency period. *Carcinogenicity:* The Department of Health and Human Services has determined that formaldehyde may reasonably be anticipated to be a carcinogen. In humans, formaldehyde exposure has been weakly associated with increased risk of nasal cancer and nasal tumors were observed in rats chronically inhaling formaldehyde. *Reproductive and Developmental Effects:* There is limited evidence that formaldehyde causes adverse reproductive effects. The TERIS database states that the risk of developmental defects to the exposed fetus ranges from none to minimal. There have been reports of menstrual disorders in women occupationally exposed to formaldehyde, but they are controversial. Studies in experimental animals have reported some effects on spermatogenesis. Formaldehyde has not been proven to be teratogenic in animals and is probably not a human teratogen at occupationally permissible levels. Formaldehyde has been shown to have genotoxic properties in human and laboratory animal studies producing sister chromatid exchange and chromosomal aberrations. Special consideration regarding the exposure of pregnant women is warranted, since formaldehyde has been shown to be a genotoxin; thus, medical counseling is recommended for the acutely exposed pregnant woman. Genetic damage from exposure has been shown in bacteria and some insects.

**Points of Attack:** Eyes and respiratory system. Cancer Site: nasal cancer.

**Medical Surveillance:** Routine laboratory studies for all exposed patients include CBC, glucose, and electrolyte determinations. Additional studies for patients exposed to formaldehyde include urinalysis (protein, casts, and red blood cells may be present), methanol level, osmolal gap, and ABG measurements (to monitor acidosis in severe toxicity). Chest radiography and pulse oximetry may be helpful in cases of inhalation exposure. Plasma formaldehyde levels are not useful.

**First Aid:** Flush liquid-exposed skin and hair with plain water for 3 to 5 minutes. Wash area thoroughly with soap and water when possible. Use caution to avoid hypothermia when decontaminating children or the elderly. Use blankets or warmers when appropriate. Irrigate exposed or irritated eyes with plain water or saline for 15 minutes. Remove contact lenses if easily removable without additional trauma to the eye. If pain or injury is evident, continue eye irrigation while transferring the victim to the Support Zone. Victims who are conscious and able to swallow should be given 4 to 8 ounces of water or milk. Gastric lavage with a small bore NG tube should be considered if it can be performed within 1 hour after ingestion. The effectiveness of activated charcoal administration is unknown, but it is suggested following lavage (administer activated charcoal at 1 gm/kg, usual adult dose 60–90 g, child dose 25–50 g). A soda can and straw may be of assistance when offering charcoal to a child. *Note to physician or authorized medical personnel:* There is no antidote for formaldehyde. Treat patients who have metabolic acidosis with sodium bicarbonate (adult dose = 1 ampule; pediatric dose = 1 Eq/kg). Further correction of acidosis should be guided by ABG measurements. Hemodialysis is effective in removing formic acid (formate) and methanol and in correcting severe metabolic acidosis. If methanol poisoning from ingestion of formalin is suspected, as indicated by a serum methanol level of greater than 20 mg/dL or elevated osmolal gap, start ethanol infusion. With 10% ethanol, the loading dose is 7.5 mL/kg body weight; maintenance dose is 1.0 to 1.5 mL/kg/hour; and maintenance dose during hemodialysis is 1.5 to 2.5 mL/kg/hour. In this setting, the target blood level of ethanol is 0.1 mg/dL.

*References:*
- EPA, Office of Pesticide Programs, Pesticide Residue Limits, "Formaldehyde," 40 CFR 180.920 and 180.1024, http://www.epa.gov/pesticides/food/viewtols.htm
- U.S. Department of Health and Human Services, Agency for Toxic Substances and Disease Registry, "ToxFAQs for Formaldehyde," (July 1999). http://www.atsdr.cdc.gov/tfacts111.html
- New Jersey Department of Health and Senior Services, "Hazardous Substance Fact Sheet: Formaldehyde," Trenton, NJ (February 1989, rev. March 2000). http://www.state.nj.us/health/eoh/rtkweb/0946.pdf
- U.S. Environmental Protection Agency, "Topical & Chemical Fact Sheets, Paraformaldehyde," Washington, DC (May 2003). http://www.epa.gov/pesticides/factsheets/chemicals/paraformaldehyde_factsheet.htm#bkmrk1
- Environmental Protection Agency, Investigation of Selected Potential Environmental Contaminants-Formaldehyde, Final Report, Office of Toxic Substances, Environmental Protection Agency, August, 1976.
- National Institute for Occupational Safety and Health, Criteria for a Recommended Standard: Occupational Exposure to Formaldehyde, NIOSH Doc. No. 77-126 (1977).
- U.S. Environmental Protection Agency, "Chemical Hazard Information Profile: Formaldehyde," Washington, DC (1979).
- U.S. Environmental Protection Agency, Formaldehyde, Health and Environmental Effects Profile No. 104, Office of Solid Waste, Washington, DC (April 30, 1980).
- Sax, N.I., Ed., "Dangerous Properties of Industrial Materials Report," 1, No. 4, 70-72 (1981) and 3, No. 3, 71-76 (1983).
- National Institute for Occupational Safety and Health, Formaldehyde: Evidence of Carcinogenicity, Current Intelligence Bulletin No. 34, DHHS (NIOSH) Publication No. 81-111, Cincinnati, Ohio (April 15, 1981).
- Clary, J.J., Gibson, J.E. and Waritz, R.S., Formaldehyde Toxicology, Epidemiology, Mechanisms, New York, Marcel Dekker, Inc. (1983).
- U.S. Environmental Protection Agency, "Chemical Profile: Formaldehyde," Washington, DC, Chemical Emergency Preparedness Program (November 30, 1987).
- Agency for Toxic Substances and Disease Registry, U.S. Department of Health and Human Services, Public Health Service, "Managing Hazardous Materials Incidents," Atlanta, GA (June 2003).
- New York State Department of Health, "Chemical Fact Sheet: Formaldehyde," Albany, NY, Bureau of Toxic Substance Assessment (March 1986).
- California Environmental Protection Agency "Chemical List of Lists," Sacramento CA (February 1997).

# Formetanate Hydrochloride

*Use Type:* Insecticide, acaricide
*CAS Number:* 23422-53-9
*Formula:* $C_{11}H_{16}ClN_3O_2$
*Synonyms:* Carbamic acid, methyl-, ester with $N'$-($m$-hydroxyphenyl)-$N,N$-dimethylformamidine, hydrochloride; $m$-[((Di-methylamino)methylene)amino] phenylcarbamate, hydrochloride; 3-Dimethylaminomethylene aminophenyl-$N$-methylcarbamate,hydrochloride; $N,N$-Dimethyl-$N'$-[((methylamino)carbonyl)oxy]phenylmethanimideamidemonohydrochloride
*Trade Names:* CARZOL®, Gowan Company (USA); CARZOL® SP, Gowan Company (USA); DICARZOL®, Gowan Company (USA); ENT 27566®; EP-332®; MORTON® EP332; NOR-AM® EP 332; SCHERING® 36056; SN 36056®
*Producers:* Gowan Company (USA)
*Chemical Class:* Carbamate
*EPA/OPP PC Code:* 097301
*California DPR Chemical Code:* 111
*RTECS Number:* FC2800000
*Uses:* An insecticide used for thrips and true bug control on fruit crops.
*Human toxicity (long-term)[77]:* Intermediate–14.00 ppb, Health Advisory
*Fish toxicity (threshold)[77]:* Very low–627.37903 ppb, MATC (Maximum Acceptable Toxicant Concentration)
*U.S. Maximum Allowable Residue Levels for Formetanate Hydrochloride (40 CFR 180.276):*

| CROP | ppm |
|---|---|
| Apple | 3 |
| Grapefruit | 4 |
| Lemon | 4 |
| Lime | 4 |
| Nectarine | 4 |

| | |
|---|---|
| Orange | 4 |
| Peach | 5 |
| Pear | 3 |
| Plum, prune, dried | 8 |
| Plum, prune, fresh | 2 |
| Tangerine | 4 |

*Carcinogen/Hazard Classifications*
**S. EPA Carcinogens:** Group E, Unlikely a carcinogen
**Label Signal Word:** DANGER
*Regulatory Authority:*
- RCRA Section 261 Hazardous Constituents
- RCRA Universal Treatment Standards: Wastewater (mg/L), 0.056; Nonwastewater (mg/kg), 1.4
- EPA Hazardous Waste Number (RCRA No.): P198
- EPCRA Section 302 Extremely Hazardous Substances: TPQ = 500/10,000 lb (227/4,540 kg)
- EPCRA Section 304 RQ: EHS, 1 lb (0.454 kg)
- FIFRA, 40CFR186: tolerances for pesticides in animal feeds
- Marine pollutant (49CFR, Subchapter 172.101, Appendix B), severe pollutant as formetanate

*Description:* Colorless to yellow, crystalline solid or white powder. Faint odor. Soluble in water; solubility = < 1 g/L. Vapor pressure = practically zero.

*Incompatibilities:* When heated to decomposition or on contact with acids or acid fumes, may produce highly toxic chloride fumes; deadly phosgene gas may be formed. May cause pitting of some metals. May form explosive materials with phosphorus pentachloride.

*Permissible Concentration in Water:* No criteria set. Runoff from spills or fire control may cause water pollution.

*Routes of Entry:* Skin absorption, ingestion and inhalation.

*Harmful Effects and Symptoms*

*Short Term Exposure:* Eye pupils are small; blurred vision; eye watering; runny nose; cough; shortness of breath; salivation; nausea, stomach cramps, diarrhea, and vomiting; increased blood pressure; profuse sweating; hypermotility, hallucinations; agitation; tingling of the skin; slow heartbeat; convulsions; fluid in lungs; loss of consciousness; incontinence; breathing stops; death. Carbamates inhibit the acetylcholinesterase enzymes and alter the way in which nervous impulses are transmitted. However, within several hours carbamates spontaneously detach from the enzymes.

*Long Term Exposure:* A potent cholinesterase inhibitor; cumulative effect is possible. This chemical may damage the nervous system with repeated exposure, resulting in convulsions, respiratory failure. May cause liver damage.

*Points of Attack:* A potent cholinesterase inhibitor; cumulative effect is possible. This chemical may damage the nervous system with repeated exposure, resulting in convulsions, respiratory failure. May cause liver damage.

*Medical Surveillance:* Medical observation is recommended for 24 to 48 hours after breathing overexposure, as pulmonary edema may be delayed. As first aid for pulmonary edema, consider administering a corticosteroid spray. Cigarette smoking may exacerbate pulmonary injury and should be discouraged for at least 72 hours following exposure. Before employment and at regular times after that, the following are recommended. Plasma and red blood cell cholinesterase levels (tests for the enzyme poisoned by this chemical). If exposure stops, plasma levels return to normal in 1-2 weeks while red blood cell levels may be reduced for 1-3 months. When acetylcholinesterase enzyme levels are reduced by 25% or more below preemployment levels, risk of poisoning is increased, even if results are in lower ranges of "normal." Reassignment to work not involving carbamate pesticides is recommended until enzyme levels recover. If symptoms develop or overexposure occurs, repeat the above tests as soon as possible and get an exam of the nervous system. Also consider complete blood count. Consider chest x-ray following acute overexposure

*First Aid:* Speed in removing material from eyes and skin is of extreme importance. *Eyes:* Eye contact can cause dangerous amounts of these chemicals to be quickly absorbed through the mucous membrane into the bloodstream. Immediately and gently flush eyes with plenty of warm or cold water (NO hot water) for at least 15 minutes, occasionally lifting the upper and lower eyelids. Get medical aid immediately. *Skin:* Get medical aid. Skin contact can cause dangerous amounts of these chemicals to be absorbed into the bloodstream. Wearing the appropriate PPE equipment and respirator for carbamate pesticides, immediately flush skin with plenty of soap and water for at least 15 minutes while removing contaminated clothing and shoes. Shampoo hair promptly if contaminated; protect eyes. *Ingestion:* Call poison control. Loosen all clothing. Never give anything by mouth to an unconscious person. Get medical aid. Do NOT induce vomiting.* If conscious, alert, and able to swallow, rinse mouth and have victim drink 4 to 8 ounces of water. Check to see if poison control instructs you to use ipecac syrup, otherwise administer slurry of activated charcoal (2 oz in 8 oz of water). If victim is *UNCONSCIOUS OR HAVING CONVULSIONS,* do nothing except keep victim warm. **In some cases you may be specifically instructed by poison control to induce vomiting by way of 2 tablespoons of syrup of ipecac (adult) washed down with a cup of water.* Do NOT give activated charcoal before or with ipecac syrup. *Inhalation:* Get medical aid. Do not contaminate yourself. Wearing the appropriate PPE equipment and respirator for carbamate pesticides, immediately remove the victim from the contaminated area to fresh air. If the victim is not breathing, administer artificial respiration. Do NOT use mouth-to-mouth resuscitation; use bag/mask apparatus. If breathing is difficult, administer oxygen through bag/mask apparatus until medical help arrives. Do not leave victim unattended. *Note to physician or authorized medical personnel.* Administer atropine, 2 mg (1/30 gr) intramuscularly or intravenously as soon as any local or systemic signs or symptoms of an intoxication are noted; repeat the administration of atropine every 3 to 8 minutes until signs of atropinization (mydriasis, dry mouth, rapid pulse, hot and dry skin) occur; initiate treatment in children with 0.05 mg mg/kg of atropine; repeat at 5 to 10 minute intervals. Watch respiration, and remove bronchial secretions if they appear to be obstructing the airway; intubate if necessary. *Medical note:* 2-PAMCI may be contraindicated in the case of some carbamate poisonings.

*References:*
- California Environmental Protection Agency "Chemical List of Lists," Sacramento CA (February 1997)
- U.S. Environmental Protection Agency, Office of Pesticide Programs, Pesticide Residue Limits, "Formetanate Hydrochloride", 40 CFR 180.276, www.epa.gov/pesticides/food/viewtols.htm

# Formic Acid

*Use Type:* Fumigant and food additive
*CAS Number:* 64-18-6
*Formula:* $CH_2O_2$; HCOOH
*Synonyms:* Acido formico (Spanish); Acide formique (French); Acido formico (Italian); Ameisensaeure (German); Aminic acid; Formylic acid; Hydrogen carboxylic acid; Methanoic acid; Mierenzur (Dutch)
*Trade Names:* ADD-F®; AI3-24237®; AMASIL®; BILORIN®; COLLO-BUEGLATT®; COLLO-DIDAX®; FORMISOTON®; MYRMICYL®
*Producers:* Ashland (USA); BASF (Germany); BP Chemicals (UK); Celanese Acetyl Products (Germany); Chongqing Chuandong Chemical (Group) (China); Cognis (Germany); DSM (Netherlands); GFS Chemicals (USA); Interstate Chemical (USA); Juhua Group Corp. (China); KCIL Group (India); Kemira Chemicals (Finland); Exxon Mobil Chemical Co. (USA); Panreac Quemica (Spain); Rashtriya Chemicals & Fertilizers (India); Samsung Fine Chemicals (South Korea); Vulcan Chemicals (USA)
*Chemical Class:* Unclassified
*EPA/OPP PC Code:* 214900
*California DPR Chemical Code:* 3208
*ICSC Number:* 0485
*RTECS Number:* LQ4900000
*EEC Number:* 607-001-00-0
*Uses:* Although formic acid is principally used in the U.S. in the manufacture of fumigants, the largest single use is found in Europe where it is used as an additive in animal feed. Formic acid is a strong reducing agent and is used as a decalcifier. It is used in dyeing color fast wool, electroplating, coagulating latex rubber, regeneration old rubber, and dehairing, plumping, and tanning leather. It is also used in the manufacture of acetic acid, airplane dope, allyl alcohol, cellulose formate, phenolic resins, and oxalate; and it is used in the laundry, textile, insecticide, refrigeration, and paper industries, as well as in drug manufacture.

*U.S. Maximum Allowable Residue Levels for Formic Acid (40 CFR 180.1178):*
The pesticide, formic acid, is exempted from the requirement of a tolerance in or on honey and beeswax when used to control tracheal mites and suppress varroa mites in bee colonies, and applied in accordance with label use directions.

*Regulatory Authority:*
- Air Pollutant Standard Set (ACGIH)[1] (DFG)[3] (HSE)[33] (OSHA)[58] (Other Countries)[35] (Several States)[60] (Several Canadian Provinces)
- Permissible Exposure Limits for Chemical Contaminants (CAL/OSHA)
- The "Director's List" (CAL/OSHA)
- Clean Water Act: Section 311 Hazardous Substances/RQ 40CFR117.3 (same as CERCLA, see below).
- Superfund/EPCRA 40CFR302.4 RQ: CERCLA, 5,000 lb (2270 kg)
- EPA Hazardous Waste Number (RCRA No.): U123
- RCRA, 40CFR261, Appendix 8 Hazardous Constituents
- EPCRA Section 313 Form R *de minimus* concentration reporting level: 1.0%
- Canada, WHMIS, Ingredients Disclosure List

*Description:* Formic acid is a colorless, fuming liquid. Pungent odor. Molecular weight = 46.02. Boiling point = 101°C. Melting/Freezing point = 8.4°C. Vapor pressure = 44 mmHg @ 25°C. Flash point = 69°C; (90% solution) 50°C. Autoignition temperature = 434°C. Explosive limits: LEL = 18%; UEL = 57%. Hazard Identification (based on NFPA-704 M Rating System): Health 3, Flammability 2, Reactivity 0. Soluble in water. Log $K_{ow}$ = Negative. Unlikely to bioaccumulate in marine organisms.

*Incompatibilities:* Forms explosive mixture with air. A medium strong acid and a strong reducing agent. Violent reaction with oxidizers, furfuryl alcohol, hydrogen peroxide, nitromethane. Incompatible with strong acids, bases, ammonia, aliphatic amines, alkanolamines, isocyanates, alkylene oxides, epichlorohydrin. Decomposes on heating and on contact with strong acids forming carbon monoxide. Attacks metals: aluminum, cast iron and steel, many plastics, rubber and coatings.

*Permissible Exposure Limits in Air:* The OSHA[2] TWA and the recommended ACGIH TWA value[1] is 5 ppm (9 mg/m³). The NIOSH[2] IDLH level = 30 ppm. The ACGIH[1] has set an STEL of 10 ppm (18 mg/m³). The 5 ppm TWA is endorsed by Argentina[35], Germany (3), the U.K.[33] and Japan[35]. The former USSR[35,43] has set an MAC in workplace air of 1.0 mg/m³. The NIOSH[2] IDLH level = 30 ppm. Several states have set guidelines or standards for formic acid in ambient air[60] ranging from 30 μg/m³ (New York) to 90 μg/m³ (Florida and North Dakota) to 150 μg/m³ (Virginia) to 180 μg/m³ (Connecticut) to 214 μg/m³ (Nevada) to 225 μg/m³ (South Carolina).

*Determination in Air:* Collection using Si gel (special); workup with water; analysis by ion chromatography; NIOSH IV, Method #2011. OSHA: #ID-112.[18]

*Permissible Concentration in Water:* No criteria set, but EPA[32] has suggested a permissible ambient goal of 124 µg/L based on health effects.

*Routes of Entry:* Inhalation of vapor, percutaneous absorption, ingestion, eye and/or skin contact.

*Harmful Effects and Symptoms*

*Short Term Exposure:* Formic acid is very corrosive to the eyes, skin, and respiratory tract. *Inhalation:* Workers exposed to 15 ppm experience nausea. Other symptoms include irritation of the nose, throat and lungs; coughing, runny nose and tearing eyes. Higher exposures can cause pulmonary edema, a medical emergency that can be delayed for several hours. This can cause death. *Skin:* Concentrated solutions may cause severe irritation, burning and blistering. Accidental exposure has resulted in death. Eyes: May cause irritation and tearing. Concentrated solutions may cause severe chemical burns. *Ingestion:* Corrosive. May affect the energy metabolism, causing acidosis. May cause salivation, vomiting, burning sensation in the mouth, vomiting of blood, diarrhea and pain. In severe cases, person may go into shock and develop difficulty in breathing. Death may result. Animal data suggest that ingestion of about 3 ounces may be fatal to a 150 pound individual.

*Long Term Exposure:* Prolonged or repeated exposure to formic acid may cause skin irritation and allergy with rash and itching. May affect the kidneys May cause genetic changes in living cells.

*Points of Attack:* Respiratory system, lungs, skin, kidneys, liver, eyes.

*Medical Surveillance:* Consideration should be given to possible irritant effects on the skin, eyes, and lungs in any placement or periodic examinations. Lung function tests. Kidney function tests checking for blood and urine. Consider chest x-ray following acute overexposure. Evaluation by a qualified allergist.

*First Aid:* If this chemical gets into the eyes, remove any contact lenses at once and irrigate immediately for at least 15 minutes, occasionally lifting upper and lower lids. Seek medical attention immediately. If this chemical contacts the skin, remove contaminated clothing and wash immediately with soap and water. Seek medical attention immediately. If this chemical has been inhaled, remove from exposure, begin rescue breathing (using universal precautions) if breathing has stopped, and CPR if heart action has stopped. Transfer promptly to a medical facility. When this chemical has been swallowed, get medical attention. If victim is conscious, administer water or milk. Do not induce vomiting.

*References:*
- EPA, Office of Pesticide Programs, Pesticide Residue Limits, "Formic Acid," 40 CFR 180.1178, http://www.epa.gov/pesticides/food/viewtols.htm
- New Jersey Department of Health and Senior Services, "Hazardous Substance Fact Sheet, Formic acid," Trenton NJ (January 1996, rev. June 2002). http://www.state.nj.us/health/eoh/rtkweb/0948.pdf
- U.S. Environmental Protection Agency, Formic Acid, Health and Environmental Effects Profile No. 105, Office of Solid Waste, Washington, DC (April 30, 1980).
- Sax, N.I., Ed., "Dangerous Properties of Industrial Materials Report," 1, No. 2, 39-41 (1980) and 3, No. 4, 53-56 (1983).
- New York State Department of Health, "Chemical Fact Sheet: Formic Acid," Albany, NY, Bureau of Toxic Substance Assessment (March 1986 and Version 2).
- California Environmental Protection Agency "Chemical List of Lists," Sacramento CA (February 1997).

# Formothion

*Use Type:* Insecticide and acaricide
*CAS Number:* 2540-82-1
*Formula:* $C_6H_{12}NO_4PS_2$; $(CH_3O)_2PSSCH_2CON(CH_3)CHO$
*Alert:* A Restricted Use Pesticide (RUP)
*Synonyms:* $O,O$-Dimethyl-$S$-($N$-formyl-$N$-methylcarbamoylmethyl)phosphorodithioate; $O,O$-Dimethyl-$S$-(3-methyl-2,4-dioxo-3-azabutyl)-dithiofosfaat (Dutch); Carbamoylmethyl phosphorodithioate; $O,O$-Dimethyldithiophosphorylacetic acid-$N$-methyl-$N$-formylamide; $O,O$-Dimethyl-$S$-(3-methyl-2,4-dioxo-3-azabutyl)-dithiophosphat (German); $O,O$-Dimethyl-$S$-($N$-methyl-$N$-formyl-carbamoylmethyl)-dithiophosphat (German); $O,O$-Dimethyl-$S$-($N$-methyl-$N$-formyl-carbamoylmethyl)-dithiophosphate; $O,O$-Dimethyl-$S$-($N$-methyl-$N$-formyl-carbamoylmethyl)-phosphorodithioate; $O,O$-Dimethylphosphorodithioate $N$-formyl-2-mercapto-$N$-methylacetamide-$S$-ester; ENT 27,257; Formotion (Spanish); $S$-(2-(Formylmethylamino)2-oxoethyl)$O,O$-dimethylphosphorodithioate; $N$-Formyl-$N$-methylcarbamoylmethyl-$O,O$-dimethylphosphorodithioate; $S$-($N$-Formyl-$N$-methylcarbamoylmethyl)-$O,O$-dimethylphosphorodithioate; $S$-($N$-formyl-$N$-methylcarbamoylmethyl)dimethylphosphorodithiolothionate

*Trade Names:* AFLIX®; ANTHIO®; ANTIO®; CP 53926®; S 6900®, Sandoz (Switzerland); SAN 244 1®, Sandoz (Switzerland); SAN 6913 1®, Sandoz (Switzerland); SAN 71071®, Sandoz (Switzerland); SPENCER S-6900®; VEL 4284®

*Producers:* Sandoz (Switzerland); Syngenta (Switzerland)
*Chemical Class:* Organophosphate
*EPA/OPP PC Code:* 366400
*RTECS Number:* TE1050000

*Uses:* Formothion is a systemic and contact insecticide used to control spider mites, aphids, psyllids, mealy bugs, whiteflies, jassids, leaf miners, ermine moths, and fruit flies.

It is used on tree fruits, vines, olives, hops, cereals, sugar cane, rice. Formothion is available as an emulsifiable concentrate and an ultra-low volume spray.

*Carcinogen/Hazard Classifications*
**Label Signal Word:** WARNING
**WHO Acute Hazard:** Class II, moderately hazardous
*Regulatory Authority:*
- Air Pollutant Standard Set (former USSR)[35, 43]
- U.S. DOT Inhalation Hazard Chemicals as organophosphates
- Superfund/EPCRA 40CFR355, Appendix B Extremely Hazardous Substances: TPQ = 100 lb (45.4 kg)
- Superfund/EPCRA 40CFR302.4 RQ: 1 lb (0.454 kg)

*Description:* Formothion is an odorless, yellowish viscous oil or crystalline mass. Practically insoluble in water. Melting/Freezing point = 25°C. Hazard Identification (based on NFPA-704 M Rating System): Health 2, Flammability 1, Reactivity 0.

*Incompatibilities:* Alkaline materials.

*Permissible Exposure Limits in Air:* The former USSR[35, 43] has set an MAC in workplace air of 0.5 mg/m$^3$ and an MAC in ambient air in residential areas of 0.01 mg/m$^3$ on a momentary basis and 0.006 mg/m$^3$ on a daily average basis.

*Determination in Air:* OSHA versatile sampler-2; Toluene/Acetone; Gas chromatography/Flame photometric detection for sulfur, nitrogen, or phosphorus; NIOSH Method IV Method #5600, Organophosphorus pesticides[18].

*Permissible Concentration in Water:* The former USSR[35, 43] has set an MAC in water bodies used for domestic purposes of 0.004 mg/L (4 µg/L).

*Routes of Entry:* Inhalation, absorbed by the skin.

*Harmful Effects and Symptoms*

*Short Term Exposure:* Early symptoms of poisoning include: headache, dizziness, weakness, perspiring, nausea, vomiting, and sensation of tightness in chest. Chronic low doses may produce symptoms similar to influenza. Formothion is one of the least toxic systemic organophosphates. Formothion is a compound of low to moderate toxicity. It causes the depression of cholinesterase leading to accumulation of acetylcholine in the nervous system, which is believed to be responsible for the symptoms. Organic phosphorus insecticides are absorbed by the skin, as well as by the respiratory and gastrointestinal tracts. They are cholinesterase inhibitors. Symptoms of exposure include headache, giddiness, blurred vision, nervousness, profound weakness, nausea, cramps, diarrhea, and discomfort in the chest. Signs include sweating, tearing, salivation, vomiting, cyanosis, convulsions, coma, loss of reflexes and loss of sphincter control. Delayed pulmonary edema may occur after inhalation.

*Long Term Exposure:* Cholinesterase inhibitor; cumulative effect is possible. This chemical may damage the nervous system with repeated exposure, resulting in convulsions, respiratory failure. May cause liver damage.

*Points of Attack:* Respiratory system, lungs, central nervous system, cardiovascular system, skin, eyes, plasma and red blood cell cholinesterase. Repeated exposure may cause changes in personality and exhibit depression, anxiety or irritability.

*Medical Surveillance:* Medical observation is recommended for 24 to 48 hours after breathing overexposure, as pulmonary edema may be delayed. Before employment and at regular times after that, the following are recommended: Plasma and red blood cell cholinesterase levels (tests for the enzyme poisoned by this chemical). If exposure stops, plasma levels return to normal in 1-2 weeks while red blood cell levels may be reduced for 1-3 months. When acetylcholinesterase enzyme levels are reduced by 25% or more below preemployment levels, risk of poisoning is increased, even if results are in lower ranges of "normal." Reassignment to work not involving organophosphate or carbamate pesticides is recommended until enzyme levels recover. If symptoms develop or overexposure occurs, repeat the above tests as soon as possible and get an exam of the nervous system. Also consider complete blood count. Consider chest x-ray following acute overexposure. Do not drink any alcoholic beverages before or during use. Alcohol promotes absorption of organophosphates.

*First Aid:* **Treatment for organophosphate poisoning consists of thorough decontamination, cardiorespiratory support, and administration of the antidotes atropine and pralidoxime. In cases of severe poisoning, diazepam, an anticonvulsant, should also be administered. Antidotes should be administered as prevention even if the diagnosis is in doubt.** Speed in removing material from eyes and skin is of extreme importance. *Eyes:* Eye contact can cause dangerous amounts of these chemicals to be quickly absorbed through the mucous membrane into the bloodstream. Immediately and gently flush eyes with plenty of warm or cold water (NO hot water) for at least 15 minutes, occasionally lifting the upper and lower eyelids. Get medical aid immediately. *Skin:* Get medical aid. Skin contact can cause dangerous amounts of these chemicals to be absorbed into the bloodstream. Wearing the appropriate PPE equipment and respirator for organophosphate pesticides, immediately flush skin with plenty of soap and water for at least 15 minutes while removing contaminated clothing and shoes. Shampoo hair promptly if contaminated. The removed, contaminated clothing and shoes should be double-bagged and left in Hot Zone for later disposal by hazardous materials experts. Skin may also be decontaminated with diluted hypochlorite solution. *Inhalation:* Get medical aid. Do not contaminate yourself. Wearing the appropriate PPE equipment and respirator for organophosphate pesticides, immediately remove the victim from the contaminated area to fresh air. If the victim is not breathing, administer artificial respiration. Do NOT use mouth-to-mouth resuscitation; use bag/mask apparatus. If breathing is difficult, administer oxygen through bag/mask apparatus until medical help arrives. Do not leave victim unattended. *Ingestion:* Call poison control. Loosen all clothing. Never give anything by mouth to an unconscious person. If victim is *unconscious or having convulsions,* do nothing except keep victim warm. Get medical aid. Transfer promptly to a medical facility. In cases of ingestion, **do not induce vomiting.** If the victim is alert and asymptomatic, administer a slurry of activated charcoal at a dose of 1 g/kg (infant, child, and adult dose). A soda can and straw may be of assistance when offering charcoal to a child. *In some cases you may be specifically instructed by poison control to induce vomiting by way of 2 tablespoons of syrup of ipecac (adult) washed down with a cup of water.* Do NOT give activated charcoal before or with ipecac syrup.

*Note to physician or authorized medical personnel:* Treat cases of respiratory compromise, coma, or excessive pulmonary secretions with respiratory support using protocols and techniques available and within the scope of training. Some cases may necessitate procedures such as endotracheal intubation or cricothyrotomy by properly trained and equipped personnel. When possible, atropine (see *Antidotes*, below) should be given under medical supervision. Patients who are comatose, hypotensive, or having seizures or cardiac arrhythmias should be treated according to advanced life support protocols. *Antidotes:* Two antidotes are administered to treat organophosphate poisoning. Atropine is a competitive antagonist of acetylcholine at muscarinic receptors and is used to control the excessive bronchial secretions which are often responsible for death. Pralidoxime relieves both the nicotinic and muscarine effects of organophosphate poisoning by regenerating acetylcholinesterase and can reduce both the bronchial secretions and the muscle weakness associated with poisoning. The initial intravenous dose of atropine in adults should be determined by the severity of symptoms: An initial adult dose of 1.0 to 2.0 mg or pediatric dose of 0.01 mg/kg (minimum 0.01 mg) should be administered intravenously. If intravenous access cannot be established, atropine may also be given intramuscularly, subcutaneously or via endotracheal tube. Doses should be repeated every 15 minutes until excessive secretions and sweating have been controlled. Once bronchial secretion has been controlled, atropine administration should be repeated whenever the secretions begin to recur. In seriously poisoned patients, very large doses may be required. Alterations of pulse rate and pupillary size should not be used as indicators of treatment adequacy. Pralidoxime should be administered as early in poisoning as possible as its efficacy may diminish when given more than 24 to 36 hours after exposure. Doses are as follows: adult 1.0 g; pediatric 25 to 50 mg/kg. The drug should be administered intravenously over 30 to 60 minutes, but in a life-threatening situation, one-half of the total dose can be given per minute for a total administration time of 2 minutes. Treatment should begin to take effect within 40 minutes with a reduction in symptoms and the amount of atropine necessary to control bronchial secretion. The initial dose can be repeated in 1 hour and then every 8 to 12 hours until the patient is clinically well and no longer requires atropine. If intravenous access cannot be established, pralidoxime may also be given intramuscularly. Early administration of diazepam in addition to the combined atropine and pralidoxime treatment may help prevent the onset of seizures and potential brain and cardiac morphologic damage following high-level organophosphate poisoning.

*References:*
- EXTOXNET, Extension Toxicology Network, "Pesticide Information Profile, Formothion," Oregon State University, Corvallis, OR (September 1995). http://extoxnet.orst.edu/pips/formothi.htm
- New Jersey Department of Health and Senior Services, "Hazardous Substance Fact Sheet, Formothion," Trenton, NJ (September 2000). http://www.state.nj.us/health/eoh/rtkweb/2439.pdf
- U.S. Environmental Protection Agency, "Chemical Profile: Formothion," Washington, DC, Chemical Emergency Preparedness Program (November 30, 1987).
- Agency for Toxic Substances and Disease Registry, U.S. Department of Health and Human Services, Public Health Service, "Managing Hazardous Materials Incidents," Atlanta, GA (June 2003).

# Fosamine Ammonium

*Use Type:* Herbicide
*CAS Number:* 25954-13-6
*Formula:* $C_3H_{11}N_2O_4P$
*Synonyms:* Ammonium-aethyl-carbamoyl-phosphonat (German); Ammonium ethyl carbamoylphosphonate; Phosphonic acid, (aminocarbonyl)-, monoethyl ester, monoammonium salt
*Trade Names:* DPX 1108®; KRENITE®, DuPont Crop Protection (USA)
*Producers:* DuPont Crop Protection (USA)
*Chemical Class:* Organophosphate
*EPA/OPP PC Code:* 106701
*California DPR Chemical Code:* 1921
*Uses:* Used to control the growth of many woody plants on non-crop areas such as rail, pipe and utility lines; right-of-ways; industrial sites; reservoirs, lakes and ponds.
*Human toxicity (long-term)*[77]*:* Low–70.00 ppb, Health Advisory
*Fish toxicity (threshold)*[77]*:* Very low–85896.50433 ppb, MATC (Maximum Acceptable Toxicant Concentration)
*Carcinogen/Hazard Classifications*
**Label Signal Word:** CAUTION
**WHO Acute Hazard:** Class U, unlikely to be hazardous

*Regulatory Authority:*
- DOT Inhalation Hazard Chemicals as organophosphates

*Description:* White crystalline solid. Sleight odor. Very soluble in water. Melting point = 175°C. Vapor pressure = $4 \times 10^{-6}$ mmHg @ 20°C.

*Determination in Air:* OSHA versatile sampler-2; Toluene/Acetone; Gas chromatography/Flame ionization detection; NIOSH IV[18], Method #5600, Organophosphorus Pesticides

*Permissible Concentration in Water:* No criteria set. Runoff from spills or fire control may cause water pollution.

*Harmful Effects and Symptoms*

*Short Term Exposure:* Eye pupils are small; blurred vision; eye watering; runny nose; cough; shortness of breath; salivation; dizziness; nausea, stomach cramps, diarrhea, and vomiting; increased blood pressure; profuse sweating; hypermotility, hallucinations; irritability; tingling of the skin; drowsiness; slow heartbeat; convulsions; fluid in lungs; loss of consciousness; incontinence; breathing stops; death. Organophosphates inhibit the action of acetylcholinesterase enzymes, and alter the way in which nervous impulses are transmitted. The effects can last for hours, days, or much longer. The action of the enzymes is reestablished after new enzymes are formed.

*Long Term Exposure:* Cholinesterase inhibitor; cumulative effect is possible. Organophosphates may damage the nervous system with repeated exposure, resulting in convulsions, respiratory failure. May cause liver damage.

*Points of Attack:* Respiratory system, central nervous system, cardiovascular system, blood cholinesterase.

*Medical Surveillance:* Medical observation is recommended for 24 to 48 hours after breathing overexposure, as pulmonary edema may be delayed. As first aid for pulmonary edema, consider administering a corticosteroid spray. Cigarette smoking may exacerbate pulmonary injury and should be discouraged for at least 72 hours following exposure.

*First Aid:* Speed in removing material from eyes and skin is of extreme importance. *Eyes:* Eye contact can cause dangerous amounts of these chemicals to be quickly absorbed through the mucous membrane into the bloodstream. Immediately and gently flush eyes with plenty of warm or cold water (NO hot water) for at least 15 minutes, occasionally lifting the upper and lower eyelids. Get medical aid immediately. *Skin:* Get medical aid. Skin contact can cause dangerous amounts of these chemicals to be absorbed into the bloodstream. Wearing the appropriate PPE equipment and respirator for organophosphate pesticides, immediately flush skin with plenty of soap and water for at least 15 minutes while removing contaminated clothing and shoes. Shampoo hair promptly if contaminated. *Ingestion:* Call poison control. Loosen all clothing. Never give anything by mouth to an unconscious person. Get medical aid. Do NOT induce vomiting.* If conscious, alert, and able to swallow, rinse mouth and have victim drink 4 to 8 ounces of water do NOT induce vomiting but immediately administer slurry of activated charcoal (2 oz in 8 oz of water). If victim is UNCONSCIOUS OR HAVING CONVULSIONS, do nothing except keep victim warm. **In some cases you may be specifically instructed by poison control to induce vomiting by way of 2 tablespoons of syrup of ipecac (adult) washed down with a cup of water.* Do NOT give activated charcoal before or with ipecac syrup. *Inhalation:* Get medical aid. Do not contaminate yourself. Wearing the appropriate PPE equipment and respirator for organophosphate pesticides, immediately remove the victim from the contaminated area to fresh air. If the victim is not breathing, administer artificial respiration. Do NOT use mouth-to-mouth resuscitation; use bag/mask apparatus. If breathing is difficult, administer oxygen through bag/mask apparatus until medical help arrives. Do not leave victim unattended. *Note to physician or authorized medical personnel.* Administer atropine, 2 mg (1/30 gr) intramuscularly or intravenously as soon as any local or systemic signs or symptoms of an intoxication are noted; repeat the administration of atropine every 3 to 8 minutes until signs of atropinization (mydriasis, dry mouth, rapid pulse, hot and dry skin) occur; initiate treatment in children with 0.05 mg mg/kg of atropine; repeat at 5 to 10 minute intervals. Watch respiration, and remove bronchial secretions if they appear to be obstructing the airway; intubate if necessary. *Notes to physician or authorized medical personnel*: N-methylpyridinium-2-aldoxime (2-PAMCI) when used in conjunction with atropine reacts with the phosphorylated cholinesterase, thereby restoring normal activity to by removing the phosphorylating group. The combination of these two chemicals is synergistic and must be administered within minutes to a few hours following exposure (depending on the specific agent) to be effective. Give 2-PAMCI (Pralidoxime; Protopam), 2.5 gm in 100 ml of sterile water or in 5% dextrose and water, intravenously, slowly, in 15-30 minutes; if sufficient fluid is not available, give 1 gm of 2-PAMCI in 3 ml of distilled water by deep intramuscular injection; repeat this every half hour if respiration weakens or if muscle fasciculation or convulsions recur. Also Diazepam, an anticonvulsant, might be considered.

*References:*
- Pesticide Management Education Program, "Fosamine ammonium (Krenite) Herbicide Profile 2/85," Cornell University, Ithaca, NY (February 1985). http://pmep.cce.cornell.edu/profiles/herb-growthreg/fatty-alcohol-monuron/fosamine-ammonium/herb-prof-fosamine-ammon.html

# Fosetyl-Al

*Use Type:* Fungicide
*CAS Number:* 39148-24-8

# Fosetyl-Al

*Formula:* $C_6H_{18}AlO_9P_3$

*Synonyms:* Aluminum tris(*O*-ethylphosphonate); Aluminum phosethyl; Fosetyl aluminum; Phosethyl aluminum; Phosphonic acid, monoethyl ester, aluminum salt (3:1); Phosethyl-Al

*Trade Names:* 32545 R®; ALIETTE®, Bayer CropScience (Germany); CHIPCO® ALIETTE WDG; EFOSITE-AL®; EFOSITE ALUMINUM®; EPAL®; LS-74783®; MIKAL®

*Producers:* Agsin (Singapore); Bayer CropScience (Germany); Epochem Co., (China)

*Chemical Class:* Organophosphate

*EPA/OPP PC Code:* 123301

*California DPR Chemical Code:* 2210

*Uses:* Used on fruits, vegetables and nut crops; also on ornamentals and greenhouse products.

*U.S. Maximum Allowable Residue Levels for Fosetyl-al (40 CFR 180.415):*

| CROP | ppm |
|---|---|
| Asparagus | 0.1 |
| Avocado | 25 |
| Banana | 3 |
| Bushberry subgroup 13b | 40 |
| Caneberry subgroup 13a | 0.1 |
| Cranberry | 0.5 |
| Fruit, citrus, group 10 | 5 |
| Fruit, pome | 10 |
| Ginseng, roots | 0.1 |
| Grape | 10 |
| Hop, dried cones | 45 |
| Juneberry | 40 |
| Lingonberry | 40 |
| Nut, macadamia | 0.2 |
| Onion, dry bulb | 0.5 |
| Pea, succulent | 0.3 |
| Pineapple | 0.1 |
| Pineapple, fodder | 0.1 |
| Pineapple, forage | 0.1 |
| Salal | 40 |
| Strawberry | 75 |
| Tomato | 3 |
| Turnip, greens | 40 |
| Turnip, roots | 15 |
| Vegetable, brassica, leafy, group 5 | 60 |
| Vegetable, cucurbit, group 9 | 15 |
| Vegetable, leafy, except brassica, group 4 | 100 |

*Carcinogen/Hazard Classifications*

**U.S. EPA Carcinogens:** Not likely a carcinogen

*Regulatory Authority:*
- Actively registered pesticide in California.
- DOT Inhalation Hazard Chemicals as organophosphates

*Description:* White, crystalline solid. Odorless. Soluble in water; solubility = 120 g/L. Molecular weight = 354.11. Vapor pressure = $1 \times 10^{-7}$ mmHg @ 20°C.

*Permissible Exposure Limits in Air:*

*Determination in Air:* OSHA versatile sampler-2; Toluene/Acetone; Gas chromatography/Flame ionization detection; NIOSH IV[18], Method #5600, Organophosphorus Pesticides.

*Permissible Concentration in Water:* No criteria set. Runoff from spills or fire control may cause water pollution.

*Harmful Effects and Symptoms*

**Short Term Exposure:** Eye pupils are small; blurred vision; eye watering; runny nose; cough; shortness of breath; salivation; dizziness; nausea, stomach cramps, diarrhea, and vomiting; increased blood pressure; profuse sweating; hypermotility, hallucinations; irritability; tingling of the skin; drowsiness; slow heartbeat; convulsions; fluid in lungs; loss of consciousness; incontinence; breathing stops; death. Organophosphates inhibit the action of acetylcholinesterase enzymes, and alter the way in which nervous impulses are transmitted. The effects can last for hours, days, or much longer. The action of the enzymes is reestablished after new enzymes are formed.

**Long Term Exposure:** Cholinesterase inhibitor; cumulative effect is possible. Organophosphates may damage the nervous system with repeated exposure, resulting in convulsions, respiratory failure. May cause liver damage.

**Points of Attack:** Respiratory system, central nervous system, cardiovascular system, blood cholinesterase.

*Medical Surveillance:* Medical observation is recommended for 24 to 48 hours after breathing overexposure, as pulmonary edema may be delayed. As first aid for pulmonary edema, consider administering a corticosteroid spray. Cigarette smoking may exacerbate pulmonary injury and should be discouraged for at least 72 hours following exposure.

*First Aid:* Speed in removing material from eyes and skin is of extreme importance. *Eyes:* Eye contact can cause dangerous amounts of these chemicals to be quickly absorbed through the mucous membrane into the bloodstream. Immediately and gently flush eyes with plenty of warm or cold water (NO hot water) for at least 15 minutes, occasionally lifting the upper and lower eyelids. Get medical aid immediately. *Skin:* Get medical aid. Skin contact can cause dangerous amounts of these chemicals to be absorbed into the bloodstream. Wearing the appropriate PPE equipment and respirator for organophosphate pesticides, immediately flush skin with plenty of soap and water for at least 15 minutes while removing contaminated clothing and shoes. Shampoo hair promptly if contaminated. *Ingestion:* Call poison control. Loosen all clothing. Never give anything by mouth to an unconscious person. Get medical aid. Do NOT induce vomiting.\* If conscious, alert, and able to swallow, rinse mouth and have victim drink 4 to 8 ounces of water do NOT induce vomiting but immediately administer slurry of activated charcoal (2 oz in 8 oz of water). If victim is *UNCONSCIOUS OR HAVING CONVULSIONS,* do nothing except keep victim warm. \**In some cases you may be specifically instructed by poison* control to induce vomiting by way of 2 tablespoons of syrup of ipecac (adult) washed down with a cup of water. Do NOT give activated charcoal before or with ipecac syrup. *Inhalation:* Get medical aid. Do not contaminate yourself. Wearing the appropriate PPE equipment and respirator for organophosphate pesticides, immediately remove the victim from the contaminated area to fresh air. If the victim is not breathing, administer artificial respiration. Do NOT use mouth-to-mouth resuscitation; use bag/mask apparatus. If breathing is difficult, administer oxygen through bag/mask apparatus until medical help arrives. Do not leave victim unattended. *Note to physician or authorized medical personnel.* Administer atropine, 2 mg (1/30 gr) intramuscularly or intravenously as soon as any local or systemic signs or symptoms of an intoxication are noted; repeat the administration of atropine every 3 to 8 minutes until signs of atropinization (mydriasis, dry mouth, rapid pulse, hot and dry skin) occur; initiate treatment in children with 0.05 mg mg/kg of atropine; repeat at 5 to 10 minute intervals. Watch respiration, and remove bronchial secretions if they appear to be obstructing the airway; intubate if necessary. *Notes to physician or authorized medical personnel:* N-methylpyridinium-2-aldoxime (2-PAMCI) when used in conjunction with atropine reacts with the phosphorylated cholinesterase, thereby restoring normal activity to by removing the phosphorylating group. The combination of these two chemicals is synergistic and must be administered within minutes to a few hours following exposure (depending on the specific agent) to be effective. Give 2-PAMCI (Pralidoxime; Protopam), 2.5 gm in 100 ml of sterile water or in 5% dextrose and water, intravenously, slowly, in 15-30 minutes; if sufficient fluid is not available, give 1 gm of 2-PAMCI in 3 ml of distilled water by deep intramuscular injection; repeat this every half hour if respiration weakens or if muscle fasciculation or convulsions recur. Also Diazepam, an anticonvulsant, might be considered.

*References:*
- U.S. Environmental Protection Agency, Office of Pesticide Programs, Pesticide Residue Limits, "Fosetyl-al", 40 CFR 180.415, www.epa.gov/pesticides/food/viewtols.htm
- California Environmental Protection Agency "Chemical List of Lists," Sacramento CA (February 1997)
- Pesticide Management Education Program, "Fosetyl-al (Aliette) Chemical Fact Sheet 10/83," Cornell University, Ithaca, NY (October 1983). http://pmep.cce.cornell.edu/profiles/fung-nemat/febuconazole-sulfur/fosetyl-al/fung-prof-fosetylal.html

# Fosthietan (ANSI)

*Use Type:* Nematicide and fumigant

*CAS Number:* 21548-32-3
*Formula:* $C_6H_{12}NO_3PS_2$
*Synonyms:* (Diethoxyphosphinylimino)-1,3-dithietane; Diethoxyphosphinylimino-2-dithietanne-1,3 (French); 1,3-Dithietan-2-ylidene phosphoramidic acid diethyl ester
*Trade Names:* AC 64475®; ACCONEM®; CL64475®; GEOFOS®; NEM-A-TAK®, BASF Corp. (Germany), canceled 9/27/1985
*Producers:* BASF Corp. (Germany)
*Chemical Class:* Organophosphate
*EPA/OPP PC Code:* 113301
*RTECS Number:* NJ6490000
*Uses:* Not registered as a pesticide in the U.S.
*Carcinogen/Hazard Classifications*
**Label Signal Word:** DANGER
*Regulatory Authority:*
- Superfund/EPCRA 40CFR355, Appendix B Extremely Hazardous Substances: TPQ = 500 lb (227kg)
- Superfund/EPCRA 40CFR302.4 RQ: EHS, 1 lb (0.454 kg)
- U.S. DOT Inhalation Hazard Chemicals as organophosphates

*Description:* Fosthietan is a pale yellow oil with a mercaptan-like odor. Moderately soluble in water.
*Incompatibilities:* Alkaline material.
*Permissible Exposure Limits in Air:* No standards set.
*Determination in Air:* OSHA versatile sampler-2; Toluene/Acetone; Gas chromatography/Flame photometric detection for sulfur, nitrogen, or phosphorus; NIOSH Method IV Method #5600, Organophosphorus pesticides[18].
*Permissible Concentration in Water:* No criteria set, but runoff from spills or fire control may cause water pollution.
*Routes of Entry:* Inhalation, ingestion, absorbed through the skin.
*Harmful Effects and Symptoms*
*Short Term Exposure:* This compound is a liquid organophosphorus insecticide. Organic phosphorus insecticides are absorbed by the skin, as well as by the respiratory and gastrointestinal tracts. Organic phosphorus insecticides are absorbed by the skin, as well as by the respiratory and gastrointestinal tracts. They are cholinesterase inhibitors. Symptoms of exposure include headache, giddiness, blurred vision, nervousness, weakness, nausea, cramps, diarrhea, and discomfort in the chest. Signs include sweating, tearing, salivation, vomiting, cyanosis, convulsions, coma, loss of reflexes and loss of sphincter control. Delayed pulmonary edema may occur after inhalation.
*Long Term Exposure:* Cholinesterase inhibitor; cumulative effect is possible. This chemical may damage the nervous system with repeated exposure, resulting in convulsions, respiratory failure. May cause liver damage.
*Points of Attack:* Respiratory system, lungs, central nervous system, cardiovascular system, skin, eyes, plasma and red blood cell cholinesterase.

*Medical Surveillance:* Medical observation is recommended for 24 to 48 hours after breathing overexposure, as pulmonary edema may be delayed. Before employment and at regular times after that, the following are recommended: Plasma and red blood cell cholinesterase levels (tests for the enzyme poisoned by this chemical). If exposure stops, plasma levels return to normal in 1-2 weeks while red blood cell levels may be reduced for 1-3 months. When acetylcholinesterase enzyme levels are reduced by 25% or more below preemployment levels, risk of poisoning is increased, even if results are in lower ranges of "normal." Reassignment to work not involving organophosphate or carbamate pesticides is recommended until enzyme levels recover. If symptoms develop or overexposure occurs, repeat the above tests as soon as possible and get an exam of the nervous system. Also consider complete blood count. Consider chest x-ray following acute overexposure. Do not drink any alcoholic beverages before or during use. Alcohol promotes absorption of organophosphates.

*First Aid:* **Treatment for organophosphate poisoning consists of thorough decontamination, cardiorespiratory support, and administration of the antidotes atropine and pralidoxime. In cases of severe poisoning, diazepam, an anticonvulsant, should also be administered. Antidotes should be administered as prevention even if the diagnosis is in doubt.** Speed in removing material from eyes and skin is of extreme importance. *Eyes:* Eye contact can cause dangerous amounts of these chemicals to be quickly absorbed through the mucous membrane into the bloodstream. Immediately and gently flush eyes with plenty of warm or cold water (NO hot water) for at least 15 minutes, occasionally lifting the upper and lower eyelids. Get medical aid immediately. *Skin:* Get medical aid. Skin contact can cause dangerous amounts of these chemicals to be absorbed into the bloodstream. Wearing the appropriate PPE equipment and respirator for organophosphate pesticides, immediately flush skin with plenty of soap and water for at least 15 minutes while removing contaminated clothing and shoes. Shampoo hair promptly if contaminated. The removed, contaminated clothing and shoes should be double-bagged and left in Hot Zone for later disposal by hazardous materials experts. Skin may also be decontaminated with diluted hypochlorite solution. *Inhalation:* Get medical aid. Do not contaminate yourself. Wearing the appropriate PPE equipment and respirator for organophosphate pesticides, immediately remove the victim from the contaminated area to fresh air. If the victim is not breathing, administer artificial respiration. Do NOT use mouth-to-mouth resuscitation; use bag/mask apparatus. If breathing is difficult, administer oxygen through bag/mask apparatus until medical help arrives. Do not leave victim unattended. *Ingestion:* Call poison control. Loosen all clothing. Never give anything by mouth to an unconscious person. If victim is *unconscious or having convulsions,* do nothing except keep victim warm. Get medical aid. Transfer promptly to a medical facility. In cases of ingestion, **do not induce vomiting**. If the victim is alert and asymptomatic, administer a slurry of activated charcoal at a dose of 1 g/kg (infant, child, and adult dose). A soda can and straw may be of assistance when offering charcoal to a child. *In some cases you may be specifically instructed by poison control to induce vomiting by way of 2 tablespoons of syrup of ipecac (adult) washed down with a cup of water.* Do NOT give activated charcoal before or with ipecac syrup.

*Note to physician or authorized medical personnel:* Treat cases of respiratory compromise, coma, or excessive pulmonary secretions with respiratory support using protocols and techniques available and within the scope of training. Some cases may necessitate procedures such as endotracheal intubation or cricothyrotomy by properly trained and equipped personnel. When possible, atropine (see *Antidotes*, below) should be given under medical supervision. Patients who are comatose, hypotensive, or having seizures or cardiac arrhythmias should be treated according to advanced life support protocols. *Antidotes:* Two antidotes are administered to treat organophosphate poisoning. Atropine is a competitive antagonist of acetylcholine at muscarinic receptors and is used to control the excessive bronchial secretions which are often responsible for death. Pralidoxime relieves both the nicotinic and muscarine effects of organophosphate poisoning by regenerating acetylcholinesterase and can reduce both the bronchial secretions and the muscle weakness associated with poisoning. The initial intravenous dose of atropine in adults should be determined by the severity of symptoms: An initial adult dose of 1.0 to 2.0 mg or pediatric dose of 0.01 mg/kg (minimum 0.01 mg) should be administered intravenously. If intravenous access cannot be established, atropine may also be given intramuscularly, subcutaneously or via endotracheal tube. Doses should be repeated every 15 minutes until excessive secretions and sweating have been controlled. Once bronchial secretion has been controlled, atropine administration should be repeated whenever the secretions begin to recur. In seriously poisoned patients, very large doses may be required. Alterations of pulse rate and pupillary size should not be used as indicators of treatment adequacy. Pralidoxime should be administered as early in poisoning as possible as its efficacy may diminish when given more than 24 to 36 hours after exposure. Doses are as follows: adult 1.0 g; pediatric 25 to 50 mg/kg. The drug should be administered intravenously over 30 to 60 minutes, but in a life-threatening situation, one-half of the total dose can be given per minute for a total administration time of 2 minutes. Treatment should begin to take effect within 40 minutes with a reduction in symptoms and the amount of atropine necessary to control bronchial secretion. The initial dose can be repeated in 1 hour and then every 8 to 12 hours until the patient is clinically well and no longer requires atropine. If intravenous access cannot be established, pralidoxime may also be given intramuscularly. Early administration of diazepam in addition to the combined atropine and pralidoxime treatment may help prevent the onset of seizures and potential brain and cardiac morphologic damage following high-level organophosphate poisoning.

*References:*
- New Jersey Department of Health and Senior Services, "Hazardous substance Fact Sheet, Fosthietan," Trenton, NJ (April 2002).
  http://www.state.nj.us/health/eoh/rtkweb/2441.pdf
- U.S. Environmental Protection Agency, "Chemical Profile: Fosthietan," Washington, DC, Chemical Emergency Preparedness Program (November 30, 1987).
- Agency for Toxic Substances and Disease Registry, U.S. Department of Health and Human Services, Public Health Service, "Managing Hazardous Materials Incidents," Atlanta, GA (June 2003).
- California Environmental Protection Agency "Chemical List of Lists," Sacramento CA (February 1997).

# Fuberidazole

*Use Type:* Fungicide for treating seeds
*CAS Number:* 3878-19-1
*Formula:* $C_{11}H_8N_2O$
*Synonyms:* Bitertanol, fuberidazole; Fuberidatol; Fuberisazol; Fubridazole; 2-(2-Furanyl)-1H-benzimidazole; 2-(2-Furyl)benzimidazole; Furidazol; Furidazole; 2-(2'-furyl)-benzimidazole
*Trade Names:* BAYCOR®, Bayer CropScience (Germany); BAYER 33172®, Bayer CropScience (Germany); BAYTAN®, Bayer CropScience (Germany); ICI BAYTAN®, Imperial Chemical Industries (UK); NEOVORONIT®; SIBUTOL®; SIBUTROL®; VORONIT®; VORONITE®; W VII/117®
*Producers:* Bayer CropScience (Germany); Imperial Chemical Industries (UK)
*Chemical Class:* Benzimidazole
*EPA/OPP PC Code:* 466200
*RTECS Number:* DD9010000
EINECS: 223-404-0
*Uses:* Uses include cereal seed dressing; and fungicidal non-mercurial seed dressing with special action against fusarium. Not registered as a pesticide in the U.S.A.
*Carcinogen/Hazard Classifications*
**WHO Acute Hazard:** Group II, moderately hazardous
*Regulatory Authority:*
- Superfund/EPCRA 40CFR355, Appendix B Extremely Hazardous Substances: TPQ = 100/10,000 lb (45.4/4,540 kg)
- Superfund/EPCRA 40CFR302.4 RQ: EHS, 1 lb (0.454 kg)

*Description:* Crystalline solid. Slightly soluble in water. Melting point = 280°C (decomposition). Hazard Identification (based on NFPA-704 M Rating System): Health 1, Flammability 1, Reactivity 0.
*Incompatibilities:* Strong oxidizers. When heated to decomposition, may form fumes of carbon monoxide and carbon dioxide.
*Permissible Exposure Limits in Air:* No standards set.
*Permissible Concentration in Water:* No criteria set, but runoff from spills or fire control may cause water pollution.
*Routes of Entry:* Inhalation, ingestion, skin contact.
*Harmful Effects and Symptoms*
*Short Term Exposure:* Fuberidazole is classified as moderately toxic. Its probable oral lethal dose in humans is 0.5–5 g/kg or between 1 ounce and 1 pint for a 70 kg (150 lb) person. The $LD_{50}$ (oral, rat) = 1100 mg/kg[9].

*First Aid:* If this chemical gets into the eyes, remove any contact lenses at once and irrigate immediately for at least 15 minutes, occasionally lifting upper and lower lids. Seek medical attention immediately. If this chemical contacts the skin, remove contaminated clothing and wash immediately with soap and water. Seek medical attention immediately. If this chemical has been inhaled, remove from exposure, begin rescue breathing (using universal precautions) if breathing has stopped, and CPR if heart action has stopped. Transfer promptly to a medical facility. When this chemical has been swallowed, get medical attention. Give large quantities of water and induce vomiting. Do not make an unconscious person vomit.

*References:*
- U.S. Environmental Protection Agency, "Chemical Profile: Fuberidazole," Washington, DC, Chemical Emergency Preparedness Program (November 30, 1987).
- California Environmental Protection Agency "Chemical List of Lists," Sacramento CA (February 1997).

# Gibberellic Acid

*Use Type:* Plant growth hormone
*CAS Number:* 77-06-5; 125-67-7 (monopotassium salt)
*Formula:* $C_{19}H_{22}O_6$; $C_{20}H_{24}O_6$ (methyl ester); $C_{19}H_{21}KO_6$ (potassium salt)
*Synonyms:* Gibb-3-ene-1,10-dicarboxylic acid, 2,4a,7-trihydroxy-1-methyl-8-methylene-, 1,4a-lactone, (1α,2β,4aα,4bβ,10β)-; NCI-C55823; 2,4a,7-Trihydroxy-1-methyl-8-methylenegibb-3-ene-1,10-dicarboxylic acid, 1,4a-lactone; GA; $GA_3$
*monopotassium salt*: Potassium gibberellate
*Trade Names:* ACTIVOL®; BERELEX®; BRELLIN®; BOLL-SET®, Micro-Flo (USA); CEKUGIB®; CROP BOOSTER®, Agriliance (USA); CYTOPLEX HMS®, DYNOGEN®; FALGRO®, Fine Agrochemicals (UK); FLORALTONE® (with 2,3,5-triiodobenzoic acid); FLORGIB®, Fine Agrochemicals (UK); FOLI-ZYME®, Stoller Enterprises (USA); GIBBEX®, Griffin (USA); GIBBERELLIN®; GIBBERELLIN $A_3$®; GIBBERELLIN X®; GIBBREL®; GIBGRO®, Nufarm (Australia); GIB-SOL®; GIB-TABS®; GIBRESCOL®; GROCEL®; KALGIBB®; MAXON®, Agriliance (USA); N-LARGE®, Stoller Enterprises (USA); NOVAGIB®, Fine Agrochemicals (UK); PGR-IV®, Micro-Flo (USA); PRO-GIBB®, Valent BioSciences (USA); REGULEX®; RELEASE®, Valent BioSciences (USA); RELAX®; RYZUP®; STIMULATE®, Stoller Enterprises (USA); VIGOR®, Stoller Enterprises (USA)
*Producers:* Agriliance (USA); Agsin (Singapore); Fine Agrochemicals (UK); Gowan (USA); Griffin (USA): ICI Group (UK); Ki-Hara Chemicals Ltd. (UK); Micro-Flo (USA); Nufarm (Australia); Stoller Enterprises (USA); Valent BioSciences (USA); Wuzhou International (China)
*Chemical Class:* Biochemical pesticide
*EPA/OPP PC Code:* 043801; 043802 (potassium salt)
*California DPR Chemical Code:* 310; 771 (potassium salt)
*RTECS Number:* LY8990000
*EINECS Number:* 201-00-0
*Uses:* Gibberellic acids are naturally occurring plant hormones that are used as plant growth regulators to stimulate both cell division and elongation that affects leaves and stems. Applications of this hormone also hastens plant maturation and seed germination. Gibberellic acids are applied to growing field crops, small fruits, vines and tree fruit, and ornamentals, shrubs and vines.
*U.S. Maximum Allowable Residue Levels for Gibberellic Acid (40 CFR 180.1098):*
*Note:* The U.S. EPA has generally exempted from tolerances many or most of the biochemical plant growth regulators. The crops enumerated in 40 CFR 180.1098 have no stated tolerances. They can be found on the web site shown in the reference section below.
*Carcinogen/Hazard Classifications*
**Label Signal Word:** WARNING or CAUTION
**WHO Acute Hazard:** Class U, unlikely to be hazardous
*Regulatory Authority:*
• Actively registered pesticide in California.
*Description:* White, crystalline solid or powder. Slightly soluble in water. Molecular weight = 346.39. Melting/Freezing point = 234°C (decomposition).
*Incompatibilities:* May react violently with strong oxidizers, bromine, 90% hydrogen peroxide, phosphorus trichloride, silver powders or dust. Incompatible with silver compounds. Mixture with some silver compounds forms explosive salts of silver oxalate.
*Permissible Concentration in Water:* No criteria set. Runoff from spills or fire control may cause water pollution.
*Harmful Effects and Symptoms*
*Short Term Exposure:* Contact with eyes or skin may cause irritation or injury. Inhalation should be avoided; use NIOSH-approved air purifying respirators for pesticides. May be harmful if swallowed.
*First Aid:* If this chemical gets into the eyes, remove any contact lenses at once and irrigate immediately for at least 15 minutes, occasionally lifting upper and lower lids. Seek medical attention immediately. If this chemical contacts the skin, remove contaminated clothing and wash immediately with soap and water. Seek medical attention immediately. If this chemical has been inhaled, remove from exposure, begin rescue breathing (using universal precautions) if breathing has stopped, and CPR if heart action has stopped. Transfer promptly to a medical facility. When this chemical has been swallowed, get medical attention. If victim is conscious and able to swallow, have victim drink 4 to 8 ounces of water. Do not induce vomiting. *Note to physician or authorized medical personnel*: Medical observation is recommended for 24 to 48 hours after breathing overexposure, as pulmonary edema may be delayed. As first aid for pulmonary edema, consider administering a corticosteroid spray. Cigarette smoking may exacerbate pulmonary injury and should be discouraged for at least 72 hours following exposure.
*References:*
• U.S. Environmental Protection Agency, "Reregistration Eligibility Decision (RED), Gibberellic Acid" Office of Prevention, Pesticides and Toxic Substances, Washington, DC (December 1995). http://www.epa.gov/REDs/4110.pdf
• Registry of Toxic Effects of Chemical Substances, "Gibberellic Acid," National Institute for Occupational

Safety and Health (NIOSH), (October 2002). http://www.cdc.gov/niosh/rtecs/ly892d30.html#L
- California Environmental Protection Agency "Chemical List of Lists," Sacramento CA (February 1997)
- U.S. Environmental Protection Agency, Office of Pesticide Programs, Pesticide Residue Limits, "Gibberellic Acid," 40 CFR 180.1098, www.epa.gov/pesticides/food/viewtols.htm

# Glufosinate-ammonium

*Use Type:* Herbicide
*CAS Number:* 77182-82-2
*Formula:* $C_5H_{15}N_2O_4P$
*Synonyms:* 2-Amino-4-(hydroxymethylphosphinyl)butanoic acid monoammonium salt; Ammonium (3-amino-3-carboxypropyl)methylphosphinate; Ammonium 2-amino-4-(hydroxymethylphosphinyl)butanoate; Ammonium-*dl*-homoalanin-4-yl (methyl) phosphinate; Ammonium (*dl*-homoalanine-4-yl)methylphosphinate; Butanoic acid, 2-amino-4-(hydroxymethylphosphinyl)-, monoammonium salt; Butanoic acid, 2-amino-4-(hydroxymethylphosphinyl)-, monoammonium salt; Monoammonium 2-amino-4-(hydroxymethylphosphinyl) butanoate; Phosphinothricin monoammonium salt
*Trade Names:* BASTA®; DERRINGER®, Bayer CropScience (Germany); FINALE®, Bayer CropScience (Germany); HOE 00661®; HOE 03986®; HOE 39866®; IGNITE®, Bayer CropScience (Germany); LIBERTY®, Bayer CropScience (Germany); RELY®, Bayer CropScience (Germany); REMOVE®, Bayer CropScience (Germany); RUBOUT®; TOTAL®
*Producers:* Bayer CropScience (Germany)
*Chemical Class:* Organophosphate
*EPA/OPP PC Code:* 128850
*California DPR Chemical Code:* 3946
*Uses:* Glufosinate-ammonium is a naturally occurring broad-spectrum contact herbicide that is used to control a wide range of weeds after the crop emerges or for total vegetation control on non-crop lands. It is used on crops that have been genetically engineered. Glufosinate herbicides are also used to desiccate crops before harvest.
*Human toxicity (long-term)*[77]*:* Very low–140.00 ppb, Health Advisory
*Fish toxicity (threshold)*[77]*:* Very low–1984.85607 ppb, MATC (Maximum Acceptable Toxicant Concentration)
*U.S. Maximum Allowable Residue Levels for Glufosinate Ammonium (40 CFR 180.473):*

| CROP | ppm |
|---|---|
| Almond, hulls | 0.50 |
| Apple | 0.05 |
| Banana | 0.30 |
| Banana, pulp | 0.20 |
| Bushberry subgroup 13b | 0.15 |
| Cattle, fat | 0.40 |
| Cattle, meat | 0.15 |
| Cattle, mbyp | 6.0 |
| Cotton, gin byproducts | 15.0 |
| Cotton, undelinted seed | 4.0 |
| Egg | 0.15 |
| Goat, fat | 0.40 |
| Goat, meat | 0.15 |
| Goat, mbyp | 6.0 |
| Grape | 0.05 |
| Hog, fat | 0.40 |
| Hog, meat | 0.15 |
| Hog, mbyp | 6.0 |
| Horse, fat | 0.40 |
| Horse, meat | 0.15 |
| Horse, mbyp | 6.0 |
| Juneberry | 0.10 |
| Lingonberry | 0.10 |
| Milk | 0.15 |
| Nut, tree, group 14 | 0.10 |
| Potato | 0.80 |
| Potato, chips | 1.60 |
| Potato granules and flakes | 2.00 |
| Poultry, fat | 0.15 |
| Poultry, meat | 0.15 |
| Poultry, mbyp | 0.60 |
| Salal | 0.10 |
| Sheep, fat | 0.40 |
| Sheep, meat | 0.15 |
| Sheep, mbyp | 6.0 |

*Carcinogen/Hazard Classifications*
**Label Signal Word:** WARNING or CAUTION
**WHO Acute Hazard:** Class III, slightly hazardous (as glufosinate)
*Regulatory Authority:*
- DOT Inhalation Hazard Chemicals as organophosphates

*Description:* Colorless, crystalline solid or powder. Soluble in water; solubility = $2 \times 10^5$ ppm @ 20°C. Molecular weight = 198.19. Melting/Freezing point = 214°C.
*Permissible Exposure Limits in Air:* OSHA versatile sampler-2; Toluene/Acetone; Gas chromatography/Flame ionization detection; NIOSH IV, Method #5600, Organophosphorus Pesticides.[18]
*Permissible Concentration in Water:* No criteria set. Runoff from spills or fire control may cause water pollution.
*Harmful Effects and Symptoms*
*Short Term Exposure:* Eye pupils are small; blurred vision; eye watering; runny nose; cough; shortness of breath; salivation; dizziness; nausea, stomach cramps, diarrhea, and vomiting; increased blood pressure; profuse sweating; hypermotility, hallucinations; irritability; tingling of the skin; drowsiness; slow heartbeat; convulsions; fluid in lungs; loss of consciousness; incontinence; breathing stops; death. Organophosphates inhibit the action of acetylcholinesterase enzymes, and alter the way in which nervous impulses are

transmitted. The effects can last for hours, days, or much longer. The action of the enzymes is reestablished after new enzymes are formed.

***Long Term Exposure:*** Cholinesterase inhibitor; cumulative effect is possible. Organophosphates may damage the nervous system with repeated exposure, resulting in convulsions, respiratory failure. May cause liver damage.

***Points of Attack:*** Respiratory system, central nervous system, cardiovascular system, blood cholinesterase.

***Medical Surveillance:*** Medical observation is recommended for 24 to 48 hours after breathing overexposure, as pulmonary edema may be delayed. As first aid for pulmonary edema, consider administering a corticosteroid spray. Cigarette smoking may exacerbate pulmonary injury and should be discouraged for at least 72 hours following exposure.

***First Aid:*** Speed in removing material from eyes and skin is of extreme importance. *Eyes:* Eye contact can cause dangerous amounts of these chemicals to be quickly absorbed through the mucous membrane into the bloodstream. Immediately and gently flush eyes with plenty of warm or cold water (NO hot water) for at least 15 minutes, occasionally lifting the upper and lower eyelids. Get medical aid immediately. *Skin:* Get medical aid. Skin contact can cause dangerous amounts of these chemicals to be absorbed into the bloodstream. Wearing the appropriate PPE equipment and respirator for organophosphate pesticides, immediately flush skin with plenty of soap and water for at least 15 minutes while removing contaminated clothing and shoes. Shampoo hair promptly if contaminated. *Ingestion:* Call poison control. Loosen all clothing. Never give anything by mouth to an unconscious person. Get medical aid. Do NOT induce vomiting.\* If conscious, alert, and able to swallow, rinse mouth and have victim drink 4 to 8 ounces of water do NOT induce vomiting but immediately administer slurry of activated charcoal (2 oz in 8 oz of water). If victim is *unconscious or having convulsions,* do nothing except keep victim warm. *\*In some cases you may be specifically instructed by poison control to induce vomiting by way of 2 tablespoons of syrup of ipecac (adult) washed down with a cup of water.* Do NOT give activated charcoal before or with ipecac syrup. *Inhalation:* Get medical aid. Do not contaminate yourself. Wearing the appropriate PPE equipment and respirator for organophosphate pesticides, immediately remove the victim from the contaminated area to fresh air. If the victim is not breathing, administer artificial respiration. Do NOT use mouth-to-mouth resuscitation; use bag/mask apparatus. If breathing is difficult, administer oxygen through bag/mask apparatus until medical help arrives. Do not leave victim unattended. *Note to physician or authorized medical personnel.* Administer atropine, 2 mg (1/30 gr) intramuscularly or intravenously as soon as any local or systemic signs or symptoms of an intoxication are noted; repeat the administration of atropine every 3 to 8 minutes until signs of atropinization (mydriasis, dry mouth, rapid pulse, hot and dry skin) occur; initiate treatment in children with 0.05 mg mg/kg of atropine; repeat at 5 to 10 minute intervals. Watch respiration, and remove bronchial secretions if they appear to be obstructing the airway; intubate if necessary. *Notes to physician or authorized medical personnel*: $N$-methylpyridinium-2-aldoxime (2-PAMCI) when used in conjunction with atropine reacts with the phosphorylated cholinesterase, thereby restoring normal activity to by removing the phosphorylating group. The combination of these two chemicals is synergistic and must be administered within minutes to a few hours following exposure (depending on the specific agent) to be effective. Give 2-PAMCI (Pralidoxime; Protopam), 2.5 gm in 100 ml of sterile water or in 5% dextrose and water, intravenously, slowly, in 15-30 minutes; if sufficient fluid is not available, give 1 gm of 2-PAMCI in 3 ml of distilled water by deep intramuscular injection; repeat this every half hour if respiration weakens or if muscle fasciculation or convulsions recur. Also Diazepam, an anticonvulsant, might be considered.

***References:***
- U.S. Environmental Protection Agency, Office of Pesticide Programs, Pesticide Residue Limits, "Glufosinate Ammonium," 40 CFR 180.473, http://www.epa.gov/fedrgstr/EPA-PEST/2003/September/Day-29/p24565.htm

# Glutamic Acid

***Use Type:*** Fungicide and plant growth regulator
***CAS Number:*** 56-86-0 (*L*-form); 617-65-2 (*dl*-form); 6893-26-1
***Formula:*** $C_5H_9NO_4$
***Synonyms:*** α-Aminoglutaric acid; l-2-Aminoglutaric acid; 2-Aminopentanedioic acid; 1-Aminopropane-1,3-dicarboxylic acid; GLU (IUPAC); l-Glutamic acid; α-Glutamic acid; Glutaminic acid; l-Glutaminic acid; α-Glutaminic acid
*dl*-form: *dl*-Glutamic acid; *dl*-α-Glutamic acid; Glutamic acid, *dl*-(synthetic racemic mix); [±]-Glutamic acid
***Trade Names:*** AUXIGRO®, Emerald BioAgriculture (USA); GLUSATE®; GLUTACID®; GLUTAMINOL®; GLUTATON®
***Producers:*** Emerald BioAgriculture (USA); Mallinckrodt Baker (USA)
***EPA/OPP PC Code:*** 374350
***California DPR Chemical Code:*** 5159
***Uses:*** This amino acid is one of the 20 building blocks for protein.
***U.S. Maximum Allowable Residue Levels for L- Glutamic Acid [40 CFR 180.1187, 1001(d)]:***

*L*-Glutamic acid is exempt from the requirement of a tolerance on all food commodities when used in accordance with good agricultural practice.

*Carcinogen/Hazard Classifications*
**Label Signal Word:** CAUTION
*Description:* White crystalline solid or powder. Moderately soluble in water; solubility = 20.54 g/L @ 25°C (*dl*-form), 8.64 g/L @ 25°C (L-form). Molecular weight 147.14. Density = 1.538 @ 4°C (L-form); 1.4601 @ 4°C (*dl*-form). Melting/Freezing point = decomposes @ 247-249°C (L-form); @ 224-227°C (*dl*-form)

*Incompatibilities:* May react violently with strong oxidizers, bromine, 90% hydrogen peroxide, phosphorus trichloride, silver powders or dust. Incompatible with silver compounds. Mixture with some silver compounds forms explosive salts of silver oxalate.

*Permissible Concentration in Water:* No criteria set. Runoff from spills or fire control may cause water pollution.

*Harmful Effects and Symptoms*
*Short Term Exposure:* Contact with eyes or skin may cause irritation or injury. Inhalation should be avoided; use NIOSH-approved air purifying respirators for pesticides. May be harmful if swallowed.

*First Aid:* If this chemical gets into the eyes, remove any contact lenses at once and irrigate immediately for at least 15 minutes, occasionally lifting upper and lower lids. Seek medical attention immediately. If this chemical contacts the skin, remove contaminated clothing and wash immediately with soap and water. Seek medical attention immediately. If this chemical has been inhaled, remove from exposure, begin rescue breathing (using universal precautions) if breathing has stopped, and CPR if heart action has stopped. Transfer promptly to a medical facility. When this chemical has been swallowed, get medical attention. Give large quantities of water and induce vomiting. Do not make an unconscious person vomit. Do not induce vomiting when formulations containing petroleum solvents are ingested.

*References:*
- U.S. Environmental Protection Agency, Office of Pesticide Programs, Pesticide Residue Limits, "L-Glutamic Acid," 40 CFR 180.1187, 1001(d), http://www.epa.gov/pesticides/food/viewtols.htm
- California Environmental Protection Agency "Chemical List of Lists," Sacramento CA (February 1997)

# Glutaraldehyde

*Use Type:* Fungicide and biocide
*CAS Number:* 111-30-8
*Formula:* $C_5H_8O_2$; $HCO(CH_2)_3CHO$
*Synonyms:;* 1,3-Diformal propane; Glutamic dialdehyde; Glutaral; Glutaraldehyd (Czech); Glutard dialdehyde; Glutaric acid dialdehyse; Glutaric dialdehyde; NCI-C55425; Pentanedial; 1,5-Pentanedial; 1,5-Pentanedione; Potentiated acid glutaraldehyde;

*Trade Names:* AQUCAR®, Dow Chemical (USA); CIDEX® (component of this product); CUDEX; ODIX (component of this product); COLDCIDE-25® microbiocide concentrate; GKN-O® microbiocide concentrate (glutaraldehyde + alkyl dimethyl benzyl ammonium chloride + alkyl dimethyl ethylbenzyl ammonium chloride); HOSPEX®; SONACIDE®

*Producers:* AMRESCO (USA); BASF (Germany); Dow Chemical (USA); Sigma-Aldrich Fine Chemicals (USA); Tokyo Kasei Kogyo (Japan); Union Carbide (USA)

*Chemical Class:*
*EPA/OPP PC Code:* 043901
*California DPR Chemical Code:* 139
*ICSC Number:* 0158
*RTECS Number:* MA2450000
*EEC Number:* 605-022-00-X

*Uses:* Used to control pollutants in water treatment plants and towers. Glutaraldehyde is also used as a cross-linking agent for protein and polyhydroxy materials. It has been used in tanning and as a fixative for tissues. It is also used as an intermediate. Buffered solutions are used as antimicrobial agents in hospitals.

*Carcinogen/Hazard Classifications*
**Label Signal Word:** DANGER
*Regulatory Authority:*
- AB 2588-Air Toxics "Hot Spots" Chemicals (CAL)
- Permissible Exposure Limits for Chemical Contaminants (CAL/OSHA)
- The "Director's List" (CAL/OSHA)
- Actively registered pesticide in California.
- California Environmental Protection Agency "Chemical List of Lists," Sacramento CA (February 1997)

*Description:* Glutaraldehyde is a colorless liquid with a pungent odor, which readily changes to a glossy polymer. The odor threshold is 0.04 ppm (NY) and 0.2 ppm (NJ). Soluble in water. Molecular weight = 100.13. Boiling point = 187-189°C (decomposes). Melting/Freezing point = –14°C. Vapor pressure = 17 mmHg @ 20°C Hazard Identification (based on NFPA-704 M Rating System): Health 2, Flammability 0, Reactivity 0. Soluble in water. Log $K_{ow}$ = –0.19. Unlikely to bioaccumulate in marine organisms.

*Incompatibilities:* May react violently with bromine, ketones. Incompatible with strong acids, azo dyes, caustics, ammonia, amines, boranes, hydrazines, strong oxidizers.

*Permissible Exposure Limits in Air:* There is no OSHA PEL. NIOSH and ACGIH recommend a ceiling value of 0.2 ppm (0.8 mg/m³) but proposed no STEL. DFG[3] and HSE[33] have adopted the same value as an 8-hour TWA and an MAK. The HSE[33] uses the same values as an STEL. Several states have set guidelines or standards for glutaraldehyde in ambient air[60] ranging from 6.0 μg/m³ (Virginia) to 7.0 μg/m³ (North Dakota) to 14.0 μg/m³ (Connecticut) to 17.0 μg/m³ (Nevada).

*Determination in Air:* Si gel; Acetonitrile; High-pressure liquid chromatography/Ultraviolet; NIOSH IV, Method #2532.

*Permissible Concentration in Water:* No criteria set. Runoff from spills or fire control may cause water pollution.

*Routes of Entry:* Inhalation, skin absorption, ingestion, skin and/or eye contact. Can be absorbed through the skin.

*Harmful Effects and Symptoms*

*Short Term Exposure:* Irritates the eyes, skin, and respiratory tract. *Inhalation:* 0.3 ppm can cause nose and throat irritation. 0.4 ppm has caused headaches. 0.5 ppm has been described as intolerably irritating. *Skin:* Can cause irritation. Contact with a 5% solution can sensitize the skin and cause an allergic response to subsequent contact of much lower concentrations. *Eyes:* Vapors of a 2% solution (0.4 ppm) have produced irritation. *Ingestion:* Can cause irritation of the mouth and stomach. The $LD_{50}$-oralratis 134 mg/kg (moderately toxic).

*Long Term Exposure:* Repeated or prolonged contact with skin may cause chemical sensitization, skin allergy and asthma. Exposure may cause liver and nervous system damage. Glutaraldehyde may cause mutations, handle with extreme caution.

*Points of Attack:* Eyes, skin, respiratory system, liver, central nervous system.

*Medical Surveillance:* If symptoms develop or overexposure has occurred, the following may be useful: Liver function tests. Evaluation by a qualified allergist, including careful exposure history and special testing, may help diagnose skin allergy.

*Note:* Testing by NIOSH has not been completed to determine the carcinogenicity of glutaraldehyde and related low-molecular-weight-aldehydes. However, the limited studies to date indicate that these substances have chemical reactivity and mutagenicity similar to acetaldehyde and malonaldehyde. NIOSH recommends that acetaldehyde and malonaldehyde be considered potential occupational carcinogens in conformance with the OSHA carcinogen policy. Therefore, NIOSH recommends that careful consideration should be given to reducing exposures to related aldehydes such as gluteraldehyde. Exposure to acetaldehyde has produced nasal tumors in rats and laryngeal tumors in hamsters, and exposure to malonaldehyde has produced thyroid gland and pancreatic islet cell tumors in rats. Further information can be found in the "NIOSH Current Intelligence Bulletin 55: Carcinogenicity of Acetaldehyde and Malonaldehyde, and Mutagenicity of Related Low-Molecular-Weight Aldehydes" [NIOSH Publication No. 91-112.]

*First Aid:* If this chemical gets into the eyes, remove any contact lenses at once and irrigate immediately for at least 15 minutes, occasionally lifting upper and lower lids. Seek medical attention immediately. If this chemical contacts the skin, remove contaminated clothing and wash immediately with soap and water. Seek medical attention immediately. If this chemical has been inhaled, remove from exposure, begin rescue breathing (using universal precautions) if breathing has stopped, and CPR if heart action has stopped. Transfer promptly to a medical facility. When this chemical has been swallowed, get medical attention. Give large quantities of water and induce vomiting. Do not make an unconscious person vomit. Do not induce vomiting when formulations containing petroleum solvents are ingested.

*References:*
- New York State Department of Health, "Chemical Fact Sheet; Glutaraldehyde, " Albany, NY, Bureau of Toxic Substance Assessment (April 1986).
- New Jersey Department of Health and Senior Services, "Hazardous Substance Fact Sheet: Glutaraldehyde," Trenton, NJ (February, 1989, rev. April 1994, January 2000). http://www.state.nj.us/health/eoh/rtkweb/0960.pdf
- Sax, N.I., Ed., "Dangerous Properties of Industrial Materials Report, " 1, No. 7, 2-4 (1981).
- California Environmental Protection Agency "Chemical List of Lists," Sacramento CA (February 1997).

# Glyodin

*Use Type:* Fungicide and algaecide

*CAS Number:* 556-22-9

*Formula:* $C_{22}H_{44}N_2O_2$

*Synonyms:* Glyodin acetate; Glyoxide; Glyoxide dry; 2-Heptadecyl-4,5-dihydro-1*H*-imidazolyl monoacetate; 2-Heptadecyl glyoxalidine acetate; 2-Heptadecyl-2-imidazoline acetate; 2-Heptadecyl-2-imidazoline acetate; 1*H*-Imidazole, 2-heptadecyl-4,5-dihydro-, monoacetate; 2-Imidazoline, 2-heptadecyl-, monoacetate

*Trade Names:* CRAG 341®

*Chemical Class:* Imidazole

*EPA/OPP PC Code:* 043601

*Uses:* Not registered in the U.S.

*Carcinogen/Hazard Classifications*

**Label Signal Word:** WARNING, CAUTION or DANGER

*Description:* Light orange crystalline solid. Insoluble in water. Melting/Freezing point = 62°C. Density = 1.032 @ 20°C.

*Incompatibilities:* May not be compatible with nitrates. Moisture may cause hydrolysis or other forms of decomposition. Forms toxic nitrogen oxides when heated to decomposition.

*Determination in Air:* Filter; none; Gravimetric; NIOSH IV [Particulates NOR; #0500 (total), #0600 (respirable)].[18]

*Permissible Concentration in Water:* No criteria set. Runoff from spills or fire control may cause water pollution.

*Harmful Effects and Symptoms*

*Short Term Exposure:* Contact with eyes or skin may cause irritation or injury. Inhalation should be avoided; use NIOSH-approved air purifying respirators for pesticides. May be harmful if swallowed.

*First Aid:* If this chemical gets into the eyes, remove any contact lenses at once and irrigate immediately for at least 15 minutes, occasionally lifting upper and lower lids. Seek medical attention immediately. If this chemical contacts the skin, remove contaminated clothing and wash immediately with soap and water. Seek medical attention immediately. If this chemical has been inhaled, remove from exposure, begin rescue breathing (using universal precautions) if breathing has stopped, and CPR if heart action has stopped. Transfer promptly to a medical facility. When this chemical has been swallowed, get medical attention. *Do not induce vomiting when formulations containing petroleum solvents are ingested.* Otherwise, give large quantities of water and induce vomiting. Do not make an unconscious person vomit.

# Glyphosate (ANSI)

*Use Type:* Herbicide
*CAS Number:* 1071-83-6
*Formula:* $C_3H_8NO_5P$; $HOCOCH_2NHCH_2PO(OH)_2$
*Synonyms:* Glyfosaat (Dutch); Glifosate (German); Glifosato (Spanish); Glycine, *N*-(phosphonomethyl)-; Phosphonomethyliminoacetic acid; *N*-Phosphonomethyl glycine; *N*-(Phosphonomethyl)-glycine
*Trade Names:* ACCORD®; AQUANEAT®, Cerexagri Inc. (USA); Nufarm (Australia); CAMPAIGN®, Monsanto (USA); COSMIC®, Calliope (France); FALLOW MASTER®, Monsanto (USA); FIELD MASTER®, Monsanto (USA); FIRE POWER®, Monsanto (USA); FLAME PLUS®, Agrochemicals del Ecuador (AGROCHEM) (Ecuador); FONT 360®, Agrochemicals del Ecuador (AGROCHEM) (Ecuador); FOZZATE®, Makhteshim-Agan Industries (Israel); GALLUP®; GLAND-UP®, Veterinary & Agricultural Products Manufacturing Co., Ltd. (VAPCO) (Jordan); GLION®, Milenia Agro Ciencias (Brazil); GLYCEL®, Excel Industries (India); GLY-FLO®, Micro Flo (USA); GLYFOCAL®, Calliope (France); GLYFOS®, Cheminova (Denmark); GLYPRO®, Dow AgroSciences (USA); GLYTEX®, Veterinary & Agricultural Products Manufacturing Co., Ltd. (VAPCO) (Jordan); GLYWEED®, Sabero Organics (India); GROUND-UP®, Veterinary & Agricultural Products Manufacturing Co., Ltd. (VAPCO) (Jordan); KEN-ROUND EXTRA®, Kenso Corp. (Malayasia); KEN-STAR PLUS®, Kenso Corp. (Malayasia); KLEERAWAY®, Monsanto (USA); LANDMASTER®, Monsanto (USA); LIDER®, Pyosa Agroquimicos (Mexico); MON 0573®, Monsanto (USA); MON 2139®, Monsanto (USA); OXALIS®, Calliope (France); PONDMASTER®; RANGER®; RAZOR®, Nufarm (Australia); READY MASTER®, Monsanto (USA); RODEO®, Monsanto (USA); ROPHOSATE®, Rotam Agrochemical (Hong Kong); ROUNDUP®, Monsanto (USA); SANOS®, Sanonda (Australia); STANDOUT®, BASF Agricultural Products Group (Germany); SWEEP®, United Phosphorus (India); TOUCHDOWN®, Syngenta (Switzerland); TREVISSIMO®, Calliope (France); TROP®, Milenia Agro Ciencias (Brazil); ZPP 1560 AS HERBICIDE®, Syngenta (Switzerland
*Producers:* Agrimor International (USA); Agsin (Singapore); Aimco Pesticides Ltd. (India); Agrochemicals del Ecuador (AGROCHEM) (Ecuador); Alcotan Laboratories (Spain); Atanor S.A. (Argentina); BASF Agricultural Products Group (Germany); Bhageria Dye-Chem (India); Biesterfeld Siemsgluess International. GmbH (Germany); Calliope (France); Chemia (Italy); Cheminova (Denmark); China Chemical (China); Dow AgroSciences (USA); DuPont Crop Protection (USA); Ehrenstorfer, Dr. (Germany); Excel Industries (India); Hindustan Insecticides (India); Jiangmen Pesticide Factory (China); Kenso Corp. (Malayasia); Ki-Hara Chemicals Ltd. (UK); Hokko Chemical Industry (Japan); Indiclay (India); Jingma Chemicals Ltd. (China); Makhteshim-Agan Industries (Israel); Micro Flo (USA); Milenia Agro Ciencias (Brazil); Monsanto (USA); Nufarm (Australia); Pyosa Agroquimicos (Mexico); Quantum Chemicals (USA); Rotam Agrochemical (Hong Kong); Sabero Organics (India); Sanonda (Australia); Scotts Company (USA); Shenzhen Guomeng Industry Co., Ltd. (China); Sinon (Taiwan); Syngenta (Switzerland); United Phosphorus (India); Veterinary & Agricultural Products Manufacturing Co., Ltd. (VAPCO) (Jordan); Whyte Agrochemicals (UK); Zago Asia Ltd. (Singapore)
*Chemical Class:* Phosphinic acid
*EPA/OPP PC Code:* 417300; 471300 use code No. 417300
*California DPR Chemical Code:* 2997
*ICSC Number:* 0160
*RTECS Number:* MC1075000
*Uses:* A General Use Pesticide (GUP). Glyphosate is a broad-spectrum, nonselective systemic herbicide used for control of annual and perennial plants including grasses, sedges, broad-leaved weeds, fruit orchards, vineyards, and woody plants. Frequently used on plantation crops such as tea, bananas, coffee, coconut, cocoa, mangoes and palms. As a pre-crop emergence control, it is used on vegetables, beet, okra, soya beans, lucerne, figs, kiwi, olives cereals, cotton, etc.) It can be used on non-cropland, aquatic weed control, and pre-harvest desiccation of cotton, cereals, peas and beans. It controls suckers on fruit trees. Glyphosate itself is an acid, but it is commonly used in salt form, most commonly the isopropylamine salt. It may also be available in acidic or trimethylsulfonium salt forms. It is generally distributed as water-soluble concentrates and powders.
*Human toxicity (long-term)*[77]: Very low–700.00 ppb, MCL (Maximum Contaminant Level)
*Fish toxicity (threshold)*[77]: Very low–26000.00 ppb, MATC (Maximum Acceptable Toxicant Concentration)

*U.S. Maximum Allowable Residue Levels for Glyphosate (40 CFR 180.364):*

| CROP | ppm |
|---|---|
| Acerola | 0.2 |
| Alfalfa, Forage | 175 |
| Alfalfa, Hay | 400 |
| Almond, Hulls | 25 |
| Aloe Vera | 0.5 |
| AmmHgella | 0.2 |
| Animal Feed, Nongrass, Group 18 | 400 |
| Artichoke, Jerusalem | 0.2 |
| Asparagus | 0.5 |
| Atemoya | 0.2 |
| Avocado | 0.2 |
| Bamboo, Shoots | 0.2 |
| Banana | 0.2 |
| Barley, Bran | 30 |
| Barley, Grain | 20 |
| Beet, Sugar, Dried Pulp | 25 |
| Beet, Sugar, Roots | 10 |
| Beet, Sugar, Tops | 10 |
| Berry Group 13 | 0.2 |
| Betelnut | 1 |
| Biriba | 0.2 |
| Borage, Seed | 0.1 |
| Breadfruit | 0.2 |
| Cacao Bean | 0.2 |
| Cactus, Fruit | 0.5 |
| Cactus, Pads | 0.5 |
| Canistel | 0.2 |
| Canola, Meal | 15 |
| Canola, Seed | 10 |
| Cattle, Liver | 0.5 |
| Chaya, Leaves | 1 |
| Cherimoya | 0.2 |
| Citrus, Dried Pulp | 1.5 |
| Coconut | 0.1 |
| Coffee, Bean | 1 |
| Corn, Field, Forage | 6 |
| Corn, Field, Grain | 1 |
| Cotton, Gin Byproducts | 100 |
| Cotton, Undelinted Seed | 15 |
| Crambe, Seed | 0.1 |
| Cranberry | 0.2 |
| Custard Apple | 0.2 |
| Date, Dried Fruit | 0.2 |
| Dokudami, Leaves | 2 |
| Durian | 0.2 |
| Egg | 0.05 |
| Epazote, Leaves | 1.3 |
| Feijoa | 0.2 |
| Fig | 0.2 |
| Fish | 0.25 |
| Flax, Meal | 8 |
| Flax, Seed | 4 |
| Fruit, Citrus, Group 10 | 0.5 |
| Fruit, Pome, Group 11 | 0.2 |
| Fruit, Stone, Group 12 | 0.2 |
| Galangal, Roots | 0.2 |
| Ginger, White, Flower | 0.2 |
| Goat, Liver | 0.5 |
| Gourd, Buffalo, Seed | 0.1 |
| Governor`s Plum | 0.2 |
| Gow Kee, Leaves | 0.2 |
| Grain, Aspirated Fractions | 100 |
| Grain, Cereal, Except Barley, Field Corn, Grain Sorghum, Oats and Wheat | 0.1 |
| Grain, Cereal, Forage, Fodder and Straw, Group 16 | 100 |
| Grain, Crops, Except Wheat, Oats, Grain Sorghum, Barley | 0.1 |
| Grape | 0.2 |
| Grass, Forage, Fodder and Hay, Group 17 | 300 |
| Guava | 0.2 |
| Herb and Spice Group 19 | 0.2 |
| Herb and Spice Group 19 | 7 |
| Hog, Kidney | 4 |
| Hog, Liver | 0.5 |
| Hop, Dried Cones | 7 |
| Horse, Liver | 0.5 |
| Ilama | 0.2 |
| Imbe | 0.2 |
| Imbu | 0.2 |
| Jaboticaba | 0.2 |
| Jackfruit | 0.2 |
| Jojoba, Seed | 0.1 |
| Juneberry | 0.2 |
| Kava, Roots | 0.2 |
| Kenaf, Forage | 200 |
| Kiwifruit | 0.2 |
| Lesquerella, Seed | 0.1 |
| Leucaena, Forage | 200 |
| Lingonberry | 0.2 |
| Longan | 0.2 |
| Lychee | 0.2 |
| Mamey Apple | 0.2 |
| Mango | 0.2 |
| Mangosteen | 0.2 |
| Marmaladebox | 0.2 |
| Meadowfoam, Seed | 0.1 |
| Mioga, Flower | 0.2 |
| Mustard, Seed | 0.1 |
| Nut, Pine | 1 |
| Nut, Tree, Group 14 | 1 |
| Oat, Grain | 20 |
| Okra | 0.5 |
| Olive | 0.2 |
| Oregano, Mexican, Leaves | 2 |
| Palm Heart, Leaves | 0.2 |
| Palm, Oil | 0.1 |

| | | | |
|---|---|---|---|
| Papaya | 0.2 | Vegetable, Brassica, Leafy, Group 5 | 0.2 |
| Papaya, Mountain | 0.2 | Vegetable, Bulb, Group 3 | 0.2 |
| Passionfruit | 0.2 | Vegetable, Cucurbit, Group 9 | 0.5 |
| Pawpaw | 0.2 | Vegetable, Foliage of Legume, Except Soybean, Subgroup 7a | 0.2 |
| Peanut | 0.1 | Vegetable, Fruiting, Group 8 | 0.1 |
| Peanut, Forage | 0.5 | Vegetable, Legume, Group 6 | 5 |
| Peanut, Hay | 0.5 | Vegetable, Root, Except Sugarbeet, Subgroup 1b | 0.2 |
| Pepper Leaf, Fresh Leaves | 0.2 | Wasabi, Roots | 0.2 |
| Peppermint, Tops | 200 | Water Spinach, Tops | 0.2 |
| Perilla, Tops | 1.8 | Watercress, Upland | 0.2 |
| Persimmon | 0.2 | Wax Jambu | 0.2 |
| Pineapple | 0.1 | Wheat, Grain | 5 |
| Pistachio | 1 | Wheat, Milled Fractions, Except Flour | 20 |
| Pomegranate | 0.2 | Yacon, Tuber | 0.2 |
| Poultry, Meat | 0.1 | | |
| Poultry, Meat Byproducts | 1 | | |
| Pulasan | 0.2 | | |
| Quinoa, Grain | 5 | | |
| Rambutan | 0.2 | | |
| Rapeseed, Meal | 15 | | |
| Rapeseed, Seed | 10 | | |
| Rose Apple | 0.2 | | |
| Safflower, Seed | 0.1 | | |
| Salal | 0.2 | | |
| Sapodilla | 0.2 | | |
| Sapote, Black | 0.2 | | |
| Sapote, Mamey | 0.2 | | |
| Sapote, White | 0.2 | | |
| Sesame, Seed | 0.1 | | |
| Sheep, Liver | 0.5 | | |
| Shellfish | 3 | | |
| Sorghum, Grain, Grain | 15 | | |
| Soursop | 0.2 | | |
| Soybean, Forage | 100 | | |
| Soybean, Hay | 200 | | |
| Soybean, Hulls | 100 | | |
| Soybean, Seed | 20 | | |
| Spanish Lime | 0.2 | | |
| Spearmint, Tops | 200 | | |
| Star Apple | 0.2 | | |
| Starfruit | 0.2 | | |
| Stevia, Dried Leaves | 1 | | |
| Strawberry | 0.2 | | |
| Sugar Apple | 0.2 | | |
| Sugarcane, Cane | 2 | | |
| Sugarcane, Molasses | 30 | | |
| Sunflower, Seed | 0.1 | | |
| Surinam Cherry | 0.2 | | |
| Tamarind | 0.2 | | |
| Tea, Dried | 1 | | |
| Tea, Instant | 7 | | |
| Teff, Grain | 5 | | |
| Ti, Leaves | 0.2 | | |
| Ti, Roots | 0.2 | | |
| Ugli Fruit | 0.5 | | |

***Carcinogen/Hazard Classifications***
**U.S. EPA Carcinogens:** Group E, Unlikely to be Carcinogen
**Label Signal Word:** WARNING–EPA Toxicity Class II
**WHO Acute Hazard:** Class U, unlikely to be hazardous
***Regulatory Authority:***
- Air Pollutant Standard Set (former USSR)[35, 43]
- AB 1803-Well Monitoring Chemical (CAL)
- MCL (Maximum Contaminants Levels) list of contaminants (CAL)
- Actively registered pesticide in California.

***Description:*** Glyphosate, is a colorless crystalline powder. Often used as a liquid in a carrier solvent which may change physical and toxicological properties. Soluble in water; solubility = $10^4$ ppm @ 25°C. Molecular weight = 169.08. Melting/Freezing point = 230°C. Log $K_{ow}$ = Negative. Unlikely to bioaccumulate in marine organisms.

***Incompatibilities:*** Solutions are corrosive to iron, unlined steel, and galvanized steel forming a highly combustible or explosive gas mixture. Do not store glyphosate in containers made from these materials.

***Permissible Exposure Limits in Air:*** No OELs have been established in the US for this chemical. The former USSR[35, 43] has set a ceiling value in workplace air of 1.5 mg/m³.

***Permissible Concentration in Water:*** The U.S. EPA has developed data on glyphosate including a no-observed-adverse effects level (NOAEL) of 10 mg/kg/day. This corresponds to a drinking water equivalent level of 3.5 mg/L from which a lifetime health advisory of 0.7 mg/L was derived. California[61] has set a guideline of 0.5 mg/L for drinking water.

***Determination in Water:*** Analysis of glyphosate is by a high-performance liquid chromatographic (HPLC) method.

***Routes of Entry:*** Inhalation, ingestion, through the skin.

***Harmful Effects and Symptoms***
***Short Term Exposure:*** Irritates the eyes, skin, and respiratory tract. Exposure to high levels can cause nausea, vomiting, diarrhea, decreased blood pressure, and convulsions. High exposures can cause arrhythmia and

possible death. The acute $LD_{50}$-oral for rats is 5600 mg/kg (insignificantly toxic).

*Long Term Exposure:* May cause liver and kidney damage. It does not seem to exhibit reproductive effects, mutagenicity or carcinogenicity in animal studies.

*Points of Attack:* Respiratory system, lungs, central nervous system, cardiovascular system, skin, eyes, plasma and red blood cell cholinesterase, liver, kidney, heart.

*Medical Surveillance:* If symptoms develop or overexposure is suspected, the following may be useful: Liver and kidney function tests. Exam of the nervous system.

*First Aid:* If this chemical gets into the eyes, remove any contact lenses at once and irrigate immediately for at least 15 minutes, occasionally lifting upper and lower lids. If this chemical contacts the skin, remove contaminated clothing and wash immediately with soap and water. When this chemical has been swallowed, get medical attention. Give large quantities of water and induce vomiting. Do not make an unconscious person vomit. If this chemical has been inhaled, remove from exposure and transfer promptly to a medical facility.

*References:*

- EXTOXNET, Extension Toxicology Network, "Pesticide Information Profile, Glyphosate," Oregon State University, Corvallis, OR (June 1996). http://extoxnet.orst.edu/pips/glyphosa.htm
- EPA, Office of Pesticide Programs, Pesticide Residue Limits, "Glyphosate," 40 CFR 180.364, http://www.epa.gov/pesticides/food/viewtols.htm
- U.S. Environmental Protection Agency, "Health Advisory: Glyphosate," Washington, DC, Office of Drinking Water (August 1987).
- California Environmental Protection Agency "Chemical List of Lists," Sacramento CA (February 1997).
- New Jersey Department of Health and Senior Services, "Hazardous Substance Fact Sheet, Glyphosate," Trenton NJ (June 1999). http://www.state.nj.us/health/eoh/rtkweb/3139.pdf

# H

## Halosulfuron-methyl

*Use Type:* Herbicide
*CAS Number:* 100784-20-1
*Formula:* $C_{13}H_{15}ClN_6O_7S$
*Synonyms:* 3-Chloro-5-[((((4,6-dimethoxy-2-pyrimidinyl)amino)carbonyl)amino)sulfonyl]-1-methyl-1*H*-pyrazole-4-carboxylic acid, methyl ester; Halosulfuron; Methyl 3-chloro-5-(4,6-dimethoxypyrimidin-2-yl carbamoylsulfamoyl)-1-methyl pyrazole-4-carboxylate; 1*H*-Pyrazole-4-carboxylic acid, 3-chloro-5-[((((4,6-dimethoxy-2-pyrimidinyl)amino)carbonyl)amino)sulfonyl]-1-methyl-, methyl ester (9CI)
*Trade Names:* ACHIVA®, Nissan Chemical Industries (Japan); BATTALION®, Monsanto (USA); F2636, FMC Agricultural Products (USA); MANAGE®, Monsanto (USA), canceled 7/9/1997; MON®, Monsanto (USA); NC-319®, Nissan Chemical Industries (Japan); PERMIT®, Monsanto (USA); SANDEA®, Gowan (USA); SEMPRA CA®, Monsanto (USA)
*Producers:* Dow AgroSciences (USA); FMC Agricultural Products (USA); Gowan (USA); Monsanto (USA); Nissan Chemical Industries (Japan)
*Chemical Class:* Sulfonylurea
*EPA/OPP PC Code:* 128721
*California DPR Chemical Code:* 3919
*Human toxicity (long-term)*[77]*:* Very low–700.00 ppb, Health Advisory
*Fish toxicity (threshold)*[77]*:* Very low–3584.60429 ppb, MATC (Maximum Acceptable Toxicant Concentration)
*U.S. Maximum Allowable Residue Levels for Halosulfuron (40 CFR 180.479):*

| CROP | ppm |
|---|---|
| Almond, hulls | 0.2 |
| Asparagus | 0.8 |
| Asparagus | 2 |
| Bean, dry, seed | 0.05 |
| Bean, snap, succulent | 0.05 |
| Cattle, mbyp | 0.1 |
| Corn, field, forage | 0.2 |
| Corn, field, grain | 0.05 |
| Corn, field, stover | 0.8 |
| Corn, pop, grain | 0.05 |
| Corn, pop, stover | 0.8 |
| Corn, sweet, forage | 0.2 |
| Corn, sweet, kernel plus cob with husks removed | 0.05 |
| Corn, sweet, stover | 0.8 |
| Cotton, gin byproducts | 0.05 |
| Cotton, undelinted seed | 0.05 |
| Goat, mbyp | 0.1 |
| Hog, mbyp | 0.1 |
| Horse, mbyp | 0.1 |
| Melon subgroup 9a | 0.1 |
| Nut, tree, group 14 | 0.05 |
| Pistachio | 0.05 |
| Rice, grain | 0.05 |
| Rice, straw | 0.2 |
| Sheep, mbyp | 0.1 |
| Sorghum, grain, forage | 0.05 |
| Sorghum, grain, grain | 0.05 |
| Sorghum, grain, stover | 0.1 |
| Squash/cucumber, subgroup 9b | 0.5 |
| Sugarcane, cane | 0.05 |
| Tomato | 0.05 |
| Vegetable, fruiting, group 8 | 0.05 |

*Carcinogen/Hazard Classifications*
**U.S. EPA Carcinogens:** Not likely a carcinogen
**Label Signal Word:** WARNING or CAUTION
*Description:* Molecular weight = 434.8105.
*Incompatibilities:* May react violently with strong oxidizers, bromine, 90% hydrogen peroxide, phosphorus trichloride, silver powders or dust. Incompatible with silver compounds. Mixture with some silver compounds forms explosive salts of silver oxalate.
*Permissible Concentration in Water:* No criteria set. Runoff from spills or fire control may cause water pollution.
*Harmful Effects and Symptoms*
*Short Term Exposure:* Contact with eyes or skin may cause irritation or injury. Inhalation should be avoided; use NIOSH-approved air purifying respirators for pesticides. May be harmful if swallowed.
*First Aid:* If this chemical gets into the eyes, remove any contact lenses at once and irrigate immediately for at least 15 minutes, occasionally lifting upper and lower lids. Seek medical attention immediately. If this chemical contacts the skin, remove contaminated clothing and wash immediately with soap and water. Seek medical attention immediately. If this chemical has been inhaled, remove from exposure, begin rescue breathing (using universal precautions) if breathing has stopped, and CPR if heart action has stopped. Transfer promptly to a medical facility. When this chemical has been swallowed, get medical attention. Give large quantities of water and induce vomiting. Do not make an unconscious person vomit. Do not induce vomiting when formulations containing petroleum solvents are ingested.
*References:*
- U.S. Environmental Protection Agency, Office of Pesticide Programs, Pesticide Residue Limits,

"Halosulfuron," 40 CFR 180.479.

http://www.epa.gov/pesticides/food/viewtols.htm

# Haloxyfop-methyl

*Use Type:* Herbicide
*CAS Number:* 69806-40-2
*Formula:* $C_{16}H_{13}ClF_3NO_4$
*Synonyms:* 2-[4-((3-Chloro-5-trifluoromethyl-2-pyridinyl)oxy)phenoxyl]propanoic acid, methyl ester; Methyl-2-[4-((3-chloro-5-(trifluoromethyl)-2-pyridinyl)oxy)phenoxy]propanoate; Propanoic acid, 2-[4-((3-chloro-5-(trifluoromethyl)-2-pyridinyl)oxy)phenoxy]-, methyl ester
*Trade Names:* BRN® 1509615; DOWCO® 453-ME (methyl ester), Dow AgroSciences (USA); DOWCO® 543 EE (haloxyfopethoxyethyl), Dow AgroSciences (USA); GALLANT®; VERDICT®; ZELLEK
*Producers:* Dow AgroSciences (USA)
*Chemical Class:* Chlorophenoxy; aryloxyphenoxy propionic acid
*EPA/OPP PC Code:* 125201
*Uses:* The common name haloxyfop is also used for haloxyfop-methyl and haloxyfop-ethoxyethyl. Both are selective pre-emergence and post-emergence herbicides used to control annual and perennial grasses. They are used on sugar beets, potatoes, oilseed, leaf vegetables, onions, strawberries, sunflowers and other crops.
*Human toxicity (long-term)[77]:* Extra high–0.04736 ppb, CHCL (Chronic Human Carcinogen Level)
*Fish toxicity (threshold)[77]:* Intermediate–43.42721 ppb, MATC (Maximum Acceptable Toxicant Concentration)
*Carcinogen/Hazard Classifications*
U.S. EPA Carcinogens: Group B2, probable carcinogen
*Regulatory Authority:*
- AB 2588-Air Toxics "Hot Spots" Chemicals (CAL) as chlorophenoxy pesticides
- The "Director's List" (CAL/OSHA) as chlorophenoxy pesticides

*Description:* Haloxyfos is a white crystal with an offensive odor. Haloxyfos methyl is an amber to straw yellow sold; it has a mild aromatic odor. Haloxyfos-ethoxyethyl is a colorless crystal which hydrolyzes to haloxyfos under acidic and alkaline conditions. Vapor pressure = $6.5 \times 10^{-7}$ mmHg @ 20°C.
*Permissible Concentration in Water:* No criteria set. Runoff from spills or fire control may cause water pollution.
*Routes of Entry:* Inhalation, ingestion, absorbed through the skin.
*Harmful Effects and Symptoms*
*Short Term Exposure:* Poisonous; may be fatal if inhaled, swallowed, or absorbed through skin. Severely irritates eyes, skin and respiratory tract, with burning sensation, pain, redness and swelling. Metabolic stimulant. If inhaled, causes coughing, dilated pupils, headache, profuse persperation, intense thirst, extreme fatigue, rapid pulse, high fever, clammy, flushed skin, rapid breathing, nausea, vomiting, cyanosis (bluish tint to skin and lips), anxiety and confusion, convulsions, risk of lung edema. If swallowed, face and lips turn bluish. Liver injury with associated jaundice, kidney failure, and cardiac arrhythmias are commonly noted. Nerve damage, which may be delayed, may include swelling of legs and feet, muscle twitch and stupor. Severe exposure can cause death from heart failure. Dust or liquid left in contact with the skin for several hours may be absorbed. This may result in severe delayed symptoms as listed above. These symptoms may last for months or years.
*Long Term Exposure:* Workers exposed to chlorophenoxy compounds such as 2,4-D (in the manufacturing process) over a five to ten year period at levels above 10 mg/m³ complained of weakness, rapid fatigue, headache and vertigo. Liver damage, low blood pressure and slowed heartbeat were also found. Based on animal tests, may affects human reproduction
*Points of Attack:* Eyes, skin, respiratory system, central nervous system, cardiovascular system, liver, kidney
*Medical Surveillance:* If symptoms develop or overexposure is suspected, the following may be useful: Liver and kidney function tests. Exam of the nervous system.
*First Aid:* If this chemical gets into the eyes, remove any contact lenses at once and irrigate immediately for at least 15 minutes, occasionally lifting upper and lower lids. Seek medical attention immediately. If this chemical contacts the skin, remove contaminated clothing and wash immediately with soap and water. Seek medical attention immediately. If this chemical has been inhaled, remove from exposure, begin rescue breathing (using universal precautions) if breathing has stopped, and CPR if heart action has stopped. Transfer promptly to a medical facility. When this chemical has been swallowed, get medical attention. *Do not induce vomiting when formulations containing petroleum solvents are ingested.* Otherwise, give large quantities of water and induce vomiting. Do not make an unconscious person vomit.
*References:*
- EXTOXNET, Extension Toxicology Network, "Pesticide Information Profile," Oregon State University, Corvallis, OR (September 1994). http://extoxnet.orst.edu/pips/haloxyfo.htm
- California Environmental Protection Agency "Chemical List of Lists," Sacramento CA (February 1997)

# Heptachlor

*Use Type:* Insecticide
*CAS Number:* 76-44-8
*Formula:* $C_{10}H_5Cl_7$

# Heptachlor

*Alert:* Registration of heptachlor-containing pesticides was canceled in 1988 by the U.S. EPA with the exception of its use for termite control outside of dwellings by in-ground (subsurface) insertion. Infants have been exposed to heptachlor and heptachlor epoxide through mothers' milk, cows' milk, and commercially prepared baby foods. It appears that infants raised on mothers' milk run a greater risk of ingesting heptachlor epoxide than if they were fedcows' milk and/or commercially prepared baby food. Persons living and working in or near heptachlor treated areas have a particularly high inhalation exposure potential. Human toxicity (long-term): Very high.

*Synonyms:* 3-Chlorochlordene; E 3314; ENT 15,152; Eptacloro (Italian); 1,4,5,6,7,8,8-Eptacloro-3a,4,7,7a-tetraidro-4,7-*endo*-metano-indene (Italian); GPKh; HEPTA; 3,4,5,6,7,8,8-Heptachlorodicyclopentadiene; 3,4,5,6,7,8,8a-Heptachlorodicyclopentadiene; Hepachloor-3a,4,7,7a-tetrahydro-4,7-*endo*-methano-indeen (Dutch); Heptachlore (French); Heptachlorane; 1,4,5,6,7,8,8-Heptachloro-3a,4,7,7a-tetrahydro-4,7-methano-1H-indene; 1,4,5,6,7,10,10-Heptachloro-4,7,8,9-tetrahydro-4,7-endomethyleneindene; 1,4,5,6,7,8,8a-Heptachloro-3a,4,7,7a-tetrahydro-4,7-methanoindene; 1,4,5,6,7,8,8-Heptachloro-3a,4,7,7a-tetrahydro-4,7-methanoindene; 1(3a),4,5,6,7,8,8-Heptachloro-3a(1),4,7,7a-tetrahydro-4,7-methanoindene; 1,4,5,6,7,8,8-Heptachloro-3a,4,7,7a-tetrahydro-4,7-methanol-1H-indene; 1,4,5,6,7,8,8-Heptachloro-3a,4,7,7a-tetrahydro-4,7-methelene Indene; 1,4,5,6,7,8,8-Heptachlor-3a,4,7,7a-tetrahydro-4,7-*endo*-methano Inden (German); Heptacloro (Spanish); 4,7-Methanoindene, 1,4,5,6,7,8,8-heptachloro-3a,4,7,7a-tetrahydro-; NCI-C00180

*Trade Names:* AAHEPTA®; AGROCERES®; ARBINEX 30TN®; BIARBINEX®; CUPINCIDA®; DRINOX®; E 3314®; FENNOTOX®; HEPTAGRAN®; HEPTAMUL®; HEPTOX®; INDENE®; RHODIACHLOR®; TERMIDE®; VELSICOL 104.®, Velsicol Chemical Corporation (USA), canceled 9/28/1987

*Producers:* Ehrenstorfer, Dr. (Germany); Sankei Chemical (Japan); Velsicol Chemical Corporation (USA)

*Chemical Class:* Orhanochlorine

*EPA/OPP PC Code:* 044801

*California DPR Chemical Code:* 317

*ICSC Number:* 0743

*RTECS Number:* PC0700000

*EEC Number:* 602-046-00-2

*Uses:* The only commercial use still permitted is for fire ant control in power transformers. Heptachlor is still available outside the U.S. Heptachlor is an organochlorine cyclodiene insecticide, first isolated from technical chlordane in 1946. During the 1960s and 1970s, it was used primarily by farmers to kill termites, ants, and soil insects in seed grains and on crops, as well as by exterminators and home owners to kill termites. Before heptachlor was banned, formulations available included dusts, wettable powders, emulsifiable concentrates, and oil solutions. It acts as a nonsystemic stomach and contact insecticide. An important metabolite of heptachlor is heptachlor epoxide, which is an oxidation product formed from heptachlor by many plant and animal species.

*Human toxicity (long-term)*[77]*:* Extra high–0.40 ppb, MCL (Maximum Contaminant Level)

*Fish toxicity (threshold)*[77]*:* Extra high–0.05276 ppb, MATC (Maximum Acceptable Toxicant Concentration)

*U.S. Maximum Allowable Levels for*

The Food and Drug Administration (FDA) limits the amount of heptachlor and heptachlor epoxide on raw food crops and on edible seafood to from 0-10 parts per billion (ppb), depending on the type of food product. The limit on edible seafood is 300 ppb, and for the fat of food-producing animals is 200 ppb.

*Carcinogen/Hazard Classifications*

**U.S. EPA Carcinogens:** Group B2, probable carcinogen
**California Prop. 65:** Carcinogen; Developmental toxin
**IARC:** Group 2B, possible carcinogen
**WHO Acute Hazard:** Class II, moderately hazardous
**Endocrine Disruptor:** Probable ED

*Regulatory Authority:*
- Air Pollutant Standard Set (ACGIH)[1] (DFG)[3] (HSE)[33] (OSHA)[58] (Argentina)[35] (former USSR)[43] (Several States)[60]
- Banned or Severely Restricted (many countries) (UN)[13]
- Carcinogen (Animal Positive) (IARC) (NCI)[9]
- AB 1803-Well Monitoring Chemical (CAL)
- MCL (Maximum Contaminants Levels) list of contaminants (CAL)
- AB 2588-Air Toxics "Hot Spots" Chemicals (CAL)
- CAL Air Resources Board/AB 1807 Toxic Air Contaminants
- Proposition 65 chemical (CAL)
- Permissible Exposure Limits for Chemical Contaminants (CAL/OSHA)
- The "Director's List" (CAL/OSHA)
- Clean Air Act: Hazardous Air Pollutants (Title I, Part A, Section 112)
- Clean Water Act: Section 311 Hazardous Substances/RQ 40CFR117.3 (same as CERCLA, see below); 40CFR423, Appendix A, Priority Pollutants; Section 313 Water Priority Chemicals (57FR41331, 9/9/92); 40CFR401.15 Section 307 Toxic Pollutants
- EPA Hazardous Waste Number (RCRA No.): P059; D031
- RCRA, 40CFR261, Appendix 8 Hazardous Constituents
- RCRA Toxicity Characteristic (Section 261.24), Maximum Concentration of Contaminants, regulatory level, 0.008 mg/L
- RCRA 40CFR268.48; 61FR15654, Universal Treatment Standards: Wastewater (mg/L), 0.0012; Nonwastewater (mg/kg), 0.066

# Heptachlor

- RCRA 40CFR264, Appendix 9; TSD Facilities Ground Water Monitoring List. Suggested test method(s) (PQL ug/L): 8080(0.05); 8270(10)
- Safe Drinking Water Act: MCL, 0.0004 mg/L; MCLG, zero.
- Superfund/EPCRA 40CFR302.4 RQ: CERCLA, 1 lb (0.454 kg)
- EPCRA Section 313 Form R *de minimus* concentration reporting level: 1.0%
- U.S. DOT Regulated Marine Pollutant (49CFR172.101, Appendix B), severe pollutant
- Canada, WHMIS, Ingredients Disclosure List

***U.S. Maximum Allowable Residue Levels for heptachlor:***

| CROP | ppm |
|---|---|
| Raw food crops and on edible seafood | 0–10 |
| Edible seafood | 300 |
| Fat of food-producing animals | 200 |

***Description:*** Heptachlor is a white, crystalline solid. Camphor-like odor. Insoluble in water. Molecular = weight 373.34. Boiling point = 135–145°C. Melting/Freezing point = 95–96°C (pure); 46–74°C (technical product). Vapor pressure = $2.9 \times 10^{-4}$ mmHg @ 25. Hazard Identification (based on NFPA-704 M Rating System): Health 3, Flammability 0, Reactivity 0. Log $K_{ow}$ = 5.41. Values at or above 3.0 are likely to bioaccumulate in marine organisms.

***Incompatibilities:*** Reacts with strong oxidizers. Attacks metal. Forms hydrogen chloride gas with iron and rust above 74°C.

***Permissible Exposure Limits in Air:*** The OSHA[2] PEL is 0.5 mg/m³ TWA, with the notation "Skin" indicating the possibility of cutaneous absorption. NIOSH[2] and ACGIH[1] recommend the same airborne limit as OSHA. This same TWA has been set by the Argentine[35] by Germany[3] and the U.K.[33], but with different STEL's in each case: 1.5 mg/m³ in the Argentine, 2.0 in the U.K. and 5.0 in Germany. The NIOSH[2] IDLH level = [Ca]35 mg/m³. The former USSR has set[35, 43] a much lower limit of 0.01 mg/m³ in workplace air and values in ambientair of residential areas of 0.001 mg/m³ on a once daily basis and 0.0002 mg/m³ on a daily average basis. A number of states have set guidelines of standards for heptachlor in ambient air[60] ranging from 0.0068 µg/m³ (Massachusetts) to 0.18 µg/m³ (Pennsylvania) to 1.19 µg/m³ (Kansas) to 1.7 µg/m³ (New York) to 2.5 µg/m³ (Connecticut and South Carolina) to 5.0 µg/m³ (Florida and North Dakota) to 8.0 µg/m³ (Virginia) to 12.0 µg/m³ (Nevada).

***Determination in Air:*** Collection by Chromosorb tube-102; Toluene; Gas chromatography/Electrochemical detection; NIOSH II(5), Method #S287[18].

***Permissible Concentration in Water:*** EPA recommends a restriction to a maximum of 2.78 ppt (parts per trillion) of heptachlor of drinking water or seafood [2.78 ppt (parts per trillion)] that you eat each day. For longer exposures, a child should not drink water with greater than 5,000 ppt heptachlor or 150 ppt (parts per trillion) heptachlor epoxide. To protect freshwater aquatic life: 0.0038 µg/L as a 24 hour average, never to exceed 0.52 µg/L. To protect saltwater aquatic life: 0.0036 µg/L as a 24 hour average, never to exceed 0.053 µg/L. To protect human health: preferably zero. An additional lifetime cancer risk of 1 in 100, 000 is imposed by a concentration of 2.78 ng/l (0.00278 µg/L).[6]. The USEPA has set health advisories[47] for heptachlor and heptachlor epoxide. The lifetime health advisory is 17.5 µg/L for heptachlor and 0.4 µg/L for heptachlor epoxide. Mexico[35] has set limits of 0.018 mg/L (18 µg/L) for both heptachlor and heptachlor epoxide in drinking water, 0.2 µg/L for heptachlor in coastal watersand 2.0 µg/L for heptachlor in estuaries. The former USSR[35, 43] has set a limit of 50 µg/L of heptachlor in water bodies used for domestic purposes. WHO[35] has set a limit of 0.1 µg/L in drinking water for heptachlor. Several states have set guidelines and standards for heptachlor and heptachlor epoxide in drinking water[61]. Illinois has set a standard of 0.1 µg/L for both heptachlor and heptachlor epoxide. Guidelines have been set for heptachlor ranging from 0.02 µg/L (California) to 0.1 µg/L (Minnesota) to 0.104 µg/L (Kansas) to 0.23 µg/L (Maine) to 0.50 µg/L (Arizona). Guidelines have been set for heptachlor epoxide in drinking water[61] ranging from 0.006 µg/L (Kansas and Minnesota) to 0.10 µg/L (California). The U.S. EPA has recently[62] proposed drinking water maximum contaminant levels for heptachlor at 0.4 µg/L and heptachlor epoxide at 0.2 µg/L.

***Determination in Water:*** Methylene chloride extraction followed by gas chromatography with electron capture or halogen specific detection (EPA Method 608) or gas chromatography plus mass spectrometry (EPA Method 625).

***Routes of Entry:*** Inhalation, skin absorption, ingestion, eye and/or skin contact.

## Harmful Effects and Symptoms

***Short Term Exposure:*** Heptachlor can cause a feeling of anxiety, headache, dizziness, weakness, a sensation of "pins and needles" on the skin, and muscle twitching. Heptachlor has been demonstrated to be highly toxic to aquatic life, to persist for prolonged periods in the environment, to bioconcentrate in organisms at various trophic levels, and to exhibit carcinogenic activity in mice. The principal metabolite of heptachlor, heptachlor epoxide is more acutely toxic than heptachlor. Most of what we know about the health effects of these pesticides comes from studies on mice and rats fed heptachlor and heptachlor epoxide in the food or water. Very high levels for short periods produce serious liver problems. Mice had trouble walking and rats developed tremors. High levels of heptachlor in the feed for several weeks damaged the livers of rats and the livers and adrenal glands of mice.

***Long Term Exposure:*** High or repeated exposure may cause brain damage with personality changes, decreased memory, difficult coordination and concentration. Higher levels can cause tremor, seizures, unconsciousness and death. This substance is possibly carcinogenic to humans. There is limited evidence that heptachlor may damage the developing fetus. May cause liver and kidney damage.

*Points of Attack:* Central nervous system, liver. Cancer site in animals: liver cancer.

*Medical Surveillance:* Consider the points of attack in preplacement and periodic physical examinations. Liver and kidney function tests. Evaluation for brain effects.

*First Aid:* If this chemical gets into the eyes, remove any contact lenses at once and irrigate immediately for at least 15 minutes, occasionally lifting upper and lower lids. Seek medical attention immediately. If this chemical contacts the skin, remove contaminated clothing and wash immediately with soap and water. Seek medical attention immediately. If this chemical has been inhaled, remove from exposure, begin rescue breathing (using universal precautions) if breathing has stopped, and CPR if heart action has stopped. Transfer promptly to a medical facility. When this chemical has been swallowed, get medical attention. Give large quantities of water and induce vomiting. Do not make an unconscious person vomit.

*References:*
- EXTOXNET, Extension Toxicology Network, "Pesticide Information Profile, Heptachlor," Oregon State University, Corvallis, OR (June 1996). http://extoxnet.orst.edu/pips/heptachl.htm
- U.S. Department of Health and Human Services; Agency for Toxic Substances and Disease Registry, "ToxFAQs for Heptachlor/Heptachlor Epoxide," Atlanta, GA (April 1993). http://www.atsdr.cdc.gov/toxprofiles/tp12.html
- New Jersey Department of Health and Senior Services, "Hazardous Substance Fact Sheet, Heptachlor," Trenton NJ (March, 1998). http://www.state.nj.us/health/eoh/rtkweb/0974.pdf
- U.S. Environmental Protection Agency, Heptachlor: Ambient Water Quality Criteria, Washington, DC (1980).
- U.S. Environmental Protection Agency, Heptachlor, Health and Environmental Effects Profile No. 108, Office of Solid Waste, Washington, DC (April 30, 1980).
- U.S. Environmental Protection Agency, Heptachlor Epoxide, Health and Environmental Effects Profile No. 109, Office of Solid Waste, Washington, DC (April 30, 1980).
- Sax, N.I., Ed., Dangerous Properties of Industrial Materials Report, 1, No. 8, 76-78 (1981) and 6, No. 5, 16-49 (1986).
- California Environmental Protection Agency "Chemical List of Lists," Sacramento CA (February 1997).

# Heptachlor Epoxide

*Use Type:* Insecticide; a breakdown product of heptachlor
*CAS Number:* 1024-57-3
*Formula:* $C_{10}H_5Cl_7O$

*Alert:* Infants have been exposed to heptachlor and heptachlor epoxide through mothers' milk, cows' milk, and commercially prepared baby foods are at risk. It appears that infants raised on mothers' milk run a greater risk of ingesting heptachlor epoxide than if they were fed cows' milk and/or commercially prepared baby food. Persons living and working in or near heptachlor treated areas have a particularly high inhalation exposure potential.

*Synonyms:* ENT 25,584; Epoxyheptachlor; HCE; 1,4,5,6,7,8,8-Heptachloro-2,3-epoxy-2,3,3a,4,7,7a-hexahydro-4,7-methanoindene; 1,4,5,6,7,8,8-Heptachloro-2,3-epoxy-3a,4,7,7a-tetrahydro-4,7-methanoindan; 2,3,5,6,7,7-Heptachloro-1a,1b,5,5a,6,6a-hexahydro-2,5-methano-2h-indeno(1,2-b)oxirene; Heptaclorepoxido (Spanish); 4,7-Methanoindan, 1,4,5,6,7,8,8-heptachloro-2,3-epoxy-3a,4,7,7a-tetrahydro

*Trade Names:* VELSICOL 53 CS 17®, Velsicol Chemical Corporation (USA)

*Producers:* Ehrenstorfer, Dr. (Germany); Sankei Chemical (Japan); Velsicol Chemical Corporation (USA)

*Chemical Class:* Chlorinated hydrocarbon
*EPA/OPP PC Code:* 044801
*California DPR Chemical Code:* 4073
*RTECS Number:* PB9450000

*Uses:* An important metabolite of heptachlor is heptachlor epoxide, which is an oxidation product formed from heptachlor by many plant and animal species. The only commercial use still permitted for heptachlor is for fire ant control in power transformers. Heptachlor is still available outside the U.S. Heptachlor is an organochlorine cyclodiene insecticide, first isolated from technical chlordane in 1946. During the 1960s and 1970s, it was used primarily by farmers to kill termites, ants, and soil insects in seed grains and on crops, as well as by exterminators and home owners to kill termites. Before heptachlor was banned, formulations available included dusts, wettable powders, emulsifiable concentrates, and oil solutions. It acts as a nonsystemic stomach and contact insecticide.

*Carcinogen/Hazard Classifications*

**U.S. EPA Carcinogens:** Group B2, probable carcinogen
**California Prop. 65:** Carcinogen
**IARC:** Group 2B, possible carcinogen as breakdown product of heptachlor
**WHO Acute Hazard:** Class II, moderately hazardous as parent compound heptachlor
**Endocrine Disruptor:** Suspected endocrine disruptor

*Regulatory Authority:*
- Air Pollutant Standard Set (ACGIH)[1] (DFG)[3] (HSE)[33] (OSHA)[58] (Argentina)[35] (former USSR)[43] (Several States)[60]
- Banned or Severely Restricted (Many, Many Countries) (UN)[13]
- Carcinogen (Animal Positive) (IARC) (NCI)[9]
- List of priority pollutants (U.S. EPA)
- AB 1803-Well Monitoring Chemical (CAL)

- MCL (Maximum Contaminants Levels) list of contaminants (CAL)
- AB 2588-Air Toxics "Hot Spots" Chemicals (CAL)
- Proposition 65 chemical (CAL)
- The "Director's List" (CAL/OSHA)
- Clean Water Act: 40CFR423, Appendix A, Priority Pollutants; 40CFR401.15 Section 307 Toxic Pollutants as hexachlorocyclohexane.
- EPA Hazardous Waste Number (RCRA No.): D031
- RCRA, 40CFR261, Appendix 8 Hazardous Constituents, waste number not listed
- RCRA Toxicity Characteristic (Section 261.24), Maximum Concentration of Contaminants, regulatory level, 0.008 mg/L
- RCRA 40CFR268.48; 61FR15654, UniversalTreatment Standards: Wastewater (mg/L), 0.016; Nonwastewater (mg/kg), 0.066
- RCRA 40CFR264, Appendix 9; TSD Facilities Ground Water Monitoring List. Suggested test method(s) (PQL ug/L): 8080[1]; 8270(10)
- Safe Drinking Water Act: MCL, 0.0002 mg/L; MCLG, zero
- Superfund/EPCRA 40CFR302.4 RQ: CERCLA, 1 lb (0.454 kg)

***U.S. Maximum Allowable Residue Levels for heptachlor epoxide:***

| CROP | ppm |
|---|---|
| Raw food crops and on edible seafood | 0-10 |
| Edible seafood | 300 |
| Fat of food-producing animals | 200 |

***Description:*** Heptachlor epoxide is a solid. It is an oxidation product of heptachlor formed by plants and animals, including humans, after exposure to heptachlor. It is also present as a contaminant in heptachlor. Practically insoluble in water; solubility = 0.350 mg/L. Melting/Freezing point = 160–162°C. Molecular weight = 389.30. Hazard Identification (based on NFPA-704 M Rating System): Health 3, Flammability 1, Reactivity 0. Soluble in water. Log $K_{ow}$ = 5.41. Values at or above 3.0 are likely to bioaccumulate in marine organisms.

***Incompatibilities:*** Melted heptachlor with iron and rust.

***Permissible Exposure Limits in Air:*** The ACGIH[1] and OSHA[2] have set a TWA of 0.5 mg/m³ with the notation "Skin" indicating the possibility of cutaneous absorption. This same TWA has been set by the Argentine[35] by Germany (3) and the U.K.[33], but with different STEL's in each case: 1.5 mg/m³ in the Argentine, 2.0 in the U.K. and 5.0 in Germany. The NIOSH[2] IDLH levelis 100 mg/m³. The former USSR has set[35, 43] a much lower limit of 0.01 mg/m³ in workplace air and values in ambientair of residential areas of 0.001 mg/m³ on a once daily basis and 0.0002 mg/m³ on a daily average basis. A number of states have set guidelines of standards for heptachlor in ambient air[60] ranging from 0.0068 µg/m³ (Massachusetts) to 0.18 µg/m³ (Pennsylvania) to 1.19 µg/m³ (Kansas) to 1.7 µg/m³ (New York) to 2.5 µg/m³ (Connecticut and South Carolina) to 5.0 µg/rn³ (Florida and North Dakota) to 8.0 µg/m³ (Virginia) to 12.0 µg/m³ (Nevada).

***Determination in Air:*** Collection by Chromosorb tube-102; Toluene; Gas chromatography/Electrochemical detection; NIOSH II(5), Method #S287[18].

***Permissible Concentration in Water:*** EPA recommends a restriction to a maximum of 2.78 ppt (parts per trillion) of heptachlor epoxide of drinking water or seafood (2.78 ppt) that you eat each day. For longer exposures, a child should not drink water with greater than 5,000 ppt (parts per trillion) heptachlor or 150 ppt (parts per trillion) heptachlor epopxide. To protect freshwater aquatic life: 0.0038 µg/L as a 24 hour average, never to exceed 0.52 µg/L. To protect saltwater aquatic life: 0.0036 µg/L as a 24 hour average, never to exceed 0.053 µg/L. To protect human health: preferably zero. An additional lifetime cancer risk of 1 in 100, 000 is imposed by a concentration of 2.78 ng/l (0.00278 µg/L).[6]. The USEPA has set health advisories[47] for heptachlor and heptachlor epoxide. The lifetime health advisory is 17.5 µg/L for heptachlor and 0.4 µg/L for heptachlor epoxide. Mexico[35] has set limits of 0.018 mg/L (18 µg/L) for both heptachlor and heptachlor epoxide in drinking water, 0.2 µg/L for heptachlor in coastal watersand 2.0 µg/L for heptachlor in estuaries. The former USSR[35, 43] has set a limit of 50 µg/L of heptachlor in water bodies used for domestic purposes. WHO[35] has set a limit of 0.1 µg/L in drinking water for heptachlor. Several states have set guidelines and standards for heptachlor and heptachlor epoxide in drinking water[61]. Illinois has set a standard of 0.1 µg/L for both heptachlor and heptachlor epoxide. Guidelines have been set for heptachlor ranging from 0.02 µg/L (California) to 0.1 µg/L (Minnesota) to 0.104 µg/L (Kansas) to 0.23 µg/L (Maine) to 0.50 µg/L (Arizona). Guidelines have been set for heptachlor epoxide in drinking water[61] ranging from 0.006 µg/L (Kansas and Minnesota) to 0.10 µg/L (California). The U.S. EPA has recently[62] proposed drinking water maximum contaminant levels for heptachlor at 0.4 µg/L and heptachlor epoxide at 0.2 µg/L.

***Routes of Entry:*** Inhalation, skin absorption, ingestion, eye and/or skin contact.

### Harmful Effects and Symptoms

***Short Term Exposure:*** Heptachlor can cause a feeling of anxiety, headache, dizziness, weakness, a sensation of "pins and needles" on the skin, and muscle twitching. Heptachlor has been demonstrated to be highly toxic to aquatic life, to persist for prolonged periods in the environment, to bioconcentrate in organisms at various trophic levels, and to exhibit carcinogenic activity in mice. The principal metabolite of heptachlor, heptachlor epoxide is more acutely toxic than heptachlor. Most of what is known about the health effects of these pesticides comes from studies on mice and rats fed heptachlor and heptachlor epoxide in the food or water. Very high levels for short periods produce serious liver problems. Mice had trouble walking and rats developed tremors. High levels of heptachlor in the feed for several weeks damaged the livers of rats and the livers and adrenal glands of mice.

*Long Term Exposure:* High or repeated exposure may cause brain damage with personality changes, decreased memory, difficult coordination and concentration. Higher levels can cause tremor, seizures, unconsciousness and death. This substance is possibly carcinogenic to humans. There is limited evidence that heptachlor may damage the developing fetus. May cause liver and kidney damage.

*Points of Attack:* Central nervous system, liver. Cancer site in animals: liver cancer.

*Medical Surveillance:* Consider the points of attack in preplacement and periodic physical examinations. Liver and kidney function tests. Evaluation for brain effects.

*First Aid:* Speed in removing material from eyes and skin is of extreme importance. If this chemical gets into the eyes, remove any contact lenses at once and irrigate immediately for at least 15 minutes, occasionally lifting upper and lower lids. Seek medical attention immediately. If this chemical contacts the skin, remove contaminated clothing and wash immediately with soap and water. Shampoo hair promptly if contaminated. Seek medical attention immediately. If this chemical has been inhaled, remove from exposure, begin rescue breathing (using universal precautions) if breathing has stopped, and CPR if heart action has stopped. Transfer promptly to a medical facility. When this chemical has been swallowed, get medical attention. Give large quantities of water and induce vomiting. Do not make an unconscious person vomit.

*References:*
- EXTOXNET, Extension Toxicology Network, "Pesticide Information Profile, Heptachlor," Oregon State University, Corvallis, OR (June 1996). http://extoxnet.orst.edu/pips/heptachl.htm
- U.S. Department of Health and Human Services; Agency for Toxic Substances and Disease Registry, "ToxFAQs for Heptachlor/Heptachlor Epoxide," Atlanta, GA (April 1993). http://www.atsdr.cdc.gov/toxprofiles/tp12.html
- U.S. Environmental Protection Agency, Heptachlor: Ambient Water Quality Criteria, Washington, DC (1980).
- U.S. Environmental Protection Agency, Heptachlor, Health and Environmental Effects Profile No. 108, Office of Solid Waste, Washington, DC (April 30, 1980).
- U.S. Environmental Protection Agency, Heptachlor Epoxide, Health and Environmental Effects Profile No. 109, Office of Solid Waste, Washington, DC (April 30, 1980).
- Sax, N.I., Ed., Dangerous Properties of Industrial Materials Report, 1, No. 8, 76-78 (1981) andfi. No. 5, 16-49 (1986).
- California Environmental Protection Agency "Chemical List of Lists," Sacramento CA (February 1997).

# Hexachlorobenzene

*Use Type:* Insecticide, fungicide and microbiocide
*CAS Number:* 118-74-1
*Formula:* $C_6Cl_6$
*Alert:* Has been banned from use in the United States. See EPA Reregistration Eligibility Decision (RED) statement on Picloram in footnote. Hexachlorobenzene (HCB) is an impurity in picloram. Human toxicity (long-term): High.
*Synonyms:* Benzene, hexachloro-; Esachlorobenzene (Italian); Granox NM; HCB; Hexa C.B; Hexachlorbenzol (German); Hexaclorobenceno (Spanish); Pentachlorophenyl chloride; Perchlorobenzene; Saatbenizfungizid (German)
*Trade Names:* AMATIN®; ANTICARIE®; BUNT-CURE®; BUNT-NO-MORE®; CEKU C.B.®; CO-OP HEXA®; GRANERO®; JULIN'S CARBON CHLORIDE®; NO BUNT®, Rhone-Poulenc Agro France (France), canceled 7/6/1984; RES-Q®, Pbi/Gordon Corporation (USA), canceled 7/6/1984; SANOCID®; SANOCIDE®; SMUT-GO®, canceled 7/6/1984; SNIECIOTOX®; THIHEX®, canceled 7/6/1984; ZAPRAWA NASIENNA SNECIOTOX®
*Producers:* Ehrenstorfer, Dr. (Germany); Pbi/Gordon Corporation (USA); Rhone-Poulenc Agro France (France); Sigma-Aldrich Fine Chemicals (USA)
*Chemical Class:* Chlorinated hydrocarbon
*EPA/OPP PC Code:* 061001
*California DPR Chemical Code:* 321
*ICSC Number:* 0895
*RTECS Number:* DA2975000
*EEC Number:* 602-065-00-6
*EINECS Number:* 204-273-9
*Uses:* Hexachlorobenzene was used widely as a pesticide to protect seeds of onions and sorghum, wheat, and other grains against fungus until 1965. It can be used with or without other seed treatments, fungicides and/or insecticides. It has fumigant action on fungal spores and is available outside the U.S. as a dry seed treatment or slurry seed treatment. This material was also used to make fireworks, ammunition for military uses, synthetic rubber, as a porosity controller in the manufacture of electrodes, as an intermediate in dye manufacture, in organic synthesis, and as a wood preservative. It is formed as a by-product of making other chemicals, in the waste streams of chloralkali and wood-preserving plants, and when burning municipal waste. Currently, there are no commercial uses of hexachlorobenzene in the United States.
*Human toxicity (long-term)*[77]*:* High–1.00 ppb, MCL (Maximum Contaminant Level)
*Fish toxicity (threshold)*[77]*:* Very low–1760.07814 ppb, MATC (Maximum Acceptable Toxicant Concentration)
*Carcinogen/Hazard Classifications*
**U.S. EPA Carcinogens:** Group 2B, probable carcinogen
**U.S. NTP Carcinogen:** Reasonably anticipated carcinogen

### Hexachlorobenzene

**California Prop. 65:** Carcinogen and developmental toxin
**IARC:** Group 2B, possible carcinogen
**Label Signal Word:** CAUTION
**WHO Acute Hazard:** Class 1 a, extremely hazardous
**Endocrine Disruptor:** Confirmed ED
*Regulatory Authority:*

- Air Pollutant Standard Set (ACGIH)[1] (former USSR)[35, 43] (Czechoslovakia)[35] (Several States)[60]
- Banned or Severely Restricted (Many Countries) (UN)[13]
- Carcinogen (Animal Positive) (IARC)[9]
- List of priority pollutants (U.S. EPA)
- AB 1803-Well Monitoring Chemical (CAL)
- MCL (Maximum Contaminants Levels) list of contaminants (CAL)
- AB 2588-Air Toxics "Hot Spots" Chemicals (CAL)
- CAL Air Resources Board/AB 1807 Toxic Air Contaminants
- Proposition 65 chemical (CAL)
- The "Director's List" (CAL/OSHA)
- Clean Air Act: Hazardous Air Pollutants (Title I, Part A, Section 112)
- Clean Water Act: Section 313 Water Priority Chemicals (57FR41331, 9/9/92)
- EPA Hazardous Waste Number (RCRA No.): U127; D032
- RCRA, 40CFR261, Appendix 8 Hazardous Constituents
- RCRA Toxicity Characteristic (Section 261.24), Maximum Concentration of Contaminants, regulatory level, 0.13 mg/L
- RCRA 40CFR268.48; 61FR15654, Universal Treatment Standards: Wastewater (mg/L), 0.055; Nonwastewater (mg/kg), 10
- RCRA 40CFR264, Appendix 9; TSD Facilities Ground Water Monitoring List. Suggested test method(s) (PQL ug/L): 8120(0.05); 8270(10)
- Safe Drinking Water Act: MCL, 0.001 mg/L; MCLG, zero
- Superfund/EPCRA 40CFR302.4 RQ: CERCLA, 10 lb (4.54 kg)
- EPCRA Section 313 Form R *de minimus* concentration reporting level: 0.1%
- Canada, WHMIS, Ingredients Disclosure List

*Description:* Hexachlorobenzene is a white crystalline solid or needles. Very slightly soluble in water; solubility = 6.0 x $10^{-3}$ ppm. Molecular weight = 284.78. Boiling point = 323–326°C. Melting/Freezing point = 228–231°C. Vapor pressure = $1.1 \times 10^{-5}$ mmHg. Flash point = 242°C. Hazard Identification (based on NFPA-704 M Rating System): Health 1, Flammability 1, Reactivity 0. Log $K_{ow}$ = above 5.4. Values at or above 3.0 are likely to bioaccumulate in marine organisms.

*Incompatibilities:* Reacts violently with oxidizers, dimethyl formamide above 65°C.

*Permissible Exposure Limits in Air:* There is not OSHA[2] PEL. ACGIH[1] recommends a TLV of 0.002 TWA [skin]; Animal Carcinogen. The former USSR[35, 43] has set an MAC in workplace air of 0.9 mg/m$^3$. They have also set[35] an MAC in ambient air of residential areas of 0.013 mg/m$^3$. Czechoslovakia[35] has set a TWA in workplace air of 1.0 mg/m$^3$ and an STEL of 2.0 mg/m$^3$. Several states have set guidelines or standards for hexachlorobenzene in ambient air[60] ranging from zero in North Dakota to 0.48 ppb (Pennsylvania) to 0.03 $\mu$g/m$^3$ (New York).

*Determination in Air:* Use NIOSH IV, Method # 1003, Halogenated hydrocarbons[18].

*Permissible Concentration in Water:* The U.S. EPA recommended that drinking water should not contain more thn 0.05 milligrams of hexachlorobenzene per liter of water (0.05 mg/L) in water that children drink and should not contain more than 0.2 mg/L in water that adults drink for longer periods (about 7 years. The U.S. EPA has set a maximum contamination level (MCL) of 0.001 mg/L in drinking water. The former USSR[35, 43] has set an MAC of 0.05 mg/L in water bodies used for domestic purposes. Several states have set guidelines for hexachlorobenzene in drinking water[61] ranging from 0.02 $\mu$g/L (Arizona) to 0.20 $\mu$g/L (Kansas) to 0.21 $\mu$g/L (Minnesota) to 5.4 $\mu$g/L (Maine). The World Health Organization (WHO)[35] has set a limit in drinking water of 0.01 $\mu$g/L.

*Determination in Water:* Methylene chloride extraction followed by concentration and gas chromatography with electron capture detection (EPA Method 612) or gas chromatography plus mass spectrometry (EPA Method 625).

*Routes of Entry:* Inhalation, ingestion, eye and skin contact.

*Harmful Effects and Symptoms*

*Short Term Exposure:* Irritates the eyes, skin, and respiratory tract. *Inhalation*: Coughing, shortness of breath and labored breathing have been reported from large, unmeasured doses or by decomposition to chlorine. *Skin:* Can cause irritation. Exposure to sunlight with (or soon after) exposure can increase effects. Following this reaction, changes in skin pigment and blistering may follow. Red or dark urine may be noticed. High doses may cause redness, pain and serious burns. *Eyes:* May cause irritation. Higher doses may cause redness, pain and blurred vision. *Ingestion*: Headache, dizziness, nausea, vomiting, numbness of hands and arms, apprehension, excitement, tremors, partial paralysis of arms and legs, loss of muscle control, loss of sensory perception, convulsions and coma may result from high doses.

*Long Term Exposure:* May affect the lungs, liver, skin, and nervous system. This substance causes cancer in laboratory animals, and may be carcinogenic to humans. May damage the developing fetus. May cause liver, thyroid, kidney and immune system damage. High, prolonged or repeated exposure may affect the nervous system. Repeated skin exposure can lead to permanent skin changes and increased hair growth. Animal tests show that this substance possibly causes toxic effects upon human reproduction Ingestion of contaminated grain, estimated at doses of 0.05-0.2 grams/day, resulted in porphyria cutanea tarda (PCT) in Turkey which is characterized by red-colored urine, skin sores, change in skin color, arthritis, and problems of the liver, nervous system, and stomach. The following symptoms were also reported: enlarged livers, porphyria in the blood, loss or appetite, weight loss and wasting of skeletal muscles. Severe and long-standing poisoning caused abnormal hair growth, loss of vision, wasting of hands, black discoloration, and skin sores which became ulcerated, healing with pigmented scars. Breast-fed children developed "pink-sore," a condition which was 95% fatal. Toxic effects on blood and active symptoms persisted up to 20 years. Studies in animals show that ingestion of this chemical can damage the liver, thyroid, nervous system, bones, kidneys, blood, and immune and endocrine system.

*Points of Attack:* Liver, skin and thyroid.

*Medical Surveillance:* Liver function tests. Thyroid function tests. Evaluation by a qualified allergist and/or dermatologist. Iron as a dietary supplement could increase liver damage. consult a physician before taking supplements. Guard against sunlight exposure to contaminated skin.

*First Aid:* If this chemical gets into the eyes, remove any contact lenses at once and irrigate immediately for at least 15 minutes, occasionally lifting upper and lower lids. Seek medical attention immediately. If this chemical contacts the skin, remove contaminated clothing and wash immediately with soap and water. Seek medical attention immediately. If this chemical has been inhaled, remove from exposure, begin rescue breathing (using universal precautions) if breathing has stopped, and CPR if heart action has stopped. Transfer promptly to a medical facility. When this chemical has been swallowed, get medical attention. Give large quantities of water and induce vomiting. Do not make an unconscious person vomit.

*Note to Physician:* For ingestions of less than 10 mg/kg body weight occurring less than an hour before treatment, induce vomiting. For ingestions of more than 10 mg/kg body weight occurring less than an hour before treatment, use gastric lavage. For ingestion occurring more than an hour before treatment, use activated charcoal. There is no specific antidote, and supervision for at least 72 hours is recommended.

*References:*
- EXTOXNET, Extension Toxicology Network, "Pesticide Information Profile, Hexachlorobenzene," Oregon State University, Corvallis, OR (June 1996). http://extoxnet.orst.edu/pips/hexachlo.htm
- U.S. Environmental Protection Agency, "Reregistration Eligibility Decision (RED), Picloram," EPA No. 738-R95-019, Washington, DC, (August 1985), http://www.epa.gov/REDs/0096.pdf
- New Jersey Department of Health and Senior Services, "Hazardous Substance Fact Sheet, Hexachlorobenzene," Trenton NJ (November 1988, rev. July 2001). http://www.state.nj.us/health/eoh/rtkweb/0978.pdf
- U.S. Environmental Protection Agency, Chlorinated Benzenes: Ambient Water Quality Criteria. Washington, DC (1980).
- U.S. Environmental Protection Agency, Status Assessment of Toxic Chemicals: Hexachlorobenzene, Report EPA-600/2-79-210g, Cincinnati, Ohio (December 1979).
- U.S. Environmental Protection Agency, Hexachlombemene, Health and Environmental Effects Profile No. 110, Office of Solid Waste, Washington, DC (April 30, 1980).
- Sax, N.I., Ed., Dangerous Properties of Industrial Materials Report, 4, No. 1, 88-92 (1984).
- New York State Department of Health, " Chemical Fact Sheet: Hexachlorobenzene (HCB), " Albany, NY, Bureau of Toxic Substance Assessment (May 1986).
- U.S. Department of Health and Human Services, "ATSDR ToxFAQs, Hexachlorobenzene," (Atlanta, GA, September 1997).
- California Environmental Protection Agency "Chemical List of Lists," Sacramento CA (February 1997).

# Hexachlorocyclohexanes

*Use Type:* Insecticide

*CAS Number:* 608-73-1 (mixed isomers; technical grade); 319-84-6 (α-isomer); 319-85-7 (β-isomer); 58-89-9 (γ-isomer) see Lindane; 319-86-8 (δ-isomer)

*Formula:* $C_6H_6Cl_6$

*Alert:* By voluntary action, the principal domestic producer of technical grade hexachlorocyclohexane (HCH), formerly known as benzene hexachloride (BHC), requested cancellations of its HCH registrations on September 1, 1976. As of July 21, 1978, all registrants of pesticide products containing HCH voluntarily canceled their registrations or switched their former HCH products to lindane formulations. Human toxicity (long-term): Very high (all isomers).

*Synonyms: Benzene hexachlorides*: *Technical grade containing 68.7% α-BHC, 6.5% β-BHC, 13.5% γ-BHC:* BHC; DBH; ENT 8,601; HCCH; Hexa; Hexaklon (in Sweden); Hexhexane; Hexachlorocyclohexane; Hexacloran (In Russia); Hexachlorocyclohexane Isomers; Hexachlorocyclohexane (Mixed Isomers); Hexaclorociclohexano (Spanish)

*α-isomer:* A13-09232; Benzene hexachloride-α-isomer; α-Benzenehexachloride; Benzene-*trans*-hexachloride; α-BHC; Cyclohexane 1,2,3,4,5,6-hexachloro-; Cyclohexane 1,2,3,4,5,6-hexachloro-(1 α,2 α,3.beta.,4 α,5β,6β)-; Cyclohexane 1,2,3,4,5,6-hexachloro-(α,dl); Cyclohexane 1,2,3,4,5,6-hexachloro-α; Cyclohexane 1,2,3,4,5,6-hexachloro-α isomer; Cyclohexane,α-1,2,3,4,5,6-

hexachloro-; ENT 9,232; α-HCH; Hexachlorcyclohexan (German); α-Hexachloran; α-Hexachlorane; 1,2,3,4,5,6-Hexachlorcyclohexane; α-1,2,3,4,5,6-Hexachlorcyclohexane; 1A,2A,3B,4A,5B,6B-Hexachlorocyclohexane; 1-α,2 α,3.β,4 α,5β,6β-Hexachlorocyclohexane; α-Hexachlorocyclohexane; Hexachlorocyclohexane; Hexachlorocyclohexan (German); 1,2,3,4,5,6-Hexaclorociclohexano (Spanish); α-Lindane

*β-isomer:* β-Benzenehexachloride; β-BHC; ENT 9,233; β-HCH; 1-α,2-β,3-α,4-β,5-α,6-β-Hexachlorocyclohexane; β-Hexachlorocyclohexane; β-1,2,3,4,5,6-Hexachlorocyclohexane; β-Lindane

*γ-isomer:* See Lindane

*δ-isomer:* δ-Benzenehexachloride; δ-BHC; ENT 9,234; δ-HCH; HCH-delta; HCH, δ-; 1-α,2-α,3-α,4-β,5-α,6-β-Hexachlorocyclohexane; δ-Hexachlorocyclohexane; δ-1,2,3,4,5,6-Hexachlorocyclohexane; δ-Lindane

*Trade Names:* BENZEX®; COMPOUND 666®; DOL®; DOLMIX®; FORLIN®; GAMAPHEX®; GAMMEXANE®; HEXABLANC®; HEXAFOR®; HEXAMUL®; HEXAPOUDRE®; HEXYCLAN®; HEXYLAN®; HCH HILBEECH®; ISOTOX®; JACUTIN®; KOTOL®; LINDACOL®; LINDAGAM®; LATKA-666®; SILVANO®

*Producers:* BASF Agricultural Products Group (Germany); Hindustan Insecticides (India); Hindustan Organic Chemicals (India); Merck (Germany); Prentiss Inc. (USA); Rallis India (India); Rhone-Poulenc Agro France (France); Syngenta (Switzerland)

*Chemical Class:* Chlorinated hydrocarbon

*EPA/OPP PC Code:* 008901 (mixed isomers)

*California DPR Chemical Code:* Hexachlorocyclohexane, technical grade (608-73-1) Carcinogen

*ICSC Number:* alpha-Hexachlorocyclohexane, 0795

*RTECS Number:* GV3150000 (mixed isomers); GV3500000 (α-isomer); GV4375000 (b-isomer)

*EEC Number:* 602-042-00-0

*EINECS Number:* 206-272-9 (*delta*-isomer)

*Uses:* The major commercial usage of *gamma*-HCH is based upon its insecticidal properties. The γ-isomer has the highest acute toxicity, but the other isomers are not without activity. It is generally advantageous to purify the γ-isomer from the less active isomers. The γ-isomer, lindane, acts on the nervous system of insects, principally at the level of the nerve ganglia. As a result, lindane has been used against insects in a wide range of applications including treatment of animals, buildings, man for ectoparasites, clothes, water for mosquitoes, living plants, fruit, vegetables, forest crops, seeds and soils. Some applications have been abandoned due to excessive residues, e.g., stored foodstuffs. HCH has not been produced in the United States since 1976. However, imported γ-HCH is available in the United States for insecticide use as a dust, powder, liquid, or concentrate. It is also available as a prescription medicine (lotion, cream, or shampoo) to treat and/or control scabies (mites) and head lice in humans.

*Human toxicity (long-term)*[77]: Extra high–0.19444 ppb, CHCL (Chronic Human Carcinogen Level)

*Carcinogen/Hazard Classifications*

**U.S. EPA Carcinogens:** Mixed isomers, Group B2, probable carcinogen

**U.S. NTP Carcinogen:** Mixed isomers and *beta*-, Reasonably anticipated carcinogen

**California Prop. 65:** Hexachlorocyclo-hexane (608-73-1) listed; Carcinogen

**IARC:** Mixed isomers and *beta*-, Group 2B, possible carcinogen

**Label Signal Word:** CAUTION

**WHO Acute Hazard:** Mixed isomers, Class II, moderately hazardous

**Endocrine Disruptor:** Mixed isomers and beta-, Suspected endocrine disruptor

*Regulatory Authority:*

- Air Pollutant Standard Set (DFG)[3] (former USSR)[43]
- Banned or Severely Restricted (Many Countries) (UN)[13]
- AB 2588-Air Toxics "Hot Spots" Chemicals (CAL) as CAS 608-73-1, mixed isomers
- Proposition 65 chemical (CAL) as CAS 608-73-1, mixed isomers
- The "Director's List" (CAL/OSHA) as CAS 608-73-1, mixed isomers

*All isomers:*

- Clean Water Act: 40CFR401.15 Section 307. Toxic Pollutants
- U.S. DOT Regulated Marine Pollutant (49CFR172.101, Appendix B), severe pollutant

*α-isomer:*

- Clean Water Act: 40CFR423, Appendix A, Priority Pollutants; 40CFR401.15 Section 307 Toxic Pollutants as hexachlorocyclohexane.
- RCRA 40CFR268.48; 61FR15654, Universal Treatment Standards: Wastewater (mg/L), 0.00014; Nonwastewater (mg/kg), 0.066
- RCRA 40CFR264, Appendix 9; TSD Facilities Ground Water Monitoring List. Suggested test method(s) (PQL $ug/L$): 8080(0.05); 8250 (10)
- Superfund/EPCRA 40CFR302.4 RQ: CERCLA, 10 lb (4.54 kg)
- EPCRA Section 313 Form R *de minimus* concentration reporting level: 1.0

*β-isomer:*

- Clean Water Act: 40CFR423, Appendix A, Priority Pollutants; 40CFR401.15 Section 307 Toxic Pollutants as hexachlorocyclohexane
- RCRA 40CFR268.48; 61FR15654, Universal Treatment Standards: Wastewater (mg/L), 0.00014; Nonwastewater (mg/kg), 0.066
- RCRA 40CFR264, Appendix 9; TSD Facilities Ground Water Monitoring List. Suggested test method(s) (PQL $ug/L$): 8080(0.05); 8250 (40)

- Superfund/EPCRA 40CFR302.4 RQ: CERCLA, 1 lb (0.454 kg)

*γ-isomer:*
See Lindane.

*δ-isomer:*
- Clean Water Act: 40CFR423, Appendix A, Priority Pollutants; 40CFR401.15 Section 307 Toxic Pollutants as hexachlorocyclohexane
- RCRA 40CFR268.48; 61FR15654, Universal Treatment Standards: Wastewater (mg/L), 0.023; Nonwastewater (mg/kg), 0.066
- RCRA 40CFR264, Appendix 9; TSD Facilities Ground Water Monitoring List. Suggested test method(s) (PQL ug/L): 8080(0.1); 8250 (30)
- Superfund/EPCRA 40CFR302.4 RQ: CERCLA, 1 lb (0.454 kg)
- Canada, WHMIS, Ingredients Disclosure List

*Description:* HCH is a white to brownish crystalline solid with a phosgene-like odor. Melting/Freezing point = 65°C. Hazard Identification (based on NFPA-704 M Rating System): Health 2, Flammability 1, Reactivity 0. It consists of eight stereoisomers of which the gamma (γ) isomer is most insecticidally active and hence most important and heavily regulated. Log $K_{ow}$ = 3.78 (*alpha-, beta-*). Values above 3.0 are likely to bioaccumulate in marine organisms. See also "Lindane." According to Prager, "*alpha-* and *beta-* isomers are carried in the food chain. The beta isomer accumulates the most, as the other isomers are either degraded, or converted to the beta isomer."

*Incompatibilities:* Decomposes on contact with powdered iron, aluminum, and zinc, and on contact with strong bases producing trichlorobenzene.

*Permissible Exposure Limits in Air:* ACGIH[1] recommends a TLV of 0.5 mg/m³ TWA [skin]. The DFG[3] has set an MAK of 0.5 mg/m³ for mixed HCH isomers. The former USSR-UNEP/IRPTC project[43] has set an MAC in workplace air of 0.1 mg/m³ and MAC value for ambient air in residential areas of 0.03 mg/m³ both on a momentary and on a daily average basis.

*Determination in Air:* Collection by filter/bubbler; workup with isooctane; analysis by gas chromatography/electrolytic conductivity detection; NIOSH IV, Method #5502[18] (recommended for Lindane).[18]

*Permissible Concentration in Water:* The U.S. EPA has set a limit in drinking water of 0.2 ppb. There are no criteria for the protection of freshwater or saltwater aquatic life from technical BHC (mixed isomers) due to insufficient data. To protect human health-preferably zero for technical product. An additional cancer risk of 1 in 100, 000 is imposed by a concentration of 0.123 μg/L[6]. The former USSR-UNEP/IRPTC project[43] has set an MAC in water bodies used for domestic purposes of 0.02 mg/L and zero in water bodies used for fishery purposes.

*Determination in Water:* Methylene chloride extraction followed by gas chromatography with electron capture or halogen specific detection (EPA Method 608) or gas chromatography plus mass spectrometry (EPA Method 625).

*Routes of Entry:* Inhalation, skin absorption, ingestion, skin and/or eye contact

*Harmful Effects and Symptoms*

*Short Term Exposure:* Irritates the eyes and respiratory tract. May affect the central nervous system, causing convulsions, respiratory failure, and collapse. Effects may be delayed. Exposure may result in death. See also below.

*Long Term Exposure:* Repeated or prolonged skin contact may cause irritation, redness. The effects of lindane and/or the α-, ß-, and δ-isomers of HCH observed in humans are lung irritation, heart disorders, blood disorders, headache, convulsions, and changes in the levels of sex hormones. These effects have occurred in workers exposed to HCH vapors during pesticide formulation and/or in individuals exposed accidentally or intentionally to large amounts of HCH. Exposure to excessive amounts of HCH can also result in death in humans and animals. Convulsions and kidney disease have been reported in animals fed lindane or γ-HCH. Liver disease has been reported in animals fed lindane and α-, ß-, or technical grade HCH. Longer exposure to lindane and α-, ß-, or technical-grade HCH has been reported to result in liver cancer. Reduced ability to fight infection was reported in animals fed lindane and injury to the ovaries and testes was reported in animals exposed to lindane or γ-HCH. In animals, there is evidence that oral exposure to lindane during pregnancy results in an increased incidence of fetuses with extra ribs. HCH is processed by the body into other chemical products, some of which probably are responsible for the harmful effects. The Department of Health and Human Services has determined that HCH may reasonably be anticipated to be carcinogenic. Liver cancer has been seen in laboratory rodents that ate HCH for long periods of time.

*Points of Attack:* Eyes, skin, respiratory system, central nervous system, blood, liver and kidneys

*Medical Surveillance:* NIOSH and OSHA recommend tests of whole blood (chemical/metabolite). See "Occupational Health Guidelines for Chemical Hazards." NIOSH Pub Nos. 81-123; 88-118, Suppls. I-IV. 1981-1995. Blood Serum. Complete Blood count.

*First Aid:* Speed in removing material from eyes and skin is of extreme importance. If this chemical gets into the eyes, remove any contact lenses at once and irrigate immediately for at least 30 minutes, occasionally lifting upper and lower lids. Seek medical attention immediately. If this chemical contacts the skin, remove contaminated clothing and wash immediately with soap and water. Shampoo hair promptly if contaminated. Seek medical attention immediately. If this chemical has been inhaled, remove from exposure, begin rescue breathing (using universal precautions) if breathing has stopped, and CPR if heart action has stopped. Transfer promptly to a medical facility. When this chemical has been swallowed, get medical attention. Give large quantities of water and induce vomiting. Do not make an unconscious person vomit.

*References:*
- New Jersey Department of Health and Senior Services, "Hazardous Substance Fact Sheet, alpha-Hexachlorocyclohexane," Trenton, NJ (October 2001). http://www.state.nj.us/health/eoh/rtkweb/0566.pdf
- New Jersey Department of Health and Senior Services, "Hazardous Substance Fact Sheet, beta-Hexachlorocyclohexane," Trenton, NJ (October 2001). http://www.state.nj.us/health/eoh/rtkweb/0567.pdf
- U.S. Department of Health and Human Services, Agency for Toxic Substances and Disease Registry, "ATSDR ToxFAQs, Hexachlorocyclohexanes," Atlanta, GA (June 1999). http://www.atsdr.cdc.gov/tfacts43.html
- U.S. Department of Health and Human Services, Agency for Toxic Substances and Disease Registry, "Public Health Statement for Hexachlorocyclohexane," Atlanta, GA (September 2003). http://www.atsdr.cdc.gov/toxprofiles/phs43.html
- U.S. Environmental Protection Agency, Hexachtorocyclohexane: Ambient Water Quality Criteria. Washington, DC (1980).
- U.S. Environmental Protection Agency, Hexachlorocyclohexane. Health and Environmental Effects Profile No. 112, Office of Solid Waste, Washington, DC (April 30, 1980).
- Sax, N.I., Ed., Dangerous Properties of Industrial Materials Report, 7, No. 4, 26-38 (1987) New York, Van Nostrand Reinhold Co. (1983).
- California Environmental Protection Agency "Chemical List of Lists," Sacramento CA (February 1997).

*Note:* See also Lindane.

# Hexachloroethane

*Use Type:* Insecticide, fungicide and in animal medicines
*CAS Number:* 67-72-1
*Formula:* $C_2Cl_6$; $CCl_3CCl_3$
*Alert:* Hexachloroethane does not occur naturally in the environment. It is no longer made in the United States, but it is formed as a by-product in the production of some chemicals. Some hexachloroethane can be formed by incinerators when materials containing chlorinated hydrocarbons are burned.
*Synonyms:* Carbon hexachloride; Ethane hexachloride; Ethylene hexachloride; Ethane, hexachloro-; HCE; Hexachloraethan (German); Hexachlorethane; 1,1,1,2,2,2-Hexachloroethane; Hexachloroethylene; Hexacloroetano (Spanish); Mottenhexe; NCI-C04604; Perchloroethane
*Trade Names:* AVLOTHANE®; DISTOKAL®; DISTOPAN®; DISTOPIN®; EGITOL®; FALKITOL®; FASCIOLIN®; PHENOHEP®

*Producers:* ATOFINA (France); ICI Group (UK); Scottish Chemical Industries (India); Seal Chemical Industries (USA)
*Chemical Class:* Halogenated organic compound
*EPA/OPP PC Code:* 045201
*ICSC Number:* 0051
*RTECS Number:* K14025000
*Uses:* In the US, about half the HCE is used by the military for smoke-producing devices. It is also used to remove air bubbles in melted aluminum. It may be present as an ingredient in some fungicides, insecticides, lubricants, and plastics. It is no longer made in the United States, but it is formed as a by-product in the production of some chemicals. Can be formed by incinerators when materials containing chlorinated hydrocarbons are burned. Some HCE can also be formed when chlorine reacts with carbon carbon compounds in drinking water. As a medicinal, HCE is used as an anthelmintic to treat fascioliasis in sheep and cattle. It is also added to the feed of ruminants, preventing methanogenesis and increasing feed efficiency. HCE is used in metal and alloy production, mainly in refining aluminum alloys. It is also used for removing impurities from molten metals, recovering metals from ores or smelting products and improving the quality of various metals and alloys. HCE is contained in pyrotechnics. It inhibits the explosiveness of methane and the combustion of ammomium perchlorate. Smoke containing HCE is used to extinguish fires. HCE has various applications as a polymer additive. It has flameproofing qualitites, increases sensitivity to radiation crosslinking, and is used as a vulcanizing agent. Added to polymer fibers, HCE acts as a swelling agent and increases affinity for dyes.

*Carcinogen/Hazard Classifications*
**U.S. EPA Carcinogens:** Group C, possible carcinogen
**U.S. NTP Carcinogen:** Reasonably anticipated carcinogen
**California Prop. 65:** Carcinogen
**IARC:** Group 2B, possible carcinogen
*Regulatory Authority:*
- Air Pollutant Standard Set (ACGIH)[1] (DFG)[3] (HSE)[33] (OSHA)[58] (Several States)[60]
- Carcinogen (Animal Positive) (IARC)[9]
- List of priority pollutants (U.S. EPA)
- AB 1803-Well Monitoring Chemical (CAL)
- AB 2588-Air Toxics "Hot Spots" Chemicals (CAL)
- CAL Air Resources Board/AB 1807 Toxic Air Contaminants
- Proposition 65 chemical (CAL)
- Permissible Exposure Limits for Chemical Contaminants (CAL/OSHA)
- The "Director's List" (CAL/OSHA)
- Clean Air Act: Hazardous Air Pollutants (Title I, Part A, Section 112)
- Clean Water Act: Section 313 Water Priority Chemicals (57FR41331, 9/9/92); Toxic Pollutant (Section 401.15) as chlorinated ethanes.

- EPA Hazardous Waste Number (RCRA No.): U131; D034
- RCRA, 40CFR261, Appendix 8 Hazardous Constituents
- RCRA Toxicity Characteristic (Section 261.24), Maximum Concentration of Contaminants, regulatory level, 3.0 mg/L
- RCRA 40CFR268.48; 61FR15654, Universal Treatment Standards: Wastewater (mg/L), 0.055; Nonwastewater (mg/kg), 30
- RCRA 40CFR264, Appendix 9; TSD Facilities Ground Water Monitoring List. Suggested test method(s) (PQL ug/L): 8120(0.5); 8270(10)
- Safe Drinking Water Act: Priority List (55 FR 1470)
- Superfund/EPCRA 40CFR355, Appendix B Extremely Hazardous Substances: TPQ = 100 lb (45.4 kg)
- Superfund/EPCRA 40CFR302.4 RQ: CERCLA, 100 lb (45.4 kg)
- EPCRA Section 313 Form R *de minimus* concentration reporting level: 1.0%
- Canada, WHMIS, Ingredients Disclosure List

*Description:* Hexachloroethane is a colorless solid. Camphor-like odor. It gradually evaporates when it is exposed to air. Practically insoluble in water; solubility = 50 ppm @ 22°C. Molecular weight = 236.76. Boiling point = (sublimes) @ 183-187°C. Melting/Freezing point = (sublimes) 187°C. Vapor pressure = 0.4 mmHg @ 20°C. Hazard Identification (based on NFPA-704 M Rating System): Health 2, Flammability 0, Reactivity 0. Log $K_{ow}$ = 3.32 to 3.89. Values at or above 3.0 are likely to bioaccumulate in marine organisms.

*Incompatibilities:* Incompatible with metals such as aluminum, cadmium, hot iron, mercury, or zinc. Alkalies forms spontaneously explosive chloroacetylene. Attacks some plastics, rubber and coatings.

*Permissible Exposure Limits in Air:* The Federal standard[58] is 1.0 ppm (10 mg/m$^3$). NIOSH[2] and ACGIH[1] recommends the same TWA. The notation "skin" is added to indicate the possibility of cutaneous absorption. The NIOSH[2] IDLH level = 300 ppm. The DFG[3] has set an MAK value of 1.0 ppm (10 mg/m$^3$) but HSE[33] has more complex limits; they are 5 ppm (50 mg/m$^3$) for vapor; 10 mg/m$^3$ for total inhalable dust and 5 mg/m$^3$ for respirable dust. Several states have set guidelines or standards for hexachloroethane in ambient air[60] ranging from 0.18 µg/m$^3$ (Massachusetts) to 50.0 µg/m$^3$ (Connecticut) to 238.095 µg/m$^3$ (Kansas) to 1000 (µg/m$^3$ (North Dakota) to 1600 µg/m$^3$ (Virginia) to 2381 µg/m$^3$ (Nevada).

*Determination in Air:* Charcoal adsorption, workup with $CS_2$, analysis by gas chromatography/flame ionization. See NIOSH Method 1003 for halogenated hydrocarbons.[18]

*Permissible Concentration in Water:* The U.S. EPA suggests (1997) that water consumed over a lifetime contain no more than 1 part HCE per billion parts of water. (1 ppb).

In an older citation, the U.S. EPA suggested the following: *To protect freshwater aquatic life:* 118,000 µg/L based on acute toxicity and 20,000 µg/L based on chronic toxicity. To protect saltwater aquatic life-113,000 µg/L based on acute toxicity. *To protect human health:* 9.4 µg/L to keep lifetime cancer risk below 10-5[6]. The former USSR[35, 43] has set an MAC in water bodies used for domestic purposes of 0.01 mg/L (10 µg/L). States which have set guidelines for hexachloroethane in drinking water[61] include Kansas (1.9 µg/L) and Minnesota (24.6 µg/L).

*Determination in Water:* Methylene chloride extraction followed by concentration, gas chromatography with electron capture detection (EPA Method 612) or gas chromatography plus mass spectrometry (EPA Method 625).

*Routes of Entry:* Inhalation, skin absorption, ingestion, eye and/or skin contact.

*Harmful Effects and Symptoms*

*Short Term Exposure:* Contact can irritate and burn the eyes and skin. Exposure can irritate the respiratory tract. High levels of exposure can cause dizziness, lightheadedness, and unconsciousness. Irritation occurs when there is an excessive amount of hexachloroethane dust in the air or when it is heated and vapors are formed. Hexachloroethane acts primarily as a central nervous system depressant, and in high concentrations it causes tremors, narcosis. It should be noted that the low vapor pressure of this compound as well as its solid state minimize its inhalation hazards.

*Long Term Exposure:* A potential occupational carcinogen. May cause kidney and liver damage.

*Points of Attack:* Eyes, skin, respiratory system and kidneys. Cancer Site in animals: liver cancer.

*Medical Surveillance:* For those with frequent or potentially high exposure (half the TLV or greater) the following are recommended before beginning work and at regular times after that: Liver and kidney function tests. More than light alcohol consumption can exacerbate liver damage. Sample of blood, urine, or feces can be tested for exposure to HCE. These tests are useful only if exposure occurred 24-48 hours prior to testing.

*First Aid:* If this chemical gets into the eyes, remove any contact lenses at once and irrigate immediately for at least 15 minutes, occasionally lifting upper and lower lids. Seek medical attention immediately. If this chemical contacts the skin, remove contaminated clothing and wash immediately with soap and water. Seek medical attention immediately. If this chemical has been inhaled, remove from exposure, begin rescue breathing (using universal precautions) if breathing has stopped, and CPR if heart action has stopped. Transfer promptly to a medical facility. When this chemical has been swallowed, get medical attention. Give large quantities of water and induce vomiting. Do not make an unconscious person vomit.

### References:

- U.S. Environmental Protection Agency, Chlorinated Ethanes: Ambient Water Quality Criteria, Washington, DC (1980).
- U.S. Environmental Protection Agency, Chemical Hazard Information Profile: Hexachloroethane, Washington, DC (1979). http://www.epa.gov/epaoswer/hazwaste/minimize/hexchlet.pdf
- New Jersey Department of Health and Senior Services, "Hazardous Substance Fact Sheet: Hexachloroethane," Trenton, NJ (May 1998). http://www.state.nj.us/health/eoh/rtkweb/0981.pdf
- U.S. Department of Health and Human Services, Agency for Toxic Substances and Disease Registry, "ATSDR ToxFAQs, Hexachloroethane," Atlanta, GA (September 1997). http://www.atsdr.cdc.gov/tfacts97.html
- U.S. Environmental Protection agency, Hexachloroethane, Health and Environmental Effects Profile No. 116, Office of Solid Waste, Washington, DC (April 30, 1980).
- Sax, N.I., Ed., "Dangerous Properties of Industrial Materials Report" 2, No. 6, 75-78 (1982) and 6, No. 4, 70-83 (1986).
- California Environmental Protection Agency "Chemical List of Lists," Sacramento CA (February 1997).

# Hexachlorophene

*Use Type:* Fungicide
*CAS Number:* 70-30-4
*Formula:* $C_{13}H_6Cl_6O_2$; $C_6H(OH)Cl_3CH_2C_6H(OH)Cl_3$
*Synonyms:* Bis(2-hydroxy-3,5,6-trichlorophenyl)methane; Bis-2,3,5-trichlor-6-hydroxyfenylmethan (Czech); Bis(3,5,6-trichlor o-2-hydroxyphenyl)methane; 2,2'-Dihydroxy-3,3',5,5',6,6'-hexachlorodiphenylmethane; 2,2'-Dihydroxy-3,5,6,3',5',6'-hexachlorodiphenylmethane; Hexaclorofeno (Spanish); HCP; 2,2',3,3',5,5'-Hex achloro-6,6'-dihydroxydiphenylmethane; Hexachlorofen (Czech); Hexachlorophane; Hexachlorophen; Hexachlorophene; Methane, Bis(2,3,5-trichloro-6-hydroxyphenyl); 2,2'-Methylenebis(3,4,6-trichlorophenol); 2,2'-Methylenebis(3,5,6-trichlorophenol); NCI-C02653; NSC 4911; Phenol, 2,2'-methylenebis(3,5,6-trichloro-; Phenol, 2,2'-methylenebis(3,4,6-trichloro)-
*Trade Names:* ACIGENA®; AI3-02372®; AT-7®; AT-17®; B 32®; BILEVON®; COTOFILM®; DISTODIN®; ESACLOROFENE®; EXOFENE®; FOMAC®; FOSTRIL®; G-11®; G-ELEVEN®; GAMOPHEN®; GAMOPHENE®; HEXAFEN®; HEXAPHENE-LV®; HEXIDE®; HEXOPHENE®; KILZOL®; NABAC®; NABAC 25 EC®
*Chemical Class:* Chlorinated phenol
*EPA/OPP PC Code:* 044901 and 600027
*California DPR Chemical Code:* 322
*ICSC Number:* 0161
*RTECS Number:* SM0700000
*EEC Number:* 604-015-00-9
*EINECS Number:* 200-733-8
*Uses:* It is used as an antifungal agent to treat various citrus fruits and vegetables. HCP also is used as an antibacterial agent in a wide variety of consumer products, including soaps, shampoos, surgical scrubs and deodorants.
*Carcinogen/Hazard Classifications*
**IARC:** Group 3, unclassifiable
**Label Signal Word:** DANGER
*Regulatory Authority:*
- Air Pollutant Standard Set (Massachusetts)[60]
- Banned or Severely Restricted (In Pharmaceuticals) (UN)[13]
- The "Director's List" (CAL/OSHA)
- Superfund/EPCRA 40CFR302.4 RQ: CERCLA, 100 lb (45.4 kg)
- EPA Hazardous Waste Number (RCRA No.): U132
- RCRA 40CFR264, Appendix 9; TSD Facilities Ground Water Monitoring List. Suggested test method(s) (PQL $ug/L$): 8270(10)
- EPCRA Section 313 Form R *de minimus* concentration reporting level: 1.0%

*Description:* Hexachlorophene is a crystalline solid. Odorless. Insoluble in water. Molecular weight = 248.74. Boiling point = 210°C. Melting/Freezing point = 165°C. Hazard Identification (based on NFPA-704 M Rating System): Health 3, Flammability 1, Reactivity 0. Log $K_{ow}$ = 6.95. Values above 3.0 are likely to bioaccumulate in marine organisms.
*Incompatibilities:* Oxidizers. Heat of decomposition forms hydrogen chloride.
*Permissible Exposure Limits in Air:* No OELs have been established. Massachusetts[61] has set a guideline for ambient air of zero.
*Permissible Concentration in Water:* A no-adverse-effect level in drinking water has been calculated by NAS/NRC as 0.008 mg/L. An ADI was calculated on the basis of the available chronic toxicity data to be 0.0012 mg/kg/day. The former USSR[35] has set an MAC in water bodies used for domestic purposes of 0.03 mg/L. Maine[61] has set a guideline for drinking water of 2.0 µg/L.
*Routes of Entry:* Inhalation, passing through the skin.
*Harmful Effects and Symptoms*
*Short Term Exposure:* Hexachlorophene may irritate the eyes and skin, and cause a skin allergy to develop. May cause permanent eye damage. Inhaling can irritate the respiratory tract. May affect the central nervous system, causing dizziness, weakness, convulsions (fits), coma, respiratory failure, or death. Exposure can cause loss of appetite, nausea, vomiting, cramps and diarrhea.
*Long Term Exposure:* Animal tests suggest an association between exposure of pregnant women to hexachlorophene and birth defects. There is limited evidence that

hexachlorophene is a teratogen in animals. Repeated or prolonged contact with skin may cause dermatitis, skin sensitization, and asthma-like allergy. May cause liver damage. Repeated exposure may cause nervous system damage, brain damage leading to paralysis and blindness.

*Points of Attack:* Nervous system, skin, liver, reproductive system.

*Medical Surveillance:* If symptoms develop or overexposure is suspected, the following may be useful: Examination of the nervous system. Eye exam. Evaluation by a qualified allergist, including careful exposure history and special testing, may help diagnose skin allergy. More than light alcohol consumption may exacerbate liver damage.

*First Aid:* If this chemical gets into the eyes, remove any contact lenses at once and irrigate immediately for at least 15 minutes, occasionally lifting upper and lower lids. Seek medical attention immediately. If this chemical contacts the skin, remove contaminated clothing and wash immediately with soap and water. Seek medical attention immediately. If this chemical has been inhaled, remove from exposure, begin rescue breathing (using universal precautions) if breathing has stopped, and CPR if heart action has stopped. Transfer promptly to a medical facility. When this chemical has been swallowed, get medical attention. Give large quantities of water and induce vomiting. Do not make an unconscious person vomit.

*References:*
- U.S. Environmental Protection Agency, Hexachlorophene, Health and Environmental Effects Profile No. 116, Office of Solid Waste, Washington, DC (April 30, 1980).
- Sax, N.I., Ed., "Dangerous Properties of Industrial Materials Report" 6, No 2, 62-66 (1986).
- California Environmental Protection Agency "Chemical List of Lists," Sacramento CA (February 1997).
- New Jersey Department of Health and Senior Services, "Hazardous Substance Fact Sheet: Hexachlorophene," Trenton, NJ (April 1999). http://www.state.nj.us/health/eoh/rtkweb/0983.pdf

# Hexaconazole (ANSI)

*Use Type:* Fungicide
*CAS Number:* 79983-71-4
*Formula:* $C_{14}H_{17}Cl_2N_3O$
*Synonyms:* α-Butyl-α-(2,4-dichlorophenyl)-1*H*-1,2,4-triazole-1-ethanol (±)-; (RS)-2-(2,4-Dichlorophenyl)-1-(1*H*-1,2,4-triazole-1-yl)hexan-2-ol; 1*H*-1,2,4-Triazole-1-ethanol, α-butyl-α-(2,4-dichlorophenyl)-, (±)-
*Trade Names:* ANVIL®; PROSEED®; PP 523®; TITAN®, Sudarshan Chemical Industries (India)

*Producers:* Hindustan Insecticides (India); Meghmani Organics (India); Rallis India (India); Sudarshan Chemical Industries (India)

*Chemical Class:* Azole
*EPA/OPP PC Code:* 128925
*Uses:* Not registered in the U.S. Used for the control of seedborne and soilborne diseases such as powdery mildew, scab, blackrot and rust on grapes, barley and wheat. Residue level is for imported bananas.

*Human toxicity (long-term)[77]:* Intermediate–35.00 ppb, Health Advisory

*Fish toxicity (threshold)[77]:* Very low–755.66000 ppb, MATC (Maximum Acceptable Toxicant Concentration)

*U.S. Maximum Allowable Residue Levels for Hexaconazole (40 CFR 180.488):*

| CROP | ppm |
|---|---|
| Imported bananas | 0.7 |

*Carcinogen/Hazard Classifications*

**U.S. EPA Carcinogens:** Group C, possible carcinogen
**WHO Acute Hazard:** Class U, unlikely to be hazardous
**Endocrine Disruptor:** Possible ED (Canada Regulatory Decision Document RDD2000-04)

*Description:* White crystalline solid. Melting point = 110-112°C.

*Permissible Concentration in Water:* No criteria set. Runoff from spills or fire control may cause water pollution.

*Routes of Entry:* Ingestion, skin contact

*Harmful Effects and Symptoms*

*Short Term Exposure:* Poisonous if swallowed. Contact may irritate skin and cause eye irritation and possible severe injury. Avoid inhalation.

*Medical Surveillance:* If this chemical gets into the eyes, remove any contact lenses at once and irrigate immediately for at least 15 minutes, occasionally lifting upper and lower lids. Seek medical attention immediately. If this chemical contacts the skin, remove contaminated clothing and wash immediately with soap and water. If this chemical has been inhaled, remove from exposure, begin rescue breathing (using universal precautions) if breathing has stopped, and CPR if heart action has stopped. Transfer promptly to a medical facility. When this chemical has been swallowed, get medical attention. Give large quantities of water and induce vomiting. Do not make an unconscious person vomit.

*First Aid:* If this chemical gets into the eyes, remove any contact lenses at once and irrigate immediately for at least 15 minutes, occasionally lifting upper and lower lids. Seek medical attention immediately. If this chemical contacts the skin, remove contaminated clothing and wash immediately with soap and water. Seek medical attention immediately. If this chemical has been inhaled, remove from exposure, begin rescue breathing (using universal precautions) if breathing has stopped, and CPR if heart action has stopped. Transfer promptly to a medical facility. When this chemical has been swallowed, get medical attention. Give large quantities of water and induce vomiting. Do not make an unconscious person vomit.

*References:*
- U.S. Environmental Protection Agency, Office of Pesticide Programs, Pesticide Residue Limits, "Bananas," 40 CFR 180.488.
  http://frwebgate.access.gpo.gov/cgi-bin/get-cfr.cgi

# Hexaflumuron

*Use Type:* Insecticide, termiticide
*CAS Number:* 86479-06-3
*Formula:* $C_{16}H_{18}Cl_2F_6N_2O_3$
*Synonyms:* AI3-29832; Benzamide, *N*-[((3,5-dichloro-4-(1,1,2,2-tetrafluoroethoxy)phenyl)amino)carbonyl]-2,6-difluoro-; 1-[3,5-Dichloro-4-(1,1,2,2-tetrafluoroethoxy)phenyl]-3-(2,6-difluorobenzoyl)urea (IUPAC); Hexafluron
*Trade Names:* CONSULT®; DE-473®; NAF-46®; RECRUIT®, Dow AgroSciences (USA); SONET; TRUENO®; XRD 473®
*Producers:* Dow AgroSciences (USA)
*Chemical Class:* Benzoylurea
*EPA/OPP PC Code:* 118202
*California DPR Chemical Code:* 3899
*ICSC Number:* 1266
*RTECS Number:* CV3800000
*Uses:* This is an insect growth regulator (IGR) that works by inhibiting the insect's growth by interfering with chitin synthesis, which termites need to form an exoskeleton. It is used to impregnate termite bait.
*Fish toxicity (threshold)*[77]*:* Intermediate–28.07482 ppb, MATC (Maximum Acceptable Toxicant Concentration)
*Carcinogen/Hazard Classifications*
**Label Signal Word:** CAUTION
**WHO Acute Hazard:** Class U, unlikely to be hazardous
*Description:* White solid or powder. Practically insoluble in water; solubility = < 0.1 ppm. Molecular weight = 461.18. Melting point = 203°C. Vapor pressure = very low/negligible. Log $K_{ow}$ = 5.7. Values above 3.0 are likely to bioaccumulate in marine organisms.
*Incompatibilities:* Irritating or toxic fumes or gases in a fire.
*Permissible Concentration in Water:* No criteria set. Runoff from spills or fire control may cause water pollution.
*Harmful Effects and Symptoms*
*Short Term Exposure:* Contact with eyes or skin may cause irritation or injury. Inhalation should be avoided; use NIOSH-approved air purifying respirators for pesticides. May be harmful if swallowed.
*Long Term Exposure:* May effect the blood; formation of methemoglobin.
*Points of Attack:* Blood.
*Medical Surveillance:* For those with frequent or potentially high exposure, the following are recommended before beginning work and at regular times after that: Lung function tests. If overexposure is suspected, also consider: Complete blood count and test for methemoglobin.
*First Aid:* If this chemical gets into the eyes, remove any contact lenses at once and irrigate immediately for at least 15 minutes, occasionally lifting upper and lower lids. Seek medical attention immediately. If this chemical contacts the skin, remove contaminated clothing and wash immediately with soap and water. Seek medical attention immediately. If this chemical has been inhaled, remove from exposure, begin rescue breathing (using universal precautions) if breathing has stopped, and CPR if heart action has stopped. Transfer promptly to a medical facility. When this chemical has been swallowed, get medical attention. Give large quantities of water and induce vomiting. Do not make an unconscious person vomit. Do not induce vomiting when formulations containing petroleum solvents are ingested.
*References:*
- National Pesticides Information Center (NPIC), "Fact Sheet, Hexaflumuron," Corvallis, OR http://www.npic.orst.edu/factsheets/hextech.pdf

# Hexaethyl Tetraphosphate

*Use Type:* Insecticide
*CAS Number:* 757-58-4
*Formula:* $C_{12}H_{30}O_{13}P_4$
*Synonyms:* Ethyl tetraphosphate; Ethyl tetraphosphate, hexa-; HET; HEPT; Hexaethyltetrafosfat; HTP; Tetrafosfato de hexaetilo (Spanish); Tetraphosphate hexaethylique (French); Tetraphosphoric acid, hexaethyl ester
*Trade Names:* BLADAN®; BLADAN® BASE
*Chemical Class:* Organophosphate
*EPA/OPP PC Code:*
*California DPR Chemical Code:*
*RTECS Number:* XF1575000
*RCRA Number:* P062
*Uses:* Contact insecticide. No longer registered with the U.S. EPA as a commercial pesticide product.
*Regulatory Authority:*
- EPA Hazardous Waste Number (RCRA No.): P062.
- RCRA Section 261 Hazardous Constituents
- EPCRA Section 304 RQ: CERCLA, 100 lb (45.4 kg)
- Marine pollutant (49CFR, Subchapter 172.101, Appendix B), solid or liquid.
- DOT Inhalation Hazard Chemicals as organophosphates
- Hexaethyl tetraphosphate and compressed gas mixture not accepted by passenger rail or air.

*Description:* Amber or yellow, mobile liquid. Odorless. Dissolves in water; hygroscopic. Molecular weight = 506.30 Density 1.2917@ 27°C. Melting/Freezing point = –75 to –90°C. Boiling point = >150°C (decomposition).
*Incompatibilities:* Hydrolyzes rapidly in aqueous solution. May react violently with antimony(V) pentafluoride. Incompatible with lead diacetate, magnesium, silver nitrate.

*Permissible Exposure Limits in Air:* NIOSH[2]/OSHA: 0.05 mg/m$^3$ as TEPP

*Determination in Air:* OSHA versatile sampler-2; Toluene/Acetone; Gas chromatography/Flame ionization detection; NIOSH IV, Method #5600, Organophosphorus Pesticides.[18]

*Permissible Concentration in Water:* No criteria set. Runoff from spills or fire control may cause water pollution. Should not exceed 0.3 ppb in surface waters [*Proposed Criteria For Water Quality, Volume I, USEPA, 10/73*].

*Routes of Entry:* Inhalation, skin.

*Determination in Water:* OSHA versatile sampler-2; Toluene/Acetone; Gas chromatography/Flame ionization detection; NIOSH IV, Method #5600, Organophosphorus Pesticides.[18]

*Harmful Effects and Symptoms*

*Short Term Exposure:* A cholinesterase inhibitor. Pin-point pupils. Eye pupils are small; blurred vision; eye watering; runny nose; cough; shortness of breath; salivation; dizziness; nausea, stomach cramps, diarrhea, and vomiting; increased blood pressure; profuse sweating; hypermotility, hallucinations; irritability; tingling of the skin; drowsiness; slow heartbeat; convulsions; fluid in lungs; loss of consciousness; incontinence; breathing stops; death. Organophosphates inhibit the action of acetylcholinesterase enzymes, and alter the way in which nervous impulses are transmitted. The effects can last for hours, days, or much longer. The action of the enzymes is reestablished after new enzymes are formed. When used medicinally, has side effects of weakness, sialorrhea, vomiting, hyperhidrosis blurring of vision, bradycardia, muscular twitching, fasciculation.

*Long Term Exposure:* Cholinesterase inhibitor; cumulative effect is possible. Organophosphates may damage the nervous system with repeated exposure, resulting in convulsions, respiratory failure. May cause liver damage.

*Points of Attack:* Respiratory system, central nervous system, cardiovascular system, blood cholinesterase.

*Medical Surveillance:* Medical observation is recommended for 24 to 48 hours after breathing overexposure, as pulmonary edema may be delayed. As first aid for pulmonary edema, consider administering a corticosteroid spray. Cigarette smoking may exacerbate pulmonary injury and should be discouraged for at least 72 hours following exposure.

*First Aid:* Speed in removing material from eyes and skin is of extreme importance. *Eyes:* Eye contact can cause dangerous amounts of these chemicals to be quickly absorbed through the mucous membrane into the bloodstream. Immediately and gently flush eyes with plenty of warm or cold water (NO hot water) for at least 15 minutes, occasionally lifting the upper and lower eyelids. Get medical aid immediately. *Skin:* Get medical aid. Skin contact can cause dangerous amounts of these chemicals to be absorbed into the bloodstream. Wearing the appropriate PPE equipment and respirator for organophosphate pesticides, immediately flush skin with plenty of soap and water for at least 15 minutes while removing contaminated clothing and shoes. Shampoo hair promptly if contaminated. *Ingestion:* Call poison control. Loosen all clothing. Never give anything by mouth to an unconscious person. Get medical aid. Do NOT induce vomiting.* If conscious, alert, and able to swallow, rinse mouth and have victim drink 4 to 8 ounces of water do NOT induce vomiting but immediately administer slurry of activated charcoal (2 oz in 8 oz of water). If victim is *unconscious or having convulsions,* do nothing except keep victim warm. **In some cases you may be specifically instructed by poison control to induce vomiting by way of 2 tablespoons of syrup of ipecac (adult) washed down with a cup of water.* Do NOT give activated charcoal before or with ipecac syrup. *Inhalation:* Get medical aid. Do not contaminate yourself. Wearing the appropriate PPE equipment and respirator for organophosphate pesticides, immediately remove the victim from the contaminated area to fresh air. If the victim is not breathing, administer artificial respiration. Do NOT use mouth-to-mouth resuscitation; use bag/mask apparatus. If breathing is difficult, administer oxygen through bag/mask apparatus until medical help arrives. Do not leave victim unattended.

*Note to physician or authorized medical personnel.* Administer atropine, 2 mg (1/30 gr) intramuscularly or intravenously as soon as any local or systemic signs or symptoms of an intoxication are noted; repeat the administration of atropine every 3 to 8 minutes until signs of atropinization (mydriasis, dry mouth, rapid pulse, hot and dry skin) occur; initiate treatment in children with 0.05 mg mg/kg of atropine; repeat at 5 to 10 minute intervals. Watch respiration, and remove bronchial secretions if they appear to be obstructing the airway; intubate if necessary. *Notes to physician or authorized medical personnel*: N-methylpyridinium-2-aldoxime (2-PAMCI) when used in conjunction with atropine reacts with the phosphorylated cholinesterase, thereby restoring normal activity to by removing the phosphorylating group. The combination of these two chemicals is synergistic and must be administered within minutes to a few hours following exposure (depending on the specific agent) to be effective. Give 2-PAMCI (Pralidoxime; Protopam), 2.5 gm in 100 ml of sterile water or in 5% dextrose and water, intravenously, slowly, in 15-30 minutes; if sufficient fluid is not available, give 1 gm of 2-PAMCI in 3 ml of distilled water by deep intramuscular injection; repeat this every half hour if respiration weakens or if muscle fasciculation or convulsions recur. Also Diazepam, an anticonvulsant, might be considered.

*References:*
- California Environmental Protection Agency "Chemical List of Lists," Sacramento CA (February 1997).

# Hexazinone (ANSI)

*Use Type:* Herbicide
*CAS Number:* 51235-04-2
*Formula:* $C_{11}H_{20}O_2N_3$
*Alert:* A General Use Pesticide (GUP)
*Synonyms:* 3-Cyclohexyl-6-(Dimethylamino)-1-methyl-*s*-triazine-2,4(1*H*,3*H*)-dione; 3-Cyclohexyl-6-dimethylamino-1-methyl-1,2,3,4-tetrahydro-1,3,5-triazine-2-,4-dione; 3-Cyclohexyl-6-(dimethylamino)-1-methyl-1,3,5-triazine-2,4(1*H*,3*H*)-dione; 3-Cyclohexyl-1-methyl-6-(dimethylamino)-*s*-trazine-2,4(1*H*,3*H*)-dione; 1,3,5-Triazine-2,4(1*H*,3*H*)-dione, 3-Cyclohexyl-6-(dimethylamino)-1-methyl-; *s*-Triazine-2,4(1*H*,3*H*)-dione, 3-Cyclohexyl-6-(dimethylamino)-1-methyl-
*Trade Names:* BO-RID®, canceled 4/25/1986; BRUSHKILLER®; DPX 3674®; K-4 HERBICIDE®, DuPont Crop Protection (USA); OUSTAR®, DuPont Crop Protection (USA); PRONONE®; VELPAR®, DuPont Crop Protection (USA); VELPAR WEED KILLER®, DuPont Crop Protection (USA)
*Producers:* Dupont Crop Protection (USA); Shenzhen Guomeng Industry Co., Ltd. (China)
*Chemical Class:* Triazine
*EPA/OPP PC Code:* 107201
*California DPR Chemical Code:* 1871
*RTECS Number:* XY7850000
*EINECS Number:* 257-074-4
*Uses:* Hexazinone is used against many annual, biennial, and perennial weeds, as well as some woody plants. It is mostly used on non-crop areas; however, it is used selectively for the control of weeds among sugar cane, pineapples, and rangeland forage. Hexazinone is a systemic herbicide that works by inhibiting photosynthesis in the target plants. Rainfall or irrigation water is needed before it becomes activated. It is available in soluble concentrate, water-soluble powder, or granular formulations.
*Human toxicity (long-term)*[77]: Very low–400.00 ppb, Health Advisory
*Fish toxicity (threshold)*[77]: Very low–24566.31408 ppb, MATC (Maximum Acceptable Toxicant Concentration)
*U.S. Maximum Allowable Residue Levels for Hexazinone (40 CFR 180.396):*

| CROP | ppm |
|---|---|
| Alfalfa, forage | 2 |
| Alfalfa, hay | 8 |
| Blueberry | 0.2 |
| Cattle, fat | 0.1 |
| Cattle, meat | 0.1 |
| Cattle, mbyp | 0.1 |
| Goat, fat | 0.1 |
| Goat, meat | 0.1 |
| Goat, mbyp | 0.1 |
| Grass, pasture | 10 |
| Grass, rangeland | 10 |
| Hog, fat | 0.1 |
| Hog, meat | 0.1 |
| Hog, mbyp | 0.1 |
| Horse, fat | 0.1 |
| Horse, meat | 0.1 |
| Horse, mbyp | 0.1 |
| Milk | 0.1 |
| Pineapple | 0.5 |
| Sheep, fat | 0.1 |
| Sheep, meat | 0.1 |
| Sheep, mbyp | 0.1 |
| Sugarcane, cane | 0.2 |
| Sugarcane, molasses | 5 |

*Carcinogen/Hazard Classifications*
**U.S. EPA Carcinogens:** Group D, unclassifiable, ambiguous data
**Label Signal Word: DANGER**
**WHO Acute Hazard:** Class III, slightly hazardous
*Regulatory Authority:*
- EPCRA Section 313 Form R *de minimus* concentration reporting level: 1.0%
- AB 1803-Well Monitoring Chemical (CAL)
- Actively registered pesticide in California.

*Description:* White crystalline solid. Practically odorless. Soluble in water; solubility = $3.4 \times 10^5$ ppm @ 25°C. Molecular weight = 252.37. Boiling point = decomposes. Melting/Freezing point = 115–117°C. Vapor Pressure = $3.0 \times 10^{-7}$ mmHg.
*Permissible Exposure Limits in Air:* No standards set.
*Permissible Concentration in Water:* The U.S. EPA has analyzed data on hexazinone and developed a no observed adverse effect level (NOAEL) of 25 mg/kg/day based on studies of dogs which resulted in a long-term health advisory of 8.75 mg/L. A NOAEL of 10 mg/kg/day was developed based on studies of rats which yielded a lifetime health advisory of 0.21 mg/L.
*Determination in Water:* Solvent extraction with methylene chloride followed by analysis by gas chromatography with a thermionic bead detector.
*Routes of Entry:* Skin, inhalation and ingestion.
*Harmful Effects and Symptoms*
*Short Term Exposure:* May cause eye and skin irritation. The acute oral $LD_{50}$ for rats is 1690 mg/kg (slightly toxic). In experience with humans, only one report was available on hexazinone. It involved a 26 year-old woman who inhaled hexazinone dust. Vomiting occurred within 24 hours.
*First Aid:* If this chemical gets into the eyes, remove any contact lenses at once and irrigate immediately for at least 15 minutes, occasionally lifting upper and lower lids. Seek medical attention immediately. If this chemical contacts the skin, remove contaminated clothing and wash immediately with soap and water. Seek medical attention immediately. If this chemical has been inhaled, remove from exposure, begin rescue breathing (using universal precautions) if

breathing has stopped, and CPR if heart action has stopped. Transfer promptly to a medical facility. When this chemical has been swallowed, get medical attention. Give large quantities of water and induce vomiting. Do not make an unconscious person vomit.

*References:*
- EXTOXNET, Extension Toxicology Network, "Pesticide Information Profile, Hexazinone," Oregon State University, Corvallis, OR (June 1996). http://extoxnet.orst.edu/pips/hexazin.htm
- U.S. Environmental Protection Agency, Office of Pesticide Programs, Pesticide Residue Limits, "Hexazinone," 40 CFR 180.396, http://www.epa.gov/pesticides/food/viewtols.htm
- U.S. Environmental Protection Agency, "Health Advisory: Hexazinone," Washington, DC, Office of Drinking Water (August 1987).
- California Environmental Protection Agency "Chemical List of Lists," Sacramento CA (February 1997).

# Hexythiazox

*Use Type:* Acaricide, insect growth regulator
*CAS Number:* 78587-05-0
*Formula:* $C_{17}H_{21}ClN_2O_2S$
*Synonyms:* trans-5-(4-Chlorophenyl)-*N*-cyclohexyl-4-methyl-2-*oxo*-3-thiazolidinecarboxamide; HTZ; trans-4-methyl-5-(4-chlorophenyl)-3-cyclohexylcarbamoyl-2-thiazolidone; 3-Thiazolidinecarboxamide, 5-(4-chlorophenyl)-*N*-cyclohexyl-4-methyl-2-*oxo*-, trans-
*Trade Names:* ACARFLOR®; ACARIFLOR®; CESAR®; DPX-Y5893®; HEXYGON® DF, Gowan Company (USA); NISSORUN®; NA 73®; ONAGER®, Gowan Company (USA); SAVEY®, Gowan Company (USA); TREVI®; ZELDOX®
*Producers:* AJE (Switzerland); Gowan Company (USA)
*Chemical Class:* Substituted urea
*EPA/OPP PC Code:* 128849
*California DPR Chemical Code:* 2303
*Human toxicity (long-term)[77]:* Intermediate–17.50 ppb, Health Advisory
*Fish toxicity (threshold)[77]:* Intermediate–62.61862 ppb, MATC (Maximum Acceptable Toxicant Concentration)
*U.S. Maximum Allowable Residue Levels for Hexythiazox (40 CFR 180.448):*

| CROP | ppm |
|---|---|
| Almond, hulls | 10 |
| Apple | 0.5 |
| Apple, wet pomace | 0.8 |
| Caneberry subgroup 13a | 1 |
| Cattle, fat | 0.02 |
| Cattle, mbyp | 0.02 |
| Cotton, gin byproducts | 3 |
| Cotton, undelinted seed | 0.2 |
| Date, dried fruit | 1 |
| Fruit, stone, group, except plum | 1 |
| Goat, fat | 0.02 |
| Goat, mbyp | 0.02 |
| Hog, fat | 0.02 |
| Hog, mbyp | 0.02 |
| Hop | 2 |
| Horse, fat | 0.02 |
| Horse, mbyp | 0.02 |
| Milk | 0.02 |
| Nut, tree, group 14 | 0.3 |
| Pear | 0.3 |
| Peppermint, tops | 2 |
| Pistachio | 0.3 |
| Plum | 0.1 |
| Plum, prune, dried | 0.4 |
| Plum, prune, fresh | 0.1 |
| Sheep, fat | 0.02 |
| Sheep, mbyp | 0.02 |
| Spearmint, tops | 2 |

*Carcinogen/Hazard Classifications*
**U.S. EPA Carcinogens:** Group C, possible carcinogen
**Label Signal Word:** CAUTION
**WHO Acute Hazard:** Class U, unlikely to be hazardous
*Regulatory Authority:*
- List of priority pollutants (U.S. EPA)
- AB 2588-Air Toxics "Hot Spots" Chemicals (CAL)
- The "Director's List" (CAL/OSHA)
- Clean Water Act: Toxic Pollutant (Section 401.15) other than those listed elsewhere; includes trichlorophenols.
- RCRA Section 261 Hazardous Constituents, waste number not listed.
- EPCRA Section 313 (as chlorophenols) Form R *de minimis* concentration reporting level: 1.0%

*Description:* White to pale yellow crystalline solid. Very slightly soluble in water; solubility = < 1 ppm. Molecular weight = 352.90. Melting point = 105–107.5°C. Vapor pressure = $2.3 \times 10^{-8}$ mmHg.
*Incompatibilities:* Slowly hydrolyzes in water, releasing ammonia and forming acetate salts.
*Permissible Concentration in Water:* No criteria set. Runoff from spills or fire control may cause water pollution.
*Routes of Entry:* Ingestion, absorbed through the unbroken skin.
*Short Term Exposure:* Inhalation can cause severe irritation, burns to the nose and throat, headache, dizziness, vomiting, lung damage, muscle twitchings, spasms, tremors, weakness, staggering and collapse. Skin contact may cause severe irritation and burns. Absorbed through the skin to cause or increase the severity of symptoms listed above. Eye contact causes severe irritation. May cause burns. Ingestion can cause irritation, burns to the mouth and throat, low blood pressure, profuse sweating, intense thirst, nausea, abdominal pain, stupor, vomiting, red blood cell damage and accumulation of fluid in the lungs followed by

pneumonia. May also cause restlessness and increased breathing rate followed by rapidly developing muscle weakness. The substance irritates the eyes, the skin and the respiratory tract.

*Long Term Exposure:* Skin sensitivity may develop. May have effects on the blood, heart, liver, lung, kidney. The state of New Jersey lists the 2-chloro- isomer as a probable carcinogen in humans, and that it causes leukemia and soft tissue cancers in humans.

*Points of Attack:* CNS, blood, bladder

*Medical Surveillance:* If symptoms develop or overexposure is suspected, the following may be useful: Liver function tests. Kidney function tests. Examination by a qualified allergist. EKG.

*First Aid:* If this chemical gets into the eyes, remove any contact lenses at once and irrigate immediately for at least 15 minutes, occasionally lifting upper and lower lids. Seek medical attention immediately. If this chemical contacts the skin, remove contaminated clothing and wash immediately with soap and water. Seek medical attention immediately. If this chemical has been inhaled, remove from exposure, begin rescue breathing (using universal precautions) if breathing has stopped, and CPR if heart action has stopped. Transfer promptly to a medical facility. When this chemical has been swallowed, get medical attention. Give large quantities of water and induce vomiting. Do not make an unconscious person vomit. Do not induce vomiting when formulations containing petroleum solvents are ingested.

*References:*
- Pesticide Management Education Program, "Hexythiazox, (Savey) Chemical Profile 4/89" Cornell University, Ithaca, NY (April 1989). http://pmep.cce.cornell.edu/profiles/insect-mite/fenitrothion-methylpara/hexythiazox/insect-prof-hexythiazox.html
- U.S. Environmental Protection Agency, Office of Pesticide Programs, Pesticide Residue Limits, "Hexythiazox," 40 CFR 180.448, http://www.epa.gov/pesticides/food/viewtols.htm
- California Environmental Protection Agency "Chemical List of Lists," Sacramento CA (February 1997)

# Hydramethylnon (ANSI)

*Use Type:* Insecticide
*CAS Number:* 67485-29-4
*Formula:* $C_{25}H_{24}F_6N_4$
*Synonyms:* AC 217300; AI3-29349; Amidinohydrazone; Caswell No. 642AB; Caswell No. 839A; Pyrimidinone; 1,4-Pentadien-3-one-1,5-bis(α,α,α-trifluoro-p-tolyl)-tetrahydro-5,5-dimethyl-2(1H)-pyrimidinylidene)hydrazone; 2(1H)-Pyrimidinone, tetrahydro-5,5-dimethyl-, [3-(4-(trifluoromethyl)phenyl]-1-[2-(4-(trifluoromethyl)phenyl] ethenyl-2-propenylidene]hydrazone; Tetrahydro-5,5-dimethyl-2(1H)-pyrimidinone[1,5-bis(α,α,α-trifluoro-p-tolyl)-1,4-pentadien-3-one]hydrazone; Tetrahydro-5,5-dimethyl-2(1H)-pyrimidinone[3-(4-(trifluoromethyl)phenyl)]-1-[2-(4-(trifluoromethyl) phenyl)ethenyl]-2-propenylidene] hydrazone

*Trade Names:* AC-217300; AMDRO®, BASF Agricultural Products Germany); CL 217,300®; COMBAT®, BASF Agricultural Products Germany); MATOX®; MAXFORCE® ANT KILLER GRANULAR BAIT, BASF Agricultural Products (Germany); MAXFORCE® ROACH GEL, BASF Agricultural Products (Germany) SENSIBLE®, BASF Agricultural Products Germany); SIEGE®, BASF Agricultural Products Germany); WIPEOUT®

*Producers:* American Cyanamid (USA); BASF Agricultural Products Germany)

*Chemical Class:* Organofluorine

*EPA/OPP PC Code:* 118401

*California DPR Chemical Code:* 2203

*Uses:* Hydramethylnon is a slow-acting toxicant used primarily to control ants in grasses and rangelands and other non-crop lands such as lawns, turf, and non-bearing nursery stock. Hydramethylnon is also registered for the control of household ant species and cockroaches in non-food use areas in and around domestic dwellings and commercial establishments. Hydramethylnon has established tolerances from use on grasses in pastures and rangeland; however hydramethylnon is almost completely metabolized within the body of ruminants and there are no detectable residues in meat, milk, or mbyp. Therefore, tolerances are not required for these commodities even though hydramethylnon is considered a food use pesticide for the purposes of reregistration and tolerance reassessment.

*Human toxicity (long-term)[77]:* Extra high–0.21 ppb, Health Advisory

*Fish toxicity (threshold)[77]:* High–8.90567 ppb, MATC (Maximum Acceptable Toxicant Concentration)

*Carcinogen/Hazard Classifications*

**U.S. EPA Carcinogens:** Group 3, possible carcinogen

*California Prop. 65:* Developmental toxin

**TRI Developmental Toxin:** Reproductive and developmental toxin

**Label Signal Word:** CAUTION

**WHO Acute Hazard:** Class III, slightly hazardous

**Endocrine Disruptor:** A developmental toxin

*Regulatory Authority:*
- Actively registered pesticide in California.
- EPCRA Section 313 Form R *de minimis* concentration reporting level: 1.0%

*Description:* Yellow to tan crystalline solid. Characteristic vegetable oil odor. Insoluble in water. Molecular weight = 494.475. Melting/Freezing point = 190°C. Vapor pressure = 2.3 x $10^{-8}$ mmHg.

*Permissible Concentration in Water:* No criteria set. Runoff from spills or fire control may cause water pollution.

### Harmful Effects and Symptoms

*Short Term Exposure:* Contact with eyes or skin may cause irritation or injury. Inhalation should be avoided; use NIOSH-approved air purifying respirators for pesticides. Toxic if inhaled, ingested, or absorbed through the skin.

*First Aid:* If this chemical gets into the eyes, remove any contact lenses at once and irrigate immediately for at least 15 minutes, occasionally lifting upper and lower lids. Seek medical attention immediately. If this chemical contacts the skin, remove contaminated clothing and wash immediately with soap and water. Seek medical attention immediately. If this chemical has been inhaled, remove from exposure, begin rescue breathing (using universal precautions) if breathing has stopped, and CPR if heart action has stopped. Transfer promptly to a medical facility. When this chemical has been swallowed, get medical attention. Give large quantities of water and induce vomiting. Do not make an unconscious person vomit. Do not induce vomiting when formulations containing petroleum solvents are ingested.

*References:*
- U.S. Environmental Protection Agency, "Reregistration Eligibility Decision (RED), Hydramethylnon," Office of Prevention, Pesticides and Toxic Substances, Washington, DC (December 1998). http://www.epa.gov/REDs/2585red.pdf
- California Environmental Protection Agency "Chemical List of Lists," Sacramento CA (February 1997)

# Hydrazine Sulfate

*Use Type:* Fungicide
*CAS Number:* 10034-93-2
*Formula:* $H_4N_2 \cdot H_2O_4S$
*Alert:* Hydrazine sulfate should be handled as a carcinogen, with extreme caution. It is used to treat some symptoms of advanced cancer.
*Synonyms:* HS; Hydrazine hydrogen; Hydrazine monosulfate; Hydrazine sulphate; Hydrazinium sulfate; Hydrazonium sulfate; Idrazina solfato (Italian); NSC-150014; Siran hydrazinu (Czech)
*Producers:* Atofina (France); Dynamit Nobel (Germany); Fairmont Chemical (USA); Hummel Croton (USA); Janssen Chimica; Mallinckrodt (USA); Otsuka Chemical; Merck (Germany); Spectrum Chemical
*RTECS Number:* MV9625000
*EINECS Number:* 233-110-4
*Uses:* Used in analysis and refining of minerals, rare metals, determination of arsenic in metals, production of rocket fuel and rust prevention products, as a catalyst and antioxidant, and in fungicides, germicides and blood tests. Used as a catalyst for making acetate fibers.

### Carcinogen/Hazard Classifications

**U.S. NTP Carcinogen:** Animal positive (8th Annual Report)
**California Prop. 65:** Carcinogen
**IARC:** Carcinogen (animal carcinogen)
*Regulatory Authority:*
- Carcinogen: (NTP) sufficient animal evidence (IARC) (OSHA)
- AB 2588-Air Toxics "Hot Spots" Chemicals (CAL)
- Proposition 65 chemical (CAL)
- EPCRA Section 313 Form R *de minimus* concentration reporting level: 0.1%.

*Description:* Hydrazine sulfate is a white or colorless, crystalline solid or powder. Soluble in water. Molecular weight = 130.16. Melting/Freezing point = 254°C (decomposition).

*Incompatibilities:* A strong reducing agent. Reacts with oxidizers, bases.

*Routes of Entry:* Inhalation, skin and/or eye contact. Absorbed through the skin.

### Harmful Effects and Symptoms

*Short Term Exposure:* Irritates the eyes and respiratory tract. Exposure can affect the brain and nervous system, causing dizziness and lightheadedness at first, followed by trembling and convulsions.

*Long Term Exposure:* Hydrazine sulfate has been shown to cause liver and lung cancers in animals. Exposure can damage the liver and kidneys. Repeated exposure can damage blood cells causing a low blood count (anemia) It can also cause methemoglobinemia with fatigue, shortness of breath, and even a bluish color to the nose, finger tips and lips. May cause skin allergy to develop.

*Points of Attack:* Liver, kidneys, blood, central nervous system, skin.

*Medical Surveillance:* Liver and kidney function tests. Complete blood count. Examination of the nervous system. Blood methemoglobin level. Evaluation by a qualified allergist.

*First Aid:* If this chemical gets into the eyes, remove any contact lenses at once and irrigate immediately for at least 15 minutes, occasionally lifting upper and lower lids. Seek medical attention immediately. If this chemical contacts the skin, remove contaminated clothing and wash immediately with soap and water. Seek medical attention immediately. If this chemical has been inhaled, remove from exposure, begin rescue breathing (using universal precautions) if breathing has stopped, and CPR if heart action has stopped. Transfer promptly to a medical facility. When this chemical has been swallowed, get medical attention. Give large quantities of water and induce vomiting. Do not make an unconscious person vomit.

*References:*
- New Jersey Department of Health and Senior Services, "Hazardous Substance Fact Sheet, Hydrazine Sulfate," Trenton, NJ (May 2001). http://www.state.nj.us/health/eoh/rtkweb/2360.pdf
- California Environmental Protection Agency "Chemical List of Lists," Sacramento CA (February 1997).

# Hydrogen Cyanide

*Use Type:* Fumigant
*CAS Number:* 74-90-8
*Formula:* CHN; HCN
*Alert:* Hydrogen cyanide is acutely toxic. Persons whose clothing or skin is contaminated with cyanide-containing solutions
can secondarily contaminate response personnel by direct contact or through offgassing vapor.
*Synonyms:* Acide cyanhydrique (French); Acido cianidrico (Italian); Blausaeure (German); Blauwzuur (Dutch); Cyaanwaterstof (Dutch); Cyanwasserstoff (German); Cyjanowodor (Polish); Formonitrile; HCN; Hydrocyanic acid; Prussic acid; Zaclon Discoids
*Trade Names:* AERO LIQUID HCN®, Degesch America (USA), canceled 7/01/1987; CYCLON®; CYCLONE B®
*Producers:* Air Products & Chemicals (USA); ATOFINA (France); BP Chemicals (UK); Ciba-Geigy (Switzerland); Cyanides & Chemicals (India); Degussa (Germany); DSM (Netherlands); Dow Chemical (USA); DuPont (USA); FMC (USA); SKW Chemicals (Germany); Sterling Chemicals (USA); Sumitomo (Japan); Rohm & Haas (USA); Varsal Instruments (USA)
*Chemical Class:* Inorganic
*EPA/OPP PC Code:* 045801
*ICSC Number:* 0492
*RTECS Number:* MW6825000
*EEC Number:* 006-006-00-X
*EINECS Number:* 200-821-6
*Uses:* Hydrogen cyanide is also used in mining, electroplating and in chemical synthesis of plastics and dyes. It may be generated in blast furnaces, gas works, and coke ovens. Cyanide salts have a wide variety of uses, including electroplating, steel hardening, fumigating, gold and silver extraction from ores, and chemical synthesis.
*U.S. Maximum Allowable Residue Levels for Hydrocyanic Acid (40 CFR 180.130):*

| CROP | ppm |
|---|---|
| Fruit, citrus | 50 |

*Carcinogen/Hazard Classifications*
**Label Signal Word:** DANGER
**WHO Acute Hazard:** Group 1b, highly hazardous as the parent chemical sodium cyanide
*Regulatory Authority:*
- Toxic Substance (World Bank)[15]
- Air Pollutant Standard Set (ACGIH)[1] (DFG)[3] (HSE)[33] (Other Countries)[35] (OSHA)[58] (Several States)[60]
- EPA/SARA 302 (EPCRA) Extremely hazardous substances
- AB 2588-Air Toxics "Hot Spots" Chemicals (CAL)
- U.S. DOT Inhalation Hazard Chemicals
- Permissible Exposure Limits for Chemical Contaminants (CAL/OSHA)
- The "Director's List" (CAL/OSHA)
- OSHA 29CFR1910.119, Appendix A, Process Safety List of Highly Hazardous Chemicals, TQ = 1000 lb (454 kg)
- Banned or Severely Restricted (Belgium, Germany, Philippines) (UN)[13]
- Clean Air Act: Accidental Release Prevention/Flammable substances, (Section 112[r], Table 3), TQ = 2,500 lb (1135 kg)
- Clean Water Act: Section 311 Hazardous Substances/RQ 40CFR117.3 (same as CERCLA, see below); Section 313 Water Priority Chemicals (57FR41331, 9/9/92)
- EPA Hazardous Waste Number (RCRA No.): P063
- U.S. DOT Regulated Marine Pollutant (49CFR172.101, Appendix B).
- Superfund/EPCRA 40CFR355, Appendix B Extremely Hazardous Substances: TPQ = 100 lb (45.4 kg)
- Superfund/EPCRA 40CFR302.4 RQ: CERCLA, 10 lb (4.54 kg)
- EPCRA Section 313 Form R de minimus concentration reporting level: 1.0%
- U.S. DOT Regulated Marine Pollutant (49CFR172.101, Appendix B)
- Canada, WHMIS, Ingredients Disclosure List

*Description:* Hydrogen cyanide is a volatile, colorless or pale-blue liquid or colorless gas with a bitter, almond-like odor. The odor threshold is 0.58 ppm. Odor is not a reliable indicator of toxic amounts of vapor; inadequate warning for acute or chronic exposure. Often used as a 96% solution in water. It is intensely poisonous, highly flammable, explosive, and is a weak acid. Specific gravity (water :1) = 0.69. Molecular weight = 27.03 daltons. Boiling point = 25.6°C. Melting/Freezing point = –13.4°C. Gas density (air: 1) = 0.94. Vapor pressure = 630 mmHg @ 20°C. Flash point = –18°C. Autoignition temperature = 538°C. Explosive limits: LEL = 5.6%; UEL = 40.0%[17]. Hazard Identification (based on NFPA-704 M Rating System): Health 4, Flammability 4, Reactivity 2. Soluble in water. Log $K_{ow}$ = 0.38. Unlikely to bioaccumulate in marine organisms.

*Incompatibilities:* Unless stabilized and maintained, samples stored more than 90 days are hazardous. The gas can form an explosive mixture with air. Material containing more than 2-5% water are less stable than dry material and can be self-reactive, forming an explosive mixture with air. Heat above 50–60°C or contact with amines or strong bases can cause polymerization. The aqueous solution is a weak acid. Violent reaction with oxidizers, acetaldehyde, hydrogen chloride in alcoholic mixtures, causing fire and explosion hazard. Incompatible with amines, strong acids, sodium hydroxide, calcium hydroxide, sodium carbonate, water, ammonia. Attacks some plastics, rubber and coatings.

*Permissible Exposure Limits in Air:* The OSHA[2] PEL is 10 ppm (11 mg/m³) STEL, not to be exceeded during any 15 minute work period. The NIOSH[2] REL is 4.7 ppm (5 mg/m³) STEL, not to be exceeded during any 15 minute work period. ACGIH[1] recommends a limit of 4.7 ppm Ceiling, not to be exceeded at any time. Sweden[35] has also set 5 mg/m³ as a ceiling value. Brazil[35] has set 8 ppm (9 mg/m³) as a TWA value and a number of countries (Argentina, Brazil, Germany, Japan) have set 10 ppm (11 mg/m³) as a TWA value. The UK[33] has set no TWA, only an STEL of 10 ppm. ACGIH[1] has set 10 ppm (10 mg/m³) as a ceiling value. Czechoslovakia and the former USSR have set 0.3 mg/m³ as an MAC in workplace air. The notation "skin" is added to indicate the possibility of cutaneous absorption. The NIOSH[2] IDLH level = 50 ppm. AIHA ERPG-2 (emergency response planning guideline) (maximum airborne concentration below which it is believed that nearly all individuals could be exposed for up to 1 hour without experiencing or developing irreversible or other serious health effects or symptoms which could impair an individual's ability to take protective action) = 10 ppm. In ambient air in residential areas, the former USSR has set an MAC of 0.01 mg/mA on a daily average basis. A number of states have set guidelines or standards for hydrogen cyanide in ambient air[60] raging from 33 µg/m³ (New York) to 80 µg/m³ (Virginia) to 100 µg/m³ (North Dakota) to 220 µg/m³ (Connecticut) to 238 µg/m³ (Nevada) to 250 µg/m³ (South Carolina) to 120-1000 µg/m³ (North Carolina).

*Determination in Air:* Soda lime; Water; Visible spectrophotometry; NIOSH IV, Method #6010[18].

*Permissible Concentration in Water:* The U.S. EPA has set a maximum contaminant level of cyanide in drinking water of 0.2 milligrams cyanide per liter of water (0.2 mg/L). A USPHS drinking water criterion for alternate source selection is 100 µg/L[32].

*Routes of Entry:* Hydrogen cyanide is highly toxic by all routes of exposure and may cause abrupt onset of profound CNS, cardiovascular, and respiratory effects, leading to death within minutes.

*Harmful Effects and Symptoms*

*Short Term Exposure:* Exposure to lower concentrations of hydrogen cyanide may produce eye irritation, headache, confusion, nausea, and vomiting followed in some cases by coma and death. Hydrogen cyanide acts as a cellular asphyxiant. By binding to mitochondrial cytochrome oxidase, it prevents the utilization of oxygen in cellular metabolism. The CNS and myocardium are particularly sensitive to the toxic effects of cyanide. Hydrogen cyanide can irritate the and burn the skin and eyes. Inhalation can irritate the respiratory tract. Lacrimation (tearing) and a burning sensation of the mouth and throat are common. Can cause dizziness, headache, weakness, anxiety, confusion, pounding heart, difficult breathing and nausea. These can rapidly lead to convulsions and death unless exposure is immediately stopped and proper first aid applied. High exposure can cause sudden death. Signs and symptoms of acute exposure to hydrocyanic acid may include hypertension (high blood pressure) and tachycardia (rapid heart rate), followed by hypotension (low blood pressure) and bradycardia (slow heart rate). Cherry red mucous membranes and blood may be noted. Cardiac arrhythmias and other cardiac abnormalities are common. Cyanosis (blue tint to the skin and mucous membranes) may be observed. Weakness, headache, vertigo (dizziness), agitation, giddiness, salivation, nausea, and vomiting, may be followed by combative behavior, convulsions, paralysis, protruding eyeballs, dilated and unreactive pupils, and coma. Tachypnea (rapid, shallow respirations) or hyperpnea (rapid, deep respirations) may be followed by respiratory depression. Lung hemorrhage and pulmonary edema may also occur. *Inhalation*: At less than 20 ppm, exposure to hydrogen cyanide may produce headache, dizziness, nausea and vomiting. Concentrations greater than 50 ppm may cause difficulty in breathing, rapid throbbing of the heart, paralysis, unconsciousness, respiratory arrest or death. 30 minutes exposure to 135 ppm may cause death. 270 ppm has caused immediate death. *Skin:* Hydrogen cyanide is readily absorbed through the skin. Symptoms are similar to above. *Eyes:* Hydrogen cyanide is irritating to the eye and rapidly absorbed. *Ingestion;* Symptoms are similar to above. Death has resulted from ingestion of 570 mg/kg of 1.4 oz for a 150 pound person.

*Long Term Exposure:* Chronically exposed workers may complain of headache, eye irritation, easy fatigue, chest discomfort, palpitations, loss of appetite, and nosebleeds. Repeated exposure can interfere with thyroid function and can cause goiter. Itching scarlet rash, red bumps, severe nose itch leading to bleeding, and possibly holes in the nose, may result from long term exposure to hydrogen cyanide. Headache, nausea, vomiting, weakness and enlarged thyroid gland have also been reported at exposures from 4 to 12 ppm. May damage the nervous system. Chronic exposure may be more serious for children because of their potential longer life span. *Carcinogenicity:* Hydrogen cyanide has not been classified for carcinogenic effects, and no carcinogenic effects have been reported for hydrogen cyanide. *Reproductive and Developmental Effects:* No reproductive or developmental effects of hydrogen cyanide have been reported in experimental animals or humans. Increased levels of thiocyanate in the umbilical cords of fetuses whose mothers smoked compared to those whose mothers were non-smokers suggests that thiocyanate, and possibly also cyanide, can cross the placenta. No data were located pertaining to hydrogen cyanide in breast milk.

*Points of Attack:* Central nervous system, cardiovascular system, thyroid, blood.

*Medical Surveillance:* Preplacement and periodic examinations should include the cardiovascular and central nervous systems, liver and kidney function, blood, history of fainting or dizzy spells. Blood cyanide test. Evaluation of thyroid function. Examination of the nervous system. Urinary thiocyanate levels have been used, but are nonspecific and are elevated in smokers.

*First Aid:* If this chemical gets into the eyes, remove any contact lenses at once and irrigate immediately for at least 15 minutes, occasionally lifting upper and lower lids. Seek medical attention immediately. If this chemical contacts the skin, remove contaminated clothing and wash immediately with soap and water. Seek medical attention immediately. If this chemical has been inhaled, remove from exposure, begin rescue breathing (using universal precautions) if breathing has stopped, and CPR if heart action has stopped. Transfer promptly to a medical facility. When this chemical has been swallowed, get medical attention. Give large quantities of water and induce vomiting. Do not make an unconscious person vomit. Use amyl nitrate capsules if symptoms develop. All area employees should be trained regularly in emergency measures for cyanide poisoning and in CPR. A cyanide antidote kit should be kept in the immediate work area and must be rapidly available. Kit ingredients should be replaced every 1-2 years to ensure freshness. Persons trained in the use of this kit, oxygen use, and CPR must be quickly available.

*Note:* Because cyanide is probably largely responsible for poisonings, antidotal measures against cyanide should be instituted promptly. Use amyl nitrate capsules if symptoms develop. All area employees should be trained regularly in emergency measures for cyanide poisoning and in CPR. A cyanide antidote kit should be kept in the immediate work area and must be rapidly available. Kit ingredients should be replaced every 1-2 years.

*Notes to physician or authorized medical personnel*: In cases of respiratory compromise secure airway and respiration via endotracheal intubation. If not possible, perform cricothyroidotomy if equipped and trained to do so. Patients who are in shock or have seizures should be treated according to advanced life support protocols. These patients or those who have arrhythmias may be seriously acidotic; consider giving, under medical supervision, each patient 1 mEq/kg intravenous sodium bicarbonate. *Antidotes* When possible, treatment with cyanide antidotes should be given under medical supervision to unconscious victims who have known or strongly suspected cyanide poisoning. Cyanide antidotes—amyl nitrite perles and intravenous infusions of sodium nitrite and sodium thiosulfate—are packaged in the cyanide antidote kit. Amyl nitrite perles should be broken onto a gauze pad and held under the nose, over the Ambu-valve intake, or placed under the lip of the face mask. Inhale for 30 seconds every minute and use a new perle every 3 minutes if sodium nitrite infusions will be delayed. If the patient has not responded to oxygen and amyl nitrite treatment, infuse sodium nitrite intravenously as soon as possible. The usual adult dose is 10 mL of a 3% solution (300 mg) infused over *absolutely no less than 5 minutes*; the average pediatric dose is 0.12 to 0.33 mL/kg body weight up to 10 mL infused as above. Monitor blood pressure during sodium nitrite administration, and slow the rate of infusion if hypotension develops. Next, infuse sodium thiosulfate intravenously. The usual adult dose is 50 mL of a 25% solution (12.5 g) infused over 10 to 20 minutes; the average pediatric dose is 1.65 mL/kg of a 25% solution. Repeat one-half of the initial dose 30 minutes later if there is an inadequate clinical response.

*References:*
- U.S. Environmental Protection Agency, Office of Pesticide Programs, Pesticide Residue Limits, "Hydrocyanic Acid," 40 CFR 180.130, http://www.epa.gov/pesticides/food/viewtols.htm
- National Institute for Occupational Safety and Health, Criteria for a Recommended Standard: Occupational Exposure to Hydrogen Cyanide, NIOSH Doc. No. 77-108 (1977).
- U.S. Environmental Protection Agency, "Chemical Profile: Hydrocyanic Acid," Washington, DC, Chemical Emergency Preparedness Program (Nov. 30, 1987).
- New York State Department of Health, "Chemical Fact Sheet: Hydrogen Cyanide, " Albany, NY, Bureau of Toxic Substance Assessment (Feb. 1986 and Version 3).
- Sax, N.I., Ed., "Dangerous Properties of Industrial Materials Report, " 1, No. 6, 61-64 (1981).
- New Jersey Department of Health and Senior Services, "Hazardous Substance Fact Sheet, Hydrogen Cyanide," Trenton NJ (June, 1998). http://www.state.nj.us/health/eoh/rtkweb/1013.pdf

# Hydrogen Fluoride

*Use Type:* Disinfectant and fumigant
*CAS Number:* 7664-39-3
*Formula:* HF

*Alert:* Victims whose clothing or skin is contaminated with hydrogen fluoride liquid or solution can secondarily contaminate response personnel by direct contact or through off-gassing vapor. Hydrogen fluoride has a strong irritating odor that is discernable at concentrations of about 0.04 ppm, which is considerably less than the OSHA[(2)] PEL. of 3 ppm.

*Synonyms:* Acido fluorhidrico (Spanish); Anhydrous hydrofluoric acid; Fluorhydric acid; Fluoric Acid; Fluoruro de hidrogeno (Spanish); Hydrogen fluoride, anhydrous; Hydrofluoric acid; Hydrofluoric acid gas

*Producers:* Abaquim (Mexico); Air Products & Chemicals (USA); Allied-Signal (USA); Ashland (USA); Bayer Chemicals (Germany); Central Glass (Japan); Derivados del Fluor (Spain); DuPont (USA); Farleyway Chem. Ltd (UK); General Chemical (USA); Great Lakes Chemical (USA); Hoechst-Celanese (USA); Honeywell Performance P & C (USA); Industrial Quimica de Mexico (Mexico); Interstate Chemical (USA); Matheson Tri-Gas (USA); Morita (Japan); Navin Fluorine Industries (India); Pelcam (South Africa); Praxair (USA); Shanghai Tian Yuan (Group) Corp. (China); Seimi Chemical Ltd. (Japan); Solvay Flour (Belgium); Stella Chemifa (Japan); Vulcan Chemicals (USA)

*Chemical Class:* Inorganic
*EPA/OPP PC Code:* 045601
*ICSC Number:* 0283
*RTECS Number:* MW7875000
*EEC Number:* 009-002-00-6
*EINECS Number:* 231-634-8

*Uses:* Hydrogen fluoride is used as a disinfectant around feeding and watering equipment and holding pens and barns for swine and cattle. Its aqueous solution hydrofluoric acid, and its salts are used in production of organic and inorganic fluorine compounds such as fluorides and plastics; as a catalyst, particularly in paraffin alkylation in the petroleum industry; as an insecticide; and to arrest the fermentation in brewing. It is utilized in the fluorination processes, especially in the aluminum industry, in separating uranium isotopes, in cleaning cast iron, copper, and brass, in removing efflorescence from brick and stone, in removing sand from metallic castings, in frosting and etching glass and enamel, in polishing crystal, in decomposing cellulose, in enameling and galvanizing iron, in working silk, in dye and analytical chemistry, and to increase the porosity of ceramics.

*Carcinogen/Hazard Classifications*
**Label Signal Word: DANGER**
*Regulatory Authority:*
- Toxic Substance (World Bank)[15]
- EPA/SARA 302 (EPCRA) Extremely hazardous substances
- AB 2588-Air Toxics "Hot Spots" Chemicals (CAL)
- Air Pollutant Standard Set (ACGIH)[1] (DFG)[3] (HSE)[33] (Other Countries)[35] (OSHA)[58] (Several States)[60]
- CAL Air Resources Board/AB 1807 Toxic Air Contaminants
- U.S. DOT Inhalation Hazard Chemicals
- Permissible Exposure Limits for Chemical Contaminants (CAL/OSHA)
- The "Director's List" (CAL/OSHA)
- OSHA 29CFR1910.119, Appendix A, Process Safety List of Highly Hazardous Chemicals, TQ = 1000 lb (454 kg)

*Hydrogen fluoride:*
- Clean Air Act: Hazardous Air Pollutants (Title I, Part A, Section 112)
- Clean Water Act: Section 311 Hazardous Substances/RQ 40CFR117.3 (same as CERCLA, see below); Section 313 Water Priority Chemicals (57FR41331, 9/9/92)
- EPA Hazardous Waste Number (RCRA No.): U134
- Superfund/EPCRA 40CFR355, Appendix B Extremely Hazardous Substances: TPQ = 100 lb (45.4 kg)
- Superfund/EPCRA 40CFR302.4 RQ: CERCLA, 100 lb (45.4 kg)
- EPCRA Section 313 Form R *de minimus* concentration reporting level: 1.0%

*Hydrofluoric acid (conc. 50% or greater):*
- Clean Air Act: Hazardous Air Pollutants (Title I, Part A, Section 112); (conc. 50% or greater, or anhydrous) Accidental Release Prevention/Flammable substances, (Section 112[r], Table 3), TQ = 1000 lb (454 kg)
- Clean Water Act: Section 311 Hazardous Substances/RQ 40CFR117.3 (same as CERCLA, see below); Section 313 Water Priority Chemicals (57FR41331, 9/9/92)
- EPA Hazardous Waste Number (RCRA No.): U134
- Superfund/EPCRA 40CFR355, Appendix B Extremely Hazardous Substances: TPQ = 100 lb (45.4 kg)
- Superfund/EPCRA 40CFR302.4 RQ: CERCLA, 100 lb (45.4 kg)
- EPCRA Section 313 Form R *de minimus* concentration reporting level: 1.0%
- Canada, WHMIS, Ingredients Disclosure List

*Description:* Hydrogen fluoride, HF, is colorless, fuming liquid or gas. *Warning properties* = Disagreeable, pungent odor at 0.04 ppm; irritation of eyes and throat at 3 ppm. Specific gravity (water:1) = 1 for liquid @ 20°C. Odor threshold = 0.03 mg/m$^3$. Miscible with water with release of heat. Molecular weight = 20.0 daltons. Gas density (air : 1) = 0.7. Boiling point =19–20°C @ 760 mmHg. Melting/Freezing point = –83°C. Vapor pressure= 783 mmHg @ 20°C; 390 mmHg @ 2.5°C. Nonflammable.

*Incompatibilities:* A super-strong acid; aqueous solutions are less strong. Reacts violently with bases. Reacts, possibly with violence, with many compounds including acetic anhydride, aliphatic amines, alcohols, alkanolamines, alkylene oxides, aromatic amines, amides, 2-aminoethanol, ammonia, ammonium hydroxide, arsenic trioxide, bismuthic acid, calcium oxide, ethylene diamine, ethyleneimine, epichlorohydrin, isocyanates, metal acetylides, nitrogen trifluoride, oleum, organic anhydrides, oxygen difluoride, phosphorous pentoxide, sulfuric acid, strong oxidizers, vinyl acetate, vinylidene fluoride. Attacks glass, concrete, ceramics, and other silicon-containing compounds. Attacks metals, some plastics, rubber and coatings.

*Permissible Exposure Limits in Air:* The OSHA[2] PEL and recommended ACGIH[1] for hydrogen fluoride (measured as fluoride) is 3 ppm TWA. The NIOSH[2] recommended REL is 3 ppm (2.5 mg/m$^3$) and a ceiling of 6 ppm (5 mg/m$^3$) [15 minute]. AIHA ERPG-2 (emergency response planning guideline) (maximum airborne concentration below which it is believed that nearly all individuals could be exposed for up to 1 hour without experiencing or developing irreversible or other serious health effects or symptoms which could impair an individual's ability to take protective action) = 20 ppm. DFG[3] and the HSE[33] have set TWA values of 3 ppm and STEL values of 6 ppm. The NIOSH[2] IDLH level = 30 ppm. Argentina, Japan and Sweden have set TWA value of 3 ppm[35]. The former USSR[35, 43] has set an MAC in workplace air of 0.05 mg/m$^3$ and Czechoslovakia has set 1.0 mg/m$^3$. Limits in ambient air in residential areas have been set by the former USSR[35] at 0.02 mg/mA on a momentary basis and 0.005 mg/m$^3$ on a daily average basis. A number of states have set guidelines or standards for hydrogen fluoride in ambient air[60] ranging from 3.4 μg/m$^3$

(Massachusetts) to 8.3 $\mu g/m^3$ (New York) to 20.0 $\mu g/m^3$ (Virginia) to 25.0 $\mu g/m^3$ (North Dakota and South Carolina) to 30.0 $\mu g/m^3$ (Rhode Island) to 50.0 $\mu g/m^3$ (Connecticut and South Dakota) to 60.0 $\mu g/m^3$ (Nevada) to 25.0-250.0 $\mu g/m^3$ (North Carolina) to 830.0 $\mu g/m^3$ (Kentucky).

*Determination in Air:* Si gel\*; $NaHCO_3/Na_2CO_3$; Ion chromatography; NIOSH IV, Method #7903[18], Inorganic Acids. See also #7902, #7906.

*Permissible Concentration in Water:* No criteria set

*Routes of Entry:* Inhalation, skin absorption, ingestion.

*Harmful Effects and Symptoms*

*Short Term Exposure:* Hydrogen fluoride is irritating to the skin, eyes, and mucous membranes, and inhalation may cause respiratory irritation or hemorrhage. Systemic effects can occur from all routes of exposure and may include nausea, vomiting, gastric pain, or cardiac arrhythmia. Symptoms may be delayed for several days, especially in the case of exposure to dilute solutions of hydrogen fluoride (less than 20%). Hydrofluoric acid is corrosive and also causes destruction of deep tissues when fluoride ions penetrate the skin. Absorption of substantial amounts of hydrogen fluoride by any route may be fatal. The systemic effects of hydrogen fluoride are due to increased fluoride concentrations in the body which can change the levels of calcium, magnesium, and potassium in the blood. Hypocalcemia can cause tetany, decreased myocardial contractility, and possible cardiovascular collapse while hyperkalemia has been suggested to cause ventricular fibrillation leading to death. Hydrogen fluoride is corrosive to the eyes, skin, and the respiratory tract. *Eye burns may not be immediately painful.* Inhalation of this gas can cause pulmonary edema, a medical emergency that can be delayed for several hours. This can cause death. High levels of exposure can cause death. Acute exposure to hydrogen fluoride will result in irritation, burns, ulcerous lesions, and necrosis of the eyes, skin, and mucous membranes. Total destruction of the eyes is possible. Other effects include nausea, vomiting, diarrhea, pneumonitis (inflammation of the lungs), and circulatory collapse. Ingestion of an estimated 1.5 grams produced sudden death without gross pathological damage. Repeated ingestion of small amounts resulted in moderately advanced hardening of the bones. Contact of skin with anhydrous liquid produces severe burns. Inhalation of anhydrous hydrogen fluoride or hydrogen fluoride mist or vapors can cause severe respiratory tract irritation that may be fatal. Hydrogen fluoride may induce hypocalcemia, causing cardiac and renal failure.

*Long Term Exposure:* Repeated ingestion of more than 6 mg of fluoride per day may result in mottling of the teeth in developing children, accumulation of fluoride in the bone, and hardening of the bone in adults and children. Long-term hydrogen fluoride exposure has been reported to damage the kidneys and liver. This chemical may cause bronchitis. Chronic exposure may be more serious for children because of their potential longer latency period. *Carcinogenicity:* Hydrogen fluoride has not been classified for carcinogenic effects. *Reproductive and Developmental Effects:* Fluoride crosses the placenta, and at low doses is thought to be essential for normal fetal development in humans. It is rarely excreted in breast milk. There have been rare cases of mottling of deciduous teeth in infants born to mothers who had high daily intakes of fluoride during pregnancy; skeletal abnormalities are considered unlikely. No reproductive effects due to hydrogen fluoride are known. The substance may cause fluorosis.

*Points of Attack:* Eyes, skin, respiratory system and bones

*Medical Surveillance:* Before beginning employment and at regular times after that, the following is recommended. Lung function tests. If symptoms develop or overexposure has occurred, the following may be useful: Liver, and kidney function tests. Consider chest x-ray after acute overexposure.

*First Aid:* If this chemical gets into the eyes, remove any contact lenses at once (contact lenses should not be worn when working with HF) and irrigate immediately for at least 30 minutes, occasionally lifting upper and lower lids. Seek medical attention immediately. If this chemical contacts the skin, remove contaminated clothing and flush immediately with large amounts of water. Immerse exposed skin area in iced 70% *ethyl alcohol.* Seek medical attention immediately. If this chemical has been inhaled, remove from exposure, begin rescue breathing (using universal precautions) if breathing has stopped, and CPR if heart action has stopped. Transfer promptly to a medical facility. When this chemical has been swallowed, get medical attention. In cases of ingestion, **do not induce vomiting**. Do not administer activated charcoal. Victims who are conscious and able to swallow should be given 4 to 8 ounces of water or milk. If available, also give 2 to 4 ounces of an antacid containing magnesium (e.g., Maalox, milk of magnesia) or calcium (e.g., Tums). Medical observation is recommended for 24 to 48 hours after breathing overexposure, as pulmonary edema may be delayed. As first aid for pulmonary edema, a doctor or authorized paramedic may consider administering a corticosteroid spray.

*Notes to physician or authorized medical personnel*: In cases of respiratory compromise secure airway and respiration via endotracheal intubation. If not possible, perform cricothyroidotomy if equipped and trained to do so. Treat patients who have bronchospasm with aerosolized bronchodilators. The use of bronchial sensitizing agents in situations of multiple chemical exposures may pose additional risks. Consider the health of the myocardium before choosing which type of bronchodilator should be administered. Cardiac sensitizing agents may be appropriate; however, the use of cardiac sensitizing agents after exposure to certain chemicals may pose enhanced risk of cardiac arrhythmias (especially in the elderly). Hydrogen cyanide poisoning is not known to pose additional risk during the use of bronchial or cardiac sensitizing agents. Consider racemic epinephrine aerosol for children who develop stridor. Dose 0.25–0.75 mL of 2.25% racemic epinephrine solution in 2.5 cc water, repeat every 20 minutes as needed, cautioning for myocardial variability. Patients who are comatose, hypotensive, or are having seizures or cardiac arrhythmias should be treated according to advanced life support protocols. Hypocalcemia (manifested by tetany and dysrhythmias) is probable after ingestion of even small amounts of hydrogen fluoride. With medical consultation, treat hypocalcemia with intravenous injections of a 10% solution of calcium gluconate. For inhalation victims, 2.5% calcium gluconate (2.5 g of calcium gluconate in 100 mL of water or 25 mL of 10% calcium gluconate diluted to 100 mL with water) administered by nebulizer with oxygen has been recommended, but the success of this therapy has not been demonstrated.

*References:*
- U.S. Department of Health and Human Services, Agency for Toxic Substances and Disease Registry, "ToxFAQs for Fluorine, Hydrogen Fluoride and Fluorides," Atlanta, GA (September 2001). http://www.atsdr.cdc.gov/tfacts11.html
- New Jersey Department of Health and Senior Services, "Hazardous Substance Fact Sheet: Hydrogen Fluoride," Trenton, NJ (October, 1998). http://www.state.nj.us/health/eoh/rtkweb/1014.pdf
- National Institute for Occupational Safety and Health, Criteria for a Recommended Standard: Occupational Exposure to Hydrogen Fluoride, NIOSH Doc. No. 76-143 (1976).
- U.S. Environmental Protection Agency, Hydrofluoric Acid, Health and Environmental Effects Profile No. 117, Office of Solid Waste, Washington, DC (April 30, 1980).
- Sax, N.I., Ed., "Dangerous Properties of Industrial Materials Report, " 1, No. 6, 64-66 (1981) and 5, No. 6, 52-56 (1985).
- U.S. Environmental Protection Agency, "Chemical Profile: Hydrogen Fluoride," Washington, DC, Chemical Emergency Preparedness Program (Now. 30, 1987).
- California Environmental Protection Agency "Chemical List of Lists," Sacramento CA (February 1997).

## Hydroprene (ANSI)

*Use Type:* Insecticide; insect growth regulator.
*CAS Number:* 41096-46-2 ($E,E$-type); 65733-18-8 ($S$-form)
*Formula:* $C_{17}H_{30}O_2$
*Synonyms:* ENT 70,459; Ethyl ($2E,4E$)-3,7,11-trimethyl-2,4-dodecadienoate; Ethyl ($E,E$)-3,7,11-trimethyl-2,4-dodecadienoate; Ethyl ($2E,4E$)-3,7,11-trimethyl-dodeca-2-4-dienoate; ($2E,4E$)-Hydroprene

*S-form:* Ethyl ($2E,4E,7S$)-trimethyl-2,4-dodecadienoate; ($7S$)-Hydroprene; 2,4-Dodecadienoic acid, 3,7,11-trimethyl-, ethyl ester, [$S$-($E,E$)]-
*Trade Names:* ALTOZAR®; ALTOZAR IGR®; GENCOR®; GENTROL®; MATOR®; OMS 1696; RAID® MAX STERILIZER DISCS; SHA 486300®; ZOECON®, Wellmark International (USA) canceled; ZR 512®; ZR 2006® ($S$-form)
*Producers:* Wellmark International (USA)
*Chemical Class:* Botanical
*EPA/OPP PC Code:* 486300 ($E,E$- type); 128966 ($S$-form)
*California DPR Chemical Code:* 2244
*Uses:* Not currently registered n the U.S. $S$-Hydroprene is used in homes, offices, hospitals, restaurants and other indoor applications as a fogger, spray and impregnated bait against cockroaches, beetles and moths. Not for direct application to food.
*Carcinogen/Hazard Classifications*
**U.S. EPA Carcinogens:** Group D, unclassifiable, insufficient data
**WHO Acute Hazard:** Class U, unlikely to be hazardous
*Description:* An amber liquid. Soluble in water. Molecular weight = 266.418.
*Permissible Concentration in Water:* No criteria set. Runoff from spills or fire control may cause water pollution.
*Routes of Entry:* Inhalation, skin absorption, ingestion.
*Harmful Effects and Symptoms*
*Short Term Exposure:* May cause irritation to the eyes, skin, and respiratory tract. Toxic if ingested or absorbed through the skin.
*Long Term Exposure:* This chemical has been shown to cause developmental effects in animals.
*First Aid:* If this chemical gets into the eyes, remove any contact lenses at once and irrigate immediately for at least 15 minutes, occasionally lifting upper and lower lids. Seek medical attention immediately. If this chemical contacts the skin, remove contaminated clothing and wash immediately with soap and water. Seek medical attention immediately. If this chemical has been inhaled, remove from exposure, begin rescue breathing (using universal precautions) if breathing has stopped, and CPR if heart action has stopped. Transfer promptly to a medical facility. When this chemical has been swallowed, get medical attention. Give large quantities of water and induce vomiting. Do not make an unconscious person vomit. Do not induce vomiting when formulations containing petroleum solvents are ingested.
*References:*
- U.S. Environmental Protection Agency, "Regulating Pesticides, Fact Sheet: Insect Growth Regulators" http://www.epa.gov/oppbppd1/biopesticides/ingredients/factsheets/factsheet_igr.htm
- National Pesticides Information Center (NPIC), "Technical Fact Sheet, Hydroprene" Corvallis, OR (April 2002). http://npic.orst.edu/factsheets/hydropretech.pdf

## Imazalil (ANSI)

*Use Type:* Fungicide
*CAS Number:* 35554-44-0
*Formula:* $C_{14}H_{14}Cl_2N_2O$
*Synonyms:* Allyl-1-(2,4-dichlorophenyl)-2-imidazol-1-ylethyl ether; (±)1-($\beta$-Allyloxy-2,4-dichlorophenethyl)imidazole; (±)-1-[$\beta$-(Allyloxy)-2,4-dichlorophenethyl]imidazole; 1-[2-(Allyloxy)-2-(2,4-dichlorophenyl)ethyl]-1H-imidazole; Caswell No. 497AB; Chloramizol; 1-(2-(2,4-Dichlorophenyl)-2-(2-propenyloxy)ethyl)-1H-imidazole; 1-[2-((2,4-Dichlorophenyl)-2-propenyloxy)-ethyl]-1H-imidazole; Enilconazole; 1H-Imidazole, 1-[2-(2,4-dichlorophenyl)-2-(2-propenyloxy)ethyl]-; 1H-Imidazole, 1-[2-(2,4-dichlorophenyl)-2-(2-propenyloxy)ethyl]-, (±)-
*Trade Names:* BAYTAN IM®, Bayer CropScience (Germany); BROMAZIL®; CEREVAX® EXTRA; CLINAFARM®, Schering-Plough Animal Health (USA); DECCOZIL®, Cerexagri Inc (France); FECUNDAL® 100EC, Janssen Pharmaceutical (USA); FF4961®; FLO-PRO IMZ®, Gustafson (USA); FRESHGARD®; FUNGAFLOR®, Janssen Pharmaceutical (USA); IMAVEROL®; MAGNET®, Makhteshim-Agan Industries (Israel); MIST-O-MATIC® liquid seed treatment; NUZONE®, Wilber Ellis (USA); RAXIL®, Gustafson (USA); RTU-VITAVAX EXTRA®, Gustafson (USA); R 23979®; VITAVAX EXTRA®, Crompton (USA)
*Producers:* Agsin (Singapore); Bayer CropScience (Germany); Cerexagri Inc (France); Crompton (USA); Gustafson (USA); Janssen Pharmacuetica (USA); Makhteshim-Agan Industries (Israel); Schering-Plough Animal Health (USA)
*Chemical Class:* Imidazole
*EPA/OPP PC Code:* 111901
*California DPR Chemical Code:* 2084
*ICSC Number:* 1303
*RTECS Number:* NI47760000
*EEC Number:* 613-042-00-5
*EINECS Number:* 252-615-0

*Uses:* Imazalil is a systemic fungicide used on fruit, vegetables and ornamentals to control powdery mildew, black spot and other fungi. Also used as a seed dressing and for post-harvest applications to bananas, citrus and other fruit. It has been shown that the use of imazalil is less likely to lead to resistant fungi strains than other fungicides.
*Human toxicity (long-term)*[77]: Intermediate–17.50 ppb, Health Advisory
*Fish toxicity (threshold)*[77]: Low–193.77060 ppb, MATC (Maximum Acceptable Toxicant Concentration)

*U.S. Maximum Allowable Residue Levels for Imazalil (40 CFR 180.413):*

| CROP | ppm |
|---|---|
| Banana | 3 |
| Banana, pulp | 0.2 |
| Barley, grain | 0.05 |
| Barley, straw | 0.5 |
| Cattle, fat | 0.01 |
| Cattle, liver | 0.5 |
| Cattle, meat | 0.01 |
| Cattle, mbyp | 0.01 |
| Citrus, dried pulp | 25 |
| Citrus, oil | 25 |
| Fruit, citrus, post-h | 10 |
| Goat, fat | 0.01 |
| Goat, liver | 0.5 |
| Goat, meat | 0.01 |
| Goat, mbyp | 0.01 |
| Hog, fat | 0.01 |
| Hog, liver | 0.5 |
| Hog, meat | 0.01 |
| Hog, mbyp | 0.01 |
| Horse, fat | 0.01 |
| Horse, liver | 0.5 |
| Horse, meat | 0.01 |
| Horse, mbyp | 0.01 |
| Milk | 0.01 |
| Sheep, fat | 0.01 |
| Sheep, liver | 0.5 |
| Sheep, meat | 0.01 |
| Sheep, mbyp | 0.01 |
| Wheat, forage | 0.5 |
| Wheat, grain | 0.05 |
| Wheat, straw | 0.5 |

*Carcinogen/Hazard Classifications*
**U.S. EPA Carcinogens:** Likely carcinogen
**TRI Developmental Toxin:** Reproductive and developmental toxin
**Label Signal Word:** CAUTION or DANGER
**WHO Acute Hazard:** Class II, moderately hazardous
*Regulatory Authority:*
- EPA 40 CFR 372.65, Specific Toxic Chemical Listings
- Actively registered pesticide in California.
- 40CFR185 tolerances in human food
- 40CFR 186 tolerances in animal foods
- EPCRA Section 313 Form R *de minimis* concentration reporting level: 1.0%

*Routes of Entry:* Inhalation, skin absorption, ingestion, eye and/or skin contact.

*Description:* Yellow to brown wax-like solidified oil. Poor solubility in water. Molecular weight = 297.18. Melting/Freezing point = 51°C. Boiling point = 320–350°C (estimated). Vapor pressure = 6.8 x $10^{-8}$ mmHg. Flash point = 192°C. Log $K_{ow}$ = 4.61. Values above 3.0 are likely to bioaccumulate in marine organisms.

*Incompatibilities:* Strong oxidizers. Forms toxic fumes of nitrogen oxides and chlorine when heated to decomposition.

*Permissible Concentration in Water:* No criteria set. Runoff from spills or fire control may cause water pollution.

*Routes of Entry:* Eyes, skin contact.

*Harmful Effects and Symptoms*

*Short Term Exposure:* Causes eye irritation with pain. May cause skin and respiratory tract irritation. May cause nausea if ingested; toxic.

*Long Term Exposure:* May cause cancer and developmental problems. May cause liver problems.

*Medical Surveillance:* Liver function tests.

*First Aid:* If this chemical gets into the eyes, remove any contact lenses at once and irrigate immediately for at least 15 minutes, occasionally lifting upper and lower lids. Seek medical attention immediately. If this chemical contacts the skin, remove contaminated clothing and wash immediately with soap and water. Seek medical attention immediately. If this chemical has been inhaled, remove from exposure, begin rescue breathing (using universal precautions) if breathing has stopped, and CPR if heart action has stopped. Transfer promptly to a medical facility. When this chemical has been swallowed, get medical attention. Give large quantities of water and induce vomiting. Do not make an unconscious person vomit. Do not induce vomiting when formulations containing petroleum solvents are ingested.

*References:*

- EXTOXNET, Extension Toxicology Network, "Pesticide Information Profile, Imazalil," Oregon State University, Corvallis, OR (June 1996). http://extoxnet.orst.edu/pips/imazalil.htm
- U.S. Environmental Protection Agency, "Reregistration Eligibility Decision (RED), Imazalil" Office of Prevention, Pesticides and Toxic Substances, Washington, DC (July 12, 2002). http://www.epa.gov/REDs/imazalil_tred.pdf
- U.S. Environmental Protection Agency, Office of Pesticide Programs, Pesticide Residue Limits, "Imazalil," 40 CFR 180.413. http://www.epa.gov/pesticides/food/viewtols.htm
- International Programme on Chemical Safety (IPCS), "Health and Safety Guide, Pesticide Residues in Food, 2000, Imazalil," Geneva, Switzerland, (2000). http://www.inchem.org/documents/jmpr/jmpmono/v00pr08.htm
- California Environmental Protection Agency "Chemical List of Lists," Sacramento CA (February 1997).

# Imazaquin

*Use Type:* Herbicide, plant growth regulator

*CAS Number:* 81335-37-7; 81335-47-9 (ammonium salt)

*Formula:* $C_{17}H_{17}N_3O_3$

*Synonyms:* 2-[4,5-Dihydro-4-methyl-4-(1-methylethyl)-5-*oxo*-1*H*-imidazol-2-yl]-3-quinolinecarboxylic acid; 2-(5-Isopropyl-5-methyl-4-*oxo*-2-imidazolin-2-yl)-3-quinolinecarboxylic acid; Quinolinecarboxylic acid, 2-(5-isopropyl-5-methyl-4-*oxo*-2-imidazolin-2-yl)-; 3-Quinolinecarboxylic acid, 2-[4,5-dihydro-4-methyl-4-(1-methylethyl)-5-*oxo*-1*H*-imidazol-2-yl]-; (ammonium salt): 3-Quinolinecarboxylic acid, 2-(4,5-dihydro-4-methyl-4-(1-methylethyl)-5-oxo-1H-imidazol-2-yl)-, monoammonium salt

*Trade Names:* AC 252,214®; ALA-SCEPT®, BASF Agricultural Products (Germany); BACKDRAFT®, BASF Agricultural Products (Germany); CL 252,214®; DETAIL®, BASF Agricultural Products (Germany); IMAGE® herbicide consumer concentrate (ammonium salt); MON-9850®, Monsanto (USA); SCEPTER®, BASF Agricultural Products (Germany); SQUADRON® (with Pendimethalin), BASF Agricultural Products (Germany); TRI-SCEPT®, BASF Agricultural Products (Germany); PARTNER®

*Producers:* American Cyanamid (USA); BASF Agricultural Products (Germany)

*Chemical Class:* Imidazolinone

*EPA/OPP PC Code:* 128848; 128840 (ammonium salt)

*California DPR Chemical Code:* 2613

*Uses:* Imazaquin is a selective, pre-emergence and post-emergence herbicide used to control grasses and broadleaf weeds. It is used on corn, wheat, soybeans, turf and ornamentals.

*Human toxicity (long-term)*[77]: Very low–1750.00 ppb, Health Advisory

*Fish toxicity (threshold)*[77]: Very low–61926.11430 ppb, MATC (Maximum Acceptable Toxicant Concentration)

*U.S. Maximum Allowable Residue Levels for Imazaquin (40 CFR 180.426):*

| CROP | ppm |
|---|---|
| Soybean | 0.05 |

*Carcinogen/Hazard Classifications*

**Label Signal Word:** WARNING, CAUTION or DANGER

**WHO Acute Hazard:** Class U, unlikely to be hazardous

*Description:* A tan crystalline solid. Pungent odor. Soluble in water; solubility = 100 mg/L. Melting/freezing point = 220°C. Molecular weight = 311.3.

*Incompatibilities:* May react violently with strong oxidizers, bromine, 90% hydrogen peroxide, phosphorus trichloride, silver powders or dust. Incompatible with silver compounds. Mixture with some silver compounds forms explosive salts of silver oxalate.

*Permissible Concentration in Water:* No criteria set. Runoff from spills or fire control may cause water pollution.

*Harmful Effects and Symptoms*
*Short Term Exposure:* May cause irritation to the eyes, skin, or respiratory tract. May be toxic if ingested or absorbed through the skin.
*First Aid:* If this chemical gets into the eyes, remove any contact lenses at once and irrigate immediately for at least 15 minutes, occasionally lifting upper and lower lids. Seek medical attention immediately. If this chemical contacts the skin, remove contaminated clothing and wash immediately with soap and water. Seek medical attention immediately. If this chemical has been inhaled, remove from exposure, begin rescue breathing (using universal precautions) if breathing has stopped, and CPR if heart action has stopped. Transfer promptly to a medical facility. When this chemical has been swallowed, get medical attention. Give large quantities of water and induce vomiting. Do not make an unconscious person vomit.
*References:*
- EXTOXNET, Extension Toxicology Network, "Pesticide Information Profile, Imazaquin," Oregon State University, Corvallis, OR. http://extoxnet.orst.edu/pips/imazaqui.htm
- U.S. Environmental Protection Agency, Office of Pesticide Programs, Pesticide Residue Limits, "Imazaquin," 40 CFR 180.426, www.epa.gov/pesticides/food/viewtols.htm
- California Environmental Protection Agency "Chemical List of Lists," Sacramento CA (February 1997)

# Imazethabenz (U.S. EPA)

*Use Type:* Herbicide
*CAS Number:* 81405-85-8
*Formula:* $C_{16}H_{20}N_2O_3$
*Synonyms:* Benzoic acid, 2-(4,5-dihydro-4-methyl-4-(1-methylethyl)-5-*xox*-1*H*-imidazol-2-yl)-4 (or 5)-methyl-, methyl ester; 2[4,5-Dihydro-4-methyl-4-(1-methylethyl)-5-*oxo*-1*H*-imidazol-2-yl]-4 (or 5)-methylbenzoic acid methyl ester; Imazamethabenz; (Methyl 6-(4-isopropyl-4-methyl-5-*oxo*-2-imidazolin-2-yl)-*m*-toluate plus; Methyl 6-(4-isopropyl-4-methyl-5-*oxo*-2-imidazolin-2-yl)-*m*-toluate); (Methyl 6-(4-isopropyl-4-methyl-5-*oxo*-2-imidazolin-2-yl)-*m*-toluate plus Methyl 6-(4-isopropyl-4-methyl-5-*oxo*-2-imidazolin-2-yl)-*m*-toluate); *m*-(or *p*-)Toluic acid, 6-(4-isopropyl-4-methyl-5-*oxo*-2-imidazolin-2-yl)-, methyl ester; *m*-(or *p*-)Toluic acid, 2-(4,5-dihydro-4-methyl-4-isopropyl-5-*oxo*-1*H*-imidazol-2-yl)-, methyl ester
*Trade Names:* ASSERT® (Benzoic acid, 2-(4,5-dihydro-5-methyl-4-(1-methylethyl) salt-5-oxo-1H-imidazol-2-yl)-5-methyl-, methyl ester and Benzoic acid, 2-(4,5-dihydro-4-methyl-4-(1-methylethyl) salt-5-oxo-1H-imidazol-2-yl)-4-methyl-, methyl ester), BASF Agricultural Products (Germany); AC 222293®; AC-293®; CL-222293®; DAGGER®

*Producers:* American Cyanamid (USA); BASF Agricultural Products (Germany); BASF Canada (Canada)
*Chemical Class:* Imidazolinone
*EPA/OPP PC Code:* 128842
*California DPR Chemical Code:* 2240
*Uses:* Generally used as a herbicide on small grain crops.
*Human toxicity (long-term)*[77]*:* Very low–441.00 ppb, Health Advisory
*Fish toxicity (threshold)*[77]*:* Very low–515.35916 ppb, MATC (Maximum Acceptable Toxicant Concentration)
*Carcinogen/Hazard Classifications*
**U.S. EPA Carcinogens:** Group D, unclassifiable
**Label Signal Word:** CAUTION or DANGER
**WHO Acute Hazard:** Class U, unlikely to be hazardous
*Regulatory Authority:*
- Actively registered pesticide in California.

*Description:* White or slightly yellow powder. Slightly soluble in water. Vapor pressure = $1.2 \times 10^{-8}$ mmHg.
*Permissible Concentration in Water:* No criteria set. Runoff from spills or fire control may cause water pollution.
*Harmful Effects and Symptoms*
*Short Term Exposure:* May cause irritation to the eyes, skin, or respiratory tract. May be toxic if ingested or absorbed through the skin.
*First Aid:* If this chemical gets into the eyes, remove any contact lenses at once and irrigate immediately for at least 15 minutes, occasionally lifting upper and lower lids. Seek medical attention immediately. If this chemical contacts the skin, remove contaminated clothing and wash immediately with soap and water. Seek medical attention immediately. If this chemical has been inhaled, remove from exposure, begin rescue breathing (using universal precautions) if breathing has stopped, and CPR if heart action has stopped. Transfer promptly to a medical facility. When this chemical has been swallowed, get medical attention. Give large quantities of water and induce vomiting. Do not make an unconscious person vomit.
*References:*
- California Environmental Protection Agency "Chemical List of Lists," Sacramento CA (February 1997)

# Imazethapyr (ANSI)

*Use Type:* Herbicide, plant growth regulator
*CAS Number:* 81335-77-5 (U.S. EPA); 101917-66-2 (ammonium salt)
*Formula:* $C_{15}H_{19}N_3O_3$; $C_{15}H_{22}N_4O_3$ (ammonium salt)
*Synonyms:* Ammonium salt of(±)-2-(4,5-dihydro-4-methyl-4-(1-methylethyl)-5-*oxo*-1H-imidazol-2-yl)-5-ethyl-3-pyridinecarboxylic acid(±)-2-(4,5-Dihydro-4-methyl-4-(1-methylethyl)-5-oxo-1H-imidazol-2-yl)-5-ethyl-3-pyridinecarboxylic acid, ammonium salt; (±)-2-[4,5-Dihydro-4-methyl-4-(1-methylethyl)-5-*oxo*-1*H*-imadazol-2-yl]-5-ethyl-3-pyridinecarboxylic acid; (±)-5-Ethyl-2-(4-

isopropyl-4-methyl-5-*oxo*-2-imidazolin-2-yl)nicotinic acid; (±)-5-Ethyl-2-(4-isopropyl-4-methyl-5-*oxo*-1H-imidazolin-2-yl)nicotinic acid (ammonium salt)
*Trade Names:* AC-263499®; CL-263499®; CONTOUR®; EXTREME®, BASF Agricultural Products Group (Germany); HAMMER®; LIGHTNING®, BASF Agricultural Products Group (Germany); ODYSSEY®, (imazamox + imazethapyr), BASF Agricultural Products Group (Germany); OVERTOP®; PATRIOT®, (atrazine + imazethapyr), BASF Canada (Canada); PASSPORT®; PIVOT®; PURSUIT,® (ammonium salt of), BASF Agricultural Products Group (Germany); PURSUIT DG® Herbicide, BASF Agricultural Products Group (Germany); RESOLVE®, BASF Agricultural Products Group (Germany); STANDOUT®, BASF Agricultural Products Group (Germany); VALOR®, (imazethapyr + pendimethalin), BASF Canada (Canada)
*Producers:* Agsin (Singapore); American Cyanamid (USA); BASF Agricultural Products Group (Germany); BASF Canada (Canada)
*Chemical Class:* Imidazolinone
*EPA/OPP PC Code:* 128922; 128923 (ammonium salt)
*California DPR Chemical Code:* 2340
*Uses:* Imazethapyr is a general use, selective pre-emergence herbicide that is used to control grasses and broadleaf weeds on a variety of field and vegetable crops including dry and edible beans, peas, soybeans, peanuts, alfalfa and corn.
*Human toxicity (long-term)*[77]: Very low–1750.00 ppb, Health Advisory
*Fish toxicity (threshold)*[77]: Very low–52267.57728 ppb, MATC (Maximum Acceptable Toxicant Concentration)
*U.S. Maximum Allowable Residue Levels for Imazethapyr (40 CFR 180.447):*

| CROP | ppm |
|---|---|
| Alfalfa, forage | 3 |
| Alfalfa, hay | 3 |
| Cattle, mbyp | 0.1 |
| Corn, field, forage | 0.1 |
| Corn, field, grain | 0.1 |
| Corn, field, stover | 0.1 |
| Crayfish | 0.1 |
| Endive | 0.1 |
| Goat, mbyp | 0.1 |
| Hog, mbyp | 0.1 |
| Horse, mbyp | 0.1 |
| Lettuce, head | 0.1 |
| Lettuce, leaf | 0.1 |
| Peanut | 0.1 |
| Rice, bran | 1.2 |
| Rice, grain | 0.2 |
| Rice, straw | 0.15 |
| Sheep, mbyp | 0.1 |
| Soybean | 0.1 |
| Vegetable, legume, group 6 | 0.1 |

*Carcinogen/Hazard Classifications*
**Label Signal Word:** WARNING, CAUTION or DANGER
**WHO Acute Hazard:** Class U, unlikely to be hazardous
*Regulatory Authority:*
- Actively registered pesticide in California.

*Description:* White to slightly yellowish crystalline solid. Pungent odor. Soluble in water; solubility = 1.42 x $10^3$ mg/L. Molecular weight = 289.33. Melting/Freezing point = 173°C.
*Incompatibilities:* May react violently with strong oxidizers, bromine, 90% hydrogen peroxide, phosphorus trichloride, silver powders or dust. Incompatible with silver compounds. Corrosive to zinc, mild steel, brass, copper and aluminum. Mixture with some silver compounds forms explosive salts of silver oxalate.
*Permissible Concentration in Water:* No criteria set. Runoff from spills or fire control may cause water pollution.
*Harmful Effects and Symptoms*
*Short Term Exposure:* May cause irritation to the eyes, skin, or respiratory tract. May be toxic if ingested or absorbed through the skin.
*First Aid:* If this chemical gets into the eyes, remove any contact lenses at once and irrigate immediately for at least 15 minutes, occasionally lifting upper and lower lids. Seek medical attention immediately. If this chemical contacts the skin, remove contaminated clothing and wash immediately with soap and water. Seek medical attention immediately. If this chemical has been inhaled, remove from exposure, begin rescue breathing (using universal precautions) if breathing has stopped, and CPR if heart action has stopped. Transfer promptly to a medical facility. When this chemical has been swallowed, get medical attention. Give large quantities of water and induce vomiting. Do not make an unconscious person vomit.
*References:*
- EXTOXNET, Extension Toxicology Network, "Pesticide Information Profile, Imazethapyr," Oregon State University, Corvallis, OR (February 1996). http://extoxnet.orst.edu/pips/imazetha.htm
- U.S. Environmental Protection Agency, Office of Pesticide Programs, Pesticide Residue Limits, "Imazethapyr," 40 CFR 180.447. http://www.epa.gov/pesticides/food/viewtols.htm
- California Environmental Protection Agency "Chemical List of Lists," Sacramento CA (February 1997)

# Imidacloprid

*Use Type:* Insecticide
*CAS Number:* 105827-78-9; 138261-41-3 (tautomeric form)
*Formula:* $C_9H_{10}ClN_5O_2$
*Synonyms:* 1-[(6-Chloro-3-pyridinyl)methyl]-*N*-nitro-2-imidazolidiniminebenzoate; 1-[(6-Chloro-3-

pyridinyl)methyl]-4,5-dihydro-*N*-nitro-1*H*-imidazol-2-amine; 1-(2-Chloro-5-pyridylmethyl)-2-(nitroamino)imidazolidine; 1*H*-Imidazol-2-amine, 1-[(6-chloro-3-pyridinyl)methyl]-4,5-dihydro-*N*-nitro-; 2-Imidazolidinimine, 1-[(6-chloro-3-pyridinyl)methyl]-*N*-nitro-benzoate

*Trade Names:* ADMIRE®, BayerCropScience (Germany); CONFIDOR® 2.5% granular; CONFIDOR® 2 flowable; ENCORE®, Bayer CropScience (Germany); GAUCHO®, Gustafson (USA); IMICIDE®, J. J. Mauget (USA); LEVERAGE®, Bayer CropScience (Germany); MARATHON®; MERIT®, BayerCropScience (Germany); NTN 33893®; PREMIER®; PREMISE®; PRESCRIBE™, Gustafson (USA); PROTREAT®, Rotam Agrochemical (Hong Kong); PROVADO®, Bayer CropScience (Germany); TRIMAX®, Bayer CropScience (Germany)

*Producers:* Agrimor International (USA); Agsin (Singapore); AJE (Switzerland); Ascot International UK); BayerCropScience (Germany); Gustafson (USA); J. J. Mauget (USA); Ki-Hara Chemicals Ltd. (UK); SuYan Agrochemical Group (China); Hindustan Insecticides (India); Rotam Agrochemical (Hong Kong); PROTREAT Rotam Agrochemical (Hong Kong); Wuzhou International (China)

*Chemical Class:* Chloronicotinyl
*EPA/OPP PC Code:* 129059; 129099
*California DPR Chemical Code:* 3849

*Uses:* A systemic insecticide used to control sucking insects in the soil, seeds and foliar environments. It is one of the most-used pesticides in the world and has a broad variety of both agriculture and non-agricultural uses, including on pets and household environments. Used on rice, cereals, maize, vegetables, fruit, sugar beets, potatoes, cotton, hops and turf. It is related to nicotine and acts on the nervous system.

*Human toxicity (long-term)[77]:* Very low–399.00 ppb, Health Advisory

*Fish toxicity (threshold)[77]:* Very low–1199.99656 ppb, MATC (Maximum Acceptable Toxicant Concentration)

*U.S. Maximum Allowable Residue Levels for Imidacloprid (40 CFR 180.472):*

*Note:* The U.S. EPA list 129 tolerances for crops on the site shown in the Reference section below.

*Carcinogen/Hazard Classifications*
**U.S. EPA Carcinogens:** Group E, Unlikely a carcinogen
**Label Signal Word:** WARNING, CAUTION or DANGER
**WHO Acute Hazard:** Class II, moderately hazardous

*Description:* Colorless, crystalline solid. Slight characteristic odor. Slightly soluble in water; solubility = 0.5 g/L. Molecular weight = 255. 63.

*Permissible Concentration in Water:* No criteria set. Runoff from spills or fire control may cause water pollution.

*Harmful Effects and Symptoms*
*Short Term Exposure:* May attack central nervous system causing apathy, incoordination. *Inhalation:* Causes labored breathing. *Ingestion:* May cause convulsions.

*Points of Attack:* Central nervous system, lungs, digestion system.

*First Aid:* If this chemical gets into the eyes, remove any contact lenses at once and irrigate immediately for at least 15 minutes, occasionally lifting upper and lower lids. Seek medical attention immediately. If this chemical contacts the skin, remove contaminated clothing and wash immediately with soap and water. Seek medical attention immediately. If this chemical has been inhaled, remove from exposure, begin rescue breathing (using universal precautions) if breathing has stopped, and CPR if heart action has stopped. Transfer promptly to a medical facility. When this chemical has been swallowed, get medical attention. Give large quantities of water and induce vomiting. Do not make an unconscious person vomit.

*References:*
- EXTOXNET, Extension Toxicology Network, "Pesticide Information Profile, Imidacloprid," Oregon State University, Corvallis, OR http://extoxnet.orst.edu/pips/imidaclo.htm
- *Journal of Pesticide Reform,* "Insecticide Factsheet, Imidacloprid," Northwest Coalition for Alternatives to Pesticides, Spring 2001, Vol. 21, No. 1. http://www.pesticide.org/imidacloprid.pdf
- U.S. Environmental Protection Agency, Office of Pesticide Programs, Pesticide Residue Limits, "Imidacloprid", 40 CFR 180.472, www.epa.gov/pesticides/food/viewtols.htm
- California Environmental Protection Agency "Chemical List of Lists," Sacramento CA (February 1997)

# Indole-3-butyric Acid

*Use Type:* Plant growth regulator
*CAS Number:* 133-32-4
*Formula:* $C_{12}H_{13}NO_2$

*Synonyms:* IBA; 1*H*-Indole-3-butanoic acid; Indole butyric; Indole butyric acid; β-Indolebutyric acid; γ-(Indole-3)-butyric acid; 3-Indolebutyric acid; γ-(Indol-3-yl)butyric acid; Indolyl-3-butyric acid; 3-Indolyl-γ-butyric acid; γ-(3-Indolyl)butyric acid; 4-(Indolyl)butyric acid; 4-(Indol-3-yl)butyric acid; 4-(3-Indolyl)butyric acid

*Trade Names:* ASSET PGR®, Helena Chemical (USA); BOLL-SET®, Micro-flo (USA); CROP BOOSTER®, Agriliance (USA); CYTOPLEX®; GOLDENGRO®; HORMEX®, Brooker Chemical (USA); HORMODIN®, E.C. Geiger (USA); JIFFY GROW®; MAXON®, Agriliance (USA); MEPEX®, Griffin (USA); PGR-IV®, Micro-flo (USA); RHIZOPON®; ROOTGRO®; ROOTONE® (with 1-Naphthaleneacetamide and 1-Naphthaleneacetic acid); SERADIX®; SNIPPER®

*Producers:* Agriliance (USA); Bonide Products (USA); Eurolabs (UK); Griffin (USA); Helena Chemical (USA); Micro-flo (USA)

*Chemical Class:* Botanical
*EPA/OPP PC Code:* 046701
*California DPR Chemical Code:* 323
*EINECS Number:* 205-101-5
*Uses:* Used as growth hormones for root cuttings.
*Carcinogen/Hazard Classifications*
**Label Signal Word:** CAUTION
*Regulatory Authority:*
• Actively registered pesticide in California.
*Description:* White or off-white crystalline solid or powder. Odorless. Insoluble in water. Molecular weight = 203.258. Melting/Freezing point = 123°C.
*Harmful Effects and Symptoms*
*Short Term Exposure:* May cause irritation of the eyes, skin, or respiratory tract. Toxic if ingested.
*First Aid:* If this chemical gets into the eyes, remove any contact lenses at once and irrigate immediately for at least 15 minutes, occasionally lifting upper and lower lids. Seek medical attention immediately. If this chemical contacts the skin, remove contaminated clothing and wash immediately with soap and water. Seek medical attention immediately. If this chemical has been inhaled, remove from exposure, begin rescue breathing (using universal precautions) if breathing has stopped, and CPR if heart action has stopped. Transfer promptly to a medical facility. When this chemical has been swallowed, get medical attention. Give large quantities of water and induce vomiting. Do not make an unconscious person vomit. Do not induce vomiting when formulations containing petroleum solvents are ingested.
*References:*
California Environmental Protection Agency "Chemical List of Lists," Sacramento CA (February 1997)

# Iprodione (ANSI)

*Use Type:* Fungicide
*CAS Number:* 36734-19-7
*Formula:* $C_{13}H_{13}Cl_2N_3O_3$
*Synonyms:* 3-(3,5-Dichlorophenyl)-*N*-(1-methylethyl)-2,4-dioxo-1-imidazolidinecarboxamide; 3-(3,5-Dichlorophenyl)-N-sopropyl-2,4-dioxo-1-imidazolidinecarboximide; Glycophen; 1-Isopropyl carbamoyl-3-(3,5-dichlorophenyl)-hydantoin
*Trade Names:* ANFOR®; CHIPCO® 26019, Bayer CropScience (Germany); DIVA®, Bayer CropScience (Germany); DOP® 26019; GLYCOPHENE®; IPRODINE®; KIDEN®; LFA 2043®; MRC 910®; PROMIDIONE®; PROTURF®, The Scotts Company (USA); ROP 500 F®; ROVRAL®, Bayer CropScience (Germany); RP-26019®; VERISAN®
*Producers:* Agriliance (USA); Agrimor International (USA); Agsin (Singapore); Bayer CropScience (Germany); Ki-Hara Chemicals Ltd. (UK); Micro-Flo (USA); The Scotts Company (USA); Wuzhou International (China)

*Chemical Class:* Dicarboximide
*EPA/OPP PC de:* 109801; (209900 old EPA code number)
*California DPR Chemical Code:* 2081
*EINECS Nuer:* 253-178-9
*Uses:* Iprodione is a contact and/or locally systemic fungicide used to control a broad range of root and stem rots, molds and mildews on a variety of field, fruit, and vegetable crops including almonds, grapes, peaches, potatoes, rice, berries, onions, peanuts and lettuce. Registration does not permit uses on turf, ornamentals and vegetable and small fruit gardens. End-uses for the formulations have been classified for outdoor use only the application methods have been restricted so to avoid undue human contact. Iprodione can also be used as a post harvest fungicide and seed treatment.
*Human toxicity (long-term)*[77]: High–7.97268 ppb, CHCL (Chronic Human Carcinogen Level)
*Fish toxicity (threshold)*[77]: Low–378.15513 ppb, MATC (Maximum Acceptable Toxicant Concentration)
*Carcinogen/Hazard Classifications*
**U.S. EPA Carcinogens:** Likely a carcinogen
**California Prop. 65:** Carcinogen
**Label Signal Word:** WARNING, CAUTION
**WHO Acute Hazard:** Class U, unlikely to be hazardous
**Endocrine Disruptor:** Suspected endocrine disruptor
*Regulatory Authority:*
• Actively registered pesticide in California.
*Description:* White, odorless crystalline solid. Melting/Freezing point = 135°C. Molecular weight = 330.18. Vapor pressure = $1.2 \times 10^{-7}$ mmHg.
*Incompatibilities:* Slowly hydrolyzes in water, releasing ammonia and forming acetate salts.
*Permissible Exposure Limits in Air:*
*Determination in Air:* Filter; none; Gravimetric; NIOSH IV [Particulates NOR; #0500 (total), #0600 (respirable)].[18]
*Permissible Concentration in Water:* No criteria set. Runoff from spills or fire control may cause water pollution.
*Harmful Effects and Symptoms*
*Short Term Exposure:* Contact with eyes or skin may cause irritation or injury. Inhalation should be avoided; use NIOSH-approved air purifying respirators for pesticides. May be harmful if swallowed.
*First Aid:* If this chemical gets into the eyes, remove any contact lenses at once and irrigate immediately for at least 15 minutes, occasionally lifting upper and lower lids. Seek medical attention immediately. If this chemical contacts the skin, remove contaminated clothing and wash immediately with soap and water. Seek medical attention immediately. If this chemical has been inhaled, remove from exposure, begin rescue breathing (using universal precautions) if breathing has stopped, and CPR if heart action has stopped. Transfer promptly to a medical facility. When this chemical has been swallowed, get medical attention. Give large quantities of water and induce vomiting. Do not make an

unconscious person vomit. Do not induce vomiting when formulations containing petroleum solvents are ingested.

*References:*
- U.S. Environmental Protection Agency, "Reregistration Eligibility Decision (RED), Iprodione" Office of Prevention, Pesticides and Toxic Substances, Washington, DC (November 1998). http://www.epa.gov/REDs/2335.pdf
- EXTOXNET, Extension Toxicology Network, "Pesticide Information Profile, Iprodione," Oregon State University, Corvallis, OR (June 1996). http://extoxnet.orst.edu/pips/iprodion.htm
- California Environmental Protection Agency "Chemical List of Lists," Sacramento CA (February 1997)

## Isobenzan

*Use Type:* Insecticide
*CAS Number:* 297-78-9
*Formula:* $C_9H_4Cl_8O$
*Alert:* This substance is highly toxic and is seldom, if ever, used today.
*Synonyms:* CP 14,957; ENT 25,545; ENT 25,545-x; Isobenzano (Spanish); Octochlorohexahydromethano isobenzofuran; 1,3,4,5,6,8,8-Octochloro-1,3,3a, 4,7,7a-hexahydro-4,7-methanoisobenzofuran; 1,3,4,5,6,7,10,10-Octochloro-4,7-endo-methylene-4,7,8,9-tetrahydrophthalan; 1,3,4,5,6,7,8,8-Octochloro-2-oxa-3a,4,7,7a-tetrahydro-4,7-methanoindene
*Trade Names:* OMTAN®; R 6700®; SD 440®, Shell Chemicals (UK); SHELL 4402®, Shell Chemicals (UK); SHELL WL 1650®, Shell Chemicals (UK); TELODRIN®; WL 1650®
*Producers:* Shell Chemicals (UK)
*Chemical Class:* Organochlorine; Halo-organics
*EPA/OPP PC Code:* 058501
*RTECS Number:* PC1225000
*Uses:* This broad spectrum insecticide was used throughout the world and manufactured from 1958 to 1965. Its use in agriculture was restricted because of its persistence and toxicity.
*Regulatory Authority:*
- Banned or Severely Restricted (Several Countries) (UN)[13]
- Very Toxic Substance (World Bank)[15]
- Superfund/EPCRA 40CFR355, Appendix B Extremely Hazardous Substances: TPQ = 100/10,000 lb (45.4/4,540 kg)
- Superfund/EPCRA 40CFR302.4 RQ: EHS, 1 lb (0.454 kg)

*Description:* Isobenzan is a white to light brown crystalline powder with a mild odor. Insoluble in water. Melting/Freezing point = 120-22°C. Hazard Identification (based on NFPA-704 M Rating System): Health 4, Flammability 0, Reactivity 0.
*Incompatibilities:* Strong oxidizers.
*Permissible Exposure Limits in Air:* No standards set.
*Permissible Concentration in Water:* No criteria set, but runoff from spills or fire control may cause water pollution.
*Routes of Entry:* Inhalation, ingestion, through the skin.
*Harmful Effects and Symptoms*
*Short Term Exposure:* This material is highly toxic. It is absorbed by the skin as well as by the respiratory and gastrointestinal tract. Symptoms may last for a long time because the material is eliminated slowly; its half-life in human blood is 2.77 years. Symptoms of exposure include headache, dizziness, drowsiness, irritability, and numbness of the legs. Convulsions may occur.
*Points of Attack:* Central nervous system
*First Aid:* Speed in removing material from eyes and skin is of extreme importance. If this chemical gets into the eyes, remove any contact lenses at once and irrigate immediately for at least 15 minutes, occasionally lifting upper and lower lids. Seek medical attention immediately. If this chemical contacts the skin, remove contaminated clothing and wash immediately with soap and water. Shampoo hair promptly if contaminated. Seek medical attention immediately. If this chemical has been inhaled, remove from exposure, begin rescue breathing (using universal precautions) if breathing has stopped, and CPR if heart action has stopped. Transfer promptly to a medical facility. When this chemical has been swallowed, get medical attention. Give large quantities of water and induce vomiting. Do not make an unconscious person vomit.
*References:*
- New Jersey Department of Health and Senior Services, "Hazardous Substance Fact Sheet, Isobenzan," Trenton, NJ (August 2000). http://www.state.nj.us/health/eoh/rtkweb/2494.pdf
- U.S. Environmental Protection Agency, "Chemical Profile: Isobenzan," Washington, DC, Chemical Emergency Preparedness Program (Nov. 30, 1987).
- California Environmental Protection Agency "Chemical List of Lists," Sacramento CA (February 1997).

## Isodrin

*Use Type:* Insecticide
*CAS Number:* 465-73-6
*Formula:* $C_{12}H_8Cl_6$
*Alert:* Isodrin must be synthesized and is no longer manufactured or used commercially in the United States.
*Synonyms:* AI3-19244; compound 711; 1,4:5,8-Dimethano naphthalene, 1,2,3,4,10,10-hexachloro-1,4,4a,5,8,8a-hexahydro-, (1α,4.α,4A.β,5 β,8 β,8A β)-; 1,4:5,8-Ddimethano naphthalene, 1,2,3,4,10,10-hexachloro-1,4,4a,5,8,8a-hexahydro-, *endo, endo*-; ENT 19,244;

(1α,4α,4aβ,5β,8β,8a.β)-1,2,3,4,10,10-Hexachloro-1,4,4a,-5,8,8a-hexahydro-1,4:5,8-dimethanonaphthalene; 1,2,3,4,10,10-Hexachloro-1,4,4a,5,8,8a-hexahydro-1,4:5,8-*endo, endo*-dimethanon aphthalene; 1,2,3,4,10,10-Hexachloro-1,4,4a,5,8,8a-hexahydro-1,4-*endo, endo*-5,8-dimethanon aphthalene; Isodrina (Spanish)

*Producers:* Sigma-Aldrich Laborchemikalien (Germany)
*Chemical Class:* Organochlorine; Halo-organics
*RTECS Number:* IO1925000
*Regulatory Authority:*
- Banned or Severely Restricted (Several Countries) (UN)[13]
- Very Toxic Substance (World Bank)[15]
- EPA Hazardous Waste Number (RCRA No.): P060
- RCRA, 40CFR261, Appendix 8 Hazardous Constituents
- RCRA 40CFR268.48; 61FR15654, Universal Treatment Standards: Wastewater (mg/L), 0.021; Nonwastewater (mg/kg), 0.066
- RCRA 40CFR264, Appendix 9; TSD Facilities Ground Water Monitoring List. Suggested test method(s) (PQL *ug*/L): 8270(10)
- Superfund/EPCRA 40CFR355, Appendix B Extremely Hazardous Substances: TPQ = 100/10,000 lb (45.4/4,540 kg)
- Superfund/EPCRA 40CFR302.4 RQ: EHS, 1 lb (0.454 kg)
- EPCRA Section 313 Form R *de minimus* concentration reporting level: 1.0%

*Description:* Isodrin, an isomer of aldrin, is a crystalline solid. Melting/Freezing point = 241°C. Decomposes above 100°C. Hazard Identification (based on NFPA-704 M Rating System): Health 3, Flammability 1, Reactivity 0. May be dissolved in flammable liquids.

*Incompatibilities:* Incompatible with concentrated mineral acids, acid catalysts, acid oxidizing agents, phenols, reactive metals.

*Permissible Exposure Limits in Air:* No standards set.

*Determination in Air:* There is no OEL established for Isodrin. However, this chemical is an isomer of aldrin. See Aldrin.

*Permissible Concentration in Water:* No criteria set, but runoff from spills or fire control may cause water pollution.

*Routes of Entry:* Inhalation, ingestion, eyes and/or skin.

*Harmful Effects and Symptoms*

*Short Term Exposure:* Isodrin is classified as extremely toxic. Probable oral lethal dose for humans is 5-50 mg/kg or between 7 drops and 2 teaspoonful for a 70 kg (150 lb) person. It causes renal damage and hyperactivity of sympathetic nervous system. Symptoms experienced are similar to poisoning by dieldrin and aldrin, including overall discomfort, headache, nausea, vomiting, dizziness, tremors, convulsions, rise in blood pressure, fever, disturbances in sleep and behavior, and rapid heartbeat. Death from respiratory arrest may occur in coma.

*Long Term Exposure:* May cause dermatitis, skin rash, and acne. May cause liver and/or kidney damage.
*Points of Attack:* Liver, kidneys and skin.
*Medical Surveillance:* Liver function tests. Kidney function tests.

*First Aid:* Speed in removing material from eyes and skin is of extreme importance. If this chemical gets into the eyes, remove any contact lenses at once and irrigate immediately for at least 15 minutes, occasionally lifting upper and lower lids. Seek medical attention immediately. If this chemical contacts the skin, remove contaminated clothing and wash immediately with soap and water. Shampoo hair promptly if contaminated. Seek medical attention immediately. If this chemical has been inhaled, remove from exposure, begin rescue breathing (using universal precautions) if breathing has stopped, and CPR if heart action has stopped. Transfer promptly to a medical facility. When this chemical has been swallowed, get medical attention. Give large quantities of water and induce vomiting. Do not make an unconscious person vomit. Effects may be delayed; keep victim under observation.

*References:*
- Sax, N.I., Ed., "Dangerous Properties of Industrial Materials Report, " 7, No. 6, 72-75 (1977).
- U.S. Environmental Protection Agency, "Chemical Profile: Isodrin," Washington, DC, Chemical Emergency Preparedness Program (Nov. 30, 1987).
- California Environmental Protection Agency "Chemical List of Lists," Sacramento CA (February 1997).

# Isofenphos

*Use Type:* Insecticide
*CAS Number:* 25311-71-1
*Formula:* $C_{15}H_{24}NO_4PS$
*Alert:* Was a Restricted Use Pesticide (RUP). Human toxicity (long-term): High
*Synonyms:* AI3-27748; Benzoic acid, 2-[(ethoxy-((1-methylethyl)amino)phosphinothioyl)oxy]-, 1-methylethyl ester; Benzoic acid, 2-[(ethoxy((1-methylethyl)amino)phosphinothioyl)oxy]-, 1-methyl ester; Caswell No. 447AB; Dipropylene glycol; 2-[(Ethoxyl((1-methylethyl)amino)phosphinothioyl)oxy]benzoic acid 1-methylethyl ester; 2-[(Ethoxy((1-methylethyl)amino)phosphinothioyl)oxy]benzoic acid 1-methylethyl ester; *O*-Ethyl *O*-(2-isopropoxycarbonyl)phenylisopropyl phosphoramidothioate; Isopropyl *O*-[ethoxy(isopropylamino)phosphinothioyl]salicylate; Isopropyl *O*-[ethoxy-*N*-isopropylamino(thiophosphoryl)]salicylate; Isopropyl salicylate *O*-ester with *O*-ethyl isopropylphosphoramidothioate; 1-Methylethyl 2-[(ethoxy((1-methylethyl)amino)phosphinothioyl)oxy]benzoate; Phosphoramidoth ioic acid, isopropyl-, *O*-ethyl *O*-(2-isopropoxycarbonylphenyl) ester;

# Isofenphos

Phosphoramidothioic acid, isopropyl-, $O$-ethyl ester, $O$-ester with isopropyl salicylate; Propanol, oxybis-; Salicylic acid, isopropyl ester, $O$-ester with $O$-ethyl isopropyl phosphoramidothioate

**Trade Names:** 40 SD®; AMAZE®, Bayer CropScience (Germany), canceled; BAY 92114®, Bayer CropScience (Germany), canceled; BAY-SRA-12869®, Bayer CropScience (Germany), canceled; OFTANOL®, Bonide Products (USA), canceled; PRYFON 6®; SRA 12869®; SRA 128691®

**Producers:** Bayer CropScience (Germany); Bonide Products (USA)

**Chemical Class:** Organophosphate

**EPA/OPP PC Code:** 109401; (512400 old EPA code number)

**California DPR Chemical Code:** 2194

**RTECS Number:** VO43955000

**EINECS Number:** 246-814-1

**Uses:** Registered products containing isofenphos were voluntarily canceled in the U.S. in 1999. Isofenphos was marketed under the basic producer's trade name Oftanol, and was used in the United States on turf and ornamental trees and shrubs to control white grubs, mole crickets, and other insects (mostly subterranean species). Isofenphos, was first registered in the United States in 1980 by Bayer Corporation for use on corn for control of the corn rootworm and was also used as a pre-emergence soil treatment in fruit crops and vegetables such as maize and carrots.

**Human toxicity (long-term)[77]:** High–3.50 ppb, Health Advisory

**Fish toxicity (threshold)[77]:** Low–116.60189 ppb, MATC (Maximum Acceptable Toxicant Concentration)

**Carcinogen/Hazard Classifications**

**U.S. EPA Carcinogens:** Not Likely a carcinogen

**Label Signal Word:** CAUTION or DANGER

**WHO Acute Hazard:** Class 1b, highly hazardous

**Regulatory Authority:**
- EPA 40 CFR 372.65, Specific Toxic Chemical Listings
- EPCRA Section 313 Form R *de minimis* concentration reporting level: 1.0%
- Marine pollutant (49CFR, Subchapter 172.101, Appendix B)
- DOT Inhalation Hazard Chemicals as organophosphates

**Description:** A colorless, oily liquid. Soluble in water; solubility = 23.8 mg/kg @ 20°C. Molecular weight = 345.4. Boiling point = 120°C. Vapor pressure = $4.2 \times 10^{-6}$ mmHg @ 20°C.

**Incompatibilities:** May react violently with oxidizers, aliphatic amines, alkalies, boranes, isocyanates, nitric acid, sulfuric acid.

**Determination in Air:** OSHA versatile sampler-2; Toluene/Acetone; Gas chromatography/Flame ionization detection; NIOSH IV, Method #5600, Organophosphorus Pesticides.[18]

**Permissible Concentration in Water:** No criteria set. Runoff from spills or fire control may cause water pollution.

**Harmful Effects and Symptoms**

**Short Term Exposure:** Eye pupils are small; blurred vision; eye watering; runny nose; cough; shortness of breath; salivation; dizziness; nausea, stomach cramps, diarrhea, and vomiting; increased blood pressure; profuse sweating; hypermotility, hallucinations; irritability; tingling of the skin; drowsiness; slow heartbeat; convulsions; fluid in lungs; loss of consciousness; incontinence; breathing stops; death. Organophosphates inhibit the action of acetylcholinesterase enzymes, and alter the way in which nervous impulses are transmitted. The effects can last for hours, days, or much longer. The action of the enzymes is reestablished after new enzymes are formed.

**Long Term Exposure:** Cholinesterase inhibitor; cumulative effect is possible. Organophosphates may damage the nervous system with repeated exposure, resulting in convulsions, respiratory failure. May cause liver damage.

**Points of Attack:** Respiratory system, central nervous system, cardiovascular system, blood cholinesterase.

**Medical Surveillance:** Medical observation is recommended for 24 to 48 hours after breathing overexposure, as pulmonary edema may be delayed. As first aid for pulmonary edema, consider administering a corticosteroid spray. Cigarette smoking may exacerbate pulmonary injury and should be discouraged for at least 72 hours following exposure.

**First Aid:** Speed in removing material from eyes and skin is of extreme importance. *Eyes:* Eye contact can cause dangerous amounts of these chemicals to be quickly absorbed through the mucous membrane into the bloodstream. Immediately and gently flush eyes with plenty of warm or cold water (NO hot water) for at least 15 minutes, occasionally lifting the upper and lower eyelids. Get medical aid immediately. *Skin:* Get medical aid. Skin contact can cause dangerous amounts of these chemicals to be absorbed into the bloodstream. Wearing the appropriate PPE equipment and respirator for organophosphate pesticides, immediately flush skin with plenty of soap and water for at least 15 minutes while removing contaminated clothing and shoes. Shampoo hair promptly if contaminated. *Ingestion:* Call poison control. Loosen all clothing. Never give anything by mouth to an unconscious person. Get medical aid. Do NOT induce vomiting.\* If conscious, alert, and able to swallow, rinse mouth and have victim drink 4 to 8 ounces of water do NOT induce vomiting but immediately administer slurry of activated charcoal (2 oz in 8 oz of water). If victim is *unconscious or having convulsions*, do nothing except keep victim warm. *\*In some cases you may be specifically instructed by poison control to induce vomiting by way of 2 tablespoons of syrup of ipecac (adult) washed down with a cup of water.* Do NOT give activated charcoal before or with ipecac syrup. *Inhalation:* Get medical aid. Do not contaminate yourself. Wearing the appropriate PPE equipment and respirator for organophosphate pesticides, immediately remove the victim from the contaminated area to fresh air. If the victim is not breathing, administer artificial respiration. Do NOT use mouth-to-mouth resuscitation; use bag/mask apparatus. If breathing is difficult, administer oxygen through bag/mask apparatus until medical help arrives. Do not leave victim unattended. *Note to physician or authorized medical personnel.* Administer atropine, 2 mg (1/30 gr) intramuscularly or intravenously as soon as any local or systemic signs or symptoms of an intoxication are noted; repeat the administration of atropine every 3 to 8 minutes until signs of atropinization (mydriasis, dry mouth, rapid pulse, hot and dry skin) occur; initiate treatment in children with 0.05 mg mg/kg of atropine; repeat at 5 to 10 minute intervals. Watch respiration, and remove bronchial secretions if they appear to be obstructing the airway; intubate if necessary. *Notes to physician or authorized medical personnel*: N-methylpyridinium-2-aldoxime (2-PAMCI) when used in conjunction with atropine reacts with the phosphorylated cholinesterase, thereby restoring normal activity to by removing the phosphorylating group. The combination of these two chemicals is synergistic and must be administered within minutes to a few hours following exposure (depending on the specific agent) to be effective. Give 2-PAMCI (Pralidoxime; Protopam), 2.5 gm in 100 ml of sterile water or in 5% dextrose and water, intravenously, slowly, in 15-30 minutes; if sufficient fluid is not available, give 1 gm of 2-PAMCI in 3 ml of distilled water by deep intramuscular injection; repeat this every half hour if respiration weakens or if muscle fasciculation or convulsions recur. Also Diazepam, an anticonvulsant, might be considered.

*References:*
- EXTOXNET, Extension Toxicology Network, "Pesticide Information Profile, Isofenphos," Oregon State University, Corvallis, OR (June 1996). http://extoxnet.orst.edu/pips/isofenph.htm
- U.S. Environmental Protection Agency, "Reregistration Eligibility Decision (RED) Facts, Isofenphos" Office of Prevention, Pesticides and Toxic Substances, Washington, DC (December 1999). http://www.epa.gov/REDs/factsheets/2345fact.pdf
- Food and Agriculture Organization of the United Nations (FAO), "FAO Plant Production and Protection Paper, Isofenphos," Geneva, Switzerland, (Noember 1981). http://www.inchem.org/documents/jmpr/jmpmono/v81pr18.htm
- California Environmental Protection Agency "Chemical List of Lists," Sacramento CA (February 1997)

# Isofluorphate

*Use Type:* An insecticide.

*CAS Number:* 55-91-4
*Formula:* $C_6H_{14}FO_3P$
*Alert:* Not in general use in the U.S.
*Synonyms:* DPF; Diflupyl; Diflurophate; Diisopropoxyphosphoryl fluoride; Diisopropylfluoro phosphate; *O,O*-Diisopropylfluoro phosphate; Diisopropylfluorophosphonate; Diisopropyl fluorophosphoric acid ester; Diisopropylfluorphosphor saeureester (German); Diisopropylphosphofluoridate; Diisopropyl phosphorofluoridate; *O,O'*-diisopropyl phosphoryl fluoride; DIPF; Dyflos; Floropryl; Fluophosphoric acid, diisopropyl ester; Fluorodiisopropyl phosphate; Fluoropryl; Fluostigmine; Isofluorphate; Isoflurophate; Isopropyl fluophosphate; Isproyl phosphorofluoridate; Neoglaucit; PF-3; Phosphorofluoridic acid, diisopropyl ester; T-1703; Tl-466
*Producers:* Molekula Fine Chemicals (UK)
*Chemical Class:* Organophosphate
*EPA/OPP PC Code:* 356100
*RTECS Number:* TE5075000
*EINECS Number:* 200-247-6
*Uses:* This material is also used as a research tool in neuroscience for its ability to inhibit cholinesterase (by phosphorylation)on an acute/sub-acute basis and to produce a delayed neuropathy. Used as a basis for "nerve gases."
*Regulatory Authority:*
- Extremely Hazardous Substance (EPA-SARA) (TPQ = 100)[7]
- U.S. DOT Inhalation Hazard Chemicals as organophosphates
- EPA Hazardous Waste Number (RCRA No.): P043
- RCRA, 40CFR261, Appendix 8 Hazardous Constituents
- Superfund/EPCRA 40CFR355, Appendix B Extremely Hazardous Substances: TPQ = 100 lb (45.4 kg)
- Superfund/EPCRA 40CFR302.4 RQ: CERCLA, 100 lb (45.4 kg)

*Description:* Isofluorphate, $C_6H_{14}FO_3P$, is an oily, colorless to faintly yellow liquid. Boiling point =62°C @ 9 mm. Melting/Freezing point = −82°C. Hazard Identification (based on NFPA-704 M Rating System): Health 4, Flammability 1, Reactivity 0. Slightly soluble in water.
*Incompatibilities:* Forms hydrofluoric acid in the presence of water. In the presence of moisture attacks metals, rubbers, plastics, coatings and silica-containing materials such as glass. When heated to decomposition, this chemical forms fumes containing toxic oxides of phosphorus and fluorine.
*Permissible Exposure Limits in Air:* No standards set. An organophosphate
*Determination in Air:* OSHA versatile sampler-2; Toluene/Acetone; Gas chromatography/Flame photometric detection for sulfur, nitrogen, or phosphorus; NIOSH Method IV Method #5600, Organophosphorus pesticides.[18]
*Permissible Concentration in Water:* No criteria set, but runoff from spills or fire control may cause water pollution.

### Harmful Effects and Symptoms

***Short Term Exposure:*** Organic phosphorus insecticides are absorbed by the skin, as well as by the respiratory and gastrointestinal tracts. They are cholinesterase inhibitors. Symptoms of exposure include headache, giddiness, blurred vision, nervousness, weakness, nausea, cramps, diarrhea, and discomfort in the chest. Signs include sweating, tearing, salivation, vomiting, cyanosis, convulsions, coma, loss of reflexes and loss of sphincter control. Isofluorphate is extremely toxic: probable oral lethal dose in humans is 5-50 mg/kg, between 7 drops and 1 teaspoonful for 70 kg person. Even traces of the vapor cause pinpoint pupils. Delayed pulmonary edema may occur after inhalation.

***Long Term Exposure:*** Cholinesterase inhibitor; cumulative effect is possible. This chemical may damage the nervous system with repeated exposure, resulting in convulsions, respiratory failure. May cause liver damage.

***Points of Attack:*** Respiratory system, lungs, central nervous system, cardiovascular system, skin, eyes, plasma and red blood cell cholinesterase.

***Medical Surveillance:*** Medical observation is recommended for 24 to 48 hours after breathing overexposure, as pulmonary edema may be delayed. Before employment and at regular times after that, the following are recommended: Plasma and red blood cell cholinesterase levels (tests for the enzyme poisoned by this chemical). If exposure stops, plasma levels return to normal in 1-2 weeks while red blood cell levels may be reduced for 1-3 months. When acetylcholinesterase enzyme levels are reduced by 25% or more below preemployment levels, risk of poisoning is increased, even if results are in lower ranges of "normal." Reassignment to work not involving organophosphate or carbamate pesticides is recommended until enzyme levels recover. If symptoms develop or overexposure occurs, repeat the above tests as soon as possible and get an exam of the nervous system. Also consider complete blood count. Consider chest x-ray following acute overexposure. Do not drink any alcoholic beverages before or during use. Alcohol promotes absorption of organophosphates.

***First Aid:* Treatment for organophosphate poisoning consists of thorough decontamination, cardiorespiratory support, and administration of the antidotes atropine and pralidoxime. In cases of severe poisoning, diazepam, an anticonvulsant, should also be administered. Antidotes should be administered as prevention even if the diagnosis is in doubt.** Speed in removing material from eyes and skin is of extreme importance. *Eyes:* Eye contact can cause dangerous amounts of these chemicals to be quickly absorbed through the mucous membrane into the bloodstream. Immediately and gently flush eyes with plenty of warm or cold water (NO hot water) for at least 15 minutes, occasionally lifting the upper and lower eyelids. Get medical aid immediately. *Skin:* Get medical aid. Skin contact can cause dangerous amounts of these chemicals to be absorbed into the bloodstream. Wearing the appropriate PPE equipment and respirator for organophosphate pesticides, immediately flush skin with plenty of soap and water for at least 15 minutes while removing contaminated clothing and shoes. Shampoo hair promptly if contaminated. The removed, contaminated clothing and shoes should be double-bagged and left in Hot Zone for later disposal by hazardous materials experts. Skin may also be decontaminated with diluted hypochlorite solution. *Inhalation:* Get medical aid. Do not contaminate yourself. Wearing the appropriate PPE equipment and respirator for organophosphate pesticides, immediately remove the victim from the contaminated area to fresh air. If the victim is not breathing, administer artificial respiration. Do NOT use mouth-to-mouth resuscitation; use bag/mask apparatus. If breathing is difficult, administer oxygen through bag/mask apparatus until medical help arrives. Do not leave victim unattended. *Ingestion:* Call poison control. Loosen all clothing. Never give anything by mouth to an unconscious person. If victim is *unconscious or having convulsions,* do nothing except keep victim warm. Get medical aid. Transfer promptly to a medical facility. In cases of ingestion, **do not induce vomiting**. If the victim is alert and asymptomatic, administer a slurry of activated charcoal at a dose of 1 g/kg (infant, child, and adult dose). A soda can and straw may be of assistance when offering charcoal to a child. *In some cases you may be specifically instructed by poison control to induce vomiting by way of 2 tablespoons of syrup of ipecac (adult) washed down with a cup of water.* Do NOT give activated charcoal before or with ipecac syrup.

*Note to physician or authorized medical personnel:* Treat cases of respiratory compromise, coma, or excessive pulmonary secretions with respiratory support using protocols and techniques available and within the scope of training. Some cases may necessitate procedures such as endotracheal intubation or cricothyrotomy by properly trained and equipped personnel. When possible, atropine (see *Antidotes*, below) should be given under medical supervision. Patients who are comatose, hypotensive, or having seizures or cardiac arrhythmias should be treated according to advanced life support protocols. *Antidotes:* Two antidotes are administered to treat organophosphate poisoning. Atropine is a competitive antagonist of acetylcholine at muscarinic receptors and is used to control the excessive bronchial secretions which are often responsible for death. Pralidoxime relieves both the nicotinic and muscarine effects of organophosphate poisoning by regenerating acetylcholinesterase and can reduce both the bronchial secretions and the muscle weakness associated with poisoning. The initial intravenous dose of atropine in adults should be determined by the severity of symptoms: An initial adult dose of 1.0 to 2.0 mg or pediatric dose of 0.01 mg/kg (minimum 0.01 mg) should be administered intravenously. If intravenous access cannot be established, atropine may also be given intramuscularly, subcutaneously or via endotracheal tube. Doses should be repeated every 15 minutes until excessive secretions and sweating have been controlled. Once bronchial secretion has been controlled, atropine administration should be repeated whenever the secretions begin to recur. In seriously poisoned patients, very large doses may be required. Alterations of pulse rate and pupillary size should not be used as indicators of treatment adequacy. Pralidoxime should be administered as early in poisoning as possible as its efficacy may diminish when given more than 24 to 36 hours after exposure. Doses are as follows: adult 1.0 g; pediatric 25 to 50 mg/kg. The drug should be administered intravenously over 30 to 60 minutes, but in a life-threatening situation, one-half of the total dose can be given per minute for a total administration time of 2 minutes. Treatment should begin to take effect within 40 minutes with a reduction in symptoms and the amount of atropine necessary to control bronchial secretion. The initial dose can be repeated in 1 hour and then every 8 to 12 hours until the patient is clinically well and no longer requires atropine. If intravenous access cannot be established, pralidoxime may also be given intramuscularly. Early administration of diazepam in addition to the combined atropine and pralidoxime treatment may help prevent the onset of seizures and potential brain and cardiac morphologic damage following high-level organophosphate poisoning.

*References:*
- U.S. Environmental Protection Agency, "Chemical Profile: Isofluorphate," Washington, DC, Chemical Emergency Preparedness Program (Nov. 30, 1987).
- Agency for Toxic Substances and Disease Registry, U.S. Department of Health and Human Services, Public Health Service, "Managing Hazardous Materials Incidents," Atlanta, GA (June 2003).
- California Environmental Protection Agency "Chemical List of Lists," Sacramento CA (February 1997).

# Isolan®

*Use Type:* Aphicide and insecticide
*CAS Number:* 119-38-0
*Formula:* $C_{10}H_{17}N_3O_2$
*Alert:* Not registered as a pesticide in the U.S. It is extremely toxic.
*Synonyms:* Dimetilcarbamato de 1-isopropil-3-metil-5-pirazolilo (Spanish); Dimethylcarbamate-d'l-isopropyl-3-methyl-5-pyrazoylle (French); Dimethylcarbamic acid 3-methyl-1-(1-methylethyl)-1H-pyrazol-5-yl ester; ENT 19,060; Isolane (French); Isopropylmethylpyrazoyl dimethylcarbamate; (1-Isopropil-3-metil-1H-pirazol-5-il)-N,N-dimetil-carbammato (Italian); (1-Isopropyl-3-methyl-1H-pyrazol-5-yl)-N,N-dimethyl-carbamaat (Dutch); (1-Isopropyl-3-methyl-1H-pyrazol-5-yl)-N,N-dimethyl-carbamat (German); Isopropylmethylpyrazoldimethyl carbamate; (1-Isopropyl-3-methyl-1H-pyrazol-5-yl)-N,N-dimethyl carbamate; 1-Isopropyl-3-methyl-5-pyrazolyl dimethyl carbamate; 1-Isopropyl-3-methylpyrazolyl-(5)-dimethylcarbamate; 5-Methyl-2-isopropyl-3-pyrazolyl dimethylcarbamate

*Trade Names:* GEIGY G-23611®, Syngenta (Switzerland), canceled; PRIMIN®; SAOLAN®
*Producers:* Syngenta (Switzerland)
*Chemical Class:* Carbamate
*EPA/OPP PC Code:* 511500
*RTECS Number:* FA2100000
*Uses:* This material is a systemic aphicide used in Europe. It is currently of little commercial interest. Not registered as a pesticide in the U.S.

*Regulatory Authority:*
- EPA Hazardous Waste Number (RCRA No.): P192
- RCRA, 40CFR261, Appendix 8 Hazardous Constituents
- RCRA 40CFR268.48; 61FR15654, Universal Treatment Standards: Wastewater (mg/L), 0.056; Nonwastewater (mg/kg), 1.4
- Superfund/EPCRA 40CFR355, Appendix B Extremely Hazardous Substances: TPQ = 500 lb (227kg)
- Superfund/EPCRA 40CFR302.4 RQ: EHS, 1 lb (0.454 kg)

*Description:* Isolan is a colorless liquid. Soluble in water. Boiling point = 103°C @ 0.7 mmHg; 117–118°C @ 2.5 mmHg.
*Permissible Exposure Limits in Air:* No standards set.
*Permissible Concentration in Water:* No criteria set, but runoff from spills or fire control may cause water pollution.

*Harmful Effects and Symptoms*
*Short Term Exposure:* In is classified as extremely toxic. Probable oral lethal dose in humans is 5-50 mg/kg or between 7 drops and 1 teaspoonful for a 150 lb person. A cholinesterase inhibitor; although it is not an organophosphate, it resembles that group in action. Can cause death due to respiratory arrest. Symptoms include cool extremities; trembling; fixed pinpoint pupils; nausea; vomiting; slight bluing of skin, lips and nailbeds; tearing; diarrhea; excessive salivation; sweating; slurring of speech; jerky movements; loss of bladder control; convulsions; coma and death. Carbamate insecticides inhibit the cholinesterase activity of enzymes, causing accumulation of acetylcholine at synapses and altering the way in which nervous impulses are transmitted. However, within several hours carbamates spontaneously detach from the enzymes.

*First Aid:* Speed in removing material from eyes and skin is of extreme importance. *Eyes:* Eye contact can cause dangerous amounts of these chemicals to be quickly absorbed through the mucous membrane into the bloodstream. Immediately and gently flush eyes with plenty of warm or cold water (NO hot water) for at least 15 minutes, occasionally lifting the upper and lower eyelids. Get medical aid immediately. *Skin:* Get medical aid. Skin contact can cause dangerous amounts of these chemicals to be absorbed into the bloodstream. Wearing the appropriate

PPE equipment and respirator for carbamate pesticides, immediately flush skin with plenty of soap and water for at least 15 minutes while removing contaminated clothing and shoes. Shampoo hair promptly if contaminated; protect eyes. *Ingestion:* Call poison control. Loosen all clothing. Never give anything by mouth to an unconscious person. Get medical aid. Do NOT induce vomiting.* If conscious, alert, and able to swallow, rinse mouth and have victim drink 4 to 8 ounces of water. Check to see if poison control instructs you to use ipecac syrup, otherwise administer slurry of activated charcoal (2 oz in 8 oz of water). If victim is *UNCONSCIOUS OR HAVING CONVULSIONS,* do nothing except keep victim warm. **In some cases you may be specifically instructed by poison control to induce vomiting by way of 2 tablespoons of syrup of ipecac (adult) washed down with a cup of water.* Do NOT give activated charcoal before or with ipecac syrup. *Inhalation:* Get medical aid. Do not contaminate yourself. Wearing the appropriate PPE equipment and respirator for carbamate pesticides, immediately remove the victim from the contaminated area to fresh air. If the victim is not breathing, administer artificial respiration. Do NOT use mouth-to-mouth resuscitation; use bag/mask apparatus. If breathing is difficult, administer oxygen through bag/mask apparatus until medical help arrives. Do not leave victim unattended. *Note to physician or authorized medical personnel.* Administer atropine, 2 mg (1/30 gr) intramuscularly or intravenously as soon as any local or systemic signs or symptoms of an intoxication are noted; repeat the administration of atropine every 3 to 8 minutes until signs of atropinization (mydriasis, dry mouth, rapid pulse, hot and dry skin) occur; initiate treatment in children with 0.05 mg mg/kg of atropine; repeat at 5 to 10 minute intervals. Watch respiration, and remove bronchial secretions if they appear to be obstructing the airway; intubate if necessary. *Medical note:* 2-PAMCI may be contraindicated in the case of some carbamate poisonings.

*References:*
U.S. Environmental Protection Agency, "Chemical Profile: Isopropylmethylpyrazolyl Dimethylcarbamate," Washington, DC, Chemical Emergency Preparedness Program (Nov. 30, 1987).
California Environmental Protection Agency "Chemical List of Lists," Sacramento CA (February 1997).

# Isopropalin (ANSI)

*Use Type:* Herbicide
*CAS Number:* 33820-53-0
*Formula:* $C_{15}H_{23}N_3O_4$
*Synonyms:* 2,6-Dinitro-*N,N*-dipropylcumidene; Isopropalin; Benzenamine, 4-(1-methylethyl)-2,6-dinitro-*N,N*-dipropyl-; 4-Isopropyl-2,6-dinitro-*N,N*-dipropylaniline; 4-(Methylethyl)-2,6-dinitro-*N,N*-dipropylbenzenamine;

*Trade Names:* EL 179®; PAARLAN®, Dow AgroSciences (USA), canceled 7/26/1987
*Producers:* Dow AgroSciences (USA)
*Chemical Class:* Dinitroaniline
*EPA/OPP PC Code:* 100201
*California DPR Chemical Code:* 1681
*Uses:* Not registered in the U.S. A pre-emergence herbicide for grasses and broadleaf weeds. Used primarily on transplanted tobacco.
*Human toxicity (long-term)*[77]*:* Very low–105.00 ppb, Health Advisory
*Fish toxicity (threshold)*[77]*:* Intermediate–98.89791 ppb, MATC (Maximum Acceptable Toxicant Concentration)
*Carcinogen/Hazard Classifications*
**Label Signal Word:** WARNING or CAUTION
*Description:* Amber, reddish-orange liquid. Very slightly soluble in water; solubility = <0.1 ppm. Molecular weight = 309.35.
*Permissible Concentration in Water:* No criteria set. Runoff from spills or fire control may cause water pollution.
*Harmful Effects and Symptoms*
*Short Term Exposure:* May cause irritation to the eyes, skin, or respiratory tract. May be toxic if ingested or upon skin contact.
*First Aid:* If this chemical gets into the eyes, remove any contact lenses at once and irrigate immediately for at least 15 minutes, occasionally lifting upper and lower lids. Seek medical attention immediately. If this chemical contacts the skin, remove contaminated clothing and wash immediately with soap and water. Seek medical attention immediately. If this chemical has been inhaled, remove from exposure, begin rescue breathing (using universal precautions) if breathing has stopped, and CPR if heart action has stopped. Transfer promptly to a medical facility. When this chemical has been swallowed, get medical attention. Give large quantities of water and induce vomiting. Do not make an unconscious person vomit.

*References:*
- California Environmental Protection Agency "Chemical List of Lists," Sacramento CA (February 1997)

# Isoproturon

*Use Type:* Herbicide
*CAS Number:* 34123-59-6
*Formula:* $C_{12}H_{18}N_2O$
*Synonyms:* 3-(4-Isopropylphenyl)-1,1-dimethylurea; Urea, *N,N*-dimethyl-*N'*-[4-(1-methylethyl)phenyl]-
*Trade Names:* ARELON® DISPERSION, Calliope (France); ATLAS FIELDGARD®, Whyte Agrochemicals (UK); CALIPURON®, Calliope (France); CGA 18731®; DAHR®, Rallis India (India); DPX 6774®; HARLEQUIN®, (isoproturon + simazine), Makhteshim-Agan Industries (Israel); HOE 16410®; JOSH®, Sudarshan Chemical

Industries (India); PROTUGAN®, Makhteshim-Agan Industries (Israel)

*Producers:* Agrimor International (USA); Agsin (Singapore); Atul (India); Bharat Rasayan (India); Calliope (France); Gharda Chemicals (India); Makhteshim-Agan Industries (Israel); Nagarjuna Agrichem (India); Rallis India (India); Sudarshan Chemical Industries (India); Whyte Agrochemicals (UK)

*Chemical Class:* Phenyl urea

*EPA/OPP PC Code:* 512200

*EINECS Number:* 251-835-4

*Uses:* Not registered in the U.S. Used as a pre-emergence and post-emergence selective herbicide for the control of grasses and broadleaf weeds in spring and winter wheat, spring and winter barley, and winter rye.

*Carcinogen/Hazard Classifications*

**Label Signal Word:** CAUTION

**WHO Acute Hazard:** Class III, slightly hazardous

*Description:* Colorless to off-white crystalline solid. Soluble in water; solubility = 70 mg/L @ 20°C. Molecular weight = 206.29. Melting/Freezing point = 152°C. Vapor pressure = $8.8 \times 10^{-6}$ mmHg @ 20°C; $3.5 \times 10^{-4}$ mmHg @ 77°C. Log $K_{ow}$ = 2.45. Unlikely to bioaccumulate in marine organisms.

*Permissible Concentration in Water:* No criteria set. Runoff from spills or fire control may cause water pollution.

*Harmful Effects and Symptoms*

*Short Term Exposure:* Contact with eyes or skin may cause irritation or injury. Inhalation should be avoided; use NIOSH-approved air purifying respirators for pesticides. May be harmful if swallowed. Skin contact may cause severe irritation or burns.

*Points of Attack:* Skin.

*First Aid:* If this chemical gets into the eyes, remove any contact lenses at once and irrigate immediately for at least 15 minutes, occasionally lifting upper and lower lids. Seek medical attention immediately. If this chemical contacts the skin, remove contaminated clothing and wash immediately with soap and water. Seek medical attention immediately. If this chemical has been inhaled, remove from exposure, begin rescue breathing (using universal precautions) if breathing has stopped, and CPR if heart action has stopped. Transfer promptly to a medical facility. When this chemical has been swallowed, get medical attention. Give large quantities of water. Do not induce vomiting when formulations containing petroleum solvents are ingested. Do not make an unconscious person vomit.

# Isoxaben (ANSI)

*Use Type:* Herbicide

*CAS Number:* 82558-50-7

*Formula:* $C_{18}H_{24}N_2O_4$

*Synonyms:* Benzamide, 2,6-dimethoxy-*N*-[3-(1-ethyl-1-methylpropyl)-5-isoxazolyl]-; Benzamide, *N*-[3-(1-ethyl-1-methylpropyl)-5-isoxazolyl]-2,6-dimethoxy-; 2,6-Dimethoxy-*N*-[3-(1-ethyl-1-methylpropyl)-5-isoxazolyl]benzamide; *N*-[3-(1-Ethyl-1-methylpropyl)-5-isoxazolyl]-2,6-dimethoxybenzamide; *N*-3-(1-Ethyl-1-methylpropyl)-5-isoxazolyl-2,6-dimethoxybenzamide; Benzamizole

*Trade Names:* EL 107®; FLEXIDOR®; GALLERY®, Dow AgroSciences (USA); NA 8318®; SNAPSHOT, Dow AgroSciences (USA)

*Producers:* Dow AgroSciences (USA); United Phosphorus (India)

*Chemical Class:* Amide

*EPA/OPP PC Code:* 125851

*California DPR Chemical Code:* 2289

*Human toxicity (long-term)*[77]: Intermediate–35.00 ppb, Health Advisory

*Fish toxicity (threshold)*[77]: Low–400.00001 ppb, MATC (Maximum Acceptable Toxicant Concentration)

*Carcinogen/Hazard Classifications*

**U.S. EPA Carcinogens:** Group C, possible carcinogen

**Label Signal Word:** CAUTION

**WHO Acute Hazard:** Class U, unlikely to be hazardous

*Description:* White crystalline solid. Molecular weight = 332.38. Practically insoluble in water; solubility = < 0.002 ppm. Vapor pressure = $4 \times 10^{-7}$ mmHg @ 20°C.

*Incompatibilities:* Slowly hydrolyzes in water, releasing ammonia and forming acetate salts. Reacts with strong oxidizers. When heated to decomposition, this chemical forms toxic oxides of nitrogen.

*Permissible Concentration in Water:* No criteria set. Runoff from spills or fire control may cause water pollution.

*Harmful Effects and Symptoms*

*Short Term Exposure:* Contact with eyes or skin may cause irritation or injury. Inhalation should be avoided; use NIOSH-approved air purifying respirators for pesticides. May be harmful if swallowed.

*First Aid:* If this chemical gets into the eyes, remove any contact lenses at once and irrigate immediately for at least 15 minutes, occasionally lifting upper and lower lids. Seek medical attention immediately. If this chemical contacts the skin, remove contaminated clothing and wash immediately with soap and water. Seek medical attention immediately. If this chemical has been inhaled, remove from exposure, begin rescue breathing (using universal precautions) if breathing has stopped, and CPR if heart action has stopped. Transfer promptly to a medical facility. When this chemical has been swallowed, get medical attention. Give large quantities of water and induce vomiting. Do not make an unconscious person vomit. Do not induce vomiting when formulations containing petroleum solvents are ingested.

*References:*
- California Environmental Protection Agency "Chemical List of Lists," Sacramento CA (February 1997)

# Isoxaflutole

*Use Type:* Herbicide
*CAS Number:* 141112-29-0
*Formula:* $C_{15}H_{12}F_3NO_4S$
*Alert:* A Restricted Use Pesticide (RUP)
*Synonyms:* 4-(2-Methylsulfonyl-4-trifluoromethyl-benzoyl)-5-cyclopropylisoxazole
*Trade Names:* BALANCE® WDG, Bayer CropScience (Germany); EPIC® (flufenacet + isoxaflutole), Bayer CropScience (Germany); RPA 201772®
*Producers:* Bayer CropScience (Germany)
*Chemical Class:* Cyclopropylisoxazola herbicide
*EPA/OPP PC Code:* 123000
*Uses:* Registered for use on field cotton.
*U.S. Maximum Allowable Residue Levels for Isoxaflutole (40 CFR 180.537):*

| CROP | ppm |
| --- | --- |
| Cattle, fat | 0.20 |
| Cattle, liver | 0.50 |
| Cattle, meat | 0.20 |
| Cattle, mbyp (except liver) | 0.10 |
| Eggs | 0.01 |
| Field corn, fodder | 0.50 |
| Field corn, forage | 1.0 |
| Field corn, grain | 0.20 |
| Goat, fat | 0.20 |
| Goat, liver | 0.50 |
| Goat, meat | 0.20 |
| Goat, mbyp (except liver) | 0.10 |
| Hog, fat | 0.20 |
| Hog, liver | 0.50 |
| Hog, meat | 0.20 |
| Hog, mbyp (except liver) | 0.10 |
| Horse, fat | 0.20 |
| Horse, liver | 0.50 |
| Horse, meat | 0.20 |
| Horse, mbyp (except liver) | 0.10 |
| Milk | 0.02 |
| Poultry, fat | 0.20 |
| Poultry, liver | 0.30 |
| Poultry, meat | 0.20 |
| Sheep, fat | 0.20 |
| Sheep, liver | 0.50 |
| Sheep, meat | 0.20 |
| Sheep, mbyp (except liver) | 0.10 |

*Carcinogen/Hazard Classifications*
**U.S. EPA Carcinogens:** Group B2, probable carcinogen; suspected developmental toxin (U.S. EPA)
*California Prop. 65:* Carcinogen
**Label Signal Word:** CAUTION
*Description:* Yellow to tan granular solid. Slight acetic acid-like odor for technical grade, no odor for end-use substance. Melting point = 135-136°C.

*Incompatibilities:* Reacts with strong oxidizers. When heated to decomposition, this chemical forms toxic fumes of nitrogen oxides, sulfur oxides, fluorine and carbon mononoxide.
*Permissible Concentration in Water:* No criteria set. Runoff from spills or fire control may cause water pollution.
*Harmful Effects and Symptoms*
*Short Term Exposure:* Contact with eyes or skin may cause irritation or injury. Inhalation should be avoided; use NIOSH-approved air purifying respirators for pesticides. May be harmful if swallowed.
*First Aid:* If this chemical gets into the eyes, remove any contact lenses at once and irrigate immediately for at least 15 minutes, occasionally lifting upper and lower lids. Seek medical attention immediately. If this chemical contacts the skin, remove contaminated clothing and wash immediately with soap and water. Seek medical attention immediately. If this chemical has been inhaled, remove from exposure, begin rescue breathing (using universal precautions) if breathing has stopped, and CPR if heart action has stopped. Transfer promptly to a medical facility. When this chemical has been swallowed, get medical attention. Give large quantities of water and induce vomiting. Do not make an unconscious person vomit. Do not induce vomiting when formulations containing petroleum solvents are ingested.
*References:*
- U.S. Environmental Protection Agency, Office of Pesticide Programs, Pesticide Residue Limits, "Isoxaflutole," 40 CFR 180.537. http://pmep.cce.cornell.edu/profiles/herb-growthreg/fatty-alcohol-monuron/isoxaflutole/Isoxaflutole_tol_998.html
- U.S. Environmental Protection Agency, Office of Pesticide Programs, "Pesticide Fact Sheet, Isoxaflutole," (September 15, 1998). http://www.epa.gov/opprd001/factsheets/isoxaflutole.pdf
- California Environmental Protection Agency "Chemical List of Lists," Sacramento CA (February 1997).

# K

## Karbutilate (ANSI)

*Use Type:* Herbicide
*CAS Number:* 4849-32-5
*Formula:* $C_{14}H_{21}N_3O_3$
*Synonyms:* tert-Butylcarbamic acid ester with 3-(m-hydroxyphenyl)-1,1-dimethylurea; Carbamic acid, 1,1-dimethylethyl-, ester with 3-(3-hydroxyphenyl)-1,1-dimethyl urea; 3-[(((Dimethylamino)carbonyl)amino)phenyl-1,1-dimethylethyl]carbamate; 1,1-Dimethyl-3-[(3-N-tert-butylcarbamyloxy)-phenyl]urea; m-(3,3-dimetylureido)phenyl-t-butylcarbamate; m-(3,3-Dimethylureido)phenyl-tert-butylcarbamate; m-(3,3-Dimethylharnstoff)-phenyl-tert-butylcarbamat (German); 3-(3-Hydroxyphenyl)-1,1-dimethylurea, tert-butylcarbamic acid ester; tert-Butylcarbamic acid, ester with 3-(m-hydroxyphenyl)-1,1-dimethylurea;
*Trade Names:* NIA 11092®; TANDEX®, Syngenta (Switzerland), canceled 7/1/1987; TANZENE®; FMC 11092®, FMC (USA), canceled
*Producers:* FMC (USA); Syngenta (Switzerland)
*Chemical Class:* Carbamate; substituted urea herbicide
*EPA/OPP PC Code:* 097401
*California DPR Chemical Code:* 691
*Uses:* Not registered in the U.S. Used to control annual and perennial broadleaf weeds and grasses, brush and vines. Used on railroad, utility and pipeline rights-of-way; drainage ditch banks; industrial plant sites; and non-crop areas.
*Carcinogen/Hazard Classifications*
**Label Signal Word:** WARNING or CAUTION
*Description:* White crystalline solid. Practically odorless. Soluble in water; solubility = 325ppm @ 20°C. Molecular weight = 279.38. Melting/Freezing point = 176°C. Vapor pressure = $<1.3 \times 10^{-4}$ mmHg @ 20°C. Log $K_{ow}$ = 1.36. Unlikely to bioaccumulate in marine organisms.
*Incompatibilities:* May form explosive materials with phosphorus pentachloride.
*Determination in Air:* Filter; none; Gravimetric; NIOSH IV [Particulates NOR; #0500 (total), #0600 (respirable)].[18]
*Permissible Concentration in Water:* No criteria set. Runoff from spills or fire control may cause water pollution.
*Routes of Entry:* Skin absorption, ingestion and inhalation.
*Harmful Effects and Symptoms*
*Short Term Exposure:* Eye pupils are small; blurred vision; eye watering; runny nose; cough; shortness of breath; salivation; nausea, stomach cramps, diarrhea, and vomiting, increased blood pressure, profuse sweating, hypermotility, hallucinations, agitation, tingling of the skin, slow heartbeat, convulsions, fluid in lungs, loss of consciousness, breathing stops, death. Carbamates inhibit the acetylcholinesterase enzymes and alter the way in which nervous impulses are transmitted. However, within several hours carbamates spontaneously detach from the enzymes.
*Long Term Exposure:* A potent cholinesterase inhibitor; cumulative effect is possible. This chemical may damage the nervous system with repeated exposure, resulting in convulsions, respiratory failure. May cause liver damage.
*Points of Attack:* Respiratory system, lungs, central nervous system, cardiovascular system, skin, eyes, plasma and red blood cell cholinesterase.
*Medical Surveillance:* Medical observation is recommended for 24 to 48 hours after breathing overexposure, as pulmonary edema may be delayed. As first aid for pulmonary edema, consider administering a corticosteroid spray. Cigarette smoking may exacerbate pulmonary injury and should be discouraged for at least 72 hours following exposure. Before employment and at regular times after that, the following are recommended: Plasma and red blood cell cholinesterase levels (tests for the enzyme poisoned by this chemical). If exposure stops, plasma levels return to normal in 1-2 weeks while red blood cell levels may be reduced for 1-3 months. When acetylcholinesterase enzyme levels are reduced by 25% or more below preemployment levels, risk of poisoning is increased, even if results are in lower ranges of "normal." Reassignment to work not involving carbamate pesticides is recommended until enzyme levels recover. If symptoms develop or overexposure occurs, repeat the above tests as soon as possible and get an exam of the nervous system. Also consider complete blood count. Consider chest x-ray following acute overexposure
*First Aid:* Speed in removing material from eyes and skin is of extreme importance. *Eyes:* Eye contact can cause dangerous amounts of these chemicals to be quickly absorbed through the mucous membrane into the bloodstream. Immediately and gently flush eyes with plenty of warm or cold water (NO hot water) for at least 15 minutes, occasionally lifting the upper and lower eyelids. Get medical aid immediately. *Skin:* Get medical aid. Skin contact can cause dangerous amounts of these chemicals to be absorbed into the bloodstream. Wearing the appropriate PPE equipment and respirator for carbamate pesticides, immediately flush skin with plenty of soap and water for at least 15 minutes while removing contaminated clothing and shoes. Shampoo hair promptly if contaminated; protect eyes. *Ingestion:* Call poison control. Loosen all clothing. Never give anything by mouth to an unconscious person. Get medical aid. Do NOT induce vomiting.* If conscious, alert,

and able to swallow, rinse mouth and have victim drink 4 to 8 ounces of water. Check to see if poison control instructs you to use ipecac syrup, otherwise administer slurry of activated charcoal (2 oz in 8 oz of water). If victim is *UNCONSCIOUS OR HAVING CONVULSIONS,* do nothing except keep victim warm. *In some cases you may be specifically instructed by poison control to induce vomiting by way of 2 tablespoons of syrup of ipecac (adult) washed down with a cup of water.* Do NOT give activated charcoal before or with ipecac syrup. *Inhalation:* Get medical aid. Do not contaminate yourself. Wearing the appropriate PPE equipment and respirator for carbamate pesticides, immediately remove the victim from the contaminated area to fresh air. If the victim is not breathing, administer artificial respiration. Do NOT use mouth-to-mouth resuscitation; use bag/mask apparatus. If breathing is difficult, administer oxygen through bag/mask apparatus until medical help arrives. Do not leave victim unattended. *Note to physician or authorized medical personnel.* Administer atropine, 2 mg (1/30 gr) intramuscularly or intravenously as soon as any local or systemic signs or symptoms of an intoxication are noted; repeat the administration of atropine every 3 to 8 minutes until signs of atropinization (mydriasis, dry mouth, rapid pulse, hot and dry skin) occur; initiate treatment in children with 0.05 mg mg/kg of atropine; repeat at 5 to 10 minute intervals. Watch respiration, and remove bronchial secretions if they appear to be obstructing the airway; intubate if necessary. *Medical note:* 2-PAMCI may be contraindicated in the case of some carbamate poisonings.

*References:*
- Pesticide Management Education Program, "Karbutilate (Tanzene, Tandex) Herbicide Profile 3/85," Cornell University, Ithaca, NY (March 1985). http://pmep.cce.cornell.edu/profiles/herb-growthreg/fatty-alcohol-monuron/karbutilate/herb-prof-karbutilate.html
- International Programme on Chemical Safety (IPCS), "International Health Criteria, Carbamate Pesticides, A General Introduction," Geneva, Switzerland (1986). http://www.inchem.org/documents/ehc/ehc/ehc64.htm
- California Environmental Protection Agency "Chemical List of Lists," Sacramento CA (February 1997)

# Kerosene

*Use Type:* Insecticide
*CAS Number:* 8008-20-6
*Formula:*
*Synonyms:* Coal oil; Deobase; Illuminating oil; Fuel oil No. 1; Jet fuel: Jp-1; Kerosine; Range oil; Lamp oil; Light petroleum; Straight run kerosene
*Trade Names:* ALCO®; DAMOIL®; Drexel Chemical (USA); TRIANGLE®, Triangle Chemical (USA); VAPONA®, Boehringer Ingelheim (Germany), canceled.

*Note:* The U.S. EPA lists 213 products containing kerosene that have been canceled or transferred; there are only four currently active kerosene insecticides.
*Producers:* Amerada-Hess; BP Amoco (UK); Bharat Petroleum (India); Boehringer Ingelheim (Germany); CITGO; Drexel Chemical (USA); DuPont; Exxon Mobil; Idemitsu Kosan (Japan); Indian Oil (India); Interstate Chemical; Koch Petroleum; ONGC (India); Sinopec Corporation (China); Shell Chemicals (UK); Total (France); and most petroleum refineries
*Chemical Class:* Petroleum derivative
*EPA/OPP PC Code:* 063501
*California DPR Chemical Code:* 2071
*ICSC Number:* 0663
*RTECS Number:* OA5500000
*EEC Number:* 649-404-00-4
*EINECS Number:* 232-366-4
*Uses:* Used as an insecticide to control larvae, eggs, etc. Kerosene is also used as a fuel for lamps, stoves, jets, and rockets. It is also used for degreasing and cleaning metals and as a vehicle for insecticides. Jet fuels JP-5 and JP-8 are used as aircraft fuels by the military. JP-8 is the primary jet fuel used by the US Navy and Air Force. Kerosene is the primary component of both JP-5 and JP-8.
*Carcinogen/Hazard Classifications*
**Label Signal Word:** CAUTION
*Regulatory Authority:*
- Air Pollutant Standard Set (former USSR)[43] (OSHA) (NIOSH)[2] (See text below)
- California: AB 1803-Well Monitoring Chemical (CAL)
- Actively registered pesticide in California.

*Description:* Kerosene is a pale yellow or clear, mobile liquid, composed of a mixture of petroleum distillates, having a characteristic odor. The odor threshold is 0.6 mg/m$^3$. The taste threshold is 0.1 mg/l in water[59]. A refined petroleum solvent (predominantly C9-C16), which typically is 25% normal paraffins, 11% branched paraffins, 30% monocycloparaffins, 12% dicycloparaffins, 1% tricycloparaffins, 16% mononuclear aromatics and 5% dinuclear aromatics. Boiling point = 150°– 301°C. Melting/Freezing point = –20°C. Flash point = 35°– 72°C. Autoignition temperature = 210°C. Explosive limits: LEL = 0.7%; UEL = 5.0% (17). Hazard Identification (based on NFPA-704 M Rating System): Health 0, Flammability 2, Reactivity 0. Insoluble in water.
*Incompatibilities:* Explosive mixture in air. Oxidizers may cause fire and explosion hazard. Incompatible with nitric acid. May accumulate static electrical charges, and may cause ignition of its vapors.
*Permissible Exposure Limits in Air:* There is no OSHA[2] PEL for kerosene, but OSHA and the Air Force Office of Safety and Health have set an exposure limit of 400 mg/m$^3$ of petroleum product for an 8-hour workday, 40-hour workweek. For kerosene NIOSH[2] recommends 100 mg/m$^3$ (approximately 14.4 ppm) as a TWA. This is due at least in

part to the variable composition of kerosene. The ACGIH® TLV recommends an 8-hour TWA of 400 mg/m³ (approximately 29 ppm). The former USSR-UNEP/IRPTC project[43] has set an MAC in workplace air of 300 mg/m³.
***Determination in Air:*** Charcoal tube; CS2; Gas chromatography/Flame ionization detection; NIOSH IV, Method #1550, Naphthas.[18]
***Permissible Concentration in Water:*** No criteria set. Runoff from spills or fire control may cause water pollution.
***Permissible Concentration in Water:*** MAC concentrations in water bodies used for domestic purposes ranging from 0.10 mg/l have been set by the former USSR-UNEP/IRPTC joint project[43]. New Jersey[59] has declined to set a maximum contaminant level for kerosene in water because of its variable composition.
***Routes of Entry:*** Inhalation, ingestion, skin and/or eye contact.
***Harmful Effects and Symptoms***
***Short Term Exposure:*** Slightly irritates the skin and respiratory tract. *Inhalation*: Does not evaporate fast enough to cause health effects except when heated or in enclosed spaces. Headache, tiredness, stupor, dizziness, nausea, coma and death, may occur with increasing exposure. *Skin*: If not promptly removed, may cause reddening, blisters, itching and an increased risk of infection. *Eyes*: Irritation may occur. *Ingestion*: Accidental ingestion of unknown amounts has caused irritation of mouth, throat and stomach, nausea, vomiting, rapid breathing, blue skin coloration, and convulsions. Death may result from as little as 1-fluid ounce. Inhalation into lungs following ingestion may result in bronchitis, chemical pneumonia, accumulation off fluid and blood in lungs, and death. As little as 1/30 oz may be fatal in this way.
***Long Term Exposure:*** Repeated or prolonged skin contact may cause defatting, itching, and rash. Absorption through skin is slow but repeated skin contact over many years has caused muscular weakness, anemia, changes in white blood cells, fever and death. Can irritate the lungs; bronchitis may develop. May cause kidney damage. A study on the use of kerosene stoves found an increase in oral cancer in men who used kerosene stoves. Skin tumors were seen in mice when their skin was exposed to jet fuel JP-5 for 60 weeks.
***Points of Attack:*** Eyes, skin, respiratory system, central nervous system
***Medical Surveillance:*** Before beginning employment and at regular times after that, for those with frequent or potentially high exposure, the following are recommended: Lung function tests. If symptoms develop or overexposure is suspected, the following may be useful: Consider chest x-ray after acute overexposure. Kidney function tests.
***First Aid:*** If this chemical gets into the eyes, remove any contact lenses at once and irrigate immediately for at least 15 minutes, occasionally lifting upper and lower lids. Seek medical attention immediately. If this chemical contacts the skin, remove contaminated clothing and wash immediately with soap and water. Seek medical attention immediately. If this chemical has been inhaled, remove from exposure, begin rescue breathing (using universal precautions) if breathing has stopped, and CPR if heart action has stopped. Transfer promptly to a medical facility. When this chemical has been swallowed, get medical attention. Give large quantities of water and induce vomiting. Do not make an unconscious person vomit. Do not induce vomiting when formulations containing petroleum solvents are ingested.
***References:***
- National Institute for Occupational Safety and Health, Criteria for a Recommended Standard: Occupational Exposure to Refined Petroleum, NIOSH Doc. No. 77-192, Washington, DC (1977).
- New Jersey Department of Health and Senior Services, "Hazardous Substance Fact Sheet, Kerosene," Trenton, NJ (January 1997). http://www.state.nj.us/health/eoh/rtkweb/1091.pdf
- New York State Department of Health, "Chemical Fact Sheet: Kerosene, " Albany, NY, Bureau of Toxic Substance Assessment (April 1986).
- U.S. Department of Health and Human Services, "ATSDR ToxFAQs, Jet Fuels JP-5 and JP-8," Atlanta, GA, August 1999. http://www.atsdr.cdc.gov/tfacts75.html
- California Environmental Protection Agency "Chemical List of Lists," Sacramento CA (February 1997)

# Kinetin (Cytokinin)

***Use Type:*** Plant growth regulator
***CAS Number:*** 525-79-1
***Formula:*** $C_{10}H_9N_5O$
***Synonyms:*** Adenine, *N*-furfuryl-; Adenine, $N^6$-furfuryl-; Cytokinin; Cytokinin, as kinetin, based on biological activity; Cytokinins (with Cytokinin B, Cytokinin R); Cytokinins (derived from aqueous extract of seaweed); $N^6$-(Furfurylamino)purine; 6-(Furfurylamino)purine; $N^6$-Furfuryladenine; 1-*H*-Purin-6-amine, *N*-(2-furanylmethyl)-; 2-Furanmethanamine, *N*-1*H*-purin-6-yl-; *N*-(2-Furanylmethyl)-1*H*-purin-6-amine; 6-Furfurylaminopurine; 1-*H*-Purin-6-amine, *N*-(2-furanylmethyl)-
***Trade Names:*** FAP®; FOLIAR TRIGGRR®; FOLI-ZYME®; GOLDENGRO®; HAPPYGRO®; MAXON®, Agriliance (USA); MEGAGRO®; MEPEX®, Griffin (USA); NITROZYME®
***Producers:*** Agriliance (USA); Griffin (USA)
***Chemical Class:*** Botanical
***EPA/OPP PC Code:*** 116801; 116802
***California DPR Chemical Code:*** 2082
***Uses:*** Cytokinin is a group of plant regulators that promote cell division, leaf expansion and retard leaf aging. Cytokinin is comprised of four naturally occurring cytokinins (derived from aqueous extract of seaweed meal). The extracts from these plant species (e.g. the naturally occurring Cytokinins)

are exempt from the requirements of tolerances when used as plant regulators in or on many raw agricultural commodities (40 CFR 180.1042). Cytokinin is applied to growing crops (field crops, vegetable crops, small fruits, vines and tree fruit), young trees, ornamental, and golf courses to increase fruit size, yield, blossoms, branching, healthy appearance, and other desirable growth effects.

*Carcinogen/Hazard Classifications*
**Label Signal Word:** CAUTION
*Regulatory Authority:*
- Actively registered pesticide in California.

*Description:* Slightly soluble in water. Molecular weight = 215.22.

*Permissible Concentration in Water:* No criteria set. Runoff from spills or fire control may cause water pollution.

*Harmful Effects and Symptoms*
*Short Term Exposure:* Contact with eyes or skin may cause irritation or injury. Inhalation should be avoided; use NIOSH-approved air purifying respirators for pesticides. May be harmful if swallowed.

*First Aid:* If this chemical gets into the eyes, remove any contact lenses at once and irrigate immediately for at least 15 minutes, occasionally lifting upper and lower lids. Seek medical attention immediately. If this chemical contacts the skin, remove contaminated clothing and wash immediately with soap and water. Seek medical attention immediately. If this chemical has been inhaled, remove from exposure, begin rescue breathing (using universal precautions) if breathing has stopped, and CPR if heart action has stopped. Transfer promptly to a medical facility. When this chemical has been swallowed, get medical attention. Give large quantities of water and induce vomiting. Do not make an unconscious person vomit. Do not induce vomiting when formulations containing petroleum solvents are ingested.

*References:*
- U.S. Environmental Protection Agency, "Reregistration Eligibility Decision (RED), Cytokinin," Office of Prevention, Pesticides and Toxic Substances, Washington, DC (December 1995). http://www.epa.gov/REDs/4107.pdf
- U.S. Environmental Protection Agency, Office of Pesticide Programs, Pesticide Residue Limits, "Cytokinin (as kinetin)," 40 CFR 180.1157. http://www.epa.gov/pesticides/food/viewtols.htm
- California Environmental Protection Agency "Chemical List of Lists," Sacramento CA (February 1997)

# Kinoprene

*Use Type:* Insect growth regulator
*CAS Number:* 42588-37-4 (U.S. EPA); 37882-31-8 (RTECS); 65733-20-2 (*S*-kinoprene)
*Formula:* $C_{18}H_{28}O_2$

*Synonyms:* 2,4-Dodecadienoic acid, 3,7,11-trimethyl-, 2-propynyl ester, (*E,E*)-; ENT 70,531; 2-Propynyl (*E,E*)-3,7,11-trimethyl-2,4-dodecadienoate; Prop-2-ynyl-3,7,11-trimethyl-2,4-dodecadienoate; 3,7,11-Trimethyl-2,4-dodecadienoic acid 2-propynyl ester

*S-kinoprene:* 2,4-Dodecadienoic acid, 3,7,11-trimethyl-, 2-propynyl ester, [*s*-(*E,E*)]-

*Trade Names:* ALTODEL®; ENSTAR®; ENSTAR II® (*s*-kinoprene); ENSTAR 5E®; ENT 70531; ZR-777®

*EPA/OPP PC Code:* 107501; (517200 old EPA code number); 107502 (*S*-kinoprene)

*California DPR Chemical Code:* 1949

*RTECS Number:* JR1760000

*Uses:* Not registered in the U.S. S-Kinoprene is a biochemical pesticide which is chemically synthesized and used as an insect juvenile hormone analog on indoor non-food/non-feed crops, including ornamental plants grown in greenhouses and interiors. S-Kinoprene, applied at a low rate, inhibits normal insect growth during the molting process causing morphogenic, ovicidal, and sterilization effects. When applied at a higher rate, *s*-Kinoprene kills the adult insects such as aphids, whiteflies, mealybugs, fungus gnats, and armored scales.

*Regulatory Authority:*
- Actively registered pesticide in California.

*Description:* Amber liquid. Soluble in water; solubility = 5 ppm. Molecular weight = 276.44.

*Permissible Concentration in Water:* No criteria set. Runoff from spills or fire control may cause water pollution.

*Harmful Effects and Symptoms*
*Short Term Exposure:* Contact with eyes or skin may cause irritation or injury. Inhalation should be avoided; use NIOSH-approved air purifying respirators for pesticides. May be harmful if swallowed.

*First Aid:* If this chemical gets into the eyes, remove any contact lenses at once and irrigate immediately for at least 15 minutes, occasionally lifting upper and lower lids. Seek medical attention immediately. If this chemical contacts the skin, remove contaminated clothing and wash immediately with soap and water. Seek medical attention immediately. If this chemical has been inhaled, remove from exposure, begin rescue breathing (using universal precautions) if breathing has stopped, and CPR if heart action has stopped. Transfer promptly to a medical facility. When this chemical has been swallowed, get medical attention. Give large quantities of water and induce vomiting. Do not make an unconscious person vomit. Do not induce vomiting when formulations containing petroleum solvents are ingested.

*References:*
- U.S. Environmental Protection Agency, Office of Pesticide Programs, "Insect Growth Regulators Fact Sheet," Washington, DC (December 2001). http://www.epa.gov/oppbppd1/biopesticides/ingredients/factsheets/factsheet_igr.htm

- U.S. Environmental Protection Agency, "Reregistration Eligibility Decision (RED), *S*-Kinoprene" Office of Prevention, Pesticides and Toxic Substances, Washington, DC (November 1996). http://www.epa.gov/REDs/4117.pdf
- California Environmental Protection Agency "Chemical List of Lists," Sacramento CA (February 1997)

## Lactofen (ANSI)

*Use Type:* Herbicide
*CAS Number:* 77501-63-4; (81362-49-4 obsolete)
*Formula:* $C_{19}H_{15}ClF_3NO_7$
*Synonyms:* Benzoic acid, 5-[2-chloro-4-(trifluoromethyl)phenoxy]-2-nitro-2-ethoxy-1-methyl-2-oxoethyl ester; 1'-(Carboethoxy)ethyl-5-[2-chloro-4-(trifluoromethyl)phenoxy]-2-nitrobenzoate; 5-[2-Chloro-4-`(trifluoromethyl)phenoxy]-2-nitrobenzoic acid 2-ethoxy-1-methyl-2-oxoethyl ester; (±)-2-Ethoxy-1-methyl-2-oxoethyl-5-[2-chloro-4-(trifluoromethyl)phenoxy]-2-nitrobenzoate; Ethyl $O$-[5-(2-chloro-$\alpha,\alpha,\alpha$-trifluoro-$p$-tolyloxy)-2-nitrobenzoyl]-*dl*-lactate
*Trade Names:* COBRA®, Valent BioSciences (USA); PPG-844®; STELLER®, Valent BioSciences (USA); V-10086®, Valent BioSciences (USA)
*Producers:* Valent BioSciences (USA)
*Chemical Class:* Diphenyl ether
*EPA/OPP PC Code:* 128888
*California DPR Chemical Code:* 3538
*Uses:* Lactofen is a broad-spectrum herbicide used for pre-emergence and post-emergent weed control on peanuts, snap beans, soybeans, cotton, and fruiting vegetables.
*Human toxicity (long-term)[77]:* High–2.05882 ppb, CHCL (Chronic Human Carcinogen Level)
*Fish toxicity (threshold)[77]:* High–1.97989 ppb, MATC (Maximum Acceptable Toxicant Concentration)
*U.S. Maximum Allowable Residue Levels for Lactofen (40 CFR 180.432):*

| CROP | ppm |
|---|---|
| Bean, snap, succulent | 0.05 |
| Soybean | 0.05 |

*Carcinogen/Hazard Classifications*
**U.S. EPA Carcinogens:** Group B2, probable carcinogen
*California Prop. 65:* Carcinogen
**Label Signal Word:** CAUTION or DANGER
*Regulatory Authority:*
- EPA 40 CFR 372.65, Specific Toxic Chemical Listings
- AB 2588-Air Toxics "Hot Spots" Chemicals (CAL)
- Safe Drinking Water Act: Priority List (55 FR 1470)
- EPCRA Section 313 Form R *de minimis* concentration reporting level: 1.0%
- Proposition 65 chemical (CAL)

*Description:* White crystalline solid in technical form; brown or tan in formulations. Melting point = 43.9–45.5°C. Molecular weight = 461.78. Vapor pressure = $7.9 \times 10^{-9}$ mmHg @ 20°C.
*Permissible Concentration in Water:* No criteria set. Runoff from spills or fire control may cause water pollution.
*Routes of Entry:* Ingestion and skin contact
*Harmful Effects and Symptoms*
**Short Term Exposure:** *Skin:* Corrosive burns. *Eyes:* Irritation. *Ingestion:* Slightly irritating
*Long Term Exposure:* Can cause permanent damage to the eyes
*Points of Attack:* Liver and kidneys, eyes and skin
*First Aid:* If this chemical gets into the eyes, remove any contact lenses at once and irrigate immediately for at least 15 minutes, occasionally lifting upper and lower lids. Seek medical attention immediately. If this chemical contacts the skin, remove contaminated clothing and wash immediately with soap and water. Seek medical attention immediately. If this chemical has been inhaled, remove from exposure, begin rescue breathing (using universal precautions) if breathing has stopped, and CPR if heart action has stopped. Transfer promptly to a medical facility. When this chemical has been swallowed, get medical attention. Give large quantities of water and induce vomiting. Do not make an unconscious person vomit. Do not induce vomiting when formulations containing petroleum solvents are ingested.
*References:*
- EXTOXNET, Extension Toxicology Network, "Pesticide Information Profile, Lactofen," Oregon State University, Corvallis, OR (July 1996). http://extoxnet.orst.edu/pips/lactofen.htm
- U.S. Environmental Protection Agency, Office of Pesticide Programs, Pesticide Residue Limits, "Lactofen," 40 CFR 180.432. http://www.ehso.com/ehso.php
- California Environmental Protection Agency "Chemical List of Lists," Sacramento CA (February 1997)

## Lead Arsenate

*Use Type:* Insecticide, rodenticide and herbicide
*CAS Number:* 7784-40-9[ $AsHO_4 \cdot Pb$]; 3687-31-8 [$Pb_3(AsO_4)_2$]; 7645-25-2 [$PbH_3AsO_4$]
*Formula:* $AsHO_4Pb$; $As_2O_8Pb_3$; $PbHAsO_4$; $Pb_3(AsO_4)_2$
*Alert:* Although lead arsenate is no longer used in the fruit industry, e.g., on apple and cherry orchards, residual contamination is present and affects humans. Lead arsenate should be handled as a carcinogen and teratogen, with extreme caution.
*Synonyms: 7645-25-2:* Arsenic acid, Lead salt; Arseniate de plomb (French); Arseniato de plomo (Spanish); Lead acetate acid; Plumbous arsenate
*Acid lead arsenate*; Acid lead arsenite; Acid lead orthoarsenate; Arsenate of lead; Arsenic acid, lead(II); Arseniato de plomo (Spanish); Arsenic acid, lead(2+);

# Lead Arsenate

Arsenic acid, lead salt; Arsinette; Dibasic lead arsenate; Gypsine; Lead acid arsenate; Salt arsenate of lead; Schultenite; Standard lead arsenate

*Acid lead arsenate*; Acid lead orthoarsenate; Arsenate of lead; Arseniato de plomo (Spanish); Arsenic acid, lead(2+) Salt; Arsinette; Dibasic lead arsenate; Gypsine; Schultenite; Standard lead arsenate; Talbot

***Trade Names:*** ARSINETTE®; GYPSINE®; NU REXFORM®; ORTHO L-10 DUST®, The Scotts Company (USA), canceled 2/21/1986; ORTHO L-40 DUST®, The Scotts Company (USA), canceled 2/21/1986; SECURITY®, canceled; SOPRABEL®, canceled; TALBOT®

***Producers:*** Molekula Fine Chemicals (UK)

***Chemical Class:*** Inorganic arsenic; inorganic lead; heavy metal

***EPA/OPP PC Code:*** 013503

***California DPR Chemical Code:*** 353

***ICSC Number:*** 0911

***RTECS Number:*** CG0980000, $PbHAsO_4$; CG0990000, $Pb_3(AsO_4)_2$; CG1000000, $Pb_xH_3AsO_4$

***EEC Number:*** 082-011-00-0

***EINECS Number:*** 3687-31-8 [ $Pb_3(AsO_4)_2$ ]; 222-979-5

***Uses:*** Insecticide and veterinary tapeworm medicine. Lead acetate was widely used in the U.S. to control pests in fruit orchards until the late 1950s, particularly apples and cherries. Its residue binds to the surface soil, exposing humans to health hazards years later when the land changes from agricultural to other uses, e.g., residential building and parkland. As late as 1986, the U.S. EPA reported lead arsenate was used as a plant growth regulator on 17% of the U.S. grapefruit crop. It was also used to control cockroaches, silverfish and crickets.

***Carcinogen/Hazard Classifications***

**U.S. EPA Carcinogens:** Not listed

**U.S. NTP Carcinogen:** Known carcinogen

**California Prop. 65:** Carcinogen

**IARC:** Group 1, known carcinogen

**Label Signal Word:** WARNING–POISON

**WHO Acute Hazard:** Group 1b, highly hazardous

***Regulatory Authority:***
- CARCINOGEN (lead compounds)
- Air Pollutant Standard Set (ACGIH)[1] (Argentina)[35] (Several States)[60]
- Banned or Severely Restricted (In Agriculture in India, Japan) (UN)[13]
- The "Director's List" (CAL/OSHA)
- Carcinogen (human positive) (DFG)[3]
- OSHA, 29CFR1910 Specifically Regulated Chemicals (See 29 CFR 1910.1025 and 1910. 1018)
- Clean Water Act: Section 311 Hazardous Substances/RQ 40CFR117.3 (same as CERCLA, see below); Section 313 Water Priority Chemicals (57FR41331, 9/9/92); 40CFR401.15 Section 307 Toxic Pollutants as lead and compounds
- RCRA, 40CFR261, Appendix 8 Hazardous Constituents, waste number not Listed as lead compounds, n.o.s.
- Superfund/EPCRA 40CFR302.4 RQ: CERCLA, 1 lb (0.454 kg)
- EPCRA Section 313 Form R *de minimus* concentration reporting level: 0.1%.
- U.S. DOT Regulated Marine Pollutant (49CFR172.101, Appendix B)
- Canada, WHMIS, Ingredients Disclosure List

*as arsenic compounds*
- Carcinogen User Register Chemical (CAL/OSHA) as inorganic arsenic compound
- AB 1803-Well Monitoring Chemical (CAL) as inorganic arsenic compound
- CAL Air Resources Board/AB 1807 Toxic Air Contaminants as inorganic arsenic compound
- Proposition 65 chemical (CAL) as inorganic arsenic compound
- The "Director's List" (CAL/OSHA) as inorganic arsenic compound
- AB 2588-Air Toxics "Hot Spots" Chemicals (CAL) as arsenic compounds as inorganic arsenic compound
- Permissible Exposure Limits for Chemical Contaminants (CAL/OSHA) as arsenic compounds as inorganic arsenic compound
- Clean Air Act: List of high risk pollutants (Section 63.74) as arsenic compounds.
- Clean water act: Section 311 Hazardous Substances/RQ (same as CERCLA); Section 313 Priority Chemicals; Toxic Pollutant (Section 401.15) as arsenic and compounds.
- RCRA Section 261 Hazardous Constituents, waste number not listed.
- EPCRA Section 304 RQ: CERCLA, 1 lbs(0.454 kgs)
- EPCRA Section 313 Form R *de minimus* concentration reporting level: 0.1%.
- U.S. DOT Regulated Marine Pollutant (49CFR, Subchapter 172.101, Appendix B). as arsenates, liquid, n.o.s; arsenates, solid, n.o.s.; arsenical pesticides liquid, toxic, flammable, n.o.s.
- Carcinogen User Register Chemical (CAL/OSHA) as inorganic arsenic compound
- AB 1803-Well Monitoring Chemical (CAL) as inorganic arsenic compound
- CAL Air Resources Board/AB 1807 Toxic Air Contaminants as inorganic arsenic compound
- Proposition 65 chemical (CAL) as inorganic arsenic compound
- The "Director's List" (CAL/OSHA) as inorganic arsenic compound
- AB 2588-Air Toxics "Hot Spots" Chemicals (CAL) as arsenic compounds as inorganic arsenic compound
- Permissible Exposure Limits for Chemical Contaminants (CAL/OSHA) as arsenic compounds as inorganic arsenic compound

# Lead Arsenate

*Description:* Lead arsenate, is an odorless, heavy, white powder, or crystals. Insoluble in cold water; soluble in hot water. Melting/Freezing point = about 280°C (decomposes); also listed @ 1042°C (decomposes). Hazard Identification (based on NFPA-704 M Rating System): Health 2, Flammability 0, Reactivity 0.

*Incompatibilities:* Violent reactions occur from contact with oxidizers, chemically active metals, strong acids. Acids and acid mists cause the release of arsine, a deadly gas. Decomposes above 270°C forming toxic fumes including arsenic and lead compounds.

*Permissible Exposure Limits in Air: For lead:* The OSHA[2] PEL is 0.05 mg(Pb)/m$^3$ TWA. NIOSH[2] recommended TWA is 0.100 mg(Pb)/m$^3$. The ACGIH[1] recommends a level of 0.15 mg/m$^3$ as a TWA for Pb(AsO$_4$)$_2$. *For arsenic:* The OSHA[2] PEL is 0.010 mg(As)/m$^3$ TWA. NIOSH recommends (inorganic arsenic) ceiling of 0.002 mg(As)/m$^3$/ 15minutes. ACGIH[1] recommends a TLV of 0.2 mg(As)/m$^3$. Argentina[35] has set 0.15 mg/m$^3$ as a TWA ambient air[60] ranging from 0.5 µg/m$^3$ (New York) to 0.75 µg/m$^3$ (South Carolina) to 0.5 µg/m$^3$ (Florida and North Dakota) to 2.5 µg/m$^3$ (Virginia) to 3.0 µg/m$^3$ (Connecticut) to 4.0 µg/m$^3$ (Nevada).

*Permissible Concentration in Water: Lead:* The U.S. EPA limits lead in drinking water to 15 µg per liter. Various organizations worldwide have set other standards for lead in drinking water as follows[35] (all in mg/L): Argentina 0.01; Czechoslovakia 0.05; Germany 0.04; EEC 0.05; Japan 0.10; Mexico 0.05; former USSR 0.03; WHO 0.10. The states of Maine and Minnesota have set guidelines for lead in drinking water[61] at the level of 20 µg/L. *Arsenic:* EPA has set a limit of 0.05 parts per million (ppm) for arsenic in drinking water. The U.S. EPA arsenic drinking water standard of 0.01 ppm (10 ppb) is based on the U.S. EPA final rule for arsenic in drinking water published in. the January 22, 2001, *Federal Register*. However, the U.S. EPA is currently reviewing the science and cost estimate supporting this rule, and, in the interim, has reverted to the previous standard for arsenic. Thus, in the US, the current EPA arsenic drinking water standard remains at 0.05 ppm (50 ppb). To protect freshwater aquatic life-total recoverable trivalent inorganic arsenic never to exceed 440 µg/L. To protect saltwater aquatic life: 508 µg/L on an acute basis. To protect human health: preferably zero. The former USSR-UNEP/IRPTC project[43] has set MAC values for inorganic arsenic compounds in water for domestic purposes at 0.05 mg/L and in water bodies for fishery purposes of 0.5 mg/L also.

*Determination in Water:* Digestion followed by atomic absorption or by colorimetric (dithizone) analysis or by inductively coupled plasma optical emission spectrometry. That gives total lead; dissolved lead may be determined by 0.45 micron filtration prior to such analysis. Collection on a filter and analysis by atomic absorption spectrometry. See NIOSH Methods #7900 and #73000, Elements[18]. See also OSHA Method ID 105 (arsenic)[58].

*Routes of Entry:* Inhalation, ingestion, skin, eye contact.

## Harmful Effects and Symptoms

*Short Term Exposure:* Lead arsenate irritates the eyes, skin, and respiratory tract. Skin contact can cause burning sensation, itching and rash. High exposure can cause poor appetite, nausea, vomiting, and muscle cramps. May affect the heart, with abnormal EKG, gastrointestinal tract, and nervous system. *Arsenic intoxication*; nausea, diarrhea; inflammation of skin and mucous membranes; *Lead intoxication:* abdominal pain, appetite loss, constipation; tiredness, weakness, nervousness; paresthesia. A rebuttable presumption against registration for pesticide uses was issued on October 18, 1978 by EPA on the basis of oncogenicity, teratogenicity and mutagenicity.

*Long Term Exposure:* Lead arsenate is a carcinogen and has been shown to cause skin, lung, and liver cancer. Lead arsenate may also affect the gastrointestinal tract, nervous system, kidneys, and blood. Lead and certain lead compounds may be teratogens and cause reproductive damage in humans. Birth defects have been observed in animals exposed to inorganic arsenic. It is likely that health effects seen in children exposed to high amounts of arsenic will be similar to the effects seen in adults.

*Points of Attack:* Several studies have shown that inorganic arsenic can increase the risk of lung cancer, skin cancer, bladder cancer, liver cancer, kidney cancer, and prostate cancer.

*Medical Surveillance:* Before first exposure and every six months thereafter, OSHA (1910. 1025) requires your employer to provide: Blood lead test. ZPP test (a special test for the effect of lead on blood cells). Examination of the nervous system. Before first exposure, and yearly for exposed person with blood lead over *40 micrograms per 100 g of whole blood*, OSHA also requires a complete medical history and exam with the above tests, and: Complete blood count. Kidney function tests. OSHA defines "exposure" for these tests as air levels which average 30 micrograms of lead or more in a cubic meter of air. OSHA requires your employer to sent the doctor a copy of the lead standard and provide one for you.

*Note:* Blood-lead level is a good indicator of total lead exposure. Current OSHA regulations require that if an individual has a blood-lead level greater than or equal to 0.050 mg lead per 100 ml blood, he or she must be removed from all exposures to lead and cannot return to the exposure environment until the blood level falls to 0.040 mg [lead]/100ml blood or less. The following tolerance levels for occupational exposures may also be useful: ACGIH[1] BEI = 50 mg/L (blood); 150 mg/g creatinine (urine). DFG BAT = 70 mg/L (blood); 30 mg/L (blood) for women < 45 years old[3]. Also seek prompt medical evaluation if health effects are noticed. With each visit, careful attention should be given to the inner nose, skin, nails and nervous system. A test for serum arsenic is recommended. NIOSH recommends urine arsenic should not be greater than 50 to 100 micrograms per liter of urine.

***First Aid:*** If this chemical gets into the eyes, remove any contact lenses at once and irrigate immediately for at least 15 minutes, occasionally lifting upper and lower lids. Seek medical attention immediately. If this chemical contacts the skin, remove contaminated clothing and wash immediately with soap and water. Seek medical attention immediately. If this chemical has been inhaled, remove from exposure, begin rescue breathing (using universal precautions) if breathing has stopped, and CPR if heart action has stopped. Transfer promptly to a medical facility. When this chemical has been swallowed, get medical attention. Give large quantities of water and induce vomiting. Do not make an unconscious person vomit. Do not induce vomiting when formulations containing petroleum solvents are ingested.

***References:***
- Pesticide Management Education Program, "Lead Arsenate EPA Pesticide Fact Sheet 12/86," Cornell University, Ithaca, NY (December 1986). http://pmep.cce.cornell.edu/profiles/insect-mite/fenitrothion-methylpara/lead-arsenate/insect-prof-leadars.html
- New Jersey Department of Health and Senior Services, "Hazardous Substance Fact Sheet: Lead Arsenate," Trenton, NJ (Feb. 3, 1988, rev. April 2002). http://www.state.nj.us/health/eoh/rtkweb/1097.pdf
- California Environmental Protection Agency "Chemical List of Lists," Sacramento CA (February 1997).

# Leptophos

***Use Type:*** Insecticide
***CAS Number:*** 21609-90-5
***Formula:*** $C_{13}H_{10}BrCl_2O_2PS$; $C_6H_5PS(OCH_3)OC_6H_2BrCl_2$
***Alert:*** Banned for use in the U.S.A.
***Synonyms:*** *O*-(4-Bromo-2,5-dichlorophenyl)*O*-methyl phenylphosphonothioate; *O*-(2,5-Dichloro-4-bromophenyl) *O*-methyl phenylthiophosphonate; MBCP; *O*-Methyl-*O*-(4-bromo-2,5-dichlorophenyl)phenyl thiophosphonate; Phenylphosphonothioic acid *O*-(4-bromo-2,5-bromo-2,5-dichlorophenyl)*O*-methyl ester; Phosphonothioic acid, phenyl-,*O*-(4-bromo-2,5-dichlorophenyl)O-methyl ester
***Trade Names:*** ABAR®; FOSVEL®; K62-105®; NK 711®; PHOSVEL®; PSL®; V.C.S®, Velsicol Chemical Corp. (USA), canceled; VCS-506®, Velsicol Chemical Corp. (USA), canceled; VELSICOL 506®, Velsicol Chemical Corp. (USA), canceled
***Producers:*** Sigma-Aldrich Laborchemikalien (Germany); Velsicol Chemical Corp. (USA)
***Chemical Class:*** Organophosphate
***EPA/OPP PC Code:*** 525300
***California DPR Chemical Code:*** 1676
***RTECS Number:*** TB1720000
***Uses:*** Its use is not permitted in the U.S.A.
***Regulatory Authority:***
- Banned or Severely Restircted (Many Countries) (UN)[13]
- Superfund/EPCRA 40CFR355, Appendix B Extremely Hazardous Substances: TPQ = 500/10,000 lb (227/4,540 kg)
- Superfund/EPCRA 40CFR302.4 RQ: EHS, 1 lb (0.454 kg)
- EPCRA Section 313 Form R *de minimus* concentration reporting level: inorganic compounds 0.1%; organic compounds 1.0%
- U.S. DOT Inhalation Hazard Chemicals as organophosphates
- U.S. DOT Regulated Marine Pollutant (49CFR172.101, Appendix B).

***Description:*** Leptophos is a tan, waxy solid. Practically insoluble in water. Molecular weight = 412.08. Melting/Freezing point = about 70°C. Hazard Identification (based on NFPA-704 M Rating System): Health 3, Flammability 1, Reactivity 0. Log $K_{ow}$ = 6.29. Values at or above 3.0 are likely to bioaccumulate in marine organisms.
***Permissible Exposure Limits in Air:*** No standards set.
***Determination in Air:*** OSHA versatile sampler-2; Toluene/Acetone; Gas chromatography/Flame photometric detection for sulfur, nitrogen, or phosphorus; NIOSH Method IV Method #5600, Organophosphorus pesticides.[18]
***Permissible Concentration in Water:*** No criteria set, but runoff from spills or fire control may cause water pollution.
***Routes of Entry:*** Inhalation, ingestion, skin contact.

***Harmful Effects and Symptoms***

***Short Term Exposure:*** Organic phosphorus insecticides are absorbed by the skin, as well as by the respiratory and gastrointestinal tracts. They are cholinesterase inhibitors. Symptoms of exposure include headache, giddiness, blurred vision, nervousness, weakness, nausea, cramps, diarrhea, and discomfort in the chest. Death may occur from failure of the respiratory center, paralysis of the respiratory muscles, intense bronchoconstriction, or all three. This material is highly toxic; it is capable of causing death or permanent injury by exposure during normal use. The $LD_{50}$-(oral, rat) = 30 mg/kg. Delayed pulmonary edema may occur after inhalation.
***Long Term Exposure:*** Cholinesterase inhibitor; cumulative effect is possible. This chemical may damage the nervous system with repeated exposure, resulting in convulsions, respiratory failure. May cause liver damage.
***Points of Attack:*** Respiratory system, lungs, central nervous system, cardiovascular system, skin, eyes, plasma and red blood cell cholinesterase.
***Medical Surveillance:*** Medical observation is recommended for 24 to 48 hours after breathing overexposure, as pulmonary edema may be delayed. Before employment and at regular times after that, the following are

recommended: Plasma and red blood cell cholinesterase levels (tests for the enzyme poisoned by this chemical). If exposure stops, plasma levels return to normal in 1-2 weeks while red blood cell levels may be reduced for 1-3 months. When acetylcholinesterase enzyme levels are reduced by 25% or more below preemployment levels, risk of poisoning is increased, even if results are in lower ranges of "normal." Reassignment to work not involving organophosphate or carbamate pesticides is recommended until enzyme levels recover. If symptoms develop or overexposure occurs, repeat the above tests as soon as possible and get an exam of the nervous system. Also consider complete blood count. Consider chest x-ray following acute overexposure. Do not drink any alcoholic beverages before or during use. Alcohol promotes absorption of organophosphates.

*First Aid:* **Treatment for organophosphate poisoning consists of thorough decontamination, cardiorespiratory support, and administration of the antidotes atropine and pralidoxime. In cases of severe poisoning, diazepam, an anticonvulsant, should also be administered. Antidotes should be administered as prevention even if the diagnosis is in doubt.** Speed in removing material from eyes and skin is of extreme importance. *Eyes:* Eye contact can cause dangerous amounts of these chemicals to be quickly absorbed through the mucous membrane into the bloodstream. Immediately and gently flush eyes with plenty of warm or cold water (NO hot water) for at least 15 minutes, occasionally lifting the upper and lower eyelids. Get medical aid immediately. *Skin:* Get medical aid. Skin contact can cause dangerous amounts of these chemicals to be absorbed into the bloodstream. Wearing the appropriate PPE equipment and respirator for organophosphate pesticides, immediately flush skin with plenty of soap and water for at least 15 minutes while removing contaminated clothing and shoes. Shampoo hair promptly if contaminated. The removed, contaminated clothing and shoes should be double-bagged and left in Hot Zone for later disposal by hazardous materials experts. Skin may also be decontaminated with diluted hypochlorite solution. *Inhalation:* Get medical aid. Do not contaminate yourself. Wearing the appropriate PPE equipment and respirator for organophosphate pesticides, immediately remove the victim from the contaminated area to fresh air. If the victim is not breathing, administer artificial respiration. Do NOT use mouth-to-mouth resuscitation; use bag/mask apparatus. If breathing is difficult, administer oxygen through bag/mask apparatus until medical help arrives. Do not leave victim unattended. *Ingestion:* Call poison control. Loosen all clothing. Never give anything by mouth to an unconscious person. If victim is *unconscious or having convulsions,* do nothing except keep victim warm. Get medical aid. Transfer promptly to a medical facility. In cases of ingestion, **do not induce vomiting**. If the victim is alert and asymptomatic, administer a slurry of activated charcoal at a dose of 1 g/kg (infant, child, and adult dose). A soda can and straw may be of assistance when offering charcoal to a child. *In some cases you may be specifically instructed by poison control to induce vomiting by way of 2 tablespoons of syrup of ipecac (adult) washed down with a cup of water.* Do NOT give activated charcoal before or with ipecac syrup.

*Note to physician or authorized medical personnel:* Treat cases of respiratory compromise, coma, or excessive pulmonary secretions with respiratory support using protocols and techniques available and within the scope of training. Some cases may necessitate procedures such as endotracheal intubation or cricothyrotomy by properly trained and equipped personnel. When possible, atropine (see *Antidotes*, below) should be given under medical supervision. Patients who are comatose, hypotensive, or having seizures or cardiac arrhythmias should be treated according to advanced life support protocols. *Antidotes:* Two antidotes are administered to treat organophosphate poisoning. Atropine is a competitive antagonist of acetylcholine at muscarinic receptors and is used to control the excessive bronchial secretions which are often responsible for death. Pralidoxime relieves both the nicotinic and muscarine effects of organophosphate poisoning by regenerating acetylcholinesterase and can reduce both the bronchial secretions and the muscle weakness associated with poisoning. The initial intravenous dose of atropine in adults should be determined by the severity of symptoms: An initial adult dose of 1.0 to 2.0 mg or pediatric dose of 0.01 mg/kg (minimum 0.01 mg) should be administered intravenously. If intravenous access cannot be established, atropine may also be given intramuscularly, subcutaneously or via endotracheal tube. Doses should be repeated every 15 minutes until excessive secretions and sweating have been controlled. Once bronchial secretion has been controlled, atropine administration should be repeated whenever the secretions begin to recur. In seriously poisoned patients, very large doses may be required. Alterations of pulse rate and pupillary size should not be used as indicators of treatment adequacy. Pralidoxime should be administered as early in poisoning as possible as its efficacy may diminish when given more than 24 to 36 hours after exposure. Doses are as follows: adult 1.0 g; pediatric 25 to 50 mg/kg. The drug should be administered intravenously over 30 to 60 minutes, but in a life-threatening situation, one-half of the total dose can be given per minute for a total administration time of 2 minutes. Treatment should begin to take effect within 40 minutes with a reduction in symptoms and the amount of atropine necessary to control bronchial secretion. The initial dose can be repeated in 1 hour and then every 8 to 12 hours until the patient is clinically well and no longer requires atropine. If intravenous access cannot be established, pralidoxime may also be given intramuscularly. Early administration of diazepam in addition to the combined atropine and pralidoxime treatment may help prevent the onset of seizures and potential brain and cardiac morphologic damage following high-level organophosphate poisoning.

*References:*
- U.S. Environmental Protection Agency, "Chemical Profile: Leptophos," Washington, DC, Chemical Emergency Preparedness Program (Nov. 30, 1987).
- California Environmental Protection Agency "Chemical List of Lists," Sacramento CA (February 1997).
- Agency for Toxic Substances and Disease Registry, U.S. Department of Health and Human Services, Public Health Service, "Managing Hazardous Materials Incidents," Atlanta, GA (June 2003).

## D-Limonene

*Use Type:* Insecticide, insect repellant, dog and cat repellant
*CAS Number:* 5989-27-5 (D-Limonene); 138-86-3 (*dl*-Limonene)
*Formula:* $C_{10}H_{16}$
*Synonyms:* Cajeputene; Carvene; Cinene; Limonene; Dipentene; *p*-Mentha-1,8-diene; 1-Methyl-4-(1-methylethenyl) cyclohexane; Terpinene; *dl-p*-mentha-1,8-diene
*Trade Names:* BUGAWAY®, canceled; BUGCHASER®, canceled; DOO-NOT®; HOLIDAY FIRE ANT KILLER®
*Producers:* Arizona Chemical (USA); Camphor & Allied Products (India); Crowley (USA); Penta Manufacturing (USA); Sanofi-Synthelabo Groupe (France); SCM Glidco (USA)
*Chemical Class:* Miscellaneous Hydrocarbon; Essential oil
*EPA/OPP PC Code:* 079701
*California DPR Chemical Code:* 979, 2531
*ICSC Number:* 0918 (D-Limonene)
*RTECS Number:* OS8100000; GW6360000 (D-Limonene)
*Uses:* Dipentene is also used as a solvent, in rubber compounding and reclamation, and to make paints, enamels, lacquer and perfumes.
*U.S. Maximum Allowable Residue Levels for Limonene (40 CFR 180.539):*
Limonene is exempt from tolerance levels when used in establishments making animal feed and precessed food.
*Carcinogen/Hazard Classifications*
**IARC:** Group 3, unclassifiable
**Label Signal Word:** CAUTION, WARNING
*Regulatory Authority:*
- Actively registered pesticide in California.

*Description:* Colorless liquid. Lemon-like odor. Insoluble in water. Molecular weight = 136.24. Boiling point = 175°C. Melting point/freezing point = –75°C. Vapor pressure = 0.4 kPa @ 14.4°C. Flash point = 45°C. Autoignition temperature = 237°C. Explosive limits: LEL = 0.7%; UEL = 6.1%, both at 150°C. Hazard Identification (based on NFPA-704 M Rating System): Health 0, Flammability 2, Reactivity 0. Log $K_{ow}$ = 4.17. Values above 3.0 are likely to bioaccumulate in marine organisms.
*Incompatibilities:* Forms explosive mixture with air. Contact with oxidizers may cause fire and explosion hazard.
*Permissible Concentration in Water:* No criteria set. Runoff from spills or fire control may cause water pollution.
*Routes of Entry:* Inhalation, passing through the skin, ingestion.
*Harmful Effects and Symptoms*
*Short Term Exposure:* Contact can irritate the eyes and skin. High exposures may damage the kidneys.
*Long Term Exposure:* Dipentene may cause a skin allergy. If allergy develops, very low future exposures can cause itching and a skin rash. There is limited evidence that dipentene causes kidney cancer in male rats (NJ).
*Points of Attack:* Skin
*Medical Surveillance:* If symptoms develop or overexposure is suspected, the following may be useful: Evaluation by a qualified allergist, including careful exposure history and special testing, may help diagnose skin allergy. Kidney function tests.
*First Aid:* If this chemical gets into the eyes, remove any contact lenses at once and irrigate immediately for at least 15 minutes, occasionally lifting upper and lower lids. Seek medical attention immediately. If this chemical contacts the skin, remove contaminated clothing and wash immediately with soap and water. Seek medical attention immediately. If this chemical has been inhaled, remove from exposure, begin rescue breathing (using universal precautions) if breathing has stopped, and CPR if heart action has stopped. Transfer promptly to a medical facility. When this chemical has been swallowed, get medical attention. Give large quantities of water and induce vomiting. Do not make an unconscious person vomit. Do not induce vomiting when formulations containing petroleum solvents are ingested.
*References:*
- Sax, N.I., Ed., "Dangerous Properties of Industrial Materials Report," 2, No. 3, 78-79 (1982).
- New Jersey Department of Health, "Hazardous Substance Fact Sheet: Dipentene," Trenton, NJ (December 1996, revised May 2003). http://www.state.nj.us/health/eoh/rtkweb/0792.pdf
- U.S. Environmental Protection Agency, Office of Pesticide Programs, Pesticide Residue Limits, "Limonene," 40 CFR 180.539, www.epa.gov/pesticides/food/viewtols.htm
- California Environmental Protection Agency "Chemical List of Lists," Sacramento CA (February 1997).

## Lindane

*Use Type:* Insecticide and rodenticide
*CAS Number:* 58-89-9
*Formula:* $C_6H_6Cl_6$

# Lindane

*Alert:* A Restricted Use Pesticide (RUP). Lindane should be handled as a carcinogen, with extreme caution. Most applications have been canceled. It has not been produced in the U.S. since 1977; however, it is still imported into this country and formulated to treat head lice, body lice and scabies. Human toxicity (long-term): Very high.

*Synonyms:* BBH; Benzene hexachloride; γ-Benzene hexachloride; Bexol; BHC; γ-BHC; Benzene hexachloride-gamma isomer; 2,5-Cyclohexane,1,2,3,4,5,6-hexachloro-, (1α,2α,3β,4α,5α,6β)-; DBH; ENT 7,796; Gammabenzene hexachlorocyclohexane (gamma isomer); gamma-BHC; Gamma-HCH; Gammahexa; Gammahexane; HCCH; HCH; γ-HCH; HCH BHC; Hexachlorocyclohexane (gamma isomer); Hexachloran; γ-Hexachloran; γ-Hexachloran; Hexachlorane; γ-Hexachlorane; gamma-Hexachlorane; γ-Hexachlorobenzene; 1α,2α,3β,4α,5α,6β-Hexachlorocyclohexane; γ-Hexachlorocyclohexane; Hexachlorocyclohexane, gamma isomer; 1,2,3,4,5,6-Hexachlor-cyclohexane; γ-1,2,3,4,5,6-Hexachlorocyclohexane; 1,2,3,4,5,6-Hexachlorocyclohexane, gamma isomer; gamma-Hexaclorobenzene; HGI; γ-Lindane; NCI-C00204

*Trade Names:* AALINDAN®; AFICIDE®; AGRISOL G-20®; AGROCIDE®; AGRONEXIT®; AMEISENATOD®; AMEISENMITTEL (MERCK)®; APARASIN®; APHTIRIA®; APLIDAL®; ARBITEX®; BEN-HEX®; BENTOX 10®; CELANEX®; CHLORESENE®; CODECHINE®; DELSANEX DAIRY FLY SPRAY®; DETMOL-EXTRAKT®; DETOX 25®; DEVORAN®; DOL GRANULE®; DRILL TOX-SPEZIAL AGLUKON®; DUAL MURGANIC RPB SEED TREATMENT®; ENTOMOXAN®; EXAGAMA®; FORLIN®; GALLOGAMA®; GAMACID®; GAMAPHEX®; GAMENE®; GAMMA-COL®; GAMMALIN®; GAMMALIN 20; GAMMALEX®; GAMMASAN 30®; GAMMATERR®; GAMMAPHEX®; GAMMEX®; GAMMEXANE®; GAMMEXENE®; GAMMOPAZ®; GEXANE®; HECLOTOX®; HEXA®; HEXAFLOW®; HEXATOX®; HEXAVERM®; HEXICIDE®; HEXYCLAN®; HORTEX®; INEXIT®; ISOTOX®; JACUTIN®; KOKOTINE®; KWELL®; LENTOX®; LINDAGRAM®; LIDENAL®; LINDAFOR®; LINDAGAM®; LINDAGRAIN®; LINDAGRANOX®; LINDAPOUDRE®; LINDATOX®; LINDOSEP®; LINTOX®; LOREXANE®; MARSTAN FLY SPRAY®; MERGAMMA 30®; MILBOL 49®; MIST-O-MATIC LINDEX®; MSZYCOL®; NEXEN FB®; NEXIT®; NEXIT-STARK®; NEXOL-E®; NICOCHLORAN®; NOVIGAM®; OMNITOX®; OVADZIAK®; OWADZIAK®; PEDRACZAK®; PFLANZOL®; QUELLADA®; RODESCO INSECT POWDER®; SANG GAMMA®; SILVANO®; SPRITZ-RAPIDIN®; SPRUEHPFLANZOL®; STREUNEX®; TAP 85®; TRI-6®; VITON®

*Producers:* Agrimor International (USA); Agsin (Singapore); Aimco Pesticides Ltd. (India); Alcotan Laboratories (Spain); BASF Agricultural Products Group (Germany); Bharat Rasayan (India); Drexel Chemical (USA); Ehrenstorfer, Dr. (Germany); Indiclay (India); Kanoria Chemicals & Industries (India); KCIL Group (India); Ki-Hara Chemicals Ltd. (UK); Merck (Germany); Prentiss (USA); Rhone-Poulenc Agro France (France); Shenzhen Guomeng Industry Co., Ltd. (China)

*Chemical Class:* Organochlorine; Halo-organics
*EPA/OPP PC Code:* 009001
*California DPR Chemical Code:* 359
*ICSC Number:* 0053
*RTECS Number:* GV4900000
*EEC Number:* 602-043-00-6
*EINECS Number:* 200-401-2

*Uses:* Lindane has been used against insects in a wide range of applications including treatment of animals, buildings, man for ectoparasites, clothes, water for mosquitoes, living plants, seeds and soils. Most applications have been canceled due to excessive residues, e.g., stored foodstuffs, that may cause cancer. Formulators, distributors and users of lindane represent a special risk group. The major use of lindane in recent years has been to pretreat seeds. Other uses include sunflowers, peas, wheat, barley and oats. Lindane is presently also used in lotions, creams and shampoos for the control of lice and mites in humans.

*Human toxicity (long-term)*[77]: Extra high–0.20 ppb, MCL (Maximum Contaminant Level)

*Fish toxicity (threshold)*[77]: Extra high–0.11310 ppb, MATC (Maximum Acceptable Toxicant Concentration)

*U.S. Maximum Allowable Residue Levels for Lindane (40 CFR 180.133):*

| CROP | ppm |
| --- | --- |
| Apple | 1 |
| Apricot | 1 |
| Asparagus | 1 |
| Avocado | 1 |
| Broccoli | 1 |
| Brussels sprouts | 1 |
| Cabbage | 1 |
| Cattle, fat | 7 |
| Cauliflower | 1 |
| Celery | 1 |
| Cherry | 1 |
| Collards | 1 |
| Cucumber | 3 |
| Eggplant | 1 |
| Goat, fat | 7 |
| Grape | 1 |
| Guava | 1 |
| Hog, fat | 4 |
| Horse, fat | 7 |
| Kale | 1 |
| Kohlrabi | 1 |
| Lettuce | 3 |
| Mango | 1 |

| | |
|---|---|
| Melon | 3 |
| Mushroom | 3 |
| Mustard greens | 1 |
| Nectarine | 1 |
| Okra | 1 |
| Onion, dry bulb | 1 |
| Peach | 1 |
| Pear | 1 |
| Pecan | 0.01 |
| Pepper | 1 |
| Pineapple | 1 |
| Plum | 1 |
| Plum, prune, fresh | 1 |
| Pumpkin | 3 |
| Quince | 1 |
| Sheep, fat | 7 |
| Spinach | 1 |
| Squash | 3 |
| Squash, summer | 3 |
| Strawberry | 1 |
| Swiss chard | 1 |
| Tomato | 3 |

*Carcinogen/Hazard Classifications*
**U.S. EPA Carcinogens:** Group B2/C, Probable and possible carcinogen
**U.S. NTP Carcinogen:** Reasonably anticipated carcinogen
**California Prop. 65:** Carcinogen
**IARC:** Group 2B, possible carcinogen
**Label Signal Word:** WARNING
**WHO Acute Hazard:** Class II, moderately hazardous
**Endocrine Disruptor:** Suspected endocrine disruptor
*Regulatory Authority:*
- Air Pollutant Standard Set (ACGIH)[1] (DFG)[3] (HSE)[33] (OSHA)[58] (Other Countries)[35] (Several States)[60]
- Carcinogen (Animal Positive) (IARC)[9]
- AB 1803-Well Monitoring Chemical (CAL)
- MCL (Maximum Contaminants Levels) list of contaminants (CAL)
- AB 2588-Air Toxics "Hot Spots" Chemicals (CAL)
- CAL Air Resources Board/AB 1807 Toxic Air Contaminants
- Proposition 65 chemical (CAL)
- Permissible Exposure Limits for Chemical Contaminants (CAL/OSHA)
- The "Director's List" (CAL/OSHA)
- Actively registered pesticide in California.
- Clean Air Act: Hazardous Air Pollutants (Title I, Part A, Section 112)
- Clean Water Act: Section 311 Hazardous Substances/RQ 40CFR117.3 (same as CERCLA, see below); 40CFR423, Appendix A, Priority Pollutants; Section 313 Water Priority Chemicals (57FR41331, 9/9/92); 40CFR401.15 Section 307 Toxic Pollutants, as hexachlorocyclohexane.
- EPA Hazardous Waste Number (RCRA No.): U129
- RCRA Toxicity Characteristic (Section 261.24), Maximum
- Concentration of Contaminants, regulatory level, 0.4 mg/L
- RCRA, 40CFR261, Appendix 8 Hazardous Constituents
- Safe Drinking Water Act: MCL, 0.0002 mg/L; MCLG, 0.0002 mg/L; Regulated chemical (47 FR 9352)
- RCRA 40CFR264, Appendix 9; TSD Facilities Ground Water Monitoring List. Suggested test method(s) (PQL $ug$/L): 8080(0.05)
- RCRA 40CFR268.48; 61FR15654, Universal Treatment Standards: Wastewater (mg/L), 0.0017; Nonwastewater (mg/kg), 0.066
- Superfund/EPCRA 40CFR355, Appendix B Extremely Hazardous Substances: TPQ = 1000/10,000 lb (454/4,540 kg)
- Superfund/EPCRA 40CFR302.4 RQ: CERCLA, 1 lb (0.454 kg)
- EPCRA Section 313 Form R *de minimis* concentration reporting level: 0.1%.
- U.S. DOT Regulated Marine Pollutant (49CFR172.101, Appendix B)
- Canada, WHMIS, Ingredients Disclosure List

*Description:* Lindane is a white to yellow, crystalline powder with a slight, musty odor (pure material is odorless). Practically insoluble in water; solubility = 17.3 ppm @ 25°C. Molecular weight = 290.83. Boiling point = 322°C. Melting/Freezing point =112°C. Hazard Identification (based on NFPA-704 M Rating System): Health 3, Flammability 0, Reactivity 0. Insoluble in water. Noncombustible solid, but may be dissolved in flammable liquids. Log $K_{ow}$ = 3.65. Values at or above 3.0 are likely to bioaccumulate in marine organisms.

*Incompatibilities:* Lindane decomposes on contact with powdered iron, aluminum, and zinc and with alkalis producing trichlorobenzene.

*Permissible Exposure Limits in Air:* The Federal standard[58] and the ACGIH[1] TWA value is 0.5 mg/m$^3$. The notation "skin" is added to indicate the possibility of cutaneous absorption. The NIOSH[2] IDLH level = 50 mg/m$^3$. Argentina[35], Germany (3, 35) and the U.K. (33, 35) have all set TWA values of 0.5 mg/m$^3$ as well. Argentina and the U.K. add STEL values of 1.5 mg/m$^3$ and Germany[35] sets an STEL of 5.0 mg/m$^3$. WHO[35] has set a lower TWA of 0.3 mg/m$^3$ and a much lower MAC value has been set by the former USSR at 0.05 mg/m$^3$ in workplace air. Several states have set guidelines or standards for lindane in ambient air[60] ranging from zero (North Dakota) to 0.068 $\mu$g/m$^3$ (Massachusetts) to 1.19$\mu$g/m$^3$ (Kansas) to 1.2 $\mu$g/m$^3$ (Pennsylvania) to 1.67 $\mu$g/m$^3$ (New York) to 5.0 $\mu$g/m$^3$ (Connecticut, Florida, South Carolina) to 8.0 $\mu$g/m$^3$ (Virginia) to 12.0 $\mu$g/m$^3$ (Nevada).

*Determination in Air:* Collection on a filter, workup with isooctane, analysis by gas chromatography/flame ionization.

*Permissible Concentration in Water:* To protect freshwater aquatic life-0.080 µg/L as a 24 hour average, never to exceed 2.0 µg/L. To protect saltwater aquatic life-never to exceed 0.16 µg/L. To protect human health-preferably zero. An additional lifetime cancer risk of 1 in 100, 000 is posed by a concentration of 0.186 µg/L[6]. Mexico[35] has set allowable limits on 0.2 µg/L in coastal waters and 2.0 µg/L in estuaries. WHO has set a limit of 3.0 µg/L in drinking water. The USEPA has set 120µg/L as a long term health advisory and 2 µg/L as a lifetime health advisory[47] for adults. The State of Maine has set a guideline for lindane in drinking water of 4.0 µg/L. EPA[62] has proposed a maximum level in drinking water of 0.2 µg/L.

*Determination in Water:* Methylene chloride extraction followed by gas chromatography with electron capture or halogen specific detection (EPA Method 608) or gas chromatography plus mass spectrometry (EPA Method 625).

*Routes of Entry:* Inhalation, skin absorption, ingestion, eye and/or skin contact.

*Harmful Effects and Symptoms*

*Short Term Exposure:* Lindane irritates the eyes and the respiratory tract and may affect the central nervous system. Symptoms of exposure include vomiting, faintness, tremor, restlessness, muscle spasms, unsteady gait, and convulsions may occur as a result of exposure. Elevated body temperature and pulmonary edema have been reported in children. Coma, respiratory failure and death can result. Exposure to vapors of this compound or its thermal decomposition products may lead to headache, nausea, vomiting, and irritation of the eyes, nose, and throat. Lindane is a stimulant of the nervous system, causing violent convulsions that are rapid in onset and generally followed by death or recovery within 24 hours. The probable human oral lethal dose in 50-500 mg/kg, or between 1 teaspoon and 1 ounce for a 150 lb (70 kg) person.

*Long Term Exposure:* Repeated or prolonged contact with skin may cause dermatitis. May damage the liver and kidneys. May damage the nerves in the arms and legs, possibly with weakness and poor coordination. May cause a serious drop in the blood cell count (aplastic anemia) or in the white blood cell count (agranulocytopenia). The Department of Health and Human Services has determined that HCH (hexachlorocyclohexanes) may reasonably be anticipated to be carcinogenic. Liver cancer has been seen in laboratory rodents that ate HCH for long periods of time. In animals, there is evidence that oral exposure to lindane during pregnancy results in an increased incidence of fetuses with extra ribs. However, ATSDR reports that animal studies have not shown birth defects in the babies of animals fed HCH during pregnancy. HCH has been detected in human breast milk.

*Points of Attack:* Eyes, central nervous system, blood, liver, kidneys and skin.

*Medical Surveillance:* Consider the points of attack in preplacement and periodic physical examinations. Examination of the nervous system. Complete blood count. Blood test for lindane (may not be accurate longer than 1 week following last exposure). Liver and kidney function tests.

*First Aid:* If this chemical gets into the eyes, remove any contact lenses at once and irrigate immediately for at least 15 minutes, occasionally lifting upper and lower lids. Seek medical attention immediately. If this chemical contacts the skin, remove contaminated clothing and wash immediately with soap and water. Seek medical attention immediately. If this chemical has been inhaled, remove from exposure, begin rescue breathing (using universal precautions) if breathing has stopped, and CPR if heart action has stopped. Transfer promptly to a medical facility. When this chemical has been swallowed, get medical attention. Give large quantities of water and induce vomiting. Do not make an unconscious person vomit.

*References:*

- EXTOXNET, Extension Toxicology Network, "Pesticide Information Profile, Lindane," Oregon State University, Corvallis, OR (June 1996). http://extoxnet.orst.edu/pips/lindane.htm
- U.S. Environmental Protection Agency, Office of Pesticide Programs, Pesticide Residue Limits, "Lindane," 40 CFR 180.133, http://www.epa.gov/pesticides/food/viewtols.htm
- U.S. Environmental Protection Agency, "Reregistration Eligibility Decision (RED), Lindane Fact Sheet," Office of Prevention, Pesticides and Toxic Substances, Washington, DC (September 2002). http://www.epa.gov/REDs/factsheets/lindane_fs.htm
- U.S. Department of Health and Human Services, Agency for Toxic Substances and Disease Registry "ToxFAQs, Hexachlorocyclohexanes," Atlanta, GA (June 1999). http://www.atsdr.cdc.gov/tfacts43.html
- New Jersey Department of Health and Senior Services, "Hazardous Substance Fact Sheet, Lindane," Trenton NJ (July 1988, rev. September 2001). http://www.state.nj.us/health/eoh/rtkweb/1117.pdf
- U.S. Environmental Protection Agency, Hexachlorocyclohexane: Ambient Water Quality Criteria, Washington, DC (1980).
- U.S. Environmental Protection Agency, gamma-Hexachloro-cyclohexane, Health and Environmental Effects Profile No. 113, Wash., DC, Office of Solid Waste (April 30, 1980).
- Sax, N.I., Ed., Dangerous Properties of Industrial Materials Report, 3, No. 1, 62-66 (1983).
- U.S. Environmental Protection Agency, "Chemical Profile: Lindane," Washington, DC, Chemical Emergency Preparedness Program (Nov. 30, 1987).
- California Environmental Protection Agency "Chemical List of Lists," Sacramento CA (February 1997).

- New York State Department of Health, "Chemical Fact Sheet: Gamma-BHC," Albany, NY, Bureau of Toxic Substance Assessment (May 1986).

# Linuron (ANSI)

*Use Type:* Herbicide
*CAS Number:* 330-55-2
*Formula:* $C_9H_{10}Cl_2N_2O_2$
*Synonyms:* Caswell No. 528; 3-(3,4-Dichlorophenyl)-1-metoxy-1-methylurea; *N*'-(3,4-Dichlorophenyl)-*N*-methoxy-*N*-methylurea; 1-(3,4-Dichlorophenyl)-3-methoxy-3-methyluree (French); 3-(3,4-Dichlorophenyl)-1-methoxymethylurea; 3-(3,4-Dichlorophenyl)-1-methoxy-1-methylurea; *N*-(3,4-Dichlorophenyl)-*N*'-methyl-*N*'-methoxyurea; 3-(3,4-dichlor-phenyl)-1-methoxy-1-methyl-harnstoff (German); 3-(4,5-Dichlorphenyl)-1-methoxy-1-methylharnstoff (German); Methoxydiuron; 1-Methoxy-1-methyl-3-(3,4-dichlorophenyl)urea; Urea, *N*'-(3,4-dichlorophenyl)-*N*-methoxy-*N*-methyl-; Urea, 3-(3,4-dichlorophenyl)-1-methoxy-1-methyl-
*Trade Names:* AFALON®, Makhteshim-Agan Industries (Israel); ALIBI®; ALISTELL®; BROADCIDE 20EC®; BRONOX®; CERTOL-LIN ONIONS®; CLOVACORN EXTRA®; CROP WEEDSTOP®; DU PONT 326®, DuPont (USA), canceled; FF6135' HERBICIDE 326®; GARNITAN®; H 326®; GEMINI®, DuPont (USA), canceled; HERBICIDE 326®; HOE 2810®; JANUS®; LANDSIDE®; LINNET®; LINEX®, Griffin (USA); LINOROX®; LINUREX®; LOREX®; LOROX®, DuPont (USA), canceled; MARKSMAN 1®; NEMINFEST®; ONSLAUGHT®; PRE-EMPT®; PREMALIN®; PROFALON®; ROTILIN®; SARCLEX®; SCARCLEX®; SINURON®; STAY KLEEN®; TEMPO®; TRIFARMON FL®; TRIFLURON®; TRILIN®; URANUS® (trifluralin + linuron), Makhteshim-Agan Industries (Israel); WARRIOR®
*Producers:* Drexel Chemical (USA); Dow AgroSciences (USA); Griffin (USA); Makhteshim-Agan Industries (Israel)
*Chemical Class:* Substituted urea
*EPA/OPP PC Code:* 035506
*California DPR Chemical Code:* 361
*ICSC Number:* 1300
*RTECS Number:* YS9100000
*EEC Number:* 006-021-00-1
*Uses:* Linuron is a selective, pre-emergence herbicide used to control grasses and broadleaf weeds in carrots, beans, peas, asparagus, maize, potatoes, soybeans, sorghum, wheat, bananas, coffee, cotton and ornamentals. It is also used for control of annual weeds in storehouses, roadsides, fence rows and other non-crop lands. Linuron is frequently used in formulations with other herbicides, insecticides and fungicides.

*Human toxicity (long-term)*[77]: High–5.60 ppb, Health Advisory
*Fish toxicity (threshold)*[77]: Intermediate–42.00007 ppb, MATC (Maximum Acceptable Toxicant Concentration)
*U.S. Maximum Allowable Residue Levels for Linuron (40 CFR 180.184):*

| CROP | ppm |
| --- | --- |
| Asparagus | 7 |
| Carrot, roots | 1 |
| Cattle, fat & meat | 1 |
| Celery | 0.5 |
| Corn, field, forage & stover | 1 |
| Corn, grain | 0.25 |
| Corn, sweet, forage | 1 |
| Corn, sweet, kernel plus Cob with husks removed | 0.25 |
| Corn, sweet, stover | 1 |
| Cotton, undelinted seed | 0.25 |
| Goat, fat & meat | 1 |
| Goat, mbyp | 1 |
| Hog, fat & meat | 1 |
| Hog, mbyp | 1 |
| Horse, fat & meat | 1 |
| Horse, mbyp | 1 |
| Parsley | 0.25 |
| Parsnip | 0.5 |
| Parsnip, roots | 0.5 |
| Potato | 1 |
| Sheep, fat & meat | 1 |
| Sheep, mbyp | 1 |
| Sorghum, forage | 1 |
| Sorghum, grain, grain | 0.25 |
| Sorghum, grain, stover | 1 |
| Soybean, dry & hay | 1 |
| Soybean, forage | 1 |
| Soybean, succulent | 1 |
| Wheat, forage & hay | 0.5 |
| Wheat, grain | 0.25 |
| Wheat, straw | 0.5 |

*Carcinogen/Hazard Classifications*
**U.S. EPA Carcinogens:** Group 3, possible carcinogen
**California Prop. 65:** Developmental toxin
**TRI Developmental Toxin:** Reproductive and developmental toxin
**Label Signal Word:** WARNING or CAUTION
**WHO Acute Hazard:** Class U, unlikely to be hazardous
**Endocrine Disruptor:** Suspected endocrine disruptor
*Regulatory Authority:*
- EPA 40 CFR 372.65, Specific Toxic Chemical Listings
- AB 1803-Well Monitoring Chemical (CAL)
- Proposition 65 chemical (CAL)
- The "Director's List" (CAL/OSHA)
- Actively registered pesticide in California.
- EPCRA Section 313 Form R *de minimis* concentration reporting level: 1.0%

***Description:*** White to colorless crystals or powder. Odorless. Melting/Freezing point = 93–94°C. Soluble in water; solubility = 75 ppm @ 25°C. Molecular weight 249.12. Vapor pressure = $1.2 \times 10^{-5}$ mmHg @ 24°C. Log $K_{ow}$ = 3.19. Values above 3.0 are likely to bioaccumulate in marine organisms.

***Determination in Air:*** Filter; none; Gravimetric; NIOSH IV [Particulates NOR; #0500 (total), #0600 (respirable)].[18]

***Permissible Concentration in Water:*** No criteria set. Runoff from spills or fire control may cause water pollution.

***Harmful Effects and Symptoms***

***Short Term Exposure:*** Contact with eyes or skin may cause irritation or injury. Inhalation should be avoided; use NIOSH-approved air purifying respirators for pesticides. May be harmful if swallowed. Skin contact may cause severe irritation or burns.

***Long Term Exposure:*** Developmental problems; mutagen.

***Points of Attack:*** Skin; reproductive system.

***First Aid:*** If this chemical gets into the eyes, remove any contact lenses at once and irrigate immediately for at least 15 minutes, occasionally lifting upper and lower lids. Seek medical attention immediately. If this chemical contacts the skin, remove contaminated clothing and wash immediately with soap and water. Seek medical attention immediately. If this chemical has been inhaled, remove from exposure, begin rescue breathing (using universal precautions) if breathing has stopped, and CPR if heart action has stopped. Transfer promptly to a medical facility. When this chemical has been swallowed, get medical attention. Give large quantities of water. Do not induce vomiting when formulations containing petroleum solvents are ingested. Do not make an unconscious person vomit.

***References:***
- EXTOXNET, Extension Toxicology Network, "Pesticide Information Profile, Linuron," Oregon State University, Corvallis, OR (June 1996). http://extoxnet.orst.edu/pips/linuron.htm
- U.S. Environmental Protection Agency, Office of Pesticide Programs, Pesticide Residue Limits, "Linuron", 40 CFR 180.184, www.epa.gov/pesticides/food/viewtols.htm
- California Environmental Protection Agency "Chemical List of Lists," Sacramento CA (February 1997)

# M

## Magnesium Chlorate

*Use Type:* Defoliant, herbicide and microbiocide
*CAS Number:* 10326-21-3
*Formula:* $Cl_2MgO_6$; $Mg(ClO_3)_2$
*Synonyms:* Chlorate de magnesium (French); Chlorate salt of magnesium; Chloric acid, magnesium; Chloric acid, magnesium salt; Magnesium dichlorate
*Trade Names:* DE-FOL-ATE®; E-Z-OFF®; MAGRON®; MC DEFOLIANT®; ORTHO MC®
*Producers:* Great Western Inorganics (USA)
*Chemical Class:* Inorganic
*EPA/OPP PC Code:* 530200
*RTECS Number:* FO0175000
*Uses:* Used as a drying agent and defoliant.
*Regulatory Authority:*
• Air Pollutant Standard Set (former USSR)[43]
*Description:* Magnesium chlorate, $Mg(ClO_3)_2$, is white crystalline solid. Soluble in water (reaction). Boiling point = 120°C. Melting/Freezing point = 35°C.
*Incompatibilities:* A strong oxidizer. Violent reactions with arsenic, carbon, charcoal, copper, phosphorus, sulfur, magnesium oxide, metal sulfides (copper sulfide, arsenic sulfide, tin sulfide), fuels, and strong acids. Reacts with moisture.
*Permissible Exposure Limits in Air:* The former USSR-UNEP/IRPTC project[43] has set an MAC in ambient air in residential areas of 0.1 mg/m$^3$ both on a momentary and a daily average basis.
*Determination in Air:* No OELs established.
*Permissible Concentration in Water:* The former USSR-UNEP/IRPTC project[43] has set an MAC in water bodies used for fishery purposes of 0.35 mg/L.
*Routes of Entry:* Inhalation, ingestion, eye and/or skin contact.
*Harmful Effects and Symptoms*
*Short Term Exposure:* Magnesium chlorate can affect you when breathed in. Contact can irritate or even burn the skin and eyes. Inhaling the dust irritates the respiratory system. Exposure can interfere with the ability of the blood to carry oxygen, causing headaches, weakness dizziness, trouble breathing, collapse and even death. Breathing the dust can irritate the air passages, causing sore throat and/or cough with phlegm.
*Long Term Exposure:* Repeated exposure can cause lung irritation bronchitis may develop with coughing, phlegm, and/or shortness of breath. May affect the kidneys.
*Points of Attack:* Lungs and blood.

*Medical Surveillance:* Before beginning employment and at regular times after that, for those with frequent or potentially high exposures, the following are recommended: Lung function tests. Kidney function tests. If symptoms develop or overexposure is suspected, the following may be useful: Blood methemoglobin level.
*First Aid:* occasionally lifting upper and lower lids. Seek medical attention immediately. If this chemical contacts the skin, remove contaminated clothing and wash immediately with soap and water. Seek medical attention immediately. If this chemical has been inhaled, remove from exposure, begin rescue breathing (using universal precautions) if breathing has stopped, and CPR if heart action has stopped. Transfer promptly to a medical facility. When this chemical has been swallowed, get medical attention. Give large quantities of water and induce vomiting. Do not make an unconscious person vomit.
*Note to physician or authorized medical personnel:* Treat for methemoglobinemia. Spectrophotometry may be required for precise determination of levels of methemoglobinemia in urine.
*References:*
• New Jersey Department of Health and Senior Services, "Hazardous Substance Fact Sheet: Magnesium Chlorate," Trenton, NJ (September 1999). http://www.state.nj.us/health/eoh/rtkweb/1139.pdf
• California Environmental Protection Agency "Chemical List of Lists," Sacramento CA (February 1997).

## Malathion (ANSI)

*Use Type:* Insecticide
*CAS Number:* 121-75-5
*Formula:* $C_{10}H_{19}O_6PS_2$
*Alert:* A General Use Pesticide (GUP) with many applications. Persons whose skin or clothing is contaminated with liquid or powdered malathion can cause secondary contamination by direct contact.
*Synonyms:* S-[1,2-Bis(aethoxy-carbonyl)-aethyl]-O,O-dimethyl-dithiophosphat (German); S-[1,2-Bis(carbethoxy)ethyl] O,O-dimethyldithiophosphate; S-[1,2-Bis(ethoxycarbonyl)ethyl] O,O-dimethyl phosphorodithioate; S-1,2-Bis(ethoxycarbonyl)ethyl-O,O-dimethylthiophosphate; S-[1,2-Bis(ethoxy-carbonyl)-ethyl]-O,O-dimethyl-dithiophosfaat (Dutch); S-[1,2-Bis(etossi-carbonil)-etil]-O,O-dimetil-ditiofosfato (Italian); Butanedioic acid, [(dimethoxyphosphinothioyl)thio]-, diethyl ester; Carbethoxy malathion; Carbofos (Russian); Carbophos (Russian); Celthion (Indian); S-(1,2-

Dicarbethoxyethyl) *O,O*-dimethylphosphorodithioate; Dicarboethoxyethyl *O,O*-dimethyl phosphorodithioate; Diethyl [(dimethoxyphosphinothioyl)thio]butanedioate; Diethyl (dimethoxyphosphinothioylthio)succinate; Diethyl (dimethoxythiophosphorylthio)succinate; Diethyl mercaptosuccinate, *O,O*-dimethyl phosphorodithioate; Diethyl mercaptosuccinate, *O,O*-dimethyl dithiophosphate, *S*-ester; Diethyl mercaptosuccinate, *O,O*-dimethyl thiophosphate; Diethyl mercaptosuccinate, *S*-ester with *O,O*-dimethyl phosphorodithioate; [(Dimethoxyphosphinothioyl) thio]butanedioic acid diethyl ester; *O,O*-Dimethyl *S*-(1,2-dicarbethoxyethyl) dithiophosphate; *O,O*-Dimethyl *S*-(1,2-dicarbaethoxyaethyl)-dithiophosphat (German); *O,O*-Dimethyl *S*-(1,2-dicarbethoxyethyl)phosphorodithioate; *O,O*-Dimethyl *S*-1,2-di(ethoxycarbamyl)ethyl phosphorodithioate; *O,O*-Dimethyldithiophosphate diethylmercaptosuccinate; *O,O*-Dimethyl dithiophosphate of diethyl mercaptosuccinate; Dithiophosphate de *O,O*-dimethyle et de *S*-(1,2-dicarboethoxyethyle) (French); Carbophos; ENT 17,034; Malathon; Malation (Spanish); Malathyl; Malatox (Indian); Maldison (Australia, New Zealand); Malmed; Malphos; Mercaptosuccinic acid diethyl ester; Mercaptothion; NCI-C00215; Oleophosphothion; Phosphothion; Succinic acid, mercapto-, Diethyl ester, *S*-ester with *O,O*-dimethyl phosphorodithioate; Vegfru (Indian)

*Trade Names:* AI3-17034®; AGRICHEM GREENFLY SPRAY®; ALCO®, Amvac Chemical (USA); ALL PURPOSE GARDEN INSECTICIDE®; AMERICAN CYANAMID 4,049®; ATRAPA 5E®, Griffin LLC (USA); BAN-MITE®; CALMATHION®; CARBETOVUR®; CARBETOX®; CELTHION®; CHEMATHION®; CIMEXAN®; COMPOUND 4049®; CROMOCIDE®; CYTHION®, Helena Chemical (USA); SPRAY CONCENTRATE®; CYTHION®; DETMOL MA®; DETMOL® 96%; DETMOL MALATHION®; DURAMITEX®; EMMATOS EXTRA®; EL 4049®; EMMATON®; EMMATOS®; ETIOL®; EVESHIELD CAPTAN/MALATHION®; EXATHIOS®; EXTERMATHION®; FYAFANON®, Helena Chemical (USA); FISONS GREENFLY AND BLACKFLY KILLER®; FOG® 3; FORMAL®; FORTHION®; FOSFOTHION®; FOSFOTION®; FYFANON®, Cheminova (Denmark); ETHIOLACAR®; GREEN DEVIL®, Drexel Chemical (USA); GREENFLY AEROSOL SPRAY®; HILTHION®; KARBOFOS®; KOP-THION®; KYPFOS®; MALACIDE®; MALAFOR®; MALAGRAN®; MALAKILL®; MALAMAR®; MALASOL®; MALASPRAY®; MALATAF®, Rallis India (India); MALATHION 60®; MALATHION E50®; MALATOL®; MALTOX®; MOSCARDA®; ORTHO MALATHION®; PBI CROP SAVER®; PRENTOX®, Prentiss Inc. (USA); PRIODERM®; PROKIL®, Gowan Company (USA); SADOFOS®; SADOPHOS®; SF® 60; SIPTOX I®; SUMITOX®; TAK®; TM-4049®; VETIOL®; ZITHIOL®

*Producers:* Agrimor International (USA); Agro-care Chemical Industry Group (China); Agsin (Singapore); Aimco Pesticides Ltd. (India); Alcotan Laboratories (Spain); Amvac Chemical (USA); BASF Agricultural Products Group (Germany); Bharat Pulverizing Mills (India); Bharat Rasayan (India); Biesterfeld Siemsgluess International. GmbH (Germany); Bonide Products (USA); Cangzhou Green Chemical Co. (China); Cheminova (Denmark); China Chemical (China); Cyanamid (USA); Dainippon Ink & Chemicals (India); Drexel Chemical (USA); Excel Industries (India); Gowan Company (USA); Griffin LLC (USA); Hebei Huafeng Chemical Group (China); Helena Chemical (USA); Hindustan Insecticides (India); Hockley International (UK); Hokko Chemical Industry (Japan); Indiclay (India); Jingma Chemicals Ltd. (China); Ki-Hara Chemicals Ltd. (UK); Luxembourg Industries (PAMOL) (Israel); Monsanto (USA); Montedison (Italy); Nissan Chemical Industries (Japan); Pbi/Gordon (USA); Prentiss Inc. (USA); Proficol (Colombia); Rallis India (India); Rhone-Poulenc Agro France (France); The Scott Company (USA); Shenzhen Guomeng Industry Co., Ltd. (China); Sigma-Aldrich Laborchemikalien (Germany); Sinon (Taiwan); Sumitomo Chemical (Japan); Tekchem (Mexico); Zago Asia Ltd. (Singapore)

*Chemical Class:* Organophosphate
*EPA/OPP PC Code:* 057701
*California DPR Chemical Code:* 367
*ICSC Number:* 0172
*RTECS Number:* WM8400000
*EEC Number:* 015-041-00-X
*EINECS Number:* 204-497-7

*Uses:* Malathion is a non-systemic, wide-spectrum organophosphate insecticide. It was one of the earliest organophosphate insecticides developed (introduced in 1950). Malathion is suited for the control of sucking and chewing insects on fruits, vegetables, citrus, cotton, corn, sorghum, ornamentals and stored products, and is also used to control mosquitoes, flies, household insects, farm and livestock parasites (ectoparasites), and head and body lice. Malathion may also be found in formulations with many other pesticides; the U.S. EPA lists 2,283 current and canceled labels of products containing malathion. Malathion is marketed as 99.6% technical grade liquid. Available formulations include wettable powders (25% and 50%), emulsifiable concentrates, dusts and aerosols.

*Human toxicity (long-term)*[77]: Very low–100.00 ppb, Health Advisory
*Fish toxicity (threshold)*[77]: Extra high–0.28991 ppb, MATC (Maximum Acceptable Toxicant Concentration)
***U.S. Maximum Allowable Residue Levels for Malathion (40 CFR 180.111):*** *Note:* The U.S. EPA allows a maximum of 0.1-135 ppm malathion per million parts of certain types of food.

| CROP | ppm |
|---|---|
| Alfalfa | 135 |

| | | | |
|---|---|---|---|
| Almond | 8 | Hog, meat | 4 |
| Almond, hulls | 50 | Hog, mbyp | 4 |
| Apple | 8 | Hop | 1 |
| Apricot | 8 | Horse, fat | 4 |
| Asparagus | 8 | Horse, meat | 4 |
| Avocado | 8 | Horse, mbyp | 4 |
| Barley, grain | 8 | Horseradish | 8 |
| Bean | 8 | Kumquat | 8 |
| Beet, garden, roots and tops | 8 | Leek | 8 |
| Beet, sugar, roots | 1 | Lemon | 8 |
| Beet, sugar, tops | 8 | Lentil | 8 |
| Blackberry | 8 | Lespedeza, hay | 135 |
| Blueberry | 8 | Lespedeza, seed | 8 |
| Boysenberry | 8 | Lespedeza, straw | 135 |
| Carrot, roots | 8 | Lime | 8 |
| Cattle, fat | 4 | Loganberry | 8 |
| Cattle, feed, concentrate, nonmedicated | 10 | Lupin, seed | 8 |
| Cattle, meat | 4 | Mango | 8 |
| Cattle, mbyp | 4 | Melon | 8 |
| Chayote | 8 | Milk, fat | 0.5 |
| Chayote, roots | 8 | Mushroom | 8 |
| Cherry | 8 | Nectarine | 8 |
| Chestnut | 1 | Nut, macadamia | 1 |
| Citrus, dried pulp | 50 | Oat, grain | 8 |
| Clover | 135 | Okra | 8 |
| Corn | 2 | Onion | 8 |
| Corn, forage | 8 | Onion, green | 8 |
| Corn, grain | 8 | Orange | 8 |
| Cotton, undelinted seed | 2 | Papaya | 1 |
| Cowpea, forage | 135 | Parsnip | 8 |
| Cowpea, hay | 135 | Passionfruit | 8 |
| Cranberry | 8 | Pea | 8 |
| Cucumber | 8 | Pea, field, hay | 8 |
| Currant | 8 | Pea, field, vines | 8 |
| Date, dried fruit | 8 | Peach | 8 |
| Dewberry | 8 | Peanut | 8 |
| Egg | 0.1 | Peanut, forage | 135 |
| Eggplant | 8 | Peanut, hay | 135 |
| Fig | 8 | Pear | 8 |
| Filbert | 1 | Pecan | 8 |
| Flax, seed | 0.1 | Pepper | 8 |
| Flax, straw | 1 | Peppermint | 8 |
| Garlic | 8 | Pineapple | 8 |
| Goat, fat | 4 | Plum | 8 |
| Goat, meat | 4 | Plum, prune | 8 |
| Goat, mbyp | 4 | Potato | 8 |
| Gooseberry | 8 | Poultry, fat | 4 |
| Grape | 8 | Poultry, meat | 4 |
| Grape, raisin | 12 | Poultry, mbyp | 4 |
| Grapefruit | 8 | Pumpkin | 8 |
| Grass | 135 | Quince | 8 |
| Grass, hay | 135 | Radish | 8 |
| Guava | 8 | Raspberry | 8 |
| Hog, fat | 4 | Rice, grain | 8 |
| | | Rice, wild | 8 |

| | |
|---|---|
| Rutabaga | 8 |
| Rye, grain | 8 |
| Safflower, refined oil | 0.6 |
| Safflower, seed | 0.2 |
| Salsify | 8 |
| Shallot, bulb | 8 |
| Sheep, fat | 4 |
| Sheep, meat | 4 |
| Sheep, mbyp | 4 |
| Sorghum, forage | 8 |
| Sorghum, grain, grain | 8 |
| Soybean, dry | 8 |
| Soybean, forage | 135 |
| Soybean, hay | 135 |
| Soybean, succulent | 8 |
| Spearmint | 8 |
| Squash, summer | 8 |
| Squash, winter | 8 |
| Strawberry | 8 |
| Sunflower, seed | 8 |
| Sweet potato | 1 |
| Tangerine | 8 |
| Tomato | 8 |
| Trefoil, birdsfoot, forage | 135 |
| Trefoil, birdsfoot, hay | 135 |
| Turnip | 8 |
| Vegetable, brassica, leafy, group 5 | 8 |
| Vegetable, leafy, except brassica, group 4 | 8 |
| Vetch, hay | 135 |
| Vetch, seed | 8 |
| Vetch, straw | 135 |
| Walnut | 8 |
| Wheat, grain | 8 |

*Carcinogen/Hazard Classifications*
**U.S. EPA Carcinogens:** Suggestive Carcinogen
**IARC:** Group 3, unclassifiable
**Label Signal Word:** CAUTION
**WHO Acute Hazard:** Class III, slightly hazardous
**Endocrine Disruptor:** Suspected endocrine disruptor
*Regulatory Authority:*
- Air Pollutant Standard Set (ACGIH)[1] (DFG)[3] (HSE)[33] (OSHA)[58] (Other Countries)[35] (Several States)[60]
- AB 1803-Well Monitoring Chemical (CAL)
- MCL (Maximum Contaminants Levels) list of contaminants (CAL)
- CAL-DHS/DHS Drinking Water Action Levels
- Permissible Exposure Limits for Chemical Contaminants (CAL/OSHA)
- The "Director's List" (CAL/OSHA)
- Actively registered pesticide in California.
- Clean Water Act: Section 311 Hazardous Substances/RQ 40CFR117.3 (same as CERCLA, see below).
- Superfund/EPCRA 40CFR302.4 RQ: CERCLA, 100 lb (45.4 kg)
- EPCRA Section 313 Form R *de minimus* concentration reporting level: 1.0%
- U.S. DOT Inhalation Hazard Chemicals as organophosphates
- U.S. DOT Regulated Marine Pollutant (49CFR172.101, Appendix B)
- Canada, WHMIS, Ingredients Disclosure List

*Description:* At room temperature, malathion is a yellow to deep brown liquid. Clear and colorless when pure. It is a solid below 37°F. It is often dissolved in a hydrocarbon solvent before use. Garlic-like odor. *Warning properties*: Garlic odor at 13.5 mg/m$^3$; inadequate warning for acute and chronic exposures. The premium grade can maintain its biological activity unchanged for approximately 2 years if stored unopened in a cool, shaded, and well aired place at 68–86 °F. Slightly soluble in water; solubility = 145 ppm @ 20°C. Specific gravity (water: 1) = 1.23. Molecular weight = 330.36 daltons. Melting/Freezing point = 2.9°C. Boiling point = 156–157°C. Vapor pressure = 8.0 x 10$^{-6}$ mmHg @ 25 °C. Flash point = >163°C. Hazard Identification (based on NFPA-704 M Rating System): Health 2, Flammability 1, Reactivity 0. Log $K_{ow}$ = 2.88. Values above 3.0 are likely to bioaccumulate in marine organisms.

*Incompatibilities:* Reacts violently with strong oxidizers, magnesium, alkaline pesticides. Attacks metals including iron, steel, tin plate, lead, copper, and some plastics, coatings, and rubbers.

*Permissible Exposure Limits in Air:* OSHA[2] and NIOSH[2] have set a TWA of 15 mg/m$^3$ with the notation "skin" to indicate the possibility of cutaneous absorption. A TWA value of 10 mg/m$^3$ has been recommended by ACGIH[1]. The NIOSH[2] IDLH level = 250 mg/m$^3$. The HSE[33] has this same standard. The DFG[3] has set an MAK of 15 mg/m$^3$. Argentina[35] and the former USSR[35, 43] have set TWA values of 0.5 mg/m$^3$ in workplace air and Argentina[35] has set STEL of 1.5 mg/m$^3$. In addition the former USSR[43] has set MAC values in the ambient air off residential areas at 0.015 mg/m$^3$ on a momentary basis and 0.006 mg/m$^3$ on a daily average basis. Several states have set guidelines or standards for malathion, in ambient air[60] ranging from 33.3 μg/m$^3$ (New York) to 100 μg/m$^3$ (Florida, North Dakota, South Carolina) to 160 μg/m$^3$ (Virginia) to 200 μg/m$^3$ (Connecticut) to 238 μg/m$^3$ (Nevada).

*Determination in Air:* OSHA versatile sampler-2; Toluene/Acetone; Gas chromatography/Flame photometric detection for sulfur, nitrogen, or phosphorus; NIOSH IV, Method #5600, Organophosphorus pesticides.[18] See also OSHA Method #62[58]

*Permissible Concentration in Water:* The U.S. EPA has established a level of 0.1 milligrams of malathion per liter of drinking water (0.1 mg/L) for lifetime exposure of adults as a level that is not expected to cause effects that are harmful to health. Some states have set guidelines for malathion in drinking water[61] ranging from 40 µg/L (Maine to 140 µg/L (Kansas) to 160 µg/L (California). The former USSR-UNEP/IRPTC project[43] has set an MAC in water bodies used for domestic purposes of 0.05 mg/L. The MAC in water bodies used for fishery purposes is zero.

*Routes of Entry:* Inhalation of vapor, skin absorption, ingestion and skin and/or eye contact.

*Harmful Effects and Symptoms*

*Short Term Exposure:* Systemic malathion toxicity due to excess cholinergic stimulation may result from all routes of exposure. Symptoms include abdominal cramps, vomiting, diarrhea, pinpoint pupils and blurred vision, excessive sweating, salivation and lacrimation, wheezing, excessive tracheobronchial secretions, agitation, seizures, bradycardia or tachycardia, muscle twitching and weakness, and urinary and fecal incontinence. Seizures are much more common in children than in adults. Death results from loss of consciousness, coma, excessive bronchial secretions, respiratory depression and cardiac irregularity. Commercial malathion products often contain impurities and hydrocarbon solvents, such as xylene or toluene, which themselves can cause toxicity. Toxicity of malathion depends on metabolic activation; thus, symptoms may appear from a few minutes to a few hours after exposure. The effects caused by many short term exposures during a week's time can be accumulated and felt as one intense response. Sometimes effects are not felt until hours or days after exposure. *Inhalation:* No effects were reported from exposures of up to 86 mg/m$^3$ for 42 days. The only effect reported due to inhalation was the reduction in activity of an important nervous system enzyme. *Skin:* Important route of exposure during formulation and usage. Prolonged contact (hours) along with poor hygiene has resulted in irritation, as well as symptoms listed under ingestion. *Eyes:* Direct contact can lead to irritation and discomfort. *Ingestion:* Swallowing of malathion has caused severe poisoning and death. Swallowing of 1-1/2 to 3 ounces of a moisture (50% malathion) has caused severe poisoning with symptoms which include nausea, vomiting, headache, abdominal pain, diarrhea, difficulty in breathing, fall in blood pressure, muscle spasms, paralysis, loss of reflexes, convulsions and coma. Between 3-1/2 to 5 ounces of a mixture (50% malathion) has caused death. Delayed pulmonary edema may occur after inhalation.

*Long Term Exposure:* High or repeated exposure may damage the nerves, causing weakness, dizziness, and poor coordination in arms and legs. Repeated exposures may cause personality changes, depression, anxiety or irritability. Persistent weakness and impaired memory have been reported to occur from low-level exposures to some organophosphates in the absence of acute cholinergic effects, but there is no reliable information on adverse health effects of chronic exposure to malathion.[ATSDR]

*Carcinogenicity:* The International Agency for Research on Cancer has determined that malathion is unclassifiable as to its carcinogenicity to humans. In animals, malathion induced liver carcinogenicity at doses that were considered excessive. *Reproductive and Developmental Effects:* Studies have been reported in which malathion induced transient testicular effects in rodents. Results from studies addressing reproductive or developmental effects in humans are inconclusive. Prolonged, daily contact with exposed areas of skin has led to skin irritation and sensitization. May cause genetic changes (mutations). Birth defects have not been observed in humans exposed to malathion, but developmental effects have been seen in the offspring of animals that ingested enough malathion while pregnant to cause health effects in the mother. Animal studies have shown that malathion can be transferred from a pregnant mother to the developing fetus and from a nursing mother to the newborns through the mother's milk.

*Points of Attack:* Eyes, skin, respiratory system, liver, blood cholinesterase, central nervous system, cardiovascular system and gastrointestinal tract.

*Medical Surveillance:* Medical observation is recommended for 24 to 48 hours after breathing overexposure, as pulmonary edema may be delayed. Before employment and at regular times after that, the following are recommended: Plasma and red blood cell cholinesterase levels (tests for the enzyme poisoned by this chemical). If exposure stops, plasma levels return to normal in 1-2 weeks while red blood cell levels may be reduced for 1-3 months. When acetylcholinesterase enzyme levels are reduced by 25% or more below preemployment levels, risk of poisoning is increased, even if results are in lower ranges of "normal." Reassignment to work not involving organophosphate or carbamate pesticides is recommended until enzyme levels recover. If symptoms develop or overexposure occurs, repeat the above tests as soon as possible and get an exam of the nervous system. Also consider complete blood count. Consider chest x-ray following acute overexposure. Preplacement and periodic medical examination shall include: Comprehensive initial or interim medical and work histories. A physical examination which shall be directed toward, but not limited to evidence of frequent headache, dizziness, nausea, tightness of the chest, dimness of vision, and difficulty in focusing the eyes. Determination, at the time of the preplacement examination, of a baseline or working baseline erythrocyte ChE activity. A judgment of the worker's physical ability to use negative or positive pressure regulators as defined in 29 CFG 1910.134. Periodic examinations shall be made available on an annual basis or at some other interval determined by the responsible physician. Medical records shall be maintained for all workers engaged in the manufacture or formulation of malathion and such records shall be kept for at least one year after termination of employment. Pertinent medical information shall be available to medical representatives of the U.S. Government, the employer and the employees. Erythrocyte cholinesterase levels should be checked as noted above and as described in detail by NIOSH Criteria Document No. 76-205.

***First Aid:*** **Treatment for organophosphate poisoning consists of thorough decontamination, cardiorespiratory support, and administration of the antidotes atropine and pralidoxime. In cases of severe poisoning, diazepam, an anticonvulsant, should also be administered. Antidotes should be administered as prevention even if the diagnosis is in doubt.** Speed in removing material from eyes and skin is of extreme importance. *Eyes:* Eye contact can cause dangerous amounts of these chemicals to be quickly absorbed through the mucous membrane into the bloodstream. Immediately and gently flush eyes with plenty of warm or cold water (NO hot water) for at least 15 minutes, occasionally lifting the upper and lower eyelids. Get medical aid immediately. *Skin:* Get medical aid. Skin contact can cause dangerous amounts of these chemicals to be absorbed into the bloodstream. Wearing the appropriate PPE equipment and respirator for organophosphate pesticides, immediately flush skin with plenty of soap and water for at least 15 minutes while removing contaminated clothing and shoes. Shampoo hair promptly if contaminated. The removed, contaminated clothing and shoes should be double-bagged and left in Hot Zone for later disposal by hazardous materials experts. Skin may also be decontaminated with diluted hypochlorite solution. *Inhalation:* Get medical aid. Do not contaminate yourself. Wearing the appropriate PPE equipment and respirator for organophosphate pesticides, immediately remove the victim from the contaminated area to fresh air. If the victim is not breathing, administer artificial respiration. Do NOT use mouth-to-mouth resuscitation; use bag/mask apparatus. If breathing is difficult, administer oxygen through bag/mask apparatus until medical help arrives. Do not leave victim unattended. *Ingestion:* Call poison control. Loosen all clothing. Never give anything by mouth to an unconscious person. If victim is *unconscious or having convulsions,* do nothing except keep victim warm. Get medical aid. Transfer promptly to a medical facility. In cases of ingestion, **do not induce vomiting**. If the victim is alert and asymptomatic, administer a slurry of activated charcoal at a dose of 1 g/kg (infant, child, and adult dose). A soda can and straw may be of assistance when offering charcoal to a child. *In some cases you may be specifically instructed by poison control to induce vomiting by way of 2 tablespoons of syrup of ipecac (adult) washed down with a cup of water.* Do NOT give activated charcoal before or with ipecac syrup.

*Note to physician or authorized medical personnel:* Treat cases of respiratory compromise, coma, or excessive pulmonary secretions with respiratory support using protocols and techniques available and within the scope of training. Some cases may necessitate procedures such as endotracheal intubation or cricothyrotomy by properly trained and equipped personnel. When possible, atropine (see *Antidotes,* below) should be given under medical supervision. Patients who are comatose, hypotensive, or having seizures or cardiac arrhythmias should be treated according to advanced life support protocols. *Antidotes:* Two antidotes are administered to treat organophosphate poisoning. Atropine is a competitive antagonist of acetylcholine at muscarinic receptors and is used to control the excessive bronchial secretions which are often responsible for death. Pralidoxime relieves both the nicotinic and muscarine effects of organophosphate poisoning by regenerating acetylcholinesterase and can reduce both the bronchial secretions and the muscle weakness associated with poisoning. The initial intravenous dose of atropine in adults should be determined by the severity of symptoms: An initial adult dose of 1.0 to 2.0 mg or pediatric dose of 0.01 mg/kg (minimum 0.01 mg) should be administered intravenously. If intravenous access cannot be established, atropine may also be given intramuscularly, subcutaneously or via endotracheal tube. Doses should be repeated every 15 minutes until excessive secretions and sweating have been controlled. Once bronchial secretion has been controlled, atropine administration should be repeated whenever the secretions begin to recur. In seriously poisoned patients, very large doses may be required. Alterations of pulse rate and pupillary size should not be used as indicators of treatment adequacy. Pralidoxime should be administered as early in poisoning as possible as its efficacy may diminish when given more than 24 to 36 hours after exposure. Doses are as follows: adult 1.0 g; pediatric 25 to 50 mg/kg. The drug should be administered intravenously over 30 to 60 minutes, but in a life-threatening situation, one-half of the total dose can be given per minute for a total administration time of 2 minutes. Treatment should begin to take effect within 40 minutes with a reduction in symptoms and the amount of atropine necessary to control bronchial secretion. The initial dose can be repeated in 1 hour and then every 8 to 12 hours until the patient is clinically well and no longer requires atropine. If intravenous access cannot be established, pralidoxime may also be given intramuscularly. Early administration of diazepam in addition to the combined atropine and pralidoxime treatment may help prevent the onset of seizures and potential brain and cardiac morphologic damage following high-level organophosphate poisoning.

***References:***
- EXTOXNET, Extension Toxicology Network, "Pesticide Information Profile, Malathion," Oregon State University, Corvallis, OR (June 1996). http://extoxnet.orst.edu/pips/malathio.htm
- U.S. Environmental Protection Agency, Office of Pesticide Programs, Pesticide Residue Limits, "Malathion," 40 CFR 180.111, http://www.epa.gov/pesticides/food/viewtols.htm
- National Institute for Occupational Safety and Health, Criteria for a Recommended Standard: Occupational

Exposure to Malathion, NIOSH Doc. No. 76-205, Wash, DC (June 1976).
- New York State Department of Health, "Chemical Fact Sheet: Malathion, " Albany, NY, Bureau of Toxic Substance Assessment (Version 2-March 1986).
- Sax, N.I., Ed., "Dangerous Properties of Industrial Materials Report, " 7, No. 5, 63-74 (1987).
- California Environmental Protection Agency "Chemical List of Lists," Sacramento CA (February 1997).
- Agency for Toxic Substances and Disease Registry, U.S. Department of Health and Human Services, Public Health Service, "Managing Hazardous Materials Incidents," Atlanta, GA (June 2003).
- New Jersey Department of Health and Senior Services, "Hazardous Substance Fact Sheet, Malathion," Trenton NJ (April, 1997).
  http://www.state.nj.us/health/eoh/rtkweb/1150.pdf

# Maleic Hydrazide

*Use Type:* Herbicide, plant growth regulator, fungicide
*CAS Number:* 123-33-1
*Formula:* $C_4H_4N_2O_2$
*Synonyms:* 1,2-Dihydropyridazine-3,6-dione; 1,2-Dihydro-3,6-pyradazinedione; 1,2-Dihydro-3,6-pyridazinedione; Hydrazid hydrazida maleica (Spanish); 6-Hydroxy-3(2*H*)-pyridazinone; KMH; MAH; Maleic acid hydrazide; Maleic hydrazine; Maleic hydrazide fungicide; Maleinsaurehydrazid (German); Mazide; *N,N*-maleoylhydrazine; Malzid; 1,2,3-tetrahydro-3,6-dioxopyridazine; MH
*Trade Names:* BH DOCK KILLER®; BOS MH®; BURTOLIN®; CHEMFORM®; DE-CUT®; DESPROUT D REXEL-SUPER P®; EC 300®; ENT 18,870®; FAIR 30®; FAIR PLUS®; FAIR PS®; MAINTAIN 3®; MALEIN 30®; MALAZIDE®; MH 30®; MH 40®; MH 36 BAYER®; PO-SAN® (with 9*H*-Fluorene-9-carboxylic acid, 2-chloro-9-hydroxy-, methyl ester)®; REGULOX®; REGULOX W®; REGULOX 50W®; RETARD®; ROYAL MH 30®; ROYAL SLO-GRO®; SLO-GRO®; SPROUT-STOP®; STUNTMAN®; SUCKER-STUFF®; SUPER DE-SPROUT®; SUPER SPROUT STOP®; VONDALDHYDE®; VONDRAX®
*Producers:* Crompton (USA); Drexel Chemical (USA)
*Chemical Class:* Unclassified
*EPA/OPP PC Code:* 051501
*California DPR Chemical Code:* 368
*RTECS Number:* UR5950000
*Uses:* Maleic hydrazide is a plant growth regulator (sprout inhibitor) and herbicide that is registered for use on tobacco, potatoes, onions, non-bearing citrus, turf, utility and highway rights-of-way, airports, industrial land, lawns, recreational areas, ornamental/shade trees and ornamental plants. Most of the use of maleic hydrazide in the U.S. is on tobacco.
*Human toxicity (long-term)*[77]*:* Very low–4000.00 ppb, Health Advisory
*Fish toxicity (threshold)*[77]*:* Very low–19952.62315 ppb, MATC (Maximum Acceptable Toxicant Concentration)
*U.S. Maximum Allowable Residue Levels for Maleic Hydrazide (40 CFR 180.175):*

| CROP | ppm |
|---|---|
| Onion, dry bulb | 15 |
| Potato | 50 |
| Potato chips | 160 |

*Carcinogen/Hazard Classifications*
**IARC:** Group 3, unclassifiable
**Label Signal Word:** WARNING or CAUTION
**WHO Acute Hazard:** Class U, unlikely to be hazardous
*Regulatory Authority:*
- FIFRA, 40CFR185: tolerances for pesticides in food
- EPA Hazardous Waste Number (RCRA No.): U148
- Actively registered pesticide in California
- Superfund/EPCRA 40CFR302.4, RQ: CERCLA, 5,000 lb (2270 kg)

*Description:* Maleic Hydrazide is a crystalline solid. Slightly soluble in cold water; more soluble in hot. Melting/Freezing point = 292°C (decomposes at 260°C); also reported as > 300°C.
*Incompatibilities:* Contact with flammable material may cause fire and explosions. Contact with combustible or oxidizable materials may form heat-, shock-, and friction-sensitive explosive mixtures. Static electricity may also cause explosions. Keep away from all acids, especially dibasic organic acids, ammonium compounds, antimony sulfide, arsenic trioxide, metal sulfides, powdered metals, calcium aluminum hydride, cyanides, manganese dioxide, phosphorus, selenium, sulfur, thiocyanates, zinc.
*Determination in Air:* Filter; none; Gravimetric; NIOSH IV [Particulates NOR; #0500 (total), #0600 (respirable)].[18]
*Permissible Concentration in Water:* No criteria set. Runoff from spills or fire control may cause water pollution.
*Determination in Water:* A lowest-observed-adverse-effect-level (LOAEL) of 500 mg/kg/day has been calculated. On the basis of this, the USEPA has calculated a lifetime health advisory for an adult of 3.5 mg/L.
*Routes of Entry:* Inhalation, ingestion, skin and/eye contact. Absorbed through the intact skin.
*Harmful Effects and Symptoms*
*Short Term Exposure:* Irritation of eyes, skin and mucous membranes, tremors, muscle spasms and skin sensitization are among the consequences of MH exposure. The oral $LD_{50}$-rat is 3800 mg/kg (slightly toxic).
*Long Term Exposure:* May cause liver damage and acute central nervous system effects. May cause mutations (genetic changes).
*Points of Attack:* Central nervous system, liver, skin.

*Medical Surveillance:* Liver function tests. Examination by a qualified allergist. Tests of the nervous system.

*First Aid: Skin:* Flood all areas of body that have contacted the substance with water. Don't wait to remove contaminated clothing; do it under the water stream. Use soap to help assure removal. Isolate contaminated clothing when removed to prevent contact by others. *Eyes:* Remove any contact lenses at once. Flush eyes well with copious quantities of water or normal saline for at least 20-23 minutes. Seek medical attention. *Inhalation:* Leave contaminated area immediately; breathe fresh air. Proper respiratory protection must be supplied to any rescuers. If coughing, difficult breathing or any other symptoms develop, seek medical attention at once, even if symptoms develop many hours after exposure. *Ingestion:* If convulsions are not present, give a glass or two of water or milk to dilute the substance. Assure that the person's airway is unobstructed and contact a hospital or poison center immediately for advice on whether or not to induce vomiting.

*References:*
- U.S. Environmental Protection Agency, "Reregistration Eligibility Decision (RED), Maleic Hydrazide," Office of Prevention, Pesticides and Toxic Substances, Washington, DC (June 1994). http://www.epa.gov/REDs/0381.pdf
- U.S. Environmental Protection Agency, "Health Advisory: Maleic Hydrazide, " Washington, DC, Office of Drinking Water (August 1987).
- U.S. Environmental Protection Agency, Office of Pesticide Programs, Pesticide Residue Limits, "Maleic Hydrazide ", 40 CFR 180.175, www.epa.gov/pesticides/food/viewtols.htm
- California Environmental Protection Agency "Chemical List of Lists," Sacramento CA (February 1997)

# Mancozeb

*Use Type:* Fungicide
*CAS Number:* 8018-01-7
*Formula:* $C_4H_6MnN_2S_4 \cdot C_4H_6N_2S_4Zn$
*Synonyms:* Carbamodithioic acid, 1,2-ethanediylbis-, manganous zinc salt; Ethylenebis(dithiocarbamic acid manganese zinc complex (8CI); Maneb-zinc; Maneb-zineb-komplex (German); Mangan-zink-aethylendiamin-bis-dithio-carbamat (German); Manganese ethylene-bis(dithiocarbamate)(polymeric) complex with zinc salt; Zinc ion and manganese ethylenebisdithiocarbamate 80%; A coordination product of manganese 16%, zinc 2%, and ethylenebisdithiocarbamate 62%

*Trade Names:* ACROBAT® (mancozeb + dimethomorph), BASF CropScience (Germany); ASHLAND SOLACE® (cymoxanil + mancozeb), Whyte Agrochemicals (UK); CARMAZINE®; CUPROFIX®, Cerexagri (France); DITHANE®, Dow AgroSciences (USA); EMCARB® (mancozeb + carbendez), Sabero Organics (India); EMTHANE M-15®, Sabero Organics (India); EVOLVE®, Gustafson (USA); F 2966®; FORE®; FORMEC®, PB/I Gordon (USA); GAUCHO® (imidacloprid + mancozeb), Bayer CropScience (Germany); GAVEL® (mancozeb + zoxamide), Dow AgroSciences (USA); GREEN-DAISEN M®; KARAMATE®; KENCOZEB®, Kenso Corp. (Malayasia); MANCOFOL®; MANOSEB®; MANTOX®, Veterinary & Agricultural Products Manufacturing Co., Ltd. (VAPCO) (Jordan); MANZATE 200®; MANZEB®; MANZIN 80®; MARZIN®; MAXIM®, Syngenta (Switzerland); MILOR®, Rotam Agrochemical (Hong Kong); NEMISPOR®; PACE® fungicide (mixture of mancozeb and metalaxyl); PENNCOZEB®;, Cerexagri (France), Whyte Agrochemicals (UK); POLICAR®; TRIZIMAN®; TRIZIMAN-D®; VONDOZEB PLUS®; ZIMANAT®; ZIMMAN-DITHANE®; ZIMANEB®

*Producers:* Agrimor International (USA); Agsin (Singapore); BASF Canada (Canada); Bayer CropScience (Germany); Cerexagri (France); Dow AgroSciences (USA); Griffin (USA); Gustafson (USA); Gujarat Pesticides (India); Hebei Huafeng Chemical Group (China); Hindustan Insecticides (India); Kenso Corp. (Malayasia); Ki-Hara Chemicals Ltd. (UK); Limin Chemical (China); Nagarjuna Agrichem (India); PB/I Gordon (USA); Rotam Agrochemical (Hong Kong); Sabero Organics (India); Simplot, J.R., Company (USA); SuYan Agrochemical Group (China); Syngenta (Switzerland); Veterinary & Agricultural Products Manufacturing Co., Ltd. (VAPCO) (Jordan); Whyte Agrochemicals (UK)

*Chemical Class:* Dithiocarbamate
*EPA/OPP PC Code:* 014504
*California DPR Chemical Code:* 211
*ICSC Number:* 0754
*RTECS Number:* ZB3200000
*EEC Number:* 006-076-00-1

*Uses:* Mancozeb is used to control a wide variety of fungal diseases, including potato blight, leaf spot, scab (on apples and pears), and rust (on roses). It is used on fruits, vegetables, nuts and field crops, and many more. It is also used as a seed treatment of cotton, potatoes, corn, safflower, sorghum, peanuts, tomatoes, flax, and cereal grains.

*Human toxicity (long-term)[77]:* High–5.82363 ppb, CHCL (Chronic Human Carcinogen Level)

*Fish toxicity (threshold)[77]:* High–3.16731 ppb, MATC (Maximum Acceptable Toxicant Concentration)

*U.S. Maximum Allowable Residue Levels for Mancozeb (40 CFR 180.176):*

| CROP | ppm |
| --- | --- |
| Apples | 7 |
| Asparagus | 0.1 |
| Banana | 4 |
| Banana, pulp | 0.5 |
| Barley, grain | 5 |

| | |
|---|---|
| Barley, milled fractions | 20 |
| Barley, straw | 25 |
| Beet, sugar | 2 |
| Beet, sugar, tops | 65 |
| Carrots | 2 |
| Celery | 5 |
| Corn, fodder | 5 |
| Corn, forage | 5 |
| Corn, grain, except popcorn | 0.1 |
| Corn, fresh, including sweet corn, kernels plus cob with husk removed | 0.5 |
| Corn, pop, grain | 0.5 |
| Cotton, undelinted seed | 0.5 |
| Crabapples | 10 |
| Cranberry | 7 |
| Cucumber | 4 |
| Fennel | 10 |
| Grape | 7 |
| Kidney | 0.5 |
| Liver | 0.5 |
| Melon | 4 |
| Oat, bran | 20 |
| Oat, grain | 5 |
| Oat, milled fractions | 20 |
| Oat, straw | 25 |
| Onion, dry bulb | 0.5 |
| Papaya | 10 |
| Peanut | 0.5 |
| Peanut, hay | 65 |
| Pear | 10 |
| Potato | 1 |
| Quince | 10 |
| Rye, grain | 5 |
| Rye, milled fractions | 20 |
| Rye, straw | 25 |
| Tomato | 4 |
| Wheat, grain | 5 |
| Wheat, milled byproducts | 20 |
| Wheat, straw | 25 |

*Carcinogen/Hazard Classifications*
**U.S. EPA Carcinogens:** Group B2, probable carcinogen
**California Prop. 65:** Carcinogen
**TRI Developmental Toxin:** Developmental toxin
**Label Signal Word:** WARNING or CAUTION
**WHO Acute Hazard:** Class U, unlikely to be hazardous
**Endocrine Disruptor:** Probable ED

*Regulatory Authority:*
- AB 1803-Well Monitoring Chemical (CAL)
- Proposition 65 chemical (CAL)
- Actively registered pesticide in California.
- AB 2588-Air Toxics "Hot Spots" Chemicals (CAL) as manganese or zinc compounds.
- Permissible Exposure Limits for Chemical Contaminants (CAL/OSHA) as manganese compounds. The "Director's List" (CAL/OSHA) as manganese or zinc compounds.
- Clean Air Act: Hazardous Air Pollutants (Title I, Part A, Section 112) as manganese compounds
- EPCRA Section 313: Form R *de minimis* concentration reporting level: 1.0% as manganese compounds.
- Clean Water Act: Section 307 Toxic Pollutants as zinc compounds
- Safe Drinking Water Act: SMCL, 0.05 mg/L as manganese.
- Safe Drinking Water Act: SMCL, 5 mg/L; Priority List (55 FR 1470) as zinc

*Description:* Grayish-yellow powder. Insoluble in water. Molecular weight = 266.31. Decomposes without melting = 192°C; also listed at 152°C. Vapor pressure = very low/negligible. Flash point = 138°C.

*Incompatibilities:* Strong oxidizers.

*Permissible Exposure Limits in Air:* ACGIH[1] TLV 0.2 mg/m$^3$ as manganese; OSHA[2] PEL ceiling limit 5 mg/m$^3$ as manganese; NIOSH[2] 1 mg/m$^3$ as manganese.

*Determination in Air:*

*Permissible Concentration in Water:* No criteria set. Runoff from spills or fire control may cause water pollution. There are a variety of maximum allowable concentrations set in various countries (35) as follows: EEC: 50.0 mg/L (drinking water); USA: 0.05 mg/L (bottled water); WHO: 0.05 mg/L (maximum desirable); WHO: 0.10 mg/L (foresthetic quality); WHO: 0.50 mg/L (maximum permissible in drinking water).

*Harmful Effects and Symptoms*

*Short Term Exposure:* Low levels of toxicity. Concentrated solutions are slightly corrosive to eyes and mucous membranes. Dust inhalation can cause irritation of the respiratory system with sneezing. Eye contact can cause irritation, watering, pain, and inflammation of the eyelids. Skin contact can cause irritation and minor ulceration. Ingestion can cause nausea, vomiting, fever, muscle twitching, seizure, rapid respiration, slow heart beat. Severe exposure may result in death.

*Points of Attack:* Respiratory system, central nervous system, cardiovascular system, skin, eyes.

*Medical Surveillance:* There are tests available to measure zinc in your blood, urine, hair, saliva, and feces. High levels of zinc in the feces can mean high recent zinc exposure. High levels of zinc in the blood can mean high zinc consumption and/or high exposure. Tests to measure zinc in hair may provide information on long-term zinc exposure; however, the relationship between levels in your hair and the amount of zinc you were exposed to is not clear. Medical observation is recommended for 24 to 48 hours after breathing overexposure, as pulmonary edema may be delayed. As first aid for pulmonary edema, consider administering a corticosteroid spray. Cigarette smoking may exacerbate pulmonary injury and should be discouraged for at least 72 hours following exposure. Before employment and at regular times after that, the following are recommended: If symptoms develop or overexposure occurs, repeat the above tests as soon as possible and get an exam of the nervous system. Also consider complete blood count. Consider chest x-ray following acute overexposure.

*First Aid:* If this chemical gets into the eyes, remove any contact lenses at once and irrigate immediately for at least 15 minutes, occasionally lifting upper and lower lids. Seek medical attention immediately. If this chemical contacts the skin, remove contaminated clothing and wash immediately with soap and water. Seek medical attention immediately. If this chemical has been inhaled, remove from exposure, begin rescue breathing (using universal precautions) if breathing has stopped, and CPR if heart action has stopped. Transfer promptly to a medical facility. When this chemical has been swallowed, get medical attention. If victim is conscious and able to swallow, have victim drink 4 to 8 ounces of water. Do not induce vomiting. *Note to physician or authorized medical personnel*: Medical observation is recommended for 24 to 48 hours after breathing overexposure, as pulmonary edema may be delayed. As first aid for pulmonary edema, consider administering a corticosteroid spray. Cigarette smoking may exacerbate pulmonary injury and should be discouraged for at least 72 hours following exposure.

*References:*
- EXTOXNET, Extension Toxicology Network, "Pesticide Information Profile, Mancozeb," Oregon State University, Corvallis, OR (June 1996). http://extoxnet.orst.edu/pips/mancozeb.htm
- U.S. Environmental Protection Agency, Office of Pesticide Programs, "Pesticide Residue Limits, Mancozeb," 40 CFR 180.176, www.epa.gov/pesticides/food/viewtols.htm
- California Environmental Protection Agency "Chemical List of Lists," Sacramento CA (February 1997)

# Maneb

*Use Type:* Fungicide
*CAS Number:* 12427-38-2
*Formula:* $C_4H_6MnN_2S_4$
*Alert:* A General Use Pesticide (GUP). Human toxicity (long-term): High
*Synonyms:* Carbamic acid, ethylenebis(dithio-), manganese salt; Carbamodithioic acid, 1,2-ethanediylbis-, manganese salt; EBDC; ENT 14,875; 1,2-Ethanediylbis (carbamodithioato)(2-)-manganese; 1,2-Ethanediylbiscarbamodithioic acid, manganese complex; 1,2-Ethanediylbiscarbamodithioic acid, manganese(2+) salt(1:1); 1,2-Ethanediylbismaneb, manganese (2+) salt (1:1); Ethylenebisdithiocarbamate manganese; $N,N'$-Ethylene bis(dithiocarbamate manganeux) (French); Ethylenebis(dithiocarbamato), manganese; Ethylenebis(dithiocarbamic acid), manganese salt; Ethylenebis(dithiocarbamic acid) manganous salt; 1,2-Ethylenediylbis(carbamodithioato)manganese; $N,N'$-Etilen-bis(ditiocarbammato) di manganese (Italian); Mangaan (II)-($N,N'$-ethyleen-bis(dithiocarbamaat)) (Dutch); Mangan (II)-[$N,N'$-aethylen-bis(dithiocarbamate)] (German); Manganese ethylene-1,2-bisdithiocarbamate; Manganese(II) ethylene di(dithiocarbamate); Manganous ethylenebis(dithiocarbamate); MEB; MNEBD

*Trade Names:* (The U.S. EPA lists 460 active and/or canceled products containing maneb) AAMANGAN®; AKZO CHEMIE MANEB®, Syngenta (Switzerland); BASF-MANEB SPRITZPULVER®, BASF Agricultural Products Group (Germany); BAVISTIN M®; COSMIC®; CHEM NEB®; CHLOROBLE M®; CLEANACRES®; CR 3029®; DELSENE M FLOWABLE®; DITHANE-22®, Dow AgroSciences (USA); F 10®; FARMANEB®; IDA MANEB®, Drexel Chemical (USA); KASCADE®; KYPMAN 80®; LONOCOL M®; MANAM®; MANEB 80®; MANEBA®; MANEBE®; MANEBGAN®; MANESAN®; MANESAN®; MANEX®, Griffin L.L.C. (USA); MANOC®; MANZATE®, DuPont Cop Protection (USA); MANZATE D®, DuPont Cop Protection (USA); MANZATE MANEB FUNGICIDE®, DuPont Cop Protection (USA); MANZEB®, Drexel Chemical (USA); MANZI®, Drexel Chemical (USA); M-DIPHAR®; MULTI-W®; NEREB®; NESPOR®; NEWSPOR®; PLANTIFOG 160M®; POLYRAM M®; REMASAN CHLOROBLE M®; RHODIANEHE®; SOPRANEBE®; SQUADRON AND QUADRANGLE MANEX®; SUPERMAN MANEB F®; SUP'R FLO®; TERSAN-LSR®; TRIMANGOL®, Whyte Agrochemicals (UK); TRIMANOC®; TRITHAC®; TUBOTHANE®; UNICROP MANEB®; VANCIDE®; VANCIDE MANEB 80®; VASSGRO MANEX®; VITAVAX®, Gustafson (USA)

*Producers:* Agriliance (USA); Agsin (Singapore); Akzo Nobel Chemicals (Netherlands); Atofina Chemicals (USA); BASF Agricultural Products Group (Germany); Biesterfeld Siemsgluess International. GmbH (Germany); Cerexagri (France); Cyanamid (USA); Dow AgroSciences (USA); Drexel Chemical (USA); DuPont Cop Protection (USA); Griffin L.L.C. (USA); Gustafson (USA); Hokko Chemical Industry (Japan); Montedison (Italy); Nissan Chemical Industries (Japan); Rhone-Poulenc Agro France (France); Rohm & Haas (USA); Sigma-Aldrich Laborchemikalien (Germany); Syngenta (Switzerland); Takeda Chemical Industries (Japan); Whyte Agrochemicals (UK)

*Chemical Class:* Dithiocarbamate
*EPA/OPP PC Code:* 014505
*California DPR Chemical Code:* 369
*ICSC Number:* 0173
*RTECS Number:* OP0700000
*EEC Number:* 006-077-00-7
*EINECS Number:* 235-654-8

*Uses:* Maneb is an ethylene(bis)dithiocarbamate fungicide used in the control of early and late blights on potatoes and tomatoes and many other diseases of fruits, vegetables, field crops, and ornamentals. Maneb controls a wider range of diseases than other fungicides. It is available as granular, wettable powder, flowable concentrate, and ready-to-use formulations. Maneb is widely used by itself and in combination with other pesticides on a variety of crops. Principal uses in California are for head and leaf lettuce, walnuts, almonds, onions, potatoes and broccoli.

*Human toxicity (long-term)*[77]: High–5.73770 ppb, CHCL (Chronic Human Carcinogen Level)

*Fish toxicity (threshold)*[77]: Extra high–0.00193 ppb, MATC (Maximum Acceptable Toxicant Concentration)

**U.S. Maximum Allowable Residue Levels for Maneb (40 CFR 180.110):**

| CROP | ppm |
|---|---|
| Almond | 0.1 |
| Apple | 2 |
| Apricot | 10 |
| Banana | 4 |
| Banana, pulp | 0.5 |
| Bean, dry | 7 |
| Bean, succulent | 10 |
| Beet, sugar, tops | 45 |
| Broccoli | 10 |
| Brussels sprouts | 10 |
| Cabbage | 10 |
| Cabbage, chinese, bok choy | 10 |
| Carrot, roots | 7 |
| Cauliflower | 10 |
| Celery | 5 |
| Collards | 10 |
| Corn, sweet, kernel plus cob with husks removed | 5 |
| Cranberry | 7 |
| Cucumber | 4 |
| Eggplant | 7 |
| Endive | 10 |
| Fig | 7 |
| Grape | 7 |
| Kale | 10 |
| Kohlrabi | 10 |
| Lettuce | 10 |
| Melon | 4 |
| Mustard greens | 10 |
| Nectarine | 10 |
| Onion | 7 |
| Papaya | 10 |
| Peach | 10 |
| Pepper | 7 |
| Potato | 0.1 |
| Pumpkin | 7 |
| Squash, summer | 4 |
| Squash, winter | 4 |
| Tomato | 4 |
| Turnip, greens | 10 |
| Turnip, roots | 7 |
| Walnut | 0.05 |

*Carcinogen/Hazard Classifications*

**U.S. EPA Carcinogens:** Group B2, probable carcinogen
**California Prop. 65:** Carcinogen
**IARC:** Group 3, unclassifiable
**Label Signal Word:** CAUTION
**WHO Acute Hazard:** Class U, unlikely to be hazardous
**Endocrine Disruptor:** Suspected endocrine disruptor

*Regulatory Authority:*
- Air Pollutant Standard Set (former USSR)[35, 43]
- Banned or Severely Restricted (In Agriculture) (former USSR)[13]
- Carcinogen (New Jersey)
- AB 1803-Well Monitoring Chemical (CAL)
- AB 2588-Air Toxics "Hot Spots" Chemicals (CAL)
- Proposition 65 chemical (CAL)
- Actively registered pesticide in California.
- EPCRA Section 313 Form R *de minimus* concentration reporting level: 1.0%
- U.S. DOT Regulated Marine Pollutant (49CFR172.101, Appendix B)
- Safe Drinking Water Act: SMCL, 0.05 mg/L as manganese.

*Description:* Yellow powder or crystalline solid with a faint odor. Moderately soluble in water. Molecular weight = 265.28. Melting/Freezing point = 130°C (decomposes below MP). Vapor pressure = $8.5 \times 10^{-8}$ mmHg @ 20°C. Unless stabilized, maneb or preparations containing 50% or more maneb may spontaneously ignite (with flare-burning effect) on contact to air. May reignite after fire is extinguished. Runoff to sewer may create fire or explosion hazard. Hazard Identification (based on NFPA-704 M Rating System): Health 0, Flammability 4, Reactivity 0. Log $K_{ow}$ = <2.5. Unlikely to bioaccumulate in marine organisms.

*Incompatibilities:* Water, acid, oxidizing materials. Heat, or contact with moisture or acids causes rapid decomposition and the generation of toxic and flammable hydrogen sulfide and carbon disulfide.

*Permissible Exposure Limits in Air:* The ACGIH[1] recommends a TWA (as Mn) of 5 mg/m³. The former USSR[35, 43] has set an MAC in workplace air of 0.5 mg/m³.

*Permissible Concentration in Water:* A no-adverse effect level in drinking water has been determined by MAS/NRC to be 0.035 mg/L. An acceptable daily intake (ADI) of 0.005 mg/kg/day has been calculated for maneb. The State of Maine has set a guideline for maneb in drinking water[61] of 10 µg/L.

*Routes of Entry:* Inhalation, ingestion, eye and/or skin contact.

*Harmful Effects and Symptoms*

*Short Term Exposure:* Maneb irritates the eyes, skin, and respiratory tract. Maneb is low in acute toxicity and does not present alarming properties during long-term administration to experimental animals, except at very high dosages. However, it is a material of concern because of evidence of mutagenic and teratogenic effects as well as the possibility of nitrosation to carcinogenic nitrosamines. Unlike carbamates the dithiocarbamates are not cholinesterase inhibitors, but some of them may react with recently ingested alcohol or alcohol-containing products including wine, medications, and cold remedies such as cough-syrups.

*Long Term Exposure:* Repeated skin contact can cause skin sensitization and rash. High or repeated exposures may interfere with thyroid function (causing goiter), damage to the central nervous system, affect liver function, or cause kidney damage.

*Points of Attack:* Skin, thyroid, liver, kidneys and central nervous system.

*Medical Surveillance:* If symptoms develop or overexposure is suspected, the following may be useful: Exam of the nervous system. Thyroid function tests. Consider kidney and liver function tests with higher or repeated exposures. Examination by a qualified allergist.

*First Aid:* If this chemical gets into the eyes, remove any contact lenses at once and irrigate immediately for at least 15 minutes, occasionally lifting upper and lower lids. Seek medical attention immediately. If this chemical contacts the skin, remove contaminated clothing and wash immediately with soap and water. Seek medical attention immediately. If this chemical has been inhaled, remove from exposure, begin rescue breathing (using universal precautions) if breathing has stopped, and CPR if heart action has stopped. Transfer promptly to a medical facility. When this chemical has been swallowed, get medical attention. Give large quantities of water and induce vomiting. Do not make an unconscious person vomit.

*References:*
- EXTOXNET, Extension Toxicology Network, "Pesticide Information Profile," Oregon State University, Corvallis, OR (June 1996). http://extoxnet.orst.edu/pips/maneb.htm
- U.S. Environmental Protection Agency, Office of Pesticide Programs, Pesticide Residue Limits, "Maneb," 40 CFR 180.110, http://www.epa.gov/pesticides/food/viewtols.htm
- New Jersey Department of Health and Senior Services, "Hazardous Substance Fact Sheet: Maneb," Trenton, NJ (November 1999). http://www.state.nj.us/health/eoh/rtkweb/1154.pdf
- California Environmental Protection Agency "Chemical List of Lists," Sacramento CA (February 1997).

# MCPA

*Use Type:* Herbicide
*CAS Number:* 94-74-6
*Formula:* $C_9H_9ClO_3$

*Alert:* A General Use Pesticide (GUP). Human toxicity (long-term): High

*Synonyms:* Acetic acid (4-chloro-2-methylphenoxy)-; Acetic acid [(4-chloro-*o*-tolyl)-oxy]-; BH MCPA; (4-Chloro-*o*-cresoxy)acetic acid; 4-Chloro-*o*-cresoxyacetic acid; (4-Chloro-2-methylphenoxy)acetic acid; 4-Chloro-2-methylphenoxyacetic acid; 4-Chloro-*o*-toloxyacetic acid; (4-Chloro-*o*-toloxy)acetic acid; [(4-Chloro-*o*-tolyl)oxy]acetic acid; MCP; Methylchlorophenoxyacetic acid; 2-Methyl-4-chlorophenoxyacetic acid; (2-Methyl-4-chlorophenoxy)acetic acid; 2-Methyl-4-chlorophenoxyessigsaeure (German)

*Trade Names:* ACME MCPA AMINE 4®; AGRITOX®; AGROXONE®; AGROZONE®; AGSCO®; ANICON KOMBI®; ANICON M®; BANLENE®; BLESEL MC®; BORDERMASTER®; BROMINAL M & PLUS®; CAMBILENE®; CHEYENNE®; CHIMAC OXY®; CHIPTOX®; CHWASTOX®; CORNOX M®; DAKOTA®; DED WEED®; DICOPUR-M®; DICOTEX®; DOW MCP AMINE WEED KILLER®, Dow AgroSciences (USA); DYVEL®, BASF Agricultural Products Group (Germany); EH1356 HERBICIDE®,; EMCEPAN®; EMPAL®; ENVOY®; HEDAPUR M 52®; HEDAREX M®; HEDONAL M®; HERBICIDE M®; HORMOTUHO®; HORNOTUHO®; KILSEM®; 4K-2M®; KVK®, A H Marks (UK); LEGUMEX DB®; LEUNA M®; LEYSPRAY®; LINORMONE®; M 40®, Pbi/Gordon (USA); 2M-4C®; 2M-4KH®; MALERBANE®; MAYCLENE®; MEPHANAC®; MIDOX®; MXL®; OKULTIN®; PHENOXYLENE 50®; PHENOXYLENE PLUS®; PHENOXYLENE SUPER®; RAZOL DOCK KILLER®; RHOMENE®; RHONOX®; SHAMOX®; B-SELEKTONON M®; SEPPIC MMD®; TILLER®; TRIMEC®, Pbi/Gordon (USA); U 46®, BASF Agricultural Products Group (Germany); VACATE®; VESAKONTUHO®; WEEDAR®, Nufarm (Australia); WEEDAR MCPA CONCENTRATE®, Nufarm (Australia); WEEDONE MCPA ESTER®, Nufarm (Australia); WEED RHAP®; ZELAN®

*Producers:* Agrimor International (USA); A H Marks (UK); Akzo Nobel Chemicals (Netherlands); Alcotan Laboratories (Spain); Atanor S.A. (Argentina); BASF Agricultural Products Group (Germany); Bayer Chemicals (Germany); Biesterfeld Siemsgluess International. GmbH (Germany); Cyanamid Agro (USA); Dow AgroSciences (USA); DuPont (USA); Hokko Chemical Industry (Japan); Ishihara Sangyo Kaisha (Japan); Ki-Hara Chemicals Ltd. (UK); Nissan Chemical Industries (Japan); Nufarm (Australia); Rhone-Poulenc Agro France (France); Shell Chemical (UK); Syngenta (Switzerland); Whyte Agrochemicals (UK); Zago Asia Ltd. (Singapore)

*Chemical Class:* Chlorophenoxy
*EPA/OPP PC Code:* 030501
*California DPR Chemical Code:* 2326
*ICSC Number:* 0054
*RTECS Number:* AG1575000

*EEC Number:* 607-051-00-3
*EINECS Number:* 202-360-6
*Uses:* MCPA is a systemic post-emergence phenoxy herbicide used to control broadleaf annual and perennial weeds (including thistle and dock) in cereals, flax, rice, vines, peas, potatoes, grasslands, forestry applications, and on rights-of-way. It is very compatible with many other compounds and may be used in formulation with many other products, including bentazone, bromoxynil, 2,4-D, dicamba, fenoxaprop, MCPB, mecoprop, thifensulfuron, and tribenuron.
*Human toxicity (long-term)*[77]: High–4.00 ppb, Health Advisory
*Fish toxicity (threshold)*[77]: Very low–17986.45312 ppb, MATC (Maximum Acceptable Toxicant Concentration)
*U.S. Maximum Allowable Residue Levels for MCPA salts and esters (40 CFR 180.339):*

| CROP | ppm |
| --- | --- |
| Alfalfa | 0.1 |
| Alfalfa, hay | 0.1 |
| Barley, grain | 0.1 |
| Barley, straw | 2 |
| Canarygrass, annual, hay | 0.1 |
| Canarygrass, annual, seed | 0.1 |
| Cattle, fat | 0.1 |
| Cattle, meat | 0.1 |
| Cattle, mbyp | 0.1 |
| Clover | 0.1 |
| Clover, hay | 0.1 |
| Flax, seed | 0.1 |
| Flax, straw | 2 |
| Goat, fat | 0.1 |
| Goat, meat | 0.1 |
| Goat, mbyp | 0.1 |
| Grass, hay | 20 |
| Grass, pasture | 300 |
| Grass, rangeland | 300 |
| Hog, fat | 0.1 |
| Hog, meat | 0.1 |
| Hog, mbyp | 0.1 |
| Horse, fat | 0.1 |
| Horse, meat | 0.1 |
| Horse, mbyp | 0.1 |
| Lespedeza | 0.1 |
| Lespedeza, hay | 0.1 |
| Milk | 0.1 |
| Oat, forage | 20 |
| Oat, grain | 0.1 |
| Oat, straw | 2 |
| Pea, field, hay | 0.1 |
| Pea, field, vines | 0.1 |
| Rice, grain | 0.1 |
| Rice, straw | 2 |
| Rye, forage | 20 |
| Rye, grain | 0.1 |
| Rye, straw | 2 |
| Sheep, fat | 0.1 |
| Sheep, meat | 0.1 |
| Sheep, mbyp | 0.1 |
| Sorghum, forage | 20 |
| Sorghum, grain, grain | 0.1 |
| Sorghum, grain, stover | 20 |
| Trefoil | 0.1 |
| Trefoil, hay | 0.1 |
| Vegetables, seed and pod | 0.1 |
| Vetch | 0.1 |
| Vetch, hay | 0.1 |
| Wheat, grain | 0.1 |
| Wheat, straw | 2 |

*Carcinogen/Hazard Classifications*
**IARC:** Group 2B, possible carcinogen
**Label Signal Word:** DANGER
**WHO Acute Hazard:** Group III, Slightly Hazardous
*Regulatory Authority:*
- EPCRA Section 313 Form R *de minimus* concentration reporting level: 0.1%
- AB 1803-Well Monitoring Chemical (CAL)
- AB 2588-Air Toxics "Hot Spots" Chemicals (CAL) as chlorophenoxy pesticides
- The "Director's List" (CAL/OSHA)

*Description:* MCPA is a colorless crystalline solid. Very slightly soluble in water; solubility = 725 ppm. Molecular weight = 200.64. Melting/Freezing point = 118–119°C. Vapor pressure = $1.5 \times 10^{-6}$ mmHg @ 20°C; 200 mPa @ 21°C. Log $K_{ow}$ = 3.31. Values above 3.0 are likely to bioaccumulate in marine organisms.
*Incompatibilities:* A weak acid. Incompatible with alkalies.
*Permissible Exposure Limits in Air:* No standards set.
*Permissible Concentration in Water:* The former USSR[35] has set an MAC in surface water of 0.25 mg/L. The USEPA has determined a no-observed-adverse effects level (NOAEL) of 1.0 mg/kg/day from which they have calculated a long-term advisory of 0.35 mg/L (350 µg/L) for an adult. They have further calculated a lifetime health advisory of 0.0036 mg/L (3.6 µg/L) for an adult. In addition, Maine has set a guideline for MCPA in drinking water[61] of 2.5 µg/L.
*Routes of Entry:* Inhalation, ingestion, skin and/or eye contact.
*Harmful Effects and Symptoms*
*Short Term Exposure:* Irritates the eyes, skin, and respiratory tract. Can cause nausea, vomiting, diarrhea and abdominal pain. This material is moderately toxic. The $LD_{50}$ (oral, rat) = 700 mg/kg. The approximate lethal dose to a 150 lb man is 3.3 tablespoonfuls (Sax).
*Long Term Exposure:* Animal tests show that this substance possibly causes toxic effects upon human reproduction. MCPA is classified as a chlorophenoxy-herbicide. These herbicides are a possible carcinogen to

humans. May cause decreased blood pressure. May cause genetic changes.
*Points of Attack:* Liver and kidneys and central nervous system.
*Medical Surveillance:* Monitor blood pressure.
*First Aid:* If this chemical gets into the eyes, remove any contact lenses at once and irrigate immediately for at least 15 minutes, occasionally lifting upper and lower lids. Seek medical attention immediately. If this chemical contacts the skin, remove contaminated clothing and wash immediately with soap and water. Seek medical attention immediately. If this chemical has been inhaled, remove from exposure, begin rescue breathing (using universal precautions) if breathing has stopped, and CPR if heart action has stopped. Transfer promptly to a medical facility. When this chemical has been swallowed, get medical attention. Give large quantities of water and induce vomiting. Do not make an unconscious person vomit.
*References:*
- EXTOXNET, Extension Toxicology Network, "Pesticide Information Profile, MCPA," Oregon State University, Corvallis, OR (June 1996). http://extoxnet.orst.edu/pips/MCPA.htm
- U.S. Environmental Protection Agency, Office of Pesticide Programs, Pesticide Residue Limits, "MCPA Salts and Esters," 40 CFR 180.339, http://www.epa.gov/pesticides/food/viewtols.htm
- New Jersey Department of Health and Senior Services, "Hazardous Substance Fact Sheet 4-Chloro-2-methyl-phenoxyacetic acid, Trenton, NJ (August 2000). http://www.state.nj.us/health/eoh/rtkweb/3094.pdf
- U.S. Environmental Protection Agency, Initial Scientific and Minieconomic Review No. 21: MCPA, Washington, DC, Office of Pesticide Programs (1975).
- U.S. Environmental Protection Agency, "Health Advisory: MCPA, " Washington, DC, Office of Drinking Water (August 1987).
- Sax, N.I., Ed., "Dangerous Properties of Industrial Materials Report, " 8, No. 6, 35-41 (1988).
- California Environmental Protection Agency "Chemical List of Lists," Sacramento CA (February 1997).

# Mecoprop

*Use Type:* Herbicide
*CAS Number:* 7085-19-0 (EPA and state of California, primary usage); 93-65-2 [International Occupational Safety and Health Information Centre (CIS), primary usage]
*Formula:* $C_{10}H_{11}ClO_3$
*Synonyms:* Acide 2-(4-chloro-2-methyl-phenoxy)propionique (French); Acido 2-(4-cloro-2-metil-fenossi)-propionico (Italian); 2-(4-Cloor-2-methyl-fenoxy)-propionzuur (Dutch); 2-(4-Chlor-2-methyl-phenoxy)-propionsaeure (German); 4-Chloro-2-methylphenoxy-α-propionic acid; 2-(4-Chloro-2-methylphenoxy)propanoic acid; (+)-α-(4-Chloro-2-methylphenoxy) propionic acid; (4-Chloro-2-methylphenoxy)propionic acid; 2-(4-Chloro-2-methylphenoxy)propionic acid; α-(4-Chloro-2-methylphenoxy)propionic acid; 2-(4-Chlorophenoxy-2-methyl)propionic acid; 2-(4-Chloro-o-tolyl)oxylpropionic acid; 2-(p-chloro-o-tolyloxy)propionic acid; CMPP; 2M-4CP; MCPP; 2-MCPP; MCPP; MCPP 2,4-D; MCPP-D-4; MCPP K-4; 2-Methyl-4-chlorophenoxy-α-propionic acid; 2-(2-Ethyl-4-chlorophenoxy)propanoic acid; α(2-Methyl-4-chlorophenoxy)propionic acid; 2-(2'-Methyl-4'-chlorophenoxy)propionic acid; 2M4KHP; NSC 60282; Propanoic acid, 2-(4-chloro-2-methylphenoxy)-; Propionic acid, 2-(4-chloro-2-methylphenoxy); Propionic acid, 2-[(4-chloro-o-tolyl)oxy]-; Propionic acid, 2-(2-methyl-4-chlorophenoxy)-
*Trade Names:* ASSASSIN®; BANVEL P®; BH MECOPROP®; CERIDOR®; CHIPCO®; CHIPCO TURF HERBICIDE MCPP®; CLEAVAL®; CLENECORN®; CLOVOTOX®; COMPITOX EXTRA®; CORNOX PLUS®; CR 205®; CRUSADER®; DOCKLENE®; EXP 419®; GRASLAM®; HARNESS®; HARRIER®; HEDONAL MCPP®; HERRISOL®; HYMEC®; IOTOX®; ISO-CORNOX®; KILPROP®; LIRANOX®; MECOBROM®; MECOMIN D®; MECOPEOP®; MECOPAR®; MECOPEX®; MECOTURF®; MEPRO®; METHOXONE®; MECHLORPROP®; N.b. MECOPROP®; MYLONE®; MUSKETEET®; POST-KITE®; PROPAL®; PROPONEX-PLUS®; RANKOTEX®; RD 4593®; RUNCATEX®; SELOXONE®; SCOTLENE®; SEL-OXONE®; SUPER GREEN AND WEED®; SUPOERTOX®; SWIPE 560 EC®; TARGET®, Syngenta (Switzerland); TERSET®; TETRALEN-PLUS®; TRIESTER II®; TRIMEC 1144 40% SP®, Pbi/Gordon Corp. (USA); TRIPLET®; U 46®; U 46 KV-ESTER®; VERDONE®; VI-PAR®; VI-PEX®; VIPEX®
*Producers:* Akzo Nobel Chemicals (Netherlands); BASF (Germany); Bayer Chemicals (Germany); Newfarm (Australia); Pbi/Gordon Corp. (USA); Scott Company (USA); Syngenta (Switzerland); Whyte Agrochemicals (UK)
*Chemical Class:* Chlorophenoxy acid or ester
*EPA/OPP PC Code:* 031501
*California DPR Chemical Code:* 374
*ICSC Number:* 0055
*RTECS Number:* UE9750000
*EEC Number:* 607-049-00-2
*EINECS Number:* 202-264-4
*Uses:* A General Use Pesticide, postemergent herbicide, most often used in combination with other pesticides. Most of the production is used on turf, including lawns, sport turf and commercial sod production for control of creeping broadleaf weeds such as clovers, chickweed, ivy, plantain and similar plants. It is also used on wheat, barley and oats. A small percentage is used in noncrop areas such as rights-of-way, drainage ditch banks and forest site preparation.

*Human toxicity (long-term)*[77]: High–7.00 ppb, Health Advisory
*Fish toxicity (threshold)*[77]: Very low–25458.68925 ppb, MATC (Maximum Acceptable Toxicant Concentration)
*Carcinogen/Hazard Classifications*
**IARC:** Group 2B, possible carcinogen
**WHO Acute Hazard:** Class III, slightly hazardous
*Regulatory Authority:*
- AB 2588-Air Toxics "Hot Spots" Chemicals (CAL) as chlorophenoxy pesticides
- The "Director's List" (CAL/OSHA) as chlorophenoxy pesticides
- EPCRA Section 313 Form R *de minimus* concentration reporting level: 0.1%
- Actively registered pesticide in California.

*Description:* Mecoprop is a colorless, crystalline solid. Odorless. Slightly soluble in water; solubility = < 650 ppm @ 20°C. Molecular weight = 214.64. Melting/Freezing point = 93–95°C. Vapor pressure = < 9.9 x $10^{-6}$ mmHg @ 20°C.
*Incompatibilities:* A weak acid. Incompatible with strong bases and oxidizers.
*Routes of Entry:* Inhalation, absorbed through the skin.
*Harmful Effects and Symptoms*
*Short Term Exposure:* This chemical can be absorbed through the skin, thereby increasing exposure. Contact irritates the eyes and skin. Irritates the respiratory tract. Exposure can cause headache, weakness, convulsions, muscle cramps, loss of coordination, unconsciousness, and death.
*Long Term Exposure:* There is limited evidence that the chemical affects human reproduction. Exposure may damage blood cells, causing anemia, and damage the kidneys. Although this chemical has not been identified as a carcinogen, several related compounds have shown limited evidence of cancer.
*Points of Attack:* Blood, kidney, nervous system.
*Medical Surveillance:* Examination of the nervous system. Complete blood count. Kidney function tests.
*First Aid:* If this chemical gets into the eyes, remove any contact lenses at once and irrigate immediately for at least 15 minutes, occasionally lifting upper and lower lids. Seek medical attention immediately. If this chemical contacts the skin, remove contaminated clothing and wash immediately with soap and water. Seek medical attention immediately. If this chemical has been inhaled, remove from exposure, begin rescue breathing (using universal precautions) if breathing has stopped, and CPR if heart action has stopped. Transfer promptly to a medical facility. When this chemical has been swallowed, get medical attention. Give large quantities of water and induce vomiting. Do not make an unconscious person vomit.
*References:*
- EXTOXNET, Extension Toxicology Network, "Pesticide Information Profile, Mecoprop," Oregon State University, Corvallis, OR (September 1995). http://extoxnet.orst.edu/pips/mecoprop.htm
- California Environmental Protection Agency "Chemical List of Lists," Sacramento CA (February 1997).
- New Jersey Department of Health and Senior Services, Hazardous Substance Fact Sheet, "2-(4-chloro-2-methylphenoxy) Propionic Acid," Trenton NJ (April 1999). http://www.state.nj.us/health/eoh/rtkweb/3093.pdf

## Mefenacet

*Use Type:* Herbicide
*CAS Number:* 73250-68-7
*Formula:* $C_{16}H_{14}N_2O_2S$
*Synonyms:* Acetamide, 2-(2-benzothiazolyloxy)-*N*-methyl-*N*-phenyl-; 2-(1,3-Benzothiazol-2-yloxy)-*N*-methylacetanilide; 2-(2-Benzothiazolyloxy)-*N*-methyl-*N*-phenylacetamide
*Trade Names:* DEFANACET®; FOE 1976®, Bayer CropScience (Germany); HINOCHLOA®, Bayer CropScience (Germany); NTN 801®; RANCHO®, Bayer CropScience (Germany)
*Producers:* Bayer CropScience (Germany); SuYan Agrochemical Group (China)
*Chemical Class:* Anilide
*Uses:* Not registered in the U.S. A selective herbicide with low toxicity; usually formulated with other herbicides. Used mainly to control barnyard grasses in irrigated paddy rice. Also used on a variety of crops including cereals, legumes, vegetables, fruits, nuts, seeds, hops and tea.
*Carcinogen/Hazard Classifications*
**WHO Acute Hazard:** Class U, unlikely to be hazardous
*Description:* White crystalline solid. Odorless. Melting point = 134.8–135°C. Molecular weight = 298.22.
*Permissible Concentration in Water:* No criteria set. Runoff from spills or fire control may cause water pollution.
*First Aid:* If this chemical gets into the eyes, remove any contact lenses at once and irrigate immediately for at least 15 minutes, occasionally lifting upper and lower lids. Seek medical attention immediately. If this chemical contacts the skin, remove contaminated clothing and wash immediately with soap and water. Seek medical attention immediately. If this chemical has been inhaled, remove from exposure, begin rescue breathing (using universal precautions) if breathing has stopped, and CPR if heart action has stopped. Transfer promptly to a medical facility. When this chemical has been swallowed, get medical attention. Give large quantities of water and induce vomiting. Do not make an unconscious person vomit. Do not induce vomiting when formulations containing petroleum solvents are ingested.

## Mefluidide (ANSI)

*Use Type:* Herbicide; plant growth regulator

*CAS Number:* 53780-34-0
*Formula:* $C_{11}H_{13}F_3N_2O_3S$
*Synonyms:* Acetamide, N-(2,4-dimethyl-5-[((trifluoromethyl)sulfonyl)amimo)phenyl]-; 5-Acetamido-2,4-dimethyltrifluoromethanesulfonanilide; N-[(2,4-Dimethyl-5-((trifluoromethyl)sulfonyl)amino)phenyl] acetamide; 2',4'-Dimethyl-5-[(trifluoromethyl)sulfonamido] acetanilide
*Trade Names:* EMmHgK®, Pbi/Gordon (USA); MBR 12325®; MBR 12325®; METHAFLUORIDAMID®; S 15733®; VEL 3973®; VISTAR® Herbicide
*Producers:* Pbi/Gordon (USA); Scotts Company (USA)
*Chemical Class:* Anilide
*EPA/OPP PC Code:* 114001; (387100 old EPA code number)
*California DPR Chemical Code:* 5082
*Uses:* Mefluidide is used to control ornamental and non-ornamental woody plants, ground cover, hedges, trees, turfgrass and broadleaf weeds.
*Human toxicity (long-term)[77]:* Very low–2100.00 ppb, Health Advisory
*Fish toxicity (threshold)[77]:* Very low–19076.48500 ppb, MATC (Maximum Acceptable Toxicant Concentration)
*Carcinogen/Hazard Classifications*
**Label Signal Word:** CAUTION
**WHO Acute Hazard:** Class III, slightly hazardous
*Regulatory Authority:*
- Actively registered pesticide in California.

*Description:* Crystalline solid. Slightly soluble in water. Melting/Freezing point = 184°C.
*Incompatibilities:* Reacts with strong oxidizers. When heated to decomposition, this chemical forms toxic fumes of nitrogen oxides, sulfur oxides, and fluorine.
*Determination in Air:* Filter; none; Gravimetric; NIOSH IV [Particulates NOR; #0500 (total), #0600 (respirable)].[18]
*Permissible Concentration in Water:* No criteria set. Runoff from spills or fire control may cause water pollution.
*First Aid:* If this chemical gets into the eyes, remove any contact lenses at once and irrigate immediately for at least 15 minutes, occasionally lifting upper and lower lids. Seek medical attention immediately. If this chemical contacts the skin, remove contaminated clothing and wash immediately with soap and water. Seek medical attention immediately. If this chemical has been inhaled, remove from exposure, begin rescue breathing (using universal precautions) if breathing has stopped, and CPR if heart action has stopped. Transfer promptly to a medical facility. When this chemical has been swallowed, get medical attention. Give large quantities of water and induce vomiting. Do not make an unconscious person vomit. Do not induce vomiting when formulations containing petroleum solvents are ingested.
*References:*
- California Environmental Protection Agency "Chemical List of Lists," Sacramento CA (February 1997).

# Mephosfolan

*Use Type:* Insecticide and acaricide
*CAS Number:* 950-10-7
*Formula:* $C_8H_{16}NO_3PS_2$
*Synonyms:* Cyclic propylene(diethoxyphosphinyl) dithioimdocarbonate; Cytrolane; p,p-Diethyl cyclic propylene ester of phosphonodithioimidocarbonic acid; Diethyl(4-methyl-1,3-dithiolan-2-ylidene)phosphoroamidate; 2-(Diethoxyphosphinylimino)-4-methyl-1,3-dithiolane; EI-47470; ENT 25,991; (4-Methyl-1,3-dithiolan-2-ylidene)phosphoramidic acid, diethyl ester
*Trade Names:* AC 47470®, American Cyanamid's Agricultural Products Group (USA); AMERICAN CYANAMID CL-47470®, American Cyanamid's Agricultural Products Group (USA); CL-47,470®
*Producers:* American Cyanamid's Agricultural Products Group (USA); Sigma-Aldrich Laborchemikalien (Germany)
*Chemical Class:* Organophosphate
*EPA/OPP PC Code:* 268310
*RTECS Number:* JP1050000
*Carcinogen/Hazard Classifications*
**WHO Acute Hazard:** Class 1 a, extremely hazardous
*Regulatory Authority:*
- Banned or Severely Restricted (In Agriculture) (India) (UN)[13]
- Superfund/EPCRA 40CFR355, Appendix B Extremely Hazardous Substances: TPQ = 500 lb (227kg)
- Superfund/EPCRA 40CFR302.4 RQ: EHS, 1 lb (0.454 kg)
- U.S. DOT Inhalation Hazard Chemicals as organophosphates
- U.S. DOT Regulated Marine Pollutant (49CFR172.101, Appendix B)

*Description:* Mephosfolan is a yellow to amber liquid. Molecular weight = 269.35. Boiling point = 120°C @ 1.0 mmHg. Hazard Identification (based on NFPA-704 M Rating System): Health 4, Flammability 1, Reactivity 0. Moderately soluble in water.
*Incompatibilities:* Strong oxidizers, strong acids.
*Permissible Exposure Limits in Air:* No standards set.
*Determination in Air:* OSHA versatile sampler-2; Toluene/Acetone; Gas chromatography/Flame photometric detection for sulfur, nitrogen, or phosphorus; NIOSH Method IV Method #5600, Organophosphorus pesticides.[18]
*Permissible Concentration in Water:* No criteria set, but runoff from spills or fire control may cause water pollution.
*Routes of Entry:* Inhalation, ingestion, skin and/or eye contact.
*Harmful Effects and Symptoms*
*Short Term Exposure:* This is a highly to extremely toxic material ($LD_{50}$ oral for rats is 9 mg/kg). Organic phosphorus insecticides are absorbed by the skin, as well as by the respiratory and gastrointestinal tracts. They are

cholinesterase inhibitors. Symptoms of exposure include headache, giddiness, blurred vision, nervousness, weakness, nausea, cramps, diarrhea, and discomfort in the chest. Signs include sweating, tearing, salivation, vomiting, cyanosis, convulsions, coma, loss of reflexes and loss of sphincter control. Delayed pulmonary edema may occur after inhalation.

*Long Term Exposure:* Cholinesterase inhibitor; cumulative effect is possible. This chemical may damage the nervous system with repeated exposure, resulting in convulsions, respiratory failure. May cause liver damage.

*Points of Attack:* Respiratory system, lungs, central nervous system, cardiovascular system, skin, eyes, plasma and red blood cell cholinesterase.

*Medical Surveillance:* Medical observation is recommended for 24 to 48 hours after breathing overexposure, as pulmonary edema may be delayed. Before employment and at regular times after that, the following are recommended: Plasma and red blood cell cholinesterase levels (tests for the enzyme poisoned by this chemical). If exposure stops, plasma levels return to normal in 1-2 weeks while red blood cell levels may be reduced for 1-3 months. When acetylcholinesterase enzyme levels are reduced by 25% or more below preemployment levels, risk of poisoning is increased, even if results are in lower ranges of "normal." Reassignment to work not involving organophosphate or carbamate pesticides is recommended until enzyme levels recover. If symptoms develop or overexposure occurs, repeat the above tests as soon as possible and get an exam of the nervous system. Also consider complete blood count. Consider chest x-ray following acute overexposure. Do not drink any alcoholic beverages before or during use. Alcohol promotes absorption of organophosphates.

*First Aid:* **Treatment for organophosphate poisoning consists of thorough decontamination, cardiorespiratory support, and administration of the antidotes atropine and pralidoxime. In cases of severe poisoning, diazepam, an anticonvulsant, should also be administered. Antidotes should be administered as prevention even if the diagnosis is in doubt.** Speed in removing material from eyes and skin is of extreme importance. *Eyes:* Eye contact can cause dangerous amounts of these chemicals to be quickly absorbed through the mucous membrane into the bloodstream. Immediately and gently flush eyes with plenty of warm or cold water (NO hot water) for at least 15 minutes, occasionally lifting the upper and lower eyelids. Get medical aid immediately. *Skin:* Get medical aid. Skin contact can cause dangerous amounts of these chemicals to be absorbed into the bloodstream. Wearing the appropriate PPE equipment and respirator for organophosphate pesticides, immediately flush skin with plenty of soap and water for at least 15 minutes while removing contaminated clothing and shoes. Shampoo hair promptly if contaminated. The removed, contaminated clothing and shoes should be double-bagged and left in Hot Zone for later disposal by hazardous materials experts. Skin may also be decontaminated with diluted hypochlorite solution. *Inhalation:* Get medical aid. Do not contaminate yourself. Wearing the appropriate PPE equipment and respirator for organophosphate pesticides, immediately remove the victim from the contaminated area to fresh air. If the victim is not breathing, administer artificial respiration. Do NOT use mouth-to-mouth resuscitation; use bag/mask apparatus. If breathing is difficult, administer oxygen through bag/mask apparatus until medical help arrives. Do not leave victim unattended. *Ingestion:* Call poison control. Loosen all clothing. Never give anything by mouth to an unconscious person. If victim is *unconscious or having convulsions,* do nothing except keep victim warm. Get medical aid. Transfer promptly to a medical facility. In cases of ingestion, **do not induce vomiting**. If the victim is alert and asymptomatic, administer a slurry of activated charcoal at a dose of 1 g/kg (infant, child, and adult dose). A soda can and straw may be of assistance when offering charcoal to a child. *In some cases you may be specifically instructed by poison control to induce vomiting by way of 2 tablespoons of syrup of ipecac (adult) washed down with a cup of water.* Do NOT give activated charcoal before or with ipecac syrup.

*Note to physician or authorized medical personnel:* Treat cases of respiratory compromise, coma, or excessive pulmonary secretions with respiratory support using protocols and techniques available and within the scope of training. Some cases may necessitate procedures such as endotracheal intubation or cricothyrotomy by properly trained and equipped personnel. When possible, atropine (see *Antidotes*, below) should be given under medical supervision. Patients who are comatose, hypotensive, or having seizures or cardiac arrhythmias should be treated according to advanced life support protocols. *Antidotes:* Two antidotes are administered to treat organophosphate poisoning. Atropine is a competitive antagonist of acetylcholine at muscarinic receptors and is used to control the excessive bronchial secretions which are often responsible for death. Pralidoxime relieves both the nicotinic and muscarine effects of organophosphate poisoning by regenerating acetylcholinesterase and can reduce both the bronchial secretions and the muscle weakness associated with poisoning. The initial intravenous dose of atropine in adults should be determined by the severity of symptoms: An initial adult dose of 1.0 to 2.0 mg or pediatric dose of 0.01 mg/kg (minimum 0.01 mg) should be administered intravenously. If intravenous access cannot be established, atropine may also be given intramuscularly, subcutaneously or via endotracheal tube. Doses should be repeated every 15 minutes until excessive secretions and sweating have been controlled. Once bronchial secretion has been controlled, atropine administration should be repeated whenever the secretions begin to recur. In seriously poisoned patients, very large doses may be required. Alterations of pulse rate and pupillary size should not be used as indicators of treatment adequacy. Pralidoxime should be administered as early in poisoning as possible as its efficacy may diminish when given more than 24 to 36 hours after exposure. Doses are as follows: adult 1.0 g; pediatric 25 to 50 mg/kg. The drug should be administered intravenously over 30 to 60 minutes, but in a life-threatening situation, one-half of the total dose can be given per minute for a total administration time of 2 minutes. Treatment should begin to take effect within 40 minutes with a reduction in symptoms and the amount of atropine necessary to control bronchial secretion. The initial dose can be repeated in 1 hour and then every 8 to 12 hours until the patient is clinically well and no longer requires atropine. If intravenous access cannot be established, pralidoxime may also be given intramuscularly. Early administration of diazepam in addition to the combined atropine and pralidoxime treatment may help prevent the onset of seizures and potential brain and cardiac morphologic damage following high-level organophosphate poisoning.

*References:*
- Sax, N.I., Ed., Dangerous Properties of Industrial Materials Report, 3, No. 1, 72-74 (1983).
- U.S. Environmental Protection Agency, "Chemical Profile: Mephosfolan, " Washington, DC, Chemical Emergency Preparedness Program (Nov. 30, 1987).
- California Environmental Protection Agency "Chemical List of Lists," Sacramento CA (February 1997).
- Agency for Toxic Substances and Disease Registry, U.S. Department of Health and Human Services, Public Health Service, "Managing Hazardous Materials Incidents," Atlanta, GA (June 2003).

# Mepiquat Chloride

*Use Type:* Herbicide, plant growth regulator
*CAS Number:* 24307-26-4
*Formula:* $C_7H_{16}ClN$
*Synonyms:* N,N-Dimethylpiperidinium chloride
*Trade Names:* BAS 083 01 W®, BASF Agricultural Products (Germany); BAS 85559X®, BASF Agricultural Products (Germany); MEPEX®, Griffin (USA); MEPICHLOR®, Micro-Flo (USA); MEPPLUS®, Micro-Flo (USA); PIX®, BASF Agricultural Products (Germany); PONNAX®, BASF Agricultural Products (Germany); ROQUAT®, Rotam Agrochemical (Hong Kong); TERPAL® (with Ethephon)
*Producers:* Agsin (Singapore); BASF Agricultural Products (Germany); Gharda Chemicals (India); Gowan (USA); Griffin (USA); Micro-Flo (USA); Rotam Agrochemical (Hong Kong)
*Chemical Class:* Quaternary ammonium compound
*EPA/OPP PC Code:* 109101
*California DPR Chemical Code:* 2075

*Uses:* Registered solely for use on cotton, to control the growth and yield.
*Human toxicity (long-term)[77]:* Very low–4200.00 ppb, Health Advisory
*Fish toxicity (threshold)[77]:* Very low–100000.00 ppb, MATC (Maximum Acceptable Toxicant Concentration)
*U.S. Maximum Allowable Residue Levels for Mepiquat Chloride (40 CFR 180.384):*

| CROP | ppm |
| --- | --- |
| Cattle, fat | 0.1 |
| Cattle, meat | 0.1 |
| Goat, fat | 0.1 |
| Goat, meat | 0.1 |
| Grape | 1 |
| Grape, raisin | 5 |
| Hog, fat | 0.1 |
| Hog, meat | 0.1 |
| Horse, fat | 0.1 |
| Horse, meat | 0.1 |
| Sheep, fat | 0.1 |
| Sheep, meat | 0.1 |

*Carcinogen/Hazard Classifications*
**U.S. EPA Carcinogens:** Group E, Unlikely a Carcinogen
**Label Signal Word:** CAUTION or WARNING
*Regulatory Authority:*
- Actively registered pesticide in California.

*Description:* Off-white powder. Slightly sweet, musty odor. Melting point = >300°C. Molecular weight = 149.63.
*Permissible Concentration in Water:* No criteria set. Runoff from spills or fire control may cause water pollution.
*First Aid:* If this chemical gets into the eyes, remove any contact lenses at once and irrigate immediately for at least 15 minutes, occasionally lifting upper and lower lids. Seek medical attention immediately. If this chemical contacts the skin, remove contaminated clothing and wash immediately with soap and water. Seek medical attention immediately. If this chemical has been inhaled, remove from exposure, begin rescue breathing (using universal precautions) if breathing has stopped, and CPR if heart action has stopped. Transfer promptly to a medical facility. When this chemical has been swallowed, get medical attention. Give large quantities of water and induce vomiting. Do not make an unconscious person vomit.

*References:*
- California Environmental Protection Agency "Chemical List of Lists," Sacramento CA (February 1997)
- U.S. Environmental Protection Agency, "Reregistration Eligibility Decision (RED), Mepiquat Chloride," Office of Prevention, Pesticides and Toxic Substances, Washington, DC (March 1997). http://www.epa.gov/REDs/2375red.pdf
- U.S. Environmental Protection Agency, Office of Pesticide Programs, Pesticide Residue Limits, "Mepiquat Chloride ", 40 CFR 180.384, www.epa.gov/pesticides/food/viewtols.htm

# Mercuric Chloride

*Use Type:* Fungicide
*CAS Number:* 7487-94-7
*Formula:* $Cl_2Hg$; $HgCl_2$
*Synonyms:* Bichloride of mercury; Bichlorure de mercure (French); Chlorid rtutnaty (Czech); Chlorure mercurique (French); Cloruro mercurico (Spanish); Cloruro di mercurio (Italian); Corrosive mercury chloride; MC; Corrosive sublimate; Mercuric bichloride; Mercury bichloride; Mercury(2+) chloride; Mercury(II) chloride; Mercury perchloride; Mercury vichloride; NCI-C60173; Quecksilber chlorid (German); Perchloride of mercury; Sulema (Russian); Sublimat (Czech)
*Trade Names:* CALO-CHLOR®; CALO-CURE®; FUNGCHEX®; TL 898®
*Producers:* GFS Chemicals (USA); United Phosphorus (India)
*Chemical Class:* Inorganic, mercury
*EPA/OPP PC Code:* 052001
*California DPR Chemical Code:* 372
*ICSC Number:* 0979
*RTECS Number:* OV9100000
*EEC Number:* 080-002-00-X
*EINECS Number:* 231-299-8
*Uses:* This compound is used in preserving wood and anatomical specimens; embalming; browning and etching steel and iron; as a catalyst for organic synthesis; disinfectant; antiseptic; tanning; textile printing aid; manufacture of dyes; agricultural chemicals and dry batteries; pharmaceuticals; and photographic chemicals.
*Carcinogen/Hazard Classifications*
**U.S. EPA Carcinogens:** Group D, unclassifiable, inadequate data
**IARC:** Group 3, unclassifiable
**California Prop. 65:** Developmental toxin
**WHO Acute Hazard:** Class 1 a, extremely hazardous
*Regulatory Authority:*
- Banned or Severely Restricted (In Agriculture) (U.K.)[13]
- Air Pollutant Standard Set (ACGIH)[1] (HSE)[33] (OSHA)[58] (former USSR)[35, 43]
- AB 2588-Air Toxics "Hot Spots" Chemicals (CAL)
- CAL Air Resources Board/AB 1807 Toxic Air Contaminants
- Proposition 65 chemical (CAL)
- Permissible Exposure Limits for Chemical Contaminants (CAL/OSHA)
- The "Director's List" (CAL/OSHA)
- Clean Water Act: Section 307 Toxic Pollutants as mercury and compounds
- Safe Drinking Water Act: MCL, 0.002 mg/L; MCLG, 0.002 mg/L as mercury
- RCRA Section 261 Hazardous Constituents, waste number not Listed as mercury compounds, n.o.s.
- EPCRA Section 302 Extremely Hazardous Substances: TPQ = 500/10,000 lb (227/4,540 kg)
- EPCRA Section 304 RQ: EHS, 1 lb (0.454 kg)
- EPCRA Section 313 (as mercury compound) Form R *de minimus* concentration reporting level: 1.0%
- U.S. DOT Regulated Marine Pollutant (49CFR172.101, Appendix B), severe pollutant as mercury(II) (mercuric) compounds (pesticides)
- Canada, WHMIS, Ingredients Disclosure List

*Description:* White crystalline solid. Odorless. Soluble in water; solubility = 7.3 g/100 cc water @ 20°C. Boiling point = 302°C. Melting/Freezing point = 276°C. Vapor pressure = 0.1 Pa @ 20°C. Hazard Identification (based on NFPA-704 M Rating System): Health 3, Flammability 0, Reactivity 0. Mercury compounds are toxic to aquatic organisms. Bioaccumulation in the food chain takes place, specifically in fish, crustacea, and birds. Bioconcentrative up to 10,000-fold. (CHRIS, U. S. Coast Guard).

*Incompatibilities:* Mercuric chloride may explode with friction or application of heat. Mixtures of mercuric chloride and sodium or potassium are shock sensitive and will explode on impact. Avoid contact with acids or acid fumes. Also avoid the presence of formats, sulfites, hypophosphites, phosphates, sulfides, albumin, gelatin, alkalies, alkaloid salts, ammonia, lime water, antimony, arsenic, bromides, borax, carbonates, reduced iron, copper, iron, lead, silver salts, infusions of cinchona, columbo, oak bark or senna, and tannic acid.

*Permissible Exposure Limits in Air:* The OSHA[2] PEL for inorganic mercury is [ceiling] is 0.1 mg/m$^3$, not to be exceeded at any time. NIOSH[2] recommends a 10-hour TWA of 0.05 mg/m$^3$ for Hg vapor and [ceiling] of 0.1 mg/m$^3$. ACGIH[1] recommends an 8-hour TWA of 0.025 mg/m$^3$. NIOSH[2] IDLH level = 10 mg/m$^3$. In addition, the former USSR-UNEP/IRTC project[43] has set an MAC in workplace air of 0.1 mg/m$^3$. Further the former USSR[35] has set an MAC for ambient air in residential areas of 0.0003 mg/m$^3$ on a daily average basis.

*Determination in Air:* Hopcalite; Acid; AA cold; NIOSH IV, Method #6009, Mercury.[18]

*Permissible Concentration in Water:* To protect freshwater aquatic life: $5.7 \times 10^{-5}$ µg/L as a 24-hour average, never to exceed 0.0017 µg/L. To protect saltwater aquatic life: 0.025 µg/L as a 24 hour average, never to exceed 3.7 µg/L. To protect human health: 0.144 µg/L (USEPA) set in 1979-80[6]. These are the limits for inorganic mercury compounds in general. In addition, the former USSR-UNEP/IRTC project[43] has set an MAC for water bodies used for domestic purposes of 0.005 µg/L.

*Determination in Water:* Total mercury is determined by flameless atomic absorption. Soluble mercury may be determined by 0.45 micron filtration followed by flameless atomic absorption.

*Routes of Entry:* Inhalation, skin absorption, ingestion, skin and/or eye contact.

### Harmful Effects and Symptoms

**Short Term Exposure:** Corrosive. The substance is corrosive to the eyes, the skin, and the respiratory tract. Corrosive on ingestion. Inhalation of its aerosol may cause lung edema. The substance may cause effects on the kidneys. Exposure far above OEL may result in death. The effects may be delayed. Medical observation is indicated. It is classified as extremely toxic. All forms of mercury are poisonous if absorbed. Probable oral lethal dose is 5-50 mg/kg; between 7 drops and 1 teaspoonful for a 150 lb person. Mercuric chloride is one of the most toxic salts of mercury. Material attacks the gastrointestinal tract and renal systems. Signs and symptoms of acute exposure or mercuric chloride may be severe and include increased salivation, foul breath, inflammation and ulceration of the mucous membranes, abdominal pain, and bloody diarrhea. Dermal exposure may result in dermatitis (red, inflamed skin) and burns. Oliguria (scanty urination), anuria (suppression of urine formation), and acute renal failure may be noted. Weak pulse, seizures, psychic disturbances, circulatory collapse, chest pain, and dyspnea (shortness of breath) may be observed.

**Long Term Exposure:** Repeated or prolonged contact with skin may result in dermatitis (red inflamed skin). Repeated or prolonged exposure may cause death by hypovolemic shock, nephrotic syndrome, or kidney failure. Mercury can cross the blood-brain and placental barriers. It is also excreted in breast milk. Children may be at increased risk for pulmonary toxicity and are more likely to develop respiratory failure.

**Points of Attack:** Eyes, skin, central nervous system, peripheral nervous system and kidneys.

**Medical Surveillance:** Before first exposure and every 6 to 12 months after, a complete medical history and exam is strongly recommended, with: Exam of the nervous system, including handwriting. Routine urine test (UA). Urine test for mercury (should be less than 0.02 mg/L). Eye exam. Consider lung function tests for persons with frequent exposure. After suspected illness or overexposure, repeat the tests above and get a blood test for mercury. Consider chest x-ray after sudden overexposure. Consider nerve conduction tests, urinary enzymes and neuro-behavioral testing.

**First Aid:** If this chemical gets into the eyes, remove any contact lenses at once and irrigate immediately for at least 15 minutes, occasionally lifting upper and lower lids. Seek medical attention immediately. If this chemical contacts the skin, remove contaminated clothing and wash immediately with soap and water. Seek medical attention immediately. If this chemical has been inhaled, remove from exposure, begin rescue breathing (using universal precautions) if breathing has stopped, and CPR if heart action has stopped. Transfer promptly to a medical facility. When this chemical has been swallowed, get medical attention. Give large quantities of water and induce vomiting. Do not make an unconscious person vomit. Medical observation is recommended for 24 to 48 hours after breathing overexposure, as pulmonary edema may be delayed. As first aid for pulmonary edema, a doctor or authorized paramedic may consider administering a corticosteroid spray.

**References:**
- New Jersey Department of Health and Senior Services, "Hazardous Substance Fact Sheet: Mercuric Chloride," Trenton, NJ (January 2000). http://www.state.nj.us/health/eoh/rtkweb/1170.pdf
- New York State Department of Health, "Chemical Fact Sheet: Mercuric Chloride," Albany, NY, Bureau of Toxic Substance Assessment (Feb. 1986 and Version 2).
- U.S. Environmental Protection Agency, "Chemical Profile: Mercuric Chloride," Washington, DC, Chemical Emergency Preparedness Program (Nov. 30, 1987).
- Agency for Toxic Substances and Disease Registry, U.S. Department of Health and Human Services, Public Health Service, "Managing Hazardous Materials Incidents," Atlanta, GA (June 2003).
- California Environmental Protection Agency "Chemical List of Lists," Sacramento CA (February 1997).

# Mercuric Oxide

**Use Type:** Fungicide
**CAS Number:** 21908-53-2
**Formula:** HgO
**Synonyms:** C.I. 77760; Mercuric oxide, red; Mercuric oxide, yellow; Mercury monoxide; Mercury oxide; Oxido mercurico rojo (Spanish); Oxido mercurico amarillo (Spanish); Oxyde de mercure (French); Red mercuric oxide; Red oxide of mercury; Red precipitate; Yellow mercuric oxide; Yellow oxide of mercury; yellow precipitate
**Trade Names:** KANKEREX®; SANTAR®
**Producers:** GFS Chemicals (USA)
**Chemical Class:** Inorganic
**EPA/OPP PC Code:** 052102
**California DPR Chemical Code:** 955
**ICSC Number:** 0981
**RTECS Number:** OW8750000
**EEC Number:** 080-002-00-6
**EINECS Number:** 244-654-7
**Uses:** Chemical intermediate for mercury salts, organic mercury compounds, and chlorine monoxide; antiseptic in pharmaceuticals; component of dry cell batteries, pigment and glass modifier; preservative in cosmetics; analytical reagent, formerly used in antifouling paints.

### Carcinogen/Hazard Classifications

**U.S. EPA Carcinogens:** Group D, unclassifiable, inadequate data
**IARC:** Group 3, unclassifiable
**California Prop. 65:** Developmental toxin
**WHO Acute Hazard:** Class 1b, highly hazardous

# Mercuric Oxide

***Regulatory Authority:***
- Banned or Severely Restricted (In Agriculture) (EEC) (UK)[13]
- Air Pollutant Standard Set (ACGIH)[1] (HSE)[33] (OSHA)[58] (former USSR)[35]
- AB 2588-Air Toxics "Hot Spots" Chemicals (CAL)
- CAL Air Resources Board/AB 1807 Toxic Air Contaminants
- Proposition 65 chemical (CAL)
- Permissible Exposure Limits for Chemical Contaminants (CAL/OSHA)
- The "Director's List" (CAL/OSHA)
- Clean Water Act: 40CFR401.15 Section 307 Toxic Pollutants as mercury and compounds
- Safe Drinking Water Act: MCL, 0.002 mg/L; MCLG, 0.002 mg/L as mercury
- RCRA, 40CFR261, Appendix 8 Hazardous Constituents, waste number not Listed as mercury compounds, n.o.s.
- Superfund/EPCRA 40CFR355, Appendix B Extremely Hazardous Substances: TPQ = 500/10,000 lb (227/4,540 kg)
- Superfund/EPCRA 40CFR302.4 RQ: EHS, 1 lb (0.454 kg)
- EPCRA Section 313 (as mercury compound): Form R *de minimis* concentration reporting level: 1.0%
- U.S. DOT Regulated Marine Pollutant (49CFR172.101, Appendix B), severe pollutant
- Canada, WHMIS, Ingredients Disclosure List

***Description:*** Mercuric oxide is a red or orange-red heavy crystalline powder; yellow when finely powdered. Odorless. Slightly soluble in water; solubility = $5 \times 10^{-3}$ g/100 ml @ 25°C. Molecular weight =216.58. Melting/Freezing point = 500°C (decomposes). Hazard Identification (based on NFPA-704 M Rating System): Health 3, Flammability 0, Reactivity 0. Mercury compounds are toxic to aquatic organisms. Bioaccumulation in the food chain takes place, specifically in fish, crustacea, and birds. Bioconcentrative up to 10,000-fold. (CHRIS, U. S. Coast Guard).

***Incompatibilities:*** A powerful oxidizer. Decomposes on exposure to light, when heated above 500°C, producing highly toxic fumes including mercury and oxygen, which will add to the intensity of an existing fire. Violent reaction with combustible materials, other oxidizers, acetyl nitrate, aluminum, diboron tetrafluoride, reducing agents, phospham., hydrogen trisulfide (on ignition), hydrazine hydrate, hydrogen peroxide, hypophosphorous acid, acetyl nitrate, chlorine, hypophosphorous acid, magnesium (when heated), disulfur dichloride, alcohols, alkali metals (i.e., lithium, sodium, potassium, rubidium, cesium, francium). Forms heat- or impact-sensitive explosive mixtures with sulfur, phosphorus and other nonmetals, potassium, magnesium, sodium, and other chemically active metals. Incompatible with strong bases and light.

***Permissible Exposure Limits in Air:*** The OSHA[2] PEL for inorganic mercury is [ceiling] is 0.1 mg/m$^3$, not to be exceeded at any time. NIOSH[2] recommends a 10-hour TWA of 0.05 mg/m$^3$ for Hg vapor and [ceiling] of 0.1 mg/m$^3$. ACGIH[1] recommends an 8-hour TWA of 0.025 mg/m$^3$. NIOSH[2] IDLH level = 10 mg/m$^3$. In addition, the former USSR has set an MAC in workplace air of 0.2 mg/m$^3$ as a ceiling and 0.05 mg/m$^3$ as a TWA. Further, the former USSR has set an MAC for ambient air in residential areas of 0.003 mg/m$^3$ on a daily average basis.

***Determination in Air:*** Hopcalite; Acid; AA cold; NIOSH IV, Method #6009, Mercury[18].

***Permissible Concentration in Water:*** To protect freshwater aquatic life: $5.7 \times 10^{-5}$ μg/L as a 24-hour average, never to exceed 0.0017 μg/L. To protect saltwater aquatic life: 0.025 μg/L as a 24 hour average, never to exceed 3.7 μg/L. To protect human health: 0.144 μg/L (USEPA) set in 1979-80[6]. These are the limits for inorganic mercury compounds in general.

***Determination in Water:*** Total mercury is determined by flameless atomic absorption. Soluble mercury may be determined by 0.45 micron filtration followed by flameless atomic absorption.

***Routes of Entry:*** Inhalation, skin absorption, ingestion, skin and/or eye contact.

***Harmful Effects and Symptoms***

***Short Term Exposure:*** Mercuric oxide dust has a corrosive effect on eyes, skin, and respiratory tract. This material is highly toxic by ingestion, inhalation, or skin absorption. Very short exposure to small quantities may cause death or permanent injury. Following ingestion, mercuric oxide is readily converted to mercuric chloride, the most dangerous mercury compounds. Signs and symptoms of acute exposure to mercuric oxide may be severe and include increased salivation, foul breath, inflammation and ulceration of the mucous membranes, abdominal pain, and bloody diarrhea. Oliguria (scanty urination), anuria (suppression of urine formation), and acute renal failure may be noted. Weak pulse, seizures, psychic disturbances, circulatory collapse, chest pain, and dyspnea (shortness of breath) may be observed.

***Long Term Exposure:*** Repeated or prolonged contact with skin may result in dermatitis and allergy. Repeated or prolonged exposure may cause brain damage and nervous system damage. Repeated or prolonged exposure may cause death by hypovolemic shock, nephrotic syndrome, and kidney failure. There is limited evidence that this chemical is a teratogen in animals. Can cause mercury to accumulate in the body and cause mercury poisoning. May cause permanent damage such as gray colored skin, brown staining of the eyes, and decreased peripheral vision. Mercury can cross the blood-brain and placental barriers. It is also excreted in breast milk. Children may be at increased risk for pulmonary toxicity and are more likely to develop respiratory failure.

***Points of Attack:*** Eyes, skin, respiratory system, central nervous system, kidneys

*Medical Surveillance:* Before first exposure and every 6 to 12 months after, a complete medical history and exam is strongly recommended, with: Exam of the nervous system, including handwriting. Routine urine test (UA), Urine test for mercury (should be less than 0.02 mg/L). Eye exam. After suspected illness or overexposure, repeat the test above and get a blood test for mercury. Consider chest x-ray after acute overexposure. Consider nerve conduction tests, urinary enzymes and neuro-behavioral testing. Evaluation by a qualified allergist. Eye examination. Consider chest x-ray following acute overexposure.

*First Aid:* Remove victims from exposure. Emergency personnel should avoid self-exposure to mercuric oxide. Evaluate vital signs including pulse and respiratory rate, and note any trauma. If no pulse is detected, provide CPR. If not breathing, provide artificial respiration. If breathing is labored, administer oxygen or other respiratory support. Remove contaminated clothing as soon as possible. If eye exposure has occurred, remove any contact lenses at once; eyes must be flushed with lukewarm water for at least 15 minutes. Wash exposed skin areas for 15 minutes with soap and water. Obtain authorization and/or further instructions from the local hospital for administration of an antidote or performance of other invasive procedures in the event if inhalation or ingestion of HgO. Rush to a health care facility.

*Antidotes and Special Procedures for medical personnel:* The drug NAP (n-acetyl penicillamine) has been used to treat mercury poisoning, with mixed success.

*References:*
- U.S. Environmental Protection Agency, "Chemical Profile: Mercuric Oxide," Washington, DC, Chemical Emergency Preparedness Program (Nov. 30, 1987).
- Agency for Toxic Substances and Disease Registry, U.S. Department of Health and Human Services, Public Health Service, "Managing Hazardous Materials Incidents," Atlanta, GA (June 2003).
- California Environmental Protection Agency "Chemical List of Lists," Sacramento CA (February 1997).
- New Jersey Department of Health and Senior Services, "Hazardous Substance Fact Sheet, Mercuric Oxide," Trenton NJ (September, 1998). http://www.state.nj.us/health/eoh/rtkweb/2537.pdf

# Mercury Alkyl Compounds

*Use Type:* Fungicide and seed disinfectant
*CAS Number:* 22967-92-6 (methyl mercury ion); 115-09-3 (methyl mercury chloride); 593-74-8 (dimethyl mercury)
*Formula:* $CH_3ClHg$; $CH_3HgCl$
*Alert:* Mercury alkyl compounds are highly toxic. Mercury and mercury compound poisoning may produce irreversible brain damage.

*Synonyms: Methyl mercury chloride:* Caspan; Chloromethylmercury; methylmercuric chloride; Methylmercury chloride; MMC; Monomethyl mercury chloride
*Dimethyl mercury:* Mercury dimethyl
*Chemical Class:* An organomercury compound
*ICSC Number:* Dimethyl mercury: 1304
*RTECS Number:* OW6320000 (methyl mercury ion); OW1225000 (methyl mercury chloride); OW301000 (dimethyl mercury)
*EEC Number:* 080-007-00-3 (dimethyl mercury)
*Uses:* Alkyl mercury compounds are also used in organic synthesis.

*Carcinogen/Hazard Classifications*
**California Prop. 65:** Listed; Reproductive toxin
**IARC:** Methyl mercury chloride: Group 2B, possible carcinogen

*Regulatory Authority:*
- Carcinogen (Chloromethyl Mercury) (RTECS)[9]
- Air Pollutant Standard Set (ACGIH)[1] (DFG)[3] (HSE)[33] (OSHA)[58] (North Dakota)[60]
- AB 2588-Air Toxics "Hot Spots" Chemicals (CAL)
- CAL Air Resources Board/AB 1807 Toxic Air Contaminants
- Proposition 65 chemical (CAL)
- Permissible Exposure Limits for Chemical Contaminants (CAL/OSHA)
- The "Director's List" (CAL/OSHA)
- Clean Air Act: Hazardous Air Pollutants (Title I, Part A, Section 112)
- Clean Water Act: Section 307 Toxic Pollutants as mercury and compounds
- Safe Drinking Water Act: MCL, 0.002 mg/L; MCLG, 0.002 mg/L as mercury
- RCRA Section 261 Hazardous Constituents, waste number not Listed as mercury compounds, n.o.s.
- EPCRA Section 302 Extremely Hazardous Substances: TPQ = 500/10,000 lb (227/4,540 kg)
- EPCRA Section 304 RQ: EHS, 1 lb (0.454 kg)
- EPCRA Section 313 (as mercury compound) Form R *de minimus* concentration reporting level: 1.0%
- U.S. DOT Regulated Marine Pollutant (49CFR172.101, Appendix B), severe pollutant as mercury based pesticides, liquid, flammable, toxic, n.o.s; mercury based pesticides, liquid, toxic, n.o.s; mercury based pesticides, solid, toxic, n.o.s; mercury compounds, liquid, n.o.s; mercury compounds, solid, n.o.s; mercury(I) (mercurous) compounds (pesticides); mercury(II) (mercuric) compounds (pesticides)

*Description:* Methyl mercury chloride is a colorless crystalline solid. Molecular weight = 251.08. Melting/Freezing point = 170°C. Dimethyl mercury is a volatile colorless liquid. Faint sweet odor. Molecular weight = 230.68. Boiling point = 96°C. Methyl mercury has a molecular weight = 215.65. Mercury compounds are toxic

## Mercury Alkyl Compounds

to aquatic organisms. Bioaccumulation in the food chain takes place, specifically in fish, crustacea, and birds. Bioconcentrative up to 10,000-fold. (CHRIS, U. S. Coast Guard).

*Incompatibilities:* Strong oxidizers such as chlorine.

*Permissible Exposure Limits in Air:* The OSHA[2] PEL for all organic mercury compounds is 0.01 mg/m$^3$ as an 8-hour TWA and ceiling of 0.04 mg/m$^3$. NIOSH[2] recommends 0.01 mg/m$^3$ as a 10-hour TWA and STEL of 0.03 mg/m$^3$ with the notation "skin" indicating the possibility of cutaneous absorption. ACGIH[1] recommends 0.01 mg/m$^3$ as an 8-hour TWA value with no STEL. The NIOSH[2] IDLH level = 2 mg/m$^3$ (as Hg). The DFG[3] has set an MAK for methyl mercury of 0.01 mg/m$^3$. They have also set[35] an STEL of 0.1 mg/m$^3$. Other countries have generally adopted the 0.01 mg/m$^3$ level. In addition, North Dakota has set guidelines for alkyl mercury compounds in ambient air[60] of 1-3 $\mu$g/m$^3$ (0.0001 to 0.0003 mg/m$^3$).

*Determination in Air:* No method available.

*Permissible Concentration in Water:* (Methylmercury): To protect freshwater aquatic life: 0.016 $\mu$g/L as a 24-hr average, never to exceed 8.8 $\mu$g/L. To protect saltwater aquatic life: 0.025 $\mu$g/L as a 24-hr average, never to exceed 2.8 $\mu$g/L. To protect human health: 0.2 $\mu$g/L[6].

*Determination in Water:* Total mercury is determined by flameless atomic absorption. Soluble mercury may be determined by 0.45 micron filtration followed by flame less atomic absorption.

*Routes of Entry:* Inhalation, ingestion, eye and/or skin contact. Absorbed through the skin.

*Harmful Effects and Symptoms*

*Short Term Exposure:* Alkyl mercury compounds can be absorbed through the skin. When deposited on the skin, they give no warning, and if contact is maintained, can cause second-degree burns. Sensitization may occur. Alkyl mercurials have very high toxicity. Systemic: The central nervous system, including the brain, is the principal target tissue for this group of toxic compounds. Severe poisoning may produce irreversible brain damage resulting in loss of higher functions. The effects of chronic poisoning with alkyl mercury compounds are progressive. In the early stages, there are fine tremors of the hands, and in some cases, of the face and arms.

*Long Term Exposure:* Repeated or prolonged contact with skin may result in dermatitis (red inflamed skin). Repeated or prolonged exposure may cause death by hypovolemic shock, nephrotic syndrome, or kidney failure. With repeated or continued exposure, tremors may become coarse and convulsive; scanning speech with moderate slurring and difficulty in pronunciation may also occur. The worker may then develop an unsteady gait of a spastic nature which can progress to severe ataxia of the arms and legs. Sensory disturbances including tunnel vision, blindness, and deafness are also common. A late symptom, constriction of the visual fields, is rarely reversible and may be associated with loss of understanding and reason which makes the victim completely out of touch with his environment. Mercury can cross the blood-brain and placental barriers. It is also excreted in breast milk. Children may be at increased risk for pulmonary toxicity and are more likely to develop respiratory failure. Severe cerebral effects have been seen in infants born to mothers who had eaten large amounts of methylmercury-contaminated fish.

*Points of Attack:* Eyes, skin, central nervous system, peripheral nervous system and kidneys.

*Medical Surveillance:* Preplacement and periodic physical examinations should be concerned particularly with the skin, vision, central nervous system, and kidneys. Consideration should be given to the possible effects on the fetus of alkyl mercury exposure in the mother. Constriction of visual fields may be a useful diagnostic sign. Blood and urine levels of mercury have been studied, especially in the case of methylmercury. A precise correlation has not been found between exposure levels and concentrations. They may be of some value in indicating that exposure has occurred, however.

*First Aid:* If this chemical gets into the eyes, remove any contact lenses at once and irrigate immediately for at least 15 minutes, occasionally lifting upper and lower lids. Seek medical attention immediately. If this chemical contacts the skin, remove contaminated clothing and wash immediately with soap and water. Seek medical attention immediately. If this chemical has been inhaled, remove from exposure, begin rescue breathing (using universal precautions) if breathing has stopped, and CPR if heart action has stopped. Transfer promptly to a medical facility. When this chemical has been swallowed, get medical attention. Give large quantities of water and induce vomiting. Do not make an unconscious person vomit.

*Antidotes and Special Procedures for medical personnel:* The drug NAP (n-acetyl penicillamine) has been used to treat mercury poisoning, with mixed success.

*References:*
- New Jersey Department of Health and Senior Services, "Hazardous Substance Fact Sheet, Dimethyl Mercury," Trenton, NJ (August 1998). http://www.state.nj.us/health/eoh/rtkweb/0763.pdf
- National Instirute for Occupational Safety and Health (NIOSH), Information Profiles on Potential Occupational Hazards: Organomercurials, pp 287-296 678, Rockville, MD (Oct. 1977).
- Agency for Toxic Substances and Disease Registry, U.S. Department of Health and Human Services, Public Health Service, "Managing Hazardous Materials Incidents," Atlanta, GA (June 2003).
- California Environmental Protection Agency "Chemical List of Lists," Sacramento CA (February 1997).
- U.S. Environmental Protection Agency, Mercury: Ambient Water Quality Criteria, Wash., DC (1979).

## Metalaxyl (ANSI)

*Use Type:* Fungicide
*CAS Number:* 57837-19-1
*Formula:* $C_{15}H_{21}NO_4$
*Synonyms:* *dl*-Alanine, *N*-(2,6-dimethylphenyl)-*N*-(methoxyacetyl)-, methyl ester; *dl*-*N*-(2,6-Dimethylphenyl)-*N*-(2'-methoxyacetyl)alaninate de methyle (French); *N*-(2,6-Dimethylphenyl)-*N*-(methoxyacetyl)alanine, methyl ester; *N*-(2,6-Dimethylphenyl)-*N*-(methoxyacetyl)-*dl*-alanine methyl ester
*Trade Names:* AGROX® PREMIERE, Agriliance (USA); ALLEGIENCE®, Gustafson (USA); APRON®, Syngenta (Switzerland); CG 117®; CGA-48988®; CHLORAXYL®, Micro-Flo (USA); COTGUARD®, Gustafson (USA); EPERON®, Syngenta (Switzerland); DELTA-COAT; FOLIO® GOLD, Syngenta (Switzerland); GAUCHO®, Drexel Chemical (USA); KODIAK®, Gustafson (USA); METALAXIL®; METAXANIN®; PACE®, Syngenta (Switzerland); PREVAIL®, Gustafson (USA); RAXIL® (tebuconazole + metalaxyl), Gustafson (USA); RIDOMIL® GOLD/BRAVO®, Syngenta (Switzerland); RIDOMIL®, Syngenta (Switzerland); RIDOMIL 2E®, Syngenta (Switzerland); SUBDUE®
*Producers:* Agriliance (USA); Agrimor International (USA); Agsin (Singapore); Drexel Chemical (USA); Gustafson (USA); Ki-Hara Chemicals Ltd. (UK); Micro-Flo (USA); Rallis India (India); Syngenta (Switzerland)
*Chemical Class:* Miscellaneous organic (Acylalanine); benzenoid
*EPA/OPP PC Code:* 113501
*California DPR Chemical Code:* 2132
*Uses:* Metalaxyl is used as a systemic fungicide on a variety of food and non-food crops including tobacco, turf and conifers, and ornamentals. Used in combination with fungicides of different mode of action as a foliar spray on tropical and subtropical crops; as a seed treatment to control downy mildew; and as a soil fumigant to control soil-borne pathogens.
*Human toxicity (long-term)[77]:* Very low–518.00 ppb, Health Advisory
*Fish toxicity (threshold)[77]:* Very low–9099.97083 ppb, MATC (Maximum Acceptable Toxicant Concentration)
*U.S. Maximum Allowable Residue Levels for Metalaxyl (40 CFR 180.408):*
The U.S. EPA lists 121 crops and their residues. They can be located at the web site indicated in the *Reference* section below.
*Carcinogen/Hazard Classifications*
**U.S. EPA Carcinogens:** Group E, Unlikely a carcinogen
**Label Signal Word:** CAUTION, WARNING or DANGER
**WHO Acute Hazard:** Class III, slightly hazardous
*Regulatory Authority:*
- Actively registered pesticide in California.
- FIFRA, 40CFR185: tolerances for pesticides in food
- FIFRA, 40CFR186: tolerances for pesticides in animal feeds.

*Description:* Flammable, white to colorless crystalline solid. Odorless. Soluble in water; solubility = 7.0 g/L @ 20°C. Molecular weight = 297.33. Practically insoluble in water. Melting/Freezing point = 70°C. Vapor pressure = $5.6 \times 10^{-6}$ mmHg @ 20°C. Hazard Identification (based on NFPA-704 M Rating System): Health 2, Flammability 2, Reactivity 0.
*Incompatibilities:* Incompatible with acids, Oxidizers (chlorates, nitrates, peroxides, permanganates, perchlorates, chlorine, bromine, fluorine, etc). May cause fires or explosions.
*Permissible Concentration in Water:* No criteria set. Runoff from spills or fire control may cause water pollution.
*Routes of Entry:* Ingestion, inhalation.
*Harmful Effects and Symptoms*
*Short Term Exposure:* Contact may burn eyes, skin, and respiratory tract. Toxic if ingested.
*Medical Surveillance:* Consult a physician if poisoning is suspected or if redness, itching, or burning of the eyes or skin develop.
*First Aid:* If this chemical gets into the eyes, remove any contact lenses at once and irrigate immediately for at least 15 minutes, occasionally lifting upper and lower lids. Seek medical attention immediately. If this chemical contacts the skin, remove contaminated clothing and wash immediately with soap and water. Seek medical attention immediately. If this chemical has been inhaled, remove from exposure, begin rescue breathing (using universal precautions) if breathing has stopped, and CPR if heart action has stopped. Transfer promptly to a medical facility. When this chemical has been swallowed, get medical attention. *Do not induce vomiting when formulations containing petroleum solvents are ingested.* Otherwise, give large quantities of water and induce vomiting. Do not make an unconscious person vomit.
*References:*
- EXTOXNET, Extension Toxicology Network, "Pesticide Information Profile, Metalaxyl," Oregon State University, Corvallis, OR (June 1996). http://extoxnet.orst.edu/pips/metalaxy.htm
- U.S. Environmental Protection Agency, Office of Pesticide Programs, Pesticide Residue Limits, "Metalaxyl", 40 CFR 180.408, http://www.epa.gov/pesticides/food/viewtols.htm
- California Environmental Protection Agency "Chemical List of Lists," Sacramento CA (February 1997)

## Metaldehyde

*Use Type:* Molluscicide
*CAS Number:* 108-62-3
*Formula:* $C_8H_{16}O_4$
*Alert:* A Restricted Use Pesticide (RUP)

*Synonyms:* Acetaldehyde, tetramer; Metacetaldehyde; Metaldehyd (German); Metaldeide (Italian); 1,3,5,7-Tetroxocane, 2,4,6,8-tetramethyl-1,3,5,7-tetraoxacyclooctane

*Trade Names:* ANTIMILACE®; ARIOTOX®; CEKUMETA®; DEADLINE®, Amvac Chemical Corp (USA); DURHAM®, Amvac Chemical Corp (USA); HALIZAN®; LIMATOR®; META®, Lonza Group (Switzerland); METASON®; NAMEKIL®; SLUG-TOX®; TRAILS END®, Amvac Chemical Corp (USA)

*Producers:* Agsin (Singapore); Amvac Chemical Corp (USA); Chevron Phillips Chemical (USA); Lonza Group (Switzerland); Sigma-Aldrich Laborchemikalien (Germany); United Agri Products (UAP)

*Chemical Class:* An aldehyde
*EPA/OPP PC Code:* 053001
*California DPR Chemical Code:* 379
*RTECS Number:* XF9900000
*EINECS Number:* 203-600-2

*Uses:* Metaldehyde is a molluscicide used in a variety of vegetable and ornamental crops in the field or greenhouse, on fruit trees, small-fruit plants, or in avocado or citrus orchards, berry plants, and banana plants. It is used to attract and kill slugs and snails. It is applied in the form of granules, sprays, and dusts, or pellets or grain bait, typically to the ground around the plants or crops. It works primarily in the stomach by producing toxic effects after it is ingested by the pest. It may be formulated with or without calcium arsenate and is also available in a mixed formulation with thiram.

*Fish toxicity (threshold)[77]:* Very low–1121.12787 ppb, MATC (Maximum Acceptable Toxicant Concentration)

*U.S. Maximum Allowable Residue Levels for Metaldehyde (40 CFR 180.523):*

| CROP | ppm |
|---|---|
| Strawberry | 0 |

*Carcinogen/Hazard Classifications*
**Label Signal Word:** CAUTION or WARNING (EPA Toxicity Class II or III)
**WHO Acute Hazard:** Class II, moderately hazardous
*Regulatory Authority:*
- TSCA: 40CFR716.120(*d*)1 as aldehydes
- Actively registered pesticide in California.

*Description:* Metaldehyde is a white crystalline solid or powder. Mild menthol odor. Molecular weight = 176.22. Boiling point = 112–116°C. Melting/Freezing point = 47°C. Flash point = 36°C. Hazard Identification (based on NFPA-704 M Rating System): Health 1, Flammability 3, Reactivity 1. Insoluble in water.

*Incompatibilities:* Strong oxidizers.
*Permissible Exposure Limits in Air:* No OELs established.
*Routes of Entry:* Inhalation, ingestion, skin and/or eye contact.

*Harmful Effects and Symptoms*

*Short Term Exposure:* Contact can irritate the eyes, skin and respiratory tract. Exposure can cause nausea, vomiting, diarrhea, abdominal pain, irritability, sleepiness, muscle twitching, convulsions, coma and death.

*Long Term Exposure:* May cause kidney and liver damage. May damage the developing fetus.

*Points of Attack:* Kidneys and liver.

*Medical Surveillance:* Kidney function tests. Liver function tests.

*First Aid:* If this chemical gets into the eyes, remove any contact lenses at once and irrigate immediately for at least 15 minutes, occasionally lifting upper and lower lids. Seek medical attention immediately. If this chemical contacts the skin, remove contaminated clothing and wash immediately with soap and water. Seek medical attention immediately. If this chemical has been inhaled, remove from exposure, begin rescue breathing (using universal precautions) if breathing has stopped, and CPR if heart action has stopped. Transfer promptly to a medical facility. When this chemical has been swallowed, get medical attention. Give large quantities of water and induce vomiting. Do not make an unconscious person vomit.

*References:*
- EXTOXNET, Extension Toxicology Network, "Pesticide Information Profile, Metaldehyde," Oregon State University, Corvallis, OR (June 1996). http://extoxnet.orst.edu/pips/metaldeh.htm
- EPA, Office of Pesticide Programs, Pesticide Residue Limits, "Metalldehyde," 40 CFR 180.523, http://www.epa.gov/pesticides/food/viewtols.htm
- New Jersey Department of Health and Senior Services, "Hazardous Substance Fact Sheet, Metaldehyde," Trenton NJ (June 1999). http://www.state.nj.us/health/eoh/rtkweb/1197.pdf
- California Environmental Protection Agency "Chemical List of Lists," Sacramento CA (February 1997).

*Note:* See the Website on META, provided by Lonza Group (Switzerland): http://www.metaldehyde.com/meta/en.html

# Metamiton

*Use Type:* Herbicide
*CAS Number:* 41394-05-2
*Formula:* $C_{10}H_{10}N_4O$
*Synonyms:* 4-Amino-3-methyl-6-phenyl-1,2,4-triazin-5(4*H*)-one; Metamitron (German); 3-Methyl-4-amino-6-phenyl-1,2,4-triazin(4*H*)-on (German)

*Trade Names:* BAY-DRW 1139®, CropScience (Germany); DRW 1139®, CropScience (Germany); GOLDBEET®, Makhteshim-Agan Industries (Israel); GOLTIX®; MM 70®, Calliope (France) and Makhteshim-Agan Industries (Israel); HERBRAK®; MARQUISE®, Makhteshim-Agan Industries (Israel); SKATER®, Makhteshim-Agan Industries (Israel);

TORERO® (metamitron + ethofumesate), Makhteshim-Agan Industries (Israel)
**Producers:** Bayer CropScience (Germany); Calliope (France); Gharda Chemicals (India); Makhteshim-Agan Industries (Israel)
**Chemical Class:** Triazinone
**California DPR Chemical Code:** 2672
**ICSC Number:** 1361
**RTECS Number:** XZ3015000
**EEC Number:** 613-129-00-8
**Uses:** Not registered in the U.S. Metamitron is a selective systemic herbicide that is absorbed through roots and leaves and inhibits photosynthesis. It is used to control grasses and broadleaf weeds in sugar beets and fodder beets.
**Carcinogen/Hazard Classifications**
**WHO Acute Hazard:** Class III, slightly hazardous
**Description:** Colorless to yellow crystalline solid. Melting/Freezing point = 166°C. Slightly soluble in water. Molecular weight = 202.24. Decomposes on heating to produce toxic fumes, including nitrogen oxides.
**Determination in Air:** Filter; none; Gravimetric; NIOSH IV [Particulates NOR; #0500 (total), #0600 (respirable)].[18]
**Permissible Concentration in Water:** No criteria set. Runoff from spills or fire control may cause water pollution.
**Routes of Entry:** Inhalation, passing through the skin, ingestion.
**Harmful Effects and Symptoms**
**Short Term Exposure:** May cause skin and severe eye irritation. Moderately poisonous if ingested or inhaled. Exposure to a triazine (simazine) has caused acute and subacute dermatitis in the former USSR, characterized by erythema, slight edema, moderate pruritus, and burning lasting 4 to 5 days. $LD_{50}$ in range (rat) 1780-7000 mg/kg.
**Long Term Exposure:** May cause lung irritation and damage. May cause skin allergy. Contact with some triazine compounds (such as atrazine) may increase risks for tumors known to be associated with hormonal factors. These have been observed in both animals and human beings, and are consistent with the known effects on the hypothalamic pituitary gonadal axis. Repeated exposure may cause weight loss and reduced red blood cell count. May be mutagenic.
**Points of Attack:** Liver, lungs, skin.
**Medical Surveillance:** Before beginning employment and at regular times after that, for those with frequent or potentially high exposures, the following is recommended: Lung function tests. Consider chest x-ray following acute overexposure. Evaluation by a qualified allergist. Examination of the nervous system.
**First Aid:** If this chemical gets into the eyes, remove any contact lenses at once and irrigate immediately for at least 15 minutes, occasionally lifting upper and lower lids. Seek medical attention immediately. If this chemical contacts the skin, remove contaminated clothing and wash immediately with soap and water. Seek medical attention immediately. If this chemical has been inhaled, remove from exposure, begin rescue breathing (using universal precautions) if breathing has stopped, and CPR if heart action has stopped. Transfer promptly to a medical facility. When this chemical has been swallowed, get medical attention. Give large quantities of water or milk and induce vomiting. Do not make an unconscious person vomit.
**References:**
- California Environmental Protection Agency "Chemical List of Lists," Sacramento CA (February 1997)

# Methanearsonic Acid

**Use Type:** Herbicide
**CAS Number:** 124-58-3
**Formula:** $CH_5AsO_3$
**Synonyms:** Arsonic acid, methyl-; MAA; Methylarsinic acid; Methylarsonic acid; Monomethylarsonic acid; MSMA
**Chemical Class:** Arsenical (organo-)
**EPA/OPP PC Code:** 128876
**ICSC Number:** 0755
**RTECS Number:** PA1575000
**EEC Number:** 033-002-00-5
**U.S. Maximum Allowable Residue Levels for Methanearsonic Acid di- and mono- sodium salts (40 CFR 180.289):**

| CROP | ppm |
|---|---|
| Citrus fruit | 0.35 |
| Cottonseed | 0.7 |
| Cottonseed hulls | 0.9 |

**Carcinogen/Hazard Classifications**
**U.S. EPA Carcinogens:** Not likely a carcinogen; Group B2, probable carcinogen (as parent chemical: cacodylic acid)
**California Prop. 65:** Carcinogen (as parent chemical: cacodylic acid)
**IARC:** Group 1, known carcinogen
**WHO Acute Hazard:** Class III, slightly hazardous
**Regulatory Authority:**
- AB 1803-Well Monitoring Chemical (CAL)
- Actively registered pesticide in California.
- Clean Air Act: Hazardous Air Pollutants (Title I, Part A, Section 112) as arsenic compounds.
- Clean Water Act: Toxic Pollutant (Section 401.15) as arsenic and compounds.
- RCRA Section 261 Hazardous Constituents, waste number not listed.
- EPCRA Section 304 RQ: CERCLA, 1 lb (0.454 kg)
- EPCRA (Section 313): Includes any unique chemical substance that contains arsenic as part of that chemical's infrastructure. Form R *de minimis* concentration reporting level: 1.0% (organic arsenic)
- Marine pollutant (49CFR, Subchapter 172.101, Appendix B) arsenical pesticides liquid, toxic, flammable, n.o.s.

- AB 2588-Air Toxics "Hot Spots" Chemicals (CAL) as arsenic compounds
- Permissible Exposure Limits for Chemical Contaminants (CAL/OSHA) as arsenic compounds
- The "Director's List" (CAL/OSHA) as arsenic compounds

*Description:* Crystalline solid. Highly soluble in water. Molecular weight = 140.01.

*Incompatibilities:* A strong acid. Incompatible with caustics, ammonia, amines, amides, organic anhydrides, isocyanates, vinyl acetate, alkylene oxides, epichlorohydrin. May not be compatible with nitrates. Moisture may cause hydrolysis or other forms of decomposition. Attacks metals in the presence of moisture.

*Permissible Exposure Limits in Air:* The OSHA[2] PEL for organic arsenic compounds is 0.5 mg/m$^3$ TWA. For an 8-hour workshift. There is no NIOSH[2] recommendation. The ACGIH[1] TLV is 0.2 mg/m$^3$ TWA.

*Determination in Air:* Filter; Reagent: Ion chromatography/hydride atomic absorption: NIOSH IV [#5022, Arsenic, organo-].[18]

*Permissible Concentration in Water:* No criteria set. Runoff from spills or fire control may cause water pollution. Toxic pollutant designated pursuant to section 307 (a) (1) of the Clean Water Act and is subject to effluent limitations (arsenic and inorganic and organic arsenic) [40 CFR 401.15 (7/1/87)]

*Determination in Water:*

*Routes of Entry:* Inhalation, eyes, ingestion, skin contact

*Harmful Effects and Symptoms*

*Short Term Exposure:* Symptoms of arsenic poisoning usually appear one-half to one hour after ingestion, but may be delayed many hours. Symptoms include a sweetish metallic taste and garlicky odor; difficulty in swallowing; abdominal pain; vomiting and painful diarrhea; dehydration, thirst, and cramps; dizziness, stupor, and delirium, rapid heart beat, headache, skin disorders, and coma.

*Long Term Exposure:* Chronic exposure to arsenic compounds can cause dermatitis and digestive disorders. Renal damage may develop.

*Points of Attack:* Skin and kidneys.

*Medical Surveillance:* Kidney function tests. Examination by a qualified allergist.

*First Aid:* If this chemical gets into the eyes, remove any contact lenses at once and irrigate immediately for at least 15 minutes, occasionally lifting upper and lower lids. Seek medical attention immediately. If this chemical contacts the skin, remove contaminated clothing and wash immediately with soap and water. Seek medical attention immediately. If this chemical has been inhaled, remove from exposure, begin rescue breathing (using universal precautions) if breathing has stopped, and CPR if heart action has stopped. Transfer promptly to a medical facility. When this chemical has been swallowed, get medical attention. *Do not induce vomiting when formulations containing petroleum solvents are ingested.* Otherwise, give large quantities of water and induce vomiting. Do not make an unconscious person vomit.

*References:*
- International Programme on Chemical Safety (IPCS), "Health and Safety Guide, Dimethylarsenic Acid, Methanearsonic Acid, and Salts," Geneva, Switzerland (1992).
- U.S. Environmental Protection Agency, Office of Pesticide Programs, Pesticide Residue Limits, "Methanearsonic Acid di- and mono-sodium salts," 40 CFR 180.289. http://www.setonresourcecenter.com/40CFR/Docs/wcd0004c/wcd04cfe.asp
- California Environmental Protection Agency "Chemical List of Lists," Sacramento CA (February 1997)

## Methamidophos (ANSI)

*Use Type:* Insecticide and miticide
*CAS Number:* 10265-92-6
*Formula:* $C_2H_8NO_2PS$; $CH_3OP(O)(NH_2)SCH_3$
*Alert:* A Restricted Use Pesticide (RUP). Human toxicity (long-term): High.

*Synonyms:* *O,S*-Dimethyl ester of amide of amidothioate; *O,S*-Dimethylphosphoramidothioate; ENT 27,396; Metamidofos (Spanish); Metamidofos estrella; NSC 190987; Thiophosphorsaeure-*O,S*-dimethylesteramid (German)

*Trade Names:* ACEPHATE-MET®; BAY 71625®, Bayer CropScience (Germany); BAYER 71628®, Bayer CropScience (Germany); CHEVRON 9006®, Chevron Phillips Chemical (USA); CHEVRON ORTHO 9006®, Chevron Phillips Chemical (USA); FILITOX®; GS-13005®; HAMIDOP®; METAFOS®, Milenia Agro Ciencias (Brazil); MONITOR®, Bayer CropScience (Germany); MTD®; NITOFOL®; NURATRON®; ORTHO 9006®, The Scotts Company (USA); PATROLE®; PILLARON®; SRA 5172®; SUPRACIDE®; SWIPE®; TAHMABON®; TAMARON®; VITARON®

*Producers:* Bayer CropScience (Germany); Biesterfeld Siemsgluess International. GmbH (Germany); Chevron Phillips Chemical (USA); China Chemical (China); Hunan Tianyu Pesticide Chemical Group (China); Jiangmen Pesticide Factory (China); Ki-Hara Chemicals Ltd. (UK); Milenia Agro Ciencias (Brazil); Saeryung Chemicals (South Korea); The Scotts Company (USA); Shandong Huayang Pesticide Group (China); Shenzhen Guomeng Industry Co., Ltd. (China); Sigma-Aldrich Laborchemikalien (Germany); Sinon (Taiwan); Valent BioSciences Corporation and Valent USA (USA); Zago Asia Ltd. (Singapore)

*Chemical Class:* Organophosphorus
*EPA/OPP PC Code:* 101201
*California DPR Chemical Code:* 1697
*ICSC Number:* 0176

## Methamidophos

*RTECS Number:* TB4970000
*EEC Number:* 015-095-00-4
*EINECS Number:* 233-606-0

*Uses:* Methamidophos is a highly active, systemic, residual organophosphate insecticide/acaricide/avicide with contact and stomach action. Its mode of action in insects and mammals is by decreasing the activity of an enzyme important for nervous system function called acetylcholinesterase. This enzyme is essential in the normal transmission of nerve impulses. Methamidophos is a potent acetylcholinesterase inhibitor. It is effective against chewing and sucking insects and is used to control aphids, flea beetles, worms, whiteflies, thrips, cabbage loopers, Colorado potato beetles, potato tubeworms, armyworms, mites, leafhoppers, and many others. Crop uses include broccoli, Brussel sprouts, cauliflower, grapes, celery, sugar beets, cotton, tobacco, and potatoes. It is used abroad for many vegetables, hops, corn, peaches, and other crops. Commercially available formulations include soluble concentrate, emulsifiable concentrate, wettable powder, granules, ultra-low volume spray and water miscible spray concentrate. Generally, methamidophos is not considered phytotoxic if used as directed, but defoliation has occurred when applied as foliar spray to deciduous fruit. It is compatible with many other pesticides, but do not use with alkaline materials. Methamidophos is slightly corrosive to mild steel and copper alloys. This compound is highly toxic to mammals, birds, and bees. Do not graze treated areas, and be sure to wear protective clothing including respirator, chemical goggles, rubber gloves, and impervious protective clothing.

*Human toxicity (long-term)$^{(77)}$:* High–7.00 ppb, Health Advisory

*Fish toxicity (threshold)$^{(77)}$:* Low–165.16992 ppb, MATC (Maximum Acceptable Toxicant Concentration)

*U.S. Maximum Allowable Residue Levels for Methamidophos (40 CFR 180.315):*

| CROP | ppm |
|---|---|
| Broccoli | 1 |
| Brussels sprouts | 1 |
| Cabbage | 1 |
| Cauliflower | 1 |
| Cotton, undelinted seed | 0.1 |
| Cucumber | 1 |
| Eggplant | 1 |
| Lettuce | 1 |
| Melon | 0.5 |
| Pepper | 1 |
| Potato | 0.1 |
| Tomato | 1 |

*Carcinogen/Hazard Classifications*
**U.S. EPA Carcinogens:** Group E, unlikely carcinogen; parent acephate is Group C, possible carcinogen
**Label Signal Word:** DANGER–POISON
**WHO Acute Hazard:** Class 1b, highly hazardous

*Regulatory Authority:*
- Superfund/EPCRA 40CFR355, Appendix B Extremely Hazardous Substances: TPQ = 100/10,000 lb (45.4/4,540 kg)
- Superfund/EPCRA 40CFR302.4 RQ: EHS, 1 lb (0.454 kg)
- AB 1803-Well Monitoring Chemical (CAL)
- EPA/SARA 302 (EPCRA) Extremely hazardous substances
- The "Director's List" (CAL/OSHA)
- Actively registered pesticide in California.
- U.S. DOT Inhalation Hazard Chemicals as organophosphates
- U.S. DOT Regulated Marine Pollutant (49CFR172.101, Appendix B)

*Description:* Off-white crystalline solid. Soluble in water; solubility = 2.5 x $10^6$ ppm @ 20°C. Molecular weight = 141.14. Melting/Freezing point = 45°C. Vapor pressure = 8 x $10^{-4}$ mmHg @ 20°C; 3.75 x $10^{-4}$ mmHg @ 30°C; 2 x $10^{-3}$ Pa @ 20°C. Hazard Identification (based on NFPA-704 M Rating System): Health 3, Flammability 1, Reactivity 0. Log $K_{ow}$= Negative. Unlikely to bioaccumulate in marine organisms. Often used as a liquid in a carrier solvent which may change physical and toxicological properties.

*Incompatibilities:* Incompatible with strong acids or alkali. Attacks mild steel and copper-containing alloys (technical grade).

*Permissible Exposure Limits in Air:* No standards set.

*Determination in Air:* OSHA versatile sampler-2; Toluene/Acetone; Gas chromatography/Flame photometric detection for sulfur, nitrogen, or phosphorus; NIOSH Method IV Method #5600, Organophosphorus pesticides.$^{(18)}$

*Permissible Concentration in Water:* No criteria set, but runoff from spills or fire control may cause water pollution.

*Routes of Entry:* Inhalation, ingestion and skin contact.

*Harmful Effects and Symptoms*

*Short Term Exposure:* Irritates the eyes. Organic phosphorus insecticides are absorbed by the skin, as well as by the respiratory and gastrointestinal tracts. They are cholinesterase inhibitors. Symptoms of exposure include headache, giddiness, blurred vision, nervousness, weakness, nausea, cramps, diarrhea, and discomfort in the chest. Signs include sweating, tearing, salivation, vomiting, cyanosis, convulsions, coma, loss of reflexes and loss of sphincter control. This material is highly toxic; the $LD_{50}$ (oral, rat) = 7.5 mg/kg. Acute exposure to methamidophos may produce the following sings an Symptoms: pinpoint pupils, blurred vision, headache, dizziness, muscle spasms, and profound weakness. Vomiting, diarrhea, abdominal pain, seizures, and coma may also occur. The heart rate may decrease following oral exposure or increase following dermal exposure. Chest pain may be noted. Hypotension (low blood pressure) may be noted, although hypertension (high blood pressure) is not uncommon. Respiratory symptoms include dyspnea (shortness of breath), respiratory depression, and respiratory paralysis. Psychosis may occur.

***Long Term Exposure:*** The substance may have effects on the nervous system, resulting in delayed neuropathy. Cholinesterase inhibitor; cumulative effect is possible. This chemical may damage the nervous system with repeated exposure, resulting in convulsions, respiratory failure. May cause liver damage.

***Points of Attack:*** Respiratory system, lungs, central nervous system, cardiovascular system, skin, eyes, plasma and red blood cell cholinesterase.

***Medical Surveillance:*** Before employment and at regular times after that, the following are recommended: Plasma and red blood cell cholinesterase levels (tests for the enzyme poisoned by this chemical). If exposure stops, plasma levels return to normal in 1-2 weeks while red blood cell levels may be reduced for 1-3 months. When acetylcholinesterase enzyme levels are reduced by 25% or more below preemployment levels, risk of poisoning is increased, even if results are in lower ranges of "normal." Reassignment to work not involving organophosphate or carbamate pesticides is recommended until enzyme levels recover. If symptoms develop or overexposure occurs, repeat the above tests as soon as possible and get an exam of the nervous system. Also consider complete blood count. Consider chest x-ray following acute overexposure. Do not drink any alcoholic beverages before or during use. Alcohol promotes absorption of organophosphates.

***First Aid:*** Speed in removing material from eyes and skin is of extreme importance. If this chemical gets into the eyes, remove any contact lenses at once and irrigate immediately for at least 15 minutes, occasionally lifting upper and lower lids. Seek medical attention immediately. If this chemical contacts the skin, remove contaminated clothing and wash immediately with soap and water. Shampoo hair promptly if contaminated. Seek medical attention immediately. If this chemical has been inhaled, remove from exposure, begin rescue breathing (using universal precautions) if breathing has stopped, and CPR if heart action has stopped. Transfer promptly to a medical facility. When this chemical has been swallowed, get medical attention. Give large quantities of water and induce vomiting. Do not make an unconscious person vomit. Obtain authorization of an antidote or performance of other invasive procedures. The effects may be delayed. Medical observation recommended.

*Note to physician or authorized medical personnel:* 1,1'-trimethylenebis(4-formylpyridinium bromide)dioxime (a.k.a TMB-4 Dibromide and TMV-4) has been used as an antidote for organophosphate poisoning.

***References:***
- EXTOXNET, Extension Toxicology Network, "Pesticide Information Profile, Methamidophos," Oregon State University, Corvallis, OR. http://extoxnet.orst.edu/pips/methamid.htm
- U.S. Environmental Protection Agency, Office of Pesticide Programs, Pesticide Residue Limits, "Methamidophos," 40 CFR 180.315, http://www.epa.gov/pesticides/food/viewtols.htm
- U.S. Environmental Protection Agency, "Chemical Profile: Methamidophos," Washington, DC, Chemical Emergency Preparedness Program (Nov. 30, 1987).

# Metham-sodium

***Use Type:*** Fungicide; nematicide, herbicide, soil fumigant, algicide
***CAS Number:*** 137-42-8; 6734-80-1 (dihydrate)
***Formula:*** $C_2H_4NNaS_2$; $C_2H_4NNaS_2 \cdot 2H_2O$ (dihydrate)
***Synonyms:*** Carbam; Carbamic acid, methyldithio-, monosodium salt; Carbamic acid, *N*-methyldithio-, monosodium salt; Carbamic acid, *N*-methyldithio-, sodium salt; Carbamodithioic acid, methyl-, monosodium salt; Carbam, sodium salt; Carbathion; Carbathione; Carbation; Carbothion; Diethylamino-2,6-acetoxylidide; Methan-sodium; *N*-Methylaminodithioformic acid sodium salt; *N*-Methylaminomethanethionothiolic acid sodium salt; Methylcarbamodithioic acid sodium salt; Methyldithiocarbamic acid, sodium salt; SMDC; Sodium metam; Sodium metham; Sodium *N*-methylaminodithioformate; Sodium *N*-methylaminomethanethionothiolate; Sodium methylcarbamodithioate; Sodium methyldithiocarbamate; Sodium *N*-methyldithiocarbamate; Sodium monomethyldithiocarbamate

***Trade Names:*** A7-VAPAM®; BASAMID-FLUID®; BUSAN®, Buckman Laboratories (USA); CHAP-FUME®; HERBATIM (dihydrate)®; KARBATION®; KARBATION (dihydrate)®; MAPOSOL®; MAPOSOL (dihydrate)®; METACIDE®, AMVAC Chemicals (USA); METAM (dihydrate)®, AMVAC Chemicals (USA); METAM-FLUID BASF®; METHAM DIHYDRATE (dihydrate)®; MONAM (dihydrate)®; N-869®; N 869 (dihydrate)®; NEMATIN®; SECTAGON®; SISTAN®; SMDC (dihydrate)®; SOLASAN 500®; SOLESAN 500®; SOMETAM®; TRAPEX®; TRIMATON (dihydrate)®; TRIMATRON®; VAPAM®; VAPAM (dihydrate)®; VAPOROOTER (dihydrate)®; VDM®; VPM (dihydrate)®; VPM® Fungicide; VPN®; WOODFUME VAPAM®

***Producers:*** AMVAC Chemicals (USA); Buckman Laboratories (USA); Drexel Chemical (USA); Limin Chemical (China); Micro-Flo (USA) J.R. Simplot (USA)
***Chemical Class:*** Dithiocarbamate
***EPA/OPP PC Code:*** 039003
***California DPR Chemical Code:*** 616
***RTECS Number:*** FC2100000
***Uses:*** A general soil biocide that is used to control weeds, weed seeds, roots, tubers, rhizomes, insects, nematodes and soil inhabiting fungi on all food and non-food crops. Also

used as a pre-planting fumigant in seed beds, vine crops, fruit trees, row crops, flowers and ornamentals. Environmental friendly; it breaks down after two weeks into carbon dioxide, water, and sodium and sulfur in small amounts.

*Human toxicity (long-term)*[77]: High–1.76768 ppb, CHCL (Chronic Human Carcinogen Level)

*Fish toxicity (threshold)*[77]: Intermediate–60.02431 ppb, MATC (Maximum Acceptable Toxicant Concentration)

***Carcinogen/Hazard Classifications***

**U.S. EPA Carcinogens:** Group B2, probable carcinogen

***California Prop. 65:*** Carcinogen and developmental toxin

**Label Signal Word:** DANGER

**WHO Acute Hazard:** Class II, moderately hazardous

***Regulatory Authority:***
- EPA 40 CFR 372.65, Specific Toxic Chemical Listings
- Actively registered pesticide in California.
- RCRA Section 261 Hazardous Constituents
- EPA Hazardous Waste Number (RCRA No.): U384
- EPCRA Section 313 Form R *de minimis* concentration reporting level: 1.0%
- Marine pollutant (49CFR, Subchapter 172.101, Appendix B).

***Description:*** Colorless to white crystalline solid. Readily soluble in water; solubility = 722 g/L @ 20°C; Unpleasant odor, similar to that of carbon disulfide. Molecular weight = 129.18, 165.21 (dihydrate). Melting/Freezing point decomposes. Boiling point = 110°C (technical product). Vapor pressure = 20 mmHg @ 20°C.

***Incompatibilities:*** Aqueous solutions are corrosive to aluminum, copper, brass, and zinc.

***Determination in Air:*** Filter; none; Gravimetric; NIOSH IV [Particulates NOR; #0500 (total), #0600 (respirable)].[18]

***Permissible Concentration in Water:*** No criteria set. Runoff from spills or fire control may cause water pollution.

***Harmful Effects and Symptoms***

*Short Term Exposure:* Low levels of toxicity. Concentrated solutions are slightly corrosive to eyes and mucous membranes. Dust inhalation can cause irritation of the respiratory system with sneezing. Eye contact can cause irritation, watering, pain, and inflammation of the eyelids. Skin contact can cause irritation and minor ulceration. Ingestion can cause nausea, vomiting, fever, muscle twitching, seizure, rapid respiration, slow heart beat. Severe exposure may result in death.

***Points of Attack:*** Respiratory system, central nervous system, cardiovascular system, skin and eyes.

***Medical Surveillance:*** Medical observation is recommended for 24 to 48 hours after breathing overexposure, as pulmonary edema may be delayed. As first aid for pulmonary edema, consider administering a corticosteroid spray. Cigarette smoking may exacerbate pulmonary injury and should be discouraged for at least 72 hours following exposure. Before employment and at regular times after that, the following are recommended: If symptoms develop or overexposure occurs, repeat the above tests as soon as possible and get an exam of the nervous system. Also consider complete blood count. Consider chest x-ray following acute overexposure.

***First Aid:*** If this chemical gets into the eyes, remove any contact lenses at once and irrigate immediately for at least 15 minutes, occasionally lifting upper and lower lids. Seek medical attention immediately. If this chemical contacts the skin, remove contaminated clothing and wash immediately with soap and water. Seek medical attention immediately. If this chemical has been inhaled, remove from exposure, begin rescue breathing (using universal precautions) if breathing has stopped, and CPR if heart action has stopped. Transfer promptly to a medical facility. When this chemical has been swallowed, get medical attention. If victim is conscious and able to swallow, have victim drink 4 to 8 ounces of water. Do not induce vomiting. *Note to physician or authorized medical personnel*: Medical observation is recommended for 24 to 48 hours after breathing overexposure, as pulmonary edema may be delayed. As first aid for pulmonary edema, consider administering a corticosteroid spray. Cigarette smoking may exacerbate pulmonary injury and should be discouraged for at least 72 hours following exposure.

***References:***
- "In-Row Sprayblade Fumigation with Metam Sodium to Control Weeds and Diseases," *Skagit Veg Trials*, Anderson, W.C. 'Andy,' and Haglund, William A., Washington State University, Mount Vernon, WA. http://www.mtvernon.wsu.edu/SkagitVegTrials/methamsodium.html
- California Environmental Protection Agency "Chemical List of Lists," Sacramento CA (February 1997)

# Methazole (ANSI)

*Use Type:* Herbicide
*CAS Number:* 20354-26-1
*Formula:* $C_9H_6Cl_2N_2O_3$
*Synonyms:* Caswell No. 549AA; 2-(3,4-Dichlorophenyl)-4-methyl-1,2,4-oxadiazolidinedione; 2-(3,4-Dichlorophenyl)-4-methyl-1,2,4-oxadiazolidine-3,5-dione; 2-(3,4-Dichlorophenyl)-4-methyl-1,2,4-oxadiazolidine-3,5-dione: 1,2,4-Oxadiazolidine-3,5-dione, 2-(3,4-dichlorophenyl)-4-methyl-; Oxydiazol
*Trade Names:* BIOXONE®; CHLORMETHAZOLE®; MEZOPUR®; PAXILON®; PROBE®, Sandoz Agro (USA), canceled; TUNIC®; VCS 438®
*EPA/OPP PC Code:* 106001; (323300 and 549200 old EPA code numbers)
*California DPR Chemical Code:* 2673
*RTECS Number:* RO0835000

*Uses:* Not registered in the U.S. Methazole is a herbicide used on cotton and was voluntarily withdrawn from the U.S. market in 1994.

*Human toxicity (long-term)*[77]: Low–52.50 ppb, Health Advisory

*Fish toxicity (threshold)*[77]: Very low–592.78286 ppb, MATC (Maximum Acceptable Toxicant Concentration)

*Carcinogen/Hazard Classifications*

**California Prop. 65:** Developmental toxin

**TRI Developmental Toxin:** Developmental toxin

**Label Signal Word:** CAUTION or WARNING

*Regulatory Authority:*
- EPA 40 CFR 372.65, Specific Toxic Chemical Listings
- EPCRA Section 313 Form R *de minimis* concentration reporting level: 1.0%
- California 65: Chemicals Known to the State to Cause Cancer or Reproductive Toxicity

*Description:* Light-tan solid. Practically insoluble in water. Melting/Freezing point = 123°C. Vapor pressure = 1 x $10^{-6}$ mmHg @ 20°C.

*Determination in Air:* Filter; none; Gravimetric; NIOSH IV [Particulates NOR; #0500 (total), #0600 (respirable)].[18]

*Permissible Concentration in Water:* No criteria set. Runoff from spills or fire control may cause water pollution.

*Harmful Effects and Symptoms*

*Short Term Exposure:* Contact with eyes or skin may cause irritation or injury. Inhalation should be avoided; use NIOSH-approved air purifying respirators for pesticides. May be harmful if swallowed.

*First Aid:* If this chemical gets into the eyes, remove any contact lenses at once and irrigate immediately for at least 15 minutes, occasionally lifting upper and lower lids. Seek medical attention immediately. If this chemical contacts the skin, remove contaminated clothing and wash immediately with soap and water. Seek medical attention immediately. If this chemical has been inhaled, remove from exposure, begin rescue breathing (using universal precautions) if breathing has stopped, and CPR if heart action has stopped. Transfer promptly to a medical facility. When this chemical has been swallowed, get medical attention. Give large quantities of water and induce vomiting. Do not make an unconscious person vomit. Do not induce vomiting when formulations containing petroleum solvents are ingested.

*References:*
- California Environmental Protection Agency "Chemical List of Lists," Sacramento CA (February 1997)

# Methidathion (ANSI)

*Use Type:* Insecticide
*CAS Number:* 950-37-8
*Formula:* $C_6H_{11}N_2O_4PS_3$

*Alert:* A Restricted Use Pesticide (RUP). There are no residential uses for methidathion. Human toxicity (long-term): High

*Synonyms:* S-(2,3-Dihydro-5-methoxy-2-oxo-1,4,4-thiadiazol-3-methyl); O,O-Dimethyl-S-(2-methoxy-1,4,4-thiadiazole-5-(4H)-onyl-(4)-methyl)-dithiophosphat (German); O,O-Dimethyl-S-(2-methoxy-1,3,4-thiadiazole-5(4H)-onyl-(4)-methyl)-phosphorodithioate; O,O-Dimethyl-S-((2-methoxy-1,3,4(4H)-thiadiazol-5-on-4-yl)-methyl)dithiofosfaat (Dutch); O,O-Dimetil-S-((2-metossoi-1,3,4(4H)-thiadiazaol-5-on-4-il)-metil)-ditifosfato (Italian); O,O-Dimethyl phosphorodithioate S-ester with 4-(mercaptomethyl)2-methoxy-.*delta*.-1,3,4-thiadiazolin-5-one; DMTP (Japan); ENT 27,193; Metidation (Spanish); S-((5-Methoxy-2-oxo-1,3,4-thiadiazol-3(2H)-yl)methyl)-O,O-dimethyl phosphordithioate

*Trade Names:* CIBA-GEIGY® GS 13005®, Ciba-Geigy (Switzerland); FISONS NC® 2964; GEIGY® 13005, Ciba-Geigy (Switzerland); GS-13005®, Ciba-Geigy (Switzerland); SOMONIC®; SOMONIL®; SURPRACIDE®, Gowan Company (USA); SUPRATHION®, Makhteshim Agan (Israel); ULTRACIDE®, Syngenta (Switzerland)

*Producers:* Agsin (Singapore); Ciba-Geigy (Switzerland); Cognis (Germany); Gowan Company (USA); Ki-Hara Chemicals Ltd. (UK); Makhteshim Agan (Israel); Nissan Chemical Industry (Japan); Shenzhen Guomeng Industry Co., Ltd. (China); Sigma-Aldrich Laborchemikalien (Germany); Syngenta (Switzerland)

*Chemical Class:* Organophosphate
*EPA/OPP PC Code:* 100301
*California DPR Chemical Code:* 1689
*RTECS Number:* TE2100000
*EINECS Number:* 213-449-4

*Uses:* Methidathion is a non-systemic organophosphate insecticide and acaricide with stomach and contact action. The compound is used to control a variety of insects and mites in many crops such as nuts, artichokes, olives, cotton, fruits, vegetables, tobacco, alfalfa, and sunflowers, and also in greenhouses and on rose cultures. It is especially useful against scale insects. In Canada, the year 2000 was the last year for any significant use of methidathion. Crops in Canada on which it had been applied were canola, mustard, sunflower, alfalfa, apple, blueberry, cherry and potato.

*Human toxicity (long-term)*[77]: High–1.05 ppb, Health Advisory

*Fish toxicity (threshold)*[77]: Extra high–0.15019 ppb, MATC (Maximum Acceptable Toxicant Concentration)

*U.S. Maximum Allowable Residue Levels for Methidathion (40 CFR 180.298):*

| CROP | ppm |
|---|---|
| Alfalfa | 12 |
| Alfalfa, hay | 12 |
| Almond, hulls | 6 |
| Artichoke, globe | 0.05 |
| Cotton, undelinted seed | 0.2 |

| | |
|---|---|
| Fruit, citrus, except mandarin | 2 |
| Fruit, pome | 0.05 |
| Fruit, stone, except cherry | 0.05 |
| Grass | 12 |
| Grass, hay | 12 |
| Kiwifruit | 0.1 |
| Longan | 0.1 |
| Mango | 0.05 |
| Nuts | 0.05 |
| Olive | 0.05 |
| Peach | 0.05 |
| Pecan | 0.05 |
| Safflower, seed | 0.5 |
| Sorghum, forage | 2 |
| Sorghum, grain, grain | 0.2 |
| Sorghum, grain, stover | 2 |
| Starfruit | 0.1 |
| Sugar apple | 0.2 |
| Sunflower, seed | 0.5 |
| Tangerine | 6 |
| Walnut | 0.05 |

*Carcinogen/Hazard Classifications*
**U.S. EPA Carcinogens:** Group C, possible carcinogen
**Label Signal Word:** DANGER or WARNING
**WHO Acute Hazard:** Class 1b, highly hazardous
*Regulatory Authority:*
- Banned or Severely Restricted (Philippines) (UN)[13]
- EPA/SARA 302 (EPCRA) Extremely hazardous substances
- The "Director's List" (CAL/OSHA)
- Actively registered pesticide in California.
- Superfund/EPCRA 40CFR355, Appendix B Extremely Hazardous Substances: TPQ = 500/10,000 lb (227/4,540 kg)
- Superfund/EPCRA 40CFR302.4 RQ: EHS, 1 lb (0.454 kg)
- U.S. DOT Regulated Marine Pollutant (49CFR172.101, Appendix B)
- Extremely Hazardous Substance (EPA-SARA) (TPQ = 500)[7]
- U.S. DOT Inhalation Hazard Chemicals as organophosphates

*Description:* Colorless crystalline solid. Very slightly soluble in water; solubility = 225 ppm @ 20°C. Molecular weight = 302.35. Melting/Freezing point = 39–40°C. Vapor pressure = $3.4 \times 10^{-6}$ mmHg @ 20°C; 0.186 mPa @ 20°C.
*Incompatibilities:* Reacts with strong oxidizers. When heated to decomposition, this chemical forms toxic oxides of nitrogen, phosphorus, and sulfur.
*Permissible Exposure Limits in Air:* No standards set.
*Permissible Concentration in Water:* No criteria set, but runoff from spills or fire control may cause water pollution.
*Routes of Entry:* Inhalation, ingestion, skin and/or eye contact. Absorbed through the skin.

*Harmful Effects and Symptoms*
*Short Term Exposure:* This material is poisonous to humans. Its toxic effects are by action on the nervous system. Organic phosphorus insecticides are absorbed by the skin, as well as by the respiratory and gastrointestinal tracts. They are cholinesterase inhibitors. Symptoms of exposure include headache, giddiness, blurred vision, nervousness, weakness, nausea, cramps, diarrhea, and discomfort in the chest. Signs include sweating, tearing, salivation, vomiting, cyanosis, convulsions, coma, loss of reflexes and loss of sphincter control. Human volunteers ingesting 0.11 mg/kg/day for 6 weeks had no clinical effects. The $LD_{50}$ (oral, rat) = 20 mg/kg (highly toxic). Symptoms are similar to parathion poisoning and may include nausea, vomiting, abdominal cramps, diarrhea, excessive salivation, headache, dizziness, giddiness, weakness, muscle twitching, difficult breathing, sensation of tightness of chest, blurring or dimness of vision, and loss of muscle coordination. Death may occur from failure of the respiratory center, paralysis of the respiratory muscles, intense bronchoconstriction, or all three. Delayed pulmonary edema may occur after inhalation.
*Long Term Exposure:* Cholinesterase inhibitor; cumulative effect is possible. This chemical may damage the nervous system with repeated exposure, resulting in convulsions, respiratory failure. May cause liver damage.
*Points of Attack:* Respiratory system, lungs, central nervous system, cardiovascular system, skin, eyes, plasma and red blood cell cholinesterase.
*Medical Surveillance:* Medical observation is recommended for 24 to 48 hours after breathing overexposure, as pulmonary edema may be delayed. Before employment and at regular times after that, the following are recommended: Plasma and red blood cell cholinesterase levels (tests for the enzyme poisoned by this chemical). If exposure stops, plasma levels return to normal in 1-2 weeks while red blood cell levels may be reduced for 1-3 months. When acetylcholinesterase enzyme levels are reduced by 25% or more below preemployment levels, risk of poisoning is increased, even if results are in lower ranges of "normal." Reassignment to work not involving organophosphate or carbamate pesticides is recommended until enzyme levels recover. If symptoms develop or overexposure occurs, repeat the above tests as soon as possible and get an exam of the nervous system. Also consider complete blood count. Consider chest x-ray following acute overexposure. Do not drink any alcoholic beverages before or during use. Alcohol promotes absorption of organophosphates.
*First Aid:* **Treatment for organophosphate poisoning consists of thorough decontamination, cardiorespiratory support, and administration of the antidotes atropine and pralidoxime. In cases of severe poisoning, diazepam, an anticonvulsant, should also be administered. Antidotes should be administered as prevention even if the diagnosis is in doubt.** *Eyes:* Speed in removing material from eyes and skin is of extreme importance. Eye

contact can cause dangerous amounts of these chemicals to be quickly absorbed through the mucous membrane into the bloodstream. Immediately and gently flush eyes with plenty of warm or cold water (NO hot water) for at least 15 minutes, occasionally lifting the upper and lower eyelids. Get medical aid immediately. *Skin:* Get medical aid. Skin contact can cause dangerous amounts of these chemicals to be absorbed into the bloodstream. Wearing the appropriate PPE equipment and respirator for organophosphate pesticides, immediately flush skin with plenty of soap and water for at least 15 minutes while removing contaminated clothing and shoes. Shampoo hair promptly if contaminated. The removed, contaminated clothing and shoes should be double-bagged and left in Hot Zone for later disposal by hazardous materials experts. Skin may also be decontaminated with diluted hypochlorite solution. *Inhalation:* Get medical aid. Do not contaminate yourself. Wearing the appropriate PPE equipment and respirator for organophosphate pesticides, immediately remove the victim from the contaminated area to fresh air. If the victim is not breathing, administer artificial respiration. Do NOT use mouth-to-mouth resuscitation; use bag/mask apparatus. If breathing is difficult, administer oxygen through bag/mask apparatus until medical help arrives. Do not leave victim unattended. *Ingestion:* Call poison control. Loosen all clothing. Never give anything by mouth to an unconscious person. If victim is *unconscious or having convulsions,* do nothing except keep victim warm. Get medical aid. Transfer promptly to a medical facility. In cases of ingestion, **do not induce vomiting**. If the victim is alert and asymptomatic, administer a slurry of activated charcoal at a dose of 1 g/kg (infant, child, and adult dose). A soda can and straw may be of assistance when offering charcoal to a child. *In some cases you may be specifically instructed by poison control to induce vomiting by way of 2 tablespoons of syrup of ipecac (adult) washed down with a cup of water.* Do NOT give activated charcoal before or with ipecac syrup.

*Note to physician or authorized medical personnel:* Treat cases of respiratory compromise, coma, or excessive pulmonary secretions with respiratory support using protocols and techniques available and within the scope of training. Some cases may necessitate procedures such as endotracheal intubation or cricothyrotomy by properly trained and equipped personnel. When possible, atropine (see *Antidotes*, below) should be given under medical supervision. Patients who are comatose, hypotensive, or having seizures or cardiac arrhythmias should be treated according to advanced life support protocols. *Antidotes:* Two antidotes are administered to treat organophosphate poisoning. Atropine is a competitive antagonist of acetylcholine at muscarinic receptors and is used to control the excessive bronchial secretions which are often responsible for death. Pralidoxime relieves both the nicotinic and muscarine effects of organophosphate poisoning by regenerating acetylcholinesterase and can reduce both the bronchial secretions and the muscle weakness associated with poisoning. The initial intravenous dose of atropine in adults should be determined by the severity of symptoms: An initial adult dose of 1.0 to 2.0 mg or pediatric dose of 0.01 mg/kg (minimum 0.01 mg) should be administered intravenously. If intravenous access cannot be established, atropine may also be given intramuscularly, subcutaneously or via endotracheal tube. Doses should be repeated every 15 minutes until excessive secretions and sweating have been controlled. Once bronchial secretion has been controlled, atropine administration should be repeated whenever the secretions begin to recur. In seriously poisoned patients, very large doses may be required. Alterations of pulse rate and pupillary size should not be used as indicators of treatment adequacy. Pralidoxime should be administered as early in poisoning as possible as its efficacy may diminish when given more than 24 to 36 hours after exposure. Doses are as follows: adult 1.0 g; pediatric 25 to 50 mg/kg. The drug should be administered intravenously over 30 to 60 minutes, but in a life-threatening situation, one-half of the total dose can be given per minute for a total administration time of 2 minutes. Treatment should begin to take effect within 40 minutes with a reduction in symptoms and the amount of atropine necessary to control bronchial secretion. The initial dose can be repeated in 1 hour and then every 8 to 12 hours until the patient is clinically well and no longer requires atropine. If intravenous access cannot be established, pralidoxime may also be given intramuscularly. Early administration of diazepam in addition to the combined atropine and pralidoxime treatment may help prevent the onset of seizures and potential brain and cardiac morphologic damage following high-level organophosphate poisoning.

***References:***

- EXTOXNET, Extension Toxicology Network, "Pesticide Information Profile, Methidation," Oregon State University, Corvallis, OR (June 1996). http://extoxnet.orst.edu/pips/methidat.htm
- U.S. Environmental Protection Agency, Office of Pesticide Programs, Pesticide Residue Limits, "Methidathion," 40 CFR 180.298, http://www.epa.gov/pesticides/food/viewtols.htm
- U.S. Environmental Protection Agency, "Pesticide Reregistration, Methidathion Facts", Washington, DC. (March 2002). http://www.epa.gov/oppsrrd1/REDs/factsheets/methidathion_fs.htm
- New Jersey Department of Health and Senior Services, "Hazardous Substance Fact Sheet, Methidathion," Trenton NJ (July 1999). http://www.state.nj.us/health/eoh/rtkweb/1206.pdf
- U.S. Environmental Protection Agency, "Chemical Profile: Methidathion," Washington, DC, Chemical Emergency Preparedness Program (Nov. 30, 1987).

- Agency for Toxic Substances and Disease Registry, U.S. Department of Health and Human Services, Public Health Service, "Managing Hazardous Materials Incidents," Atlanta, GA (June 2003).
- California Environmental Protection Agency "Chemical List of Lists," Sacramento CA (February 1997).

# Methiocarb

*Use Type:* Acaricide, molluscicide and insecticide
*CAS Number:* 2032-65-7
*Formula:* $C_{11}H_{15}NO_2S$; $C_6H_2(SCH_3)(CH_3)_2OCONHCH_3$
*Alert:* A Restricted Use Pesticide (RUP) except for residential applications.
*Synonyms:* Carbamic acid, methyl-, 3,5-dimethyl-4-(methylthio)phenyl ester; Carbamic acid, *N*-methyl-, 4-(methylthio)-3,5-xylyl ester; Carbamic acid, methyl-, 4-(methylthio)-3,5-xylyl ester; 3,5-Dimethyl-4-methyl mercaptophenyl-*N*-methyl-carbamate; 3,5-Dimethyl-4-(methylthio)phenol methylcarbamate; 3,5-Dimethyl-4-methylthiophenyl *N*-methylcarbamate; 3,5-Dimethyl-4-(methylthio)phenyl methylcarbamate; EBT 25,726; Mercaptodimethur; Methyl carbamic acid 4-(methylthio)-3,5-xylyl ester; 4-Methylmercapto-3,5-dimethylphenyl *N*-methylcarbamate; 4-Methylmercapto-3,5-xylyl methylcarbamate; 4-Methylthio-3,5-dimethylphenyl methylcarbamate; 4-(Methylthio)-3,5-xylyl-*N*-methylcarbamate; 4-(Methylthio)-3,5-xylyl methylcarbamate; Metiocarb (Spanish); Metmercapturon; Phenol, 3,5-dimethyl-4-(Methylthio)-, methylcarbamate; MXMC
*Trade Names:* AI3-25726®; ALCO SLUB"M®, Amvac Chemical (USA), canceled 9/29/1988; B 37344®, Bayer CropScience (Germany); BAY 5024®, Bayer CropScience (Germany); BAY 9026®, Bayer CropScience (Germany); BAY 37344®, Bayer CropScience (Germany); BAYER 37344®, Bayer CropScience (Germany); DCR 736®; DRAZA®; DRAZA G MICROPELLETS®, H 321®; MESUROL®, Bayer CropScience (Germany); Gowan Company (USA); METHIOCARBE®; OMS-93®; PBI SLUG GARD®; PROVADA®, Bayer CropScience (Germany); SD 9228®; SLUG-GETA®, Chevron Chemical (USA), canceled 11/30/1992
*Producers:* Bayer CropScience (Germany); Gowan Company (USA); Sigma-Aldrich Laborchemikalien (Germany)
*Chemical Class:* Carbamate
*EPA/OPP PC Code:* 100501
*California DPR Chemical Code:* 375
*RTECS Number:* FC5775000
*EINECS Number:* 217-991-2
*Uses:* Used to control slugs and snails, soil insects and spider mites in pome fruit, stone fruit, hops, strawberries, potatoes, beets, maize, vegetables and ornamentals. Also used as seed treatment to control fruit flies on maize and bird repellant on berries and cherries. Methiocarb producers deleted all food uses from their product labels between 1989-92. It is a Restricted Use Pesticide (RUP) except for residential applications.
*Human toxicity (long-term)[77]:* Intermediate–35.00 ppb, Health Advisory
*Fish toxicity (threshold)[77]:* Extra high–0.04597 ppb, MATC (Maximum Acceptable Toxicant Concentration)
*Carcinogen/Hazard Classifications*
**U.S. EPA Carcinogens:** Group D, unclassifiable, insufficient data
**Label Signal Word:** CAUTION
**WHO Acute Hazard:** Class 1b, highly hazardous
*Regulatory Authority:*
- Clean Water Act: Section 311 Hazardous Substances/RQ 40CFR117.3 (same as CERCLA, see below)
- EPA Hazardous Waste Number (RCRA No.): P199
- RCRA, 40CFR261, Appendix 8 Hazardous Constituents
- RCRA 40CFR268.48; 61FR15654, Universal Treatment Standards: Wastewater (mg/L), 0.056; Nonwastewater (mg/kg), 1.4
- Superfund/EPCRA 40CFR355, Appendix B Extremely Hazardous Substances: TPQ = 500/10,000 lb (227/4,540 kg)
- Superfund/EPCRA 40CFR302.4 RQ: CERCLA, 10 lb (4.54 kg)
- EPCRA Section 313 Form R *de minimus* concentration reporting level: 1.0%
- AB 1803-Well Monitoring Chemical (CAL)
- EPA/SARA 302 (EPCRA) Extremely hazardous substances
- The "Director's List" (CAL/OSHA)
- Actively registered pesticide in California
- U.S. DOT Regulated Marine Pollutant (49CFR172.101, Appendix B) as mercaptodimethur

*Description:* Methiocarb is a colorless crystalline powder. Vary slightly soluble in water; solubility = 25 ppm @ 20°C. Molecular weight = 225.32. Melting/Freezing point = 117–118°C. Vapor pressure = $1 \times 10^{-4}$ mmHg @ 20°C. Hazard Identification (based on NFPA-704 M Rating System): Health 3, Flammability 1, Reactivity 0. Log $K_{ow}$ = 2.94. Values at or above 3.0 are likely to bioaccumulate in marine organisms.
*Permissible Concentration in Water:* No criteria set, but runoff from spills or fire control may cause water pollution.
*Routes of Entry:* Inhalation, ingestion, skin and/or eye contact.
*Harmful Effects and Symptoms*
*Short Term Exposure:* Contact irritates the skin and eyes. Inhalation will irritate the respiratory tract. As a carbamate insecticide, this compound is a reversible cholinesterase inhibitor and acts on the nervous system. It is classified as very toxic, and the probable oral lethal dose for humans is 50-500 mg/kg or between 1 teaspoon and 1 ounce for a 150

lb adult. Symptoms include salivation, slowed heartbeat, spontaneous urination and defecation, labored breathing, headache, blurred vision, tremor, slight paralysis, and muscle twitching. Exposure to carbamate poisoning can also result in nausea, vomiting, diarrhea, and abdominal pain, convulsions, coma and death. Carbamate insecticides inhibit the cholinesterase activity of enzymes, causing accumulation of acetylcholine at synapses and altering the way in which nervous impulses are transmitted. However, within several hours carbamates spontaneously detach from the enzymes.

*Long Term Exposure:* The substance may have effects on the central nervous system and the liver.

*Points of Attack:* Central nervous system, liver, plasma and red blood cell cholinesterase.

*Medical Surveillance:* Before employment and at regular times after that, the following are recommended: Plasma and red blood cell cholinesterase levels (tests for the enzyme poisoned by this chemical). If exposure stops, plasma levels return to normal in 1-2 weeks while red blood cell levels may be reduced for 1-3 months. When acetylcholinesterase enzyme levels are reduced by 25% or more below preemployment levels, risk of poisoning is increased, even if results are in lower ranges of "normal." Reassignment to work not involving organophosphate or carbamate pesticides is recommended until enzyme levels recover. If symptoms develop or overexposure occurs, repeat the above tests as soon as possible and get an exam of the nervous system. Also consider complete blood count. Consider chest x-ray following acute overexposure. Do not drink any alcoholic beverages before or during use. Alcohol promotes absorption of organophosphates.

*First Aid:* Speed in removing material from eyes and skin is of extreme importance. *Eyes:* Eye contact can cause dangerous amounts of these chemicals to be quickly absorbed through the mucous membrane into the bloodstream. Immediately and gently flush eyes with plenty of warm or cold water (NO hot water) for at least 15 minutes, occasionally lifting the upper and lower eyelids. Get medical aid immediately. *Skin:* Get medical aid. Skin contact can cause dangerous amounts of these chemicals to be absorbed into the bloodstream. Wearing the appropriate PPE equipment and respirator for carbamate pesticides, immediately flush skin with plenty of soap and water for at least 15 minutes while removing contaminated clothing and shoes. Shampoo hair promptly if contaminated; protect eyes. *Ingestion:* Call poison control. Loosen all clothing. Never give anything by mouth to an unconscious person. Get medical aid. Do NOT induce vomiting.* If conscious, alert, and able to swallow, rinse mouth and have victim drink 4 to 8 ounces of water. Check to see if poison control instructs you to use ipecac syrup, otherwise administer slurry of activated charcoal (2 oz in 8 oz of water). If victim is UNCONSCIOUS OR HAVING CONVULSIONS, do nothing except keep victim warm. **In some cases you may be specifically instructed by poison control to induce vomiting* by way of 2 tablespoons of syrup of ipecac (adult) washed down with a cup of water. Do NOT give activated charcoal before or with ipecac syrup. *Inhalation:* Get medical aid. Do not contaminate yourself. Wearing the appropriate PPE equipment and respirator for carbamate pesticides, immediately remove the victim from the contaminated area to fresh air. If the victim is not breathing, administer artificial respiration. Do NOT use mouth-to-mouth resuscitation; use bag/mask apparatus. If breathing is difficult, administer oxygen through bag/mask apparatus until medical help arrives. Do not leave victim unattended. *Note to physician or authorized medical personnel.* Administer atropine, 2 mg (1/30 gr) intramuscularly or intravenously as soon as any local or systemic signs or symptoms of an intoxication are noted; repeat the administration of atropine every 3 to 8 minutes until signs of atropinization (mydriasis, dry mouth, rapid pulse, hot and dry skin) occur; initiate treatment in children with 0.05 mg mg/kg of atropine; repeat at 5 to 10 minute intervals. Watch respiration, and remove bronchial secretions if they appear to be obstructing the airway; intubate if necessary. *Medical note:* 2-PAMCI may be contraindicated in the case of some carbamate poisonings.

*References:*

- U.S. Environmental Protection Agency, "Reregistration Eligibility Decision (RED) Facts, Methiocarb," Washington, DC (February 1994). http://www.epa.gov/REDs/factsheets/0577fact.pdf
- U.S. Environmental Protection Agency, "Chemical Profile: Methiocarb," Washington, DC, Chemical Emergency Preparedness Program (Nov. 30, 1987).
- California Environmental Protection Agency "Chemical List of Lists," Sacramento CA (February 1997).
- New Jersey Department of Health and Senior Services, "Hazardous Substance Fact Sheet, Mercaptodimethur," Trenton NJ (November 1999). http://www.state.nj.us/health/eoh/rtkweb/1165.pdf

# Methomyl (ANSI)

*Use Type:* Insecticide and acaricide
*CAS Number:* 16752-77-5
*Formula:* $C_5H_{10}N_2O_2S$
*Alert:* A Restricted Use Pesticide (RUP)
*Synonyms:* Acetimidic acid, thio-*N*-(Mmthylcarbamoyl)oxy-,methyl ester; Acetimidothioic acid, methyl-*N*-(methylcarbamoyl) ester; ENT 27,341; Ethanimidothic acid, *N*-[(methylamino)carbonyl]; Mesomile; Methyl *N*-[methylamino(carbonyl)oxy]ethanimido)thioate; Methyl-*N*-[methyl(carbamoyl)oxy]thioacetimidate; *S*-Methyl-*N*-(methylcarbamoyloxy)thioacetimidate; 2-Methylthio-propionaldehyd-*O*-(methylcarbamoyl)oxim (German);

Metomil (Italian); Metomilo (Spanish); 3-Thiabutan-2-one,O-(methylcarbamoyl)oxime

*Trade Names:* ACINATE®; AGRINATE®; CIMETLE®, Alcotan Laboratories (Spain); DUPONT 1179®, DuPont Crop Protection (USA); FRAM FLY KILL®; FLYTEK®; IMPROVED BLUE MALRIN SUGAR BAIT®, Wellmark International (USA); IMPROVED GOLDEN MALRIN BAIT®, Wellmark International (USA); INSECTICIDE 1179®, Dupont Crop Protection (USA); KIPSIN®; KUIK®, Rotam Agrochemical (Hong Kong); LANNATE®, Dupont Crop Protection (USA); LANOX 90®; LANOX 216®; METHOMEX®, Makhteshim Agan (Israel); MEMILENE®; METHAVIN®; NU-BAIT II®; NUDRIN®; PILLARMATE®; RENTOKILL®; RENTOKIL FRAM FLY BAIT®; RIDECT®; SD 14999®; SOREX GOLDEN FLY BAIT®; WL 18236®

*Producers:* Makhteshim Agan (Israel); Agro-care Chemical Industry Group (China); Agsin (Singapore); Alcotan Laboratories (Spain); ATOFINA (France); Biesterfeld Siemsgluess International. GmbH (Germany); China Chemical (China); Dupont Crop Protection (USA); Ehrenstorfer, Dr. (Germany); Hebei Huafeng Chemical Group (China); Hokko Chemical Industry (Japan); Jingma Chemicals Ltd. (China); Ki-Hara Chemicals Ltd. (UK); Makhteshim Agan (Israel); Rhone-Poulenc Agro France (France); Rotam Agrochemical (Hong Kong); Saeryung Chemicals (South Korea); Shangdong Baoyuan Chemical (China); Shandong Huayang Pesticide Group (China); Shell Oil (UK); Shenzhen Guomeng Industry Co., Ltd. (China); Sigma-Aldrich Laborchemikalien (Germany); Sinon (Taiwan); Wellmark International (USA); Wuxj Ruize Pesticide (China)

*Chemical Class:* Carbamate
*EPA/OPP PC Code:* 090301
*California DPR Chemical Code:* 383
*ICSC Number:* 0177
*RTECS Number:* AK2975000
*EEC Number:* 006-045-00-2
*EINECS Number:* 240-815-0

*Uses:* Methomyl is broad spectrum insecticide that is particularly effective against organophosphorus resistant pests. It is used as an acaricide to control ticks and spiders. It is used for foliar treatment of vegetable, fruit and field crops, tobacco, cotton, commercial ornamentals, and in and around poultry houses and dairies. It is also used as a fly bait. Methomyl is effective as a "contact insecticide," because it kills target insects upon direct contact, and also as a "systemic insecticide" because of its capability to cause overall "systemic" poisoning in target insects, after it is absorbed and transported throughout the pests that feed on treated plants. It is capable of being absorbed by plants without being "phytotoxic" or harmful, to the plant.

*Human toxicity (long-term)[77]:* Very low–200.00 ppb, Health Advisory

*Fish toxicity (threshold)[77]:* Intermediate–80.25646 ppb, MATC (Maximum Acceptable Toxicant Concentration)

*U.S. Maximum Allowable Residue Levels for Methomyl (40 CFR 180.253 ):*

| CROP | ppm |
|---|---|
| Alfalfa | 10 |
| Apple | 1 |
| Asparagus | 2 |
| Avocado | 2 |
| Barley, grain | 1 |
| Barley, hay | 10 |
| Barley, straw | 10 |
| Bean, dry | 0.1 |
| Bean, forage | 10 |
| Bean, succulent | 2 |
| Beet, garden, tops | 6 |
| Bermudagrass | 10 |
| Bermudagrass, hay | 40 |
| Blueberry | 6 |
| Broccoli | 3 |
| Brussels sprouts | 2 |
| Cabbage | 5 |
| Cabbage, chinese, bok choy | 5 |
| Cauliflower | 2 |
| Celery | 3 |
| Collards | 6 |
| Corn, forage | 10 |
| Corn, grain | 0.1 |
| Corn, stover | 10 |
| Corn, sweet, kernel plus cob with husks removed | 0.1 |
| Cotton, undelinted seed | 0.1 |
| Cucurbits | 0.2 |
| Dandelion, leaves | 6 |
| Endive | 5 |
| Grape | 5 |
| Grapefruit | 2 |
| Hop, dried cones | 12 |
| Kale | 6 |
| Leek | 3 |
| Lemon | 2 |
| Lentil | 0.1 |
| Lettuce | 5 |
| Mint, tops | 2 |
| Mustard greens | 6 |
| Nectarine | 5 |
| Oat, forage | 10 |
| Oat, grain | 1 |
| Oat, hay | 10 |
| Oat, straw | 10 |
| Onion, green | 3 |
| Orange | 2 |
| Parsley | 6 |
| Pea | 5 |
| Pea, field, vines | 10 |

| | |
|---|---|
| Peach | 5 |
| Peanut | 0.1 |
| Pear | 4 |
| Pecan | 0.1 |
| Pepper | 2 |
| Pomegranate | 0.2 |
| Rye, forage | 10 |
| Rye, grain | 1 |
| Rye, hay | 10 |
| Rye, straw | 10 |
| Sorghum, forage | 1 |
| Sorghum, grain, grain | 0.2 |
| Soybean | 0.2 |
| Soybean, forage | 10 |
| Spinach | 6 |
| Strawberry | 2 |
| Swiss chard | 6 |
| Tangerine | 2 |
| Tomato | 1 |
| Turnip, greens | 6 |
| Vegetable, brassica, leafy, group 5 | 6 |
| Vegetable, fruiting, group 8 | 0.2 |
| Vegetable, leafy | 0.2 |
| Vegetable, root crop | 0.2 |
| Watercress | 6 |
| Wheat, forage | 10 |
| Wheat, grain | 1 |
| Wheat, hay | 10 |
| Wheat, straw | 10 |

*Carcinogen/Hazard Classifications*
**U.S. EPA Carcinogens:** Group E, unlikely carcinogen
**Label Signal Word:** DANGER
**WHO Acute Hazard:** Class 1b, highly hazardous
**Endocrine Disruptor:** Suspected endocrine disruptor
*Regulatory Authority:*
- Air Pollutant Standard Set (ACGIH)[1] (HSE)[33] (NIOSH)[2] (Several States)[60]
- AB 1803-Well Monitoring Chemical (CAL)
- EPA/SARA 302 (EPCRA) Extremely hazardous substances
- Permissible Exposure Limits for Chemical Contaminants (CAL/OSHA)
- The "Director's List" (CAL/OSHA)
- Actively registered pesticide in California
- EPA Hazardous Waste Number (RCRA No.): P066
- RCRA, 40CFR261, Appendix 8 Hazardous Constituents
- RCRA 40CFR268.48; 61FR15654, Universal Treatment Standards: Wastewater (mg/L), 0.028; Nonwastewater (mg/kg), 0.14
- Safe Drinking Water Act: Priority List (55 FR 1470)
- Superfund/EPCRA 40CFR355, Appendix B Extremely Hazardous Substances: TPQ = 500/10,000 lb (227/4,540 kg)
- Superfund/EPCRA 40CFR302.4 RQ: CERCLA, 100 lb (45.4 kg)
- U.S. DOT Regulated Marine Pollutant (49CFR172.101, Appendix B)
- Canada, WHMIS, Ingredients Disclosure List

***Description:*** Methomyl is a white crystalline solid with a slight sulfurous odor. Melting/Freezing point = 78–79°C. Vapor pressure = $6.8 \times 10^{-3}$ Pa @ 20°C. Hazard Identification (based on NFPA-704 M Rating System): Health 3, Flammability 1, Reactivity 0. Moderately soluble in water; solubility = $5.8 \times 10^4$ ppm @ 25°C. Vapor pressure = $1 \times 10^{-4}$ mmHg @ 20°C. Log $K_{ow}$ = 0.69. Unlikely to bioaccumulate in marine organisms.

***Incompatibilities:*** Strong bases. strong oxidizers. Heat causes decomposition forming toxic and irritating fumes including nitrogen oxides, sulfur oxides, hydrogen cyanide, methyl isocyanate.

***Permissible Exposure Limits in Air:*** There is no OSHA[2] PEL The NIOSH[2] 10-hour TWA and the ACGIH[1] 8-hour TWA is 2.5 mg/m$^3$. HSE[33] has set the same TWA and adds the notation "skin" indicating the possibility of Cutaneous absorption. Further, several states have set guidelines or standards for methomyl in ambient air[60] ranging from 25 µg/m$^3$ (North Dakota) to 40 µg/m$^3$ (Virginia) to 50 µg/m$^3$ (Connecticut) to 59.5 µg/m$^3$ (Nevada).

***Determination in Air:*** OSHA versatile sampler-2; Reagent; High-pressure liquid chromatography/Ultraviolet detection; NIOSH IV, Method #5601[18].

***Permissible Concentration in Water:*** The USEPA has calculated a no-observed-adverse-effects-level (NOAEL) of 2.5 mg/kg/day from which a lifetime health advisory of 175 µg/L has been calculated. The State of Maine has set a guideline for methomyl in drinking water[61] of 50 µg/L.

***Determination in Water:*** By high-performance liquid chromatography as described in EPA Health Advisory cited below.

***Routes of Entry:*** Inhalation, ingestion, skin and/or eye contact.

*Harmful Effects and Symptoms*
***Short Term Exposure:*** Cholinesterase inhibitor. Irritates the eyes. May affect the nervous system, resulting in respiratory failure, convulsions. Exposure may result in death. Methomyl has high oral toxicity, moderate inhalation toxicity and low skin toxicity. The probable oral lethal dose for humans is between 7 drops and 1 teaspoon for a 150 pound adult. Death is due to respiratory arrest. Acute exposure to methomyl usually leads to a cholinergic crisis. Signs and symptoms may include increased salivation, lacrimation (tearing), spontaneous defecation, and spontaneous urination. Pinpoint pupils, blurred vision, tremor, muscle twitching, and loss of muscle coordination may occur. Mental confusion, convulsions, and coma may also be noted. Gastrointestinal effects include nausea, vomiting, diarrhea, and abdominal pain. Bradycardia (slow heart rate) occurs frequently. Dyspnea (shortness of breath),

pulmonary edema, and respiratory arrest may also occur. Carbamate insecticides inhibit the cholinesterase activity of enzymes, causing accumulation of acetylcholine at synapses and altering the way in which nervous impulses are transmitted. However, within several hours carbamates spontaneously detach from the enzymes.
*Long Term Exposure:* Cholinesterase inhibitor; cumulative effect is possible. Methomyl may damage the nervous system with repeated exposure, resulting in convulsions, respiratory failure. May cause liver damage. May cause anemia.
*Points of Attack:* Eyes, respiratory system, central nervous system, cardiovascular system, liver, kidneys, blood cholinesterase
*Medical Surveillance:* Before employment and at regular times after that, the following are recommended: Plasma and red blood cell cholinesterase levels (tests for the enzyme poisoned by this chemical). If exposure stops, plasma levels return to normal in 1-2 weeks while red blood cell levels may be reduced for 1-3 months. When acetylcholinesterase enzyme levels are reduced by 25% or more below preemployment levels, risk of poisoning is increased, even if results are in lower ranges of "normal." Reassignment to work not involving carbamate pesticides is recommended until enzyme levels recover. If symptoms develop or overexposure occurs, repeat the above tests as soon as possible and get an exam of the nervous system. Also consider complete blood count. Consider chest x-ray following acute overexposure
*First Aid:* Speed in removing material from eyes and skin is of extreme importance. *Eyes:* Eye contact can cause dangerous amounts of these chemicals to be quickly absorbed through the mucous membrane into the bloodstream. Immediately and gently flush eyes with plenty of warm or cold water (NO hot water) for at least 15 minutes, occasionally lifting the upper and lower eyelids. Get medical aid immediately. *Skin:* Get medical aid. Skin contact can cause dangerous amounts of these chemicals to be absorbed into the bloodstream. Wearing the appropriate PPE equipment and respirator for carbamate pesticides, immediately flush skin with plenty of soap and water for at least 15 minutes while removing contaminated clothing and shoes. Shampoo hair promptly if contaminated; protect eyes. *Ingestion:* Call poison control. Loosen all clothing. Never give anything by mouth to an unconscious person. Get medical aid. Do NOT induce vomiting.* If conscious, alert, and able to swallow, rinse mouth and have victim drink 4 to 8 ounces of water. Check to see if poison control instructs you to use ipecac syrup, otherwise administer slurry of activated charcoal (2 oz in 8 oz of water). If victim is *UNCONSCIOUS OR HAVING CONVULSIONS,* do nothing except keep victim warm. **In some cases you may be specifically instructed by poison control to induce vomiting by way of 2 tablespoons of syrup of ipecac (adult) washed down with a cup of water.* Do NOT give activated charcoal before or with ipecac syrup. *Inhalation:* Get medical aid. Do not contaminate yourself. Wearing the appropriate PPE equipment and respirator for carbamate pesticides, immediately remove the victim from the contaminated area to fresh air. If the victim is not breathing, administer artificial respiration. Do NOT use mouth-to-mouth resuscitation; use bag/mask apparatus. If breathing is difficult, administer oxygen through bag/mask apparatus until medical help arrives. Do not leave victim unattended. *Note to physician or authorized medical personne:* Administer atropine, 2 mg (1/30 gr) intramuscularly or intravenously as soon as any local or systemic signs or symptoms of an intoxication are noted; repeat the administration of atropine every 3 to 8 minutes until signs of atropinization (mydriasis, dry mouth, rapid pulse, hot and dry skin) occur; initiate treatment in children with 0.05 mg mg/kg of atropine; repeat at 5 to 10 minute intervals. Watch respiration, and remove bronchial secretions if they appear to be obstructing the airway; intubate if necessary. *Medical note:* 2-PAMCI may be contraindicated in the case of some carbamate poisonings.

*References:*
- EXTOXNET, Extension Toxicology Network, "Pesticide Information Profile, Methomyl," Oregon State University, Corvallis, OR (June 1996). http://extoxnet.orst.edu/pips/methomyl.htm
- U.S. Environmental Protection Agency, Office of Pesticide Programs, Pesticide Residue Limits, "Methomyl," 180.253, http://www.epa.gov/pesticides/food/viewtols.htm
- New Jersey Department of Health and Senior Services, "Hazardous Substance Fact Sheet: Methomyl," Trenton, NJ (September 1999). http://www.state.nj.us/health/eoh/rtkweb/1208.pdf
- U.S. Environmental Protection Agency, Methomyl, Health and Environmental Effects Profile No. 125, Wash., DC, Office of Solid Waste (April 30, 1980).
- Sax, N.I., Ed., "Dangerous Properties of Industrial Materials Report" 2, No. 5, 79-81 (1982).
- U.S. Environmental Protection Agency, "Health Advisory: Methomyl," Washington, DC, Office of Drinking Water (August 1987).
- U.S. Environmental Protection Agency, "Chemical Profile: Methomyl," DC, Chemical Emergency Preparedness Program (Nov. 30, 1987).

# Methoprene

*Use Type:* Insect growth hormone
*CAS Number:* 40596-69-8
*Formula:* $C_{19}H_{34}O_3$
*Synonyms:* 2,4-Dodecadienoic acid, 11-methoxy-3,7,11-trimethyl-, ispropyl ester, (E,E)-; 2,4-Dodecadienoic acid, 11-methoxy-3,7,11-trimethyl-, 1-methylethyl ester, (E,E)-;

Isopropyl (2E,4E)-11-methoxy-3,7,11-trimethyl-2,4-dodecadienoate; Isopropyl (E,E)-11-methoxy-3,7,11-trimethyl-2,4-dodecadienoate; (E,E)-11-Methoxy-3,7,11-trimethyl-2,4-dodecandienoate; 1-Methylethyl (E,E)-11-methoxy-3,7,11-trimethyl-2,4-dodecadienoate
*Trade Names:* ALTOSID®, Wellmark International (USA); APEX®; DIACON®; DIANEX®; ENT 70,460®; EXTINGUISH®; FLEATROL®; KABAT®; MANTA®; MOORMAN'S® IGR CATTLE CONCENTRATE; OVITROL®; PHARORID®; PRECOR®; ZR-515®
*Producers:* Wellmark International (USA)
*EPA/OPP PC Code:* 105401; (549500 old EPA code number)
*California DPR Chemical Code:* 1784
*Uses:* Methoprene is an insect growth regulator (IGR) used against a variety of insects including horn flies, mosquitoes, beetles, tobacco moths, sciarid flies, fleas (eggs and larvae), fire ants, pharoah ants, midge flies and Indian meal moths. Controlling some of these insects, methoprene is used in the production of a number of foods including meat, milk, mushrooms, peanuts, rice and cereals. It also has several uses on domestic animals (pets) for controlling fleas and to control insects in wastewater, sludge beds and ponds.
*U.S. Maximum Allowable Residue Levels for Methoprene (40 CFR 180.1033):*
Methoprene is exempt from the requirement of a tolerance in or on raw agricultural commodities when used to control mosquito larvae including pastures, rice fields, vineyards, date palm orchards, nut orchards, berry orchards, and fruit orchards.
*Carcinogen/Hazard Classifications*
**Label Signal Word:** CAUTION
**WHO Acute Hazard:** Class U, unlikely to be hazardous
*Regulatory Authority:*
- Actively registered pesticide in California.
- FIFRA, 40 CFR 185: tolerances for pesticides in food
- FIFRA, 40 CFR 186: tolerances for pesticides in animal feeds.

*Description:* Amber colored liquid. Faint fruity odor. Slightly soluble in water; solubility = 1.40 ppm. Molecular weight = 310.51. Boiling point = 135 @ 0.06 mmHg.
*Permissible Concentration in Water:* No criteria set. Runoff from spills or fire control may cause water pollution.
*Harmful Effects and Symptoms*
*Short Term Exposure:* Contact with eyes or skin may cause irritation or injury. Inhalation should be avoided; use NIOSH-approved air purifying respirators for pesticides. May be harmful if swallowed.
*First Aid:* If this chemical gets into the eyes, remove any contact lenses at once and irrigate immediately for at least 15 minutes, occasionally lifting upper and lower lids. Seek medical attention immediately. If this chemical contacts the skin, remove contaminated clothing and wash immediately with soap and water. Seek medical attention immediately. If this chemical has been inhaled, remove from exposure, begin rescue breathing (using universal precautions) if breathing has stopped, and CPR if heart action has stopped. Transfer promptly to a medical facility. When this chemical has been swallowed, get medical attention. Give large quantities of water and induce vomiting. Do not make an unconscious person vomit. Do not induce vomiting when formulations containing petroleum solvents are ingested.
*References:*
- U.S. Environmental Protection Agency, Office of Pesticide Programs, "Insect Growth Regulators Fact Sheet," (December 6, 2001). http://www.epa.gov/oppbppd1/biopesticides/ingredients/factsheets/factsheet_igr.htm
- U.S. Environmental Protection Agency, "Reregistration Eligibility Decision (RED) Fact Sheet, Methoprene," Office of Prevention, Pesticides and Toxic Substances, Washington, DC (June 2001). http://www.epa.gov/oppbppd1/biopesticides/ingredients/factsheets/factsheet_105401.pdf
- U.S. Environmental Protection Agency, Office of Pesticide Programs, Pesticide Residue Limits, "Methoprene," 40 CFR 180.1033, http://www.epa.gov/pesticides/food/viewtols.htm
- EXTOXNET, Extension Toxicology Network, "Pesticide Information Profile, Methoprene," Oregon State University, Corvallis, OR (June 1996). http://extoxnet.orst.edu/pips/methopre.htm
- California Environmental Protection Agency "Chemical List of Lists," Sacramento CA (February 1997)

# Methoxychlor

*Use Type:* Insecticide
*CAS Number:* 72-43-5
*Formula:* $C_{16}H_{15}Cl_3O_2$; $H_3COC_6H_4CH(CCl_3)C_6H_4OCH_3$
*Alert:* A General Use Pesticide (GUP).
*Synonyms:* Benzene,1,1'-(2,2,2-trichloroethylidene)bis[4-methoxy-]; 2,2-bis(*p*-anisyl)-1,1,1-trichloroethane; 1,1-Bis(*p*-methoxyphenyl)-2,2,2-trichloroethane; 2,2-Bis(*p*-methoxyphenyl)-1,1,1-trichloroethane; Dianisyltrichlorethane; 2,2-Di-*p*-anisyl-1,1,1-trichloroethane; Dimethoxy-DDT; *p,p'*-Dimethoxydiphenyltrichloroethane; Dimethoxy DT; 2,2-(*p*-Methoxyphenyl)-1,1,1-trichloroethane; Di(*p*-methoxyphenyl)-trichloro methyl methane; DMDT; *p,p'*-DMDT; ENT 1,716; Methoxide; Methoxo; *p,p'*-Methoxychlor; Methoxy DDT; Metoksychlor (Polish); Metoxicloro (Spanish); NCI-C00497; 1,1,1-Trichloro-2,2-bis(4-methoxy-phenyl)aethane (German); 1,1,1-Trichloro-2,2-bis(*p*-anisyl)ethane; 1,1,1-Trichloro-2,2-bis(*p*-methoxyphenol)ethanol; 1,1,1-Trichloro-2,2-bis(*p*-methoxyphenyl)ethane; 1,1,1-Trichloro-2,2-di(4-methoxyphenyl)ethane; 1,1,1-Trichloro-2,2-di(*p*-

methoxyphenyl)ethane; 1,1-(2,2,2-Trichloroethylidene) bis(4-methoxybenzene)

*Trade Names:* (The U.S. EPA lists 826 active and/or canceled products containing methoxychlor) CHEMFORM®; HIGALMETOX®; MARLATE®; METOX®; MOXIE®; PRENTOX®, Prentiss Inc. (USA)

*Producers:* Bonide Products (USA); Drexel Chemical; DuPont; Ehrenstorfer, Dr. (Germany); Prentiss Inc; Sigma-Aldrich Laborchemikalien (Germany)

*Chemical Class:* Organochlorine; Halo-organics

*EPA/OPP PC Code:* 034001

*California DPR Chemical Code:* 384

*ICSC Number:* 1306

*RTECS Number:* KJ3675000

*EINECS Number:* 200-779-9

*Uses:* Methoxychlor was introduced as an insecticide in 1945. It is a close relative of DDT and has been increasing in use since the ban on DDT in 1972 because of its very low mammalian toxicity for home and garden, on domestic animals for fly control, for elm bark-beetle vectors of Dutch elm disease, and for blackfly larvae in streams. Methoxychlor is registered for about 87 crops such as alfalfa; nearly all fruits and vegetables, corn, wheat, rice, and other grains; beef and dairy cattle; and swine, goats and sheep, and for agricultural premises and outdoor fogging. It is available in wettable and dust powders, emulsifiable concentrates, granules, and as an aerosol. It in combined in formulations with malathion, parathion, piperonyl butoxide, and pyrethrins.

*Human toxicity (long-term)*[77]: Intermediate–40.00 ppb, MCL (Maximum Contaminant Level)

*Fish toxicity (threshold)*[77]: Extra high–0.11310 ppb, MATC (Maximum Acceptable Toxicant Concentration)

*U.S. Maximum Allowable Levels for Methoxychlor:*
The EPA limits the amount of methoxychlor present in agricultural products to 1-100 ppm. The Food and Drug Administration (FDA) limits the amount of methoxychlor in bottled water to 0.04 ppm.

*Carcinogen/Hazard Classifications*

**U.S. EPA Carcinogens:** Group D, unclassifiable, insufficient data

**IARC:** Group 3, unclassifiable

**Label Signal Word:** CAUTION

**WHO Acute Hazard:** Class U, unlikely to be hazardous

**Endocrine Disruptor:** Suspected endocrine disruptor

*Regulatory Authority:*
- Banned or Severely Restricted (In Agriculture) (Several Countries) (UN)[13]
- Air Pollutant Standard Set (ACGIH)[1] (DFG)[3] (HSE)[33] (OSHA)[58] (Argentina)[35] (Several States)[60]
- AB 1803-Well Monitoring Chemical (CAL)
- Permissible Exposure Limits for Chemical Contaminants (CAL/OSHA) as pyrethrum
- Actively registered pesticide in California. as pyrethrins and pyrethrum
- MCL (Maximum Contaminants Levels) list of contaminants (CAL)
- AB 2588-Air Toxics "Hot Spots" Chemicals (CAL)
- CAL Air Resources Board/AB 1807 Toxic Air Contaminants
- Permissible Exposure Limits for Chemical Contaminants (CAL/OSHA)
- The "Director's List" (CAL/OSHA)
- Actively registered pesticide in California.
- Clean Air Act: Hazardous Air Pollutants (Title I, Part A, Section 112)
- Clean Water Act: Section 311 Hazardous Substances/RQ 40CFR117.3 (same as CERCLA, see below); Section 313 Water Priority Chemicals (57FR41331, 9/9/92); Section 313 Water Priority Chemicals (57FR41331, 9/9/92)
- EPA Hazardous Waste Number (RCRA No.): U247
- RCRA Toxicity Characteristic (Section 261.24), Maximum
- Concentration of Contaminants, regulatory level, 10.0 mg/L
- RCRA, 40CFR261, Appendix 8 Hazardous Constituents
- RCRA 40CFR268.48; 61FR15654, Universal Treatment Standards: Wastewater (mg/L), 0.25; Nonwastewater (mg/kg), 0.18
- RCRA 40CFR264, Appendix 9; TSD Facilities Ground Water Monitoring List. Suggested test method(s) (PQL *ug*/L): 8080(2); 8270(10)
- Safe Drinking Water Act: MCL, 0.04 mg/L; MCLG, 0.04 mg/L; Regulated chemical (47 FR 9352); Priority List (55 FR 1470)
- EPCRA Section 304 RQ: CERCLA, 1 lb (0.454 kg) as pyrethrins
- Superfund/EPCRA 40CFR302.4 RQ: CERCLA, 1 lb (0.45 kg)
- EPCRA Section 313 Form R *de minimis* concentration reporting level: 1.0%
- Canada, WHMIS, Ingredients Disclosure List

The EPA limits the amount of methoxychlor present in agricultural products to 1-100 ppm.

*Description:* Methoxychlor is a colorless to tan solid. Slight fruity odor. Practically insoluble in water; solubility = 3.9 x $10^2$ ppm @ 24°C. Molecular weight = 345.64. Melting/Freezing point = 89°C; 78°C (technical product). Vapor pressure = very low. Hazard Identification (based on NFPA-704 M Rating System): Health 2, Flammability 1, Reactivity 0. Log $K_{ow}$ = 5.1. Values above 3.0 are likely to bioaccumulate in marine organisms.

*Incompatibilities:* Oxidizers

*Permissible Exposure Limits in Air:* The OSHA[2] 8-hour TWA is 15 mg/m³. HSE[33] and the ACGIH[1] set a TWA level of 10 mg/m³. The NIOSH[2] IDLH level = [Ca] 5,000 mg/m³. The German DFG[3] MAK value is 15 mg/m³. Argentina[35] has set both a TWA and an STEL at 10 mg/m³ in workplace air. Several states have set guidelines or standards for methoxychlor in ambient air[60] ranging from

23.8 $\mu g/m^3$ (Kansas) to 35.07 $\mu g/m^3$ (Pennsylvania) to 100 $\mu g/m^3$ (North Dakota) to 160 $\mu g/m^3$ (Virginia) to 200 $\mu g/m^3$ (Connecticut) to 238 $\mu g/m^3$ (Nevada).

***Determination in Air:*** Filter; Isooctane; Gas chromatography/Electrochemical detection; NIOSH II(4), Method #S371[18].

***Permissible Concentration in Water:*** The U.S. EPA limits the amount of methoxychlor that may be present in drinking water to 0.04 parts of methoxychlor per million parts of water (0.04 ppm or 400 $\mu g/L$[62]). The U.S. EPA also limits the amount of methoxychlor present in agricultural products to 1-100 ppm. The WHO[35] has recommended a limit of 30 $\mu g/L$ of methoxychlor for drinking water. A U.S. recommendation for bottled water for drinking purposes was 100 $\mu g/L$. Mexico[35] has set limits for methoxychlor of 4.0 $\mu g/L$ in coastal waters, 40.0 $\mu g/L$ in estuaries and 35 $\mu g/L$ in recovery waters used for drinking water supply. Several states have set guidelines or standards for methoxychlor in drinking water[61]. These range from a standard of 100 $\mu g/L$ in Arizona and a guideline of 100 $\mu g/L$ in Maine to a guideline of 340 $\mu g/L$ for Minnesota.

***Determination in Water:*** By liquid/liquid extraction followed by identification by gas chromatography[47].

***Routes of Entry:*** Inhalation, ingestion, skin and/or eyes. Passes through the skin.

### Harmful Effects and Symptoms

***Short Term Exposure:*** *Inhalation*: The results of accidental exposure and animal studies suggest that high levels may cause irritation to nose and throat, headache, nausea, vomiting, staggering walk, drowsiness, convulsions, coma and death. *Skin:* Absorbed in significant amounts especially when dissolved in organic solvents. Local irritation and numbing of affected area may be experienced. *Eyes:* May cause irritation. *Ingestion:* Symptoms are similar to those listed under inhalation. Ingestion of 5 ounces a day for 6 weeks resulted in no observable symptoms. The least amount causing death has been reported as one pound. Exposure can cause anxiety, fatigue, nausea, vomiting, dizziness, confusion, weakness, "pins and needles" in extremities, muscle twitching and tremor. Higher levels can cause convulsions, unconsciousness and even death. *Note:* For application, methoxychlor is dissolved in organic or petroleum distillate solvents. These solvents may have poisonous effects in addition to those above.

***Long Term Exposure:*** Experiments with animals suggest that exposure to high levels for prolonged periods may cause excess salivation, tremors, seizures and convulsions. These will generally go away when exposure stops. Methoxychlor has been shown to affect reproduction and to cause cancer at high exposure levels in some laboratory animals. Whether it does so in humans is not known. A potential occupational carcinogen. (NIOSH). May effect liver, kidneys, and central nervous system. Very high exposures may cause anemia. May be a reproductive toxin.

***Points of Attack:*** Central nervous system, liver, kidneys. Cancer site in animals: liver and ovarian cancer.

***Medical Surveillance:*** Liver and kidney function tests. Complete blood count

***First Aid:*** If this chemical gets into the eyes, remove any contact lenses at once and irrigate immediately for at least 15 minutes, occasionally lifting upper and lower lids. Seek medical attention immediately. If this chemical contacts the skin, remove contaminated clothing and wash immediately with soap and water. Seek medical attention immediately. If this chemical has been inhaled, remove from exposure, begin rescue breathing (using universal precautions) if breathing has stopped, and CPR if heart action has stopped. Transfer promptly to a medical facility. When this chemical has been swallowed, get medical attention. Give large quantities of water and induce vomiting. Do not make an unconscious person vomit.

***References:***
- EXTOXNET, Extension Toxicology Network, "Pesticide Information Profile, Methoxychlor," Oregon State University, Corvallis, OR (June 1996). http://extoxnet.orst.edu/pips/methoxyc.htm
- New Jersey Department of Health and Senior Services, "Hazardous Substance Fact Sheet, Methoxychlor," Trenton NJ (November 1999). http://www.state.nj.us/health/eoh/rtkweb/rtkhsfs.htm
- Agency for Toxic Substances and Disease Registry, "ToxFAQs for Methoxychlor," Atlanta, GA (September 2002). http://www.atsdr.cdc.gov/tfacts47.html
- Sax, N.I., Ed., "Dangerous Properties of Industrial Materials Report" 7, No. 5, 79-87 (1987).
- New York State Department of Health, "Chemical Fact Sheet: Methoxychlor," Albany, NY, Bureau of Toxic Substance Assessment (Mar. 1986 and Version 2).
- California Environmental Protection Agency "Chemical List of Lists," Sacramento CA (February 1997).

## Methoxyethylmercuric Acetate

***Use Type:*** Fungicide
***CAS Number:*** 151-38-2
***Formula:*** $C_5H_{10}HgO_3$; $CH_3OCH_2CH_2HgOOCCH_3$
***Alert:*** Not registered as a pesticide in the U.S.
***Synonyms:*** Acetato(2-methoxyethyl)mercury; Acetoxy(2-methoxyethyl)mercury; MEMA; Mercuran; Mercury, acetoxy(2-methoxyethyl)-; Methoxyethylmercury acetate; 2-Methoxyethylmerkuriacetat (German)
***Trade Names:*** CEKUSIL UNIVERSAL A®; LANDISAN®; PANOGEN®; PANOGEN® M; PANOGEN® METOX; RADOSAN®
***Chemical Class:*** Organomercury
***EPA/OPP PC Code:*** 041508
***RTECS Number:*** OV6300000
***EINECS Number:*** 205-790-2

*Uses:* Used as a pesticide in seed treatment for cotton and small grains. It is no longer approved for this use. It exhibits high fungicidal activity against leaf stripe of barley, stinking smut of wheat, snow mold of rye; against seedling diseases in beets and legumes, and for dressing "seed" potatoes, bulbs, and tubers.

*Carcinogen/Hazard Classifications*
**California Prop. 65:** Listed; Reproductive toxin
*Regulatory Authority:*
- Banned or Severely Restricted (In Agriculture) (IN U.K.)[13]
- Air Pollutant Standard Set (ACGIH)[1] (HSE)[33] (OSHA)[58] (North Dakota)[60]
- AB 2588-Air Toxics "Hot Spots" Chemicals (CAL)
- CAL Air Resources Board/AB 1807 Toxic Air Contaminants
- Proposition 65 chemical (CAL)
- Permissible Exposure Limits for Chemical Contaminants (CAL/OSHA)
- The "Director's List" (CAL/OSHA)
- Clean Water Act: Section 307 Toxic Pollutants as mercury and compounds
- Safe Drinking Water Act: MCL, 0.002 mg/L; MCLG, 0.002 mg/L as mercury
- RCRA Section 261 Hazardous Constituents, waste number not Listed as mercury compounds, n.o.s.
- EPCRA Section 302 Extremely Hazardous Substances: TPQ = 500/10,000 lb (227/4,540 kg)
- EPCRA Section 304 RQ: EHS, 1 lb (0.454 kg)
- EPCRA Section 313 (as mercury compound) Form R *de minimus* concentration reporting level: 1.0%
- U.S. DOT Regulated Marine Pollutant (49CFR172.101, Appendix B), severe pollutant as mercury based pesticides, liquid, flammable, toxic, n.o.s; mercury based pesticides, liquid, toxic, n.o.s; mercury based pesticides, solid, toxic, n.o.s; mercury compounds, liquid, n.o.s; mercury compounds, solid, n.o.s; mercury(I) (mercurous) compounds (pesticides); mercury(II) (mercuric) compounds (pesticides)
- Canada, WHMIS, Ingredients Disclosure List

*Description:* Methoxyethylmercuric acetate is a crystalline solid. Soluble in water. Melting/Freezing point = 41°C. Hazard Identification (based on NFPA-704 M Rating System): Health 3, Flammability 1, Reactivity 0. Mercury compounds are toxic to aquatic organisms. Bioaccumulation in the food chain takes place, specifically in fish, crustacea, and birds. Bioconcentrative up to 10,000-fold. (CHRIS, U. S. Coast Guard).

*Incompatibilities:* Corrosive to iron and other metals.
*Permissible Exposure Limits in Air:* The OSHA[2] PEL for all organic mercury compounds is 0.01 mg/m$^3$ as an 8-hour TWA and ceiling of 0.04 mg/m$^3$. NIOSH[2] recommends 0.01 mg/m$^3$ as a 10-hour TWA and STEL of 0.03 mg/m$^3$ with the notation "skin" indicating the possibility of cutaneous absorption. ACGIH[1] recommends 0.01 mg/m$^3$ as an 8-hour TWA value with no STEL. The NIOSH[2] IDLH level = 2 mg/m$^3$ (as Hg). HSE[33] has set 0.05 mg/m$^3$ as an 8-hour TWA and 0.15 mg/m$^3$ as an STEL. In addition, North Dakota has set guidelines for alkyl mercury compounds in ambient air[60] of 1-3 µg/m$^3$ (0.0001 to 0.0003 mg/m$^3$).

*Determination in Air:* No method available.
*Permissible Concentration in Water:* Presumably this material is covered in the Priority Toxic Pollutant Category. It is not specifically cited as an organomercurial as is methyl mercury.

*Determination in Water:* Total mercury is determined by flameless atomic absorption. Soluble mercury may be determined by 0.45 micron filtration followed by flameless atomic absorption.

*Routes of Entry:* Inhalation, ingestion skin and/or eye contact.

*Harmful Effects and Symptoms*
*Short Term Exposure:* Highly toxic. Target organs are brain and central nervous system. Inhalation can cause lung damage; ingestion can cause kidney damage. Women of childbearing age should avoid exposure. Patients complain of headache, paresthesia of tongue, lips, fingers, and toes, a metallic taste in mouth, gastrointestinal disturbances, gas, and diarrhea. Nervous system symptoms may appear first after a relatively slight exposure or have a latency period slight loss of coordination, loss of coordination of speech, writing and gait. Uncoordination may progress to loss of ability to control voluntary movements. Irritability and bad temper may progress to mania. Stupor or coma may develop. Blisters or dermatitis may be present on skin. Symptoms persist for years even in cases of mild exposure.

*Long Term Exposure:* Repeated or prolonged contact with skin may result in dermatitis and allergy. Repeated or prolonged exposure may cause brain damage amd nervous system damage. Repeated or prolonged exposure may cause death by hypovolemic shock, nephrotic syndrome, and kidney failure. Organic mercury substances have been identified as teratogen in humans. Can cause mercury to accumulate in the body and cause mercury poisoning. May cause permanent damage such as gray colored skin, brown staining of the eyes, and decreased peripheral vision.

*Points of Attack:* Eyes, skin, central nervous system, peripheral nervous system and kidneys.

*Medical Surveillance:* Before first exposure and every 6 to 12 months after, a complete medical history and exam is strongly recommended, with: Eye exam. Consider lung function tests for persons with frequent exposure. Exam of the nervous system,. Routine urine test (UA). Urine test for mercury (should be less than 0.02 mg/Liter). Consider nerve conduction tests, urinary enzymes and neurobehavioral test. After suspected illness or overexposure, repeat the tests above and get a blood test for mercury.

*First Aid:* Speed in removing material from eyes and skin is of extreme importance. If this chemical gets into the eyes, remove any contact lenses at once and irrigate immediately for at least 15 minutes, occasionally lifting upper and lower lids. Seek medical attention immediately. If this chemical contacts the skin, remove contaminated clothing and wash immediately with soap and water. Shampoo hair promptly if contaminated. Seek medical attention immediately. If this chemical has been inhaled, remove from exposure, begin rescue breathing (using universal precautions) if breathing has stopped, and CPR if heart action has stopped. Transfer promptly to a medical facility. When this chemical has been swallowed, get medical attention. Give large quantities of water and induce vomiting. Do not make an unconscious person vomit.

*Antidotes and Special Procedures for medical personnel:* The drug NAP (n-Acetyl Penicillamine) has been used to treat mercury poisoning, with mixed success.

*References:*
- U.S. Environmental Protection Agency, "Chemical Profile: Methoxyethylmercuric Acetate," Washington, DC, Chemical Emergency Preparedness Program (Nov. 30, 1987).
- California Environmental Protection Agency "Chemical List of Lists," Sacramento CA (February 1997).

## Methoxyfenozide

*Use Type:* Insecticide
*CAS Number:* 161050-58-4
*Formula:* $C_{22}H_{28}N_2O_3$
*Synonyms:* Benzoic acid, 3-methoxy-2-methyl-2-(3,5-dimethylbenzoyl)-2-(1,1-dimethylethyl)hydrazide (9CI)
*Trade Names:* INTREPID®, Dow AgroSciences (USA); PRODIGY®
*Producers:* Dow AgroSciences (USA)
*Chemical Class:* Diacylhydrazine
*EPA/OPP PC Code:* 121027
*California DPR Chemical Code:* 5698
*Uses:* Methoxyfenozide prevents insects from molting, or shedding their exoskeleton in order to grow, e.g., caterpillars and lychee webworms.

*U.S. Maximum Allowable Residue Levels for Methoxyfenozide (40 CFR 180.544):*

| CROP | ppm |
|---|---|
| Almond, hulls | 25 |
| Apple, wet pomace | 7.0 |
| Artichoke, globe | 3.0 |
| Brassica, head and stem, Subgroup | 7.0 |
| Brassica, leafy greens, subgroup | 30 |
| Cattle, fat | 0.50 |
| Cattle, meat | 0.02 |
| Corn, field, forage | 15 |
| Corn, field, grain | 0.05 |
| Corn, field, refined oil | 0.20 |
| Corn, field, stover | 125 |
| Corn, sweet, forage | 30 |
| Corn, sweet, kernal plus cob with husks removed | 0.05 |
| Corn, sweet, stover | 60 |
| Cotton, gin byproducts | 35 |
| Cotton, undelinted seed | 2.0 |
| Fruit, pome, group | 1.5 |
| Fruit, stone, group, except fresh prune plum | 3.0 |
| Goat, fat | 0.50 |
| Goat, meat | 0.02 |
| Grain, aspirated fractions | 2.0 |
| Grape | 1.0 |
| Grape, raisin | 1.5 |
| Hog, fat | 0.1 |
| Hog, meat | 0.02 |
| Horse, fat | 0.50 |
| Horse, meat | 0.02 |
| Leaf petioles subgroup | 25 |
| Leafy greens subgroup | 30 |
| Longan | 2.0 |
| Lychee | 2.0 |
| Milk | 0.10 |
| Nut, tree, group | 0.10 |
| Pistachio | 0.10 |
| Plum, prune, fresh | 0.30 |
| Poultry, fat | 0.02 |
| Poultry, meat | 0.02 |
| Pulasan | 2.0 |
| Rambutan | 2.0 |
| Sheep, fat | 0.50 |
| Sheep, meat | 0.02 |
| Spanish lime | 2.0 |
| Vegetable, fruiting, group | 2.0 |

*Carcinogen/Hazard Classifications*
**U.S. EPA Carcinogens:** Not likely a carcinogen
**Label Signal Word:** CAUTION
*Permissible Concentration in Water:* No criteria set. Runoff from spills or fire control may cause water pollution.
*Harmful Effects and Symptoms*
*Short Term Exposure:* Contact with eyes or skin may cause irritation or injury. Inhalation should be avoided; use NIOSH-approved air purifying respirators for pesticides. May be harmful if swallowed.
*First Aid:* If this chemical gets into the eyes, remove any contact lenses at once and irrigate immediately for at least 15 minutes, occasionally lifting upper and lower lids. Seek medical attention immediately. If this chemical contacts the skin, remove contaminated clothing and wash immediately with soap and water. Seek medical attention immediately. If this chemical has been inhaled, remove from exposure, begin rescue breathing (using universal precautions) if breathing has stopped, and CPR if heart action has stopped. Transfer promptly to a medical facility. When this chemical

has been swallowed, get medical attention. Give large quantities of water and induce vomiting. Do not make an unconscious person vomit.

*References:*
- U.S. Environmental Protection Agency, Office of Pesticide Programs, Pesticide Residue Limits, "Methoxyfenozide," 40 CFR 180.544. http://www.epa.gov/fedrgstr/EPA-PEST/2002/September/Day-20/p23996.htm
- California Environmental Protection Agency "Chemical List of Lists," Sacramento CA (February 1997)

# Methyl Bromide

*Use Type:* Fumigant, herbicide, insecticide and nematicide
*CAS Number:* 74-83-9
*Formula:* $CH_3Br$
*Alert:* Methyl bromide use in the U.S. will be phased out under the Montreal Protocol.
*Synonyms:* Bromomethane; Methane, bromo-; Bromure de methyle (French); Methylbromid; Metilbromid (Spenish); Monobromomethane
*Trade Names:* BROM-O-GAS®, Great Lakes Chemical (USA); BROM-O-SOL®, Great Lakes Chemical (USA); DAWSON® 100; DOWFUME®, Albemarle (USA; EDCO®; EMBAFUME®; HALON 1001®; ISCOBROME®; KAYAFUME®; MATABROM®, Dead Sea Bromine Group (Israel); METHO-GAS®, Great Lakes Chemical (USA); M-B-C FUMIGANT®, Albemarle (USA); R 40B1®; ROTOX®; TERABOL®; TERR-O-GAS®, Great Lakes Chemical (USA); ZYTOX®
*Producers:* Air Liquide Group (France); Air Products & Chemicals; Akzo Nobel Chemicals (Netherlands); Albemarle (USA); Asahi Glass (Japan); ATOFINA Chemicals (France); BOC Gases (UK); Dead Sea Bromine Group (Israel); Dow Chemical (USA); Great Lakes Chemical (USA); Laporte (UK); Messer Group (Germany); MG Industries (USA); Nippon Kayaku (Japan); Praxair (USA)
*Chemical Class:* Halogenated organic compound
*EPA/OPP PC Code:* 053201
*California DPR Chemical Code:* 385
*ICSC Number:* 0109
*RTECS Number:* PA4900000
*EEC Number:* 602-002-00-3
*EINECS Number:* 200-813-2
*Uses:* The primary use of methyl bromide is as an insect fumigant to control insects, nematodes, weeds and pathogens in more than 100 crops and for soil, grain storage, warehouses, mills, ships, etc. Use of methyl bromide in the U.S. will be phased out under the requirements of the Montreal Protocol, with some exemptions. Methyl bromide is also used as a chemical intermediate and a methylating agent, a refrigerant, a herbicide, a fire extinguishing agent, a low-boiling solvent in aniline dye manufacture, for degreasing wool, for extracting oils from nuts, seeds, and flowers, and in ionization chambers. It is used as an intermediate in the manufacture of many drugs.
*Human toxicity (long-term)*[77]*:* Intermediate–10.00 ppb, Health Advisory
*Fish toxicity (threshold)*[77]*:* Very low–1760.07814 ppb, MATC (Maximum Acceptable Toxicant Concentration)
*U.S. Maximum Allowable Residue Levels for Methyl Bromide (40 CFR 180.123):*
*Note*: 40 CFR 180.123 also lists tolerance levels for inorganic bromides resulting from fumigation with methyl bromide.

| CROP | ppm |
|---|---|
| Alfalfa, hay | 50 |
| Almond | 200 |
| Apple | 5 |
| Apricot | 20 |
| Artichoke, jerusalem | 30 |
| Asparagus | 100 |
| Avocado | 75 |
| Barley | 50 |
| Bean | 50 |
| Bean, lima | 50 |
| Bean, snap, succulent | 50 |
| Bean, succulent | 50 |
| Beet, garden, roots | 30 |
| Beet, sugar, roots | 30 |
| Blueberry | 20 |
| Butternut | 200 |
| Cabbage | 50 |
| Cacao bean | 50 |
| Cantaloupe | 20 |
| Carrot, roots | 30 |
| Cashew | 200 |
| Cherry | 20 |
| Chestnut | 200 |
| Citron, citrus | 30 |
| Coconut, copra | 100 |
| Coffee, bean | 75 |
| Corn, field, grain | 50 |
| Corn, pop | 240 |
| Corn, sweet, kernel plus cob with husks removed | 50 |
| Cotton, undelinted seed | 200 |
| Cucumber | 30 |
| Cumin | 100 |
| Eggplant | 20 |
| Filbert | 200 |
| Garlic | 50 |
| Ginger, roots | 100 |
| Grape | 20 |
| Grapefruit | 30 |
| Honeydew | 20 |
| Horseradish, roots | 30 |

| | |
|---|---|
| Kumquat | 30 |
| Lemon | 30 |
| Lime | 30 |
| Mango | 20 |
| Muskmelon | 20 |
| Nectarine | 20 |
| Nut, brazil | 200 |
| Nut, hickory | 200 |
| Nut, macadamia | 200 |
| Oat | 50 |
| Okra | 30 |
| Onion | 20 |
| Onion, cipollini, bulb | 50 |
| Orange | 30 |
| Papaya | 20 |
| Parsnip, roots | 30 |
| Pea | 50 |
| Pea, blackeyed | 50 |
| Peach | 20 |
| Peanut | 200 |
| Pear | 5 |
| Pecan | 200 |
| Pepper | 30 |
| Pepper, pimento | 30 |
| Pineapple | 20 |
| Pistachio | 200 |
| Plum, fresh prune | 20 |
| Pomegranate | 100 |
| Potato | 75 |
| Pumpkin | 20 |
| Quince | 5 |
| Radish | 30 |
| Rice | 50 |
| Rutabaga | 30 |
| Rye | 50 |
| Salsify, roots | 30 |
| Sorghum, grain, grain | 50 |
| Soybean | 200 |
| Squash, summer | 30 |
| Squash, winter | 20 |
| Squash, zucchini | 20 |
| Strawberry | 60 |
| Sweet potato, roots | 75 |
| Tangerine | 30 |
| Timothy, hay | 50 |
| Tomato | 20 |
| Turnip, roots | 30 |
| Walnut | 200 |
| Watermelon | 20 |
| Wheat | 50 |

***Carcinogen/Hazard Classifications***
**U.S. EPA Carcinogens:** Not Likely a Carcinogen
**IARC:** Group 3, unclassifiable
**California Prop. 65:** Listed; Reproductive toxin
**Label Signal Word:** DANGER

***Regulatory Authority:***
- Carcinogen (Animal Suspected) (IARC)[9] (DFG)[3] (ACGIH)[1] (suspected occupational carcinogen, NIOSH[2])
- Toxic Chemical (World Bank)[15]
- Air Pollutant Standard Set (ACGIH)[1] (DFG)[3] (HSE)[33] (OSHA)[58] (Other Countries)[35] (Several States)[60]
- List of priority pollutants (U.S. EPA)
- AB 1803-Well Monitoring Chemical (CAL)
- AB 2588-Air Toxics "Hot Spots" Chemicals (CAL)
- CAL Air Resources Board/AB 1807 Toxic Air Contaminants
- Proposition 65 chemical (CAL)
- U.S. DOT Inhalation Hazard Chemicals
- Permissible Exposure Limits for Chemical Contaminants (CAL/OSHA)
- The "Director's List" (CAL/OSHA)
- Actively registered pesticide in California.
- OSHA 29CFR1910.119, Appendix A, Process Safety List of Highly Hazardous Chemicals, TQ = 2,500 lb (1,135 kg)
- Clean Air Act: Hazardous Air Pollutants (Title I, Part A, Section 112); Stratospheric ozone protection (Title VI, Subpart A, Appendix A), Class I, Ozone Depletion Potential = 0.7
- Clean Water Act: 40CFR423, Appendix A, Priority Pollutants; Section 313 Water Priority Chemicals (57FR41331, 9/9/92)
- EPA Hazardous Waste Number (RCRA No.): U029
- RCRA, 40CFR261, Appendix 8 Hazardous Constituents
- RCRA 40CFR268.48; 61FR15654, Universal Treatment Standards: Wastewater (mg/L), 0.11; Nonwastewater (mg/kg), 15
- RCRA 40CFR264, Appendix 9; TSD Facilities Ground Water Monitoring List. Suggested test method(s) (PQL $ug/L$): 8010(20); 8240(10).
- Safe Drinking Water Act: Priority List (55 FR 1470)
- Superfund/EPCRA 40CFR355, Appendix B Extremely Hazardous Substances: TPQ = 1000 lb (454 kg)
- Superfund/EPCRA 40CFR302.4 RQ: CERCLA, 1000 lb (454 kg)
- EPCRA Section 313 Form R *de minimis* concentration reporting level: 1.0%
- U.S. DOT Regulated Marine Pollutant (49CFR172.101, Appendix B). Only as methyl bromide and ethylene dibromide mixture, liquid
- Canada, WHMIS, Ingredients Disclosure List

***Description:*** Colorless gas. A liquid below 3.3°C. Chloroform-like odor at high concentrations. Shipped as a liquefied compressed gas. *Warning properties are inadequate*; musty or fruity odor at greater than 1000 ppm; eye and throat irritation at greater than 500 ppm. Slightly soluble in water; solubility = 0.09% @ 20 °C; 13.4 g/kg @ 25°C. Molecular weight = 94.95 daltons. Gas density (air: 1) = 3.4. Boiling point = 3.6°C. Melting/Freezing point =

−93.7°C. Vapor pressure = 1824 mmHg @ 20°C. Flash point = practically non-flammable except in presence of a high energy ignition sources. Autoignition temperature = 537°C. Explosive limits: LEL = 13.5%; UEL = 14.5%. Hazard Identification (based on NFPA-704 M Rating System): Health 3, Flammability 1, Reactivity 0. Log $K_{ow}$ = 1.2. Unlikely to bioaccumulate in marine organisms.

*Incompatibilities:* Attacks aluminum to form *spontaneously* flammable aluminum trimethyl. Incompatible with strong oxidizers, aluminum, dimethylsulfoxide, ethylene oxide, water. Attacks zinc, magnesium, alkali metals and their alloys. Attacks some rubbers and coatings.

***Permissible Exposure Limits in Air:*** The OSHA[2] TWA is [ceiling] 20 ppm (80 mg/m$^3$), not to be exceeded at any time. NIOSH[2] recommends that exposure be limited to the lowest feasible concentration, ACGIH[1] recommends a TWA of 1 ppm. DFG[3] and HSE[33] ACGIH[1] is a TLV of 5 ppm (20 mg/m$^3$) as a TWA. An STEL value of 15 ppm (60 mg/m$^3$) has been set by HSE[33]. The notation "skin" is added to indicate the possibility of cutaneous absorption. The NIOSH[2] IDLH level = [Ca] 250 ppm. Sweden and Argentina[35] have set TWA values of 15 ppm (60 mg/m$^3$). Argentina has set this same level as an STEL But Sweden has set 20 ppm (80 mg/m$^3$) as an STEL. Czechoslovakia and the former USSR have set MAC values for workplace air of 1.0 mg/m$^3$. The former USSR has set MAC values for the ambient air in residential areas of 0.02 mg/m$^3$ on a momentary basis and 0.01 mg/m$^3$ on a daily average basis. Several states have set guidelines or standards for methyl bromide in ambient air[60] ranging from 2.6 $\mu$g/m$^3$ (Massachusetts) to 47.6 $\mu$g/m$^3$ (Kansas) to 100 $\mu$g/m$^3$ (South Carolina) to 200 $\mu$g/m$^3$ (North Dakota) to 350 $\mu$g/m$^3$ (Virginia) to 400 $\mu$g/m$^3$ (Connecticut) to 476 $\mu$g/m$^3$ (Nevada) to 480 $\mu$g/m$^3$ (Pennsylvania).

***Determination in Air:*** Charcoal adsorption, workup with $CS_2$, analysis by gas chromatography/flame ionization. See NIOSH Method 2520.[18]

***Permissible Concentration in Water:*** To protect human health-preferably zero. An additional lifetime cancer risk of 1 in 100,000 is posed by a concentration of 1.9 $\mu$g/L[6]. States which have set guidelines for methyl bromide in drinking water[61] include Arizona at 2.5 $\mu$g/L and Kansas at 0.19 $\mu$g/L.

***Determination in Water:*** Inert gas purge followed by gas chromatography with halide specific detection (EPA Method 601) or gas chromatography plus mass spectrometry (EPA Method 624).

***Routes of Entry:*** Methyl bromide is absorbed well by the lungs and to some degree through intact skin. Oral exposure is rare because methyl bromide is a gas at room temperature, but it may be absorbed by the gastrointestinal tract. Exposure by any route can cause systemic effects.

### Harmful Effects and Symptoms

***Short Term Exposure:*** Methyl bromide is a neurotoxic gas that can cause convulsions, coma, and long-term neuromuscular and cognitive deficits. Exposure to high concentrations of pure methyl bromide may cause inflammation of the bronchi or lungs, an accumulation of fluid in the lung, and irritation of the eyes and nose. Tearing agents added to methyl bromide to provide warning of its presence can also cause these symptoms, even at very low concentrations. Skin contact with high vapor concentrations or with liquid methyl bromide can cause systemic toxicity and may cause stinging pain and blisters. Methyl bromide irritates the respiratory tract. Inhalation of the gas can cause pulmonary edema, a medical emergency that can be delayed for several hours. This can cause death. May affect the central nervous system, causing psychological disturbances. Signs and symptoms of acute exposure to methyl bromide may be severe and include tremors, convulsions, brain hemorrhage, paralysis, coma, and permanent brain damage. Respiratory effects include cough, tachypnea (rapid respiratory rate), pulmonary edema, and respiratory collapse. Cyanosis (blue tint to the skin and mucous membranes), pallor, ventricular fibrillation, and circulatory collapse may also occur. Lethargy, profound weakness, headache, dizziness, mental confusion, slurring of speech, staggering gait, and blurred or double vision are often found. Gastrointestinal signs and symptoms include nausea, vomiting, abdominal pain, and anorexia. Oliguria (scanty urination), anuria (lack of urine formation), kidney hemorrhage, and kidney failure may occur. Contact with methyl bromide may cause dermatitis (red, inflamed skin) and conjunctivitis (red, inflamed eyes). *Inhalation:* A level of 35 ppm can cause nausea, vomiting, loss of appetite, headache, dizziness, drowsiness and dimming of vision. These effects go away soon after exposure ceases. Headaches, dizziness, and weakness can be felt at 100 ppm and can last for months after exposure. Higher levels have caused coughing, nose and throat irritation, disturbed speech and walk, visual disturbances, twitching, numbness, paralysis, convulsions and permanent nerve damage. Symptoms are often delayed 2-48 hours. Exposures of 10,000 ppm for a few minutes can cause death. Can cause abdominal cramps, respiratory failure resulting in death. *Skin*: Contact with liquid can cause burning or tingling sensation, itching, redness and swelling. Large amounts can cause blisters, numbness or aching pain. Methyl bromide can be absorbed through the skin and cause symptoms described under inhalation. Death has occurred from skin absorption. *Eye*: Can cause irritation, tearing, reddening or burning pain. *Ingestion:* Can cause throat and stomach irritation as well as symptoms described under inhalation. Delayed pulmonary edema may occur after inhalation.

*Note:* Do not wear *ordinary rubber gloves* or *adhesive bandages* while using methyl bromide. It can dissolve rapidly through rubber or adhesive tape and cause severe symptoms.

***Long Term Exposure:*** Repeated exposures have been associated with peripheral neuropathies, especially sensory neuropathy, impaired gait, behavioral changes, and mild liver and kidney dysfunction. Visual impairment secondary to atrophy of the optic nerve has been reported. Chronic exposure may be more serious for children because of their potential longer latency period. *Carcinogenicity:* The International Agency for Research on Cancer has determined that methyl bromide is not classifiable as to its carcinogenicity to humans. *Reproductive and Developmental Effects:* Methyl bromide is not considered a reproductive or developmental toxicant. No human data is available; one study of experimental animals (rats and rabbits) did not find teratogenic effects at levels below those causing maternal death. Levels between 20 and 35 ppm can cause symptoms as described under short term inhalation. Symptoms can last months or years, or can be permanent. Repeated or prolonged contact with skin may cause dermatitis, lung damage and broncho-spasms. Methyl bromide may affect the central nervous system, causing paralysis, poor vision, psychological disorders, hallucinations, numbness in the arms and legs, brain damage. May cause liver and kidney damage. Methyl bromide is a mutagen and may have a cancer risk. May damage the testes.

*Points of Attack:* Eyes, skin, respiratory system, central nervous system, brain. Cancer site in animals: lung, kidney, and forestomach tumors.

*Medical Surveillance:* Medical observation is recommended for 24 to 48 hours after breathing overexposure, as pulmonary edema may be delayed. Evaluate the central nervous system, respiratory tract, and skin in preplacement and periodic examinations. Examination of the nervous system. Blood test for bromides (unexposed persons usually have serum levels of 5 mg/L or below). Kidney function tests. Evaluation for brain effects.

*First Aid:* If this chemical gets into the eyes, remove any contact lenses at once and irrigate immediately for at least 15 minutes, occasionally lifting upper and lower lids. Seek medical attention immediately. If this chemical contacts the skin, remove contaminated clothing and wash immediately with soap and water. Seek medical attention immediately. If this chemical has been inhaled, remove from exposure, begin rescue breathing (using universal precautions) if breathing has stopped, and CPR if heart action has stopped. Transfer promptly to a medical facility. Oral exposure to methyl bromide is rare (it is a gas at temperatures above 38.5°F; however, if ingestion occurs, **do not induce vomiting**. If the victim is alert and able to swallow, administer a slurry of activated charcoal at 1 gm/kg (usual adult dose 60–90 g, child dose 25–50 g). A soda can and straw may be of assistance when offering charcoal to a child. Medical observation is recommended for 24 to 48 hours after breathing overexposure, as pulmonary edema may be delayed. As first aid for pulmonary edema, a doctor or authorized paramedic may consider administering a corticosteroid spray. If frostbite has occurred, seek medical attention immediately; do *NOT* rub the affected areas or flush them with water. In order to prevent further tissue damage, do *NOT* attempt to remove frozen clothing from frostbitten areas. If frostbite has *NOT* occurred, immediately and thoroughly wash contaminated skin with soap and water.

*Note:* Because cyanide is probably largely responsible for poisonings, antidotal measures against cyanide should be instituted promptly. Use amyl nitrate capsules if symptoms develop. All area employees should be trained regularly in emergency measures for cyanide poisoning and in CPR. A cyanide antidote kit should be kept in the immediate work area and must be rapidly available. Kit ingredients should be replaced every 1-2 years.

*Note to physician or authorized medical personnel:* In cases of respiratory compromise secure airway and respiration via endotracheal intubation. If not possible, perform cricothyroidotomy if equipped and trained to do so. Treat patients who have bronchospasm with aerosolized bronchodilators. The use of bronchial sensitizing agents in situations of multiple chemical exposures may pose additional risks. Consider the health of the myocardium before choosing which type of bronchodilator should be administered. Cardiac sensitizing agents may be appropriate; however, the use of cardiac sensitizing agents after exposure to certain chemicals may pose enhanced risk of cardiac arrhythmias (especially in the elderly). Consider racemic epinephrine aerosol for children who develop stridor. Dose 0.25–0.75 mL of 2.25% racemic epinephrine solution in 2.5 cc water, repeat every 20 minutes as needed, cautioning for myocardial variability. Patients who are comatose, hypotensive, or are having seizures or cardiac arrhythmias should be treated according to advanced life support protocols. If evidence of shock or hypotension is observed begin fluid administration. For adults, bolus 1000 mL/hour intravenous saline or lactated Ringer's solution if blood pressure is under 80 mmHg; if systolic pressure is over 90 mmHg, an infusion rate of 150 to 200 mL/hour is sufficient. For children with compromised perfusion administer a 20 mL/kg bolus of normal saline over 10 to 20 minutes, then infuse at 2 to 3 mL/kg/hour.

*References:*
- U.S. Environmental Protection Agency, Office of Pesticide Programs, Pesticide Residue Limits, "Methyl Bromide," 40 CFR 180.123, http://www.epa.gov/pesticides/food/viewtols.htm
- U.S. Department of Agriculture, Agriculture Research Service, "ARS Methyl Bromide Research," http://www.ars.usda.gov/is/mb/mebrweb.htm
- New Jersey Department of Health and Senior Services, "Hazardous Substance Fact Sheet, Methyl Bromide," Trenton NJ (June 1998). http://www.state.nj.us/health/eoh/rtkweb/1231.pdf
- U.S. Environmental Protection Agency, Halomethanes: Ambient Water quality Criteria, Wash., DC (1980).

- Agency for Toxic Substances and Disease Registry, U.S. Department of Health and Human Services, Public Health Service, "Managing Hazardous Materials Incidents," Atlanta, GA (June 2003)
- U.S. Environmental Protection Agency, Bromomethane, Health and Environmental Effects Profile No. 29, Wash., DC, Office of Solid Waste (April 30, 1980).
- U.S. Environmental Protection Agency, "Chemical Hazard Information Profile: Methyl Bromide," Wash., DC, Office of Toxic Substances (Feb. 20, 1985).
- U.S. Environmental Protection Agency, "Chemical Profile: Methyl Bromide," Wash., DC, Chemical Emergency Preparedness Program (Nov. 30, 1987).
- New York State Department of Health, "Chemical Fact Sheet: Methyl Bromide," Albany, NY, Bureau of Toxic Substance Assessment (March 1986 and Version 3).
- Sax, N.I., Ed., "Dangerous Properties of Industrial Materials Report" 5, No. 6, 37-40 (1985).
- California Environmental Protection Agency "Chemical List of Lists," Sacramento CA (February 1997).

# Methyl Formate

*Use Type:* Fumigant and larvicide
*CAS Number:* 107-31-3
*Formula:* $C_2H_4O_2$; $HCOOCH_3$
*Synonyms:* Formiate de methyle (French); Formiato de metilo (Spanish); Formic acid, methyl ester; Methyle (formiate de) (French); Methylformiaat (Dutch); Methylformiat (German); Methyl methanoate; Metil (formiato di) (Italian)
*Trade Names:* MAT 14500®, Matheson Tri-Gas (USA)
*Producers:* BASF (Germany); Celanese (Germany); ICI Group (UK); Kemira Chemicals (Finland); Matheson Tri-Gas (USA); Mitsubishi Gas Chemical (Japan); Samsung Fine Chemicals (South Korea)
*Chemical Class:* Aliphatic carboxylic acid ester
*EPA/OPP PC Code:* 053701
*California DPR Chemical Code:* 391
*ICSC Number:* 0664
*RTECS Number:* LQ8925000
*EEC Number:* 607-014-00-1
*EINECS Number:* 203-481-7
*Uses:* Methyl formate is a fumigant and larvicide for tobacco, dried fruits and cereals. It is also used as a solvent and as an intermediate in organic synthesis and in the preparation fo antileukemic agents.
*Regulatory Authority:*
- Air Pollutant Standard Set (DFG)[3] (HSE)[33] (OSHA)[58] (Several States)[60]
- Permissible Exposure Limits for Chemical Contaminants (CAL/OSHA)
- The "Director's List" (CAL/OSHA)
- Clean Air Act: Accidental Release Prevention/Flammable substances, (Section 112[r], Table 3), TQ = 10,000 lb (4540 kg)
- Canada, WHMIS, Ingredients Disclosure List

*Description:* Methyl formate is a colorless liquid with a pleasant odor. Odor threshold = 2,000 ppm. Molecular weight = 60.05. Soluble in water; solubility = $3.0 \times 10^5$ ppm @ 20°C. Boiling point = 31–32°C. Melting/Freezing point = –100°C. Vapor pressure = 586 mmHg @ 25°C. Flash point = –19°C; also listed at –9.4°C. Autoignition temperature = 449°C. Explosive limits: LEL = 4.5%; UEL = 23.0%. Hazard Identification (based on NFPA-704 M Rating System): Health 2, Flammability 4, Reactivity 0. Log $K_{ow}$ = negative. Unlikely to bioaccumulate in marine organisms.

*Incompatibilities:* Forms explosive mixture with air. Violent reaction with strong oxidizers. Reacts slowly with water to form methanol and formic acid.

*Permissible Exposure Limits in Air:* The Federal standard[58], the DFG MAK[3], the HSE[33] TWA and the ACGIH[1] TWA value is 100 ppm (250 mg/m$^3$) and ACHIH recommends a STEL of 150 ppm (375 mg/m$^3$). The NIOSH[2] IDLH level = 4,500 ppm. In addition, several states have set guidelines or standards for methyl formate in ambient air[60] ranging from 2.50-3.75 mg/m$^3$ (North Dakota) to 4.2 mg/m$^3$ (Virginia) to 5.0 mg/m$^3$ (Connecticut) to 5.952 mg/m$^3$ (Nevada).

*Determination in Air:* Collection by charcoal tube, analysis by gas liquid chromatography/flame ionization. See NIOSH Method #S291.[18]

*Permissible Concentration in Water:* No criteria set, but runoff from spills or fire control may cause water pollution.

*Routes of Entry:* Inhalation, skin absorption, ingestion, skin and/or eye contact. Passes through the skin.

*Harmful Effects and Symptoms*

*Short Term Exposure:* Methyl formate can affect you when breathed in and by passing through your skin. Exposure can irritate the eyes, nose, and throat. Higher levels can irritate the lungs and cause a build-up of fluid (pulmonary edema). This can cause death. High levels attack the nervous system and cause you to become dizzy, lightheaded, and may cause unconsciousness and death.

*Long Term Exposure:* Prolonged or repeated contact can cause cracking and drying of the skin. Repeated exposure can irritate the lungs and may cause bronchitis to develop.

*Points of Attack:* Eyes, lungs and central nervous system

*Medical Surveillance:* For those with frequent or potentially high exposure (half the TLV or greater), the following are recommended before beginning work and at regular times after that: Lung function tests. If symptoms develop or overexposure is suspected, the following may be useful: Consider chest x-ray after acute overexposure.

*First Aid:* If this chemical gets into the eyes, remove any contact lenses at once and irrigate immediately for at least

15 minutes, occasionally lifting upper and lower lids. Seek medical attention immediately. If this chemical contacts the skin, remove contaminated clothing and wash immediately with soap and water. Seek medical attention immediately. If this chemical has been inhaled, remove from exposure, begin rescue breathing (using universal precautions) if breathing has stopped, and CPR if heart action has stopped. Transfer promptly to a medical facility. When this chemical has been swallowed, get medical attention. Give large quantities of water and induce vomiting. Do not make an unconscious person vomit. Medical observation is recommended for 24 to 48 hours after breathing overexposure, as pulmonary edema may be delayed. As first aid for pulmonary edema, a doctor or authorized paramedic may consider administering a corticosteroid spray.

*References:*
- New Jersey Department of Health and Senior Services, "Hazardous Substance Fact Sheet: Methyl Formate," Trenton, NJ (March 1989). http://www.state.nj.us/health/eoh/rtkweb/1262.pdf
- California Environmental Protection Agency "Chemical List of Lists," Sacramento CA (February 1997).

# Methyl Parathion

*Use Type:* Insecticide and nematicide
*CAS Number:* 298-00-0
*Formula:* $C_8H_{10}NO_5PS$
*Alert:* A Restricted Use Pesticide (RUP). EPA has accepted voluntary cancellation of many of the most significant food crop uses of methyl parathion, one of the most toxic and most widely used organophosphate pesticides. Methyl parathion has been found to pose unacceptable dietary risks to children. Methyl parathion can no longer be used on food crops normally consumed by children. Human toxicity (long-term): High

*Synonyms:* Dimethyl *p*-nitrophenyl monothiophosphate; *O,O*-Dimethyl *O*-*p*-nitrofenylester kyseliny thiofosforecne (Czech); *O,O*-Dimethyl O-(4-nitrofenyl)-monothiofosfaat (Dutch); *O,O*-Dimethyl *O*-(4-nitrophenyl)-monothiophosphat (German); Dimethyl-*p*-nitrophenyl monothiophosphate; *O,O*-Dimethyl *O*-*p*-nitrophenyl phosphorothioate; *O,O*-Dimethyl *O*-(4-nitrophenyl)phosphorothioate; *O,O*-Dimethyl *O*-(*p*-nitrophenyl) phosphorothioate; Dimethyl *p*-nitrophenyl phosphorothionate; *O,O*-Dimethyl *O*-4-nitrophenyl phosphorothioate; Dimethyl 4-nitrophenyl phosphorothionate; *O,O*-Dimethyl *O*-(*p*-nitrophenyl) thionophosphate; Dimethyl *p*-nitrophenyl thiophosphate; *O,O*-Dimethyl *O*-*p*-nitrophenyl thiophosphate; *O,O*-Dimethyl *O*-(*p*-nitrophenyl) thiophosphate; Dimethyl parathion; *O,O*-Dimetil-*O*-(4-nitro-fenil)-monotiofosfato (Italian); ENT 17,292; Methyl fosferno; Methyl niran; Methylthiophos; Metilparationa (Spanish); Metilparation (Hungarian); Metyloparation (Polish); Metylparation (Czech); NCI-C02971; *p*-Nitrophenyldimethylthionophosphate; M-Parathion; Parathion-methyl; Parathion Metile (Spanish); Partron M; Phenol, *p*-Nitro-, *O*-ester with *O,O*-dimethyl phosphorothioate; Phosphorothioic acid, *O,O*-dimethyl *O*-(*p*-nitrophenyl) ester; Phosphorothioic acid, *O,O*-dimethyl *O*-(4-nitrophenyl) ester

*Trade Names:* A-GRO®, Arizona Agrochemical Company (USA), canceled 12/31/1987; AI3-17292®; ATOMIC®, Southern Chemical (USA), canceled 12/19/1988; AZOFOS®; AZOPHOS; BAMA BRAND®,Alabama Agriculture Service (USA), canceled 1/21/1983; BAY 11405®, Bayer CropScience (Germany); BAY E-601®, Bayer CropScience (Germany); BLADAN M®; CEKUMETHION®; CLEAN CROP®, Loveland Products (USA), canceled 7/1/1987; COTTON TOX DUST®, Griffin (USA), canceled 1/21/1983; DALF®; DECLARE®, Griffin (USA); DEVITHION®; DREXEL METHYL PARATHION 4E®, Drexel Chemical (USA), canceled; DURHAM®, Amvac Chemical (USA), canceled 9/29/1988; E 601®; EMMY®, Griffin (USA), canceled 7/1/1987; E-Z-FLO®, Bayer CropScience (Germany), canceled 7/1/1987; FALL OUT®, canceled; FMC NYNAMITE®, FMC Agricultural Products Group (USA), canceled 2/11/1983; FOLIDOC®, Bayer CropScience (Germany); FOLIDOL-80®, Bayer CropScience (Germany); FOLIDOL M®, Bayer CropScience (Germany); FOLIDOL M-40®, Bayer CropScience (Germany); FOSFERNO M 50®; GEARPHOS®; 8056HC®; KILEX PARATHION®; ME-PARATHION®; MEPTOX®; METACID 50®, Bayer CropScience (Germany); METACIDE®, Bayer CropScience (Germany); METAFOS®; METAPHOS®; METRON®; METHYL-E 605®; METRON®; NITROX®; NITROX® 80; OLEOVOFOTOX®; PARAPEST M-50®; PENNCAP M®, Cerexagri (France); PENNCAP MLS®, Cerexagri (France); QUINOPHOS®; SEIS-TRES 6-3®, Drexel Chemical (USA); SINAFID M-48®; SIXTY-THREE SPECIAL E.C. INSECTICIDE®; TEKWAISA®; THIOPHENIT®; THYLPAR M-50®; TOLL®; VERTAC METHYL PARATHION TECHNISCH 80%®; WOFATOX 50 EC®

*Producers:* Agsin (Singapore); Aimco Pesticides Ltd. (India); Alcotan Laboratories (Spain); Bayer CropScience (Germany); Bharat Rasayan (India); Cerexagri (France); Cheminova A/S (Denmark); China Chemical (China); Drexel Chemical (USA); FMC Agricultural Products Group (USA); Griffin (USA); Helena Chemical (USA); Ki-Hara Chemicals Ltd. (UK); Micro Flo (USA); Shandong Huayang Pesticide Group (China); Nagarjuna Agrichem (India); Tekchem (Mexico); YiHua Group (China)

*Chemical Class:* Organophosphate
*EPA/OPP PC Code:* 053501
*California DPR Chemical Code:* 394
*ICSC Number:* 0626
*RTECS Number:* TG175000
*EEC Number:* 015-035-00-7

# Methyl Parathion

*Uses:* This material is used as an insecticide on over 50 crops, primarily cotton, but also on walnuts, corn, dried beans and almonds and on several ornamentals. It can no longer be used on crops consumed by children.

*Human toxicity (long-term)*[77]: High–2.00 ppb, Health Advisory

*Fish toxicity (threshold)*[77]: High–5.59677 ppb, MATC (Maximum Acceptable Toxicant Concentration)

**U.S. Maximum Allowable Residue Levels for Methyl Parathion (40 CFR 180.121):** *Note:* The U.S. EPA allows a maximum of 0.1-1 part methyl parathion per million parts (ppm) of certain types of food.

| CROP | ppm |
|---|---|
| Alfalfa, forage | 1.25 |
| Alfalfa, hay | 5 |
| Almond | 0.1 |
| Almond, hulls | 3 |
| Barley | 1 |
| Bean, dry | 1 |
| Beet, sugar | 0.1 |
| Beet, sugar, tops | 0.1 |
| Cabbage | 1 |
| Clover | 1 |
| Corn | 1 |
| Corn | 1 |
| Corn, forage | 1 |
| Corn, forage | 1 |
| Cotton, undelinted seed | 0.75 |
| Cranberry | 1 |
| Cucumber | 1 |
| Currant | 1 |
| Date, dried fruit | 1 |
| Dewberry | 1 |
| Eggplant | 1 |
| Endive | 1 |
| Fig | 1 |
| Grass, forage | 1 |
| Hop | 1 |
| Hop, dried cones | 1 |
| Oat | 1 |
| Onion | 1 |
| Pea, dry | 1 |
| Pea, field, vines | 1 |
| Peanut | 1 |
| Pecan | 0.1 |
| Potato | 0.1 |
| Rapeseed, seed | 0.2 |
| Rice | 1 |
| Rye | 0.5 |
| Soybean | 0.1 |
| Soybean, hay | 1 |
| Sunflower, seed | 0.2 |
| Sweet potato | 0.1 |
| Walnut | 0.1 |
| Wheat | 1 |

*Carcinogen/Hazard Classifications*
**U.S. EPA Carcinogens:** Not likely a carcinogen
**IARC:** Group 3, unclassifiable
**Label Signal Word:** DANGER
**WHO Acute Hazard: Class 1 a, extremely hazardous**
**Endocrine Disruptor:** Suspected endocrine disruptor
*Regulatory Authority:*
- Banned or Severely Restricted (Several Countries) (UN)[13]
- Air Pollutant Standard Set (ACGIH)[1] (HSE)[33] (NIOSH)[2] (former USSR)[35] (Several States)[60]
- AB 1803-Well Monitoring Chemical (CAL)
- EPA/SARA 302 (EPCRA) Extremely hazardous substances
- MCL (Maximum Contaminants Levels) list of contaminants (CAL)
- CAL-DHS/DHS Drinking Water Action Levels
- U.S. DOT Inhalation Hazard Chemicals
- Permissible Exposure Limits for Chemical Contaminants (CAL/OSHA)
- The "Director's List" (CAL/OSHA)
- Actively registered pesticide in California.
- Clean Water Act: Section 311 Hazardous Substances/RQ 40CFR117.3 (same as CERCLA, see below).
- EPA Hazardous Waste Number (RCRA No.): P071
- RCRA, 40CFR261, Appendix 8 Hazardous Constituents
- RCRA 40CFR268.48; 61FR15654, Universal Treatment Standards: Wastewater (mg/L), 0.014; Nonwastewater (mg/kg), 4.6
- RCRA 40CFR264, Appendix 9; TSD Facilities Ground Water Monitoring List. Suggested test method(s) (PQL $ug/L$): 8140(0.5); 8270(10)
- Superfund/EPCRA 40CFR355, Appendix B Extremely Hazardous Substances: TPQ = 100/10,000 lb (45.4/4,540 kg)
- Superfund/EPCRA 40CFR302.4 RQ: CERCLA, 100 lb (45.4 kg)
- EPCRA Section 313 Form R de minimis concentration reporting level: 1.0%
- U.S. DOT Inhalation Hazard Chemicals as organophosphates
- U.S. DOT Regulated Marine Pollutant (49CFR172.101, Appendix B), severe pollutant
- Canada, WHMIS, Ingredients Disclosure List

*Description:* Methyl parathion is a white to yellow-brown, crystalline solid. The commercial product in xylene is a tan liquid (80% methyl parathion/20% xylene). Garlic odor. Very slightly soluble in water; solubility = 58 ppm. Molecular weight = 263.22. Melting/Freezing point = 35–38°C. Flash point = 46°C. Vapor pressure = low; 0.12 Pa @ 20°C. Log $K_{ow}$ = 2.7. Values above 3.0 are likely to bioaccumulate in marine organisms.

*Incompatibilities:* Incompatible with oxidizers, strong bases, heat. Mixtures with magnesium, Endrin may be violent or explosive. Slightly decomposed by acid solutions.

Rapidly decomposed by alkalies. Explosive risk when heated above 50°C. The liquid xylene solution decomposes violently at 120°C.

***Permissible Exposure Limits in Air:*** There is no OSHA[2] PEL. NIOSH[2] and ACGIH[1] recommend a TWA of 0.2 mg/m$^3$, both with the notation "skin" indicating the possibility of cutaneous adsorption. HSE[33] uses this same TWA and adds an STEL of 0.6 mg/m$^3$; Argentina[35] also has these same values. The former USSR has set an MAC in workplace air of 0.1 mg/m$^3$. The former USSR has set a MAC for ambient air in residential areas of 0.008 mg/m$^3$ on a once daily basis. Several states have set guidelines or standards for methyl parathion in ambient air[60] ranging from 2.0 μg/m$^3$ (North Dakota) to 3.5 μg/m$^3$ (Virginia) to 4.0 μg/m$^3$ (Connecticut) to 5.0 μg/m$^3$ (Nevada).

***Determination in Air:*** OSHA versatile sampler-2; Toluene/Acetone; Gas chromatography/Flame photometric detection for sulfur, nitrogen, or phosphorus; NIOSH IV, Method #5600, Organophosphorus pesticides.[18]

***Permissible Concentration in Water:*** The U.S. EPA allows 0.002 mg/L (2 μg/L) of drinking water for lifetime exposure of adults. The former USSR[35] has set an MAC in water bodies used for domestic purposes of 0.02 mg/L. Two states have set guidelines for methyl parathion in drinking water of 30 μg/L-in California and Kansas[61].

***Routes of Entry:*** Inhalation, skin absorption, ingestion, skin and/or eye contact.

*Harmful Effects and Symptoms*

**Short Term Exposure:** Methyl parathion may affect the nervous system, causing convulsion, respiratory failure, and death. A cholinesterase inhibitor. Acute exposure to parathion-methyl may produce the following symptoms: pinpoint pupils, blurred vision, headache, dizziness, muscle spasms, and profound weakness. Vomiting, diarrhea, abdominal pain, seizures, and coma may also occur. High exposure may result in death. The heart rate may decrease following oral exposure or increase following dermal exposure. Hypotension (low blood pressure) may occur although hypertension (high blood pressure) is not uncommon. Chest pain may be noted. Respiratory symptoms include dyspnea (shortness of breath), respiratory depression, and respiratory paralysis. Psychosis may occur. Because this is a mutagen, handle it as a possible carcinogen-with extreme caution. Methyl parathion may damage the developing fetus. This material is extremely toxic; the probable oral lethal dose is 5-50 mg/kg, or between 7 drops and 1 teaspoonful for a 150-lb person. Delayed pulmonary edema may occur after inhalation.

**Long Term Exposure:** Cholinesterase inhibitor; cumulative effect is possible. This chemical may damage the nervous system with repeated exposure, resulting in convulsions, respiratory failure. May cause personality changes; depression, anxiety, irritability. May cause liver damage. May damage the developing fetus.

***Points of Attack:*** Respiratory system, lungs, central nervous system, cardiovascular system, skin, eyes, plasma and red blood cell cholinesterase.

***Medical Surveillance:*** Medical observation is recommended for 24 to 48 hours after breathing overexposure, as pulmonary edema may be delayed. Before employment and at regular times after that, the following are recommended: Plasma and red blood cell cholinesterase levels (tests for the enzyme poisoned by this chemical). If exposure stops, plasma levels return to normal in 1-2 weeks while red blood cell levels may be reduced for 1-3 months. When acetylcholinesterase enzyme levels are reduced by 25% or more below preemployment levels, risk of poisoning is increased, even if results are in lower ranges of "normal." Reassignment to work not involving organophosphate or carbamate pesticides is recommended until enzyme levels recover. If symptoms develop or overexposure occurs, repeat the above tests as soon as possible and get an exam of the nervous system. Also consider complete blood count. Consider chest x-ray following acute overexposure. Do not drink any alcoholic beverages before or during use. Alcohol promotes absorption of organophosphates.

***First Aid:*** **Treatment for organophosphate poisoning consists of thorough decontamination, cardiorespiratory support, and administration of the antidotes atropine and pralidoxime. In cases of severe poisoning, diazepam, an anticonvulsant, should also be administered. Antidotes should be administered as prevention even if the diagnosis is in doubt.** Speed in removing material from eyes and skin is of extreme importance. *Eyes:* Eye contact can cause dangerous amounts of these chemicals to be quickly absorbed through the mucous membrane into the bloodstream. Immediately and gently flush eyes with plenty of warm or cold water (NO hot water) for at least 15 minutes, occasionally lifting the upper and lower eyelids. Get medical aid immediately. *Skin:* Get medical aid. Skin contact can cause dangerous amounts of these chemicals to be absorbed into the bloodstream. Wearing the appropriate PPE equipment and respirator for organophosphate pesticides, immediately flush skin with plenty of soap and water for at least 15 minutes while removing contaminated clothing and shoes. Shampoo hair promptly if contaminated. The removed, contaminated clothing and shoes should be double-bagged and left in Hot Zone for later disposal by hazardous materials experts. Skin may also be decontaminated with diluted hypochlorite solution. *Inhalation:* Get medical aid. Do not contaminate yourself. Wearing the appropriate PPE equipment and respirator for organophosphate pesticides, immediately remove the victim from the contaminated area to fresh air. If the victim is not breathing, administer artificial respiration. Do NOT use mouth-to-mouth resuscitation; use bag/mask apparatus. If breathing is difficult, administer oxygen through bag/mask apparatus until medical help arrives. Do not leave victim unattended. *Ingestion:* Call poison control. Loosen all clothing. Never give anything by mouth to an unconscious person. If victim is *unconscious or having convulsions,* do nothing except keep victim warm. Get medical aid. Transfer promptly to a medical facility. In cases of ingestion, **do not induce vomiting**. If the victim is alert and asymptomatic, administer a slurry of activated charcoal at a dose of 1 g/kg (infant, child, and adult dose). A soda can and straw may be of assistance when offering charcoal to a child. *In some cases you may be specifically instructed by poison control to induce vomiting by way of 2 tablespoons of syrup of ipecac (adult) washed down with a cup of water.* Do NOT give activated charcoal before or with ipecac syrup.

*Note to physician or authorized medical personnel:* Treat cases of respiratory compromise, coma, or excessive pulmonary secretions with respiratory support using protocols and techniques available and within the scope of training. Some cases may necessitate procedures such as endotracheal intubation or cricothyrotomy by properly trained and equipped personnel. When possible, atropine (see *Antidotes,* below) should be given under medical supervision. Patients who are comatose, hypotensive, or having seizures or cardiac arrhythmias should be treated according to advanced life support protocols. *Antidotes:* Two antidotes are administered to treat organophosphate poisoning. Atropine is a competitive antagonist of acetylcholine at muscarinic receptors and is used to control the excessive bronchial secretions which are often responsible for death. Pralidoxime relieves both the nicotinic and muscarine effects of organophosphate poisoning by regenerating acetylcholinesterase and can reduce both the bronchial secretions and the muscle weakness associated with poisoning. The initial intravenous dose of atropine in adults should be determined by the severity of symptoms: An initial adult dose of 1.0 to 2.0 mg or pediatric dose of 0.01 mg/kg (minimum 0.01 mg) should be administered intravenously. If intravenous access cannot be established, atropine may also be given intramuscularly, subcutaneously or via endotracheal tube. Doses should be repeated every 15 minutes until excessive secretions and sweating have been controlled. Once bronchial secretion has been controlled, atropine administration should be repeated whenever the secretions begin to recur. In seriously poisoned patients, very large doses may be required. Alterations of pulse rate and pupillary size should not be used as indicators of treatment adequacy. Pralidoxime should be administered as early in poisoning as possible as its efficacy may diminish when given more than 24 to 36 hours after exposure. Doses are as follows: adult 1.0 g; pediatric 25 to 50 mg/kg. The drug should be administered intravenously over 30 to 60 minutes, but in a life-threatening situation, one-half of the total dose can be given per minute for a total administration time of 2 minutes. Treatment should begin to take effect within 40 minutes with a reduction in symptoms and the amount of atropine necessary to control bronchial secretion. The initial dose can be repeated in 1 hour and then every 8 to 12 hours until the patient is clinically well and no longer requires atropine. If intravenous access cannot be established, pralidoxime may also be given intramuscularly. Early administration of diazepam in addition to the combined atropine and pralidoxime treatment may help prevent the onset of seizures and potential brain and cardiac morphologic damage following high-level organophosphate poisoning.

***References:***

- U.S. Environmental Protection Agency, "Methyl Parathion Rick Management Decision," Washington, DC (August 10, 1999). http://www.epa.gov/pesticides/factsheets/chemicals/mp factsheet.htm#action
- U.S. Environmental Protection Agency, Office of Pesticide Programs, Pesticide Residue Limits, "Methyl Parathion," 40 CFR 180.121, http://www.epa.gov/pesticides/food/viewtols.htm
- EXTOXNET, Extension Toxicology Network, "Pesticide Information Profile, Methyl Parathion," Oregon State University, Corvallis, OR (June 1996). http://extoxnet.orst.edu/pips/methylpa.htm
- Agency for Toxic substances and Disease Registry, "ToxFAQs for Methyl Parathion," Atlanta, GA (September 2001). http://www.atsdr.cdc.gov/tfacts48.html
- New Jersey Department of Health and Senior Services, "Hazardous Substance Fact Sheet: Methyl Parathion," Trenton, NJ (November 1999). http://www.state.nj.us/health/eoh/rtkweb/1283.pdf
- National Institute for Occupational Safety and Health, Criteria for a Recommended Standard: Occupational Exposure to Methyl Parathion, NIOSH Doc. No. 77-106 (1977).
- Sax, N.I., Ed., "Dangerous Properties of Industrial Materials Report" 6, No. 1, 90-97 (1986).
- U.S. Environmental Protection Agency, "Chemical Profile: Parathion-Methyl," Washington, DC, Chemical Emergency Preparedness Program (Nov. 30, 1987).
- U.S. Environmental Protection Agency, "Health Advisory: Methyl Parathion," Washington, DC, Office of Drinking Water (August 1987).
- California Environmental Protection Agency "Chemical List of Lists," Sacramento CA (February 1997).

# Methyl Phenkapton

*Use Type:* Acaricide and insecticide
*CAS Number:* 3735-23-7
*Formula:* $C_9H_{11}Cl_2O_2PS_3$
*Alert:* Not registered as a pesticide in the U.S.
*Synonyms:* (2,5-Dichlorophenylthio)methanethiol-*S*-ester with *O,O*-dimethyl phosphorodithioate; *S*-[((2,5-

# Methyl Phenkapton

Dichlorophenyl)thio)methyl] *O,O*-dimethyl phosphorodithioate; *O,O*-Dimethyl *S*-(2,5-dichlorophenylthio)methyl phosphorodithioate; *O.O*-Dimethyl *S*-(2,5-dichlorophenylthio)methyl phosphorodithioate; ENT 25,554; Methyl phencapton

*Trade Names:* GEIGY 30494®, Syngenta (Switzerland)
*Producers:* Syngenta (Switzerland)
*Chemical Class:* Organophosphate
*EPA/OPP PC Code:* 362200
*RTECS Number:* TD6125000
*Uses:* This material is an acaricide, insecticide. Not registered as a pesticide in the U.S.

*Regulatory Authority:*
- Superfund/EPCRA 40CFR355, Appendix B Extremely Hazardous Substances: TPQ = 500 lb (227kg)
- Superfund/EPCRA 40CFR302.4 RQ: EHS, 1 lb (0.454 kg)
- U.S. DOT Inhalation Hazard Chemicals as organophosphates

*Description:* Methyl phenkapton is a liquid. Hazard Identification (based on NFPA-704 M Rating System): Health 3, Flammability 1, Reactivity 0.

*Incompatibilities:* Strong oxidizers.

*Permissible Exposure Limits in Air:* No standards set.

*Permissible Concentration in Water:* No criteria set, but runoff from spills or fire control may cause water pollution.

*Routes of Entry:* Inhalation, ingestion, eye and/or skin contact.

*Harmful Effects and Symptoms*

*Short Term Exposure:* Organic phosphorus insecticides are absorbed by the skin, as well as by the respiratory and gastrointestinal tracts. They are cholinesterase inhibitors. Symptoms of exposure include headache, giddiness, blurred vision, nervousness, weakness, nausea, cramps, diarrhea, and discomfort in the chest. Signs include sweating, tearing, salivation, vomiting, cyanosis, convulsions, coma, loss of reflexes and loss of sphincter control. Delayed pulmonary edema may occur after inhalation.

*Long Term Exposure:* Cholinesterase inhibitor; cumulative effect is possible. This chemical may damage the nervous system with repeated exposure, resulting in convulsions, respiratory failure. May cause liver damage.

*Points of Attack:* Respiratory system, lungs, central nervous system, cardiovascular system, skin, eyes, plasma and red blood cell cholinesterase.

*Medical Surveillance:* Medical observation is recommended for 24 to 48 hours after breathing overexposure, as pulmonary edema may be delayed. Before employment and at regular times after that, the following are recommended: Plasma and red blood cell cholinesterase levels (tests for the enzyme poisoned by this chemical). If exposure stops, plasma levels return to normal in 1-2 weeks while red blood cell levels may be reduced for 1-3 months. When acetylcholinesterase enzyme levels are reduced by 25% or more below preemployment levels, risk of poisoning is increased, even if results are in lower ranges of "normal." Reassignment to work not involving organophosphate or carbamate pesticides is recommended until enzyme levels recover. If symptoms develop or overexposure occurs, repeat the above tests as soon as possible and get an exam of the nervous system. Also consider complete blood count. Consider chest x-ray following acute overexposure. Do not drink any alcoholic beverages before or during use. Alcohol promotes absorption of organophosphates.

*First Aid:* **Treatment for organophosphate poisoning consists of thorough decontamination, cardiorespiratory support, and administration of the antidotes atropine and pralidoxime. In cases of severe poisoning, diazepam, an anticonvulsant, should also be administered. Antidotes should be administered as prevention even if the diagnosis is in doubt.** Speed in removing material from eyes and skin is of extreme importance. *Eyes:* Eye contact can cause dangerous amounts of these chemicals to be quickly absorbed through the mucous membrane into the bloodstream. Immediately and gently flush eyes with plenty of warm or cold water (NO hot water) for at least 15 minutes, occasionally lifting the upper and lower eyelids. Get medical aid immediately. *Skin:* Get medical aid. Skin contact can cause dangerous amounts of these chemicals to be absorbed into the bloodstream. Wearing the appropriate PPE equipment and respirator for organophosphate pesticides, immediately flush skin with plenty of soap and water for at least 15 minutes while removing contaminated clothing and shoes. Shampoo hair promptly if contaminated. The removed, contaminated clothing and shoes should be double-bagged and left in Hot Zone for later disposal by hazardous materials experts. Skin may also be decontaminated with diluted hypochlorite solution. *Inhalation:* Get medical aid. Do not contaminate yourself. Wearing the appropriate PPE equipment and respirator for organophosphate pesticides, immediately remove the victim from the contaminated area to fresh air. If the victim is not breathing, administer artificial respiration. Do NOT use mouth-to-mouth resuscitation; use bag/mask apparatus. If breathing is difficult, administer oxygen through bag/mask apparatus until medical help arrives. Do not leave victim unattended. *Ingestion:* Call poison control. Loosen all clothing. Never give anything by mouth to an unconscious person. If victim is *unconscious or having convulsions,* do nothing except keep victim warm. Get medical aid. Transfer promptly to a medical facility. In cases of ingestion, **do not induce vomiting**. If the victim is alert and asymptomatic, administer a slurry of activated charcoal at a dose of 1 g/kg (infant, child, and adult dose). A soda can and straw may be of assistance when offering charcoal to a child. *In some cases you may be specifically instructed by poison control to induce vomiting by way of 2 tablespoons of syrup of ipecac (adult) washed down with a cup of water.* Do NOT give activated charcoal before or with ipecac syrup.

*Note to physician or authorized medical personnel:* Treat cases of respiratory compromise, coma, or excessive pulmonary secretions with respiratory support using protocols and techniques available and within the scope of training. Some cases may necessitate procedures such as endotracheal intubation or cricothyrotomy by properly trained and equipped personnel. When possible, atropine (see *Antidotes*, below) should be given under medical supervision. Patients who are comatose, hypotensive, or having seizures or cardiac arrhythmias should be treated according to advanced life support protocols. *Antidotes:* Two antidotes are administered to treat organophosphate poisoning. Atropine is a competitive antagonist of acetylcholine at muscarinic receptors and is used to control the excessive bronchial secretions which are often responsible for death. Pralidoxime relieves both the nicotinic and muscarine effects of organophosphate poisoning by regenerating acetylcholinesterase and can reduce both the bronchial secretions and the muscle weakness associated with poisoning. The initial intravenous dose of atropine in adults should be determined by the severity of symptoms: An initial adult dose of 1.0 to 2.0 mg or pediatric dose of 0.01 mg/kg (minimum 0.01 mg) should be administered intravenously. If intravenous access cannot be established, atropine may also be given intramuscularly, subcutaneously or via endotracheal tube. Doses should be repeated every 15 minutes until excessive secretions and sweating have been controlled. Once bronchial secretion has been controlled, atropine administration should be repeated whenever the secretions begin to recur. In seriously poisoned patients, very large doses may be required. Alterations of pulse rate and pupillary size should not be used as indicators of treatment adequacy. Pralidoxime should be administered as early in poisoning as possible as its efficacy may diminish when given more than 24 to 36 hours after exposure. Doses are as follows: adult 1.0 g; pediatric 25 to 50 mg/kg. The drug should be administered intravenously over 30 to 60 minutes, but in a life-threatening situation, one-half of the total dose can be given per minute for a total administration time of 2 minutes. Treatment should begin to take effect within 40 minutes with a reduction in symptoms and the amount of atropine necessary to control bronchial secretion. The initial dose can be repeated in 1 hour and then every 8 to 12 hours until the patient is clinically well and no longer requires atropine. If intravenous access cannot be established, pralidoxime may also be given intramuscularly. Early administration of diazepam in addition to the combined atropine and pralidoxime treatment may help prevent the onset of seizures and potential brain and cardiac morphologic damage following high-level organophosphate poisoning.

*References:*
- U.S. Environmental Protection Agency, "Chemical Profile: Methyl Phenkapton," Washington, DC, Chemical Emergency Preparedness Program (Nov. 30, 1987).
- California Environmental Protection Agency "Chemical List of Lists," Sacramento CA (February 1997).
- Agency for Toxic Substances and Disease Registry, U.S. Department of Health and Human Services, Public Health Service, "Managing Hazardous Materials Incidents," Atlanta, GA (June 2003).

# Methyl Thiocyanate

*Use Type:* Insecticide and fumigant
*CAS Number:* 556-64-9
*Formula:* $C_2H_3NS$; $CH_3CNS$
*Alert:* This is not produced in the U.S.
*Synonyms:* Methyl rhodanate; Methylrhodanid (German); Methyl sulfocyanate; Methylthiokyanat; Thiocyanic acid, methyl ester
*Producers:* Tokyo Kasei Kogyo (Japan)
*Chemical Class:* Thiocyanate (aliphatic)
*RTECS Number:* XL1575000
*Uses:* It is used as an agricultural insecticide, a fumigant, and as a research chemical. No evidence of commercial production in the U.S.
*Regulatory Authority:*
- AB 2588-Air Toxics "Hot Spots" Chemicals (CAL)
- CAL Air Resources Board/AB 1807 Toxic Air Contaminants
- Clean Air Act: Accidental Release Prevention/Flammable substances, (Section 112[r], Table 3), TQ = 20,000 lb (9080kg)
- Clean Air Act: Hazardous Air Pollutants (Title I, Part A, Section 112) as cyanide compounds
- Clean water act: Section 307 Priority Pollutants as cyanide, total; Toxic Pollutant (Section 401.15) as cyanide compounds
- EPA Hazardous Waste Number (RCRA No.): P030 as cyanide compounds
- RCRA Section 261 Hazardous Constituents as cyanide compounds
- RCRA Universal Treatment Standards: Wastewater (mg/L), 1.2 (total); 0.86 (amenable); Nonwastewater (mg/kg), 590 (total); 30 (amenable) as cyanide compounds
- RCRA Ground Water Monitoring List. Suggested test method(s) (PQL $ug$/L): 9010(40) as cyanide compounds
- Safe Drinking Water Act: MCL, 0.01 mg/L; MCLG, 0.01 mg/L;Regulated chemical (47 FR 9352) as cyanide compounds
- EPCRA Section 304 RQ: CERCLA, 10 lb (4.54 kg) as cyanide compounds
- EPCRA Section 313 Form R *de minimus* concentration reporting level: 1.0% as cyanide compounds
- U.S. DOT Regulated Marine Pollutant (49CFR, Subchapter 172.101, Appendix B) as cyanides, mixtures or solutions

- Superfund/EPCRA 40CFR355, Appendix B Extremely Hazardous Substances: TPQ = 10,000 lb (4,540 kg)
- Superfund/EPCRA 40CFR302.4 RQ: EHS, 1 lb (0.454 kg)

*Description:* Methyl thiocyanate is a colorless liquid with an onion-like odor. Boiling point = 130–133°C. Melting/Freezing point = –51°C. Very slightly soluble in water.

*Incompatibilities:* Incompatible with nitric acid. Violent reactions have occurred when mixed with chlorates, nitrates, nitric acid, peroxides, potassium chlorate, and sodium chlorate.

*Permissible Exposure Limits in Air:* Exposure limits for *hydrogen cyanide:* OSHA[2]: PEL is 10 ppm average over an 8-hour workshift; NIOSH[2]: recommended exposure limit is 4.7 ppm; ACGIH[1] : 4.7 ppm. These limits are for air levels only. When skin contact also occurs, overexposure results even when air levels are less than the limits above.

*Permissible Concentration in Water:* The U.S. EPA has set a maximum contaminant level of cyanide in drinking water of 0.2 milligrams cyanide per liter of water (0.2 mg/L). Runoff from spills or fire control may cause water pollution.

*Routes of Entry:* Inhalation, ingestion, skin and/or eye contact.

*Harmful Effects and Symptoms*

*Short Term Exposure:* Prolonged skin absorption may produce various eruptions, runny nose, dizziness, cramps, nausea, vomiting and mild or severe disturbances of the nervous system. This material is highly toxic if ingested. The ingestion of a concentrated solution may lead to vomiting. The principal systemic reaction is probably one of central nervous system depression, interrupted by periods of restlessness, abnormally fast and deep respiratory movements and convulsions. Death is usually due to respiratory arrest from paralysis of the medullary centers.

*Long Term Exposure:* May cause injury to the liver and kidneys. *Developmental Effects:* No reproductive or developmental effects of this thiocyanate have been reported in experimental animals or humans. Increased levels of thiocyanate in the umbilical cords of fetuses whose mothers smoked compared to those whose mothers were non-smokers suggests that thiocyanate, and possibly also cyanide, can cross the placenta.

*Points of Attack:* Liver and kidneys.

*Medical Surveillance:* Liver and kidney function tests. Urine thiocyanate levels. Blood cyanide levels.

*First Aid:* Treatment is as for aliphatic thiocyanates. If this chemical gets into the eyes, remove any contact lenses at once and irrigate immediately for at least 30 minutes, occasionally lifting upper and lower lids. Seek medical attention immediately. If this chemical contacts the skin, remove contaminated clothing and wash immediately with soap and water. Seek medical attention immediately. If this chemical has been inhaled, remove from exposure, begin rescue breathing (using universal precautions) if breathing has stopped, and CPR if heart action has stopped. Transfer promptly to a medical facility. Victims who are conscious and able to swallow should be given 4 to 8 ounces of water or milk. Gastric lavage with a small bore NG tube should be considered if it can be performed within 1 hour after ingestion. The effectiveness of activated charcoal administration is unknown, but it is suggested following lavage (administer activated charcoal at 1 gm/kg, usual adult dose 60–90 g, child dose 25–50 g). A soda can and straw may be of assistance when offering charcoal to a child. Medical observation is recommended for 24 to 48 hours after breathing overexposure, as pulmonary edema may be delayed. As first aid for pulmonary edema, a doctor or authorized paramedic may consider administering a corticosteroid spray.

*Note:* Because cyanide is probably largely responsible for poisonings, antidotal measures against cyanide should be instituted promptly. Use amyl nitrate capsules if symptoms develop. All area employees should be trained regularly in emergency measures for cyanide poisoning and in CPR. A cyanide antidote kit should be kept in the immediate work area and must be rapidly available. Kit ingredients should be replaced every 1-2 years to ensure freshness. Persons trained in the use of this kit, oxygen use, and CPR must be quickly available.

*References:*
- New Jersey Department of Health and Senior Services, "Hazardous Substance Fact Sheet, Methyl Thiocyanate," Trenton, NJ (June 2002). http://www.state.nj.us/health/eoh/rtkweb/2562.pdf
- U.S. Environmental Protection Agency, "Chemical Profile: Methyl Thiocyanate," Washington, DC, Chemical Emergency Preparedness Program (Nov. 30, 1987).
- California Environmental Protection Agency "Chemical List of Lists," Sacramento CA (February 1997).

## Methylmercuric Dicyanamide

*Use Type:* Fungicide
*CAS Number:* 502-39-6
*Formula:* $C_3H_6HgN_4$

*Alert:* This is a teratogen and should be handled with extreme caution. It is not registered in the U.S.

*Synonyms:* Cyanoguanidine methyl mercury derivative; Cyano(methylmercury)guanidine; Guanidine, cyano-, methylmercury deriv; MEMA; Methylmercuric cyanoguanidine; Methylmercury dicyanandimide; Methylmercury dicyandiamide; Methylmerkuridikyandiamid (German); MMD; Zaprawa nasienna plynna (Polish)

*Trade Names:* AGROSOL®; MORSODREN®; MORTON EP-227®; MORTON SOIL DRENCH®; PANDRINOX®; PANO-DRENCH®–4; PANODRIN® A-13; PANOGEN®; PANOGEN-PX®; PANOGEN TURF FUNGICIDE®;

PANOGEN TURF SPRAY; PANOSPRAY® 30; R 8®; R 8 FUNGICIDE®

*Chemical Class:* Organomercury compound

*EPA/OPP PC Code:* 051909

*California DPR Chemical Code:* 454

*RTECS Number:* OW1750000

*Uses:* This material is used as a fungicide; a seed, soil, and turf treatment especially for cereals, sorghum, sugar beets, cotton, and flax. It is not registered as a pesticide in the U.S.

*Carcinogen/Hazard Classifications*

**California Prop. 65:** Listed; Reproductive toxin

**IARC:** Group 2B, possible carcinogen

*Regulatory Authority:*
- Air Pollutant Standard Set (ACGIH)[1] (HSE)[33] (OSHA)[58]
- AB 2588-Air Toxics "Hot Spots" Chemicals (CAL)
- CAL Air Resources Board/AB 1807 Toxic Air Contaminants
- Proposition 65 chemical (CAL)
- Permissible Exposure Limits for Chemical Contaminants (CAL/OSHA)
- The "Director's List" (CAL/OSHA)
- Clean Water Act: Section 307 Toxic Pollutants as mercury and compounds
- Safe Drinking Water Act: MCL, 0.002 mg/L; MCLG, 0.002 mg/L as mercury
- RCRA Section 261 Hazardous Constituents, waste number not Listed as mercury compounds, n.o.s.
- EPCRA Section 302 Extremely Hazardous Substances: TPQ = 500/10,000 lb (227/4,540 kg)
- EPCRA Section 304 RQ: EHS, 1 lb (0.454 kg)
- EPCRA Section 313 (as mercury compound) Form R *de minimus* concentration reporting level: 1.0%
- MARINE POLLUTANT (49CFR, Subchapter 172.101, Appendix B), severe pollutant
- Canada, WHMIS, Ingredients Disclosure List

*Description:* Methylmercuric dicyanamide is a crystalline solid. Melting/Freezing point = 156°C. Hazard Identification (based on NFPA-704 M Rating System): Health 3, Flammability 1, Reactivity 0. Soluble in water.

*Permissible Exposure Limits in Air:* The OSHA[2] PEL for all organic mercury compounds is 0.01 mg/m$^3$ as an 8-hour TWA and ceiling of 0.04 mg/m$^3$. NIOSH[2] recommends 0.01 mg/m$^3$ as a 10-hour TWA and STEL of 0.03 mg/m$^3$ with the notation "skin" indicating the possibility of cutaneous absorption. ACGIH[1] recommends 0.01 mg/m$^3$ as an 8-hour TWA value with no STEL. The NIOSH[2] IDLH level = 2 mg/m$^3$ (as Hg). The DFG[3] has set an MAK for methyl mercury of 0.01 mg/m$^3$. They have also set[35] an STEL of 0.1 mg/m$^3$. Other countries have generally adopted the 0.01 mg/m$^3$ level. In addition, North Dakota has set guidelines for alkyl mercury compounds in ambient air[60] of 1-3 $\mu$g/m$^3$ (0.0001 to 0.0003 mg/m$^3$).

*Determination in Air:* No method available.

*Permissible Concentration in Water:* Methylmercury: To protect freshwater aquatic life: 0.016 $\mu$g/L as a 24-hr average, never to exceed 8.8 $\mu$g/L. To protect saltwater aquatic life: 0.025 $\mu$g/L as a 24-hr average, never to exceed 2.8 $\mu$g/L. To protect human health: 0.2 $\mu$g/L[6].

*Determination in Water:* Total mercury is determined by flameless atomic absorption. Soluble mercury may be determined by 0.45 micron filtration followed by flameless atomic absorption.

*Routes of Entry:* Inhalation, ingestion, eye, and/or skin contact.

*Harmful Effects and Symptoms*

*Short Term Exposure:* Alkyl mercury compounds are primary skin irritants and may cause dermatitis. When deposited on the skin, they give no warning, and if contact is maintained, can cause second-degree burns and blisters. Sensitization may occur. In the case of ingestion there is nausea and abdominal pain. Vomiting and diarrhea may occur. Burning or prickling of the lips, tongue, and extremities. The patient may be confused, hallucinate, be irritable, have disturbed sleep, lose muscular coordination and lose memory. Visual fields may narrow concentrically; emotional instability may occur as well as inability to concentrate, with stupor and coma. Methylmercuric dicyanamide is extremely toxic to humans. The probable lethal dose for humans is 5-50 mg/kg of body weight (between 7 drops and one teaspoon for a 150 lb person). Humans may be poisoned by feeding on the flesh of animals which have ingested this fungicide. Eating treated seeds may also cause poisoning. The poisoning may show delayed manifestations on the nervous system. Patients frequently become gradually worse after their illness is recognized and exposure is stopped.

*Long Term Exposure:* Repeated or prolonged contact with skin may result in dermatitis (red inflamed skin). Repeated or prolonged exposure may cause death by hypovolemic shock, nephrotic syndrome, or kidney failure. The central nervous system, including the brain, is the principal target tissue for this group of toxic compounds. Severe poisoning may produce irreversible brain damage resulting in loss of higher functions. The effects of chronic poisoning with alkyl mercury compounds are progressive. In the early stages, there are fine tremors of the hands, and in some cases, of the face and arms. With continued exposure, tremors may become coarse and convulsive; scanning speech with moderate slurring and difficulty in pronunciation may also occur. The worker may then develop an unsteady gait of a spastic nature which can progress to severe ataxia of the arms and legs. Sensory disturbances including tunnel vision, blindness, and deafness are also common. A late symptom, constriction of the visual fields, is rarely reversible and may be associated with loss of understanding and reason which makes the victim completely out of touch with his environment. Severe cerebral effects have been seen in infants born to mothers who had eaten large amounts of methylmercury-contaminated fish.

*Points of Attack:* Eyes, skin, central nervous system, peripheral nervous system and kidneys.

*Medical Surveillance:* Before first exposure and every 6 to 12 months after, a complete medical history and exam is strongly recommended, with: Eye exam. Consider lung function tests for persons with frequent exposure. Exam of the nervous system. Routine urine test (UA). Urine test for mercury (should be less than 0.02 mg/Liter). Consider nerve conduction tests, urinary enzymes and neurobehavioral test. After suspected illness or overexposure, repeat the tests above and get a blood test for mercury. Examination of the central nervous system, and kidneys. Consideration should be given to the possible effects on the fetus of alkyl mercury exposure in the mother. Constriction of visual fields may be a useful diagnostic sign. Blood and urine levels of mercury have been studied, especially in the case of methylmercury.

*First Aid:* If this chemical gets into the eyes, remove any contact lenses at once and irrigate immediately for at least 15 minutes, occasionally lifting upper and lower lids. Seek medical attention immediately. If this chemical contacts the skin, remove contaminated clothing and wash immediately with soap and water. Seek medical attention immediately. If this chemical has been inhaled, remove from exposure, begin rescue breathing (using universal precautions) if breathing has stopped, and CPR if heart action has stopped. Transfer promptly to a medical facility. When this chemical has been swallowed, get medical attention. Give large quantities of water and induce vomiting. Do not make an unconscious person vomit. Keep victim quiet and maintain normal body temperature. Effects may be delayed; keep victim under observation.

*References:*
- New Jersey Department of Health and Senior Services, "Hazardous Substance Fact Sheet, Methyl Mercury Dicyandiamide," Trenton NJ (May 2000). http://www.state.nj.us/health/eoh/rtkweb/1276.pdf
- U.S. Environmental Protection Agency, "Chemical Profile: Methylmercuric Dicyanamide," Washington, DC, Chemical Emergency Preparedness Program (Nov. 30, 1987).
- California Environmental Protection Agency "Chemical List of Lists," Sacramento CA (February 1997).

# Metiram

*Use Type:* Fungicide
*CAS Number:* 9006-42-2
*Formula:* $C_{16}H_{33}N_{11}S_{16}Zn_3$
*Synonyms:* Carbamic acid, 1*H*-benzimidazol-2-yl-, carbatene; Carbamodithioic acid, 1,2-ethanydiylbis-, polymer with ammonia complex of zinc EBDC; Caswell No. 041A; EBDC, polymer with ammonia complex of zinc EBDC; Ethylenebis(dithiocarbamic acid), polymer with ammonia complex of zinc EBDC; Mixture of 5.2 parts by weight (83.9%) of [ethylenebis(dithiocarbamato)] zinc with 1 part by weight (16.1%) ethylenebis(dithiocarmabic acid), bimolecular and trimolecular cyclic anhydrosulfides and disulfides; *tris*[Ammine(ethylenebis(dithiocarbamato))] zinc(2+1) (tetrahydro-1,2,4,7-dithiadiazocine-3,8-dithione)polymer; Zinc ammoniate ethylenebis(dithiocarbamate)-poly(ethylenethiuram disulfide); Zinc metiram; Zineb-ethylene thiuram disulfide adduct

*Trade Names:* ATLAS® BRAND, Helena Chemical (USA), canceled; AMAREX®; NIA 9102®, FMC Agricultural Products (USA), canceled; NIAGARA, FMC Agricultural Products (USA), canceled; POLYCARBACIN®; POLYCARBACINE®; POLYCARBAZIN®; POLYCARBAZINE®; POLYMARCIN®; POLYMARCINE®; POLYMARSIN®; POLYMARZIN®; POLYMARZINE®; POLYRAM®, BASF Agricultural Products (Germany);

*Producers:* AMVAC Chemical (USA); BASF Agricultural Products (Germany)

*Chemical Class:* Dithiocarbamate; inorganic zinc compound

*EPA/OPP PC Code:* 014601

*California DPR Chemical Code:* 493

*Uses:* Metiram is used to protect fruits, vegetables, field crops and other crops and ornamentals against many types of fungi and other foliar diseases.

*Human toxicity (long-term)[77]:* High–2.10 ppb, Health Advisory

*Fish toxicity (threshold)[77]:* Intermediate–24.87820 ppb, MATC (Maximum Acceptable Toxicant Concentration)

*U.S. Maximum Allowable Residue Levels for Metiram (40 CFR 180.217):*

| CROP | ppm |
|---|---|
| Apple | 2.0 |
| Potato | 0.5 |

*Carcinogen/Hazard Classifications*
**U.S. EPA Carcinogens:** Group B2, probable carcinogen
**California Prop. 65:** Carcinogen and developmental toxin
**TRI Developmental Toxin:** Developmental toxin
**Label Signal Word:** CAUTION
**WHO Acute Hazard:** Class U, unlikely to be hazardous
**Endocrine Disruptor:** Confirmed ED by the German Federal Environmental Agency

*Regulatory Authority:*
- EPA 40 CFR 372.65, Specific Toxic Chemical Listings
- Proposition 65 chemical (CAL) Carcinogen.
- Actively registered pesticide in California.
- EPCRA Section 313 Form R *de minimis* concentration reporting level: 1.0%
- EPA Hazardous Waste Number (RCRA No.): U114
- Clean Water Act: Section 307 Toxic Pollutants as zinc and compounds

- Safe Drinking Water Act: SMCL, 5 mg/L; Priority List (55 FR 1470) as zinc
- AB 2588-Air Toxics "Hot Spots" Chemicals (CAL) as zinc compounds
- The "Director's List" (CAL/OSHA) as zinc compounds

*Description:* Yellow powder at room temperature. Molecular weight 1088.7. Melting/Freezing point = decomposes above 140°C. Vapor pressure = $<10^{-7}$ mmHg @ 20°C.

*Permissible Concentration in Water:* No criteria set. Runoff from spills or fire control may cause water pollution. World Health Organization: 5000 µg/L as zinc in water for esthetic quality. The U.S. EPA[6] has set 5 ppm for the prevention of adverse effects due to the organoleptic properties of zinc.

*Harmful Effects and Symptoms*

*Short Term Exposure:* Low levels of toxicity. Concentrated solutions are slightly corrosive to eyes and mucous membranes. Dust inhalation can cause irritation of the respiratory system with sneezing. Eye contact can cause irritation, watering, pain, and inflammation of the eyelids. Skin contact can cause irritation and minor ulceration. Ingestion can cause nausea, vomiting, fever, muscle twitching, seizure, rapid respiration, slow heart beat. Severe exposure may result in death.

*Points of Attack:* Respiratory system, central nervous system, cardiovascular system, skin, eyes.

*Medical Surveillance:* There are tests available to measure zinc in your blood, urine, hair, saliva, and feces. High levels of zinc in the feces can mean high recent zinc exposure. High levels of zinc in the blood can mean high zinc consumption and/or high exposure. Tests to measure zinc in hair may provide information on long-term zinc exposure; however, the relationship between levels in your hair and the amount of zinc you were exposed to is not clear. Medical observation is recommended for 24 to 48 hours after breathing overexposure, as pulmonary edema may be delayed. As first aid for pulmonary edema, consider administering a corticosteroid spray. Cigarette smoking may exacerbate pulmonary injury and should be discouraged for at least 72 hours following exposure. Before employment and at regular times after that, the following are recommended: If symptoms develop or overexposure occurs, repeat the above tests as soon as possible and get an exam of the nervous system. Also consider complete blood count. Consider chest x-ray following acute overexposure.

*First Aid:* If this chemical gets into the eyes, remove any contact lenses at once and irrigate immediately for at least 15 minutes, occasionally lifting upper and lower lids. Seek medical attention immediately. If this chemical contacts the skin, remove contaminated clothing and wash immediately with soap and water. Seek medical attention immediately. If this chemical has been inhaled, remove from exposure, begin rescue breathing (using universal precautions) if breathing has stopped, and CPR if heart action has stopped. Transfer promptly to a medical facility. When this chemical has been swallowed, get medical attention. If victim is conscious and able to swallow, have victim drink 4 to 8 ounces of water. Do not induce vomiting. *Note to physician or authorized medical personnel*: Medical observation is recommended for 24 to 48 hours after breathing overexposure, as pulmonary edema may be delayed. As first aid for pulmonary edema, consider administering a corticosteroid spray. Cigarette smoking may exacerbate pulmonary injury and should be discouraged for at least 72 hours following exposure.

*References:*
- EXTOXNET, Extension Toxicology Network, "Pesticide Information Profile, Metiram," Oregon State University, Corvallis, OR (June 1996). http://extoxnet.orst.edu/pips/metiram.htm
- U.S. Environmental Protection Agency, Office of Pesticide Programs, Pesticide Residue Limits, "Metiram," 40 CFR 180.217, www.epa.gov/pesticides/food/viewtols.htm
- California Environmental Protection Agency "Chemical List of Lists," Sacramento CA (February 1997)

# Metobromuron

*Use Type:* Herbicide
*CAS Number:* 3060-89-7
*Formula:* $C_9H_{11}BrN_2O_2$
*Synonyms:* 3-(4-Bromophenyl)-1-methoxy-1-methylurea; 3-(*p*-Bromophenyl)-1-methoxy-1-methylurea; *N'*-(4-Bromophenyl)-*N*-methoxy-*N*-methylurea; 3-(*p*-Bromophenyl)-1-methyl-1-methoxyurea; 3-(4-Bromphenyl)-1-methoxyharnstoff (German); Metobromuron [ 3-(*p*-bromophenyl)-1-methoxy-1-methylurea]; Urea, *N'*-(4-bromophenyl)-*N*-methoxy-*N*-methyl-

*Trade Names:* C-3126®; CIBA 3126®; PATORAN®; PATTONEX®; PESTANAL®; Sigma-Aldrich (USA)

*Producers:* Sigma-Aldrich (USA)
*Chemical Class:* Phenyl urea
*EPA/OPP PC Code:* 035901
*Uses:* Pre-emergence herbicide.

*Carcinogen/Hazard Classifications*

**WHO Acute Hazard:** Class U, unlikely to be hazardous

*Description:* Colorless, crystalline solid. Soluble in water; solubility = 8.8 mg/L @ 20°C. Molecular weight = 259.12. Melting/Freezing point = 95°C. Vapor pressure = $3 \times 10^{-6}$ mmHg @ 20°C. Log $K_{ow}$ = 2.45. Unlikely to bioaccumulate in marine organisms.

*Incompatibilities:* Oxidizers (chlorates, nitrates, peroxides, permanganates, perchlorates, chlorine, bromine, fluorine, etc); strong acids.

*Determination in Air:* Filter; none; Gravimetric; NIOSH IV [Particulates NOR; #0500 (total), #0600 (respirable)].[18]

*Permissible Concentration in Water:* No criteria set. Runoff from spills or fire control may cause water pollution.

*Harmful Effects and Symptoms*

*Short Term Exposure:* Contact with eyes or skin may cause irritation or injury. Inhalation should be avoided; use NIOSH-approved air purifying respirators for pesticides. May be harmful if swallowed. Skin contact may cause severe irritation or burns.

*Points of Attack:* Skin.

*First Aid:* If this chemical gets into the eyes, remove any contact lenses at once and irrigate immediately for at least 15 minutes, occasionally lifting upper and lower lids. Seek medical attention immediately. If this chemical contacts the skin, remove contaminated clothing and wash immediately with soap and water. Seek medical attention immediately. If this chemical has been inhaled, remove from exposure, begin rescue breathing (using universal precautions) if breathing has stopped, and CPR if heart action has stopped. Transfer promptly to a medical facility. When this chemical has been swallowed, get medical attention. Give large quantities of water. Do not induce vomiting when formulations containing petroleum solvents are ingested. Do not make an unconscious person vomit.

## Metolachlor (ANSI)

*Use Type:* Herbicide
*CAS Number:* 51218-45-2
*Formula:* $C_{15}H_{22}ClNO_2$
*Alert:* A General Use Pesticide (GUP) in most usages. Some products may be Restricted Use Pesticides (RUP). An EPA Reregistration Eligibility Decision (RED) in April, 1995, determined that all uses of metolachlor with the exception of potatoes, soybeans and peanuts would not cause unreasonable risk to humans or the environment. Companies requested cancellation of registered uses of matolachlor effective March 22, 2002. (See Federal Register Environmental Document at http://www.epa.gov/fedrgstr/EPA-PEST/2002/March/Day-22/p6855.htm).

*Synonyms:* 2-Aethyl-6-methyl-N-(1-methyl-2-methoxyaethyl)-chloracetanilid (German); α-Chlor-6'-aethyl-N-(2-methoxy-1-methylaethyl)-acet-o-toluidin (German); 2-Chloro-6'-ethyl-N-(2-methoxy-1-methylethyl)acet-o-toluidide; α-Chloro-2'-ethyl-6'-methyl-N-(1-methyl-2-methoxyethyl)-acetanilide; 2-Chloro-N-(2-ethyl-6-methylphenyl)-N-(2-methoxy-1-methylethyl) acetamide; 2-Chloro-N-(6-ethyl-o-tolyl)-N-(2-methoxy-1-methylethyl)-acetamide; 2-Ethyl-6-methyl-1-N-(2-methoxy-1-methylethyl)chloroacetanilide; Metelilachlor

*Trade Names:* BICEP®, Syngenta (Switzerland), canceled 4/9/1998; BROADSTRIKE®, Dow AgroSciences (USA); CGA-24705®, Ciba-Geigy (Switzerland); CINCH®, DuPont Crop Protection (USA); CODAL®; COTORAN® MULTI®; CYCLE®, Syngenta (Switzerland), canceled 2/8/1996; DREXEL ME-TOO-LACHLOR®, Drexel Chemical (USA); DUAL®, Syngenta (Switzerland); DUAL MAGNUM®, Syngenta (Switzerland); DUET®, Syngenta (Switzerland), canceled 5/31/1994; INTER PLUS®, Makhteshim-Agan Industries (Israel); MEDAL®, Syngenta (Switzerland), canceled 6/27/1996; MILOCEP®; ONTRACK 8E®, Syngenta (Switzerland), canceled 10/10/1989; PENNANT®, Syngenta (Switzerland), canceled 6/27/1996; PRELUDE®, Zeneca Ag Products (USA) (now Syngenta), canceled 1/17/1996; PRIMAGRAM®; PRIMEXTRA®; TURBO®, Bayer CropScience (Germany)

*Producers:* Bayer CropScience (Germany); Ciba-Geigy (Switzerland); Dow AgroSciences (USA); Drexel Chemical (USA); DuPont Crop Protection (USA); Ehrenstorfer, Dr. (Germany); Makhteshim-Agan Industries (Israel); Shenzhen Guomeng Industry Co., Ltd. (China); Syngenta (Switzerland)

*Chemical Class:* Chloroacetanilide
*EPA/OPP PC Code:* 108801; 288700 (old EPA product code)
*California DPR Chemical Code:* 1996
*ICSC Number:* 1360
*RTECS Number:* AN3430000

*Uses:* Metolachlor is a selective herbicide that is usually applied to crops before plants emerge from the soil, and is used to control certain broadleaf and annual grassy weeds in field corn, soybeans, peanuts, grain sorghum, potatoes, pod crops, cotton, safflower, stone fruits, nut trees, highway rights-of-way and woody ornamentals. Prior to the RED of April, 1995, its primary use was on corn, soybeans and sorghum. It inhibits protein synthesis; thus, high-protein crops (e.g., soy) can be adversely affected by excessive metolachlor application. Additives may be included in product formulations to help protect sensitive crops (i.e., sorghum) from injury.

*Human toxicity (long-term)[77]:* Very low–100.00 ppb, Health Advisory

*Fish toxicity (threshold)[77]:* Very low –1117.14617 ppb, MATC (Maximum Acceptable Toxicant Concentration)

*U.S. Maximum Allowable Residue Levels for Metolachlor (40 CFR 180.368):*

*Note:* Some onion and pepper crops have regional restrictions and are not listed here.

*Carcinogen/Hazard Classifications*
**U.S. EPA Carcinogens:** Group C, possible carcinogen
**Label Signal Word:** CAUTION in most formulations, or WARNING
**WHO Acute Hazard:** Class III, slightly hazardous
**Endocrine Disruptor:** Suspected
*Regulatory Authority:*
- AB 1803-Well Monitoring Chemical (CAL)
- Actively registered pesticide in California.

*Description:* Metolachlor is a colorless or tan to brown, oily liquid. Slightly sweet odor. Very slightly soluble in water;

solubility = 510 ppm @ 20°C. Molecular weight = 283.79. Boiling point = 100°C @ 0.001 mmHg. It is stable to about 300°C. Vapor pressure = $1.5 \times 10^{-5}$ mmHg @ 20°C. Flash point = 190°C. Hazard Identification (based on NFPA-704 M Rating System): Health 1, Flammability 1, Reactivity 0. Log $K_{ow}$ = 3.0. Values above 3.0 are likely to bioaccumulate in marine organisms.

*Incompatibilities:* Oxidizers, strong acids, nitrates.

*Permissible Exposure Limits in Air:* No standards set.

*Permissible Concentration in Water:* The USEPA has set a lifetime health advisory of 10 µg/L. Several states have set guidelines for metolachlor in drinking water ranging from 1.0 µg/L (Illinois) to 17.5 µg/L (Kansas) to 25 µg/ml (Wisconsin).

*Determination in Water:* Extraction with methylene chloride followed by separation by gas chromatography and measurement using a nitrogen-phosphorus detector.

*Routes of Entry:* Inhalation

*Harmful Effects and Symptoms*

*Short Term Exposure:* Irritates the eyes and skin. The acute oral $LD_{50}$ for rats is 2780 mg/kg (slightly toxic). Signs of human intoxication from metolachlor and/or its formulations (presumably following acute deliberate or accidental exposures) include abdominal cramps, anemia, ataxia, dark urine, methemoglobinemia, cyanosis, hypothermia, collapse, convulsions, diarrhea, gastrointestinal irritation, jaundice, weakness, nausea, shock, sweating, vomiting, CNS depression, dizziness, dyspnea, liver damage, nephritis, cardiovascular failure, skin irritation, dermatitis, sensitization dermatitis, eye and mucous membrane irritation, corneal opacity and adverse reproductive effects.

*Long Term Exposure:* May cause tumors. Limited evidence of carcinogenicity animals (USEPA).

*Points of Attack:* Blood

*Medical Surveillance:* Test for methemoglobinemia. Complete blood count.

*First Aid:* If this chemical gets into the eyes, remove any contact lenses at once and irrigate immediately for at least 15 minutes, occasionally lifting upper and lower lids. Seek medical attention immediately. If this chemical contacts the skin, remove contaminated clothing and wash immediately with soap and water. Seek medical attention immediately. If this chemical has been inhaled, remove from exposure, begin rescue breathing (using universal precautions) if breathing has stopped, and CPR if heart action has stopped. Transfer promptly to a medical facility. When this chemical has been swallowed, get medical attention. Give large quantities of water and induce vomiting. Do not make an unconscious person vomit.

*Note to physician or authorized medical personnel:* Treat for methemoglobinemia. Spectrophotometry may be required for precise determination of levels of methemoglobinemia in urine.

*References:*
- EXTOXNET, Extension Toxicology Network, "Pesticide Information Profile, Metolachlor," Oregon State University, Corvallis, OR (June 1996). http://extoxnet.orst.edu/pips/metolach.htm
- U.S. Environmental Protection Agency, Office of Pesticide Programs, Pesticide Residue Limits, "Metolachlor," 40 CFR 180.368, http://www.epa.gov/pesticides/food/viewtols.htm
- U.S. Environmental Protection Agency, "Reregistration Eligibility Decision (RED), Metolachlor," Washington, DC (April 1995). http://www.epa.gov/REDs/0001.pdf
- U.S. Environmental Protection Agency, "Health Advisory," Washington, DC, Office of Drinking Water (August 1987).
- California Environmental Protection Agency "Chemical List of Lists," Sacramento CA (February 1997).

# Metolcarb

*Use Type:* Insecticide

*CAS Number:* 1129-41-5

*Formula:* $C_9H_{11}NO_2$; $C_6H_4(CH_3)OCONHCH_3$

*Alert:* Not used as a pesticide in the U.S.

*Synonyms:* Carbamic acid, methyl-, 3-methylphenyl ester; Carbamic acid, methyl-, 3-tolyl ester; Carbophen; *m*-Cresyl methylcarbamate; *m*-Cresyl ester of *N*-methylcarbamic acid; *m*-Cresyl methyl carbamate; Dicresyl; Dicresyl *N*-methylcarbamate; Metholcarb; Methylcarbamic acid *m*-toyl ester; 3-Methylphenyl-*N*-methylcarbamate; *m*-Methylphenyl methylcarbamate; MTMC; *m*-Tolyester kyseliny methyl karbaminove; *m*-Tolyl-*N*-methylcarbamate; 3-Tolyl-*N*-methylcarbamate

*Trade Names:* DRC 3341®; KUMIAI®; METACRATE®; S 1065®; SOGATOX DUST® 22; TSUMACIDE®; TSUMAUNKA®; VADEN®

*Producers:* Saeryung Chemicals (South Korea); Sigma-Aldrich Laborchemikalien (Germany)

*Chemical Class:* Carbamate

*RTECS Number:* FC8050000

*Uses:* Metolcarb is an insecticide for the control of rice green leafhoppers, plant-hoppers, codling moth, citrus mealy bug, onion thrips, fruit flies, bollworms and aphids. Not registered as a pesticide in the U.S.

*Carcinogen/Hazard Classifications*

**WHO Acute Hazard:** Class II, moderately hazardous

*Regulatory Authority:*
- EPA Hazardous Waste Number (RCRA No.): P190
- RCRA, 40CFR261, Appendix 8 Hazardous Constituents
- RCRA 40CFR268.48; 61FR15654, Universal Treatment Standards: Wastewater (mg/L), 0.056; Nonwastewater (mg/kg), 1.4

- Superfund/EPCRA 40CFR355, Appendix B Extremely Hazardous Substances: TPQ = 100/10,000 lb (45.4/4,540 kg)
- Superfund/EPCRA 40CFR302.4 RQ: EHS, 1 lb (0.454 kg)

***Description:*** Metolcarb is a colorless crystalline solid. Melting/Freezing point = 74 –75°C.

***Permissible Exposure Limits in Air:*** No standards set.

***Permissible Concentration in Water:*** No criteria set, but runoff from spills or fire control may cause water pollution.

***Routes of Entry:*** Inhalation, ingestion, skin and/or eye contact.

***Harmful Effects and Symptoms***

***Short Term Exposure:*** Metolcarb is a carbamate insecticide. Signs and symptoms of poisoning by carbamates are similar to those for organic phosphorus compounds. Symptoms of poisoning by organic phosphorus compounds include headache, giddiness, nervousness, blurred vision, weakness, nausea, cramps, diarrhea, and discomfort in the chest. Signs include sweating, myosis, tearing, salivation and other excessive respiratory tract secretion, vomiting, cyanosis, uncontrollable muscle twitches followed by muscular weakness, convulsions, coma, loss of reflexes, and loss of muscular control. Carbamate insecticides inhibit the cholinesterase activity of enzymes, causing accumulation of acetylcholine at synapses and altering the way in which nervous impulses are transmitted. However, within several hours carbamates spontaneously detach from the enzymes. Metolcarb exhibits high oral and skin toxicity, and moderate inhalation toxicity. Some carbamates appear to be carcinogenic, teratogenic, and/or mutagenic. Carbamates are cholinesterase inhibitors.

***Long Term Exposure:*** Cholinesterase inhibitor; cumulative effect is possible. This chemical may damage the nervous system with repeated exposure, resulting in convulsions, respiratory failure. May cause liver damage.

***Points of Attack:*** Respiratory system, lungs, central nervous system, cardiovascular system, skin, eyes, plasma and red blood cell cholinesterase.

***Medical Surveillance:*** Before employment and at regular times after that, the following are recommended: Plasma and red blood cell cholinesterase levels (tests for the enzyme poisoned by this chemical). If exposure stops, plasma levels return to normal in 1-2 weeks while red blood cell levels may be reduced for 1-3 months. When acetylcholinesterase enzyme levels are reduced by 25% or more below preemployment levels, risk of poisoning is increased, even if results are in lower ranges of "normal." Reassignment to work not involving organophosphate or carbamate pesticides is recommended until enzyme levels recover. If symptoms develop or overexposure occurs, repeat the above tests as soon as possible and get an exam of the nervous system. Also consider complete blood count. Consider chest x-ray following acute overexposure. Do not drink any alcoholic beverages before or during use. Alcohol promotes absorption of organophosphates.

***First Aid:*** Speed in removing material from eyes and skin is of extreme importance. *Eyes:* Eye contact can cause dangerous amounts of these chemicals to be quickly absorbed through the mucous membrane into the bloodstream. Immediately and gently flush eyes with plenty of warm or cold water (NO hot water) for at least 15 minutes, occasionally lifting the upper and lower eyelids. Get medical aid immediately. *Skin:* Get medical aid. Skin contact can cause dangerous amounts of these chemicals to be absorbed into the bloodstream. Wearing the appropriate PPE equipment and respirator for carbamate pesticides, immediately flush skin with plenty of soap and water for at least 15 minutes while removing contaminated clothing and shoes. Shampoo hair promptly if contaminated; protect eyes. *Ingestion:* Call poison control. Loosen all clothing. Never give anything by mouth to an unconscious person. Get medical aid. Do NOT induce vomiting.* If conscious, alert, and able to swallow, rinse mouth and have victim drink 4 to 8 ounces of water. Check to see if poison control instructs you to use ipecac syrup, otherwise administer slurry of activated charcoal (2 oz in 8 oz of water). If victim is *UNCONSCIOUS OR HAVING CONVULSIONS,* do nothing except keep victim warm. **In some cases you may be specifically instructed by poison control to induce vomiting by way of 2 tablespoons of syrup of ipecac (adult) washed down with a cup of water.* Do NOT give activated charcoal before or with ipecac syrup. *Inhalation:* Get medical aid. Do not contaminate yourself. Wearing the appropriate PPE equipment and respirator for carbamate pesticides, immediately remove the victim from the contaminated area to fresh air. If the victim is not breathing, administer artificial respiration. Do NOT use mouth-to-mouth resuscitation; use bag/mask apparatus. If breathing is difficult, administer oxygen through bag/mask apparatus until medical help arrives. Do not leave victim unattended. *Note to physician or authorized medical personnel.* Administer atropine, 2 mg (1/30 gr) intramuscularly or intravenously as soon as any local or systemic signs or symptoms of an intoxication are noted; repeat the administration of atropine every 3 to 8 minutes until signs of atropinization (mydriasis, dry mouth, rapid pulse, hot and dry skin) occur; initiate treatment in children with 0.05 mg mg/kg of atropine; repeat at 5 to 10 minute intervals. Watch respiration, and remove bronchial secretions if they appear to be obstructing the airway; intubate if necessary. *Medical note:* 2-PAMCI may be contraindicated in the case of some carbamate poisonings.

***References:***
- New Jersey Department of Health and Senior Services, "Hazardous Substance Fact Sheet, Metolcarb," Trenton, NJ (July 2000). http://www.state.nj.us/health/eoh/rtkweb/2563.pdf

- U.S. Environmental Protection Agency, "Chemical Profile: Metolcarb," Washington, DC, Chemical Emergency Preparedness Program (Nov. 30, 1987).
- California Environmental Protection Agency "Chemical List of Lists," Sacramento CA (February 1997).

# Metoxuron

*Use Type:* Herbicide
*CAS Number:* 19937-59-8
*Formula:* $C_{10}H_{13}ClN_2O_2$
*Synonyms:* $N'$-(3-Chlor-4-methoxy-phenyl)-$N,N$-dimethylharnstoff (German); 3-(3-Clor-4-methoxyphenyl)-1,1-dimethylharnstoff (German); $N$-(3-Chloro-4-methoxyphenyl)-$N',N'$-dimethylurea; $N'$-(3-Chloro-4-methoxyphenyl)-$N,N$-dimethylurea; 3-(3-Chloro-4-methoxyphenyl)-1,1-dimethylurea; $N,N$-Dimethyl-$N'$-(4-methoxy-3-chlorophenyl)urea; FL; Urea, $N'$-(3-chloro-4-methylphenyl)-$N,N$-dimethyl-
*Trade Names:* DEFTOR®; DOSAFLO®; DOSAGRAN®; DOSANEX®; DOSANEX FL®; DOSANEX MG®; HERBICIDE 6602®; METOX®; PURIVEL®
*Producers:* Atul (India)
*Chemical Class:* Organochlorine; Halo-organics; phenyl urea
*EPA/OPP PC Code:* 294600
*Uses:* Not registered in the U.S. A pre-emergence and post-emergence selective herbicide used to control weeds on cereals and carrots. It is particularly useful against black grass, silky bentgrass, wild oats, ryegrass and most annual broadleaf weeds.

*Carcinogen/Hazard Classifications*
**WHO Acute Hazard:** Class U, unlikely to be hazardous
*Description:* Crystalline solid or powder. Slightly soluble in water. Molecular weight = 228.72. Log $K_{ow}$ = 1.71. Unlikely to bioaccumulate in marine organisms.
*Permissible Concentration in Water:* No criteria set. Runoff from spills or fire control may cause water pollution.
*Routes of Entry:* Inhalation, ingestion. Central nervous system depressant.

*Harmful Effects and Symptoms*
*Short Term Exposure:* Symptoms include apprehension, anxiety, confusion, nervous excitation; dizziness; headache; numbness and weakness in limbs; muscle twitching, tremors; nausea and vomiting; slow, shallow respiration, bluish face; convulsions; loss of consciousness; breathing stops; death.
*Points of Attack:* CNS. May be fatal if inhaled, ingested, or absorbed through the skin
*Medical Surveillance:* Medical observation is recommended for 24 to 48 hours after breathing overexposure, as pulmonary edema may be delayed. As first aid for pulmonary edema, consider administering a corticosteroid spray. Cigarette smoking may exacerbate pulmonary injury and should be discouraged for at least 72 hours following exposure.

*First Aid:* Speed in removing material from eyes and skin is of extreme importance. *Eyes:* Eye contact can cause dangerous amounts of these chemicals to be quickly absorbed through the mucous membrane into the bloodstream. Directly, irrigate with large amounts of plain, tepid water or saline for 20 minutes, occasionally lifting the lower and upper lids. During this time, remove contact lenses, if easily removable without additional trauma to the eye. Get medical aid immediately. Have physician check for possible delayed damage. *Skin:* Get medical aid. Skin contact can cause dangerous amounts of these chemicals to be absorbed into the bloodstream. Wearing the appropriate PPE equipment and respirator for organochlorine pesticides, immediately flush exposed skin, hair, and under nails with plain, running, tepid water for 20 minutes, then wash twice with mild soap. Shampoo hair promptly if contaminated; protect eyes. **Do not scrub skin or hair**, since this can increase absorption through the skin. Rinse thoroughly with water. Victims who are able and cooperative may assist with their own decontamination. Remove and double-bag contaminated clothing and personal belongings. Leather absorbs many organochlorines; therefore, items such as leather shoes, gloves, and belts should be discarded. If the skin is swollen or inflamed, cool affected areas with cold compresses. *Ingestion:* Call poison control. Loosen all clothing. Never give anything by mouth to an unconscious person. Get medical aid. *Do not induce vomiting.** In cases of ingestion, the patient is at risk of CNS depression or seizures, which may lead to pulmonary aspiration during vomiting. If the victim is conscious and able to swallow, *administer an aqueous slurry of activated charcoal at 1 gm/kg (usual adult dose 60–90 g, child dose 25–50 g). A soda can and straw may be of assistance when offering charcoal to a child. The efficacy of activated charcoal for some organochlorine poisoning (such as chlordane) is uncertain. If victim is *unconscious or having convulsions,* do nothing except keep victim warm. **In some cases you may be specifically instructed by Poison Control to induce vomiting by way of 2 tablespoons of syrup of ipecac (adult) washed down with a cup of water. Do not give activated charcoal <u>before or with</u> ipecac syrup.* Inhalation: Get medical aid. Do not contaminate yourself. Wearing the appropriate PPE equipment and respirator for organochlorine pesticides, immediately remove the victim from the contaminated area to fresh air. For inhalation exposures, monitor for respiratory distress. If the victim is not breathing, administer artificial respiration. *Do not use mouth-to-mouth resuscitation; use bag/mask apparatus.* If cough or breathing difficulty develops, evaluate for respiratory tract irritation, bronchitis, or pneumonitis. If breathing is difficult, administer 100% humidified supplemental oxygen through bag/mask apparatus until medical help arrives. Do not leave victim unattended.

# Metribuzin

*Use Type:* Herbicide
*CAS Number:* 21087-64-9
*Formula:* $C_8H_{14}N_4OS$
*Alert:* A General Use Pesticide (GUP).
*Synonyms:* 4-Amino-6-*tert*-butyl-3-(methylthio)-1,2,4-triazin-5-one; 4-Amino-6-*tert*-butyl-3-methylthio-As-triazin-5-one; 4-Amino-6-(1,1-dimethylethyl)-3-(methylthio)-1,2,4-triazin-5-(4H)-one; Metribuzina (Spanish); 1,2,4-Triazin-5-(4H)-one, 4-Amino-6-(1,1-dimethylethyl)-3-(methylthio)-; As-triazin-5(4H)-one,4-amino-6-*tert*-butyl-3-(methylthio)-
*Trade Names:* AUTHORITY®, FMC Agricultural Products Group (USA); AXIOM®, Bayer CropScience (Germany); BAY 61597®, Bayer CropScience (Germany); BAY DIC 1468®, Bayer CropScience (Germany); BAYER 6159H®, Bayer CropScience (Germany); BAYER 6443H®, Bayer CropScience (Germany); BAYER 94337®, Bayer CropScience (Germany); BOUNDARY®, Syngenta (Switzerland); CANOPY®, DuPont Crop Protection (USA); CONQUEST®, BASF Canada (Canada); DIC 1468®; DOMAIN®, Bayer CropScience (Germany); LEXONE®, DuPont Crop Protection (USA); LEXONEEX®, DuPont Crop Protection (USA); PREVIEW®, DuPont Crop Protection (USA), canceled 3/9/2000; PYTHON®, Feinchemie Schwebda (Germany); Makhteshim-Agan Industries (Israel); SENCOR®, Bayer CropScience (Germany); SENCORAL®, Bayer CropScience (Germany); SENCOREX®, Bayer CropScience (Germany); SENCORER®, Bayer CropScience (Germany); VAPCOR®, Veterinary & Agricultural Products Manufacturing Co., Ltd. (VAPCO) (Jordan)
*Producers:* BASF Agricultural Products Group (Germany); BASF Canada (Canada); Bayer CropScience (Germany); DuPont Crop Protection (USA); Ehrenstorfer, Dr. (Germany); Feinchemie Schwebda (Germany); FMC Agricultural Products Group (USA); Hindustan Insecticides (India); Jingma Chemicals Ltd. (China); Ki-Hara Chemicals Ltd. (UK); Makhteshim-Agan Industries (Israel); Sevencontinent Agrichemical Co., Ltd. (China); Shenzhen Guomeng Industry Co., Ltd. (China); Sigma-Aldrich Laborchemikalien (Germany); Syngenta (Switzerland); Veterinary & Agricultural Products Manufacturing Co., Ltd. (VAPCO) (Jordan)
*Chemical Class:* Triazinone
*EPA/OPP PC Code:* 101101
*California DPR Chemical Code:* 1692
*ICSC Number:* 0516
*RTECS Number:* XZ2990000
*EEC Number:* 606-034-00-8
*EINECS Number:* 244-209-7
*Uses:* Metribuzin is a selective triazine herbicide which inhibits photosynthesis of susceptible plant species. It is used for control of annual grasses and numerous broadleaf weeds in asparagus, potatoes, lucerne, peas, lentils, soya beans, sugar cane, sainfoin, pineapples and cereals. It is applied to fallow lands. Metribuzin is available as liquid suspension, water dispersible granular, and dry flowable formulations.
*Human toxicity (long-term)[77]:* Very low–200.00 ppb, Health Advisory
*Fish toxicity (threshold)[77]:* Very low–7683.76758 ppb, MATC (Maximum Acceptable Toxicant Concentration)
*U.S. Maximum Allowable Residue Levels for Metribuzin (40 CFR 180.332):*

| CROP | ppm |
| --- | --- |
| Alfalfa, forage | 2 |
| Alfalfa, hay | 7 |
| Asparagus | 0.1 |
| Barley, grain | 0.75 |
| Barley, hay | 7 |
| Barley, pearled barley | 3 |
| Barley, straw | 1 |
| Carrot, roots | 0.3 |
| Cattle, fat | 0.7 |
| Cattle, meat | 0.7 |
| Cattle, mbyp | 0.7 |
| Corn, field, forage | 0.1 |
| Corn, field, grain | 0.05 |
| Corn, field, stover | 0.1 |
| Corn, pop, grain | 0.05 |
| Corn, sweet, forage | 0.1 |
| Corn, sweet, kernel plus cob with husks removed | 0.05 |
| Corn, sweet, stover | 0.1 |
| Egg | 0.01 |
| Goat, fat | 0.7 |
| Goat, meat | 0.7 |
| Goat, mbyp | 0.7 |
| Grass, forage | 2 |
| Grass, hay | 7 |
| Hog, fat | 0.7 |
| Hog, meat | 0.7 |
| Hog, mbyp | 0.7 |
| Horse, fat | 0.7 |
| Horse, meat | 0.7 |
| Horse, mbyp | 0.7 |
| Lentil | 0.05 |
| Milk | 0.05 |
| Pea | 0.1 |
| Pea, dry | 0.05 |
| Pea, dry, seed | 0.05 |
| Pea, field, hay | 4 |
| Pea, field, vines | 0.5 |
| Pea, succulent | 0.1 |
| Potato | 0.6 |
| Potato, chips | 3 |
| Poultry, fat | 0.7 |
| Poultry, meat | 0.7 |

| | |
|---|---|
| Poultry, mbyp | 0.7 |
| Sainfoin, forage | 2 |
| Sainfoin, hay | 7 |
| Sheep, fat | 0.7 |
| Sheep, meat | 0.7 |
| Sheep, mbyp | 0.7 |
| Soybean, forage | 4 |
| Soybean, hay | 4 |
| Soybean, seed | 0.3 |
| Sugarcane, cane | 0.1 |
| Sugarcane, molasses | 2 |
| Tomato | 0.1 |
| Wheat, bran | 3 |
| Wheat, forage | 2 |
| Wheat, germ | 3 |
| Wheat, grain | 0.75 |
| Wheat, hay | 7 |
| Wheat, middlings | 3 |
| Wheat, shorts | 3 |
| Wheat, straw | 1 |

*Carcinogen/Hazard Classifications*
**U.S. EPA Carcinogens:** Group D, unclassifiable, ambiguous data
**TRI Developmental Toxin:** Reproductive and developmental toxin
**Label Signal Word:** CAUTION
**WHO Acute Hazard:** Class II, moderately hazardous
**Endocrine Disruptor:** Suspected endocrine disruptor
*Regulatory Authority:*
- Air Pollutant Standard Set (ACGIH)[1] (NIOSH)[2] (Several States)[60]
- AB 1803-Well Monitoring Chemical (CAL)
- Permissible Exposure Limits for Chemical Contaminants (CAL/OSHA)
- Actively registered pesticide in California.
- EPCRA Section 313 Form R *de minimis* concentration reporting level: 1.0%
- Safe Drinking Water Act: Priority List (55 FR 1470)
- Canada, WHMIS, Ingredients Disclosure List

*Description:* Metribuzin is a colorless crystalline solid. Mild, sulfurous odor. Slightly soluble in water; solubility = $1.2 \times 10^3$ ppm. Molecular weight = 214.33. Vapor Pressure = $< 9.8 \times 10^{-4}$ mmHg @ 20°C. Melting/Freezing point = 125–127°C.

*Incompatibilities:* Strong oxidizers. When heated to decomposition, forms oxides of nitrogen and sulfur.

*Permissible Exposure Limits in Air:* There is no OSHA[2] PEL. NIOSH[2] and ACGIH[1] recommend a TWA of 5 mg/m³. Guidelines or standards for metribuzin in ambient air[60] have been set by some states ranging from 50 $\mu g/m^3$ (North Dakota) to 100 $\mu g/m^3$ (Connecticut) to 119 $\mu g/m^3$ (Nevada).

*Permissible Concentration in Water:* The USEPA has set a lifetime health advisory of 175 $\mu g/L$. Several states have set guidelines for metricuzin in drinking water[61] ranging from 1.0 $\mu g/L$ (Illinois) to 25 $\mu g/L$ (Wisconsin) to 175 $\mu g/L$ (Kansas).

*Determination in Water:* Solvent extraction with methylene chloride followed by exchange to acetone, separation by gas chromatography and measurement with a thermionic bead detector.

*Routes of Entry:* Inhalation, ingestion, skin and/or eye contact

*Harmful Effects and Symptoms*

*Short Term Exposure:* Metribuzin can affect you when breathed in and by passing through your skin. Acute poisoning can cause difficult breathing and drowsiness. High exposures may cause upset stomach, fatigue, and depression of the central nervous system, causing poor coordination, tremors and weakness.

*Long Term Exposure:* Repeated or high exposure may cause liver enzyme changes, goiter, and affect thyroid function.

*Points of Attack:* Central nervous system, thyroid and liver

*Medical Surveillance:* If symptoms develop or overexposure is suspected, the following may be useful: Thyroid function tests.

*First Aid:* If this chemical gets into the eyes, remove any contact lenses at once and irrigate immediately for at least 15 minutes, occasionally lifting upper and lower lids. Seek medical attention immediately. If this chemical contacts the skin, remove contaminated clothing and wash immediately with soap and water. Seek medical attention immediately. If this chemical has been inhaled, remove from exposure, begin rescue breathing (using universal precautions) if breathing has stopped, and CPR if heart action has stopped. Transfer promptly to a medical facility. When this chemical has been swallowed, get medical attention. Give large quantities of water and induce vomiting. Do not make an unconscious person vomit.

*References:*
- EXTOXNET, Extension Toxicology Network, "Pesticide Information Profile, Metribuzin," Oregon State University, Corvallis, OR )June 1996). http://extoxnet.orst.edu/pips/metribuz.htm
- U.S. Environmental Protection Agency, Office of Pesticide Programs, Pesticide Residue Limits, "Metribuzin," 40 CFR 180.332, http://www.epa.gov/pesticides/food/viewtols.htm
- U.S. Environmental Protection Agency, "Health Advisory: Metribuzin," Washington, DC, Office of Drinking Water (August 1987).
- California Environmental Protection Agency "Chemical List of Lists," Sacramento CA (February 1997).
- New Jersey Department of Health and Senior Services, "Hazardous Substance Fact Sheet: Metribuzin," Trenton, NJ (January 2001). http://www.state.nj.us/health/eoh/rtkweb/1302.pdf

# Metsulfuron-methyl

*Use Type:* Herbicide
*CAS Number:* 74223-64-6
*Formula:* $C_{14}H_{15}N_5O_6S$
*Synonyms:* Benzoic acid, 2-[(((((4-methoxy-6-methyl-1,3,5-triazin-2-yl)amino)carbonyl)amino)sulfonyl]-methyl ester; Methyl-2-[[(4-methoxy-6-methyl-1,3,5-triazyn-2-yl)aminocarbonyl] aminosulfonyl]benzoate; Methyl 2-[(((((4-methoxy-6-methyl-1,3,5-triazin-2-yl)amino)carbonyl)amino)sulfonyl]benzoate; Methyl 2-[3-(4-methoxy-6-methyl-1,3,5-triazin-2-yl)ureidosulphonyl]benzoate
*Trade Names:* ALLIE®; ALLY®, DuPont Crop Protection (USA); ALLY-20DF®; ANSWER®, DuPont Crop Protection (USA); BRUSH-OFF®; CANVAS®, DuPont Crop Protection (USA); CIMARRON®, DuPont Crop Protection (USA); DMC® WEED CONTROL, The Scotts Co. (USA); DPD 63760H®; DPX 6376®; DPX-T 6376®; ESCORT®, DuPont Crop Protection (USA); FINESSE®, DuPont Crop Protection (USA); GROPPER®; NUP®, Nufarm (Australia); PARTI-SAN®, Sanonda (Australia); PASTURE® MD, Nufarm (Australia); RIVERDALE®, Nufarm (Australia); ROSULFURON®, Rotam Agrochemical (Hong Kong)
*Producers:* DuPont Crop Protection (USA); Makhteshim-Agan Industries (Israel); Micro-Flo (USA); Nufarm (Australia); Rotam Agrochemical (Hong Kong); Sanonda (Australia); The Scotts Co. (USA)
*Chemical Class:* Sulfonylurea
*EPA/OPP PC Code:* 122010
*California DPR Chemical Code:* 2222
*Uses:* Metsulfuron-methyl is a pre-emergence and post-emergence herbicide used to control annual grasses, brush, woody plants and broadleaf weeds. It can be applied to cereals including barley, rye and wheat and to pastures. It is primarily used to control brush, woody plants and broadleaf weeds on rights-of-way, fence rows, storage areas, highways and other non-crop areas.
*Human toxicity (long-term)[77]:* Very low–1750.00 ppb, Health Advisory
*Fish toxicity (threshold)[77]:* Very low–31167.38603 ppb, MATC (Maximum Acceptable Toxicant Concentration)
*Carcinogen/Hazard Classifications*
**Label Signal Word:** CAUTION or WARNING
**WHO Acute Hazard:** Class U, unlikely to be hazardous
*Description:* White to pale yellow solid. Slight, sweet odor. Soluble in water; solubility = 108 mg/L @ 25°C. Molecular weight = 381.39. Melting/Freezing point = 158°C. Density = 1.47.
*Incompatibilities:* Strong oxidizers.
*Determination in Air:* Filter; none; Gravimetric; NIOSH IV [Particulates NOR; #0500 (total), #0600 (respirable)].[18]
*Permissible Concentration in Water:* No criteria set. Runoff from spills or fire control may cause water pollution.
*Routes of Entry:* Inhalation, passing through the skin, ingestion.
*Harmful Effects and Symptoms*
*Short Term Exposure:* Contact with eyes or skin may cause irritation or burns. Inhalation should be avoided; use NIOSH-approved air purifying respirators for pesticides. May be harmful if swallowed. Skin contact may cause allergic reaction.
*Long Term Exposure:* Repeated or prolonged contact may cause skin and lung sensitization, resulting in allergies.
*Points of Attack:* Skin.
*Medical Surveillance:* Evaluation by a qualified allergist, including careful exposure history and special testing, may help diagnose skin allergy.
*First Aid:* If this chemical gets into the eyes, remove any contact lenses at once and irrigate immediately for at least 15 minutes, occasionally lifting upper and lower lids. Seek medical attention immediately. If this chemical contacts the skin, remove contaminated clothing and wash immediately with soap and water. Seek medical attention immediately. If this chemical has been inhaled, remove from exposure, begin rescue breathing (using universal precautions) if breathing has stopped, and CPR if heart action has stopped. Transfer promptly to a medical facility. When this chemical has been swallowed, get medical attention. Give large quantities of water and induce vomiting. Do not make an unconscious person vomit.
*References:*
- EXTOXNET, Extension Toxicology Network, "Pesticide Information Profile, Metsulfuron-methyl," Oregon State University, Corvallis, OR (October 1996). http://extoxnet.orst.edu/pips/metsulfu.htm
- California Environmental Protection Agency "Chemical List of Lists," Sacramento CA (February 1997)

# Mevinphos

*Use Type:* Insecticide and acaricide
*CAS Number:* 7786-34-7
*Formula:* $C_7H_{13}O_6P$
*Alert:* A Restricted Use Pesticide (RUP). Amvac Chemical Corporation, the only maker of mevinphos in the U.S., voluntarily canceled all uses in the U.S. effective July 1, 1994. There are no mevinphos products in use in the U.S.
Human toxicity (long-term): High
*Synonyms:* 2-Butenoic acid, 3-[(dimethoxyphosphinyl)oxy]-, methyl ester; α-2-Carbomethoxy-1-methylvinyl dimethyl phosphate; 2-Carbomethoxy-1-methylvinyl dimethyl phosphate; α-(-2-Carbomethoxy-1-methylvinyl) dimethyl phosphate; 2-Carbomethoxy-1-methylvinyl dimethyl phosophate, α isomer; 2-Carbomethoxy-1-propen-2-yl dimethyl phosphate; CMDP; Crotonic acid, 3-hydroxy-, methyl ester, dimethyl phosphate; Crotonic acid, 3-hydroxy-methyl ester, dimethyl phosphate, (E)-; 3-[(Dimethoxy

# Mevinphos

phosphinyl)oxy]-2-butenoic acid methyl ester; *O,O*-dimethyl-*O*-(2-carbomethoxy-1-methylvinyl) phosphate; *O,O*-Dimethyl 1-carbomethoxy-1-propen-2-yl phosphate; Dimethyl-1-carbomethoxy-1-propen-2-yl phosphate; Dimethyl (2-methoxycarbonyl-1-methylvinyl) phosphate; Dimethyl methoxycarbonylpropenyl phosphate; Dimethyl (1-methoxycarboxypropen-2-yl) phosphate; *O,O*-Dimethyl *O*-(1-methyl-2-carboxyvinyl) phosphate; Dimethyl phosphate of methyl 3-hydroxy-*cis*-crotonate; Duraphos; ENT 22,374; 3-hydroxycrotonic acid methyl ester dimethyl phosphate; Menite; *cis*-2-Methoxycarbonyl-1-methylvinyl dimethylphosphate; 2-Methoxycarbonyl-1-methylvinyl dimethyl phosphate; (*cis*-2-Methoxycarbonyl-1-methylvinyl) dimethyl phosphate; 1-Methoxycarbonyl-1-propen-2-yl dimethyl phosphate; Methyl-3-((dimethoxyphosphinyl)oxy)-2-butenoate, α-isomer; Methyl 3-[(dimethoxyphosphinyl)oxy]-2-butenoate; Methyl 3-(dimethoxyphosphinyloxy) crotonate; Methyl 3-hydroxy-α-crotonate dimethyl phosphate; Methyl-3-hydroxy-α-crotonate, dimethyl phosphate ester; Methyl 3-hydroxycrotonate dimethyl phosphate ester; Mevinfos (Spanish); NSC 46470; *cis*-Phosdrin; Phosfene; Phosphene; Phosphoric acid, dimethyl ester with methyl 3-hydroxycrotonate; Phosphoric acid, (1-methoxycarboxypropen-2-yl) dimethyl ester

***Trade Names:*** AI3-22374®; APAVINPHOS®; CASWELL No. 160B®; CENTURY-CIDE®, canceled 2/21/1986; COMPOUND 2046®; DURHAM®, Amvac Chemical Corp. (USA), canceled 9/29/1988; EXCELCIDE®, canceled 7/1/1987; GESFID®; GESTID®; PD 5®; HELENA PHOSDRIN®, Helena Chemical (USA), canceled 5/31/1994

***Producers:*** Amvac Chemical Corp. (USA); Sigma-Aldrich Laborchemikalien (Germany)

***Chemical Class:*** Organophosphate
***EPA/OPP PC Code:*** 015801
***California DPR Chemical Code:*** 480
***ICSC Number:*** 0924
***RTECS Number:*** GQ5250000
***EEC Number:*** 015-020-00-5

***Uses:*** Mevinphos is an organophosphate insecticide used to control a broad spectrum of insects, including aphids, grasshoppers, leafhoppers, cutworms, caterpillars, and many other insects on a wide range of field, forage, vegetable, and fruit crops. It is also an acaricide that kills or controls mites and ticks. It acts quickly both as a contact insecticide, acting through direct contact with target pests, and as a systemic insecticide which becomes absorbed by plants on which insects feed. Not in use in the U.S.

***Human toxicity (long-term)[77]:*** High–1.75 ppb, Health Advisory

***Fish toxicity (threshold)[77]:*** Extra high–0.96183 ppb, MATC (Maximum Acceptable Toxicant Concentration)

***Carcinogen/Hazard Classifications***
**U.S. EPA Carcinogens:** Deferred classification
**Label Signal Word:** DANGER–POISON
**WHO Acute Hazard:** Class 1 a, extremely hazardous

***Regulatory Authority:***
- Banned or Severely Restricted (India, Norway) (UN)[13]
- Very Toxic Substance (World Bank)[15]
- EPA/SARA 302 (EPCRA) Extremely hazardous substances
- Permissible Exposure Limits for Chemical Contaminants (CAL/OSHA)
- The "Director's List" (CAL/OSHA)
- Actively registered pesticide in California.
- Clean Water Act: Section 311 Hazardous Substances/RQ 40CFR117.3 (same as CERCLA, see below).
- Superfund/EPCRA 40CFR355, Appendix B Extremely Hazardous Substances: TPQ = 500 lb (227kg)
- Superfund/EPCRA 40CFR302.4 RQ: CERCLA, 10 lb (4.54 kg)
- EPCRA Section 313 Form R *de minimis* concentration reporting level: 1.0%
- U.S. DOT Inhalation Hazard Chemicals as organophosphates
- U.S. DOT Regulated Marine Pollutant (49CFR, Subchapter 172.101, Appendix B), severe pollutant.

***Description:*** Mevinphos is a pale yellow to orange high-boiling liquid. Weak odor. Soluble in water. Boiling point = 99–102.5°C (decomposes). Melting/Freezing point = 7°C (*trans*-); 21°C (*cis*-). Flash point = 175°C (oc); 80°C (isomer mixture). Vapor pressure = $1 \times 10^{-4}$ mmHg @ 20°C. Hazard Identification (based on NFPA-704 M Rating System): Health 4, Flammability 1, Reactivity 0. Soluble in water. Commercial product is a mixture of the *cis*- and *trans*-isomers. Insecticide that may be absorbed on a dry carrier.

***Incompatibilities:*** Decomposes in heat producing phosphoric acid and phosphorous oxides fumes. Reacts violently with strong oxidizers. Corrosive to cast iron, some stainless steels and brass. Attacks some forms of plastics, rubber and coatings.

***Permissible Exposure Limits in Air:*** The OSHA[2] PEL is 0.1 mg/m³ TWA. DFG MAK[3], HSE[33] TWA, NIOSH[2], and ACGIH[1] recommend a TWA value of 0.01 ppm (0.1 mg/m³) and the STEL is 0.03 (0.3 mg/m³). The notation "skin" is added to indicate the possibility of cutaneous absorption. The NIOSH[2] IDLH level = 4 ppm. In addition, several states have set guidelines or standards for mevinphos in ambient air[60] ranging from 1-3 µg/m³ (North Dakota) to 1.6 µg/m³ (Virginia) to 2.0 µg/m³ (Connecticut and Nevada).

***Determination in Air:*** OSHA versatile sampler-2; Toluene/Acetone; Gas chromatography/Flame photometric detection for sulfur, nitrogen, or phosphorus; NIOSH IV, Method #5600, Organophosphorus pesticides.[18]

***Permissible Concentration in Water:*** No criteria set, but runoff from spills or fire control may cause water pollution. This chemical is highly toxic to aquatic life.

***Routes of Entry:*** Inhalation, skin absorption, ingestion, skin and/or eye contact.

*Harmful Effects and Symptoms*
*Short Term Exposure:* Cholinesterase inhibitor. Mevinphos may affect the nervous system, causing convulsions, respiratory failure. This material is super toxic; the probable oral lethal dose for humans is less than 5 mg/kg, or a taste (less than 7 drops) for a 150 lb person. It has direct and immediate effects whether it is swallowed, inhaled, or absorbed through the skin. Symptoms include nausea, vomiting, abdominal cramps, diarrhea, excessive salivation, headache, giddiness, dizziness, runny nose, tightness in the chest, blurring and dimming of vision, slurring of speech, twitching of muscles, mental confusion, disorientation, difficulty breathing, blueing of skin, convulsions, coma, and death. Delayed pulmonary edema may occur after inhalation.

*Long Term Exposure:* Cholinesterase inhibitor; cumulative effect is possible. This chemical may damage the nervous system with repeated exposure, resulting in convulsions, respiratory failure. May cause liver damage.

*Points of Attack:* Respiratory system, lungs, central nervous system, cardiovascular system, skin, eyes, plasma and red blood cell cholinesterase.

*Medical Surveillance:* Medical observation is recommended for 24 to 48 hours after breathing overexposure, as pulmonary edema may be delayed. Before employment and at regular times after that, the following are recommended: Plasma and red blood cell cholinesterase levels (tests for the enzyme poisoned by this chemical). If exposure stops, plasma levels return to normal in 1-2 weeks while red blood cell levels may be reduced for 1-3 months. When acetylcholinesterase enzyme levels are reduced by 25% or more below preemployment levels, risk of poisoning is increased, even if results are in lower ranges of "normal." Reassignment to work not involving organophosphate or carbamate pesticides is recommended until enzyme levels recover. If symptoms develop or overexposure occurs, repeat the above tests as soon as possible and get an exam of the nervous system. Also consider complete blood count. Consider chest x-ray following acute overexposure. Do not drink any alcoholic beverages before or during use. Alcohol promotes absorption of organophosphates.

*First Aid:* **Treatment for organophosphate poisoning consists of thorough decontamination, cardiorespiratory support, and administration of the antidotes atropine and pralidoxime. In cases of severe poisoning, diazepam, an anticonvulsant, should also be administered. Antidotes should be administered as prevention even if the diagnosis is in doubt.** Speed in removing material from eyes and skin is of extreme importance. *Eyes:* Eye contact can cause dangerous amounts of these chemicals to be quickly absorbed through the mucous membrane into the bloodstream. Immediately and gently flush eyes with plenty of warm or cold water (NO hot water) for at least 15 minutes, occasionally lifting the upper and lower eyelids. Get medical aid immediately. *Skin:* Get medical aid. Skin contact can cause dangerous amounts of these chemicals to be absorbed into the bloodstream. Wearing the appropriate PPE equipment and respirator for organophosphate pesticides, immediately flush skin with plenty of soap and water for at least 15 minutes while removing contaminated clothing and shoes. Shampoo hair promptly if contaminated. The removed, contaminated clothing and shoes should be double-bagged and left in Hot Zone for later disposal by hazardous materials experts. Skin may also be decontaminated with diluted hypochlorite solution. *Inhalation:* Get medical aid. Do not contaminate yourself. Wearing the appropriate PPE equipment and respirator for organophosphate pesticides, immediately remove the victim from the contaminated area to fresh air. If the victim is not breathing, administer artificial respiration. Do NOT use mouth-to-mouth resuscitation; use bag/mask apparatus. If breathing is difficult, administer oxygen through bag/mask apparatus until medical help arrives. Do not leave victim unattended. *Ingestion:* Call poison control. Loosen all clothing. Never give anything by mouth to an unconscious person. If victim is *unconscious or having convulsions,* do nothing except keep victim warm. Get medical aid. Transfer promptly to a medical facility. In cases of ingestion, **do not induce vomiting.** If the victim is alert and asymptomatic, administer a slurry of activated charcoal at a dose of 1 g/kg (infant, child, and adult dose). A soda can and straw may be of assistance when offering charcoal to a child. *In some cases you may be specifically instructed by poison control to induce vomiting by way of 2 tablespoons of syrup of ipecac (adult) washed down with a cup of water.* Do NOT give activated charcoal before or with ipecac syrup.

*Note to physician or authorized medical personnel:* Treat cases of respiratory compromise, coma, or excessive pulmonary secretions with respiratory support using protocols and techniques available and within the scope of training. Some cases may necessitate procedures such as endotracheal intubation or cricothyrotomy by properly trained and equipped personnel. When possible, atropine (see *Antidotes*, below) should be given under medical supervision. Patients who are comatose, hypotensive, or having seizures or cardiac arrhythmias should be treated according to advanced life support protocols. *Antidotes:* Two antidotes are administered to treat organophosphate poisoning. Atropine is a competitive antagonist of acetylcholine at muscarinic receptors and is used to control the excessive bronchial secretions which are often responsible for death. Pralidoxime relieves both the nicotinic and muscarine effects of organophosphate poisoning by regenerating acetylcholinesterase and can reduce both the bronchial secretions and the muscle weakness associated with poisoning. The initial intravenous dose of atropine in adults should be determined by the severity of symptoms: An initial adult dose of 1.0 to 2.0 mg or pediatric dose of 0.01 mg/kg (minimum 0.01 mg) should be administered intravenously. If intravenous access cannot be established, atropine may also be given intramuscularly, subcutaneously or via endotracheal tube. Doses should be repeated every 15 minutes until excessive secretions and sweating have been controlled. Once bronchial secretion has been controlled, atropine administration should be repeated whenever the secretions begin to recur. In seriously poisoned patients, very large doses may be required. Alterations of pulse rate and pupillary size should not be used as indicators of treatment adequacy. Pralidoxime should be administered as early in poisoning as possible as its efficacy may diminish when given more than 24 to 36 hours after exposure. Doses are as follows: adult 1.0 g; pediatric 25 to 50 mg/kg. The drug should be administered intravenously over 30 to 60 minutes, but in a life-threatening situation, one-half of the total dose can be given per minute for a total administration time of 2 minutes. Treatment should begin to take effect within 40 minutes with a reduction in symptoms and the amount of atropine necessary to control bronchial secretion. The initial dose can be repeated in 1 hour and then every 8 to 12 hours until the patient is clinically well and no longer requires atropine. If intravenous access cannot be established, pralidoxime may also be given intramuscularly. Early administration of diazepam in addition to the combined atropine and pralidoxime treatment may help prevent the onset of seizures and potential brain and cardiac morphologic damage following high-level organophosphate poisoning.

*References:*
- EXTOXNET, Extension Toxicology Network, "Pesticide Information Profile, Mevinphos," Oregon State University, Corvallis, OR (June 1996).
  http://extoxnet.orst.edu/pips/mevinpho.htm
- U.S. Environmental Protection Agency, Reregistration Eligibility Decision Facts, Mevinphos," Washington, DC (September 1994).
  http://www.epa.gov/REDs/factsheets/0250fact.pdf
- New Jersey Department of Health and Senior Services, "Hazardous Substance Fact sheet, Mevinphos," Trenton, NJ (April 2001).
  http://www.state.nj.us/health/eoh/rtkweb/1509.pdf
- Sax, N.I., Ed., "Dangerous Properties of Industrial Materials Report" 6, No. 1, 97-101 (1986).
- Pohanish, R. P. and Greene, S.A., "Hazardous Substance Resource Guide, 2nd Ed., Gale Research, Detroit MI (1977).
- U.S. Environmental Protection Agency, "Chemical Profile: Mevinphos," Washington, DC, Chemical Emergency Preparedness Program (Nov. 30, 1987).
- Agency for Toxic Substances and Disease Registry, U.S. Department of Health and Human Services, Public Health Service, "Managing Hazardous Materials Incidents," Atlanta, GA (June 2003).
- California Environmental Protection Agency "Chemical List of Lists," Sacramento CA (February 1997).

# Mexacarbate (ANSI)

*Use Type:* Insecticide and molluscicide
*CAS Number:* 315-18-4
*Formula:* $C_{12}H_{18}N_2O_2$
*Alert:* Classified as an obsolete pesticide and no longer registered in the U.S. and most other countries, with the exception of India.
*Synonyms:* Carbamate,4-dimethylamino-3,5-xylyln-methyl-; Carbamic acid, methyl-, methylcarbamate (ester); Carbamic acid, methyl-, 4-(dimethylamino)-3,5-xylyl ester; 4-(Dimethylamine)-3,5-xylyln-methylcarbamate; 4-(Dimethylamino)-3,5-dimethylphenol methylcarbamate (ester); 4-(Dimethylamino)-3,5-dimethylphenyl *N*-methylcarbamate; 4-(Dimethylamino)-3,5-xylenol,methylcarbamate (ester); 4-Dimethylamino-3,5-xylylmethylcarbamate; 4-Dimethylamino-3,5-xylyl *N*-methylcarbamate; 4-(*N,N*-Dimethylamino)-3,5-xylyl *N*-methylcarbamate; 5-Dimethylphenol methylcarbamate ester; ENT 25766; Methylcarbamic acid, 4-(dimethylamino)-3,5-xylyl ester; Methyl-4-dimethylamino-3,5-xylylcarbamate; Methyl-4-dimethylamino-3,5-xylyl ester of carbamic acid; NCI-C00544; OMS-47; Phenol, 4-(dimethylamino)-3,5-dimethyl-methylcarbamate (ester); 3,5-Xylenol, 4-(dimethylamino)-, methylcarbamate; Mexacarbato (Spanish)
*Trade Names:* DOWCO®139, Dow Chemical (USA), canceled; ZACTRAN®; ZECTANE®; ZECTRAN®; ZEXTRAN®
*Producers:* Ehrenstorfer, Dr., (Germany)
*Chemical Class:* Carbamate
*EPA/OPP PC Code:* 044201
*California DPR Chemical Code:* 623
*RTECS Number:* FC0700000
*Uses:* It is a insecticide and molluscicide for non-agricultural uses, e.g., lawn and turf, flowers, gardens, vines, forest lands, woody shrubs and trees. It is not produced or used commercially in the United States.
*Fish toxicity (threshold)*[(77)]: Intermediate–31.03658 ppb, MATC (Maximum Acceptable Toxicant Concentration)
*Carcinogen/Hazard Classifications*
**IARC:** Group 3, unclassifiable
**Label Signal Word:** WARNING or CAUTION
*Regulatory Authority:*
- Clean Water Act: Section 311 Hazardous Substances/RQ 40CFR117.3 (same as CERCLA, see below)
- EPA Hazardous Waste Number (RCRA No.): P128
- RCRA, 40CFR261, Appendix 8 Hazardous Constituents
- RCRA 40CFR268.48; 61FR15654, Universal Treatment Standards: Wastewater (mg/L), 0.056; Nonwastewater (mg/kg), 1.4
- Superfund/EPCRA 40CFR355, Appendix B Extremely Hazardous Substances: TPQ = 500/10,000 lb (227/4,540 kg)

- Superfund/EPCRA 40CFR302.4 RQ: CERCLA, 1000 lb (454 kg)
- AB 1803-Well Monitoring Chemical (CAL)
- EPA/SARA 302 (EPCRA) Extremely hazardous substances
- The "Director's List" (CAL/OSHA)
- U.S. DOT Regulated Marine Pollutant (49CFR172.101, Appendix B)

*Description:* An odorless white to tan crystalline solid. Insoluble in water. Molecular weight = 222.29. Melting/Freezing point = 85°C. Vapor pressure = 0.1 mmHg @ 20°C. Hazard Identification (based on NFPA-704 M Rating System): Health 3, Flammability 1, Reactivity 0.

*Incompatibilities:* Alkalis and strong oxidizers.

*Permissible Exposure Limits in Air:* No standards set.

*Permissible Concentration in Water:* No criteria set, but runoff from spills or fire control may cause water pollution.

*Routes of Entry:* Inhalation, ingestion, skin, and/or contact. Absorbed through the skin.

*Harmful Effects and Symptoms*

*Short Term Exposure:* Extremely toxic: probable oral lethal dose for humans is 5-50 mg/kg; between 7 drops and 1 teaspoonful for 70 kg person (150 lb). Poisonous; may be fatal if inhaled, swallowed, or absorbed through skin. Contact may cause burns to skin and eyes. Symptoms of carbamate poisoning resemble those of parathion. This material is similar to carbaryl; symptoms of carbaryl exposure include nausea, vomiting, abdominal cramps, diarrhea, excessive salivation, sweating, lassitude and weakness. Runny nose and sensation of tightness in chest may occur with inhalation exposures. Blurring or dimness of vision, tearing, eye muscle spasm, loss of muscle coordination, slurring of speech, and twitching of muscles may also occur. Carbamate insecticides inhibit the cholinesterase activity of enzymes, causing accumulation of acetylcholine at synapses and altering the way in which nervous impulses are transmitted. However, within several hours carbamates spontaneously detach from the enzymes.

*Long Term Exposure:* Cholinesterase inhibitor; cumulative effect is possible. Mexacarbate may damage the nervous system with repeated exposure, resulting in convulsions, respiratory failure. May cause liver damage.

*Points of Attack:* Respiratory system, lungs, central nervous system, cardiovascular system, skin, eyes, plasma and red blood cell cholinesterase.

*Medical Surveillance:* Before employment and at regular times after that, the following are recommended: Plasma and red blood cell cholinesterase levels (tests for the enzyme poisoned by this chemical). If exposure stops, plasma levels return to normal in 1-2 weeks while red blood cell levels may be reduced for 1-3 months.

When acetylcholinesterase enzyme levels are reduced by 25% or more below preemployment levels, risk of poisoning is increased, even if results are in lower ranges of "normal." Reassignment to work not involving organophosphate or carbamate pesticides is recommended until enzyme levels recover. If symptoms develop or overexposure occurs, repeat the above tests as soon as possible and get an exam of the nervous system. Also consider complete blood count. Consider chest x-ray following acute overexposure. Do not drink any alcoholic beverages before or during use. Alcohol promotes absorption of organophosphates.

*First Aid:* Speed in removing material from eyes and skin is of extreme importance. *Eyes:* Eye contact can cause dangerous amounts of these chemicals to be quickly absorbed through the mucous membrane into the bloodstream. Immediately and gently flush eyes with plenty of warm or cold water (NO hot water) for at least 15 minutes, occasionally lifting the upper and lower eyelids. Get medical aid immediately. *Skin:* Get medical aid. Skin contact can cause dangerous amounts of these chemicals to be absorbed into the bloodstream. Wearing the appropriate PPE equipment and respirator for carbamate pesticides, immediately flush skin with plenty of soap and water for at least 15 minutes while removing contaminated clothing and shoes. Shampoo hair promptly if contaminated; protect eyes. *Ingestion:* Call poison control. Loosen all clothing. Never give anything by mouth to an unconscious person. Get medical aid. Do NOT induce vomiting.* If conscious, alert, and able to swallow, rinse mouth and have victim drink 4 to 8 ounces of water. Check to see if poison control instructs you to use ipecac syrup, otherwise administer slurry of activated charcoal (2 oz in 8 oz of water). If victim is UNCONSCIOUS OR HAVING CONVULSIONS, do nothing except keep victim warm. **In some cases you may be specifically instructed by poison control to induce vomiting by way of 2 tablespoons of syrup of ipecac (adult) washed down with a cup of water.* Do NOT give activated charcoal before or with ipecac syrup. *Inhalation:* Get medical aid. Do not contaminate yourself. Wearing the appropriate PPE equipment and respirator for carbamate pesticides, immediately remove the victim from the contaminated area to fresh air. If the victim is not breathing, administer artificial respiration. Do NOT use mouth-to-mouth resuscitation; use bag/mask apparatus. If breathing is difficult, administer oxygen through bag/mask apparatus until medical help arrives. Do not leave victim unattended. *Note to physician or authorized medical personnel.* Administer atropine, 2 mg (1/30 gr) intramuscularly or intravenously as soon as any local or systemic signs or symptoms of an intoxication are noted; repeat the administration of atropine every 3 to 8 minutes until signs of atropinization (mydriasis, dry mouth, rapid pulse, hot and dry skin) occur; initiate treatment in children with 0.05 mg mg/kg of atropine; repeat at 5 to 10 minute intervals. Watch respiration, and remove bronchial secretions if they appear to be obstructing the airway; intubate if necessary. *Medical note:* 2-PAMCI may be contraindicated in the case of some carbamate poisonings.

*References:*
- U.S. Environmental Protection Agency, "Chemical Profile: Mexacarbate," Washington, DC, Chemical Emergency Preparedness Program (Nov. 30, 1987).
- California Environmental Protection Agency "Chemical List of Lists," Sacramento CA (February 1997).
- New Jersey Department of Health and Senior Services, "Hazardous Substance Fact Sheet, Mexacarbate." Trenton NJ (December 1999). http://www.state.nj.us/health/eoh/rtkweb/1304.pdf

# Mirex

*Use Type:* Insecticide
*CAS Number:* 2385-85-5
*Formula:* $C_{10}Cl_{12}$
*Alert:* No products in use in the U.S. since 1978.
*Synonyms:* Bichlorendo (Spanish); Dodecachlorooctahydro-1,3,4-metheno-2H-cyclobuta(c,d)pentalene; 1,1a,2,2,3,3a,4,5,5,5a,5b,6-Dodecachloroocta hydro-1,3,4-metheno-1H-cyclobuta(c,d)pentalene; Dodecachloropentacyclodecane; Ent 25,719; Hexachlorocyclopentadienedimer; 1,2,3,4,5,5-Hexachloro-1,3-cyclopentadiene dimer; NCI-C06428; Perchloro dihomocubane; Perchloropentacyclodecane
*Trade Names:* CG-1283®; DECHLORANE 4070®; FERRIAMICIDE®; HRS 1276®
*Producers:* Sigma-Aldrich Laborchemikalien (Germany)
*Chemical Class:* Organochlorine; Halo-organics
*EPA/OPP PC Code:* 039201
*California DPR Chemical Code:* 402
*RTECS Number:* PC8225000
*Uses:* Was used to control fire ants. Also used as a fire retardant in plastics. Not produced in the U.S. since 1978 but may be found in imported products.
*Fish toxicity (threshold)[77]:* High–1.98564 ppb, MATC (Maximum Acceptable Toxicant Concentration)
*U.S. Maximum Allowable Levels for Mirex:*
The U.S. EPA suggests that ingesting an amount of mirex equal to 200 picograms (pg) per kilogram (kg) of body weight per day is not likely to cause significant harmful health effects. The Food and Drug Administration (FDA) suggests that eating fish and other foods with concentrations below 100 ppt (parts per trillion) of mirex will not cause harmful health effects in people.
*Carcinogen/Hazard Classifications*
**U.S. NTP Carcinogen:** Reasonably anticipated carcinogen
**California Prop. 65:** Carcinogen
**IARC:** Group 2B, possible carcinogen
**Endocrine Disruptor:** Suspected endocrine disruptor
*Regulatory Authority:*
- Carcinogen (Animal Positive) (IARC)[9]
- AB 1803-Well Monitoring Chemical (CAL)
- AB 2588-Air Toxics "Hot Spots" Chemicals (CAL)
- Proposition 65 chemical (CAL)
- The "Director's List" (CAL/OSHA)
- Banned or Severely Restricted (Several Countries) (UN)[13]
- Air Pollutant Standard Set (Several States)[60]

*Description:* Mirex is a snow-white crystalline solid. Odorless. Practically insoluble in water; solubility = 0.20 ppm @ 24°C. Molecular weight = 545.52. Decomposes at 485°C. Vapor pressure = $8 \times 10^{-7}$ mm @ 20°C; $3 \times 10^{-7}$ mmHg @ 25°C. Hazard Identification (based on NFPA-704 M Rating System): Health 2, Flammability 1, Reactivity 0. Log $K_{ow}$ = > 5.6. Values at or above 3.0 are likely to bioaccumulate in marine organisms.
*Incompatibilities:* Strong oxidizers and dichromates.
*Permissible Exposure Limits in Air:* Several states have set guidelines or standards for mirex in ambient air[60] ranging from zero (Massachusetts) to 0.03 μg/m3 (New York) to 0.88 μg/m³ (Pennsylvania) to 4500 μg/m³ (South Carolina).
*Determination in Air:*
*Permissible Concentration in Water:* The U.S. EPA has set a limit of 1 part of mirex per trillion parts of surface water (1 ppt) to protect fish and other aquatic life from harmful effects.
*Routes of Entry:* Inhalation, ingestion, skin and/or eye contact. Passes through the skin.
*Harmful Effects and Symptoms*
*Short Term Exposure: Inhalation:* Can irritate the respiratory tract. This compound is moderately toxic (the $LD_{50}$ value for rats is 300 mg/kg). *Skin:* Can cause irritation, burning sensation and rash. *Eyes:* Can cause irritation. *Ingestion:* No cases of human toxicity reported. Possible symptoms include nausea, vomiting, restlessness, tremor, weight loss, nervous system and liver abnormalities, skin rash and reproductive failure. Exposure can cause nausea and vomiting, headache, dizziness, muscular weakness, fatigue, convulsions and may cause unconsciousness.
*Long Term Exposure:* May damage the developing fetus. May cause damage to the testes. May damage the liver and cause anemia. High exposure can cause arrhythmia (irregular heartbeat) and may cause death. Studies in mice and rats have shown that ingesting mirex and chlordecone can cause liver, adrenal gland, and kidney tumors.
*Points of Attack:* Blood, liver and central nervous system.
*Medical Surveillance:* Complete blood count, liver function tests, EKG, examination of the nervous system.
*First Aid:* Speed in removing material from eyes and skin is of extreme importance. If this chemical gets into the eyes, remove any contact lenses at once and irrigate immediately for at least 15 minutes, occasionally lifting upper and lower lids. Seek medical attention immediately. If this chemical contacts the skin, remove contaminated clothing and wash immediately with soap and water. Seek medical attention immediately. If this chemical has been inhaled, remove from exposure, begin rescue breathing (using universal precautions) if breathing has stopped, and CPR if heart

action has stopped. Transfer promptly to a medical facility. When this chemical has been swallowed, get medical attention. Give large quantities of water and induce vomiting. Do not make an unconscious person vomit.

*Note to physician or authorized medical personnel:* Gastric lavage or catharsis may be useful. High urine organic chlorine indicative of exposure, but not severity.

*References:*
- U.S. Department of Health and Human Services, Agency for Toxic Substances and Disease Registry, "ToxFAQS for Mirex and Chlordecone," Atlanta, GA (September 1996). http://www.atsdr.cdc.gov/tfacts66.html
- New Jersey Department of Health and Senior Services, "Hazardous Substance Fact Sheet, Mirex," Trenton NJ (July 1999). http://www.state.nj.us/health/eoh/rtkweb/1306.pdf
- Sax, N.I., Ed., "Dangerous Properties of Industrial Materials Report" 1, No. 2, 48 (1980) and 7, No. 5, 88-91 (1987).
- New York State Department of Health, "Chemical Fact Sheet: Mirex," Albany, NY, Bureau of Toxic Substance Assessment (March 1986).
- California Environmental Protection Agency "Chemical List of Lists," Sacramento CA (February 1997).

# Molinate

*Use Type:* Herbicide
*CAS Number:* 2212-67-1
*Formula:* $C_9H_{17}NOS$
*Synonyms:* 1*H*-Azepine-1-carbothioic acid, hexahydro-*S*-ethyl ester; Carbamic acid, hexamethylenethio-, *S*-ethyl ester; Carbamothioic acid, *N,N*-hexamethylene-, *S*-ethyl ester; Caswell No. 44; *S*-Ethyl azepane-1-carbothioate; *S*-Ethyl ester hexahydro-1*H*-azepine-1-carbothioioate; *S*-Ethyl hexahydro-1*H*-azepine-1-carbothioate; *S*-Ethyl hexahydro-1-carbothioic; ethyl 1-hexamethyleneiminecarbothiolate; *S*-Ethyl 1-hexamethyleneiminothiocarbamate; *S*-Ethyl *N,N*-hexamethyleneiminothiocarbamate; *S*-Ethyl *N*-hexamethyleneiminothiocarbamate; *S*-Ethyl perhydroazepin-1-carbothioate; *S*-Ethyl perhydroazepine-1-thiocarboxylate
*Trade Names:* ARROSOLO®, Syngenta (Switzerland); FELAN®; HIGALNATE®; HYDRAM®; JALAN®; MALERBANE-GIAVONI-L®; ORDAM®; ORDRAM®, Syngenta (Switzerland); R-4572®; RICECO, Syngenta (Switzerland); SAKKIMOL®; STAUFFER R 4,572®; YALAN®; YULAN®
*Producers:* Syngenta (Switzerland)
*Chemical Class:* Thiocarbamate
*EPA/OPP PC Code:* 041402
*California DPR Chemical Code:* 449
*RTECS Number:* CM2625000
*Uses:* Molinate is a selective herbicide used on rice for the control of water grass and other weeds. Producers in the U.S. have voluntarily withdrawn products containing molinate and sales and distribution will end August 31, 2009.

*Human toxicity (long-term)*[77]: High–1.40 ppb, Health Advisory
*Fish toxicity (threshold)*[77]: Very low–568.94463 ppb, MATC (Maximum Acceptable Toxicant Concentration)
*U.S. Maximum Allowable Residue Levels for Molinate (40 CFR 180.228):*

| CROP | ppm |
|---|---|
| Rice, grain & straw | 0.1 |

*Carcinogen/Hazard Classifications*
**U.S. EPA Carcinogens:** Group C, possible carcinogen
**TRI Developmental Toxin:** Reproductive and developmental toxin
**Label Signal Word:** CAUTION or WARNING
**WHO Acute Hazard:** Class II, moderately hazardous
*Regulatory Authority:*
- EPA 40 CFR 372.65, Specific Toxic Chemical Listings
- AB 1803-Well Monitoring Chemical (CAL)
- The "Director's List" (CAL/OSHA)
- Actively registered pesticide in California.
- EPA Hazardous Waste Number (RCRA No.): U365
- RCRA Section 261 Hazardous Constituents
- RCRA Universal Treatment Standards: Wastewater (mg/L), 0.003; Nonwastewater (mg/kg), 1.4
- RCRA Section 261 Hazardous Constituents
- EPCRA Section 313 Form R *de minimis* concentration reporting level: 1.0%

*Description:* Clear, amber liquid. Aromatic or spicy odor. Slightly soluble in water; solubility = 880 mg/L; 900 mg/L @ 21°C. Molecular weight =187.32. Density = 1.065 @ 20°C; 1.5156 @ 30°C. Vapor pressure =0.005 mm Hg at 25°C. Log $K_{ow}$ = 3.52. Values above 3.0 are likely to bioaccumulate in marine organisms. Boiling point = 202°C @ 10 mm.
*Incompatibilities:* Reacts violently with powerful oxidizers such as calcium hypochlorite.
*Permissible Exposure Limits in Air:* None listed.
*Permissible Concentration in Water:* No criteria set. Runoff from spills or fire control may cause water pollution.
*Harmful Effects and Symptoms*
*Short Term Exposure:* Low levels of toxicity. Concentrated solutions are slightly corrosive to eyes and mucous membranes. Dust inhalation can cause irritation of the respiratory system with sneezing. Eye contact can cause irritation, watering, pain, and inflammation of the eyelids. Skin contact can cause irritation and minor ulceration. Ingestion can cause nausea, vomiting, fever, muscle twitching, seizure, rapid respiration, slow heart beat. Severe exposure may result in death.
*Points of Attack:* Respiratory system, central nervous system, cardiovascular system, skin, eyes.
*Medical Surveillance:* Medical observation is recommended for 24 to 48 hours after breathing

overexposure, as pulmonary edema may be delayed. As first aid for pulmonary edema, consider administering a corticosteroid spray. Cigarette smoking may exacerbate pulmonary injury and should be discouraged for at least 72 hours following exposure. Before employment and at regular times after that, the following are recommended: If symptoms develop or overexposure occurs, repeat the above tests as soon as possible and get an exam of the nervous system. Also consider complete blood count. Consider chest x-ray following acute overexposure.

*First Aid:* If this chemical gets into the eyes, remove any contact lenses at once and irrigate immediately for at least 15 minutes, occasionally lifting upper and lower lids. Seek medical attention immediately. If this chemical contacts the skin, remove contaminated clothing and wash immediately with soap and water. Seek medical attention immediately. If this chemical has been inhaled, remove from exposure, begin rescue breathing (using universal precautions) if breathing has stopped, and CPR if heart action has stopped. Transfer promptly to a medical facility. When this chemical has been swallowed, get medical attention. If victim is conscious and able to swallow, have victim drink 4 to 8 ounces of water. Do not induce vomiting. *Note to physician or authorized medical personnel*: Medical observation is recommended for 24 to 48 hours after breathing overexposure, as pulmonary edema may be delayed. As first aid for pulmonary edema, consider administering a corticosteroid spray. Cigarette smoking may exacerbate pulmonary injury and should be discouraged for at least 72 hours following exposure.

*References:*
- U.S. Environmental Protection Agency, Office of Pesticide Programs, Pesticide Residue Limits, "Molinate," 40 CFR 180.228, www.epa.gov/pesticides/food/viewtols.htm
- EXTOXNET, Extension Toxicology Network, "Pesticide Information Profile, Molinate," Oregon State University, Corvallis, OR (June 1996). http://ace.orst.edu/cgi-bin/mfs/01/pips/molinate.htm
- California Environmental Protection Agency "Chemical List of Lists," Sacramento CA (February 1997)

# Monocarbamide Dihydrogen Sulfate

*Use Type:* Herbicide, plant growth regulator
*CAS Number:* 21351-39-3
*Formula:* $CH_4N_2O \cdot H_2O_4S$
*Synonyms:* Monourea sulfuric acid adduct; Urea sulfuric acid monoadduct; Sulfuric acid, monourea adduct; Urea dihydrogen sulfate; Urea, sulfate (1:1) (9CI)
*Trade Names:* ENQUIK®, Entek (USA); N-TAC DESSICANT®; SUPERQUIK®, Entek (USA); WILTHIN®, Entek (USA)
*Producers:* Entek (USA)

*Chemical Class:* Sulfonyl urea
*EPA/OPP PC Code:* 128961
*California DPR Chemical Code:* 2270
*Uses:* Mainly used on cotton, wine grapes, apples, asparagus and rights-of-way.
*Carcinogen/Hazard Classifications*
**Label Signal Word:** DANGER
*Permissible Concentration in Water:* No criteria set. Runoff from spills or fire control may cause water pollution.
*Harmful Effects and Symptoms*
*Short Term Exposure:* Contact with eyes or skin may cause irritation or burns. Inhalation should be avoided; use NIOSH-approved air purifying respirators for pesticides. May be harmful if swallowed. Skin contact may cause allergic reaction.
*Long Term Exposure:* Repeated or prolonged contact may cause skin and lung sensitization, resulting in allergies.
*Points of Attack:* Skin.
*Medical Surveillance:* Evaluation by a qualified allergist, including careful exposure history and special testing, may help diagnose skin allergy.
*First Aid:* If this chemical gets into the eyes, remove any contact lenses at once and irrigate immediately for at least 15 minutes, occasionally lifting upper and lower lids. Seek medical attention immediately. If this chemical contacts the skin, remove contaminated clothing and wash immediately with soap and water. Seek medical attention immediately. If this chemical has been inhaled, remove from exposure, begin rescue breathing (using universal precautions) if breathing has stopped, and CPR if heart action has stopped. Transfer promptly to a medical facility. When this chemical has been swallowed, get medical attention. Give large quantities of water and induce vomiting. Do not make an unconscious person vomit.
*References:*
- California Environmental Protection Agency "Chemical List of Lists," Sacramento CA (February 1997)

# Monocrotophos

*Use Type:* Insecticide
*CAS Number:* 6923-22-4
*Formula:* $C_7H_{14}NO_5P$
*Alert:* All uses of monocrotophos in the U.S. were discontinued in 1988. Prior to withdrawal is was a Restricted Use Pesticide (RUP). Human toxicity (long-term): Very high.
*Synonyms:* Crotonamide, 3-hydroxy-*N*-methyl-, dimethylphosphate, *cis*-; Crotonamide, 3-hydroxy-*N*-methyl-,dimethylphosphate, *(E)*-; 3-(Dimethoxyphosphinyloxy)*N*-methyl-*cis*-crotonamide; *(E)*-Dimethyl 1-methyl-3-(methylamino)-3-oxo-1-propenyl phosphate; *O,O*-Dimethyl-*O*-(2-*N*-methylcarbamoyl-1-methyl-vinyl)-fosfaat (Dutch); *O,O*-Dimethyl-*O*-(2-*N*-methylcarbamoyl-1-methyl)-vinyl-

phosphat (German); *O,O*-Dimethyl-*O*-(2-*N*-methylcarbamoyl-1-methyl-vinyl) phosphate; Dimethyl 1-methyl-2-(methylcarbamoyl)vinyl phosphate, *cis*-; Dimethyl phosphate ester of 3-hydroxy-*N*-methyl-*cis*-crotonamide; Dimethyl phosphate of 3-hydroxy-*N*-methyl-*cis*-crotonamine; *O,O*-Dimetil-*O*-(2-*N*-metilcarbamoil-1-metil-vinil)-fosfato (Italian); ENT 27,129; 3-Hydroxy-*N*-methylcrotonamide dimethyl phosphate; 3-Hydroxy-*N*-methyl-*cis*-crotonamide dimethyl phosphate; *cis*-1-Methyl-2-methyl carbamoyl vinyl phosphate; Monocron; Monocrotofos (Spanish); Monodrin; Phosphate de dimethyle et de 2-methylcarbamoyl 1-methyl vinyle (French); Phosphoric acid, dimethyl ester, ester with *cis*-3-hydroxy-*n*-methylcrotonamide

**Trade Names:** APADRIN®; AZODRIN®, Shell Chemical (UK), DuPont (USA), canceled; BILOBRAN®; BILOBORN®; C 1414®, CIBA Agriculture (Switzerland); CRISODRIN®; CIBA® 1414, CIBA Agriculture (Switzerland); CRISODIN®; GLORE PHOS 36®; MOLPHOS 36 SL®, Gujarat Pesticides (India); MONOCIL® 40; MOSUM®, Sabero Organics (India); NUVACRON®; PILLARDIN®; PLANTDRIN®; PHOSKILL®, United Phosphorus (India); SD® 9129; SHELL SD® 9129; SUFOS®, Sudarshan Chemical Industries (India); SUSVIN®; TATA MONO®, Rallis India (India); ULVAIR®

**Producers:**; Agsin (Singapore); Aimco Pesticides Ltd. (India); BEC Group (India); Bharat Pulverizing Mills (India); Cangzhou Green Chemical Co. (China); China Chemical (China); CIBA Agriculture (Switzerland); Cyanamid Agro (USA); DuPont (USA); Godavari Fertilisers and Chemicals (India); Gujarat Pesticides (India); Hindustan Insecticides (India); Hokko Chemical Industry (Japan); ICI Group (UK); Indiclay (India); Indo Gulf (India); Luxembourg Industries (PAMOL) (Israel); Makhteshim Agan (Israel); Nagarjuna Agrichem (India); Rallis India (India); Sabero Organics (India); Sankei Chemicals (Japan); Sanonda Zhengzhou Pesticide Co., Ltd. (China); Shell Chemical (UK); Shenzhen Guomeng Industry Co., Ltd. (China); Sinon (Taiwan); Sudarshan Chemical Industries (India); Syngenta (Switzerland); United Phosphorus (India); Vijayalakshmi Insecticides and Pesticides (India)

**Chemical Class:** Organophosphate
**EPA/OPP PC Code:** 058901
**California DPR Chemical Code:** 52
**ICSC Number:** 0181
**RTECS Number:** TC4375000
**EEC Number:** 015-072-00-9
**EINECS Number:** 230-042-7

**Uses:** Monocrotophos is an organophosphorus insecticide and acaricide which works systemically and on contact. It is extremely toxic to birds and is used as a bird poison. It is also very poisonous to mammals. It is used to control a variety of sucking, chewing and boring insects and spider mites on cotton, sugarcane, peanuts, ornamentals, and tobacco. Monocrotophos is available in other countries as a soluble concentrate or an ultra-low volume spray. It is also used on rice, maize, vegetables, soybeans, citrus, mangos.

**Human toxicity (long-term)**[77]**:** Extra high–0.35 ppb, Health Advisory

**Fish toxicity (threshold)**[77]**:** Very low–728.00039 ppb, MATC (Maximum Acceptable Toxicant Concentration)

**Carcinogen/Hazard Classifications**
**Label Signal Word:** DANGER
**WHO Acute Hazard:** Class 1b, highly hazardous
**Regulatory Authority:**

- Air Pollutant Standard Set (ACGIH)[1] (NIOSH)[2] (Argentina)[35] (Several States)[60]
- AB 1803-Well Monitoring Chemical (CAL)
- EPA/SARA 302 (EPCRA) Extremely hazardous substances
- Permissible Exposure Limits for Chemical Contaminants (CAL/OSHA)
- The "Director's List" (CAL/OSHA)
- U.S. DOT Inhalation Hazard Chemicals as organophosphates
- Superfund/EPCRA 40CFR355, Appendix B Extremely Hazardous Substances: TPQ = 10/10,000 lbs (4.54/4,540 kg)
- Superfund/EPCRA 40CFR302.4 RQ: EHS, 1 lbs (0.454 kg)
- U.S. DOT Regulated Marine Pollutant (49CFR172.101, Appendix B)
- Canada, WHMIS, Ingredients Disclosure List

**Description:** Monocrotophos is a colorless to reddish-brown solid with a mild ester odor. Soluble in water; solubility = 110 g/100mL @ 20°C. Boiling point = 125°C. Melting/Freezing point = 54 –55°C (pure); 25 –30°C (the reddish brown technical product). Vapor pressure = 1x $10^{-4}$ mm @ 20°C; 33 x $10^{-4}$ Pa @ 20°C. Flash point = 93°C. Hazard Identification (based on NFPA-704 M Rating System): Health 3, Flammability 1, Reactivity 0. Commercially available as a water-miscible solution.

**Incompatibilities:** Alkaline pesticides. Attacks black iron, drum steel, stainless steel, brass

**Permissible Exposure Limits in Air:** There is no OSHA[2] PEL. NIOSH[2] and ACGIH[1] recommend a TWA value of 0.25 mg/m$^3$. Argentina[35] has also set a TWA of 0.25 mg/m$^3$. In addition, some states have set guidelines or standards for monocrotophos in ambient air[60] ranging from 2.5 $\mu$g/m$^3$ (North Dakota) to 40 $\mu$g/m$^3$ (Virginia) to 5.0 $\mu$g/m$^3$ (Connecticut) to 6.0 $\mu$g/m$^3$ (Nevada). The NIOSH[2] IDLH value is 40 mg/m$^3$.

**Determination in Air:** OSHA versatile sampler-2; Toluene/Acetone; Gas chromatography/Flame photometric detection for sulfur, nitrogen, or phosphorus; NIOSH IV, Method #5600, Organophosphorus pesticides.[18]

**Permissible Concentration in Water:** No criteria set, but runoff from spills or fire control may cause water pollution.

**Routes of Entry:** Inhalation, ingestion and skin contact.

### Harmful Effects and Symptoms

***Short Term Exposure:*** Monocrotophos is a highly toxic, direct acting, water-soluble cholinesterase inhibitor which appears to be capable of penetration through the skin but which is excreted rapidly and does not accumulate in the body. Acute exposure to monocrotophos may result in the following signs and Symptoms: pinpoint pupils, blurred vision, headache, dizziness, muscle spasms, and profound weakness. Vomiting, diarrhea, abdominal pain, seizures, and coma may also occur. The heart rate may decrease following oral exposure or increase following dermal exposure. Hypotension (low blood pressure) may occur, although hypertension (high blood pressure) is not uncommon. Chest pain may be noted. Dyspnea (shortness of breath) may lead to respiratory collapse. Giddiness is common. Monocrotophos acts on the nervous system. Extremely toxic; probable oral lethal dose to humans 5-50 mg/kg or between 7 drops and 1 teaspoon for a 70 kg (150 lb) person. Repeated inhalation or skin contact with this material may, without symptoms, progressively increase susceptibility to poisoning. Monocrotophos may cause mutations. Handle with extreme caution. In animals: possible teratogenic effects. Delayed pulmonary edema may occur after inhalation.

***Long Term Exposure:*** Cholinesterase inhibitor; cumulative effect is possible. This chemical may damage the nervous system with repeated exposure, resulting in convulsions, respiratory failure. May cause liver damage. May cause personality changes with depression, anxiety, irritability.

***Points of Attack:*** Eyes, respiratory system, central nervous system, cardiovascular system, blood cholinesterase and reproductive system.

***Medical Surveillance:*** Medical observation is recommended for 24 to 48 hours after breathing overexposure, as pulmonary edema may be delayed. Before employment and at regular times after that, the following are recommended: Plasma and red blood cell cholinesterase levels (tests for the enzyme poisoned by this chemical). If exposure stops, plasma levels return to normal in 1-2 weeks while red blood cell levels may be reduced for 1-3 months. When acetylcholinesterase enzyme levels are reduced by 25% or more below preemployment levels, risk of poisoning is increased, even if results are in lower ranges of "normal." Reassignment to work not involving organophosphate or carbamate pesticides is recommended until enzyme levels recover. If symptoms develop or overexposure occurs, repeat the above tests as soon as possible and get an exam of the nervous system. Also consider complete blood count. Consider chest x-ray following acute overexposure. Do not drink any alcoholic beverages before or during use. Alcohol promotes absorption of organophosphates.

***First Aid:*** **Treatment for organophosphate poisoning consists of thorough decontamination, cardiorespiratory support, and administration of the antidotes atropine and pralidoxime. In cases of severe poisoning, diazepam, an anticonvulsant, should also be administered. Antidotes should be administered as prevention even if the diagnosis is in doubt.** Speed in removing material from eyes and skin is of extreme importance. *Eyes:* Eye contact can cause dangerous amounts of these chemicals to be quickly absorbed through the mucous membrane into the bloodstream. Immediately and gently flush eyes with plenty of warm or cold water (NO hot water) for at least 15 minutes, occasionally lifting the upper and lower eyelids. Get medical aid immediately. *Skin:* Get medical aid. Skin contact can cause dangerous amounts of these chemicals to be absorbed into the bloodstream. Wearing the appropriate PPE equipment and respirator for organophosphate pesticides, immediately flush skin with plenty of soap and water for at least 15 minutes while removing contaminated clothing and shoes. Shampoo hair promptly if contaminated. The removed, contaminated clothing and shoes should be double-bagged and left in Hot Zone for later disposal by hazardous materials experts. Skin may also be decontaminated with diluted hypochlorite solution. *Inhalation:* Get medical aid. Do not contaminate yourself. Wearing the appropriate PPE equipment and respirator for organophosphate pesticides, immediately remove the victim from the contaminated area to fresh air. If the victim is not breathing, administer artificial respiration. Do NOT use mouth-to-mouth resuscitation; use bag/mask apparatus. If breathing is difficult, administer oxygen through bag/mask apparatus until medical help arrives. Do not leave victim unattended. *Ingestion:* Call poison control. Loosen all clothing. Never give anything by mouth to an unconscious person. If victim is *unconscious or having convulsions,* do nothing except keep victim warm. Get medical aid. Transfer promptly to a medical facility. In cases of ingestion, **do not induce vomiting**. If the victim is alert and asymptomatic, administer a slurry of activated charcoal at a dose of 1 g/kg (infant, child, and adult dose). A soda can and straw may be of assistance when offering charcoal to a child. *In some cases you may be specifically instructed by poison control to induce vomiting by way of 2 tablespoons of syrup of ipecac (adult) washed down with a cup of water.* Do NOT give activated charcoal before or with ipecac syrup.

*Note to physician or authorized medical personnel:* Treat cases of respiratory compromise, coma, or excessive pulmonary secretions with respiratory support using protocols and techniques available and within the scope of training. Some cases may necessitate procedures such as endotracheal intubation or cricothyrotomy by properly trained and equipped personnel. When possible, atropine (see *Antidotes*, below) should be given under medical supervision. Patients who are comatose, hypotensive, or having seizures or cardiac arrhythmias should be treated according to advanced life support protocols. *Antidotes:* Two antidotes are administered to treat organophosphate poisoning. Atropine is a competitive antagonist of acetylcholine at muscarinic receptors and is used to control the excessive bronchial secretions which are often responsible for death. Pralidoxime relieves both the nicotinic and muscarine effects of organophosphate poisoning by regenerating acetylcholinesterase and can reduce both the bronchial secretions and the muscle weakness associated with poisoning. The initial intravenous dose of atropine in adults should be determined by the severity of symptoms: An initial adult dose of 1.0 to 2.0 mg or pediatric dose of 0.01 mg/kg (minimum 0.01 mg) should be administered intravenously. If intravenous access cannot be established, atropine may also be given intramuscularly, subcutaneously or via endotracheal tube. Doses should be repeated every 15 minutes until excessive secretions and sweating have been controlled. Once bronchial secretion has been controlled, atropine administration should be repeated whenever the secretions begin to recur. In seriously poisoned patients, very large doses may be required. Alterations of pulse rate and pupillary size should not be used as indicators of treatment adequacy. Pralidoxime should be administered as early in poisoning as possible as its efficacy may diminish when given more than 24 to 36 hours after exposure. Doses are as follows: adult 1.0 g; pediatric 25 to 50 mg/kg. The drug should be administered intravenously over 30 to 60 minutes, but in a life-threatening situation, one-half of the total dose can be given per minute for a total administration time of 2 minutes. Treatment should begin to take effect within 40 minutes with a reduction in symptoms and the amount of atropine necessary to control bronchial secretion. The initial dose can be repeated in 1 hour and then every 8 to 12 hours until the patient is clinically well and no longer requires atropine. If intravenous access cannot be established, pralidoxime may also be given intramuscularly. Early administration of diazepam in addition to the combined atropine and pralidoxime treatment may help prevent the onset of seizures and potential brain and cardiac morphologic damage following high-level organophosphate poisoning.

*References:*
- EXTOXNET, Extension Toxicology Network, "Pesticide Information Profile, Monocrotophos," Oregon State University, Corvallis, OR (September 1995). http://extoxnet.orst.edu/pips/monocrot.htm
- U.S. Environmental Protection Agency, "Chemical Profile: Monocrotophos," Washington, DC, Chemical Emergency Preparedness Program (Nov. 30, 1987).
- Agency for Toxic Substances and Disease Registry, U.S. Department of Health and Human Services, Public Health Service, "Managing Hazardous Materials Incidents," Atlanta, GA (June 2003).
- California Environmental Protection Agency "Chemical List of Lists," Sacramento CA (February 1997).
- New Jersey Department of Health and Senior Services, "Hazardous Substance Fact Sheet: Monocrotophos," Trenton, NJ (November 1999). http://www.state.nj.us/health/eoh/rtkweb/1313.pdf

# Monuron (ANSI)

*Use Type:* Herbicide; plant growth regulator
*CAS Number:* 150-68-5
*Formula:* $C_9H_{11}ClN_2O$
*Synonyms:* Caswell No. 583; Chlorea; Chlorfenidim; 3-*p*-Chlorophenyl-1,1-dimethylurea; *N*-(p-Chlorophenyl)-*N'*,*N'*-Dimethylurea; *N*-(4-Chlorophenyl)-*N'*,*N'*-dimethylurea; *N'*-(4-Chlorophenyl)-*N*,*N*-dimethylurea; 1-*p*-Chlorophenyl-3,3-dimethylurea; 3'-(4'-Chlorophenyl)-1,1-dimethylurea; 3-(4-Chlorophenyl)-1,1-dimethylurea; 3-(*p*-Chlorophenyl)-1,1-dimethylurea; 4-Chlorophenyl dimethylurea; 1-(4-Chlorophenyl)-3,3-dimethylurea; 1-(*p*-Chlorophenyl)-3,3-dimethylurea; *N-p*-Chlorophenyl-*N'*,*N'*-dimethylurea; 1,1-Dimethyl-3-(*p*-Chlorophenyl)thiourea; 1,1-Dimethyl-3-(*p*-Chlorophenyl)urea; *N*,*N*-Dimethyl-*N'*-(4-Chlorophenyl)urea; *N*-Dimethyl-*N'*-(4-chlorophenyl)urea; 1,1-Dimethyl-3-(*p*-chlorophenyl)urea; NCI-C02846; Urea, 3-(*p*-chlorophenyl)-1,1-dimethyl-; Urea, *N'*-(4-chlorophenyl)-*N*,*N*-dimethyl-
*Trade Names:* ALCO OXALIS KILLER, AMVAC Chemical (USA), canceled; CMU®; KARMEX®; KARMEX® W; LIROBETAREX®; MONUREX®, Agan Chemical (Israel), canceled; TELVAR®; TELVAR® MONURON weedkiller; TELVAR®-W MONURON weedkiller
*Chemical Class:* Substituted urea
*EPA/OPP PC Code:* 035501
*California DPR Chemical Code:* 408
*RTECS Number:* YS6300000
*Uses:* Not registered in the U.S. Used as a nonselective, broad-spectrum herbicide for the control of grasses and herbaceous weeds on non-crop areas such as rights-of-way, industrial sites and drainage ditches. Once used as a plant growth regulator on sugar cane.
*Fish toxicity (threshold)[77]:* Low–468.12877 ppb, MATC (Maximum Acceptable Toxicant Concentration)
*Carcinogen/Hazard Classifications*
**IARC:** Group 3, unclassifiable
**Label Signal Word:** CAUTION
*Regulatory Authority:*
- EPA 40 CFR 372.65, Specific Toxic Chemical Listings
- AB 1803-Well Monitoring Chemical (CAL)
- The "Director's List" (CAL/OSHA)
- EPCRA Section 313 Form R *de minimis* concentration reporting level: 1.0%

*Description:* White, crystalline solid. Odorless. Low solubility in water; solubility = 230 ppm @ 25°C. Molecular weight = 198.65. Density = 1.27 @ 20°C. Melting/Freezing point = 170-174°C. Vapor pressure = $5 \times 10^{-7}$ mmHg @ 25°C. Log $K_{ow}$ = <2.00. Values of more than 3.0 are likely to bioaccumulate in marine organisms.
*Incompatibilities:* Contact with water, bases, acids or elevated temperatures above 185°C can cause the generation of toxic 3,4-dichloraniline, and dimethylamine.

*Determination in Air:* Filter; none; Gravimetric; NIOSH IV [Particulates NOR; #0500 (total), #0600 (respirable)].[18]

*Permissible Concentration in Water:* No criteria set. Runoff from spills or fire control may cause water pollution.

*Harmful Effects and Symptoms*

*Short Term Exposure:* Contact with eyes or skin may cause irritation or injury. Inhalation should be avoided; use NIOSH-approved air purifying respirators for pesticides. May be harmful if swallowed. Skin contact may cause severe irritation or burns.

*Points of Attack:* Skin.

*First Aid:* If this chemical gets into the eyes, remove any contact lenses at once and irrigate immediately for at least 15 minutes, occasionally lifting upper and lower lids. Seek medical attention immediately. If this chemical contacts the skin, remove contaminated clothing and wash immediately with soap and water. Seek medical attention immediately. If this chemical has been inhaled, remove from exposure, begin rescue breathing (using universal precautions) if breathing has stopped, and CPR if heart action has stopped. Transfer promptly to a medical facility. When this chemical has been swallowed, get medical attention. Give large quantities of water. Do not induce vomiting when formulations containing petroleum solvents are ingested. Do not make an unconscious person vomit.

*References:*
- Pesticide Management Education Program, "Monuron (Monurex, Telvar) Herbicide Profile 2/85," Cornell University, Ithaca, NY (February 1985). http://pmep.cce.cornell.edu/profiles/herb-growthreg/fatty-alcohol-monuron/monuron/herb-prof-monuron.html
- California Environmental Protection Agency "Chemical List of Lists," Sacramento CA (February 1997)

# Myclobutanil (ANSI)

*Use Type:* Fungicide
*CAS Number:* 88671-89-0
*Formula:* $C_{15}H_{17}ClN_4$
*Synonyms:* α-Butyl-α-(4-chlorophenyl)-1*H*-1,2,4-triazole-1-propanenitrile; Caswell No. 723K; 2-(4-Chlorophenyl)-2-(1*H*-1,2,4-triazole-1-ylmethyl)hexanenitrile; 2-*p*-Chlorophenyl-2-(1*H*-1,2,4-triazole-1-ylmethyl)hexanenitrile; 1*H*-1,2,4-Triazole-1-propnenitrile, α-butyl-α-(4-chlorophenyl)
*Trade Names:* EAGLE®, Dow AgroSciences (USA); NOVA®, Dow AgroSciences (USA); NU-FLOW®, Dow AgroSciences (USA); RALLY®, Dow AgroSciences (USA); LAREDO®, Dow AgroSciences (USA); RH 3866®; SYSTHANE®, Dow AgroSciences (USA)
*Producers:* Agsin (Singapore), Dow AgroSciences (USA); Wuzhou International (China)
*Chemical Class:* Triazole
*EPA/OPP PC Code:* 128857
*California DPR Chemical Code:* 2245
*Uses:* Widely used to control powdery mildew, rust, sclerotina, spot blight, rot, black rot and similar fungi on a variety of food and non-food crops.

*U.S. Maximum Allowable Residue Levels for Myclobutanil (40 CFR 180.443):*

| CROP | ppm |
|---|---|
| Almond | 0.1 |
| Almond, hulls | 2 |
| Animal feed, nongrass, group 18 | 0.03 |
| Apple | 0.5 |
| Apple, wet pomace | 1.3 |
| Artichoke, globe | 1 |
| Banana, post-h | 4 |
| Bean, snap, succulent | 1 |
| Beet, sugar, dried pulp | 1 |
| Beet, sugar, molasses | 1 |
| Beet, sugar, refined sugar | 0.7 |
| Beet, sugar, roots | 0.05 |
| Beet, sugar, tops | 1 |
| Cattle, fat | 0.05 |
| Cattle, liver | 1 |
| Cattle, meat | 0.1 |
| Cattle, mbyp, except liver | 0.2 |
| Cherry, sweet | 5 |
| Cherry, tart | 5 |
| Cotton, undelinted seed | 0.02 |
| Currant | 3 |
| Egg | 0.02 |
| Fruit, stone, except cherry | 2 |
| Goat, fat | 0.05 |
| Goat, liver | 1 |
| Goat, meat | 0.1 |
| Goat, mbyp, except liver | 0.2 |
| Gooseberry | 2 |
| Grain, cereal, forage, fodder and straw, group 16 | 0.03 |
| Grain, cereal, group 15 | 0.03 |
| Grape | 1 |
| Grape, pomace, wet & dried | 10 |
| Grape, raisin | 10 |
| Grape, raisin, waste | 25 |
| Hog, fat | 0.05 |
| Hog, liver | 1 |
| Hog, meat | 0.1 |
| Hog, mbyp, except liver | 0.2 |
| Hop, dried cones | 5 |
| Horse, fat | 0.05 |
| Horse, liver | 1 |
| Horse, meat | 0.1 |
| Horse, mbyp, except liver | 0.2 |
| Mayhaw | 0.7 |
| Milk | 0.2 |
| Pepper | 1 |

| | |
|---|---|
| Pepper, bell | 1 |
| Pepper, nonbell | 1 |
| Plum, prune, dried | 8 |
| Poultry, fat | 0.02 |
| Poultry, meat | 0.02 |
| Poultry, mbyp | 0.02 |
| Sheep, fat | 0.05 |
| Sheep, liver | 1 |
| Sheep, meat | 0.1 |
| Sheep, mbyp, except liver | 0.2 |
| Vegetable, brassica, leafy, group 5 | 0.03 |
| Vegetable, foliage of legume, group 7 | 0.03 |
| Vegetable, fruiting, group 8 | 0.03 |
| Vegetable, leafy, except brassica, group 4 | 0.03 |
| Vegetable, leaves of root and tuber, group 2 | 0.03 |
| Vegetable, legume, group 6 | 0.03 |
| Vegetable, root and tuber, group 1 | 0.03 |

*Carcinogen/Hazard Classifications*

**U.S. EPA Carcinogens:** Group E, Unlikely a carcinogen

*California Prop. 65:* Developmental toxin; male reproductive toxin.

**TRI Developmental Toxin:** Reproductive and developmental toxin.

**U.S. TRI:** Developmental toxin and reproductive toxin

**Label Signal Word:** CAUTION or WARNING

**WHO Acute Hazard:** Class III, slightly hazardous

*Regulatory Authority:*
- Actively registered pesticide in California.
- FIFRA, 40CFR185: tolerances for pesticides in food
- FIFRA, 40CFR186: tolerances for pesticides in animal feeds.
- EPCRA Section 313 Form R *de minimis* concentration reporting level: 1.0%

*Description:* Yellow, crystalline solid. Soluble in water; solubility = 140 ppm. Melting/Freezing point = 65°C. Vapor pressure = $1.57 \times 10^{-6}$ mmHg @ 0°C.

*Permissible Concentration in Water:* No criteria set. Runoff from spills or fire control may cause water pollution.

*Routes of Entry:* Ingestion, skin contact

*Harmful Effects and Symptoms*

*Short Term Exposure:* Poisonous if swallowed. Contact may irritate skin and cause eye irritation and possible severe injury. Avoid inhalation.

*Medical Surveillance:* If this chemical gets into the eyes, remove any contact lenses at once and irrigate immediately for at least 15 minutes, occasionally lifting upper and lower lids. Seek medical attention immediately. If this chemical contacts the skin, remove contaminated clothing and wash immediately with soap and water. If this chemical has been inhaled, remove from exposure, begin rescue breathing (using universal precautions) if breathing has stopped, and CPR if heart action has stopped. Transfer promptly to a medical facility. When this chemical has been swallowed, get medical attention. Give large quantities of water and induce vomiting. Do not make an unconscious person vomit

*References:*
- U.S. Environmental Protection Agency, Office of Pesticide Programs, Pesticide Residue Limits, "Myclobutanil," 40 CFR 180.443, www.epa.gov/pesticides/food/viewtols.htm
- California Environmental Protection Agency "Chemical List of Lists," Sacramento CA (February 1997).

# N

# Nabam

*Use Type:* Fungicide, algicide, herbicide and microbiocide
*CAS Number:* 142-59-6
*Formula:* $C_4H_6N_2Na_2S_4$
*Alert:* Registered only for non-food applications.
*Synonyms:* Carbamodithioic acid, 1,2-ethanediylbis-, disodium salt; Carbamic acid, ethylenebis (dithio-), disodium salt; Disodium ethylenebis(dithiocarbamate); DSE; EBDC, disodium salt; EBDC, Sodium salt; 1,2-ethanediylbis(carbamodithioic acid), disodium salt; N,N'-Ethylenebis(dithiocarbamate de sodium) (French); Ethylenebis(dithiocarbamic acid), disodium salt; Nabame; Nabasam (obsolete)
*Trade Names:* AMA-30®, Kemira Chemical (Finland); CAMBELL'S® NABAM SOIL FUNGICIDE; CARBON D®; NALCO D-62C44®; CHEM-BAM®; DITHANE A-40®; DITHANE A-46®; DITHANE D-14®, canceled; NAFUN-IPO®; NALCO® D-62C44; PARZATE®; SPRING-BAK®
*Producers:* Bayer CropScience (Germany); Kemira Chemical (Finland); Sigma-Aldrich Laborchemikalien (Germany)
*Chemical Class:* Dithiocarbamate
*EPA/OPP PC Code:* 014503
*California DPR Chemical Code:* 417
*Uses:* Nabam is a broad spectrum fungicide/bactericide/algicide used to prevent crop damage by fungi, to protect harvested products from deterioration, and as an industrial microbiocide. As a result of the U.S. EPA review of nabam in 1989, all food uses were voluntarily canceled by the manufacturers except for one FDA-regulated food use on sugar mill grinding, crusher and/or diffuser systems, e.g., processing water systems. All other uses of nabam are for the control of algae, slime-forming bacteria and fungi in indoor non-food environments, paper mills, water cooling systems, drilling mud and packer fluids and secondary oil recovery water systems.
*Carcinogen/Hazard Classifications*
*California Prop. 65:* Developmental toxin
**TRI Developmental Toxin:** Developmental toxin
**Label Signal Word:** WARNING
*Regulatory Authority:*
- Actively registered pesticide in California.
- EPA 40 CFR 372.65, Specific Toxic Chemical Listings
- 47 FR 47669 (10/27/82); 54 FR 50020 (12/04/89); (55 FR 7935 (03) 06/90) [Environmental Protection]
- California Chemical List of Lists (February 1997)
- EPCRA Section 313 Form R *de minimis* concentration reporting level: 1.0%
- Marine pollutant (49CFR, Subchapter 172.101, Appendix B)

*Description:* Colorless crystalline solid (when pure). Light amber crystalline solid or 22% wettable powder solution. Slight odor of sulfide. Mixes with water; solubility = $>2 \times 10^5$ mg/L. Hazard Identification (based on NFPA-704 M Rating System): Health Hazards (Blue): 2; Flammability (Red): 0; Reactivity (Yellow): 0. Molecular weight = 256.34. Boiling point = decomposes. Melting/Freezing point = decomposes when heated, without melting. Specific gravity: 1.14 @ 20°C (solid). Dissociates in water; unlikely to bioaccumulate in marine organisms.
*Incompatibilities:* Dry crystalline form is unstable.
*Permissible Concentration in Water:* No criteria set. Runoff from spills or fire control may cause water pollution.
*Harmful Effects and Symptoms*
*Short Term Exposure:* Low levels of toxicity. Concentrated solutions are slightly corrosive to eyes and mucous membranes. Dust inhalation can cause irritation of the respiratory system with sneezing. Eye contact can cause irritation, watering, pain, and inflammation of the eyelids. Skin contact can cause irritation and minor ulceration. Ingestion can cause nausea, vomiting, fever, muscle twitching, seizure, rapid respiration, slow heart beat. Severe exposure may result in death.
*Points of Attack:* Respiratory system, central nervous system, cardiovascular system, skin, eyes.
*Medical Surveillance:* Medical observation is recommended for 24 to 48 hours after breathing overexposure, as pulmonary edema may be delayed. As first aid for pulmonary edema, consider administering a corticosteroid spray. Cigarette smoking may exacerbate pulmonary injury and should be discouraged for at least 72 hours following exposure. Before employment and at regular times after that, the following are recommended: If symptoms develop or overexposure occurs, repeat the above tests as soon as possible and get an exam of the nervous system. Also consider complete blood count. Consider chest x-ray following acute overexposure.
*First Aid:* If this chemical gets into the eyes, remove any contact lenses at once and irrigate immediately for at least 15 minutes, occasionally lifting upper and lower lids. Seek medical attention immediately. If this chemical contacts the skin, remove contaminated clothing and wash immediately with soap and water. Seek medical attention immediately. If this chemical has been inhaled, remove from exposure, begin rescue breathing (using universal precautions) if breathing has stopped, and CPR if heart action has stopped.

Transfer promptly to a medical facility. When this chemical has been swallowed, get medical attention. If victim is conscious and able to swallow, have victim drink 4 to 8 ounces of water. Do not induce vomiting. *Note to physician or authorized medical personnel*: Medical observation is recommended for 24 to 48 hours after breathing overexposure, as pulmonary edema may be delayed. As first aid for pulmonary edema, consider administering a corticosteroid spray. Cigarette smoking may exacerbate pulmonary injury and should be discouraged for at least 72 hours following exposure.

*References:*
- U.S. Environmental Protection Agency, "Reregistration Eligibility Decision (RED), Nabam," Office of Prevention, Pesticides and Toxic Substances, Washington, DC (January 1996). http://www.epa.gov/REDs/0641.pdf
- California Environmental Protection Agency "Chemical List of Lists," Sacramento CA (February 1997).

# Naled

*Use Type:* Insecticide, fungicide, bactericide and acaricide
*CAS Number:* 300-76-5
*Formula:* $C_4H_7Br_2Cl_2O_4P$
*Alert:* A General Use Pesticide (GUP), but not be used around the home.
*Synonyms:* BRP; *O*-(1,2-Dibrom-2,2-dichloraethyl)-*O,O*-dimethyl-phosphat (German); 1,2-Dibromo-2,2-dichloroethyl dimethyl phosphate; *O*-(1,2-Dibromo-2,2-dichloro-etil)-*O,O*-dimetil fosfato (Italian); *O,O*-dimethyl-*O*-(1,2-dibromo-2,2-dichloroethyl)phosphate; Dimethyl 1,2-dibromo-2,2-dichloroethyl phosphate; *O,O*-Dimethyl *O*-2,2-dichloro-1,2-dibromoethyl phosphate; ENT 24,988; Ethanol, 1,2-dibromo-2,2-dichloro-, dimethyl phosphate; OMS 75; Phosphate de *O, O*-dimethyle et de *O*-(1,2-dibromo-2,2-dichlorethyle) (French); Phosphoric acid, 1,2-dibromo-2,2-dichloroethyl dimethyl ester
*Trade Names:* AI3-24988®; ARTHODIBROM®; BROMCHLOPHOS®; BROMEX®; DIBROM®, Amvac Chemical Corp. (USA); FLYKILLER®, Amvac Chemical Corp. (USA); LUCANAL®; HIBROM®; ORTHO® 4355; ORTHODIBROM®, Amvac Chemical Corp. (USA); ORTHODIBROMO®, Amvac Chemical Corp. (USA); PROKIL®, Gowan Company (USA); TRUMPET®, Amvac Chemical Corp. (USA)
*Producers:* Amvac Chemical Corp. (USA); Calliope (France); Calliope (France); Chevron Phillips Chemical (USA); DuPont (USA); Ehrenstorfer, Dr. (Germany); Gowan Company (USA); Nippon Chemical Industrial (Japan)
*Chemical Class:* Organophosphate
*EPA/OPP PC Code:* 034401
*California DPR Chemical Code:* 418
*ICSC Number:* 0925
*RTECS Number:* TB9450000
*EEC Number:* 015-055-00-6
*EINECS Number:* 206-098-3
*Uses:* Naled is a fast acting, nonsystemic contact and stomach insecticide used to control aphids, mites, mosquitoes, and flies on crops and in greenhouses, mushroom houses, animal and poultry houses, kennels, food processing plants, and aquaria and in outdoor mosquito control. Liquid formulations can be applied to greenhouse heating pipes to kill insects by vapor action. It has been used by veterinarians to kill parasitic worms (other than tapeworms) in dogs. Naled may no longer be used in and around the home by residents or professional applicators. Naled is available in dust, emulsion concentrate, liquid, and ULV formulations. Also used in cooling towers, veterinary medicine, pulp and paper mill systems, hospitals, swimming pools, and bathrooms.

*U.S. Maximum Allowable Residue Levels for Naled (40 CFR 180.215):*

| CROP | ppm |
|---|---|
| Almond | 0.5 |
| Almond, hulls | 0.5 |
| Bean, dry | 0.5 |
| Bean, succulent | 0.5 |
| Beet, sugar, roots | 0.5 |
| Beet, sugar, tops | 0.5 |
| Broccoli | 1 |
| Brussels sprouts | 1 |
| Cabbage | 1 |
| Cauliflower | 1 |
| Celery | 3 |
| Collards | 3 |
| Cotton, undelinted seed | 0.5 |
| Cucumber | 0.5 |
| Eggplant | 0.5 |
| Grape | 0.5 |
| Grapefruit | 3 |
| Grass, forage | 10 |
| Hop | 0.5 |
| Kale | 3 |
| Legume, forage | 10 |
| Lemon | 3 |
| Lettuce | 1 |
| Melon | 0.5 |
| Orange | 3 |
| Pea, succulent | 0.5 |
| Peach | 0.5 |
| Pepper | 0.5 |
| Pumpkin | 0.5 |
| Raw agricultural commodities | 0.5 |
| Safflower, seed | 0.5 |
| Spinach | 3 |
| Squash, summer | 0.5 |

| | |
|---|---|
| Squash, winter | 0.5 |
| Strawberry | 1 |
| Swiss chard | 3 |
| Tangerine | 3 |
| Tomato | 0.5 |
| Turnip, greens | 3 |
| Walnut | 0.5 |

*Carcinogen/Hazard Classifications*
**U.S. EPA Carcinogens:** Group E, unlikely carcinogen
**TRI Developmental Toxin:** Reproductive toxin
**Label Signal Word:** DANGER–POISON or CAUTION
**WHO Acute Hazard:** Class II, moderately hazardous
*Regulatory Authority:*
- Air Pollutant Standard Set (ACGIH)[1] (DFG)[3] (HSE)[33] (former USSR)[35] (Several States)[60]
- AB 1803-Well Monitoring Chemical (CAL)
- Permissible Exposure Limits for Chemical Contaminants (CAL/OSHA)
- The "Director's List" (CAL/OSHA)
- Actively registered pesticide in California.
- Clean Water Act: Section 311 Hazardous Substances/RQ 40CFR117.3 (same as CERCLA, see below).
- Superfund/EPCRA 40CFR302.4 RQ: CERCLA, 10 lb (4.54 kg)
- EPCRA Section 313 Form R *de minimus* concentration reporting level: 1.0%
- U.S. DOT Inhalation Hazard Chemicals as organophosphates
- U.S. DOT Regulated Marine Pollutant (49CFR172.101, Appendix B)
- Canada, WHMIS, Ingredients Disclosure List

*Description:* Naled is a white crystalline solid (when pure) or light straw-colored liquid. Slightly pungent odor. Practically insoluble in water. Molecular weight = 380.79. Boiling point = 110°C (decomposes). Melting/Freezing point = 27°C. Vapor pressure = 2x $10^{-4}$ mm @ 20°C. Hazard Identification (based on NFPA-704 M Rating System): Health 1, Flammability 0, Reactivity 1. Log $K_{ow}$ = 1.41. Unlikely to bioaccumulate in marine organisms.

*Incompatibilities:* Incompatible with oxidizers. Hydrolyzed in presence of water. Degraded by sunlight. Decomposes when heated, on contact with acids, acid fumes, bases, producing fumes of hydrogen chloride, hydrogen bromide, phosphorous oxides. Reacts with acids, strong oxidizers in sunlight. Reacts with water. Corrosive to metals. Attacks some plastics, rubber and coatings.

*Permissible Exposure Limits in Air:* The OSHA[2] 8-hour TWA is 3 mg/m³. NIOSH[2], ACGIH[1], DFG[3] and HSE[33] set or recommend a TWA of 3 mg/m³ and HSE's STEL[33] is 6 mg/m³. The NIOSH[2] IDLH level = 200 mg/m³. The former former USSR[35] has set an MAC in workplace air of 0.5 mg/m³. Several states have set guidelines or standards for dibrom in ambient air[60] ranging from 30 µg/m³ (North Dakota) to 50 µg/m³ (Virginia) to 60 µg/m³ (Connecticut) to 71 µg/m³ (Nevada).

*Determination in Air:* No method available.
*Permissible Concentration in Water:* Mexico[35] has set maximum permissible concentration values in coastal waters of 3.0 µg/L and 0.03 mg/L (30 µg/L) in estuaries.
*Routes of Entry:* Inhalation, skin absorption, ingestion, eye and/or skin contact.
*Harmful Effects and Symptoms*
**Short Term Exposure:** Irritates the eyes, skin, and respiratory tract. May affect the nervous system, causing convulsions, respiratory failure. Organic phosphorus insecticides are absorbed by the skin, as well as by the respiratory and gastrointestinal tracts. They are cholinesterase inhibitors. Symptoms of exposure include headache, giddiness, blurred vision, nervousness, weakness, nausea, cramps, diarrhea, and discomfort in the chest. Signs include sweating, tearing, salivation, vomiting, cyanosis, convulsions, coma, loss of reflexes and loss of sphincter control. High exposure can result in death. Highly toxic; a probable human lethal dose may be between 1 teaspoon and 1 ounce. Delayed pulmonary edema may occur after inhalation.

**Long Term Exposure:** May cause skin allergy. Cholinesterase inhibitor; cumulative effect is possible. This chemical may damage the nervous system with repeated exposure, resulting in convulsions, respiratory failure. May cause liver damage.

*Points of Attack:* Respiratory system, central nervous system, cardiovascular system, skin, eyes and blood cholinesterase.

*Medical Surveillance:* Medical observation is recommended for 24 to 48 hours after breathing overexposure, as pulmonary edema may be delayed. Before employment and at regular times after that, the following are recommended: Plasma and red blood cell cholinesterase levels (tests for the enzyme poisoned by this chemical). If exposure stops, plasma levels return to normal in 1-2 weeks while red blood cell levels may be reduced for 1-3 months. When acetylcholinesterase enzyme levels are reduced by 25% or more below preemployment levels, risk of poisoning is increased, even if results are in lower ranges of "normal." Reassignment to work not involving organophosphate or carbamate pesticides is recommended until enzyme levels recover. If symptoms develop or overexposure occurs, repeat the above tests as soon as possible and get an exam of the nervous system. Also consider complete blood count. Consider chest x-ray following acute overexposure. Evaluation by a qualified allergist. Do not drink any alcoholic beverages before or during use. Alcohol promotes absorption of organophosphates.

*First Aid:* **Treatment for organophosphate poisoning consists of thorough decontamination, cardiorespiratory support, and administration of the antidotes atropine and pralidoxime. In cases of severe poisoning, diazepam, an anticonvulsant, should also be administered. Antidotes should be administered as prevention even if**

**the diagnosis is in doubt.** Speed in removing material from eyes and skin is of extreme importance. *Eyes:* Eye contact can cause dangerous amounts of these chemicals to be quickly absorbed through the mucous membrane into the bloodstream. Immediately and gently flush eyes with plenty of warm or cold water (NO hot water) for at least 15 minutes, occasionally lifting the upper and lower eyelids. Get medical aid immediately. *Skin:* Get medical aid. Skin contact can cause dangerous amounts of these chemicals to be absorbed into the bloodstream. Wearing the appropriate PPE equipment and respirator for organophosphate pesticides, immediately flush skin with plenty of soap and water for at least 15 minutes while removing contaminated clothing and shoes. Shampoo hair promptly if contaminated. The removed, contaminated clothing and shoes should be double-bagged and left in Hot Zone for later disposal by hazardous materials experts. Skin may also be decontaminated with diluted hypochlorite solution. *Inhalation:* Get medical aid. Do not contaminate yourself. Wearing the appropriate PPE equipment and respirator for organophosphate pesticides, immediately remove the victim from the contaminated area to fresh air. If the victim is not breathing, administer artificial respiration. Do NOT use mouth-to-mouth resuscitation; use bag/mask apparatus. If breathing is difficult, administer oxygen through bag/mask apparatus until medical help arrives. Do not leave victim unattended. *Ingestion:* Call poison control. Loosen all clothing. Never give anything by mouth to an unconscious person. If victim is *unconscious or having convulsions,* do nothing except keep victim warm. Get medical aid. Transfer promptly to a medical facility. In cases of ingestion, **do not induce vomiting**. If the victim is alert and asymptomatic, administer a slurry of activated charcoal at a dose of 1 g/kg (infant, child, and adult dose). A soda can and straw may be of assistance when offering charcoal to a child. *In some cases you may be specifically instructed by poison control to induce vomiting by way of 2 tablespoons of syrup of ipecac (adult) washed down with a cup of water.* Do NOT give activated charcoal before or with ipecac syrup.

*Note to physician or authorized medical personnel:* Treat cases of respiratory compromise, coma, or excessive pulmonary secretions with respiratory support using protocols and techniques available and within the scope of training. Some cases may necessitate procedures such as endotracheal intubation or cricothyrotomy by properly trained and equipped personnel. When possible, atropine (see *Antidotes*, below) should be given under medical supervision. Patients who are comatose, hypotensive, or having seizures or cardiac arrhythmias should be treated according to advanced life support protocols. *Antidotes:* Two antidotes are administered to treat organophosphate poisoning. Atropine is a competitive antagonist of acetylcholine at muscarinic receptors and is used to control the excessive bronchial secretions which are often responsible for death. Pralidoxime relieves both the nicotinic and muscarine effects of organophosphate poisoning by regenerating acetylcholinesterase and can reduce both the bronchial secretions and the muscle weakness associated with poisoning. The initial intravenous dose of atropine in adults should be determined by the severity of symptoms: An initial adult dose of 1.0 to 2.0 mg or pediatric dose of 0.01 mg/kg (minimum 0.01 mg) should be administered intravenously. If intravenous access cannot be established, atropine may also be given intramuscularly, subcutaneously or via endotracheal tube. Doses should be repeated every 15 minutes until excessive secretions and sweating have been controlled. Once bronchial secretion has been controlled, atropine administration should be repeated whenever the secretions begin to recur. In seriously poisoned patients, very large doses may be required. Alterations of pulse rate and pupillary size should not be used as indicators of treatment adequacy. Pralidoxime should be administered as early in poisoning as possible as its efficacy may diminish when given more than 24 to 36 hours after exposure. Doses are as follows: adult 1.0 g; pediatric 25 to 50 mg/kg. The drug should be administered intravenously over 30 to 60 minutes, but in a life-threatening situation, one-half of the total dose can be given per minute for a total administration time of 2 minutes. Treatment should begin to take effect within 40 minutes with a reduction in symptoms and the amount of atropine necessary to control bronchial secretion. The initial dose can be repeated in 1 hour and then every 8 to 12 hours until the patient is clinically well and no longer requires atropine. If intravenous access cannot be established, pralidoxime may also be given intramuscularly. Early administration of diazepam in addition to the combined atropine and pralidoxime treatment may help prevent the onset of seizures and potential brain and cardiac morphologic damage following high-level organophosphate poisoning.

*References:*
- EXTOXNET, Extension Toxicology Network, "Pesticide Information Profile, Naled," Oregon State University, Corvallis, OR (June 1996). http://extoxnet.orst.edu/pips/naled.htm
- U.S. Environmental Protection Agency, Office of Pesticide Programs, Pesticide Residue Limits, "Naled," 40 CFR 180.215, http://www.epa.gov/pesticides/food/viewtols.htm
- New Jersey Department of Health and Senior Services, "Hazardous Substance Fact Sheet: Dimethyl-1,2-Dibromo-2,2-Dichloroethyl Phosphate," Trenton, NJ (December 1998). http://www.state.nj.us/health/eoh/rtkweb/0751.pdf
- U.S. Environmental Protection Agency, Investigation of Selected Potential Environmental Contaminants: Haloalkyl Phosphates, Report EPA 560/2/76-007, Wash., DC (Aug. 1976).
- Agency for Toxic Substances and Disease Registry, U.S. Department of Health and Human Services, Public Health

Service, "Managing Hazardous Materials Incidents," Atlanta, GA (June 2003).
- Sax, N.I., Ed., "Dangerous Properties of Industrial Materials Report" 5, No. 3, 44-47 (1985).

# Naphthas

*Use Type:* Insecticide and fungicide
*CAS Number:* 8002-05-9 (petroleum naphtha); 64742-48-9 (petroleum naphtha, hydrotreated); 8030-30-6 & 8030-31-7 (coal tar naphtha ); 8032-32-4 (VM&P naphtha)
*Synonyms: VM&P naphtha:* Ligroin; Ligroine; Mineral spirits; Painters naphtha; Petroleum ether; Petroleum spirit; Refined solvent naphtha; Varnish makers' & painters' naphtha
*Coal tar naphtha:* Benzin; Coal tar naphtha; Crude solvent coal tar naphtha; High solvent naphtha; Naphtha
*Petroleum naphtha:* Aliphatic petroleum naphtha; Petroleum distillates; Rubber solvent; Steenkoolteerrolie-distillaat (Dutch)
*Trade Names:* (There are more than 6100 active and canceled/transferred products listed by the U.S. EPA) 415 Oil®, Drexel Chemical (USA); 435 Oil®, Drexel Chemical (USA); ACME DORMANT OIL SPRAY®, Pbi/Gordon Corp. (USA)
*Producers:* Ashland (USA); BP P.L.C. Chemical (UK); Calumet (USA); Chevron Phillips Chemical (USA); Cia. Petroquimica do Sul (COPESUL) (Brazil); CITGO (USA); Crowley (USA); Deza (Czech Republic); Drexel Chemical (USA); EniChem (Italy); Exxon Mobil Chemical (USA); Gadiv (Israel); Haldia Petrochemicals (India); Helena Chemical (USA); Koch Specialty Chemicals (USA); Lyondell (USA); Monsanto (USA); Pbi/Gordon Corp. (USA); Zeon (Japan); Pecom Energia (Argentina); PetroChina (China); Phillips Petroleum; Samsung General Chemicals (South Korea); Shell Chemicals (UK); Shin-Etsu (Japan); Sinopec Corporation (China); Statoil (Norway); Sunoco Chemicals (USA); and most petroleum refineries
*Chemical Class:* Petroleum derivatives
*EPA/OPP PC Code: VM&P naphtha:* 063506; *Petroleum naphtha:* 063503
*California DPR Chemical Code: VM&P naphtha:* 2688, 2642; *Coal tar naphtha:* 2416; *Petroleum naphtha:* 763, 1730, 2768
*RTECS Number:* SE7449000 (petroleum naphtha); DE3030000 (coal tar naphtha); OI6180000 (VM&P naphtha)
*EEC Number:* 649-327-00-6 (petroleum, hydrotreated)
*Uses:* Naphthas are used as organic solvents for dissolving or softening rubber, oils, greases, bituminous paints, varnishes, and plastics. The less flammable fractions are used in dry cleaning, the heavy naphthas serving as bases for insecticides.

*Carcinogen/Hazard Classifications*
**IARC:** *VM&P naphtha:* Group 3, unclassifiable
**Label Signal Word:** CAUTION
*Regulatory Authority:*
- Air Pollutant Standard Set (ACGIH)[1] (OSHA)[58] (Several States)[60]
- Actively registered pesticide in California. as petroleum distillates
- Permissible Exposure Limits for Chemical Contaminants (CAL/OSHA) as VM&P naphtha and Coal tar naphtha
- The "Director's List" (CAL/OSHA) as VM&P naphtha and Coal tar naphtha
- Canada, WHMIS, Ingredients Disclosure List

*Description:* Naphthas derived from both petroleum and coal tar are included in this group. *Petroleum naphtha* is a colorless liquid with a gasoline- or kerosene-like odor. A mixture of paraffins (C5 to C13) that may contain a small amount of aromatic hydrocarbons, and are termed "close-cut" fractions. "Medium-range" and "wide-range" fractions are made up of 40 to 80% aliphatic hydrocarbons, 25 to 50% naphthenic hydrocarbons 0 to 10% benzene, and 0 to 20% other aromatic hydrocarbons. Boiling point = 35–60°C. Melting/Freezing point = –73°C. Flash point = <–18 (petroleum ether); also listed as–40° to–66. Autoignition temperature = 288°C. Explosive Limits: LEL = 1.1%; UEL 5.9%. Hazard Identification (based on NFPA-704 M Rating System): Health 1, Flammability 4, Reactivity 0. Insoluble in water. *Coal tar naphtha* is a mixture of aromatic hydrocarbons, principally toluene, xylene, and cumene. Coal tar naptha is a reddish-brown, mobile liquid with an aromatic odor. Shipped as a molten solid. Benzene, is present in appreciable amounts in those coal tar naphthas with low boiling points. Boiling point = 160° to 220°C. Flash point = 38–43°C. Explosive Limits: LEL = 1.1%; UEL ?. Hazard Identification (based on NFPA-704 M Rating System): [coal tar] Health 2, Flammability 2, Reactivity 0. Insoluble in water. *VM&P naphtha* is a clear to yellowish liquid with a pleasant, aromatic odor: Boiling point = 100°-177°C. Flash points = –2 to 29°C. Autoignition temperature = 232°C. Explosive limits: vary somewhat but typical values are: LEL = 0.9%; UEL = 6.7%. Hazard Identification (based on NFPA-704 M Rating System): [vm&p] Health 1, Flammability 3, Reactivity 0. See also Stoddard Solvent.

*Incompatibilities:* Strong oxidizers.
*Permissible Exposure Limits in Air: Coal tar naphtha:* The OSHA[2] TWA is 100 ppm (400 mg/m³). *Petroleum naphtha:* The OSHA[2] TWA is 500 ppm (2,000 mg/m³). *VM&P naphtha:* There is no OSHA[2] PEL. NIOSH[2] recommends a TWA of 350 mg/m³ and a 15-minute ceiling of 1,800 mg/m³. ACGIH[1] recommends a TWA of 1,370 mg/m³. The NIOSH[2] IDLH (coal tar and petroleum naphtha) = 1000 ppm [10% LEL]. In addition, some states

have set guidelines or standards for naphtha in ambient air[60] ranging from zero (Nevada) to 60-27,000 $\mu g/m^3$ (Connecticut) to 225 $\mu g/m^3$ (Virginia).

***Determination in Air:*** Charcoal tube adsorption; workup with $CS_2$; Gas chromatography/flame ionization detection; NIOSH IV Method #1550, Naphthas[18]. See also OSHA Method #48[58].

***Permissible Concentration in Water:*** No criteria set, but runoff from spills or fire control may cause water pollution.

***Routes of Entry:*** Inhalation of vapor, ingestion, skin and/or eye contact. Percutaneous absorption of liquid is important in development of systemic effects if benzene is present.

***Harmful Effects and Symptoms***

***Short Term Exposure:*** The naphthas are irritating to the skin conjunctiva, and the mucous membranes of the upper respiratory tract. Skin chapping and photosensitivity may develop after repeated contact with the liquid. If confined against skin by clothing, the naphthas may cause skin burn. Exposure can cause dizziness, lightheadedness and unconsciousness.

*Petroleum naphtha* has a lower order of toxicity than that derived from coal tar, where the major hazard is brought about by the aromatic hydrocarbon content. Sufficient quantities of both naphthas cause central nervous system depression. Symptoms include inerriation, followed by headache and nausea. In severe cases, dizziness, convulsions, and unconsciousness occasionally result. Symptoms of anorexia and nervousness have been reported to persist for several months following an acute overexposure, but this appears to be rare. One fraction, hexane, has been reported to have been associated with peripheral neuropathy. If benzene is present, coal tar naphthas may produce blood changes such as leukopenia, aplastic anemia, or leukemia. The kidneys and spleen have also been affected in animal experiments. At vapor concentrations up to 450 ppm, petroleum naphtha inhalations may produce slight throat irritation. At 880 pm, definite throat irritation is observed. Vapors may also irritate the nose. High concentrations may produce difficulty in breathing, blue coloration of skin, excitement and dizziness. Inhalation of vapors in the absence of oxygen is immediately life-threatening. A vapor concentration up to 450 ppm mild, temporary irritation is observed; at 880 ppm more severe irritation may be experienced.

***Long Term Exposure:*** Irritates the eyes and upper respiratory system. Coal tar naphtha may contain benzene, a cancer-causing agent in humans. Exposure may cause nervous system and kidney damage. Some coal tar naphthas contain other substances that can cause blood cell damage. Long term exposure may cause drying and cracking of the skin, and make the skin sunburn more easily. Swallowing the liquid may cause chemical pneumonia.

***Points of Attack:*** Eyes, skin, respiratory system, central nervous system, liver and kidneys.

***Medical Surveillance:*** Preplacement and periodic medical examinations should include the central nervous system. If benzene exposure is present, workers should have a periodic complete blood count including hematocrits, hemoglobin, white blood cell count and differential count, mean corpuscular volume and platelet count, reticulocyte count, serum bilirubin determination, and urinary phenol in the preplacement examination and at 3-month intervals. There are no specific diagnostic tests for naphtha exposure but urinary phenols may indicate exposure to benzene and aromatic hydrocarbons. It should be noted that benzene content of vapor may be higher than predicted by content in the liquid.

***First Aid:*** If this chemical gets into the eyes, remove any contact lenses at once and irrigate immediately for at least 15 minutes, occasionally lifting upper and lower lids. Seek medical attention immediately. If this chemical contacts the skin, remove contaminated clothing and wash immediately with soap and water. Seek medical attention immediately. If this chemical has been inhaled, remove from exposure, begin rescue breathing (using universal precautions) if breathing has stopped, and CPR if heart action has stopped. Transfer promptly to a medical facility. When this chemical has been swallowed, get immediate medical attention.

*Note to physician or authorized medical personnel:* Inhalation: Bronchodilators, decongestants and oxygen may be used if necessary. Corticosteroids are useful for treating pneumonitis.

***References:***
- New Jersey Department of Health and Senior Services, "Hazardous Substance Fact Sheet: Coal Tar Naphtha," Trenton, NJ (June 1994, rev. January 2001). http://www.state.nj.us/health/eoh/rtkweb/0518.pdf
- New Jersey Department of Health and Senior Services, "Hazardous Substance Fact Sheet: VM&P Naphtha," Trenton, NJ (August 1998). http://www.state.nj.us/health/eoh/rtkweb/0206.pdf
- National Institute for Occupational Safety and Health (NIOSH), "Criteria for a Recommended Standard: Occupational Exposure to Refined Petroleum," NIOSH Doc. No. 77-192, Wash., DC (1977).
- New York State Department of Health, "Chemical Fact Sheet: V.M.& P. Naphtha," Albany, NY, Bureau of Toxic Substance Assessment (May 1986 & Version 2).
- U.S. Environmental Protection Agency, "Chemical Hazard Information Profile: Naphtha (Petroleum) Solvents," Wash., DC, Office of Toxic Substances (Sept. 28, 1984).
- California Environmental Protection Agency "Chemical List of Lists," Sacramento CA (February 1997).

# Naphthoxyacetic Acid

***Use Type:*** Plant growth regulator

## 1-Naphthaleneacetamide

*CAS Number:* 120-23-0
*Formula:* $C_{12}H_9O_4$
*Synonyms:* Acetic acid, (2-naphthyloxy)-; Acide naphthyloxyacetique (French); BNOA; 2-Naphthalenoxyacetic acid; (β-Naphthalenyloxy)acetic acid; 2-Naphthoxyacetic acid; β-Naphthoxyacetic acid; (2-Naphthoxy)acetic acid; O-(2-Naphthyl)glycolic acid; (2-Naphthyloxy)acetic acid; NOA; NOXA; 2-NOXA; β-NOXA
*Trade Names:* AMID-THIN®, K-SALT®; BETAPAL®; BETOXON®; DOXOL TOMATO LIFE®, canceled; FRUIT FIX® 200; GERLACH® 1396; PHYOMONE®; TRE-HOLD®, AMVAC Chemical (USA)
*Producers:* AMVAC Chemical (USA); Ki-Hara Chemicals Ltd. (UK)
*Chemical Class:* Botanical
*EPA/OPP PC Code:* 055601
*California DPR Chemical Code:* 432
*Uses:* Not registered in the U.S. Naphthoxyacetic Acid is an agent for promoting fruit sets, stimulating fruit enlargement, and preventing hollow fruits.
*Carcinogen/Hazard Classifications*
**Label Signal Word:** CAUTION
**WHO Acute Hazard:** Class III, slightly hazardous
*Regulatory Authority:*
- Actively registered pesticide in California.

*Description:* Crystalline solid or powder. Melting/Freezing point = 156°C. Soluble in water.
*Incompatibilities:* Keep away from oxidizers, sulfuric acid, caustics, ammonia, aliphatic amines, alkanolamines, isocyanates, alkylene oxides, epichlorohydrin.
*Permissible Concentration in Water:* No criteria set. Runoff from spills or fire control may cause water pollution.
*First Aid:* If this chemical gets into the eyes, remove any contact lenses at once and irrigate immediately for at least 15 minutes, occasionally lifting upper and lower lids. Seek medical attention immediately. If this chemical contacts the skin, remove contaminated clothing and wash immediately with soap and water. Seek medical attention immediately. If this chemical has been inhaled, remove from exposure, begin rescue breathing (using universal precautions) if breathing has stopped, and CPR if heart action has stopped. Transfer promptly to a medical facility. When this chemical has been swallowed, get medical attention. Give large quantities of water and induce vomiting. Do not make an unconscious person vomit. Do not induce vomiting when formulations containing petroleum solvents are ingested.
*References:*
- California Environmental Protection Agency "Chemical List of Lists," Sacramento CA (February 1997)

## 1-Naphthaleneacetamide

*Use Type:* Plant growth regulator

*CAS Number:* 86-86-2
*Formula:* $C_{12}H_{11}NO$
*Synonyms:* N-Acetyl-1-naphthylamine; NAAM; NAD; Naphthalene-acetamide(1-); α-Naphthaleneacetamide; α-Naphthylacetamide; 1-Naphthylacetamide
*Trade Names:* AMACTONE®; AMVAC Chemical (USA); AMID-THIN®; FRUITONE®; ROOTONE® (component, with Indole-3-butyric acid and 1-Naphthaleneacetic acid), Bayer CropScience (Germany); ROSETONE®; TRANSPLANTONE® (component, with 1-Naphthaleneacetic acid)
*Producers:* AMVAC Chemical (USA); Bayer CropScience (Germany); Ki-Hara Chemicals Ltd. (UK)
*Chemical Class:* Botanical
*EPA/OPP PC Code:* 056001
*California DPR Chemical Code:* 422
*Uses:* 1-Naphthaleneacetamide is an agent for thinning fruit sets in apples and pears.
*U.S. Maximum Allowable Residue Levels for 1-Naphthaleneacetamide (40 CFR 180.309):*

| CROP | ppm |
|---|---|
| Apple | 0.1 |
| Pear | 0.1 |

*Carcinogen/Hazard Classifications*
**Label Signal Word:** CAUTION
**WHO Acute Hazard:** Class U, unlikely to be hazardous
*Regulatory Authority:* None listed.
*Description:* White needles or crystalline solid. Odorless. Soluble in water. Molecular weight = 185.22. Melting/Freezing point = sublimes @ 181°C.
*Incompatibilities:* Slowly hydrolyzes in water, releasing ammonia and forming acetate salts.
*Determination in Air:* Filter; none; Gravimetric; NIOSH IV [Particulates NOR; #0500 (total), #0600 (respirable)].[18]
*Permissible Concentration in Water:* No criteria set. Runoff from spills or fire control may cause water pollution.
*First Aid:* If this chemical gets into the eyes, remove any contact lenses at once and irrigate immediately for at least 15 minutes, occasionally lifting upper and lower lids. Seek medical attention immediately. If this chemical contacts the skin, remove contaminated clothing and wash immediately with soap and water. Seek medical attention immediately. If this chemical has been inhaled, remove from exposure, begin rescue breathing (using universal precautions) if breathing has stopped, and CPR if heart action has stopped. Transfer promptly to a medical facility. When this chemical has been swallowed, get medical attention. Give large quantities of water and induce vomiting. Do not make an unconscious person vomit. Do not induce vomiting when formulations containing petroleum solvents are ingested.
*References:*
- U.S. Environmental Protection Agency, Office of Pesticide Programs, Pesticide Residue Limits, "1-Naphthaleneacetamide," 40 CFR 180.309, http://www.epa.gov/pesticides/food/viewtols.htm

- California Environmental Protection Agency "Chemical List of Lists," Sacramento CA (February 1997).

## 1-Naphthaleneacetic Acid

*Use Type:* Plant growth regulator
*CAS Number:* 86-87-3
*Formula:* $C_{12}H_{10}O_2$
*Synonyms:* ANA; NAA; Naphthaleneacetic acid(1-); α-Naphthaleneacetic acid; Naphthalene-1-acetic acid; α-Naphthylacetic; Naphthylacetic acid; α-Naphthylacetic acid; 1-Naphthylacetic acid; α-Naphthyleneacetic acid; α-Naphthylessigsaeure (German); Naphyl-1-essigsaeure (German)
*Trade Names:* AGRONAA®; ALCO®, AMVAC Chemical (USA); ALPHA-SPRA®; AMCOTONE®, AMVAC Chemical (USA); APPL-SET®; CELMONE®; DESTRUXOL®; DIP'N GROW®; FRUITONE®; GOLDENGRO®; HORMEX®; KLINGTITE®; LIQUI-STIK®; NAA 800®; NAFUSAKU®; NIAGARA-STIK®; NU-TONE®; PARMONE®; PHYMONE®; PIMACOL-SOL®; PLANOFIX®; PLUCKER®; PRIMACOL®; RHIZOPON B ROOTING POWDER; ROOTONE® (component, with Indole-3-butyric acid and 1-Naphthaleneacetamide); STAFAST®; STIK®; STOP-DROP®; TEKKAM®; TIPOFF®; TRANSPLANTONE® (component, with 1-Naphthaleneacetamide); TRE-HOLD®; VARDHAK®
*Producers:* Agrowchem (Canada); AMVAC Chemical (USA); Ki-Hara Chemicals Ltd. (UK)
*Chemical Class:* Botanical; naphthalene acetic acid derivative
*EPA/OPP PC Code:* 056002
*California DPR Chemical Code:* 423
*EINECS Number:* 201-705-8
*Uses:* An agent for thinning fruit sets in apples, pears, olives and some citrus. Induces root formation on cuttings and transplants. Inhibits fruit drops.
*Human toxicity (long-term)[77]:* Very low–350.00 ppb, Health Advisory
*Fish toxicity (threshold)[77]:* Very low–4918.96611 ppb, MATC (Maximum Acceptable Toxicant Concentration)
*U.S. Maximum Allowable Residue Levels for 1-Naphthaleneacetic Acid (40 CFR 180.155):*

| CROP | ppm |
|---|---|
| Apple | 1.0 |
| Cherry, sweet | 0.1 |
| Olive | 0.1 |
| Orange | 0.1 |
| Pear | 1.0 |
| Pineapple | 0.05 |
| Quince | 1.0 |
| Tangerine | 0.1 |

*Carcinogen/Hazard Classifications*
**Label Signal Word:** CAUTION, WARNING, DANGER
**WHO Acute Hazard:** Class U, unlikely to be hazardous
*Regulatory Authority:*
- Actively registered pesticide in California.

*Description:* White crystals, needles, powder, or colorless liquid. Odorless. Slight solubility in water. Molecular weight = 186.20. Melting/Freezing point = 132-134°C.
*Incompatibilities:* Keep away from oxidizers, sulfuric acid, caustics, ammonia, aliphatic amines, alkanolamines, isocyanates, alkylene oxides, epichlorohydrin.
*Permissible Concentration in Water:* No criteria set. Runoff from spills or fire control may cause water pollution.
*First Aid:* If this chemical gets into the eyes, remove any contact lenses at once and irrigate immediately for at least 15 minutes, occasionally lifting upper and lower lids. Seek medical attention immediately. If this chemical contacts the skin, remove contaminated clothing and wash immediately with soap and water. Seek medical attention immediately. If this chemical has been inhaled, remove from exposure, begin rescue breathing (using universal precautions) if breathing has stopped, and CPR if heart action has stopped. Transfer promptly to a medical facility. When this chemical has been swallowed, get medical attention. Give large quantities of water and induce vomiting. Do not make an unconscious person vomit. Do not induce vomiting when formulations containing petroleum solvents are ingested.
*References:*
- U.S. Environmental Protection Agency, Office of Pesticide Programs, Pesticide Residue Limits, "1-Naphthaleneacetic Acid," 40 CFR 180.155, www.epa.gov/pesticides/food/viewtols.htm
- Pesticide Management Education Program, "1-Naphthaleneacetic Acid (NAA) Herbicide Profile 3/85," Cornell University, Ithaca, NY (March 1985). http://pmep.cce.cornell.edu/profiles/herb-growthreg/naa-rimsulfuron/naa/herb-prof-naa.html
- California Environmental Protection Agency "Chemical List of Lists," Sacramento CA (February 1997).

## Naptalam

*Use Type:* Herbicide
*CAS Number:* 132-66-1
*Formula:* $C_{18}H_{13}NO_3$
*Alert:* Reregistration scheduled for August, 2004
*Synonyms:* Benzoic acid, 2-[(1-naphthalenylamino) carbonyl]-; Benzoic acid, 2-[(α-naphthalenylamino) carbonyl]-; Naftalame; 2-[(1-Naphthalenylamino) carbonyl]benzoic acid; N-1-Naphthylphthalamate; α-Naphthylphthalamic acid; N-1-Naphthylphthalamic acid; N-α-Naphthyl-phthalamidsaeure (German); Naptalame; NPA; Phthalamic acid, N-1-naphthyl-

*Trade Names:* ALANAP®, Crompton Corporation (USA), canceled; ALANAPE®; ANACRACK®; DYANAP®; GRELUTIN®; MOR-CRAN®; NAPTRO®; NIP-A-THIN®; PEACH-TIIN®; RESCUE®, Crompton Corporation (USA), canceled; 6Q8®; SOLO®
*Producers:* Crompton Corporation (USA)
*Chemical Class:* Amide; phthalic acid
*EPA/OPP PC Code:* 030702
*California DPR Chemical Code:* 2998
*Uses:* Not registered in the U.S. Used as a selective pre-emergence herbicide to control broadleaf weeds and grasses in soybeans, cucumbers, melons, peanuts and woody ornamentals.
*Carcinogen/Hazard Classifications*
**U.S. EPA Carcinogens:** Group D, unclassifiable, insufficient data
**WHO Acute Hazard:** Class U, unlikely to be hazardous
*Description:* Purple crystalline solid. Odorless. Molecular weight = 291.33. Density 1.40. Melting/Freezing point = 202°C.
*Incompatibilities:* Keep away from oxidizers, sulfuric acid, caustics, ammonia, aliphatic amines, alkanolamines, isocyanates, alkylene oxides, epichlorohydrin.
*Determination in Air:* Filter; none; Gravimetric; NIOSH IV [Particulates NOR; #0500 (total), #0600 (respirable)].[18]
*Permissible Concentration in Water:* No criteria set. Runoff from spills or fire control may cause water pollution.
*Routes of Entry:* Absorbed through the unbroken skin, inhalation, ingestion.
*Harmful Effects and Symptoms*
*Short Term Exposure:* May be toxic if absorbed through the skin. May cause irritation to the eyes, skin, and respiratory tract.
*First Aid:* If this chemical gets into the eyes, remove any contact lenses at once and irrigate immediately for at least 15 minutes, occasionally lifting upper and lower lids. Seek medical attention immediately. If this chemical contacts the skin, remove contaminated clothing and wash immediately with soap and water. Seek medical attention immediately. If this chemical has been inhaled, remove from exposure, begin rescue breathing (using universal precautions) if breathing has stopped, and CPR if heart action has stopped. Transfer promptly to a medical facility. When this chemical has been swallowed, get medical attention. Give large quantities of water and induce vomiting. Do not make an unconscious person vomit.
*References:*
- Pesticide Management Education Program, "Naptalam (Alanap) Herbicide Profile 3/85," Cornell University, Ithaca, NY (March 1985). http://pmep.cce.cornell.edu/profiles/herb-growthreg/naa-rimsulfuron/naptalam/herb-prof-naptalam.html
- California Environmental Protection Agency "Chemical List of Lists," Sacramento CA (February 1997)

# Napropamide

*Use Type:* Herbicide
*CAS Number:* 15299-99-7
*Formula:* $C_{17}H_{21}NO_2$
*Synonyms:* N,N-Diethyl-2-(1-naphthalenyloxy)propanamide; 2-(1-Naphthoxy)-N,N-diethylpropionamide; [2-($\alpha$-naphthoxy)-N,N-diethylpropionamide]
*Trade Names:* COLZOR TRIO® (dimethachlor + napropamide + clomazone), Syngenta (Switzerland); DEVRINOL®, United Phosphurs (India); NAPROGUARD®, Calliope (France); WAYLAY®; R-7165®
*Producers:* Calliope (France); Syngenta (Switzerland); United Phosphorus (India)
*Chemical Class:* Substituted amide
*EPA/OPP PC Code:* 103001
*California DPR Chemical Code:* 1728
*Uses:* A General Use Pesticide (GUP) that is compatible with many other fungicides and herbicides. Used to control broadleaf weeds and annual grasses on a variety of crops including vegetables, fruit trees; fruit bushes, oil seed rape, vines, sunflowers, olives, tobacco and mint.
*Carcinogen/Hazard Classifications*
**Label Signal Word:** CAUTION or DANGER
**WHO Acute Hazard:** Class U, unlikely to be hazardous
*Regulatory Authority:*
- AB 1803-Well Monitoring Chemical (CAL)
- Actively registered pesticide in California.

*Description:* Tan to brown solid in its technical form; colorless crystal in its industrial form. Soluble in water; solubility = 73 mg/L @ 20°C. Molecular weight = 271.35. Vapor pressure = $1.7 \times 10^{-7}$ mm @ 20°C. Melting/Freezing point = 70°C.
*Incompatibilities:* Slowly hydrolyzes in water, releasing ammonia and forming acetate salts.
*Permissible Exposure Limits in Air:*
*Determination in Air:* Filter; none; Gravimetric; NIOSH IV [Particulates NOR; #0500 (total), #0600 (respirable)].[18]
*Permissible Concentration in Water:* No criteria set. Runoff from spills or fire control may cause water pollution.
*First Aid:* If this chemical gets into the eyes, remove any contact lenses at once and irrigate immediately for at least 15 minutes, occasionally lifting upper and lower lids. Seek medical attention immediately. If this chemical contacts the skin, remove contaminated clothing and wash immediately with soap and water. Seek medical attention immediately. If this chemical has been inhaled, remove from exposure, begin rescue breathing (using universal precautions) if breathing has stopped, and CPR if heart action has stopped. Transfer promptly to a medical facility. When this chemical has been swallowed, get medical attention. Give large quantities of water and induce vomiting. Do not make an

unconscious person vomit. Do not induce vomiting when formulations containing petroleum solvents are ingested.

*References:*
- EXTOXNET, Extension Toxicology Network, "Pesticide Information Profile, Napropamide," Oregon State University, Corvallis, OR (June 1996). http://extoxnet.orst.edu/pips/napropam.htm
- California Environmental Protection Agency "Chemical List of Lists," Sacramento CA (February 1997).

## Nicosulfuron (ANSI)

*Use Type:* Herbicide
*CAS Number:* 111991-09-4
*Formula:* $C_{15}H_{18}N_6O_6S$
*Alert:* Some formulations may be Restricted Use Pesticides (RUP)
*Synonyms:* 2-[((((4,6-Dimethoxy-2-pyrimidinyl)amino)carbonyl)amino)sulfonyl]-*N,N*-dimethyl-3-pyridinecarboxamide; 3-Pyridinecarboxamide, 2-[((((4,6-dimethoxy-2-pyrimidinyl)amino)carbonyl)amino) sulfonyl]-*N,N*-dimethyl-
*Trade Names:* ACCENT®, Dupont Crop Protection (USA); BASIS®, Dupont Crop Protection (USA); CELEBRITY®, BASF Agricultural Products (Germany); CHALLENGER®; DASUL®, Syngenta (Switzerland); DPX 79406® Herbicide (with Rimsulfuron); DPX-V9636®, Dupont Crop Protection (USA); GHIBLI®, Syngenta (Switzerland); LAMA®; MATRIX® Herbicide (with Rimsulfuron); MILAGRO®, Syngenta (Switzerland); MISTRAL®; MOTIVEL®; NISSHIN®; SANSON®, Syngenta (Switzerland); STEADFAST®, (nicosulfuron + rimsulfuron), Dupont Crop Protection (USA)
*Producers:* BASF Agricultural Products (Germany); Dupont Crop Protection (USA); Epochem Co., (China); Syngenta (Switzerland)
*Chemical Class:* Sulfonylurea
*EPA/OPP PC Code:* 129008
*California DPR Chemical Code:* 3829
*Uses:* Used as a post-emergence herbicide to control a variety of weeds on field corn and popcorn crops.
*U.S. Maximum Allowable Residue Levels for Nicosulfuron (40 CFR 180.454):*
The following corn crops all have a residue level of 0.1 ppm: field forage, forage, grain, stover, sweet forage, sweet kernel plus cob with husks removed, and sweet stover.
*Carcinogen/Hazard Classifications*
**U.S. EPA Carcinogens:** Group E, Unlikely a carcinogen
**Label Signal Word:** CAUTION, WARNING, DANGER
**WHO Acute Hazard:** Class U, unlikely to be hazardous
*Description:* White powder or colorless crystals. Melting point = 141–144° C. Molecular weight = 410.4.
*Incompatibilities:* Slowly hydrolyzes in water, releasing ammonia and forming acetate salts.
*Determination in Air:* White powder or colorless crystals. Molecular weight = 410.4. Melting point = 141–144°C
*Permissible Concentration in Water:* No criteria set. Runoff from spills or fire control may cause water pollution.
*First Aid:* If this chemical gets into the eyes, remove any contact lenses at once and irrigate immediately for at least 15 minutes, occasionally lifting upper and lower lids. Seek medical attention immediately. If this chemical contacts the skin, remove contaminated clothing and wash immediately with soap and water. Seek medical attention immediately. If this chemical has been inhaled, remove from exposure, begin rescue breathing (using universal precautions) if breathing has stopped, and CPR if heart action has stopped. Transfer promptly to a medical facility. When this chemical has been swallowed, get medical attention. Give large quantities of water and induce vomiting. Do not make an unconscious person vomit. Do not induce vomiting when formulations containing petroleum solvents are ingested.

*References:*
- California Environmental Protection Agency "Chemical List of Lists," Sacramento CA (February 1997)
- U.S. Environmental Protection Agency, Office of Pesticide Programs, Pesticide Residue Limits, "Nicosulfuron," 40 CFR 180.454, www.epa.gov/pesticides/food/viewtols.htm
- EXTOXNET, Extension Toxicology Network, "Pesticide Information Profile, Nicosulfuron," Oregon State University, Corvallis, OR (May 1995). http://extoxnet.orst.edu/pips/nicosulf.htm

## Nicotine

*Use Type:* Insecticide
*CAS Number:* 54-11-5
*Formula:* $C_{10}H_{14}N_2$
*Alert:* Nicotine is a teratogen and should be handled with extreme caution.
*Synonyms:* ENT 3,424; 1-Methyl-2-(3-pyridyl)pyrrolidine; 3-(*N*-Methylpyrrolidino)pyridine; 3-(1-Methyl-2-pyrrolidinyl)pyridine; (s)-3-(1-Methyl-2-pyrrolidinyl)pyridine; 1-3-(1-Methyl-2-pyrrolidyl)pyridine; (-)-3-(1-Methyl-2-pyrrolidyl)pyridine; Nicotina (Italian, Spanish); 1-Nicotine; Nicotine alkaloid; Nikotin (German); Nikotyna (Polish); Pyridine, 3-(1-methyl-2-pyrrolidinyl)-; Pyridine, (s)-3-(1-methyl-2-pyrrolidinyl)-and salts; Pyridine, 3-(tetrahydro-1-methylpyrrol-2-yl); $\beta$-Pyridyl-$\alpha$-*N*-methylpyrrolidine; 3-(1-Methyl-2-pyrrolidyl) pyridine; Di-tetrahydronicotyrine
*Trade Names:* BLACK LEAF®; CAMPBELL'S NICO-SOAP®; DESTRUXOL ORCHARD SPRAY®; EMO-NIB®; FLUX MAAG®; FUMETO-TENDUST®; BAC®; MACH-NIC®; NIAGARA P.A. DUST®; NICODUST®; NICOFUME®; NICOCIDE®; ORTHO N-4 DUST®; XL ALL INSECTICIDE®

*Producers:* Infinity Industries (USA); Par Pharmaceuticals (USA); Richman Chemical (USA); Watson Laboratories (USA)

*Chemical Class:* A botanical compound

*EPA/OPP PC Code:* 056702

*California DPR Chemical Code:* 75

*ICSC Number:* 0519

*RTECS Number:* QS5250000

*EEC Number:* 614-001-00-4

*EINECS Number:* 200-193-3

*Uses:* Nicotine is used in some drugs and insecticides and in tanning.

*U.S. Maximum Allowable Residue Levels for Nicotine (40 CFR 180.167):*

| CROP | ppm |
|---|---|
| Cucumber | 2 |
| Lettuce | 2 |
| Tomato | 2 |

*Carcinogen/Hazard Classifications*

**California Prop. 65:** Listed; Reproductive toxin

**TRI Developmental Toxin:** Developmental toxin as nicotine and salts

**Label Signal Word:** CAUTION

**WHO Acute Hazard:** Class 1B Highly Hazardous

*Regulatory Authority:*

- Proposition 65 chemical (CAL)
- Banned or Severely Restricted (In Agriculture) (Germany, Hungry) (UN)[13]
- Air Pollutant Standard Set (ACGIH)[1] (DFG)[3] (HSE)[33] (OSHA)[58] (Several States)[60]
- AB 2588-Air Toxics "Hot Spots" Chemicals (CAL)
- Proposition 65 chemical (CAL)
- Permissible Exposure Limits for Chemical Contaminants (CAL/OSHA)
- The "Director's List" (CAL/OSHA)
- Actively registered pesticide in California.
- EPA Hazardous Waste Number (RCRA No.): No. P075
- RCRA, 40CFR261, Appendix 8 Hazardous Constituents
- Superfund/EPCRA 40CFR355, Appendix B Extremely Hazardous Substances: TPQ = 100 lb (45.4 kg)
- Superfund/EPCRA 40CFR302.4 RQ: CERCLA, 100 lb (45.4 kg)
- EPCRA Section 313 Form R *de minimis* concentration reporting level: 1.0%
- Canada, WHMIS, Ingredients Disclosure List

*Description:* Nicotine is a pale yellow to dark brown, oily liquid. Slight, fishy odor when warm. When pure it is colorless and odorless. Soluble in water. Boiling point =246°C. Melting/Freezing point = –80°C. Vapor pressure = $6 \times 10^{-3}$ kPa @ 20°C. Flash point = 95°C (cc). Autoignition temperature = 240°C. Explosive limits: LEL = 0.7%; UEL = 4.0%. Hazard Identification (based on NFPA-704 M Rating System): Health 4, Flammability 1, Reactivity 0. Log $K_{ow}$ = 1.18. Unlikely to bioaccumulate in marine organisms.

*Incompatibilities:* Strong oxidizers and strong acids.

*Permissible Exposure Limits in Air:* The OSHA[2] PEL, DFG MAK[3], HSE[33] TWA, and ACGIH[1] TWA value is 0.5 mg/m$^3$ (0.07 ppm) and the STEL set by HSE[33] is 1.5 mg/m$^3$. The notation "skin" indicates the possibility of cutaneous absorption. The NIOSH[2] IDLH level = 5 mg/m$^3$. Several states have set guidelines or standards for nicotine in ambient air[60] ranging from 8.0 $\mu$g/m$^3$ (Virginia) to 10.0 $\mu$g/m$^3$ (Connecticut) to 12.0 $\mu$g/m$^3$ (Nevada).

*Determination in Air:* XAD-2; workup with ethyl acetate, analysis by gas chromatography/NPD; NIOSH IV, Method #2544.[18]

*Permissible Concentration in Water:* No criteria set, but runoff from spills or fire control may cause water pollution.

*Routes of Entry:* Inhalation, ingestion, skin and/or eye contact.

*Harmful Effects and Symptoms*

*Short Term Exposure:* Irritates the eyes and skin. Even small exposures can cause increased heart rate, increased blood fat levels, and change vital hormone levels. May affect the cardiovascular system and central nervous system, resulting in convulsions and respiratory failure. Nicotine is classified as super toxic. Probable oral lethal dose in humans is less than 5 mg/kg or a taste (less than 7 drops) for a 70 kg (150 lbs) person. It may be assumed that ingestion of 40-60 mg of nicotine is lethal to humans. There is a fundamental difference between acute toxicity from use of nicotine as insecticide of from ingestion, and chronic toxicity that may be caused by prolonged exposure to small doses as occurs in smoking. Maternal smoking during pregnancy is associated with increased risk of spontaneous abortion, low birth with and still-birth. Acute exposure to nicotine may result in headache, dizziness, confusion, agitation, restlessness, lethargy, seizures, a coma. Victims may experience hypertension (high blood pressure), tachycardia (rapid heart rate), and tachypnea (rapid respirations), followed by hypotension (low blood pressure), bradycardia (slow heart rate), and respiratory depression. Cardiac arrhythmias may also occur. Gastrointestinal effects include nausea, vomiting, abdominal pain or burning sensation, and diarrhea. Increased salivation, lacrimation (tearing), and sweating may be noted. High levels, far above the OEL, may result in death.

*Long Term Exposure:* Animal tests show that this substance possibly causes toxic effects upon human reproduction. Nicotine was found as a co-carcinogen in animals.

*Points of Attack:* Central nervous system, cardiovascular system, lungs, GI tract and reproduction system. Has been shown to be a teratogen in animals; may be a teratogen in humans. Causes fat deposits in the arteries (reducing blood supply to many body organs). This increases the risk of heart attack, stroke, and many other poor circulation problems. Chronic high blood pressure can also result.

*Medical Surveillance:* Before beginning employment and at regular times after that, the following is recommended:

Blood test for nicotine (only accurate shortly after exposure); consider test to evaluate typical exposures as well as for suspected overexposure or if symptoms are present. Even those who have smoked for a long time can reduce the risk of developing health problems by stopping.

*First Aid:* If this chemical gets into the eyes, remove any contact lenses at once and irrigate immediately for at least 15 minutes, occasionally lifting upper and lower lids. Seek medical attention immediately. If this chemical contacts the skin, remove contaminated clothing and wash immediately with soap and water. Seek medical attention immediately. If this chemical has been inhaled, remove from exposure, begin rescue breathing (using universal precautions) if breathing has stopped, and CPR if heart action has stopped. Transfer promptly to a medical facility. When this chemical has been swallowed, get medical attention. Give large quantities of water and induce vomiting. Do not make an unconscious person vomit.

*Note to physician or authorized medical personnel:* The use of atropine might be considered, depending on symptoms.

*References:*
- U.S. Environmental Protection Agency, Office of Pesticide Programs, Pesticide Residue Limits, "Nicotine," 40 CFR 180.167, http://www.epa.gov/pesticides/food/viewtols.htm
- New Jersey Department of Health and Senior Services, "Hazardous Substance Fact Sheet: Nicotine," Trenton, NJ (January 1988, rev. March 2000). http://www.state.nj.us/health/eoh/rtkweb/1349.pdf
- Sax, N.I., Ed., "Dangerous Properties of Industrial Materials Report," 1, No. 8, 84-85 (1981) and 5, No. 4, 82-85 (1985).
- U.S. Environmental Protection Agency, "Chemical Profile: Nicotine," Washington, DC, Chemical Emergency Preparedness Program (Nov. 30, 1987).
- California Environmental Protection Agency "Chemical List of Lists," Sacramento CA (February 1997).

# Nicotine Sulfate

*Use Type:* Insecticide and a veterinary medicine
*CAS Number:* 65-30-5
*Formula:* $C_{10}H_{18}N_2O_8S_2$; $C_{10}H_{14}N_2 \cdot 2H_2SO_4$
*Alert:* Nicotine Sulfate is a teratogen and should be handled with extreme caution.
*Synonyms:* ENT 2,435; 1-1-Methyl-2-(3-pyridyl)-pyrrolidine sulfate; (S)-3-(1-Methyl-2-pyrrolidinyl)pyridine sulfate (2:1); 1-3-(1-Methyl-2-pyrrolidinyl)pyridine sulfate; Nicotine sulfate (2:1); Nicotine sulphate; Nicotine sulphate (2:1); Nikotinsulfat (German); Pyridine, 3-(1-methyl-2-pyrrolidinyl)-, (S)-, sulfate (2:1); Pyrrolidine, 1-methyl-2-(3-pyridyl)-, sulfate; Sulfate de nicotine (French); Sulfato de nicotina (Spanish)
*Chemical Class:* A botanical compound

*EPA/OPP PC Code:* 056703
*California DPR Chemical Code:* 430
*ICSC Number:* 0520
*RTECS Number:* QS9625000
*EEC Number:* 614-002-00-X
*EINECS Number:* 200-606-7

*Uses:* Nicotine Sulphate is an insecticide of plant origin and is effective against a wide variety of insect pests. It is also used in veterinary medicine as an anthelmintic and external parasiticide. It is used to kill aphids, thrips, bugs, worms, leaf-hoppers and similar sucking insects which attack and destroy fruit, vegetables, crops and even flowers. It is also effective against lice, mites, and ticks.

*Carcinogen/Hazard Classifications*
**Label Signal Word:** DANGER
**WHO Acute Hazard:** *As nicotine:* Class 1b, highly hazardous

*Regulatory Authority:*
- Banned or Severely Restricted (In Agriculture) (New Zealand, former USSR) (UN)[13]
- Air Pollutant Standard Set (former USSR)[35]
- Superfund/EPCRA 40CFR355, Appendix B Extremely Hazardous Substances: TPQ = 100/10,000 lb (45.4/4,540 kg)
- Superfund/EPCRA 40CFR302.4 RQ: EHS, 1 lb (0.454 kg)
- EPCRA Section 313 Form R *de minimis* concentration reporting level: 1.0%
- Canada, WHMIS, Ingredients Disclosure List

*Description:* Nicotine sulfate is a white crystalline solid. Molecular weight = 418.55. Soluble in water. Autoignition temperature = 244°C. Hazard Identification (based on NFPA-704 M Rating System): Health 4, Flammability 1, Reactivity 0.

*Incompatibilities:* Oxidizing materials.

*Permissible Exposure Limits in Air:* OSHA[2]: PEL is 0.5 mg/m$^3$ over an 8-hour shift. NIOSH[2] and ACGIH[1]: The recommended exposure limit over a 10-hour shift is 0.5 mg/m$^3$. The former USSR[35] has set an MAC in workplace air of 0.1 mg/m$^3$ and an MAC in ambient air in residential areas of 0.005 mg/m$^3$ on a once-daily basis and 0.001 mg/m$^3$ on a daily average basis.

*Permissible Concentration in Water:* No criteria set, but runoff from spills or fire control may cause water pollution.

*Routes of Entry:* Inhalation, ingestion, skin and/or eye contact.

*Harmful Effects and Symptoms*
*Short Term Exposure:* The liquid irritates the eyes and skin. Inhalation irritates nose and throat. May affect the central nervous system, causing convulsions and respiratory failure. Exposure at high concentrations may result in death. Onset of acute poisoning is rapid. Symptoms include nausea, salivation, abdominal pain, vomiting, diarrhea, cold sweat, headache, dizziness, disturbed hearing and vision, mental confusion, marked weakness, faintness and

prostration, lowered blood pressure, difficult breathing, and weak, rapid and irregular pulse. It is classified as super toxic. Probable oral lethal dose in humans is less than 5 mg/kg (less than 7 drops) for a 70 kg (150 lb) person. Death is possible from respiratory failure caused by paralysis of the respiratory muscles.

*Long Term Exposure:* Animal tests show that this substance possibly causes toxic effects upon human reproduction. $LD_{50}$ (mouse, oral) = 8.54 mg/kg.

*First Aid:* If this chemical gets into the eyes, remove any contact lenses at once and irrigate immediately for at least 15 minutes, occasionally lifting upper and lower lids. Seek medical attention immediately. If this chemical contacts the skin, remove contaminated clothing and wash immediately with soap and water. Seek medical attention immediately. If this chemical has been inhaled, remove from exposure, begin rescue breathing (using universal precautions) if breathing has stopped, and CPR if heart action has stopped. Transfer promptly to a medical facility. When this chemical has been swallowed, get medical attention. Give large quantities of water and induce vomiting. Do not make an unconscious person vomit.

*References:*
- New Jersey Department of Health and Senior Services, "Hazardous Substance Fact Sheet, Nicotine Sulfate," Trenton, NJ (April 2002). http://www.state.nj.us/health/eoh/rtkweb/1352.pdf
- Sax, N.I., Ed., "Dangerous Properties of Industrial Materials Report," 5, No. 4, 88-90 (1985).
- California Environmental Protection Agency "Chemical List of Lists," Sacramento CA (February 1997).
- U.S. Environmental Protection Agency, "Chemical Profile: Nicotine Sulfate," Washington, DC, Chemical Preparedness Program (Nov. 30, 1987).

# Nitrapyrin (ANSI)

*Use Type:* Microbiocide (bacteriostat)
*CAS Number:* 1929-82-4
*Formula:* $C_6H_3Cl_4N$
*Synonyms:* 4-Chloro-6-(Trichloromethyl)pyridine; 2-Chloro-6-trichloromethylpyridine; 2-Chloro-6-(trichloromethyl)pyridine; Pyridine, 2-chloro-6-(trichloromethyl)-
*Trade Names:* DOWCO-163®; NITRAPYRINE®; N-SERVE®, Dow AgroSciences (USA); N-SERVE NITROGEN STABILIZER®, Dow AgroSciences (USA)
*Producers:* Dow AgroSciences (USA); Ehrenstorfer, Dr. (Germany)
*Chemical Class:* Pyridines
*EPA/OPP PC Code:* 069203
*California DPR Chemical Code:* 439
*RTECS Number:* US7525000

*Uses:* Used as a nitrogen stabilizer in agricultural applications. Nitrapyrin is registered for use as a nitrogen stabilizer in corn, cotton, rice, sorghum, strawberries, and wheat.

*U.S. Maximum Allowable Residue Levels for Nitrapyrin (40 CFR 180.350):*

| CROP | ppm |
|---|---|
| Cattle, fat & meat | 0.05 |
| Cattle, mbyp | 0.05 |
| Corn, forage | 1 |
| Corn, grain | 0.1 |
| Corn, stover | 1 |
| Corn, sweet, kernel plus cob with husks removed | 0.1 |
| Goat, fat & meat | 0.05 |
| Goat, mbyp | 0.05 |
| Hog, fat & meat | 0.05 |
| Hog, mbyp | 0.05 |
| Horse, fat & meat | 0.05 |
| Horse, mbyp | 0.05 |
| Poultry, fat & meat | 0.05 |
| Poultry, mbyp | 0.05 |
| Sheep, fat & meat | 0.05 |
| Sheep, mbyp | 0.05 |
| Sorghum, forage | 0.1 |
| Sorghum, grain, grain | 0.1 |
| Sorghum, grain, stover | 0.5 |
| Strawberry | 0.2 |
| Wheat, forage & straw | 0.5 |
| Wheat, grain | 0.1 |

*Carcinogen/Hazard Classifications*
**U.S. EPA Carcinogens:** A likely carcinogen
**California Prop. 65:** Listed; Developmental toxin
**Label Signal Word:** WARNING or DANGER
**WHO Acute Hazard:** Class III, slightly hazardous
*Regulatory Authority:*
- Carcinogen (Animal Positive) (NCI)[9]
- Air Pollutant Standard Set (ACGIH)[1] (HSE)[33] (OSHA)[58] (Several States)[60]
- Permissible Exposure Limits for Chemical Contaminants (CAL/OSHA)
- The "Director's List" (CAL/OSHA)
- Proposition 65 chemical (CAL)
- Actively registered pesticide in California.
- EPCRA Section 313 Form R *de minimis* concentration reporting level: 1.0%
- Canada, WHMIS, Ingredients Disclosure List

*Description:* Nitrapyrin is a colorless crystalline solid. Mild, sweet odor. Practically insoluble in water; solubility = 43 ppm. Molecular weight = 230.88. Boiling point =–136°C. Melting/Freezing point = 62°C. Vapor pressure = $2.8 \times 10^{-3}$ mmHg @ 20°C.

*Incompatibilities:* Aluminum and magnesium.

*Permissible Exposure Limits in Air:* OSHA[2] has set a TWA of 15 mg/m³ on a total dust basis and 5 mg/m³ for the respirable fraction. The ACGIH[1] and HSE[33] have adopted a TWA value of 10 mg/m³ and set an STEL of 20 mg/m³. In addition, several states have set guidelines for nitrapyrin in ambient air[60] ranging from 100-200 $\mu$g/m³ (North Dakota) to 160 $\mu$g/m³ (Virginia) to 200 $\mu$g/m³ (Connecticut) to 238 $\mu$g/m³ (Nevada).

*Permissible Concentration in Water:* No criteria set, but runoff from spills or fire control may cause water pollution.

*Routes of Entry:* Inhalation, ingestion, skin and/or eye contact.

*Harmful Effects and Symptoms*

*Short Term Exposure:* Nitrapyrin can affect you when breathed in and by passing through your skin. Exposure can irritate the eyes, nose and throat. High levels may cause you to feel dizzy, lightheaded, and to pass out. Contact can irritate and may damage the eyes and skin. No adverse effects noted in ingestion studies with animals.

*Long Term Exposure:* There may also be damage to the liver and kidneys. Repeated exposure to nitrapyrin may cause symptoms of headaches, dizziness, loss of appetite, and trouble sleeping.

*Points of Attack:* Liver and kidneys.

*Medical Surveillance:* If symptoms develop or overexposure is suspected, the following may be useful: Liver and kidney function tests.

*First Aid:* If this chemical gets into the eyes, remove any contact lenses at once and irrigate immediately for at least 15 minutes, occasionally lifting upper and lower lids. Seek medical attention immediately. If this chemical contacts the skin, remove contaminated clothing and wash immediately with soap and water. Seek medical attention immediately. If this chemical has been inhaled, remove from exposure, begin rescue breathing (using universal precautions) if breathing has stopped, and CPR if heart action has stopped. Transfer promptly to a medical facility. When this chemical has been swallowed, get medical attention. Give large quantities of water and induce vomiting. Do not make an unconscious person vomit.

*References:*

- Pesticide Management Education Program, "Chemical Fact Sheet June, 1985, Nitrapryin," Cornell University, Ithaca, NY (June 1985).
  http://pmep.cce.cornell.edu/profiles/miscpesticides/methylchloride-xanthangum/nitrapyrin/anti-prof-nitrapyrin.html
- U.S. Environmental Protection Agency, Office of Pesticide Programs, Pesticide Residue Limits, "Nitrapyrin," 40 CFR 180.350,
  http://www.epa.gov/pesticides/food/viewtols.htm
- New Jersey Department of Health and Senior Services, "Hazardous Substance Fact Sheet: Nitrapyrin," Trenton, NJ (April 2000).
  http://www.state.nj.us/health/eoh/rtkweb/1355.pdf
- California Environmental Protection Agency "Chemical List of Lists," Sacramento CA (February 1997).

# Nitrofen

*Use Type:* Herbicide
*CAS Number:* 1836-75-5
*Formula:* $C_{12}H_7Cl_2NO_3$; $O_2NC_6H_4OC_6H_3Cl_3$
*Alert:* Nitrofen is a carcinogen and teratogen and should be handled with extreme caution. It is no longer used in the U.S.
*Synonyms:* Benzenamine, 4-ethoxy-*N*-(5-nitro-2furanyl)methylene-; Benzene, 2,4-dichloro-1-(4-nitrophenoxy)-; 2',4'-Dichloro-4'-nitrodiphenyl ether; 2,4-dichloro-1-(4-nitrophenoxy)benzene; 4-(2,4-Dichlorophenoxy)nitrobenzene; 2,4-Dichlorophenyl 4-nitrophenyl ether; 2,4-Dichlorophenyl *p*-nitrophenyl ether; 2,4-Dichlorophenyl-4-nirtophenylaether (German); Ether,2,4-dichlorophenyl *p*-nitrophenyl; Nitrochlor; 4'-Nitro-2,4-dichlorodiphenyl ether; 4-Nitro-2',4'-dichlorodiphenyl ether; NCI-C00420; Nitrofene (French); Nitrophen; Nitrophene
*Trade Names:* FW 925®; MEZOTOX®; NICLOFEN®; NIP®; PREPARATION 125®; TOK-2®, Rohm & Haas (USA), canceled 11/7/1983; TOK®, Rohm & Haas (USA), canceled 11/7/1983; TOK E®, Rohm & Haas (USA), canceled 11/7/1983; TOK E 25®, Rohm & Haas (USA), canceled 11/7/1983; TOK E 40®, Rohm & Haas (USA), canceled 11/7/1983; TOKKOM®, Rohm & Haas (USA), canceled 11/7/1983; TOKKORN®, Rohm & Haas (USA), canceled 11/7/1983; TOK WP-50®, Rohm & Haas (USA), canceled 11/7/1983; TRIZILIN®
*Producers:* Bayer CropScience (Germany); Ki-Hara Chemicals Ltd. (UK); Rhone-Poulenc Agro France (France); Shenzhen Guomeng Industry Co., Ltd. (China); Wuxj Ruize Pesticide (China)
*Chemical Class:* Diphenyl ether
*EPA/OPP PC Code:* 038201
*California DPR Chemical Code:* 592
*ICSC Number:* 0929
*RTECS Number:* KN8400000
*EEC Number:* 609-040-00-9
*EINECS Number:* 217-406-0
*Uses:* Nitrofen is a contact herbicide used for pre- and post-emergency control of annual grasses and broadleaf weeds on a variety of food and ornamental crops. It is no longer registered in the U.S.
*Carcinogen/Hazard Classifications*
**U.S. NTP Carcinogen:** Reasonably anticipated carcinogen
**California Prop. 65:** Carcinogen
**IARC:** Group 2B, possible carcinogen
**Label Signal Word:** CAUTION
**Endocrine Disruptor:** Suspected endocrine disruptor
*Regulatory Authority:*

- Carcinogen (Animal Positive) (IARC) (NCI)[9]
- Banned or Severely Restricted (Many Countries) (UN)[13]

- Air Pollutant Standard Set (former USSR)[35] (Pennsylvania)[60]
- AB 1803-Well Monitoring Chemical (CAL)
- AB 2588-Air Toxics "Hot Spots" Chemicals (CAL)
- Proposition 65 chemical (CAL)
- The "Director's List" (CAL/OSHA)
- EPCRA Section 313 Form R *de minimis* concentration reporting level: 0.1%.
- U.S. DOT Regulated Marine Pollutant (49CFR172.101, Appendix B)
- Canada, WHMIS, Ingredients Disclosure List

*Description:* Nitrofen is a crystalline solid. Very slightly soluble in water; solubility = $10^{-4}$ g/100 ml @ 22°C. Molecular weight = 284.10. Boiling point = 365°C. Vapor pressure = $10^{-3}$ Pa @ 40°C. Melting/Freezing point = 70–71°C. Vapor pressure = $1.2 \times 10^{-7}$ mm @ 20°C. Flash point = 200°C (cc). Autoignition temperature = 400°C Hazard Identification (based on NFPA-704 M Rating System): Health 2, Flammability 2, Reactivity 0. Log $K_{ow}$ = 3.5. Values above 3.0 are likely to bioaccumulate in marine organisms.

*Permissible Exposure Limits in Air:* The former USSR[35] has set an MAC in ambient air in residential areas of 0.02 mg/m$^3$ on a once-daily basis and 0.01 mg/m$^3$ on an average daily basis. Pennsylvania has set a guideline for nitrofen in ambient air[60] of 0.75 $\mu$g/m$^3$.

*Permissible Concentration in Water:* Nitrofen presumably falls under the U.S. EPA Priority Toxic Pollutant category of haloethers[6] but specific limits have not been set. The former USSR[35] has set a limit in surface water of 4.0 mg/L.

*Routes of Entry:* Inhalation, ingestion, skin and/or eye contact.

*Harmful Effects and Symptoms*

*Short Term Exposure:* Toxic by ingestion. Severe eye irritant. Causes skin irritation on contact. Inhalation can cause irritation to the respiratory tract. May cause difficult breathing, fatigue, diarrhea, loss of appetite.

*Long Term Exposure:* Long term exposure may cause damage to the blood cells, causing low white cell (leucocyte) count, reduced hemoglobin, and reduced serum cholinesterase and erythrocyte catalase activities. May cause liver damage. May affect the central nervous system.

*Points of Attack:* Blood, liver, kidneys and central nervous system.

*Medical Surveillance:* Complete blood count. Liver function tests. Examination of the nervous system and interview for brain effects.

*First Aid: Skin Contact:* Flood all areas of body that have contacted the substance with water. Don't wait to remove contaminated clothing; do it under the water stream. Use soap to help assure removal. Isolate contaminated clothing when removed to prevent contact by others. *Eye Contact:* Remove any contact lenses at once. Immediately flush eyes well with copious quantities of water or normal saline for at least 20-30 minutes. Seek medical attention. *Inhalation:* Leave contaminated area immediately; breathe fresh air. Proper respiratory protection must be supplied to any rescuers. If coughing, difficult breathing or any other symptoms develop, seek medical attention at once, even if symptoms develop many hours after exposure. *Ingestion:* Contact a physician, hospital or poison center at once. If the victim is unconscious or convulsing, do not induce vomiting or give anything by mouth. Assure that his airway is open and lay him on his side with his head lower that his body and transport immediately to a medical facility. If conscious and not convulsing, give a glass of water to dilute the substance. Vomiting should not be induced without a physician's advice.

*References:*
- New Jersey Department of Health and Senior Services, "Hazardous Substance Fact Sheet, Nitrofen," Trenton NJ (December 1988, rev. September 2001). http://www.state.nj.us/health/eoh/rtkweb/1374.pdf
- California Environmental Protection Agency "Chemical List of Lists," Sacramento CA (February 1997).

# Norflurazon (ANSI)

*Use Type:* Herbicide
*CAS Number:* 27314-13-2
*Formula:* $C_{12}H_9ClF_3N_3O$
*Synonyms:* Caswell No. 195AA; 4-Chloro-5-(methylamino)-2-[3-(trifluoromethyl)phenyl]-3(2H)-pyridazinone; 4-Chloro-5-methylamino-2-(3-trifluoromethylphenyl)pyridazin-3-one; 4-Chloro-5-(methylamino)-2-($\alpha,\alpha,\alpha$-trifluoro-m-tolyl)-3(2H)-pyridazinone; 4-Chloro-5-methylamino-2-($\alpha,\alpha,\alpha$-trifluoro-*m*-tolyl)pyridazinone-3(2H)-one; 4-Chloro-5-(methylamino)-2-[3-(trifluoromethyl)phenyl]-3(2H)-pyridazinone; Dodecylbenzenesulfonic acid, triethanolamine salt; Monomethflurazone; Norflurazon pyridazine herbicide; 3(2H)-Pyridazinone, 4-chloro-5-(methylamino)-2-[3-(trifluoromethyl)phenyl]-; 3(2H)-Pyridazinone, 4-chloro-5-(methylamino)-2-($\alpha,\alpha,\alpha$-trifluoro-*m*-tolyl)-; Triethanolamine dodecylbenzenesulfonate

*Trade Names:* EVITAL®, AMVAC Chemical (USA); H 9789®; SAN 9789 H®; SAN 97895®; SOLICAM®, Syngenta (Switzerland); TELOK®, Sandoz (USA), canceled; TRIETHANOLAMINE DBS®; ZORIAL®, Syngenta (Switzerland)

*Producers:* AMVAC Chemical (USA); Nufarm (Australia); Syngenta (Switzerland)

*Chemical Class:* Fluorinated pyridazinone
*EPA/OPP PC Code:* 105801
*California DPR Chemical Code:* 2019
*RTECS Number:* UR6150000

*Uses:* Norflurazon is a selective preemergent herbicide used to control germinating annual grasses, sedges, rushes and broadleaf weeds in fruits (cranberries, citrus, cherries, nectarines, apricots), grape vines, vegetables, nuts, cotton, peanuts, soybeans, and various nonagricultural and industrial areas.

*U.S. Maximum Allowable Residue Levels for Norflurazon (40 CFR 180.356):*

| CROP | ppm |
|---|---|
| Alfalfa, forage | 3.0 |
| Alfalfa, hay | 5.0 |
| Alfalfa, seed | 0.1 |
| Almond | 0.1 |
| Almond, hulls | 1.0 |
| Apple | 0.1 |
| Apricot | 0.1 |
| Asparagus | 0.05 |
| Avocado | 0.2 |
| Blackberry | 0.1 |
| Blueberry | 0.2 |
| Cattle, fat | 0.1 |
| Cattle, liver | 0.25 |
| Cattle, meat | 0.1 |
| Cattle, mbyp | 0.1 |
| Cattle, mbyp except liver | 0.1 |
| Cherry | 0.1 |
| Citrus, dried pulp | 0.4 |
| Citrus, molasses | 1.0 |
| Cottonseed | 0.1 |
| Cotton, undelinted seed | 0.1 |
| Cranberry | 0.1 |
| Dried citrus pulp | 0.4 |
| Filbert | 0.1 |
| Goat, fat | 0.1 |
| Goat, liver | 0.25 |
| Goat, meat | 0.1 |
| Goat, mbyp | 0.1 |
| Goat, mbyp except liver | 0.1 |
| Grape | 0.1 |
| Hog, fat | 0.1 |
| Hog, liver | 0.25 |
| Hog, meat | 0.1 |
| Hog, mbyp | 0.1 |
| Hog, mbyp except liver | 0.1 |
| Hop, fresh | 1.0 |
| Horse, fat | 0.1 |
| Horse, liver | 0.25 |
| Horse, meat | 0.1 |
| Horse, mbyp | 0.1 |
| Horse, mbyp | 0.1 |
| Milk | 0.1 |
| Nectarine | 0.1 |
| Peach | 0.1 |
| Peanut | 0.05 |
| Peanut, vines | 1.5 |
| Peanut, hay | 5.5 |
| Pear | 0.1 |
| Pecan | 0.1 |
| Plums (fresh prunes) | 0.1 |
| Poultry, fat | 0.1 |
| Poultry, meat | 0.1 |
| Poultry, mbyp | 0.1 |
| Raspberry | 0.1 |
| Sheep, fat | 0.1 |
| Sheep, liver | 0.25 |
| Sheep, meat | 0.1 |
| Sheep, mbyp | 0.1 |
| Sheep, mbyp except liver | 0.1 |
| Soybean | 0.1 |
| Soybean, forage | 1.0 |
| Soybean, hay | 1.0 |
| Walnut | 0.1 |

*Carcinogen/Hazard Classifications*
**U.S. EPA Carcinogens:** Group C, possible carcinogen
**Label Signal Word:** CAUTION or WARNING
**WHO Acute Hazard:** Class U, unlikely to be hazardous
*Regulatory Authority:*
- EPA 40 CFR 372.65, Specific Toxic Chemical Listings
- FIFRA, 40CFR185: tolerances for pesticides in food
- FIFRA, 40CFR186: tolerances for pesticides in animal feeds
- Actively registered pesticide in California.
- EPCRA Section 313 Form R *de minimis* concentration reporting level: 1.0%

*Description:* White crystalline solid. Odorless. Slightly soluble in water. Molecular weight = 303.66. Melting/Freezing point = 183°C. Vapor pressure = $2 \times 10^{-8}$ mm @ 20°C.

*Permissible Concentration in Water:* No criteria set. Runoff from spills or fire control may cause water pollution.

*Routes of Entry:* Ingestion, skin contact

*Harmful Effects and Symptoms*

*Short Term Exposure:* Contact may irritate skin and cause eye irritation and possible severe injury. Inhalation should be avoided; use NIOSH-approved air purifying respirators for pesticides. Poisonous if swallowed.

*Medical Surveillance:* If poisoning is suspected or of redness, itching, burning of skin or eyes develop.

*First Aid:* If this chemical gets into the eyes, remove any contact lenses at once and irrigate immediately for at least 15 minutes, occasionally lifting upper and lower lids. Seek medical attention immediately. If this chemical contacts the skin, remove contaminated clothing and wash immediately with soap and water. If this chemical has been inhaled, remove from exposure, begin rescue breathing (using universal precautions) if breathing has stopped, and CPR if heart action has stopped. Transfer promptly to a medical facility. When this chemical has been swallowed, get medical attention. Give large quantities of water and induce vomiting. Do not make an unconscious person vomit.

### References:

- U.S. Environmental Protection Agency, Office of Pesticide Programs, Pesticide Residue Limits, "Norflurazon," 40 CFR 180.356, www.epa.gov/pesticides/food/viewtols.htm
- U.S. Environmental Protection Agency, "Reregistration Eligibility Decision (RED), Norflurazon," Office of Prevention, Pesticides and Toxic Substances, Washington, DC
- Pesticide Management Education Program, "Norflurazon (Zorial, Solicam Herbicide profile 12/84," Cornell University, Ithaca, NY (December 1984). http://pmep.cce.cornell.edu/profiles/herb-growthreg/naa-rimsulfuron/norflurazon/herb-prof-norflurazon.html
- California Environmental Protection Agency "Chemical List of Lists," Sacramento CA (February 1997)

# Nosema Locustae

*Use Type:* Insecticide
*Synonyms:* N. locustae (Canning); Nosema Locustae Canning; Nosema Locustae Spores
*Trade Names:* LOCUSTCIDE®, canceled; NOLO-BAIT®, M & R Durango (USA); NOLO® BB CONCENTRATE, M & R Durango (USA); NOLOC®, canceled; SEMASPORE® BAIT
*Producers:* M & R Durango (USA)
*Chemical Class:* Microbial
*EPA/OPP PC Code:* 117001; 207700 (old EPA code)
*California DPR Chemical Code:* 2137
*Uses:* N. Locustae is a microbial insecticide used to control grasshoppers and crickets in crop fields, rangeland, grasses, lawns and turf, grass way drains, fence rows and hedgerows. It is made from the spores of the protozoan, N. locustae (Canning), which is infectious to certain grasshoppers and crickets. *N. locustae* must be eaten by the target insect to be effective. Although it is registered for use on crop fields, *N. locustae* is exempted from the requirement of a tolerance (or maximum limit) for residues remaining in or on all raw agricultural commodities. (40 CFR 180.1041)
*Carcinogen/Hazard Classifications*
**Label Signal Word:** CAUTION
*Regulatory Authority:*
- EPA: 40 CFR 180.1041

*Description:* Powder.
*First Aid:* If this chemical gets into the eyes, remove any contact lenses at once and irrigate immediately for at least 15 minutes, occasionally lifting upper and lower lids. Seek medical attention immediately. If this chemical contacts the skin, remove contaminated clothing and wash immediately with soap and water. Seek medical attention immediately. If this chemical has been inhaled, remove from exposure, begin rescue breathing (using universal precautions) if breathing has stopped, and CPR if heart action has stopped. Transfer promptly to a medical facility. When this chemical has been swallowed, get medical attention. Give large quantities of water and induce vomiting. Do not make an unconscious person vomit. Do not induce vomiting when formulations containing petroleum solvents are ingested.

### References:

- California Environmental Protection Agency "Chemical List of Lists," Sacramento CA (February 1997)
- U.S. Environmental Protection Agency, "Reregistration Eligibility Decision (RED) Facts, Nosema Locustae," Office of Prevention, Pesticides and Toxic Substances, Washington, DC (September 1992). http://www.epa.gov/REDs/factsheets/4104fact.pdf

# O

## Octamethyl Diphosphoramide

*Use Type:* Insecticide and acaricide.
*CAS Number:* 152-16-9
*Formula:* $C_8H_{24}N_4O_3P_2$
*Alert:* Not registered as a pesticide in the U.S.
*Synonyms:* Bis-bisdimethylaminophosphonous anhydride; Bis(bisdimethylaminophosphonous)anhydride; Bis(bisdimethylamino)phosphonousanhydride; Bis(bisdimethylamino)phosphoric anhydride; Bis-$N,N,N',N'$-Tetramethylphosphorodiamidic anhydride; ENT 17,291; Diphosphoramide, octamethyl-; Octamethyl-diforzuur-tetramide (Dutch); Octamethylpyrophosphoramide; Octamethyl-diphosphorsaeure-tetramid (German); Octamethyl pyrophosphortetramide; Octamethyl tetramido pyrophosphate; Octametilpirofosforamida (Spanish); OMPA; Ottometil-pirofosforammide (Italian); Pyrophosphoric acid octamethylteraamide; Pyrophosphorytetrakisdimethylamide; Schradan; Schradane (French); Tetrakisdimethylaminophosphoric anhydride
*Trade Names:* LETHA LAIRE G-59®; OMPACIDE®; OMPATOX®; OMPAX®; PESTOX®; PESTOX 3®; PESTOX III®; SYSTAM®; SYSTOPHOS®; SYTAM®
*Chemical Class:* Organophosphate
*EPA/OPP PC Code:* 058601
*California DPR Chemical Code:* 446
*RTECS Number:* UX5950000
*RCRA Number:* P085
*Uses:* Material is used as a systemic insecticide for plants and as an acaricide. Not registered as a pesticide in the U.S.
*Regulatory Authority:*
- Banned or Severely Restricted (In Agriculture) (former USSR, Germany, USA) (UN)[13]
- Air Pollutant Standard Set (former USSR)[35, 43]
- The "Director's List" (CAL/OSHA)
- EPA Hazardous Waste Number (RCRA No.): P085
- RCRA, 40CFR261, Appendix 8 Hazardous Constituents
- Superfund/EPCRA 40CFR355, Appendix B Extremely Hazardous Substances: TPQ = 100 lb (45.4 kg)
- Superfund/EPCRA 40CFR302.4 RQ: CERCLA, 100 lb (45.4 kg)
- U.S. DOT Inhalation Hazard Chemicals as organophosphates
- Canada, WHMIS, Ingredients Disclosure List

*Description:* Colorless liquid or in crude form, a dark brown, viscous liquid. Soluble in water. Molecular weight = 286.29. Boiling point = 120–125°C @ 0.5 mmHg; 153°C @ 2.0 mmHg. Melting/Freezing point = 14–19°C. Vapor Pressure = $1.3 \times 10^{-3}$ mmHg @ 25°C. Hazard Identification (based on NFPA-704 M Rating System): Health 3, Flammability 1, Reactivity 0.
*Incompatibilities:* Incompatible with acids. When heated to decomposition, forms toxic oxides of nitrogen and phosphorus.
*Permissible Exposure Limits in Air:* The former USSR[35, 43] has set an MAC in workplace air of 0.02 mg/m³ and an MAC for ambient air in residential areas of 0.002 mg/m³ on a momentary basis and 0.0004 mg/m³ on an average daily basis.
*Permissible Concentration in Water:* No criteria set, but runoff from spills or fire control may cause water pollution.
*Routes of Entry:* Inhalation, ingestion, skin and/or eye contact.
*Harmful Effects and Symptoms*
*Short Term Exposure:* Acute exposure to OMPA may produce the following signs and symptoms: pinpoint pupils, blurred vision, headache, dizziness, muscle spasms, and profound weakness. Vomiting, diarrhea, abdominal pain, seizures, and coma may also occur. The heart rate may decrease following oral exposure or increase following dermal exposure. Hypotension (low blood pressure) and chest pain may be noted. Hypertension (high blood pressure) is not uncommon. Respiratory symptoms include dyspnea, respiratory depression, and respiratory paralysis. Psychosis may occur. This material is extremely toxic; probable oral lethal dose in humans is 5-50 mg/kg, between 7 drops and 1 teaspoonful for a 150 lb person. It is highly toxic when inhaled. Material is a cholinesterase inhibitor. It is similar in action to other Organophosphorus pesticides in its toxicity. It is slightly less toxic than parathion. Gastrointestinal, neurologic and respiratory symptoms may accompany poisoning with this material. High doses may cause a toxic psychosis similar to acute alcoholism. Delayed pulmonary edema may occur after inhalation.
*Note:* Persons taking the following drugs may be at greater risk: Phenobarbital and phenaglycodol together, glutethimide, chlorpromazine hydrochloride, or mepromabate. These drugs appear to significantly enhance the toxicity of the material.
*Long Term Exposure:* Cholinesterase inhibitor; cumulative effect is possible. This chemical may damage the nervous system with repeated exposure, resulting in convulsions, respiratory failure. May cause liver damage.
*Points of Attack:* Respiratory system, lungs, central nervous system, cardiovascular system, skin, eyes, plasma and red blood cell cholinesterase.
*Medical Surveillance:* Medical observation is recommended for 24 to 48 hours after breathing

overexposure, as pulmonary edema may be delayed. Before employment and at regular times after that, the following are recommended: Plasma and red blood cell cholinesterase levels (tests for the enzyme poisoned by this chemical). If exposure stops, plasma levels return to normal in 1-2 weeks while red blood cell levels may be reduced for 1-3 months. When acetylcholinesterase enzyme levels are reduced by 25% or more below preemployment levels, risk of poisoning is increased, even if results are in lower ranges of "normal." Reassignment to work not involving organophosphate or carbamate pesticides is recommended until enzyme levels recover. If symptoms develop or overexposure occurs, repeat the above tests as soon as possible and get an exam of the nervous system. Also consider complete blood count. Consider chest x-ray following acute overexposure. Do not drink any alcoholic beverages before or during use. Alcohol promotes absorption of organophosphates.

*First Aid:* **Treatment for organophosphate poisoning consists of thorough decontamination, cardiorespiratory support, and administration of the antidotes atropine and pralidoxime. In cases of severe poisoning, diazepam, an anticonvulsant, should also be administered. Antidotes should be administered as prevention even if the diagnosis is in doubt.** Speed in removing material from eyes and skin is of extreme importance. *Eyes:* Eye contact can cause dangerous amounts of these chemicals to be quickly absorbed through the mucous membrane into the bloodstream. Immediately and gently flush eyes with plenty of warm or cold water (NO hot water) for at least 15 minutes, occasionally lifting the upper and lower eyelids. Get medical aid immediately. *Skin:* Get medical aid. Skin contact can cause dangerous amounts of these chemicals to be absorbed into the bloodstream. Wearing the appropriate PPE equipment and respirator for organophosphate pesticides, immediately flush skin with plenty of soap and water for at least 15 minutes while removing contaminated clothing and shoes. Shampoo hair promptly if contaminated. The removed contaminated clothing and shoes should be double-bagged and left in Hot Zone for later disposal by hazardous materials experts. Skin may also be decontaminated with diluted hypochlorite solution. *Inhalation:* Get medical aid. Do not contaminate yourself. Wearing the appropriate PPE equipment and respirator for organophosphate pesticides, immediately remove the victim from the contaminated area to fresh air. If the victim is not breathing, administer artificial respiration. Do NOT use mouth-to-mouth resuscitation; use bag/mask apparatus. If breathing is difficult, administer oxygen through bag/mask apparatus until medical help arrives. Do not leave victim unattended. *Ingestion:* Call poison control. Loosen all clothing. Never give anything by mouth to an unconscious person. If victim is *unconscious or having convulsions,* do nothing except keep victim warm. Get medical aid. Transfer promptly to a medical facility. In cases of ingestion, **do not induce vomiting.** If the victim is alert and asymptomatic, administer a slurry of activated charcoal at a dose of 1 g/kg (infant, child, and adult dose). A soda can and straw may be of assistance when offering charcoal to a child. *In some cases you may be specifically instructed by poison control to induce vomiting by way of 2 tablespoons of syrup of ipecac (adult) washed down with a cup of water.* Do NOT give activated charcoal before or with ipecac syrup.

*Note to physician or authorized medical personnel:* Treat cases of respiratory compromise, coma, or excessive pulmonary secretions with respiratory support using protocols and techniques available and within the scope of training. Some cases may necessitate procedures such as endotracheal intubation or cricothyrotomy by properly trained and equipped personnel. When possible, atropine (see *Antidotes*, below) should be given under medical supervision. Patients who are comatose, hypotensive, or having seizures or cardiac arrhythmias should be treated according to advanced life support protocols. *Antidotes:* Two antidotes are administered to treat organophosphate poisoning. Atropine is a competitive antagonist of acetylcholine at muscarinic receptors and is used to control the excessive bronchial secretions which are often responsible for death. Pralidoxime relieves both the nicotinic and muscarine effects of organophosphate poisoning by regenerating acetylcholinesterase and can reduce both the bronchial secretions and the muscle weakness associated with poisoning. The initial intravenous dose of atropine in adults should be determined by the severity of symptoms: An initial adult dose of 1.0 to 2.0 mg or pediatric dose of 0.01 mg/kg (minimum 0.01 mg) should be administered intravenously. If intravenous access cannot be established, atropine may also be given intramuscularly, subcutaneously or via endotracheal tube. Doses should be repeated every 15 minutes until excessive secretions and sweating have been controlled. Once bronchial secretion has been controlled, atropine administration should be repeated whenever the secretions begin to recur. In seriously poisoned patients, very large doses may be required. Alterations of pulse rate and pupillary size should not be used as indicators of treatment adequacy. Pralidoxime should be administered as early in poisoning as possible as its efficacy may diminish when given more than 24 to 36 hours after exposure. Doses are as follows: adult 1.0 g; pediatric 25 to 50 mg/kg. The drug should be administered intravenously over 30 to 60 minutes, but in a life-threatening situation, one-half of the total dose can be given per minute for a total administration time of 2 minutes. Treatment should begin to take effect within 40 minutes with a reduction in symptoms and the amount of atropine necessary to control bronchial secretion. The initial dose can be repeated in 1 hour and then every 8 to 12 hours until the patient is clinically well and no longer requires atropine. If intravenous access cannot be established, pralidoxime may also be given intramuscularly. Early administration of diazepam in addition to the combined atropine and pralidoxime treatment may help prevent the onset of seizures and potential brain and cardiac morphologic damage following high-level organophosphate poisoning.

*References:*
- U.S. Environmental Protection Agency, "Chemical Profile: Octamehyl Phosphoramide," Washington, DC, Chemical Emergency Preparedness Program (November 30, 1987).
  http://yosemite.epa.gov/oswer/CeppoEHS.nsf/Profiles/152-16-9
- California Environmental Protection Agency "Chemical List of Lists," Sacramento CA (February 1997).
- Agency for Toxic Substances and Disease Registry, U.S. Department of Health and Human Services, Public Health Service, "Managing Hazardous Materials Incidents," Atlanta, GA (June 2003).

# Octhilinone

*Use Type:* Fungicide, microbiocide
*CAS Number:* 26530-20-1
*Formula:* $C_{11}H_{19}NOS$
*Synonyms:* 2-Octyl-3(2H)-isothiazolone; 2-Octyl-4-isothiazolin-3-one; 3(2H)-Isothiazolone, 2-octyl-
*Trade Names:* ACTICIDE, Thor (UK); BUSAN, Buckman Laboratories (USA); CYTOX; KATHON® 893, Rohm & Haas (USA); MICRO-CHEK 11®, Ferro (USA); MICRO-CHEK SKANE®, Ferro (USA); OCTHILINONE®; PANCIL®, canceled; RH 893®; SKANE M8®, Ferro (USA); SLIMICIDE; MICROBICIDE 8®; MICROPEL
*Producers:* Buckman Laboratories (USA); Ferro (USA); Rohm & Haas (USA); Sigma-Aldrich (USA); Thor (UK)
*Chemical Class:* Isothiazolone
*EPA/OPP PC Code:* 099901
*California DPR Chemical Code:* 1881
*Uses:* Used as biocides on textiles, metalworking fluids, water thinned paints. Its use as a fungicide on cotton was canceled and the tolerances were revoked in 1998.
*Carcinogen/Hazard Classifications*
*Label Signal Word:* DANGER
*WHO Acute Hazard:* Class III, slightly hazardous
*Regulatory Authority:*
- Actively registered pesticide in California.

*Description:* Liquid. Molecular weight = 213.36
*Incompatibilities:* Oxidizers. Contact with hydrogen peroxide may form explosive material.
*Permissible Exposure Limits in Air:*
*Determination in Air:* Charcoal adsorption followed by $CS_2$ treatment and gas chromatographic analysis. See NIOSH Method 1300 [Ketones][18].
*Permissible Concentration in Water:* No criteria set. Runoff from spills or fire control may cause water pollution.
*Routes of Entry:* Eyes, skin and/or eye contact, inhalation.
*Harmful Effects and Symptoms*
*Short Term Exposure:* Contact can irritate the skin. Exposure can irritate the eyes and respiratory tract. Exposure to high concentrations may cause dizziness, lightheadedness, and unconsciousness.
*Long Term Exposure:* Repeated skin exposure can cause dryness and skin cracking. This chemical has not been adequately evaluated to determine whether brain or nerve damage could occur with repeated exposure. However, many solvents and other petroleum-based chemicals have been shown to cause such damage. Effects may include reduced memory and concentration, personality changes (withdrawal, irritability), and fatigue, sleep disturbances, reduced coordination, and/or effects on the nerves to the arms and legs (weakness, "pins and needles").
*Points of Attack:* Respiratory system, skin.
*Medical Surveillance:* Preplacement examinations should evaluate skin and respiratory conditions. Acetone can be detected in the blood, urine and expired air and can be used as an index of exposure. Evaluation for brain effects such as changes in memory, concentration, sleeping patterns and mood, as well as headaches and fatigue. Consider evaluations of the cerebellar, autonomic and peripheral nervous systems. Positive and borderline individuals should be referred for neuropsychological testing.
*First Aid:* If this chemical gets into the eyes, remove any contact lenses at once and irrigate immediately for at least 15 minutes, occasionally lifting upper and lower lids. Seek medical attention immediately. If this chemical contacts the skin, remove contaminated clothing and wash immediately with soap and water. Seek medical attention immediately. If this chemical has been inhaled, remove from exposure, begin rescue breathing (using universal precautions) if breathing has stopped, and CPR if heart action has stopped. Transfer promptly to a medical facility. When this chemical has been swallowed, get medical attention. *Do not induce vomiting when formulations containing petroleum solvents are ingested.* Otherwise, give large quantities of water and induce vomiting. Do not make an unconscious person vomit.
*References:*
- California Environmental Protection Agency "Chemical List of Lists," Sacramento CA (February 1997)

# Octylammonium Methanearsonate

*Use Type:* Herbicide and defoliant
*CAS Number:* 6379-37-9
*Synonyms:* OAMA; Methanearsonic acid, octylammonium salt; Arsonic acid, methyl-, compounded with 1-octanamine (1:1)
*Chemical Class:* Organoarsenic
*EPA/OPP PC Code:* 013804

*California DPR Chemical Code:* 27
*Uses:* Not registered in the U.S. All products were canceled in December, 1991. It was used as a herbicide on turf.
*Carcinogen/Hazard Classifications*
**U.S. EPA Carcinogens:** Group B2, probable carcinogen (as cacodylic acid)
**California Prop. 65:** Carcinogen (as cacodylic acid)
**IARC:** Group 1, known carcinogen
**Label Signal Word:** CAUTION
**WHO Acute Hazard:** Class III, slightly hazardous (as cacodylic acid)
*Regulatory Authority:*
- Clean Air Act: Hazardous Air Pollutants (Title I, Part A, Section 112) as arsenic compounds.
- Clean Water Act: Toxic Pollutant (Section 401.15) as arsenic and compounds.
- RCRA Section 261 Hazardous Constituents, waste number not listed.
- EPCRA Section 304 RQ: CERCLA, 1 lb (0.454 kg)
- EPCRA (Section 313): Includes any unique chemical substance that contains arsenic as part of that chemical's infrastructure. Form R *de minimis* concentration reporting level: (inorganics) 0.1%; organics 1.0%
- Marine pollutant (49CFR, Subchapter 172.101, Appendix B) as arsenates, liquid, n.o.s; arsenates, solid, n.o.s; arsenical pesticides liquid, toxic, flammable, n.o.s.
- AB 2588-Air Toxics "Hot Spots" Chemicals (CAL) as arsenic compounds
- Permissible Exposure Limits for Chemical Contaminants (CAL/OSHA) as arsenic compounds
- The "Director's List" (CAL/OSHA) as arsenic compounds

*Permissible Exposure Limits in Air:* The OSHA[(2)] PEL for organic arsenic compounds is 0.5 mg/m$^3$ TWA. For an 8-hour workshift. There is no NIOSH[(2)] recommendation. The ACGIH[(1)] TLV is 0.2 mg/m$^3$ TWA.
*Determination in Air:* Filter; Reagent: Ion chromatography/hydride atomic absorption: NIOSH IV [#5022, Arsenic, organo-].[(18)]
*Permissible Concentration in Water:* No criteria set. Runoff from spills or fire control may cause water pollution. Toxic pollutant designated pursuant to section 307 (a)(1) of the Clean Water Act and is subject to effluent limitations (arsenic and inorganic and organic arsenic) [40 CFR 401.15 (7/1/87)]
*Routes of Entry:* Inhalation, eyes, ingestion, skin contact
*Harmful Effects and Symptoms*
*Short Term Exposure:* Symptoms of arsenic poisoning usually appear one-half to one hour after ingestion, but may be delayed many hours. Symptoms include a sweetish metallic taste and garlicky odor; difficulty in swallowing; abdominal pain; vomiting and painful diarrhea; dehydration, thirst, and cramps; dizziness, stupor, and delirium, rapid heart beat, headache, skin disorders, and coma.

*Long Term Exposure:* Chronic exposure to arsenic compounds can cause dermatitis and digestive disorders. Renal damage may develop.
*Points of Attack:* Skin and kidneys.
*Medical Surveillance:* Kidney function tests. Examination by a qualified allergist.
*First Aid:* If this chemical gets into the eyes, remove any contact lenses at once and irrigate immediately for at least 15 minutes, occasionally lifting upper and lower lids. Seek medical attention immediately. If this chemical contacts the skin, remove contaminated clothing and wash immediately with soap and water. Seek medical attention immediately. If this chemical has been inhaled, remove from exposure, begin rescue breathing (using universal precautions) if breathing has stopped, and CPR if heart action has stopped. Transfer promptly to a medical facility. When this chemical has been swallowed, get medical attention. Give large quantities of water and induce vomiting. Do not make an unconscious person vomit. Do not induce vomiting when formulations containing petroleum solvents are ingested.
*References:*
- California Environmental Protection Agency "Chemical List of Lists," Sacramento CA (February 1997).

# Omethoate

*Use Type:* Acaricide and insecticide
*CAS Number:* 1113-02-6
*Formula:* $C_5H_{12}NO_4PS$
*Synonyms:* O-analog of Dimethoate; Dimethoate, O-analog; Dimethoate oxygen analog; Dimethoate PO-isologue; Dimethoate oxon; Dimethoxon; O,O-Dimethyl-S-(N-methyl-carbamoyl)-methyl-monothiophosphat (German); O,O-Dimethyl-S-[(methylcarbamoyl)methyl] phosphorothioate; O,O-dimethyl-S-(N-methylcarbamoylmethyl)phosphorothioate; O,O-Dimethyl S-[(methylcarbamoyl)methyl]phosphorothioate; O,O-Dimethyl-S-(N-methylcarbamoylmethyl) phosphorothiolate; Dimethyl-S-(N-methyl-carbamoyl-methyl) phosphorothiolate; O,O-Dimethyl-S-(N-methylcarbamoylmethyl)thiophosphate; O,O-Dimethyl-S-(2-*oxo*-3-azabutyl)-monothiophosphate; ENT 25,776; Ometohoat; Phosphorothioic acid, O,O-dimethyl S-[2-(methylamino)-2-oxoethyl]ester; Phosphorothioic acid, O,O-dimethyl ester, S-ester with 2-mercapto-N-methylacetamide; PO-Dimethoate; Thiophosphate de O,O-Diethyle et de s-(N-methylcarbamoyl)methyle (French)
*Trade Names:* FOLIMAT®, Bayer CropScience (Germany); BAYER® 45432, Bayer CropScience (Germany); BAYER® S-6876, Bayer CropScience (Germany); LE-MAT®, Bayer CropScience (Germany); SAFAST®, Sanonda Ltd. (Australia)
*Producers:* Agrimor International (USA); Bayer CropScience (Germany); Cangzhou Green Chemical Co.

(China); Hebei Huafeng Chemical Group (China); Sanonda Ltd. (Australia); Sanonda Zhengzhou Pesticide Co. Ltd. (China)

***Chemical Class:*** Organophosphate
***EPA/OPP PC Code:*** 035002
***California DPR Chemical Code:*** 2285
***Uses:*** Not registered in the U.S. A systemic insecticide and acaricide used to control a wide variety of sucking and biting insects, especially spider mites, which are resistant to some other organophosphorus pesticides. Used on pome, stone and citrus fruits, hops, cereals, rice, potatoes, vines, cotton, coffee, rice, sugar cane, ornamentals and other crops.
***Carcinogen/Hazard Classifications***
**U.S. EPA Carcinogens:** Group C, possible carcinogen (as dimethoate, its parent chemical)
**Label Signal Word:** DANGER
**WHO Acute Hazard:** Class 1b, highly hazardous
***Regulatory Authority:***
• DOT Inhalation Hazard Chemicals as organophosphate
***Description:*** Colorless to light yellow oily liquid. Leek- or mild, onion-like odor. Readily soluble in water. Melting point = 135°C. Molecular weight = 213.19. Density = 1.32 @ 20°C. Vapor pressure = 3.2 mPa @ 20°C. Log $K_{ow}$ = <1.0. Unlikely to bioaccumulate in marine organisms.
***Incompatibilities:*** Reacts violently with powerful oxidizers such as calcium hypochlorite. May react violently with antimony(V) pentafluoride. Incompatible with lead diacetate, magnesium, silver nitrate. Slowly hydrolyzes in water, releasing ammonia and forming acetate salts.
***Permissible Exposure Limits in Air:***
***Determination in Air:*** OSHA versatile sampler-2; Toluene/Acetone; Gas chromatography/Flame ionization detection; NIOSH IV, Method #5600, Organophosphorus Pesticides.[18]
***Permissible Concentration in Water:*** No criteria set. Runoff from spills or fire control may cause water pollution.
***Harmful Effects and Symptoms***
***Short Term Exposure:*** Eye pupils are small; blurred vision; eye watering; runny nose; cough; shortness of breath; salivation; dizziness; nausea, stomach cramps, diarrhea, and vomiting; increased blood pressure; profuse sweating; hypermotility, hallucinations; irritability; tingling of the skin; drowsiness; slow heartbeat; convulsions; fluid in lungs; loss of consciousness; incontinence; breathing stops; death. Organophosphates inhibit the action of acetylcholinesterase enzymes, and alter the way in which nervous impulses are transmitted. The effects can last for hours, days, or much longer. The action of the enzymes is reestablished after new enzymes are formed.
***Long Term Exposure:*** Cholinesterase inhibitor; cumulative effect is possible. Organophosphates may damage the nervous system with repeated exposure, resulting in convulsions, respiratory failure. May cause liver damage.
***Points of Attack:*** Respiratory system, central nervous system, cardiovascular system, blood cholinesterase.

***Medical Surveillance:*** Medical observation is recommended for 24 to 48 hours after breathing overexposure, as pulmonary edema may be delayed. As first aid for pulmonary edema, consider administering a corticosteroid spray. Cigarette smoking may exacerbate pulmonary injury and should be discouraged for at least 72 hours following exposure.
***First Aid:*** Speed in removing material from eyes and skin is of extreme importance. *Eye:* Contact can cause dangerous amounts of these chemicals to be quickly absorbed through the mucous membrane into the bloodstream. Immediately and gently flush eyes with plenty of warm or cold water (NO hot water) for at least 15 minutes, occasionally lifting the upper and lower eyelids. Get medical aid immediately. *Skin:* Get medical aid. Skin contact can cause dangerous amounts of these chemicals to be absorbed into the bloodstream. Wearing the appropriate PPE equipment and respirator for organophosphate pesticides, immediately flush skin with plenty of soap and water for at least 15 minutes while removing contaminated clothing and shoes. Shampoo hair promptly if contaminated. *Ingestion:* Call poison control. Loosen all clothing. Never give anything by mouth to an unconscious person. Get medical aid. Do NOT induce vomiting.* If conscious, alert, and able to swallow, rinse mouth and have victim drink 4 to 8 ounces of water do NOT induce vomiting but immediately administer slurry of activated charcoal (2 oz in 8 oz of water). If victim is *unconscious or having convulsions,* do nothing except keep victim warm. **In some cases you may be specifically instructed by poison control to induce vomiting by way of 2 tablespoons of syrup of ipecac (adult) washed down with a cup of water.* Do NOT give activated charcoal before or with ipecac syrup. Inhalation: Get medical aid. Do not contaminate yourself. Wearing the appropriate PPE equipment and respirator for organophosphate pesticides, immediately remove the victim from the contaminated area to fresh air. If the victim is not breathing, administer artificial respiration. Do NOT use mouth-to-mouth resuscitation; use bag/mask apparatus. If breathing is difficult, administer oxygen through bag/mask apparatus until medical help arrives. Do not leave victim unattended. *Note to physician or authorized medical personnel.* Administer atropine, 2 mg (1/30 gr) intramuscularly or intravenously as soon as any local or systemic signs or symptoms of an intoxication are noted; repeat the administration of atropine every 3 to 8 minutes until signs of atropinization (mydriasis, dry mouth, rapid pulse, hot and dry skin) occur; initiate treatment in children with 0.05 mg/kg of atropine; repeat at 5 to 10 minute intervals. Watch respiration, and remove bronchial secretions if they appear to be obstructing the airway; intubate if necessary. *Notes to physician or authorized medical personnel*: N-methylpyridinium-2-aldoxime (2-PAMCI) when used in conjunction with atropine reacts with the phosphorylated cholinesterase, thereby restoring normal activity to by

removing the phosphorylating group. The combination of these two chemicals is synergistic and must be administered within minutes to a few hours following exposure (depending on the specific agent) to be effective. Give 2-PAMCI (Pralidoxime; Protopam), 2.5 gm in 100 ml of sterile water or in 5% dextrose and water, intravenously, slowly, in 15-30 minutes; if sufficient fluid is not available, give 1 gm of 2-PAMCI in 3 ml of distilled water by deep intramuscular injection; repeat this every half hour if respiration weakens or if muscle fasciculation or convulsions recur. Also Diazepam, an anticonvulsant, might be considered.

*References:*
- International Programme on Chemical Safety (IPCS), "Pesticide Residues in Food, Omethoate," Geneva, Switzerland (1984). http://www.inchem.org/documents/jmpr/jmpmono/v84pr32.htm
- California Environmental Protection Agency "Chemical List of Lists," Sacramento CA (February 1997)

# Oryzalin (ANSI)

*Use Type:* Herbicide
*CAS Number:* 19044-88-3
*Formula:* $C_{12}H_{18}N_4O_6S$
*Synonyms:* Benzenesulfonamide, 4-(dipropylamino)-3,5-dinitro-; Caswell No. 623A; 3,5-Dinitro-$N^4,N^4$-dipropyl sulfanilamide; 3,5-Dinitro-$N^4,N^4$-dipropylsulfanilamide; 3,5-Dinitro-$N^4,N^4$-dipropyl sulphanilamide; 4-(Dipropylamino)-3,5-dinitrobenzene sulfonamide; Sulfanilamide, 3,5-dinitro-$N^4,N^4$-dipropyl-
*Trade Names:* AGVALUE®; COMPOUND 67019®; DIRIMAL®; EL-119®; EXCEL-S-PLUS®; EXPEDITE®, Monsanto (USA); FLEXLAN®, Dow AgroSciences (USA); NATIONS AG II®; ORYZA®; ORYZALIN®, Dow AgroSciences (USA); PRO-TECK®, Dow AgroSciences (USA); ROUT®, Scotts Company (USA); RYCELAN®; RYZELAN®; SNAPSHOT®, Dow AgroSciences (USA); SURFLAN®, Dow AgroSciences (USA) and United Phosphorus (India); TURF FERTILIZER®, Dow AgroSciences (USA); XL 2G®, Dow AgroSciences (USA)
*Producers:* Dow AgroSciences (USA); Monsanto (USA); Scotts Company (USA); United Phosphorus (India)
*Chemical Class:* Dinitroaniline
*EPA/OPP PC Code:* 104201
*California DPR Chemical Code:* 1868
*RTECS Number:* WO9350000
*Uses:* Oryzalin is used to control annual grasses, herbaceous plants, woody shrubs, vines and broadleaf weeds on fruit and nut trees, soya beans, peas, sweet potatoes, berries, vine and crops, cotton, Christmas tree plantations, commercial/industrial and recreation area lawns, golf course turf, residential lawns and turf, ornamental and/or shade trees, nonagricultural rights-of-way, nonagricultural uncultivated and industrial areas, power stations, paths/patios and paved areas.
*U.S. Maximum Allowable Residue Levels for Oryzalin (40 CFR 180.304):* The residual levels for the following crops are all 0.05 ppm: Almond, hulls; avocado; fig; fruit, pome; fruit, small and berry, group; fruit, stone, except cherry; fruit, stone, group 12; guava; kiwifruit; nut, tree, group 14; olive; papaya; pistachio; pomegranate.
*Carcinogen/Hazard Classifications*
**U.S. EPA Carcinogens:** Likely carcinogen
**Label Signal Word:** CAUTION or WARNING
**WHO Acute Hazard:** Class U, unlikely to be hazardous
*Regulatory Authority:*
- EPA 40 CFR 372.65, Specific Toxic Chemical Listings
- AB 1803-Well Monitoring Chemical (CAL)
- Actively registered pesticide in California.
- EPCRA Section 313 Form R *de minimis* concentration reporting level: 1.0%

*Description:* Yellow-orange crystalline solid. Odorless. Soluble in water; solubility = 2.49 mg/L. Melting/Freezing point = 138°C. Molecular weight = 346.36. Vapor pressure = $1.2 \times 10^{-8}$ mm @ 20°C.
*Incompatibilities:* Slowly hydrolyzes in water, releasing ammonia and forming acetate salts.
*Permissible Concentration in Water:* No criteria set. Runoff from spills or fire control may cause water pollution.
*Harmful Effects and Symptoms*
*Short Term Exposure:* Slight irritant to skin, eyes and mucous membranes.
*First Aid:* If this chemical gets into the eyes, remove any contact lenses at once and irrigate immediately for at least 15 minutes, occasionally lifting upper and lower lids. Seek medical attention immediately. If this chemical contacts the skin, remove contaminated clothing and wash immediately with soap and water. Seek medical attention immediately. If this chemical has been inhaled, remove from exposure, begin rescue breathing (using universal precautions) if breathing has stopped, and CPR if heart action has stopped. Transfer promptly to a medical facility. When this chemical has been swallowed, get medical attention. Give large quantities of water and induce vomiting. Do not make an unconscious person vomit.
*References:*
- U.S. Environmental Protection Agency, "Reregistration Eligibility Decision (RED), Oryzalin," Office of Prevention, Pesticides and Toxic Substances, Washington, DC (September 1994). http://www.epa.gov/REDs/0186.pdf
- U.S. Environmental Protection Agency, Office of Pesticide Programs, Pesticide Residue Limits, "Oryzalin," 40 CFR 180.304, www.epa.gov/pesticides/food/viewtols.htm
- EXTOXNET, Extension Toxicology Network, "Pesticide Information Profile, Oryzalin," Oregon State University,

University, Corvallis, OR (June 1996). http://extoxnet.orst.edu/pips/oryzalin.htm
- California Environmental Protection Agency "Chemical List of Lists," Sacramento CA (February 1997).

# Oxadiazon (ANSI)

*Use Type:* Herbicide
*CAS Number:* 19666-30-9
*Formula:* $C_{15}H_{18}Cl_2N_2O_3$
*Synonyms:* 2-*tert*-Butyl-4-(2,4-dichloro-5-isopropoxyphenyl)-δ(sup2)-1,3,4-oxadiazoline-5-one; 5-*tert*-BUTYL-3-(2,4-dichloro-5-isopropoxyphenyl)-1,3,4-oxadiazol-2(3*H*)-one; 2-*tert*-Butyl-4-(2,4-dichloro-5-isopropoxyphenyl)-δ(*sup2*)-1,3,4-oxadiazolin-5-one; 2-*tert*-Butyl-4-(2,4-dichloro-5-isopropyloxyphenyl)-1,3,4-oxadiazolin-5-one; CASWELL No. 624A; 3-(2,4-Dichloro-5-isopropyloxy-phenyl)-δ(sup4)4-5-(*tert*-butyl)-1,3,4-oxadiaz oline-2-one; 3-(2,4-Dichloro-5-(1-methylethoxy)phenyl)-5-(1,1-dimethylethyl)-1,3,4-oxadiazol-2(3*H*)-one; δ(sup2)2-1,3,4-Oxadiazolin-5-one, 2-*tert*-butyl-4-(2,4-dichloro-5-isopropyloxyphenyl)-; δ(sup2)2-1,3,4-Oxadiazolin-5-one,2-*tert*-butyl-4-(2,4-dichloro-5-isopropoxyphenyl)-; 1,3,4-Oxadiazol-2(3*H*)-one, 3-(2,4-dichloro-5-(1-methylethoxy)phenyl)-5-(1,1-dimethylethyl)-; Oxadiazon; Oxadiazone; 1,3,4-Oxazol-2(3*H*)-one, 3-[2,4-dichloro-5-(1-methylethoxy)phenyl]-5-(1,1-dimethylethyl)-
*Trade Names:* CARPETMAKER®; CHIP SHOT®, canceled; GOLD KIST®; PAR EX®, Lebanon Seaboard (USA); PRO GROW®, Scotts Company (USA); REGAL O-O®; REGALSTAR®; RONSTAR®, Bayer CropScience (Germany); RP-17623®; TURFIC®; VERTAGREEN®; WILBRO®
*Producers:* Bayer CropScience (Germany); Lebanon Seaboard (USA); The Scotts Company (USA)
*Chemical Class:* Oxadiazolinone
*EPA/OPP PC Code:* 109001; (597900 old EPA code number)
*California DPR Chemical Code:* 2017
*RTECS Number:* RO0874000
*Uses:* A pre-emergence and early post-emergence control for annual grasses, sedges and broadleaf weeds. Originally registered for use on turf and ornamentals and has wide use on golf courses.

*Carcinogen/Hazard Classifications*
**U.S. EPA Carcinogens:** Likely carcinogen
*California Prop. 65:* Carcinogen and developmental toxin
**Label Signal Word:** CAUTION, WARNING
**WHO Acute Hazard:** Class U, unlikely to be hazardous
*Regulatory Authority:*
- EPA 40 CFR 372.65, Specific Toxic Chemical Listings
- Proposition 65 chemical (CAL)
- Actively registered pesticide in California.
- EPCRA Section 313 Form R *de minimis* concentration reporting level: 1.0%

*Description:* Colorless crystalline solid. Odorless. Solubility in water; solubility = 0.7 mg/L @ 20°C. Molecular weight = 345.22. Melting/Freezing point = 89°C. Vapor pressure = $1.1 \times 10^{-6}$ mm @ 20°C.
*Determination in Air:* Filter; none; Gravimetric; NIOSH IV [Particulates NOR; #0500 (total), #0600 (respirable)].[18]
*Permissible Concentration in Water:* No criteria set. Runoff from spills or fire control may cause water pollution.
*First Aid:* If this chemical gets into the eyes, remove any contact lenses at once and irrigate immediately for at least 15 minutes, occasionally lifting upper and lower lids. Seek medical attention immediately. If this chemical contacts the skin, remove contaminated clothing and wash immediately with soap and water. Seek medical attention immediately. If this chemical has been inhaled, remove from exposure, begin rescue breathing (using universal precautions) if breathing has stopped, and CPR if heart action has stopped. Transfer promptly to a medical facility. When this chemical has been swallowed, get medical attention. Give large quantities of water and induce vomiting. Do not make an unconscious person vomit. Do not induce vomiting when formulations containing petroleum solvents are ingested.
*References:*
- Pesticide Management Education Program, "Oxadiazon (Ronstar) Herbicide Profile 2/85," Cornell University, Ithaca, NY (February 1985). http://pmep.cce.cornell.edu/profiles/herb-growthreg/naa-rimsulfuron/oxadiazon/herb-prof-oxadiazon.html
- California Environmental Protection Agency "Chemical List of Lists," Sacramento CA (February 1997)

# Oxadixyl

*Use Type:* Fungicide
*CAS Number:* 77732-09-3
*Formula:* $C_{14}H_{18}N_2O_4$
*Alert:* During 2001, U.S. registrants of oxadixyl requested voluntary cancellation of their products, which were effective September 27, 2002.
*Synonyms:* Acetamide, *N*-(2,6-dimethylphenyl)-2-methoxy-*N*-(2-*oxo*-3-oxazolininyl)-; *N*-(2,6-Dimethylphenyl)-2-methoxy-*N*-(2-*oxo*-3-oxazolidinyl)acetamide; 2-Methoxy-*N*-(2-*oxo*-1,3-oxazolidine-3-yl)-acet-2,6-xylidine
*Trade Names:* ANCHOR, Gustafson (USA), canceled 9/27/01; M 10797®; OXADIXL, Syngenta (Switzerland), canceled 9/27/02; RECOIL®; RIPOST®; SAN-371®; SAN-371F®; SANDOFAN®, Syngenta (Switzerland), canceled 2/14/97; WAKIL®
*Producers:* Gustafson (USA); Syngenta (Switzerland)
*Chemical Class:* Anilide
*EPA/OPP PC Code:* 126701

*California DPR Chemical Code:* 3539
*Uses:* Not registered in the U.S. A systemic fungicide for seed treatment. Used to control downy mildew, damping-off, seed rot and phytophthora pest species on a variety of vegetable crops, cotton, grasses, cereals, and golf course and residential turf.
*U.S. Maximum Allowable Residue Levels for Oxadixyl (40 CFR 180.456):*
Residue levels for the following crops have been set at 0.1 ppm: Grass forage, fodder and hay group.
*Carcinogen/Hazard Classifications*
**U.S. EPA Carcinogens:** Group C, possible carcinogen
**Label Signal Word:** CAUTION
**WHO Acute Hazard:** Class III, slightly hazardous
*Description:* Soluble in water.
*Incompatibilities:* Slowly hydrolyzes in water, releasing ammonia and forming acetate salts.
*Permissible Concentration in Water:* No criteria set. Runoff from spills or fire control may cause water pollution.
*First Aid:* If this chemical gets into the eyes, remove any contact lenses at once and irrigate immediately for at least 15 minutes, occasionally lifting upper and lower lids. Seek medical attention immediately. If this chemical contacts the skin, remove contaminated clothing and wash immediately with soap and water. Seek medical attention immediately. If this chemical has been inhaled, remove from exposure, begin rescue breathing (using universal precautions) if breathing has stopped, and CPR if heart action has stopped. Transfer promptly to a medical facility. When this chemical has been swallowed, get medical attention. Give large quantities of water and induce vomiting. Do not make an unconscious person vomit. Do not induce vomiting when formulations containing petroleum solvents are ingested.
*References:*
- U.S. Environmental Protection Agency, Office of Pesticide Programs, Pesticide Residue Limits, "Oxadixyl," 40 CFR 180.456, www.epa.gov/pesticides/food/viewtols.htm
- U.S. Environmental Protection Agency, "Reregistration Eligibility Decision (RED) Fact Sheet, Oxadixyl," Office of Prevention, Pesticides and Toxic Substances, Washington, DC (October 2001). http://www.epa.gov/REDs/factsheets/oxadixyl_fs.htm
- California Environmental Protection Agency "Chemical List of Lists," Sacramento CA (February 1997)

# Oxamyl (ANSI)

*Use Type:* Insecticide, nematicide and acaricide
*CAS Number:* 23135-22-0
*Formula:* $C_7H_{13}N_3O_3S$
*Alert:* Most products of oxamyl are classified as Restricted Use Pesticides (RUP)
*Synonyms:* 2-(Dimethylamino)-N[(((methylamino)carbonyl)oxy]2-oxoethanimidothioic acid methyl ester; 2-Dimethylamino-1-(methylamino)glyoxal-O-methylcarbamoylmonoxime; N,N-Dimethyl-α-methylcarbamoyloxyimino-α-(methylthio)acetamide; N,N-dimethyl-N-[(methylcarbamoyl)oxy]-1-thiooxamimidic acid methyl ester; Methyl-2-(dimethylamino)-N-[((methylamino)carbonyl)oxy]-2-oxoethanimidothioate; Methyl-1-(dimethylcarbamoyl)-N-(methylcarbamoyloxy)thioformimidate; S-Methyl-1-(dimethylcarbamoyl)-N-[(methylcarbamoyl)oxy]thioformimidate; Methyl-N,N'-dimethyl-N-[(methylcarbamoyl)oxy]-1-thiooxamimidate
*Trade Names:* BLADE®; D-1410®, DuPont Crop Protection (USA); DPX 1410®, DuPont Crop Protection (USA); INSECTICIDE-NEMACIDE 1410®, DuPont Crop Protection (USA); OXAMYL CARBAMATE INSECTICIDE®; THIOXAMYL®; VYDATE®, DuPont Crop Protection (USA); VYDATE 10G®, DuPont Crop Protection (USA); VYDATE L®, DuPont Crop Protection (USA); VYDATE INSECTICIDE/NEMATICIDE®, DuPont Crop Protection (USA); VYDATE OXAMYL INSECTICIDE/NEMATOCIDE®, DuPont Crop Protection (USA), canceled 10/10/1989
*Producers:* DuPont Crop Protection (USA)
*Chemical Class:* Carbamate
*EPA/OPP PC Code:* 103801
*California DPR Chemical Code:* 1910
*RTECS Number:* RP2300000
*EINECS Number:* 245-445-3
*Uses:* A systemic and contact insecticide/acaricide and nematicide, oxamyl is a restricted use pesticide used on apples, bananas, carrots, celery, citrus, cotton, cucumbers, eggplants, garlic, ginger, muskmelon (including cantaloupe and honeydew melon), onion (dry bulb), peanuts, pears, peppers, peppermint, pineapples, plantains, potatoes, pumpkins, soybeans, spearmint, squash, sweet potatoes, tobacco, tomatoes, watermelons, yams. Oxamyl is also used on Non-bearing apple, cherry, citrus, peach, pear, and tobacco. It is applied directly onto plants or the soil surface. It is available in both liquid and granular form, but the granular form is banned in the U.S. It has no residential use.
*U.S. Maximum Allowable Residue Levels for Oxamyl (40 CFR 180.303):*

| CROP | ppm |
| --- | --- |
| Apple | 2 |
| Banana | 0.3 |
| Cantaloupe | 2 |
| Celery | 3 |
| Cotton, undelinted seed | 0.2 |
| Cucumber | 2 |
| Eggplant | 2 |
| Fruit, stone, group 12 | 3 |
| Honeydew | 2 |

| | |
|---|---|
| Peanut | 0.2 |
| Peanut, hay | 2 |
| Pear | 2 |
| Pepper, bell | 3 |
| Pepper, nonbell | 5 |
| Peppermint, tops | 10 |
| Pineapple | 1 |
| Pineapple, bran | 6 |
| Potato | 0.1 |
| Pumpkin | 2 |
| Soybean | 0.2 |
| Spearmint, tops | 10 |
| Squash, summer | 2 |
| Squash, winter | 2 |
| Tomato | 2 |
| Vegetable, root crop | 0.1 |
| Watermelon | 2 |

*Carcinogen/Hazard Classifications*
**U.S. EPA Carcinogens:** Group E, unlikely carcinogen
**Label Signal Word:** DANGER–POISON
**WHO Acute Hazard:** Class 1B, highly hazardous
*Regulatory Authority:*
- EPA Hazardous Waste Number (RCRA No.): P194
- AB 1803-Well Monitoring Chemical (CAL)
- MCL (Maximum Contaminants Levels) list of contaminants (CAL)
- AB 1803-Well Monitoring Chemical (CAL)
- Actively registered pesticide in California.
- RCRA, 40CFR261, Appendix 8 Hazardous Constituents
- RCRA 40CFR268.48; 61FR15654, Universal Treatment Standards: Wastewater (mg/L), 0.056; Nonwastewater (mg/kg), 0.28
- Superfund/EPCRA 40CFR355, Appendix B Extremely Hazardous Substances: TPQ = 100/10,000 lb (45.4/4,540 kg)
- Superfund/EPCRA 40CFR302.4 RQ: EHS, 1 lb (0.454 kg)

*Description:* White crystalline solid. Sulfur- or garlic-like odor. Slightly soluble in water; solubility = 2.6 x $10^6$ ppm @ 25°C. Molecular weight = 219.35. Melting/Freezing point = 100°-102°C. Vapor pressure = 2.98 x $10^{-4}$ mmHg @ 25°C. Hazard Identification (based on NFPA-704 M Rating System): Health 4, Flammability 1, Reactivity 0.
*Incompatibilities:* Strong oxidizers may cause fire and explosion.
*Permissible Exposure Limits in Air:* No standards set.
*Permissible Concentration in Water:* A lifetime health advisory of 175 μg/L has been developed by EPA[47]. In addition, Massachusetts has set a guideline for oxamyl in drinking water[61] of 50 μg/L.
*Routes of Entry:* Inhalation, ingestion, skin and/or eye contact.
*Harmful Effects and Symptoms*
*Short Term Exposure:* Contact can cause skin and eye irritation. Acute exposure to oxamyl usually leads to a cholinergic crisis. Signs and symptoms may include increased salivation, lacrimation (tearing), perspiration, spontaneous defecation, and spontaneous urination. Pinpoint pupils, blurred vision, tremor, muscle twitching, mental confusion, convulsions, and coma may occur. Gastrointestinal symptoms include abdominal pain, diarrhea, nausea, and vomiting. Bradycardia (slow heart rate) is common. Dyspnea (shortness of breath) and pulmonary edema may also occur. Classified by the World Health Organization as highly hazardous. Has also been rated as extremely- to super-toxic. Acute oral exposure (ingestion) to oxamyl has caused death. Carbamate insecticides inhibit the cholinesterase activity of enzymes, causing accumulation of acetylcholine at synapses and altering the way in which nervous impulses are transmitted. However, within several hours carbamates spontaneously detach from the enzymes.
*Long Term Exposure:* Cholinesterase inhibitor; cumulative effect is possible. This chemical may damage the nervous system causing numbness and/or weakness in the hands and feet. Repeated exposure may cause personality changes with depression, anxiety, and irritability. May cause liver damage.
*Points of Attack:* Respiratory system, lungs, central nervous system, cardiovascular system, skin, eyes, liver, plasma and red blood cell cholinesterase.
*Medical Surveillance:* Before employment and at regular times after that, the following are recommended: Plasma and red blood cell cholinesterase levels (tests for the enzyme poisoned by this chemical). If exposure stops, plasma levels return to normal in 1-2 weeks while red blood cell levels may be reduced for 1-3 months. When acetylcholinesterase enzyme levels are reduced by 25% or more below preemployment levels, risk of poisoning is increased, even if results are in lower ranges of "normal." Reassignment to work not involving carbamate pesticides is recommended until enzyme levels recover. If symptoms develop or overexposure occurs, repeat the above tests as soon as possible and get an exam of the nervous system. Also consider complete blood count. Consider chest x-ray following acute overexposure. Consider liver function tests.
*First Aid:* Speed in removing material from eyes and skin is of extreme importance. *Eyes:* Eye contact can cause dangerous amounts of these chemicals to be quickly absorbed through the mucous membrane into the bloodstream. Immediately and gently flush eyes with plenty of warm or cold water (NO hot water) for at least 15 minutes, occasionally lifting the upper and lower eyelids. Get medical aid immediately. *Skin:* Get medical aid. Skin contact can cause dangerous amounts of these chemicals to be absorbed into the bloodstream. Wearing the appropriate PPE equipment and respirator for carbamate pesticides, immediately flush skin with plenty of soap and water for at least 15 minutes while removing contaminated clothing and shoes. Shampoo hair promptly if contaminated; protect

eyes. *Ingestion:* Call poison control. Loosen all clothing. Never give anything by mouth to an unconscious person. Get medical aid. Do NOT induce vomiting.* If conscious, alert, and able to swallow, rinse mouth and have victim drink 4 to 8 ounces of water. Check to see if poison control instructs you to use ipecac syrup, otherwise administer slurry of activated charcoal (2 oz in 8 oz of water). If victim is UNCONSCIOUS OR HAVING CONVULSIONS, do nothing except keep victim warm. *In some cases you may be specifically instructed by poison control to induce vomiting by way of 2 tablespoons of syrup of ipecac (adult) washed down with a cup of water.* Do NOT give activated charcoal before or with ipecac syrup. *Inhalation:* Get medical aid. Do not contaminate yourself. Wearing the appropriate PPE equipment and respirator for carbamate pesticides, immediately remove the victim from the contaminated area to fresh air. If the victim is not breathing, administer artificial respiration. Do NOT use mouth-to-mouth resuscitation; use bag/mask apparatus. If breathing is difficult, administer oxygen through bag/mask apparatus until medical help arrives. Do not leave victim unattended. *Note to physician or authorized medical personnel.* Administer atropine, 2 mg (1/30 gr) intramuscularly or intravenously as soon as any local or systemic signs or symptoms of an intoxication are noted; repeat the administration of atropine every 3 to 8 minutes until signs of atropinization (mydriasis, dry mouth, rapid pulse, hot and dry skin) occur; initiate treatment in children with 0.05 mg mg/kg of atropine; repeat at 5 to 10 minute intervals. Watch respiration, and remove bronchial secretions if they appear to be obstructing the airway; intubate if necessary. *Medical note:* 2-PAMCI may be contraindicated in the case of some carbamate poisonings.

### References:
- Environmental Protection Agency, "Interim Reregistration Eligibility Decision (IRED), Oxamyl," Washington, DC (October 2000), http://www.epa.gov/REDs/0253ired.pdf
- U.S. Environmental Protection Agency, Office of Pesticide Programs, Pesticide Residue Limits, "Oxamyl," 40 CFR 180.303, http://www.epa.gov/pesticides/food/viewtols.htm
- EXTOXNET, Extension Toxicology Network, "Pesticide Information Profile, Oxamyl," Oregon State University, Corvallis, OR (June 1996). http://extoxnet.orst.edu/pips/oxamyl.htm
- New Jersey Department of Health and Senior Services, "Hazardous Substance Fact Sheet, OXAMYL," Trenton NJ (July 1999). http://www.state.nj.us/health/eoh/rtkweb/2618.pdf
- U.S. Environmental Protection Agency, "Chemical Profile: Oxamyl," Washington, DC, Chemical Emergency Preparedness Program (Nov. 30, 1987).
- California Environmental Protection Agency "Chemical List of Lists," Sacramento CA (February 1997).

# Oxycarboxin (ANSI)

*Use Type:* Fungicide
*CAS Number:* 5259-88-1
*Formula:* $C_{12}H_{13}NO_4S$
*Synonyms:* Carboxin sulfone; DCMOD; 2,3-Dihydro-5-carboxanilido-6-methyl-1,4-oxathiin-4,4-dioxide; 5,6-Dihydro-2-methyl-3-carboxanilido-1,4-oxathiin-4,4-dioxid (German); 5,6-Dihydro-2-methyl-1,4-oxathiin-3-carboxanilide 4,4-dioxide; 5,6-Dihydro-2-methyl-N-phenyl-1,4-oxathiin-3-carboxamide-4,4-dioxide; 1,4-Oxathiin-3-carboxanilide, 5,6-dihydro-2-methyl-, 4,4-dioxide; 1,4-Oxathiin-3-carboxamide, 5,6-dihydro-2-methyl-N-phenyl-, 4, 4-dioxide; Oxycarboxine
*Trade Names:* CARBOJECT®, J.J. Mauget (USA), canceled; F461®; FUNGISOL®, J.J. Mauget (USA), canceled; PLANTVAX®, Crompton Corporation (USA); PLANT WAX®; VITAVEX®
*Producers:* Agsin (Singapore); Crompton Corporation (USA); J. J. Mauget (USA)
*Chemical Class:* Carboxamide
*EPA/OPP PC Code:* 090202
*California DPR Chemical Code:* 1434
*Uses:* Registered as a foliar systemic fungicide for use against rust on carnations and greenhouse geraniums.
*Carcinogen/Hazard Classifications*
**Label Signal Word:** CAUTION
**WHO Acute Hazard:** Class U, unlikely to be hazardous
*Description:* Colorless crystals. Melting point = 127.5–130°C. Molecular weight = 267.3.
*Permissible Concentration in Water:* No criteria set. Runoff from spills or fire control may cause water pollution.
*First Aid:* If this chemical gets into the eyes, remove any contact lenses at once and irrigate immediately for at least 15 minutes, occasionally lifting upper and lower lids. Seek medical attention immediately. If this chemical contacts the skin, remove contaminated clothing and wash immediately with soap and water. Seek medical attention immediately. If this chemical has been inhaled, remove from exposure, begin rescue breathing (using universal precautions) if breathing has stopped, and CPR if heart action has stopped. Transfer promptly to a medical facility. When this chemical has been swallowed, get medical attention. Give large quantities of water and induce vomiting. Do not make an unconscious person vomit. Do not induce vomiting when formulations containing petroleum solvents are ingested.

### References:
- Pesticide Management Education Program, "Oxycarboxin (Plantvax) Chemical Profile 2/85," Cornell University, Ithaca, NY (February 1985). http://pmep.cce.cornell.edu/profiles/fung-nemat/febuconazole-sulfur/oxycarboxin/fung-prof-oxycarboxin.html

- California Environmental Protection Agency "Chemical List of Lists," Sacramento CA (February 1997)

# Oxydisulfoton

*Use Type:* Insecticide
*CAS Number:* 2497-07-6
*Formula:* $C_8H_{19}OsPS_3$
*Alert:* Not registered in the U.S.
*Synonyms:* $O,O$-Dieyhyl-$S$-[(ethylsulfinyl)ethyl] phosphorodithioate; $O,O$-Diethyl $S$-[2-(ethylsulfinyl)ethyl] phosphorodithioate; Disulfoton disulfide; Disulfoton sulfoxide; Ethylthiomelton sulfoxide
*Trade Names:* BAY 23323®, Bayer CropScience (Germany); DISYSTON SULFOXIDE®
*Producers:* Bayer CropScience (Germany)
*Chemical Class:* Organophosphate
*EPA/OPP PC Code:* 340200
*RTECS Number:* TD8600000
*Carcinogen/Hazard Classifications*
U.S. EPA Carcinogens: As disulfoton: Group E, unlikely carcinogen
WHO Acute Hazard: As disulfoton: Class 1 a, extremely hazardous
*Regulatory Authority:*
- Very Toxic Substance (World Bank)[15]
- U.S. DOT Inhalation Hazard Chemicals as organophosphates
- Superfund/EPCRA 40CFR355, Appendix B Extremely Hazardous Substances: TPQ = 500 lb (227kg)
- Superfund/EPCRA 40CFR302.4 RQ: EHS, 1 lb (0.454 kg)

*Description:* Oxydisulfoton, $C_8H_{19}OsPS_3$ is a combustible liquid. Molecular weight = 290.43. Hazard Identification (based on NFPA-704 M Rating System): Health 4, Flammability 1, Reactivity 0.
*Incompatibilities:* Strong oxidizers.
*Permissible Exposure Limits in Air:* No standards set.
*Permissible Concentration in Water:* No criteria set, but runoff from spills or fire control may cause water pollution.
*Routes of Entry:* Inhalation, ingestion, skin and/or eye contact.
*Harmful Effects and Symptoms*
*Short Term Exposure:* Organic phosphorus insecticides are absorbed by the skin, as well as by the respiratory and gastrointestinal tracts. They are cholinesterase inhibitors. Symptoms include the following: *mild exposure*: headache, loss of appetite, nausea, dizziness; *moderate exposure*: abdominal cramps, diarrhea, salivation, excessive tearing, muscular cramps; *severe exposure*: fever, blue lips, lack of sphincter control, coma, heart shock, difficult breathing. Delayed pulmonary edema may occur after inhalation.
*Long Term Exposure:* Cholinesterase inhibitor; cumulative effect is possible. This chemical may damage the nervous system with repeated exposure, resulting in convulsions, respiratory failure. May cause liver damage.
*Points of Attack:* Respiratory system, lungs, central nervous system, cardiovascular system, skin, eyes, plasma and red blood cell cholinesterase.
*Medical Surveillance:* Medical observation is recommended for 24 to 48 hours after breathing overexposure, as pulmonary edema may be delayed. Before employment and at regular times after that, the following are recommended: Plasma and red blood cell cholinesterase levels (tests for the enzyme poisoned by this chemical). If exposure stops, plasma levels return to normal in 1-2 weeks while red blood cell levels may be reduced for 1-3 months. When acetylcholinesterase enzyme levels are reduced by 25% or more below preemployment levels, risk of poisoning is increased, even if results are in lower ranges of "normal." Reassignment to work not involving organophosphate or carbamate pesticides is recommended until enzyme levels recover. If symptoms develop or overexposure occurs, repeat the above tests as soon as possible and get an exam of the nervous system. Also consider complete blood count. Consider chest x-ray following acute overexposure. Do not drink any alcoholic beverages before or during use. Alcohol promotes absorption of organophosphates.
*First Aid:* **Treatment for organophosphate poisoning consists of thorough decontamination, cardiorespiratory support, and administration of the antidotes atropine and pralidoxime. In cases of severe poisoning, diazepam, an anticonvulsant, should also be administered. Antidotes should be administered as prevention even if the diagnosis is in doubt.** Speed in removing material from eyes and skin is of extreme importance. *Eyes:* Eye contact can cause dangerous amounts of these chemicals to be quickly absorbed through the mucous membrane into the bloodstream. Immediately and gently flush eyes with plenty of warm or cold water (NO hot water) for at least 15 minutes, occasionally lifting the upper and lower eyelids. Get medical aid immediately. *Skin:* Get medical aid. Skin contact can cause dangerous amounts of these chemicals to be absorbed into the bloodstream. Wearing the appropriate PPE equipment and respirator for organophosphate pesticides, immediately flush skin with plenty of soap and water for at least 15 minutes while removing contaminated clothing and shoes. Shampoo hair promptly if contaminated. The removed, contaminated clothing and shoes should be double-bagged and left in Hot Zone for later disposal by hazardous materials experts. Skin may also be decontaminated with diluted hypochlorite solution. *Inhalation:* Get medical aid. Do not contaminate yourself. Wearing the appropriate PPE equipment and respirator for organophosphate pesticides, immediately remove the victim from the contaminated area to fresh air. If the victim is not breathing, administer artificial respiration. Do NOT use mouth-to-mouth resuscitation; use bag/mask apparatus. If

breathing is difficult, administer oxygen through bag/mask apparatus until medical help arrives. Do not leave victim unattended. *Ingestion:* Call poison control. Loosen all clothing. Never give anything by mouth to an unconscious person. If victim is *unconscious or having convulsions,* do nothing except keep victim warm. Get medical aid. Transfer promptly to a medical facility. In cases of ingestion, **do not induce vomiting**. If the victim is alert and asymptomatic, administer a slurry of activated charcoal at a dose of 1 g/kg (infant, child, and adult dose). A soda can and straw may be of assistance when offering charcoal to a child. *In some cases you may be specifically instructed by poison control to induce vomiting by way of 2 tablespoons of syrup of ipecac (adult) washed down with a cup of water.* Do NOT give activated charcoal before or with ipecac syrup.

*Note to physician or authorized medical personnel:* Treat cases of respiratory compromise, coma, or excessive pulmonary secretions with respiratory support using protocols and techniques available and within the scope of training. Some cases may necessitate procedures such as endotracheal intubation or cricothyrotomy by properly trained and equipped personnel. When possible, atropine (see *Antidotes*, below) should be given under medical supervision. Patients who are comatose, hypotensive, or having seizures or cardiac arrhythmias should be treated according to advanced life support protocols. *Antidotes:* Two antidotes are administered to treat organophosphate poisoning. Atropine is a competitive antagonist of acetylcholine at muscarinic receptors and is used to control the excessive bronchial secretions which are often responsible for death. Pralidoxime relieves both the nicotinic and muscarine effects of organophosphate poisoning by regenerating acetylcholinesterase and can reduce both the bronchial secretions and the muscle weakness associated with poisoning. The initial intravenous dose of atropine in adults should be determined by the severity of symptoms: An initial adult dose of 1.0 to 2.0 mg or pediatric dose of 0.01 mg/kg (minimum 0.01 mg) should be administered intravenously. If intravenous access cannot be established, atropine may also be given intramuscularly, subcutaneously or via endotracheal tube. Doses should be repeated every 15 minutes until excessive secretions and sweating have been controlled. Once bronchial secretion has been controlled, atropine administration should be repeated whenever the secretions begin to recur. In seriously poisoned patients, very large doses may be required. Alterations of pulse rate and pupillary size should not be used as indicators of treatment adequacy. Pralidoxime should be administered as early in poisoning as possible as its efficacy may diminish when given more than 24 to 36 hours after exposure. Doses are as follows: adult 1.0 g; pediatric 25 to 50 mg/kg. The drug should be administered intravenously over 30 to 60 minutes, but in a life-threatening situation, one-half of the total dose can be given per minute for a total administration time of 2 minutes. Treatment should begin to take effect within 40 minutes with a reduction in symptoms and the amount of atropine necessary to control bronchial secretion. The initial dose can be repeated in 1 hour and then every 8 to 12 hours until the patient is clinically well and no longer requires atropine. If intravenous access cannot be established, pralidoxime may also be given intramuscularly. Early administration of diazepam in addition to the combined atropine and pralidoxime treatment may help prevent the onset of seizures and potential brain and cardiac morphologic damage following high-level organophosphate poisoning.

*References:*
- New Jersey Department of Health and Senior Services, "Hazardous Substance Fact Sheet, Oxydisulfoton," Trenton, NJ (December 2001). http://www.state.nj.us/health/eoh/rtkweb/2625.pdf
- U.S. Environmental Protection Agency, "Chemical Profile: Oxydisulfoton," Washington, DC, Chemical Emergency Preparedness Program (Nov. 30, 1987).
- California Environmental Protection Agency "Chemical List of Lists," Sacramento CA (February 1997).
- Agency for Toxic Substances and Disease Registry, U.S. Department of Health and Human Services, Public Health Service, "Managing Hazardous Materials Incidents," Atlanta, GA (June 2003).

# Oxyfluorfen (ANSI)

*Use Type:* Herbicide
*CAS Number:* 42874-03-3
*Formula:* $C_{15}H_{11}ClF_3NO_4$
*Synonyms:* Benzene, 2-chloro-1-(3-ethoxy-4-nitrophenoxy)-4-(trifluoromethyl)- (9CI); Caswell No. 188AAA; 2-Chloro-1-(3-ethoxy-4-nitrophenoxy)-4-(trifluoromethyl)benzene; 2-Chloro-4-trifluoromethyl-3'-ethoxy-4'-nitrodiphenyl ether; 2-Chloro-$\alpha,\alpha,\alpha$-trifluoro-$p$-tolyl-3-ethoxy-4-nitrophenyl ether; Ether, 2-chloro-$\alpha,\alpha,\alpha$-trifluoro-$p$-tolyl-3-ethoxy-4-nitro phenyl; Oxyfluorfene; Oxyfluorofen

*Trade Names:* EDGER®; FIRE POWER® (glyphosate + oxyfluorfen), Monsanto (USA); GALIGAN®, Makhteshim-Agan Industries (Israel); GOAL®, Dow AgroSciences (USA); HADAF®, Veterinary & Agricultural Products Manufacturing Co. Ltd. (VAPCO) (Jordan); KLEENUP®, Bonide (USA); KOLTAR®; MON-78095®, Monsanto (USA); RH-2915®, Rohm and Haas (USA); RH-915®, Rohm and Haas (USA); ROUT® (with oryzalin), The Scotts Company (USA); TRIOX®, Ortho Business Group (USA)

*Producers:* Agsin (Singapore); Dow AgroSciences (USA); Bonide (USA); Epochem Co., (China); Ki-Hara Chemicals Ltd. (UK), Makhteshim-Agan Industries (Israel); Monsanto (USA); Ortho Business Group (USA); Rohm and Haas (USA); The Scotts Company (USA); Veterinary &

Agricultural Products Manufacturing Co. Ltd. (VAPCO) (Jordan)
*Chemical Class:* Diphenyl ether
*EPA/OPP PC Code:* 111601; (288600 old EPA code number)
*California DPR Chemical Code:* 1973
*RTECS Number:* DV4725000
*Uses:* Oxyfluorfen is used for broad spectrum pre-emergence and post-emergent control of annual broadleaf and grassy weeds in a variety of tree fruit, nut, vine, and field crops. The largest agricultural markets are wine grapes and almonds. Also used on ornamental and forestry sites. Oxyfluorfen is also used for weed control in landscapes, patios, driveways, and similar non-crop areas in residential, highway and rights-of-way sites.
*U.S. Maximum Allowable Residue Levels for Oxyfluorfen (40 CFR 180.381):*
Residue levels for the following crops have been set at 0.1 ppm: almond hulls, peppermint and spearmint.
Residue levels for the following crops have been set at 0.05 ppm: artichokes, avocados, bananas (including plantain), broccoli, cabbage, cattle, fat, mbyp, and meat, cauliflower, cocoa beans, coffee, corn, grain, cottonseed, dates, eggs, feijoa, figs, goat, fat, mbyp and meat, grapes, Hog, fat, mbyp and meat, horseradish, Horse, fat, mbyp and meat, kiwifruit, olives, onions (dry bulb), milk, persimmons, pistachios, pome fruits group, pomegranates, poultry, fat and meat, sheep, fat, mbyp and meat, soybeans, stone fruits group, tree nuts group (except almond hulls).
*Carcinogen/Hazard Classifications*
**U.S. EPA Carcinogens:** Group C, possible carcinogen
**Label Signal Word:** CAUTION or WARNING
**WHO Acute Hazard:** Class U, unlikely to be hazardous
*Regulatory Authority:*
- EPA 40 CFR 372.65, Specific Toxic Chemical Listings
- FIFRA, 40CFR185: tolerances for pesticides in food
- Actively registered pesticide in California.
- AB 2588-Air Toxics "Hot Spots" Chemicals (CAL) as chlorophenoxy pesticides
- The "Director's List" (CAL/OSHA) as chlorophenoxy pesticides
- EPCRA Section 313 Form R *de minimis* concentration reporting level:1.0%

*Description:* White to orange crystalline solid. Smoke-like odor. Slightly soluble in water; solubility = 0.098 ppm. Molecular weight = 361.70. Melting/Freezing point = 85°C. Vapor pressure = $2.1 \times 10^{-7}$ mm @ 20°C.
*Permissible Concentration in Water:* No criteria set. Runoff from spills or fire control may cause water pollution.
*Routes of Entry:* Inhalation, ingestion, absorbed through the skin.
*Harmful Effects and Symptoms*
*Short Term Exposure:* Poisonous; may be fatal if inhaled, swallowed, or absorbed through skin. Severely irritates eyes, skin and respiratory tract, with burning sensation, pain, redness and swelling. Metabolic stimulant. If inhaled, causes coughing, dilated pupils, headache, profuse persperation, intense thirst, extreme fatigue, rapid pulse, high fever, clammy, flushed skin, rapid breathing, nausea, vomiting, cyanosis (bluish tint to skin and lips), anxiety and confusion, convulsions, risk of lung edema. If swallowed, face and lips turn bluish. Liver injury with associated jaundice, kidney failure, and cardiac arrhythmias are commonly noted. Nerve damage, which may be delayed, may include swelling of legs and feet, muscle twitch and stupor. Severe exposure can cause death from heart failure. Dust or liquid left in contact with the skin for several hours may be absorbed. This may result in severe delayed symptoms as listed above. These symptoms may last for months or years.
*Long Term Exposure:* Workers exposed to chlorophenoxy compounds such as 2,4-D (in the manufacturing process) over a five to ten year period at levels above 10 mg/m$^3$ complained of weakness, rapid fatigue, headache and vertigo. Liver damage, low blood pressure and slowed heartbeat were also found. Based on animal tests, may affects human reproduction.
*Points of Attack:* Eyes, skin, respiratory system, central nervous system, cardiovascular system, liver, kidney
*Medical Surveillance:* If symptoms develop or overexposure is suspected, the following may be useful: Liver and kidney function tests. Exam of the nervous system.
*First Aid:* If this chemical gets into the eyes, remove any contact lenses at once and irrigate immediately for at least 15 minutes, occasionally lifting upper and lower lids. Seek medical attention immediately. If this chemical contacts the skin, remove contaminated clothing and wash immediately with soap and water. Seek medical attention immediately. If this chemical has been inhaled, remove from exposure, begin rescue breathing (using universal precautions) if breathing has stopped, and CPR if heart action has stopped. Transfer promptly to a medical facility. When this chemical has been swallowed, get medical attention. *Do not induce vomiting when formulations containing petroleum solvents are ingested.* Otherwise, give large quantities of water and induce vomiting. Do not make an unconscious person vomit.
*References:*
- U.S. Environmental Protection Agency, "Reregistration Eligibility Decision (RED) Fact Sheet, Oxyfluorfen," Office of Prevention, Pesticides and Toxic Substances, Washington, DC (October 2002). http://www.epa.gov/REDs/factsheets/oxyfluorfen_red_fs.htm
- EXTOXNET, Extension Toxicology Network, "Pesticide Information Profile, Oxyfluorfen," Oregon State University, Corvallis, OR (June 1996). http://extoxnet.orst.edu/pips/oxyfluor.htm

- California Environmental Protection Agency "Chemical List of Lists," Sacramento CA (February 1997)

## Oxytetracycline Calcium

*Use Type:* Fungicide
*CAS Number:* 79-57-2 (oxytetracycline); 7179-50-2 (oxytetracycline calcium)
*Formula:* $C_{22}H_{24}N_2O_9$
*Synonyms:* 5-Hydroxytetracycline; 2-Naphthacenecarboxamide, 4-(dimethylamino)-1,4,4a,5,5a,6,11,12a-octahydro-3,5,6,10,12,12a-hexahydroxy-6-methyl-1,11-dioxo-, [4$S$-(4$\alpha$,4a$\alpha$,5$\alpha$,5a$\alpha$,6$\beta$,12a$\alpha$)]-; Oxitetracyclin; Tetracycline, 5-hydroxy
*Trade Names:* ADAMYCIN®; ANTIBIOTIC TM 25®; BERKMYCEN®; BIOSTAT® PA DABICYCLINE®; FANTERRIN®; FIREMAN®, Nufarm (Australia); GEOMYCIN®; LENOCYCLINE®; LIQUAMYCIN LA 200®; MACOCYN®; MYCOJECT®, J.J. Mauget (USA); MYCOSHIELD® AGRICULTURAL TERRAMYCIN, Syngenta (Switzerland); MYCOSHIELD TMQTHC 20®; NCI-C56473®; OKSISYKLIN®; OXYATETS®; OXYMYCIN®; OXYMYKOIN®; OXYTERRACIN®; OXYTERRACINE®; OXYTERRACYNE®; OXYTETRACYCLINE®; OXYTETRACYCLINE AMPHOTERIC®; RIOMITSIN®; RYOMYCIN®; TAOMYCIN®; TAOMYXIN®; TERRAFUNGINE®; TERRAMITSIN®; TERRAMYCIN (with oxytetracycline hydrochloride)®; TETRAN®; TREE TECH®
*Producers:* J.J. Mauget (USA); Nufarm (Australia); Syngenta (Switzerland); Pfizer (USA) drug
*Chemical Class:* Antibiotic
*EPA/OPP PC Code:* 006321 (oxytetracycline calcium); 006304 (oxytetracycline)
*California DPR Chemical Code:* 3832
*Uses:* Oxytetracycline calcium is used for preventing the growth of or killing bacteria, fungi and mycoplasma-like organisms. It is used primarily to control fire blight of pears, pear decline, bacterial spot on peaches and nectarines, lethal yellowing of coconut palm, and lethal decline of pritchardia palm. It is also used as an anti-foulant added to marine paints to prevent the growth of barnacles. Oxytetracycline is used in veterinary medicine to prevent infections in fowl, cattle and swine.
*U.S. Maximum Allowable Residue Levels for Oxytetracycline (40 CFR 180.337):*
Residue level for the following drops are 0.35 ppm: peaches and pears.
*Carcinogen/Hazard Classifications*
**U.S. EPA Carcinogens:** Group D, unclassifiable as oxytetracycline
**California Prop. 65:** Developmental toxin
**Label Signal Word:** CAUTION
*Regulatory Authority:*
- Proposition 65 chemical (CAL)

*Description:* Light yellow crystalline solid or powder. Odorless. Slightly bitter taste. Molecular weight = 460.46; 498.52. Melting/Freezing point = 180-182°C (decomposes). From *streptomyces rimosus*
*Permissible Concentration in Water:* No criteria set. Runoff from spills or fire control may cause water pollution.
*First Aid:* If this chemical gets into the eyes, remove any contact lenses at once and irrigate immediately for at least 15 minutes, occasionally lifting upper and lower lids. Seek medical attention immediately. If this chemical contacts the skin, remove contaminated clothing and wash immediately with soap and water. Seek medical attention immediately. If this chemical has been inhaled, remove from exposure, begin rescue breathing (using universal precautions) if breathing has stopped, and CPR if heart action has stopped. Transfer promptly to a medical facility. When this chemical has been swallowed, get medical attention. Give large quantities of water and induce vomiting. Do not make an unconscious person vomit. Do not induce vomiting when formulations containing petroleum solvents are ingested.
*References:*
- U.S. Environmental Protection Agency, "Reregistration Eligibility Decision (RED) Facts, Hydrotetracycline Monochloride and Oxytetracycline Calcium," Office of Prevention, Pesticides and Toxic Substances, Washington, DC (March 1993). http://www.epa.gov/REDs/factsheets/0655fact.pdf
- Pesticide Management Education Program, "Oxytetracycline EPA Pestticide Fact Sheet 12/88," Cornell University, Ithaca, NY (December 1988). http://pmep.cce.cornell.edu/profiles/fung-nemat/febuconazole-sulfur/oxytetracycline/fung-prof-oxytetracycline.html
- U.S. Environmental Protection Agency, Office of Pesticide Programs, Pesticide Residue Limits, "Oxytetracycline," 40 CFR 180.337. http://www.setonresourcecenter.com/40CFR/Docs/wcd0004d/wcd04d22.asp
- U.S. Food and Drug Administration, "Tetracycline Antibiotic Drugs, Oxytetracycline Calcium," 21 CFR 446.66. http://frwebgate.access.gpo.gov/cgi-bin/get-cfr.cgi
- California Environmental Protection Agency "Chemical List of Lists," Sacramento CA (February 1997)

## Oxythioquinox

*Use Type:* Fungicide, acaricide, insecticide and fumigant
*CAS Number:* 2439-01-2
*Formula:* $C_{10}H_6N_2OS_2$

# Oxythioquinox

*Alert:* Restricted Use Pesticide (RUP) depending upon the formulation. Bayer voluntarily requested cancellation of all products containing this substance, which became effective September 7, 1999. Toxicity: Intermediate.

*Synonyms:* A13-25606; Carbonic acid, dithio-, cyclic *S,S*-(6-methyl-2,3-quinoxalinediyl)ester; Caswell no. 576; Chinomethionat; Chinomethionate; Dithilol(4,5-*β*)quinoxalin-2-one,6-methyl-; 1,3-Dithilol(4,5-*β*)quinoxalin-2-one,6-methyl-; dithioquinox; N-Dodecylguanidine acetate; ENT 25,606; Lauryl guanidine acetate; 6-Methyldithiolo(4,5-*β*)quinoxalin-2-one; 6-Methyl-1,3-dithiolo(4,5-*β*)quinoxalin-2-one; 6-Methyl-2-oxo-1,3-dithiolo(4,5-*β*)quinoxalin; 6-Methyl-2-oxo-1,3-dithio(4,5-*β*)quinoxaline; 6-Methyl-2,3-quinoxalin dithiocarbonate; 6-Methyl-2,3-quinoxalinedithio cyclic carbonate; 6-Methyl-2,3-quinoxalinedithio cyclic *S,S*-dithiocarbonate; 6-Methyl-2,3-quinoxalinedithio cyclic dithiocarbonate; 6-Methyl-quinoxaline-2,3-dithiocyclicarbonate; *S,S*-(6-methylquinoxaline-2,3-diyl)dithiocarbonate; MQD; MSC 379587; Quinomethionate; 2,3-Quinoxalinedithiol,6-methyl-, cyclic dithiocarbonate (ester); 2,3-Quinoxalinedithiol, 6-methyl-, cyclic carbonate

*Trade Names:* BAY® 36205, Bayer CropScience (Germany); BAYER® 36205, Bayer CropScience (Germany); BAYER® 4964, Bayer CropScience (Germany); BAYER® SS2074, Bayer CropScience (Germany); ERADE®; FORSTAN®; FUNGICIDE 5223®; MELPREX®; MILPREX®; MORESTAN®, Bayer CropScience (Germany), canceled; PUREGRO®, canceled; SYLLIT®; TSITREX®; VENTUROL VONODINE®

*Producers:* Bayer CropScience (Germany)
*Chemical Class:* Dithiocarbonate
*EPA/OPP PC Code:* 054101
*California DPR Chemical Code:* 410
*RTECS Number:* FG1400000
*Uses:* Not registered in the U.S. Oxythioquinox was registered in the U.S. as an insecticide, miticide and fungicide to control mites, mite eggs, and powdery mildew. Used as a pre-bloom spray on most deciduous fruits and in both pre- and post-bloom sprays on apples and pears. Also registered for use on citrus, vegetable, walnuts, ornamentals and greenhouse fumigant.
*Human toxicity (long-term)*[77]: Intermediate–10.23392 ppb, CHCL (Chronic Human Carcinogen Level)
*Fish toxicity (threshold)*[77]: High–2.99310 ppb, MATC (Maximum Acceptable Toxicant Concentration)
*Carcinogen/Hazard Classifications*
**U.S. EPA Carcinogens:** Group B2, probable carcinogen
**California Prop. 65:** Carcinogen and developmental toxin
**Label Signal Word:** CAUTION, DANGER
**WHO Acute Hazard:** Class III, slightly hazardous
*Regulatory Authority:*
- Actively registered pesticide in California.
- EPCRA Section 313 Form R *de minimis* concentration reporting level: 1.0%

*Description:* Yellow crystalline solid. Odorless. Practically insoluble in water. Molecular weight = 234.32. Melting/Freezing point = 171-172°C. Vapor pressure = $1.98 \times 10^{-7}$ mmHg. @ 20°C. Log $K_{ow}$ = 2.87 (est). Values of less than 3 are unlikely to bioaccumulate in marine organisms.
*Incompatibilities:* Incompatible with germanium, lead diacetate, magnesium, mercurous chloride, silicon, silver nitrate, titanium. May not be compatible with nitrates. Moisture may cause hydrolysis or other forms of decomposition.
*Permissible Exposure Limits in Air:*
*Determination in Air:* Filter; none; Gravimetric; NIOSH IV [Particulates NOR; #0500 (total), #0600 (respirable)].[18]
*Permissible Concentration in Water:* No criteria set. Runoff from spills or fire control may cause water pollution.
*Harmful Effects and Symptoms*
*Short Term Exposure:* Low levels of toxicity. Concentrated solutions are slightly corrosive to eyes and mucous membranes. Dust inhalation can cause irritation of the respiratory system with sneezing. Eye contact can cause irritation, watering, pain, and inflammation of the eyelids. Skin contact can cause irritation and minor ulceration. Ingestion can cause nausea, vomiting, fever, muscle twitching, seizure, rapid respiration, slow heart beat. Severe exposure may result in death.
*Long Term Exposure:*
*Points of Attack:* Respiratory system, central nervous system, cardiovascular system, skin, eyes.
*Medical Surveillance:* Medical observation is recommended for 24 to 48 hours after breathing overexposure, as pulmonary edema may be delayed. As first aid for pulmonary edema, consider administering a corticosteroid spray. Cigarette smoking may exacerbate pulmonary injury and should be discouraged for at least 72 hours following exposure. Before employment and at regular times after that, the following are recommended: If symptoms develop or overexposure occurs, repeat the above tests as soon as possible and get an exam of the nervous system. Also consider complete blood count. Consider chest x-ray following acute overexposure.
*First Aid:* If this chemical gets into the eyes, remove any contact lenses at once and irrigate immediately for at least 15 minutes, occasionally lifting upper and lower lids. Seek medical attention immediately. If this chemical contacts the skin, remove contaminated clothing and wash immediately with soap and water. Seek medical attention immediately. If this chemical has been inhaled, remove from exposure, begin rescue breathing (using universal precautions) if breathing has stopped, and CPR if heart action has stopped. Transfer promptly to a medical facility. When this chemical has been swallowed, get medical attention. If victim is conscious and able to swallow, have victim drink 4 to 8 ounces of water. Do not induce vomiting. *Note to physician or authorized medical personnel*: Medical observation is recommended for 24 to 48 hours after breathing

overexposure, as pulmonary edema may be delayed. As first aid for pulmonary edema, consider administering a corticosteroid spray. Cigarette smoking may exacerbate pulmonary injury and should be discouraged for at least 72 hours following exposure.

*References:*
- U.S. Environmental Protection Agency, "Reregistration Eligibility Decision (RED), Oxythioquinox," Office of Prevention, Pesticides and Toxic Substances, Washington, DC (November 1999). http://www.epa.gov/REDs/factsheets/2495fact.pdf
- Pesticide Management Education Program, "Oxythioquinox (Morestan) Chemical Profile 4/85," Cornell University, Ithaca, NY (April 1985). http://pmep.cce.cornell.edu/profiles/fung-nemat/febuconazole-sulfur/oxythioquinox/insect-prof-oxythioquinox.html
- California Environmental Protection Agency "Chemical List of Lists," Sacramento CA (February 1997).

# P

## Paclobutrazol (ANSI)

*Use Type:* Plant growth regulator
*CAS Number:* 76738-62-0
*Formula:* $C_{15}H_{20}ClN_3O$
*Synonyms:* 1-*tert*-Butyl-2-(*p*-chlorobenzyl)-2-(1,2,4-triazol-1-yl)ethanol; (2RS,3RS)-1-(4-Chlorophenyl)-4,4-dimethyl-2-(1*H*-1,2,4-triazol-1-yl)pentan-3-ol; (±)-R*,R*-$\beta$-[(4-Chlorophenyl)methyl]-$\alpha$-(1,1-dimethylethyl)-1*H*-1,2,4-triazol-1-ethanol
*Trade Names:* BONZI®, Syngenta (Switzerland); CLIPPER®, Syngenta (Switzerland); CULTAR®; ICI-PP 333®; MON-7325®, Monsanto (USA), canceled; PARLAY®; PICCOLO®; PP 333®; PROFILE®; TURF MANAGER®, The Scotts Company (USA)
*Producers:* Bhageria Dye-Chem (India); Fine Agrochemicals(UK); Ki-Hara Chemicals Ltd. (UK); The Scotts Company (USA); Syngenta (Switzerland)
*Chemical Class:* Triazole
*EPA/OPP PC Code:* 125601
*California DPR Chemical Code:* 2259
*Uses:* Used by arborists to maintain deciduous and broadleaf trees near buildings, reduce the growth of trees near power lines, and extend the longevity of trees on sites with limited natural resources.
*Human toxicity (long-term):* Very low–175.00 ppb, Health Advisory
*Fish toxicity (threshold):* Very low–4075.70930 ppb, MATC (Maximum Acceptable Toxicant Concentration)
*Carcinogen/Hazard Classifications*
**U.S. EPA Carcinogens:** Group D, unclassifiable, inadequate data
**Label Signal Word:** CAUTION, WARNING or DANGER
**WHO Acute Hazard:** Class III, slightly hazardous
*Regulatory Authority:*
- Actively registered pesticide in California.

*Description:* White crystalline solid. Odorless. Soluble in water; solubility = 37 ppm. Melting/Freezing point = 165°C. Vapor pressure = $7.5 \times 10^{-9}$ mm @ 20°C.
*Incompatibilities:* Strong oxidizers.
*Permissible Concentration in Water:* No criteria set. Runoff from spills or fire control may cause water pollution.
*First Aid:* If this chemical gets into the eyes, remove any contact lenses at once and irrigate immediately for at least 15 minutes, occasionally lifting upper and lower lids. Seek medical attention immediately. If this chemical contacts the skin, remove contaminated clothing and wash immediately with soap and water. Seek medical attention immediately. If this chemical has been inhaled, remove from exposure, begin rescue breathing (using universal precautions) if breathing has stopped, and CPR if heart action has stopped. Transfer promptly to a medical facility. When this chemical has been swallowed, get medical attention. Give large quantities of water and induce vomiting. Do not make an unconscious person vomit.
*References:*
- Pesticide Management Education Program, " Paclobutrazol (Clipper 50 WP) Herbicide Profile 8/85," Cornell University, Ithaca, NY (August 1985). http://pmep.cce.cornell.edu/profiles/herb-growthreg/naa-rimsulfuron/paclobutrazol/herb-prof-paclobutrazol.html
- California Environmental Protection Agency "Chemical List of Lists," Sacramento CA (February 1997)

## Paraoxon

*Use Type:* Insecticide
*CAS Number:* 311-45-5
*Formula:* $C_{10}H_{14}NO_6P$
*Synonyms:* O,O'-Diethyl-*p*-nitrophenylphosphat (German); Diaethyl-*p*-nitrophenylphosphorsaeureester (German); Diethyl-*p*-nitrophenyl phosphate; O,O-Diethyl-*p*-nitrophenyl phosphate; O,O-Diethyl-*p*-nitrophenyl phosphate; O,O-Diethyl O-*p*-nitrophenyl phosphate; Diethyl paraoxon; O,O-Diethylphosphoric acid O-*p*-nitrophenyl ester; E 600; ENT 16,087; Ester 25; Ethyl-*p*-nitrophenyl ethylphosphate; Ethyl paraoxon; O-*p*-Nitrofenilfosfato de O,O-dietilo (Spanish); *p*-Nitrophenyl dieyhylphosphate; Oxyparathion; Phosphacol; Phosphoric acid, diethyl 4-nitrophenyl ester; Phosphoric acid, diethyl *p*-nitrophenyl ester
*Trade Names:* CHINORTA®; FOSFAKOL®; HC 2072®; MINTACO®; MINTACOL®; MIOTISAL®; MIOTISAL A®; ETICOL®; PARAOXONE®; PAROXAN®; PESTOX 101®; SOLUGLACIT®; TS 219®
*Producers:* Sigma-Aldrich Laborchemikalien (Germany)
*Chemical Class:* Organophosphate
*EPA/OPP PC Code:* 653502
*California DPR Chemical Code:* 4082
*Uses:* There are no paraoxon products registered with the U.S. EPA. Also used as a medication.
*Carcinogen/Hazard Classifications*
**U.S. EPA Carcinogens:** *As parathion:* Group C, possible carcinogen
**IARC:** *As parathion:* Group 3, unclassifiable
**WHO Acute Hazard:** *As parathion:* Class 1 a, extremely hazardous

**Endocrine Disruptor:** *As parathion:* Suspected endocrine disruptor

*Regulatory Authority:*
- EPA Hazardous Waste Number (RCRA No.): P041
- RCRA, 40CFR261, Appendix 8 Hazardous Constituents
- Superfund/EPCRA 40CFR302.4 RQ: CERCLA, 100 lb (45.4 kg)
- U.S. DOT Inhalation Hazard Chemicals as organophosphates
- U.S. DOT Regulated Marine Pollutant (49CFR172.101, Appendix B)

*Description:* Reddish-yellow oil. Odorless. Practically insoluble in water; solubility = 2.5 ppm @ 25°C. Boiling point = 170°C @ 1 mmHg. Slightly soluble in water. Paraoxon is the active metabolite of parathion. Log $K_{ow}$ = 1.61. Unlikely to bioaccumulate in marine organisms.

*Incompatibilities:* Decomposes in alkaline materials.

*Permissible Exposure Limits in Air:* No OELs have been established. However this chemical can be absorbed through the skin, thereby increasing exposure.

*Determination in Air:* OSHA versatile sampler-2; Toluene/Acetone; Gas chromatography/Flame photometric detection for sulfur, nitrogen, or phosphorus; NIOSH Method IV Method #5600, Organophosphorus pesticides.[18]

*Routes of Entry:* Inhalation, ingestion, skin and/or eye contact.

*Harmful Effects and Symptoms*

*Short Term Exposure:* Can cause rapid organophosphate poisoning. Organic phosphorus insecticides are absorbed by the skin, as well as by the respiratory and gastrointestinal tracts. They are cholinesterase inhibitors. Symptoms of exposure include headache, giddiness, blurred vision, nervousness, weakness, nausea, cramps, diarrhea, and discomfort in the chest. Signs include sweating, tearing, salivation, vomiting, cyanosis, convulsions, coma, loss of reflexes and loss of sphincter control. Delayed pulmonary edema may occur after inhalation.

*Long Term Exposure:* Cholinesterase inhibitor; cumulative effect is possible. This chemical may damage the nervous system with repeated exposure, resulting in convulsions, respiratory failure. Repeated exposure may cause personality changes, depression, anxiety, irritability. May cause liver damage.

*Points of Attack:* Respiratory system, lungs, central nervous system, cardiovascular system, skin, eyes, plasma and red blood cell cholinesterase.

*Medical Surveillance:* Medical observation is recommended for 24 to 48 hours after breathing overexposure, as pulmonary edema may be delayed. Before employment and at regular times after that, the following are recommended: Plasma and red blood cell cholinesterase levels (tests for the enzyme poisoned by this chemical). If exposure stops, plasma levels return to normal in 1-2 weeks while red blood cell levels may be reduced for 1-3 months. When acetylcholinesterase enzyme levels are reduced by 25% or more below preemployment levels, risk of poisoning is increased, even if results are in lower ranges of "normal." Reassignment to work not involving organophosphate or carbamate pesticides is recommended until enzyme levels recover. If symptoms develop or overexposure occurs, repeat the above tests as soon as possible and get an exam of the nervous system. Also consider complete blood count. Consider chest x-ray following acute overexposure. Do not drink any alcoholic beverages before or during use. Alcohol promotes absorption of organophosphates. Refer to the NIOSH Criteria Documents #78-174 and #76-147 on Manufacturing, formulating, and working safely with pesticides.

*First Aid:* **Treatment for organophosphate poisoning consists of thorough decontamination, cardiorespiratory support, and administration of the antidotes atropine and pralidoxime. In cases of severe poisoning, diazepam, an anticonvulsant, should also be administered. Antidotes should be administered as prevention even if the diagnosis is in doubt.** Speed in removing material from eyes and skin is of extreme importance. *Eyes:* Eye contact can cause dangerous amounts of these chemicals to be quickly absorbed through the mucous membrane into the bloodstream. Immediately and gently flush eyes with plenty of warm or cold water (NO hot water) for at least 15 minutes, occasionally lifting the upper and lower eyelids. Get medical aid immediately. *Skin:* Get medical aid. Skin contact can cause dangerous amounts of these chemicals to be absorbed into the bloodstream. Wearing the appropriate PPE equipment and respirator for organophosphate pesticides, immediately flush skin with plenty of soap and water for at least 15 minutes while removing contaminated clothing and shoes. Shampoo hair promptly if contaminated. The removed, contaminated clothing and shoes should be double-bagged and left in Hot Zone for later disposal by hazardous materials experts. Skin may also be decontaminated with diluted hypochlorite solution. *Inhalation:* Get medical aid. Do not contaminate yourself. Wearing the appropriate PPE equipment and respirator for organophosphate pesticides, immediately remove the victim from the contaminated area to fresh air. If the victim is not breathing, administer artificial respiration. Do NOT use mouth-to-mouth resuscitation; use bag/mask apparatus. If breathing is difficult, administer oxygen through bag/mask apparatus until medical help arrives. Do not leave victim unattended. *Ingestion:* Call poison control. Loosen all clothing. Never give anything by mouth to an unconscious person. If victim is *unconscious or having convulsions,* do nothing except keep victim warm. Get medical aid. Transfer promptly to a medical facility. In cases of ingestion, **do not induce vomiting**. If the victim is alert and asymptomatic, administer a slurry of activated charcoal at a dose of 1 g/kg (infant, child, and adult dose). A soda can and straw may be of assistance when offering charcoal to a child. *In some cases you may be specifically instructed by poison control* to induce vomiting by way of 2 tablespoons of syrup of ipecac (adult) washed down with a cup of water. Do NOT give activated charcoal before or with ipecac syrup.

*Note to physician or authorized medical personnel:* Treat cases of respiratory compromise, coma, or excessive pulmonary secretions with respiratory support using protocols and techniques available and within the scope of training. Some cases may necessitate procedures such as endotracheal intubation or cricothyrotomy by properly trained and equipped personnel. When possible, atropine (see *Antidotes*, below) should be given under medical supervision. Patients who are comatose, hypotensive, or having seizures or cardiac arrhythmias should be treated according to advanced life support protocols. *Antidotes:* Two antidotes are administered to treat organophosphate poisoning. Atropine is a competitive antagonist of acetylcholine at muscarinic receptors and is used to control the excessive bronchial secretions which are often responsible for death. Pralidoxime relieves both the nicotinic and muscarine effects of organophosphate poisoning by regenerating acetylcholinesterase and can reduce both the bronchial secretions and the muscle weakness associated with poisoning. The initial intravenous dose of atropine in adults should be determined by the severity of symptoms: An initial adult dose of 1.0 to 2.0 mg or pediatric dose of 0.01 mg/kg (minimum 0.01 mg) should be administered intravenously. If intravenous access cannot be established, atropine may also be given intramuscularly, subcutaneously or via endotracheal tube. Doses should be repeated every 15 minutes until excessive secretions and sweating have been controlled. Once bronchial secretion has been controlled, atropine administration should be repeated whenever the secretions begin to recur. In seriously poisoned patients, very large doses may be required. Alterations of pulse rate and pupillary size should not be used as indicators of treatment adequacy. Pralidoxime should be administered as early in poisoning as possible as its efficacy may diminish when given more than 24 to 36 hours after exposure. Doses are as follows: adult 1.0 g; pediatric 25 to 50 mg/kg. The drug should be administered intravenously over 30 to 60 minutes, but in a life-threatening situation, one-half of the total dose can be given per minute for a total administration time of 2 minutes. Treatment should begin to take effect within 40 minutes with a reduction in symptoms and the amount of atropine necessary to control bronchial secretion. The initial dose can be repeated in 1 hour and then every 8 to 12 hours until the patient is clinically well and no longer requires atropine. If intravenous access cannot be established, pralidoxime may also be given intramuscularly. Early administration of diazepam in addition to the combined atropine and pralidoxime treatment may help prevent the onset of seizures and potential brain and cardiac morphologic damage following high-level organophosphate poisoning.

*References:*
- New Jersey Department of Health and Senior Services, "Hazardous Substance Fact Sheet, Paraoxon," Trenton NJ (March 1989, rev. May 2000). http://www.state.nj.us/health/eoh/rtkweb/1457.pdf
- California Environmental Protection Agency "Chemical List of Lists," Sacramento CA (February 1997).
- Agency for Toxic Substances and Disease Registry, U.S. Department of Health and Human Services, Public Health Service, "Managing Hazardous Materials Incidents," Atlanta, GA (June 2003).

# Paraquat

*Use Type:* Herbicide, defoliant and desiccant
*CAS Number:* 4685-14-7 (ion); 1910-42-5 (dichloride salt)
*Formula:* $C_{12}H_{14}Cl_2N_2$; $C_{12}H_{14}N_2Cl_2$
*Alert:* A Restricted Use Pesticide (RUP). Human toxicity (long-term): High
*Synonyms:* AI3-61943; Bipyridinium, 1,1'-dimethyl-4,4'-, Dichloride; 4,4'-Bipyridinium, 1,1'-dimethyl-, Dichloride; Dimethyl violgen chloride; 1,1'-Dimethyl-4,4'-bipyridinium dichloride; *N,N'*-Dimethyl-4,4'-bipyridinium dichloride; *N,N'*-Dimethyl-4,4'-bipyridylium dichloride; 1,1'-Dimethyl-4,4'-bipyridynium dichloride; 1,1-Dimethyl-4,4-dipyridilium dichloride; 4,4'-Dimethyldipyridyl dichloride; 1,1'-Dimethyl-4,4'-dipyridylium chloride; *N,N'*-Dimethyl-4,4'-dipyridylium dichloride; 1,1'-Dimethyl-4,4'-dipyridylium dichloride; Dimethyl viologen chloride; Methyl viologen (reduced); Methyl viologen; Methyl viologen chloride; Methyl viologen dichloride; NCS 263500; NCS 88126; Paraquat dichloride; Paraquat chloride; Viologen, methyl-
*Trade Names:* AH 501®; CEKUQUAT®; CHECK-THRU®, Kenso Corp. (Malaysia); COLONEL HERBICIDE®, Zeneca Ag Products (USA) (now Syngenta), canceled; CRISQUAT®; CYCLONE®, Syngenta (Switzerland); DEXTRONE®; DEXTRONE-X®; ESGRAM®; GAMIXEL®; GOLDQUAT 276®; GRAMOXONE®, Syngenta (Switzerland); GRAMOXONE D®, Syngenta (Switzerland); GRAMOXONE DICHLORIDE®, Syngenta (Switzerland); GRAMOXONE S®, Syngenta (Switzerland); GRAMOXONE W®, Syngenta (Switzerland); HERBIKILL®, Veterinary & Agricultural Products Manufacturing Co., Ltd. (VAPCO) (Jordan); HERBOXONE®; OK 622®; ORTHO PARAQUAT CL®, Orthor Business Group (USA), canceled; PARA-COL®; PARAMINE®, Kenso Corp. (Malaysia); PARAQUAT CL®, Orthor Business Group (USA), canceled; PARAQUAT DICHLORIDE BIPYRIDYLNIUM HERBICIDE®; PATHCLEAR®; PILLARQUAT®; PILLARXONE®; PP148®; PRELUDE®, Zeneca Ag Products (USA) (now Syngenta), canceled; STARFIRE®, Zeneca Ag Products (USA) (now Syngenta); Syngenta (Switzerland);

SUREFIRE®, Syngenta (Switzerland); SWEEP®; TERRAKLENE®; TOTACOL®; TOXER TOTAL®; UNIQUAT®, United Phosphorus (India); WEEDOL®

*Producers:* Agro-care Chemical Industry Group (China); Agrochemicals del Ecuador (AGROCHEM) (Ecuador); Agsin (Singapore); Bhageria Dye-Chem (India); Biesterfeld Siemsgluess International. GmbH (Germany); ICI Group (UK); Jingma Chemicals Ltd. (China); Kenso Corp. (Malaysia); Ki-Hara Chemicals Ltd. (UK); Montedison (Italy); Otsuka Chemical (Japan); Sanonda Zhengzhou Pesticide Co., Ltd. (China); Shenzhen Guomeng Industry Co., Ltd. (China); Sigma-Aldrich Laborchemikalien (Germany); Sinon (Taiwan); Syngenta (Switzerland); United Phosphorus (India); Veterinary & Agricultural Products Manufacturing Co., Ltd. (VAPCO) (Jordan); Zeneca Ag Products (USA) (now Syngenta)

*Chemical Class:* Bipyridylium
*EPA/OPP PC Code:* 061603; 061601 (dichloride salt)
*California DPR Chemical Code:* 1601
*ICSC Number:* 0005
*RTECS Number:* DW2275000
*EEC Number:* 613-090-00-7
*EINECS Number:* 225-141-7

*Uses:* Paraquat is a quaternary nitrogen herbicide widely used for broadleaf weed control. It is a quick-acting, nonselective compound, that destroys green plant tissue on contact and by translocation within the plant. It has been employed for killing marijuana in the U.S. and in Mexico. It is also used as a crop desiccant and defoliant, and as an aquatic herbicide. It is used in many formulations with other herbicides, e.g., simazine and diquat.

Paraquat dichloride is registered for the control of weeds and grasses in agricultural and non-agricultural areas. It is used as a preplant or pre-emergence herbicide on vegetables, grains, cotton, grasses, sugarcane, peanuts, potatoes, and on areas for tree plantation establishment. Paraquat is applied as a directed spray post-emergence herbicide around fruit crops, vegetables, trees, vines, grains, soybeans, and sugarcane. It is used for dormant season applications on clover and other legumes, and for chemical fallow. It is also used as a desiccant or harvest aid on cotton, dry beans, soybeans, potatoes, sunflowers, sugarcane and as a post harvest desiccant on tomatoes. It is applied to pine trees to induce resin soaking. Paraquat dichloride is also used on non-crop areas such as public airports, electric transformer stations and around commercial buildings to control weeds.

*Carcinogen/Hazard Classifications*
**U.S. EPA Carcinogens:** Group E, unlikely carcinogen
**Label Signal Word:** DANGER
**WHO Acute Hazard:** Class II, moderately hazardous
*Regulatory Authority:*
- Banned or Severely Restricted (Several Countries) (UN)[13]
- Air Pollutant Standard Set (ACGIH)[1] (DFG)[3] (HSE)[33] (OSHA)[58] (Several States)[60]
- AB 1803-Well Monitoring Chemical (CAL)
- EPA/SARA 302 (EPCRA) Extremely hazardous substances
- Permissible Exposure Limits for Chemical Contaminants (CAL/OSHA)
- The "Director's List" (CAL/OSHA)
- Actively registered pesticide in California.
- Superfund/EPCRA 40CFR355, Appendix B Extremely Hazardous Substances: TPQ = 100/10,000 lb (45.4/4,540 kg) as paraquat dichloride
- Superfund/EPCRA 40CFR302.4 RQ: EHS, 1 lb (0.454 kg) as paraquat dichloride
- EPCRA Section 313 Form R *de minimis* concentration reporting level: 1.0% as paraquat dichloride
- Canada, WHMIS, Ingredients Disclosure List

*Description:* Paraquat dichloride is a yellow solid with a faint, ammonia-like odor. Highly soluble in water; solubility = 70 g/100mL @ 20°C. Boiling point = 175–180°C (with decomposition). Melting/Freezing point = 300°C (decomposes). Hazard Identification (based on NFPA-704 M Rating System): [*Paraquat*] Health 4, Flammability 0, Reactivity 0. Soluble in water. Log $K_{ow}$ = Negative [>–4.0]. Unlikely to bioaccumulate in marine organisms. *Paraquat dichloride*: Melting point = 300°C (decomposition).

*Incompatibilities:* Strong oxidizers, alkylaryl-sulfonate wetting agents, strong bases (hydrolysis). Decomposes in heat (see physical properties, above) and in the presence of UV light producing nitrogen oxides, hydrogen chloride; deadly phosgene gas may be formed. May cause pitting of some metals.

*Permissible Exposure Limits in Air:* The OSHA[2] PEL 8-hour TWA is 0.5 mg/m³ for the respirable fraction. The DFG MAK[3], HSE[33] TWA, and the recommended NIOSH[2] TWA is 0.1 mg/m³. ACGIH[1] recommends an 8-hour TWA of 0.1 mg/m³ for the respirable fraction and 0.5 mg/m³ for total particulates. The NIOSH[2] IDLH level = 1.5 mg/m³. Several states have set guidelines or standards for paraquat in ambient air[60] ranging from 0.33 µg/m³ (New York) to 0.50 µg/m³ (South Carolina) to 1.0 µg/m³ (Florida) to 1.6 µg/m³ (Virginia) to 2.0 µg/m³ (Connecticut and Nevada).

*Determination in Air:* Sample collection by mixed cellulose ester membrane filter; Water; High-pressure liquid chromatography/Ultraviolet detection; NIOSH IV, Method #5003.[18]

*Permissible Concentration in Water:* A lifetime health advisory of 3.0 µg/L has been derived by EPA (See Reference Below). In addition, the state of Maine[61] has set a guideline for paraquat in drinking water of 17.0 µg/L.

*Routes of Entry:* Inhalation, ingestion, skin and/or eye contact. Absorbed through the skin.

*Harmful Effects and Symptoms*
*Short Term Exposure:* Irritates the eyes, skin, and respiratory tract. Inhalation can cause pulmonary edema, a medical emergency that can be delayed for several hours.

This can cause death. Effects occur in two stages, immediate and delayed. Caution is advised. Exposure to paraquat may be fatal; there is no effective antidote. Signs and symptoms of acute exposure to paraquat may be severe and include nausea, vomiting, diarrhea, and abdominal pain. A burning sensation of the mouth and esophagus with possible ulceration may occur following ingestion. Eye exposure may result in corneal opacification (cloudiness). Dermatitis and nail atrophy may occur following dermal contact. Delayed effects include transient reversible liver injury, acute renal failure and progressive pulmonary fibrosis with associated dyspnea (shortness of breath) and pulmonary edema. Absorbed through the skin and can lead to symptoms as listed in the following paragraph. In addition, can cause finger nail discoloration and damage (which returns to normal when exposure stops), irritation, redness, swelling and burning. Exposure through ingestion may cause burning of the mouth and throat, nausea, vomiting, abdominal pain, diarrhea and damage to the kidneys, heart and liver. Lung damage, leading to death, may occur. One half ounce of a 20% solution has caused death. Ingestion can also cause lung hemorrhage and fibrosis. The substance may cause effects on the lungs, kidneys, liver, cardiovascular system and gastro intestinal tract, resulting in impaired functions, tissue lesions.

*Long Term Exposure:* Repeated or prolonged contact with skin may cause damage and possible loss of the fingernails, and can lead to dry and cracking skin with blistering. Repeated or prolonged exposure to the aerosol can cause lung irritation, lung damage; bronchitis may develop. Can cause scarring of the lungs leading to breathlessness. Can damage the liver, kidneys and affect the heart.

*Points of Attack:* Eyes, skin, respiratory system, heart, liver, kidneys and gastrointestinal tract.

*Medical Surveillance:* Consider the points of attack in preplacement and periodic physical examinations. Liver and kidney function tests. EKG. Consider chest x-ray following acute overexposure. Chemical users should be cautioned about the use of alcohol which can increase liver damage.

*First Aid:* If this chemical gets into the eyes, remove any contact lenses at once and irrigate immediately for at least 15 minutes, occasionally lifting upper and lower lids. Seek medical attention immediately. If this chemical contacts the skin, remove contaminated clothing and wash immediately with soap and water. Seek medical attention immediately. If this chemical has been inhaled, remove from exposure, begin rescue breathing (using universal precautions) if breathing has stopped, and CPR if heart action has stopped. Transfer promptly to a medical facility. When this chemical has been swallowed, get medical attention. Give large quantities of water, or bentonite clay in water, or activated charcoal in water, and induce vomiting. Do not make an unconscious person vomit. Medical observation is recommended for 24 to 48 hours after breathing overexposure, as pulmonary edema may be delayed. As first aid for pulmonary edema, a doctor or authorized paramedic may consider administering a corticosteroid spray. Obtain authorization and/or further instructions from the local hospital for performance of other invasive procedures. Rush to a health care facility.

*References:*
- EXTOXNET, Extension Toxicology Network, "Pesticide Information Profile, Paraquat," Oregon State University, Corvallis, OR (June 1996). http://extoxnet.orst.edu/pips/paraquat.htm
- Environmental Protection Agency, "Reregistration Eligibility Decision (RED), Paraquat Dichloride," Washington, DC (August 1997). http://www.epa.gov/REDs/0262red.pdf
- New Jersey Department of Health and Senior Services, "Hazardous Substance Fact Sheet, Paraquat," Trenton NJ (September 1999). http://www.state.nj.us/health/eoh/rtkweb/1458.pdf
- Pasi, A., "The Toxicology of Paraquat, Diquat and Morfamquat," Bern, Switzerland, H. Huber (1978).
- U.S. Environmental Protection Agency, "Chemical Profile: Paraquat," Washington, DC, Chemical Emergency Preparedness Program (Nov. 30, 1987).
- Sax, N.I., Ed., "Dangerous Properties of Industrial Materials Report," 8, No. 2, 67-72 (1988).
- U.S. Environmental Protection Agency, "Health Advisory: Paraquat," Washington, DC, Office of Drinking Water (August 1987).
- New York State Department of Health, "Chemical Fact Sheet: Paraquat," Albany, NY, Bureau of Toxic Substance Assessment (Version 2, Feb. 1986 and Version 3).
- California Environmental Protection Agency "Chemical List of Lists," Sacramento CA (February 1997).

# Paraquat Methosulfate

*Use Type:* Herbicide and desiccant
*CAS Number:* 2074-50-2
*Formula:* $C_{14}H_{20}N_2O_8S_2$; $C_{12}H_{14}N_2(CH_3SO_4)_2$
*Synonyms:* 4,4-Bipyridinium, 1,1'-dimethyl-, bis(methyl sulfate); 1,1'-Dimethyl-4,4'-bipyridyniumdimethylsulfate; 1,1'-Dimethyl-4,4'-dipyridynium di(methyl sulfate); Gramoxone methyl sulfate; Paraqiat-I; Paraquat bis(methyl sulfate); Paraquat dimethosulfate; Paraquat dimethyl sulphate; Paraquat dimethyl sulfate; Paraquat methsulfate bipyridylnium herbicide
*Trade Names:* LASSO EC PLUS LOROX®, Monsanto (USA), canceled 7/2/1985; PILLARQUAT®; PP 910®
*Chemical Class:* Biptridylium
*EPA/OPP PC Code:* 061602
*California DPR Chemical Code:* 458
*RTECS Number:* DW2010000
*Carcinogen/Hazard Classifications*
**Label Signal Word:** DANGER

### 670  Parathion

*Regulatory Authority:*
- Banned or Severely Restricted (Hungary) (UN)[13]
- Permissible Exposure Limits for Chemical Contaminants (CAL/OSHA)
- Superfund/EPCRA 40CFR355, Appendix B Extremely Hazardous Substances: TPQ = 100/10,000 lb (45.4/4,540 kg)
- Superfund/EPCRA 40CFR302.4 RQ: EHS, 1 lb (0.454 kg)

*Description:* Paraquat methosulfate is a white to yellow crystalline solid. Melting/Freezing point = 175–180°C (decomposition). Hazard Identification (based on NFPA-704 M Rating System): Health 4, Flammability 0, Reactivity 0. Soluble in water.

*Incompatibilities:* Strong oxidizers.

*Permissible Exposure Limits in Air:* A MAC of 0.01 mg/m$^3$ has been set in Bulgaria according to the U.S. EPA Profile (Reference below).

*Permissible Concentration in Water:* See the section in the entry on Paraquat (dichloride). There are no specific criteria for the methosulfate.

*Routes of Entry:* Inhalation, ingestion, skin and/or eye contact.

*Harmful Effects and Symptoms*

*Short Term Exposure:* Contact causes irritation. Inhalation causes nose bleeds, headaches, coughing, and a sore throat. Higher exposures can cause pulmonary edema, a medical emergency that can be delayed for several hours. This can cause death. Swallowing causes burning in mouth, throat, and abdomen, vomiting, bloody vomitus, diarrhea with bloody stools, and headaches. It can cause death by shock and/or pulmonary damage. The fatal dose is estimated to be 6 grams of paraquat ion. Exposure may cause renal tubular damage and liver dysfunction. Death may occur in 24 hours or less.

*Long Term Exposure:* Liver and kidney damage.

*Points of Attack:* Lungs, liver and kidneys.

*Medical Surveillance:* Lung function tests. Liver and kidney function tests. Consider chest x-ray following acute overexposure.

*First Aid:* If this chemical gets into the eyes, remove any contact lenses at once and irrigate immediately for at least 15 minutes, occasionally lifting upper and lower lids. Seek medical attention immediately. If this chemical contacts the skin, remove contaminated clothing and wash immediately with soap and water. Seek medical attention immediately. If this chemical has been inhaled, remove from exposure, begin rescue breathing (using universal precautions) if breathing has stopped, and CPR if heart action has stopped. Transfer promptly to a medical facility. When this chemical has been swallowed, get medical attention. Give large quantities of water and induce vomiting. Do not make an unconscious person vomit. Medical observation is recommended for 24 to 48 hours after breathing overexposure, as pulmonary edema may be delayed. As first aid for pulmonary edema, a doctor or authorized paramedic may consider administering a corticosteroid spray.

*References:*
- U.S. Environmental Protection Agency, "Chemical Profile: Paraquat Methosulfate," Washington, DC, Chemical Emergency Preparedness Program (November 30, 1987).
- California Environmental Protection Agency "Chemical List of Lists," Sacramento CA (February 1997).

## Parathion (ANSI)

*Use Type:* Insecticide, acaricide and a chemical warfare agent.
*CAS Number:* 56-38-2
*Formula:* $C_{10}H_{14}NO_5PS$
*Alert:* A Restricted Use Pesticide (RUP). Parathion is one of the most toxic pesticides. See Uses below. Persons whose skin or clothing is contaminated with liquid or powdered parathion can cause secondary contamination by direct contact.

*Synonyms:* AAT; AATP; *O,O*-Diethyl *O*-(*p*-nitrophenyl) phosphorothioate; Diethyl *p*-nitrophenyl phosphorothionate; Diethyl 4-nitrophenyl phosphorothioate; *O,O*-Diethyl *O*-(4-nitrophenyl) phosphorothioate; *O,O*-Diethyl-*O,p*-nitrophenyl phosphorothioate; Diethyl *p*-nitrophenyl thionophosphate; *O,O*-Diethyl *O*-*p*-nitrophenyl thiophosphate; Diethyl parathion; DNTP; DPP; ENT 15,108; NCI-C00226; Ethyl parathion; Nitrostigmin (German); OMS 19; Parathene; Parathion-ethyl; Parathion thiophos; Parationa (Spanish); Phosphorothioic acid, *O,O*-diethyl-*O*-(4-nitrophenyl) ester; Phosphorothioic acid, *O,O*-diethyl *O*-(*p*-nitrophenyl) ester; Phosphostigmine; RB; SNP

*Trade Names:* (There are 921 active and canceled/transferred labels registered with the U.S. EPA) ACC 3422®; ALKRON®, Kerr-McGee Chemical (USA), canceled 2/11/1983; ALLERON®; AMERICAN CYANAMID 3422®; APHAMITE®; ARALO®; B 404®, Bayer CropScience (Germany); BAY E-605®, Bayer CropScience (Germany); BAYER E-605®, Bayer CropScience (Germany); BLADAN®, Bayer CropScience (Germany); BLADAN F®, Bayer CropScience (Germany); COMPOUND 3422®; COROTHION®; CORTHION®; CORTHIONE®; DANTHION®; DREXEL PARATHION 8E®, Drexel Chemical (USA); E 605®, Bayer CropScience (Germany); E 605 F®, Bayer CropScience (Germany); ECATOX®; EKATIN WF & WF ULV®; EKATOX®; ETHLON®; ETHYL PARATHION; ETILON®; FOLIDOL®, Bayer CropScience (Germany); FOLIDOL E®, Bayer CropScience (Germany); FOLIDOL E-605®, Bayer CropScience (Germany); FOLIDOL E&E 605®, Bayer CropScience (Germany); FOLIDOL OIL®, Bayer CropScience (Germany); FOSFERMO®; FOSFERNO®; FOSFEX®; FOSFIVE®; FOSOVA®; FOSTERN®;

FOSTOX®; GEARPHOS®; GENITHION®; IDA SEIS-TRES 6-3®, Drexel Chemical (USA); KALPHOS®; KOLODUST®, FMC Agricultural Products Group (USA), canceled 2/25/1987; KYPTHION®; LETHALAIRE G-54®; LIROTHION®; MURFOS®; MURPHOS®; NIRAN®, Monsanto (USA), canceled 5/1/1987; NIUIF 100®; NITROSTIGMINE®; NOURITHION®; NOVAFOS-M®, Cheminova (Denmark); OLEOFOS 20®; OLEOPARATHENE®; OLEOPARATHION®; ORTHOPHOS®; PAC®; PACOL®; PARA-KILL®, Crompton Corporation (USA), canceled 12/31/1987; PARAMAR®; PARA-TOX®, J. R. Simplot (USA), canceled 12/31/1983; PANTHION®; PARADUST®; PARAPHOS®; PARAWET®; PENNCAP E®; PESTOX PLUS®; PETHION®; PHOSKIL®; PLEOPARAPHENE®; RHODIASOL®; RHODIATOX®; RHODIATROX®; SEIS-TRES 6-3®, Drexel Chemical (USA); SELEPHOS®; SOPRATHION®; STATHION®; SULPHOS®; SUPER RODIATOX®; T-47®; THIOMEX®; THIOPHOS®; THIOPHOS® 3422; TIOFOS®; TOX 47®; TOXOL®; VAPOPHOS®; VITREX®

*Producers:* Agrimor International (USA); Alcotan Laboratories (Spain); Bayer CropScience (Germany); Biesterfeld Siemsgluess International. GmbH (Germany); Cangzhou Green Chemical Co. (China); Cheminova (Denmark); China Chemical (China); Cognis (Germany); Drexel Chemical (USA); DuPont (USA); Ehrenstorfer, Dr. (Germany); Hebei Long Age Pesticide (China); Hunan Tianyu Pesticide Chemical Group (China); Miles (USA); Rhone-Poulenc (France); Sanonda Zhengzhou Pesticide Co., Ltd. (China); Shenzhen Guomeng Industry Co., Ltd. (China); Sigma-Aldrich Laborchemikalien (Germany); Sumitomo Chemical (Japan); Wacker-Chemie (Germany)

*Chemical Class:* Organophosphate
*EPA/OPP PC Code:* 057401, 057501, 057503
*California DPR Chemical Code:* 459
*ICSC Number:* 0006
*RTECS Number:* TF4550000
*EEC Number:* 015-034-00-1
*EINECS Number:* 200-271-7

*Uses:* The U.S. EPA announced in November, 2000, the cancellation of ethyl parathion immediately on seed corn and the eventual phase out for its use in other pesticide products by the end of 2000. By the end of October, 2003, the U.S. EPA phased out its use to control insects and mites on alfalfa, barley, corn, canola, sorghum, soybeans, sunflowers and wheat. Also used to control nuisance birds.

*U.S. Maximum Allowable Residue Levels for Parathion (40 CFR 180.121):*

| CROP | ppm |
|---|---|
| Alfalfa, forage | 1.25 |
| Alfalfa, hay | 5 |
| Almond | 0.1 |
| Almond, hulls | 3 |
| Barley | 1 |
| Bean, dry | 1 |
| Beet, sugar | 0.1 |
| Beet, sugar, tops | 0.1 |
| Cabbage | 1 |
| Clover | 1 |
| Corn | 1 |
| Corn | 1 |
| Corn, forage | 1 |
| Corn, forage | 1 |
| Cotton, undelinted seed | 0.75 |
| Cranberry | 1 |
| Cucumber | 1 |
| Currant | 1 |
| Date, dried fruit | 1 |
| Dewberry | 1 |
| Eggplant | 1 |
| Endive | 1 |
| Fig | 1 |
| Grass, forage | 1 |
| Hop | 1 |
| Hop, dried cones | 1 |
| Oat | 1 |
| Onion | 1 |
| Pea, dry | 1 |
| Pea, field, vines | 1 |
| Peanut | 1 |
| Pecan | 0.1 |
| Potato | 0.1 |
| Rapeseed, seed | 0.2 |
| Rice | 1 |
| Rye | 0.5 |
| Soybean | 0.1 |
| Soybean, hay | 1 |
| Sunflower, seed | 0.2 |
| Sweet potato | 0.1 |
| Walnut | 0.1 |
| Wheat | 1 |

*Carcinogen/Hazard Classifications*
**U.S. EPA Carcinogens:** Group 3, possible carcinogen
**IARC:** Group 3, unclassifiable
**Label Signal Word:** DANGER
**WHO Acute Hazard:** Class 1 a, extremely hazardous
**Endocrine Disruptor:** Suspected endocrine disruptor
*Regulatory Authority:*
- Banned or Severely Restricted (Many Countries) (UN)[13]
- Very Toxic Substance (World Bank)[15]
- Air Pollutant Standard Set (ACGIH)[1] (DFG)[3] (HSE)[33] (former USSR)[35] (OSHA)[58] (Several States)[60]
- AB 1803-Well Monitoring Chemical (CAL)
- MCL (Maximum Contaminants Levels) list of contaminants (CAL)
- AB 2588-Air Toxics "Hot Spots" Chemicals (CAL)
- CAL-DHS/DHS Drinking Water Action Levels

- CAL Air Resources Board/AB 1807 Toxic Air Contaminants
- Permissible Exposure Limits for Chemical Contaminants (CAL/OSHA)
- The "Director's List" (CAL/OSHA)
- Clean Water Act: Section 311 Hazardous Substances/RQ 40CFR117.3 (same as CERCLA, see below); Section 313 Water Priority Chemicals (57FR41331, 9/9/92)
- EPA Hazardous Waste Number (RCRA No.): P089
- RCRA, 40CFR261, Appendix 8 Hazardous Constituents
- RCRA 40CFR268.48; 61FR15654, Universal Treatment Standards: Wastewater (mg/L), 0.014; Nonwastewater (mg/kg), 4.6
- RCRA 40CFR264, Appendix 9; TSD Facilities Ground Water Monitoring List. Suggested test method(s) (PQL ug/L): 8270(10)
- Safe Drinking Water Act: Priority List (55 FR 1470) as parathion degradation.
- Superfund/EPCRA 40CFR355, Appendix B Extremely Hazardous Substances: TPQ = 100 lb (45.4 kg)
- Superfund/EPCRA 40CFR302.4 RQ: CERCLA, 10 lb (4.54 kg)
- EPCRA Section 313 Form R *de minimis* concentration reporting level: 1.0%
- U.S. DOT Inhalation Hazard Chemicals as organophosphates
- U.S. DOT Regulated Marine Pollutant (49CFR172.101, Appendix B), severe pollutant
- Canada, WHMIS, Ingredients Disclosure List
- Toxic Chemical (U.S. EPA)[8]

***Description:*** At room temperature, it is a combustible pale yellow to dark brown liquid that may be difficult to ignite. In commercial products, parathion is usually dissolved in hydrocarbon solvents such as toluene or xylene, which are flammable. Weak odor of garlic @ 0.47 mg/m$^3$. Practically insoluble in water; solubility = 0.001% @ 20°C. Specific gravity (water = 1) = 1.27. Boiling point = 375°C @ 760 mmHg; also listed @ 157°C for Chemical Warfare (CW) agent. Molecular weight 291.29. Melting/Freezing point = 6.1°C; also listed @ 2.9°C for CW agent. Vapor pressure = 4 x 10$^{-6}$ mmHg @ 20°C. Flash point = 120–160°C; 200°C (EPA, ATSDR). Hazard Identification (based on NFPA-704 M Rating System): Health 4, Flammability 0, Reactivity 0. Log $K_{ow}$ = 3.4 to 3.93. Values at or above 3.0 are likely to bioaccumulate in marine organisms.

***Incompatibilities:*** Reacts with strong oxidizers. Attacks some plastics, rubbers, and coatings. Rapidly hydrolyzed by alkalis.

***Permissible Exposure Limits in Air:*** The OSHA[2] PEL is 0.1 mg/m$^3$ (TWA) as is the DFG MAK[3] and the HSE[33] TWA.. The HSE[33] STEL is 0.3 mg/m$^3$. ACGIH[1] recommends a TWA of 0.1 mg/m$^3$. The notation "skin" is added to indicate the possibility of cutaneous absorption. The NIOSH[2] IDLH value is 10 mg/m$^3$. The odor threshold of parathion is five times the OSHA PEL (0.1 mg/m$^3$) and *does not provide adequate warning of hazardous concentrations.* The former USSR[35] has set 0.05 mg/m$^3$ as an MAC in workplace air. Several states have set guidelines or standards for parathion in ambient air[60] ranging from 0.238 μg/m$^3$ (Kansas) to 0.33 μg/m$^3$ (New York) to 0.5 μg/m$^3$ (South Carolina) to 1.0 μg/m$^3$ (North Dakota) to 1.6 μg/m$^3$ (Virginia) to 1.87 μg/m$^3$ (Pennsylvania) to 2.0 μg/m$^3$ (Connecticut and Nevada).

***Determination in Air:*** OSHA versatile sampler-2; Toluene/Acetone; Gas chromatography/Flame photometric detection for sulfur, nitrogen, or phosphorus; NIOSH IV, Method #5600, Organophosphorus pesticides.[18]

***Permissible Concentration in Water:*** The former USSR has set an MAC[35] in surface water of 3.0 μg/L and Mexico has set maximum permissible concentrations of 1.0 μg/L in coastal waters and 10.0 μg/L in estuaries. Some states have set guidelines for parathion in drinking water[61] ranging from 8.6 μg/L in Maine to 30.0 μg/L in California and Kansas.

***Routes of Entry:*** Inhalation, ingestion, skin and/or eye contact. Absorbed through the skin.

***Harmful Effects and Symptoms***

***Short Term Exposure:*** A cholinesterase inhibitor. Systemic toxicity due to parathion can result from all routes of exposure. Symptoms include abdominal cramps, vomiting, diarrhea, pinpoint pupils and blurred vision, excessive sweating, salivation and lacrimation, wheezing, excessive tracheobronchial secretions, agitation, seizures, bradycardia or tachycardia, muscle twitching and weakness, and urinary bladder and fecal incontinence. Seizures are much more common in children than in adults. The heart rate may decrease following oral exposure or increase following dermal exposure. Hypotension (low blood pressure) is not uncommon. Respiratory symptoms include dyspnea (shortness of breath), respiratory depression, and respiratory paralysis. Psychosis may occur. This material is extremely toxic; the probable oral lethal dose is 5-50 mg/kg, or between 7 drops and 1 teaspoonful for a 150 lb person. As little as 1 drop can endanger life if splashed in the eye. Toxicity is highest by inhalation. People at special risk are those with a history of glaucoma, cardiovascular disease, hepatic disease, renal, disease, or central nervous system abnormalities. Some additional details on short-term exposure to parathion are as follows: *Inhalation:* Occasional human exposures at concentrations of 0.1-0.8 mg/m$^3$ did not give rise to any symptoms. Occasional human exposure@ 1.5-2.0 mg/m$^3$ resulted in nausea and vomiting. Higher exposures can give rise to dizziness, blurred vision, wheezing, excessive salivation, and muscle and abdominal cramps. An estimated 10 to 20 mg (1/1500 ounce) may cause death. *Skin:* However, many human poisonings have occurred through extensive skin contact at unspecified levels. This is the greatest hazard for some workers. Symptoms of poisoning include nausea, vomiting, weakness, blurring of vision and muscle cramps. NIOSH lists the following symptoms of exposure: irritation eyes, skin, respiratory system; miosis; rhinorrhea (discharge of thin nasal mucous); headache; chest tightness, wheezing, laryngeal spasm, salivation, cyanosis; anorexia, nausea, vomiting, abdominal cramps, diarrhea; sweating; muscle fasiculation, weakness, paralysis; giddiness, confusion, ataxia; convulsions, coma; low blood pressure; cardiac irregular/irregularities. Delayed pulmonary edema may occur after inhalation.

- Death results from loss of consciousness, coma, excessive bronchial secretions, respiratory depression and cardiac irregularity.
- Commercial parathion products often contain hydrocarbon solvents, such as xylene or toluene, which themselves can cause toxicity.
- Toxicity of parathion depends on metabolic activation; thus, symptoms may be delayed for 6 to 24 hours after exposure.

*Long Term Exposure:* Persistent weakness and impaired memory have been reported to occur from low-level exposures to organophosphates in the absence of acute cholinergic effects. This chemical may damage the nervous system with repeated exposure, resulting in convulsions, respiratory failure. May cause liver damage. *Carcinogenicity:* The International Agency for Research on Cancer has determined that parathion is not classifiable as to its carcinogenicity to humans. However, EPA lists parathion as a possible human carcinogen. *Reproductive and developmental Effects:* Studies have been reported in which parathion was embryo-toxic and feto-toxic in rodents. There are no studies addressing reproductive or developmental effects in humans exposed to parathion. Cholinesterase inhibitor; cumulative effect is possible.

*Points of Attack:* Respiratory system, central nervous system, cardiovascular system, eyes, skin and blood cholinesterase.

*Medical Surveillance:* Toxicity of parathion depends on metabolic activation; thus, symptoms may be delayed for 6 to 24 hours after exposure. NIOSH recommends that medical surveillance, including preemployment and periodic examinations, shall be made available to workers who may be occupationally exposed to parathion. Biologic monitoring is also recommended as an additional safety measure. Before employment and at regular times after that, the following are recommended: Plasma and red blood cell cholinesterase levels (tests for the enzyme poisoned by this chemical). If exposure stops, plasma levels return to normal in 1-2 weeks while red blood cell levels may be reduced for 1-3 months. Do not drink any alcoholic beverages before or during use. Alcohol promotes absorption of organophosphates. When acetylcholinesterase enzyme levels are reduced by 25% or more below preemployment levels, risk of poisoning is increased, even if results are in lower ranges of "normal." Reassignment to work not involving organophosphate or carbamate pesticides is recommended until enzyme levels recover. If symptoms develop or overexposure occurs, repeat the above tests as soon as possible and get an exam of the nervous system. Also consider complete blood count. Consider chest x-ray following acute overexposure. Do not drink any alcoholic beverages before or during use. Alcohol promotes absorption of organophosphates. Medical observation is recommended for 24 to 48 hours after breathing overexposure, as pulmonary edema may be delayed.

*First Aid:* **Treatment for organophosphate poisoning consists of thorough decontamination, cardiorespiratory support, and administration of the antidotes atropine and pralidoxime. In cases of severe poisoning, diazepam, an anticonvulsant, should also be administered. Antidotes should be administered as prevention even if the diagnosis is in doubt.** Speed in removing material from eyes and skin is of extreme importance. *Eyes:* Eye contact can cause dangerous amounts of these chemicals to be quickly absorbed through the mucous membrane into the bloodstream. Immediately and gently flush eyes with plenty of warm or cold water (NO hot water) for at least 15 minutes, occasionally lifting the upper and lower eyelids. Get medical aid immediately. *Skin:* Get medical aid. Skin contact can cause dangerous amounts of these chemicals to be absorbed into the bloodstream. Wearing the appropriate PPE equipment and respirator for organophosphate pesticides, immediately flush skin with plenty of soap and water for at least 15 minutes while removing contaminated clothing and shoes. Shampoo hair promptly if contaminated. The removed, contaminated clothing and shoes should be double-bagged and left in Hot Zone for later disposal by hazardous materials experts. Skin may also be decontaminated with diluted hypochlorite solution. *Inhalation:* Get medical aid. Do not contaminate yourself. Wearing the appropriate PPE equipment and respirator for organophosphate pesticides, immediately remove the victim from the contaminated area to fresh air. If the victim is not breathing, administer artificial respiration. Do NOT use mouth-to-mouth resuscitation; use bag/mask apparatus. If breathing is difficult, administer oxygen through bag/mask apparatus until medical help arrives. Do not leave victim unattended. *Ingestion:* Call poison control. Loosen all clothing. Never give anything by mouth to an unconscious person. If victim is *unconscious or having convulsions,* do nothing except keep victim warm. Get medical aid. Transfer promptly to a medical facility. In cases of ingestion, **do not induce vomiting**. If the victim is alert and asymptomatic, administer a slurry of activated charcoal at a dose of 1 g/kg (infant, child, and adult dose). A soda can and straw may be of assistance when offering charcoal to a child. *In some cases you may be specifically instructed by poison control to induce vomiting by way of 2 tablespoons of syrup of ipecac (adult) washed down with a cup of water.* Do NOT give activated charcoal before or with ipecac syrup.

*Note to physician or authorized medical personnel:* Treat cases of respiratory compromise, coma, or excessive pulmonary secretions with respiratory support using protocols and techniques available and within the scope of training. Some cases may necessitate procedures such as endotracheal intubation or cricothyrotomy by properly trained and equipped personnel. When possible, atropine (see *Antidotes*, below) should be given under medical supervision. Patients who are comatose, hypotensive, or having seizures or cardiac arrhythmias should be treated according to advanced life support protocols. *Antidotes:* Two antidotes are administered to treat organophosphate poisoning. Atropine is a competitive antagonist of acetylcholine at muscarinic receptors and is used to control the excessive bronchial secretions which are often responsible for death. Pralidoxime relieves both the nicotinic and muscarine effects of organophosphate poisoning by regenerating acetylcholinesterase and can reduce both the bronchial secretions and the muscle weakness associated with poisoning. The initial intravenous dose of atropine in adults should be determined by the severity of symptoms: An initial adult dose of 1.0 to 2.0 mg or pediatric dose of 0.01 mg/kg (minimum 0.01 mg) should be administered intravenously. If intravenous access cannot be established, atropine may also be given intramuscularly, subcutaneously or via endotracheal tube. Doses should be repeated every 15 minutes until excessive secretions and sweating have been controlled. Once bronchial secretion has been controlled, atropine administration should be repeated whenever the secretions begin to recur. In seriously poisoned patients, very large doses may be required. Alterations of pulse rate and pupillary size should not be used as indicators of treatment adequacy. Pralidoxime should be administered as early in poisoning as possible as its efficacy may diminish when given more than 24 to 36 hours after exposure. Doses are as follows: adult 1.0 g; pediatric 25 to 50 mg/kg. The drug should be administered intravenously over 30 to 60 minutes, but in a life-threatening situation, one-half of the total dose can be given per minute for a total administration time of 2 minutes. Treatment should begin to take effect within 40 minutes with a reduction in symptoms and the amount of atropine necessary to control bronchial secretion. The initial dose can be repeated in 1 hour and then every 8 to 12 hours until the patient is clinically well and no longer requires atropine. If intravenous access cannot be established, pralidoxime may also be given intramuscularly. Early administration of diazepam in addition to the combined atropine and pralidoxime treatment may help prevent the onset of seizures and potential brain and cardiac morphologic damage following high-level organophosphate poisoning.

*References:*
- U.S. Environmental Protection Agency, Office of Pesticide Programs, Pesticide Residue Limits, "Parathion," 40 CFR 180.121, http://www.epa.gov/pesticides/food/viewtols.htm
- EXTOXNET, Extension Toxicology Network, "Pesticide Information Profile, Parathion," Oregon State University, Corvallis, OR (September 1993). http://extoxnet.orst.edu/pips/parathio.htm
- New Jersey Department of Health and Senior Services, "Hazardous Substance Fact Sheet, Parathion," Trenton, NJ (May 1989, rev. October 2000). http://www.state.nj.us/health/eoh/rtkweb/1459.pdf
- National Institute for Occupational Safety and Health, "Criteria for a Recommended Standard: Occupational Exposure to Parathion," NIOSH Doc. No. 76-190 (1976).
- Sax, N.I., Ed., "Dangerous Properties of Industrial Materials Report," 3, No. 3, 92-97 (1983).
- U.S. Environmental Protection Agency, "Chemical Profile: Parathion," Washington, DC, Chemical Emergency Preparedness Program (Nov. 30, 1987).
- Agency for Toxic Substances and Disease Registry, U.S. Department of Health and Human Services, Public Health Service, "Managing Hazardous Materials Incidents," Atlanta, GA (June 2003).
- California Environmental Protection Agency "Chemical List of Lists," Sacramento CA (February 1997).
- New York State Department of Health, "Chemical Fact Sheet: Parathion," Albany, NY, Bureau of Toxic Substance Assessment (Version 2, March 1986 and Version 3).

# Paris Green

*Use Type:* Insecticide and wood preservative
*CAS Number:* 12002-03-8
*Formula:* $C_4H_6As_6Cu_4O_{16}$
*Synonyms:* Acetoarsenite de cuivre (French); Acetoarsenito de cobre (Spanish); (Aceto) (trimrtaarsenito)dicopper; Basle green; C.I. 77410; C.I. pigment green 21 (9CI); Copper acetoarsenite; Emerald green; ENT 884; French green; Imperial green; King's green; Meadow green; Mineral green; Mitis green; Moss green; Mountain green; Neuwied green; New green; Parrot green; Patent green; Powder green; Schweinfurtergruen (German); Schweinfurt grun (German); Swedish green; Vienna green
*Trade Names:* ORTHO P-G BAIT®, The Scotts Company (USA); SOWBUG & CUTWORM BAIT®
*Producers:* Caldic Chemie (Netherlands); The Scotts Company (USA)
*Chemical Class:* Inorganic arsenic/copper compound
*EPA/OPP PC Code:* 022601
*California DPR Chemical Code:* 2485 and 460 (two codes for the same substance)
*RTECS Number:* GL6475000

*Uses:* This material is also used as a wood preservative and a pigment, particularly for ships and submarines, and also finds use as an anthelmintic.

## Carcinogen/Hazard Classifications

**U.S. NTP Carcinogen:** Known carcinogen
**California Prop. 65:** Listed
**IARC:** Group 1, known carcinogen
**Label Signal Word:** WARNING or DANGER
**WHO Acute Hazard:** Class 1b, highly hazardous
**Endocrine Disruptor:** *As arsenic:* Suspected endocrine disruptor

*Regulatory Authority:*
- Clean Air Act: Hazardous Air Pollutants (Title I, Part A, Section 112); List of high risk pollutants (Section 63.74) as arsenic compounds.
- Clean Water Act: Section 311 Hazardous Substances/RQ 1 lb (0.454 kg); Toxic Pollutant (Section 401.15) as copper and compounds; Section 313 Water Priority Chemicals (57FR41331, 9/9/92).
- Safe Drinking Water Act 47FR9352 Regulated chemical: MCL, 0.05 mg/L (Section 141.11) applies only to community water systems (arsenic)
- RCRA 40CFR264, Appendix 9; TSD Facilities Ground Water Monitoring List. Suggested test method(s) (PQL $ug/L$): 6010(60); 7210(200) Note: All species in the ground water that contain copper are included.
- Superfund/EPCRA 40CFR355, Appendix B Extremely Hazardous Substances: TPQ = 500/10,000 lb (227/4,540 kg)
- Superfund/EPCRA 40CFR302.4 RQ: EHS, 1 lb (0.454 kg)
- EPCRA Section 313 Form R *de minimis* concentration reporting level: 1.0%
- AB 2588-Air Toxics "Hot Spots" Chemicals (CAL) as copper compounds
- Permissible Exposure Limits for Chemical Contaminants (CAL/OSHA) as arsenic compounds
- The "Director's List" (CAL/OSHA)
- U.S. DOT Regulated Marine Pollutant (49CFR172.101, Appendix B) as arsenates, liquid, n.o.s.; arsenates, solid, n.o.s; arsenical pesticides liquid, toxic, flammable, n.o.s.
- Canada: Priority Substance List & Restricted Substances/Ocean Dumping Forbidden (CEPA), National Pollutant Release Inventory (NPRI) (arsenic compounds)

*Description:* Paris Green (cupric acetoarsenite), $C_4H_6As_6Cu_4O_{16}$, is an odorless emerald green crystalline powder which decomposes upon heating. Insoluble in water. Hazard Identification (based on NFPA-704 M Rating System): Health 3, Flammability 0, Reactivity 0.

*Incompatibilities:* Can react vigorously with oxidizers. Emits highly toxic arsenic fumes on contact with acid or acid fumes, and when material is heated to decomposition.

*Permissible Exposure Limits in Air: Arsenic:* The following exposure limits are for air levels only. When skin contact also occurs, overexposure is possible, even though air levels are less than the limits listed below. OSHA[2]: The legal airborne PEL is 0.010 mg/m³ averaged over an 8-hour workshift. NIOSH[2]: The recommended airborne exposure limit is 0.002 mg/m³ (ceiling), not to be exceeded during any 15 min. work period. ACGIH[1]: The recommended airborne exposure limit is 0.01 mg/m³ averaged over an 8-hour workshift. The HSE[33] (U.K.) Maximum Exposure Limit as arsenic is 0.1 mg/m³ TWA.

*Determination in Air:* Filter; Acid; Hydride generation atomic absorption spectrometry; NIOSH IV, Method #7900[18]. See also #7300, Elements (arsenic).

*Permissible Concentration in Water:* EPA has set a limit of 0.05 parts per million (ppm) for arsenic in drinking water. The U.S. EPA arsenic drinking water standard of 0.01 ppm (10 ppb) is based on the U.S. EPA final rule for arsenic in drinking water published in. the January 22, 2001, *Federal Register*. However, the U.S. EPA is currently reviewing the science and cost estimate supporting this rule, and, in the interim, has reverted to the previous standard for arsenic. Thus, in the US, the current EPA arsenic drinking water standard remains at 0.05 ppm (50 ppb). To protect freshwater aquatic life, total recoverable trivalent inorganic arsenic is never to exceed 440 µg/L. To protect saltwater aquatic life: 508 µg/L on an acute basis. To protect human health: preferably zero. The former USSR-UNEP/IRPTC project[43] has set MAC values for inorganic arsenic compounds in water for domestic purposes at 0.05 mg/L and in water bodies for fishery purposes of 0.5 mg/L also.

*Determination in Water: For arsenic:* The atomic absorption graphite furnace technique is often used for measurement of total arsenic in water. It also has been standardized by EPA. Total arsenic may be determined by digestion followed by silver diethyldithiocarbamate; an alternative is atomic absorption; another is inductively coupled plasma optical emission spectrometry. See OSHA Method #ID-105 for arsenic.[58]

*Routes of Entry:* Inhalation, ingestion, skin and/or eye contact.

## Harmful Effects and Symptoms

*Short Term Exposure:* It may cause eye and respiratory tract irritation. Industrial exposure may cause dermatitis. This material is extremely toxic; the probable oral lethal dose for humans is 5-50 mg/kg, or between 7 drops and 1 teaspoonful for a 150 lb person. Some absorption may occur through the skin and by inhalation, but most poisonings result from ingestion. Symptoms usually appear ½ to 1 hour after ingestion, but may be delayed. Causes gastric disturbance, tremors, muscular cramps, and nervous collapse which may lead to death. Symptoms of exposure also include a sweetish, metallic taste and garlicky odor; difficulty in swallowing; abdominal pain; vomiting and diarrhea; dehydration; rapid heart beat; dizziness and headache; and eventually coma, sometimes convulsions, and death.

*Long Term Exposure:* May cause liver damage. Arsenic compounds may cause blood, kidneys, and nervous system damage, and skin abnormalities may develop. Birth defects have been observed in animals exposed to inorganic arsenic. It is likely that health effects seen in children exposed to high amounts of arsenic will be similar to the effects seen in adults.

*Points of Attack:* Several studies have shown that inorganic arsenic can increase the risk of lung cancer, skin cancer, bladder cancer, liver cancer, kidney cancer, and prostate cancer.

*Medical Surveillance:* Liver and kidney function tests. Blood tests including CBC.

*First Aid:* If this chemical gets into the eyes, remove any contact lenses at once and irrigate immediately for at least 15 minutes, occasionally lifting upper and lower lids. Seek medical attention immediately. If this chemical contacts the skin, remove contaminated clothing and wash immediately with soap and water. Seek medical attention immediately. If this chemical has been inhaled, remove from exposure, begin rescue breathing (using universal precautions) if breathing has stopped, and CPR if heart action has stopped. Transfer promptly to a medical facility. When this chemical has been swallowed, get medical attention. Give large quantities of water and induce vomiting. Do not make an unconscious person vomit.

*References:*
- New Jersey Department of Health and Senior Services, "Hazardous Substance Fact Sheet, Copper Acetoarsenite," Trenton, NJ (September 1988, rev. January 1999). http://www.state.nj.us/health/eoh/rtkweb/0529.pdf
- U.S. Environmental Protection Agency, "Chemical Profile: Paris Green," Washington, DC, Chemical Emergency Preparedness Program (Nov. 30, 1987)
- California Environmental Protection Agency "Chemical List of Lists," Sacramento CA (February 1997).

# Pebulate

*Use Type:* Herbicide
*CAS Number:* 1114-71-2
*Formula:* $C_{10}H_{21}NOS$
*Synonyms:* Butylethylthiocarbamic acid *S*-propyl ester; PEBC; *S*-propyl-*N*-aethyl-*N*-butyl-thiocarbamat (German); *S*-Propyl butylethylthiocarbamate; *S*-Propyl-*N*-butyl-*N*-ethylthiolcarbamate; Propyl-ethylbutylthiocarbamate; Propyl-ethyl-*N*-butylthiocarbamate; Propyl-*N*-ethyl-*N*-butylthiocarbamate; *N*-propyl-*N*-ethyl-*N*-(*N*-butyl)thio carbamate; *S*-(*N*-Propyl)-*N*-ethyl-*N*-*N*-butyl) thiocarbamate; Propylethylbutylthiocarbamate
*Trade Names:* DASANIT, Bayer Corp (Germany), canceled 9/1/87; DEXIL, canceled; DYFONATE, canceled; GW PRE-BETA, canceled; R-2061®; STAUFFER® R-2061, Zeneca Ag (USA), canceled (now Syngenta); TILLAM®, Zeneca Ag Products (USA), canceled;
*Producers:* Bayer Corp (Germany)
*Chemical Class:* Thiocarbamate
*EPA/OPP PC Code:* 041403
*California DPR Chemical Code:* 590
*RTECS Number:* EZ0400000
*Uses:* Pebulate is used for pre-emergence control of germinating seeds of broadleaf and grassy weeds in sugar beets, tobacco, and tomatoes. There are no registered residential uses of pebulate.

*U.S. Maximum Allowable Residue Levels for Pebulate (40 CFR 180.238):*
The residue level for the following crops is 0.1 ppm: sugar beet roots and tops, and tomato.

*Carcinogen/Hazard Classifications*
**U.S. EPA Carcinogens:** Not likely a carcinogen
**Label Signal Word:** CAUTION, WARNING, DANGER
**WHO Acute Hazard:** Class II, moderately hazardous
*Regulatory Authority:*
- EPA 40 CFR 372.65, Specific Toxic Chemical Listings
- EPA Hazardous Waste Number (RCRA No.): U391
- Actively registered pesticide in California
- RCRA Section 261 Hazardous Constituents
- RCRA Universal Treatment Standards: Wastewater (mg/L), 0.003; Nonwastewater (mg/kg), 1.4
- EPCRA Section 313 Form R *de minimis* concentration reporting level: 1.0%

*Description:* Colorless, flammable liquid. Aromatic odor. Sparingly soluble in water; solubility = 92 mg/L @ 21°C. Molecular weight = 203.36. Density = 0.954 @ 30/4°C. Boiling point = 142°C @ 20 mmHg. Vapor pressure = 0.01 mm @ 20°C.

*Incompatibilities:* Oxidizers.
*Permissible Concentration in Water:* No criteria set. Runoff from spills or fire control may cause water pollution.

*Harmful Effects and Symptoms*
*Short Term Exposure:* Low levels of toxicity. Concentrated solutions are slightly corrosive to eyes and mucous membranes. Dust inhalation can cause irritation of the respiratory system with sneezing. Eye contact can cause irritation, watering, pain, and inflammation of the eyelids. Skin contact can cause irritation and minor ulceration. Ingestion can cause nausea, vomiting, fever, muscle twitching, seizure, rapid respiration, slow heart beat. Severe exposure may result in death.

*Long Term Exposure:* A reversible cholinesterase inhibitor when exposed to high doses.

*Points of Attack:* Respiratory system, central nervous system, cardiovascular system, skin, eyes.

*Medical Surveillance:* Medical observation is recommended for 24 to 48 hours after breathing overexposure, as pulmonary edema may be delayed. As first aid for pulmonary edema, consider administering a corticosteroid spray. Cigarette smoking may exacerbate pulmonary injury and should be discouraged for at least 72 hours following exposure. Before employment and at regular times after that, the following are recommended: If symptoms develop or overexposure occurs, repeat the above tests as soon as possible and get an exam of the nervous system. Also consider complete blood count. Consider chest x-ray following acute overexposure.

*First Aid:* If this chemical gets into the eyes, remove any contact lenses at once and irrigate immediately for at least 15 minutes, occasionally lifting upper and lower lids. Seek medical attention immediately. If this chemical contacts the skin, remove contaminated clothing and wash immediately with soap and water. Seek medical attention immediately. If this chemical has been inhaled, remove from exposure, begin rescue breathing (using universal precautions) if breathing has stopped, and CPR if heart action has stopped. Transfer promptly to a medical facility. When this chemical has been swallowed, get medical attention. If victim is conscious and able to swallow, have victim drink 4 to 8 ounces of water. Do not induce vomiting. *Note to physician or authorized medical personnel*: Medical observation is recommended for 24 to 48 hours after breathing overexposure, as pulmonary edema may be delayed. As first aid for pulmonary edema, consider administering a corticosteroid spray. Cigarette smoking may exacerbate pulmonary injury and should be discouraged for at least 72 hours following exposure.

*References:*
- U.S. Environmental Protection Agency, "Reregistration Eligibility Decision (RED), Pebulate," Office of Prevention, Pesticides and Toxic Substances, Washington, DC (November 1999). http://www.epa.gov/REDs/2500red.pdf
- Pesticide Management Education Program, "Pebulate (Tillam) Herbicide Profile 3/85," Cornell University, Ithaca, NY (March 1985). http://pmep.cce.cornell.edu/profiles/herb-growthreg/naa-rimsulfuron/pebulate/herb-prof-pebulate.html
- U.S. Environmental Protection Agency, Office of Pesticide Programs, Pesticide Residue Limits, "Pebulate," 40 CFR 180.238, www.epa.gov/pesticides/food/viewtols.htm
- California Environmental Protection Agency "Chemical List of Lists," Sacramento CA (February 1997)

# Pelargonic Acid

*Use Type:* Herbicide and fungicide
*CAS Number:* 112-05-0
*Formula:* $C_9H_{18}O_2$
*Synonyms:* Nonanoic acid; *n*-Nonoic acid; *n*-Nonylic acid; 1-Octanecarboxylic acid; Pelargic acid
*Trade Names:* CIRRASOL®-185A; ECONOSAN®; EMERY® 202 (mixture with *n*-octoic acid); EMFAC®-1202; HEXACID® C-9; PELARGON®; SCYTHE®, Dow AgroSciences (USA); WEST AGRO ACID SANITIZER®
*Producers:* Dow AgroSciences (USA)
*Chemical Class:* Fatty acid
*EPA/OPP PC Code:* 217500
*California DPR Chemical Code:* 2739
*Uses:* Pelargonic acid occurs naturally in many plants and animals. It is used to control the growth of weeds and as a blossom thinner for apple and pear trees. It is also used as a food additive; as an ingredient in solutions used to commercially peel fruits and vegetables; in the manufacture of lacquers, plastics and pharmaceuticals.
*Carcinogen/Hazard Classifications*
**Label Signal Word:** CAUTION or DANGER
*Regulatory Authority:*
- Actively registered pesticide in California.

*Description:* Colorless or yellowish, combustible, oily liquid. Faint odor. Slightly soluble in water. Density = 0.9055 @ 20°. Molecular weight 158.24. Melting/Freezing point = 12.4°C. Boiling point = 255.5°C. Vapor pressure = 1 mmHg @ 108°C, 100 mmHg @ 184°C.

*Incompatibilities:* May react violently with strong oxidizers, bromine, 90% hydrogen peroxide, phosphorus trichloride, silver powders or dust. Incompatible with silver compounds. Mixture with some silver compounds forms explosive salts of silver oxalate.

*Permissible Concentration in Water:* No criteria set. Runoff from spills or fire control may cause water pollution.

*First Aid:* If this chemical gets into the eyes, remove any contact lenses at once and irrigate immediately for at least 15 minutes, occasionally lifting upper and lower lids. Seek medical attention immediately. If this chemical contacts the skin, remove contaminated clothing and wash immediately with soap and water. Seek medical attention immediately. If this chemical has been inhaled, remove from exposure, begin rescue breathing (using universal precautions) if breathing has stopped, and CPR if heart action has stopped. Transfer promptly to a medical facility. When this chemical has been swallowed, get medical attention. Give large quantities of water and induce vomiting. Do not make an unconscious person vomit. Do not induce vomiting when formulations containing petroleum solvents are ingested.

*References:*
- U.S. Environmental Protection Agency, Office of Pesticide Programs, "Pelargonic Acid Fact Sheet," (April 2000). http://www.epa.gov/oppbppd1/biopesticides/ingredients/factsheets/factsheet_217500.htm
- California Environmental Protection Agency "Chemical List of Lists," Sacramento CA (February 1997)

# Penconazole

*Use Type:* Fungicide

*CAS Number:* 66246-88-6
*Formula:* $C_{13}H_{15}Cl_2N_3$
*Synonyms:* 1H-1,2,4-Triazole, 1-[2-(2,4-dichlorophenyl)pentyl]-; 1-[2-(2,4-Dichlorophenyl)pentyl]-1H-1,2,4-triazole; (RS)-1-[2-(2,4-Dichlorophenyl)-n-pentyl]-1H-1,2,4-triazole
*Trade Names:* AWARD®; CGA-71818®; ONMEX®; TOPAS®; TOPAS-C®; TOPAS-MZ®; TOPAZ®; TOPAZE®; TOPAZE-C®
*Producers:* Syngenta (Switzerland)
*Chemical Class:* Azole
*EPA/OPP PC Code:* 128999
*Uses:* Not registered in the U.S. Penconazole is a systemic fungicide used to control powdery mildew. It is used on apples and grapes and other fruits, hops, tobacco, ornamentals and on vegetables.
*Carcinogen/Hazard Classifications*
**WHO Acute Hazard:** Class U, unlikely to be hazardous
**Endocrine Disruptor:** Listed as a potential ED by the German Federal Environmental Agency
*Permissible Concentration in Water:* No criteria set. Runoff from spills or fire control may cause water pollution.
*Routes of Entry:* Ingestion, skin contact.
*Harmful Effects and Symptoms*
*Short Term Exposure:* Poisonous if swallowed. Contact may irritate skin and cause eye irritation and possible severe injury. Avoid inhalation.
*Long Term Exposure:* A potential endocrine disruptor that can affect thyroid, prostate and testes weight.
*Medical Surveillance:* If this chemical gets into the eyes, remove any contact lenses at once and irrigate immediately for at least 15 minutes, occasionally lifting upper and lower lids. Seek medical attention immediately. If this chemical contacts the skin, remove contaminated clothing and wash immediately with soap and water. If this chemical has been inhaled, remove from exposure, begin rescue breathing (using universal precautions) if breathing has stopped, and CPR if heart action has stopped. Transfer promptly to a medical facility. When this chemical has been swallowed, get medical attention. Give large quantities of water and induce vomiting. Do not make an unconscious person vomit

# Pencycuron

*Use Type:* Fungicide
*CAS Number:* 66063-05-6
*Formula:* $C_{19}H_{21}ClN_2O$
*Synonyms:* 1-(p-Chlorobenzyl)-1-cyclopentyl-3-phenylurea; N-[(4-Chlorophenyl)-methyl]-N-cyclopentyl-N'-phenylurea; Urea, N-[(4-chlorophenyl)methyl]-N-cyclopentyl-N'-phenyl-
*Trade Names:* BAY® NTN-19701, Bayer Crop Science (Germany); BAYER® NTN-19701, Bayer Crop Science (Germany); MONCEREN®; NTN-19701®, Bayer Crop Science (Germany); PENCYCURONE®
*Producers:* AgriChem (Netherlands); Bayer Crop Science (Germany); Saeryung Chemicals (South Korea); Veterinary & Agricultural Products Manufacturing Co. Ltd. (VAPCO); Yellow River Enterprise (Taiwan)
*Chemical Class:* Substituted urea
*EPA/OPP PC Code:* 128823
*Uses:* Not registered in the U.S. Pencycuron is a non-systemic fungicide with specific control against fungicidal diseases (*Rhizoctonia solani* and *Pellicularia*) in potatoes, sugar beets, rice, cotton and ornamentals.
*Carcinogen/Hazard Classifications*
**WHO Acute Hazard:** Class U, unlikely to be hazardous
*Description:* Colorless crystalline solid. Melting point = 129.5°C. Molecular weight = 328.8
*Incompatibilities:* Noncorrosive to metals.
*Permissible Concentration in Water:* No criteria set. Runoff from spills or fire control may cause water pollution.
*Harmful Effects and Symptoms*
*Short Term Exposure:* Contact with eyes or skin may cause irritation or burns. Inhalation should be avoided; use NIOSH-approved air purifying respirators for pesticides. May be harmful if swallowed. Skin contact may cause allergic reaction.
*Long Term Exposure:* Repeated or prolonged contact may cause skin and lung sensitization, resulting in allergies.
*Points of Attack:* Skin.
*Medical Surveillance:* Evaluation by a qualified allergist, including careful exposure history and special testing, may help diagnose skin allergy.
*First Aid:* If this chemical gets into the eyes, remove any contact lenses at once and irrigate immediately for at least 15 minutes, occasionally lifting upper and lower lids. Seek medical attention immediately. If this chemical contacts the skin, remove contaminated clothing and wash immediately with soap and water. Seek medical attention immediately. If this chemical has been inhaled, remove from exposure, begin rescue breathing (using universal precautions) if breathing has stopped, and CPR if heart action has stopped. Transfer promptly to a medical facility. When this chemical has been swallowed, get medical attention. Give large quantities of water and induce vomiting. Do not make an unconscious person vomit.

# Pendimethalin (ANSI)

*Use Type:* Herbicide
*CAS Number:* 40487-42-1
*Formula:* $C_{13}H_{19}N_3O_4$
*Synonyms:* N-(Aethylpropyl)-3,4-dimethyl-2,6-dinitroanilin (German); N-(1-Aethylpropyl)-2,6-dinitro-3,4-xylidin (German); Aniline, 3,4-dimethyl-2,6-dinitro-N-(1-ethylpropyl)-; Benzenamine, 3,4-dimethyl-2,6-dinitro-N-(1-ethylpropyl)-; Benzenamine, N-(1-ethylpropyl)-3,4-dimethyl-2,6-dinitro-; Caswell No. 454BB; 2,5-Dinitro-N-

(1-ethylpropyl)-3,4-xylidine; 3,4-Dimethyl-2,6-dinitro-*N*-(1-ethylpropyl)aniline; *N*-(1-Ethylpropyl)-2,6-dinitro-3,4-xylidine; *N*-(1-Ethylpropyl)-3,4-dimethyl-2,6-dinitrobenzenamine; *N*-(1-Ethylpropyl)-3,4-dimethyl-2,6-dinitroanaline; Pendimethaline; *N*-(3-Pentyl)-3,4-dimethyl-2,6-dinitroaniline; Tendimethalin; 3,4-Xylidine, 2,6-dinitro-*N*-(1-ethylpropyl)-

*Trade Names:* AC 92553®; ACCOTAB®; BULLET® (pendimethalin + cyanazine), Makhteshim-Agan Industries (Israel); GO-GO-SAN®; HERBADOX®, BASF Agricultural Products (Germany); PAY-OFF®; PENOXALIN®; PENOXALINE®; PROWL®, BASF Agricultural Products (Germany); SIPAXOL®; SQUADRON® (with imazaquin); STOMP®, BASF Agricultural Products (Germany); TATA PANIDA®, Rallis India (India); VALOR®, BASF Canada (Canada); WAY-UP®

*Producers:* Agsin (Singapore); AJE (Switzerland); Bharat Rasayan (India); BASF Agricultural Products Group (Germany); BASF Canada (Canada); Dow AgroSciences (USA); Makhteshim-Agan Industries (Israel) Rallis India (India); The Scotts Company (USA); Shandong Huayang Pesticide Group (China); Yellow River Enterprise (Taiwan); Wuzhou International (China)

*Chemical Class:* 2,6-dinitroaniline

*EPA/OPP PC Code:* 108501; (454300 old EPA code number)

*California DPR Chemical Code:* 1929

*RTECS Number:* BX5470000

*Uses:* Pendimethalin is a selective pre-emergence and post-emergence herbicide used on various agricultural and non-agricultural sites to control broadleaf weeds and grassy weeds in crops such as apricot, carrot, cherry, corn, cotton, fig, garbanzos, garlic, olive, onion, nectarine, peach, pear, pecan, plum, rice and prune, and noncrop areas. It is applied to soil preplant, pre-emergence, and post-emergence with ground and aerial equipment.

*Carcinogen/Hazard Classifications*

**U.S. EPA Carcinogens:** Group C, possible carcinogen
**Label Signal Word:** CAUTION, WARNING
**WHO Acute Hazard:** Class III, slightly hazardous
**Endocrine Disruptor:** Suspected endocrine dispruptor
*Regulatory Authority:*
- EPA 40 CFR 372.65, Specific Toxic Chemical Listings
- Actively registered pesticide in California.
- EPCRA Section 313 Form R *de minimis* concentration reporting level: 1.0%

*Description:* Yellow-orange crystalline solid. Odorless. Slightly soluble in water; solubility = 0.32 mg/L @ 20°C. Molecular weight = 281.33. Boiling point = decomposes. Melting/Freezing point = 55-58°C. Vapor pressure = 9.4 x $10^{-6}$ mmHg @ 25°C.

*Incompatibilities:* Strong oxidizers.

*Permissible Concentration in Water:* No criteria set. Runoff from spills or fire control may cause water pollution.

*Routes of Entry:* Absorbed through the skin, inhalation, ingestion.

*Harmful Effects and Symptoms*

*Short Term Exposure:* Irritates the skin, eyes, and respiratory tract. Eye contact may cause irritation, burning sensation, and damage. Harmful if ingested, inhaled or absorbed through the skin. Inhalation should be avoided; use NIOSH-approved air purifying respirators for pesticides. May be harmful if swallowed.

*Medical Surveillance:* Consult a physician if poisoning is suspected or if redness, itching, or burning of the eyes or skin develop.

*First Aid:* If this chemical gets into the eyes, remove any contact lenses at once and irrigate immediately for at least 15 minutes, occasionally lifting upper and lower lids. Seek medical attention immediately. If this chemical contacts the skin, remove contaminated clothing and wash immediately with soap and water. Seek medical attention immediately. If this chemical has been inhaled, remove from exposure, begin rescue breathing (using universal precautions) if breathing has stopped, and CPR if heart action has stopped. Transfer promptly to a medical facility. When this chemical has been swallowed, get medical attention. Give large quantities of water and induce vomiting. Do not make an unconscious person vomit. Do not induce vomiting when formulations containing petroleum solvents are ingested.

*References:*
- U.S. Environmental Protection Agency, "Reregistration Eligibility Decision (RED), Pendimethalin," Office of Prevention, Pesticides and Toxic Substances, Washington, DC (June 1997). http://www.epa.gov/REDs/0187red.pdf
- EXTOXNET, Extension Toxicology Network, "Pesticide Information Profile, Pendimethalin," Oregon State University, Corvallis, OR (June 1996). http://extoxnet.orst.edu/pips/pendimet.htm
- California Environmental Protection Agency "Chemical List of Lists," Sacramento CA (February 1997)

# Pentachlorophenol

*Use Type:* Fungicide, herbicide, slimicide and wood preservative
*CAS Number:* 87-86-5
*Formula:* $C_6HCl_5O$; $C_6Cl_5OH$
*Alert:* A Restricted Use Pesticide (RUP). It was banned as a herbicide and similar uses in 1987 and today is used as a wood preservative. Human toxicity (long-term): High
*Synonyms:* 1-Hydroxypentachlorobenzene; NCI-C54933; NCI-C55378; NCI-C56655; PCP; Penchlorol; Penta; Pentachloorfenol (Dutch); Pentachlorofenol; Pentaclorofenolo (Italian); Pentachlorophenate; 2,3,4,5,6-Pentachlorophenol; Pentachlorphenol (German);

# Pentachlorophenol

Pentachlorophenol, technical; Pentaclorofenol (Spanish); Phenol, pentachloro-

***Trade Names:*** (The U.S. EPA lists 626 active and canceled/transferred label for this chemical) CHEM-TOL®; CHLON®; CHLOROPHEN®; CRYPTOGIL OL®; DOWCIDE® 7, Dow AgroSciences LLC (USA); DOWICIDE® 7, Dow AgroSciences LLC (USA); DOW PENTACHLOROPHENOL DP-2 ANTIMICROBIAL®, Dow AgroSciences LLC (USA); DURA TREET II®; DUROTOX®; EP 30®; FORPEN-50®; FUNGIFEN®; GLAZD-PENTA®, Vulcan Chemicals (USA); GRUNDIER ARBEZOL®; LAUXTOL®; LIROPREM®; ONTRACK WE HERBICIDE®; ORTHO TRIOX®; OSMOSE WPC®; PENTACHLOROPHENOL, DOWICIDE EC-7®, Dow AgroSciences LLC (USA); PENTACHLOROPHENOL, DP-2®; PENTACON®; PENTA-KIL®; PENTA READY®; PENTASOL®; PENWAR®; PERATOX®; PERMACIDE®; PERMAGARD®; PERMASAN®; PERMATOX DP-2®; PERMATOX PENTA®; PERMITE®; POL NU®; PREVENTOL P®, Bayer CropScience (Germany); PRILTOX®; SANTOBRITE®; SANTOPHEN®; SINITUHO®; TERM-I-TROL®; THOMPSON'S WOOD FIX®; WATERSHED WP®; WEEDONE®, Nufarm Ltd. (Australia); WOODTREAT A®

***Producers:*** Dow AgroSciences LLC (USA); Ehrenstorfer, Dr. (Germany); KMG Chemicals (USA); ISK Biosciences (UK); Nufarm Ltd. (Australia); Sigma-Aldrich Laborchemikalien (Germany); Vulcan Chemicals (USA)

***Chemical Class:*** Chlorinated phenol

***EPA/OPP PC Code:*** 063001 and 600021

***California DPR Chemical Code:*** 465

***ICSC Number:*** 0069

***RTECS Number:*** SM6300000

***EEC Number:*** 604-002-00-8

***EINECS Number:*** 201-778-6

***Uses:*** Pentachlorophenol (PCP) is a commercially produced insecticide, fungicide, and slimicide. Since 1984 it has been restricted to certified applicators and is no longer available to the general public. It is primarily used to protect timber from fungal rot and wood-boring insects, but may also be used as a pre-harvest defoliant in cotton, a general pre-emergence herbicide, and as a biocide in industrial water systems. It major uses are on utility poles, pilings and railroad ties. It is available in blocks, flakes, granules, liquid concentrates, wettable powders, or ready-to-use petroleum solutions.

***Carcinogen/Hazard Classifications***
**U.S. EPA Carcinogens:** Group B2, probable carcinogen
**California Prop. 65:** Carcinogen
**IARC:** Group 2B, possible carcinogen
**WHO Acute Hazard:** Class 1b, highly hazardous
**Endocrine Disruptor:** Suspected

***Regulatory Authority:***
- Carcinogen (Human Suspected) (IARC)[9]
- Banned or Severely Restricted (Several Countries) (UN)[13]
- Air Pollutant Standard Set (ACGIH)[1] (DFG)[3] (HSE)[33] (OSHA)[58] (former USSR)[43] (Several States)[60]
- List of priority pollutants (U.S. EPA)
- AB 1803-Well Monitoring Chemical (CAL)
- MCL (Maximum Contaminants Levels) list of contaminants (CAL)
- AB 2588-Air Toxics "Hot Spots" Chemicals (CAL)
- CAL-DHS/DHS Drinking Water Action Levels
- CAL Air Resources Board/AB 1807 Toxic Air Contaminants
- Proposition 65 chemical (CAL)
- Permissible Exposure Limits for Chemical Contaminants (CAL/OSHA)
- The "Director's List" (CAL/OSHA)
- Actively registered pesticide in California.
- Clean Air Act: Hazardous Air Pollutants (Title I, Part A, Section 112)
- Clean Water Act: Section 311 Hazardous Substances/RQ 40CFR117.3 (same as CERCLA, see below); 40CFR401.15 Section 307 Toxic Pollutants; 40CFR423, Appendix A, Priority Pollutants; Section 313 Water Priority Chemicals (57FR41331, 9/9/92)
- EPA Hazardous Waste Number (RCRA No.): D037
- RCRA, 40CFR261, Appendix 8 Hazardous Constituents
- RCRA Toxicity Characteristic (Section 261.24), Maximum
- Concentration of Contaminants, regulatory level, 100 mg/L
- RCRA 40CFR268.48; 61FR15654, Universal Treatment Standards: Wastewater (mg/L), 0.089; Nonwastewater (mg/kg), 7.4
- RCRA 40CFR264, Appendix 9; TSD Facilities Ground Water Monitoring List. Suggested test method(s) (PQL $ug/L$): 8040(5); 8270(50)
- Safe Drinking Water Act: MCL, 0.001 mg/L; MCLG, zero; Regulated chemical (47 FR 9352)
- Superfund/EPCRA 40CFR302.4 RQ: CERCLA, 10 lb (4.54 kg)
- EPCRA Section 313 Form R *de minimus* concentration reporting level: 1.0%
- U.S. DOT Regulated Marine Pollutant (49CFR172.101, Appendix B), severe pollutant
- Canada, WHMIS, Ingredients Disclosure List

***Description:*** Pentachlorophenol is a colorless to white, crystalline solid. The technical grade is dark grey to brown. Benzene-like odor; pungent odor when hot. The odor threshold in water is 1600 $\mu g/L$ and the taste threshold in water is 30 $\mu g/L$. Practically insoluble in water; solubility = 15 ppm @ 20°C. Molecular weight = 266.35. Boiling point =310°C (decomposes). Melting/Freezing point = 187–189°C (anhydrous). Vapor pressure = 1.3 × 10$^{-4}$

mmHg @ 20°C. Log $K_{ow}$ = 4.98. Values at or above 3.0 are likely to bioaccumulate in marine organisms.

***Incompatibilities:*** Reacts violently with strong oxidizers, acids, alkalies and water.

***Permissible Exposure Limits in Air:*** The OSHA[2] TWA, the DFG MAK[3], the HSE[33] TWA and the recommended NIOSH[2] and ACGIH[1] TWA value is 0.5 mg/m$^3$ and the STEL value set by HSE[33] is 1.5 mg/m$^3$. The NIOSH[2] IDLH level = 2.5 mg/m$^3$. The notation "skin" is added to indicate the possibility of cutaneous absorption. The former USSR-UNEP/IRPTC project[43] has set an MAC in workplace air of 0.1 mg/m$^3$ and an MAC in ambient basis. Several states have set guidelines or standards for Pentachlorophenol in ambient air[60] ranging from zero (North Carolina) to 0.034 µg/m$^3$ (Massachusetts) to 1.67 µg/m$^3$ (New York) to 5.0 µg/m$^3$ (North Dakota and South Carolina) to 8.0 µg/m$^3$ (Virginia) to 10.0 µg/m$^3$ (Connecticut and South Dakota) to 12.0 µg/m$^3$ (Nevada and Pennsylvania) to 25.64 µg/m$^3$ (Kansas).

***Determination in Air:*** Sample collection by mixed cellulose ester membrane filter in series with ethylene glycol bubbler, analysis by high performance liquid chromatography with UV detection. NIOSH IV, Method #5512[18].

***Permissible Concentration in Water:*** The U.S. EPA has set a limit for drinking water of 1 part of pentachlorophenol per billion parts of water (1 ppb). A value of 30 µg/L is set on an organoleptic basis[6]. To protect freshwater aquatic life-55 µg/L on an acute toxicity basis and 3.2 µg/L on a chronic basis. To protect saltwater aquatic life-53µg/L on an acute basis and 34 µg/L on a chronic basis. More recently, EPA[47] has developed a lifetime health advisory of 220 µg/L. WHO[35] has set a limit on Pentachlorophenol in drinking water of 10 µg/L. The former USSR[43] has set an MAC in water bodies used for domestic purposes of 300 µg/L. More recently, EPA has set a guideline for drinking water of 200 µg/L[62]. Several states have set guidelines for Pentachlorophenol in drinking water[61] ranging from 6 µg/L (Maine) to 30 µg/L (California) to 200 µg/L (Arizona) to 220 µg/L (Kansas and Minnesota).

***Determination in Water:*** Methylene chloride extraction followed by gas chromatography with electron capture or halogen specific detection (EPA Method 608) or gas chromatography plus mass spectrometry (EPA Method 625).

***Routes of Entry:*** Inhalation, ingestion, skin and/or eye contact. Absorbed by the skin.

### Harmful Effects and Symptoms

***Short Term Exposure:*** Irritates the eyes, skin and respiratory tract. May affect the cardiovascular system. *Inhalation:* Levels of 1 mg/m$^3$ can cause severe irritation of the nose, throat and lungs. Higher exposures can cause pulmonary edema, a medical emergency that can be delayed for several hours. This can cause death. Breathing dust or particulates tainted with Pentachlorophenol can give rise to sneezing. *Skin:* A 0.04% solution can cause pain and inflammation at point of contact. Chloracne, a skin disorder, has been observed in workers in Pentachlorophenol manufacturing plants and wood preserving operations. Profuse sweating and elevated temperature are symptoms of poisoning due to prolonged contact. Excessive skin exposure has caused human death. *Eyes:* Levels of 1 mg/m$^3$ may be irritating and excessive contact can lead to loss of sight due to corneal damage. *Ingestion:* The lethal human dose is approximately equal to 1 teaspoon for a 150 lb person. Ingestion of 4 to 8 ounces followed by prompt emergency treatment still produced symptoms of poisoning which included rapid breathing followed by a decrease in breathing rate, abdominal pain, reduced blood pressure, excessive and slurred speech and weakness.

***Long Term Exposure:*** Irritation of eyes, throat, nose and upper lungs have been reported by individuals using pentachlorophenol as an insecticide for periods of a few years. Chemical acne has been associated with prolonged exposure to this compound. May affect the central nervous system, kidneys, liver, lungs. May be a carcinogen in humans. May damage the developing fetus. There is limited evidence that pentachlorophenol is a teratogen in animals. Some studies have found an increase in cancer risk in workers exposed to high levels of technical grade pentachlorophenol for a long time, but other studies have not found this. Increases in liver, adrenal gland, and nasal tumors have been found in laboratory animals exposed to high doses of pentachlorophenol. The U.S. EPA has determined that pentachlorophenol is a probable human carcinogen and the International Agency for Cancer Research (IARC) considers it possibly carcinogenic to humans. We do not know if exposure to pentachlorophenol will result in birth defects or other developmental effects in people. Death, low body weights, decreased growth, and skeletal effects have been observed in laboratory animals exposed to high levels of pentachlorophenol during development.

***Points of Attack:*** Eyes, skin, respiratory system, cardiovascular system, liver, kidneys and central nervous system. Cancer site in animals: liver.

***Medical Surveillance:*** If symptoms develop or overexposure is suspected, the following may be useful: Urine test for pentachlorophenol. Liver and kidney function tests. Refer to the NIOSH Criteria Documents #78-174 and #76-147 on Manufacturing, formulating, and working safely with pesticides.

***First Aid:*** If this chemical gets into the eyes, remove any contact lenses at once and irrigate immediately for at least 15 minutes, occasionally lifting upper and lower lids. Seek medical attention immediately. If this chemical contacts the skin, remove contaminated clothing and wash immediately with soap and water. Seek medical attention immediately. If this chemical has been inhaled, remove from exposure, begin rescue breathing (using universal precautions) if breathing has stopped, and CPR if heart action has stopped. Transfer promptly to a medical facility. When this chemical has been swallowed, get medical attention. Give large quantities of water and induce vomiting. Do not make an unconscious person vomit. Medical observation is recommended for 24 to 48 hours after breathing overexposure, as pulmonary edema may be delayed. As first aid for pulmonary edema, a doctor or authorized paramedic may consider administering a corticosteroid spray.

*References:*
- EXTOXNET, Extension Toxicology Network, "Pesticide Information Profile, Pentachlorophenol," Oregon State University, Corvallis, OR (June 1996). http://extoxnet.orst.edu/pips/pentachl.htm
- U.S. Environmental Protection Agency, "ToxFAQs for Pentachlorophenol," Washington, DC, (September 2001). http://www.atsdr.cdc.gov/tfacts51.html
- New Jersey Department of Health, "Hazardous Substance Fact Sheet: Pentachlorophenol," Trenton, NJ (September 1996, rev. August 2002). http://www.state.nj.us/health/eoh/rtkweb/1473.pdf
- U.S. Environmental Protection Agency, Pentachlorophenol: Ambient Water Quality Criteria, Washington, DC (1980).
- Rao, K.R., Ed., "Pentachlorophenol: Chemistry, Pharmacology and Environmental Toxicology," Proceedings of a Symposium, Pensacola, FL, June 1977, New York, Plenum Press (1978).
- U.S. Environmental Protection Agency, "Pentachlorophenol, Health and Environmental Effects," Profile No. 143, Office of Solid Waste, Washington, DC (April 30, 1980).
- Sax, N.I., Ed., "Dangerous Properties of Industrial Materials Report," 3, No. 4, 73-77 (1983) and 4, No. 3, 24-26 (1984).
- U.S. Public Health Service, "Toxicological Profile for Pentachlorophenol," Atlanta, Georgia, Agency for Toxic Substances and Disease Registry (Dec. 1988).
- New York State Department of Health, "Chemical Fact Sheet: Pentachlorophenol," Albany, NY, Bureau of Toxic Substance Assessment (Version 2-March 1986 and Version 3).
- California Environmental Protection Agency "Chemical List of Lists," Sacramento CA (February 1997).

# Peracetic Acid

*Use Type:* Fungicide, herbicide, rodenticide and microbiocide
*CAS Number:* 79-21-0
*Formula:* $C_2H_4O_3$; $CH_3COOOH$
*Synonyms:* Acido peracetico (Spanish); Acetic peroxide; Acetyl hydroperoxide; Acide peracetique (French); Ethaneperoxoic acid; Hydrogen peroxide and peroxyacetic acid mixture; Hydroperoxide, acetyl; Kyselina peroxyoctova (Polish); Monoperacetic acid; Peroxyacetic acid

*Trade Names:* DESOXON 1®; ESTOSTERIL®; OSBON AC®; OXYMASTER®; PROXITANE®, Solvay Group (Belgium)
*Producers:* Air Liquide Group (France); Boehringer Ingelheim (Germany); Camphor & Allied Products (India); Daicel Chemical Industries (Japan); Degussa (Germany); FMC Agricultural Products Group (USA); Laporte (UK); Mitsubishi Gas Chemical (Japan); Solvay Group (Belgium); Spectrum Chemical Mfg. (USA)
*Chemical Class:* Inorganic
*EPA/OPP PC Code:* 063201
*California DPR Chemical Code:* 2291
*ICSC Number:* 1031
*RTECS Number:* SD8750000
*EEC Number:* 607-094-00-8
*EINECS Number:* 201-186-8
*Uses:* This compound is used as bactericide and fungicide, especially in food processing, a reagent in making caprolactam and glycerol; an oxidant for preparing epoxy compounds; a bleaching agent; a sterilizing agent; and a polymerization catalyst for polyester resins.
*Carcinogen/Hazard Classifications*
**Label Signal Word: DANGER**
*Regulatory Authority:*
- Highly Reactive Substance and Explosive (World Bank)[15]
- EPA/SARA 302 (EPCRA) Extremely hazardous substances
- AB 2588-Air Toxics "Hot Spots" Chemicals (CAL)
- Actively registered pesticide in California.
- Clean Air Act: Accidental Release Prevention/Flammable substances, (Section 112[r], Table 3), TQ = 10,000 lb (4540 kg)
- Superfund/EPCRA 40CFR355, Appendix B Extremely Hazardous Substances: TPQ = 500 lb (227kg)
- Superfund/EPCRA 40CFR302.4 RQ: EHS, 1 lb (0.454 kg)
- EPCRA Section 313 Form R *de minimis* concentration reporting level: 1.0%

*Description:* Peracetic acid is a colorless liquid. Transported and stored in diluted solution to prevent explosion. Except where noted, the following data is for PAA diluted with 60% acetic acid: Boiling point = 105°C (violent decomposition @ 110°C). Melting/Freezing point = 0.1°C. Vapor Pressure = 14.8 mmHg @ 25°C. Flash point = 41°C; 56°C (32% in dilute acetic acid and < 6% hydrogen peroxide). Hazard Identification (based on NFPA-704 M Rating System): Health 3, Flammability 2, Reactivity 4 (Oxidizer). Soluble in water. Log $K_{ow}$ = Negative; < –0.9. Unlikely to bioaccumulate in marine organisms.
*Incompatibilities:* This material is a powerful oxidizer. Thermally unstable, it decomposes violently @ 110°C. Concentrated material is shock- and friction-sensitive. May explode if concentration exceeds 56% of carrier, due to evaporation. Isolate from other stored material, particularly

accelerators, oxidizers, organic or combustible materials, olefins, hydrogen peroxide, acetic anhydride, reducing substances. Keep away from acids, alkalies, heavy metals, organic materials.

*Permissible Exposure Limits in Air:* No standards set.

*Permissible Concentration in Water:* No criteria set, but runoff from spills or fire control may cause water pollution.

*Routes of Entry:* Inhalation, ingestion, skin and/or eye contact.

*Harmful Effects and Symptoms*

*Short Term Exposure:* Eye contact can cause severe irritation and burns; may cause permanent damage. Irritates the respiratory tract. Contact may burn the skin. Higher exposures can cause pulmonary edema, a medical emergency that can be delayed for several hours. This can cause death. Signs and symptoms of acute ingestion of peracetic acid may include corrosion of mucous membranes of mouth, throat, and esophagus with immediate pain and dysphagia (difficulty swallowing), ingestion may cause gastrointestinal tract irritation. This is a very toxic compound. The probable human oral lethal dose is 50-500 mg/kg, or between 1 teaspoon and 1 ounce for a 150 lb persons.

*Long Term Exposure:* There is limited evidence that PAA causes cancer in animals. It may cause cancer of the lungs. High or repeated exposure may affect the liver and kidneys.

*Points of Attack:* Liver, kidneys and lungs.

*Medical Surveillance:* Liver and kidney function tests. Consider chest x-ray following acute overexposure.

*First Aid:* If this chemical gets into the eyes, remove any contact lenses at once and irrigate immediately for at least 15 minutes, occasionally lifting upper and lower lids. Seek medical attention immediately. If this chemical contacts the skin, remove contaminated clothing and wash immediately with soap and water. Seek medical attention immediately. If this chemical has been inhaled, remove from exposure, begin rescue breathing (using universal precautions) if breathing has stopped, and CPR if heart action has stopped. Transfer promptly to a medical facility. When this chemical has been swallowed, get medical attention. If victim is <u>conscious</u>, administer water, or milk. Do not induce vomiting. Medical observation is recommended for 24 to 48 hours after breathing overexposure, as pulmonary edema may be delayed. As first aid for pulmonary edema, a doctor or authorized paramedic may consider administering a corticosteroid spray.

*References:*

- U.S. Environmental Protection Agency, "Chemical Profile: Peracetic Acid," Washington, DC, Chemical Emergency Preparedness Program (Nov. 30, 1987).
- New Jersey Department of Health and Senior Services, "Hazardous Substance Fact Sheet, Peroxyacetic acid," Trenton NJ (March 1998, revised March 1998). http://www.state.nj.us/health/eoh/rtkweb/1482.pdf
- New York State Department of Health, "Chemical Fact Sheet: Peracetic Acid," Albany, NY, Bureau of Toxic Substance Assessment (April 1986).
- California Environmental Protection Agency "Chemical List of Lists," Sacramento CA (February 1997).

# Perchloromethyl Mercaptan

*Use Type:* Used a pesticide intermediate; also used in chemical warfare.

*CAS Number:* 594-42-3

*Formula:* $CCl_4S$; $CCl_3SCl$

*Alert:* Not registered for use in the U. S.

*Synonyms:* Mercaptan methylique perchlore (French); Perchlormethylmerkaptan (Czech); Trichloromethylsulfenyl chloride; Trichloromethylsulphenyl chloride PCV; Perchloromethanethiol; PMM; Trichloromethane sulfenyl chloride; Trichloromethyl sulfur chloride

*Trade Names:* CLAIRSIT®

*Producers:* Cerexagri (France); Carbolabs (USA); Chevron Phillips Chemical (USA); Sigma-Aldrich Fine Chemicals (SA)

*EPA/OPP PC Code:* 477300

*ICSC Number:* 0311

*RTECS Number:* PB0370000

*EINECS Number:* 209-840-4

*Uses:* Perchloromethyl mercaptan is used as an intermediate in pesticide manufacture, specifically in the manufacture of Captan and Folpet. Has been considered as a warfare tear gas because of its highly irritant properties. Not registered for use as a pesticide in the U. S.

*Regulatory Authority:*

- Air Pollutant Standard Set (ACGIH)[1] (HSE)[33] (OSHA)[58] (former USSR)[43] (Several States)[60]
- EPA/SARA 302 (EPCRA) Extremely hazardous substances
- U.S. DOT Inhalation Hazard Chemicals
- Permissible Exposure Limits for Chemical Contaminants (CAL/OSHA)
- The "Director's List" (CAL/OSHA)
- OSHA 29CFR1910.119, Appendix A, Process Safety List of Highly Hazardous Chemicals, TQ = 150 lb (67.5 kg)
- Canada, WHMIS, Ingredients Disclosure List
- U.S. DOT Regulated Marine Pollutant (49CFR172.101, Appendix B), severe pollutant

*Description:* Pale yellow, oily liquid. Unbearable, foul-smelling odor. Insoluble in water. Molecular weight = 185.89. Boiling point = 147°C (decomposes). Vapor pressure = 2.8 mmHg @ 20°C; 0.39 kPa @ 20°C. Hazard Identification (based on NFPA-704 M Rating System): Health 2, Flammability 0, Reactivity 0.

*Incompatibilities:* Water contact forms hydrochloric acid, sulfur, and carbon dioxide. Reacts with alkalies, amines, hot water alcohols, oxidizers, reducing agents, iron, and steel. Attacks most metals.

*Permissible Exposure Limits in Air:* The Federal standard,[58] the HSE[33] TWA and the ACGIH[1] TWA value is 0.1 ppm (0.8 mg/m$^3$). There is no STEL value set. The NIOSH[2] IDLH level = 10 ppm. The former USSR-UNEP/IRPTC project[43] has set an MAC in workplace air of 1.0 mg/m.$^3$ Several states have set guidelines or standards for PMM in ambient air[60] ranging from 8.0 $\mu$g/m$^3$ (North Dakota) to 13.0 $\mu$g/m$^3$ (Virginia) to 16.0 $\mu$g/m$^3$ (Connecticut) to 19.0 $\mu$g/m$^3$ (Nevada).

*Determination in Air:* Sample collection by charcoal tube, analysis by gas liquid chromatography.

*Permissible Concentration in Water:* No criteria set, but runoff from spills or fire control may cause water pollution.

*Routes of Entry:* Inhalation, ingestion, skin and/or eye contact.

*Harmful Effects and Symptoms*

*Short Term Exposure:* Irritates the eyes, skin, and respiratory tract. Higher exposures can cause pulmonary edema, a medical emergency that can be delayed for several hours. This can cause death. Signs and symptoms of acute exposure to Perchloromethyl-mercaptan may lead to liver, heart, and kidney damage. Respiratory effects include coughing, dyspnea (shortness of breath), painful breathing, and lung congestion. Tachycardia (rapid heart rate) is often observed. Nausea, vomiting, abdominal cramping and diarrhea may also occur. Contact with Perchloromethyl mercaptan may result in severe dermatitis (red, inflamed skin), conjunctivitis (red, inflamed eyes), and burns with ulceration and severe pain. May cause death or permanent injury after short exposure to small quantities. Brief exposure to lower concentrations may produce central nervous system depression and lung, liver, and heart congestion. Severe exposures may be fatal. May be absorbed through the skin in quantities sufficient to cause general toxic effects. Ingestion may cause damage to mucous membranes and result in pain and burning of the mouth and throat, nausea, vomiting, cramps, and diarrhea. In severe cases, tissue ulceration and CNS depression may occur.

*First Aid:* If this chemical gets into the eyes, remove any contact lenses at once and irrigate immediately for at least 15 minutes, occasionally lifting upper and lower lids. Seek medical attention immediately. If this chemical contacts the skin, remove contaminated clothing and wash immediately with soap and water. Seek medical attention immediately. If this chemical has been inhaled, remove from exposure, begin rescue breathing (using universal precautions) if breathing has stopped, and CPR if heart action has stopped. Transfer promptly to a medical facility. When this chemical has been swallowed, get medical attention. Give large quantities of water and induce vomiting. Do not make an unconscious person vomit.

*References:*
- New Jersey Department of Health and Senior Services, "Hazardous Substance Fact sheet, Perchloromethyl Mercaptan," Trenton, NJ (June 1989, revised February 2000). http://www.state.nj.us/health/eoh/rtkweb/1480.pdf
- U. S. Environmental Protection Agency, "Chemical Profile: Perchloromethyl Mercaptan," Washington, DC, Chemical Emergency Preparedness Program (Nov. 30, 1987).
- California Environmental Protection Agency "Chemical List of Lists," Sacramento CA (February 1997).

# Perfluidone (ANSI)

*Use Type:* Herbicide
*CAS Number:* 37924-13-3
*Formula:* $C_{14}H_{12}F_3NO_4S_2$
*Alert:* PBI/Gordon voluntarily canceled the registration of Destun for use on tobacco in October, 1994. Believed to be obsolete or discontinued. The European Commission has not allowed it to be an active ingredient since July 25, 2003.
*Synonyms:* Caswell No. 903D; Methanesulfonamide, 1,1,1-trifluoro-N-(2-methyl-4-(phenylsulfonyl)phenyl)-; 2-Methyl-4-(phenylsulfonyl)trifluoromethanesulfonanilide; N-(4-Phenylsulfonyl-o-tolyl)-1,1,1-trifluoromethanesulfonamide; 1,1,1-Trifluoro-N-[2-methyl-4-(phenylsulfonyl)phenyl]methanesulfonamide
*Trade Names:* DESTUN®, Pbi/Gordon (USA) canceled; MBR 8251®; SB 1528®; WL 43423®
*Producers:* PBI/Gordon (USA)
*Chemical Class:* Sulfonanilide; sulfonamide
*EPA/OPP PC Code:* 108001
*California DPR Chemical Code:* 1895
*Uses:* Not registered in the U.S. Used to control nutsedge species, some grasses and broadleaf weeds in flue-cured tobacco crops.
*Carcinogen/Hazard Classifications*
**Label Signal Word:** CAUTION, DANGER
*Description:* Crystalline solid. Odorless. Soluble in water; solubility = 64 ppm. Melting point = 143-145°C. Molecular weight = 379.4. Vapor pressure = Very low <1 x 10$^{-5}$ mmHg.
*Incompatibilities:* Slowly hydrolyzes in water, releasing ammonia and forming acetate salts. Forms nitrogen oxides, sulfur oxides, and fluorine when heated to decomposition.
*Permissible Concentration in Water:* No criteria set. Runoff from spills or fire control may cause water pollution.
*Harmful Effects and Symptoms*
*Short Term Exposure:* Irritates eyes, skin, and respiratory tract. May be poisonous if ingested.
*Points of Attack:* Liver
*First Aid:* If this chemical gets into the eyes, remove any contact lenses at once and irrigate immediately for at least

15 minutes, occasionally lifting upper and lower lids. Seek medical attention immediately. If this chemical contacts the skin, remove contaminated clothing and wash immediately with soap and water. Seek medical attention immediately. If this chemical has been inhaled, remove from exposure, begin rescue breathing (using universal precautions) if breathing has stopped, and CPR if heart action has stopped. Transfer promptly to a medical facility. When this chemical has been swallowed, get medical attention. Give large quantities of water and induce vomiting. Do not make an unconscious person vomit.

*References:*
- Pesticide Management Education Program, "Perfluidone (Destun) Herbicide Profile 9/85," Cornell University, Ithaca, NY (September 1985). http://pmep.cce.cornell.edu/profiles/herb-growthreg/naa-rimsulfuron/perfluidone/herb-prof-perfluidone.html
- California Environmental Protection Agency "Chemical List of Lists," Sacramento CA (February 1997)

# Phenmedipham

*Use Type:* Herbicide
*CAS Number:* 13684-63-4
*Formula:* $C_{16}H_{16}N_2O_4$
*Synonyms:* Carbamic acid, (3-methylphenyl)-, 3-[(methoxycarbonyl)amino]phenyl ester; 3-[(Methoxycarbonyl)amino]phenyl N-(3-methylphenyl)carbamate; Carbanilic acid, *m*-methyl-, ester with methyl-*m*-hydroxycarbanilate (8CI); Fenmedifam; Carbanilic acid, *m*-hydroxy-, metnyl ester, *m*-methylcarbanilate (ester) (8CI); *m*-Hydroxycarbanilic acid methyl ester *m*-methylcarbanilate; Methyl *m*-hydroxycarbanilate m-methylcarbanilate; 3-(Methylphenyl)carbamic acid 3-[(methoxycarbonyl)amino]phenyl ester; Methyl 3-(*m*-tolylcarbamoyloxy)phenylcarbamate 3-Methoxycarbonylaminophenyl *N*-3'-methylphenylcarbamate; Phenmediphame
*Trade Names:* AIMSAN®; BETAMIX® (phenmedipham + desmedipham), Bayer CropScience (Germany); BETANAL®; CQ 1451® (phenmedipham + desmedipham + ethofumesate), Bayer CropScience (Germany); EC herbicide (phenmedipham + desmedipham + ethofumesate); EP 452®; KEEPER®; KEMIFAM®; MSS HERBASAN®, Whyte Agrochemicals (UK); NA 305® (phenmedipham + desmedipham + ethofumesate), Bayer CropScience (Germany); NA 308® (phenmedipham + desmedipham + ethofumesate); POWERTWIN® (phenmedipham + ethofumesate), Makhteshim-Agan Industries (Israel); PROGRESS® (phenmedipham + desmedipham + ethofumesate), Bayer CropScience (Germany); S-4075®; SCHERING 4072®; SN 38584®; SPIN-AID®, Bayer CropScience (Germany); SYNBETAN-P®; TWIN®, Makhteshim-Agan Industries (Israel); VANGARD®
*Producers:* Bayer CropScience (Germany); Calliope (France); Ki-Hara Chemicals Ltd. (UK); Makhteshim-Agan Industries (Israel); Whyte Agrochemicals (UK)
*Chemical Class:* Bis-carbamate
*EPA/OPP PC Code:* 098701
*California DPR Chemical Code:* 675
*Uses:* A post-emergence herbicide for control of annual broadleaf weeds and grasses in sugar beets, spinach, strawberries, and sunflowers.
*U.S. Maximum Allowable Residue Levels for Phenmedipham (40 CFR 180.278):*

| CROP | ppm |
|---|---|
| Beet, garden, roots | 0.2 |
| Beet, sugar, roots | 0.1 |
| Beet, sugar, tops | 0.1 |
| Spinach | 0.5 |

*Carcinogen/Hazard Classifications*
**U.S. EPA Carcinogens:** Group D, unclassifiable, inadequate data
**Label Signal Word:** WARNING or DANGER
**WHO Acute Hazard:** Class U, unlikely to be hazardous
*Regulatory Authority:*
- Actively registered pesticide in California.

*Description:* Colorless crystalline solid or needles. Odorless. Very slightly soluble in water; solubility = 1 mg/L @ 20°C; 4.7 mg/L @ 25°C. Molecular weight = 300.32. Density = 0.28 g/cm³ @ 20°C. Melting/Freezing point = 139-144°C. Vapor pressure = $1 \times 10^{-11}$ mmHg @ 25°C.
*Incompatibilities:* Noncorrosive. May form explosive materials with phosphorus pentachloride.
*Permissible Concentration in Water:* No criteria set. Runoff from spills or fire control may cause water pollution.
*Routes of Entry:* Skin absorption, ingestion, inhalation.
*Harmful Effects and Symptoms*
*Short Term Exposure:* Eye pupils are small, blurred vision, eye watering, runny nose, cough, shortness of breath, salivation, nausea, stomach cramps, diarrhea, and vomiting, increased blood pressure, profuse sweating, hypermotility, hallucinations, agitation, tingling of the skin, slow heartbeat, convulsions, fluid in lungs, loss of consciousness, incontinence, breathing stops, death. Carbamates inhibit the acetylcholinesterase enzymes and alter the way in which nervous impulses are transmitted. However, within several hours carbamates spontaneously detach from the enzymes.
*Long Term Exposure:* A potent cholinesterase inhibitor; cumulative effect is possible. This chemical may damage the nervous system with repeated exposure, resulting in convulsions, respiratory failure. May cause liver damage.
*Points of Attack:* Respiratory system, lungs, central nervous system, cardiovascular system, skin, eyes, plasma and red blood cell cholinesterase.
*Medical Surveillance:* Medical observation is recommended for 24 to 48 hours after breathing

overexposure, as pulmonary edema may be delayed. As first aid for pulmonary edema, consider administering a corticosteroid spray. Cigarette smoking may exacerbate pulmonary injury and should be discouraged for at least 72 hours following exposure. Before employment and at regular times after that, the following are recommended: Plasma and red blood cell cholinesterase levels (tests for the enzyme poisoned by this chemical). If exposure stops, plasma levels return to normal in 1-2 weeks while red blood cell levels may be reduced for 1-3 months. When acetylcholinesterase enzyme levels are reduced by 25% or more below preemployment levels, risk of poisoning is increased, even if results are in lower ranges of "normal." Reassignment to work not involving carbamate pesticides is recommended until enzyme levels recover. If symptoms develop or overexposure occurs, repeat the above tests as soon as possible and get an exam of the nervous system. Also consider complete blood count. Consider chest x-ray following acute overexposure

**First Aid:** Speed in removing material from eyes and skin is of extreme importance. *Eyes:* Eye contact can cause dangerous amounts of these chemicals to be quickly absorbed through the mucous membrane into the bloodstream. Immediately and gently flush eyes with plenty of warm or cold water (NO hot water) for at least 15 minutes, occasionally lifting the upper and lower eyelids. Get medical aid immediately. *Skin:* Get medical aid. Skin contact can cause dangerous amounts of these chemicals to be absorbed into the bloodstream. Wearing the appropriate PPE equipment and respirator for carbamate pesticides, immediately flush skin with plenty of soap and water for at least 15 minutes while removing contaminated clothing and shoes. Shampoo hair promptly if contaminated; protect eyes. *Ingestion:* Call poison control. Loosen all clothing. Never give anything by mouth to an unconscious person. Get medical aid. Do NOT induce vomiting.* If conscious, alert, and able to swallow, rinse mouth and have victim drink 4 to 8 ounces of water. Check to see if poison control instructs you to use ipecac syrup, otherwise administer slurry of activated charcoal (2 oz in 8 oz of water). If victim is *UNCONSCIOUS OR HAVING CONVULSIONS,* do nothing except keep victim warm. **In some cases you may be specifically instructed by poison control to induce vomiting by way of 2 tablespoons of syrup of ipecac (adult) washed down with a cup of water.* Do NOT give activated charcoal before or with ipecac syrup. *Inhalation:* Get medical aid. Do not contaminate yourself. Wearing the appropriate PPE equipment and respirator for carbamate pesticides, immediately remove the victim from the contaminated area to fresh air. If the victim is not breathing, administer artificial respiration. Do NOT use mouth-to-mouth resuscitation; use bag/mask apparatus. If breathing is difficult, administer oxygen through bag/mask apparatus until medical help arrives. Do not leave victim unattended.

*Note to physician or authorized medical personnel.* Administer atropine, 2 mg (1/30 gr) intramuscularly or intravenously as soon as any local or systemic signs or symptoms of an intoxication are noted; repeat the administration of atropine every 3 to 8 minutes until signs of atropinization (mydriasis, dry mouth, rapid pulse, hot and dry skin) occur; initiate treatment in children with 0.05 mg mg/kg of atropine; repeat at 5 to 10 minute intervals. Watch respiration, and remove bronchial secretions if they appear to be obstructing the airway; intubate if necessary. *Medical note:* 2-PAMCI may be contraindicated in the case of some carbamate poisonings.

**References:**
- U.S. Environmental Protection Agency, Office of Pesticide Programs, Pesticide Residue Limits, "Phenmedipham," 40 CFR 180.278, www.epa.gov/pesticides/food/viewtols.htm
- Pesticide Management Education Program, "Phenmedipham (Betanal, Spin-Aid) herbicide Profile 2/85," Cornell University, Ithaca, NY (February 1985). http://pmep.cce.cornell.edu/profiles/herb-growthreg/naa-rimsulfuron/phenmedipham/herb-prof-phenmedipham.html
- California Environmental Protection Agency "Chemical List of Lists," Sacramento CA (February 1997).

# Phenol, 3-(1-methylethyl)-, methyl carbamate

**Use Type:** Insecticide
**CAS Number:** 64-00-6
**Formula:** $C_{11}H_{15}NO_2$
**Synonyms:** Carbamic acid, methyl-, *m*-cumenyl ester; *m*-Cumenol methylcarbamate; *m*-Cumenyl methylcarbamate; ENT 25,500; ENT 25,543; HIP; 3-Isopropylphenol methylcarbamate; 3-Isopropylphenol-*N*-methylcarbamate; *m*-Isopropylphenol methylcarbamate; *m*-Isopropylphenol-*N*-methylcarbamate; 3-(1-Methylethyl)phenol methylcarbamate; Methylcarbamic acid *m*-cumenyl ester; *m*-Psopropylphenyl methylcarbamate; *m*-Isopropylphenyl-*N*-methylcarbamate; 3-Isopropylphenyl methylcarbamate; *N*-Methyl-*m*-isopropylphenyl carbamate; *N*-Methyl-3-isopropylphenyl carbamate
**Trade Names:** COMPOUND 10854®; H 5727®; H 8757®; HERCULES 5727®; HERCULES AC 5727®; OMS-15®; UC 10854®; UNION CARBIDE UC 10,854®, Union Carbide (USA)
**Producers:** Union Carbide (USA)
**Chemical Class:** Carbamate
**EPA/OPP PC Code:** 047801
**RTECS Number:** FB7875000
**Uses:** Not registered in the U.S.

*Regulatory Authority:*
- RCRA Universal Treatment Standards: Wastewater (mg/L), 0.056; Nonwastewater (mg/kg), 1.4 as *m-cumenyl methylcarbamate.*
- EPCRA Section 302 Extremely Hazardous Substances: TPQ = 500/10,000 lb (227/4,540 kg)
- EPCRA Section 304 RQ: EHS, 1 lb (0.454 kg)

*Description:* White crystalline solid. Odorless. Slightly soluble in water.

*Incompatibilities:* Incompatible with strong alkalies. May form explosive materials with phosphorus pentachloride.

*Permissible Exposure Limits in Air:* No OEL have been established.

*Permissible Concentration in Water:* No criteria set. Runoff from spills or fire control may cause water pollution.

*Routes of Entry:* Inhalation, skin and/or eye contact, ingestion. May be absorbed though the skin.

*Harmful Effects and Symptoms*

*Short Term Exposure:* Eye pupils are small, blurred vision, eye watering, runny nose, cough, shortness of breath, salivation, nausea, stomach cramps, diarrhea, and vomiting, increased blood pressure, profuse sweating, hypermotility, hallucinations, agitation, tingling of the skin, slow heartbeat, convulsions, fluid in lungs, loss of consciousness, incontinence, breathing stops, death. Carbamates inhibit the acetylcholinesterase enzymes and alter the way in which nervous impulses are transmitted. However, within several hours carbamates spontaneously detach from the enzymes.

*Long Term Exposure:* A potent cholinesterase inhibitor; cumulative effect is possible. This chemical may damage the nervous system with repeated exposure, resulting in convulsions, respiratory failure. May cause liver damage.

*Points of Attack:* Respiratory system, lungs, central nervous system, cardiovascular system, skin, eyes, plasma and red blood cell cholinesterase.

*Medical Surveillance:* Medical observation is recommended for 24 to 48 hours after breathing overexposure, as pulmonary edema may be delayed. As first aid for pulmonary edema, consider administering a corticosteroid spray. Cigarette smoking may exacerbate pulmonary injury and should be discouraged for at least 72 hours following exposure. Before employment and at regular times after that, the following are recommended: Plasma and red blood cell cholinesterase levels (tests for the enzyme poisoned by this chemical). If exposure stops, plasma levels return to normal in 1-2 weeks while red blood cell levels may be reduced for 1-3 months. When acetylcholinesterase enzyme levels are reduced by 25% or more below preemployment levels, risk of poisoning is increased, even if results are in lower ranges of "normal." Reassignment to work not involving carbamate pesticides is recommended until enzyme levels recover. If symptoms develop or overexposure occurs, repeat the above tests as soon as possible and get an exam of the nervous system. Also consider complete blood count. Consider chest x-ray following acute overexposure.

*First Aid:* Speed in removing material from skin is of extreme importance. *Eye:* Contact can cause dangerous amounts of these chemicals to be quickly absorbed through the mucous membrane into the bloodstream. Immediately and gently flush eyes with plenty of warm or cold water (NO hot water) for at least 15 minutes, occasionally lifting the upper and lower eyelids. Get medical aid immediately. *Skin:* Get medical aid. Skin and/or eye contact can cause dangerous amounts of these chemicals to be absorbed into the bloodstream. Wearing the appropriate PPE equipment and respirator for carbamate pesticides, immediately flush skin with plenty of soap and water for at least 15 minutes while removing contaminated clothing and shoes. Shampoo hair promptly if contaminated; protect eyes. *Ingestion:* Call poison control. Loosen all clothing. Never give anything by mouth to an unconscious person. Get medical aid. *Do NOT induce vomiting.\** If conscious, alert, and able to swallow, rinse mouth and have victim drink 4 to 8 ounces of water. Check to see if poison control instructs you to use ipecac syrup, otherwise administer slurry of activated charcoal (2 oz in 8 oz of water). If victim is *unconscious or having convulsions,* do nothing except keep victim warm. *\*In some cases you may be specifically instructed by poison control to induce vomiting by way of 2 tablespoons of syrup of ipecac (adult) washed down with a cup of water.* Do NOT give activated charcoal before or with ipecac syrup. *Inhalation:* Get medical aid. Do not contaminate yourself. Wearing the appropriate PPE equipment and respirator for carbamate pesticides, immediately remove the victim from the contaminated area to fresh air. If the victim is not breathing, administer artificial respiration. Do NOT use mouth-to-mouth resuscitation; use bag/mask apparatus. If breathing is difficult, administer oxygen through bag/mask apparatus until medical help arrives. Do not leave victim unattended. *Note to physician or authorized medical personnel.* Administer atropine, 2 mg (1/30 gr) intramuscularly or intravenously as soon as any local or systemic signs or symptoms of an intoxication are noted; repeat the administration of atropine every 3 to 8 minutes until signs of atropinization (mydriasis, dry mouth, rapid pulse, hot and dry skin) occur; initiate treatment in children with 0.05 mg mg/kg of atropine; repeat at 5 to 10 minute intervals. Watch respiration, and remove bronchial secretions if they appear to be obstructing the airway; intubate if necessary. *Medical note:* 2-PAMCl may be contraindicated in the case of some carbamate poisonings.

*References:*
- California Environmental Protection Agency "Chemical List of Lists," Sacramento CA (February 1997).

# Phenothiazine

*Use Type:* Insecticide
*CAS Number:* 92-84-2

# Phenothiazine

*Formula:* $C_{12}H_9NS$; $S(C_6H_4)_2NH$

*Alert:* There are no current EPA registrations for phenothiazine.

*Synonyms:* Dibenzoparathiazine; Dibenzothiazine; Dibenzo-1,4-thiazine; ENT 38; Feeno; Fenothiazine (Dutch); Fenotiazina (Italian); Phenthiazine; Thiodifenylamine (Dutch); Thiodiphenylamin (German); Tiodifenilamina (Italian)

*Trade Names:* AFI-TIAZIN®; AGRAZINE®; ANTIVERM®; BIVERM®; CONTAVERN®; FENOVERM®; FENTIAZIN®; HELMETINA®; LETHELMIN®; NEMAZENE®; NEMAZINE®; ORIMON®; PADOPHENE®; PENTHAZINE®; PHENEGIC®; PHENOSAN®; PHENOVERM®; PHENOVIS®; PHENOXUR®; PTZ®, Avecia (UK); RECONOX®; SOUFRAMINE®; VERMITIN®; WURM-THIONAL®; XL-50®

*Producers:* AstraZeneca (UK); Avecia (UK); Clariant (Switzerland); ICI Group (UK); Kawaguchi Chemical Industry (Japan); Merck (Germany); Orica (Australia); Sumitomo Chemical (Japan)

*EPA/OPP PC Code:* 064501

*California DPR Chemical Code:* 475

*ICSC Number:* 0937

*RTECS Number:* SN5075000

*Uses:* Phenothiazine is also used as anthelmintic in medicine and in veterinary medicine to rid farm animals of internal parasites; it is used widely as an intermediate in pharmaceutical manufacture; polymerization inhibitor and antioxidant. There are no pesticide uses for phenothiazine currently registered with the U.S. EPA.

*Regulatory Authority:*
- Air Pollutant Standard Set (ACGIH)[1] (NIOSH)[2] (Several States)[60]
- Permissible Exposure Limits for Chemical Contaminants (CAL/OSHA)
- The "Director's List" (CAL/OSHA)
- Canada, WHMIS, Ingredients Disclosure List

*Description:* Phenothiazine is a greenish-yellow to greenish-gray crystalline solid. Slight odor and taste. Practically insoluble in water. Boiling point = 371°C (decomposes). Melting/Freezing point = 182°C (sublimes @130°C @ 1mmHg). Log $K_{ow}$ = 4.18. Values above 3.0 are likely to bioaccumulate in marine organisms.

*Incompatibilities:* Contact with strong acids, acid fumes, or elevated heat produces sulfur oxides and nitrogen oxides.

*Permissible Exposure Limits in Air:* There is no OSHA[2] PEL. NIOSH[2] and ACGIH[1] recommend a TWA of 5 mg/m³ with the notation "skin" indicating the possibility of cutaneous absorption. The STEL value is 10 mg/m³. Several states have set guidelines or standards for Phenothiazine is ambient air[60] ranging from 50 μg/m³ (North Dakota) to 80 μg/m³ (Virginia) to 100 μg/m³ (Connecticut) to 119 μg/m³ (Nevada).

*Permissible Concentration in Water:* No criteria set, but runoff from spills or fire control may cause water pollution.

*Routes of Entry:* Inhalation, ingestion, skin and/or eye contact. Skin absorption.

*Harmful Effects and Symptoms*

*Short Term Exposure:* Phenothiazine can affect you when breathed in and by passing through your skin. Exposure can irritate the skin and eyes. Exposure can cause an inflammation in the eye (keratitis). This can also be made worse by sunlight (photosensitization) and cause a severe skin reaction with rash and color changes. Can cause a severe allergic liver reaction. High levels of exposure may affect the blood cells causing hemolytic anemia, and toxic liver degeneration. Exposure may affect the nervous system, causing muscle twitching and shaking. May affect heart rhythm, causing irregular heartbeat.

*Long Term Exposure:* Repeated or prolonged contact with skin may cause dermatitis and allergy. Can cause kidney and liver damage. Repeated or prolonged contact may cause skin sensitization as well as skin photophobia (abnormal visual intolerance to light). There is limited evidence that this chemical may damage the developing fetus. Several related phenothiazine compounds have been associated with human tertogenic effects.

*Points of Attack:* Skin, cardiovascular system, liver, kidneys, and heart.

*Medical Surveillance:* For those with frequent or potentially high exposure (half the TLV or greater, or significant skin contact), the following are recommended before beginning work and at regular times after that: Exam of the nervous system and eyes. Liver function tests especially bile salts. Complete blood count. Evaluation by a qualified allergist, including careful exposure history and special testing, may help diagnose skin allergy. EKG.

*First Aid:* If this chemical gets into the eyes, remove any contact lenses at once and irrigate immediately for at least 15 minutes, occasionally lifting upper and lower lids. Seek medical attention immediately. If this chemical contacts the skin, remove contaminated clothing and wash immediately with soap and water. Seek medical attention immediately. If this chemical has been inhaled, remove from exposure, begin rescue breathing (using universal precautions) if breathing has stopped, and CPR if heart action has stopped. Transfer promptly to a medical facility. When this chemical has been swallowed, get medical attention. Give large quantities of water and induce vomiting. Do not make an unconscious person vomit.

*References:*
- New Jersey Department of Health, "Hazardous Substance Fact Sheet: Phenothiazine," Trenton, NJ (June 1986, revised May 2000). http://www.state.nj.us/health/eoh/rtkweb/1489.pdf
- California Environmental Protection Agency "Chemical List of Lists," Sacramento CA (February 1997).

# D-Phenothrin

*Use Type:* Insecticide
*CAS Number:* 26002-80-2
*Formula:* $C_{23}H_{26}O_3$
*Synonyms:* A13-29062; Caswell No. 652B; Cyclopropanecarboxylic acid, 2,2-dimethyl-3-(2-methyl-1-propenyl)-,(3-phenoxyphenyl) methyl ester; Cyclopropanecarboxylic acid, 2,2-dimethyl-3-(2-methylpropenyl)-, *m*-phenoxybenzyl ester; Cyclopropanecarboxylic acid, 2,2-dimethyl-3-(2-methyl-1-propenyl)-, (3-phenoxyphenyl)methyl ester; 2,2-Dimethyl-3-(2-methyl-1-propenyl)cyclopropanecarboxylic acid (3-phenoxyphenyl)methyl ester; ENT 27972; EPA Fenotrina (Spanish); Fenothrin, (±)-; (+)-*cis,trans*-Fenothrin; Fenothrin, (+)-*cis,trans*-; Fenothrin, Forte; (+)-*cis,trans*-phenothrin; Phenothrin; 3-Phenoxybenzyl *d-Z/E* chrysanthemate; 3-Phenoxybenzyl *d-Z/E* chrysanthemate; 3-Phenoxybenzyl D-*cis,trans*-chrysanthemate; 3-Phenoxybenzyl *cis,trans*-chrysanthemate; 3-Phenoxybenzyl (1RS)-*cis,trans*-chrysanthemate; 3-Phenoxybenzyl (±)-*cis,trans*-chrysanthemate; *m*-Phenoxybenzyl 2,2-dimethyl-3-(2-methylpropenyl)cyclopropanecarboxylate; 3-Phenoxybenzyl 2-dimethyl-3-(methylpropenyl) cyclopropanecarboxylate; 3-Phenoxybenzyl (1RS,3RS;1RS,3SR)-2,2-dimethyl-3-(2-methylprop-1-enyl)cyclop ropanecarboxylate; 3-Phenoxybenzyl(1RS)-*cis,trans*-2,2-dimethyl-3-(2-methylprop-1-enyl)cyclopropanecarboxylate; 3-Phenoxybenzyl(1RS)-(Z),(E)-2,2-dimethyl-3-(2-methylprop-1-enyl)cyclopropanecarboxylate
*Trade Names:* FORTE®; MULTICIDE-2154®; OMS 1809®; OMS 1810®; PHENOXYTHRIN®; PT-515; S-2539®; SUMETHRIN®; SUMITHRIN®; WELLCIDE®
*Note:* The U.S. EPA currently lists 724 active or canceled products, 242 of which are active. Most products are for residential flying insect control, e.g., fleas, wasps, flies, mosquitoes.
*Producers:* K-Hara Chemicals Ltd. (UK); Changzhou Kangmei Chemical Industry Co., Ltd. (China)
*Chemical Class:* Pyrethroid
*EPA/OPP PC Code:* 069005
*California DPR Chemical Code:* 2093
*ICSC Number:* 0313
*RTECS Number:* GZ1975000
*Uses:* Used world-wide to control household insects and to protect stored grain. It is frequently formulated with other insecticides and is formulated in a number of carriers (aerosols, oil, dust, emulsifiable concentrates), and in powders, shampoos and lotions.
*Carcinogen/Hazard Classifications*
**Label Signal Word:** CAUTION
**WHO Acute Hazard:** Class U, unlikely to be hazardous
**Endocrine Disruptor:** Suspected endocrine disruptor

*Regulatory Authority:*
- EPA 40 CFR 372.65, Specific Toxic Chemical Listings
- Actively registered pesticide in California.
- EPCRA Section 313 Form R *de minimis* concentration reporting level: 1.0%

*Description:* Pale yellow to yellow-brown liquid. Insoluble in water. Molecular weight = 350.5. Vapor pressure = 10 Pa @ 20°C. May be highly toxic to marine life.

*Incompatibilities:* May react violently with strong oxidizers, bromine, 90% hydrogen peroxide, phosphorus trichloride, silver powders or dust. Incompatible with silver compounds. Mixture with some silver compounds forms explosive salts of silver oxalate.

*Harmful Effects and Symptoms*

*Short Term Exposure:* Pyrethroids can affect you when breathed in and by passing through your skin. Irritates the eyes and respiratory tract. High exposure can affect the nervous system causing headache, nausea, vomiting, fatigue, and restlessness, rhinorrhea (discharge of thin nasal mucous).

*Long Term Exposure:* High or repeated exposure can cause lung allergy (with cough, wheezing and/or shortness of breath) or hay fever symptoms (sneezing, runny or stuffy nose). Allergic "pneumonia" can also occur with cough, chest pain, breathing difficulty and abnormal chest x-ray. Repeated attacks may lead to permanent scarring. Skin allergy may also develop with rash and itching, even with lower exposures. Skin contact can cause rash with redness, blisters and intense itching. A severe generalized allergy can occur with weakness and collapse.

*Points of Attack:* Respiratory system, skin, central nervous system, liver and kidneys.

*Medical Surveillance:* Before beginning employment and at regular times after that, the following are recommended: Lung function tests. These may be normal if the person is not having an attack at the time of the test. Consider chest x-ray if lung symptoms are present. Evaluation by a qualified allergist, including careful exposure history and special testing, may help diagnose skin allergy.

*First Aid:* If this chemical gets into the eyes, remove any contact lenses at once and irrigate immediately for at least 15 minutes, occasionally lifting upper and lower lids. Seek medical attention immediately. If this chemical contacts the skin, remove contaminated clothing and wash immediately with soap and water. Seek medical attention immediately. If this chemical has been inhaled, remove from exposure, begin rescue breathing (using universal precautions) if breathing has stopped, and CPR if heart action has stopped. Transfer promptly to a medical facility. When this chemical has been swallowed, get medical attention. Give large quantities of water and induce vomiting. Do not make an unconscious person vomit.

*References:*
- International Programme on Chemical Safety (IPCS), "Environmental Health Criteria, d-Phenothrin," Geneva,

"Environmental Health Criteria, d-Phenothrin," Geneva, Switzerland. (1990). http://www.inchem.org/documents/ehc/ehc/ehc96.htm
* *Journal of Pesticide Reform*, "Insecticide Factsheet, Sumithrin (d-Phenothrin)," Northwest Coalition for Alternatives to Pesticides, Eugene, OR, Summer 2003, Vol 23, No. 2. http://www.pesticide.org/sumithrin.pdf
* California Environmental Protection Agency "Chemical List of Lists," Sacramento CA (February 1997)

# Phenthoate

*Use Type:* Insecticide and acaricide
*CAS Number:* 2597-03-7
*Formula:* $C_{12}H_{17}O_4PS_2$
*Synonyms:* Acetic acid, (*O,O*-dimethyldithiophosphoryl phenyl)-, ethyl ester; Dimephenthioate; dimephenthoate; Dimethenthoate; *O,O*-Dimethyl-*S*-(1-carboethoxybenzyl) dithiophosphate; *O,O*-Dimethyl *S*-α-(ethoxycarbonyl)benzyl phosphorothiolothionate; *O,O*-Dimethyl-*S*-αethoxycarbonylbenzyl phosphorodithioate; *O,O*-Dimethyl-*S*-(phenylacetic acid ethyl ester) phosphorodithioate; *O,O*-Dimethyl-*S*-(phenyl)(carboethoxy)methyl phosphoro dithioate; [Dimethyl-*S*-(phenylethoxycarbonylmethyl) phosphorothiolothionate]; ENT 23,438; ENT 27,386GC; *S*-[α-(Ethoxycarbonyl)benzyl] *O,O*-dimethyl phosphoro dithioate; *S*-α-Ethoxycarbonylbenzyl-*O,O*-dimethyl phosphorodithioate; *S*-α-Ethoxycarbonylbenzyl dimethyl phosphorothiolothionate; Ethyl-α-[(dimethoxy phosphenothioyl)thio]benzeneacetate; Ethyl-*O,O*-dimethyl phosphorodithioylphenyl acetate; Ethyl ester of *O,O*-dimethyldithiophosphoryl α-phenyl acetate acid; *S*-Ethylmercaptophenylacetate-*O,O*-dimethyl phosphorodithioate; (*O,O*-Dimethyldithiophosphorylphenyl) acetic acid ethyl ester; Ethyl mercaptophenylacetate-*O,O*-dimethyl phosphorocithioate; Fenthoate; Phosphorodithioic acid, *O,O*-dimethyl ester, *S*-ester with ethylmercapto phenylacetate
*Trade Names:* AIMSAN®; BAY 33051®, Bayer CropScience (Germany); BAYER-18510®, Bayer CropScience (Germany); CIDEMUL®; CIDIAL®, canceled; ELSAN®; ERUCIN®; ERUSAN®; L-561®; MONTECATINI L-561®; NSC-190978®; OMS 1075®; PAP®; PAPTHION®; PHENDAL®; ROGODIAL®; S 2940®; TANONE®; TH 346-1®; TSIDIAL®
*Producers:* Agsin (Singapore); Bharat Rasayan (India); Bayer CropScience (Germany)
*Chemical Class:* Organophosphate
*EPA/OPP PC Code:* 104901
*Uses:* Not registered in the U.S. A broad spectrum insecticide used to control pests such as aphids, mosquito larvae, houseflies, blowflies and scales on cotton, rice, fruit and vegetable crops and ornamentals. Also used around agricultural buildings.

*Carcinogen/Hazard Classifications*
**WHO Acute Hazard:** Class II, moderately hazardous
*Regulatory Authority:*
* DOT Inhalation Hazard Chemicals

*Description:* Light yellow oily liquid. Odorless. Soluble in water; solubility = 11 ppm @ 20°C. Molecular weight 320.39. Melting/Freezing point = 17.8°C. Density = 1.22. Vapor pressure = $2.6 \times 10^{-6}$ mmHg @ 20°C; $5 \times 10^{-5}$ mmHg @ 40°C.

*Incompatibilities:* May react violently with antimony(V) pentafluoride. Incompatible with lead diacetate, magnesium, silver nitrate. Keep away from oxidizers, sulfuric acid, caustics, ammonia, aliphatic amines, alkanolamines, isocyanates, alkylene oxides, epichlorohydrin. Moisture may cause hydrolysis or other forms of decomposition.

*Permissible Exposure Limits in Air:*
*Determination in Air:* OSHA versatile sampler-2; Toluene/Acetone; Gas chromatography/Flame ionization detection; NIOSH IV, Method #5600, Organophosphorus Pesticides.[18]

*Permissible Concentration in Water:* No criteria set. Runoff from spills or fire control may cause water pollution.

*Harmful Effects and Symptoms*
*Short Term Exposure:* Eye pupils are small, blurred vision, eye watering, runny nose, cough, shortness of breath, salivation, dizziness, nausea, stomach cramps, diarrhea, and vomiting, increased blood pressure, profuse sweating, hypermotility, hallucinations, irritability, tingling of the skin, drowsiness, slow heartbeat, convulsions, fluid in lungs, loss of consciousness, incontinence, breathing stops, death. Organophosphates inhibit the action of acetylcholinesterase enzymes, and alter the way in which nervous impulses are transmitted. The effects can last for hours, days, or much longer. The action of the enzymes is reestablished after new enzymes are formed.

*Long Term Exposure:* Cholinesterase inhibitor; cumulative effect is possible. Organophosphates may damage the nervous system with repeated exposure, resulting in convulsions, respiratory failure. May cause liver damage.

*Points of Attack:* Respiratory system, central nervous system, cardiovascular system, blood cholinesterase.

*Medical Surveillance:* Medical observation is recommended for 24 to 48 hours after breathing overexposure, as pulmonary edema may be delayed. As first aid for pulmonary edema, consider administering a corticosteroid spray. Cigarette smoking may exacerbate pulmonary injury and should be discouraged for at least 72 hours following exposure.

*First Aid:* Speed in removing material from eyes and skin is of extreme importance. Eye contact can cause dangerous amounts of these chemicals to be quickly absorbed through the mucous membrane into the bloodstream. Immediately and gently flush eyes with plenty of warm or cold water

(NO hot water) for at least 15 minutes, occasionally lifting the upper and lower eyelids. Get medical aid immediately. *Skin:* Get medical aid. Skin contact can cause dangerous amounts of these chemicals to be absorbed into the bloodstream. Wearing the appropriate PPE equipment and respirator for organophosphate pesticides, immediately flush skin with plenty of soap and water for at least 15 minutes while removing contaminated clothing and shoes. Shampoo hair promptly if contaminated. *Ingestion:* Call poison control. Loosen all clothing. Never give anything by mouth to an unconscious person. Get medical aid. Do NOT induce vomiting.* If conscious, alert, and able to swallow, rinse mouth and have victim drink 4 to 8 ounces of water do NOT induce vomiting but immediately administer slurry of activated charcoal (2 oz in 8 oz of water). If victim is *unconscious or having convulsions,* do nothing except keep victim warm. **In some cases you may be specifically instructed by poison control to induce vomiting by way of 2 tablespoons of syrup of ipecac (adult) washed down with a cup of water.* Do NOT give activated charcoal before or with ipecac syrup. *Inhalation:* Get medical aid. Do not contaminate yourself. Wearing the appropriate PPE equipment and respirator for organophosphate pesticides, immediately remove the victim from the contaminated area to fresh air. If the victim is not breathing, administer artificial respiration. Do NOT use mouth-to-mouth resuscitation; use bag/mask apparatus. If breathing is difficult, administer oxygen through bag/mask apparatus until medical help arrives. Do not leave victim unattended. *Note to physician or authorized medical personnel.* Administer atropine, 2 mg (1/30 gr) intramuscularly or intravenously as soon as any local or systemic signs or symptoms of an intoxication are noted; repeat the administration of atropine every 3 to 8 minutes until signs of atropinization (mydriasis, dry mouth, rapid pulse, hot and dry skin) occur; initiate treatment in children with 0.05 mg mg/kg of atropine; repeat at 5 to 10 minute intervals. Watch respiration, and remove bronchial secretions if they appear to be obstructing the airway; intubate if necessary. *Notes to physician or authorized medical personnel*: N-methylpyridinium-2-aldoxime (2-PAMCI) when used in conjunction with atropine reacts with the phosphorylated cholinesterase, thereby restoring normal activity to by removing the phosphorylating group. The combination of these two chemicals is synergistic and must be administered within minutes to a few hours following exposure (depending on the specific agent) to be effective. Give 2-PAMCI (Pralidoxime; Protopam), 2.5 gm in 100 ml of sterile water or in 5% dextrose and water, intravenously, slowly, in 15-30 minutes; if sufficient fluid is not available, give 1 gm of 2-PAMCI in 3 ml of distilled water by deep intramuscular injection; repeat this every half hour if respiration weakens or if muscle fasciculation or convulsions recur. Also Diazepam, an anticonvulsant, might be considered.

*References:*
- International Programme on Chemical Safety (IPCS), "Data Sheets on Pesticides, Phenthoate," Geneva, Switzerland, http://www.inchem.org/documents/pds/pds/pest48_e.htm

# Phenylmercury Acetate

*Use Type:* Fungicide and herbicide
*CAS Number:* 62-38-4
*Formula:* $C_8H_8HgO_2$; $C_6H_5HgOOCCH_3$
*Alert:* There are no pesticide applications currently registered with the U.S. EPA. Phenylmercury acetate is a teratogen and should be handled with extreme caution.
*Synonyms:* Acetate phenylmercurique (French); (Aceato)phenylmercury; acetato fenilmercurio (Spanish); Acetic acid, phenylmercury derivitive; Benzene, (acetoxymercuri)-; Benzene, (acetoxymercurio); Femma; Fenylmercuriacetat (Czech); FMA; Mercuriphenyl acetate; Mercury(II) acetate, phenyl; Mercury, (acetoxy)phenyl-; Octan fenylrtutnaty (Czech); Phenomercury acetate; Phenylmurcuriacetate; Phenylmercuric acetate; Phenylquecksilberacetat (German); PHIX; PMA; PMAC; Pmacetate; PMAL; PMAS
*Trade Names:* AGROSAN®; AGROSAND®; AGROSAN GN 5®; ALGIMYCIN®; ANTIMUCIN WDR®; BUFEN®; CEKUSIL®; CELMER®; CERESAN®; CERESOL®; CONTRA CREME®; COSAN®; DYNACIDE®; FUNGITOX OR®; GALLOTOX®; HL-331®; HONG KIEN®; HOSTAQUICK®; INTERCIDE®, Akzo Nobel (Netherlands), canceled 8/13/1990; KWIKSAN®; LEYTOSAN®; LIQUIPHENE®; MERGAMMA®; MERSOLITE®, Troy Chemical (USA), canceled 10/10/1989; METASOL 30®; NORFORMS®; NUODEX PMA®, Creanova (USA), canceled 8/13/1990; NYMERATE®; PAMISAN®; PHENMAD®; PURASAN-SC-10®; PURATURF 10®; RIOGEN®; QUICKSAN®; SANITIZED SPG®; SC-110®; SCUTL®; SEEDTOX®; SHIMMEREX®; SPOR-KIL®; TAG®; TAG-HL-33®; 1 TRIGOSAN®; ZIARNIK®
*Producers:* Akzo Nobel (Netherlands)
*Chemical Class:* Organomercury
*EPA/OPP PC Code:* 066003
*California DPR Chemical Code:* 491
*ICSC Number:* 0540
*RTECS Number:* OV6475000
*EEC Number:* 080-011-00-5
*EINECS Number:* 200-532-5
*Uses:* Used as an antiseptic, fungicide, herbicide; mildewcide for paints; slimicide in paper mills. It was also used in contraceptive gels and foams. There are no pesticide applications currently registered with the U.S. EPA.

### Phenylmercury Acetate

*Carcinogen/Hazard Classifications*
**California Prop. 65:** Reproductive toxin as mercury and mercury compounds.
**Label Signal Word:** DANGER
**WHO Acute Hazard:** Class 1 a, extremely hazardous
*Regulatory Authority:*
- Banned or Severely Restricted (Several Countries) (UN)[13]
- Air Pollutant Standard Set (ACGIH)[1] (OSHA)[58] (Argentina)[35]
- AB 2588-Air Toxics "Hot Spots" Chemicals (CAL)
- CAL Air Resources Board/AB 1807 Toxic Air Contaminants
- Proposition 65 chemical (CAL)
- Permissible Exposure Limits for Chemical Contaminants (CAL/OSHA)
- The "Director's List" (CAL/OSHA)
- EPA Hazardous Waste Number (RCRA No.): P092
- RCRA, 40CFR261, Appendix 8 Hazardous Constituents
- Superfund/EPCRA 40CFR355, Appendix B Extremely Hazardous Substances: TPQ = 500/10,000 lb (227/4,540 kg)
- Superfund/EPCRA 40CFR302.4 RQ: CERCLA, 100 lb (45.4 kg)
- Safe Drinking Water Act: MCL, 0.002 mg/L; MCLG, 0.002 mg/L as mercury
- U.S. DOT Regulated Marine Pollutant (49CFR172.101, Appendix B), severe pollutant.
- Canada, WHMIS, Ingredients Disclosure List

*Description:* Phenylmercury acetate is a white or yellow crystalline solid. Slightly soluble in water; solubility = 1500 ppm @ 15°C; 0.44 g/100 mL @ 20°C. Melting/Freezing point =151°C. Vapor pressure = $8.5 \times 10^{-6}$ mmHg @ 35°C. Flash point = 37.8°C. Hazard Identification (based on NFPA-704 M Rating System): Health 3, Flammability 1, Reactivity 0. Mercury compounds are toxic to aquatic organisms. Bioaccumulation in the food chain takes place, specifically in fish, crustacea, and birds. Bioconcentrative up to 10,000-fold. (CHRIS, U. S. Coast Guard).

*Incompatibilities:* Strong oxidizers and halogens. The dust mixed with air is explosive.

*Permissible Exposure Limits in Air:* The OSHA[2] PEL for all organo mercury compounds is 0.01 mg/m$^3$ as an 8-hour TWA and ceiling of 0.04 mg/m$^3$. NIOSH[2] recommends 0.01 mg/m$^3$ as a 10-hour TWA and STEL of 0.03 mg/m$^3$ with the notation "skin" indicating the possibility of cutaneous absorption. ACGIH[1] recommends 0.01 mg/m$^3$ as an 8-hour TWA value. The NIOSH[2] IDLH level = 2 mg/m$^3$ (as Hg). HSE[33] has set 0.05 mg/m$^3$ as an 8-hour TWA and 0.15 mg/m$^3$ as an STEL. For mercury aryl compounds: Germany: 0.1 mg/m$^3$; Czechoslovakia: 0.05 mg/m$^3$; Sweden: 0.05 mg/m$^3$; former USSR: 0.01 mg/m$^3$.

*Determination in Air:* No test available.

*Permissible Concentration in Water:* To protect freshwater aquatic life: 0.00057 µg/L as a 24-hour average, never to exceed 0.0017 µg/L. To protect saltwater aquatic life: 0.025 µg/L as a 24 hour average, never to exceed 3.7 µg/L. To protect human health: 0.144µg/L (USEPA) set in 1979-80[6]. These are the limits for inorganic mercury compounds in general.

*Determination in Water:* Total mercury is determined by flameless atomic absorption. Soluble mercury may be determined by 0.45 micron filtration followed by flameless atomic absorption.

*Routes of Entry:* Inhalation, ingestion, skin and/or eye contact. Absorbed through the skin.

*Harmful Effects and Symptoms*
*Short Term Exposure:* Irritates the eyes, skin, and respiratory tract. Overexposure affect the kidneys, causing renal function failure. Extremely toxic. The probable oral lethal dose for humans is 5-50 mg/kg, between 7 drops and 1 teaspoonful for a 70 kg (150 lb) person. Symptoms arising from acute exposure may occur at varying intervals up to several weeks following exposure. Ingestion of mercurial fungicide treated grain resulted in gastrointestinal irritation with nausea, vomiting, abdominal pain, and diarrhea. Alkylmercurials produce severe neurologic toxicity, such as loss of feeling in lips, tongue, and extremities, confusion, hallucinations, irritability, sleep disturbances, staggering walk, memory loss, slurred speech, auditory defects, emotional instability, and inability to concentrate. It is also a strong skin irritant, erythema and blistering may result 6-12 hours after exposure. Phenylmercury acetate, at sufficient concentration, is expected to be injurious to the eye externally. Mercury poisoning can cause "shakes," irritability, sore gums, increased saliva, personality change and brain damage. Skin contact can cause burns, skin allergy and a gray skin color. Heating or contact with acid or acid "fumes" releases toxic mercury vapors.

*Long Term Exposure:* Mercury accumulates in the body. Repeated or prolonged contact with skin may cause dermatitis. May affect the nervous system, causing nervous disorders. Based on animal tests, phenylmercuric acetate should be handled as a teratogen-with extreme caution. It also may cause mutations.

*Points of Attack:* Eyes, skin, central nervous system, peripheral nervous system and kidneys.

*Medical Surveillance:* Before first exposure and every 6 to 12 months after, a complete medical history and exam is strongly recommended with: Exam of the nervous system, including handwriting. Routine urine test (UA). Urine test for mercury (should be less than 0.02 mg/L). Consider lung function tests for persons with frequent exposures. After suspected illness or overexposure, repeat the above tests and get a blood test for mercury. Consider chest x-ray after acute overexposure.

*First Aid:* If this chemical gets into the eyes, remove any contact lenses at once and irrigate immediately for at least 15 minutes, occasionally lifting upper and lower lids. Seek medical attention immediately. If this chemical contacts the skin, remove contaminated clothing and wash immediately with soap and water. Seek medical attention immediately. If this chemical has been inhaled, remove from exposure, begin rescue breathing (using universal precautions) if breathing has stopped, and CPR if heart action has stopped. Transfer promptly to a medical facility. When this chemical has been swallowed, get medical attention. Give large quantities of water and induce vomiting. Do not make an unconscious person vomit. Keep victim quiet and maintain normal body temperature. Effects may be delayed; keep victim under observation.

*Antidotes and Special Procedures for medical personnel:* The drug NAP (n-Acetyl Penicillamine) has been used to treat mercury poisoning, with mixed success.

*References:*
- U.S. Environmental Protection Agency, "Chemical Profile: Phenylmercury Acetate," Washington, DC, Chemical Emergency Preparedness Program (Nov. 30, 1987).
- California Environmental Protection Agency "Chemical List of Lists," Sacramento CA (February 1997).
- New Jersey Department of Health, "Hazardous Substance Fact Sheet: Phenylmercuric Acetate," Trenton, NJ (January 1987, rev. February 2000). http://www.state.nj.us/health/eoh/rtkweb/1502.pdf

## o-Phenylphenol

*Use Type:* Fungicide and microbiocide
*CAS Number:* 90-43-7
*Formula:* $C_{12}H_{10}O$
*Synonyms:* (1,1'-Biphenyl)-2-ol; 2-Biphenylol; o-Biphenylol; (1,1'-biphenyl)-2-ol; o-Diphenylol; o-Hydroxybiphenyl; 2-Hydroxybiphenyl; o-Hydroxydiphenyl; 2-Hydroxydiphenyl; 2-Hydroxy-1,1'-biphenyl; NCI-C50351; OPP; Orthohydroxydiphenyl; Orthophenylphenol; Orthoxenol; 2-Phenylphenol; o-Xenol
*Trade Names:* ANTHRAPOLE 73®; DOWCIDE-1®, Dow Chemical (USA); Invalon OP®; KIWI LUSTR-277®; NECTRYL®; PREVENTOL-O Extra®, Bayer CropScience (Germany); REMOL TRF®; TETROSIN OE®; TETROSIN OE-N®; TORSITE®; TUMESCAL OPE®
*Producers:* Bayer Chemicals (Germany); Coalite Chemicals (UK); Clariant (Switzerland); Dow Chemical (USA); R.S.A. Corp (USA)
*Chemical Class:* Phenol
*EPA/OPP PC Code:* 064103
*California DPR Chemical Code:* 448
*ICSC Number:* 0669
*RTECS Number:* DV5775000
*EEC Number:* 604-020-00-6
*Uses:* Used to make fungicides. Also used to make dye stuffs and rubber chemicals, but used primarily as a disinfectant cleaner.

*U.S. Maximum Allowable Residue Level o-Phenylphenol (40 CFR 180.129):*
Tolerances are established for combined residues of the fungicide o-phenylphenol and sodium o-phenylphenate, each expressed as o-phenylphenol, from post-harvest application on the following crops:

| CROP | ppm |
|---|---|
| Apple | 25 |
| Cantaloupe | 125 |
| Cantaloupe, pulp | 10 |
| Carrot, roots | 20 |
| Cherry | 5 |
| Citron, citrus | 10 |
| Cucumber | 10 |
| Fruit, citrus | 10 |
| Grapefruit | 10 |
| Kiwifruit | 20 |
| Kumquat | 10 |
| Lemon | 10 |
| Lime | 10 |
| Nectarine | 5 |
| Orange | 10 |
| Peach | 20 |
| Pear | 25 |
| Pepper, bell | 10 |
| Pineapple | 10 |
| Plum, fresh prune | 20 |
| Sweet potato, roots | 15 |
| Tangerine | 10 |
| Tomato | 10 |

*Carcinogen/Hazard Classifications*
**U.S. EPA Carcinogens:** Group B2, probable carcinogen
**California Prop. 65:** Carcinogen
**U.S. TRI:** Carcinogen and developmental toxin
**IARC:** Group 3, unclassifiable
**Label Signal Word:** DANGER
**WHO Acute Hazard:** Class U, unlikely to be hazardous
*Regulatory Authority:*
- Actively registered pesticide in California.
- AB 2588-Air Toxics "Hot Spots" Chemicals (CAL)
- EPCRA Section 313 Form R *de minimis* concentration reporting level: 1.0%

*Description:* Nearly white or light-buff crystalline solid. Practically insoluble in water; solubility = 700 ppm @ 25°C. Molecular weight = 170.21. Density = 1.22 @ 25°C. Melting/Freezing point = 56°C. Boiling point = 279°C. Vapor pressure = 100 mmHg @ 205°C. Flash point = 125
*Incompatibilities:* Reacts with boranes, alkalies, aliphatic amines, amides, nitric acid, sulfuric acid. Keep away from oxidizers (chlorates, nitrates, peroxides, permanganates, perchlorates, chlorine, bromine, fluorine, etc).
*Permissible Concentration in Water:* No criteria set. Runoff from spills or fire control may cause water pollution.
*Routes of Entry:* Inhalation, skin, eye contact, ingestion.

*Harmful Effects and Symptoms*
*Short Term Exposure:* Skin irritation. Eye irritation, burning.
*Long Term Exposure:* Possible eye damage.
*Points of Attack:* Eyes
*First Aid:* If this chemical gets into the eyes, remove any contact lenses at once and irrigate immediately for at least 15 minutes, occasionally lifting upper and lower lids. Seek medical attention immediately. If this chemical contacts the skin, remove contaminated clothing and wash immediately with soap and water. Seek medical attention immediately. If this chemical has been inhaled, remove from exposure, begin rescue breathing (using universal precautions) if breathing has stopped, and CPR if heart action has stopped. Transfer promptly to a medical facility. When this chemical has been swallowed, get medical attention. Give large quantities of water and induce vomiting. Do not make an unconscious person vomit. Do not induce vomiting when formulations containing petroleum solvents are ingested.
*References:*
- U.S. Environmental Protection Agency, Office of Pesticide Programs, Pesticide Residue Limits, "o-Phenylphenol," 40 CFR 180.129. www.epa.gov/pesticides/food/viewtols.htm
- New Jersey Department of Health and Senior Services, "Hazardous Substance Fact Sheet, o-Phenylphenol," Trenton NJ (September 1986, rev. December 2000). http://www.state.nj.us/health/eoh/rtkweb/1439.pdf
- California Environmental Protection Agency "Chemical List of Lists," Sacramento CA (February 1997).

# Phenylthiourea

*Use Type:* Rodenticide
*CAS Number:* 103-85-5
*Formula:* $C_7H_8N_2S$; $C_6H_5NCHCSNH_2$
*Alert:* There are no phenylthiourea products registered with the U.S. EPA.
*Synonyms:* NCI-C02017; Phenylthiocarbamide; N-Phenylthiourea; α-Phenylthiourea; Phenyl-2-thiourea; 1-Phenylthiourea; PTC; PTU, U 6324
*Producers:* Degussa (Germany)
*RTECS Number:* YU1400000
*EINECS Number:* 203-151-2
*Uses:* Used as a repellent for rats, rabbits, and weasels; in the manufacture of rodenticides and in medical genetics.
*Regulatory Authority:*
- EPA Hazardous Waste Number (RCRA No.): P093
- RCRA, 40CFR261, Appendix 8 Hazardous Constituents
- Superfund/EPCRA 40CFR355, Appendix B Extremely Hazardous Substances: TPQ = 100/10,000 lb (45.4/4,540 kg)
- Superfund/EPCRA 40CFR302.4 RQ: CERCLA, 100 lb (45.4 kg)

*Description:* N-phenylthiourea is a colorless crystalline solid. Melting/Freezing point = 148–154°C. Soluble in water.
*Incompatibilities:* Incompatible with oxidizers, strong bases, and acids. Contact with acids or acid fumes produces toxic fumes of sulfur oxide.
*Permissible Exposure Limits in Air:* No standards set.
*Permissible Concentration in Water:* No criteria set, but runoff from spills or fire control may cause water pollution.
*Routes of Entry:* Inhalation, ingestion, skin and/or eye contact.
*Harmful Effects and Symptoms*
*Short Term Exposure:* Irritates the eyes, skin and respiratory tract. High exposures can cause lung irritation, coughing and/or shortness of breath. Higher exposures can cause pulmonary edema, a medical emergency that can be delayed for several hours. This can cause death. Exposure may result in vomiting, difficult breathing, noisy breathing, cyanosis, and low body temperature. It is classified as extremely toxic. The probable oral lethal dose is 5-50 mg/kg or between 7 drops and 1 teaspoon for a 70 kg (150 lb) person.
*Long Term Exposure:* Not tested for long term health effects. May cause methemoglobinemia, cyanosis, and anemia. Phenylthiourea is reported to be similar to ANTU.
*Points of Attack:* Lungs
*Medical Surveillance:* Lung function tests. Blood methemogloblin level. Completed blood count. Consider chest x-ray following acute overexposure.
*First Aid:* If this chemical gets into the eyes, remove any contact lenses at once and irrigate immediately for at least 15 minutes, occasionally lifting upper and lower lids. Seek medical attention immediately. If this chemical contacts the skin, remove contaminated clothing and wash immediately with soap and water. Seek medical attention immediately. If this chemical has been inhaled, remove from exposure, begin rescue breathing (using universal precautions) if breathing has stopped, and CPR if heart action has stopped. Transfer promptly to a medical facility. When this chemical has been swallowed, get medical attention. Give large quantities of water and induce vomiting. Do not make an unconscious person vomit.Medical observation is recommended for 24 to 48 hours after breathing overexposure, as pulmonary edema may be delayed. As first aid for pulmonary edema, a doctor or authorized paramedic may consider administering a corticosteroid spray.
*Note to physician or authorized medical personnel:* Treat for methemoglobinemia. Spectrophotometry may be required for precise determination of levels of methemoglobinemia in urine.
*References:*
- U.S. Environmental Protection Agency, "Chemical Profile: Phenylthiourea," Washington, DC, Chemical Emergency Preparedness Program (November 30, 1987).

- California Environmental Protection Agency "Chemical List of Lists," Sacramento CA (February 1997).
- New Jersey Department of Health and Senior Services, "Hazardous Substance Fact Sheet, Phenylthiourea," Trenton NJ (August 1999).
  http://www.state.nj.us/health/eoh/rtkweb/2664.pdf

# Phorate (ANSI)

*Use Type:* Insecticide, acaricide and nematicide
*CAS Number:* 298-02-2
*Formula:* $C_7H_{17}O_2PS_3$
*Alert:* A Restricted Use Pesticide (RUP). Human toxicity (long-term): High
*Synonyms:* O,O-Diaethyl-S-(aethylthio-methyl)-dithiophosphat (German); O,O-Diethyl S-ethylmercapto methyl dithiophosphonate; O,O-Diethyl-S-(ethylthio-methyl)-dithiofosfaat (Dutch); O,O-Diethyl S-ethylthio methyldithiophosphonate; O,O-Diethylethylthiomethyl phosphorodithioate; O,O-Diethyl S[(ethylthio)methyl] phosphorodithioate; O,O-Diethyl S-(ethylthio)methyl phosphorodithioate; O,O-Diethyl S-ethylthiomethyl thiothionophosphate; O,O-Dietil-S-(etiltio-metil)-ditiofosfato (Italian); Dithiophosphatede O,O-diethyle et d'ethylthio methyle (French); ENT 24,042; Foraat (Dutch); Forato (Spanish); Methanethiol, (ethylthio)-, S-ester with O,O-diethylphosphorodithioate; Phorat (German); Timet (Russia)
*Trade Names:* AASTAR®, BASF Agricultural Products Group (Germany), canceled 1/22/1991; AC 3911®; AGRIMET®; AMERICAN CYANAMID 3,911®, Cyanamid Agricultural Products Div. (USA); EL 3911®; EXPERIMENTAL INSECTICIDE 3911®; GEOMET®; GRAMTOX®; GRANUTOX®; L 11/6®; METAPHOR®, Sudarshan Chemical Industries (India); PHORATE-10G®; PHORIL®, Rallis India (India); RAMPART®, Loveland Products (USA); TERRACLOR®, Crompton Corporation (USA); TERRATHION GRANULES®; THIMENOX®; THIMET®, BASF Agricultural Products Group (Germany); THEMET®, BASF Agricultural Products Group (Germany); UMET®, United Phosphorus (India); VEGFRU®; VERGFRU FORATOX®
*Producers:* Agrimor International (USA); Aimco Pesticides Ltd. (India); BASF Agricultural Products Group (Germany); Crompton Corporation (USA); Cyanamid Agricultural Products Div. (USA); Ehrenstorfer, Dr. (Germany); Gujarat Pesticides (India); Hebei Huafeng Chemical Group (China); Hebei Long Age Pesticide (China); Hindustan Insecticides (India); ICI Group (UK); Nagarjuna Agrichem (India); Orica (Australia); Rallis India (India); Shenzhen Guomeng Industry Co., Ltd. (China); Sudarshan Chemical Industries (India); United Agri Products (UAP) (Loveland Products) (USA); United Phosphorus (India)
*Chemical Class:* Organophosphate
*EPA/OPP PC Code:* 057201
*California DPR Chemical Code:* 478
*ICSC Number:* 1060
*RTECS Number:* TD9450000
*EEC Number:* 015-033-00-6
*EINECS Number:* 206-052-2
*Uses:* Phorate is an organophosphorus insecticide and acaricide used to control a wide variety of sucking and chewing insects, leafhoppers, leafminers, mites, some nematodes, and rootworms. It is used on many crops, including root and field crops such as corn, cotton, coffee, potatoes, sugar beets, beans, peanuts, wheat, some ornamental and herbaceous plants, and bulb. In the U.S., 80% of the annual use of phorate is applied to corn, potatoes and cotton. It is available in granular and emulsifiable concentrate formulations. Phorate has been shown to be responsible for a large number of bird kills and it is extremely toxic to mammals.

*U.S. Maximum Allowable Residue Levels for Phorate (40 CFR 180.206):*

| CROP | ppm |
| --- | --- |
| Bean | 0.1 |
| Beet, sugar, roots | 0.3 |
| Beet, sugar, tops | 3 |
| Coffee, bean | 0.02 |
| Corn, forage | 0.5 |
| Corn, grain | 0.1 |
| Corn, sweet, kernel plus cob with husks removed | 0.1 |
| Cotton, undelinted seed | 0.05 |
| Hop | 0.5 |
| Peanut | 0.1 |
| Potato | 0.5 |
| Sorghum, grain, grain | 0.1 |
| Sorghum, grain, stover | 0.1 |
| Soybean | 0.1 |
| Sugarcane, cane | 0.1 |
| Wheat, grain | 0.05 |
| Wheat, hay | 1.5 |
| Wheat, straw | 0.05 |

*Carcinogen/Hazard Classifications*
**U.S. EPA Carcinogens:** Group E, unlikely carcinogen
**Label Signal Word:** DANGER–POISON
**WHO Acute Hazard:** Class 1 a, extremely hazardous
*Regulatory Authority:*
- Banned or Severely Restricted (Malaysia) (UN)[13]
- Very Toxic Substance (World Bank)[15]
- Air Pollutant Standard Set (ACGIH)[1] (HSE)[33] (NIOSH)[2] (Several States)[60]
- AB 1803-Well Monitoring Chemical (CAL)
- Permissible Exposure Limits for Chemical Contaminants (CAL/OSHA)
- The "Director's List" (CAL/OSHA)
- Actively registered pesticide in California.
- EPA Hazardous Waste Number (RCRA No.): P094
- RCRA, 40CFR261, Appendix 8 Hazardous Constituents

- RCRA 40CFR268.48; 61FR15654, Universal Treatment Standards: Wastewater (mg/L), 0.021; Nonwastewater (mg/kg), 4.6
- RCRA 40CFR264, Appendix 9; TSD Facilities Ground Water Monitoring List. Suggested test method(s) (PQL $ug/L$): 8140(2); 8270(10)
- Superfund/EPCRA 40CFR355, Appendix B Extremely Hazardous Substances: TPQ = 10 lb (4.54 kg)
- Superfund/EPCRA 40CFR302.4 RQ: CERCLA, 10 lb (4.54 kg)
- U.S. DOT Inhalation Hazard Chemicals as organophosphates
- U.S. DOT Regulated Marine Pollutant (49CFR172.101, Appendix B), severe pollutant
- Canada, WHMIS, Ingredients Disclosure List

*Description:* Phorate is a clear, pale yellow mobile liquid. Skunk-like odor. Practically insoluble in water; solubility 40–50 ppm. Molecular weight = 260.39. Boiling point = 118–120°C @ 0.8 mm. Melting/Freezing point = –43°C. Vapor pressure = $8.8 \times 10^{-4}$ mmHg @ 20 °C. Flash point = 160°C. Hazard Identification (based on NFPA-704 M Rating System): Health 4, Flammability 1, Reactivity 0. Log $K_{ow}$ = 3.89. Values at or above 3.0 are likely to bioaccumulate in marine organisms.

*Incompatibilities:* Water, alkalis. Hydrolyzed in the presence of moisture and by alkalis; may produce toxic oxides of phosphorus and sulfur.

*Permissible Exposure Limits in Air:* There is no OSHA[2] PEL. NIOSH[2] and ACGIH[1] recommend a TWA of 0.05 mg/m$^3$ and STEL of 0.2 mg/m$^3$. HSE[33] set the same limits. The notation "skin" is added to indicate the possibility of cutaneous absorption. A number of states have set guidelines or standards for phorate in ambient air[60] ranging from 0.5 to 2.0 $\mu g/m^3$ (North Dakota) to 0.8 $\mu g/m^3$ (Virginia) to 1.0 $\mu g/m^3$ (Connecticut and Nevada).

*Determination in Air:* OSHA versatile sampler-2; Toluene/Acetone; Gas chromatography/Flame photometric detection for sulfur, nitrogen, or phosphorus; NIOSH IV, Method #5600, Organophosphorus pesticides.[18]

*Permissible Concentration in Water:* Maine[61] has set a guideline for phorate in drinking water of 0.2 $\mu g/L$.

*Determination in Water:*

*Routes of Entry:* Inhalation, ingestion, skin and/or eye contact. Absorbed through the skin.

*Harmful Effects and Symptoms*

*Short Term Exposure:* Acute exposure to phorate may produce the following signs and symptoms: pinpoint pupils, blurred vision, headache, dizziness, muscle spasms, and profound weakness. Vomiting, diarrhea, abdominal pain, seizures, and coma may also occur. The heart rate may decrease following oral exposure or increase following dermal exposure. Chest pain may be noted. Hypotension (low blood pressure) may occur, although hypertension (high blood pressure) is not uncommon. Dyspnea (shortness of breath) may be followed by respiratory collapse. Giddiness is common. This material is one of the more toxic organophosphorus insecticides. It is a cholinesterase inhibitor that acts on the nervous system, and produces toxicity similar to parathion. The probable oral lethal dose for humans is less than 5 mg/kg, i.e. a taste (less than 7 drops) for a 70 kg (150 lb) person. Delayed pulmonary edema may occur after inhalation.

*Long Term Exposure:* Cholinesterase inhibitor; cumulative effect is possible. This chemical may damage the nervous system with repeated exposure, resulting in convulsions, respiratory failure. May cause liver damage.

*Points of Attack:* Respiratory system, lungs, central nervous system, cardiovascular system, skin, eyes, plasma and red blood cell cholinesterase.

*Medical Surveillance:* Medical observation is recommended for 24 to 48 hours after breathing overexposure, as pulmonary edema may be delayed. Before employment and at regular times after that, the following are recommended: Plasma and red blood cell cholinesterase levels (tests for the enzyme poisoned by this chemical). If exposure stops, plasma levels return to normal in 1-2 weeks while red blood cell levels may be reduced for 1-3 months. Do not drink any alcoholic beverages before or during use. Alcohol promotes absorption of organophosphates. When acetylcholinesterase enzyme levels are reduced by 25% or more below preemployment levels, risk of poisoning is increased, even if results are in lower ranges of "normal." Reassignment to work not involving organophosphate or carbamate pesticides is recommended until enzyme levels recover. If symptoms develop or overexposure occurs, repeat the above tests as soon as possible and get an exam of the nervous system. Also consider complete blood count. Consider chest x-ray following acute overexposure. Do not drink any alcoholic beverages before or during use. Alcohol promotes absorption of organophosphates.

*First Aid:* **Treatment for organophosphate poisoning consists of thorough decontamination, cardiorespiratory support, and administration of the antidotes atropine and pralidoxime. In cases of severe poisoning, diazepam, an anticonvulsant, should also be administered. Antidotes should be administered as prevention even if the diagnosis is in doubt.** Speed in removing material from eyes and skin is of extreme importance. *Eyes:* Eye contact can cause dangerous amounts of these chemicals to be quickly absorbed through the mucous membrane into the bloodstream. Immediately and gently flush eyes with plenty of warm or cold water (NO hot water) for at least 15 minutes, occasionally lifting the upper and lower eyelids. Get medical aid immediately. *Skin:* Get medical aid. Skin contact can cause dangerous amounts of these chemicals to be absorbed into the bloodstream. Wearing the appropriate PPE equipment and respirator for organophosphate pesticides, immediately flush skin with plenty of soap and water for at least 15 minutes while removing contaminated clothing and shoes. Shampoo hair promptly if contaminated.

The removed, contaminated clothing and shoes should be double-bagged and left in Hot Zone for later disposal by hazardous materials experts. Skin may also be decontaminated with diluted hypochlorite solution. *Inhalation:* Get medical aid. Do not contaminate yourself. Wearing the appropriate PPE equipment and respirator for organophosphate pesticides, immediately remove the victim from the contaminated area to fresh air. If the victim is not breathing, administer artificial respiration. Do NOT use mouth-to-mouth resuscitation; use bag/mask apparatus. If breathing is difficult, administer oxygen through bag/mask apparatus until medical help arrives. Do not leave victim unattended. *Ingestion:* Call poison control. Loosen all clothing. Never give anything by mouth to an unconscious person. If victim is *unconscious or having convulsions,* do nothing except keep victim warm. Get medical aid. Transfer promptly to a medical facility. In cases of ingestion, **do not induce vomiting**. If the victim is alert and asymptomatic, administer a slurry of activated charcoal at a dose of 1 g/kg (infant, child, and adult dose). A soda can and straw may be of assistance when offering charcoal to a child. *In some cases you may be specifically instructed by poison control to induce vomiting by way of 2 tablespoons of syrup of ipecac (adult) washed down with a cup of water.* Do NOT give activated charcoal before or with ipecac syrup.

*Note to physician or authorized medical personnel:* Treat cases of respiratory compromise, coma, or excessive pulmonary secretions with respiratory support using protocols and techniques available and within the scope of training. Some cases may necessitate procedures such as endotracheal intubation or cricothyrotomy by properly trained and equipped personnel. When possible, atropine (see *Antidotes*, below) should be given under medical supervision. Patients who are comatose, hypotensive, or having seizures or cardiac arrhythmias should be treated according to advanced life support protocols. *Antidotes:* Two antidotes are administered to treat organophosphate poisoning. Atropine is a competitive antagonist of acetylcholine at muscarinic receptors and is used to control the excessive bronchial secretions which are often responsible for death. Pralidoxime relieves both the nicotinic and muscarine effects of organophosphate poisoning by regenerating acetylcholinesterase and can reduce both the bronchial secretions and the muscle weakness associated with poisoning. The initial intravenous dose of atropine in adults should be determined by the severity of symptoms: An initial adult dose of 1.0 to 2.0 mg or pediatric dose of 0.01 mg/kg (minimum 0.01 mg) should be administered intravenously. If intravenous access cannot be established, atropine may also be given intramuscularly, subcutaneously or via endotracheal tube. Doses should be repeated every 15 minutes until excessive secretions and sweating have been controlled. Once bronchial secretion has been controlled, atropine administration should be repeated whenever the secretions begin to recur. In seriously poisoned patients, very large doses may be required. Alterations of pulse rate and pupillary size should not be used as indicators of treatment adequacy. Pralidoxime should be administered as early in poisoning as possible as its efficacy may diminish when given more than 24 to 36 hours after exposure. Doses are as follows: adult 1.0 g; pediatric 25 to 50 mg/kg. The drug should be administered intravenously over 30 to 60 minutes, but in a life-threatening situation, one-half of the total dose can be given per minute for a total administration time of 2 minutes. Treatment should begin to take effect within 40 minutes with a reduction in symptoms and the amount of atropine necessary to control bronchial secretion. The initial dose can be repeated in 1 hour and then every 8 to 12 hours until the patient is clinically well and no longer requires atropine. If intravenous access cannot be established, pralidoxime may also be given intramuscularly. Early administration of diazepam in addition to the combined atropine and pralidoxime treatment may help prevent the onset of seizures and potential brain and cardiac morphologic damage following high-level organophosphate poisoning.

*References:*
- EXTOXNET, Extension Toxicology Network, "Pesticide Information Profile, Phorate," Oregon State University, Corvallis, OR (June 1996). http://extoxnet.orst.edu/pips/phorate.htm
- U.S. Environmental Protection Agency, Office of Pesticide Programs, Pesticide Residue Limits, "Phorate," 40 CFR 180.206, http://www.epa.gov/pesticides/food/viewtols.htm
- New Jersey Department of Health, "Hazardous Substance Fact Sheet: Phorate," Trenton, NJ (April 1986, rev. September 2001). http://www.state.nj.us/health/eoh/rtkweb/1508.pdf
- U.S. Environmental Protection Agency, "Phorate, Health and Environmental Effects," Profile No. 145, Office of Solid Waste, Washington, DC (April 30, 1980).
- U.S. Environmental Protection Agency, "Chemical Profile: Phorate," Washington, DC, Chemical Emergency Preparedness Program (November 30, 1987).
- Agency for Toxic Substances and Disease Registry, U.S. Department of Health and Human Services, Public Health Service, "Managing Hazardous Materials Incidents," Atlanta, GA (June 2003).
- California Environmental Protection Agency "Chemical List of Lists," Sacramento CA (February 1997).

# Phosacetim

*Use Type:* Rodenticide
*CAS Number:* 4104-14-7
*Formula:* $C_{14}H_{13}Cl_2N_2O_2PS$

## Phosacetim

***Synonyms:*** Acetimidoylphosphoramidothioic acid *O,O*-bis(*p*-chlorophenyl)ester; *O,O*-Bis(*p*-chlorophenyl) acetimidoylphosphoramidothioate; *O,O*-Bis(4-chlorophenyl)(1-iminoethyl)phosphoramidothioate; *O,O*-Bis(4-chlorophenyl)*N*-acetimidoylphosphoramidothioate; *O,O*-Bis(4-chlorophenyl) (1-iminoethyl) phosphoramidothioic acid; (1-Iminoethyl) phosphoramidothioic acid, *O,O*-bis(4-chlorophenyl) ester; Phosazetim; Phosphonodithioimidocarbonic acid, (1-iminoethyl)-*O,O*-bis(*p*-chlorophenyl) ester; Phosphonodithioimidocarbonic acid, acetimidoyl-, *O,O*-bis(p-chlorophenyl) ester

***Trade Names:*** BAY 33819®, Bayer CropScience (Germany); BAYER 33819®, Bayer CropScience (Germany); DRC-714; GOPHACIDE®, Bayer CropScience (Germany), canceled 1/22/1991

***Producers:*** Bayer CropScience (Germany)

***Chemical Class:*** Organophosphate

***EPA/OPP PC Code:*** 018501

***California DPR Chemical Code:*** 1523

***RTECS Number:*** TB4725000

***Uses:*** Used as a rodenticide on gophers, rats and mice. No products currently registered with the U.S. EPA.

***Carcinogen/Hazard Classifications***

**Label Signal Word:** WARNING

***Regulatory Authority:***
- Banned or Severely Restricted (Philippines) (UN)[13]
- Very Toxic Substance (World Bank)[15]
- U.S. DOT Inhalation Hazard Chemicals as organophosphates
- The "Director's List" (CAL/OSHA)
- Superfund/EPCRA 40CFR355, Appendix B Extremely Hazardous Substances: TPQ = 100/10,000 lb (45.4/4,540 kg)
- Superfund/EPCRA 40CFR302.4 RQ: EHS, 1 lb (0.454 kg)

***Description:*** Phosacetim is a crystalline solid. Hazard Identification (based on NFPA-704 M Rating System): Health 4 Flammability 1, Reactivity 0.

***Incompatibilities:*** Strong oxidizers, nitrates. May hydrolyze on contact with moisture.

***Permissible Exposure Limits in Air:*** No standards set.

***Determination in Air:*** OSHA versatile sampler-2 Toluene/Acetone Gas chromatography/Flame photometric detection for sulfur, nitrogen, or phosphorus NIOSH IV, Method #5600, Organophosphorus pesticides.[18]

***Permissible Concentration in Water:*** No criteria set, but runoff from spills or fire control may cause water pollution.

***Routes of Entry:*** Inhalation, ingestion, skin and/or eye contact. Absorbed by the skin.

***Harmful Effects and Symptoms***

***Short Term Exposure:*** Organic phosphorus insecticides are absorbed by the skin, as well as by the respiratory and gastrointestinal tracts. They are cholinesterase inhibitors. Symptoms of exposure include headache, giddiness, blurred vision, nervousness, weakness, nausea, cramps, diarrhea, and discomfort in the chest. Signs include sweating, tearing, salivation, vomiting, cyanosis, convulsions, coma, loss of reflexes and loss of sphincter control. Highly toxic. $LD_{50}$-(oral, rat) = 3.7 mg/kg. Delayed pulmonary edema may occur after inhalation.

***Long Term Exposure:*** Cholinesterase inhibitor; cumulative effect is possible. This chemical may damage the nervous system with repeated exposure, resulting in convulsions, respiratory failure. May cause liver damage.

***Points of Attack:*** Respiratory system, lungs, central nervous system, cardiovascular system, skin, eyes, plasma and red blood cell cholinesterase.

***Medical Surveillance:*** Medical observation is recommended for 24 to 48 hours after breathing overexposure, as pulmonary edema may be delayed. Before employment and at regular times after that, the following are recommended: Plasma and red blood cell cholinesterase levels (tests for the enzyme poisoned by this chemical). If exposure stops, plasma levels return to normal in 1-2 weeks while red blood cell levels may be reduced for 1-3 months. When acetylcholinesterase enzyme levels are reduced by 25% or more below preemployment levels, risk of poisoning is increased, even if results are in lower ranges of "normal." Reassignment to work not involving organophosphate or carbamate pesticides is recommended until enzyme levels recover. If symptoms develop or overexposure occurs, repeat the above tests as soon as possible and get an exam of the nervous system. Also consider complete blood count. Consider chest x-ray following acute overexposure. Do not drink any alcoholic beverages before or during use. Alcohol promotes absorption of organophosphates.

***First Aid:*** **Treatment for organophosphate poisoning consists of thorough decontamination, cardiorespiratory support, and administration of the antidotes atropine and pralidoxime. In cases of severe poisoning, diazepam, an anticonvulsant, should also be administered. Antidotes should be administered as prevention even if the diagnosis is in doubt.** Speed in removing material from eyes and skin is of extreme importance. *Eyes:* Eye contact can cause dangerous amounts of these chemicals to be quickly absorbed through the mucous membrane into the bloodstream. Immediately and gently flush eyes with plenty of warm or cold water (NO hot water) for at least 15 minutes, occasionally lifting the upper and lower eyelids. Get medical aid immediately. *Skin:* Get medical aid. Skin contact can cause dangerous amounts of these chemicals to be absorbed into the bloodstream. Wearing the appropriate PPE equipment and respirator for organophosphate pesticides, immediately flush skin with plenty of soap and water for at least 15 minutes while removing contaminated clothing and shoes. Shampoo hair promptly if contaminated. The removed, contaminated clothing and shoes should be double-bagged and left in Hot Zone for later disposal by hazardous materials experts. Skin may also be decontaminated with diluted hypochlorite solution. *Inhalation:* Get medical aid. Do not contaminate yourself. Wearing the appropriate PPE equipment and respirator for organophosphate pesticides, immediately remove the victim from the contaminated area to fresh air. If the victim is not breathing, administer artificial respiration. Do NOT use mouth-to-mouth resuscitation; use bag/mask apparatus. If breathing is difficult, administer oxygen through bag/mask apparatus until medical help arrives. Do not leave victim unattended. *Ingestion:* Call poison control. Loosen all clothing. Never give anything by mouth to an unconscious person. If victim is *unconscious or having convulsions,* do nothing except keep victim warm. Get medical aid. Transfer promptly to a medical facility. In cases of ingestion, **do not induce vomiting**. If the victim is alert and asymptomatic, administer a slurry of activated charcoal at a dose of 1 g/kg (infant, child, and adult dose). A soda can and straw may be of assistance when offering charcoal to a child. *In some cases you may be specifically instructed by poison control to induce vomiting by way of 2 tablespoons of syrup of ipecac (adult) washed down with a cup of water.* Do NOT give activated charcoal before or with ipecac syrup.

*Note to physician or authorized medical personnel:* Treat cases of respiratory compromise, coma, or excessive pulmonary secretions with respiratory support using protocols and techniques available and within the scope of training. Some cases may necessitate procedures such as endotracheal intubation or cricothyrotomy by properly trained and equipped personnel. When possible, atropine (see *Antidotes,* below) should be given under medical supervision. Patients who are comatose, hypotensive, or having seizures or cardiac arrhythmias should be treated according to advanced life support protocols. *Antidotes:* Two antidotes are administered to treat organophosphate poisoning. Atropine is a competitive antagonist of acetylcholine at muscarinic receptors and is used to control the excessive bronchial secretions which are often responsible for death. Pralidoxime relieves both the nicotinic and muscarine effects of organophosphate poisoning by regenerating acetylcholinesterase and can reduce both the bronchial secretions and the muscle weakness associated with poisoning. The initial intravenous dose of atropine in adults should be determined by the severity of symptoms: An initial adult dose of 1.0 to 2.0 mg or pediatric dose of 0.01 mg/kg (minimum 0.01 mg) should be administered intravenously. If intravenous access cannot be established, atropine may also be given intramuscularly, subcutaneously or via endotracheal tube. Doses should be repeated every 15 minutes until excessive secretions and sweating have been controlled. Once bronchial secretion has been controlled, atropine administration should be repeated whenever the secretions begin to recur. In seriously poisoned patients, very large doses may be required. Alterations of pulse rate and pupillary size should not be used as indicators of treatment adequacy. Pralidoxime should be administered as early in poisoning as possible as its efficacy may diminish when given more than 24 to 36 hours after exposure. Doses are as follows: adult 1.0 g; pediatric 25 to 50 mg/kg. The drug should be administered intravenously over 30 to 60 minutes, but in a life-threatening situation, one-half of the total dose can be given per minute for a total administration time of 2 minutes. Treatment should begin to take effect within 40 minutes with a reduction in symptoms and the amount of atropine necessary to control bronchial secretion. The initial dose can be repeated in 1 hour and then every 8 to 12 hours until the patient is clinically well and no longer requires atropine. If intravenous access cannot be established, pralidoxime may also be given intramuscularly. Early administration of diazepam in addition to the combined atropine and pralidoxime treatment may help prevent the onset of seizures and potential brain and cardiac morphologic damage following high-level organophosphate poisoning.

*References:*
- New Jersey Department of Heaht and Senior Services, "Hazardous Substance Fact Sheet, Phosacetim," Trenton, NJ (April 2002). http://www.state.nj.us/health/eoh/rtkweb/2669.pdf
- U.S. Environmental Protection Agency, "Chemical Profile: Phosacetim," Washington, DC, Chemical Emergency Preparedness Program (November 30, 1987). http://www.state.nj.us/health/eoh/rtkweb/2669.pdf
- California Environmental Protection Agency "Chemical List of Lists," Sacramento CA (February 1997).
- Agency for Toxic Substances and Disease Registry, U.S. Department of Health and Human Services, Public Health Service, "Managing Hazardous Materials Incidents," Atlanta, GA (June 2003).

# Phosalone (ANSI)

*Use Type:* Insecticide, molluscicide, acaricide
*CAS Number:* 2310-17-0
*Formula:* $C_{12}H_{15}ClNO_4PS_2$
*Synonyms:* Benzophosphate; S-[(6-Chloro-2-oxo-3(2H)-benzoxazolyl)methyl] O,O-diethylphosphordithioate; S-[6-Chloro-3-(mercaptomethyl)-2-benzoxazolinone]-O,O-diethylphosphorodithioate; 3-(6-Chloro-2-oxobenzoxazolin-3-yl)methyl-O,O-diethyl phosphorothiolothionate; O,O-Diaethyl-S-(6-chlor-2-oxo-ben($\beta$)-1,3-oxalin-3-yl)-methyl-dithiophosphat (German); O,O-Diethyl-S-(6-chlorobenzoxazolinyl-3-methyl)dithiophosphate; O,O-Diethyl-S-[(6-chloro-2-oxobenzoxazolin-3-yl)methyl]phosphorodithioate; O,O-Diethyl S-[6-chloro-3-(mercaptomethyl)-2-benzoxazolinone]phosphorodithioate; 3-Diethyldithiophosphorylmethyl-6-chlorobenzoxazolone-

2; ENT 27,163; Phosphorodithioic acid, S-[(6-chloro-3-(mercaptomethyl)-2-benzoxazolinone] O,O-diethyl ester
**Trade Names:** AZOFENE®; BENZPHOS®; CHIPMAN®-11974; FOZALON®; NIA-9241®; NIAGARA®-9241; NPH-1091®; PHASOLON®; PHOSALON®; PHOSALONE®, Bayer CropScience (Germany), canceled; PHOZALON®; RHODIA-RP-11974®; RP 11974®; RUBITOX®; ZOLON®, Rhone-Poulenc (France), canceled; ZOOLON®
**Producers:** Chromos Agro (Croatia); Agsin (Singapore)
**Chemical Class:** Organophosphate
**EPA/OPP PC Code:** 097701; (294500 old EPA code number)
**California DPR Chemical Code:** 479
**ICSC Number:** 0797
**RTECS Number:** TD5175000
**EEC Number:** 015-067-00-1
**Uses:** Not registered in the U.S. It is a broad spectrum insecticide that was used in the U.S. to control chewing and sucking insects on deciduous fruit trees, garden crops, potatoes, rape and cotton.
*Carcinogen/Hazard Classifications*
**U.S. EPA Carcinogens:** Not likely a carcinogen
**WHO Acute Hazard:** Class II, moderately hazardous
**Regulatory Authority:**
- AB 1803-Well Monitoring Chemical (CAL)
- DOT Inhalation Hazard Chemicals
- FIFRA, 180.3(4); class of chlorinated organic pesticide
- FIFRA, 40 CFR 186: tolerances for pesticides in animal feeds
- U.S. DOT Regulated Marine Pollutant (49CFR172.101, Appendix B), severe pollutant

**Description:** Colorless crystalline solid. Garlic-like odor. Practically insoluble in water; solubility = < 100 ppm. Molecular weight = 367.82. Melting/Freezing point = 47.5°C. Vapor pressure = $5 \times 10^{-7}$ mmHg @ 25°C. Log $K_{ow}$ = 4.33. Values above 3.0 are likely to bioaccumulate in marine organisms.
**Incompatibilities:** May react violently with antimony(V) pentafluoride. Incompatible with lead diacetate, magnesium, silver nitrate.
**Permissible Exposure Limits in Air:**
**Determination in Air:** OSHA versatile sampler-2; Toluene/Acetone; Gas chromatography/Flame ionization detection; NIOSH IV, Method #5600, Organophosphorus Pesticides.[18]
**Permissible Concentration in Water:** No criteria set. Runoff from spills or fire control may cause water pollution.
**Routes of Entry:** Inhalation, ingestion, through the skin.
*Harmful Effects and Symptoms*
**Short Term Exposure:** Eye pupils are small; blurred vision; eye watering; runny nose; cough; shortness of breath; salivation; dizziness; nausea, stomach cramps, diarrhea, and vomiting; increased blood pressure; profuse sweating; hypermotility, hallucinations; irritability; tingling of the skin; drowsiness; slow heartbeat; convulsions; fluid in lungs; loss of consciousness; incontinence; breathing stops; death. Organophosphates inhibit the action of acetylcholinesterase enzymes, and alter the way in which nervous impulses are transmitted. The effects can last for hours, days, or much longer. The action of the enzymes is reestablished after new enzymes are formed.
**Long Term Exposure:** Cholinesterase inhibitor; cumulative effect is possible. Organophosphates may damage the nervous system with repeated exposure, resulting in convulsions, respiratory failure. May cause liver damage.
**Points of Attack:** Respiratory system, central nervous system, cardiovascular system, blood cholinesterase.
**Medical Surveillance:** Medical observation is recommended for 24 to 48 hours after breathing overexposure, as pulmonary edema may be delayed. As first aid for pulmonary edema, consider administering a corticosteroid spray. Cigarette smoking may exacerbate pulmonary injury and should be discouraged for at least 72 hours following exposure.
**First Aid:** Speed in removing material from eyes and skin is of extreme importance. Eye contact can cause dangerous amounts of these chemicals to be quickly absorbed through the mucous membrane into the bloodstream. Immediately and gently flush eyes with plenty of warm or cold water (NO hot water) for at least 15 minutes, occasionally lifting the upper and lower eyelids. Get medical aid immediately. *Skin:* Get medical aid. Skin contact can cause dangerous amounts of these chemicals to be absorbed into the bloodstream. Wearing the appropriate PPE equipment and respirator for organophosphate pesticides, immediately flush skin with plenty of soap and water for at least 15 minutes while removing contaminated clothing and shoes. Shampoo hair promptly if contaminated. *Ingestion:* Call poison control. Loosen all clothing. Never give anything by mouth to an unconscious person. Get medical aid. Do NOT induce vomiting.* If conscious, alert, and able to swallow, rinse mouth and have victim drink 4 to 8 ounces of water do NOT induce vomiting but immediately administer slurry of activated charcoal (2 oz in 8 oz of water). If victim is *unconscious or having convulsions,* do nothing except keep victim warm. *\*In some cases you may be specifically instructed by poison control to induce vomiting by way of 2 tablespoons of syrup of ipecac (adult) washed down with a cup of water.* Do NOT give activated charcoal before or with ipecac syrup. *Inhalation:* Get medical aid. Do not contaminate yourself. Wearing the appropriate PPE equipment and respirator for organophosphate pesticides, immediately remove the victim from the contaminated area to fresh air. If the victim is not breathing, administer artificial respiration. Do NOT use mouth-to-mouth resuscitation; use bag/mask apparatus. If breathing is difficult, administer oxygen through bag/mask apparatus until medical help arrives. Do not leave victim unattended. *Note to physician or authorized medical personnel.* Administer atropine, 2 mg (1/30 gr) intramuscularly or

intravenously as soon as any local or systemic signs or symptoms of an intoxication are noted; repeat the administration of atropine every 3 to 8 minutes until signs of atropinization (mydriasis, dry mouth, rapid pulse, hot and dry skin) occur; initiate treatment in children with 0.05 mg mg/kg of atropine; repeat at 5 to 10 minute intervals. Watch respiration, and remove bronchial secretions if they appear to be obstructing the airway; intubate if necessary. *Notes to physician or authorized medical personnel*: N-methylpyridinium-2-aldoxime (2-PAMCI) when used in conjunction with atropine reacts with the phosphorylated cholinesterase, thereby restoring normal activity to by removing the phosphorylating group. The combination of these two chemicals is synergistic and must be administered within minutes to a few hours following exposure (depending on the specific agent) to be effective. Give 2-PAMCI (Pralidoxime; Protopam), 2.5 gm in 100 ml of sterile water or in 5% dextrose and water, intravenously, slowly, in 15-30 minutes; if sufficient fluid is not available, give 1 gm of 2-PAMCI in 3 ml of distilled water by deep intramuscular injection; repeat this every half hour if respiration weakens or if muscle fasciculation or convulsions recur. Also Diazepam, an anticonvulsant, might be considered.

*References:*
- U.S. Environmental Protection Agency, Office of Pesticide Programs, "Phosalone Facts," (January 2001). http://www.epa.gov/REDs/phosalonetred.pdf
- EXTOXNET, Extension Toxicology Network, "Pesticide Information Profile, Phosalone," Oregon State University, Corvallis, OR (June 1996). http://extoxnet.orst.edu/pips/phosalon.htm
- California Environmental Protection Agency "Chemical List of Lists," Sacramento CA (February 1997)

# Phosfolan

*Use Type:* Insecticide
*CAS Number:* 947-02-4
*Formula:* $C_7H_{14}NO_3PS_2$
*Alert:* There are no products containing phosfolan currently registered with the U.S. EPA.
*Synonyms:* C.I. 47031; Cyclic ethylene (diethoxyphosphinothioyl)dithioimidocarbonate; Cyclic ethylene *p,p*-diethylphosphono dithioimidocarbonate; (Diethoxyphosphinyl)dithioimidocarbonic acid cyclic ethylene ester; 2-(Diethoxyphosphinylimino)-1,3-dithiolan; 2-(Diethoxyphosphinylimino)-1,3-dithiolane; *p,p*-Diethyl cyclic ethylene ester of phosphonodithioimidocarbonate; *p,p*-Diethyl cyclic ethylene ester of phosphonodithioimido carbonic acid; Diethyl 1,3-dithiolan-2-ylidene phosphoramidate; ENT 25,830; 1,2-Ethanedithiol, cyclic ester with *p,p*-diethyl phosphonodithioimidocarbonate; 1,2-Ethanedithiol, cyclic ester with phosphonodithioimido carbonic acid *p,p*-diethyl ester; Imidocarbonic acid, phosphonodithio-, cyclic ethylene *p,p*-diethyl ester; Phosphoroamidic acid, 1,3-dithiolan-2-ylidene-, diethyl ester

*Trade Names:* AC 47031®, BASF Agricultural Products Group (Germany); AMERICAN CYANAMID 47031®, BASF Agricultural Products Group (Germany); CYLAN®; CYOLANE®; CYOLANE INSECTICIDE®; EI 47031®
*Producers:* BASF Agricultural Products Group (Germany)
*Chemical Class:* Organophosphate
*EPA/OPP PC Code:* 268300
*RTECS Number:* NJ6475000
*Uses:* A systemic insecticide. There are no products containing phosfolan currently registered with the U.S. EPA.
*Regulatory Authority:*
- Superfund/EPCRA 40CFR355, Appendix B Extremely Hazardous Substances: TPQ = 100/10,000 lb (45.4/4,540 kg)
- Superfund/EPCRA 40CFR302.4 RQ: EHS, 1 lb (0.454 kg)
- U.S. DOT Inhalation Hazard Chemicals as organophosphates

*Description:* Phosfolan is a colorless to yellow solid. Boiling point = 115°-118°C @ 0.001 mm. Melting/Freezing point =37°-45°C. Hazard Identification (based on NFPA-704 M Rating System): Health 4, Flammability 1, Reactivity 0. Soluble in water.
*Incompatibilities:* Incompatible with nitrates and water. May hydrolyze upon contact with water, steam and moisture and produce toxic oxides of phosphorus, nitrogen, sulfur, and chlorine.
*Permissible Exposure Limits in Air:* No standards set.
*Determination in Air:* OSHA versatile sampler-2 Toluene/Acetone Gas chromatography/Flame photometric detection for sulfur, nitrogen, or phosphorus NIOSH IV, Method #5600, Organophosphorus pesticides.[18]
*Permissible Concentration in Water:* No criteria set, but runoff from spills or fire control may cause water pollution.
*Routes of Entry:* Inhalation, ingestion, skin and/or eye contact. Absorbed through the skin.
*Harmful Effects and Symptoms*
*Short Term Exposure:* Similar to parathion in health hazards. Death may result due to respiratory arrest as a result of paralysis of respiratory muscles and intense bronchoconstriction. Also considered a cholinesterase inhibitor. Symptoms similar to parathion include nausea, vomiting, abdominal cramps, diarrhea, excessive salivation, headache, giddiness, dizziness, tightness in the chest, blurring or dimness of vision, tearing, loss of muscle coordination, slurring of speech, twitching of muscles, drowsiness, difficulty in breathing, respiratory rales, and random jerky movements. Delayed pulmonary edema may occur after inhalation.

***Long Term Exposure:*** Cholinesterase inhibitor; cumulative effect is possible. This chemical may damage the nervous system with repeated exposure, resulting in convulsions, respiratory failure. May cause liver damage.

***Points of Attack:*** Respiratory system, lungs, central nervous system, cardiovascular system, skin, eyes, plasma and red blood cell cholinesterase.

***Medical Surveillance:*** Medical observation is recommended for 24 to 48 hours after breathing overexposure, as pulmonary edema may be delayed. Before employment and at regular times after that, the following are recommended: Plasma and red blood cell cholinesterase levels (tests for the enzyme poisoned by this chemical). If exposure stops, plasma levels return to normal in 1-2 weeks while red blood cell levels may be reduced for 1-3 months. When acetylcholinesterase enzyme levels are reduced by 25% or more below preemployment levels, risk of poisoning is increased, even if results are in lower ranges of "normal." Reassignment to work not involving organophosphate or carbamate pesticides is recommended until enzyme levels recover. If symptoms develop or overexposure occurs, repeat the above tests as soon as possible and get an exam of the nervous system. Also consider complete blood count. Consider chest x-ray following acute overexposure. Do not drink any alcoholic beverages before or during use. Alcohol promotes absorption of organophosphates.

***First Aid:* Treatment for organophosphate poisoning consists of thorough decontamination, cardiorespiratory support, and administration of the antidotes atropine and pralidoxime. In cases of severe poisoning, diazepam, an anticonvulsant, should also be administered. Antidotes should be administered as prevention even if the diagnosis is in doubt.** Speed in removing material from eyes and skin is of extreme importance. *Eyes:* Eye contact can cause dangerous amounts of these chemicals to be quickly absorbed through the mucous membrane into the bloodstream. Immediately and gently flush eyes with plenty of warm or cold water (NO hot water) for at least 15 minutes, occasionally lifting the upper and lower eyelids. Get medical aid immediately. *Skin:* Get medical aid. Skin contact can cause dangerous amounts of these chemicals to be absorbed into the bloodstream. Wearing the appropriate PPE equipment and respirator for organophosphate pesticides, immediately flush skin with plenty of soap and water for at least 15 minutes while removing contaminated clothing and shoes. Shampoo hair promptly if contaminated. The removed, contaminated clothing and shoes should be double-bagged and left in Hot Zone for later disposal by hazardous materials experts. Skin may also be decontaminated with diluted hypochlorite solution. *Inhalation:* Get medical aid. Do not contaminate yourself. Wearing the appropriate PPE equipment and respirator for organophosphate pesticides, immediately remove the victim from the contaminated area to fresh air. If the victim is not breathing, administer artificial respiration. Do NOT use mouth-to-mouth resuscitation; use bag/mask apparatus. If breathing is difficult, administer oxygen through bag/mask apparatus until medical help arrives. Do not leave victim unattended. *Ingestion:* Call poison control. Loosen all clothing. Never give anything by mouth to an unconscious person. If victim is *unconscious or having convulsions,* do nothing except keep victim warm. Get medical aid. Transfer promptly to a medical facility. In cases of ingestion, **do not induce vomiting.** If the victim is alert and asymptomatic, administer a slurry of activated charcoal at a dose of 1 g/kg (infant, child, and adult dose). A soda can and straw may be of assistance when offering charcoal to a child. *In some cases you may be specifically instructed by poison control to induce vomiting by way of 2 tablespoons of syrup of ipecac (adult) washed down with a cup of water.* Do NOT give activated charcoal before or with ipecac syrup.

*Note to physician or authorized medical personnel:* Treat cases of respiratory compromise, coma, or excessive pulmonary secretions with respiratory support using protocols and techniques available and within the scope of training. Some cases may necessitate procedures such as endotracheal intubation or cricothyrotomy by properly trained and equipped personnel. When possible, atropine (see *Antidotes,* below) should be given under medical supervision. Patients who are comatose, hypotensive, or having seizures or cardiac arrhythmias should be treated according to advanced life support protocols. *Antidotes:* Two antidotes are administered to treat organophosphate poisoning. Atropine is a competitive antagonist of acetylcholine at muscarinic receptors and is used to control the excessive bronchial secretions which are often responsible for death. Pralidoxime relieves both the nicotinic and muscarine effects of organophosphate poisoning by regenerating acetylcholinesterase and can reduce both the bronchial secretions and the muscle weakness associated with poisoning. The initial intravenous dose of atropine in adults should be determined by the severity of symptoms: An initial adult dose of 1.0 to 2.0 mg or pediatric dose of 0.01 mg/kg (minimum 0.01 mg) should be administered intravenously. If intravenous access cannot be established, atropine may also be given intramuscularly, subcutaneously or via endotracheal tube. Doses should be repeated every 15 minutes until excessive secretions and sweating have been controlled. Once bronchial secretion has been controlled, atropine administration should be repeated whenever the secretions begin to recur. In seriously poisoned patients, very large doses may be required. Alterations of pulse rate and pupillary size should not be used as indicators of treatment adequacy. Pralidoxime should be administered as early in poisoning as possible as its efficacy may diminish when given more than 24 to 36 hours after exposure. Doses are as follows: adult 1.0 g; pediatric 25 to 50 mg/kg. The drug should be administered intravenously over 30 to 60 minutes, but in a life-threatening situation, one-half of the total dose can be given per minute for a total administration time of 2 minutes. Treatment should begin to take effect within 40 minutes with a reduction in symptoms and the amount of atropine necessary to control bronchial secretion. The initial dose can be repeated in 1 hour and then every 8 to 12 hours until the patient is clinically well and no longer requires atropine. If intravenous access cannot be established, pralidoxime may also be given intramuscularly. Early administration of diazepam in addition to the combined atropine and pralidoxime treatment may help prevent the onset of seizures and potential brain and cardiac morphologic damage following high-level organophosphate poisoning.

*References:*
- New Jersey Department of Health and Senior Services, "Hazardous Substance Fact Sheet, Phosfolan," Trenton, NJ (August 2000) http://www.state.nj.us/health/eoh/rtkweb/2670.pdf
- U.S. Environmental Protection Agency, "Chemical Profile: Phosfolan," Washington, DC, Chemical Emergency Preparedness Program (November 30, 1987).
- California Environmental Protection Agency "Chemical List of Lists," Sacramento CA (February 1997).
- Agency for Toxic Substances and Disease Registry, U.S. Department of Health and Human Services, Public Health Service, "Managing Hazardous Materials Incidents," Atlanta, GA (June 2003).

# Phosmet

*Use Type:* Insecticide and acaricide
*CAS Number:* 732-11-6
*Formula:* $C_{11}H_{12}NO_4PS_2$
*Alert:* A General Use Pesticide (GUP). Human toxicity (long-term): High. Effective 12/27, 2004, deleted from EHS list.
*Synonyms:* (O,O-Dimethyl-phthalimidiomethyl-dithiophosphate); O,O-Dimethyl S-(N-phthalimidomethyl)dithiophosphate; O,O-Dimethyl S-phthalimidomethylphosphorodithioate; ENT 25,705; Fosmet (Spanish); N-(Mercaptomethyl)phthalimide S-(O,O-dimethyl phosphorodithioate); Phosphorodithioic acid, S-[(1,3-dihydro-1,3-dioxo-isoindol-2-yl)methyl] O,O-dimethyl ester; Phosphorodithioic acid, O,O-dimethyl ester, S-ester with N-(mercaptomethyl)phthalimide; Phthalimide, N-(mercaptomethyl)-, S-ester with O,O-dimethylphosphorodithioate; Phthalimido O,O-dimethyl phosphorodithioate; Phthalimidomethyl O,O-dimethyl phosphorodithioate; Phthalophos (USSR); PMP (Japan)
*Trade Names:* APPA®; DECEMTHION®; DEL-PHOS®, Schering-Plough Animal Health (USA); FESDAN®; FIREBAN®, Gowan Company (USA); FTALOPHOS®; IMIDAN®, Gowan Company (USA); KEMOLATE®; PERCOLATE®; PMC®; PROLATE®, canceled; R 1504®; SAFIDON®; SMIDAN®; STARBAR CATTLE DUST®, Wellmark International (USA), canceled; STAUFFER R 1504®; VET-KEM®, Wellmark International (USA); ZEOCON®, Wellmark International (USA)
*Producers:* Alcotan Laboratories (Spain); Cyanamid Agro (USA); Dainippon Ink & Chemical (Japan); Ehrenstorfer, Dr. (Germany); General Quimica (Spain); Gowan Company (USA); Hokko Chemical Industry (Japan); Schering-Plough Animal Health (USA); Shenzhen Guomeng Industry Co., Ltd. (China); Takeda Chemical Industries (Japan); Wellmark International (USA)
*Chemical Class:* Organophosphate
*EPA/OPP PC Code:* 059201
*California DPR Chemical Code:* 335
*ICSC Number:* 0543
*RTECS Number:* TE2275000
*EEC Number:* 015-101-00-5
*EINECS Number:* 211-987-4
*Uses:* Phosmet is a non-systemic insecticide used on both plants and animals. It is mainly used on apple trees for control of coddling moth, though it is also used on a wide range of crops including alfalfa, nuts, grapes, blueberries, peas, potatoes, fruit crops, ornamentals, and vines for the control of aphids, suckers, mites, fire ants and fruit flies. The compound is also an active ingredient in some dog collars. It is used as an insecticide on swine and cattle.
*U.S. Maximum Allowable Residue Levels for Phosmet (40 CFR 180.261):*

| CROP | ppm |
|---|---|
| Alfalfa | 40 |
| Almond, hulls | 10 |
| Apple | 10 |
| Apricot | 5 |
| Blueberry | 10 |
| Cattle, fat | 0.2 |
| Cattle, meat | 0.2 |
| Cattle, mbyp | 0.2 |
| Cherry | 10 |
| Cotton, undelinted seed | 0.1 |
| Crabapple | 20 |
| Cranberry | 10 |
| Fruit, stone, group 12 | 5 |
| Goat, fat | 0.2 |
| Goat, meat | 0.2 |
| Goat, mbyp | 0.2 |
| Grape | 10 |
| Hog, fat | 0.2 |
| Hog, meat | 0.2 |
| Hog, mbyp | 0.2 |
| Horse, fat | 0.2 |
| Horse, meat | 0.2 |
| Horse, mbyp | 0.2 |
| Kiwifruit | 25 |
| Nectarine | 5 |
| Nuts | 0.1 |
| Pea | 0.5 |

| | |
|---|---|
| Pea, field, hay | 10 |
| Pea, field, vines | 10 |
| Peach | 10 |
| Pear | 10 |
| Pistachio | 0.1 |
| Plum, prune, fresh | 5 |
| Potato | 0.1 |
| Sheep, fat | 0.2 |
| Sheep, meat | 0.2 |
| Sheep, mbyp | 0.2 |
| Sweet potato, roots | 10 |

*Carcinogen/Hazard Classifications*
**U.S. EPA Carcinogens:** Suspected Carcinogen
**Label Signal Word:** DANGER or WARNING
**WHO Acute Hazard:** Class II, moderately hazardous
*Regulatory Authority:*
- Air Pollutant Standard Set (former USSR)[35]
- AB 1803-Well Monitoring Chemical (CAL)
- Actively registered pesticide in California.
- Superfund/EPCRA 40CFR355, Appendix B Extremely Hazardous Substances: TPQ = 10/10,000 lb (4.54/4,540 kg), removed December 27, 2004.
- Superfund/EPCRA 40CFR302.4 RQ: EHS, 1 lb (0.454 kg). See above.
- U.S. DOT Inhalation Hazard Chemicals as organophosphates
- MARINE POLLUTANT (49CFR, Subchapter 172.101, Appendix B)
- Canada, WHMIS, Ingredients Disclosure List

*Description:* Phosmet, $C_{11}H_{12}NO_4PS_2$ is a white crystalline solid. Practically insoluble in water; solubility = 25 ppm @ 25°C. Molecular weight = 317.34. Boiling point = decomposes >100°C. Melting/Freezing point = 72°C. Vapor pressure = 4.9 x$10^{-7}$ mmHg @ 20°C. Hazard Identification (based on NFPA-704 M Rating System): Health 3, Flammability 1, Reactivity 0. Log $K_{ow}$ = 2.82. Values at or above 3.0 are likely to bioaccumulate in marine organisms.

*Incompatibilities:* Not compatible with other pesticides under alkaline conditions. Contact with water, steam or moisture forms phthalic acids. Slightly corrosive to metals in the presence of moisture. When heated to decomposition, forms toxic oxides of nitrogen, phosphorus, and sulfur.

*Permissible Exposure Limits in Air:* The former USSR[35] has set a ceiling value in workplace air of 0.3 mg/m³. The former USSR has also set an MAC in ambient air in residential area of 0.009 mg/m³ on a once-daily basis and 0.004 mg/m³ on an average daily basis.

*Determination in Air:* OSHA versatile sampler-2; Toluene/Acetone; Gas chromatography/Flame photometric detection for sulfur, nitrogen, or phosphorus; NIOSH Method IV Method #5600, Organophosphorus pesticides.[18]

*Permissible Concentration in Water:* The former USSR[36] has set an MAC in surface water of 0.2 mg/L

*Routes of Entry:* Inhalation, ingestion, skin and/or eye contact. Absorbed by the skin.

*Harmful Effects and Symptoms*
*Short Term Exposure:* Irritates the eyes and skin on contact. This material is a highly toxic organophosphate; the probable oral lethal dose for humans is 50-500 mg/kg, or between 1 teaspoon and 1 oz for a 150 lb person. It is a cholinesterase inhibitor and has central nervous system effects. Oral lethal doses in humans have been reported at 50 mg/kg. Acute exposure to phosmet may produce the following signs and symptoms: pinpoint pupils, blurred vision, headache, dizziness, muscle spasms, and profound weakness. Vomiting, diarrhea, abdominal pain, seizures, and coma may also occur. The heart rate may decrease following oral exposure or increase following dermal exposure. Chest pain may be noted. Hypotension (low blood pressure) may occur, although hypertension (high blood pressure) is not uncommon. Dyspnea (shortness of breath) may be followed by respiratory collapse. Giddiness is common. Delayed pulmonary edema may occur after inhalation.

*Long Term Exposure:* Cholinesterase inhibitor; cumulative effect is possible. This chemical may damage the nervous system with repeated exposure, resulting in convulsions, respiratory failure. May cause liver damage.

*Points of Attack:* Respiratory system, lungs, central nervous system, cardiovascular system, skin, eyes, plasma and red blood cell cholinesterase.

*Medical Surveillance:* Medical observation is recommended for 24 to 48 hours after breathing overexposure, as pulmonary edema may be delayed. Before employment and at regular times after that, the following are recommended: Plasma and red blood cell cholinesterase levels (tests for the enzyme poisoned by this chemical). If exposure stops, plasma levels return to normal in 1-2 weeks while red blood cell levels may be reduced for 1-3 months. When acetylcholinesterase enzyme levels are reduced by 25% or more below preemployment levels, risk of poisoning is increased, even if results are in lower ranges of "normal." Reassignment to work not involving organophosphate or carbamate pesticides is recommended until enzyme levels recover. If symptoms develop or overexposure occurs, repeat the above tests as soon as possible and get an exam of the nervous system. Also consider complete blood count. Consider chest x-ray following acute overexposure. Liver function tests. Do not drink any alcoholic beverages before or during use. Alcohol promotes absorption of organophosphates.

*First Aid:* **Treatment for organophosphate poisoning consists of thorough decontamination, cardiorespiratory support, and administration of the antidotes atropine and pralidoxime. In cases of severe poisoning, diazepam, an anticonvulsant, should also be administered. Antidotes should be administered as prevention even if the diagnosis is in doubt.** Speed in removing material from eyes and skin is of extreme importance. *Eyes:* Eye contact can cause dangerous amounts of these chemicals to be quickly absorbed through the mucous membrane into the bloodstream. Immediately and gently flush eyes with plenty of warm or cold water (NO hot water) for at least 15 minutes, occasionally lifting the upper and lower eyelids. Get medical aid immediately. *Skin:* Get medical aid. Skin contact can cause dangerous amounts of these chemicals to be absorbed into the bloodstream. Wearing the appropriate PPE equipment and respirator for organophosphate pesticides, immediately flush skin with plenty of soap and water for at least 15 minutes while removing contaminated clothing and shoes. Shampoo hair promptly if contaminated. The removed, contaminated clothing and shoes should be double-bagged and left in Hot Zone for later disposal by hazardous materials experts. Skin may also be decontaminated with diluted hypochlorite solution. *Inhalation:* Get medical aid. Do not contaminate yourself. Wearing the appropriate PPE equipment and respirator for organophosphate pesticides, immediately remove the victim from the contaminated area to fresh air. If the victim is not breathing, administer artificial respiration. Do NOT use mouth-to-mouth resuscitation; use bag/mask apparatus. If breathing is difficult, administer oxygen through bag/mask apparatus until medical help arrives. Do not leave victim unattended. *Ingestion:* Call poison control. Loosen all clothing. Never give anything by mouth to an unconscious person. If victim is *unconscious or having convulsions,* do nothing except keep victim warm. Get medical aid. Transfer promptly to a medical facility. In cases of ingestion, **do not induce vomiting**. If the victim is alert and asymptomatic, administer a slurry of activated charcoal at a dose of 1 g/kg (infant, child, and adult dose). A soda can and straw may be of assistance when offering charcoal to a child. *In some cases you may be specifically instructed by poison control to induce vomiting by way of 2 tablespoons of syrup of ipecac (adult) washed down with a cup of water.* Do NOT give activated charcoal before or with ipecac syrup.

*Note to physician or authorized medical personnel:* Treat cases of respiratory compromise, coma, or excessive pulmonary secretions with respiratory support using protocols and techniques available and within the scope of training. Some cases may necessitate procedures such as endotracheal intubation or cricothyrotomy by properly trained and equipped personnel. When possible, atropine (see *Antidotes*, below) should be given under medical supervision. Patients who are comatose, hypotensive, or having seizures or cardiac arrhythmias should be treated according to advanced life support protocols. *Antidotes:* Two antidotes are administered to treat organophosphate poisoning. Atropine is a competitive antagonist of acetylcholine at muscarinic receptors and is used to control the excessive bronchial secretions which are often responsible for death. Pralidoxime relieves both the nicotinic and muscarine effects of organophosphate poisoning by regenerating acetylcholinesterase and can reduce both the bronchial secretions and the muscle weakness associated with poisoning. The initial intravenous dose of atropine in adults should be determined by the severity of symptoms: An initial adult dose of 1.0 to 2.0 mg or pediatric dose of 0.01 mg/kg (minimum 0.01 mg) should be administered intravenously. If intravenous access cannot be established, atropine may also be given intramuscularly, subcutaneously or via endotracheal tube. Doses should be repeated every 15 minutes until excessive secretions and sweating have been controlled. Once bronchial secretion has been controlled, atropine administration should be repeated whenever the secretions begin to recur. In seriously poisoned patients, very large doses may be required. Alterations of pulse rate and pupillary size should not be used as indicators of treatment adequacy. Pralidoxime should be administered as early in poisoning as possible as its efficacy may diminish when given more than 24 to 36 hours after exposure. Doses are as follows: adult 1.0 g; pediatric 25 to 50 mg/kg. The drug should be administered intravenously over 30 to 60 minutes, but in a life-threatening situation, one-half of the total dose can be given per minute for a total administration time of 2 minutes. Treatment should begin to take effect within 40 minutes with a reduction in symptoms and the amount of atropine necessary to control bronchial secretion. The initial dose can be repeated in 1 hour and then every 8 to 12 hours until the patient is clinically well and no longer requires atropine. If intravenous access cannot be established, pralidoxime may also be given intramuscularly. Early administration of diazepam in addition to the combined atropine and pralidoxime treatment may help prevent the onset of seizures and potential brain and cardiac morphologic damage following high-level organophosphate poisoning.

***References:***
- EXTOXNET, Extension Toxicology Network, "Pesticide Information Profile, Phosmet," Oregon State University, Corvallis, OR (June 1996). http://extoxnet.orst.edu/pips/phosmet.htm
- U.S. Environmental Protection Agency, Office of Pesticide Programs, Pesticide Residue Limits, "Phosmet," 40 CFR 180.261, http://www.epa.gov/pesticides/food/viewtols.htm
- U.S. Environmental Protection Agency, "Chemical Profile: Phosmet," Washington, DC, Chemical Emergency Preparedness Program (November 30, 1987).
- Agency for Toxic Substances and Disease Registry, U.S. Department of Health and Human Services, Public Health Service, "Managing Hazardous Materials Incidents," Atlanta, GA (June 2003).
- California Environmental Protection Agency "Chemical List of Lists," Sacramento CA (February 1997).
- New Jersey Department of Health and Senior Services, "Hazardous Substance Fact Sheet, Decemthion," Trenton NJ (March 1999). http://www.state.nj.us/health/eoh/rtkweb/0603.pdf

# Phosphamidon (ANSI)

*Use Type:* Insecticide and acaricide
*CAS Number:* 13171-21-6
*Formula:* $C_{10}H_{19}ClNO_5P$
*Alert:* No products containing phosphamidon are currently registered with the U.S. EPA.
*Synonyms:* (2-Chlor-3-diaethylamino-methyl-3-oxo-prop-1-en-yl)-dimethylphosphat (German); 2-Chloro-2-diethylcarbamoyl-1-methylvinyldimethylphosphate; 1-Chloro-diethylcarbamoyl-1-propen-2-yl dimethyl phosphate; (2-Chloro-3-dietilamino-1-metil-3-oxo prop-1-en-il)-dimetilfosfato (Italian); 2-Chloro-3-(diethylamino)-1-methyl-3-oxo-1-propenyldimethyl phosphate; Dimethyl 2-chloro-2-diethylcarbamoyl-1-methylvinylphosphate; O,O-Dimethyl O-(2-chloro-2-(N,N-diethylcarbamoyl)-1-methylvinyl)phosphate; Dimethyl diethylamido-1-chlorocrotonyl (2)phosphate; O,O-Dimethyl-O-(1-methyl-2-chlor-2-N,N-diethyl-carbamoyl)-vinyl-phosphat (German); (O,O-Dimethyl-O-(1-methyl-2-chloro-2-diethylcarbamoyl-vinyl)-phosphate; Dimethyl phosphate ester with 2-chloro-N,N-diethyl-3-hydroxycrotonamide; Dimethyl phosphate of 2-chloro-N,N-diethyl-3-hydroxycrotonamide; ENT 25,515; Fosfamidon (Spanish); Fosfamidone; Foszfamidon; NCI-C00588; OMS 1325 or 1191; Phosphate de dimethyle et de (2-chloro-2-diethylcarbamoyl-1-methyl-vinyle) (French); Phosphoric acid, 2-chloro-3-(diethylamino)-1-methyl-3-oxo-1-propenyl dimethyl ester; Phosphoric acid, dimethyl ester with 2-chloro-N,N-diethyl-3-hydroxycrotonamide
*Trade Names:* APAMIDON®; C 570®, Syngenta (Switzerland); CIBA 570®, Syngenta (Switzerland); CROPHOSPHATE®; DIMECRON®; DIMONEX®; DIXON®; MERKON PHOSPHAMIDONE®; ML 97®; SWAT INSECTICIDE–MITICIDE®, Syngenta (Switzerland), canceled 5/28/1991; TOXO®, Loveland Products (USA), canceled
*Producers:* Bharat Pulverizing Mills (India); DuPont (USA); Eli Lilly (USA); Gujarat Pesticides (India); Hindustan Insecticides (India); Nagarjuna Agrichem (India); Sudarshan Chemical Industries (India); Syngenta (Switzerland)
*Chemical Class:* Organophosphate
*EPA/OPP PC Code:* 018201
*California DPR Chemical Code:* 482
*ICSC Number:* 0189
*RTECS Number:* TC2800000
*EEC Number:* 015-022-00-6
*EINECS Number:* 236-116-5
*Uses:* This material is an insecticide to control mites, beetles, aphids and other plant pests on citrus, cotton, sugar cane, rice and deciduous fruit and nut crops. There are no products containing phosphamidon currently registered with the U.S. EPA.
**U.S. EPA Carcinogens:** Group C, possible carcinogen
**Label Signal Word:** DANGER
**WHO Acute Hazard:** Class 1 a, extremely hazardous
*Regulatory Authority:*
- Very Toxic Substance (World Bank)[15]
- AB 1803-Well Monitoring Chemical (CAL)
- The "Director's List" (CAL/OSHA)
- Superfund/EPCRA 40CFR355, Appendix B Extremely Hazardous Substances: TPQ = 100 lb (45.4 kg)
- Superfund/EPCRA 40CFR302.4 RQ: EHS, 1 lb (0.454 kg)
- U.S. DOT Inhalation Hazard Chemicals as organophosphates
- U.S. DOT Regulated Marine Pollutant (49CFR172.101, Appendix B), severe pollutant.

*Description:* Phosphamidon is a pale yellow oily liquid. Soluble in water. Boiling point = 162°C @ 1.5 mm. Melting/Freezing point = –45°C. Vapor Pressure = 2.5 x $10^{-5}$ mmHg @ 20°C. Hazard Identification (based on NFPA-704 M Rating System): Health 3, Flammability 1, Reactivity 0. Log $K_{ow}$ = 0.82. Unlikely to bioaccumulate in marine organisms.
*Incompatibilities:* Reacts with alkalis (hydrolysis). Attacks metals such as aluminum, iron, tin. When heated to decomposition, this chemical forms toxic oxides of nitrogen and phosphorus and fumes of chlorine.
*Permissible Exposure Limits in Air:* No standards set.
*Permissible Concentration in Water:* No criteria set, but runoff from spills or fire control may cause water pollution.
*Routes of Entry:* Inhalation, ingestion, skin and/or eye contact.
*Harmful Effects and Symptoms*
*Short Term Exposure:* Irritates the eyes. may affect the nervous system, causing convulsions, respiratory failure and death. Higher exposures can cause pulmonary edema, a medical emergency that can be delayed for several hours. This can cause death. This material is extremely toxic; the probable oral lethal dose for humans if 5-50 mg/kg, or between 7 drops and 1 teaspoonful for a 150 lb person. It is a cholinesterase inhibitor. Acute exposure to phosphamidon may produce pinpoint pupils, blurred vision, headache, dizziness, muscle spasms, and profound weakness. Vomiting, diarrhea, abdominal pain, seizures, and coma may also occur. The heart rate may decrease following oral exposure or increase following dermal exposure. Hypotension (low blood pressure) may occur, although hypertension (high blood pressure) is not uncommon. Chest pain may be noted. Respiratory effects include dyspnea (shortness of breath), respiratory depression, and respiratory paralysis. Psychosis may occur. Delayed pulmonary edema may occur after inhalation.
*Long Term Exposure:* Cholinesterase inhibitor; cumulative effect is possible. This chemical may damage the nervous system with repeated exposure, resulting in convulsions, respiratory failure. May cause liver damage.

*Points of Attack:* Respiratory system, lungs, central nervous system, cardiovascular system, skin, eyes, plasma and red blood cell cholinesterase.

*Medical Surveillance:* Medical observation is recommended for 24 to 48 hours after breathing overexposure, as pulmonary edema may be delayed. Before employment and at regular times after that, the following are recommended: Plasma and red blood cell cholinesterase levels (tests for the enzyme poisoned by this chemical). If exposure stops, plasma levels return to normal in 1-2 weeks while red blood cell levels may be reduced for 1-3 months. When acetylcholinesterase enzyme levels are reduced by 25% or more below preemployment levels, risk of poisoning is increased, even if results are in lower ranges of "normal." Reassignment to work not involving organophosphate or carbamate pesticides is recommended until enzyme levels recover. If symptoms develop or overexposure occurs, repeat the above tests as soon as possible and get an exam of the nervous system. Also consider complete blood count. Consider chest x-ray following acute overexposure. Do not drink any alcoholic beverages before or during use. Alcohol promotes absorption of organophosphates.

*First Aid:* **Treatment for organophosphate poisoning consists of thorough decontamination, cardiorespiratory support, and administration of the antidotes atropine and pralidoxime. In cases of severe poisoning, diazepam, an anticonvulsant, should also be administered. Antidotes should be administered as prevention even if the diagnosis is in doubt.** Speed in removing material from eyes and skin is of extreme importance. *Eyes:* Eye contact can cause dangerous amounts of these chemicals to be quickly absorbed through the mucous membrane into the bloodstream. Immediately and gently flush eyes with plenty of warm or cold water (NO hot water) for at least 15 minutes, occasionally lifting the upper and lower eyelids. Get medical aid immediately. *Skin:* Get medical aid. Skin contact can cause dangerous amounts of these chemicals to be absorbed into the bloodstream. Wearing the appropriate PPE equipment and respirator for organophosphate pesticides, immediately flush skin with plenty of soap and water for at least 15 minutes while removing contaminated clothing and shoes. Shampoo hair promptly if contaminated. The removed, contaminated clothing and shoes should be double-bagged and left in Hot Zone for later disposal by hazardous materials experts. Skin may also be decontaminated with diluted hypochlorite solution. *Inhalation:* Get medical aid. Do not contaminate yourself. Wearing the appropriate PPE equipment and respirator for organophosphate pesticides, immediately remove the victim from the contaminated area to fresh air. If the victim is not breathing, administer artificial respiration. Do NOT use mouth-to-mouth resuscitation; use bag/mask apparatus. If breathing is difficult, administer oxygen through bag/mask apparatus until medical help arrives. Do not leave victim unattended. *Ingestion:* Call poison control. Loosen all clothing. Never give anything by mouth to an unconscious person. If victim is *unconscious or having convulsions,* do nothing except keep victim warm. Get medical aid. Transfer promptly to a medical facility. In cases of ingestion, **do not induce vomiting**. If the victim is alert and asymptomatic, administer a slurry of activated charcoal at a dose of 1 g/kg (infant, child, and adult dose). A soda can and straw may be of assistance when offering charcoal to a child. *In some cases you may be specifically instructed by poison control to induce vomiting by way of 2 tablespoons of syrup of ipecac (adult) washed down with a cup of water.* Do NOT give activated charcoal before or with ipecac syrup.

*Note to physician or authorized medical personnel:* Treat cases of respiratory compromise, coma, or excessive pulmonary secretions with respiratory support using protocols and techniques available and within the scope of training. Some cases may necessitate procedures such as endotracheal intubation or cricothyrotomy by properly trained and equipped personnel. When possible, atropine (see *Antidotes*, below) should be given under medical supervision. Patients who are comatose, hypotensive, or having seizures or cardiac arrhythmias should be treated according to advanced life support protocols. *Antidotes:* Two antidotes are administered to treat organophosphate poisoning. Atropine is a competitive antagonist of acetylcholine at muscarinic receptors and is used to control the excessive bronchial secretions which are often responsible for death. Pralidoxime relieves both the nicotinic and muscarine effects of organophosphate poisoning by regenerating acetylcholinesterase and can reduce both the bronchial secretions and the muscle weakness associated with poisoning. The initial intravenous dose of atropine in adults should be determined by the severity of symptoms: An initial adult dose of 1.0 to 2.0 mg or pediatric dose of 0.01 mg/kg (minimum 0.01 mg) should be administered intravenously. If intravenous access cannot be established, atropine may also be given intramuscularly, subcutaneously or via endotracheal tube. Doses should be repeated every 15 minutes until excessive secretions and sweating have been controlled. Once bronchial secretion has been controlled, atropine administration should be repeated whenever the secretions begin to recur. In seriously poisoned patients, very large doses may be required. Alterations of pulse rate and pupillary size should not be used as indicators of treatment adequacy. Pralidoxime should be administered as early in poisoning as possible as its efficacy may diminish when given more than 24 to 36 hours after exposure. Doses are as follows: adult 1.0 g; pediatric 25 to 50 mg/kg. The drug should be administered intravenously over 30 to 60 minutes, but in a life-threatening situation, one-half of the total dose can be given per minute for a total administration time of 2 minutes. Treatment should begin to take effect within 40 minutes with a reduction in symptoms and the amount of atropine necessary to control bronchial secretion. The initial dose can be repeated in 1 hour and then every 8 to 12 hours until the patient is clinically well and no longer requires atropine. If intravenous access cannot be established, pralidoxime may also be given intramuscularly. Early administration of diazepam in addition to the combined atropine and pralidoxime treatment may help prevent the onset of seizures and potential brain and cardiac morphologic damage following high-level organophosphate poisoning.

*References:*
- New Jersey Department of Health and Senior Services, "Hazardous Substance Fact Sheet, Phosphamidon," Trenton NJ (September 1999). http://www.state.nj.us/health/eoh/rtkweb/1513.pdf
- U.S. Environmental Protection Agency, "Chemical Profile: Phosphamidon," Washington, DC, Chemical Emergency Preparedness Program (November 30, 1987).
- California Environmental Protection Agency "Chemical List of Lists," Sacramento CA (February 1997).
- Agency for Toxic Substances and Disease Registry, U.S. Department of Health and Human Services, Public Health Service, "Managing Hazardous Materials Incidents," Atlanta, GA (June 2003).

# Phosphine

*Use Type:* Fumigant and insecticide
*CAS Number:* 7803-51-2
*Formula:* $H_3P$; $PH_3$
*Alert:* Phosphine gas is a Restricted Use Pesticide (RUP)
*Synonyms:* Celphos; Delicia; Detia Gas-EX-B; Fosfamia (Spanish); Fosforowodor (Polish); Hydrogen phosphide; Phosphine gas; Phosphorous trihydride; Phosphorous hydride; Phosphorated hydrogen; Phosphorwasserstoff (German); Phostoxin
*Trade names:* $ECO_2FUME\ TM^®$, Cytec Industries (USA); $VAPORPH_3OS^®$, Cytec Industries (USA)
*Producers:* Air Liquide Group (France); Air Products & Chemicals (USA); Aldrich Chemical (USA); Argus Chemicals (Italy); Avecia (UK); BOC Gases (UK); Cognis (Germany); Cytec Industries (USA); Matheson Tri-Gas (USA); Messer Group (Germany); Nippon Sanso (Japan); Praxair (USA); Showa Denko (Japan); Sigma-Aldrich Fine Chemicals (USA); Solkatronic (USA); Sumitomo Chemical (Japan); UCB Group (Belgium)
*Chemical Class:* Inorganic phosphine
*EPA/OPP PC Code:* 066500
*California DPR Chemical Code:* 3541
*ICSC Number:* 0694
*RTECS Number:* SY7525000
*Uses:* Phosphine gas is used indoors to control a broad spectrum of insects for non-food/non-feed commodities in sealed containers or structures. There are no homeowner or agricultural row crop uses for this product. The end-use product is a poisonous liquefied gas under pressure, and is a Restricted Use Pesticide (RUP) due to the acute inhalation toxicity of phosphine gas. Phosphine is only occasionally used in industry, and exposure usually results accidentally as a by-product of various processes. Exposures may occur when acid or water comes in contact with metallic phosphides (aluminum Phosphide, calcium Phosphide). These two phosphides are used as insecticides or rodenticides for grain, and phosphine is generated during grain fumigation. Phosphine may also evolve during the generation of acetylene from impure calcium carbide, as well as during metal shaving, sulfuric acid tank cleaning, rustproofing, and ferrosilicon, phosphoric acid and yellow phosphorus explosive handling.

*U.S. Maximum Allowable Residue Levels for Phosphine (40 CFR 180.225 ):*

| CROP | ppm |
|---|---|
| Almond | 0.1 |
| Animal feed | 0.1 |
| Avocado | 0.01 |
| Banana | 0.01 |
| Barley, grain | 0.1 |
| Cabbage, Chinese, bok choy | 0.01 |
| Cacao bean, dried bean | 0.1 |
| Cashew | 0.1 |
| Citron, citrus | 0.01 |
| Coffee, bean | 0.1 |
| Corn, field, grain | 0.1 |
| Corn, pop, grain | 0.1 |
| Cotton, undelinted seed | 0.1 |
| Date, dried fruit | 0.1 |
| Dill, seed | 0.01 |
| Eggplant | 0.01 |
| Endive | 0.01 |
| Filbert | 0.1 |
| Grapefruit | 0.01 |
| Kumquat | 0.01 |
| Lemon | 0.01 |
| Lettuce | 0.01 |
| Lime | 0.01 |
| Mango | 0.01 |
| Millet, grain | 0.1 |
| Mushroom | 0.01 |
| Nut, brazil | 0.1 |
| Oat, grain | 0.1 |
| Okra | 0.01 |
| Orange | 0.01 |
| Papaya | 0.01 |
| Peanut | 0.1 |
| Pecan | 0.1 |
| Pepper, black, post-h | 0.01 |
| Pepper, pimento | 0.01 |

| | |
|---|---|
| Pepper, red, post-h | 0.01 |
| Pepper, white, post-h | 0.01 |
| Persimmon | 0.01 |
| Pistachio | 0.1 |
| Processed food | 0.01 |
| Raw agricultural commodities | 0.01 |
| Rice, grain | 0.1 |
| Rye, grain | 0.1 |
| Safflower, seed | 0.1 |
| Salsify, tops | 0.01 |
| Sesame, post-h | 0.1 |
| Sorghum, grain | 0.1 |
| Soybean, seed | 0.1 |
| Sunflower, seed | 0.1 |
| Sweet potato | 0.01 |
| Tangelo | 0.01 |
| Tangerine | 0.01 |
| Tomato | 0.01 |
| Vegetable, legume, group 6 | 0.01 |
| Walnut | 0.1 |
| Wheat, grain | 0.1 |

*Carcinogen/Hazard Classifications*

**U.S. EPA Carcinogens:** Group D, unclassifiable, inadequate data

**Label Signal Word:** DANGER

*Regulatory Authority:*
- Banned or Severely Restricted (Several Countries) (UN)[13]
- Very Toxic Substance (World Bank)[15]
- Air Pollutant Standard Set (ACGIH)[1] (DFG)[3] (HSE)[33] (former USSR)[35] (OSHA)[58] (Several States)[60]
- CLEAN AIR ACT: Hazardous Air Pollutants (Title I, Part A, Section 112); Accidental Release Prevention/Flammable substances, (Section 112[r], Table 3), TQ = 5,000 lb (2270 kg)
- EPA HAZARDOUS WASTE NUMBER (RCRA No.): P096
- RCRA, 40CFR261, Appendix 8 Hazardous Constituents
- SUPERFUND/EPCRA 40CFR355, Appendix B Extremely Hazardous Substances: TPQ = 500 lb (227kg)
- SUPERFUND/EPCRA 40CFR302.4 RQ: CERCLA, 100 lb (45.4 kg)
- EPCRA Section 313 Form R de minimus concentration reporting level: 1.0%

*Description:* Phosphine is a colorless gas with the foul odor of decaying fish. *Odor is not an adequate indicator of phosphine's presence and may not provide reliable warning of hazardous concentrations.* The odor threshold is 0.14 ppm. Phosphine presents an additional hazard in that it ignites at very low temperatures. Shipped as a liquefied compressed gas. The pure compound is odorless. Slightly soluble in water; solubility = 26 ml/100 ml @ 17°C. Molecular weight = 34.00. Boiling point = –88°C. Melting/Freezing point = –134°C. Vapor pressure = >760 mm Hg @ 20°C. Flash point = Flammable Gas. Autoignition temperature = 100°C. Explosive limits: LEL= 1.8%. Hazard Identification (based on NFPA-704 M Rating System): Health 4, Flammability 4, Reactivity 2.

*Incompatibilities:* Reacts with water*, air, acids, moisture, oxidizers, oxygen, chlorine, nitrogen oxides, metal nitrates, halogens, halogenated hydrocarbons, copper and many other substances, causing fire and explosion hazard. May ignite spontaneously on contact with air at or about 100°C. Attacks many metals.

*Note:* When phosphine burns it produces a dense white cloud of phosphorus pentoxide ($P_2O_5$) fume. This fume is a severe respiratory tract irritant due to the rapid formation of orthophosphoric acid($H_3PO_4$) on contact with water.

*Permissible Exposure Limits in Air:* The OSHA TWA is 0.3 ppm (0.4 mg/m$^3$). NIOSH and ACGIH recommend the same TWA and STEL value is 1 ppm (1.4 mg/m$^3$). The NIOSH IDLH level is 50 ppm. HSE[33] has set the same values as OSHA and ACGIH. DFG[3] has however set a lower MAK of 0.1 ppm (0.15 mg/m$^3$). Czechoslovakia[35] has set a TWA of 0.1 mg/m$^3$ and a ceiling of 0.2 mg/m$^3$. The former USSR has set a ceiling value in workplace air of 0.1 mg/m$^3$. The former USSR has also set an MAC for phosphine in the ambient air in residential areas of 0.01 mg/m$^3$ on a once-daily basis and 0.001 mg/m$^3$ on a daily average basis. In addition, several states have set guidelines or standards for phosphine in ambient air[60] ranging from 1.33 µg/m$^3$ (New York) to 4.0 µg/m$^3$ (Florida and North Dakota) to 6.7 µg/m$^3$ (Virginia) to 8.0 µg/m$^3$ (Connecticut) to 10.0 µg/m$^3$ (Nevada and North Dakota) to 130 µg/m$^3$ (North Carolina). *ERPG-2 (Emergency Response Planning Guideline):* (maximum airborne concentration below which it is believed nearly all individuals could be exposed for up to 1 hour without experiencing or developing irreversible or other serious adverse health effects or symptoms that could impair an individuals's ability to take protective action) = 0.5 ppm

*Determination in Air:* Beaded carbon*; $H_2O_2$/buffer; Ion chromatography; OSHA Method #ID180. Collection on silver nitrate impregnated cellulose pad to give free silver which is converted to the nitrate, then to the sulfide which is analyzed colorimetrically; See also NIOSH #332, Phosphine.

*Permissible Concentration in Water:* No criteria set, but EPA (32) has suggested a permissible ambient goal of 5.5 µg/L based on health effects.

*Routes of Entry:* Inhalation, skin and/or eye contact with the liquid. Most phosphine exposures occur by inhalation of the gas or ingestion of metallic phosphides, but dermal exposure to phosphides can also cause systemic effects.

*Harmful Effects and Symptoms*

*Short Term Exposure: Inhalation:* Inhalation is the major route of phosphine toxicity. Phosphine is a respiratory tract irritant that attacks primarily the cardiovascular and respiratory systems causing cardiac arrhythmias and peripheral vascular collapse, cardiac arrest and failure, and pulmonary edema, a medical emergency that can be delayed for several hours. This can cause death. *Skin/Eye Contact:* Phosphides may be absorbed dermally, especially through broken skin, and can cause systemic toxicity by this route. Phosphine gas produces no adverse effects on the skin or eyes, and contact does not result in systemic toxicity. Phosphine is a super-toxic gas with a probable oral lethal dose of 5 mg/kg or 7 drops for a 150 lb person. An air concentration of 3 ppm is safe for long term exposure, 500 ppm is lethal in 30 minutes, and concentration of 1000 ppm is lethal after a few breaths. Contact with liquefied or compressed phosphine gas may cause frostbite. *Ingestion* Ingestion of phosphine is unlikely because it is a gas at room temperature. Ingestion of metallic phosphides can produce phosphine intoxication when the solid phosphide contacts gastric acid. Contact with the liquid may cause frostbite. Phosphine interferes with enzymes and protein synthesis, primarily in the mitochondria of heart and lung cells. As a result, effects may include hypotension, reduction in cardiac output, tachycardia, oliguria, anuria, cyanosis, pulmonary edema, tachypnea, jaundice, hepatosplenomegaly, ileus, seizures, and diminished reflexes. Acute exposure to phosphine usually results in headache, cough, tightness and pain in the chest, shortness of breath, dizziness, lethargy, and stupor. Fatigue, muscle pain, chills, tremors, loss of coordination, seizures, and coma may be seen. Gastrointestinal symptoms include nausea, vomiting, abdominal pain, and diarrhea. Renal (kidney) damage, hepatic (liver) damage, and jaundice may also occur.

*Long Term Exposure:* Chronic exposure to very low concentrations may result in anemia, bronchitis, gastrointestinal disturbances, and visual, speech, and motor disturbances. Chronic exposure may be more serious for children because of their potential longer latency period. Chronic poisoning may cause toothache, swelling of the jaw, spontaneous fractures of bones. May cause anemia. May damage the liver and kidneys. The effects are cumulative. Although most survivors of acute phosphine exposure show no permanent disabilities, damage due to insufficient blood supply to the heart and brain have been reported. Subacute poisoning resulting from exposure for a few days may cause reactive airways dysfunction syndrome (RADS) months later.

*Points of Attack:* Liver, kidney, and respiratory system

*Medical Surveillance:* For those with frequent or potentially high exposure (half the TLV or greater), the following are recommended before beginning work and at regular times after that: Lung function tests. If symptoms develop or overexposure is suspected, the following may be useful: Consider chest x-ray after acute overexposure. Liver function tests. Typically, liver injury does not become evident until 48 to 72 hours after exposure. Findings may include jaundice, enlarged liver, elevated serum transaminases, and increased bilirubin in the blood. Analysis of blood gases may reveal combined respiratory and metabolic acidosis. Also, there have been reports of significant hypomagnesemia and hypermagnesemia associated with massive focal myocardial damage.

*First Aid:* Brush all visible particles from clothes, skin, and hair. Remove and double-bag contaminated clothing and personal belongings. Thoroughly flush exposed skin and hair with water for 3 to 5 minutes, then wash with mild soap. Rinse thoroughly with water. Use caution to avoid hypothermia when decontaminating children or the elderly. Use blankets or similar warmers when appropriate. If phosphides have been ingested, *do not induce vomiting.* Phosphides will release phosphine in the stomach; therefore, watch for signs similar to those produced by phosphine inhalation. Administer a slurry of activated charcoal at 1 gm/kg (usual adult dose 60–90 g, child dose 25–50 g). A soda can and a straw may be of assistance when offering charcoal to a child.

*Advanced Treatment* In cases of respiratory compromise, secure airway and respiration via endotracheal intubation. If not possible, perform cricothyroidotomy if equipped and trained to do so. Treat patients who have bronchospasm with aerosolized bronchodilators. The use of bronchial sensitizing agents in situations of multiple chemical exposures may pose additional risks. Consider the health of the myocardium before choosing which type of bronchodilator should be administered. Cardiac sensitizing agents may be appropriate; however, the use of cardiac sensitizing agents after exposure to certain chemicals may pose enhanced risk of cardiac arrhythmias (especially in the elderly). Consider racemic epinephrine aerosol for children who develop stridor. Dose 0.25–0.75 mL of 2.25% racemic epinephrine solution in 2.5 cc water, repeat every 20 minutes as needed, cautioning for myocardial variability. Patients who are comatose, hypotensive, or having seizures or cardiac arrhythmias should be treated according to advanced life support protocols. If evidence of shock or hypotension is observed, begin fluid administration. For adults, bolus 1000 mL/hour intravenous saline or lactated Ringer's solution if blood pressure is under 80 mm Hg; if systolic pressure is over 90 mm Hg, an infusion rate of 150 to 200 mL/hour is sufficient. For children with compromised perfusion, administer a 20 mL/kg bolus of normal saline over 10 to 20 minutes, then infuse at 2 to 3 mL/kg/hour.

*References:*
- U.S. Environmental Protection Agency, "Pesticide Fact Sheet, Phosphine," Washington, DC, Office of Prevention, Pesticides and Toxic Substances, (December 1999).
  http://www.epa.gov/opprd001/factsheets/phosphine.pdf
- U.S. Environmental Protection Agency, Office of Pesticide Programs, Pesticide Residue Limits,

- New Jersey Department of Health, "Hazardous Substance Fact Sheet: Phosphine," Trenton, NJ (May 1997, rev. April 2004). http://www.state.nj.us/health/eoh/rtkweb/1514.pdf
- Sax, N.I., Ed., "Dangerous Properties of Industrial Materials Report," 6, No. 2, 103-107 (1986).

# Phosphoric Acid

*Use Type:* Fertilizer (pH adjustment), fungicide, herbicide and microbiocide
*CAS Number:* 7664-38-2
*Formula:* $H_3O_4P$; $H_3PO_4$
*Alert:* A General Use Pesticide (GUP)
*Synonyms:* Acide phodphorique (French); Acido fosforico (Italian, Spanish); Fosforzuuroplossingen (Dutch); Orthophosphoric acid; o-Phosphoric acid; Phosphorsaeureloesungen (German); White phosphoric acid
*Trade Names:* (There are 403 active and canceled/transferred products registered with the U.S. EPA) DECON® 4512; EVITS®; RAIMONT®, Agrochemicals del Ecuador (AGROCHEM) (Ecuador); SONAC®; WC-REINIGER®
*Producers:* Abaquim (Mexico); Agrium (Canada); Agrochemicals del Ecuador (AGROCHEM) (Ecuador); Albright & Wilson Pty. (UK); Aldrich Chemical (USA); Ashland (USA); ATOFINA Chemicals (France); Central Glass (Japan); Chemical Company (USA); Chemische Fabrik Budenheim (Germany); Chongqing Chuandong Chemical (Group) (China); Eastman Chemical (USA); FMC (USA); Fosbrasil (Brazil); IMC Agrico (USA); Indo Gulf (India); Interstate Chemical (USA); Israel Chemicals (Israel); Jiangmen Pesticide Factory (China); Kunming Baoyi Phosphate Chemicals (China); Kynoch (South Africa); Mitsui Chemicals (Japan); OxyChem (USA); Rhodia Group (France); Rhodia Specialty Phosphates (USA); Petroquimica de Venezuela (Pequiven); Sichuan Chuanxi Xingda Chemical Plant (China); Simplot, J.R. (USA); Tekchem (Mexico); Solaris ChemTech (India); Tessenderlo Chemie (Belgium); Thermphos (Netherlands); Vulcan Chemicals (USA); WMC (Australia)
*Chemical Class:* Inorganic
*EPA/OPP PC Code:* 076001
*California DPR Chemical Code:* 871
*ICSC Number:* 1008
*RTECS Number:* TB6300000
*EEC Number:* 015-011-00-6
*EINECS Number:* 231-633-2
*Uses:* Most of the currently registered farm uses of phosphoric acid are in sanitizing products for farm animals. Phosphoric acid is used in the manufacture of fertilizers, phosphate salts, polyphosphates, detergents, activated carbon, animal feed, ceramics, dental cement, pharmaceuticals, soft drinks, gelatin, rust inhibitors, wax, and rubber latex. Also used in electropolishing, engraving, photoengraving, lithographing, metal cleaning, sugar refining, and water-treating.
*U.S. Maximum Allowable Residue Levels for Phosphoric acid:*
Exempt from the requirement of tolerances
*Carcinogen/Hazard Classifications*
**Label Signal Word: DANGER**
*Regulatory Authority:*
- Air Pollutant Standard Set (ACGIH)[1] (OSHA)[58] (Other Countries)[35] (Several States)[60]
- AB 2588-Air Toxics "Hot Spots" Chemicals (CAL)
- Permissible Exposure Limits for Chemical Contaminants (CAL/OSHA)
- The "Director's List" (CAL/OSHA)
- Actively registered pesticide in California.
- Clean Water Act: Section 311 Hazardous Substances/RQ 40CFR117.3 (same as CERCLA, see below); Section 313 Water Priority Chemicals (57FR41331, 9/9/92)
- Superfund/EPCRA 40CFR302.4 RQ: CERCLA, 5,000 lb (2270 kg)
- EPCRA Section 313 Form R *de minimis* concentration reporting level: 1.0%
- Canada, WHMIS, Ingredients Disclosure List

*Description:* Phosphoric acid is a colorless, crystalline solid, or thick syrupy liquid. Odorless. Physical state is strength and temperature dependent. Highly soluble in water. Boiling Point = (decomposes below BP) 215°C (loses 50% water). Melting/Freezing point = 41.8°C. Hazard Identification (based on NFPA-704 M Rating System): Health 3, Flammability 0, Reactivity 0.
*Incompatibilities:* The substance is a medium strong acid. Incompatible with strong caustics, most metals. Readily attacks and reacts with metals forming flammable hydrogen gas. DO NOT mix with solutions containing bleach or ammonia. Violently polymerizes on contact with azo compounds, epoxides, and other polymerizable compounds. Decomposes on contact with metals, alcohols, aldehydes, cyanides, ketones, phenols, esters, sulfides, halogenated organics producing toxic fumes. Becomes anhydrous at 66°C; very corrosive to ferrous metals and alloys particularly at temperatures above 82°C. At 93°C, changes to pyrophosphoric acid. Above 148°C, changes to metaphosphoric acid. Corrosive to glass and pottery above 199°C.
*Permissible Exposure Limits in Air:* The OSHA[2] PEL is 1 mg/m³ TWA. The NIOSH[2] and ACGIH[1] TWA value is 1 mg/m³ and the STEL is 3 mg/m³ and the NIOSH[2] STEL is not to be exceeded during any 15 minute work period. The NIOSH[2] IDLH level = 1000 mg/m³. Argentina[35] has set a TWA of 1 ppm and an STEL of 3 ppm. Japan and Sweden have set 1 mg/m³. The former USSR has set of 20 µg/m³ on a once-daily basis. Further, several states have set guidelines or standards for phosphoric acid in ambient air[60] ranging from 1.4 µg/m³ (Massachusetts) to 10-30 µg/m³

(North Dakota) to 10-33 $\mu g/m^3$ (Virginia) to 20.0 $\mu g/m^3$ (Connecticut) to 24.0 $\mu g/m^3$ (Nevada) to 25.0 $\mu g/m^3$ (South Carolina).

***Determination in Air:*** Si gel (special); $NaHCO_3/Na_2CO_3$; Ion chromatography; NIOSH IV, Method #7903, Inorganic Acids[18]. Collection on a cellulose membrane filter, workup with water, colorimetric determination. See OSHA Methods ID-111 and ID-165-5G[58].

***Permissible Concentration in Water:*** No criteria set, but runoff from spills or fire control may cause water pollution.

***Routes of Entry:*** Inhalation, ingestion, skin and/or eye contact.

*Harmful Effects and Symptoms*

***Short Term Exposure:*** Corrosive to the eyes, skin, and respiratory tract. Eye contact may cause permanent damage. Inhalation can cause pulmonary edema, a medical emergency that can be delayed for several hours. This can cause death. Solid is especially irritating to skin in the presence of moisture. Corrosive if swallowed. May cause pain in the throat and stomach, nausea, vomiting and intense thirst. Severe exposures may result in shock with clammy skin, weak and rapid pulse, shallow breathing, reduced urine output and death. 1-5 $mg/m^3$ may cause irritation of nose and throat. 4-11 $mg/m^3$ may cause coughing. Inhalation of acid mist can cause lung irritation. 1-5 $mg/m^3$ may cause irritation of nose and throat. 4-11 $mg/m^3$ may cause coughing.

***Long Term Exposure:*** Repeated or prolonged skin exposure may cause irritation, drying, cracking, and dermatitis. Can cause bronchitis to develop.

***Points of Attack:*** Eyes, skin and respiratory system.

***Medical Surveillance:*** Before beginning employment and at regular times after that, the following is recommended: Lung function tests. If symptoms develop or overexposure is suspected, the following may be useful: Consider chest x-ray after acute overexposure.

***First Aid:*** If this chemical gets into the eyes, remove any contact lenses at once and irrigate immediately for at least 15 minutes, occasionally lifting upper and lower lids. Seek medical attention immediately. If this chemical contacts the skin, remove contaminated clothing and wash immediately with soap and water. Seek medical attention immediately. If this chemical has been inhaled, remove from exposure, begin rescue breathing (using universal precautions) if breathing has stopped, and CPR if heart action has stopped. Transfer promptly to a medical facility. When this chemical has been swallowed, get medical attention. If victim is <u>conscious</u>, administer water, or milk. Do not induce vomiting. Medical observation is recommended for 24 to 48 hours after breathing overexposure, as pulmonary edema may be delayed. As first aid for pulmonary edema, a doctor or authorized paramedic may consider administering a corticosteroid spray.

*References:*
- Sax, N.I., Ed., "Dangerous Properties of Industrial Materials Report," 3, No. 4, 84-87 (1983).
- New Jersey Department of Health, "Hazardous Substance Fact Sheet: Phosphoric Acid," Trenton, NJ (April 2004). http://www.state.nj.us/health/eoh/rtkweb/1516.pdf
- New York State Department of Health, "Chemical Fact Sheet: Phosphoric Acid," Albany, NY, Bureau of Toxic Substance Assessment (April 1986 and Version 2).
- California Environmental Protection Agency "Chemical List of Lists," Sacramento CA (February 1997).

# Phosphorus

***Use Type:*** Fertilizers and rodenticide
***CAS Number:*** 7723-14-0
***Formula:*** P
***Synonyms:*** Fosforo blanco (Spanish); Fosforo bianco (Italian); Gelber phosphor (German); Phosphore blanc (French); Phosphorus elemental, white; Phosphorous yellow; Phosphorus-31; Red phosphorus; Tetrafosfor (Dutch); Tetraphosphor (German); Weiss phosphor (German); White phosphorus; Yellow phosphorus
***Trade Names:*** BONIDE BLUE DEATH RAT KILLER®; COMMON SENSE COCKROACH AND RAT PREPARATIONS®; EXOLITE® 405; EXOLIT LPKN® 275; EXOLIT VPK-N® 361
***Producers:*** Akzo Nobel Chemicals (Netherlands); Albright & Wilson Pty. (UK); Archimica (UK); BEC Group (India); Chemische Fabrik Budenheim (Germany); Chongqing Chuandong Chemical (Group) (China); Excel Industries (India); Faci Europe (Italy); FMC Agricultural Products Group (USA); Great Western Inorganics (USA); Kemira GrowHow (Finland); Kunming Baoyi Phosphate Chemicals (China); Mississippi Chemical (USA); Monsanto (USA); OxyChem (USA); Pechiney (France); Potash Corporation (Canada); Rasa Industries (Japan); Rhodia (France); Search Chem Industries (India); Sichuan Chuanxi Xingda Chemical Plant (China); Thermphos (Netherlands); United Phosphorus (India)
***Chemical Class:*** Inorganic
***EPA/OPP PC Code:*** 066502
***California DPR Chemical Code:*** 513
***ICSC Number:*** 0628 (Yellow Phosphorus)
***RTECS Number:*** TH3500000
***EEC Number:*** 015-001-00-1
***EINECS Number:*** 231-768-7
***Uses:*** Yellow phosphorus is handled away from air so that exposure is usually limited. Phosphorus was at one time used for the production of matches or "lucifers" but has long since been replaced due to its chronic toxicity. Phosphorus is used in the manufacture of munitions, pyrotechnics, explosives, smoke bombs, and other incendiaries, in

artificial fertilizers, rodenticides, phosphor bronze alloy, semiconductors, electro-luminescent coating, and chemicals, such as phosphoric and metallic phosphides.

*Carcinogen/Hazard Classifications*

**U.S. EPA Carcinogens:** Group D, unclassifiable, insufficient data

*Regulatory Authority:*
- Air Pollutant Standard Set (ACGIH)[1] (DFG)[3] (HSE)[33] (OSHA)[58] (former USSR)[35] (Several States)[60]
- AB 2588-Air Toxics "Hot Spots" Chemicals (CAL)
- CAL Air Resources Board/AB 1807 Toxic Air Contaminants
- Permissible Exposure Limits for Chemical Contaminants (CAL/OSHA)
- The "Director's List" (CAL/OSHA)
- Actively registered pesticide in California
- Canada, WHMIS, Ingredients Disclosure List
- Clean Air Act: Hazardous Air Pollutants (Title I, Part A, Section 112)
- Clean Water Act: Section 311 Hazardous Substances/RQ 40CFR117.3 (same as CERCLA, see below); Section 313 Water Priority Chemicals (57FR41331, 9/9/92)
- Superfund/EPCRA 40CFR355, Appendix B Extremely Hazardous Substances: TPQ = 100 lb (45.4 kg)
- Superfund/EPCRA 40CFR302.4 RQ: CERCLA, 1 lb (0.454 kg)
- EPCRA Section 313 (yellow or white) Form R *de minimis* concentration reporting level: 1.0%
- U.S. DOT Regulated Marine Pollutant (49CFR172.101, Appendix B), severe pollutant, white, yellow dry, molten or in solution

*Description:* Phosphorus is a white to yellow, soft, waxy solid with acrid fumes in air.

*White/yellow phosphorus* is either a yellow or colorless, volatile, crystalline solid which darkens when exposed to light and ignited in air to form white fumes and greenish light. It has a garlic-like odor. Usually shipped or stored in water. Boiling point = 280°C. Melting/Freezing point = 44°C (decomposes). Autoignition temperature = 30°C. Vapor Pressure = $2.6 \times 10^{-2}$ mmHg @ 20°C (yellow). Hazard Identification (based on NFPA-704 M Rating System): [black] Health 1, Flammability 1, Reactivity 0. Insoluble in water.

*Red phosphorus* is a brick red, reddish-brown, or violet amorphous powder, frequently contaminated with a small amount of the yellow. Boiling point = 280°C (with ignition @ 200°C). Melting/Freezing point = 416°C (sublimes). Flash point = 260°C. Autoignition temperature = 260°C. Hazard Identification (based on NFPA-704 M Rating System): Health 1, Flammability 1, Reactivity 1. Insoluble in water.

*Incompatibilities:* Phosphorus is incompatible with air, all oxidizing agents, including elemental sulfur, strong caustics. *White/yellow* reacts with air, halogens, halides, sulfur, oxidizers, alkali hydroxides (forming phosphine gas), and metals (forming reactive phosphides). *Red* is a combustible solid. Friction or contact with oxidizers can cause ignition. Incompatible with many other substances. Forms phosphine gas and phosphoric acid on contact with moisture. Opened packages of red phosphorus should be stored under inert gas blanket.

*Permissible Exposure Limits in Air:* The OSHA[2] PEL is 0.1 mg/m³ TWA. The HSE[33] TWA, DFG MAK[3], and the recommended ACGIH[1] TWA value[1] is 0.1 mg/m³ and the STEL is 0.3 mg/m³. The NIOSH[2] IDLH level = 5 mg/m³ level has been set. Czechoslovakia[35] has set 0.03 mg/m³ as a TWA and 0.06 mg/m³ as an STEL in workplace air. The former USSR[35, 43] has set 0.03 mg/m³ as a ceiling value in workplace air. Several states have set guidelines or standards for yellow phosphorus in ambient air[60] ranging from 0.33 µg/m³ (New York) to 1.0 µg/m³ (Florida) to 1.6 µg/m³ (Virginia) to 2.0 µg/m³ (Connecticut and Nevada) to 10.0-30.0 µg/m³ (North Dakota).

*Determination in Air:* Tenax Gas chromatography; Xylene; Gas chromatography/Flame photometric detection for sulfur, nitrogen, or phosphorus; NIOSH IV, Method #7905.[18]

*Permissible Concentration in Water:* EPA[32] has suggested a permissible ambient goal of 1.4 µg/L based on health effects. The former USSR[35, 43] has set a MAC in water bodies used for domestic purposes of 0.1 µg/L and in surface water used for fishery purposes of zero.

*Routes of Entry:* Inhalation, ingestion, skin and/or eye contact. Yellow phosphorus can be absorbed through the skin.

*Harmful Effects and Symptoms*

*Short Term Exposure:* Phosphorus is corrosive to the eyes, skin, and respiratory tract. Eye contact may lead to a total destruction of the eyes. Victims may experience spontaneous hemorrhaging of phosphorus-contaminated skin and mucous membranes. Sudden death, possibly due to irregular heartbeat, may occur after relatively minor (10-15%) burns. *Yellow:* Fumes are irritating to the respiratory tract and cause severe ocular irritation. On contact with the skin it may ignite and produce severe skin burns with blistering. Very high exposure may cause severe or fatal poisoning. *Red:* Irritates eyes. Corrosive if ingested. Inhalation can cause pulmonary edema, a medical emergency that can be delayed for several hours. This can cause death. May affect the kidneys, liver. Exposure may result in death. Phosphorus is classified as super toxic. The probable lethal dose is less than 5 mg/kg (a taste or less than 7 drops) for 70 kg (150 lb) person. Signs and symptoms of acute exposure to phosphorus may be severe and occur in three stages. The first stage will involve burns, pain, shock, intense thirst, nausea, vomiting, diarrhea, severe abdominal pain, and "smoking stools." The breath and feces may have garlicky odor. The second stage will be a symptom-free period of several day in which the patient appears to be recovering. The third stage may be severe and include nausea, bloody vomitus, diarrhea (may be bloody), jaundice, liver enlargement with tenderness, renal damage, hematuria (bloody urine), and either oliguria (little urine formation) or anuria (no urine formation). Headache, convulsions, delirium, coma, cardiac arrhythmias, and cardiovascular collapse may also occur. If phosphorus contacts the eyes, then severe irritation and burns, blepharospasm (spasmodic winking), lacrimation (tearing), and photophobia (heightened sensitivity to light) may occur.

***Long Term Exposure:*** Phosphorus may affect the bones, causing bone degeneration (especially the jaw bone, known as "phossy" jaw), dental pain, salivation, jaw pain and swelling. This process can extend into one or both eye sockets. Repeated low exposure can cause low blood count (anemia), weight loss, and bronchitis. May cause jaundice; liver and kidney damage; cachexia. May cause nervous system damage.

***Points of Attack:*** Respiratory system, liver, kidneys, jaw, teeth, blood, eyes and skin.

***Medical Surveillance:*** Complete blood count with differential. X-ray of teeth and jaw and dental exam. Lung function tests. EKG. Liver function tests. Special consideration should be given to the skin, eyes, jaws, teeth, respiratory tract, and liver. Preplacement medical and dental examination with x-ray of teeth is highly recommended in the case of yellow phosphorus exposure. Poor dental hygiene may increase the risk in yellow phosphorus exposures, and any required dental work should be completed before workers are assigned to areas of possible exposure. Workers experiencing any jaw injury, tooth extraction, or any abnormal dental conditions should be removed from areas of exposure and observed. X-ray examinations may show necrosis; however, in order to prevent full development of sequestra, the disease should be diagnosed in earlier stages. Liver function should be evaluated periodically.

***First Aid:*** If this chemical gets into the eyes, remove any contact lenses at once and irrigate immediately for at least 30 minutes, occasionally lifting upper and lower lids. Seek medical attention immediately. If this chemical contacts the skin, remove contaminated clothing and brush all traces of dry chemical from skin. Submerge burning phosphorus (yellow) in water or 1% copper sulfate solution if embedded in skin, or wash exposed area with large amounts of water. Seek medical attention immediately. Skin burns from yellow phosphorus should be observed for 1-3 days for possible delayed effects. If this chemical has been inhaled, remove from exposure, begin rescue breathing (using universal precautions) if breathing has stopped, and CPR if heart action has stopped. Transfer promptly to a medical facility. When this chemical has been swallowed, get medical attention. Give large quantities of water and induce vomiting. Do not make an unconscious person vomit. Medical observation is recommended for 24 to 48 hours after breathing overexposure, as pulmonary edema may be delayed. As first aid for pulmonary edema, a doctor or authorized paramedic may consider administering a corticosteroid spray.

***References:***
- New Jersey Department of Health, "Hazardous Substance Fact Sheet: Phosphorus," Trenton, NJ (October 2002). http://www.state.nj.us/health/eoh/rtkweb/1520.pdf
- Sax, N.I., Ed., "Dangerous Properties of Industrial Materials Report," 3, No. 4, 90-93 (1983).
- U.S. Environmental Protection Agency, "Chemical Profile: Phosphorus," Washington, DC, Chemical Emergency Preparedness Program (November 30, 1987).
- California Environmental Protection Agency "Chemical List of Lists," Sacramento CA (February 1997).

# Phoxim

***Use Type:*** Insecticide, miticide
***CAS Number:*** 14816-18-3
***Formula:*** $C_{12}H_{15}N_2O_3PS$
***Synonyms:*** Benzoyl cyanide-*o*-(diethoxyphosphinothioyl) oxime; *O,O*-Diaethyl-*o*-(α-cyanbenzyliden-amino)-thionphosphat (German); *O,O*-Diaethyl-*O*-(α-cyano-benzylidenamino)-monothiophosphat (German); α-[((Diethoxyphosphinothioyl)oxy)imino]benzeneacetonitrile; (Diethoxy-thiophosphoryloxyimino)-phenyl acetonitrile; *O,O*-Diethyl phosphorothioate, *O*-ester with phenyl glyoxylonitrile oxime; ENT 27,488; 4-Ethoxy-7-phenyl-3,5-dioxa-6-aza-4-phosphaoct-6-ene-8-nitrile-4-sulfide; Glyoxylonitrile, phenyl-, oxime, *O,O*-diethyl phosphorothioate; Phenylglyoxylonitrile oxime *O,O*-diethyl phosphorothioate; Phoxime

***Trade Names:*** BAY®-5621, Bayer CropScience (Germany); BAY®-77488, Bayer CropScience (Germany); BAY SRA 7502, Bayer CropScience (Germany); BAYER®-77488, Bayer CropScience (Germany); BAYTHION®, Bayer CropScience (Germany); SEBACIL®; VALEXONE®; VOLATON®

***Producers:*** Agrimor International (USA); Bayer CropScience (Germany); Hebei Huafeng Chemical Group (China)

***Chemical Class:*** Organophosphate
***EPA/OPP PC Code:*** 598800
***Uses:*** Not registered in the U.S. Phoxim is an organophosphorus insecticide used for topical treatment of cattle, sheep, goats and swine. Also used to control insects that infest stored grains, peas, beans, rice, tobacco, skins and dried fruit.

***Carcinogen/Hazard Classifications***
**WHO Acute Hazard:** Class II, moderately hazardous
***Regulatory Authority:*** as cyanides and cyanide compounds
- AB 2588-Air Toxics "Hot Spots" Chemicals (CAL)
- CAL Air Resources Board/AB 1807 Toxic Air Contaminants

- Clean Air Act: Hazardous Air Pollutants (Title I, Part A, Section 112)
- Clean Water Act: Section 307 Priority Pollutants as cyanide, total; Toxic Pollutant (Section 401.15)
- RCRA Section 261 Hazardous Constituents
- RCRA Universal Treatment Standards: Wastewater (mg/L), 1.2 (total); 0.86 (amenable); Nonwastewater (mg/kg), 590 (total); 30 (amenable)
- RCRA Ground Water Monitoring List. Suggested test method(s) (PQL ug/L): 9010(40)
- EPA Hazardous Waste Number (RCRA No.): P030 as cyanides
- Safe Drinking Water Act: MCL, 0.2 mg/L; MCLG, 0.2 mg/L; Regulated chemical (47 FR 9352)
- EPCRA Section 304 RQ: CERCLA, 10 lb (4.54 kg)
- EPCRA Section 313 Form R *de minimis* concentration reporting level: 1.0%
- DOT Inhalation Hazard Chemicals as organophosphate
- Marine pollutant (49CFR, Subchapter 172.101, Appendix B) as cyanides, mixtures or solutions

*Description:* Pale yellow to yellow-brown, oily liquid. Molecular weight = 298.31. Density = 1.176°C @ 20 mmHg. Melting/Freezing point = 5°C. Boiling point = 102°C @ 0.01 mmHg. (decomposition).

*Incompatibilities:* May react violently with strong oxidizers, chlorobenzenediazonium salts, mercurous chloride. Incompatible with cadmium bromide, zinc acetate.

*Permissible Exposure Limits in Air:*

*Determination in Air:* OSHA versatile sampler-2; Toluene/Acetone; Gas chromatography/Flame ionization detection; NIOSH IV, Method #5600, Organophosphorus Pesticides.[18]

*Permissible Concentration in Water:* No criteria set. Runoff from spills or fire control may cause water pollution.

*Harmful Effects and Symptoms*

*Short Term Exposure:* Eye pupils are small, blurred vision, eye watering, runny nose, cough, shortness of breath, salivation, dizziness, nausea, stomach cramps, diarrhea, and vomiting, increased blood pressure, profuse sweating, hypermotility, hallucinations, irritability, tingling of the skin, drowsiness, slow heartbeat, convulsions, fluid in lungs, loss of consciousness, incontinence, breathing stops, death. Organophosphates inhibit the action of acetylcholinesterase enzymes, and alter the way in which nervous impulses are transmitted. The effects can last for hours, days, or much longer. The action of the enzymes is reestablished after new enzymes are formed.

*Long Term Exposure:* Cholinesterase inhibitor; cumulative effect is possible. Organophosphates may damage the nervous system with repeated exposure, resulting in convulsions, respiratory failure. May cause liver damage.

*Points of Attack:* Respiratory system, central nervous system, cardiovascular system, blood cholinesterase.

*Medical Surveillance:* Medical observation is recommended for 24 to 48 hours after breathing overexposure, as pulmonary edema may be delayed. As first aid for pulmonary edema, consider administering a corticosteroid spray. Cigarette smoking may exacerbate pulmonary injury and should be discouraged for at least 72 hours following exposure.

*First Aid:* Speed in removing material from eyes and skin is of extreme importance. Eye contact can cause dangerous amounts of these chemicals to be quickly absorbed through the mucous membrane into the bloodstream. Immediately and gently flush eyes with plenty of warm or cold water (NO hot water) for at least 15 minutes, occasionally lifting the upper and lower eyelids. Get medical aid immediately. *Skin:* Get medical aid. Skin contact can cause dangerous amounts of these chemicals to be absorbed into the bloodstream. Wearing the appropriate PPE equipment and respirator for organophosphate pesticides, immediately flush skin with plenty of soap and water for at least 15 minutes while removing contaminated clothing and shoes. Shampoo hair promptly if contaminated. *Ingestion:* Call poison control. Loosen all clothing. Never give anything by mouth to an unconscious person. Get medical aid. Do NOT induce vomiting.* If conscious, alert, and able to swallow, rinse mouth and have victim drink 4 to 8 ounces of water do NOT induce vomiting but immediately administer slurry of activated charcoal (2 oz in 8 oz of water). If victim is *unconscious or having convulsions,* do nothing except keep victim warm. *In some cases you may be specifically instructed by poison control to induce vomiting by way of 2 tablespoons of syrup of ipecac (adult) washed down with a cup of water.* Do NOT give activated charcoal before or with ipecac syrup. *Inhalation:* Get medical aid. Do not contaminate yourself. Wearing the appropriate PPE equipment and respirator for organophosphate pesticides, immediately remove the victim from the contaminated area to fresh air. If the victim is not breathing, administer artificial respiration. Do NOT use mouth-to-mouth resuscitation; use bag/mask apparatus. If breathing is difficult, administer oxygen through bag/mask apparatus until medical help arrives. Do not leave victim unattended. *Note to physician or authorized medical personnel.* Administer atropine, 2 mg (1/30 gr) intramuscularly or intravenously as soon as any local or systemic signs or symptoms of an intoxication are noted; repeat the administration of atropine every 3 to 8 minutes until signs of atropinization (mydriasis, dry mouth, rapid pulse, hot and dry skin) occur; initiate treatment in children with 0.05 mg mg/kg of atropine; repeat at 5 to 10 minute intervals. Watch respiration, and remove bronchial secretions if they appear to be obstructing the airway; intubate if necessary. *Notes to physician or authorized medical personnel*: N-methylpyridinium-2-aldoxime (2-PAMCl) when used in conjunction with atropine reacts with the phosphorylated cholinesterase, thereby restoring normal activity to by removing the phosphorylating group. The combination of these two chemicals is synergistic and must be administered within minutes to a few hours following exposure (depending on the specific agent) to be effective. Give 2-PAMCI (Pralidoxime; Protopam), 2.5 gm in 100 ml of sterile water or in 5% dextrose and water, intravenously, slowly, in 15-30 minutes; if sufficient fluid is not available, give 1 gm of 2-PAMCI in 3 ml of distilled water by deep intramuscular injection; repeat this every half hour if respiration weakens or if muscle fasciculation or convulsions recur. Also Diazepam, an anticonvulsant, might be considered.

*References:*
- International Programme on Chemical Safety (IPCS), "Data Sheets on Pesticides, Phoxim," Geneva, Switzerland (2000). http://www.inchem.org/documents/pds/pds/pest31_e.htm

# Physostigmine

*Use Type:* Insecticide
*CAS Number:* 57-47-6
*Formula:* $C_{15}H_{21}N_3O_2$
*Alert:* Not registered in the U.S.
*Synonyms:* Calabarine; Erserine; Eserine; Eserolein; Fisostigmina (Spanish); Methylcarbamate (Ester); Methylcarbamic acid, ester with eseroline; Physostol
*Producers:* Boehringer Ingelheim (Germany); Sandoz (Switzerland)
*Chemical Class:* Carbamate
*RTECS Number:* TJ2100000
*EINECS Number:* 200-332-8
*Uses:* Material is used as a cholinergic (anticholinesterase) agent and as a veterinary medication. It is used to enhance memory in Alzheimer patients. Listed as a carbamate pesticide; however, physostigmine is not registered in the U.S.
*Regulatory Authority:*
- EPA Hazardous Waste Number (RCRA No.): P204
- RCRA, 40CFR261, Appendix 8 Hazardous Constituents
- RCRA 40CFR268.48; 61FR15654, Universal Treatment Standards: Wastewater (mg/L), 0.056; Nonwastewater (mg/kg), 1.4
- Superfund/EPCRA 40CFR355, Appendix B Extremely Hazardous Substances: TPQ = 100/10,000 lb (45.4/4,540 kg)
- Superfund/EPCRA 40CFR302.4 RQ: EHS, 1 lb (0.454 kg)

*Description:* Physostigmine, $C_{15}H_{21}N_3O_2$, is an odorless white crystalline solid. Slightly soluble in water. Melting/Freezing point = 86–87°C (unstable form); 105–106°C (stable form).
*Incompatibilities:* Light and heat. When heated to decomposition, this chemical forms toxic oxides of nitrogen.
*Permissible Exposure Limits in Air:* No standards set.
*Permissible Concentration in Water:* No criteria set, but runoff from spills or fire control may cause water pollution.
*Routes of Entry:* Inhalation, ingestion, skin and/or eye contact.
*Harmful Effects and Symptoms*
*Short Term Exposure:* Super toxic. Probable oral lethal dose is less than 5 mg/kg for a 70 kg (150 lb) person. Material is a cholinesterase inhibitor. Effects of exposure may involve the respiratory, gastrointestinal, cardiovascular and central nervous systems. Death occurs due to respiratory paralysis or impaired cardiac function. Time to death may vary from 5 minutes to 24 hours, in severely poisoned patients, depending on factors such as the dose and route. General symptoms include: increased secretions, fatigue and generalized weakness, involuntary twitching, severe weakness of skeletal muscles. Symptoms of exposure to material by major organ system: gastrointestinal: lack of appetite, nausea and vomiting, abdominal cramps and diarrhea. Central nervous system: confusion, uncoordination, slurred speech, loss of reflexes, rapid irregular breathing, generalized convulsions, and coma. Cardiovascular: slowed heart beat resulting in Hypotension and fall in cardiac output. Carbamate insecticides inhibit the cholinesterase activity of enzymes, causing accumulation of acetylcholine at synapses and altering the way in which nervous impulses are transmitted. However, within several hours carbamates spontaneously detach from the enzymes.
*Long Term Exposure:* Cholinesterase inhibitor; cumulative effect is possible. This chemical may damage the nervous system with repeated exposure, resulting in convulsions, respiratory failure. May cause liver damage.
*Points of Attack:* Respiratory system, lungs, central nervous system, cardiovascular system, skin, eyes, plasma and red blood cell cholinesterase.
*Medical Surveillance:* Before employment and at regular times after that, the following are recommended: Plasma and red blood cell cholinesterase levels (tests for the enzyme poisoned by this chemical). If exposure stops, plasma levels return to normal in 1-2 weeks while red blood cell levels may be reduced for 1-3 months.
When acetylcholinesterase enzyme levels are reduced by 25% or more below preemployment levels, risk of poisoning is increased, even if results are in lower ranges of "normal." Reassignment to work not involving carbamate pesticides is recommended until enzyme levels recover. If symptoms develop or overexposure occurs, repeat the above tests as soon as possible and get an exam of the nervous system. Also consider complete blood count. Consider chest x-ray following acute overexposure
*First Aid:* Speed in removing material from eyes and skin is of extreme importance. *Eyes:* Eye contact can cause dangerous amounts of these chemicals to be quickly absorbed through the mucous membrane into the bloodstream. Immediately and gently flush eyes with plenty

of warm or cold water (NO hot water) for at least 15 minutes, occasionally lifting the upper and lower eyelids. Get medical aid immediately. *Skin:* Get medical aid. Skin contact can cause dangerous amounts of these chemicals to be absorbed into the bloodstream. Wearing the appropriate PPE equipment and respirator for carbamate pesticides, immediately flush skin with plenty of soap and water for at least 15 minutes while removing contaminated clothing and shoes. Shampoo hair promptly if contaminated; protect eyes. *Ingestion:* Call poison control. Loosen all clothing. Never give anything by mouth to an unconscious person. Get medical aid. Do NOT induce vomiting.* If conscious, alert, and able to swallow, rinse mouth and have victim drink 4 to 8 ounces of water. Check to see if poison control instructs you to use ipecac syrup, otherwise administer slurry of activated charcoal (2 oz in 8 oz of water). If victim is *UNCONSCIOUS OR HAVING CONVULSIONS,* do nothing except keep victim warm. **In some cases you may be specifically instructed by poison control to induce vomiting by way of 2 tablespoons of syrup of ipecac (adult) washed down with a cup of water.* Do NOT give activated charcoal before or with ipecac syrup. *Inhalation:* Get medical aid. Do not contaminate yourself. Wearing the appropriate PPE equipment and respirator for carbamate pesticides, immediately remove the victim from the contaminated area to fresh air. If the victim is not breathing, administer artificial respiration. Do NOT use mouth-to-mouth resuscitation; use bag/mask apparatus. If breathing is difficult, administer oxygen through bag/mask apparatus until medical help arrives. Do not leave victim unattended. *Note to physician or authorized medical personnel.* Administer atropine, 2 mg (1/30 gr) intramuscularly or intravenously as soon as any local or systemic signs or symptoms of an intoxication are noted; repeat the administration of atropine every 3 to 8 minutes until signs of atropinization (mydriasis, dry mouth, rapid pulse, hot and dry skin) occur; initiate treatment in children with 0.05 mg mg/kg of atropine; repeat at 5 to 10 minute intervals. Watch respiration, and remove bronchial secretions if they appear to be obstructing the airway; intubate if necessary. *Medical note:* 2-PAMCI may be contraindicated in the case of some carbamate poisonings.

*References:*
- U.S. Environmental Protection Agency, "Chemical Profile: Physostigmine," Washington, DC, Chemical Emergency Preparedness Program (November 30, 1987).
- California Environmental Protection Agency "Chemical List of Lists," Sacramento CA (February 1997).

# Picloram (ANSI)

*Use Type:* Herbicide
*CAS Number:* 1918-02-1
*Formula:* $C_6H_3Cl_3N_2O_2$

*Alert:* A Restricted Use Pesticide (RUP). A carcinogen, handle with extreme caution.

*Synonyms:* 4-aminotrichloropicolinic acid; 4-Amino-3,5,6-trichloro-2-picolinic acid; 4-Amino-3,5,6-trichloropicolinic acid; 4-Amino-3,5,6-trichloro-2-pyridinecarboxylic acid; 4-Amino-3,5,6-trichloropyridine-2-carboxylic acid; 4-Amino-3,5,6-trichlorpicolinsaeure (German); ATCP; Chloramp (Russian); NCI-C00237; NSC 233899; Picolinic acid, 4-Amino-3,5,6-trichloro-; 2-Pyridine carboxylic acid, 4-amino-3,5,6-trichloro-; 3,5,6-Trichloro-4-aminopicolinic acid

*Trade Names:* ACCESS®; AMDON®; AMDON GRAZON®; BOROLIN®; GRAZON®; K-PIN®; PATHWAY®; TORDON®, Dow Chemical (USA), canceled 1/11/1980

*Producers:* Dow AgroSciences (USA); Ehrenstorfer, Dr. (Germany); Micro Flo Company (USA); Rhone-Poulenc Agro France (France); Shell Oil (Netherlands); Wuzhou International Co. (China)

*Chemical Class:* Pyridine
*EPA/OPP PC Code:* 005101
*California DPR Chemical Code:* 593
*ICSC Number:* 1246
*RTECS Number:* TJ7525000
*EINECS Number:* 217-636-1

*Uses:* Picloram is a systemic herbicide used for control of woody plants and a wide range of broad-leaved weeds along roads, power lines and long right-of-ways. Most grasses are resistant to picloram, so it is used in range management programs to control noxious weeds and brush. It is used to prepare sites for tree planting. Picloram is formulated either as an acid (technical product), a potassium or triisopropanolamine salt, or an isooctyl ester, and is available as either soluble concentrates, pellets, or granular formulations. During the Vietnam war, a herbicide named *Agent White* was used to control vegetation. It was a mixture of 2,4-D, triisopropanolamine salt and picloram.

*U.S. Maximum Allowable Residue Levels for Picloram (40 CFR 180.292):*

| CROP | ppm |
|---|---|
| Barley, grain | 0.5 |
| Barley, milled fractions, except flour | 3 |
| Barley, straw | 1 |
| Cattle, fat | 0.2 |
| Cattle, kidney | 5 |
| Cattle, liver | 0.5 |
| Cattle, meat | 0.2 |
| Cattle, mbyp, except kidney and liver | 0.2 |
| Egg | 0.05 |
| Goat, fat | 0.2 |
| Goat, kidney | 5 |
| Goat, liver | 0.5 |
| Goat, meat | 0.2 |

| | |
|---|---|
| Goat, mbyp, except kidney and liver | 0.2 |
| Grass, forage | 80 |
| Hog, fat | 0.2 |
| Hog, kidney | 5 |
| Hog, liver | 0.5 |
| Hog, meat | 0.2 |
| Hog, mbyp, except kidney and liver | 0.2 |
| Horse, fat | 0.2 |
| Horse, kidney | 5 |
| Horse, liver | 0.5 |
| Horse, meat | 0.2 |
| Horse, mbyp, except kidney and liver | 0.2 |
| Milk | 0.05 |
| Oat, forage and hay | 1 |
| Oat, grain | 0.5 |
| Oat, milled fractions, except flour | 3 |
| Oat, straw | 1 |
| Poultry, fat | 0.05 |
| Poultry, meat | 0.05 |
| Poultry, mbyp | 0.05 |
| Sheep, fat | 0.2 |
| Sheep, kidney | 5 |
| Sheep, liver | 0.5 |
| Sheep, meat | 0.2 |
| Sheep, mbyp, except kidney and liver | 0.2 |
| Wheat, forage | 1 |
| Wheat, grain | 0.5 |
| Wheat, milled fractions, except flour | 3 |
| Wheat, straw | 1 |

*Carcinogen/Hazard Classifications*
**U.S. EPA Carcinogens:** Group E, unlikely carcinogen
**IARC:** Group 3, unclassifiable
**Label Signal Word:** DANGER
**WHO Acute Hazard:** Class U, unlikely to be hazardous
*Regulatory Authority:*
- Carcinogen (Positive, Rat; Negative, Mouse) (NcI)[9]
- Banned or Severely Restricted (Sweden) (UN)[13]
- Air Pollutant Standard Set (ACGIH)[1] (HSE)[33] (NIOSH)[58] (former USSR)[35, 43] (Several States)[60]
- AB 1803-Well Monitoring Chemical (CAL)
- MCL (Maximum Contaminants Levels) list of contaminants (CAL)
- Permissible Exposure Limits for Chemical Contaminants (CAL/OSHA)
- The "Director's List" (CAL/OSHA)
- Safe Drinking Water Act: MCL, 0.5 mg/L; MCLG, 0.5 mg/L; Regulated chemical (47 FR 9352) as pichloram
- EPCRA Section 313 Form R *de minimis* concentration reporting level: 1.0%
- Canada, WHMIS, Ingredients Disclosure List

*Description:* Colorless crystalline solid or powder. Chlorine-like odor. Very slight solubility in water; solubility = 425 ppm @ 25°C. Molecular weight = 241.47. Melting/Freezing point = 218–219°C; decomposes above 190°C. Vapor pressure = $6.15 \times 10^{-7}$ mmHg @ 35°C.

*Incompatibilities:* This material is acidic. Reacts with hot concentrated alkali (hydrolyzes), strong bases. Attacks some metals.

*Permissible Exposure Limits in Air:* The NIOSH[2] recommended 10-hour TWA is 15 mg/m$^3$ based on total dust and 5 mg/m$^3$ based on the respirable fraction. HSE[33] and ACGIH[1] have adopted a TWA value of 10 mg/m$^3$ and the STEL is 20 mg/m$^3$. The former USSR has set an MAC in workplace air of 2.0 mg/m$^{3(35)}$ and a tentative safe exposure level in the workplace of 5 mg/m$^{3(43)}$. The former USSR[35] has also set a MAC for ambient air of 0.03 mg/m$^3$ on a once-daily basis and 0.02 mg/m$^3$ on an average daily basis. Several states have set guidelines or standards for picloram in ambient air[60] ranging from 0.1-0.2 mg/m$^3$ (North Dakota) to 0.16 mg/m$^3$ (Virginia) to 0.2 mg/m$^3$ (Connecticut) to 0.238 mg/m$^3$ (Nevada).

*Determination in Air:* Filter; none; Gravimetric; NIOSH IV, Particulates NOR: Method #0500 (total); Method #0600 (respirable).[18]

*Permissible Concentration in Water:* The USEPA has set a lifetime health advisory of 0.49 mg/L. The former USSR[35] has set an MAC in surface water of 10.0 mg/L. States which have set guidelines for picloram in drinking water[61] include Kansas at 0.175 mg/L and Maine at 0.3 mg/L.

*Routes of Entry:* Inhalation, ingestion, skin and/or eye contact.

*Harmful Effects and Symptoms*
*Short Term Exposure:* Irritates the eyes, skin, and respiratory tract. Exposure can cause nausea.
*Long Term Exposure:* Picloram should be handled as a carcinogen with extreme caution. It may damage the testes. May affect the kidneys and liver. In animals: liver, kidney changes.

*Points of Attack:* Eyes, skin, respiratory system, liver and kidneys. Cancer site in animals: liver, uterus, pituitary gland.

*Medical Surveillance:* Liver and kidney function tests.

*First Aid:* If this chemical gets into the eyes, remove any contact lenses at once and irrigate immediately for at least 15 minutes, occasionally lifting upper and lower lids. Seek medical attention immediately. If this chemical contacts the skin, remove contaminated clothing and wash immediately with soap and water. Seek medical attention immediately. If this chemical has been inhaled, remove from exposure, begin rescue breathing (using universal precautions) if breathing has stopped, and CPR if heart action has stopped. Transfer promptly to a medical facility. When this chemical has been swallowed, get medical attention. Give large quantities of water and induce vomiting. Do not make an unconscious person vomit.

*References:*
- EXTOXNET, Extension Toxicology Network, "Pesticide Information Profile, Picloram," Oregon State University, Corvallis, OR (June 1996). http://extoxnet.orst.edu/pips/picloram.htm
- EPA, Office of Pesticide Programs, Pesticide Residue Limits, "Picloram," 40 CFR 180.292, http://www.epa.gov/pesticides/food/viewtols.htm
- U.S. Environmental Protection Agency, "Health Advisory: Picloram," Washington, DC, Office of Drinking Water (August 1987).
- California Environmental Protection Agency "Chemical List of Lists," Sacramento CA (February 1997).
- New Jersey Department of Health, "Hazardous Substance Fact Sheet: Picloram," Trenton, NJ (May 2001). http://www.state.nj.us/health/eoh/rtkweb/1536.pdf

# Pindone

*Use Type:* Rodenticide
*CAS Number:* 83-26-1
*Formula:* $C_{14}H_{14}O_3$; $C_6H_4(CO)_2CHCOC(CH_3)_3$
*Alert:* No products currently registered with the U.S. EPA.
*Synonyms:* *tert*-Butyl valone; 1,3-Dioxo-2-pivaloy-lindane; Pivalyl; 2-Pivalyl-1,3-indandione; Pivaldione (French); 2-(Trimetil-acetil)-indan-1,3-dione (Italian)
*Trade Names:* CHEMRAT®; PIVAL®; PIVALYL VALONE®; TRI-BAN®
*Producers:* R.S.A. Corp (USA)
*Chemical Class:* Indandione
*EPA/OPP PC Code:* 067703
*California DPR Chemical Code:* 489
*RTECS Number:* NK6300000
*Uses:* Pindone is used to control Norway rats, roof rats, house mice and insects around stored grain and often used as a rabbit poison. It is used over a period of 6 to 10 days of continuous feeding. When it was used in the U.S., it was a Restricted Use Pesticide (RUP) when used in concentrations above 3%.
*Carcinogen/Hazard Classifications*
*Label Signal Word:* CAUTION or DANGER
*WHO Acute Hazard:* Class 1b, highly hazardous
*Regulatory Authority:*
- Air Pollutant Standard Set (ACGIH)[1] (OSHA)[58] (Several States)[60]
- Permissible Exposure Limits for Chemical Contaminants (CAL/OSHA)
- The "Director's List" (CAL/OSHA)
- Actively registered pesticide in California.
- Canada, WHMIS, Ingredients Disclosure List

*Description:* Bright yellow crystalline solid or powder. May turn yellow-brown on contact with air. Nearly odorless. Practically insoluble in water; 15 ppm @ 25°C. Boiling point = 178°C (decomposes). Melting/Freezing point = 105–110°C; 205–210°C (Na salt).
*Incompatibilities:* None reported.
*Permissible Exposure Limits in Air:* The OSHA[2], NIOSH[2], and ACGIH[1] TWA value is 0.1 mg/m³. The NIOSH[2] IDLH level = 100 mg/m³. Several states have set guidelines or standards for pindone in ambient air[60] ranging from 1.0-3.0 $\mu g/m^3$ (North Dakota) to 1.6 $\mu g/m^3$ (Virginia) to 2.0 $\mu g/m^3$ (Connecticut and Nevada).
*Determination in Air:* No method available.
*Permissible Concentration in Water:* No criteria set, but runoff from spills or fire control may cause water pollution.
*Routes of Entry:* Inhalation and ingestion.
*Harmful Effects and Symptoms*
*Short Term Exposure:* Nosebleeds (epostaxis), excessive bleeding of minor cuts and bruises, smoky urine, black tarry stools, abdominal and back pain. Reduced blood-clotting which leads to hemorrhaging; symptoms resembling warfarin: depressed formation of prothrombin and capillary fragility leading to hemorrhages.
*Points of Attack:* Blood prothrombin.
*Medical Surveillance:* Consider the points of attack in preplacement and periodic physical examinations.
*First Aid:* If this chemical gets into the eyes, remove any contact lenses at once and irrigate immediately for at least 15 minutes, occasionally lifting upper and lower lids. Seek medical attention immediately. If this chemical contacts the skin, remove contaminated clothing and wash immediately with soap and water. Seek medical attention immediately. If this chemical has been inhaled, remove from exposure, begin rescue breathing (using universal precautions) if breathing has stopped, and CPR if heart action has stopped. Transfer promptly to a medical facility. When this chemical has been swallowed, get medical attention. Give large quantities of water and induce vomiting. Do not make an unconscious person vomit.
*References:*
- Pesticide Education Management Program, "Pindone Chemical Profile 1/85," Cornel University, Ithaca, NY (January 1985). http://pmep.cce.cornell.edu/profiles/rodent/pindone/rod-prof-pindone.ht
- California Environmental Protection Agency "Chemical List of Lists," Sacramento CA (February 1997).

# Piperidine

*Use Type:* Insecticide
*CAS Number:* 110-89-4
*Formula:* $C_5H_{11}N$
*Alert:* Not registered for use in the U. S.

*Synonyms:* Azacyclohexane; Cyclopentimine; Cypentil; Hexahydropyridine; Hexazane; Pentamethyleneimine; Peperidin (German)

*Producers:* BASF (Germany); Bayer Chemicals (Germany); Daicel Chemical Industries (Japan); DSM Fine Chemicals (Netherlands); Koei Chemical (Japan); Nepera (USA); Nippon Steel Chemical (Japan); Penta Manufacturing (USA); Reilly Industries (USA); Rutherford Chemicals (USA); Whyte Chemicals (UK)

*ICSC Number:* 0317

*RTECS Number:* TM3500000

*EEC Number:* 613-027-00-3

*EINECS Number:* 203-813-0

*Uses:* Piperidine is used as a solvent and intermediate; as a curing agent for rubber and epoxy resins; catalyst for condensation reactions; as an ingredient in oils and fuels; complexing agent; manufacture of local anesthetics, in analgesics, pharmaceuticals, wetting agents, and germicides; synthetic flavoring. Not registered as a pesticide in the U.S.

*Regulatory Authority:*
- Air Pollutant Standard Set (former USSR)[43]
- Clean Air Act: Accidental Release Prevention/Flammable Substances, (Section 112[r], Table 3), TQ = 15,000 lb (6810 kg)
- Superfund/EPCRA 40CFR355, Appendix B Extremely Hazardous Substances: TPQ = 1000 lb (454 kg)
- Superfund/EPCRA 40CFR302.4 RQ: EHS, 1 lb (0.454 kg)

*Description:* Clear, colorless liquid. Amine odor. Soluble in water. Molecular weight = 85.15. Boiling point = 105.8°C. Melting/Freezing point = –6.9°C. Flash point = 16°C. Vapor pressure = 48 mmHg @ 29°C; 5.2 kPa @ 29°C. Log $K_{ow}$ = 0.8. Unlikely to bioaccumulate in marine organisms.

*Incompatibilities:* Piperidine is a medium-strong base. Reacts violently with acids, oxidizers.

*Permissible Exposure Limits in Air:* The former USSR-UNEP/IRPTC project[43] has set an MAC in workplace air of 0.2 mg/m$^3$.

*Permissible Concentration in Water:* The former USSR-UNEP/IRPTC project[43] has set an MAC in water bodies used for domestic purposes of 0.06 mg/L. Runoff from spills or fire control may cause water pollution.

*Routes of Entry:* Inhalation, ingestion, skin and/or eye contact.

*Harmful Effects and Symptoms*

*Short Term Exposure:* Corrosive to the eyes, skin and respiratory tract. Sore throat, coughing, labored breathing, and dizziness occur after inhalation. Higher exposures can cause pulmonary edema, a medical emergency that can be delayed for several hours. This can cause death. Exposure may cause increased blood pressure. May cause permanent injury after short exposure to small amounts. Ingestion may involve both irreversible and reversible changes. 30 to 60 mg/kg may cause symptoms in humans. Symptoms upon oral administration include weakness, nausea, vomiting, salivation, labored respiration, muscular paralysis, and asphyxiation. Redness, pain, and burns occur upon contact with skin.

*Long Term Exposure:* Irritating substances may cause lung irritation; bronchitis may develop. May affect the liver and kidneys.

*Points of Attack:* Lungs, blood, liver and kidneys.

*Medical Surveillance:* Monitor blood pressure. Lung function tests. Liver and kidney function tests. Consider chest x-ray following acute overexposure.

*First Aid:* If this chemical gets into the eyes, remove any contact lenses at once and irrigate immediately for at least 15 minutes, occasionally lifting upper and lower lids. Seek medical attention immediately. If this chemical contacts the skin, remove contaminated clothing and wash immediately with soap and water. Seek medical attention immediately. If this chemical has been inhaled, remove from exposure, begin rescue breathing (using universal precautions) if breathing has stopped, and CPR if heart action has stopped. Transfer promptly to a medical facility. When this chemical has been swallowed, get medical attention. Give large quantities of water and induce vomiting. Do not make an unconscious person vomit. Medical observation is recommended for 24 to 48 hours after breathing overexposure, as pulmonary edema may be delayed. As first aid for pulmonary edema, a doctor or authorized paramedic may consider administering a corticosteroid spray.

*References:*
- U.S. Environmental Protection Agency, "Chemical Profile: Piperidine,: Washington, DC, Chemical Emergency Preparedness Program (November 30, 1987).
- California Environmental Protection Agency "Chemical List of Lists," Sacramento CA (February 1997).
- New Jersey Department of Health and Senior Services, "Hazardous Substance Fact Sheet, Piperidine," Trenton NJ (September, 1999).
  http://www.state.nj.us/health/eoh/rtkweb/1543.pdf

# Piperonyl Butoxide

*Use Type:* Insecticide synergist

*CAS Number:* 51-03-6

*Formula:* $C_{19}H_{30}O_5$

*Synonyms:* AI3-14250; 1,3-Benzodioxole, 5-[(2-(2-butoxyethoxy)ethoxy)methyl]-6-propyl-; Butoxide; butoxido de piperonilo (Spanish); α-[2-(2-N-Butoxyethoxy)ethoxy]-4,5-methylenedioxy-2-propyltoluene; α-[2-(2-Butoxyethoxy)ethoxy]-4,5-Methylenedioxy-2-propyltoluene; 5-[(2-(2-Butoxyethoxy)ethoxy)methyl]-6-propyl-1,3-benzodioxole; 2-(2-Butoxyethoxy)ethyl 6-propylpiperonyl ether; Butyl carbitol 6-propylpiperonyl ether; Butyl-carbityl (6-propylpiperonyl) ether; (Butylcarbityl)(6-propylpiperonyl) ether 80% and related compounds 20%; Caswell No. 670; ENT 14,250; Ethanol butoxide; 3,4-Methylendioxy-6-propylbenzyl-*N*-butyl-diaethylenglykolaether (German); (3,4-Methylenedioxy-6-propylbenzyl)(butyl)diethylene glycol ether; 3,4-Methylenedioxy-6-propylbenzyl *N*-butyl diethyleneglycol ether; 4,5-Methylenedioxy-2-propylbenzyldiethylene glycol butyl ether; NCI-C02813; PB; Piperonyl butoxyde; 6-(Propylpiperonyl)-butyl carbityl ether; 6-Propylpiperonyl butyl diethylene glycol ether; 5-Propyl-4-(2,5,8-trioxa-dodecyl)-1,3-benzodioxol (German); Toluene, α-[2-(2-butoxyethoxy)ethoxy]-4,5-(methylenedioxy)-2-propyl-

*Trade Names:* The U.S. EPA lists 7,340 products containing this substance, 1,622 of which are active.

*Chemical Class:* Glycol ether

*EPA/OPP PC Code:* 067501

*California DPR Chemical Code:* 486

*ICSC Number:* 1347

*RTECS Number:* XS8050000

*EINECS Number:* 200-076-7

*Uses:* Piperonyl butoxide is a synergist, i.e., not a pesticide itself, but enhances the properties of other chemicals. It is used with other pesticides such as pyrethrins, pyrethroids, rotenone and carbamates in food and non-food agricultural products, home and garden products, termite and mosquito products, and veterinary pesticide products. It inhibits the insect's ability to break down an insecticide before it takes effect, thereby prolonging the action and reduces the necessity for using a stronger formulation.

*U.S. Maximum Allowable Residue Levels Piperonyl Butoxide (40 CFR 180.127): Note:* 180.127 gives tolerance levels, but 180.905 says its exempt under certain circumstances.

| CROP | ppm |
|---|---|
| Almond, post-h | 8 |
| Apple, post-h | 8 |
| Barley, post-h | 20 |
| Bean, post-h | 8 |
| Birdseed, mixtures, post-h | 20 |
| Blackberry, post-h | 8 |
| Blueberry, post-h | 8 |
| Boysenberry, post-h | 8 |
| Buckwheat, post-h | 20 |
| Cacao bean, post-h | 8 |
| Cattle, fat | 0.1 |
| Cattle, meat | 0.1 |
| Cattle, mbyp | 0.1 |
| Cherry, post-h | 8 |
| Coconut, copra, post-h | 8 |
| Corn, field, grain, post-h | 20 |
| Corn, pop | 20 |
| Cotton, undelinted seed, post-h | 8 |
| Crabapple, post-h | 8 |
| Currant, post-h | 8 |
| Dewberry, post-h | 8 |
| Egg | 1 |
| Fig, post-h | 8 |
| Flax, seed, post-h | 8 |
| Goat, fat | 0.1 |
| Goat, meat | 0.1 |
| Goat, mbyp | 0.1 |
| Gooseberry, post-h | 8 |
| Grain, cereal, milled fractions, except flour | 10 |
| Grape, post-h | 8 |
| Guava, post-h | 8 |
| Hog, fat | 0.1 |
| Hog, meat | 0.1 |
| Hog, mbyp | 0.1 |
| Horse, fat | 0.1 |
| Horse, meat | 0.1 |
| Horse, mbyp | 0.1 |
| Loganberry, post-h | 8 |
| Mango, post-h | 8 |
| Milk, fat | 0.25 |
| Muskmelon, post-h | 8 |
| Oat, post-h | 8 |
| Orange, post-h | 8 |
| Pea, post-h | 8 |
| Peach, post-h | 8 |
| Peanut, post-h | 8 |
| Pear, post-h | 8 |
| Pineapple, post-h | 8 |
| Plum, fresh prune, post-h | 8 |
| Potato, post-h | 0.25 |
| Poultry, fat | 3 |
| Poultry, meat | 3 |
| Poultry, mbyp | 3 |
| Processed food | 10 |
| Raspberry, post-h | 8 |
| Rice, post-h | 20 |
| Rye, post-h | 20 |
| Sheep, fat | 0.1 |
| Sheep, meat | 0.1 |
| Sheep, mbyp | 0.1 |
| Sorghum, grain, post-h | 8 |
| Stored feed, dried, 4% fat or less, from cotton bags | 10 |
| Stored feed, dried, 4% fat or less, from paper bags | 10 |
| Sweet potato, roots, post-h | 0.25 |
| Tomato, post-h | 8 |
| Walnut, post-h | 8 |
| Wheat, post-h | 20 |

*Carcinogen/Hazard Classifications*

**U.S. EPA Carcinogens:** Group 3, possible carcinogen

**IARC:** Group 3, unclassifiable

**WHO Acute Hazard:** Class U, unlikely to be hazardous

*Regulatory Authority:*
- EPA 40 CFR 372.65, Specific Toxic Chemical Listings
- Actively registered pesticide in California.
- FIFRA, 40 CFR 185: tolerances for pesticides in food

- FIFRA, 40CFR186: tolerances for pesticides in animal feeds
- EPCRA Section 313 Form R *de minimis* concentration reporting level: 1.0%
- Clean Air Act: Hazardous Air Pollutants (Title I, Part A, Section 112) as glycol ethers
- CAL Air Resources Board/AB 1807 Toxic Air Contaminants as glycol ethers
- AB 2588-Air Toxics "Hot Spots" Chemicals (CAL) as glycol ethers
- Actively registered pesticide in California. as glycol ethers

*Description:* Light yellow to light brown liquid. Nearly odorless. Very slightly soluble in water. Molecular weight = 338.49. Boiling point = 180°C @ 1 mmHg. Flash point = 171°C. Density = 1.06 @ 25°C. Log $K_{ow}$ = 4.30. Values above 3.0 are likely to bioaccumulate in marine organisms.

*Incompatibilities:* Contact with oxidizers (chlorates, nitrates, peroxides, permanganates, perchlorates, chlorine, bromine, fluorine, etc) may cause fire and explosions. May react violently with aliphatic amines, alkalies, boranes, isocyanates, nitric acid, sulfuric acid.

*Permissible Concentration in Water:* No criteria set. Runoff from spills or fire control may cause water pollution.

*Routes of Entry:* Inhalation, ingestion, skin and/or eye contact. Absorbed through the skin.

*Harmful Effects and Symptoms*

*Short Term Exposure:* May affect you when breathed in and by passing through your skin. Contact can irritate the eyes and skin. Exposure can irritate the nose and throat. High levels of glycol ether vapor may cause central nervous system depression causing dizziness, lightheadedness, and unconsciousness. Very high levels of exposure may cause lung, liver, and kidney damage.

*Long Term Exposure:* Causes skin dryness; dermatitis. May cause liver and kidney damage. May irritate the lungs; bronchitis may develop.

*Points of Attack:* Eyes, skin, respiratory system, central nervous system.

*Medical Surveillance:* If symptoms develop or overexposure is suspected, the following may be useful: Liver and kidney function tests. Lung function tests.

*First Aid:* If this chemical gets into the eyes, remove any contact lenses at once and irrigate immediately for at least 15 minutes, occasionally lifting upper and lower lids. Seek medical attention immediately. If this chemical contacts the skin, remove contaminated clothing and wash immediately with soap and water. Seek medical attention immediately. If this chemical has been inhaled, remove from exposure, begin rescue breathing (using universal precautions) if breathing has stopped, and CPR if heart action has stopped. Transfer promptly to a medical facility. When this chemical has been swallowed, get medical attention. Give large quantities of water and induce vomiting. Do not make an unconscious person vomit.

*References:*
- National Pesticides Information Center (NPIC), "Fact Sheet, Piperonyl Butoxide," Corvallis, OR (November 2000). http://npic.orst.edu/factsheets/pbogen.pdf
- California Environmental Protection Agency "Chemical List of Lists," Sacramento CA (February 1997)
- U.S. Environmental Protection Agency, Office of Pesticide Programs, Pesticide Residue Limits, "Piperonyl Butoxide," 40 CFR 180.127 and 40CFR 180.905, www.epa.gov/pesticides/food/viewtols.htm

# Pirimicarb (ANSI)

*Use Type:* Insecticide
*CAS Number:* 23103-98-2
*Formula:* $C_{11}H_{18}N_4O_2$
*Synonyms:* Carbamic acid, dimethyl-, 2-(dimethylamino)-5,6-dimethyl-4-pyrimidinyl ester; 2-Dimethylamino-5,6-dimethyl-4-pyrimidinyl dimethylcarbamate; 2-(Dimethylamino)-5,6-dimethyl-4-pyrimidinyldimethylcarbamate; 2-Dimethylamino-5,6-dimethylpryimidin-4-yl *N,N*-dimethylcarbamate; Dimethylcarbamic acid 2-(dimethylamino)-5,6-dimethyl-4-pyrimidinyl ester; 5,6-Dimethyl-2-dimethylamino-4-pyrimidinyldimethylcarbamate; ENT 27,766
*Trade Names:* ABOL®; AFICIDA®; APHOX®; FERNOS®; PIRIMIKARB; PIRIMOR®, Syngenta (Switzerland); PP 062®; PYRIMOR®; RAPID®
*Producers:* Agsin (Singapore); Ki-Hara Chemicals Ltd. (UK); Syngenta (Switzerland)
*Chemical Class:* Methyl carbamate
*EPA/OPP PC Code:* 106101
*California DPR Chemical Code:* 1875
*Uses:* Originally registered in the U.S. for non-food use on alfalfa grown for seed in selected. Elsewhere it is used on a wide range of cereals, potatoes, fruits, vegetables and other crops.
*Carcinogen/Hazard Classifications*
**Label Signal Word:** WARNING
**WHO Acute Hazard:** Class II, moderately hazardous
*Description:* Colorless crystalline solid. Odorless. Soluble in water. Melting point = 90.5°C. Molecular weight = 238.3.
*Incompatibilities:* May form explosive materials with phosphorus pentachloride.
*Permissible Concentration in Water:* No criteria set. Runoff from spills or fire control may cause water pollution.
*Routes of Entry:* Skin absorption, ingestion and inhalation.
*Harmful Effects and Symptoms*
*Short Term Exposure:* Eye pupils are small, blurred vision, eye watering, runny nose, cough, shortness of breath, salivation, nausea, stomach cramps, diarrhea, and vomiting, increased blood pressure, profuse sweating, hypermotility, hallucinations, agitation, tingling of the skin, slow heartbeat, convulsions, fluid in lungs, loss of consciousness,

incontinence, breathing stops, death. Carbamates inhibit the acetylcholinesterase enzymes and alter the way in which nervous impulses are transmitted. However, within several hours carbamates spontaneously detach from the enzymes.

*Long Term Exposure:* A potent cholinesterase inhibitor; cumulative effect is possible. This chemical may damage the nervous system with repeated exposure, resulting in convulsions, respiratory failure. May cause liver damage.

*Points of Attack:* Respiratory system, lungs, central nervous system, cardiovascular system, skin, eyes, plasma and red blood cell cholinesterase.

*Medical Surveillance:* Medical observation is recommended for 24 to 48 hours after breathing overexposure, as pulmonary edema may be delayed. As first aid for pulmonary edema, consider administering a corticosteroid spray. Cigarette smoking may exacerbate pulmonary injury and should be discouraged for at least 72 hours following exposure. Before employment and at regular times after that, the following are recommended: Plasma and red blood cell cholinesterase levels (tests for the enzyme poisoned by this chemical). If exposure stops, plasma levels return to normal in 1-2 weeks while red blood cell levels may be reduced for 1-3 months. When acetylcholinesterase enzyme levels are reduced by 25% or more below preemployment levels, risk of poisoning is increased, even if results are in lower ranges of "normal." Reassignment to work not involving carbamate pesticides is recommended until enzyme levels recover. If symptoms develop or overexposure occurs, repeat the above tests as soon as possible and get an exam of the nervous system. Also consider complete blood count. Consider chest x-ray following acute overexposure

*First Aid:* Speed in removing material from eyes and skin is of extreme importance. *Eyes:* Eye contact can cause dangerous amounts of these chemicals to be quickly absorbed through the mucous membrane into the bloodstream. Immediately and gently flush eyes with plenty of warm or cold water (NO hot water) for at least 15 minutes, occasionally lifting the upper and lower eyelids. Get medical aid immediately. *Skin:* Get medical aid. Skin contact can cause dangerous amounts of these chemicals to be absorbed into the bloodstream. Wearing the appropriate PPE equipment and respirator for carbamate pesticides, immediately flush skin with plenty of soap and water for at least 15 minutes while removing contaminated clothing and shoes. Shampoo hair promptly if contaminated; protect eyes. *Ingestion:* Call poison control. Loosen all clothing. Never give anything by mouth to an unconscious person. Get medical aid. Do NOT induce vomiting.* If conscious, alert, and able to swallow, rinse mouth and have victim drink 4 to 8 ounces of water. Check to see if poison control instructs you to use ipecac syrup, otherwise administer slurry of activated charcoal (2 oz in 8 oz of water). If victim is *UNCONSCIOUS OR HAVING CONVULSIONS,* do nothing except keep victim warm. **In some cases you may be* *specifically instructed by poison control to induce vomiting by way of 2 tablespoons of syrup of ipecac (adult) washed down with a cup of water. Do NOT give activated charcoal before or with ipecac syrup. Inhalation: Get medical aid. Do not contaminate yourself. Wearing the appropriate PPE equipment and respirator for carbamate pesticides, immediately remove the victim from the contaminated area to fresh air. If the victim is not breathing, administer artificial respiration. Do NOT use mouth-to-mouth resuscitation; use bag/mask apparatus. If breathing is difficult, administer oxygen through bag/mask apparatus until medical help arrives. Do not leave victim unattended. Note to physician or authorized medical personnel.* Administer atropine, 2 mg (1/30 gr) intramuscularly or intravenously as soon as any local or systemic signs or symptoms of an intoxication are noted; repeat the administration of atropine every 3 to 8 minutes until signs of atropinization (mydriasis, dry mouth, rapid pulse, hot and dry skin) occur; initiate treatment in children with 0.05 mg mg/kg of atropine; repeat at 5 to 10 minute intervals. Watch respiration, and remove bronchial secretions if they appear to be obstructing the airway; intubate if necessary. *Medical note:* 2-PAMCI may be contraindicated in the case of some carbamate poisonings.

*References:*
- U.S. Environmental Protection Agency, Office of Pesticide Programs, "Pesticide Fact Sheet, Pirimicarb," http://www.epa.gov/opprd001/factsheets/pirimicarb.pdf
- Pesticide Management Education Program, "Pirimicarb (Pirimor) Chemical Profile 4/85," Cornell University, Ithaca, NY (April 1985). http://pmep.cce.cornell.edu/profiles/insect-mite/mevinphos-propargite/pirimicarb/insect-prof-pirimicarb.html
- New Jersey Department of Health and Senior Services, "Hazardous Substance Fact Sheet, Pirimicarb," Trenton NJ (March 1989, rev. January 2001). http://www.state.nj.us/health/eoh/rtkweb/1544.pdf
- California Environmental Protection Agency "Chemical List of Lists," Sacramento CA (February 1997).

# Pirimiphos-ethyl (ANSI)

*Use Type:* Insecticide
*CAS Number:* 23505-41-1
*Formula:* $C_{13}H_{24}N_3O_3PS$
*Alert:* There are no currently registered products with the U.S. EPA.
*Synonyms:* $O$-(2-(Diethylamino)-6-methyl-4-pyrimidinyl)$O,O$-diethyl phosphorothioate; $O$-$O$-Diethyl-$O$-(2-diethylamino-6-methyl-4-pyrimidinyl)phosphorothioate; Diethyl $O$-(2-diethylamino-6-methyl-4-pyrimidinyl)phosphorothioate; 2-Diethylamino-6-methylpyrimidin-4-yl diethylphosphorothioate; Diethyl 2-

dimethylamino-4-methylpyrimidin-6-yl phosphorothionate; Ethyl pirimiphos; Phosphorothioic acid, O-[2-(diethylamino)-6-methyl-4-pyrimidinyl] O,O-diethyl ester; Primifosethyl

*Trade Names:* FERNEX®; PP 211®; PRIMICID®, ICI Group (Imperial Chemical Industries Plc) (UK); PRIMOTEC®; PRINICID®, canceled; R 42211®; SOLGARD®

*Producers:* Cognis (Germany); ICI Group (Imperial Chemical Industries Plc) (UK); Shenzhen Guomeng Industry Co., Ltd. (China); Sigma-Aldrich Laborchemikalien (Germany)

*Chemical Class:* Organophosphate

*EPA/OPP PC Code:* 108101

*California DPR Chemical Code:* 2781

*EINECS Number:* 245-704-0

*Carcinogen/Hazard Classifications*

**WHO Acute Hazard:** Class 1b, highly hazardous

*Regulatory Authority:*

- Superfund/EPCRA 40CFR355, Appendix B Extremely Hazardous Substances: TPQ = 1000 lb (454 kg)
- U.S. DOT Inhalation Hazard Chemicals as organophosphates
- U.S. DOT Regulated Marine Pollutant (49CFR172.101, Appendix B), severe pollutant

*Description:* Pirimfos-ethyl is a straw-colored liquid. Decomposes in water. Decomposes @ 130°C; no boiling point can be determined.

*Permissible Exposure Limits in Air:* No standards set.

*Permissible Concentration in Water:* No criteria set, but runoff from spills or fire control may cause water pollution.

*Routes of Entry:* Inhalation, ingestion, skin and/or eye contact. Absorbed through the skin.

*Harmful Effects and Symptoms*

**Short Term Exposure:** As with other Organophosphorus pesticides,[18], symptoms are secondary to cholinesterase inhibition: headache, giddiness, blurred vision, nervousness, weakness, nausea, cramps, diarrhea, and discomfort in the chest. Other signs include sweating, tearing, salivation, vomiting, cyanosis, convulsions, coma, loss of reflexes and loss of sphincter control. Delayed pulmonary edema may occur after inhalation.

*Long Term Exposure:* Cholinesterase inhibitor; cumulative effect is possible. This chemical may damage the nervous system with repeated exposure, resulting in convulsions, respiratory failure. May cause liver damage.

*Points of Attack:* Respiratory system, lungs, central nervous system, cardiovascular system, skin, eyes, plasma and red blood cell cholinesterase.

*Medical Surveillance:* Medical observation is recommended for 24 to 48 hours after breathing overexposure, as pulmonary edema may be delayed. Before employment and at regular times after that, the following are recommended: Plasma and red blood cell cholinesterase levels (tests for the enzyme poisoned by this chemical). If exposure stops, plasma levels return to normal in 1-2 weeks while red blood cell levels may be reduced for 1-3 months. When acetylcholinesterase enzyme levels are reduced by 25% or more below preemployment levels, risk of poisoning is increased, even if results are in lower ranges of "normal." Reassignment to work not involving organophosphate or carbamate pesticides is recommended until enzyme levels recover. If symptoms develop or overexposure occurs, repeat the above tests as soon as possible and get an exam of the nervous system. Also consider complete blood count. Consider chest x-ray following acute overexposure. Do not drink any alcoholic beverages before or during use. Alcohol promotes absorption of organophosphates. Do not drink any alcoholic beverages before or during use. Alcohol promotes absorption of organophosphates.

*First Aid:* **Treatment for organophosphate poisoning consists of thorough decontamination, cardiorespiratory support, and administration of the antidotes atropine and pralidoxime. In cases of severe poisoning, diazepam, an anticonvulsant, should also be administered. Antidotes should be administered as prevention even if the diagnosis is in doubt.** Speed in removing material from eyes and skin is of extreme importance. *Eyes:* Eye contact can cause dangerous amounts of these chemicals to be quickly absorbed through the mucous membrane into the bloodstream. Immediately and gently flush eyes with plenty of warm or cold water (NO hot water) for at least 15 minutes, occasionally lifting the upper and lower eyelids. Get medical aid immediately. *Skin:* Get medical aid. Skin contact can cause dangerous amounts of these chemicals to be absorbed into the bloodstream. Wearing the appropriate PPE equipment and respirator for organophosphate pesticides, immediately flush skin with plenty of soap and water for at least 15 minutes while removing contaminated clothing and shoes. Shampoo hair promptly if contaminated. The removed, contaminated clothing and shoes should be double-bagged and left in Hot Zone for later disposal by hazardous materials experts. Skin may also be decontaminated with diluted hypochlorite solution. *Inhalation:* Get medical aid. Do not contaminate yourself. Wearing the appropriate PPE equipment and respirator for organophosphate pesticides, immediately remove the victim from the contaminated area to fresh air. If the victim is not breathing, administer artificial respiration. Do NOT use mouth-to-mouth resuscitation; use bag/mask apparatus. If breathing is difficult, administer oxygen through bag/mask apparatus until medical help arrives. Do not leave victim unattended. *Ingestion:* Call poison control. Loosen all clothing. Never give anything by mouth to an unconscious person. If victim is *unconscious or having convulsions,* do nothing except keep victim warm. Get medical aid. Transfer promptly to a medical facility. In cases of ingestion, **do not induce vomiting**. If the victim is alert and asymptomatic, administer a slurry of activated charcoal at a dose of 1 g/kg (infant, child, and adult dose). A soda can and straw may be of assistance when offering charcoal to a child. *In some cases you may be specifically instructed by poison control to induce vomiting by way of 2 tablespoons of syrup of ipecac (adult) washed down with a cup of water.* Do NOT give activated charcoal before or with ipecac syrup.

*Note to physician or authorized medical personnel:* Treat cases of respiratory compromise, coma, or excessive pulmonary secretions with respiratory support using protocols and techniques available and within the scope of training. Some cases may necessitate procedures such as endotracheal intubation or cricothyrotomy by properly trained and equipped personnel. When possible, atropine (see *Antidotes*, below) should be given under medical supervision. Patients who are comatose, hypotensive, or having seizures or cardiac arrhythmias should be treated according to advanced life support protocols. *Antidotes:* Two antidotes are administered to treat organophosphate poisoning. Atropine is a competitive antagonist of acetylcholine at muscarinic receptors and is used to control the excessive bronchial secretions which are often responsible for death. Pralidoxime relieves both the nicotinic and muscarine effects of organophosphate poisoning by regenerating acetylcholinesterase and can reduce both the bronchial secretions and the muscle weakness associated with poisoning. The initial intravenous dose of atropine in adults should be determined by the severity of symptoms: An initial adult dose of 1.0 to 2.0 mg or pediatric dose of 0.01 mg/kg (minimum 0.01 mg) should be administered intravenously. If intravenous access cannot be established, atropine may also be given intramuscularly, subcutaneously or via endotracheal tube. Doses should be repeated every 15 minutes until excessive secretions and sweating have been controlled. Once bronchial secretion has been controlled, atropine administration should be repeated whenever the secretions begin to recur. In seriously poisoned patients, very large doses may be required. Alterations of pulse rate and pupillary size should not be used as indicators of treatment adequacy. Pralidoxime should be administered as early in poisoning as possible as its efficacy may diminish when given more than 24 to 36 hours after exposure. Doses are as follows: adult 1.0 g; pediatric 25 to 50 mg/kg. The drug should be administered intravenously over 30 to 60 minutes, but in a life-threatening situation, one-half of the total dose can be given per minute for a total administration time of 2 minutes. Treatment should begin to take effect within 40 minutes with a reduction in symptoms and the amount of atropine necessary to control bronchial secretion. The initial dose can be repeated in 1 hour and then every 8 to 12 hours until the patient is clinically well and no longer requires atropine. If intravenous access cannot be established, pralidoxime may also be given intramuscularly. Early administration of diazepam in addition to the combined atropine and pralidoxime treatment may help prevent the onset of seizures and potential brain and cardiac morphologic damage following high-level organophosphate poisoning.

*References:*
- U.S. Environmental Protection Agency, "Chemical Profile: Pirimiphos-Ethyl," Washington, DC, Chemical Emergency Preparedness Program (November 30, 1987).
- California Environmental Protection Agency "Chemical List of Lists," Sacramento CA (February 1997).
- Agency for Toxic Substances and Disease Registry, U.S. Department of Health and Human Services, Public Health Service, "Managing Hazardous Materials Incidents," Atlanta, GA (June 2003).

## Pirimiphos-methyl (ANSI)

*Use Type:* Insecticide and acaricide.
*CAS Number:* 29232-93-7
*Formula:* $C_{11}H_{20}N_3O_3PS$
*Synonyms:* AI3-27699; Caswell No. 334B; *O*-[2-(Diethylamino]-6-methyl-4-pyrimidinyl) *O,O*-dimethyl phosphorothioate; *O*-(2-Diethylamino-6-methylpyrimidin-4-yl) *O,O*-dimethyl phosphorothioate; *O*-[2-(Diethylamino]-6-methyl-4-pyrimidinyl]-*O,O*-dimethyl phosphorothioate; 2-diethylamino-6-methylpyrimidin-4-yl dimethyl phosphorothionate; *O,O*-dimethyl-*O*-[2-(diethylamino)-6-methyl-4-pyrimidinyl]; *O,O*-Dimethyl *O*-[2-(diethylamino)-6-methyl-4-pyrimidinyl]phosphorothioate; ENT 27699GC; Methylpirimiphos; Methylpyrimiphos; Phosphorothioic acid, *O*-[2-(diethylamino)-6-methyl-4-pyrimidinyl] *O,O*-dimethyl ester; 4-Pyrimidinol, 2-(diethylamino)-6-methyl-, *O*-ester with *O,O*-dimethyl phosphorothioate; Pyridimine phosphate

*Trade Names:* ACTELLIC®, Agriliance (USA); ACTELLIFOG®, Agriliance (USA); BLEX®; ENT 27699Gc®; DOMINATOR® EAR TAG, Schering-Plough Animal Health (USA); DOUBLE BARREL® EAR TAG, Schering-Plough Animal Health (USA); PLANT PROTECTION PP511®; PP511®; SILOSAN®; SYBOL®; TOMAHAWK®, canceled

*Producers:* Agriliance (USA); Agsin (Singapore); Ascot International UK); Biesterfeld Siemsgluess International GmbH (Germany); ICI Group (UK); Ki-Hara Chemicals Ltd. (UK); Schering-Plough Animal Health (USA); Shenzhen Guomeng Industry Co., Ltd. (China); Sigma-Aldrich Laborchemikalien (Germany); Syngenta (Switzerland); Zeneca Ag Products (USA) (now Syngenta)

*Chemical Class:* Organophosphate
*EPA/OPP PC Code:* 108102; (334300 old EPA code number)
*California DPR Chemical Code:* 2217
*RTECS Number:* TF1410000
*Uses:* Pirimiphos-methyl is a post-harvest insecticide used to control a variety of insects in stored grain products and

seed such as corn, rice, wheat and sorghum. It is also incorporated into cattle ear tags, and used for the fogging treatment of iris bulbs and pre-harvest clean-up of fruits and vegetables.

***U.S. Maximum Allowable Residue Levels Pirimiphos-methyl (40 CFR 180.409):***

| CROP | ppm |
|---|---|
| Cattle, fat | 0.2 |
| Cattle, kidney | 2.0 |
| Cattle, liver | 2.0 |
| Cattle, meat | 0.2 |
| Cattle, mbyp | 0.2 |
| Corn | 8.0 |
| Goat, fat | 0.2 |
| Goat, kidney | 2.0 |
| Goat, liver | 2.0 |
| Goat, mbyp | 0.2 |
| Hog, fat | 0.2 |
| Hog, kidney | 2.0 |
| Hog, liver | 2.0 |
| Hog, mbyp | 0.2 |
| Horse, fat | 0.2 |
| Horse, kidney | 2.0 |
| Horse, liver | 2.0 |
| Horse, mbyp | 0.2 |
| Kiwifruit | 5.0 |
| Poultry, fat | 0.2 |
| Sheep, fat | 0.2 |
| Sheep, kidney | 2.0 |
| Sheep, liver | 2.0 |
| Sheep, mbyp | 0.2 |
| Sorghum, grain | 8.0 |
| Wheat, flour | 8.0 |

***Carcinogen/Hazard Classifications***
**U.S. EPA Carcinogens:** Unclassifiable
**Label Signal Word:** CAUTION or WARNING
**WHO Acute Hazard:** Class III, slightly hazardous
***Regulatory Authority:***
- EPA 40 CFR 372.65, Specific Toxic Chemical Listings
- FIFRA, 180.3(5); class of cholinesterase-inhibiting pesticide
- FIFRA, 40CFR185: tolerances for pesticides in food
- FIFRA, 40CFR186: tolerances for pesticides in animal feeds
- DOT Inhalation Hazard Chemicals as organophosphates
- EPCRA Section 313 Form R *de minimis* concentration reporting level: 1.0%

***Description:*** Light yellow or amber oily liquid. Odorless when pure. Slightly soluble in water. Melting point = 15-18°C. Boiling point = decomposes above 100°C. Molecular weight = 274.4. Density = 1.157 @ 30 mmHg.

***Incompatibilities:*** May react violently with antimony(V) pentafluoride. Incompatible with lead diacetate, magnesium, silver nitrate. Contact with oxidizers (chlorates, nitrates, peroxides, permanganates, perchlorates, chlorine, bromine, fluorine, etc) may cause fire and explosions.

***Determination in Air:*** OSHA versatile sampler-2; Toluene/Acetone; Gas chromatography/Flame ionization detection; NIOSH IV, Method #5600, Organophosphorus Pesticides.[18]

***Permissible Concentration in Water:*** No criteria set. Runoff from spills or fire control may cause water pollution.

***Routes of Entry:*** Inhalation, ingestion, through the skin.

***Harmful Effects and Symptoms***

***Short Term Exposure:*** Eye pupils are small; blurred vision; eye watering; runny nose; cough; shortness of breath; salivation; dizziness; nausea, stomach cramps, diarrhea, and vomiting; increased blood pressure; profuse sweating; hypermotility, hallucinations; irritability; tingling of the skin; drowsiness; slow heartbeat; convulsions; fluid in lungs; loss of consciousness; incontinence; breathing stops; death. Organophosphates inhibit the action of acetylcholinesterase enzymes, and alter the way in which nervous impulses are transmitted. The effects can last for hours, days, or much longer. The action of the enzymes is reestablished after new enzymes are formed.

***Long Term Exposure:*** Cholinesterase inhibitor; cumulative effect is possible. Organophosphates may damage the nervous system with repeated exposure, resulting in convulsions, respiratory failure. May cause liver damage.

***Points of Attack:*** Respiratory system, central nervous system, cardiovascular system, blood cholinesterase.

***Medical Surveillance:*** Medical observation is recommended for 24 to 48 hours after breathing overexposure, as pulmonary edema may be delayed. As first aid for pulmonary edema, consider administering a corticosteroid spray. Cigarette smoking may exacerbate pulmonary injury and should be discouraged for at least 72 hours following exposure.

***First Aid:*** Speed in removing material from eyes and skin is of extreme importance. Eye contact can cause dangerous amounts of these chemicals to be quickly absorbed through the mucous membrane into the bloodstream. Immediately and gently flush eyes with plenty of warm or cold water (NO hot water) for at least 15 minutes, occasionally lifting the upper and lower eyelids. Get medical aid immediately. *Skin:* Get medical aid. Skin contact can cause dangerous amounts of these chemicals to be absorbed into the bloodstream. Wearing the appropriate PPE equipment and respirator for organophosphate pesticides, immediately flush skin with plenty of soap and water for at least 15 minutes while removing contaminated clothing and shoes. Shampoo hair promptly if contaminated. *Ingestion:* Call poison control. Loosen all clothing. Never give anything by mouth to an unconscious person. Get medical aid. Do NOT induce vomiting.* If conscious, alert, and able to swallow, rinse mouth and have victim drink 4 to 8 ounces of water do NOT induce vomiting but immediately administer slurry of activated charcoal (2 oz in 8 oz of water). If victim is *unconscious or having convulsions,* do nothing except keep victim warm. *\*In some cases you may be specifically instructed by poison control to induce vomiting by way of 2 tablespoons of syrup of ipecac (adult) washed down with a cup of water.* Do NOT give activated charcoal before or with ipecac syrup. *Inhalation:* Get medical aid. Do not contaminate yourself. Wearing the appropriate PPE equipment and respirator for organophosphate pesticides, immediately remove the victim from the contaminated area to fresh air. If the victim is not breathing, administer artificial respiration. Do NOT use mouth-to-mouth resuscitation; use bag/mask apparatus. If breathing is difficult, administer oxygen through bag/mask apparatus until medical help arrives. Do not leave victim unattended. *Note to physician or authorized medical personnel.* Administer atropine, 2 mg (1/30 gr) intramuscularly or intravenously as soon as any local or systemic signs or symptoms of an intoxication are noted; repeat the administration of atropine every 3 to 8 minutes until signs of atropinization (mydriasis, dry mouth, rapid pulse, hot and dry skin) occur; initiate treatment in children with 0.05 mg mg/kg of atropine; repeat at 5 to 10 minute intervals. Watch respiration, and remove bronchial secretions if they appear to be obstructing the airway; intubate if necessary. *Notes to physician or authorized medical personnel*: N-methylpyridinium-2-aldoxime (2-PAMCI) when used in conjunction with atropine reacts with the phosphorylated cholinesterase, thereby restoring normal activity to by removing the phosphorylating group. The combination of these two chemicals is synergistic and must be administered within minutes to a few hours following exposure (depending on the specific agent) to be effective. Give 2-PAMCI (Pralidoxime; Protopam), 2.5 gm in 100 ml of sterile water or in 5% dextrose and water, intravenously, slowly, in 15-30 minutes; if sufficient fluid is not available, give 1 gm of 2-PAMCI in 3 ml of distilled water by deep intramuscular injection; repeat this every half hour if respiration weakens or if muscle fasciculation or convulsions recur. Also Diazepam, an anticonvulsant, might be considered.

*References:*
- U.S. Environmental Protection Agency, Office of Pesticide Programs, Pesticide Residue Limits, "Pirimiphos-methyl," 40 CFR 180.409, www.epa.gov/pesticides/food/viewtols.htm
- International Programme on Chemical Safety (IPCS), "Data Sheets on Pesticides, Pirimiphos-methyl," Geneva, Switzerland (January 1983). http://www.inchem.org/documents/pds/pds/pest49_e.htm
- Pesticide Management Education Program, "Pirimiphos-methyl (Actellic) Chemical Fact Sheet 6/85," Cornell University, Ithaca, NY (June 1985), http://pmep.cce.cornell.edu/profiles/insect-mite/mevinphos-propargite/pirimiphos-methyl/insect-prof-actellic.html
- U.S. Environmental Protection Agency, "Interim Reregistration Eligibility Decision (IRED) Facts, Pirimiphos-Methyl," Office of Prevention, Pesticides and Toxic Substances, Washington, DC (January 2003). http://www.epa.gov/REDs/factsheets/pirimiphosmethyl_ired_fs.htm
- California Environmental Protection Agency "Chemical List of Lists," Sacramento CA (February 1997)

# Polyoxin D, Zinc salt

*Use Type:* Fungicide
*CAS Number:* 146659-78-1
*Synonyms:* β-D-Allofuranuronic acid, 5-[(2-amino-5-*O*-(aminocarbonyl)-2-deoxy-L-xylonoyl)amino]-1-(5-carboxy-3,4-dihydro-2,4-dioxo-1(2*H*)-pyrimidinyl)-1,5-dideoxy-, zinc salt (1:1)
*Trade Names:* ENDORSE® WP Turf fungicide, Arvesta (USA)
*Producers:* Arvesta (USA)
*Chemical Class:* Inorganic zinc
*EPA/OPP PC Code:* 230000
*California DPR Chemical Code:* 5788
*Uses:* Used on turf on golf courses, parks, residential, industrial and commercial sites. Not intended for use on sod or seed production sites. Lack of toxicity and harmless.
*Carcinogen/Hazard Classifications*
**U.S. TRI:** Developmental and reproductive toxin
**Label Signal Word:** CAUTION
*Regulatory Authority:*
- Clean water act: Section 307 Toxic Pollutants as zinc and compounds
- AB 2588-Air Toxics "Hot Spots" Chemicals (CAL) as zinc compounds
- The "Director's List" (CAL/OSHA) as zinc compounds
- EPCRA Section 313: Form R *de minimis* concentration reporting level: 1.0% as zinc compounds

*Incompatibilities:* When heated to decomposition, this material forms toxic fumes of zinc oxide.
*Harmful Effects and Symptoms*
*Short Term Exposure:* Eye irritation.
*Medical Surveillance:* There are tests available to measure zinc in your blood, urine, hair, saliva, and feces. High levels of zinc in the feces can mean high recent zinc exposure. High levels of zinc in the blood can mean high zinc consumption and/or high exposure. Tests to measure zinc in hair may provide information on long-term zinc exposure; however, the relationship between levels in your hair and the amount of zinc you were exposed to is not clear.
*First Aid:* If this chemical gets into the eyes, remove any contact lenses at once and irrigate immediately for at least 15 minutes, occasionally lifting upper and lower lids. Seek

medical attention immediately. If this chemical contacts the skin, remove contaminated clothing and wash immediately with soap and water. Seek medical attention immediately. If this chemical has been inhaled, remove from exposure, begin rescue breathing (using universal precautions) if breathing has stopped, and CPR if heart action has stopped. Transfer promptly to a medical facility. When this chemical has been swallowed, get medical attention. Give large quantities of water and induce vomiting. Do not make an unconscious person vomit.

*References:*
- U.S. Environmental Protection Agency, Office of Pesticide Programs, "Polyoxin D Zinc Salt Fact Sheet," (July 2004). http://www.epa.gov/oppbppd1/biopesticides/ingredients/factsheets/factsheet_230000.htm
- California Environmental Protection Agency "Chemical List of Lists," Sacramento CA (February 1997)

# Potassium Arsenate

*Use Type:* Arsenates are used in agriculture as insecticides, herbicides, larvicides, and rodenticides
*CAS Number:* 7784-41-0
*Formula:* $AsH_2KO_4$; $KH_2AsO_4$
*Alert:* Arsenic compounds are generally considered carcinogens and should be handled with extreme caution.
*Synonyms:* Arsenic acid, monopotassium salt; Arseniato potasico (Spanish); Macquer's salt; Monopotassium arsenate; Monopotassium dihydrogen arsenate; Potassium acid arsenate; Potassium arsenate, monobasic; Potassium dihydrogen arsenate; Potassium hydrogen arsenate
*Trade Names:*
*Producers:* Air Products & Chemicals (USA); Aldrich Chemical (USA); ATOFINA N.A. (USA); ASARCO (USA); Cia. Universal de Industrias (Mexico); Degussa (Germany); Great Western Inorganics (USA); GFS Chemicals (USA); Mining & Chemical Products Ltd. (MCP) (UK); Mitsubishi Materials (Japan); Newmont Koch (UK); PPM Pure Metals (Germany); Union Miniere (Belgium)
*Chemical Class:* Inorganic arsenic compound
*ICSC Number:* 1210
*RTECS Number:* CG1100000
*EEC Number:* 033-005-00-1
*EINECS Number:* 232-065-8
*Uses:* Potassium arsenate is used in the textile, tanning, and paper industries and as an insecticide, especially for fly paper, and other pesticide applications.
*Carcinogen/Hazard Classifications*
**U.S. EPA Carcinogens:** Group A, known carcinogen, as an arsenic compound
**U.S. NTP Carcinogen:** Known carcinogen
**California Prop. 65:** Carcinogen

**IARC:** Group 1, known carcinogen (human positive)
*Regulatory Authority:*
- Air Pollutant Standard Set (ACGIH)[1] (HSE)[33] (OSHA)[63]
- Clean Water Act: Section 311 Hazardous Substances/RQ 40CFR117.3 (same as CERCLA, see below); Section 313 Water Priority Chemicals (57FR41331, 9/9/92)
- Superfund/EPCRA 40CFR302.4 RQ: CERCLA, 1 lb (0.454 kg)
- Canada, WHMIS, Ingredients Disclosure List
- U.S. DOT Regulated Marine Pollutant (49CFR172.101, Appendix B)

*As arsenic compounds:*
- Clean Air Act: Hazardous Air Pollutants (Title I, Part A, Section 112) as arsenic compounds
- Clean Water Act: Toxic Pollutant (Section 401.15) as arsenic and compounds
- RCRA, 40CFR261, Appendix 8 Hazardous Constituents, waste number not listed
- Superfund/EPCRA 40CFR302.4 RQ: CERCLA, 1 lb (0.454 kg)
- EPCRA (Section 313): Includes any unique chemical substance that contains arsenic as part of that chemical's infrastructure. Form R *de minimus* concentration reporting level: inorganics 0.1%;organics 1.0%
- U.S. DOT Regulated Marine Pollutant (49CFR172.101, Appendix B) as arsenates, liquid, n.o.s; arsenates, solid, n.o.s; arsenical pesticides liquid, toxic, flammable, n.o.s.
- Carcinogen User Register Chemical (CAL/OSHA) as inorganic arsenic compound
- AB 1803-Well Monitoring Chemical (CAL) as inorganic arsenic compound
- CAL Air Resources Board/AB 1807 Toxic Air Contaminants as inorganic arsenic compound
- Proposition 65 chemical (CAL) as inorganic arsenic compound
- The "Director's List" (CAL/OSHA) as inorganic arsenic compound
- AB 2588-Air Toxics "Hot Spots" Chemicals (CAL) as inorganic arsenic compound
- Permissible Exposure Limits for Chemical Contaminants (CAL/OSHA) as inorganic arsenic compound
- Canada: Priority Substance List & Restricted Substances/Ocean Dumping Forbidden (CEPA), National Pollutant Release Inventory (NPRI) (arsenic compounds)

*Description:* Potassium arsenate is a colorless to white crystalline solid. Slightly soluble in water; solubility = 20 g/100 mL @ 20°C. Boiling point = (decomposes). Melting/Freezing point = 288°C.

*Incompatibilities:* A weak base. Reacts with strong oxidizers, bromine azide, acids, and decomposes on contact with strong acids producing acetic acid fumes. Arsine, a very deadly gas, can be released in the presence of acid, acid mists, or hydrogen gas.

*Permissible Exposure Limits in Air:* OSHA[2] has set a PEL of 0.010 mg/m³ for inorganic arsenic. NIOSH[2] recommends a limit of 0.002 mg/m³ not to be exceeded at any time. ACGIH[1] recommends a TWA of 0.01 mg/m³. The NIOSH[2] IDLH level = [Ca] 5 mg/m³ as As. (See also Arsenic and Inorganic Arsenic Compounds)

*Determination in Air:* Filter; Acid; Hydride generation atomic absorption spectrometry; NIOSH IV, Method #7900. See also #7300, Elements[18].

*Permissible Concentration in Water:* EPA has set a limit of 0.05 parts per million (ppm) for arsenic in drinking water. The U.S. EPA arsenic drinking water standard of 0.01 ppm (10 ppb) is based on the U.S. EPA final rule for arsenic in drinking water published in. the January 22, 2001, *Federal Register*. However, the U.S. EPA is currently reviewing the science and cost estimate supporting this rule, and, in the interim, has reverted to the previous standard for arsenic. Thus, in the US, the current EPA arsenic drinking water standard remains at 0.05 ppm (50 ppb). To protect freshwater aquatic life-total recoverable trivalent inorganic arsenic never to exceed 440 µg/L. To protect saltwater aquatic life: 508 µg/L on an acute basis. To protect human health: preferably zero. The former USSR-UNEP/IRPTC project[43] has set MAC values for inorganic arsenic compounds in water for domestic purposes at 0.05 mg/L and in water bodies for fishery purposes of 0.5 mg/L also.

*Determination in Water:* See OSHA Method ID-105 for arsenic.[58]

*Routes of Entry:* Inhalation, ingestion, skin and/or eye contact. Absorbed through the skin.

*Harmful Effects and Symptoms* (See also Arsenic and Inorganic Arsenic Compounds)

*Short Term Exposure:* Potassium arsenate can affect you when breathed, and may enter through skin. Potassium arsenate is a carcinogen handle with extreme caution. Eye contact causes irritation, burns and red, watery eyes. Skin contact can cause burning, itching, and rash. Breathing can irritation with sneezing and coughing. High or repeated exposures can cause nerve damage, with numbness and weakness of arms and legs, and can cause poor appetite, nausea, cramps and if severe, vomiting and diarrhea. $LD_{50}$ (rat, oral) = 14.0 mg/kg.

*Long Term Exposure:* Long term exposure can cause ulcer or hole in the nasal septum; hoarseness and sore eyes also occur. Repeated exposure can cause nervous system, bone marrow, kidney, and liver damage. Repeated skin contact can cause thickened skin and/or patchy area of darkening and loss of pigment. Animal tests suggest this substance may be a reproductive toxin.

*Points of Attack:* Central nervous system and skin. Several studies have shown that inorganic arsenic can increase the risk of lung cancer, skin cancer, bladder cancer, liver cancer, kidney cancer, and prostate cancer.

*Medical Surveillance:* Before beginning employment and at regular times after that, the following are recommended: Exam of the nose, skin, eyes, nails, nervous system. Test for urine arsenic (may not be accurate within 2 days of eating shellfish or fish; most accurate at the end of a work-day). At NIOSH recommended exposure levels, urine arsenic should not be greater than 50 to 100 micrograms per liter of urine. After suspected overexposure, repeat these tests. Also examine your skin periodically for abnormal growths. Skin cancer from arsenic is easily cured when detected early. Kidney and liver function tests. Complete blood count.

*First Aid:* If this chemical gets into the eyes, remove any contact lenses at once and irrigate immediately for at least 15 minutes, occasionally lifting upper and lower lids. Seek medical attention immediately. If this chemical contacts the skin, remove contaminated clothing and wash immediately with soap and water. Seek medical attention immediately. If this chemical has been inhaled, remove from exposure, begin rescue breathing (using universal precautions) if breathing has stopped, and CPR if heart action has stopped. Transfer promptly to a medical facility. When this chemical has been swallowed, get medical attention. Give large quantities of water and induce vomiting. Do not make an unconscious person vomit.

*Note to physician or authorized medical personnel:* For severe poisoning BAL has been used. For milder poisoning penicillamine (not penicillin) has been used, both with mixed success. Side effects occur with such treatment and it is never a substitute for controlling exposures. It can only be done under strict medical care.

*References:*
- Sax, N.I., Ed., "Dangerous Properties of Industrial Materials Report," 3, No. 4, 101-103 (1983).
- California Environmental Protection Agency "Chemical List of Lists," Sacramento CA (February 1997).
- New Jersey Department of Health, "Hazardous Substance Fact Sheet: Potassium Arsenate," Trenton, NJ (March 1998). http://www.state.nj.us/health/eoh/rtkweb/1556.pdf

# Potassium Cyanide

*Use Type:* Rodenticide
*CAS Number:* 151-50-8
*Formula:* CN·K; KCN
*Synonyms:* Cianuro potasico (Spanish); Cyanide of potassium; Cyanure de potassium (French); Hydrocyanic acid, potassium salt; Kalium-cyanid (German)
*Producers:* Alquimia Mexicana (Mexico); Ashland Chemical (USA); Bayer Corp. (Germany); Cia. Quimica Universal de Industrias (Mexico); CP/PhibroChem (USA); Cyanides and Chemicals Company (India); Degussa (Germany); DuPont (USA); EniChem (Italy); ICI Group (UK); Harcros Chemicals (USA); Mitsui Chemical (Japan); Nippon Soda (Japan); Orica (Australia); Quantum Chemicals (Australia); Whyte Chemicals (UK)
*Chemical Class:* Inorganic

*EPA/OPP PC Code:* 599600
*California DPR Chemical Code:* 497
*ICSC Number:* 0671
*RTECS Number:* TS8750000
*EEC Number:* 006-007-00-5
*EINECS Number:* 205-793-3

*Uses:* Used to control rodents, possums, and other varmints. Used in electroplating, steel hardening, extraction of precious metals form ores; a reagent in analytical chemistry.

*Carcinogen/Hazard Classifications*
**Label Signal Word: DANGER**
**WHO Acute Hazard:** *As sodium cyanide:* Class 1b, highly hazardous

*Regulatory Authority:*
- Clean Water Act: Section 311 Hazardous Substances/RQ 40CFR117.3 (same as CERCLA, see below); Section 313 Water Priority Chemicals (57FR41331, 9/9/92)
- EPA Hazardous Waste Number (RCRA No.): P098
- RCRA, 40CFR261, Appendix 8 Hazardous Constituents
- Superfund/EPCRA 40CFR355, Appendix B Extremely Hazardous Substances: TPQ = 100 lb (45.4 kg)
- Superfund/EPCRA 40CFR302.4 RQ: CERCLA, 10 lb (4.54 kg)
- The "Director's List" (CAL/OSHA)
- U.S. DOT Regulated Marine Pollutant (49CFR172.101, Appendix B)
- Canada, WHMIS, Ingredients Disclosure List; National Pollutant Release Inventory (NPRI); CEPA Priority Substance List, Ocean dumping prohibited

*As cyanide compounds:*
- Clean Air Act: Hazardous Air Pollutants (Title I, Part A, Section 112)
- Clean Water Act: 40CFR423, Appendix A, Priority Pollutants, as cyanide, total
- EPA Hazardous Waste Number (RCRA No.): P030 as cyanides soluble salts and complexes, n.o.s.
- RCRA, 40CFR261, Appendix 8 Hazardous Constituents as cyanides, soluble salts and complexes, n.o.s.
- AB 2588-Air Toxics "Hot Spots" Chemicals (CAL)
- CAL Air Resources Board/AB 1807 Toxic Air Contaminants
- EPA Hazardous Waste Number (RCRA No.): P030 as cyanide compounds
- RCRA Section 261 Hazardous Constituents as cyanide compounds
- RCRA Universal Treatment Standards: Wastewater (mg/L), 1.2 (total); 0.86 (amenable); Nonwastewater (mg/kg), 590 (total); 30 (amenable) as cyanide compounds
- RCRA Ground Water Monitoring List. Suggested test method(s) (PQL $ug$/L): 9010(40) as cyanide compounds
- Safe Drinking Water Act: MCL, 0.01 mg/L; MCLG, 0.01 mg/L;Regulated chemical (47 FR 9352) as cyanide compounds
- EPCRA Section 304 RQ: CERCLA, 10 lb (4.54 kg) as cyanide compounds
- EPCRA Section 313 Form R *de minimus* concentration reporting level: 1.0% as cyanide compounds
- U.S. DOT Regulated Marine Pollutant (49CFR, Subchapter 172.101, Appendix B) as cyanides, mixtures or solutions
- Canada, WHMIS, Ingredients Disclosure List (cyanide compounds, inorganic, n.o.s)

*Description:* Potassium cyanide is a white lump, granular powder or colorless solution. Faint almond odor. Highly soluble in water; solubility = 80 g/100 mL @ 25°C. Boiling point = 1625°C. Melting/Freezing point = 634°C. Explosive limits: LEL = %; UEL = %. Hazard Identification (based on NFPA-704 M Rating System): Health 3, Flammability 0, Reactivity 0.

*Incompatibilities:* Incompatible with strong acids, acid salts, organic anhydrides, isocyanates, alkylene oxides, epichlorohydrin, aldehydes, alcohols, glycols, phenols, cresols, caprolactum, strong oxidizers, nitrogen trichloride, sodium chlorate. Violent reaction with strong acids and acid salts (such as hydrochloric, sulfuric, and nitric acids) producing flammable hydrogen cyanide. Reacts with water releasing hydrogen cyanide. Attacks aluminum, copper, zinc in the presence of moisture.

*Permissible Exposure Limits in Air:* The OSHA[2] PEL 8-hour TWA is 5 mg/m$^3$ (4.7 ppm). NIOSH[2] recommended ceiling (10 minute) is also 5 mg/m$^3$ (4.7 ppm). ACGIH[1], DFG[3], HSE[33] have recommended 5 mg/m$^3$ as a TWA. The NIOSH[2] IDLH level = 25 mg/m$^3$. Regulatory authorities add the notation "skin" indicating the possibility off cutaneous absorption. The former USSR-UNEP/IRPTC project[43] has set an MAC in workplace air of 0.3 mg/m$^3$. Further, they have set MAC values for ambient air in residential areas of 0.009 mg/m$^3$ on a momentary basis and 0.004 mg/m$^3$ on an average daily basis. Several states have set guidelines or standards for cyanides in ambient air[60] ranging from 16.7 $\mu$g/m$^3$ (New York) to 50.0 $\mu$g/m$^3$ (Florida and North Dakota) to 80.0 $\mu$g/m$^3$ (Virginia) to 100 $\mu$g/m$^3$ (Connecticut and South Dakota) to 125 $\mu$g/m$^3$ (South Carolina) to 119.0 $\mu$g/m$^3$ (Nevada).

*Determination in Air:* Filter/Bubbler; Potassium hydroxide; Ion-specific electrode; NIOSH IV Method #7904, Cyanides. See also Method #6010, Hydrogen Cyanide.[18]

*Permissible Concentration in Water:* The U.S. EPA has set a maximum contaminant level of cyanide in drinking water of 0.2 milligrams cyanide per liter of water (0.2 mg/L). As of 1980, the criteria are: To protect freshwater aquatic life: 3.5 $\mu$g/L as a 24-hour average, never to exceed 52.0 $\mu$g/L. To protect saltwater aquatic life: 30.0 $\mu$g/L on an acute toxicity basis; 2.0 $\mu$g/L on a chronic toxicity basis. Internationally, the South African Bureau of Standards has set 10 $\mu$g/L, the World Health Organization (WHO) 10 $\mu$g/L and the Federal Republic of Germany 50 $\mu$g/L as drinking water standards. Other international limits[35] include an

EEC limit of 50 µg/L; Mexican limits of 200 µg/L in drinking water and 1.0 µg/L in coastal waters and a Swedish limit of 100 µg/L. The former USSR-UNEP/IRPTC project[43] has set an MAC of 100 µg/L in water bodies used for domestic purposes and 50 µg/L in water for fishery purposes. The U.S. EPA[49] has determined a no-observed-adverse-effect-level (NOAEL) of 10.8 mg/kg/day which yields a lifetime health advisory of 154 µg/L. States which have set guidelines for cyanides in drinking water[61] include Arizona at 160 µg/L and Kansas at 220 µg/L.

*Determination in Water:* Distillation followed by silver nitrate titration or colorimetric analysis using pyridine pyrazolone (or barbituric acid).

*Routes of Entry:* Inhalation, ingestion, skin and/or eye contact. Absorbed through the skin.

*Harmful Effects and Symptoms*

*Short Term Exposure:* Potassium cyanide is corrosive to the eyes, the skin and the respiratory tract. Contact can cause skin and eye burns, and possible permanent eye damage. Inhalation can cause lung irritation with coughing, sneezing, and difficult breathing, slow gasping respiration. Corrosive if swallowed. These substances may affect the central nervous system. Symptoms include headaches, confusion, nausea, pounding heart, weakness, unconsciousness and death.

*Long Term Exposure:* Repeated or prolonged contact with potassium cyanide may cause thyroid gland enlargement and interfere with thyroid function. May cause nosebleeds and sores in the nose; changes in blood cell count. May cause central nervous system damage with headache, dizziness, confusion, nausea, vomiting, pounding heart, weakness in the arms and legs, unconsciousness and death. may affect liver and kidney function. *Developmental Effects:* No reproductive or developmental effects of this cyanate have been reported in experimental animals or humans. Increased levels of thiocyanate in the umbilical cords of fetuses whose mothers smoked compared to those whose mothers were non-smokers suggests that thiocyanate, and possibly also cyanide, can cross the placenta.

*Points of Attack:* Liver, kidneys, skin, cardiovascular system, central nervous system and the thyroid gland.

*Medical Surveillance:* Consider the points of attack in preplacement and periodic physical examinations. Urine thiocyanate levels. Blood cyanide levels. Complete blood count. Evaluation of thyroid function. Liver function tests. Kidney function tests. Central nervous system tests. EKG.

*First Aid:* Speed in removing material from eyes and skin is of extreme importance. If this chemical gets into the eyes, remove any contact lenses at once and irrigate immediately for at least 15 minutes, occasionally lifting upper and lower lids. Seek medical attention immediately. If this chemical contacts the skin, remove contaminated clothing and wash immediately with soap and water. Shampoo hair promptly if contaminated. Seek medical attention immediately. If this chemical has been inhaled, remove from exposure, begin rescue breathing (using universal precautions) if breathing has stopped, and CPR if heart action has stopped. Transfer promptly to a medical facility. When this chemical has been swallowed, get medical attention. Give large quantities of water and induce vomiting. Do not make an unconscious person vomit. Keep under observation for 24 to 48 hours as symptoms may return. Use amyl nitrate capsules if symptoms develop. All area employees should be trained regularly in emergency measures for cyanide poisoning and in CPR. A cyanide antidote kit should be kept in the immediate work area and must be rapidly available. Kit ingredients should be replaced every 1-2 years to ensure freshness. Persons trained in the use of this kit, oxygen use, and CPR must be quickly available.

*References:*
- New Jersey Department of Health and Senior Services, "Hazardous Substance Fact Sheet, Potassium cyanide," Trenton NJ (May 1998). http://www.state.nj.us/health/eoh/rtkweb/1562.pdf
- California Environmental Protection Agency "Chemical List of Lists," Sacramento CA (February 1997).

## Potassium Fluoride

*Use Type:* Insecticide and intermediate
*CAS Number:* 7789-23-3
*Formula:* FK; KF
*Synonyms:* Fluorure de potassium (French); Potassium fluorure (French)
*Producers:* Air Products & Chemicals (USA); Bayer Chemicals (Germany); ATOFINA Chemicals (France); Clariant (Switzerland); Derivados del Fluor (Spain); Dharamsi Morarji (India); Harcros Chemicals (USA); Morita Chemical Industries (Japan); Navin Fluorine Industries (India); Rhodia (France); Scottish Chemical Industries (India); Solvay Group (Belgium); Stella Chemifa (Japan)
*Chemical Class:* Inorganic
*RTECS Number:* TT0700000
*Uses:* Potassium fluoride is used as an insecticide and also in etching glass and in fluorination, flux for solder, and as a preservative.
*Regulatory Authority:*
- Air Pollutant Standard Set (ACGIH)[1] (DFG)[3] (HSE)[33] (former USSR)[43] (Several States)[60]
- Canada, WHMIS, Ingredients Disclosure List

*Description:* Potassium Fluoride is a white crystalline solid. Soluble in water; solubility will increase with temperature. Boiling point = 1505°C. Melting/Freezing point = 860°C. Hazard Identification (based on NFPA-704 M Rating System): Health 3, Flammability 0, Reactivity 0.
*Incompatibilities:* Strong acids.
*Permissible Exposure Limits in Air:* The applicable regulations are those for the fluoride ion. The OSHA[2] PEL,

and NIOSH[2] REL, and ACGIH[1] TLV is 2.5 mg/m³ TWA as fluoride. HSE[33] and DFG[3] set the same values. NIOSH IDLH = 250 mg F/m³ as Fluorides (F). The former USSR-UNEP/IRPTC project[43] has set MAC values for soluble fluorides in ambient air in residential areas of 0.03 mg/m³ on a momentary basis and 0.01 mg/m³ on an average daily basis. A number of states have set limits for fluoride in ambient air[60] ranging from as low as 2.85 µg/m3 (Iowa) to as high as 60,000 µg/m3 (Kentucky).

*Permissible Concentration in Water:* As with air, the applicable regulations are those for the fluoride ion. The values which have been set for drinking water[61] are a standard of 4.0 mg/L set by EPA and a guideline of 2.4 mg/L set by the State of Maine.

*Routes of Entry:* Inhalation, ingestion, skin and/or eye contact.

*Harmful Effects and Symptoms*

*Short Term Exposure:* Potassium fluoride can affect you when breathed in. Inhalation of dust or mist can cause severe irritation and burns of the eyes and skin. May cause permanent eye damage. Inhalation can cause irritation of the nose and throat causing sneezing, coughing and sore throat. High exposure can irritate the lungs causing a build-up of fluid in the lungs. This can cause death.

*Long Term Exposure:* These effects do not occur at the levels of fluorides used in water to prevent cavities. Repeated exposure can cause fluoride to build-up in the body. Can irritate the lungs; bronchitis may develop. Repeated exposure can cause fluoride to buildup in the body causing stiffness, brittle bones, and crippling. Prolonged contact can cause sores in the nose and perforated septum.

*Points of Attack:* Lungs and skin.

*Medical Surveillance:* For those with frequent or potentially high exposure (half the TLV or greater), the following are recommended before beginning work and at regular times after that: Lung function tests. Fluoride levels in urine (levels higher than 4 mg/L may indicate overexposure). If symptoms develop or overexposure is suspected, the following may be useful: Kidney function test. Consider chest x-ray after acute overexposure.

*First Aid:* If this chemical gets into the eyes, remove any contact lenses at once and irrigate immediately for at least 15 minutes, occasionally lifting upper and lower lids. Seek medical attention immediately. If this chemical contacts the skin, remove contaminated clothing and wash immediately with soap and water. Seek medical attention immediately. If this chemical has been inhaled, remove from exposure, begin rescue breathing (using universal precautions) if breathing has stopped, and CPR if heart action has stopped. Transfer promptly to a medical facility. If victim is *conscious*, administer water, or milk. Do not induce vomiting. Medical observation is recommended for 24 to 48 hours after breathing overexposure, as pulmonary edema may be delayed. As first aid for pulmonary edema, a doctor or authorized paramedic may consider administering a corticosteroid spray.

*References:*
- New Jersey Department of Health, "Hazardous Substance Fact Sheet, Potassium Fluoride," Trenton, NJ (March 1998). http://www.state.nj.us/health/eoh/rtkweb/1565.pdf
- California Environmental Protection Agency "Chemical List of Lists," Sacramento CA (February 1997).

# Potassium Nitrate

*Use Type:* Rodenticide, microbiocide, fertilizer
*CAS Number:* 7757-79-1
*Formula:* $KNO_3$
*Synonyms:* Kaliumnitrat (German); Niter; Nitrate de potassium (France); Nitrato potasico (Spanish); Nitre; Nitric acid, Potassium salt; Saltpeter; Saltpetre; Vicknite
*Trade Names:* HAWK T-A®, Pic Corp. (USA), canceled; REVENGE®, Roxide International (USA); SMOKE'EM®, Nott Products (USA)
*Producers:* Allipo Chemicals (India); Alquimia Mexicana (Mexico); a.m.p.e.r.e. Industrie SA Dept Chimie (France); Caldic Chemie (Netherlands); Celtic Chemicals (UK); GFS Chemicals (USA); Grande Paroisse (France); Haldor Topsoe (Denmark); Holvoet Chimie (France); Honeywell Performance P & C (USA); Kemira GrowHow (Finland); Rhodia (France); Rhone-Poulenc (France); Showa Denko Chemicals Group (Japan); S.K. Chemical Industries (India); SQM (Chile); UBE Industries Agri-Materials (Japan)
*Chemical Class:* Inorganic
*EPA/OPP PC Code:* 076103
*California DPR Chemical Code:* 726
*ICSC Number:* 0184
*EINECS Number:* 231-818-8
*Uses:* It is used as a fertilizer on high value crops. Used to make explosives, gun powder, matches, fireworks, glass and rocket fuel, and also used in rodenticides and fertilizers. Used in medicine as a diuretic.
*Carcinogen/Hazard Classifications*
**Label Signal Word:** DANGER
*Regulatory Authority:*
- Actively registered pesticide in California.

*Description:* White or colorless crystalline powder. Odorless. Salty taste. Highly soluble in water; solubility = 40 g/100mL @ 25°C. Molecular weight = 101.11. Boiling point = 400°C (decomposition). Melting/Freezing point = 334°C.

*Incompatibilities:* A powerful oxidizer. Dangerously reactive and friction- and shock-sensitive when mixed with organic materials and many materials. Violent reactions with reducing agents, chemically active metals, charcoal and trichloroethylene.

*Permissible Exposure Limits in Air:* None established.

*Routes of Entry:* Inhalation, ingestion, skin and/or eye contact.

*Harmful Effects and Symptoms*

*Short Term Exposure:* Contact can cause eye and skin irritation. Inhalation can cause respiratory tract irritation, coughing and wheezing. High levels of exposure can interfere with the blood's ability to carry oxygen, causing headache, dizziness, cyanosis, methemoglobineamia, with blue color to the skin and lips. Higher levels can cause breathing difficulty, collapse and death.

*Long Term Exposure:* There is limited evidence that this chemical can damage the developing fetus.

*Points of Attack:* Blood

*Medical Surveillance:* Blood test for methemoglobin.

*First Aid:* If this chemical gets into the eyes, remove any contact lenses at once and irrigate immediately for at least 15 minutes, occasionally lifting upper and lower lids. Seek medical attention immediately. If this chemical contacts the skin, remove contaminated clothing and wash immediately with soap and water. Seek medical attention immediately. If this chemical has been inhaled, remove from exposure, begin rescue breathing (using universal precautions) if breathing has stopped, and CPR if heart action has stopped. Transfer promptly to a medical facility. When this chemical has been swallowed, get medical attention. Give large quantities of water and induce vomiting. Do not make an unconscious person vomit.

*References:*
- New Jersey Department of Health and Senior Services, "Hazardous Substance Fact Sheet, Potassium Nitrate," Trenton NJ (March 1998).
http://www.state.nj.us/health/eoh/rtkweb/1574.pdf

# Potassium Nitrite

*Use Type:* Rodenticide and microbiocide
*CAS Number:* 7758-09-0
*Formula:* $KNO_2$
*Synonyms:* Kaliumnitrat (German); Niter; Nitre; Nitrous Acid, Potassium Salt; Saltpeter
*Producers:* Allipo Chemicals (India); Alquimia Mexicana (Mexico); Celtic Chemicals (UK); General Chemical (USA); GFS Chemicals (USA); Grande Paroisse (France); Haldor Topsoe (Denmark); Holvoet Chimie (Belgium); Honeywell Performance P & C (USA); Humco (USA); Kemira Chemicals (Finland); Monterrey Chemical Products (Mexico); Rhodia (France); Rhone-Poulenc (France); Showa Denko Chemicals Group (Japan); SQM (Chile); UBE Industries (Japan)
*Chemical Class:* Inorganic
*EPA/OPP PC Code:* 076203
*California DPR Chemical Code:* 3356
*ICSC Number:* 1069
*RTECS Number:* TT3750000
*EEC Number:* 007-011-00-X
*EINECS Number:* 231-832-4

*Uses:* Potassium nitrite is used in chemical analysis, heat transfer salts, as a food additive, in medications, fertilizers and rodenticides.

*Description:* Potassium nitrite is a white to yellowish crystalline solid. Highly soluble in water; 275 g/100mL @ 25°C. Molecular weight = 85.12. Boiling point = (decomposition) 350°C. Melting/Freezing point = 441°C.

*Incompatibilities:* A strong oxidizer. Reacts violently with combustible and reducing materials. Heat above 530°C may cause explosion. Incompatible with cyanide salts, boron, ammonium sulfate, potassium amide, and acids. Decomposes on contact with even weak acids producing toxic nitrogen oxide fumes.

*Permissible Exposure Limits in Air:* No standards set.

*Permissible Concentration in Water:* No criteria set, but runoff from spills or fire control may cause water pollution.

*Routes of Entry:* Inhalation, ingestion, skin and/or eye contact.

*Harmful Effects and Symptoms*

*Short Term Exposure:* Potassium nitrite can affect you when breathed in. Contact can cause eye and skin burns. Breathing the dust or mist can irritate the nose, throat and lungs, and may cause cough with phlegm. Higher exposures can cause pulmonary edema, a medical emergency that can be delayed for several hours. This can cause death. High levels can affect the vascular system and interfere with the ability of the blood to carry oxygen (methemoglobinemia), causing headaches, weakness, dizziness and cyanosis, a bluish color to the skin and lips. Higher levels can cause trouble breathing, collapse and even death.

*Long Term Exposure:* Repeated skin contact causes dermatitis; drying and cracking. May cause lung irritation; bronchitis may develop. There is limited evidence that potassium nitrite may damage the developing fetus.

*Points of Attack:* Eyes, skin, blood and lungs.

*Medical Surveillance:* If symptoms develop or overexposure is suspected, the following may be useful: Blood test for methemoglobin. Lung function tests. Consider chest x-ray after acute overexposure.

*First Aid:* If this chemical gets into the eyes, remove any contact lenses at once and irrigate immediately for at least 15 minutes, occasionally lifting upper and lower lids. Seek medical attention immediately. If this chemical contacts the skin, remove contaminated clothing and wash immediately with soap and water. Seek medical attention immediately. If this chemical has been inhaled, remove from exposure, begin rescue breathing (using universal precautions) if breathing has stopped, and CPR if heart action has stopped. Transfer promptly to a medical facility. When this chemical has been swallowed, get medical attention. Give large quantities of water and induce vomiting. Do not make an unconscious person vomit. Medical observation is recommended for 24 to 48 hours after breathing overexposure, as pulmonary edema may be delayed. As first aid for pulmonary edema, a doctor or authorized paramedic may consider administering a corticosteroid spray.

*Note to physician or authorized medical personnel:* Treat for methemoglobinemia. Spectrophotometry may be required for precise determination of levels of methemoglobinemia in urine.

*References:*
- New Jersey Department of Health, "Hazardous Substance Fact Sheet: Potassium Nitrite," Trenton, NJ (March 1998). http://www.state.nj.us/health/eoh/rtkweb/1575.pdf
- California Environmental Protection Agency "Chemical List of Lists," Sacramento CA (February 1997).

# Primisulfuron-methyl

*Use Type:* Herbicide
*CAS Number:* 86209-51-0
*Formula:* $C_{15}H_{12}F_4N_4O_7S$
*Synonyms:* Benzoic acid, 2-[(((((4,6-bis(difluoromethoxy)-2-pyrimidinyl)amino)carbonyl)amino)sulfonyl]-, methyl ester; 3-[4,6-Bis(difluoromethoxy)-pyrimidin-2-yl]-1-(methoxycarbonylphenylsulfonyl)urea; N-(2-Methoxycarbonylphenylsulfonyl)-N-[4,6-bis(difluoromethoxy)pyrimidin-2-yl]urea; Methyl-2-[(((4,6-bis(difluoromethoxy)-2-pyrimidinyl)amino)carbonyl)amino)sulfonyl]benzoate
*Trade Names:* BEACON®, Syngenta (Switzerland); CGA 136,872®, Syngenta (Switzerland); EXCEED®, Syngenta (Switzerland); NORTHSTAR®, Syngenta (Switzerland); RIFLE®, Syngenta (Switzerland), canceled; SPIRIT®, Syngenta (Switzerland); TELL®
*Producers:* Syngenta (Switzerland)
*Chemical Class:* Sulfonylurea
*EPA/OPP PC Code:* 128973
*California DPR Chemical Code:* 5103
*Uses:* A broad spectrum herbicide for use on corn. EPA announced in 2002 that tolerances for sweet corn will be revoked and it will establish tolerances for Kentucky bluegrass grown for seed for use as forage and hay.
*U.S. Maximum Allowable Residue Levels for Primisulfuron-methyl (40 CFR 180.452):*
Corn, grain and milk: 0.02 ppm: Cattle, Fat; Cattle, Meat; Cattle, mbyp; Corn, Forage; Corn, Stover; Corn, Sweet, Kernel plus Cob with Husks Removed; Egg; Goat, Fat; Goat, Meat; Goat, mbyp; Hog, Fat; Hog, Meat; Hog, mbyp; Horse, Fat; Horse, Meat; Horse, mbyp; Poultry, Fat; Poultry, Meat; Poultry, mbyp; Sheep, Fat; Sheep, Meat; Sheep, mbyp: 0.1 ppm.
*Carcinogen/Hazard Classifications*
**U.S. EPA Carcinogens:** Group D, unclassifiable
**Label Signal Word:** CAUTION
**WHO Acute Hazard:** Class U, unlikely to be hazardous

*Description:* Colorless crystalline solid. Very slightly soluble in water. Melting point = 203.1°C. Molecular weight = 468.35.
*Permissible Concentration in Water:* No criteria set. Runoff from spills or fire control may cause water pollution.
*Harmful Effects and Symptoms*
*Short Term Exposure:* Contact with eyes or skin may cause irritation or burns. Inhalation should be avoided; use NIOSH-approved air purifying respirators for pesticides. May be harmful if swallowed. Skin contact may cause allergic reaction.
*Long Term Exposure:* Repeated or prolonged contact may cause skin and lung sensitization, resulting in allergies.
*Points of Attack:* Skin. Adverse effects have been found in the testicular and thyroid glands.
*Medical Surveillance:* Evaluation by a qualified allergist, including careful exposure history and special testing, may help diagnose skin allergy.
*First Aid:* If this chemical gets into the eyes, remove any contact lenses at once and irrigate immediately for at least 15 minutes, occasionally lifting upper and lower lids. Seek medical attention immediately. If this chemical contacts the skin, remove contaminated clothing and wash immediately with soap and water. Seek medical attention immediately. If this chemical has been inhaled, remove from exposure, begin rescue breathing (using universal precautions) if breathing has stopped, and CPR if heart action has stopped. Transfer promptly to a medical facility. When this chemical has been swallowed, get medical attention. Give large quantities of water and induce vomiting. Do not make an unconscious person vomit.

*References:*
- U.S. Environmental Protection Agency, Office of Pesticide Programs, Pesticide Residue Limits, "Primisulfuron-methyl," 40 CFR 180. www.epa.gov/pesticides/food/viewtols.htm
- EXTOXNET, Extension Toxicology Network, "Pesticide Information Profile, Primisulfuron-methyl," Oregon State University, Corvallis, OR (June 1996). http://extoxnet.orst.edu/pips/primisul.htm
- U.S. Environmental Protection Agency, FQPA Tolerance Reassessment Progress and Interim Risk Management Decision (TRED) for Primisulfuron-methyl. (July 2002). http://www.epa.gov/REDs/factsheets/primisulfuron_tred_fs.htm
- California Environmental Protection Agency "Chemical List of Lists," Sacramento CA (February 1997)

# Procymidone

*Use Type:* Fungicide
*CAS Number:* 32809-16-8
*Formula:* $C_{13}H_{11}Cl_2NO_2$

*Synonyms:* 1,2-Cyclopropanedicarboximide, *N*-(3,5-dichlorophenyl)-1,2-dimethyl-; 3-(3,5-Dichlorophenyl)-1,5-dimethyl-3-azabicyclo(3.1.0)hexane-2,4-dione; *N*-(3,5-Dichlorophenyl)-1,2-dimethylcyclopropane-1,2-dicarboxamide; 3-(3,5-Dichlorophenyl)-1,5-dimethyl-3-azabicyclo(3.1.0)hexane-2,4-dione
*Trade Names:* CAMPILEX®; CYON®; FORTRESS®; SALITHIEX®; SIALEX®; SUMIBOTO®, Sumitomo Chemical Japan); SUMICROS®, Sumitomo Chemical Japan); SUMILEX®, Sumitomo Chemical Japan); SUMISCLEX®, Sumitomo Chemical Japan); S-7131®Sumitomo Chemical Japan)
*Producers:* Agsin (Singapore); Ki-Hara Chemicals Ltd. (UK); Sumitomo Chemical Japan); Wuzhou International (China)
*Chemical Class:* Dicarboximide
*EPA/OPP PC Code:* 129044
*California DPR Chemical Code:* 2241
*Uses:* Not registered in the U.S. Used on wine grapes.
*Carcinogen/Hazard Classifications*
**U.S. EPA Carcinogens:** Group B2, probable carcinogen
**California Prop. 65:** Carcinogen
**WHO Acute Hazard:** Class U, unlikely to be hazardous
**Endocrine Disruptor:** Suspected endocrine disruptor
*Regulatory Authority:*
- Proposition 65 chemical (CAL)

*Description:* Colorless, crystalline solid. Soluble in water; solubility = 4.5ppm @ 25°C. Molecular weight 284.12. Melting/Freezing point = 165°C. Vapor pressure = $1.79 \times 10^{-4}$ mmHg @ 25°C.
*Incompatibilities:* Slowly hydrolyzes in water, releasing ammonia and forming acetate salts.
*Permissible Exposure Limits in Air:*
*Determination in Air:* Filter: none. Gravimetric; NIOSH IV [Particulates NOR; #0500 (total), #0600 (respirable)].[18]
*Permissible Concentration in Water:* No criteria set. Runoff from spills or fire control may cause water pollution.
*First Aid:* If this chemical gets into the eyes, remove any contact lenses at once and irrigate immediately for at least 15 minutes, occasionally lifting upper and lower lids. Seek medical attention immediately. If this chemical contacts the skin, remove contaminated clothing and wash immediately with soap and water. Seek medical attention immediately. If this chemical has been inhaled, remove from exposure, begin rescue breathing (using universal precautions) if breathing has stopped, and CPR if heart action has stopped. Transfer promptly to a medical facility. When this chemical has been swallowed, get medical attention. Give large quantities of water and induce vomiting. Do not make an unconscious person vomit.
*References:*
- California Environmental Protection Agency "Chemical List of Lists," Sacramento CA (February 1997).

# Prodiamine (ANSI)

*Use Type:* Herbicide
*CAS Number:* 29091-21-2
*Formula:* $C_{13}H_{17}F_3N_4O_4$
*Synonyms:* 2,4-Dinitro-$N^3$,$N^3$-dipropyl-6-(trifluoromethyl)-1,3-benzenediamine; $N^3$,$N^3$-Dipropyl-2,4-dinitro-6-(trifluoromethyl)-*m*-phenylenediamine; 1,3-Benzenediamine, 2,6-dinitro-$N^1$,$N^1$-dipropyl-4-(trifluoromethyl)-
*Trade Names:* BLOCKADE®, Syngenta (Switzerland); CARPETMAKER X-X-X®; CN-11-2936®; ENDURANCE®; FACTOR®; MARATHON®; PAR EX®, Lebanon Seaboard (USA); PRO-MATE BARRICADE®, Helena Chemical (USA); REGALKADE®; RYDEX®; SIPCAM®, Sipcam Agro (USA); USB-3153®
*Producers:* Bayer CropScience (Germany); Helena Chemical (USA); Lebanon Seaboard (USA); Simplot, J.R. (USA); Sipcam Agro (USA); Syngenta (Switzerland)
*Chemical Class:* Dinitroaniline
*EPA/OPP PC Code:* 110201; (391200 old EPA code number)
*California DPR Chemical Code:* 2236
*Uses:* A pre-emergence herbicide used to control broadleaf weeds and annual grasses on a large variety of crops and ornamentals.
*Carcinogen/Hazard Classifications*
**U.S. EPA Carcinogens:** Group C, possible carcinogen
**Label Signal Word:** CAUTION
**WHO Acute Hazard:** Class U, unlikely to be hazardous
*Description:* Orange crystalline solid. Odorless. Molecular weight = 350.28. Vapor pressure = $2.4 \times 10^{-8}$ mmHg @ 25°C.
*Incompatibilities:* Strong oxidizers. When heated to decomposition this material forms toxic oxides of nitrogen and fluoride fumes.
*Permissible Concentration in Water:* No criteria set. Runoff from spills or fire control may cause water pollution.
*First Aid:* If this chemical gets into the eyes, remove any contact lenses at once and irrigate immediately for at least 15 minutes, occasionally lifting upper and lower lids. Seek medical attention immediately. If this chemical contacts the skin, remove contaminated clothing and wash immediately with soap and water. Seek medical attention immediately. If this chemical has been inhaled, remove from exposure, begin rescue breathing (using universal precautions) if breathing has stopped, and CPR if heart action has stopped. Transfer promptly to a medical facility. When this chemical has been swallowed, get medical attention. Give large quantities of water and induce vomiting. Do not make an unconscious person vomit. Do not induce vomiting when formulations containing petroleum solvents are ingested.

## Profenofos (ANSI)

*Use Type:* Insecticide miticide
*CAS Number:* 41198-08-7
*Formula:* $C_{11}H_{15}BrClO_3PS$
*Alert:* Some formulations may be a Restricted Use Pesticide (RUP).
*Synonyms:* A13-29236; *O*-(4-Bromo-2-chlorophenyl)-*O*-ethyl-*S*-propylphosphorothioate; Caswell No. 266AA; CGA 15,324; Phosphorothioic acid, *O*-(4-bromo-2-chlorophenyl)-*O*-ethyl-*S*-propyl ester
*Trade Names:* CGA-15324®; CURACRON®, Syngenta (Switzerland); POLYCRON®; SELECRON®
*Producers:* Agsin (Singapore); Epochem Co., (China); Hindustan Insecticides (India); Syngenta (Switzerland); Nagarjuna Agrichem (India); Vijayalakshmi Insecticides and Pesticides (India)
*Chemical Class:* Organophosphate
*EPA/OPP PC Code:* 111401; (210700 old EPA code number)
*California DPR Chemical Code:* 2042
*RTECS Number:* TE9675000
*Uses:* Profenofos is a Restricted Use Pesticide (RUP) used solely on cotton to control a number of pests including tobacco budworm, cotton aphid and bollworm, armyworm, fleahopper and whiteflies. It is not registered for residential use.
*U.S. Maximum Allowable Residue Levels for Profenofos (40 CFR 180.404):*

| CROP | ppm |
| --- | --- |
| Cattle, fat, meat, mbyp | 0.05 |
| Cotton, undelinted seed | 3.0 |
| Goat, fat, meat, mbyp | 0.05 |
| Horse, fat, meat, mbyp | 0.05 |
| Milk | 0.01 |
| Sheep, fat, meat, mbyp | 0.05 |

*Carcinogen/Hazard Classifications*
**U.S. EPA Carcinogens:** Group E, Unlikely carcinogen
**Label Signal Word:** WARNING or DANGER
**WHO Acute Hazard:** Class II, moderately hazardous
*Regulatory Authority:*
- EPA 40 CFR 372.65, Specific Toxic Chemical Listings
- FIFRA, 40CFR186: tolerances for pesticides in animal feeds
- Actively registered pesticide in California
- EPCRA Section 313 Form R *de minimis* concentration reporting level: 1.0%
- DOT Inhalation Hazard Chemicals as organophosphates

*Description:* Pale yellow liquid (technical profenofos). Boiling point = 100°C. Limited solubility in water. Molecular weight = 373.65. Vapor pressure = $9 \times 10^{-7}$ mmHg @ 20°C.
*Incompatibilities:* Contact with flammable material may cause fire and explosions. Contact with combustible or oxidizable materials may form heat-, shock-, and friction-sensitive explosive mixtures. Static electricity may also cause explosions. Keep away from all acids, especially dibasic organic acids, ammonium compounds, antimony sulfide, arsenic trioxide, metal sulfides, powdered metals, calcium aluminum hydride, cyanides, manganese dioxide, phosphorus, selenium, sulfur, thiocyanates, zinc.
*Permissible Exposure Limits in Air:*
*Determination in Air:* OSHA versatile sampler-2; Toluene/Acetone; Gas chromatography/Flame ionization detection; NIOSH IV, Method #5600, Organophosphorus Pesticides.[18]
*Permissible Concentration in Water:* No criteria set. Runoff from spills or fire control may cause water pollution.
*Determination in Water:*
*Routes of Entry:*
*Harmful Effects and Symptoms*
*Short Term Exposure:* Eye pupils are small; blurred vision; eye watering; runny nose; cough; shortness of breath; salivation; dizziness; nausea, stomach cramps, diarrhea, and vomiting; increased blood pressure; profuse sweating; hypermotility, hallucinations; irritability; tingling of the skin; drowsiness; slow heartbeat; convulsions; fluid in lungs; loss of consciousness; incontinence; breathing stops; death. Organophosphates inhibit the action of acetylcholinesterase enzymes, and alter the way in which nervous impulses are transmitted. The effects can last for hours, days, or much longer. The action of the enzymes is reestablished after new enzymes are formed.
*Long Term Exposure:* Cholinesterase inhibitor; cumulative effect is possible. Organophosphates may damage the nervous system with repeated exposure, resulting in convulsions, respiratory failure. May cause liver damage.
*Points of Attack:* Respiratory system, central nervous system, cardiovascular system, blood cholinesterase.
*Medical Surveillance:* Medical observation is recommended for 24 to 48 hours after breathing overexposure, as pulmonary edema may be delayed. As first aid for pulmonary edema, consider administering a corticosteroid spray. Cigarette smoking may exacerbate pulmonary injury and should be discouraged for at least 72 hours following exposure.
*First Aid:* Speed in removing material from eyes and skin is of extreme importance. *Eyes:* Eye contact can cause dangerous amounts of these chemicals to be quickly absorbed through the mucous membrane into the bloodstream. Immediately and gently flush eyes with plenty of warm or cold water (NO hot water) for at least 15 minutes, occasionally lifting the upper and lower eyelids. Get medical aid immediately. *Skin:* Get medical aid. Skin contact can cause dangerous amounts of these chemicals to

be absorbed into the bloodstream. Wearing the appropriate PPE equipment and respirator for organophosphate pesticides, immediately flush skin with plenty of soap and water for at least 15 minutes while removing contaminated clothing and shoes. Shampoo hair promptly if contaminated. *Ingestion:* Call poison control. Loosen all clothing. Never give anything by mouth to an unconscious person. Get medical aid. Do NOT induce vomiting.* If conscious, alert, and able to swallow, rinse mouth and have victim drink 4 to 8 ounces of water do NOT induce vomiting but immediately administer slurry of activated charcoal (2 oz in 8 oz of water). If victim is *unconscious or having convulsions,* do nothing except keep victim warm. *In some cases you may be specifically instructed by poison control to induce vomiting by way of 2 tablespoons of syrup of ipecac (adult) washed down with a cup of water.* Do NOT give activated charcoal before or with ipecac syrup. *Inhalation:* Get medical aid. Do not contaminate yourself. Wearing the appropriate PPE equipment and respirator for organophosphate pesticides, immediately remove the victim from the contaminated area to fresh air. If the victim is not breathing, administer artificial respiration. Do NOT use mouth-to-mouth resuscitation; use bag/mask apparatus. If breathing is difficult, administer oxygen through bag/mask apparatus until medical help arrives. Do not leave victim unattended. *Note to physician or authorized medical personnel.* Administer atropine, 2 mg (1/30 gr) intramuscularly or intravenously as soon as any local or systemic signs or symptoms of an intoxication are noted; repeat the administration of atropine every 3 to 8 minutes until signs of atropinization (mydriasis, dry mouth, rapid pulse, hot and dry skin) occur; initiate treatment in children with 0.05 mg mg/kg of atropine; repeat at 5 to 10 minute intervals. Watch respiration, and remove bronchial secretions if they appear to be obstructing the airway; intubate if necessary. *Notes to physician or authorized medical personnel*: N-methylpyridinium-2-aldoxime (2-PAMCI) when used in conjunction with atropine reacts with the phosphorylated cholinesterase, thereby restoring normal activity to by removing the phosphorylating group. The combination of these two chemicals is synergistic and must be administered within minutes to a few hours following exposure (depending on the specific agent) to be effective. Give 2-PAMCI (Pralidoxime; Protopam), 2.5 gm in 100 ml of sterile water or in 5% dextrose and water, intravenously, slowly, in 15-30 minutes; if sufficient fluid is not available, give 1 gm of 2-PAMCI in 3 ml of distilled water by deep intramuscular injection; repeat this every half hour if respiration weakens or if muscle fasciculation or convulsions recur. Also Diazepam, an anticonvulsant, might be considered.

*References:*
- U.S. Environmental Protection Agency, Office of Pesticide Programs, Pesticide Residue Limits, "Profenofos," 40 CFR 180.404, www.epa.gov/pesticides/food/viewtols.htm
- U.S. Environmental Protection Agency, "Interim Reregistration Eligibility Decision (IRED), Profenofos," Office of Prevention, Pesticides and Toxic Substances, Washington, DC (August 2000). http://www.epa.gov/REDs/2540ired.pdf
- California Environmental Protection Agency "Chemical List of Lists," Sacramento CA (February 1997)

# Profluralin (ANSI)

*Use Type:* Herbicide
*CAS Number:* 26399-36-0
*Formula:* $C_{14}H_{16}F_3N_3O_4$
*Synonyms:* Benzenamine, N-(cyclopropylmethyl)-2,6-dinitro-N-propyl-4-(trifluoromethyl)-; N-(Cyclopropylmethyl)-α,α,α-trifluoro-2,6-dinitro-N-propyl-p-toluidine
*Trade Names:* CGA-10832®, Syngenta (Switzerland); ER-5461®; GA-10832®, Syngenta (Switzerland); PREGARD®; TOLBAN®, Syngenta (Switzerland), canceled 4/20/84
*Producers:* Ciba-Geigy (Switzerland); Syngenta (Switzerland)
*Chemical Class:* Dinitroaniline
*EPA/OPP PC Code:* 106601; (304300 old EPA code number)
*California DPR Chemical Code:* 1897
*Uses:* Not registered in the U.S. Believed to be obsolete or discontinued (WHO). Was used to control a broad spectrum of insects and used in a variety of crops and established turf.
*Regulatory Authority:*
• AB 1803-Well Monitoring Chemical (CAL)
*Description:* Yellow-orange crystalline solid. Odorless. Very slightly soluble in water; solubility = < 0.2 ppm. Molecular weight 347.32 Melting/Freezing point = 32–34°C. Vapor pressure = $1 \times 10^{-4}$ mmHg @ 20°C.
*Determination in Air:* Filter; none; Gravimetric; NIOSH IV [Particulates NOR; #0500 (total), #0600 (respirable)].[18]
*Permissible Concentration in Water:* No criteria set. Runoff from spills or fire control may cause water pollution.
*First Aid:* If this chemical gets into the eyes, remove any contact lenses at once and irrigate immediately for at least 15 minutes, occasionally lifting upper and lower lids. Seek medical attention immediately. If this chemical contacts the skin, remove contaminated clothing and wash immediately with soap and water. Seek medical attention immediately. If this chemical has been inhaled, remove from exposure, begin rescue breathing (using universal precautions) if breathing has stopped, and CPR if heart action has stopped. Transfer promptly to a medical facility. When this chemical has been swallowed, get medical attention. Give large quantities of water and induce vomiting. Do not make an

unconscious person vomit. Do not induce vomiting when formulations containing petroleum solvents are ingested.

*References:*
- Pesticide Management Education Program, "Profluralin (Tolban) Herbicide Profile 3/85," Cornell University, Ithaca, NY (March 1985). http://pmep.cce.cornell.edu/profiles/herb-growthreg/naa-rimsulfuron/profluralin/herb-prof-profluralin.html
- California Environmental Protection Agency "Chemical List of Lists," Sacramento CA (February 1997).

# Prohexadione Calcium

*Use Type:* Plant growth regulator
*CAS Number:* 127277-53-6
*Formula:* $C_{10}H_{11}CaO_5$
*Synonyms:* Calcium 3-oxido-5-*oxo*-4-propionylcyclohex-3-enecarboxylate; Cyclohexanecarboxylic acid, 3,5-dioxo-4-(1-oxopropyl)-, ion$^{(1-)}$, calcium, calcium salt; 3,5-Dioxo-4-(1-oxopropyl)cyclohexanecarboxylic acid ion(1-) calcium salt
*Trade Names:* APOGEE® PLANT GROWTH REGULATOR, BASF Agricultural Products (Germany); BASELINE® PLANT REGULATOR, BASF Agricultural Products (Germany); BX 112®; K-I CHEMICAL, BASF Agricultural Products (Germany); KIM-112®; KUH-833®; VIVIFUL®
*Producers:* BASF Agricultural Products (Germany)
*EPA/OPP PC Code:* 112600
*California DPR Chemical Code:* 5497
*Uses:* Used to inhibit foliar growth on apples and pears, reduces maturation of fruit and foliar covering. Also used on turf grass.
*U.S. Maximum Allowable Residue Levels for Prohexadione Calcium (40 CFR 180.547):*

| CROP | ppm |
| --- | --- |
| Cattle, kidney | 0.10 |
| Cattle, mbyp (except kidney) | 0.05 |
| Goat, kidney | 0.10 |
| Goat, mbyp (except kidney) | 0.05 |
| Grass, forage | 0.10 |
| Grass, hay | 0.10 |
| Grass, seed screenings | 3.5 |
| Grass, straw | 1.2 |
| Hog, kidney | 0.10 |
| Hog, mbyp (except kidney) | 0.05 |
| Horse, kidney | 0.10 |
| Horse, mbyp (except kidney) | 0.05 |
| Peanuts | 1.0 |
| Peanut hay | 0.60 |
| Fruit, pome, group | 3.0 |
| Sheep, kidney | 0.10 |
| Sheep, mbyp (except kidney) | 0.05 |

*Carcinogen/Hazard Classifications*
**U.S. EPA Carcinogens:** Group E, unlikely carcinogen
**Label Signal Word:** CAUTION
*Description:* Pale yellow-brown fine powder.
*Incompatibilities:* May react violently with strong oxidizers, bromine, 90% hydrogen peroxide, phosphorus trichloride, silver powders or dust. Incompatible with silver compounds. Mixture with some silver compounds form explosive salts of silver oxalate.
*First Aid:* If this chemical gets into the eyes, remove any contact lenses at once and irrigate immediately for at least 15 minutes, occasionally lifting upper and lower lids. Seek medical attention immediately. If this chemical contacts the skin, remove contaminated clothing and wash immediately with soap and water. Seek medical attention immediately. If this chemical has been inhaled, remove from exposure, begin rescue breathing (using universal precautions) if breathing has stopped, and CPR if heart action has stopped. Transfer promptly to a medical facility. When this chemical has been swallowed, get medical attention. Give large quantities of water and induce vomiting. Do not make an unconscious person vomit. Do not induce vomiting when formulations containing petroleum solvents are ingested.

*References:*
- U.S. Environmental Protection Agency, Office of Pesticide Programs, "Pesticide Fact Sheet, Prohexadione Calcium," (April 26, 2000). http://www.epa.gov/opprd001/factsheets/prohexadione.pdf
- U.S. Environmental Protection Agency, Office of Pesticide Programs, Pesticide Residue Limits, "Prohexadione Calcium," 40 CFR 180.547, http://www.setonresourcecenter.com/40CFR/Docs/wcd0004d/wcd04dd8.asp
- California Environmental Protection Agency "Chemical List of Lists," Sacramento CA (February 1997).

# Promecarb

*Use Type:* Insecticide
*CAS Number:* 2631-37-0
*Formula:* $C_{12}H_{17}NO_2$; $C_6H_3(CH_3)(OCONHCH_3)CH(CH_3)_2$
*Alert:* No products currently registered by the U.S. EPA.
*Synonyms:* Carbamic acid, methyl-, *m*-cym-5-yl ester; Carbamic acid, (3-methyl-5-(1-methylethyl)phenyl-, methyl ester; Carbamic acid, *N*-methyl-, 3-methyl-5-isopropylphenyl ester; Carbanilic acid, 3-isopropyl-5-methyl-, methyl ester; 3-Isopropyl-5-methylcarbamic acid methyl ester; *m*-Cym-5-yl methylcarbamate; ENT 27,300; ENT 27,300-a; EP 316; 3-Isopropyl-5-methylphenyl-*N*-methylcarbamate; 5-Isopropyl-*m*-tolyl methyl-carbamate; Methylcarbamic acid-*m*-cym-5-yl ester; 3-Methyl-5-isopropyl-*N*-methyl carbamate; 3-Methyl-5-(1-methylethyl)phenyl-carbamic acid methyl ester; 5-Methyl *m*-

cumenyl methylcarbamate; 3-Methyl-5-isopropylphenyl-*N*-methyl carbamate; *N*-Methylcarbamic acid 3-methyl-5-isopropylphenyl ester; (3-Methyl-5-isopropylphenyl)-*N*-methylcarbamat (German); 3-Methyl-5-(1-methylethyl)phenol methylcarbamate; Phenol, 3-methyl-5-(1-methylethyl)-, methylcarbamate

*Trade Names:* CARBAMULT®, canceled; MINACIDE®, canceled; MORTON EP-316®, canceled; OMS® 716, canceled; SCHERING® 34615, Schering-Plough (USA), canceled; SN® 316, Schering-Plough (USA), canceled; SN® 34615, Schering-Plough (USA), canceled; UC® 9880, Union Carbide (USA), canceled; UNION CARBIDE UC-9880®, Union Carbide (USA), canceled

*Producers:* Sigma-Aldrich Laborchemikalien (Germany)
*Chemical Class:* Carbamate
*EPA/OPP PC Code:* 271400
*California DPR Chemical Code:* 4092
*RTECS Number:* FB8050000

*Uses:* A non-systemic contact insecticide. Promecarb is used to combat foliage and fruit eating insects because it is not readily metabolized by plants. No products currently registered in the U.S.

*Regulatory Authority:*
- EPA Hazardous Waste Number (RCRA No.): P201
- RCRA, 40CFR261, Appendix 8 Hazardous Constituents
- RCRA 40CFR268.48; 61FR15654, Universal Treatment Standards: Wastewater (mg/L), 0.056; Nonwastewater (mg/kg), 1.4
- Superfund/EPCRA 40CFR355, Appendix B Extremely Hazardous Substances: TPQ = 500/10,000 lb (227/4,540 kg)
- Superfund/EPCRA 40CFR302.4 RQ: EHS, 1 lb (0.454 kg)
- U.S. DOT Regulated Marine Pollutant (49CFR172.101, Appendix B)

*Description:* Promecarb is a colorless, odorless, sand-like solid. Slightly soluble in water. Melting/Freezing point = 87-88°C. Hazard Identification (based on NFPA-704 M Rating System): Health 3, Flammability 1, Reactivity 0.

*Incompatibilities:* Alkalis
*Permissible Exposure Limits in Air:* No standard set.
*Determination in Air:*
*Permissible Concentration in Water:* No criteria set, but runoff from spills or fire control may cause water pollution.
*Determination in Water:*
*Routes of Entry:* Inhalation, ingestion, skin and/or eye contact.

*Harmful Effects and Symptoms*
*Short Term Exposure:* Promecarb is highly toxic by ingestion and is absorbed through the intact skin. It is a reversible cholinesterase inhibitor and its effects are related to action on the central nervous system. Symptoms of exposure include diarrhea, nausea, vomiting, excessive salivation, headache, pinpoint pupils and uncoordinated muscle movements. These are all common symptoms of exposure to carbamate insecticides.

*Long Term Exposure:* Cholinesterase inhibitor; cumulative effect is possible. This chemical may damage the nervous system with repeated exposure, resulting in convulsions, respiratory failure. May cause liver damage.

*Points of Attack:* Respiratory system, lungs, central nervous system, cardiovascular system, skin, eyes, plasma and red blood cell cholinesterase.

*Medical Surveillance:* Before employment and at regular times after that, the following are recommended: Plasma and red blood cell cholinesterase levels (tests for the enzyme poisoned by this chemical). If exposure stops, plasma levels return to normal in 1-2 weeks while red blood cell levels may be reduced for 1-3 months. When acetylcholinesterase enzyme levels are reduced by 25% or more below preemployment levels, risk of poisoning is increased, even if results are in lower ranges of "normal." Reassignment to work not involving carbamate pesticides is recommended until enzyme levels recover. If symptoms develop or overexposure occurs, repeat the above tests as soon as possible and get an exam of the nervous system. Also consider complete blood count. Consider chest x-ray following acute overexposure

*First Aid:* If this chemical gets into the eyes, remove any contact lenses at once and irrigate immediately for at least 15 minutes, occasionally lifting upper and lower lids. Seek medical attention immediately. If this chemical contacts the skin, remove contaminated clothing and wash immediately with soap and water. Seek medical attention immediately. If this chemical has been inhaled, remove from exposure, begin rescue breathing (using universal precautions) if breathing has stopped, and CPR if heart action has stopped. Transfer promptly to a medical facility. When this chemical has been swallowed, get medical attention. Give large quantities of water and induce vomiting. Do not make an unconscious person vomit.

*References:*
- New Jersey Department of Health and Senior Services, "Hazardous Substance Fact Sheet, Promecarb," Trenton, NJ (April 2002). http://www.state.nj.us/health/eoh/rtkweb/2710.pdf
- U.S. Environmental Protection Agency, "Chemical Profile: Promercarb," Washington, DC, Chemical Emergency Preparedness Program (November 30, 1987).
- California Environmental Protection Agency "Chemical List of Lists," Sacramento CA (February 1997).

# Prometon (ANSI)

*Use Type:* Herbicide
*CAS Number:* 1610-18-0
*Formula:* $C_{10}H_{19}N_5O$

***Synonyms:*** 2,4-Bis(isopropylamino)-6-methoxy-s-triazine; 2,6-Diisopropylamino-4-methoxytriazine; N,N'-Diisopropyl-6-methoxy-1,3,5-triazine-2,4-diyldiamine; N,N'-Diisopropyl-6-methoxy-1,3,5-triazine-2,4-diamine; 2-Methoxy-4,6-bis(isopropylamino)-s-triazine; 2-Methoxy-4,6-bis(isopropylamino)-1,3,5-triazine; 6-Methoxy-N,N'-bis(1-methylethyl)-1,3,5-triazine-2,4-diamine; Methoxypropazine; s-Triazine, 2,4-bis(isopropylamino)-6-methoxy-

***Trade Names:*** ACME®, Pbi/Gordon (USA); G-31435®; GESAFRAM® 50; GESAFRAM®; GESAGRAM®; GROUND ZERO®; KLEENWALK®; NIX®; NOXALL®; ONTRACK®; PRIMATOL®; PROMETONE®; WEED-GO®

***Producers:*** Agan Chemical (Israel); PBI/Gordon (USA); Bonide Products (USA); Makhteshim-Agan (Israel); Syngenta (Switzerland)

***Chemical Class:*** Triazine

***EPA/OPP PC Code:*** 080804

***California DPR Chemical Code:*** 499

***Uses:*** A nonselective pre-emergence and post-emergence herbicide. Use around buildings, storage areas, industrial sites, fences, recreational areas, rights-of-way, railroads, pipelines, lumberyards, tank farms, and similar areas. Controls broadleaf weeds and grasses over an extended period of time.

***Carcinogen/Hazard Classifications***

**U.S. EPA Carcinogens:** Group D, unclassifiable

**Label Signal Word:** WARNING or DANGER

**WHO Acute Hazard:** Class U, unlikely to be hazardous

***Regulatory Authority:***
- AB 1803-Well Monitoring Chemical (CAL)
- Actively registered pesticide in California.

***Description:*** Colorless crystalline solid or white powder. Odorless. Slightly soluble in water; solubility = 750 mg/L @ 20°C. Molecular weight 225.92. Melting/Freezing point = 91°C. Vapor pressure = $7.7 \times 10^{-6}$ mmHg @ 20°C. Log $K_{ow}$ = 2.98. Values above 3.0 May bioaccumulate in marine organisms.

***Permissible Concentration in Water:*** No criteria set. Runoff from spills or fire control may cause water pollution.

***Routes of Entry:*** Inhalation, passing through the skin, ingestion.

***Harmful Effects and Symptoms***

***Short Term Exposure:*** May cause skin and severe eye irritation. Moderately poisonous if ingested or inhaled. Exposure to a triazine (simazine) has caused acute and subacute dermatitis in the former USSR, characterized by erythema, slight edema, moderate pruritus, and burning lasting 4 to 5 days. $LD_{50}$ in range (rat) 1780-7000 mg/kg.

***Long Term Exposure:*** May cause lung irritation and damage. May cause skin allergy. Contact with some triazine compounds (such as atrazine) may increase risks for tumors known to be associated with hormonal factors. These have been observed in both animals and human beings, and are consistent with the known effects on the hypothalamic pituitary gonadal axis. Repeated exposure may cause weight loss and reduced red blood cell count. May be mutagenic.

***Points of Attack:*** Liver, lungs, skin.

***Medical Surveillance:*** Before beginning employment and at regular times after that, for those with frequent or potentially high exposures, the following is recommended: Lung function tests. Consider chest x-ray following acute overexposure. Evaluation by a qualified allergist. Examination of the nervous system.

***First Aid:*** If this chemical gets into the eyes, remove any contact lenses at once and irrigate immediately for at least 15 minutes, occasionally lifting upper and lower lids. Seek medical attention immediately. If this chemical contacts the skin, remove contaminated clothing and wash immediately with soap and water. Seek medical attention immediately. If this chemical has been inhaled, remove from exposure, begin rescue breathing (using universal precautions) if breathing has stopped, and CPR if heart action has stopped. Transfer promptly to a medical facility. When this chemical has been swallowed, get medical attention. Give large quantities of water or milk and induce vomiting. Do not make an unconscious person vomit.

***References:***
- Pesticide Management Education Program, "Prometon (Pramitol) Herbicide Profile 2/85," Cornell University, Ithaca, NY (February 1985). http://pmep.cce.cornell.edu/profiles/herb-growthreg/naa-rimsulfuron/prometon/herb-prof-prometon.html
- California Environmental Protection Agency "Chemical List of Lists," Sacramento CA (February 1997)

# Prometryn (ANSI)

***Use Type:*** Herbicide

***CAS Number:*** 7287-19-6

***Formula:*** $C_{10}H_{19}N_5S$

***Synonyms:*** A13-60366; 2,4-Bis(isopropylamino)-6-(methylmercapto)-S-triazine; 2,4-Bis(isopropylamino)-6-(methylthio)-S-triazine; 2,4-Bis(isopropylamino)-6-(methylthio)-1,3,5-triazine; N,N'-Bis(1-methylethyl)-6-methylthio-1,3,5-triazine-2,4-diamine; N,N'-Bis(1-methylethyl)-6-(methylthio)-1,3,5-triazine-2,4-diamine; Caswell No. 097; N,N'-Di-isopropyl-6-methylthio-1,3,5-triazine-2,4-diamine; N,N'-Di-isopropyl-6-methylthio-1,3,5-triazine-2,4-diyldiamine; 2-(Methylmercapto)-4,6-bis(isopropylamino)-S-triazine; 2-(Methylthio)-4,6-bis(isopropylamino)-S-triazine; NSC 163049; Prometrin; Prometrene; Prometryne (USDA); S-triazine,4,6-bis(isopropylamino)-2-(methylmercapto)-; S-triazine, 2,4-bis(isopropylamino)-6-(methylthio)-; 1,3,5-Triazine-2,4-diamine, N,N'-bis(1-methylethyl)-6-(methylthio)-

***Trade Names:*** A-1114®; CAPAROL®, Syngenta (Switzerland); COTTON PRO®, Griffin (USA); G 34161®; GESAGARD®, Syngenta (Switzerland); MERCASIN®;

MERCAZIN®; MERKAZIN®; POLISIN®; PRIMAPIN®; PRIMATOL-Q®; PRIMAZE®, Syngenta (Switzerland), canceled; PROKIL®, Gowan (USA), canceled; PROMET®; PROMETREX®; SELECTIN®; SELECTIN-50®; SELEKTIN®; SESAGARD®; SUPREND®, Syngenta (Switzerland); UVON®

*Producers:* Agriliance (USA); Griffin (USA); Ki-Hara Chemicals Ltd. (UK); Makhteshim-Agan (Israel); OXON Italia S.p.A. (Italy); Syngenta (Switzerland); United Agri Products (UAP)

*Chemical Class:* Triazine
*EPA/OPP PC Code:* 080805
*California DPR Chemical Code:* 502
*RTECS Number:* XY4390000
*Uses:* Prometryn is used to control several annual grasses and broadleaf weeds in terrestrial food and feed crops. Its major uses are on cotton and celery, and is often used on dill and pigeon peas.

*U.S. Maximum Allowable Residue Levels for Prometryn (40 CFR 180.222):*

| CROP | ppm |
|---|---|
| Carrots[1] | 0.1 |
| Celery | 0.5 |
| Corn, grain | 0.25 |
| Cottonseed | 0.25 |
| Pigeon peas | 0.25 |

[1] There are no U.S. registrations as of April 10, 1998 for use on carrots.

*Carcinogen/Hazard Classifications*
**U.S. EPA Carcinogens:** Group E, unlikely carcinogen
**TRI Developmental Toxin:** Reproductive and developmental toxin
**Label Signal Word:** CAUTION
**WHO Acute Hazard:** Class U, unlikely to be hazardous
*Regulatory Authority:*
- EPA 40 CFR 372.65, Specific Toxic Chemical Listings
- Actively registered pesticide in California.
- AB 1803-Well Monitoring Chemical (CAL)
- EPCRA Section 313 Form R *de minimis* concentration reporting level: 1.0%

*Description:* Colorless crystalline solid. Slightly soluble in water; solubility = 48 ppm @ 20°C. Molecular weight = 241.385. Melting/Freezing point = 119°C. Vapor pressure = $1.24 \times 10^{-6}$ mmHg @ 20°C. Log $K_{ow}$ = 3.33. Values > 3.0 may bioaccumulate in marine organisms.
*Determination in Air:* Filter; none; Gravimetric; NIOSH IV [Particulates NOR; #0500 (total), #0600 (respirable)].[18]
*Permissible Concentration in Water:* No criteria set. Runoff from spills or fire control may cause water pollution.
*Routes of Entry:* Inhalation, passing through the skin, ingestion.
*Harmful Effects and Symptoms*
**Short Term Exposure:** May cause skin and severe eye irritation. Moderately poisonous if ingested or inhaled. Exposure to a triazine (simazine) has caused acute and subacute dermatitis in the former USSR, characterized by erythema, slight edema, moderate pruritus, and burning lasting 4 to 5 days.
**Long Term Exposure:** May cause lung irritation and damage. May cause skin allergy. Contact with some triazine compounds (such as atrazine) may increase risks for tumors known to be associated with hormonal factors. These have been observed in both animals and human beings, and are consistent with the known effects on the hypothalamic pituitary gonadal axis. Repeated exposure may cause weight loss and reduced red blood cell count. May be mutagenic.
*Points of Attack:* Liver, lungs, skin.
*Medical Surveillance:* Before beginning employment and at regular times after that, for those with frequent or potentially high exposures, the following is recommended: Lung function tests. Consider chest x-ray following acute overexposure. Evaluation by a qualified allergist. Examination of the nervous system.
*First Aid:* If this chemical gets into the eyes, remove any contact lenses at once and irrigate immediately for at least 15 minutes, occasionally lifting upper and lower lids. Seek medical attention immediately. If this chemical contacts the skin, remove contaminated clothing and wash immediately with soap and water. Seek medical attention immediately. If this chemical has been inhaled, remove from exposure, begin rescue breathing (using universal precautions) if breathing has stopped, and CPR if heart action has stopped. Transfer promptly to a medical facility. When this chemical has been swallowed, get medical attention. Give large quantities of water or milk and induce vomiting. Do not make an unconscious person vomit.
*References:*
- U.S. Environmental Protection Agency, "Reregistration Eligibility Decision (RED), Prometryn," Office of Prevention, Pesticides and Toxic Substances, Washington, DC (February 1996). http://www.epa.gov/REDs/0467.pdf
- EXTOXNET, Extension Toxicology Network, "Pesticide Information Profile, Prometryn," Oregon State University, Corvallis, OR (June 1996). http://extoxnet.orst.edu/pips/prometry.htm
- U.S. Environmental Protection Agency, Office of Pesticide Programs, Pesticide Residue Limits, "Prometryn," 40 CFR 180.222. http://www.setonresourcecenter.com/40CFR/Docs/wcd0004c/wcd04cd7.asp
- California Environmental Protection Agency "Chemical List of Lists," Sacramento CA (February 1997)

# Pronamide

*Use Type:* Herbicide
*CAS Number:* 23950-58-5
*Formula:* $C_{12}H_{11}Cl_2NO$

*Alert:* A Restricted Use Pesticide (RUP).
*Synonyms:* Benzamide, 3,5-dichloro-*N*-(1,1-dimethyl-2-propynyl); Caswell No. 306A; 3,5-Dichloro-*N*-(1,1-dimethyl-2-propynyl)benzamide; 3,5-Dichloro-*N*-(1,1-dimethylpropynyl)benzamide; 3,5-Dichloro-*N*-(1,1-dimethylprop-2-ynyl)benzamide; *N*-(1,1-Dimethylpropynyl)-3,5-dichlorobenzamide; Propyzamide
*Trade Names:* BENZAMIDE®; CAMPBELL'S RAPIER®; CLANEX®; KERB®, Dow AgroSciences (USA); KERB 50W®, Dow AgroSciences (USA); KERB® PROPYZAMIDE 50, Dow AgroSciences (USA); RH-315 RAPIER®; RONAMID®
*Producers:* Dow AgroSciences (USA)
*Chemical Class:* Amide
*EPA/OPP PC Code:* 101701
*California DPR Chemical Code:* 694
*RTECS Number:* CV3460000
*Uses:* Pronamide is a selective herbicide used either before weeds emerge (pre-emergence), and/or after weeds come up (post-emergence). It controls a wide range of annual and perennial grasses, as well as certain annual broadleaf weeds. It is used primarily on lettuce and alfalfa crops, as well as on blueberries, ornamentals, fruit trees, forage legumes, and on pastures and rangelands. Pronamide is usually incorporated into the soil by cultivation, irrigation, or rain immediately following application. It is available in wettable powder and granular formulations.
*Carcinogen/Hazard Classifications*
**U.S. EPA Carcinogens:** Group B2, probable carcinogen
**California Prop. 65:** Carcinogen
**Label Signal Word:** CAUTION
**WHO Acute Hazard:** Class U, unlikely to be hazardous
*Regulatory Authority:*
- Carcinogen (U.S. EPA) (See Reference Below)
- Banned or Severely Restricted (USA) (UN)[13]
- Proposition 65 chemical (CAL)
- Actively registered pesticide in California.
- EPA Hazardous Waste Number (RCRA No.): U192
- RCRA, 40CFR261, Appendix 8 Hazardous Constituents
- RCRA 40CFR268.48; 61FR15654, Universal Treatment Standards: Wastewater (mg/L), 0.093; Nonwastewater (mg/kg), 1.5
- RCRA 40CFR264, Appendix 9; TSD Facilities Ground Water Monitoring List. Suggested test method(s) (PQL *u*g/L): 8270(10)
- EPCRA Section 313 Form R *de minimus* concentration reporting level: 1.0%

*Description:* Pronamide is a colorless, crystalline solid or powder. Odorless. Practically insoluble in water; solubility = 15 ppm @ 25°C. Molecular weight = 256.12. Melting/Freezing point = 155–156°C. Vapor pressure = $1.3 \times 10^{-4}$ mbar @ 25°C.
*Permissible Exposure Limits in Air:* No standards set.
*Determination in Air:* No tests

*Permissible Concentration in Water:* The U.S. EPA has derived a lifetime health advisory of 0.052 mg/L (52 µg/L).
*Determination in Water:* Extraction with Methylene chloride, separation by capillary-column gas chromatography, then measurement using a nitrogen-phosphorus detector.
*Routes of Entry:* Inhalation, ingestion, skin and/or eye contact.
*Harmful Effects and Symptoms*
*Short Term Exposure:* Eye contact can cause irritation. Inhalation can cause irritation of the respiratory tract with cough, phlegm, and/or chest tightness. The acute oral $LD_{50}$ for male rats is 8350 mg/kg and for female rats is 5620 mg/kg (insignificantly toxic in both cases).
*Long Term Exposure:* Applying the criteria described in EPA's final guidelines for assessment of carcinogenic risk, pronamide has tentatively been classified in Group C: possible human carcinogen. This category is for substances with limited evidence of carcinogenicity in animals in the absence of human data. There is limited animal evidence of liver cancer.
*Points of Attack:* Cancer site in animals: liver.
*First Aid:* If this chemical gets into the eyes, remove any contact lenses at once and irrigate immediately for at least 15 minutes, occasionally lifting upper and lower lids. Seek medical attention immediately. If this chemical contacts the skin, remove contaminated clothing and wash immediately with soap and water. Seek medical attention immediately. If this chemical has been inhaled, remove from exposure, begin rescue breathing (using universal precautions) if breathing has stopped, and CPR if heart action has stopped. Transfer promptly to a medical facility. When this chemical has been swallowed, get medical attention. Give large quantities of water and induce vomiting. Do not make an unconscious person vomit.
*References:*
- EXTOXNET, Extension Toxicology Network, "Pesticide Information Profile, Pronamide," Oregon State University, Corvallis, OR (June 1996). http://extoxnet.orst.edu/pips/pronamid.htm
- U.S. Environmental Protection Agency, "Health Advisory: Pronamide," Washington, DC, Office of Drinking Water (August 1987).
- California Environmental Protection Agency "Chemical List of Lists," Sacramento CA (February 1997).
- New Jersey Department of Health and Senior Services, "Hazardous Substance Fact Sheet, Pronamide," Trenton NJ (May 1998). http://www.state.nj.us/health/eoh/rtkweb/1592.pdf

# Propachlor

*Use Type:* Herbicide
*CAS Number:* 1918-16-7

*Formula:* $C_{11}H_{14}ClNO$; $(ClCH_2CO)N(C_6H_5)CH(CH_3)_2$

*Synonyms:* Acetamide, 2-chloro-N-(1-methylethyl)-N-phenyl-; Acetamide, 2-chloro-N-isopropyl-; α-Chloro-N-isopropylacetanilide; Chloressigsaeure-n-isopropylanilid (German); α-Chloro-N-isopropylacetanilide; 2-Chloro-N-isopropylacetanilide; 2-Chloro-N-isopropyl-N-phenylacetamide; 2-Chloro-N-(1-methylethyl)-N-phenylacetamide; N-Isopropyl-α-chloroacetanilide; N-Isopropyl-2-chloroacetanilide; Propachlore; Propacloro (Spanish)

*Trade Names:* AATRAM®, Syngenta (Switzerland), canceled 2/17/1984; ACLID®; AI3-51503®; ALBRASS®; BEXTON®, Dow Chemical (USA), canceled; CIPA®; CP 31393®; KARTEX A®; NITICID®; RAMROD®, Monsanto (USA); RAMROD® 65, Monsanto (USA); SATECID®; WALLOP®, Monsanto (USA), canceled

*Producers:* Agrichem (International) Limited (UK); Crop Care Australasia (Australia); Drexel Chemical (USA); Makhteshim Agan (Israel); Monsanto (USA); Shenzhen Guomeng Industry Co., Ltd. (China); Sigma-Aldrich Laborchemikalien (Germany)

*Chemical Class:* Chloroacetanilide

*EPA/OPP PC Code:* 019101

*California DPR Chemical Code:* 511

*RTECS Number:* AE1575000

*EINECS Number:* 217-638-2

*Uses:* A pre-emergence herbicide used to combat annual grasses and broad-leaved weeds in corn, sorghum, soybeans, cotton, sugar cane, sugar beets, vegetable crops, forage crops, pasture land and range land. Also used to control weeds in groundnuts, leeks, onions, peas, maize, roses and ornamental trees and shrubs.

*U.S. Maximum Allowable Residue Levels for* Propachlor (40 CFR 180.211):

| CROP | ppm |
| --- | --- |
| Cattle, fat | 0.02 |
| Cattle, meat | 0.02 |
| Cattle, mbyp | 0.02 |
| Corn, forage | 1.5 |
| Corn, grain | 0.1 |
| Eggs | 0.02 |
| Goats, fat | 0.02 |
| Goats, meat | 0.02 |
| Goats, mbyp | 0.02 |
| Hogs, fat | 0.02 |
| Hogs, meat | 0.02 |
| Hogs, mbyp | 0.02 |
| Horses, fat | 0.02 |
| Horses, meat | 0.02 |
| Horses, mbyp | 0.02 |
| Milk | 0.02 |
| Poultry, fat | 0.02 |
| Poultry, meat | 0.02 |
| Poultry, mbyp | 0.02 |
| Sheep, fat | 0.02 |
| Sheep, meat | 0.02 |
| Sheep, mbyp | 0.02 |
| Sorghum, fodder & forage | 5 |
| Sorghum, grain | 0.25 |

*Carcinogen/Hazard Classifications*

**U.S. EPA Carcinogens:** Likely carcinogen

**California Prop. 65:** Carcinogen

**TRI Developmental Toxin:** Listed Reproductive and developmental toxin

**Label Signal Word:** WARNING or DANGER

**WHO Acute Hazard:** Class III, slightly hazardous

*Regulatory Authority:*
- Air Pollutant Standard Set (former USSR)[35, 43]
- AB 1803-Well Monitoring Chemical (CAL)
- Proposition 65 chemical (CAL)
- EPCRA Section 313 Form R *de minimis* concentration reporting level: 1.0%

*Description:* Propachlor is a light tan or white crystalline solid. Slightly soluble in water. oiling point = 110°C @ 0.03 mm. Melting/Freezing point = 67–76°C. Flash point = about 316°C. Vapor Pressure = $2.5 \times 10^{-4}$ mbar @ 25°C.

*Incompatibilities:* Incompatible with alkaline materials, strong acids, strong oxidizers. Attacks carbon steel.

*Permissible Exposure Limits in Air:* The former USSR[35, 43] has set an MAC in workplace air of 0.5 mg/m$^3$.

*Permissible Concentration in Water:* The former USSR[35, 43] has set an MAC in water bodies used for domestic purposes of 0.01 mg/L and in water for fishery purposes of zero. The U.S. EPA has set a lifetime health advisory of 0.092 mg/L (92 μg/L). States which have set guidelines for propachlor in drinking water include Kansas at 700 μg/L and Maine @ 200 μg/L.

*Routes of Entry:* Inhalation, ingestion, skin and/or eye contact.

*Harmful Effects and Symptoms*

*Short Term Exposure:* The maximal tolerated dosage of propachlor without adverse effect is reported as 133.3 mg/kg/day in both rats and dogs. Other workers reported slight organ pathology in rats, mice, and rabbits at 100 mg/kg/day or higher; this agrees approximately with the former data. Apparently, no long-term toxicity studies have been completed that would contribute information on reproductive effects or carcinogenic potential of propachlor or its degradation products, which include aniline derivatives. These studies are needed.

*Long Term Exposure:* May be a mutagen.

*First Aid:* If this chemical gets into the eyes, remove any contact lenses at once and irrigate immediately for at least 15 minutes, occasionally lifting upper and lower lids. Seek medical attention immediately. If this chemical contacts the skin, remove contaminated clothing and wash immediately with soap and water. Seek medical attention immediately. If this chemical has been inhaled, remove from exposure, begin rescue breathing (using universal precautions) if breathing has stopped, and CPR if heart action has stopped.

Transfer promptly to a medical facility. When this chemical has been swallowed, get medical attention. Give large quantities of water and induce vomiting. Do not make an unconscious person vomit.

*References:*
- EXTOXNET, Extension Toxicology Network, "Pesticide Information Profile," Oregon State University, Corvallis, OR.
- EPA, Office of Pesticide Programs, Pesticide Residue Limits, "Propachlor," 40 CFR 180.211, http://www.epa.gov/pesticides/food/viewtols.htm
- U.S. Environmental Protection Agency, "Health Advisory: Propachlor," Washington, DC, Office of Drinking Water (August 1987).
- California Environmental Protection Agency "Chemical List of Lists," Sacramento CA (February 1997).

# Propanil

*Use Type:* A contact herbicide
*CAS Number:* 709-98-8
*Formula:* $C_9H_9Cl_2NO$
*Alert:* A General Use Pesticide (GUP).
*Synonyms:* Cekupropanil; DCPA; N-(3,4-Dichlorophenyl)propanamide; 3',4'-Dichlorophenyl propionanilide; 3,4-Dichloropropionanilide; 3',4'-Dichloropropionanilide; Dichloropropionanilide; Dipram; DPA; NSC 31312; Propanamide, N-(3,4-Dichlorophenyl)-; Propanide; Propionanilide, 3',4'-Dichloro-; Propionic acid, 3,4-dichloroanilide
*Trade Names:* AI3-31382®; ARROSOLO®, Syngenta (Switzerland); ATLAS®, Helena Chemical (USA); BAY® 30130, Bayer CropScience (Germany); CHEM RICE®, Tifa Lts. (USA), canceled 9/30/1991; CRYSTAL PROPANIL-4®, Rice Company, The (TRC) (USA), canceled 8/25/2000; DREXEL PROP-JOB®, Drexel Chemical (USA); DROPAVEN®; DUET®, Rice Company, The (TRC) (USA); ERBAN®; FARMCO PROPANIL®; FW-734®; HERBANIL 368®, Milenia Agro Ciencias (Brazil); HERBAX TECHNICAL®; IDA PROP-A-NEL®, Drexel Chemical (USA); KENSOLO®, Kenso Corp. (Malaysia); LONDAX PRO-PACK BNB®, Agriliance (USA); MONTROSE PROPANIL®, International Minerals (USA), canceled 6/15/1984; PANTOX®, Pyosa Agroquimicos (Mexico); PROPANEX®, Cumberland Internationqal (USA), canceled 10/10/1989; PROP-JOB®, Drexel Chemical (USA); PROPANIL MILENIA®, Milenia Agro Ciencias (Brazil); RISELECT®; ROGUE®, Monsanto (USA), canceled 6/15/1984; ROSANIL®; RICECO TOUCHE®, Syngenta (Switzerland); S-10165®; SETRA PROWL®, Helena Chemical (USA); STAM®, Dow AgroSciences (USA); STAM®F-34, Dow AgroSciences (USA); STAM® LV 10, Dow AgroSciences (USA); STAMPEDE® 3E, Dow AgroSciences (USA); STAM SUPERNOX®, Dow AgroSciences (USA); STREL®; SURCOPUR®; SURPUR®; SUPERNOX®, Dow AgroSciences (USA); SYNPRAN N®; TURF! EZ®, Makhteshim-Agan Industries (Israel); VERTAC®; WHAM! EZ®, Rice Company, The (TRC) (USA)

*Producers:* Agriliance (USA); Bayer CropScience (Germany); Bhageria Dye-Chem (India); Biesterfeld Siemsgluess International. GmbH (Germany); Drexel Chemical (USA); Chemia (Italy); Dow AgroSciences (USA); Ehrenstorfer, Dr. (Germany); Epochem Co., (China); Helena Chemical (USA); Hokko Chemical Industry (Japan); Kenso Corp. (Malaysia); Makhteshim-Agan Industries (Israel); Micro Flo Company (USA); Milenia Agro Ciencias (Brazil); Nufarm Ltd. (Australia); Nissan Chemical Industries (Japan); Proficol (Colombia); Pyosa Agroquimicos (Mexico); Rice Company, The (TRC) (USA); Shenzhen Guomeng Industry Co., Ltd. (China); Sigma-Aldrich Laborchemikalien (Germany); Syngenta (Switzerland); Veterinary & Agricultural Products Manufacturing Co., Ltd. (VAPCO) (Jordan); Zeneca Ag Products (USA) (now Syngenta)

*Chemical Class:* Acetanilide
*EPA/OPP PC Code:* 028201
*California DPR Chemical Code:* 503
*ICSC Number:* 0552
*RTECS Number:* UE4900000
*EEC Number:* 616-009-00-3
*EINECS Number:* 211-914-6

*Uses:* Propanil is a post-emergence herbicide with no residual effect. It is used against numerous grasses and broad-leaved weeds in rice, potatoes, and wheat. Mixing with carbamates or organophosphorous compounds is not recommended. It is also used on wheat in a mixture with MCPA. With carbaryl, it is used in citrus crops grown in sod culture.

*Human toxicity (long-term)[77]:* Intermediate–35.00 ppb, Health Advisory

*Fish toxicity (threshold)[77]:* Extra high–0.48990 ppb, MATC (Maximum Acceptable Toxicant Concentration)

*U.S. Maximum Allowable Residue Levels for* Propanil (40 CFR 180.274):

| CROP | ppm |
|---|---|
| Barley, grain | 0.2 |
| Barley, straw | 0.75 |
| Cattle, fat | 0.1 |
| Cattle, meat | 0.1 |
| Cattle, mbyp | 0.1 |
| Egg | 0.05 |
| Goat, fat | 0.1 |
| Goat, meat | 0.1 |
| Goat, mbyp | 0.1 |
| Hog, fat | 0.1 |
| Hog, meat | 0.1 |
| Hog, mbyp | 0.1 |
| Horse, fat | 0.1 |

| | |
|---|---|
| Horse, meat | 0.1 |
| Horse, mbyp | 0.1 |
| Milk | 0.05 |
| Oat, grain | 0.2 |
| Oat, straw | 0.75 |
| Poultry, fat | 0.1 |
| Poultry, meat | 0.1 |
| Poultry, mbyp | 0.1 |
| Rice | 2 |
| Rice, bran | 10 |
| Rice, hulls | 10 |
| Rice, milled fractions | 10 |
| Rice, polished rice | 10 |
| Rice, straw | 75 |
| Sheep, fat | 0.1 |
| Sheep, meat | 0.1 |
| Sheep, mbyp | 0.1 |
| Wheat, grain | 0.2 |
| Wheat, straw | 0.75 |

*Carcinogen/Hazard Classifications*
**U.S. EPA Carcinogens:** Suggested Carcinogen
**Label Signal Word:** WARNING or CAUTION
**WHO Acute Hazard:** Class III, slightly hazardous
*Regulatory Authority:*
- Air Pollutant Standard Set (former USSR)[43]
- AB 1803-Well Monitoring Chemical (CAL)
- The "Director's List" (CAL/OSHA)
- Actively registered pesticide in California.
- Safe Drinking Water Act: Priority List (55 FR 1470) as DCPA (and its acid metabolites)
- EPCRA Section 313 Form R *de minimis* concentration reporting level: 1.0%
- Canada, WHMIS, Ingredients Disclosure List

*Description:* Propanil is a medium to dark gray crystalline solid. The technical product is a brown crystalline solid. Slightly soluble in water; solubility =125 ppm @ 22°C. Molecular weight 218.1. Melting/Freezing point = 92–93°C (pure); 87–89°C (technical grade). Melting/Freezing point (pure) = 89–92°C (pure); 85–89°C (technical grade). Vapor pressure = $2.6 \times 10^{-7}$ mmHg @ 30°C. Log $K_{ow}$ = 3.12. Values > 3.0. May bioaccumulate in marine organisms.

*Permissible Exposure Limits in Air:* The former USSR-UNEP/IRPTC project[43] has set an MAC in workplace air of 0.1 mg/m³ and has set MAC values for ambient air in residential areas of 0.005 mg/m³ on a once-daily basis and 0.001 mg/m³ on a daily average basis.

*Permissible Concentration in Water:* The former USSR-UNEP/IRPTC project[43] has set an MAC in water bodies used for domestic purposes of 0.1 mg/L.

*Routes of Entry:* Inhalation, ingestion, skin and/or eye contact.

*Harmful Effects and Symptoms*
*Short Term Exposure:* Causes blue lips or fingernails, blue skin, headaches and dizziness. Propanil is well tolerated by experimental animals on a chronic basis, and there is little or no indication of mutagenic or oncogenic properties of the compound. The highest no-adverse-effect concentration of propanil based on reproduction in the rat and acute, subchronic, and chronic studies in rats and dogs is 400 pm in the diet. Based on this data, ADI was calculated at 0.02 mg/kg/day.

*Points of Attack:* Digestive system, skin, nails, central nervous system.

*First Aid:* If this chemical gets into the eyes, remove any contact lenses at once and irrigate immediately for at least 15 minutes, occasionally lifting upper and lower lids. Seek medical attention immediately. If this chemical contacts the skin, remove contaminated clothing and wash immediately with soap and water. Seek medical attention immediately. If this chemical has been inhaled, remove from exposure, begin rescue breathing (using universal precautions) if breathing has stopped, and CPR if heart action has stopped. Transfer promptly to a medical facility. When this chemical has been swallowed, get medical attention. Give large quantities of water and induce vomiting. Do not make an unconscious person vomit.

*References:*
- EXTOXNET, Extension Toxicology Network, "Pesticide Information Profile, Propanil," Oregon State University, Corvallis, OR (June 1986)
- U.S. Environmental Protection Agency, Office of Pesticide Programs, Pesticide Residue Limits, "Propanil," 40 CFR 180.274, http://www.epa.gov/pesticides/food/viewtols.htm
- California Environmental Protection Agency "Chemical List of Lists," Sacramento CA (February 1997).

# Propargyl Bromide

*Use Type:* A soil fumigant and nematicide
*CAS Number:* 106-96-7
*Formula:* $C_3H_3Br$; $BrCH_2C\equiv CH$
*Alert:* Not registered as a pesticide in the U.S.
*Synonyms:* γ-Bromoallylene; 3-Bromopropyne; 3-Bromo-1-propyne; Bromuro de propargilo (Spanish)
*Producers:* Advances Synthesis Technologies (USA); Dhruv Chemical Industries (India); GFS Chemicals(USA); Sigma-Aldrich Fine Chemicals (USA)
*Chemical Class:* Halogenated organic compound
*EPA/OPP PC Code:* 068701
*California DPR Chemical Code:*
*RTECS Number:* UK4375000
*EINECS Number:* 203-447-1

*Uses:* This material is used as a soil fumigant but is not registered as a pesticide in the U.S. It is being investigated as a possible replacement for methyl bromide in agricultural applications. It is also used as an intermediate in organic synthesis and as a corrosion inhibitor.

## Propazine

*U.S. Maximum Allowable Residue Levels:*
*Note*: inorganic bromides resulting from soil treatment with combinations of chloropicrin, methyl bromide and propargyl bromide (40 CFR 180.199):

| CROP | ppm |
|---|---|
| Asparagus | 300 |
| Broccoli | 25 |
| Cauliflower | 25 |
| Eggplant | 60 |
| Ginger, roots | 100 |
| Lettuce | 300 |
| Muskmelon | 40 |
| Onion, dry bulb | 300 |
| Pepper | 25 |
| Pineapple | 25 |
| Strawberry | 25 |
| Tomato | 40 |

*Regulatory Authority:*
- Superfund/EPCRA 40CFR355, Appendix B Extremely Hazardous Substances: TPQ = 10 lb (4.54 kg)
- Superfund/EPCRA 40CFR302.4 RQ: EHS, 1 lb (0.454 kg)

*Description:* Propargyl bromide is a colorless liquid. Sharp odor; a lacrimator. Molecular weight = 118.98. Boiling point = 89–90°C. Melting/Freezing point = –61°C. Flash point = 10°C (oc). Autoignition temperature = 324°C. Explosive limits: LEL = 3.0%; UEL = ?. Hazard Identification (based on NFPA-704 M Rating System): Health 3, Flammability 3, Reactivity 4.

*Incompatibilities:* Violent reaction with oxidizers. Becomes shock- or heat-sensitive when mixed with trichloronitromethane or chloropicrin. Detonates when heated to 220°C, or when heated while confined. May explode on contact with copper, copper alloys, mercury, silver.

*Permissible Exposure Limits in Air:* No standards set.

*Permissible Concentration in Water:* No criteria set, but runoff from spills or fire control may cause water pollution.

*Routes of Entry:* Inhalation, ingestion, skin and/or eye contact.

*Harmful Effects and Symptoms*

**Short Term Exposure:** This material is very toxic via the oral route; may be harmful if inhaled. Contact may cause burns to skin and eyes. Symptoms of exposure include skin irritation and tearing of the eyes.

*First Aid:* If this chemical gets into the eyes, remove any contact lenses at once and irrigate immediately for at least 15 minutes, occasionally lifting upper and lower lids. Seek medical attention immediately. If this chemical contacts the skin, remove contaminated clothing and wash immediately with soap and water. Seek medical attention immediately. If this chemical has been inhaled, remove from exposure, begin rescue breathing (using universal precautions) if breathing has stopped, and CPR if heart action has stopped. Transfer promptly to a medical facility. When this chemical has been swallowed, get medical attention. Give large quantities of water and induce vomiting. Do not make an unconscious person vomit. Keep victim quiet and maintain normal body temperature.

*References:*
- U.S. Environmental Protection Agency, Office of Pesticide Programs, "Pesticide Residue Limits," 40 CFR 180.199, http://www.epa.gov/pesticides/food/viewtols.htm
- U.S. Environmental Protection Agency, "Chemical Profile: Propargyl Bromide," Washington, DC, Chemical Emergency Preparedness Program (November 30, 1987).
- California Environmental Protection Agency "Chemical List of Lists," Sacramento CA (February 1997).

## Propazine (ANSI)

*Use Type:* Pre-emergence herbicide
*CAS Number:* 139-40-2
*Formula:* $C_9H_{16}ClN_5$
*Alert:* Some formulations are Restricted Use Pesticides (RUP).
*Synonyms:* 2,4-Bis(isopropylamino)-6-chloro-*s*-triazine; 2,4-Bis(propylamino)-6-chlor-1,3,5-triazin (German); 2-Chloro-4,6-bis(isopropylamino)-*s*-triazine; Propasin
*Trade Names:* GEIGY® 30,028; GESAMIL®; MAXX-90®; MILOCEP®; MILOGARD®, Syngenta (Switzerland), canceled 12/23/1988; MILO-PRO®; PLANTULIN®; PRIMATOL P®; PROPAZIN®; PROPINEX®; PROZINEX®
*Producers:* Ciba-Geigy (Switzerland); Ehrenstorfer, Dr. (Germany); Griffin L.L.C. (USA); Makhteshim Agan (Israel); Nissan Chemical Industries (Japan); Sigma-Aldrich Laborchemikalien (Germany); Syngenta (Switzerland)
*Chemical Class:* Triazine
*EPA/OPP PC Code:* 080808
*California DPR Chemical Code:* 504
*ICSC Number:* 0697
*RTECS Number:* XY5300000
*EINECS Number:* 205-359-9
*Uses:* Atrazine, simazine, and propazine and their common chlorinated degradates have a common mechanism of toxicity. They have similar applications. Propazine is used for control of broadleaf weeds and annual grasses in sweet sorghum. It is applied as a spray at the time of planting or immediately following planting, but prior to weed or sorghum emergence. It is also used as a post-emergence selective herbicide on carrots, celery and fennel.

*U.S. Maximum Allowable Residue Levels for Propazine (40 CFR 180.243):*

| CROP | ppm |
|---|---|
| Sorghum, forage | 0.25 |
| Sorghum, grain | 0.25 |

| | |
|---|---|
| Sorghum, grain, stover | 0.25 |
| Sorghum, sweet | 0.25 |

*Carcinogen/Hazard Classifications*
**U.S. EPA Carcinogens:** Group C, possible carcinogen
**Label Signal Word:** CAUTION
**WHO Acute Hazard:** Class U, unlikely to be hazardous
*Regulatory Authority:*
- Air Pollutant Standard Set (former USSR)[43]
- AB 1803-Well Monitoring Chemical (CAL)

*Description:* Propazine is a colorless, crystalline solid or powder. Slightly soluble in water; solubility = 9.0 ppm @ 20°C; 4.5 x $10^{-4}$ g/100 mL @ 20°C. Molecular weight = 230.09. Melting/Freezing point = 212–214°C. Vapor pressure = 1.3 x $10^{-7}$ mmHg @ 20°C. Log $K_{ow}$ = ~3.0. Values at or above 3.0 are likely to bioaccumulate in marine organisms.

*Permissible Exposure Limits in Air:* The former USSR-UNEP/IRPTC project[43] has set a MAC in workplace air of 5.0 mg/m³ and an MAC in ambient air in residential areas of 0.04 mg/m³ both on a momentary and an average daily basis.

*Permissible Concentration in Water:* The former USSR-UNEP/IRPTC project[43] has set an MAC in water bodies used for domestic purposes of 1.0 mg/L. The USEPA has determined a lifetime health advisory of 0.014 mg/L (14 µg/L) (see Reference below). States which have set guidelines for propazine in drinking water[61] include Kansas at 325 µg/L and Maine at 93 µg/L.

*Determination in Water:* Analysis of propazine is by a gas chromatographic (GC) method applicable to the determination of certain nitrogen-phosphorus containing pesticides in water samples. In this method, approximately 1 liter of sample is extracted with Methylene chloride. The extract is concentrated and the compounds are separated using capillary column GC. Measurement is made using a nitrogen-phosphorus detector. The method detection limit has not been determined for propazine, but it is estimated that the detection limits for analytes included in this method are in the range of 0.1 to 2 ug/L.

*Routes of Entry:* Inhalation, ingestion, skin and/or eye contact. May be absorbed by the skin.

*Harmful Effects and Symptoms*
*Short Term Exposure:* May cause eye irritation. Contact dermatitis was reported in workers involved in propazine manufacturing. Poisonous if ingested. $LD_{50}$ in range (rat) 1780-7000 mg/kg. No other information on the health effects of propazine in humans was found in the available literature.
*Long Term Exposure:* May cause skin allergy.
*Points of Attack:* Skin
*Medical Surveillance:* Examination by a qualified allergist.
*First Aid:* If this chemical gets into the eyes, remove any contact lenses at once and irrigate immediately for at least 15 minutes, occasionally lifting upper and lower lids. Seek medical attention immediately. If this chemical contacts the skin, remove contaminated clothing and wash immediately with soap and water. Seek medical attention immediately. If this chemical has been inhaled, remove from exposure, begin rescue breathing (using universal precautions) if breathing has stopped, and CPR if heart action has stopped. Transfer promptly to a medical facility. When this chemical has been swallowed, get medical attention. Give large quantities of water and induce vomiting. Do not make an unconscious person vomit.

*References:*
- EXTOXNET, Extension Toxicology Network, "Pesticide Information Profile, Propazine," Oregon State University, Corvallis, OR (June 1996). http://extoxnet.orst.edu/pips/propazin.htm
- EPA, Office of Pesticide Programs, Pesticide Residue Limits, "Propazine," 40 CFR 180.243, http://www.epa.gov/pesticides/food/viewtols.htm
- U.S. Environmental Protection Agency, "Pesticide Fact Sheet: Propazine," Washington, DC, Office of Prevention, Pesticides and Toxic Substances (September 1998). http://www.epa.gov/opprd001/factsheets/propazine.pdf
- California Environmental Protection Agency "Chemical List of Lists," Sacramento CA (February 1997).

# Propamocarb Hydrochloride

*Use Type:* Fungicide
*CAS Number:* 25606-41-1
*Formula:* $C_9H_{21}ClN_2O_2$
*Synonyms:* Carbamic acid, [3-(dimethylamino)propyl]-, propyl ester, monohydrochloride; N-(γ-Dimethylaminopropyl)carbamic acid propyl ester, monohydrochloride; Propyl [3-(dimethylamino)propyl]carbamate monohydrochloride
*Trade Names:* BANOL®, Bayer CropScience (Germany); PREVEX®; PREVICUR®, Bayer CropScience (Germany); SH-66752®; TATTOO®, Bayer CropScience (Germany)
*Producers:* Agsin (Singapore); Bayer CropScience (Germany)
*Chemical Class:* Carbamates
*EPA/OPP PC Code:* 119302
*California DPR Chemical Code:* 4022
*Uses:* Propamocarb hydrochloride is used to the plant disease "damping-off" and has fungicidal activity against *Pythium* spp. and *Phytophthora* spp. It is used on non-food sites such as ornamental lawns and turf, sod farms, plants, vines and woody plants.
*U.S. Maximum Allowable Residue Levels for Propamocarb Hydrochloride (40 CFR 180.499):*

| CROP | ppm |
|---|---|
| Tomato | 2.0 |

## 748  Propamocarb Hydrochloride

Tomato, paste      5.0
Tomato, puree      1.0

***Carcinogen/Hazard Classifications***
**U.S. EPA Carcinogens:** Not likely a Carcinogen
**Label Signal Word:** CAUTION or DANGER
***Description:*** Colorless to yellow granules or liquid. Odorless. Highly soluble in water. Molecular weight = 224.73.
***Incompatibilities:*** May form explosive materials with phosphorus pentachloride. When heated to decomposition or on contact with acids or acid fumes, it may produce highly toxic chloride fumes; deadly phosgene gas may be formed. May cause pitting of some metals.
***Permissible Concentration in Water:*** No criteria set. Runoff from spills or fire control may cause water pollution.
***Routes of Entry:*** Skin absorption, ingestion, inhalation.
***Harmful Effects and Symptoms***
***Short Term Exposure:*** Eye pupils are small; blurred vision; eye watering; runny nose; cough; shortness of breath; salivation; nausea, stomach cramps, diarrhea, and vomiting; increased blood pressure; profuse sweating; hypermotility, hallucinations; agitation; tingling of the skin; slow heartbeat; convulsions; fluid in lungs; loss of consciousness; incontinence; breathing stops; death. Carbamates inhibit the acetylcholinesterase enzymes and alter the way in which nervous impulses are transmitted. However, within several hours carbamates spontaneously detach from the enzymes.
***Long Term Exposure:*** A potent cholinesterase inhibitor; cumulative effect is possible. This chemical may damage the nervous system with repeated exposure, resulting in convulsions, respiratory failure. May cause liver damage.
***Points of Attack:*** Respiratory system, lungs, central nervous system, cardiovascular system, skin, eyes, plasma and red blood cell cholinesterase.
***Medical Surveillance:*** Medical observation is recommended for 24 to 48 hours after breathing overexposure, as pulmonary edema may be delayed. As first aid for pulmonary edema, consider administering a corticosteroid spray. Cigarette smoking may exacerbate pulmonary injury and should be discouraged for at least 72 hours following exposure. Before employment and at regular times after that, the following are recommended: Plasma and red blood cell cholinesterase levels (tests for the enzyme poisoned by this chemical). If exposure stops, plasma levels return to normal in 1-2 weeks while red blood cell levels may be reduced for 1-3 months. When acetylcholinesterase enzyme levels are reduced by 25% or more below preemployment levels, risk of poisoning is increased, even if results are in lower ranges of "normal." Reassignment to work not involving carbamate pesticides is recommended until enzyme levels recover. If symptoms develop or overexposure occurs, repeat the above tests as soon as possible and get an exam of the nervous system. Also consider complete blood count. Consider chest x-ray following acute overexposure.

***First Aid:*** Speed in removing material from eyes and skin is of extreme importance. *Eyes:* Eye contact can cause dangerous amounts of these chemicals to be quickly absorbed through the mucous membrane into the bloodstream. Immediately and gently flush eyes with plenty of warm or cold water (NO hot water) for at least 15 minutes, occasionally lifting the upper and lower eyelids. Get medical aid immediately. *Skin:* Get medical aid. Skin contact can cause dangerous amounts of these chemicals to be absorbed into the bloodstream. Wearing the appropriate PPE equipment and respirator for carbamate pesticides, immediately flush skin with plenty of soap and water for at least 15 minutes while removing contaminated clothing and shoes. Shampoo hair promptly if contaminated; protect eyes. *Ingestion:* Call poison control. Loosen all clothing. Never give anything by mouth to an unconscious person. Get medical aid. *Do NOT induce vomiting.\** If conscious, alert, and able to swallow, rinse mouth and have victim drink 4 to 8 ounces of water. Check to see if poison control instructs you to use ipecac syrup, otherwise administer slurry of activated charcoal (2 oz in 8 oz of water). If victim is *unconscious or having convulsions,* do nothing except keep victim warm. *\*In some cases you may be specifically instructed by poison control to induce vomiting by way of 2 tablespoons of syrup of ipecac (adult) washed down with a cup of water.* Do NOT give activated charcoal before or with ipecac syrup. *Inhalation:* Get medical aid. Do not contaminate yourself. Wearing the appropriate PPE equipment and respirator for carbamate pesticides, immediately remove the victim from the contaminated area to fresh air. If the victim is not breathing, administer artificial respiration. Do NOT use mouth-to-mouth resuscitation; use bag/mask apparatus. If breathing is difficult, administer oxygen through bag/mask apparatus until medical help arrives. Do not leave victim unattended. *Note to physician or authorized medical personnel.* Administer atropine, 2 mg (1/30 gr) intramuscularly or intravenously as soon as any local or systemic signs or symptoms of an intoxication are noted; repeat the administration of atropine every 3 to 8 minutes until signs of atropinization (mydriasis, dry mouth, rapid pulse, hot and dry skin) occur; initiate treatment in children with 0.05 mg mg/kg of atropine; repeat at 5 to 10 minute intervals. Watch respiration, and remove bronchial secretions if they appear to be obstructing the airway; intubate if necessary. *Medical note:* 2-PAMCI may be contraindicated in the case of some carbamate poisonings.

***References:***
- U.S. Environmental Protection Agency, Office of Pesticide Programs, Pesticide Residue Limits, "Propamocarb Hydrochloride," 40 CFR 180.499, www.epa.gov/pesticides/food/viewtols.htm

- U.S. Environmental Protection Agency, "Reregistration Eligibility Decision (RED), Propamocarb Hydrochloride," Office of Prevention, Pesticides and Toxic Substances, Washington, DC http://www.epa.gov/REDs/3124red.pdf
- California Environmental Protection Agency "Chemical List of Lists," Sacramento CA (February 1997)

# Propargite (ANSI)

*Use Type:* Miticide and acaricide
*CAS Number:* 2312-35-8
*Formula:* $C_{19}H_{26}O_4S$
*Alert:* Propargite is a flammable liquid and a fire hazard.
*Synonyms:* A13-27226; BPPS; 2-(*p-tert*-Butylphenoxy)cyclohexyl propargyl sulfite; Caswell No. 1301; Cyclosulfyne; 2-[4-(1,1-dimethylethyl)phenoxy]cyclohexyl 2-propynyl sulfite; 2-[4-(1,1-Dimethylethyl)phenoxy]cyclohexyl 2-propynyl sulfurous acid; ENT 27,226; Propargil; Propargita (Spanish); Sulfurous acid, 2-[4-(1,1-dimethylethyl)phenoxy]cyclohexyl 2-propynyl ester; Sulfurous acid, 2-(*p-tert*-butylphenoxy)cyclohexyl-2-propynyl ester; 2-(*p-tert*-Butylphenoxy)cyclohexyl 2-propynyl sulfite; 2-(*p-tert*-Butylphenoxy)cyclohexyl propargyl sulfite; 2-(4-*tert*-Butylphenoxy)cyclohexyl prop-2-ynyl sulfite
*Trade Names:* COMITE II®, Crompton Corporation (USA); COMITE® AGRICULTURAL MITICIDE, Crompton Corporation (USA); D-014®, Crompton Corporation (USA); DIBROM OMITE®, Ortho Business Group (USA); DICTATOR®, Veterinary & Agricultural Products Manufacturing Co. Ltd. (VAPCO) (Jordan); NAUGATUCK® D-014, Crompton Corporation (USA); OMAIT®; OMITE®, Crompton Corporation (USA); RED-TOP®, canceled; UNIROYAL® D-014, Crompton Corporation (USA); U.S. RUBBER D-014®, Crompton Corporation (USA)
*Producers:* Agrimor International (USA); Agsin (Singapore); AJE (Switzerland); Crompton Corporation (USA); Ortho Business Group (USA); Veterinary & Agricultural Products Manufacturing Co. Ltd. (VAPCO) (Jordan); Wuzhou International (China)
*Chemical Class:* Organosulphite
*EPA/OPP PC Code:* 097601
*California DPR Chemical Code:* 445
*RTECS Number:* WT2900000
*Uses:* Used on a variety of bearing and non-bearing food crops and non-food sites. Grapes, walnuts, almonds, nectarines and mint are the most highest treated crops. Other crops include alfalfa, avocados, beans, boysenberries, carrots, cherries, citrus, corn, currants, dates, filberts, grapefruit, jojoba, grapes, hops, peanuts, sugar beets, cotton and ornamentals.

*Carcinogen/Hazard Classifications*
**U.S. EPA Carcinogens:** Group B2, probable carcinogen
***California Prop. 65:*** Carcinogen and developmental toxin
**U.S. TRI:** Developmental toxin
**Label Signal Word:** CAUTION, WARNING or DANGER
**WHO Acute Hazard:** Class III, slightly hazardous
*Regulatory Authority:*
- EPA 40 CFR 372.65, Specific Toxic Chemical Listings
- AB 1803-Well Monitoring Chemical (CAL)
- Proposition 65 chemical (CAL)
- The "Director's List" (CAL/OSHA)
- Actively registered pesticide in California.
- Clean Water Act: Section 311 Hazardous Substances/RQ (same as CERCLA)
- EPCRA Section 304 RQ: CERCLA, 10 lb (4.54 kg)
- EPCRA Section 313 Form R *de minimis* concentration reporting level: 1.0%

*Description:* Flammable, dark amber viscous liquid. Faint solvent odor. A strong odor of sulfur dioxide may indicate that partial decomposition has occurred. Practically insoluble in water; sinks slowly in water. Molecular weight = 350.472. Density = 1.085-1.115 @ 25°C. Boiling point = Decomposes. Vapor pressure = $3 \times 10^{-3}$ mmHg @ 20°C. Flash point = 28°C (oc). Specific gravity: 1.085-1.115 @ 20°C. Melting/Freezing point = 70°C. Hazard Identification (based on NFPA-704 M Rating System): Health 2, Flammability 2, Reactivity 0.
*Incompatibilities:* Keep away from oxidizers, sulfuric acid, caustics, ammonia, aliphatic amines, alkanolamines, isocyanates, alkylene oxides, epichlorohydrin. May react with water, steam, or acids forming corrosive acid solution and sulfur oxide fumes. Incompatible with lead diacetate, mercury(I) chloride.
*Permissible Concentration in Water:* No criteria set. Runoff from spills or fire control may cause water pollution.
*Harmful Effects and Symptoms*
*Short Term Exposure:* Contact may burn eyes, skin, and respiratory tract. Toxic if ingested. If spilled on clothing and allowed to remain may cause smarting and reddening of skin. Grade 2; $LD_{50}$ = 0.5 to 5 g/kg
*Points of Attack:* Ingestion, skin contact.
*Medical Surveillance:* Consult a physician if poisoning is suspected or if redness, itching, or burning of the eyes or skin develop.
*First Aid:* If this chemical gets into the eyes, remove any contact lenses at once and irrigate immediately for at least 15 minutes, occasionally lifting upper and lower lids. Seek medical attention immediately. If this chemical contacts the skin, remove contaminated clothing and wash immediately with soap and water. Seek medical attention immediately. If this chemical has been inhaled, remove from exposure, begin rescue breathing (using universal precautions) if breathing has stopped, and CPR if heart action has stopped. Transfer promptly to a medical facility. When this chemical

has been swallowed, get medical attention. *Do not induce vomiting when formulations containing petroleum solvents are ingested.* Otherwise, give large quantities of water and induce vomiting. Do not make an unconscious person vomit.

*References:*
- U.S. Environmental Protection Agency, "Reregistration Eligibility Decision (RED), Propargite, Case No.0243," Office of Prevention, Pesticides and Toxic Substances, Washington, DC http://www.epa.gov/REDs/propargite_red.pdf
- New Jersey Department of Health and Senior Services, "Hazardous Substance Fact Sheet, Propargite," Trenton NJ (January 2002). http://www.state.nj.us/health/eoh/rtkweb/1596.pdf
- International Programme on Chemical Safety (IPCS), "Propargite," Geneva, Switzerland (1978). http://www.inchem.org/documents/jmpr/jmpmono/v078pr26.htm
- California Environmental Protection Agency "Chemical List of Lists," Sacramento CA (February 1997)

## Propetamphos (ANSI)

*Use Type:* Fungicide, insecticide
*CAS Number:* 31218-83-4
*Formula:* $C_{10}H_{20}NO_4PS$
*Alert:* Some formulations may be Restricted Use Pesticides (RUP).
*Synonyms:* 2-Butenoic acid, 3-[((ethylamino)methoxyphosphinothioyl)oxy]-, isopropylester, *(E)*-; 2-Butenoic acid, 3-[((ethylamino)methoxyphosphinothioyl)oxy]-,1-methylethyl ester, *(E)*-; Caswell No. 706A; Crotonic acid, 3-hydroxy-, isopropyl ester, O-ester with O-methyl ethylphosphoramidothioate, *(E)*-; ENT 27989; 3-[(Ethylamino)methoxyphosphinothioyl)oxy]-2-butenoic acid, 1-methylethyl ester; *(E)*-O-2-Isopropoxycarbonyl-1-methylvinyl O-methylethylphosphoramidothioate; O-(1-Isopropoxycarbonyl-1-propen-2-yl)-O-methyl-ethyl-phosphoramidothionate; Isopropyl 3-[((ethylamino)methoxyphosphinothioyl]oxy)crotonate; Isopropyl 3-(ethylamino(methoxy)phosphinothioyloxy) isocrotonate; 1-Methylethyl*(E)*-3-[((ethylamino)methoxyphosphinothioyl)oxy]-2-butenoate; *(E)*-1-Methylethyl 3-[((ethylamino)methoxy phosphinothioyl)oxy]-2-butenoate; Z-O-2-iso-Propoxycarbonyl-1-methylvinyl O-methyl ethyl phosphoramidothioate; Phosphoramidothioic acid, *N*-ethyl-, *(E)*-O-(2-isopropoxycarbonyl-1-methylvinyl) O-methyl ester
*Trade Names:* BLOTIC®; OVIDIP®; SAFROTIN®, Wellmark (USA), canceled; SAN-52139®, Syngenta (Switzerland); SANDOZ®- 52139, Syngenta (Switzerland); SERAPHOS®; TSAR®; VEL-4283®; ZOECON®, Wellmark (USA)

*Producers:* Syngenta (Switzerland); Wellmark (USA)
*Chemical Class:* Organophosphate
*EPA/OPP PC Code:* 113601; (216800 old EPA code number)
*California DPR Chemical Code:* 2122
*RTECS Number:* GQ4750000
*Uses:* Propetamphos is used indoors for the control of structural insects, e.g., ants, cockroaches, fleas, and termites. It is applied at indoor residential, medical, commercial, and industrial buildings, and in food service establishments where there is no contact with food, and where no processing, packing, or warehousing of food occurs. It is also used in veterinary practice to control ticks, lice, mites and other parasites in livestock.
*Carcinogen/Hazard Classifications*
**U.S. EPA Carcinogens:** Unlikely carcinogen
**Label Signal Word:** CAUTION or WARNING
**WHO Acute Hazard:** Class 1b, highly hazardous
*Regulatory Authority:*
- EPA 40 CFR 372.65, Specific Toxic Chemical Listings
- Actively registered pesticide in California.
- EPCRA Section 313 Form R *de minimis* concentration reporting level: 1.0%
- DOT Inhalation Hazard Chemicals

*Description:* Yellowish, oily liquid at room temperature. Practically insoluble in water. Molecular weight = 281.30. Density = 1.13 @ 20°C. Boiling point = 88°C @ 0.005 mmHg.
*Determination in Air:* OSHA versatile sampler-2; Toluene/Acetone; Gas chromatography/Flame ionization detection; NIOSH IV, Method #5600, Organophosphorus Pesticides.[18]
*Permissible Concentration in Water:* No criteria set. Runoff from spills or fire control may cause water pollution.
*Harmful Effects and Symptoms*
*Short Term Exposure:* Eye pupils are small; blurred vision; eye watering; runny nose; cough; shortness of breath; salivation; dizziness; nausea, stomach cramps, diarrhea, and vomiting; increased blood pressure; profuse sweating; hypermotility, hallucinations; irritability; tingling of the skin; drowsiness; slow heartbeat; convulsions; fluid in lungs; loss of consciousness; incontinence; breathing stops; death. Organophosphates inhibit the action of acetylcholinesterase enzymes, and alter the way in which nervous impulses are transmitted. The effects can last for hours, days, or much longer. The action of the enzymes is reestablished after new enzymes are formed.
*Long Term Exposure:* Cholinesterase inhibitor; cumulative effect is possible. Organophosphates may damage the nervous system with repeated exposure, resulting in convulsions, respiratory failure. May cause liver damage.
*Points of Attack:* Respiratory system, central nervous system, cardiovascular system, blood cholinesterase.

*Medical Surveillance:* Medical observation is recommended for 24 to 48 hours after breathing overexposure, as pulmonary edema may be delayed. As first aid for pulmonary edema, consider administering a corticosteroid spray. Cigarette smoking may exacerbate pulmonary injury and should be discouraged for at least 72 hours following exposure.

*First Aid:* Speed in removing material from eyes and skin is of extreme importance. *Eyes:* Eye contact can cause dangerous amounts of these chemicals to be quickly absorbed through the mucous membrane into the bloodstream. Immediately and gently flush eyes with plenty of warm or cold water (NO hot water) for at least 15 minutes, occasionally lifting the upper and lower eyelids. Get medical aid immediately. *Skin:* Get medical aid. Skin contact can cause dangerous amounts of these chemicals to be absorbed into the bloodstream. Wearing the appropriate PPE equipment and respirator for organophosphate pesticides, immediately flush skin with plenty of soap and water for at least 15 minutes while removing contaminated clothing and shoes. Shampoo hair promptly if contaminated. *Ingestion:* Call poison control. Loosen all clothing. Never give anything by mouth to an unconscious person. Get medical aid. Do NOT induce vomiting.* If conscious, alert, and able to swallow, rinse mouth and have victim drink 4 to 8 ounces of water do NOT induce vomiting but immediately administer slurry of activated charcoal (2 oz in 8 oz of water). If victim is *unconscious or having convulsions,* do nothing except keep victim warm. **In some cases you may be specifically instructed by poison control to induce vomiting by way of 2 tablespoons of syrup of ipecac (adult) washed down with a cup of water.* Do NOT give activated charcoal before or with ipecac syrup. *Inhalation:* Get medical aid. Do not contaminate yourself. Wearing the appropriate PPE equipment and respirator for organophosphate pesticides, immediately remove the victim from the contaminated area to fresh air. If the victim is not breathing, administer artificial respiration. Do NOT use mouth-to-mouth resuscitation; use bag/mask apparatus. If breathing is difficult, administer oxygen through bag/mask apparatus until medical help arrives. Do not leave victim unattended. *Note to physician or authorized medical personnel.* Administer atropine, 2 mg (1/30 gr) intramuscularly or intravenously as soon as any local or systemic signs or symptoms of an intoxication are noted; repeat the administration of atropine every 3 to 8 minutes until signs of atropinization (mydriasis, dry mouth, rapid pulse, hot and dry skin) occur; initiate treatment in children with 0.05 mg mg/kg of atropine; repeat at 5 to 10 minute intervals. Watch respiration, and remove bronchial secretions if they appear to be obstructing the airway; intubate if necessary. *Notes to physician or authorized medical personnel*: $N$-methylpyridinium-2-aldoxime (2-PAMCI) when used in conjunction with atropine reacts with the phosphorylated cholinesterase, thereby restoring normal activity to by removing the phosphorylating group. The combination of these two chemicals is synergistic and must be administered within minutes to a few hours following exposure (depending on the specific agent) to be effective. Give 2-PAMCI (Pralidoxime; Protopam), 2.5 gm in 100 ml of sterile water or in 5% dextrose and water, intravenously, slowly, in 15-30 minutes; if sufficient fluid is not available, give 1 gm of 2-PAMCI in 3 ml of distilled water by deep intramuscular injection; repeat this every half hour if respiration weakens or if muscle fasciculation or convulsions recur. Also Diazepam, an anticonvulsant, might be considered.

*References:*
- U.S. Environmental Protection Agency, "Interim Reregistration Eligibility Decision (IRED), Propetamphos," Office of Prevention, Pesticides and Toxic Substances, Washington, DC (October 2000). http://www.epa.gov/REDs/2550ired.pdf
- EXTOXNET, Extension Toxicology Network, "Pesticide Information Profile, Propetamphos," Oregon State University, Corvallis, OR (June 1996). http://extoxnet.orst.edu/pips/propetam.htm
- California Environmental Protection Agency "Chemical List of Lists," Sacramento CA (February 1997).

# Propham

*Use Type:* Herbicide and plant growth regulator
*CAS Number:* 122-42-9
*Formula:* $C_{10}H_{13}NO_2$; $C_6H_5NHCOOCH(CH_3)_2$
*Alert:* There are no propham products currently registered in the U.S.
*Synonyms:* Carbanilic acid, isopropyl ester; IFC (Russia); IFC; IFK; IPC; IPPC; Isopropil-$N$-fenil-carbammato (Italian); Isopropyl carbanilate; Isopropyl carbanilic acid ester; Isopropyl-$N$-fenyl-carbamaat (Dutch); Isopropyl-$N$-phenyl-carbamat (German); Isopropyl phenylcarbamate; Isopropyl-$N$-phenylcarbamate; $o$-Isopropyl-$N$-phenyl carbamate; Isopropyl-$N$-phenyl carbamate; Isopropyl-$N$-phenyurethan (German); $N$-Phenylcarbamate d'isopropyle (French); Phenyl carbamic acid-1-methylethyl ester; $N$-Phenyl isopropyl carbamate
*Trade Names:* AGERMIN®; BAN-HOE®; BEET-KLEEN®; BIRGIN®; CHEM-HOE®, PPG Industries (USA), canceled 10/10/1989; COLLAVIN®; ORTHO GRASS KILLER®; PREMALOX®; PROFAM®; PROFOS®; TIXIT®; TRIHERBIDE®; TRIHERBIDE-IPC®; TUBERIT®; TUBERITE®
*Producers:* Bayer CropScience (Germany); Ishihara Sangyo Kaisha (Japan); Nissan Chemical Industries (Japan); Sigma-Aldrich Laborchemikalien (Germany)
*Chemical Class:* Isopropyl Carbanilate

*EPA/OPP PC Code:* 047601
*California DPR Chemical Code:* 339
*RTECS Number:* ED9100000
*EINECS Number:* 204-542-0
*Uses:* Propham is the active substance in products used as herbicides and to prevent potato sprouting while potatoes are in storage. It is a pre- and post-emergent herbicide to control annual broadleaf weeds and grasses in alfalfa, white clover, red or crimson clover, flax, lettuce, safflower, spinach, sugar beets, lentils, peas and fallow land. There are no propham products currently registered in the U.S.
*Carcinogen/Hazard Classifications*
**IARC:** Group 3, unclassifiable
**Label Signal Word:** WARNING or DANGER
*Regulatory Authority:*
- Air Pollutant Standard Set (former USSR)[43]
- AB 1803-Well Monitoring Chemical (CAL)
- The "Director's List" (CAL/OSHA)
- EPA Hazardous Waste Number (RCRA No.): U363
- RCRA, 40CFR261, Appendix 8 Hazardous Constituents
- Superfund/EPCRA 40CFR302.4 RQ: EHS, 1 lb (0.454 kg)

*Description:* Propham is a colorless crystalline solid. Practically insoluble in water. Melting/Freezing point = 87–88°C; 84°C (technical grade). Boiling point = Decomposition above 150°C. Log $K_{ow}$ = negative. Unlikely to bioaccumulate in marine organisms.

*Permissible Exposure Limits in Air:* The former USSR-UNEP/IRPTC project[43] has set an MAC in workplace air of 2.0 mg/m$^3$ and an MAC in ambient air of residential areas of 0.02 mg/m$^3$, both on a momentary and an average daily basis.

*Permissible Concentration in Water:* The former USSR-UNEP/IRPTC project[43] has set an MAC in water bodies used for domestic purposes of 0.2 mg/L. The U.S. EPA has set a lifetime health advisory of 0.12 mg/L. (See Reference Below).

*Determination in Water:* Analysis of propham is by a high-performance liquid chromatographic (HPLC) method applicable to the determination of certain carbamate and urea pesticides in water samples. This method requires a solvent extraction of approximately 1 liter of sample with methylene chloride using a separatory funnel. The Methylene chloride extract dried and concentrated to a volume of 10 ml or less. Compounds are separated by HPLC, and measurement is conducted with a UV detector. The method detection limit has not been determined for propham, but it is estimated that the detection limits for analytes included in this method are in the range of 1 to 5 $u$g/L.

*Routes of Entry:* Inhalation, ingestion, skin and/or eye contact.

*Harmful Effects and Symptoms*
*Short Term Exposure:* Doses to rats of 2,000 mg/kg produced loss of righting reflex, ptosis, piloerection, decreased locomotor activity, chronic pulmonary disease and rugation, and irregular thickening of the stomach. The acute oral $LD_{50}$ values in male and female rats were reported to be 3,000 ±232 mg/kg and 2,360 ±118 mg/kg, respectively. Carbamates are cholinesterase inhibitors. Symptoms of exposure include headache, giddiness, blurred vision, nervousness, weakness, nausea, cramps, diarrhea, and discomfort in the chest. Signs include sweating, tearing, salivation, vomiting, cyanosis, convulsions, coma, loss of reflexes and loss of sphincter control.

*Long Term Exposure:* Cholinesterase inhibitor; cumulative effect is possible. This chemical may damage the nervous system with repeated exposure, resulting in convulsions, respiratory failure. May cause liver damage.

*Points of Attack:* Respiratory system, lungs, central nervous system, cardiovascular system, skin, eyes, plasma and red blood cell cholinesterase.

*Medical Surveillance:* Before employment and at regular times after that, the following are recommended: Plasma and red blood cell cholinesterase levels (tests for the enzyme poisoned by this chemical). If exposure stops, plasma levels return to normal in 1-2 weeks while red blood cell levels may be reduced for 1-3 months. When acetylcholinesterase enzyme levels are reduced by 25% or more below preemployment levels, risk of poisoning is increased, even if results are in lower ranges of "normal." Reassignment to work not involving carbamate pesticides is recommended until enzyme levels recover. If symptoms develop or overexposure occurs, repeat the above tests as soon as possible and get an exam of the nervous system. Also consider complete blood count. Consider chest x-ray following acute overexposure.

*First Aid:* Speed in removing material from eyes and skin is of extreme importance. If this chemical gets into the eyes, remove any contact lenses at once and irrigate immediately for at least 15 minutes, occasionally lifting upper and lower lids. Seek medical attention immediately. If this chemical contacts the skin, remove contaminated clothing and wash immediately with soap and water. Shampoo hair promptly if contaminated. Seek medical attention immediately. If this chemical has been inhaled, remove from exposure, begin rescue breathing (using universal precautions) if breathing has stopped, and CPR if heart action has stopped. Transfer promptly to a medical facility. When this chemical has been swallowed, get medical attention. Give large quantities of water and induce vomiting. Do not make an unconscious person vomit.

*References:*
- Pesticide Management Education Program, "Propham (Chem Hoe) Herbicide Profile 2/85," Cornell University, Ithaca, NY (February 1985).

http://pmep.cce.cornell.edu/profiles/herb-growthreg/naa-rimsulfuron/propham/herb-prof-propham.html
- U.S. Environmental Protection Agency, "Health Advisory: Propham," Washington, DC, Office of Drinking Water (August 1987).
- California Environmental Protection Agency "Chemical List of Lists," Sacramento CA (February 1997).

# Propiconazole

*Use Type:* Fungicide
*CAS Number:* 60207-90-1
*Formula:* $C_{15}H_{17}Cl_2N_3O_2$
*Synonyms:* Caswell No. 323EE; Desmel; (±)-1-[2-(2,4-dichlorophenyl)-4-propyl-1,3-dioxolan-2-ylmethyl]-1*H*-1,2,4-triazole; 1-[(2-(2,4-Dichlorophenyl)-4-propyl-1,3-dioxolan-2-yl)methyl]-1*H*-1,2,4-triazole; 1-[2-(2,4-Dichlorophenyl)-4-propyl-1,3-dioxolan-2-yl]-methyl-1*H*-1,2,4,-triazole; Proconazole; 1*H*-1,2,4-Triazole, 1-[(2-(2,4-dichlorophenyl)-4-propyl-1,3-dioxolan-2-yl)methyl]-
*Trade Names:* ALAMO®; BANNER®, Syngenta (Switzerland); BENIT®; BREAK®; BUMPER®, Makhteshim-Agan Industries (Israel); CGA-64250®, Syngenta (Switzerland); CGD-92710F®, Syngenta (Switzerland); DESMEL®; FIDIS®; JUNO®; MANTI® S; MAXX®; NOVEL®; ORBIT®, Syngenta (Switzerland); PRACTIS®; PROPIMAX®, Dow AgroSciences (USA); RADAR®; RESTORE®; SPIRE®; STRATEGO® (trifloxystrobin + propiconazole), Bayer CropScience (Germany); TASPA®; TILT®, Syngenta (Switzerland); WOCOSIN®
*Producers:* Bayer CropScience (Germany); Bharat Rasayan (India); Dow AgroSciences (USA); Epochem (China); Ki-Hara Chemicals Ltd. (UK); Makhteshim-Agan Industries (Israel); Sipcam Agro (USA); Syngenta (Switzerland); Yellow River Enterprise (Taiwan)
*Chemical Class:* Triazole
*EPA/OPP PC Code:* 122101
*California DPR Chemical Code:* 2276
*Uses:* Used to control fungi on a broad range of crops and turf. Used on ornamentals, range land and rights-of-way to prevent and control powdery mildew and fungi on hardwoods and conifers.
*U.S. Maximum Allowable Residue Levels for Propiconazole (40 CFR 180.434):*

| CROP | ppm |
|---|---|
| Banana | 0.2 |
| Barley, grain | 0.1 |
| Barley, straw | 1.5 |
| Bean, dry | 0.5 |
| Bean, dry, forage | 8.0 |
| Bean, dry, hay | 8.0 |
| Blueberry | 1.0 |
| Cattle, fat | 0.1 |
| Cattle, kidney | 2.0 |
| Cattle, liver | 2.0 |
| Cattle, meat | 0.1 |
| Cattle, mbyp, except kidney and liver | 0.1 |
| Celery | 5.0 |
| Corn, field, forage | 12.0 |
| Corn, field, grain | 0.1 |
| Corn, field, stover | 12.0 |
| Corn, sweet, kernel plus cob with husks removed | 0.1 |
| Cranberry | 1.0 |
| Fruit, stone, group 12 | 1.0 |
| Goat, fat | 0.1 |
| Goat, kidney | 2.0 |
| Goat, liver | 2.0 |
| Goat, meat | 0.1 |
| Goat, mbyp, except kidney and liver | 0.1 |
| Grain, aspirated fractions | 20.0 |
| Grass, forage | 0.5 |
| Grass, hay | 40.0 |
| Hog, fat | 0.1 |
| Hog, kidney | 2.0 |
| Hog, liver | 2.0 |
| Hog, meat | 0.1 |
| Hog, mbyp, except kidney and liver | 0.1 |
| Horse, fat | 0.1 |
| Horse, kidney | 2.0 |
| Horse, liver | 2.0 |
| Horse, meat | 0.1 |
| Horse, mbyp, except kidney and liver | 0.1 |
| Milk | 0.05 |
| Mint | 0.3 |
| Mushroom | 0.1 |
| Oat, forage | 10.0 |
| Oat, grain | 0.1 |
| Oat, hay | 30.0 |
| Oat, straw | 1.0 |
| Peanut | 0.2 |
| Peanut, hay | 20 |
| Pecan | 0.1 |
| Pineapple | 0.1 |
| Pineapple, fodder | 0.1 |
| Rice, grain | 0.1 |
| Rice, straw | 3.0 |
| Rice, wild | 0.5 |
| Rye, grain | 0.1 |
| Rye, straw | 1.5 |
| Sheep, fat | 0.1 |
| Sheep, kidney | 2.0 |

| | |
|---|---|
| Sheep, liver | 2.0 |
| Sheep, meat | 0.1 |
| Sheep, mbyp, except kidney and liver | 0.1 |
| Sorghum, grain, grain | 0.2 |
| Sorghum, grain, stover | 1.5 |
| Wheat, grain | 0.1 |
| Wheat, straw | 1.5 |

*Carcinogen/Hazard Classifications*

**U.S. EPA Carcinogens:** Group C, possible carcinogen

**U.S. TRI:** Developmental toxin

**Label Signal Word:** CAUTION, WARNING or DANGER

**WHO Acute Hazard:** Class II, moderately hazardous

*Regulatory Authority:*
- EPA 40 CFR 372.65, Specific Toxic Chemical Listings
- Actively registered pesticide in California.
- EPCRA Section 313 Form R *de minimis* concentration reporting level: 1.0%

*Description:* Thick yellow liquid. Odorless. Soluble in water; solubility = 112 ppm. Molecular weight = 342.23. Boiling point = 178°C. Vapor pressure = 4.2 x $10^{-7}$ mmHg @ 20°C.

*Incompatibilities:* Strong oxidizers

*Permissible Concentration in Water:* No criteria set. Runoff from spills or fire control may cause water pollution.

*Routes of Entry:* Inhalation, passing through the skin and ingestion.

*Harmful Effects and Symptoms*

*Short Term Exposure:* Irritates the skin, eyes, and respiratory tract. Eye contact may cause irritation, burning sensation, and damage. Harmful if ingested, inhaled or absorbed through the skin. Inhalation should be avoided; use NIOSH-approved air purifying respirators for pesticides. May be harmful if swallowed.

*Medical Surveillance:* Consult a physician if poisoning is suspected or if redness, itching, or burning of the eyes or skin develop.

*First Aid:* If this chemical gets into the eyes, remove any contact lenses at once and irrigate immediately for at least 15 minutes, occasionally lifting upper and lower lids. Seek medical attention immediately. If this chemical contacts the skin, remove contaminated clothing and wash immediately with soap and water. Seek medical attention immediately. If this chemical has been inhaled, remove from exposure, begin rescue breathing (using universal precautions) if breathing has stopped, and CPR if heart action has stopped. Transfer promptly to a medical facility. When this chemical has been swallowed, get medical attention. Give large quantities of water and induce vomiting. Do not make an unconscious person vomit. Do not induce vomiting when formulations containing petroleum solvents are ingested.

*References:*
- U.S. Environmental Protection Agency, Office of Pesticide Programs, Pesticide Residue Limits, "Propiconazole," 40 CFR 180.434, www.epa.gov/pesticides/food/viewtols.htm
- EXTOXNET, Extension Toxicology Network, "Pesticide Information Profile, Propiconazole," Oregon State University, Corvallis, OR (October 1997). http://pmep.cce.cornell.edu/profiles/extoxnet/metiram-propoxur/propiconazole-ext.html
- California Environmental Protection Agency "Chemical List of Lists," Sacramento CA (February 1997)

# Propineb

*Use Type:* Fungicide

*CAS Number:* 12071-83-9

*Formula:* $(C_5H_8N_2S_4Zn)_x$

*Synonyms:* [((1-Methyl-1,2-ethanediyl)bis(carbamodithioato)](2-)zinc homopolymer; Methyl zineb; Zinc,(1-methyl-1,2-ethanediyl)bis(carbamodithioato)(2-)-; Propylenebis(dithiocarbamato)zinc; Zinc-1,2-propylene-bisdithiocarbamate; Zinc propylenebis(dithiocarbamate); Zinc [$N,N'$-propylene-1,2-bis(dithiocarbamate)]; Zink-[$N,N'$-propylen-1,2-bis(dithiocarbamat)] (German)

*Trade Names:* AIRONE®; ANTRACOL®; BAY®-46131, Bayer Crop Science (Germany); BAYER®-46131, Bayer Crop Science (Germany); LH 3012®; LH 30/Z®; MEZINEB®; PROPINEBE®; ZIPROMAT®

*Producers:* Bayer Crop Science (Germany); Limin Chemical (China); Yellow River Enterprise (Taiwan)

*Chemical Class:* Dithiocarbamate; inorganic zinc compound

*EPA/OPP PC Code:* 522200

*Uses:* Not registered in the U.S. Used to control downy mildew, black rot, brown rot and other fungi on vines, apples, pears, citrus, berry fruit, stone fruit, tomatoes, potatoes, tobacco, vegetables, rice, ornamentals and tea.

*Carcinogen/Hazard Classifications*

**U.S. TRI:** Developmental and reproductive toxin

**WHO Acute Hazard:** Class U, unlikely to be hazardous

*Regulatory Authority:*
- Clean Water Act: Section 307 Toxic Pollutants as zinc and compounds
- Safe Drinking Water Act: SMCL, 5 mg/L; Priority List (55 FR 1470) as zinc
- AB 2588-Air Toxics "Hot Spots" Chemicals (CAL) as zinc compounds
- The "Director's List" (CAL/OSHA) as zinc compounds
- EPCRA Section 313: Form R *de minimis* concentration reporting level: 1.0% as zinc compounds

*Description:* Yellowish to white crystalline solid or powder. Slight characteristic odor. Insoluble in water. Molecular weight = 289.8

*Permissible Concentration in Water:* No criteria set. Runoff from spills or fire control may cause water pollution.

## Harmful Effects and Symptoms

*Short Term Exposure:* Low levels of toxicity. Concentrated solutions are slightly corrosive to eyes and mucous membranes. Dust inhalation can cause irritation of the respiratory system with sneezing. Eye contact can cause irritation, watering, pain, and inflammation of the eyelids. Skin contact can cause irritation and minor ulceration. Ingestion can cause nausea, vomiting, fever, muscle twitching, seizure, rapid respiration, slow heart beat. Severe exposure may result in death.

*Points of Attack:* Respiratory system, central nervous system, cardiovascular system, skin, eyes.

*Medical Surveillance:* There are tests available to measure zinc in your blood, urine, hair, saliva, and feces. High levels of zinc in the feces can mean high recent zinc exposure. High levels of zinc in the blood can mean high zinc consumption and/or high exposure. Tests to measure zinc in hair may provide information on long-term zinc exposure; however, the relationship between levels in your hair and the amount of zinc you were exposed to is not clear. Medical observation is recommended for 24 to 48 hours after breathing overexposure, as pulmonary edema may be delayed. As first aid for pulmonary edema, consider administering a corticosteroid spray. Cigarette smoking may exacerbate pulmonary injury and should be discouraged for at least 72 hours following exposure. Before employment and at regular times after that, the following are recommended: If symptoms develop or overexposure occurs, repeat the above tests as soon as possible and get an exam of the nervous system. Also consider complete blood count. Consider chest x-ray following acute overexposure.

*First Aid:* If this chemical gets into the eyes, remove any contact lenses at once and irrigate immediately for at least 15 minutes, occasionally lifting upper and lower lids. Seek medical attention immediately. If this chemical contacts the skin, remove contaminated clothing and wash immediately with soap and water. Seek medical attention immediately. If this chemical has been inhaled, remove from exposure, begin rescue breathing (using universal precautions) if breathing has stopped, and CPR if heart action has stopped. Transfer promptly to a medical facility. When this chemical has been swallowed, get medical attention. If victim is conscious and able to swallow, have victim drink 4 to 8 ounces of water. Do not induce vomiting. *Note to physician or authorized medical personnel*: Medical observation is recommended for 24 to 48 hours after breathing overexposure, as pulmonary edema may be delayed. As first aid for pulmonary edema, consider administering a corticosteroid spray. Cigarette smoking may exacerbate pulmonary injury and should be discouraged for at least 72 hours following exposure.

*References:*
- International Programme on Chemical Safety (IPCS), "Environmental Health Criteria, Dithiocarbamate Pesticides," Geneva, Switzerland (1988). http://www.inchem.org/documents/ehc/ehc/ehc78.htm
- International Programme on Chemical Safety (IPCS), "Propineb," Geneva, Switzerland (1977). http://www.inchem.org/documents/jmpr/jmpmono/v077pr41.htm

# Propoxur

*Use Type:* Insecticide and molluscicide
*CAS Number:* 114-26-1
*Formula:* $C_{11}H_{15}NO_3$
*Alert:* a General Use Pesticide (GUP)
*Synonyms:* Carbamic acid, methyl-, *o*-isopropoxyphenyl ester; ENT 25,671; IPMC; 2-Isopropoxyphenyl *N*-methylcarbamate; 2-Isopropoxyphenyl methylcarbamate; *o*-isopropoxyphenyl methylcarbamate; *o*-(2-Isopropoxyphenyl) *N*-methylcarbamate; *o*-Isopropoxyphenyl *N*-methylcarbamate; 2-(1-Methylethoxy)phenyl *N*-methylcarbamate; Mrowkozol (Polish); OMS 33; PHC; Phenol, 2-(1-methylethoxy)-, methylcarbamate

*Trade Names:* (There are currently 695 registered active and/or canceled or transferred products in the U.S.)
ARPROCARB®; BAY® 39007, Bayer CropScience Corp. (Germany); BAY® 5122, Bayer CropScience Corp. (Germany); BAYER® 39007, Bayer CropScience Corp. (Germany); BAYER® B 5122, Bayer CropScience Corp. (Germany); BAYGON®, Bayer CropScience Corp. (Germany); BIFEX®; BLATTANEX®; BLATTOSEP®; BOLFO®; BO Q 58-12-315®; BORUHO®; BORUHO® 50; BRIFUR®; BRYGOU®; CHEMAGRO® 9010; COMPOUND 39007; DALF DUST®; INVISI-GARD®; PILLARGON®; PRENTOX CARBAMATE®; PROPOGON®; PROPOTOX®; PROPOXYLOR®; PROPYON®; RHODEN®; SENDRAN®; SUNCIDE®; TENDEX®; TUGEN®; UNDEN®; UNDENE®

*Producers:* Agrimor International (USA); Agsin (Singapore); Atul (India); Bayer CropScience Corp. (Germany); Borregaard (Norway); Ehrenstorfer, Dr. (Germany); EniChem (Italy); Jingma Chemicals Ltd. (China); Ki-Hara Chemicals Ltd. (UK); Shenzhen Guomeng Industry Co., Ltd. (China); Sigma-Aldrich Laborchemikalien (Germany)

*Chemical Class:* Carbamate
*EPA/OPP PC Code:* 047802
*California DPR Chemical Code:* 62
*ICSC Number:* 0191
*RTECS Number:* FC3150000
*EEC Number:* 006-16-00-4
*EINECS Number:* 204-043-8

*Uses:* A non-systemic insecticide compatible with most fungicides and insecticides except those that are alkaline. It is often used in combination with azinphosmethyl, chlorpyrifos, cyfluthrin, dichlorvos, disulfoton or methocarb. It is used on sugar cane, cocoa, pome and stone fruit, grapes, maize, hops, rice, sugar beets, vegetables, cotton, and forestry and ornamentals to control pests such as chewing and sucking insects, ants, crickets, flies, mosquitoes, millepedes, jassids and cockroaches.

*Carcinogen/Hazard Classifications*

**U.S. EPA Carcinogens:** Group B2, probable carcinogen
**Label Signal Word:** CAUTION, WARNING or DANGER depending on formulation
**WHO Acute Hazard:** Class II, moderately hazardous

*Regulatory Authority:*
- Air Pollutant Standard Set (ACGIH)[1] (DFG)[3] (HSE)[33] (NIOSH)[2] (Several States)[60]
- AB 1803-Well Monitoring Chemical (CAL)
- MCL (Maximum Contaminants Levels) list of contaminants (CAL)
- AB 2588-Air Toxics "Hot Spots" Chemicals (CAL)
- CAL-DHS/DHS Drinking Water Action Levels
- CAL Air Resources Board/AB 1807 Toxic Air Contaminants
- Permissible Exposure Limits for Chemical Contaminants (CAL/OSHA)
- The "Director's List" (CAL/OSHA)
- Actively registered pesticide in California.
- Clean Air Act: Hazardous Air Pollutants (Title I, Part A, Section 112)
- EPA Hazardous Waste Number (RCRA No.): U411
- RCRA, 40CFR261, Appendix 8 Hazardous Constituents
- RCRA 40CFR268.48; 61FR15654, Universal Treatment Standards: Wastewater (mg/L), 0.056; Nonwastewater (mg/kg), 1.4
- Superfund/EPCRA 40CFR302.4 RQ: CERCLA, 1 lb (0.454 kg)
- EPCRA Section 313 Form R *de minimus* concentration reporting level: 1.0%
- U.S. DOT Regulated Marine Pollutant (49CFR172.101, Appendix B)
- Canada, WHMIS, Ingredients Disclosure List

*Description:* Propoxur is a white to tan crystalline solid or powder. Faint characteristic odor. Slightly soluble in water; solubility = 1800 ppm @ 20°C. Boiling point = (decomposes). Melting/Freezing point = 91°C. Molecular weight = 209.27. Vapor pressure = $9.7 \times 10^{-6}$ mmHg @ 20°C. Flash point = >149°C. Hazard Identification (based on NFPA-704 M Rating System): Health 2, Flammability 1, Reactivity 0. Log $K_{ow}$ = 1.54. Unlikely to bioaccumulate in marine organisms.

*Incompatibilities:* Strong oxidizers, alkalis, heat, and moisture. Emits highly toxic methyl isocyanate fumes when heated to decomposition. When heated to decomposition, this material forms toxic oxides of nitrogen.

*Permissible Exposure Limits in Air:* There is no OSHA[2] PEL. NIOSH[2] and ACGIH[1] recommend a TWA value of 0.5 mg/m³. HSE[33] has set the same TWA as NIOSH[2] and a STEL of 2.0 mg/m³. The DFG[3] has set an MAK of 2.0 mg/m³. Several states have set guidelines or standards for Baygon in ambient air[60] ranging from 5-20 µg/m³ (North Dakota) to 8 µg/m³ (Virginia) to 10 µg/m³ (Connecticut) to 12 µg/m³ (Nevada).

*Determination in Air:* No method available.

*Permissible Concentration in Water:* No criteria set, but runoff from spills or fire control may cause water pollution.

*Routes of Entry:* Inhalation, ingestion, skin and/or eye contact.

*Harmful Effects and Symptoms*

*Short Term Exposure:* Propoxur can affect you when breathed in and quickly enters the body by passing through the skin. Severe poisoning can occur from skin contact. It is a moderately toxic carbamate chemical. Exposure can cause severe carbamate poisoning, with symptoms of headaches, sweating, nausea and vomiting, diarrhea, muscle twitching, loss of coordination and even death. May affect the nervous system, liver, kidneys. Carbamate insecticides inhibit the cholinesterase activity of enzymes, causing accumulation of acetylcholine at synapses and altering the way in which nervous impulses are transmitted. However, within several hours carbamates spontaneously detach from the enzymes.

*Long Term Exposure:* Propoxur may cause mutations. Handle with extreme caution. It may damage the developing fetus. Cholinesterase inhibitor; cumulative effect is possible. This chemical may damage the nervous system with repeated exposure, resulting in convulsions, respiratory failure. May cause liver damage.

*Points of Attack:* Central nervous system, liver, kidneys, gastrointestinal tract, blood cholinesterase.

*Medical Surveillance:* If symptoms develop or overexposure is suspected, the following may be useful: Serum and RBC cholinesterase levels (a test for the enzyme in the body affected by Propoxur). These tests are only useful if done 1 to 2 hours after exposure and can return to normal before the person feels well. Before employment and at regular times after that, the following are recommended: Plasma and red blood cell cholinesterase levels (tests for the enzyme poisoned by this chemical). If exposure stops, plasma levels return to normal in 1-2 weeks while red blood cell levels may be reduced for 1-3 months. When acetylcholinesterase enzyme levels are reduced by 25% or more below preemployment levels, risk of poisoning is increased, even if results are in lower ranges of "normal." Reassignment to work not involving carbamate pesticides is recommended until enzyme levels recover. If symptoms develop or overexposure occurs, repeat the above tests as soon as possible and get an exam of the nervous system. Also consider complete blood count. Consider chest x-ray following acute overexposure.

*First Aid:* Speed in removing material from eyes and skin is of extreme importance. *Eyes:* Eye contact can cause dangerous amounts of these chemicals to be quickly absorbed through the mucous membrane into the bloodstream. Immediately and gently flush eyes with plenty of warm or cold water (NO hot water) for at least 15 minutes, occasionally lifting the upper and lower eyelids. Get medical aid immediately. *Skin:* Get medical aid. Skin contact can cause dangerous amounts of these chemicals to be absorbed into the bloodstream. Wearing the appropriate PPE equipment and respirator for carbamate pesticides, immediately flush skin with plenty of soap and water for at least 15 minutes while removing contaminated clothing and shoes. Shampoo hair promptly if contaminated; protect eyes. *Ingestion:* Call poison control. Loosen all clothing. Never give anything by mouth to an unconscious person. Get medical aid. Do NOT induce vomiting.* If conscious, alert, and able to swallow, rinse mouth and have victim drink 4 to 8 ounces of water. Check to see if poison control instructs you to use ipecac syrup, otherwise administer slurry of activated charcoal (2 oz in 8 oz of water). If victim is UNCONSCIOUS OR HAVING CONVULSIONS, do nothing except keep victim warm. *In some cases you may be specifically instructed by poison control to induce vomiting by way of 2 tablespoons of syrup of ipecac (adult) washed down with a cup of water.* Do NOT give activated charcoal before or with ipecac syrup. *Inhalation:* Get medical aid. Do not contaminate yourself. Wearing the appropriate PPE equipment and respirator for carbamate pesticides, immediately remove the victim from the contaminated area to fresh air. If the victim is not breathing, administer artificial respiration. Do NOT use mouth-to-mouth resuscitation; use bag/mask apparatus. If breathing is difficult, administer oxygen through bag/mask apparatus until medical help arrives. Do not leave victim unattended. *Note to physician or authorized medical personnel.* Administer atropine, 2 mg (1/30 gr) intramuscularly or intravenously as soon as any local or systemic signs or symptoms of an intoxication are noted; repeat the administration of atropine every 3 to 8 minutes until signs of atropinization (mydriasis, dry mouth, rapid pulse, hot and dry skin) occur; initiate treatment in children with 0.05 mg mg/kg of atropine; repeat at 5 to 10 minute intervals. Watch respiration, and remove bronchial secretions if they appear to be obstructing the airway; intubate if necessary. *Medical note:* 2-PAMCI may be contraindicated in the case of some carbamate poisonings.

*References:*
- EXTOXNET, Extension Toxicology Network, "Pesticide Information Profile, Propoxur," Oregon State University, Corvallis, OR (June 1996). http://extoxnet.orst.edu/pips/propoxur.htm
- New Jersey Department of Health, "Hazardous Substance Fact Sheet: Propoxur," Trenton, NY (May 1998). http://www.state.nj.us/health/eoh/rtkweb/1604.pdf
- California Environmental Protection Agency "Chemical List of Lists," Sacramento CA (February 1997).

# Prosulfuron

*Use Type:* Herbicide
*CAS Number:* 94125-34-5
*Formula:* $C_{15}H_{16}F_3N_5O_4S$
*Synonyms:* Benzenesulfonamide, N-[((4-methoxy-6-methyl-1,3,5-triazin-2-yl)amino)carbonyl]-2-(3,3,3-trifluoropropyl)-; 1-(4-Methoxy-6-methyl-triazin-2-yl)-3-[2-(3,3,3-trifluoropropyl)phenylsuulfonyl]urea; N-[((4-Methoxy-6-methyl-1,3,5-triazin-2-yl)amino)carbonyl]-2-(3,3,3-trifluoropropyl)benzenesulfonamide
*Trade Names:* CGA®-152005, Syngenta (Switzerland); EXCEED®, Syngenta (Switzerland); PEAK®, Syngenta (Switzerland); SPIRIT®, Syngenta (Switzerland)
*Producers:* Ciba-Geigy (Switzerland); Syngenta (Switzerland)
*Chemical Class:* Sulfonylurea
*EPA/OPP PC Code:* 129031
*California DPR Chemical Code:* 5115
*Uses:* Tolerances expired 12/31/1999. A petition for new tolerances under 40 CFR 180.481 was filed by Syngenta and reported in the Federal Register December 31, 2002. Used as a post-emergence herbicide on corn and cereals such as barley, millet, oats, rye, sorghum and wheat. Used on sugar cane in some countries.
*Carcinogen/Hazard Classifications*
**U.S. EPA Carcinogens:** Group D, unclassifiable
**Label Signal Word:** CAUTION
*Description:* Colorless crystalline solid. Odorless. Highly soluble in water. Molecular weight = 419.40. Log $K_{ow}$ = < 1.0 @ pH 7.0. Unlikely to bioaccumulate in marine organisms.
*Permissible Concentration in Water:* No criteria set. Runoff from spills or fire control may cause water pollution.
*Routes of Entry:* Inhalation, passing through the skin and ingestion.
*Harmful Effects and Symptoms*
*Short Term Exposure:* Contact with eyes or skin may cause irritation or burns. Inhalation should be avoided; use NIOSH-approved air purifying respirators for pesticides. May be harmful if swallowed. Skin contact may cause allergic reaction.
*Long Term Exposure:* Repeated or prolonged contact may cause skin and lung sensitization, resulting in allergies. May affect breast, testicles and uterus.
*Points of Attack:* Skin
*Medical Surveillance:* Evaluation by a qualified allergist, including careful exposure history and special testing, may help diagnose skin allergy.

*First Aid:* If this chemical gets into the eyes, remove any contact lenses at once and irrigate immediately for at least 15 minutes, occasionally lifting upper and lower lids. Seek medical attention immediately. If this chemical contacts the skin, remove contaminated clothing and wash immediately with soap and water. Seek medical attention immediately. If this chemical has been inhaled, remove from exposure, begin rescue breathing (using universal precautions) if breathing has stopped, and CPR if heart action has stopped. Transfer promptly to a medical facility. When this chemical has been swallowed, get medical attention. Give large quantities of water and induce vomiting. Do not make an unconscious person vomit.

*References:*
- California Environmental Protection Agency "Chemical List of Lists," Sacramento CA (February 1997)

# Prothoate

*Use Type:* Acaricide and insecticide
*CAS Number:* 2275-18-5
*Formula:* $C_9H_{20}NO_3PS_2$
*Synonyms:* O,O-Diethyldithiophosphorylacetic acid-N-monoisopropylamide; O,O-Diethyl-S-(N-isopropyl carbamoylmethyl)dithiophosphate; O,O-Diethyl-S-isopropylcarbamoylmethylphosphorodithioate; O,O-Diethyl-S-(N-isopropylcarbamoylmethyl) phosphorodithioate; ENT 24,652; Isopropyl diethyldithiophosphorylacetamide; N-Isopropyl-2-mercaptoacetamide-S-ester with O,O-diethyl phosphorodithioate; L 343; N-Monoisopropylamide of O,O-diethyldithiophosphorylacetic acid; Phosphorodithioic acid O,O-diethylesters-ester with N-isopropyl-2-mercaptoacetamide; Phosphorodithioic acid, O,O-diethyl S-[2-((1-methylethyl]amino)]-2-oxoethyl]ester; Trimethoate
*Trade Names:* AC® 18682®, American Cyanamid's Agricultural Products Group (USA); AMERICAN CYANAMID 18682®, American Cyanamid's Agricultural Products Group (USA); FAC®; FAC® 20; FOSTION®; OLEOFAC®; TELEFOS®
*Producers:* American Cyanamid's Agricultural Products Group (USA); BASF Agricultural Products Group (Germany)
*Chemical Class:* Organophosphate
*EPA/OPP PC Code:* 344300
*RTECS Number:* TD8225000
*EINECS Number:* 218-893-2
*Regulatory Authority:*
- Banned or Severely Restricted (Malaysia) (UN)[13]
- Superfund/EPCRA 40CFR355, Appendix B Extremely Hazardous Substances: TPQ = 100/10,000 lb (45.4/4,540 kg)
- Superfund/EPCRA 40CFR302.4 RQ: EHS, 1 lb (0.454 kg)
- U.S. DOT Inhalation Hazard Chemicals as organophosphates
- U.S. DOT Regulated Marine Pollutant (49CFR172.101, Appendix B)

*Description:* Prothoate is an amber to yellow crystalline solid. Camphor-like odor. Molecular weight = 285.40. Boiling point = 135°C @ 0.1 mm. Melting/Freezing point = 29°C. Hazard Identification (based on NFPA-704 M Rating System): Health 4, Flammability 1, Reactivity 0. Slightly soluble in water.

*Permissible Exposure Limits in Air:* No standards set.

*Determination in Air:* OSHA versatile sampler-2 Toluene/Acetone Gas chromatography/Flame photometric detection for sulfur, nitrogen, or phosphorus NIOSH Method IV Method #5600, Organophosphorus pesticides.[18]

*Permissible Concentration in Water:* No criteria set, but runoff from spills or fire control may cause water pollution.

*Routes of Entry:* Inhalation, ingestion, skin and/or eye contact.

## Harmful Effects and Symptoms

*Short Term Exposure:* This is a highly toxic material capable of causing death or permanent injury due to exposure during normal use. Small doses at frequent intervals are additive. Similar to parathion. Symptoms may include nausea, vomiting, abdominal cramps, diarrhea, excessive salivation, headache, giddiness, dizziness, weakness, muscle twitching, difficult breathing, blurring or dimness of vision, and loss of muscle coordination. Death may occur from failure of the respiratory center, paralysis of the respiratory muscles, intense bronchoconstriction, or all three. Delayed pulmonary edema may occur after inhalation.

*Long Term Exposure:* Cholinesterase inhibitor cumulative effect is possible. This chemical may damage the nervous system with repeated exposure, resulting in convulsions, respiratory failure. May cause liver damage.

*Points of Attack:* Respiratory system, lungs, central nervous system, cardiovascular system, skin, eyes, plasma and red blood cell cholinesterase.

*Medical Surveillance:* Medical observation is recommended for 24 to 48 hours after breathing overexposure, as pulmonary edema may be delayed. Before employment and at regular times after that, the following are recommended: Plasma and red blood cell cholinesterase levels (tests for the enzyme poisoned by this chemical). If exposure stops, plasma levels return to normal in 1-2 weeks while red blood cell levels may be reduced for 1-3 months. When acetylcholinesterase enzyme levels are reduced by 25% or more below preemployment levels, risk of poisoning is increased, even if results are in lower ranges of "normal." Reassignment to work not involving organophosphate or carbamate pesticides is recommended until enzyme levels recover. If symptoms develop or overexposure occurs, repeat the above tests as soon as possible and get an exam of the nervous system. Also consider complete blood count.

Consider chest x-ray following acute overexposure. Do not drink any alcoholic beverages before or during use. Alcohol promotes absorption of organophosphates.

***First Aid:*** **Treatment for organophosphate poisoning consists of thorough decontamination, cardiorespiratory support, and administration of the antidotes atropine and pralidoxime. In cases of severe poisoning, diazepam, an anticonvulsant, should also be administered. Antidotes should be administered as prevention even if the diagnosis is in doubt.** Speed in removing material from eyes and skin is of extreme importance. *Eyes:* Eye contact can cause dangerous amounts of these chemicals to be quickly absorbed through the mucous membrane into the bloodstream. Immediately and gently flush eyes with plenty of warm or cold water (NO hot water) for at least 15 minutes, occasionally lifting the upper and lower eyelids. Get medical aid immediately. *Skin:* Get medical aid. Skin contact can cause dangerous amounts of these chemicals to be absorbed into the bloodstream. Wearing the appropriate PPE equipment and respirator for organophosphate pesticides, immediately flush skin with plenty of soap and water for at least 15 minutes while removing contaminated clothing and shoes. Shampoo hair promptly if contaminated. The removed, contaminated clothing and shoes should be double-bagged and left in Hot Zone for later disposal by hazardous materials experts. Skin may also be decontaminated with diluted hypochlorite solution. *Inhalation:* Get medical aid. Do not contaminate yourself. Wearing the appropriate PPE equipment and respirator for organophosphate pesticides, immediately remove the victim from the contaminated area to fresh air. If the victim is not breathing, administer artificial respiration. Do NOT use mouth-to-mouth resuscitation; use bag/mask apparatus. If breathing is difficult, administer oxygen through bag/mask apparatus until medical help arrives. Do not leave victim unattended. *Ingestion:* Call poison control. Loosen all clothing. Never give anything by mouth to an unconscious person. If victim is *unconscious or having convulsions,* do nothing except keep victim warm. Get medical aid. Transfer promptly to a medical facility. In cases of ingestion, **do not induce vomiting**. If the victim is alert and asymptomatic, administer a slurry of activated charcoal at a dose of 1 g/kg (infant, child, and adult dose). A soda can and straw may be of assistance when offering charcoal to a child. *In some cases you may be specifically instructed by poison control to induce vomiting by way of 2 tablespoons of syrup of ipecac (adult) washed down with a cup of water.* Do NOT give activated charcoal before or with ipecac syrup.

*Note to physician or authorized medical personnel:* Treat cases of respiratory compromise, coma, or excessive pulmonary secretions with respiratory support using protocols and techniques available and within the scope of training. Some cases may necessitate procedures such as endotracheal intubation or cricothyrotomy by properly trained and equipped personnel. When possible, atropine (see *Antidotes,* below) should be given under medical supervision. Patients who are comatose, hypotensive, or having seizures or cardiac arrhythmias should be treated according to advanced life support protocols. *Antidotes:* Two antidotes are administered to treat organophosphate poisoning. Atropine is a competitive antagonist of acetylcholine at muscarinic receptors and is used to control the excessive bronchial secretions which are often responsible for death. Pralidoxime relieves both the nicotinic and muscarine effects of organophosphate poisoning by regenerating acetylcholinesterase and can reduce both the bronchial secretions and the muscle weakness associated with poisoning. The initial intravenous dose of atropine in adults should be determined by the severity of symptoms: An initial adult dose of 1.0 to 2.0 mg or pediatric dose of 0.01 mg/kg (minimum 0.01 mg) should be administered intravenously. If intravenous access cannot be established, atropine may also be given intramuscularly, subcutaneously or via endotracheal tube. Doses should be repeated every 15 minutes until excessive secretions and sweating have been controlled. Once bronchial secretion has been controlled, atropine administration should be repeated whenever the secretions begin to recur. In seriously poisoned patients, very large doses may be required. Alterations of pulse rate and pupillary size should not be used as indicators of treatment adequacy. Pralidoxime should be administered as early in poisoning as possible as its efficacy may diminish when given more than 24 to 36 hours after exposure. Doses are as follows: adult 1.0 g; pediatric 25 to 50 mg/kg. The drug should be administered intravenously over 30 to 60 minutes, but in a life-threatening situation, one-half of the total dose can be given per minute for a total administration time of 2 minutes. Treatment should begin to take effect within 40 minutes with a reduction in symptoms and the amount of atropine necessary to control bronchial secretion. The initial dose can be repeated in 1 hour and then every 8 to 12 hours until the patient is clinically well and no longer requires atropine. If intravenous access cannot be established, pralidoxime may also be given intramuscularly. Early administration of diazepam in addition to the combined atropine and pralidoxime treatment may help prevent the onset of seizures and potential brain and cardiac morphologic damage following high-level organophosphate poisoning.

***References:***
- New Jersey Department of Health and Senior Services, "Hazardous substance Fact Sheet, Prothoate," Trenton, NJ (May 2002).
 http://www.state.nj.us/health/eoh/rtkweb/2715.pdf
- U.S. Environmental Protection Agency, "Chemical Profile: Prothoate," Washington, DC, Chemical Emergency Preparedness Program (November 30, 1987).

- California Environmental Protection Agency "Chemical List of Lists," Sacramento CA (February 1997).
- Agency for Toxic Substances and Disease Registry, U.S. Department of Health and Human Services, Public Health Service, "Managing Hazardous Materials Incidents," Atlanta, GA (June 2003).


## Pymetrozine

***Use Type:*** Insecticide
***CAS Number:*** 123312-89-0
***Formula:*** $C_{10}H_{11}N_5O$
***Synonyms:*** *(E)*-4,5-Dihydro-6-methyl-4-[(3-pyridinylmethylene)amino]-1,2,4-triazin-3(2*H*)-one; 1,2,4-Triazin-3(2*H*)-one, 4,5-dihydro-6-methyl-4-[(3-pyridinylmethylene)amino]-, *(E)*- (9CI)
***Trade Names:*** CGA-215944®, Syngenta (Switzerland); CHESS®, Syngenta (Switzerland); ENDEAVOR®, Syngenta (Switzerland); FULFILL®, Syngenta (Switzerland); PLENUM®, Syngenta (Switzerland); STERLING®
***Producers:*** Syngenta (Switzerland)
***Chemical Class:*** Pyridine azomethine; triazine
***EPA/OPP PC Code:*** 101103
***California DPR Chemical Code:*** 5232
***Uses:*** The U.S. EPA considers pymetrozine a replacement fo organophosphate pesticides when used in the same use patterns. Used on pecans, foliar ground plants and ornamentals, potatoes and other tuberous vegetables, leafy and fruiting vegetables, hops, cotton and tobacco.
***U.S. Maximum Allowable Residue Levels for Pymetrozine (40 CFR 180.556):***

| CROP | ppm |
|---|---|
| Brassica, head and stem, subgroup (crop subgroup 5-a) | 0.5 |
| Brassica, leafy greens, Subgroup (crop subgroup 5-b) | 0.25 |
| Cotton gin byproducts | 2.0 |
| Cotton, undelinted seed | 0.3 |
| Hops, dried cones | 6.0 |
| Pecans | 0.02 |
| Turnip, greens | 0.25 |
| Vegetable, fruiting, group (crop group 8) | 0.2 |
| Vegetable, cucurbit, group (crop group 9) | 0.1 |
| Vegetable, leafy, execpt brassica, Group (crop group 4) | 0.6 |
| Vegetable, tuberous and corm, subgroup (crop subgroup 1-c) | 0.02 |

***Carcinogen/Hazard Classifications***
**U.S. EPA Carcinogens:** Likely carcinogen
**Label Signal Word:** CAUTION
***Description:*** White to beige crystalline solid. Slightly sweet odor. Slightly soluble in water. Melting point = 217° C. Log $K_{ow}$ = < 1.0. Unlikely to bioaccumulate in marine organisms.
***Permissible Concentration in Water:*** No criteria set. Runoff from spills or fire control may cause water pollution.
***First Aid:*** If this chemical gets into the eyes, remove any contact lenses at once and irrigate immediately for at least 15 minutes, occasionally lifting upper and lower lids. Seek medical attention immediately. If this chemical contacts the skin, remove contaminated clothing and wash immediately with soap and water. Seek medical attention immediately. If this chemical has been inhaled, remove from exposure, begin rescue breathing (using universal precautions) if breathing has stopped, and CPR if heart action has stopped. Transfer promptly to a medical facility. When this chemical has been swallowed, get medical attention. Give large quantities of water and induce vomiting. Do not make an unconscious person vomit.
***References:***


- U.S. Environmental Protection Agency, Office of Pesticide Programs, "Pesticide Fact Sheet, Pymetrozine." (August 2000). http://www.epa.gov/opprd001/factsheets/pymetrozine.pdf
- U.S. Environmental Protection Agency, Office of Pesticide Programs, Pesticide Residue Limits, "Pymetrozine," 40 CFR 180.556. http://www.setonresourcecenter.com/40CFR/Docs/wcd0004d/wcd04de1.asp
- California Environmental Protection Agency "Chemical List of Lists," Sacramento CA (February 1997)


## Pyrazon (ANSI)

***Use Type:*** Herbicide
***CAS Number:*** 1698-60-8
***Formula:*** $C_{10}H_8ClN_3O$
***Synonyms:*** 5-Amino-4-chloro-2,3-dihydro-3-*oxo*-2-phenylpyridazine; 5-Amino-4-chloro-2-phenyl-3(2H)-pyridazinone; 1-Phenyl-4-amino-5-chloropyridazon-(6) (German); 1-Phenyl-4-amino-5-chloropyridazone-6; 1-Phenyl-4-amino-5-chloro-6-pyridazone; 1-Phenyl-4-amino-5-chlorpyridaz-6-one
***Trade Names:*** BUREX®; CHLORIDAZON®; HS-119-1®; PCA®; PHENOSANE®; PYRAMIN®, BASF Agricultural Products (Germany); PYRAMINE®; PYRAZON®; PYRAZONE®; PYRAZONL®
***Producers:*** BASF Agricultural Products (Germany)
***Chemical Class:*** Pyridazine
***EPA/OPP PC Code:*** 069601
***California DPR Chemical Code:*** 509
***Uses:*** Used on sugar beets, red beets and fodder beets for pre-emergence and post-emergence weed control.

*U.S. Maximum Allowable Residue Levels for Pyrazon (40 CFR 180.316):*

| CROP | ppm |
|---|---|
| Beet tops | 1.0 |
| Sugar beet tops | 1.0 |
| in or on beets | 0.1 |
| in or on sugar beets | 0.1 |

*Carcinogen/Hazard Classifications*
**Label Signal Word:** CAUTION
**WHO Acute Hazard:** Class U, unlikely to be hazardous
*Description:* Tan to brown powder as pure compound; dark brown solid in technical state. Odorless. Practically insoluble in water. Molecular weight = 221.63. Melting/Freezing point = 205°C. Vapor pressure = 5 x $10^{-2}$ mmHg @ 20°C.
*Determination in Air:* Filter; none; Gravimetric; NIOSH IV [Particulates NOR; #0500 (total), #0600 (respirable)].[18]
*Permissible Concentration in Water:* No criteria set. Runoff from spills or fire control may cause water pollution.
*Routes of Entry:* Inhalation and ingestion.
*Harmful Effects and Symptoms*
*Short Term Exposure:* Contact may irritate skin and cause eye irritation and possible severe injury. Inhalation should be avoided; use NIOSH-approved air purifying respirators for pesticides. Poisonous if swallowed.
*Medical Surveillance:* If poisoning is suspected or of redness, itching, burning of skin or eyes develop.
*First Aid:* If this chemical gets into the eyes, remove any contact lenses at once and irrigate immediately for at least 15 minutes, occasionally lifting upper and lower lids. Seek medical attention immediately. If this chemical contacts the skin, remove contaminated clothing and wash immediately with soap and water. If this chemical has been inhaled, remove from exposure, begin rescue breathing (using universal precautions) if breathing has stopped, and CPR if heart action has stopped. Transfer promptly to a medical facility. When this chemical has been swallowed, get medical attention. Give large quantities of water and induce vomiting. Do not make an unconscious person vomit
*References:*
- Pesticide Management Education Program, "Pyrazon (Pyramin) Herbicide Profile 2/85," Cornell University, Ithaca, NY (February 1985). http://pmep.cce.cornell.edu/profiles/herb-growthreg/naa-rimsulfuron/pyrazon/herb-prof-pyrazon.html
- U.S. Environmental Protection Agency, Office of Pesticide Programs, Pesticide Residue Limits, "Pyrazon," 40 CFR 180.316. http://www.setonresourcecenter.com/40CFR/Docs/wcd0004d/wcd04d16.asp
- California Environmental Protection Agency "Chemical List of Lists," Sacramento CA (February 1997)

# Pyrethrins or Pyrethrum

*Use Type:* Insecticide
*CAS Number:* 8003-34-7; 121-21-1 (I); 121-29-9 (II)
*Formula:* $C_{20-21}H_{28-30}O_{3-5}$
*Alert:* These substances are General Use Pesticides (GUP).
*Synonyms:* Chrysanthemum cinerariefolium; Cinerin I; Cinerin II; Dalmatian insect flowers; Jasmolin I; Jasmolin II; Persian insect powder; Piretrina (Spanish); Pyrethrin I; Pyrethrin II; Pyrethrum flowers; Trieste flowers
*Trade Names:* (The U.S. EPA currently registers 7,295 products, including 1,441 active ones and those that have been canceled or transferred.) BUHACH®; CHRYSANTHEMUM CINERAREAEFOLIUM®; DALMATION INSECT FLOWERS®; FIRMOTOX®; INSECT POWDER®; OFIRMOTOX®; PAREXAN®; PRENTOX®, Prentiss, Inc. (USA); PYRETRINER®; PYRETHRUM INSECTICIDE®
*Producers:* Agropharm Ltd. (UK); Bayer CropScience (Germany); Botanical Resources Australia (Australia); Ki-Hara Chemicals Ltd. (UK); Pbi/Gordon Corporation (USA); Prentiss Inc. (USA); Sigma-Aldrich Laborchemikalien (Germany); SuYan Agrochemical Group (China); Vijayalakshmi Insecticides and Pesticides (India); Zeneca Ag Products (USA) (now Syngenta)
*Chemical Class:* Botanical
*EPA/OPP PC Code:* 069000 (Pyrethrum); 069001 (Mixed Pyrethrins I and Pyrethrins II); 069002 [Ground Pyrethrum (other than pyrethrins)]. Sprays may be dissolved in xylene or kerosene.
*California DPR Chemical Code:* 510
*RTECS Number:* UR4200000
*EINECS Number:* 232-319-8
*Uses:* Pyrethrum is toxic to human lice, mosquitoes, fleas, roaches, ants, beetles, silverfish and many other insects. Pyrethrum is derived from the dried flowers of Chrysanthemum Cinerariaefolium. Pyrethrins is the name given to the active insecticidal components of the dried flowers–Pyrethrin I and II–which, in turn, have four different ingredients–Cinerin I and II and Jasmolin I and II.
*U.S. Maximum Allowable Residue Levels for Pyrethrum powder other than pyrethrins (40 CFR 180.905):*
Pyrethrins or pyrethrum powders when applied to growing crops in accordance with good agricultural practice, are exempt from the requirements of tolerance. These pesticides are not exempted from the requirement of a tolerance when applied to a crop at the time of, or after, the harvest.
*Carcinogen/Hazard Classifications*
**U.S. EPA Carcinogens:** A Likely carcinogen
**Label Signal Word:** CAUTION
**WHO Acute Hazard:** Class II, moderately hazardous
*Regulatory Authority:*
- Air Pollutant Standard Set (ACGIH)[1] (DFG)[3] (HSE)[33] (OSHA)[58] (Several States)[60]

- AB 1803-Well Monitoring Chemical (CAL) as pyrethrins
- Permissible Exposure Limits for Chemical Contaminants (CAL/OSHA) as pyrethrum
- Actively registered pesticide in California. as pyrethrins

*Pyrethrin I:*
- Clean Water Act: Section 311 Hazardous Substances/RQ 40CFR117.3 (same as CERCLA, see below)
- EPA Hazardous Waste Number (RCRA No.): P008
- RCRA, 40CFR261, Appendix 8 Hazardous Constituents
- Superfund/EPCRA 40CFR302.4 RQ: CERCLA, 1 lb (0.454 kg)

*Pyrethrin II:*
- Clean Water Act: Section 311 Hazardous Substances/RQ 40CFR117.3 (same as CERCLA, see below)
- Superfund/EPCRA 40CFR302.4 RQ: CERCLA, 1 lb (0.454 kg)
- Canada, WHMIS, Ingredients Disclosure List

***Description:*** The pyrethrins are a variable mixture of compounds which are found in pyrethrum flowers: Cinerin, Jasmolin, and Pyrethrin. Brown, viscous oil or solid. Insoluble in water. Sprays may be dissolved in xylene or kerosene. Boiling point = (I) 170°C @ 0.1 mm (decomposition). Vapor pressure = (estimate) $1 \times 10^{-8}$ mmHg @ 20°C. Flash point = 82–88°C. Hazard Identification (based on NFPA-704 M Rating System): Health 2, Flammability 1, Reactivity 0.

***Incompatibilities:*** Violent reaction with strong oxidizers and alkaline materials.

***Permissible Exposure Limits in Air:*** The OSHA[(2)] PEL, the DFG MAK[(3)] and the recommended ACGIH[(1)] TWA value[(1)] is 5 mg/m³. HSE[(33)] set the same TWA level and the STEL set by HSE[(33)] is 10 mg/m³. The NIOSH[(2)] IDLH level = 5,000 ppm. Several states have set guidelines or standards for pyrethrum in ambient air[(60)] ranging from 16.7 µg/m³ (New York) to 50.0 µg/m³ (Florida and South Carolina) to 50.0–100.0 µg/m³ (North Dakota) to 80.0 µg/m³ (Virginia) to 100.0 µg/m³ (Connecticut) to 119.0 µg/m³ (Nevada).

***Determination in Air:*** Collection by impinger or fritted bubbler, analysis by gas liquid chromatography/ultraviolet. See NIOSH IV, Method #5008[(18)].

***Permissible Concentration in Water:*** No criteria set, but runoff from spills or fire control may cause water pollution.

***Routes of Entry:*** Inhalation, ingestion, skin and/or eye contact.

***Harmful Effects and Symptoms***

***Short Term Exposure:*** Pyrethrum can affect you when breathed in and by passing through your skin. Irritates the eyes and respiratory tract. High exposure can affect the nervous system causing headache, nausea, vomiting, fatigue, and restlessness, rhinorrhea (discharge of thin nasal mucous).

***Long Term Exposure:*** High or repeated exposure can cause lung allergy (with cough, wheezing and/or shortness of breath) or hay fever symptoms (sneezing, runny or stuffy nose). Allergic "pneumonia" can also occur with cough, chest pain, breathing difficulty and abnormal chest x-ray. Repeated attacks may lead to permanent scarring. Skin allergy may also develop with rash and itching, even with lower exposures. Skin contact can cause rash with redness, blisters and intense itching. A severe generalized allergy can occur with weakness and collapse.

***Points of Attack:*** Respiratory system, skin and central nervous system.

***Medical Surveillance:*** Before beginning employment and at regular times after that, the following are recommended: Lung function tests. These may be normal if the person is not having an attack at the time of the test. Consider chest x-ray if lung symptoms are present. Evaluation by a qualified allergist, including careful exposure history and special testing, may help diagnose skin allergy.

***First Aid:*** If this chemical gets into the eyes, remove any contact lenses at once and irrigate immediately for at least 15 minutes, occasionally lifting upper and lower lids. Seek medical attention immediately. If this chemical contacts the skin, remove contaminated clothing and wash immediately with soap and water. Seek medical attention immediately. If this chemical has been inhaled, remove from exposure, begin rescue breathing (using universal precautions) if breathing has stopped, and CPR if heart action has stopped. Transfer promptly to a medical facility. When this chemical has been swallowed, get medical attention. Give large quantities of water and induce vomiting. Do not make an unconscious person vomit.

***References:***
- EXTOXNET, Extension Toxicology Network, "Pesticide Information Profile," Oregon State University, Corvallis, OR (March 1994). http://extoxnet.orst.edu/pips/pyrethri.htm
- EPA, Office of Pesticide Programs, Pesticide Residue Limits, "Pyrethrum powder other than pyrethrins," 40 CFR 180. 905. http://www.epa.gov/pesticides/food/viewtols.htm
- New Jersey Department of Health and Senior Services, "Hazardous Substance Fact sheet, Pyrethrum," Trenton, NJ (August 2002). http://www.state.nj.us/health/eoh/rtkweb/1623.pdf
- California Environmental Protection Agency "Chemical List of Lists," Sacramento CA (February 1997).

# Pyridaben

***Use Type:*** Insecticide and acaricide
***CAS Number:*** 96489-71-3
***Formula:*** $C_{19}H_{25}ClN_2OS$
***Synonyms:*** 2-(*tert*-Butyl)-5-(4-*tert*-butyl-benzylthio)-4-chloropyridazin-3-(2*H*)one; 3(2*H*)-Pyridazinone, 4-chloro-2-(1,1-dimethylethyl)-5-[((4-(1,1-dimethylethyl) phenyl)methyl)thio]-

*Trade Names:* BAS®-300, BASF Agricultural Products (Germany); NCI®-129, Nissan Chemical Industries (Japan); NESTER®; PYRAMITE®, BASF Agricultural Products (Germany); SANMITE®, BASF Agricultural Products (Germany)

*Producers:* Agrimor International (USA); Agsin (Singapore); BASF Agricultural Products (Germany); Hebei Huafeng Chemical Group (China); Ki-Hara Chemicals Ltd. (UK); Nissan Chemical Industries (Japan); Sanonda Zhengzhou Pesticide Co. Ltd. (China); Sheyang Pesticides and Chemical Industry Co.(China); SuYan Agrochemical Group (China)

*EPA/OPP PC Code:* 129105

*California DPR Chemical Code:* 3959

*Uses:* Used to control mites, whiteflies, leafhoppers and psyllids on fruit trees, vegetables, ornamentals and other field crops.

*U.S. Maximum Allowable Residue Levels for Pyridaben (40 CFR 180.494):*

| CROP | ppm |
|---|---|
| Almond hulls | 4.0 |
| Apple | 0.5 |
| Apple, wet pomace | 0.75 |
| Apricot | 0.05 |
| Cattle, fat | 0.05 |
| Cattle, meat | 0.05 |
| Cattle, mbyp | 0.05 |
| Cherry, sweet | 0.05 |
| Cherry, tart | 0.05 |
| Citrus | 0.5 |
| Citrus, dried pulp | 1.5 |
| Citrus, oil | 10.0 |
| Goat, fat | 0.0 |
| Goat, meat | 0.05 |
| Goat mbyp | 0.05 |
| Grape | 1.5 |
| Hog, fat | 0.05 |
| Hog, meat | 0.05 |
| Hog mbyp | 0.05 |
| Horse, fat | 0.05 |
| Horse meat | 0.05 |
| Horse, mbyp | 0.05 |
| Milk | 0.01 |
| Nectarine | 2.5 |
| Nut, tree crop group | 0.05 |
| Peach | 2.5 |
| Pear | 0.75 |
| Pistachio | 0.05 |
| Plum | 2.5 |
| Prune | 2.5 |
| Sheep, fat | 0.05 |
| Sheep, meat | 0.05 |
| Sheep, mbyp | 0.05 |

*Carcinogen/Hazard Classifications*

*U.S. EPA Carcinogens:* Group E, Unlikely Carcinogen

*Label Signal Word:* CAUTION or WARNING

*WHO Acute Hazard:* Class III, slightly hazardous

*Description:* White to light brown crystalline solid. Melting point = 111–112°C.

*Permissible Concentration in Water:* No criteria set. Runoff from spills or fire control may cause water pollution.

*First Aid:* If this chemical gets into the eyes, remove any contact lenses at once and irrigate immediately for at least 15 minutes, occasionally lifting upper and lower lids. Seek medical attention immediately. If this chemical contacts the skin, remove contaminated clothing and wash immediately with soap and water. Seek medical attention immediately. If this chemical has been inhaled, remove from exposure, begin rescue breathing (using universal precautions) if breathing has stopped, and CPR if heart action has stopped. Transfer promptly to a medical facility. When this chemical has been swallowed, get medical attention. Give large quantities of water and induce vomiting. Do not make an unconscious person vomit. Do not induce vomiting when formulations containing petroleum solvents are ingested.

*References:*
- U.S. Environmental Protection Agency, Office of Pesticide Programs, Pesticide Residue Limits, "Pyridaben," 40 CFR 180.494, www.epa.gov/pesticides/food/viewtols.htm
- California Environmental Protection Agency "Chemical List of Lists," Sacramento CA (February 1997)

# Pyridate

*Use Type:* Herbicide

*CAS Number:* 55512-33-9

*Formula:* $C_{19}H_{23}ClN_2O_2S$

*Synonyms:* Carbonothioic acid, *O*-(6-chloro-3-phenyl-4-pyridizinyl) *S*-octyl ester; *O*-(6-Chloro-3-phenyl-4-pyridazinyl) *S*-octyl carbonothioate; 6-Chloro-3-phenylpridazin-4-yl-*S*-octyl-thiocarbonate; *O*-(6-Chloro-3-phenyl-4-pyridazinyl)-carbonothioic acid *S*-octyl ester; Fenpyrate; Octyl-*O*-(6-chloro-3-phenylpyridazin-4-yl)carbothioate (BSI)

*Trade Names:* CL-11344®; LENTAGRAN®, Syngenta (Switzerland); ST-9551®; TOUGH®, Syngenta (Switzerland)

*Producers:* Syngenta (Switzerland)

*Chemical Class:* Thiocarbamate

*EPA/OPP PC Code:* 128834

*California DPR Chemical Code:* 3939

*U.S. Maximum Allowable Residue Levels for Pyridate (40 CFR 180.462):*

| CROP | ppm |
|---|---|
| Brassica, head and stem, subgroup 5a | 0.03 |
| Cabbage | 0.03 |

| | |
|---|---|
| Chickpea, seed | 0.1 |
| Collards | 0.03 |
| Corn, field, forage | 0.03 |
| Corn, forage, grain, stover | 0.03 |
| Peanut | 0.03 |
| Peppermint, tops | 0.2 |
| Spearmint, tops | 0.2 |

*Carcinogen/Hazard Classifications*
**Label Signal Word:** CAUTION or WARNING
**WHO Acute Hazard:** Class III, slightly hazardous
*Description:* Brown oily liquid. very slightly soluble in water; solubility = 85 ppm. Melting/Freezing point = 27–28°C.
*Incompatibilities:* Incompatible with germanium, lead diacetate, magnesium, mercurous chloride, silicon, silver nitrate, titanium. When heated to decomposition, this chemical forms toxic oxides of nitrogen, sulfur, and fumes of chlorine.
*Permissible Concentration in Water:* No criteria set. Runoff from spills or fire control may cause water pollution.
*Harmful Effects and Symptoms*
*Short Term Exposure:* Low levels of toxicity. Concentrated solutions are slightly corrosive to eyes and mucous membranes. Dust inhalation can cause irritation of the respiratory system with sneezing. Eye contact can cause irritation, watering, pain, and inflammation of the eyelids. Skin contact can cause irritation and minor ulceration. Ingestion can cause nausea, vomiting, fever, muscle twitching, seizure, rapid respiration, slow heart beat. Severe exposure may result in death.
*Long Term Exposure:*
*Points of Attack:* Respiratory system, central nervous system, cardiovascular system, skin, eyes.
*Medical Surveillance:* Medical observation is recommended for 24 to 48 hours after breathing overexposure, as pulmonary edema may be delayed. As first aid for pulmonary edema, consider administering a corticosteroid spray. Cigarette smoking may exacerbate pulmonary injury and should be discouraged for at least 72 hours following exposure. Before employment and at regular times after that, the following are recommended: If symptoms develop or overexposure occurs, repeat the above tests as soon as possible and get an exam of the nervous system. Also consider complete blood count. Consider chest x-ray following acute overexposure.
*First Aid:* If this chemical gets into the eyes, remove any contact lenses at once and irrigate immediately for at least 15 minutes, occasionally lifting upper and lower lids. Seek medical attention immediately. If this chemical contacts the skin, remove contaminated clothing and wash immediately with soap and water. Seek medical attention immediately. If this chemical has been inhaled, remove from exposure, begin rescue breathing (using universal precautions) if breathing has stopped, and CPR if heart action has stopped. Transfer promptly to a medical facility. When this chemical has been swallowed, get medical attention. If victim is conscious and able to swallow, have victim drink 4 to 8 ounces of water. Do not induce vomiting.
*Note to physician or authorized medical personnel:* Medical observation is recommended for 24 to 48 hours after breathing overexposure, as pulmonary edema may be delayed. As first aid for pulmonary edema, consider administering a corticosteroid spray. Cigarette smoking may exacerbate pulmonary injury and should be discouraged for at least 72 hours following exposure.
*References:*
- U.S. Environmental Protection Agency, Office of Pesticide Programs, Pesticide Residue Limits, "Pyridate," 40 CFR 180.462, www.epa.gov/pesticides/food/viewtols.htm
- California Environmental Protection Agency "Chemical List of Lists," Sacramento CA (February 1997).

# Pyrimethanil

*Use Type:* Fungicide
*CAS Number:* 53112-28-0
*Formula:* $C_{12}H_{13}N_3$
*Synonyms:* 4,6-Dimethyl-*N*-phenyl-2-pyrimidinamine
*Trade Names:* SCALA®, Bayer CropScience (Germany); SN 100309®
*Producers:* Bayer CropScience (Germany); Zhejiang Heben Pesticide & Chemicals (China)
*Chemical Class:* Pyrimidine
*EPA/OPP PC Code:* 288201
*Uses:* Used on grapes, strawberries, tomatoes, onions, beans, cucumbers, eggplant, and ornamentals.
*Carcinogen/Hazard Classifications*
**U.S. EPA Carcinogens:** Group C, possible carcinogen
**WHO Acute Hazard:** Class U, unlikely to be hazardous
**Endocrine Disruptor:** Suspected endocrine disruptor
*Description:* Pyrimethanil is a white crystalline powder. Melting point = 93.3°C. Molecular weight = 199.26
*Incompatibilities:* Strong oxidizers. When heated to decomposition, this material forms toxic oxides of nitrogen.
*Permissible Concentration in Water:* No criteria set. Runoff from spills or fire control may cause water pollution.
*First Aid:* If this chemical gets into the eyes, remove any contact lenses at once and irrigate immediately for at least 15 minutes, occasionally lifting upper and lower lids. Seek medical attention immediately. If this chemical contacts the skin, remove contaminated clothing and wash immediately with soap and water. Seek medical attention immediately. If this chemical has been inhaled, remove from exposure, begin rescue breathing (using universal precautions) if breathing has stopped, and CPR if heart action has stopped. Transfer promptly to a medical facility. When this chemical

has been swallowed, get medical attention. Give large quantities of water and induce vomiting. Do not make an unconscious person vomit.

# Pyriproxyfen

*Use Type:* Insect growth regulator, insecticide
*CAS Number:* 95737-68-1
*Formula:* $C_{20}H_{19}NO_3$
*Synonyms:* 2-[1-Methyl-2-(4-phenoxyphenoxy)ethoxy]pyridine; 4-Phenoxyphenyl (*RS*)-2-(2-pyridyloxy)propyl ether; Pyridine, 2-[1-methyl-2-(4-phenoxyphenoxy)ethoxy]-
*Trade Names:* ARCHER®, Syngenta (Switzerland); DALAR®, Mclaughlin Gormley King (USA); DISTANCE®, Valent BioSciences (USA); ESTEEM®, Valent BioSciences (USA); NYLAR®; S-9318®; S 31183®; SUMILARV®
*Producers:* Mclaughlin Gormley King (USA); Syngenta (Switzerland); Valent BioSciences (USA)
*Chemical Class:* Unclassified
*EPA/OPP PC Code:* 129032
*California DPR Chemical Code:* 4019
*ICSC Number:* 1269
*RTECS Number:* UT5804000
*Uses:* Pyriproxyfen is found in a number of household products as sprays, powders, baits, mists and shampoos for the control of fleas, ticks, mites and flying insects on pets, in the air, and in carpets and rugs.

*U.S. Maximum Allowable Residue Levels for Pyriproxyfen (40 CFR 180.510):*

| CROP | ppm |
|---|---|
| Almond hulls | 2.0 |
| Apple, pomace, wet | 0.8 |
| Citrus fruits | 0.3 |
| Citrus oil | 20.0 |
| Citrus pulp, dried | 2.0 |
| Cotton, gin byproducts | 2.0 |
| Cottonseed | 0.05 |
| Fruiting vegetables, except cucurbits | 0.2 |
| Pistachio | 0.02 |
| Pome fruits | 0.2 |
| Tree nuts | 0.02 |
| Walnuts | 0.02 |

*Carcinogen/Hazard Classifications*
**U.S. EPA Carcinogens:** Group E, Unlikely Carcinogen
**Label Signal Word:** CAUTION or WARNING
**WHO Acute Hazard:** Class U, unlikely to be hazardous
*Description:* Pale yellow waxy solid or liquid. Molecular weight = 321.41.
*Incompatibilities:* Reacts with strong acids and strong oxidizers. When heated to decomposition, this material forms toxic oxides of nitrogen and carbon monoxide.

*Permissible Concentration in Water:* No criteria set. Runoff from spills or fire control may cause water pollution.
*First Aid:* If this chemical gets into the eyes, remove any contact lenses at once and irrigate immediately for at least 15 minutes, occasionally lifting upper and lower lids. Seek medical attention immediately. If this chemical contacts the skin, remove contaminated clothing and wash immediately with soap and water. Seek medical attention immediately. If this chemical has been inhaled, remove from exposure, begin rescue breathing (using universal precautions) if breathing has stopped, and CPR if heart action has stopped. Transfer promptly to a medical facility. When this chemical has been swallowed, get medical attention. Give large quantities of water and induce vomiting. Do not make an unconscious person vomit.
*References:*
- U.S. Environmental Protection Agency, Office of Pesticide Programs, Pesticide Residue Limits, "Pyriproxyfen," 40 CFR 180.510, www.epa.gov/pesticides/food/viewtols.htm
- California Environmental Protection Agency "Chemical List of Lists," Sacramento CA (February 1997).

# Pyriminil

*Use Type:* Rodenticide
*CAS Number:* 53558-25-1
*Formula:* $C_{13}H_{12}N_4O_3$
*Alert:* Not registered as a pesticide in the U.S.
*Synonyms:* *N*-(4-Nitrophenyl)-*N'*-(3-pyridinylmethyl)urea; 1-(4-Nitrophenyl)-3-(3-pyridinylmethyl)urea; *N*-3-Pyridylmethyl-*N'*-*p*-nitrophenylurea; Pyridylmethyl-*N'*-*para*-nitrophenyl urea; 1-(3-Pyridylmethyl)-3-(4-nitrophenyl)urea; Urea, *N*-(4-nitrophenyl)-*N'*-(3-pyridinylmethyl)-
*Trade Names:* DPL-87®; DLP 787®; PMP-787®; PYRINURON®; RH-787®; VACOR®
*EPA/OPP PC Code:* 104501
*California DPR Chemical Code:* 1916
*RTECS Number:* YI9690000
*Uses:* Used to control Norway rats, roof rats and house mice. Not registered as a pesticide in the U.S.
*Regulatory Authority:*
- Superfund/EPCRA 40CFR355, Appendix B Extremely Hazardous Substances: TPQ = 100/10,000 lb (45.4/4,540 kg)
- Superfund/EPCRA 40CFR302.4 RQ: EHS, 1 lb (0.454 kg)

*Description:* Pyriminil is a yellow crystalline solid resembling corn meal. Melting/Freezing point = 223°C (decomposition). Hazard Identification (based on NFPA-704 M Rating System): Health 3, Flammability 1, Reactivity 0.

*Permissible Exposure Limits in Air:* No standards set.
*Permissible Concentration in Water:* No criteria set, but runoff from spills or fire control may cause water pollution.
*Routes of Entry:* Inhalation, ingestion, skin and/or eye contact.

### Harmful Effects and Symptoms

*Short Term Exposure:* This chemical may cause death by cardiovascular collapse and respiratory failure. Symptoms include nausea, vomiting, abdominal pains, chills, mental confusion, anorexia, aching, dilated pupils, dehydration, chest pain, urinary retention, irregular heartbeat, and muscular weakness. Exposure may also result in visual disturbances, central nervous system depression and tremors.

*Long Term Exposure:* It may damage the pancreas, causing diabetes. Human survivors regularly develop an insulin-deficient, ketosis-prone form of diabetes mellitus. It also affects the central nervous system.

*Points of Attack:* Central nervous system.

*Medical Surveillance:* Blood sugar. Examination of the central nervous system.

*First Aid:* If this chemical gets into the eyes, remove any contact lenses at once and irrigate immediately for at least 15 minutes, occasionally lifting upper and lower lids. Seek medical attention immediately. If this chemical contacts the skin, remove contaminated clothing and wash immediately with soap and water. Seek medical attention immediately. If this chemical has been inhaled, remove from exposure, begin rescue breathing (using universal precautions) if breathing has stopped, and CPR if heart action has stopped. Transfer promptly to a medical facility. When this chemical has been swallowed, get medical attention. Give large quantities of water and induce vomiting. Do not make an unconscious person vomit. Keep victim quiet and maintain normal body temperature. Effects may be delayed; keep victim under observation.

*References:*
- New Jersey Department of Health and Senior Services, "Hazardous Substance Fact Sheet, Pyriminil," Trenton, NJ (May 2002).
  http://www.state.nj.us/health/eoh/rtkweb/2719.pdf
- U.S. Environmental Protection Agency, "Chemical Profile: Pyriminil," Washington, DC, Chemical Emergency Preparedness Program (November 30, 1987).
- California Environmental Protection Agency "Chemical List of Lists," Sacramento CA (February 1997).

# Pyrithiobac-sodium (ANSI)

*Use Type:* Herbicide
*CAS Number:* 123343-16-8
*Formula:* $C_{13}H_{10}ClN_2NaO_4S$

*Synonyms:* Benzoic acid, 2-chloro-6-[(4,6-dimethoxy-2-pyrimidinyl)thio]-, sodium salt (9CI); KIH 2031; Sodium 2-chloro-6-(4,6-dimethoxypyrimidin-2-ylthio)benzoate

*Trade Names:* STAPLE®, DuPont Crop Protection (USA), FMC Agricultural Products Group (USA); DPX-PE350®, DuPont Crop Protection (USA); KIH® 2031, DuPont Crop Protection (USA)

*Producers:* DuPont Crop Protection (USA); FMC Agricultural Product Group (USA)

*Chemical Class:* Unclassified
*EPA/OPP PC Code:* 078905
*California DPR Chemical Code:* 3940

*Uses:* Used to control a wide range of broadleaf weeds and grasses, pre-emergence and post-emergence in cotton.

*U.S. Maximum Allowable Residue Levels for Pyrithiobac-sodium (40 CFR 180.487):*

| CROP | ppm |
| --- | --- |
| Cotton gin by-products | 0.15 |
| Cotton, undelinted seed | 0.02 |

### Carcinogen/Hazard Classifications

*U.S. EPA Carcinogens:* Group C, possible carcinogen
*Label Signal Word:* CAUTION or WARNING
*WHO Acute Hazard:* Class U, unlikely to be hazardous
*Description:* White solid. Soluble in water. Melting point = 223.8–234.2°C.

*Incompatibilities:* Reacts with strong oxidizers. When heated to decomposition, this material forms toxic oxides of sulfur and nitrogen; carbon monoxide and fumes of chlorine.

*Permissible Concentration in Air:* No criteria set.
*Permissible Concentration in Water:* No criteria set. Runoff from spills or fire control may cause water pollution.

*First Aid:* If this chemical gets into the eyes, remove any contact lenses at once and irrigate immediately for at least 15 minutes, occasionally lifting upper and lower lids. Seek medical attention immediately. If this chemical contacts the skin, remove contaminated clothing and wash immediately with soap and water. Seek medical attention immediately. If this chemical has been inhaled, remove from exposure, begin rescue breathing (using universal precautions) if breathing has stopped, and CPR if heart action has stopped. Transfer promptly to a medical facility. When this chemical has been swallowed, get medical attention. Give large quantities of water and induce vomiting. Do not make an unconscious person vomit.

*References:*
- U.S. Environmental Protection Agency, Office of Pesticide Programs, Pesticide Residue Limits, "Pyrithiobac-sodium," 40 CFR 180.487, www.epa.gov/pesticides/food/viewtols.htm
- California Environmental Protection Agency "Chemical List of Lists," Sacramento CA (February 1997)

# Q

## Quinalphos

*Use Type:* Insecticide, acaricide, plant growth regulator
*CAS Number:* 13593-03-8
*Formula:* $C_{12}H_{15}N_2O_3PS$
*Synonyms:* O,O-Diaethyl-O-[chinoxalyl-(2)]-monothiophosphat (German); O,O-Diethyl-O-(2-chinoxalyl)phosphorothioate; Diethquinalphion; Diethquinalphione; O,O-Diethyl O-2-quinoxalinyl phosphorothioate; O,O-Diethyl-O-quinoxalin-2-yl phosphorothioate; O,O-Diethyl O-(quinoxalin-2-yl)thiophosphate; O,O-Diethyl O-quinoxalin-2-yl thionophosphate; O,O-Diethyl-O-(2-quinoxalinyl)phosphorothioate; O,O-Diethyl-O-(2-quinoxalyl)phosphorothioate; ENT 27,394; NSC
*Trade Names:* BAYRUSIL®; CHINALPHOS®; EKALUX®, BEC Group (India); EKAQUIN®, BEC Group (India); DHANULUX®, Dhanuka Group (India); QUINALTAF®, Rallis India (India); SAN® 6538 I, Syngenta (Switzerland); SAN® 6626 I, Syngenta (Switzerland); SANDOZ® 6538, Syngenta (Switzerland); SPENCER® S-6538; SRA-7312®; SUQUIN®, Sudarshan Chemical Industries (India); WIE OBEN®
*Producers:* Agrimor International (USA); Agsin (Singapore); Bayer CropScience (Germany); BEC Group (India); Dhanuka Group (India); Gharda Chemicals (India); Gujarat Pesticides (India); Indiclay (India); Nagarjuna Agrichem (India); Rallis India (India); Sudarshan Chemical Industries (India); Syngenta (Switzerland)
*Chemical Class:* Organophosphate
*EPA/OPP PC Code:* 381400
*Uses:* Not registered in the U.S. Used to control a broad variety of insect pests; for caterpillars on vegetables, ground nuts and cotton; scales and caterpillars on fruit trees.
*Carcinogen/Hazard Classifications*
**Label Signal Word:** WARNING
**WHO Acute Hazard:** Class II, moderately hazardous
*Regulatory Authority:*
- DOT Inhalation Hazard Chemicals as organophosphate compound

*Description:* Colorless, crystalline solid. Slightly soluble in water; solubility = 25 mg/L @ 20°C. Molecular weight 298.31. Density = 1.22 @ 20°C. Boiling point = 142°C @ $3 \times 10^{-4}$ mmHg. (decomposes). Melting/Freezing point = 31°C. Vapor pressure = $3.5 \times 10^{-6}$ mmHg @ 20°C.
*Incompatibilities:* May react violently with antimony(V) pentafluoride. Incompatible with lead diacetate, magnesium, silver nitrate.
*Permissible Exposure Limits in Air:* No standards set.

*Determination in Air:* OSHA versatile sampler-2; Toluene/Acetone; Gas chromatography/Flame ionization detection; NIOSH IV, Method #5600, Organophosphorus Pesticides.[18]
*Permissible Concentration in Water:* No criteria set. Runoff from spills or fire control may cause water pollution.
*Harmful Effects and Symptoms*
*Short Term Exposure:* Eye pupils are small; blurred vision; eye watering; runny nose; cough; shortness of breath; salivation; dizziness; nausea, stomach cramps, diarrhea, and vomiting; increased blood pressure; profuse sweating; hypermotility, hallucinations; irritability; tingling of the skin; drowsiness; slow heartbeat; convulsions; fluid in lungs; loss of consciousness; incontinence; breathing stops; death. Organophosphates inhibit the action of acetylcholinesterase enzymes, and alter the way in which nervous impulses are transmitted. The effects can last for hours, days, or much longer. The action of the enzymes is reestablished after new enzymes are formed.
*Long Term Exposure:* Cholinesterase inhibitor; cumulative effect is possible. Organophosphates may damage the nervous system with repeated exposure, resulting in convulsions, respiratory failure. May cause liver damage.
*Points of Attack:* Respiratory system, central nervous system, cardiovascular system, blood cholinesterase.
*Medical Surveillance:* Medical observation is recommended for 24 to 48 hours after breathing overexposure, as pulmonary edema may be delayed. As first aid for pulmonary edema, consider administering a corticosteroid spray. Cigarette smoking may exacerbate pulmonary injury and should be discouraged for at least 72 hours following exposure.
*First Aid:* Speed in removing material from eyes and skin is of extreme importance. Eye contact can cause dangerous amounts of these chemicals to be quickly absorbed through the mucous membrane into the bloodstream. Immediately and gently flush eyes with plenty of warm or cold water (NO hot water) for at least 15 minutes, occasionally lifting the upper and lower eyelids. Get medical aid immediately. *Skin:* Get medical aid. Skin contact can cause dangerous amounts of these chemicals to be absorbed into the bloodstream. Wearing the appropriate PPE equipment and respirator for organophosphate pesticides, immediately flush skin with plenty of soap and water for at least 15 minutes while removing contaminated clothing and shoes. Shampoo hair promptly if contaminated. *Ingestion:* Call poison control. Loosen all clothing. Never give anything by mouth to an unconscious person. Get medical aid. Do NOT induce vomiting.* If conscious, alert, and able to swallow, rinse mouth and have victim drink 4 to 8 ounces of water do NOT

induce vomiting but immediately administer slurry of activated charcoal (2 oz in 8 oz of water). If victim is *unconscious or having convulsions,* do nothing except keep victim warm.**In some cases you may be specifically instructed by poison control to induce vomiting by way of 2 tablespoons of syrup of ipecac (adult) washed down with a cup of water.* Do NOT give activated charcoal before or with ipecac syrup. *Inhalation:* Get medical aid. Do not contaminate yourself. Wearing the appropriate PPE equipment and respirator for organophosphate pesticides, immediately remove the victim from the contaminated area to fresh air. If the victim is not breathing, administer artificial respiration. Do NOT use mouth-to-mouth resuscitation; use bag/mask apparatus. If breathing is difficult, administer oxygen through bag/mask apparatus until medical help arrives. Do not leave victim unattended. *Note to physician or authorized medical personnel.* Administer atropine, 2 mg (1/30 gr) intramuscularly or intravenously as soon as any local or systemic signs or symptoms of an intoxication are noted; repeat the administration of atropine every 3 to 8 minutes until signs of atropinization (mydriasis, dry mouth, rapid pulse, hot and dry skin) occur; initiate treatment in children with 0.05 mg mg/kg of atropine; repeat at 5 to 10 minute intervals. Watch respiration, and remove bronchial secretions if they appear to be obstructing the airway; intubate if necessary. *Notes to physician or authorized medical personnel*: N-methylpyridinium-2-aldoxime (2-PAMCl) when used in conjunction with atropine reacts with the phosphorylated cholinesterase, thereby restoring normal activity to by removing the phosphorylating group. The combination of these two chemicals is synergistic and must be administered within minutes to a few hours following exposure (depending on the specific agent) to be effective. Give 2-PAMCI (Pralidoxime; Protopam), 2.5 gm in 100 ml of sterile water or in 5% dextrose and water, intravenously, slowly, in 15-30 minutes; if sufficient fluid is not available, give 1 gm of 2-PAMCI in 3 ml of distilled water by deep intramuscular injection; repeat this every half hour if respiration weakens or if muscle fasciculation or convulsions recur. Also Diazepam, an anticonvulsant, might be considered.

# Quinclorac

*Use Type:* Herbicide
*CAS Number:* 84087-01-4
*Formula:* $C_{10}H_5Cl_2NO_2$
*Synonyms:* 3,7-Dichloroquinoline-8-carboxylic acid; 3,7-Dichloro-8-quinolinecarboxylic acid; Quinclorac tech; 8-Quinolinecarboxylic acid(8-), 3,7-dichloro-
*Trade Names:* BAS-514 00H®, BASF Agricultural Products (Germany); CAO-NENG®; DRIVE 75®, BASF Agricultural Products (Germany); FACET®, BASF Agricultural Products (Germany); FAS-NOX®; PARAMOUNT®, BASF Agricultural Products (Germany); PROPACET®; TURF BUILDER WITH WEED CONTROL®, Scotts Company (USA)
*Producers:* BASF Agricultural Products (Germany); Scotts Company (USA)
*Chemical Class:* Unclassified
*EPA/OPP PC Code:* 128974
*California DPR Chemical Code:* 5104
*U.S. Maximum Allowable Residue Levels for Quinclorac (40 CFR 180.463):*

| CROP | ppm |
|---|---|
| Cattle, meat | 0.05 |
| Egg | 0.05 |
| Goat, meat | 0.05 |
| Hog, meat | 0.05 |
| Horse, meat | 0.05 |
| Milk | 0.05 |
| Poultry, meat | 0.05 |
| Poultry, mbyp | 0.1 |
| Rice, bran | 15 |
| Rice, grain | 5 |
| Rice, straw | 12 |
| Sheep, meat | 0.05 |
| Wheat, hay | 0.5 |

*Carcinogen/Hazard Classifications*
**U.S. EPA Carcinogens:** Group D, unclassifiable
**Label Signal Word:** CAUTION or DANGER
**WHO Acute Hazard:** Class U, unlikely to be hazardous
*Regulatory Authority:*
- FIFRA, 40CFR186: Tolerances for pesticides in animal foods

*Incompatibilities:* Keep away from oxidizers, sulfuric acid, caustics, ammonia, aliphatic amines, alkanolamines, isocyanates, alkylene oxides, epichlorohydrin. Incompatible with silver compounds. Mixture with some silver compounds forms explosive salts of silver oxalate.
*Permissible Concentration in Water:* No criteria set. Runoff from spills or fire control may cause water pollution.
*Routes of Entry:* Inhalation, ingestion.
*First Aid:* If this chemical gets into the eyes, remove any contact lenses at once and irrigate immediately for at least 15 minutes, occasionally lifting upper and lower lids. Seek medical attention immediately. If this chemical contacts the skin, remove contaminated clothing and wash immediately with soap and water. Seek medical attention immediately. If this chemical has been inhaled, remove from exposure, begin rescue breathing (using universal precautions) if breathing has stopped, and CPR if heart action has stopped. Transfer promptly to a medical facility. When this chemical has been swallowed, get medical attention. Give large quantities of water and induce vomiting. Do not make an unconscious person vomit. Do not induce vomiting when formulations containing petroleum solvents are ingested.

*References:*
- U.S. Environmental Protection Agency, Office of Pesticide Programs, Pesticide Residue Limits, "Quinclorac," 40 CFR 180.463. www.epa.gov/pesticides/food/viewtols.htm

# Quintozene

*Use Type:* Soil fungicide, nematicide and seed treatment
*CAS Number:* 82-68-8
*Formula:* $C_6Cl_5NO_2$
*Alert:* Most uses of this pesticide, normally called quintozene or PCNB, have been canceled in the U.S. Human toxicity (long-term): High
*Synonyms:* Benzene, pentachloronitro-; NCI-C00419; Nitropentachlorobenzene; Olipsan; Olpisan; PCNB; Pentachloronitrobenzene; Pentachlornirtobenzol (German); Quinosan; Quintocene; Quintoceno (Spanish)
*Trade Names:* (EPA lists 290 active and canceled or transferred products) AVICOL (PESTICIDE)®; BOTRILEX®; BLOCKER 4F®, Amvac Chemical Corp. (USA); BOTRILEX®; BRASSICOL®; BRASSICOL EARTHCIDE®; BRASSICOL 75®; BRASSICOL SUPER®; CHINOZAN®; EARTHCIDE®; FARTOX®; FOLOSAN®; FOMAC 2®; FUNGICHLOR®; GC 3944-3-4®; KOBU®; KOBUTOL®; KODIAK A-T FUNGICIDE®, Gustafson (USA); KP 2®; MARISAN FORTE®; MEFENOXAM®, Syngenta (Switzerland); PARFLO®, Amvac Chemical Corp. (USA); PENTAGEN®; PHOMASAN®; PKhNB®; RTU 1010®; SANICLOR 30®; TERRACHLOR®, Gustafson (USA); TERRACLOR®, Gustafson (USA); TERRACLOR 30 G®, Gustafson (USA); TERRA-COAT®, Gustafson (USA); TERRAFUN®; TERRAZAN®; TILCAREX®; TRIPCNB®; TRIQUINTAM®; TRITISAN®; TUBERGRAN®; TURFCIDE®, Crompton Corporation (USA); VITAVAX®, Gustafson (USA)
*Producers:* Amvac Chemical Corp. (USA); Crompton Corporation (USA); Dainippon Ink & Chemical (Japan); Drexel Chemical (USA); Gowan Company (USA); Gustafson (USA); Hokkio Chemical Industry (Japan); Luxembourg Industries (PAMOL) (Israel); Montedison (Italy); Nippon Kayaku (Japan); Shenzhen Guomeng Industry Co., Ltd. (China); Simplot, J.R. Company (USA); Syngenta (Switzerland); Uniroyal Chemical (USA)
*Chemical Class:* Organochlorine; Halo-organics
*EPA/OPP PC Code:* 056502
*California DPR Chemical Code:* 464
*ICSC Number:* 0745
*RTECS Number:* DA6650000
*EEC Number:* 609-043-00-5
*EINECS Number:* 201-435-0
*Uses:* Quintozene, the common name for PCNB or pentachlkoronitrobenzene, is a organochlorine fungicide used as a seed dressing or soil treatment to control a wide range of fungi species in such crops as potatoes, wheat, onions, lettuce, tomatoes, tulips, garlic, and others. Depending on the producer and the manufacturing procedure, PCNB impurities can include hexachlorobenzene, pentachlorobenzene, and tetrachloronitrobenzene. The fungicide is often used in combination with insecticides and fungicides including carbaryl, imazalil, tridimenol, etridiazole, and fuberidazole. It is available as a dustable or wettable powder, in granular form, emulsifiable concentrate, and seed treatment.
*Carcinogen/Hazard Classifications*
**U.S. EPA Carcinogens:** Group C, Possible carcinogen
**IARC:** Group 3, animal sufficient evidence
**Label Signal Word:** CAUTION
**WHO Acute Hazard:** Class U, unlikely to be hazardous
**Endocrine Disruptor:** Suspected endocrine disruptor
*Regulatory Authority:*
- Carcinogen (Animal Positive) (IARC)[9]; (Animal Negative, rat, mouse) (NCI)[9]
- Banned or Severely Restricted (Germany, U.S.) (UN)[13]
- Air Pollutant Standard Set (former USSR)[43)(35] (Pennsylvania)[60]
- AB 1803-Well Monitoring Chemical (CAL)
- MCL (Maximum Contaminants Levels) list of contaminants (CAL)
- AB 2588-Air Toxics "Hot Spots" Chemicals (CAL)
- CAL-DHS/DHS Drinking Water Action Levels
- CAL Air Resources Board/AB 1807 Toxic Air Contaminants
- The "Director's List" (CAL/OSHA)
- Actively registered pesticide in California.
- EPA Hazardous Waste Number (RCRA No.): U185
- RCRA, 40CFR261, Appendix 8 Hazardous Constituents
- RCRA 40CFR268.48; 61FR15654, Universal Treatment Standards: Wastewater (mg/L), 0.055; Nonwastewater (mg/kg), 4.8
- RCRA 40CFR264, Appendix 9; TSD Facilities Ground Water Monitoring List. Suggested test method(s) (PQL $ug/L$): 8270(10)
- Superfund/EPCRA 40CFR302.4 RQ: CERCLA, 100 lb (45.4 kg)
- EPCRA Section 313 Form R *de minimis* concentration reporting level: 1.0%
- Canada, WHMIS, Ingredients Disclosure List

*Description:* PCNB, $C_6Cl_5NO_2$, forms colorless needles. Technical-grade PCNB contains an average of 97.8% PCNB, 1.8% hexachlorobenzene (HCB), 0.4% 2,3,4,5-tetrachloronitrobenzene (TCNB), and less than 0.1% pentachlorobenzene. Practically insoluble in water. Molecular weight 295.33. Boiling point = 328°C. Melting/Freezing point = 146°C. Vapor pressure = $1 \times 10^{-4}$ mbar @ 25°C (pentachloronitrobenzene). Log $K_{ow}$ = 4.5 to 5.42. Values at or above 3.0 are likely to bioaccumulate in marine organisms.

*Incompatibilities:* Reacts with alkalies. When heated to decomposition, this material forms toxic oxides of nitrogen and fumes of chlorine.

*Permissible Exposure Limits in Air:* The former USSR[35, 43] has set an MAC in workplace air of 0.5 mg/m$^3$ and MAC values for ambient air in residential areas of 0.01 mg/m$^3$ on a momentary basis and 0.006 mg/m$^3$ on a daily average basis. A guideline for PCNB in ambient air has been set[60] in Pennsylvania at 2.47 $\mu$g/m$^3$.

*Permissible Concentration in Water:* No criteria set, but runoff from spills or fire control may cause water pollution.

*Routes of Entry:* Inhalation, ingestion, skin and/or eye contact.

*Harmful Effects and Symptoms*

*Short Term Exposure:* May cause skin and eye irritation, sensitization with erythema, itching and edema. A rebuttable presumption against registration of PCNB for pesticidal uses was issued on October 13, 1977 by EPA on the basis of oncogenicity.

*Long Term Exposure:* There is limited evidence that this compound is an animal carcinogen.

*First Aid: Skin Contact:* Flood all areas of body that have contacted the substance with water. Don't wait to remove contaminated clothing; do it under the water stream. Use soap to help assure removal. Isolate contaminated clothing when removed to prevent contact by others. *Eye Contact:* Remove any contact lenses at once. Flush eyes well with copious quantities of water or normal saline for at least 20-30 minutes. Seek medical attention. *Inhalation:* Leave contaminated area immediately; breathe fresh air. Proper respiratory protection must be supplied to any rescuers. If coughing, difficult breathing or any other symptoms develop, seek medical attention at once, even if symptoms develop many hours after exposure. *Ingestion:* If convulsions are not present, give a glass or two of water or milk to dilute the substance. Assure that the person's airway is unobstructed and contact a hospital or poison center immediately for advice on whether or not to induce vomiting.

*References:*
- EXTOXNET, Extension Toxicology Network, "Pesticide Information Profile, Quintozene," Oregon State University, Corvallis, OR (June 1986). http://extoxnet.orst.edu/pips/quintoze.htm
- U.S. Environmental Protection Agency, "Pentachloronitrobenzene, Health and Environmental Effects Profile," No. 142, Office of Solid Waste, Washington, DC (April 30, 1980).
- Lee, C.C., "Environmental Law Index to Chemicals," Government Institutes, Rockville, MD (1996).
- Sax, N.I., Ed., "Dangerous Properties of Industrial Materials Report," 5, No. 3, 11-16 (1985).
- California Environmental Protection Agency "Chemical List of Lists," Sacramento CA (February 1997).

# Quizalofop-ethyl

*Use Type:* Herbicide
*CAS Number:* 76578-14-8
*Formula:* $C_{19}H_{17}ClN_2O_4$
*Synonyms:* Caswell No. 215D; 2-[4-((6-Chloro-2-quinoxalinyl)oxy)phenoxy]ethyl propionate; 2-[4-((6-Chloro-2-quinoxalinyl)oxy)phenoxy]propionic acid, ethyl ester; Ethyl 2-[4-(6-chloro-2-quinoxalyloxy)phenoxy] propionate; NCI-C99983; Propanoic acid, 2-[4-((6-chloro-2-quinoxalinyl)oxy)phenoxy]-, ethyl ester; Quinofop-ethyl; Xylofop-ethyl

*Trade Names:* ASSURE®, DuPont Crop Protection (USA), canceled 3/14/1994; DPX-Y 6202®; FBC®; 32197; INY-6202; MATADOR®, FMC Agricultural Products Group (USA); MON® 78746, Monsanto (USA); NC-302®; PILOT®; TARGA®, Nissan Chemical Industries (Japan)

*Producers:* DuPont Crop Protection (USA); Epochem Co., (China); FMC Agricultural Products Group (USA); Monsanto (USA); Nissan Chemical Industries (Japan)

*Chemical Class:* Chlorophenoxy
*EPA/OPP PC Code:* 128711; (128201 old EPA code number)
*California DPR Chemical Code:* 2226
*RTECS Number:* GW71910000
*U.S. Maximum Allowable Residue Levels for Quizalofop-Ethyl (40 CFR 180.441):*

| CROP | ppm |
| --- | --- |
| Cattle, fat | 0.05 |
| Cattle, meat | 0.02 |
| Cattle, mbyp | 0.05 |
| Eggs | 0.02 |
| Goat, fat | 0.05 |
| Goat, meat | 0.02 |
| Goat, mbyp | 0.05 |
| Hog, fat | 0.05 |
| Hog, meat | 0.02 |
| Hog, mbyp | 0.05 |
| Horse, fat | 0.05 |
| Horse, meat | 0.02 |
| Horse, mbyp | 0.05 |
| Milk | 0.01 |
| Milk, fat | 0.05 |
| Poultry, fat | 0.05 |
| Poultry, meat | 0.02 |
| Poultry, mbyp | 0.05 |
| Sheep, fat | 0.05 |
| Sheep, meat | 0.02 |
| Sheep, mbyp | 0.05 |
| Soybean flour | 0.5 |
| Soybean hulls | 0.02 |
| Soybean meal | 0.5 |
| Soybean soapstock | 1.0 |
| Soybeans | 0.05 |

## Quizalofop-ethyl

*Carcinogen/Hazard Classifications*
**U.S. EPA Carcinogens:** Group D, unclassifiable
**California Prop. 65:** Male reproductive toxin
**U.S. TRI:** Developmental toxin; male reproductive toxin
**Label Signal Word:** CAUTION or DANGER
**WHO Acute Hazard:** Class III, slightly hazardous as quizalofop (parent element)
*Regulatory Authority:*
- EPA 40 CFR 372.65, Specific Toxic Chemical Listings
- FIFRA, 40CFR185: tolerances for pesticides in food
- FIFRA, 40CFR186: tolerances for pesticides in animal feeds
- Actively registered pesticide in California.
- AB 2588-Air Toxics "Hot Spots" Chemicals (CAL) as chlorophenoxy pesticides
- The "Director's List" (CAL/OSHA) as chlorophenoxy pesticides
- EPCRA Section 313 Form R *de minimis* concentration reporting level: 1.0%

*Description:* White crystalline solid. Practically insoluble in water; solubility = $<2 \times 10^{-8}$ ppm. Molecular weight = 372.78. Vapor pressure = $3 \times 10^{-7}$ mmHg @ 20°C.

*Permissible Concentration in Water:* No criteria set. Runoff from spills or fire control may cause water pollution.

*Routes of Entry:* Inhalation, ingestion, absorbtion through the skin.

*Harmful Effects and Symptoms*

*Short Term Exposure:* Poisonous; may be fatal if inhaled, swallowed, or absorbed through skin. Severely irritates eyes, skin and respiratory tract, with burning sensation, pain, redness and swelling. Metabolic stimulant. If inhaled, causes coughing, dilated pupils, headache, profuse persperation, intense thirst, extreme fatigue, rapid pulse, high fever, clammy, flushed skin, rapid breathing, nausea, vomiting, cyanosis (bluish tint to skin and lips), anxiety and confusion, convulsions, risk of lung edema. If swallowed, face and lips turn bluish. Liver injury with associated jaundice, kidney failure, and cardiac arrhythmias are commonly noted. Nerve damage, which may be delayed, may include swelling of legs and feet, muscle twitch and stupor. Severe exposure can cause death from heart failure. Dust or liquid left in contact with the skin for several hours may be absorbed. This may result in severe delayed symptoms as listed above. These symptoms may last for months or years.

*Long Term Exposure:* Workers exposed to chlorophenoxy compounds such as 2,4-D (in the manufacturing process) over a five to ten year period at levels above 10 mg/m$^3$ complained of weakness, rapid fatigue, headache and vertigo. Liver damage, low blood pressure and slowed heartbeat were also found. Based on animal tests, may affects human reproduction.

*Points of Attack:* Eyes, skin, respiratory system, central nervous system, cardiovascular system, liver, kidney.

*Medical Surveillance:* If symptoms develop or overexposure is suspected, the following may be useful: Liver and kidney function tests. Exam of the nervous system.

*First Aid:* If this chemical gets into the eyes, remove any contact lenses at once and irrigate immediately for at least 15 minutes, occasionally lifting upper and lower lids. Seek medical attention immediately. If this chemical contacts the skin, remove contaminated clothing and wash immediately with soap and water. Seek medical attention immediately. If this chemical has been inhaled, remove from exposure, begin rescue breathing (using universal precautions) if breathing has stopped, and CPR if heart action has stopped. Transfer promptly to a medical facility. When this chemical has been swallowed, get medical attention. *Do not induce vomiting when formulations containing petroleum solvents are ingested.* Otherwise, give large quantities of water and induce vomiting. Do not make an unconscious person vomit.

*References:*
- U.S. Environmental Protection Agency, Office of Pesticide Programs, Pesticide Residue Limits, "Quizalofop-Ethyl," 40 CFR 180.441, www.epa.gov/pesticides/food/viewtols.htm
- California Environmental Protection Agency "Chemical List of Lists," Sacramento CA (February 1997)

# R

## Resmethrin (ANSI); (+)-*d-trans*-Resmethrin; (+)-*cis*-Resmethrin; (−)-*trans*-Resmethrin

*Use Type:* Insecticide
*CAS Number:* 10453-86-8; 28434-01-7 (*d-trans*-isomer); 35764-59-1 (*cis*-isomer); 33911-28-3 (−)-(*trans*-isomer)
*Formula:* $C_{22}H_{26}O_3$
*Alert:* Restricted Use Pesticide (RUP) when formulated for use in mosquito abatement and pest control treatments at non-agricultural sites.
*Synonyms:* Al3-27474; Benzofuroline; Benzyfuroline; 5-Benzylfurfuryl chrysanthemate; (5-Benzyl-3-furyl)methyl chrysanthemate; 5-Benzyl-3-furylmethyl (±)-*cis-trans*-chrysanthemate; 5-Benzyl-3-furylmethyl(+)-*trans*-chrysanthemate; *d-trans*(5-Benzyl-3-furyl)methyl 2,2-dimethyl-3-(2-methylpropenyl) cyclopropanecarboxylate; 5-Benzyl-3-furylmethyl(1RS)-*cis,trans*-2,2-dimethyl-3-(2-methylprop-1-enyl)cyclopropanecarboxylate; 5-Benzyl-3-furylmethyl(1RS)-*(Z),(E)*-2,2-dimethyl-3-(2-methylprop-1-enyl)cyclopropanecarboxylate; 5-Benzyl-3-furylmethyl (1RS,3RS;1RS,3SR)-2,2-dimethyl-3-(2-methylprop-1-enyl)cyclopropanecarboxylate; (5-Benzyl-3-furyl)methyl 2,2-dimethyl-3-(2-methylpropenyl) cyclopropanecarboxylate; Bioresmethrin (*d-trans* isomer); Caswell No. 083E; Cyclopropanecar boxylic acid, 2,2-dimethyl-3-(2-methylpropenyl)-, (4-(2-benzyl)furyl) methyl ester; Cyclopropanecarboxylic acid, 2,2-dimethyl-3-(2-methylpropenyl)-, (5-benzyl-3-furyl)methyl ester (8CI); Cyclopropanecarboxylic acid, 2,2-dimethyl-3-(2-methyl-1-propenyl)-, (5-(phenylmethyl)-3-furanyl)methyl ester, *cis,trans*-(±)-; Cyclopropanecarboxylic acid, 2,2-dimethyl-3-(2-methyl-1-propenyl)-, [5-(phenylmethyl)-3-furanyl]methyl ester, *(Z),(E)*-(±)-; Cyclopropanecarboxylic acid, 2,2-dimethyl-3-(2-methyl-1-propenyl)-, [5-(phenylmethyl)-3-furanyl]methyl ester (9CI); Cyclopropanecarboxylic acid, 2,2-dimethyl-3-(2-methyl-1-propenyl)-(5-(phenylmethyl)-3-furanyl)methyl ester, (1R-*cis*)-; Dimethyl 3-(2-methyl-1-propenyl) cyclopropanecarboxylate; 2,2-dimethyl-3-(2-methyl-1-propenyl)cyclopropanecarboxylic acid; ENT 27474; NSC 195022; [5-(Phenylmethyl)-3-furanyl]methyl 2,2-dimethyl-3-furylmethyl-2,2-dimethyl-3-(2-methylpropenyl) cyclopropanecarboxylate; [5-(Phenylmethyl)-3-furanyl] methyl 2,2-dimethyl-3-(2-methyl-1-propenyl)cyclopropanecarboxylate); 5-(Phenylmethyl)-3-furanyl)methyl 2,2-dimethyl-3-(2-methyl-1-propenyl)cyclopropanecarboxylate; [5-(Phenylmethyl)-3-furanyl]methyl ester; Resmethrin, (±)-; Resmethrin, (+)-*trans,cis*-; Resmethrin, (+)-*(E),(Z)*-
*Trade Names:* BIORESMETHRIN® (*d-trans*-isomer); CHRYSRON®; CISMETHRIN® (*cis*-isomer); CROSSFIRE®; DERRINGER®; FMC® 17370, FMC Agricultural Products (USA); ISATHRIN®; NRDC® 107 (*d-trans*-isomer); NIA 26021® (*cis*-isomer), FMC Agricultural Products (USA); NIA-17370®, FMC Agricultural Products (USA); NIAGARA® 18739 (*d-trans*-isomer), FMC Agricultural Products (USA); NIAGARA® 26021 (*cis*-isomer), FMC Agricultural Products (USA); NRDC® 119 (*cis*-isomer); OBLIQUE®; PYNOSECT®; PYRETHERM®; RAID®; RESPOND®; RU-11484® (*d-trans*-isomer); SBP® 1382 (*d-trans*-isomer); *d-trans*-SBP® 1382 (*d-trans*-isomer); SBP®-1390; S.B. PENICK 1382®; SCOURGE®, Bayer CropScience (Germany); SUN-BUGGER®; SYNTHRIN®; SYNTOX®; VECTRIN®; WHITMIRE® PT-110
*Producers:* Bayer CropScience (Germany); FMC Agricultural Products (USA)
*Chemical Class:* Pyrethroid
*EPA/OPP PC Code:* 097801; 097802 (*d-trans*-isomer); 097804 (*cis*-isomer)
*California DPR Chemical Code:* 2119
*ICSC Number:* 0324
*RTECS Number:* GZ1310000
*Uses:* Currently, there are 287 registered products in the U.S., primarily used for household sprays to control flying and crawling insects. Also used in mushroom houses, stored products and for mosquito control by aerial application. It is applied to horses and in stables, to fabrics, and in pet sprays and shampoos. The U.S. EPA has records of another 1000 products that have been canceled and/or transferred.
*Carcinogen/Hazard Classifications*
**California Prop. 65:** Developmental toxin
**U.S. TRI:** Developmental and reproductive toxin
**Label Signal Word:** CAUTION
**WHO Acute Hazard:** Class III, slightly hazardous
**Endocrine Disruptor:** Suspected endocrine disruptor
*Regulatory Authority:*
- EPA 40 CFR 372.65, Specific Toxic Chemical Listings
- Actively registered pesticide in California.
- AB 1803-Well Monitoring Chemical (CAL) as pyrethrins
- Permissible Exposure Limits for Chemical Contaminants (CAL/OSHA) as pyrethrum
- Clean Water Act: Section 311 Hazardous Substances/RQ (same as CERCLA) as pyrethrins
- EPCRA Section 304 RQ: CERCLA, 1 lb (0.454 kg) as pyrethrins
- EPCRA Section 313 Form R *de minimis* concentration reporting level: 1.0%

*Description:* Off-white to tan waxy solid, and colorless crystals. Chrysanthemum-like odor. Molecular weight = 338.48 (all isomers). Boiling point = 175° @ $8 \times 10^{-4}$ mmHg. Vapor pressure = $1 \times 10^{-8}$ mmHg @ 20°C.

*Incompatibilities:* May react violently with strong oxidizers, bromine, 90% hydrogen peroxide, phosphorus trichloride, silver powders or dust. Incompatible with silver compounds. Mixture with some silver compounds forms explosive salts of silver oxalate.

*Permissible Concentration in Water:* No criteria set. Runoff from spills or fire control may cause water pollution.

*Harmful Effects and Symptoms*

*Short Term Exposure:* Pyrethrins can affect you when breathed in and by passing through your skin. Irritates the eyes and respiratory tract. High exposure can affect the nervous system causing headache, nausea, vomiting, fatigue, and restlessness, rhinorrhea (discharge of thin nasal mucous).

*Long Term Exposure:* High or repeated exposure can cause lung allergy (with cough, wheezing and/or shortness of breath) or hay fever symptoms (sneezing, runny or stuffy nose). Allergic "pneumonia" can also occur with cough, chest pain, breathing difficulty and abnormal chest x-ray. Repeated attacks may lead to permanent scarring. Skin allergy may also develop with rash and itching, even with lower exposures. Skin contact can cause rash with redness, blisters and intense itching. A severe generalized allergy can occur with weakness and collapse.

*Points of Attack:* Respiratory system, skin, central nervous system.

*Medical Surveillance:* Before beginning employment and at regular times after that, the following are recommended: Lung function tests. These may be normal if the person is not having an attack at the time of the test. Consider chest x-ray if lung symptoms are present. Evaluation by a qualified allergist, including careful exposure history and special testing, may help diagnose skin allergy.

*First Aid:* If this chemical gets into the eyes, remove any contact lenses at once and irrigate immediately for at least 15 minutes, occasionally lifting upper and lower lids. Seek medical attention immediately. If this chemical contacts the skin, remove contaminated clothing and wash immediately with soap and water. Seek medical attention immediately. If this chemical has been inhaled, remove from exposure, begin rescue breathing (using universal precautions) if breathing has stopped, and CPR if heart action has stopped. Transfer promptly to a medical facility. When this chemical has been swallowed, get medical attention. *Do not induce vomiting when formulations containing petroleum solvents are ingested.* Otherwise, give large quantities of water and induce vomiting. Do not make an unconscious person vomit.

*References:*
- EXTOXNET, Extension Toxicology Network, "Pesticide Information Profile, Resmethrin," Oregon State University, Corvallis, OR (June 1996). http://extoxnet.orst.edu/pips/resmethr.htm
- California Environmental Protection Agency "Chemical List of Lists," Sacramento CA (February 1997)

# Rimsulfuron (ANSI)

*Use Type:* Herbicide
*CAS Number:* 122931-48-0
*Formula:* $C_{14}H_{17}N_5O_7S_2$
*Alert:* A Restricted Use Pesticide (RUP) for some formulations.
*Synonyms:* N-[(4,6-Dimethoxypyrimidin-2-yl)aminocarbonyl]-3-(ethylsulfonyl)-2-pyridinesulfonamide; 1-(4,6-Dimethoxypyrimidin-2-yl)-3-(3-ethylsulfonyl-2-pyridylsulfonyl)urea; N-[((4,6-Dimethoxy-2-pyrimidinyl)amino)carbonyl]-3-(ethylsulfonyl)-2-pyridinesulfonamide; 2-Pyridinesulfonamide, N-[((4,6-dimethoxy-2-pyrimidinyl)amino)carbonyl]-3-(ethylsulfonyl)-
*Trade Names:* ACCENT®, DuPont Crop Protection (USA); BASIS® (rimsulfuron + thifensulfuron methyl), DuPont Crop Protection (USA); DPX-E9636®, DuPont Crop Protection (USA); DPX 79406® (nicosulfuron + rimsulfuron), DuPont Crop Protection (USA); Matrix® (nicosulfuron + rimsulfuron), DuPont Crop Protection (USA); SHADEOUT®, DuPont Crop Protection (USA); STEADFAST®, (nicosulfuron + rimsulfuron),Dupont Crop Protection (USA); TRANXIT®, Griffin (USA)
*Producers:* DuPont Crop Protection (USA); Griffin (USA)
*Chemical Class:* Sulfonylurea
*EPA/OPP PC Code:* 129009; (129024 old EPA code number)
*California DPR Chemical Code:* 3835
*Uses:* For controlling broadleaf weeds.
*U.S. Maximum Allowable Residue Levels for Rimsulfuron (40 CFR 180.478):*

| CROP | ppm |
|---|---|
| Corn, field, forage | 0.1 |
| Corn, field, grain | 0.1 |
| Corn, field, stover | 0.1 |
| Potato | 0.1 |
| Tomato | 0.05 |

*Carcinogen/Hazard Classifications*
**U.S. EPA Carcinogens:** Group E, Unlikely Carcinogen
**Label Signal Word:** CAUTION, WARNING or DANGER
**WHO Acute Hazard:** Class U, unlikely to be hazardous
*Description:* Soluble in water.
*Incompatibilities:* Slowly hydrolyzes in water, releasing ammonia and forming acetate salts.
*Permissible Concentration in Water:* No criteria set. Runoff from spills or fire control may cause water pollution.

*Harmful Effects and Symptoms*
*Short Term Exposure:* Contact with eyes or skin may cause irritation or burns. Inhalation should be avoided; use NIOSH-approved air purifying respirators for pesticides. May be harmful if swallowed. Skin contact may cause allergic reaction.
*Long Term Exposure:* Repeated or prolonged contact may cause skin and lung sensitization, resulting in allergies.
*Points of Attack:* Skin, inhalation.
*Medical Surveillance:* Evaluation by a qualified allergist, including careful exposure history and special testing, may help diagnose skin allergy.
*First Aid:* If this chemical gets into the eyes, remove any contact lenses at once and irrigate immediately for at least 15 minutes, occasionally lifting upper and lower lids. Seek medical attention immediately. If this chemical contacts the skin, remove contaminated clothing and wash immediately with soap and water. Seek medical attention immediately. If this chemical has been inhaled, remove from exposure, begin rescue breathing (using universal precautions) if breathing has stopped, and CPR if heart action has stopped. Transfer promptly to a medical facility. When this chemical has been swallowed, get medical attention. Give large quantities of water and induce vomiting. Do not make an unconscious person vomit.
*References:*
- U.S. Environmental Protection Agency, Office of Pesticide Programs, Pesticide Residue Limits, "Rimsulfuron," 40 CFR 180.478, www.epa.gov/pesticides/food/viewtols.htm
- California Environmental Protection Agency "Chemical List of Lists," Sacramento CA (February 1997)

# Ronnel (ANSI)

*Use Type:* Insecticide
*CAS Number:* 299-84-3
*Formula:* $C_8H_8Cl_3O_3PS$; $Cl_3C_6H_2OP(S)(OCH_3)_2$
*Alert:* There are no registered ronnel products in the U.S. All uses in the United States were canceled by 1991.
*Synonyms:* Dermafosu (Polish); *O,O*-Dimethyl-*O*-2,4,5-trichlorophenyl phosphorothioate; Dimethyl trichlorophenyl thiophosphate; *O,O*-Dimethyl-*O*-(2,4,5-trichlorophenyl)thiophosphate; *O,O*-Dimethyl-*O*-(2,4,5-trichlorphenyl)-thionophosphat (German); ENT 23,284; Fenchloorfos (Dutch); Fenchlorfos; Fenchlorfosu (Polish); Fenchlorophos; Fenchlorphos; Phosphorothioic acid, *O,O*-dimethyl *O*-(2,4,5-trichlorophenyl) ester; Thiophosphate de *O,O*-dimethyle et de *O*-(2,4,5-trichlorophenyle) (French); *O*-(2,4,5-trichloor-fenyl)-*O,O*-dimethyl-monothiofosfaat (Dutch); Trichlorometafos; 2,4,5-Trichlorophenol, *O*-ester with *O,O*-dimethyl phosphorothioate; *O*-(2,4,5-Trichlorphenyl)-*O,O*-dimethyl-monothiophosphat (German); *O*-(2,4,5-tricloro-fenil)-*O,O*-dimetil-monotiofosfato (Italian)

*Trade Names:* BLITEX®; DERMAPHOS®; DOW ET 14®, Dow AgroSciences (USA), canceled; DOW ET 57®, Dow AgroSciences (USA), canceled; ECTORAL®; ET 14®, Dow AgroSciences (USA), canceled; ET 57®, Dow AgroSciences (USA), canceled; ETROLENE®; KARLAN®; KORLAN®, Dow AgroSciences (USA), canceled; KORLANE®, Dow AgroSciences (USA), canceled; NANCHOR®; NANKER®; NANKOR®; REMELT®; ROVAN®; TROLEN®; TROLENE®; VIOZENE®
*Producers:* Dow AgroSciences (USA)
*Chemical Class:* Organophosphate
*EPA/OPP PC Code:* 058301
*California DPR Chemical Code:* 517
*ICSC Number:* 0975
*RTECS Number:* TG0525000
*EEC Number:* 015-052-00-X
*Uses:* Used to control flies, cockroaches and other insects in housing for beef and dairy cattle, goats, hogs, and chickens. Can be orally administered to livestock to control cattle grubs, lice, horn flies, face flies, screw-worms, ticks, sheep ked and wool maggots. Also used to control household pests that are attracted to cats and dogs. The use of ronnel was canceled by the U.S. EPA by 1991 and all tolerances were revoked as of March 23, 1994.
*Regulatory Authority:*
- Air Pollutant Standard Set (ACGIH)[1] (HSE)[33] (OSHA)[58] (Other Countries)[35] (Several States)[60]
- AB 1803-Well Monitoring Chemical (CAL)
- Permissible Exposure Limits for Chemical Contaminants (CAL/OSHA)
- The "Director's List" (CAL/OSHA)
- U.S. DOT Inhalation Hazard Chemicals as organophosphates
- Canada, WHMIS, Ingredients Disclosure List

*Description:* Ronnel is a white-to light tan crystalline solid. Practically insoluble in water; solubility = 35 ppm. Molecular weight = 321.54. Boiling point = (decomposes). Melting/Freezing point = 41°C. Vapor pressure = $7.8 \times 10^{-4}$ mbar @ 25°C. Log $K_{ow}$ = 4.9. Values at or above 3.0 are likely to bioaccumulate in marine organisms.
*Incompatibilities:* Strong oxidizers. Store below 25–30°C.
*Permissible Exposure Limits in Air:* The OSHA[2] TWA is 15 mg/m³. The NIOSH[2], HSE[33] TWA, and ACGIH[1] TWA value[1] is 10 mg/m³. The NIOSH[2] IDLH level = 300 mg/m³. Argentina[35] has set 10 mg/m³ both as a TWA and an STEL. The former USSR[35] has set an MAC of 0.3 mg/m³ in workplace air. Several states have set guidelines or standards for ronnel in ambient air[60] ranging from 100 μg/m³ (North Dakota) to 160 μg/m³ (Virginia) to 200 μg/m³ (Connecticut) to 238 μg/m³ (Virginia).
*Determination in Air:* OSHA versatile sampler-2; Toluene/Acetone; Gas chromatography/Flame photometric detection for sulfur, nitrogen, or phosphorus; NIOSH Method IV Method #5600, Organophosphorus pesticides.[18]

***Permissible Concentration in Water:*** Mexico[35] has set maximum permissible concentration of 50 µg/L in estuaries and 5 µg/L in coastal waters.
***Routes of Entry:*** Inhalation, ingestion, skin and/or eye contact. Absorbed through the skin.
***Harmful Effects and Symptoms***
***Short Term Exposure:*** Irritates the eyes. Organic phosphorus insecticides are absorbed by the skin, as well as by the respiratory and gastrointestinal tracts. They are cholinesterase inhibitors. Symptoms of exposure include headache, giddiness, blurred vision, nervousness, weakness, nausea, cramps, diarrhea, and discomfort in the chest. Signs include sweating, tearing, salivation, vomiting, cyanosis, convulsions, coma, loss of reflexes and loss of sphincter control. Delayed pulmonary edema may occur after inhalation.
***Long Term Exposure:*** Cholinesterase inhibitor; cumulative effect is possible. This chemical may damage the nervous system with repeated exposure, resulting in convulsions, respiratory failure. There is limited evidence that ronnel may damage the developing fetus. May cause personality changes, depression, anxiety, irritability. High or repeated exposure may cause nerve damage, causing weakness, a feeling of "pins and needles" in the arms and legs, and poor coordination.
***Points of Attack:*** Skin, central nervous system and blood plasma.
***Medical Surveillance:*** Medical observation is recommended for 24 to 48 hours after breathing overexposure, as pulmonary edema may be delayed. Before employment and at regular times after that, the following are recommended: Plasma and red blood cell cholinesterase levels (tests for the enzyme poisoned by this chemical). If exposure stops, plasma levels return to normal in 1-2 weeks while red blood cell levels may be reduced for 1-3 months. When acetylcholinesterase enzyme levels are reduced by 25% or more below preemployment levels, risk of poisoning is increased, even if results are in lower ranges of "normal." Reassignment to work not involving organophosphate or carbamate pesticides is recommended until enzyme levels recover. If symptoms develop or overexposure occurs, repeat the above tests as soon as possible and get an exam of the nervous system. Also consider complete blood count. Consider chest x-ray following acute overexposure. Do not drink any alcoholic beverages before or during use. Alcohol promotes absorption of organophosphates. Examination of the nervous system.
***First Aid:* Treatment for organophosphate poisoning consists of thorough decontamination, cardiorespiratory support, and administration of the antidotes atropine and pralidoxime. In cases of severe poisoning, diazepam, an anticonvulsant, should also be administered. Antidotes should be administered as prevention even if the diagnosis is in doubt.** Speed in removing material from eyes and skin is of extreme importance. *Eyes:* Eye contact can cause dangerous amounts of these chemicals to be quickly absorbed through the mucous membrane into the bloodstream. Immediately and gently flush eyes with plenty of warm or cold water (NO hot water) for at least 15 minutes, occasionally lifting the upper and lower eyelids. Get medical aid immediately. *Skin:* Get medical aid. Skin contact can cause dangerous amounts of these chemicals to be absorbed into the bloodstream. Wearing the appropriate PPE equipment and respirator for organophosphate pesticides, immediately flush skin with plenty of soap and water for at least 15 minutes while removing contaminated clothing and shoes. Shampoo hair promptly if contaminated. The removed, contaminated clothing and shoes should be double-bagged and left in Hot Zone for later disposal by hazardous materials experts. Skin may also be decontaminated with diluted hypochlorite solution. *Inhalation:* Get medical aid. Do not contaminate yourself. Wearing the appropriate PPE equipment and respirator for organophosphate pesticides, immediately remove the victim from the contaminated area to fresh air. If the victim is not breathing, administer artificial respiration. Do NOT use mouth-to-mouth resuscitation; use bag/mask apparatus. If breathing is difficult, administer oxygen through bag/mask apparatus until medical help arrives. Do not leave victim unattended. *Ingestion:* Call poison control. Loosen all clothing. Never give anything by mouth to an unconscious person. If victim is *unconscious or having convulsions,* do nothing except keep victim warm. Get medical aid. Transfer promptly to a medical facility. In cases of ingestion, **do not induce vomiting**. If the victim is alert and asymptomatic, administer a slurry of activated charcoal at a dose of 1 g/kg (infant, child, and adult dose). A soda can and straw may be of assistance when offering charcoal to a child. *In some cases you may be specifically instructed by poison control to induce vomiting by way of 2 tablespoons of syrup of ipecac (adult) washed down with a cup of water.* Do NOT give activated charcoal before or with ipecac syrup.
*Note to physician or authorized medical personnel:* Treat cases of respiratory compromise, coma, or excessive pulmonary secretions with respiratory support using protocols and techniques available and within the scope of training. Some cases may necessitate procedures such as endotracheal intubation or cricothyrotomy by properly trained and equipped personnel. When possible, atropine (see *Antidotes*, below) should be given under medical supervision. Patients who are comatose, hypotensive, or having seizures or cardiac arrhythmias should be treated according to advanced life support protocols. *Antidotes:* Two antidotes are administered to treat organophosphate poisoning. Atropine is a competitive antagonist of acetylcholine at muscarinic receptors and is used to control the excessive bronchial secretions which are often responsible for death. Pralidoxime relieves both the nicotinic and muscarine effects of organophosphate poisoning by regenerating acetylcholinesterase and can

reduce both the bronchial secretions and the muscle weakness associated with poisoning. The initial intravenous dose of atropine in adults should be determined by the severity of symptoms: An initial adult dose of 1.0 to 2.0 mg or pediatric dose of 0.01 mg/kg (minimum 0.01 mg) should be administered intravenously. If intravenous access cannot be established, atropine may also be given intramuscularly, subcutaneously or via endotracheal tube. Doses should be repeated every 15 minutes until excessive secretions and sweating have been controlled. Once bronchial secretion has been controlled, atropine administration should be repeated whenever the secretions begin to recur. In seriously poisoned patients, very large doses may be required. Alterations of pulse rate and pupillary size should not be used as indicators of treatment adequacy. Pralidoxime should be administered as early in poisoning as possible as its efficacy may diminish when given more than 24 to 36 hours after exposure. Doses are as follows: adult 1.0 g; pediatric 25 to 50 mg/kg. The drug should be administered intravenously over 30 to 60 minutes, but in a life-threatening situation, one-half of the total dose can be given per minute for a total administration time of 2 minutes. Treatment should begin to take effect within 40 minutes with a reduction in symptoms and the amount of atropine necessary to control bronchial secretion. The initial dose can be repeated in 1 hour and then every 8 to 12 hours until the patient is clinically well and no longer requires atropine. If intravenous access cannot be established, pralidoxime may also be given intramuscularly. Early administration of diazepam in addition to the combined atropine and pralidoxime treatment may help prevent the onset of seizures and potential brain and cardiac morphologic damage following high-level organophosphate poisoning.

*References:*
- Pesticide Management Education Program, "Ronnel (Ectoral, Korlan) Chemical Profile 4/85," Cornell University, Ithaca, NY (April, 1985). http://pmep.cce.cornell.edu/profiles/insect-mite/propetamphos-zetacyperm/ronnel/insect-prof-ronnel.html
- California Environmental Protection Agency "Chemical List of Lists," Sacramento CA (February 1997).
- Agency for Toxic Substances and Disease Registry, U.S. Department of Health and Human Services, Public Health Service, "Managing Hazardous Materials Incidents," Atlanta, GA (June 2003).
- New Jersey Department of Health and Senior Services, "Hazardous Substance Fact Sheet, Ronnel," Trenton, NJ (May 1998). http://www.state.nj.us/health/eoh/rtkweb/1637.pdf

# Rotenone

*Use Type:* Insecticide and acaracide

*CAS Number:* 83-79-4
*Formula:* $C_{23}H_{22}O_6$
*Alert:* A General Use Pesticide (GUP). Because of its low toxicity, rotenone has been used in combination with pyrethrum and DDT.
*Synonyms:* (1)Benzopyrano(3,4-b)furo(2,3-H)(1)benzopyrano-6(6a$H$)-one,1,2,12,12a-tetrahydro-8,9-dimethoxy-2-(1-methylethenyl)-,[2R(-2α,6aα,12aα)]; ENT 133; NCI-C55210; Hydrogenated rotenone; Rotenon; Rotenona (Spanish); Rotenone, hydrogenated; [2r-(2a,6aa,12aa)]-1,2,12,12a-Tetrahydro-8,9-dimethoxy-2-(1-methylethenyl)[1]-benzopyrano[3,4-b]furo[2,3-H][1]-benzopyran-6(6a$H$)one
*Trade Names:* ACME®, Pbi/Gordon (USA); AROL GORDON DUST®; BARBASCO®; BONIDE CUKE ANDMELON DUST®, Bonide Products (USA); CENOL GARDEN DUST®; CHEM FISH®; CHEM-MITE®; CUBE®; CUBE EXTRACT®; CUBE-PULVER®; CUBEROL®; CUBE ROOT®; CUBOR®; CUREX FLEA DUSTER®; DACTINOL®; DERIL®; DERRIN®; DERRIS®; DRI-KIL®; ENT-133®; EXTRAX®; FISH-TOX®; GREEN CROSS WARBLE POWDER®; HAIARI®; LIQUID DERRIS®; MEXIDE®; NICOULINE®; NOXFIRE®; NOXFISH®; PARADERIL®; POWDER AND ROOT®; PRENTOX®, Prentiss Inc. (USA); PRO-NOX FISH®; RO-KO®; RONONE®; ROTACIDE®; ROTEFIVE®; ROTEFOUR®; ROTESSENOL®; SINID®; TOX-R®; TUBATOXIN®
*Producers:* Bonide Products (USA); Pbi/Gordon (USA); Prentiss Inc. (USA); Rhone-Poulenc Agro (France)
*Chemical Class:* Botanical
*EPA/OPP PC Code:* 071003
*California DPR Chemical Code:* 518
*ICSC Number:* 0944
*RTECS Number:* DJ2800000
*EEC Number:* 650-005-00-2
*EINECS Number:* 201-501-9
*Uses:* Rotenone is a selective, non-specific botanical insecticide with some acaricidal properties, and used in agriculture to control insects on vine fruit, flowers and vegetables. Rotenone is used in home gardens for insect control, for lice and tick control on pets, and for fish eradications as part of water body management. Rotenone is a rotenoid plant extract obtained from such species as barbasco, cub, haiari, nekoe, and timbo, members of the pea (Leguminosae) family.
*Carcinogen/Hazard Classifications*
**U.S. EPA Carcinogens:** Group E, unlikely carcinogen
**Label Signal Word:** CAUTION or DANGER, depending on the formulation.
**WHO Acute Hazard:** Class II, moderately hazardous
*Regulatory Authority:*
- Air Pollutant Standard Set (ACGIH)[1] (DFG)[3] (HSE)[33] (OSHA)[58] (Several States)[60]

- Permissible Exposure Limits for Chemical Contaminants (CAL/OSHA)
- The "Director's List" (CAL/OSHA)
- Actively registered pesticide in California.
- Canada, WHMIS, Ingredients Disclosure List
- U.S. DOT Regulated Marine Pollutant (49CFR172.101, Appendix B)

***Description:*** Rotenone is a white crystalline solid when pure. Oxidation will cause yellowing to bright red coloring. Odorless. Sprays may be dissolved in xylene or kerosene. Practically insoluble in water; solubility = 10–12 ppm @ 100°C. Molecular weight = 394.43. Boiling point = 210–220°C @ 0.5 mm (decomposes). Melting/Freezing point = 164°C; also reported @178°C. Vapor pressure = 1 x $10^{-6}$ mmHg @ 20°C. Log $K_{ow}$ = 4.12. Values at or above 3.0 are likely to bioaccumulate in marine organisms.

***Incompatibilities:*** Strong oxidizers and alkalies.

***Permissible Exposure Limits in Air:*** The OSHA/NIOSH TWA[2], the DFG MAK[3], the HSE[33] TWA and ACGIH[1] TWA value is 5 mg/m³. The STEL set by HSE[33] is 10 mg/m³. The NIOSH[2] IDLH level = 2,500 mg/m³. Several states have set guidelines or standards for rotenone in ambient[60] air ranging from 1.67 µg/m³ (New York) to 50.0 µg/m³ (Florida and South Carolina) to 50.0-100.0 µg/m³ (North Dakota) to 80.0 µg/m³ (Virginia) to 100.0 µg/m³ (Connecticut) to 119.0 µg/m³ (Nevada).

***Determination in Air:*** Filter; $CH_3CN$; High-pressure liquid chromatography/Ultraviolet detection; NIOSH IV, Method #5007[18].

***Permissible Concentration in Water:*** The state of Maine has set a guideline for rotenone in drinking water of 4.0 µg/L[61].

***Routes of Entry:*** Inhalation, ingestion, skin and/or eye contact.

***Harmful Effects and Symptoms***

***Short Term Exposure:*** Irritates the eyes, skin, and respiratory tract. Eye contact can cause severe irritation and permanent damage. Exposure can cause numbness of the mucous membrane; nausea, vomiting, abdominal pain; muscular tremors, incoherence, clonic convulsions; stupor. May cause a severe drop in blood sugar. Higher exposures can cause pulmonary edema, a medical emergency that can be delayed for several hours. This can cause death.

***Long Term Exposure:*** May affect the liver and kidneys. There is limited evidence that rotenone causes cancer of the liver and breast in animals, and damage to the developing fetus. There is limited evidence that this chemical is stored in breast milk and passed on to nursing infants. Repeated skin contact can cause severe rash.

***Points of Attack:*** Central nervous system, eyes, respiratory system; liver and kidneys.

***Medical Surveillance:*** Consider the points of attack in preplacement and periodic physical examinations. Lung function tests. Blood sugar. Liver or function tests.

***First Aid:*** Speed in removing material from eyes and skin is of extreme importance. If this chemical gets into the eyes, remove any contact lenses at once and irrigate immediately for at least 15 minutes, occasionally lifting upper and lower lids. Seek medical attention immediately. If this chemical contacts the skin, remove contaminated clothing and wash immediately with soap and water. Seek medical attention immediately. If this chemical has been inhaled, remove from exposure, begin rescue breathing (using universal precautions) if breathing has stopped, and CPR if heart action has stopped. Transfer promptly to a medical facility. When this chemical has been swallowed, get medical attention. Give large quantities of water and induce vomiting. Do not make an unconscious person vomit. Medical observation is recommended for 24 to 48 hours after breathing overexposure, as pulmonary edema may be delayed. As first aid for pulmonary edema, a doctor or authorized paramedic may consider administering a corticosteroid spray.

***References:***

- EXTOXNET, Extension Toxicology Network, "Pesticide Information Profile, Rotenone," Oregon State University, Corvallis, OR (June 1996). http://extoxnet.orst.edu/pips/rotenone.htm
- New Jersey Department of Health and Senior Services, "Hazardous Substance Fact Sheet, Rotenone," Trenton NJ (March, 1989, rev. October 2000). http://www.state.nj.us/health/eoh/rtkweb/1638.pdf
- California Environmental Protection Agency "Chemical List of Lists," Sacramento CA (February 1997).

# S

## Salicylic Acid

*Use Type:* Fungicide, insecticide and microbiocide
*CAS Number:* 69-72-7
*Formula:* $C_7H_6O_3$; $C_6H_4(OH)COOH$
*Alert:* There are no salicylic acid products currently registered with the U.S. EPA.
*Synonyms:* Acide salicilique (French); Acido salicilico (Italian); Benzoic acid, 2-hydroxy-; *o*-Hydroxybenzoic acid; 2-Hydroxybenzoic acid; Keralyt; Orthohydroxybenzoic acid
*Trade Names:* RETARDER W®; SA®; SAX®
*Producers:* Alquimia Mexicana (Mexico); Alta Laboratories (India); Archimica (UK); Atofina (France); Bayer Chemicals (Germany); Cia. Universal de Industrias S.A. de C.V. (Mexico); Clariant (Switzerland); GFS Chemicals (USA); Merck (Germany); Mitsui Chemicals (Japan); Rhodia Pharmaceutical (France); Rhone-Poulenc (France); Total Specialty Chemicals (USA)
*Chemical Class:* Benzoic acid
*EPA/OPP PC Code:* 076602
*California DPR Chemical Code:* 1170
*ICSC Number:* 0563
*RTECS Number:* VO0525000
*EINECS Number:* 200-712-3
*Uses:* Used in manufacture of aspirin, salicylates, resins, as a dyestuff intermediate, pre-vulcanization inhibitor, analytical reagent, antiseptic and food preservative. No pesticide uses registered with the U.S. EPA.
*Description:* White crystalline solid; needles. Practically insoluble in water; solubility = 1700 ppm @ 20°C; 2,059 ppm @ 25°C. Molecular weight = 138.14. Boiling point = 211°C. Melting/Freezing point = 157°C. Vapor Pressure = 7.9 x$10^{-5}$ mbar @ 24°C. Flash point = 157°C. Autoignition temperature = 540°C. Explosive limits: LEL = 1.1% @ 200°C; UEL = ?. Hazard Identification (based on NFPA-704 M Rating System): Health 0, Flammability 1, Reactivity 0. Log $K_{ow}$ = 2.25 (est). Unlikely to bioaccumulate in marine organisms.
*Incompatibilities:* Iron salts, lead acetate, iodine. Forms an explosive mixture in air.
*Permissible Exposure Limits in Air:* No standards set.
*Permissible Concentration in Water:* No criteria set, but runoff from spills or fire control may cause water pollution.
*Routes of Entry:* Ingestion, inhalation, eyes and/or skin contact. Absorbed through the skin.
*Harmful Effects and Symptoms*
*Short Term Exposure:* Overexposure may affect the central nervous system and the body's acid-base balance, causing delirium and tremors. *Inhalation:* May cause ringing in the ears, confusion, rapid pulse and breathing, headache, dizziness, nausea and vomiting. *Skin:* May be very irritating and cause skin sores. May act as a systemic poison if applied to large areas of the skin. *Eyes:* Causes irritation; may be severe. *Ingestion:* 10 grams may cause headache, dizziness, nausea and vomiting. Death may occur from ingestion of about 1 ounce.
*Long Term Exposure:* Repeated large doses may cause, in addition to the symptoms listed above, abdominal pain, loss of appetite, heartburn, poor digestion, stomach ulcers, bleeding of the stomach, iron-deficiency anemia, restlessness, incoherent speech, tremor, kidney damage, coma, convulsions and death. Repeated or prolonged contact with skin may cause acne-like skin sores.
*Points of Attack:* Skin, blood and kidneys.
*Medical Surveillance:* Complete blood count. Kidney function tests. Evaluation by a qualified allergist.
*First Aid:* If this chemical gets into the eyes, remove any contact lenses at once and irrigate immediately for at least 15 minutes, occasionally lifting upper and lower lids. Seek medical attention immediately. If this chemical contacts the skin, remove contaminated clothing and wash immediately with soap and water. Seek medical attention immediately. If this chemical has been inhaled, remove from exposure, begin rescue breathing (using universal precautions) if breathing has stopped, and CPR if heart action has stopped. Transfer promptly to a medical facility. When this chemical has been swallowed, get medical attention. Give large quantities of water and induce vomiting. Do not make an unconscious person vomit. *Note to physician or authorized medical personnel:* Induced vomiting, gastric lavage, activated charcoal, or a combination of these, may be necessary to clear the gastrointestinal tract. Sodium bicarbonate (IV) with added potassium may be necessary for blood acidosis.
*References:*
- New York State Department of Health, "Chemical Fact Sheet: Salicylic Acid," Albany, NY, Bureau of Toxic Substance Assessment (April 1986).
- California Environmental Protection Agency "Chemical List of Lists," Sacramento CA (February 1997).

## Sethoxydim

*Use Type:* Herbicide
*CAS Number:* 74051-80-2
*Formula:* $C_{17}H_{29}NO_3S$
*Synonyms:* Caswell No. 072A; 2-Cyclohexen-1-one, 2-[1-(ethoxyimino)butyl]-5-[2-(ethylthio)propyl]-3-hydroxy-; 2-

[1-(Ethoxyimino) butyl]-5-[2-(ethylthio)propyl]-3-hydroxyl-2-cyclohexen-1-one; Cyethoxydim; (±)-2-[1-(ethoxyimino)butyl]-5-[2-(ethylthio)propyl]-3-hydroxy-2-cyclohexen-1-one; 2-[1-(Ethoxyimino)butyl]-5-[2-(ethylthio)propyl]-3-hydroxy-2-cyclohexen-1-one; 2-[1-(Ethoxyimino)butyl]-5-[2-(ethylthio)propyl]-3-hydroxyl-2-cyclohexen-1-one; (ZE)-2-(1-Ethoxyiminobutyl)-5-[2-(ethylthio)propyl]-3-hydroxycyclohex-2-enone; Sethoxydim cyclohexanone herbicide

*Trade Names:* ALJADEN®; ALLOXOL® S; ARD 34/02®; BASF 9052®, BASF Agricultural Products (Germany); CHECKMATE®; CONCLUDE®, BASF Agricultural Products (Germany); EXPAND®; FERVINAL®; GRASIDIM®; GRASSOUT®, BASF Agricultural Products (Germany); NABU®; NP-55®; POAST®, BASF Agricultural Products (Germany); TRITEX-EXTRA®; ULTIMA®, BASF Agricultural Products (Germany); VANTAGE®

*Producers:* BASF Agricultural Products (Germany); Cangzhou Green Chemical Co. (China)

*Chemical Class:* Cyclohexanone derivative

*EPA/OPP PC Code:* 121001

*California DPR Chemical Code:* 2177

*RTECS Number:* GW7191000

*Uses:* A selective post-emergence herbicide used to control annual and perennial grass weeds in broad-leaved vegetables, field crops, fruit, ornamentals and indoor applications.

*U.S. Maximum Allowable Residue Levels for Sethoxydim (40 CFR 180.412):*

| CROP | ppm | CFR |
|---|---|---|
| Alfalfa, forage | 40 | |
| Alfalfa, hay | 40 | |
| Almond, hulls | 2 | |
| Apple, pomace, wet & dry | 0.8 | |
| Apricot | 0.2 | |
| Artichoke, globe | 5 | |
| Asparagus | 4 | |
| Bean, dry | 20 | |
| Bean, forage | 15 | |
| Bean, hay | 50 | |
| Bean, succulent | 15 | |
| Beet, garden | 1 | |
| Beet, sugar, molasses | 10 | |
| Beet, sugar, roots | 1 | |
| Beet, sugar, tops | 3 | |
| Blueberry | 4 | |
| Caneberry subgroup 13a | 5 | |
| Canola/rapeseed | 35 | |
| Canola/rapeseed, meal | 40 | |
| Carrot, roots | 1 | |
| Cattle, fat | 0.2 | |
| Cattle, meat | 0.2 | |
| Cattle, mbyp | 1 | |
| Cherry, sweet | 0.2 | |
| Cherry, tart | 0.2 | |
| Citrus, dried pulp | 1.5 | |
| Citrus, molasses | 1.5 | |
| Clover, forage | 35 | |
| Clover, hay | 50 | |
| Coriander, leaves | 4 | |
| Corn, field, grain | 0.5 | |
| Corn, forage | 2 | |
| Corn, stover | 2.5 | |
| Corn, sweet, forage | 3 | |
| Corn; sweet, kernel plus cob with husks removed | 0.4 | |
| Corn, sweet, stover | 3.5 | |
| Cotton, seed, soapstock | 15 | |
| Cotton, undelinted seed | 5 | |
| Cranberry | 2 | |
| Egg | 2 | |
| Flax, meal | 7 | |
| Flax, seed | 5 | |
| Flax, straw | 2 | |
| Fruit, pome | 0.2 | |
| Goat, fat | 0.2 | |
| Goat, meat | 0.2 | |
| Goat, mbyp | 1 | |
| Grape | 1 | |
| Grape, raisin | 2 | |
| Hog, fat | 0.2 | |
| Hog, meat | 0.2 | |
| Hog, mbyp | 1 | |
| Horse, fat | 0.2 | |
| Horse, meat | 0.2 | |
| Horse, mbyp | 1 | |
| Horseradish | 4 | |
| Juneberry | 5 | |
| Lentil | 30 | |
| Lingonberry | 5 | |
| Milk | 0.5 | |
| Nectarine | 0.2 | |
| Nut, tree, group 14 | 0.2 | |
| Pea, dry | 40 | |
| Pea, field, hay | 40 | |
| Pea, field, vines | 20 | |
| Pea, succulent | 10 | |
| Peach | 0.2 | |
| Peanut | 25 | |
| Peanut, soapstock | 75 | |
| Peppermint | 30 | |
| Pistachio | 0.2 | |
| Potato, granules/flakes | 8 | |
| Potato, processed potato waste | 8 | |
| Poultry, fat | 0.2 | |
| Poultry, meat | 0.2 | |
| Poultry, mbyp | 2 | |
| Safflower | 15 | |
| Salal | 5 | |

| | |
|---|---|
| Sheep, fat | 0.2 |
| Sheep, meat | 0.2 |
| Sheep, mbyp | 0.2 |
| Soybean | 16 |
| Soybean, hay | 10 |
| Spearmint | 30 |
| Strawberry | 10 |
| Sunflower, meal | 20 |
| Sunflower, seed | 7 |
| Tomato, concentrated products | 24 |
| Tomato, dried pomace | 12 |
| Vegetable, brassica, leafy, group5 | 5 |
| Vegetable, bulb | 1 |
| Vegetable, cucurbit, group 9 | 4 |
| Vegetable, fruiting, group 8 | 4 |
| Vegetable, leafy, except brassica, group 4 | 4 |
| Vegetable, tuberous and corm, subgroup 1c | 4 |

*Carcinogen/Hazard Classifications*
**Label Signal Word:** WARNING or DANGER
**WHO Acute Hazard:** Class III, slightly hazardous
*Regulatory Authority:*
- EPA 40 CFR 372.65, Specific Toxic Chemical Listings
- FIFRA, 40CFR185: tolerances for pesticides in food
- FIFRA, 40CFR186: tolerances for pesticides in animal feeds
- Actively registered pesticide in California.
- EPCRA Section 313 Form R *de minimis* concentration reporting level: 1.0%

*Description:* Amber color, oily liquid. Odorless. Flammable. Molecular weight = 327.47. Practically insoluble in water. Vapor pressure = $1.6 \times 10^{-7}$ mmHg @ 20°C. Hazard Identification (based on NFPA-704 M Rating System): Health 2, Flammability 2, Reactivity 0.

*Incompatibilities:* Incompatible with strong acids, oxidizers (chlorates, nitrates, peroxides, permanganates, perchlorates, chlorine, bromine, fluorine, etc). Contact may cause fire and explosions.

*Permissible Concentration in Water:* No criteria set. Runoff from spills or fire control may cause water pollution.

*Routes of Entry:* Ingestion, inhalation, skin contact.

*Harmful Effects and Symptoms*

*Short Term Exposure:* Contact may burn eyes, skin, and respiratory tract. Toxic if ingested.

*Points of Attack:* Consult a physician if poisoning is suspected or if redness, itching, or burning of the eyes or skin develop.

*First Aid:* If this chemical gets into the eyes, remove any contact lenses at once and irrigate immediately for at least 15 minutes, occasionally lifting upper and lower lids. Seek medical attention immediately. If this chemical contacts the skin, remove contaminated clothing and wash immediately with soap and water. Seek medical attention immediately. If this chemical has been inhaled, remove from exposure, begin rescue breathing (using universal precautions) if breathing has stopped, and CPR if heart action has stopped. Transfer promptly to a medical facility. When this chemical has been swallowed, get medical attention. *Do not induce vomiting when formulations containing petroleum solvents are ingested.* Otherwise, give large quantities of water and induce vomiting. Do not make an unconscious person vomit.

*References:*
- U.S. Environmental Protection Agency, Office of Pesticide Programs, Pesticide Residue Limits, "*Sethoxydim*," 40 CFR 180.412, http://www.epa.gov/pesticides/food/viewtols.htm
- EXTOXNET, Extension Toxicology Network, "Pesticide Information Profile, Sethoxydim," Oregon State University, Corvallis, OR (June 1996). http://extoxnet.orst.edu/pips/sethoxyd.htm
- California Environmental Protection Agency "Chemical List of Lists," Sacramento CA (February 1997)

# Siduron (ANSI)

*Use Type:* Herbicide
*CAS Number:* 1982-49-6
*Formula:* $C_{14}H_{20}N_2O$
*Synonyms:* 1-(2-Methylcyclohexyl)-3-phenylurea; *N*-(2-Methylcyclohexyl)-*N'*-phenylurea; Urea, *N*-(2-methylcyclohexyl)-*N'*-phenyl-
*Trade Names:* GREENFIELD®, Lebanon Seaboard (USA); GRO-TONE®, Vigoro (Canada), canceled; H-1318®, DuPont Crop Protection (USA); TREY®; TUPERSAN®, Gowan Company (USA)
*Producers:* Bonide Products (USA); DuPont Crop Protection (USA); Gowan Company (USA); Lebanon Seaboard (USA); Vigoro (Canada)
*Chemical Class:* Substituted urea
*EPA/OPP PC Code:* 035509
*California DPR Chemical Code:* 603
*Uses:* A pre-emergence herbicide used to control annual grasses, e.g., crabgrass, foxtail, and barnyard grass. Used on newly seeded or established turf and lawn grasses and as pre-emergence treatment to bare soil following spring seeding.

*Carcinogen/Hazard Classifications*
**Label Signal Word:** CAUTION
*Regulatory Authority:*
- AB 1803-Well Monitoring Chemical (CAL)
- The "Director's List" (CAL/OSHA)

*Description:* Colorless crystalline solid. Odorless. Soluble in water; solubility = 18.3 ppm @ 25°C. Molecular weight = 232.35. Melting/Freezing point = 135°C. Vapor pressure = $4 \times 10^{-9}$ mmHg @ 20°C. Log $K_{ow}$ = 3.1. Values above 3.0 are likely to bioaccumulate in marine organisms.

*Determination in Air:* Filter; none; Gravimetric; NIOSH IV [Particulates NOR; #0500 (total), #0600 (respirable)].[18]
*Permissible Concentration in Water:* No criteria set. Runoff from spills or fire control may cause water pollution.
*Harmful Effects and Symptoms*
*Short Term Exposure:* Contact with eyes or skin may cause irritation or injury. Inhalation should be avoided; use NIOSH-approved air purifying respirators for pesticides. May be harmful if swallowed. Skin contact may cause severe irritation or burns.
*Points of Attack:* Skin
*First Aid:* If this chemical gets into the eyes, remove any contact lenses at once and irrigate immediately for at least 15 minutes, occasionally lifting upper and lower lids. Seek medical attention immediately. If this chemical contacts the skin, remove contaminated clothing and wash immediately with soap and water. Seek medical attention immediately. If this chemical has been inhaled, remove from exposure, begin rescue breathing (using universal precautions) if breathing has stopped, and CPR if heart action has stopped. Transfer promptly to a medical facility. When this chemical has been swallowed, get medical attention. Give large quantities of water. Do not induce vomiting when formulations containing petroleum solvents are ingested. Do not make an unconscious person vomit.
*References:*
- Pesticide Management Education Program, "Siduron (Tupersan) Herbicide Profile 2/85," Cornell University, Ithaca, NY (February 1985). http://pmep.cce.cornell.edu/profiles/herb-growthreg/sethoxydim-vernolate/siduron/herb-prof-siduron.html
- California Environmental Protection Agency "Chemical List of Lists," Sacramento CA (February 1997)

# Silvex (ANSI)

*Use Type:* Herbicide and plant growth regulator
*CAS Number:* 93-72-1
*Formula:* $C_9H_7Cl_3O_3$, $Cl_3C_6H_2OCH(CH_3)COOH$
*Alert:* All registered uses of silvex in the U.S. were canceled in 1979 and all tolerances revoked effective June 16, 1993.
*Synonyms:* Acide, 2-(2,4,5-trichloro-phenoxy) propionique (French); Acido 2-(2,4,5-triclorofenossi)-propionico (Italian); Fenoprop; 2,4,5-TC; 2,4,5-TCPPA; 2,4,5-TP; 2(2,4,5-trichloor-fenoxy)-propionzuur (Dutch); 2-(2,4,5-Trichlorophenoxy)propanoic acid; α-(2,4,5-Trichlorophenoxy)propanoic acid; 2,4,5-Trichlorophenoxy-α-; 2-(2,4,5-Trichlor-phenoxy)propionsaeure (German)
*Trade Names:* AMCHEN 2,4,5-TP®; AQUA-VEX®; COLOR-SET®; DED-WEED®; DOUBLR STRENGTH®; FENOPROP®; FENOMORE®; FRUIT-O-NET®; FRUITONE T®; KURAN®; KURON®; KUROSAL®; KUROSALG®; MILLER NU SET®; PROPON®; SILVI-RHAP®; STA-FAST®; SILVEX HERBICIDE®
*Chemical Class:* Chlorophenoxy acid or ester
*EPA/OPP PC Code:* 082501
*California DPR Chemical Code:* 530
*RTECS Number:* UF8225000
*EINECS Number:* 202-271-2
*Uses:* There are no silvex products currently registered by the U.S. EPA. It is applied to woody plants, primarily on fence lines around pasture land.
*Carcinogen/Hazard Classifications*
**U.S. EPA Carcinogens:** Group D, unclassifiable, inadequate date
**IARC:** Group 2b, possible carcinogen
**Label Signal Word:** CAUTION
*Regulatory Authority:*
- Carcinogen (Human Suspected) (IARC)[9]
- Banned or Severely Restricted (Several Countries) (UN)[13]
- AB 1803-Well Monitoring Chemical (CAL)
- MCL (Maximum Contaminants Levels) list of contaminants (CAL)
- The "Director's List" (CAL/OSHA)
- Clean Water Act: Section 311 Hazardous Substances/RQ 40CFR117.3 (same as CERCLA, see below) as 2,4,5-tp acid.
- EPA Hazardous Waste Number (RCRA No.): U233
- RCRA, 40CFR261, Appendix 8 Hazardous Constituents
- RCRA 40CFR264, Appendix 9; TSD Facilities Ground Water Monitoring List. Suggested test method(s) (PQL $ug/L$): 8150(2)
- Safe Drinking Water Act: MCL, 0.01 mg/L; MCLG, 0.01 mg/L; Regulated chemical (47 FR 9352) as 2,4,5-TP
- RCRA 40CFR268.48; 61FR15654, Universal Treatment Standards: Wastewater (mg/L), 0.72; Nonwastewater (mg/kg), 7.9
- Superfund/EPCRA 40CFR302.4 RQ: CERCLA, 100 lb (45.4 kg)

*Description:* Colorless crystalline solid or powder. Practically insoluble in water; solubility = 150 ppm @ 25°C. Boiling point = @ >149°C @ 1 atm. Melting/Freezing point = 182°C. Vapor pressure = $1 \times 10^{-7}$ mmHg @ 20°C. Hazard Identification (based on NFPA-704 M Rating System): Health 2, Flammability 1, Reactivity 0.
*Incompatibilities:* Strong oxidizers.
*Permissible Exposure Limits in Air:* No standards set.
*Permissible Concentration in Water:* Surface water levels should never exceed 2.5 ppb silvex (butoxyethyl ester); 2.0 ppb (propyleneglycolbutylether ester) (U.S. EPA). Runoff from spills or fire control may cause water pollution.
*Routes of Entry:* Inhalation, ingestion, skin and/or eye contact.

### Simazine

*Harmful Effects and Symptoms*
*Short Term Exposure:* May cause skin and eye irritation. Poisonous by ingestion. Approximate lethal dose = 2.4 tablespoonful/150 lb man.
*Long Term Exposure:* Silvex has caused liver and kidney damage in experimental animals.
*Points of Attack:* Liver, kidneys, skin and eyes.
*Medical Surveillance:* Liver and kidney function tests.
*First Aid:* If this chemical gets into the eyes, remove any contact lenses at once and irrigate immediately for at least 15 minutes, occasionally lifting upper and lower lids. Seek medical attention immediately. If this chemical contacts the skin, remove contaminated clothing and wash immediately with soap and water. Seek medical attention immediately. If this chemical has been inhaled, remove from exposure, begin rescue breathing (using universal precautions) if breathing has stopped, and CPR if heart action has stopped. Transfer promptly to a medical facility. When this chemical has been swallowed, get medical attention. Give large quantities of water and induce vomiting. Do not make an unconscious person vomit.

*References:*
- New Jersey Department of Health and Senior Services, "Hazardous Substance Fact Sheet, Trichlorophenoxypropionic Acid," Trenton, NJ (November, 2001). http://www.state.nj.us/health/eoh/rtkweb/1899.pdf
- California Environmental Protection Agency "Chemical List of Lists," Sacramento CA (February 1997).

## Simazine (ANSI)

*Use Type:* Pre-emergence herbicide and algicide
*CAS Number:* 122-34-9
*Formula:* $C_7H_{12}ClN_5$
*Alert:* A Restricted Use Pesticide (RUP) for all land uses because of its potential to contaminate ground water. Human toxicity (long-term): High
*Synonyms:* A 2079; AI3-51142; 2,4-Bis(aethylamino)-6-chlor-1,3,5-triazin (German); 2,4-Bis(ethylamino)-6-chloro-s-triazine; Caswell No. 740; 2-Chloro-4,6-bis(ethylamino)-1,3,5-triazine; 1-Chloro-3,5-bis(ethylamino)-2,4,6-triazine; 2-Chloro-4,6-bis(ethylamino)-s-triazine; 6-chloro-$N,N'$-diethyl-1,3,5-triazine-2,4-diamine; 6-Chloro-$N2,N4$-Diethyl-1,3,5-triazine-2,4-diamine; 6-chloro-$N,N'$-diethyl-1,3,5-triazine-2,4-diyldiamine; NSC 25999; Simazina (Spanish); s-Triazine, 2-chloro-4,6-bis(ethylamino)-; 1,3,5-Triazine-2,4-diamine, 6-chloro-$N,N'$-diethyl-
*Trade Names:* AKTINIT S®; ALCO®, Amvac Chemical (USA); AQUAZINE®; ATLAS SIMAZINE®, Whyte Agrochemicals (UK); BATAZINA®; BITEMOL®; CALIBER®; CDT®; CEKUSAN®; CEKUZINA-S®; FRAMED®; G 27692®, Ciba-Geigy (Switzerland); GEIGY 27692®, Ciba-Geigy (Switzerland); GESARAN®; GESATOP®; GESATOP-50®; H 1803®; HARLEQUIN®, Makhteshim-Agan Industries (Israel); HERBAZIN® 500 BR, Milenia Agro Ciencias (Brazil); HERBAZIN® 50, Milenia Agro Ciencias (Brazil); HERBEX®; HERBOXY®; HUNGAZIN DT®; OXON ITALIA SIM-TROL®, OXON Italia S.p.A. (Italy); PREMAZINE®; PRIMATEL S®; PRIMATOL S®, Makhteshim-Agan Industries (Israel); PRINCEP®, Syngenta (Switzerland); PRINCEP® 80W, Syngenta (Switzerland); SIMADEX®; SIMANEX®; SIMAZINE® 80W; SIMAZAT®, Drexel Chemical (USA); SIM-TROL®, Sipcam Agro USA (USA); TAFAZINE®; TAFAZINE® 50-W; TANZINE®; TAPHAZINE®; TOTAZINE®; TRIAZINE A 384®; W 6658®; WEEDEX®; ZEAPUR®

*Producers:* Agan Chemical Manufacturers Ltd. (Israel); Amvac Chemical (USA); Atanor S.A. (Argentina); Bharat Pulverizing Mills (India); Biesterfeld Siemsgluess International. GmbH (Germany); Calliope (France); Ciba-Geigy (Switzerland); Cyanamid (USA); Drexel Chemical (USA); Ehrenstorfer, Dr. (Germany); Makhteshim-Agan Industries (Israel); Milenia Agro Ciencias (Brazil); Monsanto (USA); Montedison (Italy); Nihon Nohyaku (Japan); Nippon Kayaku (Japan); Nissan Chemical Industries (Japan); OXON Italia S.p.A. (Italy); Rhone-Poulenc (France); Rallis India (India); Shenzhen Guomeng Industry Co., Ltd. (China); Sigma-Aldrich Laborchemikalien (Germany); Sipcam Agro USA (USA); Syngenta (Switzerland); United Agri Products (UAP); Vijayalakshmi Insecticides and Pesticides (India); Whyte Agrochemicals (UK); Zhejiang Changxing Zhongshan Chemical Industry (China)

*Chemical Class:* Triazine
*EPA/OPP PC Code:* 080807
*California DPR Chemical Code:* 531
*ICSC Number:* 0699
*RTECS Number:* XY5250000
*EEC Number:* 612-088-00-3
*EINECS Number:* 204-535-2

*Uses:* Simazine, a pre-emergence selective herbicide, is severely regulated because of effect on humans through ground and drinking water. It is used to control broad-leaved weeds and annual grasses in corn, sorghum, sugar cane and other field crops, artichokes, asparagus, berry fruit, nuts, citrus crops, coffee, hops, oil palms, olives, vegetables and ornamental crops, turf grass, orchards, and vineyards. At higher rates, it is used for non-selective weed control in industrial areas. Before 1992, simazine was used to control submerged weeds and algae in large aquariums, farm ponds, fish hatcheries, swimming pools, ornamental ponds, and cooling towers. Simazine is available in wettable powder, water dispersible granule, liquid, and granular formulations. It may be soil-applied.

*U.S. Maximum Allowable Residue Levels for Simazine (40 CFR 180.213):*

| CROP | ppm |
|---|---|
| Alfalfa, forage & hay | 15 |
| Almond and hulls | 0.25 |
| Apple | 0.25 |
| Artichoke, globe | 0.5 |
| Asparagus | 10 |
| Avocado | 0.25 |
| Banana | 0.2 |
| Bermuda grass, forage & hay | 15 |
| Blackberry | 0.25 |
| Blueberry | 0.25 |
| Boysenberry | 0.25 |
| Cattle, fat | 0.02 |
| Cattle, meat & byproducts | 0.02 |
| Cherry | 0.25 |
| Corn, forage, grain & stover | 0.25 |
| Corn, sweet, kernel plus cob with husks removed | 0.25 |
| Cranberry | 0.25 |
| Currant | 0.25 |
| Dewberry | 0.25 |
| Egg | 0.02 |
| Filbert | 0.25 |
| Fish | 12 |
| Goat, fat | 0.02 |
| Goat, meat & byproducts | 0.02 |
| Grape | 0.25 |
| Grapefruit | 0.25 |
| Grass | 15 |
| Grass, forage & hay | 15 |
| Hog, fat | 0.02 |
| Hog, meat & byproducts | 0.02 |
| Horse, fat | 0.02 |
| Horse, meat & byproducts | 0.02 |
| Lemon | 0.25 |
| Loganberry | 0.25 |
| Milk | 0.02 |
| Nut, Macadamia | 0.25 |
| Olive | 0.25 |
| Orange | 0.25 |
| Peach | 0.25 |
| Pear | 0.25 |
| Pecan | 0.1 |
| Plum | 0.25 |
| Poultry, fat | 0.02 |
| Poultry, meat & mbyp | 0.02 |
| Raspberry | 0.25 |
| Sheep, fat | 0.02 |
| Sheep, mbyp | 0.02 |
| Strawberry | 0.25 |
| Sugarcane, cane | 0.25 |
| Sugarcane, molasses | 1 |
| Walnut | 0.2 |

*Carcinogen/Hazard Classifications*
**U.S. EPA Carcinogens:** Group C, possible carcinogen
**IARC:** Group 3, unclassifiable
**Label Signal Word:** CAUTION
**WHO Acute Hazard:** Class U, unlikely to be hazardous
**Endocrine Disruptor:** Suspected endocrine disruptor
*Regulatory Authority:*
- Air Pollutant Standard Set (former USSR)[35)(43]
- AB 1803-Well Monitoring Chemical (CAL)
- MCL (Maximum Contaminants Levels) list of contaminants (CAL)
- Actively registered pesticide in California.
- Safe Drinking Water Act: MCL, 0.004 mg/L; MCLG, 0.004 mg/L; Regulated chemical (47 FR 9352)
- EPCRA Section 313 Form R *de minimis* concentration reporting level: 1.0%
- Canada, WHMIS, Ingredients Disclosure List

*Description:* Simazine is a combustible, white crystalline solid. Practically odorless. Practically insoluble in water; solubility = 4 ppm @ 20°C. Molecular weight = 201.68. Melting/Freezing point = 226°C. Vapor pressure = 2.2 x $10^{-8}$ mmHg @ 20°C. Hazard Identification (based on NFPA-704 M Rating System): Health 1, Flammability 1, Reactivity 0. Log $K_{ow}$ = 2.3. Unlikely to bioaccumulate in marine organisms.

*Incompatibilities:* A combustible solid. Incompatible with strong oxidizers.

*Permissible Exposure Limits in Air:* The former USSR has set an MAC for simazine in workplace air[35] of 2.0 mg/m³. It has also set an MAC for ambient air in residential areas of 0.02 mg/m³ both on a momentary and a daily average basis.

*Permissible Concentration in Water:* The former USSR[35] has set an MAC in water bodies used for fishery purposes of 2.4 µg/L and a limit in water bodies used for domestic purposes of zero. A lifetime health advisory of 35 µg/L has been developed by EPA (see reference below). Various states have developed guidelines for simazine in drinking water[61] ranging from 150 µg/L (California) to 430 µg/L (Maine) to 2,150 µg/L (Wisconsin).

*Determination in Water:* Analysis of simazine is by a gas chromatographic (GC) method applicable to the determination of certain nitrogen-phosphorus-containing pesticides in water samples. In this method, approximately 1 liter of sample is extracted with methylene chloride. The extract is concentrated and the compounds are separated using capillary column GC. Measurement is made using a nitrogen-phosphorus detector. The method detection limit has not been determined for this compound but it is estimated that the detection limits for the method analytes are in the range of 0.1 to 2 µg/L.

*Routes of Entry:* Inhalation, ingestion, skin and/or eye contact.

*Harmful Effects and Symptoms*
*Short Term Exposure:* May cause skin or eye irritation. Moderately poisonous if ingested. Approximate lethal dose

= 1.5 cupful/150 lb man. $LD_{50}$ in range (rat) 1780-7000 mg/kg. No case of poisoning in man from simazine has been reported, although exposure to simazine has caused acute and Subacute dermatitis in the former USSR, characterized by erythema, slight edema, moderate pruritus, and burning lasting 4 to 5 days.

*Long Term Exposure:* Repeated exposure may cause weight loss and reduced red blood cell count. Simazine fed to rats for 2 years @ 1.0, 10, and 100 ppm produced no difference between treated and control animals in gross appearance or behavior. The rats fed 100 ppm had approximately twice as many thyroid and mammary tumors as the control animals, but it was stated that these were not attributable to simazine. A 2-year chronic-feeding study of simazine in dogs with simazine 80W fed at 15, 150, and 1,500 ppm showed only a slight thyroid hyperplasia at 1,500 ppm and slight increases in serum alkaline phosphatase and serum glutamic oxalacetic transaminase in several of the dogs fed 1,500 ppm.

*Points of Attack:* Blood

*Medical Surveillance:* Complete blood count.

*First Aid:* If this chemical gets into the eyes, remove any contact lenses at once and irrigate immediately for at least 15 minutes, occasionally lifting upper and lower lids. Seek medical attention immediately. If this chemical contacts the skin, remove contaminated clothing and wash immediately with soap and water. Seek medical attention immediately. If this chemical has been inhaled, remove from exposure, begin rescue breathing (using universal precautions) if breathing has stopped, and CPR if heart action has stopped. Transfer promptly to a medical facility. When this chemical has been swallowed, get medical attention. Give large quantities of water and induce vomiting. Do not make an unconscious person vomit.

*References:*
- EXTOXNET, Extension Toxicology Network, "Pesticide Information Profile, Simazine," Oregon State University, Corvallis, OR (June 1996). http://extoxnet.orst.edu/pips/simazine.htm
- U.S. Environmental Protection Agency, Office of Pesticide Programs, Pesticide Residue Limits, "Simazine," 40 CFR 180.213, http://www.epa.gov/pesticides/food/viewtols.htm
- Sax, N.I., Ed., "Dangerous Properties of Industrial Materials Report," 7, No. 4, 109-113 (1987).
- U.S. Environmental Protection Agency, "Health Advisory: Simazine," Washington, DC, Office of Drinking Water (August 1987).
- California Environmental Protection Agency "Chemical List of Lists," Sacramento CA (February 1997).

# Sodium Aluminum Fluoride

*Use Type:* Insecticide

*CAS Number:* 15096-52-3

*Formula:* $AlF_6Na_3$; $Na_3AlF_6$

*Synonyms:* Aluminum sodium fluoride; Cryolite; ENT 24,984; Greenland spar; Hexafluoroaluminato de trisodio (Spanish); Ice Spar; Kryolith (German); Mumfluorid (German); Natriumhexafluoroaluminate (German); Sodium aluminofluoride; Sodium fluoaluminate; Sodium hexafluoroaluminate; Trinatriumhexafluoroaluminat (German); Villiaumite

*Trade Names:* KRYOCIDE®, Cerexagri (France); CHIOLITE®; CRYOLITE 93®, AMVAC Chemical (USA); ICE-SPAR®; ICETONE®; KOYOSIDE®; KRIOLIT®; PROKIL®, Gowan (USA); VILLIAUMITE®

*Producers:* Amvac Chemical (USA); Bayer CropScience (Germany); Cerexagri Inc (France); Gowan Company (USA); Quantum Chemicals (Australia)

*Chemical Class:* Inorganic fluorine compound; Inorganic, aluminofluoride salt

*EPA/OPP PC Code:* 075101

*California DPR Chemical Code:* 173

*RTECS Number:* WA9625000

*EINECS Number:* 239-148-8

*Uses:* Sodium aluminum fluoride is used as an insecticide on food crops and ornamentals, and also in aluminum refining and making ceramics, glass and polishes. Cryolite is used to control a variety of pests on cucurbits (melons, cantaloupe, water melon, pumpkins, all types of squash), fruiting vegetables (eggplant, pepper, broccoli, Brussels sprouts, cabbage, cauliflower, collards, head and leaf lettuce, kohlrabi), kiwi (in California only), pears, radish, cranberry and peaches, grapefruit, lemon, lime, orange, tangelo, tangerines, tomatoes, apples, potatoes, beans and grapes. Also on ornamental plants, woody shrubs and vines.

*Fish toxicity (threshold)[77]:* Very low–7784.44797 ppb, MATC (Maximum Acceptable Toxicant Concentration)

*U.S. Maximum Allowable Residue Levels for Cryolite (40 CFR 180.145):* The following crops have residue levels of 7 ppm: Apricot, Blackberry, Blueberry, Boysenberry, Broccoli, Brussels sprouts, Cabbage, Cauliflower, Collards, Cranberry, Cucumber, Dewberry, Eggplant, Fruit, stone, group 12, Grape, Kale, Kohlrabi, Lettuce, Loganberry, Melon, Nectarine, Peach, Pepper, Plum, prune, fresh, Pumpkin, Raspberry, Squash, Squash, summer, Strawberry, Tomato, Youngberry. The following crop has a residue level of 15 ppm: Kiwifruit.

*Carcinogen/Hazard Classifications*

**U.S. EPA Carcinogens:** Group D, unclassifiable, inadequate data

**Label Signal Word:** CAUTION

**WHO Acute Hazard:** Class U, unlikely to be hazardous

*Regulatory Authority:*
- Air Pollutant Standard Set (ACGIH)[1] (HSE)[33] (OSHA)[58]
- Actively registered pesticide in California.
- Canada, WHMIS, Ingredients Disclosure List

*Description:* Snow-white solid or vitreous mass. Crystalline solid (natural product may be colored reddish or brown or even black but loses this discoloration on heating); synthetic product is an amorphous powder. Odorless. Very slightly soluble in water. Molecular weight = 209.97. Density = 2.95. Melting/Freezing point = 995°C. Boiling point = (decomposes).

*Incompatibilities:* Strong acids and strong oxidizers.

*Permissible Exposure Limits in Air:* OSHA: 8-hr time-weighted average: 2.5 mg/m$^3$ [fluoride (as (F)]; NIOSH[(2)]: 10-hr time-weighted average: 2.5 mg/m$^3$ [fluorides (as (F)]; threshold-limit 8-hr time-weighted average (TWA) 2.5 mg/m$^3$ [fluorides (as (F)]. NIOSH IDLH = 250 mg F/m$^3$ as Fluorides (F).

*Permissible Concentration in Water:* As regards drinking water, *aluminum* guidelines[(61)] are 1.43 mg/L in Massachusetts and 5.0 mg/L in Kansas. *Fluoride* guidelines are 1.8 mg/L in Arizona, 2.4 mg/L in Maine and 4.0 mg/L according to EPA.

*Routes of Entry:* Inhalation, ingestion, skin and/or eye contact.

*Harmful Effects and Symptoms*

*Short Term Exposure:* Sodium aluminum fluoride can affect you when breathed in. Eye contact can cause severe irritation, burns with possible loss of vision. Skin contact can cause irritation and even burns, especially if prolonged. Inhalation can irritate the nose, throat and air passages. Higher exposures can cause pulmonary edema, a medical emergency that can be delayed for several hours. This can cause death. Exposure can cause nausea, abdominal pain, diarrhea; salivation, thirst, sweating

*Long Term Exposure:* Repeated exposure can cause stiff spine; calcification of ligaments of ribs, pelvis. Repeated or high exposures may cause permanent lung damage.

*Points of Attack:* Eyes, skin, lungs and skeletal system.

*Medical Surveillance:* Before beginning employment and at regular times after that, the following are recommended: Lung function tests. Urine fluoride test (levels above 3 to 4 mg/L at the end of exposure represent increased exposure).

*First Aid:* Speed in removing material from eyes and skin is of extreme importance. If this chemical gets into the eyes, remove any contact lenses at once and irrigate immediately for at least 15 minutes, occasionally lifting upper and lower lids. Seek medical attention immediately. If this chemical contacts the skin, remove contaminated clothing and wash immediately with soap and water. Seek medical attention immediately. If this chemical has been inhaled, remove from exposure, begin rescue breathing (using universal precautions) if breathing has stopped, and CPR if heart action has stopped. Transfer promptly to a medical facility. When this chemical has been swallowed, get medical attention. If victim is _conscious_, administer water, or milk. Do not induce vomiting. Medical observation is recommended for 24 to 48 hours after breathing overexposure, as pulmonary edema may be delayed. As first aid for pulmonary edema, a doctor or authorized paramedic may consider administering a corticosteroid spray.

*References:*
- U.S. Environmental Protection Agency, Office of Pesticide Programs, Pesticide Residue Limits, "Cryolite," 40 CFR 180.145, http://www.epa.gov/pesticides/food/viewtols.htm
- U.S. Environmental Protection Agency, "Reregistration Eligibility Decision (RED), Cryolite", Office of Prevention, Pesticides and Toxic Substances, Washington, DC (August 1996). http://www.epa.gov/REDs/0087.pdf
- Pesticide Management Education Program, "Chemical Fact Sheet, Cryolite (Kryocide)," Cornell University, Ithaca, NY (June 1983). http://pmep.cce.cornell.edu/profiles/insect-mite/cadusafos-cyromazine/cryolite/insect-prof-cryolite.html
- California Environmental Protection Agency "Chemical List of Lists," Sacramento CA (February 1997).
- New Jersey Department of Health and Senior Services, "Hazardous Substance Fact Sheet: Sodium Aluminum Fluoride," Trenton, NJ (September 1987, rev. April 2000). http://www.state.nj.us/health/eoh/rtkweb/1676.pdf

# Sodium Arsenite

*Use Type:* Insecticide, herbicide, fungicide and rodenticide
*CAS Number:* 7784-46-5
*Formula:* $AsO_2Na$; $NaAsO_2$
*Alert:* This is a carcinogen and teratogen and should be handled with extreme caution.
*Synonyms:* Arsenito sodico (Spanish); Arsenous acid, sodium salt; Arsenite de sodium (French); Disodium arsenate heptahydrate; Sodium dioxoarsenate; Sodium metaarsenite
*Trade Names:* ATLAS A®; CHEM PELS C®; CHEM-SEN 56®; KILL-ALL®; PENITE®; PRODALUMNOL DOUBLE®; SODANIT®
*Producers:* Great Western Inorganics (USA); Luxembourg Industries (Israel); William Blythe (UK)
*Chemical Class:* Inorganic arsenic compound
*EPA/OPP PC Code:* 013603
*California DPR Chemical Code:* 534
*RTECS Number:* CG3675000
*EINECS Number:* 232-070-5
*Uses:* This material is used in manufacturing of arsenical soap for use on skin; treating vines against certain scale diseases; wood preservation; as a reagent in preparation of methylene iodide; corrosion inhibitor; and as a non-selective herbicide, insecticides and other pesticides. Sodium arsenite is used for the treatment of fungus disease on vines, clearing tropical plantation weeds and trees, and in sheep and cattle

# Sodium Arsenite

dips. However, there are no sodium arsenite products currently registered with the U.S. EPA. Sodium arsenite has been linked to an increase risk of testicular cancer.

## Carcinogen/Hazard Classifications

**U.S. EPA Carcinogens:** Not listed
**U.S. NTP Carcinogen:** Known carcinogen
**California Prop. 65:** Carcinogen
**IARC:** Group 1, known carcinogen
**Label Signal Word:** DANGER
**WHO Acute Hazard:** Class 1b, highly hazardous

## Regulatory Authority:

- Banned or Severely Restricted (Several Countries) (UN)[13]
- Carcinogen User Register Chemical (CAL/OSHA) as inorganic arsenic compound
- AB 1803-Well Monitoring Chemical (CAL) as inorganic arsenic compound
- CAL Air Resources Board/AB 1807 Toxic Air Contaminants as inorganic arsenic compound
- Proposition 65 chemical (CAL) as inorganic arsenic compound
- AB 2588-Air Toxics "Hot Spots" Chemicals (CAL) as arsenic compounds as inorganic arsenic compound
- Permissible Exposure Limits for Chemical Contaminants (CAL/OSHA) as arsenic compounds as inorganic arsenic compound
- The "Director's List" (CAL/OSHA)
- Clean Air Act: List of high risk pollutants (Section 63.74) as arsenic compounds.
- Clean Water Act: Section 311 Hazardous Substances/RQ 40CFR117.3 (same as CERCLA, see below); Section 313 Water Priority Chemicals (57FR41331, 9/9/92)
- Superfund/EPCRA 40CFR355, Appendix B Extremely Hazardous Substances: TPQ = 500/10,000 lb (227/4,540 kg)
- Superfund/EPCRA 40CFR302.4 RQ: CERCLA, 1 lb (0.454 kg)
- Air Pollutant Standard Set (ACGIH)[1] (OSHA)[63]
- EPCRA Section 313 Form R *de minimus* concentration reporting level: 0.1%
- U.S. DOT Regulated Marine Pollutant (49CFR, Subchapter 172.101, Appendix B) as arsenates, liquid, n.o.s; arsenates, solid, n.o.s; arsenical pesticides liquid, toxic, flammable, n.o.s.
- Canada: Priority Substance List & Restricted Substances/Ocean Dumping Forbidden (CEPA), National Pollutant Release Inventory (NPRI) (arsenic compounds)

**Description:** Sodium arsenite is a white or grayish-white powder or flakes. Highly soluble in water. Melting/Freezing point = 615°C. Hazard Identification (based on NFPA-704 M Rating System): Health 3, Flammability 0, Reactivity 0.

**Incompatibilities:** Chemically active metals. Arsine, a very deadly gas, can be released in the presence of acid, acid mists, or hydrogen gas.

**Permissible Exposure Limits in Air:** OSHA[2] has set a PEL of 0.010 mg/m$^3$ for inorganic arsenic. NIOSH[2] recommends a limit of 0.002 mg/m$^3$ not to be exceeded at any time. ACGIH[1] recommends a TWA of 0.01 mg/m$^3$. The NIOSH[2] IDLH level = [Ca] 5 mg/m$^3$ as arsenic.

**Determination in Air:** Filter; Acid; Hydride generation atomic absorption spectrometry; NIOSH IV Method #7900[18].

**Permissible Concentration in Water:** EPA has set a limit of 0.05 parts per million (ppm) for arsenic in drinking water. The U.S. EPA arsenic drinking water standard of 0.01 ppm (10 ppb) is based on the U.S. EPA final rule for arsenic in drinking water published in. the January 22, 2001 *Federal Register*. However, the U.S. EPA is currently reviewing the science and cost estimate supporting this rule, and, in the interim, has reverted to the previous standard for arsenic. Thus, in the US, the current EPA arsenic drinking water standard remains at 0.05 ppm (50 ppb). To protect freshwater aquatic life, total recoverable trivalent inorganic arsenic never to exceed 440 µg/L. To protect saltwater aquatic life: 508 µg/L on an acute basis. To protect human health: preferably zero. The former USSR-UNEP/IRPTC project[43] has set MAC values for inorganic arsenic compounds in water for domestic purposes at 0.05 mg/L and in water bodies for fishery purposes of 0.5 mg/L also.

**Determination in Water:** *For arsenic:* The atomic absorption graphite furnace technique is often used for measurement of total arsenic in water. It also has been standardized by EPA. Total arsenic may be determined by digestion followed by silver diethyldithiocarbamate; an alternative is atomic absorption; another is inductively coupled plasma optical emission spectrometry. See OSHA Method #ID-105 for arsenic[58].

**Routes of Entry:** Inhalation, ingestion, skin and/or eye contact. Absorbed by the skin.

## Harmful Effects and Symptoms

**Short Term Exposure:** Sodium arsenite is extremely toxic. Poisonous if swallowed or inhaled. Probable lethal dose (oral, human) = 5-50 mg/kg. Between 7 drops and one teaspoon for 70 kg person (150 lb). May irritate or burn the skin, eyes, and mucous membranes. Skin contact can cause burning sensation, itching, and rash. Signs and symptoms of acute exposure to sodium arsenite may be severe and include headache, vomiting, stomach pain, vomiting, cough, dyspnea (shortness of breath), hypotension (low blood pressure), and chest pain. Gastrointestinal effects include difficulty swallowing, intense thirst, generalized abdominal pain, vomiting, and painful diarrhea; blood may be noted in the vomitus and feces. A weak pulse, cyanosis (blue tint to the skin and mucous membranes), and cold extremities may also be observed. Neurological effects include giddiness, delirium, mania, stupor, weakness, headache, dizziness, and fainting. Convulsions, paralysis, and coma may occur.

**Long Term Exposure:** Repeated or prolonged contact may cause skin sensitization and dermatitis. May affect the peripheral nervous system, skin, mucous membranes, causing neuropathy, skin thickening and pigmentation disorders, ulcers and perforation of nasal septum, and liver cirrhosis. Nerve damage may include "pins and needles," burning, numbness, and later weakness in the limbs. This substance contains 50 to 75% arsenic and is carcinogenic to humans. Birth defects have been observed in animals exposed to inorganic arsenic. It is likely that health effects seen in children exposed to high amounts of arsenic will be similar to the effects seen in adults. A probable teratogen in humans.

*Points of Attack:* Lymphatic system. Several studies have shown that inorganic arsenic can increase the risk of lung cancer, skin cancer, bladder cancer, liver cancer, kidney cancer, and prostate cancer.

*Medical Surveillance:* For those with frequent or potentially high exposure (half the TLV or greater, or significant skin contact), the following are recommended before beginning work and at regular times after that: Exam of the nose, skin, eyes, nails and nervous system. Test for urine arsenic (may not be accurate within 2 days of eating shellfish or fish; most accurate at the end of a workday). At NIOSH recommended exposure levels, urine arsenic should not be greater that 100 micrograms per gram creatinine in the urine.

*First Aid:* If this chemical gets into the eyes, remove any contact lenses at once and irrigate immediately for at least 15 minutes, occasionally lifting upper and lower lids. Seek medical attention immediately. If this chemical contacts the skin, remove contaminated clothing and wash immediately with soap and water. Seek medical attention immediately. If this chemical has been inhaled, remove from exposure, begin rescue breathing (using universal precautions) if breathing has stopped, and CPR if heart action has stopped. Transfer promptly to a medical facility. When this chemical has been swallowed, get medical attention. Give a slurry of activated charcoal in water to drink and induce vomiting. Do not make an unconscious person vomit.

*Note to physician or authorized medical personnel:* For severe poisoning BAL has been used. For milder poisoning penicillamine (not penicillin) has been used, both with mixed success. Side effects occur with such treatment and it is never a substitute for controlling exposure. It can only be done under strict medical care.

*References:*
- U.S. Environmental Protection Agency, "Chemical Profile: Sodium Arsenite," Washington, DC, Chemical Emergency Preparedness Program (November 30, 1987).
- California Environmental Protection Agency "Chemical List of Lists," Sacramento CA (February 1997).
- New Jersey Department of Health and Senior Services, "Hazardous Substance Fact Sheet: Sodium Arsenite," Trenton, NJ (January 1996, rev. March 2002). http://www.state.nj.us/health/eoh/rtkweb/1683.pdf

# Sodium Azide

*Use Type:* Fungicide and nematocide
*CAS Number:* 26628-22-8
*Formula:* $N_3Na$; $NaN_3$
*Alert:* No products currently registered by the U.S. EPA.
*Synonyms:* Azida sodico (Spanish); Azoture de sodium (French); Caswell No. 744A; Hydrazoic acid, sodium salt; Natriumazid (German); Natriummazide (Dutch); NCI-C06462
*Trade Names:* AI3-50436®; AXIUM®; AZIDE®; AZIUM®; DAZOE®; KAZOE®; SMITE®, PPG Industries (USA), canceled; U-3886®
*Producers:* American Pacific (USA); ATOFINA Chemicals (France); Degussa (Germany); Eurolabs Ltd. (UK); Merck (Germany); Mitsui Chemicals (Japan); Tokyo Kasei Kogyo (Japan)
*Chemical Class:* Inorganic compound
*EPA/OPP PC Code:* 107701; 245100 use code No. 107701
*California DPR Chemical Code:*
*ICSC Number:* 0950
*RTECS Number:* VY8050000
*EEC Number:* 011-004-00-7
*EINECS Number:* 247-852-1

*Uses:* Sodium azide is used to replace methyl bromide in pesticide applications. It has been used for a wide variety of military, laboratory, medical and commercial purposes. It is used extensively as an intermediate in the production of lead azide, commonly used in detonators and other explosives. It is used in automobile airbag inflation devices. One of the largest potential exposures is that to automotive workers, repairmen and wreckers if sodium azide is used as the inflation chemical. Commercial applications include use as a fungicide, nematocide, and soil sterilizing agent and as a preservative for seeds and wine. The lumber industry has used sodium azide to limit the growth of enzymes responsible for formation of brown stain on sugar pine, while the Japanese beer industry used it to prevent the growth of a fungus which darkens its product. The chemical industry has used sodium azide as a retarder in the manufacture of sponge rubber, to prevent coagulation of styrene and butadiene latexes stored in contact with metals, and to decompose nitrites in the presence of nitrates. There are no currently registered products by the U.S. EPA.

*Regulatory Authority:*
- Air Pollutant Standard Set (ACGIH)[1] (DFG)[3] (HSE)[33] (NIOSH)[2] (Several States)[60]
- Permissible Exposure Limits for Chemical Contaminants (CAL/OSHA)
- The "Director's List" (CAL/OSHA)
- EPA Hazardous Waste Number (RCRA No.): P105
- RCRA, 40CFR261, Appendix 8 Hazardous Constituents
- Superfund/EPCRA 40CFR355, Appendix B Extremely Hazardous Substances: TPQ = 500 lb (227kg)

- Superfund/EPCRA 40CFR302.4 RQ: CERCLA, 1000 lb (454 kg)
- EPCRA Section 313 Form R *de minimis* concentration reporting level: 1.0%
- Canada, WHMIS, Ingredients Disclosure List

***Description:*** Sodium azide is a colorless to white, odorless crystalline solid. Soluble in water; reaction; solubility = 40% @ 10°C; 42 g/100 mL @ 18°C. Molecular weight = 65.03. Boiling point=(decomposes above 280°C). Melting/Freezing point = 275° (the solid crystals decompose and may explode below the MP with the evolution of nitrogen gas, leaving a residue of sodium oxide).

***Incompatibilities:*** Contact with water forms hydrazoic acid. May explode when heated above the Melting/Freezing point, especially on rapid heating. Reacts with acids, producing toxic, shock-sensitive, and explosive hydrogen azide. Reacts with benzoyl chloride and potassium hydroxide, bromine, carbon disulfide, copper, lead, nitric acid, barium carbonate, sulfuric acid, chromium (II) hypochlorite, dimethyl sulfate, dibromomalononitrile, silver, mercury. Over a period of time, sodium azide may react with copper, lead, brass, or solder in plumbing systems to form an accumulation of the HIGHLY EXPLOSIVE and shock-sensitive compounds of lead azide and copper azide.

***Permissible Exposure Limits in Air:*** There is no OSHA[2] PEL. NIOSH[2] recommends a ceiling of 0.11 ppm as hydrazoic acid or a ceiling of 0.3 mg/m$^3$ as sodium azide. ACGIH[1] recommends a ceiling value of 0.1 ppm (0.3 mg/m$^3$). The HSE[33] has set a STEL of 0.3 mg/m$^3$. The DFG[3] has set an MAK of 0.07 ppm (0.2 mg/m$^3$). Several states have set guidelines or standards for sodium azide in ambient air[60] ranging from 0.7 $\mu$g/m$^3$ (Nevada) to 2.5 $\mu$g/m$^3$ (Virginia) to 3.0 $\mu$g/m$^3$ (North Dakota).

***Determination in Air:*** Filter/Si gel (special coating must be added); NaHCO$_3$/Na$_2$CO$_3$; Ion chromatography/Ultraviolet-Visible spectrophotometry; OSHA Method #ID121.[58]

***Permissible Concentration in Water:*** No criteria set, but runoff from spills or fire control may cause water pollution.

***Routes of Entry:*** Inhalation, ingestion, skin contact. Absorbed through the skin.

### Harmful Effects and Symptoms

***Short Term Exposure:*** Severely irritates the eyes, skin, and respiratory tract. Contact with skin causes redness and pain. Contact with eyes causes redness, pain, and blurred vision; may cause loss of vision. Inhalation or ingestion causes dizziness, weakness, blurred vision, slight shortness of breath, Hypotension, slowed heart rate, abdominal pain and spasms. Serious cases of exposure may result in convulsions, unconsciousness and death. Exposure slightly above the exposure limits in air can cause death by affecting the central nervous system. Azides can cause blood pressure to drop and thus have action similar to cyanides and nitrites. Higher exposures can cause pulmonary edema, a medical emergency that can be delayed for several hours. This can cause death. Sodium azide is a broad-spectrum, metabolic poison that interferes with oxidation enzymes and inhibits nuclear phosphorylation. Although the effects in these systems are complex, there is general agreement that azide causes a dissociation of phosphorylation and cellular respiration. For this reason parallels have been drawn to other metabolic inhibitors such as cyanide, malonitrile, and fluoride.

***Long Term Exposure:*** May cause kidney damage. Sodium azide is a potent mutagen in barley, peas, rice, and soybeans. It is also a very effective mutagen in bacteria. Its potency as a mutagen is comparable to the nitrosamines as a class. For these reasons sodium azide has been suspected of being a carcinogen. However, several studies have been performed to determine whether it is. In each instance the results were negative.

***Points of Attack:*** Eyes, skin, central nervous system, cardiovascular system and kidneys.

***Medical Surveillance:*** If symptoms develop or overexposure is suspected, the following may be useful: Exam of the nervous system and vision (including visual fields). Lung function tests. Consider chest x-ray after acute overexposure.

***First Aid:*** If this chemical gets into the eyes, remove any contact lenses at once and irrigate immediately for at least 15 minutes, occasionally lifting upper and lower lids. Seek medical attention immediately. If this chemical contacts the skin, remove contaminated clothing and wash immediately with soap and water. Seek medical attention immediately. If this chemical has been inhaled, remove from exposure, begin rescue breathing (using universal precautions) if breathing has stopped, and CPR if heart action has stopped. Transfer promptly to a medical facility. When this chemical has been swallowed, get medical attention. If victim is <u>conscious</u>, administer water, or milk. Do not induce vomiting.Medical observation is recommended for 24 to 48 hours after breathing overexposure, as pulmonary edema may be delayed. As first aid for pulmonary edema, a doctor or authorized paramedic may consider administering a corticosteroid spray.

***References:***
- New Jersey Department of Health and Senior Services, "Hazardous Substance Fact Sheet: Sodium Azide," Trenton, NJ (October 1998). http://www.state.nj.us/health/eoh/rtkweb/1684.pdf
- National Institute for Occupational Safety and Health, "Profiles on Occupational Hazards for Criteria Document Priorities: Sodium Azide," Report PB-274,073, Cincinnati, Ohio, pp. 306-308 (1977).
- U.S. Environmental Protection Agency, "Chemical Hazard Information Profile: Sodium Azide," Washington, DC (August 1, 1977).
- Sax, N.I., Ed., "Dangerous Properties of Industrial Materials Report," 2, No. 6, 74-78 (1982).
- California Environmental Protection Agency "Chemical List of Lists," Sacramento CA (February 1997).

- U.S. Environmental Protection Agency, "Chemical Profile: Sodium Azide," Washington, DC, Chemical Emergency Preparedness Program (November 30, 1987).

# Sodium Benzoate

*Use Type:* Fungicide and insecticide
*CAS Number:* 532-32-1
*Formula:* $C_7H_5NaO_2$; $C_6H_5COONa$
*Synonyms:* Antimol; Benzoate of soda; Benzoate sodium; Benzoesaeure (Na-salz) (German); Benzoic acid, sodium salt; Sobenate; Sodium benzoic acid
*Producers:* Abaquim (Mexico); Ashland (USA); Bayer Group (Germany); DSM Fine Chemicals (Netherlands); Gwalior Chemicals Industries (India); Harcros Chemicals (USA); Merck (Germany); Total Speciality Chemicals (USA); Velsicol Chemical Corporation (USA)
*EPA/OPP PC Code:* 009103
*California DPR Chemical Code:* 535
*RTECS Number:* DH6650000
*EINECS Number:* 208-534-8
*Uses:* This material is used as a preservative in food and beverage products, tobacco and cosmetic products. It is an anti-fungal agent, antiseptic, rust and mildew inhibitor and intermediate in the manufacture of dyes.
*U.S. Maximum Allowable Residue Levels for Sodium Benzoate (40 CFR 180.1001 and 180.2):* Sodium benzoate is exempted from the requirement of a tolerance when used in accordance with good agricultural practice as inert (or occasionally active) ingredients in pesticide formulations applied to growing crops or to raw agricultural commodities after harvest.
*Carcinogen/Hazard Classifications*
U.S. EPA Carcinogens: Group D, unclassifiable, inadequate date (as benzoic acid)
*Regulatory Authority:*
- Actively registered pesticide in California.

*Description:* Sodium benzoate is a white crystalline solid. Odorless and nonflammable. Soluble in water. Decomposes @ 120°C.
*Incompatibilities:* Strong oxidizers.
*Permissible Exposure Limits in Air:* No standards set.
*Permissible Concentration in Water:* No criteria set, but runoff from spills or fire control may cause water pollution.
*Routes of Entry:* Inhalation, ingestion, skin and/or eye contact.
*Harmful Effects and Symptoms*
*Short Term Exposure:* An eye and skin irritant. The accumulation of dust in the eyes, ears, nose, throat and lungs may be sufficiently unpleasant and distracting to make work near machinery hazardous. Irritation may result from abrasion or chemical action. Sodium benzoate can cause allergic reactions. Sodium benzoate has been given GRAS (Generally Recognized As Safe) status by the Food and Drug Administration at the levels currently being used as a food preservative. Ingestion of 8 to 10 grams (1/3 oz) may cause nausea and vomiting. 12 grams has caused gastric pain and loss of appetite. These symptoms disappear when exposure stops.
*Long Term Exposure:* Sodium benzoate may produce an allergic reaction and, in addition, may intensify the symptoms of allergies to other substances.
*Points of Attack:* Skin and lungs.
*Medical Surveillance:* Examination by a qualified allergist. Lung function tests.
*First Aid:* If this chemical gets into the eyes, remove any contact lenses at once and irrigate immediately for at least 15 minutes, occasionally lifting upper and lower lids. Seek medical attention immediately. If this chemical contacts the skin, remove contaminated clothing and wash immediately with soap and water. Seek medical attention immediately. If this chemical has been inhaled, remove from exposure, begin rescue breathing (using universal precautions) if breathing has stopped, and CPR if heart action has stopped. Transfer promptly to a medical facility. When this chemical has been swallowed, get medical attention. Give large quantities of water and induce vomiting. Do not make an unconscious person vomit.
*References:*
- U.S. Environmental Protection Agency, Office of Pesticide Programs, Pesticide Residue Limits, "Sodium benzoate," 40 CFR 180.910, http://www.epa.gov/pesticides/food/viewtols.htm
- New York State Department of Health, "Chemical Fact Sheet: Sodium Benzoate," Albany, NY, Bureau of Toxic Substance Assessment (March 1986).
- California Environmental Protection Agency "Chemical List of Lists," Sacramento CA (February 1997).

# Sodium Bromide

*Use Type:* Herbicide, insecticide, molluscicide and microbiocide
*CAS Number:* 7647-15-6
*Formula:* BrNa
*Synonyms:* Bromide salt of sodium; Bromnatrium (German); Sedoneural
*Trade Names:* BROMIDE PLUS®, Dead Sea Bromine Group (Israel)
*Producers:* Dead Sea Bromine Group (Israel); Ethyl Corp. (USA); Great Lakes Chemical (USA); Morton Intl. (USA)
*Chemical Class:* Inorganic
*EPA/OPP PC Code:* 013907
*California DPR Chemical Code:* 1103
*EINECS Number:* 231-599-9
*Uses:* Sodium bromide is used as a microbiocide to control algae, bacteria and fungi in pasteurizers and cannery cooling water recirculation systems, pulp and paper mill water

systems, and ornamental ponds and aquaria. It also is an active ingredient in pesticide products used to repel moths from clothing, and fleas from pets and their sleeping quarters. Sodium bromide has been used for many years in medicine as a sedative.

***Carcinogen/Hazard Classifications***
**Label Signal Word:** CAUTION or DANGER
***Regulatory Authority:***
- Actively registered pesticide in California.
- FIFRA, 40CFR186: tolerances for pesticides in animal feeds. as inorganic bromides
- AB 2588-Air Toxics "Hot Spots" Chemicals (CAL) as bromine-containing (inorganic compounds)

***Description:*** Hygroscopic, white or colorless, crystalline granules or powder. Odorless. Bitter, salty taste. Soluble in water; solubility = 0.92 g/mL. Molecular weight = 102.92. Density = 3.208. Melting/Freezing point = 756°C. Boiling point = 1390°C.

***Incompatibilities:*** Incompatible with acids, Oxidizers (chlorates, nitrates, peroxides, permanganates, perchlorates, chlorine, bromine, fluorine, etc); salts of heavy metals.

***Permissible Concentration in Water:*** No criteria set. Runoff from spills or fire control may cause water pollution.

***First Aid:*** If this chemical gets into the eyes, remove any contact lenses at once and irrigate immediately for at least 15 minutes, occasionally lifting upper and lower lids. Seek medical attention immediately. If this chemical contacts the skin, remove contaminated clothing and wash immediately with soap and water. Seek medical attention immediately. If this chemical has been inhaled, remove from exposure, begin rescue breathing (using universal precautions) if breathing has stopped, and CPR if heart action has stopped. Transfer promptly to a medical facility. When this chemical has been swallowed, get medical attention. Give large quantities of water and induce vomiting. Do not make an unconscious person vomit. Do not induce vomiting when formulations containing petroleum solvents are ingested.

***References:***
- U.S. Environmental Protection Agency, "Reregistration Eligibility Decision (RED), Inorganic Halides," Office of Prevention, Pesticides and Toxic Substances, Washington, DC (1993). http://www.epa.gov/REDs/factsheets/4051fact.pdf
- California Environmental Protection Agency "Chemical List of Lists," Sacramento CA (February 1997)

# Sodium Cacodylate

***Use Type:*** Herbicide, rodenticide and defoliant
***CAS Number:*** 124-65-2
***Formula:*** $C_2H_6AsNaO_2$; $(CH_3)_2AsOONa$
***Synonyms:*** Cacodilato sodico (Spanish); Cacodylate de sodium (French); Cacodylic acid sodium salt; Dimethylarsinic acid, sodium salt; [(Dimethylarsino)oxy]sodium-As-oxide; [(Dimethylarsino)oxy]sodium-arsenic-oxide; Hydrodimethylarsine oxide, sodium salt; Kakodylan dodny; Sodium cacodylate trihydrate; Sodium dimethylarsinate; Sodium dimethyl arsonate; Sodium salt of cacodylic acid

***Trade Names:*** ACME®, Pbi/Gordon Corporation (USA); ALKARSODYL®; ANSAR 160®; ARSECODILE®; ARSYCODILE®; BOLLS-EYE®; CHEMAID®; DREXEL EZY-PICKIN COTTON DEFOLIANT®, Drexel Chemical (USA); DREXEL KACK HERBICIDE®, Drexel Chemical (USA); DUTCH-TREAT®; HERB-ALL®, Luxembourg Industries (Israel); PHYTAR 560® (with Cacodylic acid); RAD-E-CATE®; SILVISAR®

***Producers:*** BASF Agricultural Products Group (Germany); Drexel Chemical (USA); Luxembourg Industries (Israel); Merck (Germany); Pbi/Gordon Corporation (USA); Rhone-Poulenc (France)

***Chemical Class:*** Organoarsenic compound
***EPA/OPP PC Code:*** 012502
***California DPR Chemical Code:*** 1673
***RTECS Number:*** CH7890000
***EINECS Number:*** 204-708-2

***Uses:*** This material has been used as a non-selective herbicide for general weed control.

***Carcinogen/Hazard Classifications***
**IARC:** Group 1, Known carcinogen
**California Prop. 65:** Carcinogen
**Label Signal Word:** CAUTION
***Regulatory Authority:***
- Banned or Severely Restricted (Portugal) (UN)[13]
- Air Pollutant Standard Set (ACGIH)[1] (OSHA)[63]
- Carcinogen User Register Chemical (CAL/OSHA) as inorganic arsenic compound
- AB 1803-Well Monitoring Chemical (CAL) as inorganic arsenic compound
- CAL Air Resources Board/AB 1807 Toxic Air Contaminants as inorganic arsenic compound
- Proposition 65 chemical (CAL) as inorganic arsenic compound
- The "Director's List" (CAL/OSHA) as inorganic arsenic compound
- AB 2588-Air Toxics "Hot Spots" Chemicals (CAL) as arsenic compounds as inorganic arsenic compound
- Permissible Exposure Limits for Chemical Contaminants (CAL/OSHA) as arsenic compounds as inorganic arsenic compound
- Actively registered pesticide in California.
- Superfund/EPCRA 40CFR355, Appendix B Extremely Hazardous Substances: TPQ = 100/10,000 lb (45.4/4,540 kg)
- Superfund/EPCRA 40CFR302.4 RQ: EHS, 1 lb (0.454 kg) as arsenic compounds
- Clean Air Act: Hazardous Air Pollutants (Title I, Part A, Section 112) as arsenic compounds

- Clean Water Act: Toxic Pollutant (Section 401.15) as arsenic and compounds
- RCRA, 40CFR261, Appendix 8, Hazardous Constituents, waste number not listed
- EPCRA (Section 313): Includes any unique chemical substance that contains arsenic as part of that chemical's infrastructure. Form R *de minimis* concentration reporting level: organics 1.0%
- U.S. DOT Regulated Marine Pollutant (49CFR172.101, Appendix B) as arsenates, liquid, n.o.s; arsenates, solid, n.o.s; arsenical pesticides liquid, toxic, flammable, n.o.s.
- Canada: Priority Substance List & Restricted Substances/Ocean Dumping Forbidden (CEPA), National Pollutant Release Inventory (NPRI) (arsenic compounds)
- Canada, WHMIS, Ingredients Disclosure List

*Description:* Sodium Cacodylate is a white crystalline solid which occurs as the trihydrate. It liquifies in the water of hydration at 60°C and becomes anhydrous at 120°C. Molecular weight = 160.01. Hazard Identification (based on NFPA-704 M Rating System): Health 4, Flammability 0, Reactivity 0. Soluble in water.

*Incompatibilities:* Incompatible with oxidizers, strong bases, acids, active metals (iron, aluminum, zinc). Contact with acids react to form highly toxic dimethylarsine gas. Attacks some metals.

*Permissible Exposure Limits in Air:* The OSHA 8-hour TWA for organic arsenic compounds is 0.5 mg/m$^3$. There is no NIOSH[2] REL.

*Determination in Air:* Filter; Reagent; Ion chromatography/Hydride generation atomic absorption spectrometry; NIOSH IV, Method #5022, Arsenic, Organo-[18].

*Permissible Concentration in Water:* EPA has set a limit of 0.05 parts per million (ppm) for arsenic in drinking water. The U.S. EPA arsenic drinking water standard of 0.01 ppm (10 ppb) is based on the U.S. EPA final rule for arsenic in drinking water published in the January 22, 2001 *Federal Register*. However, the U.S. EPA is currently reviewing the science and cost estimate supporting this rule, and, in the interim, has reverted to the previous standard for arsenic. Thus, in the US, the current EPA arsenic drinking water standard remains at 0.05 ppm (50 ppb). To protect freshwater aquatic life, total recoverable trivalent inorganic arsenic is never to exceed 440 µg/L. To protect saltwater aquatic life: 508 µg/L on an acute basis. To protect human health: preferably zero. The former USSR-UNEP/IRPTC project[43] has set MAC values for inorganic arsenic compounds in water for domestic purposes at 0.05 mg/L and in water bodies for fishery purposes of 0.5 mg/L also.

*Determination in Water:* For *arsenic:* The atomic absorption graphite furnace technique is often used for measurement of total arsenic in water. It also has been standardized by the EPA. Total arsenic may be determined by digestion followed by silver diethyldithiocarbamate; an alternative is atomic absorption; another is inductively coupled plasma optical emission spectrometry. See OSHA Method #ID-105 for arsenic[58].

*Routes of Entry:* Inhalation, ingestion, skin and/or eye contact.

*Harmful Effects and Symptoms*

*Short Term Exposure:* Sodium cacodylate is corrosive to the skin, eyes, and mucous membranes. Moderately toxic; probable oral lethal dose in humans is 0.5-5 g/kg or between 1 ounce and 1 pint (or 1 lb) for a 70 kg (150 lb) person. It may cause disturbances of the blood, kidneys, and nervous system. Acute exposure to sodium cacodylate may be fatal. Headache, red-stained eyes, and a garlicky odor of the breath may be the first effects noticed. Other signs and symptoms include generalized weakness, intense thirst, muscle cramping, seizures, toxic delirium, and shock. Nausea, vomiting, anorexia, abdominal pain, and diarrhea may occur. Hypotension (low blood pressure), tachycardia (rapid heart rate), pulmonary edema, ventricular fibrillation, and other cardiac abnormalities are usually found following severe exposure.

*Long Term Exposure:* Repeated exposure may cause ulcers and hole in the nasal septum. Hoarseness and sore eyes also occur. Repeated contact may cause thickened skin, pigmentation changes. May cause liver damage and nerve damage causing sensation of "pins and needles," weakness and loss of coordination in the limbs. May cause gastrointestinal tract and reproductive effects. Repeated exposures can cause metallic taste, poor appetite, nausea, vomiting, diarrhea and stomach pain, seizures and death. Birth defects have been observed in animals exposed to inorganic arsenic. It is likely that health effects seen in children exposed to high amounts of arsenic will be similar to the effects seen in adults.

*Points of Attack:* Several studies have shown that inorganic arsenic can increase the risk of lung cancer, skin cancer, bladder cancer, liver cancer, kidney cancer, and prostate cancer. May also attack central nervous system, gastrointestinal tract, and reproductive system.

*Medical Surveillance:* Examination of the nose, skin, eyes and nails. Examination of the nervous system. Liver function tests. Test for urine arsenic. NIOSH recommended exposure levels should not exceed 100 micrograms per liter of creatinine in the urine. Results may be accurate within 2 days of eating shellfish or fish (which may increase arsenic levels); they are most accurate at the end of a workday.

*First Aid:* If this chemical gets into the eyes, remove any contact lenses at once and irrigate immediately for at least 15 minutes, occasionally lifting upper and lower lids. Seek medical attention immediately. If this chemical contacts the skin, remove contaminated clothing and wash immediately with soap and water. Seek medical attention immediately. If this chemical has been inhaled, remove from exposure, begin rescue breathing (using universal precautions) if breathing has stopped, and CPR if heart action has stopped. Transfer promptly to a medical facility. When this chemical

has been swallowed, get medical attention. Give a slurry of activated charcoal in water to drink and induce vomiting. Do not make an unconscious person vomit. Obtain authorization and/or further instructions from the local hospital for administration of an antidote or performance of other invasive procedures. Rush to health care facility.

*References:*
- U.S. Environmental Protection Agency, "Chemical Profile: Sodium Cacodylate," Washington, DC, Chemical Emergency Preparedness Program (November 30, 1987).
- California Environmental Protection Agency "Chemical List of Lists," Sacramento CA (February 1997).
- New Jersey Department of Health and Senior Services, "Hazardous Substance Fact Sheet, Sodium Cacodylate," Trenton NJ (August 1999).
  http://www.state.nj.us/health/eoh/rtkweb/1687.pdf

## Sodium Chlorate

*Use Type:* Herbicide, defoliant and microbiocide
*CAS Number:* 7775-09-9
*Formula:* ClNaO3; NaClO3
*Synonyms:* Chlorate of soda; Chlorate salt of sodium; Chloric acid, sodium salt; Chlorsaure (German); Natriumchloraat (Dutch); Natriumchlorat (German); Soda chlorate; Sodio (clorato di) (Italian); Sodium (chlorate de) (French)
*Trade Names:* ASEX®; ATLACIDE®, Rhone-Poulenc Agro France (France), canceled; ATRATOL B-HERBATOX®, Syngenta (Switzerland), canceled; BAREGROUND®, Pro-Serve (USA); BEST MAG-CHLOR DEFOLIANT®, Simplot, J.R. Company (USA); CHAPMAN WEED FREE®, Drexel Chemical (USA); CHLORAX®; D-LEAF COTTON DEFOLIANT®, Gowan (USA); DE-FOL-ATE®; DESOLET®; DREXEL DEFOL®, Drexel Chemical (USA); DROP LEAF®; EVAU-SUPERFALL®; FALL®; GRAIN SORGHUM HARVEST AID®; GRANEX OK®; HARVEST-AID®; KLOREX®; KM SODIUM CHLORATE®, Kerr-McGee Chemical (USA); KUSA-TOHRUKUSATOL®; LOREX®; ORTHO C-1 DEFOLIANT & WEED KILLER®, Scotts Company, The (USA); OXYCIL®; RASIKAL®; SHED-A-LEAF®; TRAVEX®; TUMBLEAF®; 20 MULE TEAM HIBOR®, canceled; UNITED CHEMICAL DEFOLIANT NO. 1®; VAL-DROP®
*Producers:* ATOFINA Chemicals (France); Nexen Chemicals (Canada); Daiso (Japan); Drexel Chemical (USA); Eka Chemicals (Netherlands); Finnish Chemicals (Finland); Georgia Gulf (USA); Gowan (USA); Grupo Aragonesas (Spain); Harcros Chemicals (USA); Helena Chemical (USA); Kerr-McGee Chemical (USA); OxyChem (USA); Pro-Serve (USA; Showa Denko Chemicals Group (Japan); Simplot, J.R., Company (USA); Solvay Group (Belgium); Sterling Chemicals (USA); Syngenta (Switzerland); YiHua Group (China)
*Chemical Class:* Inorganic
*EPA/OPP PC Code:* 073301
*California DPR Chemical Code:* 536
*ICSC Number:* 1117
*RTECS Number:* FO0525000
*EEC Number:* 017-005-00-9
*EINECS Number:* 231-887-4
*Uses:* Sodium chlorate is a non-selective herbicide that is used on cotton, corn, flax, soybeans, peppers, grain sorghum, peas, greens, sunflower crops, sudan grass forage, rice and safflower crops to control, among others, Canadian thistle, johnson grass, morning glory and St. Johnswort. It is also used on non-crop lands, ditches, fences and roadways for vegetation control. It has a soil-sterilant effect. Sodium chlorate is also used to manufacture dyes, explosives, and in paper pulp processing.
*U.S. Maximum Allowable Residue Levels for Sodium Chlorate (40 CFR 180.1020):* Sodium chlorate is exempted from the requirement of a tolerance for residues in or on the following raw agricultural commodities when used as a defoliant, desiccant, or fungicide in accordance with good agricultural practice: Bean, dry, Corn, forage, Corn, grain, Corn, stover, Cotton, undelinted seed, Flax, seed, Flax, straw, Guar, forage, Pea, southern, succulent, Pepper, chili, Potato, Rice, Rice, straw, Safflower, Sorghum, forage, Sorghum, grain, Sorghum, grain, stover, Soybean, Sunflower, seed, Wheat.
*Carcinogen/Hazard Classifications*
**Label Signal Word:** WARNING
**WHO Acute Hazard:** Class III, slightly hazardous
*Regulatory Authority:*
- Highly Reactive Substance and Explosive (World Bank)[15]
- AB 1803-Well Monitoring Chemical (CAL)
- Actively registered pesticide in California.

*Description:* Sodium chlorate is a white crystalline solid. Soluble in water; solubility = 95 g/100mL @ 20°C. Melting/Freezing point = 248°C; also listed at 255°C and 264°C. Decomposes below boiling point @ 300°C. Hazard Identification (based on NFPA-704 M Rating System): Health 1, Flammability 0, Reactivity 2.

*Incompatibilities:* A strong oxidizer. Reacts violently with combustibles, sulfuric acid, and reducing materials. Reacts with strong acids, giving off carbon dioxide. Explosions may be caused by contact with ammonia salts, ammonium thiosulfate, antimony sulfide, arsenic, carbon, charcoal, organic matter, organic acids, thiocyanates, chemically active metals, oils, metal sulfides, nitrobenzene, powdered metals, sugar. Reacts with organic contaminants, forming shock-sensitive mixtures. Decomposes on heating above 300°C or on burning, producing oxygen and toxic chlorine fumes. Attacks zinc, magnesium, and steel.

*Permissible Exposure Limits in Air:* No standards set.
*Permissible Concentration in Water:* The former USSR-UNEP/IRPTC project[43] has set an MAC in water bodies used for domestic purposes of 20 mg/L.
*Routes of Entry:* Inhalation, ingestion, skin and/or eye contact.
*Harmful Effects and Symptoms*
*Short Term Exposure:* Sodium chlorate can affect you when breathed in. Eye or skin contact can cause severe irritation and even burns. Breathing sodium chlorate, especially dust or mist, can irritate the nose and throat. It can also cause cyanosis causing the skin to turn blue (methemoglobinemia) because it interferes with the blood's ability to carry oxygen. Damage to red blood cells (hemolytic anemia) can also occur. If severe or repeated, this can cause kidney damage. Ingestion can cause kidney damage. The effects may be delayed. Delayed pulmonary edema may occur after inhalation.
*Long Term Exposure:* Repeated or prolonged contact with skin may cause dermatitis. Kidney damage can occur from severe or repeated damage to red blood cells resulting from exposure. Very irritating substances may cause lung damage.
*Points of Attack:* Kidneys, lungs and skin.
*Medical Surveillance:* Medical observation is recommended for 24 to 48 hours after breathing overexposure, as pulmonary edema may be delayed. Before beginning employment and at regular times after that, for those with frequent or potentially high exposures, the following are recommended: Lung function tests. If symptoms develop or overexposure is suspected, the following may be useful: Complete blood count. Test for methemoglobin if skin is blue. *Note:* Do not use Methylene blue treatment to treat methemoglobinemia as it can increase toxicity.
*First Aid:* If this chemical gets into the eyes, remove any contact lenses at once and irrigate immediately for at least 15 minutes, occasionally lifting upper and lower lids. Seek medical attention immediately. If this chemical contacts the skin, remove contaminated clothing and wash immediately with soap and water. Seek medical attention immediately. If this chemical has been inhaled, remove from exposure, begin rescue breathing (using universal precautions) if breathing has stopped, and CPR if heart action has stopped. Transfer promptly to a medical facility. When this chemical has been swallowed, get medical attention. Give large quantities of water and induce vomiting. Do not make an unconscious person vomit. Keep victim under medical observation.
*Antidotes and Special Procedures:* Do not use *Methylene blue* to treat methemoglobinemia from sodium chlorate as it can cause increased toxicity. *Note to physician or authorized medical personnel:* Treat for methemoglobinemia. Spectrophotometry may be required for precise determination of levels of methemoglobinemia in urine.
*References:*
- Pesticide Management Education Program, "Pesticide Information Profile, Sodium Chlorate," Cornell University, Ithaca, NY, (September 1995). http://pmep.cce.cornell.edu/profiles/extoxnet/pyrethrins-ziram/sodium-chlorate-ext.html
- U.S. Environmental Protection Agency, Office of Pesticide Programs, Pesticide Residue Limits, "Sodium Chlorate," 40 CFR 180.1020, http://www.epa.gov/pesticides/food/viewtols.htm
- Sax, N.I., Ed., "Dangerous Properties of Industrial Materials Report," 3, No. 1, 28-32 (1983).
- New Jersey Department of Health and Senior Services, "Hazardous Substance Fact Sheet: Sodium Chlorate," Trenton, NJ (May 1986, rev. February 2001). http://www.state.nj.us/health/eoh/rtkweb/1688.pdf

## Sodium Cyanide

*Use Type:* Rodenticide
*CAS Number:* 143-33-9; 592-01-8 (U.S. EPA)
*Formula:* NaCN
*Alert:* A Restricted Use Pesticide (RUP)
*Synonyms:* Cianuro sodico (Spanish); Cianuro di sodio (Italian); Cyanide of sodium; Cyanobrik; Cyanogran; Cyanure de sodium (French); Cymag; Hydrocyanic acid, sodium salt; Kyanid sodny (Czech); Natriumcyanid (Dutch); Prussiate of soda; Sodium cyanide, solid; Sodium cyanide, solution
*Producers:* Abaquim (Mexico); Alquimia Mexicana (Mexico); Bayer Chemicals (Germany); Coogee Chemicals (Australia); CP/PhilbroChem (USA); Cyanides & Chemicals (India); Degussa (Germany); Dow Chemical (USA); DSM (Netherlands); DuPont (USA); EniChem (Italy); Grupo Aragonesas (Spain); Harcros Chemicals (USA); ICI Group (UK); Mitsui Chemicals (Japan); Nippon Soda (Japan); Orica (Australia); Philips Brothers (USA); Polifin (South Africa); Rhodia (France); Showa Denko Chemicals Group (Japan); Sterling Chemicals (USA); Univertical (USA); Westfarmers CSBP (Australia)
*Chemical Class:* Inorganic
*EPA/OPP PC Code:* 074002
*California DPR Chemical Code:* 688
*ICSC Number:* 1118
*RTECS Number:* VZ7530000
*RCRA Number:* P106
*EEC Number:* 006-007-00-5
*EINECS Number:* 205-599-4
*Uses:* Sodium cyanide is a Restricted Use Pesticide used in rodenticides. It is also used as a solid or in solution to extract metal ores, in electroplating and metal cleaning baths, in metal hardening, and in such applications as dyes

and pharmaceuticals. In gold mining, it is used to extract gold from gold bearing ore.

***Carcinogen/Hazard Classifications***
**Label Signal Word:** DANGER
**WHO Acute Hazard:** Class 1b, highly hazardous
***Regulatory Authority:***
- Air Pollutant Standard Set (ACGIH)[1] (DFG)[3] (HSE)[33] (UNEP)[43] (OSHA)[58]
- AB 2588-Air Toxics "Hot Spots" Chemicals (CAL)
- CAL Air Resources Board/AB 1807 Toxic Air Contaminants
- Clean Water Act: Section 311 Hazardous Substances/RQ 40CFR117.3 (same as CERCLA, see below); Section 313 Water Priority Chemicals (57FR41331, 9/9/92)
- RCRA, 40CFR261, Appendix 8 Hazardous Constituents
- Superfund/EPCRA 40CFR355, Appendix B Extremely Hazardous Substances: TPQ = 100 lb (45.4 kg)
- Superfund/EPCRA 40CFR302.4 RQ: CERCLA, 10 lb (4.54 kg)
- EPCRA Section 313: See Cyanide Compounds
- U.S. DOT Regulated Marine Pollutant (49CFR172.101, Appendix B)
- Canada, WHMIS, Ingredients Disclosure List; National Pollutant Release Inventory (NPRI); CEPA Priority Substance List, Ocean dumping prohibited

*As cyanide compounds:*
- Clean Air Act: Hazardous Air Pollutants (Title I, Part A, Section 112) as cyanide compounds
- Clean water act: Section 307 Priority Pollutants as cyanide, total; Toxic Pollutant (Section 401.15) as cyanide compounds
- EPA Hazardous Waste Number (RCRA No.): P030 as cyanide compounds
- RCRA Section 261 Hazardous Constituents as cyanide compounds
- RCRA Universal Treatment Standards: Wastewater (mg/L), 1.2 (total); 0.86 (amenable); Nonwastewater (mg/kg), 590 (total); 30 (amenable) as cyanide compounds
- RCRA Ground Water Monitoring List. Suggested test method(s) (PQL ug/L): 9010(40) as cyanide compounds
- Safe Drinking Water Act: MCL, 0.01 mg/L; MCLG, 0.01 mg/L; Regulated chemical (47 FR 9352) as cyanide compounds
- EPCRA Section 304 RQ: CERCLA, 10 lb (4.54 kg) as cyanide compounds
- EPCRA Section 313 Form R *de minimus* concentration reporting level: 1.0% as cyanide compounds
- U.S. DOT Regulated Marine Pollutant (49CFR, Subchapter 172.101, Appendix B) as cyanides, mixtures or solutions
- Canada, WHMIS, Ingredients Disclosure List (cyanide compounds, inorganic, n.o.s)

**Description:** Sodium cyanide, NaCN, is found as white granules, flakes or lumps. Faint, almond-like odor. Soluble in water; solubility = ~50 g/100 cc water @ 10°C. Boiling point = 1496°C. Melting/Freezing point = 564°C. Vapor Pressure = 1 mbar @ 817°C.

***Incompatibilities:*** Strong oxidizers such as acids, acid salts, chlorates, and nitrates. Contact with acids and acid salts releases highly toxic and flammable hydrogen cyanide. A strong base. Reacts violently with acid, strong oxidizers such as nitrates and chlorates. Decomposes in the presence of air, moisture, or carbon dioxide producing highly toxic and flammable hydrogen cyanide gas. Absorbs moisture from the air forming a syrup. Corrosive to active metals such as aluminum, copper, and zinc.

***Permissible Exposure Limits in Air:*** The OSHA[2] PEL 8-hour TWA is 5 mg/m$^3$ (4.7 ppm). NIOSH[2] recommended ceiling is 5 mg/m$^3$ (4.7 ppm), not to be exceeded during any 10 minute work period. ACGIH[1], DFG[3], HSE[33] have recommended 5 mg/m$^3$ as a TWA. The NIOSH[2] IDLH level = 25 mg/m$^3$. Regulatory authorities add the notation "skin" indicating the possibility off cutaneous absorption. The former USSR-UNEP/IRPTC project[43] has set an MAC in workplace air of 0.3 mg/m$^3$. Further, they have set MAC values for ambient air in residential areas of 0.009 mg/m$^3$ on a momentary basis and 0.004 mg/m$^3$ on an average daily basis. Several states have set guidelines or standards for cyanides in ambient air[60] ranging from 16.7 $\mu$g/m$^3$ (New York) to 50.0 $\mu$g/m$^3$ (Florida and North Dakota) to 80.0 $\mu$g/m$^3$ (Virginia) to 100 $\mu$g/m$^3$ (Connecticut and South Dakota) to 125 $\mu$g/m$^3$ (South Carolina) to 119.0 $\mu$g/m$^3$ (Nevada).

***Determination in Air:*** Filter/Bubbler; Potassium hydroxide; Ion-specific electrode; NIOSH IV Method #7904, Cyanides[18]. See also Method #6010, Hydrogen Cyanide.[18]

***Permissible Concentration in Water:*** The U.S. EPA has set a maximum contaminant level of cyanide in drinking water of 0.2 milligrams cyanide per liter of water (0.2 mg/L). The allowable daily intake for man is 8.4 mg/day[6]. On the international scene, the South African Bureau of Standards has set 10 $\mu$g/L, the World Health Organization (WHO) 10 $\mu$g/L and the Federal Republic of Germany 50 $\mu$g/L as drinking water standards. Other international limits[35] include an EEC limit of 50 $\mu$g/L; Mexican limits of 200 $\mu$g/L in drinking water and 1.0 $\mu$g/L in coastal waters and a Swedish limit of 100 $\mu$g/L. The former USSR-UNEP/IRPTC project[43] has set an MAC of 100 $\mu$g/L in water bodies used for domestic purposes and 50 $\mu$g/L in water for fishery purposes. The U.S. EPA[49] has determined a no-observed-adverse-effect-level (NOAEL) of 10.8 mg/kg/day which yields a lifetime health advisory of 154 $\mu$g/L. States which have set guidelines for cyanides in drinking water[61] include Arizona at 160 $\mu$g/L and Kansas at 220 $\mu$g/L.

***Determination in Water:*** Distillation followed by silver nitrate titration or colorimetric analysis using pyridine pyrazolone (or barbituric acid).

*Routes of Entry:* Inhalation, skin absorption, ingestion, skin and/or eye contact.

*Harmful Effects and Symptoms*

*Short Term Exposure:* Sodium cyanide can be absorbed through the skin, thereby increasing exposure. Sodium cyanide is corrosive to the eyes, skin, and respiratory tract. Contact can cause skin and eye burns, and possible permanent eye damage. Inhalation can cause lung irritation with coughing, sneezing, and difficult breathing, slow gasping respiration. Corrosive if swallowed. These substances may affect the central nervous system. Symptoms include headaches, confusion; nausea, pounding heart, weakness, unconsciousness and possible death.

*Long Term Exposure:* Repeated or prolonged contact with sodium cyanide may cause thyroid gland enlargement and interfere with thyroid function. May cause nosebleed and sores in the nose; changes in blood cell count. May cause central nervous system damage with headache, dizziness, confusion; nausea, vomiting, pounding heart, weakness in the arms and legs, unconsciousness and death. May affect liver and kidney function. Repeated lower exposures can cause sores in the nose with nosebleeds.

*Points of Attack:* Liver, kidneys, skin, cardiovascular system, central nervous system and thyroid gland.

*Medical Surveillance:* Consider the points of attack in preplacement and periodic physical examinations. Urine thiocyanate levels. Blood cyanide levels. Complete blood count. Evaluation of thyroid function. Liver function tests. Kidney function tests. Central nervous system tests. EKG. Smokers may have somewhat higher blood cyanide and urine thiocyanate levels.

*First Aid:* If this chemical gets into the eyes, remove any contact lenses at once and irrigate immediately for at least 15 minutes, occasionally lifting upper and lower lids. Seek medical attention immediately. If this chemical contacts the skin, remove contaminated clothing and wash immediately with soap and water. Seek medical attention immediately. If this chemical has been inhaled, remove from exposure, begin rescue breathing (using universal precautions) if breathing has stopped, and CPR if heart action has stopped. Transfer promptly to a medical facility. When this chemical has been swallowed, get medical attention. Give large quantities of water and induce vomiting. Do not make an unconscious person vomit. Use amyl nitrate capsules if symptoms develop. All area employees should be trained regularly in emergency measures for cyanide poisoning and in CPR. A cyanide antidote kit should be kept in the immediate work area and must be rapidly available. Kit ingredients should be replaced every 1-2 years to ensure freshness. Persons trained in the use of this kit, oxygen use, and CPR must be quickly available.

*References:*
- New Jersey Department of Health and Senior Services and Senior Services, "Hazardous Substance Fact Sheet, Sodium Cyanide," Trenton NJ (June 1998). http://www.state.nj.us/health/eoh/rtkweb/1693.pdf
- California Environmental Protection Agency "Chemical List of Lists," Sacramento CA (February 1997).

# Sodium Diacetate

*Use Type:* Fungicide, pH adjustment
*CAS Number:* 126-96-5
*Formula:* $CH_3COONa \cdot CH_3COOH$
*Synonyms:* Acetic acid, sodium salt (2:1); Sodium acid acetate
*Trade Names:* CROP CURE®; DYKON®; GRAIN CURE®, canceled; SENTRY SODIUM ACETATE®, Union Carbide (USA), canceled
*EPA/OPP PC Code:* 044008
*California DPR Chemical Code:* 5065
*Uses:* Sodium diacetate is a fungicide and bactericide registered to control molds and bacteria, thereby preventing spoilage in stored grains. The pesticide is applied to hay as a dust or soluble concentrate (liquid spray) during the baling process. It is applied to silage as an "aid" in fermentation, to preserve the quality of field corn, alfalfa, sorghum, oats and grasses, stored in silos. Sodium diacetate is composed of acetic acid and sodium acetate. It dissociates to acetate, sodium and hydrogen ions, normal components of plants and animals, and of human foods. Acetates are formed in living organisms during the metabolism of food. Acetates and acetic acid have long been used in both human and animal foods, without significant adverse effects. Sodium diacetate is also used as an antimicrobial agent in baked goods and other foods, and in medicine and cosmetics manufacturing as a freshener, flavoring and pH regulator.

*U.S. Maximum Allowable Residue Levels for Sodium Diacetate (40 CFR 180.1058):*

Sodium diacetate, when used post-harvest as a fungicide, is exempt from the requirement of a tolerance for residues in or on alfalfa hay, Bermuda grass hay, bluegrass hay, brome grass hay, clover hay, corn grain, oat grain, orchard grass hay, sorghum grain, sudan grass hay, rye grass hay, and timothy hay.

*Carcinogen/Hazard Classifications*
**Label Signal Word:** WARNING

*Description:* White solid. Soluble in water. Boiling point = (decomposes) > 150°C.

*Incompatibilities:* Contact with water liberates acetic acid. May not be compatible with nitrates. Moisture may cause hydrolysis or other forms of decomposition.

*First Aid:* If this chemical gets into the eyes, remove any contact lenses at once and irrigate immediately for at least 15 minutes, occasionally lifting upper and lower lids. Seek medical attention immediately. If this chemical contacts the skin, remove contaminated clothing and wash immediately with soap and water. Seek medical attention immediately. If this chemical has been inhaled, remove from exposure, begin rescue breathing (using universal precautions) if breathing has stopped, and CPR if heart action has stopped. Transfer promptly to a medical facility. When this chemical has been swallowed, get medical attention. *Do not induce vomiting when formulations containing petroleum solvents are ingested.* Otherwise, give large quantities of water and induce vomiting. Do not make an unconscious person vomit.

*References:*
- U.S. Environmental Protection Agency, Office of Pesticide Programs, Pesticide Residue Limits, "Sodium Diacetate," 40 CFR 180.1058, www.epa.gov/pesticides/food/viewtols.htm
- U.S. Environmental Protection Agency, "Reregistration Eligibility Decision Facts (RED), Sodium Diacetate," Office of Prevention, Pesticides and Toxic Substances, Washington, DC (September 1991). http://www.epa.gov/REDs/factsheets/4001fact.pdf
- California Environmental Protection Agency "Chemical List of Lists," Sacramento CA (February 1997)

# Sodium Dicamba

*Use Type:* Herbicide
*CAS Number:* 1982-69-0
*Formula:* $C_8H_6Cl_2O_3$ (Dicamba)
*Synonyms:* o-Anisic acid, 3,6-dichloro-, sodium salt; Benzoic acid, 3,6-dichloro-2-methoxy-, sodium salt (9CI); Dicamba-sodium; Dicamba, sodium salt; 3,6-Dichloro-o-anisic acid, sodium salt; 3,6-dichloro-2-methoxybenzoic acid, sodium salt; 2-Methoxy-3,6-dichlorobenzoic acid sodium salt; Sodium dicambate; Sodium 3,6-dichloro-o-anisate; Sodium 3,6-dichloro-2-methoxybenzoate; Sodium 2-methoxy-3,6-dichlorobenzoate; Sodium salt of 3,6-dichloro-o-anisic acid
*Trade Names:* AC 513,995 DG®, BASF Agricultural Products (Germany); Banvel® II/SGF, BASF Agricultural Products (Germany); CELEBRITY®, BASF Agricultural Products (Germany); DISTINCT®, BASF Agricultural Products (Germany); DMA 2# AG®, Micro-Flo (USA); NC-398 WG®, Nissan Chemical Industries (Japan); NORTHSTAR®, Syngenta (Switzerland); RAVE®, Syngenta (Switzerland); RESOLVE®, BASF Agricultural Products (Germany); SAN845H®, BASF Agricultural Products (Germany); YUKON®, Monsanto (USA)
*Producers:* BASF Agricultural Products (Germany); Micro-Flo (USA); Monsanto (USA); Nissan Chemical Industries (Japan); Syngenta (Switzerland)
*Chemical Class:* Chlorophenoxy; benzoic acid
*EPA/OPP PC Code:* 029806
*California DPR Chemical Code:* 5057
*Uses:* A herbicide used for the control of various annual and perennial broadleaf weeds, brush and vines on field corn, corn grown for seed and grain sorghum, and in range land and non-cropland areas.

*Carcinogen/Hazard Classifications*
**U.S. EPA Carcinogens:** Group D, unclassifiable (as parent chemical dicamba)
**U.S. TRI:** Developmental toxin
**Label Signal Word:** CAUTION, WARNING, DANGER
**WHO Acute Hazard:** Class III, slightly hazardous (as parent chemical dicamba)

*Regulatory Authority:*
- EPA 40 CFR 372.65, Specific Toxic Chemical Listings
- AB 2588-Air Toxics "Hot Spots" Chemicals (CAL) as chlorophenoxy pesticides
- The "Director's List" (CAL/OSHA) as chlorophenoxy pesticides
- EPCRA Section 313 Form R *de minimis* concentration reporting level: 1.0%

*Permissible Exposure Limits in Air:* Although no U.S. exposure limits have been established, this chemical can be absorbed through the skin, thereby increasing exposure.
*Permissible Concentration in Water:* No criteria set. Runoff from spills or fire control may cause water pollution.
*Routes of Entry:* Inhalation, ingestion, absorbed through the skin.

*Harmful Effects and Symptoms*
*Short Term Exposure:* Poisonous; may be fatal if inhaled, swallowed, or absorbed through skin. Severely irritates eyes, skin and respiratory tract, with burning sensation, pain, redness and swelling. Metabolic stimulant. If inhaled, causes coughing, dilated pupils, headache, profuse persperation, intense thirst, extreme fatigue, rapid pulse, high fever; clammy, flushed skin; rapid breathing, nausea, vomiting, cyanosis (bluish tint to skin and lips), anxiety and confusion, convulsions, risk of lung edema. If swallowed, face and lips turn bluish. Liver injury with associated jaundice, kidney failure, and cardiac arrhythmias are commonly noted. Nerve damage, which may be delayed, may include swelling of legs and feet, muscle twitch and stupor. Severe exposure can cause death from heart failure. Dust or liquid left in contact with the skin for several hours may be absorbed. This may result in severe delayed symptoms as listed above. These symptoms may last for months or years.
*Long Term Exposure:* Workers exposed to chlorophenoxy compounds such as 2,4-D (in the manufacturing process) over a five to ten year period at levels above 10 mg/m³ complained of weakness, rapid fatigue, headache and vertigo. Liver damage, low blood pressure and slowed heartbeat were also found. Based on animal tests, may affect human reproduction
*Points of Attack:* Eyes, skin, respiratory system, central nervous system, cardiovascular system, liver, kidney.
*Medical Surveillance:* If symptoms develop or overexposure is suspected, liver or kidney function tests may be useful.

*First Aid:* If this chemical gets into the eyes, remove any contact lenses at once and irrigate immediately for at least 15 minutes, occasionally lifting upper and lower lids. Seek medical attention immediately. If this chemical contacts the skin, remove contaminated clothing and wash immediately with soap and water. Seek medical attention immediately. If this chemical has been inhaled, remove from exposure, begin rescue breathing (using universal precautions) if breathing has stopped, and CPR if heart action has stopped. Transfer promptly to a medical facility. When this chemical has been swallowed, get medical attention. *Do not induce vomiting when formulations containing petroleum solvents are ingested.* Otherwise, give large quantities of water and induce vomiting. Do not make an unconscious person vomit. *Note to Physician:* If ingested, remove by lavage or vomiting. Use general supportive measures for CNS depression. Consider the use of quinidine for myotonia.

*References:*
- California Environmental Protection Agency "Chemical List of Lists," Sacramento CA (February 1997)

# Sodium Dichromate

*Use Type:* Wood preservative
*CAS Number:* 10588-01-9 (anhydrous); 7789-12-0 (dihydrate)
*Formula:* $Na_2Cr_2O_7$; $Cr_2Na_2O_7 \cdot 2H_2O$ (dihydrate)
*Alert:* A Restricted Use Pesticide (RUP)
*Synonyms:* Chromic acid ($H_2Cr_2O_7$), disodium salt (monohyrate); Dichromic acid, disodium salt; Dichromic acid heptaoxide; Disodium dichromate(VI); Disodium dichromium heptaoxide
*Trade Names:* CELCURE®, canceled; COMPOSITION NO. 155®, canceled; CSI® 70%; OSMOSALTS®, canceled; PYRESOTE®, canceled; WOLMAN SALTS®, canceled
*Producers:* Bayer Group (Germany); Honeywell (USA); Lords Chemicals (India); Mallinckrodt Baker (USA); Oxychem (USA)
*Chemical Class:* Inorganic chromium(VI) compound
*EPA/OPP PC Code:* 068304
*California DPR Chemical Code:* 3395
*ICSC Number:* 1369
*RTECS Number:* HX7700000
*EEC Number:* 024-004-00-7
*EINECS Number:* 234-190-3
*Uses:* Used as a wood preservative, pigments, leather tanning, metal treating, and pharmaceuticals.
*Carcinogen/Hazard Classifications*
**U.S. EPA Carcinogens:** Group A, known carcinogen
**U.S. NTP Carcinogen:** Known carcinogen
**California Prop. 65:** Carcinogen
**IARC:** Group 1, known carcinogen
**Label Signal Word:** DANGER

*Regulatory Authority:*
- AB 2588-Air Toxics "Hot Spots" Chemicals (CAL)
- The "Director's List" (CAL/OSHA)
- Actively registered pesticide in California.
- EPCRA Section 304 RQ: CERCLA, 10 lb (4.54 kg)
- Clean Air Act: Hazardous Air Pollutants (Title I, Part A, Section 112)
- Clean Water Act: Toxic Pollutant (Section 401.15); Section 307 Toxic Pollutants as chromium and compounds
- Safe Drinking Water Act: MCL, 0.05 mg/L as chromium, hexavalent
- RCRA Section 261 Hazardous Constituents, waste number not listed
- EPCRA (Section 313): Includes any unique chemical substances that contains chromium as part of that chemical's infrastructure. Form R *de minimis* concentration reporting level: Chromium(VI) compounds: 0.1%

*Description:* Red-orange hygroscopic crystalline solid. Soluble in water; solubility = 236 g/100 mL @ 20°C. Molecular weight = 262.0. Boiling point (decomposes) = 395°C; Melting/Freezing point = 357°C.

*Incompatibilities:* A strong oxidizer; violent reaction with reducing agents, combustible materials. Reacts with fluorine, hydrazine (violent), zirconium dusts, potassium iodide, sodium tetraborate, sodium tetraborate decahydrate, sodium borohydride.

*Permissible Exposure Limits in Air:* OSHA (ceiling): 0.1 mg/m$^3$ (as $CrO_3$); ACGIH[1] : 0.5 mg/m$^3$ (as Cr); NIOSH[2]: 0.001 mg/m$^3$ (as Cr); MAK (Germany): Class 2.

*Determination in Air:* Hexavalent chromium may be determined by filtration followed by visible absorption spectrophotometry according to NIOSH Method #7600[18]. Also, Filter (5.0-$\mu$m PVC membrane); Ion chromatography, conductivity detection; NIOSH Method #7604[18].

*Permissible Concentration in Water:* To protect human health, hexavalent chromium should be held below 0.05 mg/L according to EPA[6] in studies on priority toxic pollutants. This is also a WHO recommendation for total chromium in drinking water.

*Determination in Water:* Chromium(VI) may be determined by extraction and atomic absorption or colorimetry (using diphenylhydrazide).

*Routes of Entry:* Inhalation, ingestion, skin and eye contact.

*Harmful Effects and Symptoms*
*Short Term Exposure: Eyes:* Redness, pain, blurred vision. Highly corrosive; can cause severe and deep burns. *Skin:* Burning sensation and redness. *Inhalation:* Cough, sore throat, wheezing and a burning sensation. *Ingestion:* Abdominal pain and burning sensation, nausea and vomiting. Victim of exposure may go into shock or collapse. IDLH level = 15 mg/m$^3$ as chromium(VI); potential carcinogen. May cause liver and kidney damage.

**Long Term Exposure:** Repeated or prolonged skin contact may cause sensitization. Repeated or prolonged inhalation exposure may cause asthma. May affect the respiratory tract, kidneys, liver. May cause nasal septum perforation. Carcinogenic to humans.

**Points of Attack:** Kidneys, liver, respiratory tract.

**Medical Surveillance:** If symptoms develop or overexposure is suspected, the following may be useful: Examination of the respiratory tract. Liver and kidney function tests. Evaluation by a qualified allergist.

**First Aid:** If this chemical gets into the eyes, remove any contact lenses at once and irrigate immediately for at least 15 minutes, occasionally lifting upper and lower lids. Seek medical attention immediately. If this chemical contacts the skin, remove contaminated clothing and wash immediately with soap and water. Seek medical attention immediately. If this chemical has been inhaled, remove from exposure, begin rescue breathing (using universal precautions) if breathing has stopped, and CPR if heart action has stopped. Transfer promptly to a medical facility. When this chemical has been swallowed, get medical attention. Give large quantities of water and induce vomiting. Do not make an unconscious person vomit. Do not induce vomiting when formulations containing petroleum solvents are ingested.

**References:**
- California Environmental Protection Agency "Chemical List of Lists," Sacramento CA (February 1997)

# Sodium Dimethyldithiocarbamate

**Use Type:** Fungicide
**CAS Number:** 128-04-1
**Formula:** $C_3H_6NNaS_2$
**Synonyms:** Aceto SDD 40; AI3-14673; Carbamic acid, dimethyldithio-, sodium salt; Carbamodithioic acid, dimethyl-, sodium salt; Caswell No. 762; DDC; *N,N*-Dimethyldithiocarbamate, Sodium salt; dimethyldithio carbamate sodium salt; *N,N*-Dimethyldithio carbamic acid, sodium salt; Dimethyldithio carbamic acid, sodium salt; Dimetilditio carbamato sodico (Spanish); DMDK; Methyl namate; MSL; NSC 85566; SDDC; Sodium dimethylaminecarbodithioate; Sodium dimethylamino carbodithioate; Sodium dimethylcarbamodithioate; Sodium dimethyldithiocarbamate; Sodium *N,N*-dimethyl dithiocarbamate

**Trade Names:** ALCOBAM NM®; ALGEX, Kemira Chemical (Finland); ALGATROL-30, Kemira Chemical (Finland); AMA-30, Kemira Chemical (Finland); BIOQUEST, Kemira Chemical (Finland); BIOTREAT, Kemira Chemical (Finland); BROGDEX 555®; CARBON-S®; DIBAM®; FRESHGARD 40®; SHARSTOP 204®; STA-FRESH 615®; STERISEAL LIQUID-40®; THIOSTOP-N®; VULNOPOL® NM; WING STOP-B®

**Producers:** Huntsman (USA); Kemira Chemical (Finland); Tessenderlo Chemie (Belgium)
**Chemical Class:** Dithiocarbamate
**EPA/OPP PC Code:** 034804
**California DPR Chemical Code:** 548
**RTECS Number:** FD3500000
**EINECS Number:** 204-876-7
**Uses:** Used in wastewater treatment and for the removal of heavy metals.
**U.S. Maximum Allowable Residue Levels for Sodium Dimethyldithiocarbamate (40 CFR 180.152):**

| CROP | ppm |
|---|---|
| Melon | 25.0 |

**Carcinogen/Hazard Classifications**
**U.S. EPA Carcinogens:** Likely carcinogen
**California Prop. 65:** Developmental toxin
**U.S. TRI:** Developmental toxin
**Label Signal Word:** WARNING
**Regulatory Authority:**
- EPA 40 CFR 372.65, Specific Toxic Chemical Listings
- Actively registered pesticide in California
- RCRA Section 261 Hazardous Constituents
- EPA Hazardous Waste Number (RCRA No.): U382
- EPCRA Section 313 Form R *de minimis* concentration reporting level: 1.0%

**Description:** Crystalline solid. Molecular weight = 143.22.
**Determination in Air:** Filter; none; Gravimetric; NIOSH IV [Particulates NOR; #0500 (total), #0600 (respirable)].[18]
**Permissible Concentration in Water:** No criteria set. Runoff from spills or fire control may cause water pollution.
**Harmful Effects and Symptoms**
**Short Term Exposure:** Concentrated solutions are slightly corrosive to eyes and mucous membranes. Dust inhalation can cause irritation of the respiratory system with sneezing. Eye contact can cause irritation, watering, pain, and inflammation of the eyelids. Skin contact can cause irritation and minor ulceration. Ingestion can cause nausea, vomiting, fever, muscle twitching, seizure, rapid respiration, slow heart beat. Severe exposure may result in death.

**Points of Attack:** Respiratory system, central nervous system, cardiovascular system, skin, eyes.

**Medical Surveillance:** Medical observation is recommended for 24 to 48 hours after breathing overexposure, as pulmonary edema may be delayed. As first aid for pulmonary edema, consider administering a corticosteroid spray. Cigarette smoking may exacerbate pulmonary injury and should be discouraged for at least 72 hours following exposure. Before employment and at regular times after that, the following are recommended: If symptoms develop or overexposure occurs, repeat the above tests as soon as possible and get an exam of the nervous system. Also consider complete blood count. Consider chest x-ray following acute overexposure.

**First Aid:** If this chemical gets into the eyes, remove any contact lenses at once and irrigate immediately for at least

15 minutes, occasionally lifting upper and lower lids. Seek medical attention immediately. If this chemical contacts the skin, remove contaminated clothing and wash immediately with soap and water. Seek medical attention immediately. If this chemical has been inhaled, remove from exposure, begin rescue breathing (using universal precautions) if breathing has stopped, and CPR if heart action has stopped. Transfer promptly to a medical facility. When this chemical has been swallowed, get medical attention. If victim is conscious and able to swallow, have victim drink 4 to 8 ounces of water. Do not induce vomiting. *Note to physician or authorized medical personnel*: Medical observation is recommended for 24 to 48 hours after breathing overexposure, as pulmonary edema may be delayed. As first aid for pulmonary edema, consider administering a corticosteroid spray. Cigarette smoking may exacerbate pulmonary injury and should be discouraged for at least 72 hours following exposure.

*References:*
- U.S. Environmental Protection Agency, Office of Pesticide Programs, Pesticide Residue Limits, "Sodium Dimethyldithiocarbamate," 40 CFR 180.152, www.epa.gov/pesticides/food/viewtols.htm
- California Environmental Protection Agency "Chemical List of Lists," Sacramento CA (February 1997)

# Sodium Fluoride

*Use Type:* Insecticide and wood preservative
*CAS Number:* 7681-49-4
*Formula:* NaF
*Synonyms:* Disodium difluoride; FDA 0101; Fluorid sodny (Czech); Fluorure de sodium (French); Fluoruro sodico (Spanish); Natrium fluoride; NCI-C55221; Sodium hydrofluoride; Sodium monofluoride; Trisodium trifluoride
*Trade Names:* (Note: Many of these trade names are for roach control or wood preservatives and used in combination with other compounds) ALCOA SODIUM FLUORIDE®, Alcoa Inc (USA); ANTIBULIT®; CAVTROL®; CHECKMATE®; CHEMIFLOUR®; CREDO®; F1-TABS®; FLORIDINE®; FLOROCID®; FLOZENGES®; FLUORAL®; FLUORIDENT®; FLUORIGARD®; FLUORINEED®; FLUORINSE®; FLUORITAB®; FLOUR-O-KOTE®; FLURA-GEL®; FLURCARE®; FUNGOL B®; GEL II®; GELUTION®; IRADICAV®; KARIDIUM®; KARIGEL®; KARI-RINSE®; LEA-COV®; LEMOFLUR®; LURIDE®; NAFEEN®; NUFLOUR®; OSSALIN®; OSSIN®; PEDIAFLOR®; PEDIDENT®; PENNWHITE®; PERGANTENE®; PHOS-FLUR®; POINT TWO®; PROPORTION®; RAFLUOR®; RESCUE SQUAD®; ROACH SALT®; SO-FLO®; STAY-FLO®; STUDAFLOUR®; SUPER-DENT®; T-FLUORIDE®; THERA-FLUR-N®; VILLIAUMITE®; ZENDIUM®; ZYMAFLUOR®

*Producers:* Air Products & Chemicals (USA); Alcoa Inc (USA); ATOFINA Chemicals (France); Bayer Chemicals (Germany); Central Glass (Japan); Derivados del Fluor (Spain); Morita (Japan); Navin Fluorine Industries (India); Olin Corp. (USA); Philipp Brothers Chemicals (USA); Solvay Group (Belgium); Stella Chemifa (Japan)
*Chemical Class:* Inorganic
*EPA/OPP PC Code:* 075202
*California DPR Chemical Code:* 537
*ICSC Number:* 0951
*RTECS Number:* WB0350000
*EEC Number:* 009-004-00-7
*EINECS Number:* 231-667-8
*Uses:* Widely used in the chemical industry; in water treatment; as an insecticide, primarily for roach control; chemical cleaning, electroplating, glass manufacture, vitreous enamels, preservative for adhesives, toothpastes, disinfectant, dental prophylaxis. It is used as a wood preservative, particularly to preserve railroad ties. Also used orally in the treatment of various bone diseases to increase bone density and to relieve bone pain.
*Carcinogen/Hazard Classifications*
**IARC:** Group 3, unclassifiable
**Label Signal Word:** DANGER or WARNING
**WHO Acute Hazard:** Class II, moderately hazardous
*Regulatory Authority:*
- Clean Water Act: Section 311 Hazardous Substances/RQ 40CFR117.3 (same as CERCLA, see below)
- Superfund/EPCRA 40CFR302.4 RQ: CERCLA, 1000 lb (454 kg)
- The "Director's List" (CAL/OSHA)
- Actively registered pesticide in California

*Description:* Odorless, white powder or colorless crystals. Pesticide grade is often dyed blue. Often used in a solution. Boiling point = 1695–1704°C. Melting/Freezing point = 988–993°C. Vapor pressure = 1 mbar @ 1080°C. Hazard Identification (based on NFPA-704 M Rating System): Health 3, Flammability 0, Reactivity 0. Slightly soluble in water.
*Incompatibilities:* Strong oxidizers and acids.
*Permissible Exposure Limits in Air:* The applicable regulations are those for the fluoride ion. The OSHA[2] PEL, and NIOSH[2] REL, and ACGIH[1] TLV is 2.5 mg/m$^3$ TWA as fluoride. HSE[33] and DFG[3] set the same values. The NIOSH[2] IDLH is 250 mg/m$^3$ as Fluorides (F). The former USSR-UNEP/IRPTC project[43] has set MAC values for soluble fluorides in ambient air in residential areas of 0.03 mg/m$^3$ on a momentary basis and 0.01 mg/m$^3$ on an average daily basis. A number of states have set limits for fluoride in ambient air[60] ranging from as low as 2.85 $\mu$g/m$^3$ (Iowa) to as high as 60,000 $\mu$g/m$^3$ (Kentucky).
*Determination in Air:* Treated pad (special coating)/Pre-Filter; Sodium hydroxide; Ion-specific electrode; NIOSH IV, Method #7902, Fluorides[18]. See also #7906.

## Sodium Fluoroacetate

*Permissible Concentration in Water:* As with air, the applicable regulations are those for the fluoride ion. The values which have been set for drinking water[61] are a standard of 4.0 mg/L set by EPA and a guideline of 2.4 mg/L set by the State of Maine.

*Routes of Entry:* Inhalation, ingestion, skin and/or eye contact. Liquid can be absorbed through the skin.

*Harmful Effects and Symptoms*

*Short Term Exposure:* Sodium fluoride can affect you when breathed in. Inhalation of dust or mist can cause severe irritation and burns of the eyes and skin. Irritates the eyes and respiratory system. May cause permanent eye damage. Exposure can cause nausea, abdominal pain, diarrhea, salivation, thirst, sweating.

*Long Term Exposure:* Repeated or prolonged industrial contact can cause dermatitis. Repeated exposure can cause fluoride to build-up in the body. Can irritate the lungs; bronchitis may develop. Repeated exposure can cause fluoride to build up in the body causing stiffness, brittle bones, stiff spine, calcification of ligaments of ribs, pelvis; and crippling. Repeated exposures can cause weakness, and muscle twitching, tremors, convulsions, coma and even death. May cause kidney damage. Prolonged contact can cause sores in the nose and perforated septum. High concentrations can damage the developing fetus. These effects DO NOT occur when sodium fluoride is used in drinking water for dental cavity prevention.

*Points of Attack:* Eyes, skin, respiratory system, central nervous system, skeleton and kidneys.

*Medical Surveillance:* Lung function tests. Urine fluorine levels (above 4 mg/dl indicates overexposure). Kidney function tests.

*First Aid:* If this chemical gets into the eyes, remove any contact lenses at once and irrigate immediately for at least 15 minutes, occasionally lifting upper and lower lids. Seek medical attention immediately. If this chemical contacts the skin, remove contaminated clothing and wash immediately with soap and water. Seek medical attention immediately. If this chemical has been inhaled, remove from exposure, begin rescue breathing (using universal precautions) if breathing has stopped, and CPR if heart action has stopped. Transfer promptly to a medical facility. When this chemical has been swallowed, get medical attention. Give large quantities of water and induce vomiting. Do not make an unconscious person vomit.

*References:*
- New Jersey Department of Health and Senior Services, "Hazardous Substance Fact Sheet, Sodium Fluoride," Trenton NJ (March 1998).
  http://www.state.nj.us/health/eoh/rtkweb/1699.pdf
- Lee, C.C., "Environmental Law Index to Chemicals," Government Institutes, Rockville, MD (1996).
- California Environmental Protection Agency "Chemical List of Lists," Sacramento CA (February 1997).

# Sodium Fluoroacetate

*Use Type:* Insecticide and rodenticide (rat poison).
*CAS Number:* 62-74-8
*Formula:* $C_2H_2FNaO_2$; $FCH_2COONa$
*Alert:* A Restricted Use Pesticide (RUP). It is seldom used as an insecticide.
*Synonyms:* 1080; Acetic acid, fluoro-, sodium salt; Fluoroacetic acid, sodium salt; Compound 1080; Fluoacetato sodico (Spanish); Fluoracetato di sodio (Italian); Fluoressigsaeure (German); Monofluoressigsaure, natrium (German); Natriumfluoracetaat (Dutch); Natriumfluoracetat (German); NSC 77690; SMFA; Sodio, fluoracetato di (Italian); Sodium fluoacetate; Sodium fluoacetic acid; Sodium fluoracetate; Sodium fluoroacetate de (French); Sodium monofluoroacetate
*Trade Names:* AI3-08434®; FLUORAKIL® 3; FRATOL®; FURATOL®; RATBANE 1080®; TEN-EIGHTY®; TL 869®; YASOKNOCK®
*Producers:* (*Note:* 1080 and Compound 1080 are currently active and manufactured for use in the United States by the U.S. Department of Agriculture and various state agricultural departments and used as bait for rodents and large predators.) Fluorochem Ltd. (UK); Tull Chemical (USA)
*Chemical Class:* Unclassified
*EPA/OPP PC Code:* 075003
*California DPR Chemical Code:* 633
*ICSC Number:* 0484
*RTECS Number:* AH9100000
*EEC Number:* 607-082-00-2
*Uses:* Sodium Fluoroacetate is primarily used around sewers, ships and warehouses, and in agriculture to control rats, mice, squirrels, prairie dogs, coyotes, rabbits and other small pests. It is seldom used as an insecticide. It is very toxic to birds, domestic animals and wildlife either by consuming the bait or eating poisoned carcasses. It is sometimes used in 1% solutions which are injected into collars which are strapped to the necks of sheep, goats and other livestock that predators are attracted to. Coyotes that puncture the collars are likely to be fatally poisoned by the sodium fluoroacetate as a result.
*Carcinogen/Hazard Classifications*
**TRI Developmental Toxin:** Reproductive toxin
**Label Signal Word:** DANGER
**WHO Acute Hazard:** Class 1 a, extremely hazardous
*Regulatory Authority:*
- Banned or Severely Restricted (Several Countries) (UN)[13]
- Air Pollutant Standard Set (ACGIH)[1] (DFG)[3] (HSE)[33] (OSHA)[58]
- Permissible Exposure Limits for Chemical Contaminants (CAL/OSHA)
- The "Director's List" (CAL/OSHA)

- EPA Hazardous Waste Number (RCRA No.): P058
- RCRA, 40CFR261, Appendix 8 Hazardous Constituents
- Superfund/EPCRA 40CFR355, Appendix B Extremely Hazardous Substances: TPQ = 10/10,000 lb (4.54/4,540 kg)
- Superfund/EPCRA 40CFR302.4 RQ: CERCLA, 10 lb (4.54 kg)
- EPCRA Section 313 Form R *de minimis* concentration reporting level: 1.0%
- Canada, WHMIS, Ingredients Disclosure List

*Description:* Sodium fluoroacetate is a fluffy, colorless, odorless, hygroscopic solid (sometimes dyed black). Soluble in water; solubility = 110 g/100 cc @ 20°C. Boiling point = (decomposes). Non-volatile. Melting/Freezing point = 200°C (decomposes below MP). Hazard Identification (based on NFPA-704 M Rating System): Health 4, Flammability 0, Reactivity 0.

*Incompatibilities:* Alkaline metals and carbon disulfide.

*Permissible Exposure Limits in Air:* The OSHA[2] TWA is 0.05 mg/m$^3$. The NIOSH[2] TWA is the same and the NIOSH[2] STEL is 0.15 mg/m$^3$. The HSE[33] TWA, the DFG MAK[3] and the recommended ACGIH[1] TWA value is 0.05 mg/m$^3$. The STEL value is 0.15 mg/m$^3$. The notation "skin" is added to indicate the possibility of cutaneous absorption. The NIOSH[2] IDLH level = 2.5 mg/m$^3$.

*Determination in Air:* Filter; Water; Ion chromatography; NIOSH II(5) Method #S30[18].

*Permissible Concentration in Water:* No criteria set, but runoff from spills or fire control may cause water pollution.

*Routes of Entry:* Inhalation, skin absorption, ingestion, eye and/or skin contact.

*Harmful Effects and Symptoms*

*Short Term Exposure:* May affect the cardiovascular system and central nervous system, causing cardiac disorders and respiratory failure. Exposure may result in death This material is super toxic. Higher exposures can cause pulmonary edema, a medical emergency that can be delayed for several hours. This can cause death. The probable oral lethal dose in humans is less that 5 mg/kg, or a taste (less than 7 drops) for a 150 lb person. Symptoms include nausea, vomiting, apprehension, auditory hallucinations, facial paresthesia, twitching face muscle, pulsus altenans, ectopic heartbeat, tacard, ventricular fibrillation. Symptoms are usually seen within one-half hour of exposure, but severe effects may be delayed as long as 20 hours. A rebuttable presumption against registration of sodium fluoroacetate for pesticidal uses was issued on December 1, 1976 by the U.S. EPA. on the basis of reductions in nontarget and endangered species and because there is no human antidote.

*Long Term Exposure:* May cause liver and kidney damage. Affects the central nervous system causing epileptiform convulsive seizures that may be followed by severe depression.

*Points of Attack:* Cardiovascular system, lungs, kidneys, liver and central nervous system.

*Medical Surveillance:* Consider the points of attack in preplacement and periodic physical examinations. Liver and kidney function tests. Lung function tests. Consider chest x-ray following acute overexposure. Examination of the nervous system.

*First Aid:* Speed in removing material from eyes and skin is of extreme importance. If this chemical gets into the eyes, remove any contact lenses at once and irrigate immediately for at least 15 minutes, occasionally lifting upper and lower lids. Seek medical attention immediately. If this chemical contacts the skin, remove contaminated clothing and wash immediately with soap and water. Seek medical attention immediately. If this chemical has been inhaled, remove from exposure, begin rescue breathing (using universal precautions) if breathing has stopped, and CPR if heart action has stopped. Transfer promptly to a medical facility. When this chemical has been swallowed, get medical attention. Give large quantities of water and induce vomiting. Do not make an unconscious person vomit.

*References:*
- Environmental Protection Agency, "Reregistration Eligibility Decision (RED), Sodium Fluoroacetate," Office of Prevention, Pesticides and Toxic Substances, Washington, DC (September 1995), http://www.epa.gov/REDs/3073.pdf
- New Jersey Department of Health and Senior Services, "Hazardous Substance Fact Sheet, Sodium Fluoroacetate," Trenton, NJ (December 1997, rev. April 2000). http://www.state.nj.us/health/eoh/rtkweb/1700.pdf
- U.S. Environmental Protection Agency, "Chemical Profile: Sodium Fluoroacetate," Washington, DC, Chemical Emergency Preparedness Program (November 30, 1987).
- California Environmental Protection Agency "Chemical List of Lists," Sacramento CA (February 1997).

# Sodium Metaborate

*Use Type:* Herbicide
*CAS Number:* 7775-19-1
*Formula:* $NaBO_2$
*Synonyms:* Sodium metaborate ($NaBO_2$); Monosodium metaborate; Boric acid ($HBO_2$), sodium salt; Sodium borate ($NaBO_2$) (6CI)
*Trade Names:* ALLPRO BARACIDE®; ATRATOL®, canceled; BAREGROUND®; MONOBOR-CHLORATE®; PRAMITOL®, Makhteshim-Agan Industries (Israel); TRI-KILL®; UREABOR®
*Producers:* Ashland (USA); Makhteshim-Agan Industries (Israel); Rio Tinto (UK)
*Chemical Class:* Inorganic
*EPA/OPP PC Code:* 011104

*California DPR Chemical Code:* 689

*Uses:* Used in photographic chemicals, adhesives, textile processing, detergent, cleaners, and corrosion inhibition.

*Carcinogen/Hazard Classifications*

**Label Signal Word:** CAUTION, WARNING, DANGER

*Regulatory Authority:*
- Actively registered pesticide in California.

*Description:* White lumpy solid or powder. Soluble in water. Molecular weight = 65.83. Density 2.463. Melting point = 90°C. Fuses to clear glass = 965°C. Boiling point = 1434°C.

*Incompatibilities:* Contact with moisture forms a basic mixture. Keep away from acids.

*Permissible Concentration in Water:* No criteria set. Runoff from spills or fire control may cause water pollution.

*Harmful Effects and Symptoms*

*Short Term Exposure:* Corrosive. Severe eye irritant. Irritates the skin and respiratory tract.

*Medical Surveillance:*

*First Aid:* If this chemical gets into the eyes, remove any contact lenses at once and irrigate immediately for at least 15 minutes, occasionally lifting upper and lower lids. Seek medical attention immediately. If this chemical contacts the skin, remove contaminated clothing and wash immediately with soap and water. Seek medical attention immediately. If this chemical has been inhaled, remove from exposure, begin rescue breathing (using universal precautions) if breathing has stopped, and CPR if heart action has stopped. Transfer promptly to a medical facility. When this chemical has been swallowed, get medical attention. If victim is conscious and able to swallow, have victim drink 4 to 8 ounces of water. Do not induce vomiting. *Note to physician or authorized medical personnel*: Medical observation is recommended for 24 to 48 hours after breathing overexposure, as pulmonary edema may be delayed. As first aid for pulmonary edema, consider administering a corticosteroid spray. Cigarette smoking may exacerbate pulmonary injury and should be discouraged for at least 72 hours following exposure.

*References:*
- California Environmental Protection Agency "Chemical List of Lists," Sacramento CA (February 1997)

# Sodium Methanearsonate

*Use Type:* Herbicide and defoliant

*CAS Number:* 2163-80-6

*Formula:* $CH_4AsNaO_3$

*Synonyms:* Arsonic acid, methyl-, monosodium salt; Disodium methane arsonate; DSMA; Disodium methyl arsonate; Methanearsonic acid, monosodium salt; Methylarsonic acid, monosodium salt; Monosodium acid methanearsonate; Monosodium methane arsonate; MSMA; Monosodium methyl arsonate; Superarsonate

*Trade Names:* ANSAR®, KMG Chemicals (USA); ARSANOTE® Liquid, OxyChem (USA), canceled; ASAZOL BUENO®, KMG Chemicals (USA); DACONATE®, KMG Chemicals (USA); DICONATE 6®; DAL-E-RAD®, canceled; DIUMATE, Drexel Chemical (USA); DREXAR®-530, Drexel Chemical (USA); EH 1143, PBI/Gordon (USA); FERTI-LOME; HERB-ALL®; HERBAN M®; IDA BRANDS, Drexel Chemical (USA); KACK, Drexel Chemical (USA); MAGMA, Luxembourg Industries (Israel); MERGE®; MESAMATE®; MONATE®; MONATE® Merge 823; NUTGRASS KILLER®, PBI/Gordon (USA); PHYBAN®; PHYBAN H.C.®; QUADMEC, PBI/Gordon (USA); RED PANTHER®; RIVERDALE®, Nufarm (Australia); RIVERSIDE®, Agriliance (USA); SELECT-KIL®; SILVISAR®; SILVISAR-550®; TARGET MSMA®; TRANS-VERT®; VERSAR®, Drexel Chemical (USA); WEED-108®; WEED-BROOM® (DSMA + Bomacil + 2,4-D); WEED-E-RAD®; WEED-S-RAD®; WEED HOE®, Albaugh (USA)

*Producers:* Agriliance (USA); Bayer CropScience (Germany); Dow AgroSciences (USA); Drexel Chemical (USA); Helena Chemical (USA); KMG Chemicals (USA); Luxembourg Industries (Israel); Nufarm (Australia); PBI/Gordon (USA); The Scotts Company (USA)

*Chemical Class:* Arsenical (organo-)

*EPA/OPP PC Code:* 013803

*California DPR Chemical Code:* 34

*Uses:* MSMA is used extensively as a selective post-emergence herbicide to control grassy weeds in cotton crops, in non-agricultural areas and in industrial applications.

*Human toxicity (long-term)*[77]: Low–70.00 ppb, Health Advisory

*Fish toxicity (threshold)*[77]: Very low–1936.86508 ppb, MATC (Maximum Acceptable Toxicant Concentration)

*Carcinogen/Hazard Classifications*

**U.S. EPA Carcinogens:** Group B2, probable carcinogen (as cacodylic acid, the parent chemical)

**California Prop. 65:** Carcinogen (as cacodylic acid, the parent chemical)

**IARC:** Group 1, known carcinogen

**Label Signal Word:** CAUTION or WARNING

**WHO Acute Hazard:** Class III, slightly hazardous (as cacodylic acid, the parent chemical)

*Regulatory Authority:*
- AB 2588-Air Toxics "Hot Spots" Chemicals (CAL)
- Permissible Exposure Limits for Chemical Contaminants (CAL/OSHA)
- The "Director's List" (CAL/OSHA)
- Clean Air Act: Hazardous Air Pollutants (Title I, Part A, Section 112) as arsenic compounds
- Clean Water Act: Toxic Pollutant (Section 401.15) as arsenic and compounds
- RCRA Section 261 Hazardous Constituents, waste number not listed

- EPCRA Section 304 RQ: CERCLA, 1 lb (0.454 kg)
- EPCRA (Section 313): Includes any unique chemical substance that contains arsenic as part of that chemical's infrastructure. Form R *de minimis* concentration reporting level: as organic arsenic compounds, 1.0%
- Marine pollutant (49CFR, Subchapter 172.101, Appendix B) as arsenates

*Description:* Colorless crystalline solid; solution may be dyed red or green. Odorless. Solid may float or sink in water; solid and solution are highly soluble in water. Molecular weight = 162; 292 (hexahydrate). Boiling point = Decomposes. Melting/Freezing point = 117°C; 58°C (hexahydrate); Specific gravity: 1.0 20°C (solid); 1.4–1.6 @ 20°C (solutions). Hazard Identification (based on NFPA-704 M Rating System): Health Hazards (Blue): 1; Flammability (Red): 1; Reactivity (Yellow): 0.

*Permissible Exposure Limits in Air:* The OSHA[2] PEL for organic arsenic compounds is 0.5 mg/m$^3$ TWA for an 8-hour workshift. There is no NIOSH[2] recommendation. The ACGIH[1] TLV is 0.2 mg/m$^3$ TWA.

*Determination in Air:* Filter; Reagent; Ion chromatography/Hydride generation atomic absorption spectrometry; NIOSH IV, Method #5022, Arsenic, organo-[18]

*Permissible Concentration in Water:* No criteria set. Runoff from spills or fire control may cause water pollution. EPA[6] recommends a zero concentration of arsenic for human health reasons but has set a guideline of 50 $\mu$g/L[61] for drinking water.

*Determination in Water:* See OSHA Method ID-105 for arsenic.

*Routes of Entry:* Inhalation, ingestion, skin and/or eye contact.

*Harmful Effects and Symptoms*
*Short Term Exposure:* Grade 2; LD$_{50}$ = 0.5 to 5 g/kg (rat).
*Long Term Exposure:* Repeated contact may cause skin sensitivity. Chronic exposure to arsenic compounds can cause dermatitis and digestive disorders. Renal damage may develop. In animals: kidney damage, muscle tremor, seizure, possible gastrointestinal tract, reproductive effects, possible liver damage.

*Points of Attack:* Skin, kidneys, liver.

*Medical Surveillance:* Kidney function tests. Examination by a qualified allergist. Kidney, liver, lung function tests. Consider chest x-ray following acute overexposure.

*First Aid:* If artificial respiration is administered, *avoid mouth-to-mouth resuscitation; use bag/mask apparatus.* Solid or solution: Irritating to skin and eyes. If swallowed, will cause nausea, vomiting, or loss of consciousness. Remove contaminated clothing and shoes. Flush affected areas with plenty of water. If in eyes, hold eyelids open and flush with plenty of water. If swallowed and victim is conscious and able to swallow, have victim drink 4 to 8 ounces of water and have victim induce vomiting. If swallowed and victim is unconscious or having convulsions, do nothing except keep victim warm.

*References:*
- International Programme on Chemical Safety (IPCS), "Health and Safety Guide, Dimethyarseinic Acid, Methanearsonic Acid, and Salts," Geneva, Switzerland (1992). http://www.inchem.org/documents/hsg/hsg/hsg69.htm#PartNumber:1
- California Environmental Protection Agency "Chemical List of Lists," Sacramento CA (February 1997)

# Sodium Pentachlorophenate

*Use Type:* Fungicide, algaecide, molluscicide, microbiocide and as a wood preservative
*CAS Number:* 131-52-2
*Formula:* $C_6Cl_5NaO$; $C_6Cl_5ONa$
*Alert:* There are no registered products of sodium pentachlorophenate in the U.S.
*Synonyms:* PCP-sodium; PCP sodium salt; Pentachlorophenate sodium; Pentachlorophenol, sodium salt; Pentachlorophenoxy sodium; Pentaclorofenato sodico (Spanish); Pentaphenate; Phenol, pentachloro-, sodium salt; Phenol, pentachloro-, sodium salt, monohydrate; PKHFN; sodium PCP; Sodium pentachlorophenol; Sodium pentachlorophenolate; Sodium pentachlorophenoxide; Sodium, (pentachlorophenoxy)-; Sodium pentachlorphenate
*Trade Names:* AI3-16418®; DOW DORMANT FUNGICIDE®, Dow AgroSciences (USA), canceled; DOWICIDE G®, Dow AgroSciences (USA), canceled; DOWICIDE G-ST®, Dow AgroSciences (USA), canceled; GR 48-11PS®; GR 48-32S®; NAPCLOR-G®; SANTOBRITE®; SANTOBRITE D®; WEEDBEADS®
*Producers:* Dow AgroSciences (USA); Fluorochem (UK)
*Chemical Class:* Chlorinated phenol; chlorophenoxy
*EPA/OPP PC Code:* 063003
*California DPR Chemical Code:* 540
*ICSC Number:* 0532
*RTECS Number:* SM6490000
*EEC Number:* 604-003-00-3
*EINECS Number:* 205-025-2

*Uses:* Used as a wood preservative and a fungicide in cellulose products, textiles, water-based latex paints, adhesives, leather, pulp, paper and industrial waste systems; general disinfectant and control of the intermediate snail host off schistosomiasis. In 1993, all manufacturers discontinued their registrations of products containing sodium pentachlorophenate.

*Carcinogen/Hazard Classifications*
**U.S. EPA Carcinogens:** Group B2, probable carcinogen (as pentachlorophenate)
**California Prop. 65:** Listed (as pentachlorophenate)
**IARC:** Group 2B, possible carcinogen

**Label Signal Word:** WARNING or DANGER
**WHO Acute Hazard:** Class 1b, highly hazardous (as pentachlorophenate)
**Endocrine Disruptor:** Suspected endocrine disruptor (as pentachlorophenate)

*Regulatory Authority:*
- Air Pollutant Standard Set (former USSR)[43]
- EPA Hazardous Waste Number (RCRA No.): listed as "None"
- AB 2588-Air Toxics "Hot Spots" Chemicals (CAL) as chlorophenoxy pesticides
- The "Director's List" (CAL/OSHA) as chlorophenoxy pesticides
- RCRA, 40CFR261, Appendix 8 Hazardous Constituents
- EPCRA Section 313 Form R *de minimis* concentration reporting level: 1.0%
- U.S. DOT Regulated Marine Pollutant (49CFR172.101, Appendix B), severe pollutant
- Canada, WHMIS, Ingredients Disclosure List

*Description:* White or tan crystalline solid or powder. Phenolic odor. Soluble in water; solubility = 33% (wt/vol) @ 25°C; 33 g/100 ml water at 25°C.

*Incompatibilities:* Oxidizers

*Permissible Exposure Limits in Air:* The former USSR-UNEP/IRPTC project[43] has set an MAC of 0.1 mg/m$^3$ in workplace air and MAC values for ambient air in residential areas as follows: 0.005 mg/m$^3$ on a momentary basis and 0.001 mg/m$^3$ on a daily average basis.

*Determination in Air:* No OELs have been established.

*Permissible Concentration in Water:* The former USSR-UNEP/IRPTC project[43] set a MAC in water bodies used for domestic purposes of 5 mg/m$^3$ and in water bodies used for fishery purposes of 0.0005 mg/L.

*Routes of Entry:* Inhalation, ingestion, skin and/or eye contact.

*Harmful Effects and Symptoms*

*Short Term Exposure:* Exposure to fine dusts or sprays cause burning in eyes and painful irritation in upper respiratory tract. If inhaled, it will induce violent coughing and sneezing. Skin irritation results from brief exposures, causing a burning sensation or rash. Symptoms of severe systemic intoxication include loss of appetite, respiratory difficulties, anesthesia, fever, sweating, difficulty in breathing and rapidly progressive coma. Severe intoxications, including fatalities, have been reported from uncontrolled use. This compound causes inflamed gastric mucosa, congestion of the lungs, edema in the brain, cardiac dilatation, degeneration of the liver and kidneys. Individuals suffering from kidney and liver diseases have a lowered resistance and should not be exposed.

*Long Term Exposure:* May cause skin allergy. May cause anemia. May damage the liver and kidneys. Repeated exposure can cause headache, weakness, sweating, fever, muscle twitching, dizziness, confusion, and death.

*Points of Attack:* Skin, blood, liver, and kidneys.

*Medical Surveillance:* Completed blood count. Liver and kidney function tests. Evaluation by a qualified allergist.

*First Aid:* If this chemical gets into the eyes, remove any contact lenses at once and irrigate immediately for at least 15 minutes, occasionally lifting upper and lower lids. Seek medical attention immediately. If this chemical contacts the skin, remove contaminated clothing and wash immediately with soap and water. Seek medical attention immediately. If this chemical has been inhaled, remove from exposure, begin rescue breathing (using universal precautions) if breathing has stopped, and CPR if heart action has stopped. Transfer promptly to a medical facility. When this chemical has been swallowed, get medical attention. Give large quantities of water and induce vomiting. Do not make an unconscious person vomit.

*References:*
- Sax, N.I., Ed., "Dangerous Properties of Industrial Materials Report," 6, N0. 2, 5-30 (1986).
- U.S. Environmental Protection Agency, "Chemical Profile: Sodium Pentachlorophenate," Washington, DC, Chemical Emergency Preparedness Program (November 30, 1987).
- California Environmental Protection Agency "Chemical List of Lists," Sacramento CA (February 1997).
- New Jersey Department of Health and Senior Services, "Hazardous Substance Fact Sheet, Sodium Pentachlorophenate," Trenton NJ (August 1999). http://www.state.nj.us/health/eoh/rtkweb/1712.pdf

# Sodium Tellurite

*Use Type:* Pesticide
*CAS Number:* 10102-20-2
*Formula:* $Na_2O_3Te$; $Na_2TeO_3$
*Alert:* No longer registered as a pesticide in the U.S.
*Synonyms:* Sodium tellurate(IV); Telluric acid, disodium salt; Tellurous acid, disodium salt; Telurito sodico (Spanish)
*Producers:* Great Western Inorganics (USA); Molekula Fine Chemicals (UK)
*RTECS Number:* WY24500000
*EINECS Number:* 233-268-4
*Uses:* Used in bacteriology and medicine. Formerly used as pesticide.

*Regulatory Authority:*
- Air Pollutant Standard Set (ACGIH)[1] (DFG)[3] (HSE)[33] (OSHA)[58] (former USSR)[43]
- Superfund/EPCRA 40CFR355, Appendix B Extremely Hazardous Substances: TPQ = 500/10,000 lb (227/4,540 kg)
- Superfund/EPCRA 40CFR302.4 RQ: EHS, 1 lb (0.454 kg)
- Canada, WHMIS, Ingredients Disclosure List

*Description:* Sodium tellurite is a white crystalline solid. Hazard Identification (based on NFPA-704 M Rating

System): Health 2, Flammability 0, Reactivity 0. Slightly soluble in water.

***Incompatibilities:*** Cadmium, nitric acid, halogens and oxidizers.

***Permissible Exposure Limits in Air:*** The OSHA[2] PEL for tellurium compounds of 0.1 mg/m$^3$. NIOSH[2] and ACGIH[1] recommend the same level as does HSE[33] and DFG[3].

***Permissible Concentration in Water:*** The former USSR[43] has set an MAC for tellurium in water bodies used for domestic purposes of 0.01 mg/L.

***Routes of Entry:*** Inhalation, ingestion, skin and/or eye contact. Absorbed through the skin.

*Harmful Effects and Symptoms*

***Short Term Exposure:*** Irritates the skin and eyes. Irritates the respiratory tract causing cough and wheezing. The material is both an oral and dermal toxic hazard. The material is toxic by ingestion. Oral ingestion of tellurium compounds is generally regarded as extremely toxic. The probable oral lethal dose is 5-50 mg/kg or between 7 drops and 1 teaspoonful for a 70 kg (150 lb) person. Tellurium compounds are regarded as super toxic for skin exposures. Symptoms of exposure are sleepiness, fatigue, stupor; loss of appetite, nausea, vomiting, stomach pain, metallic taste, garlic odor of the breath and sweat, dryness of the mouth or excessive salivation, renal pain, bronchitis, irregular breathing; cyanosis, fatty degeneration of the liver, and unconsciousness.

***Long Term Exposure:*** Tellurium compounds cause reproductive damage. May cause liver and kidney damage. Can cause lung irritation; bronchitis may develop with cough, phlegm, and/or shortness of breath. Repeated exposure can cause weakness, fatigue, dizziness, and loss of consciousness.

***Points of Attack:*** Liver, kidneys and lungs.

***Medical Surveillance:*** Lung function tests. Liver and kidney function tests.

***First Aid:*** *Skin Contact:* Flood all areas of body that have contacted the substance with water. Don't wait to remove contaminated clothing; do it under the water stream. Use soap to help assure removal. Isolate contaminated clothing when removed to prevent contact by others. *Eye Contact:* Remove any contact lenses at once. Immediately flush eyes well with copious quantities of water or normal saline for at least 20-30 minutes. Seek medical attention. *Inhalation:* Leave contaminated area immediately; breathe fresh air. Proper respiratory protection must be supplied to any rescuers. If coughing, difficult breathing or any other symptoms develop, seek medical attention at once, even if symptoms develop many hours after exposure. *Ingestion:* If unconscious or convulsing, do not induce vomiting or give anything by mouth. Assure that victim's airway is open and lay him on his side with his head lower than his body and transport at once to a medical facility. If conscious and not convulsing, give a glass of water to dilute the substance. If medical advice is not readily available, consider inducing vomiting of this toxic material. Transport at once to a medical facility.

***References:***
- U.S. Environmental Protection Agency, "Chemical Profile: Sodium Tellurite," Washington, DC, Chemical Emergency Preparedness Program (November 30, 1987).
- California Environmental Protection Agency "Chemical List of Lists," Sacramento CA (February 1997).
- New Jersey Department of Health and Senior Services, "Hazardous Substance Fact Sheet, Sodium Tellurite," Trenton NJ (August 1999). http://www.state.nj.us/health/eoh/rtkweb/2783.pdf

## Sodium Tetraborate

***Use Type:*** Insecticide, herbicide and molluscicide

***CAS Number:*** 1303-96-4; 1330-43-4 as borates, tetra, sodium salts (anhydrous)

***Formula:*** $Na_2B_4O_7$

***Synonyms:*** Borates, tetra, sodium salt, anhydrous; Borax ($B_4Na_2O_7 \cdot 10H_2O$); Caswell No. 109; Disodium tetraborate decahydrate; Sodium tetraborate decahydrate; Sodium borate anhydrous; Sodium borate decahydrate

***Trade Names:*** BORATEEM®, U.S. Borax (USA), canceled; BORICIN®; BURA®; GERSTLEY® BORATE®; GROTAN®, canceled; JAIKIN®; NEOBOR®; POLYBOR®; RASORITE 65®; SOLUBOR®; THREE ELEPHANT V-BOR®, IMC Chemicals (USA); 20 MULE TEAM®, U.S. Borax (USA)

***Producers:*** Agrium (Canada); American Borate (USA); Allipo Chemicals (India); Chemettal (Germany); Cia. Quimica Universal de Industrias (Mexico); Eagle-Picher Minerals (USA); Holvoet Chimie (Belgium); IMC Chemicals (USA); Kerr-McGee (USA); OxyChem (USA); Quantum Chemicals (Australia); Rohm & Haas (USA); SQM (Chile); Textron Systems (USA); SB Boron (USA); TCI America (USA); U.S. Borax (USA); Varsal Instruments (USA)

***Chemical Class:*** Inorganic

***EPA/OPP PC Code:*** 011102; 011112 as borates, tetra, sodium salts (anhydrous)

***California DPR Chemical Code:*** 79; 5054 as borates, tetra, sodium salts (anhydrous)

***ICSC Number:*** 0567; 1229 as borates, tetra, sodium salts (anhydrous)

***RTECS Number:*** VZ2275000; ED4588000 as borates, tetra, sodium salts (anhydrous)

***Uses:*** Pesticide products containing boric acid and its sodium salts are registered in the U.S. for use as insecticides, fungicides and herbicides. They are used on several agricultural and many non-agricultural sites including residential, commercial, medical, veterinary, industrial, forestry and food/feed handling areas. They are marketed in many formulations including liquids, soluble and emulsifiable concentrates, granulars, powders, dusts, pellets, tablets, solids, pastes, baits, and crystalline rods.

*U.S. Maximum Allowable Residue Levels for Borax $(B_4Na_2O_7 \cdot 10H_2O)$, (40 CFR 180.1121):*
Borax along with boric acid and its salts are exempt from the requirements of a tolerance when used in or on raw agricultural commodities.

*Carcinogen/Hazard Classifications*
**U.S. EPA Carcinogens:** Group E, Unlikely carcinogen
**Label Signal Word:** CAUTION, WARNING, DANGER
**WHO Acute Hazard:** Class U, unlikely to be hazardous
*Regulatory Authority:*
- Actively registered pesticide in California.

*Description:* White to gray crystalline solid, granules, or powder. Becomes colorless on exposure to air. Odorless. Soluble in water; solubility = 2.56 g/100 mL @ 20°C. Molecular weight = 201.25. Density = 1.73. Melting/Freezing point = 756°C. Boiling point = 1575°C (decomposition).

*Incompatibilities:* Moisture, acids, metallic salts. Forms partial hydrate in moist air.

*Permissible Exposure Limits in Air:* NIOSH[2]: 1 mg/m$_3$ TWA; ACGIH[1] TLV: 1 mg/m$_3$ TWA as Borates. tetra, sodium salts.

*Determination in Air:* Collection on a filter and gravimetric analysis. See NIOSH Method #0500 Particulates NOR (total)[18].

*Permissible Concentration in Water:* No criteria set but EPA has suggested[32] an ambient water limit of 138 µg/L based on health effects. Runoff from spills or fire control may cause water pollution.

*Routes of Entry:* Inhalation, skin, eyes, ingestion.

*Harmful Effects and Symptoms*
*Short Term Exposure:* Irritates the eyes, skin, and respiratory tract. Ingestion in high dose or through damaged skin, may affect the central nervous system, kidneys, and gastrointestinal tract.

*Long Term Exposure:* Repeated or prolonged contact with skin may cause dermatitis. Repeated or prolonged inhalation may affect the respiratory tract.

*Points of Attack:* Eyes, skin, respiratory system.

*Medical Surveillance:* Consider the points of attack in preplacement and periodic physical examinations.

*First Aid:* If this chemical gets into the eyes, remove any contact lenses at once and irrigate immediately for at least 15 minutes, occasionally lifting upper and lower lids. Seek medical attention immediately. If this chemical contacts the skin, remove contaminated clothing and wash immediately with soap and water. Seek medical attention immediately. If this chemical has been inhaled, remove from exposure, begin rescue breathing (using universal precautions) if breathing has stopped, and CPR if heart action has stopped. Transfer promptly to a medical facility. When this chemical has been swallowed, get medical attention. *Do not induce vomiting when formulations containing petroleum solvents are ingested.* Otherwise, give large quantities of water and induce vomiting. Do not make an unconscious person vomit.

*References:*
- U.S. Environmental Protection Agency, "Reregistration Eligibility Decision Facts (RED), Boric Acid," Office of Prevention, Pesticides and Toxic Substances, Washington, DC (September 1993). http://www.epa.gov/REDs/factsheets/0024fact.pdf
- U.S. Environmental Protection Agency, Office of Pesticide Programs, Pesticide Residue Limits, "Borax $(B_4Na_2O_7 \cdot 10H_2O)$," 40 CFR 180.1121, www.epa.gov/pesticides/food/viewtols.htm
- California Environmental Protection Agency "Chemical List of Lists," Sacramento CA (February 1997)

# Stibine

*Use Type:* Fumigant
*CAS Number:* 7803-52-3
*Formula:* $H_3Sb$; $SbH_3$
*Alert:* No products of Stibine registered in the U.S.
*Synonyms:* Antimonwasserstoffes (German); Antimony hydride; Antimony trihydride; Antymonowodor (Polish); Hydrogen antimonide; Stibnite
*ICSC Number:* 0776
*RTECS Number:* WJ0700000
*EEC Number:* 051-003-00-9
*Uses:* Stibine is used as a fumigating agent. It is used in metallurgy, welding or cutting with blow torches, soldering, filling of hydrogen balloons, etching of zinc, and chemical processes. No products registered in the U.S.

*Carcinogen/Hazard Classifications*
**TRI Developmental Toxin:** Reproductive and developmental toxin as antimony compounds
*Regulatory Authority:*
- Very Toxic Substance (World Bank)[15]
- U.S. DOT Inhalation Hazard Chemicals
- Permissible Exposure Limits for Chemical Contaminants (CAL/OSHA)
- The "Director's List" (CAL/OSHA)
- OSHA 29CFR1910.119, Appendix A, Process Safety List of Highly Hazardous Chemicals, TQ = 500 lb (227 kg)
- Air Pollutant Standard Set (ACGIH)[1] (DFG)[3] (HSE)[33] (OSHA)[58] (Several States)[60]

*As antimony compounds*:
- Banned or Severely Restricted (New Zealand)[13] (Many countries, especially in food) (UN)[35]
- Air Pollutant Standard Set (ACGIH)[1] (OSHA)[2] (California) (HSE) (Ontario, Quebec)
- AB 2588-Air Toxics "Hot Spots" Chemicals (CAL) as antimony compounds
- CAL Air Resources Board/AB 1807 Toxic Air Contaminants as antimony compounds

- Permissible Exposure Limits for Chemical Contaminants (CAL/OSHA) as antimony compounds
- The "Director's List" (CAL/OSHA) as antimony compounds
- Clean Air Act, 42USC7412; Title I, Part A,§112 hazardous pollutants (as antimony compounds)
- Clean Water Act: 40CFR116.4 Hazardous Substances; 40CFR117.3, RQ (same as CERCLA); 40CFR423, Appendix A, Priority Pollutants; Section 313 Water Priority Chemicals (57FR41331, 9/9/92); 40CFR401.15 Section 307 Toxic Pollutants, as antimony compounds.
- RCRA, 40CFR261, Appendix 8 Hazardous Constituents, waste number not listed (as antimony compounds, n.o.s.)
- Safe Drinking Water Act, MCL, treatment technique; MCL, 0.006 mg/L; MCLG, 0.006 mg/L; Regulated Chemical (47FR9352)
- Superfund/EPCRA 40CFR302.4 RQ: CERCLA, 100 lb (45.4 kg)
- EPCRA Section 313: Includes any unique chemical substance that contains antimony as part of that chemical's infrastructure. Form R *de minimus* concentration reporting level: 0.1%
- Canada, WHMIS, Ingredients Disclosure List; National Pollutant Release Inventory (as antimony compounds)

***Description:*** Stibine is a colorless gas with a characteristic disagreeable odor. It is produced by dissolving zinc/antimony or magnesium-antimony in hydrochlorid acid. Poor solubility in water. Boiling point = –18°C. Melting/Freezing point = –88°C. Flash point = flammable gas. Hazard Identification (based on NFPA-704 M Rating System): Health 4, Flammability 4, Reactivity 2.

***Incompatibilities:*** A flammable gas. Incompatible with acids, halogenated hydrocarbons, oxidizers, moisture, chlorine, ammonia. Reacts violently with chlorine, concentrated nitric acid and ozone. Decomposes in air. Thermally unstable: quick decomposition @ 200°C producing metallic antimony and explosive hydrogen gas.

***Permissible Exposure Limits in Air:*** The OSHA[2] TWA, the DFG MAK[3] and ACGIH[1] TWA value is 0.1 ppm (0.5 mg/m$^3$). The NIOSH[2] IDLH level = 5 ppm. Several states have set guidelines or standards for Stibine in ambient air[60] ranging from 5.0 µg/m$^3$ (North Dakota) to 8.0 µg/m$^3$ (Virginia) to 10.0 µg/m$^3$ (Connecticut) to 12.0 µg/m$^3$ (Nevada).

***Determination in Air:*** Si gel (special coating); Hydrochloric acid; Visible spectrophotometry; NIOSH IV, Method #6008[18].

***Permissible Concentration in Water:*** The U.S. EPA allows 0.006 ppm of antimony per million parts of drinking water.

***Routes of Entry:*** Inhalation of gas.

***Harmful Effects and Symptoms***

***Short Term Exposure:*** Stibine can affect you when breathed in. May be fatal if absorbed through the skin or inhaled. A strong sensitizer. May cause severe allergic respiratory reaction. Exposure can cause rapid, fatal poisoning, with symptoms of headaches, nausea, dark or bloody urine, pain in the back and abdomen, slowed breathing and death. Exposure can also irritate the lungs and may lead to a build-up of fluid (pulmonary edema) a medical emergency that can be delayed for several hours. This can cause death. Stibine destroys red blood cells and can also cause liver and kidney damage.

***Long Term Exposure:*** Stibine destroys red blood cells(hemolysis). May affect the kidneys, liver; hemoglobinuria, hematuria (blood in the urine), hemolytic anemia, jaundice. May affect the central nervous system.

***Points of Attack:*** Blood, liver, kidneys, respiratory system and central nervous system. Lung cancer has been observed in some studies of rats that breathed high levels of antimony. No human studies are available. It is unknown whether antimony will cause cancer in people.

***Medical Surveillance:*** Antimony can be measured in the urine, feces, and blood for several days after exposure. For those with frequent or potentially high exposure (half the TLV or greater), the following are recommended before beginning work and at regular times after that: Lung function tests. If symptoms develop or overexposure is suspected, the following may be useful: Consider chest x-ray after acute overexposure. Blood and urine hemoglobin levels. Liver and kidney function tests. Examination of the central nervous system. Examination by a qualified allergist.

***First Aid:*** If this chemical gets into the eyes, remove any contact lenses at once and irrigate immediately for at least 15 minutes, occasionally lifting upper *and* lower lids. Seek medical attention immediately. If this chemical contacts the skin, remove contaminated clothing and wash immediately with soap and water. Seek medical attention immediately. If this chemical has been inhaled, remove from exposure, begin rescue breathing (using universal precautions) if breathing has stopped, and CPR if heart action has stopped. Transfer promptly to a medical facility. When this chemical has been swallowed, get medical attention. Give large quantities of water and induce vomiting. Do not make an unconscious person vomit. Medical observation is recommended for 24 to 48 hours after breathing overexposure, as pulmonary edema may be delayed. As first aid for pulmonary edema, a doctor or authorized paramedic may consider administering a corticosteroid spray.

***References:***
- Sax, N.I., Ed., "Dangerous Properties of Industrial Materials Report," 2, No. 4, 17-18 (1982).
- Pohanish. R.P., "Rapid Guide to Hazardous Chemicals in the Environment," Van Nostrand Reinhold, NY (1997).
- California Environmental Protection Agency "Chemical List of Lists," Sacramento CA (February 1997).
- New Jersey Department of Health and Senior Services, "Hazardous Substance Fact Sheet: Stibine," Trenton, NJ (January 1987, rev. August 2001).
  http://www.state.nj.us/health/eoh/rtkweb/1735.pdf

# Stoddard Solvent

*Use Type:* Insecticide and herbicide
*CAS Number:* 8052-41-3
*Formula:* $C_9H_{20}$
*Alert:* There are no Stoddard Solvent pesticide products currently registered in the U.S.
*Synonyms:* Cleaning solvent; Dry cleaner naphtha; Mineral spirits; Naphtha safety solvent; Petroleum solvent; Spotting solvent; Varnoline; White spirits
*Chemical Class:* Unclassified (miscellaneous hydrocarbon)
*EPA/OPP PC Code:* 063504
*ICSC Number:* 0361
*RTECS Number:* WJ8925000
*EEC Number:* 649-345-00-4
*EINECS Number:* 232-489-3
*Uses:* Stoddard Solvent is used in dry cleaning, in degreasing of metal parts and as a paint thinner. It is also an insecticide and herbicide; however, there are no pesticide products currently registered in the U.S.
*Regulatory Authority:*
- Air Pollutant Standard Set (ACGIH)[1] (OSHA)[58] (Several States)[60]
- Permissible Exposure Limits for Chemical Contaminants (CAL/OSHA)
- The "Director's List" (CAL/OSHA)
- Canada, WHMIS, Ingredients Disclosure List

*Description:* A mixture of saturated various hydrocarbons ($C_7$–$C_{12}$) that may include benzene. A refined petroleum solvent containing >65% $C_{10}$ or higher hydrocarbons. Colorless liquid. Kerosene-like odor; odor threshold = 1 ppm (NY) to 30 ppm (NJ). Insoluble in water. Boiling point = 154–202°C; also 130–230°C. Melting/Freezing point = <-30°C. Flash point = 39–60°C; also listed at 21°C and >38°C. Vapor pressure = 0.1-1.4 kPa @ 20°C. Autoignition temperature = 229°C; also 230-240°C (ICSC). Explosive limits: LEL = 0.6%; UEL = 8.0%. Hazard Identification (based on NFPA-704 M Rating System): Health 0, Flammability 2, Reactivity 0. Log $K_{ow}$ = 3.16-7.06. Values above 3.0 are likely to bioaccumulate in marine organisms. (ICSC)
*Incompatibilities:* Forms explosive mixture with air. Strong oxidizers. Attacks some forms of plastics, rubber, and coatings.
*Permissible Exposure Limits in Air:* The OSHA[2] PEL is 500 ppm (2,900 mg/m³) TWA. NIOSH[2] recommends a TWA 350 mg/m³ and a ceiling of 1800 mg/m³ not to be exceeded during any 15 minute work period. ACGIH[1] recommends a TWA of 100 ppm (525 mg/m³). The NIOSH[2] IDLH level = 20,000 mg/m³. Several states have set guidelines or standards for Stoddard Solvent in ambient air[60] ranging from 5.25 to 10.50 mg/m³ (North Dakota) to 7.0 mg/m³ (Connecticut) to 12.5 mg/m³ (Nevada).

*Determination in Air:* Adsorption on charcoal, workup with $CS_2$, analysis by gas chromatography/flame ionizaton. See NIOSH Method 1550 for Naphthas.[18]
*Permissible Concentration in Water:* No criteria set, but runoff from spills or fire control may cause water pollution.
*Routes of Entry:* Inhalation, ingestion, skin and/or eye contact.
*Harmful Effects and Symptoms*
*Short Term Exposure: Inhalation:* Causes irritation of the eyes and respiratory tract. Exposure to levels above 2400 mg/m³ may cause headache, dizziness and nose and throat irritation. More severe exposures may cause nausea and vomiting, a feeling of intoxication, weakness, muscle twitches and in extreme cases convulsions, unconsciousness and death. *Skin:* Contact with liquid may cause irritation and drying of skin. This can result in dermatitis. *Eyes:* Contact with liquid or vapor levels of 900 mg/m³ to 2400 mg/m³ may cause irritation and tearing. *Ingestion:* Small amounts may cause headache, dizziness, nausea, vomiting, intoxication, weakness, muscle twitches, convulsions and unconsciousness. May cause aspiration into the lungs and chemical pneumonia. As little as 3 ounces may be fatal. If liquid is breathed into the lungs as little as 1 ounce may cause death due to respiratory failure.
*Long Term Exposure:* Prolonged or repeated contact with liquid may cause defatting of the skin with drying, irritation, and skin ulcers. Exposure to vapor may cause eye, nose and throat irritation, fatigue, headache, anemia, jaundice, and damage to the liver and bone marrow. In animals: kidney damage. Repeated exposure may cause a rare reaction in some people that destroys blood cells (aplastic anemia). This can be fatal. Many petroleum-based solvents have been shown to cause brain and/or nerve damage. Effects may include reduced memory and concentration, personality changes, fatigue, sleep disturbances, reduced coordination, effects on the autonomic nerves and/or nerves to the limbs.
*Points of Attack:* Eyes, skin, respiratory system, central nervous system, liver and kidneys.
*Medical Surveillance:* If symptoms develop or overexposure is suspected, the following may be useful: Complete blood count. Evaluation for brain effects. Liver and kidney function tests.
*First Aid:* If this chemical gets into the eyes, remove any contact lenses at once and irrigate immediately for at least 15 minutes, occasionally lifting upper and lower lids. Seek medical attention immediately. If this chemical contacts the skin, remove contaminated clothing and wash immediately with soap and water. Seek medical attention immediately. If this chemical has been inhaled, remove from exposure, begin rescue breathing (using universal precautions) if breathing has stopped, and CPR if heart action has stopped. Transfer promptly to a medical facility. When this chemical has been swallowed, get medical attention. Do not induce vomiting.

*Note to physician or authorized medical personnel:* Treat symptomatically for central nervous system depression. Supportive treatment for pulmonary edema using oxygen may be needed when aspiration of liquids or massive exposure to vapors has occurred.

*References:*
- National Institute for Occupational Safety and Health, Criteria for a Recommended Standard: Occupational Exposure to Refined Petroleum, NIOSH Doc. No. 77-192, Washington, DC (1977).
- New Jersey Department of Health and Senior Services, "Hazardous Substance Fact Sheet: Stoddard Solvent," Trenton, NJ (August 1998). www.state.nj.us/health/eoh/rtkweb/1736.pdf
- California Environmental Protection Agency "Chemical List of Lists," Sacramento CA (February 1997).
- New York State Department of Health, "Chemical Fact Sheet: Stoddard Solvent," Albany, NY, Bureau of Toxic Substance Assessment (March 1986 and Version 2).

# Streptomycin

*Use Type:* Fungicide and microbiocide
*CAS Number:* 57-92-1
*Formula:* $C_{21}H_{39}N_7O_{12}$
*Synonyms:* NSC-14083; Streptomycin A; D-Streptamine, $O$-2-deoxy-2-(methylamino)-$\alpha$-L-glucopyranosyl-(1->2)-$O$-5-deoxy-3-C-formyl-$\alpha$-L-lyxofuranosyl-(1->4)-$N$, $N'$-bis(aminomethyl)-; Streptomycine; Streptomycinum; Streptomyzin (German)
*Trade Names:* AGRIMYCIN 17®; AGREPT®; AGRI-STREP®; AGRO STREP®, Zeneca Agro (UK) (now Syngenta), canceled; AS-50®; CHEMFORM®; GEROX®; HOKKO-MYCIN®; PLANTOMYCIN®; RIMOSIDEN®; REPAR® STREPTOMYCIN; STREPCEN®
*Producers:* Zeneca Agro (UK) (now Syngenta)
*Chemical Class:* Unclassified
*EPA/OPP PC Code:* 006306
*California DPR Chemical Code:* 1217
*Uses:* Streptomycin is a human antibiotic drug which also is used as a pesticide, to control bacteria, fungi and algae in crops. Streptomycin controls bacterial and fungal diseases of certain fruit, vegetables, seed, and ornamental crops, and controls algae in ornamental ponds and aquaria. The use of streptomycin to control fireblight on apples and pears accounts for 58% of its total use. Other significant uses are on nursery stock and in landscape maintenance (17% of use), and on tobacco (7% of use).
*Carcinogen/Hazard Classifications*
*Label Signal Word:* CAUTION, WARNING, DANGER
*Regulatory Authority:*
- Actively registered pesticide in California.
- AB 1803-Well Monitoring Chemical (CAL)
- FIFRA, 40CFR180.102-1147: tolerances and/or tolerance exemptions for pesticides in or on raw agricultural commodities

*Description:* Off-white powder derived from *Streptomyces griseus* bacteria. Soluble in water. Molecular weight = 581.6
*Incompatibilities:* Unstable in strong acids and alkalis. Incompatible with alkaline materials.
*Permissible Concentration in Water:* No criteria set. Runoff from spills or fire control may cause water pollution.
*First Aid:* If this chemical gets into the eyes, remove any contact lenses at once and irrigate immediately for at least 15 minutes, occasionally lifting upper and lower lids. Seek medical attention immediately. If this chemical contacts the skin, remove contaminated clothing and wash immediately with soap and water. Seek medical attention immediately. If this chemical has been inhaled, remove from exposure, begin rescue breathing (using universal precautions) if breathing has stopped, and CPR if heart action has stopped. Transfer promptly to a medical facility. When this chemical has been swallowed, get medical attention. Give large quantities of water and induce vomiting. Do not make an unconscious person vomit.
*References:*
- EXTOXNET, Extension Toxicology Network, "Pesticide Information Profile, Streptomycin," Oregon State University, Corvallis, OR (September 1995). http://extoxnet.orst.edu/pips/streptom.htm
- U.S. Environmental Protection Agency, "Reregistration Eligibility Decision (RED), Streptomycin and Streptomycin Sulfate," Office of Prevention, Pesticides and Toxic Substances, Washington, DC, (September 1992). http://www.epa.gov/REDs/factsheets/0169fact.pdf
- California Environmental Protection Agency "Chemical List of Lots," Sacramento CA (February 1997).

# Streptomycin Sulfate

*Use Type:* Fungicide
*CAS Number:* 3810-74-0
*Formula:* $C_{42}H_{78}N_{14}O_{24} \cdot H_6O_{12}S_3$
*Synonyms:* Streptomycin sesquisulfate; D-Streptamine, $O$-2-deoxy-2-(methylamino)-$\alpha$-L-glucopyranosyl-(1->2)-$O$-5-deoxy-3-C-formyl-$\alpha$-L-lyxofuranosyl-(1->4)-$N,N'$-bis(aminoiminomethyl)-, sulfate (2:3) (salt)
*Trade Names:* AGRI-MYCIN® Syngenta (Switzerland); AGRISTREP®; AS-50®, Syngenta (Switzerland); PHYTOMYCIN®; PLANTOMYCIN®; STREPCIN®; STREP-GRAN®; STREPSULFAT®; STREPTOMYCIN SULFATE®; STREPTOMYCIN SULPHATE B.P.®; STREPTOREX®; STREPVET®; VETSTREP®
*Producers:* Gustafson (USA); Nufarm (Australia); Syngenta (Switzerland)
*EPA/OPP PC Code:* 006310
*California DPR Chemical Code:* 3834

*Uses:* Streptomycin is a human antibiotic drug which also is used as a pesticide to control bacteria, fungi and algae in crops. It controls bacterial and fungal diseases of certain fruit, vegetables, seed, and ornamental crops, and controls algae in ornamental ponds and aquaria. The use of streptomycin to control fireblight on apples and pears accounts for 58% of its total use. Other significant uses are on nursery stock and in landscape maintenance (17% of use), and on tobacco (7% of use). [U.S. EPA RED Facts]

*Carcinogen/Hazard Classifications*
**California Prop. 65:** Developmental toxin
**Label Signal Word:** CAUTION, WARNING
*Regulatory Authority:*
- Proposition 65 chemical (CAL)
- Actively registered pesticide in California.

*Description:* Off-white powder derived from *Streptomyces griseus* bacteria. Soluble in water. Molecular weight = 1457.38

*Incompatibilities:* May react violently with carbon dust, finely divided aluminum, magnesium, potassium and alkaline materials.

*Permissible Concentration in Water:* No criteria set. Runoff from spills or fire control may cause water pollution.

*First Aid:* If this chemical gets into the eyes, remove any contact lenses at once and irrigate immediately for at least 15 minutes, occasionally lifting upper and lower lids. Seek medical attention immediately. If this chemical contacts the skin, remove contaminated clothing and wash immediately with soap and water. Seek medical attention immediately. If this chemical has been inhaled, remove from exposure, begin rescue breathing (using universal precautions) if breathing has stopped, and CPR if heart action has stopped. Transfer promptly to a medical facility. When this chemical has been swallowed, get medical attention. Give large quantities of water and induce vomiting. Do not make an unconscious person vomit. Do not induce vomiting when formulations containing petroleum solvents are ingested.

*References:*
- U.S. Environmental Protection Agency, "Reregistration Eligibility Decision Facts (RED), Streptomycin and Streptomycin Sulfate," Office of Prevention, Pesticides and Toxic Substances, Washington, DC (September 1992). http://www.epa.gov/REDs/factsheets/0169fact.pdf
- EXTOXNET, Extension Toxicology Network, "Pesticide Information Profile, Streptomycin," Oregon State University, Corvallis, OR (September 1995). http://extoxnet.orst.edu/pips/streptom.htm
- California Environmental Protection Agency "Chemical List of Lists," Sacramento CA (February 1997)

# Strychnine

*Use Type:* Rodenticide and avicide
*CAS Number:* 57-24-9; 60-41-3 (sulfate)
*Formula:* $C_{21}H_{22}N_2O_2$
*Alert:* A Restricted Use Pesticide (RUP)
*Synonyms:* Estricnina (Spanish); Stricnina (Italian); Strychnidin-10-One; Strychnin (German); Strychnos
*Trade Names:* BOOMER-RID®; CERTOX®; DOLCO MOUSE CEREAL®; GOPHER BAIT®; GOPHER-GETTER®, Elston Manufacturing (USA); GOPHER-GO AG BAIT®, Southwest Chemical Co. (USA); HARE-RID®; KWIK-KIL®; MOLE DEATH®; MOUSE-NOTS®; MOUSE-RID®; MOUSE-TOX®; NUX VOMICA®; PIED PIPER MOUSE SEED®; RO-DEX®; SANASEED®
*Producers:* Elston Manufacturing (USA); Kothari Phytochemicals International (India); Southwest Chemical Co. (USA)
*Chemical Class:* Botanical
*EPA/OPP PC Code:* 076901
*California DPR Chemical Code:* 554
*ICSC Number:* 0197
*RTECS Number:* WL2275000; WL255000 (sulfate)
*EEC Number:* 614-003-00-5; 614-004-00-0 (sulfate)
*EINECS Number:* 200-319-7
*Uses:* Strychnine products are allowed for use only below ground where exposure to food and feed crops is not expected, and it is a Restricted Use Pesticide (RUP). It can be used in orchards, feed crop sites, pastures, range land, alfalfa fields, irrigation systems, nonagriculture rights-of-way, forests, and residential sites. Pocket gophers are primary targets.

*Carcinogen/Hazard Classifications*
**Label Signal Word:** DANGER
**WHO Acute Hazard:** Class 1b, highly hazardous
*Regulatory Authority:*
- Banned or Severely Restricted (Several Countries) (UN)[13]
- Air Pollutant Standard Set (ACGIH)[1] (DFG)[3] (HSE)[33] (OSHA)[58] (Several States)[60]
- Permissible Exposure Limits for Chemical Contaminants (CAL/OSHA)
- The "Director's List" (CAL/OSHA)
- Actively registered pesticide in California.
- Clean Water Act: Section 311 Hazardous Substances/RQ 40CFR117.3 (same as CERCLA, see below)
- EPA Hazardous Waste Number (RCRA No.): P108
- RCRA, 40CFR261, Appendix 8 Hazardous Constituents
- Superfund/EPCRA 40CFR355, Appendix B Extremely Hazardous Substances: TPQ = 100/10,000 lb (45.4/4,540 kg)
- Superfund/EPCRA 40CFR302.4 RQ: CERCLA, 10 lb (4.54 kg)
- EPCRA Section 313 Form R *de minimis* concentration reporting level: 1.0%
- U.S. DOT Regulated Marine Pollutant (49CFR172.101, Appendix B)
- Canada, WHMIS, Ingredients Disclosure List

*Description:* White crystalline prisms or white powder. Sprays may be dissolved in xylene or kerosene. Very bitter taste. Practically insoluble in water; solubility = 140 ppm @ 22°C. The sulfate is slightly more soluble. Molecular weight = 334.44; 383.49 (sulfate). Boiling point = 270°C @ 5 mbar (decomposes). Melting/Freezing point = 268–285°C (decomposes). Log $K_{ow}$ = 1.62. Unlikely to bioaccumulate in marine organisms.

*Incompatibilities:* Strong oxidizers. Dangerous when heated.

*Permissible Exposure Limits in Air:* The OSHA[2] PEL, DFG MAK[3], HSE[33] TWA, and the ACGIH[1] TWA value is 0.15 mg/m³. The STEL set by HSE[33] is 0.45 mg/m³. The NIOSH[2] IDLH value is 3.0 mg/m³. Several states have set guidelines or standards for strychnine in ambient air[60] ranging from 1.5 µg/m³ (North Dakota) to 2.5 µg/m³ (Virginia) to 3.0 µg/m³ (Connecticut) to 4.0 µg/m³ (Nevada).

*Determination in Air:* Filter; Reagent; High-pressure liquid chromatography/Ultraviolet detection; NIOSH IV, Method #5016.[18]

*Permissible Concentration in Water:* No criteria set, but runoff from spills or fire control may cause water pollution.

*Routes of Entry:* Inhalation of dust, ingestion, skin and/or eye contact.

### Harmful Effects and Symptoms

*Short Term Exposure:* Affects the central nervous system, causing convulsions, muscle contractions, and respiratory failure. Super toxic; probable oral lethal dose in humans is less than 5 mg/kg, a taste (less than 7 drops) for a 70 kg (150 lb) person. It causes violent generalized convulsions. Death results from respiratory arrest as the respiratory muscles are in sustained spasm. The lowest lethal oral dose reported for humans if 30 mg/kg. Respiratory paralysis and arrest are likely to occur following severe exposure to strychnine. Signs and symptoms of acute exposure generally involve excitation of all portions of the central nervous system. Convulsions, bilateral horizontal nystagmus (rapid, synchronous, horizontal, oscillations of the eyeballs), agitation, restlessness, apprehension, and abrupt, jerking movements of the extremities may occur. Victims may also experience stiffness, painful muscle cramping (especially in the legs), and opisthotonos (spasm in which the spine and extremities are bent with convexity forward, the body resting on the head and heels). Vomiting and renal failure, as well as cyanosis (blue tint to skin and mucous membranes) and rhabdomyolysis (destruction of skeletal muscle), may be found.

*Long Term Exposure:* Chronic allergen if inhaled or ingested.

*Points of Attack:* Central nervous system.

*Medical Surveillance:* Consider the points of attack in preplacement and periodic physical examinations. Examination by a qualified allergist.

*First Aid:* Remove victims from exposure. Emergency personnel should avoid self-exposure to strychnine. *Warning:* Any unnecessary sensory input may induce seizures. Isolate the victims from any avoidable distractions. Rush to a health care facility! Evaluate vital signs including pulse and respiratory rate, and note any trauma. If no pulse is detected, provide CPR. If not breathing, provide artificial respiration. If breathing is labored, administer oxygen or other respiratory support. Remove contaminated clothing as soon as possible. If eye exposure has occurred, remove any contact lenses at once; eyes must be flushed with lukewarm water for at least 15 minutes. Wash exposed skin areas thoroughly with soap and water. Obtain authorization and/or further instructions from the local hospital for administration of an antidote or performance of other invasive procedures.

*References:*

- U.S. Environmental Protection Agency, "Reregistration Eligibility Decision (RED), Strychnine," Office of Prevention, Pesticides and Toxic Substances, Washington, DC (July 1996). http://www.epa.gov/REDs/3133.pdf
- New Jersey Department of Health and Senior Services, "Hazardous Substance Fact sheet, Strychnine," Trenton, NJ (April 2002). http://www.state.nj.us/health/eoh/rtkweb/1747.pdf
- Sax, N.I., Ed., "Dangerous Properties of Industrial Materials Report," 2, No. 2, 63-65 (1982) and 8, No. 1, 78-83 (1988).
- U.S. Environmental Protection Agency, "Chemical Profile: Strychnine," Washington, DC, Chemical Emergency Preparedness Program (November 30, 1987).
- U.S. Environmental Protection Agency, "Chemical Profile: Strychnine Sulfate," Washington, DC, Chemical Emergency Preparedness Program (November 30, 1987).
- California Environmental Protection Agency "Chemical List of Lists," Sacramento CA (February 1997).

## Sulfallate

*Use Type:* Herbicide
*CAS Number:* 95-06-7
*Formula:* $C_8H_{14}ClNS_2$
*Alert:* No products registered with the U.S. EPA. It is a carcinogen, use with extreme caution.
*Synonyms:* CDEC; Chlorallyl diethyldithiocarbamate; 2-Chlorallyl diethyldithiocarbamate; 2-Chlorallyl-*N,N*-diethyldithiocarbamate; 2-Chloroallyl-*N,N*-diethyldithiocarbamate; 2-Chloro-2-propene-1-thiol diethyldithiocarbamate; 2-Chloro-2-propenyldiethylcarbamodithioate; CP 4572; Diethylcarbamodithioic acid 2-Chloro-2-propenyl ester; Diethyldithiocarbamic acid-2-chloroallyl ester; NCI-C00453; Thioallate

*Trade Names:* VEGADEX®, Monsanto (USA), canceled 8/17/1984; VEGADEX SUPER®, Monsanto (USA), canceled 8/17/1984; VEGA-RAND®, Helena Chemical (USA), canceled 6/1/1984
*Chemical Class:* Dithiocarbamate
*EPA/OPP PC Code:* 039001
*California DPR Chemical Code:* 115
*RTECS Number:* EZ5075000
*Uses:* The major use for sulfallate is as a pre-emergent selective herbicide to control certain annual grasses and broadleaf weeds around vegetable and fruit crops. Sulfallate has also been used for weed control among shrubbery and ornamental plants. There are no currently registered products in the U.S.
*Carcinogen/Hazard Classifications*
**U.S. NTP Carcinogen:** Reasonably anticipated carcinogen
**California Prop. 65:** Carcinogen
**IARC:** Group 2B, possible carcinogen
*Regulatory Authority:*
- Carcinogen (Animal Positive) (IARC, NCI)[9]
- AB 2588-Air Toxics "Hot Spots" Chemicals (CAL)
- Proposition 65 chemical (CAL)
- The "Director's List" (CAL/OSHA)

*Description:* Sulfallate is an amber liquid. Molecular weight = 223.79. Boiling point = 128–130°C under 1.0 mm pressure. Flash point = 88°C. Hazard Identification (based on NFPA-704 M Rating System): Health 1, Flammability 1, Reactivity 0. Slightly soluble in water.
*Incompatibilities:* Strong oxidizers.
*Permissible Exposure Limits in Air:* No standards set.
*Permissible Concentration in Water:* No criteria set, but runoff from spills or fire control may cause water pollution.
*Routes of Entry:* Inhalation, ingestion, skin and/or eye contact. May be absorbed by the skin.
*Harmful Effects and Symptoms*
*Short Term Exposure:* Irritates the eyes and skin. High exposure may cause fatigue, sleepiness, headache, dizziness, upset stomach, severe rash and personality changes; muscle weakness and collapse may result. The $LD_{50}$ (oral, rat) = 850 mg/kg (slightly toxic). Unlike carbamates the dithiocarbamates are not cholinesterase inhibitors, but some of them may react with recently ingested alcohol or alcohol-containing products including wine, medications, and cold remedies such as cough-syrups.
*Long Term Exposure:* Repeated or prolonged skin contact may cause rash from irritation. A probable carcinogen in humans. Repeated exposure may cause kidney damage. Sulfallate, a chlorinated dithiocarbamate, administered in the feed, was carcinogenic to Osborne-Mendel rats and to mice, inducing mammary gland tumors in females of both species, tumors of the fore stomach in male rats, and lung tumors in male mice. May be a cholinesterase inhibitor.
*Points of Attack:* Skin, eyes, kidneys, plasma and red blood cell cholinesterase. Cancer site in animals: stomach, lungs.

*Medical Surveillance:* Before employment and at regular times after that, the following are recommended: Plasma and red blood cell cholinesterase levels (tests for the enzyme poisoned by this chemical). If exposure stops, plasma levels return to normal in 1-2 weeks while red blood cell levels may be reduced for 1-3 months. When acetylcholinesterase enzyme levels are reduced by 25% or more below preemployment levels, risk of poisoning is increased, even if results are in lower ranges of "normal." Reassignment to work not involving carbamate pesticides is recommended until enzyme levels recover. If symptoms develop or overexposure occurs, repeat the above tests as soon as possible and get an exam of the nervous system. Also consider complete blood count. Kidney function tests. Consider chest x-ray following acute overexposure
*First Aid: Skin Contact:* Flood all areas of body that have contacted the substance with water. Don't wait to remove contaminated clothing; do it under the water stream. Use soap to help assure removal. Isolate contaminated clothing when removed to prevent contact by others. *Eye Contact:* Remove any contact lenses at once. Flush eyes well with copious quantities of water or normal saline for at least 20-30 minutes. Seek medical attention. *Inhalation:* Leave contaminated area immediately; breathe fresh air. Proper respiratory protection must be supplied to any rescuers. If coughing, difficult breathing or any other symptoms develop, seek medical attention at once, even if symptoms develop many hours after exposure. *Ingestion:* If convulsions are not present, give a glass or two of water or milk to dilute the substance. Assure that the person's airway is unobstructed and contact a hospital or poison center immediately for advice on whether or not to induce vomiting.
*References:*
- New Jersey Department of Health and Senior Services, "Hazardous Substance Fact Sheet, Sulfallate," Trenton NJ (June 1988, rev. October 2001). http://www.state.nj.us/health/eoh/rtkweb/1753.pdf
- California Environmental Protection Agency "Chemical List of Lists," Sacramento CA (February 1997).

# Sulfentrazone (ANSI)

*Use Type:* Herbicide
*CAS Number:* 122836-35-5
*Formula:* $C_{11}H_{10}Cl_2F_2N_4O_3S$
*Synonyms:* 1-(2,4-Dichloro-5-methylsulfonylamidophenyl)-4-difluoromethyl-4,5-dihydro-3-methyl-1$H$-1,2,4-triazol-5-one; 2-(2,4-Dichloro-5-methylsulfonylamidophenyl)-4-difluoromethyl-2,4-dihydro-5-methyl-3$H$-1,2,4-triazol-3-one; 1-[2,4-Dichloro-5-($N^2$-methylsulfonylamino)phenyl]-3-methyl-4-difluoromethyl-$\Delta^2$-1,2,4-triazolin-5-one; Methanesulfonamide, $N$-[2,4-dichloro-5-(4-

(difluoromethyl)-4,5-dihydro-3-methyl-5-*oxo*-1*H*-1,2,4-triazol-1-yl)phenyl]-

*Trade Names:* AUTHORITY®, FMC Agricultural Products Group (USA); CANOPY XL®, DuPont Crop Protection (USA); COVER®, DuPont Crop Protection (USA); F6285®, FMC Agricultural Products (USA); FMC® 97285, FMC Agricultural Products (USA); GAUNTLET®, FMC Agricultural Products (USA); SPARTAN®, FMC Agricultural Products (USA); SULFENTRAZONE® (F6285) 4F, FMC Agricultural Products (USA); SULFENTRAZONE® (F6285) 75DF, FMC Agricultural Products (USA)

*Producers:* DuPont Crop Protection (USA); FMC Agricultural Products (USA); PBI/Gordon (USA)

*Chemical Class:* Triazolone

*EPA/OPP PC Code:* 129081

*Uses:* Used for pre-plant weed control on a variety of products.

*U.S. Maximum Allowable Residue Levels for Sulfentrazone (40 CFR 180.498):*

| CROP | ppm |
|---|---|
| Asparagus | 0.15 |
| Bean, lima, succulent | 0.1 |
| Bean, lima, succulent | 0.15 |
| Cabbage | 0.2 |
| Chickpea, seed | 0.1 |
| Corn, field, forage | 0.2 |
| Corn, field, grain | 0.15 |
| Corn, field, stover | 0.3 |
| Cowpea, succulent | 0.1 |
| Flax, seed | 0.2 |
| Grain, cereal, bran, except sweet corn | 0.15 |
| Grain, cereal, forage, except sweet corn | 0.2 |
| Grain, cereal, grain, except sweet corn | 0.1 |
| Grain, cereal, hay, except sweet corn | 0.2 |
| Grain, cereal, hulls, except sweet corn | 0.3 |
| Grain, cereal, stover, except sweet corn | 0.1 |
| Grain, cereal, straw, except sweet corn | 0.6 |
| Horseradish, roots | 0.2 |
| Pea and bean, dried shelled, except soybean, subgroup 6c | 0.15 |
| Peanut | 0.2 |
| Peanut, meal | 0.4 |
| Peppermint, tops | 0.3 |
| Potato | 0.15 |
| Soybean, seed | 0.05 |
| Spearmint, tops | 0.3 |
| Strawberry | 0.6 |
| Sugarcane, cane | 0.15 |
| Sugarcane, molasses | 0.2 |

*Carcinogen/Hazard Classifications*

**U.S. EPA Carcinogens:** Group E, Unlikely carcinogen

**Label Signal Word:** CAUTION or WARNING

*Description:* Tan solid. Faint sulfur-like odor. Melting point = 120-122°C.

*Permissible Concentration in Water:* No criteria set. Runoff from spills or fire control may cause water pollution.

*Routes of Entry:* Inhalation, passing through the skin, ingestion.

*Harmful Effects and Symptoms*

*Short Term Exposure:* May cause skin and severe eye irritation. Moderately poisonous if ingested or inhaled. Exposure to a triazine (simazine) has caused acute and subacute dermatitis in the former USSR, characterized by erythema, slight edema, moderate pruritus, and burning lasting 4 to 5 days.

*Long Term Exposure:* May cause lung irritation and damage. May cause skin allergy. Contact with some triazine compounds (such as atrazine) may increase risks for tumors known to be associated with hormonal factors. These have been observed in both animals and human beings, and are consistent with the known effects on the hypothalamic pituitary gonadal axis. Repeated exposure may cause weight loss and reduced red blood cell count. May be mutagenic.

*Points of Attack:* Liver, lungs, skin.

*Medical Surveillance:* Before beginning employment and at regular times after that, for those with frequent or potentially high exposures, the following is recommended: Lung function tests. Consider chest x-ray following acute overexposure. Evaluation by a qualified allergist. Examination of the nervous system.

*First Aid:* If this chemical gets into the eyes, remove any contact lenses at once and irrigate immediately for at least 15 minutes, occasionally lifting upper and lower lids. Seek medical attention immediately. If this chemical contacts the skin, remove contaminated clothing and wash immediately with soap and water. Seek medical attention immediately. If this chemical has been inhaled, remove from exposure, begin rescue breathing (using universal precautions) if breathing has stopped, and CPR if heart action has stopped. Transfer promptly to a medical facility. When this chemical has been swallowed, get medical attention. Give large quantities of water or milk and induce vomiting. Do not make an unconscious person vomit.

*References:*

- U.S. Environmental Protection Agency, Office of Pesticide Programs, Pesticide Residue Limits, "Sulfentrazone," 40 CFR 180.498, www.epa.gov/pesticides/food/viewtols.htm
- U.S. Environmental Protection Agency, Office of Pesticide Programs, "Pesticide Fact Sheet, Sulfentrazone." (February 27, 1997). http://www.epa.gov/opprd001/factsheets/sulfentrazone.pdf
- California Environmental Protection Agency "Chemical List of Lists," Sacramento CA (February 1997).
- Pohanish, R.P., "Rapid Guide to Hazardous Chemicals in the Environment," Van Nostrand Reinhold, New York, NY (1997).

## Sulfluramid

*Use Type:* Insecticide, acaricide
*CAS Number:* 4151-50-2
*Formula:* $C_{10}H_6F_{17}NO_2S$
*Synonyms:* N-Ethylperfluorooctanesulfonamide; 1-Octane sulfonamide, N-ethyl-1,1,2,2,3,3,4,4,5,5,6,6,7,7,8,8,8-heptadecafluoro-
*Trade Names:* FINITRON®, Griffin (USA); FIRSTLINE®, FMC Agricultural Products (USA); FLUORGUARD®, FMC Agricultural Products (USA); GX-071®, Griffin (USA); MICRO-GEN ANT REACTOR®, Whitnire Micro-Gen Research Laboratories (USA); VOLCANO®, Griffin (USA)
*Producers:* FMC Agricultural Products (USA); Griffin (USA); Whitnire Micro-Gen Research Laboratories (USA)
*Chemical Class:* Fluorinated sulfonamide
*EPA/OPP PC Code:* 128992
*California DPR Chemical Code:* 2314
*Uses:* Used as ant, roach and termite trap bait.
*Carcinogen/Hazard Classifications*
**Label Signal Word:** CAUTION
**WHO Acute Hazard:** Class III, slightly hazardous
*Description:* White crystalline solid. Odorless. Melting point = 96°C. Insoluble in water.
*Incompatibilities:* Slowly hydrolyzes in water, releasing ammonia and forming acetate salts.
*Permissible Concentration in Water:* No criteria set. Runoff from spills or fire control may cause water pollution.
*Routes of Entry:* Inhalation and skin contact.
*Harmful Effects and Symptoms*
*Short Term Exposure:* Will cause skin burns and respiratory insufficiency. Eye contamination.
*Points of Attack:* Skin, eyes, lungs.
*First Aid:* If this chemical gets into the eyes, remove any contact lenses at once and irrigate immediately for at least 15 minutes, occasionally lifting upper and lower lids. Seek medical attention immediately. If this chemical contacts the skin, remove contaminated clothing and wash immediately with soap and water. Seek medical attention immediately. If this chemical has been inhaled, remove from exposure, begin rescue breathing (using universal precautions) if breathing has stopped, and CPR if heart action has stopped. Transfer promptly to a medical facility. When this chemical has been swallowed, get medical attention. Give large quantities of water and induce vomiting. Do not make an unconscious person vomit.
*References:*
- Pesticide Management Education Program, "Sulfluramid (GX-071) EPA Pesticide Fact Sheet 3/89," Cornell University, Ithaca, NY (March 1989). http://pmep.cce.cornell.edu/profiles/insect-mite/propetamphos-zetacyperm/sulfluramid/insect-prof-sulfluramid.html
- Fluoride Action Network, Hazardous Substances Data Bank, "Sulfluramid," October 22, 2003. http://www.fluoridealert.org/pesticides/sulfluramid.hsdb.oct.2003.htm
- California Environmental Protection Agency "Chemical List of Lists," Sacramento CA (February 1997)

## Sulfometuron-methyl

*Use Type:* Herbicide
*CAS Number:* 74222-97-2
*Formula:* $C_{15}H_{16}N_4O_5S$
*Synonyms:* Benzoic acid, 2-[((((4,6-dimethyl-2-pyrimidinyl)amino)carbonyl)amino)sulfonyl]-, methyl ester; Benzoic acid, o-[(3-(4,6-dimethyl-2-pyrimidinyl)ureido)sulfonyl]-, methyl ester; Methyl 2-[((((4,6-dimethyl-2-pyrimidinyl)amino)carbonyl)amino)sulfonyl]benzoate; Sulfometuron methyl; Methyl 2-[((((4,6-dimethyl-2-pyrimidinyl)amino)carbonyl)amino)sulfonyl]benzoate
*Trade Names:* DPX-T5648®, DuPont Crop Protection (USA); KNOCKOUT®; LANDMARK® MP, (sulfometuron-methyl + chlorsulfuron), DuPont Crop Protection (USA); OUST®, DuPont Crop Protection (USA); OUSTAR®, DuPont Crop Protection (USA); RIVERDALE®, Nufarm (Australia); STAMPRO®, Dow AgroSciences (USA)
*Producers:* Dow AgroSciences (USA); DuPont Crop Protection (USA); Micro-flo (USA); Nufarm (Australia)
*Chemical Class:* Sulfonylurea
*EPA/OPP PC Code:* 122001; (122002 old EPA code number)
*California DPR Chemical Code:* 2149
*Uses:* Used to control annual and perennial grasses and broadleaf weeds on landscapes, rights-of-ways, fence rows, forests, industrial structure areas and non-crop lands.
*Carcinogen/Hazard Classifications*
**Label Signal Word:** CAUTION, DANGER
**WHO Acute Hazard:** Class U, unlikely to be hazardous (as sulfometuron, parent compound)
*Regulatory Authority:*
- AB 1803-Well Monitoring Chemical (CAL)
- Actively registered pesticide in California.

*Description:* White to colorless solid. Odorless. Slightly soluble in water. Melting/Freezing point = 203-205°C. Molecular weight = 364.4.
*Determination in Air:* Filter; none; Gravimetric; NIOSH IV [Particulates NOR; #0500 (total), #0600 (respirable)].[18]
*Permissible Concentration in Water:* No criteria set. Runoff from spills or fire control may cause water pollution.
*Harmful Effects and Symptoms*
*Short Term Exposure:* Contact with eyes or skin may cause irritation or burns. Inhalation should be avoided; use NIOSH-approved air purifying respirators for pesticides.

May be harmful if swallowed. Skin contact may cause allergic reaction.

*Long Term Exposure:* Repeated or prolonged contact may cause skin and lung sensitization, resulting in allergies.

*Points of Attack:* Skin.

*Medical Surveillance:* Evaluation by a qualified allergist, including careful exposure history and special testing, may help diagnose skin allergy.

*First Aid:* If this chemical gets into the eyes, remove any contact lenses at once and irrigate immediately for at least 15 minutes, occasionally lifting upper and lower lids. Seek medical attention immediately. If this chemical contacts the skin, remove contaminated clothing and wash immediately with soap and water. Seek medical attention immediately. If this chemical has been inhaled, remove from exposure, begin rescue breathing (using universal precautions) if breathing has stopped, and CPR if heart action has stopped. Transfer promptly to a medical facility. When this chemical has been swallowed, get medical attention. Give large quantities of water and induce vomiting. Do not make an unconscious person vomit.

*References:*
- EXTOXNET, Extension Toxicology Network, "Pesticide Information Profile, Sulfometuron-methyl," Oregon State University, Corvallis, OR (June 1996). http://extoxnet.orst.edu/pips/sulfomet.htm
- California Environmental Protection Agency "Chemical List of Lists," Sacramento CA (February 1997)

# Sulfosate

*Use Type:* Herbicide
*CAS Number:* 81591-81-3
*Formula:* $C_6H_{16}NO_5PS$
*Synonyms:* Glycine, *N*-(phosphonomethyl)-, ion(1-), trimethylsulfonium; Glyphosate-trimesium; Sulfonium, trimethyl-, salt with *N*-(phosphonomethyl)glycine (1:1); Trimethylsulfonium carboxymethylaminomethyl phosphonate; Trimethylsulfonium *N*-phosphono methylglycine; Trimethylsulfonium salt of *N*-(phosphonomethyl)glycine
*Trade Names:* SC-0224®; TOUCHDOWN®, Syngenta (Switzerland)
*Producers:* Syngenta (Switzerland)
*Chemical Class:* Glyphosate; Phosphonoglycine
*EPA/OPP PC Code:* 128501
*California DPR Chemical Code:* 2327
*Uses:* Used as a non-selective herbicide for the control of grasses and broadleaf weeds.
*U.S. Maximum Allowable Residue Levels for Sulfosate (40 CFR 180.489):*

| CROP | ppm |
|---|---|
| Almond, hulls | 1 |
| Banana | 0.05 |
| Cattle, fat | 0.5 |
| Cattle, kidney | 6 |
| Cattle, meat | 1 |
| Cattle, mbyp, except kidney | 1.5 |
| Corn, field, forage | 0.1 |
| Corn, field, grain | 0.2 |
| Corn, field, stover | 0.3 |
| Corn, pop, grain | 0.2 |
| Corn, pop, stover | 0.3 |
| Corn, sweet, forage | 20 |
| Corn, sweet, kernel plus cob with husks removed | 0.15 |
| Corn, sweet, stover | 170 |
| Cotton, gin byproducts | 120 |
| Cotton, undelinted seed | 40 |
| Egg | 0.05 |
| Fruit, pome | 0.05 |
| Fruit, stone, except cherry | 0.05 |
| Fruit, stone, group 12 | 0.05 |
| Goat, fat | 0.5 |
| Goat, kidney | 6 |
| Goat, meat | 1 |
| Goat, mbyp, except kidney | 1.5 |
| Grain, aspirated fractions | 1300 |
| Grape | 0.1 |
| Grape, raisin | 0.2 |
| Hog, fat | 0.5 |
| Hog, kidney | 6 |
| Hog, meat | 1 |
| Hog, mbyp, except kidney | 1.5 |
| Horse, fat | 0.5 |
| Horse, kidney | 6 |
| Horse, meat | 1 |
| Horse, mbyp, except kidney | 1.5 |
| Milk | 1.5 |
| Nut, tree, group 14 | 0.05 |
| Pea and bean, dried shelled, except soybean, subgroup 6c | 6 |
| Pea and bean, succulent shelled, subgroup 6b | 0.2 |
| Pistachio | 0.05 |
| Plum, prune | 0.2 |
| Poultry, fat | 0.05 |
| Poultry, meat | 0.05 |
| Poultry, mbyp | 0.5 |
| Radish, roots | 16 |
| Radish, tops | 10 |
| Sheep, fat | 0.5 |
| Sheep, kidney | 6 |
| Sheep, meat | 1 |
| Sheep, mbyp, except kidney | 1.5 |
| Sorghum, grain, forage | 0.2 |
| Sorghum, grain, grain | 35 |

| | |
|---|---|
| Sorghum, grain, stover | 140 |
| Soybean, forage | 2 |
| Soybean, hay | 5 |
| Soybean, hulls | 45 |
| Soybean, seed | 21 |
| Vegetable, fruiting, group 8 | 0.05 |
| Vegetable, legume, edible podded, subgroup 6a | 0.5 |
| Vegetable, root and tuber, Group 1, except radish | 0.3 |
| Vegetable, root, except radish | 0.15 |
| Vegetable, tuberous and corm, subgroup 1c | 1 |
| Wheat, bran | 30 |
| Wheat, forage | 35 |
| Wheat, grain | 10 |
| Wheat, hay | 1 |
| Wheat, milled byproducts | 1.5 |
| Wheat, shorts | 20 |
| Wheat, straw | 90 |

*Carcinogen/Hazard Classifications*
**U.S. EPA Carcinogens:** Group E, Unlikely a carcinogen
**Label Signal Word:** CAUTION
**WHO Acute Hazard:** Class U, unlikely to be hazardous (as glyphosate, the parent compound)
*Incompatibilities:* Solutions are corrosive to iron, unlined steel, and galvanized steel forming a highly combustible or explosive gas mixture. Do not store glyphosate compounds in containers made from these materials.
*Permissible Exposure Limits in Air:* No OELs have been established in the US for this chemical. For glyphosate (the parent compound), the former USSR[35, 43] has set a ceiling value in workplace air of 1.5 mg/m$^3$.
*Permissible Concentration in Water:* The U.S. EPA has developed data on glyphosate including a no-observed-adverse effects level (NOAEL) of 10 mg/kg/day. This corresponds to a drinking water equivalent level of 3.5 mg/L from which a lifetime health advisory of 0.7 mg/L was derived. California[61] has set a guideline of 0.5 mg/L for drinking water.
*Determination in Water:* Analysis of glyphosate is by a high-performance liquid chromatographic (HPLC) method.
*Routes of Entry:* Inhalation, ingestion, skin and/or eye contact. Absorbed through the skin.
*Harmful Effects and Symptoms*
*Short Term Exposure:* Irritates the eyes and skin. Nausea is often the first symptom, followed by vomiting, abdominal cramps, diarrhea, excessive salivation, headache, giddiness, dizziness, weakness, tightness in the chest, loss of muscle coordination, slurring of speech, muscle twitching (particularly the tongue and eyelid), respiratory difficulty, blurring or dimness of vision, pinpoint pupils, profound weakness, mental confusion, disorientation and drowsiness.
*Long Term Exposure:* May cause liver and kidney damage.

It does not seem to exhibit reproductive effects, mutagenicity, or carcinogenicity in animal studies.
*Points of Attack:* Respiratory system, central nervous system, cardiovascular system, eyes, skin and blood cholinesterase.
*Medical Surveillance:* If symptoms develop or overexposure is suspected, the following may be useful: Liver and kidney function tests. Exam of the nervous system.
*First Aid:* Speed in removing material from eyes and skin is of extreme importance. If this chemical gets into the eyes, remove any contact lenses at once and irrigate immediately for at least 15 minutes, occasionally lifting upper and lower lids. Seek medical attention immediately. If this chemical contacts the skin, remove contaminated clothing and wash immediately with soap and water. Shampoo hair promptly if contaminated. Seek medical attention immediately. If this chemical has been inhaled, remove from exposure, begin rescue breathing (using universal precautions) if breathing has stopped, and CPR if heart action has stopped. Transfer promptly to a medical facility. When this chemical has been swallowed, get medical attention. Give large quantities of water and induce vomiting. Do not make an unconscious person vomit.
*References:*
- U.S. Environmental Protection Agency, Office of Pesticide Programs, Pesticide Residue Limits, "Sulfosate," 40 CFR 180.498, www.epa.gov/pesticides/food/viewtols.htm
- New Jersey Department of Health and Senior Services, "Hazardous Substance Fact Sheet, Glyphosate," Trenton NJ (June 1999).
- U.S. Environmental Protection Agency, "Health Advisory: Glyphosate," Washington, DC, Office of Drinking Water (August 1987).
- California Environmental Protection Agency "Chemical List of Lists," Sacramento CA (February 1997)

# Sulfotepp

*Use Type:* Insecticide
*CAS Number:* 3689-24-5
*Formula:* $C_8H_{20}O_5P_2S_2$
*Alert:* A Restricted Use Pesticide (RUP). Not registered in the U.S.
*Synonyms:* Bis-$O,O$-Diethylphosphorothionic anhydride; Dithio; Dithiodiphosphoric acid, tetraethyl ester; Dithiotep; Dithiofos; Dithion; Dithione; Dithiophos; Di(thiophosphoric) acid, tetraethyl ester; Dithiopyrophosphate de tetraethyle (French); ENT 16,273; Ethyl thiopyrophosphate; Pirofos; Pyrophosphorodithioic acid, tetraethyl ester; Pyrophosphorodithioic acid,$O,O,O,O$-tetraethyl ester; Sulfatep; TEDP; TEDTP; $O,O,O,O$-Tetraethyl-dithio-difosfaat (Dutch);

# Sulfotepp

Tetraethyldithiopyrophosphate; *O,O,O,O-*Tetraethyldithiopyrophosphate; Tetraethyl dithio pyrophosphate; *O,O,O,O-*Tetraetil-di tio-pirofosfato (Italian); Thiotepp

***Trade Names:*** ASP® 47; BAY E-393®, Bayer CropScience (Germany), canceled; BAYER-E-393®, Bayer CropScience (Germany), canceled; BLADAFUM®; BLADAFUME®; BLADAFUN®; E393®; LETHALAIRE G-57®; PLANT DITHIO AEROSOL®; PLANTFUME 103 SMOKE GENERATOR®

***Producers:*** Bayer CropScience (Germany); Rallis India (India); Shenzhen Guomeng Industry Co., Ltd. (China)

***Chemical Class:*** Organophosphate

***EPA/OPP PC Code:*** 079501

***California DPR Chemical Code:*** 558

***ICSC Number:*** 0895

***RTECS Number:*** XN4375000

***EEC Number:*** 015-027-00-3

***EINECS Number:*** 222-995-2

***Uses:*** Sulfotepp is a restricted use insecticide/acaricide used on greenhouse ornamentals, including carnations, chrysanthemums, geraniums, gladiolus, poinsettias, snapdragons, azaleas, and roses. Sulfotepp is used to control aphids, spider mites, thrips, and whiteflies prior to shipment of plants, which is important for ensuring that plants are pest-free when shipped as mandated by intrastate, interstate, and international requirements. There are no sulfotepp products currently registered in th U.S.

***Carcinogen/Hazard Classifications***

**Label Signal Word:** DANGER

**WHO Acute Hazard:** Class 1 a, extremely hazardous

***Regulatory Authority:***

- Banned or Severely Restricted (in former USSR) (UN)[13]
- Very Toxic Substance (World Bank)[15]
- Air Pollutant Standard Set (ACGIH)[1] (DFG)[3] (HSE)[33] (OSHA)[58] (Several States)[6]
- U.S. DOT Inhalation Hazard Chemicals
- Permissible Exposure Limits for Chemical Contaminants (CAL/OSHA)
- The "Director's List" (CAL/OSHA)
- Actively registered pesticide in California.
- EPA Hazardous Waste Number (RCRA No.): P109
- RCRA, 40CFR261, Appendix 8 Hazardous Constituents
- RCRA 40CFR264, Appendix 9; TSD Facilities Ground Water Monitoring List. Suggested test method(s) (PQL ug/L): 8270(10)
- Superfund/EPCRA 40CFR355, Appendix B Extremely Hazardous Substances: TPQ = 500 lb (227kg)
- Superfund/EPCRA 40CFR302.4 RQ: CERCLA, 100 lb (45.4 kg)
- U.S. DOT Inhalation Hazard Chemicals as organophosphates
- U.S. DOT Regulated Marine Pollutant (49CFR172.101, Appendix B)
- Canada, WHMIS, Ingredients Disclosure List

***Description:*** Sulfotepp is a pale yellow, mobile liquid. Garlic-like odor. A pesticide that may be absorbed on a solid carrier or mixed in a more flammable liquid. Practically insoluble in water; solubility = 50 ppm @ 20°C; 1:1500 @ 22. Molecular weight = 322. 35. Boiling point = 131–135°C @ 2 mmHg. Vapor pressure = $1.7 \times 10^{-4}$ mbar @ 20°C. Hazard Identification (based on NFPA-704 M Rating System): Health 4, Flammability 1, Reactivity 1. Log $K_{ow}$ = 4.1. Values above 3.0 are likely to bioaccumulate in marine organisms.

***Incompatibilities:*** Strong oxidizers. Hydrolyzes very slowly in aqueous solution. Attacks some forms of plastic, rubber and, coating. Corrosive to iron.

***Permissible Exposure Limits in Air:*** The OSHA/NIOSH TWA[2] and the recommended ACGIH[1] TWA is 0.2 mg/m$^3$. The HSE[33] has set a TWA of 0.2 mg/m$^3$ and an STEL of 0.6 mg/m$^3$. The notation "skin" is added to indicate the possibility of cutaneous absorption. The NIOSH[2] IDLH level = 10 mg/m$^3$. The DFG[3] has set an MAK of 0.015 ppm (0.2 mg/m$^3$). Several states have set guidelines or standards for sulfotep in ambient air ranging from 3.5 $\mu$g/m$^3$ (Virginia) to 4.0 $\mu$g/m$^3$ (Connecticut) to 5.0 $\mu$g/m$^3$ (Nevada) to 20.0 $\mu$g/m$^3$ (North Dakota).

***Determination in Air:*** OSHA versatile sampler-2; Toluene/Acetone; Gas chromatography/Flame photometric detection for sulfur, nitrogen, or phosphorus; NIOSH Method IV Method #5600, Organophosphorus pesticides.[18]

***Permissible Concentration in Water:*** No criteria set, but runoff from spills or fire control may cause water pollution.

***Routes of Entry:*** Inhalation, skin absorption, ingestion, skin and/or eye contact.

***Harmful Effects and Symptoms***

***Short Term Exposure:*** Irritates the eyes and skin. Can cause rapid, fatal, organophosphate poisoning. Contact may cause eye pain, blurred vision. May affect the nervous system. Symptoms of exposure include, lacrimation (discharge of tears), rhinorrhea (discharge of thin nasal mucous), headache, cyanosis, anorexia, nausea, vomiting, diarrhea, localized sweating, weakness, twitching, paralysis, Cheyne-Stokes respiration, convulsions, low blood pressure, cardiac irregular/irregularities, respiratory failure and death. Super toxic, probable oral lethal dose in humans is less than 5 mg/kg, or a taste (less than 7 drops) for a 70 kg (150 lb) person. It is a cholinesterase inhibitor. Material is similar to parathion in symptomatology, including nausea followed by vomiting, abdominal cramps, diarrhea, excessive salivation, headache, giddiness, dizziness, weakness, tightness in chest, blurring of vision, tearing, slurring of speech, confusion, difficulty breathing, convulsions, coma and even death. Delayed pulmonary edema may occur after inhalation.

***Long Term Exposure:*** A cholinesterase inhibitor; cumulative effect is possible. This chemical may damage the nervous system with repeated exposure, resulting in convulsions, respiratory failure. Repeated exposure may cause personality changes of depression, anxiety, or irritability.

***Points of Attack:*** Eyes, skin, respiratory system, central nervous system, cardiovascular system, blood cholinesterase.

***Medical Surveillance:*** Medical observation is recommended for 24 to 48 hours after breathing overexposure, as pulmonary edema may be delayed. Before employment and at regular times after that, the following are recommended: Plasma and red blood cell cholinesterase levels (tests for the enzyme poisoned by this chemical). If exposure stops, plasma levels return to normal in 1-2 weeks while red blood cell levels may be reduced for 1-3 months. When acetylcholinesterase enzyme levels are reduced by 25% or more below preemployment levels, risk of poisoning is increased, even if results are in lower ranges of "normal." Reassignment to work not involving organophosphate or carbamate pesticides is recommended until enzyme levels recover. If symptoms develop or overexposure occurs, repeat the above test as soon as possible and get an exam of the nervous system.

***First Aid:*** **Treatment for organophosphate poisoning consists of thorough decontamination, cardiorespiratory support, and administration of the antidotes atropine and pralidoxime. In cases of severe poisoning, diazepam, an anticonvulsant, should also be administered. Antidotes should be administered as prevention even if the diagnosis is in doubt.** Speed in removing material from eyes and skin is of extreme importance. *Eyes:* Eye contact can cause dangerous amounts of these chemicals to be quickly absorbed through the mucous membrane into the bloodstream. Immediately and gently flush eyes with plenty of warm or cold water (NO hot water) for at least 15 minutes, occasionally lifting the upper and lower eyelids. Get medical aid immediately. *Skin:* Get medical aid. Skin contact can cause dangerous amounts of these chemicals to be absorbed into the bloodstream. Wearing the appropriate PPE equipment and respirator for organophosphate pesticides, immediately flush skin with plenty of soap and water for at least 15 minutes while removing contaminated clothing and shoes. Shampoo hair promptly if contaminated. The removed, contaminated clothing and shoes should be double-bagged and left in Hot Zone for later disposal by hazardous materials experts. Skin may also be decontaminated with diluted hypochlorite solution. *Inhalation:* Get medical aid. Do not contaminate yourself. Wearing the appropriate PPE equipment and respirator for organophosphate pesticides, immediately remove the victim from the contaminated area to fresh air. If the victim is not breathing, administer artificial respiration. Do NOT use mouth-to-mouth resuscitation; use bag/mask apparatus. If breathing is difficult, administer oxygen through bag/mask apparatus until medical help arrives. Do not leave victim unattended. *Ingestion:* Call poison control. Loosen all clothing. Never give anything by mouth to an unconscious person. If victim is *unconscious or having convulsions,* do nothing except keep victim warm. Get medical aid. Transfer promptly to a medical facility. In cases of ingestion, **do not induce vomiting**. If the victim is alert and asymptomatic, administer a slurry of activated charcoal at a dose of 1 g/kg (infant, child, and adult dose). A soda can and straw may be of assistance when offering charcoal to a child. *In some cases you may be specifically instructed by poison control to induce vomiting by way of 2 tablespoons of syrup of ipecac (adult) washed down with a cup of water.* Do NOT give activated charcoal before or with ipecac syrup.

*Note to physician or authorized medical personnel:* Treat cases of respiratory compromise, coma, or excessive pulmonary secretions with respiratory support using protocols and techniques available and within the scope of training. Some cases may necessitate procedures such as endotracheal intubation or cricothyrotomy by properly trained and equipped personnel. When possible, atropine (see *Antidotes*, below) should be given under medical supervision. Patients who are comatose, hypotensive, or having seizures or cardiac arrhythmias should be treated according to advanced life support protocols. *Antidotes:* Two antidotes are administered to treat organophosphate poisoning. Atropine is a competitive antagonist of acetylcholine at muscarinic receptors and is used to control the excessive bronchial secretions which are often responsible for death. Pralidoxime relieves both the nicotinic and muscarine effects of organophosphate poisoning by regenerating acetylcholinesterase and can reduce both the bronchial secretions and the muscle weakness associated with poisoning. The initial intravenous dose of atropine in adults should be determined by the severity of symptoms: An initial adult dose of 1.0 to 2.0 mg or pediatric dose of 0.01 mg/kg (minimum 0.01 mg) should be administered intravenously. If intravenous access cannot be established, atropine may also be given intramuscularly, subcutaneously or via endotracheal tube. Doses should be repeated every 15 minutes until excessive secretions and sweating have been controlled. Once bronchial secretion has been controlled, atropine administration should be repeated whenever the secretions begin to recur. In seriously poisoned patients, very large doses may be required. Alterations of pulse rate and pupillary size should not be used as indicators of treatment adequacy. Pralidoxime should be administered as early in poisoning as possible as its efficacy may diminish when given more than 24 to 36 hours after exposure. Doses are as follows: adult 1.0 g; pediatric 25 to 50 mg/kg. The drug should be administered intravenously over 30 to 60 minutes, but in a life-threatening situation, one-half of the total dose can be given per minute for a total administration time of 2 minutes. Treatment should begin to take effect within 40 minutes with a reduction in symptoms and the amount of atropine necessary to control bronchial secretion. The initial dose can be repeated in 1 hour and then every 8 to 12 hours until the patient is clinically well and no longer requires atropine. If intravenous access cannot be established, pralidoxime may also be given intramuscularly. Early administration of diazepam in addition to the combined atropine and pralidoxime treatment may help prevent the onset of seizures and potential brain and cardiac morphologic damage following high-level organophosphate poisoning.

*References:*
- Environmental Protection Agency, "Reregistration Eligibility Decision (RED), Sulfotepp," Office of Prevention, Pesticides and Toxic Substances, Washington, DC (September 1999). http://www.epa.gov/REDs/0338red.pdf
- Pesticide Management Education Program, "EPA Pesticide Fact Sheet 9/88, Sulfotepp (Bladafum, Plantfum)" Cornell University, Ithaca, NY (September 1988). http://pmep.cce.cornell.edu/profiles/insect-mite/propetamphos-zetacyperm/sulfotepp/insect-prof-sulfotepp.html
- New Jersey Department of Health and Senior Services, "Hazardous Substance Fact Sheet: Sulfotepp," Trenton, NJ (September 7, 1987, rev. December 2000). http://www.state.nj.us/health/eoh/rtkweb/1756.pdf
- U.S. Environmental Protection Agency, "Chemical Profile: Sulfotepp," Washington, DC, Chemical Emergency Preparedness Program (November 30, 1987).
- Agency for Toxic Substances and Disease Registry, U.S. Department of Health and Human Services, Public Health Service, "Managing Hazardous Materials Incidents," Atlanta, GA (June 2003).
- California Environmental Protection Agency "Chemical List of Lists," Sacramento CA (February 1997).

# Sulfoxide

*Use Type:* Insecticide and synergist
*CAS Number:* 120-62-7
*Formula:* $C_{18}H_{28}O_3S$
*Synonyms:* ENT 16,634; Isosafrole-*n*-octylsulfoxide; Isosafrole, octyl sulfoxide; 1,2-(Methylenedioxy)-4-[2-(octylsulfinyl)propyl]benzene; 1,2-(methylenedioxy)-4-[2-(octylsulfinyl)propyl]benzene; 1-Methyl-2-(3,4-methylenedioxyphenyl)ethyl octyl sulfoxide; NCI-C02824; *n*-Octylisosafrole sulfoxide; *n*-Octyl sulfoxide isosafrole; Piperonyl sulfoxide; Sulfoxyl; Sulphoxide
*Trade Names:* SULFOX-CIDE®, Prentiss (USA), canceled
*Producers:* Bayer CropScience (Germany); Prentiss (USA)
*EPA/OPP PC Code:* 057101
*California DPR Chemical Code:* 559
*Uses:* Not registered in the U.S. Primary uses are in household and pet sprays.
*Carcinogen/Hazard Classifications*
*Label Signal Word:* CAUTION or WARNING
*Regulatory Authority:*
- Actively registered pesticide in California.

*Description:* Light yellow to brownish, oily liquid. Practically insoluble in water. Molecular weight = 324.49.
*Incompatibilities:* Violent reaction with perchloric acid.
*Permissible Concentration in Water:* No criteria set. Runoff from spills or fire control may cause water pollution.
*Routes of Entry:* Absorbed through the unbroken skin. Ingestion.

*Harmful Effects and Symptoms*
*Short Term Exposure:* May cause irritation to skin, eyes, and respiratory tract. Slightly to moderate toxic effects by skin contact, or ingestion.
*Long Term Exposure:* May be carcinogenic.
*First Aid:* If this chemical gets into the eyes, remove any contact lenses at once and irrigate immediately for at least 15 minutes, occasionally lifting upper and lower lids. Seek medical attention immediately. If this chemical contacts the skin, remove contaminated clothing and wash immediately with soap and water. Seek medical attention immediately. If this chemical has been inhaled, remove from exposure, begin rescue breathing (using universal precautions) if breathing has stopped, and CPR if heart action has stopped. Transfer promptly to a medical facility. When this chemical has been swallowed, get medical attention. Give large quantities of water and induce vomiting. Do not make an unconscious person vomit. Do not induce vomiting when formulations containing petroleum solvents are ingested.

*References:*
- California Environmental Protection Agency "Chemical List of Lists," Sacramento CA (February 1997)

# Sulfur

*Use Type:* Insecticide, fungicide and fertilizer
*CAS Number:* 7704-34-9
*Formula:* S
*Alert:* A General Use Pesticide (GUP)
*Synonyms:* Azufre (Spanish); Brimstone; Bensulfoid; Colloidal sulfur; Colloidal-S; Crystex; Dusting sulphur; Elemental sulfur; Elosal; Enxofre; Flour sulphur; Flowers of sulphur; Ground vocle sulphur; Schwefel (German); Sofril; Soufre (French); Soufre micronise (French); Soufre sublime (French); Spersul thiovit; Thiovit sulphur; Zwavel
*Trade Names:* (There are 131 sulfur products registered in the U.S. and 1,339 active and/or canceled or transferred products). ACME WETTABLE DUSTING SULFUR®, Pbi/Gordon (USA); COLLOKIT®; COLSUL®; COROSAL D AND S®; COSAN®; COSAN® 80; CRISAZUFRE®; HEXASUL®; KOLOFOG®; KOLOSPRAY®; KUMULUS® 5, BASF (Germany); LACCO MAGIC SULPHUR®; MAGNETIC 70®; MICROFLOTOX®; MICROTHIOL® DISPERSS®, Cerexagri (France); POPCORN®, Agrium (Canada); SPERLOX-S®; SUL-CIDE®; SULFIDAL®; SULFLOX®; SULFORON®; SULKOL®; SULSOL®;

SVOVEL®; SVOVL®; TECHNETIUM TC 99M®; TESULOID®; THIOLUX®; THION® 80; THION® 95; THIORIT®; TIOLENE®; ZOLVIS®

*Producers:* Agrium (Canada); Akzo Nobel (Netherlands); ASARCO (USA); BASF (Germany); Bonide Products (USA); Boliden (Sweden); Cerexagri (France); Chemical Products (USA); Cia. Quimica Universal de Industrias (Mexico); Coogee Chemicals (Australia); DuPont (USA); Esseco (Italy); Exxon Mobil Chemical (USA); Georgia Gulf Sulfur (USA); Goldschmidt (Germany); Gujarat Pesticides (India); Incitec Pivot (Australia); International Sulfur (USA); Juhua Group Corp. (China); Montana Sulphur (USA); OxyChem (USA); Pbi/Gordon (USA); Rhodia (France); Sasol Chemical (South Africa); Sinopec Corporation (China); Solvay Barium Strontium (Germany); Synthetic Products (USA); Teck Cominco (Canada); Tessenderlo (Belgium); United Agri Products (UAP)

*Chemical Class:* Inorganic
*EPA/OPP PC Code:* 077501
*California DPR Chemical Code:* 560
*ICSC Number:* 1166
*RTECS Number:* WS4250000
*EINECS Number:* 231-722-6

*Uses:* Widely used in the manufacture of drugs, fungicides, gunpowder, wood pulp, rubber, and various chemical products including sulfuric acid, and carbon bisulfide. Sulfur is used to control powdery mildew, rust, brown rot, cotton root rot, black spot, leaf rot, scab, mites; thirps and flea hopper. As a fertilizer it increases crude protein of forages, improves drought resistance and winter hardiness and general plant hardiness, color and uniformity.

*U.S. Maximum Allowable Residue Levels for Sulfur (40 CFR 180.1001 and 180.930): Note:* There are 229 crops with residue levels for sulfur and compounds of sulfur. Sulfur is exempt from residue tolerances.

*Carcinogen/Hazard Classifications*
Label Signal Word: CAUTION
WHO Acute Hazard: Class U, unlikely to be hazardous
*Regulatory Authority:*
- Air Pollutant Standard Set (former USSR)[35]
- Actively registered pesticide in California.
- Canada, WHMIS, Ingredients Disclosure List

*Description:* Sulfur is a yellow crystalline solid or powder. Often transported in the molten state. Insoluble in water. Boiling point = 445°C. Melting/Freezing point = 113°-120°C. Flash point = 207°C. Autoignition temperature = 232°C. Hazard Identification (based on NFPA-704 M Rating System): Health 2, Flammability 1, Reactivity 0.

*Incompatibilities:* Combustible solid. Liquid forms sulfur dioxide with air. Violent reaction with strong oxidizers, halogen compounds, phosphorus, sodium, tin, uranium, metal carbides and other compounds. Forms explosive, shock-sensitive or pyrophoric mixtures with ammonia, ammonium nitrate, bromates, calcium carbide, charcoal, chlorates, hydrocarbons, iodates, iron. Attacks steel when moist.

*Permissible Exposure Limits in Air:* The former USSR[35] has set an MAC for workplace air of 6.0 mg/m³ of elemental sulfur.

*Determination in Air:* Filter; Acid; Inductively coupled plasma; NIOSH IV, Method #7300, Elements.[18]

*Permissible Concentration in Water:* Mexico[35] has set an MAC of 0.5 mg/L of sulfides in estuaries.

*Routes of Entry:* Inhalation, ingestion, skin and/or eye contact.

*Harmful Effects and Symptoms*

*Short Term Exposure:* Sulfur can affect you when breathed in. Irritates the eyes, skin and respiratory tract. Exposure can cause inflammation of the nose and irritate the lungs.

*Long Term Exposure:* Repeated exposures may cause chronic bronchitis to develop with cough, phlegm, and/or shortness of breath. Contact can irritate the skin and may cause a skin allergy. Repeated exposure to sulfur dust may cause permanent eye damage (clouding of the eye lens and chronic irritation).

*Points of Attack:* Skin and respiratory tract.

*Medical Surveillance:* Before beginning employment and at regular times after that, the following are recommended: Lung function tests. Eye examination. If symptoms develop or overexposure is suspected, the following may be useful: Evaluation by a qualified allergist, including careful exposure history and special testing, may help diagnose skin allergy.

*First Aid:* If this chemical gets into the eyes, remove any contact lenses at once and irrigate immediately for at least 15 minutes, occasionally lifting upper and lower lids. Seek medical attention immediately. If this chemical contacts the skin, remove contaminated clothing and wash immediately with soap and water. Seek medical attention immediately. If this chemical has been inhaled, remove from exposure, begin rescue breathing (using universal precautions) if breathing has stopped, and CPR if heart action has stopped. Transfer promptly to a medical facility. When this chemical has been swallowed, get medical attention. Give large quantities of water and induce vomiting. Do not make an unconscious person vomit.

*References:*
- U.S. Environmental Protection Agency, Office of Pesticide Programs, Pesticide Residue Limits, "Sulfur," 40 CFR 188.930, http://www.epa.gov/pesticides/food/viewtols.htm
- EXTOXNET, Extension Toxicology Network, "Pesticide Information Profile, Sulfur," Oregon State University, Corvallis, OR (September 1995). http://extoxnet.orst.edu/pips/sulfur.htm
- New Jersey Department of Health and Senior Services, "Hazardous Substance Fact Sheet: Sulfur," Trenton, NJ (September 1986, rev. August 2002). http://www.state.nj.us/health/eoh/rtkweb/1757.pdf

- Sax, N.I., Ed., "Dangerous Properties of Industrial Materials Report," 2, N0. 2, 65-68 (1982), New York, Van Nostrand Reinhold Co. (1982).

# Sulfuryl Fluoride

*Use Type:* Fumigant
*CAS Number:* 2699-79-8
*Formula:* $F_2O_2S$; $SO_2F_2$
*Alert:* A Restricted Use Pesticide (RUP).
*Synonyms:* Fluorure de sulfuryle (French); Fluoruro de sulfurilo (Spanish); Sulfonyl fluoride; Sulfur difluoride dioxide; Sulfuric oxyfluoride; Sulphuryl difluoride; Sulphuryl fluoride
*Trade Names:* TERMAFUME®; VIKANE®, Dow AgroSciences (USA); VIKANE FUMIGANT®, Dow AgroSciences (USA)
*Producers:* Dow AgroSciences (USA); Fluorochem Ltd. (USA); Ozark Fluorine Specialties (USA)
*Chemical Class:* Inorganic gas
*EPA/OPP PC Code:* 078003
*California DPR Chemical Code:* 618
*ICSC Number:* 1402
*RTECS Number:* WT5075000
*EEC Number:* 009-015-00-7
*EINECS Number:* 220-281-5
*Uses:* Sulfuryl fluoride is used to fumigate closed structures and their contents such as domestic dwellings, garages, barns, storage buildings, commercial warehouses, ships in port, and railroad cars. It controls numerous insect pests including termites, powder post beetles, old house borers, bedbugs, carpet beetles, clothes moths and cockroaches, as well as rats and mice. It is also used in organic synthesis of drugs and dyes.

*U.S. Maximum Allowable Residue Levels for Sulfuryl Fluoride (40 CFR 180.575):*

| CROP | ppm |
|---|---|
| Barley, bran, post-h | 0.05 |
| Barley, flour, post-h | 0.05 |
| Barley, grain, post-h | 0.1 |
| Barley, pearled barley | 0.05 |
| Corn, field, flour, post-h | 0.01 |
| Corn, field, grain, post-h | 0.05 |
| Corn, field, grits, post-h | 15 |
| Corn, field, meal, post-h | 0.01 |
| Corn, pop, grain, post-h | 0.05 |
| Grape, raisin | 0.004 |
| Grape, raisin (40 CFR 180.145) | 30 |
| Millet, grain, post-h | 0.1 |
| Nut, tree, group 14, post-h | 3 |
| Oat, flour, post-h | 0.05 |
| Oat, grain, post-h | 0.1 |
| Oat, groats/rolled oats, post-h | 0.1 |
| Pistachio, post-h | 3 |
| Rice, bran, post-h | 0.01 |
| Rice, grain, post-h | 0.04 |
| Rice, hulls, post-h | 0.1 |
| Rice, polished rice, post-h | 0.01 |
| Rice, wild, grain, post-h | 0.05 |
| Sorghum, grain, grain, post-h | 0.1 |
| Triticale, grain, post-h | 0.1 |
| Walnut | 2 |
| Walnut (40 CFR 180.145) | 12 |
| Wheat, bran, post-h | 0.05 |
| Wheat, flour, post-h | 0.05 |
| Wheat, germ, post-h | 0.02 |
| Wheat, grain, post-h | 0.1 |
| Wheat, milled byproducts, post-h | 0.05 |
| Wheat, shorts, post-h | 0.05 |

*Carcinogen/Hazard Classifications*
**U.S. EPA Carcinogens:** Unlikely Carcinogen
*Regulatory Authority:*
- Air Pollutant Standard Set (ACGIH)[1] (HSE)[33] (OSHA)[58] (Several States)[60]
- U.S. DOT Inhalation Hazard Chemicals
- Permissible Exposure Limits for Chemical Contaminants (CAL/OSHA)
- The "Director's List" (CAL/OSHA)
- Actively registered pesticide in California.
- EPCRA Section 313 Form R *de minimis* concentration reporting level: 1.0%
- Canada, WHMIS, Ingredients Disclosure List

*Description:* Sulfuryl fluoride is a colorless, odorless gas. Boiling point = –55°C. Melting/Freezing point = –137°C. Slightly soluble in water.
*Incompatibilities:* Can react with water, steam. Fluorides form explosive gases on contact with strong acids or acid fumes.
*Permissible Exposure Limits in Air:* The OSHA[2] TWA, the HSE[33] and ACGIH[1] TWA value is 5 ppm (20 mg/m$^3$). The HSE[33] STEL is 10 ppm (40 mg/m$^3$). The NIOSH[2] IDLH level = 200 ppm. Several states have set guidelines or standards for sulfuryl fluoride in ambient air[60] ranging from 200-400 µg/m$^3$ (North Dakota) to 350 µg/m$^3$ (Virginia) to 400 µg/m$^3$ (Connecticut) to 476 µg/m$^3$ (Nevada).
*Determination in Air:* Charcoal tube; Sodium hydroxide; Ion chromatography; NIOSH IV, Method #6012[18].
*Permissible Concentration in Water:* No criteria set, but runoff from spills or fire control may cause water pollution.
*Routes of Entry:* Inhalation, eye and/or skin contact (liquid).

*Harmful Effects and Symptoms*
*Short Term Exposure:* May cause conjunctivitis, rhinitis, pharyngitis, paresthesia. Contact with the liquid may cause frostbite. High exposures can cause pulmonary edema, a medical emergency that can be delayed for several hours. This can cause death. Overexposure can cause nausea, vomiting, itching, muscle twitching, tremors and seizures.

*Long Term Exposure:* May cause kidney damage. Repeated high exposures can cause deposits of fluorides in the bones (fluorosis) that may cause pain, disability and mottling of the teeth.

*Points of Attack:* Eyes, skin, respiratory system, central nervous system and kidneys.

*Medical Surveillance:* Consider the points of attack in preplacement and periodic physical examinations. Fluoride level in urine. (for fluoride in urine use NIOSH #8308). Levels higher than 4 mg/L may indicate overexposure. If symptoms develop or overexposure is suspected, the following may be useful: Consider chest x-ray after acute overexposure. Kidney function tests. Examination of the nervous system. Kidney function tests.

*First Aid:* If this chemical gets into the eyes, remove any contact lenses at once and irrigate immediately for at least 15 minutes, occasionally lifting upper and lower lids. Seek medical attention immediately. If this chemical contacts the skin, remove contaminated clothing and wash immediately with soap and water. Seek medical attention immediately. If this chemical has been inhaled, remove from exposure, begin rescue breathing (using universal precautions) if breathing has stopped, and CPR if heart action has stopped. Transfer promptly to a medical facility. Medical observation is recommended for 24 to 48 hours after breathing overexposure, as pulmonary edema may be delayed. As first aid for pulmonary edema, a doctor or authorized paramedic may consider administering a corticosteroid spray. If frostbite has occurred, seek medical attention immediately; do *NOT* rub the affected areas or flush them with water. In order to prevent further tissue damage, do *NOT* attempt to remove frozen clothing from frostbitten areas. If frostbite has *NOT* occurred, immediately and thoroughly wash contaminated skin with soap and water.

*References:*
- U.S. Environmental Protection Agency, Office of Pesticide Programs, Pesticide Residue Limits, "Sulfuryl Fluoride," 40 CFR 180.575. http://www.epa.gov/pesticides/food/viewtols.htm
- Environmental Protection Agency, "Reregistration Eligibility Decision (RED) Fact Sheet, Sulfuryl Fluoride," Office of Prevention, Pesticides and Toxic Substances, Washington DC (September 1993). http://www.epa.gov/REDs/factsheets/0176fact.pdf
- New Jersey Department of Health and Senior Services, "Hazardous Substance Fact Sheet, Sulfuryl Fluoride," Trenton NJ (March 1989, rev. May 2000). http://www.state.nj.us/health/eoh/rtkweb/1769.pdf
- EXTOXNET, Extension Toxicology Network, "Pesticide Information Profile, Sulfuryl Fluoride," Oregon State University, Corvallis, OR (June 1996). http://extoxnet.orst.edu/pips/sulfuryl.htm
- California Environmental Protection Agency "Chemical List of Lists," Sacramento CA (February 1997).

# Sulprofos

*Use Type:* Insecticide
*CAS Number:* 35400-43-2
*Formula:* $C_{12}H_{19}O_2PS_3$
*Alert:* There are no sulprofos products registered with the U.S. EPA.
*Synonyms:* *O*-Ethyl *O*-(4-(methylmercapto)phenyl)-*S*-*N*-propylphosphorothionothiolate; *O*-Ethyl *O*-(4-(methylthio)phenyl)phosphorodithioic acid *S*-propyl ester; *O*-Ethyl *O*-(4-(methylthio)phenyl)phosphorodithioic acid *S*-propyl ester; *O*-Ethyl *O*-(4-methylthiophenyl) *S*-propyl dithiophosphate; *O*-Ethyl *O*-(4-(methylthio)phenyl) *S*-propyl phosphorodithioate; Phosphorodithioic acid, *O*-ethyl *O*-(4-(methylthio)phenyl) *S*-propyl ester; Phosphorothioic acid, *O*-ethyl *O*-(4-(methylthio)phenyl) *S*-propyl ester
*Trade Names:* AI3-29149®; BAYER NTN 9306®, Bayer CropScience (Germany), canceled; BAY-NTN-9306®, Bayer CropScience (Germany), canceled; BOLSTAR®, Bayer CropScience (Germany), canceled; HELOTHION®, Bayer CropScience (Germany), canceled; MERCAPROFOS®; MERCAPROPHOS®
*Producers:* Bayer CropScience (Germany)
*Chemical Class:* Organophosphate
*EPA/OPP PC Code:* 111501
*California DPR Chemical Code:* 2006
*ICSC Number:* 1248
*RTECS Number:* TE4165000
*EINECS Number:* 252-545-0
*Uses:* No products registered in the U.S.
*Carcinogen/Hazard Classifications*
**U.S. EPA Carcinogens:** Group E, unlikely carcinogen
**TRI Developmental Toxin:** Developmental toxin
**Label Signal Word:** DANGER
**WHO Acute Hazard:** Class II, moderately hazardous
*Regulatory Authority:*
- Banned or Severely Restricted (Germany, Malaysia) (UN)[13]
- Air Pollutant Standard Set (ACGIH)[1] (NIOSH)[2] (Several States)[60]
- Permissible Exposure Limits for Chemical Contaminants (CAL/OSHA)
- The "Director's List" (CAL/OSHA)
- Actively registered pesticide in California.
- EPCRA Section 313 Form R *de minimis* concentration reporting level: 1.0%
- U.S. DOT Inhalation Hazard Chemicals as organophosphates
- U.S. DOT Regulated Marine Pollutant (49CFR172.101, Appendix B), severe pollutant
- Canada, WHMIS, Ingredients Disclosure List

*Description:* Sulprofos is a colorless to tan-colored oily liquid. Poor solubility in water; solubility = <50 ppm.

Molecular weight = 322.45. Boiling point = 155–158°C @ 0.1 mbar. Vapor pressure = 63 mmHg @ 20°C.

*Incompatibilities:* Strong oxidizers. When heated to decomposition, this material forms toxic oxides of phosphorus and sulfur.

*Permissible Exposure Limits in Air:* There is no OSHA[2] PEL. NIOSH[2] and ACGIH[1] recommends a TWA of 1.0 mg/m$^3$. Several states have set guidelines of standards for sulprofos in ambient air[60] ranging from 16.0 µg/m$^3$ (Virginia) to 20.0 µg/m$^3$ (Connecticut) to 24.0 µg/m$^3$ (Nevada) to 100.0 µg/m$^3$ (North Dakota).

*Determination in Air:* OSHA versatile sampler-2; Toluene/Acetone; Gas chromatography/Flame photometric detection for sulfur, nitrogen, or phosphorus; NIOSH IV, Method #5600, Organophosphorus pesticides.[18]

*Permissible Concentration in Water:* No criteria set, but runoff from spills or fire control may cause water pollution.

*Routes of Entry:* Inhalation, ingestion, skin and/or eye contact. Absorbed by the skin.

*Harmful Effects and Symptoms*

*Short Term Exposure:* May affect the nervous system, causing convulsions and respiratory failure. Shows typical anticholinesterase effects. Sulprofos can affect you when breathed in and quickly enters the body by passing through the skin. Severe poisoning can occur from skin contact. It is an organophosphate pesticide. Exposure can cause rapid severe poisoning with headache, sweating, nausea and vomiting, diarrhea, loss of coordination, convulsions, respiratory failure, and death. This is considered a moderately toxic compound (LD$_{50}$ for rats is 65 mg/kg). Delayed pulmonary edema may occur after inhalation.

*Long Term Exposure:* Cholinesterase inhibitor; cumulative effect is possible. This chemical may damage the nervous system with repeated exposure, resulting in convulsions, respiratory failure.

*Points of Attack:* Respiratory system, central nervous system, cardiovascular system and blood cholinesterase

*Medical Surveillance:* Medical observation is recommended for 24 to 48 hours after breathing overexposure, as pulmonary edema may be delayed. Before employment and at regular times after that, the following are recommended: Plasma and red blood cell cholinesterase levels (tests for the enzyme poisoned by this chemical). If exposure stops, plasma levels return to normal in 1-2 weeks wile red blood cell levels may be reduced for 1-3 months. When acetylcholinesterase enzyme levels are reduced by 25% or more below preemployment levels, risk of poisoning is increased, even if results are in lower ranges of "normal". Reassignment to work not involving organophosphate or carbamate pesticides is recommended until enzyme levels recover. If symptoms develop or overexposure occurs, repeat the above tests as soon as possible and get an exam of the nervous system.

*First Aid:* **Treatment for organophosphate poisoning consists of thorough decontamination, cardiorespiratory support, and administration of the antidotes atropine and pralidoxime. In cases of severe poisoning, diazepam, an anticonvulsant, should also be administered. Antidotes should be administered as prevention even if the diagnosis is in doubt.** Speed in removing material from eyes and skin is of extreme importance. *Eyes:* Eye contact can cause dangerous amounts of these chemicals to be quickly absorbed through the mucous membrane into the bloodstream. Immediately and gently flush eyes with plenty of warm or cold water (NO hot water) for at least 15 minutes, occasionally lifting the upper and lower eyelids. Get medical aid immediately. *Skin:* Get medical aid. Skin contact can cause dangerous amounts of these chemicals to be absorbed into the bloodstream. Wearing the appropriate PPE equipment and respirator for organophosphate pesticides, immediately flush skin with plenty of soap and water for at least 15 minutes while removing contaminated clothing and shoes. Shampoo hair promptly if contaminated. The removed, contaminated clothing and shoes should be double-bagged and left in Hot Zone for later disposal by hazardous materials experts. Skin may also be decontaminated with diluted hypochlorite solution. *Inhalation:* Get medical aid. Do not contaminate yourself. Wearing the appropriate PPE equipment and respirator for organophosphate pesticides, immediately remove the victim from the contaminated area to fresh air. If the victim is not breathing, administer artificial respiration. Do NOT use mouth-to-mouth resuscitation; use bag/mask apparatus. If breathing is difficult, administer oxygen through bag/mask apparatus until medical help arrives. Do not leave victim unattended. *Ingestion:* Call poison control. Loosen all clothing. Never give anything by mouth to an unconscious person. If victim is *unconscious or having convulsions,* do nothing except keep victim warm. Get medical aid. Transfer promptly to a medical facility. In cases of ingestion, **do not induce vomiting.** If the victim is alert and asymptomatic, administer a slurry of activated charcoal at a dose of 1 g/kg (infant, child, and adult dose). A soda can and straw may be of assistance when offering charcoal to a child. *In some cases you may be specifically instructed by poison control to induce vomiting by way of 2 tablespoons of syrup of ipecac (adult) washed down with a cup of water.* Do NOT give activated charcoal before or with ipecac syrup.

*Note to physician or authorized medical personnel:* Treat cases of respiratory compromise, coma, or excessive pulmonary secretions with respiratory support using protocols and techniques available and within the scope of training. Some cases may necessitate procedures such as endotracheal intubation or cricothyrotomy by properly trained and equipped personnel. When possible, atropine (see *Antidotes*, below) should be given under medical supervision. Patients who are comatose, hypotensive, or having seizures or cardiac arrhythmias should be treated according to advanced life support protocols. *Antidotes:* Two antidotes are administered to treat organophosphate poisoning. Atropine is a competitive antagonist of acetylcholine at muscarinic receptors and is used to control the excessive bronchial secretions which are often responsible for death. Pralidoxime relieves both the nicotinic and muscarine effects of organophosphate poisoning by regenerating acetylcholinesterase and can reduce both the bronchial secretions and the muscle weakness associated with poisoning. The initial intravenous dose of atropine in adults should be determined by the severity of symptoms: An initial adult dose of 1.0 to 2.0 mg or pediatric dose of 0.01 mg/kg (minimum 0.01 mg) should be administered intravenously. If intravenous access cannot be established, atropine may also be given intramuscularly, subcutaneously or via endotracheal tube. Doses should be repeated every 15 minutes until excessive secretions and sweating have been controlled. Once bronchial secretion has been controlled, atropine administration should be repeated whenever the secretions begin to recur. In seriously poisoned patients, very large doses may be required. Alterations of pulse rate and pupillary size should not be used as indicators of treatment adequacy. Pralidoxime should be administered as early in poisoning as possible as its efficacy may diminish when given more than 24 to 36 hours after exposure. Doses are as follows: adult 1.0 g; pediatric 25 to 50 mg/kg. The drug should be administered intravenously over 30 to 60 minutes, but in a life-threatening situation, one-half of the total dose can be given per minute for a total administration time of 2 minutes. Treatment should begin to take effect within 40 minutes with a reduction in symptoms and the amount of atropine necessary to control bronchial secretion. The initial dose can be repeated in 1 hour and then every 8 to 12 hours until the patient is clinically well and no longer requires atropine. If intravenous access cannot be established, pralidoxime may also be given intramuscularly. Early administration of diazepam in addition to the combined atropine and pralidoxime treatment may help prevent the onset of seizures and potential brain and cardiac morphologic damage following high-level organophosphate poisoning.

*References:*
- New Jersey Department of Health and Senior Services, "Hazardous Substance Fact Sheet: Sulprofos," Trenton, NJ (October 1998).
  http://www.state.nj.us/health/eoh/rtkweb/1771.pdf
- California Environmental Protection Agency "Chemical List of Lists," Sacramento CA (February 1997).
- Agency for Toxic Substances and Disease Registry, U.S. Department of Health and Human Services, Public Health Service, "Managing Hazardous Materials Incidents," Atlanta, GA (June 2003).

# T

## 2,4,5-T

*Use Type:* Herbicide and plant growth regulator
*CAS Number:* 93-76-5
*Formula:* $C_8H_5Cl_3O_3$; $Cl_3C_6H_2OCH_2COOH$
*Alert:* Use of this chemical in the U.S. was canceled in 1985.
*Synonyms:* Acetic acid, (2,4,5-T)-; Acetic acid, (2,4,5-trichlorophenoxy)-; Acide 2,4,5-trichlorophenoxyacetique (French); Acido (2,4,5-tricloro-fenossi)-acetico (Italian); Acido 2,4,5-triclorofenoxiacetico (Spanish); Amine; (2,4,5-Trichloor-fenoxy)-azijnzuur (Dutch); 2,4,5-Trichlorophenoxyacetic acid; (2,4,5-Trichlor-phenoxy)-essigsaeure (German)
*Trade Names:* (Note: all of the following products that contain 2,4,5-T have been canceled in the U.S.) BCF-BUSHKILLER®; BRUSH-OFF 445 LOW VOLATILE BRUSH KILLER®; BRUSH RHAP®; BRUSHTOX®; DACAMINE®; DEBROUSSAILLANT CONCENTRE®; DEBROUSSAILLANT SUPER CONCENTRE®; DECAMINE 4T®; DED-WEED BRUSHKILLER®; DED-WEED LV-6 BRUSH KIL®; T-5 BRUSH KIL®; DINOXOL®; ENVERT-T®; ESTERCIDE T-2 AND T-245®; ESTERON®; ESTERON 245®; ESTERON BRUSH KILLER®; FENCE RIDER®; FORRON®; FORST U 46®; FORTEX®; FRUITONEA®; INVERTON 245®; LINE RIDER®; PHORTOX®; REDDON®; REDDOX®; SPONTOX®; SUPER D WEEDONE®; TIPPON®; T-NOX®; TORMONA®; TRANSAMINE®; TRIBUTON®; TRINOXOL®; TRIOXON®; TRIOXONE®; TRIOXAL®; VEON®; VEON 245®; VERTON2T®; VISKO RHAP LOW VOLATILE ESTER®; WEEDAR®; WEEDONE
*Chemical Class:* Chlorophenoxy acid
*EPA/OPP PC Code:* 082001
*California DPR Chemical Code:* 639
*ICSC Number:* 0075
*RTECS Number:* AJ8400000
*EEC Number:* 607-041-00-9
*EINECS Number:* 202-273-3
*Uses:* 2,4,5-T products in the U.S. were canceled in 1985. It was used as a post-emergence herbicide to control broadleaf weeds, shrubs and woody plants on range land, rice, vacant areas, turf, residential areas and food crops. It can also act as a growth regulator to control fruit drops and increase the yield of citrus crops.
*Human toxicity (long-term)[77]:* Low–70.00 ppb, Health Advisory
*Fish toxicity (threshold)[77]:* Very low–79154.36150 ppb, MATC (Maximum Acceptable Toxicant Concentration)
*Carcinogen/Hazard Classifications*
**California Prop. 65:** Listed
**IARC:** Group 2B, possible carcinogen
**Label Signal Word:** CAUTION
**Endocrine Disruptor:** Probable ED
*Regulatory Authority:*
- Carcinogen (Human Suspected) (IARC)[9]
- AB 2588-Air Toxics "Hot Spots" Chemicals (CAL) as chlorophenoxy pesticides
- The "Director's List" (CAL/OSHA) as chlorophenoxy pesticides
- Banned or Severely Restricted (Many Countries) (UN)[13]
- Air Pollutant Standard Set (ACGIH)[1] (DFG)[3] (HSE)[33] (OSHA)[58] (Several States)[60]
- Clean Water Act: Section 311 Hazardous Substances/RQ 40CFR117.3 (same as CERCLA, see below)
- EPA Hazardous Waste Number (RCRA No.): U232
- RCRA, 40CFR261, Appendix 8 Hazardous Constituents
- RCRA 40CFR268.48; 61FR15654, Universal Treatment Standards: Wastewater (mg/L), 0.72; Nonwastewater (mg/kg), 7.9
- RCRA 40CFR264, Appendix 9; TSD Facilities Ground Water Monitoring List. Suggested test method(s) (PQL $ug/L$): 8150(2)
- Safe Drinking Water Act: Priority List (55FR1470)
- Superfund/EPCRA 40CFR302.4 RQ: CERCLA, 1000 lb (454 kg)

*Description:* Colorless to tan crystalline solid. Slightly soluble in water; solubility = 278 mg/L @ 25°C. Molecular weight 255.49. Melting/Freezing point = 154–158°C. Vapor pressure = very low/negligible. Log $K_{ow}$ = 3.3/3.9. Also reported at 4.0 (Kenaga, E.E., & C.A. Goring. *Aquatic Toxicology*, 1980). Values at or above 3.0 are likely to bioaccumulate in marine organisms.
*Incompatibilities:* The aqueous solution is a weak acid. Incompatible with sulfuric acid, bases, ammonia, aliphatic amines, alkanolamines, isocyanates, alkylene oxides, epichlorohydrin; strong oxidizers such as chlorine, bromine, fluorine, and strong bases.
*Permissible Exposure Limits in Air:* The OSHA/NIOSH TWA[2], the DFG MAK[3], the HSE[33] TWA, and the ACGIH[1] TWA value is 10 mg/m³. The HSE[33] STEL value is 20 mg/m³. The NIOSH[2] IDLH level = 250 mg/m³. Several states have set guidelines or standards for 2,4,5-T in ambient air[60] ranging from 1.0 μg/m³ (Pennsylvania) to 100.0 μg/m³ (North Dakota) to 160 μg/m³ (Virginia) to 238.0 μg/m³ (Nevada).
*Determination in Air:* Filter; Methanol; High-pressure liquid chromatography/Ultraviolet detection; NIOSH IV, Method #5001[18].

*Permissible Concentration in Water:* The U.S. EPA (see reference below) has set a lifetime health advisory of 21.0 μg/L. Mexico[35] has set limits of 100 μg/L in estuaries and 10 μg/L in coastal waters. The State of Kansas[61] has set a guideline for drinking water of 700.0 μg/L.

*Determination in Water:* Liquid extraction and gas chromatography (See EPA reference below)

*Routes of Entry:* Inhalation, ingestion, skin and/or eye contact. Absorbed through the skin.

*Harmful Effects and Symptoms*

*Short Term Exposure:* Irritates the eyes, skin, and respiratory tract. *Inhalation:* Increasingly severe symptoms may include nose an throat irritation, weakness, tiredness, metallic taste in mouth, loss of appetite, diarrhea, heart problems, heart failure and death. *Skin:* Reddening and itching may develop. absorption is slow, but may contribute significantly to total exposure. *Eyes:* Irritation may develop. *Ingestion:* 350 mg (0.01 ounce) produces only a metallic taste in the mouth, lasting about 2 hours. Approximate 4 teaspoonfuls (150 lb man) may cause weakness, tiredness, loss of appetite, diarrhea, heart problems, heart failure, and death. *Note:* Reported effects of 2,4,5-T are due to accidental exposures, often at unknown levels or duration. In addition, 2,4,5-T may be contaminated with very small amounts of another more toxic compound 2,3,7,8-tetrachlorodibenzo-p-dioxin (Dioxin). Therefore, some of the symptoms off exposure to 2,4,5,-T may be due to contaminants.

*Long Term Exposure:* Levels above the standard may produce skin irritation, acne-like skin sores, loss of skin coloration in small patches, GI tract ulcer, and nerve disorders resulting in difficulty controlling muscles. Animal studies also indicate the possibility of an increased susceptibility to infection. Changes in generic material and birth defects have been reported in laboratory studies and may be due to 2,4,5-T or its contaminant. Whether these effects are produced in humans is unknown. In animals: ataxia; skin irritation, acne-like rash; liver damage.

*Points of Attack:* Skin, liver and gastrointestinal tract.

*Medical Surveillance:* If symptoms develop or overexposure is suspected, liver or kidney function tests may be useful.

*First Aid:* If this chemical gets into the eyes, remove any contact lenses at once and irrigate immediately for at least 15 minutes, occasionally lifting upper and lower lids. Seek medical attention immediately. If this chemical contacts the skin, remove contaminated clothing and wash immediately with soap and water. Seek medical attention immediately. If this chemical has been inhaled, remove from exposure, begin rescue breathing (using universal precautions) if breathing has stopped, and CPR if heart action has stopped. Transfer promptly to a medical facility. When this chemical has been swallowed, get medical attention. Give large quantities of water and induce vomiting. Do not make an unconscious person vomit. *Note to physician or authorized medical personnel:* If ingested, remove by lavage or vomiting. Use general supportive measures for CNS depression. Use quinidine for myotonia.

*References:*

- U.S. Environmental Protection Agency, "Health Advisory: 2,4,5-Trichlorphenoxy-Acetic Acid," Washington, DC, Office of Drinking Water (August 1987).
- Sax, N.I., Ed., "Dangerous Properties of Industrial Materials Report," 3, No. 5, 20-21 (1983).
- New Jersey Department of Health and Senior Services, "Hazardous Substance Fact Sheet: 2,4,5-(Trichlorphenoxy) Acetic Acid," Trenton, NJ (January 1988, rev. August 2001).
  http://www.state.nj.us/health/eoh/rtkweb/1896.pdf
- California Environmental Protection Agency "Chemical List of Lists," Sacramento CA (February 1997).
- New York State Department of Health, "Chemical Fact Sheet: 2,4,5-T," Albany, NY, Bureau of Toxic Substance Assessment (March 1980).

## Tebuconazole

*Use Type:* Fungicide
*CAS Number:* 107534-96-3
*Formula:* $C_{16}H_{23}ClN_3O$
*Synonyms:* (RS)-1-(4-Chlorophenyl)-4,4-dimethyl-3-(1*H*-1,2,4-triazol-1-ylmethyl)pentan-3-ol; (RS)-1-*p*-Chlorophenyl-4,4-dimethyl-3-(1*H*-1,2,4-triazol-1-ylmethyl)pentan-3-ol; (±)-α-[2-(4-Chlorophenyl)ethyl]-α-(1,1-dimethylethyl)-1*H*-1,2,4-triazole-1-ethanol; Ethyltrianol; Fenetrazole; Terbutrazole; 1*H*-1,2,4-Triazole-1-ethanol, α-[2-(4-chlorophenyl)ethyl]-α-(1,1-dimethylethyl)-, (±)-

*Trade Names:* BAY®-HWG 1608, BayerCropScience (Germany); BAYER®-HWG-1608, BayerCropScience (Germany); CORAIL®; ELITE®, BayerCropScience (Germany); FOLICUR®, BayerCropScience (Germany); FOLITRAZOLE®; GAUCHO®, Gustafson (USA); HORIZON®; HWG 1608®; LYNX-1.2®; PREVENTOL®, BayerCropScience (Germany); RAXIL® (tebuconazole + metalaxyl), Gustafson (USA); SILVACUR®; TEBUJECT®, J. J. Mauget (USA); WOLMAN®; WOODLIFE®

*Producers:* Ascot International (UK); Bayer CropScience (Germany); Bharat Rasayan (India); Gustafson (USA); J. J. Mauget (USA); Shandong Huayang Pesticide Group (China); SuYan Agrochemical Group (China)

*Chemical Class:* Azole
*EPA/OPP PC Code:* 128997
*California DPR Chemical Code:* 3850
*Uses:* Used as a seed treatment against smuts and bunts of cereals; as a foliar spray against diseases of cereal, peanuts, oilseed rape, grapes, bananas, stone fruit, and pome fruit.

*U.S. Maximum Allowable Residue Levels for Tebuconazole (40 CFR 180.474):*

| CROP | ppm |
| --- | --- |
| Banana | 0.05 |
| Barley, grain | 0.05 |
| Barley, hay & straw | 0.1 |
| Cattle mbyp | 0.2 |
| Cherry | 4.0 |
| Goat, mbyp | 0.2 |
| Grape | 5.0 |
| Grass, forage | 8.0 |
| Grass, hay | 25.0 |
| Grass, seed screening | 55.0 |
| Grass, straw | 30.0 |
| Horse, mbyp | 0.2 |
| Milk | 0.1 |
| Oat, forage | 0.1 |
| Oat, grain | 0.05 |
| Oat, hay | 0.1 |
| Oat, straw | 0.1 |
| Peachea, including nectarenes | 1.0 |
| Peanuts | 0.1 |
| Sheep, mbyp | 0.2 |
| Wheat, forage | 0.1 |
| Wheat, grain | 0.05 |
| Wheat, hay & straw | 0.1 |

*Carcinogen/Hazard Classifications*
**U.S. EPA Carcinogens:** Group C, possible carcinogen
**Label Signal Word:** CAUTION, WARNING or DANGER
**WHO Acute Hazard:** Class III, slightly hazardous
*Description:* Clear, crystalline solid. Slightly soluble in water. Molecular weight = 307.87. Melting/Freezing point = 105-106°C.
*Permissible Concentration in Water:* No criteria set. Runoff from spills or fire control may cause water pollution.
*Routes of Entry:* Ingestion, skin contact.
*Harmful Effects and Symptoms*
*Short Term Exposure:* Poisonous if swallowed. Contact may irritate skin and cause eye irritation and possible severe injury. Avoid inhalation.
*Medical Surveillance:* If this chemical gets into the eyes, remove any contact lenses at once and irrigate immediately for at least 15 minutes, occasionally lifting upper and lower lids. Seek medical attention immediately. If this chemical contacts the skin, remove contaminated clothing and wash immediately with soap and water. If this chemical has been inhaled, remove from exposure, begin rescue breathing (using universal precautions) if breathing has stopped, and CPR if heart action has stopped. Transfer promptly to a medical facility. When this chemical has been swallowed, get medical attention. Give large quantities of water and induce vomiting. Do not make an unconscious person vomit
*References:*
- U.S. Environmental Protection Agency, Office of Pesticide Programs, Pesticide Residue Limits, "Tebuconazole," 40 CFR 180.474, www.epa.gov/pesticides/food/viewtols.htm
- California Environmental Protection Agency "Chemical List of Lists," Sacramento CA (February 1997)

## Tebufenozide (ANSI)

*Use Type:* Insecticide, insect growth regulator
*CAS Number:* 112410-23-8
*Formula:* $C_{22}H_{28}N_2O_2$
*Synonyms:* Benzoic acid, 3,5-dimethyl-, 1-(1,1-dimethylethyl)-2-(4-ethylbenzoyl)hydrazide; *N-tert*-Butyl-*N'*-(4-ethylbenzoyl)-3,5-dimethylbenzoylhydrazide; 3,5-Dimethylbenzoic acid 1-(1,1-dimethylethyl)-2-(4-ethylbenzoyl)hydrazine
*Trade Names:* CONFIRM®, Dow AgroSciences (USA); MIMIC®, Dow AgroSciences (USA); RH-5992®
*Producers:* Dow AgroSciences (USA)
*Chemical Class:* Diacylhydrazine
*EPA/OPP PC Code:* 129026
*California DPR Chemical Code:* 3957
*Uses:* Tebufenozide is an insect growth regulator that interferes with molting of Lepidopteran larvae. It is used on fruitworm, fireworms, false armyworm, gypsy moth, and spanworms, and is applied pre-harvest.

*U.S. Maximum Allowable Residue Levels for Tebufenozide (40 CFR 180.482):*

| CROP | ppm |
| --- | --- |
| Almond, hulls | 25 |
| Animal feed, nongrass, group 18 | 0.5 |
| Beet, garden, roots | 0.3 |
| Beet, garden, tops | 9 |
| Berry group 13 | 3 |
| Brassica, head and stem, subgroup 5a | 5 |
| Brassica, leafy greens, subgroup 5b | 10 |
| Canola, refined oil | 4 |
| Canola, seed | 2 |
| Egg | 0.01 |
| Fruit, pome | 1.5 |
| Grain, cereal, forage | 0.5 |
| Grain, cereal, hay | 0.5 |
| Grain, cereal, stover, group | 0.5 |
| Grape | 3 |
| Grape, wine | 0.5 |
| Grass, forage | 5 |
| Grass, forage, fodder and hay, group 17 | 0.5 |
| Grass, hay | 18 |
| Hog, liver | 1 |
| Hog, mbyp | 0.1 |
| Kiwifruit | 0.5 |
| Leaf petioles subgroup 4b | 2 |
| Longan | 1 |
| Lychee | 1 |

| | |
|---|---|
| Nut, tree, group 14 and pistachios | 0.1 |
| Peanut | 0.05 |
| Peanut, hay | 5 |
| Peanut, meal | 0.15 |
| Peanut, refined oil | 0.15 |
| Peppermint | 10 |
| Poultry, fat | 0.1 |
| Poultry, meat | 0.01 |
| Poultry, mbyp | 0.05 |
| Spearmint | 10 |
| Sweet potato | 0.25 |
| Turnip, roots | 0.3 |
| Vegetable, foliage of legume, group 7 | 0.1 |
| Vegetable, fruiting, group 8 | 1 |

*Carcinogen/Hazard Classifications*
**U.S. EPA Carcinogens:** Group E, Unlikely a carcinogen
**Label Signal Word:** CAUTION
*Description:* Practically insoluble in water. Melting/Freezing point = 189°C. Log $K_{ow}$ = > 3.0. Likely to bioaccumulate in marine organisms.
*Incompatibilities:* Contact with flammable material may cause fire and explosions. Contact with combustible or oxidizable materials may form heat-, shock-, and friction-sensitive explosive mixtures. Static electricity may also cause explosions. Keep away from all acids, especially dibasic organic acids, ammonium compounds, antimony sulfide, arsenic trioxide, metal sulfides, powdered metals, calcium aluminum hydride, cyanides, manganese dioxide, phosphorus, selenium, sulfur, thiocyanates, zinc.
*Permissible Concentration in Water:* No criteria set. Runoff from spills or fire control may cause water pollution.
*Routes of Entry:* Ingestion
*Harmful Effects and Symptoms*
*Short Term Exposure:* Toxic if ingested.
*First Aid:* If this chemical gets into the eyes, remove any contact lenses at once and irrigate immediately for at least 15 minutes, occasionally lifting upper and lower lids. Seek medical attention immediately. If this chemical contacts the skin, remove contaminated clothing and wash immediately with soap and water. Seek medical attention immediately. If this chemical has been inhaled, remove from exposure, begin rescue breathing (using universal precautions) if breathing has stopped, and CPR if heart action has stopped. Transfer promptly to a medical facility. When this chemical has been swallowed, get medical attention. Give large quantities of water and induce vomiting. Do not make an unconscious person vomit.
*References:*
- U.S. Environmental Protection Agency, Office of Pesticide Programs, Pesticide Residue Limits, "Tebufenozide," 40 CFR 180.482, www.epa.gov/pesticides/food/viewtols.htm
- California Environmental Protection Agency "Chemical List of Lists," Sacramento CA (February 1997)

# Tebuthiuron (ANSI)

*Use Type:* Herbicide
*CAS Number:* 34014-18-1
*Formula:* $C_9H_{16}N_4OS$
*Synonyms:* 1-(5-*tert*-Butyl-1,3,4-thiadiazol-2-yl)-3-dimethylharnstoff (German); Caswell No. 366AA; *N*-[5-(1,1-Dimethylaethyl)-1,3,4-thiadiazol-2-yl]-*N,N'*-dimethylharnstoff (German); *N*-[5-(1,1-Dimethylethyl)-1,3,4-thiadiazol-2-yl]-*N,N'*-dimethylurea; *N*-[5-(1,1-Dimethylethyl)-1,3,4-thiadiazol-2-yl]-*N,N'*-dimethylurea; 1-(5-*tert*-Butyl-1,3,4-thiadiazol-2-yl)-1,3-dimethylurea; Urea, *N*-(5-(1,1-dimethylethyl)-1,3,4-thiadiazol-2-yl)-*N,N'*-dimethy l-; Urea, 2-(5-*tert*-butyl-1,3,4-thiadiazol-2-yl)-1,3-dimethyl-; Urea, 1-(5-*tert*-butyl-1,3,4-thiadiazol-2-yl)-1,3-dimethyl-; Urea, *N*-[5-(1,1-dimethylethyl)-1,3,4-thiadiazol-2-yl]-*N,N'*-dimethyl-
*Trade Names:* BRULAN®; BRUSH-BULLET®; E-103®, Dow AgroSciences (USA); GRASLAN®; HERBEC®; HERBIC®; PERFLAN®; RECLAIM; SHA-105501®; SPIKE® Dow AgroSciences (USA) SPRAKIL®, SSI Maxim (USA); TEBULAN®; TIUROLAN®
*Producers:* Dow AgroSciences (USA)
*Chemical Class:* Sulfonylurea
*EPA/OPP PC Code:* 105501
*California DPR Chemical Code:* 1810
*RTECS Number:* YS4250000
*Uses:* Tebuthiuron is a broad spectrum herbicide that is used to control weeds in non-cropland areas, range lands, rights-of-way and industrial sites. In grasslands and sugar cane it controls woody and herbaceous plants and weeds such as alfalfa, bluegrasses, chickweed, clover, dock, goldenrod and mullein.
*U.S. Maximum Allowable Residue Levels for Tebuthiuron (40 CFR 180.390):*

| CROP | ppm |
|---|---|
| Cattle, fat | 2 |
| Cattle, meat | 2 |
| Cattle, mbyp | 2 |
| Goat, fat | 2 |
| Goat, meat | 2 |
| Goat, mbyp | 2 |
| Grass, forage | 10 |
| Grass, hay | 10 |
| Horse, fat | 2 |
| Horse, meat | 2 |
| Horse, mbyp | 2 |
| Milk | 0.3 |
| Sheep, fat | 2 |
| Sheep, meat | 2 |
| Sheep, mbyp | 2 |

*Carcinogen/Hazard Classifications*
**U.S. EPA Carcinogens:** Group D, unclassifiable
**U.S. TRI:** Developmental toxin

**Label Signal Word:** CAUTION or WARNING
**WHO Acute Hazard:** Class III, slightly hazardous
*Regulatory Authority:*
- EPA 40 CFR 372.65, Specific Toxic Chemical Listings
- AB 1803-Well Monitoring Chemical (CAL)
- Actively registered pesticide in California.
- EPCRA Section 313 Form R *de minimis* concentration reporting level: 1.0%

*Description:* Grayish to dark-brown pellets, granules, or powder; Technical grade is colorless. Musty odor. Slightly soluble in water: solubility: 2,575 ppm @ 25°C (powder). Molecular weight = 228.35. Melting/Freezing point = 163°C. Vapor pressure = 2.2 x $10^{-6}$ mmHg @ 20°C; 0.27 mPa @ 25°C. Log $K_{ow}$ = 1.78. Unlikely to bioaccumulate in marine organisms.

*Permissible Concentration in Water:* No criteria set. Runoff from spills or fire control may cause water pollution.

*Routes of Entry:* Ingestion. Skin contact.

*Harmful Effects and Symptoms*

*Short Term Exposure:* Contact with eyes or skin may cause irritation or burns. Inhalation should be avoided; use NIOSH-approved air purifying respirators for pesticides. May be harmful if swallowed. Skin contact may cause allergic reaction.

*Long Term Exposure:* Repeated or prolonged contact may cause skin and lung sensitization, resulting in allergies.

*Points of Attack:* Skin.

*Medical Surveillance:* Evaluation by a qualified allergist, including careful exposure history and special testing, may help diagnose skin allergy.

*First Aid:* If this chemical gets into the eyes, remove any contact lenses at once and irrigate immediately for at least 15 minutes, occasionally lifting upper and lower lids. Seek medical attention immediately. If this chemical contacts the skin, remove contaminated clothing and wash immediately with soap and water. Seek medical attention immediately. If this chemical has been inhaled, remove from exposure, begin rescue breathing (using universal precautions) if breathing has stopped, and CPR if heart action has stopped. Transfer promptly to a medical facility. When this chemical has been swallowed, get medical attention. Give large quantities of water and induce vomiting. Do not make an unconscious person vomit.

*References:*
- U.S. Environmental Protection Agency, Office of Pesticide Programs, Pesticide Residue Limits, "Tebuthiuron," 40 CFR 180.390, www.epa.gov/pesticides/food/viewtols.htm
- EXTOXNET, Extension Toxicology Network, "Pesticide Information Profile, Tebuthiuron," Oregon State University, Corvallis, OR (September 1993). http://pmep.cce.cornell.edu/profiles/extoxnet/pyrethrins-ziram/tebuthiuron-ext.html
- U.S. Environmental Protection Agency, "Reregistration Eligibility Decision Fact Sheet (RED), Tebuthiuron," Office of Prevention, Pesticides and Toxic Substances, Washington, DC (April 1994). http://www.epa.gov/REDs/factsheets/0054fact.pdf
- California Environmental Protection Agency "Chemical List of Lists," Sacramento CA (February 1997)

# Tecnazene

*Use Type:* Fungicide and plant growth regulator
*CAS Number:* 117-18-0
*Formula:* $C_6HCl_4NO_2$; $NO_2-C_6H-Cl_4$
*Synonyms:* 3-Nitro-1,2,4,5-tetrachlorobenzene; TCNB; Tecnazen (German); Tecnazene; 1,2,4,5-Tetrachloro-3-nitrobenzene; 2,3,5,6-Tetrachloronitrobenzene; 2,3,5,6-Tetrachlor-3-nitrobenzol (German)
*Trade Names:* CHIPMAN 3,142®; FOLOSAN®; FUSAREX®; FUMITE®; FOLOSAN®
*Chemical Class:* Nitro compound of an aromatic compound
*EPA/OPP PC Code:* 055201
*California DPR Chemical Code:* 2918
*Uses:* Not registered in the U.S. Used as a sprout inhibitor on stored potatoes; as a fungicide mainly as a smoke formulation in greenhouses.

*Carcinogen/Hazard Classifications*
**Label Signal Word:** CAUTION or WARNING
**WHO Acute Hazard:** Class U, unlikely to be hazardous

*Description:* A colorless crystalline solid. Odorless. Slightly soluble in water. Molecular weight = 260.87. Boiling point = 304°C. Melting/Freezing point = 99°C; Log $K_{ow}$ = >3.85. Values above 3.0 are likely to bioaccumulate in marine organisms.

*Incompatibilities:* Contact with alkalies may form explosive metal salts.

*Permissible Concentration in Water:* No criteria set. Runoff from spills or fire control may cause water pollution.

*Routes of Entry:* Skin

*First Aid:* If this chemical gets into the eyes, remove any contact lenses at once and irrigate immediately for at least 15 minutes, occasionally lifting upper and lower lids. Seek medical attention immediately. If this chemical contacts the skin, remove contaminated clothing and wash immediately with soap and water. Seek medical attention immediately. If this chemical has been inhaled, remove from exposure, begin rescue breathing (using universal precautions) if breathing has stopped, and CPR if heart action has stopped. Transfer promptly to a medical facility. When this chemical has been swallowed, get medical attention. Give large quantities of water and induce vomiting. Do not make an unconscious person vomit. Do not induce vomiting when formulations containing petroleum solvents are ingested.

*References:*
International Programme on Chemical Safety (IPCS), "Environmental Health Criteria #42, Tecnazene," Geneva,

Switzerland, (1984).
http://www.inchem.org/documents/ehc/ehc/ehc42.htm
- California Environmental Protection Agency "Chemical List of Lists," Sacramento CA (February 1997).

# Tefluthrin

*Use Type:* Insecticide, miticide
*CAS Number:* 79538-32-2
*Formula:* $C_{17}H_{14}ClF_7O_2$
**Alert:** A Restricted Use Pesticide (RUP)
*Synonyms:* Cyclopropanecarboxylic acid, 3-(2-chloro-3,3,3-trifluoro-1-propenyl)-2,2-dimethyl-, (2,3,5,6-tetrafluoro-4-methylphenyl)methyl ester, $(1\alpha,3\alpha(Z))$-($\pm$)-; (2,3,5,6-Tetrafluoro-4-methylphenyl)methyl *cis*-3-(2-chloro-3,3,3-trifluoro-1-propenyl)-2,2-dimethylcyclopropane carboxylate; Tefluthrine
*Trade Names:* FORCE® Syngenta (Switzerland); FORZA®; JF 6064®; KOMET-RP®; PP 993®; R 151993®
*Producers:* Syngenta (Switzerland)
*Chemical Class:* Pyrethroid
*EPA/OPP PC Code:* 128912
*California DPR Chemical Code:* 3839
*Uses:* Registered in the U.S. for use on corn. Elsewhere it is registered for use on a variety of crops, e.g., peanuts, sweet potato, sugarcane, cabbage, radish, Brussels sprouts, and strawberries.
*U.S. Maximum Allowable Residue Levels Tefluthrin (40 CFR 180.440):*

| CROP | ppm |
|---|---|
| Corn, field, forage | 0.06 |
| Corn, field, grain | 0.06 |
| Corn, field, stover | 0.06 |
| Corn, pop, forage | 0.06 |
| Corn, pop, grain | 0.06 |
| Corn, pop, stover | 0.06 |
| Corn, sweet, forage | 0.06 |
| Corn, sweet, kernel plus cob with husks removed | 0.06 |
| Corn, sweet, stover | 0.06 |

*Carcinogen/Hazard Classifications*
**Label Signal Word:** CAUTION or DANGER
**WHO Acute Hazard:** Class 1b, highly hazardous
**Endocrine Disruptor:** Suspected endocrine disruptor
*Description:* Crystalline solid. Slightly soluble in water; solubility = < 0.002 ppm. Molecular weight = 418.71.
*Incompatibilities:* May react violently with strong oxidizers, bromine, 90% hydrogen peroxide, phosphorus trichloride, silver powders or dust. Incompatible with silver compounds. Mixture with some silver compounds forms explosive salts of silver oxalate.
*Permissible Concentration in Water:* No criteria set. Runoff from spills or fire control may cause water pollution.

*First Aid:* If this chemical gets into the eyes, remove any contact lenses at once and irrigate immediately for at least 15 minutes, occasionally lifting upper and lower lids. Seek medical attention immediately. If this chemical contacts the skin, remove contaminated clothing and wash immediately with soap and water. Seek medical attention immediately. If this chemical has been inhaled, remove from exposure, begin rescue breathing (using universal precautions) if breathing has stopped, and CPR if heart action has stopped. Transfer promptly to a medical facility. When this chemical has been swallowed, get medical attention. Give large quantities of water and induce vomiting. Do not make an unconscious person vomit. Do not induce vomiting when formulations containing petroleum solvents are ingested.
*References:*
- U.S. Environmental Protection Agency, Office of Pesticide Programs, Pesticide Residue Limits, "Tefluthrin," 40 CFR 180.440, www.epa.gov/pesticides/food/viewtols.htm
- California Environmental Protection Agency "Chemical List of Lists," Sacramento CA (February 1997).

# Temephos (ANSI)

*Use Type:* Insecticide and larvicide
*CAS Number:* 3383-96-8
*Formula:* $C_{16}H_{20}O_6P_2S_3$
*Synonyms:* Bis-*p*-(*O*,*O*-dimethyl *O*-phenyl phosphorothioate)sulfide; *O*,*O*-Dimethylphosphorothioate *O*,*O*-diester with 4,4'-thiodiphenol; ENT 27,165; Phenol,4,4'-thiodi-, *O*,*O*-diester with *O*,*O*-dimethyl phosphorothioate; Phosphorothioic acid, *O*,*O*'-dimethyl ester, *O*,*O*-Diester with 4,4'-thiodiphenol; Phosphorothioic acid, *O*,*O*'-(thiodi-4,1-phenylene)*O*,*O*,*O*',*O*'-tetramethyl ester; Phosphorothioic acid, *O*,*O*'-(thiodi-*p*-phenylene)*O*,*O*,*O*',*O*'-tetramethyl ester; Swebate; Temefos (Spanish); Temophos; Tetrafenphos; Tetramethyl-*O*,*O*'-thiodi-*p*-phenylene phosphorothioate; *O*,*O*,*O*',*O*'-Tetramethyl *O*,*O*'-thiodi-*p*-phenylene Bis(phosphorothioate); *O*,*O*,*O*',*O*'-Tetramethyl *O*,*O*'-thiodi-*p*-phenylenephosphorothioate; *O*,*O*'-(Thiodi-4,1-phenylene)bis(*O*,*O*-dimethylphosphorothioate); *O*,*O*'-(Thiodi-4,1-phenylene)phosphorothioic acid *O*,*O*,*O*',*O*'-tetramethyl ester; *O*,*O*'-(Thiodi-*p*-phenylene)*O*,*O*,*O*',*O*'-tetramethylbis(phosphorothioate)
*Trade Names:* 27165®; ABAT®; ABATE®, Clarke Mosquito Control Products (USA); ABATE® 1-SG, American Cyanamid's Agricultural Products Group (USA); ABATE® 2-CG, American Cyanamid's Agricultural Products Group (USA); ABATE® 4-E, American Cyanamid's Agricultural Products Group (USA); ABATE® 5CG, American Cyanamid's Agricultural Products Group (USA); ABATHION®; AI3-27165®; AC 52160®, American Cyanamid's Agricultural Products Group (USA);

# Temephos

AMERICAN CYANAMID AC-52,160®, American Cyanamid's Agricultural Products Group (USA); AMERICAN CYANAMID CL-52160®, American Cyanamid's Agricultural Products Group (USA); AMERICAN CYANAMID E.I. 52,160®, American Cyanamid's Agricultural Products Group (USA); BIOTHION®; BITHION®; CL 52160®; DIFENPHOS®; DIFENTHOS®; DIFOS®; DIPHOS®; ECOPRO®; ECOPRO® 1707; EI 52160®; NEPHIS®; NEPHIS® 1G; NIMITEX®; NIMITOX®

*Producers:* Agrimor International (USA); Agsin (Singapore); BASF Agricultural Products Group (Germany); Biesterfeld Siemsgluess International. GmbH (Germany); American Cyanamid's Agricultural Products Group (USA); Ehrenstorfer, Dr. (Germany); Gharda Chemicals (India); Ki-Hara Chemicals Ltd. (UK); Merck (Germany); Sigma-Aldrich Laborchemikalien (Germany)

*Chemical Class:* Organophosphate
*EPA/OPP PC Code:* 059001
*California DPR Chemical Code:* 1
*ICSC Number:* 0199
*RTECS Number:* TF6890000
*EINECS Number:* 222-191-1

*Uses:* Temephos is one of the few organophosphates that ise registered to control mosquito larvae. It is a non-systemic insecticide for use on wetlands, ponds, lakes and other moist areas to control mosquito, gnat, black fly, midge, pinkie and sandfly larvae.

*Carcinogen/Hazard Classifications*
**Label Signal Word:** CAUTION, DANGER, WARNING, depending on the formulation
**WHO Acute Hazard:** Class U, unlikely to be hazardous
*Regulatory Authority:*
- Air Pollutant Standard Set (ACGIH)[1] (OSHA)[58] (former USSR)[35, 43] (Several States)[60]
- Permissible Exposure Limits for Chemical Contaminants (CAL/OSHA)
- The "Director's List" (CAL/OSHA)
- EPCRA Section 313 Form R *de minimis* concentration reporting level: 1.0%
- U.S. DOT Inhalation Hazard Chemicals as organophosphates
- U.S. DOT Regulated Marine Pollutant (49CFR172.101, Appendix B)
- Canada, WHMIS, Ingredients Disclosure List

*Description:* Temephos is a crystalline solid. The technical product is an amber viscous liquid. Molecular weight = 466.47. Melting/Freezing point = 31°C. Practically insoluble in water. Combustible.

*Incompatibilities:* Strong acids and bases. Gives off toxic oxides of sulfur and phosphorus in fire.

*Permissible Exposure Limits in Air:* OSHA[2] has set a TWA of 15 $mg/m^3$ for total dust and 5 $mg/m^3$ for the respirable fraction. NIOSH[2] recommends a TWA of 10 $mg/m^3$ for total dust and 5 $mg/m^3$ for the respirable fraction. ACGIH[1] recommends a TWA value of 10 $mg/m^3$. The former USSR[35, 43] has set a MAC in workplace air of 0.5 $mg/m^3$. Several states have set guidelines or standards for Temephos in ambient air[60] ranging from 100 $\mu g/m^3$ (North Dakota) to 160 $\mu g/m^3$ (Virginia) to 200 $\mu g/m^3$ (Connecticut) to 238 $\mu g/m^3$ (Nevada).

*Determination in Air:* Collection on a filter and gravimetric analysis.

*Permissible Concentration in Water:* No criteria set. Experience in the field for a period of more than one year has shown, however, that 1 mg/L in drinking water is without effect.

*Determination in Water:* Techniques used for residue determination include colorimetry and gas liquid chromatography and may be applicable to water analysis.

*Routes of Entry:* Inhalation, skin absorption, ingestion, skin and/or eye contact

*Harmful Effects and Symptoms*
*Short Term Exposure:* Temephos can affect you when breathed in and quickly enters the body by passing through the skin. Severe poisoning can occur from skin contact. It is a moderately toxic organophosphate chemical. Exposure can cause rapid severe poisoning with headache, sweating, nausea and vomiting, diarrhea, loss of coordination, and death. Delayed pulmonary edema may occur after inhalation.

*Long Term Exposure:* Cholinesterase inhibitor; cumulative effect is possible. This chemical may damage the nervous system with repeated exposure, resulting in convulsions, respiratory failure. May cause liver damage.

*Points of Attack:* Respiratory system, lungs, central nervous system, cardiovascular system, skin, eyes, plasma and red blood cell cholinesterase.

*Medical Surveillance:* Medical observation is recommended for 24 to 48 hours after breathing overexposure, as pulmonary edema may be delayed. Before employment and at regular times after that, the following are recommended: Plasma and red blood cell cholinesterase levels (tests for the enzyme poisoned by this chemical). If exposure stops, plasma levels return to normal in 1-2 weeks while red blood cell levels may be reduced for 1-3 months. When acetylcholinesterase enzyme levels are reduced by 25% or more below preemployment levels, risk of poisoning is increased, even if results are in lower ranges of "normal." Reassignment to work not involving organophosphate or carbamate pesticides is recommended until enzyme levels recover. If symptoms develop or overexposure occurs, repeat the above tests as soon as possible and get an exam of the nervous system. Do not drink any alcoholic beverages before or during use. Alcohol promotes absorption of organophosphates.

*First Aid:* **Treatment for organophosphate poisoning consists of thorough decontamination, cardiorespiratory support, and administration of the antidotes atropine and pralidoxime. In cases of severe poisoning, diazepam,** an anticonvulsant, should also be administered. **Antidotes should be administered as prevention even if the diagnosis is in doubt.** Speed in removing material from eyes and skin is of extreme importance. *Eyes:* Eye contact can cause dangerous amounts of these chemicals to be quickly absorbed through the mucous membrane into the bloodstream. Immediately and gently flush eyes with plenty of warm or cold water (NO hot water) for at least 15 minutes, occasionally lifting the upper and lower eyelids. Get medical aid immediately. *Skin:* Get medical aid. Skin contact can cause dangerous amounts of these chemicals to be absorbed into the bloodstream. Wearing the appropriate PPE equipment and respirator for organophosphate pesticides, immediately flush skin with plenty of soap and water for at least 15 minutes while removing contaminated clothing and shoes. Shampoo hair promptly if contaminated. The removed, contaminated clothing and shoes should be double-bagged and left in Hot Zone for later disposal by hazardous materials experts. Skin may also be decontaminated with diluted hypochlorite solution. *Inhalation:* Get medical aid. Do not contaminate yourself. Wearing the appropriate PPE equipment and respirator for organophosphate pesticides, immediately remove the victim from the contaminated area to fresh air. If the victim is not breathing, administer artificial respiration. Do NOT use mouth-to-mouth resuscitation; use bag/mask apparatus. If breathing is difficult, administer oxygen through bag/mask apparatus until medical help arrives. Do not leave victim unattended. *Ingestion:* Call poison control. Loosen all clothing. Never give anything by mouth to an unconscious person. If victim is *unconscious or having convulsions,* do nothing except keep victim warm. Get medical aid. Transfer promptly to a medical facility. In cases of ingestion, **do not induce vomiting.** If the victim is alert and asymptomatic, administer a slurry of activated charcoal at a dose of 1 g/kg (infant, child, and adult dose). A soda can and straw may be of assistance when offering charcoal to a child. *In some cases you may be specifically instructed by poison control to induce vomiting by way of 2 tablespoons of syrup of ipecac (adult) washed down with a cup of water.* Do NOT give activated charcoal before or with ipecac syrup.

*Note to physician or authorized medical personnel:* Treat cases of respiratory compromise, coma, or excessive pulmonary secretions with respiratory support using protocols and techniques available and within the scope of training. Some cases may necessitate procedures such as endotracheal intubation or cricothyrotomy by properly trained and equipped personnel. When possible, atropine (see *Antidotes,* below) should be given under medical supervision. Patients who are comatose, hypotensive, or having seizures or cardiac arrhythmias should be treated according to advanced life support protocols. *Antidotes:* Two antidotes are administered to treat organophosphate poisoning. Atropine is a competitive antagonist of acetylcholine at muscarinic receptors and is used to control the excessive bronchial secretions which are often responsible for death. Pralidoxime relieves both the nicotinic and muscarine effects of organophosphate poisoning by regenerating acetylcholinesterase and can reduce both the bronchial secretions and the muscle weakness associated with poisoning. The initial intravenous dose of atropine in adults should be determined by the severity of symptoms: An initial adult dose of 1.0 to 2.0 mg or pediatric dose of 0.01 mg/kg (minimum 0.01 mg) should be administered intravenously. If intravenous access cannot be established, atropine may also be given intramuscularly, subcutaneously or via endotracheal tube. Doses should be repeated every 15 minutes until excessive secretions and sweating have been controlled. Once bronchial secretion has been controlled, atropine administration should be repeated whenever the secretions begin to recur. In seriously poisoned patients, very large doses may be required. Alterations of pulse rate and pupillary size should not be used as indicators of treatment adequacy. Pralidoxime should be administered as early in poisoning as possible as its efficacy may diminish when given more than 24 to 36 hours after exposure. Doses are as follows: adult 1.0 g; pediatric 25 to 50 mg/kg. The drug should be administered intravenously over 30 to 60 minutes, but in a life-threatening situation, one-half of the total dose can be given per minute for a total administration time of 2 minutes. Treatment should begin to take effect within 40 minutes with a reduction in symptoms and the amount of atropine necessary to control bronchial secretion. The initial dose can be repeated in 1 hour and then every 8 to 12 hours until the patient is clinically well and no longer requires atropine. If intravenous access cannot be established, pralidoxime may also be given intramuscularly. Early administration of diazepam in addition to the combined atropine and pralidoxime treatment may help prevent the onset of seizures and potential brain and cardiac morphologic damage following high-level organophosphate poisoning.

*References:*
- EXTOXNET, Extension Toxicology Network, "Pesticide Information Profile, Temephos," Oregon State University, Corvallis, OR (June 1996). http://extoxnet.orst.edu/pips/temephos.htm
- New Jersey Department of Health and Senior Services, "Hazardous Substance Fact Sheet: Temephos," Trenton, NJ (April 1986, rev. May 2000). http://www.state.nj.us/health/eoh/rtkweb/1780.pdf
- California Environmental Protection Agency "Chemical List of Lists," Sacramento CA (February 1997).
- Pohanish, R.P., "Rapid Guide to Hazardous Chemicals in the Environment," Van Nostrand Reinhold, New York, NY (1997).
- Agency for Toxic Substances and Disease Registry, U.S. Department of Health and Human Services, Public Health Service, "Managing Hazardous Materials Incidents," Atlanta, GA (June 2003).

# TEPP

*Use Type:* Aphicide and acaricide
*CAS Number:* 107-49-3
*Formula:* $C_8H_{20}O_7P_2$
*Alert:* There are no TEPP products currently registered in the U.S. It is a Restricted Use Pesticide (RUP)
*Synonyms:* Bis-*O,O*-diethylphosphoric anhydride; Diphosphoric acid, tetraethyl ester; ENT 18,771; Ethyl pyrophosphate, tetra-; Phosphoric acid, tetraethyl ester; Commercial 40%; Pyrophosphate de tetraethyle (French); TEP; *O,O,O,O*-tetraaethyl-diphosphat, Bis(*O,O*-diaethylphosphorsaeure-anhydrid (German); *O,O,O,O*-Tetraethyl-difosfaat (Dutch); *O,O,O,O*-Tetraetil-pirofosfato (Italian); Tetraethyl pyrofosfaat (Belgian); Tetraethyl pyrophosphate; Tetraethyl pyrophosphate, liquid; Tetrastigmine
*Trade Names:* BLADAN®; BLADON®; FOSVEX®; GRISOL®; HEPT®; HEXAMITE;KILLAX®; KILMITE 40®; LETHALAIRE® G-52; LIROHEX®; MORTOPAL®; MOTOPAL®; NIFOS®; NIFOS T®; NIFROST®; TETRON®; TETRON-100®; VAPOTONE®
*Chemical Class:* Organophosphate
*EPA/OPP PC Code:* 079601
*California DPR Chemical Code:* 577
*ICSC Number:* 1158
*RTECS Number:* UX6825000
*EEC Number:* 015-025-00-2
*Uses:* Used as an insecticide and also, in medicine, as a parasympathic nervous system stimulant.
*Regulatory Authority:*
- Very Toxic Substance (World Bank)[15]
- Air Pollutant Standard Set (ACGIH)[1] (DFG)[3] (HSE)[33] (OSHA)[58] (Several States)[60]
- AB 1803-Well Monitoring Chemical (CAL)
- Permissible Exposure Limits for Chemical Contaminants (CAL/OSHA)
- The "Director's List" (CAL/OSHA)
- Clean Water Act: Section 311 Hazardous Substances/RQ 40CFR117.3 (same as CERCLA, see below)
- EPA Hazardous Waste Number (RCRA No.): P111
- Superfund/EPCRA 40CFR355, Appendix B Extremely Hazardous Substances: TPQ = 100 lb (45.4 kg)
- Superfund/EPCRA 40CFR302.4 RQ: CERCLA, 10 lb (4.54 kg)
- U.S. DOT Inhalation Hazard Chemicals as organophosphates
- U.S. DOT Regulated Marine Pollutant (49CFR172.101, Appendix B)
- Canada, WHMIS, Ingredients Disclosure List

*Description:* TEPP is a colorless to amber liquid. Faint, fruity odor. Combustible; flash point not found. Soluble in water. Molecular weight = 290.21 Boiling point = 138°C @ 2.3 mbar. Decomposes below boiling point @170°C. Melting/Freezing point = 0°C. Vapor pressure = $1.48 \times 10^{-4}$ mbar @ 20°C; $4.5 \times 10^{-4}$ mbar @ 30°C. Log $K_{ow}$ = 2.96. Values at or above 3.0 are likely to bioaccumulate in marine organisms.

*Incompatibilities:* Strong oxidizers, alkalis, water. Hydrolyzes quickly in water, forming pyrophosphoric acid and highly flammable ethylene gas.

*Permissible Exposure Limits in Air:* The OSHA[2] TWA, the DFG MAK[3], the HSE,[33] as well as ACGIH[1] has set a TWA of 0.004 ppm (0.05 mg/m$^3$). HSE[33] adds an STEL of 0.01 ppm (0.2 mg/m$^3$). The notation "skin" is added to indicate the possibility of cutaneous absorption. The NIOSH[2] IDLH level = 5 mg/m$^3$. Several states have set guidelines or standards for TEPP in ambient air[60] ranging from 1.0 $\mu g/m^3$ (Connecticut) to 5.0 $\mu g/m^3$ (North Dakota) to 80,000 $\mu g/m^3$ (Virginia) to a much higher value for Nevada.

*Determination in Air:* Chromosorb tube-102(2); Toluene; Gas chromatography/Flame photometric detection for sulfur, nitrogen, or phosphorus; NIOSH IV Method #2504[18], Tetraethyl Pyrophosphate. OSHA versatile sampler-2; Toluene/Acetone; Gas chromatography/Flame photometric detection for sulfur, nitrogen, or phosphorus; NIOSH Method IV Method #5600, Organophosphorus pesticides.[18]

*Permissible Concentration in Water:* No criteria set, but runoff from spills or fire control may cause water pollution.

*Routes of Entry:* Inhalation, skin absorption, ingestion, skin and/or eye contact.

*Harmful Effects and Symptoms*
*Short Term Exposure:* Symptoms of exposure include eye pain, blurred vision, lacrimation (discharge of tears), rhinorrhea (discharge of thin nasal mucous), headache, chest tightness, cyanosis; anorexia, nausea, vomiting, diarrhea, weakness, twitching, paralysis, Cheyne-Stokes respiration, convulsions, low blood pressure, cardiac irregular/irregularities; sweating. TEPP is classified as super toxic. Probable oral lethal dose in humans is less than 5 mg/kg (a taste) for a 150 lb person. A small drop in the eye may cause death. Small doses at frequent intervals are additive. Poisonings always develop at a rapid rate. It is a cholinesterase inhibitor. Delayed pulmonary edema may occur after inhalation.

*Long Term Exposure:* Cholinesterase inhibitor; cumulative effect is possible. This chemical may damage the nervous system with repeated exposure, resulting in convulsions, respiratory failure. May cause liver damage.

*Points of Attack:* Eyes, respiratory system, central nervous system, cardiovascular system, gastrointestinal tract and blood cholinesterase.

*Medical Surveillance:* Medical observation is recommended for 24 to 48 hours after breathing overexposure, as pulmonary edema may be delayed. Before employment and at regular times after that, the following are recommended: Plasma and red blood cell cholinesterase levels (tests for the enzyme poisoned by this chemical). If exposure stops, plasma levels return to normal in 1-2 weeks while red blood cell levels may be reduced for 1-3 months. When acetylcholinesterase enzyme levels are reduced by 25% or more below preemployment levels, risk of poisoning is increased, even if results are in lower ranges of "normal." Reassignment to work not involving organophosphate or carbamate pesticides is recommended until enzyme levels recover. If symptoms develop or overexposure occurs, repeat the above tests as soon as possible and get an exam of the nervous system. Also consider complete blood count. Consider chest x-ray following acute overexposure. Do not drink any alcoholic beverages before or during use. Alcohol promotes absorption of organophosphates.

*First Aid:* **Treatment for organophosphate poisoning consists of thorough decontamination, cardiorespiratory support, and administration of the antidotes atropine and pralidoxime. In cases of severe poisoning, diazepam, an anticonvulsant, should also be administered. Antidotes should be administered as prevention even if the diagnosis is in doubt.** Speed in removing material from eyes and skin is of extreme importance. *Eyes:* Eye contact can cause dangerous amounts of these chemicals to be quickly absorbed through the mucous membrane into the bloodstream. Immediately and gently flush eyes with plenty of warm or cold water (NO hot water) for at least 15 minutes, occasionally lifting the upper and lower eyelids. Get medical aid immediately. *Skin:* Get medical aid. Skin contact can cause dangerous amounts of these chemicals to be absorbed into the bloodstream. Wearing the appropriate PPE equipment and respirator for organophosphate pesticides, immediately flush skin with plenty of soap and water for at least 15 minutes while removing contaminated clothing and shoes. Shampoo hair promptly if contaminated. The removed, contaminated clothing and shoes should be double-bagged and left in Hot Zone for later disposal by hazardous materials experts. Skin may also be decontaminated with diluted hypochlorite solution. *Inhalation:* Get medical aid. Do not contaminate yourself. Wearing the appropriate PPE equipment and respirator for organophosphate pesticides, immediately remove the victim from the contaminated area to fresh air. If the victim is not breathing, administer artificial respiration. Do NOT use mouth-to-mouth resuscitation; use bag/mask apparatus. If breathing is difficult, administer oxygen through bag/mask apparatus until medical help arrives. Do not leave victim unattended. *Ingestion:* Call poison control. Loosen all clothing. Never give anything by mouth to an unconscious person. If victim is *unconscious or having convulsions,* do nothing except keep victim warm. Get medical aid. Transfer promptly to a medical facility. In cases of ingestion, **do not induce vomiting.** If the victim is alert and asymptomatic, administer a slurry of activated charcoal at a dose of 1 g/kg (infant, child, and adult dose). A soda can and straw may be of assistance when offering charcoal to a child. *In some cases you may be specifically instructed by poison control to induce vomiting by way of 2 tablespoons of syrup of ipecac (adult) washed down with a cup of water.* Do NOT give activated charcoal before or with ipecac syrup.

*Note to physician or authorized medical personnel:* Treat cases of respiratory compromise, coma, or excessive pulmonary secretions with respiratory support using protocols and techniques available and within the scope of training. Some cases may necessitate procedures such as endotracheal intubation or cricothyrotomy by properly trained and equipped personnel. When possible, atropine (see *Antidotes,* below) should be given under medical supervision. Patients who are comatose, hypotensive, or having seizures or cardiac arrhythmias should be treated according to advanced life support protocols. *Antidotes:* Two antidotes are administered to treat organophosphate poisoning. Atropine is a competitive antagonist of acetylcholine at muscarinic receptors and is used to control the excessive bronchial secretions which are often responsible for death. Pralidoxime relieves both the nicotinic and muscarine effects of organophosphate poisoning by regenerating acetylcholinesterase and can reduce both the bronchial secretions and the muscle weakness associated with poisoning. The initial intravenous dose of atropine in adults should be determined by the severity of symptoms: An initial adult dose of 1.0 to 2.0 mg or pediatric dose of 0.01 mg/kg (minimum 0.01 mg) should be administered intravenously. If intravenous access cannot be established, atropine may also be given intramuscularly, subcutaneously or via endotracheal tube. Doses should be repeated every 15 minutes until excessive secretions and sweating have been controlled. Once bronchial secretion has been controlled, atropine administration should be repeated whenever the secretions begin to recur. In seriously poisoned patients, very large doses may be required. Alterations of pulse rate and pupillary size should not be used as indicators of treatment adequacy. Pralidoxime should be administered as early in poisoning as possible as its efficacy may diminish when given more than 24 to 36 hours after exposure. Doses are as follows: adult 1.0 g; pediatric 25 to 50 mg/kg. The drug should be administered intravenously over 30 to 60 minutes, but in a life-threatening situation, one-half of the total dose can be given per minute for a total administration time of 2 minutes. Treatment should begin to take effect within 40 minutes with a reduction in symptoms and the amount of atropine necessary to control bronchial secretion. The initial dose can be repeated in 1 hour and then every 8 to 12 hours until the patient is clinically well and no longer requires atropine. If intravenous access cannot be established, pralidoxime may also be given intramuscularly. Early administration of diazepam in addition to the combined atropine and pralidoxime treatment may help prevent the onset of seizures and potential brain and cardiac morphologic damage following high-level organophosphate poisoning.

*References:*
- EXTOXNET, Extension Toxicology Network, "Pesticide Information Profile, Cholinesterase Inhibitor," Oregon State University, Corvallis, OR (September 1993). http://extoxnet.orst.edu/tibs/cholines.htm
- U.S. Environmental Protection Agency, "Chemical Profile: TEPP," Washington, DC, Chemical Emergency Preparedness Program (November 30, 1987).
- California Environmental Protection Agency "Chemical List of Lists," Sacramento CA (February 1997).
- Agency for Toxic Substances and Disease Registry, U.S. Department of Health and Human Services, Public Health Service, "Managing Hazardous Materials Incidents," Atlanta, GA (June 2003).
- New Jersey Department of Health and Senior Services, "Hazardous Substance Fact Sheet: TEPP," Trenton, NJ (January 1987, rev. January 2001). http://www.state.nj.us/health/eoh/rtkweb/1781.pdf

# Terbacil (ANSI)

*Use Type:* Herbicide
*CAS Number:* 5902-51-2
*Formula:* $C_9H_{13}ClN_2O_2$
*Synonyms:* 3-*tert*-Butyl-5-chlor-6-methyluracil (German); 3-*tert*-Butyl-5-chloro-6-methyluracil; 5-Chloro-3-*tert*-butyl-6-methyluracil; 5-Chloro-3-(1,1-dimethylethyl)-6-methyl-2,4(1$H$,3$H$)-pyrimidinedione; 5-Chloro-3-(1,1-dimethyl)-6-methyl-2,4(1$H$,3$H$)-pyrimidinedione; 5-Chloro-3-(1,1-dimethylethyl)-6-methyl-2,4(1H,3H)-pyrimidinedione; 5-Chloro-3-*tert*-butyl-6-methyluracil; 2,4(1$H$,3$H$)-Pyrimidinedione, 5-chloro-3-(1,1-dimethylethy)-6-methyl-; Uracil, 3-*tert*-butyl-5-chloro-6-methyl-
*Trade Names:* COMPOUND® 732, DuPont Crop Protection (USA); DUPONT® 732, DuPont Crop Protection (USA); EXPERIMENTAL HERBICIDE 732, DuPont Crop Protection (USA); GEONTER®; SINBAR®, DuPont Crop Protection (USA); TURBSVIL®; ZOBAR®, DuPont Crop Protection (USA), canceled 12/16/1991
*Producers:* DuPont Crop Protection (USA)
*Chemical Class:* Substituted uracil
*EPA/OPP PC Code:* 012701
*California DPR Chemical Code:* 532
*RTECS Number:* YQ9360000
*Uses:* Terbacil is a selective herbicide that treats a broad spectrum of broadleaf weeds and grasses. It is formulated as a wettable powder and is applied by aircraft or ground equipment on terrestrial food and feed crops (e.g., apples, mint/peppermint/spearmint, sugarcane, and ornamentals), forestry [e.g., cottonwood (forest/shelterbelt)], terrestrial food (e.g., asparagus, blackberry, boysenberry, dewberry, loganberry, peach, raspberry, youngberry and strawberry), and terrestrial feed (e.g., alfalfa, sainfoin (hay and fodder), and forage).

*U.S. Maximum Allowable Residue Levels for Terbacil (40 CFR 180.209):*

| CROP | ppm |
|---|---|
| Alfalfa, forage | 1.0 |
| Alfalfa, hay | 2.0 |
| Apple | 0.3 |
| Asparagus | 0.4 |
| Blueberry | 0.2 |
| Caneberry subgroup 13a | 0.2 |
| Peach | 0.2 |
| Peppermint, tops | 2.0 |
| Spearmint, tops | 2.0 |
| Strawberry | 0.1 |
| Sugarcane, cane | 0.4 |
| Watermelon | 0.4 |

*Carcinogen/Hazard Classifications*
**U.S. EPA Carcinogens:** Group E, Unlikely a carcinogen
**California Prop. 65:** Developmental toxin
**U.S. TRI:** Developmental toxin
**Label Signal Word:** CAUTION
**WHO Acute Hazard:** Class U, unlikely to be hazardous
*Regulatory Authority:*
- EPCRA Section 313 Form R *de minimis* concentration reporting level: 1.0%

*Description:* White crystalline powder; colorless crystalline solid. Odorless. Slightly soluble in water; solubility = 712 ppm @ 25°C. Melting/Freezing point = 176°C. Molecular weight = 216.69. Density = 1.34 @ 25°C. Vapor pressure = 4.5 x $10^{-7}$ mmHg @ 29.5°C.
*Determination in Air:* Filter; none; Gravimetric; NIOSH IV [Particulates NOR; #0500 (total), #0600 (respirable)].[18]
*Permissible Concentration in Water:* No criteria set. Runoff from spills or fire control may cause water pollution.
*First Aid:* If this chemical gets into the eyes, remove any contact lenses at once and irrigate immediately for at least 15 minutes, occasionally lifting upper and lower lids. Seek medical attention immediately. If this chemical contacts the skin, remove contaminated clothing and wash immediately with soap and water. Seek medical attention immediately. If this chemical has been inhaled, remove from exposure, begin rescue breathing (using universal precautions) if breathing has stopped, and CPR if heart action has stopped. Transfer promptly to a medical facility. When this chemical has been swallowed, get medical attention. Give large quantities of water and induce vomiting. Do not make an unconscious person vomit. Do not induce vomiting when formulations containing petroleum solvents are ingested.
*References:*
- U.S. Environmental Protection Agency, Office of Pesticide Programs, Pesticide Residue Limits, "Terbacil," 40 CFR 180.209, www.epa.gov/pesticides/food/viewtols.htm
- U.S. Environmental Protection Agency, "Reregistration Eligibility Decision (RED), Terbacil," Office of

Prevention, Pesticides and Toxic Substances, Washington, DC (January 1998). http://www.epa.gov/REDs/0039red.pdf
- EXTOXNET, Extension Toxicology Network, "Pesticide Information Profile, Terbacil," Oregon State University, Corvallis, OR (June 1996). http://extoxnet.orst.edu/pips/terbacil.htm
- California Environmental Protection Agency "Chemical List of Lists," Sacramento CA (February 1997)

# Terbufos (ANSI)

*Use Type:* Insecticide and nematicide
*CAS Number:* 13071-79-9
*Formula:* $C_9H_{21}O_2PS_3$; $(C_2H_5O)_2PSSCH_2SC(CH_3)_3$
*Alert:* A Restricted Use Pesticide (RUP).
*Synonyms:* S-[(Tert-butylthio)methyl]-O,O-diethylphosphorodithioate; S-[((1,1-Dimethylethyl)thio)methyl]-O,O-diethylphosphorodithioate; Phosphorodithioic acid S-[(tert-butylthio)methyl]-O,O-diethylester; Phosphorodithioic acid S-[((1,1-dimethylethyl)thio)methyl]-O,O-diethyl ester
*Trade Names:* AC 921000®; ARAGRAN®; CONTRAVEN®; COUNTER®, BASF Agricultural Products Group (Germany); COUNTER 15G SYSTEMIC INSECTICIDE®, BASF Agricultural Products Group (Germany); PLYDOX®; TERBUROX®, Rotam Agrochemical (Hong Kong)
*Producers:* Biesterfeld Siemsgluess International. GmbH (Germany); BASF Agricultural Products Group (Germany); Hebei Long Age Pesticide (China); Ki-Hara Chemicals Ltd. (UK); Rotam Agrochemical (Hong Kong); Shenzhen Guomeng Industry Co., Ltd. (China); Sigma-Aldrich Laborchemikalien (Germany)
*Chemical Class:* Organophosphate
*EPA/OPP PC Code:* 105001
*California DPR Chemical Code:* 2925
*RTECS Number:* TD7200000
*EINECS Number:* 235-963-8
*Uses:* This insecticide and nematicide is applied at planting time to corn, sugar beets, sorghum, maize, cotton, bananas and cabbage. It controls corn rootworms, wireworms, white grubs, maggots, billbugs and nematodes. Some above ground pests can be controlled when soil has been treated with terbufos. Terbufos has no residential use.
*U.S. Maximum Allowable Residue Levels for Terbufos (40 CFR 180.352):*

| CROP | ppm |
|---|---|
| Banana | 0.025 |
| Beet, sugar, roots | 0.05 |
| Beet, sugar, tops | 0.1 |
| Corn, field, forage | 0.05 |
| Corn, field, stover | 0.05 |
| Corn, grain | 0.05 |
| Corn, pop, forage | 0.5 |
| Corn, pop, stover | 0.5 |
| Corn, sweet, forage | 0.5 |
| Corn, sweet, kernel plus cob with husks removed | 0.05 |
| Corn, sweet, stover | 0.5 |
| Sorghum, forage | 0.5 |
| Sorghum, grain | 0.05 |
| Sorghum, grain, stover | 0.5 |

*Carcinogen/Hazard Classifications*
**U.S. EPA Carcinogens:** Group E, unlikely carcinogen
**Label Signal Word:** DANGER
**WHO Acute Hazard:** Class Ia, Extremely Hazardous
*Regulatory Authority:*
- Superfund/EPCRA 40CFR355, Appendix B Extremely Hazardous Substances: TPQ = 100 lb (45.4 kg)
- Superfund/EPCRA 40CFR302.4 RQ: EHS, 1 lb (0.454 kg)
- U.S. DOT Inhalation Hazard Chemicals as organophosphates
- U.S. DOT Regulated Marine Pollutant (49CFR172.101, Appendix B), severe pollutant

*Description:* Colorless to pale yellow liquid. Slightly soluble in water. Molecular weight = 288.45. Boiling point = 70°C @ 0.01 mm. Melting/Freezing point = –29°C. Vapor pressure = $3 \times 10^{-4}$ mmHg @ 20°C. Flash point = 88°C (oc). Hazard Identification (based on NFPA-704 M Rating System): Health 4, Flammability 3, Reactivity 0.
*Incompatibilities:* Strong oxidizers.
*Permissible Exposure Limits in Air:* No standards set.
*Determination in Air:* OSHA versatile sampler-2; Toluene/Acetone; Gas chromatography/Flame photometric detection for sulfur, nitrogen, or phosphorus; NIOSH Method IV Method #5600, Organophosphorus pesticides.[18]
*Permissible Concentration in Water:* The U.S. EPA (see reference below) has developed a lifetime health advisory of 0.18 μg/L. No other criteria set, but runoff from spills or fire control may cause water pollution.
*Determination in Water:* Analysis of terbufos is by a gas chromatographic (GC) method applicable to the determination of certain nitrogen-phosphorus containing pesticides in water samples. In this method, approximately 1 liter of sample is extracted with Methylene chloride. The extract is concentrated and the compounds are separated using capillary column GC. Measurement is made using a nitrogen-phosphorus detector.
*Routes of Entry:* Inhalation, ingestion and skin contact.
*Harmful Effects and Symptoms*
*Short Term Exposure:* This material may be fatal if swallowed, inhaled, or absorbed through the skin. Repeated inhalation or skin contact may progressively increase susceptibility to poisoning. Acute exposure to terbufos may produce the following signs and symptoms: pinpoint pupils, blurred vision, headache, dizziness, muscle spasms, and profound weakness. Vomiting, diarrhea, abdominal pain,

seizures, and coma may also occur. The heart rate may decrease following oral exposure or increase following dermal exposure. Chest pain may be noted. Hypotension (low blood pressure) may be noted, although hypertension (high blood pressure) is not uncommon. Respiratory symptoms include dyspnea (shortness of breath), respiratory depression, and respiratory paralysis. Psychosis may occur. Delayed pulmonary edema may occur after inhalation.

*Long Term Exposure:* Cholinesterase inhibitor; cumulative effect is possible. This chemical may damage the nervous system with repeated exposure, resulting in convulsions, respiratory failure. May cause liver damage.

*Points of Attack:* Respiratory system, lungs, central nervous system, cardiovascular system, skin, eyes, plasma and red blood cell cholinesterase.

*Medical Surveillance:* Medical observation is recommended for 24 to 48 hours after breathing overexposure, as pulmonary edema may be delayed. Before employment and at regular times after that, the following are recommended: Plasma and red blood cell cholinesterase levels (tests for the enzyme poisoned by this chemical). If exposure stops, plasma levels return to normal in 1-2 weeks while red blood cell levels may be reduced for 1-3 months. When acetylcholinesterase enzyme levels are reduced by 25% or more below preemployment levels, risk of poisoning is increased, even if results are in lower ranges of "normal." Reassignment to work not involving organophosphate or carbamate pesticides is recommended until enzyme levels recover. If symptoms develop or overexposure occurs, repeat the above tests as soon as possible and get an exam of the nervous system. Also consider complete blood count. Consider chest x-ray following acute overexposure. Do not drink any alcoholic beverages before or during use. Alcohol promotes absorption of organophosphates.

*First Aid:* **Treatment for organophosphate poisoning consists of thorough decontamination, cardiorespiratory support, and administration of the antidotes atropine and pralidoxime. In cases of severe poisoning, diazepam, an anticonvulsant, should also be administered. Antidotes should be administered as prevention even if the diagnosis is in doubt.** Speed in removing material from eyes and skin is of extreme importance. *Eyes:* Eye contact can cause dangerous amounts of these chemicals to be quickly absorbed through the mucous membrane into the bloodstream. Immediately and gently flush eyes with plenty of warm or cold water (NO hot water) for at least 15 minutes, occasionally lifting the upper and lower eyelids. Get medical aid immediately. *Skin:* Get medical aid. Skin contact can cause dangerous amounts of these chemicals to be absorbed into the bloodstream. Wearing the appropriate PPE equipment and respirator for organophosphate pesticides, immediately flush skin with plenty of soap and water for at least 15 minutes while removing contaminated clothing and shoes. Shampoo hair promptly if contaminated. The removed, contaminated clothing and shoes should be double-bagged and left in Hot Zone for later disposal by hazardous materials experts. Skin may also be decontaminated with diluted hypochlorite solution. *Inhalation:* Get medical aid. Do not contaminate yourself. Wearing the appropriate PPE equipment and respirator for organophosphate pesticides, immediately remove the victim from the contaminated area to fresh air. If the victim is not breathing, administer artificial respiration. Do NOT use mouth-to-mouth resuscitation; use bag/mask apparatus. If breathing is difficult, administer oxygen through bag/mask apparatus until medical help arrives. Do not leave victim unattended. *Ingestion:* Call poison control. Loosen all clothing. Never give anything by mouth to an unconscious person. If victim is *unconscious or having convulsions,* do nothing except keep victim warm. Get medical aid. Transfer promptly to a medical facility. In cases of ingestion, **do not induce vomiting.** If the victim is alert and asymptomatic, administer a slurry of activated charcoal at a dose of 1 g/kg (infant, child, and adult dose). A soda can and straw may be of assistance when offering charcoal to a child. *In some cases you may be specifically instructed by poison control to induce vomiting by way of 2 tablespoons of syrup of ipecac (adult) washed down with a cup of water.* Do NOT give activated charcoal before or with ipecac syrup.

*Note to physician or authorized medical personnel:* Treat cases of respiratory compromise, coma, or excessive pulmonary secretions with respiratory support using protocols and techniques available and within the scope of training. Some cases may necessitate procedures such as endotracheal intubation or cricothyrotomy by properly trained and equipped personnel. When possible, atropine (see *Antidotes,* below) should be given under medical supervision. Patients who are comatose, hypotensive, or having seizures or cardiac arrhythmias should be treated according to advanced life support protocols. *Antidotes:* Two antidotes are administered to treat organophosphate poisoning. Atropine is a competitive antagonist of acetylcholine at muscarinic receptors and is used to control the excessive bronchial secretions which are often responsible for death. Pralidoxime relieves both the nicotinic and muscarine effects of organophosphate poisoning by regenerating acetylcholinesterase and can reduce both the bronchial secretions and the muscle weakness associated with poisoning. The initial intravenous dose of atropine in adults should be determined by the severity of symptoms: An initial adult dose of 1.0 to 2.0 mg or pediatric dose of 0.01 mg/kg (minimum 0.01 mg) should be administered intravenously. If intravenous access cannot be established, atropine may also be given intramuscularly, subcutaneously or via endotracheal tube. Doses should be repeated every 15 minutes until excessive secretions and sweating have been controlled. Once bronchial secretion has been controlled, atropine administration should be repeated whenever the secretions begin to recur. In seriously poisoned patients, very large doses may be required.

Alterations of pulse rate and pupillary size should not be used as indicators of treatment adequacy. Pralidoxime should be administered as early in poisoning as possible as its efficacy may diminish when given more than 24 to 36 hours after exposure. Doses are as follows: adult 1.0 g; pediatric 25 to 50 mg/kg. The drug should be administered intravenously over 30 to 60 minutes, but in a life-threatening situation, one-half of the total dose can be given per minute for a total administration time of 2 minutes. Treatment should begin to take effect within 40 minutes with a reduction in symptoms and the amount of atropine necessary to control bronchial secretion. The initial dose can be repeated in 1 hour and then every 8 to 12 hours until the patient is clinically well and no longer requires atropine. If intravenous access cannot be established, pralidoxime may also be given intramuscularly. Early administration of diazepam in addition to the combined atropine and pralidoxime treatment may help prevent the onset of seizures and potential brain and cardiac morphologic damage following high-level organophosphate poisoning.

*References:*
- U.S. Environmental Protection Agency, Office of Pesticide Programs, Pesticide Residue Limits, "Terbufos," 40 CFR 180.352, http://www.epa.gov/pesticides/food/viewtols.htm
- EXTOXNET, Extension Toxicology Network, "Pesticide Information Profile, Terbufos," Oregon State University, Corvallis, OR (June 1996). http://extoxnet.orst.edu/pips/terbufos.htm
- U.S. Environmental Protection Agency, "Interim Reregistration Eligibility Decision (IRED) Fact Sheet, Terbufos," Washington, DC (October 2001). http://www.epa.gov/REDs/factsheets/terbufos_ired_fs.htm
- New Jersey Department of Health and Senior Services, "Hazardous Substance Fact Sheet, Terbufos," Trenton, NJ (May 2002). http://www.state.nj.us/health/eoh/rtkweb/2801.pdf
- U.S. Environmental Protection Agency, "Health Advisory: Terbufos," Washington, DC, Office of Drinking Water (August 1987).
- California Environmental Protection Agency "Chemical List of Lists," Sacramento CA (February 1997).
- Pohanish, R.P., "Rapid Guide to Hazardous Chemicals in the Environment," Van Nostrand Reinhold, New York, NY (1997).
- Agency for Toxic Substances and Disease Registry, U.S. Department of Health and Human Services, Public Health Service, "Managing Hazardous Materials Incidents," Atlanta, GA (June 2003).
- U.S. Environmental Protection Agency, "Chemical Profile: Terbufos," Washington, DC, Chemical Emergency Preparedness Program (November 30, 1987).

# Terbutryn (ANSI)

*Use Type:* Herbicide
*CAS Number:* 886-50-0
*Formula:* $C_{10}H_{19}N_5S$
*Synonyms:* 4-Aethylamino-2-*tert*-butylamino-6-methylthio-*S*-triazin (German); 2-*tert*-Butylamino-4-ethylamino-6-methylmercapto-*S*-triazine; 2-*tert*-Butylamino-4-ethylamino-6-methylthio-*S*-triazine; Caswell No. 125-D; 2-Methylthio-4-ethylamino-6-*tert*-butylamino-*S*-triazine; Terbutryne; *S*-Triazine, 2-(*tert*-butylamino)-4-(ethylamino)-6-(methylthio)-; 1,3,5-Triazine-2,4-diamine, *N*-(1,1-dimethylethyl)-*N'*-ethyl-6-(methylthio)-
*Trade Names:* A-1866®; BATTALION®, Makhteshim-Agan Industries (Israel); CLAROSAN®; GESAPRIM®; GS-14260®; HS-14260®; IGRAN®, Syngenta (Switzerland), canceled; IGRATER®; PLANTONIT®; PREBANE®; PROKIL, Gowan (USA), canceled; SENATE®; SHORT-STOP®; SHORTSTOP®; TERBUTREX®
*Producers:* Gowan (USA); Makhteshim-Agan Industries (Israel); Syngenta (Switzerland)
*Chemical Class:* Triazine
*EPA/OPP PC Code:* 080813
*California DPR Chemical Code:* 1691
*Uses:* Not registered in the U.S. Terbutryn is a selective herbicide for pre-emergence and post-emergence control of most grasses and annual broadleaf weeds. It is used in winter wheat, winter barley, sorghum, sugarcane, sunflowers, peas and potatoes. It is also used as an aquatic herbicide for control of algae and floating weeds in waterways, reservoirs, and fish ponds.
*Human toxicity (long-term)*[77]: Extra high–0.70 ppb, Health Advisory
*Fish toxicity (threshold)*[77]: Low–101.20343 ppb, MATC (Maximum Acceptable Toxicant Concentration)
*Carcinogen/Hazard Classifications*
**U.S. EPA Carcinogens:** Group C, possible carcinogen
**Label Signal Word:** CAUTION or WARNING
**WHO Acute Hazard:** Class U, unlikely to be hazardous
*Description:* Colorless crystalline solid or white powder. Very slightly soluble in water; solubility = 26 mg/L @ 20°C. Molecular weight = 241.39. Melting/Freezing point = 104°C. Boiling point = 154–160°C @ 0.06 mmHg. Vapor pressure = 2.3 x $10^{-6}$ mbar @ 20°C; also listed at 9.6 x $10^{-6}$ mmHg @ 20°C. Log $K_{ow}$ = 3.55. Values above 3.0 are likely to bioaccumulate in marine organisms.
*Incompatibilities:* Hydrolized in strong acids or basic solutions and is decomposed by ultraviolet exposure. It is not corrosive.
*Permissible Concentration in Water:* No criteria set. Runoff from spills or fire control may cause water pollution.
*Routes of Entry:* Inhalation, passing through the skin, ingestion.

## Harmful Effects and Symptoms

**Short Term Exposure:** May cause skin and severe eye irritation. Moderately poisonous if ingested or inhaled. Exposure to a triazine (simazine) has caused acute and subacute dermatitis in the former USSR, characterized by erythema, slight edema, moderate pruritus, and burning lasting 4 to 5 days.

**Long Term Exposure:** May cause lung irritation and damage. May cause skin allergy. Contact with some triazine compounds (such as atrazine) may increase risks for tumors known to be associated with hormonal factors. These have been observed in both animals and human beings, and are consistent with the known effects on the hypothalamic pituitary gonadal axis. Repeated exposure may cause weight loss and reduced red blood cell count. May be mutagenic.

**Points of Attack:** Liver, lungs and skin.

**Medical Surveillance:** Before beginning employment and at regular times after that, for those with frequent or potentially high exposures, the following is recommended: Lung function tests. Consider chest x-ray following acute overexposure. Evaluation by a qualified allergist. Examination of the nervous system.

**First Aid:** If this chemical gets into the eyes, remove any contact lenses at once and irrigate immediately for at least 15 minutes, occasionally lifting upper and lower lids. Seek medical attention immediately. If this chemical contacts the skin, remove contaminated clothing and wash immediately with soap and water. Seek medical attention immediately. If this chemical has been inhaled, remove from exposure, begin rescue breathing (using universal precautions) if breathing has stopped, and CPR if heart action has stopped. Transfer promptly to a medical facility. When this chemical has been swallowed, get medical attention. Give large quantities of water or milk and induce vomiting. Do not make an unconscious person vomit.

*References:*
- EXTOXNET, Extension Toxicology Network, "Pesticide Information Profile, Terbutyrn," Oregon State University, Corvallis, OR (September 1995). http://extoxnet.orst.edu/pips/terbutry.htm
- California Environmental Protection Agency "Chemical List of Lists," Sacramento CA (February 1997)

# Tetrachlorvinphos

**Use Type:** Insecticide
**CAS Number:** 961-11-5 (E-isomer); 22248-79-9 (Z-isomer)
**Formula:** $C_{10}H_9Cl_4O_4P$
**Synonyms:** Benzyl alcohol, 2,4,5-trichloro-α-(chloromethylene)-, dimethyl phosphate; 2-Chloro-1-(2,4,5-trichlorophenyl)ethenyl dimethyl phosphate; 2-Chloro-1-(2,4,5-trichlorophenyl)vinyl dimethyl phosphate; 2-Chloro-1-(2,4,5-trichlorophenyl)vinyl phosphoric acid dimethyl ester; CVMP; *O,O*-Dimethyl-*O*-2-chloro-1-(2,4,5-trichlorophenyl)vinyl phosphate; *O,O*-Dimethyl-*O*-2-chlor-1-(2,4,5-trichlorophenyl)vinyl phosphat (German); Dimethyl 2,4,5-trichloro-α-(chloromethylene)benzyl phosphate; ENT 25841; NCI-C00168; OMS 595; Phosphoric acid, 2-chloro-1-(2,3,5-trichlorophenyl)ethenyl dimethyl ester; Phosphoric acid, 2-chloro-1-(2,4,5-trichlorophenyl)vinyl dimethyl ester; Phosphoric acid, 2-chloro-1-(2,4,5-trichlorophenyl)ethenyldimethyl ester; Phosphoric acid, 2-chloro-1-(2,3,5-trichlorophenyl)ethenyl dimethyl ester; Tetraclorvinfos (Spanish); 2,4,5-Trichloro-α-(chloromethylene)benzyl phosphate

**Trade Names:** AMERICARE RABON®, KMG Chemicals (USA); CLEAN CROP®; EQUI-FLY® ORAL LARVICIDE; EQUITROL®; FLY PATROL®; GARDONA®; IPO 8®; RABON®, KMG Chemicals (USA); RABOND®; SD 8447®; STIROFOS®

**Producers:** BASF Agricultural Products (Germany); KMG Chemicals (USA)

**Chemical Class:** Organophosphate

**EPA/OPP PC Code:** 083702 (E-isomer); 083701 (Z-isomer)

**California DPR Chemical Code:** 305

**RTECS Number:** TB9100000

**Uses:** Tetrachlorvinphos is applied dermally to livestock to control flies and mites. It is used as an oral larvicide in cattle, Hog, goats and horses; in cattle ear tags to control flies; in cattle feedlots; in poultry dust boxes to control poultry mites; and in poultry houses. Tetrachlorvinphos also is used in pet sleeping areas and pet flea collars and to control flies around refuse sites, recreational areas, and for general outdoor treatment. Currently, there are 83 products registered with the U.S. EPA. There are a total 471 products that are active, canceled or transferred.

**Human toxicity (long-term)**[77]**:** Very low–875.00 ppb, Health Advisory

**Fish toxicity (threshold)**[77]**:** Intermediate–49.75254 ppb, MATC (Maximum Acceptable Toxicant Concentration)

**U.S. Maximum Allowable Residue Levels for Tetrachlorvinphos (40 CFR 180.252):**

| CROP | ppm |
|---|---|
| Cattle | – |
| Cattle, fat | 1.5 |
| Cattle, meat & mbyp | - |
| Egg | 0.1 |
| Goat, fat | 0.5 |
| Hog | – |
| Horse | – |
| Horse, fat | 0.5 |
| Milk, fat | 0.5 |
| Poultry, fat | 0.75 |

**Carcinogen/Hazard Classifications**
**U.S. EPA Carcinogens:** Group C, possible carcinogen
**IARC:** Group 3, unclassifiable
**Label Signal Word:** CAUTION
**WHO Acute Hazard:** Class U, unlikely to be hazardous

***Regulatory Authority:***
- EPA 40 CFR 372.65, Specific Toxic Chemical Listings
- AB 2588-Air Toxics "Hot Spots" Chemicals (CAL)
- The "Director's List" (CAL/OSHA)
- Actively registered pesticide in California.
- EPCRA Section 313 Form R *de minimis* concentration reporting level: 1.0%
- DOT Inhalation Hazard Chemicals as organophosphates

***Description:*** Off-white powder. Slightly soluble in water. Melting/Freezing point = 98°C. Vapor pressure = $4.2 \times 10^{-8}$ mmHg @ 20°C.

***Incompatibilities:*** May react violently with antimony(V) pentafluoride. Incompatible with lead diacetate, magnesium, silver nitrate. When heated to decomposition, this material forms toxic oxides of phosphorus and fumes of chlorine.

***Determination in Air:*** OSHA versatile sampler-2; Toluene/Acetone; Gas chromatography/Flame ionization detection; NIOSH IV, Method #5600, Organophosphorus Pesticides.[18]

***Permissible Concentration in Water:*** No criteria set. Runoff from spills or fire control may cause water pollution.

***Routes of Entry:*** Inhalation, ingestion, skin and eye contact.

***Harmful Effects and Symptoms***

***Short Term Exposure:*** Eye pupils are small, blurred vision, eye watering, runny nose, cough, shortness of breath, salivation, dizziness, nausea, stomach cramps, diarrhea, and vomiting, increased blood pressure, profuse sweating, hypermotility, hallucinations, irritability, tingling of the skin, drowsiness, slow heartbeat, convulsions, fluid in lungs, loss of consciousness, incontinence, breathing stops, death. Organophosphates inhibit the action of acetylcholinesterase enzymes, and alter the way in which nervous impulses are transmitted. The effects can last for hours, days, or much longer. The action of the enzymes is reestablished after new enzymes are formed.

***Long Term Exposure:*** Cholinesterase inhibitor; cumulative effect is possible. Organophosphates may damage the nervous system with repeated exposure, resulting in convulsions, respiratory failure. May cause liver damage.

***Points of Attack:*** Respiratory system, central nervous system, cardiovascular system, blood cholinesterase, kidneys

***Medical Surveillance:*** Medical observation is recommended for 24 to 48 hours after breathing overexposure, as pulmonary edema may be delayed. As first aid for pulmonary edema, consider administering a corticosteroid spray. Cigarette smoking may exacerbate pulmonary injury and should be discouraged for at least 72 hours following exposure.

***First Aid:*** Speed in removing material from eyes and skin is of extreme importance. Eye contact can cause dangerous amounts of these chemicals to be quickly absorbed through the mucous membrane into the bloodstream. Immediately and gently flush eyes with plenty of warm or cold water (NO hot water) for at least 15 minutes, occasionally lifting the upper and lower eyelids. Get medical aid immediately. *Skin:* Get medical aid. Skin contact can cause dangerous amounts of these chemicals to be absorbed into the bloodstream. Wearing the appropriate PPE equipment and respirator for organophosphate pesticides, immediately flush skin with plenty of soap and water for at least 15 minutes while removing contaminated clothing and shoes. Shampoo hair promptly if contaminated. *Ingestion:* Call poison control. Loosen all clothing. Never give anything by mouth to an unconscious person. Get medical aid. Do NOT induce vomiting.* If conscious, alert, and able to swallow, rinse mouth and have victim drink 4 to 8 ounces of water do NOT induce vomiting but immediately administer slurry of activated charcoal (2 oz in 8 oz of water). If victim is *unconscious or having convulsions,* do nothing except keep victim warm. **In some cases you may be specifically instructed by poison control to induce vomiting by way of 2 tablespoons of syrup of ipecac (adult) washed down with a cup of water.* Do NOT give activated charcoal before or with ipecac syrup. *Inhalation:* Get medical aid. Do not contaminate yourself. Wearing the appropriate PPE equipment and respirator for organophosphate pesticides, immediately remove the victim from the contaminated area to fresh air. If the victim is not breathing, administer artificial respiration. Do NOT use mouth-to-mouth resuscitation; use bag/mask apparatus. If breathing is difficult, administer oxygen through bag/mask apparatus until medical help arrives. Do not leave victim unattended. *Note to physician or authorized medical personnel.* Administer atropine, 2 mg (1/30 gr) intramuscularly or intravenously as soon as any local or systemic signs or symptoms of an intoxication are noted; repeat the administration of atropine every 3 to 8 minutes until signs of atropinization (mydriasis, dry mouth, rapid pulse, hot and dry skin) occur; initiate treatment in children with 0.05 mg mg/kg of atropine; repeat at 5 to 10 minute intervals. Watch respiration, and remove bronchial secretions if they appear to be obstructing the airway; intubate if necessary. *Notes to physician or authorized medical personnel*: N-methylpyridinium-2-aldoxime (2-PAMCl) when used in conjunction with atropine reacts with the phosphorylated cholinesterase, thereby restoring normal activity to by removing the phosphorylating group. The combination of these two chemicals is synergistic and must be administered within minutes to a few hours following exposure (depending on the specific agent) to be effective. Give 2-PAMCI (Pralidoxime; Protopam), 2.5 gm in 100 ml of sterile water or in 5% dextrose and water, intravenously, slowly, in 15-30 minutes; if sufficient fluid is not available, give 1 gm of 2-PAMCI in 3 ml of distilled water by deep intramuscular injection; repeat this every half hour if respiration weakens or if muscle fasciculation or convulsions recur. Also Diazepam, an anticonvulsant, might be considered.

*References:*
- U.S. Environmental Protection Agency, Office of Pesticide Programs, Pesticide Residue Limits, "Tetrachlorvinphos," 40 CFR 180., www.epa.gov/pesticides/food/viewtols.htm
- U.S. Environmental Protection Agency, "Reregistration Eligibility Decision (RED), Tetrachlorvinphos," Office of Prevention, Pesticides and Toxic Substances, Washington, DC (September 1995). http://www.epa.gov/REDs/0321red.pdf
- California Environmental Protection Agency "Chemical List of Lists," Sacramento CA (February 1997)

## Tetradifon (ANSI)

*Use Type:* Insecticide, acaricide, ovicide
*CAS Number:* 116-29-0
*Formula:* $C_{12}H_6Cl_4O_2S$
*Synonyms:* 4-Chlorophenyl 2,4,5-trichlorophenyl sulfone; *p*-Chlorophenyl 2,4,5-trichlorophenyl sulfone; ENT 23,737; Sulfone-2,4,4',5-tetrachlorodiphenyl sulfone; Sulfone, *p*-chlorophenyl 2,4,5-trichlorophenyl; 2,4,4',5-Tetrachlordiphenyl-sulfon (German); 2,4,4',5-Tetrachlorodiphenyl sulfone; 2,4,5,4'-Tetrachlorodiphenylsulphone; Tetradiphon; 1,2,4-trichloro-5-[( 4-chlorophenyl)sulfonyl]-benzene
*Trade Names:* ACAROIL TD®; ACARVIN®; AGREX T-7.5®; AKARITOX®; ARACNOL K®; AREDION®; DE PESTER TEDION®, Crompton (USA), canceled; DICTATOR-PLUS®, VAPCO® (Jordan); DUPHAR®; FMC 5488®, FMC Agricultural Products (USA), canceled; MITION®; NIAGARA 5488®, FMC Agricultural Products (USA), canceled; MITIFON®, NOX-MITE®, canceled; POLACARITOX®; ROZTOCZOL®, ROZTOZOL®; TEDION®; TEDION V-18®; V-18®; TETRADICHLONE®, TETRANOL V18®; VAPCOTHION®, VAPCO (Jordan)
*Producers:* Crompton (USA); FMC Agricultural Products (USA); VAPCO (Jordan); Wuzhou International (China)
*Chemical Class:* Organochlorine
*EPA/OPP PC Code:* 079202
*California DPR Chemical Code:* 581
*ICSC Number:* 0747
*RTECS Number:* WR5850000
*Uses:* Used to control eggs and young active stages of phytophagous mites on deciduous fruits, citrus, cotton, vines, vegetables, ornamentals, cotton, hops, coffee, tea, and rice. It is also a food additive permitted in food for human consumption.
*U.S. Maximum Allowable Residue Levels for Tetradifon (40 CFR 180.174):*

| CROP | ppm |
|---|---|
| Apple | 5 |
| Apricot | 5 |
| Cherry | 5 |
| Citron, citrus | 2 |
| Crabapple | 5 |
| Cucumber | 1 |
| Fig | 6 |
| Fig, dried fruit | 10 |
| Grape | 5 |
| Grapefruit | 2 |
| Hop, dried cones | 120 |
| Hop, fresh | 30 |
| Lemon | 2 |
| Lime | 2 |
| Meat | 0 |
| Melon | 1 |
| Milk | 0 |
| Nectarine | 5 |
| Orange | 2 |
| Peach | 5 |
| Pear | 5 |
| Peppermint | 100 |
| Plum, prune, fresh | 5 |
| Pumpkin | 1 |
| Quince | 5 |
| Spearmint | 100 |
| Squash, winter | 1 |
| Strawberry | 5 |
| Tangerine | 2 |
| Tea, dried | 8 |
| Tomato | 1 |

*Carcinogen/Hazard Classifications*
**Label Signal Word:** CAUTION or DANGER
**WHO Acute Hazard:** Class U, unlikely to be hazardous
*Regulatory Authority:*
- FIFRA, 180.3(4); class of chlorinated organic pesticide
- FIFRA, 40CFR185: tolerances for pesticides in food.

*Description:* Colorless to white crystalline solid or whitish to slightly yellow powder. Liquid may contain a flammable solvent. Odorless. Practically insoluble in water; solubility = 50 µg/L @ 20°C. Molecular weight = 356.04. Melting/Freezing point = 145-148°C. Vapor pressure = 0.032 mPa @ 20°C. Log $K_{ow}$ = 4.58. Values above 3.0 are likely to bioaccumulate in marine organisms.
*Incompatibilities:* Strong oxidizers. Sulfur oxides and hydrogen chloride are produced in the heat of fire.
*Permissible Concentration in Water:* No criteria set. Runoff from spills or fire control may cause water pollution.
*Routes of Entry:* Inhalation of dust.
*Harmful Effects and Symptoms*
*Short Term Exposure:* May cause irritation of the eyes, skin, and respiratory tract.
*Long Term Exposure:* May cause kidney and liver damage.
*Points of Attack:* Kidneys and liver.
*Medical Surveillance:* Kidney and liver function tests.
*First Aid:* If this chemical gets into the eyes, remove any contact lenses at once and irrigate immediately for at least 15 minutes, occasionally lifting upper and lower lids. Seek medical attention immediately. If this chemical contacts the

skin, remove contaminated clothing and wash immediately with soap and water. Seek medical attention immediately. If this chemical has been inhaled, remove from exposure, begin rescue breathing (using universal precautions) if breathing has stopped, and CPR if heart action has stopped. Transfer promptly to a medical facility. When this chemical has been swallowed, get medical attention. Give large quantities of water and induce vomiting. Do not make an unconscious person vomit. Do not induce vomiting when formulations containing petroleum solvents are ingested.

*References:*
- International Programme on Chemical Safety (IPCS), "Environmental Health Criteria #67, Tetradifon," Geneva, Switzerland (1986). http://www.inchem.org/documents/ehc/ehc/ehc67.htm
- U.S. Environmental Protection Agency, Office of Pesticide Programs, Pesticide Residue Limits, "Tetradifon," 40 CFR 180. 174. http://www.epa.gov/pesticides/food/viewtols.htm
- California Environmental Protection Agency "Chemical List of Lists," Sacramento CA (February 1997).

# Tetramethrin (ANSI)

*Use Type:* Insecticide
*CAS Number:* 7696-12-0
*Formula:* $C_{19}H_{25}NO_4$
*Synonyms:* AI3-27339; Bioneopynamin; Caswell No. 844; (1-Cyclohexane-1,2-dicarboximido)methyl chrysanthemumate; Cyclohex-1-ene-1,2-dicarboximidomethyl (±)-*cis-trans*-chrysanthemate; (1-Cyclohexene-1,2-dicarboximido)methyl 2,2-dimethyl-3-(2-methylpropenyl)cyclopropanecarboxylate; Cyclopropanecarboxylic acid, 2,2-dimethyl-3-(2-methyl-1-propenyl)-, (1,3,4,5,6,7-hexahydro-1,3-dioxo-2*H*-isoindol-2-yl)methyl ester (9CI); 2,2-Dimethyl-3-(2-methyl-1-propenyl)cyclopropanecarboxylic acid (1,3,4,5,6,7-hexahydro-1,3-dioxo-2*H*-isoindol-2-yl)methyl ester; ENT 27339; NSC 190939; *d*-Phthalthrin; 2,3,4,5-Tetrahydrophthalimidomethylchrysanthemate; 3,4,5,6-Tetrahydrophthalimidomethyl (±)-*cis-trans*-chrysanthemate; 3,4,5,6-Tetrahydrophthalimidomethyl (±)-*(Z)-(E)*-chrysanthemate; 3,4,5,6-Tetrahydrophthalimidomethyl *cis and trans* dlchrysanthemummonocarboxylic acid; *N*-(3,4,5,6-Tetra hydrophthalimido)-methyl *dl-cis,trans*-chrysanthemate; *N*-(3,4,5,6-Tetrahydrophthalimido)-methyl *dl-(Z),(E)*-chrysanthemate; Tetramethrin, (±)-; Tetramethrine; Tetramethrin, racemic; (±)-*cis/trans*-Phthalthrin
*Trade Names:* FMC 9260®, FMC Agricultural Products (USA); ENT-27339; EVERCIDE INTERMEDIATE® 2265 (tetramethrin + fenvalerate); MULTICIDE®; NEO-PYNAMIN®; NEOPYNAMINE®; NEOPYNAMIN FORTE®; NIAGARA®-9260, FMC Agricultural Products (USA); NIA®-9260; PHTHALTHRIN®; SP-1103; SUMITOMO® SP-1103, Sumitomo Chemical (Japan)
*Producers:* Agsin (Singapore); Ascot International UK); Bonide Products (USA); Changzhou Kangmei Chemical Industry Co., Ltd. (China); FMC Agricultural Products (USA); Ki-Hara Chemicals Ltd. (UK); Sumitomo Chemical (Japan); SuYan Agrochemical Group (China); Valent BioSciences (USA)
*Chemical Class:* Pyrethroid
*EPA/OPP PC Code:* 069003
*California DPR Chemical Code:* 1695
*ICSC Number:* 0334
*RTECS Number:* GZ1730000
*Uses:* Tetramethrin is formulated as an aerosol and used primarily for indoor pest control or in mosquito coils. It is also used in shampoos to control fleas and ticks on pets. It is often formulated with other insecticides and synergists. Currently, there are more than 319 registered products containing tetramethrin. Another 500 products have either been canceled or transferred in the U.S.
*Human toxicity (long-term)*[77]*:* Very low–8750.00 ppb, Health Advisory
*Fish toxicity (threshold)*[77]*:* Extra high–0.26608 ppb, MATC (Maximum Acceptable Toxicant Concentration)
*Carcinogen/Hazard Classifications*
**U.S. EPA Carcinogens:** Group C, possible carcinogen
**Label Signal Word:** CAUTION
**WHO Acute Hazard:** Class U, unlikely to be hazardous
**Endocrine Disruptor:** Suspected in Colborn list
*Regulatory Authority:*
- EPA 40 CFR 372.65, Specific Toxic Chemical Listings
- Actively registered pesticide in California.
- EPCRA Section 313 Form R *de minimis* concentration reporting level: 1.0%

*Description:* White to colorless crystalline solid. Practically insoluble in water. Molecular weight = 331.39. Melting/Freezing point = 69-74°C. Vapor pressure = 10 Pa @ 20°C.
*Incompatibilities:* May react violently with strong oxidizers, bromine, 90% hydrogen peroxide, phosphorus trichloride, silver powders or dust. Incompatible with silver compounds. Mixture with some silver compounds forms explosive salts of silver oxalate.
*Permissible Concentration in Water:* No criteria set. Runoff from spills or fire control may cause water pollution.
*Harmful Effects and Symptoms*
*Short Term Exposure:* Pyrethroids can affect you when breathed in and by passing through your skin. Irritates the eyes and respiratory tract. High exposure can affect the nervous system causing headache, nausea, vomiting, fatigue, and restlessness, rhinorrhea (discharge of thin nasal mucous).
*Long Term Exposure:* High or repeated exposure can cause lung allergy (with cough, wheezing and/or shortness of breath) or hay fever symptoms (sneezing, runny or stuffy

nose). Allergic "pneumonia" can also occur with cough, chest pain, breathing difficulty and abnormal chest x-ray. Repeated attacks may lead to permanent scarring. Skin allergy may also develop with rash and itching, even with lower exposures. Skin contact can cause rash with redness, blisters and intense itching. A severe generalized allergy can occur with weakness and collapse.

*Points of Attack:* Respiratory system, skin, central nervous system.

*Medical Surveillance:* Before beginning employment and at regular times after that, the following are recommended: Lung function tests. These may be normal if the person is not having an attack at the time of the test. Consider chest x-ray if lung symptoms are present. Evaluation by a qualified allergist, including careful exposure history and special testing, may help diagnose skin allergy.

*First Aid:* If this chemical gets into the eyes, remove any contact lenses at once and irrigate immediately for at least 15 minutes, occasionally lifting upper and lower lids. Seek medical attention immediately. If this chemical contacts the skin, remove contaminated clothing and wash immediately with soap and water. Seek medical attention immediately. If this chemical has been inhaled, remove from exposure, begin rescue breathing (using universal precautions) if breathing has stopped, and CPR if heart action has stopped. Transfer promptly to a medical facility. When this chemical has been swallowed, get medical attention. Give large quantities of water and induce vomiting. Do not make an unconscious person vomit.

*References:*
- International Programme on Chemical Safety (IPCS), "Environmental Health Criteria, Tetramethrin," Geneva, Switzerland (1990). http://www.inchem.org/documents/ehc/ehc/ehc98.htm
- California Environmental Protection Agency "Chemical List of Lists," Sacramento CA (February 1997)

# Thallium Sulfate

*Use Type:* Rodenticide and insecticide
*CAS Number:* 10031-59-1(sulfate); 7446-18-6 [sulfate(I)]
*Formula:* $Tl_2SO_4$ (CAS 7446-18-6)
*Alert:* No longer registered in the U.S.
*Synonyms: Thallium Sulfate:* Sulfato de talio (Spanish); Sulfuric acid, thallium salt; Thallium sulfate; Thallium sulphate
*Thallium Sulfate(I):* C.F.S; CFS-Giftweizen (German); Dithallium sulfate; Dithallium(1+) sulfate; Dithallium(I) sulfate; Eccothal; M7-Giftkoerner; Rattengiftkonserv (German); Sulfato de talio (Spanish); Sulfuric acid, dithallium (+1) salt; Sulfuric acid, dithallium (I) salt(8CI,9CI); Sulfuric acid, thallium(1+) salt(1:2); Sulfuric acid, thallium(I) salt(1:2); Thallium(1+) sulfate (2:1); Thallium(I) Sulfate (2:1); Thallous Sulfate

*Trade Names:* RATOX®; ZELIO®
*Producers:* American Elements (USA); Fluorochem (UK); GFS Chemicals (USA); Great Western Inorganics (USA); Newmont Koch (UK); PPM Pure Metals (Germany); UMICORE (Belgium)
*Chemical Class:* Inorganic, heavy metal salt
*EPA/OPP PC Code:* 080001
*California DPR Chemical Code:*
*ICSC Number:* 0336 [thallium(I) sulfate]
*RTECS Number:* XG6600000 (sulfate); XG7800000 [sulfate(I)]
*EEC Number:* 081-002-00-9 [sulfate(I)]
*EINECS Number:* 231-201-3 [thallium(I) sulfate]
*Uses:* Thallium sulfate is used mainly to control rats, squirrels, prairie dogs, mice, house mice and moles. It is also used to control ants and cockroaches. It is readily absorbed from ingestion and also absorbed through the skin. Its unintended effects is to poison livestock, wildlife and domestic animals. It is no longer registered for pesticide use in the U.S.

*Carcinogen/Hazard Classifications*
**U.S. EPA Carcinogens:** Group D, unclassifiable
**WHO Acute Hazard:** Class Ib, highly hazardous
*Regulatory Authority:*
- Banned or Severely Restricted (Many Countries) (UN)[13]
- Air Pollutant Standard Set (ACGIH)[1] (DFG)[3] (HSE)[33] (OSHA)[58] (former USSR)[43] (Several States)[60]

*Thallium(I) sulfate (7446-18-6):*
- Clean Water Act: Section 311 Hazardous Substances/RQ 40CFR117.3 (same as CERCLA, see below); 40CFR401.15 Section 307 Toxic Pollutants
- EPA Hazardous Waste Number (RCRA No.): P115
- RCRA, 40CFR261, Appendix 8 Hazardous Constituents
- Superfund/EPCRA 40CFR302.4 RQ: CERCLA, 100 lb (45.4 kg)
- EPCRA Section 313 Form R *de minimis* concentration reporting level: 1.0%
- Superfund/EPCRA 40CFR355, Appendix B Extremely Hazardous Substances: TPQ = 100/10,000 lb (45.4/4,540 kg)
- The "Director's List" (CAL/OSHA)
- U.S. DOT Regulated Marine Pollutant (49CFR172.101, Appendix B)

*Thallium sulfate (10031-59-1):*
- Clean Water Act: Section 311 Hazardous Substances/RQ 40CFR117.3 (same as CERCLA, see below); 40CFR401.15 Section 307 Toxic Pollutants
- Superfund/EPCRA 40CFR355, Appendix B Extremely Hazardous Substances: TPQ = 100/10,000 lb (45.4/4,540 kg)
- Superfund/EPCRA 40CFR302.4 RQ: CERCLA, 100 lb (45.4 kg)
- EPCRA Section 313 Form R *de minimis* concentration reporting level: 1.0%

- Appendix B Extremely Hazardous Substances: TPQ = 100/10,000 lb (45.4/4,540 kg)
- U.S. DOT Regulated Marine Pollutant (49CFR172.101, Appendix B)

*Thallium compounds:*
- Clean Water Act: 40CFR401.15 Section 307 Toxic Pollutants as thallium and compounds.
- RCRA, 40CFR261, Appendix 8 Hazardous Constituents, waste number not listed.
- PCRA Section 313 Form R *de minimis* concentration reporting level: 1.0%
- U.S. DOT Regulated Marine Pollutant (49CFR172.101, Appendix B) as thallium compounds, n.o.s; thallium compounds (pesticides)
- Canada, WHMIS, Ingredients Disclosure List
- AB 2588-Air Toxics "Hot Spots" Chemicals (CAL)
- Permissible Exposure Limits for Chemical Contaminants (CAL/OSHA)
- The "Director's List" (CAL/OSHA)

***Description:*** Thallium sulfate(I) is a white or colorless, crystalline solid. Odorless. Slightly soluble in water; solubility 5 g/100mL @ 20°C. Melting/Freezing point = 632°C.

***Incompatibilities:*** Varies. Thallium metal reacts violently with strong acids (such as hydrochloric, sulfuric and nitric) and strong oxidizers (such as chlorine, bromine and fluorine). Reacts with other halogens at room temperature.

***Permissible Exposure Limits in Air:*** The OSHA[2] PEL, the DFG MAK[3], the HSE[33] and the ACGIH[1] TWA value for thallium (soluble compounds) is 0.1 mg Tl/m$^3$. The NIOSH[2] IDLH level = 15 mg/m$^3$. The notation "skin" is added to indicate the possibility of cutaneous absorption. The former USSR-UNEP/IRPTC project[43] has set a MAC in workplace air for thallium bromide and iodide of 0.01 mg/m$^3$. Several states have set guidelines or standards for thallium soluble compounds in ambient air[60] ranging from 0.238 $\mu$g/m$^3$ (Kansas) to 0.33 $\mu$g/m$^3$ (New York) to 1.0 $\mu$g/m$^3$ (Florida, North Dakota) to 1.6 $\mu$g/m$^3$ (Virginia) to 2.0 $\mu$g/m$^3$ (Connecticut and Nevada) to 2.47 $\mu$g/m$^3$ (Pennsylvania).

***Determination in Air:*** Filter; Acid; Inductively coupled plasma; NIOSH IV, Method #7300, Element. See also OSHA Method ID-121.[58]

***Permissible Concentration in Water:*** To protect freshwater aquatic life, 1400 $\mu$g/L on an acute toxicity basis and 40 $\mu$g/L on a chronic basis. To protect saltwater aquatic life, 2,130 $\mu$g/L on an acute toxicity basis. For the protection of human health from the toxic properties of thallium ingested through water and contaminated aquatic organisms, the ambient water criterion is 13.0 $\mu$g/L[6]. Kansas[61] has set a guideline for thallium in drinking water of 13.0 $\mu$g/L. The former USSR[35] has set an MAC in water bodies used for domestic purposes of 0.1 $\mu$g/L.

***Determination in Water:*** Digestion followed by atomic absorption measurement gives total thallium. Dissolved thallium may be determined by the same procedure preceded by 0.45 micron filtration.

***Routes of Entry:*** Ingestion and percutaneous absorption of dust, eye/skin contact.

## Harmful Effects and Symptoms

***Short Term Exposure:*** Thallium salts may be eye and skin irritants and skin sensitizers. Exposure can cause fatigue, weakness, poor appetite, insomnia, and mood changes. Acute poisoning rarely occurs in industry, and is usually due to ingestion of thallium. When it occurs, gastrointestinal symptoms, abdominal colic, loss of kidney function, peripheral neuritis, strabismus, disorientation, convulsions, joint pain, and alopecia develop rapidly. The symptoms of acute thallium poisoning (except for gastrointestinal symptoms) do not become manifest until 12 hours to 4 days after exposure. Death is due to damage to the central nervous system. Thallium may affect the peripheral and the central nervous system, liver and kidneys, the gastrointestinal tract, skin (hair) and the cardiovascular system, resulting in polyneuritis, optic nerve atrophy, encephalopathy, cardiac disturbances, liver and kidney damage, alopecia. Exposure may result in death. The nitrate can irritate and burn the skin and eyes. The nitrate can damage the nervous system causing headache, weakness, irritability, pain, "pins and needles" in arms and legs, convulsions, coma, and death. The sulfate(I) irritates the eyes and the skin. May affect the nervous system, cardiovascular system, kidneys and gastrointestinal tract. Exposure may result in death. Exposure may result in hair loss. Delayed pulmonary edema may occur after inhalation.

***Long Term Exposure:*** Thallium is an extremely toxic and cumulative poison. In non-fatal occupational cases of moderate or long-term exposure, early symptoms usually include fatigue, limb pain, metallic taste in the mouth and loss of hair, although loss of hair is not always present as an early symptom. Later, peripheral neuritis, proteinuria, and joint pains occur. Occasionally, neurological signs are the presenting factor, especially in more severe poisonings. Long-term exposure may produce optic atrophy, paresthesia, and changes in papillary and superficial tendon reflexes (slowed responses). Some thallium compounds are teratogens in animals.

***Points of Attack:*** Eyes, central nervous system, lungs, liver, kidneys, gastrointestinal tract and body hair.

***Medical Surveillance:*** Medical observation is recommended for 24 to 48 hours after breathing overexposure, as pulmonary edema may be delayed. Preplacement and periodic examinations should give special consideration to the eyes, central nervous system, gastrointestinal symptoms, and liver and kidney function. Hair loss may be a significant sign. Urine examinations may be helpful.

***First Aid:*** If this chemical gets into the eyes, remove any contact lenses at once and irrigate immediately for at least 15 minutes, occasionally lifting upper and lower lids. Seek medical attention immediately. If this chemical contacts the skin, remove contaminated clothing and wash immediately with soap and water. Seek medical attention immediately. If this chemical has been inhaled, remove from exposure, begin rescue breathing (using universal precautions) if breathing has stopped, and CPR if heart action has stopped. Transfer promptly to a medical facility. When this chemical has been swallowed, get medical attention. Give a slurry of activated charcoal in water to drink and induce vomiting. Do not make an unconscious person vomit. The symptoms of acute thallium poisoning (except for gastrointestinal symptoms) do not become manifest until 12 hours to 4 days after exposure.

*References:*
- New Jersey Department of Health and Senior Services, "Hazardous Substance Fact Sheet: Thallium Sulfate," Trenton, NJ (November 2000). http://www.state.nj.us/health/eoh/rtkweb/1842.pdf
- U.S. Environmental Protection Agency, Thallium: Ambient Water Quality Criteria, Washington, DC (1980).
- U.S. Environmental Protection Agency, Thallium, Health and Environmental Effects Profile No. 159, Washington, DC, Office of Solid Waste (April 30, 1980).Sax, N.I., Ed., "Dangerous Properties of Industrial Materials Report," 4, No. 1, 94-97 (1984) (Sulfate) 7, No. 2, 92-94 (1987) (Acetate) and 8, no. 4, 13-22 (1988) (Nitrate).
- U.S. Environmental Protection Agency, "Chemical Profile: Thallium Sulfate," Washington, DC, Chemical Emergency Preparedness Program (November 30, 1987).
- U.S. Environmental Protection Agency, "Chemical Profile: Thallous Sulfate," Washington, DC, Chemical Emergency Preparedness Program (November 30, 1987).
- California Environmental Protection Agency "Chemical List of Lists," Sacramento CA (February 1997).

# Thiabendazole

*Use Type:* Fungicide
*CAS Number:* 148-79-8
*Formula:* $C_{10}H_7N_3S$
*Synonyms:* AI3-50598; 1*H*-Benzimidazole, 2-(4-thiazolyl)-; Benzimidazole, 2-(4-thiazolyl)-; 4-(2-Benzimidazolyl)thiazole; Caswell No. 849A; TBDZ; TBZ; 2-Thiazole-4-ylbenzimidazole; 2-(Thiazol-4-yl)benzimidazole; 2-(1,3-Thiazol-4-yl)benzimidazole; 2-(4'-Thiazolyl)benzimidazole; 2-(4-Thiazolyl)benzimidazole; 2-(4-Thiazolyl)-1*H*-benzimidazole; Tiabendazol (Spanish)
*Trade Names:* AGROSOL®, Agriliance (USA); AGROSOL®-T, (with thiram); APL-LUSTER®, Cerexagri (France); ARBOTECT®; BOVIZOLE®; BRODEX®; CHEM-TEK®; CITRUS LUSTR®, Cerexagri (France); DECCO SALT NO. 19®, Cerexagri (France); E-Z-EX®; EPROFIL®; EQUIVET TZ®; EQUIZOLE®; FRESHGARD®, FMC Agricultural Products (USA); FUNGICIDE 4T®; GRANOX®, Agriliance (USA); IRGAGUARD®, Syngenta (Switzerland); LOMBRISTOP®; MERTEC®; MERTECT 160®; METASOL TK-100®, BASF Agricultural Products (Germany); MINTEZOL®; MINZOLUM®; MK-360®; MYCOZOL®; NEMAPAN®; NSC 525040®; OMNIZOLE®; POLIVAL®; RIVAL®, (captan + PCNB + thiabendazole), Gustafson (USA); RPH®; RTU-VITAVAX-EXTRA®, Gustafson (USA); STA-FRESH®, FMC Agricultural Products (USA); TBZ 6®; TECTO®; TECTO RPH®; TECTO 10P®; TECTO 40F®; TESTO®; THIABEN®; THIABENDAZOLUM®; THIABENZAZOLE®; THIABENZOLE®; THIBENZOL®; THIBENZOLE®; THIBENZOLE 200®; THIBENZOLE ATT®; TIABENDAZOLE®; TOBAZ®; TOP FORM WORMER®; VITAVAX®, Crompton (USA);
*Producers:* Agriliance (USA); Agsin (Singapore); BASF Agricultural Products (Germany); Cerexagri (France); Crompton (USA); FMC Agricultural Products (USA); Gustafson (USA); Kawaguchi Chemical Industry (Japan); Syngenta (Switzerland)
*Chemical Class:* Benzimidazole
*EPA/OPP PC Code:* 060101
*California DPR Chemical Code:* 587
*RTECS Number:* DE0700000
*Uses:* Thiabendazole is a fungicide used in diseases such as blight, mold, stain and rot that are found on fruit and vegetables; Dutch Elm disease; and diseases found in food storage. It is also used to treat roundworms and similar conditions in livestock and humans. Thiabendazole is used medicinally as a chelating agent to bind metals.
*Human toxicity (long-term)*[77]*:* Very low–700.00 ppb, Health Advisory
*Fish toxicity (threshold)*[77]*:* Intermediate–18.65477 ppb, MATC (Maximum Acceptable Toxicant Concentration)
*U.S. Maximum Allowable Residue Levels for Thiabendazole (40 CFR 180.242):*

| CROP | ppm |
|---|---|
| Apple, post-h | 10 |
| Avocado | 10 |
| Banana, post-h | 3 |
| Banana, pulp | 0.4 |
| Banana, pulp, post-h | 0.4 |
| Bean, dry | 0.1 |
| Beet, sugar, dried pulp | 3.5 |
| Beet, sugar, roots | 0.25 |
| Beet, sugar, tops | 10 |
| Cantaloupe, post-h | 15 |
| Carrot, roots, post-h | 10 |
| Cattle, fat | 0.1 |
| Cattle, meat | 0.1 |
| Cattle, mbyp | 0.1 |
| Citrus, dried pulp, post-h | 35 |
| Egg | 0.1 |
| Fruit, citrus, post-h | 10 |
| Goat, fat | 0.1 |

| | |
|---|---|
| Goat, meat | 0.1 |
| Goat, mbyp | 0.1 |
| Hog, fat | 0.1 |
| Hog, meat | 0.1 |
| Hog, mbyp | 0.1 |
| Horse, fat | 0.1 |
| Horse, meat | 0.1 |
| Horse, mbyp | 0.1 |
| Lentil, seed | 0.1 |
| Mango | 10 |
| Milk | 0.4 |
| Mushroom | 40 |
| Papaya, post-h | 5 |
| Pear, post-h | 10 |
| Potato, post-h | 10 |
| Potato, processed, potato waste | 30 |
| Poultry | 0.1 |
| Poultry, meat | 0.1 |
| Poultry, mbyp | 0.1 |
| Rice, grain | 3 |
| Rice, hulls | 8 |
| Rice, straw | 10 |
| Sheep, fat | 0.1 |
| Sheep, meat | 0.1 |
| Sheep, mbyp | 0.1 |
| Soybean | 0.1 |
| Squash, hubbard | 1 |
| Strawberry | 5 |
| Sweet potato, roots, post-h | 0.02 |
| Wheat, grain | 1 |
| Wheat, milled fractions, except flour | 3 |
| Wheat, straw | 0.1 |

***Carcinogen/Hazard Classifications***
**U.S. EPA Carcinogens:** Likely carcinogen in high doses; Unlikely Carcinogen in low doses.
**U.S. TRI:** Developmental toxin
**Label Signal Word:** CAUTION, WARNING or DANGER
**WHO Acute Hazard:** Class U, unlikely to be hazardous
***Regulatory Authority:***
- EPA 40 CFR 372.65, Specific Toxic Chemical Listings
- Actively registered pesticide in California.
- FIFRA, 40CFR185: tolerances for pesticides in food
- FIFRA, 40CFR186: tolerances for pesticides in animal feeds
- EPCRA Section 313 Form R *de minimis* concentration reporting level: 1.0%

***Description:*** White to tan crystalline solid. Odorless. Soluble in water. Melting/Freezing point = 304°C. Molecular weight = 201.26. Vapor pressure = $4.1 \times 10^{-8}$ mmHg @ 20°C.

***Incompatibilities:*** When heated to decomposition this material forms toxic oxides of nitrogen and sulfur.

***Determination in Air:*** Filter; none; Gravimetric; NIOSH IV [Particulates NOR; #0500 (total), #0600 (respirable)].[18]

***Permissible Concentration in Water:*** No criteria set. Runoff from spills or fire control may cause water pollution.

***First Aid:*** If this chemical gets into the eyes, remove any contact lenses at once and irrigate immediately for at least 15 minutes, occasionally lifting upper and lower lids. Seek medical attention immediately. If this chemical contacts the skin, remove contaminated clothing and wash immediately with soap and water. Seek medical attention immediately. If this chemical has been inhaled, remove from exposure, begin rescue breathing (using universal precautions) if breathing has stopped, and CPR if heart action has stopped. Transfer promptly to a medical facility. When this chemical has been swallowed, get medical attention. Give large quantities of water and induce vomiting. Do not make an unconscious person vomit.

***References:***
- EXTOXNET, Extension Toxicology Network, "Pesticide Information Profile, Thiabendazole," Oregon State University, Corvallis, OR (June 1996). http://extoxnet.orst.edu/pips/thiabend.htm
- U.S. Environmental Protection Agency, Office of Pesticide Programs, Pesticide Residue Limits, "Thiabendazole," 40 CFR 180.242, www.epa.gov/pesticides/food/viewtols.htm
- Pohanish, R.P., "Rapid Guide to Hazardous Chemicals in the Environment," Van Nostrand Reinhold, New York, NY (1997).
- California Environmental Protection Agency "Chemical List of Lists," Sacramento CA (February 1997).

## Thiafluamide (Flufenacet)

***Use Type:*** Herbicide
***CAS Number:*** 142459-58-3
***Formula:*** $C_{14}H_{12}O_2N_3SF_4$
***Synonyms:*** Acetamide, *N*-(4-fluorophenyl)-*N*-(1-methylethyl)-2-[(5-(trifluoromethyl)-1,3,4-thiadiazol-2-yl)oxy]-; Flufenacet; N-(4-Fluorophenyl)-*N*-(1-methylethyl)-2[(5-(trifluoromethyl)-1,3,4-thiadiazol-2-yl)oxy]acetamide benzoate
***Trade Names:*** AXIOM®, Bayer CropScience (Germany); DOMAIN®, Bayer CropScience (Germany); EPIC®, Bayer CropScience (Germany); FOE 5043® technical, Bayer CropScience (Germany)
***Producers:*** Bayer CropScience (Germany)
***Chemical Class:*** Thiadiazole (EPA); Anilide
***EPA/OPP PC Code:*** 121903
***California DPR Chemical Code:*** 5293
***Uses:*** Flufenacet is applied to the soil surface or incorporated pre-emergence in field corn, corn grown for silage, or soybeans to control certain annual grasses and broadleaf weeds.

*Human toxicity (long-term)*[77]*:* Intermediate–28.00 ppb, Health Advisory

*Fish toxicity (threshold)*[77]*:* Low–245.00 ppb, MATC (Maximum Acceptable Toxicant Concentration)

*Carcinogen/Hazard Classifications*

**U.S. EPA Carcinogens:** Not Likely a carcinogen

**Label Signal Word:** CAUTION, WARNING or DANGER

**WHO Acute Hazard:** Class III, slightly hazardous

*Description:* Tan solid. Mercaptan-like odor. Slightly soluble in water. Melting point = 75.5–77.0°C. Molecular weight = 362.

*Permissible Concentration in Water:* No criteria set. Runoff from spills or fire control may cause water pollution.

*Routes of Entry:* Inhalation, ingestion.

*Harmful Effects and Symptoms*

*Short Term Exposure:* Contact with eyes or skin may cause irritation or injury. Inhalation should be avoided; use NIOSH-approved air purifying respirators for pesticides. May be harmful if swallowed.

*First Aid:* If this chemical gets into the eyes, remove any contact lenses at once and irrigate immediately for at least 15 minutes, occasionally lifting upper and lower lids. Seek medical attention immediately. If this chemical contacts the skin, remove contaminated clothing and wash immediately with soap and water. Seek medical attention immediately. If this chemical has been inhaled, remove from exposure, begin rescue breathing (using universal precautions) if breathing has stopped, and CPR if heart action has stopped. Transfer promptly to a medical facility. When this chemical has been swallowed, get medical attention. Give large quantities of water and induce vomiting. Do not make an unconscious person vomit. Do not induce vomiting when formulations containing petroleum solvents are ingested.

*References:*
- U.S. Environmental Protection Agency, Office of Pesticide Programs, Pesticide Fact Sheet, "Flufenacet." (April 1998). http://www.epa.gov/opprd001/factsheets/flufenacet.pdf
- California Environmental Protection Agency "Chemical List of Lists," Sacramento CA (February 1997)

# Thiazopyr (ANSI)

*Use Type:* Herbicide

*CAS Number:* 117718-60-2

*Formula:* $C_{16}H_{17}F_5N_2O_2S$

*Alert:* Not permitted as an active ingredient by the European Commission after July 25, 2003. Human toxicity (long-term): High

*Synonyms:* 3-Pyridinecarboxylic acid, 2-(difluoromethyl)-5-(4,5-dihydro-2-thiazolyl)-4-(2-methylpropyl)-6-(trifluoromethyl)-, methyl ester

*Trade Names:* MANDATE®, Dow AgroSciences (USA); MON-13200®, Dow AgroSciences (USA); VISOR®, Dow AgroSciences (USA)

*Producers:* Dow AgroSciences (USA); Rohm & Haas (USA)

*Chemical Class:* Pyridine

*California DPR Chemical Code:* 3984

*Uses:* Used as a preemergent herbicide to control annual grasses and certain broadleaf weeds on citrus crops.

*Human toxicity (long-term)*[77]*:* High–5.60 ppb, Health Advisory

*Fish toxicity (threshold)*[77]*:* Low–406.10450 ppb, MATC (Maximum Acceptable Toxicant Concentration)

*U.S. Maximum Allowable Residue Levels for Thiazopyr (40 CFR 180.496):*
The residue limit is 0.05 ppm for the following: grapefruit and oranges.

*Carcinogen/Hazard Classifications*

**U.S. EPA Carcinogens:** Group C, possible carcinogen

**Label Signal Word:** CAUTION or WARNING

**Endocrine Disruptor:** Suspected endocrine disruptor

*Description:* Light tan granular solid. Sulfurous odor. Low solubility in water. Melting point = 77.3–79.1°C.

*Incompatibilities:* May react violently with strong oxidizers, bromine, 90% hydrogen peroxide, phosphorus trichloride, silver powders or dust. Incompatible with silver compounds. Mixture with some silver compounds forms explosive salts of silver oxalate.

*Permissible Concentration in Water:* No criteria set. Runoff from spills or fire control may cause water pollution.

*Routes of Entry:* Inhalation, ingestion.

*Harmful Effects and Symptoms*

*Short Term Exposure:* Skin: moderately irritating. Eyes: substantially irritating. High toxicity.

*First Aid:* If this chemical gets into the eyes, remove any contact lenses at once and irrigate immediately for at least 15 minutes, occasionally lifting upper and lower lids. Seek medical attention immediately. If this chemical contacts the skin, remove contaminated clothing and wash immediately with soap and water. Seek medical attention immediately. If this chemical has been inhaled, remove from exposure, begin rescue breathing (using universal precautions) if breathing has stopped, and CPR if heart action has stopped. Transfer promptly to a medical facility. When this chemical has been swallowed, get medical attention. Give large quantities of water and induce vomiting. Do not make an unconscious person vomit. Do not induce vomiting when formulations containing petroleum solvents are ingested.

*References:*
- U.S. Environmental Protection Agency, Office of Pesticide Programs, Pesticide Fact Sheet, "Thiazopyr." (February 20, 1997). http://www.epa.gov/opprd001/factsheets/thiazopyr.pdf
- U.S. Environmental Protection Agency, Office of

Pesticide Programs, Pesticide Residue Limits, "Thiazopyr," 40 CFR 180.496, www.epa.gov/pesticides/food/viewtols.htm
- California Environmental Protection Agency "Chemical List of Lists," Sacramento CA (February 1997)

## Thidiazuron

*Use Type:* Defoliant, plant growth regulator
*CAS Number:* 51707-55-2
*Formula:* $C_9H_8N_4OS$
*Synonyms:* 5-*N*-Phenylcarbamoylamino-1,2,3-thiadiazole; 1-Phenyl-3-(1,2,3-thiadiazol-5-yl)urea; *N*-Phenyl-*N'*-1,2,3-thiadiazol-5-yl-urea; *N*-Phenyl-*N'*-(1,2,3-thiadiazyl)urea; TDZ; (N-1,2,3-Thiadiazolyl-5)-*N'*-phenylurea; Urea, *N*-phenyl-*N'*-1,2,3-thiadiazol-5-yl (9CI)
*Trade Names:* DAZE®, FMC Agricultural Products (USA); DEFOLIT®; DROPP®, Bayer Crop Science (Germany); GINSTAR EC® (thidiazuron + diuron), Bayer Crop Science (Germany); LEAFLESS®, Compton (USA); SN 49537®;
*Producers:* AJE (Switzerland) Bayer Crop Science (Germany); Crompton (USA); FMC Agricultural Products (USA); Micro-Flo (USA); Wuzhou International (China)
*Chemical Class:* Sulfonylurea
*EPA/OPP PC Code:* 120301 (208100 and 208800 are old EPA code numbers)
*California DPR Chemical Code:* 2162
*Uses:* Used primarily as a cotton defoliant in order to increase the harvest yield. Not applied to food crops.
*U.S. Maximum Allowable Residue Levels for Thidiazuron (40 CFR 180.403):*

| CROP | ppm |
|---|---|
| Cattle, fat, meat & mbyp | 0.2 |
| Cotton, hulls | 0.8 |
| Cotton, undelinted seed | 0.4 |
| Egg | 0.1 |
| Goat, fat, meat & mbyp | 0.2 |
| Hog, fat, meat & mbyp | 0.2 |
| Horse, fat, meat & mbyp | 0.2 |
| Milk | 0.05 |
| Poultry, fat, meat & mbyp | 0.2 |
| Sheep, fat, meat & mbyp | 0.2 |

*Carcinogen/Hazard Classifications*
**Label Signal Word:** CAUTION, DANGER
**WHO Acute Hazard:** Class U, unlikely to be hazardous
*Regulatory Authority:*
- FIFRA, 40CFR186: tolerances for pesticides in animal feeds

*Description:* Crystalline solid. Slightly soluble in water. Molecular weight = 220.29.
*Incompatibilities:* Heat of decomposition can form toxic nitrogen oxides and sulfur oxides.
*Permissible Concentration in Water:* No criteria set. Runoff from spills or fire control may cause water pollution.

*Routes of Entry:* Inhalation, ingestion.
*Harmful Effects and Symptoms*
*Short Term Exposure:* Contact with eyes or skin may cause irritation or burns. Inhalation should be avoided; use NIOSH-approved air purifying respirators for pesticides. May be harmful if swallowed. Skin contact may cause allergic reaction.
*Long Term Exposure:* Repeated or prolonged contact may cause skin and lung sensitization, resulting in allergies.
*Points of Attack:* Skin
*Medical Surveillance:* Evaluation by a qualified allergist, including careful exposure history and special testing, may help diagnose skin allergy.
*First Aid:* If this chemical gets into the eyes, remove any contact lenses at once and irrigate immediately for at least 15 minutes, occasionally lifting upper and lower lids. Seek medical attention immediately. If this chemical contacts the skin, remove contaminated clothing and wash immediately with soap and water. Seek medical attention immediately. If this chemical has been inhaled, remove from exposure, begin rescue breathing (using universal precautions) if breathing has stopped, and CPR if heart action has stopped. Transfer promptly to a medical facility. When this chemical has been swallowed, get medical attention. Give large quantities of water and induce vomiting. Do not make an unconscious person vomit.
*References:*
- U.S. Environmental Protection Agency, Office of Pesticide Programs, Pesticide Residue Limits, "Thidiazuron," 40 CFR 180.403, www.epa.gov/pesticides/food/viewtols.htm
- California Environmental Protection Agency "Chemical List of Lists," Sacramento CA (February 1997).

## Thifensulfuron Methyl

*Use Type:* Herbicide
*CAS Number:* 79277-27-3
*Formula:* $C_{11}H_{11}N_5O_6S_2$
*Synonyms:* Methyl 3-[((((4-methoxy-6-methyl-1,3,5-triazin-2-yl)amino)carbonyl)amino)sulfonyl]-2-thiophenecarboxylate; 2-Thiophenecarboxylic acid, 3-[((((4-methoxy-6-methyl-1,3,5-triazin-2-yl)amino)carbonyl)amino)sulfonyl]-, methyl ester; Thiameturon-methyl
*Trade Names:* ALLY®, DuPont Crop Protection (USA); BASIS®, (rimsulfuron + thifensulfuron methyl), DuPont Crop Protection (USA); CANVAS®, (thifensulfuron methyl + tribenuron methyl + metsulfuron-methyl), DuPont Crop Protection (USA); DPX-M6316®, DuPont Crop Protection (USA); EXPRESS®, DuPont Crop Protection (USA); HARMONY® Extra, (thifensulfuron methyl + tribenuron methyl), DuPont Crop Protection (USA); INM-6316®; PINNACLE®, DuPont Crop Protection (USA);

PROSPECT®, Whyte Agrochemicals (UK); RELIANCE®, DuPont Crop Protection (USA); SYNCHRONY®, (chlorimuron-ethyl + thifensulfuron methyl), DuPont Crop Protection (USA)
*Producers:* DuPont Crop Protection (USA); Epochem Co., (China); Whyte Agrochemicals (UK)
*Chemical Class:* Sulfonylurea; triazine
*EPA/OPP PC Code:* 128845
*California DPR Chemical Code:* 2237
*Human toxicity (long-term)[77]:* Low–91.00 ppb, Health Advisory
*Fish toxicity (threshold)[77]:* Very low–19952.62315 ppb, MATC (Maximum Acceptable Toxicant Concentration)
*Carcinogen/Hazard Classifications*
*Label Signal Word:* CAUTION or WARNING
*WHO Acute Hazard:* Class U, unlikely to be hazardous
*Description:* Vapor pressure = $1.3 \times 10^{-10}$ mmHg @ 20°C. Human toxicity (long-term): High
*Incompatibilities:* May react violently with strong oxidizers, bromine, 90% hydrogen peroxide, phosphorus trichloride, silver powders or dust. Incompatible with silver compounds. Mixture with some silver compounds forms explosive salts of silver oxalate.
*Permissible Concentration in Water:* No criteria set. Runoff from spills or fire control may cause water pollution.
*Routes of Entry:* Inhalation, passing through the skin, ingestion.
*Harmful Effects and Symptoms*
*Short Term Exposure:* May cause skin and severe eye irritation. Moderately poisonous if ingested or inhaled. Exposure to a triazine (simazine) has caused acute and subacute dermatitis in the former USSR, characterized by erythema, slight edema, moderate pruritus, and burning lasting 4 to 5 days.
*Long Term Exposure:* May cause lung irritation and damage. May cause skin allergy. Contact with some triazine compounds (such as atrazine) may increase risks for tumors known to be associated with hormonal factors. These have been observed in both animals and human beings, and are consistent with the known effects on the hypothalamic pituitary gonadal axis. Repeated exposure may cause weight loss and reduced red blood cell count.
*Points of Attack:* Liver, lungs, skin.
*Medical Surveillance:* Before beginning employment and at regular times after that, for those with frequent or potentially high exposures, the following is recommended: Lung function tests. Consider chest x-ray following acute overexposure. Evaluation by a qualified allergist. Examination of the nervous system.
*First Aid:* If this chemical gets into the eyes, remove any contact lenses at once and irrigate immediately for at least 15 minutes, occasionally lifting upper and lower lids. Seek medical attention immediately. If this chemical contacts the skin, remove contaminated clothing and wash immediately with soap and water. Seek medical attention immediately. If this chemical has been inhaled, remove from exposure, begin rescue breathing (using universal precautions) if breathing has stopped, and CPR if heart action has stopped. Transfer promptly to a medical facility. When this chemical has been swallowed, get medical attention. Give large quantities of water or milk and induce vomiting. Do not make an unconscious person vomit.
*References:*
- California Environmental Protection Agency "Chemical List of Lists," Sacramento CA (February 1997)

## Thiobencarb (ANSI)

*Use Type:* Herbicide
*CAS Number:* 28249-77-6
*Formula:* $C_{12}H_{16}ClNOS$
*Synonyms:* Carbamic acid, diethyl-, S-(4-chlorobenzyl)ester; Carbamothioic acid, diethyl-, S-[(4-chlorophenyl)methyl] ester; S-(4-Chlorobenzyl) N,N-diethylthiocarbamate; S-(p-Chlorobenzyl)diethylthio carbamate; p-Chlorobenzyl N,N-diethylthiolcarbamate; S-[(4-Chlorophenyl)methyl] diethylcarbamothioate; S-[(4-Chlorophenyl)methyl] N,N-diethylthiocarbamate; Carbamic acid, diethylthio-S-(p-chlorobenzyl) ester; carbamothioic acid, diethyl-,S-[chlorophenyl)methyl]ester; Caswell No. 207DA; S-(p-Chlorobenzyl)diethylthiocarbamate; S-4-Chlorobenzyl diethylthiocarbamate; S-(4-Chlorobenzyl) N,N-diethylthiocarbamate; S-[(4-chlorophenyl)methyl] diethylcarbamothiote; thiobencarbe; α-Toluenethiol, p-chloro-, diethylcarbamate
*Trade Names:* B-3015®; BENCARB®; BENTHIOCARB®; BOLERO®, Valent USA (USA); IMC 3950®; SATURN®; SATURNO®; SIACARB®; TAMARIZ®
*Producers:* Valent USA (USA)
*Chemical Class:* Thiocarbamate
*EPA/OPP PC Code:* 108401; (283500 old EPA code number)
*California DPR Chemical Code:* 1933
*RTECS Number:* EZ7260000
*Uses:* Thiobencarb is a thiocarbamate herbicide that is applied primarily to rice as well as to lettuce, celery, and endive to control grasses and broadleaf weeds.
*Human toxicity (long-term)[77]:* Low–70.00 ppb, Health Advisory
*Fish toxicity (threshold)[77]:* Low–185.74195 ppb, MATC (Maximum Acceptable Toxicant Concentration)
*U.S. Maximum Allowable Residue Levels for Thiobencarb (40 CFR 180.401):*
The maximum allowable residue level is 0.05 ppm for milk. The maximum allowable residue level is 1.0 for rice, straw. The maximum allowable residue level is 0.2 for the following: Cattle, fat; Cattle, meat; Cattle, mbyp; Celery; Egg; Endive; Goat, fat; Goat, meat; Goat, mbyp; Hog, fat; Hog, meat; Hog, mbyp; Horse, fat; Horse, meat; Horse,

mbyp; Lettuce; Poultry, Fat; Poultry, meat; Poultry, meat byproducts; Rice, Grain; Sheep, fat; Sheep, meat; Sheep, mbyp

*Carcinogen/Hazard Classifications*
**U.S. EPA Carcinogens:** Group D, unclassifiable, inadequate data
**Label Signal Word:** CAUTION
**WHO Acute Hazard:** Class II, moderately hazardous
*Regulatory Authority:*
- EPA 40 CFR 372.65, Specific Toxic Chemical Listings
- AB 1803-Well Monitoring Chemical (CAL)
- The "Director's List" (CAL/OSHA)
- MCL (Maximum Contaminants Levels) list of contaminants (CAL)
- Actively registered pesticide in California.
- FIFRA, 40CFR186: tolerances for pesticides in animal feeds
- EPCRA Section 313 Form R *de minimis* concentration reporting level: 1.0%

*Description:* Light yellow to brownish-yellow liquid. Very sparingly soluble in water; solubility = 30 mg/L @ 20°C. Molecular weight 257.8. Density = 1.145–1.180 @ 20°C. Melting/Freezing point = 3.3°C. Boiling point = 126–129°C @ 0.008 mmHg. Vapor pressure = $1.9 \times 10^{-6}$ mbar @ 20°C. Log $K_{ow}$ = 3.42. Values above 3.0 are likely to bioaccumulate in marine organisms.

*Permissible Concentration in Water:* No criteria set. Runoff from spills or fire control may cause water pollution.

*Harmful Effects and Symptoms*

*Short Term Exposure:* Low levels of toxicity. Concentrated solutions are slightly corrosive to eyes and mucous membranes. Dust inhalation can cause irritation of the respiratory system with sneezing. Eye contact can cause irritation, watering, pain, and inflammation of the eyelids. Skin contact can cause irritation and minor ulceration. Ingestion can cause nausea, vomiting, fever, muscle twitching, seizure, rapid respiration, slow heart beat. Severe exposure may result in death.

*Points of Attack:* Respiratory system, central nervous system, cardiovascular system, skin, eyes.

*Medical Surveillance:* Medical observation is recommended for 24 to 48 hours after breathing overexposure, as pulmonary edema may be delayed. As first aid for pulmonary edema, consider administering a corticosteroid spray. Cigarette smoking may exacerbate pulmonary injury and should be discouraged for at least 72 hours following exposure. Before employment and at regular times after that, the following are recommended: If symptoms develop or overexposure occurs, get an exam of the nervous system. Also consider complete blood count. Consider chest x-ray following acute overexposure.

*First Aid:* If this chemical gets into the eyes, remove any contact lenses at once and irrigate immediately for at least 15 minutes, occasionally lifting upper and lower lids. Seek medical attention immediately. If this chemical contacts the skin, remove contaminated clothing and wash immediately with soap and water. Seek medical attention immediately. If this chemical has been inhaled, remove from exposure, begin rescue breathing (using universal precautions) if breathing has stopped, and CPR if heart action has stopped. Transfer promptly to a medical facility. When this chemical has been swallowed, get medical attention. If victim is conscious and able to swallow, have victim drink 4 to 8 ounces of water. Do not induce vomiting. *Note to physician or authorized medical personnel*: Medical observation is recommended for 24 to 48 hours after breathing overexposure, as pulmonary edema may be delayed. As first aid for pulmonary edema, consider administering a corticosteroid spray. Cigarette smoking may exacerbate pulmonary injury and should be discouraged for at least 72 hours following exposure.

*References:*
- U.S. Environmental Protection Agency, "Reregistration Eligibility Decision (RED), Thiobencarb," Office of Prevention, Pesticides and Toxic Substances, Washington, DC (December 1997). http://www.epa.gov/REDs/2665red.pdf
- U.S. Environmental Protection Agency, Office of Pesticide Programs, Pesticide Residue Limits, "Thiobencarb," 40 CFR 180.401, www.epa.gov/pesticides/food/viewtols.htm
- California Environmental Protection Agency "Chemical List of Lists," Sacramento CA (February 1997).

## Thiodicarb (ANSI)

*Use Type:* Insecticide, molluscicide
*CAS Number:* 59669-26-0
*Formula:* $C_{10}H_{18}N_4O_4S_3$
*Synonyms:* AI3-29311; Bismethomylthioether; Bis-(*O*-1-methylthioethylimino)-*N*-methylcarbamic acid)-*N,N'*-sulfide; [Carbamic acid, *N*-methyl-, compounded with (2-methylthio)acetaldoxime]bis-, thioether; Caswell No. 900AA; Dimethyl-*N,N'*-[thiobis((methylimino) carbonyl)oxy)]bis(ethanimidothioate); Dimethyl *N,N'*-[thiobis((methylimino)carbonyloxy)]bis(thioimidoacetate); Dimethyl *N,N'*-[thiobis((methylimino) carbonyloxy)]bis(ethanimidothioate); Ethanimidothioic acid, *N'N'*-[thiobis((methylimino)carbonyloxy)]bis-, dimethyl ester; 3,7,9,13-Tetramethyl-5,11-dioxa-2,8,14-trithia-4,7,9,12-tetra-azapentadeca-3,12-diene-6,10-dione; *N,N'*-[Thiobis((methylimino)carbonyloxy)]bis dimethylester

*Trade Names:* CGA® 45156; CHIPCO, Bayer CropScience (Germany), canceled; DICARBOSULF®; DICARBASULF®; LARVIN®, Bayer CropScience (Germany); LEPICRON®; SEMEVIN®; NIVRAL®; UC-51762®; UC 51769®; UC 80502®

*Producers:* Agriliance (USA); Agsin (Singapore); Bayer CropScience (Germany); Saeryung Chemicals (South Korea); Wuzhou International (China)
*Chemical Class:* Carbamate
*EPA/OPP PC Code:* 114501
*California DPR Chemical Code:* 2202
*RTECS Number:* KJ4301050
*Uses:* Thiodicarb is used primarily on cotton, sweet corn, and soybeans. The remaining usage is spread among leafy vegetables, cole crops, ornamentals, and other minor use sites. Thiodicarb acts as an ovicide against cotton bollworms and budworms.
*Human toxicity (long-term)*[77]*:* Intermediate–18.61702 ppb, CHCL (Chronic Human Carcinogen Level)
*Fish toxicity (threshold)*[77]*:* Low–353.55339 ppb, MATC (Maximum Acceptable Toxicant Concentration)
*U.S. Maximum Allowable Residue Levels for Methomyl (40 CFR 180.253):* The U.S. EPA has determined that methomyl is a degradate of thiodicarb, which is a registered pesticide. Therefore, methomyl residues resulting from applications of both thiodicarb and methomyl have been considered in an aggregate risk assessment and compared to appropriate toxicological endpoints for methomyl.
*Carcinogen/Hazard Classifications*
**U.S. EPA Carcinogens:** Group B2, probable carcinogen
*California Prop. 65:* Carcinogen
**Label Signal Word:** CAUTION, WARNING
**WHO Acute Hazard:** Class II, moderately hazardous
*Regulatory Authority:*
- EPA 40 CFR 372.65, Specific Toxic Chemical Listings
- Actively registered pesticide in California.
- RCRA Section 261 Hazardous Constituents
- EPA Hazardous Waste Number (RCRA No.): U410
- RCRA Universal Treatment Standards: Wastewater (mg/L), 0.019; Nonwastewater (mg/kg), 1.4
- Safe Drinking Water Act: Priority List (55 FR 1470)
- EPCRA Section 313 Form R *de minimis* concentration reporting level: 1.0%

*Description:* White to light tan crystalline powder. Slight sulfurous odor. Soluble in water. Molecular weight = 354.52. Melting/freezing point = 168-175°C. Vapor pressure = $1.1 \times 10^{-7}$ mmHg @ 20°C.
*Incompatibilities:* May react violently with strong oxidizers, chlorobenzenediazonium salts, mercurous chloride. Incompatible with cadmium bromide, zinc acetate. May form explosive materials with phosphorus pentachloride. May not be compatible with nitrates. Moisture may cause hydrolysis or other forms of decomposition.
*Permissible Concentration in Water:* No criteria set. Runoff from spills or fire control may cause water pollution.
*Routes of Entry:* Skin contact, ingestion and inhalation.
*Harmful Effects and Symptoms*
*Short Term Exposure:* Eye pupils are small, blurred vision, eye watering, runny nose, cough, shortness of breath, salivation, nausea, stomach cramps, diarrhea, and vomiting, increased blood pressure, profuse sweating, hypermotility, hallucinations, agitation, tingling of the skin, slow heartbeat, convulsions, fluid in lungs, loss of consciousness, incontinence, breathing stops, death. Carbamates inhibit the acetylcholinesterase enzymes and alter the way in which nervous impulses are transmitted. However, within several hours carbamates spontaneously detach from the enzymes.
*Long Term Exposure:* A potent cholinesterase inhibitor; cumulative effect is possible. This chemical may damage the nervous system with repeated exposure, resulting in convulsions, respiratory failure. May cause liver damage.
*Points of Attack:* Respiratory system, lungs, central nervous system, cardiovascular system, skin, eyes, plasma and red blood cell cholinesterase.
*Medical Surveillance:* Medical observation is recommended for 24 to 48 hours after breathing overexposure, as pulmonary edema may be delayed. As first aid for pulmonary edema, consider administering a corticosteroid spray. Cigarette smoking may exacerbate pulmonary injury and should be discouraged for at least 72 hours following exposure. Before employment and at regular times after that, the following are recommended: Plasma and red blood cell cholinesterase levels (tests for the enzyme poisoned by this chemical). If exposure stops, plasma levels return to normal in 1-2 weeks while red blood cell levels may be reduced for 1-3 months. When acetylcholinesterase enzyme levels are reduced by 25% or more below preemployment levels, risk of poisoning is increased, even if results are in lower ranges of "normal." Reassignment to work not involving carbamate pesticides is recommended until enzyme levels recover. If symptoms develop or overexposure occurs, repeat the above tests as soon as possible and get an exam of the nervous system. Also consider complete blood count. Consider chest x-ray following acute overexposure.
*First Aid:* Speed in removing material from eyes and skin is of extreme importance. *Eyes:* Eye contact can cause dangerous amounts of these chemicals to be quickly absorbed through the mucous membrane into the bloodstream. Immediately and gently flush eyes with plenty of warm or cold water (NO hot water) for at least 15 minutes, occasionally lifting the upper and lower eyelids. Get medical aid immediately. Skin: Get medical aid. Skin contact can cause dangerous amounts of these chemicals to be absorbed into the bloodstream. Wearing the appropriate PPE equipment and respirator for carbamate pesticides, immediately flush skin with plenty of soap and water for at least 15 minutes while removing contaminated clothing and shoes. Shampoo hair promptly if contaminated; protect eyes. *Ingestion:* Call poison control. Loosen all clothing. Never give anything by mouth to an unconscious person. Get medical aid. *Do NOT induce vomiting.*\* If conscious, alert, and able to swallow, rinse mouth and have victim drink 4 to 8 ounces of water. Check to see if poison control instructs you to use ipecac syrup, otherwise administer slurry of

activated charcoal (2 oz in 8 oz of water). If victim is *unconscious or having convulsions,* do nothing except keep victim warm. **In some cases you may be specifically instructed by poison control to induce vomiting by way of 2 tablespoons of syrup of ipecac (adult) washed down with a cup of water.* Do NOT give activated charcoal before or with ipecac syrup. *Inhalation:* Get medical aid. Do not contaminate yourself. Wearing the appropriate PPE equipment and respirator for carbamate pesticides, immediately remove the victim from the contaminated area to fresh air. If the victim is not breathing, administer artificial respiration. Do NOT use mouth-to-mouth resuscitation; use bag/mask apparatus. If breathing is difficult, administer oxygen through bag/mask apparatus until medical help arrives. Do not leave victim unattended. *Note to physician or authorized medical personnel.* Administer atropine, 2 mg (1/30 gr) intramuscularly or intravenously as soon as any local or systemic signs or symptoms of an intoxication are noted; repeat the administration of atropine every 3 to 8 minutes until signs of atropinization (mydriasis, dry mouth, rapid pulse, hot and dry skin) occur; initiate treatment in children with 0.05 mg mg/kg of atropine; repeat at 5 to 10 minute intervals. Watch respiration, and remove bronchial secretions if they appear to be obstructing the airway; intubate if necessary. *Medical note:* 2-PAMCI may be contraindicated in the case of some carbamate poisonings.

*References:*
- U.S. Environmental Protection Agency, "Reregistration Eligibility Decision (RED), Thiodicarb," Office of Prevention, Pesticides and Toxic Substances," Washington, DC. (December 1998). http://www.epa.gov/REDs/2675red.pdf
- Pesticide Management Education Program, "Thiodicarb (Larvin) Chemical Fact Sheet 8/84," Cornell University, Ithaca, NY (August 1984). http://pmep.cce.cornell.edu/profiles/insect-mite/propetamphos-zetacyperm/thiodicarb/insect-prof-thiodicarb.html
- California Environmental Protection Agency "Chemical List of Lists," Sacramento CA (February 1997)

# Thiofanox

*Use Type:* A systemic insecticide and acaricide.
*CAS Number:* 39196-18-4
*Formula:* $C_9H_{18}N_2O_2S$; $(CH_3)_3CC(CH_2SCH_3)=NOCONHCH_3$
*Synonyms:* 3,3-Dimethyl-1-(methylthio)-2-butanone-*O*-[(methylamino)carbonyl]oxime; ENT 27,851; Thiofanocarb (South Africa); Thiophanox
*Trade Names:* DACAMOX®; DIAMOND SHAMROCK DS-15647®; DS-15647®
*Producers:* Rhone-Poulenc Agro France (France)

*Chemical Class:* Methyl carbamate
*EPA/OPP PC Code:* 109201
*California DPR Chemical Code:* 2938
*RTECS Number:* EL8200000
*EINECS Number:* 254-346-4
*Uses:* Has been used to control pests on cotton and mango crops.
*Carcinogen/Hazard Classifications*
**WHO Acute Hazard:** Class 1b, highly hazardous
*Regulatory Authority:*
- EPA Hazardous Waste Number (RCRA No.): P045
- RCRA, 40CFR261, Appendix 8 Hazardous Constituents
- Superfund/EPCRA 40CFR355, Appendix B Extremely Hazardous Substances: TPQ = 100/10,000 lb (45.4/4,540 kg)
- Superfund/EPCRA 40CFR302.4 RQ: CERCLA, 100 lb (45.4 kg)

*Description:* Colorless solid with a pungent odor. Molecular weight = 218.33. Melting/Freezing point = 57°C.
*Permissible Exposure Limits in Air:* No standards set.
*Permissible Concentration in Water:* No criteria set, but runoff from spills or fire control may cause water pollution.
*Routes of Entry:* Inhalation, ingestion, eye and/or skin contact.
*Harmful Effects and Symptoms*
*Short Term Exposure:* This material is moderately to highly toxic. It is a cholinesterase inhibitor. Symptoms of exposure include nausea, vomiting, abdominal cramps, diarrhea, excessive salivation, sweating, weakness, runny nose, tightness of chest (inhalation exposure), blurred vision, tearing, muscle spasm, loss of eye coordination, ocular pain, extreme dilation of the pupil, loss of muscle coordination, slurring of speech, difficulty in breathing, excessive respiratory tract mucous, skin discoloration, and hypertension. High exposures can cause pulmonary edema, a medical emergency that can be delayed for several hours. This can cause death.
*Long Term Exposure:* Cholinesterase inhibitor; cumulative effect is possible. This chemical may damage the nervous system with repeated exposure, resulting in convulsions, respiratory failure.
*Points of Attack:* Blood, eyes and lungs.
*Medical Surveillance:* Before employment and at regular times after that, the following are recommended: Plasma and red blood cell cholinesterase levels (tests for the enzyme poisoned by this chemical). If exposure stops, plasma levels return to normal in 1-2 weeks while red blood cell levels may be reduced for 1-3 months. When acetylcholinesterase enzyme levels are reduced by 25% or more below preemployment levels, risk of poisoning is increased, even if results are in lower ranges of "normal." Reassignment to work not involving carbamate or organophosphate pesticides is recommended until enzyme levels recover. If symptoms develop or overexposure occurs, repeat the above tests as soon as possible and get an exam of the nervous

system. Also consider complete blood count. Consider chest x-ray following acute overexposure. Do not drink any alcoholic beverages before or during use. Eye examination. Lung function tests. Consider chest x-ray following acute overexposure.

*First Aid:* Speed in removing material from eyes and skin is of extreme importance. If this chemical gets into the eyes, remove any contact lenses at once and irrigate immediately for at least 15 minutes, occasionally lifting upper and lower lids. Seek medical attention immediately. If this chemical contacts the skin, remove contaminated clothing and wash immediately with soap and water. Shampoo hair promptly if contaminated. Seek medical attention immediately. If this chemical has been inhaled, remove from exposure, begin rescue breathing (using universal precautions) if breathing has stopped, and CPR if heart action has stopped. Transfer promptly to a medical facility. When this chemical has been swallowed, get medical attention. Give large quantities of water and induce vomiting. Do not make an unconscious person vomit. Medical observation is recommended for 24 to 48 hours after breathing overexposure, as pulmonary edema may be delayed. As first aid for pulmonary edema, a doctor or authorized paramedic may consider administering a corticosteroid spray.

*References:*
- New Jersey Department of Health and Senior Services, "Hazardous Substance Fact Sheet, Thiofanox," Trenton, NJ (May 2002). http://www.state.nj.us/health/eoh/rtkweb/2820.pdf
- U.S. Environmental Protection Agency, "Chemical Profile: Thiofanox," Washington, DC, Chemical Emergency Preparedness Program (November 30, 1987).
- California Environmental Protection Agency "Chemical List of Lists," Sacramento CA (February 1997).

# Thionazin

*Use Type:* Insecticide, fungicide and nematicide
*CAS Number:* 297-97-2
*Formula:* $C_8H_{13}N_2O_3PS$
*Synonyms:* AC 18133; *O,O*-Diaethyl-*O*-(pyrazin-2yl)-monothiophosphat (German); *O,O*-Diaethyl-*O*-(2-pyrazinyl)-thionophosphat (German); *O,O*-Diethyl-*O*,2-pyrazinyl phosphorothioate; Diethyl-*O*-2-pyrazinyl phosphorothionate; *O,O*-Diethyl-*O*-2-pyrazinyl phosphothionate; *O,O*-Diethyl-*O*-pyrazinyl thiophosphate; EN 18133; ENT 25 580; Phosphorothioic Acid-*O,O*-diethyl-*O*-2-pyrazinyl ester; Pyrazinol-*O*-ester with *O,O*-diethyl phosphorothioate pyrazinol *O* ester
*Trade Names:* AMERCIAN CYANAMID® 18133; CL®18133®; CYNEM®; EXPERIMENTAL NEMATOCIDE 18,133®; NEMAFOS®; NEMAPHOS®; NEMATOCIDE®; ZINOPHOS®

*Producers:* Diachem (Italy); Schering (Germany); Shell Chemicals (UK)
*Chemical Class:* Organophosphate
*EPA/OPP PC Code:* 032401
*California DPR Chemical Code:* 2939
*RTECS Number:* TF5775000
*RCRA Number:* P040
*EINECS Number:* 206-97-2
*Uses:* Discontinued use as a pesticide.
*Regulatory Authority:*
- Very Toxic Substance (World Bank)[15]
- RCRA, 40CFR261, Appendix 8 Hazardous Constituents
- RCRA 40CFR264, Appendix 9; TSD Facilities Ground Water Monitoring List. Suggested test method(s) (PQL *ug/L*): 8270(10)
- Superfund/EPCRA 40CFR355, Appendix B Extremely Hazardous Substances: TPQ = 500 lb (227kg)
- Superfund/EPCRA 40CFR302.4 RQ: CERCLA, 100 lb (45.4 kg)
- U.S. DOT Inhalation Hazard Chemicals as organophosphates

*Description:* Thionazin, $C_8H_{13}N_2O_3PS$, is an amber to colorless liquid. Very slightly soluble in water; solubility = 1150 ppm @ 20°C. Molecular weight = 284.25. Melting/Freezing point = −1.7°C. Boiling point = 80°C @ 0.001 mmHg. Vapor pressure = $3.8 \times 10^{-3}$ mbar @ 39°C. Hazard Identification (based on NFPA-704 M Rating System): Health 4, Flammability 1, Reactivity 0.
*Incompatibilities:* Strong alkalies.
*Permissible Exposure Limits in Air:* No standards set. However it is an organophosphate pesticide.
*Determination in Air:* OSHA versatile sampler-2; Toluene/Acetone; Gas chromatography/Flame photometric detection for sulfur, nitrogen, or phosphorus; NIOSH Method IV Method #5600, Organophosphorus pesticides.[18]
*Permissible Concentration in Water:* No criteria set, but runoff from spills or fire control may cause water pollution.
*Routes of Entry:* Inhalation, ingestion, eye and/or skin contact. Absorbed through the skin.
*Harmful Effects and Symptoms*
*Short Term Exposure:* Irritates the eyes, skin, and respiratory tract. Contact may cause burns to skin and eyes. Poisonous; may be fatal if inhaled, swallowed or absorbed through skin. Acute effects include loss of appetite, nausea, vomiting, diarrhea, excessive salivation, papillary constriction, bronchoconstriction, muscle twitching, convulsions, and coma. Organic phosphorus insecticides are absorbed by the skin, as well as by the respiratory and gastrointestinal tracts. They are cholinesterase inhibitors. Symptoms of exposure include headache, giddiness, blurred vision, nervousness, weakness, nausea, cramps, diarrhea, and discomfort in the chest. Signs include sweating, tearing, salivation, vomiting, cyanosis, convulsions, coma, loss of reflexes and loss of sphincter control. Delayed pulmonary edema may occur after inhalation.

*Long Term Exposure:* Cholinesterase inhibitor; cumulative effect is possible. This chemical may damage the nervous system with repeated exposure, resulting in convulsions, respiratory failure. May cause liver damage. May cause dermatitis.

*Points of Attack:* Respiratory system, lungs, central nervous system, cardiovascular system, skin, eyes, plasma and red blood cell cholinesterase.

*Medical Surveillance:* Medical observation is recommended for 24 to 48 hours after breathing overexposure, as pulmonary edema may be delayed. Before employment and at regular times after that, the following are recommended: Plasma and red blood cell cholinesterase levels (tests for the enzyme poisoned by this chemical). If exposure stops, plasma levels return to normal in 1-2 weeks while red blood cell levels may be reduced for 1-3 months. When acetylcholinesterase enzyme levels are reduced by 25% or more below preemployment levels, risk of poisoning is increased, even if results are in lower ranges of "normal." Reassignment to work not involving organophosphate or carbamate pesticides is recommended until enzyme levels recover. If symptoms develop or overexposure occurs, repeat the above tests as soon as possible and get an exam of the nervous system. Also consider complete blood count. Consider chest x-ray following acute overexposure. Do not drink any alcoholic beverages before or during use. Alcohol promotes absorption of organophosphates.

*First Aid:* **Treatment for organophosphate poisoning consists of thorough decontamination, cardiorespiratory support, and administration of the antidotes atropine and pralidoxime. In cases of severe poisoning, diazepam, an anticonvulsant, should also be administered. Antidotes should be administered as prevention even if the diagnosis is in doubt.** Speed in removing material from eyes and skin is of extreme importance. *Eyes:* Eye contact can cause dangerous amounts of these chemicals to be quickly absorbed through the mucous membrane into the bloodstream. Immediately and gently flush eyes with plenty of warm or cold water (NO hot water) for at least 15 minutes, occasionally lifting the upper and lower eyelids. Get medical aid immediately. *Skin:* Get medical aid. Skin contact can cause dangerous amounts of these chemicals to be absorbed into the bloodstream. Wearing the appropriate PPE equipment and respirator for organophosphate pesticides, immediately flush skin with plenty of soap and water for at least 15 minutes while removing contaminated clothing and shoes. Shampoo hair promptly if contaminated. The removed, contaminated clothing and shoes should be double-bagged and left in Hot Zone for later disposal by hazardous materials experts. Skin may also be decontaminated with diluted hypochlorite solution. *Inhalation:* Get medical aid. Do not contaminate yourself. Wearing the appropriate PPE equipment and respirator for organophosphate pesticides, immediately remove the victim from the contaminated area to fresh air. If the victim is not breathing, administer artificial respiration. Do NOT use mouth-to-mouth resuscitation; use bag/mask apparatus. If breathing is difficult, administer oxygen through bag/mask apparatus until medical help arrives. Do not leave victim unattended. *Ingestion:* Call poison control. Loosen all clothing. Never give anything by mouth to an unconscious person. If victim is *unconscious or having convulsions,* do nothing except keep victim warm. Get medical aid. Transfer promptly to a medical facility. In cases of ingestion, **do not induce vomiting**. If the victim is alert and asymptomatic, administer a slurry of activated charcoal at a dose of 1 g/kg (infant, child, and adult dose). A soda can and straw may be of assistance when offering charcoal to a child. *In some cases you may be specifically instructed by poison control to induce vomiting by way of 2 tablespoons of syrup of ipecac (adult) washed down with a cup of water.* Do NOT give activated charcoal before or with ipecac syrup.

*Note to physician or authorized medical personnel:* Treat cases of respiratory compromise, coma, or excessive pulmonary secretions with respiratory support using protocols and techniques available and within the scope of training. Some cases may necessitate procedures such as endotracheal intubation or cricothyrotomy by properly trained and equipped personnel. When possible, atropine (see *Antidotes*, below) should be given under medical supervision. Patients who are comatose, hypotensive, or having seizures or cardiac arrhythmias should be treated according to advanced life support protocols. *Antidotes:* Two antidotes are administered to treat organophosphate poisoning. Atropine is a competitive antagonist of acetylcholine at muscarinic receptors and is used to control the excessive bronchial secretions which are often responsible for death. Pralidoxime relieves both the nicotinic and muscarine effects of organophosphate poisoning by regenerating acetylcholinesterase and can reduce both the bronchial secretions and the muscle weakness associated with poisoning. The initial intravenous dose of atropine in adults should be determined by the severity of symptoms: An initial adult dose of 1.0 to 2.0 mg or pediatric dose of 0.01 mg/kg (minimum 0.01 mg) should be administered intravenously. If intravenous access cannot be established, atropine may also be given intramuscularly, subcutaneously or via endotracheal tube. Doses should be repeated every 15 minutes until excessive secretions and sweating have been controlled. Once bronchial secretion has been controlled, atropine administration should be repeated whenever the secretions begin to recur. In seriously poisoned patients, very large doses may be required. Alterations of pulse rate and pupillary size should not be used as indicators of treatment adequacy. Pralidoxime should be administered as early in poisoning as possible as its efficacy may diminish when given more than 24 to 36 hours after exposure. Doses are as follows: adult 1.0 g; pediatric 25 to 50 mg/kg. The drug should be administered intravenously over 30 to 60 minutes, but in a life-threatening situation, one-half of the total dose can be given per minute for a total administration time of 2 minutes. Treatment should begin to take effect within 40 minutes with a reduction in symptoms and the amount of atropine necessary to control bronchial secretion. The initial dose can be repeated in 1 hour and then every 8 to 12 hours until the patient is clinically well and no longer requires atropine. If intravenous access cannot be established, pralidoxime may also be given intramuscularly. Early administration of diazepam in addition to the combined atropine and pralidoxime treatment may help prevent the onset of seizures and potential brain and cardiac morphologic damage following high-level organophosphate poisoning.

*References:*
- U.S. Environmental Protection Agency, "Chemical Profile: Thionazin," Washington, DC, Chemical Emergency Preparedness Program (November 30, 1987).
- California Environmental Protection Agency "Chemical List of Lists," Sacramento CA (February 1997).
- Agency for Toxic Substances and Disease Registry, U.S. Department of Health and Human Services, Public Health Service, "Managing Hazardous Materials Incidents," Atlanta, GA (June 2003).

# Thiophanate-methyl

*Use Type:* Fungicide
*CAS Number:* 23564-05-8 (thiophanate-methyl); 23564-06-9 (thiophanate)
*Formula:* $C_{12}H_{14}N_4O_4S_2$
*Synonyms:* Allophanic acid, 4,4'-*O*-phenylenebis(3-thio-, diethyl ester; 1,2-Bis(ethoxycarbonylthioureido)benzene; 1,2-Bis(3-(ethoxycarbonyl)-2-thioureido)benzene; 1,2-Bis[3-(ethoxycarbonyl)thioureido]benzene; Carbamic acid, [1,2-phenylenebis(iminocarbonothioyl)]bis-, diethyl ester; Caswell No. 344A; Cercobin-methyl; 1,2-Di-(3-ethoxycarbonyl-2-thioureido)benzene; Diethyl [1,2-phenylenebis(iminocarbonothioyl)]bis(carbamate); Diethyl [(1,2-phenylene)bis(iminocarbonothioyl)]bis(carbamate); Diethyl 4,4'-*O*-phenylenebis(3-thioallophanate); Diethyl 4,4'-(*O*-phenylene)bis(3-thioallophanate); Ethyl thiophanate; 4,4'-*O*-Phenylenebis(ethyl 3-thioallophanate); [1,2-Methyl thiophanate; Phenylenebis(iminocarbonothioyl)]biscarbamic acid diethyl ester; (1,2-Phenylenebis[iminocarbonothioyl]) biscarbamic acid diethyl ester; 4,4'-*O*-Phenylenebis(3-thioallophanic acid)dimethyl ester; Thiofanate; Thiophanate; Thiophenite
*Trade Names:* BASF® 32500F, BASF Agricultural Products (Germany); BASF® 32500 Fungicide, BASF Agricultural Products (Germany); CERCOBIN®; CLEARY® 3336, Cleary Chemical (USA); CONSYST®; DITEK®, Sandoz Agro ((USA), canceled; DOMAIN®, The Scotts Co. (USA); DOUSAN®, The Scotts Co. (USA), canceled; ENOVIT®; EVOLVE®, Gustafson (USA); FANATE®; FUNGITOX®; FUNGO®, The Scotts Co. (USA); NEOTOPSIN®; NF-35®; NF-44®; NSC 170810®; PELT®; PRO-TURF®, The Scotts Co. (USA); SIPCAVIT®; SPECTRO®, Cleary Chemical (USA); SYSTEC®; SYSTEMIC® FUNGICIDE, The Scotts Co. (USA); TD 1771®; TOPSIN®, Cerexagri (France); TOPSIN-WP METHYL®, Cerexagri (France); 3336 TURF FUNGICIDE®; ZYBAN®, The Scotts Co. (USA)
*Producers:* Agrimor International (USA); Agsin (Singapore); BASF Agricultural Products (Germany); Cerexagri (France); Cleary Chemical (USA); Gowan (USA); Gustafson (USA); Hebei Huafeng Chemical Group (China); Hindustan Insecticides (India); Micro-Flo (USA); Nagarjuna Agrichem (India); Nufarm (Australia); Saeryung Chemicals (South Korea); The Scotts Co. (USA); Shandong Huayang Pesticide Group (China); Yellow River Enterprise (Taiwan)
*Chemical Class:* Carbamate; benzimidazole group
*EPA/OPP PC Code:* 103401 (thiophanate); 102001 (thiophanate-methyl)
*California DPR Chemical Code:* 1684 (thiophanate); 1696 (thiophanate-methyl)
*Uses:* Thiophanate-methyl is a systemic fungicide used to control a broad spectrum of fungal diseases on fruits, vegetables, turf and ornamentals, including shade trees, and diseases in the field, nurseries, and in greenhouses.
*U.S. Maximum Allowable Residue Levels for Thiophanate-Methyl (40 CFR 180.371):*

| CROP | ppm |
| --- | --- |
| Almond | 0.2 |
| Almond, hulls | 1 |
| Apple, post-h | 7 |
| Apple, dried pomace | 40 |
| Apricot, post-h | 15 |
| Banana | 2 |
| Banana, pulp | 0.2 |
| Bean, dry | 2 |
| Bean, forage | 50 |
| Bean, hay | 50 |
| Bean, snap, succulent | 2 |
| Beet, sugar, roots | 0.2 |
| Beet, sugar, tops | 15 |
| Blueberry | 1.5 |
| Canola, seed | 0.1 |
| Cattle, fat | 0.1 |
| Cattle, kidney | 0.2 |
| Cattle, liver | 2.5 |
| Cattle, meat | 0.1 |
| Cattle, mbyp, except kidney and liver | 0.1 |
| Celery | 3 |
| Cherry, post-h | 15 |
| Citrus | 0.5 |
| Cucumber | 1 |
| Egg | 0.1 |

| | |
|---|---|
| Goat, fat | 0.1 |
| Goat, kidney | 0.2 |
| Goat, liver | 2.5 |
| Goat, meat | 0.1 |
| Goat, mbyp, except kidney and liver | 0.1 |
| Grape | 5 |
| Horse, fat | 0.1 |
| Horse, liver | 1 |
| Horse, meat | 0.1 |
| Horse, mbyp, except liver | 0.1 |
| Melon | 1 |
| Milk | 1 |
| Mushroom | 0.01 |
| Nectarine, post-h | 15 |
| Onion, dry bulb | 3 |
| Onion, green | 3 |
| Peach, post-h | 15 |
| Peanut | 0.2 |
| Peanut, forage | 15 |
| Peanut, hay | 15 |
| Pear | 3 |
| Pecan | 0.2 |
| Pistachio | 0.1 |
| Plum, post-h | 15 |
| Plum, prune, post-h | 15 |
| Potato | 0.05 |
| Potato | 0.1 |
| Pumpkin | 1 |
| Sheep, fat | 0.1 |
| Sheep, kidney | 0.2 |
| Sheep, liver | 2.5 |
| Sheep, meat | 0.1 |
| Sheep, mbyp, except kidney and liver | 0.1 |
| Soybean | 0.2 |
| Squash | 1 |
| Strawberry | 5 |
| Sugarcane | 0.1 |
| Vegetable, fruiting, group 8 | 0.5 |
| Wheat, grain | 0.05 |
| Wheat, hay | 0.1 |
| Wheat, straw | 0.1 |

***Carcinogen/Hazard Classifications***
**U.S. EPA Carcinogens:** Likely carcinogen
***California Prop. 65:*** Listed female and male reproductive toxin
**U.S. TRI:** Reproductive toxin
**Label Signal Word:** CAUTION, WARNING, or DANGER
**WHO Acute Hazard:** Class U, unlikely to be hazardous
***Regulatory Authority:***
- EPA 40 CFR 372.65, Specific Toxic Chemical Listings
- AB 1803-Well Monitoring Chemical (CAL)
- AB 2588-Air Toxics "Hot Spots" Chemicals (CAL)
- Proposition 65 chemical (CAL)
- Actively registered pesticide in California.
- EPA Hazardous Waste Number (RCRA No.): U409
- RCRA Section 261 Hazardous Constituents
- EPCRA Section 313 Form R *de minimis* concentration reporting level: 1.0%

***Description:*** Colorless crystalline solid. Low solubility in water; solubility = 3.58 ppm @ 20°C. Molecular weight = 342.41. Melting/Freezing point = 177.6°C.

***Incompatibilities:*** May form explosive materials with phosphorus pentachloride.

***Determination in Air:*** Filter; none; Gravimetric; NIOSH IV [Particulates NOR; #0500 (total), #0600 (respirable)].[18]

***Permissible Concentration in Water:*** No criteria set. Runoff from spills or fire control may cause water pollution.

***Routes of Entry:*** Absorbed through the skin, inhalation, ingestion.

***Harmful Effects and Symptoms***

***Short Term Exposure:*** Eye pupils are small, blurred vision, eye watering, runny nose, cough, shortness of breath, salivation, nausea, stomach cramps, diarrhea, and vomiting, increased blood pressure, profuse sweating, hypermotility, hallucinations, agitation, tingling of the skin, slow heartbeat, convulsions, fluid in lungs, loss of consciousness, incontinence, breathing stops, death. Carbamates inhibit the acetylcholinesterase enzymes and alter the way in which nervous impulses are transmitted. However, within several hours carbamates spontaneously detach from the enzymes.

***Long Term Exposure:*** A potent cholinesterase inhibitor; cumulative effect is possible. This chemical may damage the nervous system with repeated exposure, resulting in convulsions, respiratory failure. May cause liver damage.

***Points of Attack:*** Respiratory system, lungs, central nervous system, cardiovascular system, skin, eyes, plasma and red blood cell cholinesterase.

***Medical Surveillance:*** Medical observation is recommended for 24 to 48 hours after breathing overexposure, as pulmonary edema may be delayed. As first aid for pulmonary edema, consider administering a corticosteroid spray. Cigarette smoking may exacerbate pulmonary injury and should be discouraged for at least 72 hours following exposure. Before employment and at regular times after that, the following are recommended: Plasma and red blood cell cholinesterase levels (tests for the enzyme poisoned by this chemical). If exposure stops, plasma levels return to normal in 1-2 weeks while red blood cell levels may be reduced for 1-3 months. When acetylcholinesterase enzyme levels are reduced by 25% or more below preemployment levels, risk of poisoning is increased, even if results are in lower ranges of "normal." Reassignment to work not involving carbamate pesticides is recommended until enzyme levels recover. If symptoms develop or overexposure occurs, repeat the above tests as soon as possible and get an exam of the nervous system. Also consider complete blood count. Consider chest x-ray following acute overexposure.

*First Aid:* Speed in removing material from eyes and skin is of extreme importance. *Eyes:* Eye contact can cause dangerous amounts of these chemicals to be quickly absorbed through the mucous membrane into the bloodstream. Immediately and gently flush eyes with plenty of warm or cold water (NO hot water) for at least 15 minutes, occasionally lifting the upper and lower eyelids. Get medical aid immediately. *Skin:* Get medical aid. Skin contact can cause dangerous amounts of these chemicals to be absorbed into the bloodstream. Wearing the appropriate PPE equipment and respirator for carbamate pesticides, immediately flush skin with plenty of soap and water for at least 15 minutes while removing contaminated clothing and shoes. Shampoo hair promptly if contaminated; protect eyes. *Ingestion:* Call poison control. Loosen all clothing. Never give anything by mouth to an unconscious person. Get medical aid. *Do NOT induce vomiting.*\* If conscious, alert, and able to swallow, rinse mouth and have victim drink 4 to 8 ounces of water. Check to see if poison control instructs you to use ipecac syrup, otherwise administer slurry of activated charcoal (2 oz in 8 oz of water). If victim is *unconscious or having convulsions,* do nothing except keep victim warm. *\*In some cases you may be specifically instructed by poison control to induce vomiting by way of 2 tablespoons of syrup of ipecac (adult) washed down with a cup of water.* Do NOT give activated charcoal before or with ipecac syrup. *Inhalation:* Get medical aid. Do not contaminate yourself. Wearing the appropriate PPE equipment and respirator for carbamate pesticides, immediately remove the victim from the contaminated area to fresh air. If the victim is not breathing, administer artificial respiration. Do NOT use mouth-to-mouth resuscitation; use bag/mask apparatus. If breathing is difficult, administer oxygen through bag/mask apparatus until medical help arrives. Do not leave victim unattended. *Note to physician or authorized medical personnel.* Administer atropine, 2 mg (1/30 gr) intramuscularly or intravenously as soon as any local or systemic signs or symptoms of an intoxication are noted; repeat the administration of atropine every 3 to 8 minutes until signs of atropinization (mydriasis, dry mouth, rapid pulse, hot and dry skin) occur; initiate treatment in children with 0.05 mg mg/kg of atropine; repeat at 5 to 10 minute intervals. Watch respiration, and remove bronchial secretions if they appear to be obstructing the airway; intubate if necessary. *Medical note:* 2-PAMCI may be contraindicated in the case of some carbamate poisonings.

*References:*
- U.S. Environmental Protection Agency, Office of Pesticide Programs, Pesticide Residue Limits, "Thiophanate-Methyl," 40 CFR 180.371, www.epa.gov/pesticides/food/viewtols.htm
- Pesticide Management Education Program, "Thiophanate-methyl (Topspin M) Chemical Profile 2/85," Cornell University, Ithaca, NY (February 1985). http://pmep.cce.cornell.edu/profiles/fung-nemat/tcmtb-ziram/thiophanate-methyl/fung-prof-thiophanate.html
- California Environmental Protection Agency "Chemical List of Lists," Sacramento CA (February 1997)

## Thiosemicarbazide

*Use Type:* Rodenticide
*CAS Number:* 79-19-6
*Formula:* $CH_5N_3S$; $H_2NNHCSNH_2$
*Synonyms:* AI3-16319; Aminothiourea; 1-Aminothiourea; *N*-Aminothiourea; 1-Amino-2-thiourea; Hydrazinecarbothioamide; Isothiosemicarbazide; Semicarbazide, 3-Thio-; Semicarbazide, thio-; Thiocarbamoylhydrazine; Thiocarbamylhydrazine; 2-Thiosemicarbazide; 3-Thiosemicarbazide; Tiosemicarbazida (Spanish); TSC; TSZ
*Producers:* Atofina (France); Eurolabs Ltd. (UK); Fairmount Chemical (USA); Fluorchem (UK); GFS Chemicals (USA); OxonItalia (Italy); TCI America (USA)
*RTECS Number:* VT4200000
*EINECS Number:* 201-184-7
*Uses:* This compound is used as a reagent for ketones and certain metals, for photography and as a rodenticide. It is also effective for control of bacterial leaf blight of rice.
*Regulatory Authority:*
- EPA Hazardous Waste Number (RCRA No.): P116
- RCRA, 40CFR261, Appendix 8 Hazardous Constituents
- Superfund/EPCRA 40CFR355, Appendix B Extremely Hazardous Substances: TPQ = 100/10,000 lb (45.4/4,540 kg)
- Superfund/EPCRA 40CFR302.4 RQ: CERCLA, 100 lb (45.4 kg)
- EPCRA Section 313 Form R *de minimis* concentration reporting level: 1.0%

*Description:* White crystalline powder. Odorless. Soluble in water. Molecular weight = 91.14. Melting/Freezing point = (decomposes) 182°C. Hazard Identification (based on NFPA-704 M Rating System): Health 3, Flammability 1, Reactivity 0.
*Incompatibilities:* Strong oxidizers, nitrates.
*Permissible Exposure Limits in Air:* No standards set.
*Permissible Concentration in Water:* No criteria set, but runoff from spills or fire control may cause water pollution.
*Routes of Entry:* Inhalation, ingestion, eye and/or skin contact. Absorbed through the skin.
*Harmful Effects and Symptoms*
*Short Term Exposure:* Azides can cause decreased blood pressure and consequently have action similar to cyanides and nitrites. This material is highly toxic by ingestion. May cause delayed toxic effects in blood and skin. May be

mutagenic in human cells. Thiosemicarbazide may induce goiter and has also been reported to cause bone marrow depression with accompanying decreases in white blood cells and platelets. It may also cause skin irritation.

*Long Term Exposure:* May cause delayed toxic effects in blood and skin. May be mutagenic in human cells. May be a cholinesterase inhibitor; cumulative effect is possible. This chemical may damage the nervous system with repeated exposure, resulting in convulsions, respiratory failure.

*Points of Attack:* Respiratory system, lungs, central nervous system, cardiovascular system, skin, eyes, plasma and red blood cell cholinesterase.

*Medical Surveillance:* Before employment and at regular times after that, the following are recommended: Plasma and red blood cell cholinesterase levels (tests for the enzyme poisoned by this chemical). If exposure stops, plasma levels return to normal in 1-2 weeks while red blood cell levels may be reduced for 1-3 months. When acetylcholinesterase enzyme levels are reduced by 25% or more below preemployment levels, risk of poisoning is increased, even if results are in lower ranges of "normal." Reassignment to work not involving carbamate pesticides (or organophosphates) is recommended until enzyme levels recover. If symptoms develop or overexposure occurs, repeat the above tests as soon as possible and get an exam of the nervous system. Also consider complete blood count. Consider chest x-ray following acute overexposure. Before employment and at regular times after that, the following are recommended: Plasma and red blood cell cholinesterase levels (tests for the enzyme poisoned by this chemical). If exposure stops, plasma levels return to normal in 1-2 weeks while red blood cell levels may be reduced for 1-3 months. When acetylcholinesterase enzyme levels are reduced by 25% or more below preemployment levels, risk of poisoning is increased, even if results are in lower ranges of "normal." Reassignment to work not involving carbamate pesticides (or organophosphates) is recommended until enzyme levels recover. If symptoms develop or overexposure occurs, repeat the above tests as soon as possible and get an exam of the nervous system. Also consider complete blood count. Consider chest x-ray following acute overexposure. Do not drink any alcoholic beverages before or during use.

*First Aid:* Speed in removing material from eyes and skin is of extreme importance. If this chemical gets into the eyes, remove any contact lenses at once and irrigate immediately for at least 15 minutes, occasionally lifting upper and lower lids. Seek medical attention immediately. If this chemical contacts the skin, remove contaminated clothing and wash immediately with soap and water. Shampoo hair promptly if contaminated. Seek medical attention immediately. If this chemical has been inhaled, remove from exposure, begin rescue breathing (using universal precautions) if breathing has stopped, and CPR if heart action has stopped. Transfer promptly to a medical facility. When this chemical has been swallowed, get medical attention. Give large quantities of water and induce vomiting. Do not make an unconscious person vomit.

*References:*
- New Jersey Department of Health and Senior Services, Hazardous Substance Fact Sheet, Thiosemicabazide," Trenton, NJ (June 2002). http://www.state.nj.us/health/eoh/rtkweb/2823.pdf
- U.S. Environmental Protection Agency, "Chemical Profile: Thiosemicarbazide," Washington, DC, Chemical Emergency Preparedness Program (November 30, 1987).
- California Environmental Protection Agency "Chemical List of Lists," Sacramento CA (February 1997).

# Thiram

*Use Type:* Fungicide and rodenticide
*CAS Number:* 137-26-8
*Formula:* $C_6H_{12}N_2S_4$
*Alert:* A General Use Pesticide (GUP). Human toxicity (long-term): Very High.
*Synonyms:* Aceto TETD; A13-00987; Bis(diethylthiocarbamoyl) sulfide; Bis[(dimethylamino)carbonothioyl] disulphide; Bis[(dimethylamino)carbonothioyl] disulfide; Bis(dimethylthiocarbamoyl) disulfide; Bis(dimethylthiocarbamoyl) disulphide; Disulfide, bis(dimethylthiocarbamoyl); α,α'-Dithiobis(dimethylthio)formamide; N,N-(Dithiodicarbonothioyl)bis(N-methylmethanamine); ENT 987; Formamide, 1,1'-dithiobis(N,N-dimethylthio-; Methyl thiram; Methylthiuram disulfide; Methyl tuads; NSC 1771; Teramethylthiuram disulfide; Tetramethyldiurane sulphite; Tetramethylenethiuram disulfide; Tetramethylenethiuram disulphide; Tetramethylthiocarbamoyldisulphide; Tetramethylthioperoxydicarbonic diamide; Tetramethylthiuram; Tetramethylthiuram bisulfide; Tetramethylthiuram bisulphide; Tetramethylthiuram disulfide; N,N-Tetramethylthiuram disulfide; N,N,N',N'-Tetramethylthiuram disulfide; Tetramethylthiuram disulphide; Tetramethylthiuran disulphide; Tetramethylthiurane disulfide; Tetramethyl thiurane disulphide; Tetramethylthiurum disulfide; Tetramethylthiurum disulphide; Tetrathiuram disulfide; Tetrathiuram disulphide; Thioperoxydicarbo NIC diamide, Tetramethyl-; Thirame (French); Thiuram (Japan); Tiram (Spanish); Tiuram (Polish); Tiuramyl; TMTD (former USSR); TMTDS; TTD

*Trade Names:* AAPIROL®; AATACK®; AATIRAM®; ACCELERATOR T®; ACCELERATOR THIURAM®; ACCEL TMT®; AGROSOL POUR-ON®, Agriliance (USA); ANLES®; ARASAN®, DuPont (USA), canceled 2/21/1986; ATIRAM®; ATTACK®; AULES®; CHIPCO THIRAM 75®; CRYLCOAT®; CUNITEX®; CYURAM DS®; DELSAN®; EBECRYL®; EKAGOM TB®;

EVERSHIELD T SEED PROTECTORANT®, Gustafson LLC (USA); FALITIRAM®; FERMIDE®; FERNACOL®; FERNASAN®; FERNIDE®; FLO PRO T SEED PROTECTANT®, Gustafson LLC (USA), canceled 7/1/1987; FMC 2070®, FMC (USA), canceled; FORMALSOL®; HERMAL®; HERYL®; HEXATHIR®; HY-VIC®; KODIAK T®, Gustafson LLC (USA); KREGASAN®; LIQUID MOLY-CO-THI®, Gustafson LLC (USA); MERCURAM®; METIURAC®; MOLY-T®, Gustafson LLC (USA); NA2771®; NOBECUTAN®; NOMERSAN®; NORMERSAN®; OPTIMA®, Crompton Corporation (USA); PANORAM 75®; POLYRAM ULTRA®; POMARSOL®; POMARSOL FORTE®; POMASOL®; PRO-GRO®, Crompton Corporation (USA); PURALIN®; RAXIL®, Gustafson LLC (USA); REZIFILM®; ROOTONE®, Bayer CropScience (Germany); ROYAL TMTD®; RTU-BAYTAN-THIRAM®, Gustafson LLC (USA); RTU FLOWABLE SOYBEAN FUNGICIDE®, Gustafson LLC (USA); SADOPLON®; SOLUCRYL®; SPOTRETE®, Taminco (Belgium); SPOTRETE-F®, Taminco (Belgium); SQ 1489®; SRANAN-SF-X®; TERSAN 75®, DuPpont (USA), canceled 2/21/1986; TERSANTETRAMETHYL DIURANE SULFIDE®; TETRAPOM®; TETRASIPTON®; THIANOSAN®; THILLATE®; THIMAR®; THIMER®; THIOKNOCK®; THIOSAN®; THIOSCABIN®; THIOTEX®; THIOTOX®; THIRAM 75®; THIRAM 80®; THIRAMAD®; THIRAM B®; THIRAMPA®; THIRASAN®; THIULIN®; THIULIX®; THIURAD®; THIURAMIN®; THIURAMYL®; THYLATE®; TIRAMPA®; TITAN FL®, Gustafson LLC (USA); TRAMETAN®; TRIDIPAM®; TRIPOMOL®; TUADS®; TUEX®; TULISAN®; UCECOAT®; UCECRYL®; UVECRYL®; VANCIDA TM-95®; VANCIDE TM®; VITAFLO 280®, Gustafson LLC (USA); VITAVAX®, Gustafson LLC (USA); VITAVAX-T®, Crompton Corporation (USA); VUAGT-1-4®; VULCAFOR TMTD®; VULKACIT MTIC®; VULKACIT THIURAM®; VULKACIT THIURAM/C®

**Producers:** Agriliance (USA); Agrimor International (USA); Akzo Nobel Chemicals (Netherlands); American Cyanamid's Agricultural Products Group (USA); ATOFINA (France); Bayer CropScience (Germany); Bonide Products (USA); China Chemicals (China); Crompton Corporation (USA); Gustafson LLC (USA); Hebei Huafeng Chemical Group (China); Hokko Chemical Industry (Japan); Kawaguchi Chemical Industry (Japan); Micro Flo Company (USA); Rhone-Poulenc (France); Shenzhen Guomeng Industry Co., Ltd. (China); Sumitomo Chemicals (Japan); Taminco (Belgium); UCB Group (Belgium)

**Chemical Class:** Dithiocarbamate
**EPA/OPP PC Code:** 079801
**California DPR Chemical Code:** 589
**ICSC Number:** 0757
**RTECS Number:** JO14000000
**EEC Number:** 006-005-00-4

**EINECS Number:** 205-286-2

**Uses:** Thiram is used as a fungicide to prevent crop damage in the field and to prevent crops from deterioration in storage or transport. Thiram is also used as a seed, nut, fruit, and mushroom disinfectant from a variety of fungal diseases. In addition, it is used as an animal repellent to protect fruit trees and ornamentals from damage by rabbits, rodents, and deer. Thiram has been used in the treatment of human scabies, as a sun screen, and as a bactericide applied directly to the skin or incorporated into soap. Thiram is used as a rubber accelerator and vulcanizer and as a bacteriostat for edible oils and fats. It is also used as a rodent repellent, wood preservative, and may be used in the blending of lubricant oils.

**Human toxicity (long-term)[77]:** Low–56.00 ppb, Health Advisory

**Fish toxicity (threshold)[77]:** Extra high–0.53654 ppb, MATC (Maximum Acceptable Toxicant Concentration)

**U.S. Maximum Allowable Residue Levels for Thiram (40 CFR 180.132):**

| CROP | ppm |
|---|---|
| Apple | 7 |
| Banana, pulp | 1 |
| Peach | 7 |
| Strawberry | 7 |

**Carcinogen/Hazard Classifications**
**TRI Developmental Toxin:** Reproductive and developmental toxin
**IARC:** Group 3, unclassifiable
**WHO Acute Hazard:** Class III, slightly hazardous
**Endocrine Disruptor:** Suspected endocrine disruptor
**Regulatory Authority:**
- Air Pollutant Standard Set (ACGIH)[1] (DFG)[3] (HSE)[33] (OSHA)[58] (former USSR)[35, 43] (Several States)[60]
- Permissible Exposure Limits for Chemical Contaminants (CAL/OSHA)
- The "Director's List" (CAL/OSHA)
- Actively registered pesticide in California.
- EPA Hazardous Waste Number (RCRA No.): U244
- RCRA, 40CFR261, Appendix 8 Hazardous Constituents
- Superfund/EPCRA 40CFR302.4 RQ: CERCLA, 10 lb (4.54 kg)
- EPCRA Section 313 Form R *de minimis* concentration reporting level: 1.0%
- Canada, WHMIS, Ingredients Disclosure List

**Description:** Thiram is a colorless to yellow, crystalline solid. The methyl analog of disulfiram.* Characteristic odor. Commercial pesticide products may be dyed blue. Practically insoluble in water; solubility = 26 ppm @ 20°C; 30 ppm @ 25°C. Boiling point = 129°C. Melting/Freezing point = 156°C. Flash point = 148°C. Hazard Identification (based on NFPA-704 M Rating System): Health 2, Flammability 1, Reactivity. Log $K_{ow}$ = 1.79 to 2.0. Unlikely to bioaccumulate in marine organisms. *See also disulfiram.

*Incompatibilities:* Strong oxidizers, strong acids and oxidizable materials.

*Permissible Exposure Limits in Air:* The OSHA/NIOSH[2] TWA, the DFG MAK[3] and the HSE[33] TWA value for Thiram is 5 mg/m$^3$. The STEL value set by HSE[33] is 10 mg/m$^3$. The NIOSH[2] IDLH level = 100 mg/m$^3$. The ACGIH TWA is 1.0 mg/m$^{3(1)}$. The former USSR[35,43] has set a MAC for workplace air of 0.5 mg/m$^3$. The former USSR[35,43] has also set MAC values for ambient air in residential areas of 0.01 mg/m$^3$ on a momentary basis and 0.006 mg/m$^3$ on a daily average basis. Several states have set guidelines or standards for Thiram in ambient air[60] ranging from 50 $\mu$g/m$^3$ (North Dakota) to 80 $\mu$g/m$^3$ (Virginia) to 100 $\mu$g/m$^3$ (Connecticut) to 119 $\mu$g/m$^3$ (Nevada).

*Determination in Air:* Filter; CH$_3$CN; High-pressure liquid chromatography/Ultraviolet detection; NIOSH IV, Method #5005.[18]

*Permissible Concentration in Water:* The former USSR-UNEP/IRPTC project[43] has set an MAC in water bodies used for domestic purposes of 1.0 $\mu$g/L. Further, it has set a MAC in water bodies used for fishery purposes of zero. The State of Maine[61] has set a guideline for Thiram in drinking water of 10 $\mu$g/L.

*Routes of Entry:* Inhalation, ingestion, skin and/or eye contact.

*Harmful Effects and Symptoms*

*Short Term Exposure:* Irritates the eyes, skin, and respiratory tract. High exposures can cause kidney and liver damage. Brain and nerve damage can also occur. *Inhalation:* Inhalation can cause irritation of the respiratory tract with stuffy nose, nosebleeds, hoarseness, cough and/or phlegm. Animal studies indicate that irritation of the nose and throat may occur at levels above 5 mg/m$^3$. *Skin:* Exposure to spray containing 45% Thiram resulted in irritation and skin sensitization. Skin irritation can lead to rash, and allergy. *Eyes:* May cause irritation, tearing and sensitivity to light. *Ingestion:* No information available on human exposure. In animal studies, 38 ppm in food caused nausea, vomiting, diarrhea, hyperexcitability, weakness and loss of muscle control. Death may occur from ingestion of approximately one teaspoonful. *Note:* Unlike carbamates the dithiocarbamates are not cholinesterase inhibitors, but some of them may react with recently ingested alcohol or alcohol-containing products including wine, medications, and cold remedies such as cough-syrups. See also disulfiram.

*Long Term Exposure:* Repeated or prolonged contact may cause skin sensitization. Prolonged contact has caused eye irritation, tearing, increased sensitivity to light, reduced night vision and blurred vision. Occupational exposures to 0.03 mg/m$^3$ over a 5 year period has caused mild irritation of the nose and throat. Whether it has this effect in humans is not known. May affect the thyroid and liver. Thiram has caused birth defects in laboratory animals and has been shown to be a teratogen in animals.

*Points of Attack:* Eyes, skin, respiratory system and central nervous system.

*Medical Surveillance:* Preplacement and periodic medical examinations should give special attention to history of skin allergy, eye irritation, and significant respiratory, liver, or kidney disease. Workers should be aware of the potentiating action of alcoholic beverages when working with tetramethylthiuram disulfide.

*First Aid:* If this chemical gets into the eyes, remove any contact lenses at once and irrigate immediately for at least 15 minutes, occasionally lifting upper and lower lids. Seek medical attention immediately. If this chemical contacts the skin, remove contaminated clothing and wash immediately with soap and water. Seek medical attention immediately. If this chemical has been inhaled, remove from exposure, begin rescue breathing (using universal precautions) if breathing has stopped, and CPR if heart action has stopped. Transfer promptly to a medical facility. When this chemical has been swallowed, get medical attention. Give large quantities of water and induce vomiting. Do not make an unconscious person vomit.

*References:*

- EXTOXNET, Extension Toxicology Network, "Pesticide Information Profile, Thiram," Oregon State University, Corvallis, OR (June 1996). http://extoxnet.orst.edu/pips/thiram.htm
- U.S. Environmental Protection Agency, Office of Pesticide Programs, Pesticide Residue Limits, "Thiram," 40 CFR 180.132, http://www.epa.gov/pesticides/food/viewtols.htm
- Sax, N.I., Ed., "Dangerous Properties of Industrial Materials Report," 1, No. 5, 41-42 (1981).
- New York State Department of Health, "Chemical Fact Sheet: Thiram," Albany, NY, Bureau of Toxic Substance Assessment (April 1986).
- New Jersey Department of Health and Senior Services, "Hazardous Substance Fact Sheet, Thiram," Trenton NJ (March 1989, rev. June 2000). http://www.state.nj.us/health/eoh/rtkweb/1854.pdf

# Thymol

*Use Type:* Dye and biocide
*CAS Number:* 89-83-8
*Formula:* $C_{10}H_{14}O$
*Synonyms:* p-Cymen-3-ol; 3-p-Cymenol; 3-Hydroxy-4-cymene; 3-Hydroxy-p-cymene; 3-Hydroxy-1-methyl-4-isopropylbenzene; Isopropyl cresol; 6-Isopropyl-3-cresol; 6-Isopropyl-m-cresol; 2-Isopropyl-5-methylphenol; 1-Methyl-3-hydroxy-4-isopropylbenzene; 5-Methyl-2-isopropyl-1-phenol; 5-Methyl-2-(1-methylethyl)phenol; Thyme camphor; Thymic acid; m-Thymol
*Trade Names:* BENEFECT®; ECOPAC®; RO-PE® L; THYMO-CIDE®; TOPPS®

*Chemical Class:* Phenol
*EPA/OPP PC Code:* 080402
*California DPR Chemical Code:* 991
*Uses:* Thymol is a constituent of oil of thyme, a naturally occurring mixture of compounds in the plant Thymus vulgaris L, or thyme. Thymol is an active ingredient in pesticide products registered for use as animal repellents, fungicides/fungistats, medical disinfectants, tuberculocides, and virucides. These products are used on a variety of indoor and outdoor sites, to control target pests including animal pathogenic bacteria and fungi, several viruses including HIV-I, and birds, squirrels, beavers, rats, mice, dogs, cats and deer. Products are liquids applied by spray, mop, brush-on, wipe-on dip, aerosol, immersion and spot treatment. Thymol also has many non-pesticidal uses, including use in perfumes, food flavorings, mouthwashes, pharmaceutical preparations and cosmetics.
*U.S. Maximum Allowable Residue Levels for Thymol (40 CFR 180.1240):*
Thymol has been exempt from residue levels on honey and honeycomb.
*Carcinogen/Hazard Classifications*
**Label Signal Word:** CAUTION or DANGER
*Regulatory Authority:*
- Actively registered pesticide in California.

*Description:* White, crystalline solid or colorless plates. Slightly soluble in water; solubility = 850 ppm @ 20°C. Molecular weight 150.2. Density = 0.972. Boiling point = 234°C. Melting/Freezing point = 51–53°C. Vapor pressure = 1 mmHg @ 64°. Log $K_{ow}$ = 3.32. Values above 3.0 are likely to bioaccumulate in marine organisms.
*Incompatibilities:* React with boranes, alkalies, aliphatic amines, amides, nitric acid. Keep away from oxidizers, sulfuric acid, caustics, ammonia, aliphatic amines, alkanolamines, isocyanates, alkylene oxides, epichlorohydrin.
*Permissible Concentration in Water:* No criteria set. Runoff from spills or fire control may cause water pollution.
*First Aid:* If this chemical gets into the eyes, remove any contact lenses at once and irrigate immediately for at least 15 minutes, occasionally lifting upper and lower lids. Seek medical attention immediately. If this chemical contacts the skin, remove contaminated clothing and wash immediately with soap and water. Seek medical attention immediately. If this chemical has been inhaled, remove from exposure, begin rescue breathing (using universal precautions) if breathing has stopped, and CPR if heart action has stopped. Transfer promptly to a medical facility. When this chemical has been swallowed, get medical attention. Give large quantities of water and induce vomiting. Do not make an unconscious person vomit. Do not induce vomiting when formulations containing petroleum solvents are ingested.
*References:*
- U.S. Environmental Protection Agency, "Reregistration Eligibility Decision Facts (RED), Thymol," Office of Prevention, Pesticides and Toxic Substances, Washington, DC (September 1993). http://www.epa.gov/REDs/factsheets/3143fact.pdf
- U.S. Environmental Protection Agency, Office of Pesticide Programs, Pesticide Residue Limits, "Thymol," 40 CFR 180.1240, www.epa.gov/pesticides/food/viewtols.htm
- California Environmental Protection Agency "Chemical List of Lists," Sacramento CA (February 1997)

# Toxaphene

*Use Type:* Insecticide
*CAS Number:* 8001-35-2
*Formula:* $C_{10}H_{10}Cl_8$
*Alert:* Toxaphene is a carcinogen and teratogen and should be handled with extreme caution. There are no currently registered products of toxaphene in the U.S; all uses were banned in 1982. Human toxicity (long-term): High.
*Synonyms:* Camphechlor; Chem-Phene; Camphene, octachloro-; Camphochlor; Camphoclor; Camphofene chlorinated camphene; Chlorocamphene; ENT 9735; NCI-C00259; Octachlorocamphene; PCC; PCHK; Penphene; Polychlorcamphene; Polychlorinated camphene; Polychlorocamphene; Technical chlorinated camphene, 67-69% chlorine; Toxafeen (Dutch); Toxafeno (Spanish); Toxaphen (German)
*Trade Names:* AGRICIDE MAGGOT KILLER (F)®; ALLTEX®; ALLTOX®; ANATOX®; ATTAC-2®; ATTAC 6®; ATTAC 6-3®; HUILEUX®; CANFECLOR®; CLOR CHEM T-590®; COMPOUND 3956®; COTTON TOX MP82®; CRESTOXO®; CRISTOXO 90®; DR. ROGER'S TOXENE®; ESTONOX®; FASCO-TERPENE®; GENIPHENE®; GY-PHENE®; HERCULES® 3956; HERCULES TOXAPHENE®; KAMFOCHLOR®; M 5055®; MELIPAX®; MOTOX®; PHENACIDE®; PHENATOX®; ROYAL BRAND BEAN TOX 82®; SECURITY TOX-SOL-6®; STROBANE T®; STROBANE T 90®; SYNTHETIC 3956®; TOXADUST®; TOXAKIL®; TOXASPRAY®; TOXON 63®; TOXYPHEN®; VERTAC 90%®; VERTAC TOXAPHENE 90®
*Producers:* DuPont Crop Protection (USA)
*Chemical Class:* Organochlorine; Halo-organics
*EPA/OPP PC Code:* 080501
*California DPR Chemical Code:* 594
*ICSC Number:* 0843
*RTECS Number:* XW5250000
*EEC Number:* 602-044-00-1
*EINECS Number:* 232-283-3
*Uses:* Toxaphene is a broad spectrum, chlorinated hydrocarbon pesticide that contains over 670 chemicals. All uses of toxaphene were banned in the U.S. in 1982. It was used primarily in the southern United States to control insect

## 862 Toxaphene

pests on cotton and other crops. It was also used to control insect pests on livestock and to kill unwanted fish in lakes.

***Human toxicity (long-term)***[77]: High–3.00 ppb, MCL (Maximum Contaminant Level)

***Fish toxicity (threshold)***[77]: Extra high–0.03900 ppb, MATC (Maximum Acceptable Toxicant Concentration)

***Carcinogen/Hazard Classifications***

**U.S. EPA Carcinogens:** Group B2, probable carcinogen
**U.S. NTP Carcinogen:** Reasonably anticipated carcinogen
**California Prop. 65:** Carcinogen
**IARC:** Group 2B, possible carcinogen
**Label Signal Word:** DANGER, WARNING, CAUTION depending on formulation.
**Endocrine Disruptor:** Suspected endocrine disruptor

***Regulatory Authority:***

- Carcinogen (Animal Positive) (IARC) (NCI)[9]
- Banned or Severely Restricted (Many Countries) (UN)[13]
- Air Pollutant Standard Set (ACGIH)[1] (DFG)[3] (former USSR)[35] (OSHA)[58] (Several States)[60]
- List of priority pollutants (U.S. EPA)
- AB 1803-Well Monitoring Chemical (CAL)
- EPA/SARA 302 (EPCRA) Extremely Hazardous Substances
- MCL (Maximum Contaminants Levels) list of contaminants (CAL)
- AB 2588-Air Toxics "Hot Spots" Chemicals (CAL)
- CAL Air Resources Board/AB 1807 Toxic Air Contaminants
- Proposition 65 chemical (CAL)
- Permissible Exposure Limits for Chemical Contaminants (CAL/OSHA)
- The "Director's List" (CAL/OSHA)
- Clean Air Act: Hazardous Air Pollutants (Title I, Part A, Section 112)
- Clean Water Act: Section 311 Hazardous Substances/RQ 40CFR117.3 (same as CERCLA, see below); Toxic Pollutant (Section 401.15); 40CFR423, Appendix A, Priority Pollutants; Section 313 Water Priority Chemicals (57FR41331, 9/9/92)
- EPA Hazardous Waste Number (RCRA No.): P123
- RCRA, 40CFR261, Appendix 8 Hazardous Constituents
- RCRA Toxicity Characteristic (Section 261.24), Maximum Concentration of Contaminants, regulatory level, 0.5 mg/L
- RCRA 40CFR268.48; 61FR15654, Universal Treatment Standards: Wastewater (mg/L), 0.0095; Nonwastewater (mg/kg), 2.6
- RCRA 40CFR264, Appendix 9; TSD Facilities Ground Water Monitoring List. Suggested test method(s) (PQL ug/L): 8080(2); 8250(10)
- Safe Drinking Water Act: MCL, 0.003 mg/L; MCLG, zero; Regulated chemical (47 FR 9352)
- Superfund/EPCRA 40CFR355, Appendix B Extremely Hazardous Substances: TPQ = 500/10,000 lb (227/4,540 kg)
- Superfund/EPCRA 40CFR302.4 RQ: CERCLA, 1 lb (0.454 kg)
- EPCRA Section 313 Form R *de minimis* concentration reporting level: 0.1%
- U.S. DOT Regulated Marine Pollutant (49CFR172.101, Appendix B)
- Canada, WHMIS, Ingredients Disclosure List

***Description:*** Amber, waxy solid. Mild, piney, chlorine- and camphor-like odor; somewhat like turpentine. Usually dissolved in a flammable solvent. Flammability depends on the solvent used. Practically insoluble in water; solubility = 2.8 ppm @ 20°C. It sticks to the soil and settles into lakes, stream sediments and mud. Molecular weight = 413.84. Melting/Freezing point = 65–90°C. Vapor pressure = 7.1 x $10^{-6}$ mmHg @ 20°C. Flash point = 135°C (solid); 29°C (solution). Autoignition temperature= 530°C. Explosive limits: LEL = 1.1%; UEL = 6.4% (solvents only). Hazard Identification (based on NFPA-704 M Rating System): Health 3, Flammability 0, Reactivity 0. Log $K_{ow}$ = 3.5; also reported at 5.25. Values at or above 3.0 are likely to bioaccumulate in marine organisms.

***Incompatibilities:*** Contact with strong oxidizers may cause fire and explosion hazard. Decomposes, producing fumes of hydrogen chloride and chlorine in heat above 155°C, on contact with strong bases, strong sunlight, and catalysts such as iron. Slightly corrosive to metals in the presence of moisture.

***Permissible Exposure Limits in Air:*** The OSHA[2] PEL, DFG MAK[3], and ACGIH[1] TWA value is 0.5 mg/m$^3$, and the ACGIH STEL[1] is 1.0 mg/m$^3$. The notation "skin" is added to indicate the possibility of cutaneous absorption. The NIOSH[2] IDLH level = [Ca] 200 mg/m$^3$. The former USSR[35] has set an MAC for ambient air in residential areas of 7.0 µg/m$^3$ on a once-daily basis. A number of states have set guidelines or standards for toxaphene in ambient air[60] ranging from 1.19 µg/m$^3$ (Kansas) to 1.2 µg/m$^3$ (Pennsylvania) to 1.67 µg/m$^3$ (New York) to 2.5 µg/m$^3$ (Connecticut and South Carolina) to 5.0 µg/m$^3$ (Florida) to 5.0-10.0 µg/m$^3$ (North Dakota) to 8.0 µg/m$^3$ (Virginia) to 12.0 µg/m$^3$ (Nevada).

***Determination in Air:*** Filtration from air, working up with petroleum ether and analysis by gas chromatography. See NIOSH Method S-67.

***Permissible Concentration in Water:*** The U.S. EPA has set a drinking water standard of 0.003 milligrams of toxaphene per liter of drinking water (0.003 mg/L). To protect freshwater aquatic life-0.013 µg/L as a 24-hour average is never to exceed 1.6 µg/L. To protect saltwater aquatic life-never to exceed 0.07 µg/L. To protect human health, preferably zero. An additional lifetime cancer risk of 1 in 100,000 is presented by a concentration of 0.0071 µ/L.[6]. The International Joint Commission of the United States and Canada has recommended a water standard of 0.008 ppm for protection of aquatic life. States which have set drinking water guidelines include Maine at 5.0 µg/L and Minnesota at 0.3 μg/L[61]. Mexico[35] has set MPC values of 30 μg/L for estuaries, 3.0 μg/L for coastal waters and 5.0 μg/L for receiving waters used for drinking water supply.

***Determination in Water:*** Gas chromatography (EPA Method 608) or gas chromatography plus mass spectrometry (EPA Method 625).

***Routes of Entry:*** Inhalation, skin absorption, ingestion, eye and/or skin contact.

***Harmful Effects and Symptoms***

***Short Term Exposure:*** High concentrations can irritate the skin and eyes. A nervous system depressant causing tremors, weakness, dizziness, increased saliva, convulsions, unconsciousness, and possible death. High exposures can cause pulmonary edema, a medical emergency that can be delayed for several hours. This can cause death. Inhalation of spray at unknown levels has caused bronchitis and inflammation of the lungs. If spilled on clothing and allowed to remain, can cause pain and reddening. May be absorbed through skin, contributing to symptoms described under ingestion. Absorbed through the stomach and intestines. 0.6 grams (about 1/50 of an oz) has caused convulsions. Other symptoms include nausea, vomiting, bluish coloration of the skin, and coma. Estimated lethal dose for an adult is between 2 and 7 grams (1/15-1/4 ounce).

***Long Term Exposure:*** Aplastic anemia (low blood count) is an uncommon but serious reaction to toxaphene. High or repeated exposure may cause liver and kidney damage. Toxaphene is suspected of causing brain damage. Changes in genetic material have been observed in workers exposed to toxaphene. An animal study reported that toxaphene caused cancer of the thyroid gland when the animals were exposed to high levels over their lifetimes. It is not known whether toxaphene can affect reproduction or cause birth defects in people. Animal studies have reported that toxaphene affects the development of newborn animals when their mothers are exposed during pregnancy.

***Points of Attack:*** Central nervous system, skin. Cancer site in animals: liver cancer; thyroid tumors.

***Medical Surveillance:*** Consider the points of attack in preplacement and periodic physical examinations. Complete blood count. Complete examination of the nervous system. Liver and kidney function tests. Consider chest x-ray following acute overexposure.

***First Aid:*** If this chemical gets into the eyes, remove any contact lenses at once and irrigate immediately for at least 15 minutes, occasionally lifting upper and lower lids. Seek medical attention immediately. If this chemical contacts the skin, remove contaminated clothing and wash immediately with soap and water. Seek medical attention immediately. If this chemical has been inhaled, remove from exposure, begin rescue breathing (using universal precautions) if breathing has stopped, and CPR if heart action has stopped. Transfer promptly to a medical facility. When this chemical has been swallowed, get medical attention. Give large quantities of water and induce vomiting. Do not make an unconscious person vomit. Medical observation is recommended for 24 to 48 hours after breathing overexposure, as pulmonary edema may be delayed. As first aid for pulmonary edema, a doctor or authorized paramedic may consider administering a corticosteroid spray.

***References:***

- New Jersey Department of Health and Senior Services, "Hazardous Substance Fact Sheet, Toxaphene," Trenton NJ (December 1994, rev. May 2001). http://www.state.nj.us/health/eoh/rtkweb/1871.pdf
- Agency for Toxic Substances and Disease Registry, "ToxFAQs for Toxaphene," Atlanta, GA (September 1997). http://www.atsdr.cdc.gov/tfacts94.html
- U.S. Environmental Protection Agency, "Consumer Factsheet on Toxaphene," Washington, DC (November 2002). http://www.epa.gov/OGWDW/dwh/c-soc/toxaphen.html
- U.S. Environmental Protection Agency, "Reviews of the Environmental Effects of Pollutants: X. Toxaphene", Report No. EPA-600/1-79-044 (1979).
- U.S. Environmental Protection Agency, " Toxaphene, Health and Environmental Effects" Profile No. 163," Washington, DC, Office of Solid Waste (April 30, 1980).
- U.S. Environmental Protection Agency, "Chemical Profile: Camphechlor," Washington, DC, Chemical Emergency Preparedness Program (November 30, 1987).
- Sax, N.I., Ed., "Dangerous Properties of Industrial Materials Report," 2, No. 2, 68-70 (1982), 4, No. 1, 27-28 (1984) and 7, No. 5, 100-107 (1987).
- New York State Department of Health, "Chemical Fact Sheet: Toxaphene," Albany, NY, Bureau of Toxic Substance Assessment (April 1986).
- California Environmental Protection Agency "Chemical List of Lists," Sacramento CA (February 1997).
- USEPA Ambient Water Quality Criteria Document, "Toxaphene," 1980 [EPA 440/5-80-076]. Originally appeared in publication by the International Joint Commission, United States, and Canada, "Toxaphene" Vol. 2 (1977).

# Tralkoxydim

***Use Type:*** Herbicide
***CAS Number:*** 87820-88-0
***Formula:*** $C_{20}H_{27}NO_3$
***Synonyms:*** 2-Cyclohexen-1-one, 2-[1-(ethoxyimino)propyl]-3-hydroxy-5-(2,4,6-trimethylphenyl)-(9CI); 2-[1-(Ethoxyimino)propyl]-3-hydroxy-5-mesitylcyclohex-2-en-one; 2-[1-(Ethoxyimino)propyl]-3-hydroxy-5-(2,4,6-trimethylphenyl)-2-cyclohexen-1-one; Tralkoxydime
***Trade Names:*** ACHIEVE®, Syngenta (Switzerland); ACHIEVE®-40DG, Syngenta (Switzerland); GRASP®,

Syngenta (Switzerland); ICI-A 604®; PP 604®; SPLENDOR®, Syngenta (Switzerland)
*Producers:* Syngenta (Switzerland)
*Chemical Class:* Cyclohexanedione
*EPA/OPP PC Code:* 121000
*California DPR Chemical Code:* 5457
*Uses:* Tralkoxydim is applied to actively growing weeds in wheat, barley, triticale and cereal rye to control wild oats, green foxtail, yellow foxtail, annual ryegrass (Italian) and Persian darnel.
*Human toxicity (long-term)*[77]*:* Intermediate–20.83 ppb, CHCL (Chronic Human Carcinogen Level)
*Fish toxicity (threshold)*[77]*:* Very low–1154.96119 ppb, MATC (Maximum Acceptable Toxicant Concentration)
*U.S. Maximum Allowable Residue Levels for Tralkoxydim (40 CFR 180.548):*
The revocation date for the following residue levels was February 28, 2003.

| CROP | ppm |
| --- | --- |
| Barley, grain | 0.02 |
| Barley, hay | 0.02 |
| Barley, straw | 0.05 |
| Wheat, forage | 0.05 |
| Wheat, grain | 0.02 |
| Wheat, hay | 0.02 |
| Wheat, straw | 0.05 |

*Carcinogen/Hazard Classifications*
**U.S. EPA Carcinogens:** Likely carcinogen
**Label Signal Word:** CAUTION
**WHO Acute Hazard:** Class III, slightly hazardous
*Description:* Off white to pale pink crystalline solid. Faint burnt odor. Soluble in water. Molecular weight = 329.47. Melting/Freezing point = 110°C. Vapor pressure = Extremely low; $3.5 \times 10^{-10}$. Non-corrosive, stable at normal warehouse temperatures.
*Permissible Concentration in Water:* No criteria set. Runoff from spills or fire control may cause water pollution.
*First Aid:* If this chemical gets into the eyes, remove any contact lenses at once and irrigate immediately for at least 15 minutes, occasionally lifting upper and lower lids. Seek medical attention immediately. If this chemical contacts the skin, remove contaminated clothing and wash immediately with soap and water. Seek medical attention immediately. If this chemical has been inhaled, remove from exposure, begin rescue breathing (using universal precautions) if breathing has stopped, and CPR if heart action has stopped. Transfer promptly to a medical facility. When this chemical has been swallowed, get medical attention. Give large quantities of water and induce vomiting. Do not make an unconscious person vomit. Do not induce vomiting when formulations containing petroleum solvents are ingested.
*References:*
- U.S. Environmental Protection Agency, Office of Pesticide Programs, Pesticide Fact Sheet, "Tralkoxydim." (December 4, 1998).
- http://www.epa.gov/opprd001/factsheets/tralkoxystrobin.pdf
- California Environmental Protection Agency "Chemical List of Lists," Sacramento CA (February 1997).

# Tralomethrin

*Use Type:* Insecticide
*CAS Number:* 66841-25-6
*Formula:* $C_{22}H_{19}Br_4NO_3$
*Synonyms:* (S)-α-Cyano-3-phenoxybenzyl (1R,3R)-3-(2,2–dibromovinyl-2,2-dimethylcyclopropanecarboxylate [metabolate of tralomethrin]; (S)-α-Cyano-3-phenoxybenzyl (1S,3R)-3-(2,2–dibromovinyl)-2,2-dimethylcyclopropanecarboxylate [metabolate of tralomethrin]; Cyano(3-phenoxyphenyl)methyl; Cyclopropanecarboxylic acid, 2,2-dimethyl-3-(1,2,2,2-tetrabromoethyl)-, cyano(3-phenoxyphenyl)methyl ester (9CI); 2,2-Dimethyl-3-(1,2,2,2-tetrabromoethyl)cyclopropropanecarboxlic acid, cyano(3-phenoxyphenyl)methyl ester; 2,2-Dimethyl-3-(1,2,2,2-tetrabromoethyl)cyclopropanecarboxylate; (1R,3S)3[(1'RS)(1',2',2',2'-Tetrabromoethyl)]-2,2-dimethylcyclopropanecarboxylic acid, (S)-α-cyano-3-phenoxybenzyl ester; Tralomethrine
*Trade Names:* DETHMOR®; HAG-107®; RU-25472®; RU-25474®; SCOUT®; SCOUT® X-TRA Gel insecticide
*Chemical Class:* Pyrethroid
*EPA/OPP PC Code:* 121501; (128822 old EPA code number)
*California DPR Chemical Code:* 2329
*Human toxicity (long-term)*[77]*:* Low–52.50 ppb, Health Advisory
*Fish toxicity (threshold)*[77]*:* Extra high–0.06714 ppb, MATC (Maximum Acceptable Toxicant Concentration)
*U.S. Maximum Allowable Residue Levels for Tralomethrin (40 CFR 180.422):*

| CROP | ppm |
| --- | --- |
| Broccoli | 0.5 |
| Cotton, refined oil | 0.2 |
| Cotton, undelinted seed | 0.02 |
| Feed processing and storage areas | 0.02 |
| Lettuce, head | 1.0 |
| Lettuce, leaf | 3.0 |
| Processed food | 0.02 |
| Soybean | 0.05 |
| Sunflower, seed | 0.05 |

*Carcinogen/Hazard Classifications*
**Label Signal Word:** CAUTION
**Endocrine Disruptor:** Suspected endocrine disruptor
*Regulatory Authority:*
- FIFRA, 40CFR185: tolerances for pesticides in food

*Description:* Yellow-orange resinous solid. Soluble in water; solubility = 70 mg/kg. Molecular weight = 665.05. Density = 1.70 @ 20°C. Boiling point = 138°C. Vapor pressure = 17 pPa

*Incompatibilities:* May react violently with strong oxidizers, bromine, 90% hydrogen peroxide, phosphorus trichloride, silver powders or dust. Incompatible with silver compounds. Mixture with some silver compounds forms explosive salts of silver oxalate.

*Permissible Exposure Limits in Air:*

*Determination in Air:* Collection by impinger or fritted bubbler, analysis by gas liquid chromatography/ultraviolet. See NIOSH IV, Method #5008[18] (pyrethrum).

*Permissible Concentration in Water:* No criteria set. Runoff from spills or fire control may cause water pollution.

*Determination in Water:* Collection by impinger or fritted bubbler, analysis by gas liquid chromatography/ultraviolet. See NIOSH IV, Method #5008.[18]

*Routes of Entry:* Inhalation, ingestion, absorbed through skin.

*Harmful Effects and Symptoms*

*Short Term Exposure:* Contact may cause burns to skin and eyes. Poisonous; may be fatal if inhaled, swallowed or absorbed through skin. Contact may cause burns to skin and eyes.

*First Aid:* If this chemical gets into the eyes, remove any contact lenses at once and irrigate immediately for at least 15 minutes, occasionally lifting upper and lower lids. Seek medical attention immediately. If this chemical contacts the skin, remove contaminated clothing and wash immediately with soap and water. Seek medical attention immediately. If this chemical has been inhaled, remove from exposure, begin rescue breathing (using universal precautions) if breathing has stopped, and CPR if heart action has stopped. Transfer promptly to a medical facility. When this chemical has been swallowed, get medical attention. Give large quantities of water and induce vomiting. Do not make an unconscious person vomit. Do not induce vomiting when formulations containing petroleum solvents are ingested.

*References:*
- U.S. Environmental Protection Agency, Office of Pesticide Programs, Pesticide Residue Limits, "Tralomethrin," 40 CFR 180.422, www.epa.gov/pesticides/food/viewtols.htm
- California Environmental Protection Agency "Chemical List of Lists," Sacramento CA (February 1997)

# Triacontanol

*Use Type:* Plant growth regulator
*CAS Number:* 593-50-0
*Formula:* $C_{30}H_{62}O$
*Synonyms:* 1-Hydroxytriacontane; Melissyl alcohol; Myricyl alcohol; 1-Triacontanol

*Trade Names:* M 8164®; TRIACON-10®, canceled; ULTRIA®, canceled; UREKA®, Sudarshan Chemical Industries (India)
*Producers:* Sudarshan Chemical Industries (India)
*Chemical Class:* Naturally occurring plant hormone
*EPA/OPP PC Code:* 116201
*Uses:* Not registered in the U.S. Triacontanol raises plant yield by improving cell division and photosynthesis.
*Description:* White crystalline solid. Practically insoluble in water. Melting/Freezing point = 91°C. Molecular weight = 438.82

*Permissible Exposure Limits in Air:*

*Determination in Air:* Collection by charcoal tube, 2-Butanol/$CS_2$; analysis by gas chromatography/flame ionization detection; NIOSH (IV) [Method #1400, Alcohols I].

*Permissible Concentration in Water:* No criteria set. Runoff from spills or fire control may cause water pollution.

*First Aid:* If this chemical gets into the eyes, remove any contact lenses at once and irrigate immediately for at least 15 minutes, occasionally lifting upper and lower lids. Seek medical attention immediately. If this chemical contacts the skin, remove contaminated clothing and wash immediately with soap and water. Seek medical attention immediately. If this chemical has been inhaled, remove from exposure, begin rescue breathing (using universal precautions) if breathing has stopped, and CPR if heart action has stopped. Transfer promptly to a medical facility. When this chemical has been swallowed, get medical attention. Give large quantities of water and induce vomiting. Do not make an unconscious person vomit.

# Triadimefon

*Use Type:* Fungicide
*CAS Number:* 43121-43-3
*Formula:* $C_{14}H_{16}ClN_3O_2$
*Synonyms:* 2-Butanone, 1-(4-chlorophenoxy)-3,3-dimethyl-1-(1-*H*-1,2,4-triazol-1-yl)-; 2-Butanone, 1-(4-chlorophenoxy)-3,3-dimethyl-1-(1,2,4-triazol-1-yl)-; 1-[(*tert*-Butylcarbonyl-4-chlorophenoxy)methyl]-1*H*-1,2,4-triazole; Caswell No. 862AA; 1-(4-Chlorophenoxy)-3,3-dimethyl-1-(1*H*-1,2,4-triazol-1-yl)-2-butanone; 1-(4-Chlorophenoxy)-3,3-dimethyl-1-(1,2,4-triazol-1-yl)-butan-2-one; 1-(4-Chlorophenoxy)-3,3-dimethyl-1-(1*H*-1,2,4-triazol-1-yl)butanone-; 1-(4-Chlorophenoxy)-3,3-dimethyl-1-(1,2,4-triazol-1-yl)-2-butan-2-one; 1-(4-Chlorophenoxy)-3,3-dimethyl-1-(1*H*-1,2,4-triazol-1-yl)-2-butanone; 1-(4-Chlorophenoxy)-3,3-dimethyl-1-(1,2,4-triazol-1-yl)butanone-; NSC 303303; Triadimefon triazole fungicide; Triadimefone; Triadimeform; 1*H*-1,2,4-Triazole, 1-[(*tert*-butylcarbonyl-4-chlorophenoxy)methyl]-; 1-(1,2,4-Triazoyl-1)-1-(4-chloro-phenoxy)3,3-dimethylbutanone; 1-[(*tert*-Butylcarbonyl-4-chlorophenoxy)methyl]-

# Triadimefon

*Trade Names:* ACCOST®; ACIZOL®; AMIRAL®; BAY® 6681-F, Bayer CropScience (Germany); BAYLETON®, Bayer CropScience (Germany); BAY®-MEB-6447, Bayer CropScience (Germany); BAYER® 6681-F, Bayer CropScience (Germany); BAYER® MEB-6447, Bayer CropScience (Germany); MEB 6447®; PRO-TEK®; ROFON®

*Producers:* Agrimor International (USA); Agsin (Singapore); Bayer CropScience (Germany); Bonide (USA)

*Chemical Class:* Chlorophenoxy; azole; triazole

*EPA/OPP PC Code:* 109901

*California DPR Chemical Code:* 2133

*RTECS Number:* EL7100000

*Uses:* Triadimefon is a systemic fungicide that is used to control powdery mildews, rusts, and other fungi on coffee, seed grasses, cereals, fruits, grapes, vegetables, vines, pineapple, sugar cane, sugar beets, turf, shrubs, and trees.

*Human toxicity (long-term)[77]:* Intermediate–28.00 ppb, Health Advisory

*Fish toxicity (threshold)[77]:* Intermediate–68.11770 ppb, MATC (Maximum Acceptable Toxicant Concentration)

*U.S. Maximum Allowable Residue Levels for Triadimefon (40 CFR 180.410):*

| CROP | ppm |
| --- | --- |
| Apple | 1.0 |
| Apple, pomace, wet & dry | 4.0 |
| Barley, milled fractions, except flour | 4.0 |
| Beet, sugar | 0.5 |
| Beet, sugar, tops | 3.0 |
| Cattle, fat | 1.0 |
| Cattle, meat | 1.0 |
| Cattle, mbyp | 1.0 |
| Chickpea, seed | 0.1 |
| Cucurbits | 0.3 |
| Egg | 0.04 |
| Goat, fat | 1.0 |
| Goat, meat | 1.0 |
| Goat, mbyp | 1.0 |
| Grape | 1.0 |
| Grape, pomace, wet & dried | 3.0 |
| Grape, raisin, waste | 7.0 |
| Grass, forage | 0.2 |
| Grass, seed screenings | 145.0 |
| Grass, straw, grown for seed | 105.0 |
| Hog, fat | 0.04 |
| Hog, meat | 0.04 |
| Hog, mbyp | 0.04 |
| Horse, fat | 1.0 |
| Horse, meat | 1.0 |
| Horse, mbyp | 1.0 |
| Milk | 0.04 |
| Nectarine | 4.0 |
| Pear | 1.0 |
| Pineapple | 3.0 |
| Poultry, fat | 0.04 |
| Poultry, meat | 0.04 |
| Poultry, mbyp | 0.04 |
| Raspberry | 2.0 |
| Sheep, fat | 1.0 |
| Sheep, meat | 1.0 |
| Sheep, mbyp | 1.0 |
| Wheat, forage | 15.0 |
| Wheat, grain | 1.0 |
| Wheat, milled fractions, except flour | 4.0 |
| Wheat, straw | 5.0 |

*Carcinogen/Hazard Classifications*

**U.S. EPA Carcinogens:** Group 3, possible carcinogen

**California Prop. 65:** Developmental toxin; female and male reproductive toxin

**TRI Developmental Toxin:** Reproductive and developmental toxin

**Label Signal Word:** CAUTION, WARNING or DANGER

**WHO Acute Hazard:** Class III, slightly hazardous

**Endocrine Disruptor:** Suspected endocrine disruptor

*Regulatory Authority:*

- EPA 40 CFR 372.65, Specific Toxic Chemical Listings
- FIFRA, 40CFR185: tolerances for pesticides in food
- FIFRA, 40CFR186: tolerances for pesticides in animal feeds
- AB 2588-Air Toxics "Hot Spots" Chemicals (CAL) as chlorophenoxy pesticides
- The "Director's List" (CAL/OSHA) as chlorophenoxy pesticides
- Actively registered pesticide in California.
- EPCRA Section 313 Form R *de minimis* concentration reporting level: 1.0%

*Description:* Colorless crystalline solid. Soluble in water; solubility = 250 ppm. Melting point = 82.3°C. Molecular weight = 293.78

*Permissible Concentration in Water:* No criteria set. Runoff from spills or fire control may cause water pollution.

*Routes of Entry:* Inhalation, ingestion, absorbed through the skin.

*Harmful Effects and Symptoms*

**Short Term Exposure:** Poisonous; may be fatal if inhaled, swallowed, or absorbed through skin. Severely irritates eyes, skin and respiratory tract, with burning sensation, pain, redness and swelling. Metabolic stimulant. If inhaled, causes coughing, dilated pupils, headache, profuse persperation, intense thirst, extreme fatigue, rapid pulse, high fever, clammy, flushed skin, rapid breathing, nausea, vomiting, cyanosis (bluish tint to skin and lips), anxiety and confusion, convulsions, risk of lung edema. If swallowed, face and lips turn bluish. Liver injury with associated jaundice, kidney failure, and cardiac arrhythmias are commonly noted. Nerve damage, which may be delayed, may include swelling of legs and feet, muscle twitch and stupor. Severe exposure can cause death from heart failure. Dust or liquid left in contact with the skin for several hours may be absorbed. This may result in severe delayed symptoms as listed above. These symptoms may last for months or years.

*Long Term Exposure:* Workers exposed to chlorophenoxy compounds such as 2,4-D (in the manufacturing process) over a five to ten year period at levels above 10 mg/m$^3$ complained of weakness, rapid fatigue, headache and vertigo. Liver damage, low blood pressure and slowed heartbeat were also found. Based on animal tests, may affect human reproduction.

*Points of Attack:* Eyes, skin, respiratory system, central nervous system, cardiovascular system, liver, kidney.

*Medical Surveillance:* If symptoms develop or overexposure is suspected, the following may be useful: Liver and kidney function tests. Exam of the nervous system.

*First Aid:* If this chemical gets into the eyes, remove any contact lenses at once and irrigate immediately for at least 15 minutes, occasionally lifting upper and lower lids. Seek medical attention immediately. If this chemical contacts the skin, remove contaminated clothing and wash immediately with soap and water. Seek medical attention immediately. If this chemical has been inhaled, remove from exposure, begin rescue breathing (using universal precautions) if breathing has stopped, and CPR if heart action has stopped. Transfer promptly to a medical facility. When this chemical has been swallowed, get medical attention. *Do not induce vomiting when formulations containing petroleum solvents are ingested.* Otherwise, give large quantities of water and induce vomiting. Do not make an unconscious person vomit.

*References:*
- EXTOXNET, Extension Toxicology Network, "Pesticide Information Profile, Triadimefon," Oregon State University, Corvallis, OR (June 1996). http://extoxnet.orst.edu/pips/triadime.htm
- U.S. Environmental Protection Agency, Office of Pesticide Programs, Pesticide ResidueLimits, "Triadimefon," 40 CFR 180.410, www.epa.gov/pesticides/food/viewtols.htm
- California Environmental Protection Agency "Chemical List of Lists," Sacramento CA (February 1997)

# Triadimenol

*Use Type:* Fungicide; breakdown product
*CAS Number:* 55219-65-3
*Formula:* $C_{14}H_{18}ClN_3O_2$
*Synonyms:* β-(4-Chlorophenoxy)-α-(1,1-dimethylethyl)-1*H*-1,2,4-triazole-1-ethanol; (1RS,2RS,1RS,2SR)-1-(4-Chlorophenoxy)-3,3-dimethyl-1-(1*H*-1,2,4-triazol-1-yl)butan-2-ol; 2-(4-Chlorophenoxy)-1-tert-butyl-2-(1*H*-1,2,4-triazole-1-yl)ethanol; Ethanol, 2-(4-Chlorophenoxy)-1-*tert*-butyl-2-(1*H*-1,2,4-triazole-1-yl)-; 1*H*-1,2,4-Triazole-1-ethanol, β-(4-chlorophenoxy)-α-(1,1-dimethylethyl)-

*Trade Names:* BAYFIDAN®, Bayer CropScience (Germany); BAYFRDAN EW®, Bayer CropScience (Germany); BAY KWG 0519®, Bayer CropScience (Germany); BAYTAN® SEED TREATMENT, Bayer CropScience (Germany); BAYTAN 30® FUNGICIDE, Gustafson (USA); PROTEGE ALLEGIANCE BAYTAN®, Gustafson (USA); SPINNAKER®; SUMMIT®; TRIADIMENOL®; TRIAFOL®; TRIAPHOL®

*Producers:* Agsin (Singapore); Bayer CropScience (Germany); Gustafson (USA)

*Chemical Class:* Chlorophenoxy; azole
*EPA/OPP PC Code:* 127201
*California DPR Chemical Code:* 2307
*EINECS Number:* 259-537-6

*Uses:* Triadimenol is used to control seed- and soil-borne diseases and to provide early season control of foliar diseases. It is applied to seeds of barley, corn, oats, rye, sorghum and wheat and also to fruits, vegetables and ornamentals.

*Human toxicity (long-term)*[77]*:* Intermediate–26.600 ppb, Health Advisory

*Fish toxicity (threshold)*[77]*:* Very low–2294.778860 ppb, MATC (Maximum Acceptable Toxicant Concentration)

*U.S. Maximum Allowable Residue Levels for Triadimenol (40 CFR 180.450):*

| CROP | ppm |
|---|---|
| Banana | 0.2 |
| Barley, grain | 0.05 |
| Barley, straw | 0.2 |
| Cattle, fat | 0.1 |
| Cattle, meat | 0.1 |
| Cattle, mbyp | 0.1 |
| Corn, forage | 0.05 |
| Corn, grain | 0.05 |
| Corn, stover | 0.05 |
| Corn, sweet, kernel plus cob with husks removed | 0.05 |
| Cotton, forage | 0.02 |
| Cotton, undelinted seed | 0.02 |
| Egg | 0.01 |
| Goat, fat | 0.1 |
| Goat, meat | 0.1 |
| Goat, mbyp | 0.1 |
| Hog, fat | 0.1 |
| Hog, meat | 0.1 |
| Hog, mbyp | 0.1 |
| Horse, fat | 0.1 |
| Horse, meat | 0.1 |
| Horse, mbyp | 0.1 |
| Milk | 0.01 |
| Oat, forage and hay | 2.5 |
| Oat, grain | 0.05 |
| Oat, straw | 0.2 |
| Poultry, fat | 0.01 |
| Poultry, meat | 0.01 |

| | |
|---|---|
| Poultry, mbyp | 0.01 |
| Rye, forage | 2.5 |
| Rye, grain | 0.05 |
| Rye, straw | 0.1 |
| Sheep, fat | 0.1 |
| Sheep, meat | 0.1 |
| Sheep, mbyp | 0.1 |
| Sorghum, forage | 0.05 |
| Sorghum, grain | 0.01 |
| Sorghum, grain, stover | 0.01 |
| Wheat, forage | 2.5 |
| Wheat, grain | 0.05 |
| Wheat, straw | 0.2 |

*Carcinogen/Hazard Classifications*
**U.S. EPA Carcinogens:** Group C, possible carcinogen
**Label Signal Word:** CAUTION or WARNING
**WHO Acute Hazard:** Class III, slightly hazardous
**Endocrine Disruptor:** Suspected endocrine disruptor
*Regulatory Authority:*
- AB 2588-Air Toxics "Hot Spots" Chemicals (CAL) as chlorophenoxy pesticides
- The "Director's List" (CAL/OSHA) as chlorophenoxy pesticides
- Actively registered pesticide in California.

*Description:* Colorless solid and white to tan powder. Melting point = 82°C. Molecular weight = 293.75. Vapor pressure = $3.12 \times 10^{-10}$ mmHg @ 20°C.
*Permissible Concentration in Water:* No criteria set. Runoff from spills or fire control may cause water pollution.
*Routes of Entry:* Inhalation, ingestion, absorbed through the skin.
*Harmful Effects and Symptoms*
*Short Term Exposure:* Poisonous; may be fatal if inhaled, swallowed, or absorbed through skin. Severely irritates eyes, skin and respiratory tract, with burning sensation, pain, redness and swelling. Metabolic stimulant. If inhaled, causes coughing, dilated pupils, headache, profuse persperation, intense thirst, extreme fatigue, rapid pulse, high fever, clammy, flushed skin, rapid breathing, nausea, vomiting, cyanosis (bluish tint to skin and lips), anxiety and confusion, convulsions, risk of lung edema. If swallowed, face and lips turn bluish. Liver injury with associated jaundice, kidney failure, and cardiac arrhythmias are commonly noted. Nerve damage, which may be delayed, may include swelling of legs and feet, muscle twitch and stupor. Severe exposure can cause death from heart failure. Dust or liquid left in contact with the skin for several hours may be absorbed. This may result in severe delayed symptoms as listed above. These symptoms may last for months or years.
*Long Term Exposure:* Workers exposed to chlorophenoxy compounds such as 2,4-D (in the manufacturing process) over a five to ten year period at levels above 10 mg/m³ complained of weakness, rapid fatigue, headache and vertigo. Liver damage, low blood pressure and slowed heartbeat were also found. Based on animal tests, may affects human reproduction.
*Points of Attack:* Eyes, skin, respiratory system, central nervous system, cardiovascular system, liver, kidney.
*Medical Surveillance:* If symptoms develop or overexposure is suspected, liver or kidney function tests may be useful. Liver function tests.
*First Aid:* If this chemical gets into the eyes, remove any contact lenses at once and irrigate immediately for at least 15 minutes, occasionally lifting upper and lower lids. Seek medical attention immediately. If this chemical contacts the skin, remove contaminated clothing and wash immediately with soap and water. Seek medical attention immediately. If this chemical has been inhaled, remove from exposure, begin rescue breathing (using universal precautions) if breathing has stopped, and CPR if heart action has stopped. Transfer promptly to a medical facility. When this chemical has been swallowed, get medical attention. Give large quantities of water and induce vomiting. Do not make an unconscious person drink or vomit. *Note to Physician:* If ingested, remove by lavage or vomiting. Use general supportive measures for CNS depression. Use quinidine for myotonia.
*References:*
- Pesticide Management Education Program, "Triadimenol (Baytan) Chemical Fact sheet 7/89," Cornell University, Ithaca, NY (July 1989). http://pmep.cce.cornell.edu/profiles/fung-nemat/tcmtb-ziram/triadimenol/fung-prof-triadimenol.html
- U.S. Environmental Protection Agency, Office of Pesticide Programs, Pesticide Residue Limits, "Triadimenol,", 40 CFR 180.450, www.epa.gov/pesticides/food/viewtols.htm
- California Environmental Protection Agency "Chemical List of Lists," Sacramento CA (February 1997)

# Triallate

*Use Type:* Herbicide
*CAS Number:* 2303-17-5
*Formula:* $C_{10}H_{16}Cl_3NOS$
*Synonyms:* Carbamic acid, diisopropylthio-, S-(2,3,3-trichloroallyl) ester; Bis(1-methylethyl)carbamothioic acid S-(2,3,3-trichloro-2-propenyl) ester; Carbamothioic acid, bis(1-methylethyl)-, S-(2,3,3-trichloro-2-propenyl) ester; Caswell No. 870A; N-Diisopropylthiocarbamic acid S-2,3,3-trichloro-2-propenyl ester; Diisopropyltrichloroallyl thiocarbamate; NSC 379698; 2-Propene-1-thiol, 2,3,3-trichloro-, diisopropylcarbamate; thiocarbamic acid, N-diisopropyl-, S-2,3,3-Trichloroallyl ester; Tri-allate; 2,3,3-Trichloroallyl N,N-diisopropylthiocarbamate; S-2,3,3-Trichloroallyl N,N-diisopropylthiocarbamate; S-(2,3,3-Trichloroallyl) diisopropylthiocarbamate; 2,3,3-

Trichloroallyl diisopropylthiocarbamate; S-(2,3,3-Trichloro-2-propenyl)bis(1-methylethyl)carbamothioate

***Trade Names:*** AVADEX BW®, Monsanto (USA); BUCKLE®, (triallate + trifluralin), Monsanto (USA); CP-23426®; DIPTHAL®; FAR-GO®, Monsanto (USA), canceled; FAR-GO®; FORTRESS®, Monsanto (USA); OVADEX BW®

***Producers:*** Monsanto (USA); Gowan Company (USA)

***Chemical Class:*** Thiocarbamate

***EPA/OPP PC Code:*** 078802

***California DPR Chemical Code:*** 49

***ICSC Number:*** 0201

***RTECS Number:*** EZ8575000

***RCRA Number:*** U389

***EEC Number:*** 006-039-00-X

***Uses:*** Triallate is a pre-emergent or post-emergent herbicide used to control a variety of annual grasses and wild oats on several grains, oilseed and vegetable crops. Its use has been restricted to use in CO, ID, KS, MN, MT, NE, NV, ND, OR, SD, UT, WA, and WY. Its use on canary grass has been revoked. Monsanto announced in March, 2004, that its brands of triallate products have been sold to Gowan Company.

***Human toxicity (long-term)[77]:*** High–9.10 ppb, Health Advisory

***Fish toxicity (threshold)[77]:*** Intermediate–54.442630 ppb, MATC (Maximum Acceptable Toxicant Concentration)

***U.S. Maximum Allowable Residue Levels: for Triallate (40 CFR 180.314):***

| CROP | ppm |
|---|---|
| Barley, grain | 0.05 |
| Barley, straw | 0.05 |
| Beet, sugar, dried pulp | 0.2 |
| Beet, sugar, roots | 0.1 |
| Beet, sugar, tops | 0.5 |
| Lentil | 0.05 |
| Lentil, hay | 0.05 |
| Pea | 0.05 |
| Pea, field, hay | 0.05 |
| Pea, field, vines | 0.05 |
| Wheat, grain | 0.05 |
| Wheat, straw | 0.05 |

***Carcinogen/Hazard Classifications***

***U.S. EPA Carcinogens:*** Group C, possible carcinogen

***Label Signal Word:*** CAUTION

***WHO Acute Hazard:*** Class III, slightly hazardous

***Regulatory Authority:***
- EPA 40 CFR 372.65, Specific Toxic Chemical Listings
- RCRA Section 261 Hazardous Constituents
- RCRA Universal Treatment Standards: Wastewater (mg/L), 0.003; Nonwastewater (mg/kg), 1.4
- EPCRA Section 313 Form R *de minimis* concentration reporting level: 1.0%

***Description:*** Oily, amber liquid (technical) or colorless, crystalline solid. May be dissolved in a flammable solvent. Practically insoluble in water; solubility = 4 ppm @ 25°C. Molecular weight = 304.66. Density = 1.273 @ 25°C; also reported 1.04 @ 25°C. Boiling point = 117°C at 0.4 mmHg. Melting/Freezing point = 29°C. Vapor pressure = $1.6 \times 10^{-4}$ mbar @ 25°C. Log $K_{ow}$ = 3.96. Values above 3.0 are likely to bioaccumulate in marine organisms.

***Determination in Air:*** Filter; none; Gravimetric; NIOSH IV[18] [Particulates NOR; #0500 (total), #0600 (respirable)]

***Permissible Concentration in Water:*** No criteria set. Runoff from spills or fire control may cause water pollution.

***Harmful Effects and Symptoms***

***Short Term Exposure:*** Low levels of toxicity. Concentrated solutions are slightly corrosive to eyes and mucous membranes. Dust inhalation can cause irritation of the respiratory system with sneezing. Eye contact can cause irritation, watering, pain, and inflammation of the eyelids. Skin contact can cause irritation and minor ulceration. Ingestion can cause nausea, vomiting, fever, muscle twitching, seizure, rapid respiration, slow heart beat. Severe exposure may result in death.

***Points of Attack:*** Respiratory system, central nervous system, cardiovascular system, skin, eyes.

***Medical Surveillance:*** Medical observation is recommended for 24 to 48 hours after breathing overexposure, as pulmonary edema may be delayed.

***First Aid:*** If this chemical gets into the eyes, remove any contact lenses at once and irrigate immediately for at least 15 minutes, occasionally lifting upper and lower lids. Seek medical attention immediately. If this chemical contacts the skin, remove contaminated clothing and wash immediately with soap and water. Seek medical attention immediately. If this chemical has been inhaled, remove from exposure, begin rescue breathing (using universal precautions) if breathing has stopped, and CPR if heart action has stopped. Transfer promptly to a medical facility. When this chemical has been swallowed, get medical attention. If victim is conscious and able to swallow, have victim drink 4 to 8 ounces of water. Do not induce vomiting. *Note to physician or authorized medical personnel*: Medical observation is recommended for 24 to 48 hours after breathing overexposure, as pulmonary edema may be delayed. As first aid for pulmonary edema, consider administering a corticosteroid spray. Cigarette smoking may exacerbate pulmonary injury and should be discouraged for at least 72 hours following exposure.

***References:***
- U.S. Environmental Protection Agency, "Reregistration Eligibility Decision (RED) Facts, Triallate," Office of Prevention, Pesticides and Toxic Substances, Washington, DC (March 2001). http://www.epa.gov/REDs/factsheets/triallatefact.pdf
- U.S. Environmental Protection Agency, Office of Pesticide Programs, Pesticide Residue Limits, "Triallate," 40 CFR 180.314, www.epa.gov/pesticides/food/viewtols.htm

# Triasulfuron

*Use Type:* Herbicide
*CAS Number:* 82097-50-5
*Formula:* $C_{14}H_{16}ClN_5O_5S$
*Synonyms:* Benzenesulfonamide, 2-(2-chloroethoxy)-*N*-[((4-methoxy-6-methyl-1,3,5-triazin-2-yl)amino)carbonyl]-; 2-(2-Chloroethoxy)-*N*-[((4-methoxy-6-methyl-1,3,5-triazin-2-yl)amino)carbonyl]benzenesulfonamide; 1-[2-(2-Chloroethoxy)phenylsulfonyl]-3-(4-methoxy-6-methyl-1,3,5-triazin-2-yl)urea; 3-(6-Methoxy-4-methyl-1,3,5-triazin-2-yl)-1-[2-(2-chloroethoxy)phenylsulfonyl]urea; Urea, *N*-[2-(2-chloroethoxy)phenylsulfonyl]-*N'*-(6-methoxy-4-methyl-1,3,5-triazinyl-2-yl)-
*Trade Names:* AMBER®, Syngenta (Switzerland); CGA 131036®, Syngenta (Switzerland); LOGRAN®; RAVE®, Syngenta (Switzerland)
*Producers:* Syngenta (Switzerland)
*Chemical Class:* Sulfonyl urea
*EPA/OPP PC Code:* 128969 (128985 old EPA code number)
*California DPR Chemical Code:* 5100
*Uses:* Used for the control of annual ryegrass, paradoxa grass and a wide range of broadleaf weeds in wheat and the post-emergence control of wild radishes in wheat, oats and barley.
*Human toxicity (long-term)[77]:* Low–70.00 ppb, Health Advisory
*Fish toxicity (threshold)[77]:* Very low–68600.000680 ppb, MATC (Maximum Acceptable Toxicant Concentration)
*U.S. Maximum Allowable Residue Levels for Triasulfuron (40 CFR 180.459):*

| CROP | ppm |
| --- | --- |
| Barley, grain | 0.02 |
| Barley, straw | 2.0 |
| Cattle, fat | 0.1 |
| Cattle, kidney | 0.5 |
| Cattle, meat | 0.1 |
| Cattle, mbyp, except kidney | 0.1 |
| Goat, fat | 0.1 |
| Goat, kidney | 0.5 |
| Goat, meat | 0.1 |
| Goat, mbyp, except kidney | 0.1 |
| Grass, forage | 7.0 |
| Grass, hay | 2.0 |
| Hog, fat | 0.1 |
| Hog, kidney | 0.5 |
| Hog, meat | 0.1 |
| Hog, mbyp, except kidney | 0.1 |
| Horse, fat | 0.1 |
| Horse, kidney | 0.5 |
| Horse, meat | 0.1 |
| Horse, mbyp, except kidney | 0.1 |
| Milk | 0.02 |
| Sheep, fat | 0.1 |
| Sheep, kidney | 0.5 |
| Sheep, meat | 0.1 |
| Sheep, mbyp, except kidney | 0.1 |
| Wheat, forage | 5.0 |
| Wheat, grain | 0.02 |
| Wheat, straw | 2.0 |

*Carcinogen/Hazard Classifications*
**U.S. EPA Carcinogens:** Group E, Unlikely a carcinogen
**Label Signal Word:** CAUTION
**WHO Acute Hazard:** Class U, unlikely to be hazardous
*Description:* White crystalline solid. Soluble in water. Molecular weight = 401.85. Melting point = 178.1°C.
*Incompatibilities:* Slowly hydrolyzes in water, releasing ammonia and forming acetate salts.
*Permissible Concentration in Water:* No criteria set. Runoff from spills or fire control may cause water pollution.
*Routes of Entry:* Inhalation, passing through the skin, ingestion.
*Harmful Effects and Symptoms*
*Short Term Exposure:* Contact with eyes or skin may cause irritation or burns. Inhalation should be avoided; use NIOSH-approved air purifying respirators for pesticides. May be harmful if swallowed. Skin contact may cause allergic reaction.
*Long Term Exposure:* Repeated or prolonged contact may cause skin and lung sensitization, resulting in allergies.
*Points of Attack:* Skin.
*Medical Surveillance:* Evaluation by a qualified allergist, including careful exposure history and special testing, may help diagnose skin allergy.
*First Aid:* If this chemical gets into the eyes, remove any contact lenses at once and irrigate immediately for at least 15 minutes, occasionally lifting upper and lower lids. Seek medical attention immediately. If this chemical contacts the skin, remove contaminated clothing and wash immediately with soap and water. Seek medical attention immediately. If this chemical has been inhaled, remove from exposure, begin rescue breathing (using universal precautions) if breathing has stopped, and CPR if heart action has stopped. Transfer promptly to a medical facility. When this chemical has been swallowed, get medical attention. Give large quantities of water and induce vomiting. Do not make an unconscious person vomit.
*References:*
- U.S. Environmental Protection Agency, Office of Pesticide Programs, Pesticide Residue Limits, "Triasulfuron," 40 CFR 180.459, www.epa.gov/pesticides/food/viewtols.htm

## Triazamate

*Use Type:* Insecticide
*CAS Number:* 112143-82-5
*Formula:* $C_{13}H_{22}N_4O_3S$
*Synonyms:* Acetic acid, [(1-((dimethylamino)carbonyl)-3-(1,1-dimethylethyl)-1*H*-1,2,4-triazol-5-yl)thio]-, ethyl ester (9CI); Ethyl(3-*tert*-butyl-1-dimethylcarbamoyl-1*H*-1,2,4-triazol-5-ylthio)acetate
*Trade Names:* AZTEC®, BASF Agricultural Products (Germany); RH-7988®; RH-7988-25W® experimental aphicide in water-soluble pouches
*Producers:* BASF Agricultural Products (Germany); Omya Agro (Switzerland)
*Chemical Class:* Unclassified
*EPA/OPP PC Code:* 128100
*California DPR Chemical Code:* 5517
*Uses:* Not registered in the U.S. Used to control root aphids on fir Christmas trees and apple trees.
*Carcinogen/Hazard Classifications*
U.S. EPA Carcinogens: Not likely a carcinogen
WHO Acute Hazard: Class II, moderately hazardous
*Incompatibilities:* May not be compatible with nitrates. Moisture may cause hydrolysis or other forms of decomposition.
*Determination in Air:* Ionization. NIOSH IV[(18)] Method#1450, Esters
*Permissible Concentration in Water:* No criteria set. Runoff from spills or fire control may cause water pollution.
*Routes of Entry:* Inhalation, passing through the skin, ingestion.
*Harmful Effects and Symptoms*
*Short Term Exposure:* May cause skin and severe eye irritation. Moderately poisonous if ingested or inhaled. Exposure to a triazine (simazine) has caused acute and subacute dermatitis in the former USSR, characterized by erythema, slight edema, moderate pruritus, and burning lasting 4 to 5 days.
*Long Term Exposure:* May cause lung irritation and damage. May cause skin allergy. Contact with some triazine compounds (such as atrazine) may increase risks for tumors known to be associated with hormonal factors. These have been observed in both animals and human beings, and are consistent with the known effects on the hypothalamic pituitary gonadal axis. Repeated exposure may cause weight loss and reduced red blood cell count. May be mutagenic.
*Points of Attack:* Liver, lungs, skin.
*Medical Surveillance:* Before beginning employment and at regular times after that, for those with frequent or potentially high exposures, the following is recommended: Lung function tests. Consider chest x-ray following acute overexposure. Evaluation by a qualified allergist. Examination of the nervous system.
*First Aid:* If this chemical gets into the eyes, remove any contact lenses at once and irrigate immediately for at least 15 minutes, occasionally lifting upper and lower lids. Seek medical attention immediately. If this chemical contacts the skin, remove contaminated clothing and wash immediately with soap and water. Seek medical attention immediately. If this chemical has been inhaled, remove from exposure, begin rescue breathing (using universal precautions) if breathing has stopped, and CPR if heart action has stopped. Transfer promptly to a medical facility. When this chemical has been swallowed, get medical attention. Give large quantities of water or milk and induce vomiting. Do not make an unconscious person vomit.
*References:*
- California Environmental Protection Agency "Chemical List of Lists," Sacramento CA (February 1997)

## Triazophos

*Use Type:* Insecticide
*CAS Number:* 24017-47-8
*Formula:* $C_{12}H_{16}N_3O_3PS$
*Synonyms:* *O,O*-Diethyl *O*-(1-phenyl-1*H*-1,2,4-triazol-3-yl) phosphorothioate; 1-Phenyl-3-(*O,O*-diethylthionophophoryl)-1,2,4-triazole; 1-Phenyl-1,2,4-triazolyl-3-(*O,O*-diethylthionophosphate); Phosphorothioic acid, *O,O*-diethyl *O*-(1-phenyl-1*H*-1,2,4-triazol-3-yl) ester
*Trade Names:* HILAZPOPHOS®, Hindustan Insecticides (India); HOE 2960 OJ®; HOE 002960®; HOSTATHION®, AgrEvo (Germany), Bayer CropScience (Germany); SUTATHION®, Sudarshan Chemical Industries (India)
*Producers:* Agsin (Singapore); AgrEvo (Germany), Bayer CropScience (Germany); Biesterfeld Siemsgluess International. GmbH (Germany); Gharda Chemicals (India); Hindustan Insecticides (India); Hunan Tianyu Pesticide Chemical Group (China); Sanonda Ltd. (Australia); Sanonda Zhengzhou Pesticide Co. Ltd. (China); Shenzhen Guomeng Industry Co., Ltd. (China); Sudarshan Chemical Industries (India)
*Chemical Class:* Organophosphate
*EPA/OPP PC Code:* 344600
*California DPR Chemical Code:* 3543
*RTECS Number:* TF5635000
*Uses:* Not registered in the U.S. Triazophos is a broad spectrum insecticide and acaricide used to control sucking and chewing pests on a variety of crops, including cotton, rice, corn, beets and fruit trees.
*Carcinogen/Hazard Classifications*
WHO Acute Hazard: Class 1b, highly hazardous
*Regulatory Authority:*
- EPCRA Section 302 Extremely Hazardous Substances: TPQ = 500 lb (227 kg)

- EPCRA Section 304 RQ: EHS, 1 lb (0.454 kg)
- DOT Inhalation Hazard Chemicals
- U.S. DOT Regulated Marine Pollutant (49CFR172.101, Appendix B)

*Description:* Tan, oily liquid. Slightly soluble in water; solubility = 40 mg/L @ 25°C. Molecular weight = 313.30.

*Determination in Air:* OSHA versatile sampler-2; Toluene/Acetone; Gas chromatography/Flame ionization detection; NIOSH IV[18], Method #5600, Organophosphorus Pesticides.

*Permissible Concentration in Water:* No criteria set. Runoff from spills or fire control may cause water pollution.

*Harmful Effects and Symptoms*

*Short Term Exposure:* Eye pupils are small; blurred vision; eye watering; runny nose; cough; shortness of breath; salivation; dizziness; nausea, stomach cramps, diarrhea, and vomiting; increased blood pressure; profuse sweating; hypermotility, hallucinations; irritability; tingling of the skin; drowsiness; slow heartbeat; convulsions; fluid in lungs; loss of consciousness; incontinence; breathing stops; death. Organophosphates inhibit the action of acetylcholinesterase enzymes, and alter the way in which nervous impulses are transmitted. The effects can last for hours, days, or much longer. The action of the enzymes is reestablished after new enzymes are formed.

*Long Term Exposure:* Cholinesterase inhibitor; cumulative effect is possible. Organophosphates may damage the nervous system with repeated exposure, resulting in convulsions, respiratory failure. May cause liver damage.

*Points of Attack:* Respiratory system, central nervous system, cardiovascular system, blood cholinesterase.

*Medical Surveillance:* Medical observation is recommended for 24 to 48 hours after breathing overexposure, as pulmonary edema may be delayed. As first aid for pulmonary edema, consider administering a corticosteroid spray. Cigarette smoking may exacerbate pulmonary injury and should be discouraged for at least 72 hours following exposure.

*First Aid: Eyes:* Speed in removing material from skin is of extreme importance. Eye contact can cause dangerous amounts of these chemicals to be quickly absorbed through the mucous membrane into the bloodstream. Immediately and gently flush eyes with plenty of warm or cold water (NO hot water) for at least 15 minutes, occasionally lifting the upper and lower eyelids. Get medical aid immediately. *Skin:* Get medical aid. Skin and/or eye contact can cause dangerous amounts of these chemicals to be absorbed into the bloodstream. Wearing the appropriate PPE equipment and respirator for organophosphate pesticides, immediately flush skin with plenty of soap and water for at least 15 minutes while removing contaminated clothing and shoes. Shampoo hair promptly if contaminated. *Ingestion:* Call poison control. Loosen all clothing. Never give anything by mouth to an unconscious person. Get medical aid. Do NOT induce vomiting.* If conscious, alert, and able to swallow, rinse mouth and have victim drink 4 to 8 ounces of water do NOT induce vomiting but immediately administer slurry of activated charcoal (2 oz in 8 oz of water). If victim is *unconscious or having convulsions,* do nothing except keep victim warm. *\*In some cases you may be specifically instructed by poison control to induce vomiting by way of 2 tablespoons of syrup of ipecac (adult) washed down with a cup of water.* Do NOT give activated charcoal before or with ipecac syrup. *Inhalation:* Get medical aid. Do not contaminate yourself. Wearing the appropriate PPE equipment and respirator for organophosphate pesticides, immediately remove the victim from the contaminated area to fresh air. If the victim is not breathing, administer artificial respiration. Do NOT use mouth-to-mouth resuscitation; use bag/mask apparatus. If breathing is difficult, administer oxygen through bag/mask apparatus until medical help arrives. Do not leave victim unattended. *Note to physician or authorized medical personnel.* Administer atropine, 2 mg (1/30 gr) intramuscularly or intravenously as soon as any local or systemic signs or symptoms of an intoxication are noted; repeat the administration of atropine every 3 to 8 minutes until signs of atropinization (mydriasis, dry mouth, rapid pulse, hot and dry skin) occur; initiate treatment in children with 0.05 mg mg/kg of atropine; repeat at 5 to 10 minute intervals. Watch respiration, and remove bronchial secretions if they appear to be obstructing the airway; intubate if necessary. *Notes to physician or authorized medical personnel:* N-methylpyridinium-2-aldoxime (2-PAMCl) when used in conjunction with atropine reacts with the phosphorylated cholinesterase, thereby restoring normal activity to by removing the phosphorylating group. The combination of these two chemicals is synergistic and must be administered within minutes to a few hours following exposure (depending on the specific agent) to be effective. Give 2-PAMCI (Pralidoxime; Protopam), 2.5 gm in 100 ml of sterile water or in 5% dextrose and water, intravenously, slowly, in 15-30 minutes; if sufficient fluid is not available, give 1 gm of 2-PAMCI in 3 ml of distilled water by deep intramuscular injection; repeat this every half hour if respiration weakens or if muscle fasciculation or convulsions recur. Also Diazepam, an anticonvulsant, might be considered.

*References:*
- California Environmental Protection Agency "Chemical List of Lists," Sacramento CA (February 1997)

# Tribenuron-methyl

*Use Type:* Herbicide
*CAS Number:* 101200-48-0
*Formula:* $C_{15}H_{17}N_5O_6S$
*Synonyms:* Benzoic acid, 2-[(((((4-methoxy-6-methyl-1,3,5-triazin-2-yl)methylamino)carbonyl)amino)sulfonyl]-, methyl

ester; Benzoic acid, 2-[(((((4-methoxy-6-methyl-1,3,5-triazin-2-yl)-*N*-methylamino)carbonyl)amino)sulfonyl]-, methyl ester; 2-[(((((4-Methoxy-6-methyl-1,3,5-triazin-2-yl)methylamino)carbonyl)amino)sulfonyl)-, methyl ester; Methyl 2-[(((((4-methoxy-6-methyl-1,3,5-triazin-2-yl)methylamino)carbonyl)amino)sulfonyl]benzoate; Sulfmethmeton-methyl

*Trade Names:* ALLY®, DuPont Crop Protection (USA); CANVAS®, DuPont Crop Protection (USA); DPX-L-5300®, DuPont Crop Protection (USA); EXPRESS®, DuPont Crop Protection (USA); EXPRESS®-75 DF, DuPont Crop Protection (USA); HARMONY EXTRA®, DuPont Crop Protection (USA); INL-5300®, DuPont Crop Protection (USA); L 5300®, DuPont Crop Protection (USA); MATRIX®

*Producers:* DuPont Crop Protection (USA); Epochem Co., (China); Ki-Hara Chemicals Ltd. (UK)

*Chemical Class:* Triazinyl sulfonylurea

*EPA/OPP PC Code:* 128887

*California DPR Chemical Code:* 2338

*ICSC Number:* 1359

*RTECS Number:* DH3565000

*EEC Number:* 607-177-00-9

*Uses:* Used to control annual grasses and broadleaf weeds on barley, blueberries, oats, wheat, flax and rape seed (canola).

*Human toxicity (long-term)*[77]*:* High–5.60 ppb, Health Advisory

*Fish toxicity (threshold)*[77]*:* Very low–251188.64315 ppb, MATC (Maximum Acceptable Toxicant Concentration)

*Carcinogen/Hazard Classifications*

**U.S. EPA Carcinogens:** Group C, possible carcinogen

**Label Signal Word:** CAUTION or WARNING

*Regulatory Authority:*
• EPCRA Section 313 Form R *de minimis* concentration reporting level: 1.0%

*Description:* Off-white to brown crystalline solid. Slight pungent odor. Soluble in water. Melting point = 141°C. Molecular weight = 395.40. Log $K_{ow}$ = –0.439. Unlikely to bioaccumulate in marine organisms.

*Permissible Concentration in Water:* No criteria set. Runoff from spills or fire control may cause water pollution.

*Routes of Entry:* Inhalation, passing through the skin and ingestion.

*Harmful Effects and Symptoms*

*Short Term Exposure:* May cause skin and severe eye irritation. Moderately poisonous if ingested or inhaled. Exposure to a triazine (simazine) has caused acute and subacute dermatitis in the former USSR, characterized by erythema, slight edema, moderate pruritus, and burning lasting 4 to 5 days.

*Long Term Exposure:* May cause lung irritation and damage. May cause skin allergy. Contact with some triazine compounds (such as atrazine) may increase risks for tumors known to be associated with hormonal factors. These have been observed in both animals and human beings, and are consistent with the known effects on the hypothalamic pituitary gonadal axis. Repeated exposure may cause weight loss and reduced red blood cell count. May be mutagenic.

*Points of Attack:* Liver, lungs, skin.

*Medical Surveillance:* Before beginning employment and at regular times after that, for those with frequent or potentially high exposures, the following is recommended: Lung function tests. Consider chest x-ray following acute overexposure. Evaluation by a qualified allergist. Examination of the nervous system.

*First Aid:* If this chemical gets into the eyes, remove any contact lenses at once and irrigate immediately for at least 15 minutes, occasionally lifting upper and lower lids. Seek medical attention immediately. If this chemical contacts the skin, remove contaminated clothing and wash immediately with soap and water. Seek medical attention immediately. If this chemical has been inhaled, remove from exposure, begin rescue breathing (using universal precautions) if breathing has stopped, and CPR if heart action has stopped. Transfer promptly to a medical facility. When this chemical has been swallowed, get medical attention. Give large quantities of water or milk and induce vomiting. Do not make an unconscious person vomit.

*References:*
• Pesticide Management Education Program, "Tribenuron methyl (Express) Herbicide Profile 6/89," Cornell University, Ithaca, NY (June 1989). http://pmep.cce.cornell.edu/profiles/herb-growthreg/sethoxydim-vernolate/tribenuron-methyl/herb-prof-tribenuron-meth.html
• California Environmental Protection Agency "Chemical List of Lists," Sacramento CA (February 1997)

# Tribufos

*Use Type:* Herbicide, defoliant, plant growth regulator

*CAS Number:* 78-48-8

*Formula:* $C_{12}H_{27}OPS_3$

*Synonyms:* A13-25812; Butyl phosphorotrithioate; Caswell No. 864; Fosforotritioato de *S,S,S*-tributilo (Spanish); Merphos-oxide; Phosphorotrithioic acid, *S,S,S*-tributyl ester; TBTP; Tribufos; *S,S,S*-Tributyl phosphorotrithioate; *S,S,S*-Tributyl trithiophosphate

*Trade Names:* B-1776®; BUTIFOS®; BUTIPHOS®, canceled; CHEMAGRO® 1776; CHEMAGRO® B-1776; DE-GREEN®; DEF®, Bayer CropScience (Germany); DEF DEFOLIANT®, Bayer CropScience (Germany); DELEAF DEFOLIANT®; EASY OFF®-D; E-Z-OFF® D; FOLEX® 6EC, AMVAC Chemical (USA); FOS-FALL® A; ORTHO® phosphate defoliant

*Producers:* AMVAC Chemical (USA); Bayer CropScience (Germany); Cerexagri (France); Micro-Flo (USA)

*Chemical Class:* Organophosphate

# Tribufos

*EPA/OPP PC Code:* 074801
*California DPR Chemical Code:* 190
*RTECS Number:* TG5425000

**Uses:** Tribufos is an organophosphate defoliant used for cotton crops. It is specifically used to defoliate cotton in preparation for machine harvesting. It was first registered in the United States in 1961. Tolerances for tribufos were revoked in July, 2002, because the specific tolerances are either no longer needed or are associated with food uses that are no longer registered in the United States.

***Human toxicity (long-term)***[77]***:*** High–1.46444 ppb, CHCL (Chronic Human Carcinogen Level)

***Fish toxicity (threshold)***[77]***:*** Intermediate–13.345590 ppb, MATC (Maximum Acceptable Toxicant Concentration)

***Carcinogen/Hazard Classifications***

**U.S. EPA Carcinogens:** Likely carcinogen in high doses; not likely carcinogen in low doses.

**Label Signal Word:** CAUTION, WARNING or DANGER

***Regulatory Authority:***
- Actively registered pesticide in California.
- FIFRA, 180.3(5); class of cholinesterase-inhibiting pesticide
- DOT Inhalation Hazard Chemicals
- EPCRA Section 313 Form R *de minimis* concentration reporting level: 1.0%

***Description:*** Clear, colorless to light yellow liquid. Mercaptan-like odor. Insoluble in water. Molecular weight 314.54. Density = 1.06 @ 20°C. Boiling point = 150°C @ 0.3 @ mmHg. Melting/Freezing point = < –25°C. Vapor pressure = $1.6 \times 10^{-6}$ mmHg @ 20°C. Log $K_{ow}$ = 5.68. Values above 3.0 are likely to bioaccumulate in marine organisms.

***Incompatibilities:*** Alkaline materials.

***Determination in Air:*** OSHA versatile sampler-2; Toluene/Acetone; Gas chromatography/Flame ionization detection; NIOSH IV[18], Method #5600, Organophosphorus Pesticides.

***Permissible Concentration in Water:*** No criteria set. Runoff from spills or fire control may cause water pollution.

***Harmful Effects and Symptoms***

***Short Term Exposure:*** Eye pupils are small; blurred vision; eye watering; runny nose; cough; shortness of breath; salivation; dizziness; nausea, stomach cramps, diarrhea, and vomiting; increased blood pressure; profuse sweating; hypermotility, hallucinations; irritability; tingling of the skin; drowsiness; slow heartbeat; convulsions; fluid in lungs; loss of consciousness; incontinence; breathing stops; death. Organophosphates inhibit the action of acetylcholinesterase enzymes, and alter the way in which nervous impulses are transmitted. The effects can last for hours, days, or much longer. The action of the enzymes is reestablished after new enzymes are formed.

***Long Term Exposure:*** Cholinesterase inhibitor; cumulative effect is possible. Organophosphates may damage the nervous system with repeated exposure, resulting in convulsions, respiratory failure. May cause liver damage.

***Points of Attack:*** Respiratory system, central nervous system, cardiovascular system, blood cholinesterase.

***Medical Surveillance:*** Medical observation is recommended for 24 to 48 hours after breathing overexposure, as pulmonary edema may be delayed. As first aid for pulmonary edema, consider administering a corticosteroid spray. Cigarette smoking may exacerbate pulmonary injury and should be discouraged for at least 72 hours following exposure.

***First Aid:*** Speed in removing material from skin is of extreme importance. *Eye:* Contact can cause dangerous amounts of these chemicals to be quickly absorbed through the mucous membrane into the bloodstream. Immediately and gently flush eyes with plenty of warm or cold water (NO hot water) for at least 15 minutes, occasionally lifting the upper and lower eyelids. Get medical aid immediately. *Skin:* Get medical aid. Skin and/or eye contact can cause dangerous amounts of these chemicals to be absorbed into the bloodstream. Wearing the appropriate PPE equipment and respirator for organophosphate pesticides, immediately flush skin with plenty of soap and water for at least 15 minutes while removing contaminated clothing and shoes. Shampoo hair promptly if contaminated. *Ingestion:* Call poison control. Loosen all clothing. Never give anything by mouth to an unconscious person. Get medical aid. Do NOT induce vomiting.\* If conscious, alert, and able to swallow, rinse mouth and have victim drink 4 to 8 ounces of water do NOT induce vomiting but immediately administer slurry of activated charcoal (2 oz in 8 oz of water). If victim is *unconscious or having convulsions,* do nothing except keep victim warm. \**In some cases you may be specifically instructed by poison control to induce vomiting by way of 2 tablespoons of syrup of ipecac (adult) washed down with a cup of water.* Do NOT give activated charcoal before or with ipecac syrup. *Inhalation:* Get medical aid. Do not contaminate yourself. Wearing the appropriate PPE equipment and respirator for organophosphate pesticides, immediately remove the victim from the contaminated area to fresh air. If the victim is not breathing, administer artificial respiration. Do NOT use mouth-to-mouth resuscitation; use bag/mask apparatus. If breathing is difficult, administer oxygen through bag/mask apparatus until medical help arrives. Do not leave victim unattended. *Note to physician or authorized medical personnel.* Administer atropine, 2 mg (1/30 gr) intramuscularly or intravenously as soon as any local or systemic signs or symptoms of an intoxication are noted; repeat the administration of atropine every 3 to 8 minutes until signs of atropinization (mydriasis, dry mouth, rapid pulse, hot and dry skin) occur; initiate treatment in children with 0.05 mg mg/kg of atropine; repeat at 5 to 10 minute intervals. Watch respiration, and remove bronchial secretions if they appear to be obstructing the airway; intubate if necessary. *Notes to physician or authorized medical personnel*: N-methylpyridinium-2-aldoxime (2-PAMCI) when used in conjunction with atropine reacts with the phosphorylated cholinesterase, thereby restoring normal activity to by removing the phosphorylating group. The combination of these two chemicals is synergistic and must be administered within minutes to a few hours following exposure (depending on the specific agent) to be effective. Give 2-PAMCI (Pralidoxime; Protopam), 2.5 gm in 100 ml of sterile water or in 5% dextrose and water, intravenously, slowly, in 15-30 minutes; if sufficient fluid is not available, give 1 gm of 2-PAMCI in 3 ml of distilled water by deep intramuscular injection; repeat this every half hour if respiration weakens or if muscle fasciculation or convulsions recur. Also Diazepam, an anticonvulsant, might be considered.

*References:*
- U.S. Environmental Protection Agency, "Interim Reregistration Eligibility Decision (IRED), Tribufos," Office of Prevention, Pesticides and Toxic Substances, Washington, DC (September 2000). http://www.epa.gov/REDs/2145ired.pdf
- U.S. Environmental Protection Agency, Office of Pesticide Programs, "Tribufos Facts." (October 2000). http://www.epa.gov/REDs/factsheets/2145iredfact.pdf
- California Environmental Protection Agency "Chemical List of Lists," Sacramento CA (February 1997)

# Tributyltin Chloride

*Use Type:* Rodent repellant, antifoulant, and microbiocide
*CAS Number:* 1461-22-9
*Formula:* $C_{12}H_{27}ClSn$
*Synonyms:* Chlorotributylstannane; O,O-Dimethyl dithiobis(thioformate); Dimexan; Dimexano; Formic acid, dithiobis(thio-, O,O-dimethyl ester; Thioperoxydicarbonic acid [[(HO)C(S)]$_2$S$_2$], dimethyl ester; Stannane, tributylchloro-; Tributylchlorostannane; Tri-*n*-butyltin chloride; Tri-*n*-butylzinn-chlorid (German)
*Trade Names:* BIOMET® 12 RODENT REPELLENT, Atofina Chemicals (USA), canceled 7/1/1987
*Producers:* Atofina Chemicals (USA)
*Chemical Class:* Organotin
*EPA/OPP PC Code:* 083107
*California DPR Chemical Code:* 1891
*Uses:* Tributyltin compounds have been registered as molluscicides, as antifoulants on boats, ships, quays, buoys, crabpots, fish nets, and cages, as wood preservatives, as slimicides on masonry, as disinfectants, and as biocides for cooling systems, power station cooling towers, pulp and paper mills, breweries, leather processing, and textile mills.
*Carcinogen/Hazard Classifications*
**Label Signal Word:** WARNING

**Endocrine Disruptor:** Known Endocrine Disruptor.
*Regulatory Authority:*
- Actively registered pesticide in California.
- Marine pollutant (49CFR, Subchapter 172.101, Appendix B) as tributyltin compounds
- The Food and Drug Administration (FDA) regulates the use of some organic tin compounds in coatings and plastic food packaging.

*Incompatibilities:* Violent reaction with oxidizers, barium, potassium. and sodium. When heated to decomposition or on contact with acids or acid fumes, may produce highly toxic chloride fumes; deadly phosgene gas may be formed. May cause pitting of some metals.
*Permissible Exposure Limits in Air:* OSHA/NIOSH, and ACGIH (1) TWA is 0.1 mg/m³ (for tin organic compounds) and ACGIH STEL is 0.2 mg/m³. The notation "skin" is added indicating the possibility of cutaneous absorption. The NIOSH[2] IDLH level = 25 mg/m³ (as Sn).
*Determination in Air:* Filter/XAD-2® (tube); Acetic acid/$CH_3CN$; High-pressure liquid chromatography/Graphite furnace atomic absorption spectrometry; NIOSH IV[18], Method #5504, Organotin compounds.
*Permissible Concentration in Water:* No criteria set. Runoff from spills or fire control may cause water pollution. No criteria set, but EPA[32] has suggested a permissible ambient goal of 1.4 µg/L based on health effects.
*Harmful Effects and Symptoms*
*Short Term Exposure:* Irritates the eyes, skin, and respiratory tract. Contact may cause skin burns. Inhalation can cause coughing, wheezing and/or shortness of breath. Toxic hazard rating is high for oral, intravenous, intraperitoneal administration. This material causes swelling of the brain and spinal cord. Exposure may result in muscular weakness and paralysis, leading to respiratory failure, convulsive movements, closure of eyelids and sensitivity to light, headaches, and EEG changes, headache, dizziness, psychological and neurological disturbances, vertigo (an illusion of movement), sore throat, cough, abdominal pain, nausea, vomiting, diarrhea, urine retention, paresis, focal anesthesia, pruritus. Higher levels can cause unconsciousness, collapse and death.
*Long Term Exposure:* Repeated or prolonged contact can cause dermatitis; dry and cracked skin. May cause brain damage, hepatic necrosis; kidney damage.
*Points of Attack:* Skin, brain, kidneys.
*Medical Surveillance:* Tests to measure total tin and specific organotin compounds in blood, urine, feces, and body tissues. Normally, small amounts of tin can be found in the body because of the daily exposure to small amounts in the food. Therefore, the available tests cannot tell you when you were exposed or the exact amount of tin to which you were exposed, but can help determine if you were exposed to an unusually high amount of tin in the near past.

Kidney function tests. Psychological testing. Examination of the nervous system. EEG

*First Aid:* If this chemical gets into the eyes, remove any contact lenses at once and irrigate immediately for at least 15 minutes, occasionally lifting upper and lower lids. Seek medical attention immediately. If this chemical contacts the skin, remove contaminated clothing and wash immediately with soap and water. Seek medical attention immediately. If this chemical has been inhaled, remove from exposure, begin rescue breathing (using universal precautions) if breathing has stopped, and CPR if heart action has stopped. Transfer promptly to a medical facility. When this chemical has been swallowed, get medical attention. *Do not induce vomiting when formulations containing petroleum solvents are ingested.* Otherwise, give large quantities of water and induce vomiting. Do not make an unconscious person vomit.

*References:*
- EXTOXNET, Extension Toxicology Network, "Pesticide Information Profile, Tributyltin," Oregon State University, Corvallis, OR (September 1993). http://pmep.cce.cornell.edu/profiles/extoxnet/pyrethrins-ziram/tributyltin-ext.html
- International Programme on Chemical Safety (IPCS), "Environmental Health Criteria, Tributyltin Compounds," Geneva, Switzerland (1990). http://www.inchem.org/documents/ehc/ehc/ehc116.htm#SectionNumber:1.1
- California Environmental Protection Agency "Chemical List of Lists," Sacramento CA (February 1997)

# Tributyltin Fluoride

*Use Type:* Fungicide, antifoulant, and microbiocide
*CAS Number:* 1983-10-4
*Formula:* $C_{12}H_{27}FSn$
*Synonyms:* Fluorotributylstannane; Stannane, tributylfluoro-; Tri-*n*-butyltin flouride; Tributylflourostannane
*Chemical Class:* Organotin
*EPA/OPP PC Code:* 083112
*California DPR Chemical Code:* 1030
*Uses:* Not registered in the U.S. Most uses are for antifouling products for boat hulls, paints, fishing gear and other marine applications.
*Regulatory Authority:*
- Actively registered pesticide in California.
- EPCRA Section 313 Form R *de minimis* concentration reporting level: 1.0%
- Marine pollutant (49CFR, Subchapter 172.101, Appendix B) as tributyltin compounds

*Incompatibilities:* Violent reaction with oxidizers, barium, potassium. and sodium.
*Permissible Exposure Limits in Air:* OSHA/NIOSH, and ACGIH (1) TWA is 0.1 mg/m³ (for tin organic compounds) and ACGIH STEL is 0.2 mg/m³. The notation "skin" is added indicating the possibility of cutaneous absorption. The NIOSH[2] IDLH level = 25 mg/m³ (as Sn).

*Determination in Air:* Filter/XAD-2® (tube); Acetic acid/$CH_3CN$; High-pressure liquid chromatography/Graphite furnace atomic absorption spectrometry; NIOSH IV[18], Method #5504, Organotin compounds.

*Permissible Concentration in Water:* No criteria set. Runoff from spills or fire control may cause water pollution. No criteria set, but EPA[32] has suggested a permissible ambient goal of 1.4 µg/L based on health effects.

*Harmful Effects and Symptoms*

*Short Term Exposure:* Irritates the eyes, skin, and respiratory tract. Contact may cause skin burns. Inhalation can cause coughing, wheezing and/or shortness of breath. Toxic hazard rating is high for oral, intravenous, intraperitoneal administration. This material causes swelling of the brain and spinal cord. Exposure may result in muscular weakness and paralysis, leading to respiratory failure; convulsive movements; closure of eyelids and sensitivity to light; headaches, and EEG changes, headache, dizziness, psychological and neurological disturbances, vertigo (an illusion of movement), sore throat, cough, abdominal pain, nausea, vomiting, diarrhea, urine retention, paresis, focal anesthesia, pruritus. Higher levels can cause unconsciousness, collapse and death.

*Long Term Exposure:* Repeated or prolonged contact can cause dermatitis; dry and cracked skin. May cause brain damage, hepatic necrosis, kidney damage.

*Points of Attack:* Skin, brain, kidneys.

*First Aid:* If this chemical gets into the eyes, remove any contact lenses at once and irrigate immediately for at least 15 minutes, occasionally lifting upper and lower lids. Seek medical attention immediately. If this chemical contacts the skin, remove contaminated clothing and wash immediately with soap and water. Seek medical attention immediately. If this chemical has been inhaled, remove from exposure, begin rescue breathing (using universal precautions) if breathing has stopped, and CPR if heart action has stopped. Transfer promptly to a medical facility. When this chemical has been swallowed, get medical attention. *Do not induce vomiting when formulations containing petroleum solvents are ingested.* Otherwise, give large quantities of water and induce vomiting. Do not make an unconscious person vomit.

*References:*
- EXTOXNET, Extension Toxicology Network, "Pesticide Information Profile, Tributyltin," Oregon State University, Corvallis, OR (September 1993). http://pmep.cce.cornell.edu/profiles/extoxnet/pyrethrins-ziram/tributyltin-ext.html
- International Programme on Chemical Safety (IPCS), "Environmental Health Criteria, Tributyltin Compounds," Geneva, Switzerland (1990). http://www.inchem.org/documents/ehc/ehc/ehc116.htm#SectionNumber:1.1

- California Environmental Protection Agency "Chemical List of Lists," Sacramento CA (February 1997).

# Trichlorfon

*Use Type:* Insecticide
*CAS Number:* 52-68-6
*Formula:* $C_4H_8Cl_3O_4P$; $(CH_3O)_2POCHOHCCl_3$
*Alert:* Use of trichlorfon in the U.S. for food and feed applications was canceled in 1995.
*Synonyms:* Chlorophos; Chlorophthalm; Chloroxyphos; Clorofos (Russian); DETF; Dimethoxy-2,2,2-trichloro-1-hydroxy-ethylphosphine oxide; *O,O*-Dimethyl (1-hydroxy-2,2,2-trichloraethyl)phosphat (German); *O,O*-Dimethyl (1-hydroxy-2,2,2-trichloraethyl)phosphonsaeure ester (German); *O,O*-Dimethyl (1-hydroxy-2,2,2-trichloroethyl)phosphonate; Dimethyl 1-hydroxy-2,2,2-trichloroethylphosphonate; *O,O*-Dimethyl (2,2,2-trichloro-1-hydroxyethyl)phosphonate; Dimethyl (2,2,2-trichloro-1-hydroxyethyl)phosphonate; *O,O*-Dimetil-(2,2,2-tricloro-1-idrossi-etil)-fosfonato (Italian); ENT 19,763; Foschlorem (Polish); 1-Hydroxy-2,2,2-trichloroethylphosphonic acid dimethyl ester; Hypodermacid; Methyl chlorophos; Metifonate; Metrifonate; Metriphonate; NCI-C54831; Phoschlor; Phosphonic acid, (2,2,2-trichloro-1-hydroxyethyl)-, dimethyl ester; Polfoschlor; Trichlorofon (Dutch); 2,2,2-Trichloro-1-hydroxyethyl-phosphonate, dimethyl ester; (2,2,2-Trichloro-1-hydroxyethyl)phosphonic acid dimethyl ester; Trichlorophene; Trichlorphon
*Trade Names:* AEROL 1 (PESTICIDE)®; AGROFOROTOX®; ANTHON®; BAY 15922®, Bayer CropScience (Germany); BAYER 15922®, Bayer CropScience (Germany); BAYER L 13/59®, Bayer CropScience (Germany); BILARCIL®; BOVINOX®, canceled; BRITON®; BRITTEN®; CEKUFON®; CHLORAK®; CHLOROFTALM®; CICLO-SOM®; COMBOT®; COMBOT EQUINE®; DANEX®, canceled; DEP (PESTICIDE)®; DEPTHON®; DIMETOX®; DIPTEREX®, Bayer CropScience (Germany); DIPTEREX® 50, Bayer CropScience (Germany); DIPTEVU®; DITRIFON®; DYLOX®, Bayer CropScience (Germany); DYLOX-METASYSTOX-R®, Bayer CropScience (Germany); DYREX®; DYVON®; EQUINO-ACID®; EQUINO-AID®; FLIBOL E®; FLIEGENTELLE®; FOROTOX®; FOSCHLOR®; FOSCHLOR R®; FOSCHLOR R-50®; LEIVASOM®; LOISOL®; MASOTEN®, Bayer Healthcare (Germany), canceled 8/11/1987; MAZOTEN®; NEGUVON®, Bayer CropScience (Germany); NEGUVON A®, Bayer CropScience (Germany); PHOSCHLOR R50®; PROXOL®, Bayer CropScience (Germany); RICIFON®; RITSIFON®; SATOX 20WSC®; SOLDEP®; SOTIPOX®; TRICHLORPHON FN®; TRINEX®; TUGON®; TUGON FLY BAIT®; TUGON STABLE SPRAY®; VERMICIDE BAYER 2349®; VOLFARTOL®; VOTEXIT®; WEC 50®; WOTEXIT®
*Producers:* Agsin (Singapore); Alcotan Laboratories (Spain); Bayer CropScience (Germany); Calliope (France); Epochem Co., (China); Hokko Chemical Industry (Japan); Hunan Tianyu Pesticide Chemical Group (China); I.N.D.I.A. Industrie Chimiche (Italy); Miles (USA); Saeryung Chemicals (South Korea); Sankei Chemical (Japan); Sanonda Zhengzhou Pesticide Co. Ltd. (China); Shenzhen Guomeng Industry Co., Ltd. (China); Sinon Corporation (Taiwan); Takeda Chemical Industries (Japan)
*Chemical Class:* Organophosphate
*EPA/OPP PC Code:* 057901
*California DPR Chemical Code:* 88
*ICSC Number:* 0585
*RTECS Number:* AO700000
*EEC Number:* 015-021-00-0
*EINECS Number:* 200-149-3
*Uses:* Trichlorfon has non-agriculture uses on golf course turf, home lawns and similar venues, and in non-food contact areas of food and meat processing plants. Also on ornamental shrubs and plants, and ornamental and bait fish ponds. Overseas, trichlorfon is used as cattle pour-on, which is classified as a food-use. It is used against insects such as lepidopteran larvae (caterpillars), white grubs, mole crickets, cattle lice, sod webworms, leaf miners, stink bugs, flies, ants, cockroaches, earwigs, crickets, diving beetles, water scavenger beetles, water boatman, backswimmers, water scorpions, giant water bugs and pillbugs. Use of all food and feed uses in the U.S. were voluntarily canceled November 21, 1995. It was used on Brussels sprouts, barley, beets, blueberries, beans (dry and snap), corn, field corn, popcorn, sweet corn, cotton, cow peas, lima beans, tomatoes, cabbage, carrots (including tops), cauliflower, collards, cowpeas, southern peas, black-eyed peas, crowder peas, pumpkins, collards, lettuce and alfalfa, cotton, peanuts, peppers, pumpkins, tobacco, soybeans and treatment to manure.
*Human toxicity (long-term)*[77]: Intermediate–14.00 ppb, Health Advisory
*Fish toxicity (threshold)*[77]: Intermediate–24.99773 ppb, MATC (Maximum Acceptable Toxicant Concentration)
*U.S. Maximum Allowable Residue Levels for Trichlorfon (40 CFR 180.198):* The Maximum Residue levels for cattle fat, cattle meat, and cattle meat byproducts (mbyp) are 0.1 ppm.
*Carcinogen/Hazard Classifications*
**U.S. EPA Carcinogens:** Likely carcinogen in high doses; Unlikely carcinogen in low doses
**IARC:** Group 3, unclassifiable
**Label Signal Word:** WARNING or CAUTION
**WHO Acute Hazard:** Class II, moderately hazardous
*Regulatory Authority:*
- Air Pollutant Standard Set (former USSR)[35, 43]

- AB 2588-Air Toxics "Hot Spots" Chemicals (CAL)
- The "Director's List" (CAL/OSHA)
- Actively registered pesticide in California
- Clean Water Act: Section 311 Hazardous Substances/RQ 40CFR117.3 (same as CERCLA, see below)
- Superfund/EPCRA 40CFR302.4 RQ: CERCLA, 100 lb (45.4 kg)
- Dropped from Extremely Hazardous Substance (EPA-SARA) in 1988
- EPCRA Section 313 Form R *de minimis* concentration reporting level: 1.0%
- U.S. DOT Inhalation Hazard Chemicals as organophosphates
- U.S. DOT Regulated Marine Pollutant (49CFR172.101, Appendix B)

***Description:*** White to pale yellow crystalline solid. Ethyl ether odor. Slightly soluble in water; rapid hydrolysis. Molecular weight = 257.42. Boiling point = 100°C @ 1mmHg. Melting/Freezing point = 83–84°C. Vapor Pressure = $7.6 \times 10^{-6}$ mbar @ 20°C. Hazard Identification (based on NFPA-704 M Rating System): Health 2, Flammability 1, Reactivity 0.

***Incompatibilities:*** Alkaline materials, e.g., lime, lime sulfur, etc. Corrosive to iron and steel.

***Permissible Exposure Limits in Air:*** The former USSR[35, 43] has set an MAC in workplace air of 0.5 mg/m$^3$ and in ambient air in residential areas of 0.04 mg/m$^3$ on a momentary basis and 0.02 mg/m$^3$ on a daily average basis.

***Determination in Air:*** OSHA versatile sampler-2; Toluene/Acetone; Gas chromatography/Flame photometric detection for sulfur, nitrogen, or phosphorus; NIOSH Method IV Method #5600, Organophosphorus pesticides.[18]

***Permissible Concentration in Water:*** The former USSR[35, 43] has set an MAC in water bodies used for domestic purposes of 0.05 mg/L and in water bodies used for fishery purposes of zero.

***Routes of Entry:*** Inhalation, ingestion and skin contact.

***Harmful Effects and Symptoms***

***Short Term Exposure:*** Very toxic. Probable oral lethal dose (human) 50-500 mg/kg, between 1 teaspoon and 1 ounce for 150 lb (70 kg) person. Toxicity relatively low among organophosphate insecticides, although a potent cholinesterase inhibitor. Skin sensitivity has been reported. Symptoms of exposure: muscle weakness, twitching, respiratory depression, sweating, vomiting, diarrhea, chest and abdominal distress, sometimes pulmonary edema, excessive salivation, headache, giddiness, vertigo and weakness, runny nose and sensation of tightness in chest (inhalation), blurring of vision, tearing, ocular pain, loss of muscle coordination, and slurring of speech. Delayed pulmonary edema may occur after inhalation.

***Long Term Exposure:*** Cholinesterase inhibitor; cumulative effect is possible. This chemical may damage the nervous system with repeated exposure, resulting in convulsions, respiratory failure. May cause liver damage.

***Points of Attack:*** Respiratory system, lungs, central nervous system, cardiovascular system, skin, eyes, plasma and red blood cell cholinesterase.

***Medical Surveillance:*** Medical observation is recommended for 24 to 48 hours after breathing overexposure, as pulmonary edema may be delayed. Before employment and at regular times after that, the following are recommended: Plasma and red blood cell cholinesterase levels (tests for the enzyme poisoned by this chemical). If exposure stops, plasma levels return to normal in 1-2 weeks while red blood cell levels may be reduced for 1-3 months. When acetylcholinesterase enzyme levels are reduced by 25% or more below preemployment levels, risk of poisoning is increased, even if results are in lower ranges of "normal." Reassignment to work not involving organophosphate or carbamate pesticides is recommended until enzyme levels recover. If symptoms develop or overexposure occurs, repeat the above tests as soon as possible and get an exam of the nervous system. Also consider complete blood count. Consider chest x-ray following acute overexposure. Do not drink any alcoholic beverages before or during use. Alcohol promotes absorption of organophosphates.

***First Aid:*** **Treatment for organophosphate poisoning consists of thorough decontamination, cardiorespiratory support, and administration of the antidotes atropine and pralidoxime. In cases of severe poisoning, diazepam, an anticonvulsant, should also be administered. Antidotes should be administered as prevention even if the diagnosis is in doubt.** Speed in removing material from eyes and skin is of extreme importance. *Eyes:* Eye contact can cause dangerous amounts of these chemicals to be quickly absorbed through the mucous membrane into the bloodstream. Immediately and gently flush eyes with plenty of warm or cold water (NO hot water) for at least 15 minutes, occasionally lifting the upper and lower eyelids. Get medical aid immediately. *Skin:* Get medical aid. Skin contact can cause dangerous amounts of these chemicals to be absorbed into the bloodstream. Wearing the appropriate PPE equipment and respirator for organophosphate pesticides, immediately flush skin with plenty of soap and water for at least 15 minutes while removing contaminated clothing and shoes. Shampoo hair promptly if contaminated. The removed, contaminated clothing and shoes should be double-bagged and left in Hot Zone for later disposal by hazardous materials experts. Skin may also be decontaminated with diluted hypochlorite solution. *Inhalation:* Get medical aid. Do not contaminate yourself. Wearing the appropriate PPE equipment and respirator for organophosphate pesticides, immediately remove the victim from the contaminated area to fresh air. If the victim is not breathing, administer artificial respiration. Do NOT use mouth-to-mouth resuscitation; use bag/mask apparatus. If breathing is difficult, administer oxygen through bag/mask apparatus until medical help arrives. Do not leave victim unattended. *Ingestion:* Call poison control. Loosen all clothing. Never give anything by mouth to an unconscious person. If victim is *unconscious or having convulsions,* do nothing except keep victim warm. Get medical aid. Transfer promptly to a medical facility. In cases of ingestion, **do not induce vomiting**. If the victim is alert and asymptomatic, administer a slurry of activated charcoal at a dose of 1 g/kg (infant, child, and adult dose). A soda can and straw may be of assistance when offering charcoal to a child. *In some cases you may be specifically instructed by poison control to induce vomiting by way of 2 tablespoons of syrup of ipecac (adult) washed down with a cup of water.* Do NOT give activated charcoal before or with ipecac syrup.

*Note to physician or authorized medical personnel:* Treat cases of respiratory compromise, coma, or excessive pulmonary secretions with respiratory support using protocols and techniques available and within the scope of training. Some cases may necessitate procedures such as endotracheal intubation or cricothyrotomy by properly trained and equipped personnel. When possible, atropine (see *Antidotes*, below) should be given under medical supervision. Patients who are comatose, hypotensive, or having seizures or cardiac arrhythmias should be treated according to advanced life support protocols. *Antidotes:* Two antidotes are administered to treat organophosphate poisoning. Atropine is a competitive antagonist of acetylcholine at muscarinic receptors and is used to control the excessive bronchial secretions which are often responsible for death. Pralidoxime relieves both the nicotinic and muscarine effects of organophosphate poisoning by regenerating acetylcholinesterase and can reduce both the bronchial secretions and the muscle weakness associated with poisoning. The initial intravenous dose of atropine in adults should be determined by the severity of symptoms: An initial adult dose of 1.0 to 2.0 mg or pediatric dose of 0.01 mg/kg (minimum 0.01 mg) should be administered intravenously. If intravenous access cannot be established, atropine may also be given intramuscularly, subcutaneously or via endotracheal tube. Doses should be repeated every 15 minutes until excessive secretions and sweating have been controlled. Once bronchial secretion has been controlled, atropine administration should be repeated whenever the secretions begin to recur. In seriously poisoned patients, very large doses may be required. Alterations of pulse rate and pupillary size should not be used as indicators of treatment adequacy. Pralidoxime should be administered as early in poisoning as possible as its efficacy may diminish when given more than 24 to 36 hours after exposure. Doses are as follows: adult 1.0 g; pediatric 25 to 50 mg/kg. The drug should be administered intravenously over 30 to 60 minutes, but in a life-threatening situation, one-half of the total dose can be given per minute for a total administration time of 2 minutes. Treatment should begin to take effect within 40 minutes with a reduction in symptoms and the amount of atropine necessary to control bronchial secretion. The initial dose can be repeated in 1 hour and then every 8 to 12 hours until the patient is clinically well and no longer requires atropine. If intravenous access cannot be established, pralidoxime may also be given intramuscularly. Early administration of diazepam in addition to the combined atropine and pralidoxime treatment may help prevent the onset of seizures and potential brain and cardiac morphologic damage following high-level organophosphate poisoning.

*References:*
- U.S. Environmental Protection Agency, "Reregistration Eligibility Decision (RED) for Trichlorfon," Office of Prevention, Pesticides and Toxic Substances, Washington, DC (January 1997). http://www.epa.gov/REDs/0104.pdf
- New Jersey Department of Health and Senior Services, "Hazardous Substance Fact Sheet, Trichlorfon," Trenton NJ (March 1998). http://www.state.nj.us/health/eoh/rtkweb/1882.pdf
- Sax, N.I., Ed., "Dangerous Properties of Industrial Materials Report," 7, No. 2, 95-101 (1987).
- U.S. Environmental Protection Agency, "Chemical Profile: Trichlorophon," Washington, DC, Chemical Emergency Preparedness Program (October 31, 1985).
- Agency for Toxic Substances and Disease Registry, U.S. Department of Health and Human Services, Public Health Service, "Managing Hazardous Materials Incidents," Atlanta, GA (June 2003).
- California Environmental Protection Agency "Chemical List of Lists," Sacramento CA (February 1997).

## Trichlorobenzoic Acid

*Use Type:* Herbicide
*CAS Number:* 50-31-7
*Formula:* $C_7H_3Cl_3O_2$
*Synonyms:* NCI-C60242; NSC 21914; 2,3,6-TBA; 2,3,6-TCB; 2,3,6-TCBA; 2,3,6-TrCB acid; 2,3,6-Trichlorbenzoesaeure (German); 2,3,6-Trichlorobenzoic acid
*Trade Names:* BENZABAR®; BENZAC®; BENZOBOR®, U.S. Borax (USA), canceled; FEN-ALL®; HC 1281®; T-2®; TRIBAC®; TRYBEN®; TRYSBEN 200®; UREABOR®, U.S. Borax (USA), canceled; ZOBAR®, DuPont (USA)
*Producers:* DuPont (USA); U.S. Borax (USA)
*Chemical Class:* Organochlorine
*EPA/OPP PC Code:* 017302
*California DPR Chemical Code:* 602
*Uses:* Not registered in the U.S.
*Carcinogen/Hazard Classifications*
**Label Signal Word:** CAUTION
**WHO Acute Hazard:** Class III, slightly hazardous
*Regulatory Authority:*
- FIFRA, 180.3(4); class of chlorinated organic pesticide.

*Incompatibilities:* Keep away from oxidizers, sulfuric acid, caustics, ammonia, aliphatic amines, alkanolamines, isocyanates, alkylene oxides, epichlorohydrin. When heated to decomposition this material forms toxic fumes of chlorine.

*Permissible Concentration in Water:* No criteria set. Runoff from spills or fire control may cause water pollution.

*Routes of Entry:* Inhalation, ingestion. Absorbed through the intact skin.

*Harmful Effects and Symptoms*

*Short Term Exposure:* Symptoms include apprehension, anxiety, confusion, nervous excitation, dizziness, headache, numbness and weakness in limbs, muscle twitching, tremors, nausea and vomiting, slow, shallow respiration, bluish face, convulsions, loss of consciousness, breathing stops, death.

*Long Term Exposure:*

*Points of Attack:* CNS. May be fatal if inhaled, ingested, or absorbed through the skin

*Medical Surveillance:* Medical observation is recommended for 24 to 48 hours after breathing overexposure, as pulmonary edema may be delayed. As first aid for pulmonary edema, consider administering a corticosteroid spray. Cigarette smoking may exacerbate pulmonary injury and should be discouraged for at least 72 hours following exposure.

*First Aid:* **Eyes:** *Eyes:* Speed in removing material from skin is of extreme importance. Eye contact can cause dangerous amounts of these chemicals to be quickly absorbed through the mucous membrane into the bloodstream. Directly, irrigate with large amounts of plain, tepid water or saline for 20 minutes, occasionally lifting the lower and upper lids. During this time, remove contact lenses, if easily removable without additional trauma to the eye. Get medical aid immediately. Have physician check for possible delayed damage. *Skin:* Get medical aid. Skin and/or eye contact can cause dangerous amounts of these chemicals to be absorbed into the bloodstream. Wearing the appropriate PPE equipment and respirator for organochlorine pesticides, immediately flush exposed skin, hair, and under nails with plain, running, tepid water for 20 minutes, then wash twice with mild soap. Shampoo hair promptly if contaminated; protect eyes. **Do not scrub skin or hair**, since this can increase absorption through the skin. Rinse thoroughly with water. Victims who are able and cooperative may assist with their own decontamination. Remove and double-bag contaminated clothing and personal belongings. Leather absorbs many organochlorines; therefore, items such as leather shoes, gloves, and belts should be discarded. If the skin is swollen or inflamed, cool affected areas with cold compresses. *Ingestion:* Call poison control. Loosen all clothing. Never give anything by mouth to an unconscious person. Get medical aid. *Do not induce vomiting.\** In cases of ingestion, the patient is at risk of CNS depression or seizures, which may lead to pulmonary aspiration during vomiting. If the victim is conscious and able to swallow, \*administer an aqueous slurry of activated charcoal at 1 gm/kg (usual adult dose 60–90 g, child dose 25–50 g). A soda can and straw may be of assistance when offering charcoal to a child. The efficacy of activated charcoal for some organochlorine poisoning (such as chlordane) is uncertain. If victim is *unconscious or having convulsions,* do nothing except keep victim warm. *\*In some cases you may be specifically instructed by Poison Control to induce vomiting by way of 2 tablespoons of syrup of ipecac (adult) washed down with a cup of water. Do not give activated charcoal before or with ipecac syrup. Inhalation:* Get medical aid. Do not contaminate yourself. Wearing the appropriate PPE equipment and respirator for organochlorine pesticides, immediately remove the victim from the contaminated area to fresh air. For inhalation exposures, monitor for respiratory distress. If the victim is not breathing, administer artificial respiration. *Do not use mouth-to-mouth resuscitation; use bag/mask apparatus.* If cough or breathing difficulty develops, evaluate for respiratory tract irritation, bronchitis, or pneumonitis. If breathing is difficult, administer 100% humidified supplemental oxygen through bag/mask apparatus until medical help arrives. Do not leave victim unattended.

*References:*
- California Environmental Protection Agency "Chemical List of Lists," Sacramento CA (February 1997)

# Trichloronate

*Use Type:* Insecticide
*CAS Number:* 327-98-0
*Formula:* $C_{10}H_{12}Cl_3O_2PS$; $C_6H_2(Cl_3)$-O-$P(S)(CH_2CH_3)OCH_2CH_3$
*Synonyms:* O-Aethyl-O-(2,4,5-trichlorphenyl)-aethylthionophosphonat (German); ENT 25,712; O-Ethyl-O-2,4,5-trichlorophenyl ethyl-phosphonothioate; Ethyl trichlorophenylethylphosphonothioate; Phosphonothioic acid, ethyl-, O-ethyl O-(2,4,5-trichlorophenyl) ester; Trichloronat; 2,4,5-Trichlorophenol-O-ester with O-ethyl ethylphosphonothioate

*Trade Names:* AGRISIL®; AGRITOX®; BAY® 37289, Bayer CropScience (Germany); BAYER® 37289, Bayer CropScience (Germany); BAYER S® 4400, Bayer CropScience (Germany); CHEMAGRO® 37289; FENOPHOSPHON®; FITOSOL®; PHYTOSOL®; RICHLORONATE®; STAUFFER® N-3049; WIRKSTOFF® 37289

*Producers:* Bayer CropScience (Germany); Ehrenstorfer, Dr. (Germany)

*Chemical Class:* Organophosphate
*EPA/OPP PC Code:* 456200
*California DPR Chemical Code:* 5001
*RTECS Number:* TB0700000
*EINECS Number:* 206-326-1

*Uses:* This is a non-systemic, organophosphate insecticide which is used for the control of soil insects. It is particularly harmful to honey bees. There are no registered products in the U.S.

*Fish toxicity (threshold)[77]:* Intermediate–14.47908 ppb, MATC (Maximum Acceptable Toxicant Concentration)

*Regulatory Authority:*
- Superfund/EPCRA 40CFR355, Appendix B Extremely Hazardous Substances: TPQ = 500 lb (227kg)
- Superfund/EPCRA 40CFR302.4 RQ: EHS, 1 lb (0.454 kg)
- AB 1803-Well Monitoring Chemical (CAL)
- The "Director's List" (CAL/OSHA) as a trichlorophenol
- U.S. DOT Inhalation Hazard Chemicals as organophosphates
- U.S. DOT Regulated Marine Pollutant (49CFR172.101, Appendix B) as trichloronat

*Description:* Yellow to amber liquid. Practically insoluble in water; solubility = 50ppm @ 20°C. Molecular weight = 333.59. Boiling point = 108°C @ 0.012 mbar. Vapor pressure = $1.7 \times 10^{-5}$ mbar @ 20°C. Hazard Identification (based on NFPA-704 M Rating System): Health 3, Flammability 1, Reactivity 0.

*Incompatibilities:* Strong bases.

*Permissible Exposure Limits in Air:* No standards set.

*Determination in Air:* OSHA versatile sampler-2; Toluene/Acetone; Gas chromatography/Flame photometric detection for sulfur, nitrogen, or phosphorus; NIOSH Method IV Method #5600, Organophosphorus pesticides.[18]

*Permissible Concentration in Water:* No criteria set, but runoff from spills or fire control may cause water pollution.

*Routes of Entry:* Inhalation, skin absorption, ingestion, skin and/or eye contact.

*Harmful Effects and Symptoms*

*Short Term Exposure:* Toxic effects are due to action on the nervous system. It has high oral toxicity and death can occur in acute poisonings. Delayed neurotoxicity has been reported. Symptoms of exposure include headache, dizziness, nausea, salivation, vomiting, abdominal pain, diarrhea, chest pain, decreased heart rate, excessive discharge of mucous from the air passages, difficult breathing, contraction of the pupil, blurred vision, profuse perspiration, muscle twitching and spasms, profound weakness, psychotic behavior, uncoordination, unconsciousness, rarely convulsions. Low level absorption syndrome is similar to influenza. High dosage may cause toxic psychosis similar to alcoholism. Exposures may be misdiagnosed as asthma and heart failure. Organic phosphorus insecticides are absorbed by the skin, as well as by the respiratory and gastrointestinal tracts. They are cholinesterase inhibitors. Symptoms of exposure include headache, giddiness, blurred vision, nervousness, weakness, nausea, cramps, diarrhea, and discomfort in the chest. Signs include sweating, tearing, salivation, vomiting, cyanosis, convulsions, coma, loss of reflexes and loss of sphincter control. Delayed pulmonary edema may occur after inhalation.

*Long Term Exposure:* Cholinesterase inhibitor; cumulative effect is possible. This chemical may damage the nervous system with repeated exposure, resulting in convulsions, respiratory failure. May cause liver damage.

*Points of Attack:* Respiratory system, lungs, central nervous system, cardiovascular system, skin, eyes, plasma and red blood cell cholinesterase.

*Medical Surveillance:* Medical observation is recommended for 24 to 48 hours after breathing overexposure, as pulmonary edema may be delayed. Before employment and at regular times after that, the following are recommended: Plasma and red blood cell cholinesterase levels (tests for the enzyme poisoned by this chemical). If exposure stops, plasma levels return to normal in 1-2 weeks while red blood cell levels may be reduced for 1-3 months. When acetylcholinesterase enzyme levels are reduced by 25% or more below preemployment levels, risk of poisoning is increased, even if results are in lower ranges of "normal." Reassignment to work not involving organophosphate or carbamate pesticides is recommended until enzyme levels recover. If symptoms develop or overexposure occurs, repeat the above tests as soon as possible and get an exam of the nervous system. Also consider complete blood count. Consider chest x-ray following acute overexposure. Do not drink any alcoholic beverages before or during use. Alcohol promotes absorption of organophosphates.

*First Aid:* **Treatment for organophosphate poisoning consists of thorough decontamination, cardiorespiratory support, and administration of the antidotes atropine and pralidoxime. In cases of severe poisoning, diazepam, an anticonvulsant, should also be administered. Antidotes should be administered as prevention even if the diagnosis is in doubt.** Speed in removing material from eyes and skin is of extreme importance. *Eyes:* Eye contact can cause dangerous amounts of these chemicals to be quickly absorbed through the mucous membrane into the bloodstream. Immediately and gently flush eyes with plenty of warm or cold water (NO hot water) for at least 15 minutes, occasionally lifting the upper and lower eyelids. Get medical aid immediately. *Skin:* Get medical aid. Skin contact can cause dangerous amounts of these chemicals to be absorbed into the bloodstream. Wearing the appropriate PPE equipment and respirator for organophosphate pesticides, immediately flush skin with plenty of soap and water for at least 15 minutes while removing contaminated clothing and shoes. Shampoo hair promptly if contaminated. The removed, contaminated clothing and shoes should be double-bagged and left in Hot Zone for later disposal by hazardous materials experts. Skin may also be decontaminated with diluted hypochlorite solution. *Inhalation:* Get medical aid. Do not contaminate yourself. Wearing the appropriate PPE equipment and respirator for organophosphate pesticides, immediately remove the victim from the contaminated area to fresh air. If the victim is not breathing, administer artificial respiration. Do NOT use mouth-to-mouth resuscitation; use bag/mask apparatus. If breathing is difficult, administer oxygen through bag/mask apparatus until medical help arrives. Do not leave victim unattended. *Ingestion:* Call poison control. Loosen all clothing. Never give anything by mouth to an unconscious person. If victim is *unconscious or having convulsions,* do nothing except keep victim warm. Get medical aid. Transfer promptly to a medical facility. In cases of ingestion, **do not induce vomiting**. If the victim is alert and asymptomatic, administer a slurry of activated charcoal at a dose of 1 g/kg (infant, child, and adult dose). A soda can and straw may be of assistance when offering charcoal to a child. *In some cases you may be specifically instructed by poison control to induce vomiting by way of 2 tablespoons of syrup of ipecac (adult) washed down with a cup of water.* Do NOT give activated charcoal before or with ipecac syrup.

Note to physician or authorized medical personnel: Treat cases of respiratory compromise, coma, or excessive pulmonary secretions with respiratory support using protocols and techniques available and within the scope of training. Some cases may necessitate procedures such as endotracheal intubation or cricothyrotomy by properly trained and equipped personnel. When possible, atropine (see *Antidotes*, below) should be given under medical supervision. Patients who are comatose, hypotensive, or having seizures or cardiac arrhythmias should be treated according to advanced life support protocols. *Antidotes:* Two antidotes are administered to treat organophosphate poisoning. Atropine is a competitive antagonist of acetylcholine at muscarinic receptors and is used to control the excessive bronchial secretions which are often responsible for death. Pralidoxime relieves both the nicotinic and muscarine effects of organophosphate poisoning by regenerating acetylcholinesterase and can reduce both the bronchial secretions and the muscle weakness associated with poisoning. The initial intravenous dose of atropine in adults should be determined by the severity of symptoms: An initial adult dose of 1.0 to 2.0 mg or pediatric dose of 0.01 mg/kg (minimum 0.01 mg) should be administered intravenously. If intravenous access cannot be established, atropine may also be given intramuscularly, subcutaneously or via endotracheal tube. Doses should be repeated every 15 minutes until excessive secretions and sweating have been controlled. Once bronchial secretion has been controlled, atropine administration should be repeated whenever the secretions begin to recur. In seriously poisoned patients, very large doses may be required. Alterations of pulse rate and pupillary size should not be used as indicators of treatment adequacy. Pralidoxime should be administered as early in poisoning as possible as its efficacy may diminish when given more than 24 to 36 hours after exposure. Doses are as follows: adult 1.0 g; pediatric 25 to 50 mg/kg. The drug should be administered intravenously over 30 to 60 minutes, but in a life-threatening situation, one-half of the total dose can be given per minute for a total administration time of 2 minutes. Treatment should begin to take effect within 40 minutes with a reduction in symptoms and the amount of atropine necessary to control bronchial secretion. The initial dose can be repeated in 1 hour and then every 8 to 12 hours until the patient is clinically well and no longer requires atropine. If intravenous access cannot be established, pralidoxime may also be given intramuscularly. Early administration of diazepam in addition to the combined atropine and pralidoxime treatment may help prevent the onset of seizures and potential brain and cardiac morphologic damage following high-level organophosphate poisoning.

*References:*
- New Jersey Department of Health and Senior Services, "Hazardous Substance Fact Sheet, Trichloronate," Trenton, NJ (June 1999). http://www.state.nj.us/health/eoh/rtkweb/2837.pdf
- U.S. Environmental Protection Agency, "Chemical Profile: Trichloronate," Washington, DC, Chemical Emergency Preparedness Program (November 30, 1987).
- Agency for Toxic Substances and Disease Registry, U.S. Department of Health and Human Services, Public Health Service, "Managing Hazardous Materials Incidents," Atlanta, GA (June 2003).
- California Environmental Protection Agency "Chemical List of Lists," Sacramento CA (February 1997).

## Trichlorophenols

*Use Type:* Defoliant, fungicide and algaecide
*CAS Number:* 25167-82-2 (mixed isomers); 15950-66-0 (2,3,4-); 933-78-8 (2,3,5-); 933-75-5 (2,3,6-); 95-95-4 (2,4,5-); 88-06-2 (2,4,6-); 609-19-8 (3,4,5-)
*Formula:* $C_6H_3Cl_3O$; $HOC_6H_2Cl_3$
*Alert:* 2,4,6-Trichlorophenol is a carcinogen and should be handled with extreme caution. No pesticide uses are registered in the U.S.
*Synonyms: 25167-82-2:* Phenol, trichloro-; Trichlorofenol (Czech, Spanish); Triclorofenol
*15950-66-0:* Phenol, 2,3,4-trichloro-; 2,3,4-Trichlorofenol (Czech, Spanish); 2,3,4-Trichlorophenol; Trichlorophenol, 2,3,4-
*933-78-8:* Phenol, 2,3,5-trichloro-; 2,3,5-Trichlorofenol (Czech, Spanish); 2,3,5-Trichlorophenol; Trichlorophenol, 2,3,5-
*933-75-5:* Phenol, 2,3,6-trichloro-; 2,3,6-Trichlorofenol (Czech, Spanish); 2,3,6-Trichlorophenol; Trichlorophenol, 2,3,6-
*88-06-2:* NCI-CO2904; Phenachlor; phenol, 2,4,6-Trichloro-; 2,4,6-Trichlorfenol (Czech, Spanish); 2,4,6-Trichlorophenol; Trichlorophenol, 2,4,6-; 1,3,5-Trichloro-2-hydroxybenzene

*609-19-8:* Phenol, 3,4,5-trichloro-; 3,4,5-Trichlorofenol (Czech, Spanish); 3,4,5-Trichlorophenol; Trichlorophenol, 3,4,5-

*Trade Names: 25167-82-2:* OMAL®; *88-06-2:* DOWICIDE 2S®, Dow AgroSciences (USA); OMAL®; *95-95-4;* PREVENTOL 1®, Bayer CropScience (Germany)

*Producers:* Coalite Chemicals (UK); Dow AgroSciences (USA); Ehrenstorfer, Dr. (Germany); Excel Industries (India); Fluorochem (UK); Nippon Kayaku (Japan); Rhodia (France)

*Chemical Class:* Chlorinated phenols

*EPA/OPP PC Code:* 064210 (2,4,5-isomer); 064212 (*2,4,6-isomer*)

*California DPR Chemical Code:* 1189 (Trichlorophenols); 1382 (*2,4,5-isomer*); 640 (*2,4,6-isomer*)

*ICSC Number:* 0879 (2,4,5-isomer); 1122 (*2,4,6-isomer*)

*RTECS Number:* SN1400000 (*2,4,5-isomer*); SN1575000 (*2,4,6-isomer*); SN1650000 (*3,4,5-isomer*); SN1300000 (*2,3,6-isomer*)

*EEC Number:* 604-017-00-X (*2,4,5-isomer*); 604-012-00-2 (*2,4,6-isomer*)

*EINECS Number:* 202-467-8 (*2,4,5-isomer*)

*Uses:* 2,4,5-TCP is used to produce defoliant 2,4,5-T and related products. Also used directly as a fungicide, antimildew and preservative agent, algicide, bactericide. 2,4,6-TCP is used to produce 2,3,4,6-TCP and PCP. Used directly as germicide, bactericide, glue and wood preservative, and anti-mildew treatment. No pesticide uses are registered ion the U.S.

*Carcinogen/Hazard Classifications*

**U.S. EPA Carcinogens:** Group B2, probable carcinogen (*2,4,6-isomer*)

**U.S. NTP Carcinogen:** Reasonably anticipated carcinogen (*2,4,6-isomer*)

**California Prop. 65:** Listed (*2,4,6-isomer*)

**IARC:** Group 2B, possible carcinogen (*2,4,5-isomer*; *2,4,6-isomer*) *Note:* In animal studies, one chlorophenol, 2,4,6-trichlorophenol, caused leukemia in rats and liver cancer in mice. The Department of Health and Human Services has determined that 2,4,6-trichlorophenol may reasonably be anticipated to be a carcinogen.

*Regulatory Authority:*
- Carcinogen (Human Limited Evidence) (IARC)[9]
- Air Pollutant Standard Set (Sweden, former USSR)[35] (Several States)[61]

*Mixed isomers:*
- Clean Water Act: Section 311 Hazardous Substances/RQ 40CFR117.3 (same as CERCLA, see below)
- Superfund/EPCRA 40CFR302.4 RQ: CERCLA, 10 lb (4.54 kg)
- The "Director's List" (CAL/OSHA) as trichlorophenols *2,3,4-; 2,3,5-; 2,3,6-; 3,4,5-isomers:*
- Superfund/EPCRA 40CFR302.4 RQ: CERCLA, 10 lb (4.54 kg)

*2,4,5-isomers*
- Clean Air Act: Hazardous Air Pollutants (Title I, Part A, Section 112)
- EPA Hazardous Waste Number (RCRA No.): U230
- RCRA Toxicity Characteristic (Section 261.24), Maximum
- Concentration of Contaminants, regulatory level, 400.0 mg/L
- RCRA, 40CFR261, Appendix 8 Hazardous Constituents
- RCRA 40CFR268.48; 61FR15654, Universal Treatment Standards: Wastewater (mg/L), 0.18; Nonwastewater (mg/kg), 7.4
- RCRA 40CFR264, Appendix 9; TSD Facilities Ground Water Monitoring List. Suggested test method(s) (PQL ug/L): 8270(10)
- Superfund/EPCRA 40CFR302.4 RQ: CERCLA, 10 lb (4.54 kg)
- AB 2588-Air Toxics "Hot Spots" Chemicals (CAL)
- CAL Air Resources Board/AB 1807 Toxic Air Contaminants
- EPCRA Section 313 Form R *de minimis* concentration reporting level: 1.0%

*2,4,6-isomer:*
- Clean Air Act: Hazardous Air Pollutants (Title I, Part A, Section 112)
- EPA Hazardous Waste Number (RCRA No.): U231
- RCRA Toxicity Characteristic (Section 261.24), Maximum
- Concentration of Contaminants, regulatory level, 2.0 mg/L
- RCRA, 40CFR261, Appendix 8 Hazardous Constituents
- RCRA 40CFR268.48; 61FR15654, Universal Treatment Standards: Wastewater (mg/L), 0.035; Nonwastewater (mg/kg), 7.4
- RCRA 40CFR264, Appendix 9; TSD Facilities Ground Water Monitoring List. Suggested test method(s) (PQL ug/L): 8040(5); 8270(10)
- Superfund/EPCRA 40CFR302.4 RQ: CERCLA, 10 lb (4.54 kg)
- List of priority pollutants (U.S. EPA)
- AB 1803-Well Monitoring Chemical (CAL)
- AB 2588-Air Toxics "Hot Spots" Chemicals (CAL)
- CAL Air Resources Board/AB 1807 Toxic Air Contaminants
- Proposition 65 chemical (CAL)
- The "Director's List" (CAL/OSHA)
- EPCRA Section 313 Form R *de minimis* concentration reporting level: 0.1%
- Canada, WHMIS, Ingredients Disclosure List

***Description:*** Trichlorophenols exists as 6 isomers (*2,4,5-; 3,4,5-; 2,4,6-; 2,3,4-; 2,3,5-, and 2,3,6-*). The most important (heavily regulated) are the *2,4,5-* and *2,4,6-isomers*. The *2,4,5-isomer* is white powder or needles; the *2,3,5-* and *2,3,6-isomers* are colorless crystals; the *2,4,5-*

*isomer* is a grey crystalline solid or flakes; the *2,4,6-isomer* is a colorless to light yellow crystalline solid. They have a phenolic odor. Boiling point = 248–253°C (*2,3,5-isomer*); 253°C (*2,3,6-isomer*); 253°C (*2,4,5-isomer*); 246°C *(2,4,6-isomer)*. Melting/Freezing point = 84°C (*2,3,4-isomer*); 62°C (*2,3,5-isomer*); 58°C (*2,3,6-isomer*); 67°C (*2,4,5-isomer*); 70°C (*2,4,6-isomer*). Flash point = 78°C (*2,3,6-isomer*); 99°C (*2,4,6-isomer*). Hazard Identification (based on NFPA-704 M Rating System): [*2,4,6-isomer*] Health 2, Flammability 1, Reactivity 0. All isomers are slightly soluble or practically insoluble in water. Log $K_{ow}$ = 4.6. (*3,4,5-isomer*); 3.6 (*2,3,6-isomer*); 3.7 (*2,4,5-isomer*); 3.9 (*2,4,6-isomer*). Values above 3.0 are likely to bioaccumulate in marine organisms.

*Incompatibilities:* Perhaps the most notable incompatibility is the reaction of 2,4,5-trichlorophenol in alkaline medium at high temperatures to produce dioxin. (*2,3,4-isomer*) reacts with oxidizers, acid anhydrides, and acid chlorides. (*2,3,5-isomer*) Decomposes on heating, on burning and on contact with strong oxidants producing toxic and corrosive fumes of hydrogen chloride. The substance is a weak acid. (*2,3,6-isomer*) The substance is a weak acid. pH= 4.8/4.2. (*2,4,6-isomer*) Reacts violently with strong oxidants and is incompatible with acid chlorides and acid anhydrides.

*Permissible Exposure Limits in Air:* Sweden[35] has set an MAC in workplace air of 0.5 mg/m³ and an STEL of 1.5 mg/m³. The former USSR[35] has set a MAC in ambient air in residential areas of 3.0 μg/m³ for the 2,4,6-isomer on a once-daily basis. Some states have set guidelines or standards for the trichlorophenols in ambient air[60]. Massachusetts has set zero for the 2,4,6-isomer and 1.6 μg/m³ for the 2,4,5-isomer. Pennsylvania has set 3,500 μg/m³ for the 2,4,5-isomer on a one-year exposure basis.

*Determination in Air:* Use NIOSH: (*o*-chlorophenol) P&CAM Method #337, Chlorophenols.[18]

*Permissible Concentration in Water:* For 2,4,6-trichlorophenol, to protect freshwater aquatic life, 970 μg/L on a chronic toxicity basis. To protect saltwater aquatic life- no criteria developed due to insufficient data. To protect human health, for 2,4,5-TCP, 2,600 μg/L; for 2,4,6-TCP, preferably zero. An additional lifetime cancer risk of 1 in 100,000 occurs at a level of 12 μg/L. These are based on organoleptic effects. A limit based on toxicological effects for 2,4,5-TCP would be 1,600 μg/L[6]. Kansas[61] has set a guideline for the 2,4,5-isomer in drinking water of 1.0 μg/L. Values for the 2,4,6-isomer have been set by Kansas at 17.0 μg/L, by Minnesota at 17.5 μg/L and by Maine at 700.0 μg/L. The WHO[35] has set a limit for the 2,4,6-isomer in drinking water of 10.0 μg/L. EPA recommends that drinking water contain no more than 0.04 mg/L of 2-chlorophenol for a lifetime exposure for an adult, and 0.05 mg/L for a 1-day, 10-day, or longer exposure for a child. For 2,4-dichlorophenol, EPA recommends that drinking water contain no more than 0.03 mg/L for a 1-day, 10-day, or longer exposure for a child.

*Determination in Water:* Methylene chloride extraction followed by gas chromatography with flame ionization or electron capture detection (EPA Method 604) or gas chromatography plus mass spectrometry (EPA Method 625).

*Routes of Entry:* Inhalation, ingestion, skin and/or eye contact. May be absorbed through the skin.

*Harmful Effects and Symptoms*

*Short Term Exposure:* Trichlorophenols irritate the eyes, the skin and the respiratory tract. A central nerveous system depressant. High exposures can cause weakness, difficulty in breathing, tremors, convulsions, coma, and possible death. See also chlorophenols. (*2,3,5-iosmer*) A mixture of trichlorophenols may cause irritation of the skin, eyes and respiratory tract. These substances may cause acute metabolic effects resulting in damage in several organs notably the CNS. Some technical products may contain highly toxic impurities including polychlorinated dibenzo-*p*-dioxins and -furans. (*2,4,5-isomer*)[52]: Irritation of the skin, eyes, nose, and pharynx, redness and edema of the skin, dermatitis, corneal injury, iritis, sweating, thirst, nausea, vomiting, diarrhea, abdominal pain, cyanosis, hyperactivity, stupor, decreased activity and motor weakness, increase followed by decrease in respiratory rate and urinary output, fever, increased bowel action, lung, liver or kidney damage, convulsions, collapse, and coma.

*Long Term Exposure:* Repeated or prolonged contact with skin may cause dermatitis; drying and cracking. May affect the liver and kidneys. A related chemical, *phenol*, can cause liver and kidney damage. May be carcinogenic to humans. If any of the trichlorophenols is contaminated with 2,3,7,8-tetrachlorodibenzo-*p*-dioxin, the following effects may occur: acne-like skin rash, liver damage, nervous system damage with symptoms of weakness, pain in the legs, and numbness. Chlorophenols leave the body quickly, so they are not likely to accumulate in the mother's tissues or breast milk. There are no human studies on the effects of chlorophenols on developing fetuses. Studies in rats showed that chlorophenols can pass through the placenta and produce toxic effects to the developing fetuses. The most common problems in children are delayed hardening of the bones of the breastbone, spine, and skull.

*Points of Attack:* Cancer site in animals (*2,4,6-isomer*): liver and leukemia.

*Medical Surveillance:* Liver and kidney function tests. Complete blood count.

*First Aid: Skin Contact:* Flood all areas of body that have contacted the substance with water. Don't wait to remove contaminated clothing; do it under the water stream. Use soap to help assure removal. Isolate contaminated clothing when removed to prevent contact by others. *Eye Contact:* Remove any contact lenses at once. Immediately flush eyes well with copious quantities of water or normal saline for at least 20 to 30 minutes. Seek medical attention. *Inhalation:* Leave contaminated area immediately; breathe fresh air.

Proper respiratory protection must be supplied to any rescuers. If coughing, difficult breathing or any other symptoms develop, seek medical attention at once, even if symptoms develop many hours after exposure. *Ingestion:* If unconscious or convulsing, do not induce vomiting or give anything by mouth. Assure that victim's airway is open and lay him on his side with his head lower than his body and transport at once to a medical facility. If conscious and not convulsing, give a slurry of activated charcoal in water. If medical advice is not readily available, do not induce vomiting, and rush the victim to the nearest medical facility.

*References:*
- New Jersey Department of Health and Senior Services, "Hazardous Substance Fact Sheet, 2,4,6-Trichlorophenol," Trenton NJ (June 1988, rev. August 2002). http://www.state.nj.us/health/eoh/rtkweb/1894.pdf
- U.S. Environmental Protection Agency, "Chlorinated Phenols: Ambient Water Quality Criteria," Washington, DC (1980).
- U.S. Environmental Protection Agency, "2,4,6-Trichlorophenol, Health and Environmental Effects," Profile No. 168, Washington, DC, Office of Solid Waste (April 30, 1980).
- Sax, N.I., Ed., "Dangerous Properties of Industrial Materials Report," 3, No. 6, 79-81 (1983) and 5, No. 1, 87-99 (1985) (2,4,5-Isomer) and 4, No. 5, 46-58 (1984) (2,4,6-Isomer).
- California Environmental Protection Agency "Chemical List of Lists," Sacramento CA (February 1997).
- Agency for Toxic Substances and Disease Registry "Toxicological Profile for Chlorophenols," 1999, Atlanta, GA.

# 2,4,5-Trichlorophenoxyacetic Acid, Esters

*Use Type:* Herbicide, defoliant, plant growth stimulant
*CAS Number:* 93-76-5; 93-78-7 (isopropyl ester); 93-79-8 (butyl ester); 1928-47-8 (ethylhexyl ester); 2545-59-7 (butoxyethyl ester); 25168-15-4 (isooctyl ester); 61792-07-2 (propyl ester)
*Formula:* $C_8H_5Cl_3O_3$; $C_{11}H_{11}Cl_3O_3$ (isopropyl ester); $C_{12}H_{13}Cl_3O_3$ (butyl ester); $C_{14}H_{17}Cl_3O_4$ (butoxyethyl ester); $C_{16}H_{21}Cl_3O_3$ (isooctyl ester)
*Alert:* 2,4,5-T, when combined with 2,4-D, is the chemical known as *Agent Orange* which received notoriety after it was extensively used as a defoliant in the Vietnam war. It is a plant growth stimulant, closely related to a number of other herbicides such as 2,4-D and MCPA, which cannot be metabolized by plants. When applied in high concentrations they cause uncontrollable and grossly distorted growth.
*Synonyms: 93-76-5:* Phenoxyacetic acid, 2,4,5-trichloro-; 2,4,5-Trichlorophenoxyacetic acid
*93-78-7:* 2,4,5-Trichlorophenoxyacetic acid, isopropyl ester; Isopropyl 2,4,5-trichlorophenoxyacetate; Acetic acid, (2,4,5-trichlorophenoxy)-, isopropyl ester; 2,4,5-T, isopropyl ester
*93-79-8:* Acetic acid, (2,4,5-T)-, butyl ester; Acetic acid, (2,4,5-trichlorophenoxy)-, butyl ester; Butyl-2,4,5-T; Butylate-2,4,5-T; Butyl 2,4,5-trichlorophenoxyacetate; n-Butyl(2,4,5-trichlorophenoxy)acetate; 2,4,5-T-n-Butyl ester; 2,4,5-Trichlorophenoxyacetic acid, butyl ester
*1928-47-8:* Acetic acid, (2,4,5-trichlorophenoxy)-, 2-ethylhexyl ester; Ethylhexyl-2,4,5-T; 2,4,5-T Ethylhexyl ester; 2-Ethylhexyl 2,4,5-trichlorophenoxyacetate; 2,4,5-Trichlorophenoxyacetic acid, 2-ethylhexyl ester
*2545-59-7:* Acetic acid, (2,4,5-trichlorophenoxy)-, 2-butoxyethyl ester; Butoxyethyl 2,4,5-T; Butoxyethyl 2,4,5-trichlorophenoxyacetate; Butoxyethanol 2,4,5-trichlorophenoxyacetate; 2,4,5-T Butoxyethyl ester; 2,4,5-Trichlorophenoxyacetic acid, butoxyethanol ester (25168-15-4) Acetic acid, (2,4,5-trichlorophenoxy)-, isooctyl ester; 2,4,5-Isooctyl ester; Isooctyl 2,4,5-trichlorophenoxyacetate; 2,4,5-Trichlorophenoxyacetic acid, isooctyl ester
*61792-07-2:* Acetic acid, (2,4,5-trichlorophenoxy)-, 1-methyl propyl ester
*Trade Names:* ARBORICID® (butyl ester); BCF-BUSHKILLER®; BLADEX-H® (butoxyethyl ester); BRUSH RHAP®; BRUSH-OFF® 445 mild volatile brush killer; BRUSHTOX®; DACAMINE®; DEBROUSSAILLANT® CONCENTRE (French); DEBROUSSAILLANT® SUPER-CONCENTRE (French); DECAMINE-4T®; DED-WEED® LV-6 BRUSH KIL; DED-WEED® T-5 BRUSH KIL; DED-WEED® BRUSH KILLER; DINOXOL®; ENVERT-T®; ESTERCIDE® T-2; ESTERCIDE® T-245; ESTERON® 245 BE; ESTERON® BRUSH KILLER; FARMCO FENCE RIDER®; FLOMORE® (butyl ester); FORRON®; FORST® U 46; FORTEX®; FRUITONE®-A; HORMOSLYR® 500T (butoxyethyl ester); INVERTON® 245; KILEX®-3 (butyl ester); KRZEWOTOKS® (butyl ester); LINE RIDER®; LO VOL®; PHORTOX®; REDDON®; REDDOX®; SPONTOX®; SUPER D WEEDONE®; TIPPON®; TORMONA®; TORMONA® (butyl ester); TRANSAMINE®; TRIBUTON®; TRINOXOL®; TRIOXON®; TRIOXONE®; TRIOXONE® (butyl ester); U 46®; U 46T® (isooctyl ester); U46KW® (butyl ester); VEON® 245; VERTON® 2T; VISKO RHAP® low volatile ester; WEEDAR®; WEEDONE®
*Producers:* Agriliance (USA); Dow Chemical (USA); Bayer CropScience (Germany); Occidental Chemical (USA);
*Chemical Class:* Chlorophenoxy
*EPA/OPP PC Code:* 082001; 082056 (butyl ester); 082066 (isopropyl ester); 082063 (isooctyl ester); 082053 (butoxy ethanol ester) (082072 use code No. 082053)
*California DPR Chemical Code:* 639
*ICSC Number:* 0075
*RTECS Number:* AJ8400000; AJ8485000 (butyl ester)

## 2,4,5-Trichlorophenoxyacetic Acid, Esters

*EEC Number:* 607-041-00-9

*Uses:* Use of 2,4,5-T in the United States has been canceled since 1985. Some or all applications may be classified by the U.S. EPA as Restricted Use Pesticides (RUP). Used as a herbicide around industrial and residential sites, lumber yards, and vacant lots, range land, aquatic sites, turf, and food crops. Has been used as a growth regulator to increase size of citrus fruits and reduce drop of deciduous fruit.

*Human toxicity (long-term)*[77]: (CAS: 93-76-5) Low–70.00 ppb, Health Advisory

*Fish toxicity (threshold)*[77]: (CAS: 93-76-5) Very low–79154.36150 ppb, MATC (Maximum Acceptable Toxicant Concentration)

*Carcinogen/Hazard Classifications*

*California Prop. 65:* Carcinogen

**IARC:** Group 2B, possible carcinogen

**Endocrine Disruptor:** Known ED

*Regulatory Authority:*
- AB 2588-Air Toxics "Hot Spots" Chemicals (CAL) as chlorophenoxy pesticides
- The "Director's List" (CAL/OSHA) as chlorophenoxy pesticides
- Clean Water Act: Section 311 Hazardous Substances/RQ (same as CERCLA)
- EPA Hazardous Waste Number (RCRA No.): F027; U232
- EPCRA Section 304 RQ: CERCLA, 1000 lb (454 kg)

*Description:* Colorless to tan crystalline solid. Odorless. Slightly soluble in water; solubility = 265 mg/L @ 25°C. Melting/Freezing point = 154–158°C. Vapor pressure = < 1 x $10^{-7}$ mm.Hg; < 0.01 mPa @ 20°C. Log $K_{ow}$ = 3.95. Values above 3.0 are likely to bioaccumulate in marine organisms.

*Incompatibilities:* The aqueous solution is a weak acid. Incompatible with sulfuric acid, bases, ammonia, aliphatic amines, alkanolamines, isocyanates, alkylene oxides, epichlorohydrin; oxidizers (chlorates, nitrates, peroxides, permanganates, perchlorates, chlorine, bromine, fluorine, etc). May not be compatible with nitrates. Moisture may cause hydrolysis or other forms of decomposition. When heated to decomposition, this material forms toxic fumes of chlorine.

*Permissible Exposure Limits in Air:* The OSHA/NIOSH TWA (58), and the ACGIH TWA value is 10 mg/m³. The NIOSH[2] IDLH level = 250 mg/m³.

*Determination in Air:* Filter; Methanol; High-pressure liquid chromatography/Ultraviolet detection; NIOSH IV[18], Method #5001.

*Permissible Concentration in Water:* No criteria set. Runoff from spills or fire control may cause water pollution. The U.S. EPA (see reference below) has set a lifetime health advisory of 21.0 μg/L.

*Determination in Water:* Liquid extraction and gas chromatography (See EPA reference below)

*Routes of Entry:* Inhalation, ingestion, skin and/or eye contact. Absorbed through the skin.

*Harmful Effects and Symptoms*

*Short Term Exposure:* Poisonous; may be fatal if inhaled, swallowed, or absorbed through skin. Severely irritates eyes, skin and respiratory tract, with burning sensation, pain, redness and swelling. Metabolic stimulant. If inhaled, causes coughing, dilated pupils, headache, profuse persperation, intense thirst, extreme fatigue, rapid pulse, high fever, clammy, flushed skin, rapid breathing, nausea, vomiting, cyanosis (bluish tint to skin and lips), anxiety and confusion, convulsions, risk of lung edema. If swallowed, face and lips turn bluish. Liver injury with associated jaundice, kidney failure, and cardiac arrhythmias are commonly noted. Nerve damage, which may be delayed, may include swelling of legs and feet, muscle twitch and stupor. Severe exposure can cause death from heart failure. Dust or liquid left in contact with the skin for several hours may be absorbed. This may result in severe delayed symptoms as listed above. These symptoms may last for months or years. *Skin:* Reddening and itching may develop. absorption is slow, but may contribute significantly to total exposure. *Eyes:* Irritation may develop. *Ingestion:* 350 mg (0.01 ounce) produces only a metallic taste in the mouth, lasting about 2 hours. Approximate 4 teaspoonfuls (150 lb man) may cause weakness, tiredness, loss of appetite, diarrhea, heart problems, heart failure, and death. *Note:* Reported effects of 2,4,5-T are due to accidental exposures, often at unknown levels or duration. In addition, 2,4,5-T may be contaminated with very small amounts of another more toxic compound 2,3,7,8-Tetrachlorodibenzo-*p*-dioxin (TCDD, Dioxin)*. Therefore, some of the symptoms off exposure to 2,4,5,-T may be due to contaminants. * Said to be the most toxic, chlorine-containing compound.

*Long Term Exposure:* Levels above the standard may produce skin irritation, acne-like skin sores, loss of skin coloration in small patches, GI tract ulcer, and nerve disorders resulting in difficulty controlling muscles. Animal studies also indicate the possibility of an increased susceptibility to infection. Changes in generic material and birth defects have been reported in laboratory studies and may be due to 2,4,5-T or its contaminant. Whether these effects are produced in humans is unknown. In animals: ataxia, skin irritation, acne-like rash, liver damage.

*Points of Attack:* If symptoms develop or overexposure is suspected, liver or kidney function tests may be useful. Liver function tests. Test gastrointestinal tract.

*Medical Surveillance:* If symptoms develop or overexposure is suspected, liver or kidney function tests may be useful. Liver function tests.

*First Aid:* If this chemical gets into the eyes, remove any contact lenses at once and irrigate immediately for at least 15 minutes, occasionally lifting upper and lower lids. Seek medical attention immediately. If this chemical contacts the skin, remove contaminated clothing and wash immediately with soap and water. Seek medical attention immediately. If this chemical has been inhaled, remove from exposure, begin rescue breathing (using universal precautions) if breathing has stopped, and CPR if heart action has stopped. Transfer promptly to a medical facility. When this chemical has been swallowed, get medical attention. *Do not induce vomiting when formulations containing petroleum solvents are ingested.* Otherwise, give large quantities of water and induce vomiting. Do not make an unconscious person vomit. *Note to Physician:* If ingested, remove by lavage or vomiting. Use general supportive measures for CNS depression. Consider the use of quinidine for myotonia.

*References:*
- California Environmental Protection Agency "Chemical List of Lists," Sacramento CA (February 1997)

# Triclocarban

*Use Type:* Fungicide and antiseptic
*CAS Number:* 101-20-2
*Formula:* $C_{13}H_9Cl_3N_2O$
*Synonyms:* N-(4-Chlorophenyl)-N'-(3,4-dichlorophenyl)urea; N-(3,4-Dichlorophenyl)-N'-(4-chlorophenyl)urea; ENT 26,925; NSC-72005; TCC; Trichlorocarbanilide; 3,4,4'-Trichlorocarbanilide; 3,4,4'-Trichlorodiphenylurea; Urea, N-(4-chlorophenyl)-N'-(3,4-dichlorophenyl)-(9CI)
*Trade Names:* CUSITER®; CUTISAN®; GENOFACE®; NOBACTER®; PROCUTENE®; SOLUBACTER®
*Chemical Class:* Anilide
*EPA/OPP PC Code:* 027901
*California DPR Chemical Code:* 844
*Uses:* Not registered in the U.S. Triclocarban is used as an anti-bacterial and anti-fungal agent in disinfectants, shampoos, shower cream, cosmetics, soaps, and other household products. It is used to kill bacteria and many other diseases.
*Regulatory Authority:*
- Actively registered pesticide in California.

*Description:* White crystals or powder. Molecular weight = 315.60. Melting point = 250-256°C.
*Incompatibilities:* When heated to decomposition this material forms toxic oxides of nitrogen and fumes of chlorine.
*First Aid:* If this chemical gets into the eyes, remove any contact lenses at once and irrigate immediately for at least 15 minutes, occasionally lifting upper and lower lids. Seek medical attention immediately. If this chemical contacts the skin, remove contaminated clothing and wash immediately with soap and water. Seek medical attention immediately. If this chemical has been inhaled, remove from exposure, begin rescue breathing (using universal precautions) if breathing has stopped, and CPR if heart action has stopped. Transfer promptly to a medical facility. When this chemical has been swallowed, get medical attention. Give large quantities of water and induce vomiting. Do not make an unconscious person vomit.

*References:*
- California Environmental Protection Agency "Chemical List of Lists," Sacramento CA (February 1997)

# Triclopyr (ANSI)

*Use Type:* Herbicide
*CAS Number:* 55335-06-3
*Formula:* $C_7H_4Cl_3NO_3$
*Alert:* Some products may be classified as Restricted Use Pesticides (RUP)
*Synonyms:* Acetic acid, [(3,5,6-trichloro-2-pyridinyl)oxy]-; Triclopyr, triethylamine salt; 3,5,6-Trichloro-2-pyridinyloxyacetic acid
*Trade Names:* [*Note:* See the following record, Triclopry, triethylamine salt, for trade names containing the salt of triclopry]. ACCESS®; CROSSBOW®, (triclopyr + 2,4-D ester), Dow AgroSciences (USA);ET®; GARLON®, Dow AgroSciences (USA); GRAZON®; PATHFINDER®; REDEEM®, Dow AgroSciences (USA); RELY®; REMEDY®; RIVERDALE TAHOE® Nufarm (Australia); TURFLON®, Dow AgroSciences (USA)
*Producers:* Dow AgroSciences (USA); Gharda Chemicals (India); Micro-Flo (USA); Nufarm (Australia)
*Chemical Class:* Chloropyridinyl; pyridine compound
*EPA/OPP PC Code:* 116001
*Uses:* Triclopry is a systemic herbicide used on rice, range land and pasture, rights-of-way, forestry and grasslands, including home lawns, for control of broadleaf weeds and woody plants. Triclopry is usually available as a triclopyr butoxyethyl ester (BEE) or as a triclopry triethylamine salt (TEA).
*Human toxicity (long-term)[77]:* Very low–350.00 ppb, Health Advisory
*Fish toxicity (threshold)[77]:* Very low–23713.98000 ppb, MATC (Maximum Acceptable Toxicant Concentration)
*Carcinogen/Hazard Classifications*
**U.S. EPA Carcinogens:** Group D, unclassifiable, ambiguous data
**Label Signal Word:** CAUTION, DANGER
**WHO Acute Hazard:** Class III, slightly hazardous
*Regulatory Authority:*
- Actively registered pesticide in California.

*Description:* White to colorless, feathery solid. Molecular weight = 256.48. Soluble in water; solubility = 440 mg/L @ 24°C. Melting/Freezing point = 148–150°C. Boiling point = decomposes @ 290°C. Vapor pressure = $1.26 \times 10^{-6}$ mmHg @ 25°C.
*Permissible Concentration in Water:* No criteria set. Runoff from spills or fire control may cause water pollution.
*First Aid:* If this chemical gets into the eyes, remove any contact lenses at once and irrigate immediately for at least

15 minutes, occasionally lifting upper and lower lids. Seek medical attention immediately. If this chemical contacts the skin, remove contaminated clothing and wash immediately with soap and water. Seek medical attention immediately. If this chemical has been inhaled, remove from exposure, begin rescue breathing (using universal precautions) if breathing has stopped, and CPR if heart action has stopped. Transfer promptly to a medical facility. When this chemical has been swallowed, get medical attention. Give large quantities of water and induce vomiting. Do not make an unconscious person vomit. Do not induce vomiting when formulations containing petroleum solvents are ingested.

*References:*
- U.S. Environmental Protection Agency, "Reregistration Eligibility Decision (RED), Triclopry," Office of Prevention, Pesticides and Toxic Substances, Washington, DC (October 1998). http://www.epa.gov/REDs/2710red.pdf
- EXTOXNET, Extension Toxicology Network, "Pesticide Information Profile, Triclopry," Oregon State University, Corvallis, OR (June 1996). http://extoxnet.orst.edu/pips/triclopy.htm
- California Environmental Protection Agency "Chemical List of Lists," Sacramento CA (February 1997)

## Triclopyr, triethylammonium salt

*Use Type:* Herbicide
*CAS Number:* 57213-69-1
*Formula:* $C_7H_4Cl_3NO_3$
*Synonyms:* Acetic acid, [(3,5,6-trichloro-2-pyridinyl)oxy]-, compounded with *N,N*-diethylethanamine (1:1); Caswell No. 882J; *N,N*-Diethyletha namine compounded with [(3,5,6-trichloro-2-pyridinyl)oxy]acetic acid (1:1); (3,5,6-Trichloro-2-pyridinyl)oxyacetic acid, triethylamine salt; 3,5,6-Trichloro-2-pyridinyloxyacetic acid, TEA salt; [(3,5,6-Trichloro-2-pyridinyl)oxy]acetic acid compounded with *N,N*-diethylethanamine (1:1); [(3,5,6-Trichloro-2-pyridyl)oxy]acetic acid, compound with triethylamine (1:1); Triclopyr triethylamine; Triethylammonium triclopyr; Triethylamine triclopyr; TTEA
*Trade Names:* BRUSH-B-GON®, Ortho Business Group (USA); CONFRONT®, Dow AgroSciences (USA); CROSSBOW®, Dow AgroSciences (USA); DOWELANCO® BRUSH AND WEED, Dow AgroSciences (USA); DTDA/DMA-TEA-DMA® SELECTIVE HERBICIDE, Nufarm (Australia); GARLON-3A®, Dow AgroSciences (USA); GRANDSTAND®, Dow AgroSciences (USA); MON® 78736, Monsanto (USA); REDEEM® R & P, Dow AgroSciences (USA); RENOVATE®, Dow AgroSciences (USA); RIVERDALE DTDA® SELECTIVE HERBICIDE, Nufarm (Australia); RIVERDALE HORSEPOWER®, Nufarm (Australia); TRICLOPRY-EZ-JECT®; TURFLON® AMINE, Dow AgroSciences (USA); XRM-5202®, Dow AgroSciences (USA); WEEDEX®
*Producers:* Bayer CropScience (Germany); Bonide Products (USA); Dow AgroSciences (USA); Lebanon Seaboard (USA); Monsanto (USA); Nufarm (Australia); Ortho Business Group (USA)
*Chemical Class:* Chloropyridinyl
*EPA/OPP PC Code:* 116002
*California DPR Chemical Code:* 2131
*RTECS Number:* AJ8974000
*Uses:* A systemic herbicide used on rice, range land and pasture, rights-of-way, forestry and grasslands, including home lawns, for control of broadleaf weeds and woody plants.
*Carcinogen/Hazard Classifications*
**U.S. EPA Carcinogens:** Group D, unclassifiable, ambiguous data
**Label Signal Word:** CAUTION, WARNING or DANGER
**WHO Acute Hazard:** Class III, slightly hazardous as triclopry, the parent chemical
*Regulatory Authority:*
- Actively registered pesticide in California.
- EPCRA Section 313 Form R *de minimis* concentration reporting level: 1.0%

*Permissible Concentration in Water:* No criteria set. Runoff from spills or fire control may cause water pollution.
*First Aid:* If this chemical gets into the eyes, remove any contact lenses at once and irrigate immediately for at least 15 minutes, occasionally lifting upper and lower lids. Seek medical attention immediately. If this chemical contacts the skin, remove contaminated clothing and wash immediately with soap and water. Seek medical attention immediately. If this chemical has been inhaled, remove from exposure, begin rescue breathing (using universal precautions) if breathing has stopped, and CPR if heart action has stopped. Transfer promptly to a medical facility. When this chemical has been swallowed, get medical attention. Give large quantities of water and induce vomiting. Do not make an unconscious person vomit. Do not induce vomiting when formulations containing petroleum solvents are ingested.
*References:*
- California Environmental Protection Agency "Chemical List of Lists," Sacramento CA (February 1997)

## Triclosan

*Use Type:* Microbiocide, pesticide
*CAS Number:* 3380-34-5
*Formula:* $C_{12}H_7Cl_3O_2$
*Synonyms:* 5-Chloro-2-(2,4-dichlorophenoxy)phenol; 5-Chloro-2-(2,4-dichlorophenoxyphenol); 2'-Hydroxy-2,4,4'-trichloro-phenylether; Phenol, 5-chloro-2-(2,4-dichlorophenoxy)-; TCC; 2,4,4'-Trichloro-2'-hydroxy diphenyl ether

*Trade Names:* BAC-TEX; CH-3565®; ITGAGUARD®, Ciba Specialty Chemicals (Switzerland); IRGASAN®, Ciba Specialty Chemicals (Switzerland); IRGASAN® DP-300, Ciba Specialty Chemicals (Switzerland); LEXOL® 300; MICROBAN® ADDITIVE; SANITIZED® BRAND; TRICLOSAN®; ULTRA FRESH®; VIKOL®; VINYZENE®, Rohm and Haas (USA)

*Producers:* Ciba Specialty Chemicals (Switzerland); Novartis (Switzerland) (now Syngenta); Rohm and Haas (USA); Schering-Plough Health Care Products (USA); Syngenta (Switzerland)

*Chemical Class:* Chlorophenol

*EPA/OPP PC Code:* 054901

*California DPR Chemical Code:* 1371

*Uses:* Triclosan is a broad-spectrum antibacterial and antimicrobial agent used to make creams, lotions, deodorants, detergents, dish soaps, laundry soaps, cosmetics, toothpastes, acne creams and mouthwashes.

*Carcinogen/Hazard Classifications*

**U.S. EPA Carcinogens:** Unclassifiable

**IARC:** Group 2B, possible carcinogen

**Label Signal Word:** CAUTION, WARNING or DANGER

*Regulatory Authority:*
- AB 2588-Air Toxics "Hot Spots" Chemicals (CAL) as chlorophenoxy pesticides
- The "Director's List" (CAL/OSHA) as chlorophenoxy pesticides
- Actively registered pesticide in California.

*Description:* Off-white, crystalline solid or powder. Insoluble in water. Molecular weight = 289.52. Melting/Freezing point = 55°C.

*Incompatibilities:* Reacts with boranes, alkalies, aliphatic amines, amides, nitric acid, sulfuric acid.

*Permissible Concentration in Water:* No criteria set. Runoff from spills or fire control may cause water pollution.

*Routes of Entry:* Inhalation, ingestion, absorbed through the skin.

*Harmful Effects and Symptoms*

*Short Term Exposure:* Poisonous; may be fatal if inhaled, swallowed, or absorbed through skin. Severely irritates eyes, skin and respiratory tract, with burning sensation, pain, redness and swelling. Metabolic stimulant. If inhaled, causes coughing, dilated pupils, headache, profuse persperation, intense thirst, extreme fatigue, rapid pulse, high fever, clammy, flushed skin, rapid breathing, nausea, vomiting, cyanosis (bluish tint to skin and lips), anxiety and confusion, convulsions, risk of lung edema. If swallowed, face and lips turn bluish. Liver injury with associated jaundice, kidney failure, and cardiac arrhythmias are commonly noted. Nerve damage, which may be delayed, may include swelling of legs and feet, muscle twitch and stupor. Severe exposure can cause death from heart failure. Dust or liquid left in contact with the skin for several hours may be absorbed. This may result in severe delayed symptoms as listed above. These symptoms may last for months or years.

*Long Term Exposure:* Workers exposed to chlorophenoxy compounds such as 2,4-D (in the manufacturing process) over a five to ten year period at levels above 10 mg/m³ complained of weakness, rapid fatigue, headache and vertigo. Liver damage, low blood pressure and slowed heartbeat were also found. Based on animal tests, may affects human reproduction.

*Points of Attack:* Eyes, skin, respiratory system, central nervous system, cardiovascular system, liver, kidney

*Medical Surveillance:* If symptoms develop or overexposure is suspected, liver or kidney function tests may be useful. Liver function tests.

*First Aid:* If this chemical gets into the eyes, remove any contact lenses at once and irrigate immediately for at least 15 minutes, occasionally lifting upper and lower lids. Seek medical attention immediately. If this chemical contacts the skin, remove contaminated clothing and wash immediately with soap and water. Seek medical attention immediately. If this chemical has been inhaled, remove from exposure, begin rescue breathing (using universal precautions) if breathing has stopped, and CPR if heart action has stopped. Transfer promptly to a medical facility. When this chemical has been swallowed, get medical attention. Give large quantities of water and induce vomiting. Do not make an unconscious person drink or vomit. *Note to Physician:* If ingested, remove by lavage or vomiting. Use general supportive measures for CNS depression. Consider the use of quinidine for myotonia.

*References:*
- California Environmental Protection Agency "Chemical List of Lists," Sacramento CA (February 1997)

# Triethanolamine Dodecylbenzene Sulfonate

*Use Type:* Fungicide, insecticide, microbiocide and adjuvant

*CAS Number:* 27323-41-7

*Formula:* $C_{18}H_{20}O_3S \cdot C_6H_{15}NO_3$

*Synonyms:* AI3-26730-X; Benzenesulfonic acid, dodecyl-, compounded with 2,2',2"-nitrilotris[ethanol](1:1); Dodecylbenzenesulfonic acid, triethanolamine salt; Dodecylbenzenesulfonic acid, compounded with 2,2',2"-nitrilotris(ethanol) (1:1); Dodecylbenzenesulfonic acid triethanolamine salt; 2-2',2"-Nitrilotris-dodecylbenzene sulfonate (Salt); Triethanolamine DBS; Triethanolamine dodecylbenzene sulfonate

*Trade Names:* WITCONATE®-60L; WITCONATE®-60T; WITCONATE®-79S; WITCONATE®-5725; WITCONATE® S-1280; WITCONATE® TAB

*Chemical Class:* Soap

*EPA/OPP PC Code:* 079020

*California DPR Chemical Code:* 984

*RTECS Number:* DB6700000

*Uses:* Not registered in the U.S. Widely used as a surfactant.
*Carcinogen/Hazard Classifications*
**Label Signal Word:** CAUTION, WARNING or DANGER
*Regulatory Authority:*
- The "Director's List" (CAL/OSHA)
- Actively registered pesticide in California.
- Clean Water Act: Section 311 Hazardous Substances/RQ (same as CERCLA)
- EPCRA Section 304 RQ: CERCLA, 1000 lb (454 kg)

*Description:* Yellowish brown or amber liquid. Sinks and mixes with water. Molecular weight = 475.6 (solute). Hazard Identification (based on NFPA-704 M Rating System): Health Hazards (Blue): 1; Flammability (Red): 1; Reactivity (Yellow): 0.

*Incompatibilities:* Acids, acid fumes, oxidizers (chlorates, nitrates, peroxides, permanganates, perchlorates, chlorine, bromine, fluorine, etc).

*Permissible Exposure Limits in Air:* Not established.

*Permissible Concentration in Water:* No criteria set. Runoff from spills or fire control may cause water pollution. Not established.

*Harmful Effects and Symptoms*

*Short Term Exposure:* Contact with eyes or prolonged contact with skin may cause irritation. May cause irritation of mouth and stomach.

*First Aid:* If this chemical gets into the eyes, remove any contact lenses at once and irrigate immediately for at least 15 minutes, occasionally lifting upper and lower lids. Seek medical attention immediately. If this chemical contacts the skin, remove contaminated clothing and wash immediately with soap and water. Seek medical attention immediately. If this chemical has been inhaled, remove from exposure, begin rescue breathing (using universal precautions) if breathing has stopped, and CPR if heart action has stopped. Transfer promptly to a medical facility. When this chemical has been swallowed, get medical attention. Give large quantities of water and induce vomiting. Do not make an unconscious person vomit.

*References:*
- California Environmental Protection Agency "Chemical List of Lists," Sacramento CA (February 1997)

# Triflumizole

*Use Type:* Fungicide
*CAS Number:* 68694-11-1
*Formula:* $C_{15}H_{15}ClF_3N_3O$
*Synonyms:* 1-[1-((4-Chloro-2-(trifluoromethyl)phenyl)imino)-2-propoxyethyl]-1*H*-imidazole; 1-[*N*-(4-Chloro-2-trifluoromethylphenyl)-propoxyacetimidoyl]-imidazole; *(E)*-4-Chloro-α,α,α-trifluoro-*N*-(1-(1*H*-imidazol-1-yl)-2-propoxyethylidene)-*O*-toluidine; 1*H*-Imidazole, 1-[1-((4-Chloro-2-(trifluoromethyl)phenyl)imino]-2-propoxyethyl)-, *(E)*-

*Trade Names:* A815®, Crompton Corporation (USA); CONDOR®; DUO TOP®; NF-114®; PROCURE®, Crompton (USA); TERRAGUARD®, Crompton Corporation (USA); TRIFMINE®

*Producers:* Crompton Corporation (USA)
*Chemical Class:* Imidazole; conazole fungicide
*EPA/OPP PC Code:* 128879
*California DPR Chemical Code:* 2260
*ICSC Number:* 1252
*RTECS Number:* NI4490000

*Uses:* Triflumizole is used in control of Cylindrocladium root and petiole rot on Spathiphyllum.

*Human toxicity (long-term)*[77]: Very low–105.00 ppb, Health Advisory

*Fish toxicity (threshold)*[77]: Intermediate–47.02080 ppb, MATC (Maximum Acceptable Toxicant Concentration)

*U.S. Maximum Allowable Residue Levels for Triflumizole (40 CFR 180.476):*

| CROP | ppm |
|---|---|
| Apple | 0.5 |
| Apple, pomace, wet & dry | 2.0 |
| Cattle, fat | 0.5 |
| Cattle, meat | 0.05 |
| Cattle, mbyp | 0.5 |
| Cherry, sweet | 1.5 |
| Cherry, tart | 1.5 |
| Egg | 0.05 |
| Filbert | 0.05 |
| Goat, fat | 0.5 |
| Goat, meat | 0.05 |
| Goat, mbyp | 0.5 |
| Grape | 2.5 |
| Grape, pomace, wet & dried | 15.0 |
| Grape, raisin, waste | 10.0 |
| Hog, fat | 0.5 |
| Hog, meat | 0.05 |
| Hog, mbyp | 0.5 |
| Horse, fat | 0.5 |
| Horse, meat | 0.05 |
| Horse, mbyp | 0.5 |
| Milk | 0.05 |
| Pear | 0.5 |
| Poultry, fat | 0.05 |
| Poultry, meat | 0.05 |
| Poultry, mbyp | 0.1 |
| Sheep, fat | 0.5 |
| Sheep, meat | 0.05 |
| Sheep, mbyp | 0.5 |
| Strawberry | 2.0 |
| Vegetable, cucurbit, group 9 | 0.5 |

*Carcinogen/Hazard Classifications*
**U.S. EPA Carcinogens:** Group E, Unlikely carcinogen
**Label Signal Word:** CAUTION
**WHO Acute Hazard:** Class III, slightly hazardous

*Description:* Colorless crystals. Moderately soluble in water; solubility = 1.253 g/100 ml @ 20°C. Molecular weight = 345.7. Melting/Freezing point = 63°C. Vapor pressure = 1.1 x $10^{-8}$ mmHg @ 20°C; 1.4 x $10^{-3}$ Pa at 25°C. Log $K_{ow}$ = 1.43. Unlikely to bioaccumulate in marine organisms.

*Incompatibilities:* Strong oxidizers. Decomposes in fire forming toxic and corrosive oxides of nitrogen, hydrogen fluoride, and hydrogen chloride.

*Permissible Concentration in Water:* No criteria set. Runoff from spills or fire control may cause water pollution.

*Routes of Entry:* Inhalation or ingestion.

*Harmful Effects and Symptoms*

*Long Term Exposure:* Skin sensitization. Liver damage and decreased hemoglobin levels.

*Points of Attack:* Liver and blood.

*Medical Surveillance:* Liver function tests. Hemoglobin level. Complete blood count.

*First Aid:* If this chemical gets into the eyes, remove any contact lenses at once and irrigate immediately for at least 15 minutes, occasionally lifting upper and lower lids. Seek medical attention immediately. If this chemical contacts the skin, remove contaminated clothing and wash immediately with soap and water. Seek medical attention immediately. If this chemical has been inhaled, remove from exposure, begin rescue breathing (using universal precautions) if breathing has stopped, and CPR if heart action has stopped. Transfer promptly to a medical facility. When this chemical has been swallowed, get medical attention. Give large quantities of water and induce vomiting. Do not make an unconscious person vomit.

*References:*
- Pesticide Management Education Program, "Triflumizole (Terraguard, Procure) Pesticide Fact Sheet 10/91," Cornell University, Ithaca, NY (October 24, 1991). http://pmep.cce.cornell.edu/profiles/fung-nemat/tcmtb-ziram/triflumizole/fung-prof-triflumizole.html
- U.S. Environmental Protection Agency, Office of Pesticide Programs, Pesticide Residue Limits, "Triflumizole," 40 CFR 180.476, www.epa.gov/pesticides/food/viewtols.htm
- California Environmental Protection Agency "Chemical List of Lists," Sacramento CA (February 1997)

# Trifluralin (ANSI)

*Use Type:* Herbicide
*CAS Number:* 1582-09-8
*Formula:* $C_{10}H_9F_3N_3O_4$; $C_3H_7N-C_6H_2(NO_2)_2CF_3$
*Synonyms:* Benzenamine, 2,6-dinitro-*N,N*-dipropyl-4-(trifluoromethyl-); Benzeneamine, 2,6-dinitro-*N,N*-dipropyl-4-(trifluoromethylaniline); 2,6-Dinitro-*N,N*-dipropyl-4-(trifluoromethyl)benzenamine; 2,6-Dinitro-*N,N*-di-*N*-propyl-α,α,α-trifluro-*p*-toluidine; 4-(Di-*N*-propylamino)-3,5-dinitro-1-trifluoromethylbenzene; *N,N*-Di-*N*-propyl-2,6-dinitro-4-trifluoromethylaniline; 2,6-Dinitro-4-trifluoromethyl-*N,N*-dipropylanilin (German); *N,N*-Dipropyl-4-trifluoromethyl-2,6-dinitroaniline; Ethane, trifluoro-; 2,6-dinitro-*N,N*-dipropyl-4-(trifluoromethyl)aniline; NCI-C00442; Synfloran; *p*-Toluidine, α,α,α-trifluoro-2,6-dinitro-*N,N*-dipropyl-; α,α,α-Trifluoro-2,6-dinitro-*N,N*-dipropyl-4-toluidine; α,α,α-Trifluoro-2,6-dinitro-*N,N*-dipropyl-*p*-toluidine; Trifluralina (Spanish); Trifluraline

*Trade Names:* AGREFLAN®; AGRIFLAN® 24; ASHLADE TRIMARAN®, Whyte Agrochemicals (UK); AUTUMN KITE®; BROADSTRIKE®, Dow AgroSciences (USA); BUCKLE®, Monsanto (USA); CALLIFORT®, Calliope (France); CAMPBELL'S TRIFLURON®; CANNON HERBICIDE®, Monsanto (USA), canceled 8/31/1994; CHANDOR®; COMMENCE®, FMC Agricultural Products Group (USA); CRISALIN®; DEVRINOL T®; DIGERMIN®; ELANCOLAN®; FLEXLAN®, Dow AgroSciences (USA), canceled 6/15/1987; FLINT®; FLORA®, United Phosphorus (India); FLURENE SE®; FLUTRIX®; FREEDOM®, Monsanto (USA); GORDON'S WEEDER®, Pbi/Gordon (USA); HERBIFLURIN®, Veterinary & Agricultural Products Manufacturing Co., Ltd. (VAPCO) (Jordan); IPERSAN®; JANUS®; L-36352®; LILLY 36,352®; LINNET®; MARKSMAN®; MARKSMAN 2, TRIGARD®; M. T.F®; NITRAN®; OLITREF®; ONSLAUGHT®; PREMERLIN 600 CE®, Milenia Agro Ciencias (Brazil); PROKIL®, Gowan Company (USA), canceled 9/29/1988; SEDAGRI TRIFLURALIN® 480, Bayer CropScience (Germany), canceled 9/4/2002; SINFLOWAN®; SOLO®; SU SEGURO CARPIDOR®; TEAM®; TREFANOCIDE®; TREFICON®; TREFLAN®, Dow AgroSciences (USA); TREFLANOCIDE®; TREFMID 50W®, Dow AgroSciences (USA), canceled 2/21/1986; TRIFARMON®; TRIFLURALINA® 600; TRIFLUREX®; TRIFUREX®; TRIGARD®; TRIKEPIN®; TRILIN®; TRILIN® 10G; TRIM®; TRIMARAN®; TRIPART TRIFLURALIN 48 EC; TRISTAR®; TRUST®; TURFLAN®; URANUS®, Makhteshim-Agan Industries (Israel)

*Producers:* Agan Chemical Manufacturers Ltd. (Israel); Agriliance (USA); Atanor S.A. (Argentina); Bayer CropScience (Germany); Biesterfeld Siemsgluess International. GmbH (Germany); Calliope (France); Dow AgroSciences (USA); Drexel Chemical (USA); FMC Agricultural Products Group (USA); Gowan Company (USA); Makhteshim-Agan Industries (Israel); Milenia Agro Ciencias (Brazil); Monsanto (USA); Montedison (Italy); Pbi/Gordon (USA); Pyosa Agroquimicos (Mexico); Rhone-Poulenc (France); Shenzhen Guomeng Industry Co., Ltd. (China); United Agri Products (UAP); United Phosphorus (India); Veterinary & Agricultural Products Manufacturing Co., Ltd. (VAPCO) (Jordan); Whyte Agrochemicals (UK)

*Chemical Class:* Organofluorine
*EPA/OPP PC Code:* 036101

# Trifluralin

*California DPR Chemical Code:* 597
*ICSC Number:* 0205
*RTECS Number:* XU9275000
*EEC Number:* 609-046-00-1
*EINECS Number:* 216-428-8

*Uses:* Trifluralin is a selective, pre-emergence herbicide used to control many annual grasses and broadleaf weeds in a large variety of tree fruit, nut, vegetable, and grain crops, including soybeans, sunflowers, cotton, and alfalfa. It is used on winter wheat and barley, oilseed rape, carrots, lettuce, sugar beets and beans. It is also used on set-aside land, i.e., arable land temporarily taken out of use. Trifluralin should be incorporated into the soil by mechanical means within 24 hours of application. Granular formulations may be incorporated by overhead irrigation.

*Human toxicity (long-term)*[77]: High–5.00 ppb, Health Advisory

*Fish toxicity (threshold)*[77]: High–1.57645 ppb, MATC (Maximum Acceptable Toxicant Concentration)

**U.S. Maximum Allowable Residue Levels for Trifluralin (40 CFR 180.207):**

| CROP | ppm |
|---|---|
| Alfalfa, hay | 0.2 |
| Asparagus | 0.05 |
| Barley, hay | 0.05 |
| Barley, straw | 0.05 |
| Bean, mung, sprouts | 2 |
| Carrot, roots | 1 |
| Corn, field, forage | 0.05 |
| Corn, field, grain | 0.05 |
| Corn, field, stover | 0.05 |
| Cotton, undelinted seed | 0.05 |
| Cress, upland | 0.05 |
| Flax, seed | 0.05 |
| Fruit, citrus, group 10 | 0.05 |
| Fruit, stone, group 12 | 0.05 |
| Grain, crops, except corn (sweet) and rice (grain) | 0.05 |
| Grape | 0.05 |
| Hop | 0.05 |
| Legume, forage | 0.05 |
| Nut, tree, group 14 | 0.05 |
| Peanut | 0.05 |
| Peppermint, tops | 0.05 |
| Rape seed, seed | 0.05 |
| Safflower, seed | 0.05 |
| Sorghum, forage | 0.05 |
| Sorghum, grain, stover | 0.05 |
| Spearmint, tops | 0.05 |
| Sugarcane, cane | 0.05 |
| Sunflower, seed | 0.05 |
| Vegetable, cucurbit, group 9 | 0.05 |
| Vegetable, fruiting, group 8 | 0.05 |
| Vegetable, leafy | 0.05 |
| Vegetable, root crop, except carrot | 0.05 |
| Vegetables, seed and pod | 0.05 |
| Wheat, grain | 0.05 |
| Wheat, straw | 0.05 |

*Carcinogen/Hazard Classifications*
**U.S. EPA Carcinogens:** Group C, possible carcinogen
**IARC:** Group 3, unclassifiable
**Label Signal Word:** WARNING or CAUTION
**WHO Acute Hazard:** Class U, unlikely to be hazardous
**Endocrine Disruptor:** Suspected endocrine disruptor

*Regulatory Authority:*
- Banned or Severely Restricted (in U.S.) (UN)[13]
- AB 1803-Well Monitoring Chemical (CAL)
- AB 2588-Air Toxics "Hot Spots" Chemicals (CAL)
- CAL Air Resources Board/AB 1807 Toxic Air Contaminants
- The "Director's List" (CAL/OSHA)
- Actively registered pesticide in California.
- Clean Air Act: Hazardous Air Pollutants (Title I, Part A, Section 112)
- Safe Drinking Water Act: Priority List (55 FR 1470)
- EPCRA Section 313 Form R *de minimis* concentration reporting level: 1.0%
- Canada, WHMIS, Ingredients Disclosure List

*Description:* Yellow-orange to orange crystalline solid. Practically insoluble in water; solubility = 3.9 ppm @ 27°C. Molecular weight = 335.31. Boiling point = 139°C @ 4.2 mbar. Melting/Freezing point =49°C. Vapor pressure = 2.11 x $10^{-4}$ mmHg @ 29°C. Hazard Identification (based on NFPA-704 M Rating System): Health 1, Flammability 1, Reactivity 0. Log $K_{ow}$ = 5.1. Values above 3.0 are likely to bioaccumulate in marine organisms.

*Incompatibilities:* Store in temperatures above 4.4°C. Fluorocarbons can react violently with barium, potassium, sodium.

*Permissible Exposure Limits in Air:* The state of Pennsylvania has set a guideline for trifluralin in ambient air[60] of 1,150 $\mu g/m^3$.

*Permissible Concentration in Water:* EPA has set a lifetime health advisory of 2.0 $\mu g/L$. The state of Maine has set a guideline for drinking water of 200.0 $\mu g/L$[61].

*Routes of Entry:* Inhalation, skin and/or eye contact.

*Harmful Effects and Symptoms*

*Short Term Exposure:* Inhalation can cause irritation of the respiratory tract with cough, phlegm, and/or tightness in the chest. The vapor can cause eye and skin irritation. Skin contact can cause irritation and rash which can be exacerbated by sunlight. The majority of reported trifluralin exposure cases were occupational in nature. Other reported symptoms include respiratory involvement, abdominal cramps, nausea, diarrhea, headache, lethargy and parasthesia following dermal or inhalation exposure.

*Long Term Exposure:* May cause skin sensitization. High or repeated exposure may affect the liver and kidneys and/or cause anemia. There is some dispute about the actual carcinogenic effect of trifluralin. NCI[9] reports clear

evidence of carcinogenicity in mice but not in rats. Some authorities feel that dipropyl nitrosamine formed in trifluralin manufacture and contained in the technical material might be the actual culprit and the purified trifluralin might not have this problem.
*Points of Attack:* Skin, eyes, liver, kidneys, and blood.
*Medical Surveillance:* Kidney and liver function tests. Complete blood count.
*First Aid: Skin Contact:* Flood all areas of body that have contacted the substance with water. Don't wait to remove contaminated clothing; do it under the water stream. Use soap to help assure removal. Isolate contaminated clothing when removed to prevent to prevent contact by others. *Eye Contact:* Remove any contact lenses at once. Flush eyes well with copious quantities of water or normal saline for at least 20 to 30 minutes. Seek medical attention. *Inhalation:* Leave contaminated area immediately; breathe fresh air. Proper respiratory protection must be supplied to any rescuers. If coughing, difficult breathing or any other symptoms develop, seek medical attention at once, even if symptoms develop many hours after exposure. *Ingestion:* If convulsions are not present, give a glass or two of water or milk to dilute the substance. Assure that the person's airway is unobstructed and contact a hospital or poison center immediately for advice on whether or not to induce vomiting.
*References:*
- EXTOXNET, Extension Toxicology Network, "Pesticide Information Profile, Trifluralin," Oregon State University, Corvallis, OR (June 1996). http://extoxnet.orst.edu/pips/triflura.htm
- U.S. Environmental Protection Agency, Office of Pesticide Programs, Pesticide Residue Limits, "Trifluralin," 40 CFR 180.207, http://www.epa.gov/pesticides/food/viewtols.htm
- U.S. Environmental Protection Agency, "Reregistration Eligibility Decision (RED), Trifluralin," Office of Prevention, Pesticides and Toxic Substances, Washington, DC (April 1996). http://www.epa.gov/REDs/0179.pdf
- New Jersey Department of Health and Senior Services, "Hazardous Substance Fact Sheet, Trifluralin," Trenton, NJ (March 1998). http://www.state.nj.us/health/eoh/rtkweb/1918.pdf
- Sax, N.I., Ed., "Dangerous Properties of Industrial Materials Report," 1, No. 2, 70-71 (1980).
- Pohanish, R.P., "Rapid Guide to Hazardous Chemicals in the Environment," Van Nostrand Reinhold, New York, NY (1997).
- U.S. Environmental Protection Agency, "Health Advisory: Trifluralin," Washington, DC, Office of Drinking Water (August 1987).
- California Environmental Protection Agency "Chemical List of Lists," Sacramento CA (February 1997).

# Trimethacarb (ANSI)

*Use Type:* Insecticide, molluscicide, dog and cat repellant
*CAS Number:* 12407-86-2
*Formula:* $C_{11}H_{15}NO_2$
*Synonyms:* Carbamic acid, methyl-, mixed 3,4,5- and 2,3,5-triphenylmethyl esters (4:1); Methylcarbamic acid, trimethylphenyl ester; Trimethacarb; 3,4,5-Trimethylphenyl methylcarbamate and 2,3,5-trimethylphenyl methylcarbamate
*Trade Names:* BROOT®, Union Carbide (USA), canceled; LANDRIN® [3,4,5- and 2,3,5-Trimethylphenyl methylcarbamate isomers (4:1)]; OMS 597®; SD 8530® (mixture of 3,4,5- and 2,3,5-Trimethylphenyl methylcarbamate isomers); UC 27867®, Union Carbide (USA), canceled
*Producers:* Drexel Chemical (USA); Union Carbide (USA)
*Chemical Class:* Carbamate
*EPA/OPP PC Code:* 102400
*California DPR Chemical Code:* 2962
*Uses:* Not registered in the U.S. Tolerances were revoked by the U.S. EPA October 15, 2002. A general use pesticide used against the corn rootworm.
*Carcinogen/Hazard Classifications*
**Label Signal Word:** WARNING
*Description:* Off-white granules. Molecular weight = 193.24. Does not melt.
*Incompatibilities:* May form explosive materials with phosphorus pentachloride. Avoid contact with acids and alkalis.
*Permissible Concentration in Water:* No criteria set. Runoff from spills or fire control may cause water pollution.
*Routes of Entry:* Inhalation, skin contact, ingestion.
*Harmful Effects and Symptoms*
*Short Term Exposure:* Eye pupils are small; blurred vision; eye watering; runny nose; cough; shortness of breath; salivation; nausea, stomach cramps, diarrhea, and vomiting; increased blood pressure; profuse sweating; hypermotility; hallucinations; agitation; tingling of the skin; slow heartbeat; convulsions; fluid in lungs; loss of consciousness; incontinence; breathing stops; death. Carbamates inhibit the acetylcholinesterase enzymes and alter the way in which nervous impulses are transmitted. However, within several hours carbamates spontaneously detach from the enzymes.
*Long Term Exposure:* A potent cholinesterase inhibitor; cumulative effect is possible. This chemical may damage the nervous system with repeated exposure, resulting in convulsions, respiratory failure. May cause liver damage.
*Points of Attack:* Respiratory system, lungs, central nervous system, cardiovascular system, skin, eyes, plasma and red blood cell cholinesterase.
*Medical Surveillance:* Medical observation is recommended for 24 to 48 hours after breathing

overexposure, as pulmonary edema may be delayed. As first aid for pulmonary edema, consider administering a corticosteroid spray. Cigarette smoking may exacerbate pulmonary injury and should be discouraged for at least 72 hours following exposure. Before employment and at regular times after that, the following are recommended: Plasma and red blood cell cholinesterase levels (tests for the enzyme poisoned by this chemical). If exposure stops, plasma levels return to normal in 1-2 weeks while red blood cell levels may be reduced for 1-3 months. When acetylcholinesterase enzyme levels are reduced by 25% or more below preemployment levels, risk of poisoning is increased, even if results are in lower ranges of "normal." Reassignment to work not involving carbamate pesticides is recommended until enzyme levels recover. If symptoms develop or overexposure occurs, repeat the above tests as soon as possible and get an exam of the nervous system. Also consider complete blood count. Consider chest x-ray following acute overexposure.

*First Aid: Eyes:* Speed in removing material from skin is of extreme importance. Eye contact can cause dangerous amounts of these chemicals to be quickly absorbed through the mucous membrane into the bloodstream. Immediately and gently flush eyes with plenty of warm or cold water (NO hot water) for at least 15 minutes, occasionally lifting the upper and lower eyelids. Get medical aid immediately. *Skin:* Get medical aid. Skin and/or eye contact can cause dangerous amounts of these chemicals to be absorbed into the bloodstream. Wearing the appropriate PPE equipment and respirator for carbamate pesticides, immediately flush skin with plenty of soap and water for at least 15 minutes while removing contaminated clothing and shoes. Shampoo hair promptly if contaminated; protect eyes. *Ingestion:* Call poison control. Loosen all clothing. Never give anything by mouth to an unconscious person. Get medical aid. *Do NOT induce vomiting.*\* If conscious, alert, and able to swallow, rinse mouth and have victim drink 4 to 8 ounces of water. Check to see if poison control instructs you to use ipecac syrup, otherwise administer slurry of activated charcoal (2 oz in 8 oz of water). If victim is *unconscious or having convulsions,* do nothing except keep victim warm. *\*In some cases you may be specifically instructed by poison control to induce vomiting by way of 2 tablespoons of syrup of ipecac (adult) washed down with a cup of water.* Do NOT give activated charcoal before or with ipecac syrup. *Inhalation:* Get medical aid. Do not contaminate yourself. Wearing the appropriate PPE equipment and respirator for carbamate pesticides, immediately remove the victim from the contaminated area to fresh air. If the victim is not breathing, administer artificial respiration. Do NOT use mouth-to-mouth resuscitation; use bag/mask apparatus. If breathing is difficult, administer oxygen through bag/mask apparatus until medical help arrives. Do not leave victim unattended. *Note to physician or authorized medical personnel.* Administer atropine, 2 mg (1/30 gr) intramuscularly or intravenously as soon as any local or systemic signs or symptoms of an intoxication are noted; repeat the administration of atropine every 3 to 8 minutes until signs of atropinization (mydriasis, dry mouth, rapid pulse, hot and dry skin) occur; initiate treatment in children with 0.05 mg mg/kg of atropine; repeat at 5 to 10 minute intervals. Watch respiration, and remove bronchial secretions if they appear to be obstructing the airway; intubate if necessary. *Medical note:* 2-PAMCI may be contraindicated in the case of some carbamate poisonings.

*References:*
- Pesticide Management Education Program, "Trimethacarb (Broot) Chemical Profile 3/85," Cornell University, Ithaca, NY (March 1985). http://pmep.cce.cornell.edu/profiles/insect-mite/propetamphos-zetacyperm/trimethacarb/insect-prof-trimethacarb.html
- California Environmental Protection Agency "Chemical List of Lists," Sacramento CA (February 1997).

# Triphenyltin compounds

*Use Type:* Fungicide, herbicide, algicides and molluscicide
*CAS Number:* 900-95-8 (*acetate*); 639-58-7 (*chloride*); 76-87-9 (*hydroxide*); 892-20-6 (*hydride*); 379-52-2 (*fluoride*); 2155-70-6 (*methacrylate*)
*Formula:* $C_{18}H_{16}Sn$(*hydride*); $C_{20}H_{18}O_2Sn$ (*acetate*); $C_{18}H_{15}ClSn$ (chloride); $C_{18}H_{16}OSn$ (*hydroxide*); $C_{18}H_{15}FSn$ (*fluoride*)
*Alert:* A Restricted Use Pesticide (RUP), depending on the formulation
*Synonyms: Acetate:* Acetate de triphenyl-etain (French); Aceto di stagno trifenile (Italian); Acetotriphenylstannine; Acetoxy-triphenyl-stannan (German); Acetoxytriphenylstannane; Acetoxytriphenyltin; (Acetyloxy)triphenyl-stannane (9CI); ENT 25,208; Fenolovo acetate; Fentin acetaat (Dutch); Fentin acetat (German); Fentin acetate; Fentine acetate (French); Fintin acetato (Italian); Phentin acetate; Phentinoacetate; Stannane, acetoxytriphenyl- (U.S. EPA); Tin triphenyl acetate; TPTA; TPZA; Trifenyltinacetaat (Dutch); Triphenylaceto stannane; Triphenyltin acetate; Triphenyl-zinnacetat (German)
*Chloride:* Chlorotriphenylstannane; Chlorotriphenyltin; Cloruro de trifenilesta (Spanish); Fentin chloride; NSC 43675; Phenostat-C; Stannane, chlorotriphenyl-; TPTC; Triphenylchlorostannane; Triphenylchlorotin; Triphenyltin chloride (U.S. EPA)
*Hydroxide:* ENT 28,009; Fentin; Fentin hydroxide; Fintine hydroxyde (French); Fintin hydroxid (German); Fintin hydroxyde (Dutch); Fintin idrossido (Italian); Hydroxyde de triphenyl-etain (French); Hydroxytriphenylstannane; Hydroxytriphenyltin; Idrossido di stagno trifenile (Italian); NCI-C00260; NSC 113243; Phenostat-H; Stannane, hydroxytriphenyl-; Stannol, triphenyl-; Tin,

hydroxytriphenyl-; TN IV; TPTH; TPTH technical; TPTOH; Trifenyl-tin-hydroxyde (Dutch); Triphenyltin hydroxide (U.S. EPA); Triphenylstannanol; Triphenylstannium hydroxide; Triphenyltin(IV) hydroxide; Triphenyltin hydroxide organotin fungicide; Triphenyltin oxide; Triphenyl-zinnhydroxid (German)

***Trade Names:*** *Acetate:* BATASAN®; BRESTAN®; GC 6936®; HOE-2824®; LIROMATIN®; LIROSTANOL®; SUZI®; TINESTAN®; TINESTAN 60 WP®; TUBOTIN®

*Chloride:* AI3-25207®; AQUATIN 20 EC®; BRESTANOL®; GC 8993®; GENERAL CHEMICALS 8993®; HOE 2872®; LS 4442®; TINMATE®

*Hydroxide:* AGRI TIN®, Nufarm Ltd. (Australia); AGTROL®, Nufarm Ltd. (Australia); AI3-28009®; BRESTAN H 47.5 WP FUNGICIDE®, Clariant International (Switzerland), canceled 9/17/1994; DOWCO 186®, Dow Chemical (USA), canceled; DUTER®, Harcros Chemicals (USA), canceled 8/8/1985; DU-TER®, Harcros Chemicals (USA); DUTER EXTRA®, Griffin (USA), canceled; DU-TER FUNGICIDE®, Griffin (USA), canceled; DU-TER FUNGICIDE WETTABLE POWDER®, Griffin (USA), canceled; DU-TER PB-47 FUNGICIDE®, Griffin (USA), canceled; DU-TER W-50®, Griffin (USA), canceled; DU-TUR FLOWABLE-30®, Griffin (USA), canceled 9/8/1993; ENABLE®, Nufarm Ltd. (Australia); FLO TIN 4L®; HAITIN®; HAITIN® WP 20 (FENTIN HYDROXIDE 20%); HAITIN® WP 60 (FENTIN HYDROXIDE 60%); IDA, IMC FLO-TIN 4L®, Nufarm Ltd. (Australia), canceled 6/27/1989; K-19®; ORBIT®, Griffin (USA); PHOTON FUNGICIDE®, Bayer CropScience (Germany), canceled 2/27/2003; PRO-TEX®, Griffin (USA), canceled 4/13/1992; SUPER TIN 4L®, Griffin (USA); SUZU H®; TRIPLE-TIN®, Griffin (USA), canceled 9/30/1991; TRIPLE TIN 4L®, Mid America Chemical (USA), canceled 8/20/1983; VANCIDE KS®, canceled 10/12/1983; VITO SPOT FUNGICIDE®, canceled 12/18/1990; WESLEY TRIPLE TIN 4L®, Griffin (USA), canceled 9/30/1991

***Producers:*** ATOFINA (France); Bayer CropScience (Germany); Cerexagri (France); Clariant International (Switzerland); Fluorochem (UK); Griffin (USA); Harcros Chemicals (USA); Hokko Chemical Industry (Japan); Nufarm Ltd. (Australia); Sankyo Organic Chemicals (Japan); TCI America (USA)

***Chemical Class:*** Organotin, heavy metal

***EPA/OPP PC Code:*** 496700 (*acetate*); 496500 (*chloride*); 083601 (*hydroxide)*; 083602 (*fluoride*); 496900 use code No. 083602 (*fluoride*)

***California DPR Chemical Code:*** 599 (*hydroxide*)

***RTECS Number:*** WH6650000 (*acetate*); WH6860000 (*chloride*); WH8575000 (*hydroxide*)

***EINECS Number:*** 211-358-4 (*chloride*); 200-990-6 (*hydroxide*)

***Uses:*** The U.S. EPA terminated the special review of Triphenyltin hydroxide in September, 2000, which put into question the tolerance revocation of the compound in pesticides. Triphenyltin compounds are used as biocides to prevent fouling of boats, to preserve wood, kill mollusks and as anti-fouling agents in paint. It has been used to control blight on potatoes, leaf spots on sugar beets and peanuts, and scab and other diseases on pecans. Also used on carrots. Use of organotin compounds in antifouling paint has been restricted in many countries because of their adverse effects on the oyster industry and aquatic ecosystems.

***Carcinogen/Hazard Classifications***

**U.S. EPA Carcinogens:** Group B2, probable carcinogen (*hydroxide*)

**California Prop. 65:** Carcinogen (*hydroxide*); Developmental toxin (*hydroxide, methacrylate*)

**Label Signal Word:** DANGER

**WHO Acute Hazard:** *Acetate:* Class II, moderately hazardous; *(hydroxide):* Class II, moderately hazardous

**Endocrine Disruptor:** Suspected (*hydroxide*)

***Regulatory Authority:***
- Air Pollutant Standard Set (See text below)
- U.S. DOT Regulated Marine Pollutant (49CFR, Subchapter 172.101, Appendix B), severe pollutant, as organotin pesticide compounds.

*Acetate:*
- Superfund/EPCRA 40CFR355, Appendix B Extremely Hazardous Substances: TPQ = 500/10,000 lb (227/4,540 kg)
- Superfund/EPCRA 40CFR302.4 RQ: EHS, 1 lb (0.454 kg)
- U.S. DOT Regulated Marine Pollutant (49CFR172.101, Appendix B) severe pollutant, as triphenyltin compounds

*Chloride:*
- Superfund/EPCRA 40CFR355, Appendix B Extremely Hazardous Substances: TPQ = 500/10,000 lb (227/4,540 kg)
- Superfund/EPCRA 40CFR302.4 RQ: EHS, 1 lb (0.454 kg)
- EPCRA Section 313 Form R *de minimis* concentration reporting level: 1.0%
- U.S. DOT Regulated Marine Pollutant (49CFR172.101, Appendix B) severe pollutant, as triphenyltin compounds

*Hydroxide:*
- EPCRA Section 313 Form R *de minimis* concentration reporting level: 1.0%
- Proposition 65 chemical (CAL)
- U.S. DOT Regulated Marine Pollutant (49CFR172.101, Appendix B) severe pollutant, as triphenyltin compounds
- Canada, WHMIS, Ingredients Disclosure List

***Description:*** *Triphenyltin acetate*: $(C_6H_5)_3SnOOCCH_3$, is a white solid. Melting/Freezing point = 122°C. Practically insoluble in water. *Triphenyltin chloride:* $(C_6H_5)_3SnCl$ is a colorless to yellow crystalline solid with a characteristic odor. Boiling point = 240° @ 13.5 mmHg. Melting/Freezing point = 106°C. Hazard Identification (based on NFPA-704

M Rating System): Health 3, Flammability 2, Reactivity 0. Insoluble in water. *Triphenyltin hydroxide*: $(C_6H_5)_3SnOH$ is a white crystalline solid or powder. Practically insoluble in water; solubility = 1 ppm @ 20°C. Molecular weight 276.42. Melting/Freezing point =119°C (decomposes). Vapor pressure = $4.3 \times 10^{-7}$ mbar @ 50°C.

*Incompatibilities: Triphenyltin chloride:* violent reaction with strong oxidizers. Keep away from moisture.

*Permissible Exposure Limits in Air:* OSHA/NIOSH[2], and ACGIH[1] TWA is 0.1 mg/m³ (for tin organic compounds) and ACGIH STEL[1] is 0.2 mg/m³. The notation "skin" is added indicating the possibility of cutaneous absorption. The NIOSH[2] IDLH level = 25 mg/m³ (as Sn).

*Determination in Air:* Filter/XAD-2® (tube); Acetic acid/$CH_3CN$; High-pressure liquid chromatography/Graphite furnace atomic absorption spectrometry; NIOSH IV, Method #5504, Organotin compounds.[18]

*Permissible Concentration in Water:* No criteria set, but EPA[32] has suggested a permissible ambient goal of 1.4 μg/L based on health effects.

*Routes of Entry:* Inhalation, skin absorption, ingestion, skin and/or eye contact.

### Harmful Effects and Symptoms

*Short Term Exposure:* These chemicals are strong poisons that can cause neurologic emergencies. Breathing, swallowing, or skin contact with some organotins, such as trimethyltin and triethyltin compounds, can interfere with the way the brain and nervous system work. In severe cases, it can cause death. Toxic and irritating to the eyes, skin, and respiratory system. Dermal exposure may lead to severe skin burns as well as renal failure and possible death in the case of the chloride. Symptom of exposure include headache, vertigo (an illusion of movement), psycho-neurologic disturbance, sore throat, cough, abdominal pain, vomiting, urine retention, paresis, focal anesthesia.

*Long Term Exposure:* Exposure may affect the nervous system causing headache, nausea, vomiting, dizziness, decreased coordination, muscle weakness, and visual changes. Triphenyltin chloride can irritate the lungs; bronchitis may develop. In animals: hemolysis, hepatic necrosis, kidney damage. Tributyltins, have been shown to affect the immune system in animals, but this has not been examined in people. Studies in animals also have shown that some organotins, such as tributyltins can affect the reproductive system. This, also, has not been examined in people.

*Points of Attack:* Kidneys and liver.

*Medical Surveillance:* Kidney and liver function tests. Evaluation of the nervous system. Lung function tests.

*First Aid:* (triphenyltin hydroxide) Speed in removing material from skin and eyes is of extreme importance. *Skin Contact:* Flood all areas of body that have contacted the substance with water. Don't wait to remove contaminated clothing; do it under the water stream. Use soap to help assure removal. Shampoo hair promptly if contaminated. Isolate contaminated clothing when removed to prevent contact by others. *Eye Contact:* Remove any contact lenses at once. Immediately flush eyes well with copious quantities of water or normal saline for at least 20 to 30 minutes. Seek medical attention. *Inhalation:* Leave contaminated area immediately; breathe fresh air. Proper respiratory protection must be supplied to any rescuers. If coughing, difficult breathing or any other symptoms develop, seek medical attention at once, even if symptoms develop many hours after exposure. *Ingestion:* Contact a physician, hospital or poison center at once. If the victim is unconscious or convulsing, do not induce vomiting or give anything by mouth. Assure that his airway is open and lay him on his side with his head lower than his body and transport immediately to a medical facility. If conscious and not convulsing, give a glass of water to dilute the substance. Vomiting should not be induced without a physician's advice.

### References:

- Pesticide Management Education Program, "Chemical Fact Sheet 9/84, Triphenyltin Hydroxide," Cornell University, Ithaca, NY (September 1984). http://pmep.cce.cornell.edu/profiles/fung-nemat/tcmtb-ziram/triphenyltin/fung-prof-tpth.html
- New Jersey Department of Health and Senior Services, "Hazardous Substance Fact Sheet, Triphenyltin Chloride," Trenton, NJ (March 1998). http://www.state.nj.us/health/eoh/rtkweb/1952.pdf
- New Jersey Department of Health and Senior Services, "Hazardous Substance Fact Sheet, Triphenyltin Hydroxide," Trenton, NJ (November 2001). http://www.state.nj.us/health/eoh/rtkweb/1953.pdf
- U.S. Environmental Protection Agency, "Chemical Profile: Acetoxytriphenyl Stannane," Washington, DC, Chemical Emergency Preparedness Program (November 30, 1987).
- U.S. Environmental Protection Agency, "Chemical Profile: Triphenyltin Chloride," Washington, DC, Chemical Emergency Preparedness Program (November 30, 1987).
- Sax, N.I., Ed., "Dangerous Properties of Industrial Materials Report," 2, No. 4, 92-94 (1982).
- California Environmental Protection Agency "Chemical List of Lists," Sacramento CA (February 1997).

# Triflusulfuron-methyl

*Use Type:* Herbicide
*CAS Number:* 126535-15-7
*Formula:* $C_{16}H_{17}F_3N_6O_6S$ (triflusulfuron)
*Synonyms:* Benzoic acid, 2-[((((4-(dimethylamino)-6-(2,2,2-trifluoroethoxy)-1,3,5-triazin-2-yl)amino)carbonyl)amino)sulfonyl]-3-methyl-, methyl ester; Methyl 2-[((((4-

(dimethylamino)-6-(2,2,2-trifluofoethoxy)-1,3,5-triazin-2-yl)amino)carbonyl)amino)sulfonyl]-3-, methylbenzoate
*Trade Names:* DPX-66037®, DEBUT®, DuPont Crop Protection (USA); SAFARI®, DuPont Crop Protection (USA); UPBEET®, DuPont Crop Protection (USA)
*Producers:* DuPont Crop Protection (USA)
*Chemical Class:* Sulfonylurea
*EPA/OPP PC Code:* 129002
*California DPR Chemical Code:* 3875
*Uses:* Registered for use on sugar beets and chicory.
*Human toxicity (long-term)[77]:* Intermediate–16.80 ppb, Health Advisory
*Fish toxicity (threshold)[77]:* Very low–185735.513150 ppb, MATC (Maximum Acceptable Toxicant Concentration)
*U.S. Maximum Allowable Residue Levels for Triflusulfuron methyl (40 CFR 180.492):*
The maximum allowable residue level for the following products is 0.05 ppm: beet sugar, roots and tops); chicory roots.
*Carcinogen/Hazard Classifications*
*U.S. EPA Carcinogens:* Group C, possible carcinogen
*Label Signal Word:* CAUTION
*WHO Acute Hazard:* Class U, unlikely to be hazardous
*Permissible Concentration in Water:* No criteria set. Runoff from spills or fire control may cause water pollution.
*Routes of Entry:* Inhalation, passing through the skin, ingestion.
*Harmful Effects and Symptoms*
*Short Term Exposure:* May cause skin and severe eye irritation. Moderately poisonous if ingested or inhaled. Exposure to a triazine (simazine) has caused acute and subacute dermatitis in the former USSR, characterized by erythema, slight edema, moderate pruritus, and burning lasting 4 to 5 days.
*Long Term Exposure:* May cause lung irritation and damage. May cause skin allergy. Contact with some triazine compounds (such as atrazine) may increase risks for tumors known to be associated with hormonal factors. These have been observed in both animals and human beings, and are consistent with the known effects on the hypothalamic pituitary gonadal axis. Repeated exposure may cause weight loss and reduced red blood cell count. May be mutagenic.
*Points of Attack:* Liver, lungs, skin.
*Medical Surveillance:* Before beginning employment and at regular times after that, for those with frequent or potentially high exposures, the following is recommended: Lung function tests. Consider chest x-ray following acute overexposure. Evaluation by a qualified allergist. Examination of the nervous system.
*First Aid:* If this chemical gets into the eyes, remove any contact lenses at once and irrigate immediately for at least 15 minutes, occasionally lifting upper and lower lids. Seek medical attention immediately. If this chemical contacts the skin, remove contaminated clothing and wash immediately with soap and water. Seek medical attention immediately. If this chemical has been inhaled, remove from exposure, begin rescue breathing (using universal precautions) if breathing has stopped, and CPR if heart action has stopped. Transfer promptly to a medical facility. When this chemical has been swallowed, get medical attention. Give large quantities of water or milk and induce vomiting. Do not make an unconscious person vomit.
*References:*
- U.S. Environmental Protection Agency, Office of Pesticide Programs, Pesticide Residue Limits, "Truflusulfuron methyl," 40 CFR 180.492. http://a257.g.akamaitech.net/7/257/2422/08aug20031600/edocket.access.gpo.gov/cfr_2003/julqtr/40cfr180.492.htm
- California Environmental Protection Agency "Chemical List of Lists," Sacramento CA (February 1997)

## Tris(hydroxymethyl)-nitromethane

*Use Type:* Microbiocide
*CAS Number:* 126-11-4
*Formula:* $C_4H_9NO_5$
*Synonyms:* 2-(Hydroxymethyl)-2-nitro-1,3-propanediol; 2-Hydroxymethyl-2-nitropropane-1,3-diol; Isobutylglycerol, nitro-; Methane, trimethylolnitro-; Nitroisobutylglycerol; Nitrotris(hydroxymethyl)methane; 2-Nitro-2-(hydroxymethyl)-1,3-propanediol; Trihydroxymethylnitromethane; Trimethylolnitromethane; Tris(hydroxymethyl)nitromethane
*Trade Names:* CIMCOOL®; DC & R®; KNOCK OUT®; TRIS NITRO®, Dow Chemical (USA)
*Producers:* Dow Chemical (USA)
*Chemical Class:* Unclassified
*EPA/OPP PC Code:* 083902
*California DPR Chemical Code:* 1105
*Uses:* Used as a biocide in water treatment facilities
*Carcinogen/Hazard Classifications*
*Label Signal Word:* CAUTION
*Regulatory Authority:*
- Actively registered pesticide in California.

*Description:* White crystalline solid. Soluble in water. Molecular weight = 151.16. Boiling point = decomposes. Melting/Freezing point = 214°C (pure); 175°C (decomposes).
*Permissible Exposure Limits in Air:* No OELs established.
*Determination in Air:* Filter; none; Gravimetric; NIOSH IV[18] [Particulates NOR; #0500 (total), #0600 (respirable)]
*Permissible Concentration in Water:* No criteria set. Runoff from spills or fire control may cause water pollution.
*Harmful Effects and Symptoms*
*Short Term Exposure:* Vapor caused eye irritation and redness. Inhalation may cause dizziness, rapid breathing, headache. May be narcotic in high concentrations. Ingestion

is poisonous and may cause gastrointestinal irritation with dizziness, and fatigue.
***Long Term Exposure:*** May cause skin dryness and cracking.
***First Aid:*** If this chemical gets into the eyes, remove any contact lenses at once and irrigate immediately for at least 15 minutes, occasionally lifting upper and lower lids. Seek medical attention immediately. If this chemical contacts the skin, remove contaminated clothing and wash immediately with soap and water. Seek medical attention immediately. If this chemical has been inhaled, remove from exposure, begin rescue breathing (using universal precautions) if breathing has stopped, and CPR if heart action has stopped. Transfer promptly to a medical facility. When this chemical has been swallowed, get medical attention. Give large quantities of water and induce vomiting. Do not make an unconscious person vomit.
***References:***
- California Environmental Protection Agency "Chemical List of Lists," Sacramento CA (February 1997)

# Trisodium Phosphate

***Use Type:*** Fungicide, herbicide, pH adjustment, microbiocide
***CAS Number:*** 7601-54-9
***Formula:*** $O_4P \cdot 3Na$
***Synonyms:*** Disodium dihydrogen pyrophosphate; Fosfato sodico (Spanish); Monosodium dihydrogen phosphate; Monosodium phosphate; MSP; Phosphoric acid, trisodium salt; Sodium acid pyrophosphate; Sodium dihydrogen phosphate; Sodium phosphate; Sodium phosphate, anhydrous; Sodium phosphate, monobasic; Sodium phosphate dibasic; Sodium phosphate, tribasic; Tertiary sodium phosphate; Tribasic sodium phosphate; Trinatriumphosphat (German); Trisodium orthophosphate; TSP
***Trade Names:*** ANTISAL®-4; DRI-TRI®; EMULSIPHOS®; EMULSIPHOS® 440/660; NUTRIFOS® STP; OAKITE®; TROMETE®
***Producers:*** Abaquim (Mexico); Alquimia Mexicana (Mexico); AMRESCO (USA); Chongqing Chuandong Chemical (Group) (China); EM Industries (USA); FMC (USA); Harcros Chemicals (USA); Hindustan Chemical (India); Kemira Chemicals (Finland); Montedison (Italy); Pacific Century (USA); Rhodia (France); Solutia (USA)
***Chemical Class:*** Phosphate
***EPA/OPP PC Code:*** 076406
***California DPR Chemical Code:*** 1579
***ICSC Number:*** 1178
***RTECS Number:*** TC9490000
***EINECS Number:*** 231-509-8
***Uses:*** Used to remove insecticide residues from fruit and inhibiting mold; as an emulsifier in processed cheese; to inhibit potato virus X, and tobacco mosaic virus; pH value adjustment. Also used in photographic developers; removing boiler scale, in water softeners, aluminum polishing, sugar refining, cleaning compounds, and in animal feeds.
***U.S. Maximum Allowable Residue Levels:*** Residues of trisodium phosphate are exempted from the requirement of a tolerance when used as a surfactant, emulsifier, or wetting agent in pesticide formulations applied to growing crops or to raw agricultural commodities after harvest.
***Carcinogen/Hazard Classifications***
**Label Signal Word:** CAUTION
***Regulatory Authority:***
- The "Director's List" (CAL/OSHA)
- Actively registered pesticide in California.

***Description:*** Colorless or white granular powder. Odorless. Soluble in water; solubility = 8.8 g/100 ml; 1 part boiling water; the aqueous solution is strongly alkaline. Molecular weight = 163.9. Density = 1.6. Melting/Freezing point = about 75°C (rapid heating); Log $K_{ow}$ = negative. Unlikely to bioaccumulate in marine organisms.
***Incompatibilities:*** Contact with water forms a caustic solution. Vigorous reaction with strong acids. In the presence of moisture, mild steel, copper, or brass may be corroded by TSP. May react violently with antimony(V) pentafluoride. Incompatible with lead diacetate, magnesium, silver nitrate.
***Permissible Exposure Limits in Air:*** No OELs established.
***Determination in Air:*** Filter; Hydrochloric acid; Titration; NIOSH IV[18], Method #7401, Alkaline Dust. See also OSHA Method ID-121.
***Permissible Concentration in Water:*** No criteria set. Runoff from spills or fire control may cause water pollution. There are no criteria for TSP *per se*. Based on similar materials, EPA guidelines recommended criteria for pH as follows: to protect freshwater aquatic life-pH 6.5 to 9.0; to protect saltwater aquatic life-pH 6.5 to 8.5; and to protect humans' drinking water-pH 5 to 9.
***Routes of Entry:*** Inhalation, ingestion, skin contact.
***Harmful Effects and Symptoms***
***Short Term Exposure:*** Caustic material. Contact can cause skin and severe eye irritation. Prolonged contact can cause burns, eye damage and possible blindness. Dust will irritate respiratory tract. Exposure may cause lung edema. Ingestion will damage gastrointestinal tract.
***Points of Attack:*** Eyes, skin, respiratory system
***Medical Surveillance:*** Skin irritation may develop from repeated exposure to the solid or low concentrations of the liquid. Irritation to the lungs, nose, throat and mouth may occur if exposed to low levels for long periods of time. May cause temporary loss of hair.
***First Aid:*** If this chemical gets into the eyes, remove any contact lenses at once and irrigate immediately for at least 15 minutes, occasionally lifting upper and lower lids. Seek medical attention immediately. If this chemical contacts the

skin, remove contaminated clothing and wash immediately with soap and water. Seek medical attention immediately. If this chemical has been inhaled, remove from exposure, begin rescue breathing (using universal precautions) if breathing has stopped, and CPR if heart action has stopped. Transfer promptly to a medical facility. When this chemical has been swallowed, get medical attention. If victim is conscious and able to swallow, have victim drink 4 to 8 ounces of water. *Do not induce vomiting. Note to physician or authorized medical personnel*: Medical observation is recommended for 24 to 48 hours after breathing overexposure, as pulmonary edema may be delayed. As first aid for pulmonary edema, consider administering a corticosteroid spray. Cigarette smoking may exacerbate pulmonary injury and should be discouraged for at least 72 hours following exposure.

### *References:*
- California Environmental Protection Agency "Chemical List of Lists," Sacramento CA (February 1997).

# U

## Urea

*Use Type:* Fertilizer and fungicide
*CAS Number:* 57-13-6
*Formula:* $CH_4N_2O$; $H_2NCONH_2$
*Synonyms:* Carbamide; Carbamide resin; Carbamimidic acid; Carbonyl diamide; Carbonyldiamine; Isourea; NCI-C02119; Pseudourea
*Trade Names:* PRESPERSION, 75 UREA®; SUPERCEL 3000®; UREAPHIL®; UREOPHIL®; UREVERT®; VARIOFORM II®
*Producers:* Abaquim (Mexico); Achema (Lithuania); Agrium (Canada); Alaska Nitrogen (USA); Borden Chemical (USA); BP Chemicals (UK); Cargill Crop Nutrition (USA); CF Industries (USA); Chambal Fertilisers and Chemicals (India); Chemopetrol (Czech Republic); Chimco (Bulgaria); DSCL (India); DSM (Netherlands); EniChem S.p.A. (Italy); Grande Paroisse (France); Hydro Agri Chemicals (Norway); ICI Group (UK); Incitec Pivot (Australia); Indo Gulf (India); Jilin Chemical (China); Juhua Group Corp. (China); Kemira GrowHow Oy (Finland); Kynoch (South Africa); Messer Group (Germany); Monsanto (USA); Nagarjuna Fertilizers and Chemicals (India); Neste Chemicals (Finland); Nissan Chemicals (Japan); Petroquimica de Venezuela (Pequiven); Potash Corporation of Saskatchewan Inc. (Canada); Prodica (USA); Qatar Fertiliser (Qatar); Saudi Basic Industries Corp. (Saudi Arabia); Southern Petrochemical Industries Corporation Ltd. (SPIC) (India); Terra Nitrogen (USA); Ube Agri-Materials, (Japan); Wesfarmers CSBP Ltd. (Australia)
*Chemical Class:* Inorganic
*EPA/OPP PC Code:* 085702
*California DPR Chemical Code:* 662
*ICSC Number:* 0595
*RTECS Number:* YR6250000
*EINECS Number:* 200-315-5
*Uses:* Used in fertilizers and animal feeds, as a fungicide, in the manufacture of resins and plastics, as a stabilizer in explosives and in medicines, and others. Urea is used to protect against frost and is used in some pesticides as an inert ingredient as a stabilizer, as an inhibitor and as an intensifier for herbicides.
*U.S. Maximum Allowable Residue Levels for Urea (40 CFR 180.920 and 40CFR 180.950):* Urea is exempt from requirements for a tolerance.
*Regulatory Authority:*
- Air Pollutant Standard Set (New York)[60] (former USSR)[35]
- Actively registered pesticide in California.
- Canada, WHMIS, Ingredients Disclosure List

*Description:* White crystalline solid or powder. Soluble in water; solubility = $1.2 \times 10^6$ ppm @ 25°C. Molecular weight = 60.08. Boiling point = (decomposes). Melting/Freezing point = 133°C. Log $K_{ow}$ = negative. Unlikely to bioaccumulate in marine organisms.
*Incompatibilities:* Violent with strong oxidizers, chlorine, permanganates, dichromates, nitrites, inorganic chlorides, chlorites, and perchlorates Contact with hypochlorites can result in the formation of explosive compounds.
*Permissible Exposure Limits in Air:* The state of New York[61] has set a guidelines for urea in ambient air of 0.03 $\mu g/m^3$. The former USSR has set an MAC in ambient air in residential areas of 0.2 $mg/m^3$ on a daily average basis.
*Permissible Concentration in Water:* The former USSR-UNEP/IRPTC project[43] has set an MAC in water bodies used for domestic purposes of 80.0 mg/L.
*Routes of Entry:* Inhalation, ingestion, eye and/or skin contact.
*Harmful Effects and Symptoms*
*Short Term Exposure: Inhalation:* Causes irritation of the respiratory tract. Dust may cause difficult breathing especially if the person has asthma. *Skin:* May cause irritation, burning or stinging. *Eyes:* Causes irritation. *Ingestion:* There have been no reported cases of human toxicity. However, some toxic effects have been seen in sheep with impaired liver function.
*Long Term Exposure:* Prolonged skin contact may cause dermatitis.
*Points of Attack:* Skin
*First Aid:* If this chemical gets into the eyes, remove any contact lenses at once and irrigate immediately for at least 15 minutes, occasionally lifting upper and lower lids. Seek medical attention immediately. If this chemical contacts the skin, remove contaminated clothing and wash immediately with soap and water. Seek medical attention immediately. If this chemical has been inhaled, remove from exposure, begin rescue breathing (using universal precautions) if breathing has stopped, and CPR if heart action has stopped. Transfer promptly to a medical facility. When this chemical has been swallowed, get medical attention. Give large quantities of water and induce vomiting. Do not make an unconscious person vomit.
*References:*
- U.S. Environmental Protection Agency, Office of Pesticide Programs, Pesticide Residue Limits, "Urea," 40 CFR 180.920 and 40CFR 180.950, http://www.epa.gov/pesticides/food/viewtols.htm

- New York State Department of Health, "Chemical Fact Sheet: Urea," Albany, NY, Bureau of Toxic Substance Assessment (April 1986).
- California Environmental Protection Agency "Chemical List of Lists," Sacramento CA (February 1997).

## Validamycin

*Use Type:* Fungicide
*CAS Number:* 37248-47-8
*Formula:* $C_{20}H_{35}NO_{13}$; $C_{20}H_{35}NO_{13} \cdot HCl$ (hydrochloride)
*Synonyms:* [1S(1α,4α,5β, 6α)]-1,5,6-Trideoxy-4-O-β-D-glucopyranosyl-5-(hydromethyl)-1-[(4,5,6-trihydroxy-3-(hydroxymethyl)-2-cyclohexen-1-yl)amino]-D-chiroinositol; NSC 122023; Validamycin A
*Trade Names:* ANTIBIOTIC N-329 B®; SOLACOL®; VALIDACIN®; VALIMON®; VALINOMICIN®
*Producers:* Agrimor International (USA); Epochem Co., (China); Hebei Huafeng Chemical Group (China); Nagarjuna Agrichem (India); Wuhan kernel Bio-pesticide (China)
*Chemical Class:* Carbohydrate
*RTECS Number:* YV9468000
*Uses:* Not registered in the U.S. Validamycin is a non-systemic plant antibiotic that controls plant fungi that attacks roots. It is used against soil borne diseases for the control of Rhizoctonia solani in rice, potatoes, vegetables, and others as well as damping off diseases in vegetable seedlings, cotton, sugar beets, rice and other plants. It is a general use fungicide and also mixed with a variety of pesticides.
*Carcinogen/Hazard Classifications*
**WHO Acute Hazard:** Class U, unlikely to be hazardous
*Regulatory Authority:*
- EPCRA Section 302 Extremely Hazardous Substances: TPQ = 1000/10,000 lb (454/4,540 kg)
- EPCRA Section 304 RQ: EHS, 1 lb (0.454 kg)

*Description:* Colorless powder or crystalline solid. Odorless. Soluble in water. Molecular weight = 497.5. Melting point = 130-135°C.
*Incompatibilities:* Slightly unstable in acidic media.
*Permissible Concentration in Water:* No criteria set. Runoff from spills or fire control may cause water pollution.
*First Aid:* If this chemical gets into the eyes, remove any contact lenses at once and irrigate immediately for at least 15 minutes, occasionally lifting upper and lower lids. Seek medical attention immediately. If this chemical contacts the skin, remove contaminated clothing and wash immediately with soap and water. Seek medical attention immediately. If this chemical has been inhaled, remove from exposure, begin rescue breathing (using universal precautions) if breathing has stopped, and CPR if heart action has stopped. Transfer promptly to a medical facility. When this chemical has been swallowed, get medical attention. Give large quantities of water and induce vomiting. Do not make an unconscious person vomit.
*References:*
- EXTOXNET, Extension Toxicology Network, "Pesticide Information Profile, Validamycin," Oregon State University, Corvallis, OR (September 1995). http://extoxnet.orst.edu/pips/validamy.htm

## Vernolate

*Use Type:* Herbicide
*CAS Number:* 1929-77-7
*Formula:* $C_{10}H_{21}NOS$
*Synonyms:* Carbamic acid, dipropylthio-, S-propyl ester; Carbamothioic acid, dipropyl-, S-propyl ester; Dipropylthiocarbamic acid-S-propyl ester; PPTC; Propyl-N,N-dipropylthiolcarbamate; S-Propyl dipropylthiocarbamate; S-Propyl dipropyl (thiocarbamate); N-Propyl-di-N-propylthiolcarbamate
*Trade Names:* R-1607®; REWARD®, Zeneca Ag Products (USA) (now Syngenta), canceled; SURPASS®, Zeneca Ag Products (USA) (now Syngenta), canceled; SURPASS®-E, Zeneca Ag Products (USA) (now Syngenta), canceled; VERNAM®, Drexel Chemical (USA), canceled; VERNAM®-E, Drexel Chemical (USA), canceled; VERNAM®-G, Drexel Chemical (USA), canceled; VOMZLATE®
*Producers:* Drexel Chemical (USA); Zeneca Ag Products (USA) (now Syngenta)
*Chemical Class:* Thiocarbamate
*EPA/OPP PC Code:* 041404
*California DPR Chemical Code:* 1987
*Uses:* Not registered in the U.S. Vernolate is a selective herbicide used to control a variety of weeds as their seeds germinate. These weeds include annual broadleaf weeds, annual grasses, perennial grasses, nut sedges and seedling Johnson grass. Agricultural crop use sites include soybeans, peanuts, groundnuts, soya beans, maize, tobacco, and sweet potatoes.
*Human toxicity (long-term)[77]:* High–7.00 ppb, Health Advisory
*Fish toxicity (threshold)[77]:* Low–344.93242 ppb, MATC (Maximum Acceptable Toxicant Concentration)
*Carcinogen/Hazard Classifications*
**Label Signal Word:** CAUTION
**WHO Acute Hazard:** Class II, moderately hazardous
*Regulatory Authority:*
- AB 1803-Well Monitoring Chemical (CAL)
- EPA Hazardous Waste Number (RCRA No.): U385

- RCRA 40CFR268.48; 61FR15654, Universal Treatment Standards: Wastewater (mg/L), 0.003; Nonwastewater (mg/kg), 1.4

*Description:* Clear, colorless liquid. Faint, aromatic odor. Soluble in water; solubility = 95 ppm @ 20°C; 107 mg/L @ 25°C. Molecular weight 203.37. Density = 0.953 @ 20°C Boiling point = 149°C @ 30 mmHg; 150° @ 30 mmHg. Vapor pressure = $1.04 \times 10^{-2}$ mmHg @ 25°C. Log $K_{ow}$ = 3.49. Values above 3.0 may be able to bioaccumulate in marine organisms.

*Permissible Concentration in Water:* No criteria set. Runoff from spills or fire control may cause water pollution.

*Harmful Effects and Symptoms*

*Short Term Exposure:* Low levels of toxicity. Concentrated solutions are slightly corrosive to eyes and mucous membranes. Dust inhalation can cause irritation of the respiratory system with sneezing. Eye contact can cause irritation, watering, pain, and inflammation of the eyelids. Skin contact can cause irritation and minor ulceration. Ingestion can cause nausea, vomiting, fever, muscle twitching, seizure, rapid respiration, slow heart beat. Severe exposure may result in death.

*Points of Attack:* Respiratory system, central nervous system, cardiovascular system, skin, eyes.

*Medical Surveillance:* Medical observation is recommended for 24 to 48 hours after breathing overexposure, as pulmonary edema may be delayed. As first aid for pulmonary edema, consider administering a corticosteroid spray. Cigarette smoking may exacerbate pulmonary injury and should be discouraged for at least 72 hours following exposure. Before employment and at regular times after that, the following are recommended: If symptoms develop or overexposure occurs, repeat the above tests as soon as possible and get an exam of the nervous system. Also consider complete blood count. Consider chest x-ray following acute overexposure.

*First Aid:* If this chemical gets into the eyes, remove any contact lenses at once and irrigate immediately for at least 15 minutes, occasionally lifting upper and lower lids. Seek medical attention immediately. If this chemical contacts the skin, remove contaminated clothing and wash immediately with soap and water. Seek medical attention immediately. If this chemical has been inhaled, remove from exposure, begin rescue breathing (using universal precautions) if breathing has stopped, and CPR if heart action has stopped. Transfer promptly to a medical facility. When this chemical has been swallowed, get medical attention. If victim is conscious and able to swallow, have victim drink 4 to 8 ounces of water. Do not induce vomiting. *Note to physician or authorized medical personnel*: Medical observation is recommended for 24 to 48 hours after breathing overexposure, as pulmonary edema may be delayed. As first aid for pulmonary edema, consider administering a corticosteroid spray. Cigarette smoking may exacerbate pulmonary injury and should be discouraged for at least 72 hours following exposure.

*References:*

- U.S. Environmental Protection Agency, "Reregistration Eligibility Decision (RED) Facts, Vernolate," Office of Prevention, Pesticides and Toxic Substances, Washington, DC (March 1999). http://www.epa.gov/REDs/factsheets/2735fact.pdf
- EXTOXNET, Extension Toxicology Network, "Pesticide Information Profile, Vernolate," Oregon State University, Corvallis, OR (September 1995). http://extoxnet.orst.edu/pips/vernolat.htm
- California Environmental Protection Agency "Chemical List of Lists," Sacramento CA (February 1997)

# Vinclozolin

*Use Type:* Fungicide
*CAS Number:* 50471-44-8
*Formula:* $C_{12}H_9Cl_2NO_3$
*Synonyms:* Caswell No. 323C; (RS)-3-(3,5-Dichlorophenyl)-5-ethenyl-5-methyl-2,4-oxazolidinedione; 3-(3,5-Dichlorophenyl)-5-ethenyl-5-methyl-2,4-oxazolidinedione; 3-(3,5-Dichlorophenyl)-5-ethenyl-5-methyl-2,4-oxazolidinedione; (RS)-3-(3,5-Dichlorophenyl)-5-methyl-5-vinyl-1,3-oxazolidine-2,4-dione; *N*-3,5-Dichlorophenyl-5-methyl-5-vinyl-1,3-oxazolidine-2,4-dione; 2,4-Oxazolidinedione, 3-(3,5-dichlorophenyl)-5-ethenyl-5-methyl-; 2,4-Oxazolidinedione, 3-(3,5-dichlorophenyl)-5-methyl-5-vinyl-

*Trade Names:* BAS-352-F®, BASF Agricultural Products (Germany); BAS-35204-F®, BASF Agricultural Products (Germany); CURALAN® BASF Agricultural Products (Germany); FLOTILLA®, BASF Agricultural Products (Germany); FUMITE RONALIN®, BASF Agricultural Products (Germany); MASCOT® contact turf fungicide; ORNALIN®, canceled; POWERDRIVE®; RONILAN®, BASF Agricultural Products (Germany); RONILAN-DF®, BASF Agricultural Products (Germany); RONALINE-FL®, BASF Agricultural Products (Germany); TOUCHE®, BASF Agricultural Products (Germany); VINCHLOZOLINE®; VINCLOZOLINE®; VORLAND®, BASF Agricultural Products (Germany)

*Producers:* BASF Agricultural Products (Germany); BASF Canada (Canada)

*Chemical Class:* Dicarboximide
*EPA/OPP PC Code:* 113201
*California DPR Chemical Code:* 2129
*RTECS Number:* RP8530000

*Uses:* Vinclozolin is a non-systemic fungicide currently registered in the United States for use on vines (such as grapes), strawberries, raspberries, chicory grown for Belgian endive, lettuce, kiwi, canola, succulent beans, and dry bulb onions. Import tolerances have been established to permit importation of vinclozolin-treated cucumbers, sweet peppers, and wine (from treated grapes), but there are no U.S. registrations for these uses. Vinclozolin is also registered for use on ornamentals and turf. BASF, the manufacturer of vinclozolin, requested the phase-out of the following uses: onions, raspberries and ornamentals immediately; kiwi and chicory in December, 2001; and lettuce and snap beans in July, 2004. BASF also requested revocation of the import tolerances to cover residues in/on peppers and cucumbers in January, 2001. After 2004, only use on canola, non-domestic wine grapes, and turf will remain, Labels have been recently amended to prohibit use on turf except for golf courses and industrial park landscapes and to prohibit use on sod except for transplant onto golf courses.

*Fish toxicity (threshold)[77]*: Low–396.87174000 ppb, MATC (Maximum Acceptable Toxicant Concentration)

*U.S. Maximum Allowable Residue Levels for Vinclozolin (40 CFR 180.380)*:

| CROP | ppm |
|---|---|
| Bean, succulent | 2.0 |
| Canola, seed | 1.0 |
| Cattle, fat | 0.05 |
| Cattle, meat | 0.05 |
| Cattle, mbyp | 0.05 |
| Egg | 0.05 |
| Endive, belgium | 5.0 |
| Goat, fat | 0.05 |
| Goat, meat | 0.05 |
| Goat, mbyp | 0.05 |
| Grape, wine | 6.0 |
| Hog, fat | 0.05 |
| Hog, meat | 0.05 |
| Hog, mbyp | 0.05 |
| Horse, fat | 0.05 |
| Horse, meat | 0.05 |
| Horse, mbyp | 0.05 |
| Kiwifruit | 10.0 |
| Lettuce, head | 10.0 |
| Lettuce, leaf | 10.0 |
| Milk | 0.05 |
| Onion, dry bulb | 1.0 |
| Poultry, fat | 0.1 |
| Poultry, meat | 0.1 |
| Poultry, mbyp | 0.1 |
| Raspberry | 10.0 |
| Sheep, fat | 0.05 |
| Sheep, meat | 0.05 |
| Sheep, mbyp | 0.05 |

*Carcinogen/Hazard Classifications*
**U.S. EPA Carcinogens:** Group C, possible carcinogen
**California Prop. 65:** Carcinogen and developmental toxin
**TRI Developmental Toxin:** Reproductive and developmental toxin
**Label Signal Word:** CAUTION

**WHO Acute Hazard:** Class U, unlikely to be hazardous
**Endocrine Disruptor:** Suspected endocrine disruptor
*Regulatory Authority:*
- Actively registered pesticide in California.
- EPCRA Section 313 Form R *de minimis* concentration reporting level: 1.0%
- EPA 40 CFR 372.65, Specific Toxic Chemical Listings

*Description:* Colorless, crystalline solid. Slight aromatic odor. Practically insoluble in water. Molecular weight = 286.11. Melting point = 108°C. Vapor pressure = 1.25 x $10^{-7}$ mmHg @ 20°C.

*Determination in Air:* Filter; none; Gravimetric; NIOSH IV[18] [Particulates NOR; #0500 (total), #0600 (respirable)]

*Permissible Concentration in Water:* No criteria set. Runoff from spills or fire control may cause water pollution.

*Routes of Entry:* Inhalation of dust or aerosols, ingestion.

*First Aid:* If this chemical gets into the eyes, remove any contact lenses at once and irrigate immediately for at least 15 minutes, occasionally lifting upper and lower lids. Seek medical attention immediately. If this chemical contacts the skin, remove contaminated clothing and wash immediately with soap and water. Seek medical attention immediately. If this chemical has been inhaled, remove from exposure, begin rescue breathing (using universal precautions) if breathing has stopped, and CPR if heart action has stopped. Transfer promptly to a medical facility. When this chemical has been swallowed, get medical attention. Give large quantities of water and induce vomiting. Do not make an unconscious person vomit.

*References:*
- U.S. Environmental Protection Agency, Office of Pesticide Programs, Pesticide Residue Limits, "Vinclozolin," 40 CFR 180.380, www.epa.gov/pesticides/food/viewtols.htm
- U.S. Environmental Protection Agency, "Reregistration Eligibility Decision (RED),Vinclozolin," Office of Prevention, Pesticides and Toxic Substances, Washington, DC (October 2000). http://www.epa.gov/REDs/2740red.pdf
- EXTOXNET, Extension Toxicology Network, "Pesticide Information Profile, Vinclozolin," Oregon State University, Corvallis, OR (June 1996).
- California Environmental Protection Agency "Chemical List of Lists," Sacramento CA (February 1997)

# W

## Warfarin

*Use Type:* Rodenticide
*CAS Number:* 81-81-2 (warfarin); 129-06-6 (sodium salt)
*Formula:* $C_{19}H_{16}O_4$
*Synonyms:* 3-(α-Acetonylbenzyl)-4-hydroxycoumarin; 2H-1-Benzopyran-2-one,4-hydroxy-3-(3-oxo-1-phenylbutyl)-; Coumafene (French); Coumadin; Coumarin (Japan); Coumarin, 3-(α-acetonylbenzyl)-4-hydroxy-; 4-Hydroxy-3-(3-oxo-1-phenylbutyl)coumarin; Kumander; Kypfarin; 3-(α-Phenyl-β-acetylethyl)-4-hydroxycoumarin; 3-(1'-Phenyl-2'-acetylethyl)-4-hydroxycoumarin; (Phenyl-1 acetyl-2-ethyl)-3-hydroxy-4 coumarine (French); Prothromadin; Toxic chemical category code, N874; Warfarine (French); Zoocoumarin (Dutch and Russian); Zoocoumaring (Russian)
*Sodium:* 3-(α-Acetonylbenzyl)-4-hydroxy-coumarin sodium salt; Coumadin sodium; 4-Hydroxy-3-(3-oxo-1-phenylbutyl)-2H-1-benzopyran-2-one sodium salt (9CI); Marevan (sodium salt); Prothrombin; Sodium coumadin; Sodium warfarin
*Trade Names:* ARAB RAT DETH®; ATROMBINE-K®; BRUMIN®; COMPOUND 42®; D-CON®, Reckitt Benckiser Plc (UK); CO-RAX®, Prentiss Inc. (USA); DETHMORE®; EAGLES-7®; EASTERN STATES DUOCIDE®; GROVEX SEWER BAIT®; HOPKINS BAR BAIR®, Neogen Corporation (USA); HOPKINS COV-R-TOX®, Neogen Corporation (USA); HOPKINS RODEX®, Neogen Corporation (USA); KILLGERM SEWARIN P®; KILMOL®; LIQUA-TOX®; MAR-FIN®; MOUSE PAK®; PLUSBAIT®; RAT-A-WAY®; RAT-B-GON®; RAT-O-CIDE®; RAT-GARD®; RAT & MICE BAIT®; RATRON®; RATS-NO-MORE®; RATTUNAL®; RAX®; RCR SQUIRREL KILLER®; RENTOKIL®; RENTOKIL BIOTROL®; RODEX BLOX®, Neogen Corporation (USA); RODENTEX®; RO-DETH®; RODEX®, Neogen Corporation (USA); ROUGH & READY MOUSE MIX®; SAKARAT®; SOLFARIN®; SOREXA PLUS®; SOREX CR1®; SEWARIN®; SPRAY-TROL BRANCH®; TWIN LIGHT RAT AWAY®; RODEN-TROL®; WARFARAT®; WARF COMPOUND®; VAMPIRINIP®;
*Sodium Salt:* ATHROMBIN®; LIQUA-TOX®, Bell Laboratories (USA); PANWARFIN®; RATSUL SOLUBLE®; TINTORANE®; VARFINE®; WARAN®; WARCOUMIN®; WARFILONE®
*Producers:* Agrimor International (USA); Bell Laboratories (USA); Diachem (Italy); DuPont (USA); I.N.D.I.A. Industrie Chimiche (Italy); Neogen Corporation (USA); Prentiss Inc. (USA); Rhone-Poulenc Jardin (France); Reckitt Benckiser Plc (UK); Shacco (USA); Shell Chemical (Netherlands); Shenzhen Guomeng Industry Co., Ltd. (China); Whyte Agrochemical (UK)
*Chemical Class:* Coumarin
*EPA/OPP PC Code:* 086002 (warfarin); 086003 (sodium salt)
*California DPR Chemical Code:* 621 (warfarin); 1184 (sodium salt)
*ICSC Number:* 0821 (warfarin)
*RTECS Number:* GN4550000; GN4725000 (sodium)
*EEC Number:* 607-056-00-0 (warfarin)
*EINECS Number:* 201-377-6 (warfarin)
*Uses:* Warfarin and its sodium salt is an anticoagulant rodenticide used for controlling rats and house mice in and around homes, animal and agricultural premises, and commercial and industrial sites. It is effective in very low dosages. About a week is required before a marked reduction in the rodent population is noticeable. Rodents do not become bait-shy after once tasting warfarin; they continue to consume it until its anti-clotting properties have produced death through internal hemorrhaging. It can be used year-after-year wherever a rodent problem exists. Warfarin and its sodium salt is only slightly dangerous to humans and domestic animals when used as directed, but care must be taken with young pigs, which are especially susceptible. The sodium salt is also used to treat people with blood hyper-coagulation problems.
*Carcinogen/Hazard Classifications*
**California Prop. 65:** Listed; Reproductive toxin
**TRI Developmental Toxin:** Developmental toxin as warfarin and salts
**Label Signal Word:** DANGER or CAUTION, depending on the concentration
**WHO Acute Hazard:** Class 1b, highly hazardous (warfarin)
**Endocrine Disruptor:** Suspected endocrine disruptor (warfarin)
*Regulatory Authority:*
- Very Toxic Substance (World Bank)[15]
- Air Pollutant Standard Set (ACGIH)[1] (DFG)[3] (HSE)[33] (OSHA)[58] (former USSR)[35] (Several States)[60]
- AB 2588-Air Toxics "Hot Spots" Chemicals (CAL)
- Proposition 65 chemical (CAL)
- Permissible Exposure Limits for Chemical Contaminants (CAL/OSHA)
- The "Director's List" (CAL/OSHA)
- Actively registered pesticide in California.
- EPA Hazardous Waste Number (RCRA No.): U248, warfarin salts when present at concentrations less than 0.3%; P001, warfarin salts when present at concentrations greater than 0.3%.

- RCRA, 40CFR261, Appendix 8 Hazardous Constituents
- Superfund/EPCRA 40CFR355, Appendix B Extremely Hazardous Substances: TPQ = 500/10,000 lb (227/4,540 kg) warfarin and warfarin sodium
- Superfund/EPCRA 40CFR302.4 RQ: CERCLA, 100 lb (45.4 kg)
- EPCRA Section 313 Form R *de minimis* concentration reporting level: 1.0%
- U.S. DOT Regulated Marine Pollutant (49CFR172.101, Appendix B)
- Canada, WHMIS, Ingredients Disclosure List

*Sodium salt:*

- EPCRA Section 313 Form R *de minimis* concentration reporting level: 1.0%
- Superfund/EPCRA 40CFR355, Appendix B Extremely Hazardous Substances: TPQ = 100/10,000 lb (45.4/4,540 kg)
- EPA Hazardous Waste Number (RCRA No.): U248, warfarin salts when present at concentrations less than 0.3%; P001, warfarin salts when present at concentrations greater than 0.3%
- U.S. DOT Regulated Marine Pollutant (49CFR172.101, Appendix B)

***Description:*** Warfarin is a colorless, odorless crystalline solid. Although warfarin is usually available commercially as the sodium salt, the following physical properties refer to the pure substance. Practically insoluble in water; solubility = 15 ppm @ 20°C. Molecular weight = 308.36. Melting/Freezing point =161°C (decomposes below BP). Vapor pressure = $8.7 \times 10^{-2}$ mbar @ 20°C. Log $K_{ow}$ = <2.4. Unlikely to bioaccumulate in marine organisms. The sodium salt is soluble in water and is even less likely to bioaccumulate in marine organisms.

***Incompatibilities:*** Strong oxidizers, strong acids and strong bases.

***Permissible Exposure Limits in Air:*** The OSHA/NIOSH[2] TWA, the HSE[33] TWA, and the ACGIH TWA value[1] is 0.1 mg/m³. The STEL value added by HSE[33] is 0.3 mg/m³. The NIOSH[2] IDLH level = 100 mg/m³. The DFG MAK value is 0.5 mg/m³[3]. The former USSR has set an MAC of 1.0 µg/m³[35]. Several states have set guidelines or standards for warfarin in ambient air[60] ranging from 0.016 µg/m³ (Virginia) to 1.0–3.0 µg/m³ (North Dakota) to 2.0 µg/m³ (Connecticut and Nevada).

***Determination in Air:*** Filter; Methanol; High-pressure liquid chromatography/Ultraviolet detection; NIOSH IV[18], Method #5002.[18]

***Permissible Concentration in Water:*** No criteria set, but runoff from spills or fire control may cause water pollution.

***Routes of Entry:*** Ingestion, inhalation, skin and/or eye contact. Skin contact should be avoided.

***Harmful Effects and Symptoms***

***Short Term Exposure:*** Warfarin is classified as very toxic, and may cause hemorrhage at even low levels. Probable oral lethal dose in humans is 50-500 mg/kg, between 1 teaspoon and 1 ounce for a 150 lb person. Warfarin sodium is a powerful anticoagulant. Hemorrhage is the most common sign and may be manifested by hemorrhagic skin rashes and lip, nose, and upper airway bleeding. Upper airway pain, difficulty in speaking and swallowing, and dyspnea (shortness of breath) may occur. Back pain may be noted. Other symptoms of warfarin exposure begin a few days or weeks after ingestion. They include epistaxis (nose bleed), bleeding gums, pallor, and sometimes hematomas around joints and on buttocks, blood in urine and feces; hematoma arms, legs; bleeding lips, mucous membrane hemorrhage; petechial rash; abnormal/abnormalities hematologic indices. Later, paralysis due to cerebral hemorrhage, and finally hemorrhagic shock and death may occur.

***Long Term Exposure:*** Anemia can result from severe or repeated bleeding. Repeated exposure may affect the liver and kidneys. Material is believed to be teratogenic in humans. There is limited evidence that warfarin may decrease fertility in females. Animal tests indicates that warfarin may cause malformations in human babies.

***Points of Attack:*** Blood and cardiovascular system.

***Medical Surveillance:*** Consider the points of attack in preplacement and periodic physical examinations. Blood test for prothrombin time. Complete blood count. Persons taking "blood thinning" medications are at increased risk.

***First Aid:*** If this chemical gets into the eyes, remove any contact lenses at once and irrigate immediately for at least 15 minutes, occasionally lifting upper and lower lids. Seek medical attention immediately. If this chemical contacts the skin, remove contaminated clothing and wash immediately with soap and water. Seek medical attention immediately. If this chemical has been inhaled, remove from exposure, begin rescue breathing (using universal precautions) if breathing has stopped, and CPR if heart action has stopped. Transfer promptly to a medical facility. When this chemical has been swallowed, get medical attention. Give large quantities of water and induce vomiting. Do not make an unconscious person vomit. Medical observation is recommended.

***References:***

- EXTOXNET, Extension Toxicology Network, "Pesticide Information Profile, Warfarin," Oregon State University, Corvallis, OR (September 1995). http://extoxnet.orst.edu/pips/warfarin.htm
- U.S. Environmental Protection Agency, "Reregistration Eligibility Decision (RED) Fact Sheet, Warfarin," Office of Pesticides and Toxic Substances, Washington, DC (June 1991). http://www.epa.gov/REDs/factsheets/0011fact.pdf
- New Jersey Department of Health and Senior Services, "Hazardous Substance Fact Sheet, Warfarin," Trenton, NJ (May 1998). http://www.state.nj.us/health/eoh/rtkweb/2012.pdf

- U.S. Environmental Protection Agency, "Chemical Profile: Warfarin," Washington, DC, Chemical Emergency Preparedness Program (November 30, 1987).
- U.S. Environmental Protection Agency, "Chemical Profile: Warfarin Sodium," Washington, DC, Chemical Emergency Preparedness Program (November 30, 1987).
- California Environmental Protection Agency "Chemical List of Lists," Sacramento CA (February 1997).

# Z

## Zilkonium Chloride

*Use Type:* Herbicide, algaecide, antiseptic, antimicrobial agent.
*CAS Number:* 8001-54-5; 8045-22-5
*Synonyms:* Alkyl dimethyl benzyl ammonium chloride [50% ($C_{12}$), 30% ($C_{14}$),17% ($C_{16}$), 3% ($C_{18}$)]; Alkyl dimethylbenzyl ammonium chloride; Alkyldimethyl(phenylmethyl)quaternary ammonium chlorides; Benzalkonium chloride; BTC; Pheneene germicidal solution and tincture; Quaternary ammonium compounds, alkylbenzyldimethyl, chlorides; Zephiran chloride
*Trade Names:* AMMONYX®; ARQUAD DMMCB-75®; BARQUAT® MB-50; BARQUAT® MB-80; BAYCLEAN®; BIO-QUAT 50-24®, Lonza Group (Switzerland); CATAMINE® AB; CONSAN®; DRAPOLENE®; GARDIQUAT®-1450; HYAMINE®-3500; INTEXAN® LB-50; KATAMINE® AB; NEO GERM-I-TOL®; ONYX BTC; OSVAN; RODALON®; SENTINEL®; TRITON® K-60; VIKROL® RQ
*Producers:* Lonza Group (Switzerland); Onyx Oil & Chem Co
*Chemical Class:* Quaternary ammonium compound
*EPA/OPP PC Code:* 069106, 069109*
*Uses:* Used as an algicide to control slime mold, algae, fish pathogens, and mollusks in ponds, canals and bodies of water. Also widely used in deodorants, detergents sanitizers and germicides for applications in food plants, laundries, and operating rooms.
*Carcinogen/Hazard Classifications*
**Label Signal Word:** DANGER
*Regulatory Authority:*
- Actively registered pesticide in California.

*Description:* White to yellowish-white amorphous powder. Colorless. Aromatic odor, bitter taste. Very soluble in water; forms an alkaline solution
*Incompatibilities:* Keep aqueous solution away from acids. When heated to decomposition or on contact with acids or acid fumes, may produce highly toxic chloride fumes; deadly phosgene gas may be formed. May cause pitting of some metals.
*Harmful Effects and Symptoms*
*Short Term Exposure:* Irritates skin and eyes. Central nervous system (CNS) poison.
*First Aid:* If this chemical gets into the eyes, remove any contact lenses at once and irrigate immediately for at least 15 minutes, occasionally lifting upper and lower lids. Seek medical attention immediately. If this chemical contacts the skin, remove contaminated clothing and wash immediately with soap and water. Seek medical attention immediately. If this chemical has been inhaled, remove from exposure, begin rescue breathing (using universal precautions) if breathing has stopped, and CPR if heart action has stopped. Transfer promptly to a medical facility. When this chemical has been swallowed, get medical attention. Give large quantities of water and induce vomiting. Do not make an unconscious person vomit. Do not induce vomiting when formulations containing petroleum solvents are ingested.
*References:*
- California Environmental Protection Agency "Chemical List of Lists," Sacramento CA (February 1997)

## Zineb

*Use Type:* Fungicide, insecticide
*CAS Number:* 12122-67-7
*Formula:* $C_4H_6N_2S_4Zn$
*Alert:* Zineb was formerly registered in the U.S. as a General Use Pesticide (GUP) and was rated as a pesticide of low toxicity–EPA toxicity class IV. Products containing zineb were required to carry the Signal Word CAUTION on the label. Following an EPA Special Review of all the ethylene(bis)dithiocarbamate pesticides (EDBCs), including zineb, all registrations for zineb were voluntarily canceled by the manufacturer. All tolerances for zineb in agricultural commodities in the U.S. (except grapes used in winemaking) were revoked, effective 12/31/94. The tolerance for grapes used in winemaking was revoked, effective 12/31/97. (EXTOXNET)
*Synonyms:* Carbamodithioic acid, 1,2-ethanediylbis-, zinc complex; Carbamodithioic acid, 1,2-ethanediylbis-, zinc salt; EBDC, zinc salt; ENT 14,874; 1,2-Ethanediylbis(carbamodithioato)zinc; [1,2-Ethanediylbis(carbamodithioato)](2-)zinc; 1,2-Ethanediylbis(carbamodithioato)(2-)-S,S'-zinc; 1,2-Ethanediylbiscarbamodithioic acid, zinc complex; 1,2-Ethanediylbiscarbamothioic acid, zinc salt; Ethylenebis(dithiocarbamato)zinc; Ethylenebis(dithiocarbamic acid), zinc salt; Ethyl zimate; Zinc ethylenebisdithiocarbamate; Zinc ethylenebis (dithiocarbamate); Zinc ethylene-1,2-bisdithiocarbamate; Zinc, [ethylenebis(dithiocarbamato)]-; Zinc ethylenebis(dithiocarbamate); Zink-[N,N'-aethylen-bis(dithiocarbamat)] (German)
*Trade Names:* ASPOR®; ASPORUM®; BERCEMA®; BLIGHTOX®; BLITEX®; BLIZENE®; CARBADINE®; CHEM ZINEB®; CINEB®; CRITTOX®; CYNKOTOX®; DAISEN®; DEVIZEB®; DIPHER®; DITHANE® 7-78;

DITHANE® Z; DITHANE® Z-78; DITIAMINA®; FBC® PROTECTANT FUNGICIDE; HEXATHANE®; KUPRATSIN®; KYPZIN®; LIROTAN®; LONACOL®; MICIDE®; MILTOX®; MILTOX® SPECIAL; NOVOSIR® N; NOVOZIN® N 50; NOVOZIR®; NOVOZIR® N; NOVOZIR® N 50; PAMOSOL® 2 FORTE PARZATE®; PARZATE® ZINEB; PEROSIN®; PEROSIN® 75B; PEROZIN®; PEROZINE®; POLYRAM® Z; POLYGRAM® Z; SPERLOX®-Z; THIODOW®; TRITHAC®; TURBAIR® ZINEB; TIEZENE®; TRIPART®; TRIPART® BLUE; TITOFTOROL®; TRIMANZONE®; TRITOFTOROL®; TSINEB (Russian); TURBAIR® DICAMATE Z-78®; ZEBENIDE®; ZEBTOX®; ZIDAN®; ZIMATE®; ZINOSAN®; ZEBTOX®

**Producers:** Agrimor International (USA); Barmac Industries (Australia); Limin Chemical (China)
**Chemical Class:** Dithiocarbamate
**EPA/OPP PC Code:** 014506
**California DPR Chemical Code:** 627
**ICSC Number:** 0350
**RTECS Number:** ZH3325000
**EEC Number:** 006-078-00-2
*Uses:* Not registered in the U.S. Ethylene(bis)dithiocarbamate pesticides (EDBCs), including zineb, are dithiocarbamate fungicides used to prevent crop damage in the field and to protect harvested crops from deterioration during storage or transport. Zineb was used to protect fruit and vegetable crops from a wide range of foliar and other diseases. Zineb can be formed by combining nabam and zinc sulfate in the spray tank.
*Human toxicity (long-term)[77]:* Very low–350.00 ppb, Health Advisory
*Fish toxicity (threshold)[77]:* Very low–3547.0692700 ppb, MATC (Maximum Acceptable Toxicant Concentration)
*Carcinogen/Hazard Classifications*
**U.S. TRI:** Developmental and reproductive toxin
**California Prop. 65:** Delisted as a carcinogen Oct. 29, 1999
**IARC:** Group 3, unclassifiable
**Label Signal Word:** CAUTION, WARNING or DANGER
**WHO Acute Hazard:** Class U, unlikely to be hazardous
**Endocrine Disruptor:** Suspected endocrine disruptor
*Regulatory Authority:*
- EPCRA Section 313 Form R *de minimis* concentration reporting level: 1.0%
- AB 1803-Well Monitoring Chemical (CAL)
- EPA 40 CFR 372.65, Specific Toxic Chemical Listings
- AB 2588-Air Toxics "Hot Spots" Chemicals (CAL)
- The "Director's List" (CAL/OSHA)
- Clean Water Act: Section 307 Toxic Pollutants as zinc and compounds
- Safe Drinking Water Act: SMCL, 5 mg/L; Priority List (55 FR 1470) as zinc
- FIFRA, 40CFR180.102-1147: tolerances and/or tolerance exemptions for pesticides in or on raw agricultural commodities

*Description:* Yellow to light-tan crystalline solid or powder. Slightly soluble in water; solubility = 12 ppm. Molecular weight = 275.73. Melting/Freezing point = Decomposes below melting point at 157°C. Vapor pressure = $7.5 \times 10^{-8}$ mmHg @ 20°C. Combustible. Forms explosive mixture with air. Flash point = 880°C. Autoignition temperature = 149°C. Log $K_{ow}$ = $20^{(ICSC)}$. Values above 3.0 may be able to bioaccumulate in marine organisms.
*Incompatibilities:* Unstable in moisture and light. Decomposition products in fire includes oxides of nitrogen and sulfur.
*Determination in Air:* Filter; none; Gravimetric; NIOSH IV[18] [Particulates NOR; #0500 (total), #0600 (respirable)]
*Permissible Concentration in Water:* No criteria set. Runoff from spills or fire control may cause water pollution.
*Routes of Entry:* Inhalation of dust or aerosols, ingestion.
*Harmful Effects and Symptoms*
*Short Term Exposure:* Low levels of toxicity. Concentrated solutions are slightly corrosive to eyes and mucous membranes. Dust inhalation can cause irritation of the respiratory system with sneezing. Eye contact can cause irritation, watering, pain, and inflammation of the eyelids. Skin contact can cause irritation and minor ulceration. Ingestion can cause nausea, vomiting, fever, muscle twitching, seizure, rapid respiration, slow heart beat. Severe exposure may result in death.
*Long Term Exposure:* Repeated or prolonged contact may cause dermatitis and skin sensitization. May affect the blood, central nervous system, and liver.
*Points of Attack:* Respiratory system, central nervous system, cardiovascular system, skin, eyes.
*Medical Surveillance:* There are tests available to measure zinc in your blood, urine, hair, saliva, and feces. High levels of zinc in the feces can mean high recent zinc exposure. High levels of zinc in the blood can mean high zinc consumption and/or high exposure. Tests to measure zinc in hair may provide information on long-term zinc exposure; however, the relationship between levels in your hair and the amount of zinc you were exposed to is not clear. Before employment and at regular times after that, the following are recommended: If symptoms develop or overexposure occurs, repeat the above tests as soon as possible and get an exam of the nervous system. Also consider complete blood count and liver function tests. Consider chest x-ray following acute overexposure. Medical observation is recommended for 24 to 48 hours after breathing overexposure, as pulmonary edema may be delayed. As first aid for pulmonary edema, consider administering a corticosteroid spray. Cigarette smoking may exacerbate pulmonary injury and should be discouraged for at least 72 hours following exposure.

*First Aid:* If this chemical gets into the eyes, remove any contact lenses at once and irrigate immediately for at least 15 minutes, occasionally lifting upper and lower lids. Seek medical attention immediately. If this chemical contacts the skin, remove contaminated clothing and wash immediately with soap and water. Seek medical attention immediately. If this chemical has been inhaled, remove from exposure, begin rescue breathing (using universal precautions) if breathing has stopped, and CPR if heart action has stopped. Transfer promptly to a medical facility. When this chemical has been swallowed, get medical attention. If victim is conscious and able to swallow, have victim drink 4 to 8 ounces of water. Do not induce vomiting. *Note to physician or authorized medical personnel*: Medical observation is recommended for 24 to 48 hours after breathing overexposure, as pulmonary edema may be delayed. As first aid for pulmonary edema, consider administering a corticosteroid spray. Cigarette smoking may exacerbate pulmonary injury and should be discouraged for at least 72 hours following exposure.

*References:*
- EXTOXNET, Extension Toxicology Network, "Pesticide Information Profile, Zineb," Oregon State University, Corvallis, OR (June 1996). http://extoxnet.orst.edu/pips/zineb.htm
- California Environmental Protection Agency "Chemical List of Lists," Sacramento CA (February 1997)

# Zinc Phosphide

*Use Type:* Rodenticide
*CAS Number:* 1314-84-7
*Formula:* $P_2Zn_3$; $Zn_3P_2$
*Alert:* Metallic phosphides on clothes, skin, or hair can react with water or moisture to generate phosphine gas. Vomitus containing phosphides can also off-gas phosphine. Phosphine is extremely flammable and explosive; it may ignite spontaneously on contact, with air.
*Synonyms:* Fosfuro de zinc (Spanish); Phosphure de zinc (French); Phosvin; Zinc fosfid; Zinkfosfide (Dutch); Zinco(fosfuro di) (Italian); Zinc(phosphure de) (French); Zinkphosphid (German)
*Trade Names:* BAKER BRAND®, canceled 10/10/1989; BLUE-OX®; E-Z FLO®, Bayer CropScience (Germany), canceled 12/2/1985; GOPHA-RID®, Bell Laboratories (USA); HOPKINS®, Neogen Corp. (USA); KILRAT®; MOLETOX II®, Bonide Products (USA); MOUS-CON®; MR. KILL RAT®; MR RAT GUARD®; NOTT ZINC PHOSPHIDE 93®, Neogen Corp. (USA); RATOL®; ROBAN II AG®, canceled 7/11/2001; RUMETAN®; ZINC-TOX®; ZP®, Bell Laboratories (USA)
*Producers:* Agrimor International (USA); Bayer CropScience (Germany); Bell Laboratories (USA); Bharat Pulverizing Mills (India); Bonide Products (USA); Excel Industries (India); Fluorochem (UK); Neogen Corp. (USA); Jingma Chemicals Ltd. (China); Shenzhen Guomeng Industry Co., Ltd. (China); United Phosphorus (India)
*Chemical Class:* Inorganic zinc compound
*EPA/OPP PC Code:* 088601
*California DPR Chemical Code:* 626
*ICSC Number:* 0602
*RTECS Number:* ZH4900000
*EEC Number:* 015-006-00-9
*EINECS Number:* 215-244-5
*Uses:* Zinc phosphide reacts with the acidic conditions in the gut to form phosphine gas, which interferes with cell respiration. The rodenticide may be used to control many species of rodents, including mice, ground squirrels, prairie dogs, voles, moles, rats, muskrats, nutria and gophers. It may be used as an indoor or outdoor spot treatment for rodents as well as around burrows or underground in orchards, vineyards, various food crops, range lands, and non-crop areas. Zinc phosphide is formulated as a bait/solid, dust, granular, pellet/tablet or wettable powder and is also applied as a broadcast treatment by ground or aerial applications.

*U.S. Maximum Allowable Residue Levels for Zinc Phosphide (40 CFR 180.225):*
*Note:* The following residue limits are for phosphine compounds that produce phosphine gas.

| CROP | ppm |
|---|---|
| Almond | 0.1 |
| Animal feed | 0.1 |
| Avocado | 0.01 |
| Banana, including plantains | 0.01 |
| Barley, grain | 0.1 |
| Brazil nuts | 0.1 |
| Cabbage, chinese, bok choy | 0.01 |
| Cacao bean, dried | 0.1 |
| Cashew | 0.1 |
| Citron, citrus | 0.01 |
| Coffee, bean | 0.1 |
| Corn, field, grain | 0.1 |
| Corn, pop, grain | 0.1 |
| Cotton, undelinted seed | 0.1 |
| Date, dried | 0.1 |
| Dill, seed | 0.01 |
| Eggplant | 0.01 |
| Endive | 0.01 |
| Filbert | 0.1 |
| Grapefruit | 0.01 |
| Kumquat | 0.01 |
| Lemon | 0.01 |
| Lettuce | 0.01 |
| Lime | 0.01 |
| Mango | 0.01 |
| Millet, grain | 0.1 |
| Mushroom | 0.01 |
| Oat, grain | 0.1 |

| | |
|---|---|
| Okra | 0.01 |
| Orange | 0.01 |
| Papaya | 0.01 |
| Peanut | 0.1 |
| Pecan | 0.1 |
| Pepper, black, post-h | 0.01 |
| Pepper, red, post-h | 0.01 |
| Pepper, white, post-h | 0.01 |
| Persimmon | 0.01 |
| Pimentos | 0.01 |
| Pistachio | 0.1 |
| Processed food | 0.01 |
| Raw agricultural commodities | 0.01 |
| Rice, grain | 0.1 |
| Rye, grain | 0.1 |
| Safflower, seed | 0.1 |
| Salsify, tops | 0.01 |
| Sesame, post-h | 0.1 |
| Sorghum, grain, grain | 0.1 |
| Soybean, seed | 0.1 |
| Sunflower, seed | 0.1 |
| Sweet potato | 0.01 |
| Tangelo | 0.01 |
| Tangerine | 0.01 |
| Tomato | 0.01 |
| Vegetable, legume (crop group 6), exc. soybeans | 0.01 |
| Walnut | 0.1 |
| Wheat, grain | 0.1 |

***Carcinogen/Hazard Classifications***
**Label Signal Word:** DANGER, WARNING or CAUTION
**WHO Acute Hazard:** Class 1b, highly hazardous
***Regulatory Authority:***
- Clean Water Act: Section 311 Hazardous Substances/RQ 40CFR117.3 (same as CERCLA, see below); 40CFR401.15 Section 307 Toxic Pollutants; Section 313 Water Priority Chemicals (57FR41331, 9/9/92)
- Safe Drinking Water Act: SMCL, 5 mg/L; Priority List (55 FR 1470) as zinc
- EPA Hazardous Waste Number (RCRA No.): P122; U249 (concentrations of 10% or less)
- FIFRA, 40CFR152.175: restricted use pesticide
- FIFRA, 40CFR180.102-1147: tolerances and/or tolerance exemptions for pesticides in or on raw agricultural commodities
- RCRA, 40CFR261, Appendix 8 Hazardous Constituents
- Superfund/EPCRA 40CFR355, Appendix B Extremely Hazardous Substances: TPQ = 500 lb (227kg)
- Superfund/EPCRA 40CFR302.4 RQ: CERCLA, 100 lb (45.4 kg), when present at concentrations greater than 10%
- AB 2588-Air Toxics "Hot Spots" Chemicals (CAL) as zinc compounds
- The "Director's List" (CAL/OSHA)
- Actively registered pesticide in California
- EPCRA Section 313 Form R *de minimis* concentration reporting level: 1.0%
- Canada, WHMIS, Ingredients Disclosure List

***Description:*** Zinc phosphide is a gray crystalline powder. Mild garlic odor. Insoluble in cold water; slowly decomposes. Boiling point = 1100°C. Melting/Freezing point = 420°C.

***Incompatibilities:*** Decomposes in water or on contact with strong acids evolution of spontaneously flammable phosphine. Decomposes on heating and on contact water producing toxic and flammable fumes of phosphorus, zinc oxides, and phosphine. Incompatible with carbon dioxide, halogenated agents.

***Permissible Exposure Limits in Air:*** OSHA[2] PEL = 0.3 ppm (averaged over an 8-hour workshift). NIOSH[2] IDLH (immediately dangerous to life or health) = 50 ppm. ERPG-2 (Emergency Response Planning Guideline) (maximum airborne concentration below which it is believed nearly all individuals could be exposed for up to 1 hour without experiencing or developing irreversible or other serious adverse health effects or symptoms that could impair an individuals's ability to take protective action) = 0.5 ppm.

***Permissible Concentration in Water:*** There are a number of standards for *zinc* in water set around the world[35]:

| | | |
|---|---|---|
| EEC | 100-500 µg/L | for drinking water |
| Germany | 2000 µg/L | for drinking water |
| Mexico | 10,000 µg/L | for estuaries |
| Mexico | 10 µg/L | for coastal waters |
| former USSR | 5000 µg/L | for drinking water |
| former USSR | 1000 µg/L | for surface water |
| former USSR | 10 µg/L | in water for fishery purposes |
| WHO | 5000 µg/L | in water for esthetic quality |

*Note*: The USEPA[6] recommends that drinking water should contain no more than 5 milligrams per liter of water (5 mg/L) because of taste. The state of Kansas has set a drinking water limit of 5 mg/L also[61].

***Routes of Entry:*** Inhalation, ingestion, eye and/or skin contact. Absorbed through the skin.

***Harmful Effects and Symptoms***
***Short Term Exposure:*** Irritates the respiratory tract. Contact with the eyes can cause severe irritation, burns and permanent damage. Skin contact causes irritation. This chemical is a CNS depressant. Inhalation of zinc phosphide dust is followed in several hours by vomiting, diarrhea, cyanosis (bluing of skin), rapid pulse, fever and shock. The breath smells of phosphine. The compound is very caustic and may cause closing of the esophagus. Inhalation of phosphine (formed when zinc phosphide is exposed to flame, water or acids) can cause pulmonary edema, a medical emergency that can be delayed for several hours. This can cause death. Zinc phosphide is very caustic when ingested and forms phosphine. The probable oral lethal dose is 5-50 mg/kg, or between 7 drops and 1 teaspoonful for a 70 kg (150 lb) person. Most patients die after about 30 hours from peripheral vascular collapse secondary to the

compound's direct effects. Extensive liver damage and kidney damage can also occur. Ingestion of 4-5 grams has produced death in human adults, but also doses of 25 to 50 grams have been survived. The lowest oral lethal dose reported for women is 80 mg/kg. Symptoms of oral ingestion include nausea, abdominal pain, vomiting, tightness in chest, excitement, agitation and chills; faintness, weakness, dyspnea, fall in blood pressure, change in pulse rate, diarrhea, intense thirst, convulsions, paralysis, and coma. Early labored breathing, shock, halted urinary output, metabolic acidosis, muscle cramps and convulsions are grave prognostic signs.

*Long Term Exposure:* The substance may cause effects on the liver, kidneys, heart and nervous system. Repeated exposure of low exposures causes chronic poisoning, anemia, bronchitis, and gastrointestinal, visual, speech, and motor disturbances.

*Points of Attack:* Lungs, liver, kidneys, heart, blood and central nervous system.

*Medical Surveillance:* There are tests available to measure zinc in your blood, urine, hair, saliva, and feces. High levels of zinc in the feces can mean high recent zinc exposure. High levels of zinc in the blood can mean high zinc consumption and/or high exposure. Tests to measure zinc in hair may provide information on long-term zinc exposure; however, the relationship between levels in your hair and the amount of zinc you were exposed to is not clear. Liver and kidney function tests. EKG. Lung function tests. Consider chest x-ray following acute overexposure. Complete blood count.

*First Aid:* If this chemical gets into the eyes, remove any contact lenses at once and irrigate immediately for at least 15 minutes, occasionally lifting upper and lower lids. Seek medical attention immediately. If this chemical contacts the skin, remove contaminated clothing and wash immediately with soap and water. Seek medical attention immediately. If this chemical has been inhaled get medical attention for phosphine (IDLH level = 50 ppm), remove from exposure, begin rescue breathing (using universal precautions) if breathing has stopped, and CPR if heart action has stopped. Transfer promptly to a medical facility. If the zinc salt is ingested, get medical attention for phosphine poisoning. Give one tablespoonful of mustard in a glass of warm water; repeat until vomit fluid is clear; avoid use of all oils. Do not make an unconscious person vomit. Medical observation is recommended for at least 24 to 48 hours after breathing overexposure, as pulmonary edema may be delayed. Death from phosphine may occur following convulsions, which may occur suddenly after apparent recovery of patient.

*Note to physician or authorized medical personnel:* In case of zinc salt ingestion, empty stomach with emetrics or gastric lavage with 1:5000 solution of potassium permanganate. As first aid for pulmonary edema, a physician or authorized paramedic may consider administering a corticosteroid spray.

*Advanced Treatment for phosphine exposure:* In cases of respiratory compromise secure airway and respiration via endotracheal intubation. If not possible, perform cricothyroidotomy if equipped and trained to do so. Treat patients who have bronchospasm with aerosolized bronchodilators. The use of bronchial sensitizing agents in situations of multiple chemical exposures may pose additional risks. Consider the health of the myocardium before choosing which type of bronchodilator should be administered. Cardiac sensitizing agents may be appropriate; however, the use of cardiac sensitizing agents after exposure to certain chemicals may pose enhanced risk of cardiac arrhythmias (especially in the elderly). Consider racemic epinephrine aerosol for children who develop stridor. Dose 0.25–0.75 mL of 2.25% racemic epinephrine solution in 2.5 cc water, repeat every 20 minutes as needed, cautioning for myocardial variability. Patients who are comatose, hypotensive, or having seizures or cardiac arrhythmias should be treated according to advanced life support protocols. If evidence of shock or hypotension is observed begin fluid administration. For adults, bolus 1000 mL/hour intravenous saline or lactated Ringer's solution if blood pressure is under 80 mm Hg; if systolic pressure is over 90 mm Hg, an infusion rate of 150 to 200 mL/hour is sufficient. For children with compromised perfusion administer a 20 mL/kg bolus of normal saline over 10 to 20 minutes, then infuse at 2 to 3 mL/kg/hour.

*References:*
- EXTOXNET, Extension Toxicology Network, "Pesticide Information Profile, Zinc Phosphide," Oregon State University, Corvallis, OR (June 1996). http://extoxnet.orst.edu/pips/zincphos.htm
- U.S. Environmental Protection Agency, "Reregistration Eligibility Decision (RED) Fact Sheet, Zinc Phosphide," Office of Pesticides and Toxic Substances, Washington, DC (July 1998). http://www.epa.gov/REDs/0026red.pdf
- New Jersey Department of Health and Senior Services, "Hazardous Substance Fact Sheet, Zinc Phosphide," Trenton, NJ (May 2002). http://www.state.nj.us/health/eoh/rtkweb/2041.pdf
- Sax, N.I., Ed., "Dangerous Properties of Industrial Materials Report," 5, No. 5, 103-106 (1985).
- U.S. Environmental Protection Agency, "Chemical Profile: Zinc Phosphide," Washington, DC, Chemical Emergency Preparedness Program (November 30, 1987).
- Pohanish, R.P., "Rapid Guide to Hazardous Chemicals in the Environment," Van Nostrand Reinhold, New York, NY (1997).
- California Environmental Protection Agency "Chemical List of Lists," Sacramento CA (February 1997).
- Agency for Toxic Substances and Disease Registry, U.S. Department of Health and Human Services, Public Health Service, "Managing Hazardous Materials Incidents," Atlanta, GA (June 2003).

# Zinc Sulfate Heptahydrate

*Use Type:* Fungicide, herbicide, miticide
*CAS Number:* 7733-02-0 (heptahydrate); 7446-19-7 (monohydrate)
*Formula:* $ZnSO_4 \cdot 7H_2O$ (heptahydrate); $ZnSO_4 \cdot H_2O$ (monohydrate)
*Synonyms:* AI3-03967; Caswell No. 927; Sulfate de zinc (French); Sulfuric acid zinc salt; White-vitriol; White-copperas; Zinc sulphate monohydrate; Zincum-sulfuricum; Zinc sulphate heptahydrate
*Trade Names:* BUFOPTO-ZINC-SULFATE®; MEDIZINC®; MOSS-B-WARE®; NEOZIN®; OP-THAL-ZIN®; OP-THAL-ZIN®; OPTISED®; OPTRAEX®; ORAZINC®; PREFRIN-Z®; SOLVEZINC®; SOLVEZINK®; VERAZINC®; VISINE-AC®; ZINC-200®; ZINC-VITRIOL®; ZINCATE®; ZINCFRIN®; ZINCI-SULFAS®; ZINCOMED®; ZINK-GROB®; ZINKLET®; ZINKOSITE®
*Producers of zinc and zinc compounds:* ASARCO (USA); Anglo American (UK); Arcon International Resources (Ireland); Billiton (UK); Boliden (Sweden); CP/Philbrochem (USA); Dowa Mining (Japan); Faci (Italy); Grillo-Werke (Germany); Hall, C.P. (USA); Industrias Penoles (Mexico); Jost; Kemira (Finland); Korea Zinc (South Korea); Maranda Mining (South Africa); Max Atotech (India); TotalMetorex (South Africa); Mitsubishi Materials (Japan); Mitsui Mining (Japan); Nippon Mining & Metals (Japan); Noranda (Canada); Numinor (Israel); Pasminco (Australia); PPM Pure Metals (Germany); Rio Tinto (UK); Shepherd Chemical (USA); Synthetic Products (USA); Teck Cominco (Canada); UMICORE (Belgium); Zinc Corp. of America (USA); Zinc Corp. of South Africa (South Africa)
*Chemical Class:* Inorganic zinc
*EPA/OPP PC Code:* 089001 (heptahydrate and anhydrous); 527200 (monohydrate)
*California DPR Chemical Code:* 667 (heptahydrate); 1852 (anhydrous); 2995 (monohydrate)
*RTECS Number:* ZH5260000
*EINECS Number:* 231-793-3 (heptahydrate)
*Uses:* Zinc sulfate anhydrous and zinc sulfate heptahydrate are not registered as pesticides in the U.S. Zinc sulfate is used in small amounts to increase the yield of cotton, cereals, vegetables and fruit and crops such as pecan, deciduous fruit, peanuts, corn, and many citrus. Zinc sulfate is used in the production of water treatment chemicals, the production of rayon, and in animal feed supplements.
*Carcinogen/Hazard Classifications*
*U.S. TRI:* Developmental and reproductive toxins
*Label Signal Word:* DANGER (heptahydrate)
*Regulatory Authority:*
- Clean Water Act: Section 307 Toxic Pollutants as zinc and compounds
- AB 2588-Air Toxics "Hot Spots" Chemicals (CAL) as zinc compounds
- The "Director's List" (CAL/OSHA) as zinc compounds
- Clean Water Act: 40CFR116.4 Hazardous Substances; 40CFR117.3, RQ (same as CERCLA) as zinc sulfate
- Safe Drinking Water Act: SMCL, 5 mg/L; Priority List (55 FR 1470) as zinc
- Superfund/EPCRA 40CFR302.4 RQ: CERCLA, 1000 lb (454 kg) as zinc sulfate
- EPCRA Section 313: Form R *de minimis* concentration reporting level: 1.0% as zinc compounds
- Actively registered pesticide in California as CAS: 7733-02-0

*Description:* Colorless granules or crystalline solid. Odorless. Soluble in water; solubility 54 g/100 ml @ x20°C. Molecular weight =161.43. Density = 3.54 @ 25. Decomposes below boiling point @ 500°C. Melting/Freezing point = 100°C (heptahydrate); monohydrate loses water above 250°C. Log $K_{ow}$ = > 3.0. Values above 3.0 are likely to bioaccumulate in marine organisms.
*Permissible Concentration in Water:* No criteria set. Runoff from spills or fire control may cause water pollution. If powder is inhaled will cause coughing or labored breathing. Solid is irritating to skin and eyes. Swallowing will cause nausea and vomiting.
*Routes of Entry:* Inhalation, ingestion.
*Harmful Effects and Symptoms*
*Short Term Exposure:* Irritates the eyes, skin, and respiratory tract.
*Medical Surveillance:* There are tests available to measure zinc in your blood, urine, hair, saliva, and feces. High levels of zinc in the feces can mean high recent zinc exposure. High levels of zinc in the blood can mean high zinc consumption and/or high exposure. Tests to measure zinc in hair may provide information on long-term zinc exposure; however, the relationship between levels in your hair and the amount of zinc you were exposed to is not clear.
*First Aid:* If this chemical gets into the eyes, remove any contact lenses at once and irrigate immediately for at least 15 minutes, occasionally lifting upper and lower lids. Seek medical attention immediately. If this chemical contacts the skin, remove contaminated clothing and wash immediately with soap and water. Seek medical attention immediately. If this chemical has been inhaled, remove from exposure, begin rescue breathing (using universal precautions) if breathing has stopped, and CPR if heart action has stopped. Transfer promptly to a medical facility. When this chemical has been swallowed, get medical attention. Give large quantities of water and induce vomiting. Do not make an unconscious person vomit.
*References:*
- California Environmental Protection Agency "Chemical List of Lists," Sacramento CA (February 1997)

# Ziram

*Use Type:* Fungicide, microbiocide, cat and dog repellant
*CAS Number:* 137-30-4
*Formula:* $C_6H_{12}N_2S_4Zn$
*Synonyms:* Amyl zimate; Antrene bis(dimethyl carbamodithioato-*s,s'*)zinc; Bis(dimethyldithio carbamato)zinc; Bis(dimethyldithio carbamate de zinc) (French); Carbamic acid, dimethyldithio-, zinc salt; Carbazinc; Carbamodithioic acid, dimethyl-, zinc salt; Dimethylcarbamodithioic acid, zinc complex; Dimethylcarbamodithioic acid, zinc salt; Dimethylcarbamate, zinc salt; Dimethylcarbamodithio carbamic acid, zinc salt; ENT 988; Methyl zimate; Methyl ziram; NCI-C50442; Zinc bis(dimethyldithiocarbamate); Zinc bis(dimethyldithiocarbamoyl)disulphide; Zinc dimethyldithiocarbamate; Zinc, bis(dimethyl carbamodithioato-*s,s'*)-, (T-4)-; Zink-bis(*N,N*-dimethyl-dithiocarbamat) (German); Zinkcarbamate; Zinc *N,N*-dimethyldithiocarbamate; ZnDMDC
*Trade Names:* AAPROTECT®; AAVOLEX®; AAZIRA®; ACCELERATOR®-L; ACCELERATOR® MZ® Powder; ACETO ZDED®; ACETO ZDMD®; ALCOBAM ZM®; ANCANZATE ME®; CARBAZINC®; CIRAM®; CORONA COROZATE®; COROZATE®; CUMAN®; CUMAN L®; CYMATE®; DRUPINA® 90; EPTAC-1®; FUCLASIN®; FUCLASIN® ULTRA; FUKLASIN®; FUNGOSTOP®; HERMAT ZDM®; HEXAZIR®; KARBAM WHITE®; KYPZIN®; METHASAN®; METHAZATE®; MEXENE®; MEZENE®; MILBAM®; MILBAN®; MOLURAME®; MYCRONIL®; OCTOCURE ZDM-50®; ORCHARD® BRAND ZIRAM; PERKACIT ZDMC®; POMARSOL® Z FORTE; PRODARAM®; PROKIL®, Gowan (USA); RHODIACID®; SOXINAL®-PZ; SOXINOL®-PZ; TRICARBAMIX Z®; TSIMAT®; TSIRAM® (Russian); ULTRA ZINC DMC®; VANCIDE® MZ-96; VANCIDE® 51Z Dispersion (with Zinc 2-mercaptobenzothiazolate); VANCIDE® 51Z Dispersion (with ziram); ZERLATE®; ZINCMATE®; ZIMATE®; ZIMATE®; METHYL®; ZIRAMVIS®; ZIRASAN®; ZIRBERK®; ZIREX 90®; ZIRIDE®; ZIRTHANE®; ZITOX®
*Producers:* Bonide Products (USA); Cerexagri (France); Drexel Chemical (USA); Gowan (USA); Hebei Huafeng Chemical Group (China); Taminco (Belgium)
*Chemical Class:* Dithiocarbamate
*EPA/OPP PC Code:* 034805
*California DPR Chemical Code:* 629
*ICSC Number:* 0348
*RTECS Number:* ZH0525000
*EEC Number:* 006-012-00-2
*Uses:* Ziram is an agricultural fungicide registered to control fungal diseases on a wide range of crops including stone fruits, pome fruits, nut crops, vegetables and commercially grown ornamentals, and as a soil and seed treatment. In addition, it is formulated as a bird and rabbit repellent for outdoor foliar applications to ornamentals and as an additive in industrial adhesives, caulking, and latex paints.
*Human toxicity (long-term)*[77]: Very low–140.00 ppb, Health Advisory
*Fish toxicity (threshold)*[77]: Extra high–0.6214300 ppb, MATC (Maximum Acceptable Toxicant Concentration)
*U.S. Maximum Allowable Residue Levels for Ziram (40 CFR 180.116):*
The maximum allowable residue level is 0.1 ppm for almonds and pecans. The maximum allowable residue level is 7.0 ppm for the following crops: Apple, Apricot, Bean; Beet, garden, roots; Beet, garden, roots and tops; Beet, garden, tops; Blackberry, Blueberry, Boysenberry, Broccoli, Brussels sprouts, Cabbage; Carrot, roots; Cauliflower, Celery, Cherry, Collards, Cranberry, Cucumber, Dewberry, Eggplant, Gooseberry, Grape, Huckleberry, Kale, Kohlrabi, Lettuce, Loganberry, Melon, Nectarine, Onion, Pea, Peach, Peanut, Pear, Pepper, Pumpkin, Quince, Radish; Radish, roots; Radish, tops; Raspberry, Rutabaga; Rutabaga, roots; Rutabaga, tops; Spinach, Squash; Squash, summer; Strawberry, Tomato, Turnip; Turnip, greens; Turnip, roots; Youngberry
*Carcinogen/Hazard Classifications*
**U.S. EPA Carcinogens:** Likely carcinogen
**U.S. TRI:** Developmental and reproductive toxin
**IARC:** Group 3, unclassifiable
**Label Signal Word:** CAUTION, WARNING or DANGER
**WHO Acute Hazard:** Class III, slightly hazardous
**Endocrine Disruptor:** Suspected endocrine disruptor
*Regulatory Authority:*
- AB 1803-Well Monitoring Chemical (CAL)
- Actively registered pesticide in California.
- Clean Water Act: Section 307 Toxic Pollutants as zinc and compounds
- Safe Drinking Water Act: SMCL, 5 mg/L; Priority List (55 FR 1470) as zinc
- AB 2588-Air Toxics "Hot Spots" Chemicals (CAL) as zinc compounds
- EPA Hazardous Waste Number (RCRA No.): P205
- RCRA, 40CFR261, Appendix 8 Hazardous Constituents
- Superfund/EPCRA 40CFR302.4 RQ: CERCLA, 1 lb (0.454 kg) as Zinc, bis(dimethylcarbamodithioato-*S,S'*)-
- FIFRA,180.3[3] dithiocarbamates
- The "Director's List" (CAL/OSHA) as zinc compounds
- EPCRA Section 313: Form R *de minimis* concentration reporting level: 1.0% as zinc compounds
- Regulated Marine Pollutant
- Canada, WHMIS, Ingredients Disclosure List

*Description:* White powder (pure). Odorless. Practically insoluble in water; solubility = 72 ppm @ 25°C. Molecular weight = 305.81. Density: 1.71 1.65 @ 20°C. Melting/Freezing point = 248°C. Vapor pressure = 1 x 10$^{-7}$ mmHg @ 20°C. Combustible. Dust forms explosive mixture with air. Log $K_{ow}$ = 1.1. Highly toxic to marine organisms, but unlikely to bioaccumulate.

*Incompatibilities:* Contact with acids can cause decomposition. Decomposition products in fire includes oxides of nitrogen and sulfur.

*Permissible Concentration in Water:* No criteria set. Runoff from spills or fire control may cause water pollution.

*Routes of Entry:* Inhalation of dust or aerosol, ingestion.

*Harmful Effects and Symptoms*

*Short Term Exposure:* Low levels of toxicity. Concentrated solutions are slightly corrosive to eyes and mucous membranes. Dust inhalation can cause irritation of the respiratory system with sneezing. Eye contact can cause irritation, watering, pain, and inflammation of the eyelids. Skin contact can cause irritation and minor ulceration. Ingestion can cause nausea, vomiting, fever, muscle twitching, seizure, rapid respiration, slow heart beat. Severe exposure may result in death.

*Long Term Exposure:* May affect the central nervous system

*Points of Attack:* Respiratory system, central nervous system, cardiovascular system, skin, eyes.

*Medical Surveillance:* There are tests available to measure zinc in your blood, urine, hair, saliva, and feces. High levels of zinc in the feces can mean high recent zinc exposure. High levels of zinc in the blood can mean high zinc consumption and/or high exposure. Tests to measure zinc in hair may provide information on long-term zinc exposure; however, the relationship between levels in your hair and the amount of zinc you were exposed to is not clear. Medical observation is recommended for 24 to 48 hours after breathing overexposure, as pulmonary edema may be delayed. As first aid for pulmonary edema, consider administering a corticosteroid spray. Cigarette smoking may exacerbate pulmonary injury and should be discouraged for at least 72 hours following exposure. Before employment and at regular times after that, the following are recommended: If symptoms develop or overexposure occurs, repeat the above tests as soon as possible and get an exam of the nervous system. Also consider complete blood count. Consider chest x-ray following acute overexposure.

*First Aid:* If this chemical gets into the eyes, remove any contact lenses at once and irrigate immediately for at least 15 minutes, occasionally lifting upper and lower lids. Seek medical attention immediately. If this chemical contacts the skin, remove contaminated clothing and wash immediately with soap and water. Seek medical attention immediately. If this chemical has been inhaled, remove from exposure, begin rescue breathing (using universal precautions) if breathing has stopped, and CPR if heart action has stopped. Transfer promptly to a medical facility. When this chemical has been swallowed, get medical attention. If victim is conscious and able to swallow, have victim drink 4 to 8 ounces of water. Do not induce vomiting. *Note to physician or authorized medical personnel*: Medical observation is recommended for 24 to 48 hours after breathing overexposure, as pulmonary edema may be delayed. As first aid for pulmonary edema, consider administering a corticosteroid spray. Cigarette smoking may exacerbate pulmonary injury and should be discouraged for at least 72 hours following exposure.

*References:*

- U.S. Environmental Protection Agency, "Reregistration Eligibility Decision (RED) Fact Sheet, Ziram," Office of Prevention, Pesticides and Toxic Substances, Washington, DC (July 2004).
http://www.epa.gov/REDs/factsheets/ziram_red_fs.pdf
- EXTOXNET, Extension Toxicology Network, "Pesticide Information Profile, Ziram," Oregon State University, Corvallis, OR (July 1996).
http://extoxnet.orst.edu/pips/ziram.htm
- U.S. Environmental Protection Agency, Office of Pesticide Programs, Pesticide Residue Limits, "Ziram," 40 CFR 180.116,
www.epa.gov/pesticides/food/viewtols.htm
- International Programme on Chemical Safety (IPCS), "Data Sheets on Pesticides, Ziram," Geneva, Switzerland.
http://www.inchem.org/documents/pds/pds/pest73_e.htm
- California Environmental Protection Agency "Chemical List of Lists," Sacramento CA (February 1997).

# Bibliography

(1) American Conference of Governmental Industrial Hygienists, *Threshold Limit Values for Chemical Substances and Physical Agents in the Workroom Environment with Intended Changes*, Cincinnati, Ohio, ACGIH (2003/2004).

(2) National Institute for Occupational Safety and Health, *NIOSH/OSHA Pocket Guide to Chemical Hazards.* DHHS (NIOSH) Publication No. 97-140, Washington, DC (June 1997).

(3) Deutsche Forschungsgemeinschaft (DFG), "*List of MAK and BAT Values 1999*," Wiley-VCH Publishers, New York, NY (1999).

(4) U.S. Environmental Protection Agency, "Water Programs: Hazardous Substances," *Federal Register,* 43, No. 49, 10474-10508 (March 13, 1978).

(5) U.S. Environmental Protection Agency, "Identification and Testing of Hazardous Waste," *Federal Register,* 45, No. 98, 33084-33133, Washington, DC (May 19, 1980).

(6) U.S. Environmental Protection Agency, *Federal Register,* 43, 4109, Washington, DC (January 31, 1978). See also *Federal Register,* 44, 44501 (July 30, 1979) and also *Federal Register,* 45, 79318-79379 (November 28, 1980).

(7) U.S. Environmental Protection Agency, Emergency Planning and Community Right-to-Know Programs, *Federal Register,* 51, No. 221, 41570-41594 (November 17, 1986).

(8) U.S. Environmental Protection Agency, "*Toxic Chemical Release Reporting; Community Right-to-Know,*" *Federal Register,* 52, 107, 21152-21208, Washington, DC (June 4, 1987).

(9) National Institute for Occupational Safety and Health, *Registry of Toxic Effects of Chemical Substances,* DHEW (NIOSH) Publication No. 87-114, Cincinnati, OH (1985-86).

(10) U.S. Department of Health and Human Services, *Fourth Annual Report on Carcinogens,* National Toxicology Program, Research Triangle Park, NC, (1985).

(11) Verscheuren, K., "*Handbook of Environmental Data on Organic Chemicals,*" 3rd Edition, New York, Van Nostrand Reinhold Co, New York, NY (1998).

(12) International Agency for Research on Cancer, "*IARC Monographs on the Carcinogenic Risks of Chemicals to Humans,*" Lyon, France.

(13) United Nations, "*Consolidated List of Products Whose Consumption and/or Sale Have Been Banned, Withdrawn, Severely Restricted or Not Approved by Governments,*" Second Issue, U.N. Sales No. E.87.IV.1, United Nations, Geneva, (1987).

(14) U.S. Environmental Protection Agency, "*Report on the Status of Chemicals in the Special Review Program, Registration Program and Data Call-in Program,*" Office of Pesticide Programs, Washington, DC, (Annual).

(15) World Bank, "*Manual of Industrial Hazard Assessment Techniques,*" Office of Environmental and Scientific Affairs, Washington, DC (1985).

(16) National Research Council, "*Drinking Water and Health,*" National Academy Press, Washington, DC, (1980). See also Reference 46.

(17) National Fire Protection Association, "*Fire Protection Guide on Hazardous Materials,*" 11th Edition, Quincy, MA, (1994).

(18) National Institute of Occupational Safety and Health, "*NIOSH Manual of Analytical Methods,*" 1985 Supplement to 3rd. Edition, NIOSH Publication No. 84-100, Cincinnati, OH (1985).

(19) U.S. Department of Transportation, "*Performance-Oriented Packaging Standards,*" 49 CFR 17179, *Federal Register,* 52, No. 215, 42772-43000 incl., Washington, DC (November 6, 1987).

(20) United Nations, "*Recommendations on the Transport of Dangerous Goods,*" Fourth Revised Edition, U.N. Sales No. E.85.VIII.3, United Nations, New York, NY (1986).

(21) United Nations, "*Recommendations on the Transport of Dangerous Goods; Tests and Criteria,*" First Edition, U.N. Sale No. E.85.VIII.2, United Nations, New York, NY (1986).

(22) International Register of Potentially Toxic Chemicals, "*Treatment and Disposal Methods for Waste Chemicals,*" U.N. Sale No. E.85.111.2, United Nations Environment Programme, Geneva, Switzerland (1985).

(23) Worthing, C.R. and Walker, S.B., Eds., "*The Pesticide Manual,*" 8th Edition, The British Crop Protection Council, Thornton Heath, U.K. (1987).

(24) The International Technical Information Institute, "*Toxic and Hazardous Industrial Chemicals Safety Manual for Handling and Disposal with Toxicity and Hazard Data,*" Tokyo, Japan (1986).

(25) U.S. Environmental Protection Agency, "*Guidelines Establishing Test Procedures for the Analysis of Pollutants; Proposed Regulations,*" *Federal Register,* 44, No. 233, 69464-69575 (December 3, 1979) and also a corrected version in *Federal Register,* 44, No. 244, 7502875052 Washington, DC (December 18, 1979).

(26) Sittig, M., Editor, "*Priority Toxic Pollutants: Health Impacts and Allowable Limits*," Noyes Data Corp, Park Ridge, NJ, (1980).

(27) National Institute for Occupational Safety and Health, *Occupational Diseases: A Guide to Their Recognition*, DHEW (NIOSH) Publication No. 77-181, Washington, DC (June 1977).

(28) Proctor, N.H., Hughes, J.P., and Fischman, M.L, *Chemical Hazards of the Workplace*, 3rd Ed. Van Nostrand Reinhold, New York, NY (1991).

(29) Plunkett, E.R., *Handbook of Industrial Toxicology*, Chemical Publishing Co., Inc. Third Edition, New York, NY (1987).

(30) Parmeggiani, L., Ed., *Encyclopedia of Occupational Health and Safety*, Third Edition, International Labor Office (ILO) Geneva, Switzerland (1983).

(31) Research and Special Programs Administration, U.S. Department of Transportation, *2000 North American Emergency Response Guidebook*, Washington, DC (2000). See also Reference 56.

(32) U.S. Environmental Protection Agency, "*Multimedia Environmental Goals for Environmental Assessment*," Report EPA-600/7-77-136, Research Triangle Park, NC (November 1977).

(33) Health and Safety Executive, "*Occupational Exposure Limits 1988*," Guidance Note EH 40/88, London, England (1988).

(34) Health and Safety Executive, "*Monitoring Strategies for Toxic Substances*," Guidance Note EH 42, London, England (November 1984).

(35) International Register of Potentially Toxic Chemicals (IRPTC), "*IRPTC Legal File (1986),*" U.N. Sales No. E.87.III.D5, United Nations Environment Programme, Geneva, Switzerland (1987).

(36) U.S. Environmental Protection Agency, "*Drinking Water; Proposed Substitution of Contaminants and Proposed List of Additional Substances Which May Require Regulation Under the Safe Drinking Water Act*," *Federal Register*, 52, No. 130, 25720-25734, Washington, DC (July 8, 1987).

(37) U.S. Environmental Protection Agency, "*Hazardous Waste Management System; Identification and Testing of Hazardous Waste; Notification Requirements; Reportable Quantity Adjustments; Proposed Rule*," *Federal Register*, 51, No. 114, 21648-21693 incl., Washington, DC (June 13, 1986).

(38) U.S. Environmental Protection Agency, "*Land Disposal Restrictions for Certain California List Hazardous Wastes and Modifications to the Framework; Final Rule*," *Federal Register*, 52, No. 130, 25760-792 incl., Washington, DC (July 8, 1987).

(39) U.S. Environmental Protection Agency, "*Notice of the First Priority List of Hazardous Substances That Will Be the Subject of Toxicological Profiles and Guidelines for the Development of Toxicological Profiles*," *Federal Register*, 52, No. 74, 12868-12874 incl., Washington, DC (April 17, 1987).

(40) U.S. Environmental Protection Agency, "*National Primary Drinking Water Regulations-Synthetic Organic Chemicals; Monitoring for Unregulated Contaminants; Final Rule*," *Federal Register*, 52, No. 130, 25690-25717 incl., Washington, DC (July 8, 1987).

(41) Pohanish, R.P. and Greene, S.A., "*Hazardous Materials Handbook*," Van Nostrand Reinhold, New York, NY (1996).

(42) U.S. Environmental Protection Agency, "*Guidelines Establishing Test Procedures for the Analysis of Pollutants; Interim Final Rule and Request for Comments and Proposed Regulation*," *Federal Register*, 52, No. 171, 33542-557 incl. Washington, DC (September 3, 1987).

(43) United Nations Environment Program, "*Maximum Allowable Concentrations and Tentative Safe Exposure Levels of Harmful Substances in the Environmental Media (Hygiene Standards Officially Approved in the former USSR)*," Moscow, Centre of International Projects, Moscow, Russia (1984).

(44) Lewis, R.J., Sr., "*Hazardous Chemicals Desk Reference*," 5th Edition, Van Nostrand Reinhold Co, New York, NY (2001).

(45) National Institute for Occupational Safety and Health, "*NIOSH Recommendations for Occupational Safety and Health Standards*," Supplement to Morbidity and Mortality Weekly Report, Centers for Disease Control, Atlanta Georgia (September 26, 1986).

(46) National Research Council, "*Drinking Water and Health*," National Academy of Sciences, Washington, DC (1977). See also Reference 16.

(47) U.S. Environmental Protection Agency, "*Health Advisories for 16 Pesticides*," Office of Drinking Water, Report PB-87-200176, Washington, DC (March 1987).

(48) U.S. Environmental Protection Agency, "*Health Advisories for 25 Organics*," Office of Drinking Water, Report PB 87-235578, Washington, DC (March 1987).

(49) U.S. Environmental Protection Agency, "*Health Advisories for Legionella and Seven Inorganics*," Office of Drinking Water, Report PB 87-235586, Washington, DC (March 1987).

(50) Lewis, R.J., Sr., "*Rapid Guide to Hazardous Chemicals in the Workplace*," 3rd Edition, Van Nostrand Reinhold Co, New York, NY (1994).

(51) U.S. Environmental Protection Agency, "*Organic Chemicals and Plastics and Synthetic Fibers Category Effluent Limitations Guidelines,*

(52) Keith, L.H. and Walters. D.B., editors, "*Compendium of Safety Data Sheets for Research and Industrial Chemicals,*" VCH Publishers, Inc.,(Vols. I-III, 1985 and Vols. IV-VI, 1987), New York, NY .

Pretreatment Standards and New Source Performance Standards," *Federal Register,* 52, No. 214, 42522-42584, Washington, DC (November 5, 1987).

(53) American Conference of Governmental Industrial Hygienists, "*Documentation of the Threshold Limit Values and Biological Exposure Indices, 1998-1999,*" ACGIH, Cincinnati, OH (1999)

(54) U.S. Congress, Office of Technology Assessment, " *Identifying and Regulating Carcinogens,*" Report OTA-BP-H-42, U.S. Government Printing Office, Washington, DC (November 1987).

(55) U.S. Environmental Protection Agency, "*Pesticide Fact Handbook,*" Noyes Data Corp, Park Ridge, NJ (1988).

(56) U.S. Department of Transportation, "*2000 Emergency Response Guidebook,*" Washington, DC (2000). See also Reference 31.

(57) Dutch Association of Safety Experts, Dutch Chemical Industry Association and Dutch Safety Institute, "*Handling Chemicals Safely*," 2nd. Ed., Amsterdam, Holland (1980).

(58) U.S. Department of Labor, "*Air Contaminants-Final Rule,*" 29 CFR Part 1910, *Federal Register,* 54, No. 12, pp 2332-2983 incl. Occupational Safety and Health Administration, Washington, DC (January 19, 1989).

(59) New Jersey Drinking Water Institute, "*Maximum Contaminant Level Recommendations for Hazardous Contaminants in Drinking Water,*" Appendix B: Health-Based Maximum Contaminant Level Support Documents, Trenton, NJ (March 26, 1987).

(60) U.S. Environmental Protection Agency, "*NATICH Data Base Report on State, Local and EPA Air Toxics Activities,*" Office of Air Quality Planning and Standards, Research Triangle Park, NC (July 1988).

(61) U.S. Environmental Protection Agency, "*Summary of State and Federal Drinking Water Standards and Guidelines,*" Federal-State Toxicology and Regulatory Alliance Committee (FSTRAC), Office of Drinking Water, Washington, DC (March 1988).

(62) U.S. Environmental Protection Agency, "*National Primary and Secondary Drinking Water Regulations,*" *Federal Register,* 54, No. 97, 22062-22160 incl., Washington, DC (May 22, 1989).

(63) U.S. Department of Labor, "*Occupational Safety and Health Standards,*" 29CFR1910, Washington, DC (July 1, 1988).

(64) U.S. Department of Transportation, "*Chemical Data Guide for Bulk Shipment by Water,*" United States Coast Guard, Washington, DC (1990).

(65) New York State Department of Health, "*Chemical Fact Sheets,*" Bureau of Toxic Substance Assessment, Albany, NY, various issues and dates.

(66) U.S. Environmental Protection Agency, "*Consolidated List of Chemicals Subject to the Emergency Planning and Community Right-to-Know Act (EPCRA) and Section 112 (r) of The Clean Air Act, As Amended,*" Washington, DC, 1998 (EPA 550-B-98-017).

(67) Bomgardner, Paul M, ed., "*Handling Hazardous Materials,*" American Trucking Association, Alexandria, VA (1997).

(68) Lewis, Richard J, Sr., "*Hawley's Condensed Chemical Dictionary,*" 13th ed., Van Nostrand Reinhold, New York, NY (1998).

(69) Pohanish, Richard P. and Greene, Stanley A., "*Hazardous Substance Resource Guide*", 2nd ed. Gale Research, Inc., Detroit, MI (1997).

(70) New Jersey Department of Health and Senior Services, Right-to-Know Project, "*Hazardous Substance Fact Sheets,*" Trenton, NJ (various dates from 1985-2005). See also http://www.state.nj.us/health/eoh/rtkweb/rtkhsfs.htm

(71) Scott, R. Stricoff and Partridge, Lawrence J. Jr., eds., "*NIOSH/OSHA Occupational Health Guidelines for Chemical Hazards,*" A.D. Little, Inc., United States Department of Health and Human Services, United States Department of Labor, Occupational Safety and Health Administration, Cincinnati, OH (1997).

(72) U.S. Environmental Protection Agency, "*Pollution Prevention Fact Sheets: Chemical Production, FREG-1 (PPIC)*", United States Environmental Protection Agency, Washington, DC.

(73) U.S. Environmental Protection Agency, "*Polychlorinated Biphenyl (PWB) Information Package,*" TSCA Information Service, Washington, DC (April, 1993).

(74) Pohanish, Richard P. and Stanley A. Greene, "*Rapid Guide to Chemical Incompatibilities,*" Van Nostrand Reinhold, New York (1997).

(75) Lewis, Richard J., Sr., "*Sax's Dangerous Properties of Industrial Materials,*" 9th ed., Van Nostrand Reinhold, New York, NY (1996).

(76) Agency for Toxic Substances and Disease Registry, "*Toxicological Fact Sheets,*" U.S. Department of Health and Human Services, Public Health Service, Atlanta, Ga (various dates).

(77) U.S. Department of Health and Human Services/National Institute for Occupational Safety and Health, *CD-ROM NIOSH/OSHA Pocket Guide to Chemical Hazards and other Databases.* DHHS (NIOSH) Publication No. 99-115, Washington, DC (April 1999).

(78) Federal Emergency Management Agency/United States Fire Administration, "*Hazardous Materials for First Responders*," Washington DC (1999).

(79) Pohanish, Richard P., "*Rapid Guide to Hazardous Chemicals in the Environment*," Van Nostrand Reinhold, New York, NY (1997).

(76) Agency for Toxic Substances and Disease Registry (ATSDR), U.S. Department of Health and Human Services, Public Health Service, "*Managing Hazardous Materials Incidents*," Atlanta, GA (June 2003).

(77) WIN-PST Database, U.S. Department of Agriculture, Natural Resources Conservation Service, National Water and Climate Center, Portland, OR (2004)

# Appendix A: List of Companies Cited

## - A -

A H Marks & Co., Ltd., United Kingdom
A.M.P.E.R.E. Industrie SA Dept Chimie, France
Abaquim, MexicoAcido de Mexico S.A. de C.V., Mexico
ABCR GmbH & Co. KG, Germany
Acetex Corporation, Canada
Achema SC, Lithuania
Acros Organics N.V., Belgium
Adheswara Group of Companies, India
Adrian Resources Ltd., Canada
Advanced Synthesis Technologies, United States
Aero Agro Chemical Industries, Ltd., India
AGA Gas, see Linde Gas AG, (Germany), Spain
Agan Chemical Manufacturers Ltd., Israel
AgraQuest, United States
AgrEvo, see Bayer CropScience (Germany), Germany
Agrichem (International) Limited, United Kingdom
AgriChem B.V., The Netherlands
Agrides S.A., Spain
Agriliance, United States
Agrimor International Co., United States
Agrium, Canada
Agro-care Chemical Industry Group, China
Agrochemicals del Ecuador (AGROCHEM), Ecuador
Agropharm Ltd., United Kingdom
Agrowchem Inc., Canada
Agsin Pte. Ltd., Singapore
Agway Inc., United States
Aimco Pesticides Ltd., India
Air Liquide Group, France
Air Products & Chemicals, United States
AJE GmbH, Switzerland
Akzo Nobel Functional Chemicals BV, The Netherlands
Akzo Nobel, The Netherlands
Alaska Nitrogen Products LLC, United States
Albemarle Corporation, United States
Albright & Wilson Pty., United Kingdom
Alcide Corporation, United States
Alcoa Inc, United States
Alcotan Laboratories S.A., Spain
Aldrich Chemical Co., United States
Alfa Rio Quimica Ltda, Brazil
Alfa Aesar, United States
Alkyl Amines Chemicals, India
Allipo Chemicals, India
Alquimia Mexicana, Mexico
Alta Laboratories, India
Amchem Products, see Dow AgroScience (USA), United States
American Cyanamid's Agricultural Products Group, see BASF Agricultural Products Group (Germany), United States
American Elements Inc, United States
American Gas Group, United States
Amomex Mexicana, S.A. de C.V., Mexico
AMVAC Chemical, United States
Anhui Huaxing Chemical Industry Co., Ltd., China
Anhui Fengle Agrochemical Co., Ltd., China
ANWIL S. A., Poland
Apache Nitrogen Products, United States
Aquashade Inc., United States
Arab Potash Company, Jordan
Archimica Ltd., United Kingdom
Aristech Chemical Corporation, see Sunoco Chemicals Unit of Sunoco Inc. (USA), United States
Arkema Group S.A., France
Arkema Group Inc., see Arkema Group S.A. (France), United States
Arvesta Corporation, United States
Aryan Pesticides Ltd., India
Asahi Glass Chemicals Division, Japan
ASARCO, United States
Ascot International Ltd., United Kingdom
Ashland Inc., United States
Asia Pacific Resources Ltd., Thailand
Atanor S.A., Argentina
Atofina Chemicals Inc., see Arkema Inc., (USA), United States
ATOFINA, see Arkema Group S.A. (France), France
Atul Ltd., India
Aventis CropScience, see Bayer CropScience (Germany), France
Avitrol Corporation, United States
Azot Association, Ukraine

## - B -

Baker Petrolite (USA), United States
Barmac Industries, Australia
BASF Agricultural Products Group, United States
BASF Group AG, Germany
BASF Agricultural Products Group, Germany
BASF Canada Inc, Canada
BASF Corporation, United States
Bayer Group, Germany
Bayer CropScience AG, Germany
BEC Group Ltd. (Bhilai Engineering Corporation), India
Becker Underwood Inc., United States

## List of Companies Cited

Bell Laboratories, Inc., United States
Bhageria Dye-Chem Ltd., India
Bharat Rasayan Ltd., India
Bharat Pulverizing Mills Pvt. Ltd., India
Bhp Billiton Ltd., Australia
BHP Billiton Plc, United Kingdom
Biesterfeld Siemsgluess International. GmbH, Germany
Bilbaina de Alquitranes, Spain
Bilt Chemicals, see Solaris ChemTech Ltd. (India), India
BOC Gases, United Kingdom
Boehringer Ingelheim, Germany
Boliden AB, Spain
Bombay Ammonia and Chemical Company, India
Bonide Products, Inc., United States
Borden Chemical, United States
Borregaard, Norway
BP Chemicals, United Kingdom
Brotherton Specialty Products, United Kingdom
Buckman Laboratories, Inc., United States

- C -

C.P. Hall, United States
Caffaro S.p.A, Italy, Italy
Caldic Chemie B.V., The Netherlands
Calliope SA, France
Cangzhou Green Chemical Co., Ltd., China
Carburos Metalicos, Spain
Cargill Crop Nutrition, United States
CECA, France
Cedar Chemical Corporation, United States
Celanese AG, Germany
Celite Corp., United States
Celtic Chemicals Ltd., United Kingdom
Central Glass, Japan
CEPSA, Spain
Cerexagri Inc. - North America, United States
Cerexagri Inc. - Europe/International, France
Certis USA, LLC, United States
CF Industries Inc., United States
Changfeng Chemical, China
Changzhou Fengdeng Pesticide Factory, China
Changzhou Wujin Linchuan Chemical Factory, China
Changzhou Kangmei Chemical Industry Co., Ltd., China
Chemada Fine Chemicals, Israel
Chemical Company, The, United States
Chemical Lime Co., United States
Chemical Products Corporation, United States
Cheminova A/S, Denmark
Chevron Phillips Chemical Company LP, United States
ChevronTexaco Corp, United States
Chimco AD, Bulgaria
China Chemicals, China

China Petrochemical Development Corporation (CPDC), Taiwan
Chongqing Chuandong Chemical (Group), China
Chromos Agro d.d., Croatia
Cia. Petroquimica do Sul (COPESUL), Brazil
Cia. Universal de Industrias S.A. de C.V., Mexico
Ciba Specialty Chemicals, Switzerland
Ciba-Geigy, now Sygenta, Switzerland
Clariant International, Switzerland
Clariant International, Germany
Cleary Chemical Corp, United States
Cleveland Potash, United Kingdom
Coalite Chemicals Ltd., United Kingdom
Cognis GmbH, Germany
CONDEA GmbH, Germany
Continental Lime, United States
Coogee Chemicals, Australia
Copene Petroquimica do Nordeste SA, Brazil
Coromandel Fertilisers Ltd., India
Creanova, United States
Crompton Corporation, United States
Crop Care Australasia Pty Ltd., Australia
Crowley Chemical, United States
Crown Technology, United States
CTC Organics, United States
Cyanides & Chemicals Company, India
Cytec Industries, United States

- D -

Daicel Chemical Industries, Japan
Dainippon Ink and Chemicals, Japan
Dankalk, Denmark
Dead Sea Bromine Group, Israel
Deepak Fertilizers and Petrochemicals Corp. Ltd., India
Degesch America, United States
Degussa AG, Germany
Degussa-Huls AG, see Degussa AG (Germany), Germany
Delta Chemical, United States
Denki Kagaku Kogyo Kabushiki Kaisha (Denka), Japan
Derivados del Fluor SA, Spain
Deza a. s., Czech Republic
Dhanuka Group, India
Dharamsi Morarji Chemicals, India
Dhruv Chemical Industries, India
Diachem S.p.A., Italy
Dow AgroSciences LLC, United States
Dow Chemical, United States
Drexel Chemical Company, United States
DSM Agro BV, The Netherlands
DSM Fine Chemicals, The Netherlands
DuPont, United States
DuPont Crop Protection, United States
Dynamit Nobel AG, Germany

## - E -

E.I.D. Parry (India) Ltd., see Coromandel Fertilisers (India), India
EaglePicher Industries, United States
Eastman Chemical Co., United States
Ecogen Inc., United States
EDEN Bioscience Corp., United States
Ehrenstorfer, Dr., GmbH, Germany
El Dorado Chemical Company, United States
Elston Manufacturing, United States
EM Industries, United States
Emerald BioAgriculture Corp, United States
EniChem S.p.A., Italy
Entek Corporation, United States
Epochem Co., Ltd., China
Equistar Chemicals, United States
Ercros S.A., Spain
Esseco Group S.p.A., Italy
Eurolabs Ltd., United Kingdom
Excel Industries Ltd., India
Exxon Mobil Chemical Company, United States

## - F -

Fairmount Chemical Company, United States
Feinchemie Schwebda GmbH, Germany
Ferro Corporation, United States
Fine Agrochemicals Ltd., United Kingdom
Fluorochem Ltd., United Kingdom
FMC Agricultural Products Group, United States
Forward (Beihai) Hepu Pesticide Co., Ltd., China
Fulon Chemical Industrial Co., Ltd., Taiwan

## - G -

Gadiv Petrochemical Industries Ltd., Israel
Gayatri Minerals & Chemicals, India
General Quimica S.A., Spain
Georgia Gulf Sulfur, United States
Georgia-Pacific Corporation, United States
GFS Chemicals Inc., United States
Gharda Chemicals Ltd., India
GlaxoSmithKline Plc, United Kingdom
Godavari Fertilisers and Chemicals, India
Goldschmidt Chemicals, see Degussa Group (Germany), Germany
Gowan Company, United States
Grande Paroisse SA, France
Great Western Inorganics, United States
Great Lakes Chemical Corporation, United States
Griffin L.L.C., United States
Grupo Aragonesas S. A., Spain
Grupo Mexico SA de C.V., Mexico

Gujarat Pesticides, India
Gustafson L LC, United States
Gwalior Chemical Industries, India

## - H -

Haifa Chemicals Ltd., Israel
Haldor Topsoe A/S, Denmark
Halocarbon Products Corporation, United States
Harcros Chemicals Inc, United States
Hebei Huafeng Chemical Group, China
Hebei Long Age Pesticide Co., Ltd., China
Helena Chemical Company, United States
Hellenic Corundum SA, Greece
Hercules Inc., United States
Hindustan Organic Chemicals, India
Hindustan Insecticides Ltd., India
Hockley International Ltd., United Kingdom
Hoek Loos BV, The Netherlands
Hokko Chemical Industry Co., Ltd., Japan
Holvoet Chimie, France
Honeywell Performance Polymers & Chemicals, United States
Honshu Chemical Industry, Japan
Hummel Croton Inc., United States
Hunan Tianyu Pesticide Chemical Group Co., Ltd., China
Hunan Darong Chemical & Pesticide Co., Ltd., China
Huntsman, United States
Hydro Agri Chemicals, Norway

## - I -

I.N.D.I.A. Industrie Chimiche S.p.A., Italy
ICC Industries, United States
ICI Group (Imperial Chemical Industries Plc), United Kingdom
IMC Chemicals, United States
IMC Vigoro, see Vigoro (Canada), Canada
Incitec Industrial Chemicals, see Incitec Pivot, Australia
Indian Oil Corporation, India
Indiclay Co., Ltd., India
Indo Gulf Corporation, India
Indofine Chemical Company, United States
Industria Quimica Loser S.A. de C.V., Mexico
INEOS Phenols, Germany
Ingenieria Industrial S.A. de C.V. (Bravo), Mexico
International Specialty Products, United States
International Sulfur Inc., United States
Interstate Chemical, United States
ISOCHEM, France
Israel Chemicals Ltd., Israel

## - J -

J. J. Mauget Co., United States
Janssen Pharmacuetica Products, United States
Japan Aldehyde Co., Ltd., Japan
Jiangmen Pesticide Factory, China
Jiangsu Wujin Zhenhua Chemical Plant, China
Jilin Chemical Industrial Co., China
Jingma Chemicals Ltd., China
Juhua Group Corporation, China
Junsei Chemical, Japan

## - K -

Kali und Salz Group AG (K+S), Germany
Kanoria Chemicals & Industries Ltd. (KCIL), India
Kawaguchi Chemical Industry Co., Ltd., Japan
Kawasaki Kasei Chemicals, Japan
Kemira Chemical Oy, Finland
Kemira GrowHow Oy, Finland
Kenso Corporation, Malaysia
Kerr-McGee Chemical LLC, United States
Ki-Hara Chemicals Ltd., United Kingdom
KMG Chemicals Inc, United States
KMG Chemicals, United States
Koei Chemical, Japan
Kothari Phytochemicals International, India
Kunshan Chemical Group (Industries) Corp., China
Kynoch Fertilizer, South Africa
Kyowa Hakko Kogyo Co., Ltd., Japan

## - L -

Lancaster Synthesis, United Kingdom
Laporte Performance Chemicals Plc, United Kingdom
LaRoche Industries, United States
Lebanon Seaboard Corp., United States
Lhoist Group, France
Lime Industries, Australia
Limin Chemical Co., Ltd., China
Linde Gas Group, Germany
Liphatech, United States
Lonza Group AG, Switzerland
Lords Chemicals Ltd., India
Lukoil Oil Company, Russia
Luxembourg Industries Ltd., Israel
Lyondell Chemical Company, United States

## - M -

M & R Durango, Inc. Insectary, United States
Makhteshim-Agan Industries Ltd., Israel
Mallinckrodt Baker, United States
Matheson Tri-Gas Corporation, United States
Matheson Gas Products, see Matheson Tri-Gas (USA), United States
Mclaughlin Gormley King Co., United States
Meghmani Organics Ltd., India
Merck KgaA, Germany
Messer Group GmbH, Germany
MFA Inc., United States
MG Industries, United States
Micro-Flo Company, United States
Milenia Agro Ciencias SA, Brazil
Millennium Chemical, see Lyondell Chemical (USA), United States
Minerals Research & Development, United States
Mining & Chemical Products Ltd. (MCP), United Kingdom
Mississippi Lime Company, United States
Mississippi Chemical Corporation, United States
Mitsubishi Materials Corporation, Japan
Mitsubishi Chemical Corporation, Japan
Mitsubishi Gas Chemical Company, Japan
Mitsui Chemicals Inc. (MCI), Japan
Mitsui Mining & Smelting, Japan
Molekula Fine Chemicals, United Kingdom
Monsanto Company, see Dow AgroSciences (USA), United States
Montana Sulphur and Chemical, United States
Morita Chemical Industries Co., Ltd., Japan
Morton International, see Rohm & Haas Inc. (USA), United States

## - N -

Nagarjuna Group, India
Nagarjuna Fertilizer and Chemicals Ltd. (NFCL) Ltd., India
Nagarjuna Agrichem, India
Nanjing Agrochemicals Co., Ltd., sub. of AgroDragon, China
Navin Fluorine Industries, India
Navy Brand Manufacturing Co., United States
Neogen Corporation, United States
Nepera, United States
Neste Chemicals Oy, Finland
Newmont Koch, United Kingdom
Niacet Corporation, United States
Nihon Kagaku Sangyo, Japan
Nihon Nohyaku Co., Ltd., Japan
Nippon Steel Chemical Co., Ltd., Japan
Nippon Chemical Industrial Co., Ltd., Japan
Nippon Mining & Metals, Japan
Nippon Kayaku Co., Ltd., Japan
Nippon Carbide Industries Co., Inc., Japan
Nippon Soda, Japan
Nissan Chemical Industries Ltd., Japan
Nitrokemia 2000 Rt., Hungary

## List of Companies Cited

NOR-AM Chemical Company, see Bayer CropScience (Germany), United States
Noranda DuPont LLC, see NorFalco (USA), United States
Noranda Inc., Canada
Nova Biogenetics, Inc., United States
Novartis, now Syngenta, Switzerland
Nufarm Ltd., Australia

### - O -

Occidental Petroleum Corporation, United States
Ocean Chemicals Group, United Kingdom
Olin Corporation, United States
Omnia Group, South Africa
Sasol Ltd., South Africa
Omya AG, Switzerland
Orica Ltd., Australia
Ortho Business Group, see Scotts Company, The, United States
Osmose, Inc, United States
Otsuka Chemical Co., Ltd., United States
Oxon Italia S.p.A., Italy
OxyChem, see Occidental Chemical Corporation, United States
Ozark Fluorine Specialties, United States

### - P -

P.D. Industries Ltd., India
Pazchem Ltd., Israel
Pbi/Gordon Corporation, United States
PCF Chimie, France
Pechiney SA, France
Penta Manufacturing Company, United States
Pestcon Systems, United States
Petrobas Energia S.a, Argentina
Petroleo Brasileiro SA (PETROBRAS), Brazil
Petroquimica Uniao SA (PQU), Brazil
Petroquimica de Venezuela (Pequiven), Venezula
Pharm - Chem Manufacturing Co., Ltd, China
Pharmacia Animal Health, United States
Phelps Dodge Corporation, United States
Phenolchemie GmbH & Co. KG, see INEOS Phenols, Germany
Philbro-Tech Inc, United States
Philipp Brothers Chemicals, United States
Phosphate Resource Partners Limited Partnership, United States
Phosphoric Fertilizers Industry SA (PFI), Greece
PI Industries Ltd., India
Pilarquim Corp., Taiwan
Pioneer Enterprise, India
Potash Corporation of Saskatchewan Inc., Canada
PPM Pure Metals GmbH, Germany

Praxair, United States
Prentiss Inc., United States
Prince Manufacturing Company, United States
Pro-Outdoors, United States
Pro-Serve Inc., United States
Prodica LLC, United States
Proficol S.A., Colombia
Pursell Technologies Inc., United States
Pyosa Agroquimicos S.A. de C.V., Mexico

### - Q -

Qatar Fertiliser Company S.A.Q., Qatar
Quantum Chemicals, Australia

### - R -

R.S.A. Corporation, United States
Rallis India Ltd., India
Rashtriya Chemicals & Fertilizers Ltd. (RCF), India
Reckitt Benckiser Plc, United Kingdom
Recochem Inc., International Division, Canada
Reilly Industries, United States
Reliance Industries, India
Rentokil Initial Pty. Ltd., Australia
Rhodia Specialty Phosphates, United States
Rhodia Eco Services, France
Rhodia Group SA, France
Rhone-Poulenc Agro France, see Aventis CropScience (France), France
Rice Company, The (TRC), United States
Richman Chemical, United States
Rio Tinto Borax (U.S. Borax), United States
Rio Tinto Plc, United Kingdom
Rohm and Haas, United States
Rotam Agrochemical (HK) Co., Ltd., Hong Kong
Roussel Uclaf, see Akzo Nobel Pharma Unit (Netherlands), France
Rutherford Chemicals, United States

### - S -

Sabero Organics Gujarat Ltd., India
Saeryung Chemicals Co., Ltd., South Korea
Salvi Chemical Industries, India
Samsung Atofina Co., Ltd., South Korea
Sandoz, now Sygenta, Switzerland
Sankei Chemical Co., Ltd, Japan
Sanonda Ltd., Australia
Sanonda Zhengzhou Pesticide Co., Ltd., China
Saudi Basic Industries Corporation (SABIC), Saudi Arabia
Schenectady Herdillia Chemicals Ltd., India
Schenectady International, United States
Schering AG, Germany

Schering-Plough Animal Health Corp., United States
Scott Specialty Gases, United States
Scottish Chemical Industries, India
Scotts Company, The, United States
Sepro Corporation, United States
Sevencontinent Agrichemical Co., Ltd., China
Shacco Inc., United States
Shandong Changyi Salt Chemicals (Group) Co., Ltd., China
Shandong Huayang Pesticide Group, China
Shangdong Baoyuan Chemical, China
Shanghai Agricultural Chemical Industry Corp., China
Shangyu Chemical Industry Corp., China
Shanxi Friends Union Chemicals (China), China
Shell Chemicals, United Kingdom
Shenzhen Guomeng Industry Co., Ltd., China
Sheyang Pesticides and Chemical Industry Co., Ltd., China
Shield Industries, Inc., United States
Showa Denko K.K., Japan
Shyam Chemicals, India
Sigma-Aldrich Laborchemikalien GmbH, Germany
Sigma-Aldrich Co., United States
Simplot, J.R., Company, United States
Sinon Corporation, Taiwan
Sinopec Corporation, China
Sipcam Agro USA, Inc., United States
Sloss Industries, United States
SNPE Agro, see ISOCHEM (France), France
Sociedad Quimica y Minera de Chile SA, see SQM, Chile
Societe Commerciale des Potasses et de l'Azote (SPCA), France
Solaris ChemTech Ltd., India
Solvay Barium Strontium, Germany
Solvay Group SA, Belgium
Southern Petrochemical Industries Corporation Ltd. (SPIC), India
Southern Peru Copper Corporation, Peru
Southwest Chemical Co., United States
Spectrum Chemical, United States
Spiess-Urania Chemicals GmbH, Germany
SQM SA (Sociedad Quimica y Minera de Chile SA), Chili
Stella Chemifa, Japan
Stepan Company, United States
Sterling Chemicals, United States
Stoller Enterprises, United States
Sudarshan Chemical Industries Ltd., India
Sulphur Mills Ltd., India
Sumika Agro Manufacturing Co., Ltd., Japan
Sumika Fine Chemicals Co., Ltd, Japan
Sumika Takeda Agrochemical Co., Ltd., now Sumika Agro Manufacturing Co., Ltd., Japan
Sumitomo Chemical Co., Ltd, Japan
Sundat Pte. Ltd., Singapore
Sunoco Chemicals, United States
SuYan Agrochemical Group, China

Sybron Chemicals, Inc., see Bayer AG (Germany, United States
Syngenta Crop Protection AG, Switzerland

- T -

Taminco, Belgium
Taoka Chemical, Japan
TCI America, United States
Teck Cominco Ltd., Canada
Terra Industries, United States
Terra Nitrogen Company, United States
Tessenderlo Chemie, Belgium
TETRA Technologies, United States
Thor GmbH, United Kingdom
Timminco Ltd., Canada
Tokuyama Group, Japan
Tokyo Kasei Kogyo Co., Ltd. (TKK), see Tokyo Chemical Industry Co., Ltd. (Japan), Japan
Tokyo Chemical Industry Co., Ltd. (TKK), Japan
Tosoh Corporation, Japan
Total Speciality Chemicals, Inc., see Atofina (France), United States
Total SA., France
Tull Chemical, United States

- U -

U.S. Borax Inc., see Rio Tinto Plc (United Kingdom), United States
Ube Agri-Materials, Ltd., sub. of Industries Ltd., Japan
UCB Group SA, Belgium
UMICORE N.V., Belgium
Union Miniere N.V., see UMICORE (Belgium), Belgium
Union Carbide, see Dow AgroSciences LLC, United States
Uniqema Ltd, United Kingdom
UNIROYAL Crop Protectioin, see Crompton Corporation, United States
United Agri Products (UAP), United States
United Agro Industries, India
United Phosphorus Ltd., India
Univertical Corporation, United States
Upjohn Inc. (US), see Pharmacia Animal Health (USA), United States

- V -

Valent USA, United States
Valent BioSciences Corporation, United States
Vani Chemicals & Intermediates Ltd., India
Varsal Instruments, United States
Velsicol Chemical Corporation, United States
Veterinary & Agricultural Products Manufacturing Co., Ltd. (VAPCO), Jordan

Vigoro, Canada
Vijayalakshmi Insecticides and Pesticides Pvt. Ltd., India
Vulcan Chemicals, United States

## - W -

W. Neudorff GmbH KG, Germany
Wangs Ltd., China
Webcot, Australia
Wellmark International, United States
Wesfarmers Csbp Ltd., Australia
West Agro, Inc, United States
White Mountain Natural Products, Inc, United States
Whitnire Micro-Gen Research Laboratories, Inc., United States
Whyte Agrochemical, United Kingdom
William Blythe Ltd., United Kingdom
Witco, see Crompton (USA), United States
World Minerals Inc., United States
World Metal LLC, United States
Wuhan kernel Bio-pesticide Co., Ltd., India
Wuhan Kernel Bio-pesticide Co., Ltd., China
Wuxj Ruize Pesticide Co., Ltd., China
Wuzhou International Co., Ltd., China

## - Y -

Yellow River Enterprise Co., Ltd., Taiwan
YiHua Group, China

## - Z -

Zagro Asia Ltd., Singapore
Zeneca Ag Products, see Syngenta (Switzerland), United States
Zeneca Agro (UK), see Syngenta Crop Protection (Switzerland), United Kingdom
Zeon Corporation, Japan
Zhejiang Chem-tech Group Co., Ltd., China
Zhejiang Heben Pesticide & Chemicals Co., Ltd. , China
Zhejiang Yifan Chemical Co., China

# Appendix B: Directory of Agrochemical Manufacturers

## ARGENTINA

**Company:** Atanor S.A.
**Address:** San Martin 140, Buenos Aires, Argentina
**E-mail:** osune@atanor.com.ar
**Web site:** www.atanor.com.ar

**Company:** Pecom Energia SA, now Petrobas Energia S.S. (Argentina)

**Company:** Petrobas Energia SA
**Address:** Perez Companc Building, Maipu 1 22$^{nd}$ Floor, C1084ABA, Buenos Aires, Argentina
**Phone:** +54-11 4344 6000
**Fax:** +54-11 4344-6315
**Web site:** www.petrobrasenergia.com
**Stock listing:** Sao Paulo SE: POB; NYSE: PBR
**Parent Company:** Petroleo Brasileiro SA (PETROBRAS) (Brazil)

## AUSTRALIA

**Company:** Barmac Industries
**Address:** Box Flat Estate, Swanbank Rd., Swanbank QLD 4306, Australia
**Phone:** +61 7-3280-3000
**Fax:** +61 7-3280-3030
**E-mail:** sales@barmac.com.au
**Web site:** www.barmac.com.au

**Company:** BHP Billiton Ltd.
**Address:** BHP Billiton Centre, 180 Lonsdale St., Melbourne, Victoria 3000, Australia
**Phone:** +61 1300-55-47-57
**Fax:** +61 3-9609- 3015
**Web site:** www.bhpbilliton.com
**Stock listing:** Australia, London, NYSE SE: BHP

**Company:** Coogee Chemicals Pty. Ltd.
**Address:** P.O. Box 5051, Rockingham Beach, Western Australia 6969, Australia
**Phone:** +61 8 9439 8200
**Fax:** +61 8 9439 8300
**Web site:** www.coogee.com.au

**Company:** Crop Care Australasia Pty Ltd.
**Address:** 77 Tingira St., Pinkenba, Queensland 4008, Australia
**Phone:** +61 7-3867-9100
**Fax:** +61 7-3867-9110
**Web site:** www.cropcare.com

**Company:** Incitec Industrial Chemicals, now Incitec Pivot, Ltd. (Australia)

**Company:** Incitec Pivot, Ltd.
**Address:** GPO Box 1322L, Melbourne, Victoria 3001, Australia
**Phone:** +61 3 8695 4400
**Fax:** +613 8695 4419
**E-mail:** contactus@incitecpivot.com.au
**Web site:** www.incitecpivot.com.au
**Parent Company:** Orica Ltd. (Australia)
**Note:** Result of merger between Incitec Fertilizers and Pivot Ltd., June 1, 2003.

**Company:** Lime Industries Pty Ltd
**Address:** P.O. Box 1544, Osborne Park W.A. 6916, Australia
**Phone:** +61 8 94468644
**Fax:** +61 8 92442071
**E-mail:** general@limeinductries.com.au
**Web site:** www.limeindustries.com.au

**Company:** Nufarm Ltd.
**Address:** 103-105 Pipe Road, Laverton North, Victoria 3026, Australia
**Phone:** +61 3 9282 1000
**Fax:** +61 3 9282 1001
**E-mail:** corporate.information@au.nufarm.com
**Web site:** www.nufarm.com
**Stock listing:** Australian SE: NUF
**Parent Company:** Fernz Corporation (New Zealand)

**Company:** Orica Ltd.
**Address:** 1 Nicholson St., Melbourne, Victoria 3000, Australia
**Phone:** +61 3 9665 7111
**Fax:** +61 3 9665 7937
**E-mail:** companyinfo@orica.com
**Web site:** www.orica.com.au
**Stock listing:** Australian SE: ORI

**Company:** Quantum Chemicals Pty Ltd.
**Address:** 70 Quantum Close, Quantum Industrial Par., Dandenong South, Victoria 3175, Australia
**Phone:** +61 3 8795 8000
**Fax:** +61 3 8795 8099
**E-mail:** sales@quantum-group.com.au
**Web site:** www.quantum-group.com.au
**Parent Company:** Quantum Group (Australia)

**Company:** Rentokil Initial Pty. Ltd.

Address: Level 2, 150 Mowbray Rd., Willougby, NSW 2068, Australia
Phone: +61 2 9370 9300
Fax: +61 2 9958 0776
E-mail: rentokiul@rentokilinitial.com.au
Web site: www.rentokilinitial.com.au

Company: Sanonda (Australia) Pty Ltd.
Address: P.O. Box 943, Tewantin Qld 4565, Australia
Phone: +61 7 5473 0344
Fax: +61 7 5473 0544
Web site: www.sanondaoz.com

Company: Webcot
Address: APP House, 14 Rodborough Rd., Frenches Forest, NSW 2086, Australia
Phone: +61 2 9984 2255
Fax: +61 2 9984 2222
E-mail: mail@webcot.com.au
Web site: www.webcot.com.au

Company: Wesfarmers CSBP Ltd.
Address: Kwinana Beach Rd, P.O. Box 345, Kwinana, Western Australia 6167, Australia
Phone: +61 8 9411 8777
Fax: +61 8 9411 8425
E-mail:: corinne_hawke@csbp.wesfarmers.com.au
Web site: www.csbp.com.au

Company: WMC Limited
Address: IBM Centre, Level 16, 60 City Rd., Southbank, Victoria 3006, Australia
Phone: +61 3-9685-6000
Fax: +61 3-9686-3569
E-mail: info@wmc.com
Web site: www.wmc.com

# BELGIUM

Company: Acros Organics N.V.
Address: Geel West Zone 2, Janssen Pharmaceuticalaan 3A, B-2440 Geel, Belgium
Phone: +32 14 575211
Fax: +32 14 593434
E-mail: infodesk@be.acros.com
Web site: www.be.acros.com
Parent Company: Fisher Scientific (USA)

Company: Chimac-Agriphar S.A. (AGRIPHAR)
Address: Rue de Renory, 26, B-4102 Ougree, Belgium
Web site: www.agriphar.com

Company: Taminco
Address: Pantserschipstraat 207, 9000 Gent, Belgium
Phone: +32 (0)9 254 14 11
Fax: +32 (0)9 254 14 10
Web site: www.tamico.com

Company: Solvay Group SA
Address: Rue du Prince Albert 33, B-1050 Brussels, Belgium
Phone: +32 (0) 2 509 60 16 (Corporate Communications)
Fax: +32 (0) 2 509 72 40
E-mail: solvay.international@solvay.com
Web site: www.solvay.com

Company: Tessenderlo Chemie N.V.
Address: Rue du Trône 130, 1050 Brussels, Belgium
Phone: +32 2 639 18 11
Fax: +32 2 639 19 99
E-mail: info@tessenderlo.be
Web site: www.tessenderlo.be
Stock listing: Brussels SE: TES
Parent Company: Tessenderlo Group (Belgium)

Company: UCB Group SA
Address: Allee de la Recherche 60, B-1070 Brussels, Belgium
Phone: +32 2 559 99 99
Fax: +32 2 559 95 71
E-mail: Maryse.Delbart@UCB-group.com
Web site: www.ucb.be

Company: UMICORE N.V.
Address: Rue du Marais 31, B-1000 Brussels, Belgium
Phone: +32 (0)2 227 71 11
Fax: +32 (0)2 227 79 00
Web site: www.um.be
Stock listing: Brussels SE: UNIM
*Note:* Formerly known as Union Miniere (Belgium)

Company: Union Miniere N.V., now UMICORE (Belgium)

# BRAZIL

Company: Alfa Rio Quimica Ltda.
Address: Estrada Rio/Teresopolis, 6.410 Km 140, Imbane, 25271-970 Duque de Caxias, RJ, Brazil
Phone: +55 021-776-1997
Fax: +55 021-776-3621
E-mail: alfario@netscape.net
Web site: http://www.cibadbostongear.com/pqs/bhsx.htm

Company: Cia. Petroquimica do Sul (COPESUL)
Address: Br. 386 Rodovia Tabai Canoas-Km 419, III Polo Petroq 95853-000 Triunfo RS, Brazil
Phone: +55 51 457 6356

E-mail: comunicacao@copesul.com.br
Web site: www.copesul.com.br
Stock listing: Rio SE: CPS3

**Company: Copene Petroquimica do Nordeste SA**
Address: Rua Eteno 1561, Polo Petroquimico de Camacari, Camacari, Bahia-CEP, Brazil
Phone: +55 71 832 5504
Fax: +55 71 832 1733
E-mail: info@copene.com
Web site: www.copene.com.br
Stock listing: NYSE: PNE; Sao Paulo SE: CP5

**Company: Milenia Agro Ciencias SA**
Address: Rua Pedro Antonio De Souza 400, Parque Rui Barbosa, Londrina Parana 86031-610, Brazil
Phone: +55 43 371-9000
Fax: +55 43 371-9014
E-mail: evicente@milenia.br
Web site: www.milenia.com.br
Parent Company: Makteshim-Agan Industries (Israel)

**Company: Petrobras,** see Petroleo Brasileiro (Brazil)

**Company: Petroleo Brasileiro SA (PETROBRAS)**
Address: Avenida Republica de Chile 65, 20035-900 Rio de Janeiro, Brazil
Phone: +55 21 534 4477
Fax: +55 21 534 3247
Web site: www.petrobras.com.br
Stock listing: Sao Paulo SE: POB; NYSE: PBR

**Company: Petroquimica Uniao SA (PQU)**
Address: Av. Brigadeiro Faria Lima, 2020, 5 an, cj 51, Sao Paulo SP, Brazil
Phone: +55 11 816 7333
Web site: www.pqu.com.br

## BULGARIA

**Company: Chimco AD**
Address: 3037 Vratza, Bulgaria
Phone: +359 92 61071
Fax: +259 92 61118
E-mail: info@chimco.bg
Web site: www.chimco.bg

## CANADA

**Company: Acetex Corporation**
Address: 750 World Trade Center, 999 Canada Place, Vancouver, British Columbia, Canada V6C 3E1
Phone: +1 604-688-9600
Fax: +1 604-688-9620
E-mail: info@acetex.com
Web site: www.acetex.com
Stock listing: Toronto SE: ATX

**Company: Adrian Resources Ltd.**
Address: 2000-1055 W. Hastings St., Vancouver, British Columbia, Canada V6E 2E9
Phone: +1 604-331-8772
Fax: +1 604-331-8773
Web site: www.adrian.com
Stock listing: Toronto SE: ADL; Nasdaq: ADLRF

**Company: Agrium Inc.**
Address: 13131 Lake Fraser Dr. SE., Calgary, Alberta, Canada T2J 7E8
Phone: +1 403-225-7000
Fax: +1 403-225-7609
E-mail: indprod@argium.com
Web site: www.agrium.com
Stock listing: Toronto SE: AGU; NYSE: AGU
Parent Company: Unocal Inc. (USA)

**Company: Agrowchem Inc.**
Address: 620 Cataraqui Woods Dr., Unit 2, Kingston, ON K7P 1TB, Canada
Phone: +1 613-384-4957
Fax: +1 613-384-0662
E-mail: info@agrowchem.com
Web site: www.agrowchem.com

**Company: BASF Canada Inc.**
Address: 345 Carlingview Dr., Toronto, Ontario M9W 6N9, Canada
Phone: +1 416-675-3611
Fax: +1 416-674-2588
Web site: www.basf.ca
Stock listing: Frankfurt SE: BAS; NYSE: BF
Parent Company: BASF Group AG (Germany)

**Company: First Quantum Minerals Ltd.**
Address: 543 Granville St., Vancouver, British Columbia, Canada V6C 1X8
Phone: +1 604-688-6577
Fax: +1 604-688-3818
E-mail: info@first-quantum.com
Web site: www.first-quantum.com
Stock listing: Toronto SE: FM

**Company: IMC Vigoro,** see Vigoro (Canada)

**Company: Noranda Inc.**
Address: 181 Bay St., Ste. 200, BCE Place, Toronto, Ontario, Canada M5J 2T3
Phone: +1 416-982-7111

Fax: +1 416-982-7423
Web site: www.noranda.com
Stock listing: Toronto SE: NOR

**Company: Potash Corporation of Saskatchewan Inc.**
Address: 121 1st Avenue South, Saskatoon Suite 500, Saskatoon, Saskatchewan, Canada S7K 7G3
Phone: +1 306 933 8500
Fax: +1 306 933 8877
E-mail: corporate.relations@potashcorp.com
Web site: www.potashcorp.com
Stock listing: Toronto & NYSE: POT

**Company: Recochem Inc., International Division**
Address: 850-T Montee De Liesse, Montreal Quebec, Canada H4T 1P4
Phone: +1 514-341-3550
Fax: +1 514-341-1292
E-mail: internationaldivision@recochem.com
Web site: www.recochem.com

**Company: Teck Cominco Ltd.**
Address: 600 - 200 Burrard St., Vancouver, British Columbia, Canada V6C 3L9
Phone: +1 604-687-1117
E-mail: info@teckcominco.com
Web site: www.teckcominco.com
Stock listing: Toronto and American SE: TEK.B

**Company: Timminco Ltd.**
Address: WaterPark Place, 10 Bay St., Ste. 901, Toronto, Ontario, Canada M5J 2R8
Phone: +1 416 364 5171
Fax: +1 416 364 3451
E-mail: info@timminco.com
Web site: www.timminco.com
Stock listing: Toronto SE: TIM

**Company: Vigoro**
Address: 10 Craig St., Brantford, Ontario, Canada N3R 7J1
Phone: +1 519-757-0077
Fax: +1 519-757-0080
E-mail: products@nu-gro.ca
Web site: www.vigoro.on.ca
Parent Company: Nu-Grow Corp. (Canada)

# CHILE

**Company: Codelco SA (Corporacion Nacional del Cobre)**
Address: Huerfanos 1270, Santiago, Chile
Phone: +56 2 690 3000
Fax: +56 2 690 3059
Web site: www.codelco.com

**Company: Corporacion Nacional del Cobre SA,** see Codelco SA (Chile)

**Company: Minera Escondida Ltda.**
Address: Avda. Americo Vespucio Sur 100, Las Condes, Santiago, Chile
Phone: +56 2-330-50-00
Fax: +56 2-207-65-20
Web site: www.escondida.cl
Parent Company: BHP Billiton (Australia) and Rio Tinto (UK)

**Company: Sociedad Quimica y Minera de Chile SA,** see SQM (Chile)

**Company: SQM SA (Sociedad Quimica y Minera de Chile SA)**
Address: El Trovador 4285, Las Condes, Santiago, Chile
Phone: +56 2 425 2000
Fax: +56 2 425 2268
Web site: www.sqm.com
Stock listing: NYSE: SQM; Santiago SE: SQM-A

# CHINA

**Company: Agro-care Chemical Industry Group Limited**
Address: 369 Jiangsu Rd., Shanghai 200050, China
Phone: +86 21-5240-0067
Fax: +86 21-5240-1089
E-mail: sales@agrocare.com.cn
Web site: www.agrocare.com.cn

**Company: Anhui Fengle Agrochemical Co., Ltd.**
Address: Heyo Rd., Hefei, Anhui 230011, China
Phone: +86 551-453-1880
Fax: +86 551-453-0643
E-mail: fengle@fengle-pesticide.com
Web site: www.fengle-agrochem.com

**Company: Anhui Huaxing Chemical Industry Co., Ltd.**
Address: Wujiang, Hexian, Anhui, China
Phone: +86 565-539 3988
Fax: +86 551-283 3421
E-mail: hxtrade@huaxingchem.com
Web site: www.huaxingchem.com

**Company: Cangzhou Green Chemical Co., Ltd.**
Address: Xingji Town Industrial Park, Cangzhou City, Hebei Province 061021, China
Phone: +86 317-4851-690
Fax: +86 317-4851-707
E-mail: market@green-chem.com
Web site: www.green-chem.com

**Company:** Changfeng Chemical Co., Ltd.
**Address:** 29 W. Hua Huiyuan Rd, Rm. 28-2, YuBei District, Changqing 401147, China
**Phone:** +86 23-6791-9006
**Fax:** +86 23-6791-7791
**E-mail:** changfeng@public.cta.cq.cn
**Web site:** www.changfengchem.com

**Company:** Changzhou Fengdeng Pesticide Factory
**Address:** Dengguan County, Jintan City, Jiangsu Province 213252, China
**Phone:** +86 519-242-2752
**E-mail:** jlx@fdpesticide.com
**Web site:** www.fdpesticide.com

**Company:** Changzhou Kangmei Chemical Industry Co., Ltd.
**Address:** Rulin Town, Jintan City, Jinagsu Province 213225, China
**Phone:** +86 519-283-5448
**Fax:** +86 519-256-1700
**Web site:** www.kangmei.com

**Company:** Changzhou Wujin Linchuan Chemical Plant
**Address:** Miaoqiao Town, Wujin City, Jiang-su Province, China
**Phone:** +86 519-646-1196
**Fax:** +86 519-646-3703
**E-mail:** czmqgxx@public.cz.js.cn
**Web site:** www.xingcheng.com.cn

**Company:** China Chemicals
**Address:** Luxun Mansion, Ste. 12-G, 568 Ou Yang Rd., Shanghai 200 081, China
**Phone:** +86 21-5671-9141
**Fax:** +86 21-5671-9140
**Web site:** www.china-chemicals.com

**Company:** China Shenghua Group Agrochemical Company
**Address:** 12th West Fl., Fancy Garden, No. 6 Xibahe South Rd., Chaoyang District, Beijing 100028, China
**Phone:** +86 10-51666078
**Fax:** +86 10-51666079
**Web site:** www.agrofar.com

**Company:** Chongqing Chuandong Chemical (Group) Co., Ltd.
**Address:** 20 Zhongyao Street, Baishatuo Nanan, Municipality of Chongqing 400 063, China
**Phone:** +86 23-6295-1148
**Fax:** +86 23-6295-0365
**E-mail:** ggf@hi2000.com
**Web site:** www.cqchemgroup.com

**Company:** Epochem Co., Ltd.
**Address:** 168 Changdong Rd., Ste. 679, Shanghai 201612, China
**Phone:** +86 21 6764-3383
**Fax:** +86 21 6764-3350
**Web site:** www.epochem.com

**Company:** Forward (Beihai) Hepu Pesticide Co., Ltd.
**Address:** No. 1 Chinshaichung, Lianzhou, Hepu, Beihai, Guangxi, China
**Web site:** www.forwardinter.com
**Parent Company:** Forward International Ltd. (China)

**Company:** Hebei Huafeng Chemical Group
**Address:** 19 Changxing Rd., Qing Liang Dian Town, Wuyi County, Hengshui City 053000, China
**Phone:** +86 318-581-5276
**Fax:** +86 318-581-5858
**E-mail:** info@huafeng-chemical.com
**Web site:** www.huafeng-chemical.com

**Company:** Hebei Long Age Pesticide Co., Ltd.
**Address:** No. 999, Fuqiang Rd., Zaoqiang County, Hengshui City, Hebei Province 053 100, China
**Phone:** +86 318-822-8383
**Fax:** +86 318-822-8015
**E-mail:** sjny@163.com
**Web site:** www.shiji-pesticide.com

**Company:** Hunan Darong Chemical & Pesticide Co., Ltd.
**Address:** Cha Shan Ao, Hengyang City, Hunan 421173, China
**Phone:** +86 734-871-0121
**Fax:** +86 734-871-0122
**E-mail:** hndrcm@163.net
**Web site:** www.hndarong.com

**Company:** Hunan Tianyu Pesticide Chemical Group Co., Ltd.
**Address:** XiangJiang South, R.D. # 83, Hengyang City, Hunan 421001, China
**Phone:** +86 734-820-2592
**Fax:** +86 734-826-0705
**E-mail:** typgs@mail.hy.hn.cn
**Web site:** www.tianyugroup.com

**Company:** Jiangmen Pesticide Factory
**Address:** 69, Jianghui Rd., Jiangmen, Guangdong, China
**Phone:** +86 750-353-4351
**Fax:** +86 750-355-3048
**E-mail:** jimnyc@pub.jiangmen.gd.cn
**Web site:** www.china-jmpesticide.com

**Company:** Jiangsu Wujin Zhenhua Chemical Plant
**Address:** Henlin Town, Wujin, Changzhou, Jiangsu 213101, China
**Phone:** +86 519-8781340
**Fax:** +86 519-8781339
**E-mail:** zhuachem@public.cz.js.cn
**Web site:** www.zhuachem.com

**Company:** Jilin Chemical Industrial Co., Ltd.
**Address:** 31 East Zunyi Road, Longtan District, Jilin City, Jilin Province 132021, China
**Phone:** +86 432-397-6445
**Fax:** +86 432-302-8126
**E-mail:** webmaster@jcic.com
**Web site:** www.jcic.com
**Stock listing:** Hong Kong SE: 0368, NYSE: JCC
**Parent Company:** Jilin Chemical Group Corporation, State owned by Jilin Provincial Government (China)

**Company:** Jingma Chemicals Ltd.
**Address:** 104 Yu Gu Road, Hangzhou, China
**Phone:** +86 571-798-0695
**Fax:** +86 571-798-0702
**E-mail:** jmc@mail.hz.zj.cn
**Web site:** www.jmcchem.com

**Company:** Juhua Group Corporation
**Address:** West Lake Mansions No. 202, Wenter Rd., Hangzhou, Zhejiang, China
**Phone:** +86 571-884-1942
**Fax:** +86 571-884-1950
**Web site:** www.juhua.com.cn
**Stock listing:** Hong Kong SE

**Company:** Kunshan Chemical Group (Industries) Corp.
**Address:** No. 21 Chao Yang Subroad, Kunshan, Jiangsu Province 215300, China
**Phone:** +86 520-730-2401-888
**Fax:** +86 520-730-2401
**E-mail:** kh306764@publicl.sz.js.cn
**Web site:** www.js.cei.gov.cn/JSfamous/0504005/eshenchq.htm

**Company:** Limin Chemical Co., Ltd.
**Address:** Xinhua Rd., Xinyi City, Jiangsu Province, China
**Phone:** +86 516-892-3527
**Fax:** +86 516-893-7719
**Web site:** www.limin-chemical.com.cn

**Company:** Nanjing Agrochemicals Co., Ltd. of AgroDragon
**Address:** Rm. 822, East Bldg., No. 1500, Century Ave., Pudong New Area, Shanghai 200122, China
**Phone:** +86 21-2890-1828

**Fax:** +86 21-2890-1829
**E-mail:** adragon@online.sh.cn
**Web site:** www.pesticide-china.com
**Parent Company:** AgroDragon (China)

**Company:** Pharm - Chem Manufacturing Co., Ltd
**Address:** No. 3 East of Da Yuan Rd., Bai Yun Hill, Guangzhou, China
**Phone:** +86 20-875-30037
**Fax:** +86 20-875-10853
**E-mail:** abc123@public.guangzhou.gd.cn
**Web site:** www.chmes.com
**Parent Company:** Sino-US Joint Venture Sheng Fa Wang Group (China)

**Company:** Sanonda Zhengzhou Pesticide Co., Ltd.
**Address:** No. 57 Chengdongnan Rd., Zhengzhou, Henan, 450009, China
**Phone:** +86 371-6818-876
**E-mail:** sndzzpco@public2.zz.ha.cn
**Web site:** www.zzpesticide.com

**Company:** Sevencontinent Agrichemical Co., Ltd.
**Address:** 28 Chengbei Rd., Zhangjiagang, 215600, China
**Phone:** +86 512-586-78398
**Fax:** +86 512 -586-86955
**E-mail:** info@sevencontinent.com
**Web site:** www.sevencontinent.com

**Company:** Shangdong Baoyuan Chemical Co.
**Address:** Huantai, Zibo, Shangdong 256401, China
**Phone:** +86 533-8514528
**Fax:** +86 533-8514368
**E-mail:** ilyandimy@sina.com
**Web site:** www.baoyuanchem.com

**Company:** Shandong Changyi Salt Chemicals (Group) Co., Ltd.
**Address:** 10 Beihai Rd, Changyi City 261300, Shandong Province, China
**Phone:** +86 536-721-1854
**Fax:** +86 536-721-1455
**E-mail:** cyyygs@public.wfptt.sd.cn
**Web site:** www.sdcyyh.com

**Company:** Shandong Huayang Pesticide Group
**Address:** Ciyao, Ningyang, Shandong, China
**Phone:** +86 538-582-6155
**Fax:** +86 538-582-6258
**E-mail:** huayang@public.taptt.sd.cn
**Web site:** http://tay.51.net

**Company:** Shanghai Agricultural Chemical Industry Corp.

Address: 10th Floor E, 818 Dongfang Rd., Pu Dong New Area, Shanghai 200 122, China
Phone: +86 21-582-00369
Fax: +86 21-508-12742
E-mail: info@shanghai-pesticides.com
Web site: www.shanghai-agrochem.com

**Company: Shangyu Chemical Industry Corp.**
Address: Sanpeng Bridge, Baiguan Town, Shangyu City, Zhejiang 312351, China
Phone: +86 575-2014544
Fax: +86 575-2011434
E-mail: ingo@shangyuchem.com
Web site: www.shangyuchem.com

**Company: Shanxi Friends Union Chemicals Co., Ltd.**
Address: Hengxin Bldg. 7F, Str. No. 18, Jincheng, Shanxi 048000, China
Phone: +86 356-205-5722
Fax: +86 356-205-5664
E-mail: nfu@public.jc.sx.cn
Web site: www.thiocyanate.com

**Company: Shenyang Harvest Agrochemical Co., Ltd.**
Address: 100 Jidong Rd., Linsheng Town, Sujiatun District, Shengyang 110108, China
Fax: +86 24 89487788
Web site: www.agrochemcn.com

**Company: Shenzhen Guomeng Industry Co., Ltd.**
Address: 14/F No. 2 Ruipeng Bldg., No. 22 South Dongmeng Rd., Shenzhen, China
Phone: +86 755-822-84398
Fax: +86 755-822-83498
E-mail: guomeng@public.szptt.net.cn
Web site: www.gmagrochem.com

**Company: Sheyang Pesticides and Chemical Industry Co., Ltd.**
Address: No. 32 East Renmin Rd., Shenyang County, Jiansu Province 224300, China
Phone: +86 515-2324359
Fax: +86 515-2325430
E-mail: fychem@fengyangchem.com
Web site: www.fengyangchem.com

**Company: Sinopec Corporation (China Petroleum & Chemical Corp.)**
Address: No. A6 Hui xin East St., Chaoyang District, Beijing, 100029, China
Phone: +86 10-6499-8828
E-mail: webmaster@sinopec.com.cn
Web site: www.sinopec.com

Stock listing: NYSE: SNP; London: SNP; Hong Kong: 0386; Shanghai: 600028

**Company: SuYan Agrochemical Group**
Address: No. 80, Luosiqiano St., Nanjing, China
Phone: +86 25-6301-856
Fax: +86 25-6301-702
E-mail: suyan@china-pesticide.com
Web site: www.china-pesticide.com

**Company: Wangs Ltd.**
Address: 16 Rongtai Villas, Xiangzhou, Zhuhai 519001, China
Phone: +86 756-252-2501
Fax: +86 756-226-8793
E-mail: info@ewangs.com
Web site: www.ewangs.com

**Company: Wuhan kernel Bio-pesticide Co., Ltd.**
Phone: +86 27-87514086
Fax: +86 27-87514063
E-mail: admin@kenuo.com.cn
Web site: www.e-kernel.net

**Company: Wuxj Ruize Pesticide Co., Ltd.**
Address: 354 North St., Yunting Town, Jiangyin, Jiangsu 214422, China
Phone: +86 510-615-1215
Fax: +86 510-615-1137
E-mail: wxiz.jy@public1.wx.js.cn
Web site: www.ruizepesticide.com
Parent Company: Ministry of Chemical of Jiangsu Province (China)

**Company: Wuzhou International Co., Ltd.**
Address: No. 6/F, West Block of Financial Bldg., 182 Quigchun Rd., Hangzhou, China
Phone: +86 571-8724-4967
Fax: +86 571-8724-4707
E-mail: josephq@mail.hz.zj.cn
Web site: www.chinax.com/pages/wuzhou

**Company: YiHua Group**
Address: 164 Fuma Rd., Fuzhou, Fujian 350 011, China
Phone: +86 591 366 0186
Fax: +86 591 366 0140
E-mail: fyh@pub2.fz.fj.cn
Web site: www.chinachlorate.com
Parent Company: Fuzhou YiHua Group Co., Ltd. (China)

**Company: Zhejiang Heben Chemicals Co., Ltd.**
Address: No. 3-1 Jujiang East Rd., Yangfushan, Wenzhou Zhejiang Province 325 003, China
Phone: +86 577-889-17785

Fax: +86 577-889-17329
E-mail: hb-p@hb-p.com
Web site: www.hb-p.com

**Company: Zhejiang Yifan Chemical Co., Ltd**
Address: No. 2, Zone 4, Fuping, Wenzhou, Zhejiang 325013, China
Phone: +86 577-8663-0101
Fax: +86 577-8663-6638
E-mail: yifan@chinayifan.com
Web site: www.chinayifan.com

## COLOMBIA

**Company: Proficol S.A.**
Address: Carrera 11 No.87-51 Piso 4, Bogota D.C., Colombia
Phone: +91 644-6730
Fax: +91 640-1210
E-mail: contact.us@proficol.com
Web site: www.proficol.com
Parent Company: Makheteshim-Agan Industries (Israel)

## CROATIA

**Company: Chromos Agro d.d.**
Address: Zitnjak b.b., Zagreb, Croatia
Phone: +385 1 2404 188
Fax: +385 1 2404 420
Web site: www.chromos-agro.hr

## CZECH REPUBLIC

**Company: Deza a. s.**
Address: Masarykova 753, 757 28 Valasske Mezioiei, Czech Republic
Phone: +42 65 169 11 11
Fax: +42 65 161 15 46
E-mail: info@deza.cz
Web site: www.deza.cz

## DENMARK

**Company: Cheminova A/S**
Address: P.O. Box 9, DK-7620 Lemvig, Denmark
Phone: +45 96 90 96 90
Fax: +45 96 90 96 91
E-mail: info@cheminova.dk
Web site: www.cheminova.com
Parent Company: Auriga Industries A/S (Denmark)

**Company: Dankalk**
Address: Aggersundvej 50, DK 9670 Logstor, Denmark
Phone: +45 98 67 31 55
Fax: +45 98 67 14 16
E-mail: prc@dlq.dk
Web site: www.dankalk.dk
Parent Company: DLG - Danish Co-operative Farm Supply

**Company: Haldor Topsoe A/S**
Address: P.O. Box 213, Nymollvej 55, DK-2800 Lyngby, Denmark
Phone: +45 4527 2000
Fax: +45 4527 2999
E-mail: webmaster@topsoe.dk
Web site: www.haldortopsoe.com

## ECUADOR

**Company: Agrochemicals del Ecuador (AGROCHEM)**
Address: Av. Juan Tanca Marengo y Jaime Roldos, Guayaquil, Ecuador
Phone: +593 4 354 157
Fax: +593 4 357 773
E-mail: agrochem@email.com
Web site: www.agrochem.com.ec

## FINLAND

**Company: Kemira Chemical Oy**
Address: Porkkalankatu 3, P.O. Box 330, FIN-00101 Helsinki, Finland
Phone: +358 10 86 1211
Fax: +358 10 862 1124
Web site: kc.kemira.com
Stock listing: Helsinki SE: KRA
Parent Company: Kemira Group Oy (Finland)

**Company: Kemira GrowHow Oy**
Address: P.O. Box 900, FIN-00181, Helsinki, Finland
Phone: +358 1021 5111
Fax: +358 102 152 126
E-mail: infr@kimera-growhow.com
Web site: www.kemira-grohow.com
Stock listing: Helsinki SE: KRA
Parent Company: Kemira Group Oy (Finland)

**Company: Neste Chemicals Oy**
Address: Snellmaninkatu 13, 00170 Helsinki, Finland
Phone: +358 10 585 2000
Fax: +358 10 585 2001
E-mail: nordkemi.communications@neste.com
Web site: www.nestechemicals.com
Parent Company: Dynea International Oy

## FRANCE

**Company:** Air Liquide Group
**Address:** 75 quai d'Orsay, 75007 Paris, France
**Phone:** +33 1 40 62 55 55
**Fax:** +33 1 40 62 56 92
**Web site:** www.airliquide.com
**Stock listing:** Paris Bourse: AL

**Company:** a.m.p.e.r.e. Industrie SA Dept Chimie
**Address:** BP 9125, 18 , Av de I'tle-de-France, Saint-Ouen l'Aumone, 95074 Cergy-Pontoise CEDEX, France
**Phone:** +33 134401280
**Fax:** +33 134645699
**E-mail:** chimie@ampere.com

**Company:** Arkema Group S.A.
**Address:** 4-8, cours Michelet, La Defense 10, 92091 Paris La Defense Dedex, France
**Phone:** +33 1 49 00 80 80
**Fax:** +33 1 49 00 83 96
**E-mail:** info@atofina.com
**Web site:** www.arkemagroup.com
**Parent Company:** Total (France)
**Note:** Arkema Group is former Atofina Chemical subsidiary of Total (France)

**Company:** ATOFINA, now Arkema Group S.A. (France)

**Company:** Aventis CropScience, now Bayer CropScience, Germany

**Company:** Calliope SA
**Address:** Route d'Artix BP 80, 64150 Nogueres, France
**Phone:** +33 5 59 60 92 92
**Fax:** +33 5 59 60 92 99
**E-mail:** calliope-nogueres@calliope-sa.com
**Web site:** www.calliope-sa.com
**Parent Company:** Arysta LifeScience Corp. (Japan)

**Company:** CECA
**Address:** Immeuble Iris - La Défense 2, 92062 Paris, France
**Phone:** +33 1 47 96 90 90
**Fax:** +33 1 47 96 91 91
**Web site:** www.ceca.fr
**Parent Company:** ATOFINA, Subsidiary of Total (France)

**Company:** Cerexagri Inc. - Europe/International
**Address:** 1, Rue des Freres, Lumiere, 78370 Plaisir, France
**Phone:** +33 1-30-81-7300
**Fax:** +33 1-30-81-7250
**E-mail:** contact@cerexagri.com
**Web site:** www.cerexagri.com
**Parent Company:** ATOFINA (France), Subsidiary of Total (France)
*Note:* As of January 1, 2001, the Agrichemicals Division of Elf Atochem became Cerexagri, Inc.

**Company:** Elf Aquitaine, now Arkema, Subsidiary of Total (France)

**Company:** Elf Atochem, now Arkema, Subsidiary of Total (France)

**Company:** Fina, now Total (France)

**Company:** Fina Oil, now Total (France)

**Company:** Grande Paroisse SA
**Address:** 12 place de l'Iris, La Defense 2, 92062 Paris La Defense Cedex, France
**Phone:** +33 1 47 96 97 66
**Fax:** +33 1 47 78 11 60
**Web site:** www.grande-paroisse.fr/gp/e
**Stock listing:** Paris Bourse: 12027; NYSE: TOT
**Parent company:** ATOFINA (France), Subsidiary of Total (France)

**Company:** Groupe SPCA, see Societe Commerciale des Potasses et de l'Azote

**Company:** Holvoet Chimie
**Address:** Industrial area of Tourni Quest II, Rue des Sablieres, 1, 7522 Blandain, France
**Phone:** +32 69 88 10 00
**Fax:** +32 69 88 10 01
**E-mail:** info@holvoet-chimie.com
**Web site:** www.holvoet-chimie.com

**Company:** ISOCHEM
**Address:** Chemin de la Loge, 31078 Toulouse Cedex 4, France
**Phone:** +33 5 62 25 72 40
**Fax:** +33 5 62 25 72 15
**E-mail:** agro@snpe.com
**Web site:** www.isochem.biz/Iso_home_uk
**Parent Company:** Groupe SNPE SA (France)

**Company:** Lhoist Group
**Address:** Saint-Jean-des-Bois, B-1342 Limelette, France
**Phone:** +33 32 10 23 07 11
**Fax:** +33 32 10 23 09 50
**E-mail:** info@lhoist.com
**Web site:** www.lhoist.com

**Company:** Novartis agriculture units, now Syngenta AG (Switzerland)

**Company:** PCF Chimie
**Address:** Quai Jean-Jaures, 07800 La Voulte Sur Rhone, France
**Phone:** +33 4 75 85 88 00
**Fax:** +33 4 75 85 31 38
**E-mail:** contact@pcf-chimie.com
**Web site:** www.pcf-chimie.com
**Parent Company:** Pharmacie Central de France SA (France)

**Company:** Pechiney SA
**Address:** 10, Place des Vosges, La Defense 5, 92400 Coubevoie, France
**Phone:** +33 01 15 62 82 00 0
**Fax:** +33 01 15 62 83 38 2
**E-mail:** investor.relations@pechiney.com
**Web site:** www.pechiney.com
**Stock listing:** NYSE: AL
**Parent Company:** Alcan (Canada)

**Company:** Rhodia Eco Services
**Address:** Les Bureaux du Lac II Immeuble P 39, rue Robert Caumont, 33049 Bordeaux Cedex, France
**Phone:** +33 5 56 69 68 90
**Fax:** +33 5 56 50 67 34
**Web site:** www.rhodia-eco-services.com
**Stock listing:** Paris Bourse & NYSE: RHA
**Parent Company:** Rhodia Group SA (France)

**Company:** Rhodia Group SA
**Address:** 26, quai AlphonsenLe Gallo, 92512 Boulogne-Billancourt Cedex 92512, France
**Phone:** +33 1 55 38 40 00
**Fax:** +33 1 55 38 44 71
**E-mail:** produit.info@eu.rhodia.com
**Web site:** www.rhodia.com
**Stock listing:** Paris Bourse & NYSE: RHA

**Company:** Rhone-Poulenc, see Aventis SA (France)

**Company:** Rhone-Poulenc Agro France, now Bayer CropScience (Germany)

**Company:** Roussel Uclaf, see Akzo Nobel (Netherlands)

**Company:** Societe Commerciale des Potasses et de l'Azote (SCPA)
**Address:** 2, place du General de Gaulle, BP 1170, 68053 Mulhouse Cedex, France
**Phone:** +33 3-89-36-36-00
**Fax:** +33 3-89-45-79-17
**Web site:** www.scpa.fr

**Company:** Total
**Address:** 2, Place de la Coulope, 92400 Courbevoie La Defense 6, France
**Phone:** +33 01 47 44 45 46
**Fax:** +33 01 47 44 78 78
**E-mail:** webmaster@total.com
**Web site:** www.total.com
**Stock listing:** Paris Bourse: 12027; NYSE: TOT
*Note:* Formerly TotalFinaElf (France)

## GERMANY

**Company:** ABCR GmbH & Co. KG
**Address:** Im Schlehert 10, 76187 Karlsruhe, Germany
**Phone:** +49 721-95061-16
**Fax:** +49 721-95061-80
**E-mail:** kiso@abcr.de
**Web site:** www.abcr.de

**Company:** AgrEvo, now Bayer CropScience (Germany)

**Company:** BASF Group AG
**Address:** Carl-Bosch Strasse 38, 67056 Ludwigshafen, Germany
**Phone:** +49 621 60-0
**Fax:** +49 621 60-42525
**E-mail:** info.service@basf-ag.de
**Web site:** www.basf.com
**Stock listing:** Frankfurt SE: BAS; NYSE: BF

**Company:** BASF Agricultural Products Group
**Address:** Carl-Bosch Strasse 38, 67056 Ludwigshafen, Germany
**Phone:** +49 621 60-0
**Fax:** +49 621 60-42525
**E-mail:** info.service@basf-ag.de
**Web site:** www.basf-ag.de
**Stock listing:** Frankfurt SE: BAS; NYSE: BF
**Parent Company:** BASF Group AG (Germany)
*Note:* A result of the merger of American Cyanamid's agricultural products group with BASF agricultural products.

**Company:** Bayer Group AG
**Address:** 51368 Leverkusen, Germany
**Phone:** +49 214-30-1
**Fax:** +49 214-30-7
**Web site:** www.bayer.de
**Stock listing:** Nasdaq: BAYZY

**Company:** Bayer CropScience AG
**Address:** Alfred-Nobel-Str. 50, D-40789 Monheim am Rhein, Germany

Web site: www.bayercropscience.com
**Parent Company:** Bayer Group AG, Germany
*Note:* Bayer CropScience is the result of 2002 merger with Aventis CropScience (formerly Rhone-Poulenc Agro France) into the Bayer Agricultural Group, a subsidiary of Bayer Group AG (Germany)

**Company: Biesterfeld Siemsgluess International. GmbH**
**Address:** Ferdinandstrasse 41, D-20095 Hamburg, Germany
**Phone:** +49 40-320-08-608
**Fax:** +49 40-320-08-608
**E-mail:** international@biesterfeld.com
**Web site:** www.biesterfeld-siemsgluess.com
**Parent Company:** Biesterfield Group (Germany)

**Company: Boehringer Ingelheim GmbH**
**Address:** Binger Strasse 173, 55216 Ingelheim, Germany
**Phone:** +49 61-32-77 0
**Fax:** +49 6-32-72 0
**Web site:** www.boehringer-ingelheim.com

**Company: Celanese AG**
**Address:** Corporate Center, Frankfurt Str. 111, D-61476 Kronberg/Ts, Germany
**Phone:** +49 69-305-160-00
**Fax:** +49 69-305-160-09
**E-mail:** m.buss@celanese.com
**Web site:** www.celanese.com
**Stock listing:** NYSE: CZ, Frankfurt SE

**Company: Cognis GmbH**
**Address:** Henkel Strasse 67, 40589 Dusseldorf, Germany
**Phone:** +49 211-794-00
**Fax:** +49 211-798-40-08
**Web site:** www.cognis.com
**Stock listing:** German SE: HNKG.F

**Company: CONDEA GmbH**
**Address:** Uberseering 40, P.O. Box 60 04 49, 22297 Hamburg, Germany
**Phone:** +49 40-637-50
**Fax:** +49 40 637-535-88
**Web site:** www.sasol.com
**Parent Company:** Sasol Ltd. (South Africa)

**Company: Degussa AG**
**Address:** Degussa Feed Additives Division, Rodenbacher Chaussee 4, D-63457 Hanau-Wolfgang, Germany
**Phone:** +49 6181-59-6782
**Fax:** +49 69 6181-59-6734
**E-mail:** feed.additives@degussa.com
**Web site:** www.degussa.com
**Stock listing:** Frankfurt SE: 542 190

*Note:* Degussa is the result of the merger between Degussa-Huls AG and SKW Trostberg AG in February, 2001.

**Company: Degussa-Huls AG,** see Degussa AG (Germany)

**Company: Dynamit Nobel AG**
**Address:** Kaiserstrasse 1, P.O. Box 12 61, 53839 Troisdorf, Germany
**Phone:** +49 22-41-89-0
**Fax:** +49 22-41-89-15 40
**E-mail:** info@dynamit-nobel.com
**Web site:** www.dynamitnobel.com
**Parent Company:** MG Technologies AG (Germany)

**Company: Ehrenstorfer, Dr., GmbH**
**Address:** Bgm.-Schlosser - Str. 6A, D-86199 Augsburg, Germany
**Phone:** +49 821-90-60-80
**Fax:** +49 821-90-60-888
**E-mail:** info@analytical-standards.com
**Web site:** 213.73.6.114/eqbin/eqweb.dll?subid=10

**Company: Feinchemie Schwebda GmbH**
**Address:** Strassburger Strasse 5, D-37269 Eschwege, Germany
**Phone:** +49 56-51-92-37-0
**Fax:** +49 56 51-92-37-55
**E-mail:** info@fcs-feinchemie.com
**Web site:** www.fcs-feinchemie.com

**Company: Goldschmidt Chemicals AG**
**Address:** Konzernzentrale, Goldschmidtstrasse 100, 45127 Essen, Germany
**Phone:** +49 201-173-01
**Fax:** +49 201-173-3000
**E-mail:** info@goldschmidt.com
**Web site:** www.goldschmidt.com
**Parent Company:** Degussa Group (Germany)

**Company: INEOS Phenol**
**Address:** Dechenstrasse 3, 45966 Gladbeck, Germany
**Phone:** +49 20-43-958-302
**Fax:** +49 20-43-958-947
**E-mail:** marketing@phenolchemie.de
**Web site:** www.phenolchemie.com
**Stock listing:** Frankfurt SE: 542 190
**Parent Company:** Degussa Group (Germany

**Company: Kali und Salz Group AG (K+S)**
**Address:** Bertha-von-Suttner, Strasse 7, 34111 Kassel, Germany
**Phone:** +49 561-9301-2322
**Fax:** +49 561-9301-1186

E-mail: hans-dieter.wolter@k-plus-s.com
Web site: www.kalisalz.de
Stock listing: Frankfurt SE: SDF

Company: Linde Gas AG
Address: Seitnerstrasse 70, D-82049 Hollriegelskreuth, Germany
Phone: +49 89-74-46-0
Fax: +49 89 74-46-11-44
E-mail: info@linde.de
Web site: www.linde.de/linde-gas
Parent Company: Linde Group (Germany)

Company: Maag Agro (Germany), now Syngenta (Switzerland)

Company: Merck KGaA
Address: Frankfurter Strasse 250, 64293 Darmstadt, Germany
Phone: +49 6151-720
Fax: +49 6151-722-000
E-mail: service@merck.de
Web site: www.merck.de
Stock listing: Frankfurt SE: MRK, Zurich SE

Company: Messer Group GmbH
Address: Futingweg 34, 47805 Krefeld, Germany
Phone: +49 21-51-379-0
Fax: +49 21-51-379-9115
E-mail: communication@messer.de
Web site: www.messergroup.com
Parent Company: Allianz Capital Partners + Goldman Sachs (USA)

Company: W. Neudorff GmbH KG
Address: An der Muhle 3, 31860 Emmerthal, Germany
Phone: +49 0180-563-8367
Fax: +49 0180-05-155-6010
E-mail: info@neudorff.de
Web site: www.neudorff.de

Company: Phenolchemie GmbH & Co. KG, now INEOS Phenol (Germany)

Company: PPM Pure Metals GmbH
Address: Am Bahnhof 1, 38685 Langelsheim, Germany
Phone: +49 53-26-507-0
Fax: +49 53-26-507-151
E-mail: ppm@ppmpuremetals.de
Web site: www.ppmpuremetals.de
Parent company: Metaleurop SA (France)

Company: Schering AG
Address: Muellerstrasse 178, 13353 Berlin, Germany
Phone: +49 30-46811-11
Fax: +49 30-46815-305
E-mail: info@schering.de
Web site: www.schering.de
Stock listing: NYSE: SHR; Frankfurt: SCH

Company: Sigma-Aldrich Laborchemikalien GmbH
Address: P.O. Box 10 02 62, 30918 Seelze, Germany
Phone: +49 51-37-8238-0
Fax: +49 51-37-8238-120
E-mail: riedel@sial.com
Web site: www.rdh-lab.de
Parent Company: Sigma-Aldrich (USA)

Company: Solvay Barium Strontium GmbH
Address: Hans-Bockler Allee20, D-30173 Hannover, Germany
Phone: +49 5 11/857 27 25
Fax: +49 5 11/857 21 22
Web site: www.solvay.com/bs/en
Parent Company: Solvay Group SA (Belgium)

Company: Spiess-Urania Chemicals GmbH
Address: Heidenkampsweg 77, 200 97 Hamburg, P.O. Box 10 62 200, 200 42 Hamburg, Germany
Phone: +49 40 23 65 20
Fax: +49 40 23 65 22 55
E-mail: mail@speiss-urania.com
Web site: www.urania.de

## GREECE

Company: Hellenic Corundum SA
Address: 3 Kallifrona St., 112 Athens, Greece
Phone: +30 18 647 274
Fax: +30 18 616 473

Company: Phosphoric Fertilizers Industry SA (PFI)
Address: 97 Syngrou Ave., 117 45 Athens, P.O. Box 101 83, 541 10 Thessaloniki, Greece
Phone: +30 1 922 7391
Fax: +30 1 922 8087
E-mail: headoffice@pfi.gr
Web site: www.add.gr/pfi

## HONG KONG

Company: Rotam Agrochemical (HK) Co., Ltd.
Address: 7/F Cheung Tat Center, 18 Cheung Lee St., Chai Wan, Hong Kong
Phone: +852 2896-5608
Fax: +852 2896-5730
E-mail: general@rotamhk.com

Web site: www.rotam.com/agroc-what.asp
Parent Company: Rotam Group (Canada)

## HUNGARY

Company: Nitrokemia 2000 Rt.
Address: 1012 Budapest I. Palya u 9, Hungary
Phone: +36 1 487 5310, 88 352 011
Fax: +36 1 487 5311, 88 451 423
E-mail: nitrokemia2000@nitrokemia.hu
Web site: www.nitrokemia.hu

## INDIA

Company: Adheswara Group of Companies
Address: 26, Royapattah High Rd., Chennai, Tamil Nadu, 600 014, India
Phone: +91 44-826-7019
Fax: +91 44-823-3321
E-mail: adheswara@epages.webindia.com
Web site: epages.webindia.com/india/adheswara

Company: Aero Agro Chemical Industries, Ltd.
Address: 28, Strand Road, Calcutta 700 001, West Bengal, India
Phone: +91 33-2221-4090
Fax: +91 33-2221-4089
E-mail: aerochem@cal2.vsnl.net.in
Web site: www.aacil.com

Company: Aimco Pesticides Ltd.
Address: P.O. Box 6822, 8th Rd., East Santacruz, Bombay 400 055, India
Phone: +91 22-616-3744
Fax: +91 22-611-6736
E-mail: aimco@vsnl.com
Web site: www.aimcopesticides.com

Company: Alkyl Amines Chemicals Ltd.
Address: 401-407, Nirman Vyapar, Kendra, Plot No. 10, Sector 17, Vashi, Navi Bombay 400 703, India
Phone: +91 22-789-0632
Fax: +91 22-789-0631
E-mail: corporate@alkylamines.com
Web site: www.alkylamines.com
Stock listing: Bombay SE: 6767

Company: Allipo Chemicals
Address: 49 GIDC, Makarpura, Vadodara 390010, Gujarat, India
Phone: +91 265-642798
Fax: +91 265-642708
E-mail: info@allipochem.com
Web site: www.allipochem.com

Company: Alta Laboratories Ltd.
Address: Alta Bhavan, 532 Senapati Bapat Marg, Dadar, Bombay 400 028, India
Phone: +91 22-2430-7441
Fax: +91 22-2430-8707
E-mail: altaind@vsnl.com
Web site: www.altaind.com

Company: Aryan Pesticides Ltd.
Address: B-206/212, 2nd Floor, Arjun Centre, B.S. Devshi Marg, Govandi (E), Bombay 400 088, India
Phone: +91 22-555-4484
Fax: +91 22-556-1786
E-mail: aryan@bom3.vsnl.net.in
Web site: www.aryanindia.com

Company: Atul Ltd.
Address: Ashoka Chambers, Rasala Marg, Mithakhali Cross Road, Ellisbridge, Ahmedabad 380 006, Gujarat, India
Phone: +91 79-646-294/ 0520/ 3706
Fax: +91 79-640-4111
E-mail: ahd@atul.co.in
Web site: www.atul.co.in
Parent Company: Lalbhai Group (India)

Company: BEC Group Ltd. (Bhilai Engineering Corporation)
Address: 31, Maker Chambers III, Nariman Point, Bombay 400 021, India
Phone: +91 22-287-1331
Fax: +91 22-287-35611
E-mail: bec@bec-group.com
Web site: www.bec-group.com

Company: Bhageria Dye-Chem Ltd.
Address: A/101, Virwani Industrial Estate, W.E. Highway, Goregaon (E), Bombay 400 063, India
Phone: +91 22-876-0936
Fax: +91 22-874-8389
E-mail: info@bhageriagroup.com
Web site: www.bhageriagroup.com
Parent Company: Bhageria Group of Companies (India)

Company: Bharat Pulverizing Mills Pvt. Ltd.
Address: Hexamer House, Sayani Rd., Bombay Maharashtra 400028, India
Phone: +91 22-200-6155
Fax: +91 22-206-2751
Web site: www.cflindia.com/fproducts.html
Parent Company: E.I.D. Parry (India) Ltd., now Coromandel Fertilisers (India)

**Company:** Bharat Rasayan Ltd.
**Address:** 211, Shivlok House - 1, Karampura Commercial Complex, New Delhi 100 015, India
**Phone:** +91 11-545-0727
**Fax:** +91 11-544-8264
**E-mail:** bharat.brl@gems.vsnl.net.in
**Web site:** www.bharatrasayan.com

**Company:** Bilt Chemicals, now Solaris ChemTech Ltd. (India)

**Company:** Bombay Ammonia and Chemical Company
**Address:** 204-B, Neelam Centre, Behind Glaxo, Worli, Bombay 400 025, India
**Phone:** +91 22-249-48652/ +91 22-249-48659
**Fax:** +91 22-249-52186
**E-mail:** aroraj@bom5.vsnl.net.in
**Web site:** www.bomammonia.com

**Company:** Coromandel Fertilisers Ltd.
**Address:** Coromandel House, 1-2-10, Sardar Patel Rd., Sceunderabad - 500 003, Andhra Pradesh, India
**Phone:** +91 40-784-2034
**Fax:** +91 40-784-4117
**Web site:** www.cflindia.com/fproducts.html
**Parent Company:** Murugappa Group (India)

**Company:** Cyanides & Chemicals Company
**Address:** 65 Free Press House, 215 Free Press Journal Road, Nariman Point, Bombay 400 021, India
**Phone:** +91 22-285-3669
**Fax:** +91 22-202-9430
**E-mail:** cyanides@bom2.vsnl.net.in
**Web site:** www.cyanides-chemicals.com
**Parent Company:** Hindustan Engineering & Industries Ltd. (India)

**Company:** Deepak Fertilizers and Petrochemicals Corp. Ltd.
**Address:** Opp. Golf Course, Shastri Nagar, Yerwada, Pune 411 066, India
**Phone:** +91 20-668-4155
**Fax:** +91 20-668-3727
**Web site:** www.deepakgroup.com
**Parent Company:** Deepak Group (India)

**Company:** Dhanuka Group
**Address:** Dhanuka House, 861-62, Joshi Rd., Karol Bagh, New Delhi 110 005, India
**Phone:** +91 11-351-9461-63
**Fax:** +91 (011) 351-8981
**E-mail:** dhanuka@bol.net.in
**Web site:** www.dhanuka.com

**Company:** Dharamsi Morarji Chemicals Co., Ltd.
**Address:** 12, Ring Road, NDSE Part 1, New Delhi 110 049, Delhi, India
**Phone:** +91 11-692-097
**Fax:** +91 11-462-5170
**Web site:** www.dmccl.com
**Stock listing:** Bombay SE: DHRC.BO, 6405
**Parent Company:** R M Goculdas Group (India)

**Company:** Dhruv Chemical Industries
**Address:** Basement, Ratnamani Complex, N.H. No. 8,Nr. Thakkarnagar Char Rasta, Ahmedabad 382 350, Gujarat, India
**E-mail:** sendyourenquiry@yahoo.co.in
**Web site:** www.needsinfo.com/dhruvchemindustries.htm

**Company:** E.I.D. Parry (India) Ltd., now Coromandel Fertilisers (India)

**Company:** Excel Industries Ltd.
**Address:** 184/87, S.V. Road, Jogeshwari (W), Bombay 400 102, India
**Phone:** +91 22-678-8258
**Fax:** +91 22-678-4522
**E-mail:** excelmumbai@excelind.com
**Web site:** www.excelind.com
**Stock listing:** Bombay SE: EXCI.BO; 500650

**Company:** Gayatri Minerals & Chemicals
**Address:** 34-A Shastrinagar, Nizampura, Baroda 390 002, India
**Phone:** +91 265-278-1284
**Fax:** +91 265-278-0228
**E-mail:** info@gayatrionline.com
**Web site:** www.gayatrionline.com

**Company:** Gharda Chemicals Ltd.
**Address:** 5/6, Jer Mansion, W.P. Warde Road, Bandra West, Bombay 400 052, India
**Phone:** +91 22-6452492
**Fax:** +91 22-6404224
**E-mail:** ghardaho@gharda.com
**Web site:** www.gharda.com

**Company:** Godavari Fertilisers and Chemicals Ltd.
**Address:** 50, Sebastian Rd., Vani Nilayam, Secunderabad 500 003, India
**E-mail:** hyd_lgfcl@sancharnet.in
**Web site:** www.gfcl.co.in

**Company:** Gujarat Pesticides
**Address:** F-15, G.I.D.C. Estate, Phase 2, Naroda. Ahmedabnad 382 330, India
**Phone:** +91 79-22811126

Fax: +91 79-22818067
E-mail: gujpesticide@icenet.net
Web site: www.gujaratpesticides.com

**Company: Gwalior Chemical Industries Ltd.**
Address: 29, Bank St., 1st Floor, Fort, Bombay 400 023, India
Phone: +91 22-2266-2843
Fax: +91 22-2266-2453
E-mail: enquiry@gwaliorchemicals.com
Web site: www.gwaliorchemicals.com

**Company: Hindustan Insecticides Ltd.**
Address: Core 6, 2nd Floor, SCOPE Complex, 7, Lodi Road, New Delhi 110 003, India
Phone: +91 11-2436-2165
Fax: +91 11-2436-2116
E-mail: hilhq@nde.vsnl.net.in
Web site: www.hil-india.com
Parent Company: Government of India

**Company: Hindustan Organic Chemicals Ltd.**
Address: Harchandrai House, 81, Maharshi Karve Marg, Bombay 400 002, India
Phone: +91 22-220-14269
Fax: +91 22-220-59533
Web site: www.hoclindia.com
Stock listing: Bombay SE: 500449
Parent Company: Government of India

**Company: Indian Oil Corporation Ltd.**
Address: Core II, 7, Scope Complex, Institutional Area Lodhi Road, New Delhi 110 003, India
Phone: +91 11-436-2896
Fax: +91 11-436-4602
E-mail: info@iocl.co.in
Web site: www.iocl.com
Stock listing: Bombay SE: IOC; 530965

**Company: Indiclay Co., Ltd.**
Address: Plot No. 2, Udyog Nagar, Goregaon (West), Bombay 400 062, India
Phone: +91 22-2874-3823
Fax: +91 22-2874-2986
E-mail: info@indiclay.com
Web site: www.indiclay.com

**Company: Indo Gulf Corporation Ltd.**
Address: P.O. Jagdishpur Industrial Area, Sultanpur 227 817, India
Phone: +91 5361-270032
E-mail: igfl@adityabirla.com
Web site: www.indo-gulf.com
Parent Company: The Aditya Group Ltd. (India)

**Company: Kanoria Chemicals & Industries Ltd.**
Address: Indraprakash, 21 Barakhamba Road, New Delhi 110 001, India
Phone: +91 11-2371-6580
Fax: +91 11-2371-7203
E-mail: info@kanoriachem.com
Web site: www.kanoriachem.com
Stock listing: Bombay SE: 6525
Parent Company: Kanoria Chemicals & Industries Ltd. (KCIL) Group

**Company: KCIL Group (India)**, see Kanoria Chemicals & Industries Ltd. (India)

**Company: Kothari Phytochemicals International**
Address: 766, Anna Nagar, Madurai 625 020, Tamil Nadu, India
Phone: +91 452-535807
Fax: +91 452-534138
E-mail: kothariphyto@eth.net
Web site: www.chemicals-india.com
Parent Company: Kothari Plantatioins & Industries Ltd. (India)

**Company: Lords Chemicals Ltd.**
Address: 5C, Electronic Centre, 1/1A, Biplabi Anukul, Chandra St., Calcutta, West Bengal 700 072, India
Phone: +91 33-26 8903 4240
Fax: +91 30-215-1297
E-mail: nkjain@cal.vsnl.net.in
Web site: www.dialindia.com

Company: Meghmani Organics Ltd.
Address: Meghmani House, Sree Nivas Society, Off New Vikas Gruh Rd., Paldi Ahmedabad 380 007, India
Phone: +91 79-664-0668
Fax: +91 79-664-0670
E-mail: vshekhar@meghmani.com
Web site: www.meghmani.com

**Company: Nagarjuna Agrichem Ltd. (NACL)**
Address: Nagarjuna Hills, Panjagutta, Hyderabad, Andhra Pradesh 500 082, India
Phone: +91 40-335-7204
Fax: +91 40-335-4788
E-mail: mail@nagarjunagroup.com
Web site: www.nagarjunagroup.com
Stock listing: Bombay SE: 530064
Parent Company: Nagarjuna Group (India)

**Company: Nagarjuna Fertilizer and Chemicals Ltd.**
Address: Nagarjuna Hills, Panjagutta, Hyderabad, Andhra Pradesh 500 082, India
Phone: +91 40-335-7204

Fax: +91 40-335-4788
E-mail: mail@nagarjunagroup.com
Web site: www.nagarjunagroup.com
Stock listing: Bombay SE: 530064
Parent Company: Nagarjuna Group (India)

**Company: Nagarjuna Group**
Address: Nagarjuna Hills, Panjagutta, Hyderabad, Andhra Pradesh 500 082, India
Phone: +91 40-335-7204
Fax: +91 40-335-4788
E-mail: mail@nagarjunagroup.com
Web site: www.nagarjunagroup.com
Stock listing: Bombay SE: 530064

**Company: Navin Fluorine Industries**
Address: 1st Floor, Sakhi Naka, Corporate Park, Sion-Trombay Rd., Chembur, Bombay 400071, India
Phone: +91 22 527 4003 / 6
Fax: +91 22 524 0421
E-mail: info@mafnav.com
Web site: www.mafnav.com/fluorin
Parent Company: Arvind Mafatlal Group (AMG), Chemical Division

**Company: P.D. Industries Ltd.**
Address: 504, Sahakar Bhavan, 340/48, Narsi Natha St., Bombay 400 009, India
Phone: +91 22-2344-9061
Fax: +91 22-2342-1235
E-mail: neelwale@vsnl.com
Web site: www.pdindustries.com

**Company: PI Industries Ltd.**
Address: P.O. Box 20, Udaisager Rd., Udaipur, Rajasthan 313 001, India
Phone: +91 294-492-541
Fax: +91 294-491-946
Stock listing: Bombay SE: 523642
*Note:* Formerly Pesticides India Ltd.

**Company: Pioneer Enterprise**
Address: 101, Raudat Tehera St., Bombay 400 003, India
Phone: +91 22-347 25 34
Fax: +91 22 347 03 25
E-mail: pionlace@vsnl.com
Web site: www.pioneerherbs.com

**Company: Rallis India Ltd.**
Address: Ralli House, 21 D Sukhadwala Marg, Bombay 400 001, India
Phone: +91 22-207-8221
Fax: +91 22-287-5980
E-mail: admn.ho@rallis.co.in
Web site: www.rallis.co.in
Stock listing: Bombay SE: 355
Parent Company: Tata Group (India)

**Company: Rashtriya Chemicals & Fertilizers Ltd. (RCF)**
Address: Priyadarshini, Eastern Express Highway, Sion, Bombay, Maharashtra 400 022, India
Phone: +91 22-2404-3644
Fax: +91 22-2404-5111
E-mail: sbalan@rcfltd.com
Web site: www.rcfltd.com
Stock listing: Bombay SE: RSTC.BO; 524230

**Company: Reliance Industries Ltd.**
Address: Maker Chambers IV, Nariman Point, Bombay 400 021, India
Phone: +91 22-2283-1633
Fax: +91 22-2204-2268
E-mail: investor@ril.com
Web site: www.ril.com
Stock listing: Bombay SE: RIL; 500325

**Company: Sabero Organics Gujarat Ltd.**
Address: A-302 Phoenix House, 3rd Floor, 462 Senapati Bapat Marg, Worli (E), Bombay 400 013, India
Phone: +91 22-496-0979
Fax: +91 22-495-3727
E-mail: sabero@vsnl.com
Web site: www.sabero.com
Stock listing: Bombay SE: 24446

**Company: Salvi Chemical Industries Ltd.**
Address: B-108 Ashoka Towers, Kulupwadi Road, Borivli (East), Bombay 400 066 Maharashtra, India
Phone: +91 22-2861-9276
Fax: +91 22-2805-9274
E-mail: info@salvichem.com
Web site: www.salvichem.com

**Company: Schenectady Herdillia Chemicals Ltd.**
Address: Air India Building, Nariman Point, Bombay 400 021, India
Phone: +91 22-202-4224
Fax: +91 22-204-2379
E-mail: herdillia@herdillia.com
Web site: www.herdillia.com
Parent Company: Schenectady International Group, Inc. (USA)

**Company: Scottish Chemical Industries**
Address: 407-412 Span Centre, 4th Fl., South Avenue, Santacruz (W), Bombay 400 054, India
Phone: +91 22-2605-6666

Fax: +91 22-2605-6060
E-mail: scottish@vsnl.com
Web site: www.scottish-chem.com

**Company: Shyam Chemicals Pvt. Ltd.**
Address: Jash Chambers, 5th Floor, Sir P.M. Road, Fort, Bombay 400 001 Maharashtra, India
Phone: +91 22-266-0015
Fax: +91 22-266-6973
E-mail: shyamchemicals@vsnl.com
Web site: www.shyamchemicals.com

**Company: Solaris ChemTech Ltd.**
Address: First India Place, Tower-C, Mehrauli-Gurgaon Rd., Gurgaon 122022,,Haryana, India
Phone: +91 124-680-4242

Fax: +91 124-680-4263
Web site: www.solarischemtech.com

**Company: Southern Petrochemical Industries Corporation Ltd. (SPIC)**
Address: SPIC House, 88, Mount Road, Guiny, Chennai 600 032, TamilNadu, India
Phone: +91 44 235 0245
Fax: +91 44 235 2163
E-mail: spiccorp@giasmd01@vsnl.net.in
Web site: www.spic-india.com
Stock listing: Bombay SE: 758
Parent Company: SPIC Group (India)

**Company: Sudarshan Chemical Industries Ltd.**
Address: 162, Wellesley Road, Pune 411 001, India
Phone: +91 20-612-7334
Fax: +91 20-612-5900
E-mail: contact@sudarshan.com
Web site: www.sudarshan.com
Stock listing: Bombay SE: 6655

**Company: Sulphur Mills Ltd.**
Address: 303/304 T.V. Estate, S.K. Ahire Marg, Worli, Bombay 400 025, India
Phone: +91 22493-7685
Fax: +91 22-493-9586

**Company: United Agro Industries**
Address: 9/70, Sakthi Nagar, Seelannaickenapatti, Salem 636 201, Tamil Nadu, India
Phone: +91 427-246-1188
Fax: +91 427-424-61198
E-mail: nimbex@eth.net
Web site: www.indiamert.com/united-agro/

**Company: United Phosphorus Ltd.**

Address: Uniphos House, Madhu Park Centre, Opp Madhu Park, Chitrakar Dhurandar Marg, Khar (West), Bombay 400 053, India
Phone: +91 22-604-1111
Fax: +91 22-604-1010
E-mail: info@uniphos.com
Web site: www.uplonline.com
Stock listing: Bombay SE: 500429

**Company: Vani Chemicals & Intermediates Ltd.**
Address: Divya Shakti Commercial Complex, Ameerpet, Hyderabad 500 016, Andhra Pradesh, India
Phone: +91 40-23731475
Fax: +91 40-23730099
E-mail: vni@vanigroup.com
Web site: www.vanigroup.com

**Company: Vijayalakshmi Insecticides Pesticides Ltd. (VIPL)**
Address: Plot 61, Nagarjuna Hills, Hyderabad 500 082, India
Phone: +91 40-335-0235
Fax: +91 40-335-8062
E-mail: mail@nagarjunagroup.com
Web site: www.nagarjunagroup.com
Parent Company: Nagarjuna Group (India)

## ISRAEL

**Company: Agan Chemical Manufacturers Ltd.**
Address: New Industrial Zone, P.O. Box 262, Ashdod 77102, Israel
Phone: +972 8-8515211
Fax: +972 8-8522806
E-mail: agan@agan.co.il
Web site: www.agan.co.il
Stock listing: Tel Aviv SE: MAIN
Parent Company: Makhteshim-Agan Industries Ltd., sub. of Koor Industries Ltd. (Israel)

**Company: Chemada Fine Chemicals**
Address: Kibbutz Nir Itzhak D.N., HaNegev 85455, Israel
Phone: +972 8-9983421
Fax: +972 8-9983598
E-mail: sauliko@chemada.com
Web site: www.chemada.com

**Company: Dead Sea Bromine Group**
Address: Makleff House, 12 Kroitzer St., P.O. Box 180, Beer Sheva 84101, Israel
Phone: +972 8-6297209
Fax: +972 8-6297848
E-mail: golanb@dsbg.com
Web site: www.dsbg.com

Parent Company: Israel Chemicals Ltd. (Israel)

Company: **Gadiv Petrochemical Industries Ltd.**
Address: P.O. Box 32, Haifa, Israel 31000
Phone: +972 4-8788020
Fax: +972 4-8788018
E-mail: gadiv@orl.co.il
Web site: www.gadiv.com
Parent Company: Oil Refineries Ltd. (Israel)

Company: **Haifa Chemicals Ltd.**
Address: P.O.Box 10809, Haifa Bay 261-20, Israel
Phone: +972-4-8469610/4
Fax: +972-4-8450588
E-mail: info@haifachem.co.il
Web site: www.haifachem.com
Parent Company: Trans-Resources Inc. (USA)

Company: **Israel Chemicals Ltd.**
Address: Millenium Tower, 23 Aranha St., Tel-Aviv 61202, Israel
Phone: +972 3-6844401
Fax: +972 3-6844428
E-mail: info@israelchemicals.com
Web site: www.israelchemicals.co.il

Company: **Luxembourg Industries Ltd.**
Address: P.O. Box 13, Tel Aviv 61000, Israel
Phone: +972 3-796-4300
Fax: +972 3-510-0474
E-mail: main@luxpam.com
Web site: www.luxembourg.co.il

Company: **Makhteshim Chemical Work Ltd.**, see Makhteshim-Agan Industries Ltd. (Israel)

Company: **Makhteshim-Agan Industries Ltd.**
Address: Yahav Bldg., 9 Omarim St., P.O. Box 1646, Omer 84965, Israel
Phone: +972 7-6469837
Fax: +972 7-6469846
E-mail: main@main.co.il
Web site: www.main.co.il
Parent Company: Koor Industries Ltd. (Israel)

Company: **Pazchem Ltd.**
Address: P.O. Box 100, 78100 Ashkelon, Israel
Phone: +972-3-6888016
Fax: +972-3-6889619
E-mail: zvifeler@attglobal.net
Web site: www.pazchem.coml
Parent Company: Rallis India Ltd. (India)
*Note:* Rallis India suspended operations of Pazchem in 2001.

## ITALY

Company: **Caffaro S.p.A**
Address: Via Friuli, 55, 20031 Cesano Maderno, Milan, Italy
Phone: +39 0362-514-1
Fax: +39 0362-514-889
E-mail: caffaro.chem@caffaro.it
Web site: www.caffarochem.com
Parent Company: Snia S.p.A. (Italy)

Company: **Diachem S.p.A.**
Address: Via Tonale, 15-24061 Albano SA, Italy
Phone: +39 35 581228
Fax: +39 35 581357
Web site: www.diachemagro.it

Company: **EniChem S.p.A.**
Address: Piazza Boldrini, 1, 20097 San Donato, Milan, Italy
Phone: +39 02-52-01
Fax: +39 02-31-42-15
Web site: www.enichemnet.com
Stock listing: Italian SE: ENI.MI; NYSE: E
Parent Company: Eni Group S.p.A. (Italy)

Company: **Esseco Group S.p.A.**
Address: Via S. Cassiano, 99, 28069 Trecate NO, Italy
Phone: +39 03 21 79 01
Fax: +39 03 21 79 02 89
E-mail: esseco@esseco.it
Web site: www.esseco.it

Company: **I.N.D.I.A. Industrie Chimiche S.p.A.**
Address: Nona Strada, 57, Zona Industriale, 35129 Padova, Italy
Phone: +39 49-807-61-44
Fax: +39 49-807-61-46
E-mail: export@indiapesticides.com
Web site: www.indiapesticides.com

Company: **Oxon Italia S.p.A.**
Address: Via Sempione, 195-20016 Pero, Milano, Italy
Phone: +39 2-353-784.1
Fax: +39 2-339-0275
E-mail: Epallucca@oxon.it
Web site: www.oxon.it
Parent Company: Sipcam Oxon Group (Italy)

## JAPAN

Company: **Asahi Glass Chemicals Division**
Address: 12-1, Yurakucho 1-chome, Chiyoda-ku, Tokyo 100-8405, Japan

Phone: +81 3-3218-5700
Fax: +81 3-3214-1574
Web site: www.agc.co.jp/english/chemicals
Parent Company: Asahi Glass Co., Ltd., sub. of Mitsubishi Corporation (Japan)

Company: Central Glass Co., Ltd.
Address: Kowa-Hitotsubashi Building, 7-1, Kanda-Nishikicho, 3-chome, Chiyoda-ku, Tokyo 101-0054, Japan
Phone: +81 3-3259-7031
E-mail: info@cgco.co.jp
Web site: www.cgco.co.jp
Stock listing: Tokyo SE: 4044

Company: Daicel Chemical Industries Ltd.
Address: Osaka Head Office: 1, Teppo-cho, Sakai-shi, Osaka 590-8501. Tokyo Head Office: 2-5, Kasumigaseki 3-chome, Chiyoda-ku, Tokyo 100-6077, Japan
Phone: Osaka: +81 722-27-3111. Tokyo: +81 3-3507-3111
Fax: Osaka: +81 722-27-3000. Tokyo: +81 3-3507-3139
Web site: www.daicel.co.jp
Stock listing: Tokyo SE: 4202

Company: Dainippon Ink and Chemicals, Inc. (DIC)
Address: DIC Bldg., 7-20, Nihonbashi 3-chome, Chuo-ku, Tokyo 103-8233, Japan
Phone: +81 3-3272-4511
Fax: +81 3-3278-8558
E-mail: info@dic.co.jp
Web site: www.dic.co.jp
Stock listing: Tokyo SE: 4631

Company: Denki Kagaku Kogyo Kabushiki Kaisha (Denka)
Address: Sanshin Bldg., 4-1, Yuraku-cho 1-chome, Chiyoda-ku, Tokyo 100-8455, Japan
Phone: +81 3-3507-5055
Fax: +81 3-3507-5059
Web site: www.denka.co.jp
Stock listing: Tokyo SE: 4061
Parent Company: Denka Group (Japan)

Company: Dowa Mining Co., Ltd.
Address: 1-8-2, Marunouchi, Chiyoda-ku, Tokyo 100-8282, Japan
Phone: +81 3-3201-1061
Fax: +81 3-3201-1259
E-mail: info@dowa.co.jp
Web site: www.dowa.co.jp
Stock listing: Tokyo SE: 5714

Company: Hokko Chemical Industry Co., Ltd.
Address: Mitsui Building No. 2, 4-20, Nihonbashi Hongoku-cho 4-chome, Chuo-ku, Tokyo 103-8341, Japan
Phone: +81 3-3279-5831
Fax: +81 3-3279-5067
E-mail: finc@hokkochem.co.jp
Web site: www.hokkochem.co.jp
Stock listing: Tokyo SE: 4992

Company: Honshu Chemical Industry Co., Ltd.
Address: 1-1, Kyobashi 1-crome, Chuo-ku, Tokyo 104 003, Japan
Phone: +81 3-3272-1481
Fax: +81 3-3274-3870
Stock listing: Tokyo SE: 4115
Parent Company: Mitsui Group (Japan)

Company: Japan Aldehyde Co., Ltd.
Address: Tokyo Sumitomo Twin Bldg., 2-27-1, Shinkawa, Chuo-ku, Tokyo 104, Japan
Phone: +81 3-5543-5304
Fax: +81 3-5543-5911
Web site: www.sumitomo-chem.co.jp
Stock listing: Tokyo SE: 4005
Parent Company: Sumitomo Chemical Co., Ltd (Japan)

Company: Junsei Chemical Co., Ltd.
Address: 4-416, Nihonbashi Hon-cho, Chuo-ku, Tokyo 103, Japan
Phone: +81 3-3270-5414
Fax: +81 3-3241-8298
E-mail: jouhou@junsei.co.jp
Web site: www.junsei.co.jp

Company: Kawaguchi Chemical Industry Co., Ltd.
Address: 8-4, Uchi-Kanda 2-chome, Chiyoda-ku, Tokyo 101 0047, Japan
Phone: +81 3-3254-8481
Fax: +81 3-325-4-8497
Web site: www.kawachem.co.jp
Stock listing: Tokyo SE: 4361

Company: Kawasaki Kasei Chemicals Ltd.
Address: Kawasakiekimae TowerRiverK Bldg., 17F, 12-1, Ekimaehoncho, Kawasaki-ku, Kawasaki City, Kanagawa 210 0007, Japan
Phone: +81 44-246-7100
Fax: +81 44-246-7462
E-mail: kkcpr@kk-chem.co.jp
Web site: www.kk-chem.co.jp

Company: Koei Chemical Co., Ltd.
Address: 12-13, Hanaten-nishi 2-chome, Joto-ku, Osaka 536 0011, Japan, Japan
Phone: +81 6-6961-0252
Fax: +81 6-6961-0498
Web site: www.koeichem.com

Stock listing: Tokyo SE: 4367
Parent Company: Sumitomo Chemical Co., Ltd. (Japan)

Company: Kyowa Hakko Kogyo Co., Ltd.
Address: 1-6-1, Ohtemachi, Chiyoda-ku, Tokyo 100 8185, Japan
Phone: +81 3-3282-0007
Fax: +81 3-3284-1968
Web site: www.kyowa.co.jp
Stock listing: Tokyo SE: 4151

Company: Mitsubishi Chemical Corporation
Address: 5-2, Marunouchi 2-chome, Chiyoda-ku, Tokyo 100 0005, Japan
Phone: +81 3-3283-6111
Fax: +81 3-3283-5874
E-mail: mccpr@cc.m-kagaku.co.jp
Web site: www.m-kagaku.co.jp
Stock listing: Tokyo SE: 4010; Frankfurt SE
Parent company: Mitsubishi Corporation (Japan)

Company: Mitsubishi Gas Chemical Company Inc
Address: Mitsubishi Bldg., 2-5-2, Marunouchi, Chiyoda-ku, Tokyo 100 8324, Japan
Phone: +81 3-3283-5000
Fax: +81 3-3283-5120
E-mail: info@mgc.co.jp
Web site: www.mgc.co.jp
Stock listing: Tokyo SE: 4182
Parent Company: Mitsubishi Corporation (Japan)

Company: Mitsubishi Materials Corporation
Address: 1-5-1, Ohtemachi, Chiyoda-ku, Tokyo 100 8117, Japan
Phone: +81 3-5252-5206
Fax: +81 3-5252-5272
E-mail: www.adm@mmc.co.jp
Web site: www.mmc.co.jp
Stock listing: Tokyo SE: 5711
Parent Company: Mitsubishi Corporation (Japan)

Company: Mitsui Chemicals Inc. (MCI)
Address: 2-5, Kasumigaseki 3-chome, Chiyoda-ku, Tokyo 100 6070, Japan
Phone: +81 3-3592-4060
Fax: +81 3-3592-4211
Web site: www.mitsui-chem.co.jp
Stock listing: Tokyo SE: 418

Company: Mitsui Mining & Smelting Co., Ltd.
Address: 11-1, Osaki 1-chome, Shinagawa-ku, Tokyo 141-8584, Japan
Phone: +81 3 5437 8031
Fax: +81 3 5437 8033
E-mail: kouhou@mitsui-mining.co.jp
Web site: http://www.mitsui-kinzoku.co.jp/en/index.htm
Stock listing: Tokyo SE: 5706

Company: Morita Chemical Industries Co., Ltd.
Address: 19-3, Toyosaki 3, Kita-ku, Osaka, Japan
Phone: +81 6 376 3501
Fax: +81 6 376 3579
E-mail: info@morita-kagaku.co.jp
Web site: www.morita-kagaku.co.jp

Company: Nihon Kagaku Sangyo Co., Ltd.
Address: 20-5, Shitaya 2-chome, Taito-ku, Tokyo 110 0004, Japan
Phone: +81 3 38739223
Fax: +81 3 38766963
Web site: www.nihonkagakusangyo.co.jp

Company: Nihon Nohyaku Co., Ltd.
Address: 1-2-5 Nihonbashi, Eitaro Bldg., Chuo-ku, Tokyo 103-8236, Japan
Phone: +81 3 3274 3374
Fax: +81 3 3282 5462
E-mail: nnc-overseasdiv@nichino.co.jp
Web site: www.nichino.co.jp/eng/index
Note: Acquired the Agrochemical Division of Mitsubishi Chemical in 2002.

Company: Nippon Carbide Industries Co., Inc.
Address: 2-11-19 Kohnan, Minato-ku. Tokyo 108 8466, Japan
Phone: +81 3-5462-8207
Fax: +81 3-5462-8273
Web site: www.carbide.co.jp
Stock listing: Tokyo SE: 4064
Parent Company: Asahi Glass Group (Japan)

Company: Nippon Chemical Industrial Co., Ltd.
Address: 11-1, Kameido 9-chome, Koto-ku, Tokyo 136 8515, Japan
Phone: +81 3-3636-8111
Fax: +81 3-3636-6817
E-mail: comm.sales@nippon-chem.co.jp
Web site: www.nippon-chem.com
Stock listing: Tokyo SE: 4092

Company: Nippon Kayaku Co., Ltd.
Address: Tokyo Fujimi Bldg., 11-2, Fujimi 1-chome, Chiyoda-ku, Tokyo 102 8172, Japan
Phone: +81 3-3237-5046
E-mail: nk-kouho@magical2.egg.or.jp
Web site: www.nipponkayaku.co.jp/english
Stock listing: Tokyo SE: 4272

**Company:** Nippon Mining & Metals Co., Ltd.
**Address:** 10-1, Toranomon 2-chome, Minato-ku, Tokyo 105 0001, Japan
**Phone:** +81 3-5573-7148
**Web site:** www.nikko-metal.co.jp
**Stock listing:** Tokyo SE: 5716
**Parent Company:** Japan Energy Corp. (Japan)

**Company:** Nippon Soda Co., Ltd.
**Address:** 2-1, 2-chome, Ohtemachi, Chiyoda-ku, Tokyo 100 8165, Japan
**Phone:** +81 3-3245-6054
**Fax:** +81 3-3245-6238
**Web site:** www.nippon-soda.co.jp
**Stock listing:** Tokyo SE:

**Company:** Nippon Steel Chemical Co., Ltd.
**Address:** 7-21-11, Nishi Gotanda Shinagawa-ku, Tokyo 141 0031, Japan
**Phone:** +81 3-5759-2741
**Fax:** +81 3-5759-2777
**Web site:** www.nscc.co.jp
**Stock listing:** Tokyo SE: 4363
**Parent Company:** Nippon Steel Coporation (Japan)

**Company:** Nissan Chemical Industries Ltd.
**Address:** 7-1, Kanda-Nishiki-cho 3-chome, Chiyoda-ku, Tokyo 101 0054, Japan
**Phone:** +81 3-3296-8320
**Fax:** +81 3-3296-8210
**E-mail:** info@nissanchem.co.jp
**Web site:** www.nissanchem.co.jp
**Stock listing:** Tokyo SE: 4021

**Company:** Sankei Chemical Co., Ltd
**Address:** 9 Nanei 2-chome, Kagoshima 981-0122, Kagoshima, Japan
**Phone:** +81 99 268 7588
**Fax:** +81 99 269 6121
**Web site:** www.sankei-chem.com

**Company:** Showa Denko K.K.
**Address:** 13-9, Shiba Daimon 1-chome, Minato-ku, Tokyo 105 8518, Japan
**Phone:** +81 3-5470-3235
**Fax:** +81 3-3436-2625
**E-mail:** shodex@hq.sdk.co.jp
**Web site:** www.sdk.co.jp/index_e
**Stock listing:** Tokyo SE: 4004
**Parent Company:** Fuyo Group (Japan)

**Company:** Stella Chemifa
**Address:** NM Plaza Midosuji 3F, 3-6-3 Awajimachi, Chuo-ku, Osaka 541 0047, Japan
**Phone:** +81 6 4707 1511
**Fax:** +81 6 4707 1521
**E-mail:** somu@stella-chemifa.co.jp
**Web site:** www.stella-chemifa.co.jp

**Company:** Sumika Agro Manufacturing Co., Ltd.
**Address:** 2-12-10 Nihonbashi, Chuo-ku, Tokyo, Japan
**Parent Companies:** Sumitomo Chemical (Japan)
*Note:* Formerly Sumika-Takeda Agro Manufacturing Co., a subsidiary of Sumitomo Chemical Co. (Japan)

**Company:** Sumika Fine Chemicals Co., Ltd
**Address:** 1-21, Utajima 3-chome, Nishiyodogawa-ku, Osaka 555-0021, Japan
**Phone:** +81 6-7473-0331
**Fax:** +81 6-6474-2468
**E-mail:** www.@sumika-fine-chem.co.jp
**Web site:** www.sumika-fine-chem.co.jp

Company: Sumika-Takeda Agro Manufacturing Co., now Sumika Agro Manufacturing Co., Ltd.

**Company:** Sumitomo Chemical Co., Ltd
**Address:** Osaka Head Office: Sumitomo Bldg., 5-33, Kitahama 4-chome, Chuo-ku, Osaka 541-8550. Tokyo Head Office: Tokyo Sumitomo Twin Bldg. East, 27-1, Shinkawa 2-chome, Chuo-ku, Tokyo 104-8260, Japan
**Phone:** Osaka: +81 6-6220-3891. Tokyo: +81 3-5543-5500
**Fax:** Osaka: +81 6-6220-3345. Tokyo: +81 3-5543-5901
**Web site:** www.sumitomo-chem.co.jp
**Stock listing:** Tokyo SE: 4005

**Company:** Taoka Chemical Co., Ltd.
**Address:** 4-2-11, Nishi-mikuni 4-chome, Yodogawa-ku, Osaka 532 0006, Japan
**Phone:** +81 6-6394-1221
**Fax:** +81 6-6394-1658
**Web site:** www.taoka-chem.co.jp
**Stock listing:** Osaka SE: 4113

**Company:** Tokuyama Group
**Address:** Shibuya Konno Bldg., 3-1, Shibuya 3-chome, Shibuya-ku, Tokyo 150 8383, Japan
**Phone:** +81 3-3499-8937
**Fax:** +81 3-3499-8967
**Web site:** www.tokuyama.co.jp
**Stock listing:** Tokyo SE: 4043
**Parent Company:** Tokuyama Corporation (Japan)

**Company:** Tokyo Chemical Industry Co., Ltd. (TKK)
**Address:** 4-10-2, Nihonbashi-honcho, Chuo-ku, Tokyo 103-0023, Japan
**Phone:** +81 3-5640-8851
**Fax:** +81 3-5640-8865

E-mail: international@tokyokasei.co.jp
Web site: www.tokyokasei.co.jp

**Company:** Tokyo Kasei Kogyo Co., Ltd. (TKK), see Tokyo Chemical Industry Co., Ltd. (Japan)

**Company:** Tosoh Corporation
Address: 3-8-2, Shiba, Minato-ku, Tokyo 105 8623, Japan
Phone: +81 3-5427-5118
Fax: +81 3-5427-5198
E-mail: info@tosoh.co.jp
Web site: www.tosoh.com
Stock listing: Tokyo SE: 4042

**Company:** Ube Agri-Materials, Ltd., sub. of Industries Ltd.
Address: 1978-96, Kogushi, Ube, Yamaguchi 755 8633, Japan
Phone: +81 3-5419-5350
Fax: +81 3-5419-6352
Web site: www.ube-ind.co.jp
Stock listing: Tokyo SE: 4208

**Company:** Zeon Corporation
Address: Furukawa Sogo Bldg., 2-6-1, Marunouchi, Chiyoda-ku, Tokyo 100 8323, Japan
Phone: +81 3-3216-1772
Fax: +81 3-3216-0501
Web site: www.zeon.co.jp
Stock listing: Tokyo SE

## JORDAN

**Company:** Arab Potash Company (APC)
Address: 6 Jaheth St., Shmeisani, Amman 11118, Jordan
Phone: +962 6-666-165
Fax: +962 6-674-416
Web site: www.arabpotash.com

**Company:** Veterinary & Agricultural Products Manufacturing Co., Ltd. (VAPCO)
Address: P.O. Box 17058, Shaker bin Zaid 12, Bldg. 7, Amman 11195, Jordan
Phone: +962 6-5694991
Fax: +962 6-5694998
E-mail: vapco@vapco.net
Web site: www.vapco.net

## LITHUANIA

**Company:** Achema SC
Address: Taurostos 26, 5000 Jonava, Lithuania
Phone: +370 (8-349) 56626
Fax: +370 (8-349) 56619

E-mail: rvs@achema.com
Web site: www.achema.com

## MALAYSIA

**Company:** Kenso Corporation
Address: Unit 601, Menara PJ, 18 Persiaran Barat, 46050 Petaling Jaya, Selangor DE, Malaysia
Phone: +60 3-7958-8828
Fax: +60 3-7956-6818
E-mail: enquiry@kenso.com.my
Web site: www.kensocorp.com

## MEXICO

**Company:** Abaquim, SA
Address: Cerrada de Colima No. 4, Col. Roma, 06700 Mexico D.F., Mexico
Phone: +52/5 525-8420
Fax: +52/5 207-7907
E-mail: ventas@abaquim.com.mx
Web site: www.abaquim.com.mx

**Company:** Acido de Mexico S.A. de C.V. (Acimex)
Address: Lago Chapala No. 58, Col. Anahuac, 11320 Mexico D.F., Mexico
Phone: +52/5 396-8997
Fax: +52/5 888-0384
E-mail: info@acimex.com.mx
Web site: www.acimex.com.mx

**Company:** Alquimia Mexicana, S. de R. L.
Address: Cerrada de Colima No. 2-2, P.O. Box 7-843, Col Roma, Deleg. Cuahtemco, 06700 Mexico D.F., Mexico
Phone: +55 5533-3964
Fax: +55 5511-8970
E-mail: alquimia@mexred.net.mx
Web site: www.alquimiamex.com.mx

**Company:** Amomex Mexicana, SA de C.V.
Address: Calle Progreso No. 8, Col. Industrial Puente de Vigas, 54070 Tlalnepantla, Mexico
Phone: +52/5 565-5241
Fax: +52/5 565-4795
E-mail: amomex@mpsnet.com.mx
Web site: www.cosmos.com.mx/chem/amomex

**Company:** Cia. Universal de Industrias S.A. de C.V.
Address: Calle Flor de Maria No.20, Col. Atlamaya San Angel I, 01760 Mexico D.F., Mexico
Phone: +55 5683-6166
Fax: +55 5683-6750
E-mail: cuisa@netservice.com.mx
Web site: www.cosmos.com.mx/chem/universal

**Address:** Avenida Baja California 200, Col. Roma Sur, 06760 Mexico D.F., Mexico
**Phone:** +52 55-5080-0050
**Web site:** www.grupomexico.com
**Stock listing:** NASDAQ: GMBXF.PK

**Company: Industria Quimica Loser S.A. de C.V.**
**Address:** Av. Revolucion No. 1875 P.H., Col. San Angel, Deleg. Alvaro Obregon, 01000 Mexico D.F., Mexico
**Phone:** +52/5 616-1902
**Fax:** +52/5 227-7979
**E-mail:** polinala@mpsnet.com.mx
**Web site:** www.cosmos.com.mx/chem/loser

**Company: Ingenieria Industrial S.A. de C.V. (Bravo)**
**Address:** Av. Coyoacan No. 1878-403, Col. Del Valle, 03100 Mexico D.F., Mexico
**Phone:** +52/81 8625-5600
**Fax:** +52/55 5524-8270
**Web site:** www.bravoag.com.mx

**Company: Pyosa Agroquimicos, S.A. de C.V.**
**Address:** Industrias #1200 Pte. Ave., Col. Bella Vista, Monterrey N.L. 64410, Mexico
**Phone:** +52 81 8625-5600
**Fax:** +52 81 376-9398
**E-mail:** info@pyosa.com.mx
**Web site:** www.pyosa.com

## NETHERLANDS

**Company: AgriChem B.V.**
**Address:** Weststad Industrial Estate, Koopvaardijweg 9, 4906 CV Oosterhout, The Netherlands
**Phone:** +31 (0) 162 431931
**Fax:** +31 (0) 162 456797
**E-mail:** info@agrichem.nl
**Web site:** www.agrichem.net

**Company: Akzo Nobel NV**
**Address:** Velperweg 76, 6824 BM Arnhem, P.O. Box 9300, 6800 SB Arnhem, Netherlands
**Phone:** +31 26 366 44 33
**Fax:** +31 26 366 32 50
**E-mail:** acc@akzonobel.com
**Web site:** www.akzonobel.com
**Stock listing:** Amsterdam SE; NASDAQ: AKZOY; all major exchanges

**Company: Akzo Nobel Functional Chemicals BV**
**Address:** Barchman Wuijtierslaan 10, P.O. Box 247, 3800 AE Amersfoort, Netherlands
**Phone:** +31 33 467 67 67
**Fax:** +31 33 467 61 46
**Web site:** www.functionalchemicals.com
**Stock listing:** Amsterdam SE; NASDAQ: AKZOY; all major exchanges
**Parent Company:** Akzo Nobel NV (Netherlands)

**Company: Caldic Chemie B.V.**
**Address:** Blaak 22, 3011 TA Rotterdam, 3011 AA Rotterdam, The Netherlands
**Phone:** +31 10-4136420
**Fax:** +31 10-4047458
**E-mail:** mailbox@caldic.nl
**Web site:** www.caldic.com
**Parent Company:** Caldic Group (Netherlands)

**Company: DSM Fine Chemicals**
**Address:** Poststraat 1, Sittard, P.O. Box 43, 66130 AA Sittard, Netherlands
**Phone:** +31 46 477 34 87
**Fax:** +31 46 477 31 72
**E-mail:** info@dsm.com
**Web site:** www.dsm.com
**Stock listing:** Amsterdam Exchanges, German Bourse, Swiss Exchange
**Parent Company:** DSM NV (Netherlands)

**Company: DSM Agro BV**
**Address:** Poststraat 1, 6135 KR Sittard, P.O. Box 43, 6130 AA Sittard, Netherlands
**Phone:** +31 46 477 03 20
**Fax:** +31 46 452 86 15
**E-mail:** dsm.agro@dsm.com
**Web site:** www.dsm.com/dag
**Stock listing:** Amsterdam Exchanges, German Bourse, Swiss Exchange
**Parent Company:** DSM NV (Netherlands)

**Company: Hoek Loos BV**
**Address:** Havenstraat 1, 3100 AB Schiedam, Netherlands
**Phone:** +31 10 246 16 00
**Fax:** +31 10 24 6 16 00
**E-mail:** info@hoekloos.nl
**Web site:** www.hoekloos.nl
**Parent Company:** Linde Gas Group (Germany)

## NORWAY

**Company: Borregaard**
**Address:** P.O. Box 162 N-1701 Sarpsborg, Norway
**Phone:** +47 69 11 80 00
**Fax:** +47 69 11 87 70
**E-mail:** borregaard@borregaard.com
**Web site:** www.orkla.com
**Parent Company:** Orkla Group ASA (Norway)

Web site: www.orkla.com
Parent Company: Orkla Group ASA (Norway)

**Company: Hydro Agri Chemicals**
Address: Drammensveien 264, Vaekero 0240, Oslo, Norway
Phone: +47 22 53 81 00
Fax: +47 22 53 30 18
E-mail: agri@hydro.com
Web site: www.hydroagro.com
Stock listing: Oslo SE and NYSE: NHY
Parent Company: Norsk Hydro (Norway)

## PERU

**Company: Southern Peru Copper Corporation**
Address: Av. Caminos del Inca 171, Chacarilla del Estanque Santiago de Surco, Lima 33, Peru
Phone: +51 1-372-1414
Fax: +51 1-372-0077
Web site: www.southernperu.com
Stock listing: NYSE: PCU
Parent Company: Grupo Mexico SA de C.V. (Mexico)

## POLAND

**Company: ANWIL S. A.**
Address: ul. Torunska 222, 87-805 Wloclawek, Poland
Phone: +48 54 236 30 91
Fax: +48 54 236 19 83
E-mail: anwil@anwil.com.pl
Web site: www.anwil.com.pl
Parent Company: Grupa Orlen (Poland)

## QATAR

**Company: Qatar Fertiliser Company S.A.Q.**
Address: P.O. Box 50001, Mesaieed, Qatar
Phone: +974 4779779
Fax: +974 4770347
E-mail: admin@qafco.com
Web site: www.qafco.com
Parent Company: Qatar General Petroleum (Qatar)

## RUSSIA

**Company: Lukoil Oil Company**
Address: 11 Sretenski Boulevard, 101000 Moscow, Russia
Phone: +7 95 927 4444
Fax: +7 95 928 9841
E-mail: webmaster@lukoil.com
Web site: www.lukoil.com
Stock listing: Russian Trading System (RTS) and Moscow SE: LUKOY

## SAUDI ARABIA

**Company: Saudi Basic Industries Corporation (SABIC)**
Address: P.O. Box 5101, Riyadh 11422, Saudi Arabia
Phone: +966 1 225 8000, Ext. 9812
Fax: +966 1 225 9000
Web site: www.sabic.com

## SINGAPORE

**Company: Agsin Pte. Ltd.**
Address: Zagro Asia Bldg., 5 Woodlands Terrace, Woodland East Industrial Estate, Singapore 738 430
Phone: +65 6759-1811
Fax: +65 6758-7118
E-mail: contact@agsin.com
Web site: www.agsin.com
Parent Company: Zagro Asia Ltd.

**Company: Sundat Pte. Ltd.**
Address: 26 Gul Crescent, Singapore 629 532
Phone: +65 6861-2460
Fax: +65 6862-0287
E-mail: sundat@singnet.com.sg

**Company: Zagro Asia Ltd.**
Address: Zagro Asia Bldg., 5 Woodlands Terrace, Woodlands East Industrial Estate, Singapore 738 430
Phone: +65 6759-1811
Fax: +65 6759-1855
E-mail: zsingapore@zagro.com
Web site: www.zagro.com

## SOUTH AFRICA

**Company: Kynoch Fertilizer**
Address: P.O. Box 3836, Randburg 2125, South Africa
Phone: +27 11 293 6800
Fax: +27 11 789 1669
E-mail: info@aeci.co.za
Web site: www.aeci.co.za
Parent Company: AECI Ltd. (South Africa)

**Company: Metorex (Pty)Ltd.**
Address: 2nd Floor, Cradock Heights, 21 Cradock Ave., Rosebank, Johannesburg, P.O. Box 2814, Saxonwold, 2132, South Africa
Phone: +27 11-880-3155
Fax: +27 11-880-3322
E-mail: info@metorexgroup.com
Web site: www.metorexgroup.com
Stock listing: Johannesburg SE: MTX; London SE: MTX

**Company:** Omnia Group
**Address:** Omnia House, 13 Sloane St, Epsom Downs, Bryanston 2021, Gauteng, South Africa
**Phone:** +22 11-709-8888
**Fax:** +22 11-706-4022
**E-mail:** group@omnia.co.za
**Web site:** www.omnia.co.za

**Company:** Sasol Ltd.
**Address:** 1 Sturdee Ave., Rosebank, Johannesburg, Gauteng 2196, South Africa, P. O. Box 5486, Rosebank, Johannesburg Gauteng 2000, South Africa
**Phone:** +27 11-441-3111
**Fax:** +27 11-788-5092
**E-mail:** sasolinternet@sasol.com
**Web site:** www.sasol.com

## SOUTH KOREA

**Company:** Saeryung Chemicals Co., Ltd.
**Address:** 543-6, Kajwa3-dong, Seo-Ku, Inchon 404-253, South Korea
**Phone:** +82 32-578-5171
**Fax:** +82 32-582-5173
**E-mail:** saeryung@kotis.net
**Web site:** http://gov.ec21.com/incheon/inc_saeryung

**Company:** Samsung Atofina Co., Ltd. (SGC)
**Address:** 11F, Samsung Life Bldg., 150, 2-Ga Taepyung-ro, Chung-gu, Seoul, South Korea
**Phone:** +82 2-772-6619
**Fax:** +82 2-772-6614
**Web site:** www.samsungatofina.com
**Parent Company:** Arkema (France)

**Company:** Samsung General Chemicals Co., Ltd., see Samsung Atofina Co., Ltd. (South Korea)

## SPAIN

**Company:** Agrides S.A. (Sapec Agro, S.A.)
**Address:** Crta. de Constanti, km 3 - Poligono Nirsa, 43206 REUS (Tarragona), Spain
**Phone:** +34 977-77-02-11
**Fax:** +34 977-77-14-19
**E-mail:** agrides@readysoft.es
**Web site:** www.agrides.com

**Company:** Alcotan Laboratories SA
**Address:** Poligono Industrial Center, La Isla. Avenida Oeste, percela 63, P.O. Box 2204 1700, Dos Hermanas, Seville, Spain
**Phone:** +34 954- 93-00-01
**Fax:** +34 954-93-00-33
**E-mail:** informacion@alcotan-lab.com
**Web site:** www.alcotan-lab.com

**Company:** Bilbaina de Alquitranes
**Address:** Obispo Olaechea, 49, 48903 Luchana, Baracaldo, Spain
**Phone:** +34 94-499-31-78
**Fax:** +34 944-99-97-21
**E-mail:** nathalie@bilbaina.com
**Web site:** www.bilbaina.com

**Company:** Carburos Metalicos
**Address:** Aragon 300, 08009 Barcelona, Spain
**Phone:** +34 932-90-26-00
**Fax:** +34 932-90-26-03
**E-mail:** info@carburos.com
**Web site:** www.carburos.com
**Parent Company:** Air Products & Chemicals (USA)

**Company:** CEPSA (Spanish Petrol Company SA)
**Address:** Campo de las Naciones, Avda. Del Partenon, 12, 28042 Madrid, Spain
**Phone:** +34 900-10-12-82
**Web site:** www.cepsa.com
**Stock listing:** Madrid SE
**Parent Company:** Grupo Cepsa (Spain)

**Company:** Derivados del Fluor SA
**Address:** 39708 Onton, Castro Urdiales, Cantabria, Spain
**Phone:** +34 942-87-94 00
**Fax:** +34 942-87-92-46
**E-mail:** ddf@ceoecant.es
**Web site:** www.ceoecant.es/derivadosdelfluor

**Company:** Ercros SA
**Address:** Avda. Diagonal, 595, No. 10 pl., 08014 Barcelona, Spain
**Phone:** +34 934 39 30 09
**Fax:** +34 934 19 66 52
**Web site:** www.ercros.es
**Stock listing:** Barcelona and Madrid SE: ECR

**Company:** General Quimica S.A.
**Address:** Apartado 13 - 09200 Miranda de Ebro, Zubillaga - Lantaron (Alava), Spain
**Phone:** +945-332-145
**Fax:** +945-332-888
**Web site:** www.gequisa.es

**Company:** Grupo Aragonesas S. A.
**Address:** Paseo de Recoletos, 27, 28004 Madrid, Spain
**Phone:** +34 91-585 38 00
**Fax:** +34 91-585 23 00
**E-mail:** aragonesas@eiasa.es

Web site: www.grupoaragonesas.com
Stock listing: Spain SE: ARA

## SWEDEN

**Company:** AGA Gas AB, now Linde Gas AG (Germany)

**Company:** Boliden AB
**Address:** Kanalvagen 18, InfraCity, Box 500, SE-194 05 Upplands, Vasby, Sweden1
**Phone:** +46 8 610 15 00
**Fax:** +46 8 31 55 45
**Web site:** www.boliden.com
**Stock listing:** Sweden and London SE

**Company:** Perstorp Chemicals Division
**Address:** Perstop AB, SE-284 80 Perstorp, Sweden
**Phone:** +46 435 38000
**Fax:** +46 435 38100
**E-mail:** perstorp@perstorp.com
**Web site:** www.perstorp.com
**Stock listing:** Swiss Exchange, London SE
**Parent Company:** Perstorp Group (Sweden)

## SWITZERLAND

**Company:** AJE GmbH
**Address:** Sihleggstrasse 23, CH-8832 Wollerau, Switzerland
**Phone:** +41 43-888-20-10
**Fax:** +41 43-888-20-19
**E-mail:** info@aje.cc
**Web site:** www.aje.cc

**Company:** Ciba Specialty Chemicals Holding AG
**Address:** Klybeckstrasse 141, Schwartzwaldallee 215, 4002 Basel, Switzerland
**Phone:** +41 61 636 11 11
**Fax:** +41 61 636 12 12
**Web site:** www.cibasc.com
**Stock listing:** NYSE: CSB; Swiss Exchange

**Company:** Clariant International Ltd.
**Address:** Rothausstrasse 61, CH-4132 Muttenz 1, Switzerland
**Phone:** +41 61 469 5111
**Fax:** +41 61 469 6512
**Web site:** www.clariant.com
**Stock listing:** Swiss Exchange: CLN

**Company:** Lonza Group AG
**Address:** Muenchensteinerstrasse 38, 4002 Basel, Switzerland
**Phone:** +41 61 316 81 11
**Fax:** +41 61 316 91 11
**E-mail:** contact.bs@lonzagroup.com
**Web site:** www.lonza.com
**Stock listing:** Swiss Exchange: LONN

**Company:** Omya AG
**Address:** CH-4665 Oftringer, Switzerland
**Phone:** +41 62 789 29 29
**Fax:** +41 62 789 20 77
**E-mail:** info.ch@omya.com
**Web site:** www.omya.com

**Company:** Syngenta Crop Protection AG
**Address:** Schwarzwaldalle 215, P.O. Box CH-4002, 4058 Basel, Switzerland
**Phone:** +41 61 323 1111
**Web site:** www.syngenta.com
**Stock listing:** NYSE: SYT; London SE: SYA; Swiss Exchange
*Note:* Formed from the agriculture units of Novartis (formerly Ciba and Sandoz) and AstraZeneca (Zeneca Agro units).

## TAIWAN

**Company:** China Petrochemical Development Corporation (CPDC)
**Address:** 8-11 Floor, No. 12 Tunghsing Rd., Taipei, Taiwan
**Phone:** +886 2-8787-8187
**Fax:** +886 2-8787-8111
**Web site:** www.cpdc.com.tw

**Company:** Fulon Chemical Industrial Co., Ltd.
**Address:** 51-10, No. 9 Lin, Paochang Tsun, Kuaninshiang, Taoyuan County, Taiwan
**Phone:** +886 3-4830-960
**Fax:** +886 3-4830-625

**Company:** Pilarquim Corp.
**Address:** 9F 332, Chien Kuo, S. Rd., Sec. 2, Taipei, Taiwan
**Phone:** +886 2-2362-2222
**Fax:** +886 2-2362-0000
**E-mail:** tpe@pilarquim.com

**Company:** Sinon Corporation
**Address:** Agrochemical Department, 45 Wu Chuan Center St., Taichung, Taiwan
**Phone:** +886 4-693-4261
**Fax:** +886 4-693-4265
**E-mail:** service@sinon.com
**Web site:** www.sinon.com

**Company:** Yellow River Enterprise Co., Ltd.
**Address:** 100 Chung Cheng Rd., Hsing Ying City, Tainan County, Taiwan
**Phone:** +886 6 635 3326
**Fax:** +886 6 635 3313
**E-mail:** service@yelori.com
**Web site:** www.yelori.com

## THAILAND

**Company:** Asia Pacific Resources Ltd.
**Address:** Ste. 2002, 20 Fl., Abdulrahim Place, 990, Rama IV Rd, Silom, Bangrak, Bangkok 10500, Thailand
**Phone:** +662 636-1600
**Fax:** +662 636-1599
**Web site:** www.apq-potash.com
**Stock listing:** OTC: APQC

## UKRAINE

**Company:** Azot Association
**Address:** 93403, Severodonetsk, Luganskaya oblastj, ul. Pivovarova, 5, Ukraine
**Phone:** +380 6452 9 26 80
**Fax:** +380 6452 2 97 02
**E-mail:** efc@azot.lg.ua
**Web site:** www.azot.lg.ua
**Parent Company:** Severodonetsk State Manufacturing Enterprise

## UNITED KINGDOM

**Company:** Agrichem (International) Limited
**Address:** Industrial Estate, Station Rd., Whittlesey, Cambridgeshire PE7 2EY, UK
**Phone:** +44 1733-204019
**Fax:** +44 1733-204162
**E-mail:** info@agrichem.co.uk
**Web site:** www.agrichem.co.uk

**Company:** Agropharm Ltd.
**Address:** Buckingham House, Church Rd., Penn, High Wycombe HP10 8LN, UK
**Phone:** +44 1952-740333
**Fax:** +44 1952-740207
**E-mail:** sales@agropharm.co.uk
**Web site:** www.agropharm.co.uk

**Company:** A H Marks & Co., Ltd.
**Address:** Wyke, Bradford, West Yorkshire BD12 9EJ, UK
**Phone:** +44 1274 691234
**Fax:** +44 1274 691176
**E-mail:** postmaster@ahmarks.com
**Web site:** www.ahmarks.com

**Company:** Albright & Wilson Pty.
**Address:** P.O. Box, 210-222 Hagley Rd. West, Oldbury, West Midlands B68 ONN, UK
**Phone:** +44 (0) 121 429 4942
**Fax:** +44 (0) 121 420 5151
**Web site:** www.albriw.com
**Parent Company:** Rhodia Group (France)

**Company:** Anglo American Plc
**Address:** 20 Carlton House Terrace, London SW1Y 5AN, UK
**Phone:** +44 207-698-8500
**Fax:** +44 207-698-8888
**Web site:** www.angloamerican.co.uk
**Stock listing:** London SE: AGM; Munich SE; France Bourse: AAL; Nasdaq: AAUK

**Company:** Antofagasta Plc
**Address:** 5 Princes Gate, London SW7 1QJ, UK
**Phone:** +44 20-7808-0988
**Fax:** +44 20-7807-0986
**Web site:** www.antofagasta.co.uk
**Stock listing:** London SE: ANTO

**Company:** Archimica Ltd.
**Address:** Sandycroft, Deeside, Flintshire CH5 2PX, United Kigdom
**Phone:** +44 (0) 1244-520-777
**Fax:** +44 (0) 1244-537-216
**E-mail:** uk@archimica.com
**Web site:** www.archimica.com
**Parent Company:** BTP Plc (UK)

**Company:** Ascot International (1996) Ltd.
**Address:** Welcroft St., Stockport SK1 3DF, UK
**Phone:** +44 0161-476-6161
**Fax:** +44 0161-476-5775
**E-mail:** melissa@ascot1.com
**Web site:** www.ascot1.com

**Company:** BHP Billiton Plc.
**Address:** Neathhouose Place, Victoria, London SW1V 1BH, UK
**Phone:** +44 20-7802-4000
**Fax:** +44 20-7802-4111
**Web site:** www.bhpbilliton.com
**Stock listing:** London, Australia, NYSE: BHP
**Parent Company:** BHP Billiton Ltd. (Australia)

**Company:** BOC Gases
**Address:** The Priestley Centre, 10 Priestley Road, Surrey Research Park, Guildford, Surrey GU2 5XY, UK
**Phone:** +44 1483 579 857, 0800 111333 (customer service toll free in UK)

Web site: www.boc.com/gases
Parent Company: BOC Group (UK)

**Company:** BP Chemical Company Plc
**Address:** Britannic House, 1 Finsbury Circus, London EC2M 7BA, UK
**Phone:** +44 20 7496 4000
**Fax:** +44 20 7496 4630
**E-mail:** chem_americas@bp.com
**Web site:** www.bpchemicals.com
**Stock listing:** NYSE: BP; London SE: BP
**Parent Company:** BP Plc (UK)

**Company:** Brotherton Specialty Products ltd.
**Address:** Calder Vale Rd., Wakefield, West Yorkshire WF1 5PH, UK
**Phone:** +44 19 24 371 919
**Fax:** +44 19 24 290 408
**Web site:** www.brotherton.co.uk
**Parent Company:** Church & Dwight Inc. (US)

**Company:** Celtic Chemicals Ltd.
**Address:** Gas Works Estate, Victoria Road, Port Talbot, West Glamorgan SA12 6DB, UK
**Phone:** +44 1639 886236
**Fax:** +44 1639 893147
**Web site:** www.celticchemicals.co.uk

**Company:** Cleveland Potash Ltd.
**Address:** Boulby Mine, Loftus, Satlburn-by-the-Sea, Cleveland TS13 4UZ, UK
**Phone:** +44 (0) 1287 640 934
**Fax:** +44 (0) 1287 640 935
**E-mail:** enquiries@clevelandpotash.co.uk
**Web site:** www.clevelandpotash.ltd.uk

**Company:** Coalite Chemicals Ltd.
**Address:** P.O. Box 152, Buttermilk Lane, Bolsover, Chesterfield, Derbyshire S44 6AZ, UK
**Phone:** +44 1246-826816
**Fax:** +44 1246-240309
**E-mail:** chemicals@coalitechemicals.com
**Web site:** www.coalitechemicals.com

**Company:** Eurolabs Ltd.
**Address:** London House, London Road South, Paynton, Cheshire Sk12 1YP: UK
**Phone:** +44 1625-850089
**Fax:** +44 1625-858854
**Web site:** www.eurolabs.co.uk

**Company:** Fine Agrochemicals Ltd.
**Address:** Hill End House, Whittington, Worcester WR5 2RQ, UK
**Phone:** +44 1905 361 800
**Fax:** +44 1905 361 810
**E-mail:** enquire@fine-agrochemicals.com
**Web site:** www.fine-agrochemicals.com

**Company:** Fluorochem Ltd.
**Address:** Wesley Street, Old Glossop, Derbyshire SK13 7RY, UK
**Phone:** +44 14 57 868921
**Fax:** +44 14 57 869360
**E-mail:** enquiries@fluorochem.co.uk
**Web site:** www.fluorochem.co.uk

**Company:** GlaxoSmithKline Plc
**Address:** Glaxo Wellcome House, Berkley Ave., Greenford, Middlesex UB6 ONN, UK
**Phone:** +44 171 493 4060
**Fax:** +44 181 966 8330
**Web site:** www.gsk.com
**Stock listing:** NYSE: GSK; London SE: GSK

**Company:** Hockley International Ltd.
**Address:** Hockley House, 354 Park Lane, Poynton, Stockport, Cheshire SK12 1RL, UK
**Phone:** +44 1625 878 590
**Fax:** +44 1625 877 285
**E-mail:** mail@hockley.co.uk
**Web site:** www.hockley.co.uk

**Company:** ICI Group (Imperial Chemical Industries Plc)
**Address:** Imperial Chemical House, Milbank, London SW1P 3JF, UK
**Phone:** +44 (0) 20 7834 4444
**Fax:** +44 (0) 20 7834 2042
**E-mail:** ici@ici.com
**Web site:** www.ici.com
**Stock listing:** London and NYSE: ICI

**Company:** Ki-Hara Chemicals Ltd.
**Address:** 20-22 Harborne Rd., Egdbaston, Birmingham B15 3AA, UK
**Phone:** +44 121-693-5900
**Fax:** +44 121-693-5901
**Web site:** www.ki-hara.co.uk

**Company:** Lancaster Synthesis
**Address:** Newgate, White Lund, Morecambe, Lancashire LA3 3PT, UK
**Phone:** +44 1524 36101
**Fax:** +44 1524 39727
**E-mail:** lancaster_UKsales@clariant.com
**Web site:** www.lancastersynthesis.com
**Parent Company:** Clariant International (Switzerland)

**Company:** Laporte Performance Chemicals Plc
**Address:** Charleston Industrial Estate, Hardley, Hythe, Southampton SO45 3ZG, UK
**Phone:** +44 2380 894666
**Fax:** +44 2380 243113
**Web site:** www.inspec.co.uk

**Company:** Mining & Chemical Products Ltd. (MCP)
**Address:** 1-4 Nielson Rd, Finedon Road Industrial Estate, Wellingborough, Northants NN8 4PE, UK
**Phone:** +44 1933-225-766
**Fax:** +44 1933-27-814
**E-mail:** info@mcp-group.co.uk
**Web site:** www.mcp-group.com

**Company:** Molekula Fine Chemicals Ltd.
**Address:** Technology Rd., Poole, Dorset BH177DA, UK
**Phone:** +44 (0)1202 330066
**Fax:** +44 (0)1202 330055
**E-mail:** info@molekula.co.uk
**Web site:** www.molekula.co.uk

**Company:** Newmont Koch
**Address:** Newmet House, Rue de St. Lawrence, Waltham Abbey, Essex EN9 1PF, UK
**Phone:** +44 1992-711-111
**Fax:** +44 1992-768-393
**E-mail:** materials@newmet.co.uk
**Web site:** www.newmet.co.uk
**Parent Company:** New Metals and Chemicals Ltd. (Newmet)

**Company:** Ocean Chemicals Group
**Address:** Wesley St., Old Glossop, Derbyshire SK12 7RY, UK
**Phone:** +44 1457-867-708
**Fax:** +44 1457-860-927
**E-mail:** enquiries@oceanchemicals.co.uk
**Web site:** www.oceanchemicals.co.uk

**Company:** Reckitt Benckiser Plc
**Address:** 103-105 Bath Rd., Slough, Berks SL1 3UH, UK
**Phone:** +44 1753-217800
**Fax:** +44 1753-217899
**Web site:** www.reckitt.com

**Company:** Rio Tinto Plc
**Address:** 6 St. James's Square, London SW1Y 4LD, UK
**Phone:** +44 (071) 930 2399
**Fax:** +44 (071) 930 3249
**Web site:** www.riotinto.com
**Stock listing:** London SE: RIO; NYSE: RTP
**Parent Company:** Rio Tinto Group (Australia & United Kingdom)

**Company:** Shell Chemicals
**Address:** Cheshire Innovation Park, Chester CH1 3SH, UK
**Phone:** +44 12 44 68 5000
**Fax:** +44 12 44 68 5 010
**Web site:** www.shellchemicals.com
**Parent Company:** Royal Dutch Petroleum Company (Netherlands)

**Company:** Sorex Ltd.
**Address:** St. Michael's Industrial Estate, Widnes, Cheshire WA8 8TJ, UK
**Phone:** +44 01510-420-7151
**Fax:** +44 0150-495-1163
**Web site:** http://www.sorex.com

**Company:** Thor GmbH
**Address:** Ramsgate Rd., Margate, Kent CT9 4JY, UK
**Phone:** +44 (0) 1843-222404
**Fax:** +44 (0) 1843-229413
**E-mail:** webmaster@thor.com
**Web site:** www.thor.com

**Company:** Uniqema Ltd
**Address:** Bebington, Wirral, Merseyside CH62 4UF, UK
**Phone:** +44 151643 3200
**Fax:** +44 151 645 9197
**Web site:** www.uniqema.com
**Parent Company:** ICI Group (Imperial Chemical Industries Plc) (UK)

**Company:** Whyte Agrochemicals Ltd.
**Address:** Denaby Lane Industrial Estate, Denaby Lane, Old Denaby, Doncaster, South Yorkshire DN12 4LQ, UK
**Phone:** +44 01709-772-200
**Fax:** +44 01709-772-201
**E-mail:** sales@whytechemicals.co.uk
**Web site:** www.whytechemicals.co.uk
**Parent Company:** Whyte Chemicals Ltd., (UK)

**Company:** William Blythe Ltd.
**Address:** Church, Accrington, Lancashire BB5 4PD, UK
**Phone:** +44 1254-320-000
**Fax:** +44 1254-320-001
**E-mail:** info@wm-blythe.co.uk
**Web site:** www.wm-blythe.co.uk
**Stock listing:** London SE: YULC
**Parent Company:** Yule Catto & Co. Plc (UK)

**Company:** Zeneca Agro (UK), now Syngenta (Switzerland)

# UNITED STATES

**Company:** Advanced Synthesis Technologies, SA

Address: P.O. Box 437920, San Ysidro, CA 92173, USA
Phone: +1 619-423-7821
Fax: +1 619-43-7793
E-mail: advancedsynthesis@yahoo.com
Web site: www.advancedsynthesis.com

Company: AgraQuest
Address: 1530 Drew Ave., Davis, CA 95616, USA
Phone: +1 530-750-0150
Fax: +1 530-750-0153
Web site: www.agraquest.com

Company: Agriliance LLC
Address: P.O. Box 64089, St. Paul, MN 55164, USA
Phone: +1 651-451-5000, 800-535-4635
Fax: +1 651-451-5405
Web site: www.agriliance.com

Company: Agrimor International Co.
Address: 210 174th St, S-1819, Sunny Isles Beach, FL 33160, USA
Phone: +1 305-682-1211
Fax: +1 305-682-1196
E-mail: box@agrimore.com
Web site: www.agrimor.com

Company: Agway Inc.
Address: P.O. Box 4933, Syracuse, NY 13221-4933, USA
Phone: +1 315-449-7061
Web site: www.agway.com
Stock listing: Farmer-owned coop.

Company: Air Products and Chemicals Inc.
Address: 7201 Hamilton Boulevard, Allentown, PA 18195-1501, USA
Phone: +1 610-481-4911
Fax: +1 610-481-5900
E-mail: info@apci.com
Web site: www.airproducts.com
Stock listing: NYSE: APD

Company: Alaska Nitrogen Products LLC
Address: P.O. Box 575, Kenai, AK 99611-0575, USA
Phone: +1 907-776-8121
Fax: +1 907-776-5579
Parent Company: Agrium (Canada)

Company: Albemarle Corporation
Address: 451 Florida St., Baton Rouge, LA 70801, USA
Phone: +1 225-388-7402
Fax: +1 225-388-7848
Web site: www.albemarle.com
Stock listing: NYSE: ALB

Company: Alcide Corporation
Address: 8561 154th Ave. NE, Redmond, WA 98052, USA
Phone: +1 425-882-2555
Fax: +1 425-861-0173
E-mail: info@alcide.com
Web site: www.alcide.com

Company: Alcoa Inc.
Address: Alcoa Corporate Center, 201 Isabella St., Pittsburgh, PA 15212, USA
Phone: +1 412-553-3042
Fax: +1 412-553-3129
Web site: www.alcoa.com
Stock listing: NYSE: AA

Company: Aldrich Chemical Co., Inc
Address: P.O. Box 355, Milwaukee, WI 53201, USA
Phone: +1 414-273-3850, 800-558-9160
Fax: +1 414-273-4979
Web site: www.sigma-aldrich.com
Stock listing: Nasdaq: SIAL
Parent Company: Sigma-Aldrich Co. (USA)

Company: Alfa Aesar
Address: 30 Bond St., Ward Hill, MA 01835, USA
Phone: +1 978 521 6300
Fax: ++1 978 521 6350
E-mail: info@alfa.com
Web site: www.alfa.com
Parent Company: John Matthey Ltd. (UK)

Company: Amchem Products, now Union Carbide (USA) and then Dow AgroScience (USA)

Company: American Cyanamid's Agricultural Products Group, now BASF Agricultural Products Group (Germany).

Company: American Elements Inc.
Address: 1093 Broxton Ave., Ste. 2000, Los Angeles, CA 90024, USA
Phone: +1 310-208-0551
Fax: +1 310-208-0351
E-mail: customerservice@americanelements.com
Web site: www.americanelements.com

Company: American Gas Group
Address: 6055 Brent Dr., Toledo, OH 43611, USA
Phone: +1 419-729-7732, 800-471-7013
Fax: +1 419-729-2411
Web site: www.americangasgroup.com

Company: AMVAC Chemical Corporation
Address: 4100 E. Washington Boulevard, Los Angeles,

CA 90023, USA
Phone: +1 323-264-3910
Fax: +1 323-268-1028
Web site: www.amvac-chemical.com
Parent Company: American Vanguard Corporation (USA)

Company: Apache Nitrogen Products Inc.
Address: P.O. Box 700, Benson, AZ 85602, USA
Phone: +1 520-720-2217
Fax: +1 520-720-4158
E-mail: anpi@apachenitro.com
Web site: www.apachenitro.com

Company: Aquashade Inc.
Address: W175 N11163 Stonewood Dr., Ste. 234, Germantown, WI 53022, USA
Phone: +1 800-558-5106
Web site: www.aquashade.com

Company: Arvesta Corporation
Address: 100 First St., Ste. 1700, San Francisco, CA 94105, USA
Phone: +1 415-536-3480
E-mail: : hkaufmann@arvesta.com
Web site: www.arvesta.com
Parent Company: Arysta LifeScience Corporation, Japan

Company: ASARCO
Address: 2575 E. Camelback Rd., Ste. 500, Phoenix, AZ 85016, USA
Phone: +1 602-977-6500
Fax: +1 602-977-6700
Web site: www.asarco.com
Parent Company: Grupo Mexico SA de C.V. (Mexico)

Company: Aristech Chemical Corporation, now Sunoco Chemicals (US)

Company: Ashland Inc.
Address: 50 E. River Center Boulevard, P.O. Box 391, Covington, KY 41012-0391, USA
Phone: +1 859-815-3333
Fax: +1 859-815-5053
E-mail: info@ashland.com
Web site: www.ashland.com
Stock listing: NYSE: ASH

Company: Atofina Chemicals Inc.
Address: 2000 Market St., Philadelphia, PA 19103-3222, USA
Phone: +1 215-419-7000, +1 800-225-7788
Fax: +1 215-419-7591
E-mail: webmaster@ato.com
Web site: www.arkemagroup.com
Parent Company: Arkema Group (France), Subsidiary of Total (France)

Company: Avitrol Corporation
Address: 7644 E 46th St., Tulsa, OK 74145, USA
Web site: www.avitrol.com

Company: Baker Petrolite Corporation
Address: 12645 West Airport Boulevard, Sugar Land, TX 77478, USA
Phone: +1 281-276-5400
Web site: www.bakerhughes.com/bakerpetrolite
Parent Company: Baker Hughes (USA)

Company: BASF Agricultural Products Group
Address: 3000 Continental Dr., North, Mount Olive, NJ 07828-1234, USA
Phone: +1 973-426-2600
Fax: +1 973-426-2610
Web site: www.cyanamid.com, www.agro.basf.com
Note: Merger of American Cyanamid with BASF AG (Germany)

Company: BASF Corporation
Address: 3000 Continental Dr., North, Mount Olive, NJ 07828-1234, USA
Phone: +1 973-426-2600
Fax: +1 973-426-2610
Web site: www.cyanamid.com, www.agro.basf.com
Note: Merger of American Cyanamid with BASF AG (Germany)

Company: Becker Underwood Inc.
Address: 801 Dayton Ave., Ames, IA 50010, USA
Phone: +1 515-232-5907
Web site: www.beckerunderwood.com

Company: Bell Laboratories, Inc.
Address: 3699 Kinsman Boulevard, Madison, WI 53704, USA
Phone: +1 608-241-0202
Fax: +1 608-241-9631
E-mail: emea@belllabs.com
Web site: www.belllabs.com

Company: Bonide Products, Inc.
Address: 6301 Sutliff Rd., Oriskany, NY 13424, USA
Phone: +1 315-736-8231
Fax: +1 315-736-7582
Web site: www.bonideproducts.com

Company: Borden Chemical Inc.
Address: 180 East Broad St., Columbus, OH 43215-3799, USA

Phone: +1 614-225-4000
Fax: +1 502-560-5260
Web site: www.bordenchem.com
Parent Company: Borden Family of Companies, Kohlberg Kravis Roberts & Co.

Company: Buckman Laboratories, Inc.
Address: 1256 No. Mclean Boulevard, Memphis, TN 38108, USA
Phone: +1 901-278-0330
Fax: +1 901-276-5343
Web site: www.buckman.com
Parent Company: Bulab Holdings, Inc. (USA)

Company: Cargill Crop Nutrition
Address: 12105 Lynn Ave. South, Savage, MN 55378, USA
Phone: +1 800-527-7491
Web site: www.cargillfertilizer.com

Company: Cedar Chemical Corporation went into bankruptcy March, 2002

Company: Celite Corporation
Address: 137 West Central Ave., Lompac, CA 93436, USA
Phone: +1 805 737 2455, 800 893 4445
Fax: +1 805 737 5699
E-mail: info@worldminerals.com
Web site: www.worldminerals.com
Stock listing: NYSE: Y
Parent Company: **World Minerals, sub. of Allegheny Corporation (US)**

Company: Cerexagri Inc. - North America
Address: 630 Freedom Business Center, Ste. 402, King of Prussia, PA 19406, USA
Phone: +1 610-491-2800, +1 800-438-6071
Fax: +1 610-491-2801
Web site: www.cerexagri.com
Parent Company: ATOFINA (France), Subsidiary of Total (France)
*Note:* As of January 1, 2001, the Agrichemicals Division of Elf Atochem became Cerexagri, Inc.

Company: Certis USA, LLC
Address: 9145 Guilford Rd., Ste. 175, Columbia, MD 21046, USA
Phone: +1 800-847-5620
Web site: www.certisusa.com

Company: CF Industries Inc.
Address: One Salem Lake Dr., Long Grove, IL 60047, USA
Phone: +1 847-438-9500
Fax: +1 847-438-0211

E-mail: contact@cfindustries.com
Web site: www.cfindustries.com

Company: Chemical Company, The
Address: 19 Narragansett Ave., P. O. Box 436, Jamestown, RI 02835, USA
Phone: +1 401-423-3100
Fax: +1 401-423-3102
Web site: www.thhechemco.com

Company: Chemical Lime Co.
Address: P.O. Box 98504, Fort Worth, TX 76185, USA
Phone: +1 817-732-8164
Fax: +1 817-732-8564
E-mail: solution@chemicallime.com
Web site: www.chemicallime.com
Parent Company: Lhoist Group (Belgium)

Company: Chemical Products Corporation
Address: 102 Old Mill Rd., P.O. Box 2470, Cartersville, GA 30120-1692, USA
Phone: +1 770-382-2144
Fax: +1 770-386-6053
E-mail: cpcsales@mindspring.com
Web site: www.chemicalproductscorp.com

Company: Chevron Phillips Chemical Company LP
Address: 1301 McKinney St., Houston, TX 77010-3030, P.O. Box 3766, Houston, TX 77253-3766, USA
Phone: +1 713-289-4100
E-mail: products@cpchem.com
Web site: www.cpchem.com
Parent Company: Joint venture of Phillips Petroleum and ChevronTexaco (US)

Company: ChevronTexaco Corp.
Address: 575 Market St., San Francisco, CA 94105, USA
Phone: +1 415-894-7700
E-mail: comment@chevrontexaco.com
Web site: www.chevrontexaco.com
Stock listing: NYSE: CVX

Company: Cleary Chemical Corp.
Address: 178 Ridge Rd., Dayton, OH 08810, USA
Phone: +1 800-524-1662
E-mail: info@clearychemical.com
Web site: www.clearychemical.com

Company: Continental Lime Inc.
Address: 670 East 3900 South, Ste. 205, Salt Lake City, UT 84107, USA
Phone: +1 801-262-3942
Fax: +1 801-264-8039
Web site: www.continentallime.com

**Company: C. P. Hall Company, The**
Address: 311 S. Wacker Dr., Ste. 4700, Chicago, IL 60606, USA
Phone: +1 312-554-7400
Fax: +1 312-554-7499
Web site: www.cphall.com

**Company: Creanova Inc.**
Address: 220 Davidson Ave., Somerset, NJ 08873, P.O. Box 6821, Somerset, NJ 08875-6821, USA
Phone: +1 732-560-6800
Fax: +1 732-560-6958
Web site: www.creanovainc.com
Parent Company: Degussa AG (Germany)

**Company: Crompton Corporation**
Address: One American Lane, Greenwich, CT 06831-2559, USA
Phone: +1 203-552-2000
Fax: +1 203-353-5424
E-mail: cromptonlit@mmcweb.com
Web site: www.cromptoncorp.com
Stock listing: NYSE: CK
*Note:* Formed September 1, 1999, by the merger of Crompton & Knowles Corporation (USA) and the Witco Corporation (USA).

**Company: Crowley Chemical Co.**
Address: 261 Madison Ave., New York, NY 10016, USA
Phone: +1 212-682-1200
Fax: +1 212-953-3487
E-mail: info@crowleychemical.com
Web site: www.crowleychemical.com

**Company: Crown Technology Inc.**
Address: 7513 E. 96th St., P.O. Box 50426, Indianapolis, IN 46250-0426, USA
Phone: +1 317-845-0045
Fax: +1 317-845-9086
E-mail: info@crowntech.com
Web site: www.crowntech.com

**Company: CTC Organics**
Address: 792 Windsor St. SW, P.O. Box 6933, Atlanta, GA 30315, USA
Phone: +1 404-524-6744

**Company: Cytec Industries Inc.**
Address: 5 Garret Mountain Plaza, West Paterson, NJ 07424, USA
Phone: +1 973-357-3100, 800-652-6013
Fax: +1 973-357-3065
E-mail: info@gm.cytec.com
Web site: www.cytec.com

**Company: Degesch America Inc.**
Address: P.O. Box 116, Weyers Cave, VA 24486, USA
Phone: +1 800-330-2525, 540-234-9281
Fax: +1 540-234-8225
E-mail: info@degeschamerica.com
Web site: www.degeschamerica.com

**Company: Delta Chemical Corporation**
Address: 2601 Cannery Ave., Baltimore, MD 21226-1595, USA
Phone: +1 410-354-0100, 800-282-5322
Fax: +1 410-354-1021
Web site: www.deltachemical.com

**Company: Dow AgroSciences LLC**
Address: 9330 Zionsville Rd., Indianapolis, IN 46268-1054, USA
Phone: +1 800-891-9157
Fax: +1 800-905-7326
E-mail: info@dowagro.com
Web site: www.dowagro.com
Stock listing: NYSE: DOW
Parent Company: Dow Chemical Co. (US)

**Company: Dow Chemical Company**
Address: 47 Building, Midland, MI 48674, USA
Phone: +1 989-636-1000
Fax: +1 989-636-7238
Web site: www.dow.com
Stock listing: NYSE: DOW

**Company: Drexel Chemical Company**
Address: P.O. Box 13327, Memphis, TN 38113-0327, USA
Phone: +1 901 774 4370
Fax: +1 901 774 4666
E-mail: drexchem@bellsouth.net
Web site: www.drexchem.com

**Company: DuPont (E.I. du Pont de Nemours & Co.)**
Address: DuPont Building, 1007 Market St., Wilmington, DE 19898, USA
Phone: +1 302-774-1000, US only 800-441-7515
Fax: +1 302-774-7321
E-mail: info@dupont.com
Web site: www.dupont.com
Stock listing: NYSE: DD

**Company: DuPont Crop Protection**
Address: 1007 Market St., Wilmington, DE 19898, USA
Phone: +1 888-638-7668
Web site: www.cropprotection.dupont.com
Stock listing: NYSE: DD

**Company:** Eagle Picher Industries Inc.
**Address:** Eagle Picher Filtration & Minerals, Inc, 9785 Gateway Dr., Ste. 1000, Reno, NV 89521, USA
**Phone:** +1 775-824-7600
**Fax:** +1 775-824-76-1
**E-mail:** minerals@eaglepicher.com
**Web site:** www.epcorp.com
**Parent Company:** Eagle Picher Inc. (USA)

**Company:** Eastman Chemical Co.
**Address:** 100 North Eastman Rd., P.O. Box 511, Kingsport, TN 37662-5075, USA
**Phone:** +1 423-229-2000
**Fax:** +1 423-229-1351
**E-mail:** info@eastman.com
**Web site:** www.eastman.com
**Stock listing:** NYSE: EMN

**Company:** Ecogen Inc.
**Address:** 2005 West Cabot Boulevard, Langhorne, PA 19047, USA
**Phone:** +1 215-757-1590
**Fax:** +1 215-757-2956

**Company:** EDEN Bioscience Corp.
**Address:** 3830 Monte Villa Pkwy., Ste. 100, Bothell, WA 98021, USA
**Phone:** +1 425-806-7300, 888-879-2420
**Fax:** +1 425-806-7400
**E-mail:** messenger@edenbio.com
**Web site:** www.edenbio.com
**Stock listing:** NASDAQ: EDEN

**Company:** El Dorado Chemical Company
**Address:** 1950 Alpha Rd., Ste. 100, Rockwell, TX 75087, USA
**Phone:** +1 800-264-2853
**E-mail:** information@eldoradochemical.com
**Web site:** www.eldoradochemical.com
**Parent Company:** LSB Industries (US)

**Company:** Elston Manufacturing Co.
**Address:** 706 No. Weber, Sioux Falls, SD 57103, USA
**Phone:** +1 800-845-1385
**Web site:** www.elstonmfg.com

**Company:** Emerald BioAgriculture Corp.
**Address:** 3125 Sovereign Dr., Ste. B, Lansing, MI 48911-4240, USA
**Phone:** +1 517-882-7370
**Fax:** +1 517-882-7560
**Web site:** www.emeraldbio.com

**Company:** EM Industries Inc.
**Address:** 7 Skyline Drive, Hawthorne, NY 10532, USA
**Phone:** +1 914-592-4660
**Fax:** +1 914-592-9469
**Web site:** www.emindustries.com
**Stock listing:** Frankfurt SE: MRK; Zurich SE
**Parent Company:** Merck KgaA (Germany)

**Company:** Entek Corporation
**Address:** 6835 Deerpath Rd., Ste. E, Elkridge, MD 21075, USA
**Phone:** +1 800-760-7150
**Fax:** +1 410-579-1633
**E-mail:** entek@entekcorp.com
**Web site:** www.entekcorp.com

**Company:** Equistar Chemicals LP
**Address:** 1221 McKinney St., Houston, TX 77010, P.O. Box 2583, Houston, TX 77252-2583, USA
**Phone:** +1 713-652-7300
**Fax:** +1 713-652-4151
**Web site:** www.equistarchem.com
**Parent Company:** 70% owned by Lyondell Chemical Company (USA), 30% owned by Millennium Chemicals (USA)

**Company:** Exxon Mobil Chemical Company
**Address:** 13501 Katy Freeway, Houston, TX 77079-1306, USA
**Phone:** +1 281-584-7600
**Fax:** +1 281-870-6661
**E-mail:** info@exxonmobilchemical.com
**Web site:** www.exxonmobilchemical.com
**Parent Company:** Exxon Mobil Corporation (USA)

**Company:** Fairmount Chemical Company
**Address:** 117 Banchard St., Newark, NJ 07105, USA
**Phone:** +1 973-344-5790
**Fax:** +1 973-690-5298
**Stock listing:** OTC: FMTC.OB

**Company:** Ferro Corporation
**Address:** 100 Lakeside Ave., Cleveland, OH 44114, USA
**Phone:** +1 216-641-8580
**Web site:** www.ferro.com

**Company:** FMC Agricultural Products Group
**Address:** 1735 Market St., Philadelphia, PA 19103, USA
**Phone:** +1 215-299-6000
**Fax:** +1 215-299-6568
**E-mail:** info@fmc.com
**Web site:** www.fmc.com
**Stock listing:** NYSE: FMC

**Company:** Georgia Gulf Corporation
**Address:** 400 Perimeter Center Terrace, Suite 595, P.O. Box 105197, Atlanta, Georgia 30346, USA
**Phone:** +1 770-395-4500
**Fax:** +1 770-395-4529
**Web site:** www.georgiagulf.com
**Stock listing:** NYSE: GGC

**Company:** Georgia-Pacific Corporation
**Address:** 133 Peachtree St., N.E., Atlanta, GA 30303, USA
**Phone:** +1 404-6521-4000
**Fax:** +1 404-230-1674
**Web site:** www.gp.com
**Stock listing:** NYSE: GP
**Parent Company:** Georgia Pacific Group (US)

**Company:** GFS Chemicals Inc.
**Address:** P.O. Box 245, Powell, OH 43065, USA
**Phone:** +1 877-534-0795 (US and Canada), 740-881-5501 (International)
**Fax:** +1 740-881-5989
**E-mail:** gfschem@gfschemicals.com
**Web site:** www.gfschemicals.com

**Company:** Gowan Company L.L.C.
**Address:** P.O. Box 5569, Yuma, AZ 85366-5569, USA
**Phone:** +1 520-783-8844, 800-883-1844
**Web site:** www.gowanco.com

**Company:** Great Lakes Chemical Corporation
**Address:** 9025 No. river Rd., Ste. 400, Indianapolis, IN 46240, USA
**Phone:** +1 317-715-3000
**Fax:** +1 317-715-3050
**E-mail:** info@glcc.com
**Web site:** www.greatlakeschem.com
**Stock listing:** NYSE: GLK

**Company:** Great Western Inorganics
**Address:** 17400 Highway 72, Arvada, CO 80007, USA
**Phone:** +1 303-423-9770
**Fax:** +1 303-423-9772
**E-mail:** gwi46@cs.com
**Web site:** www.greatwesterninorganics.com

**Company:** Griffin L.L.C.
**Address:** 2509 Rocky Ford Rd., P.O. Box 1847, Valdosta, GA 31601, USA
**Phone:** +1 229-242-8635
**Fax:** ++1 229-244-5813
**Web site:** www.griffinllc.com

**Company:** Gustafson LLC
**Address:** 1400 Preston Rd., Ste. 400, Plano, TX 75093, USA
**Phone:** +1 972-985-8877, 800-248-6907
**Fax:** +1 972-985-1696
**Web site:** www.gustafson.com

**Company:** Halocarbon Products Corporation
**Address:** P.O. Box 661, River Edge, NJ 07661, USA
**Phone:** +1 201-262-8899
**Fax:** +1 201-262-0019
**E-mail:** info@halocarbon.com
**Web site:** www.halocarbon.com

**Company:** Harcros Chemicals Inc.
**Address:** 5200 Speaker Rd., P.O. Box 2930, Kansas City, KS 66110-2930, USA
**Phone:** +1 913-621-7747
**Fax:** +1 913-621-7746
**E-mail:** custserv@harcroschem.com
**Web site:** www.harcroschem.com

**Company:** Helena Chemical Company
**Address:** 225 Schilling Blvd, Collierville, TN 38017, USA
**Phone:** +1 901-761-0050
**Fax:** +1 901-761-5754
**E-mail:** hccsales@helenachemical.com
**Web site:** www.helenachemical.com

**Company:** Hercules Inc.
**Address:** 1313 North Market St., Wilmington, DE 19894-0001, USA
**Phone:** +1 302-594-5000
**Fax:** +1 302-594-5400
**Web site:** www.herc.com
**Stock listing:** NYSE: HPC

**Company:** Honeywell Performance Polymers & Chemicals Inc.
**Address:** Honeywell Specialty Materials, 101 Columbia Rd., P.O. Box 1057, Morristown, NJ 07962, USA
**Phone:** +1 800-322-2766
**Fax:** +1 973-455-5000
**Web site:** www.specialtychem.com
**Stock listing:** NYSE: HON
**Parent Company:** Honeywell International Inc. (US)

**Company:** Hummel Croton Inc.
**Address:** 10 Harmich Rd., South Plainfield, NJ 07080-4899, USA
**Phone:** +1 908 754 1800
**Fax:** +1 908 754 1815
**E-mail:** sales@hummelcroton.com
**Web site:** www.hummelcroton.com

**Company:** Huntsman Corporation
**Address:** 500 Huntsman Way, Salt Lake City, UT 84108, USA
**Phone:** +1 801-584-5700, 800-421-2411
**Fax:** +1 801-584-5781
**E-mail:** info@huntsman.com
**Web site:** www.huntsman.com

**Company:** ICC Industries Inc.
**Address:** 460 Park Ave., New York, NY 10022, USA
**Phone:** +1 212-521-1700
**Fax:** +1 212-521-1970
**E-mail:** trading@iccchem.com
**Web site:** www.iccchem.com

**Company:** IMC Chemicals
**Address:** 9401 Indian Creek Pkwy., Ste. 1000, Overland Park, KS, USA
**Phone:** +1 800-837-2775
**Web site:** www.imcchemicals.com
**Parent Company:** IMC Global Inc. (USA)

**Company:** Indofine Chemical Company Inc.
**Address:** P.O. Box 473, Somerville, NJ 08876, USA
**Phone:** +1 908-359-6778, 888-463-6346
**Fax:** +1 908-359-1179
**E-mail:** chemical@indofinechemical.com
**Web site:** www.indofinechemical.com

**Company:** International Specialty Products
**Address:** 1361 Alps Rd., Wayne, NJ 07470 (USA)
**Phone:** +1 973-628-4000
**Fax:** +1 973-628-4001
**Web site:** www.ispcorp.com

**Company:** International Sulfur Inc.
**Address:** P.O. Box 611, Mt. Pleasant, TX 75456-0611, USA
**Phone:** +1 903-577-5500, 800-828-7857
**Fax:** +1 903-577-5540
**E-mail:** isulphur@internationalsulphur.com
**Web site:** www.internationalsulphur.com

**Company:** Interstate Chemical Corporation
**Address:** 2797 Freedland Rd., P.O. Box 1600, Hermitage, PA 16148-0600, USA
**Phone:** +1 724-981-3771
**Fax:** +1 724-981-3675
**Web site:** www.interstatechemical.com

**Company:** Janssen Pharmacuetica Products, L.P.
**Address:** 1125 Trenton-Harbourton Rd., Titusville, NJ 08560, USA
**Phone:** +1 609-730-2000
**Fax:** +1 609-730-2323
**Web site:** www.janssen.com
**Stock listing:** NYSE: JNJ
**Parent Company:** Johnson & Johnson (USA)

**Company:** J. J. Mauget Company
**Address:** 5435 Peck Rd., Arcadia, CA 91006, USA
**Phone:** +1 626-444-1057
**Web site:** www.mauget.com

**Company:** Kerr-McGee Chemical LLC
**Address:** Kerr-McGee Center, 123 Robert S. Kerr Ave., Oklahoma City, OK 73102, USA
**Phone:** +1 405-270-1313, 800-786-2556
**Fax:** +1 405-270-3029
**Web site:** www.kerr-mcgee.com
**Stock listing:** NYSE: KMG
**Parent Company:** Kerr-McGee Corporation (USA)

**Company:** KMG Chemicals Inc.
**Address:** 10611 Harwin Dr., Ste. 402, Houston, TX 77036, USA
**Phone:** +1 713-988-9252
**Fax:** +1 713-988-9298
**Web site:** www.kmgb.com
**Stock listing:** Nasdaq Small Cap: KMG

**Company:** LaRoche Industries Inc.
**Address:** 1100 Johnson Ferry Rd., NE, Atlanta, GA 30342, USA
**Phone:** +1 404-851-0300
**Fax:** +1 404-851-0317
**Web site:** www.larocheind.com

**Company:** Lebanon Seaboard Corp.
**Address:** 1600 E. Cumberland St., Lebanon, PA 17042, USA
**Phone:** +1 800-233-0626
**Fax:** +1 717-273-9466
**E-mail:** customersercvice@lebsea.com
**Web site:** www.lebsea.com

**Company:** Liphatech, Inc.
**Address:** 3600 West Elm St., Milwaukee 53209, Wisconsin, USA
**Phone:** +1 888-331-7900
**Fax:** +1 414-247-8166
**Web site:** www.liphatech.com

**Company:** Lyondell Chemical Company
**Address:** 1221 McKinney, Houston, TX 77010, P.O. Box 3646, Houston, TX 77253-3646, USA
**Phone:** +1 713 652 7200
**E-mail:** info@lyondell.com

Web site: www.lyondell.com
Stock listing: NYSE: LYO

**Company:** Mallinckrodt Baker Inc.
Address: 222 Red School Lane, Phillipsburg, NJ 08865, USA
Phone: +1 908 859 2151, In USA: 800 582 2537
Fax: +1 908 859 6905
E-mail: infombi@mkg.com
Web site: www.mallbaker.com

**Company:** M & R Durango, Inc. Insectary
Address: P.O. Box 886, Bayfield, CO 81122, USA
Phone: +1 970-259-3521
Fax: +1 970-259-3857
E-mail: mail@goodbug.com
Web site: www.goodbug.com

**Company:** Matheson Gas Products, see Matheson Tri-Gas (US)

**Company:** Matheson Tri-Gas Corporation
Address: 959 Route 46 East, P.O. Box 624, Parsippany, NJ 07054-0624, USA
Phone: +1 973 257 1100
Fax: +1 973 257 9393
E-mail: info@matheson-trigas.com
Web site: www.mathesongas.com
Parent Company: Nippon Sanso Corporation (Japan)

**Company:** Mclaughlin Gormley King Co.
Address: 8810 Tenth Ave. North, Minneapolis, MN 55427-4372, USA
Phone: +1 763-544-0341
Web site: www.mgk.com

**Company:** MFA Inc.
Address: 201 Ray Young Dr., Columbia, MO 65201, USA
Phone: +1 573-874-5111
Web site: www.mfa-inc.com

**Company:** MG Industries
Address: 3 Great Valley Parkway, Malvern, PA 19355, USA
Phone: +1 610-695-7400, +1 800-869-6644
Fax: +1 610-695-7600
Web site: www.mgindustries.com
Parent Company: Messer Group (Germany), a Hoechst Subsidiary of Aventis S. A. (France)

**Company:** Micro Flo Company
Address: P.O. Box 772099, Memphis, TN 38117, USA
Phone: +1 800-451-8461
Fax: +1 901-432-5100
E-mail: sales@microflocompany.com
Web site: www.microflocompany.com
Parent Company: BASF Corp. (US)

**Company:** Millennium Chemicals, Inc.
Address: 230 Half Mile Rd., Red Bank, NJ 07701, USA
Phone: +1 732-933-5000
Fax: +1 732-933-5200
E-mail: feedback@millenniumchem.com
Web site: www.millenniumchem.com
Stock listing: NYSE: MCH
Parent Company: Lyondell Chemical Co. (USA)

**Company:** Minerals Research & Development
Address: One Woodlawn Green, Ste. 250, 200 E. Woodlawn Rd., Charlotte, NC 28217, USA
Phone: +1 704-525-2771, 800-334-0417 (toll free)
Fax: +1 704-527-8232
E-mail: mrdc@mrdc.com
Web site: www.mrdc.com
Parent Company: Chemical Specialties Inc., Subsidiary of Rockwood Specialties Inc. (US)

**Company:** Mississippi Chemical Corporation
Address: 3622 Highway 49 East, P.O. Box 388, Yazoo City, MS 39194-0388, USA
Phone: +1 662-746-4131
Fax: +1 662-746-9158
E-mail: corpcomm@misschem.com
Web site: www.misschem.com
Stock listing: NYSE: GRO
Parent Company: Terra Industries (USA)

**Company:** Mississippi Lime Company
Address: 7 Alby St., P.O. Box 2247, Alton, IL 62002-9004, USA
Phone: +1 618-465-7741, 800-437-5463
Fax: +1 618-465-7786
E-mail: info@mississippilime.com
Web site: www.mississippilime.com

**Company:** Monsanto Company
Address: 800 N. Lindbergh Boulevard, St. Louis, MO 63167, USA
Phone: +1 314-694-1000
Web site: www.monsanto.com
Stock listing: NYSE: MON
*Note:* This is now a pure agricultural chemical and seed business. Monsanto's chemical businesses were spun off to Solutia and the remainder purchased by Pharmacia. Pharmacia distributed its remaining shares in Monsanto to its shareholders on August 13, 2002, making Monsanto a completely independent agrichemical company.

**Company:** Montana Sulphur and Chemical
**Address:** P.O. Box 31118, Billings, MT 59107-1118, USA
**Phone:** +1 406-252-9324
**Fax:** +1 406-252-8250
**Web site:** montanasulphur.com

**Company:** Morton International, now Rohm & Haas Inc. (US)

**Company:** Navy Brand Manufacturing Co.
**Address:** 3670 Scarlet Oak Industrial Boulevard, St. Louis, MO 63122, USA
**Phone:** +1 636-861-5500, 800-325-3312
**Fax:** +1 636-861-5509
**Web site:** www.navybrand.com

**Company:** Neogen Corporation
**Address:** 620 Lesher Place, Lansing, MI 48912, USA
**Phone:** +1 800-234-5333
**Fax:** +1 517-372-0108
**E-mail:** neogencorp@aol.com
**Web site:** www.neogen.com

**Company:** Nepera
**Address:** 41 Arden House Rd., Harriman, N.Y. 10926, USA
**Phone:** +1 845-782-1200
**Fax:** +1 845-783-9713
**Web site:** www.rutherfordchemicals.com/nepera
**Parent Company:** Rutherford Chemicals (USA)

**Company:** Niacet Corporation
**Address:** 400 47th St., Niagara Falls, NY 14304, USA
**Phone:** +1 716-285-1474, 800-828-1207
**Fax:** +1 716-285-1497
**Web site:** www.niacet.com

**Company:** NOR-AM Chemical Company, now Bayer CropScience (Germany)

**Company:** Noranda DuPont LLC, now NorFalco LLC (USA)

**Company:** NorFalco LLC
**Address:** 6050 Oak Tree Boulevard, Ste. 190, Independence, OH 44131, USA
**Phone:** +1 216-642-7342
**Fax:** +1 216-642-9169
**Web site:** www.sulfuricacid.com
**Parent Company:** Noranda Corp. (Canada) and Falconbridge Corp. (US)

**Company:** Nova Biogenetics, Inc.
**Address:** 3353 Peachtree Rd., Ste. 942A, Atlanta, GA 30326, USA
**Phone:** +1 404-239-6087
**Web site:** www.novabiogenetics.com

**Company:** Occidental Chemical Corporation (OxyChem)
**Address:** Occidental Tower, 5005 LBJ Freeway, Dallas, TX 75244, USA
**Phone:** +1 972-404-3800
**Fax:** +1 972-404-3669
**E-mail:** info@oxychem.com
**Web site:** www.oxychem.com
**Stock listing:** NYSE: OXY
**Parent Company:** Occidental Petroleum Corp., (USA)

**Company:** Occidental Petroleum Corporation
**Address:** 10889 Wilshire Boulevard, Los Angeles, CA 90024-4201, USA
**Phone:** +1 310-208-8800
**Fax:** +1 310-443-6694
**E-mail:** oxyweb@oxy.com
**Web site:** www.oxy.com
**Stock listing:** NYSE: OXY

**Company:** Olin Corporation
**Address:** 501 Merritt Seven, Norwalk, CT 06856-4500, USA
**Phone:** +1 203-750-3000
**Fax:** +1 203-750-3292
**E-mail:** info@olin.com
**Web site:** www.olin.com
**Stock listing:** NYSE: OLN

**Company:** Ortho Business Group, see Scotts Company, The

**Company:** Osmose, Inc.
**Address:** 980 Ellicott St., Buffalo, NY 14209, USA
**Phone:** +1 716-882-5905
**Fax:** +1 716-882-5134
**E-mail:** info@osmose.com
**Web site:** www.osmose.com

**Company:** OxyChem, see Occidental Chemical Corporation

**Company:** Ozark Fluorine Specialties
**Address:** 1830 Columbia Ave., Folcroft, PA 19032, USA
**Phone:** +1 610-522-5960
**Fax:** +1 610-522-5969
**E-mail:** sales@ozarkfluorine.com
**Web site:** www.ozarkfluorine.com
**Parent Company:** LithChem International Div. of Toxco, Inc. (USA)

**Company:** Pbi/Gordon Corporation
**Address:** 1217 West 12th St., Kansas City, MO 64101, USA
**Phone:** +1 816-421-4070, 800-821-7925
**Fax:** +1 816-474-0462
**E-mail:** webmaster@pbigordon.com
**Web site:** www.pbigordon.com

**Company:** Penta Manufacturing Company
**Address:** 50 Okner Parkway, Livingston, NJ 07039-1604, USA
**Phone:** +1 973-740-2300
**Fax:** +1 973-740-1839
**Web site:** www.pentamfg.com
**Parent Company:** Penta International Corp. (US)

**Company:** Pestcon Systems, Inc.
**Address:** 1808 Firestone Prkwy., Wilson, NC 27893, USA
**Phone:** +1 800-548-2778
**Fax:** +1 252-243-1832
**E-mail:** khamm@pestcon.com
**Web site:** www.pestcon.com

**Company:** Pharmacia Animal Health
**Address:** 7000 Portage Rd., Kalamazoo, MI 49001-0199, USA
**Phone:** +1 800-793-0596
**Fax:** +1 800-984-9647
**Web site:** www.pharmaciaah.com
**Parent Company:** Pfizer (USA) [absorbed Pharmacia Corporation (USA) 2003]

**Company:** Phelps Dodge Corporation
**Address:** 2600 N. Central Ave., Phoenix, AZ 85004, USA
**Phone:** +1 602-366-8100
**Fax:** +1 602-366-8337
**Web site:** www.phelpsdodge.com
**Stock listing:** NYSE: PD

**Company:** Philbro-Tech Inc.
**Address:** One Parker Plaza, Fort Lee, NJ 07024, USA
**Phone:** +1 201-944-6000
**Fax:** +1 201-944-7916
**Web site:** http://www.phibro-tech.com

**Company:** Philipp Brothers Chemicals Inc.
**Address:** One Parker Plaza, 400 Kelby St., Fort Lee, NJ 07024, USA
**Phone:** +1 201-944-6020
**Fax:** +1 201-944-6245
**E-mail:** info@phibro-tech.com
**Web site:** www.philipp-brothers.com

**Company:** Phosphate Resource Partners Limited Partnership
**Address:** 100 S. Saunders Rd., Ste. 300, Lake Forest, IL 60045, USA
**Phone:** +1 847-739-1200
**Fax:** +1 847-739-1617
**Web site:** www.phosplp.com
**Stock listing:** NYSE: PLP
**Parent Company:** IMC Global (USA)

**Company:** Praxair Inc.
**Address:** 39 Old Ridgebury Rd., Danbury, CT 06810-5113, USA
**Phone:** +1 800-772-9247, +1 716-879-4077
**Fax:** +1 800-772-9985, +1 716-879-2040
**Web site:** www.praxair.com
**Stock listing:** NYSE: PX

**Company:** Prentiss Inc.
**Address:** 21 Vernon St., C.B. 2000, Floral Park, NY 11001, USA
**Phone:** +1 516-326-1919
**Fax:** +1 516-326-2312
**E-mail:** info@prentiss.com
**Web site:** www.prentiss.com

**Company:** Prince Manufacturing Company
**Address:** One Prince Plaza, P.O. Box 1009, Quincy, IL 62306, USA
**Phone:** +1 217-222-8854
**Fax:** +1 217-222-5098
**E-mail:** prince@princemfg.com
**Web site:** www.princemfg.com
**Parent Company:** Philipp Brothers Chemicals Inc. (USA)

**Company:** Pro-Outdoors
**Address:** Corners Circle, Norcross, GA 30092, USA
**Phone:** +1 800-297-7947, +1 770-446-1983
**Fax:** +1 770-446-6823
**Web site:** www.pro-outdoors.com

**Company:** Prodica LLC
**Address:** 376 S. Valencia Ave., P.O. Box 2347, Brea, CA 92823, USA
**Phone:** +1 877-PRODICA
**E-mail:** info@prodica.com
**Web site:** www.unocal.com
**Stock listing:** NYSE: UCL
**Parent Company:** Unocal Corp. (USA)

**Company:** Pro-Serve Inc.
**Address:** P.O. Box 161059, 400 E. Brooks Rd., Memphis, TN 38109, USA
**Phone:** +1 877-776-7375

Fax: +1 901-346-7157
Web site: www.pro-serveinc.com

Company: Pursell Technologies Inc.
Address: 201 W. Fourth St., Sylacauga, AL 35150, USA
Phone: +1 703-305-7448
E-mail: answers@pursell.com
Web site: www.fertilizer.com

Company: Reilly Industries Inc.
Address: 300 N. Meridian St., Ste. 1500, Indianapolis, IN 46204, USA
Phone: +1 317-247-8141
Fax: +1 317-248-6472
E-mail: webmaster@reillyind.com
Web site: www.reillyind.com

Company: Rhodia Specialty Phosphates
Address: 259 Prospect Plains Rd. CN 7500, Cranbury, NJ 08512-7500, USA
Phone: +1 609-860-3511
Fax: +1 609-855-8704
E-mail: brenda.watkinson@us.rhodia.com
Web site: ww.rhodia-phosphates.com
Parent company: Rhodia Group SA (France)

Company: Rice Company, The (TRC)
Address: 1624 Santa Clara, Ste. 145, Roseville, CA 95661, USA
Phone: +1 916-784-7745
Fax: +1 916-784-7681
Web site: www.riceco.com

Company: Richman Chemical Inc.
Address: 768 N. Bethlehem Pike, Lower Gwynedd, PA 19002, USA
Phone: +1 215-628-2946
Fax: +1 215-628-4262
Web site: www.richmanchemical.com

Company: Rio Tinto Borax (U.S. Borax)
Address: 26877 Tourney Rd., Valencia, CA 91355-1847, USA
Phone: +1 661 287 5400
Fax: +1 661 287 5495
Web site: www.borax.com
Parent Company: Rio Tinto Group (Australia & UK)

Company: Rohm and Haas Company
Address: 100 Independence Mall West, Philadelphia, PA 19106-2399, USA
Phone: +1 215-592-3000
Fax: +1 215-592-3377
E-mail: info@rohmhaas.com

Web site: www.rohmhaas.com
Stock listing: NYSE: ROH

Company: R.S.A. Corporation
Address: 36 Old Sherman Turnpike, Danbury, CT 06810, USA
Phone: +1 203 790 8100
Fax: +1 203 790 1709
E-mail: RSASteph@cs.com
Web site: www.rsa-corporation.com

Company: Rutherford Chemicals
Address: C/o Cambrex Corporation, One Meadowlands Plaza, East Rutherford, NJ 07073, USA
Phone: +1 888-283-4022
E-mail: susan.ostrowski@rutherfordchemicals.com
Web site: www.rutherfordchemicals.com
Stock listing: NYSE: CBM
Parent Company: Cambrex Corporation (US)

Company: Schenectady International Inc.
Address: P.O. Box 1046, Schenectady, NY 12301, USA
Phone: +1 518-370-4200
Fax: +1 518-346-3111
Web site: www.siigroup.com

Company: Schering-Plough Animal Health Corp.
Address: 10409 I St., Omaha, NE 68127, USA
Phone: +1 800-521-5767
Fax: +1 800-462-3720
Web site: http://usa.spah.com
Parent Company: Schering-Plough Inc., USA

Company: Scotts Company, The
Address: 41 S. High St., Ste. 3500, Columbus, OH 43215, USA
Phone: +1 614 719 5500
Fax: +1 614 719 5754
Web site: www.scottscompany.com
Stock listing: NYSE: SMG

Company: Scott Specialty Gases Inc.
Address: 6141 Easton Rd., Box 310, Plumsteadville, PA 18949, USA
Phone: +1 215-766-8861
Fax: +1 215-766-0320
Web site: www.scottgas.com

Company: Sepro Corporation
Address: 11550 N. Meridian St., Ste. 600, Carmel, IN 46032, USA
Phone: +1 317-580-8282
Web site: www.sepro.com

**Company:** Shacco Inc.
**Address:** 537 Atlas Ave., Madison, WI 53714, P.O. Box 7190, Madison, WI 53707, USA
**Phone:** +1 608-221-6200
**Fax:** +1 608-221-6208

**Company:** Shield Industries, Inc.
**Address:** 131 Smokehill Lane, Woodstock, GA 30188, USA
**Phone:** +1 770-517-6869
**Web site:** www.bugspray.com

**Company:** Sigma-Aldrich Co.
**Address:** 3050 Spruce Street, St. Louis, MO 63103, USA
**Phone:** +1 314-771-5765, 800-521-8956
**Fax:** +1 314-771-5757
**E-mail:** custserv@sial.com
**Web site:** www.sigma-aldrich.com
**Stock listing:** Nasdaq: SIAL

**Company:** Sigma-Aldrich Fine Chemicals Division
**Address:** 3050 Spruce Street, St. Louis, MO 63103, USA
**Phone:** +1 800-336-9719, 800-521-8956
**Fax:** +1 800-368-4661
**E-mail:** SAFcust@sial.com
**Web site:** www.sigma-aldrich.com
**Parent Company:** Sigma-Aldrich Co. (USA)

**Company:** Simplot, J.R., Company
**Address:** 999 Main St., Ste. 1300, Boise, ID 83702, P.O. Box 27, Boise, ID 83707, USA
**Phone:** +1 208-336-2110
**E-mail:** jrs_info@simplot.com
**Web site:** www.simplot.com

**Company:** Sipcam Agro USA, Inc.
**Address:** 300 Colonial Center Parkway, Ste. 300, Roswell, GA 30076, USA
**Phone:** +1 800-295-0733
**Fax:** +1 770-587-1115
**Web site:** www.sipcamagrousa.com
**Parent Company:** Sipcam Oxon Group (Italy)

**Company:** Sloss Industries Corporation
**Address:** Chemical Division, 3500 35th Ave. North, P.O. Box 5327, Birmingham, AL 35207, USA
**Phone:** +1 205-808-7911
**Fax:** +1 205-808-7948
**Web site:** www.sloss.com
**Parent Company:** Walter Industries Inc. (USA)

**Company:** Spectrum Chemical Mfg. Corp.
**Address:** 14422 South San Pedro St., Gardena, CA 90248-2027, USA
**Phone:** +1 310-516-8000, 800-772-8786
**Fax:** +1 310-516-7512, 800-525-2299
**E-mail:** sales@spectrumchemical.com
**Web site:** www.spectrumchemical.com

**Company:** Stauffer Chemical Co., now Imperial Chemical Industry (ICI) (UK)

**Company:** Stepan Company
**Address:** 22 W. Frontage Rd., Northfield, IL 60093, USA
**Phone:** +1 847-446-7500
**Fax:** +1 847-501-2100
**Web site:** www.stepan.com

**Company:** Sterling Chemicals Inc.
**Address:** 1200 Smith St., Ste. 1900, Houston, TX 77002-4312, USA
**Phone:** +1 713-650-3700
**Fax:** +1 713-654-9551
**Web site:** www.sterlingchemicals.com
**Stock listing:** Nasdaq: STXX.OB
**Parent Company:** Sterling Chemicals Holdings Inc. (US)

**Company:** Stoller Enterprises, Inc.
**Address:** 4001 W. Sam Houston Pkwy., North, Ste. 100, Houston, TX 77043, USA
**Phone:** +1 800-539-5283
**Web site:** www.stollerusa.com

**Company:** Sunoco Chemicals
**Address:** Ten Penn Center, 1801 Market St., Philadelphia, PA 19103, USA
**Phone:** +1 215-977-3321, 800-481-7840
**Fax:** +1 215-977-3470
**E-mail:** info@sunocochem.com
**Web site:** www.sunocochem.com
**Stock listing:** NYSE: SUN
**Parent Company:** Sunoco Inc. (USA)

**Company:** Sybron Chemicals, Inc., now Bayer AG (Germany)

**Company:** TCI America
**Address:** 9211 North Harborgate St., Portland, OR 97203, USA
**Phone:** +1 800-423-8616
**Fax:** +1 503-283-1987
**E-mail:** webmaster@tciamerica.com
**Web site:** www.tciamerica.com
**Parent Company:** Tokyo Chemical Industry. Ltd. (TKK) (Japan)

**Company:** Terra Industries Inc.
**Address:** 600 Fourth St., Box 6000, Sioux City, IA 51101

Phone: +1 712-277-1340
Fax: +1 712-2770-7364
E-mail: webmanager@terraindustries.com
Web site: www.terraindustries.com
Stock listing: NYSE: TRA

Company: Terra Nitrogen Company L.P.
Address: 600 Fourth St., P.O. Box 6000, Sioux City, IA 51101, USA
Phone: +1 712-277-1340
Fax: +1 712-277-7364
E-mail: webmanager@terraindustries.com
Web site: www.terranitrogen.com
Stock listing: NYSE: TNH
Parent company: Terra Industries Inc. (USA)

Company: TETRA Technologies Inc.
Address: 25025 Route I-45 North, The Woodlands, TX 77380, USA
Phone: +1 281-367-1983, 800-327-7817
Fax: +1 281-367-4306
Web site: www.tetratec.com

Company: Total Speciality Chemicals, Inc., now Atofina, subsidiary of Total (France)

Company: Tull Chemical Company
Address: 130 Burton St, P.O. Box 3246, Oxford, AL 36203, USA
Phone: +1 205 831 3845
Fax: +1 205 831 1154

Company: Union Carbide, now Dow AgroSciences LLC (USA)

Company: Uniroyal Crop Protection, now Crompton Corporation (USA)

Company: United Agri Products (UAP)
Address: ConAgra Foods, Inc., One ConAgra Dr., Omaha, NE 68102-5001, USA
Phone: +1 402-595-4000
Web site: www.uap.com
Parent Company: Apollo Management (USA)

Company: Univertical Corporation
Address: 203 Weatherhead St., Angola, IN 46703, USA
Phone: +1 260-665-1500
Fax: +1 260- 665-1400
E-mail: phwalker@msn.com
Web site: www.univertical.com

Company: Upjohn Inc. , now Pharmacia Animal Health (US)

Company: U.S. Borax, now Rio Tinto Borax (US)

Company: Valent U.S.A./Valent BioSciences Corporation
Address: 870 Technology Way, Ste. 100, Libertyville, IL 60048, USA
Phone: +1 800-323-9597
E-mail: vbcwebmaster@valent.com
Web site: www.valent.com
Parent Company: Formerly the Agriculture Division of Abbott Laboratories. Now owned by Sumitomo Chemical (Japan). In the US, products are sold by Valent U.S.A. Corp.

Company: Velsicol Chemical Corporation
Address: 10400 w. Higgins Rd., Ste. 600, Rosemont, IL 60018, USA
Phone: +1 847-298-9000
Fax: +1 847-298-9018
E-mail: expertworld@velsicol.com
Web site: www.velsicol.com

Company: Vulcan Chemicals
Address: 1200 Urban Center Dr., Birmingham, 35242, P.O. Box 385014, Birmingham, AL 35238-5014, USA
Phone: +1 800-633-8280
Fax: +1 800-933-6039
E-mail: contactus@vul.com
Web site: www.vulcanmaterials.com
Stock listing: NYSE: VMC
Parent Company: Vulcan Materials Company (USA)

Company: Wellmark International
Address: 1100 East Woodfield Rd., Ste. 500, Schumburg, IL 60173, USA
Phone: +1 800-877-6374
Web site: http://wellmarkinternational.com
Parent Company: Sandoz Agro Inc. (USA)

Company: West Agro, Inc.
Address: 11100 N. Congress Ave., Kansas City, MO 64153, USA
Phone: +1 816-891-1600
Fax: +1 816-891-1595
Web site: www.westagro.com

Company: White Mountain Natural Products, Inc.
Address: 2047 100$^{th}$ St., Paton, IA 50217, USA
Phone: +1 515-968-4341
E-mail: info@whitemountainnatural.com
Web site: www.whitemountainnatural.com

Company: Whitmire Micro-Gen Research Laboratories, Inc.

**Address:** 3568 Tree Court Industrial Boulevard, St. Louis, MO 63122, USA
**Phone:** +1 800-777-8570
**Web site:** www.wmmg.com

**Company: Witco, now Crompton Corporation (USA)**

**Company: World Metal LLC**
**Address:** 104 Industrial Boulevard, Ste. 202, Sugar Land, TX 77478, USA
**Phone:** +1 281-491-7474, 877-363-8257
**Fax:** +1 281-491-7979
**E-mail:** wmetal@aol.com
**Web site:** www.worldmetalllc.com

**Company: World Minerals Inc.**
**Address:** 130 Castilian Dr., Santa Barbara, CA 93117, USA
**Phone:** +1 805 562 0200
**Fax:** +1 805 562 0288
**E-mail:** info@worldminerals.com
**Web site:** www.worldminerals.com
**Stock listing:** NYSE: Y
**Parent Company:** Allegheny Corporation (US)

**Company: Zeneca Ag Products, now Syngenta (Switzerland)**

## VENEZUELA

**Company: Petroquimica de Venezuela (Pequiven)**
**Address:** Centro Corporativo, Torre Este, La Campina, Caracas, Venezuela
**Web site:** www.pequiven.pdv.com
**Parent Company:** Petroleos de Venezuela (PDVSA) (Venezuela)

# Appendix C: Directory of Federal and International Regulatory Agencies: Environment and Pesticides

## ARGENTINA

**Argentina Environmental Agency**
**Address:** Ministerio de Desarrolloo Social y Medio Ambiente
Av. 9 de Julio 1925 piso 18 of. 1807, Buenos Aires, Argentina
**Phone:** +54 1 4381-5960
**Fax:** +54 1 4383-6017
**Web site:** www.desarrollosocial.gov.ar

## AUSTRALIA

**Australia National Registration Authority (NRA)**
**Address:** John Curtin House, 22 Brisbane Ave, Barton, ACT 2600 Australia
PO Box E240, Kingston, ACT 2604 Australia
**Phone:** +61 2 6272 5852
**Fax:** +61 2 6272 4753
**E-mail:** nra.contact@nra.gov.au
**Web site:** www.nra.gov.au
The NRA operates the Australian system which evaluates, registers and regulates agricultural and veterinary chemicals. Before an agricultural or veterinary chemical product can enter the Australian market, it must go through the NRA's rigorous assessment process to ensure that it meets high standards of safety and effectiveness.

**Australia Department of Agriculture, Fisheries and Forestry**
**Address:** Edmund Barton Bldg., Broughton St., Barton, Canberra ACT 2601, Australia
**Phone:** +61 2 6272 3933
**E-mail:** plant.protection@affa.gov.au, for plant protection information
**Web site:** www.affa.gov.au
Has the dual roles of providing customer services to the agricultural, food, fisheries and forest industries, and addressing the challenges of natural resource management. It also helps build and promote the whole food and fibre chain from paddock to plate for domestic and international markets.

**Australian Pesticides & Veterinary Medicines Authority**
**Address:** John Curtain House, 22 Brisbane Ave, Barton, ACT 2600, Australia
**Phone:** +61 2 6272 5852
**Fax:** +61 2 6272 4753
**Web site:** www.apvma.gov.au
Responsible for the assessment and registration of pesticides and veterinary medicines and for their regulation up to and including the point of retail sale.

**Environment Australia**
**Address:** Department of the Environment and Heritage
John Gorton Building, King Edward Terrace, Parkes ACT 2600, GPO Box 787, Canberra ACT 2601, Australia
**Phone:** +61 2 6274 1111
**Fax:** +61 2 6274 1123
**E-mail:** comments/quiries@environment.gov.au
**Web site:** www.erin.gov.au
Environment Australia advises the Commonwealth Government on policies and programs for the protection and conservation of the environment, including both natural and cultural heritage places.

## AUSTRIA

**Austria Federal Environment Agency Ltd.**
**Address:** Spittelauer Lände 5, A-1090 Wien, Austria
**Phone:** +43 1 31304-0
**Fax:** +43 1 31304-5400
**Web site:** www.ubavie.gv.at

**Austria Federal Ministry of Agriculture Forestry Environment and Water Management**
**Address:** Stubenring 1, 1012 Vienna, Austria
**Phone:** +43 1 711 00-0
**Fax:** +43 1 711 00-0
**Web site:** www.lebensministerium.at

## BELGIUM

**Belgium Environmental Protection Agency**
**Address:** Vlaamse Milieumaatschappij (VMM)
A. Van De Maelestraat 96, 9320 Erembodegem, Belgium
**Phone:** +32 053 72 62 11
**Fax:** +32 053 77 71 68
**Web site:** www.vmm.be

**Belgian Interregional Cell for the Environment**
**Address:** Avenue des Arts 10-11, 1210 Brussels, Belgium
**Phone:** +32 2 227 57 01
**Fax:** +32 2 227 56 99
**E-mail:** celinair@irdeline.be
**Web site:** www.irceline.be
An agreement between the three Belgian regions (Flanders, Brussels and Wallonia) to deal with national and international environmental problems.

**Belgium Department for the Environment**
Address: Vesalius Bldg., 7th Fl, Pacheco Lane 19 Box 5, B-1010 Brussels, Belgium
Phone: +32 2 210 45 32
Fax: +32 2 210 48 52
E-mail: environment@health.fgov.be
Web site: www.environment.fgov.be

**FYTO WEB**
Web site: www.phytoweb.fgov.be
A government data base for legal pesticides in Belgium

## BRAZIL

**Brazil Institute of Environment and Renewable Natural Resources**
Address: SAS Q. 05 Lt. 5 B1.H, Ed. Subhevea, 70800-200 Brasilia DF, Brazil
Phone: +55 61 226 6851
Fax: +55 61 226 6851
Web site: www.ibama.gov.br

## BULGARIA

**Bulgarian Ministry of Environment and Water**
Address: Environmental Executive Agency
136 Tsar Boris III Boulevard, 1618 Sofia, Bulgaria
Phone: +359 2 955 90 11
Fax: +359 2 944 90 15
Web site: www.moew.govrn.bg

## CANADA

**Canadian Centre for Occupational Health and Safety**
Address: 250 Main St. East, Hamilton, Ontario, L8N 1H6, Canada
Phone: +1 905-570-8094, +1 800-668-4284, (toll-free in Canada and USA)
Fax: +1 905-572-2206
E-mail: custserv@ccohs.ca
Web site: www.ccohs.ca
Provides information and advice about occupational safety and health matters in Canada and internationally. Offers telephone and mail inquiry services and print and electronic products on health and safety issues.

**Canadian Pollution Prevention Information Clearinghouse**
Address: The National Office of Pollution Prevention
Phone: +1 800-667-9790
Fax: +1 519-337-3486
Web site: www.ec.gc.ca/cppic

**Environment Canada**
Address: 351 St. Joseph Boulevard, Hull, Quebec K1A 0H3, Canada
Phone: +1 819-997-2800, 800-668-6767
Fax: +1 819-953-2225
E-mail: enviroinfo@ec.gc.ca
Web site: www.ec.gc.ca

**Canadian Pest Management Regulatory Agency**
Address: 2720 Riverside Dr., Ottawa, Ontario K1A 0K9, Canada
Fax: +1 613-736-3798
E-mail: pmra_infoserv@hc-sc.gc.ca
Web site: www.hc-sc.gc.ca/pmra-arla
All products designed to manage, destroy, attract or repel pests that are used, sold or imported into Canada are regulated by the PMRA, a unit in Health Canada.

## CHILE

**Chile Environmental Protection Commission**
Address: Obispo Donoso N° 6, Providencia, Santiago, Chile
Phone: +56 2 405 600
Web site: www.conama.cl

## CHINA

**China State Environmental Protection Administration**
Address: No. 115 Xizhimennei Nanxiaojie, Beijing 100035, China
Phone: +86 10 6615 3366
Fax: +86 10 6615 1768
Web site: www.zhb.gov.cn/english

## CZECH REPUBLIC

**Czech Republic Ministry of the Environment**
Address: Nabrezi Edvarda Benese 4, Prague 1 -Mala Strana, Czech Republic 118 01
Phone: +420 2 2400 2111
Web site: www.env.cz

## DENMARK

**European Environmental Agency**
Address: Kongens Nytorv 6, 1050 Copenhagen K, Denmark
Phone: +45 3336 7100
Fax: +45 3336 7199
E-mail: eea@eea.eu.int
Web site: www.eea.eu.int
The European Environmental Agency aims to support sustainable development and to help achieve significant

and measurable improvement in Europe's environment by providing timely, targeted, relevant and reliable information to policy making agents and the public. The home page of EEA has links to environmental agencies in 26 European countries.

**Danish Environmental Protection Agency (MST)**
Danish Ministry of the Environment
29 Strandgade, DK-1401 Kubenhavn K, Denmark
Phone: +45 33 66 01 00
Fax: +45 33 32 22 27
Web site: http://mst.dk
The Danish Environmental Protection Agency spheres of activity are concentrated on preventing and combating water, soil and air pollution. The Agency belongs under the Danish Ministry of the Environment and has some 360 employees.

**Danish Ministry of Food, Agriculture and Fisheries**
Address: Holbergsgade 2, DK-1057 Copenhagen K, Denmark
Phone: +45 33 92 33 01
Fax: +45 33 14 50 42
E-mail: fvm@fvm.dk
Web site: www.fvm.dk

## ESTONIA

**Estonia Ministry of Agriculture**
Address: 39/41 Lai St Tallium, 15056, Estonia
E-mail: pm@agri.ee
Web site: www.agri.ee

**Estonian Ministry of Environment**
Address: Keskkonnaministeerium, Toompuiestee 24, 15172 Tallinn, Estonia
Phone: +372 2 626 2800
Fax: 372 2 626 2801
E-mail: info@ekm.envir.ee
Web site: www.envir.ee

## FINLAND

Finnish Ministry of Agriculture and Forestry
Address: P.O. Box 30, FIN-00023 Government, Helsinki, Finland
Phone: +358-9-16001
Web Site: www.mmm.fi

**Finnish Ministry of the Environment**
Address: Kasarmikatu 25, P.O. Box 33, FIN- 00023 Government, Finland
Tel.: +358 9-160 07
Fax.: +358 9 1603 9545

E-mail: kirjaamo.ym@vmparisto.fi
Web site: www.vyh.fi

## FRANCE

**French Agency for Food Safety**
Address: Agence françaisee sécurité des aliments
27/31, avenue du général Leclerc, BP 19-94701 Maisons-Alfort, France
Phone: +33 1 49 77 13 50
Fax: +33 1 49 77 26 12
E-mail: eb.internet@afssa.fr
Web site: www.afssa.fr

**French Environmental Institute**
Address: Institut Français de l'Environnement, 61, blvd Alexandre Martin, 45058 Orleans Cedex 1, France
Phone: +33 2 38 79 78 78
Fax: +33 2 38 79 78 70
E-mail: ifen@ifen.fr
Web site: www.ifen.fr
A public administrative body under the authority of the French Ministry of the Environment.

French Ministry of the Environment
Address: 18, ave. Canot, 94234 Cachan CEDEX, France
Phone: +33 1 41 24 18 00
Fax: +33 1 41 24 18 55
Web site: www.environnement.gouv.fr
Its mission is to monitor the quality of the environment, protect nature, prevent, reduce or totally eliminate pollution and other nuisances, and enhance the quality of life.

**World Organization for Animal Health**
Address: Office International des Epizooties
12, rue de Prony, 75017 Paris, France
Phone: +33 (0)1 44 15 18 88
Fax: +33 (0)1 42 67 09 87
E-mail: oie@oie.int
Web site: www.oie.int
The OIE is an intergovernmental organization that collects and analyzes the latest scientific information on animal disease control. This information is then made available to member countries to enable them to improve the methods used to control and eradicate these diseases. The OIE also provides technical support with animal disease control and eradication operations, including diseases transmissible to humans.

## GERMANY

**Germany Federal Biological Research Centre for Agriculture and Forestry (BBA)**

**Address:** Konigin-Luise-Strasse 19, 14195 Berlin, Germany
**Phone:** +30 83 04-1
**Fax:** +30 83 04 20 02
**E-mail:** pressestelle@bba.de
**Web site:** http://www.bba.de
The BBA is engaged in plant pathology, entomology, plant protection and related fields. The BBA concerns itself with the effect of pesticides on humans, animals and the environment.

**Germany Federal Environmental Agency**
**Address:** Bismarckplatz 1, 14193 Berlin, Germany
**Phone:** +30 89 03 0
**Fax:** +30 89 03 2285
**Web site:** www.umweltbundesamt.de
Provide scientific and technical support for the Federal Environment Ministry, especially with the preparation of legal and administrative regulations in the fields of air quality control, noise abatement, waste management, water resources management, soil conservation, environmental chemicals, and health-related environmental issues.

**Germany Federal Environment Ministry (BMU)**
**Address:** Alexanderplatz 6, D-10178 Berlin, Germany
**Phone:** +49 1 888 305 0
**Fax:** +49 1 888 305 4375
**E-mail:** oea-100@bmu.de
**Web site:** www.bmu.de

**Germany Ministry of Consumer Protectioin, Food and Agriculture**
**Address:** Rochusstr, 1, 53123 Bonn, Germany
Mailbox 14 02 70, 53107 Bonn, Germany
**Phone:** +30 0228/529-0
**Fax:** +30 0228/529-4262
**E-mail:** internet@bmvel.bund.de
**Web site:** www.verbraucherministerium.de

## GREECE

**Hellenic Ministry for the Environment**
**Address:** Web site: www.minenv.gr

## HONG KONG

**Hong Kong Environmental Protection Department**
**Address:** 28th Floor, Southorn Centre, 130 Hennessy Road, Wan Chai, Hong Kong
**Phone:** +852 2573 7746
**E-mail:** enquiry@epd.gov.hk
**Web site:** www.info.gov.hk/epd

## HUNGARY

**Hungarian Ministry for the Environment**
**Address:** H-1011 Budapest Foutca 44-50, Hungary
**Phone:** +36 1 457 3369
**Fax:** +36 1 201 4361
**E-mail:** gridbp@kik.ktm.hu
**Web site:** www.gridbp.meh.hu

## ICELAND

**Iceland Ministry of Environmental Affairs**
**Address:** Vonarstraeti 4 - 150, Reykjavik, Iceland
**Phone:** +354 1 560 9600
**Fax:** +354 1 562 4566
**E-mail:** postur@umh.stjr.is
**Web site:** http://brunnur.stjr.is

## INDIA

**India Ministry of Environment & Forests**
**Address:** Paryavaran Bhavan, CGO Complex, Lodhi Road, New Delhi 110003, India
**Phone:** +91 11 436 1896
**E-mail:** secy@envfor.delhi.nic.in
**Web site:** http://envfor.nic.in

## IRELAND

**Ireland Pesticide Control Service**
**Address:** Department of Agriculture, Food and Rural Development
Abbotstown, Snugborough Rd., Castleknock, Dublin 15, Ireland
**Phone:** +353-1-6072655
**Fax:** +353-1-8204260
**Web site:** www.pcs.agriculture.gov.ie
PCS is responsible for implementing the regulatory system for plant protection products and biocidal products. Biocidal products include disinfectants, preservatives, pest control products and antifouling products for use in industry and the home, as well as taxidermist and embalming fluids, etc.

**Irish Environmental Protection Agency**
**Address:** P.O.Box 3000, Johnstown Castle Estate, County Wexford, Ireland.
**Phone:** +353 53 606 00
**Fax:** +353 53 606 99
**E-mail:** info@epa.ie
**Web site:** www.epa.ie

# ISRAEL

## Israel Agro-Ecology Division, Ministry of the Environment
**Address:** Phone: +972-2 655 3845
**Fax:** +972-2 655 3848
**Web site:** www1.sviva.gov.il

The Agro-Ecology Division deals with the prevention and treatment of environmental degradation arising from improper agricultural practices in Israel's rural sector, especially in the areas of pesticides, aerial and ground spraying, fertilizer contamination, agricultural effluents from dairy farms, and agricultural wastes. The Ministries of Agriculture, Health and the Environment are responsible for monitoring pesticide use and chemical control for agriculture, which is applied to approximately 95% of Israeli crops.

## Israel Ministry of Environment
**Address:** Phone: +972 2 655 3802
**Fax:** +972 2 655 3817
**Web site:** www1.sviva.gov.il

# ITALY

## Codex Alimentarius Commission
**Address:** Secretariat, FAO/WHO Food Standards Programme, Food and Agriculture Organization of the United National
Viale delle Terme di Caracalla, 00100 Rome, Italy
**Phone:** +39 6 5705-1
**Fax:** +39 6 5705-4593
**E-mail:** codex@fao.org
**Web site:** www.codexalimentarius.net

The Codex Alimentarius Commission was created in 1963 by FAO and WHO to develop food standards, guidelines and related texts such as codes of practice under the Joint FAO/WHO Food Standards Programme. The main purposes of this Programme are protecting health of the consumers and ensuring fair trade practices in the food trade, and promoting coordination of all food standards work undertaken by international governmental and non-governmental organizations.

## European Chemical Bureau (ECB)
**Address:** Institute for Health and Consumer Protection
Joint Research Centre
Via Fermi 1, I-21020 Ispra (VA), Italy
**E-mail:** remi.allanou@jrd.it
**Web site:** http://ecb.jrd.it

The ECB is the focal point for the assessment of dangerous chemicals for the European Community.

## Food and Agricultural Organization of the United Nations
**Address:** Viale delle Terme di Caracalla, 00100 Rome, Italy
**Phone:** +39 06 5705 1
**Fax:** +39 06 5705 3152
**Cable Address:** FOODAGRI ROME
**E-mail:** FAO-HQ@fao.org
**Web site:** www.fao.org

FAO is one of the largest specialized agencies in the United Nations system and the lead agency for agriculture, forestry, fisheries and rural development. An intergovernmental organization, FAO has 183 member countries plus one member organization, the European Community.

## Food and Agriculture Pesticide Management
**Address:** Food and Agriculture Organization of the United Nations (FAO)
Viale delle Terme di Caracella, 00100 Rome, Italy
**Phone:** +39 6 5705-1
**Fax:** +39 6 5705 3152
**E-mail:** FAO-HQ@fao.org
**Web site:** www.fao.org

Pesticide Management is an activity carried out within the overall framework of the Plant Protection Service of FAO. It is designed to work together with member countries as a partner to introduce sustainable and environmentally sound agricultural practices which reduce the health hazard associated with the use of pesticides.

## International Plant Protection Convention
**Address:** Secretariat, IPPC
Food and Agriculture Organization of the United Nations (FAO)
Viale delle Terme di Caracella, 00100 Rome, Italy
**Phone:** +39 6 5705 4812
**Fax:** +39 6 5705 6347
**E-mail:** IPPC@fao.org
**Web site:** www.ippc.int/IPP/En/default

The purpose of the International Plant Protection Convention is to secure a common and effective action to prevent the spread and introduction of pests of plants and plant products, and to promote appropriate measures for their control. It also includes both direct and indirect damage by pests, thus including weeds. The provisions extend to cover conveyances, containers, storage places, soil and other objects or material capable of harboring plant pests. National Plant Protection Organizations and Regional Plant Protection Organizations work together to help contracting parties meet their IPPC obligations.

## Italian Department for the Environment
**Address:** Web site: wwwamb.casaccia.enea.it

Department for the Environment promotes and performs studies, research, and experimental and demonstration work in the environmental field and transfers know-how and results to the public administration, economic operators and the community.

**Italian National Environmental Protection Agency**; Agenzia Nazionale per la Protezione dell'Ambiente
Address: Via Vitaliano Brancati, 48, 00144 Rome, Italy
Phone: +39 6 50071
E-mail: websinanet@anpa.it
Web site: www.sinanet.anpa.it

## JAPAN

**Japan Agricultural Chemicals Inspection Station**
Address: 2-772, Suzuki-cho, Kodaira-shi, Tokyo 187-0011, Japan
Phone: +81-42-383-2151
Fax: +81-42-385-3361
Web site: www.acis.go.jp

**Japan Environment Corporation**
Address: Nittochi Bldg., 1-4-1 Kasumigaseki, Chiyoda-ku, Tokyo 100-0013, Japan
Phone: +81-3-5251-1017
Fax: +81-3-3592-5056
E-mail: jec@jec.go.jp
Web site: http://www.jec.go.jp

**Japan Ministry of Health, Labour and Welfare**
Address: Department of Food Sanitation
1-2-2 Kasumigaseki Chiyoda-ku, Tokyo 100-8916, Japan
Phone: +81-3-5253-1111
E-mail: www-admin@mhlw.go.jp
Web site: www.mhlw.go.jp

**Japan Ministry of the Environment**
Address: No. 5 Godochosha, 1-2-2 Kasumigaseki, Chiyoda-ku, Tokyo 100-8975, Japan
Phone: +81-3-3581-3351
E-mail: moe@env.go.jp
Web site: www.env.go.jp

**Japan National Institute for Environmental Studies**
Address: 16-2 Onogawa, Tsukuba-shi, Ibaraki 305-8506, Japan
E-mail: www@nies.go.jp
Web site: www.nies.go.jp
NIES is the main research branch of the Environment Agency (EA) of the Government of Japan.

## LATVIA

**Latvian Ministry of Environmental Protection and Regional Development**
Address: Peldu St. 25, Riga, LV-1494, Latvia
Phone: +371 702 6470
Fax: +371 782 0442
E-mail: abava@varam.gov.lv
Web site: www.varam.gov.lv

## LITHUANIA

**Lithuania Ministry of Environment of Republic**
Address: Lietuvos Respublikos Aplinkos Ministerija, Jakôto 4/9, Vilnius 2694, Lithuania
Phone: +370 61 0558
Fax: +370 22 0847
E-mail: info@nt.gamta.lt
Web site: www.gamta.lt

## LUXEMBOURG

**Luxembourg Ministry of the Environment**
Address: 18, Montee de la Petrusse, L-2918 Luxembourg
Fax: +352 400 410
Web site: www.mev.etat.lu

## MACEDONIA REPUBLIC

**Republic of Macedonia Ministry of Environment and Physical Planning**
Address: Drezdenska 52, 91000 Skopje, Macedonia
Phone: +389 91 366 930
Fax: +389 91 366 931
E-mail: info@moe.gov.mk
Web site: www.mia.com.mk/moe

## MEXICO

**Mexico Secretariat of Environment and Natural Resources**
Address: Boulevard Adolfo Ruiz Corrines, 4209 Col. Jardines en la Montana, Deleg. Tlalpan C.P., 14210 Mexico D.F., Mexico
E-mail: pagina@semarnat.gob.mx
Web site: www.semarnat.gob.mx

## THE NETHERLANDS

**Netherlands National Institute of Public Health and the Environment**
Address: Rijksinstituut voor Volksgezondheid en Milieu, P.O. Box 1, 3720 BA Bilthoven, The Netherlands
Phone: +70 339 5050

E-mail: info@rivm.nl
Web site: www.rivm.nl

**Netherlands Ministry of Housing, Spatial Planning and the Environment**
Web site: www.minvrom.nl

## NEW ZEALAND

Agricultural Compounds and Veterinary Medicines Group
Address: New Zealand Food Safety Authority
P.O. Box 2835, Wellington, New Zealand
Phone: +64 4 463-2550
Web site: www.nzfsa.govt.nz

**New Zealand Ministry for the Environment**
Address: Grand Annexe Building, 84 Boulcott St., PO Box 10362, Wellington, New Zealand
Phone: +64 4 917 7400
Fax: +64 4 917 7523
E-mail: library@mfe.govt.nz
Web site: www.mfe.govt.nz

## NORWAY

Norway Ministry of the Environment

**Norsk Plantevern Forening (NPF)**
Address: Myntgata 2, P.O. Box 8013 Dep, 0030 Oslo, Norway
Phone: +47 22 24 90 90
Fax: +47 22 24 95 60
E-mail: postmottak@md.dep.no
Web site: http://odin.dep.no

## REPUBLIC OF THE PHILIPPINES

**Philippines Department of Environment and Natural Resources**
Address: Visayas Ave., Diliman, Quezon City, The Philippines
Phone: +63 9 26 70 31
Fax: +63 9 24 25 40
E-mail: misd@denr.gov.ph
Web site: www.denr.gov.ph

## POLAND

**Poland Chief Inspectorate for Environmental Protection**
Address: Inspekcja Ochrony |Rodowiska (IOS), ul.Wawelska 52/54, 00-922 Warsaw, Poland
Phone: (0 22) 825 00 01 do 09
Web site: www.pios.gov.pl

**Poland Ministry of the Environment**
Address: Wawelska 52/54, 00-922 Warsaw, Poland
Phone: +48 22 825-00-01
E-mail: webmaster@mos.gov.pl
Web site: www.mos.gov.pl

## PORTUGAL

**Portugal General Directorate for the Environment**
Address: Rua da Murgueira - Bairro do Zambujal, 2721-865 Amadora, Portugal
Phone: +351 1 472 82 00
Fax:: +351 1 471 90 74
E-mail: dga@dga.min-amb.pt
Web site: www.dga.min-amb.pt

## SLOVAK REPUBLIC

**Slovak Environmental Agency**
Address: Tajovskeho 28, 975 90 Banska Bystrica, Slovak Republic
Phone: +42 88 413 5131
Fax: +42 88 423 0409
E-mail: toncik@sazp.sk
Web site: www.sazp.sk

## SLOVENIA

**Slovenia Ministry of the Environment and Spatial Planning**
Address: Dunajska cesta 48, 1000 Ljubljana, Slovenia
Phone: +386 1 478 7400
Fax: +386 1 478 7422
E-mail: info.mop@gov.si
Web site: www.sigov.si

## SOUTH AFRICA

**South Africa Department of Environmental Affairs and Tourism**
Address: Fedsure Forum Building, North Tower, Cnr van der Walt and Pretorius Sts., Pretoria, 0001, South Africa
Phone: +27 12 310 3911
Fax:: +27 12 322 2682
E-mail: bes@ozone.pwv.gov.za
Web site: www.environment.gov.za

## SPAIN

**Spanish Ministry of the Environment**
Address: Pza. San Juan de la Cruz, s/n, 28046 Madrid, Spain
Phone: +34 91 597 70 00

E-mail: webmaster@mma.es
Web site: www.mma.es

## SWEDEN

**Swedish Chemicals Inspectorate - KemI**
Address: P.O. Box 2, SE-172 13 Sundbyberg, Sweden
Phone: +46 8 519 411 00
E-mail: kemi@kemi.se
Web site: www.kemi.se

**Swedish Environmental Protection Agency**
Address: SE-106 48 Stockholm, Sweden
Phone: +46 8 698 10 00
Fax: +46 8 20 29 25
E-mail: natur@environ.se
Web site: www.internat.environ.se

**Swedish National Food Administration**
Address: Box 622, 751 26 Uppsala, Sweden
Phone: +46 18 17 55 00
Fax: +46 18 10 58 48
E-mail: livsmedelsverket@slv.se
Web site: www.slv.se

The Swedish National Food Administration, an agency under the Ministry of Agriculture, is the central administrative authority for matters concerning food. The Chemistry Division develops methods of analysis for the control of residues of pesticides and other substances in food.

## SWITZERLAND

**Swiss Registration Authority for Plant Protectioin Products**
Address: Mattenhofstrasse 5, 3003 Bern, Switzerland
Phone: +41 31 322 25 11
Fax: +41 31 322 26 34
E-mail: info@blw.admin.ch
Web site: www.blw.admin.ch

**Food Safety Programme**
Address: World Health Organization, Department of Protection of the Human Environment Cluster on Sustainable Development and Healthy Environments Avenue Appia 20, 1211 Geneva 27, Switzerland
Phone: +41 22 791 21 11
Fax: +41 22 791 31 13
E-mail: foodsafety@who.int
Web site: www.who.int/fsf

World Health Organization's work towards the improvement of food safety involves both technical cooperation with Member States to strengthen national food safety programs and functions such as developing the scientific basis for managing food safety programs and food safety-related issues.

**World Health Organization of the United Nations**
Address: Avenue Appia 20, 1211 Geneva 27, Switzerland
Phone: +41 22 791 21 11
Fax: +41 22 791 31 13
Telegraph: UNISANTE GENEVA
E-mail: info@who.int
Web site: www.who.int

World Health Organization's work towards the improvement of food safety involves both technical cooperation with member states to strengthen national food safety programs and functions such as developing the scientific basis for managing food safety programs and food safety-related issues. WHO's food safety work is coordinated and implemented at headquarters by Food Safety Programme, Department of Protection of the Human environment, Cluster on Sustainable Development and Healthy Environments (FOS/PHE/SDE) and, at the regional and country level, by regional advisers.

## UNITED KINGDOM

**Pesticides Safety Directorate**
Address: Mallard House, Kings Pool, 3 Peasholme Green, York Y01 7PX, UK
Phone: +44 1904 640500
Fax: +44 1904 455733
E-mail: information@psd.defra.gsi.gov.uk
Web site: www.pesticides.gov.uk

The principal functions of PSD, an executive agency of DEFRA, are to evaluate and process applications for approval of pesticide products for use in Great Britain and provide advice to the Government on pesticides policy.

**United Kingdom Department for Environment, Food & Rural Affairs (DEFRA)**
Address: Nobel House, 17 Smith Square, London, SW1P 3JR, UK
Phone: +44 8459 33 55 77
Fax: +44 20 7238 6591
E-mail: helpline@defra.gsi.gov.uk
Web site: www.defra.gov.uk

Created in June, 2001, to combine the various United Kingdom programs on the environment, food, farming and rural affairs.

**United Kingdom Environmental Agency**
Address: Rio House, Waterside Drive, Aztec West, Almonsbury, Bristol BS32 4UD, UK
Phone: +44 01 845 933 3111
Fax: +44 01 454 624 409

Emergency Hotline: +44 0800 807 060
E-mail: enquiries@environment-agency.gov.uk
Web site: www.environment-agency.gov.uk
The leading public body for protecting and improving the environment in England and Wales.

## UNITED STATES

**Agency for Toxic Substances and Disease Registry**
Address: U.S. Department of Health and Human Services, ATSDR, U.S. Public Health Service, Centers for Disease Control and Prevention, 1600 Clifton Rd. Atlanta, GA 30333, USA
Phone: +1 888-422-8737
Fax: +1 404-498-0093
Hotline: +1 404-639-6360, 24-hour emergency number.
E-mail: atsdric@cdc.gov
Web site: www.atsdr.cdc.gov
ATSDR's mission is to prevent exposure and adverse human health effects and diminished quality of life associated with exposure to hazardous substances from waste sites, unplanned releases, and other sources of pollution present in the environment. The ATSDR ToxFAQs™ is a series of summaries about hazardous substances developed by the ATSDR Division of Toxicology. Information for this series is excerpted from the ATSDR Toxicological Profiles and Public Health Statements. Each fact sheet serves as a quick and easy to understand guide. Answers are provided to the most frequently asked questions about exposure to hazardous substances found around hazardous waste sites and the effects of exposure on human health.

**Agriculture Research Service**
Address: U.S. Department of Agriculture
Jamie L. Whitten Bldg., 302-A, 14th & Independence Ave., S.W., Washington, DC 20250, USA
Phone: +1 202-720-3656
Fax: +1 202-720-5437
E-mail: eknipling@ars.usda.gov
Web site: www.ars.usda.gov
Agriculture Research Service is the in-house research agency of the U.S. Department of Agriculture. It is one of the Research, Education and Economics agencies charged with extending the nation's scientific knowledge across a broad range of program areas that affect the American people on a daily basis. The Agricultural Research Service is one of the largest agencies of USDA. ARS research covers a broad range scientific disciplines, and is carried out across the United States in over 120 labs within 9 Area Offices.

**Biopesticides and Pollution Prevention Division**
Address: Office of Pesticide Programs

U. S. Environmental Protection Agency
Ariel Rios Building, 1200 Pennsylvania Ave., N.W., Washington, D.C. 20460 , USA
Phone: +1 703-305-8098 (Ombudsman )
Web site: www.epa.gov/pesticides/biopesticides
The biopesticides program seeks to expedite the registration of biopesticides and to encourage the use of safer pest management practices including biopesticides. BPPD is responsible for risk/benefit assessment and risk management functions for microbial pesticides, tolerance reassessment, biochemical pesticides, plant pesticides, and the Pesticide Environmental Stewardship Program (PESP).

**Center for Environmental Research Information**
Address: U.S. Environmental Protection Agency
Office of Research and Development (ORD)

26 W. Martin Luther King Dr., P.O. Box 12505, Cincinnati, OH 45268, USA
Phone: +1 513-569-7562
Fax: +1 513-569-7566
CERI is the focal point for the exchange of scientific and technical environmental information produced by the Office of Research and Development, its 12 research laboratories, and associated programs nationwide. Hours: 8:00 a.m. - 4:30 p.m., Monday - Friday EST.

**Food Safety and Inspection Service**
Address: U. S. Department of Agriculture
Washington, D.C. 20250-3700
Phone: (Meat and Poultry Hotline) +1 800-535-4555
Medical Inquiries: +1 202-720-9113
E-mail: MPHotline.fsis@usda.gov
Web site: www.fsis.usda.gov
The Food Safety and Inspection Service is the public health agency in the U.S. Department of Agriculture responsible for ensuring that the nation's commercial supply of meat, poultry, and egg products is safe, wholesome, and correctly labeled and packaged, as required by the Federal Meat Inspection Act, the Poultry Products Inspection Act, and the Egg Products Inspection Act. The FSIS Web site has a complete list of e-mail contacts within the service.

**Food Safety Research Information Office**
National Agricultural Library, U.S. Department of Agriculture, Agriculture Research Service
Address: 10301 Baltimore Ave., Room 303, Beltsville, MD 20705-2351, USA
Phone: +1 301-504-7374
E-mail: yalonso@nal.usda.gov
Wed site: www.nal.usda.gov/fsrio
Suppports the food safety research community with an online database.

**Headquarters Information Resources Center**
Address: U.S. Environmental Protection Agency
Ariel Rios Building, 1200 Pennsylvania Ave., N.W.,
Washington, D.C. 20460 , USA
Phone: +1 202-260-5922
Fax: +1 202-260-5153
E-mail: library-hq@epamail.epa.gov
Web site: www.epa.gov/natlibra/hqirc
Provides access to EPA information for United States and International requests, and has a range of information services consisting of environmental and related subjects of interest to EPA staff, including online searching of commercial databases. The focus of the IRC collection is on environmental regulations, policy, planning, and administration. Hours: 8 a.m. - 5 p.m., Monday - Friday

**National Center for Toxicological Research**
Address: U.S. Food and Drug Administration
3900 NCTR Road, Jefferson, AR 72079, USA
Phone: +1 870-543-7130
E-mail: rhuber@nctr.fda.gov
Web site: www.fda.gov/nctr
Provides fundamental and applied research specifically designed to define biological mechanisms of action underlying the toxicity of products regulated by the FDA. This research is aimed at understanding critical biological events in the expression of toxicity and at developing methods to improve assessment of human exposure, susceptibility and risk.

**National Health Information Center**
Address: U.S. Public Health Service
Office of Disease Prevention and Health Promotion (ODPHP)
Address: P.O. Box 1133, Washington, DC 20013-1133, USA
Phone: +1 301-565-4167, 800-336-4797
Fax: +1 301-984-4256
E-mail: nhicinfo@health.org
Web site: www.health.gov/nhic
NHIC puts health professionals and consumers in touch with those organizations that are best able to provide answers.

**National Institute of Environmental Health Sciences**
Address: National Institutes of Health
Address: P.O. Box 12233, Research Triangle Park, NC 27709, USA
Phone: +1 919-541-3201, 800-643-4794, Communications Office, 919-541-2605
Fax: +1 919-541-2260
E-mail: environmental@niehs.nih.gov
Web site: www.niehs.nih.gov
Supports and conducts basic research focusing on the interaction between humans and potentially toxic or harmful agents in the environment. Its research is the basis for preventive programs for environment-related disease and for action by regulatory agencies. Has a basic program to research problems of hazardous wastes. Has one of the world's largest research libraries. Operates the Environmental Health Clearinghouse and the National Toxicology Program. Provides research grants to Environmental Health Sciences Centers at universities and medical schools.

**National Service Center for Environmental Publications**
Address: U.S. Environmental Protection Agency
P.O. Box 42419, Cincinnati, OH 45242-0419, USA
Phone: +1 513-489-8190, 800-490-9198
Fax: +1 513-489-8695
E-mail: ncepimal@one.net
Web site: www.epa.gov/ncepi
The central clearinghouse for the exchange of scientific, technical and public-oriented environmental information published by the Environmental Protection Agency. Provides copies of EPA documents and multimedia products upon request. It operates Monday through Friday, 7:30 a.m. - 5:30 p.m. EST.

**National Toxicology Program**
Address: National Institute of Environmental Health Science, P.O. Box 12233, 111 Alexander Dr., Research Triangle Park, NC 27709, USA
Phone: +1 919-541-3345
E-mail: liaison@starbase.niehs.nih.gov
Web site: http://ntp-server.niehs.nih.gov
The NTP is an interagency program consisting of relevant toxicology activities of the National Institutes of Health's National Institute of Environmental Health Sciences , the Centers for Disease Control and Prevention's National Institute for Occupational Safety and Health , and the Food and Drug Administration's National Center for Toxicological Research.

**Occupational Safety and Health Administration**
Address: U.S. Department of Labor
200 Constitution Ave. NW, Washington, DC 20210, USA
Phone: Public Information: +1 202-693-1999
Emergency phone: +1 800-321-OSHA, 800-321-6742
Fax: +1 202-219-5986
Web site: www.osha.gov
The mission of the Occupational Safety and Health Administration is to save lives, prevent injuries and protect the health of America's workers.

## Office of Air and Radiation
**Address:** U.S. Environmental Protection Agency
Ariel Rios Building, 1200 Pennsylvania Ave., N.W., Washington, D.C. 20460 , USA
**Phone:** +1 202-564-7400
**E-mail:**
http://joshua.epa.gov/oar/task.nsf/Comments?OpenForm
**Web site:** www.epa.gov/oar
Deals with issues that affect the quality of our air and protection from exposure to harmful radiation. OAR develops national programs, technical policies, and regulations for controlling air pollution and radiation exposure. Areas of concern to OAR include indoor and outdoor air quality, stationary and mobile sources of air pollution, radon, acid rain, stratospheric ozone depletion, radiation protection, and pollution prevention.

## Office of Air Quality Planning and Standards
**Address:** Office of Air and Radiation
U.S. Environmental Protection Agency
Mail Drop 10, Research Triangle Park, NC 27711, USA
**Phone:** +1 919-541-5618
**Fax:** +1 919-541-0501
**E-mail:**
http://joshua.epa.gov/oar/task.nsf/Comments?OpenForm
**Web site:** www.epa.gov/oar/oaqps
Directs national efforts to meet air quality goals, particularly for smog, air toxics, carbon monoxide, lead, particulate matter (soot and dust), sulfur dioxide, and nitrogen dioxide. The office is responsible for more than half of the guidance documents, regulations, and regulatory activities required by the Clean Air Act Amendments of 1990.

## Office of Emergency and Remedial Response (Superfund)
**Address:** Emergency Response Division
U.S. Environmental Protection Agency
Crystal Gateway One, 13th Fl., 1235 Jefferson Davis Hwy., Arlington, VA 22202, USA
**Phone:** +1 703-603-8760
**National Response Center:** +1 800-424-8802
**Web site:** www.epa.gov/superfund
The Comprehensive Environmental Response, Compensation, and Liability Act ( CERCLA), also known as the 'Superfund' Act, requires EPA to create new processes, policies, and procedures, and develop new technical capabilities for treating and containing hazardous substances.

## Office of Environmental Information
**Address:** U.S. Environmental Protection Agency
Ariel Rios Building, 1200 Pennsylvania Ave., N.W., Washington, D.C. 20460 , USA
**Phone:** +1 202-260-2090 - EPA Locator, 202-260-2080 - Public Information Center

**E-mail:** public-access@epa.gov
**Web site:** www.epa.gov/eq
A one-stop source of data and information on environmental quality, status and trends, including environmental profiles, the Digital Library of Environmental Quality, and the environmental Atlas.

## Office of Pesticide Programs
**Address:** U. S. Environmental Protection Agency
Ariel Rios Building, 1200 Pennsylvania Ave., N.W., Washington, D.C. 20460 , USA
**Phone:** +1 703-305-5805
**Web site:** www.epa.gov/pesticides
The OPP regulates the use of all pesticides in the U.S. and establishes maximum levels for pesticide residue in food.

## Office of Prevention, Pesticides and Toxic Substances
**Address:** U.S. Environmental Protection Agency
Ariel Rios Building, 1200 Pennsylvania Ave., N.W., Washington, D.C. 20460 , USA
**Phone:** +1 202-260-9262
**Web site:** www.epa.gov/oppts/asstadmn.htm
Promotes pollution prevention and evaluates pesticides.

## Office of Solid Waste and Emergency Response
**Address:** U.S. Environmental Protection Agency
Ariel Rios Bldg., 1200 Pennsylvania Ave., N.W., Washington, DC 20460, USA
**E-mail:** www.epa.gov/swerrims/contact
**Web site:** www.epa.gov/swerrims

## Office of Water
**Address:** Office of Groundwater and Drinking Water
U.S. Environmental Protection Agency
Ariel Rios Bldg., 1200 Pennsylvania Ave., N.W., Washington, DC 20460, USA
**Phone:** +1 202-260-5543
**E-mail:** ow-general@epamail.epa.gov
**Web site:** www.epa.gov/ow

## Pesticide Data Program
**Address:** U.S. Department of Agriculture (USDA), Agriculture Marketing Service (AMS), Science & Technology, Monitoring Programs Office, 8609 Sudley Rd., Ste. 206, Manassas, VA 20110
**Phone:** +1 703-330-2300
**Fax:** +1 703-369-0678
**Web site:** www.ams.usda.gov/science/pdp/quick.htm
Established in 1991 to test commodities in the U.S. food supply for pesticide residues.

## U.S. Department of Agriculture
**Address:** 1400 Independence Ave., S.W., MS-9100, Washington, D.C. 20250-9100, USA
**Phone:** +1 202-0906
**Fax:** +1 202-401-4770
**Web site:** www.usda.gov/agriculture

**U.S. Department of Transportation**
Address: 400 7th St., S.W., Washington, DC 20590, USA
Phone: +1 202-366-5580
E-mail: dot.comments@ost.dot.gov
Web site: www.dot.gov

The Department of Transportation is responsible for setting safety standards for rail, highway, air and water transportation and providing law enforcement and traffic management for airspace and waterways. Also regulates manufacturers of containers and transporters of hazardous materials.

**U.S. Environmental Protection Agency**
Address: Ariel Rios Bldg., 1200 Pennsylvania Ave., NW, Washington, DC 20460, USA
Phone: +1 202-260-2080
Fax: +1 202-260-7883
Web site: www.epa.gov

The Agency's mission is to control and abate pollution in the areas of air, water, solid waste, pesticides, radiation, and toxic substances. The EPA coordinates and supports research and anti-pollution activities by state and local governments, private and public groups, individuals, and educational institutions. The Pesticides Page is located at www.epa.gov/pesticides

**U.S. Food and Drug Administration**
Address: U.S. Department of Health and Human Services 5600 Fishers Lane, Rockville, MD 20857-0001, USA
Phone: +1 888-463-6332
Web site: www.fda.gov

The FDA is a public health agency, charged with enforcing the Federal Food, Drug, and Cosmetic Act and several related public health laws. The FDA has some investigators and inspectors who cover the country's almost 95,000 FDA-regulated businesses. Feed and drugs for pets and farm animals also come under FDA scrutiny.

## VENEZUELA

**Venezuelan Ministry of Environment and Renewable Natural Resources**
Address: Web site: www.marnr.gov.ve

# Appendix D: Directory of State Regulatory Agencies: Environment and Pesticide Management

## AMERICAN SAMOA

**American Samoa Environmental Protection Agency**
Address: P.O. Box 2609, Pago Pago, AS 97699
Phone: +1-415-972-3767
E-mail: goldstein.carl@epa.gov
Web site:
www.epa.gov/region09/cross_pr/islands/samoa.html

## GUAM

**Guam Environmental Protection Agency**
Address: 17-3304 Mariner Ave., Tiyan 96913, P.O. Box 22439 GMF, Barrigada 96921, Guam
Phone: +671-475-1658
Fax: +671-477-9402
E-mail: guamepa@mail.gov.gu
Web site:
www.epa.gov/region09/cross_pr/islands/guam.html

## NORTHERN MARIANA ISLANDS

**Northern Mariana Islands Division of Environmental Quality**
Address: P.O. Box 1304 CK&N bsp, Saipan, MP 96950
Phone: +670-664-8500
Fax: +670-664-8540
E-mail: deq@saipan.com
Web site: www.deq.gov.mp

## U.S. VIRGIN ISLANDS

**Virgin Islands Department of Agriculture**
Address: Estate Lower Love, Kingshill, St. Croix, Virgin Islands 00850
Phone: +340-778-0997
Fax: +340-744-1823
E-mail: agriculture@usvi.org
Web site: www.usvi.org/agriculture

## ALABAMA

**Alabama Department of Agriculture and Industries**
Address: Plant Protection and Pesticide Management Division, Richard Beard Building, 1445 Federal Dr., P.O. Box 3336, Montgomery, AL 36109-0336, USA
Phone: +1 334-240-7239
Fax: +1 334-240-7168
E-mail: alagicom01@agri-ind.state.al.us
Web site: www.agri-ind.state.al.us/pppm
Includes Plant Protection, Pesticide Management, and Pesticide Residue Laboratory.

**Alabama Department of Environmental Management**
Address: 1400 Coliseum Boulevard, Montgomery, AL 36110-2059, P.O. Box 301463, Montgomery, AL 36130-1463, USA
Phone: Ombudsman: +1 800-533-ADEM, +1 800-533-2336
Fax: +1 334 -394-4383
E-mail: Gle@adem.state.al.us
Web site: www.adem.state.al.us

## ALASKA

**Alaska Department of Environmental Conservation**
Address: Division of Environmental Health, Pesticide Services, 410 Willoughby Ave., Ste. 303, Juneau, AK 99801-1795, USA
Phone: +1 907-745-3236
Pesticide Use and Disposal Information: +1 800-478-2577
Fax: +1 907-745-8125
E-mail: website@envircon.state.ak.us
Web site: www.state.ak.us/dec

## ARIZONA

**Arizona Department of Agriculture**
Address: Environmental Services Division
1688 W. Adams, Phoenix, AZ 85007, USA
Phone: +1 602-542-4373, +1 602-542-3578
Emergency line: Pesticide Hotline: +1 800-423-8876
Fax: +1 602-542-0466
Web site: http://agriculture.state.az.us/esd/esd

**Arizona Department of Environmental Quality**
Address: 1110 W. Washington St., Phoenix, AZ 85007, USA
Phone: +1 602-771-2300, +1 800-234-5677 outside Phoenix area, TDD 602-771-4829
Emergency Response: +1 602-771-2330, +1 800-234-5677, extension 2330 toll free in Arizona
Fax: +1 602-207-2218
E-mail: munday.staci@ev.state.az.us
Web site: www.adeq.state.az.us

## ARKANSAS

**Arkansas Department of Environmental Quality**
Address: 8001 National Dr., Little Rock, AR 72209, USA
Phone: +1 501-562-0744
Fax: +1 501-562-0297
E-mail: help-custsvs@adeq.state.ar.us
Web site: www.adeq.state.ar.us

**Arkansas State Plant Board**
Address: Division of Feed, Fertilizer and Pesticides
Pesticide Information and Registration

P.O. Box 1069, Little Rock, AR 72203, USA
Phone: +1 501-225-1598
Web site: www.plantboard.org

## CALIFORNIA

**California Department of Pesticide Regulation**
Address: 1001 I St., P.O. Box 4015, Sacramento, CA 95812-4015, USA
Phone: +1 916-445-4300
Fax: +1 916-324-1452
E-mail: webmaster@pestreg.cdpr.ca.gov
Web site: www.cdpr.ca.gov
California Environmental Protection Agency
Address: 1001 I St., Sacramento, CA 95814, USA
Phone: +1 916-445-3846, Help Desk, +1 800-808-8058 or +1 916-327-1848.
Emergency line: Hazardous Materials Spill Release, +1 800-852-7550
Fax: +1 916 445-6401
E-mail: epasecty@calepa.ca.gov
Web site: www.calepa.ca.gov

## COLORADO

**Colorado Department of Agriculture**
Address: Division of Plant Industry, Pesticide Applicator Section, 700 Kipling St., Ste. 4000, Lakewood, CO 80215-8000, USA
Phone: +1 303-239-4140
Fax: +1 303-239-4177
E-mail: stacy.romero@ag.state.co.us
Web site: www.ag.state.co.us/DPI/home.html

**Colorado Department of Public Health and Environment**
Address: 4300 Cherry Creek Dr. South, Denver, CO 80246-1530, USA
Phone: +1 303-692-2035, +1 800-886-7689
Fax: +1 303-782-4969
E-mail: cdphe.information@state.co.us
Web site: www.cdphe.state.co.us

## CONNECTICUT

**Connecticut Department of Environmental Protection**
Address: Pesticide Management Division
79 Elm St., Hartford, CT 06106-5127, USA
Phone: +1 860-424-3369
Fax: +1 860-424-4153
Emergency lines: Spill reporting, +1 860-424-3338
Fax: +1 860-424-4051
E-mail: dep.webmaster@po.state.ct.us
Web site: http://dep.state.ct.us

## DELAWARE

**Delaware Department of Agriculture**
Address: Pesticide Compliance Section
2320 S. DuPont Hwy., Dover, DE 19901, USA
Phone: +1 302-698-4571
Fax: +1 302-697-4483
Web site: www.state.de.us/deptagri/pesticides

**Delaware Department of Agriculture**
Address: Plant Industries Section
2320 S. DuPont Hwy., Dover, DE 19901, USA
Phone: +1 302-698-4500
Fax: +1 302-697-6287
Web site: www.state.de.us/deptagri/plantind

**Delaware Department of Natural Resources and Environmental Control**
Address: 89 Kings Highway, Richardson and Robbins Bldg., Dover, DE 19901, USA
Phone: +1 302-739-4403
Fax: +1 302-739-6242
E-mail: mpolo@state.de.us
Web site: www.dnrec.state.de.us

## FLORIDA

**Florida Department of Agriculture and Consumer Services**
Address: Division of Agricultural Environmental Services, Bureau of Pesticides, 3125 Conner Boulevard, Bldg. 6, Tallahassee, FL 32399-1650, USA
Phone: +1 850-488-0532
Fax: +1 850-488-8497
E-mail: rutzs@doacs.state.fl.us
Web site: http://doacs.state.fl.us/~aes/pesticides

**Florida Department of Environmental Protection**
Address: Twin Towers Office Bldg., 3900 Commonwealth Boulevard, Tallahassee, FL 32399-3000, USA
Phone: +1 850-245-2086
Fax: +1 850-921-3267 (Ombudsman)
Emergency Response: +1 850 488-2974, 800 413-9911
Web site: www.dep.state.fl.us

## GEORGIA

**Georgia Department of Agriculture**
Address: Pesticide Division
19 Martin Luther King Dr., S.W., Atlanta, GA 30334, USA
Phone: +1 404-656-3685
Fax: +1 404 651-7957
Web site: www.agr.state.ga.us

**Georgia Department of Natural Resources**
Address: Environmental Protection Division
2 Martin Luther King Jr., Dr., 1152 East Tower, Atlanta, GA 30334, USA
Phone: +1 404-657-5947, 888 373-5947
Emergency Reporting: +1 400-4656-4300
Fax: +1 404-651-5778
Web site: www.dnr.state.ga.us/dnr/environ

## HAWAII

**Hawaii Department of Agriculture**
Address: Plant Industry Division, Pesticides Branch
P.O. Box 22159, Honolulu, HI 96823-2159
Phone: +1 808-973-9401
Fax: +1 808-973-9533
E-mail: hdoa.info@hawaii.gov
Web site: www.hawaiiag.org/hdoa/pi

**Hawaii Department of Health**
Address: Environmental Quality Control
1250 Punchbowl St., P.O. Box 3378, Honolulu, HI 96801, USA
Phone: +1 808 586-4185
Fax: +1 808 586-4186
Web site: www.state.hi.us/health

## IDAHO

**Idaho Department of Agriculture**
Address: P.O. Box 790, Boise, ID 83712, USA
Phone: +1 208-331-8605
Web site: www.agri.idaho.gov/agresource

## ILLINOIS

**Illinois Department of Agriculture**
Address: Agriculture Industry Regulations Division
Environmental Protection Bureau
State Fair Ground, P.O. Box 19281, Springfield, IL 62794-9281, USA
Phone: +1 217-782-2172
Fax: +1 217-785-4505
Web site: www.agr.state.il.us

**Illinois Environmental Protection Agency**
Address: 1021 North Grand Ave. East, P.O. Box 19276, Springfield, IL 62702, USA
Phone: +1 217-782-3397
Web site: www.epa.state.il.us

## INDIANA

**Indiana Office of the State Chemist**
Address: Pesticide Administrator
1154 Biochemistry Bldg., Purdue University, West Lafayette, IN 47907-1154, USA
Phone: +1 765-494-1492
Fax: +1 765494-4331
Web site: www.isco.purdue.edu

## IOWA

**Iowa Department of Agriculture and Land Stewardship**
Address: Pesticide Bureau
Henry A. Wallace Bldg., E. 9th St. and Grand Ave., Des Moines, IA 50319, USA
Phone: +1 515-281-8591
E-mail: chuck.eckermann@idals.state.ia.us
Web site: www2.state.ia.us/agriculture

## KANSAS

**Kansas Department of Agriculture**
Address: Pesticide & Fertilizer Program
109 SW 9th St., Topeka, KS 66612-1281, USA
Phone: +1 785-296-3786
Fax: +1 785-296-0673
Web site: www.accesskansas.org/kda/pest&fert/

**Kansas Department of Health and Environment**
Address: Environment Division
Forbes Field, Bldg. 740, Topeka, KS 66620-0001, USA
Phone: +1 785 296-1535
Fax: +1 785 296-8464
Web site: www.kdhe.state.ks.us/environment

## KENTUCKY

**Kentucky Department of Agriculture**
Address: Division of Pesticide Regulation
100 Fair Oaks Lane, 2nd Fl., Frankfort, KY 40601, USA
Phone: +1 502-564-7274
Web site: www.kyagr.com

**Kentucky Natural Resources and Environmental Protection Cabinet**
Department for Environmental Protection
14 Reilly Rd., Frankfort, KY 40601, USA
Phone: +1 502 564-2150
Fax: +1 502 564-4245
Web site: www.nr.state.ky.us

## LOUISIANA

**Louisiana Department of Agriculture and Forestry**
Address: Office of Agricultural and Environmental Sciences
Pesticide and Environmental Programs
P.O. Box 631, Baton Rouge, LA 70821-0631, USA
Phone: +1 225-922-1235
Fax: +1 225-922-1253
Web site: www.ldaf.state.la.us/aes/pesticide

**Louisiana Department of Environmental Quality**
Address: P.O. Box 4301, Baton Rouge, LA 70821, USA
Phone: +1 225-219-3953
Fax: +1 225-219-3971
Web site: www.deq.state.la.us

## MAINE

**Maine Department of Agriculture**
Address: Board of Pesticide Control
28 State House Station, Augusta, ME 04333-0028, USA
Phone: +1 207-287-2731

Fax: +1 207-287-7548
E-mail: pesticides@maine.gov
Web site: www.state.me.us/agriculture/pesticides

**Maine Department of Environmental Protection**
Address: 17 State House Station, Augusta, ME 04333-0017, USA
Phone: +1 207-287-7688, +1 800-452-1942
Emergency Response: +1 800-482-0777
Fax: +1 207-287-2812
Web site: www.state.me.us/dep

## MARYLAND

**Maryland Department of Agriculture**
Address: Office of Plant Industries and Pest Management
Pesticide Regulation Section
50 Harry S Truman Pkwy., Annapolis, MD 21401-8960, USA
Phone: +1 410-841-5700
Fax: +1 410 841-5914
E-mail: settingm@mda.state.md.us
Web site: www.mda.state.md.us

**Maryland Department of the Environment**
Address: 1800 Washington Boulevard, Baltimore, MD 21230, USA
Phone: +1 410-537-3000
Fax: +1 410-537-3888
Web site: www.mde.state.md.us

## MASSACHUSETTS

**Massachusetts Department of Agricultural Resources**
Address: Pesticide Bureau
251 Causeway St., Ste. 500, Boston, MA 02114, USA
Phone: +1 617-626-1700
Fax: +1 617-626-1850
Web site: www.state.ma.us/dfa/

## MICHIGAN

**Michigan Department of Agriculture**
Address: Pesticide and Plant Pest Management Division
Ottawa Bldg., North Tower, 4th Fl., 611 W. Ottawa St., Lansing, MI 48909, USA
Phone: +1 517-373-1087
Fax: +1 517-335-1423
Web site: www.michigan.gov/mda/0,1607,7-125-1572_2875-8324--,00.html

**Michigan Department of Environmental Quality**
Address: P.O. Box 30457, Lansing, MI 48909-7957, USA
Phone: +1 800-662-9278
Fax: +1 517-335-4729
Emergency Response: +1 800 292-4706
Web site: www.michigan.gov/deq

## MINNESOTA

**Minnesota Department of Agriculture**
Address: Agronomy and Plant Protection Division
90 W. Plato Boulevard, St. Paul, MN 55107, USA
Phone: +1 651-297-2200, 800-967-2474
E-mail: webinfo@mda.state.mn.us
Web site: www.mda.state.mn.us

## MISSISSIPPI

**Mississippi Department of Agriculture and Commerce**
Address: Pesticide Programs
P.O. Box 5207, Mississippi State, MS 39762, USA
Phone: +1 662-325-7763
Fax: +1 662-325-0397
Web site: www.mdac.state.ms.us

## MISSOURI

**Missouri Department of Agriculture**
Address: Plant Industries Division, Bureau of Pesticide Control
P.O. Box 630, Jefferson City, MO 65102, USA
Phone: +1 573-751-2462
Fax: +1 573-751-0005
E-mail: aginfo@mail.state.mo.us
Web site: www.mda.state.mo.us/Pest/d7h.htm

**Missouri Department of Natural Resources**
Address: Division of Environmental Quality
205 Jefferson St., P.O. Box 176, Jefferson City, MO 65102-0176, USA
Phone: +1 800-361-4827
E-mail: oac@mail.dnr.mo.gov
Web site: www.dnr.state.mo.us

## MONTANA

**Montana Department of Agriculture**
Address: Pesticides & Fertilizers
P.O. Box 20021, Helena, MT 59620-0201, USA
Phone: +1 406-444-3144
Fax: +1 406-444-5409
E-mail: agr@state.mt.us
Web site: http://agr.state.mt.us

**Montana Department of Environmental Quality**
Address: Metcalf Bldg., 1520 E. Sixth St., P.O. Box 200901, Helena, MT 59620-0901, USA
Phone: +1 406 444-2544
Web site: www.deq.state.mt.us

## NEBRASKA

**Nebraska Department of Agriculture**
Address: Bureau of Plant Industry
301 Centennial Mall S., 4th Fl., P.O. Box 94756, Lincoln,

NE 68509-4756, USA
Phone: +1 402-471-2394
Fax: +1 402-471-6892
Web site: www.agr.state.ne.us/division/bpi/bpi

## NEVADA

**Nevada Department of Conservation and Natural Resources**
Address: Division of Environmental Protection
333 W. Nye Lane, Rm. 138, Carson City, NV 89706-0851, USA
Phone: +1 775-687-4670
Fax: +1 775-687-5856
Web site: http://ndep.state.nv.us

**Nevada Department of Agriculture**
Address: Division of Plant Industry
350 Capitol Hill Ave., Reno, NV 89520, USA
Phone: +1 702-688-1180
Web site: http://agri.state.nv.us/pl_ind2

## NEW HAMPSHIRE

**New Hampshire Department of Agriculture, Markets and Food**
Address: Division of Pesticides Control
25 Capitol St., Concord, NH 03301, USA
Phone: +1 603-271-3550
Fax: +1 603 271-1109
E-mail: pesticides@agr.state.nh.us
Web site: www.agriclture.nh.gov/about/pesticide_control

## NEW JERSEY

**New Jersey Department of Environmental Protection**
Address: Pesticide Control Program
401 E. State St., P.O. Box 401, Trenton, NJ 08625-0402, USA
Phone: +1 609-530-4070
Fax: +1 609-292-7695
E-mail: askDEP@dep.state.nj.us
Web site: www.nj.gov/dep

**New Jersey Department of Health and Senior Services**
Address: Division of Epidemiology, Environmental and Occupational Health
Right to Know Program
P.O. Box 360, Trenton, NJ 08625-0360, USA
Phone: +1 609-984-1863
Fax: +1 609-292-5677
Web site: www.state.nj.us/health/eoh/rtkweb/rtkhsfs.htm#A
This agency produces "Right to Know Hazardous Substance Fact Sheets" that summarize information from many sources. Industrial hygienists are available to answer questions regarding the control of chemical exposures.

## NEW MEXICO

**New Mexico Department of Agriculture**
Address: Agricultural and Environmental Services Division
Pesticide Product Registration
MSC 3189, Corner of Gregg & Espina, Box 30005, Las Cruces, NM 88003-88005, USA
Phone: +1 505-646-3007
Fax: +1 505-646-3300
Web site: http://nmdaweb.nmsu.edu

## NEW YORK

**New York Department of Environmental Conservation**
Address: Bureau of Pesticides Management, Division of Solid and Hazardous Materials
625 Broadway, Albany, NY 12233, USA
Phone: +1 518-402-8788
E-mail: pestmgt@gw.dec.state.ny.us
Web site: www.dec.state.ny.us/website/dshm/pesticid/pesticid.htm

## NORTH CAROLINA

**North Carolina Department of Agriculture and Consumer Services**
Address: Food and Drug Protection Division, Pesticide Section
2109 Blue Ridge Ave., Raleigh 27607, 1090 Mail Service Center, Raleigh, NC 27699-1090, USA
Phone: +1 919-733-3556
Fax: +1 919-733-9796
Web site: www.ncagr.com/fooddrug/pesticid

## NORTH DAKOTA

**North Dakota Department of Agriculture**
Address: Pesticide Programs
600 East Boulevard Ave., Dept. 602, Bismarck, ND 58505-0020, USA
Phone: +1 7011328-2231, 800-242-7535
Fax: +1 701-328-4567
E-mail: ndda@state.nd.us
Web site: www.agdepartment.com

## OHIO

**Ohio Department of Agriculture**
Address: Division of Plant Industry, Pesticide Regulation
8995 E. Main St., Reynoldsburg, OH 43068-3399, USA
Phone: +1 614-728-6987
E-mail: plantpest@mail.agri.state.oh.us
Web site: www.ohioagriculture.gov/pubs/divs/plnt/curr/pr/plnt-pr-index.stm

## OKLAHOMA

**Oklahoma Department of Agriculture, Food & Forestry**
Address: Plant Industry and Consumer Services Division, Pest Management Section
2800 N. Lincoln Boulevard, Oklahoma City, OK 73105-4298, USA
Phone: +1 40- 271-1400
Fax: +1 405-521-3864
Web site: www.state.ok.us/~okag

## OREGON

**Oregon Department of Agriculture**
Address: Pesticides Division
635 Capitol St., NE, Salem, OR 97301-2532
Phone: +1 503-986-4635
Fax: +1 503-986-4735
E-mail: pestx@oda.state.or.us
Web site: www.oda.state.or.us/pesticide

## PENNSYLVANIA

**Pennsylvania Department of Agriculture**
Address: Bureau of Plant Industry, Agronomic Services Division
2301 N. Cameron St., Harrisburg, PA 17110-9408
Phone: +1 717-787-4843
Fax: +1 717-783-3275
Web site: www.agriculture.state.pa.us/plantindystry

## PUERTO RICO

**Puerto Rico Department of Economics and Agriculture**
Address: Analysis and Registration of Agricultural Materials, Laboratory Division
P.O. Box 10163, Santurce, PR 00908
Phone: +1 787 721-7000
Fax: +1 787 721-6634
E-mail: hramirez@fortaleza.gobierno.pr
Web site: www.fortaleza.gobierno.pr

**Puerto Rico Department of Planning, Natural Resources and Environmental Protection**
Address: Environmental Quality Board
Sernades Juncos Station, P.O. Box 11488, Santurce, PR 00910-1488
Phone: +1 787 721-7000
Fax: +1 787 724-5963
E-mail: ivila@fortaleza.gobierno.pr
Web site: www.fortaleza.gobierno.pr

## RHODE ISLAND

**Rhode Island Bureau of Natural Resources**
Address: Division of Agriculture
235 Promenade St., Providence, RI 02908-5767, USA
Phone: +1 401-222-2781
Fax: +1 401-222-6047
Web site: www.state.ri.us/dem

## SOUTH CAROLINA

**South Carolina Department of Plant Industry**
Address: Department of Pesticide Regulation
511 Westinghouse Rd., Pendleton, SC 29670, USA
Phone: +1 86- 646-2150
Fax: +1 864-646-2179
Web site: www.dpr.clemson.edu

## SOUTH DAKOTA

**South Dakota Department of Agriculture**
Address: Division of Agricultural Services, Pesticide Program
523 E. Capitol Ave., Pierre, SD 57501-3182, USA
Phone: +1 605-773-3375, 800-228-5254
Fax: +1 605-773-3481
E-mail: agmail@state.sd.us
Web site: www.state.sd.us/doa/das

## TENNESSEE

**Tennessee Department of Agriculture**
Address: Plant Industries Division
Ellington Agriculture Center, Nashville, TN 37204, USA
Phone: +1 615-837-5103
Fax: +1 615-837-5333
Web site: www.state.tn.us/agriculture

## TEXAS

**Texas Department of Agriculture**
Address: Licencing/Regulations
P.O. Box 12847, Austin, TX 78711, USA
Phone: +1 800-TELL-TDA
Fax: +1 512-463-1104
E-mail: contact@agr.state.tx.us
Web site: http://www.agr.state.tx.us/

## UTAH

**Utah Department of Agriculture and Food**
Address: Plant Industry Division
350 N. Redwood Rd., P.O. Box 146500, Salt Lake City, UT 84114-6500, USA
Phone: +1 801-538-7100
Fax: +1 801-538-7189
E-mail: agmain.lmlewis@email.state.ut.us
Web site: www.ag.state.ut.us

## VERMONT

**Vermont Agency of Agriculture, Food & Markets**
Address: Plant Industry Division

116 State St., State Office Bldg., Montpelier, VT 05620-2901, USA
Phone: +1 802-828-3472
Fax: +1 802-828-3831
E-mail: phil@agr.state.vt.us
Web site: www.vermontagriculture.com/pid.htm

## VIRGINIA

**Virginia Department of Agriculture and Consumer Service**
Address: Office of Pesticide Services
P.O. Box 1163, Richmond, VA 23218, USA
Phone: +1 804-371-6558
Fax: +1 804-371-8598
Web site: www.vdacs.state.va.us/pesticides

## WASHINGTON

**Washington State Department of Agriculture**
Address: Pesticide Management Division
1111 Washington St., P.O. Box 42589, Olympia, WA 98504-2589, USA
Phone: +1 360-902-2040, 877-301-4555
Fax: +1 360-902-2093
Web site: www.wa.gov/pestfert/default.htm

## WEST VIRGINIA

**West Virginia Department of Agriculture**
Address: Plant Industries Division
1900 Kanawha Boulevard., E., Charleston, WV 25305-0191, USA
Phone: +1 304-558-2212
Fax: +1 304-558-0451
Web site: www.wvagriculture.org

## WISCONSIN

**Wisconsin Department of Agriculture, Trade and Consumer Protection**
Address: Agricultural Resource Management Division
2811 Agriculture Dr., P.O. Box 8911, Madison, WI 53708-8911, USA
Phone: +1 608-224-4500
Fax: +1 608-224-4656
Web site: www.datcp.state.wi.us/core/agriculture/pest-fert/index.html

## WYOMING

**Wyoming Department of Agriculture**
Address: Plant Industry Programs
2219 Carey Ave., Cheyenne, WY 82002-0100, USA
Phone: +1 307-777-7321
Fax: +1 307-777-6593
E-mail: wda@state.wy.us
Web site: http://wyagric.state.wy.us

# Appendix E: Directory of Industrial & Professional Agrochemical & Food-related Organizations

*Note:* The national chemical societies have sections on agrochemicals.

## ALBANIA

**Society of Albanian Chemists**
Address: Faculty of Natural Sciences
University of Tirana, Tirana, Albania
Phone: +355 42 27 669
Fax: +355 42 26 724
E-mail: imalo@fshn.tirana.al

## ARGENTINA

**Argentina Chemical Association**
Address: Asociacion Quimica Argentina
Sanchez de Bustamante 1749, Buenos Aires 1425, Argentina
Phone: +54 1 822 4886
Fax: +54 1 822 4886
E-mail: info@aqa.org.ar

**Argentina Crop Protection Association**
Address: Cámara de Sanidad Agropecuaria y Fertilizantes
Rivadavia 1367-piso 7-"B", 1033 Buenos Aires, Argentina
Phone: +54 11 4381 2742
Fax: +54 11 4383 1562
E-mail: casafe@casafe.org
Web site: www.casafe.org

**Chemical and Petrochemical Industry Association of Argentina**
Address: Camera de la Industria Quimica y Petroquimica de Argentina, Av. Cordoba 629, piso 4 - (C1054AAF), Buenos Aires, Argentina
Phone: +54 11-4313-1000
Fax: +54 11-4313-1059
E-mail: informacion@ciqyp.org.ar
Web site: www.ciqyp.org.ar

**Latin American Petrochemical and Chemical Association**
Address: Esmeralda 351 piso 3 Of. B, C1025ABG Buenos Aires, Argentina
Phone: +54-11-4325-1422
E-mail: info@apla.com.ar
Web site: www.apla.com.ar
APLA seeks to boost and promote business in the chemical and petrochemical sectors in Latin America.

## ARMENIA

**Armenian National Academy of Sciences**
Address: 375019 Yerevan, Pr Marshala Bagramyana 24, Armenia
Phone: +374 2 527 031
Fax: +374 7 885 39 0 6867
E-mail: academy@sci.am

## AUSTRALIA

**Australian Mineral Foundation Inc.**
Address: 63 Conyngham St., Glenside 5065, South Australia, Australia
Phone: +61 8 8379 0444
Fax: +61 8 8379 4634
E-mail: amf@amf.com.au
Web site: www.amf.com.au

**Minerals Council of Australia**
Address: Mining Industry House, 216 Northbourne Ave., Braddon, ACT 2612, Australia
P.O. Box 363, Dickson ACR 2602, Australia
Phone: +61 2 6279 3600
Fax: +61 2 6279 3699
Web site: www.minerals.org.au

**Royal Australian Chemical Institute, Division of Cereal Chemistry**
Address: 1/21 Vale St., North Melbourne, Victoria 3051, Australia
Phone: +61 3 9328 2033
Fax: +61 3 9328 2670
E-mail: member@raci.org.au
Web site: www.raci.org.au
The RACI is both the qualifying body in Australia for professional chemists and a learned society promoting the science and practice of chemistry in all its branches. The Division of Cereal Chemistry concentrates on chemical aspects of the cereal-grain industry such as bread and other products from wheat, beer, and malt from barley.

## AUSTRIA

**Association of the Chemical Industry of Austria**
Address: Fachverband der Chemischen Industrie Osterreichs
Wiedner Hauptstraße 63, A-1045 Vienna, Austria
Phone: +43 1 501 05-3340
Fax: +43 1 501 05-280
E-mail: office@fcio.wko.at
Web site: http://fcio.at/home

**The Austrian Chemical Society**
Address: Gesellschaft Osterreichischer Chemiker (GOC)
Nibelungengasse 11/6 , A-1010 Vienna, Austria
Phone: +43 1 587 42 49 or 587 39 80
Fax: +43 1 587 89 66
E-mail: office@goech.at

## BANGLADESH

**Bangladesh Chemical Society**
Address: 10/11 Eastern Plaza, Sonargaon Road, Hatirpool, Dhaka - 1205, Bangladesh
E-mail: bcsir@bangla.net

## BELARUS

**Belarus Academy of Sciences**
Address: 220072 Minsk, Pr Skaryny 66, Belarus
Phone: +375 (17) 284 18 01
Fax: +375 (17) 239 31 63
E-mail: academy@mserv.bas-net.by

## BELGIUM

**Association of Petrochemicals Producers in Europe ,**
see European Chemical Industry Council

**Belgium Association of Plant Protection Products**
Address: Phytofar ASBL, Square Marie-Louise 49, 1000 Brussels, Belgium
Phone: +32 2 238.97.72
Fax: +32 2 280.03.48
E-mail: phytosec@fedichem.be
Web site: www.phytofar.be

**Belgium Association of the Biocides Industry**
Address: Secretary-General, Square Marie-Louise 49, B-1000 Brussels, Belgium
Phone: +1 32 2 238 97 72
Fax: +1 32 2 280 03 80
E-mail: bioplus@fedichem.be
Web site: www.fedichem.be

**Belgian Association of the Industry of Nitrogen Products**
Address: Secretary-General, Square Marie-Louise 49, B-1000 Brussels, Belgium
Phone: +1 32 2 238 97 72
Fax: +1 32 2 280 03 80
E-mail: abipa-bvsi@fedichem.be
Web site: www.fedichem.be

**Belgium Association of the Phosphate Fertilizers Industry**
Address: Secretary-General, Square Marie-Louise 49, B-1000 Brussels, Belgium
Phone: +1 32 2 238 97 72
Fax: +1 32 2 280 03 80
E-mail: abiphos@fedichem.be
Web site: www.fedichem.be

**Bromine Science and Environmental Forum**
Address: 118 Avenue de Cortenbergh, 1000 Brussels, Belgium
Phone: +32 2 733.93.70
Fax: +32 2 735.60.63
E-mail: mail@BSEFsecr.com
Web site: www.bsef.com.be

**CropLife International**
Address: 143 Avenue Louise, B-1000 Brussels, Belgium
Phone: +32 2 542 04 10
Fax: +32 2 542 04 19
E-mail: croplife@croplife.org
Web site: www.croplife.org
CropLife International is the global federation representing the plant science industry. It supports a network of 75 regional and national associations and their member companies worldwide, in six Regional Associations: Africa/Middle-East, Asia-Pacific, Europe, Japan, Latin America and North America. CropLife International was formerly Global Crop Protection Association.

**European Association for Bioindustries, The**
Address: Avenue de l'Armee 6, 1040 Brussels, Belgium
Phone: +32 2 735.03.13
Fax: +32 2 735.49.60
E-mail: mail@europabio.org
Web site: www.europabio.org
EuropaBio, the European Association for Bioindustries, represents nearly 40 member companies operating worldwide and 18 national biotechnology associations. Members are involved in a wide range of activities: human and animal health care, diagnostics, bio-informatics, chemicals, crop protection, agriculture, food and environmental products and services. EuropaBio also has associate members such as international commercial, financial, asset management and other service-providing companies, regional biotechnology development organizations and scientific institutes.

**European Centre for Ecotoxicology and Toxicology of Chemicals**
Address: Avenue E. Van Nieuwenuye 4, Box 6, B-1160 Brussels, Belgium
Phone: +32 2 675 3600
Fax: +32 2 675 3625
Web site: www.ecetoc.org
A scientific, non-profit making, non-commercial association, financed by 50 of the leading companies with interests in the manufacture and use of chemicals. It provides a scientific forum through which the European chemical industry can research, review, assess and publish studies on the ecotoxicology and toxicology of chemicals.

**European Chemical Industry Council**
Address: Avenue Van Nieuwenhuyse, 4 Box 1, B-1160 Brussels, Belgium
Phone: +32 2 676 7211
Fax: +32 2 676 7300
E-mail: mail@cefic.be
Web site: www.cefic..be
CEFIC is made up of the national chemical industry federations of 22 countries in Europe and large international

companies which are members in their own right. Web site for individual federations and their members is www.cefic.org/product_families.

**European Chemical Society**
Address: Department of Chemistry, Universite Catholique de Louvain, Batiments Lavoisier, Place Louis Pasteur, 1, B-1348 Louvain-la-Neuve, Belgium
Fax: +32-0-47 80 33
E-mail: ecs@chim.ucl.ac.be
Web site: www.chim.ucl.ac.be/CHIM/ECS/
The European Chemical Society was founded in 1995. It seeks to promote European chemistry both inside and outside Europe, to network and unite chemists at the European level, to circulate information among European chemists, to provide a forum for European chemists, to promote public awareness and appreciation of chemistry, and to encourage the young research students to become actively involved in the Society.

**European Crop Protection Association**
Address: 6 Avenue E van Nieuwenhuyse, B-1160 Brussels, Belgium
Phone: +32 2 663 15 50
Fax: +32 2 663 15 60
E-mail: ecpa@ecpa.be
Web site: www.ecpa.be
ECPA is the pan-European voice of the crop protection industry. The membership includes both national associations and companies throughout Europe, including Central and Eastern Europe. In addition to traditional crop protection products such as herbicides, fungicides and insecticides, members also provide biopesticides and genetically modified crops. ECPA is a member of CropLife International (formerly Global Crop Protection Association).

**European Fertilizer Manufacturers Association**
Address: Avenue E, Van Nieuwenhuyse 4, B-1160 Brussels, Belgium
Phone: +32 2 675 3550
Fax: +32 2 675 3961
E-mail: main@efma.be
Web site: www.efma.org
EFMA is a trade association to promote role of mineral fertilizers in European agriculture and horticulture.

**FaBeChim - VeBeVeChem**
Address: Avenue Eduoard de Thibault 51, B-1040 Brussels, Belgium
Fax: +32 2 735 5681
Web site: http://curie.sc.ucl.ac.be/curievh/acl/FABECHIM

**Federation of the Belgium Chemical Industry**
Address: Square Marie-Louise 49, B-1000 Brussels, Belgium
Phone: +32 2 238 97 11
Fax: +32 2 231 13 01
E-mail: info@fedichem.be
Web site: www.fedichem.be

**Global Crop Protection Federation**, See CropLife International (Belgium)

**Royal Flemish Chemical Society**
Address: Koninklijke Vlaamse Chemische Vereniging Groot Begijnhof 6, B-3000 Leuven, Belgium
Phone: +32 16 29 32 14
Fax: +32 16 22 68 92
E-mail: Robben@uia.ua.ac.be
Web site: www.kvcv.be

**Society of Environmental Toxicology and Chemistry**
Address: Avenue E. Mounier 83, Box 31, 1200 Brussels, Belgium
Phone: +32 2 772 72 81
Fax: +32 2 770 53 86
E-mail: setac@ping.be
Web site: www.setac.org
An independent, international, nonprofit professional society that provides a forum for individuals and institutions engaged in the study of environmental issues, management and conservation of natural resources, environmental education, and environmental research and development. Offices in Belgium, Australia and the United States.

## BOLIVIA

**Bolivian Chemical Society**
Address: Socuedad Boliviana de Quimica Casilla 13514, La Paz, Bolivia
Phone: +591 2 794 164
Fax: +591 2 359 491

## BRAZIL

**Brazilian Chemical Association**
Address: Asociacion Brasilena de Quimica (ABQ) Rua Alcino Guanabara 24 Conj 1606, 13° andar, CEP 20031 Rio de Janiero, RJ, Brazil
Phone: +55 21 262 1837
Fax: +55 21 262 6044

**Brazilian Chemical Industry Associatio**
Address: razilian Chemical Manufacturers Association (ABIQUIM)

**Associação Brasileira da Indústria Química**
Address: Rua Santo Antonio, 184 - 17° andar, 01314-900 São Paulo, SP, Brazil
Fax: +55 11 232 0919
E-mail: abiquim@abiquim.org.br
Web site: www.brazchemicals.org.br

**Brazilian Chemical Society**
Address: Sociedade Brasileira de Química
Instituto de Química da USP
26037, 05513-970, São Paulo - SP, Brazil
Phone: +55 11 210 2299
Fax: +55 11 814 3602
E-mail: paula@qt.dq.ufscar.br
Web site: www.sbq.org.br

## CANADA

**Canadian Chemical Producers Association**
Address: Suite 805, 350 Sparks Street, Ottawa, Ontario, Canada K1R 7S8
Phone: +1 613 237 6215
Fax: +1 613 237 4061
E-mail: info@ccpa.ca
Web site: www.ccpa.ca
A trade association representing 75 chemical manufacturing industries in Canada.

**Canadian Institute of Food Science and Technology**
Address: P.O. Box 152, Apple Hill, Ontario, Canada K0C 1B0
Phone: +1 613-525-2833
Fax: +1 613-525-4328
E-mail: cifst@cifst.ca
Web site: www.cifst.ca
CIFST is the national association for food industry professionals. Its membership of more than 1500 is comprised of scientists and technologists in industry, government and academia who are committed to advancing food science and technology.

**Chemical Institute of Canada**
Address: 130 Slater St., Ste. 550, Ottawa, Ontario, Canada K1P 6E2
Phone: +1 613 232 6252
Fax: +1 613 232 5862
E-mail: info@cheminst.ca
Web site: www.cic.ca

**CropLife Canada**
Address: 21 Four Seasons Place, Suite 627, Etobicoke, Ontario, Canada M9B 6J8
Phone: +1 416-622-9771
Fax: +1 416-622-6764
E-mail: pilbeaml@croplife.ca
Web site: www.croplife.ca
CropLife Canada represents the manufacturers, developers and distributors of plant life science solutions for agriculture, forestry and pest management in Canada.. CropLife Canada was formerly the Crop Protection Institute.

**Crop Protection Institute,** see CropLife Canada

**International Union of Food Science and Technology**
Address: P.O. Box 61021, 511 Maplegrove Road, Ste. 19, Oakville, Ontario, Canada L6J 6X0
Phone: +1 905 815 1926
Fax: +1 905 815 1574
E-mail: secretariat@iufost.org
Web site: www.iufost.org
The International Union of Food Science and Technology, a country-membership organization, is the sole global food science and technology organization. It is a voluntary, non-profit association of national food science organizations linking the world's food scientists and technologists. It promotes world-wide exchange of knowledge in those scientific disciplines and technologies that relate to the expansion, improvement, distribution and conservation of the world's food supply through its sponsorship of international and regional congresses, conferences, workshops and symposia, and through its encouragement of appropriate educational programs.

## CHILE

**Association of Chilean Chemical Manufacturers**
Address: Asociacion Gremial de Industriales Quimicos de Chile
Av. Andres Bello 2777 Of. 501, Las Condes, Santiago, Chile
Phone: +56 2 203 3350
Fax: +56 2 203 3351
E-mail: asiquim@asiquim.cl
Web site: www.asiquim.cl

**Chilean Chemical Society**
Address: Sociedad Chilena de Quimica (SCQ)
Bernardo O'Higgins 1061-G, Casilla de Correo 2613, Concepcion, Chile
Phone: +56 41 235-819
Fax: +56 41 240-280

## CHINA

**Shanghai Pesticide Research Institute**
Address: 2356 Xietu Rd., Shanghai 200032, China
Phone: +86 21-6469-0260
Fax: +86 21-6469-0261
E-mail: spri@hi2000.com
Web site: http://www.chinachemnet.com/spri

## COSTA RICA

**Latin American Crop Protection Association**
Address: Apartado 94-2020, Centro Postal Zapote, San José, Costa Rica
Phone. +506 272 0716
Fax: +506 272 5304
E-mail: lacpacr@sol.racsa.co.cr
Web site: www.lacpa.org
The Latin America Crop Protection Association (LACPA), represents the Crop Protection Industry through nineteen

national associations in Argentina, Belize, Bolivia, Brazil, Chile, Colombia, Costa Rica, Dominican Republic, Ecuador, El Salvador, Guatemala, Honduras, Mexico, Nicaragua, Panama, Paraguay, Peru, Uruguay, and Venezuela.

## CZECH REPUBLIC

### Association of Chemical Industry of the Czech Republic
Address: Svaz Chemického Prumyslu Ceské Republiky
Kodanská 46, CR - 100 10 Prague 10, Czech Republic
Phone: +42 2 6715 4131
Fax: +42 2 6715 4130
E-mail: oinchzs@mbox.vol.cz
Web site: http://www.schp.cz

### Czech Chemical Society
Address: Novotneho lavka 5, CZ-116 68 Prague 1, Czech Republic
Phone: +42 2 210 82 382
Fax: +420 2 222 20 184
E-mail: csch@csch.cz
Web site: www.csch.cz

## DENMARK

### Danish Chemical Society
Address: Kemisk Forening, Universitetsparken 5, DK-2100 Copenhagen, Denmark
Phone: +45 3537 0850
Fax: +45 3537 5376
Web site: www.rub.ruc.dk/dis/chem/kemfor

### Danish Crop Protection Association
Address: Dansk Plantevaern, Amalievej 20, 1875 Frederiksber C, Denmark
Phone: +33 24 42 66
Fax: +33 25 84 16
E-mail: dp@plantevaern.dk
Web site: http://www.plantevaern.dk

## ETHIOPIA

### National Chemical Corporation
Address: P.O. Box 5747, Addis Ababa, Ethiopia
Phone: +251 1 611 3111
Fax: +251 1 610 296

## FINLAND

### Association of Finnish Chemical Societies, The
Address: Kemiska Sallskapet I Finland
Hietanie menkatu 2, FIN-00100, Helsinki, Finland
Phone: +358 9 454 2040
Fax: +358 9 408 780
E-mail: skks@kemia.pp
Web site: www.kemiaseura.fi

### Finnish Chemical Society
Address: Kemianteollisuus, RY
Etelaranta 10, Pl 4, 00130 Helsinki, Finland
Phone: +358 9 172 841
Fax: +358 9 630 225
Web site: www.chemind.fi

### Finnish Crop Protection Association
Address: Etelaranta 10, P.O. Box 4, FIN-00131 Helsinki, Finland
Phone: +358 9 172 841
Fax: +358 9 630 225
E-mail: maija.pohjakallio@kaste.net
Web site: www.kaste.net
The Finnish Crop Protection Association is a member of the Chemical Industry Federation of Finland. The membership of European Crop Protection Association (ECPA) is shared with the sister associations in Sweden and Norway.

## FRANCE

### French Union of Industrial Chemistry
Web site: www.uic.fr
The UIC is the professional body representing French chemical companies. It provides them with exchange and meeting structures, and encourages their development. It represents and defends them in its various fields of activities, such as social, economic, technical, fiscal and legal affairs.

### French Chemical Society
Address: Société Française de Chimie (SFC)
250, rue Saint-Jacques, F-75005 Paris, France
Phone: +33 1 40 46 71 60
Fax: +33 1 40 46 71 61
E-mail: sfc@sfc.fr
Web site: www.sfc.fr

### French Crop Protection Association
Address: Union des Industries de la Protection des Plantes (UIPP)
E-mail: uipp@uipp.net
Web site: www.uipp.org
The UIPP (French Crop Protection Association) is a professional association committed to explaining the role of the industry in modern agriculture and the benefits of its products to the community. It was created in 1918 and has 29 member companies which represent 96% of the French market, the premier market in the European Union. It is a member of the European Crop Protection Association (ECPA).

### French Society of Industrial Chemistry
Address: Societe de Chimie Industrielle (SCI)
28, rue Saint Dominique, F-75007 Paris, France
Phone: +33 1 53 59 02 10
Fax: +33 1 45 55 40 33
E-mail: sci.fr@wanadoo.fr
Web site: www.scifrance.org/sci/default.html

**International Agency for Research on Cancer**
Address: 150 Cours Albert Thomas, 69372 Lyon CEDEX 08, France
Phone: +33 (0)4 72 73 84 85
Fax: +33 (0)4 72 73 85 75
Web site: www.iarc.fr
The International Agency for Research on Cancer coordinates and conducts both epidemiological and laboratory research into the causes of cancer.

**International Fertilizer Industry Association**
Address: 28, rue Marbeuf, 75008 Paris, France
Phone: +33 1 53 93 05 00
Fax: +33 1 53 93 05 45/47
E-mail: ifa@fertilizer.org
Web site: www.fertilizer.org
IFA's mission is to promote efficient and responsible production and use of plant nutrients to maintain and increase agricultural production worldwide in a sustainable manner, to improve the operating environment of the fertilizer industry, to collect, compile and disseminate information, and to provide a discussion forum for its members and others on all aspects of the production, distribution and consumption of fertilizers, their intermediates and raw materials.

## GERMANY

**German Society for Chemical Engineering, Biotechnology and Environmental Protection**
Address: Gesellschaft für Chemische Technik und Biotechnologie e.V
Theodor-Heuss-Allee 25, D-60486 Frankfurt am Main, Germany
Phone: +49 69 7564-0
Fax: +49 69 7564-201
E-mail: internetinfo@dechema.de
Web site: www.dechema.de

**German Chemical Industry Association**
Address: Verband der Chemischen Industrie e.V. (VCI)
Karlstr. 21, 60329 Frankfurt am Main, Germany
Fax: +49 69 25 56 16 12
Web site: www.chemische-industrie.de

**German Chemical Society**
Address: Gesellschaft Deutscher Chemiker (GDCh)
Varrentrappstraße 40-42, D-60486 Frankfurt am Main, Germany
Phone: +49 69 7917 326
E-mail: pr@gdch.de
Web site: www.gdch.de

**German Crop Protection Association**
Address: Industrieverband Agrar eV
Web site: www.iva.de

## GREECE

**Hellenic Association of Chemical Industries**
Address: 23, Lagoumitzi Ave., 176 71 Kallithea, Athens, Greece
Phone: +30 1 55 74 501-5
Fax +30 1 55 74 910
E-mail: haci@biznet.com.gr
Web site: www.biznet.com.gr/industrial/haci

**Hellenic Crop Protection Association**
Address: 53 Patission Street, 104 33, Athens, Greece
Phone: +30 10 52 29 786
Fax: +33 10 52 21 542
E-mail: info@esyf.gr
Web site: www.hcpa.gr
The Hellenic Crop Protection Association companies involved in the crop protection industry and other associated members, namely distributors of crop protection products. The HCPA works in close collaboration with the Cyprus Crop Protection Association. HCPA is a member of the European Crop Protection Association (ECPA) and a member of CropLife International.

## HONG KONG

**Association of International Chemical Manufacturers**
Address: G.P.O. Box 1607, Central, Hong Kong
Phone: +852 2866 2131
Fax: +852 2866 2131
E-mail: pmcdoug@attglobal.net
Web site: www.aicmasia.com
A chemical industry association with focus on China. Its members comprise more than 50 international companies engaged in the chemical industry in China, Hong Kong SAR and the Asia Pacific region.

## HUNGARY

**Hungarian Chemical Industry Association**
Address: Magyar Vegyipari Szovetseg (MAVESZ), Pesticide Industry Professional Association
1146 Budapest, Erzsébet királyné útja 1/C, Hungary
Phone: +36 1 343-8920
Fax: +36 1 343-0980
E-mail: kozpont@mavesz.hu
Web site: www.mavesz.hu

## INDIA

Indian Chemical Manufacturers Association (ICMA)
Address: Sir Vithaldas Chambers, 16, Bombay Samachar Marg, Bombay 400 023, India
Phone: +91 22 204 7649
Fax: +91 22 204 8057
E-mail: mail@icmaindia.com
Web site: www.icmaindia.com

**Indian Crop Protection Association**
Address: 102, Creative Industrial Bldg., Sundernagar, Road No. 2, Kalina, Santacruz (East), Bombay 400 098, India
Phone: +91 22 455 238
Web site: www.kisan.net/icpa

**Pesticides Association of India**
Address: 1202, New Delhi House, 27, Barakhamba Road, New Delhi 110 001, India
E-mail: pest@satyam.net.in

**Pesticides Manufacturers & Formulators Association of India**
Address: B-4, Anand Co-operative Hsg. Society, Sitadevi Temple Road, Mahim, Bombay 400 016, India
Phone: +91 22 437 5279
Fax: +91 22 437 6856
E-mail: pmfai@bom4.vsnl.net.in
Web site: www.pmfai.org
The Pesticides Manufacturers & Formulators Association of India mission is to include basic technical grade manufacturers of pesticides in order to foster the interest of the general public and PMFAI's members, promote innovations and environmentally sound use of crop protection / public health products so as to ensure high quality, abundant food, fiber and maintenance for growing population in the Indian sub-continental.

## IRELAND

**Ireland Animal and Plant Health Association**
Address: 8, Woodbine Park, Blackrock, Co. Dublin, Ireland
Phone: +353 1 2603050
Fax: +353 1 2603021
E-mail: info@alpha.ie
Web site: www.apha.ie
The Animal and Plant Health Association (APHA) is the representative body for manufacturers and sole distributors of animal health (veterinary medicines) and plant health (plant protection/agrochemical) products in Ireland. APHA is a member of FEDESA (European Federation of Animal Health) and the European Crop Protection Association.

**Ireland Pharmaceutical and Chemical Manufacturers' Federation**
Address: Confederation House, 84/86 Lower Baggot St., Dublin 2, Ireland
Phone: +353 1 660 1011
Fax: +353 1 660 1717
E-mail: matt.moran@ibec.ie

## ISRAEL

**Manufacturers Association of Israel - MAI, Chemical & Pharmaceutical Division**
Address: Industry House, 29 Hamered St., P.O. Box 50022, Tel Aviv 61500, Israel
Phone: +972 (3) 519 87 87
Fax: +972 (3) 516 20 26
E-mail: chemical@industry.org.il
Web site: www.ifpma.org/ifpma3.html#IS

## ITALY

**Agrofarma**
Address: Associazione Nazionale Imprese Prodotti Fitosanitari
Web site: agrofarma.federchimica.it

**Codex Alimentarius**
Address: Viale della Teme di Caracella, 00100 Roma, Italy
Phone: +39 06 5705.1
Fax: +39 06 5705.4593
E-mail: Codex@fao.org
Web site: www.codexalimentarius.net
The Codex Alimentarius Commission was created in 1963 by the Food and Agricultural Organization of the United Nations and the World Health Organization to develop food standards, guidelines and related texts such as codes of practice under the Joint FAO/WHO Food Standards Program. The main purposes of this Program are protecting health of the consumers and ensuring fair trade practices in the food trade, and promoting coordination of all food standards work undertaken by international governmental and non-governmental organizations.

**Food and Agricultural Organization of the United Nations**
Address: Viale della Terme di Caracalla, 00100 Roma, Italy
Phone: +39 06 57051
Fax: +39 06 570 53152
E-mail: FAO-HQ@fao.org
Web site: www.fao.org
Mandate is to raise levels of nutrition, improve agricultural productivity, better the lives of rural populations and contribute to the growth of the world's economy.

**Italian Chemical Manufacturers Association**
Address: Federazione Nazionale dell'Industria Chimica Via Accademia 33, 20131 Milano, Italy
Phone: +39 2 268 10.1
Fax: +39 2 268 10.310
E-mail: federchimica@federchimica.it
Web site: www.federchimica.it

**Italian Plant Protection Association**
Address: Associazione Italiana per la Protezione della Piante, Agostino Brunelli, Department of Agricultural Science, University of Bologna, Via Filippo Re, 8, 40126 Bologna, Italy
Phone: +39-051-2091352
Fax: +39-051-2091363
E-mail: brunelli@agrsci.unibo.it

**Web site:** pwhux.tin.it/vegetale/aipp2-1.htm

AIPP, a non-profit association, aims at developing the research, education and training in plant protection, at fostering the spreading of information and know-how related to new, environmentally sound strategies for the management of diseases and pests of crops and commodities, in order to avoid the contamination of the environment as well as the adoption of inadequate agricultural practices.

## JAPAN

### Chemical Society of Japan
**Address:** 1-5, Kanda-Surugadai, Chiyoda-ku, Tokyo 101-0062, Japan
**Phone:** +81 3 3292-6161
**Fax:** +81 3 3292-6318
**E-mail:** library@chemistry.or.jp
**Web site:** www.soc.nacsis.ac.jp/csj

### Federation of Asian Chemical Societies /The Chemical Society of Japan
**Address:** c/o Prof. Yoshito Takeuchi
1-5 Kanda-Surugadai, Chiyoda-ku, Tokyo, 101-0062, Japan
**Phone:** +81 3 3292-6161
**Fax:** +81 3 3292-6318
**E-mail:** facs@chemistry.or.jp
**Web site:** www.ozchemnet.adfa.oz.au/FACS

### Japan Chemical Industry Association
**Address:** Kazan Bldg., 3-2-4 Kasumigaseki, Chiyoda-Ku, Tokyo, 100-0013, Japan
**Phone:** +81 3 3580 0751
**Fax:** +81 3 3580 0970
**E-mail:** chemical@jcia-net.or.jp
**Web site:** www.nikkakyo.org

### Japan Crop Protection Association
**Address:** Nihonbashi Club Bldg., 5-8, 1-chome Muromachi, Nihonbashi, Chuo-ku, Tokyo 103, Japan
**Phone:** +81 3 3241 02 30
**Fax:** +81 3 3241 31 49
**Web site:** www.jcpa.or.jp

The Japan Crop Protection Association is the nonprofit organization of Japanese manufacturers, formulators and distributors of crop protection products. JCPA represents 90 companies that manufacture, sell and distribute more than 95% of Japan's crop protection products. It is a member of CropLife International.

### Pesticides Manufacturers & Formulators Association of India
**Address:** Nippon Noyaku Gakkai
1-43-11 Komagome, Toshima-ku, Tokyo 170-8484, Japan
**Fax:** +81 3 3943 6086
**Web site:** www.soc.nii.ac.pk

Provides a forum for the exchange of ideas on developing pesticides and their contribution to environmental and life sciences.

## JORDAN

### Africa/Middle East Working Group (AMEWG), see CropLife Africa Middle East (Jordan)

### CropLife Africa Middle East
**Address:** P.O. Box 961810 Sport City, 11196 Amman, Jordan
**Phone:** +962 79 567 409
**Fax:** +962 6 5671 878

The Africa/Middle-East Working Group represents Crop Protection Industry Associations in Cameroon, Ivory Coast, Ghana, Kenya, Malawi, Nigeria, South Africa, Tanzania, Zambia, Zimbabwe, Algeria, Egypt, Jordan, Israel, Lebanon, Morocco and Syria.

CropLife Africa & Middle East promotes the benefits of crop protection and biotechnology products and their importance to sustainable agriculture and food production. CropLife Africa & Middle East is a member of CropLife International.

## LATVIA

### Latvian Institute of Organic Synthesis
**Address:** Aizkraukles iela 21, LV-1006 Riga, Latvia
**Phone:** +371 7 551 822
**Fax:** +371 7 553 493
**Web site:** www.osi.lanet.lv/osi-info

## MACEDONIA

### Society of Chemists and Technologists of Macedonia
**Address:** Institute of Chemistry, Arhimedova 5, P.O.B. 161, Skopje, Macedonia
**Phone:** +389 91 117-055
**Fax:** +389 91 226-865
**Web site:** www.pmf.ukim.edu.mk

## MEXICO

### CIQUIM
**Address:** Consultoria en Ingenieria Quimica
Boyaca No. 557-2, Col. Guadalupe Tepeyec, 07740 Mexico D.F., Mexico
**Phone:** +52 5 587-00-52
**Fax:** +52 5 587-00-52
**E-mail:** egc@dec5500.ind.imp.mx

### National Association of the Chemical Industry
**Address:** Asociacion Nacional de la Industria Química, Providencia 1118 Col. del Valle C.P., 03100 México, D.F., Mexico
**Phone:** +55 52-30-51-00
**Fax:** +52 52-30-51-07

E-mail: aniq@aniq.org.mx
Web site: www.aniq.org.mx

## THE NETHERLANDS

**Association of the Dutch Chemical Industry**
Address: Vereniging van de Nederlandse Chemische Industrie (VNCI)
P. O. Box 443, 2260 AK, Leidschendam, The Netherlands
Phone: +31 70 337 8787
Fax: +31 70 320 3903
E-mail: info@vnci.nl
Web site: www.vnci.nl

**Netherlands Crop Protection Association**
Address: Nederlandse Stichting voor Fytofarmacie (NEFYTO)
P.B. 80523, 2508 Gm den Haag, The Netherlands
Phone: +31 70-351 48 51
Fax: +31 70-354 97 66
Web site: www.nefyto.nl
Nefyto is the trade association of the Dutch agrochemical industry. Affiliated to Nefyto are companies that manufacture and/or market crop protection products in the Netherlands. Nefyto has 15 participants, together representing 90% of the turnover of crop protection products in the Netherlands.

**Organotin Environmental Programme Association**
Address: Drs. J.A. Jonker
Wachttoren 11, 4336 KL Middelburg, The Netherlands
Phone: +31 118 617 063
Fax: +31 118 617 349
Promotes and fosters the dissemination of scientific and technical information on the environmental effects of organic compounds.

**Royal Netherlands Chemical Society**
Address: Koninklijke Nederlandse Chemische Vereniging
Burnierstraat 1, 2596 HV The Hague, The Netherlands
Phone: +31 70 346 9406
Fax: +31 70 362 5197
E-mail: kncv@kncv.nl
Web site: www.kncv.nl/kncven
The professional organization for chemists and chemical engineers in The Netherlands.

## NEW ZEALAND

**New Zealand Chemical Industry Council**
Address: P. O. Box 5069, Wellington, New Zealand
Phone: +64 4 499 4311
Fax: +64 4 472 7100
E-mail: nzcic@ibm.net
Web site: www.Webnz.com/nzcic

## NORWAY

**Processing Industry Association (Norway)**
Address: Prosessindustriens Landsforening (PIL)
P.O. Box 5487 Majorstua, 0305 Oslo, Norway
Phone: +47 23 08 78 00
Fax: +47 23 08 78 99
E-mail: pil@pil.no
Web site: Web@pil.no

## PERU

**Chemical Society of Peru**
Address: Sociedad Quimica del Peru (SQP)
Av. Nicolas e Aranibar 696, Santa Beatriz, Lima 1, Peru
Phone: +51 1 472-3925
Fax: +51 1 265-9049
E-mail: olock@pucp.edu.pe
Web site: www.pucp.edu.pe/~quimica/sqp

## POLAND

**Pesticides Producers Association**
Address: Polish Chamber of Chemical Industryul, Czackiego 15/17, Rm. 320, 00-043 Warsaw, Poland
Phone: +48 22 829 73 35
Fax: +48 22 829 73 39
E-mail: pipc@pipc.org.pl
Web site: http://www.pipc.org.pl

**Polish Chamber of Chemical Industry**
Address: ul. Czackiego 15/17 room 320, 00-043 Warsaw, Poland
Phone: +48 22 829 73 35
Fax: +48 22 829 73 39
E-mail: pipc@pipc.org.pl
Web site: www.pipc.org.pl

## PORTUGAL

**Portugese Chemical Manufacturing Association;** Associacao Portuguesa das Empresas Quimicas
Address: Avenida D. Carlos I, n°45-3, P - 1200-646 Lisbon, Portugal
Phone: +351 1 390 6796
Fax: +351 1 396 3052
E-mail: apeqassociacao@mail.telepac.pt
Web site: www.apequimica.pt

## SLOVAK REPUBLIC

**Slovakia Association of Chemical and Pharmaceutical Industry**
Address: Zväz Chemického a Farmaceutického Priemyslu Slovenskej Republiky (ZCHFP)
Drienova 24, SK - 826 03 Bratislava, Slovak Republic

Phone: +421 7 235 226
Fax: +421 7 235 226
E-mail: zchfp@isnet.sk

## SLOVENIA

**Slovenia Chemical and Rubber Industry Association**
Address: Chamber of Commerce and Industry of Slovenia
Dimiceva 9, SI-1504 Ljubljana, Republic of Slovenia
Phone: +386 1 5898 000
Fax: +386 1 5898 100
E-mail: infolink@gzs.si
Web site: www.gzs.si

## SOUTH AFRICA

**Agricultural & Veterinary Chemicals Association**
Address: P. O. Box 1995, Halfway House, 1685 South Africa
Phone: +27 11 805 2079
Fax: +27 11 805 2222
Web site: www.mbendi.co.za

**Chemical and Allied Industries Association**
Address: P. O. Box 91415, 15th Fl., Metal Box Centre, 25 Owl St., Auckland Park 2600, Johannesburg, South Africa
Phone: +27 11 482 1671
Fax: +27 11 726 8310
E-mail: caia@iafrica.com
Web site: www.caia.co.za

## SOUTH KOREA

**Korea Agricultural Chemicals Industrial Association**
Address: 1358-9, Seocho 4-dong, Seocho-gu, Seoul, Korea
Phone: +82 2-3-3474-1590 4
Fax: +82 2-3472-4134
Web site: www.kacia.or.kr

## SPAIN

**Spanish Chemical Association**
Address: Asociacion Nacional de Quimicos de Espana (ANQUE)
Apartado de correos 6049, 47080 Valladolid, Spain
E-mail: polyfemo@luna.gui.uva.es
Web site: www.gui.uva.es

**Spanish Crop Protection Association**
Address: Asociación Empresarial para la Protección de las Plantas
c/ Almagro 44, 4°, 28010 Madrid, Spain
Phone: +91 3 1002 38
Fax: +91 3 1977 34
Web site: www.aepla.es

**Spanish Federation of Chemical Industries**
Address: Federación Empresarial de la Industria Química Española
C/ Hermosilla, 31 - 1° Ext. Dcha., 28001 Madrid, Spain
Phone: +34 91 431 79 64
Fax: +34 91 576 33 81
E-mail: feique@interbook.net
Web site: www.feique.org

## SWEDEN

**National Chemicals Inspectorate**
Address: P.O. Box 1384, SE - 171 27 Solna, Sweden
Phone: +46 8 783 11 00
Fax: +46 8 735 76 98
E-mail: kemi@kemi.se
Web site: www.kemi.se

**Swedish Chemical Industries Association**
Address: Kemikontoret, Box 5501, SE-114 85 Stockholm, Sweden
Phone: +46 8 783 80 00
Fax: +46 8 663 63 23
Web site: www.chemind.se

**Swedish Chemical Society**
Address: Svenska Kemistsamfundet, Wallingatan 24 3 tr, 111 24 Stockholm, Sweden
Phone: +46 8 411 52 60/80
Fax: +46 8 10 66 78
Web site: www.chemsoc.se

## SWITZERLAND

**Collaborative International Pesticides Analytical Council**
Address: Mr. Markus D. Muller, Chairman
Swiss Federal Research Station, CH-8820, Wadenswil, Switzerland
Phone: +41 1 783 64 12
Fax: +41 1 783 64-39
E-mail: markus.mueller@faw.admin.ch
Web site: www.cipac.org
An international, non-profit, non-governmental organization that promotes international agreement on methods for pesticides analysis and tests methods of formulations, promotes programs for the evaluation of test methods.

**International Potash Institute**
Address: Schneidergasse 27, PO Box 1609, CH-4001 Basel, Switzerland
Phone: +41 61 261 29 22
Fax: +41 61 261 29 25
E-mail: ipi@IPIpotash.org
Web site: http://www.ipipotash.org
IPI deals with the nutrient potassium and its effect on soil and plants.

**International Register of Potentially Toxic Chemicals**
Address: P.O. Box 356, 15 chemin des Anemones, Chalelaine CH-1219, Geneva, Switzerland
Phone: +41 22 979 9111
Fax: +41 22 797 34 60
E-mail: irptc@unep.ch
Web site: www.irptc.unep.ch/irptc

**Nouvelle Société Suisse de Chimie**; New Swiss Chemical Society; Neue Schweizerische Chemische Gesellschaft
Address: c/o Novartis, WKL-24.1.09, CH-4002 Basel, Switzerland
Phone: +41 61 696.67.96
Fax: +41 61 696.69.85
E-mail: nscg.darms@group.novartis.com
Web site: www.nscs.ch

**Swiss Society of Chemical Industries; Société Suisse des Industries Chimiques**; Schweizerische Gesellschaft für Chemische Industrie
Address: Nordstrasse 15, P.O. 8035, Zürich, Switzerland
Phone: +41 1 368 17 11
Fax: +41 1 368 17 70
E-mail: mailbox@sgci.ch
Web site: www.sgci.de

**United Nations Environmental Program, Chemical Unit**
Address: 11-13, chemin des Anemones, 1219 chatelaine, Geneva, Switzerland
Phone: +41 22 917 81 11
Fax: +41 22 797 34 60
E-mail: chemicals@unep.ch
Web site: www.chem.unep.ch/irptc
UNEP Chemicals is the center for all chemicals-related activities of the United Nations Environment Programme. The unit works towards making the world a safer place from toxic chemicals by helping governments take needed global actions for the sound management of chemicals, by promoting the exchange of information on chemicals, and by helping to build the capacities of countries around the world to use chemicals safely.

## THAILAND

**Asia-Pacific Crop Protection Association**, see CropLife Asia

**CropLife Asia**
Address: 28th floor, Rasa Tower Building, 555, Pahonyothin Road, Chatuchak, Bangkok 10900, Thailand.
Phone: +66 2937-0487
Fax: +66 2937-0491
E-mail: info@croplifeasia.org
Web site: www.croplifeasia.org
CropLife Asia (formerly the Asia-Pacific Crop Protection Association) provides regional leadership and representation for the plant science industry. It promotes and supports the safe and responsible use of crop production technologies, and their role in the development of a sustainable agriculture system in the Asia-Pacific region.

**Fertilizer Advisory, Development and Information Network for Asia and the Pacific**
Address: FADINAP, Rural Development Section, Population and rural and Urban Development Division
United Nations Building, Bangkok 10200, Thailand
Web site: www.fadinap.org
*Note:* Since 1991, FADINAP has been promoting environmentally-friendly fertilization in an effort to promote a better environment while ensuring food security. FADINAP works closely with national and international fertilizer organizations as well as with the fertilizer industry.

## TURKEY

**Turkish Chemical Manufacturers Association**
Address: Turkiye Kimya Sanayicileri Dernegi
Deðirmen Sokak, Þaþmaz Sitesi, No:19, Duranbey Apt. K:3
D.9, Kozyataðý / Erenköy, 81090 Istanbul, Turkey
Phone: +90 216 416 76 44
Fax: +90 216 416 92 18
E-mail: tksd@turk,net
Web site: www.tksd.ork.tr

## UNITED KINGDOM

**British Crop Protection Council-BCPC**
Address: 49 Downing St., Farnham, Surrey, GU9 7PH, UK
Phone: +44 1252 7330721
Fax: +44 1252 727194
E-mail: gensec@bcpc.org
Web site: www.bcpc.org
BCPC promotes and encourages the science and practice of crop protection. Formerly called the British Crop Protection Council.

**British Biochemical Society, The**
Address: 59 Portland Place, London W1N 3AJ, UK
Phone: +44 207 580 5530
Fax: +44 207 637 3626
E-mail: genadmin@biochemistry.org
Web site: www.biochemsoc.org.uk

**British Agrochemicals Association**, see Crop Protection Association (UK)

**British Calcium Carbonates Federation**
Address: Omya UL Ltd.
Curtis Rd., Dorking, Surrey RH4 1XA, UK
Phone: +44 1 3068 86688
Fax: +44 1 3067 47444
E-mail: lindsay.watson@omya.com.uk

**CABI Biosciences**
Address: Bakeham Lane, Egham, Surrey TW20 9TY, UK
Phone: +44 1491-829080

Fax: +44 1491-829100
E-mail: bioscienceUK@cabi.org
Web site: www.cabi-bioscience.org
CABI *Bioscience* integrates four former international biological institutes, the International Institute of Biological Control (IIBC), the International Institute of Entomology (IIE), the International Institute of Parasitology (IIP) and the International Mycological Institute (IMI). Today it forms a group of scientists working in the fields of agricultural sustainability and biological diversity. It operates from six centers worldwide, in Kenya, Malaysia, Trinidad, Pakistan, Switzerland and the UK..

**Chemical Industries Association**
Address: Kings Buildings, Smith Square, London SW1P 3JJ, UK
Phone: +44 171 834 3399
Fax: +44 171 834 4469
E-mail: enquiries@cia.org.uk
Web site: www.cia.org.uk
The Chemical Industries Association is the UK chemical industry's leading trade and employer organization and embraces all the industry's trade sectors, product types and business activities.

**Crop Protection Association**
Address: 4 Lincoln Court, Lincoln Road, Peterborough PE1 2RP, UK
Phone: +44 17 33 349 225
Fax: +44 17 33 562 523
E-mail: info@cropprotection.org.uk
Web site: www.cropprotection.org.uk
The Crop Protection Association is the United Kingdom trade body representing companies engaged in manufacture, formulation and distribution of pesticide products for agriculture, forestry, horticulture, gardening, industrial, amenity and local authority uses. Formerly the British Agrochemicals Association.

**Institute of Food Science and Technology**
Address: 5 Cambridge Court, 210 Shepherd's Bush Road, London W6 7NJ, UK
Phone: +44 (0)20-7603 6316
Fax: +44 (0)20-7602 9936
E-mail: ifst@ifst.org
Web site: www.ifst.org
The IFST seeks to further the application of science and technology to all aspects of the supply of safe, wholesome, nutritious and attractive food, nationally and internationally.

**Pesticide Action Network UK**
Address: Development House
56-64 Leonard St., London 7065 0907, UK
E-mail: admin@pan-uk.org
Web site: www.pan-uk.org
Mission is to eliminate the hazards of pesticides; reduce dependence on pesticides and prevent unnecessary expansion of their use; and increase sustainable and ecological alternatives to chemical pest control.

**Royal Society of Chemistry**
Address: Burlington House, Piccadilly, London W1V 0BN, UK
Phone: +44 207 440 3312
Fax: +44 207 437 8883
E-mail: franklin@rsc.org.uk
Web site: www.rsc.org.uk
The learned society for chemistry and professional chemists in the United Kingdom.

**Society of Chemical Industry**
Address: 14/15 Belgrave Square, London SW1X 8PS, UK
Phone: +44 207 598 1500
Fax: +44 207 598 1545
E-mail: members@chemind.demon.co.uk
Web site: www.sci.mond.org
A global interdisciplinary network with deep roots in business, manufacturing, consumer affairs, research and education at all levels. Has particular strengths in the agricultural, food, pharmaceutical, water, construction, energy and environmental product and service areas.

## UNITED STATES

**American Association of Cereal Chemists**
Address: 3340 Pilot Knob Road, St. Paul, MN 55121, USA
Phone: +1 651-454-7250
Fax: +1 651-454-0766
E-mail: aacc@scisoc.org
Web site: www.aaccnet.org
The American Association of Cereal Chemists members are specialists in the use of cereal grains in foods. AACC gathers and disseminates scientific and technical information to professionals in the grain-based foods industry worldwide.

**American Chemistry Council**
Address: 1300 Wilson Boulevard, Arlington, VA 22209, USA
Phone: +1 703 741-5000
Hotline: +1 800 424-9300, Non-emergency, 800 262-8200
Fax: +1 703 741-6000
E-mail: nicole_naylor@americanchemistry.com
Web site: www.americanchemistry.com
Formerly known as Chemical Manufacturers Association (CMA). A trade association of chemical manufacturers, representing more than 90 percent of the production for basic industrial chemicals in the U.S. Administers research in areas significant to chemical manufacturing such as air and water pollution control, operates Chemical Transportation Emergency Center (CHEMTREC) to control and report chemical accidents. ACC is organized by industrial groups and also by issue groups that serve as coordinators and advocates in their fields of specializations.

**American Chemical Society**
Address: 1155 16th Street, NW, Washington, D.C. 20036, USA
Phone: +1 202 872-4600, 800 227-5558
Fax: +1 202 872-4615
E-mail: nbyerly@vt.edu
Web site: www.acs.org
A professional organization of chemists and chemical engineers. Conducts studies and surveys, professional and student conferences and programs, administers grants and fellowship programs, maintains extensive chemical databanks, online services, and more than 33 divisions and hundreds of subcommittees covering all aspects of fundamental and applied chemistry. Helps interpret technical data and refers citizens to local scientists.

**American Council on Science and Health**
Address: 1995 Broadway, 2nd Floor, New York, NY 10023-5860, USA
Phone: +1 212-362-7044
Fax: +1 212-362-4919
Web site: www.acsh.org
ACSH is a nonprofit, consumer education organization dedicated to providing the public with mainstream scientific information on issues related to food, nutrition, chemicals, pharmaceuticals, lifestyle, the environment and health.

**American Crop Protection Association**, see CropLife America (US)

**American Meat Science Association**
Address: 1111 N. Dunlap Ave, Savoy, IL 61874, USA
Phone: +1 217 356-5368
Fax: +1 217 398-4119
Web site: www.meatscience.org
The Association seeks to promote the application of science and technology in the production, processing, packaging, distribution, preparation, evaluation, and utilization of all types of meat and meat related products from all animal species.

**Association of American Plant Food Control Officials**
Address: D.L. Terry, Secretary, University of Kentucky, 103 Regulatory Services Building, Lexington, KY 40546-0275, USA
Phone: +1 859-257-2668
Fax: +1 859-257-9478
E-mail: dterry@uky.edu
Web site: http://www.aapfco.org
The AAPFCO is an organization of fertilizer control officials from each state in the United States, from Canada and from Puerto Rico who are actively engaged in the administration of fertilizer laws and regulations. Also, research workers employed by these governments who are engaged in any investigation concerning mixed fertilizers, fertilizer materials, their effect, and/or their component parts.

**Association of Natural Bio-control Producers**
Address: Executive Director, 2230 Martin Dr., Tustin Ranch, CA 92782, USA
Phone: +1 714-544-8295
Fax: +1 714-544-8295
E-mail: maclayb2@aol.com
Web site: www.anbp.org
Represents the commercial biocontrol industry.

**Biotechnical Industry Organization**
Address: 1255 Eye St., N.W., Ste. 400, Washington, DC 20005, USA
Phone: 202-962-9200
E-mail: ldry@bio.org
Web site: www.bio.org
This organization unites two biotechnology trade organizations, comprising 503 companies, under one umbrella. The organizations are the Biotechnology Association (IBA) and the Association of Biotechnology Companies (ABC), which represented emerging companies and universities, and focused on technology transfer issues, meetings and other business development activities. Food and agriculture issues are a subdivision of BIO.

**Center for Integrated Pest Management**
Address: College of Agriculture and Life Sciences
North Carolina State University
Raleigh, NC, USA
Phone: +1 919-513-1432
E-mail: cipm@ncsu.edu
Web site: http://ipmwww.ncsu.edu/cipm
The Center for Integrated Pest Management was established in 1991 to serve a lead role in technology development, program implementation, training, and public awareness for IPM at the state, regional, and national levels. The CIPM is an organizational unit within the College of Agriculture and Life Sciences at North Carolina State University. It is composed of faculty members from all academic departments in the College and involves all relevant disciplines impacting on IPM. The CIPM also involves scientists from other universities across the nation through grants, contracts, or other formal working relationships. It is part of the National Science Foundation.

**Center for Science in the Public Interest**
Address: 1875 Connecticut Ave, N.W., Ste. 300, Washington, DC 20009, USA
Phone: +1 202-332-9110
Fax: +1 202-265-4954
E-mail: cspi@cspinet.org
Web site: www.cspinet.org
The Center for Science in the Public Interest (CSPI) is a nonprofit education and advocacy organization that focuses on improving the safety and nutritional quality of our food supply and on reducing the damage caused by alcoholic beverages. CSPI seeks to promote health through educating the public about nutrition and alcohol. It represents citizens' interests before legislative, regulatory, and judicial bodies, and it works to ensure advances in science are used for the public good.

**Chemical Abstract Services (CAS)**
**Address:** 2540 Olentangy River Rd., P.O. Box 3012, Columbus, Ohio 43210, USA
**Phone:** +1 614 447-3600, 800 753-4227
**Fax:** +1 614 447-3713
**E-mail:** help@cas.org
**Web site:** www.cas.org
Chemical Abstracts Service produces the world's largest and most comprehensive databases of chemical information. Principal databases, CHEMICAL ABSTRACTS and REGISTRY, include nearly 16 million abstracts of chemistry-related literature and patents and more than 30 million substance records respectively.

**Chemical Industry Institute of Toxicology**
**Address:** 6 Davis Dr., P.O. Box 12137, Research Triangle Park, NC 27709-2137, USA
**Phone:** +1 919 558-1200
**Fax:** +1 919 558-1300
**E-mail:** CIITinfo@ciit.org
**Web site:** www.ciit.org
A not-for-profit toxicology research institute that provides an improved scientific basis for understanding and assessing the potential adverse effects of chemicals, pharmaceuticals, and consumer products on human health.

**Chemical Manufacturers Association (CMA)**, see American Chemistry Council (ACC)

**Chemical Transportation Emergency Center (CHEMTREC)**
**Address:** American Chemistry Council, 1300 Wilson Boulevard, Arlington, VA 22209-2380, USA
**Phone:** +1 800 262-8200
Continental US emergency phone only: +1 800 424-9300, non-emergency, 800 262-8200
International emergencies only, +1 703 527-3887, non-emergency, 703 741-5516
**Fax:** +1 703 741-6037
**E-mail:** chemtrec@americanchemistry.com
**Web site:** www.chemtrec.com
Handles emergency calls regarding chemical spills, leaks, fires, exposures, or accidents. Chemical emergency number open 24 hours, every day. CHEMTREC provides emergency response information for incidents involving hazardous materials, serving the chemical and transportation industries and emergency services who may be called upon as first responders. Also provides referrals to manufacturers, regulatory agencies, and research institutions. A public service of the American Chemistry Council.

**Chlorine Chemistry Council**
**Address:** 1300 Wilson Boulevard, Arlington, VA 22209, USA
**Phone:** +1 703 741-5000
**Web site:** http://c3.org
Comprised of chlorine and chlorinated product manufacturers, CCC is a business council of the American Chemistry Council. It strives to achieve policies that promote the continuing, responsible uses of chlorine and chlorine-based products.

**Chlorine Institute, Inc., The**
**Address:** 2001 L Street N.W., Suite 506, Washington, DC 20036
**Phone:** +1 202 775-2790
**Fax:** +1 202 223-7225
**E-mail:** tkerns@cl2.com
**Web site:** www.cl2.com
A trade association of companies that are involved or interested in the safe production, distribution and use of chlorine, sodium and potassium hydroxides, and sodium hypochlorite, and the distribution and use of hydrogen chloride.

**Consortium for International Crop Protection**
**Address:** Dr. Richard E. Ford, Executive Director
Oregon State University, c/o Integrated Plant Protection Center, 2040 Cordley Hall, Corvallis, OR 97330, USA
**E-mail:** R-FORD1@uiuc.edu
**Web site:** www.ipmnet.org
A non-profit organization formed in 1978 by a group of U.S. universities. Its principal purpose is to assist developing nations reduce food crop losses caused by pests while also safe-guarding the environment.

**Council for Biotechnology Information**
**Address:** P.O. Box 34280, Washington, DC, 20043-0380, USA
**Phone:** 202-467-6565
**Web site:** www.whybiotech.com
The council was launched in April of 2000 by seven leading biotechnology companies and two trade associations with a vision to create a new communications initiative built on a mix of research, advertising, media relations and constituency relations. CBI's founding companies are BASF, Bayer CropScience, Dow Chemical, DuPont, Monsanto and Syngenta. Two trade associations, the Biotechnology Industry Organization and CropLife America, also are members.

**CropLife America**
**Address:** 1156 15th Street, N.W., Ste. 400, Washington, DC 20005, USA
**Phone:** +1 202-296-1585
**Fax:** +1 202-463-0474
**E-mail:** member-service@croplifeamerica.org
**Web site:** www.croplifeamerica.org
CropLife America represents the developers, manufacturers, formulators and distributors of plant science solutions for agriculture and pest management in the United States. It promotes the environmentally sound use of crop protection products for the economical production of safe, high quality, abundant food, fiber and other crops. CropLife America was formerly the American Crop Protection Association.

**Environmental Working Group**
Address: 1436 U Street NW, Ste. 100, Washington DC 20009, USA
Phone: +1 202-667-6982
Fax: +1 202-232-2592
E-mail: info@ewg.org
Web site: www.ewg.org
This organization investigates issues affecting the environment, particularly issues involving the use of agricultural chemicals.

**Fertilizer Institute, The**
Address: Union Center Plaza, 820 First Street, N.E., Suite 430, Washington, D.C. 20002, USA
Phone: +1 202-962-0490
Fax: +1 202-962-0577
E-mail: kmathers@tfi.org
Web site: http://www.tfi.org
Represent the makers, transporters and providers of fertilizer.

**Insecticide Resistance Action Committee**
Address: P.O. Box 413708, Kansas City, MO 64141-3708, USA
Web site: www.plantprotection.org/IRAC
Provides a coordinated crop protection industry response to the development of resistance in insect and mite pests. During the last decade, IRAC has formed several international working groups to provide practical solutions to mite and insect resistance problems within major crops and pesticide groups.

**Institute of Food Technologists**
Address: 525 West Van Buren, Ste. 1000, Chicago, IL 60607, USA
Phone: +1 312 782-8424
Fax: +1 312 782-8348
E-mail: info@ift.org
Web site: www.ift.org
The Institute of Food Technologists is a nonprofit scientific society with 28,000 members working in food science, food technology, and related professions in industry, academia and government.

**National Lime Association (NLA)**
Address: 200 North Glebe Road, Suite 800, Arlington, Virginia 22203, USA
Phone: +1 703 243-5463
Fax: +1 703 243-5489
E-mail: natlime@lime.org
Web site: www.lime.org
The trade association for U. S. and Canadian manufacturers of high calcium quicklime, dolomitic quicklime, and hydrated lime, collectively referred to as "lime."

**Natural Resources Defense Council (NRDC)**
Address: 40 West 20th St., New York, NY 10011, USA
Phone: +1 212-727-2700
Fax: +1 212-727-1773
E-mail: nrdcinfo@nrdc.org
Web site: www.nrdc.org
NRDC uses law, science, and the support of more than 500,000 members nationwide to protect the planet's wildlife and wild places and to ensure a safe and healthy environment for all living things. Maintains offices in Washington, Los Angeles, and San Francisco.

**Pesticide Action Network (PAN); Pesticide Action Network North America (PANNA)**
Address: 49 Powell St., Ste. 500, San Francisco, CA 94102, USA
Phone: +1 415-981-1771
Fax: +1 415-981-1991
E-mail: panna@panna.org
Web site: www.panna.org
Pesticide Action Network North America is the regional center for PAN International. PANNA has campaigned to replace pesticides with ecologically sound alternatives since 1982. The organization links over 100 affiliated health, consumer, labor, environment, progressive agriculture and public interest groups in Canada, Mexico and the U.S. with more than 600 partners worldwide to promote healthier, more effective pest management through research, policy development, education, media, demonstrations of alternatives and international advocacy campaigns. This activist group has a wealth of information useful to anyone in the industry.

**Society of Environmental Toxicology and Chemistry**
Address: 1010 N. 12th Ave., Pensacola, FL 32501-3370, USA
Phone: +1 904 469 1500
Fax: +1 904 469 9778
E-mail: setac@setac.org
Web site: www.setac.org
An independent, international, nonprofit professional society that provides a forum for individuals and institutions engaged in the study of environmental issues, management and conservation of natural resources, environmental education, and environmental research and development. Offices in Belgium, Australia and the United States.

**Sulphur Institute, The**
Address: 1140 Connecticut Ave., N.W., Ste. 612, Washington, DC 20036, USA
Phone: +1 202-293-9660
Fax: +1 202-293-2940
E-mail: sulphur@sulphurinstitute.org
Web site: www.sulphurinstitute.org
TSI is an international, non-profit organization supported by the world's sulphur industry, dedicated to promoting its consumption in established and new markets, as well as its safe handling and transport

## URUGUAY

**Latin American and Caribbean Food Science and Technology Association (ALACCTA)**
**Address:** Asociacion Latinoamericana y del Caribe de Ciencia y Tecnologia de Alimentos (ALACCTA)
Ave. Centenario 3143 Ap. 302, 11600 Montevideo, Uruguay
**Fax:** +582 2 480 3932
**E-mail:** mtaranto@adinet.com.uy
**Web site:** http://acd.ufrj.br/consumo/alaccta
This association has links to member associations in Latin America and the Caribbean.

**Uruguayan Chemical and Pharmaceutical Association**
**Address:** Asociacion de Quimica y Farmacia del Uruguay (AQFU)
Ejido 1589, Montevideo, Uruguay
**Phone:** +598 2 900 0711
**Fax:** +598 2 900 6340
**E-mail:** info@aqfu.org.uy
**Web site:** www.aqfu.org.uy

**Uruguayan Chemical Industry Association**
**Address:** Associaciòn de la Industria Química de Uruguay
Sudy Lever SA, Treinta y Tres 1269, Casilla de Correo 1385, U – 11000 Montevideo, Uruguay

## VENEZUELA

**Venezuelan Chemical Society**
**Address:** Sociedad Venezolana de Quimica (SVQ)
Central University of Venezuela, Institute of Chemistry, Apartado 3895, Caracas 1040, Venezuela
**Fax:** +582 962-1025
**Web site:** http://socvenquim.org

# Appendix F: Directory of Hotlines, Databases, and Web Sites

**Air Risk Information Support Center**
Address: Office of Air Quality Planning and Standards
U.S. Environmental Protection Agency
Research Triangle Park, NC 27711, USA
Hotline: +1 919 541-0888
Fax: +1 919 541-1818
Web site: www.epa.gov/ttn/atw/hepindex.html
Assists state and local air pollution control agencies and EPA regional offices with technical matters pertaining to health, exposure, and risk assessment of air pollutants. Services are also extended to the general public, small businesses and international agencies. Hours: Thursday 8:00 a.m. to 5:00 p.m., EST; Friday 8:00 a.m. to 4 p.m., EST.

**Aerometric Information Retrieval System**
Address: Office of Air and Radiation, Office of Air Quality Planning and Standards, U.S. Environmental Agency, Research Triangle Park, NC 27711, USA
Hotline: +1 800 334-7909
Fax: +1 919 541-0028
Web site: www.epa.gov/airs/
AIRS is a computer-based repository of information about airborne pollution in the United States and various World Health Organization member countries. The system is administered by the U.S. Environmental Protection Agency, Office of Air Quality Planning and Standards (OAQPS), Information Transfer and Program Integration Division (ITPID), located in Research Triangle Park, North Carolina, and provides a national repository for air pollution data that is reported to the EPA by states and local agencies. Call for other network addresses and available linkage types. Hours: Voice, 8:30 a.m. to 5:00 p.m. EST, Monday through Friday. Electronic: 24 hours, everyday.

**Aquatic Toxicity Information Retrieval Database**
Address: Mid-Continent Ecology Division Laboratory
National Health and Environmental Effects Research Laboratory
U.S. Environmental Protection Agency
6201 Congdon Boulevard, Duluth, MN 55804-2595, USA
Phone: +1 218 529-5225
Fax: +1 218 529-5003
E-mail: ecotox.support@epa.gov
Web site: www.epa.gov/ecotox
Contains information on the toxic effects of 5,600 chemicals on more than 2,800 aquatic species of animals and plants, excluding birds, aquatic mammals, and bacteria. Has now been incorporated into ECOTOX Data System. Hours: 8:00 a.m. to 4:30 p.m. CST, Monday - Friday.

**Canadian Centre for Occupational Health and Safety - CCINFO MSDS Series**
Address: Canadian Centre for Occupational Health and Safety, 250 Main St. East, Hamilton, Ontario, L8N 1H6, Canada
Phone: +1 905 570-8094, 800 668-4284, (toll-free in Canada and USA)
Fax: +1 905 572-2206
E-mail: custserv@ccohs.ca
Web site: www.ccohs.ca
A bundle of subscription services of more than 50 databases on occupational health and safety information, are fully operable in both English and French. Contains over 70,000 MSDSs as well as CHEMINFO database of hazardous information on over 1,044 chemicals. Available online through CCINFOline; on CD-ROM through CCINFOdisc.

**Canadian Pollution Prevention Information Clearinghouse**
Address: The National Office of Pollution Prevention
351 St. Joseph Boulevard, Hull, Quebec K1A 0H3, Canada
Phone: +1 819 997-2800
Hotline: +1 800 668-6767
Fax: +1 819 953-2225
E-mail: enviroinfo@ec.qc.ca
Fax: +1 519 337 3486
Web site: www.ec.gc.ca/cppic
Provides pollution information from 1,200 references.

**Center for Environmental and Regulatory Information Systems**
Address: Entomology Department, Purdue University
1231 Cumberland Ave., Ste. A, West Lafayette, IN 47906-1317, USA
Phone: +1 765 494-6616
Fax: +1 765 494-9727
E-mail: info@ceris.purdue.edu
Web site: www.ceris.purdue.edu
The Center for Environmental and Regulatory Information Systems is home to a collection of databases of information on pesticides, plant export/import, and exotic pest tracking. Data are received or derived principally from the US Environmental Protection Agency, US Department of Agriculture, and USA state agencies. CERIS is a center within the Entomology Department at Purdue University. Provides current information on EPA product registration and tolerance data for pesticides and hazardous chemicals. It uses the National Pesticides Information Retrieval System (NPIRS), EPA, U.S. Department of Agriculture, and state information to provide pesticide product information, registration guidelines, and descriptions of studies in the field. This a membership clearinghouse operated for the Office of Pesticide Programs, U. S. Environmental Protection Agency.

**CHEMDEX**
Web site: www.chemdex.org
A directory of more than 5800 international chemistry links on the World Wide Web. Maintained at the Department of Chemistry, University of Sheffield, UK.

**The Chemical Alliance**
Web site: www.chemalliance.org
Provides regulatory information for the chemical process industry. The ChemAlliance site was made possible in large part due to funding provided by the United States Environmental Protection Agency. ChemAlliance is a partnership between the Chemical Industry, EPA's Office of Enforcement and Compliance Assurance, and the ChemAlliance staff, who reside at Michigan Technological University, Pacific Northwest National Laboratory, and University of Wisconsin.

**Chemical Carcinogenicity Research Information System**
**Address:** National Library of Medicine
8600 Rockville Pike, Bethesda, MD 20894, USA
**Phone:** +1 800 272-4787
**Fax:** +1 301 496-0822
**Web site:** www.toxnet.nlm.nih.gov/cgi-bin/sis/htmlgen?ccris
A scientifically evaluated and fully referenced data bank, developed and maintained by the National Cancer Institute (NCI). It contains some 8,000 chemical records with carcinogenicity, mutagenicity, tumor promotion, and tumor inhibition test results. Data are derived from studies cited in primary journals, current awareness tools, NCI reports, and other special sources. Test results have been reviewed by experts in carcinogenesis and mutagenesis.

**ChemConnect**
Web site: http://my.chemconnect.com
A membership site for daily, world-wide news and resources for the chemical industry. Also a buyers' trading forum.

**Chemcyclopedia**
Web site: www.chemcyclopedia.ims.ca
A buyer's guide of commercially available chemicals in the United States, as submitted by the suppliers. Sponsored by the American Chemical Society.

**Chemical Evaluation Search & Retrieval System**
**Address:** Michigan State Department of Natural Resources and the Ontario Ministry of the Environment
Great Lakes and Environmental Assessment Section
Knapp's Office Centre, P.O. Box 30273, Lansing, MI 48909, USA
**Phone:** +1 800 668-4284
**Fax:** +1 905 572-2200
**Web site:** www.ccohs.ca/products/databases/cesars.html
The CESARS database contains comprehensive environmental and health information on chemicals. It provides detailed descriptions of chemical toxicity to humans, mammals, aquatic and plant life, as well as data on physical chemical properties, and environmental fate and persistence. Each record consists of chemical identification information and provides descriptive data on up to 23 topic areas, ranging from chemical properties to toxicity to environmental transport and fate. Records are in English. Available online through CCINFOline from the Canadian Centre For Occupational Health and Safety (CCOHS) and Chemical Information System (CIS); on CD-ROM through CCINFOdisc.

**ChemExper Chemical Directory**
Web site: www.chemexper.com
A Belgian site giving details of over 70,000 chemicals, as submitted by the suppliers.

**ChemExtra**
Web site: www.chemextra.com
A British site that lists over 28,000 companies from 135 countries, with 100,000 chemicals, products and services as submitted by the suppliers. Sponsored by the British Chemical Industries Association. Very extensive and easy to use, but discontinued in August, 2004. May be reinstated in the future.

**ChemFinder.com**
Web site:
http://chemfinder.cambridgesoft.com/siteslist.asp
Database and Internet searching.

**Chemical Hazard Response Information System - CHRIS**
**Address:** Office of Marine Environmental Protection Division
U.S. Coast Guard
2100 2nd St., SW, Washington, DC 20593, USA
**Phone:** +1 202 267-2611
**Fax:** +1 202 426-7881
**E-mail:** stinet@dtic.mil
**Web site:** www.chrismanual.com
Provides emergency response and chemical handling information for more than 1,200 chemical substances that would be involved in chemical transport accidents, particularly by water. Contains information on labeling, physical and chemical properties, health and fire hazards and hazard classifications, chemical reactivity, and water pollution. Includes safety procedures for preventing emergency situations. Available online through Chemical Information System (CIS) and Canadian Centre for Occupational Health and Safety (COOHS); on CD-ROM through CCINFOdisc, SilverPlatter CHEM-BANK, as part of TOMES Plus System; on magnetic tape through TOMES Plus System; and from National Information Service Corporation.

**ChemIndustry.com**
Web site: http://chemindustry.com
A marvelous chemical directory to industry sectors, manufacturers, services, careers, equipment and software, portals, events, organizations, institutions, software, and much, much more. Has links to other search sites.

**Chemical Registry System**
**Address:** Office of Environmental Information
U.S. Environmental Protection Agency
**Web site:** www.epa.gov/crs

CRS provides information on chemical substances and how they are represented in the Environmental Protection Agency regulations and data systems. A search engine for chemicals by CAS number, name, molecular formula, chemical type, definition, or other data identifiers.

**CHEMTREC** (24-hour emergency number for chemical accidents and spills)
**Address:** Chemical Transportation Emergency Center
American Chemistry Council (ACC)
1300 Wilson Boulevard, Arlington, VA 22209, USA
**Phone:** +1 703 741-5525; Continental US emergency phone only: +1 800 424-9300; Non-emergency, +1 800 262-8200; International emergencies only, +1 703 527-3887; Non-emergency, +1 703 741-5516
**Fax:** Emergency only, +1 703 741-6090; Non-emergency, +1 703 741-6089
**E-mail:** chemtrec@cmahq.org
**Web site:** www.chemtrec.com
Handles emergency calls regarding chemical spills, leaks, fires, exposures, or accidents. Chemical emergency number open 24 hours, every day. CHEMTREC provides emergency response information for incidents involving hazardous materials, serving the chemical and transportation industries and emergency services who may be called upon as first responders. Also provides referrals to manufacturers, regulatory agencies, and research institutions. A public service of the American Chemistry Council.

**Clean Air Technology Center**
**Address:** Office of Air Quality, Planning and Standards (AOQPS)
U.S. Environmental Protection Agency
**Hotline:** +1 919 541-0800 (English); 919 541-1800 (Spanish)
**Fax:** +1 919 541-0242
**E-mail:** catcmail@epamail.epa.gov
**Web site:** www.epa.gov/ttn/catc
Serves as a resource on all areas of emerging and existing air pollution prevention and control technologies, and provides public access to data and information on their use, effectiveness and cost. In addition, CATC provides technical support, including access to EPA's knowledge base, to government agencies and others, as resources allow, related to the technical and economic feasibility, operation and maintenance of these technologies.

**Clean Water Act**
**Address:** Office of Water
U.S. Environmental Protection Agency
**Hotline:** +1 202 260-5700
**E-mail:** OW-general@epamail.epa.gov
**Web site:** www.epa.gov/OW
EPA's Office of Water directs callers with questions about the Clean Water Act to the appropriate EPA office. EPA also maintains a bibliographic database of Office of Water publications. The Safe Drinking Water Hotline number is +1 800 426-4791.

**ChemWeb.com**
**Address:** Web site: www.chemweb.com
A home page for chemical information.

**China Chemical Network**
**Address:** Web site: www.hi2000.net
A one-stop site for contacts and information about the chemical industry in China.

**Compliance Resource Center for CFR Information**
**Web site:**
www.setonresourcecenter.com/40CFR/Docs/wcd00000/wcd0008d.asp
A quick-find instrument to CFR titles 29 (OSHA), 40 and 49 (Transportation). Its index to 40 CFR 180 is an easy-to-use reference to "Tolerances and Exemptions from Tolerances for Pesticide Chemicals in Food." Presented by the Seton Company. A similar web site is maintained by the EPA but not in as concise format. The EPA site can be located at
http://www.epa.gov/pesticides/food/viewtols.htm.

**Comprehensive Chemical Contaminant Series**
Electronic version of Lewis "Sax's Dangerous Properties of Industrial Materials," 9$^{th}$ ed., "Hawley's Condensed Chemical Dictionary, " Prager's "Environmental Contaminant Reference Databook, Vols, I, II & III," Verschueren's "Handbook of Environmental Data on Organic Chemicals, 3$^{rd}$ ed., " and Pohanish and Greene's "Hazardous Materials Handbook," on CD-ROM, available from John Wiley and Sons, New York, NY, 1998.

**Comprehensive Environmental Response, Compensation, and Liability Information System**
**Address:** Office of Solid Waste and Emergency Response (OSW), U.S. Environmental Protection Agency
401 M St., SW, Washington, DC 20460, USA
**Phone:** +1 202 260-8321; 800 775-5037
**Fax:** +1 703 603-9133
**Web site:** www.epa.gov/superfund/sites/cursites
The Superfund database containing information on all aspects of hazardous waste sites from initial discovery to listing on the National Priorities List. Magnetic tapes are available quarterly from NTIS. Summary data under the Freedom of Information Act is available free by calling the Superfund Automated Phone System +1 800 775-5037.

**COSMOS Online**
**Web site:** www.cosmos.com.mx
The online directory to chemicals and chemical companies in Mexico and South America. Includes links to companies.

**ECOTOX Database System**
**Address:** Office of Research and Development, National Health and Environmental Effects Research Laboratory
Mid-Continent Ecology Division, U.S. Environmental Protection Agency, 6201 Congdon Boulevard, Duluth, MN 55804-2595, USA.

Phone: +1 218 529-5225
Fax: +1 218 529-5003
E-mail: ecotox.support@epa.gov
Web site: www.epa.gov/ecotox
The ECOTOXicology database is a source for locating single chemical toxicity data for aquatic life, terrestrial plants and wildlife. ECOTOX integrates three toxicology effects databases: AQUIRE (aquatic life), PHYTOTOX (terrestrial plants), and TERRETOX (terrestrial wildlife). These databases were created by the U.S. EPA, Office of Research and Development, and the National Health and Environmental Effects Research Laboratory (NHEERL), Mid-Continent Ecology Division. Hours: 8:00 a.m. to 4:30 p.m. CST, Monday - Friday.

**Emergency Planning and Community Right-to-Know Act (EPCRA)**
Address: Office of Solid Waste, U.S. Environmental Protection Agency
401 M St., SW, Washington, DC 20460, USA
Phone: +1 703 412-9810; TDD 703 412-3323
Hotline: +1 800 424-9346; TDD 800 535-7672
E-mail: tri.us@epamail.gov
Web site: www.epa.gov/epaoswer/hotline
Answers questions and distributes guidance regarding the emergency planning and community right-to-know regulations. Programs include the Resource Conservation and Recovery Act (RCRA), Underground Storage Tank program, the Risk Management Program (RMP), the Comprehensive Environmental Response, Compensation and Liability Act (CERCLA or Superfund), and the EPA's Oil Program. The EPCRA Hotline operates weekdays from 9:00 a.m. to 6:00 p.m., EST, excluding federal holidays. Services are also available in Spanish.

**Envirofacts Master Chemical Integrator (EMCI)**
Address: Web site:
www.epa.gov/enviro/html/emci/chemref/
Lists chemicals monitored by the EPA major programs: Air (AFS), Water (PCS), Hazardous Waste (RCRIS), Superfund (CERCLIS), and Toxic Release Inventory (TRIS).

**Environmental RouteNet**
Address: Cambridge Scientific Abstracts
7200 Wisconsin Ave., Ste. 601 NW, Bethesda, MD 20814, USA
Phone: +1 301 961-6700; 800 843-7751
Fax: +1 301 961-6720
E-mail: fred@csa.com
Web site: www.csa.com/routenet
Environmental RouteNet provides a single gateway to the world's foremost databases and information sources available on the Internet. The service includes searchable links to hundreds of carefully-screened environmentally-related resources, selected and indexed by the editors at Cambridge Scientific Abstracts. In addition, the site provides access to proprietary environmentally-related databases and to daily updates of environmentally-related news stories, regulations and legislation, plus much, much more.

**European Environmental Information and Observation Network**
Address: Kongens Nytorv 6, DK-1050 Copenhagen K, Denmark
Phone: +45 33 367 100
Fax: +45 333 671 99
Web site: www.eionet.eu.int
EIONET is a collaborative network of the European Environment Agency and its member countries, connecting national agencies in the European Union, European reference centers, and principal organizations. These organizations jointly provide the information that is used for making decisions for improving the state of the environment in Europe and making EU policies more effective. EIONET is both a network of organizations and a electronic network (e-EIONET).

**India Chemical Manufacturers Directory**
Web site:
http://epages.webindia.com/bycategory/chemicals
Links to Indian chemical manufacturers.

Indian Mart
Address: Web site: www.indiamart.com
Indian Chemical Manufacturers and Exporters Directory.

**International Council of Chemical Associations**
Web site: www.icca-chem.org
This site has articles and papers on issues facing the chemical industry throughout the world.

**Inventory of Information Sources on Chemicals**
Address: UNEP Chemicals
11-13, chemin des Anemones, 1219 chatelaine, Geneva, Switzerland
Phone: +41 22 917 81 11
Fax: +41 22 797 34 60
E-mail: chemicals@unep.ch
Web site: www.chem.unep.ch/irptc
Acts as a central point from which the existence and whereabouts of international chemical information can be obtained.

**Integrated Risk Information Hotline**
Address: Office of Research and Development, U.S. Environmental Protection Agency
26 W. Martin Luther King Dr., Cincinnati, OH 45268, USA
Phone: +1 513 569-7254
Hotline: +1 513 569-7159
E-mail: Hotline.IRIS@epa.gov
Web site: www.epa.gov/iris
Provides information on how levels of exposure of hazardous chemicals affect human health. Covers levels of exposure to hazardous chemicals below which no adverse health effects are expected to occur in various segments of

the human population. The reference dose and carcinogenicity assessments on IRIS can serve as guides in evaluating potential health hazards and selecting response to alleviate a potential risk to human health. Hours: 8:00 a.m. to 4:40 p.m. EST, Monday - Friday.

**Malaysia Company Directory**
Web site: www.malaysiacompany.com
A directory of Malaysia industry.

**MBendi Information for Africa**
Web site: http://mbendi.co.za
National overviews and product finder for the African chemical industry. Includes links to chemical companies.

**MDL Information Systems, Inc.**
Address: 14600 Catalina St., San Leandro, CA 94577, USA
Phone: +1 510 895-1313
Fax: +1 510 614-3608
E-mail: info@mdli.com
Web site: www.mdli.com
Provides information for the life science and chemical industries, including an enterprise framework for identifying successful new products.

**MEDLINE**
Address: National Center for Biotechnology Information, National Library of Medicine
Bldg. 38A, Rm. 8N805, Bethesda, MD 20894, USA
Phone: +1 301 496-2475; 800 272-4787
Fax: +1 301 480-9241; +1 301 496-0822
Web site: www.nlm.nih.gov
MEDLINE is the National Library of Medicine's premier bibliographic database covering the fields of medicine, nursing, dentistry, veterinary medicine, the health care system, and the preclinical sciences. MEDLINE contains bibliographic citations and author abstracts from more than 4,000 biomedical journals published in the United States and 70 other countries. The file contains over 11 million citations dating back to the mid-1960's. Coverage is worldwide, but most records are from English-language sources or have English abstracts. Especially useful for researching toxicology of a particular chemical; for biological monitoring of a particular chemical; for linking a disease to chemical exposure, and similar medical and health matters. PubMed is the electronic service feature of MEDLINE, linking users directly to publishers' sites and full texts.

**National Pesticide Information Center**
Address: Oregon State University, Department of Agricultural Chemistry Extension
Weniger, Room 333, Corvallis, OR 97331-6502, USA
Hotline: +1 800 858-7378 (general public) or 800 858-7377 (medical and government personnel)
Fax: +1 503 737-0761
E-mail: nptn@npic.orst.edu
Web site: http://npic.orst.edu
Provides access to detailed information on all categories of pesticides including herbicides, fungicides, insecticides, and rodenticides. Included is information on pesticide toxicity, health effects, residual data, efficacy, and other information. NPIC is a cooperative effort of the U.S. EPA and the Oregon State University Department of Agricultural Chemistry. NPIC is staffed from 6:30 a.m to 4:30 p.m, Pacific Standard Time.

**National Response Center**
Address: U.S. Coast Guard Headquarters
2100 2nd Street, S.W., Room 2611, Washington, DC 20593, USA
Hotline: +1 800 424-8802
Fax: +1 202-267-2675
Web site: www.nrc.uscg.mil
The federal government's national communication center that receives all reports of releases involving hazardous substances and oil. Hours: 24 hours, everyday.

**National Technical Information Service Information Center**
Address: U.S. Department of Commerce
5285 Port Royal Road, Springfield, VA 22161, USA
Phone: +1 703 605-6000
Fax: +1 703 605-6900
Web site: www.ntis.gov
The federal government's central source for the sale of scientific, technical, engineering and related business information produced by or for the U.S. government. Information on more than 600,000 information products covering over 350 subject areas from more than 200 federal agencies.

**New Jersey Right to Know Hazardous Substance Fact Sheets**
Address: New Jersey Department of Health and Senior Services, Division of Epidemiology, Environmental and Occupational Health
Right to Know Program
P.O. Box 360, Trenton, NJ 08625-0360, USA
Phone: +1 609-984-1863
Fax: +1 609-292-5677
Web site:
www.state.nj.us/health/eoh/rtkweb/rtkhsfs.htm#A
This agency produces "Right to Know Hazardous Substance Fact Sheets" that summarize information from many sources. Industrial hygienists are available to answer questions regarding the control of chemical exposures.

**NIOSH Technical Information Center Database - NIOSHTIC**
Address: Standards Development and Technology Transfer Division, Technical Information Branch
U.S. National Institute for Occupational Safety and Health
4676 Columbia Pkwy., Cincinnati, OH 45226-1998, USA
Phone: +1 513 533-8328; 1+800 356-4674
Fax: +1 513 533-8573

E-mail: pubstaff@cdc.gov
Web site: www.cdc.gov/niosh/nioshtic.html
The electronic, bibliographic database of more than 2,000 journals and 70,000 monographs on all aspects of occupational safety and health. Available through DIALOG Information Services, MEDLARS as part of TOXLINE, Silver Platter and other commercial systems.

**Online Library System**
Address: Public Information Center, U.S. Environmental Protection Agency
406 M. St., SW, Washington, DC 20460, USA
Phone: +1 919 541-7862; 800 334-2405
Fax: +1 202 260-6257
E-mail: public-access@epamail.epa.gov
Web site: www.epa.gov/natlibra/ols.htm
OLS is the Online Library System for the Library Network of the United States Environmental Protection Agency. It consists of several related databases that can be used to locate books, reports, and audiovisual materials on a variety of topics.

**Pesticides and Foods - Residue Limits on Foods**
Address: U.S. Environmental Protection Agency
401 M St., SW, Washington, DC 20460, USA
Web-site: www.epa.gov/pesticides/food/viewtols.htm
EPA database of tolerance limits for pesticide residue on foods. The EPA tolerance database is being updated; meanwhile this site leads to 40 CFR 180 published July 1, 2004. See above: Compliance Resource Center for CFR Information.

**Pollution Prevention Information Clearinghouse**
Address: Office of Pollution Prevention, Pesticides and Toxic Substances Chemical Library (OPPTS Chemical Library)
U. S. Environmental Protection Agency
401 M St., SW, Washington, DC 20460, USA
Hotline: +1 202 260-1023
Fax: +1 202 260-4659
E-mail: ppic@epa.gov
Web site: www.epa.gov/oppt/library/ppicindex.htm
A free, non-regulatory service of the EPA dedicated to reducing or eliminating industrial pollutants through technology transfer, education, and public awareness. Hours: 8:30 a.m. - 4:00 p.m. EST; 24-hour-a-day voice mail.

**RCRA, Superfund & EPCRA Hotline**
Address: Office of Solid Waste
U. S. Environmental Protection Agency
Hotline: +1 800 424-9346; +1 703 412-9810
E-mail: tri.us@epamail.gov
Web site: www.epa.gov/epaoswer/hotline
Provides information about all RCRA regulations and programs including the Resource Conservation and Recovery Act (RCRA); Comprehensive Environmental Response Compensation and Liability Act (CERCLA, or Superfund); and Emergency Planning and Community Right-to-Know Act (EPCRA)/Superfund Amendments Reauthorization Act
(SARA) Title III. Operates weekdays from 9:00 a.m. to 6:00 p.m., EST, excluding federal holidays. Services are also available in Spanish.

**Registry of Toxic Effects of Chemical Substances**
Address: Standards Development and Technology Transfer Division, Technical Information Branch
U.S. National Institute for Occupational Safety and Health
4676 Columbia Pkwy., Cincinnati, OH 45226-1998, USA
Phone: +513 533-8328; +1 800 356-4674
Fax: +1 513 533-8573
E-mail: pubstaff@cdc.gov
Web site: www.cdc.gov/niosh/rtecs.html
RTECS provides toxicology data for more than 130,000 chemicals in four categories: substance identification; toxicity/biomedical effects; toxicology and carcinogenicity review; and exposure standards and regulations. Built and maintained by the National Institute for Occupational Safety and Health (NIOSH) as a segment of the Toxic Release Inventory database. Available online, CD-ROM and computer tape from NIOSH and from commercial database vendors.

**Safe Drinking Water Data & Databases**
Address: Office of Water (OW), Ground Water and Drinking Water
U.S. Environmental Protection Agency
401 M St., SW, Washington, DC 20460, USA
Hotline: +1 703 285-1098; +1 800 426-4791
Fax: +1 703 285-1101
E-mail: hotline-sdwa@epamail.epa.gov
Web site: www.epa.gov/safewater/databases.html
Provides information about drinking water regulations and related topics. Hours: 9:00 a.m. through 5:30 p.m. EST, weekdays except federal holidays.

**TOMES PLUS Information System.**
Address: Micromedex Inc
6200 S. Syracuse Way, Ste. 300, Greenwood Village, CO 80111-4740, USA
Phone: +1 303 486-6400; 800 525-9083
Fax: +1 303 486-6464
E-mail: info@mdx.com
Web site: www.micromedex.com
Contains a collection of proprietary and government databases on toxicology, environment, and industrial medicine and the hazardous effects related to chemical exposure in manufacturing and transportation, and gives the proper response in chemical emergency situations. Contains several proprietary files as well as files of major international journals, standard reference sources, professional specialists, chemical manufacturers, government agencies, and poison centers. Databases include Hazardtext Hazards Management, SARAtext System, HSDB (Hazardous Substance Data Bank) from NLM, CHRIS from the U.S. Coast Guard, IRIS, and New Jersey Fact Sheets.

**Toxic Substances Control Act (TSCA) Hotline**
Address: TSCA Assistance Information Service,
Office of Prevention, Pesticides, and Toxic Substances (OPPTS), Office of Pollution Prevention and Toxics,
U. S. Environmental Protection Agency
Phone: +1 202 554-1404; TDD 202 554-0551
Fax: +1 202 554-5603
E-mail: tsca-hotline@epa.gov
Web site: www.epa.gov/Region5/defs/html/tsca.htm
Operating under contract to EPA, the TCSA Hotline provides technical assistance and information about programs under the Toxic Substances Control Act (TSCA), including the Asbestos School Hazard Abatement Act, the Asbestos Hazard Emergency Response Act, and the Lead Exposure Reduction Act. Hours: 8:30 a.m. - 5:00 p.m. EST, weekdays.

TOXNET - Toxicology Data Network
Address: National Library of Medicine, Specialized Information Services
8600 Rockville Pike, Bethesda, MD 20894, USA
Phone: +1 800 272-4787
Fax: +1 301 480-3537
E-mail: tehip@teh.nlm.nih.gov
Web site: http://toxnet.nlm.nih.gov
A cluster of databases on toxicology, hazardous chemicals and related areas, i.e. the pharmacological, biochemical, physiological, and toxicological effects of drugs and other chemicals.

**University of Iowa Center for Global and Regional Environmental Research**
Address: University of Iowa, 204 IATL, Iowa City, IA 52242, USA
E-mail: webadm@cgrer.uiowa.edu
Web site: www.cgrer.uiowa.edu
Has a world-wide database of federal and non-profit environmental research organizations that focus on the multiple aspects of global environmental change, including the regional effects on natural ecosystems, environments and resources as well as on human health, culture and social systems.

**U.S. Federal Government Agencies Directory**
Web site: www.lib.lsu.edu/gov/fedgov
A list of federal agencies on the Internet. Maintained by Louisiana State University.

**Vermont SIRI - Safety Information Resources, Inc.**
Address: University of Vermont
E-mail: dan@siri.org
Web site: http://siri.uvm.edu/msds
An exhaustive source of environmental health and occupational safety information: MSDSs; links to other MSDS sites; links to safety sites on the Internet; links to occupational safety and health and environmental organizations and research programs; OSHA and EPA regulations; NIOSH databases; discussion boards; and much more.

**Yahoo Chemistry Web Directory**
Web site: http://dir.yahoo.com/science/chemistry
An exhaustive directory of resources on the web for chemists in varied fields: industrial, research, and academia.

# Appendix G: Agrochemical Web Sites
## Sources of Information about Agrochemicals and Food Safety

**AGRICOLA**
Web site: www.nal.usda.gov/ag98
AGRICOLA (**AGRI**Cultural **OnL**ine **A**ccess) is a bibliographic database of citations to the agricultural literature created by the National Agricultural Library and its cooperators. Production of these records in electronic form began in 1970, but the database covers materials in all formats, including printed works from the 15th century. The records describe publications and resources encompassing all aspects of agriculture and allied disciplines, including animal and veterinary sciences, entomology, plant sciences, forestry, aquaculture and fisheries, farming and farming systems, agricultural economics, extension and education, food and human nutrition, and earth and environmental sciences. Although AGRICOLA does not contain the materials, thousands of AGRICOLA records are linked to online full-text documents, with new links being added every day. AGRICOLA is searchable on the World Wide Web at http://www.nal.usda.gov/ag98. For information on how to obtain library materials from NAL, see NAL's Document Delivery Services
Web site: http://www.nal.usda.gov/ddsb/
The AGRICOLA database is organized into two bibliographic data sets, which must be searched separately. One data set is the Online Public Access Catalog, known as "Books, etc.," that contains citations to books, audiovisual materials, serial publications, and other materials in the NAL collection. AGRICOLA also contains bibliographic records for items cataloged by other libraries and not held in NAL's collection. The other data set is the Journal Article Citation Index, known as "Articles, etc." It includes citations, many with abstracts, to journal articles (see "List of Journals Indexed"), book chapters, reports, and reprints, selected primarily from the materials cataloged in Books, etc. Both data sets are updated daily with newly cataloged and indexed materials.
AGRICOLA is searchable via the Web using the VTLS Web Gateway. For those who prefer a text-oriented interface for searching the two data sets, Telnet access to ISIS (Integrated System for Information Services) is also available. AGRICOLA and ISIS are the same database, but ISIS is accessed from the production side of the system, via the VTLS99 interface. AGRICOLA can be accessed for a fee through several commercial vendors, both online and on CD-ROM. Users can also lease the AGRICOLA file from the National Technical Information Service (NTIS).

**Agriculture Network Information Center (AgNIC)**
Web site: http://www.agnic.org/

**AgSearch - a service of Ceres OnLine**
Web site: www.ceresgroup.com/col/agsearch/index
A service of Ceres Online, AgSearch is a gateway to over seventy search engines within the field of agriculture. Users are provided subject categories within the main menu and also able to specify searches through a series of simple commands. AgSearch provides searches in the following categories: news (publications); research; genetics; production, economics, pest management, Internet and computer subjects; trade; government subjects; and water and irrigation.

**Australia Agrifood Awareness**
Address: P.O. Box E10, Kingston, ACT, 2604, Australia
Phone: +2 6273-9535
Fax: +2 6273-2331
E-mail : info@afaa.com.au
Web site: www.afaa.com.au
Agrifood Awareness Australia is an industry initiative, established to increase public awareness of, and encourage informed debate about, gene technology.

**Australian and Asian Agriculture Links**
Web site: http://www.sanondaoz.com
A directory of more than 125 links to agriculture and pesticide matters in the Pacific rim. Part of the Sanonda Ltd. company web site. Includes links to herbicides, insecticides, crop production and control, literature, weather, and much more.

**California Department of Pesticide Regulation (DPR) USEPA/OPP Database Queries**
Web site: www.cdpr.ca.gov/docs/epa/epamenu
The California Department of Pesticide Regulation works closely with the USEPA Office of Pesticide Programs to develop internet access to data that are of significant value to the general public, chemical and agricultural industries. Brief registration information on approximately 90,000 products is currently online. The data include product number and name, company number and name, registration date, cancellation date, existing stocks date, and reason (if canceled), and product manager name and telephone number.
In addition, OPP's databases contain chemical ingredient and firm information. The chemical data is searchable by common, technical, synonym, CAS number, or trade names. The firm data is searchable by firm number, name, or portions thereof. These chemical, firm, and product databases have complementary links and are searchable by multiple variables. More information about the data and structure of EPA's PPIS can be found at: Pesticide Product Information System
With the advent of online access to PPLS (a collection of product label images in TIF format) has been linked directly to the product reports, enabling a user to look for products by any of the above criteria and then view the EPA Label image. *Note:* these label images do not reflect the California Registered label. For information regarding California registered labels contact CDPR's Label Resource Center at 916-324-0399.

**Chemical Carcinogenicity Research Information System**
Address: National Library of Medicine
8600 Rockville Pike, Bethesda, MD 20894, USA
Phone: +1 800 272-4787
Fax: +1 301 496-0822
Web site: www.toxnet.nlm.nih.gov/cgi-bin/sis/htmlgen?ccris
A scientifically evaluated and fully referenced data bank, developed and maintained by the National Cancer Institute (NCI). It contains some 8,000 chemical records with carcinogenicity, mutagenicity, tumor promotion, and tumor inhibition test results. Data are derived from studies cited in primary journals, current awareness tools, NCI reports, and other special sources. Test results have been reviewed by experts in carcinogenesis and mutagenesis.

**Chemical Contaminants in Food**
Web site: www.who.int/fsf/Chemicalcontaminants/index2.htm
An excellent entry point to world-wide information about food safety. Includes access to GEMS/Food - the Global Environment Monitoring system/Food Contamination Monitoring and Assessment Programme. Includes links to documents and position papers on chemical hazards in the food supply chain, monitoring chemical contaminants, acute hazards exposure assessments, mycotoxins, and industrial and environmental contaminants.

**Compliance Resource Center for CFR Information**
Web site: www.setonresourcecenter.com/40CFR/Docs/wcd00000/wcd0008d.asp
A quick-find instrument to CFR titles 29 (OSHA), 40 and 49 (Transportation). Its index to 180 CFR 180 is an easy-to-use reference to "Tolerances and Exemptions from Tolerances for Pesticide Chemicals in Food." Presented by the Seton Company. A similar web site is maintained by the EPA but not in as concise format. The EPA site can be located at http://www.epa.gov/pesticides/food/viewtols.htm

**ENVIRO-ONE - Environmental News Information Service**
Web site: http://enviroone.com/searchresults.php?searchtext=pesticides
A nifty source for all things related to environmental information. The Pesticide site, alone, lists 109 resources for U.S. federal and state statistics, foreign information, a many, many resources.

**EPA Envirofacts Warehouse**
Web site: www.epa.gov/enviro
Public access to databases maintained by the U.S. Environmental Protection Agency. Databases such as Air, Chemicals, Facility Information, Releases, Water Permits, Drinking Water Contaminant Occurrence, Maps, and more.

**EXTOXNET - Extension Toxicology Network**
Web site: http://extoxnet.orst.edu/
This is a comprehensive entry gate for information on individual pesticides and data sources. EXTOXNET is a cooperative effort of University of California-Davis, Oregon State University, Michigan State University, Cornell University, and the University of Idaho. Primary files are maintained and archived at Oregon State University.

**FASonline**
International Maximum Residue Limit Database
USDA Foreign Agricultural Service
Web site: www.mrldatabase.com
This residue data base allows one to search for residue limits in nearly 90 countries. Searches can be made by specifying the commodity, pesticide, or pesticide type. It is managed by the Foreign Agricultural Service. The FAS web site provides resources of agricultural data bases and search engines, export data, and reports, statistics and contact information.

**Food Safety Web Site**
Web site: www.foodsafety.gov
The gateway to food safety information. Managed by the U.S. Department of Agriculture. Topics include consumer safety alerts and advice, industry assistance, foodborne pathogens, chemical contaminants, and food additives. This is the place to report illnesses and product complaints.

**National Agricultural Library**
Web site: www.nal.usda.gov
The National Agricultural Library is one of four National Libraries in the United States. NAL is a major international source for agriculture and related information. This Web site provides access to NAL's many resources and a gateway to its associated institutions.

**National Agriculture Safety Database (NASD)**
Address: Robert A. Taft Laboratories, 4676 Columbia Pkwy., Cincinnati, OH 45226, USA
Phone: +1 800-35-NIOSH; +1 800-356-4674
Web site: www.cdc.gov/nasd
The National Agriculture Safety Database (NASD) is a collection of information about health, safety and injury prevention in agriculture. The information in the database was contributed by safety professionals and organizations from across the nation in an effort to promote safety in agriculture. The database contains agriculture health and safety publications from 32 states, 4 federal agencies and 5 national organizations. The collection includes OSHA and EPA Standards; extension publications; abstracts and ordering information for agriculture safety-related videos; a NIOSH bibliography with abstracts for over 500 scientific publications; training materials; posters; sample new releases; and public service announcement scripts. Materials are categorized into topical, organizational, language, and state menus. Information in the database can be accessed on-screen and/or printed on demand.

**National Integrated Pest Management Centers Information System, U.S. Department of Agriculture**
Web site: http://www.ipmcenters.org
The National IPM Network is a cooperating group of universities, government agencies, and other organizations that provides up-to-date information for pest management. There are four regional centers whose purpose are to identify, prioritize and coordinate a national pest management research, extension, and education program that is implemented on a regional basis.

**National Library of Medicine, Specialized Information Services**
Web site: http://sis.nlm.nih.gov
A super list of world-wide sources of information about drugs, pesticides, environmental pollutants and other potential toxins. Ties into ASDTR, EPA, the National Toxicology Program Carcinogen List (NTPA), and other governmental sources.

**National Pesticide Information Center**
Address: Oregon State University
Department of Agricultural Chemistry Extension
Weniger, Room 333, Corvallis, OR 97331-6502, USA
Hotline: +1 800 858-7378 (general public) or 800 858-7377 (medical and government personnel)
Fax: +1 503 737-0761
E-mail: nptn@npic.orst.edu
Web site: http://npic.orst.edu
Provides access to detailed information on all categories of pesticides including herbicides, fungicides, insecticides, and rodenticides. Included is information on pesticide toxicity, health effects, residual data, efficacy, and other information. NPIC is a cooperative effort of the U.S. EPA and the Oregon State University Department of Agricultural Chemistry. NPIC is staffed from 6:30 a.m to 4:30 p.m, Pacific Standard Time.

**National Pesticide Information Retrieval System**
Address: Center for Environmental and Regulatory Information Systems (CERIS)
Purdue University
1231 Cumberland Ave., Ste. A, West Lafayette, IN 47907, USA
Phone: +1 765 494-5249
Fax: +1 765 494-0535
E-mail: vcassens@purdue.edu
Web site: www.ceris.purdue.edu/npirs
A collection of pesticide-related databases available by subscription which provides current information on products registered with the EPA and tolerance data for pesticides and hazardous chemicals. It also provides state registration information, registration guideline information, and descriptions of studies on pesticides and hazardous chemicals. It is administered by CERIS at Purdue University.

**Office of Pesticide Programs**
Address: U. S. Environmental Protection Agency
Ariel Rios Building, 1200 Pennsylvania Ave., N.W., Washington, D.C. 20460, USA
Phone: +1 703-305-5805
Web site: www.epa.gov/pesticides
This site leads you into everything you need to know about the federal pesticide programs, including notices about open comment periods, various pesticide uses, restricted use products, labeling, sources of information, safety programs, tolerance index files, and much more.

**Office of Pesticide Programs, Pesticide Residue Limits**
Web site: www.epa.gov/pesticides/food/viewtols.htm
EPA sets limits on how much of a pesticide residue can remain on food. These pesticide residue limits are known as tolerances. Inspectors from the Food and Drug Administration and the United States Department of Agriculture monitor food in interstate commerce to ensure that these limits are not exceeded. The site can be searched by crop or by chemical.

**Pesticide Action Network (PAN); Pesticide Action Network North America (PANNA)**
Address: 49 Powell St., Ste. 500, San Francisco, CA 94102, USA
Phone: +1 415-981-1771
Fax: +1 415-981-1991
E-mail: panna@panna.org
Web site: www.pesticideinfo.org
Pesticide Action Network North America is the regional center for PAN International. PANNA has campaigned to replace pesticides with ecologically sound alternatives since 1982. The PAN Pesticide Database brings together a diverse array of information on pesticides from many different sources, providing human toxicity (chronic and acute), ecotoxicity and regulatory information for about 5,400 pesticide active ingredients and their transformation products, as well as adjuvants and solvents used in pesticide products. This database of active ingredients has been integrated with the U.S. EPA product databases, which provide information on formulated products (the form of the pesticide that growers and consumers purchase for use) containing the active ingredients. The information is most complete for pesticides registered for use in the United States and provides a wealth of information useful to anyone in the industry. There are five regional networks which can be used for more specific information world-wide: PAN North America, PAN Europe, PAN Asia and the Pacific, PAN Africa, and PAN Latin America.

**Pesticide.net**
Address: Wright & Sielaty, P.C.
1990 Old Bridge Rd., Ste. 202, Lake Ridge, VA 22192, USA
Phone: +1 703-492-0055
Fax: +1 703-492-0066
Web site: www.pestlaw.com

A comprehensive source of pesticide related news, laws, and regulatory information. Full text documents and other resources. Subscription needed for full access.

**Pesticides and Food: Pesticide Residue Limits**
Web site:
http://www.epa.gov/pesticides/food/viewtols.htm
Pesticide residue tolerances limits for crops as reported by the EPA with references to EPA data bases and the Code of federal Regulations.

**Pesticide Product Information System**
Address: Office of Pesticide Programs
U. S. Environmental Protection Agency
Ariel Rios Building, 1200 Pennsylvania Ave., N.W., Washington, D.C. 20460, USA
Phone: +1 703-305-5805
Web site: www.epa.gov/opppmsd1/PPISdata
The Pesticide Product Information System contains information concerning all pesticide products registered in the United States. The files located in this download area are presented in ASCII to enable interested parties to access them using a variety of database and spreadsheet software.

**Specific Pesticides Fact Sheets**
Web site:
http://www.epa.gov/pesticides/factsheets/chemical_fs.htm
Each Fact Sheet, prepared by the EPA, contains information about pesticides such as their physical properties, how they are used, scientific findings and pertinent regulatory activity. This site will also lead the user to a broader spectrum of regulatory information. They are most comprehensive and should be incorporated in any pesticide investigation.

**Toxicology Information Response Center**
Address: Toxicology and Risk Assessment Section (TARA)
Oak Ridge National Laboratory
1060 Commerce Park, MS 6480, Oak Ridge, TN 37830, USA
Phone: +1 865 576-1746
Fax: +1 865 574-9888
Web site:
www.ornl.gov/sci/techresources/tirc/hmepg.shtml
Offers direct access to virtually all of the world's scientific and data bases for toxicology and related information. Covers chemicals, pesticides, food additives, industrial chemicals, heavy metals, environmental pollutants, and pharmaceuticals. The Center is online to more than 400 computerized databases, including DIALOG, MEDLARS, STN International, ITIS, and DROLS. It performs searches for outside users for a fee.

**Tolerance Reassessment & Reregistration, EPA**
Web site: www.epa.gov/oppsrrd1/reregistration
EPA is reviewing older pesticides (those initially registered prior to November 1984) under the federal Insecticide, Fungicide, and Rodenticide Act to ensure that they meet current scientific and regulatory standards. This process, called reregistration, considers the human health and ecological effects of pesticides and results in actions to reduce risks that are of concern. EPA also is reassessing tolerances (pesticide residue limits in food) to ensure that they meet the safety standard established by the Food Quality Protection Act of 1996. EPA has integrated reregistration and tolerance reassessment to most effectively accomplish the goals of both programs. They reregistration eligibility statements are the most comprehensive studies available from the EPA.

**University of Florida Institute of Food and Agricultural Science (AGRIGATOR)**
Web site: http://agrigator.ifas.ufl.edu/

**Web-agri**
Address: Hyltel Multimedia
12a Rue de Brest, Rennes 35000, France
Web site: www.web-agri.com
This is a French site that offers an agricultural search engine proporting to search 859,512 agricultural web pages.

**World Health Organization Food Safety Page**
Web site: http://www.who.int/fsf
An excellent entry point to world-wide information about food safety. Includes links to documents and position papers on chemical hazards in the food supply chain, monitoring chemical contaminants, acute hazards exposure assessments, mycotoxins, and industrial and environmental contaminants.

# Index 1: Synonym and Trade Name-Cross Index

4-WAY® see...Etridiazole
6Q8® see...Naptalam
40 SD® see...Isofenphos
88-R® see...Aramite
20 MULE TEAM® see...Sodium Tetraborate
20 MULE TEAM HIBOR® see...Sodium Chlorate
60-CS-16® see...Chlormequat Chloride
75 SP® see...Acephate
415 Oil® see...Naphthas
435 Oil® see...Naphthas
869® see...Metham-Sodium
1080 see...Sodium Fluoroacetate
3336 TURF FUNGICIDE® see...Thiophanate-Methyl
8056 HC® see...Methyl Parathion
8057 HC® see...Fenitrothion
11561 RP® see...Carbetamide
27165® see...Temephos
32545 R® see...Fosetyl-Al
330541® see...Diuron

- A -

A-42® see...Aspon®
A-361 see...Atrazine
A-820 see...Butralin
A-980 see...Barban
A-1114 see...Prometryn
A-1866 see...Terbutryn
A-2079 see...Simazine
A-7881 see...Ethametsulfuron-methyl
A13-09232 see...Hexachlorocyclohexanes
A13-25606 see...Oxythioquinox
A13-25812 see...Tribufos
A13-27093 see...Aldicarb
A13-27164 see...Carbofuran
A13-27967 see...Amitraz
A13-29235 see...Fenvalerate
A13-29236 see...Profenfos
A13-60366 see...Prometryn
72-A34® see...Butralin
A7-VAPAM® see...Metham-Sodium
A815 see...Triflumizole
AA see...Allyl Alcohol
AACAPTAN® see...Captan
AADIBROOM® see...Ethylene Dibromide
AAF see...Acetylaminofluorene
2-AAF see...Acetylaminofluorene
AAFERTIS® see...Ferbam
AAHEPTA® see...Heptachlor
AALINDAN® see...Lindane
AAM see...Acrylamide
AAMANGAN® see...Maneb
AAPIROL® see...Thiram
AAPROTECT® see...Ziram
AASTAR see...Flucythrinate

AASTAR® see...Phorate
AAT see...Parathion
AATACK® see...Thiram
AATERRA® see...Etridiazole
AATIRAM® see...Thiram
AATOX® see...Dinoseb
AATP see...Parathion
AATRAM® see...Propachlor
AATRAM® see...Atrazine
AATREX® see...Atrazine
AAVOLEX® see...Ziram
AAZDIENO® see...Amitraz
AAZIRA® see...Ziram
ABACIDE® see...Abamectin
ABACOL® see...Carbendazim
ABAR® see...Leptophos
ABAT® see...Temephos
ABATE® see...Temephos
ABATE® 1-SG see...Temephos
ABATE® 2-CG see...Temephos
ABATE® 4-E see...Temephos
ABATE® 5-CG see...Temephos
ABATHION® see...Temephos
ABG-3034® see...6-Benzaldenine
ABG-3097® see...Aminoethoxyvinylglycine Hydrochloride
ABG-6215® see...Fenoxycarb
ABMINTHIC® see...Dithiazanine Iodide
ABOL® see...Pirimicarb
ABORTRINE® see...Benomyl
ABOUND® see...Azaxystrobin
ABSTENSIL® see...Disulfiram
ABSTINYL® see...Disulfiram
AC-293® see...Imazethabenz
AC-263499® see...Imazethapyr
AC 528® see...Dioxathion
AC 3422® see...Ethion
AC 3911® see...Phorate
AC 5223® see...Dodine
AC 18133® see...Thionazin
AC 18682® see...Prothoate
AC 18737® see...Endothion
AC 22234® see...Diethatyl-ethyl
AC 38023® see...Famphur
AC 38555® see...Chlormequat Chloride
AC 47031® see...Phosfolan
AC 47470® see...Mephosfolan
AC 52160® see...Temephos
AC 64475® see...Fosthietan
AC 84777® see...Difenzoquat
AC 92553® see...Pendimethalin
AC 217300® see...Hydramethylnon
AC 222293® see...Imazethabenz
AC 222705® see...Flucythrinate
AC 303630® see...Chlorfenapyr

AC 513995 DG® see...Sodium Dicamba
AC 921000® see...Terbufos
ACADREX® see...Amitraz
ACARABEN 4E® see...Chlorobenzilate
ACARABEN® see...Chlorobenzilate
ACARACIDE® see...Aramite
ACARAC® see...Amitraz
ACARFLOR® see...Hexythiazox
ACARIFLOR® see...Hexythiazox
ACARIN® see...Dicofol
ACARITHION® see...Carbophanothion
ACAROIL TD® see...Tetradifon
ACARON® see...Chlordimeform
ACARSTIN® see...Cyhexatin
ACARVIN® see...Tetradifon
ACAR® see...Chlorobenzilate
ACC 3422® see...Parathion
ACCEL 22® see...Ethylene Thiourea
ACCEL TMT® see...Thiram
ACCELERATE® see...Endothall
ACCELERATOR THIURAM® see...Thiram
ACCELERATOR T® see...Thiram
ACCELERATOR® -L see...Ziram
ACCELERATOR® MZ® see...Ziram
ACCEL® see...6-Benzaldenine
ACCENT® see...Flumetsulam
ACCENT® see...Nicosulfuron
ACCENT® see...Clopyralid
ACCESS® see...Triclopry
ACCESS® see...Picloram
ACCLAIM® see...Fenoxaprop-ethyl
ACCOMPLISH® see...Dodecylbenzenesulfonic Acid
ACCONEM® see...Fosthietan
ACCORD® see...Glyphosate
ACCOST® see...Triadimefon
ACCOTAB® see...Pendimethalin
ACCOTHION® see...Fenitrothion
ACCUSPIN ASX-10 SPIN-ON DOPANT® see..Arsenic and Inorganic Arsenic Compounds
(Aceato)phenylmercury see...Phenylmercury Acetate
ACECAP SYSTEMIC INSECTICIDE IMPLANTS® see. Acephate
ACEFAL 75 PS® see...Acephate
ACEHERO® see...Acephate
Acenafeno (Spanish) see...Acenaphthene
Acenaphthylene, 1,2-Dihydro- see...Acenaphthene
ACENIT® see...Acetochlor
ACEOTHION® see...Fenitrothion
Acephat (German) see...Acephate
ACEPHATE 97 EG® see...Acephate
ACEPHATE PCO SP INSECTICIDE® see...Acephate
ACEPHATE-MET® see...Methamidophos
ACEPHATE 75SP® see...Acephate
ACESUL® see...Acephate
Acetaldehyde, chloro- see...Chloroacetaldehyde
Acetaldehyde, tetramer see...Metaldehyde
Acetamide, 2-(2-benzothiazolyloxy)-N-methyl-N-phenyl- see...Mefenacet
Acetamide, N-(butoxymethyl)-2-chloro-N-(2,6-diethylphenyl)- see...Butachlor
Acetamide, 2-chloro-n-(2,6-diethylphenyl)-N-(Methoxymethyl)- see...Alachlor
Acetamide, 2-chloro-N,N-di-2-propenyl- see..Allidochlor
Acetamide, 2-chloro-N-(ethoxymethyl)-N-(2-ethyl-6-methylphenyl)- see...Acetochlor
Acetamide, 2-chloro-N-(1-methylethyl)-N-phenyl- see. Propachlor
Acetamide, 2-chloro-N-isopropyl- see...Propachlor
Acetamide, 2-chloro-N-(2,4-dimethyl-3-thienyl)-N-(2-methoxy-1-methylethyl)- see...Dimethenamid
Acetamide, 2-chloro-N-(2,4-dimethyl-3-thienyl)-N-(2-methoxy-1-methylethyl)- see...Dimethenamid
Acetamide, 2-cyano-N-[(ethylamino)carbonyl]-2-(methoxyimino)- see...Cymoxanil
Acetamide, N,N-dimethyl-2,2-diphenyl- see. Diphenamid
Acetamide, N-(2,4-dimethyl-5-[((trifluoromethyl)sulfonyl)amimo)phenyl]- see. Mefluidide
Acetamide,N-fluoren-2-yl- see...Acetylaminofluorene
Acetamide,N-9H-fluoren-2-yl see...Acetylaminofluorene
Acetamide, N-(4-fluorophenyl)-N-(1-methylethyl)-2-[(5-(trifluoromethyl)-1,3,4-thiadiazol-2-yl)oxy]- see.. Thiafluamide
5-Acetamido-2,4-dimethyltrifluoromethanesulfonanilide see...Mefluidide
2-Acetamidofluorene see...Acetylaminofluorene
Acetanilide, 2-chloro-2',6'-diethyl-N-(butoxymethyl)- see. Butachlor
Acetanilide, 2-chloro-2',6'-diethyl-N-methoxymethyl)- see...Alachlor
Acetate de cuivre (French) see...Cupric Acetate
Acetate phenylmercurique (French) see...Phenylmercury Acetate
Acetato de cobre (Spanish) see...Cupric Acetate
Acetate de triphenyl-etain (French) see...Triphenyltin Compounds
Acetato fenilmercurio (Spanish) see...Phenylmercury Acetate
Acetato(2-methoxyethyl)mercury see. Methoxyethylmercuric Acetate
Acetato(trimetaarsenito)dicopper see...Copper Acetoarsenite
Acetene see...Ethylene
Acetic acid see...Dichloroacetic Acid
Acetic acid, (2,4,5-t)- see...2,4,5-T
Acetic acid, [(4-amino-3,5-dichloro-6-fluoro-2-pyridinyl)oxy]-,1-methylheptyl ester, see...Fluroxypyr 1-methylheptyl Ester
Acetic acid, (2,4,5-t)-, butyl ester see...2,4,5-Trichlorophenoxyacetic Acid, Esters
Acetic acid, chloro- see...Chloroacetic Acid
Acetic acid (4-chloro-2-methylphenoxy)- see...MCPA
Acetic acid [(4-chloro-o-tolyl)-oxy]- see...MCPA
Acetic acid, copper(II) salt see..Cupric Acetate
Acetic acid, copper(2+) salt see...Cupric Acetate
Acetic acid, cupric salt see...Cupric Acetate

Acetic acid, dichloro- see...Dichloroacetic Acid
Acetic acid (2,4-dichlorophenoxy)- see...2,4-D
Acetic acid, (2,4-dichlorophenoxy)-, 2-butoxyethyl ester see...2,4-D, butoxyethyl ester
Acetic acid, (2,4-dichlorophenoxy)-, isopropyl ester see. 2,4-D, isopropyl ester
Acetic acid, (2,4-dichlorophenoxy)-,1-methylethyl ester see...2,4-D, isopropyl ester
Acetic acid, [(1-((dimethylamino)carbonyl)-3-(1,1-dimethylethyl)-1*H*-1,2,4-triazol-5-yl)thio]-, ethyl ester see...Triazamate
Acetic acid, *O,O*-dimethyldithiophosphoryl-, *N*-monomethylamide salt see...Dimethoate
Acetic acid, (*O,O*-dimethyldithiophosphorylphenyl)-, ethyl ester see...Phenthoate
Acetic acid, diphenyl-, 2-fluoroethyl ester see...Fluenetil
Acetic acid, fluoro-, sodium salt see...Sodium Fluoroacetate
Acetic acid, (2-naphthyloxy)- see...Naphthoxyacetic Acid
Acetic acid, phenylmercury derivitive see. .Phenylmercury Acetate
Acetic acid, sodium salt (2:1) see...Sodium Diacetate
Acetic acid, (2,4,5-trichlorophenoxy)- see...2,4,5-T
Acetic acid, (2,4,5-trichlorophenoxy)-, 2-butoxyethyl ester see...2,4,5-Trichlorophenoxyacetic Acid, Esters
Acetic acid, (2,4,5-trichlorophenoxy)-, butyl ester see. 2,4,5-Trichlorophenoxyacetic Acid, Esters
Acetic acid, (2,4,5-trichlorophenoxy)-, 2-ethylhexyl ester see...2,4,5-Trichlorophenoxyacetic Acid, Esters
Acetic acid, (2,4,5-trichlorophenoxy)-, isooctyl ester see. 2,4,5-Trichlorophenoxyacetic Acid, Esters
Acetic acid, (2,4,5-trichlorophenoxy)-, isopropyl ester see...2,4,5-Trichlorophenoxyacetic Acid, Esters
Acetic acid, (2,4,5-trichlorophenoxy)-, 1-methyl propyl ester, 2,4,5-Trichlorophenoxyacetic Acid, Esters
Acetic acid, [(3,5,6-trichloro-2-pyridinyl)oxy]- see. . Triclopry
Acetic acid, [(3,5,6-trichloro-2-pyridinyl)oxy]-, compounded with *N,N*-diethylethanamine (1:1) see. Triclopyr, Triethylammonium Salt
Acetic peroxide see...Peracetic Acid
Acetimidic acid, thio-*N*-(Mmthylcarbamoyl)oxy-,methyl ester see...Methomyl
Acetimidothioic acid, methyl-*N*-(methylcarbamoyl) ester see...Methomyl
Acetimidoylphosphoramidothioic acid *O,O*-bis(*p*-chlorophenyl)ester see...Phosacetim
Aceto de *N*-dodecilguanidina (Spanish) see...Dodine
3-(α-Acetonylbenzyl)-4-hydroxycoumarin see. .Warfarin
3-(α-Acetonylbenzyl)-4-hydroxy-coumarin sodium salt see...Warfarin
3-(α-Acetonylfurfuryl)-4-hydroxycoumarin see. . Coumafuryl
Aceto SDD 40 see...Sodium Dimethyldithiocarbamate
Aceto TETD see...Thiram
Aceto di stagno trifenile (Italian) see...Triphenyltin Compounds
(Aceto)(trimrtaarsenito)dicopper see...Paris Green
ACE-TOX® see...Acephate
ACETO ZDED® see...Ziram
ACETO ZDMD® see...Ziram
Acetoarsenite de cuivre (French) see...Paris Green
Acetoarsenite de cuivre (French) see...Copper Acetoarsenite
Acetoarsenito de cobre (Spanish) see...Copper Acetoarsenite
Acetoarsenito de cobre (Spanish) see...Paris Green
*O*-Acetotoluidide, 2-chloro-*N*-(ethoxymethyl)-6'-ethyl- see...Acetochlor
Acetotriphenylstannine see...Triphenyltin Compounds
Acetoxy(2-methoxyethyl)mercury see. .Methoxyethylmercuric Acetate
Acetoxy-triphenyl-stannan (German) see...Triphenyltin Compounds
Acetoxytriphenylstannane see...Triphenyltin Compounds
Acetoxytriphenyltin see...Triphenyltin Compounds
2-Acetylamino-fluoren (German) see. . Acetylaminofluorene
2-Acetylaminofluorene see...Acetylaminofluorene
2-2-Acetylamidofluorene see...Acetylaminofluorene
*N*-Acetyl-2-aminofluorene see...Acetylaminofluorene
Acetyl hydroperoxide see...Peracetic Acid
*N*-Acetyl-1-naphthylamine see...1-Naphthaleneacetamide
(Acetyloxy)triphenyl-stannane (9CI) see...Triphenyltin Compounds
Acetylphosphoramidothioic acid, *O,S*-dimethyl ester see. . Acephate
ACHERO® see...Acephate
ACHIEVE® see...Tralkoxydim
ACHIEVE®-40DG see...Tralkoxydim
ACHIVA® see...Halosulfuron-methyl
Acide arsenieux (French) see...Arsenous Oxide
Acide benzoique (French) see...Benzoic Acid
Acide cacodylique (French) see...Cacodylic Acid
Acide chloracetique (French) see...Chloroacetic Acid
Acide 2-(4-chloro-2-methyl-phenoxy)propionique (French) see...Mecoprop
Acide cyanhydrique (French) see...Hydrogen Cyanide
Acide 2,4-dichloro phenoxyacetique (French) see... 2,4-D
Acide dimethylarsinique (French) see...Cacodylic Acid
Acide formique (French) see...Formic Acid
Acide monochloracetique (French) see...Chloroacetic Acid
Acide monofluoracetique (French) see...Fluoroacetic Acid
Acide naphthyloxyacetique (French) see. . Naphthoxyacetic Acid
Acide peracetique (French) see...Peracetic Acid
Acide phodphorique (French) see...Phosphoric Acid
Acide salicilique (French) see...Salicylic Acid
Acide 2,4,5-trichlorophenoxyacetique (French) see. 2,4,5-T

Acide, 2-(2,4,5-trichloro-phenoxy) propionique (French) see...Silvex
Acide-2-(2,4-dichloro-phenoxy)propionique (French) see...Dichlorprop
ACIDET® see...Dodecylbenzenesulfonic Acid
ACIDISOL® see...Dodecylbenzenesulfonic Acid
Acid lead arsenite see...Lead Arsenate
Acid lead orthoarsenate see...Lead Arsenate
Acido arsenico (Spanish) see...Arsenic Acid
Acido benzoico (Spanish) see...Benzoic Acid
Acido cacodilico (Spanish) see...Cacodylic Acid
Acido 2-(4-cloro-2-metil-fenossi)-propionico (Italian) see...Mecoprop
Acido cianidrico (Italian) see...Hydrogen Cyanide
Acido cloroacetico (Spanish) see...Chloroacetic Acid
Acido decanoico (Spanish) see...Decanoic Acid
Acido (2,4-dicloro-fenossi)-acetico (Italian) see...2,4-D
Acido-2-(2,4-dicloro-fenossi)propionico (Italian) see. .Dichlorprop
Acido 2,4-diclorofenoxiacetico (Spanish) see...2,4-D
Acido 2,4-diclorofenoxibutirico (Spanish) see...2,4-DB
Acido 2-(2,4-diclorofenoxi)propionico (Spanish) see. .Dichlorprop
Acido (3,6-dichloro-2-metossi)-benzoico (Italian) see. . Dicamba
Acido 2,2-dicloropropionico (Spanish) see...Dalapon
Acido dodecilbencenosulfonico (Spanish) see. .Dodecylbenzenesulfonic Acid
Acido formico (Italian) see...Formic Acid
Acido formico (Spanish) see...Formic Acid
Acido fosforico (Italian, Spanish) see...Phosphoric Acid
Acido fluoroacetico (Spanish) see...Fluoroacetic Acid
Acido fluorhidrico (Spanish) see...Hydrogen Fluoride
Acidomonocloroacetico (Italian) see...Chloroacetic Acid
Acido monofluoroacetico (Italian) see...Fluoroacetic Acid
Acid orthoarsenite see...Copper Arsenite
Acid oxalate see...Amiton Oxalate
Acido peracetico (Spanish) see...Peracetic Acid
Acido salicilico (Italian) see...Salicylic Acid
Acido (2,4,5-tricloro-fenossi)-acetico (Italian) see. .2,4,5-T
Acido 2,4,5-triclorofenoxiacetico (Spanish) see...2,4,5-T
Acido 2-(2,4,5-triclorofenossi)-propionico (Italian) see. Silvex
ACIFAT® see...Acephate
Acifluorfene see...Acifluorfen
ACIFON® see...Azinphos-methyl
ACIGENA® see...Hexachlorophene
ACINATE® see...Methomyl
ACIZOL® see...Triadimefon
ACL 70® see...Dichloroisocyanuric Acid
ACLID® see...Propachlor
ACME® see...2,4-D
ACME® see...DCPA
ACME® see...Prometon

ACME® see...Rotenone
ACME® see...Sodium Cacodylate
ACME DORMANT OIL SPRAY® see...Naphthas
ACP-M-728® see...Chloramben
ACME MCPA AMINE 4® see...MCPA
ACME WETTABLE DUSTING SULFUR® see...Sulfur
ACQ® see...Copper Ammonium Carbonate
ACQUINITE® see...Acrolein
ACQUINITE® see...Chloropicrin
Acrehyde see...Acrolein
ACRICID® see...Binapacryl
Acrilamida (Spanish) see...Acrylamide
Acrilonitrilo (Spanish) see...Acrylonitrile
ACRITET® see...Acrylonitrile
ACRITET® see...Carbon Tetrachloride
ACROBAT® see...Mancozeb
ACROBAT® WP see...Dimethomorph
Acroleina (Italian) see...Acrolein
Acroleine (Dutch, French) see...Acrolein
ACRYLAGEL® see...Acrylamide
Acrylaldehyde see...Acrolein
Acrylamide, 30% see...Acrylamide
Acrylamide,5 0% see...Acrylamide
Acrylamide monomer see...Acrylamide
Acrylehyd (German) see...Acrolein
Acrylehyde see...Acrolein
Acrylic acid amide, (50%) see...Acrylamide
Acrylic aldehyde see...Acrolein
Acrylic amide see...Acrylamide
Acrylic amide 30% see...Acrylamide
Acrylic amide 50% see...Acrylamide
Acrylnitril (Dutch, German) see...Acrylonitrile
Acrylonitrile monomer see...Acrylonitrile
ACRYLON® see...Acrylonitrile
ACR® 2913 see...Fenoxycarb
ACR® 2984F see...Fenoxycarb
ACTAMER® see...Bithionol
ACTELLIC® see...Pirimiphos-Methyl
ACTELLIFOG® see...Pirimiphos-Methyl
ACTI-AID® see...Cycloheximide
ACTICIDE see...Octhilinone
ACTIDIONE® TGF see...Cycloheximide
ACTIDIONE® see...Cycloheximide
ACTIDONE® see...Cycloheximide
ACTINIT® see...Atrazine
ACTINITE P® see...Atrazine
ACTISPRAY see...Cycloheximide
ACTIVOL® see...Gibberellic Acid
ACTOX® see...Allidochlor
ACTUAL DINOCAP® see...Dinocap
ADAMS® see...Dipropyl Isocinchomeronate
ADAMYCIN® see...Oxytetracycline Calcium
ADD-F® see...Formic Acid
ADDRESS® see...Acephate
Adenine, N-benzyl- see...6-Benzaldenine
Adenine, N-furfuryl- see...Kinetin (Cytokinin)
Adenine, $N^6$-furfuryl- see...Kinetin (Cytokinin)
ADEPT® see...Diflubenzuron

ADIOS® see...Carbaryl
ADIOS® see...Cinnamaldehyde
ADJUST® see...Chlormequat Chloride
ADMIRE® see...Imidacloprid
ADVANTAGE® see...Carbosulfan
AERO® see...Calcium Cyanide
AERO-CYANAMID® see...Calcium Cyanamide
AERO-FLYING INSECT SPRAY® see...Dipropyl Isocinchomeronate
AEROL 1 (PESTICIDE)® see...Trichlorfon
AERO LIQUID HCN® see...Hydrogen Cyanide
3-(Aethoxycarbonylaminophenyl)-N-phenyl-carbamat (German) see...Desmedipham
5-Aethoxy-3-trichlormethyl-1,2,4-thiadiazol (German) see...Etridiazole
4-Aethylamino-2-*tert*-butylamino-6-methylthio-*S*-triazin (German) see...Terbutryn
2-Aethylamino-4-chlor-6-isopropylamino-1,3,5-triazin (German) see...Atrazine
*S*-Aethyl-*N,N*-dipropylthiocarbamat (German) see..EPTC
1,1'-Aethylen-2,2'-bipyridinium-dibromid (German) see..Diquat Dibromide
Aethylenbromid (German) see...Ethylene Dibromide
Aethylenchlorid (German) see...Ethylene Dichloride
Aethylenoxid (German) see...Ethylene Oxide
Aethylformiat (German) see...Ethyl Formate
2-Aethyl-6-methyl-*N*-(1-methyl-2-methoxyaethyl)-chloracetanilid (German) see...Metolachlor
*O*-Aethyl-*O*-(3-methyl-4-methylthiophenyl)-isopropylamido-phosphorsaeure ester (German) see..Fenamiphos
*O*-Aethyl-*O*-*N*(4-nitrophenyl)-phenylmonothiophosphonat (German) see...EPN
*O*-Aethyl-*S*-phenyl-aethyl-dithiophosphonat (German) see...Fonofos
*N*-(Aethylpropyl)-3,4-dimethyl-2,6-dinitroanilin (German) see...Pendimethalin
*N*-(1-Aethylpropyl)-2,6-dinitro-3,4-xylidin (German) see..Pendimethalin
Aethylrhodanid (German) see...Ethylthiocyanate
*O*-Aethyl-*O*-(2,4,5-trichlorphenyl)-aethylthionophosphonat (German) see...Trichloronate
AF 101® see...Diuron
AFALON® see...Linuron
AFFIRM® see...Abamectin
AFFIRM® see...Emamectin Benzoate
AFICIDA® see...Pirimicarb
AFICIDE® see...Lindane
AFIDEN® see...Endosulfan
AFI-TIAZIN® see...Phenothiazine
AFL 1081® see...Fluoroacetamide
AFLIX® see...Formothion
AFNOR® see...Chlorphacinone
AG-500® see...Diazinon
Agent Blue see...Cacodylic Acid
AGERMIN® see...Propham
AGIMIX® see...Alachlor
AGIMIX® see...Atrazine
AGLIME® see...Calcium Carbonate
AGNIQUE MMF MOSQUITO LARVICIDE & PUPICIDE® see...Arosurf® MSF
AGRAZINE® see...Phenothiazine
AGREFLAN® see...Trifluralin
AGREPT® see...Streptomycin
AGREX T-7.5® see...Tetradifon
AGRIA 1050® see...Fenitrothion
AGRI-MEK® see...Abamectin
AGRI-MYCIN® see...Streptomycin Sulfate
AGRI-STREP® see...Streptomycin
AGRI-TIN® see...Fentin Hydroxide
AGRI TIN® see...Triphenyltin Compounds
AGRICHEM GREENFLY SPRAY® see...Malathion
AGRICIDE MAGGOT KILLER (F)® see...Toxaphene
Agricultural limestone see...Calcium Carbonate
AGRIDIP® see...Coumaphos
AGRIFLAN® 24 see...Trifluralin
AGRIGARD® see...Capsaicin
AGRIMET® see...Phorate
AGRIMYCIN 17® see...Streptomycin
AGRINATE® see...Methomyl
AGRISIL® see...Trichloronate
AGRISOL G-20® see...Lindane
AGRISTREP® see...Streptomycin Sulfate
AGRITAN® see...DDT
AGRITOX® see...Copper Sulfate
AGRITOX® see...MCPA
AGRITOX® see...Trichloronate
AGRIYA 1050® see...Fenitrothion
AGRIZAN® see...Copper Oxychloride
A-GRO® see...Methyl Parathion
AGROCER COMPLEX® see...2,4-D
AGROCERES® see...Heptachlor
AGROCIDE® see...Lindane
AGROCITE® see...Benomyl
AGROFOROTOX® see...Trichlorfon
AGRONAA® see...1-Naphthaleneacetic Acid
AGRONEXIT® see...Lindane
AGROSAN® see...Phenylmercury Acetate
AGROSAND® see...Phenylmercury Acetate
AGROSAN GN 5® see...Phenylmercury Acetate
AGROSOL® see...Methylmercuric Dicyanamide
AGROSOL® see...Thiabendazole
AGROSOL POUR-ON® see...Thiram
AGROSOL S® see...Captan
AGROSOL®-T see...Thiabendazole
AGRO STREP® see...Streptomycin
AGROTECT® see...2,4-D
AGROTHION® see...Fenitrothion
AGROTHRIN® see...Cypermethrin
AGROX® 2-WAY and 3-WAY see...Captan
AGROXONE® see...MCPA
AGROX® PREMIERE see...Metalaxyl
AGROZONE® see...MCPA
AGSCO® see...MCPA
AGSTONE® see...Calcium Carbonate

AGTROL® see...6-Benzaldenine
AGTROL® see...Fentin Hydroxide
AGTROL® see...Triphenyltin Compounds
AGVALUE® see...Oryzalin
AGWAY® see...Dichlorvos
AGWAY FOOD PLANT FOGGING SPRAY® see...Allethrins
AH 501® see...Paraquat
AI 50 see...Dichloran
AI3-00027 see...Chloropicrin
AI3-00987 see...Thiram
AI3-01122 see...Dinoseb
AI3-02370 see...Dichlorophene
AI3-02372 see...Hexachlorophene
AI3-2824 see...Atrazine
AI3-03967 see...Zinc Sulfate Heptahydrate
AI3-08434 see...Sodium Fluoroacetate
AI3-08870 see...Dichloran
AI3-14250 see...Piperonyl Butoxide
AI3-14673 see...Sodium Dimethyldithiocarbamate
AI3-14689 see...Ferbam
AI3-16319 see...Thiosemicarbazide
AI3-16418 see...Sodium Pentachlorophenate
AI3-16667 see...2,4-D, isopropyl ester
AI3-17034 see...Malathion
AI3-17292 see...Methyl Parathion
AI3-17591 see...Dipropyl Isocinchomeronate
AI3-19244 see...Isodrin
AI3-19507 see...Diazinon
AI3-22374 see...Mevinphos
AI3-24237 see...Formic Acid
AI3-24964 see...Demeton-methyl
AI3-24988 see...Naled
AI3-25207 see...Triphenyltin Compounds
AI3-25644 see...Famphur
AI3-25726 see...Methiocarb
AI3-26730-X see...Triethanolamine Dodecylbenzene Sulfonate
AI3-27165 see...Temephos
AI3-27226 see...Propargite
AI3-27318 see...Ethoprop
AI3-27318 see...Ethoprop
AI3-27339 see...Tetramethrin
AI3-27474 see...Resmethrin
AI3-27556 see...Dicamba
AI3-27695 see...Bendiocarb
AI3-27699 see...Pirimiphos-Methyl
AI3-27738 see...Fenbutatin Oxide
AI3-27748 see...Isofenphos
AI3-28009 see...Fentin Hydroxide
AI3-28009 see...Triphenyltin Compounds
AI3-29054 see...Diflubenzuron
AI3-29062 see...D-Phenothrin
AI3-29149 see...Sulprofos
AI3-29234 see...Fenpropathrin
AI3-29311 see...Thiodicarb
AI3-29349 see...Hydramethylnon
AI3-29426 see...Fluvalinate

AI3-29460 see...Fenoxycarb
AI3-29832 see...Hexaflumuron
AI3-31382 see...Propanil
AI3-50436 see...Sodium Azide
AI3-50598 see...Thiabendazole
AI3-51142 see...Simazine
AI3-51503 see...Propachlor
AI3-51506 see...Alachlor
AI3-61438 see...Diuron
AI3-61943 see...Paraquat
AIMCO SYSTOX® see...Demeton-methyl
AIMCOZIM see...Carbendazim
AIMSAN® see...Phenmedipham
AIMSAN® see...Phenthoate
AIMTHENE® see...Acephate
AIP see...Aluminum Phosphide
AIRONE® see...Propineb
AITC see...Allyl Isothiocyanate
AKAR 338® see...Chlorobenzilate
AKARI® see...Fenpyroximate
Akarithion see...Carbophanothion
AKARITOX® see...Tetradifon
AKROCHEM ETU-22® see...Ethylene Thiourea
Akrolein (Czech) see...Acrolein
Akroleina (Polish) see...Acrolein
Akrylamid (Czech) see...Acrylamide
Akrylonitryl (Polish) see...Acrylonitrile
AKTIKON® see...Atrazine
AKTIKON PK® see...Atrazine
AKTINIT A® see...Atrazine
AKTINIT S® see...Simazine
AKZO CHEMIE MANEB® see...Maneb
AL-50® see...Dichloran
Alachlore see...Alachlor
ALAGAM® see...Alachlor
ALAGAN® see...Alachlor
ALAMO® see...Propiconazole
ALANAP® see...Naptalam
ALANAPE® see...Naptalam
ALANEX® see...Alachlor
β-Alanine, N-[((((2,3-dihydro-2,2-dimethyl-7-benzofuranyl)oxy)carbonyl)methylamino)thio]-N-(1-methylethyl)-, ethyl ester see...Benfuracarb
ALAPAZ® see...Alachlor
ALAR® see...Daminozide
ALAR-85® see...Daminozide
ALATEX® see...Dalapon
ALATOX 480® see...Alachlor
ALAZINE® see...Alachlor
ALAZINE® see...Atrazine
ALBRASS® see...Propachlor
ALCLOR 48 LE® see...Alachlor
ALCO® see...Dichlorvos
ALCO® see...Dipropyl Isocinchomeronate
ALCO® see...Kerosene
ALCO® see...Malathion
ALCO® see...1-Naphthaleneacetic Acid
ALCO® see...Simazine

ALCOA SODIUM FLUORIDE® see...Sodium Fluoride
ALCOBAM NM® see...Sodium Dimethyldithiocarbamate
ALCOBAM ZM® see...Ziram
ALCO CITRUS FIX® see...2,4-D, isopropyl ester
Alcohol bencilico (Spanish) see...Benzyl Alcohol
Alcool allilco (Italian) see...Allyl Alcohol
Alcool allylique (French) see...Allyl Alcohol
ALCO® OXALIS KILLER see...Monuron
ALCOPHOBIN® see...Disulfiram
ALCO SLUB"M® see...Methiocarb
ALCOV® see...Dioxathion
Aldecarb see...Aldicarb
Aldecarbe (French) see...Aldicarb
Aldehyde acrylique (French) see...Acrolein
Aldehyde formique (French) see...Formaldehyde
Aldeide acrilica (Italian) see...Acrolein
Aldeide formica (Italian) see...Formaldehyde
Aldicarb sulfure (French) see...Aldoxycarb
ALDICARB SULFONE® see...Aldoxycarb
ALDOCIT® see...Aldrin
ALDREC® see...Aldrin
ALDREX® see...Aldrin
ALDREX-30® see...Aldrin
ALDREX-40® see...Aldrin
ALDRIN 37 EQUIVALENT SOLUTION® see..Aldrin
Aldrina (Spanish) see...Aldrin
Aldrine (French) see...Aldrin
ALDRITE® see...Aldrin
ALDRON® see...Aldrin
ALDROSOL® see...Aldrin
ALFA-TOX® see...Diazinon
ALGATROL-30 see...Sodium Dimethyldithiocarbamate
ALGEX see...Sodium Dimethyldithiocarbamate
ALGIMYCIN® see...Phenylmercury Acetate
ALGISTAT® see...Dichlone
ALGRAN® see...Aldrin
ALIBI® see...Linuron
Alidochlor see...Allidochlor
ALIETTE® see...Fosetyl-Al
ALIGN® see...Azadirachtin
Alilico Alcohol (Spanish) see...Allyl Alcohol
Aliphatic petroleum naphtha see...Naphthas
ALIROX® see...EPTC
ALISTELL® see...Linuron
ALJADEN® see...Sethoxydim
ALK-AUBS® see...Disulfiram
ALKARSODYL® see...Sodium Cacodylate
ALKRON® see...Parathion
Alkyl dimethyl benzyl ammonium chloride see.Zilkonium Chloride
Alkyldimethyl(phenylmethyl)quaternary ammonium chlorides see...Zilkonium Chloride
ALLBRI NATURAL COPPER® see...Copper and Copper Compounds
ALL BUG® see...Ammonium Hexafluorosilicate
ALLEGIENCE® see...Metalaxyl
(+)-Allelrethonyl see...Allethrins

ALLERON® see...Parathion
Allethrin I see...Allethrins
d-Allethrin see...Allethrins
ALLEVIATE® see...Allethrins
ALLIE® see...Metsulfuron-methyl
Allilowy alkohol (Polish) see...Allyl Alcohol
ALLISAN® see...Dichloran
$\beta$-D-Allofuranuronic acid, 5-[(2-amino-5-$O$-(aminocarbonyl)-2-deoxy-L-xylonoyl)amino]-1-(5-carboxy-3,4-dihydro-2,4-dioxo-1(2$H$)-pyrimidinyl)-1,5-dideoxy-, zinc salt (1:1) see...Polyoxin D, Zinc Salt
Allophanic acid, 4,4'-$O$-phenylenebis(3-thio-, diethyl ester see...Thiophanate-Methyl
ALLOXOL® S see...Sethoxydim
1-[2-(Allyloxy)-2-(2,4-dichlorophenyl)ethyl]-1$H$-imidazole see...Imazalil
ALLPRO BARACIDE® see..Sodium Metaborate
ALL PURPOSE GARDEN INSECTICIDE® see..Malathion
ALLTEX® see...Toxaphene
ALLTOX® see...Toxaphene
ALLY® see...Metsulfuron-methyl
ALLY® see...Thifensulfuron Methyl
ALLY® see...Tribenuron-Methyl
ALLY-20DF® see...Metsulfuron-methyl
Allyl Al see...Allyl Alcohol
Allylaldehyde see...Acrolein
Allylalkohol (German) see...Allyl Alcohol
Allyl cinerin see...Allethrins
Allyl-1-(2,4-dichlorophenyl)-2-imidazol-1-ylethyl ether see...Imazalil
Allyl homolog of cinerin I see...Allethrins
Allylic Alcohol see...Allyl Alcohol
Allylidene diacetate see...Acrolein diacetate
Allyl isorhodanide see...Allyl Isothiocyanate
Allyl isosulfocyanate see...Allyl Isothiocyanate
Allyl isothiocyanate see...stabilized see...Allyl Isothiocyanate
3-Allyl-4-keto-2-methylcyclopentenyl chrysanthemummonocarboxylate 3 see...Allethrins
3-Allyl-2-methyl-4-oxo-2-cyclopenten-1-yl chrysanthemate see...Allethrins
Allyl mustard oil see...Allyl Isothiocyanate
($\pm$)-1-[$\beta$-(Allyloxy)-2,4-dichlorophenethyl]imidazole see.Imazalil
Allylrethronyl $dl$-$cis$-$trans$-chrysanthemate see.Allethrins
Allylsenfoel (German) see...Allyl Isothiocyanate
Allyl sevenolum see...Allyl Isothiocyanate
Allyl thiocarbonimide see...Allyl Isothiocyanate
Alphacypermethrin see...$alpha$-Cypermethrin
(+)-Alphamethrin see...$alpha$-Cypermethrin
Alphanaphtyl thiourea (French) see...ANTU
ALPHASET IPE see...2,4-D, isopropyl ester
ALPHA-SPRA® see...1-Naphthaleneacetic Acid
AL-PHOS® see...Aluminum Phosphide
ALRATO® see...ANTU
ALTO® see...Cyproconazole
ALTODEL® see...Kinoprene

ALTOSID® see...Methoprene
ALTOX® see...Aldrin
ALTOZAR® see...Hydroprene
ALTOZAR IGR® see...Hydroprene
ALUDOR® see...Chlorpyrifos
Alum see...Aluminum Sulfate
Aluminum alum see...Aluminum Sulfate
Aluminum fosfide (Dutch) see...Aluminum Phosphide
Aluminum monoPhosphide see...Aluminum Phosphide
Aluminum phosethyl see...Fosetyl-Al
Aluminum sodium fluoride see...Sodium Aluminum fluoride
Aluminum sulphate see...Aluminum Sulfate
Aluminum tris(O-ethylphosphonate) see...Fosetyl-Al
Aluminum trisulfate see...Aluminum Sulfate
ALVIT® see...Dieldrin
ALZODEF® see...Calcium Cyanamide
AMA-20® see...Dazomet
AMA-30® see...Nabam
AMA-30 see...Sodium Dimethyldithiocarbamate
AMACTONE® see...1-Naphthaleneacetamide
AMASIL® see...Formic Acid
AMATIN® see...Hexachlorobenzene
AMAZE® see...Isofenphos
AMAZIN® see...Azadirachtin
AMBEN® see...Chloramben
AMBER® see...Triasulfuron
AMBIBEN® see...Chloramben
AMBOX® see...Binapacryl
AMCHEM® see...Amitrole
AMCHEM A-280® see...Butralin
AMCHEM® 68-250 see...Ethephon
AMCHEM 70-25® see...Butralin
AMCHEM® WEED KILLER 650 see...2,4-D, isopropyl ester
AMCHEN 2,4,5-TP® see...Silvex
AMCIDE® see...Ammonium Sulfamate
AMCOTHENE® see...Acephate
AMCOTONE® see...1-Naphthaleneacetic Acid
AMDON® see...Picloram
AMDON GRAZON® see...Picloram
AMDRO® see...Hydramethylnon
AMEISENATOD® see...Lindane
AMEISENMITTEL (MERCK)® see...Lindane
Ameisensaeure (German) see...Formic Acid
AMERICAN CYANAMID 3422® see...Parathion
AMERICAN CYANAMID 3,911® see...Phorate
AMERICAN CYANAMID 4,049® see...Malathion
AMERICAN CYANAMID 5223® see...Dodine
AMERCIAN CYANAMID 18133® see...Thionazin
AMERICAN CYANAMID 18682® see...Prothoate
AMERICAN CYANAMID 38023® see...Famphur
AMERICAN CYANAMID 47031® see...Phosfolan
AMERICAN CYANAMID AC-52,160® see..Temephos
AMERICAN CYANAMID CL-52,160® see...Temephos
AMERICAN CYANAMID CL-47,300® see..Fenitrothion
AMERICAN CYANAMID CL-47470® see..Mephosfolan
AMERICAN CYANAMID E.I. 52,160® see..Temephos
AMERICARE RABON® see...Tetrachlorvinphos
AMERICARE® see...Esfenvalerate
AMERCIDE® see...Captan
AMERESCO ACRYL-40® see...Acrylamide
AMEROL® see...Amitrole
AMERTREX® see...Ametryn
AMESIP® see...Ametryn
AMETRON SC® see...Ametryn
AMETRON SC® see...Diuron
AMETRYNE 2E® see...Ametryn
AMETRYNE 80W HERBICIDE® see...Ametryn
AMETRYNE TECHNICAL® see...Ametryn
AMEX 820® see...Butralin
AMEXINE® see...Butralin
AMEX® see...Butralin
AM-FOL see...Ammonia
AMIBEN® see...Chloramben
AMIBIN® see...Chloramben
AMICIDE® see...Ammonium Sulfamate
AMID-THIN® see...1-Naphthaleneacetamide
AMID-THIN® see...Naphthoxyacetic Acid
Amidinohydrazone see...Hydramethylnon
Amidocyanogen see...Cyanamide
AMIDOSULFATE® see...Ammonium Sulfamate
AMIDOX® see...2,4-D
AMIGAN® see...Ametryn
Amine see...2,4,5-T
Aminic acid see...Formic Acid
trans-L-2-Amino-4-(2-aminoethoxy)3-butenoic acid hydrochloride see...Aminoethoxyvinylglycine Hydrochloride
(s)-trans-2-Amino-4-(2-aminoethyoxy)-3-butenoic acid hydrochloride see...Aminoethoxyvinylglycine Hydrochloride
4-Amino-benzolsulfonyl-methylcarbamat (German) see. Asulam
4-Amino-6-tert-butyl-3-methylthio-As-triazin-5-one see. Metribuzin
4-Amino-6-tert-butyl-3-(methylthio)-1,2,4-triazin-5-one see...Metribuzin
4-Amino-6-tert-butyl-3-methylthio-As-triazin-5-one see.. Metribuzin
5-Amino-4-chloro-2,3-dihydro-3-oxo-2-phenylpyridazine see...Pyrazon
5-Amino-4-chloro-2-phenyl-3(2H)-pyridazinone see.. Pyrazon
5-Amino-3-cyano-1-(2,6-dichloro-4-trifluoromethylphenyl)-4-trifluoromethylsulfinylpyrazole see...Fipronil
3-Amino-2,5-dichlorobenzoic acid see...Chloramben
3-Amino-2,6-dichlorobenzoic acid see...Chloramben
[(4-Amino-3,5-dichloro-6-fluoro-2-pyridinyl)oxy]acetic acid, 1-methylheptyl ester

5-Amino-1-(2,6-dichloro-4-(trifluoromethyl)phenyl)-4-(1,R,S)-(trifluoromethyl)sulfinyl)-1H-pyrazole-3-carbonitrile  see...Fipronil
(±)-5-Amino-1-(2,6-dichloro-α,α,α-trifluoro-p-tolyl)-4-trifluoromethylsulfinylpyrazole-3-carbonitrile  see..Fipronil
4-Amino-6-(1,1-dimethylethyl)-3-(methylthio)-1,2,4-triazin-5-(4H)-one  see...Metribuzin
2-Aminoethanol salt of 2',5-dichloro-4'-nitrosalicylanilide  see...Clonitralid
2-Aminoethanol salt of 5-chloro-N-(2-chloro-4-nitrophenyl)-2-hydroxybenzamide  see...Clonitralid
L-α-(2-Aminoethoxyvinyl)glycine hydrochloride  see..Aminoethoxyvinylglycine Hydrochloride
Aminofuracarb  see...Benfuracarb
l-2-Aminoglutaric acid  see...Glutamic Acid
α-Aminoglutaric acid  see...Glutamic Acid
AMINOL 806® see...2,4-D
4-Amino-3-methyl-6-phenyl-1,2,4-triazin-5(4H)-one  see. Metamiton
2-Aminopentanedioic acid  see...Glutamic Acid
4-(Aminophenylsulfonyl)carbamate, methyl ester  see..Asulam
4-Aminopiridina (Spanish)  see...4-Aminopyridine
1-Aminopropane-1,3-dicarboxylic acid  see...Glutamic Acid
γ-Aminopyridine  see...4-Aminopyridine
p-Aminopyridine  see...4-Aminopyridine
Amino-4-pyridine  see...4-Aminopyridine
Aminosulfulan  see...Benfuracarb
Aminothiourea  see...Thiosemicarbazide
1-Aminothiourea  see...Thiosemicarbazide
N-Aminothiourea  see...Thiosemicarbazide
1-Amino-2-thiourea  see...Thiosemicarbazide
AMINOTRIAZOLE BAYER® see...Amitrole
AMINO TRIAZOLE WEEDKILLER 90® see..Amitrole
Aminotriazole  see...Amitrole
2-Aminotriazole  see...Amitrole
3-Aminotriazole  see...Amitrole
2-Amino-1,3,4-triazole  see...Amitrole
3-Amino-1,2,4-triazole  see...Amitrole
3-Amino-1H-1,2,4-triazole  see...Amitrole
3-Amino-S-triazole  see...Amitrole
4-Aminotrichloropicolinic acid  see...Picloram
4-Amino-3,5,6-trichloropicolinic acid  see...Picloram
4-Amino-3,5,6-trichloro-2-picolinic acid  see..Picloram
4-Amino-3,5,6-trichlorpicolinsaeure (German)  see..Picloram
4-Amino-3,5,6-trichloro-2-pyridinecarboxylic acid  see. Picloram
4-Amino-3,5,6-trichloropyridine-2-carboxylic acid  see. Picloram
AMINOZ® see...2,4-D
AMINOZID® see...Daminozide
AMINOZIDE® see...Daminozide
AMIPAZ® see...Amitraz
AMIRAL® see...Triadimefon
Amisia-mottenschutz  see...para-Dichlorobenzene

AMISTAR® see...Azaxystrobin
AMITOL® see...Amitrole
Amitraz estrella  see...Amitraz
Amitraze  see...Amitraz
AMITRIL® see...Amitrole
AMITROL 90  see...Amitrole
AMITROL-T® see...Amitrole
AMIZOL® see...Amitrole
AMIZOL DP NAU® see...Amitrole
AMMAT® see...Ammonium Sulfamate
AMMATE  see...Ammonium Sulfamate
AMMO® see...Cypermethrin
Ammonia, anhydrous  see...Ammonia
Ammoniac (French)  see...Ammonia
Ammoniaca (Italian)  see...Ammonia
Ammonia gas  see...Ammonia
Ammoniale (German)  see...Ammonia
Ammonium acid sulfite  see...Ammonium Sulfite
Ammonium-aethyl-carbamoyl-phosphonat (German)  see..Fosamine Ammonium
Ammonium amide  see...Ammonia
Ammonium amidosulfonate  see...Ammonium Sulfamate
Ammonium amidosulphate  see...Ammonium Sulfamate
Ammonium aminoformate  see...Ammonium Carbamate
Ammonium aminosulfonate  see...Ammonium Sulfamate
Ammonium, (2-chloroethyl)trimethyl-, Chloride 2-chloro-N,N,N-trimethylethanaminium chloride  see..Chlormequat Chloride
Ammonium chromate(VI)  see...Ammonium Chromate
Ammonium ethyl carbamoylphosphonate  see...Fosamine Ammonium
Ammonium fluorosilicate  see...Ammonium Hexafluorosilicate
Ammonium hexafluorosilicate  see...Ammonium Hexafluorosilicate
Ammonium hydrogen sulfite  see...Ammonium Sulfite
Ammonium hydrosulfite  see...Ammonium Sulfite
Ammonium hydroxide  see...Ammonia
Ammonium hyposulfite  see...Ammonium Thiosulfate
Ammonium monosulfite  see...Ammonium Sulfite
Ammoniumnitrat (German)  see...Ammonium Nitrate
Ammonium(i) Nitrate(1:1)  see...Ammonium Nitrate
Ammonium orthophosphate, dibasic  see...Ammonium Phosphate
Ammonium orthophosphate, monohydrogen  see..Ammonium Phosphate
Ammonium phosphate, dibasic  see...Ammonium Phosphate
Ammonium phosphate, hydrogen  see...Ammonium Phosphate
Ammonium phosphate secondary  see...Ammonium Phosphate
Ammonium salt  see...Ammonium Nitrate
Ammonium Saltpeter  see...Ammonium Nitrate
Ammonium salz der amidosulfonsaure (German)  see..Ammonium Sulfamate
Ammonium silicofluoride  see...Ammonium Hexafluorosilicate

Ammonium silicon fluoride see...Ammonium Hexafluorosilicate
Ammonium sulfite, hydrogen see...Ammonium Sulfite
Ammonium sulphamate see...Ammonium Sulfamate
Ammonium salt of (±)-2-(4,5-dihydro-4-methyl-4-(1-methylethyl)-5-oxo-1H-imidazol-2-yl)-5-ethyl-3-pyridinecarboxylic acid(±)-2-(4,5-Dihydro-4-methyl-4-(1-methylethyl)-5-oxo-1H-imidazol-2-yl)-5-ethyl-3-pyridinecarboxylic acid, ammonium salt see...Imazethapyr
AMMONYX® see...Zilkonium Chloride
AMOBEN® see...Chloramben
Amoniaco (Spanish) see...Ammonia
Amoniaco anhidro (Spanish) see...Ammonia
Amoniak (Polish) see...Ammonia
Amorphous Silica see...Diatomaceous Earth
AMOXONE® see...2,4-D
AMPLIFY™ see...Cloransulam-methyl
AMPROLENE® see...Ethylene Oxide
AMS see...Ammonium Sulfamate
AMS® AMMONIUM SULFAMATE WEED & BRUSH KILLER see...Ammonium Sulfamate
Amthio see...Ammonium Thiosulfate
AMTRATE® see...Ammonium Nitrate
Amyl zimate see...Ziram
Amyphyt see...Ametryn
AN see...Acrylonitrile
ANA see...1-Naphthaleneacetic Acid
ANAC 110® see...Copper and Copper Compounds
ANACRACK® see...Naptalam
Anagrapha falcifera MNPV PIB's in aqueous suspension see...Anagrapha Falcifera
Anagrapha falcifera multi-nuclear polyhedrosis virus (AfMNPV) see...Anagrapha Falcifera
O-Analog of Dimethoate see...Omethoate
ANATOX® see...Toxaphene
ANCANZATE ME® see...Ziram
Androst-5-en-3-ol, 17-[(((3-(dimethylamino)propyl)methyl)amino]-,dihydrochloride, (3β,17β)- see...Azacosterol Dihydrochloride
ANELDA PLUS® see...Butylate
ANELDAZIN® see...Atrazine
ANELDAZIN® see...Butylate
ANELIROX® see...Butylate
ANELMID® see...Dithiazanine Iodide
ANFOR® see...Iprodione
ANGUIFUGAN® see...Dithiazanine Iodide
Anhydride of ammonium carbonate see...Ammonium Carbamate
Anhydride arsenique (French) see...Arsenic Pentoxide
Anhydride arsenieux (French) see...Arsenous Oxide
Anhydrous ammonia see...Ammonia
Anhydrous hydrofluoric acid see...Hydrogen Fluoride
ANICON KOMBI® see...MCPA
ANICON M® see...MCPA
Anidrino arsenioso (Italian) see...Arsenic Pentoxide
ANILAZIN® see...Anilazine

Aniline, N-sec-butyl-4-tert-butyl-2,6-dinitro- see...Butralin
Aniline, 2,6-dichloro-4-nitro- see...Dichloran
Aniline, 3,4-dimethyl-2,6-dinitro-N-(1-ethylpropyl)- see...Pendimethalin
o-Anisic acid, 3,6-dichloro- see...Dicamba
o-Anisic acid, 3,6-dichloro-, sodium salt see...Sodium Dicamba
ANLES® see...Thiram
ANOFEX® see...DDT
ANPROLENE® see...Ethylene Oxide
ANPROLINE® see...Ethylene Oxide
ANSAR® see...Sodium Methanearsonate (MSMA)
ANSAR® see...Cacodylic Acid
ANSAR 160® see...Sodium Cacodylate
Ansax see...Ammonium Nitrate
ANSWER® see...Metsulfuron-methyl
ANTADIX® see...Disulfiram
ANTAENYL® see...Disulfiram
ANTAETHAN® see...Disulfiram
ANTAETHYL® see...Disulfiram
ANTAETIL® see...Disulfiram
ANTALCOL® see...Disulfiram
ANTETAN® see...Disulfiram
ANTETHYL® see...Disulfiram
ANTETIL® see...Disulfiram
ANTEYL® see...Disulfiram
ANTHIO® see...Formothion
ANTHIPHEN® see...Dichlorophene
ANTHON® see...Trichlorfon
Anthracen (German) see...Anthracene
9,10-Anthracenedione see...Anthraquinone
Anthracene oil see...Anthracene
Anthracene polycyclic aromatic compound see...Anthracene
Anthracin see...Anthracene
Anthradione see...Anthraquinone
ANTHRAPOLE 73® see...o-Phenylphenol
(p)ANTHRAPEL® see...Anthraquinone
9,10-Anthraquinone see...Anthraquinone
ANTIAETHAN® see...Disulfiram
ANTIBIOTIC N-329 B® see...Validamycin
ANTIBIOTIC TM 25® see...Oxytetracycline Calcium
ANTIBULIT® see...Sodium Fluoride
ANTICARIE® see...Hexachlorobenzene
ANTIETANOL® see...Disulfiram
ANTI-ETHYL® see...Disulfiram
ANTIETIL® see...Disulfiram
ANTIGAL® see...Diazinon
ANTIKOL® see...Disulfiram
Antimicina A (Spanish) see...Antimycin A
ANTIMILACE® see...Metaldehyde
Antimol see...Sodium Benzoate
Antimonate (2-), bis μ-2,3-dihydroxybutanedioata (4-)-01,02:03,04di-, dipotassium, trihydrate, stereoisomer see...Antimony Potassium Tartrate
Antimony hydride see...Stibine
Antimony trihydride see...Stibine

Antimonyl potassium tartrate see...Antimony Potassium Tartrate
Antimonwasserstoffes (German) see...Stibine
ANTIMUCIN WDR® see...Phenylmercury Acetate
Antimycin A see...Antimycin A
ANTINONIN® see...Dinitro-o-cresol (DNOC)
ANTINONNIN® see...Dinitro-o-cresol (DNOC)
ANTIO® see...Formothion
Antipiricullin see...Antimycin A
ANTISAL®-4 see...Trisodium Phosphate
ANTIVERM® see...Phenothiazine
ANTIVITIUM® see...Disulfiram
ANTLAK® see...Diazinon
ANTOR® see...Diethatyl-ethyl
Antraceno (Spanish) see...Anthracene
ANTRACOL® see...Propineb
Antraquinona (Spanish) see...Anthraquinone
Antrene bis(dimethylcarbamodithioato-s,s')zinc see...Ziram
ANTURAT® see...ANTU
Antymonowodor (Polish) see...Stibine
ANVIL® see...Hexaconazole
APACHE® see...Cadusafos
APACHLOR® see...Chlorfenvinphos
APADODINE® see...Dodine
APADRIN® see...Monocrotophos
APAMIDON® see...Phosphamidon
APARASIN® see...Lindane
APAVAP® see...Dichlorvos
APAVINPHOS® see...Mevinphos
APEX® see...Methoprene
APHAMITE® see...Parathion
APHOX® see...Pirimicarb
[2-(α-Aphthoxy)-N,N-diethylpropionamide] see...Napropamide
APHTIRIA® see...Lindane
APISTAN® see...Fluvalinate
APLIDAL® see...Lindane
APL-LUSTER® see...Thiabendazole
APOGEE® PLANT GROWTH REGULATOR see...Prohexadione Calcium
APOLLO® see...Clofentezine
APPA® see...Phosmet
APPLAUD® see...Buprofezin
APPLE DUST No. 1® see...Ferbam
APPL-SET® see...1-Naphthaleneacetic Acid
APRON® see...Captan
APRON® see...Metalaxyl
APTAL® see...p-Chloro-m-Cresol
Apyonine auramarine base see...Auramine
4-AP® see...4-Aminopyridine
AQ-10 Biofungicide® see...Ampelomyces Quisqualis isolate M10
AQ-10 Technical Powder® see...Ampelomyces Quisqualis isolate M10
Aqua ammonia see...Ammonia
AQUA-CLEAR® see...Diquat
AQUA-KLEEN® see...2,4-D

AQUA-KLEEN® see...2,4-D, butoxyethyl ester
AQUA-VEX® see...Silvex
AQUACIDE® see...Diquat
AQUACIDE® see...Diquat Dibromide
AQUAKILL® see...Diquat
AQUAKILL® see...Diquat Dibromide
Aqualine see...Acrolein
AQUALIN® see...Acrolein
AQUANEAT® see...Glyphosate
AQUATHOL® see...Endothall
AQUATIN 20 EC® see...Triphenyltin Compounds
AQUAZINE® see...Simazine
AQUCAR® see...Glutaraldehyde
ARAB RAT DETH® see...Warfarin
ARACIDE® see...Aramite
ARACNOL K® see...Tetradifon
Aragonite (mineral) see...Calcium Carbonate
ARAGRAN® see....Terbufos
ARALO® see...Parathion
ARARAMITE-15W® see...Aramite
ARASAN® see...Thiram
ARATHANE® see...Dinocap
ARATRON® see...Aramite
ARBINEX 30TN® see...Heptachlor
ARBITEX® see...Lindane
ARBOGAL® see...Fenitrothion
ARBORICID® see...2,4,5-Trichlorophenoxyacetic Acid, Esters
ARBOROL® see...Dinitro-o-cresol (DNOC)
ARBOTECT® see...Thiabendazole
ARCHER® see...Pyriproxyfen
ARD 34/02® see...Sethoxydim
ARDAP® see...Cypermethrin
ARDENT® see...Diflufenican
AREDION® see...Tetradifon
Areginal see...Ethyl Formate
ARELON® DISPERSION see...Isoproturon
A-REST® see...Ancymidol
ARETIT® see...Dinoseb
ARGEZIN® see...Atrazine
ARILAT® see...Carbaryl
ARILATE® see...Benomyl
ARILATE® see...Carbaryl
ARIOTOX® see...Metaldehyde
ARKOTINE® see...DDT
ARMOR® see...Cyromazine
ARMY® see...Amitraz
AROL GORDON DUST® see...Rotenone
AROSURF® 66ES see...Arosurf® MSF
AROSURF® 66E2 see...Arosurf® MSF
ARPROCARB® see...Propoxur
ARQUAD DMMCB-75® see...Zilkonium Chloride
ARRIVO® see...Cypermethrin
ARROSOLO® see...Propanil
ARROSOLO® see...Molinate
ARSANOTE Liquid® see...Sodium Methanearsonate (MSMA)
ARSAN® see...Cacodylic Acid

ARSECODILE® see...Sodium Cacodylate
Arsen (German, Polish) see...Arsenic and Inorganic Arsenic Compounds
Arsenate see...Arsenic Acid
Arsenate de calcium (French) see...Calcium Arsenate
Arsenate of lead see...Lead Arsenate
Arseniate de plomb (French) see...Lead Arsenate
Arseniato calcico (Spanish) see...Calcium Arsenate
Arseniato de plomo (Spanish) see...Lead Arsenate
Arseniato potasico (Spanish) see...Potassium Arsenate
Arsenic-75 see...Arsenic and Inorganic Arsenic Compounds
o-Arsenic Acid see...Arsenic Acid
Arsenic acid anhydride see...Arsenic Pentoxide
Arsenic acid, calcium salt (2:3) see...Calcium Arsenate
Arsenic acid, lead(2+) see...Lead Arsenate
Arsenic acid, lead(II) see...Lead Arsenate
Arsenic acid, lead salt see...Lead Arsenate
Arsenic acid, lead(2+) salt see...Lead Arsenate
Arsenic acid, monopotassium salt see...Potassium Arsenate
Arsenic anhydride see...Arsenic Pentoxide
Arsenic black see...Arsenic and Inorganic Arsenic Compounds
Arsenic blanc (French) see...Arsenous Oxide
Arsenic, metallic see...Arsenic and Inorganic Arsenic Compounds
Arsenic oxide see...Arsenic Pentoxide
Arsenic(III) oxide see...Arsenous Oxide
Arsenic(V) oxide see...Arsenic Pentoxide
Arsenic pentaoxide see...Arsenic Pentoxide
Arsenic pentoxide see...Arsenic Acid
Arsenic sesquioxide see...Arsenous Oxide
Arsenic, solid see...Arsenic and Inorganic Arsenic Compounds
Arsenic trioxide see...Arsenous Oxide
Arsenic trioxide, solid see...Arsenous Oxide
Arsenicals see...Arsenic and Inorganic Arsenic Compounds
Arsenico (Spanish) see...Arsenic and Inorganic Arsenic Compounds
Arsenicum album see...Arsenous Oxide
Arsenigen saure (German) see...Arsenous Oxide
Arsenious acid see...Arsenous Oxide
Arsenious acid, copper(2+) salt (1:1) see...Copper Arsenite
Arsenious oxide see...Arsenous Oxide
Arsenious trioxide see...Arsenous Oxide
Arsenite see...Arsenous Oxide
Arsenite de sodium (French) see...Sodium Arsenite
Arsenito calcico (Spanish) see...Calcium Arsenite
Arsenito de cobre (Spanish) see...Copper Arsenite
Arsenito sodico (Spanish) see...Sodium Arsenite
Arsenolite see...Arsenous Oxide
Arsenous acid see...Arsenous Oxide
Arsenous acid anhydride see...Arsenous Oxide
Arsenous acid, calcium salt see...Calcium Arsenite
Arsenous acid, sodium salt see...Sodium Arsenite

Arsenous anhydride see...Arsenous Oxide
Arsenous oxide anhydride see...Arsenous Oxide
Arsinette see...Lead Arsenate
ARSINETTE® see...Lead Arsenate
Arsinic acid, dimethyl- see...Cacodylic Acid
Arsodent see...Arsenous Oxide
Arsonic acid, calcium salt (1:1) see...Calcium Arsenite
Arsonic acid, copper(2+) salt (1:1) see...Copper Arsenite
Arsonic acid, methyl- see...Methanearsonic Acid
Arsonic acid, methyl-, calcium salt (2:1) see...Calcium Methanearsonate
Arsonic acid, methyl-, compounded with 1-octanamine see...Octylammonium Methanearsonate
Arsonic acid, methyl-, monosodium salt see...Sodium Methanearsonate (MSMA)
ARSYCODILE® see...Sodium Cacodylate
ARTESIAN® see...Flutolanil
ARTHODIBROM® see...Naled
Artificial mustard oil see...Allyl Isothiocyanate
ARVEST® see...Ethephon
ARWOOD COPPER® see...Copper and Copper Compounds
ARYLAM® see...Carbaryl
AS-50® see...Streptomycin
AS-50® see...Streptomycin Sulfate
AS-120® see...Arsenic and Inorganic Arsenic Compounds
AS-217® see...Arsenic and Inorganic Arsenic Compounds
ASAGIO® see...Bentazon
ASANA® see...Esfenvalerate
ASANA® DPX-YB656-84 see...Esfenvalerate
ASANA-XL® see...Esfenvalerate
ASATAF® see...Acephate
ASAZOL BUENO® see...Sodium Methanearsonate (MSMA)
ASEX® see...Sodium Chlorate
ASHLAND SOLACE® see...Mancozeb
ASHLAND TRIMARAN® see...Trifluralin
ASIF® see...Acifluorfen
ASIFY® see...Acephate
ASILAN® see...Asulam
ASP47® see...Sulfotepp
ASP 51® see...Aspon®
ASPON-CHLORDANE® see...Chlordane
ASPOR® see...Zineb
ASPORUM® see...Zineb
ASSASSIN® see...Mecoprop
ASSERT® see...Imazethabenz
ASSET PGR® see...Indole-3-Butyric Acid
ASSURE® see...Quizalofop-Ethyl
ASTONEX® see...Diflubenzuron
As-triazin-5(4H)-one,4-amino-6-tert-butyl-3-(methylthio)- see...Metribuzin
ASTROBOT® see...Dichlorvos
ASULFOX F® see...Asulam
ASULOX® see...Asulam

ASULOX 40® see...Asulam
ASUNTOL® see...Coumaphos
AT-7® see...Hexachlorophene
AT-17® see...Hexachlorophene
AT-90® see...Amitrole
ATA see...Amitrole
ATABRON® see...Chlorfluazuron
ATAZINAX® see...Atrazine
ATCP see...Picloram
ATEMI® see...Cyproconazole
ATEMI-50-SL® see...Cyproconazole
ATERBUTEX® see...Atrazine
ATERBUTOX® see...Atrazine
ATGARD® see...Dichlorvos
ATHROMBIN® see...Warfarin
ATHYL-GUSATHION® see...Azinphos-ethyl
Athylen (German) see...Ethylene
ATIRAM® see...Thiram
ATLACIDE® see...Sodium Chlorate
ATLAS® see...Propanil
ATLAS A® see...Sodium Arsenite
ATLAS ATRAZINE® see...Atrazine
ATLAS CHLORMEQUAT® see...Chlormequat Chloride
ATLAS CROPGARD® see...Chlorothalonil
ATLAS LIGNUM (FORMULATION)® see...Dalapon
ATLAS FIELDGARD® see...Isoproturon
ATLAS SIMAZINE® see...Simazine
ATLAZIN® see...Amitrole
ATLAZIN D-WEED® see...Atrazine
ATLAZINE® FLOWABLE see...Amitrole
ATOMIC® see...Methyl Parathion
ATOMIT® see...Calcium Carbonate
ATRA-BUTE® see...Butylate
ATRAFLOW PLUS® see...Amitrole
ATRAPA 5E® see...Malathion
ATRASINE® see...Atrazine
ATRATAF® see...Atrazine
ATRATOL® see...Atrazine
ATRATOL® see...Sodium Metaborate
ATRATOL B-HERBATOX® see...Sodium Chlorate
Atrazin (German) see...Atrazine
Atrazina (Spanish) see...Atrazine
ATRAZINE 90DF® see...Atrazine
ATRAZINEK® see...Atrazine
ATRED® see...Atrazine
ATREX® see...Atrazine
ATROMBINE-K® see...Warfarin
AT® see...Amitrole
3-AT® see...Amitrole
AT® see...Amitrole
3-AT® see...Amitrole
ATTAC 2® see...Toxaphene
ATTAC 6® see...Toxaphene
ATTAC 6-3® see...Toxaphene
ATTACK® see...Acephate
ATTACK® see...Thiram
ATTATOX® see...Cyfluthrin

AU'ULTRAMICIN® see...Carbofuran
AULES® see...Thiram
Auramina (Spanish) see...Auramine
Auramine base see...Auramine
Auramine N base see...Auramine
Auramine OAF see...Auramine
Auramine O base see...Auramine
Auramine SS see...Auramine
AUTHORITY® see...Chlorimuron-ethyl
AUTHORITY® see...Metribuzin
AUTHORITY® see...Sulfentrazone
AUTUMN KITE® see...Trifluralin
AUXIGRO® see...Glutamic Acid
AVADEX® see...Diallate
AVADEX BW® see...Triallate
AVAST® see...Fluridone
AVENGE® see...Difenzoquat
Avermectin see...Abamectin
Avermectin $B_1$ see...Abamectin
Avermectin $B_{1a}$ +Avermectin $B_{1b}$ mixture see...Abamectin
Avermectin $B_1$, 4"-deoxy-4"-(methylamino)-, (4"R)-benzoate (salt) see...Emamectin Benzoate
AVERSAN® see...Disulfiram
AVERZAN® see...Disulfiram
AVG see...Aminoethoxyvinylglycine Hydrochloride
AVICADE® see...Cypermethrin
AVICOL (PESTICIDE)® see...Quintozene
AVID® see...Abamectin
AVITROL® see...4-Aminopyridine
AVITROL 200® see...4-Aminopyridine
AVLOTHANE® see...Hexachloroethane
AVOMEC® see...Abamectin
AWARD see...Fenoxycarb
AWARD® see...Penconazole
AXIOM® see...Atrazine
AXIOM® see...Metribuzin
AXIOM® see...Thiafluamide
AXIUM® see...Sodium Azide
Azacosterol hydrochloride see...Azacosterol Dihydrochloride
Azacyclohexane see...Piperidine
Azadirachtin A see...Azadirachtin
AZAPLANT® see...Amitrole
AZAPLANT KOMBI® see...Amitrole
AZASTEROL® see...Azacosterol Dihydrochloride
Azasterol HCL see...Azacosterol Dihydrochloride
AZATIN EC® see...Azadirachtin
AZATIN®-XL PLUS see...Azadirachtin
AZATROL EC® see...Azadirachtin
1H-Azepine-1-carbothioic acid, hexahydro-S-ethyl ester see...Molinate
Azetochlor see...Acetochlor
Azetylaminofluoren see...Acetylaminofluorene
Azida sodico (Spanish) see...Sodium Azide
AZIDE® see...Sodium Azide
Azinfos-ethyl (Dutch) see...Azinphos-ethyl
Azinfos-methyl (Dutch) see...Azinphos-methyl

Azinos see...Azinphos-ethyl
AZINOTOX® see...Atrazine
Azinphos-aethyl (German) see...Azinphos-ethyl
Azinphos etile (Italian) see...Azinphos-ethyl
AZINPHOS-METHYL GUTHION see...Azinphos-methyl
Azinphosmetile (Italian) see...Azinphos-methyl
AZIUM® see...Sodium Azide
AZODIENO® see...Amitraz
AZODRIN® see...Monocrotophos
AZOFENE® see...Phosalone
AZOFOS® see...Methyl Parathion
Azoksystrobin see...Azaxystrobin
AZOLAN® see...Amitrole
AZOLE® see...Amitrole
AZOPHOS see...Methyl Parathion
AZOTOX® see...DDT
Azoture de sodium (French) see...Sodium Azide
Azoxistrobin see...Azaxystrobin
Azoxystrolin see...Azaxystrobin
AZTEC® see...Cyfluthrin
AZTEC® see...Triazamate
Azufre (Spanish) see...Sulfur
AZUNTHOL® see...Coumaphos

- B -

B-9® see...Daminozide
B-32® see...Hexachlorophene
B-404® see...Parathion
B-622® see...Anilazine
B-995® see...Daminozide
B-1776® see...Tribufos
B-3015® see...Thiobencarb
B-37344® see...Methiocarb
BA® see...6-Benzaldenine
6-BA® see...6-Benzaldenine
BAAM® see...Amitraz
BAC® see...Nicotine
BACARA® see...Diflufenican
BAC-TEX® see...Triclosan
BAKER BRAND® see...Zinc Phosphide
BAKTOL® see...*p*-Chloro-*m*-Cresol
BAKTOLAN® see...*p*-Chloro-*m*-Cresol
BALAN® see...Benefin
BALANCE® WDG see...Ioxaflutole
BALFIN® see...Benefin
BAMA BRAND® see...Methyl Parathion
BANEX® see...Dicamba
BANGTON® see...Captan
BAN-HOE® see...Propham
BANKIT® see...Azaxystrobin
BANLEN® see...Dicamba
BANLENE® see...MCPA
BAN-MITE® see...Malathion
BANNER® see...Propiconazole
BANOL® see...Propamocarb Hydrochloride
BANOL C® see...Chlorothalonil

BANROT® see...Etridiazole
BANTU® see...ANTU
BANVEL® see...Dicamba
BANVEL CST® see...Dicamba
BANVEL 4WS® see...Dicamba
BANVEL P® see...Mecoprop
BANVEL 4S® see...Dicamba
BANVEL® II/SGF see...Sodium Dicamba
BANVEL®-520 see...2,4-D, isooctyl ester
BANVEL HERBICIDE® see...Dicamba
BANVEL II HERBICIDE® see...Dicamba
BAP see...6-Benzaldenine
6-BAP see...6-Benzaldenine
BAR 500 EC® see...Chlorpyrifos
BARBAMATE® see...Barban
BARBANE® see...Barban
BARBASCO® see...Rotenone
BARBER'S® WEED KILLER (ESTER FORMULATION) see...2,4-D, isopropyl ester
BARDIKE® see...Cuprous Thiocyanate
BAREGROUND® see...Sodium Chlorate
BAREGROUND® see...Sodium Metaborate
BAR-FUNGAL PLUS® see...4-Chloro-3,5-xylenol
Bario (Spanish) see...Barium and Barium Compounds
Barium metal see...Barium and Barium Compounds
Barium, elemental see...Barium and Barium Compounds
BARIZON® see...BPMC
BARQUAT® MB-50 see...Zilkonium Chloride
BARQUAT® MB-80 see...Zilkonium Chloride
BARRAGE® see...2,4-D
BARRICADE® see...Cypermethrin
BARRIER® see...Dichlobenil
BAS 083 01 W® see...Mepiquat Chloride
BAS 351-H® see...Bentazon
BAS 352-F® see...Vinclozolin
BAS 392-H® see...Fluchloralin
BAS 514 00H® see...Quinclorac
BAS 530 04® see...Fomesafen
BAS 35204-F® see...Vinclozolin
BAS 85559X® see...Mepiquat Chloride
BASAGRAN® see...Bentazon
BASALIN® see...Fluchloralin
BASAMID® see...Dazomet
BASAMID-FLUID® see...Metham-Sodium
BASAMID® G see...Dazomet
BASAMID®-GRANULAR see...Dazomet
BASAMID® P see...Dazomet
BASAMID-PUDER® see...Dazomet
BASANITE® see...Dinoseb
BASAPON® see...Dalapon
BASAPON B® see...Dalapon
BASAPON N® see...Dalapon
BASELINE® PLANT REGULATOR see...Prohexadione Calcium
BASFAPON® see...Dalapon
BASFAPON® B see...Dalapon
BASFAPON® N see...Dalapon
BASF® 250 see...Ethephon

BASF® 9052  see...Sethoxydim
BASF® 32500 Fungicide  see...Thiophanate-Methyl
BASF® GRUNKUPFER  see...Copper Oxychloride
BASF-MANEB SPRITZPULVER®  see...Maneb
BASICOP®  see...Copper Sulfate
Basic copper chloride  see...Copper Oxychloride
Basic cupric carbonate  see...Copper Carbonate, Basic
Basic cupric chloride  see...Copper Oxychloride
Basic Yellow 2  see...Auramine
BASINEX®  see...Dalapon
BASIS®  see...Nicosulfuron
BASIS®  see...Thifensulfuron Methyl
Basle green  see...Copper Acetoarsenite
Basle green  see...Paris Green
Baso Yellow 124  see...Auramine
BASSA®  see...BPMC
BASUDIN®  see...Diazinon
BASUS®  see...Fenoxycarb
BAS® 300  see...Pyridaben
BAS® 3460  see...Carbendazim
BAS® 67054  see...Carbendazim
BAS® 32500F  see...Thiophanate-Methyl
BATAMIX® PROGRESS  see...Ethofumesate
BATASAN®  see...Triphenyltin Compounds
BATAZINA®  see...Simazine
BATTALION®  see...Halosulfuron-methyl
BATTALION®  see...Terbutryn
BATTAL®  see...Carbendazim
BAVISTIN®  see...Carbendazim
BAVISTIN M®  see...Maneb
BAY 21/199®  see...Coumaphos
BAY 73®  see...Clonitralid
BAY 5024®  see...Methiocarb
BAY 5122®  see...Propoxur
BAY 5621®  see...Phoxim
BAY 6076®  see...Clonitralid
BAY 6681-F®  see...Triadimefon
BAY 9026®  see...Methiocarb
BAY 9027®  see...Azinphos-methyl
BAY 11405®  see...Methyl Parathion
BAY 15203®  see...Demeton-methyl
BAY 15922®  see...Trichlorfon
BAY 16225®  see...Azinphos-ethyl
BAY 18436®  see...Demeton-methyl
BAY 18436®  see...Demeton
BAY 18436®  see...Demeton
BAY 19149®  see...Dichlorvos
BAY 19639®  see...Disulfoton
BAY 21097®  see...Demeton-methyl
BAY 22555®  see...Fenaminosulf
BAY 23323®  see...Oxydisulfoton
BAY 25634®  see...Coumatetralyl
BAY 30130®  see...Propanil
BAY 33051®  see...Phenthoate
BAY 33819®  see...Phosacetim
BAY 34727®  see...Cyanofos
BAY 36205®  see...Oxythioquinox
BAY 37289®  see...Trichloronate

BAY 37344®  see...Methiocarb
BAY 39007®  see...Propoxur
BAY 41637-C®  see...BPMC
BAY 41637®  see...BPMC
BAY 41831®  see...Fenitrothion
BAY 46131®  see...Propineb
BAY 61597®  see...Metribuzin
BAY 68138®  see...Fenamiphos
BAY 71625®  see...Methamidophos
BAY 77488®  see...Phoxim
BAY 78537®  see...Carbofuran
BAY 92114®  see...Isofenphos
BAY 704143®  see...Carbofuran
BAY CARB®  see...BPMC
BAY DIC 1468®  see...Metribuzin
BAY DRW 1139®  see...Metamiton
BAY E-393®  see...Sulfotepp
BAY E-601®  see...Methyl Parathion
BAY E-605®  see...Parathion
BAY ENE® 11183B  see...Coumatetralyl
BAY FCR 1272®  see...Cyfluthrin
BAY HWG 1608®  see...Tebuconazole
BAY KWG 0519®  see...Triadimenol
BAY KWG 0599®  see...Bitertanol
BAY MEB 6447®  see...Triadimefon
BAY NAK 1654®  see...Fenfluthrin
BAY NTN 6867®  see...Diethatyl-ethyl
BAY NTN 9306®  see...Sulprofos
BAY NTN 19701®  see...Pencycuron
BAY SRA 7502  see...Phoxim
BAY SRA-12869®  see...Isofenphos
BAYCLEAN®  see...Zilkonium Chloride
BAYCOR®  see...Bitertanol
BAYCOR®  see...Fuberidazole
BAYER 21/116®  see...Demeton-methyl
BAYER 21/199®  see...Coumaphos
BAYER 25-154®  see...Demeton
BAYER 25-154®  see...Demeton-methyl
BAYER 73®  see...Clonitralid
BAYER 4964®  see...Oxythioquinox
BAYER 5072®  see...Fenaminosulf
BAYER 6159H®  see...Metribuzin
BAYER 6443H®  see...Metribuzin
BAYER 6681-F®  see...Triadimefon
BAYER 8169®  see...Demeton
BAYER 10756®  see...Demeton
BAYER 15922®  see...Trichlorfon
BAYER 16259®  see...Azinphos-ethyl
BAYER 17147®  see...Azinphos-methyl
BAYER 18510®  see...Phenthoate
BAYER 19149®  see...Dichlorvos
BAYER 19639®  see...Disulfoton
BAYER 21097®  see...Demeton-methyl
BAYER 25,634®  see...Coumatetralyl
BAYER 25648®  see...Clonitralid
BAYER 33172®  see...Fuberidazole
BAYER 33819®  see...Phosacetim
BAYER 34727®  see...Cyanofos

BAYER 36205® see...Oxythioquinox
BAYER 37289® see...Trichloronate
BAYER 37344® see...Methiocarb
BAYER 39007® see...Propoxur
BAYER 41367C® see...BPMC
BAYER 41831® see...Fenitrothion
BAYER 45432® see...Omethoate
BAYER 46131® see...Propineb
BAYER 68138® see...Fenamiphos
BAYER 71628® see...Methamidophos
BAYER 77488® see...Phoxim
BAYER 94337® see...Metribuzin
BAYER B 5122® see...Propoxur
BAYER-E-393® see...Sulfotepp
BAYER E-605® see...Parathion
BAYER-HWG-1608® see...Tebuconazole
BAYER L 13/59® see...Trichlorfon
BAYER MEB-6447® see...Triadimefon
BAYER NTN 9306® see...Sulprofos
BAYER NTN-19701® see...Pencycuron
BAYER S 4400® see...Trichloronate
BAYER S 5660® see...Fenitrothion
BAYER S-6876® see...Omethoate
BAYER SS2074® see...Oxythioquinox
BAYFIDAN EW® see...Triadimenol
BAYFIDAN® see...Triadimenol
BAYGON® see...Propoxur
BAYLETON® see...Triadimefon
BAYLUSCID® see...Clonitralid
BAYLUSCIDE® see...Clonitralid
BAYMAT-SPRAY® see...Bitertanol
BAYMIX® see...Coumaphos
BAYMIX® 50 see...Coumaphos
BAYNAC® see...Fenfluthrin
BAYRUSIL® see...Quinalphos
BAYTAN® see...Fuberidazole
BAYTAN 30® FUNGICIDE see...Triadimenol
BAYTAN IM® see...Imazalil
BAYTAN® SEED TREATMENT see...Triadimenol
BAYTHION® see...Phoxim
BAYTHROID® see...Cyfluthrin
BAYTHROID® H see...Cyfluthrin
BAYTHROID® TECHNICAL see...Cyfluthrin
BAZUDEN® see...Diazinon
BB CHLOROTHALONIL® see...Chlorothalonil
BBC 12® see...Dibromochloropropane
BBC 6597® see...Benomyl
BBC see...Benomyl
BBH see...Lindane
BCEE see...Dichloroethyl Ether
BCF-BUSHKILLER® see...2,4,5-T
BCF-BUSHKILLER® see...2,4,5-Trichlorophenoxyacetic Acid, Esters
BCM see...Carbendazim
BCS COPPER FUNGICIDE® see...Copper Sulfate
BEACON® see...Primisulfuron-Methyl
BEAN SEED PROTECTANT® see...Captan
BEET-KLEEN® see...Propham

BEISTERGARD® see...Captan
BELCO® see...Dieldrin
Bell mine see...Calcium Hydroxide
BELL MINE PULVERIZED LIMESTONE® see...Calcium Carbonate
BELMARK® see...Fenvalerate
BELT® see...Chlordane
BENATOL® see...Benzyl Alcohol
BENCARB® see...Thiobencarb
Bencarbate see...Bendiocarb
BENDAZIM® see...Carbendazim
BENDEX® see...Fenbutatin Oxide
Bendiocarbe see...Bendiocarb
BENDIOXIDE® see...Bentazon
BENEFECT® see...Thymol
BENEFEX® see...Benefin
BENEX® see...Benomyl
Benfluralin see...Benefin
Benfluraline see...Benefin
BENFOS® see...Dichlorvos
Benfuracarb see...Benfuracarb
BEN-HEX® see...Lindane
BENIT® see...Propiconazole
BENLAT® see...Benomyl
BENLATE® see...Benomyl
Benomilo (Spanish) see...Benomyl
BENOMYL® 50W see...Benomyl
BENOSAN® see...Benomyl
Bensonitrile, 3,5-dibromo-4-hydroxy- see...Bromoxynil
Bensulfoid see...Sulfur
BENSUMEC® see...Bensulide
BENTALOL® see...Benzyl Alcohol
BENTAZONE® see...Bentazon
BENTA® see...Bentazon
BENTHIOCARB® see...Thiobencarb
BENTOX 10® see...Lindane
Benz-o-chlor see...Chlorobenzilate
BENZ-O-CHLOR® see...Chlorobenzilate
BENZABAR® see...Trichlorobenzoic Acid
BENZAC® see...Trichlorobenzoic Acid
Benzalkonium chloride see...Zilkonium Chloride
Benzamide see...Flutolanil
BENZAMIDE® see...Pronamide
Benzamide, N-[((4-chlorophenyl)amino)carbonyl]-2,6-difluoro see...Diflubenzuron
Benzamide, 5-[2-chloro-4-(trifluoromethyl)phenoxy]-N-(methylsulfonyl)-2-nitro- see...Fomesafen
Benzamide, N-[((3,5-dichloro-4-{{3-chloro-5-(trifluoromethyl)-2-pyridinyl}oxy}phenyl)amino)carbonyl]-2,6-difluoro- see...Chlorfluazuron
Benzamide, 3,5-dichloro-N-(1,1-dimethyl-2-propynyl) see...Pronamide
Benzamide, N-[((3,5-dichloro-4-(1,1,2,2-tetrafluoroethoxy)phenyl)amino)carbonyl]-2,6-difluoro- see...Hexaflumuron
Benzamide, 2,6-dimethoxy-N-[3-(1-ethyl-1-methylpropyl)-5-isoxazolyl]- see...Isoxaben

Benzamide, N-[3-(1-ethyl-1-methylpropyl)-5-isoxazolyl]-2,6-dimethoxy- see...Isoxaben
Benzamizole see...Isoxaben
Benzenamine, N-butyl-N-ethyl-2,6-dinitro-4-(trifluoromethyl)- see...Benefin
Benzenamine, N-(2-chloroethyl)-2,6-dinitro-N-propyl-4-(trifluoromethyl)- see...Fluchloralin
Benzenamine, N-(cyclopropylmethyl)-2,6-dinitro-N-propyl-4-(trifluoromethyl)- see...Profluralin
Benzenamine, 2,6-dichloro-4-nitro- see...Dichloran
Benzenamine, 3,4-dimethyl-2,6-dinitro-N-(1-ethylpropyl)- see...Pendimethalin
Benzenamine, 4-(1,1-dimethylethyl)-N-(1-methylpropyl)-2,6-dinitro- see...Butralin
Benzenamine, 2,6-dinitro-N,N-dipropyl-4-(trifluoromethyl-) see...Trifluralin
Benzenamine, 4-ethoxy-N-(5-nitro-2furanyl)methylene- see...Nitrofen
Benzenamine, N-ethyl-N-(2-methyl-2-propenyl)-2,6-dinitro-4-(trifluoromethyl)- see...Lindane
Benzenamine, N-(1-ethylpropyl)-3,4-dimethyl-2,6-dinitro- see...Pendimethalin
Benzenamine, 4-(1-methylethyl)-2,6-dinitro-N,N-dipropyl- see...Isopropalin
1,3-Benzenedicarbonitrile,2,4,6,6-tetrachloro- see...Chlorothalonil
Benzeneacetamide, N,N-dimethyl-α-phenyl- see...Diphenamid
Benzeneacetic acid, 4-chloro-α-(4-chlorophenyl)-α-hydroxy-,ethyl ester see...Chlorobenzilate
Benzeneacetic acid, 4-chloro-α-(1-methylethyl)-, cyano(3-phenoxyphenyl)methyl ester see...Fenvalerate
Benzeneacetic acid, 4-chloro-α-(1-methylethyl)-, cyano(3-phenoxyphenyl)methyl ester, [s-(R*,R*)]- see...Esfenvalerate
Benzeneacetic acid, 4-(difluoromethoxy)-α-(1-methylethyl)-, cyano(3-phenoxyphenyl)methyl ester see...Flucythrinate
Benzeneacetic acid, 2,3,6-trichloro- see...Fenac
Benzene, (acetoxymercuri)- see...Phenylmercury Acetate
Benzene, (acetoxymercurio) see...Phenylmercury Acetate
Benzeneamine, 4,4'-cabonimidoylbis[n-dimethyl- see...Auramine
Benzeneamine, 2,6-dinitro-N,N-dipropyl-4-(trifluoromethylaniline) see...Trifluralin
Benzenecarbinol see...Benzyl Alcohol
Benzenecarboxylic acid see...Benzoic Acid
Benzene, 2-chloro-1-(3-ethoxy-4-nitrophenoxy)-4-(trifluoromethyl)- (9CI) see...Oxyfluorfen
Benzene, 1-chloro-2-methyl- see...o-Chlorotoluenel
1,3-Benzenediamine, 2,6-dinitro-N¹,N¹-dipropyl-4-(trifluoromethyl)- see...Prodiamine
Benzene, dichloro- see...para-Dichlorobenzene
Benzene, 1,4-dichloro- see...para-Dichlorobenzene
Benzene, p-dichloro- see...para-Dichlorobenzene
Benzene, 1,1'-(2,2-dichloroethylidene)bis(4-ethyl- see...Ethylan
Benzene, 2,4-dichloro-1-(4-nitrophenoxy)- see...Nitrofen
Benzene, 1-[(2-(4-ethoxyphenyl)-2-methylpropoxy)methyl]-3-phenoxy- see...Ethofenprox
Benzene fluoride see...Fluorobenzene
Benzene, fluoro- see...Fluorobenzene
Benzeneformic acid see...Benzoic Acid
Benzene hexachloride see...Lindane
α-Benzenehexachloride see...Hexachlorocyclohexanes
β-Benzenehexachloride see...Hexachlorocyclohexanes
γ-Benzene hexachloride see...Lindane
δ-3-Benzenehexachloride see...Hexachlorocyclohexanes
Benzene hexachloride-α-isomer see. Hexachlorocyclohexanes
Benzene hexachloride-gamma isomer see...Lindane
Benzene-trans-hexachloride see. .Hexachlorocyclohexanes
Benzene, hexachloro- see...Hexachlorobenzene
Benzenemethanoic acid see...Benzoic Acid
Benzenemethanol see...Benzyl Alcohol
Benzenemethanol, 4-chloro-α-(4-chlorophenyl)-α-(trichloromethyl)- see...Dicofol
Benzene, methoxy see...Anisole
Benzene, pentachloronitro- see...Quintozene
Benzenesulfonamide, 2-(2-chloroethoxy)-N-[((4-methoxy-6-methyl-1,3,5-triazin-2-yl)amino)carbonyl]- see...Triasulfuron
Benzenesulfonamide, 2-chloro-N-[((4-methoxy-6-methyl-1,3,5-triazin-2-yl)amino)carbonyl]- see...Chlorsulfuron
Benzenesulfonamide, 4-(dipropylamino)-3,5-dinitro- see. Oryzalin
Benzenesulfonamide, p-hydroxy-N,N-dimethyl-, O-ester with O,O-dimethyl phosphorothioate see...Famphur
Benzene sulfonic acid, dodecyl- see...Dodecylbenzenesulfonic Acid
Benzene sulphonic acid, dodecyl- see. .Dodecylbenzenesulfonic Acid
Benzenesulfonic acid, dodecyl-, compounded with 2,2',2"-nitrilotris[ethanol](1:1) see...Triethanolamine Dodecylsenzene Sulfonate
Benzene sulfonic acid, dodecyl ester see...Dodecylbenzenesulfonic Acid
Benzene sulphonic acid, dodecyl ester see...Dodecylbenzenesulfonic Acid
Benzene, 1,1'-(2,2,2-trichloroethylidene)bis(4-chloro) see...DDT
Benzene,1,1'-(2,2,2-trichloroethylidene)bis[4-methoxy-] see...Methoxychlor
BENZEX® see...Hexachlorocyclohexanes
Benzhydrol, 4,4'-dichloro-.alpha.-(trichloromethyl)- see...Dicofol
Benzilan see...Chlorobenzilate
BENZILAN® see...Chlorobenzilate
Benzilic acid, 4,4'-dichloro-,ethyl ester see...Chlorobenzilate
2-Benzimidazolecarbamic acid, 1-(butylcarbamoyl)-, methyl ester see...Benomyl

Benzimidazole-2-carbamic acid, methyl ester see.. Carbendazim
Benzimidazole, 2-(4-thiazolyl)- see...Thiabendazole
1*H*-Benzimidazole, 2-(4-thiazolyl)- see...Thiabendazole
*N*-2-(Benzimidazolyl) carbamate see...Carbendazim
1*H*-Benzimidazol-2-ylcarbamic acid methyl ester see.. Carbendazim
4-(2-Benzimidazolyl)thiazole see...Thiabendazole
Benzin see...Naphthas
Benzinoform see...Carbon Tetrachloride
Benzoate see...Benzoic Acid
Benzoate of soda see...Sodium Benzoate
Benzoate sodium see...Sodium Benzoate
BENZOBOR® see...Trichlorobenzoic Acid
1,3-Benzodioxole, 5-[(2-(2-butoxyethoxy)ethoxy)methyl]-6-propyl- see.. Piperonyl Butoxide
1,3-Benzodioxole, 2,2-dimethyl-1,3-benzodioxol-4-ol methylcarbamate see...Bendiocarb
1,3-Benzodioxole, 2,2-dimethyl-4-(*N*-methylcarbamato)- see...Bendiocarb
1,3-Benzodioxol-4-ol, 2,2-dimethyl-,methylcrbamate see..Bendiocarb
α-Benzoepin see...Endosulfan
β-Benzoepin see...Endosulfan
Benzoesaeure (German) see...Benzoic Acid
Benzoesaeure (Na-salz) (German) see...Sodium Benzoate
7-Benzofuranol, 2,3-dihydro-2,2-dimethyl-, methylcarbamate see...Carbofuran
5-Benzofuranol, 2-ethoxy-2,3-dihydro-3,3-dimethyl-, methanesulfonate (+)- see...Ethofumesate
Benzofuroline see...Resmethrin
Benzoic acid, 3-amino-2,5-dichloro- see...Chloramben
Benzoic acid, 2-[((((4,6-bis(difluoromethoxy)-2-pyrimidinyl)amino)carbonyl)amino)sulfonyl]-, methyl ester see...Primisulfuron-Methyl
Benzoic acid, 2-chloro-6-[(4,6-dimethoxy-2-pyrimidinyl)thio]-, sodium salt (9CI) see...Pyrithiobac-Sodium
Benzoic acid, 3-chloro-2-[((5-ethoxy-7-fluoro(1,2,4)triazolo(1,5-c)pyrimidin-2-yl)sulfonyl)amino]-, methyl ester see...Cloransulam-methyl
Benzoic acid,2-[((((4-chloro-6-methoxy-2-pyrimidinyl)amino)carbonyl)amino)sulfonyl]-,ethyl ester see...Chlorimuron-ethyl
Benzoic acid, 5-(2-chloro-4-(trifluoromethyl)phenoxy)-2-nitro- see...Acifluorfen
Benzoic acid, 5-[2-chloro-4-(trifluoromethyl)phenoxy]-2-nitro-2-ethoxy-1-methyl-2-oxoethyl ester see.. Lactofen
Benzoic acid, 3,6-dichloro-2-methoxy- see...Dicamba
Benzoic acid, 3,6-dichloro-2-methoxy-, sodium salt (9CI) see...Sodium Dicamba
Benzoic acid, 5-(2,4-dichlorophenyl)-2-nitro-, methyl ester see...Bifenox
Benzoic acid, 2-(4,5-dihydro-4-methyl-4-(1-methylethyl)-5-xox-1*H*-imidazol-2-yl)-4 (or 5)-methyl-, methyl ester see...Imazethabenz

Benzoic acid, 2-[(((4-(dimethylamino)-6-(2,2,2-trifluoroethoxy)-1,3,5-triazin-2-yl)amino)carbonyl)amino)sulfonyl]-3-methyl-, methyl ester see...Triflusulfuron-Methyl
Benzoic acid, 3,5-dimethyl-, 1-(1,1-dimethylethyl)-2-(4-ethylbenzoyl)hydrazide see...Tebufenozide
Benzoic acid, 4-[((((1,3-dimethyl-5-phenoxy-1*H*-pyrazol-4-yl)methylene)amino)oxy)methyl]-, 1,1-dimethylethyl ester, (E)- see...Fenpyroximate
Benzoic acid, 2-[(((((4,6-dimethyl-2-pyrimidinyl)amino)carbonyl)amino)sulfonyl]-, methyl ester see...Sulfometuron-Methyl
Benzoic acid, *o*-[(3-(4,6-dimethyl-2-pyrimidinyl)ureido)sulfonyl]-, methyl ester see.. Sulfometuron-Methyl
Benzoic acid, 2-[(((((4,6-dimethoxy-2-pyrimidinyl)amino)carbonyl)amino)sulfonyl)methyl]-, methyl ester see...Bensulfuron-methyl
Benzoic acid, 2-[((((4-ethoxy-6-(methylamino)-1,3,5-triazin-2-yl)amino)carbonyl)amino)sulfonyl]-, methyl ester see...Ethametsulfuron-methyl
Benzoic acid, 2-[(ethoxy-((1-methylethyl)amino)phosphinothioyl)oxy]-, 1-methylethyl ester see...Isofenphos
Benzoic acid, 2-[(ethoxy((1-methylethyl)amino)phosphinothioyl)oxy]-, 1-methyl ester see...Isofenphos
Benzoic acid, 2-hydroxy- see...Salicylic Acid
Benzoic acid, 3- methoxy-2-methyl-2-(3,5-nimethylbenzoyl)-2-(1,1-nimethylethyl)hydrazide (9CI) see...Methoxyfenozide
Benzoic acid, 2-[((((4-methoxy-6-methyl-1,3,5-triazin-2-yl)amino)carbonyl)amino)sulfonyl]- methyl ester see..Metsulfuron-methyl
Benzoic acid, 2-[((((4-methoxy-6-methyl-1,3,5-triazin-2-yl)methylamino)car bonyl)amino)sulfonyl]-, methyl ester see...Tribenuron-Methyl
Benzoic acid, 2-[((((4-methoxy-6-methyl-1,3,5-triazin-2-yl)-*N*-methylamino)carbonyl)amino)sulfonyl]-, methyl ester see...Tribenuron-Methyl
Benzoic acid, 2-[(1-naphthalenylamino)carbonyl]- see.. Naptalam
Benzoic acid, 2-[(α-naphthalenylamino)carbonyl]- see. Naptalam
Benzoic acid, sodium salt see...Sodium Benzoate
Benzonitrile, 2,6-dichloro- see...Dichlobenil
Benzonitrile, 3,5-dibromo-4-hydroxy- see...Bromoxynil
Benzophosphate see...Phosalone
2*H*-1-Benzopyran-2-one, 3-(3-[4'-bromo(1,1'-biphenyl)-4-yl]-3-hydroxy-1-phenylpropyl)-4-hydroxy- see..Bromadiolone
Benzopyrano(3,4-b)furo(2,3-H)(1)benzopyrano-6(6a*H*)-one,1,2,12,12a-tetrahydro-8,9-dimethoxy-2-(1-methylethenyl)-, [2R-(2α,6aα,12aα)] see...Rotenone
2*H*-1-Benzopyran-2-one,4-hydroxy-3-(3-oxo-1-phenylbutyl)- see...Warfarin
2*H*-1-Benzopyran-2-one, 4-hydroxy-3-(1,2,3,4-tetrahydro-1-naphthalenyl)- see...Coumatetralyl

1H-2,1,3-Benzothiadiazin-4(3H)-one, 3-(1-methylethyl)-, 2,2-dioxide see...Bentazon
3(2H)-Benzothiazoleacetic acid, 4-chloro-2-oxo-, ethyl ester see...Benazolin Ethyl
2-(2-Benzothiazolyloxy)-N-methyl-N-phenylacetamide see...Mefenacet
2-(1,3-Benzothiazol-2-yloxy)-N-methylacetanilide see. Mefenacet
Benzotriazine derivative of an ethyl dithiophosphate see. Azinphos-ethyl
Benzotriazine derivative of a methyl dithiophosphate see. .Azinphos-methyl
Benzotriazinedithiophosphoric acid dimethoxy ester see. Azinphos-methyl
Benzoyl cyanide-o-(diethoxyphosphinothioyl)oxime see. Phoxim
BENZPHOS® see...Phosalone
Benzulfide see...Bensulide
Benzyfuroline see...Resmethrin
Benzyladenine see...6-Benzaldenine
N-Benzyladenine see...6-Benzaldenine
$N^6$-Benzyladenine see...6-Benzaldenine
Benzyl alcohol,2,4-dichloro-α-(chloromethylene)-, Diethyl phosphate see...Chlorfenvinphos
Benzyl alcohol, 2,4,5-trichloro-α-(chloromethylene)-, dimethyl phosphate see...Tetrachlorvinphos
Benzylaminopurine see...6-Benzaldenine
6-(Benzylamino)purine see...6-Benzaldenine
6-(N-Benzylamino)purine see...6-Benzaldenine
$N^6$-(Benzylamino)purine see...6-Benzaldenine
5-Benzylfurfuryl chrysanthemate see...Resmethrin
(5-Benzyl-3-furyl)methyl chrysanthemate see.. Resmethrin
5-Benzyl-3-furylmethyl(+)-trans-chrysanthemate see. Resmethrin
5-Benzyl-3-furylmethyl (±)-cis-trans-chrysanthemate see. Resmethrin
(5-Benzyl-3-furyl)methyl 2,2-dimethyl-3-(2-methylpropenyl)cyclopropanecarboxylate see. Resmethrin
5-Benzyl-3-furylmethyl(1RS)-cis,trans-2,2-dimethyl-3-(2-methylprop-1-enyl)cyclopropanecarboxylate see. Resmethrin
d-trans(5-Benzyl-3-furyl)methyl 2,2-dimethyl-3-(2-methylpropenyl) cyclopropanecarboxylate see. Resmethrin
5-Benzyl-3-furylmethyl(1RS,3RS;1RS,3SR)-2,2-dimethyl-3-(2-methylprop-1-enyl)cyclopropanecarboxylate see...Resmethrin
5-Benzyl-3-furylmethyl(1RS)-(Z),(E)-2,2-dimethyl-3-(2-methylprop-1-enyl)cyclopropanecarboxylate see. Resmethrin
BENZYLICUM® see...Benzyl Alcohol
Benzylideneacetaldehyde see...Cinnamaldehyde
Benzytol see...4-Chloro-3,5-xylenol
BEOSIT® see...Endosulfan
BERCEMA® see...Zineb
BERCEMA FERTAM 50® see...Ferbam

BERCEMA NMC50® see...Carbaryl
BERELEX® see...Gibberellic Acid
BERKMYCEN® see...Oxytetracycline Calcium
BERMAT® see...Chlordimeform
Bernsteinsaeure-2,2-dimethylhydrazid (German) see..Daminozide
BEST MAG-CHLOR DEFOLIANT® see...Sodium Chlorate
BESTOX® see...alpha-Cypermethrin
BES® 602 see...Fipronil
BETAMEC® see...Bensulide
BETAMIX® see...Phenmedipham
BETAMIX® 70 WP see...Desmedipham
BETANAL® see...Ethofumesate
BETANAL® see...Phenmedipham
BETANAL®-475 see...Desmedipham
BETANAL® AM see...Desmedipham
BETANEX® see...Desmedipham
BETANEX® 70 WP see...Desmedipham
BETAPAL® see...Naphthoxyacetic Acid
BETASAN® see...Bensulide
BETASAN-E® see...Bensulide
BETASAN-G® see...Bensulide
BETHRODINE® see...Benefin
BETOXON® see...Naphthoxyacetic Acid
Bexol see...Lindane
BEXTON® see...Propachlor
BFPO see...Dimefox
BFV see...Formaldehyde
BHC see...Hexachlorocyclohexanes
BHC see...Lindane
α-BHC see...Hexachlorocyclohexanes
β-BHC see...Hexachlorocyclohexanes
γ-BHC see...Lindane
δ-BHC see...Hexachlorocyclohexanes
gamma-BHC see...Lindane
BH 2,4-D® see...2,4-D
BH 2,4-DP® see...Dichlorprop
BH DALAPON see...Dalapon
BH DOCK KILLER® see...Maleic Hydrazide
BH MCPA see...MCPA
BH MECOPROP® see...Mecoprop
BH Prefix D® see...Dichlobenil
BH RASINOX R® see...Dalapon
BH TOTAL (FORMULATION)® see...Dalapon
BHULAN® see...Benefin
BIARBINEX® see...Heptachlor
Bibenzene see...Biphenyl
BIBESOL® see...Dichlorvos
Bicam ULV see...Bendiocarb
Bicarburretted hydrogen see...Ethylene
BICEP® see...Atrazine
BICEP® see...Metolachlor
Bichloro- see...Dichloroacetic Acid
Bichlorendo (Spanish) see...Mirex
Bichloride of mercury see...Mercuric Chloride
Bichloroacetic acid see...Dichloroacetic Acid
1,2-Bichloroethane see...Ethylene Dichloride

Bichlorure d'ethylene (French) see...Ethylene Dichloride
Bichlorure de mercure (French) see...Mercuric Chloride
Bichlorure de propylene (French) see...1,2-Dichlorophene
BIDIPHEN® see...Bithionol
Bidiphenbis(2-hydroxy-3,5-dichlorophenyl) sulfide see. Bithionol
BIDIRL® see...Dicrotophos
BIDRIN® see...Dicrotophos
BIDRIN-R® see...Dicrotophos
BIDRIN (SHELL)® see...Dicrotophos
BIFEX® see...Propoxur
BIFLEX® see...Bifenthrin
BILARCIL® see...Trichlorfon
BILEVON® see...Hexachlorophene
BILOBORN® see...Monocrotophos
BILOBRAN® see...Monocrotophos
BILORIN® see...Formic Acid
BILOXAZOL® see...Bitertanol
BINNELL® see...Benefin
BIO 5,462® see...Endosulfan
BIOALLETHRIN® see...Allethrins
BIOALLETHRIN TECHNICAL® see...Allethrins
Bioaltrina see...Allethrins
BIOCHEK® see...1,2-Dibromo-2,4-dicyanobutane
Biocide see...Acrolein
BIOCLEAR® see...1,2-Dibromo-2,4-dicyanobutane
BIODAC® see...Ethylene Oxide
BIOMET® 12 RODENT REPELLENT see...Tributyltin Chloride
Bioneopynamin see...Tetramethrin
BIO-QUAT 50-24® see...Zilkonium Chloride
BIOQUEST see...Sodium Dimethyldithiocarbamate
BIOQUIN® see...Copper(II)-8-hydroxyquinoline
BIOQUIN®-1 see...Copper(II)-8-hydroxyquinoline
Bioresmethrin (d-trans isomer) see...Resmethrin
BIORESMETHRIN® see...Resmethrin
BIO SLIME® see...4-Chloro-3,5-xylenol
BIOSTAT® PA see...Oxytetracycline Calcium
BIOTHION® see...Temephos
BIOTREAT see...Sodium Dimethyldithiocarbamate
BIOXONE® see...Methazole
Biphenthrin see...Bifenthrin
1,1'-Biphenyl see...Biphenyl
4-Biphenylacetic acid,2-fluoroethyl ester see...Fluenetil
(1,1'-Biphenyl)-4-acetic acid, 2-fluoroethyl ester see.. Fluenetil
[1,1'-Biphenyl]-2,2'-diyl oxide see...Dibenzofuran
2,2'-Biphenylene oxide see...Dibenzofuran
2-Biphenylol see...o-Phenylphenol
o-Biphenylol see...o-Phenylphenol
(1,1'-Biphenyl)-2-ol see...o-Phenylphenol
2,2'-Biphenylyleme oxide see...Dibenzofuran
β-[(1,1'-Biphenyl)-4-yloxy]-α-(1,1-dimethylethyl)-1H-1,2,4-triazole-1-ethanol see...Bitertanol
3-(3,1,1'-Biphenyl-4-yl-1,2,3,4-tetrahydro-1-napthalenyl)-4-hydroxy-1(2H)-benzopyran-2-one see...Difenacoum

Bipyridinium, 1,1'-dimethyl-4,4'-, dichloride see.. Paraquat
4,4'-Bipyridinium, 1,1'-dimethyl-, dichloride see.. Paraquat
4,4-Bipyridinium, 1,1'-dimethyl-, bis(methyl sulfate) see. Paraquat Methosulfate
BIRGIN® see...Propham
BIRLANE® see...Chlorfenvinphos
BIRLANE LIQUID® see...Chlorfenvinphos
S-[1,2-Bis(aethoxy-carbonyl)-aethyl]-O,O-dimethyl-dithiophosphat (German) see...Malathion
2,4-Bis(aethylamino)-6-chlor-1,3,5-triazin (German) see. Simazine
2,2-Bis(p-anisyl)-1,1,1-trichloroethane see.. Methoxychlor
BISTAR® see...Bifenthrin
Bis(bisdimethylamino)phosphonousanhydride see.. Octamethyl Diphosphoramide
Bis(bisdimethylamino)phosphoric anhydride see.. Octamethyl Diphosphoramide
Bis(bisdimethylaminophosphonous)anhydride see.. Octamethyl Diphosphoramide
Bis-bisdimethylaminophosphonous anhydride see.. Octamethyl Diphosphoramide
1,3-Bis(carbamoylthio)-2-(N,N-dimethylamino)propane hydrochloride see...Cartap Hydrochloride
S-[1,2-Bis(carbethoxy)ethyl] O,O-dimethyldithiophosphate see...Malathion
Bis(2-chloroethyl) ether see...Dichloroethyl Ether
Bis(β-chloroethyl) ether see...Dichloroethyl Ether
Bis(chlorohydroxyphenyl)methane see...Dichlorophene
Bis(5-chloro-2-hydroxyphenyl)methane see..Dichlorophene
O,O,-Bis(p-chlorophenyl)acetimidoylphosphoramidothioate see.. Phosacetim
O,O,-Bis(4-chlorophenyl)N-acetimidoylphosphoramidothioate see...Phosacetim
O,O,-Bis(4-chlorophenyl) (1-iminoethyl)phosphoramidothioic acid see...Phosacetim
O,O,-Bis(4-chlorophenyl)(1-iminoethyl)phosphoramidothioate see...Phosacetim
3,6-Bis(2-chlorophenyl)-1,2,4,5-tetrazine see.. Clofentezine
1,1-Bis-(p-chlorophenyl)-2,2,2-trichloroethane see.. DDT
2,2-Bis(p-chlorophenyl)-1,1-trichloroethane see...DDT
α,α-Bis(p-chlorophenyl)-β,β,β-trichlorethane see...DDT
1,1-Bis(4-chlorophenyl)-2,2,2-trichloroethanol see..Dicofol
1,1-Bis(p-chlorophenyl)-2,2,2-trichloroethanol see.. Dicofol
Bisclofentezin see...Clofentezine
Bis(2-cloroetil)eter (Spanish) see...Dichloroethyl Ether
Bis(S-(diethoxyphosphinothioyl)mercapto)methane see.. Ethion
(Bis(diethylamino)thioxomethyl) disulphide see.. Disulfiram

Bis-*O,O*-diethylphosphoric anhydride  see. . .TEPP
Bis-*O,O*-diethylphosphorothionic anhydride  see. . Sulfotepp
Bis(diethylthiocarbamoyl) disulfide  see. . .Disulfiram
Bis(*N,N*-diethylthiocarbamoyl) disulfide  see. . Disulfiram
Bis(*N,N*-diethylthiocarbamoyl) disulphide  see. . Disulfiram
Bis(diethylthiocarbamoyl) sulfide  see. . .Thiram
3-[4,6-Bis(difluoromethoxy)-pyrimidin-2-yl]-1-(methoxycarbonylphenylsulfonyl)urea  see. Primisulfuron-Methyl
Bis(dimethylamido)fluorophosphate  see. . .Dimefox
Bis(dimethylamido)fluorophosphine oxide  see. .Dimefox
Bis(dimethylamido)phosphoryl fluoride  see. . .Dimefox
Bis[(dimethylamino)carbonothioyl] disulfide  see. Thiram
Bis[(dimethylamino)carbonothioyl] disulphide  see. Thiram
Bis(dimethylamino)fluorophosphate  see. . .Dimefox
Bisdimethylaminofluorophosphine oxide  see. . .Dimefox
Bis(dimethyldithiocarbamate de zinc) (French)  see. Ziram
Bis(dimethyldithiocarbamato)zinc  see. . .Ziram
Bis-*p*-(*O,O*-dimethyl *O*-phenylphosphorothioate)sulfide  see. . .Temephos
Bis(dimethylthiocarbamoyl) disulfide  see. . .Thiram
Bis(dimethylthiocarbamoyl) disulphide  see. . .Thiram
Bis-*O,O*-di-*n*-propylphosphorothionic anhydride  see. Aspon®
Bis (dithiophosphatede *O,O*-diethyle) de *S,S*'-methylene (French)  see. . .Ethion
Bis(dithiophospate de *O,O*-diethyle) de *S,S*'-(1,4-dioxanne-2,3-diyle) (French)  see. . .Dioxathion
*S*-[1,2-Bis(ethoxy-carbonyl)-ethyl]-*O,O*-dimethyl-dithiophosfaat (Dutch)  see. . .Malathion
1,2-Bis(3-(ethoxycarbonyl)-2-thioureido)benzene  see. Thiophanate-Methyl
*S*-[1,2-Bis(ethoxycarbonyl)ethyl] *O,O*-dimethyl phosphorodithioate  see. . .Malathion
*S*-1,2-Bis(ethoxycarbonyl)ethyl-*O,O*-dimethylthiophosphate  see. . .Malathion
1,2-Bis[3-(ethoxycarbonyl)thioureido]benzene  see. Thiophanate-Methyl
1,2-Bis(ethoxycarbonylthioureido)benzene  see. .Thiophanate-Methyl
2,4-Bis(ethylamino)-6-chloro-*s*-triazine  see. . .Simazine
1,1-Bis(*p*-ethylphenyl)-2,2-dichloroethane  see. . .Ethylan
*S*-[1,2-Bis(etossi-carbonil)-etil]-*O,O*-dimetil-ditiofosfato (Italian)  see. . .Malathion
Bis(2-hydroxy-5-chlorophenyl)methane  see. .Dichlorophene
Bis(2-hydroxy-3,5, 6-trichlorophenyl)methane  see. Hexachlorophene
2,4-Bis(isopropylamino)-6-chloro-*s*-triazine  see. . Propazine
2,4-Bis(isopropylamino)-6-ethylthio-*S*-triazine  see. . Dipropetryn
2,4-Bis(isopropylamino)-6-methoxy-*s*-triazine  see. . Prometon

2,4-Bis(isopropylamino)-6-(methylmercapto)-*S*-triazine  see. . .Prometryn
2,4-Bis(isopropylamino)-6-(methylthio)-*S*-triazine  see. . Prometryn
2,4-Bis(isopropylamino)-6-(methylthio)-1,3,5-triazine  see. . .Prometryn
1,4-Bis(methanesulfonoxy)butane  see. . .Busulfan
[1,4-Bis(methanesulfonyloxy)butane]  see. . .Busulfan
1,1-Bis(*p*-methoxyphenyl)-2,2,2-trichloroethane  see. . Methoxychlor
2,2-Bis(*p*-methoxyphenyl)-1,1,1-trichloroethane  see. . Methoxychlor
Bis(1-methylethyl) carbamothioic acid, *S*-(2,3-dichloro-2-propenyl)ester  see. . .Diallate
*N,N*'-Bis(1-methylethyl)-6-(methylthio)-1,3,5-triazine-2,4-diamine  see. . .Prometryn
*N,N*'-Bis(1-methylethyl)-6-methylthio-1,3,5-triazine-2,4-diamine  see. . .Prometryn
*O,O,*-Bis(1-methylethyl)-*S*-[2-((phenylsulfonyl) amino)ethyl]pheosphorodithioate  see. . .Bensulide
Bismethomylthioether  see. . .Thiodicarb
Bis(2-methylpropyl)carbamothioic acid-*S*-ethyl ester  see. Butylate
Bis-(*O*-1-methylthioethylimino)-*N*-methylcarbamic acid)-*N,N*'-sulfide  see. . .Thiodicarb
Bis(8-oxyquinoline)copper  see. . .Copper(II)-8-hydroxyquinoline
Bis(pentachlor-2,4-cyclopentadien-1-yl)  see. .Dienochlor
Bis(pentachlorocyclopentadienyl)  see. . .Dienochlor
Bis(pentachloro-2,4-cyclopentadien-1-yl)  see. Dienochlor
2,4-Bis(propylamino)-6-chlor-1,3,5-triazin (German)  see. Propazine
Bis(8-quinolinato)copper  see. . .Copper(II)-8-hydroxyquinoline
Bis(8-quinolinolato)copper  see. . .Copper(II)-8-hydroxyquinoline
Bis(8-quinolinolato-N1,O8)-copper  see. . .Copper(II)-8-hydroxyquinoline
Bisulfan  see. . .Busulfan
Bisulphane  see. . .Busulfan
Bis-*N,N,N',N'*-tetramethylphosphorodiamidic anhydride  see. . .Octamethyl Diphosphoramide
Bis-2,3,5-trichlor-6-hydroxyfenylmethan (Czech)  see. Hexachlorophene
Bis(3,5,6-trichlor *o*-2-hydroxyphenyl)methane  see. .Hexachlorophene
Bis(trineophyltin) oxide  see. . .Fenbutatin Oxide
Bis[tris(*β,β*-dimethylphenethyl)tin]oxide  see. .Fenbutatin Oxide
Bis[tris(2-methyl-2-phenylpropyl)tin]oxide  see. Fenbutatin Oxide
*N,N*-Bis(2,4-xyliminomethyl)methylamine  see. . Amitraz
BITEMOL® see. . .Simazine
Bitertanol, fuberidazole  see. . .Fuberidazole
BITERTANOL® see. . .Bitertanol
BITHION® see. . .Temephos
BITHIONOL® see. . .Bithionol

BITIN® see...Bithionol
BIVERM® see...Phenothiazine
BLACK LEAF® see...Nicotine
BLADAFUM® see...Sulfotepp
BLADAFUME® see...Sulfotepp
BLADAFUN® see...Sulfotepp
BLADAN® see...Hexaethyl Tetraphosphate
BLADAN® see...Parathion
BLADAN® see...TEPP
BLADAN® BASE see...Hexaethyl Tetraphosphate
BLADAN F® see...Parathion
BLADAN M® see...Methyl Parathion
BLADE® see...Oxamyl
BLADEX® see...Cyanazine
BLADEX® 80WP see...Cyanazine
BLADEX/ATRAZINE (2:1) 80W® see...Atrazine
BLADEX-H® (butoxyethyl ester) see...2,4,5-Trichlorophenoxyacetic Acid, Esters
BLADON® see...TEPP
BLAST® see...Bentazon
BLATTANEX® see...Propoxur
BLATTOSEP® see...Propoxur
Blausaeure (German) see...Hydrogen Cyanide
Blauwzuur (Dutch) see...Hydrogen Cyanide
BLAZER® see...Acifluorfen
BLESEL MC® see...MCPA
BLEX® see...Pirimiphos-Methyl
BLIGHTOX® see...Zineb
BLITEX® see...Ronnel
BLITEX® see...Zineb
BLITOX® see...Copper Oxychloride
BLITOX® 50 see...Copper Oxychloride
BLIZENE® see...Zineb
BLOCKADE® see...Prodiamine
BLOCKER 4F® see...Quintozene
BLOC® see...Fenarimol
BLOTIC® see...Propetamphos
BLUE CONTROL® see...Copper(II)-8-hydroxyquinoline
Blue copper see...Copper Oxychloride
Blue copper see...Copper Sulfate
BLUE COPPER-50® see...Copper Oxychloride
BLUE-OX® see...Zinc Phosphide
Blue stone see...Copper Sulfate
Blue vitriol see...Copper Sulfate
BMC see...Carbendazim
BNM see...Benomyl
BNOA see...Naphthoxyacetic Acid
BNP 20® see...Dinoseb
BNP 30® see...Dinoseb
BO-ANA® see...Famphur
BOLDO® see...Bromadiolone
BOLERO® see...Thiobencarb
BOLFO® see...Propoxur
BOLL'D® see...Ethephon
BOLL-SET® see...Gibberellic Acid
BOLL-SET® see...Indole-3-Butyric Acid
BOLLS-EYE® see...Cacodylic Acid

BOLLS-EYE® see...Sodium Cacodylate
BOLSTAR® see...Sulprofos
BOMmHgDIER® see...Chlorothalonil
BONALAN® see...Benefin
BONIBAL® see...Disulfiram
BONIDE CUKE ANDMELON DUST® see...Rotenone
BONIDE BLUE DEATH RAT KILLER® see...Phosphorus
BONZI® see...Paclobutrazole
BOOMER-RID® see...Strychnine
BOOT HILL® see...Bromadiolone
BOOTS® see...Amitraz
BO Q 58-12-315® see...Propoxur
BO-RID® see...Hexazinone
Boracic acid see...Borax and Boric Acid
BORASCU® see...Borax and Boric Acid
BORATEEM® see...Sodium Tetraborate
Borates, tetra, sodium salt, anhydrous see...Sodium Tetraborate
Borax ($B_4Na_2O_7 \cdot 10H_2O$) see...Sodium Tetraborate
BORDERMASTER® see...MCPA
BOREA® see...Bromacil
BORER SOL® see...Ethylene Dichloride
Boric acid ($HBO_2$), sodium salt see...Sodium Metaborate
BORICIN® see...Borax and Boric Acid
BORICIN® see...Sodium Tetraborate
BOROCIL EXTRA® see...Bromacil
BOROFAX® see...Borax and Boric Acid
BOROFLOW® A/ATA see...Amitrole
BOROLIN® see...Picloram
BORTRAN® see...Dichloran
BORTRYSAN® see...Anilazine
BORUHO® see...Propoxur
BORUHO® 50 see...Propoxur
BOSAN SUPRA® see...DDT
BOS MH® see...Maleic Hydrazide
BOTRAN® see...Dichloran
BOTRILEX® see...Quintozene
BOUNDARY® see...Metribuzin
BOUNDRY® see...Diuron
BOVIDERMOL® see...DDT
BOVINOX® see...Trichlorfon
BOVIZOLE® see...Thiabendazole
BP 736® see...Binapacryl
BP 855® see...Binapacryl
BPF see...Dimefox
BPPS see...Propargite
BRASSICOL® see...Quintozene
BRASSICOL 75® see...Quintozene
BRASSICOL EARTHCIDE® see...Quintozene
BRASSICOL SUPER® see...Quintozene
BRAVO® see...Chlorothalonil
BRAVO® 6F see...Chlorothalonil
BRAVO® 500 see...Chlorothalonil
BRAVO D® see...Fenaminosulf
BRAVO ULTREX® see...Chlorothalonil
BRAVO-W-75® see...Chlorothalonil

BREAK® see...Propiconazole
BRELLIN® see...Gibberellic Acid
BRESTAN H 47.5 WP FUNGICIDE® see..Triphenyltin Compounds
BRESTANOL® see...Triphenyltin Compounds
BRESTAN® see...Triphenyltin Compounds
BRESTAN® H see...Fentin Hydroxide
BREVINYL® see...Dichlorvos
BREVINYL E 50® see...Dichlorvos
BRIDGEPORT® SPOT WEED KILLER see...2,4-D, isopropyl ester
BRIFUR® see...Carbofuran
BRIFUR® see...Propoxur
BRIGADE® see...Bifenthrin
Brilliant green see...C.I. Basic Green 1
Brilliant Oil Yellow see...Auramine
Brimstone see...Sulfur
BRIOTRIL® see...Bromoxynil
BRITON® see...Trichlorfon
BRITTEN® see...Trichlorfon
BRITTOX® see...Bromoxynil
BRN® 1509615 see...Haloxyfop-methyl
BROADSIDE® see...Cacodylic Acid
BROADCIDE 20EC® see...Linuron
BROADSTRIKE® see...Flumetsulam
BROADSTRIKE® see...Metolachlor
BROADSTRIKE® see...Trifluralin
BROCIDE® see...Ethylene Dichloride
BRODAN® see...Chlorpyrifos
BRODEX® see...Thiabendazole
BROFENE® see...Bromophos
BROGDEX 555® see...Sodium Dimethyldithiocarbamate
O-(4-Brom-2,5-dichlor-phenyl)-O,O-dimethyl-monothiophosphat (German) see...Bromophos
BROM-O-GAS® see...Chloropicrin
BROM-O-GAS® see...Methyl Bromide
BROM-O-SOL® see...Methyl Bromide
Bromacil 1.5 see...Bromacil
α-BROMACIL 80 WP® see...Bromacil
Bromadialone see...Bromadiolone
Bromallylene see...Allyl Bromide
BROMAX® see...Bromacil
Bromazil see...Bromacil
BROMAZIL® see...Imazalil
BROMCHLOPHOS® see...Naled
BROMEFLOR® see...Ethephon
BROMEX® see...Naled
BROMIDE PLUS® see...Sodium Bromide
Bromide salt of sodium see...Sodium Bromide
BROMINAL® see...Bromoxynil
BROMINAL M & PLUS® see...MCPA
BROMINAL ME-4® see...Bromoxynil
Bromine cyanide see...Cyanogen Bromide
BROMINEX® see...Bromoxynil
BROMINIL® see...Bromoxynil
Bromnatrium (German) see...Sodium Bromide
γ-Bromoallylene see...Propargyl Bromide

3-[3-(4'-Bromo(1,1'-biphenyl)-4-yl)3-hydroxy-1(phenylpropyl)-4-hydroxy-2H-1(benzopyran-2-one see. Bromadiolone
2-Bromo-2-(bromomethyl)glutaronitrile see...1,2-Dibromo-2,4-dicyanobutane
1-Bromo-1-(bromomethyl)-1,3(propanedicarbonitrile see. 1,2-Dibromo-2,4-dicyanobutane
5-Bromo-3-sec(butyl-6(methyl see...Bromacil
5-Bromo-3-sec(butyl-6(methyluracil see...Bromacil
4-Bromo-2-(4(chlorophenyl)-1-(ethoxymethyl)-5-(trifluoromethyl)pyrrole-3(carbonitrile (IUPAC) see..Chlorfenapyr
O-(4-Bromo-2(chlorophenyl)-O-ethyl-S(propylphosphorothioate see...Profenfos
4-Bromo-2,5(dichlorophenol-O-ester with O,O(diethyl phosphorothioate see...Bromophos-ethyl
4-Bromo-2,5(dichlorophenyl dimethyl phosphorothionate see...Bromophos
O-(4-Bromo-2,5(dichlorophenyl) O,O(diethyl phosphorothioate see...Bromophos-ethyl
O-(4-Bromo-2,5(dichlorophenyl) O,O(dimethylphosphorothioic acid see...Bromophos
O-(4-Bromo-2,5(dichlorophenyl)O(methyl phenylphosphonothioate see...Leptophos
5-Bromo-6(methyl-3-(1(methylpropyl)- see...Bromacil
5-Bromo-6(methyl-3-(1(methylpropyl)-2,4-(1H,3H)(pyrimidinedione see...Bromacil
5-Bromo-6(methyl-3-(1(methylpropyl)-2,4(1H,3H)(pyrimidinedione see...Bromacil
3-[3-(4'-Bromobiphenyl)-4-yl]3-hydroxy-1(phenylpropyl)-4-hydroxy(coumarin see.. Bromadiolone
Bromocyan see...Cyanogen Bromide
Bromocyanide see...Cyanogen Bromide
Bromocyanogen see...Cyanogen Bromide
BROMOFLOR® see...Ethephon
Bromofos see...Bromophos
Bromofos methyl see...Bromophos
BROMOFUME® see...Ethylene Dibromide
Bromomethane see...Methyl Bromide
BROMONE® see...Bromadiolone
3-[α-(ρ-(ρ-Bromophenyl)-β-hydroxyphenethyl)benzyl]-4-hydroxy(coumarin see...Bromadiolone
3-(4-Bromophenyl)-1(methoxy-1(methylurea see..Metobromuron
3-(p-Bromophenyl)-1(methoxy-1(methylurea see..Metobromuron
N'-(4-Bromophenyl)-N(methoxy-N(methylurea see.. Metobromuron
3-(p-Bromophenyl)-1(methyl-1(methoxyurea see.. Metobromuron
3-Bromopropene see...Allyl Bromide
1-Bromo, 2(propene see...Allyl Bromide
3-Bromopropeno (Spanish) see...Allyl Bromide
3-Bromopropylene see...Allyl Bromide
3-Bromopropylene see...Allyl Bromide
3-Bromopropyne see...Propargyl Bromide
3-Bromo-1(propyne see...Propargyl Bromide

BROMOTRIL® see...Bromoxynil
Bromox 2E see...Bromoxynil
BROMOXYNIL NITRILE HERBICIDE® see...Bromoxynil
3-(4-Bromphenyl)-1(methoxyharnstoff (German) see...Metobromuron
Bromuro de alilo (Spanish) see...Allyl Bromide
Bromuro de cianogeno (Spanish) see...Cyanogen Bromide
Bromure de cyanogen (French) see...Cyanogen Bromide
Bromure de methyle (French) see...Methyl Bromide
Bromuro de propargilo (Spanish) see...Propargyl Bromide
Bromuro di etile (Italian) see...Ethylene Dibromide
BRONATE® see...Bromoxynil
BRONCO® see...Alachlor
BRONOX® see...Linuron
Bronze powder see...Copper and Copper Compounds
BROOT® see...Trimethacarb
BROPHENE® see...Bromophos
Brown copper oxide see...Cuprous Oxide
BROXYNIL® see...Bromoxynil
BROZONE® see...Chloropicrin
BRP see...Naled
BRUCIL® see...Bromoxynil
Brucina (Italian, Spanish) see...Brucine
(-)Brucine see...Brucine
(-)Brucine dihydrate see...Brucine
Brucine hydrate see...Brucine
BRULAN® see...Tebuthiuron
BRUMIN® see...Warfarin
Bruomophos (Russian) see...Bromophos
BRUSH-B-GON® see...Triclopyr, Triethylammonium Salt
BRUSH-BULLET® see...Tebuthiuron
BRUSH BUSTER® see...Dicamba
BRUSHKILLER® see...Hexazinone
BRUSH-OFF® see...Metsulfuron-methyl
BRUSH-OFF® AMMONIUM SULFAMATE BRUSH WEED KILLER see...Ammonium Sulfamate
BRUSH-OFF® 445 LOW VOLATILE BRUSH KILLER see...2,4,5-T
BRUSH-OFF® 445 MILD VOLATILE BRUSH KILLER see...2,4,5-Trichlorophenoxyacetic Acid, Esters
BRUSH-RHAP® see...2,4-D
BRUSH RHAP® see...2,4,5-T
BRUSH RHAP® see...2,4,5-Trichlorophenoxyacetic Acid, Esters
BRUSHTOX® see...2,4,5-T
BRUSHTOX® see...2,4,5-Trichlorophenoxyacetic Acid, Esters
BRUS KILLER 64® see...2,4-D, butoxyethyl ester
BRYGOU® see...Propoxur
BSC FLOWABLE® see...Copper Sulfate
B-SELEKTONON® see...2,4-D
B-SELEKTONON M® see...MCPA
BTC see...Zilkonium Chloride
BTS 27,419® see...Amitraz

BUCKLE® see...Triallate
BUCKLE® see...Trifluralin
BUCTRIL® see...Bromoxynil
BUCTRIL + ATRAZINE GEL® see...Atrazine
BUCTRIL® 4EC GELBUCTRIL INDUSTRIAL® see...Bromoxynil
BUCTRIL® GEL HERBICIDE see...Bromoxynil
BUFEN® see...Phenylmercury Acetate
BUFOPTO-ZINC-SULFATE® see...Zinc Sulfate Heptahydrate
BUGAWAY® see...D-Limonene
BUG BAN PLUS® see...Dipropyl Isocinchomeronate
BUG-B-GON® see...Cyfluthrin
BUGCHASER® see...D-Limonene
BUGMASTER® see...Carbaryl
BUHACH® see...Pyrethrins or Pyrethrum
BULLET® see...Cyanazine
BULLET® see...Pendimethalin
BUMETRAN® see...Amitraz
BUMPER® see...Propiconazole
BUNT-CURE® see...Hexachlorobenzene
BUNT-NO-MORE® see...Hexachlorobenzene
BURA® see...Sodium Tetraborate
BUREX® see...Pyrazon
BURTOLIN® see...Maleic Hydrazide
BUSAN see...Octhilinone
BUSAN® see...Metham-Sodium
BUSH KILLER® see...2,4-D
BUSHWHACKER® see...Dicamba
BUSULFEX® see...Busulfan
Butalin see...Butralin
Butanedioic acid, [(dimethoxyphosphinothioyl)thio]-, diethyl ester see...Malathion
Butanedioic acid mono(2,2-dimethylhydrazide) see...Daminozide
1,4-Butanediol dimethanesulphonate see...Busulfan
1,4-Butanediol dimethyl sulfonate see...Busulfan
BUTANEX® see...Butachlor
Butanoic acid, 4-(2,4-dichlorophenoxy)- see...2,4-DB
2-Butanone, 1-(4-chlorophenoxy)-3,3-dimethyl-1-(1,2,4-triazol-1-yl)- see...Triadimefon
2-Butanone, 1-(4-chlorophenoxy)-3,3-dimethyl-1-(1-$H$-1,2,4-triazol-1-yl)- see...Triadimefon
BUTANOX® see...Butachlor
BUTAPHENE® see...Dinoseb
3-Butenoic acid, 2-amino-4-(2-aminoethoxy)-, monohydrochloride, [$s$-($E$)]- see...Aminoethoxyvinylglycine Hydrochloride
2-Butenoic acid, 3-((dimethoxyphosphinyl)oxy)-, methyl ester see...Mevinphos
2-Butenoic acid, 3-[((ethylamino)methoxyphosphinothioyl)oxy]-, isopropylester, ($E$)- see...Propetamphos
2-Butenoic acid, 3-[((ethylamino)methoxyphosphinothioyl)oxy]-1-methylethyl ester, ($E$)- see...Propetamphos
2-Butenoic acid, 2-isooctyl-4,6-dinitrophenyl ester see...Dinocap

2-Butenoic acid, 4-isooctyl-2,6-dinitrophenyl ester see...Dinocap
2-Butenoic acid 2-(1-methylheptyl)-4,6-dinitrophenyl ester see...Dinocap
BUTIFOS® see...Tribufos
Butilate see...Butylate
BUTILATE® see...Butylate
Butilchlorofos see...Bromoxynil
BUTIPHOS® see...Tribufos
BUTIREX® see...2,4-DB
BUTOFLIN® see...Deltamethrin
BUTORMONE® see...2,4-DB
BUTOSS® see...Deltamethrin
Butoxide see...Piperonyl Butoxide
Butoxido de piperonilo (Spanish) see...Piperonyl Butoxide
BUTOXON® see...2,4-DB
BUTOXONE® see...2,4-DB
BUTOXONE® AMINE see...2,4-DB
BUTOXONE® ESTER see...2,4-DB
Butoxy-D3 see...2,4-D, butoxyethyl ester
Butoxyethanol ester of 2,4-dichlorophenoxyacetic acid see...2,4-D, butoxyethyl ester
Butoxyethanol 2,4,5-trichlorophenoxyacetate see...2,4,5-Trichlorophenoxyacetic Acid, Esters
α-[2-(2-Butoxyethoxy)ethoxy]-4,5-methylenedioxy-2-propyltoluene see...Piperonyl Butoxide
α-[2-(2-N-Butoxyethoxy)ethoxy]-4,5-methylenedioxy-2-propyltoluene see...Piperonyl Butoxide
5-[(2-(2-Butoxyethoxy)ethoxy)methyl]-6-propyl-1,3-benzodioxole see...Piperonyl Butoxide
2-(2-Butoxyethoxy)ethyl 6-propylpiperonyl ether see...Piperonyl Butoxide
Butoxyethyl 2,4-dichlorophenoxyacetate see...2,4-D, butoxyethyl ester
2-Butoxyethyl 2,4-dichlorophenoxyacetate see...2,4-D, butoxyethyl ester
2,4,5-T Butoxyethyl ester see...2,4,5-Trichlorophenoxyacetic Acid, Esters
Butoxyethyl 2,4,5-trichlorophenoxyacetate see...2,4,5-Trichlorophenoxyacetic Acid, Esters
Butoxyethyl 2,4,5-T see...2,4,5-Trichlorophenoxyacetic Acid, Esters
Butoxymethyl see...Butachlor
N-(Butoxymethyl)-2-chloro-2',6'-diethylacetanilide see...Butachlor
N-(Butoxymethyl)-2-chloro-N-(2,6-diethylphenyl)acetamide see...Butachlor
BUTOX® see...Deltamethrin
Butraline see...Butralin
Butter of Arsenic see...Arsenic and Inorganic Arsenic Compounds
Butyl-2,4,5-T see...2,4,5-Trichlorophenoxyacetic Acid, Esters
1-(Butylamino)carbonyl-1H-benzimidazol-2-yl-, methyl ester see...Benomyl
2-tert-Butylamino-4-ethylamino-6-methylmercapto-S-triazine see...Terbutryn

2-tert-Butylamino-4-ethylamino-6-methylthio-S-triazine see...Terbutryn
Butylate-2,4,5-T see...2,4,5-Trichlorophenoxyacetic Acid, Esters
3-sek-Butyl-5-brom-6-methyluracil (German) see...Bromacil
2-(tert-Butyl)-5-(4-tert-butyl-benzylthio)-4-chloropyridazin-3-(2H)one see...Pyridaben
N-sec-Butyl-4-tert-butyl-2,6-dinitroaniline see...Butralin
tert-Butylcarbamic acid, ester with 3-(m-hydroxyphenyl)-1,1-dimethylurea see...Karbutilate
1-(Butylcarbamoyl)-2-benzimidazolecarbamic acid, methyl ester see...Benomyl
1-(–Butylcarbamoyl)-2-(methoxy-carboxamido)-benzamidazol (German) see...Benomyl
1-(n-Butylcarbamoyl)-2-(methoxy-carboxamido)-benzimidazol (German) see...Benomyl
Butyl carbitol 6-propylpiperonyl ether see...Piperonyl Butoxide
Butyl-carbityl (6-propylpiperonyl) ether see...Piperonyl Butoxide
1-[(tert-Butylcarbonyl-4-chlorophenoxy)methyl]- see...Triadimefon
1-[(tert-Butylcarbonyl-4-chlorophenoxy)methyl]-1H-1,2,4-triazole see...Triadimefon
(Butylcarbityl)(6-propylpiperonyl) ether 80% and related compounds 20% see...Piperonyl Butoxide
1-tert-Butyl-2-(p-chlorobenzyl)-2-(1,2,4-triazol-1-yl)ethanol see...Paclobutrazole
2-Butynyl-4-chloro-m-chlorocarbanilate see...Barban
3-tert-Butyl-5-chlor-6-methyluracil (German) see...Terbacil
3-tert-Butyl-5-chloro-6-methyluracil see...Terbacil
O-(4-tert-Butyl-2-chlor-phenyl)-O-methyl-phosphorsaeure-N-methylamid (German) see...Crufomate
4-t-Butyl-2-chlorophenyl methyl methylphosphoramidate see...Crufomate
4-tert-Butyl 2-chlorophenyl methylphosphoramidate de methyle (French) see...Crufomate
α-Butyl-α-(4-chlorophenyl)-1H-1,2,4-triazole-1-propanenitrile see...Myclobutanil
2-tert-Butyl-4-(2,4-dichloro-5-isopropoxyphenyl-$\delta$(sup2)-1,3,4-oxadiazoline-5-one see...Oxadiazon
2-tert-Butyl-4-(2,4-dichloro-5-isopropyloxyphenyl)-1,3,4-oxadiazolin-5-one see...Oxadiazon
5-tert-Butyl-3-(2,4-dichloro-5-isopropoxyphenyl)-1,3,4-oxadiazol-2(3H)-one see...Oxadiazon
α-Butyl-α-(2,4-dichlorophenyl)-1H-1,2,4-triazole-1-ethanol (±)- see...Hexaconazole
tert-Butyl (E)-4-[((((1,3-dimethyl-5-phenoxy-1H-pyrazol-4-yl)methylene)amino)oxy)methyl]benzoate see...Fenpyroximate
tert-Butyl (E)α-(1,3-dimethyl-5-phenoxypyrazol-4-methyleneaminooxy)-p-toluate see...Fenpyroximate
N-Butyl-2,6-dinitro-N-ethyl-4-trifluoromethylaniline see...Benefin
2-sec-Butyl-4,6-dinitrophenol see...Dinoseb
o-tert-Butyl-4,6-dinitrophenol see...Dinoterb

2-*sec*-Butyl-4,6-dinitrophenol 6-*sec* see...Dinoseb
2-*sec*-Butyl-4,6-dinitrophenyl-3,3-dimethylacrylate see..Binapacryl
2-*sec*-Butyl-4,6-dinitrophenyl-3-methyl-2-butenoate see..Binapacryl
2-*sec*-Butyl-4,6-dinitrophenyl-3-methylcrotonate see. Binapacryl
2-*sec*-Butyl-4,5-dinitrophenyl senecioate see..Binapacryl
2,4,5-T-n-Butyl ester see...2,4,5-Trichlorophenoxyacetic Acid, Esters
*N*-Butyl-*N*-ethyl-2,6-dinitro-4-trifluoromethylaniline see. Benefin
*N*-Butyl-*N*-ethyl-2,6-dinitro-4-trifluoromethylbenzenamine see...Benefin
*N*-Butyl-*N*-ethyl-2,6-dinitro-4-(trifluromethyl)benzeneamine see...Benefin
*N*-*tert*-Butyl-*N*'-(4-ethylbenzoyl)-3,5-dimethylbenzoyl hydrazide see...Tebufenozide
Butylethylthiocarbamic acid *S*-propyl ester see..Pebulate
*N*-Butyl-*N*-ethyl-α,α,α-trifluoro-2,6-dinitro-*p*-toluidine see...Benefin
2-*tert*-Butylimino-3-isopropyl-5-phenylperhydro-1,3,5-thidiazin-4-one see...Buprofezin
Butylphen see...Butylphenols
2-*n*-Butylphenol see...Butylphenols
2-*sec*-Butylphenol see...Butylphenols
2-*tert*-Butylphenol see...Butylphenols
4-*sec*-Butylphenol see...Butylphenols
4-*tert*-Butylphenol see...Butylphenols
*o*-*sec*-Butylphenol see...Butylphenols
*p*-*sec*-Butylphenol see...Butylphenols
*p*-*tert*-Butylphenol see...Butylphenols
2-(*p*-*tert*-Butylphenoxy)cyclohexyl propargyl sulfite see. Propargite
2-(*p*-*tert*-Butylphenoxy)cyclohexyl 2-propynyl sulfite see...Propargite
2-(4-*tert*-Butylphenoxy)cyclohexyl prop-2-ynyl sulfite see...Propargite
Butylphenoxyisopropyl chloroethyl sulfite see...Aramite
2-(*p*-Butylphenoxy)isopropyl 2-chloroethyl sulfite see. .Aramite
2-(4-*tert*-Butylphenoxy)isopropyl-2-chloroethyl sulfite see...Aramite
2-(*p*-*tert*-Butylphenoxy)isopropyl 2'-chloroethyl sulphite see...Aramite
2-(*p*-Butylphenoxy)-1-methylethyl 2-chloroethyl sulfite see...Aramite
2-(*p*-*tert*-Butylphenoxy)-1-methylethyl 2-chloroethyl ester of sulphurous acid see...Aramite
2-(*p*-*tert*-Butylphenoxy)-1-methylethyl-2-chloroethyl sulfite ester see...Aramite
2-(*p*-*tert*-Butylphenoxy)-1-methylethyl 2'-chloroethyl sulphite see...Aramite
2-(*p*-*tert*-Butylphenoxy)-1-methylethyl sulphite of 2-chloroethanol see...Aramite
1-(*p*-*tert*-Butylphenoxy)-2-propanol-2-chloroethyl sulfite see...Aramite
2-*sec*-Butylphenyl *n*-methylcarbamate see...BPMC

*o*-*sec*-Butylphenyl methylcarbamate see...BPMC
Butyl phosphorotrithioate see...Tribufos
1-(5-*tert*-Butyl-1,3,4-thiadiazol-2-yl)-3-dimethylharnstoff (German) see...Tebuthiuron
1-(5-*tert*-Butyl-1,3,4-thiadiazol-2-yl)-1,3-dimethylurea see...Tebuthiuron
Butyl 2,4,5-trichlorophenoxyacetate see...2,4,5-Trichlorophenoxyacetic Acid, Esters
*n*-Butyl(2,4,5-trichlorophenoxy)acetate see...2,4,5-Trichlorophenoxyacetic Acid, Esters
Butyl(*RS*)-2-[4-((5-(trifluoromethyl)-2-pyridinyl)oxy)phenoxy]propanoate see...Fluazifop-butyl
(±)-Butyl-2-[4-(((5-trifluoro-methyl)-2-pyridinyl)oxy)phenoxy]propanoate see...Fluazifop-butyl
Butyl 2-[4-((5-(trifluoromethyl)-2-pyridyl)oxy)phenoxy]propionate see...Fluazifop-butyl
*tert*-Butyl valone see...Pindone
BUTYRAC® see...2,4-DB
BUTYRAC® 118 see...2,4-DB
BUTYRAC® 200 see...2,4-DB
BUTYRAC® ESTER see...2,4-DB
Butyric acid, 4-(2,4-dichlorophenoxy)- see...2,4-DB
BUX see...Brufencarb
BUX-TEN® see...Brufencarb
Buzulfan see...Busulfan
BX 112® see...Prohexadione Calcium
BYE BUGS® see...Ammonium Hexafluorosilicate

- C -

C 570® see...Phosphamidon
C 709® see...Dicrotophos
C 1414® see...Monocrotophos
C 1983® see...Chloroxuron
C 2059® see...Fluometuron
C 3126® see...Metobromuron
C 8514® see...Chlordimeform
C 8949® see...Chlorfenvinphos
C 10015® see...Chlorfenvinphos
C.B. 2041 see...Busulfan
C.F.S see...Thallium Sulfate
C.I. 41000B see...Auramine
C.I. 47031 see...Phosfolan
C.I. 77180 see...Cadmium
C.I. 77400 see...Copper and Copper Compounds
C.I. 77402 see...Cuprous Oxide
C.I. 77410 see...Paris Green
C.I. 77760 see...Mercuric Oxide
C.I. Basic Yellow 2, free base see...Auramine
C.I. Pigment Green 21 see...Copper Acetoarsenite
C.I. Pigment Green 21 (9CI) see...Paris Green
C.I. Pigment Metal 2 see...Copper and Copper Compounds
C.I. Solvent Yellow 34 see...Auramine
Cacodilato sodico (Spanish) see...Sodium Cacodylate
Cacodylate de sodium (French) see...Sodium Cacodylate
Cacodylic acid sodium salt see...Sodium Cacodylate

CADAN® see...Carboxin
CADAN® see...Cartap Hydrochloride
CADDY® see...Cadmium Chloride
CADENCE® see...Dicamba
Cadmio (Spanish) see...Cadmium
Cadmium dichloride see...Cadmium Chloride
Cadmium monosulfate see...Cadmium Sulfate
Cadmium sulphate see...Cadmium Sulfate
CAID® see...Chlorophacinone
Cajeputene see...D-Limonene
Cake alum see...Aluminum Sulfate
Calabarine see...Physostigmine
CALAR® see...Calcium Methanearsonate
Calcid see...Calcium Cyanide
Calcite (mineral) see...Calcium Carbonate
Calcium acid methanearsonate see...Calcium Methanearsonate
Calciumarsenat (German) see...Calcium Arsenate
Calcium carbimide see...Calcium Cyanamide
Calcium(II)carbonate (1:1) see...Calcium Carbonate
Calcium cyanamid see...Calcium Cyanamide
Calcium hydrate see...Calcium Hydroxide
Calcium hydrogen methanearsonate see...Calcium Methanearsonate
Calcium meta-arsenate see...Calcium Arsenite
Calcium methanearsonate see...Calcium Methanearsonate
Calcium (II) nitrate (1:2) see...Calcium Nitrate
Calcium orthoarsenate see...Calcium Arsenate
Calcium 3-oxido-5-*oxo*-4-propionylcyclohex-3-enecarboxylate see...Prohexadione Calcium
Calcium phosphid see...Calcium Phosphide
Calcyan see...Calcium Cyanide
Calcyanide see...Calcium Cyanide
CALDAN® see...Cartap Hydrochloride
CALDON® see...Dinoseb
CALIBER® see...Simazine
CALIPURON® see...Isoproturon
CALLIFOL® see...Dicofol
CALLIFORT® see...Trifluralin
CALMATHION® see...Malathion
CAL-META® see...Calcium Arsenate
CALO-CHLOR® see...Mercuric Chloride
CALO-CURE® see...Mercuric Chloride
CALSOFT LAS 99® see...Dodecylbenzenesulfonic Acid
CAMA® see...Calcium Methanearsonate
CAMBELL'S® NABAM SOIL FUNGICIDE see...Nabam
CAMBILENE® see...MCPA
CAMPAIGN® see...2,4-D
CAMPAIGN® see...Glyphosate
CAMPAPRIM® A 1544 see...Amitrole
CAMPBELL'S® DB STRAIGHT see...2,4-DB
CAMPBELL'S NICO-SOAP® see...Nicotine
CAMPBELL'S RAPIER® see...Pronamide
CAMPBELL'S® REDIPON see...Dichlorprop
CAMPBELL'S® REDLEGOR see...2,4-DB
CAMPBELL'S TRIFLURON® see...Trifluralin

Camphechlor see...Toxaphene
Camphene, octachloro- see...Toxaphene
Camphochlor see...Toxaphene
Camphoclor see...Toxaphene
Camphofene chlorinated camphene see...Toxaphene
CAMPILEX® see...Procymidone
Campilit see...Cyanogen Bromide
CAMPOSAN® see...Ethephon
CANADIEN 2000® see...Bromadiolone
CANDASEPTIC® see...*p*-Chloro-*m*-Cresol
CANDEX® see...Atrazine
CANFECLOR® see...Toxaphene
CANNON HERBICIDE® see...Alachlor
CANNON HERBICIDE® see...Trifluralin
CANOGARD® see...Dichlorvos
CANOPY® see...Chlorimuron-ethyl
CANOPY® see...Metribuzin
CANOPY XL® see...Sulfentrazone
CANVAS® see...Metsulfuron-methyl
CANVAS® see...Thifensulfuron Methyl
CANVAS® see...Tribenuron-Methyl
CAO-NENG® see...Quinclorac
CAPAROL® see...Prometryn
CAPFOS® see...Fonofos
CAPRANE® see...Dinocap
Capric acid see...Decanoic Acid
*n*-Capric acid see...Decanoic Acid
Caprinic acid see...Decanoic Acid
CAPROLIN® see...Carbaryl
Capryldinitrophenyl crotonate see...Dinocap
2-Capryl-4,6-dinitrophenyl crotonate see...Dinocap
Caprynic acid see...Decanoic Acid
Capsaicine see...Capsaicin
CAPSYN® see...Capsaicin
CAPTAF® see...Captan
CAPTAFOL® see...Captafol
CAPTAN® 50W see...Captan
CAPTANCAPTENEET® 26,538 see...Captan
CAPTAN SC® see...Captan
CAPTATOL® see...Captafol
CAPTEX® see...Captan
CAPTOFOL® see...Captafol
CAPTURE® see...Bifenthrin
CAPTURE® see...Diflufenican
Carbacryl see...Acrylonitrile
CARBADINE® see...Zineb
Carbam see...Metham-Sodium
Carbam, sodium salt see...Metham-Sodium
Carbamate,4-dimethylamino-3,5-xylyln-methyl- see...Mexacarbate
Carbamato amonico (Spanish) see...Ammonium Carbamate
CARBAMEC® see...Carbaryl
Carbamic acid see...Aldoxycarb
Carbamic acid, aimethyldithio-, iron salt see...Ferbam
Carbamic acid, [(4-aminophenyl)sulfonyl]-, methyl ester see...Asulam

Carbamic acid, ammonium salt see...Ammonium Carbamate
Carbamic acid, 1*H*-benzimidazole-2-yl-, Methyl ester see...Carbendazim
Carbamic acid, 1-(butylamino)carbonyl- 1H-benzimidazol-2yl, methyl ester see...Benomyl
Carbamic acid, 3-chlorophenyl-, 4-chloro-2-butynyl ester see...Barban
Carbamic acid, [(dibutylamino)thio]methyl-, 2,3-dihydro-2,2-dimethyl-7-benzofuranyl ester see...Carbosulfan
Carbamic acid,[(dibutylamino)thio]methyl-, 2,2-dimethyl-2,3-dihydro-7-benzofuranyl ester see...Carbosulfan
Carbamic acid, diethyl-, *S*-(4-chlorobenzyl)ester see..Thiobencarb
Carbamic acid, diethylthio-*S*-(*p*-chlorobenzyl) ester see..Thiobencarb
Carbamic acid, diisopropylthio-, *S*-(2,3,3-trichloroallyl) ester see...Triallate
Carbamic acid, [3-(dimethylamino)propyl]-, propyl ester, monohydrochloride see...Propamocarb Hydrochloride
Carbamic acid, dimethyl-, 1-[(dimethylamino)carbonyl]-5-methyl-1*H*-pyrazol-2-yl ester see...Dimetilan
Carbamic acid, dimethyl-, 2-(dimethylamino)-5,6-dimethyl-4-pyrimidinyl ester see...Pirimicarb
Carbamic acid, dimethyldithio-, sodium salt see..Sodium Dimethyldithiocarbamate
Carbamic acid, dimethyldithio-, zinc salt see...Ziram
Carbamic acid, dimethyl-, ester with 3-hydroxy-*N,N*-5-trimethylpyrazole-1-carboxamide see...Dimetilan
Carbamic acid, 1,1-dimethylethyl-, ester with 3-(3-hydroxyphenyl)-1,1-dimethyl urea see...Karbutilate
Carbamic acid, dipropylthio-, *S*-ethyl ester see...EPTC
Carbamic acid, dipropylthio-, *S*-propyl ester see..Vernolate
Carbamic acid, ethylenebis (dithio-), disodium salt see..Nabam
Carbamic acid, ethylenebis(dithio-), manganese salt see.Maneb
Carbamic acid, hexamethylenethio-, *S*-ethyl ester see..Molinate
[Carbamic acid, –methyl-, compounded with (2-methylthio)acetaldoxime]bis-, thioether see...Thiodicarb
Carbamic acid, methyl-, *m*-cumenyl ester, Phenol, 3-(1-methyethyl)-, methylcarbamate see..Promecarb
Carbamic acid, methyl-, *m*-cym-5-yl ester see..Promecarb
Carbamic acid, methyl-, 4-(dimethylamino)-3,5-xylyl ester see...Mexacarbate
Carbamic acid, methyl-, 2,2-dimethyl-2,3-dihydrobenzofuran-7-yl ester see...Carbofuran
Carbamic acid, methyl-, 2,3-(dimethylmethylenedioxy)phenyl ester see...Bendiocarb
Carbamic acid, methyl-, 3,5-dimethyl-4-(methylthio)phenyl ester see...Methiocarb
Carbamic acid, methyldithio-, monosodium salt see.Metham-Sodium
Carbamic acid, *N*-methyldithio-, monosodium salt see..Metham-Sodium

Carbamic acid, *N*-methyldithio-, sodium salt see..Metham-Sodium
Carbamic acid, methyl-, ester with *N'*-(*m*-hydroxyphenyl)-*N,N*-dimethylformamidine, hydrochloride see..Formetanate Hydrochloride
Carbamic acid, methyl-, *o*-isopropoxyphenyl ester see..Propoxur
Carbamic acid, methyl-, 2,3-(isopropylidenedioxy)phenyl ester see...Bendiocarb
Carbamic acid, methyl-, methylcarbamate (ester) see..Mexacarbate
Carbamic acid, (3-methyl-5-(1-methylethyl)phenyl-, methyl ester see...Promecarb
Carbamic acid, *N*-methyl-, 3-methyl-5-isopropylphenyl ester see...Promecarb
Carbamic acid, methyl -,*O* -((2 -methyl -2 -(methylthio)propylidene)amino) deriv. see...Aldicarb
Carbamic acid, methyl-, 3-methylphenyl ester see..Metolcarb
Carbamic acid, (3-methylphenyl)-, 3-[(methoxycarbonyl)amino]phenyl ester see..Phenmedipham
Carbamic acid, methyl-, 4-(methylthio)-3,5-xylyl ester see...Methiocarb
Carbamic acid, *N*-methyl-, 4-(methylthio)-3,5-xylyl ester see...Methiocarb
Carbamic acid, methyl-, mixed (1-methylbutyl)phenyl and (1-ethylpropyl)phenyl esters, Brufencarb
Carbamic acid, methyl-, mixed 3,4,5- and 2,3,5-triphenylmethyl esters (4:1) see...Trimethacarb
Carbamic acid, methyl-, 1-naphthyl ester see...Carbaryl
Carbamic acid, methyl-, 3-tolyl ester see...Metolcarb
Carbamic acid, monoammonium salt see...Ammonium Carbamate
Carbamic acid, [2-(4-phenoxyphenoxy)ethyl]-, ethyl ester see...Fenoxycarb
Carbamic acid, [2-4(-phenoxyphenoxy)ethyl]-, ethyl ester see...Fenoxycarb
Carbamic acid, [1,2-phenylenebis(iminocarbonothioyl)]bis-, diethyl ester see.Thiophanate-Methyl
Carbamic acid, *N*-phenyl-, 3-[(ethoxycarbonyl)amino]phenyl ester see..Desmedipham
Carbamic acid, sulfanilyl-, methyl ester see...Asulam
Carbamide see...Urea
Carbamide resin see...Urea
Carbamimidic acid see...Urea
CARBAMINE® see...Carbaryl
Carbamodithioic acid, dimethyl-, sodium salt see.Sodium Dimethyldithiocarbamate
Carbamodithioic acid, dimethyl-, zinc salt see...Ziram
Carbamodithioic acid, 1,2-ethanediylbis-, disodium salt see...Nabam
Carbamodithioic acid, 1,2-ethanediylbis-, manganese salt see...Maneb
Carbamodithioic acid, 1,2-ethanediylbis-, manganous zinc salt see...Mancozeb

Carbamodithioic acid, 1,2-ethanediylbis-, zinc complex see...Zineb
Carbamodithioic acid, 1,2-ethanediylbis-, zinc salt see..Zineb
Carbamodithioic acid, methyl-, monosodium salt see..Metham-Sodium
Carbamonitrile see...Cyanamide
Carbamothioic acid, bis(1-methylethyl)-S-(2,3-dichloro-2-propenyl) ester see...Diallate
Carbamothioic acid, bis(1-methylethyl)-, S-(2,3,3-trichloro-2-propenyl) ester see...Triallate
Carbamothioic acid, diethyl-,S-[chlorophenyl)methyl] ester see...Thiobencarb
Carbamothioic acid, diethyl-, S-[(4-chlorophenyl)methyl] ester see...Thiobencarb
Carbamothioic acid-S,S'-[2-(dimethylamino)-1,3-propanediyl]ester monohydrochloride see...Cartap Hydrochloride
Carbamothioic acid, dipropyl-, S-ethyl ester see...EPTC
Carbamothioic acid, dipropyl-, S-propyl ester see..Vernolate
Carbamothioic acid, N,N-hexamethylene-, S-ethyl ester see...Molinate
Carbamoylmethyl phosphorodithioate see...Formothion
CARBAMULT® see...Promecarb
Carbanilic acid, 3-chloro-, 4-chloro-2-butynyl ester see..Barban
d(-)-Carbanilic acid(1-ethylcarbamoyl)ethyl ester see..Carbetamide
Carbanilic acid, m-hydroxy-, metnyl ester, m-methylcarbanilate (ester) (8Cl) see...Phenmedipham
Carbanilic acid, isopropyl ester see...Propham
Carbanilic acid, 3-isopropyl-5-methyl-, methyl ester see..Promecarb
Carbanilic acid, m-methyl-, ester with methyl-m-hydroxycarbanilate (8Cl) see...Phenmedipham
m-Carbaniloyloxycarbanilic acid ethyl ester see..Desmedipham
Carbanolate see...Aldicarb
Carbaril (Italian) see...Carbaryl
Carbaryl, NAC see...Carbaryl
CARBATE® see...Carbendazim
Carbathrin see...Carboxin
Carbathion see...Metham-Sodium
Carbathione see...Metham-Sodium
Carbation see...Metham-Sodium
CARBATOX® see...Carbaryl
CARBAVUR® see...Carbaryl
CARBAX® see...Dicofol
Carbazinc see...Ziram
CARBAZINC® see...Ziram
CARBENDAZIME® see...Carbendazim
CARBENDAZOL® see...Carbendazim
CARBENDAZOLE® see...Carbendazim
CARBENDAZYM® see...Carbendazim
CARBENDOR® see...Carbendazim
CARBETAMEX® see...Carbetamide
Carbetamid (German) see...Carbetamide

Carbethoxy malathion see...Malathion
CARBETOVUR® see...Malathion
CARBETOX® see...Malathion
CARBICRIN® see...Dicrotophos
CARBICRON® see...Dicrotophos
Carbimide see...Cyanamide
CARBIN® see...Barban
CARBODAN® see...Carbofuran
Carbodiimide see...Cyanamide
1'-(Carboethoxy)ethyl-5-[2-chloro-4-(trifluoromethyl)phenoxy]-2-nitrobenzoate see..Lactofen
Carbofenothion (Dutch) see...Carbophanothion
Carbofos (Russian) see...Malathion
CARBOFUORFEN® see...Acifluorfen
Carbofurano (Spanish) see...Carbofuran
CARBOJECT® see...Oxycarboxin
Carbolineun see...Anthracene
CARBOMATE® see...Carbaryl
2-Carbomethoxy-1-methylvinyl dimethyl phosphate see..Mevinphos
2-Carbomethoxy-1-methylvinyl dimethyl phosophate, α isomer see...Mevinphos
α-2-Carbomethoxy-1-methylvinyl dimethyl phosphate see...Mevinphos
α-(2-Carbomethoxy-1-methylvinyl) dimethyl phosphate see...Mevinphos
2-Carbomethoxy-1-propen-2-yl dimethyl phosphate see..Mevinphos
CARBOMICRON® see...Dicrotophos
Carbon bisulfide see...Carbon Disulfide
Carbon bisulphide see...Carbon Disulfide
Carbon chloride see...Carbon Tetrachloride
CARBON D® see...Nabam
Carbon disulphide see...Carbon Disulfide
Carbon hexachloride see...Hexachloroethane
CARBON-S® see...Sodium Dimethyldithiocarbamate
Carbon sulfide see...Carbon Disulfide
Carbon tet see...Carbon Tetrachloride
Carbona see...Carbon Tetrachloride
(Carbonato)dihydroxydicopper see...Copper Carbonate, Basic
Carbone (sufure de) (French) see...Carbon Disulfide
Carbonic acid, ammonium copper salt see...Copper Ammonium Carbonate
Carbonic acid, calcium salt (1:1) see...Calcium Carbonate
Carbonic acid, dithio-, cyclic S,S-(6-methyl-2,3-quinoxalinediyl)ester see...Oxythioquinox
4,4'-Carbonimidoylbis(n,n-dimethylbenzenamine) see..Auramine
Carbonio (solfuro di) (Italian) see...Carbon Disulfide
Carbonothioic acid, O-(6-chloro-3-phenyl-4-pyridizinyl) S-octyl ester see...Pyridate
Carbonyldiamine see...Urea
Carbophen see...Metolcarb
Carbophos see...Malathion
CARBOSIP 5G® see...Carbofuran
CARBOSPOL® see...Allyl Isothiocyanate

CARBOTHIALDINE® see...Dazomet
CARBOTHIALDIN® see...Dazomet
Carbothion see...Metham-Sodium
5-Carboxanilido-2,3-dihydro-6-methyl-1,4-oxathiin see...Carboxin
Carboxide see...Calcium Hydroxide
CARBOXIN OXATHION PESTICIDE® see...Carboxin
Carboxin sulfone see...Oxycarboxin
Carboxine see...Carboxin
Carboxybenzene see...Benzoic Acid
Carboxylbenzene see...Benzoic Acid
N-(2-Carboxymethyl-6-chlorophenyl)-5-ethoxy-7-fluoro(1,2,4)triazolo-(1,5-c)pyrimidine-2-sulfonamide see...Cloransulam-methyl
CARBYNE® see...Barban
CARFENE® see...Azinphos-methyl
CARMAZINE® see...Mancozeb
CARPENE® see...Dodine
CARPETMAKER® see...Oxadiazon
CARPETMAKER X-X-X® see...Prodiamine
CARPIDOR® see...Benefin
CARPOLIN® see...Carbaryl
CARSORON® see...Dichlobenil
CARSORON® G see...Dichlobenil
CARSORON® G4 see...Dichlobenil
CARSORON® G20-SR see...Dichlobenil
Cartap monohydrochloride see...Cartap Hydrochloride
Carvene see...D-Limonene
CARVIL® see...BPMC
CARYLDERM® see...Carbaryl
CARYNE® see...Barban
CARZOL® see...Chlordimeform
CARZOL® see...Formetanate Hydrochloride
CARZOL® SP see...Formetanate Hydrochloride
CASORON® 133 see...Dichlobenil
Caspan see...Mercury Alkyl Compounds
Cassia aldehyde see...Cinnamaldehyde
CASTRIX® see...Crimidine
Caswell No. 011A see...Aldicarb
Caswell No. 040 see...Amitrole
Caswell No. 44 see...Molinate
Caswell No. 045 see...Ammonium Nitrate
Caswell No. 072A see...Sethoxydim
Caswell No. 077A see...Fenvalerate
Caswell No. 083E see...Resmethrin
Caswell No. 097 see...Prometryn
Caswell No. 109 see...Sodium Tetraborate
Caswell No. 119 see...Bromoxynil
Caswell No. 125-D see...Terbutryn
Caswell No. 130I see...Propargite
Caswell No. 130 see...Benefin
Caswell No. 160B see...Mevinphos
Caswell No. 165 A see...Carboxin
Caswell No. 179 see...Diethatyl-ethyl
Caswell No. 188AAA see...Oxyfluorfen
Caswell No. 193B see...Chlorimuron-ethyl
Caswell No. 195AA see...Norflurazon
Caswell No. 207AA see...Fenarimol

Caswell No. 207DA see...Thiobencarb
Caswell No. 214 see...Chloropicrin
Caswell No. 215D see...Quizalofop-Ethyl
Caswell No. 266AA see...Profenfos
Caswell No. 273H see...Fenpropathrin
Caswell No. 295 see...Dicamba
Caswell No. 306A see...Pronamide
Caswell No. 311 see...Dichloran
Caswell No. 315AI see...2,4-D, butoxyethyl ester
Caswell No. 315AV see...2,4-D, isopropyl ester
Caswell No. 316 see...2,4-DB
Caswell No. 319A see...Diclofop-methyl
Caswell No. 320 see...Dichlorprop
Caswell No. 323C see...Vinclozolin
Caswell No. 323EE see...Propiconazole
Caswell No. 334B see...Pirimiphos-Methyl
Caswell No. 342 see...Diazinon
Caswell No. 344A see...Thiophanate-Methyl
Caswell No. 366AA see...Tebuthiuron
Caswell No. 391D see...Dinocap
Caswell No. 392DD see...Dinoseb
Caswell No. 400 see...Dipropyl Isocinchomeronate
Caswell No. 410 see...Diuron
Caswell No. 419 see...Dodine
Caswell No. 431C see...Fenoxaprop-ethyl
Caswell No. 434C see...Ethoprop
Caswell No. 435 see...EPTC
Caswell No. 447AB see...Isofenphos
Caswell No. 454BB see...Pendimethalin
Caswell No. 455 see...Demeton-methyl
Caswell No. 456D see...Famphur
Caswell No. 458 see...Ferbam
Caswell No. 460C see...Fluazifop-butyl
Caswell No. 463F see...Bifenthrin
Caswell No. 472AA see...Dimethipin
Caswell No. 481DD see...Fenbutatin Oxide
Caswell No. 497AB see...Imazalil
Caswell No. 528 see...Linuron
Caswell No. 549AA see...Methazole
Caswell No. 563 see...Dichlorophene
Caswell No. 576 see...Oxythioquinox
Caswell No. 583 see...Monuron
Caswell No. 623A see...Oryzalin
Caswell No. 624A see...Oxadiazon
Caswell No. 642AB see...Hydramethylnon
Caswell No. 652B see...D-Phenothrin
Caswell No. 652C see...Fenoxycarb
Caswell No. 670 see...Piperonyl Butoxide
Caswell No. 706A see...Propetamphos
Caswell No. 723K see...Myclobutanil
Caswell No. 740 see...Simazine
Caswell No. 744A see...Sodium Azide
Caswell No. 762 see...Sodium Dimethyldithiocarbamate
Caswell No. 839A see...Hydramethylnon
Caswell No. 840 see...Dazomet
Caswell No. 844 see...Tetramethrin
Caswell No. 849A see...Thiabendazole
Caswell No. 862AA see...Triadimefon

Caswell No. 864 see...Tribufos
Caswell No. 870A see...Triallate
Caswell No. 882J see...Triclopyr Triethylammonium Salt
Caswell No. 896E see...Fentin Hydroxide
Caswell No. 900AA see...Thiodicarb
Caswell No. 903D see...Perfluidone
Caswell No. 927 see...Zinc Sulfate Heptahydrate
Caswell No. 934 see...Fluvalinate
CATAMINE® AB see...Zilkonium Chloride
CAV-TROL® see...Sodium Fluoride
CBN see...Barban
CCC see...Calcium Cyanamide
CCC PLANT GROWTH REGULANT® see..Chlormequat Chloride
CCN52® see...Cypermethrin
CD 68® see...Chlordane
CDA 101® see...Copper and Copper Compounds
CDA 122® see...Copper and Copper Compounds
CDA SIMFLOW PLUS® see...Amitrole
CDA 102® see...Copper and Copper Compounds
CDA 110® see...Copper and Copper Compounds
CDAA see...Allidochlor
CDAAT see...Allidochlor
CDB 60® see...Dichloroisocyanuric Acid
CDEC see...Sulfallate
CDM see...Chlordimeform
CDNA see...Dichloran
CDT® see...Simazine
CEASEFIRE® see...Fipronil
CEKIURON® see...Diuron
CEKUCAP® 25 WP see...Dinocap
CEKU C.B.® see...Hexachlorobenzene
CEKUDAZIM® see...Carbendazim
CEKUDIFOL® see...Dicofol
CEKUFON® see...Trichlorfon
CEKUGIB® see...Gibberellic Acid
CEKUMETA® see...Metaldehyde
CEKUMETHION® see...Methyl Parathion
Cekupropanil see...Propanil
CEKUQUAT® see...Paraquat
CEKUSAN® see...Dichlorvos
CEKUSAN® see...Simazine
CEKUSIL® see...Phenylmercury Acetate
CEKUSIL UNIVERSAL A® see..Methoxyethylmercuric Acetate
CEKUTHOATE® see...Dimethoate
CEKUTROTHION® see...Fenitrothion
CEKUZINA-S® see...Simazine
CEKUZINA-T® see...Atrazine
CELA S-1942® see...Bromophos
CELA S-2957® see...Chlorthiophos
CELA S-2225® see...Bromophos-ethyl
CELAMERCK S-2957® see...Chlorthiophos
CELANEX® see...Lindane
CELATHION® see...Chlorthiophos
CELATOX-DP® see...Dichlorprop
CELCURE® see...Sodium Dichromate

CELEBRITY® see...Nicosulfuron
CELEBRITY® see...Sodium Dicamba
Celery looper moth NPV see...Anagrapha Falcifera
CELITE® see...Diatomaceous Earth
CELLU-QUIN® see...Copper(II)-8-hydroxyquinoline
CELMER® see...Phenylmercury Acetate
Celmide see...Ethylene Dibromide
CELMONE® see...1-Naphthaleneacetic Acid
CELPHIDE® see...Aluminum Phosphide
CELPHOS® see...Aluminum Phosphide
Celphos see...Phosphine
CELTHION® see...Malathion
Celthion (Indian) see...Malathion
Cemerim see...Ametryn
CENOL GARDEN DUST® see...Rotenone
CENTURY-CIDE® see...Mevinphos
CEP see...Ethephon
2-CEPA see...Ethephon
CEPHA® see...Ethephon
CEPHA® 10LS see...Ethephon
CERANO® see...Clomazone
Cercobin- methyl see...Thiophanate-Methyl
CERCOBIN® see...Thiophanate-Methyl
Ceresan see...Ethyl Mercuric Chloride
CERESAN® see...Ethyl Mercuric Chloride
CERESAN® see...Phenylmercury Acetate
CERESOL® see...Phenylmercury Acetate
CEREVAX® EXTRA see...Imazalil
CERIDOR® see...Mecoprop
CERONE® see...Ethephon
CERTOL-LIN ONIONS® see...Linuron
CERTOX® see...Strychnine
CES see...Aramite
CESAR® see...Hexythiazox
CESAREX® see...DDT
CFS-Giftweizen (German) see...Thallium Sulfate
CFV® see...Chlorfenvinphos
CFZ see...Chlorfluazuron
CG 117® see...Metalaxyl
CG 1283® see...Mirex
CGA 10832® see...Profluralin
CGA 15324® see...Profenfos
CGA 18731® see...Isoproturon
CGA 24705® see...Metolachlor
CGA 26351® see...Chlorfenvinphos
CGA 45156® see...Thiodicarb
CGA 48988® see...Metalaxyl
CGA 64250® see...Propiconazole
CGA 71818® see...Penconazole
CGA 72662® see...Cyromazine
CGA 112913® see...Chlorfluazuron
CGA 131036® see...Triasulfuron
CGA 136,872® see...Primisulfuron-Methyl
CGA 163935® see...Cimectacarb
CGA 169374® see...Difenoconazole
CGA 169374® see...Difenoconazole
CGA 215944® see...Pymetrozine
CGA 219417® see...Cyprodinil

CGD 92710F® see...Propiconazole
CH-3565® see...Triclosan
Chalk see...Calcium Carbonate
CHALLENGER® see...Nicosulfuron
CHAMPION® see...Copper Hydroxide
CHAMPMAN PQ-8 see...Copper(II)-8-hydroxyquinoline
CHANDOR® see...Trifluralin
CHAP-FUME® see...Metham-Sodium
CHAPCO® Cu-NAP see...Copper Naphthenate
CHAPMAN WEED FREE® see...Sodium Chlorate
CHARGE® see...*lamda*-Cyhalothrin
CHECK-MATE® see...Cacodylic Acid
CHECK-THRU® see...Paraquat
CHECKMATE® see...Sethoxydim
CHECKMATE® see...Sodium Fluoride
CHECKMITE® see...Coumaphos
CHEMAGRO® 1776 see...Tribufos
CHEMAGRO® 9010 see...Propoxur
CHEMAGRO® 37289 see...Trichloronate
CHEMAGRO® B-1776 see...Tribufos
CHEMAID® see...Sodium Cacodylate
CHEMATHION® see...Malathion
CHEM-AX® see...Cacodylic Acid
CHEM-BAM® see...Nabam
CHEM FISH® see...Rotenone
CHEMFORM® see...Streptomycin
CHEMFORM® see...Maleic Hydrazide
CHEMFORM® see...Methoxychlor
CHEM-HOE® see...Propham
Chemical 109 see...ANTU
CHEMIFLOUR® see...Sodium Fluoride
CHEMIURON® see...Diuron
CHEMLEY® see...Ethoxyquin
CHEM-MITE® see...Rotenone
CHEM NEB® see...Maneb
CHEMOCIN® see...Copper Oxychloride
CHEMOX® see...Dinoseb
CHEMOX P.E.® see...Dinoseb
CHEMOX GENERAL® see...Dinoseb
CHEMPAR® see...Copper Oxychloride
CHEM PELS C® see...Sodium Arsenite
Chem-Phene see...Toxaphene
CHEMRAT® see...Pindone
CHEM RICE® see...Propanil
CHEMSECT DNBP® see...Dinoseb
CHEM-SEN 56® see...Sodium Arsenite
CHEM-TEK® see...Thiabendazole
CHEM-TOL® see...Pentachlorophenol
CHEM ZINEB® see...Zineb
CHESS® see...Pymetrozine
Chestnut compound see...Copper Carbonate, Basic
CHEVRON 9006® see...Methamidophos
CHEVRON ORTHO 9006® see...Methamidophos
CHEVRON RE 12420® see...Acephate
CHEYENNE® see...MCPA
CHILTERN KOCIDE® 101 see...Copper Hydroxide
CHILTERN OLE® see...Chlorothalonil

CHIMAC OXY® see...MCPA
CHIMICHLOR® see...Alachlor
CHIMIGOR 40® see...Dimethoate
CHINALPHOS® see...Quinalphos
Chinomethionat see...Oxythioquinox
Chinomethionate see...Oxythioquinox
CHINORTA® see...Paraoxon
CHINOZAN® see...Quintozene
CHINUFUR® see...Carbofuran
CHIOLITE® see...Sodium Aluminum fluoride
CHIP-CAL® see...Calcium Arsenate
CHIPCO® see...Fipronil
CHIPCO® see...Mecoprop
CHIPCO® see...Thiodicarb
CHIPCO® 26019 see...Iprodione
CHIPCO® ALIETTE WDG see...Fosetyl-Al
CHIPCO BUCTRIL® see...Bromoxynil
CHIPCO CRAB-KLEEN® see...Bromoxynil
CHIPCO FLOREL PRO see...Ethephon
CHIPCO THIRAM 75® see...Thiram
CHIPCO TURF HERBICIDE "D"® see...2,4-D
CHIPCO TURF HERBICIDE MCPP® see...Mecoprop
CHIPMAN 3,142® see...Tecnazene
CHIPMAN 6199® see...Amiton Oxalate
CHIPMAN 6200® see...Amiton
CHIPMAN 11974® see...Phosalone
CHIPMAN® PATH WEEDKILLER see...Amitrole
CHIPMAN® R-6,199 see...Amiton Oxalate
CHIP SHOT® see...Oxadiazon
CHIPTOX® see...MCPA
CHLON® see...Pentachlorophenol
Chloordaan (Dutch) see...Chlordane
*O*-2-Chloor-1-(2,4-dichloor-fenyl)-vinyl-*O,O*-diethylfosfaat (Dutch) see...Chlorfenvinphos
2-Chloor-4-dimethylamino-6-methyl-pyrimidine (Dutch) see...Crimidine
Chloorpikrine (Dutch) see...Chloropicrin
α-Chlor-6'-aethyl-*N*-(2-methoxy-1-methylaethyl)-acet-*o*-toluidin (German) see...Metolachlor
(4-Chlor-but-2-in-yl)-*n*-(3-chlor-phenyl)-carbamat (German) see...Barban
(2-Chlor-3-diethylamino-methyl-3-oxo-prop-1-en-yl)-dimethylphosphat (German) see...Phosphamidon
*O*-2-Chlor-1-(2,4-dichlor-phenyl)-vinyl-*O,O*-diaethylphosphat (German) see...Chlorfenvinphos
CHLOR KIL® see...Chlordane
*N*'-(3-Chlor-4-methoxy-phenyl)-*N,N*-dimethylharnstoff (German) see...Metoxuron
2-(4-Chlor-2-methyl-phenoxy)-propionsaeure (German) see...Mecoprop
CHLOR-O-PIC® see...Chloropicrin
Chloracetic acid see...Chloroacetic Acid
2-Chloraethyl phosphonsaeure (German) see...Ethephon
2-Chloraethyl-trimethylammoniumchlorid (German) see...Chlormequat Chloride
CHLORAK® see...Trichlorfon
2-Chlorallyl diethyldithiocarbamate see...Sulfallate
Chlorallyl diethyldithiocarbamate see...Sulfallate

2-Chlorallyl-*N,N*-diethyldithiocarbamate see...Sulfallate
Chlorambed see...Chloramben
Chloramben, Aromatic carboxylic acid see..Chloramben
Chloramben benzoic acid herbicide see...Chloramben
Chlorambene see...Chloramben
Chloramizol see...Imazalil
Chloramp (Russia) see...Picloram
2-(2-Chloranilin)-4,6-dichlor-1,3,5-triazin (German) see. Anilazine
Chlorate salt of magnesium see...Magnesium Chlorate
Chlorate of soda see...Sodium Chlorate
Chlorate salt of sodium see...Sodium Chlorate
Chlorate de magnesium (French) see...Magnesium Chlorate
CHLORAX® see...Sodium Chlorate
CHLORAXYL® see...Metalaxyl
CHLORAXYL® SEED TREATER see...Chloroneb
CHLORBAN® see...Chlorpyrifos
Chlorbenzalate see...Chlorobenzilate
Chlorbenzilat see...Chlorobenzilate
Chlorcholinchlorid see...Chlormequat Chloride
Chlorcholine chloride see...Chlormequat Chloride
Chlorcyan see...Cyanogen Chloride
Chlordan see...Chlordane
Chlorea see...Monuron
CHLORESENE® see...Lindane
Chloressigsaeure-*n*-isopropylanilid (German) see.. Propachlor
Chlorethephon see...Ethephon
Chlorethoxyphos see...Chlorethoxyfos
2-Chlorethylphosphonic acid see...Ethephon
Chlorex see...Dichloroethyl Ether
Chlorfacinon (German) see...Chlorophacinone
CHLORFENAC® see...Fenac
Chlorfenamidine see...Chlordimeform
Chlorfenidim see...Monuron
Chloric acid, magnesium see...Magnesium Chlorate
Chloric acid, magnesium salt see...Magnesium Chlorate
Chloric acid, sodium salt see...Sodium Chlorate
Chlorid rtutnaty (Czech) see...Mercuric Chloride
CHLORIDAZON® see...Pyrazon
Chlorimuronethyl ester see...Chlorimuron-ethyl
Chlorimuronethyl [Ethyl-2-[[[(4-chloro-6-methoxyprimidin-2-yl)-carbonyl]-amino]sulfonyl]benzoate] see...Chlorimuron-ethyl
Chlorinat see...Barban
CHLORINDAN® see...Chlordane
Chlorine cyanide see...Cyanogen Chloride
Chlormequat see...Chlormequat Chloride
CHLORMETHAZOLE® see...Methazole
Chlormethylfos see...Chlorpyrifos-methyl
2-Chloroacetaldehyde see...Chloroacetaldehyde
Chloroacetaldehyde monomer see...Chloroacetaldehyde
*N*-Chloroacetyl-*N*-(2,6-diethylphenyl)glycine, ethyl ester see...Diethatyl-ethyl
β-Chloroallyl chloride see...1,3-Dichloropropene
2-Chloroallyl-*N,N*-diethyldithiocarbamate see..Sulfallate
(*o*-Chloroanilino)dichlorotriazine see...Anilazine
4-Chloro-2-*oxo*-3(2*H*)-benzothiazoleacetic acid, ethyl ester see...Benazolin Ethyl
3-(6-Chloro-2-*oxo*-benzoxazolin-3-yl)methyl-O,O-diethyl phosphorothiolothionate see...Phosalone
*S*-[(6-Chloro-2-*oxo*-3(2H)-benzoxazolyl)methyl] O,O-diethylphosphordithioate see...Phosalone
2-[4-((6-Chloro-2-benzoxazolyl)oxy)phenoxy]propionic acid, ethyl ester, (±)- see...Fenoxaprop-ethyl
1-(*p*-Chlorobenzyl)-1-cyclopentyl-3-phenylurea see. Pencycuron
*p*-Chlorobenzyl *N,N*-diethylthiolcarbamate see.. Thiobencarb
*S*-(*p*-Chlorobenzyl)diethylthiocarbamate see.. Thiobencarb
*S*-(4-Chlorobenzyl) *N,N*-diethylthiocarbamate see.. Thiobencarb
2-(2-Chlorobenzyl)-4,4-dimethyl-1,2-oxazolidin-3-one see...Clomazone
1-Chloro-3,5-bis(ethylamino)-2,4,6-triazine see.. Simazine
2-Chloro-4,6-bis(ethylamino)-1,3,5-triazine see.. Simazine
2-Chloro-4,6-bis(ethylamino)-*s*-triazine see...Simazine
2-Chloro-4,6-bis(isopropylamino)-*s*-triazine see. Propazine
CHLOROBLE M® see...Maneb
4-Chlorobut-2-ynyl-*m*-chlorocarbanilate see...Barban
4-Chlorobut-2-ynyl-3-chlorophenylcarbamate see. Barban
5-Chloro-3-*tert*-butyl-6-methyluracil see...Terbacil
Chloro-2-butynyl-*m*-chlorocarbamate see...Barban
4-Chloro-2-butynyl-*m*-chlorocarbanilate see...Barban
4-Chloro-2-butynyl-(3-chlorophenyl)carbamate see.. Barban
4-Chloro-2-butynyl-*n*-(3-chlorophenyl)carbamate see.. Barban
Chlorocamphene see...Toxaphene
*m*-Chloro carbanilic acid-4-chloro-2-butynyl ester see.. Barban
Chlorocholine chloride see...Chlormequat Chloride
3-Chloro-*N*-[3-chloro-2,6-dinitro-4-(trifluoromethyl)phenyl]-5-(trifluoromethyl)-2-pyridinamine see...Fluazinam®
1-Chloro-2-(ß-chloroethoxy)ethane see...Dichloroethyl Ether
5-Chloro-*N*-(2-chloro-4-nitrophenyl)-2-hydroxybenzamide, 2-aminoethanol salt see...Clonitralid
5-Chloro-*N*-(2-chloro-4-nitrophenyl)-2-hydroxybenzamide with 2-aminoethanol (1:1) see. Clonitralid
4-Chloro-α-(4-chlorophenyl)-α-(trichloromethyl)benzene methanol see...Dicofol
3-Chlorochlordene see...Heptachlor
3-Chloro-*N*-(3-chloro-5-trifluoromethyl-2-pyridinyl)-α,α,α-trifluoro-2,6-dinitro-*p*-toluidine see...Fluazinam®
Chlorocresol see...*p*-chloro-*m*-cresol
*p*-Chlorocresol see...*p*-chloro-*m*-cresol
4-Chloro-*m*-cresol see...*p*-chloro-*m*-cresol

6-Chloro-*m*-cresol see...*p*-chloro-*m*-cresol
4-Chloro-*o*-cresoxyacetic acid see...MCPA
(4-Chloro-*o*-cresoxy)acetic acid see...MCPA
Chlorocyan see...Cyanogen Chloride
Chlorocyanide see...Cyanogen Chloride
Chlorocyanogen see...Cyanogen Chloride
2-Chloro-4-((1-cyano-1-methylethyl)amino)-6-(ethylamino)-*s*-triazine, Cyanazine 2-chloro-4-(1-cyano-1-methylethylamino)-6-ethylamino-1,3,5-triazine see...Cyanazine
CHLORODANE® see...Chlordane
2-Chloro-*N,N*-diallylacetamide see...Allidochlor
α-Chloro-*N,N*-diallylacetamide see...Allidochlor
1-Chloro-2,3-dibromopropane see...Dibromochloropropane
3-Chloro-1,2-dibromopropane see...Dibromochloropropane
5-Chloro-2-(2,4-dichlorophenoxy)phenol see...Triclosan
5-Chloro-2-(2,4-dichlorophenoxyphenol) see...Triclosan
β-2-Chloro-1-(2',4'-dichlorophenyl) vinyl dethylphosphate see...Chlorfenvinphos
*N*-(4-Chlorophenyl)-*N*'-(3,4-dichlorophenyl)urea see...Triclocarban
2-Chloro-1-(2,4-dichlorophenyl)vinyl diethyl phosphate see...Chlorfenvinphos
*N*-(2-Chloro-1-(diethoxyphosphinpthioylthio)ethyl) phthalimide see...Dialifor
2-Chloro-3-(diethylamino)-1-methyl-3-*oxo*-1-propenyldimethyl phosphate see...Phosphamidon
2-Chloro-2',6'-diethyl-*N*-(butoxymethyl)acetanalide see...Butachlor
2-Chloro-2-diethylcarbamoyl-1-methylvinyldimethyl phosphate see...Phosphamidon
1-Chloro-diethylcarbamoyl-1-propen-2-yl dimethyl phosphate see...Phosphamidon
α-Chloro-2',6'-diethyl-*N*-(methoxymethyl)acetanilide see...Alachlor
2-Chloro-*N*-(2,6-diethylphenyl)-n-(methoxymethyl)acetamide see...Alachlor
S-4-Chlorobenzyl diethylthiocarbamate see...Thiobencarb
6-Chloro-*N,N*'-diethyl-1,3,5-triazine-2,4-diamine see...Simazine
6-Chloro-*N2,N4*-diethyl-1,3,5-triazine-2,4-diamine see...Simazine
6-Chloro-*N,N*'-diethyl-1,3,5-triazine-2,4-diyldiamine see...Simazine
(2-Chloro-3-dietilamino-1-metil-3-oxo-prop-1-en-il)-dimetil-fosfato (Italian) see...Phosphamidon
5-Chloro-3-(1,1-dimethyl)-6-methyl-2,4(1*H*,3*H*)-pyrimidinedione see...Terbacil
5-Chloro-3-(1,1-dimethylethyl)-6-methyl-2,4(1*H*,3*H*)-pyrimidinedione see...Terbacil
4-Chloro-3,5-dimethylphenol see...4-chloro-3,5-xylenol
2-Chloro-*N*-(2,4-dimethyl-3-thienyl)-*N*-(2-methoxy-1-methylethyl)acetamide see...Dimethenamid
2-Chloro-*N,N*-di-2-propenylacetamide see...Allidochlor
2-Chloroethanal see...Chloroacetaldehyde

2-Chloro-1-ethanal see...Chloroacetaldehyde
2-Chloroethanephosphonic acid see...Ethephon
Chloroethanoic acid see...Chloroacetic Acid
2-Chloroethanol-2-(*p-tert*-butylphenoxy)-1-methylethyl sulfite see...Aramite
2-Chloroethanol ester with 2-(*p-tert*-butylphenoxy)-1-methylethyl sulfite see...Aramite
2-(2-Chloroethoxy)-*N*-[((4-methoxy-6-methyl-1,3,5-triazin-2-yl)amino)carbonyl]benzenesulfonamide see...Triasulfuron
2-Chloro-*N*-(ethoxymethyl)-6'-ethyl-*O*-acetotoluidide see...Acetochlor
2-Chloro-*N*-(ethoxymethyl)-*N*-(2-ethyl-6-methylphenyl)acetamide see...Acetochlor
2-Chloro-1-(3-ethoxy-4-nitrophenoxy)-4-(trifluoromethyl)benzene see...Oxyfluorfen
2-Chloro-4-ethylamineisopropylamine-s-triazine see...Atrazine
2-Chloro-4-ethylamino-6-(1-cyano-1-methyl)ethylamino-*s*-triazine see...Cyanazine
1-Chloro-3-ethylamino-5-isopropylamino-s-triazine see...Atrazine
1-Chloro-3-ethylamino-5-isopropylamino-2,4,6-triazine see...Atrazine
2-Chloro-4-ethylamono-6-isopropylamino- see...Atrazine
2-Chloro-4-ethylamino-6-isopropylamino-1,3,5-triazine see...Atrazine
2-Chloro-4-ethylamino-6-isopropylamino-s-triazine see...Atrazine
2-[(4-Chloro-6-(ethylamino)-1,3,5-triazin-2-yl)amino]-2-methylpropanenitrile see...Cyanazine
2-[(4-Chloro-6-(ethylamino)-*s*-triazin-2-yl)amino]-2-methylpropionitrile see...Cyanazine
2-((4-Chloro-6-(ethylamino)-*s*-triazin-2-yl)amino)-2-methylpropanenitrile see...Cyanazine
2-(4-Chloro-6-ethylamino-1,3,5-triazin-2-ylamino)-2-methylpropionitrile see...Cyanazine
β-Chloroethyl-β'-(*p-tert*-butylphenoxy)-α'-methylethyl sulfite see...Aramite
β-Chloroethyl-β-(*p-tert*-butylphenoxy)-α-methylethyl sulphite see...Aramite
*N*-(2-Chloroethyl)-2,6-dinitro-*n*-propyl-4-(trifluoromethyl)aniline see...Fluchloralin
*N*-(2-Chloroethyl)-2,6-dinitro-*n*-propyl-4-(trifluoromethyl)benzenamide see...Fluchloralin
Chloroethyl ether (DOT) see...Dichloroethyl Ether
6-Chloro-*N*-ethyl-*n*-isopropyl-1,3,5-triazinediyl-2,4-diamine see...Atrazine
Chloroethyl mercury see...Ethyl Mercuric Chloride
2-Chloro-6'-ethyl-*N*-(2-methoxy-1-methylethyl)acet-*o*-toluidide see...Metolachlor
2-Chloroethyl 1-methyl-2-(*p-tert*-butylphenoxy)ethyl sulphite see...Aramite
6-Chloro-*N*-ethyl-*N*'-(1-methylethyl)-1,3,5-triazine-2,4-diamine see...Atrazine
α-Chloro-2'-ethyl-6'-methyl-*N*-(1-methyl-2-methoxyethyl)-acetanilide see...Metolachlor

2-Chloro-*N*-(2-ethyl-6-methylphenyl)-*N*-(2-methoxy-1-methylethyl) acetamide  see. . .Metolachlor
(2-Chloroethyl)phosphonic acid  see. . .Ethephon
2-Chloroethyl sulfurous acid-2-(4-(1,1-dimethylethyl)phenoxy)-1-methylethyl ester  see. . Aramite
2-Chloroethyl sulphite of 1-(*p-tert*-butylphenoxy)-2-propanol  see. . .Aramite
2-Chloro-*N*, (6-ethyl-*o*-tolyl)-*N*-(2-methoxy-1-methylethyl)-acetamide  see. . .Metolachlor
*N*-(2-Chloroethyl)-α,α,α-trifluoro-2,6-dinitro-*N*-propyl-*p*-toluidine  see. . .Fluchloralin
(E)-4-Chloro-α,α,α-trifluoro-*N*-(1-(1*H*-imidazol-1-yl)-2-propoxyethylidene)-*O*-toluidine  see. . .Triflumizole
2-Chloroethyl trimethylammonium chloride  see. . Chlormequat Chloride
(2-Chloroethyl)trimethylammonium chloride  see. . Chlormequat Chloride
(β-Chloroethyl)trimethylammonium chloride  see. . Chlormequat Chloride
2-Chloro-*N,N,N*-ethyl)trimethylethanaminium chloride  see. . .Chlormequat Chloride
Chlorofenvinphos  see. . .Chlorfenvinphos
Chloroform, nitro-  see. . .Chloropicrin
CHLOROFTALM®  see. . .Trichlorfon
3-Chloro-7-hydroxy-4-methyl-coumarin *O,O*-diethyl phosphorothioate  see. . .Coumaphos
3-Chloro-7-hydroxy-4-methyl-coumarin *O*-ester with *O,O*-diethylphosphorothioate  see. . .Coumaphos
2-Chloro-hydroxytoluene  see. . .*p*-Chloro-*m*-cresol
6-Chloro-3-hydroxytoluene  see. . .*p*-Chloro-*m*-cresol
2-Chloro-*N*-isopropylacetanilide  see. . .Propachlor
α-Chloro-*N*-isopropylacetanilide  see. . .Propachlor
2-Chloro-*N*-isopropyl-*N*-phenylacetamide  see. . .Propachlor
*S*-[6-Chloro-3-(mercaptomethyl)-2-benzoxazolinone]-*O,O*-diethylphosphorodithioate  see. . .Phosalone
2-Chloro-*N*-[((4-methoxy-6-methyl-1,3,5-triazin-2-yl)amino)carbonyl]benzenesulfonamide  see. . Chlorsulfuron
2-Chloro-*N*-[(4-methoxy-6-methyl-1,3,5-triazin-2-yl)aminocarbonyl]-benzenesulfonamide  see. . Chlorsulfuron
1-[2-(2-Chloroethoxy)phenylsulfonyl]-3-(4-methoxy-6-methyl-1,3,5-triazin-2-yl)urea  see. . .Triasulfuron
3-(3-Chloro-4-methoxyphenyl)-1,1-dimethylurea  see. . Metoxuron
*N*-(3-Chloro-4-methoxyphenyl)-*N',N'*-dimethylurea  see. . Metoxuron
*N'*-(3-Chloro-4-methoxyphenyl)-*N,N*-dimethylurea  see. . Metoxuron
2-[(((4-Chloro-6-methoxy-2-pyrimidinyl)amino)carbonyl)amino)sulfonyl]benzoic acid  see. . .ethyl ester  see. . .Chlorimuron-ethyl
*S*-Chloromethyl-*O,O*-diethyl phosphorodithioate  see. . Chlormephos
*S*-(Chloromethyl)-*O,O*-diethyl phosphorodithioate  see. . Chlormephos
*S*-(Chloromethyl)-*O,O*-diethyl phosphorodithioic acid  see. . .Chlormephos
*S*-Chloromethyl-*O,O*-diethyl phosphorodithiothiolothionate  see. . .Chlormephos
*S*-(2-Chloro-1-(1,3-dihydro-1,3-dioxo-2*H*-isoindol-2-yl)ethyl)-*O,O*-diethyl phosphorodithioate  see. . .Dialifor
3-Chloro-5-[((((4,6-dimethoxy-2-pyrimidinyl)amino)carbonyl)amino)sulfonyl]-1-methyl-1*H*-pyrazole-4-carboxylic acid  see. . .Halosulfuron-methyl
4-Chloro-5-methylamino-2-(3-trifluoromethylphenyl)pyridazin-3-one  see. . .Norflurazon
4-Chloro-5-(methylamino)-2-[3-(trifluoromethyl)phenyl]-3(2H)-pyridazinone  see. . .Norflurazon
4-Chloro-5-methylamino-2-(α, α,α-trifluoro-*m*-tolyl)pyridazinone-3(2H)-one  see. . .Norflurazon
4-Chloro-5-(methylamino)-2-(α,α,α-trifluoro-*m*-tolyl)-3(2H)-pyridazinone  see. . .Norflurazon
2-Chloro-1-methylbenzene  see. . .*o*-Chlorotoluenel
[*s*-(R*,R*)]-4-Chloro-α-(1-methylethyl)benzeneacetic acid, cyano(3-phenoxyphenyl)methyl ester  see. .Esfenvalerate
2-Chloro-*N*-(1-methylethyl)-*N*-phenylacetamide  see. . Propachlor
*S*-(Chloromethyl) *O,O*-diethyl ester phosphorodithioic acid  see. . .Chlormephos
4-Chloro-3-methylphenol  see. . .*p*-chloro-*m*-cresol
4-Chloro-2-methylphenoxyacetic acid  see. . .MCPA
(4-Chloro-2-methylphenoxy)acetic acid  see. . .MCPA
2-(4-Chloro-2-methylphenoxy)propanoic acid  see. . Mecoprop
(4-Chloro-2-methylphenoxy)propionic acid  see. . Mecoprop
α-(4-Chloro-2-methylphenoxy)propionic acid  see. . Mecoprop
4-Chloro-α-(1-methylethyl)benzeneacetic acid cyano(3-phenoxyphenyl)methyl ester  see. . .Fenvalerate
3-Chloro-4-methyl-7-coumarinyldiethyl phosphorothioate  see. . .Coumaphos
*O*-3-Chloro-4-methyl-7-coumarinyl *O,O*-diethyl phosphorothioate  see. . .Coumaphos
2-Chloro-4-methyl-6-dimethylaminopyrimidine  see. . Crimidine
Chloromethylmercury  see. . .Mercury Alkyl Compounds
*N'*-(4-Chloro-2-methylphenyl)-*N,N*-dimethylmethanimidamide  see. . .Chlordimeform
3-Chloro-4-methyl-7-hydroxycoumarindiethyl thiophosphoric acid ester  see. . .Coumaphos
2-Chloro-*N*-[(1-methyl-2-methoxy)-ethyl]-*N*-(2,4-dimethyl-thien-3-yl)acetamide  see. . .Dimethenamid
4-Chloro-2-methylphenoxy-α-propionic acid  see. . Mecoprop
(+)-α-(4-Chloro-2-methylphenoxy) propionic acid  see. . Mecoprop
3-Chloro-4-methylumbelliferoneo-ester with *O,O*-diethyl phosphorothioate  see. . .Coumaphos
Chloronebe (French)  see. . .Chloroneb
CHLORONIL®  see. . .Chlorothalonil

1-Chloro-1-nitropropano (Spanish) see...1-Chloro-1-Nitropropane
Chloronitropropane see...1-Chloro-1-Nitropropane
1,1-Chloronitropropane see...1-Chloro-1-Nitropropane
CHLOROPHEN® see...Pentachlorophenol
Chlorophenothan see...DDT
Chlorophenothane see...DDT
α.-Chlorophenothane see...DDT
Chlorophenotoxum see...DDT
2-(4-Chlorophenoxy)-1-*tert*-butyl-2-(1*H*-1,2,4-triazole-1-yl)ethanol see...Triadimenol
β-(4-Chlorophenoxy)-α-(1,1-dimethylethyl)-1*H*-1,2,4-triazole-1-ethanol see...Triadimenol
1-(4-Chlorophenoxy)-3,3-dimethyl-1-(1,2,4-triazol-1-yl)butanone- see...Triadimefon
1-(4-Chlorophenoxy)-3,3-dimethyl-1-(1,2,4-triazol-1-yl)-butan-2-one see...Triadimefon
1-(4-Chlorophenoxy)-3,3-dimethyl-1-(1,2,4-triazol-1-yl)-2-butan-2-one see...Triadimefon
1-(4-Chlorophenoxy)-3,3-dimethyl-1-(1*H*-1,2,4-triazol-1-yl)-2-butanone see...Triadimefon
1-(4-Chlorophenoxy)-3,3-dimethyl-1-(1*H*-1,2,4-triazol-1-yl)-2-butanone see...Triadimefon
1-(4-Chlorophenoxy)-3,3-dimethyl-1-(1*H*-1,2,4-triazol-1-yl)butanone- see...Triadimefon
(1RS,2RS,1RS,2SR)-1-(4-Chlorophenoxy)-3,3-dimethyl-1-(1*H*-1,2,4-triazol-1-yl)butan-2-ol see...Triadimenol
2-(4-Chlorophenoxy-2-methyl)propionic acid see. Mecoprop
3-[4-(4-Chlorophenoxy)phenyl]-1,1-dimethylurea see. Chloroxuron
3-[*p*-(*p*-Chlorophenoxy)phenyl-1,1]-dimethylurea see. Chloroxuron
*N*'-[4-(4-Chlorophenoxy)phenyl]-*N*,*N*-dimethylurea see. Chloroxuron
*N*-[((4-Chlorophenyl)amino)carbonyl]-2,6-difluorobenzamide see...Diflubenzuron
*N*-(3-Chlorophenyl)carbamate de 4-chloro 2-butynyle (French) see...Barban
(3-Chlorophenyl)carbamic acid 4-chloro-2-butynyl ester see...Barban
*p*-Chlorophenyl chloride see...*para*-Dichlorobenzene
α-(2-Chlorophenyl)-α-(4-chlorophenyl)-5-pyridinemethanol see...Fenarimol
α-(2-Chlorophenyl)-α-(4-chlorophenyl)-5-pyrimidinemethanol see...Fenarimol
(2-Chlorophenyl)-α-(4-chlorophenyl)-5-pyrimidinemethanol see...Fenarimol
*trans*-5-(4-Chlorophenyl)-*N*-cyclohexyl-4-methyl-2-*oxo*-3-thiazolidinecarboxamide see...Hexythiazox
α-(4-Chlorophenyl)-α-(1-cyclopropylethyl)-1*H*-1,2,4-triazole-1-ethanol see...Cyproconazole
(2RS,3RS)-2-(4-Chlorophenyl)-3-cyclopropyl-1-(1*H*-1,2,4-triazol-1-yl)butan-2-ol see...Cyproconazole
1-(4-Chlorophenyl)-3-(2,6-difluorobenzoyl)urea see. Diflubenzuron
*S*-[(4-Chlorophenyl)methyl] diethylcarbamothioate see. Thiobencarb
2-(α-*p*-Chlorophenylacetyl)indane-1,3-dione see. Chlorophacinone
4-[3-(4-Chlorophenyl)-3-(3,4-dimethoxyphenyl)-1-*oxo*-2-propenyl]morpholine see...Dimethomorph
(*E*,*Z*)-4-[3-(4-Chlorophenyl)-3-(3,4-dimethoxyphenyl)acryloyl]morpholine see. Dimethomorph
3-(4-Chlorophenyl)-3-(3,4-dimethyphenyl)acrylic acid morpholide see...Dimethomorph
(2RS,3RS)-1-(4-Chlorophenyl)-4,4-dimethyl-2-(1*H*-1,2,4-triazol-1-yl)pentan-3-ol see...Paclobutrazole
(RS)-1-*p*-Chlorophenyl-4,4-dimethyl-3-(1*H*-1,2,4-triazol-1-ylmethyl)pentan-3-ol see...Tebuconazole
(RS)-1-(4-Chlorophenyl)-4,4-dimethyl-3-(1*H*-1,2,4-triazol-1-ylmethyl)pentan-3-ol see...Tebuconazole
(RS)-1-*p*-Chlorophenyl-4,4-dimethyl-3-(1*H*-1,2,4-triazol-1-ylmethyl)pentan-3-ol see...Tebuconazole
4-Chlorophenyl dimethylurea see...Monuron
1-(4-Chlorophenyl)-3,3-dimethylurea see...Monuron
1-*p*-Chlorophenyl-3,3-dimethylurea see...Monuron
3'-(4'-Chlorophenyl)-1,1-dimethylurea see...Monuron
3'-(4'-Chlorophenyl)-1,1-dimethylurea see...Monuron
3-(4-Chlorophenyl)-1,1-dimethylurea see...Monuron
3-(*p*-Chlorophenyl)-1,1-dimethylurea see...Monuron
3-*p*-Chlorophenyl-1,1-dimethylurea see...Monuron
*N*'-(4-Chlorophenyl)-*N*,*N*-dimethylurea see...Monuron
*N*-(4-Chlorophenyl)-*N*',*N*'-dimethylurea see...Monuron
*N*-(p-Chlorophenyl)-*N*',*N*'-dimethylurea see...Monuron
N-p-Chlorophenyl-*N*',*N*'-dimethylurea see...Monuron
(±)-α-[2-(4-Chlorophenyl)ethyl]-α-(1,1-dimethylethyl)-1*H*-1,2,4-triazole-1-ethanol see...Tebuconazole
-α-(4-Chlorophenyl)-α-hydroxybenzeneacetic acid ethyl ether see...Chlorobenzilate
(±)-R*,R*-β-[(4-Chlorophenyl)methyl]-α-(1,1-dimethylethyl)-1*H*-1,2,4-triazol-1-ethanol see. Paclobutrazole
2-[(2-Chlorophenyl)methyl]-4,4-dimethyl-3-isoxazolidinone see...Clomazone
2-[(*p*-Chlorophenyl)phenylacetyl]-1,3-indandione see. Chlorophacinone
2[(4-Chlorophenyl)phenylacetyl]-1*H*-indene-1,3(2*H*)-dione see...Chlorophacinone
[(4-Chlorophenyl)-1-phenyl]-acetyl-1,3-indandion (German) see...Chlorophacinone
*S*-[(*p*-Chlorophenylthio)methyl]-*O*,*O*-diethyl phosphorodithioate see...Carbophanothion
*S*-[(4-Chlorophenyl)methyl] *N*,*N*-diethylthiocarbamate see...Thiobencarb
α-[2-(4-Chlorophenyl)ethyl]-α-phenyl-1*H*-1,2,4-triazole-1-propane nitrile see...Fenbuconazole
α-[2-(4-Chlorophenyl)ethyl]-α-phenyl-3-(1*H*-1,2,4-triazole)-1-propanenitrile see...Fenbuconazole
*N*-[(4-Chlorophenyl)-methyl]-*N*-cyclopentyl-*N*'-phenylurea see...Pencycuron
*S*-[(4-Chlorophenyl)methyl]diethylcarbamothiote see. Thiobencarb
4-(4-Chlorophenyl)-2-phenyl-2-(1*H*-1,2,4-triazol-1-ylmethyl)butyronitrile see...Fenbuconazole

*O*-(6-Chloro-3-phenyl-4-pyridazinyl)-carbonothioic acid *S*-octyl ester **see**...Pyridate
*O*-(6-Chloro-3-phenyl-4-pyridazinyl) *S*-octyl carbonothioate **see**...Pyridate
2[2-(4-Chlorophenyl]-2-phenylacetyl)indan-1,3-dione **see**...Chlorophacinone
6-Chloro-3-phenylpridazin-4-yl-*S*-octyl-thiocarbonate **see**...Pyridate
1-[(*O*-Chlorophenyl)sulfonyl]-3-(4-methoxy-6-methyl-*S*-triazin-2-yl)urea **see**...Chlorsulfuron
*S*-(4-Chlorophenylthiomethyl)diethyl Phosphorothiolothionate **see**...Carbophanothion
2-*p*-Chlorophenyl-2-(1*H*-1,2,4-triazole-1-ylmethyl)hexanenitrile **see**...Myclobutanil
2-(4-Chlorophenyl)-2-(1*H*-1,2,4-triazole-1-ylmethyl)hexanenitrile **see**...Myclobutanil
2-Chloro-6-trichloromethylpyridine **see**...Nitrapyrin
Chlorophos **see**...Trichlorfon
*S*-(2-Chloro-1-phthalimidoethyl)-*O,O*-diethylphosphorodithioate **see**...Dialifor
Chlorophthalm **see**...Trichlorfon
Chloropicrine (French) **see**...Chloropicrin
2-Chloro-2-propene-1-thiol diethyldithiocarbamate **see**...Sulfallate
3-Chloropropenyl chloride **see**...1,3-Dichloropropene
(E)-2-[1-(((3-Chloro-2-propenyl)oxy)imino)propyl]-5-[2-(ethylthio)propyl]-3-hydroxy-2-cyclohexen-1-one **see**...Clethodim
2-Chloro-2-propenyldiethylcarbamodithioate **see**...Sulfallate
1-[1-((4-Chloro-2-(trifluoromethyl)phenyl)imino)-2-propoxyethyl]-1*H*-imidazole **see**...Triflumizole
2-Chloro-4-(2-propylamino)-6-ethylamino-s-triazine **see**...Atrazine
*N*-(2-Chloro-4-pyridinyl)-*N*'-phenylurea **see**...Forchlorfenuron
1-[(6-Chloro-3-pyridinyl)methyl]-4,5-dihydro-*N*-nitro-1*H*-imidazol-2-amine **see**...Imidacloprid
1-[(6-Chloro-3-pyridinyl)methyl]-*N*-nitro-2-imidazolidinimine benzoate **see**...Imidacloprid
1-(2-Chloro-4-pyridyl)-3-phenylurea **see**...Forchlorfenuron
1-(2-Chloro-5-pyridylmethyl)-2-(nitroamino)imidazolidine **see**...Imidacloprid
2-[4-((6-Chloro-2-quinoxalinyl)oxy)phenoxy]ethyl propionate **see**...Quizalofop-Ethyl
2-[4-((6-Chloro-2-quinoxalinyl)oxy)phenoxy]propionic acid ethyl ester **see**...Quizalofop-Ethyl
4-Chloro-*o*-toloxyacetic acid **see**...MCPA
(4-Chloro-*o*-toloxy)acetic acid **see**...MCPA
[(4-Chloro-*o*-tolyl)oxy]acetic acid **see**...MCPA
*N*'-(4-Chloro-*o*-tolyl)-*N,N*-dimethylformamidine **see**...Chlordimeform
2-(4-Chloro-*o*-tolyl)oxylpropionic acid **see**...Mecoprop
2-(*p*-Chloro-*o*-tolyloxy)propionic acid **see**...Mecoprop
2-Chlorotoluene **see**...*o*-chlorotoluenel
Chlorotributylstannane **see**...Tributyltin Chloride
2-Chloro-6-trichloromethylpyridine **see**...Nitrapyrin

4-Chloro-6-(trichloromethyl)pyridine **see**...Nitrapyrin
2-Chloro-1-(2,4,5-trichlorophenyl)ethenyl dimethyl phosphate **see**...Tetrachlorvinphos
*p*-Chlorophenyl 2,4,5-trichlorophenyl sulfone **see**...Tetradifon
4-Chlorophenyl 2,4,5-trichlorophenyl sulfone **see**...Tetradifon
2-Chloro-1-(2,4,5-trichlorophenyl)vinyl phosphoric acid dimethyl ester **see**...Tetrachlorvinphos
2-Chloro-1-(2,4,5-trichlorophenyl)vinyl dimethyl phosphate **see**...Tetrachlorvinphos
2-Chloro-4-trifluoromethyl-3'-ethoxy-4'-nitrodiphenyl ether **see**...Oxyfluorfen
*N*-(2-Chloro-4-(trifluoromethyl)phenyl]-*dl*-valinecyano(3-phenoxylphenyl)methyl ester] **see**...Fluvalinate
5-[2-Chloro-4-(trifluoromethyl)phenoxy]-*N*-methylsulfonyl)-2-nitrobenzamide **see**...Fomesafen
5-(2-Chloro-4-(trifluoromethyl)phenoxy)-2-nitrobenzoic acid **see**...Acifluorfen
5-[2-Chloro-4-(trifluoromethyl)phenoxy]-*N*-(methylsulphonyl)-2-nitrobenzamide **see**...Fomesafen
5-[2-Chloro-4-(trifluoromethyl)phenoxy]-2-nitrobenzoic acid 2-ethoxy-1-methyl-2-oxoethyl ester **see**...Lactofen
*N*-[2-Chloro-4-(trifluoromethyl)phenyl]-*dl*-valine(±)-cyano(3-phenoxylphenyl)methyl ester] **see**...Fluvalinate
1-[*N*-(4-Chloro-2-trifluoromethylphenyl)-propoxyacetimidoyl]-imidazole **see**...Triflumizole
*N*-[4-(3-Chloro-5-trifluoromethyl-2-pyridinyl-oxy)-3,5-dichloro-phenyl-aminocarbonyl]-2,6-difluorobenzamide **see**...Chlorfluazuron
2-[4-((3-Chloro-5-trifluoromethyl-2-pyridinyl)oxy)phenoxyl]propanoic acid, methyl ester **see**...Haloxyfop-methyl
2-*N*-(3-Chloro-5-trifluoromethyl-2-pyridyl)-2,6-dinitro-3-chloro-4-trifluoromethylaniline **see**...Fluazinam®
(1*S*+1*R*)-*cis*-3-(*Z*-2-Chloro-3,3,3-trifluoroprop-1-enyl)-2,2-dimethylcyclopropanecarboxylate **see**...*lamda*-Cyhalothrin
5-(2-Chloro-α,α,α-trifluoro-*p*-tolyloxy)-*N*-methylsulfonyl-2-nitrobenzamide **see**...Fomesafen
5-(2-Chloro-α-α-α-trifluoro-*p*-tolyloxy)-2-nitrobenzoic acid **see**...Acifluorfen
2-Chloro-α,α,α-trifluoro-*p*-tolyl-3-ethoxy-4-nitrophenyl ether **see**...Oxyfluorfen
*N*-[2-Chloro-α,α,α-(trifluoro-*p*-tolyl)-*dl*-valine*alpha*-cyano-phenoxybenzyl ester **see**...Fluvalinate
Chlorotriphenylstannane **see**...Triphenyltin Compounds
Chlorotriphenyltin **see**...Triphenyltin Compounds
Chloroxifenidum **see**...Chloroxuron
CHLOROXONE® **see**...2,4-D
Chloroxylenol **see**...4-Chloro-3,5-xylenol
Chloro-*m*-xylenol **see**...4-Chloro-3,5-xylenol
*p*-Chloro-*m*-xylenol **see**...4-Chloro-3,5-xylenol
Chloroxyphos **see**...Trichlorfon
Chlrphacinon (Italian) **see**...Chlorophacinone
Chlrphenamidine **see**...Chlordimeform
Chlrphenvinfos **see**...Chlorfenvinphos
Chlrphenvinphos **see**...Chlorfenvinphos

1-(4-Chlorphenyl)-1-phenyl-acetyl indan-1,3-dion (German) see...Chlorophacinone
Chlorpikrin (German) see...Chloropicrin
CHLORPIRIFOS 480 CE MILENIA® see..Chlorpyrifos
α-Chlorpyrifos 48EC (α) see...Chlorpyrifos
Chlorpyrifos-ethyl see...Chlorpyrifos
Chlorsaure (German) see...Sodium Chlorate
Chlorthal dimethyl see...DCPA
Chlorthal-dimethyl see...DCPA
Chlorthal-methyl see...DCPA
Chlorthalonil (German) see...Chlorothalonil
CHLORTHIEPIN® see...Endosulfan
CHLORTOX® see...Chlordane
Chlorure de cyanogene (French) see...Cyanogen Chloride
Chlorure mercurique (French) see...Mercuric Chloride
Chlorure d'ethylene (French) see...Ethylene Dichloride
Chlorvinphos see...Dichlorvos
CHOIR® see...Chlorpyrifos
Choline dichloride see...Chlormequat Chloride
CHORUS® see...Cyprodinil
Chromic acid ($H_2Cr_2O_7$), disodium salt (monohyrate) see. Sodium Dichromate
Chromic acid, diammonium salt see...Ammonium Chromate
CHROMOZIN® see...Atrazine
CHRYSAL BVB® see...6-Benzaldenine
(+)-Cis,trans-Chrysanthemate see...Allethrins
Chrysanthemum cinerariefolium see...Pyrethrins or Pyrethrum
CHRYSANTHEMUM CINEAREAEFOLIUM® see..Pyrethrins or Pyrethrum
CHRYSRON® see...Resmethrin
CHWASTOX® see...MCPA
Ciafos see...Cyanofos
Cianamida calcica (Spanish) see...Calcium Cyanamide
Cianofos (Spanish) see...Cyanofos
Cianuro di sodio (Italian) see...Sodium Cyanide
Cianuro di vinile (Italian) see...Acrylonitrile
Cianuro de cobre (Spanish) see...Copper Cyanide
Cianuro calcico (Spanish) see...Calcium Cyanide
Cianuro potasico (Spanish) see...Potassium Cyanide
Cianuro sodico (Spanish) see...Sodium Cyanide
CIBA 570® see...Phosphamidon
CIBA 709® see...Dicrotophos
CIBA 1414® see...Monocrotophos
CIBA 1983® see...Chloroxuron
CIBA 2059® see...Fluometuron
CIBA 3126® see...Metobromuron
CIBA 8514® see...Chlordecone (Kepone®)
CIBA 8514® see...Chlordimeform
CIBA-GEIGY GS 13005® see...Methidathion
4,4'-Cichlorbenzilsaeureaethylester (German) see..Chlorobenzilate
CICLO-SOM® see...Trichlorfon
CIDEMUL® see...Phenthoate
CIDEX® see...Glutaraldehyde
CIDIAL® see...Phenthoate

Cihexatin see...Cyhexatin
CIMARRON® see...Metsulfuron-methyl
CIMCOOL® see...Tris(hydroxymethyl)nitromethane
CIMETLE® see...Methomyl
CIMEXAN® see...Malathion
CINCH® see...Metolachlor
CINEB® see...Zineb
Cinene see...D-Limonene
Cinerin I see...Pyrethrins or Pyrethrum
Cinerin I allyl homolog see...Allethrins
Cinerin II see...Pyrethrins or Pyrethrum
Cinnamal see...Cinnamaldehyde
trans-Cinnamaldehyde see...Cinnamaldehyde
(E)-Cinnamaldehyde see...Cinnamaldehyde
trans-Cinnamic aldehyde see...Cinnamaldehyde
Cinnamyl aldehyde see...Cinnamaldehyde
trans-Cinnamylaldehyde see...Cinnamaldehyde
Cinnimic aldehyde see...Cinnamaldehyde
CIODRIN® VINYL PHOSPHATE see...Crotoxyphos
CIODRIN® see...Crotoxyphos
CIOVAP® see...Crotoxyphos
CIPA® see...Propachlor
CIRAM® see...Ziram
CIRRASOL®-185A see...Pelargonic Acid
d-Cisallethrin see...Allethrins
d-CISALLETHRIN® see...Allethrins
CISLIN® see...Deltamethrin
CISMETHRIN® see...Resmethrin
CITATION® see...Cyromazine
Citosulfan see...Busulfan
CITOX® see...DDT
CITRAM® see...Amiton Oxalate
CITRAM® see...Amiton
CITRUS FIX® see...2,4-D
CITRUS LUSTR® see...Thiabendazole
CL 11344® see...Pyridate
CL 18133® see...Thionazin
CL 38023 see...Famphur
CL 47300® see...Fenitrothion
CL 47,470® see...Mephosfolan
CL 52160® see...Temephos
CL 64475® see...Fosthietan
CL 217,300® see...Hydramethylnon
CL 222293® see...Imazethabenz
CL 263499® see...Imazethapyr
CL 303630® see...Chlorfenapyr
CLAIRSIT® see...Perchloromethyl Mercaptan
CLANEX® see...Pronamide
CLARITY® see...Dicamba
CLAROSAN® see...Terbutryn
CLASSIC® see...Chlorimuron-ethyl
Claudelite see...Arsenous Oxide
CLEANACRES® see...Maneb
CLEAN CROP® see...Endosulfan
CLEAN CROP® see...Methyl Parathion
CLEAN CROP® see...Tetrachlorvinphos
CLEAN CROP ACEPHATE 80 DF SEED PROTECTORANT® see...Acephate

CLEAN KILL® see...2,4-D, isooctyl ester
Cleaning solvent see...Stoddard Solvent
CLEANSWEEP® see...Diquat
CLEANSWEEP® see...Diquat Dibromide
CLEAN-UP® see...Cacodylic Acid
CLEARWAY® see...Amitrole
CLEARY® 3336 see...Thiophanate-Methyl
CLEAVAL® see...Mecoprop
CLENECORN® see...Mecoprop
CLETODIME® see...Clethodim
CLINAFARM® see...Imazalil
CLIPPER® see...Paclobutrazole
CLOFENOTANE® see...DDT
Clofenvinfos see...Chlorfenvinphos
Clonitarlid see...Clonitralid
2-(4-Cloor-2-methyl-fenoxy)-propionzuur (Dutch) see...Mecoprop
CLOR CHEM T-590® see...Toxaphene
3-(3-Clor-4-methoxyphenyl)-1,1-dimethylharnstoff (German) see...Metoxuron
Clordan (Italian) see...Chlordane
Clordano (Spanish) see...Chlordane
Clorex see...Dichloroethyl Ether
Clorfenvinfos (Spanish) see...Chlorfenvinphos
Clormecuato de cloroacetilo (Spanish) see...Chlormequat Chloride
O-2-Cloro-1-(2,4-dicloro-fenil)-vinil-O,O-di etilfosfato (Italian) see...Chlorfenvinphos
4-Cloro-3-metilfenol (Spanish) see...p-Chloro-m-Cresol
Cloroacetaldehido (Spanish) see...Chloroacetaldehyde
Clorofacinona (Spanish) see...Chlorophacinone
Clorofos (Russian) see...Trichlorfon
Cloroxuron (Spanish) see...Chloroxuron
Clorphacinon (Italian) see...Chlorophacinone
Clorpicrina (Italian, Spanish) see...Chloropicrin
Clorpirifos metil (Spanish) see...Chlorpyrifos-methyl
Clorpirifos (Spanish) see...Chlorpyrifos
Cloruro de cadmio (Spanish) see...Cadmium Chloride
Cloruro de cianogeno (Spanish) see...Cyanogen Chloride
Cloruro de trifenilesta (Spanish) see...Triphenyltin Compounds
Cloruro di ethene (Italian) see...Ethylene Dichloride
Cloruro di mercuri (Italian) see...Mercuric Chloride
Cloruro mercurico (Spanish) see...Mercuric Chloride
CLOVACORN EXTRA® see...Linuron
CLOVOTOX® see...Mecoprop
CMDP see...Mevinphos
CME 151® see...Dimethomorph
CMK® see...p-Chloro-m-Cresol
CMPP see...Mecoprop
CM S 2957® see...Chlorthiophos
CMU® see...Monuron
CN-11-2936® see...Prodiamine
CN-11-3183® see...Forchlorfenuron
CNA see...Dichloran
CNC see...Copper Naphthenate
CNN 52® see...Cypermethrin

Coal oil see...Kerosene
Coal tar distillate (boiling beween 270-300° C) see...Anthracene
Coal tar naphtha see...Naphthas
COBEX® see...Ethalfluralin
COBOX BLUE® see...Copper Oxychloride
COBOX® see...Copper Oxychloride
COBRA® see...Lactofen
Cobre (Spanish) see...Copper and Copper Compounds
CODAL® see...Metolachlor
CODE H 133® see...Dichlobenil
CODECHINE® see...Lindane
COLDCIDE-25® see...Glutaraldehyde
Colecalciferol see...Cholecalciferol
COLLAVIN® see...Propham
COLLO-BUEGLATT® see...Formic Acid
COLLO-DIDAX® see...Formic Acid
Colloidal arsenic see...Arsenic and Inorganic Arsenic Compounds
Colloidal cadmium see...Cadmium
Colloidal-S see...Sulfur
Colloidal sulfur see...Sulfur
COLLOIDOX® see...Copper Oxychloride
COLLOKIT® see...Sulfur
COLONEL HERBICIDE® see...Paraquat
COLOR-SET® see...Silvex
COLSUL® see...Sulfur
COLZOR TRIO® see...Clomazone
COLZOR TRIO® see...Napropamide
COMAC PARASOL® see...Copper Hydroxide
COMBAT® see...Fipronil
COMBAT® see...Hydramethylnon
COMBOT® see...Trichlorfon
COMBOT EQUINE® see...Trichlorfon
COMITE II® see...Propargite
COMITE® AGRICULTURAL MITICIDE see...Propargite
COMMAND® see...Clomazone
COMMENCE® see...Trifluralin
Commercial 40% see...TEPP
COMMODORE® see...lamda-Cyhalothrin
COMMON SENSE COCKROACH AND RAT PREPARATIONS® see...Phosphorus
COMMON SENSE DRIONE 79700® see...Ammonium Hexafluorosilicate
COMPITOX EXTRA® see...Mecoprop
COMPLY® see...Fenoxycarb
COMPO® see...Difenacoum
COMPOSITION NO. 155® see...Sodium Dichromate
COMPOUND 42® see...Warfarin
COMPOUND 88R® see...Aramite
COMPOUND 118® see...Aldrin
COMPOUND 269® see...Endrin
COMPOUND 338® see...Chlorobenzilate
COMPOUND 497® see...Dieldrin
COMPOUND 604® see...Dichlone
COMPOUND 666® see...Hexachlorocyclohexanes
COMPOUND 711 see...Isodrin

COMPOUND 732® see...Terbacil
COMPOUND 1080 see...Sodium Fluoroacetate
COMPOUND 1081® see...Fluoroacetamide
COMPOUND 1189® see...Chlordecone (Kepone®)
COMPOUND 01748® see...Dithiazanine Iodide
COMPOUND 1809® see...Fluoroacetic Acid
COMPOUND 1861® see...4-Aminopyridine
COMPOUND 2046® see...Mevinphos
COMPOUND 3422® see...Parathion
COMPOUND 3956® see...Toxaphene
COMPOUND 4049® see...Malathion
COMPOUND 4072® see...Chlorfenvinphos
COMPOUND 7744® see...Carbaryl
COMPOUND 10854® see...Phenol, 3-(1-methyethyl)-, methylcarbamate
COMPOUND 39007 see...Propoxur
COMPOUND 56722® see...Fenarimol
COMPOUND 67019® see...Oryzalin
COMPOUND 72500® see...Flurprimidol
COMPOUND 94961® see...Ethalfluralin
COMPOUND B DICAMBA® see...Dicamba
CONCERN® see...Copper Octanoate
CONCERT® see...Chlorimuron-ethyl
CONCLUDE® see...Sethoxydim
CONCORD® see...*alpha*-Cypermethrin
CONDOR® see...Triflumizole
CONFIDOR® see...Imidacloprid
CONFIRM® see...Tebufenozide
CONFRONT® see...Clopyralid
CONFRONT® see...Triclopyr, Triethylammonium Salt
CONKILL® see...Diquat Dibromide
CONQUEST® see...Metribuzin
CONSAN® see...Zilkonium Chloride
CONSULT® see...Hexaflumuron
CONSYST® see...Thiophanate-Methyl
CONTACT® 75 see...Chlorothalonil
CONTAVERN® see...Phenothiazine
CONTOUR® see...Imazethapyr
CONTRA CREME® see...Phenylmercury Acetate
CONTRAC® see...Bromadiolone
CONTRALIN® see...Disulfiram
CONTRAPOT® see...Disulfiram
CONTRAVEN® see...Terbufos
CONTUR® see...Cyfluthrin
CO-OP ATRAZINE® see...Atrazine
CO-OP HEXA® see...Hexachlorobenzene
CO-RAL® see...Coumaphos
CO-RAX® see...Warfarin
COOP RTU® CATTLE SPRAY see...Crotoxyphos
COP-TOX® see...Copper Oxychloride
COPOX® see...Cuprous Oxide
Copper-8 see...Copper(II)-8-hydroxyquinoline
Copper-8-hydroxyquinolate see...Copper(II)-8-hydroxyquinoline
Copper-8-hydroxyquinolinate see...Copper(II)-8-hydroxyquinoline
Copper-8-hydroxyquinoline see...Copper(II)-8-hydroxyquinoline
Copper-8-quinolate see...Copper(II)-8-hydroxyquinoline
Copper-8-quinolinol see...Copper(II)-8-hydroxyquinoline
Copper-8-quinolinolate see...Copper(II)-8-hydroxyquinoline
Copper acetate see...Cupric Acetate
Copper(2+) acetate see...Cupric Acetate
Copper(II) acetate see...Cupric Acetate
Copper aceto-arsenite see...Copper Acetoarsenite
Copper acetoarsenite see...Paris Green
Copper bronze see...Copper and Copper Compounds
Copper carbonate hydroxide see...Copper Carbonate, Basic
Copper(I) chloride see...Cuprous Chloride
Copper chloride, basic see...Copper Oxychloride
Copper chloride, mixed with copper oxide, hydrate see...Copper Oxychloride
Copper chloride oxide see...Copper Oxychloride
Copper chloride oxide, hydrate see...Copper Oxychloride
Copper chloroxide see...Copper Oxychloride
Copper Count N see...Copper Ammonium Carbonate
Copper(1+) cyanide see...Copper Cyanide
Copper(I) cyanide see...Copper Cyanide
Copper(II) cyanide see...Copper Cyanide
Copper cynanamide see...Copper Cyanide
Copper diacetate see...Cupric Acetate
Copper(2+) diacetate see...Cupric Acetate
Copper(II) diacetate see...Cupric Acetate
Copper dihydroxide see...Copper Hydroxide
Copper dinitrate see...Cupric Nitrate
Copper hydrate see...Copper Hydroxide
Copper(II) hydroxide see...Copper Hydroxide
Copper hydroxyquinolate see...Copper(II)-8-hydroxyquinoline
Copper monochloride see...Cuprous Chloride
Copper monosulfate see...Copper Sulfate
Copper(2+) nitrate see...Cupric Nitrate
Copper(II) nitrate see...Cupric Nitrate
COPPER NORDOX® see...Cuprous Oxide
Copper OC fungicide see...Copper Oxychloride
Copper orthoarsenite see...Copper Arsenite
Copper oxide ($Cu_2O$) see...Cuprous Oxide
Copper oxide hydrated see...Copper Hydroxide
Copper(I) oxide see...Cuprous Oxide
Copper oxinate see...Copper(II)-8-hydroxyquinoline
Copper (II) oxinate see...Copper(II)-8-hydroxyquinoline
Copper oxine see...Copper(II)-8-hydroxyquinoline
Copper oxychloride see...Copper Oxychloride
Copper oxyquinolate see...Copper(II)-8-hydroxyquinoline
Copper oxyquinoline see...Copper(II)-8-hydroxyquinoline
Copper quinolate see...Copper(II)-8-hydroxyquinoline
Copper quinolinolate see...Copper(II)-8-hydroxyquinoline
COPPER SARDEX® see...Cuprous Oxide

Copper suboxide  see...Cuprous Oxide
Copper(2+) sulfate  see...Copper Sulfate
Copper(2+) sulfate (1:1)  see...Copper Sulfate
Copper(II) sulfate  see...Copper Sulfate
Copper sulfate pentahydrate  see...Copper Sulfate
Copper sulfate (1:1)  see...Copper Sulfate
Copper(I) thiocyanate  see...Cuprous Thiocyanate
Copper uversol  see...Copper Naphthenate
COPPERAS®  see...Ferrous Sulfate
COPPESAN BLUE®  see...Copper Oxychloride
COPPESAN®  see...Copper Oxychloride
COPRANTOL®  see...Copper Oxychloride
COPREX®  see...Copper Oxychloride
COPROSAN BLUE®  see...Copper Oxychloride
COPSIN®  see...Copper Sulfate
COPSOL®  see...Copper Ammonium Carbonate
CORAIL®  see...Tebuconazole
CORBEL®  see...Carbendazim
CORNOX M®  see...MCPA
CORNOX PLUS®  see...Mecoprop
CORNOX RD®  see...Dichlorprop
COROBAN®  see...Chlorpyrifos
CORODANE®  see...Chlordane
CORONA COROZATE®  see...Ziram
COROSAL D AND S®  see...Sulfur
COROTHION®  see...Parathion
COROZATE®  see...Ziram
Corrosive mercury chloride  see...Mercuric Chloride
Corrosive sublimate  see...Mercuric Chloride
CORTHION®  see...Parathion
CORTHIONE®  see...Parathion
Cortilan-Neu  see...Chlordane
COSAN®  see...Phenylmercury Acetate
COSAN®  see...Sulfur
COSAN® 80  see...Sulfur
COSMIC®  see...Glyphosate
COSMIC®  see...Maneb
COTGUARD®  see...Metalaxyl
COTNION-ETHYL®  see...Azinphos-ethyl
COTNION-METHYL®  see...Azinphos-methyl
COTOFILM®  see...Hexachlorophene
COTOFOR®  see...Dipropetryn
COTORAN MULTI 50WP®  see...Fluometuron
COTORAN®  see...Fluometuron
COTORAN® MULTI®  see...Metolachlor
COTOREX®  see...Fluometuron
COTTON AIDE HC®  see...Cacodylic Acid
COTTONEX®  see...Fluometuron
COTTON PRO®  see...Prometryn
COTTON TOX DUST®  see...Methyl Parathion
COTTON TOX MP82®  see...Toxaphene
COUGAR®  see...Diflufenican
Coumadin  see...Warfarin
Coumadin sodium  see...Warfarin
Coumafene (French)  see...Warfarin
Coumafos  see...Coumaphos
Coumarin (Japan)  see...Warfarin

Coumarin, 3-(α-acetonylbenzyl)-4-hydroxy-  see..Warfarin
Coumarin, 3-[3-(4'-bromo-1,1'-biphenyl-4-yl)-3-hydroxy-1-phenylpropyl]-4-hydroxy-  see...Bromadiolone
Coumarin, 3-[α-(p-(p-bromophenyl)-β-hydroxyphenethyl)benzyl]-4-hydroxy-  see..Bromadiolone
Coumarin, 4-hydroxy-3-(1,2,3,4-tetrahydro-1-naphthyl)-  see...Coumatetralyl
COUNTER®  see...Terbufos
COUNTER 15G SYSTEMIC INSECTICIDE®  see..Terbufos
COVER®  see...Sulfentrazone
COXYSAN®  see...Copper Oxychloride
COYOTE®  see...Amitraz
CP 3438®  see...Bithionol
CP 4572®  see...Sulfallate
CP 6343®  see...Allidochlor
CP 14,957®  see...Isobenzan
CP 15,336®  see...Diallate
CP 23426®  see...Triallate
CP 31393®  see...Propachlor
CP 47114®  see...Fenitrothion
CP 53619®  see...Butachlor
CP 53926®  see...Formothion
CP 55097®  see...Acetochlor
CP BASIC SULFATE®  see...Copper Sulfate
CPC  see...Dinocap
CPCA  see...Dicofol
CPPU  see...Forchlorfenuron
CQ 1451®  see...Phenmedipham
CR 205®  see...Mecoprop
CR 409  see...Dimefox
CR 1639®  see...Dinocap
CR 3029®  see...Maneb
CRACKDOWN®  see...Deltamethrin
CRAG®  see...Dazomet
CRAG 341®  see...Glyodin
CRAG® FUNGICIDE 974  see...Dazomet
CRAG HERBICIDE 1®  see...2,4-DES-Sodium
CRAG® NEMACIDE  see...Dazomet
CRAG SESONE®  see...2,4-DES-Sodium
CRAG SEVIN®  see...Carbaryl
CREDO®  see...Sodium Fluoride
o-Cresol, 4,6-dinitro-  see...Dinitro-o-cresol (DNOC)
CRESTOXO®  see...Toxaphene
m-Cresyl ester of N-methylcarbamic acid  see..Metolcarb
m-Cresyl methylcarbamate  see...Metolcarb
m-Cresyl methyl carbamate  see...Metolcarb
Crimidin (German)  see...Crimidine
Crimidina (Italian)  see...Crimidine
CRIMITOX®  see...Crimidine
CRIPTAN®  see...Captan
CRISALIN®  see...Trifluralin
CRISAPON®  see...Dalapon
CRISATRINA®  see...Atrazine
CRISATRINA®  see...Ametryn
CRISAZINE®  see...Atrazine

CRISAZUFRE® see...Sulfur
CRISCOBRE® see...Copper Hydroxide
CRISFOLATAN® see...Captafol
CRISFURAN® see...Carbofuran
CRISODIN® see...Monocrotophos
CRISODRIN® see...Monocrotophos
CRISQUAT® see...Paraquat
CRISTOXO 90® see...Toxaphene
CRISUFAN® see...Endosulfan
CRISURON® see...Diuron
CRITTOX® see...Zineb
CROLEAN® see...Acrolein
Cromato amonico (Spanish) see...Ammonium Chromate
CROMOCIDE® see...Malathion
CRONETAL® see...Disulfiram
CROP BOOSTER® see...Gibberellic Acid
CROP BOOSTER® see...Indole-3-Butyric Acid
CROP CURE® see...Sodium Diacetate
CROP GUARD® see...Diatomaceous Earth
CROPHOSPHATE® see...Phosphamidon
CROP RIDER® see...2,4-D
CROP RIDER® 3.34D see...2,4-D, isopropyl ester
CROP RIDER® 3-34D-2 see...2,4-D, isopropyl ester
CROP STAR® see...Alachlor
CROPTEX ONYX® see...Bromacil
CROP WEEDSTOP® see...Linuron
CROSSBOW® see...Triclopry
CROSSBOW® see...Triclopyr, Triethylammonium Salt
CROSSBOW® see...2,4-D, butoxyethyl ester
CROSSFIRE® see...Resmethrin
CROTILIN® see...2,4-D
Crotonamide, 3-hydroxy-*N*-*N*-dimethyl-, dimethylphosphate, *(E)*- see...Dicrotophos
Crotonamide, 3-hydroxy-*N*-methyl-,dimethylphosphate, *(E)*- see...Monocrotophos
Crotonamide, 3-hydroxy-*N*-*N*,-dimethyl-, *cis*-, Dimethyl phosphate see...Dicrotophos
Crotonamide, 3-hydroxy-*N*-methyl-, dimethylphosphate, *cis*- see...Monocrotophos
Crotonamide, 3-hydroxy-*N*-*N*-dimethyl-, dimethylphosphate, *cis*- see...Dicrotophos
Crotonic acid 2,4-dinitro-6-(1-methylheptyl)phenyl ester see...Dinocap
Crotonic acid 2,4-dinitro-6-(2-octyl)phenyl ester see..Dinocap
Crotonic acid, 3-hydroxy-, isopropyl ester, O-ester with *O*-methylethylphosphoramidothioate, *(E)*- see..Propetamphos
Crotonic acid, 3-hydroxy-, α-methylbenzyl ester, dimethylphosphate, *(E)*- see...Crotoxyphos
Crotonic acid, 3-hydroxy-, methyl ester, dimethyl phosphate, *(E)*- see...Mevinphos
Crotonic acid, 3-hydroxy-, methyl ester, dimethyl phosphate see...Mevinphos
Crotonic acid 2-(1-methylheptyl)-4,6-dinitrophenyl ester see...Dinocap
Crotonic acid, 4-(1-methylheptyl)-2,6-dinitrophenyl ester see...Dinocap

CROTOTHANE® see...Dinocap
Crude arsenic see...Arsenous Oxide
Crude solvent coal tar naphtha see...Naphthas
CRUNCH® see...Carbaryl
CRUSADER® see...Mecoprop
CRYLCOAT® see...Thiram
Cryolite see...Sodium Aluminum Fluoride
CRYOLITE 93® see...Sodium Aluminum fluoride
CRYPTOGIL OL® see...Pentachlorophenol
CRYSTAL PROPANIL-4® see...Propanil
Crystallized verdigris see...Cupric Acetate
Crystals of Venus see...Cupric Acetate
Crystex see...Sulfur
CRYSTHION® see...Azinphos-ethyl
CRYSTHION 2L® see...Azinphos-methyl
CRYSTHYON® see...Azinphos-ethyl
CRYSTHYON® see...Azinphos-methyl
Crytophthalite see...Ammonium Hexafluorosilicate
CS-847® see...Barban
CSI® 70% see...Sodium Dichromate
CTR® 6669 see...Carbendazim
CU-56® see...Copper Oxychloride
CU-75® see...Cuprous Oxide
Cu basic sulfate see...Copper Sulfate
CUB® see...Diflufenican
CUBE® see...Rotenone
CUBE EXTRACT® see...Rotenone
CUBE-PULVER® see...Rotenone
CUBEROL® see...Rotenone
CUBE ROOT® see...Rotenone
CUBOR® see...Rotenone
CUCUMBER DUST® see...Calcium Arsenate
CUDEX see...Glutaraldehyde
CUDGEL® see...Fonofos
CULTAR® see...Paclobutrazole
Cumafos (Dutch, Spanish) see...Coumaphos
Cumafuryl (German) see...Coumafuryl
CUMAN® see...Ziram
CUMAN L® see...Ziram
Cumatetralyl (German, Dutch) see...Coumatetralyl
*m*- see...Cumenol methylcarbamate see...Phenol, 3-(1-methylethyl)-, methylcarbamate
*m*- see...Cumenyl methylcarbamate see...Phenol, 3-(1-methylethyl)-, methylcarbamate
CUNAPSOL® see...Copper Naphthenate
CUNILATE® see...Copper(II)-8-hydroxyquinoline
CUNILATE®-2472 see...Copper(II)-8-hydroxyquinoline
CUNITEX® see...Thiram
CUPINCIDA® see...Heptachlor
CUPRAL 45® see...Copper Oxychloride
CUPRAMAR® see...Copper Oxychloride
CUPRAMER® see...Copper Oxychloride
CUPRANTOL® see...Copper Oxychloride
CUPRAVET® see...Copper Oxychloride
Cupravit blue see...Copper Hydroxide
Cupravit blau (German) see...Copper Hydroxide
CUPRAVIT® see...Copper Oxychloride
CUPRAVIT® FORTE see...Copper Oxychloride

CUPRAVIT GREEN® see...Copper Oxychloride
Cupric-8-hydroxyquinolate see...Copper(II)-8-hydroxyquinoline
Cupric-8-quinolinolate see...Copper(II)-8-hydroxyquinoline
Cupric acetoarsenite see...Copper Acetoarsenite
Cupric arsenite see...Copper Arsenite
Cupric carbonate see...Copper Carbonate, Basic
Cupric diacetate see...Cupric Acetate
Cupric dinitrate see...Cupric Nitrate
Cupric green see...Copper Arsenite
Cupric hydroxide see...Copper Hydroxide
Cupric oxide chloride see...Copper Oxychloride
Cupric sulfate anhydrous see...Copper Sulfate
Cupric sulphate see...Copper Sulfate
Cupricin see...Copper Cyanide
CUPRINOL® see...Copper Naphthenate
CUPRICOL® see...Copper Oxychloride
CUPRITOX® see...Copper Oxychloride
CUPROFIX® see...Copper Sulfate
CUPROFIX® see...Mancozeb
CUPROID® see...Cuprous Chloride
CUPROKYLT® see...Copper Oxychloride
CUPROL® see...Copper Oxychloride
CUPROSAN® see...Copper Oxychloride
CUPROSANA® see...Copper Oxychloride
CUPROSAN BLUE® see...Copper Oxychloride
Cuprous sulfocyanate see...Cuprous Thiocyanate
Cuprous sulfocyanide see...Cuprous Thiocyanate
Cuprous dichloride see...Cuprous Chloride
CUPROVINOL® see...Copper Oxychloride
CUPROXOL® see...Copper Oxychloride
CUPROX® see...Copper Oxychloride
CURACRON® see...Profenfos
CURALAN® see...Vinclozolin
CURETERR® see...Carbofuran
CUREX FLEA DUSTER® see...Rotenone
CURIGNA® see...Chlorpyrifos
CURITAN® see...Dodine
CURTAIL® see...Clopyralid
CURTAIL® see...2,4-D
CURTAIL M® see...Clopyralid
CURZATE® see...Cymoxanil
CUSITER® see...Triclocarban
CUSTOS® see...Carbendazim
CUTISAN® see...Triclocarban
CUTLESS® see...Flurprimidol
CVMP see...Tetrachlorvinphos
CVP see...Chlorfenvinphos
Cy-L 500 see...Calcium Cyanamide
CY-PRO® see...Cyanazine
Cyaanwaterstof (Dutch) see...Hydrogen Cyanide
Cyanamid see...Calcium Cyanamide
Cyanamide see...Calcium Cyanamide
Cyanamide calcique (French) see...Calcium Cyanamide
Cyanamide, calcium salt (1:1) see...Calcium Cyanamide
Cyanamid granular see...Calcium Cyanamide
Cyanamid special grade see...Calcium Cyanamide
Cyanide of potassium see...Potassium Cyanide
Cyanide of sodium see...Sodium Cyanide
Cyano (3-phenoxyphenyl]-methyl 3-(2,2-dichlorovinyl-2,2-dimethylcyclopropanecarboxylate see...Cypermethrin
s-(R*,R*)-Cyano (3-phenoxyphenyl) methyl 4-chloro-2-(1-methylethyl) benzene- see...Esfenvalerate
Cyano(methylmercury)guanidine see...Methylmercuric Dicyanamide
Cyano(3-phenoxyphenyl)methyl ester of 4-chloro-α-(1-methylethyl)benzeneacetic acid see...Fenvalerate
Cyano(3-phenoxyphenyl)methyl 3-(2,2-dichlorovinyl)-2,2-dimethylcyclopropanecarboxylate, (±)-cis isomer see...alpha-Cypermethrin
Cyano(3-phenoxyphenyl)methyl N-[((2-chloro-4-trifluoromethyl)phenyl)]-d-valinate see...Fluvalinate
Cyano(3-phenoxyphenyl)methyl 3-(2,2-dichloroethenyl)-2,2-dimethylcyclopropanecarboxylate see...Cypermethrin
Cyano(3-phenoxyphenyl)methyl 3-(2,2-dichloroethenyl)-2,2-dimethylcyclopropanecarboxylate see...Cypermethrin
Cyano(3-phenoxyphenyl)methyl 4-chloro-α-(1-methylethyl)benzeneacetate see...Fenvalerate
Cyano(3-phenoxyphenyl)methyl see...Tralomethrin
1R-[1-α(S*),3-α)]-cyano(3-phenoxyphenyl)methyl-3-(2,2-dibromovinyl)-2,2-dimethylcyclopropanecarboxylate see...Deltamethrin
(±)-Cyano(3-phenoxyphenyl)methyl(+)-4-(difluoromethoxy)-α(1-methylethyl)benzeneacetate see...Flucythrinate
s-Cyano(3-phenoxyphenyl)methyl (±)-cis/trans-3-(2,2-dichloethenyl)-2,2-dimethylcyclopropanecarboxylate see...alpha-Cypermethrin
Cyano(4-fluoro-3-phenoxyphenyl)methyl 3-(2,2-dichloroethenyl)-2,2-dimethylcyclopropanecarboxylate see...Cyfluthrin
Cyano-(3-phenoxybenzyl)-methyl 2-(4-chlorophenyl)-3-methylbutyrate see...Fenvalerate
(RS)-Cyano-(3-phenoxyphenyl)methyl (S)-4-(difluoromethoxy)-α-(1-methylethyl)-benzeneacetate see...Flucythrinate
(+)-Cyano-(3-phenoxyphenyl)methyl (+)-4-(difluoromethoxy)-α-(1-methylethyl) benzene acetate see...Flucythrinate
Cyano-(3-phenoxyphenyl)methyl 4-chloro-α-(1-methylethyl)benzeneacetate see...Fenvalerate
2-Cyano-2-phenyl-2-(β-p-chlorophenethyl)ethyl-1H-1,2,4-triazole see...Fenbuconazole
(RS)-α-(Cyano-3-phenoxybenzyl n-(2-chloro-α,α,α-trifluoro-p-tolyl)-d-valinate see...Fluvalinate
(RS)-α-Cyano-3-phenoxybenzyl (1RS)-cis, trans-3-(2,2-dichlorovinyl)-2,2-dimethylcyclopropanecarboxylate see...Cypermethrin
(IRS)-α-Cyano-3-phenoxybenzyl (RS)-2-(4-chlorophenyl)-3-methylbutyrate see...Fenvalerate

(RS)-α-Cyano-3-phenoxybenzyl (R)-2-[2-chloro-4-(trifluoromethyl)anilino]-3-methylbutanoate see. .Fluvalinate
(R+S)-α-Cyano-3-phenoxybenzyl
(S)–α-Cyano-3-phenoxybenzyl (1S,3R)-3-(2,2–dibromovinyl)-2,2-dimethylcyclopropanecarboxylate [metabolate of tralomethrin] see. . .Tralomethrin
(S)-α-Cyano-3-phenoxybenzyl (S)-2-(4-chlorophenyl)isovalerate see. . .Esfenvalerate
(RS)-α-Cyano-3-phenoxybenzyl (S)-2-(4-difluoromethoxyphenyl)-3-methylbutyrate see. .Flucythrinate
(±)-α-Cyano-3-phenoxybenzyl 2,2-dimethyl-3-(2,2-dichlorovinyl)cyclopropanecarboxylate see. .Cypermethrin
Cyano-3-phenoxybenzyl-2,2,3,3-tetramethylcyclopropanecarboxylate see. . .Fenpropathrin
[(RS-α-Cyano-3-phenoxybenzyl (1R)-cis,trans-crysanthemate see. . .Cyphenothrin
α-Cyano-3-phenoxybenzyl 3-(2,2-dichlorovinyl)-2,2-dimethylcyclopropanecarboxylate, (±)-cis isomer see.alpha-Cypermethrin
(S)–α-Cyano-3-phenoxybenzyl (1R,3R)-3-(2,2–dibromovinyl-2,2-dimethylcyclopropanecarboxylate [metabolate of tralomethrin] see. . .Tralomethrin
α-Cyano-3-phenoxybenzyl-2-(4-chlorophenyl)-3-methybutyrate see. . .Fenvalerate
α-Cyano-3-phenoxybenzyl 2-(4-chlorophenyl)isovalerate see. . .Fenvalerate
α-Cyano-3-phenoxybenzyl 2,2-dimethyl-3-(2-methylpropenyl)cyclopropanecarboxylate see. .Cyphenothrin
α-Cyano-3-phenoxybenzyl 2,2,3,3-tetramethyl-1-cyclopropanecarboxylate see. . .Fenpropathrin
α-Cyano-3-phenoxybenzyl 2,2,3,3-tetramethylcyclopropanecarboxylate see. . .Fenpropathrin
(+)-α-Cyano-m-phenoxybenzyl alcohol ester of (+)-2-(p-difluoromethoxy)phenyl-3-methylbutyric acid see. .Flucythrinate
α-Cyano-m-phenoxybenzyl 2-(p-chlorophenyl)-3-methylbutyrate see. . .Fenvalerate
(S)-α-Cyano-m-phenoxybenzyl (1R,3R)-3-(2,2-dibromovinyl)-2,2-dimethylcyclopropanecarboxylate see. Deltamethrin
2-Cyano-N-ethylcarbamoyl-2-methoxyiminoacetamide see. . .Cymoxanil
2-Cyano-N-[(ethylamino)carbonyl]-2-(methoxyimino)acetamide see. . .Cymoxanil
Cyanobrik see. . .Sodium Cyanide
Cyanobromide see. . .Cyanogen Bromide
Cyanoethylene see. . .Acrylonitrile
Cyanogas see. . .Calcium Cyanide
Cyanogen nitride see. . .Cyanamide
Cyanogen monobromide see. . .Cyanogen Bromide
Cyanogenamide see. . .Cyanamide
Cyanogran see. . .Sodium Cyanide
Cyanoguanidine methyl mercury derivative see. .Methylmercuric Dicyanamide

O-p-Cyanophenyl O,O-dimethyl phosphorothioate see. .Cyanofos
O-(4-Cyanophenyl) O,O-dimethyl Phosphorothioate see. Cyanofos
Cyanophos organophosphate compound see. . .Cyanofos
Cyanophos see. . .Dichlorvos
CYANOX® see. . .Cyanofos
Cyanure de calcium (French) see. . .Calcium Cyanide
Cyanure de vinyle (French) see. . .Acrylonitrile
Cyanure de potassium (French) see. . .Potassium Cyanide
Cyanure de cuivre (French) see. . .Copper Cyanide
Cyanure de sodium (French) see. . .Sodium Cyanide
Cyanwasserstoff (German) see. . .Hydrogen Cyanide
CYAP® see. . .Cyanofos
CYAZIN® see. . .Atrazine
CYBOLT® see. . .Flucythrinate
CYCLE® see. . .Cyanazine
CYCLE® see. . .Metolachlor
Cyclic propylene(diethoxyphosphinyl) dithioimdocarbonate see. . .Mephosfolan
Cyclic ethylene p,p-diethylphosphono dithioimidocarbonate see. . .Phosfolan
Cyclic ethylene(diethoxyphosphinothioyl)dithioimidocarbonate see. . .Phosfolan
4-(Cyclo-α-hydroxymethylene)-3,5-dioxocyclohexanecarboxylic acid ethyl ester see. .Cimectacarb
CYCLOCEL® see. . .Chlormequat Chloride
CYCLODAN® see. . .Endosulfan
Cyclododecyl-2,6-dimethylmorpholine acetate see. .Dodemorph Acetate
Cyclododecyl(4)-2,6-dimethylmorpholine acetate see. .Dodemorph Acetate
4-Cyclododecyl-2,6-dimethylmorpholine acetate see. .Dodemorph Acetate
N-Cyclododecyl-2,6-dimethylmorpholinium acetate see. .Dodemorph Acetate
Cyclohex-1-ene-1,2-dicarboximidomethyl (±)-cis-trans-chrysanthemate see. . .Tetramethrin
Cyclohexane 1,2,3,4,5,6-hexachloro- see. .Hexachlorocyclohexanes
Cyclohexane 1,2,3,4,5,6-hexachloro-(1 α,2 α,3.beta.,4 α,5β,6β)- see. . .Hexachlorocyclohexanes
Cyclohexane 1,2,3,4,5,6-hexachloro-α isomer see. .Hexachlorocyclohexanes
Cyclohexane 1,2,3,4,5,6-hexachloro-α see. .Hexachlorocyclohexanes
Cyclohexane 1,2,3,4,5,6-hexachloro-(α,dl) see. .Hexachlorocyclohexanes
(1- Cyclohexane-1,2-dicarboximido)methyl chrysanthemumate see. . .Tetramethrin
Cyclohexane,α-1,2,3,4,5,6-hexachloro- see. .Hexachlorocyclohexanes
2,5-Cyclohexane,1,2,3,4,5,6-hexachloro-, (1α,2α,3β,4 α,5α,6β)- see. . .Lindane
Cyclohexanecarboxylic acid, 3,5-dioxo-4-(1-oxopropyl)-, ion$^{(1-)}$, calcium salt see. . .Prohexadione Calcium

Cyclohexanecarboxylic acid, 4-(cyclopropylhydroxymethylene)-3,5-dioxo-, ethyl ester see...Cimectacarb
2-Cyclohexen-1-one, 2-[1-(ethoxyimino)butyl]-5-[2-(ethylthio)propyl]-3-hydroxy- see...Sethoxydim
2-Cyclohexen-1-one, 2-[1-(ethoxyimino)propyl]-3-hydroxy-5-(2,4,6-trimethylphenyl)- (9CI) see...Tralkoxydim
2-Cyclohexen-1-one, 2-[1-(((3-chloro-2-propenyl)oxy)imino)propyl]-5-[2-(ethylthio)propyl]-3-hydroxy- see...Clethodim
(1- Cyclohexene-1,2-dicarboximido)methyl 2,2-dimethyl-3-(2-methylpropyl)cyclopropanecarboxylate see...Tetramethrin
4-Cyclohexene-1,2-dicarboximide,N-[(trichloromethyl)mercapto see...Captan
4-Cyclohexene-1,2-dicarboximide, N-(1,1,2,2-Tetrachloroethyl)thiol-1H-isoindole-1,3(2H)-dione,3A,4,7,7A-tetrahydro-2-(1,1,2,2-tetrachloroethyl)thio- see...Captafol
2-Cyclohexyl-4,6-dinitrophenol see...Dinex
2-Cyclohexyl-4,6-dinitrophenol see...Dinex
6-Cyclohexyl-2,4-dinitrophenol see...Dinex
3-Cyclohexyl-6-(dimethylamino)-1-methyl-1,3,5-triazine-2,4(1H,3H)-dione see...Hexazinone
3-Cyclohexyl-6-(dimethylamino)-1-methyl-s-triazine-2,4(1H,3H)-dione see...Hexazinone
3-Cyclohexyl-6-dimethylamino-1-methyl-1,2,3,4-tetrahydro-1,3,5-triazine-2-,4-dione see...Hexazinone
Cyclohexylethylcarbamothioic acid-S-ethyl ester see...Cycloate
Cyclohexylethylthiocarbamic acid-S-ethyl ester see...Cycloate
3-Cyclohexyl-1-methyl-6-(dimethylamino)-s-trazine-2,4(1H,3H)-dione see...Hexazinone
CYCLOMORPH® see...Dodemorph Acetate
CYCLONE® see...Paraquat
CYCLONE B® see...Hydrogen Cyanide
CYCLON® see...Hydrogen Cyanide
Cyclopentimine see...Piperidine
Cyclopropanecar boxylic acid, 2,2-dimethyl-3-(2-methylpropenyl)-, (4-(2-benzyl)furyl) methyl ester see...Resmethrin
Cyclopropanecarboxamide, 1-carboxy-, N-(2,4-dichlorophenyl)- see...Cyclanilide
Cyclopropanecarboxylic acid, 2,2-dimethyl-3-(2-methylpropenyl)-, (5-benzyl-3-furyl)methyl ester (8CI) see...Resmethrin
Cyclopropanecarboxylic acid, 2,2-dimethyl-3-(2-methyl-1-propenyl)-(5-(phenylmethyl)-3-furanyl)methyl ester, (1R-cis)- see...Resmethrin
Cyclopropanecarboxylic acid, 2,2-dimethyl-3-(1,2,2,2-tetrabromoethyl)-, cyano(3-phenoxyphenyl)methyl ester (9CI) see...Tralomethrin
Cyclopropanecarboxylic acid, 2,2-dimethyl-3-(2-methyl-1-propenyl)-, [5-(phenylmethyl)-3-furanyl]methyl ester, (Z),(E)-(±)- see...Resmethrin
Cyclopropanecarboxylic acid, 2,2-dimethyl-3-(2-methyl-1-propenyl)-, (1,3,4,5,6,7-hexahydro-1,3-dioxo-2H-isoindol-2-yl)methyl ester (9CI) see...Tetramethrin
Cyclopropanecarboxylic acid, 2,2-dimethyl-3-(2-methyl-1-propenyl)-, 5-(phenylmethyl)-3-furanyl]methyl ester (9CI) see...Resmethrin
Cyclopropanecarboxylic acid, 3-(2-chloro-3,3,3-trifluoro-1-propenyl)-2,2-dimethyl-, (2,3,5,6-tetrafluoro-4-methylphenyl)methyl ester, [1α,3α(Z)]-(±)- see...Tefluthrin
Cyclopropanecarboxylic acid, 2,2-dimethyl-3-(2-methyl-1-propenyl)-, (3-phenoxyphenyl)methyl ester see...D-Phenothrin
Cyclopropanecarboxylic acid, 2,2-dimethyl-3-(2-methylpropenyl)-, m-phenoxybenzyl ester see...D-Phenothrin
Cyclopropanecarboxylic acid, 2,2-dimethyl-3-(2-methyl-1-propenyl)-,(3-phenoxyphenyl) methyl ester see...D-Phenothrin
Cyclopropanecarboxylic acid, 2,2,3,3-tetramethyl-,cyano(3-phenoxyphenyl)methyl ester see...Fenpropathrin
Cyclopropanecarboxylic acid, 3-(2,2-dichloroethenyl)-2,2-dimethyl-, (pentafluorophenyl)methyl ester, (1R-trans)- see...Fenfluthrin
Cyclopropanecarboxylic acid, 3-(2,2-dibromoethenyl)-2,2-dimethyl-, cyano(3-phenoxyphenyl)methyl ester, 1R-[1α(S*,3α)]- see...Deltamethrin
Cyclopropanecarboxylic acid, 3-(2,2-dichloroethenyl)-2,2-dimethyl-, cyano(3-phenoxyphenyl)methyl ester, [1α(S*), 3α]-(+)- see...alpha-Cypermethrin
Cyclopropanecarboxylic acid, 3-(2-chloro-3,3,3-trifluoro-1-propenyl)-2,2-dimethyl-, cyano(3-phenoxyphenyl)methyl ester, [1α(S*),3α(Z)]-(+)- see.lamda-Cyhalothrin
Cyclopropanecarboxylic acid, 1-[((2,4-dichlorophenyl)amino)carbonyl]- see...Cyclanilide
Cyclopropanecarboxylic acid, 3-(2,2-dichloroethenyl)-2,2-dimethyl-, cyano(4-fluoro-3-phenoxyphenyl)methyl ester see...Cyfluthrin
Cyclopropanecarboxylic acid, 2,2-dimethyl-3-(2-methyl-1-propenyl)-, cyano(3-phenoxyphenyl)methyl ester see...Cyphenothrin
Cyclopropanecarboxylic acid, 3-(2,2-dichloroethenyl)-2,2-dimethyl-, cyano(3-phenoxyphenyl)methyl ester, (S)- see...alpha-Cypermethrin
Cyclopropanecarboxylic acid,3-(2-chloro-3,3,3-trifluoro-1-propenyl)-2,2-dimethyl-,[2-methyl(1,1'-biphenyl)3-yl]methyl ester,(Z)- see...Bifenthrin
Cyclopropanecarboxylic acid, 2-(2,2-dichlorovinyl)-3,3-dimethyl-, ester with (4-fluoro-3-phenoxyphenyl)hydroxyacetonitrile see...Cyfluthrin
Cyclopropanecarboxylic acid, 3-(2,2-dichloroethenyl)-2,2-dimethyl-, cyano(3-phenoxyphenyl)methyl ester see...Cypermethrin
Cyclopropanecarboxylic acid, 2,2-dimethyl-3-(2-methyl-1-propenyl)-, [5-(phenylmethyl)-3-furanyl]methyl ester, cis,trans-(±)- see...Resmethrin

1,2-Cyclopropanedicarboximide, N-(3,5-dichlorophenyl)-1,2-dimethyl- see...Procymidone
α-Cyclopropyl-α-(4-methoxyphenyl)-5-pyrimidinemethanol- see...Ancymidol
N-Cyclopropyl-1,3,5-triazine-2,4,6-triamine see...Cyromazine
α-Cyclopropyl-4-methoxy-α-(pyrimidin-5-yl)benzyl alcohol see...Ancymidol
4-Cyclopropyl-6-methyl-N-phenyl-2-pyrimidinamine see. Cyprodinil
N-(4-Cyclopropyl-6-methyl-pyrimidin-2-yl)- see...Cyprodinil
2-Cyclopropylamino-4,6-diamino-s-triazine see...Cyromazine
Cyclopropylmelamine see...Cyromazine
N-(Cyclopropylmethyl)-α,α,α-trifluoro-2,6-dinitro-N-propyl-p-toluidine see...Profluralin
Cyclosulfyne see...Propargite
CYCOCEL® see...Chlormequat Chloride
CYCOCEL-EXTRA® see...Chlormequat Chloride
CYCOGAN® see...Chlormequat Chloride
CYCOGAN EXTRA® see...Chlormequat Chloride
Cyethoxydim see...Sethoxydim
CYFEN® see...Fenitrothion
CYFLEE® see...Famphur
Cyfluthin see...Cyfluthrin
Cyfluthrine see...Cyfluthrin
Cyfoxylate see...Cyfluthrin
CYGON 400® see...Dimethoate
Cyhalothrin-K see...lamda-Cyhalothrin
Cyjanowodor (Polish) see...Hydrogen Cyanide
CYLAN® see...Phosfolan
CYLATHRIN® see...Cyfluthrin
m- see...Cym-5-yl methylcarbamate see...Promecarb
Cymag see...Sodium Cyanide
CYMATE® see...Ziram
CYMBUSH® 2E see...Cypermethrin
CYMBUSH® 3E see...Cypermethrin
p-Cymen-3-ol see...Thymol
3-p-Cymenol see...Thymol
Cymonic acid see...Fluoroacetic Acid
CYMPERATOR® see...Cypermethrin
CYNEM® see...Thionazin
CYNEX® 41 see...Cyanazine
CYNKOTOX® see...Zineb
CYNOFF® see...Cypermethrin
Cynogan see...Bromacil
CYNOGEN® see...Bromacil
CYOCEL® see...Chlormequat Chloride
CYODRIN® see...Crotoxyphos
CYOLANE INSECTICIDE® see...Phosfolan
CYOLANE® see...Phosfolan
CYON® see...Procymidone
Cypentil see...Piperidine
CYPERCARE® see...Cypermethrin
CYPERKILL® see...Cypermethrin
zeta-Cypermethrin see...alpha-Cypermethrin
Cypermethrin-minus see...alpha-Cypermethrin

CYPERSECT® see...Cypermethrin
CYPONA E.C.® see...Crotoxyphos
CYPONA® see...Dichlorvos
CYPREX® see...Dodine
CYPREX® 65W see...Dodine
CYREN® see...Chlorpyrifos
CYRUX® see...Cypermethrin
CYTAC® see...Amitraz
CYTEL® see...Fenitrothion
CYTEN® see...Fenitrothion
CYTHION® see...Malathion
CYTHRIN® see...Flucythrinate
Cytokinin see...Kinetin (Cytokinin)
CYTOPLEX HMS® see...Gibberellic Acid
CYTOPLEX® see...Indole-3-Butyric Acid
CYTOX® see...Octhilinone
CYTOX® 2160 see...Dodine
Cytrolane see...Mephosfolan
CYTROL® see...Amitrole
CYTROLE® see...Amitrole
CYURAM DS® see...Thiram
Czterochlorek wegla (Polish) see...Carbon Tetrachloride

- D -

D3-VIGANTOL® see...Cholecalciferol
D 014® see...Propargite
D 31® see...Dieldrin
D 50® see...2,4-D
D 735® see...Carboxin
D 1221® see...Carbofuran
D 1410® see...Oxamyl
D 1991® see...Benomyl
2,4-D acid see...2,4-D
2,4-D-BEE® see...2,4-D, butoxyethyl ester
2,4-D, butoxyethanol ester see...2,4-D, butoxyethyl ester
2,4-D, butoxyethyl ester see...2,4-D, butoxyethyl ester
2,4-D (butoxyethyl) see...2,4-D, butoxyethyl ester
2,4-D 2-butoxyethyl ester see...2,4-D, butoxyethyl ester
2,4-D butyric see...2,4-DB
D-CON® see...Warfarin
1,4:5,8-Dimethano naphthalene, 1,2,3,4,10,10-hexachloro-1,4,4a,5,8,8a-hexahydro-, endo,endo-, Isodrin
D-D SOIL FUMIGANT® see...1,2-Dichloropropane
2,4-D esters see...2,4-D, isooctyl ester
2,4-D esters see...2,4-D, isopropyl ester
2,4-D (IOE) see...2,4-D, isooctyl ester
2,4-D-isopropyl see...2,4-D, isopropyl ester
D-LEAF COTTON DEFOLIANT® see...Sodium Chlorate
2,4-D L.V. 4 ESTER® see...2,4-D, isooctyl ester
2,4-D PHENOXY PESTICIDE® see...2,4-D
D-Phthalthrin see...Tetramethrin
D REXEL-SUPER P® see...Maleic Hydrazide
2,4-D, salts and esters see...2,4-D
D-Streptamine,O-2-deoxy-2-(methylamino)-α-L-glucopyranosyl-(1->2)-O-5-deoxy-3-C-formyl-α-L-

lyxofuranosyl-(1->4)-*N,N'*-bis(aminoiminomethyl)-, sulfate (2:3) (salt) **see**...Streptomycin Sulfate
D-Streptamine, *O*-2-deoxy-2-(methylamino)-α-L-glucopyranosyl-(1->2)-*O*-5-deoxy-3-*C*-formyl-α-L-lyxofuranosyl-(1->4)-*N, N'*-bis(aminomethyl)- **see**. Streptomycin
D-Valine, *N*-(2-chloro-4-(trifluoromethyl)phenyl)-, cyano(3-phenoxyphenyl)methyl ester **see**.. Fluvalinate
D.Z.N.® **see**...Diazinon
DABICYCLINE® **see**...Oxytetracycline Calcium
DAC 893® **see**...DCPA
DACAMID® **see**...2,4-D
DACAMINE® **see**...2,4,5-Trichlorophenoxyacetic Acid, Esters
DACAMINE® **see**...2,4-D
DACAMINE® **see**...2,4,5-T
DACAMOX® **see**...Thiofanox
DACONATE® **see**...Sodium Methanearsonate (MSMA)
DACONIL® **see**...Chlorothalonil
DACONIL® 2787 FUNGICIDE **see**...Chlorothalonil
DACONIL® 2787 W **see**...Chlorothalonil
DACONIL® F **see**...Chlorothalonil
DACONIL® M **see**...Chlorothalonil
DACONIL® TURF **see**...Chlorothalonil
DACOSOIL® **see**...Chlorothalonil
DACTHAL® **see**...DCPA
DACTHAL® W-75 **see**...DCPA
DACTINOL® **see**...Rotenone
DACUTOX® **see**...Diazinon
DAC® 2787 **see**...Chlorothalonil
DAGADIP® **see**...Carbophanothion
DAGGER® **see**...Imazethabenz
DAHR® **see**...Isoproturon
DAILON® **see**...Diuron
DAISEN® **see**...Zineb
DAKOTA® **see**...MCPA
DALAPON 85® **see**...Dalapon
Dalapon aliphatic acid herbicide **see**...Dalapon
DALAR® **see**...Pyriproxyfen
DAL-E-RAD® **see**...Sodium Methanearsonate (MSMA)
DALF® **see**...Methyl Parathion
DALF DUST® **see**...Propoxur
DALMATION INSECT FLOWERS® **see**...Pyrethrins or Pyrethrum
DAMOIL® **see**...Kerosene
DANADIM® **see**...Dimethoate
DANEX® **see**...Trichlorfon
DANICUT® **see**...Amitraz
DANITOL® **see**...Fenpropathrin
DANTHION® **see**...Parathion
DAPA® **see**...Fenaminosulf
DAPACRYL® **see**...Binapacryl
DAP-DIAMMONIUM PHOSPHATE **see**...Ammonium Phosphate
DAPHENE® **see**...Dimethoate
DAS® **see**...Fenaminosulf
DASANIT **see**...Pebulate
DASSITOX® **see**...Diazinon

DASUL® **see**...Nicosulfuron
DATC **see**...Diallate
DAVCO® **see**...Aldrin
DAWSON® 100 **see**...Methyl Bromide
DAXAD-32s® **see**...Ammonia
DAZA® **see**...Dihydroazadirachtin
DAZE® **see**...Thidiazuron
DAZIDE® **see**...Daminozide
DAZIDE® ENHANCE **see**...Daminozide
DAZOE® **see**...Sodium Azide
DAZOMET®-POWDER **see**...Dazomet
DAZZEL® **see**...Diazinon
4(2,4-DB) **see**...2,4-DB
DBCP **see**...Dibromochloropropane
DBD® **see**...Azinphos-methyl
DBE **see**...Ethylene Dibromide
DBH **see**...Hexachlorocyclohexanes
DBH **see**...Lindane
DBN **see**...Dichlobenil
2,6-DBN **see**...Dichlobenil
DBNF **see**...Dinoseb
DC & R® **see**...Tris(hydroxymethyl)nitromethane
DCA **see**...Dichloroacetic Acid
DCB **see**...Dichlobenil
DCB **see**...*para*-Dichlorobenzene
*p*-DCB **see**...Benzene
*p*-1,4-DCB **see**...Benzene
DCBN **see**...Dichlobenil
2,3-DCDT **see**...Diallate
DCEE **see**...Dichloroethyl Ether
DCMO **see**...Carboxin
DCMOD **see**...Oxycarboxin
DCMU (In Japan) **see**...Diuron
DCNA **see**...Dichloran
3,6-DCP **see**...Clopyralid
DCPA **see**...Propanil
DCPC **see**...Dinocap
DCR 736® **see**...Methiocarb
DDB **see**...Diquat Dibromide
DDBSA **see**...Dodecylbenzenesulfonic Acid
DDC **see**...Sodium Dimethyldithiocarbamate
DDM **see**...Dichlorophene
DDDM **see**...Dichlorophene
4,4'DDT **see**...DDT
*p,p*'-DDT **see**...DDT
DDVF **see**...Dichlorvos
DDVP (Insecticide) **see**...Dichlorvos
DE-473® **see**...Hexaflumuron
DE-498® **see**...Flumetsulam
DE-CUT® **see**...Maleic Hydrazide
DE-FEND® **see**...Dimethoate
DE-FOL-ATE® **see**...Magnesium Chlorate
DE-FOL-ATE® **see**...Sodium Chlorate
DE-GREEN® **see**...Tribufos
DE-PESTER® **see**...Diphacione
DE-PESTER® **see**...Endosulfan
DE-PESTER FUMIGANT® **see**...Carbon Disulfide
DE PESTER TEDION® **see**...Tetradifon

DEADLINE® see...Metaldehyde
DEBROUSSAILLANT 600® see...2,4-D
DEBROUSSAILLANT CONCENTRE® see...2,4,5-T
DEBROUSSAILLANT® CONCENTRE (French) see.2,4,5-Trichlorophenoxyacetic Acid, Esters
DEBROUSSAILLANT SUPER CONCENTRE® see.2,4,5-T
DEBROUSSAILLANT® SUPER-CONCENTRE (French) see...2,4,5-Trichlorophenoxyacetic Acid, Esters
DEBUT® see...Triflusulfuron-Methyl
Decabane see...Dichlobenil
DECABANE® see...Dichlobenil
Decachlor see...Dienochlor
Decachlorobis(2,4-cyclopentadiene-1-yl) see..Dienochlor
1,1',2,2',3,3',4,4',5,5'-Decachloro-bis(2,4-cyclopentadien-1-yl) see...Dienochlor
1,2,3,5,6,7,8,9,10-Decachloro(5.2.2.0$^{2.6}$.0$^{3.9}$.0$^{5.8}$)decano-4-one see. Chlordecone (Kepone®)
Decachloroketone see...Chlordecone (Kepone®)
1,1a,3,3a,4,5,5,5a,5b,6-Decachloro-octahydro-1,3,4-metheno-2$H$-cyclobuta[c,d]pentalen-2-one see. Chlordecone (Kepone®)
Decachlorooctahydro-1,3,4-metheno-2$H$-cyclobuta[c,d]pentalen-2-one see...Chlordecone (Kepone®)
Decachlorooctahydrokepone-2-one see...Chlordecone (Kepone®)
1,1a, 3, 3a, 4,5,5,5a, 5b see..6-Decachlorooctahydro-1,3,4-metheno-2$H$-cyclobuta[c,d]pentalen-2-one see...Chlordecone (Kepone®)
Decachlorooctahydro-1,3,4-metheno-2$H$-cyclobuta[c,d]-pentalen-2-one see...Chlordecone (Kepone®)
Decachlorotetracyclodecanone see...Chlordecone (Kepone®)
Decachlorotetrahydro-4,7-methanoindeneone see. Chlordecone (Kepone®)
Decamethrin see...Deltamethrin
DECAMINE® see...2,4-D
DECAMINE 4T® see...2,4,5-T
DECAMINE-4T® see...2,4,5-Trichlorophenoxyacetic Acid, Esters
$n$-Decanoic acid see...Decanoic Acid
DECCO SALT NO. 19® see...Thiabendazole
DECCOQUIN 305® see...Ethoxyquin
DECCOZIL® see...Imazalil
DECEMTHION®
DECHLORANE 4070® see...Mirex
DECIMATE® see...DCPA
DECIS® see...Deltamethrin
DECLARE® see...Methyl Parathion
DECOFOL® see...Dicofol
$n$-Decoic acid see...Decanoic Acid
DECON® 4512 see...Phosphoric Acid
DECROTOX® see...Crotoxyphos
Decylic acid see...Decanoic Acid
$n$-Decylic acid see...Decanoic Acid

DED-WEED® see...2,4-D
DED-WEED® see...Dalapon
DED WEED® see...MCPA
DED-WEED® see...Silvex
DED-WEED BRUSHKILLER® see...2,4,5-T
DED-WEED® BRUSH KILLER see...2,4,5-Trichlorophenoxyacetic Acid, Esters
DED-WEED LV-6 BRUSH KIL® see...2,4,5-T
DED-WEED® LV-6 BRUSH KIL see...2,4,5-Trichlorophenoxyacetic Acid, Esters
DED-WEED® T-5 BRUSH KIL see...2,4,5-Trichlorophenoxyacetic Acid, Esters
DEDELO® see...DDT
DEDEVAP® see...Dichlorvos
DEER-OFF® see...Capsaicin
DEF® see...Tribufos
DEFANACET® see...Mefenacet
DEF DEFOLIANT® see...Tribufos
DEFEND® see...Dimethoate
DEFENSOR® see...Carbendazim
DEFOLIT® see...Thidiazuron
DEFTOR® see...Metoxuron
DEGRASSAN® see...Dinitro-o-cresol (DNOC)
DEGREE® see...Acetochlor
DEHERBAN® see...2,4-D
7-Dehydrochloesterol see...Cholecalciferol
2,3-Dehydro-2,3-dimethyl-,tetroxide see...Dimethipin
DEIQUAT® see...Diquat
DEIQUAT® see...Diquat Dibromide
DEJO® see...Dithiazanine Iodide
DEKRYSIL® see...Dinitro-o-cresol (DNOC)
DEKSONAL® see...Fenaminosulf
DELEAF DEFOLIANT® see...Tribufos
DELICE® see...Coumaphos
Delicia see...Phosphine
DELICIA® see...Aluminum Phosphide
DELNATEX® see...Dioxathion
DELNAV® see...Dioxathion
DEL-PHOS®
DELSAN® see...Thiram
DELSANEX DAIRY FLY SPRAY® see...Lindane
DELSENE® see...Carbendazim
DELSENE M FLOWABLE® see...Maneb
DELSTEROL® see...Cholecalciferol
DELTA® see...Chlorophacinone
DELTA® see...Deltamethrin
DELTA-COAT® see...Metalaxyl
DELTA-COAT® II see...Chloroneb
DELTAGUARD® see...Deltamethrin
DELVEX® see...Dithiazanine Iodide
DEMAND® see...*lamda*-Cyhalothrin
DEMAND CS® see...*lamda*-Cyhalothrin
Demethon-methyl see...Demeton-methyl
Demetona (Spanish) see...Demeton
Demeton-$O$+demeton-$S$ mixture see...Demeton
Demeton-methyl sulphoxide see...Demeton-methyl
Demeton-$O$-methyl sulfoxide see...Demeton-methyl
Demeton-$S$-methyl sulfoxide see...Demeton-methyl

DEMON® see...Cypermethrin
DEMOS NF® see...Dimethoate
DEMOSAN® see...Chloroneb
DEMOX® see...Demeton
DENAPON® see...Carbaryl
DENIM® see...Emamectin Benzoate
DENOX® see...Demeton
Deobase see...Kerosene
DEP (PESTICIDE)® see...Trichlorfon
Depallethrin see...Allethrins
DEPARAL® see...Cholecalciferol
DEPON® see...Fenoxaprop-ethyl
DEPTHON® see...Trichlorfon
DERIL® see...Rotenone
Dermafosu (Polish) see...Ronnel
DERMAPHOS® see...Ronnel
DERMATON® see...Chlorfenvinphos
DEROSAL® see...Carbendazim
DERRIBAN® see...Dichlorvos
DERRIBANTE® see...Dichlorvos
DERRINGER® see...Resmethrin
DERRIN® see...Rotenone
DERRIS® see...Rotenone
DERROPRENE® see...Carbendazim
DES® see...Dichlorvos
DES-I-CATE® see...Endothall
DESICCANT L-10® see...Arsenic Acid
DESICOIL® see...Dinoseb
Desmel see...Propiconazole
DESMEL® see...Propiconazole
2,4-DES-Na see...2,4-DES-Sodium
2,4-DES-NATRIUM (German) see...2,4-DES-Sodium
DESOLET® see...Sodium Chlorate
DESORMONE® see...2,4-D
DESORMONE® see...2,4-DB
DESORMONE® see...Dichlorprop
DESOXON 1® see...Peracetic Acid
DESPROUT® see...Maleic Hydrazide
DESSON® see...4-Chloro-3,5-xylenol
DESTRAL® see...Dalapon
DESTROY® see...Endosulfan
DESTRUXOL® see...1-Naphthaleneacetic Acid
DESTRUXOL BORER-SOL® see...Ethylene Dichloride
DESTRUXOL ORCHARD SPRAY® see...Nicotine
DESTUN® see...Perfluidone
DETAIL® see...Dimethenamid
DETAL® see...Dinitro-o-cresol (DNOC)
DETF see...Trichlorfon
DETHMOR® see...Tralomethrin
DETHMORE® see...Warfarin
DETIA-GAS-EX see...Aluminum Phosphide
DETIA GAS-EX-B® see...Phosphine
DETMOL® 96% see...Malathion
DETMOL-EXTRAKT® see...Lindane
DETMOL MA® see...Malathion
DETMOL MALATHION® see...Malathion
DETMOL U.A.® see...Chlorpyrifos
DETOX 25® see...Lindane

DETTOL® see...4-Chloro-3,5-xylenol
DEVICOPPER® see...Copper Oxychloride
DEVIGON® see...Dimethoate
DEVIKOL® see...Dichlorvos
DEVIPON® see...Dalapon
DEVISULPHAN® see...Endosulfan
DEVITHION® see...Methyl Parathion
DEVIZEB® see...Zineb
DEVORAN® see...Lindane
DEVRINOL® see...Napropamide
DEVRINOL T® see...Trifluralin
DEXIL see...Pebulate
DEXOL EARWIG BAIT® see...Ammonium Hexafluorosilicate
DEXON® see...Fenaminosulf
DEXTRONE® see...Diquat
DEXTRONE® see...Diquat Dibromide
DEXTRONE® see...Paraquat
DEXTRONE-X® see...Paraquat
DFF see...Diflufenican
DHANULUX® see...Quinalphos
Diaaluminum trisulfate see...Aluminum Sulfate
Diacetoxypropene see...Acrolein diacetate
1,1-Diacetoxy-2-propene see...Acrolein diacetate
1,1-Diacetoxypropene-2 see...Acrolein diacetate
3,3-Diacetoxypropene see...Acrolein diacetate
DIACON® see...Methoprene
DIACTIV® see...Diatomaceous Earth
$O,O$,-Diaethy-S-[(4-chlor-phenyl-thio)-methyl]dithiophosphat (German) see...Carbophanothion
$O,O$,-Diaethyl-$o$-(α-cyanbenzyliden-amino)-thionphosphat (German) see...Phoxim
$O,O$,-Diaethyl-$O$-(α-cyano-benzylidenamino)-monothiophosphat (German) see...Phoxim
$O,O$,-Diaethyl-$O$-(2-isopropyl-4-methyl-6-pyrimidyl)-thionophosphat (German) see...Diazinon
$O,O$,-Diaethyl-$O$-(2-isopropyl-4-methyl-pyrimidin-6-yl)-monothiophosphat (German) see...Diazinon
$O,O$,-Diaethyl-$O$-(2-pyrazinyl)-thionophosphat (German) see...Thionazin
$O,O$,-Diaethyl-$O$-(2,5-dichlor-4-bromphenyl)-thionophosphat (German) see...Bromophos-ethyl
$O,O$,-Diaethyl-$O$-(3-chlor-4-methyl-cu marin-7-yl)-monothiophosphat (German) see...Coumaphos
$O,O$,-Diaethyl-$O$-(4-brom-2,5-dichlor)-phenyl-monothiophosphat (German) see...Bromophos-ethyl
$O,O$,-Diaethyl-$O$-(pyrazin-2yl)-monothiophosphat (German) see...Thionazin
$O,O$,-Diaethyl-$O$-3,5,6-trichlor-2-pyridylmonothiophosphat (German) see...Chlorpyrifos
$O,O$,-Diaethyl-$O$-[chinoxalyl-(2)]-monothiophosphat (German) see...Quinalphos
Diaethyl-$p$-nitrophenylphosphorsaeureester (German) see...Paraoxon
$O,O$,-Diaethyl-$S$(2-aethyltio-aethyl)monothiophosphat (Russia) see...Demeton
$O,O$,-Diaethyl-$S$-(2-aethylthio-aethyl)-dithiophosphat (German) see...Disulfoton

O,O,-Diaethyl-S-(3-thia-pentyl)-dithiophosphat (German) see...Disulfoton
O,O,-Diaethyl-S-(6-chlor-2-oxo-ben(β)-1,3-oxalin-3-yl)-methyl-dithiophosphat (German) see...Phosalone
O,O,-Diaethyl-S-(aethylthio-methyl)-dithiophosphat (German) see...Phorate
DIAFIL® see...Diatomaceous Earth
DIAFURAN® see...Carbofuran
DIAGRAN® see...Diazinon
Dialifos see...Dialifor
Diallaat (Dutch) see...Diallate
Diallat (German) see...Diallate
Di-allate see...Diallate
Diallate carbamate herbicide see...Diallate
Diallylchloroacetamide see...Allidochlor
N,N-Diallyl-α-chloroacetamide see...Allidochlor
N,N-Diallyl-2-chloroacetamide see...Allidochlor
N,N-Diallylchloroacetamide see...Allidochlor
Dialuminum sulfate see...Aluminum Sulfate
2,4-Diamino-6-(cyclopropylamino)-s-triazine see..Cyromazine
18-46-0DI-AMMONIUM PHOSPHATE® see..Ammonium Phosphate
Diammonium chromate see...Ammonium Chromate
Diammonium fluosilicate see...Ammonium Hexafluorosilicate
Diammonium hydrogen phosphate see...Ammonium Phosphate
Diammonium orthophosphate see...Ammonium Phosphate
Diammonium orthophosphate, see..Ammonium Phosphate
Diammonium phosphate see...Ammonium Phosphate
Diammonium phosphate, see..Ammonium Phosphate
Diammonium phosphate see...monohydrogen see..Ammonium Phosphate
Diammonium silicon hexafluoride see...Ammonium Hexafluorosilicate
Diammonium sulfite see...Ammonium Sulfite
Diammonium thiosulfate see...Ammonium Thiosulfate
DIAMOND SHAMROCK DS-15647® see...Thiofanox
Dianat (Russian) see...Dicamba
DIANATE® see...Dicamba
DIANEX® see...Methoprene
Dianisyltrichlorethane see...Methoxychlor
2,2-Di-p-anisyl-1,1,1-trichloroethane see..Methoxychlor
DIANON® see...Diazinon
DIAPADRIN® see...Dicrotophos
Diarsenic trioxide see...Arsenous Oxide
Diarsenic pentoxide see...Arsenic Pentoxide
DIATER® see...Diuron
DIATERR-FOS® see...Diazinon
Diatomaceous silica see...Diatomaceous Earth
DI-ATOMATE® see...Diatomaceous Earth
Diatomite see...uncalcined see...Diatomaceous Earth
Diazacholesterol dihydrochloride see...Azacosterol Dihydrochloride

20,25-Diazacosterol hydrochloride see...Azacosterol Dihydrochloride
DIAZAJET® see...Diazinon
DIAZATOL® see...Diazinon
DIAZIDE® see...Diazinon
DIAZINON AG 500 WBC® see...Diazinon
DIAZINONE® see...Diazinon
DIAZITOL® see...Diazinon
DIAZOBEN® see...Fenaminosulf
DIAZOL® see...Diazinon
DIBAM® see...Sodium Dimethyldithiocarbamate
Dibasic ammonium phosphate see...Ammonium Phosphate
Dibasic lead arsenate see...Lead Arsenate
Dibenzene see...Biphenyl
Dibenzo[b,d]furan see...Dibenzofuran
Dibenzofurano (Spanish) see...Dibenzofuran
Dibenzoparathiazine see...Phenothiazine
Dibenzothiazine see...Phenothiazine
Dibenzo-1,4-thiazine see...Phenothiazine
Dibenzothiazine see...Phenothiazine
DIBROM® see...Naled
DIBROM OMITE® see...Propargite
1,2-Dibromaethan (German) see...Ethylene Dibromide
Dibromchlorpropan (German) see..Dibromochloropropane
1,2-Dibromo-3-chloro- see...Dibromochloropropane
1,2-Dibrom-3-chlor-propan (German) see..Dibromochloropropane
1,2-Dibromo-3-cloropropano (Spanish) see..Dibromochloropropane
1,2-Dibromo-3-cloro-propano (Italian) see..Dibromochloropropane
2,6-Dibromo-4-cyanophenol see...Bromoxynil
Dibromodicyanobutane see...1,2-dibromo-2,4-dicyanobutane
1,2-Dibromoetano (Italian, Spanish) see..Ethylene Dibromide
Dibromoethane see...Ethylene Dibromide
1,2-Dibromoethane see...Ethylene Dibromide
α,β-Dibromoethane see...Ethylene Dibromide
sym-Dibromoethane see...Ethylene Dibromide
2,6-Dibromo-4-hydroxybenzonitrile see...Bromoxynil
3,5-Dibromo-4-hydroxybenzonitrile see...Bromoxynil
3,5-Dibromo-4-hydroxyphenyl cyanide see..Bromoxynil
3,5-Dibromo-4-octanoyloxybenzonitrile see..Bromoxynil
2,6-Dibromo-4-phenylcyanide see...Bromoxynil
1,2-Dibroom-3-chloorpropaan (Dutch) see..Dibromochloropropane
1,2-Dibroomethaan (Dutch) see...Ethylene Dibromide
Dibromure d'ethylene (French) see...Ethylene Dibromide
Dibromuro de etileno (Spanish) see...Ethylene Dibromide
O-(1,2-Dibrom-2,2-dichloraethyl)-O,O-dimethyl-phosphat (German) see...Naled

O-(1,2-Dibromo-2,2-dichloro-etil)-O,O-dimetil fosfato (Italian) see...Naled
1,2-Dibromo-2,2-dichloroethyl dimethyl phosphate see...Naled
(1R,3R)-3-(2,2-Dibromovinyl)-2,2-dimethylcyclopropane carboxylic acid, (S)-α-cyano-3-phenoxybenzyl ester see...Deltamethrin
Dibutalin see...Butralin
DIBUTOX® see...Dinoseb
[(Dibutylamino)thio]methylcarbamic acid see...2,2-dimethyl-2,3-dihydro-7-benzofuranyl ester see...Carbosulfan
DIC 1468® see...Metribuzin
Dicamba benzoic acid herbicide see...Dicamba
Dicamba-sodium see...Sodium Dicamba
Dicamba see...sodium salt see...Sodium Dicamba
DICAP® see...Dimethoate
DICAP® see...Dinocap
DICARBAM® see...Carbaryl
DICARBASULF® see...Thiodicarb
Dicarboethoxyethyl O,O-dimethyl phosphorodithioate see...Malathion
S-(1,2-Dicarbethoxyethyl) O,O-dimethylphosphorodithioate see...Malathion
DICARBOSULF® see...Thiodicarb
1,2-Dicarboxy 3,6-endoxocyclohexane see...Endothall
Dicarburetted hydrogen see...Ethylene
DICARZOL® see...Formetanate Hydrochloride
DICATHION® see...Fenitrothion
Dichlofos see...Dichlorvos
1,2-Dichlor-aethan (German) see...Ethylene Dichloride
Di(β-chloroethyl)ether see...Dichloroethyl Ether
Di(2-chloroethyl) ether see...Dichloroethyl Ether
Di-chloricide see...para-Dichlorobenzene
2,2'-Dichlor-diaethylaether (German) see..Dichloroethyl Ether
Dichlordiphenprop see...Diclofop-methyl
Dichloremulsion see...Ethylene Dichloride
Dichlorethanoic acid see...Dichloroacetic Acid
2,2'-Dichlorethyl Ether (DOT) see...Dichloroethyl Ether
Dichlorfenidim see...Diuron
Dichlorfop methyl ester see...Diclofop-methyl
Dichlorfos (Polish) see...Dichlorvos
Dichlorman see...Dichlorvos
3,6-Dichlor-3-methoxy-benzoesaeure (German) see..Dicamba
Di-chlor-mulsion see...Ethylene Dichloride
2,3-Dichlor-1,4-naphthochinon (German) see...Dichlone
Dichloroallyldiisopropylthiocarbamate see...Diallate
S-2,3-Dichloroallyl diisopropylthiocarbamate see..Diallate
2,3-Dichloroallyl N,N-diisopropylthiolcarbamate see..Diallate
S-2,3-Dichloroallyl di-isopropyl(thiocarbamate) see..Diallate
S-(2,3-Dichloroallyl) diisopropylthiocarbamate see..Diallate
2,5-Dichloro-3-aminobenzoic acid see...Chloramben

3,6-Dichloro-o-anisic acid see...Dicamba
3,6-Dichloro-o-anisic acid see...sodium salt see..Sodium Dicamba
p-Dichlorobenzene see...para-Dichlorobenzene
Dichlorobenzene (Mixed Isomers) see...para-Dichlorobenzene
4,4'-Dichlorobenzilate see...Chlorobenzilate
4,4'-Dichlorobenzilic acid ethyl ester see..Chlorobenzilate
2,6-Dichlorobenzonitrile see...Dichlobenil
α,α-Dichloro-2,2-bis(p-ethylphenyl)ethane see...Ethylan
1,1-Dichloro-2,2-bis(4-ethylphenyl)ethane see...Ethylan
1,1-Dichloro-2,2-bis(p-ethylphenyl)ethane see...Ethylan
2,2-Dichloro-1,1-bis(p-ethylphenyl)ethane see...Ethylan
O-(2,5-Dichloro-4-bromophenyl) O-methyl phenylthiophosphonate see...Leptophos
Dichlorocadmium see...Cadmium Chloride
2,4-Dichloro-6-o-chloranilino-s-triazine see...Anilazine
Dichlorochlordene see...Chlordane
2,4-Dichloro-6-(2-chloroanilino)-1,3,5-triazine see..Anilazine
2,4-Dichloro-6-(o-chloroanilino)-s-triazine see..Anilazine
4,6-Dichloro-n-(2-chlorophenyl)-1,3,5-triazin-2-amine see...Anilazine
2,6-Dichlorocyanobenzene see...Dichlobenil
ß,ß'-Dichlorodiethyl ether see...Dichloroethyl Ether
2,2'-Dichloro-diethylether see...Dichloroethyl Ether
5,5'-Dichloro-2,2'-dihydroxydiphenylmethane see..Dichlorophene
1,4-Dichloro-2,5-dimethoxybenzene see...Chloroneb
3,5-Dichloro-n-(1,1-dimethylpropynyl)benzamide see..Pronamide
3,5-Dichloro-n-(1,1-dimethyl-2-propynyl)benzamide see...Pronamide
3,5-Dichloro-n-(1,1-dimethylprop-2-ynyl)benzamide see..Pronamide
Dichlorodiphenyltrichloroethane see...DDT
4,4'-Dichlorodiphenyltrichloroethane see...DDT
p,p'-Dichlorodiphenyltrichloroethane see...DDT
Dichloro-1,2-ethane (French) see...Ethylene Dichloride
Dichlorodiphenyl trichloroethane 2,2-bis(p-chlorophenyl)-1,1,1-trichloroethane see...DDT
α,β-Dichloroethane see...Ethylene Dichloride
1,2-Dichloroethane see...Ethylene Dichloride
1,2-Dichloroethane see...Ethylene Dichloride
sym-Dichloroethane see...Ethylene Dichloride
2,2-Dichloroethenol dimethyl phosphate see..Dichlorvos
2,2-Dichloroethenyl dimethyl phosphate see..Dichlorvos
Dichloroether see...Dichloroethyl Ether
Dichloroethyl ether see...Dichloroethyl Ether
ß,ß'-Dichloroethyl ether see...Dichloroethyl Ether
2,2'-Dichloroethyl ether see...Dichloroethyl Ether
sym-Dichloroethyl ether see...Dichloroethyl Ether
Dichloroethyl oxide see...Dichloroethyl Ether
Dichloroethylene see...Ethylene Dichloride
Dichlorofen see...Dichlorophene

Di(5-chloro-2-hydroxyphenyl)methane see...Dichlorophene
Dichloroisocyanurate see...Dichloroisocyanuric Acid
Dichloroisocyanuric acid see...Dichloroisocyanuric Acid
Dichlorokelthane see...Dicofol
2,5-Dichloro-6-methoxybenzoic acid see...Dicamba
3,6-Dichloro-2-methoxybenzoic acid see...Dicamba
3,6-Dichloro-2-methoxybenzoic acid see...sodium salt see...Sodium Dicamba
4,4'-Dichloro-2,2'-methylenediphenol see...Dichlorophene
3-(2,4-Dichloro-5-(1-methylethoxy)phenyl)-5-(1,1-dimethylethyl)-1,3,4-oxadiazol-2(3$H$)-one see...Oxadiazon
1-(2,4-Dichloro-5-methylsulfonylamidophenyl)-4-difluoromethyl-4,5-dihydro-3-methyl-1$H$-1,2,4-triazol-5-one see...Sulfentrazone
2-(2,4-Dichloro-5-methylsulfonylamidophenyl)-4-difluoromethyl-2,4-dihydro-5-methyl-3$H$-1,2,4-triazol-3-one see...Sulfentrazone
1-[2,4-Dichloro-5-($N^2$-methylsulfonylamino)phenyl]-3-methyl-4-difluoromethyl-$\Delta^2$-1,2,4-triazolin-5-one see...Sulfentrazone
$O$-[Dichloro(methylthio)phenyl] $O,O$-diethyl phosphorothioate (3 isomers) see...Chlorthiophos
2,3-Dichloro-1,4-naphthalenedione see...Dichlone
Dichloronaphthoquinone see...Dichlone
2,3-Dichloronaphthoquinone see...Dichlone
2.3-Dichloro-α-naphthoquinone see...Dichlone
2,3-Dichloronaphthoquinone-1,4 see...Dichlone
2,3-Dichloro-1,4-naphthoquinone see...Dichlone
2,6-Dichloro-4-nitroaniline see...Dichloran
2,6-Dichloro-4-nitrobenzenamine see...Dichloran
2',4'-Dichloro-4'-nitrodiphenyl ether see...Nitrofen
2,4-Dichloro-1-(4-nitrophenoxy)benzene see...Nitrofen
2',5-Dichloro-4'-nitrosalicylanilide see...2-aminoethanol salt see...Clonitralid
5,2-Dichloro-4-nitrosalicylicanilide-2-aminoethanol salt see...Clonitralid
5,2'-Dichloro-4'-nitrosalicylanilide ethanolamine salt see. Clonitralid
2',5-Dichloro-4'-nitrosalicyloylanilide ethanolamine salt see...Clonitralid
1,2-Dichloorethaan (Dutch) see...Ethylene Dichloride
2,2'-Dichloorethylether (Dutch) see...Dichloroethyl Ether
(2,4-Dichloor-fenoxy)-azijnzuur (Dutch) see...2,4-D
2(2,4-Dichloor-fenoxy)propionzuur (Dutch) see...Dichlorprop
3-(3,4-Dichloor-fenyl)-1,1-dimethylureum (Dutch) see...Diuron
3,6-Dichloor-2-methoxy-benzoeizuur (Dutch) see...Dicamba
(2,2-Dichloor-vinyl)-dimethyl-fosfaat (Dutch) see...Dichlorvos
Dichloorvo (Dutch) see...Dichlorvos
DICHLOROPHEN® see...Dichlorophene

DICHLOROPHEN B® see...Dichlorophene
DICHLOROPHENE 10® see...Dichlorophene
3-(3,4-Dichlorophenol)-1,1-dimethylurea see...Diuron
Dichlorophenoxyacetic acid see...2,4-D
2,4-Dichlorphenoxyacetic acid see...2,4-D
2,4-Dichlorophenoxyacetic acid see...salts and esters see...2,4-D
2,4-Dichlorophenoxyacetic acid isooctyl ester see...2,4-D, isooctyl ester
2,4-Dichlorophenoxyacetic acid butoxyethanol ester see. 2,4-D, butoxyethyl ester
(2,4-Dichlorophenoxy)acetic acid butoxyethyl ester see...2,4-D, butoxyethyl ester
2,4-Dichlorophenoxyacetic acid 2-butoxyethyl ester see...2,4-D, butoxyethyl ester
2,4-Dichlorophenoxyacetic acid butoxyethyl ester see...2,4-D, butoxyethyl ester
2,4-Dichlorophenoxyacetic acid ethylene glycol butyl ether ester see...2,4-D, butoxyethyl ester
2,4-Dichlorophenoxyacetic acid isopropyl ester see. 2,4-D, isopropyl ester
4-(2,4-Dichlorophenoxy)butyric acid see...2,4-DB
γ-(2,4-Dichlorophenoxy)butyric acid see...2,4-DB
(2,4-Dichlor-phenoxy)-essigsaeure (German) see...2,4-D
2-(2,4-Dichlorophenoxy)ethanol hydrogen sulfate sodium salt see...2,4-DES-Sodium
2,4-Dichlorophenoxyethyl sulfate, sodium salt see. 2,4-DES-Sodium
4-(2,4-Dichlorophenoxy)nitrobenzene see...Nitrofen
5-(2,4-Dichlorophenoxy)-nitrobenzoic acid, methyl ester see...Bifenox
2-[4-(2,4-Dichlorophenoxy)phenoxy]-methyl-propionate see...Diclofop-methyl
2-[4-(2,4-Dichlorophenoxy)phenoxy]propanoic acid methyl ester see...Diclofop-methyl
2-(2,4-Dichlor-phenoxy)-propionsaeure (German) see...Dichlorprop
2-(2,4-Dichlorophenoxy)propionic acid see...Dichlorprop
(±)-2-(2,4-Dichlorophenoxy)propionic acid see. .Dichlorprop
α-(2,4-Dichlorophenoxy)propionic acid see...Dichlorprop
2,4-Dichlorophenoxy-α-propionic acid see...Dichlorprop
2,4-Dichlorophenoxypropionic acid see...Dichlorprop
1-(2,4-Dichlorophenylaminocarbonyl)cyclopropanecarboxylic acid see...Cyclanilide
$N$-(3,4-Dichlorophenyl)-$n$'-(4-chlorophenyl)urea see...Triclocarban
3-(3,5-Dichlorophenyl)-1,5-dimethyl-3-azabicyclo(3.1.0)hexane-2,4-dione see...Procymidone
$N$-(3,5-Dichlorophenyl)-1,2-dimethylcyclopropane-1,2-dicarboxamide see...Procymidone
3-(3,4-Dichlor-phenyl)-1,1-dimethylharnstoff (German) see...Diuron

[(R-(E)]-1-(2,4-Dichlorophenyl-4,4-dimethyl-2-(1*H*-1,2,4-triazol-1-yl)pent-1-en-3-ol  see...Diniconazole
1-(3,4-Dichlorophenyl)-3,3-dimethylurea  see...Diuron
3-(3,4-Dichlorophenyl)-1,1-dimethylurea  see...Diuron
*N*'-(3,4-Dichlorophenyl)-*n*,*N*-dimethylurea
*N*-(3,4-Dichlorophenyl)-*n*',*N*-dimethylurea
1(3,4-Dichlorophenyl)-3,3-dimethyluree (French)
3-(3,5-Dichlorophenyl)-5-ethenyl-5-methyl-2,4-oxazolidinedione  see...Vinclozolin
(RS)-3-(3,5-Dichlorophenyl)-5-ethenyl-5-methyl-2,4-oxazolidinedione  see...Vinclozolin
2,4-Dichlorophenyl 3-(methoxycarbonyl)-4-nitrophenyl ether  see...Bifenox
3-(3,4-Dichlor-phenyl)-1-methoxy-1-methyl-harnstoff (German)  see...Linuron
3-(3,4-Dichlorophenyl)-1-methoxymethylurea  see. Linuron
3-(3,4-Dichlorophenyl)-1-methoxy-1-methylurea  see.. Linuron
*N*'-(3,4-Dichlorophenyl)-*n*-methoxy-*n*-methylurea  see.. Linuron
1-(3,4-Dichlorophenyl)-3-methoxy-3-methyluree (French)  see...Linuron
Dichlorophenyl)-4-methyl-  see...Methazole
3-(3,5-Dichlorophenyl)-*n*-(1-methylethyl)-2,4-dioxo-1-imidazolidinecarboxamide  see...Iprodione
*N*-(3,4-Dichlorophenyl)-*n*'-methyl-*n*'-methoxyurea  see.. Linuron
2-(3,4-Dichlorophenyl)-4-methyl-1,2,4-oxadiazolidine-3,5-dione  see...Methazole
2-(3,4-Dichlorophenyl)-4-methyl-1,2,4-oxadiazolidinedione  see...Methazole
*N*-3,5-Dichlorophenyl-5-methyl-5-vinyl-1,3-oxazolidine-2,4-dione  see...Vinclozolin
(RS)-3-(3,5-2,4-Dichlorophenyl-4-nirtophenylaether (German)  see...Nitrofen
2,4-Dichlorophenyl 4-nitrophenyl ether  see...Nitrofen
2,4-Dichlorophenyl *p*-nitrophenyl ether  see...Nitrofen
1-[2-(2,4-Dichlorophenyl)pentyl]-1*H*-1,2,4-triazole  see. .Penconazole
(RS)-1-[2-(2,4-Dichlorophenyl)-*n*-pentyl]-1*H*-1,2,4-triazole  see...Penconazole
*N*-(3,4-Dichlorophenyl)propanamide  see...Propanil
1-(2-(2,4-Dichlorophenyl)-2-(2-propenyloxy)ethyl)-1*H*-imidazole  see...Imazalil
1-[2-((2,4-Dichlorophenyl)-2-propenyloxy)-ethyl]-1*H*-imidazole  see...Imazalil
3',4'-Dichlorophenylpropionanilide  see...Propanil
1-[(2-(2,4-Dichlorophenyl)-4-propyl-1,3-dioxolan-2-yl)methyl]-1*H*-1,2,4-triazole  see...Propiconazole
1-[2-(2,4-Dichlorophenyl)-4-propyl-1,3-dioxolan-2-yl]-methyl-1*H*-1,2,4,-triazole  see...Propiconazole
(±)-1-[2-(2,4-Dichlorophenyl)-4-propyl-1,3-dioxolan-2-ylmethyl]-1*H*-1,2,4-triazole  see...Propiconazole
3-(3,5-Dichlorophenyl)-n-sopropyl-2,4-dioxo-1-imidazolidinecarboximide  see...Iprodione
S-(((2,5-Dichlorophenyl)thio)methyl) *O*,*O*-dimethyl phosphorodithioate  see...Methyl Phenkapton

(2,5-Dichlorophenylthio)methanethiol-*S*-ester with *O*,*O*-dimethyl phosphorodithioate  see...Methyl Phenkapton
(RS)-2-(2,4-Dichlorophenyl)-1-(1*H*-1,2,4-triazole-1-yl)hexan-2-ol  see...Hexaconazole
Di(*p*-chlorophenyl)trichloromethyl carbinol  see.. Dicofol
3,6-Dichloropicolinic acid  see...Clopyralid
3,6-Dichloro-2-picolinic acid  see...Clopyralid
Dichloroprop  see...Dichlorprop
α,β-Dichloropropane  see...1,2-dichlorophene
1,3-Dichloro-1-propene  see...1,3-Dichloropropene
1,3-Dichloro-2-propene  see...1,3-Dichloropropene
1,3-d  see...1,3-Dichloropropene  see...1,3-Dichloropropene
1,3-Dichloropropene and 1,2-dichloropropane mixture  see...D-D mixture
1,3-Dichloro-1-propene  see...mixture with 1,2-dichloropropane  see...D-D mixture
2,3-Dichloro-2-propene-1-thiol  see..
Iisopropylcarbamate  see...Diallate
S-(2,3-Dichloro-2-propenyl)bis(1-methylethtl)carbamothioate  see...Diallate
Dichloropropionanilide  see...Propanil
3,4-Dichloropropionanilide  see...Propanil
3',4'-Dichloropropionanilide  see...Propanil
α-Dichloropropionic acid  see...Dalapon
α,α-Dichloropropionic acid  see...Dalapon
2,2-Dichloropropionic acid  see...Dalapon
α,ϐ-Dichloropropylene  see...1,3-Dichloropropene
1,3-Dichloropropylene  see...1,3-Dichloropropene
3,6-Dichloro-2-pyridinecarboxylic acid  see...Clopyralid
3,7-Dichloroquinoline-8-carboxylic acid  see..
Quinclorac
3,7-Dichloro-8-quinolinecarboxylic acid  see..
Quinclorac
1-[3,5-Dichloro-4-(1,1,2,2-tetrafluoroethoxy)phenyl]-3-(2,6-difluorobenzoyl)urea (IUPAC)  see...Hexaflumuron
1,3-Dichloro-*s*-triazine-2,4,6-(1*H*,3*H*,5*H*)-trione  see..
Dichloroisocyanuric Acid
Dichloro-*s*-triazinetrione  see...Dichloroisocyanuric Acid
2,2-Dichlorovinyl dimethyl phosphate  see...Dichlorvos
(2,2-Dichlorvinyl)-dimethyl-phosphat (German)  see..
Dichlorvos
(1*R*)-*trans*-(2,2-Dichlorovinyl)-2,2-dimethylcyclopropanecarboxylate  see...Fenfluthrin
Dichlorovos  see...Dichlorvos
DICHLORPHEN®  see...Dichlorophene
Dichlorpropan-Dichlorpropengemisch (German)  see. .D-D mixture
*O*-(2,2-Dichlorvinyl)-*O*,*O*-dimethylphosphat (German)  see...Dichlorvos
(±)-2,4'-Dichloro-α-(pyrimidin-5-yl)benzhydryl alcohol  see...Fenarimol
2,4'-Dichloro-α-(pyrimidin-5-yl)benzhydryl alcohol  see..
Fenarimol
4,4'-Dichloro-α-(trichloromethyl)benzhydrol  see..
Dicofol
Dichromic acid  see...disodium salt  see...Sodium Dichromate

Dichromic acid heptaoxide see...Sodium Dichromate
DICID® see...Diazinon
Diclofop methyl ester see...Diclofop-methyl
Diclona (spanish) see...Dichlone
Diclone see...Dichlone
Dicloran see...Dichloran
DICLORCAL 50® see...Dichlorvos
Diclorobenceno (Spanish) see...*para*-Dichlorobenzene
1,4-Diclorobenceno (Spanish) see...*para*-Dichlorobenzene
*p*-Diclorobenceno (Spanish) see...*para*-Dichlorobenzene
Diclorodifeniltricloroetano (Spanish) see...DDT
1,2-Dicloroetano (Italian, Spanish) see...Ethylene Dichloride
2,2'-Dicloroetiletere (Italian) see...Dichloroethyl Ether
3-(3,4-Diclorofenil)-1,1-dimetilurea (Spanish) see...
Diclorofeno (Spanish) see...Dichlorophene
3-(3,4-Dicloro-fenyl)-1,1-dimetil-urea (Italian)
1,2-Dicloropropano (Spanish) see...1,2-Dichloropropane
1,3-Dicloropropeno (Spanish) see...1,3-Dichloropropene
(2,2-Dicloro-vinil)dimetilfosfato (Italian) see...Dichlorvos
DICOFEN® see...Fenitrothion
DICOMITE® see...Dicofol
DICONATE 6® see...Sodium Methanearsonate (MSMA)
DICOPHANE® see...DDT
Dicopper dichloride see...Cuprous Chloride
Dicopper dihydroxycarbonate see...Copper Carbonate Basic
Dicopper monoxide see...Cuprous Oxide
DICOPUR® see...2,4-D
DICOPUR-M® see...MCPA
DICOTEX® see...MCPA
DICOTOX® see...2,4-D
Dicresyl see...Metolcarb
Dicresyl *N*-methylcarbamate see...Metolcarb
DICRON® see...Dicrotophos
Dicroptophos see...Dicrotophos
Dicrotofos (Dutch) see...Dicrotophos
DICTATOR® see...Propargite
DICTATOR-PLUS® see...Tetradifon
DICUPRAL® see...Disulfiram
1,3-Dicyanotetrachlorobenzene see...Chlorothalonil
DIDANDIN® see...Diphacione
DIDIGAM® see...DDT
DIDIMAC® see...DDT
1,5-Di-(2,4dimethylphenyl)-3-methyl-1,3,5-triazapenta-1,4-diene see...Amitraz
DIDIVANE see...Dichlorvos
DIDRIN® see...Dicrotophos
DIDROXANE® see...Dichlorophene
DIELDREX® see...Dieldrin
Dieldrina (Spanish) see...Dieldrin
Dieldrine (French) see...Dieldrin

DIELDRITE® see...Dieldrin
Dienochlor see...Dienochlor
Diethion (France) see...Ethion
α-[((Diethoxyphosphinothioyl)oxy)imino]benzene acetonitrile see...Phoxim
(Diethoxyphosphinyl)dithioimidocarbonic acid cyclic ethylene ester see...Phosfolan
(Diethoxyphosphinylimino)-1,3-dithietane see...Fosthietan
Diethoxyphosphinylimino-2-dithietanne-1,3 (French) see...Fosthietan
2-(Diethoxyphosphinylimino)-1,3-dithiolane see...Phosfolan
2-(3 see...Diethoxyphosphinylimino)-1,3-dithiolan see...Phosfolan
2-(Diethoxyphosphinylimino)-4-methyl-1,3-dithiolane see...Mephosfolan
Diethoxy thiophosphoric acid ester of 2-ethylmercaptoethanol see...Demeton
(Diethoxy-thiophosphoryloxyimino)-phenyl acetonitrile see...Phoxim
Diethquinalphion see...Quinalphos
Diethquinalphione see...Quinalphos
*O,O,*-Diethyl see...Chlorpyrifos
Diethylamino-2,6-aceto xylidide see...Metham-Sodium
2-(2-Diethylamino)ethyl-*O,O*-diethyl ester oxalate see...Amiton Oxalate
*S*-(2-Diethylaminoethyl)-*O,O*-diethylphosphorothioate hydrogen oxalate see...Amiton Oxalate
*S*-(2-Diethylamino) ethyl phosphorothioic acid-*O,O*-diethyl ester see...Amiton
*O*-(2-(Diethylamino)-6-methyl-4(pyrimidinyl)*O,O*-diethyl phosphorothioate see...Pirimiphos-Ethyl
2-Diethylamino-6-methylpyrimidin-4-yl diethylphosphorothionate see...Pirimiphos-Ethyl
2-Diethylamino-6-methylpyrimidin-4-yl dimethyl phosphorothionate see...Pirimiphos-Methyl
*O*-(2-(Diethylamino)-6-methyl-4(pyrimidinyl) *O,O*-dimethyl phosphorothioate see...Pirimiphos-Methyl
*O*-(2-Diethylamino-6-methylpyrimidin-4-yl) *O,O*-dimethyl phosphorothioate see...Pirimiphos-Methyl
Diethylcarbamodithioic acid 2-chloro-2-propenyl ester see...Sulfallate
*O,O,*-Diethyl-*O*-(2-chinoxalyl)phosphorothioate see...Quinalphos
*O,O,*-Diethyl-*O*-(3-chloor-4-methyl-cumarin-7-yl)monothiofosfaat (Dutch) see...Coumaphos
*O,O,*-Diethy-S-[(4-chloor-fenyl-thio)-methyl]dithiofosfaat (Dutch) see...Carbophanothion
*O,O,*-Diethyl-*S*-(6-chlorobenzoxazolinyl-3-methyl)dithiophosphate see...Phosalone
*O,O,*-Diethylo-(2-chloro-1-(2',4'-dichlorophenyl)vinyl) phosphate see...Chlorfenvinphos
*O,O,*-Diethylo-(3-chloro-4-methylcoumarinyl-7)thiophosphate see...Coumaphos
*O,O,*-Diethylo-(3-chloro-4-methyl-7-coumarinyl)phosphorothioate see...Coumaphos

$O,O,$-Diethylo-(3-chloro-4-methyl-2-oxo-2H-benzopyran-7-yl)phosphorothioate  see...Coumaphos
$O,O,$-Diethylo-(3-chloro-4-methylumbelliferyl)phosphorothioate  see...Coumaphos
$O,O,$-Diethyl $S$-[6-chloro-3-(mercaptomethyl)-2-benzoxazolinone]phosphorodithioate  see...Phosalone
$O,O,$-Diethyl-3-chloro-4-methyl-7-umbelliferone thiophosphate  see...Coumaphos
Diethyl 3-chloro-4-methylumbelliferyl thionophosphate  see...Coumaphos
$O,O,$-Diethyl-$S$-[(6-chloro-2-oxobenzoxazolin-3-yl)methyl]phosphorodithioate  see...Phosalone
$O,O,$-Diethyl-4-chlorophenylmercaptomethyl dithiophosphate  see...Carbophanothion
$O,O,$-Diethyl-$p$-chlorophenylmercaptomethyl dithiophosphate  see...Carbophanothion
$O,O,$-Diethy-$S$-$p$-chlorophenylthiomethyl dithiophosphate  see...Carbophanothion
$O,O,$-Diethy-$S$-($p$-chlorophenylthiomethyl)phosphoro dithioate  see..Carbophanothion
$O,O,$-Diethyl-$S$-(2-chloro-1-phthalimidoethyl)phosphorodithioate  see...Dialifor
$p,p$-Diethyl cyclic ethylene ester of phosphonodithioimidocarbonate  see...Phosfolan
$p,p$-Diethyl cyclic ethylene ester of phosphonodithioimidocarbonic acid  see...Phosfolan
$p,p$-Diethyl cyclic propylene ester of phosphonodithioimidocarbonic acid  see...Mephosfolan
$O,O,$-Diethyl $O$-2,5-dichloro-4-bromophenyl-phosphorothioate  see...Bromophos-ethyl
$O,O,$-Diethyl $O$-(2,5-dichloro-4-bromophenyl)thiophosphate  see...Bromophos-ethyl
$O,O,$-[Diethyl-$O$-2,4,5-dichloro(methylthio)phenyl]thionophosphate  see..Chlorthiophos
Diethyl-(2,4-dichlorophenyl)-2-chlorovinyl phosphate  see...Chlorfenvinphos
Diethyl $O$-(2-diethylamino-6-methyl-4-pyrimidinyl)phosphorothioate  see...Pirimiphos-Ethyl
Diethyl (dimethoxyphosphinothioylthio)succinate  see...Malathion
Diethyl [(dimethoxyphosphinothioyl)thio]butanedioate  see...Malathion
Diethyl (dimethoxythiophosphorylthio)succinate  see...Malathion
Diethyl 2-dimethylamino-4-methylpyrimidin-6-yl phosphorothionate  see...Pirimiphos-Ethyl
Diethyldiphenyl  see...Ethylan
Diethyldithiocarbamic acid-2-chloroallyl ester  see..Sulfallate
Diethyl 1,3-dithiolan-2-ylidenephosphoramidate  see..Phosfolan
$O,O,$-Diethy-dithiophosphoric acid, $p$-chlorophenylthiomethyl ester  see...Carbophanothion
$O,O,$-Diethyldithiophosphorylacetic acid-$N$-monoisopropylamide  see...Prothoate
3-Diethyldithiophosphorylmethyl-6-chlorobenzoxazolone-2  see...Phosalone

$N,N$-Diethyletha namine compounded with [(3,5,6-trichloro-2-pyridinyl)oxy]acetic acid (1:1)  see. Triclopyr, Triethylammonium Salt
$O,O,$-Diethyl $S$-(2-ethioethyl)phosphorothioate  see...Demeton
$O,O,$-Diethyl $S$-(2-eththioethyl)phosphorodithioate  see...Disulfoton
1,2-Di-(3-ethoxycarbonyl-2-thioureido)benzene  see...Thiophanate-Methyl
$O,O,$-Diethyl-$S$-(2-ethyl-$N,N$-diethylamino)ethylphosphorothioate hydrogen oxalate  see...Amiton Oxalate
$O,O,$-Diethyl $S$-(2-ethylmercaptoethyl)dithiophosphate  see...Disulfoton
$O,O,$-Diethyl-2-ethylmercaptoethyl thiophosphate diethoxythiophosphoric acid  see...Demeton
$O,O,$-Diethyl $S$-ethylmercaptomethyl dithiophosphonate  see...Phorate
$O,O,$-Diethyl-$S$-ethyl-2-ethylmercaptophosphorothiolate  see...Demeton
$O,O,$-Diethyl-$S$-(2-ethylthio-ethyl)-monothiofosfaat  see..Demeton
Di($p$-ethylphenyl)dichloroethane  see...Ethylan
$O,O,$-Diethyl $S$-[2-(ethylsulfinyl)ethyl]phosphorodithioate  see...Oxydisulfoton
$O,O,$-Diethyl-$S$-[(ethylsulfinyl)ethyl]phosphorodithioate  see...Oxydisulfoton
$O,O,$-Diethyl 2-ethylthioethylphosphorodithioate  see...Disulfoton
$O,O,$-Diethyl-$S$-(2-ethylthio-ethyl)-dithiofosfaat (Dutch)  see...Disulfoton
$O,O,$-Diethyl $S$-(2-(ethylthio)ethyl) phosphorothioate  see. Demeton
$O,O,$-Diethyl $O$-(2-(ethylthio)ethyl) phosphorothioate  see...Demeton
Diethyl-$S$-(2-(ethylthio)ethyl)phosphorothiolate  see...Demeton
$O,O,$-Diethyl 2-ethylthioethylphosphorodithioate  see...Disulfoton
$O,O,$-Diethyl $S$-(2-eththioethyl)thiothionophosphate  see...Disulfoton
$O,O,$-Diethyl-$S$-(ethylthio-methyl)-dithiofosfaat (Dutch)  see...Phorate
$O,O,$-Diethylethylthiomethyl phosphorodithioate  see..Phorate
$O,O,$-Diethyl $S$((ethylthio)methy)phosphorodithioate  see. Phorate
$O,O,$-Diethyl $S$-(ethylthio)methylphosphorodithioate  see. Phorate
$O,O,$-Diethyl $S$-ethylthiomethylthionophosphate  see.. Phorate
$O,O,$-Diethyl $S$-ethylthiomethyldithiophosphonate  see.. Phorate
$O,O,$-Diethyls-2-(ethylthio)ethylphosphorodithioate  see...Disulfoton
$O,O,$-Diethyl $S$-2-(ethylthio)ethyl phosphorothioate mixed with phosphorothioic acid,$O,O,$-diethyl $O$-2-(ethylthio) ethyl ester  see...Demeton

*O,O,*-Diethyl-*S*-isopropylcarbamoylmethyl phosphorodithioate see. .Prothoate
*O,O,*-Diethyl-*S*-(*N*-isopropylcarbamoylmethyl) phosphorodithioate see. . Prothoate
*O,O,*-Diethyl-*S*-(*N*-isopropylcarbamoylmethyl) dithiophosphate see. . Prothoate
*O,O,*-Diethyl-*O*-(2-isopropyl-4-methyl-pyrimidin-6-yl)-monothiofospaat (Dutch) see. . .Diazinon
Diethyl 2-isopropyl-4-methyl-6-pyrimidinl phosphorothionate see. . .Diazinon
Diethyl 2-isopropyl-4-methyl-6-pyrimidylthionophosphate see. . .Diazinon
Diethyl 4-(2-isopropyl-6-methylpyrimidinl) phosphorothionate see. . .Diazinon
*O,O,*-Diethyl 2-isopropyl-4-methylpyrimidyl-6-thiophosphate see. . .Diazinon
*O,O,*-Diethyl *O*-(2-isopropyl-4-methyl-6-pyrimidyl)thionophosphate see. . .Diazinon
*O,O,*-Diethyl-*O*-(2-isopropyl-4-methyl-6-pyrimidyl)phosphorothionate see. . .Diazinon
*O,O,*-Diethyl *O*-2-isopropyl-6-methylpyrimidin-4-ylphosphorothionate see. . .Diazinon
Diethyl mercaptosuccinate, *O,O*-dimethyl thiophosphate see. . .Malathion
Diethyl mercaptosuccinate, *O,O*-dimethyl dithiophosphate *S*-ester see. . .Malathion
Diethyl mercaptosuccinate, *O,O*-dimethyl phosphorodithioate see. . .Malathion
Diethyl mercaptosuccinate *S*-ester with *O,O*-dimethyl phosphorodithioate see. . .Malathion
Diethyl(4-methyl-1,3-dithiolan-2-ylidene)phosphoroamidate see. . .Mephosfolan
*O,O,*-Diethyl *O*-6-methyl-2-isopropyl-4-pyrimidinyl phosphorthioate see. . .Diazinon
*O,O,*-Diethyl *O*-(6-methyl-2-(1-methylethyl)-4-pyrimidinyl) phosphorthioate see. . .Diazinon
*N,N*-Diethyl-2-(1-naphthalenyloxy)propanamide see. . Napropamide
*O,O'*-Diethyl-*p*-nitrophenylphosphat (German) see. . Paraoxon
Diethyl-*p*-nitrophenyl phosphate see. . .Paraoxon
*O,O,*-Diethyl-*p*-nitrophenyl phosphate see. . .Paraoxon
*O,O,*-Diethyl *O*-*p*-nitrophenyl phosphate see. . .Paraoxon
Diethyl *p*-nitrophenyl phosphorothionate see. . .Parathion
Diethyl 4-nitrophenyl phosphorothionate see. . .Parathion
*O,O,*-Diethyl *O*-(4-nitrophenyl) phosphorothioate see. . Parathion
*O,O,*-Diethyl *O*-(*p*-nitrophenyl) phosphorothioate see. . Parathion
*O,O,*-Diethyl-*O*,*p*-nitrophenyl phosphorothioate see. . Parathion
Diethyl *p*-nitrophenyl thionophosphate see. . .Parathion
*O,O,*-Diethyl *O*-*p*-nitrophenyl thiophosphate see. . Parathion
*O,O,*-Diethyl-*S*-(4-oxobezotriazin-3-methyl)-dithiophosphat (German) see. . .Azinphos-ethyl
*O,O,*-Diethyl-*S*-(4-oxobezotriazino-3-methyl)-phosphorodithioate see. . .Azinphos-ethyl

*O,O,*-Diethyl-*S*-[(4-oxo-3H-1,2,3-bezotriazin-3yl)methyl]-dithio fosfaat (Dutch) see. . .Azinphos-ethyl
*O,O,*-Diethyl-*S*-[(4-oxo-3H-1,2,3-bezotriazin-3-yl)-methyl]-dithiophosphat (German) see. . .Azinphos-ethyl
Diethyl paraoxon see. . .Paraoxon
Diethyl parathion see. . .Parathion
Diethyl [(1,2-phenylene)bis(iminocarbonothioyl)] bis(carbamate) see. . Thiophanate-Methyl
Diethyl [1,2-phenylenebis(iminocarbonothioyl)]bis(carbamate) see. . Thiophanate-Methyl
Diethyl 4,4'-*O*-phenylenebis(3-thioallophanate) see. . Thiophanate-Methyl
Diethyl 4,4'-(*O*-phenylene)bis(3-thioallophanate) see. . Thiophanate-Methyl
*O,O,*-Diethyl *O*-(1-phenyl-1*H*-1,2,4-triazol-3-yl) phosphorothioate see. . .Triazophos
*O,O,*-Diethylphosphoric acid *O*-*p*-nitrophenyl ester see. . Paraoxon
*O,O,*-Diethyl phosphorodithioate *S*-ester with *N*-(2-Chloro-1-mercaptoethyl)phthalimide see. . .Dialifor
*O,O,*-Diethylphosphorodithioate-ester with 3-(mercaptomethyl)-1,2,3-benzotriazin-4(3H)-one see. . Azinphos-ethyl
*O,O,*-Diethyl phosphorothioat, *O*-ester with phenylglyoxylonitrile oxime see. . .Phoxim
Diethyl-*O*-2-pyrazinyl phosphorothionate see. . Thionazin
*O,O,*-Diethyl-*O*,2(pyrazinyl phosphorothioate see. . Thionazin
*O,O,*-Diethyl-*O*-2-pyrazinyl phosphothionate see. . Thionazin
*O,O,*-Diethyl-*O*(pyrazinyl thiophosphate see. . .Thionazin
*O,O,*-Diethyl-*O*(quinoxalin-2-yl phosphorothioate see. . Quinalphos
*O,O,*-Diethyl *O*-2-quinoxalinyl phosphorothioate see. . Quinalphos
*O,O,*-Diethyl *O*-(quinoxalin-2-yl)thiophosphate see. . Quinalphos
*O,O,*-Diethyl *O*(quinoxalin-2-yl thionophosphate see. . Quinalphos
*O,O,*-Diethyl-*O*-(2-quinoxalinyl)phosphorothioate see. . Quinalphos
*O,O,*-Diethyl-*O*-(2-quinoxalyl)phosphorothioate see. . Quinalphos
*O,O,*-Diethyl *O*-(1,2,2,2-tetrachloroethyl) phosphorothioate see. . .Chlorethoxyfos
*O,O,*-Diethyl *O*-(1,2,2,2-tetrachloroethyl) thionophosphate see. . .Chlorethoxyfos
Diethylthiadicarbocyanine iodide see. . .Dithiazanine Iodide
3,3'-Diethylthiadicarbocyanine iodide see. . .Dithiazanine Iodide
Diethylthiophosphoric acid ester of 3-chloro-4-methyl-7-hydroxycoumarin see. . .Coumaphos
*O,O,*-Dietil-*S*-[(4-clorofenil-tio)-metile]-ditiofosfato (Italian) see. . .Carbophanothion
*O,O,*-Dietil-*S*-[(*p*-clorofenil-tio)-metile]-ditiofosfato (Italian) see. . .Carbophanothion

*O,O,*-Dietil-*O*-(3-cloro-4-metil-cumarin-7-il-monotiofosfato) (Italian) see...Coumaphos
*O,O,*-Dietil-*S*-(2-etiltio-etil) monotiofosfato see...Demeton
*O,O,*-Dietil-*S*-(etiltio-metil)-ditiofosfato (Italian) see...Phorate
*O,O,*-Dietil-*S*-(2-etiltio-metil)-ditiofosfato (Italian) see...Disulfoton
*O,O,*-Dietil-*S*-((4-oxo-3H-1,2,3-bezotriazin-3il)metil)-ditiofosfato (Italian) see...Azinphos-ethyl
DIF 4® see...Diphenamid
Difenamid (Spanish) see...Diphenamid
DIFENPHOS® see...Temephos
DIFENTHOS® see...Temephos
Difenzoquat methyl sulfate see...Difenzoquat
Diflubenzuron (Spanish) see...Diflubenzuron
Diflufenicanil (French) see...Diflufenican
2',4'-Difluoro-2-(α-α-α-trifluoro-*m*-tolyloxy)nicotinanilide see...Diflufenican
*N*-(2,4-Difluorophenyl)-2-[3-(trifluoromethyl)phenoxy]-3-pyridinecarboxamide see...Diflufenican
*N*-(2,6-Difluorophenyl)-5-methyl-(1,2,4)triazolo-(1,5-α)pyrimidine-2-sulfonamide see...Flumetsulam
Diflupyl see...Isofluorphate
DIFLURON® see...Diflubenzuron
Diflurophate see...Isofluorphate
DIFO see...Dimefox
DIFOCAP® see...Captafol
DIFOL® see...Dicofol
DIFOLATAN® see...Captafol
DIFONATE® see...Fonofos
1,3-Diformal propane see...Glutaraldehyde
DIFOS® see...Temephos
DIFOSAN® see...Captafol
DIGERMIN® see...Trifluralin
1,2-Dihydroacenaphthylene see...Acenaphthene
1,8-Dihydroacenaphthylene see...Acenaphthene
22,23-Dihydroazadirachtin see...Dihydroazadirachtin
2,3-Dihydro-5-carboxanilido-6-methyl-1,4-oxathiin see...Carboxin
2,3-Dihydro-5-carboxanilido-6-methyl-1,4-oxathiin-4,4-dioxide see...Oxycarboxin
9,10-Dihydro-8a,10,-diazoniaphenanthrene dibromide see...Diquat
9,10-Dihydro-8A,10A,-diazoniaphenanthrene dibromide see...Diquat Dibromide
9,10-Dihydro-8a,10a-diazoniaphenanthrene(1,1'-ethylene-2,2'-bipyridylium)dibromide see...Diquat
9,10-Dihydro-8A,10A-diazoniaphenanthrene(1,1'-ethylene-2,2'-bipyridylium)dibromide see...Diquat Dibromide
2,3-Dihydro-2,2-dimethyl-7-benzofuranol-*N*-methylcarbamate see...Carbofuran
2,3-Dihydro-2,2-dimethyl-7-benzofuranolmethylcarbamate see...Carbofuran
2,3-Dihydro-2,2-dimethyl-7-benzofuranyl(di-*N*-butylaminosulfenyl)methylcarbamate see...Carbosulfan
2,2-Dihydro-2,2-dimethyl-7-benzofuranyl[(dibutylamino)thio]methylcarbamate see...Carbosulfan
2,3-Dihydro-2,2-dimethyl-7-benzofuranyl *N*-(*N*-2-(ethoxycarbonyl)ethyl-*N*-isopropylaminosulfenyl)-*N*-methylcarbamate see...Benfuracarb
2,3-Dihydro-2,2-dimethylbenzofuran-7-yl methylcarbamate see...Carbofuran
2,3-Dihydro-2,2-dimethylbenzofuranyl-7-*N*-methylcarbamate see...Carbofuran
2,3-Dihydro-5,6-dimethyl-1,4-dithiin 1,1,4,4-tetraoxide see...Dimethipin
5,6-Dihydro-dipyrido(1,2A,2,1C)pyrazinium dibromide see...Diquat Dibromide
5,6-Dihydro-dipyrido(1,2-a:2,1'-c)pyrazinium dibromide see...Diquat
6,7-Dihydrodipyrido(1,2-A:2',1'-C)pyrazinediium dibromide see...Diquat Dibromide
5,6-Dihydro-dipyrido(1,2-A:2,1'-C)pyrazinium dibromide see...Diquat Dibromide
5,6-Dihydro-dipyrido(1,2a,2,1c)pyrazinium dibromide see...Diquat
1,2-Dihydro-6-3 see...ethoxy-2,2,4-trimethylquinoline see...Ethoxyquin
4,5-Dihydroimidazole-2(3*H*)-thione see...Ethylene Thiourea
4,5-Dihydro-2-mercaptoimidazole see...Ethylene Thiourea
*S*-(2,3-Dihydro-5-methoxy-2-oxo-1,4,4-thiadiazol-3-methyl) see...Methidathion
5,6-Dihydro-2-methyl-3-carboxanilido-1,4-oxathiin-4,4-dioxid (German) see...Oxycarboxin
2[4,5-Dihydro-4-methyl-4-(1-methylethyl)-5-*oxo*-1*H*-imidazol-2-yl]-4 (or 5)-methylbenzoic acid methyl ester see...Imazethabenz
(±)-2-[4,5-Dihydro-4-methyl-4-(1-methylethyl)-5-*oxo*-1*H*-imidazol-2-yl]-5-ethyl-3-pyridinecarboxylic acid see...Imazethapyr
2,3-Dihydro-6-methyl-1,4-oxathiin-5-carboxanilide see...Carboxin
5,6-Dihydro-2-methyl-1,4-oxathiin-3-carboxanilide see...Carboxin
5,6-Dihydro-2-methyl-1,4-oxathiin-3-carboxanilide 4,4-dioxide see...Oxycarboxin
Dihydrooxirene see...Ethylene Oxide
2,3-Dihydro-6-methyl-5-phenylcarbamoyl-1,4-oxathiin see...Carboxin
5,6-Dihydro-2-methyl-*n*-phenyl-1,4-oxathiin-3-carboxamide see...Carboxin
5,6-Dihydro-2-methyl-*N*-phenyl-1,4-oxathiin-3-carboxamide-4,4-dioxide see...Oxycarboxin
(*E*)-4,5-Dihydro-6-methyl-4-[(3-pyridinylmethylene)amino]-1,2,4-triazin-3(2*H*)-one see...Pymetrozine
3,4-Dihydro-4-oxo-3-benzotriazinylmethyl *O,O*-diethyl phosphorodithioate see...Azinphos-ethyl

S-(3,4-Dihydro-4-oxo-1,2,3-benzotriazin-3-ylmethyl) O,O-dimethyl phosphorodithioate see...Azinphos-methyl
S-(3,4-Dihydro-4-oxobenzo[a][1,2,3]triazin-3-ylmethyl) O,O-dimethyl phosphorodithioate see...Azinphos-methyl
S-(3,4-Dihydro-4-oxobenzol[d][1,2,3]triazin-3-ylmethyl) O,O-dimethyl phosphorodithioate see...Azinphos-methyl
1,2-Dihydro-3,6-pyradazinedione see...Maleic Hydrazide
1,2-Dihydropyridazine-3,6-dione see...Maleic Hydrazide
1,2-Dihydro-3,6-pyridazinedione see...Maleic Hydrazide
6,7-Dihydropyrido(1,2-a:2',1'-c)pyrazinedium dibromide see...Diquat
6,7-Dihydropyridol(1,2-a:2',1'-c)pyrazinedium dibromide see...Diquat
6,7-Dihydropyridol(1,2-A:2',1'-C)pyrazinedium dibromide see...Diquat Dibromide
Dihyrosamidin see...Antimycin A
1,2-Dihydro-2,2,4-trimethyl-6-ethoxyquinoline see. Ethoxyquin
2,2'-Dihydroxy-5,5'-dichlorodiphenylmethane see.. Dichlorophene
2,2'-Dihydroxy-3,3',5,5',6,6'-hexachlorodiphenylmethane see...Hexachlorophene
2,2'-Dihydroxy-3,5,6,3',5',6'-hexachlorodiphenylmethane see...Hexachlorophene
Dihydroxydichlorodiphenylmethane see..Dichlorophene
[(Dihydroxydichloro)diphenyl]methane see. Dichlorophene
2,2'-Dihydroxy-3,3',5,5'-tetrachlorodiphenylsulfide see.Bithionol
Diiron trisulfate see...Ferric Sulfate
Diisobutylthiocarbamic acid-S-ethyl ester see...Butylate
Diisocarb see...Butylate
Diisopropoxyphosphoryl fluoride see...Isofluorphate
2,6-Diisopropylamino-4-methoxytriazine see...Prometon
N-(β-O,O-Diisopropyldithiophosphorylethyl)bezenesulfonamide see...Bensulide
N-[2-(O,O-Diisopropyldithiophosphoryl)ethyl]benzene sulfonamide see...Bensulide
Diisopropylfluorophosphate see...Isofluorphate
O,O-Diisopropylfluorophosphate see...Isofluorphate
Diisopropylfluorophosphonate see...Isofluorphate
Diisopropylfluorophosphoric acid ester see.Isofluorphate
Diisopropylfluorphosphorsaeureester (German) see.Isofluorphate
N,N-Di-isopropyl-6-methylthio-1,3,5-triazine-2,4-diamine see...Prometryn
N,N'-Di-isopropyl-6-methylthio-1,3,5-triazine-2,4-diyldiamine see...Prometryn
N,N-Diisopropyl-6-methoxy-1,3,5-triazine-2,4-diamine see...Prometon
N,N'-Diisopropyl-6-methoxy-1,3,5-triazine-2,4-diyldiamine see...Prometon

Diisopropylphosphofluoridate see...Isofluorphate
S-(O,O-Diisopropyl phosphorodithioate) ester of N-(2-mercaptoethyl)benzenesulfonamide see...Bensulide
Diisopropyl phosphorofluoridate see...Isofluorphate
O,O'-Diisopropyl phosphoryl fluoride see..Isofluorphate
N-Diisopropylthiocarbamic acid S-2,3,3-trichloro-2-propenyl ester see...Triallate
Diisopropylthiocarbamic acid,-(2,3-dichloroallyl) ester see...Diallate
Di-isopropylthiolocarbamate des-(2,3-dichloro allyle) (French) see...Diallate
Diisopropyltrichloroallylthiocarbamate see...Triallate
DIKAMIN® see...2,4-D
DIKAR® see..EZENOAN® see...Dinocap
DIKONIRT® see...2,4-D
DILATIN DBI® see...para-Dichlorobenzene
DILIC® see...Cacodylic Acid
DILLEX® see..Dinitro-o-cresol (DNOC)
Dilombrin see...Dithiazanine Iodide
DIMAS® see...Daminozide
DIMATE 267® see...Dimethoate
DIMAZ® see...Disulfoton
DIMECRON® see...Phosphamidon
DIMENSION® see...Dithiopyr
Dimephenthioate see...Phenthoate
Dimephenthoate see...Phenthoate
1,4-Dimesyloxybutane see...Busulfan
1,4-Dimethanesulfonoxbutane see...Busulfan
1,4-Di(methanesulfonyloxy)butane see...Busulfan
1,4-Dimethanesulphonyloxybutan see...Busulfan
1,4:5,8-Dimethano naphthalene, 1,2,3,4,10,10-hexachloro-1,4,4a,5,8,8a-hexahydro-, (1α,4.α,4A.β.5 β,8 β,8A β)- see...Isodrin
2,7:3,6-Dimethanonaphth(2, 3-b)oxirene see.3,4,5,6,9,9-Hexachloro-1a,2,2a,3,6,6a,7,7a-octahydro-, (aα,2.β,2aβ,2aβ,3α,6α,6aβ,7β,7aα)- see...Endrin
1,4:5,8-Dimethanonaphthalene,1,2,3,4,10,10-hexachloro-1,4,4a,5,8,8a-hexahydro-(1α,4α,4β,5α,8α,8β)- see.Aldrin
1,4:5,8-Dimethanonaphthalene,1,2,3,4,10,10-hexachloro-1,4,4a,5,8,8a-hexahydro-, endo-exo- see...Aldrin
2,7:3,6-Dimethanonaphtha[2,3B]oxirene,3,4,5,6,9,9-hexachloro-1a,2,2a,3,6,6a,7,7a-octahydro-(1a α,2.β,2A.α,3β,6.β,6Aα,7β,7Aα) see...Dieldrin
Dimethazone see...Clomazone
Dimethenthoate see...Phenthoate
Dimethoate see...O-analog see...Omethoate
Dimethoate PO isologue see...Omethoate
Dimethoate oxon see...Omethoate
Dimethoate oxygen analog see...Omethoate
DIMETHOPGAN® see...Dimethoate
Dimethoxon see...Omethoate
Dimethoxy-DDT see...Methoxychlor
p,p'-Dimethoxydiphenyltrichloroethane see.Methoxychlor
Dimethoxy DT see...Methoxychlor
2,6-Dimethoxy-N-[3-(1-ethyl-1-methylpropyl)-5-isoxazolyl]benzamide see...Isoxaben

Di(*p*-methoxyphenyl)-trichloro methyl methane see.Methoxychlor
[(Dimethoxyphosphinothioyl)thio]butanedioic acid diethyl ester see. . .Malathion
3-[(Dimethoxyphosphinyl)oxy]-2-butenoic acid methyl ester see. . .Mevinphos
(E)-3-[(Dimethoxyphosphinyl)oxy]-2-butenoic acid 1-phenylethyl ester see. . .Crotoxyphos
3-Dimethoxyphosphinyloxy)-*N* see. . .*N*-dimethyl-*(E)*-crotonamide see. . .Dicrotophos
3-Dimethoxyphosphinyloxy)-*N,N*-dimethyl-*cis*-crotonamide see. . .Dicrotophos
3-Dimethoxyphosphinyloxy)-*N,N*-dimethylisocrotonamide see. . .Dicrotophos
3-Dimethoxyphosphinyloxy)*N*-methyl-*cis*-crotonamide see. . .Monocrotophos
2-[(((((4,6-Dimethoxy-2-pyrimidinyl)amino)carbonyl)amino)sulfonyl]-*N,N*-dimethyl-3-pyridinecarboxamide see. . .Nicosulfuron
2-[(((((4,6-Dimethoxy-2-pyrimidinyl)amino)carbonyl)amino)sulfonyl)methyl]benzoic acid, methyl ester see. . .Bensulfuron-methyl
2,3-Dimethoxystrichnidin-10-one see. . .Brucine
Dimethoxy strychnine see. . .Brucine
2,3-Dimethoxystrychnine see. . .Brucine
10,11-Dimethoxystrychnine see. . .Brucine
Dimethoxy-2,2,2-trichloro-1-hydroxy-ethylphosphine oxide see. . .Trichlorfon
3,3-Dimethyl-acrylate de 2,4-dinitro-6-(1-methylpropyle) phenyle (French) see. . .Binapacryl
Dimethylarsenic acid see. . .Cacodylic Acid
Dimethylarsinic acid see. . .Cacodylic Acid
Dimethylarsinic acid, sodium salt see. . .Sodium Cacodylate
3,3-Dimethylacrylic acid 2-*sec*-butyl-4,5-dinitrophenyl ester see. . .Binapacryl
*O,O,*-Dimethyl-*S*-(2-aethtyl-thio-aethyl)-monothio phosphat (German) see. . .Demeton-methyl
*N*-[5-(1,1-Dimethylaethyl)-1,3,4-thiadiazol-2-yl]-*N,N'*-dimethylharnstoff (German) see. . .Tebuthiuron
4-(Dimethylamine)-3,5-xylyln-methylcarbamate see.Mexacarbate
*p*-Dimethylaminobenzene diazo sodium sulfonate see.Fenaminosulf
*p*-(Dimethylamino)benzenediazosulfonate see.Fenaminosulf
*p*-(Dimethylamino)benzenediazosulphonate see.Fenaminosulf
4-Dimethylaminobenzenediazosulfonic acid see. .sodium salt see. . .Fenaminosulf
*p*-Dimethylaminobenzenediazosulfonic acid see. .sodium salt see. . .Fenaminosulf
4-Dimethylaminobenzenediazosulphonic acid see.sodium salt see. . .Fenaminosulf
*p*-(Dimethylamino)benzenediazosulphonic acid see.sodium salt see. . .Fenaminosulf
*p*-Dimethylaminobenzoldiazosulfonat (Natriumsalz) (German) see. . .Fenaminosulf

4,4'-Dimethylaminobenzophenonimide see. . .Auramine
*N*-Dimethyl amino-β-carbamyl propionic acid see.Daminozide
3-[(((Dimethylamino)carbonyl)amino)phenyl-1,1-dimethylethyl]carbamate see. . .Karbutilate
4-(Dimethylamino)-3,5-dimethylphenol methylcarbamate (ester) see. . .Mexacarbate
4-(Dimethylamino)-3,5-dimethylphenyl *N*-methylcarbamate see. . .Mexacarbate
2-Dimethylamino-5,6-dimethylpryimidin-4-yl *N,N*-dimethylcarbamate see. . .Pirimicarb
2-Dimethylamino-5,6-dimethyl-4-pyrimidinyl dimethylcarbamate see. . .Pirimicarb
2-(Dimethylamino)-5,6-dimethyl-4-pyrimidinyldimethylcarbamate see. . .Pirimicarb
3-(Dimethylamino)-1-methyl-3-oxo-1-propenyl dimethyl phosphate see. . .Dicrotophos
2-(Dimethylamino)-*N*[(((-methylamino)carbonyl)oxy]2-oxoethanimidothioic acid methyl ester see. . .Oxamyl
2-Dimethylamino-1-(-methylamino)glyoxal-*O*-methylcarbamoylmonoxime, Oxamyl
*m*-[((Di-methylamino)methylene)amino]phenylcarbamate, hydrochloride see. . .Formetanate Hydrochloride
*m*-[((Di-methylamino)methylene)amino]phenylcarbamate, hydrochloride see. . .Formetanate Hydrochloride
3-Dimethylaminomethyleneaminophenyl-*N*-methyl carbamate,hydrochloride see. . .Formetanate Hydrochloride
*p*-(Dimethylamino)-phenyldiazo-natriumsulfonat (German) see. . .Fenaminosulf
4-[(Dimethylamino)phenyl]diazenesulfonic acid, sodium salt see. . .Fenaminosulf
*N*-(γ-Dimethylaminopropyl)carbamic acid propyl ester , monohydrochloride see. . .Propamocarb Hydrochloride
17-β-[(3-(Dimethylamino)-propyl)methylamino]androst-5-en-3-β-ol dihydrochloride see. . .Azacosterol
5,6-Dimethyl-2-dimethylamino-4-pyrimidinyldimethylcarbamate see. . .Pirimicarb
*N*-(Dimethylamino)succinamic acid see. . .Daminozide
*N*-Dimethylamino-succinamidsaeure (German) DMASA see. . .Daminozide
*O*-[4-((Dimethylamino)sulfonyl)phenyl]*O,O*-dimethyl phosphorothioate see. . .Famphur
*O*-[4-((Dimethylamino)sulphonyl)phenyl]*O,O*-dimethyl thiophosphate see. . .Famphur
*S,S'*-[2-(Dimethylamino)trimethylene]bis(thiocarbamate) hydrochloride see. . .Cartap Hydrochloride
4-(Dimethylamino)-3,5-xylenol,methylcarbamate (ester) see. . .Mexacarbate
4-Dimethylamino-3,5-xylylmethylcarbamate see.Mexacarbate
4-(*N,N*-Dimethylamino)-3,5-xylyl *N*-methylcarbamate see. . .Mexacarbate
*N,N*-Dimethyl-*p*-anilinediazosulfonic acid sodium salt see. . .Fenaminosulf
Dimethylarsinic arsinic acid see. . .Cacodylic Acid
[(Dimethylarsino)oxy]sodium-As-oxide see. . .Sodium Cacodylate

[(Dimethylarsino)oxy]sodium-arsenic-oxide see. .Sodium Cacodylate
[(Dimethylarsino)oxy]sodium-arsenic-oxide see. .Sodium Cacodylate
[(Dimethylarsino)oxy]sodium-As-oxide see. . .Sodium Cacodylate
2,2-Dimethylbenzo-1,3-benzodioxol-4-yl-*N*-methylcarbamate see. . .Bendiocarb
2,2-Dimethylbenzo-1,3-dioxol-4-yl methylcarbamate see. Bendiocarb
2,2-Dimethyl-1,3-benzodioxol-4-yl-*N*-methylcarbamate see. . .Bendiocarb
3,5-Dimethylbenzoic acid 1-(1,1-dimethylethyl)-2-(4-ethylbenzoyl)hydrazine see. . .Tebufenozide
*O,O,*-Dimethyl-*S*-(1,2,3-bezotriazinyl-4-keto) methylphosphorodithioate see. . .Azinphos-methyl
*N,N'*-Dimethyl-4,4'-bipyridinium dichloride see.Paraquat
1,1'-Dimethyl-4,4'-bipyridynium dichloride see. .Paraquat
1,1'-Dimethyl-4,4'-bipyridyniumdimethylsulfate see. Paraquat Methosulfate
*O,O,*-Dimethyl-*O*-(4-bromo-2,5-dichlorophenyl) phosphorothioate see. . .Bromophos
*m*-(3,3-Dimethylureido)phenyl-*tert*-butylcarbamate see. Karbutilate
1,1-Dimethyl-3-[(3-*N*-*tert*-butylcarbamyloxy)-phenyl]urea see. . .Karbutilate
Dimethylcarbamate-d'l-isopropyl-3-methyl-5-pyrazoylle (French) see. . .Isolan®
Dimethyl carbamate ester of 3-hydroxy-*N,N*-5-trimethylpyrazole-1-carboxamide see. . .Dimetilan
Dimethylcarbamate, zinc salt see. . .Ziram
Dimethylcarbamic acid-1-[(-dimethylamino)carbonyl]-5-methyl-1*H*-pyrazol-3-yl ester see. . .Dimetilan
Dimethylcarbamic acid 2-(-dimethylamino)-5,6-dimethyl-4-pyrimidinyl ester see. . .Pirimicarb
Dimethylcarbamic acid ester with 3-hydroxy-*N,N*,5-trimethylpyrazole-1-carboxamide see. . .Dimetilan
Dimethylcarbamic acid-5-methyl-1*H*-carboxamine see.Dimetilan
Dimethylcarbamic acid 3-methyl-1-(1-methylethyl)-1*H*-pyrazol-5-yl ester see. . .Isolan®
Dimethylcarbamic acid-5-methyl-1*H*-pyrazol-3-yl ester see. . .Dimetilan
Dimethylcarbamo dithioic acid, iron complex see.Ferbam
Dimethylcarbamodithioic acid, iron(3+) salt see.Ferbam
Dimethylcarbamodithioic acid see. . .zinc complex see.Ziram
Dimethylcarbamodithioic acid, zinc salt see.Ziram
Dimethylcarbamodithiocarbamic acid, zinc salt see. .Ziram
(OC-6-11)-tris(Dimethylcarbamodithioato-*S,S'*)iron see.Ferbam
1-Dimethylcarbamoyl-5-methylpyrazol-3-yl dimethylcarbamate see. . .Dimetilan
2-Dimethylcarbamoyl-3-methylpyrazolyl-(5)-*N,N*-dimethylcarbamat see. . .Dimetilan
Dimethylcarbamoyl-3-methyl-5-pyrazolyldimethylcarbamate see. . .Dimetilan

*(E)*-2-Dimethylcarbamoyl-1-methylvinyl dimethylphosphate see. . .Dicrotophos
*cis*-2-Dimethylcarbamoyl-1-methylvinyl dimethylphosphate see. . .Dicrotophos
*O,O,*-Dimethyl-*S*-(1-carboethoxybenzyl) dithiophosphate see. . .Phenthoate
*O,O,*-Dimethyl-*O*-(2-carbomethoxy-1-methylvinyl) phosphate see. . .Mevinphos
Dimethyl-1-carbomethoxy-1-propen-2-yl phosphate see. . Mevinphos
*O,O,*-Dimethyl *O*-(2-chloro-2-(*N,N*-diethylcarbamoyl)-1-methylvinyl)phosphate see. . .Phosphamidon
3,5-Dimethyl-4-chlorophenol see. . .4-Chloro-3,5-xylenol
1,1-Dimethyl-3-(*p*-chlorophenyl)thiourea see. . .Monuron
1,1-Dimethyl-3-(*p*-chlorophenyl)urea see. . .Monuron
*N*-Dimethyl-*N'*-(4-chlorophenyl)urea see. . .Monuron
*N,N*-Dimethyl-*N'*-(4-chlorophenyl)urea see. . .Monuron
*O,O,*-Dimethyl-*O*-2-chloro-1-(2,4,5-trichlorophenyl)vinyl phosphate see. . .Tetrachlorvinphos
*O,O,*-Dimethyl-*O*-2-chlor-1-(2,4,5-trichlorophenyl)vinyl phosphat (German) see. . .Tetrachlorvinphos
2,2-Dimethyl-7-coumaranyl *N*-methylcarbamate see.Carbofuran
*O,O,*-Dimethyl-*O*-(4-cyano-phenyl)-monothiophosphat (German) see. . .Cyanofos
*O,O,*-Dimethyl-*O*-4-cyanophenyl-phosphorothioate see.Cyanofos
*O,O,*-Dimethyl-*O*-*p*-cyanophenyl-phosphorothioate see.Cyanofos
*O,O,*-Dimethyl-*O*-(1,2-dibromo-2,2-dichloroethyl)phosphate see. . .Naled
*O,O,*-Dimethyl *S*-(1,2-dicarbaethoxyaethyl)-dithiophosphat (German) see. . .Malathion
*O,O,*-Dimethyl *S*-(1,2-dicarbethoxyethyl) dithiophosphate see. . .Malathion
*O,O,*-Dimethyl *S*-(1,2-dicarbethoxyethyl)phosphorodithioate see. . .Malathion
*O,O,*-Dimethyl-*O*-(2,5-dichlor-4-bromphenyl)-thionophosphat (German) see. . .Bromophos
*O,O,*-Dimethyl-*O*-(2,5-dichloro-4-bromophenyl) thiophosphate see. . .Bromophos
*O,O,*-Dimethyl-*O*-(2,5-dichloro-4-bromophenyl)phosphorothioate see. . .Bromophos
*O,O,*-Dimethyl *O*-2,2-dichloro-1,2-dibromoethyl phosphate see. . .Naled
Dimethyl 2,2-dichloroethenyl phosphate see. .Dichlorvos
*O,O,*-Dimethyl *S*-(2,5-dichlorophenylthio)methyl phosphorodithioate see. . .Methyl Phenkapton
1,1-Dimethyl-3-(3,4-dichlorophenyl)urea
Dimethyl dichlorovinyl phosphate see. . .Dichlorvos
Dimethyl 2,2-dichlorovinyl phosphate see. . .Dichlorvos
*O,O,*-Dimethyl *S*-1,2-di(ethoxycarbamyl)ethyl phosphorodithioate see. . .Malathion
Dimethyl diethylamido-1-chlorocrotonyl (2)phosphate see. . .Phosphamidon
*O,O,*-Dimethyl-*O*-(2-(-diethylamino)-6-methyl-4-pyrimidinyl) see. . .Pirimiphos-Methyl

*O,O,*-Dimethyl *O*-(2-(-diethylamino)-6-methyl-4-pyrimidinyl)phosphorothioate see. . .Pirimiphos-Methyl
2,2-Dimethyl-2,3-dihydro-7-benzofuranyl-*N*-methylcarbamate see. . .Carbofuran
2,2-Dimethyl-2,2-dihydrobenzofuranyl-7 *N*-methylcarbamate see. . .Carbofuran
*O,O,*-Dimethyl-*S*-(3,4-dihydro-4-keto-1,2,3-bezotriazinyl-3-methyl) dithiophosphate see. . .Azinphos-methyl
*O,O,*-Dimethyl *S*-1,2-dikarbetoxyethylditiofosfat (Czech) see. . .Malathion
*O,O,*-Dimethyl-*O*-(2-dimethyl-carbamoyl-1-methyl-vinyl)phosphat (German) see. . .Dicrotophos
*O,O,*-Dimethylo-(*N,N*-dimethylcarbamoyl-1-methylvinyl) phosphate see. . .Dicrotophos
*O,O,*-Dimethyl-*O*-(1,4-dimethyl-3-oxo-4-aza-pent-1-enyl)phosphate see. . .Dicrotophos
*O,O,*-Dimethyl-*O*-(1,4-dimethyl-3-oxo-4-aza-pent-1-enyl)fosfaat (Dutch) see. . .Dicrotophos
*O,O,*-Dimethyl *O*-[*p*-(-dimethylsulfamoyl)phenyl]phosphorothioate see. . .Famphur
*O,O,*-Dimethyl *O*-[*p*-(*N,N*-dimethylsulfamoyl)phenyl]phosphorothioate see. . .Famphur
3,4-Dimethyl-2,6-dinitro-*N*-(1-ethylpropyl)aniline see. . .Pendimethalin
*N,N*-Dimethyldiphenylacetamide see. . .Diphenamid
*N,N*-Dimethyl-α,α-diphenylacetamide see. . .Diphenamid
*N,N*-Dimethyl-2,2-diphenylacetamide see. . .Diphenamid
1,2-Dimethyl-3,5-diphenyl-1*H*-pyrazolium methyl sulfate see. . .Difenzoquat
1,1-Dimethyl-4,4-dipyridilium dichloride see. . .Paraquat
4,4'-Dimethyldipyridyl dichloride see. . .Paraquat
1,1'-Dimethyl-4,4'-dipyridylium chloride see. . .Paraquat
1,1'-Dimethyl-4,4'-dipyridylium dichloride see. . .Paraquat
1,1'-Dimethyl-4,4'-dipyridynium di(-methyl sulfate) see. . .Paraquat Methosulfate
*O,O,*-Dimethyl dithiobis(thioformate) see. . .Tributyltin Chloride
*N,N*-Dimethyldithiocarbamate, sodium salt see.Sodium Dimethyldithiocarbamate
Dimethyldithiocarbamate sodium salt see. . .Sodium Dimethyldithiocarbamate
Dimethyldithiocarbamic acid, iron salt see.Ferbam
Dimethyldithiocarbamic acid, iron(3+) salt see.Ferbam
*N,N*-Dimethyldithiocarbamic acid, sodium salt see. Sodium Dimethyldithiocarbamate
Dimethyldithiocarbamic acid, sodium salt see.Sodium Dimethyldithiocarbamate
*O,O,*-Dimethyldithiophosphate diethylmercaptosuccinate see. . .Malathion
*O,O,*-Dimethyl dithiophosphate of diethyl mercaptosuccinate see. . .Malathion
Dimethyldithiophosphoric acid *N*-methylbenzazimide ester see. . .Azinphos-methyl
*O,O,*-Dimethyldithiophosphorylacetic acid-*N*-methyl-*N*-formylamide see. . .Formothion
(*O,O*-Dimethyldithiophosphorylphenyl)acetic acid ethyl ester see. . .Phenthoate
Dimethylene oxide see. . .Ethylene Oxide
*O,O,*-Dimethyl-*S*-α-ethoxy-carbonylbenzyl phosphorodithioate see. . .Phenthoate
*O,O,*-Dimethyl *S*-α-(ethoxycarbonyl)benzyl phosphorothiolothionate see. . .Phenthoate
*O,O,*-Dimethyl-*S*-(2-(eththio)ethyl)phosphorthioate see. Demeton-methyl
*O,O,*-Dimethyl-*S*-(2-eththioethyl)phosphorothioate see. Demeton-methyl
Dimethyl-*S*-(2-eththioethyl)thiophosphate see. .Demeton-methyl
*O,O,*-Dimethyl *S*-(2-eththionylethyl) phosphorothioate see. . .Demeton-methyl
Dimethyl *S*-(2-eththionylethyl) thiophosphate see. Demeton-methyl
2-(1,1-Dimethylethyl)-4,6-dinitrophenol see. . .Dinoterb
*O,O,*-Dimethyl-*S*-ethylmercaptoethyl thiophosphate see. Demeton-methyl
*O,O,*-Dimethyl-*S*-ethylmercaptoethyl thiophosphate, thiolo isomer see. . .Demeton-methyl
Dimethyl-ethyl-*n*-(1-methylpropyl)-2,6-dinitrobenzeneamine[4-(1,1-)] see. . .Butralin
4-(1,1-Dimethylethyl)-*N*-(1-methylpropyl)-2,6-dinitrobenzenamine see. . .Butralin
4-(1,1-Dimethylethyl)phenol see. . .Butylphenols
2-[4-(1,1-Dimethylethyl)phenoxy]cyclohexyl 2-propynyl sulfurous acid see. . .Propargite
2-[4-(1,1-Dimethylethyl)phenoxy]cyclohexyl 2-propynyl sulfite see. . .Propargite
*O,O,*-Dimethyl *S*-[2-(-ethylsulfinyl)ethyl]monothiophosphate see. . .Demeton-methyl
*O,O,*-Dimethyl *S*-2-(-ethyl sulfinylethyl)phosphorothioate see. . .Demeton-methyl
*O,O,*-Dimethyl *S*-[2-(-ethylsulfinyl)ethyl]phosphorothioate see. . .Demeton-methyl
*O,O,*-Dimethyl *S*-[2-(-ethylsulfinyl)ethyl]thiophosphate see. . .Demeton-methyl
*O,O,*-Dimethyl *S*-ethylsulphinylethyl phosphorothiolate see. . .Demeton-methyl
*O,O,*-Dimethyl *S*-(2-ethylsulfinyl)ethyl thiophosphate see. . .Demeton-methyl
*N*-[5-(1,1-Dimethylethyl)-1,3,4-thiadiazol-2-yl]-*N,N*'-dimethylurea see. . .Tebuthiuron
*S*-[((1,1-Dimethylethyl)thio)methyl]-*O,O*-diethylphosphorodithioate see. . .Terbufos
Dimethylformamidine *N*'-(4-chlor-*o*-tolyl)-*N,N*-dimethylformamidin (German) see. . .Chlordimeform
Dimethylformocarbothialdine see. . .Dazomet
*O,O,*-Dimethyl-*S*-(*N*-formyl-*N*-methyl carbamoylmethyl)phosphorodithioate see. . .Formothion
*m*-(3,3-Dimethylharnstoff)-phenyl-*tert*-butylcarbamat (German) see. . .Karbutilate
*O,O,*-Dimethyl-*S*-[(5-methoxy-pyron-2-yl)-methyl]-thiolphosphat (German) see. . .Endothion

*O,O*,-Dimethyl-*S*-(5-methoxypyronyl-2-methyl)thiolphosphate see...Endothion

*O,O*,-Dimethyl-*S*-[(2-methoxy-1,3,4(4H)-thiadiazol-5-on-4-yl)-methyl]dithiofosfaat (Dutch) see...Methidathion

*O,O*,-Dimethyl)-*S*-(2-methoxy-1,3,4-thiadiazole-5(4H)-onyl-(4)-methyl)-phosphorodithioate see...Methidathion

*O,O*,-Dimethyl)-*S*-(2-methoxy-1,4,4-thiadiazole-5-(4H)-onyl-(4)-methyl)-dithiophosphat (German) see. Methidathion

N,N-Dimethyl-*N'*-(4-methoxy-3-chlorophenyl)urea see. Metoxuron

*O,O*,-Dimethyl-*S*-(5-methoxy-4-oxo-4H-pyran-2-yl)phosphorothioate see...Endothion

Dimethyl methoxycarbonylpropenyl phosphate see. Mevinphos

2,2-Dimethyl-4-(*N*-methyl aminocarboxylato)- see. Bendiocarb

2,2-Dimethyl-4-(*N*-methyl aminocarboxylato)-1,3-benxodioxole see...Bendiocarb

*N,N*-Dimethyl-*N'*-[((methylamino)carbonyl)oxy]phenylmethanimideamidemonohydrochloride see...Formetanate Hydrochloride

*(E)*-Dimethyl 1-methyl-3-(methylamino)-3-oxo-1-propenyl phosphate see...Monocrotophos

*O,O*,-Dimethyl *S*-(*N*-methylcarbamoylmethyl) dithio phosphate see...Dimethoate

*O,O*,-Dimethyl-*S*-(*N*-methyl-carbamoyl)-methyl-monothio phosphat (German) see...Omethoate

*O,O*,-Dimethyl-*S*-[(methyl carbamoyl)methyl] phosphorothioate see...Omethoate

Dimethyl-*S*-(*N*-methyl-carbamoyl-methyl) phosphorothiolate see...Omethoate

*O,O*,-Dimethyl-*O*-(2-*N*-methyl carbamoyl-1-methyl-vinyl) phosphate see...Monocrotophos

*O,O*,-Dimethyl-*O*-(2-*N*-methyl carbamoyl-1-methyl)-vinyl-phosphat (German) see...Monocrotophos

*O,O*,-Dimethyl-*O*-(2-*N*-methyl carbamoyl-1-methyl-vinyl)-fosfaat (Dutch) see...Monocrotophos

*O,O*,-Dimethyl-*S*-(*N*-methyl carbamoylmethyl) thiophosphate see...Omethoate

*O,O*,-Dimethyl-*S*-(*N*-methyl carbamoylmethyl) phosphorothiolate see...Omethoate

*O,O*,-Dimethyl-*S*-(*N*-methyl carbamoylmethyl)phosphorothioate see...Omethoate

*N,N*-Dimethyl-α-methyl carbamoyloxyimino-α-(methyl thio)acetamide see...Oxamyl

*N,N*-Dimethyl-*N*-[(methylcarbamoyl)oxy]-1-thiooxamimidic acid methyl ester see...Oxamyl

*O,O*,-Dimethyl *S*-[(methylcarbamoyl)methyl]phosphorothioate see. Omethoate

*O,O*,-Dimethyl-*O*-(1-methyl-2-carboxy-α-phenylethyl)vinyl phosphate see...Crotoxyphos

*O,O*,-Dimethyl *O*-(1-methyl-2-carboxyvinyl) phosphate see...Mevinphos

*O,O*,-Dimethyl-*O*-(1-methyl-2-chlor-2-*N,N*-diethyl-carbamoyl)-vinyl-phosphat (German) see.Phosphamidon

(*O,O*-Dimethyl-*O*-(1-methyl-2-chloro-2-diethylcarbamoyl-vinyl)-phosphate see...Phosphamidon

N,N-Dimethyl-*N'*-(2-methyl-4-chlorophenyl)formamidine see...Chlordimeform

N,N-Dimethyl-*N'*-(2-methyl-4-chlorphenyl)-formadin (German) see...Chlordimeform

*O,O*,-Dimethyl-*S*-(3-methyl-2,4-dioxo-3-aza-butyl)-dithiophosphat (German) see...Formothion

*O,O*,-Dimethyl-*S*-(3-methyl-2,4-dioxo-3-azabutyl)-dithiofosfaat (Dutch) see...Formothion

*O,O*,-Dimethyl-*S*-(*N*-methyl-*N*-formyl-carbamoylmethyl)-dithiophosphate see...Formothion

*O,O*,-Dimethyl-*S*-(*N*-methyl-*N*-formyl-carbamoylmethyl)-dithiophosphat (German) see...Formothion

*O,O*,-Dimethyl-*S*-(*N*-methyl-*N*-formyl-carbamoylmethyl)-phosphorodithioate see...Formothion

3,5-Dimethyl-4-methyl mercaptophenyl-*N*-methyl-carbamate see...Methiocarb

*O,O*,-Dimethyl-*O*-(3-methyl-4-nitrofenyl)-monothiofosfaat (Dutch) see...Fenitrothion

*O,O*,-Dimethyl-*O*-(3-methyl-4-nitrophenyl)-monothiophosphat (German) see...Fenitrothion

*O,O*,-Dimethyl-*O*-(3-methyl-4-nitrophenyl)-phosphorothioate see...Fenitrothion

*O,O*,-Dimethyl-*O*-(3-methyl-4-nitrophenyl)-thiophosphate see...Fenitrothion

Dimethyl-*cis*-1-methyl-2-(1-phenylethoxycarbonyl)vinyl phosphate see...Crotoxyphos

*O,O*,-Dimethyl-*O*-(3-methyl)phosphorothioate see. Fenitrothion

2,2-Dimethyl-3-(2-methyl-1-propenyl)cyclopropanecarboxylic acid (3-phenoxyphenyl)methyl ester see...D-Phenothrin

2,2-Dimethyl-3-(2-methyl-1-propenyl)cyclopropane carboxylic acid see...Resmethrin

2,2-Dimethyl-3-(2-methyl-1-propenyl)cyclopropane carboxylic acid (1,3,4,5,6,7-hexahydro-1,3-dioxo-2*H*-isoindol-2-yl)methyl ester see.Tetramethrin

3,3-Dimethyl-1-(methyl thio)-2-butanone-*O*-[(methyl amino)carbonyl]oxime see...Thiofanox

3,5-Dimethyl-4-(methyl thio)phenol methyl carbamate see...Methiocarb

3,5-Dimethyl-4-(methyl thio)phenyl methyl carbamate see...Methiocarb

3,5-Dimethyl -4-methyl thiophenyl *N*-methyl carbamate see...Methiocarb

*O,O*,-Dimethyl *O*-(4-nitrol fenyl)-monothiofosfaat (Dutch) see...Methyl Parathion

*O,O*,-Dimethyl *O*-(4-nitrol phenyl)-monothiophosphat (German) see...Methyl Parathion

*O,O*,-Dimethyl *O*-*p*-nitrol phenyl phosphorothioate see. Methyl Parathion

*O,O*,-Dimethyl *O*-(*p*-nitrol phenyl) phosphorothioate see. Methyl Parathion

*O,O*,-Dimethyl *O*-4-nitrol phenyl phosphorothioate see. Methyl Parathion

*O,O*,-Dimethyl *O*-(4-nitrol phenyl)phosphorothioate see. Methyl Parathion

*O,O*,-Dimethyl *O-p*-nitrol fenylester kyseliny thiofosforecne (Czech) see...Methyl Parathion
*O,O*,-Dimethyl *O*-(*p*-nitrol phenyl) thionophosphate see..Methyl Parathion
*O,O*,-Dimethyl *O-p*-nitrol phenyl thiophosphate see..Methyl Parathion
*O,O*,-Dimethyl *O*-(*p*-nitrol phenyl) thiophosphate see. Methyl Parathion
*O,O*,-Dimethyl-*O*-(4-nitro-3-methylphenyl)thiophosphate see...Fenitrothion
Dimethyl-*p*-nitrophenyl monothiophosphate see..Methyl Parathion
Dimethyl 4-nitrol phenyl phosphorothionate see..Methyl Parathion
*O,O*,-Dimethyl-*O*-4-nitro-*m*-toylphosphorothioate see. Fenitrothion
*O,O*,-Dimethyl *O*-(3,5,6-trichloro-2-pyridinyl)phosphorothioate see...Chlorpyrifos
*O,O*,-Dimethyl O-(3,5,6-trichloro-2-pyridyl)phosophorothioate see...Chlorpyrifos-methyl
*O,O*-Dimethyl-*S*-(2-*oxo*-3-azabutyl)-monothiophosphate see...Omethoate
*O,O*,-Dimethyl-*S*-(4-*oxo*-benzotriazino-3-methyl) phosphorodithioate see...Azinphos-methyl
*O,O*,-Dimethyl-*S*-(4-oxo-1,2,3-benzotriazino(3)-methyl) thiophosphorodithioate see...Azinphos-methyl
*O,O*,-Dimethyl-*S*-(4-oxo-1,2,3-bezotriazin-3(4H)-yl methyl)phosphorodithioate see...Azinphos-methyl
*O,O*,-Dimethyl-*S*-oxo-1,2,3-benzotriazin-3-(4H)-yl-methyl) phosphodithioate see...Azinphos-methyl
*O,O*,-Dimethyl-*S*-((4-oxo-3H-1,2,3-benzotriazin-3-yl)-methyl)dithiophosphat (German) see...Azinphos-methyl
*O,O*,-Dimethyl-*S*-(4-oxo-3H-1,2,3-benzotriazin-3-yl)-methyl)dithiofosfaat (Dutch) see...Azinphos-methyl
*O,O*,-Dimethyl-*S*-(4-oxo-3H-1,2,3-benzotriazine-3-methyl) phosphorodithioate see...Azinphos-methyl
3[2-(3,5-Dimethyl-2-oxocyclohexyl)-2-hydroxyethyl]glutarimide see...Cycloheximide
Dimethyl *p*-nitrol phenyl thiophosphate see...Methyl Parathion
Dimethyl *p*-nitrol phenyl monothiophosphate see.Methyl Parathion
Dimethyl *p*-nitrol phenyl phosphorothionate see..Methyl Parathion
Dimethyl parathion see...Methyl Parathion
3,5-Dimethylperhydro-1,3,5-thiadiazin-2-thion (Czech, German) see...Dazomet
5-Dimethylphenol methylcarbamate ester see. Mexacarbate
4-[(((((1,3-Dimethyl-5-phenoxy-1*H*-pyrazol-4-yl)methylene)amino)oxy)methyl]benzoic acid see. Fenpyroximate
*O,O*,-Dimethyl-*S*-(phenylacetic acid ethyl ester) phosphorodithioate see...Phenthoate
*N,N*-Dimethyl-α-phenylbenzeneacetamide see. Diphenamid

*N'*-(2,4-Dimethylphenyl)-*N*-(((2,4-dimethylphenyl)imino)methyl)-*N*-methylmethanimidamide see...Amitraz
*N*-(2,6-Dimethylphenyl)-*N*-(-methoxyacetyl)alanine, methyl ester see...Metalaxyl
*N*-(2,6-Dimethylphenyl)-*N*-(-methoxyacetyl)-*dl*-alanine methyl ester see...Metalaxyl
*N'*-(2,4-Dimethylphenyl)-3-methyl-1,3,5-triazapenta-1,4-diene see...Amitraz
4,6-Dimethyl-*N*-phenyl-2-pyrimidinamine see. Pyrimethanil
(Dimethyl-*S*-(phenylethoxycarbonylmethyl)phosphorothiolothionate) see...Phenthoate
Dimethyl phosphate of 2-chloro-*N,N*-diethyl-3-hydroxycrotonamide see...Phosphamidon
Dimethyl phosphate ester with 2-chloro-*N*,N-diethyl-3-hydroxycrotonamide see...Phosphamidon
Dimethyl phosphate ester of 3-hydroxy-*N*-methyl-*cis*-crotonamide see...Monocrotophos
Dimethyl phosphate of 3-hydroxy-*N*-methyl-*cis*-crotonamine see...Monocrotophos
Dimethyl phosphate ester with 3-hydroxy-*N,N*-dimethyl-*cis*-crotonamide see...Dicrotophos
Dimethyl phosphate ester of α-methylbenzyl-3-hydroxy-*cis*-crotonate see...Crotoxyphos
Dimethyl phosphate of methyl 3-hydroxy-*cis*-crotonate see...Mevinphos
*O,S*-Dimethylphosphoramidothioate see.Methamidophos
*O,O*,-Dimethylphosphorodithioate *N*-formyl-2-mercapto-*N*-methylacetamide-*S*-ester see...Formothion
*O,O*,-Dimethyl-*S*-(phenyl)(carboethoxy)methyl phosphorodithioate see...Phenthoate
*O,O*,-Dimethyl phosphorothioate *O*-ester with *p*-hydroxy-N,N-dimethylbenzenesulfonamide see...Famphur
*O,O*,-Dimethyl S-(*N*-phthalimidomethyl)dithiophosphate (*O,O*-Dimethyl-phthalimidiomethyl-dithiophosphate)
*O,O*,-Dimethyl S-phthalimidomethylphosphorodithioate
*N,N*-Dimethylpiperidinium chloride see...Mepiquat Chloride
*N*-(1,1-Dimethylpropynyl)-3,5-dichlorobenzamide see.Pronamide
10,11-Dimethylstrychnine see...Brucine
*O*-4-Dimethylsulfam oylphenyl *O,O*-dimethyl phosphorothioate see...Famphur
1,4-Dimethylsulfonoxybutane see...Busulfan
*O*-4-Dimethylsulpha moylphenyl *O,O*-dimethylphosphorothioate see...Famphur
2,2-Dimethyl-3-(1,2,2,2-tetrabromoethyl)cyclopropanecarboxylate see. Tralomethrin
2,2-Dimethyl-3-(1,2,2,2-tetrabromoethyl)cyclopropanecarboxlic acid, cyano(3-phenoxyphenyl)methyl ester see..Tralomethrin
Dimethyl tetrachloroterephthalate see...DCPA
3,5-Dimethyltetrahydro-1,3,5-thiadiazine-2-thione see. Dazomet

3,5-Dimethyl-1,2,3,5-tetrahydro-1,3,5-thiadiazinethione-2 see...Dazomet
3,5-Dimethyltetrahydro-1,3,5-2*H*-thiadiazine-2-thione see...Dazomet
*O,O,*-Dimethyl-*S*-(3-thia-pentyl)-monothiophosphat (German) see...Demeton-methyl
3,5-Dimethyl-1,3,5-thiadiazinane-2-thione see...Dazomet
Dimethyl-*N,N'*-[thiobis(((methylimino)carbonyl)oxy)]bis(ethanimidothioate) see...Thiodicarb
Dimethyl *N,N'*-[thiobis((methylimino)carbonyloxy)]bis(thioimidoacetate) see...Thiodicarb
Dimethyl *N,N'*-[thiobis((methylimino)carbonyloxy)]bis(ethanimidothioate) see...Thiodicarb
3,5-Dimethyl-2-thionotetrahydro-1,3,5-thiadiazine see. Dazomet
Dimethyl trichlorophenyl thiophosphate see...Ronnel
*O,O,*-Dimethyl-*O*-2,4,5-trichlorophenyl phosphorothioate see...Ronnel
*O,O,*-Dimethyl-*O*-(2,4,5-trichlorophenyl)thiophosphate see...Ronnel
*O,O,*-Dimethyl-*O*-(2,4,5-trichlorphenyl)-thionophosphat (German) see...Ronnel
2',4'-Dimethyl-5-[(trifluoromethyl)sulfonamido]acetanilide see.Mefluidide
*N*-(2,4-Dimethyl-5-[((trifluoromethyl)sulfonyl)amino)phenyl]acetamide see.Mefluidide
1,1-Dimethyl-3-(3-trifluoromethylphenyl)urea see.Fluometuron
Dimethyl violgen chloride see...Paraquat
Dimethyl viologen chloride see...Paraquat
*O,O,*-Dimetil-(2,2,2-tricloro-1-idrossi-etil)-fosfonato (Italian) see...Trichlorfon
*O,O,*-Dimetil-*O*-(1,4-dimetil-3-oxo-4-aza-pent-1-enil)-fosfato (Italian) see...Dicrotophos
*O,O,*-Dimetil-*O*-(2-*N*-metilcarbamoil-1-metil-vinil)-fosfato (Italian) see...Monocrotophos
*O,O,*-Dimetil-*O*-(4-nitro-fenil)-monotiofosfato (Italian) see...Methyl Parathion
*O,O,*-Dimetil-*S*-((2-metossoi-1,3,4(4H)-thiadiazaol-5-on-4-il)-metil)-ditifosfato (Italian) see...Methidathion
*O,O,*-Dimetil-*S*-((4-oxo-3H-1,2,3-benzotriazin-3-il-metil)-ditiofosfato (Italian) see...Azinphos-methyl
Dimetilane see...Dimetilan
Dimetilcarbamato de 1-isopropil-3-metil-5-pirazolilo (Spanish) see...Isolan®
Dimetilditiocarbamato sodico (Spanish) see...Sodium Dimethyldithiocarbamate
DIMET® see...Dimethoate
DIMETOX® see...Trichlorfon
*m*-(3,3-Dimetylureido)phenyl-*t*-butylcarbamate see. Karbutilate
Dimexan see...Tributyltin Chloride
Dimexano see...Tributyltin Chloride
Dimid see...Diphenamid
DIMILIN® see...Diflubenzuron
DIMONEX® see...Phosphamidon
DIMPYLATE® see...Diazinon

DINAPACRYL® see...Binapacryl
Diniconazole M see...Diniconazole
Dinitrall see...Dinoseb
DINITRALL® see...Dinoseb
4,6-Dinitro-2-*sec*-butylfenol (Czech) see...Dinoseb
2,4-Dinitro-6-*tert*-butylphenol see...Dinoterb
2,4-Dinitro-6-*sec*-butylphenol see...Dinoseb
2,4-Dinitro-6-*sec*-butylphenol see...Dinoseb
4,6-Dinitro-*o-sec*-butylphenol see...Dinoseb
4,6-Dinitro-*o-sec*-butylphenol see...Dinoseb
Dinitro-*ortho-sec*-butylphenol see...Dinoseb
Dinitrobutylphenol see...Dinoseb
4,6-Dinitro-2-*sec*-butylphenol see...Dinoseb
4,6-Dinitro-2-*sec*-butylphenol see...Dinoseb
4,6-Dinitro-2-*sec*-butylphenyl β,β-dimethylacrylate see. Binapacryl
2,4-Dinitro-6-*sec*-butylphenyl-2-methylcrotonate see. Binapacryl
4,6-Dinitro-2-(2-capryl)phenyl crotonate see...Dinocap
4,6-Dinitro-2-caprylphenyl crotonate see...Dinocap
Dinitrocaprylphenyl crotonate see...Dinocap
Dinitrocresol see...Dinitro-o-cresol (DNOC)
3,5-Dinitro-*o*-cresol see...Dinitro-o-cresol (DNOC)
4,6-Dinitro-o-cresol and salts see...Dinitro-o-cresol (DNOC)
2,4-Dinitro-6-cyclohexylphenol see...Dinex
4,6-Dinitro-*o*-cyclohexylphenol see...Dinex
Dinitrocyclohexylphenol see...Dinex
4,6-Dinitro-*o*-cyclohexylphenol see...Dinex
Dinitro-*o*-cyclohexylphenol see...Dinex
Dinitrocyclophenol see...Dinex
Dinitrodendtroxal see...Dinitro-o-cresol (DNOC)
2,6-Dinitro-*N,N*-di-*N*-propyl-α,α,α-trifluro-*p*-toluidine see...Trifluralin
DINITRO® see...Dinoseb
Dinitrol see...Dinitro-o-cresol (DNOC)
2,6-Dinitro-*N,N*-dipropylcumidene see...Isopropalin
3,5-Dinitro-*N⁴,N⁴*-dipropylsulfanilamide see...Oryzalin
3,5-Dinitro-*N⁴,N⁴*-dipropylsulphanilamide see...Oryzalin
2,6-Dinitro-*N,N*-dipropyl-4-(trifluoromethyl)aniline see. Trifluralin
2,6-Dinitro-*N,N*-dipropyl-4-(trifluoromethyl)benzenamine see...Trifluralin
2,4-Dinitro-*N³,N³*-dipropyl-6-(trifluoromethyl)-1,3-benzenediamine see...Prodiamine
2,5-Dinitro-*N*-(1-ethylpropyl)-3,4-xylidine see. Pendimethalin
3,5-Dinitro-2-hydroxytoluene see...Dinitro-o-cresol (DNOC)
Dinitromethylheptyphenyl crotonate see...Dinocap
Dinitro(1-methylheptyl)phenyl crotonate see...Dinocap
2,4-Dinitro-6-(1-methylheptyl)phenyl crotonate see. Dinocap
4,6-Dinitro-2-(1-methylheptyl)phenyl crotonate see.Dinocap
2,4-Dinitro-6-methylphenol see...Dinitro-o-cresol (DNOC)

4,6-Dinitro-2-methylphenol see...Dinitro-o-cresol (DNOC)
2,4-Dinitro-6-(1-methylpropyl)phenol see...Dinoseb
4,6-Dinitro-2-(1-methyl-propyl)phenol see...Dinoseb
4,6-Dinitro-2-(1-methyl-N-propyl)phenol see...Dinoseb
2,6-Dinitro-4-octyl-phenyl crotonate see...Dinocap
2,4-Dinitro-6-octyl-phenyl crotonate see...Dinocap
2,4-Dinitro-6-(2-octyl)phenyl crotonate see...Dinocap
2,4-Dinitro-6-octyl* phenyl crotonate see...2,6-dinitro-4-octyl* phenylcrotonate see...and nitrooctylphenols (principally dinitro) see...Dinocap
4,6-Dinitrophenyl-2-sec-butyl-3-methyl-2-butenonate see...Binapacryl
Dinitroterb see...Dinoterb
2,6-Dinitro-4-trifluormethyl-N,N-dipropylanilin (German) see...Trifluralin
DINOC® see...Dinitro-o-cresol (DNOC)
DINOCIDE® see...DDT
Dinoseb methacrylate see...Binapacryl
DINOXOL® see...2,4-D
DINOXOL® see...2,4,5-T
DINOXOL® see...2,4,5-Trichlorophenoxyacetic Acid, Esters
DINURANIA® see...Dinitro-o-cresol (DNOC)
DIOLICE® see...Coumaphos
DI-ON® see...Diuron
1,4-Dioxan-2,3-diyl S,S-di(O,O-diethyl phosphorodithioate) see...Dioxathion
2,3-Dioxanedithiol S,S-bis(O,O-diethylphosphorodithioate) see...Dioxathion
S,S'-para-Dioxane-2,3-diyl bis(O,O-diethylphosphorodithioate) see...Dioxathion
S,S'-1,4-Dioxane-2,3-diyl bis(O,O-diethyl phosphorodithioate) see...Dioxathion
S,S'-(1,4-Dioxane-2,3-diyl) O,O,O',O'-tetraethylbis(phosphorodithioate) see...Dioxathion
9,10-Dioxoanthracene see...Anthraquinone
3,5-Dioxo-4-(1-oxopropyl)cyclohexanecarboxylic acid ion(1-) calcium calcium salt see...Prohexadione Calcium
1,3-Dioxo-2-pivaloy-lindane see...Pindone
DIPAXIN® see...Diphacione
Dipazin see...Diphacione
Dipentene see...D-Limonene
DIPF see...Isofluorphate
Diphacin (Italy and Turkey) see...Diphacione
Diphacinon see...Diphacione
Diphenacin see...Diphacione
Diphenadion see...Diphacione
Diphenadione see...Diphacione
Diphenamide see...Diphenamid
DIPHENTANE 70® see...Dichlorophene
DIPHENTHANE 70® see...Dichlorophene
Diphenyl see...Biphenyl
1,1'-Diphenyl see...Biphenyl
2-Diphenylacetyl-1,3-diketohydrindene see..Diphacione
2-(Diphenylacetyl)indan-1,3-indandione see..Diphacione
2-(3,Diphenylacetyl)-1H-indene-1,3(2h)-dione see...Diphacione

Diphenylamide see...Diphenamid
2,2-Diphenyl-N,N-dimethylacetamide see...Diphenamid
Diphenylene oxide see...Dibenzofuran
o-Diphenylol see...o-Phenylphenol
Diphenyltrichloroethane see...DDT
DIPHER® see...Zineb
Diphosphoramide, octamethyl- see...Octamethyl Diphosphoramide
Diphosphoric acid, tetraethyl ester see...TEPP
DIPHOS® see...Temephos
DIP'N GROW® see...1-Naphthaleneacetic Acid
DIPOFENE® see...Diazinon
Dipram see...Propanil
Dipropetryne see...Dipropetryn
Dipropetryn [2-(ethylthio)-4,6-bis(isopropylamino)-S-triazine] see...Dipropetryn
4-(Dipropylamino)-3,5-dinitrobenzenesulfonamide see. Oryzalin
4-(Di-N-propylamino)-3,5-dinitro-1-trifluoromethylbenzene see...Trifluralin
Dipropylcarbamothioic acid S-ethyl ester see...EPTC
N,N-Di-N-propyl-2,6-dinitro-4-trifluoromethylaniline see. Trifluralin
$N^3,N^3$-Dipropyl-2,4-dinitro-6-(trifluoromethyl)-m-phenylenediamine see...Prodiamine
Dipropylene glycol see...Isofenphos
Di-propylisocinchomeronate see...Dipropyl Isocinchomeronate
Dipropyl isocincnomeronate see...Dipropyl Isocinchomeronate
Di-N-propyl isocinchomeronate see...Dipropyl Isocinchomeronate
Dipropyl 2,5-pyridinedicarboxylate see...Dipropyl Isocinchomeronate
Di-N-propyl 2,5-pyridinedicarboxylate see...Dipropyl Isocinchomeronate
Dipropyl pyridine-2,5-dicarboxylate see...Dipropyl Isocinchomeronate
N,N-Dipropylthiocarbamic acid S-ethyl ester see...EPTC
Dipropylthiocarbamic acid-S-propyl ester see.Vernolate
N,N-Dipropyl-4-trifluoromethyl-2,6-dinitroaniline see. Trifluralin
DIPTEREX® see...Trichlorfon
DIPTEREX® 50 see...Trichlorfon
DIPTEVU® see...Trichlorfon
DIPTHAL® see...Triallate
Dipyrido(1,2-a:2',1'-c)pyrazinediium, 6,7-dihydro-, dibromide see...Diquat
Dipyrido(1,2-A:2',1'-C)pyrazinediium, 6,7-dihydro-, dibromide see...Diquat Dibromide
o-Diquat see...Diquat
Diquat bromide see...Diquat Dibromide
DIQUAT WEED KILLER® see...Diquat Dibromide
DIRAX® see...ANTU
DIREX® see...Diuron
DIREZ® see...Anilazine
DIRIMAL® see...Oryzalin
DISAN® see...Bensulide

DISCIPLINE® see...Bifenthrin
DIE-SECTICIDE® see...Diatomaceous Earth
DISETIL® see...Disulfiram
Disodium arsenate heptahydrate see...Sodium Arsenite
Disodium dichromate(VI) see...Sodium Dichromate
Disodium dichromium heptaoxide see...Sodium Dichromate
Disodium difluoride see...Sodium Fluoride
Disodium dihydrogen pyrophosphate see...Trisodium Phosphate
Disodium ethylenebis(dithiocarbamate) see...Nabam
Disodium methane arsonate see...Sodium Methanearsonate (MSMA)
Disodium methyl arsonate see...Sodium Methanearsonate (MSMA)
Disodium tetraborate see...Borax and Boric Acid
Disodium tetraborate decahydrate see...Sodium Tetraborate
DISSULFAN CE® see...Endosulfan
DISTANCE® see...Pyriproxyfen
Distannoxane see...hexakis($\beta,\beta$-dimethylphenethyl)- see...Fenbutatin Oxide
Distannoxane see...hexakis(2-methyl-2-phenylpropyl)- see...Fenbutatin Oxide
DISTINCT® see...Dicamba
DISTINCT® see...Sodium Dicamba
DISTODIN® see...Hexachlorophene
DISTOKAL® see...Hexachloroethane
DISTOPAN® see...Hexachloroethane
DISTOPIN® see...Hexachloroethane
Disul see...2,4-DES-Sodium
Disulfan see...Disulfiram
DISULFATON® see...Disulfoton
Disulfide, bis(dimethylthiocarbamoyl) see...Thiram
Disulfoton disulfide see...Oxydisulfoton
Disulfoton sulfoxide see...Oxydisulfoton
Disulfuram see...Disulfiram
Disul-Na see...2,4-DES-Sodium
Disul-sodium see...2,4-DES-Sodium
Disulphuram see...Disulfiram
DISYSTON SULFOXIDE® see...Oxydisulfoton
DI-SEPTON® see...Demeton
DISYSTON® see...Demeton
DISYSTON® see...Disulfoton
DI-SYSTON® see...Disulfoton
DISYSTOX® see...Disulfoton
DITEK® see...Thiophanate-Methyl
Di-tetrahydronicotyrine see...Nicotine
Dithallium sulfate see...Thallium Sulfate
Dithallium(1+) sulfate see...Thallium Sulfate
Dithallium(I) sulfate see...Thallium Sulfate
DITHANE® see...Mancozeb
DITHANE® 7-78 see...Zineb
DITHANE-22® see...Maneb
DITHANE A-40® see...Nabam
DITHANE A-46® see...Nabam
DITHANE D-14® see...Nabam
DITHANE® Z see...Zineb

DITHANE® Z-78 see...Zineb
2,3-$p$-Dithiane see...2,3-dehydro-2,3-dimethyl-, tetroxide see...Dimethipin
$p$-Dithiane see...dimethipin [2,3,-dihydro-5,6-dimethyl-1,4-dithiin-1,1,4,4-tetraoxide] see...Dimethipin
Dithiazanin iodide see...Dithiazanine Iodide
Dithiazanine iodide see...Dithiazanine Iodide
Dithiazinine see...Dithiazanine Iodide
1,3-Dithietan-2-ylidene phosphoramidic acid diethyl ester see...Fosthietan
1,4-Dithiin see...2,3-dihydro-5,6-dimethyl-,1,1,4,4-tetraoxide see...Dimethipin
Dithilol(4,5-$\beta$)quinoxalin-2-one,6-methyl- see. Oxythioquinox
1,3-Dithilol(4,5-$\beta$)quinoxalin-2-one,6-methyl- see. Oxythioquinox
Dithio see...Sulfotepp
1,1'-Dithiobis($N,N$-diethylthioformamide) see.Disulfiram
$\alpha,\alpha'$-Dithiobis(dimethylthio)formamide see...Thiram
Dithiocarbonic anhydride see...Carbon Disulfide
DITHIODEMETON® see...Disulfoton
$N,N$-(Dithiodicarbonothioyl)bis($N$-methylmethanamine) see...Thiram
Dithiodiphosphoric acid, tetraethyl ester see.Sulfotepp
Dithiofos see...Sulfotepp
Dithion see...Sulfotepp
Dithione see...Sulfotepp
Dithiophos see...Sulfotepp
Dithiophosphate de $O,O$-diethyle et de (4-chlorophenyl) thiomethyle (French) see...Carbophanothion
Dithiophosphate de $O,O$-diethyle etde $S$-(2-ethylthio-ethyle) (French) see...Disulfoton
Dithiophosphatede $O,O$-diethyle et d'ethylthiomethyle (French) see...Phorate
Dithiophosphate de $O,O$-dimethyle et de $S$-(1,2-dicarboethoxyethyle) (French) see...Malathion
Di(thiophosphoric) acid, tetraethyl ester see.Sulfotepp
Dithiopyrophosphate de tetraethyle (French) see..Sulfotepp
Dithioquinox see...Oxythioquinox
DITHIOSYSTOX® see...Disulfoton
Dithiotep see...Sulfotepp
DITIAMINA® see...Zineb
DITOX® see...Diuron
DITRAC® see...Diphacione
DITRANIL® see...Dichloran
DITRIFON® see...Trichlorfon
DITROSOL® see...Dinitro-$o$-cresol (DNOC)
DIUMATE® see...Diuron
DIUMATE® see...Sodium Methanearsonate (MSMA)
DIUREX® see...Diuron
DIUROL® see...Amitrole
DIUROL® see...Diuron
DIURON 4L® see...Diuron
DIVA® see...Iprodione
DIVA FUNGICIDE® see...Chlorothalonil
DIVIDEND® see...Difenoconazole

DIVIDEND® EXTREME FUNGICIDE see.Difenoconazole
DIVIPAN® see...Dichlorvos
DIXON® see...Phosphamidon
*N*,*N*-Di-(2,4-xylyliminomethyl)methylamine see.Amitraz
DIZIKTOL® see...Diazinon
DIZINON® see...Diazinon
Di[tri(2,2-dimethyl-2-phenylethyl)tin]oxide see. Fenbutatin Oxide
*dl*-Alanine, *N*-(2,6-dimethylphenyl)-*N*-(methoxyacetyl)-, methyl ester see...Metalaxyl
*dl*-2-allyl-4-hydroxy-3-methyl-2-cyclopenten-1-one-d,l-chrysanthemum monocarboxylate see...Allethrins
*dl*-3-allyl-2-methyl-4-oxocyclopent-2-enyl *dl*-cis trans chrysanthemate see...Allethrins
*dl*-*N*-(2,6-Dimethylphenyl)-*N*-(2'-methoxyacetyl)alaninate de methyle (French) see...Metalaxyl
*dl*-Glutamic acid see...Glutamic Acid
*dl*-α-Glutamic acid see...Glutamic Acid
*dl*-*p*-mentha-1,8-diene see...D-Limonene
*dl*-Valine,n-[2-chloro-4-(trifluoromethyl)phenyl]-cyano(3-phenoxylphenyl)methyl ester see...Fluvalinate
DLP 787® see...Pyriminil
2,4-DM see...2,4-DB
DMA 2AG® see...Sodium Dicamba
DMA-4® see...2,4-D
DMAA see...Cacodylic Acid
DMC® WEED CONTROL see...Metsulfuron-methyl
DMDK see...Sodium Dimethyldithiocarbamate
DMDT see...Methoxychlor
*p*,*p*'-DMDT see...Methoxychlor
DMF see...Dimefox
DMOC see...Carboxin
DMSA see...Daminozide
DMTP (Japan) see...Methidathion
DMTT see...Dazomet
DMU® see...Diuron
DN® see...Dinex
DN 111® see...Dinex
DN 289® see...Dinoseb
DNBP see...Dinoseb
DN DUST No. 12® see...Dinex
DN DRY MIX see...Dinex
DNOC® see...Dinitro-o-cresol (DNOC)
DNOCHP see...Dinex
DNOCP see...Dinocap
DNOPC see...Dinocap
DNOSBP see...Dinoseb
DNPB see...Dinoseb
DNSBP see...Dinoseb
DNTBP see...Dinoterb
DNTP see...Parathion
DOCKLENE® see...Mecoprop
Dodanic acid 83 see...Dodecylbenzenesulfonic Acid
DODAT® see...DDT
Dodecachlorooctahydro-1,3,4-metheno-2H-cyclobuta(c,d)pentalene see...Mirex

1,1a,2,2,3,3a,4,5,5,5a,5b,6-Dodecachlorooctahydro-1,3,4-metheno-1H-cyclobuta(c,d)pentalene see...Mirex
Dodecachloropentacyclodecane see...Mirex
2,4-Dodecadienoic acid, 11-methoxy-3,7,11-trimethyl-, ispropyl ester, (E,E)- see.Methoprene
2,4-Dodecadienoic acid, 11-methoxy-3,7,11-trimethyl-, 1-methylethyl ester, (E,E)- see.Methoprene
2,4-Dodecadienoic acid, 3,7,11-trimethyl-, .ethyl ester, [*S*-(*E*,*E*)]- see...Hydroprene
2,4-Dodecadienoic acid, 3,7,11-trimethyl-, 2-propynyl ester, (E,E)- see...Kinoprene
2,4-Dodecadienoic acid, 3,7,11-trimethyl-,2-propynyl ester, [*s*-(E,E)]- see...Kinoprene
*N*-Dodecyl benzenesulfonic acid see. Dodecylbenzenesulfonic Acid
Dodecyl benzenesulfonate see..Dodecylbenzenesulfonic Acid
Dodecyl benzenesulphonate see.Dodecylbenzenesulfonic Acid
*N*-Dodecyl benzenesulphonic acid see. Dodecylbenzene sulfonic Acid
Dodecylbenzenesulfonic acid, compounded with 2,2',2"-nitrilotris(ethanol) (1:1) see...Triethanolamine Dodecyl benzene Sulfonate
Dodecylbenzenesulfonic acid, triethanolamine salt see...Norflurazon
Dodecylbenzenesulfonic acid, triethanolamine salt see..Triethanolamine Dodecylbenzene Sulfonate
Dodecylbenzenesulfonic acid triethanolamine salt see. Triethanolamine Dodecylbenzene Sulfonate
Dodecylbenzenesulphonic acid see. Dodecylbenzenesulfonic Acid
*N*-Dodecylguanidine acetate see...Dodine
*N*-Dodecylguanidineacetat (German) see...Dodine
Dodecylguanidine acetate see...Dodine
*N*-Dodecylguanidine acetate see...Oxythioquinox
Dodecylguanidine monoacetate see...Dodine
1-Dodecylguanidinium acetate see...Dodine
Dodemorfe (French) see...Dodemorph Acetate
Dodguadine see...Dodine
Dodin see...Dodine
Dodine acetate see...Dodine
Dodine see...mixture with glyodin see...Dodine
Doguadine see...Dodine
DOKIRIN® see...Copper(II)-8-hydroxyquinoline
DOL® see...Hexachlorocyclohexanes
DOLCO MOUSE CEREAL® see...Strychnine
DOLCO MOUSE CEREAL® see...Brucine
DOL GRANULE® see...Lindane
DOLMIX® see...Hexachlorocyclohexanes
DOMAIN® see...Metribuzin
DOMAIN® see...Thiafluamide
DOMAIN® see...Thiophanate-Methyl
DOMATOL® see...Amitrole
DOMINATOR® EAR TAG see...Pirimiphos-Methyl
DOMINEX® see...*alpha*-Cypermethrin
Domolite see...Calcium Carbonate
DOOM® see...Dichlorvos

DOO-NOT® see...D-Limonene
DOP® 26019 see...Iprodione
DOQUADINE® see...Dodine
DORMEX® see...Cyanamide
DORMONE® see...2,4-D
DORSAN® see...Chlorpyrifos
DORSAN®-C see...Chlorpyrifos
DORSAN-C® see...Cypermethrin
DORUPLANT® see...Ametryn
DOSAFLO® see...Metoxuron
DOSAGRAN® see...Metoxuron
DOSANEX® see...Metoxuron
DOSANEX FL® see...Metoxuron
DOSANEX MG® see...Metoxuron
DOTAN® see...Chlormephos
DOUBLE BARREL® see...*lamda*-Cyhalothrin
DOUBLE BARREL® EAR TAG see...Pirimiphos-Methyl
DOUBLE DOWN® see...Fonofos
DOUBLE STRENGTH® see...Silvex
DOUBLE THREAT® see...Bifenthrin
DOUSAN® see...Thiophanate-Methyl
DOVIP® see...Famphur
DOW ATRAZINE 80W HERBICIDE® see...Atrazine
DOWCHLOR® see...Chlordane
DOWCIDE 1® see...*o*-Phenylphenol
DOWCIDE 7® see...Pentachlorophenol
DOWCO 132® see...Crufomate
DOWCO 139® see...Mexacarbate
DOWCO 163® see...Nitrapyrin
DOWCO 179® see...Chlorpyrifos
DOWCO 186® see...Triphenyltin Compounds
DOWCO 186® see...Fentin Hydroxide
DOWCO 213® see...Cyhexatin
DOWCO 217® see...Chlorpyrifos-methyl
DOWCO 290® see...Clopyralid
DOWCO 433® MHE see...Fluroxypyr 1-methylheptyl Ester
DOWCO 453®-ME see...Haloxyfop-methyl
DOWCO 543® EE see...Haloxyfop-methyl
DOW DORMANT FUNGICIDE® see...Sodium Pentachlorophenate
DOWELANCO® BRUSH AND WEED see...Triclopyr, Triethylammonium Salt
DOW ET 14® see...Ronnel
DOW ET 57® see...Ronnel
DOWFUME® see...Carbon Disulfide
DOWFUME® see...Carbon Tetrachloride
DOWFUME® see...Chloropicrin
DOWFUME®-N see...D-D mixture
DOWFUME® see...Ethylene Dibromide
DOWFUME® see...Ethylene Dichloride
DOWFUME® see...Methyl Bromide
DOW GENERAL® see...Dinoseb
DOW GENERAL WEED KILLER® see...Dinoseb
DOWICIDE 2S® see...Trichlorophenols
DOWICIDE® 7 see...Pentachlorophenol
DOWICIDE EC-7® see...Pentachlorophenol
DOWICIDE G® see...Sodium Pentachlorophenate
DOWICIDE G-ST® see...Sodium Pentachlorophenate
DOW-KLOR® see...Chlordane
DOW MCP AMINE WEED KILLER® see...MCPA
DOW PENTACHLOROPHENOL DP-2 ANTI MICROBIAL® see...Pentachlorophenol
DOWPON® see...Dalapon
DOWPON M® see...Dalapon
DOWPON®-RAE see...Dalapon
DOW SELECTIVE WEED KILLER® see...Dinoseb
DOW SPRAY®-17 see...Dinex
DOWTHERM® see...*para*-Dichlorobenzene
DOXOL TOMATO LIFE® see...Naphthoxyacetic Acid
2,4-DP see...Dichlorprop
DP-2® see...Pentachlorophenol
DP-FLUID® see...Dichlorprop
DPA see...Propanil
DPC see...Dinocap
DPD 63760H® see...Metsulfuron-methyl
DPF see...Isofluorphate
DPL-87® see...Pyriminil
DPP see...Parathion
DPX-A 7881® see...Ethametsulfuron-methyl
DPX-F5384® see...Bensulfuron-methyl
DPX-F6025® see...Chlorimuron-ethyl
DPX-L-5300® see...Tribenuron-Methyl
DPX-M6316® see...Thifensulfuron Methyl
DPX-PE350® see...Pyrithiobac-Sodium
DPX-PM082® see...Dimethenamid
DPX-T3217® see...Cymoxanil
DPX-T5648® see...Sulfometuron-Methyl
DPX-T 6376® see...Metsulfuron-methyl
DPX-V9636® see...Nicosulfuron
DPX-Y5893® see...Hexythiazox
DPX-Y 6202® see...Quizalofop-Ethyl
DPX 1108® see...Fosamine Ammonium
DPX 1410® see...Oxamyl
DPX 3217® see...Cymoxanil
DPX 3217M® see...Cymoxanil
DPX 3674® see...Hexazinone
DPX 4189® see...Chlorsulfuron
DPX 6376® see...Metsulfuron-methyl
DPX 6774® see...Isoproturon
DPX 43898® Chlorethoxyfos
DPX 66037® see...Triflusulfuron-Methyl
DPX 79406® see...Nicosulfuron
DQUIGARD® see...Dichlorvos
DR. ROGER'S TOXENE® see...Toxaphene
Dracyclic acid see...Benzoic Acid
DRAPOLENE® see...Zilkonium Chloride
DRAT® see...Chlorophacinone
DRAWIZON® see...Diazinon
DRAZA G MICROPELLETS® see...Methiocarb
DRAZA® see...Methiocarb
DRC 3341® see...Metolcarb
DRC-714 see...Phosacetim
DRENCH® see...Crufomate

DREXAR®-530 see...Sodium Methanearsonate (MSMA)
DREXEL ACEPHATE 75 WSP® see...Acephate
DREXEL CROAK® see...Fluometuron
DREXEL DEFOL® see...Sodium Chlorate
DREXEL DIURON 4L® see...Diuron
DREXEL EZY-PICKIN COTTON DEFOLIANT® see. Sodium Cacodylate
DREXEL KACK HERBICIDE® see...Sodium Cacodylate
DREXEL METHYL PARATHION 4E® see...Methyl Parathion
DREXEL ME-TOO-LACHLOR® see...Metolachlor
DREXEL PARATHION 8E® see...Parathion
DREXEL PROP-JOB® see...Propanil
DRI-DYE® see...Ammonium Hexafluorosilicate
DRI-KIL® see...Rotenone
DRILL TOX-SPEZIAL AGLUKON® see...Lindane
DRINOX® see...Aldrin
DRINOX® see...Heptachlor
DRI-TRI® see...Trisodium Phosphate
DRIVE 75® see...Quinclorac
DROPAVEN® see...Propanil
DROP LEAF® see...Sodium Chlorate
DROPP® see...Thidiazuron
DROPP ULTRA® see...Diuron
DRUPINA® 90 see...Ziram
DRW 1139® see...Metamiton
Dry cleaner naphtha see...Stoddard Solvent
DRY MIX No. 1® see...Dinex
DS-15647® see...Thiofanox
DSDP see...Amiton
DSE see...Nabam
DSM® see...Demeton-methyl
DSMA see...Sodium Methanearsonate (MSMA)
DTDA/DMA-TEA-DMA® SELECTIVE HERBICIDE see...Triclopyr, Triethylammonium Salt
DTMC see...Dicofol
DU 112307® see...Diflubenzuron
DUAL® see...Metolachlor
DUAL MAGNUM® see...Metolachlor
DUAL MURGANIC RPB SEED TREATMENT® see. Lindane
DUET® see...Bensulfuron-methyl
DUET® see...Metolachlor
DUET® see...Propanil
DUO-KILL® see...Crotoxyphos
DUO-KILL® see...Dichlorvos
DUO TOP® see...Triflumizole
DUPONT 326® see...Linuron
DUPONT® 732 see...Terbacil
DUPONT 1179® see...Methomyl
DUPONT 1991® see...Benomyl
DUPONT HERBICIDE 976® see...Bromacil
DU-SPREX® see...Dichlobenil
DU-TER® see...Fentin Hydroxide
DU-TER® see...Triphenyltin Compounds

DU-TUR FLOWABLE-30® see...Triphenyltin Compounds
DU-TER FUNGICIDE® see...Triphenyltin Compounds
DU-TER FUNGICIDE WETTABLE POWDER® see.Triphenyltin Compounds
DU-TER PB-47 FUNGICIDE® see...Triphenyltin Compounds
DU-TER W-50® see...Triphenyltin Compounds
DUPHAR® see...Tetradifon
DUPHAR® PH 60-40 see...Diflubenzuron
DURAMITEX® see...Malathion
DURAN® see...Diuron
Duraphos see...Mevinphos
DURATOX® see...Demeton
DURATOX® see...Demeton-methyl
DURA TREET II® see...Pentachlorophenol
DURAVOS® see...Dichlorvos
DURETTER® see...Ferrous Sulfate
DURHAM® see...Metaldehyde
DURHAM® see...Methyl Parathion
DURHAM® see...Mevinphos
DURHAM NEMATOCIDE® see...1,3-Dichloropropene
DUROFERON® see...Ferrous Sulfate
DUROTOX® see...Pentachlorophenol
DURSBAN® see...Chlorpyrifos
Dursban methyl see...Chlorpyrifos-methyl
Dusting sulphur see...Sulfur
DUTER EXTRA® see...Triphenyltin Compounds
DUTCH LIQUID® see...Ethylene Dichloride
DUTCH OIL® see...Ethylene Dichloride
DUTCH-TREAT® see...Sodium Cacodylate
DUTER® see...Fentin Hydroxide
DUTER® see...Triphenyltin Compounds
DW 3418® see...Cyanazine
DWELL® see...Etridiazole
Dwubromoetan (Polish) see...Ethylene Dibromide
2,4-Dwuchlorofenoksyoctowy kwas (Polish) see...2,4-D
Dwuchloropropan (Polish) see...1,2-Dichloropropane
DYANAP® see...Naptalam
DYBAR® see...Fenitrothion
DYCARB® see...Bendiocarb
DYCLOMEC® see...Dichlobenil
Dyflos see...Isofluorphate
DYFONATE® see...Fonofos
DYFONATE® see...Pebulate
DYKOL® see...DDT
DYKON® see...Sodium Diacetate
DYLOX® see...Trichlorfon
DYLOX-METASYSTOX-R® see...Trichlorfon
DYMEC® see...2,4-D
DYMET® see...Diazinon
DYMID® see...Diphenamid
DYNA-CARBYL® see...Carbaryl
DYNACIDE® see...Phenylmercury Acetate
DYNAMYTE® see...Dinoseb
DYNEX® see...Diuron
DYNOFORM® see...Formaldehyde
DYPHONATE® see...Fonofos

DYRENE® see...Anilazine
DYRENE 50W® see...Anilazine
DYREX® see...Trichlorfon
DYSECT® see...Cypermethrin
DYTOP® see...Dinoseb
DYVEL® see...Dicamba
DYVEL® see...MCPA
DYVON® see...Trichlorfon
DYZOL® see...Diazinon

- E -

E 2® see...Ammonium Nitrate
E 99® see...2,4-D, butoxyethyl ester
E 103® see...Tebuthiuron
E 393® see...Sulfotepp
E 600® see...Paraoxon
E 601® see...Methyl Parathion
E 605 F® see...Parathion
E 605® see...Parathion
E 965® see...Carbendazim
E 1059® see...Demeton
E 1059® see...Demeton
E 3314® see...Heptachlor
E 3314® see...Heptachlor
E 7256® see...Dodecylbenzenesulfonic Acid
EACITHION® see...Ethion
EAGLE® see...Myclobutanil
EAGLES-7® see...Warfarin
EAQUA ETHION® see...Ethion
EARTHCIDE® see...Quintozene
EASIGRAZE® see...Ammonium Nitrate
EASTERN STATES DUOCIDE® see...Warfarin
EASTERON® 99 CONCENTRATE see...2,4-D
see.butoxyethyl ester
EASTMAN 7663® see...Dithiazanine Iodide
EASY OFF®-D see...Tribufos
EBDC see...Maneb
EBDC see...disodium salt see...Nabam
EBDC, sodium salt see...Nabam
EBDC, zinc salt see...Zineb
EBECRYL® see...Thiram
EBLADAN® see...Ethion
EBT see...Allethrins
EBT 25,726 see...Methiocarb
EBUFOS® see...Cadusafos
EC 300® see...Maleic Hydrazide
EC HERBICIDE see...Phenmedipham
ECATOX® see...Parathion
ECB see...Chlorobenzilate
Eccothal see...Thallium Sulfate
Echlomezole (Japan) see...Etridiazole
ECHO® see...Chlorothalonil
ECO₂FUME TM® see...Phosphine
ECOMMANDO INSECTICIDE CATTLE EAR TAG®
see...Ethion
ECONOSAN® see...Decanoic Acid
ECONOSAN® see...Pelargonic Acid

ECOPAC® see...Thymol
ECOPRO® see...Temephos
ECOPRO® 1707 see...Temephos
ECOTRU® see...4-Chloro-3,5-xylenol
ECOZIN® see...Azadirachtin
ECTIN® see...Fenvalerate
ECTODEX® see...Amitraz
ECTOGARD® see...Fenoxycarb
ECTORAL® see...Ronnel
EDB® see...Ethylene Dibromide
E-D-BEE® see...Ethylene Dibromide
EDB-85® see...Ethylene Dibromide
EDC see...Ethylene Dichloride
EDCO® see...Methyl Bromide
EDGER® see...Oxyfluorfen
EDRASTIC® see...Ethion
EDRIZAN® see...Amitraz
EDRIZAR® see...Amitraz
EEMBATHION® see...Ethion
EEREX® see...Bromacil
EETHANOX® see...Ethion
EETHIOL® see...Ethion
EETHODAN® see...Ethion
EETHOPAZ® see...Ethion
EF 121® see...Chlorpyrifos
EFFUSAN® see...Dinitro-o-cresol (DNOC)
EFFUSAN 3436® see...Dinitro-o-cresol (DNOC)
EFMC-1240® see...Ethion
EFOSFATOXE® see...Ethion
EFOSFONO 50® see...Ethion
EFOSITE ALUMINUM® see...Fosetyl-Al
EFOSITE-AL® see...Fosetyl-Al
EFUZIN® see...Dodine
EGITOL® see...Hexachloroethane
EH 1143® see...Sodium Methanearsonate (MSMA)
EH1356 HERBICIDE®, see...MCPA
EH-1400® see...Dithiopyr
EH-YAN-KU® see...Difenzoquat
EHYLEMOX® see...Ethion
EI 783® see...Azadirachtin
EI 38555® see...Chlormequat Chloride
EI 47031® see...Phosfolan
EI 47300® see...Fenitrothion
EI 47470® see...Mephosfolan
EI 52160® see...Temephos
Eisendimethyldithiocarbamat (German) see...Ferbam
Eisen(III)-tris(N,N-dimethyldithiocarbamat) (German)
see...Ferbam
EITOPAZ® see...Ethion
EKAGOM TB® see...Thiram
EKAGOM TEDS® see...Disulfiram
EKALUX® see...Quinalphos
EKAQUIN® see...Quinalphos
EKATIN WF & WF ULV® see...Parathion
EKATIN TD® see...Disulfoton
EKATOX® see...Parathion
EKTAFOS® see...Dicrotophos
EKTOFOS® see...Dicrotophos

EKWIT® see...Ethion
EL 107® see...Isoxaben
EL 110® see...Benefin
EL 119® see...Oryzalin
EL 161® see...Ethalfluralin
EL 171® see...Fluridone
EL 179® see...Isopropalin
EL 222® see...Fenarimol
EL 400® see...Bromophos
EL 500® see...Flurprimidol
EL 531® see...Ancymidol
EL 3911® see...Phorate
EL 4049® see...Malathion
ELANCOLAN® see...Trifluralin
ELASTREL® see...Dichlorvos
Elayl see...Ethylene
Elemental cadmium see...Cadmium
Elemental copper see...Copper and Copper Compounds
Elemental sulfur see...Sulfur
ELFAN WA SULPHONIC ACID®
see.Dodecylbenzenesulfonic Acid
ELGETOL® see...Dinitro-o-cresol (DNOC)
ELGETOL 30® see...Dinitro-o-cresol (DNOC)
ELGETOL 318® see...Dinoseb
ELIMINATOR® see...Fenoxycarb
ELIPOL® see...Dinitro-o-cresol (DNOC)
ELITE® see...Tebuconazole
ELMASIL® see...Amitrole
Elosal see...Sulfur
ELSAN® see...Phenthoate
EMBAFUME® see...Methyl Bromide
EMBASSADOR® see...Diniconazole
EMBLEM® see...Benefin
EMBUTONE® see...2,4-DB
EMBUTOX® see...2,4-DB
EMBUTOX KLEAN-UP® see...2,4-DB
EMBUTOX® see...Dichlorprop
EMC see...Ethyl Mercuric Chloride
EMCARB® see...Mancozeb
EMCEPAN® see...MCPA
Emerald green see...C.I. Basic Green 1
Emerald green see...Copper Acetoarsenite
Emerald green see...Paris Green
EMERY® 202 see...Pelargonic Acid
Emetique (French) see...Antimony Potassium Tartrate
EMFAC®-1202 see...Pelargonic Acid
EMISOL® see...Amitrole
EMITKILL® see...Ethion
EMMATON® see...Malathion
EMMATOS® see...Malathion
EMMATOS EXTRA® see...Malathion
EMmHgK® see...Mefluidide
EMMY® see...Methyl Parathion
EMO-NIB® see...Nicotine
EMPAL® see...MCPA
EMPIRE® see...Chlorpyrifos
EMPOWER® see...Bifenthrin
EMQ® see...Ethoxyquin

EMTHANE M-15® see...Mancozeb
EMULSAMINE BK® see...2,4-D
EMULSAMINE E-3® see...2,4-D
EMULSIPHOS® see...Trisodium Phosphate
EMULSIPHOS® 440/660 see...Trisodium Phosphate
EN 57® see...Endrin
EN 18133® see...Thionazin
ENABLE® see...Fenbuconazole
ENABLE® see...Fentin Hydroxide
ENABLE® see...Triphenyltin Compounds
ENAGATA® see...Ethion
ENCORE® see...Imidacloprid
ENDEAVOR® see...Pymetrozine
ENDOCEL® see...Endosulfan
ENDOCID® see...Endothion
ENDOCIDE® see...Endosulfan
ENDOCIDE® see...Endothion
3,6-Endo-epoxy-1,2-cyclohexanedicarboxylic acid
see.Endothall
3,6-Endooxohexahydrophthalic acid see...Endothall
ENDORSE® WP Turf Fungicide see...Polyoxin D, Zinc Salt
ENDOSAN® see...Binapacryl
ENDOSOL® see...Endosulfan
END-O-SULFAN® see...Endosulfan
Endosulphan see...Endosulfan
Endosulfan-1 see...Endosulfan
Endosulfan-2 see...Endosulfan
Endosulfan-α see...Endosulfan
Endosulfan-A see...Endosulfan
beta Endosulfan see...Endosulfan
α-Endosulfan see...Endosulfan
β-Endosulfan see...Endosulfan
ENDOTAF® see...Endosulfan
Endothal (Great Britian) see...Endothall
Endothal chlorophenoxy herbicide see...Endothall
Endothall technical see...Endothall
Endotiona (Spanish) see...Endothion
3,6-Endoxohexahydrophthalic acid see...Endothall
ENDOX® see...Coumatetralyl
ENDOX® see...Endosulfan
ENDREX® see...Endrin
ENDRICOL® see...Endrin
Endrina (Spanish) see...Endrin
ENDRIN CHLORINATED HYDROCARBON INSECTICIDE® see...Endrin
Endrine (French) see...Endrin
ENDROCID® see...Coumatetralyl
ENDROCIDE® see...Coumatetralyl
ENDURANCE® see...Prodiamine
ENDYL® see...Carbophanothion
ENE 11183 see...Coumatetralyl
ENFORCER® see...Diquat Dibromide
ENFORCER® see...Esfenvalerate
ENIA 1240® see...Ethion
ENIAGARA 1240® see...Ethion
ENIALATE® see...Ethion
ENIDE® see...Diphenamid

Enilconazole  see...Imazalil
ENOVIT® see...Thiophanate-Methyl
ENQUIK® see...Monocarbamide Dihydrogen Sulfate
Ensodulfan (Spanish)  see...Endosulfan
β-Ensodulfan (Spanish)  see...Endosulfan
ENSTAR® see...Kinoprene
ENSTAR II® see...Kinoprene
ENSTAR 5E® see...Kinoprene
ENSURE® see...Endosulfan
ENT 38  see...Phenothiazine
ENT 54  see...Acrylonitrile
ENT 133  see...Rotenone
ENT 133  see...Rotenone
ENT 157  see...Dinex
ENT 884  see...Paris Green
ENT 987  see...Thiram
ENT 988  see...Ziram
ENT 1,122  see...Dinoseb
ENT 1,506  see...DDT
ENT 1,656  see...Ethylene Dichloride
ENT 1,716  see...Methoxychlor
ENT 2,435  see...Nicotine Sulfate
ENT 3,424  see...Nicotine
ENT 3,776  see...Dichlone
ENT 4,504  see...Dichloroethyl Ether
ENT 4705  see...Carbon Tetrachloride
ENT 7,796  see...Lindane
ENT 8,420  see...D-D mixture
ENT 8,538  see...2,4-D
ENT 8,601  see...Hexachlorocyclohexanes
ENT 9,232  see...Hexachlorocyclohexanes
ENT 9,233  see...Hexachlorocyclohexanes
ENT 9,234  see...Hexachlorocyclohexanes
ENT 9735  see...Toxaphene
ENT 9,932  see...Chlordane
ENT 14,250  see...Piperonyl Butoxide
ENT 14,689  see...Ferbam
ENT 14,874  see...Zineb
ENT 14,875  see...Maneb
ENT 15,108  see...Parathion
ENT 15,152  see...Heptachlor
ENT 15,349  see...Ethylene Dibromide
ENT 15,406  see...1,2-Dichloropropane
ENT 15949  see...Aldrin
ENT 16,087  see...Paraoxon
ENT 16,225  see...Dieldrin
ENT 16,273  see...Sulfotepp
ENT 16,391  see...Chlordecone (Kepone®)
ENT 16,436  see...Dodine
ENT 16,519  see...Aramite
ENT 16,634  see...Sulfoxide
ENT 16,894  see...Aspon®
ENT 17,034  see...Malathion
ENT 17,251  see...Endrin
ENT 17,291  see...Octamethyl Diphosphoramide
ENT 17,292  see...Methyl Parathion
ENT 17295  see...Demeton
ENT 17,510  see...Allethrins
ENT 17591  see...Dipropyl Isocinchomeronate
ENT 17591  see...Dipropyl Isocinchomeronate
ENT 17,798  see...EPN
ENT 17,957  see...Coumaphos
ENT 18,596  see...Chlorobenzilate
ENT 18,771  see...TEPP
ENT 18,862  see...Demeton-methyl
ENT 18,870® see...Maleic Hydrazide
ENT 19,060  see...Isolan®
ENT 19,109  see...Dimefox
ENT 19,244  see...Isodrin
ENT 19,507  see...Diazinon
ENT 19,763  see...Trichlorfon
ENT 20,738  see...Dichlorvos
ENT 20,852  see...Bromoxynil
ENT 22,014  see...Azinphos-ethyl
ENT 22,374  see...Mevinphos
ENT 22879  see...Dioxathion
ENT 23,233  see...Azinphos-methyl
ENT 23,284  see...Ronnel
ENT 23,437  see...Disulfoton
ENT 23,438  see...Phenthoate
ENT 23,648  see...Dicofol
ENT 23,708  see...Carbophanothion
ENT 23,737  see...Tetradifon
ENT 23969  see...Carbaryl
ENT 23,979  see...Endosulfan
ENT 24,042  see...Phorate
ENT 24,105  see...Ethion
ENT 24,482  see...Dicrotophos
ENT 24,652  see...Prothoate
ENT 24,653  see...Endothion
ENT 24,717  see...Crotoxyphos
ENT 24727  see...Dinocap
ENT 24,964  see...Demeton-methyl
ENT 24969  see...Chlorfenvinphos
ENT 24,980-X  see...Amiton
ENT 24,984  see...Sodium Aluminum fluoride
ENT 24,988  see...Naled
ENT 25,208  see...Triphenyltin Compounds
ENT 25445  see...Amitrole
ENT 25,500  see...Phenol, 3-(1-methyethyl)-, methylcarbamate
ENT 25,515  see...Phosphamidon
ENT 25,543  see...Phenol, 3-(1-methyethyl)-, methylcarbamate
ENT 25,545  see...Isobenzan
ENT 25,545-x  see...Isobenzan
ENT 25,552-X  see...Chlordane
ENT 25,554  see...Methyl Phenkapton
ENT 25,580  see...Thionazin
ENT 25,584  see...Heptachlor Epoxide
ENT 25,595-X  see...Dimetilan
ENT 25,602-X  see...Crufomate
ENT 25,606  see...Oxythioquinox
ENT 25,644  see...Famphur
ENT 25,671  see...Propoxur
ENT 25,675  see...Cyanofos

ENT 25,705  see...Phosmet
ENT 25,712  see...Trichloronate
ENT 25,715  see...Fenitrothion
ENT 25,718  see...Dienochlor
ENT 25,719  see...Mirex
ENT 25766  see...Mexacarbate
ENT 25,776  see...Omethoate
ENT 25,793  see...Binapacryl
ENT 25,796  see...Fonofos
ENT 25,830  see...Phosfolan
ENT 25841  see...Tetrachlorvinphos
ENT 25,922  see...Dimetilan
ENT 25,991  see...Mephosfolan
ENT 26,058  see...Anilazine
ENT 26,263  see...Ethylene Oxide
ENT 26538  see...Captan
ENT 26,925  see...Triclocarban
ENT 27093  see...Aldicarb
ENT 27,129  see...Monocrotophos
ENT 27,162  see...Bromophos
ENT 27,163  see...Phosalone
ENT 27,164  see...Carbofuran
ENT 27,165  see...Temephos
ENT 27,193  see...Methidathion
ENT 27,226  see...Propargite
ENT 27,257  see...Formothion
ENT 27,258  see...Bromophos-ethyl
ENT 27,300  see...Promecarb
ENT 27,300-a  see...Promecarb
ENT 27311  see...Chlorpyrifos
ENT 27,318  see...Ethoprop
ENT 27,320  see...Dialifor
ENT 27,335  see...Chlordimeform
ENT 27339  see...Tetramethrin
ENT 27339  see...Tetramethrin
ENT 27,341  see...Methomyl
ENT 27,386GC  see...Phenthoate
ENT 27,394  see...Quinalphos
ENT 27,396  see...Methamidophos
ENT 27474  see...Resmethrin
ENT 27,488  see...Phoxim
ENT 27,520  see...Chlorpyrifos-methyl
ENT 27,566  see...Formetanate Hydrochloride
ENT 27,567  see...Chlordimeform
ENT 27,572  see...Fenamiphos
ENT 27,635  see...Chlorthiophos
ENT 27699Gc  see...Pirimiphos-Methyl
ENT 27699GC  see...Pirimiphos-Methyl
ENT 27738  see...Fenbutatin Oxide
ENT 27,766  see...Pirimicarb
ENT 27822  see...Acephate
ENT 27,851  see...Thiofanox
ENT 27967  see...Amitraz
ENT 27972  see...D-Phenothrin
ENT 27989  see...Propetamphos
ENT 28,009  see...Fentin Hydroxide
ENT 28,009  see...Triphenyltin Compounds
ENT 29,054 see... Diflubenzuron

ENT 50,434  see...Antimony Potassium Tartrate
ENT 70,459  see...Hydroprene
ENT 70,460  see...Methoprene
ENT 70,531  see...Kinoprene
ENT AI 3-29261  see...Aldoxycarb
ENTOMOXAN®  see...Lindane
ENTRY®  see...Bentazon
ENVERT-T®  see...2,4,5-T
ENVERT-T®  see...2,4,5-Trichlorophenoxyacetic Acid, Esters
ENVERT®  see...2,4-D
ENVOY®  see...MCPA
Enxofre  see...Sulfur
E.O  see...Ethylene Oxide
EP 30®  see...Pentachlorophenol
EP 316®  see...Promecarb
EP 332®  see...Formetanate Hydrochloride
EP 333®  see...Chlordimeform
EP 452®  see...Phenmedipham
EP 475®  see...Desmedipham
EPA Fenotrina (Spanish)  see...D-Phenothrin
EPAL®  see...Fosetyl-Al
EPERON®  see...Metalaxyl
EPHORRAN®  see...Disulfiram
EPHOSPHOTOX E®  see...Ethion
EPIC®  see...Isoxaflutole
EPIC®  see...Thiafluamide
4"-Epimethylamino-4"-deoxyavermectin $B_{1a}$ and $B_{1b}$ benzoates  see...Emamectin Benzoate
1,2-Epoxyaethan (German)  see...Ethylene Oxide
3,6-Epoxycyclohexane-1,2-dicarboxylic acid  see.Endothall
Epoxyethane  see...Ethylene Oxide
1,2-Epoxyethane  see...Ethylene Oxide
Epoxyheptachlor  see...Heptachlor Epoxide
EPROFIL®  see...Thiabendazole
EPROKIL®  see...Ethion
EPTAC-1®  see...Ziram
Eptacloro (Italian)  see...Heptachlor
1,4,5,6,7,8,8-Eptacloro-3a,4,7,7a-tetraidro-4,7-*endo*-metano-indene (Italian)  see...Heptachlor
EPTAM®  see...EPTC
EPTAM® 6E  see...EPTC
EPTAM 10G  see...EPTC
EPTAM 2.3G  see...EPTC
EPTC®  see...EPTC
EQUI-FLY® ORAL LARVICIDE  see.Tetrachlorvinphos
EQUIGARD®  see...Dichlorvos
EQUIGEL®  see...Dichlorvos
EQUINO-ACID®  see...Trichlorfon
EQUINO-AID®  see...Trichlorfon
EQUITDAZIN®  see...Carbendazim
EQUITROL  see...Tetrachlorvinphos
EQUIVET TZ®  see...Thiabendazole
EQUIZOLE®  see...Thiabendazole
EQ®  see...Ethoxyquin
ER-5461®  see...Profluralin
ERADE®  see...Oxythioquinox

ERADEX® see...Chlorpyrifos
ERADICANE® see...EPTC
ERASE® see...Cacodylic Acid
ERBAN® see...Propanil
ERHODIACIDE® see...Ethion
ERHODOCIDE® see...Ethion
ERITHANE® see...Fentin Hydroxide
ERODOCID® see...Ethion
ERP-THION® see...Ethion
Erserine see...Physostigmine
ERUCIN® see...Phenthoate
ERUNIT® see...Acetochlor
ERUNIT 500 FW® see...Atrazine
ERUSAN® see...Phenthoate
Esachlorobenzene (Italian) see...Hexachlorobenzene
ESACLOROFENE® see...Hexachlorophene
ESBECYTHRIN® see...Deltamethrin
ESBIOTHRIN® see...Allethrin
ESCORT® see...Metsulfuron-methyl
ESENTRY® see...Ethion
Eserine see...Physostigmine
Eserolein see...Physostigmine
ESGRAM® see...Paraquat
ESOPRATHION® see...Ethion
ESPADOL® see...4-Chloro-3,5-xylenol
ESPENAL® see...Disulfiram
Esperal (France) see...Disulfiram
ESSO® FUNGICIDE 406 see...Captan
ESTEEM® see...Pyriproxyfen
Ester 25 see...Paraoxon
O-Ester-p-nitrophenol with O-ethylphenyl phosphonothioate see...EPN
ESTERCIDE T-2® AND T-245® see...2,4,5-T
ESTERCIDE® T-245 see...2,4,5-Trichlorophenoxyacetic Acid, Esters
ESTERCIDE® T-2 see...2,4,5-Trichlorophenoxyacetic Acid, Esters
ESTERON® see...2,4-D
ESTERON® see...2,4,5-T
ESTERON 245® see...2,4,5-T
ESTERON® 245 BE see...2,4,5-Trichlorophenoxyacetic Acid, Esters
ESTERON 44 WEED KILLER® see...2,4-D
ESTERON® 44 see...2,4-D, isopropyl ester
ESTERON 99 CONCENTRATE® see...2,4-D
ESTERON BRUSH KILLER® see...2,4-D
ESTERON® BRUSH KILLER see...2,4,5-T
ESTERON® BRUSH KILLER see...2,4,5-Trichlorophenoxyacetic Acid, Esters
ESTERONE FOUR® see...2,4-D
ESTONATE® see...DDT
ESTONE® see...2,4-D
ESTONOX® see...Toxaphene
ESTOSTERIL® see...Peracetic Acid
Estricnina (Spanish) see...Strychnine
ESTROSEL® see...Dichlorvos
ESTROSOL® see...Dichlorvos
ET® see...Triclopry

ET 14® see...Ronnel
ET 57® see...Ronnel
Etabus see...Disulfiram
ETAFETHION® see...Ethion
Eteno (Spanish) see...Ethylene
ETHALFLURALIN® see...Ethalfluralin
ETHALFLURLIN® see...Ethalfluralin
Ethametsulfuron-methyl see...Ethametsulfuron-methyl
Ethanaminium, 2-chloro-$N,N,N$-trimethyl-, Chloride (9Cl) see...Chlormequat Chloride
Ethane, 1,2-dibromo- see...Ethylene Dibromide
Ethane dichloride see...Ethylene Dichloride
Ethane, 1,2-dichloro- see...Ethylene Dichloride
Ethane, 1,1-dichloro-2,2-bis($p$-ethylphenyl)- see. Ethylan
1,2-Ethanedithiol, cyclic ester with $p,p$-diethyl phosphonodithioimidocarbonate see...Phosfolan
1,2-Ethanedithiol see...cyclic ester with phosphonodithioimidocarbonic acid $p,p$-diethyl ester see. Phosfolan
1,2-Ethanediylbis(carbamodithioato)(2-)-manganese see. Maneb
1,2-Ethanediylbis(carbamodithioato)zinc see...Zineb
[1,2-Ethanediylbis(carbamodithioato)](2-)zinc see.Zineb
1,2-Ethanediylbis(carbamodithioato)(2-)-$S,S'$-zinc see. Zineb
1,2-Ethanediylbis(carbamodithioic acid), disodium salt see...Nabam
1,2-Ethanediylbiscarbamodithioic acid, manganese complex see...Maneb
1,2-Ethanediylbiscarbamodithioic acid, manganese(2+) salt(1:1) see...Maneb
1,2-Ethanediylbiscarbamodithioic acid, zinc complex see...Zineb
1,2-Ethanediylbiscarbamothioic acid, zinc salt see.Zineb
1,2-Ethanediylbismaneb, manganese (2+) salt (1:1) see...Maneb
Ethane hexachloride see...Hexachloroethane
Ethane, hexachloro- see...Hexachloroethane
Ethane, 1,1'-oxybis 2-chloro- see...Dichloroethyl Ether
Ethaneperoxoic acid see...Peracetic Acid
Ethane, thiocyanato- (Italian) see. Ethylthiocyanate
Ethanethiol, 2-(ethylsulfinyl)-, $S$-ester with $O,O$-dimethylphosphorothioate see...Demeton-methyl
Ethanethiol, 2-(ethylthio)-, $S$-ester with $O$ see...salt $O$-diethylphosphorothioate see...Demeton
Ethane, 1,1,1-trichloro-2,2-bis($p$-chlorophenyl)- see...DDT
Ethane, trifluoro- see...Trifluralin
Ethanimidothic acid, $N$-[(methylamino)carbonyl] see...Methomyl
Ethanimidothioic acid, $N'N'$-[thiobis((methylimino)carbonyloxy)]bis-, dimethyl ester see...Thiodicarb
Ethanolamine salt of 5,2'-dichloro-4'-nitrosalicyclicanilide see...Clonitralid
Ethanol butoxide see...Piperonyl Butoxide
Ethanol, 2-(4-chlorophenoxy)-1-tert-butyl-2-(1H-1,2,4-triazole-1-yl)- see...Triadimenol

Ethanol, 1,2-dibromo-2,2-dichloro- see...dimethyl phosphate see...Naled
Ethanol, 2,2,2-trichloro-1,1-bis(4-chlorophenyl)- see...Dicofol
Ethazole see...Etridiazole
ETHAZOLE® see...Etridiazole
Ethefon see...Ethephon
ETHEL® see...Ethephon
Ethene see...Ethylene
Ethene oxide see...Ethylene Oxide
Ethenol, 2,2-dichloro-, dimethyl phosphate see...Dichlorvos
ETHEPON® see...Ethephon
Ether, 2-chloro-α,α,α-trifluoro-p-tolyl-3-ethoxy-4-nitro phenyl see...Oxyfluorfen
Ether dichlore (French) see...Dichloroethyl Ether
Ether, 2,4-dichlorophenyl p-nitrophenyl see...Nitrofen
Ether, methyl phenyl see...Anisole
Etherin see...Ethylene
ETHEVERSE® see...Ethephon
ETHIOLACAR® see...Malathion
ETHLON® see...Parathion
Ethoprophos see...Ethoprop
ETHOSAT® 500 see...Ethofumesate
3-[(Ethoxycarbonyl)amino]phenyl N-phenylcarbamate see...Desmedipham
S-α-Ethoxycarbonylbenzyl dimethyl phosphorothiolothionate see...Phenthoate
S-α-Ethoxycarbonylbenzyl-O,O-dimethyl phosphorodithioate see...Phenthoate
S-(α-(Ethoxycarbonyl)benzyl) O,O-dimethyl phosphorodithioate see...Phenthoate
2-Ethoxy-2,3-dihydro-3,3-dimethyl-5-benzofuranyl methanesulfonate,(+)- see...Ethofumesate
2-Ethoxy-2,3-dihydro-3,3-dimethylbenzofuran-5-yl methanesulfonate see...Ethofumesate
6-Ethoxy-1,2-dihydro-2,2,4-trimethyl quinoline see. Ethoxyquin
2-[1-(Ethoxyimino)butyl]-5-[2-(ethylthio)propyl]-3-hydroxy-2-cyclohexen-1-one see...Sethoxydim
2-[1-(Ethoxyimino)butyl]-5-[2-(ethylthio)propyl]-3-hydroxyl-2-cyclohexen-1-one see...Sethoxydim
(±)-2-[1-(Ethoxyimino)butyl]-5-[2-(ethylthio)propyl]-3-hydroxy-2-cyclohexen-1-one see...Sethoxydim
(ZE)-2-(1-Ethoxyiminobutyl)-5-[2-(ethylthio)propyl]-3-hydroxycyclohex-2-enone see...Sethoxydim
2-[1-(Ethoxyimino)propyl]-3-hydroxy-5-mesitylcyclohex-2-en-one see...Tralkoxydim
2-[1-(Ethoxyimino)propyl]-3-hydroxy-5-(2,4,6-trimethylphenyl)-2-cyclohexen-1-one see...Tralkoxydim
2-[(Ethoxy((1-methylethyl)amino)phosphinothioyl)oxy]benzoic acid 1-methylethyl ester see...Isofenphos
2-[(Ethoxyl((1-methylethyl)amino]phosphinothioyl) oxy]benzoic acid 1-methylethyl ester see...Isofenphos
(±)-2-Ethoxy-1-methyl-2-oxoethyl-5-[2-chloro-4-(trifluoromethyl)phenoxy]-2-nitrobenzoate see..Lactofen
Ethoxy-4-nitrophenoxyphenylphosphine sulfide see.EPN

4-Ethoxy-7-phenyl-3,5-dioxa-6-aza-4-phosphaoct-6-ene-8-nitrile-4-sulfide see...Phoxim
2-(4-Ethoxyphenyl)-2-methylpropyl 3-phenoxybenzyl ether see...Ethofenprox
Ethoxyquine see...Ethoxyquin
5-Ethoxy-3-(trichloromethyl)-1,2,4-thiadiazole see. Etridiazole
6-Ethoxy-2,2,4-trimethyl-1,2-dihydroquinoline see. Ethoxyquin
ETHREL® see...Ethephon
d-N-Ethylacetamide carbanilate see...Carbetamide
2-Ethylamino-4-isopropylamino-6-chloro-S-triazine see. Atrazine
2-Ethylamino-4-isopropylamino-6-methylmercarpo-s-triazine see...Ametryn
2-Ethylamino-4-isopropylamino-6-methylthio triazine see...Ametryn
2-Ethylamino-4-isopropylamino-6-methylthio-1,3,5-triazine see...Ametryn
3-[(Ethylamino)m ethoxyphosphinothioyl)oxy]-2-butenoic acid, 1-methylethyl ester see...Propetamphos
S-Ethyl azepane-1-carbothioate see...Molinate
Ethyl azinphos see...Azinphos-ethyl
S-Ethyl bis(2-methylpropyl)carbamothioate see..Butylate
O-Ethyl S,S-[bis(1-methylpropyl)]phosphorodithioate see...Cadusafos
Ethyl bromophos see...Bromophos-ethyl
Ethyl(3-tert-butyl-1-dimethylcarbamoyl-1H-1,2,4-triazol-5-ylthio)acetate d-(-)-1-(Ethylcarbamoyl)ethyl phenylcarbamate see.Carbetamide
Ethyl N-(chloroacetyl)-N-(2,6-diethylphenyl)glycinate see...Diethatyl-ethyl
Ethyl-2-[4-((6-chlorobenzoxazol-2-yl)oxy)phenoxy] propionate see...Fenoxaprop-ethyl
Ethyl (D+)-2-[4-(6-chlor-2-benzoxazolyloxy)phenoxy] propanoate see..Fenoxaprop-ethyl
(±)-Ethyl 2-[4-((6-chloro-2-benzoxazolyl)oxy)phenoxy] propanoate see..Fenoxaprop-ethyl
(±)Ethyl-2-[-((6-chloro-2-benzoxazolyl)oxy)phenoxy] ropionate see...Fenoxaprop-ethyl
Ethyl-2-[(4-(6-chloro-2-benzoxazolyloxy))-phenoxy]propionate see...Fenoxaprop-ethyl
Ethyl 4-chloro-α-(4-chlorophenyl)-α-hydroxybenzene acetate see...Chlorobenzilate
Ethyl 2-[(((4-chloro-6-methoxypyrimidine-2-yl)aminocarbonyl)aminosulfonyl]benzoate see. Chlorimuron-ethyl
Ethyl 2-[((((4-chloro-6-methoxy-2-pyrimidinyl)amino)carbonyl)amino)sulfonyl]benzoate see...Chlorimuron-ethyl
Ethyl-2-[(((4-chloro-6-methoxyprimidin-2-yl)-carbonyl)-amino]sulfonyl]benzoate see...Chlorimuron-ethyl
Ethyl 4-chloro-2-oxo-3(2H)-benzothiazoleacetate see. Benazolin Ethyl
2-(2-Ethyl-4-chlorophenoxy)propanoic acid see. Mecoprop

Ethyl 2-[4-(6-chloro-2-quinoxalyloxy)phenoxy]propionate see...Quizalofop-Ethyl
Ethyl O-[5-(2-chloro-α,α,α-trifluoro-p-tolyloxy)-2-nitrobenzoyl]-dl-lactate see...Lactofen
S-Ethyl cyclohexylethylthiocarbamate see...Cycloate
Ethyl 4-(cyclopropylhydroxymethylene)-3,5-dioxocyclohexanecarboxylate see...Cimectacarb
p,p-Ethyl DDD see...Ethylan
p,p'-Ethyl-DDD see...Ethylan
Ethyl 4,4'-dichlorobenzilate see...Chlorobenzilate
Ethyl-p,p'-dichlorobenzilate see...Chlorobenzilate
Ethyl-4,4'-dichlorodiphenyl glycollate see. Chlorobenzilate
Ethyl-4,4'-dichlorophenyl glycollate see..Chlorobenzilate
Ethyl [((((2,3-dihydro-2,2-dimethyl-7-benzofuranyl)oxy)carbonyl)methylamino)thio]-N-(1-methylethyl)-β-alanine see...Benfuracarb
Ethyl N-(2,3-dihydro-2,2-dimethylbenzofuraN-7-yloxycarbonyl(methyl)aminothio)-N-isopropyl-β-alaninate see...Benfuracarb
S-Ethyldiisobutyl thiocarbamate see...Butylate
S-Ethyl N,N-diisobutylthiocarbamate see...Butylate
Ethyl-N,N-diisobutyl thiolcarbamate see...Butylate
Ethyl-α-[(dimethoxyphosphenothioyl]thio)benzeneacetate see...Phenthoate
Ethyl-O,O-dimethyl phosphorodithioylphenyl acetate see. Phenthoate
S-Ethyl dipropylcarbamothioate see...EPTC
Ethyl dipropylthiocarbamate see...EPTC
S-Ethyl dipropylthiocarbamate see...EPTC
S-Ethyl-N,N-di-N-propylthiocarbamate see...EPTC
Ethyl N,N-dipropylthiocarbamate see...EPTC
Ethyl di-N-propylthiolcarbamate see...EPTC
Ethyl N,N-dipropylthiolcarbamate see...EPTC
Ethyl N,N-di-N-propylthiolcarbamate see...EPTC
O-Ethyl S,S-[di(sec-butyl)]phosphorodithioate see. Cadusafos
Ethyldithiourame see...Disulfiram
Ethyldithiurame see...Disulfiram
Ethyleendichloride (Dutch) see...Ethylene Dichloride
Ethyleenoxide (Dutch) see...Ethylene Oxide
Ethyle (formiate D') (French) see...Ethyl Formate
1,1-Ethylene 2,2-dipyridylium dibromide see...Diquat
Ethylene aldehyde see...Acrolein
1,1'-Ethylene-2,2'-bipyridyliumdibromide see...Diquat
1,1'-Ethylene-2,2'-bipyridylium dibromide see...Diquat Dibromide
Ethylenebis(dithiocarbamato), manganese see.Maneb
Ethylenebis(dithiocarbamato)zinc see...Zineb
Ethylenebis(dithiocarbamic acid), manganese salt see...Maneb
Ethylenebis(dithiocarbamic acid manganese zinc complex (8CI) see...Mancozeb
Ethylenebis(dithiocarbamic acid), zinc salt see.Zineb
N,N'-Ethylene bis(dithiocarbamate de sodium) (French) see...Nabam
Ethylenebisdithiocarbamate manganese see...Maneb

N,N'-Ethylene bis(dithiocarbamate manganeux) (French) see...Maneb
Ethylenebis(dithiocarbamic acid) see...disodium salt see...Nabam
Ethylenebis(dithiocarbamic acid) manganous salt see. Maneb
Ethylene bromide see...Ethylene Dibromide
Ethylenecarboxamide see...Acrylamide
Ethylene chloride see...Ethylene Dichloride
1,2-Ethylene dibromide see...Ethylene Dibromide
1,2-Ethylene dichloride see...Ethylene Dichloride
Ethylene dipyridylium dibromide see...Diquat
Ethylene dipyridylium dibromide see...Diquat Dibromide
1,1'-Ethylene-2,2'-dipyridylium dibromide see...Diquat Dibromide
1,1'-Ethylene-2,2'-dipyridylium dibromide see...Diquat Dibromide
1,2-Ethylenediylbis(carbamodithioato)manganese see. Maneb
Ethylene hexachloride see...Hexachloroethane
Ethylene monoclinic tablets carboxamide see.Acrylamide
Ethylenenaphthalene see...Acenaphthene
1,8-Ethylenenaphthalene see...Acenaphthene
peri-Ethylenenaphthalene see...Acenaphthene
Ethylene (oxyde d') (French) see...Ethylene Oxide
1,3-Ethylenethiourea see...Ethylene Thiourea
N,N'-Ethylenethiourea see...Ethylene Thiourea
Ethyl ester of 4,4'-dichlorobenzilic acid see. Chlorobenzilate
Ethyl ester of O,O-dimethyldithiophosphoryl α-phenyl acetate acid see...Phenthoate
S-Ethyl ester hexahydro-1H-azepine-1-carbothioioate see...Molinate
3-Ethyl-2-(5-(3-ethyl-2-benzothiazolinylidene)-1,3-pentadienyl)benzothiazolium iodide see...Dithiazanine Iodide
S-Ethyl-N-ethyl-N-cyclohexylthiolcarbamate see. Cycloate
O,O,-Ethyl S-2(ethylthio)ethylphosphorodithioate see. Disulfoton
O-Ethyl S,S-dipropyl phosphorodithioate see...Ethoprop
O-Ethyl S,S-dipropyl dithiophosphate see...Ethoprop
Ethylformiaat (Dutch) see...Ethyl Formate
Ethyl formic ester see...Ethyl Formate
Ethyl green see...C.I. Basic Green 1
Ethyl guthion see...Azinphos-ethyl
S-Ethyl hexahydro-1H-azepine-1-carbothioate see. Molinate
S-Ethyl hexahydro-1-carbothioic see...Molinate
Ethyl 1-hexamethyleneiminecarbothiolate see...Molinate
S-Ethyl 1-hexamethyleneiminothiocarbamate see.Molinate
S-Ethyl N-hexamethyleneiminothiocarbamate see. Molinate
S-Ethyl N,N-hexamethyleneiminothiocarbamate see. Molinate
2,4,5-t Ethylhexyl ester see...2,4,5-Trichloro phenoxyacetic Acid, Esters

Ethylhexyl-2,4,5-t see...2,4,5-Trichlorophenoxyacetic Acid, Esters
2-Ethylhexyl 2,4,5-trichlorophenoxyacetate see...2,4,5-Trichlorophenoxyacetic Acid, Esters
Ethyl 2-hydroxy-2,2-bis(4-chlorophenyl)acetate see. Chlorobenzilate
Ethyl m-hydroxycarbanilate carbanilate see. Desmedipham
O-Ethyl O-(2-isopropoxycarbonyl) phenylisopropylphosphoramidothioate see...Isofenphos
(±)-5-Ethyl-2-(4-isopropyl-4-methyl-5-oxo-1H-imidazolin-2-yl)nicotinic acid (ammonium salt) see. Imazethapyr
(±)-5-Ethyl-2-(4-isopropyl-4-methyl-5-oxo-2-imidazolin-2-yl)nicotinic acid see...Imazethapyr
d-N-Ethyllactamide carbanilate (ester) see. Carbetamide
Ethyl mercaptophenylacetate-O,O-dimethyl phosphorocithioate see...Phenthoate
S-Ethylmercaptophenylacetate-O,O-dimethylphosphorodithioate see...Phenthoate
O-Ethyl O-(4-(methylmercapto)phenyl)-S-N-propylphosphorothionothiolate see...Sulprofos
Ethylmercuric chloride see...Ethyl Mercuric Chloride
Ethylmercury chloride see...Ethyl Mercuric Chloride
Ethyl methanoate see...Ethyl Formate
Ethyl methylene phosphorodithioate see...Ethion
2'-Ethyl-6'-methyl-N-(ethoxymethyl)-2-chloroacetanilide see...Acetochlor
2-Ethyl-6-methyl-1-N-(2-methoxy-1-methylethyl)chloroacetanilide see...Metolachlor
N-Ethyl-N-(2-methyl-2-propenyl)-2,6-dinitro-4-(trifluoromethyl)benzenamine see...Ethalfluralin
N-3-(1-Ethyl-1-methylpropyl)-5-isoxazolyl-2,6-dimethoxybenzamide see...Isoxaben
N-[3-(1-Ethyl-1-methylpropyl)-5-isoxazolyl]-2,6-dimethoxybenzamide see...Isoxaben
Ethyl 3-methyl-4-(methylthio)phenyl(1-methylethyl)phosphoramidate see...Fenamiphos
O-Ethyl O-(4-(methylthio)phenyl) S-propyl phosphorodithioate see...Sulprofos
O-Ethyl O-(4-(methylthio)phenyl)phosphorodithioic acid S-propyl ester see...Sulprofos
O-Ethyl O-(4-methylthiophenyl) S-propyl dithiophosphate see...Sulprofos
Ethyl 4-(methylthio)-m-tolylisopropylphosphoramidate see...Fenamiphos
O-Ethyl-O-[(4-nitrofenyl)-fenyl]monothiofosfonaat (Dutch) see...EPN
O-Ethyl-O-[(4-nitrofenyl)-fenyl]monothiofosfonaat (Dutch) see...EPN
Ethyl-p-nitrophenyl benzenethionophosphate see...EPN
Ethyl-p-nitrophenyl benzenethiophosphonate see...EPN
Ethyl-p-nitrophenyl ethylphosphate see...Paraoxon
O-Ethyl-O-p-nitrophenyl phenylphosphonothioate see. EPN
O-Ethyl-O-(4-nitrophenyl phenyl)phenylphosphonothioate see...EPN
Ethyl-p-nitrophenyl phenylphosphonothioate see...EPN
O-Ethyl-O-p-nitrophenyl phenylphosphonothioate see. EPN
Ethyl-p-nitrophenyl thionobenzenephosphate see...EPN
O-Ethyl-O-(4-nitrophenyl)-benzenethionophosphonate see...EPN
Ethyl-p-nitrophenylbenzenethionophosphonate see..EPN
Ethyl paraoxon see...Paraoxon
Ethyl parathion see...Parathion
ETHYL PARATHION see...Parathion
N-Ethylperfluorooctanesulfonamide see...Sulfluamid
S-Ethyl perhydroazepin-1-carbothioate see...Molinate
S-Ethyl perhydroazepine-1-thiocarboxylate see.Molinate
Ethyl [2-(4-phenoxyphenoxy)ethyl]carbamate see. Fenoxycarb
Ethyl[2-(p-phenoxyphenoxy)ethyl]carbamate see. Fenoxycarb
Ethyl 2-(p-phenoxyphenoxy)ethyl carbamate see. Fenoxycarb
N-Ethyl-2-[((phenylamino)carbonyl)oxy]propanamide, (d) isomer see...Carbetamide
(R)-N-Ethyl-2-[((phenylamino)carbonyl)oxy]propanamide see...Carbetamide
Ethyl phenylcarbamoyloxyphenylcarbamate see. Desmedipham
O-Ethyl-S-phenyl ethyldithiophosphonate see...Fonofos
O-Ethyl-S-phenyl ethylphosphonodithioate see..Fonofos
O-Ethyl-S-phenyl(RS)-ethylphosphonodithioate see. Fonofos
O-Ethyl phenyl-p-nitrophenylthiophosphonate see..EPN
Ethyl pirimiphos see...Pirimiphos-ethyl
N-(1-Ethylpropyl)-3,4-dimethyl-2,6-dinitroanaline see. Pendimethalin
N-(1-Ethylpropyl)-3,4-dimethyl-2,6-dinitrobenzenamine see...Pendimethalin
N-(1-Ethylpropyl)-2,6-dinitro-3,4-xylidine see. Pendimethalin
Ethyl pyrophosphate, tetra- see...TEPP
Ethyl rhodanate see...Ethylthiocyanate
S-[2-(Ethylsulfinyl)ethyl] O,O-dimethyl ester phosphorothioic acid see...Demeton-methyl
S-2-Ethylsulfinylethyl O,O-dimethyl phosphorothioate see...Demeton-methyl
S-[2-(Ethylsulfinyl)ethyl]O,O-dimethyl phosphorothioate see...Demeton-methyl
S-(2-(Ethylsulfinyl)ethyl) O,O-dimethyl phosphorothioate see...Demeton-methyl
Ethyl sulfocyanate see...Ethylthiocyanate
S-2-Ethylsulphinylethyl O,O-dimethyl phosphorothioate see...Demeton-methyl
Ethyl tetraphosphate see...Hexaethyl Tetraphosphate
Ethyl tetraphosphate, hexa- see...Hexaethyl Tetraphosphate
2-(Ethylthio)-4,6-bis(isopropylamino)-S-triazine see. Dipropetryn
Ethylthiodemeton see...Disulfoton
S-(2-(Ethylthio)ethyl)-O,O-dimethylphosphorothioate see...Demeton-methyl

*O*-(2-(Ethylthio)ethyl) *O,O*-dimethyl phosphorothioate see...Demeton-methyl
*S* (and *O*)-2-(Ethylthio)ethyl-*O,O*-dimethyl phosphorothioate see...Demeton-methyl
*S*-(2-(Ethylthio)ethyl)dimethyl phosphorothiolate see. Demeton-methyl
*S*-2-(Ethylthio)ethyl *O,O*-diethylester of phosphorodithioic acid see...Disulfoton
*S*-(2-(Ethylthio)ethyl phosphoric acid see...*O,O*-diethyl ester see...Demeton
*S*-(2-(Ethylthio)ethyl *O,O*-diethyl thiophosphate see. Demeton
*O*-(2-(Ethylthio)ethyl) *O,O*-diethyl thiophosphate see. Demeton
*S*-(2-(Ethylthio)ethyl)-*O,O*-dimethyl thiophosphosphate see...Demeton-methyl
Ethylthiomelton sulfoxide see...Oxydisulfoton
Ethyl thiophanate see...Thiophanate-methyl
Ethyl thiopyrophosphate see...Sulfotepp
Ethyl thiram see...Disulfiram
Ethyl thiudad see...Disulfiram
Ethyl thiurad see...Disulfiram
Ethyltrianol see...Tebuconazole
Ethyl trichlorophenylethylphosphonothioate see. Trichloronate
*O*-Ethyl-*O*-2,4,5-trichlorophenyl ethyl-phosphonothioate see...Trichloronate
Ethyl (*E,E*)-3,7,11-trimethyl-2,4-dodecadienoate see. Hydroprene
Ethyl (2*E*,4*E*)-3,7,11-trimethyl-dodeca-2-4-dienoate see. Hydroprene
Ethyl (2*E*,4*E*)-3,7,11-trimethyl-2,4-dodecadienoate see. Hydroprene
Ethyl (2*E*,4*E*,7*S*)-trimethyl-2,4-dodecadienoate see. Hydroprene
Ethyl tuads see...Disulfiram
Ethyl tuex see...Disulfiram
Ethyl zimate see...Zineb
ETICOL® see...Paraoxon
Etil azinfos (Spanish) see...Azinphos-ethyl
Etile (Formiato di) (Italian) see...Ethyl Formate
*N,N*'-Etilen-bis(ditiocarbammato) di manganese (Italian) see...Maneb
Etilene (ossido di) (Italian) see...Ethylene Oxide
Etilentiourea (Spanish) see...Ethylene Thiourea
ETILON® see...Parathion
Etiltriazotion see...Azinphos-ethyl see...Azinphos-ethyl
ETIOL® see...Malathion
Etion (Spanish) see...Ethion
ETMT see...Etridiazole
ETO see...Ethylene Oxide
Etofenprox see...Ethofenprox
Etridiazole see...Etridiazole
ETROLENE® see...Ronnel
ETSAN® see...Cycloate
ETU see...Ethylene Thiourea
Etylenu Tlenek (Polish) see...Ethylene Oxide

EULAN SP® see...Cyfluthrin
EUREX® see...Cycloate
EUXYL®-K-100 see...Benzyl Alcohol
EVAU-SUPERFALL® see...Sodium Chlorate
EVEGFRU FOSMITE® see...Ethion
EVERCIDE® see...Esfenvalerate
EVERCIDE® see...Fenvalerate
EVERCIDE INTERMEDIATE® 2265 see...Tetramethrin
EVERSHIELD CAPTAN/MALATHION® see.Malathion
EVERSHIELD T SEED PROTECTORANT® see.Thiram
EVIK® see...Ametryn
EVIPOL® see...Cyproconazole
EVITAL® see...Norflurazon
EVITS® see...Phosphoric Acid
EVOLA® see...*para*-Dichlorobenzene
EVOLVE® see...Cymoxanil
EVOLVE® see...Mancozeb
EVOLVE® see...Thiophanate-Methyl
EXAGAMA® see...Lindane
EXATHIOS® see...Malathion
EXCALIBER® see...*lamda*-Cyhalothrin
EXCEED® see...Primisulfuron-Methyl
EXCEL® see...Fenoxaprop-ethyl
EXCELCIDE® see...Mevinphos
EXCEL-S-PLUS® see...Oryzalin
EXHORAN® see...Disulfiram
EXHORRAN® see...Disulfiram
EXILIS® see...6-Benzaldenine
EXODIN® see...Diazinon
EXOFENE® see...Hexachlorophene
EXOLIT LPKN® 275 see...Phosphorus
EXOLIT VPK-N® 361 see...Phosphorus
EXOLITE® 405 see...Phosphorus
EXOTHERM® see...Chlorothalonil
EXOTHERM TERMIL® see...Chlorothalonil
EXOTHION® see...Endothion
EXP 419® see...Mecoprop
EXP 31039B® see...Cyclanilide
EXPAND® see...Sethoxydim
EXPEDITE® see...Oryzalin
EXPERIMENTAL HERBICIDE 732® see...Terbacil
EXPERIMENTAL INSECTICIDE 3911® see...Phorate
EXPERIMENTAL INSECTICIDE 7744® see...Carbaryl
EXPERIMENTAL NEMATOCIDE 18,133 ® see. Thionazin
EXPORSAN® see...Bensulide
EXPRESS® see...Thifensulfuron Methyl
EXPRESS® see...Tribenuron-Methyl
EXPRESS®-75 DF see...Tribenuron-Methyl
Exsiccated ferrous sulfate see...Ferrous Sulfate
Exsiccated ferrous sulphate see...Ferrous Sulfate
EXTERMATHION® see...Malathion
EXTERRA REQUIEM TERMITE BAIT® see. Chlorfluazuron
EXTHRIN FMC 249® see...Allethrin
EXTINGUISH® see...Methoprene
EXTRAR® see...Dinitro-o-cresol (DNOC)
EXTRAX® see...Rotenone

EXTRAZINE® see...Cyanazine
EXTREME® see...Imazethapyr
E-Z-EX® see...Thiabendazole
E-Z FLO® see...Endosulfan
E-Z-FLO® see...Methyl Parathion
E-Z FLO® see...Zinc Phosphide
E-Z-OFF® see...Magnesium Chlorate
E-Z-OFF® D see...Tribufos
EZY-PICKIN' COTTON DEFOLIANT® see..Cacodylic Acid

- F -

F 10® see...Maneb
F 461® see...Oxycarboxin
F 735® see...Carboxin
F 1991® see...Benomyl
F 2636 see...Halosulfuron-methyl
F 2966® see...Mancozeb
F 5384® see...Bensulfuron-methyl
F 6285® see...Sulfentrazone
FA see...Formaldehyde
FAA see...Acetylaminofluorene
2-FAA see...Acetylaminofluorene
FAA see...Fluoroacetamide
FAA see...Fluoroacetic Acid
FAC® see...Prothoate
FAC® 20 see...Prothoate
FACET® see...Quinclorac
FACTOR® see...Prodiamine
FAIR 30® see...Maleic Hydrazide
FAIR PLUS® see...Maleic Hydrazide
FAIR PS® see...Maleic Hydrazide
FALGRO® see...Gibberellic Acid
FALIGRUEN® see...Copper Oxychloride
FALITHION® see...Fenitrothion
FALITIRAM® see...Thiram
FALKITOL® see...Hexachloroethane
FALL OUT® see...Methyl Parathion
FALLOWMASTER® see...Dicamba
FALLOW MASTER® see...Glyphosate
FALL® see...Sodium Chlorate
FAMFOS® see...Famphur
Famfur (Spanish) see...Famphur
FAMIX® see...Famphur
FAMOPHOS® see...Famphur
FAMOPHOS WARBEX® see...Famphur
FAMPHOS® see...Famphur
FANATE® see...Thiophanate-Methyl
FANFOS® see...Famphur
FANICIDE® see...Dinoseb
FANNOFORM® see...Formaldehyde
FANTERRIN® see...Oxytetracycline Calcium
FAP® see...Kinetin (Cytokinin)
FAR-GO® see...Triallate
FARMANEB® see...Maneb
FARMCO DIURON® see...Diuron
FARMCO® see...Amitrole

FARMCO® see...2,4-D
FARMCO® ATRAZINE see...Atrazine
FARMCO FENCE RIDER® see...2,4,5-Trichlorophenoxyacetic Acid, Esters
FARMCO PROPANIL® see...Propanil
FARMON PDQ® see...Diquat
FARMON PDQ® see...Diquat Dibromide
FARMOZ® see...Aluminum Phosphide
FARTOX® see...Quintozene
Fasciolin see...Carbon Tetrachloride
FASCIOLIN® see...Hexachloroethane
FASCO PARIS GREEN® see...Copper Acetoarsenite
FASCO-TERPENE® see...Toxaphene
FAS-NOX® see...Quinclorac
FASTAC® see...*alpha*-Cypermethrin
FASTAC® see...Cypermethrin
FASTER® see...Fomesafen
FATAL® see...DCPA
FATEL® see...Acephate
FB/2® see...Diquat
FB/2® see...Diquat Dibromide
FBC® see...Quizalofop-Ethyl
FBC® PROTECTANT FUNGICIDE see...Zineb
F.C.® FORMULAS see...Carbon Tetrachloride
FCR 1272® see...Cyfluthrin
FDA 1446 see...Allethrins
FDA 1541 see...EPTC
FDA 0101 see...Sodium Fluoride
FDN® see...Diphenamid
FECAMA® see...Dichlorvos
FECUNDAL® 100EC see...Imazalil
Feeno see...Phenothiazine
FEGLOX® see...Diquat
FEGLOX® see...Diquat Dibromide
FEKAMA® see...Dichlorvos
FELAN® see...Molinate
FEMA No. 2097 see...Anisole
Femma see...Phenylmercury Acetate
FENAB® see...Fenac
FEN-ALL® see...Trichlorobenzoic Acid
FENAM® see...Diphenamid
FENAMIN® see...Atrazine
FENAMINE® see...Amitrole
FENAMINE® FENATROL® see...Atrazine
FENAMINOSULF® see...Fenaminosulf
FENATROL® see...Fenac
FENAVAR® see...Amitrole
Fenbutatin-oxyde see...Fenbutatin Oxide
FENCAL® see...Calcium Arsenate
FENCE RIDER® see...2,4,5-T
Fenchloorfos (Dutch) see...Ronnel
Fenchlorfos see...Ronnel
Fenchlorfosu (Polish) see...Ronnel
Fenchlorophos see...Ronnel
Fenchlorphos see...Ronnel
FENDONA® see...*alpha*-Cypermethrin
FENETHANIL® see...Fenbuconazole
Fenetrazole see...Tebuconazole

FENITEX® see...Fenitrothion
FENITOX® see...Fenitrothion
Fenitrotion (Hungarian) see...Fenitrothion
FENKILL® see...Fenvalerate
Fenmedifam see...Phenmedipham
FENNOSAN® B 100 see...Dazomet
FENNOTOX® see...Heptachlor
FENOBCARB® see...BPMC
FENOBUCARB® see...BPMC
FENOCIL® see...Bromacil
Fenolovo acetate see...Triphenyltin Compounds
FENOLOVO® see...Fentin Hydroxide
FENOMORE® see...Silvex
FENOPHOSPHON® see...Trichloronate
Fenoprop see...Silvex
FENOPROP® see...Silvex
Fenothiazine (Dutch) see...Phenothiazine
Fenothrin see...D-Phenothrin
Fenothrin, (±)- see...D-Phenothrin
Fenothrin, (+)-cis,trans- see...D-Phenothrin
(+)-cis,trans-Fenothrin see...D-Phenothrin
Fenotiazina (Italian) see...Phenothiazine
FENOVERM® see...Phenothiazine
FENOXYPROP® see...Fenoxaprop-ethyl
FENPROPANATE® see...Fenpropathrin
Fenpyrate see...Pyridate
FENSTAN® see...Fenitrothion
Fenthoate see...Phenthoate
FENTIAZIN® see...Phenothiazine
Fentin see...Fentin Hydroxide
Fentin see...Triphenyltin Compounds
Fentin acetaat (Dutch) see...Triphenyltin Compounds
Fentin acetat (German) see...Triphenyltin Compounds
Fentin acetate see...Triphenyltin Compounds
Fentin chloride see...Triphenyltin Compounds
Fentine acetate (French) see...Triphenyltin Compounds
Fentin hydroxide see...Triphenyltin Compounds
Fenvalerate A-α see...Esfenvalerate
s-Fenvalerate (S)-α-cyano-3-phenoxybenzyl (S)-2-(4-chlorophenyl)-3-methylbutyrate see...Esfenvalerate
Fenvaleriato (Spanish) see...Fenvalerate
Fenylmercuriacetat (Czech) see...Phenylmercury Acetate
FEOSOL® see...Ferrous Sulfate
FEOSPAN® see...Ferrous Sulfate
Ferbam, iron salt see...Ferbam
FERBAM 50® see...Ferbam
FERBECK® see...Ferbam
FER-IN-SOL® see...Ferrous Sulfate
FERKETHION® see...Dimethoate
FERMATE FERBAM FUNGICIDE® see...Ferbam
FERMIDE® see...Thiram
FERMOCIDE® see...Ferbam
FERNACOL® see...Thiram
FERNASAN® see...Thiram
FERNESTA® see...2,4-D
FERNEX® see...Pirimiphos-Ethyl
FERNIDE® see...Thiram

FERNIMINE® see...2,4-D
FERNOS® see...Pirimicarb
FERNOXONE® see...2,4-D
FERRADOUR® see...Ferbam
FERRADOW® see...Ferbam
FERRALYN® see...Ferrous Sulfate
FERRIAMICIDE® see...Mirex
Ferric dimethyl dithiocarbamate see...Ferbam
FERRO-GRADUMET® see...Ferrous Sulfate
Ferro-theron see...Ferrous Sulfate
Ferrosulfat (German) see...Ferrous Sulfate
Ferrosulfate see...Ferrous Sulfate
Ferrosulphate see...Ferrous Sulfate
Ferrous sulphate (1:1) see...Ferrous Sulfate
Fersolate see...Ferrous Sulfate
FERTI-LOME® see...Sodium Methanearsonate (MSMA)
FERVINAL® see...Sethoxydim
FERXONE® see...2,4-D
FESDAN®
FEZUDIN® see...Diazinon
FF4961® see...Imazalil
FF6135' HERBICIDE 326® see...Linuron
F.i.a. GRAIN FUMIGANT® see...Carbon Tetrachloride
FI CLOR 71® see...Dichloroisocyanuric Acid
FICAM® see...Bendiocarb
FIDIS® see...Propiconazole
FIELD MASTER® see...Atrazine
FIELD MASTER® see...Glyphosate
FILARIOL® see...Bromophos-ethyl
FILITOX® see...Methamidophos
FINAVEN® see...Difenzoquat
FINESSE® see...Chlorsulfuron
FINESSE® see...Metsulfuron-methyl
FINISH® see...Cyclanilide
FINISH® see...Ethephon
FINITRON® see...Sulfluamid
Fintin acetato (Italian) see...Triphenyltin Compounds
Fintin hydroxid (German) see...Fentin Hydroxide
Fintin hydroxid (German) see...Triphenyltin Compounds
Fintine hydroxyde (French) see...Fentin Hydroxide
Fintine hydroxyde (French) see...Triphenyltin Compounds
Fintin hydroxyde (Dutch) see...Triphenyltin Compounds
Fintin idrossido (Italian) see...Triphenyltin Compounds
FINTROL® see...Antimycin A
FIREBAN® see...Phosmet
FIREMAN® see...Oxytetracycline Calcium
FIRE POWER® see...Glyphosate
FIRE POWER® see...Oxyfluorfen
FIRMOTOX® see...Pyrethrins or Pyrethrum
FIRST CUT No. 8® see...Ammonium Nitrate
FIRSTLINE® see...Sulfluamid
FIRSTRATE® see...Cloransulam-methyl
FISH-TOX® see...Rotenone
FISONS B25® see...Barban

FISONS GREENFLY AND BLACKFLY KILLER® see.Malathion
FISONS NC® 2964  see...Methidathion
Fisostigmina (Spanish) see...Physostigmine
FITOSOL® see...Trichloronate
FL see...Metoxuron
FLAC® see...Calcium Arsenate
FLAGON® see...Bromoxynil
FLAME PLUS® see...Glyphosate
FLAVIN-SANDOZ® see...Dinitro-o-cresol (DNOC)
FLEATROL® see...Methoprene
FLECTRON® see...Cypermethrin
FLEXIDOR® see...Isoxaben
FLEXLAN® see...Oryzalin
FLEXLAN® see...Trifluralin
FLEXSTAR see...Fomesafen
FLEX® see...Fomesafen
FLIBOL E® see...Trichlorfon
FLIEGENTELLE® see...Trichlorfon
FLINT® see...Cyproconazole
FLINT® see...Trifluralin
FLIT® 406 see...Captan
FLO-MET® see...Fluometuron
FLOMORE® (butyl ester) see...2,4,5-Trichlorophenoxyacetic Acid, Esters
FLO-PRO IMZ® see...Imazalil
FLO PRO T SEED PROTECTANT® see...Thiram
FLO PRO V SEED PROTECTANT® see...Carboxin
FLORA® see...Trifluralin
FLORALTONE® (with 2,3,5-triiodobenzoic acid) see.Gibberellic Acid
FLORDIMEX® see...Ethephon
FLOREL® see...Ethephon
FLORGIB® see...Gibberellic Acid
FLORIDINE® see...Sodium Fluoride
FLOROCID® see...Sodium Fluoride
Floropryl see...Isofluorphate
FLOTILLA® see...Vinclozolin
FLO TIN 4L® see...Fentin Hydroxide
FLO TIN 4L® see...Triphenyltin Compounds
FLOUR-O-KOTE® see...Sodium Fluoride
Flour sulphur see...Sulfur
FLOWABLE ATRAZINE® see...Atrazine
Flowers of sulphur see...Sulfur
FLOWMASTER® see...Dicamba
FLOZENGES® see...Sodium Fluoride
FL-TABS® see...Sodium Fluoride
FLUAZINAM 50 WP® see...Fluazinam®
FLUBALEX® see...Benefin
Fluenethyl see...Fluenetil
Fluenyl see...Fluenetil
Flufenacet see...Thiafluamide
Flukoids see...Carbon Tetrachloride
Fluoacetato sodico (Spanish) see...Sodium Fluoroacetate
2-3 see...Fluoetanol (Spanish) see...Ethylene Fluorohydrin
Fluophosphoric acid di(dimethylamide) see...Dimefox

Fluophosphoric acid see...diisopropyl ester see.Isofluorphate
Fluoracetato di sodio (Italian) see...Sodium Fluoroacetate
FLUORAKIL® 3 see...Sodium Fluoroacetate
FLUORAKIL 100® see...Fluoroacetamide
FLUORAL® see...Sodium Fluoride
2-Fluorenylacetamide see...Acetylaminofluorene
N-2-Fluoren-2-yl acetamide see...Acetylaminofluorene
Fluoressigsaeure (German) see...Sodium Fluoroacetate
FLUORGUARD® see...Sulfluamid
Fluorhydric acid see...Hydrogen Fluoride
Fluoric Acid see...Hydrogen Fluoride
Fluorid sodny (Czech) see...Sodium Fluoride
FLUORIDENT® see...Sodium Fluoride
FLUORIGARD® see...Sodium Fluoride
FLUORINEED® see...Sodium Fluoride
FLUORINSE® see...Sodium Fluoride
FLUORITAB® see...Sodium Fluoride
2-Fluoroacetamide see...Fluoroacetamide
Fluoroacetate see...Fluoroacetic Acid
2-Fluoroacetic acid see...Fluoroacetic Acid
Fluoroacetic acid amide see...Fluoroacetamide
Fluoroacetic acid, sodium salt see...Sodium Fluoroacetate
Fluorocythrin see...Flucythrinate
Fluorodiisopropyl phosphate see...Isofluorphate
Fluoroethanoic acid see...Fluoroacetic Acid
β-Fluoroethanol see...Ethylene Fluorohydrin
β-Fluoroethyl-4-biphenylacetate see...Fluenetil
Fluoroethylic ester of xenylacetic acid see...Fluenetil
N-(4-Fluorophenyl)-N-(1-methylethyl)-2[(5-(trifluoromethyl)-1,3,4-thiadiazol-2-yl)oxy]acetamide benzoate see...Thiafluamide
Fluoropryl see...Isofluorphate
Fluorotributylstannane see...Tributyltin Fluoride
Fluoruro de hidrogeno (Spanish) see...Hydrogen Fluoride
Fluorure de potassium (French) see...Potassium Fluoride
Fluorure de sodium (French) see...Sodium Fluoride
Fluoruro de sulfurilo (Spanish) see...Sulfur Fluoride
Fluorure de sulfuryle (French) see...Sulfur Fluoride
Fluorure de N,N,N',N'-tetramethyle phosphoro-diamide (French) see...Dimefox
Fluoruro sodico (Spanish) see...Sodium Fluoride
Fluosilicate de ammonium (French) see...Ammonium Hexafluorosilicate
Fluosilicato amonico (Spanish) see...Ammonium Hexafluorosilicate
Fluostigmine see...Isofluorphate
FLURA-GEL® see...Sodium Fluoride
FLURCARE® see...Sodium Fluoride
FLURENE SE® see...Trifluralin
2-Fluroethanol see...Ethylene Fluorohydrin
FLUTRIX® see...Trifluralin
FLUX MAAG® see...Nicotine
FLYBANDS® see...Dimetilan

FLY-DIE® see...Dichlorvos
FLY FIGHTER® see...Dichlorvos
FLYKILLER® see...Naled
FLY PATROL® see...Tetrachlorvinphos
FLYTEK® see...Methomyl
FLYTROL® see...Diazinon
FMA see...Phenylmercury Acetate
FMC 2070® see...Thiram
FMC 5462 see...Endosulfan
FMC 5488® see...Tetradifon
FMC 9044® see...Binapacryl
FMC 9260® see...Tetramethrin
FMC 10242® see...Carbofuran
FMC 11092® see...Karbutilate
FMC 17370® see...Resmethrin
FMC 30980® see...Cypermethrin
FMC 35001® see...Carbosulfan
FMC 45497® see...*alpha*-Cypermethrin
FMC 45497® see...Cypermethrin
FMC 45498® see...Deltamethrin
FMC 45806® see...Cypermethrin
FMC 56701® see...*alpha*-Cypermethrin
FMC 57020® see...Clomazone
FMC 58000® see...Bifenthrin
FMC 67825® see...Cadusafos
FMC 97285® see...Sulfentrazone
FMC NYNAMITE® see...Methyl Parathion
FOE 1976® see...Mefenacet
FOE 5043® see...Thiafluamide
FOG® 3 see...Malathion
FOLBEX® see...Chlorobenzilate
FOLBEX SMOKE STRIPS® see...Chlorobenzilate
FOLCID® see...Captafol
FOLCORD® see...Cypermethrin
FOLETHION® see...Fenitrothion
FOLEX® 6EC see...Tribufos
FOLI-ZYME® see...Gibberellic Acid
FOLI-ZYME® see...Kinetin (Cytokinin)
FOLIAR TRIGGRR® see...Kinetin (Cytokinin)
FOLICUR® see...Tebuconazole
FOLIDOC® see...Methyl Parathion
FOLIDOL® see...Parathion
FOLIDOL-80® see...Methyl Parathion
FOLIDOL E® see...Parathion
FOLIDOL E-605® see...Parathion
FOLIDOL E&E 605® see...Parathion
FOLIDOL M® see...Methyl Parathion
FOLIDOL M-40® see...Methyl Parathion
FOLIDOL OIL® see...Parathion
FOLIMAT® see...Omethoate
FOLIO® GOLD see...Metalaxyl
FOLISTAR® see...Flutolanil
FOLITHION® see...Fenitrothion
FOLITRAZOLE® see...Tebuconazole
FOLOSAN® see...Quintozene
FOLOSAN® see...Tecnazene
FOMAC® see...Hexachlorophene
FOMAC 2® see...Quintozene

FOMESAFEN® SODIUM see...Fomesafen
FONT 360® see...Glyphosate
Foraat (Dutch) see...Phorate
Forato (Spanish) see...Phorate
FORCE® see...Tefluthrin
FOREDEX 75® see...2,4-D
FORE® see...Mancozeb
FORLIN® see...Hexachlorocyclohexanes
FORLIN® see...Lindane
Formaldehido (Spanish) see...Formaldehyde
Formaldehyd (Czech) see...Formaldehyde
Formaldehyd (Polish) see...Formaldehyde
Formalin see...Formaldehyde
Formalin 40 see...Formaldehyde
Formalina (Italian, Spanish) see...Formaldehyde
Formaline (German) see...Formaldehyde
FORMAL® see...Malathion
FORMALITH® see...Formaldehyde
Formalin-loesungen (German) see...Formaldehyde
FORMALSOL® see...Thiram
Formamide see...1,1'-dithiobis(*N,N*-dimethylthio-
see.Thiram
Formamidine, *N'*-(4-chloro-*o*-tolyl)-*N*-dimethyl- see...Chlordimeform
Formamidine, *N*-methyl-*N'*-2,4-xylyl-*N*-(*N*-2,4-xylylformimidoyl)- see...Amitraz
FORMEC® see...Mancozeb
Formiate de methyle (French) see...Methyl Formate
Formiato de metilo (Spanish) see...Methyl Formate
Formic acid, dithiobis(thio-, O,O-dimethyl ester see...Tributyltin Chloride
Formic acid, ethyl ester see...Ethyl Formate
Formic acid, methyl ester see...Methyl Formate
Formic aldehyde see...Formaldehyde
Formic ether see...Ethyl Formate
FORMISOTON® see...Formic Acid
FORMOL® see...Formaldehyde
Formonitrile see...Hydrogen Cyanide
Formotion (Spanish) see...Formothion
FORMULA 40® see...2,4-D
Formylic acid see...Formic Acid
*S*-(2-(Formylmethylamino)2-oxoethyl)*O,O*-dimethylphosphorodithioate see...Formothion
*S*-(*N*-Formyl-*N*-methylcarbamoylmethyl)dimethylphosphorodithiolothionate see...Formothion
*N*-Formyl-*N*-methylcarbamoylmethyl-*O,O*-dimethylphosphorodithioate see...Formothion
*S*-(*N*-Formyl-*N*-methylcarbamoylmethyl)-*O,O*-dimethylphosphorodithioate see...Formothion
FOROTOX® see...Trichlorfon
FORPEN-50® see...Pentachlorophenol
FORPHATE® see...Acephate
FORRON® see...2,4,5-T
FORRON® see...2,4,5-Trichlorophenoxyacetic Acid, Esters
FORSTAN® see...Oxythioquinox
FORST U 46® see...2,4,5-T

FORST® U 46  see...2,4,5-Trichlorophenoxyacetic Acid, Esters
Forte  see...D-Phenothrin
FORTE®  see...D-Phenothrin
FORTEX®  see...2,4,5-T
FORTEX®  see...2,4,5-Trichlorophenoxyacetic Acid, Esters
FORTEX SC®  see...Diuron
FORTHION®  see...Malathion
FORTRESS®  see...Chlorethoxyfos
FORTRESS®  see...Procymidone
FORTRESS®  see...Triallate
FORTROL®  see...Cyanazine
FORTURF®  see...Chlorothalonil
FORUM DC®  see...Dimethomorph
FORZA®  see...Tefluthrin
FOSCHLOR®  see...Trichlorfon
FOSCHLOR R®  see...Trichlorfon
FOSCHLOR R-50®  see...Trichlorfon
Foschlorem (Polish)  see...Trichlorfon
Fosetyl aluminum  see...Fosetyl-Al
FOSFAKOL®  see...Paraoxon
FOS-FALL® A  see...Tribufos
Fosfamia (Spanish)  see...Phosphine
Fosfamidon (Spanish)  see...Phosphamidon
Fosfamidone  see...Phosphamidon
Fosfato sodico (Spanish)  see...Trisodium Phosphate
FOSFERMO®  see...Parathion
FOSFERNO®  see...Parathion
FOSFERNO M 50®  see...Methyl Parathion
FOSFEX®  see...Parathion
FOSFIVE®  see...Parathion
Fosforo bianco (Italian)  see...Phosphorus
Fosforo blanco (Spanish)  see...Phosphorus
Fosforotritioato de S,S,S-tributilo (Spanish)  see..Tribufos
Fosforowodor (Polish)  see...Phosphine
Fosforzuuroplossingen (Dutch)  see...Phosphoric Acid
FOSFOTHION®  see...Malathion
FOSFOTION®  see...Malathion
Fosfuri di alluminio (Italian)  see...Aluminum Phosphide
Fosfuro aluminico (Spanish)  see...Aluminum Phosphide
Fosfuro de zinc (Spanish)  see...Zinc Phosphide
Fosmet (Spanish)
FOSOVA®  see...Parathion
FOSTERN®  see...Parathion
FOSTION®  see...Prothoate
FOSTION MM®  see...Dimethoate
FOSTOX®  see...Parathion
FOSTRIL®  see...Hexachlorophene
FOSVEL®  see...Leptophos
FOSVEX®  see...TEPP
Foszfamidon  see...Phosphamidon
Fotox  see...Arsenic Pentoxide
FOUMARIN®  see...Coumafuryl
FOZALON®  see...Phosalone
FOZZATE®  see...Glyphosate
FRAMED®  see...Simazine
FRAM FLY KILL®  see...Methomyl

FRANKLIN®  see...Calcium Carbonate
FRATOL®  see...Sodium Fluoroacetate
FREEDOM®  see...Trifluralin
FREEFLO®  see...Diuron
French green  see...Copper Acetoarsenite
French green  see...Paris Green
Freon 10  see...Carbon Tetrachloride
Freon 150  see...Ethylene Dichloride
FRESHGARD®  see...Imazalil
FRESHGARD®  see...Thiabendazole
FRESHGARD 40®  see...Sodium Dimethyldithiocarbamate
FRONTIER®  see...Dimethenamid
FRONTIERSMAN®  see...Capsaicin
FRONTLINE  see...Fipronil
FRONTROW®  see...Cloransulam-methyl
FRONTROW®  see...Flumetsulam
FROWNCIDE®  see...Fluazinam®
FRUITDO®  see...Copper(II)-8-hydroxyquinoline
FRUIT FIX® 200  see...Naphthoxyacetic Acid
FRUITONE®  see...1-Naphthaleneacetic Acid
FRUITONE®  see...1-Naphthaleneacetamide
FRUITONEA®  see...2,4,5-T
FRUITONE-A®  see...2,4,5-Trichlorophenoxyacetic Acid, Esters
FRUITONE T®  see...Silvex
FRUIT-O-NET®  see...Silvex
FRUMIN-AL®  see...Disulfoton
FRUMIN G®  see...Disulfoton
FTALOPHOS®
FUAM®  see...Bendiocarb
Fuberidatol  see...Fuberidazole
Fuberisazol  see...Fuberidazole
Fubridazole  see...Fuberidazole
FUCHING JUJR®  see...Flucythrinate
FUCLASIN®  see...Ziram
FUCLASIN® ULTRA  see...Ziram
Fuel oil No. 1  see...Kerosene
FUKLASIN®  see...Ziram
FUKLASIN ULTRA®  see...Ferbam
FULFILL®  see...Pymetrozine
FULTIME®  see...Acetochlor
FUM-A-CIDE®  see...Ethylene Dibromide
FUM-A-CIDE® 15  see...Chloropicrin
FUMAGONE®  see...Dibromochloropropane
FUMARIN®  see...Coumafuryl
FUMAZONE®  see...Dibromochloropropane
FUMAZONE®  see...1,3-Dichloropropene
FUMETO-TENDUST®  see...Nicotine
FUMIGRAIN®  see...Acrylonitrile
FUMITE®  see...Tecnazene
FUMITE DICOFOL®  see...Dicofol
FUMITE DICLORAN SMOKE ACARICIDE®  see...Dichloran
FUMITE RONALIN®  see...Vinclozolin
FUMITOXIN®  see...Aluminum Phosphide
FUMO-GAS®  see...Ethylene Dibromide
FUNDAL®  see...Chlordimeform

FUNDAL® 500 see...Chlordimeform
FUNDAZOL® see...Benomyl
FUNDEX® see...Chlordimeform
FUNGACIDE D-1991® see...Benomyl
FUNGAFLOR® see...Imazalil
FUNGCHEX® see...Mercuric Chloride
FUNGICHLOR® see...Quintozene
FUNGICIDE 4T® see...Thiabendazole
FUNGICIDE 406® see...Captan
FUNGICIDE 1991® see...Benomyl
FUNGICIDE 5223® see...Oxythioquinox
FUNGICIDE F® see...Dichlorophene
FUNGICIDE GM® see...Dichlorophene
FUNGICIDE M® see...Dichlorophene
FUNGIFEN® see...Pentachlorophenol
FUNGIMAR® see...Cuprous Oxide
FUNGINIL® see...Chlorothalonil
FUNGI-RHAP® see...Cuprous Oxide
FUNGISOL® see...Carbendazim
FUNGISOL® see...Oxycarboxin
FUNGI-SPERSE II see...Copper Sulfate
FUNGITOX® see...Thiophanate-Methyl
FUNGITOX OR® see...Phenylmercury Acetate
FUNGO® see...Thiophanate-Methyl
FUNGOCHROM® (USA) see...Benomyl
FUNGOL B® see...Sodium Fluoride
FUNGOSTOP® see...Ziram
FUNGUS BAN® TYPE II see...Captan
FURACARB® see...Carbofuran
FURACON® see...Benfuracarb
FURADAN® see...Carbofuran
FURAN® see...Carbofuran
2-Furanmethanamine see...$N$-1$H$-purin-6-yl- see.
Kinetin (Cytokinin)
2-(2-Furanyl)-1H-benzimidazole see...Fuberidazole
$N$-(2-Furanylmethyl)-1$H$-purin-6-amine see...Kinetin
(Cytokinin)
FURATOL® see...Sodium Fluoroacetate
$N^6$-Furfuryladenine see...Kinetin (Cytokinin)
6-Furfurylaminopurine see...Kinetin (Cytokinin)
6-(Furfurylamino)purine see...Kinetin (Cytokinin)
$N^6$-(Furfurylamino)purine see...Kinetin (Cytokinin)
Furidazol see...Fuberidazole
Furidazole see...Fuberidazole
FURODAN® see...Carbofuran
FURORE® see...Fenoxaprop-ethyl
2-(2'-Furyl)-benzimidazole see...Fuberidazole
2-(2-3-Furyl)benzimidazole see...Fuberidazole
3-(1-Furyl-3-acetylethyl)-4-hydroxycoumarin see.
Coumafuryl
3-($\alpha$-Furyl-b-acetylaethyl)-4-hydroxycumarin (German)
see...Coumafuryl
FURY® see...*alpha*-Cypermethrin
FUSAREX® see...Tecnazene
FUSILADE® see...Fluazifop-butyl
FUSION® see...Fluazifop-butyl
FUSSOL® see...Fluoroacetamide
FW 293® see...Dicofol

FW 734® see...Propanil
FW 925® see...Nitrofen
FYAFANON® see...Malathion
FYCOL 8® see...Copper Oxychloride
FYDE® see...Formaldehyde
FYDULAN® see...Dichlobenil
FYDULAN® see...Dalapon
FYDUMAS® see...Dichlobenil
FYDUSIT® see...Dichlobenil
FYRAN 206K® see...Ammonium Sulfamate
FYFANON® see...Malathion
FYTOLAN® see...Copper Oxychloride

- G -

G 4® see...Dichlorophene
G 11® see...Hexachlorophene
G 301® see...Diazinon
G 338® see...Chlorobenzilate
G 996® see...Ethephon
G 23992® see...Chlorobenzilate
G 24480® see...Diazinon
G 27692® see...Simazine
G 30027® see...Atrazine
G 31435® see...Prometon
G 34161® see...Prometryn
G 34162® see...Ametryn
GA see...Gibberellic Acid
GA$_3$ see...Gibberellic Acid
GA 10832® see...Profluralin
GALAXY® see...Acifluorfen
GALECRON® see...Chlordimeform
GALESAN® see...Diazinon
GALIGAN® see...Oxyfluorfen
GALLANT® see...Haloxyfop-methyl
GALLERY® see...Isoxaben
GALLOGAMA® see...Lindane
GALLOTOX® see...Phenylmercury Acetate
GALLUP® see...Glyphosate
GAMACID® see...Lindane
GAMAPHEX® see...Hexachlorocyclohexanes
GAMAPHEX® see...Lindane
GAMBIT® see...Clomazone
GAMENE® see...Lindane
GAMIXEL® see...Paraquat
Gammabenzene hexachlorocyclohexane (gamma isomer)
see...Lindane
GAMMA-COL® see...Lindane
Gammahexa see...Lindane
Gammahexane see...Lindane
GAMMALEX® see...Lindane
GAMMALIN® see...Lindane
GAMMALIN 20 see...Lindane
GAMMAPHEX® see...Lindane
GAMMASAN 30® see...Lindane
GAMMATERR® see...Lindane
GAMMEX® see...Lindane
GAMMEXANE® see...Hexachlorocyclohexanes

GAMMEXANE® see...Lindane
GAMMEXENE® see...Lindane
GAMMOPAZ® see...Lindane
GAMONIL® see...Carbaryl
GAMOPHEN® see...Hexachlorophene
GAMOPHENE® see...Hexachlorophene
GARDENTOX® see...Diazinon
GARDIQUAT®-1450 see...Zilkonium Chloride
GARDONA® see...Tetrachlorvinphos
GARIAL® see...Amitraz
GARLON® see...Triclopry
GARLON-3A® see...Triclopyr, Triethylammonium Salt
GARNITAN® see...Linuron
GARRATHION® see...Carbophanothion
GARVOX® see...Bendiocarb
GAUCHO® see...Imidacloprid
GAUCHO® see...Mancozeb
GAUCHO® see...Metalaxyl
GAUCHO® see...Tebuconazole
GAUNTLET® see...Cloransulam-methyl
GAUNTLET® see...Sulfentrazone
GAVEL® see...Mancozeb
GC 1189® see...Chlordecone (Kepone®)
GC 3944-3-4® see...Quintozene
GC 4072® see...Chlorfenvinphos
GC 6936® see...Triphenyltin Compounds
GC 8993® see...Triphenyltin Compounds
GEARPHOS® see...Methyl Parathion
GEARPHOS® see...Parathion
GEBUTOX® see...Dinoseb
GEFIR® see...Dichlorophene
GEIGY 338® see...Chlorobenzilate
GEIGY 13005® see...Methidathion
GEIGY 22870® see...Dimetilan
GEIGY 24480® see...Diazinon
GEIGY 27692® see...Simazine
GEIGY 30,027® see...Atrazine
GEIGY 30,028® see...Propazine
GEIGY 30494® see...Methyl Phenkapton
GEIGY G-23611® see...Isolan®
GEIGY GS-13332® see...Dimetilan
GEL II® see...Sodium Fluoride
Gelber phosphor (German) see...Phosphorus
G-ELEVEN® see...Hexachlorophene
GELUTION® see...Sodium Fluoride
GEMINI® see...Chlorimuron-ethyl
GEMINI® see...Linuron
GENATE® see...Butylate
GENCOR® see...Hydroprene
GENEP® EPTC see...EPTC
GENERAL CHEMICALS 1189® see...Chlordecone
GENERAL CHEMICALS 8993® see...Triphenyltin Compounds
GENIPHENE® see...Toxaphene
GENITHION® see...Parathion
GENITOX® see...DDT
GENOFACE® see...Triclocarban
GENPROPATHRIN® see...Fenpropathrin

GENTROL® see...Hydroprene
GEOCARB-50EC® see...BPMC
GEOFOS® see...Fosthietan
GEOMET® see...Phorate
GEOMYCIN® see...Oxytetracycline Calcium
GEONTER® see...Terbacil
GERLACH® 1396 see...Naphthoxyacetic Acid
GERMAIN'S® see...Carbaryl
German Saltpeter see...Ammonium Nitrate
GEROX® see...Streptomycin
GERSTLEY BORATE® see...Borax and Boric Acid
GERSTLEY BORATE® see...Sodium Tetraborate
GESAFID® see...DDT
GESAFRAM® see...Prometon
GESAFRAM® 50 see...Prometon
GESAGARD® see...Prometryn
GESAGRAM® see...Prometon
GESAMIL® see...Propazine
GESAPAX® see...Ametryn
GESAPON® see...DDT
GESAPRIM® see...Atrazine
GESAPRIM® see...Terbutryn
GESARAN®, see...Simazine
GESAREX® see...DDT
GESAROL® see...DDT
GESATOP® see...Simazine
GESATOP-50® see...Simazine
GESFID® see...Mevinphos
GESOPRIM® see...Atrazine
GESTID® see...Mevinphos
GEXANE® see...Lindane
GEXAREX® see...DDT
GH see...Dichlorophene
GHIBLI® see...Nicosulfuron
Gibb-3-ene-1,10-dicarboxylic acid, 2,4a,7-trihydroxy-1-methyl-8-methylene-, 1,4a-lactone, (1$\alpha$,2$\beta$,4a$\alpha$,4b$\beta$,10$\beta$)- see...Gibberellic Acid
GIBBERELLIN® see...Gibberellic Acid
GIBBERELLIN A$_3$® see...Gibberellic Acid
GIBBERELLIN X® see...Gibberellic Acid
GIBBEX® see...Gibberellic Acid
GIBBREL® see...Gibberellic Acid
GIBGRO® see...Gibberellic Acid
GIBRESCOL® see...Gibberellic Acid
GIB-SOL® see...Gibberellic Acid
GIB-TABS® see...Gibberellic Acid
Gifblaar Poison see...Fluoroacetic Acid\
GINSTAR® see...Diuron
GINSTAR EC® see...Thidiazuron
GKN-O® see...Glutaraldehyde
GLAND-UP® see...Glyphosate
Glauramine see...Auramine
GLAZD-PENTA® see...Pentachlorophenol
GLEAN® see...Chlorsulfuron
GLEAN 20DF® see...Chlorsulfuron
GLEBOFOS® see...Disulfoton
Glifosate (German) see...Glyphosate
Glifosato (Spanish) see...Glyphosate

GLION® see...Glyphosate
GLOBAL CRAWLING INSECT BAIT® see. Chlorpyrifos
GLORE PHOS 36® see...Monocrotophos
GLU (IUPAC) see...Glutamic Acid
GLUSATE® see...Glutamic Acid
GLUTACID® see...Glutamic Acid
Glutaminic acid see...Glutamic Acid
l-Glutamic acid see...Glutamic Acid
α-Glutamic acid see...Glutamic Acid
[±]-Glutamic acid see...Glutamic Acid
Glutamic acid, *dl*-(synthetic racemic mix) see. Glutamic Acid
Glutamic dialdehyde see...Glutaraldehyde
l-Glutaminic acid see...Glutamic Acid
GLUTAMINOL® see...Glutamic Acid
Glutaral see...Glutaraldehyde
Glutaraldehyd (Czech) see...Glutaraldehyde
Glutard dialdehyde see...Glutaraldehyde
Glutaric acid dialdehyse see...Glutaraldehyde
Glutaric dialdehyde see...Glutaraldehyde
Glutaronitrile, 2-bromo-2-(bromomethyl)- see.1,2-Dibromo-2,4-dicyanobutane
GLUTATON® see...Glutamic Acid
GLYCEL® see...Glyphosate
Glycine, *N*-(chloroacetyl)-*N*-(2,6-diethylphenyl)-, ethyl ester see...Diethatyl-ethyl
Glycine, *N*-(phosphonomethyl)- see...Glyphosate
Glycin, *N*-(phosphonomethyl)-, ion(1-), trimethylsulfonium see...Sulfosate
Glycol bromide see...Ethylene Dibromide
Glycol dibromide see...Ethylene Dibromide
Glycol dichloride see...Ethylene Dichloride
Glycophen see...Iprodione
GLYCOPHENE® see...Iprodione
GLY-FLO® see...Glyphosate
GLYFOCAL® see...Glyphosate
Glyfosaat (Dutch) see...Glyphosate
GLYFOS® see...Glyphosate
GLYODEX® 37-22 see...Captan
Glyodin acetate see...Glyodin
Glyoxide see...Glyodin
Glyoxide dry see...Glyodin
Glyoxylonitrile, phenyl-,.oxime see...*O,O*-diethyl phosphorothioate see...Phoxim
Glyphosate isopropylamine salt see...Alachlor
Glyphosate-trimesium see...Sulfosate
GLYPRO® see...Glyphosate
GLYTEX® see...Glyphosate
GLYWEED® see...Glyphosate
GOAL® see...Oxyfluorfen
GO-GO-SAN® see...Pendimethalin
GOKILAHT® see...Cyphenothrin
1721 Gold see...Copper and Copper Compounds
Gold bronze see...Copper and Copper Compounds
GOLDBEET® see...Metamiton
GOLD CREST® see...Chlordane
GOLD CREST® see...Diphacione

GOLDENGRO® see...Kinetin (Cytokinin)
GOLDENGRO® see...Indole-3-Butyric Acid
GOLDENGRO® see...1-Naphthaleneacetic Acid
GOLD KIST® see...Oxadiazon
GOLD ORANGE MP® see...Fenaminosulf
GOLDQUAT 276® see...Paraquat
GOLTIX® see...Metamiton
GOPHACIDE® see...Phosacetim
GOPHA-RID® see...Zinc Phosphide
GOPHER BAIT® see...Strychnine
GOPHER-GETTER® see...Strychnine
GOPHER-GO AG BAIT® see...Strychnine
GORDON'S TRIGUARD® see...Dicamba
GORDON'S TRI-MEC® see...Dicamba
GORDON'S WEEDER® see...Trifluralin
GOTHNION® see...Azinphos-methyl
GPKh see...Heptachlor
GR 48-11PS® see...Sodium Pentachlorophenate
GR 48-32S® see...Sodium Pentachlorophenate
GRAIN CURE® see...Sodium Diacetate
GRAIN SORGHUM HARVEST AID® see...Sodium Chlorate
GRAMEVIN® see...Dalapon
GRAMINON-PLUS® see...Dichlorprop
GRAMOXONE® see...Paraquat
GRAMOXONE D® see...Paraquat
GRAMOXONE DICHLORIDE® see...Paraquat
Gramoxone methyl sulfate see...Paraquat Methosulfate
GRAMOXONE S® see...Paraquat
GRAMOXONE W® see...Paraquat
GRAMTOX® see...Phorate
GRANDSTAND® see...Triclopyr, Triethylammonium Salt
GRANERO® see...Hexachlorobenzene
GRANEX OK® see...Sodium Chlorate
GRANOSAN® see...Ethyl Mercuric Chloride
GRANOX® see...Thiabendazole
Granox NM see...Hexachlorobenzene
GRANOX PFM® see...Captan
GRANOZAN® see...Ethyl Mercuric Chloride
GRANUTOX® see...Phorate
GRASIDIM® see...Sethoxydim
GRASLAM® see...Mecoprop
GRASLAN® see...Tebuthiuron
GRASP® see...Tralkoxydim
GRASS-B-GONE® see...Fluazifop-butyl
GRASSOUT® see...Sethoxydim
GRAZON® see...Picloram
GRAZON® see...Triclopry
GREEN CROSS WARBLE POWDER® see...Rotenone
GREEN-DAISEN M® see...Mancozeb
GREEN DEVIL® see...Malathion
GREENFIELD® see...Siduron
GREENFLY AEROSOL SPRAY® see...Malathion
Greenland spar see...Sodium Aluminum fluoride
GREENMASTER AUTUMN® see...Ferric Sulfate
Green Oil see...Anthracene
Green vitriol iron monosulfate see...Ferrous Sulfate

GRELUTIN® see...Naptalam
GRENADE® see...*lamda*-Cyhalothrin
GRENADIER® see...Diflufenican
Grey arsenic see...Arsenic and Inorganic Arsenic Compounds
GRIFFEX® see...Atrazine
GRIFFIN® ATRAZINE 90 DRY FLOWABLE HERBICIDE® see...Atrazine
Griffin super Cu see...Copper Sulfate
GRISOL® see...TEPP
GROCEL® see...Gibberellic Acid
GROPPER® see...Metsulfuron-methyl
GROTAN® see...Sodium Tetraborate
GRO-TONE® see...Siduron
GROUNDHOG SOLTAIR® see...Diquat
GROUNDHOG SOLTAIR® see...Diquat Dibromide
GROUND-UP® see...Glyphosate
Ground vocle sulphur see...Sulfur
GROUND ZERO® see...Prometon
GROVEX SEWER BAIT® see...Warfarin
GRUNDIER ARBEZOL® see...Pentachlorophenol
GS-13005® see...Methamidophos
GS-13005® see...Methidathion
GS-14260® see...Terbutryn
GS-16068® see...Dipropetryn
GT-41® see...Busulfan
GT-2041® see...Busulfan
Guanidine, cyano-, methylmercury deriv. see Methylmercuric Dicyanamide
Guanidine, dodecyl-, acetate see...Dodine
Guanidine, dodecyl-, monoacetate see.Dodine
GUARDIAN® see...Acetochlor
GUARDIAN® see...Flucythrinate
GUARDSMAN® see...Dimethenamid
GUESAPON® see...DDT
GUESAROL® see...DDT
GUSATAFSON ACEPHATE 90 SEED PROTECTORANT® see...Acephate
GUSATHION® see...Azinphos-methyl
GUSATHION A® see...Azinphos-ethyl
GUSATHION A INSECTICIDE® see...Azinphos-ethyl
GUSATHION ETHYL® see...Azinphos-ethyl
GUSATHION M® see...Azinphos-methyl
GUSTAFSON CAPTAN 30-DD see...Captan
GUTHION® see...Azinphos-methyl
GUTHION ETHYL® see...Azinphos-ethyl
GUTHION INSECTICIDE® see...Azinphos-ethyl
GX-071® see...Sulfluamid
GY-PHENE® see...Toxaphene
Gypsine see...Lead Arsenate
GYPSINE® see...Lead Arsenate
GYRON® see...DDT

- H -

H 10® see...Arsenic Acid
H 35-F 87 (BVM)® see...Fenitrothion
H 133® see...Dichlobenil
H 321® see...Methiocarb
H 326® see...Linuron
H 1313® see...Dichlobenil
H 1318® see...Siduron
H 1803® see...Simazine
H 5727® see...Phenol, 3-(1-methyethyl)-, methylcarbamate
H 8757® see...Phenol, 3-(1-methyethyl)-, methylcarbamate
H 9789® see...Norflurazon
H 22234® see...Diethatyl-ethyl
H&G® see...Fipronil
HACHE UNO SUPER® see...Fluazifop-butyl
HADAF® see...Oxyfluorfen
HAG-107® see...Tralomethrin
HAIARI® see...Rotenone
HAIPEN® see...Captafol
HAITIN® see...Fentin Hydroxide
HAITIN® see...Triphenyltin Compounds
HAITIN® WP 20 (FENTIN HYDROXIDE 20%) see.Triphenyltin Compounds
HAITIN® WP 60 see...Fentin Hydroxide
HAITIN® WP 60 (FENTIN HYDROXIDE 60%) see.Triphenyltin Compounds
HALIZAN® see...Metaldehyde
HALLMARK® see...*lamda*-Cyhalothrin
HALMARK® see...Esfenvalerate
Halon 104 see...Carbon Tetrachloride
HALON 1001® see...Methyl Bromide
Halosulfuron see...Halosulfuron-methyl
HALT® see...Capsaicin
HAMIDOP® see...Methamidophos
HAMMER® see...Imazethapyr
HANANE® see...Dimefox
HAPPYGRO® see...Kinetin (Cytokinin)
HARE-RID® see...Strychnine
HARLEQUIN® see...Isoproturon
HARLEQUIN® see...Simazine
HARMONY® EXTRA see...Thifensulfuron Methyl
HARMONY EXTRA® see...Tribenuron-Methyl
HARNESS® see...Acetochlor
HARNESS® see...Mecoprop
HARRIER® see...Mecoprop
HARVADE® see...Dimethipin
HARVADE-5F® see...Dimethipin
HARVEST-AID® see...Sodium Chlorate
HAVERO-EXTRA® see...DDT
HAVILAND® ATRAZINE LINURON WEED KILLER see...Atrazine
HAWK® see...Bromadiolone
HAWK T-A® see...Potassium Nitrate
HC 1281® see...Trichlorobenzoic Acid
HC 2072® see...Paraoxon
HCB see...Hexachlorobenzene
HCCH see...Hexachlorocyclohexanes
HCCH see...Lindane
HCE see...Heptachlor Epoxide
HCE see...Hexachloroethane

HCH see...Lindane
α-HCH see...Hexachlorocyclohexanes
β-HCH see...Hexachlorocyclohexanes
γ-HCH see...Lindane
δ-HCH see...Hexachlorocyclohexanes
gamma-HCH see...Lindane
HCH-delta see...Hexachlorocyclohexanes
HCH, δ- see...Hexachlorocyclohexanes
HCH BHC see...Lindane
HCH HILBEECH® see...Hexachlorocyclohexanes
HCN see...Hydrogen Cyanide
HCP see...Hexachlorophene
HCS 3260® see...Chlordane
Heavy carburetted hydrogen see...Ethylene
HECLOTOX® see...Lindane
HEDAPUR M 52® see...MCPA
HEDAREX M® see...MCPA
HEDOLIT® see...Dinitro-o-cresol (DNOC)
HEDOLITE® see...Dinitro-o-cresol (DNOC)
HEDONAL® see...2,4-D
HEDONAL® see...Dichlorprop
HEDONAL M® see...MCPA
HEDONAL MCPP® see...Mecoprop
HELENA ATRAZINE TECHNICAL® see...Atrazine
HELENA BRAND ATRAZINE® see...Atrazine
HELENA PHOSDRIN® see...Mevinphos
HEL-FIRE® see...Dinoseb
HELIX® see...Chlorfluazuron
HELIX® see...Difenoconazole
HELMETINA® see...Phenothiazine
HELMSMAN® see...Carbetamide
HELOTHION® see...Sulprofos
HEOD see...Dieldrin
Hepachloor-3a,4,7,7a-tetrahydro-4,7-endo-methano-indeen (Dutch) see...Heptachlor
HEPT see...Hexaethyl Tetraphosphate
HEPTA see...Heptachlor
Heptachlorane see...Heptachlor
Heptachlore (French) see...Heptachlor
3,4,5,6,7,8,8-Heptachlorodicyclopentadiene see. Heptachlor
3,4,5,6,7,8,8a-Heptachlorodicyclopentadiene see.Heptachlor
1,4,5,6,7,8,8-Heptachloro-2,3-epoxy-2,3,3a,4,7,7a-hexahydro-4,7-methanoindene see...Heptachlor Epoxide
1,4,5,6,7,8,8-Heptachloro-2,3-epoxy-3a,4,7,7a-tetrahydro-4,7-methanoindan see...Heptachlor Epoxide
2,3,5,6,7,7-Heptachloro-1a,1b,5,5a,6,6a-hexahydro-2,5-methano-2h-indeno(1,2-b)oxirene see...Heptachlor Epoxide
1,4,5,6,7,10,10-Heptachloro-4,7,8,9-tetrahydro-4,7-endomethyleneindene see...Heptachlor
1,4,5,6,7,8,8-Heptachlor-3a,4,7,7a-tetrahydro-4,7-endo-methano Inden (German) see...Heptachlor
1,4,5,6,7,8,8-Heptachloro-3a,4,7,7a-tetrahydro-4,7-methanoindene see...Heptachlor
1,4,5,6,7,8,8-Heptachloro-3a,4,7,7a-tetrahydro-4,7-methano-1H-indene see...Heptachlor
1(3a),4,5,6,7,8,8-Heptachloro-3a(1),4,7,7a-tetrahydro-4,7-methanoindene see...Heptachlor
1,4,5,6,7,8,8a-Heptachloro-3a,4,7,7a-tetrahydro-4,7-methanoindene see...Heptachlor
1,4,5,6,7,8,8-Heptachloro-3a,4,7,7a-tetrahydro-4,7-methelene Indene see...Heptachlor
1,4,5,6,7,8,8-Heptachloro-3a,4,7,7a-tetrahydro-4,7-methanol-1H-indene see...Heptachlor
Heptacloro (Spanish) see...Heptachlor
Heptaclorepoxido (Spanish) see...Heptachlor Epoxide
2-Heptadecyl-4,5-dihydro-1H-imidazolyl monoacetate see...Glyodin
2-Heptadecyl glyoxalidine acetate see...Glyodin
2-Heptadecyl-2-imidazoline acetate see...Glyodin
HEPTAGRAN® see...Heptachlor
HEPTAMUL® see...Heptachlor
HEPTOX® see...Heptachlor
HEPT® see...TEPP
HERALD® see...Fenpropathrin
HERB-ALL® see...Cacodylic Acid
HERB-ALL® see...Sodium Cacodylate
HERB-ALL® see...Sodium Methanearsonate (MSMA)
HERBADOX® see...Pendimethalin
HERBAN M® see...Sodium Methanearsonate (MSMA)
HERBANIL® see...2,4-D
HERBANIL 368® see...Propanil
HERBATIM (dihydrate)® see...Metham-Sodium
HERBATOXOL® see...Atrazine
HERBAX TECHNICAL® see...Propanil
HERBAZIN® 50 see...Simazine
HERBAZIN® 500 BR see...Simazine
HERBAZIN PLUS SC® see...Amitrole
HERBEC® see...Tebuthiuron
HERBEX® see...Simazine
HERBIC® see...Tebuthiuron
HERBICIDE 326® see...Linuron
HERBICIDE 976® see...Bromacil
HERBICIDE 6602® see...Metoxuron
HERBICIDE C-2059® see...Fluometuron
HERBICIDE M® see...MCPA
HERBICIDE TOTAL® see...Amitrole
HERBI D-480® see...2,4-D
HERBIDAL® see...2,4-D
HERBIFLURIN® see...Trifluralin
HERBIKILL® see...Paraquat
HERBIMIX SC® see...Atrazine
HERBIPAK® see...Ametryn
HERBITRIN 500 BR® see...Atrazine
HERBIZID DP® see...Dichlorprop
HERBIZOLE® see...Amitrole
HERBOGIL® see...Dinoterb
HERBOXONE® see...2,4-D
HERBOXONE® see...Paraquat
HERBOXY® see...Simazine
HERBRAK® see...Metamiton
HERBURON 500 BR® see...Diuron
Herco Prills see...Ammonium Nitrate
HERCULES 37M6-8® see...Formaldehyde

HERCULES 3956® see...Toxaphene
HERCULES 5727® see...Phenol, 3-(1-methyethyl)-, methylcarbamate
HERCULES 14503® see...Dialifor
HERCULES 22234® see...Diethatyl-ethyl
HERCULES AC 528® see...Dioxathion
HERCULES AC 5727® see...Phenol, 3-(1-methyethyl)-, methylcarbamate
HERCULES TOXAPHENE® see...Toxaphene
HERITAGE® see...Azaxystrobin
HERKOL® see...Dichlorvos
HERMAL® see...Thiram
HERMAT ZDM® see...Ziram
HERRISOL® see...Mecoprop
HERYL® see...Thiram
HET see...Hexaethyl Tetraphosphate
Hexa see...Hexachlorocyclohexanes
HEXABLANC® see...Hexachlorocyclohexanes
HEXACAP® see...Captan
Hexa C.B see...Hexachlorobenzene
Hexachloraethan (German) see...Hexachloroethane
Hexachloran see...Lindane
α-Hexachloran see...Hexachlorocyclohexanes
γ-Hexachloran see...Lindane
Hexachlorane see...Lindane
α-Hexachlorane see...Hexachlorocyclohexanes
γ-Hexachlorane see...Lindane
gamma-Hexachlorane see...Lindane
Hexachlorbenzol (German) see...Hexachlorobenzene
Hexachlorcyclohexan (German) see. Hexachlorocyclohexanes
1,2,3,4,5,6-Hexachlor-cyclohexane see...Lindane
α-1,2,3,4,5,6-Hexachlorcyclohexane see. Hexachlorocyclohexanes
Hexachlorethane see...Hexachloroethane
γ-Hexachlorobenzene see...Lindane
A,β-1,2,3,4,7,7-Hexachlorobiclo(2,2,1)hepten-5,6-bioxymethylenesulfite see...Endosulfan
1,2,3,4,7,7-Hexachlorobiclo(2,2,1)hepten-5,6-bioxymethylenesulfite see...Endosulfan
Hexachlorocyclohexan (German) see. Hexachlorocyclohexanes
Hexachlorocyclohexane see...Hexachlorocyclohexanes
Hexachlorocyclohexane Isomers see...Hexachlor cyclohexanes
Hexachlorocyclohexane (Mixed Isomers) see. Hexachlorocyclohexanes
Hexachlorocyclohexane, gamma isomer see.Lindane
1,2,3,4,5,6-Hexachlorcyclohexane see.Hexachlorocyclohexanes
1,2,3,4,5,6-3, Hexachlorocyclohexane, gamma isomer see...Lindane
1A,2A,3B,4A,5B,6B-Hexachlorocyclohexane see.Hexachlorocyclohexanes
1-α,2,α,3,β,4α,5,α,6β-Hexachlorocyclohexane see. Lindane
1-α,2,α,3,β,4α,5β,6β-Hexachlorocyclohexane see. Hexachlorocyclohexanes
1-α,2,α,3,α,4,β,5,α,6,β-Hexachlorocyclohexane see. Hexachlorocyclohexanes
1-α,2-β,3-α,4-β,5-α,6-β-Hexachlorocyclohexane see. Hexachlorocyclohexanes
α-Hexachlorocyclohexane see..Hexachlorocyclohexanes
β-Hexachlorocyclohexane see..Hexachlorocyclohexanes
γ-Hexachlorocyclohexane see...Lindane
δ-Hexachlorocyclohexane see..Hexachlorocyclohexanes
β-1,2,3,4,5,6-Hexachlorocyclohexane see. Hexachlorocyclohexanes
γ-1,2,3,4,5,6-Hexachlorocyclohexane see...Lindane
δ-1,2,3,4,5,6-Hexachlorocyclohexane see. Hexachlorocyclohexanes
1,2,3,4,5,5-Hexachloro-1,3-cyclopentadiene dimer see. Mirex
Hexachlorocyclopentadienedimer see...Mirex
2,2',3,3',5,5'-Hexachloro-6,6'-dihydroxydiphenylmethane see... Hexachlorophene
Hexachloroepoxyoctahydro-endo,endo-dimethanonapthalene see...Endrin
Hexachloroepoxyoctahydro-endo,exo-dimethanonaphthalene see...Dieldrin
1,2,3,4,10,10-Hexachloro-6,7-epoxy-1,4,4a,5,6,7,8,8a-octahydro-1,4-endo,exo-5,8-di-methanonaphthaleNE see. Dieldrin
1,2,3,4,10,10-Hexachloro-6,7-epoxy-1,4,4a,5,6,7,8,8a-octahydro-1,4-endo-endo-1,4,5,8-dimethanonaphthalene see... Endrin
1,1,1,2,2,2-Hexachloroethane see...Hexachloroethane
Hexachloroethylene see...Hexachloroethane
Hexachlorofen (Czech) see...Hexachlorophene
Hexachlorohexahydro-endo-exo-dimethanonaphthalene see...Aldrin
1,2,3,4,10,10-Hexachloro-1,4,4a,5,8,8a-hexahydro-1,4,5,8-dimethanonaphthalene see...Aldrin
1,2,3,4,10,10-Hexachloro-1,4,4a,5,8,8a-hexahydro-exo-1,4-endo-5,8-dimethanonaphthalene see...Aldrin
1,2,3,4,10,10-Hexachloro-1,4,4A,5,8,8A-hexahydro-1,4-endo-exo-5,8-dimethanonaphthalene see...Aldrin
1,2,3,4,10-10-Hexachloro-1,4,4a,5,8,8a-hexahydro-1,4,5,8-endo-exo-dimethanonaphthalene see...Aldrin
(1α,4α,4aβ,5β,8β,8a.β)-1,2,3,4,10,10-Hexachloro-1,4,4a,-5,8,8a-hexahydro-1,4,5,8-dimethanonaphthalene see.Isodrin
1,2,3,4,10,10-Hexachloro-1,4,4a,5,8,8a-hexahydro-1,4:5,8-endo, endo-dimethanon aphthalene see...Isodrin
1,2,3,4,10,10-Hexachloro-1,4,4a,5,8,8a-hexahydro-1,4-endo, endo-5,8-dimethanon aphthalene see.Isodrin
Hexachlorohexahydromethano 2,4,3-benzodioxathiepin-3-oxide see...Endosulfan
6,7,8,9,10,10-Hexachloro-1,5,5a,6,9,9a-hexahydro-6,9-methano-2,4,3-benzodioxathiepin-3-oxide see. Endosulfan
gamma-Hexaclorobenzene see...Lindane
1,4,5,6,7,7-Hexachloro-5-norborene-2,3-dimethanol cyclic sulfite see...Endosulfan
1,4,5,6,7,7-Hexachloro-5-norbornene-2,3-dimethanol, cyclic sulfite, exo- see...Endosulfan

1,4,5,6,7,7-Hexachloro-5-norbornene-2,3-dimethanol, cyclic sulfite, *endo-* **see**...Endosulfan
3,4,5,6,9,9-Hexachloro-1a,2,2a,3,6,6a,7,7a-octahydro-2,7:3,6-dimethano **see**...Dieldrin
3,4,5,6,9,9-Hexachloro-1a,2,2a,3,6,6a,7,7a-octahydro-2,7:3,6-dimethanonaphth(2,3-b)oxirene **see**...Dieldrin
(1r,4s,4as,ss,7r,8r,8ar)-1,2,3,4,10-Hexachloro-1,4,4a,5,6,7,8,8a-octahydro-6,7-epoxy-1,4:5,8- dimethano naphthalene **see**...Endrin
(1R,4S,4AS,5R,6R,7S,8S,8AR)1,2,3,4,10,10-Hexachloro-1,4,4a,5,6,7,8,8a-octahydro-6,7-epoxy-1,4:5,8-dimethanonaphthalene **see**...Dieldrin
Hexachlorophane **see**...Hexachlorophene
Hexachlorophen **see**...Hexachlorophene
Hexachlorophene **see**...Hexachlorophene
c,c'-(1,4,5,6,7,7-Hexachloro-8,9,10-trinorborn-5-en-2,3-ylene)(dimethylsulphite)6,7,8,9,10,10-hexachloro-1,5,5a,6,9,9a-hexahydro-6,9-methano-2,4,3-benzodioxathiepin 3-oxide **see**...Endosulfan
HEXACID®-1095 **see**...Decanoic Acid
HEXACID® C-9 **see**...Pelargonic Acid
Hexacloran (In Russia) **see**...Hexachlorocyclohexanes
Hexaclorobenceno (Spanish) **see**...Hexachlorobenzene
Hexaclorociclohexano (Spanish) **see**. Hexachlorocyclohexanes
1,2,3,4,5,6-Hexaclorociclohexano (Spanish) **see**.Hexachlorocyclohexanes
Hexacloroetano (Spanish) **see**...Hexachloroethane
Hexaclorofeno (Spanish) **see**...Hexachlorophene
HEXADRIN® **see**...Endrin
Hexaethyltetrafosfat **see**...Hexaethyl Tetraphosphate
HEXAFEN® **see**...Hexachlorophene
HEXAFERB® **see**...Ferbam
HEXAFLOW® **see**...Lindane
Hexafluoroaluminato de trisodio (Spanish) **see**...Sodium Aluminum Fluoride
Hexafluron **see**...Hexaflumuron
HEXAFOR® **see**...Hexachlorocyclohexanes
Hexahydro-3,6-*endo*-oxyphthalic acid **see**...Endothall
Hexahydropyridine **see**...Piperidine
Hexakis **see**...Fenbutatin Oxide
Hexakis(β,β-dimethylphenethyl)distannoxane **see**. Fenbutatin Oxide
Hexakis(2-methyl-2-phenylpropyl)distannoxane **see**. Fenbutatin Oxide
Hexaklon (in Sweden) **see**...Hexachlorocyclohexanes
HEXAMITE **see**...TEPP
HEXAMUL® **see**...Hexachlorocyclohexanes
HEXAPHENE-LV® **see**...Hexachlorophene
HEXAPOUDRE® **see**...Hexachlorocyclohexanes
HEXASAN® **see**...Ethyl Mercuric Chloride
HEXASUL® **see**...Sulfur
HEXASULFAN® **see**...Endosulfan
HEXATHANE® **see**...Zineb
HEXATHION® **see**...Carbophanothion
HEXATHIR® **see**...Thiram
HEXATOX® **see**...Lindane
HEXAVERM® **see**...Lindane

HEXAVIN® **see**...Carbaryl
Hexazane **see**...Piperidine
HEXAZIR® **see**...Ziram
HEXA® **see**...Lindane
Hexhexane **see**...Hexachlorocyclohexanes
HEXICIDE® **see**...Lindane
HEXIDE® **see**...Hexachlorophene
HEXOPHENE® **see**...Hexachlorophene
HEXYCLAN® **see**...Hexachlorocyclohexanes
HEXYCLAN® **see**...Lindane
HEXYGON® DF **see**...Hexythiazox
HEXYLAN® **see**...Hexachlorocyclohexanes
Hexylthiocarbam **see**...Cycloate
HFA **see**...Fluoroacetic Acid
HGI **see**...Lindane
HHDM **see**...Aldrin
HHDN **see**...Aldrin
HHPN **see**...Aldrin
HI-DEP® **see**...2,4-D
HI-YIELD DESICCANT H-10® **see**...Arsenic Acid
HIBOR **see**...Bromacil
HIBROM® **see**...Naled
HICO CCC® **see**...Chlormequat Chloride
HIFOL® **see**...Dicofol
HIGALCOTON® **see**...Fluometuron
HIGALMETOX® **see**...Methoxychlor
HIGALNATE® **see**...Molinate
High solvent naphtha **see**...Naphthas
HILAZPOPHOS® **see**...Triazophos
HILDAN® **see**...Endosulfan
HILDIT® **see**...DDT
HILITE 60® **see**...Dichloroisocyanuric Acid
HILTACHLOR® **see**...Butachlor
HILTHION® **see**...Malathion
HINOCHLOA® **see**...Mefenacet
HIP® **see**...Phenol, 3-(1-methyethyl)-, methylcarbamate
HIVERTOX® **see**...Dinoseb
HIVOL 44® **see**...2,4-D, isopropyl ester
HIZAROCIN® **see**...Cycloheximide
HL-331® **see**...Phenylmercury Acetate
HOBANE® **see**...Bromoxynil
HOCA® **see**...Disulfiram
HOCH® **see**...Formaldehyde
HOE 2671® **see**...Endosulfan
HOE 2784® **see**...Binapacryl
HOE 2810® **see**...Linuron
HOE 2824® **see**...Triphenyltin Compounds
HOE 2872® **see**...Triphenyltin Compounds
HOE 002960® **see**...Triazophos
HOE 2960 OJ® **see**...Triazophos
HOE 16410® **see**...Isoproturon
HOE 17411® **see**...Carbendazim
HOE 23408® **see**...Diclofop-methyl
HOE 26150® **see**...Dinoseb
HOE 033171® **see**...Fenoxaprop-ethyl
HOE 555-02A® **see**...Fenpyroximate
HOE-A 25-01® **see**...Fenoxaprop-ethyl
HOE-GRASS® **see**...Diclofop-methyl

HOELON® see...Diclofop-methyl
HOELON® 3EC see...Diclofop-methyl
HOKKO-MYCIN® see...Streptomycin
HOKMATE® see...Ferbam
HOLIDAY FIRE ANT KILLER® see...D-Limonene
HONG KIEN® see...Phenylmercury Acetate
HOOKER® HRS-16 see...Dienochlor
HOOKER® HRS 1654 see...Dienochlor
HOPCIN® see...BPMC
HOPKINS® see...Zinc Phosphide
HOPKINS BAR BAIR® see...Warfarin
HOPKINS COV-R-TOX® see...Warfarin
HOPKINS RODEX® see...Warfarin
HORIZON® see...Fluazifop-butyl
HORIZON® see...Tebuconazole
HORMATOX® see...Dichlorprop
HORMEX® see...Indole-3-Butyric Acid
HORMEX® see...1-Naphthaleneacetic Acid
HORMOCEL-2CCC® see...Chlormequat Chloride
HORMODIN® see...Indole-3-Butyric Acid
HORMOSLYR® 500T (butoxyethyl ester) see...2,4,5-Trichlorophenoxyacetic Acid, Esters
HORMOTOX® see...2,4-D
HORMOTUHO® see...MCPA
HORNET® see...Clopyralid
HORNET® see...Flumetsulam
HORNOTUHO® see...MCPA
HORTEX® see...Lindane
HOSPEX® see...Glutaraldehyde
HOSTAQUICK® see...Phenylmercury Acetate
HOSTATHION® see...Triazophos
HRS 16® see...Dienochlor
HRS 16A® see...Dienochlor
HRS 1276® see...Mirex
HRS 1654® see...Dienochlor
HS see...Hydrazine Sulfate
HS-119-1® see...Pyrazon
HS-14260® see...Terbutryn
HTP see...Hexaethyl Tetraphosphate
HTZ see...Hexythiazox
HUILEUX® see...Toxaphene
HUNGAZIN® see...Atrazine
HUNGAZIN DT® see...Simazine
HUSEPT® see...4-Chloro-3,5-xylenol
HUSEPT EXTRA® see...4-Chloro-3,5-xylenol
HW 920® see...Diuron
HWG 1608® see...Tebuconazole
HYAMINE®-3500 see...Zilkonium Chloride
HYDON® see...Bromacil
HYDOUT® see...Endothall
HYDRAM® see...Molinate
Hydrated lime see...Calcium Hydroxide
Hydrated kemikal see...Calcium Hydroxide
Hydrazid hydrazida maleica (Spanish) see...Maleic Hydrazide
Hydrazinecarbothioamide see...Thiosemicarbazide
Hydrazine hydrogen see...Hydrazine Sulfate
Hydrazine monosulfate see...Hydrazine Sulfate

Hydrazine sulphate see...Hydrazine Sulfate
Hydrazinium sulfate see...Hydrazine Sulfate
Hydrazoic acid, sodium salt see...Sodium Azide
Hydrazonium sulfate see...Hydrazine Sulfate
Hydrocyanic acid see...Hydrogen Cyanide
Hydrocyanic acid, potassium salt see...Potassium Cyanide
Hydrocyanic acid, sodium salt see...Sodium Cyanide
Hydrodimethylarsine oxide, sodium salt see.Sodium Cacodylate
Hydrofluoric acid see...Hydrogen Fluoride
Hydrofluoric acid gas see...Hydrogen Fluoride
Hydrogen antimonide see...Stibine
Hydrogenated rotenone see...Rotenone
Hydrogen carboxylic acid see...Formic Acid
Hydrogen cyanamide see...Cyanamide
Hydrogen fluoride, anhydrous see...Hydrogen Fluoride
Hydrogen oxalate of amiton see...Amiton Oxalate
Hydrogen peroxide and peroxyacetic acid mixture see. Peracetic Acid
Hydrogen phosphide see...Phosphine
Hydroperoxide, acetyl see...Peracetic Acid
(7$S$)-Hydroprene see...Hydroprene
(2$E$,4$E$)-Hydroprene see...Hydroprene
HYDROTHAL-47® see...Endothall
HYDROTHOL® see...Endothall
2-Hydroxybenzoic acid see...Salicylic Acid
$o$-Hydroxybenzoic acid see...Salicylic Acid
2-Hydroxybiphenyl see...$o$-Phenylphenol
$o$-Hydroxybiphenyl see...$o$-Phenylphenol
2-Hydroxy-1,1'-biphenyl see...$o$-Phenylphenol
$m$-Hydroxycarbanilic acid methyl ester $m$-methylcarbanilate see...Phenmedipham
(Hydroxy-4-coumarinyl 3)-3 phenyl-3(bromo-4 biphenyl-4)-1 propanol-1 (French) see...Bromadiolone
($E$)-3-Hydroxy-crotonic acid, $\alpha$-methylbenzyl ester, dimethyl phosphate see...Crotoxyphos
3-Hydroxycrotonic acid methyl ester dimethyl phosphate see...Mevinphos
3-Hydroxy-4-cymene see...Thymol
3-Hydroxy-$p$-cymene see...Thymol
Hydroxyde de triphenyl-etain (French) see..Triphenyltin Compounds
Hydroxyde de triphenyl-etain (French) see...Fentin Hydroxide
4-Hydroxy-3,5-dibromobenzonitrile see...Bromoxynil
2-Hydroxy-3,5-dichlorophenyl sulphide see...Bithionol
$p$-Hydroxy-$N,N$-dime thylbenzenesulfonamide ester with phosphorothioic acid $O,O$-dimethyl ester see...Famphur
3-Hydroxydimethyl crotonamide dimethyl phosphate see. Dicrotophos
3-Hydroxy-$N,N$-dimethyl-($E$)-crotonamide dimethyl phosphate see...Dicrotophos
3-Hydroxy-$N,N$-dimethyl-$cis$-crotonamide dimethyl phosphate see...Dicrotophos
Hydroxydimethylarsine oxide see...Cacodylic Acid
2-Hydroxydiphenyl see...$o$-Phenylphenol
$o$-Hydroxydiphenyl see...$o$-Phenylphenol

*trans*-N-[(4-Hydroxy-3-methoxyphenyl)methyl]-8-methyl-6-noneamide  see...Capsaicin
N-[(4-Hydroxy-3-methoxyphenyl)methyl]-8-methyl-6-nonenamide  see...Capsaicin
3-Hydroxy-N-methylcrotonamide dimethyl phosphate  see...Monocrotophos
3-Hydroxy-N-methyl-*cis*-crotonamide dimethyl phosphate  see...Monocrotophos
3-Hydroxy-1-methyl-4-isopropylbenzene  see...Thymol
2-Hydroxymethyl-2-nitropropane-1,3-diol  see. Tris(hydroxymethyl)nitromethane
2-(Hydroxymethyl)-2-nitro-1,3-propanediol  see. Tris(hydroxymethyl)nitromethane
4-Hydroxy-3-(3-oxo-1-phenylbutyl)-2H-1-benzopyran-2-one sodium salt (9CI)  see...Warfarin
4-Hydroxy-3-(3-oxo-1-phenylbutyl)coumarin  see. Warfarin
1-Hydroxypentachlorobenzene  see...Pentachlorophenol
3-(3-Hydroxyphenyl)-1,1-dimethylurea, *tert*-butylcarbamic acid ester  see...Karbutilate
3-Hydroxypropene  see...Allyl Alcohol
6-Hydroxy-3(2H)-pyridazinone  see...Maleic Hydrazide
8-Hydroxyquinoline copper complex  see...Copper (II)-8-hydroxyquinoline
1-Hydroxy-4-*tert*-butylbenzene  see...Butylphenols
5-Hydroxytetracycline  see...Oxytetracycline Calcium
4-Hydroxy-3-(1,2,3,4-tetrahydro-1-naftyl)-4-cumarine (Dutch)  see...Coumatetralyl
4-Hydroxy-3-(1,2,3,4-tetrahydro-1-napthalenyl)-2H-1-benzopyran-2-one  see...Coumatetralyl
4-Hydroxy-3-(1,2,3,4-tetrahydro-1-napthyl)cumarin  see. .Coumatetralyl
α-Hydroxytoluene  see...Benzyl Alcohol
1-Hydroxytriacontane  see...Triacontanol
1-Hydroxy-2,2,2-trichloroethylphosphonic acid dimethyl ester  see...Trichlorfon
2'-Hydroxy-2,4,4'-trichloro-phenylether  see...Triclosan
3-Hydroxy-N,N,5-trimethylpyrazole-1-carboxamidedimethylcarbamate (ester)  see...Dimetilan
Hydroxytriphenylstannane  see...Fentin Hydroxide
Hydroxytriphenylstannane  see...Triphenyltin Compounds
Hydroxytriphenyltin  see...Fentin Hydroxide
Hydroxytriphenyltin  see...Triphenyltin Compounds
HYMEC®  see...Mecoprop
HYOSAN  see...Dichlorophene
Hypodermacid  see...Trichlorfon
HYVAR®  see...Bromacil
HYVAR X BROMACIL®  see...Bromacil
HYVAR X-L®  see...Bromacil
HYVAR X WEED KILLER®  see...Bromacil
HYVAR X-WS®  see...Bromacil
HYVAR-X®  see...Bromacil
HY-VIC®  see...Thiram

- I -

IBA  see...Indole-3-Butyric Acid

ICE-SPAR®  see...Sodium Aluminum fluoride
ICETONE®  see...Sodium Aluminum fluoride
ICI-A0009®  see...Fluazifop-butyl
ICI-A 604®  see...Tralkoxydim
ICI BAYTAN®  see...Fuberidazole
ICIA-192®  see...Fluazinam®
ICIA5504 80WG®  see...Azaystrobin
ICI-PP 333®  see...Paclobutrazole
ICON®  see...Fipronil
ICON®  see...*lamda*-Cyhalothrin
IDA  see...Triphenyltin Compounds
IDA BRANDS®  see...Sodium Methanearsonate (MSMA)
IDA FLO-TIN 4L®  see...Fentin Hydroxide
IDA MANEB®  see...Maneb
IDA PROP-A-NEL®  see...Propanil
IDA SEIS-TRES 6-3®  see...Parathion
Idrazina solfato (Italian)  see...Hydrazine Sulfate
Idrossido di stagno trifenile (Italian)  see...Triphenyltin Compounds
IFC (Russia)  see...Propham
IFC  see...Propham
IFK  see...Propham
IGRAN®  see...Terbutryn
IGRATER®  see...Terbutryn
IKF-1216®  see...Fluazinam®
IKI-7899®  see...Chlorfluazuron
IKURIN®  see...Ammonium Sulfamate
ILLOXAN®  see...Diclofop-methyl
ILLOXOL®  see...Dieldrin
Illuminating oil  see...Kerosene
ILOXAN®  see...Diclofop-methyl
IMAVEROL®  see...Imazalil
Imazamethabenz  see...Imazethabenz
IMC 3950®  see...Thiobencarb
IMD-760®  see...Azacosterol Dihydrochloride
IMICIDE®  see...Imidacloprid
IMIDAN®
1H-Imidazol-2-amine, 1-[(6-chloro-3-pyridinyl)methyl]-4,5-dihydro-N-nitro-  see...Imidacloprid
1H-Imidazole, 1-[1-((4-chloro-2-(trifluoromethyl)phenyl)imino]-2-propoxyethyl]-,(E)-  see...Triflumizole
1H-Imidazole, 1-[2-(2,4-dichlorophenyl)-2-(2-propenyloxy)ethyl]-  see...Imazalil
1H-Imidazole, 1-[2-(2,4-dichlorophenyl)-2-(2-propenyloxy)ethyl]-,(±)-  see...Imazalil
1H-Imidazole, 2-heptadecyl-4,5-dihydro-,monoacetate  see...Glyodin
Imidazolidinethione  see...Ethylene Thiourea
2-Imidazolidinethione  see...Ethylene Thiourea
2-Imidazolidinimine, 1-[(6-chloro-3-pyridinyl)methyl]-N-nitro-benzoate  see...Imidacloprid
2-Imidazoline, 2-heptadecyl-, monoacetate  see. .Glyodin
Imidazoline-2-thiol  see...Ethylene Thiourea
2-(Imidazoline-2-thiol  see...Ethylene Thiourea
Imidazoline-2(3H)-thione  see...Ethylene Thiourea

Imidocarbonic acid, phosphonodithio-, cyclic ethylene *p,p*-diethyl ester  see. . .Phosfolan
4,4-(Imidocarbonyl)bis(n,n-dimethylaniline)  see. . .Auramine
(1-Iminoethyl)phosphoramidothioic acid, *O,O*-bis(4-chlorophenyl) ester  see. . .Phosacetim
IMISOL® see. . .Carbendazim
IMPACT EXCEL® see. . .Chlorothalonil
IMPASSE® see. . .*lamda*-Cyhalothrin
IMPERATOR® see. . .Cypermethrin
Imperial green  see. . .Copper Acetoarsenite
Imperial green  see. . .Paris Green
IMPROVED BLUE MALRIN SUGAR BAIT® see. . .Methomyl
IMPROVED GOLDEN MALRIN BAIT® see.Methomyl
INAKOR® see. . .Atrazine
INCRECEL® see. . .Chlormequat Chloride
1,3-Indandione, 2-[(*p*-chlorophenyl)phenylacetyl]-  see. . .Chlorophacinone
INDAR® see. . .Fenbuconazole
INDENE® see. . .Heptachlor
1*H*-Indene-1,3(2*H*)-dione, 2-[(4-chlorophenyl)phenylacetyl]-  see. . .Chlorophacinone
1*H*-Indole-3-butanoic acid  see. . .Indole-3-Butyric Acid
Indole butyric  see. . .Indole-3-Butyric Acid
Indole butyric acid  see. . .Indole-3-Butyric Acid
Indolyl-3-butyric acid  see. . .Indole-3-Butyric Acid
3-Indolebutyric acid  see. . .Indole-3-Butyric Acid
3-Indolyl-γ-butyric acid  see. . .Indole-3-Butyric Acid
4-(Indolyl)butyric acid  see. . .Indole-3-Butyric Acid
4-(Indol-3-yl)butyric acid  see. . .Indole-3-Butyric Acid
4-(3-Indolyl)butyric acid  see. . .Indole-3-Butyric Acid
β-Indolebutyric acid  see. . .Indole-3-Butyric Acid
γ-(Indole-3)-butyric acid  see. . .Indole-3-Butyric Acid
γ-(Indol-3-yl)butyric acid  see. . .Indole-3-Butyric Acid
γ-(3-Indolyl)butyric acid  see. . .Indole-3-Butyric Acid
INEXIT® see. . .Lindane
INFERNO® see. . .Amiton
INJECT-A-CIDE AV® see. . .Abamectin
INL-5300® see. . .Tribenuron-Methyl
INM-6316® see. . .Thifensulfuron Methyl
INSECT POWDER® see. . .Pyrethrins or Pyrethrum
856 INSECT REPELLENT® see. . .Dipropyl Isocinchomeronate
INSECTICIDE 1179® see. . .Methomyl
INSECTICIDE-NEMACIDE 1410® see. . .Oxamyl
INSECTIGAS D® see. . .Dichlorvos
INSECTOPHENE® see. . .Endosulfan
INSECTO® see. . .Endosulfan
INSEGAR® see. . .Fenoxycarb
INTERCIDE® see. . .Phenylmercury Acetate
INTER PLUS® see. . .Metolachlor
INTEXAN® LB-50  see. . .Zilkonium Chloride
INTRACEL-15® see. . .Arsenic Acid
INTREPID® see. . .Methoxyfenozide
INTUDER® see. . .Cyfluthrin
INTUDER HPX® see. . .Cyfluthrin
Invalon OP® see. . .*o*-Phenylphenol
INVERTON 245® see. . .2,4,5-T
INVERTON 245® see. . .2,4,5-Trichlorophenoxyacetic Acid, Esters
INVISI-GARD® see. . .Propoxur
INY-6202  see. . .Quizalofop-Ethyl
IONIZ® see. . .Diflufenican
IOTOX® see. . .Mecoprop
IPANER® see. . .2,4-D
IPC  see. . .Propham
IPERSAN® see. . .Trifluralin
IPMC  see. . .Propoxur
IPO 8® see. . .Tetrachlorvinphos
IPPC  see. . .Propham
IPRODINE® see. . .Iprodione
IRADICAV® see. . .Sodium Fluoride
IRGAGUARD® see. . .Thiabendazole
IRGASAN® see. . .Triclosan
IRGASAN® DP-300  see. . .Triclosan
Iron dimethyldithiocarbamate  see. . .Ferbam
Iron(III) dimethyldithiocarbamate  see. . .Ferbam
Iron persulfate  see. . .Ferric Sulfate
Iron protosulfate  see. . .Ferrous Sulfate
Iron sesquisulfate  see. . .Ferric Sulfate
Iron sulfate (1:1)  see. . .Ferrous Sulfate
Iron sulfate (2:3)  see. . .Ferric Sulfate
Iron(2+) sulfate  see. . .Ferrous Sulfate
Iron(2+) sulfate (1:1)  see. . .Ferrous Sulfate
Iron (3+) sulfate  see. . .Ferric Sulfate
Iron(II) sulfate  see. . .Ferrous Sulfate
Iron(III) sulfate  see. . .Ferric Sulfate
Iron tersulfate  see. . .Ferric Sulfate
Iron, tris(dimethylcarbamodithioato-*S,S'*-)-  see. . .Ferbam
Iron, tris(dimethylcarbamodithioato-*S,S'*)-, (OC-6-11)-  see. . .Ferbam
Iron, tris(dimethyldithiocarbamato)-  see. . .Ferbam
Iron tris(dimethyldithiocarbamate)  see. . .Ferbam
Iron vitriol  see. . .Ferrous Sulfate
IROSPAN® see. . .Ferrous Sulfate
IROSUL® see. . .Ferrous Sulfate
ISA-20E® see. . .Arosurf® MSF
ISATHRIN® see. . .Resmethrin
ISCOBROME® see. . .Methyl Bromide
ISCOBROME D® see. . .Ethylene Dibromide
ISCOTHANE® see. . .Dinocap
Isobenzano (Spanish)  see. . .Isobenzan
Isobutylglycerol, nitro-  see. . .Tris(hydroxymethyl)nitromethane
ISOCIL® see. . .Bromacil
Isocinchomeronic acid, dipropyl ester  see. . .Dipropyl Isocinchomeronate
Isocinchomeronyl dipropylester  see. . .Dipropyl Isocinchomeronate
ISO-CORNOX® see. . .Mecoprop
ISOCOTHANE® see. . .Dinocap
Isocyanuric acid, dichloro-  see. . .Dichloroisocyanuric Acid
Isocyanuric dichloride  see. . .Dichloroisocyanuric Acid

α-Isodecyl-ω-hydroxypoly(oxy-1,2-ethanediyl) see. Arosurf® MSF
Isodemeton see...Demeton
Isodrina (Spanish) see...Isodrin
Isofluorphate see...Isofluorphate
Isoflurophate see...Isofluorphate
1H-Isoindole-1,3(2H)-dione,3a,4,7,7a-tetrahydro-2-[(trichloromethyl)thiol]- see...Captan
Isolane (French) see...Isolan®
ISOMETASYSTOX® see...Demeton
ISOMETASYSTOX® see...Demeton-methyl
Isomethylsystox sulfoxide see...Demeton-methyl
ISOMETHYLSYSTOX® see...Demeton
ISOMETHYLSYSTOX® see...Demeton-methyl
Isooctyl alcohol (2,4-dichlorophenoxy)acetate see...2,4-D, isooctyl ester
Isooctyl 2,4-dichlorophenoxyacetate see...2,4-D, isooctyl ester
Isooctyl ester of dichloro 2,4-chloroacetic acid see...2,4-D, isooctyl ester
2,4,5-Isooctyl ester see...2,4,5-Trichlorophenoxyacetic Acid, Esters
Isooctyl 2,4,5-trichlorophenoxyacetate see...2,4,5-Trichlorophenoxyacetic Acid, Esters
Isophthalonitrile, tetrachloro see...Chlorothalonil
2-Isopr 3-isopropyl-5-methylphenyl-N-methylcarbamate see...Promecarb
Isopropalin see...Isopropalin
Isopropil-N-fenil-carbammato (Italian) see...Propham
(1-Isopropil-3-metil-1H-pirazol-5-il)-N,N-dimetil-carbammato (Italian) see...Isolan®
(E)-O-2-Isopropoxycarbonyl-1-methylvinyl O-methylethylphosphoramidothioate see...Propetamphos
O-(1-Isopropoxycarbonyl-1-propen-2-yl)-O-methyl-ethyl-phosphoramidothionate see...Propetamphos
2-Isopropoxyphenyl methylcarbamate see...Propoxur
2-Isopropoxyphenyl N-methylcarbamate see...Propoxur
o-Isopropoxyphenyl methylcarbamate see...Propoxur
o-Isopropoxyphenyl N-methylcarbamate see...Propoxur
o-(2-Isopropoxyphenyl) N-methylcarbamate see...Propoxur
Isopropylamino-O-ethyl-(4-methylmer capto-3-methylphenyl)phosphate see...Fenamiphos
3-Isopropyl-2,1,3-benzothiadiazinon-(4)-2,2-dioxid (German) see...Bentazon
3-Isopropyl-1H-2,1,3-benzothiadiazin-4(3H)-one-2,2-dioxide see...Bentazon
1-Isopropyl carbamoyl-3-(3,5-dichlorophenyl)-hydantoin see...Iprodione
Isopropyl carbanilate see...Propham
Isopropyl carbanilic acid ester see...Propham
N-Isopropyl-α-chloroacetanilide see...Propachlor
N-Isopropyl-2-chloroacetanilide see...Propachlor
Isopropyl cresol see...Thymol
6-Isopropyl-3-cresol see...Thymol
6-Isopropyl-m-cresol see...Thymol
Isopropyl 2,4-dichlorophenoxyacetate see...2,4-D, isopropyl ester
Isopropyl (2,4-dichlorophenoxy)acetate see...2,4-D, isopropyl ester
Isopropyl diethyldithiophosphorylacetamide see...Prothoate
4-Isopropyl-2,6-dinitro-N,N-dipropylaniline see...Isopropalin
2,4,5-T, Isopropyl ester see...2,4,5-Trichlorophenoxyacetic Acid, Esters
Isopropyl O-[ethoxy-N-isopropylamino(thiophosphoryl)]salicylate see...Isofenphos
Isopropyl O-[ethoxy(isopropylamino)phosphinothioyl]salicylate see...Isofenphos
Isopropyl 3-(ethylamino(methoxy)phosphinothioyloxy)isocrotonate see...Propetamphos
Isopropyl 3-[((ethylamino)methoxyphosphinothioyl]oxy)crotonate see...Propetamphos
Isopropyl-N-fenyl-carbamaat (Dutch) see...Propham
Isopropyl fluophosphate see...Isofluorphate
2,3-Isopropylidene-dioxyphenyl methylcarbamate see...Bendiocarb
N-Isopropyl-2-mercaptoacetamide-S-ester with O,O-diethyl phosphorodithioate see...Prothoate
Isopropyl (2E,4E)-11-methoxy-3,7,11-trimethyl-2,4-dodecadienoate see...Methoprene
Isopropyl (E,E)-11-methoxy-3,7,11-trimethyl-2,4-dodecadienoate see...Methoprene
3-Isopropyl-5-methylcarbamic acid methyl ester see..Promecarb
(1-Isopropyl-3-methyl-1H-pyrazol-5-yl)-N,N-dimethyl-carbamaat (Dutch) see...Isolan®
(1-Isopropyl-3-methyl-1H-pyrazol-5-yl)-N,N-dimethyl-carbamat (German) see...Isolan®
Isopropylmethylpyrazoldimethylcarbamate see...Isolan®
Isopropylmethylpyrazoyl dimethylcarbamate see..Isolan®
1-Isopropyl-3-methylpyrazolyl-(5)-dimethylcarbamate see...Isolan®
1-Isopropyl-3-methyl-5-pyrazolyl dimethyl carbamate see...Isolan®
(1-Isopropyl-3-methyl-1H-pyrazol-5-yl)-N,N-dimethyl carbamate see...Isolan®
Isopropylmethylpyrimidyl diethyl thiophosphate see...Diazinon
O-2-Isopropyl-4-methylpyrimyl-O,O-diethyl phosphorothioate see...Diazinon
m-Isopropylphenol methylcarbamate see...Phenol, 3-(1-methyethyl)-, methylcarbamate
m-Isopropylphenol-N-methylcarbamate see...Phenol, 3-(1-methyethyl)-, methylcarbamate
3-Isopropylphenol methylcarbamate see...Phenol, 3-(1-methyethyl)-, methylcarbamate
3-Isopropylphenol-N-methylcarbamate see...Phenol, 3-(1-methyethyl)-, methylcarbamate
Isopropyl-N-phenyl-carbamat (German) see...Propham
Isopropyl phenylcarbamate see...Propham
Isopropyl-N-phenylcarbamate see...Propham

Isopropyl-*N*-phenyl carbamate  see. . .Propham
*o*-Isopropyl-*N*-phenyl carbamate  see. . .Propham
3-(4-Isopropylphenyl)-1,1-dimethylurea  see. .Isoproturon
*m*-Isopropylphenyl-*N*-methylcarbamate  see. . .Phenol, 3-(1-methyethyl)-, methylcarbamate
3-Isopropylphenyl methylcarbamate  see. . .Phenol, 3-(1-methyethyl)-, methylcarbamate
Isopropyl-*N*-phenyurethan (German)  see. . .Propham
Isopropylphosphoramidic acid ethyl 4-(methylthio)-*m*-toyl ester  see. . .Fenamiphos
Isopropyl salicylate O-ester with O-ethyl isopropylphosphoramidothioate  see. . .Isofenphos
5-Isopropyl-*m*-tolyl methyl-carbamate  see. . .Promecarb
Isopropyl 2,4,5-trichlorophenoxyacetate  see. . .2,4,5-Trichlorophenoxyacetic Acid, Esters
α-Isopropyl-α-[*p*-(trifluoromethoxy)phenyl]-5-pyrimidinemethanol  see. . .Flurprimidol
Isoproyl phosphorofluoridate  see. . .Isofluorphate
Isopto carbachol  see. . .Captan
Isosafrole, octyl sulfoxide  see. . .Sulfoxide
Isosafrole-*n*-octylsulfoxide  see. . .Sulfoxide
Isosteareth-2  see. . .Arosurf® MSF
ISOSYSTOX®  see. . .Demeton
3(2*H*)-Isothiazolone, 2-octyl-  see. . .Octhilinone
Isothiocyanate d'allyle (French)  see. . .Allyl Isothiocyanate
3-Isothiocyanato-1-propene  see. . .Allyl Isothiocyanate
Isothiocyanic acid, allyl ester  see. . .Allyl Isothiocyanate
Isothiosemicarbazide  see. . .Thiosemicarbazide
ISOTOX®  see. . .Hexachlorocyclohexanes
ISOTOX®  see. . .Lindane
ISOTOX SEED TREATER® "D" and "F"  see. . .Captan
Isourea  see. . .Urea
Isovaleric acid-8-ester with 3-formamido-*N*-(7-hexyl-8-hydroxy-4,9-dimethyl-2,6-dioxo-1,5-dioxonan-3-yl)salicylamide isovaleric acid 8 ester  see. . .Antimycin A
ISTAMBUL®  see. . .Amitraz
ITGAGUARD®  see. . .Triclosan
IVALON®  see. . .Formaldehyde
IVORAN®  see. . .DDT
IVOSIT®  see. . .Dinoseb
IXODEX®  see. . .DDT
JACUTIN®  see. . .Hexachlorocyclohexanes
JACUTIN®  see. . .Lindane
JAIKIN®  see. . .Sodium Tetraborate
JALAN®  see. . .Molinate
JANUS®  see. . .Linuron
JANUS®  see. . .Trifluralin
Jasmolin I  see. . .Pyrethrins or Pyrethrum
Jasmolin II  see. . .Pyrethrins or Pyrethrum
JAVELIN®  see. . .Diflufenican
Jet fuel: Jp-1  see. . .Kerosene
JF 5705F®  see. . .Cypermethrin
JF 6064®  see. . .Tefluthrin
JIFFY GROW®  see. . .Indole-3-Butyric Acid
JMC 45498®  see. . .Deltamethrin
JOLT®  see. . .Ethoprop
JONNIX®  see. . .Asulam

JOSH®  see. . .Isoproturon
JULIN'S CARBON CHLORIDE®  see.Hexachlorobenzene
JUNO®  see. . .Propiconazole
JUPITAL®  see. . .Chlorothalonil

- K -

4K-2M®  see. . .MCPA
K-4®  see. . .Diuron
K-4 HERBICIDE®  see. . .Hexazinone
K-19®  see. . .Triphenyltin Compounds
K-19®  see. .Fentin Hydroxide
K62-105®  see. . .Leptophos
K III®  see. . .Dinitro-o-cresol (DNOC)
K IV®  see. . .Dinitro-o-cresol (DNOC)
KABAT®  see. . .Methoprene
KACK®  see. . .Cacodylic Acid
KACK®  see. . .Sodium Methanearsonate (MSMA)
Kadmium (German)  see. . .Cadmium
Kadmiumchlorid (Germany)  see. . .Cadmium Chloride
Kadmu (Polish)  see. . .Cadmium
Kafar copper  see. . .Copper and Copper Compounds
KAFIL® SUPER  see. . .Cypermethrin
KAKEN®  see. . .Cycloheximide
Kakodylan dodny  see. . .Sodium Cacodylate
KALEIT®  see. . .Fenitrothion
KALGIBB®  see. . .Gibberellic Acid
Kalium-cyanid (German)  see. . .Potassium Cyanide
Kaliumnitrat (German)  see. . .Potassium Nitrate
Kaliumnitrat (German)  see. . .Potassium Nitrite
KALO®  see. . .Calcium Arsenate
KALPHOS®  see. . .Parathion
Kalziumarseniat (German)  see. . .Calcium Arsenate
KAMFOCHLOR®  see. . .Toxaphene
KAMPOSAN®  see. . .Ethephon
KANEPAR®  see. . .Fenac
KANKEREX®  see. . .Mercuric Oxide
KAPTAN®  see. . .Captan
KARAMATE®  see. . .Mancozeb
KARATE®  see. . .*lamda*-Cyhalothrin
KARATHANE® WD  see. . .Dinocap
KARATHANE®  see. . .Dinocap
KARATHENE®  see. . .Dinocap
KARBAM BLACK®  see. . .Ferbam
KARBAM CARBAMATE®  see. . .Ferbam
KARBAM WHITE®  see. . .Ziram
KARBASPRAY®  see. . .Carbaryl
KARBATION®  see. . .Metham-Sodium
KARBATION (dihydrate)®  see. . .Metham-Sodium
KARBATOX®  see. . .Carbaryl
KARBICRON®  see. . .Dicrotophos
KARBOFOS®  see. . .Malathion
KARBOSEP®  see. . .Carbaryl
KARIDIUM®  see. . .Sodium Fluoride
KARIGEL®  see. . .Sodium Fluoride
KARI-RINSE®  see. . .Sodium Fluoride
KARLAN®  see. . .Ronnel

KARMEX® see...Diuron
KARMEX® see...Monuron
KARMEX DIURON HERBICIDE® see...Diuron
KARMEX DW® see...Diuron
KARMEX® W see...Monuron
KARPEN® see...Dodine
KARSAN® see...Formaldehyde
KARTAN® see...Fluvalinate
KARTEX A® see...Propachlor
KASCADE® see...Maneb
KATAMINE® AB see...Zilkonium Chloride
Katharin see...Carbon Tetrachloride
KATHON® 893 see...Octhilinone
KAURITIL® see...Copper Oxychloride
KAVADEL® see...Dioxathion
KAYAFUME® see...Methyl Bromide
KAYAPHENONE® see...Bensulide
KAYAZINON® see...Diazinon
KAYAZOL® see...Diazinon
KAYNITRO® see...Ammonium Nitrate
KAZOE® see...Sodium Azide
KEEN SUPERKILL ANT AND ROACH EXTERMINATOR® see...Fenitrothion
KEEPER® see...Phenmedipham
KEEPOUT® see...Capsaicin
KELTANE® see...Dicofol
KELTHANE® see...Dicofol
KELTHANETHANOL® see...Dicofol
KEMATE® see...Anilazine
KEMDAZIN® see...Carbendazim
KEMIFAM® see...Phenmedipham
KEMIKER® see...Carboxin
KEMIRON® see...Ethofumesate
KEMOLATE® see...Phosmet
KENAPON® see...Dalapon
KENCIS® see...Cypermethrin
KENCOZEB® see...Mancozeb
KENDAN® see...Endosulfan
KENFURAN® see...Carbofuran
KENITE® see...Diatomaceous Earth
KENLOGO® see...Dimethoate
KENOFOL® see...Captafol
KENOFURAN® see...Carbofuran
KEN-ROUND EXTRA® see...Glyphosate
KENSBAN® see...Chlorpyrifos
KENSOLO® see...Propanil
KEN-STAR PLUS® see...Glyphosate
KEPONE® see...Chlordecone (Kepone®)
Keralyt see...Salicylic Acid
KERB® see...Pronamide
KERB 50W® see...Pronamide
KERB® PROPYZAMIDE 50 see...Pronamide
KERNTOX® see...Endosulfan
Kerosine see...Kerosene
KEYSTONE LA® see...Acetochlor
K-I CHEMICAL see...Prohexadione Calcium
KIDEN® see...Iprodione
Kieselguhr (German) see...Diatomaceous Earth

KIH 2031 see...Pyrithiobac-Sodium
KIH® 2031 see...Pyrithiobac-Sodium
KILDIP® see...Dichlorprop
KILEX LINDANE® see...Chlordane
KILEX PARATHION® see...Methyl Parathion
KILEX® see...Copper Oxychloride
KILEX®-3 (butyl ester) see...2,4,5-Trichlorophenoxyacetic Acid, Esters
KILL-ALL® see...Sodium Arsenite
KILLAX® see...TEPP
KILLGERM DETHLAC INSECTICIDAL LAQUER® see...Dieldrin
KILLGERM SEWARIN P® see...Warfarin
KILLGERM TETRACIDE INSECTICIDAL SPRAY® see...Fenitrothion
KILL KANTZ® see...ANTU
KILL-OFF® see...Borax and Boric Acid
KILL-RO RAT KILLER® see...Diphacione
KILMAG® see...Calcium Arsenate
KILMITE 40® see...TEPP
KILMOL® see...Warfarin
KILOSEB® see...Dinoseb
KILPROP® see...Mecoprop
KILRAT® see...Zinc Phosphide
KILSEM® see...MCPA
KILZOL® see...Hexachlorophene
KIM-112® see...Prohexadione Calcium
King's green see...Copper Acetoarsenite
King's green see...Paris Green
KIPSIN® see...Methomyl
KISVAX® see...Carboxin
KITRON® see...Acephate
KIWI LUSTR-277® see...o-Phenylphenol
KLARTAN® see...Fluvalinate
KLEENUP® see...Acifluorfen
KLEENUP® see...Diquat Dibromide
KLEENUP® see...Oxyfluorfen
KLEENWALK® see...Prometon
KLEERAWAY® see...Acifluorfen
KLEERAWAY® see...Glyphosate
KLEER-LOT® see...Amitrole
KLINGTITE® see...1-Naphthaleneacetic Acid
KLOP® see...Chloropicrin
KLOREX® see...Sodium Chlorate
KMH see...Maleic Hydrazide
KM SODIUM CHLORATE® see...Sodium Chlorate
KNOCKMATE® see...Ferbam
KNOCKOUT® see...Sulfometuron-Methyl
KNOCK OUT® see...Tris(hydroxymethyl)nitromethane
KNOWX-WEED® see...Dinoseb
KNOX-WEED® see...Dinoseb
KOBAN® see...Etridiazole
KOBU® see...Quintozene
KOBUTOL® see...Quintozene
KOCIDE® 101 see...Copper Hydroxide
KOCIDE® 2000 see...Copper Arsenite
KODIAK® see...Metalaxyl
KODIAK A-T FUNGICIDE® see...Quintozene

KODIAK T® see...Thiram
Kohlendisulfid (German) see...Carbon Disulfide
KOKOTINE® see...Lindane
KOLODUST® see...Parathion
KOLOFOG® see...Sulfur
KOLOSPRAY® see...Sulfur
KOLTAR® see...Oxyfluorfen
KOMET-RP® see...Tefluthrin
KOPFUME® see...Ethylene Dibromide
KOP KARB® see...Copper Carbonate see...Basic
KOP MITE® see...Chlorobenzilate
KOPSOL® see...DDT
KOP-THIODAN® see...Endosulfan
KOP-THION® see...Malathion
KORANDA® see...Acephate
KORANDA® see...Fenvalerate
KORAX® see...1-Chloro-1-Nitropropane
KORAX 6® see...1-Chloro-1-Nitropropane
KORIUM® see...Dichlorophene
KORLAN® see...Ronnel
KORLANE® see...Ronnel
KORTOFIN® see...Aldrin
K-OTHRINE® see...Deltamethrin
KOTION® see...Fenitrothion
KOTOL® see...Hexachlorocyclohexanes
KOYOSIDE® see...Sodium Aluminum fluoride
KOZINC® see...Copper Hydroxide
KP 2® see...Quintozene
K-PIN® see...Picloram
K-SALT® see...Naphthoxyacetic Acid
KRECALVIN® see...Dichlorvos
KREGASAN® see...Thiram
KRENITE® see...Fosamine Ammonium
KREOZAN® see...Dinitro-o-cresol (DNOC)
KREZOTOL 50® see...Dinitro-o-cresol (DNOC)
KRIOLIT® see...Sodium Aluminum fluoride
KRISMAT® see...Ametryn
KRITAP® see...Cartap Hydrochloride
NTD 2® see...Cartap Hydrochloride
KROTENAL® see...Disulfiram
KROTILINE® see...2,4-D
KROVAR IDF® see...Diuron
KROVAR® see...Bromacil
KRUMKIL® see...Coumafuryl
KRYOCIDE® see...Sodium Aluminum Fluoride
Kryolith (German) see...Sodium Aluminum fluoride
KRYSID® see...ANTU
KRYSID PI® see...ANTU
KRZEWOTOKS® (butyl ester) see...2,4,5-Trichlorophenoxyacetic Acid, Esters
KT 30® see...Forchlorfenuron
KT 35® see...Copper Oxychloride
KUH-833® see...Prohexadione Calcium
KUIK® see...Methomyl
Kumander see...Warfarin
KUMIAI® see...Metolcarb
KUMULUS® 5 see...Sulfur

Kupfercarbonat (German) see...Copper Carbonate see.Basic
Kupferoxychlorid (German) see...Copper Oxychloride
Kupferoxydul (German) see...Cuprous Oxide
Kupfersulfat (German) see...Copper Sulfate
Kuprablau (German) see...Copper Hydroxide
KUPRATSIN® see...Zineb
KUPRICOL® see...Copper Oxychloride
KUPRIKOL® see...Copper Oxychloride
KURAN® see...Silvex
KURON® see...Silvex
KUROSALG® see...Silvex
KUROSAL® see...Silvex
KUSA-TOHRUKUSATOL® see...Sodium Chlorate
KVK® see...MCPA
KWARC® see...Diflufenican
Kwasu 2,4-dwuchlorofenoksoctowego see...2,4-D
Kwas 2,4-dwuchlorofenoksyoctowy see...2,4-D
KWELL® see...Lindane
KWG 0599® see...Bitertanol
KWIK-KIL® see...Strychnine
KWIKSAN® see...Phenylmercury Acetate
Kyanid sodny (Czech) see...Sodium Cyanide
KYLAR® see...Daminozide
KYPCHLOR® see...Chlordane
Kypfarin see...Warfarin
KYPFOS® see...Malathion
KYPMAN 80® see...Maneb
KYPTHION® see...Parathion
KYPZIN® see...Ziram
KYPZIN® see...Zineb
Kyselina benzoova (Czech) see...Benzoic Acid
Kyselina 2,4-dichlorfenoxyoctova (Czech) see...2,4-D
Kyselina peroxyoctova (Polish) see...Peracetic Acid

- L -

L 11/6® see...Phorate
L 343® see...Prothoate
L 561® see...Phenthoate
L 01748® see...Dithiazanine Iodide
L 5300® see...Tribenuron-Methyl
L 34314® see...Diphenamid
L 36352® see...Trifluralin
L 676,863® ($B_{1a}$) see...Abamectin
LABUCTRIL® see...Bromoxynil
LACCO MAGIC SULPHUR® see...Sulfur
LACCO PARIS GREEN® see...Los Angeles Chemical Co. (USA) see...Copper Acetoarsenite
LADDOK® see...Atrazine
LADDOK® see...Bentazon
LADOB® see...Dinoseb
LAMA® see...Nicosulfuron
LAMBAST® see...Butachlor
LAMBROL® see...Fluenetil
Lamp oil see...Kerosene
LANCER® see...Acephate
LANCO ATRAZINE® see...Atrazine

LAND MASTER® see...2,4-D
LANDISAN® see...Methoxyethylmercuric Acetate
LANDMARK® MP see...Chlorsulfuron
LANDMARK® MP see...Sulfometuron-Methyl
LANDMASTER® see...Glyphosate
LANDRIN® see...Trimethacarb
LANDSIDE® see...Linuron
LANEX® see...Fluometuron
LANNATE® see...Methomyl
LANOX 90® see...Methomyl
LANOX 216® see...Methomyl
LANSTAN® see...1-Chloro-1-Nitropropane
LAREDO® see...Myclobutanil
LARGON® see...Diflubenzuron
LARIAT® see...Alachlor
LARIAT® see...Atrazine
LARVACIDE® see...Chloropicrin
LARVACIDE 100® see...Chloropicrin
LARVADEX® see...Cyromazine
LARVAKIL® see...Diflubenzuron
LARVIN® see...Thiodicarb
LASAGRIN® see...Alachlor
LASEB® see...Dinoseb
LASER® see...Cyfluthrin
LASHER® see...Chlorsulfuron
LASSAGRIN® see...Alachlor
LASSO® see...Alachlor
LASSO® EC PLUS LOROX® see...Paraquat Methosulfate
LASSO MICRO-TECH® see...Alachlor
LATKA-666® see...Hexachlorocyclohexanes
Laurylbenzenesulfonate see...Dodecylbenzenesulfonic Acid
Laurylbenzenesulphonate see...Dodecylbenzenesulfonic Acid
Laurylbenzenesulfonic acid see.Dodecylbenzenesulfonic Acid
Laurylbenzenesulphonic acid see.Dodecylbenzenesulfonic Acid
Lauryl guanidine acetate see...Oxythioquinox
LAUXTOL® see...Pentachlorophenol
LAWN-KEEP® see...2,4-D
LAZO® see...Alachlor
le Captane (French) see...Captan
LEA-COV® see...Sodium Fluoride
Lead acetate acid see...Lead Arsenate
Lead acid arsenate see...Lead Arsenate
LEADER® see...Bentazon
LEADOFF® see...Atrazine
LEADOFF® see...Dimethenamid
LEAFLESS® see...Dimethipin
LEAFLESS see...Thidiazuron
LEGUMEX® D see...2,4-DB
LEGUMEX DB® see...MCPA
LEGURAME® see...Carbetamide
LEIVASOM® see...Trichlorfon
LE-MAT® see...Omethoate
LEMOFLUR® see...Sodium Fluoride

LEMONENE® see...Biphenyl
LENOCYCLINE® see...Oxytetracycline Calcium
LENTAGRAN® see...Pyridate
LENTOX® see...Lindane
LEPICRON® see...Thiodicarb
LESAN® see...Fenaminosulf
LETHALAIRE® G-52 see...TEPP
LETHALAIRE G-54® see...Parathion
LETHALAIRE G-57® see...Sulfotepp
LETHALAIRE G-59® see...Octamethyl Diphosphoramide
LETHELMIN® see...Phenothiazine
LETHOX® see...Carbophanothion
LEUCOSULFAN® see...Busulfan
LEUNA M® see...MCPA
LEVERAGE® see...Imidacloprid
LEXOL® 300 see...Triclosan
LEXONE® see...Metribuzin
LEXONEEX® see...Metribuzin
LEYSPRAY® see...MCPA
LEYTOSAN® see...Phenylmercury Acetate
LFA 2043® see...Iprodione
LH 30/Z® see...Propineb
LH 3012® see...Propineb
LIDENAL® see...Lindane
LIDER® see...Glyphosate
Light petroleum see...Kerosene
LIGHTNING® see...Imazethapyr
LIGNASAN® see...Carbendazim
Ligroin see...Naphthas
Ligroine see...Naphthas
LIHOCIN® see...Chlormequat Chloride
LILLY 34,314® see...Diphenamid
LILLY 36,352® see...Trifluralin
LIMATOR® see...Metaldehyde
Lime nitrogen see...Calcium Cyanamide
Lime saltpeter see...Calcium Nitrate
Limestone see...Calcium Carbonate
Lime water see...Calcium Hydroxide
Limonene see...D-Limonene
LINDACOL® see...Hexachlorocyclohexanes
LINDAFOR® see...Lindane
LINDAGAM® see...Hexachlorocyclohexanes
LINDAGAM® see...Lindane
LINDAGRAIN® see...Lindane
LINDAGRAM® see...Lindane
LINDAGRANOX® see...Lindane
LINDAN® see...Dichlorvos
α-Lindane see...Hexachlorocyclohexanes
β-Lindane see...Hexachlorocyclohexanes
γ-Lindane see...Lindane
δ-Lindane see...Hexachlorocyclohexanes
LINDAPOUDRE® see...Lindane
LINDATOX® see...Lindane
LINDOSEP® see...Lindane
LINE RIDER® see...2,4,5-T
LINE RIDER® see...2,4,5-Trichlorophenoxyacetic Acid, Esters

LINEX® see...Linuron
LINNET® see...Linuron
LINNET® see...Trifluralin
LINORMONE® see...MCPA
LINOROX® see...Linuron
LINTOX® see...Lindane
LINUREX® see...Linuron
LIPAN® see...Dinitro-*o*-cresol (DNOC)
LIPHADIONE® see...Chlorophacinone
LIQUAMYCIN LA 200® see...Oxytetracycline Calcium
LIQUA-TOX® see...Diphacione
LIQUA-TOX® see...Warfarin
LIQUA-TOX® see...Warfarin
LIQUI-STIK® see...1-Naphthaleneacetic Acid
Liquid ammonia see...Ammonia
LIQUID DERRIS® see...Rotenone
LIQUID MOLY-CO-THI® see...Thiram
LIQUIPHENE® see...Phenylmercury Acetate
LIRANOX® see...Mecoprop
LIROBETAREX® see...Monuron
LIRO DNBP® see...Dinoseb
LIROHEX® see...TEPP
LIROMATIN® see...Triphenyltin Compounds
LIROPON® see...Dalapon
LIROPREM® see...Pentachlorophenol
LIROSTANOL® see...Triphenyltin Compounds
LIROTAN® see...Zineb
LIROTHION® see...Parathion
LITAROL® see...Bromoxynil
Lithographic stone see...Calcium Carbonate
LM 91® see...Chlorophacinone
LM-637® see...Bromadiolone
LOCUSTCIDE® see...Nosema Locustae
LO-ESTASOL® see...2,4-D, butoxyethyl ester
LOGRAN® see...Triasulfuron
LOISOL® see...Trichlorfon
LOMBRISTOP® see...Thiabendazole
LONACOL® see...Zineb
LONDAX® see...Bensulfuron-methyl
LONDAX PRO-PACK BNB® see...Propanil
LONOCOL M® see...Maneb
LONTREL® see...Clopyralid
LONTREL® 3 see...Clopyralid
LONTRIL® F see...Clopyralid
LONTRIL® T see...Clopyralid
LOREXANE® see...Lindane
LOREX® see...Linuron
LOREX® see...Sodium Chlorate
LOROTHIDOL® see...Bithionol
LOROX® see...Chlorimuron-ethyl
LOROX® see...Linuron
LORSBAN® see...Chlorpyrifos
LO VOL® see...2,4,5-Trichlorophenoxyacetic Acid, Esters
LS 4442® see...Triphenyltin Compounds
LS-74783® see...Fosetyl-Al
LUCANAL® see...Naled
LUFOX® see...Fenoxycarb

LURIDE® see...Sodium Fluoride
LYNX-1.2® see...Tebuconazole
LYSOFORM® see...Formaldehyde

- M -

M2 Copper see...Copper and Copper Compounds
2M-4C® see...MCPA
2M-4CP see...Mecoprop
2M-4KH® see...MCPA
2M-4KHP see...Mecoprop
M7-Giftkoerner see...Thallium Sulfate
M 40® see...MCPA
M 73® see...Clonitralid
M-74® see...Disulfoton
M 140® see...Chlordane
M 410® see...Chlordane
M 2060® see...Fluenetil
M 5055® see...Toxaphene
M 8164® see...Triacontanol
M&B 10731® see...Bromoxynil
M&B 10064® see...Bromoxynil
M&B 38544® see...Diflufenican
MAA see...Methanearsonic Acid
MABLIN® see...Busulfan
MACHETE® see...Butachlor
MACHETTE® see...Butachlor
MACH-NIC® see...Nicotine
MACOCYN® see...Oxytetracycline Calcium
Macquer's salt see...Potassium Arsenate
MACRONDRAY® see...2,4-D
MAFU® see...Dichlorvos
MAGIC CARPET FERTILIZER WITH ATRAZINE® see...Atrazine
MAGISTER® see...Clomazone
MAGMA® see...Sodium Methanearsonate (MSMA)
MAGNACIDE® see...Acrolein
MAGNACIDE H® see...Acrolein diacetate
Magnesium dichlorate see...Magnesium Chlorate
MAGNET® see...Imazalil
MAGNETIC 70® see...Sulfur
MAGNIFLOC 156C FLOCCULANT® see.Formaldehyde
MAGRON® see...Magnesium Chlorate
MAH see...Maleic Hydrazide
MAINTAIN 3® see...Maleic Hydrazide
Major Capsaicinoids see...Capsaicin
MAKI® see...Bromadiolone
Malachite see...Copper Carbonate see...Basic
Malachite green G see...C.I. Basic Green 1
MALACIDE® see...Malathion
MALAFOR® see...Malathion
MALAGRAN® see...Malathion
MALAKILL® see...Malathion
MALAMAR® see...Malathion
MALASOL® see...Malathion
MALASPRAY® see...Malathion
MALATAF® see...Malathion

MALATHION E50® see...Malathion
MALATHION 60® see...Malathion
Malathon see...Malathion
Malathyl see...Malathion
Malation (Spanish) see...Malathion
MALATOL® see...Malathion
Malatox (Indian) see...Malathion
MALAZIDE® see...Maleic Hydrazide
Maldison (Australia see...New Zealand) see..Malathion
Maleic acid hydrazide see...Maleic Hydrazide
Maleic hydrazide fungicide see...Maleic Hydrazide
Maleic hydrazine see...Maleic Hydrazide
MALEIN 30® see...Maleic Hydrazide
Maleinsaurehydrazid (German) see...Maleic Hydrazide
$N,N$-Maleoylhydrazine see...Maleic Hydrazide
MALERBANE® see...2,4-D
MALERBANE® see...MCPA
MALERBANE-GIAVONI-L® see...Molinate
MALIPUR® see...Captan
MALIX® see...Endosulfan
MALLET PM BROMOXYNIL® see...Atrazine
Malmed see...Malathion
MAL-O-CHLOR® see...Chlorobenzilate
Malphos see...Malathion
MALTOX® see...Malathion
Malzid see...Maleic Hydrazide
MANAGE® see...Halosulfuron-methyl
MANAM® see...Maneb
MANCOFOL® see...Mancozeb
MANDATE® see...Thiazopry
MANEB 80® see...Maneb
MANEBA® see...Maneb
MANEBE® see...Maneb
MANEBGAN® see...Maneb
Maneb-zinc see...Mancozeb
Maneb-zineb-komplex (German) see...Mancozeb
MANESAN® see...Maneb
MANEX® see...Maneb
Mangaan (II)-[$N,N$'-ethyleen-bis(dithiocarbamaat)] (Dutch) see...Maneb
Mangan (II)-[$N,N$'-aethylen-bis(dithiocarbamate)] (German) see...Maneb
Manganese ethylene-1,2-bisdithiocarbamate see..Maneb
Manganese ethylene-bis(dithiocarbamate)(polymeric) complex with zinc salt see...Mancozeb
Manganese(II) ethylene di(dithiocarbamate) see..Maneb
Manganous ethylenebis(dithiocarbamate) see...Maneb
Mangan-zink-aethylendiamin-bis-dithio-carbamat (German) see...Mancozeb
MANOC® see...Maneb
MANOSEB® see...Mancozeb
MANTA® see...Methoprene
MANTI® S see..Propiconazole
MANTOX® see...Mancozeb
MANZATE® see...Maneb
MANZATE 200® see..Mancozeb
MANZATE D® see..Maneb
MANZATE MANEB FUNGICIDE® see...Maneb

MANZEB® see...Maneb
MANZEB® see...Mancozeb
MANZIN 80® see...Mancozeb
MANZI® see...Maneb
MAPOSOL® see...Metham-Sodium
MAPOSOL® see...Metham-Sodium
MAR-FIN® see...Warfarin
MARATHON® see...Imidacloprid
MARATHON® see...Prodiamine
Marble see...Calcium Carbonate
Marevan (sodium salt) see...Warfarin
MARGOSAN-O® see...Azadirachtin
MARISAN FORTE® see...Quintozene
MARKSMAN® see...Atrazine
MARKSMAN® see...Dicamba
MARKSMAN® see...Trifluralin
MARKSMAN 1® see...Linuron
MARKSMAN 2® see...Trifluralin
MARLATE® see...Methoxychlor
MARMER® see...Diuron
MARQUISE® see...Metamiton
MARSHAL® see...Carbosulfan
MARSHALL® see...Carbosulfan
MARSTAN FLY SPRAY® see...Lindane
MARVEX® see...Dichlorvos
MARZIN® see...Mancozeb
MARZONE ATRAZINE® see...Atrazine
MASCOT® see...Vinclozolin
MASCOT HIGHWAY® see...Amitrole
MASOTEN® see...Trichlorfon
MASTER BRAND® see...Aldrin
MAT 14500® see...Methyl Formate
MATABROM® see...Methyl Bromide
MATADOR® see...*lamda*-Cyhalothrin
MATADOR® see...Quizalofop-Ethyl
MATA-SYSTOX® see...Demeton-methyl
MATAVEN® see...Difenzoquat
MATCH® see...Cyanazine
MATON® see...2,4-D
MATOR® see...Hydroprene
MATOX® see...Hydramethylnon
MATRAK® see...Difenacoum
MATRIGON® see...Clopyralid
MATRIX® see...Nicosulfuron
MATRIX® see...Tribenuron-Methyl
MAUX® see...Endosulfan
MAVRIK AQUAFLOW® see...Fluvalinate
MAVRIK®,Wellmark International (USA) see. Fluvalinate
MAXFORCE® ANT KILLER GRANULAR BAIT see. Hydramethylnon
MAXFORCE® ANT STATION see...Fipronil
MAXFORCE® ROACH GEL see...Hydramethylnon
MAXFORCE® ROACH STATION see...Fipronil
MAXICROP MOSS KILLER® see...Ferric Sulfate
MAXIM® see...Mancozeb
MAXON® see...Indole-3-Butyric Acid
MAXON® see...Gibberellic Acid

MAXON® see...Kinetin (Cytokinin)
MAXX® see...Propiconazole
MAXX-90® see...Propazine
MAYCLENE® see...MCPA
Mazide see...Maleic Hydrazide
MAZOTEN® see...Trichlorfon
MB 9057® see...Asulam
MB 10064® see...Bromoxynil
MB 10731® see...Bromoxynil
MB 38183® see...Diflufenican
MB 46030® see...Fipronil
MBC see...Benomyl
MBC see...Carbendazim
M-B-C FUMIGANT® see...Methyl Bromide
MBCP see...Leptophos
MBR 8251® see...Perfluidone
MBR 12325® see...Mefluidide
MC see...Mercuric Chloride
MC 2188® see...Chlormephos
MC 4379® see...Bifenox
MC 6897® see...Bendiocarb
MCA see...Chloroacetic Acid
MC DEFOLIANT® see...Magnesium Chlorate
MCP see...MCPA
2-MCPP see...Mecoprop
MCPP see...Mecoprop
MCPP 2,4-D see...Mecoprop
MCPP-D-4 see...Mecoprop
MCPP K-4 see...Mecoprop
MDBA see...Dicamba
M-DIPHAR® see...Maneb
ME4 BROMINAL® see...Bromoxynil
Meadow green see...Copper Acetoarsenite
Meadow green see...Paris Green
MEB see...Maneb
MEB 6447® see...Triadimefon
MECHLORPROP® see...Mecoprop
MECOBROM® see...Mecoprop
MECOMIN D® see...Mecoprop
MECOPAR® see...Mecoprop
MECOPEOP® see...Mecoprop
MECOPEX® see...Mecoprop
MECOTURF® see...Mecoprop
MEDAL® see...Metolachlor
MEDIBEN® see...Dicamba
MEDIZINC® see...Zinc Sulfate Heptahydrate
MEFENOXAM® see...Quintozene
MEFENOXAM/COPPER® see...Copper Hydroxide
MEGAGRO® see...Kinetin (Cytokinin)
MEGATOX® see...Fluoroacetamide
MELDANE® see...Coumaphos
MELDONE® see...Coumaphos
MELIPAX® see...Toxaphene
Melissyl alcohol see...Triacontanol
MELPREX® see...Dodine
MELPREX® see...Oxythioquinox
MELPREX® 65 see...Dodine
MELTATOX® see...Dodemorph Acetate

MEMA see...Methoxyethylmercuric Acetate
MEMA see...Methylmercuric Dicyanamide
MEMATOCIDE® see...Dibromochloropropane
MEMILENE® see...Methomyl
MENAPHAM® see...Carbaryl
MENAP® see...Ethoprop
MENDRIN® see...Endrin
Menite see...Mevinphos
p-Mentha-1,8-diene see...D-Limonene
ME-PARATHION® see...Methyl Parathion
MEP (PESTICIDE)® see...Fenitrothion
MEPEX® see...Indole-3-Butyric Acid
MEPEX® see...Kinetin (Cytokinin)
MEPEX® see...Mepiquat Chloride
MEPHANAC® see...MCPA
MEPICHLOR® see...Mepiquat Chloride
MEPPLUS® see...Mepiquat Chloride
MEPRO® see...Mecoprop
MEPTOX® see...Methyl Parathion
MERCAPROFOS® see...Sulprofos
MERCAPROPHOS® see...Sulprofos
Mercaptan methylique perchlore (French) see. Perchloromethyl Mercaptan
Mercaptodimethur see...Methiocarb
N-(2-Mercaptoethylbenzenesulfonamide)-S-($O,O$-diisopropyl phosphorodithioate) see...Bensulide
Mercaptofos (in former USSR) see...Demeton
Mercaptofos teolery see...Demeton
Mercaptoimidazoline see...Ethylene Thiourea
2-Mercaptoimidazoline see...Ethylene Thiourea
2-Mercapto-2-imidazoline see...Ethylene Thiourea
3-(Mercaptomethyl)-1,2,3-benzotriazin-4(3H)-one-$O,O$-dimethyl phosphorodithioate see...Azinphos-methyl
3-(Mercaptomethyl)-1,2,3-benzotriazin-4(3H)-one-$O,O$-dimethyl phos-phorodithioate-$S$-ester see...Azinphos-methyl
N-(Mercaptomethyl)phthalimide $S$-($O,O$-dimethyl phosphorodithioate)
Mercaptosuccinic acid diethyl ester see...Malathion
Mercaptothion see...Malathion
MERCASIN® see...Prometryn
MERCAZIN® see...Prometryn
MERCAZIN I® see...Ethylene Thiourea
MERCK® 48051 see...1,2-Dibromo-2,4-dicyanobutane
Merco Prills see...Ammonium Nitrate
MERCURAM® see...Thiram
Mercuran see...Methoxyethylmercuric Acetate
Mercuric bichloride see...Mercuric Chloride
Mercuric oxide, red see...Mercuric Oxide
Mercuric oxide, yellow see...Mercuric Oxide
Mercuriphenyl acetate see...Phenylmercury Acetate
Mercury(II) acetate, phenyl see...Phenylmercury Acetate
Mercury, acetoxy(2-methoxyethyl)- see. Methoxyethylmercuric Acetate
Mercury, (acetoxy)phenyl- see...Phenylmercury Acetate
Mercury bichloride see...Mercuric Chloride
Mercury(II) chloride see...Mercuric Chloride

Mercury(2+) chloride see...Mercuric Chloride
Mercury dimethyl see...Mercury Alkyl Compounds
Mercury monoxide see...Mercuric Oxide
Mercury oxide see...Mercuric Oxide
Mercury perchloride see...Mercuric Chloride
Mercury vichloride see...Mercuric Chloride
MEREX® see...Chlordecone (Kepone®)
MERGAL® see...Carbendazim
MERGAMMA® see...Phenylmercury Acetate
MERGAMMA 30® see...Lindane
MERGE® see...Sodium Methanearsonate (MSMA)
MERIT® see...Bromoxynil
MERIT® see...Clomazone
MERIT® see...Imidacloprid
2-Merkaptoimidazolin (Czech) see...Ethylene Thiourea
MERKAZIN® see...Prometryn
MERKON PHOSPHAMIDONE® see...Phosphamidon
MERPAFOL® see...Captafol
MERPAN® see...Captan
Merphos-oxide see...Tribufos
MERPOL® see...Ethylene Oxide
MERSOLITE® see...Phenylmercury Acetate
MERTEC® see...Thiabendazole
MERTECT 160® see...Thiabendazole
MESAMATE® see...Sodium Methanearsonate (MSMA)
Mesomile see...Methomyl
MESUROL® see...Methiocarb
META® see...Metaldehyde
Metaarsenic Acid see...Arsenic Acid
Metacetaldehyde see...Metaldehyde
Metachlor see...Alachlor
METACHLOR® see...Alachlor
METACIDE® see...Metham-Sodium
METACIDE® see...Methyl Parathion
METACIDE 38® see...1,2-Dibromo-2,4-dicyanobutane
METACID 50® see...Methyl Parathion
METACRATE® see...Metolcarb
METAFOS® see...Methamidophos
METAFOS® see...Methyl Parathion
METAG® see...Calcium Arsenate
METAISOSEPTOX® see...Demeton
METAISOSEPTOX® see...Demeton-methyl
Metaisosystox sulfoxide see...Demeton-methyl
METAISOSYSTOX® see...Demeton
METAISOSYSTOX® see...Demeton-methyl
METALAXIL® see...Metalaxyl
Metaldehyd (German) see...Metaldehyde
Metaldeide (Italian) see...Metaldehyde
METALKAMATE® see...Brufencarb
Metallic arsenic see...Arsenic and Inorganic Arsenic Compounds
METAM® see...Metham-Sodium
METAM-FLUID BASF® see...Metham-Sodium
Metamidofos (Spanish) see...Methamidophos
Metamidofos estrella see...Methamidophos
Metamitron (German) see...Metamiton
Metam sodium see...Metham-Sodium
METAPHOR® see...Phorate

METAPHOS® see...Methyl Parathion
METAPICRIN® see...Chloropicrin
METASOL® see...1,2-Dibromo-2,4-dicyanobutane
METASOL 30® see...Phenylmercury Acetate
METASOL TK-100® see...Thiabendazole
METASON® see...Metaldehyde
Metasystemox see...Demeton-methyl
Metasystemox R see...Demeton-methyl
METASYSTOX FORTE® see...Demeton
METASYSTOX FORTE® see...Demeton-methyl
Metasystox R see...Demeton-methyl
Metatetrachlorophthalodinitrile see...Chlorothalonil
METATHION® see...Fenitrothion
METATHIONE® see...Fenitrothion
METATION® see...Fenitrothion
METAXANIN® see...Metalaxyl
Metelilachlor see...Metolachlor
Methachlor see...Alachlor
METHAFLUORIDAMID® see...Mefluidide
METHAM DIHYDRATE® see...Metham-Sodium
Methanal see...Formaldehyde
Methanearsonic acid see...Calcium salt (2:1) see.Calcium Methanearsonate
Methanearsonic acid, monosodium salt see.Sodium Methanearsonate (MSMA)
Methanearsonic acid, octylammonium salt see.Octylammonium Methanearsonate
Methane see...bis(2,3,5-trichloro-6-hydroxyphenyl) see.Hexachlorophene
Methane, bromo- see...Methyl Bromide
Methanedithiol see...$S,S$-diester with $O,O$-diethyl phosphorodithioate acid see...Ethion
Methanesulfonamide see...$N$-[2,4-dichloro-5-(4-(difluoromethyl)-4,5-dihydro-3-methyl-5-*oxo*-1$H$-1,2,4-triazol-1-yl)phenyl]- see...Sulfentrazone
Methanesulfonamide see...1,1,1-trifluoro-$N$-(2-methyl-4-(phenylsulfonyl)phenyl)- see...Perfluidone
Methanesulfonic acid tetramethylene ester see..Busulfan
Methane tetrachloride see...Carbon Tetrachloride
Methane, tetrachloro- see...Carbon Tetrachloride
Methanethiol, (ethylthio)- see...$S$-ester with $O,O$-diethylphosphorodithioate see...Phorate
Methane, trichloronitro- see...Chloropicrin
Methane, trimethylolnitro- see.Tris(hydroxymethyl)nitromethane
Methanimidamide see...$N'$-(4-chloro-2-methylphenyl)-$N,N$-dimethyl- see...Chlordimeform
6,9-Methano-2,4,3-benzodioxathiepin see..6,7,8,9,10,10-hexachloro-1,5,5a,6,9,9a-hexahydro-, 3-oxide see..Endosulfan
6,9-Methano-2,4,3-benzodioxathiepin, 6,7,8,9,10,10-hexachloro-1,5,5α,6,9,9α-hexahydro-, 3-oxide, (3,5aβ,6,9,9a.β)- see...Endosulfan
6,9-Methano-2,4,3-benzodioxathiepin, 6,7,8,9,10,10-hexachloro-1,5,5A,6,9,9A-hexahydro-, 3-oxide, (3α,5Aα,6β,9β,9Aα)- see...Endosulfan
Methanoic acid see...Formic Acid

4,7-Methanoindan see...1,4,5,6,7,8,8-heptachloro-2,3-epoxy-3a,4,7,7a-tetrahydro see...Heptachlor Epoxide
4,7-Methanoindene see...1,4,5,6,7,8,8-heptachloro-3a,4,7,7a-tetrahydro- see...Heptachlor
4,7-Methanoindan see...1,2,4,5,6,8,8-octachloro 3a,4,7,7a-tetrahydro see...Chlordane
4,7-Methanoindan see...1,2,3,4,5,6,7,8,8-octachloro-2,3,3a,4,7,7a-hexahydro- see...Chlordane
4,7-Methano-1H-indene,1,2,4,5,6,7,8,8-octachloro-2,3,3a,4,7,7a-hexahydro- see...Chlordane
Methanol see...phenyl- see...Benzyl Alcohol
Methan-sodium see...Metham-Sodium
METHASAN® see...Ziram
METHAVIN® see...Methomyl
METHAZATE® see...Ziram
1,3,4-Metheno-2H-cyclobuta[c,d]pentalen-2-one,1,1a,3,3a,4,5,5a,5b,6-decachloro-octahydro-Kepone® see...Chlordecone (Kepone®)
Methidathion
METHIOCARBE® see...Methiocarb
METHOFAN® see...Endosulfan
METHO-GAS® see...Methyl Bromide
Metholcarb see...Metolcarb
METHOMEX® see...Methomyl
Methoxide see...Methoxychlor
Methoxo see...Methoxychlor
METHOXONE® see...Mecoprop
2-[(((((4-Methoxy-6-methyl-1,3,5-triazin-2-yl)-methylamino)carbonyl)amino)sulfonyl)-, methyl ester see...Tribenuron-Methyl
Methoxybenzene see...Anisole
6-Methoxy-N,N'-bis(1-methylethyl)-1,3,5-triazine-2,4-diamine see...Prometon
2-Methoxy-4,6-bis(isopropylamino)-s-triazine see...Prometon
2-Methoxy-4,6-bis(isopropylamino)-1,3,5-triazine see...Prometon
2-(Methoxy-carbonylamino)-benzimidazol see...Carbendazim
2-(Methoxycarbonylamino)-benzimidazole see...Carbendazim
3-[(Methoxycarbonyl)amino]phenyl N-(3-methylphenyl)carbamate see...Phenmedipham
2-Methoxycarbonyl-1-methylvinyl dimethyl phosphate see...Mevinphos
cis-2-Methoxycarbonyl-1-methylvinyl dimethylphosphate see...Mevinphos
(cis-2-Methoxycarbonyl-1-methylvinyl) dimethyl phosphate see...Mevinphos
1-Methoxycarbonyl-1-propen-2-yl dimethyl phosphate see...Mevinphos
N-(2-Methoxycarbonylphenylsulfonyl)-N-[4,6-bis(difluoromethoxy)pyrimidin-2-yl]urea see...Primisulfuron-Methyl
p,p'-Methoxychlor see...Methoxychlor
Methoxy DDT see...Methoxychlor
2-Methoxy-3,6-dichlorobenzoic acid see...Dicamba

2-Methoxy-3,6-dichlorobenzoic acid sodium salt see. Sodium Dicamba
5-Methoxy-2-(dimethoxyphosphinylthiomethyl)pyrone-4 see...Endothion
Methoxydiuron see...Linuron
Methoxyethylmercury acetate see.Methoxyethylmercuric Acetate
2-Methoxyethylmerkuriacetat (German) see. Methoxyethylmercuric Acetate
1-Methoxy-1-methyl-3-(3,4-dichlorophenyl)urea see. Linuron
N-(Methoxymethyl)2,6-diethylchloroacetamide see. Alachlor
3-(6-Methoxy-4-methyl-1,3,5-triazin-2-yl)-1-[2-(2-chloroethoxy)phenylsulfonyl]urea see...Triasulfuron
S-5-Methoxy-4-oxopyran-2-ylmethyl dimethyl phosphorothioate see...Endothion
S-[(5-Methoxy-2-oxo-1,3,4-thiadiazol-3(2H)-yl)methyl]-O,O-dimethyl phosphordithioate see...Methidathion
2,2-(p-Methoxyphenyl)-1,1,1-trichloroethane see. Methoxychlor
Methoxypropazine see...Prometon
S-((5-Methoxy-4H-pyron-2-yl)-methyl)-O,O-dimethyl-monothiofosfaat (Dutch) see...Endothion
S-((5-Methoxy-4H-pyron-2-yl)-methyl)-O,O-dimethyl-monothiophosphat (German) see...Endothion
S-(5-Methoxy-4-pyron-2-ylmethyl)dimethylphosphorothiolate see...Endothion
(E,E)-11-Methoxy-3,7,11-trimethyl-2,4-dodecandienoate see...Methoprene
Methyl aldehyde see...Formaldehyde
Methyl-4-aminobenzenesulphonyl-carbamate see.Asulam
Methyl-N-(4-aminobenzenesulfonyl)carbamate see. Asulam
N-Methylaminodithioformic acid sodium salt see. Metham-Sodium
N-Methylaminomethanethionothiolic acid sodium salt see...Metham-Sodium
Methyl [(4-aminophenyl)sulfonyl] carbamate see.Asulam
3-Methyl-4-amino-6-phenyl-1,2,4-triazin(4H)-on (German) see...Metamiton
Methylarsinic acid see...Methanearsonic Acid
Methylarsonic acid see...Methanearsonic Acid
Methylarsonic acid see...Calcium salt (2:1) see.Calcium Methanearsonate
Methylarsonic acid, monosodium salt see. .Sodium Methanearsonate (MSMA)
Methyl azinphos see...Azinphos-methyl
N-Methylbenzazimide see...dimethyldithiophosphoric acid ester see...Azinphos-methyl
Methyl 1H-benzemedazol-2-yl carbamate see. Carbendazim
Methyl 2-benzimidazolecarbamate see...Carbendazim
Methyl benzimidazole-2-yl carbamate see. .Carbendazim
α-Methylbenzyl 3-(dimethoxyphosphinoxy)-cis-crotonate see...Crotoxyphos
1-Methylbenzyl-3-(dimethoxyphosphinyloxo) isocrotonate see...Crotoxyphos

α-Methylbenzyl-3-hydroxy-crotonate dimethyl phosphate see...Crotoxyphos
Methyl-2-[(((((4,6-bis(difluoromethoxy)-2-pyrimidinyl)amino)carbonyl)amino)sulfonyl]benzoate see...Primisulfuron-Methyl
N-Methylbis(2,4-xylyliminomethyl)amine see...Amitraz
Methylbromid see...Methyl Bromide
O-Methyl-O-(4-bromo-2,5-dichlorophenyl)phenyl thiophosphonate see...Leptophos
Methyl bromofos see...Bromophos
Methyl bromophos see...Bromophos
Methyl 1-(butylcarbamoyl)-2-benzimidazolyl carbamate see...Benomyl
3-(1-Methylbutyl)phenyl methylcarbamate and 3-(1-ethylpropyl)phenyl methylcarbamate, mixed esters (3:1) see...Brufencarb
N-Methylcarbamate de 1-naphtyle (French) see..Carbaryl
Methylcarbamate (Ester) see...Physostigmine
Methylcarbamate 1-naphthalenol see...Carbaryl
Methylcarbamic acid o-sec-butylphenyl ester see..BPMC
Methylcarbamic acid m-cumenyl ester see...Phenol, 3-(1-methyethyl)-, methylcarbamate
Methylcarbamic acid-m-cym-5-yl ester see...Promecarb
Methyl carbamic acid 2,3-dihydro-2,2-dimethyl-7-benzofuranyl ester see...Carbofuran
Methylcarbamic acid, 4-(dimethylamino)-3,5-xylyl ester see...Mexacarbate
Methyl-carbamic acid, ester with eseroline see. Physostigmine
Methylcarbamic acid 2,3-(isopropylidenedioxy)phenyl ester see...Bendiocarb
Methylcarbamic acid-m-[(1-methyl)butyl]phenyl ester mixed with carbamic acid, methyl-m-(1-ethylpropyl)phenyl ester (3:1) see...Brufencarb
N-Methylcarbamic acid 3-methyl-5-isopropylphenyl ester see...Promecarb
Methyl carbamic acid 4-(methylthio)-3,5-xylyl ester see. Methiocarb
Methylcarbamic acid, 1-naphthyl ester see.Carbaryl
Methylcarbamic acid, toyl ester see...Metolcarb
Methylcarbamic acid, trimethylphenyl ester see. Trimethacarb
Methylcarbamodithioic acid sodium salt see...Metham-Sodium
2-Methylchlorobenzene see...o-Chlorotoluenel
1-Methyl-2-chlorobenzene see...o-Chlorotoluenel
Methyl 3-chloro-5-(4,6-dimethoxypyrimidin-2-yl carbamoylsulfamoyl)-1-methyl pyrazole-4-carboxylate see...Halosulfuron-methyl
3-Methyl-4-chlorophenol see...p-Chloro-m-cresol
Methylchlorophenoxyacetic acid see...MCPA
2-Methyl-4-chlorophenoxyacetic acid see...MCPA
(2-Methyl-4-chlorophenoxy)acetic acid see...MCPA
2-Methyl-4-chlorophenoxyessigsaeure (German) see. MCPA
α(2-Methyl-4-chlorophenoxy)propionic acid see. Mecoprop
2-(2'-Methyl-4'-chlorophenoxy)propionic acid see. Mecoprop
2-Methyl-4-chlorophenoxy-α-propionic acid see. Mecoprop
trans-4-Methyl-5-(4-chlorophenyl)-3-cyclohexylcarbamoyl-2-thiazolidone see...Hexythiazox
N'-(2-Methyl-4-chlorophenyl)-N,N-dimethylformamidine see...Chlordimeform
Methyl chlorophos see...Trichlorfon
N'-(2-Methyl-4-chlorphenyl)-formamidin-hydrochlorid (German) see...Chlordimeform
Methyl chlorpyrifos see...Chlorpyrifos-methyl
3-Methylcrotonic acid 2-sec-butyl-4,6-dinitrophenyl ester see...Binapacryl
5-Methyl m-cumenyl methylcarbamate see...Promecarb
1-(2-Methylcyclohexyl)-3-phenylurea see...Siduron
N-(2-Methylcyclohexyl)-N'-phenylurea see...Siduron
Methyl-2-[4-((3-chloro-5-(trifluoromethyl)-2-pyridinyl)oxy)phenoxy]propanoate see...Haloxyfop-methyl
Methyl demeton see...Demeton-methyl
Methyl-O-demeton see...Demeton-methyl
Methyl demeton-O see...Demeton-methyl
Methyl demeton-O-sulfoxide see...Demeton-methyl
Methyl demeton thioester see...Demeton-methyl
Methyl 5-(2,4-dichlorophenoxy)-2-nitrobenzoate see. Bifenox
Methyl 2-[2-(2,4-dichlorophenoxy)phenoxy]propanoate see...Diclofop-methyl
Methyl 3-((dimethoxyphosphinyl)oxy)-2-butenoate see. Mevinphos
Methyl-3-((dimethoxyphosphinyl)oxy)-2-butenoate, α isomer see...Mevinphos
Methyl 3-(dimethoxyphosphinyloxy)crotonate see. Mevinphos
Methyl 2-[(((((4,6-dimethoxy-2-pyrimidinyl)amino)carbonyl)amino)sulfonyl)methyl]benzoate see...Bensulfuron-methyl
Methyl-2-(dimethylamino)-N-[((methylamino)carbonyl)oxy]-2-oxoethanimidothioate see...Oxamyl
Methyl 2-[((((4-(dimethylamino)-6-(2,2,2-trifluoethoxy)-1,3,5-triazin-2-yl)amino)carbonyl)amino)sulfonyl]-3-, methylbenzoate see...Triflusulfuron-Methyl
Methyl-4-dimethylamino-3,5-xylylcarbamate see. Mexacarbate
Methyl-4-dimethylamino-3,5-xylyl ester of carbamic acid see...Mexacarbate
Methyl-1-(dimethylcarbamoyl)-N-(methylcarbamoyloxy) thioformimidate see...Oxamyl
S-Methyl-1-(dimethylcarbamoyl)-N-[(methylcarbamoyl) oxy]thioformimidate see...Oxamyl
Methyl-N,N'-dimethyl-N-((methylcarbamoyl)oxy)-1-thiooxamimidate see...Oxamyl
Methyl 2-[(((4,6-dimethyl-2-pyrimidinyl)amino)carbonyl)amino)sulfonyl]benzoate see...Sulfometuron-Methyl

2-Methyl-4,6-dinitrophenol see...Dinitro-o-cresol (DNOC)
6-Methyl-2,4-dinitrophenol see...Dinitro-o-cresol (DNOC)
Methyldithiocarbamic acid, sodium salt see. Metham-Sodium
(4-Methyl-1,3-dithiolan-2-ylidene)phosphoramidic acid, diethyl ester see...Mephosfolan
6-Methyl-1,3-dithiolo(4,5-β)quinoxalin-2-one see. Oxythioquinox
6-Methyldithiolo(4,5-β)quinoxalin-2-one see. Oxythioquinox
2-Methyl-1,3-di(2,4-xylylimino)-2-azapropane see. Amitraz
Methyl dursban see...Chlorpyrifos-methyl
METHYL-E 605® see...Methyl Parathion
Methyle (formiate de) (French) see...Methyl Formate
Methyleen-S,S'-bis(O,O-diethyl-dith iofosfaat) (Dutch) see...Ethion
3,4-Methylendioxy-6-propylbenzyl-N-butyl-diaethylenglykolaether (German) see...Piperonyl Butoxide
2,2'-Methylenebis(4-chlorophenol) see...Dichlorophene
Methylene-S,S'-bis(O,O-diethyl-dithiophosphat) (German) see...Ethion
2,2'-Methylenebis(3,4,6-trichlorophenol) see. .Hexachlorophene
2,2'-Methylenebis(3,5,6-trichlorophenol) see. .Hexachlorophene
1,2-(Methylenedioxy)-4-[2-(octylsulfinyl)propyl]benzene see...Sulfoxide
3,4-Methylenedioxy-6-propylbenzyl N-butyl diethyleneglycol ether see...Piperonyl Butoxide
(3,4-Methylenedioxy-6-propylbenzyl)(butyl)diethylene glycol ether see...Piperonyl Butoxide
4,5-Methylenedioxy-2-propylbenzyldiethylene glycol butyl ether see...Piperonyl Butoxide
Methylene glycol see...Formaldehyde
Methylene oxide see..Formaldehyde
S,S'-Methylene O,O,O',O'-tetraethyl phosphorodithioate see...Ethion
a,S'-Methylene O,O,O',O'-tetraethyl ester phosphorodithioic acid see...Ethion
Methyl ester of 2-[4-(2,4-dichlorophenoxy)phenoxy] propanoic acid see..Diclofop-methyl
[((1-Methyl-1,2-ethanediyl)bis(carbamodithioato)](2-)zinc homopolymer see...Propineb
Methyl 2-[(4-ethoxy-6-methylamino-1,3,5-triazin-2-yl)carbamoylsulfamoyl]benzoate see...Ethametsulfuron-methyl
Methyl 2-[((((4-ethoxy-6-(methylamino)-1,3,5-triazin-2-yl)amino)carbonyl)amino)sulfonyl]benzoate see. Ethametsulfuron-methyl
2-(1-Methylethoxy)phenyl N-methylcarbamate see. Propoxur
N[3-(1-Methylethoxy)phenyl]-2-(trifluoromethyl)benzamide see...Flutolanil

(E)-1-Methylethyl 3-[((ethylamino)methoxyphosphinothioyl)oxy]-2-butenoate see...Propetamphos
3-(1-Methylethyl)-1H-2,1,3-benzothiazain-4(3H)-one-2,2-dioxide see...Bentazon
4-(Methylethyl)-2,6-dinitro-N,N-dipropylbenzenamine see...Isopropalin
1-Methylethyl 2-[(ethoxy((1-methylethyl)amino)phosphinothioyl)oxy]benzoate see. Isofenphos
1-Methylethyl(E)-3-[((ethylamino)methoxyphosphinothioyl)oxy]-2-butenoate see...Propetamphos
1-(Methylethyl)-ethyl 3-methyl-4-(methylthio)phenyl phosphoramidate see...Fenamiphos
1-Methylethyl (E,E)-11-methoxy-3,7,11-trimethyl-2,4-dodecadienoate see...Methoprene
3-(1-Methylethyl)phenol methylcarbamate see...Phenol, 3-(1-methylethy)-, methylcarbamate
(1-Methylethyl) phosphoramidic acid ethyl 3-methyl-4-(methylthio)phenyl ester see...Fenamiphos
α-(1-Methylethyl)-α-[4-(trifluoromethoxy)phenyl]-5-pyrimidinemethanol see...Flurprimidol
Methylformiaat (Dutch) see...Methyl Formate
Methylformiat (German) see...Methyl Formate
Methyl fosferno see...Methyl Parathion
METHYL GUTHION® see...Azinphos-methyl
1-Methylheptyl [(4-amino-3,5-dichloro-6-fluoro-2-pyridinyl)oxy]acetate[6-(1-Methyl-heptyl)-2,3-dinitro-phenyl]-crotonat (German) see...Dinocap
2-(1-Methylheptyl)-4,6-dinitrophenylcrotonate see. Dinocap
Methyl m-hydroxycarbanilate m-methylcarbanilate see. Phenmedipham
Methyl 3-hydroxycrotonate dimethyl phosphate ester see. Mevinphos
Methyl-3-hydroxy-α-crotonate see...dimethyl phosphate ester see...Mevinphos
1-Methyl-3-hydroxy-4-isopropylbenzene see...Thymol
N,N-((Methylimino)dimethylidyne)bis(2,4-xylidine) see. Amitraz
N,N-((Methylimino)dimethylidyne)d-2,4-xylidine see. Amitraz
3-Methyl-5-isopropyl-N-methyl carbamate see. Promecarb
Methyl 6-(4-isopropyl-4-methyl-5-oxo-2-imidazolin-2-yl)-m-toluate) see...Imazethabenz
(Methyl 6-(4-isopropyl-4-methyl-5-oxo-2-imidazolin-2-yl)-m-toluate plus see...Imazethabenz
(Methyl 6-(4-isopropyl-4-methyl-5-oxo-2-imidazolin-2-yl)-m-toluate plus Methyl 6-(4-isopropyl-4-methyl-5-oxo-2-imidazolin-2-yl)-m-toluate) see...Imazethabenz
5-Methyl-2-isopropyl-1-phenol see...Thymol
N-Methyl-m-isopropylphenyl carbamate see...Phenol, 3-(1-methylethy)-, methylcarbamate
N-Methyl-3-isopropylphenyl carbamate see...Phenol, 3-(1-methylethy)-, methylcarbamate

3-Methyl-5-isopropylphenyl-*N*-methyl carbamate see. Promecarb
(3-Methyl-5-isopropylphenyl)-*N*-methylcarbamat (German) see...Promecarb
5-Methyl-2-isopropyl-3-pyrazolyl dimethylcarbamate see...Isolan®
METHYL ISOSYSTOX® see...Demeton
Methyl isosystox see...Demeton-methyl
2-(Methylmercapto)-4,6-bis(isopropylamino)-*S*-triazine see...Prometryn
4-Methylmercapto-3,5-dimethylphenyl *N*-methylcarbamate see...Methiocarb
2-Methylmercapto-4-ethylamino-6-isopropylamino-*s*-triazine see...Ametryn
2-Methylmercapto-4-isopropylamino-6-ethylamino-*s*-triazine see...Ametryn
Methyl mercaptophos see...Demeton-methyl
4-Methylmercapto-3,5-xylyl methylcarbamate see. Methiocarb
Methylmercuric chloride see...Mercury Alkyl Compounds
Methylmercuric cyanoguanidine see...Methylmercuric Dicyanamide
Methylmercury chloride see...Mercury Alkyl Compounds
Methylmercury dicyanandimide see...Methylmercuric Dicyanamide
Methylmercury dicyandiamide see...Methylmercuric Dicyanamide
Methylmerkuridikyandiamid (German) see. Methylmercuric Dicyanamide
Methyl methanoate see...Methyl Formate
Methyl-2-[[(4-methoxy-6-methyl-1,3,5-triazyn-2-yl)aminocarbonyl] aminosulfonyl]benzoate see. Metsulfuron-methyl
Methyl 2-[((((4-methoxy-6-methyl-1,3,5-triazin-2-yl)amino)carbonyl)amino)sulfonyl]benzoate see. Metsulfuron-methyl
Methyl 3-[((((4-methoxy-6-methyl-1,3,5-triazin-2-yl)amino)carbonyl)amino)sulfonyl]-2-thiophene carboxylate see...Thifensulfuron Methyl
Methyl 2-[((((4-methoxy-6-methyl-1,3,5-triazin-2-yl)methylamino)carbonyl)amino)sulfonyl]benzoate see. Tribenuron-Methyl
Methyl 2-[3-(4-methoxy-6-methyl-1,3,5-triazin-2-yl)ureidosulphonyl]benzoate see...Metsulfuron-methyl
Methyl *N*-[methylamino(carbonyl)oxy]ethanimido)thioate see...Methomyl
*cis*-1-Methyl-2-methyl carbamoyl vinyl phosphate see. Monocrotophos
Methyl-*N*-[methyl(carbamoyl)oxy]thioacetimidate see. Methomyl
*S*-Methyl-(methylcarbamoyloxy)thioacetimidate see. Methomyl
1-Methyl-2-(3,4-methylenedioxyphenyl)ethyl octyl sulfoxide see...Sulfoxide
1-Methyl-4-(1-methylethenyl) cyclohexane see...D-Limonene

3-Methyl-5-(1-methylethyl)phenol methylcarbamate see. Promecarb
5-Methyl-2-(1-methylethyl)phenol see...Thymol
3-Methyl-5-(1-methylethyl)phenyl-carbamic acid methyl ester see...Promecarb
2-Methyl-2-(methylsulfonyl)propanal-*O*-[(methylamino)carbonyl]oxime see...Aldoxycarb
2-Methyl-2-(methylsulfonyl)propionaldehyde-*O*-(methylcarbamoyl)oxime see...Aldoxycarb
2-Methyl-2-(methylthio)propanal, *O*-[(methylamino)carbonyl]oxime see...Aldicarb
2-Methyl-2-(methylthio)propanaldehyde, *O*-(methylcarbamoyl)oxime see...Aldicarb
2-Methyl-2-methylthio-propionaldehyd-*O*-(*N*-methyl-carbamoyl)-oxim (German) see...Aldicarb
Methyl namate see...Sodium Dimethyldithiocarbamate
*N*-Methyl-1-naphthyl-carbamat (German) see...Carbaryl
*N*-Methyl-α-naphthylcarbamate see...Carbaryl
*N*-Methyl-1-naphthyl carbamate see...Carbaryl
*N*-Methyl-α-naphthylurethan see...Carbaryl
Methyl niran see...Methyl Parathion
Methylnitrophos see...Fenitrothion
6-Methyl-2-oxo-1,3-dithiolo(4,5-β)quinoxalin see. Oxythioquinox
6-Methyl-2-oxo-1,3-dithio(4,5-β)quinoxaline see. Oxythioquinox
Methyl oxydemeton *S* see...Demeton-methyl
Methyl phencapton see...Methyl Phenkapton
2-[1-Methyl-2-(4-phenoxyphenoxy)ethoxy]pyridine see. Pyriproxyfen
3-(Methylphenyl)carbamic acid 3-[(methoxycarbonyl)amino]phenyl ester see. Phenmedipham
Methyl phenyl ether see...Anisole
*m*-Methylphenyl methylcarbamate see...Metolcarb
3-Methylphenyl-*N*-methylcarbamate see...Metolcarb
2-(Methyl-2-phenylpropyl)distannoxane see...Fenbutatin Oxide
2-Methyl-4-(phenylsulfonyl)trifluoromethane sulfonanalide see. Perfluidone
1-Methyl-3-phenyl-5-[3-(trifluoromethyl)phenyl]-4(1*H*)-pyridinone see...Fluridone
Methylphosphoramidic acid,4-*t*-butyl-2-chlorophenyl methyl ester see...Crufomate
Methylpirimiphos see...Pirimiphos-Methyl
6-(1-Methyl-propyl)-2,4-dinitrofenol (Dutch) see. Dinoseb
2-(1-Methylpropyl)-4,6-dinitrophenol see...Dinoseb
(6-(1-Methyl-propyl)-2,4-dinitro-phenyl)-3,3-dimethyl acrylat (German) see...Binapacryl
2-(1-Methylpropyl)-4,6-dinitrophenyl-β,β-dimethacrylate see...Binapacryl
2-(1-Methylpropyl)-4,6-dinitrophenyl ester 3-methyl-2-butenoic acid see...Binapacryl
2-(1-Methylpropyl)phenyl methylcarbamate see..BPMC
5-Methyl-1*H*-pyrazol-3-yl dimethylcarbamate see. Dimetilan
1-Methyl-2-(3-pyridyl)pyrrolidine see...Nicotine

1-1-Methyl-2-(3-pyridyl)-pyrrolidine sulfate see. Nicotine Sulfate
Methylpyrimiphos see...Pirimiphos-Methyl
3-(N-Methylpyrrolidino)pyridine see...Nicotine
3-(1-Methyl-2-pyrrolidinyl)pyridine see...Nicotine
(S)-3-(1-Methyl-2-pyrrolidinyl)pyridine see...Nicotine
1-3-(1-Methyl-2-pyrrolidinyl)pyridine sulfate see. Nicotine Sulfate
(S)-3-(1-Methyl-2-pyrrolidinyl)pyridine sulfate (2:1) see. Nicotine Sulfate
1-3-(1-Methyl-2-pyrrolidyl)pyridine see...Nicotine
3-(1-Methyl-2-pyrrolidyl) pyridine see...Nicotine
(-)-3-(1-Methyl-2-pyrrolidyl)pyridine see...Nicotine
6-Methyl-quinoxaline-2,3-dithiocyclicarbonate see. Oxythioquinox
6-Methyl-2,3-quinoxalin dithiocarbonate see. Oxythioquinox
S,S-(6-Methylquinoxaline-2,3-diyl)dithiocarbonate see. Oxythioquinox
6-Methyl-2,3-quinoxalinedithio cyclic carbonate see. Oxythioquinox
6-Methyl-2,3-quinoxalinedithio cyclic dithiocarbonate see...Oxythioquinox
6-Methyl-2,3-quinoxalinedithio cyclic S,S-dithiocarbonate see...Oxythioquinox
Methyl rhodanate see...Methyl Thiocyanate
Methylrhodanid (German) see...Methyl Thiocyanate
Methyl sulfanilyl carbamate see...Asulam
Methyl sulfanilylcarbamate see...Asulam
Methyl sulfocyanate see...Methyl Thiocyanate
4-(2-Methylsulfonyl-4-trifluoromethyl-benzoyl)-5-cyclopropylisoxazole see...Isoxaflutole
METHYL SYSTOX® see...Demeton-methyl
2-(Methylthio)-4,6-bis(isopropylamino)-S-triazine see. Prometryn
4-Methylthio-3,5-dimethylphenyl methylcarbamate see. Methiocarb
2-Methylthio-4-ethylamino-6-tert-butylamino-S-triazine see...Terbutryn
2-Methylthio-4-ethylamino-6-isopropylamino-s-triazine see...Ametryn
Methylthiokyanat see...Methyl Thiocyanate
[1,2-Methyl thiophanate see...Thiophanate-Methyl
Methylthiophos see...Methyl Parathion
2-Methylthio-propionaldehyd-O-(methylcarbamoyl)oxim (German) see...Methomyl
4-(Methylthio)-3,5-xylyl methylcarbamate see. Methiocarb
4-(Methylthio)-3,5-xylyl-N-methylcarbamate see. Methiocarb
Methyl thiram see...Thiram
Methylthiuram disulfide see...Thiram
Methyl 3-(tolylcarbamoyloxy)phenylcarbamate 3-methoxycarbonylaminophenyl N-3'-methylphenylcarbamate see...Phenmedipham
Methyl tuads see...Thiram
trans-8-Methyl-N-vanillyl-6-noneamide see...Capsaicin

8-Methyl-N-vanillyl-6-nonenamide, (E)- (8CI) see. Capsaicin
Methyl viologen see...Paraquat
Methyl viologen chloride see...Paraquat
Methyl viologen dichloride see...Paraquat
Methyl viologen (reduced) see...Paraquat
Methyl zimate see...Ziram
Methyl zineb see...Propineb
Methyl ziram see...Ziram
Metidation (Spanish) see...Methidathion
Metifonate see...Trichlorfon
Metil azinfos (Spanish) see...Azinphos-methyl
Metilbromid (Spenish) see...Methyl Bromide
Metilen-S,S'-bis(O,O-dietil-ditiofosfato) (Italian) see. Ethion
Metil fenil eter (Spanish) see...Anisole
Metil (formiato di) (Italian) see...Methyl Formate
N-Metil-1-naftil-carbammato (Italian) see...Carbaryl
Metilparation (Hungarian) see...Methyl Parathion
Metilparationa (Spanish) see...Methyl Parathion
6-(1-Metil-propil)-2,4-dinitrnolo (Italian) see...Dinoseb
2-Metil-2-tiometil-propionaldeid-O-(n-metil-carbamoil)-ossima (Italian) see...Aldicarb
Metiltriazotion (Russian) see...Azinphos-methyl
Metiocarb (Spanish) see...Methiocarb
METIURAC® see...Thiram
Metmercapturon see...Methiocarb
Metobromuron [ 3-(p-bromophenyl)-1-methoxy-1-methylurea] see...Metobromuron
Metoksychlor (Polish) see...Methoxychlor
Metomil (Italian) see...Methomyl
Metomilo (Spanish) see...Methomyl
Metoxicloro (Spanish) see...Methoxychlor
METOX® see...Methoxychlor
METOX® see...Metoxuron
METRAMAC® see...Amiton
METRAMAK® see...Amiton
Metribuzina (Spanish) see...Metribuzin
Metrifonate see...Trichlorfon
Metriphonate see...Trichlorfon
METRON® see...Methyl Parathion
METURON 80 DF® see...Fluometuron
Meturone see...Fluometuron
Metyloparation (Polish) see...Methyl Parathion
Metylparation (Czech) see...Methyl Parathion
Mevinfos (Spanish) see...Mevinphos
Mexacarbato (Spanish) see...Mexacarbate
MEXENE® see...Ziram
MEXIDE® see...Rotenone
MEXTROL-BIOX® see...Bromoxynil
Mezcla de dicloropropeno y dicloropropano (Spanish) see...D-D mixture
MEZENE® see...Ziram
MEZINEB® see...Propineb
MEZOPUR® see...Methazole
MEZOTOX® see...Nitrofen
MF-344® see...Etridiazole
MFA see...Fluoroacetic Acid

MFB  see...Fluorobenzene
MG 02® see...Acetochlor
MGK 264® see...Allethrin
MGK 326® see...Dipropyl Isocinchomeronate
MGK INTERMEDIATE 10® see...Allethrin
MGK REPELLENT-326® see...Dipropyl Isocinchomeronate
MH see...Maleic Hydrazide
MH 30® see...Maleic Hydrazide
MH 36 BAYER® see...Maleic Hydrazide
MH 40® see...Maleic Hydrazide
MI (Copper) see...Copper and Copper Compounds
MICIDE® see...Zineb
MICROBAN® ADDITIVE see...Triclosan
MICROBICIDE 8® see...Octhilinone
MICROCARB® see...Carbaryl
MICRO-CHEK 11® see...Octhilinone
MICRO-CHECK® 12 see...Captan
MICRO-CHEK SKANE® see...Octhilinone
MICROCOP® see...Copper Oxychloride
MICROFLOTOX® see...Sulfur
MICO-FUME® see...Dazomet
MICRO-GEN ANT REACTOR® see...Sulfluamid
MICROMITE® see...Diflubenzuron
MICROMITE® see...Fenitrothion
MICRON® see...Cuprous Thiocyanate
MICROPEL® see...Octhilinone
MICROTHIOL® DISPERSS® see...Sulfur
MICROZUL® see...Chlorophacinone
MIDOX® see...MCPA
MIEDZIAN® see...Copper Oxychloride
MIEDZIAN 50® see...Copper Oxychloride
MIELUCIN® see...Busulfan
Mierenzur (Dutch) see...Formic Acid
MIFATOX® see...Demeton-methyl
MIKAL® see...Fosetyl-Al
MILAGRO® see...Nicosulfuron
MILBAM® see...Ziram
MILBAN® see...Dodemorph Acetate
MILBAN® see...Ziram
MILBOL® see...Dicofol
MILBOL 49® see...Lindane
MILDANE® see...Dinocap
MILDEX® see...Dinocap
MILLENNIUM® see...Clopyralid
MILLER NU SET® see...Silvex
MILLER'S FUMIGRAIN® see...Acrylonitrile
MILMER® see...Copper(II)-8-hydroxyquinoline
MILOCEP® see...Metolachlor
MILOCEP® see...Propazine
MILOGARD® see...Propazine
MILO-PRO® see...Propazine
MILOR® see...Mancozeb
MILPREX® see...Dodine
MILPREX® see...Oxythioquinox
MILTOX® see...Zineb
MILTOX® SPECIAL see...Zineb
MIMIC® see...Tebufenozide

MINACIDE® see...Promecarb
Mineral green see...Paris Green
Mineral spirits see...Naphthas
Mineral spirits see...Stoddard Solvent
MINTACO® see...Paraoxon
MINTACOL® see...Paraoxon
MINTEZOL® see...Thiabendazole
MINZOLUM® see...Thiabendazole
MIOSTAT® see...Captan
MIOTISAL® see...Paraoxon
MIOTISAL A® see...Paraoxon
MIRACLE® see...2,4-D
MIST-O-MATIC® see...Imazalil
MIST-O-MATIC LINDEX® see...Lindane
MISTRAL® see...Nicosulfuron
MISULBAN® see...Busulfan
MITAC® see...Atrazine
MITIFON® see...Tetradifon
MITIGAN® see...Dicofol
MITION® see...Tetradifon
Mitis green see...Copper Acetoarsenite
Mitis green see...Paris Green
MITOL® see...Demeton
MITOSTAN® see...Busulfan
M ITROL® see...Chlorobenzilate
MIX No. 1® see...Dinex
MK 244® see...Emamectin Benzoate
MK-0244® see...Emamectin Benzoate
MK-360® see...Thiabendazole
MK 936® see...Abamectin
MK 936® ($B_{1a}$) see...Abamectin
ML 97® see...Phosphamidon
MM 70® see...Metamiton
MMC see...Mercury Alkyl Compounds
MMD see...Methylmercuric Dicyanamide
MNEBD see...Maneb
MOBIL V-C 9-104® see...Ethoprop
MOCAP® see...Ethoprop
MOCAP 10G® see...Ethoprop
MODOWN® see...Bifenox
MOLE DEATH® see...Strychnine
MOLETOX II® see...Zinc Phosphide
MOLLUSCICIDE BAYER 73® see...Clonitralid
MOLPHOS 36 SL® see...Monocrotophos
MOLURAME® see...Ziram
MOLY-T® see...Thiram
MON® see...Halosulfuron-methyl
MON 097® see...Acetochlor
MON 0573® see...Glyphosate
MON 2139® see...Glyphosate
MON 7200® see...Dithiopyr
MON 7325® see...Paclobutrazole
MON 13200® see...Thiazopry
MON 15100® see...Dithiopyr
MON 58420® see...Acetochlor
MON 78095® see...Oxyfluorfen
MON 78567® see...Diquat Dibromide
MON 78736® see...Triclopyr, Triethylammonium Salt

MON 78746® see...Quizalofop-Ethyl
MONAM (dihydrate)® see...Metham-Sodium
MONATE® see...Sodium Methanearsonate (MSMA)
MONATE® Merge 823 see...Sodium Methanearsonate (MSMA)
MONCEREN® see...Pencycuron
MONCUT® see...Flutolanil
MONITOR® see...Methamidophos
Monoammonium salt of sulfamic acid see...Ammonium Sulfamate
Monoammonium sulfamate see...Ammonium Sulfamate
MONOBOR-CHLORATE® see...Sodium Metaborate
Monobromomethane see...Methyl Bromide
Monocalcium arsenite see...Calcium Arsenite
Monochloorazijnzuur (Dutch) see...Chloroacetic Acid
Monochloracetic acid see...Chloroacetic Acid
Monochloressigsaeure (German) see...Chloroacetic Acid
Monochloroacetaldehyde see...Chloroacetaldehyde
Monochloroacetic acid see...Chloroacetic Acid
Monochloroethanoic acid see...Chloroacetic Acid
MONOCIDE® see...Cacodylic Acid
MONOCIL® 40 see...Monocrotophos
Monocron see...Monocrotophos
Monocrotofos (Spanish) see...Monocrotophos
Monodrin see...Monocrotophos
Monofluorazijnzuur (Dutch) see...Fluoroacetic Acid
Monofluoressigsaeure (German) see...Fluoroacetic Acid
Monofluoressigsaure see...natrium (German) see...Sodium Fluoroacetate
Monofluoroacetamide see...Fluoroacetamide
Monofluoroacetate see...Fluoroacetic Acid
Monofluoroacetic acid see...Fluoroacetic Acid
Monofluorobenzene see...Fluorobenzene
$N$-Monoisopropylamide of $O,O$-diethyldithiophosphorylacetic acid see...Prothoate
Monomethflurazone see...Norflurazon
Monomethyl mercury chloride see...Mercury Alkyl Compounds
Monomethylarsonic acid see...Methanearsonic Acid
Monoperacetic acid see...Peracetic Acid
Monopotassium arsenate see...Potassium Arsenate
Monopotassium dihydrogen arsenate see...Potassium Arsenate
MONOSAN® see...2,4-D
Monosodium acid methanearsonate see...Sodium Methanearsonate (MSMA)
Monosodium dihydrogen phosphate see...Trisodium Phosphate
Monosodium metaborate see...Sodium Metaborate
Monosodium methane arsonate see...Sodium Methanearsonate (MSMA)
Monosodium methyl arsonate see...Sodium Methanearsonate (MSMA)
Monosodium phosphate see...Trisodium Phosphate
Monosodium sulfite see...Ammonium Sulfite
Monourea sulfuric acid adduct see...Monocarbamide Dihydrogen Sulfate

MONSANTO CP 47114® see...Fenitrothion
MONSANTO®-CP 51969 see...Bromophos
MONSANTO® 2,4-D ISOPROPYL ESTER see...2,4-D, isopropyl ester
MONTAR® see...Cacodylic Acid
MONTECATINI L-561® see...Phenthoate
MONTREL® see...Crufomate
MONTROSE PROPANIL® see...Propanil
MONUREX® see...Monuron
MOORMAN'S® IGR CATTLE CONCENTRATE see. Methoprene
MOPARI® see...Dichlorvos
MORBICID® see...Formaldehyde
MOR-CRAN® see...Naptalam
MORESTAN® see...Oxythioquinox
MORKIT® see...Anthraquinone
MOROCIDE® see...Binapacryl
Morpholine see...3-[3-(4-chlorophenyl)-3-(3,4-dimethoxyphenyl)-1-oxo-2-propenyl]- see. Dimethomorph
Morpholine see...$N$-cyclododecyl-2,6-dimethyl-, acetate see...Dodemorph Acetate
MORROCID® see...Binapacryl
MORSODREN® see...Methylmercuric Dicyanamide
MORTON EP-227® see...Methylmercuric Dicyanamide
MORTON EP-316® see...Promecarb
MORTON EP-332® see...Formetanate Hydrochloride
MORTON SOIL DRENCH® see...Methylmercuric Dicyanamide
MORTOPAL® see...TEPP
MOS-570 see...Endosulfan
MOSCARDA® see...Malathion
MOSS-B-WARE® see...Zinc Sulfate Heptahydrate
Moss green see...Copper Acetoarsenite
Moss green see...Paris Green
MOSUM® see...Monocrotophos
MOTA MASKROS® see...2,4-D
MOTIVEL® see...Nicosulfuron
MOTOPAL® see...TEPP
MOTOX® see...Toxaphene
Mottenhexe see...Hexachloroethane
MOTT-EX® see...para-Dichlorobenzene
MOTTENSCHUTZMITTEL EVAU P® see...para-Dichlorobenzene
Mountain green see...Copper Acetoarsenite
Mountain green see...Paris Green
MOUS-CON® see...Zinc Phosphide
MOUSE-NOTS® see...Strychnine
MOUSE PAK® see...Warfarin
MOUSE-RID® see...Strychnine
MOUSE-TOX® see...Strychnine
MOXIE® see...Methoxychlor
MOXONE® see...2,4-D
MOXY 2E® see...Bromoxynil
M-Parathion see...Methyl Parathion
MQD see...Oxythioquinox
MRC 910® see...Iprodione
MR. KILL RAT® see...Zinc Phosphide

Mrowczan etylu (Polish) see...Ethyl Formate
Mrowkozol (Polish) see...Propoxur
MR RAT GUARD® see...Zinc Phosphide
MSC 379587® see...Oxythioquinox
MSL see...Sodium Dimethyldithiocarbamate
MSMA see...Methanearsonic Acid
MSMA see...Sodium Methanearsonate (MSMA)
MSP see...Trisodium Phosphate
MSR® see...Demeton-methyl
MSS AMINOTRIAZOLE® see...Amitrole
MSS FLOTIN® see...Fentin Hydroxide
MSS HERBASAN® see...Phenmedipham
MSS SIMAZINE® see...Amitrole
MSZYCOL® see...Lindane
MTD® see...Methamidophos
M.T.F® see...Trifluralin
MTI 500® see...Ethofenprox
MTMC see...Metolcarb
MUGAN® see...Carbaryl
MULTAMAT® see...Bendiocarb
MULTICIDE® see...Cyphenothrin
MULTICIDE® see...Tetramethrin
MULTICIDE-2154® see...D-Phenothrin
MULTIMET® see...Bendiocarb
MULTI-W® see...Maneb
Mumfluorid (German) see...Sodium Aluminum Fluoride
MURFOS® see...Parathion
MURIOL® see...Chlorophacinone
MURPHOS® see...Parathion
MURPHY SUPER ROOT GUARD® see...Chlorpyrifos
MURVIN® see...Carbaryl
MUSCATOX® see...Coumaphos
MUSKETEET® see...Mecoprop
Mustard oil see...Allyl Isothiocyanate
MUSTER® see...Ethametsulfuron-methyl
MUTOXIN® see...DDT
MXL® see...MCPA
MXMC see...Methiocarb
MYCOJECT® see...Oxytetracycline Calcium
MYCOSHIELD® AGRICULTURAL TERRAMYCIN see...Oxytetracycline Calcium
MYCOSHIELD TMQTHC 20® see...Oxytetracycline Calcium
MYCOZOL® see...Thiabendazole
Mycrolysin see...Chloropicrin
MYCRONIL® see...Ziram
MYELOLEUKON® see...Busulfan
MYLERAN® see...Busulfan
MYLON® see...Dazomet
MYLONE® see...Dazomet
MYLONE® see...Mecoprop
MYLONE® 85 see...Dazomet
Myricyl alcohol see...Triacontanol
MYRMICYL® see...Formic Acid
MYTROL® see...Fluenetil
MZ-CURZATE® see...Cymoxanil

- N -

N 252® see...Dimethipin
N 521® see...Dazomet
N 869 (dihydrate)® see...Metham-Sodium
NA 22® see...Ethylene Thiourea
NA 73® see...Hexythiazox
NA 305® see...Phenmedipham
NA 308® see...Phenmedipham
NA 2761® see...DDT
NA 2771® see...Thiram
NA 2783® (DOT) see...Diazinon
NA 8318® see...Isoxaben
NAA see...1-Naphthaleneacetic Acid
NAA 800® see...1-Naphthaleneacetic Acid
NAAM see...1-Naphthaleneacetamide
NABAC® see...Hexachlorophene
NABAC 25 EC® see...Hexachlorophene
Nabame see...Nabam
Nabasam (obsolete) see...Nabam
NABU® see...Sethoxydim
NAC® see...Carbaryl
NACCONOL 98 SA® see...Dodecylbenzenesulfonic Acid
NAD see...1-Naphthaleneacetamide
NAF-2® see...Flumetsulam
NAF-9® see...Flumetsulam
NAF-46® see...Hexaflumuron
NAF-280® see...Clopyralid
NAFEEN® see...Sodium Fluoride
Naftalame see...Naptalam
1-Naftil-tiourea (Italian) see...ANTU
1-Naftylthioureum (Dutch) see...ANTU
NAFUN-IPO® see...Nabam
NAFUSAKU® see...1-Naphthaleneacetic Acid
NAGATA® see...Cypermethrin
NALCO D-62C44® see...Nabam
NALCON 243® see...Dazomet
NALCO® D-62C44 see...Nabam
NALKIL® see...Bromacil
NAMEKIL® see...Metaldehyde
NAMFUME® see...Chloropicrin
NANCHOR® see...Ronnel
NANKER® see...Ronnel
NANKOR® see...Ronnel
NANSA SSA see...Dodecylbenzenesulfonic Acid
NAPCLOR-G® see...Sodium Pentachlorophenate
Naphtha see...Naphthas
2-Naphthacenecarboxamide see...4-(dimethylamino)-1,4,4a,5,5a,6,11,12a-octahydro-3,5,6,10,12,12a-hexahydroxy-6-methyl-1,11-dioxo- see...[4S-(4α,4aα,5α,5aα,6β,12aα)]- see...Oxytetracycline Calcium
Naphthalene-acetamide(1-) see...1-Naphthaleneacetamide
α-Naphthaleneacetamide see...1-Naphthaleneacetamide
Naphthalene-1-acetic acid see...1-Naphthaleneacetic Acid

Naphthaleneacetic acid(1-) see...1-Naphthaleneacetic Acid
α-Naphthaleneacetic acid see...1-Naphthaleneacetic Acid
2-Naphthalenoxyacetic acid see...Naphthoxyacetic Acid
2-[(1-Naphthalenylamino)carbonyl]benzoic acid see. Naptalam
(β-Naphthalenyloxy)acetic acid see...Naphthoxyacetic Acid
Naphtha safety solvent see...Stoddard Solvent
Naphthenic acids, copper salts see...Copper Naphthenate
1-Naphthol see...Carbaryl
α-Naphthothiourea see...ANTU
β-Naphthoxyacetic acid see...Naphthoxyacetic Acid
2-Naphthoxyacetic acid see...Naphthoxyacetic Acid
(2-Naphthoxy)acetic acid see...Naphthoxyacetic Acid
2-(1-Naphthoxy)-N,N-diethylpropionamide see. Napropamide
α-Naphthylacetamide see...1-Naphthaleneacetamide
1-Naphthylacetamide see...1-Naphthaleneacetamide
Naphthylacetic acid see...1-Naphthaleneacetic Acid
α-Naphthylacetic see...1-Naphthaleneacetic Acid
α-Naphthylacetic acid see...1-Naphthaleneacetic Acid
1-Naphthylacetic acid see...1-Naphthaleneacetic Acid
α-Naphthyleneacetic acid see...1-Naphthaleneacetic Acid
Naphthyleneethylene see...Acenaphthene
α-Naphthylessigsaeure (German) see...1-Naphthaleneacetic Acid
Naphyl-1-essigsaeure (German) see...1-Naphthaleneacetic Acid
O-(2-Naphthyl)glycolic acid see...Naphthoxyacetic Acid
1-Naphthylmethylcarbamate see...Carbaryl
α-Naphthyl N-methylcarbamate see...Carbaryl
1-Naphthyl N-methylcarbamate see...Carbaryl
(2-Naphthyloxy)acetic acid see...Naphthoxyacetic Acid
N-1-Naphthylphthalamate see...Naptalam
N-1-Naphthylphthalamic acid see...Naptalam
α-Naphthylphthalamic acid see...Naptalam
N-α-Naphthyl-phthalamidsaeure (German) see.Naptalam
α-Naphthylthiocarbamide see...ANTU
1-Naphthyl-thioharnstoff (German) see...ANTU
α-Naphthyl thiourea see...ANTU
1-Naphthylthiourea see...ANTU
1-(1-Naphthyl)-2-thiourea see...ANTU
N-(1-Naphthyl)-2-thiourea see...ANTU
1-Naphthyl-thiouree (French) see...ANTU
NAPHTOX® see...ANTU
NAPROGUARD® see...Napropamide
Naptalame see...Naptalam
1,4-Napthalenedione, 2,3-dichloro- see...Dichlone
NAPTRO® see...Naptalam
NARAMYCIN A® see...Cycloheximide
NARAMYCIN® see...Cycloheximide
NATIONS AG II® see...Oryzalin
Natriumaluminumfluorid (German) see...Sodium Aluminum fluoride

Natriumazid (German) see...Sodium Azide
Natriumchloraat (Dutch) see...Sodium Chlorate
Natriumchlorat (German) see...Sodium Chlorate
Natriumcyanid (Dutch) see...Sodium Cyanide
Natrium-2,4-dichlorphenoxyathylsulfat (German) see. 2,4-DES-Sodium
Natriumfluoracetaat (Dutch) see...Sodium Fluoroacetate
Natriumfluoracetat (German) see...Sodium Fluoroacetate
Natrium fluoride see...Sodium Fluoride
Natriumhexafluoroaluminate (German) see...Sodium Aluminum Fluoride
Natriummazide (Dutch) see...Sodium Azide
NAUGATUCK® D-014 see...Propargite
NAVADEL® see...Dioxathion
NAVRON® see...Fluoroacetamide
N.b. MECOPROP® see...Mecoprop
NC,-Clofentezine
NC-302® see...Quizalofop-Ethyl
NC-319® see...Halosulfuron-methyl
NC-398 WG® see...Sodium Dicamba
NC-6897® see...Bendiocarb
NC-8438® see...Ethofumesate
NC-21314 see...Clofentezine
NCI-C00055 see...Chloramben
NCI-C00066 see...Azinphos-methyl
NCI-0077 see...Captan
NCI-C00099 see...Chlordane
NCI-C00102 see...Chlorothalonil
NCI-C00113 see...Dichlorvos
NCI-C00124 see...Dieldrin
NCI-129 see...Pyridaben
NCI-C00157 see...Endrin
NCI-C00168 see...Tetrachlorvinphos
NCI-C00180 see...Heptachlor
NCI-C00191 see...Chlordecone (Kepone®)
NCI-C00204 see...Lindane
NCI-C00215 see...Malathion
NCI-C00226 see...Parathion
NCI-C00237 see...Picloram
NCI-C00259 see...Toxaphene
NCI-C00260 see...Triphenyltin Compounds
NCI-C00260 see...Fentin Hydroxide
NCI-C00395 see...Dioxathion
NCI-C00408 see...Chlorobenzilate
NCI-C00419 see...Quintozene
NCI-C00420 see...Nitrofen
NCI-C00431 see...Clonitralid
NCI-C00442 see...Trifluralin
NCI-C00453 see...Sulfallate
NCI-C00464 see...DDT
NCI-C00486 see...Dicofol
NCI-C00497 see...Methoxychlor
NCI-C00500 see...Dibromochloropropane
NCI-C00511 see...Ethylene Dichloride
NCI-C00522 see...Ethylene Dibromide
NCI-C00533 see...Chloropicrin
NCI-C00544 see...Mexacarbate

NCI-C00566 see...Endosulfan
NCI-C00588 see...Phosphamidon
NCI-C01592 see...Busulfan
NCI-C02017 see...Phenylthiourea
NCI-C02119 see...Urea
NCI-C02653 see...Hexachlorophene
NCI-C02799 see...Formaldehyde
NCI-C02813 see...Piperonyl Butoxide
NCI-C02824 see...Sulfoxide
NCI-C02846 see...Monuron
NCI-C02868 see...Ethylan
NCI-CO2904 see...Trichlorophenols
NCI-C02937 see...Calcium Cyanamide
NCI-C02959 see...Disulfiram
NCI-C02960 see...Chlormequat Chloride
NCI-C02971 see...Methyl Parathion
NCI-C03010 see...Fenaminosulf
NCI-C03372 see...Ethylene Thiourea
NCI-C03827 see...Daminozide
NCI-C04035 see...Allidochlor
NCI-C04591 see...Carbon Disulfide
NCI-C04604 see...Hexachloroethane
NCI-C06111 see...Benzyl Alcohol
NCI-C06428 see...Mirex
NCI-C06462 see...Sodium Azide
NCI-08640 see...Aldicarb
NCI-C08662 see...Coumaphos
NCI-C08673 see...Diazinon
NCI-C08684 see...Anilazine
NCI-C08695 see...Fluometuron
NCI-C50088 see...Ethylene Oxide
NCI-C50215 see...Acrylonitrile
NCI-C50351 see...o-Phenylphenol
NCI-C50442 see...Ziram
NCI-C50464 see...Allyl Isothiocyanate
NCI-C54831 see...Trichlorfon
NCI-C54933 see...Pentachlorophenol
NCI-C55141 see...1,2-Dichloropropane
NCI-C55210 see...Rotenone
NCI-C55221 see...Sodium Fluoride
NCI-C55378 see...Pentachlorophenol
NCI-C55425 see...Glutaraldehyde
NCI-C55823 see...Gibberellic Acid
NCI-C56111 see...Cinnamaldehyde
NCI-C56473 see...Oxytetracycline Calcium
NCI-C56655 see...Pentachlorophenol
NCI-C60173 see...Mercuric Chloride
NCI-C60231 see...Chloroacetic Acid
NCI-C60242 see...Trichlorobenzoic Acid
NCI-C60413 see...Chlorobenzilate
NCI-C60628 see...Bithionol
NCI-C99983 see...Quizalofop-Ethyl
NCS 88126 see...Paraquat
NCS 263500 see...Paraquat
NCR CE EE DOV7® see...Bromoxynil
Necarboxylic acid see...Allethrins
Necatorina see...Carbon Tetrachloride
Necatorine see...Carbon Tetrachloride

NECTRYL® see...o-Phenylphenol
NEEMAZAL® see...Azadirachtin
NEEM® see...Azadirachtin
NEFIS® see...Ethylene Dibromide
NEFRAFOS® see...Dichlorvos
NEGASHUNT® see...Coumaphos
NEGUVON A® see...Trichlorfon
NEGUVON® see...Trichlorfon
NEM-A-TAK® see...Fosthietan
NEMABROM® see...Dibromochloropropane
NEMACUR® see...Famphur
NEMACUR® see...Fenamiphos
NEMACURP® see...Fenamiphos
NEMAFENE® see...D-D mixture
NEMAFOS® see...Thionazin
NEMAFUM® see...Dibromochloropropane
NEMAGON® see...Dibromochloropropane
NEMAGON® see...1,3-Dichloropropene
NEMANAX® see...Dibromochloropropane
NEMAPAN® see...Thiabendazole
NEMAPAZ® see...Dibromochloropropane
NEMAPHOS® see...Thionazin
NEMASET® see...Dibromochloropropane
NEMATIN® see...Metham-Sodium
NEMATOCIDE® see...Dibromochloropropane
NEMATOCIDE® see...Thionazin
NEMATOX® see...Dibromochloropropane
NEMAX® see...Chloropicrin
NEMAZENE® see...Phenothiazine
NEMAZINE® see...Phenothiazine
NEMAZON® see...Dibromochloropropane
NEMEX® see...1,3-Dichloropropene
NEMINFEST® see...Linuron
NEMISPOR® see...Mancozeb
NENDRIN® see...Endrin
NEOBAN® see...Barban
NEOBOR® see...Sodium Tetraborate
NEOCID® see...DDT
NEOCIDOL® see...Diazinon
NEOCIDOL® (OIL) see...Diazinon
NEOCYCLOHEXIMIDE® see...Cycloheximide
NEO GERM-I-TOL® see...Zilkonium Chloride
Neoglaucit see...Isofluorphate
NEO-FAT 10® see...Decanoic Acid
NEOPELLIS® see...Bithionol
NEOPYNAMINE® see...Tetramethrin
NEO-PYNAMIN® see...Tetramethrin
NEOPYNAMIN FORTE® see...Tetramethrin
NEORAM BLU® see...Copper Oxychloride
NEOSOREXA see...Difenacoum
NEOSOREXA PP580® see...Difenacoum
NEOSTANOX® see...Fenbutatin Oxide
NEOTOPSIN® see...Thiophanate-Methyl
NEOVORONIT® see...Fuberidazole
NEOZIN® see...Zinc Sulfate Heptahydrate
NEPHIS® see...Ethylene Dibromide
NEPHIS® see...Temephos
NEPHIS® 1G see...Temephos

NEPHOCARP® see...Carbophanothion
NERACID® see...Captan
NEREB® see...Maneb
NERKOL® see...Dichlorvos
NESPOR® see...Maneb
NESTER® see...Pyridaben
NETAGRONE® see...2,4-D
NETAL® see...Bromophos
NETOCYD® see...Dithiazanine Iodide
NEU 1140F® see...Copper Octanoate
NEUDORFF DN 50® see...Dinitro-o-cresol (DNOC)
Neutral ammonium chromate see...Ammonium Chromate
Neutral verdigris see...Cupric Acetate
Neuwied green see...Paris Green
NEVIREX® see...Acetochlor
NEW CHLOREA® see...Atrazine
New green see...Paris Green
NEWSPOR® see...Maneb
NEX® see...Carbofuran
NEXAGAN® see...Bromophos-ethyl
NEXEN FB® see...Lindane
NEXION® see...Bromophos
NEXION®-40 see...Bromophos
NEXIT® see...Lindane
NEXIT-STARK® see...Lindane
NEXOL-E® see...Lindane
NF-35® see...Thiophanate-Methyl
NF-44® see...Thiophanate-Methyl
NF-114® see...Triflumizole
NIA 249® see...Allethrin
NIA 5462® see...Endosulfan
NIA 5767® see...Endothion
NIA 5996® see...Dichlobenil
NIA 9044® see...Binapacryl
NIA 9241® see...Phosalone
NIA 10242® see...Carbofuran
NIA 11092® see...Karbutilate
NIA 17370® see...Resmethrin
NIA 26021® see...Resmethrin
NIACIDE® see...Ferbam
NIAGARA MALACHLOR LIVESTOCK SPRAY CODE 983® see...Alachlor
NIAGARA 5488® see...Tetradifon
NIAGARA KOLO MALACHLOR DUST® see.Alachlor
NIAGARA 10242® see...Carbofuran
NIAGARA ZINEB® see...DDT
NIAGARA SOILFUME 85® see...Ethylene Dibromide
NIAGARA-STIK® see...1-Naphthaleneacetic Acid
NIAGARAMITE® see...Aramite
NIAGARA® 26021 see...Resmethrin
NIAGARA® 9044 see...Binapacryl
NIAGARA® 5,996 see...Dichlobenil
NIAGARA® 5006 see...Dichlobenil
NIAGARA® ESTASOL see...2,4-D, Isopropyl Ester
NIAGARA® 18739 see...Resmethrin
NIAGARA® see...EPN
NIAGARA® see...Calcium Arsenate

NIAGARA®-9241 see...Phosalone
NIAGARA®-9260 see...Tetramethrin
NIAGRA 5767® see...Endothion
NIAGRA P.A. DUST® see...Nicotine
NIAGRA 10242 see...Carbofuran
NIAGRA NIA-10242 see...Carbofuran
NIAGRA 5,462® see...Endosulfan
NIAGRATHOL® see...Endothall
NIA® ESTASOL see...2,4-D, Isopropyl Ester
NIA®-9260 see...Tetramethrin
NICLOFEN® see...Nitrofen
Niclosamide see...Clonitralid
NICLOSAMIDE® see...Clonitralid
NICOCHLORAN® see...Lindane
NICOCIDE® see...Nicotine
NICODUST® see...Nicotine
NICOFUME® see...Nicotine
Nicotina (Italian, Spanish) see...Nicotine
Nicotine sulphate see...Nicotine Sulfate
Nicotine sulphate (2:1) see...Nicotine Sulfate
1-Nicotine see...Nicotine
Nicotine sulfate (2:1) see...Nicotine Sulfate
Nicotine alkaloid see...Nicotine
NICOULINE® see...Rotenone
NIFLEX® see...Ethoxyquin
NIFOS T® see...TEPP
NIFOS® see...TEPP
NIFROST® see...TEPP
Nikotin (German) see...Nicotine
Nikotinsulfat (German) see...Nicotine Sulfate
Nikotyna (Polish) see...Nicotine
NIMITEX® see...Temephos
NIMITOX® see...Temephos
B-NINE® see...Daminozide
NINJA® see...lamda-Cyhalothrin
NIOMIL® see...Bendiocarb
NIP-A-THIN® see...Naptalam
NIPACIDE PCMC® see...p-Chloro-m-Cresol
NIPACIDE® see...4-Chloro-3,5-xylenol
NIPSAN® see...Diazinon
NIP® see...Nitrofen
NIRAN® see...Parathion
NIRAN® see...Chlordane
NISSHIN® see...Nicosulfuron
NISSORUN® see...Hexythiazox
Niter see...Potassium Nitrite
Niter see...Potassium Nitrate
NITICID® see...Propachlor
NITOFOL® see...Methamidophos
Nitram see...Ammonium Nitrate
NITRAM® see...Ammonium Nitrate
NITRAN® see...Trifluralin
NITRAPYRINE® see...Nitrapyrin
Nitrate de potassium (France) see...Potassium Nitrate
Nitrate d'ammonium (French) see...Ammonium Nitrate
Nitrato potasico (Spanish) see...Potassium Nitrate
Nitrato amonico (Spanish) see...Ammonium Nitrate
Nitrato de cobre (Spanish) see...Cupric Nitrate

Nitre see...Potassium Nitrite
Nitre see...Potassium Nitrate
Nitric acid see...Calcium salt, see...Calcium Nitrate
Nitric acid see...Potassium salt see...Potassium Nitrate
Nitric acid see...copper(II) salt see...Cupric Nitrate
Nitric acid see...copper(2+) salt see...Cupric Nitrate
Nitric acid see...Ammonium Nitrate
Nitric acid see...Calcium salt, see...Calcium Nitrate
Nitrile acrilico (Italian) see...Acrylonitrile
Nitrile acrylique (French) see...Acrylonitrile
2-2',2"-Nitrilotris-dodecylbenzenesulfonate (Salt) see. Triethanolamine Dodecylbenzene Sulfonate
3-Nitro-1,2,4,5-tetrachlorobenzene see...Tecnazene
4-Nitro-2',4'-dichlorodiphenyl ether see...Nitrofen
2-Nitro-2-(hydroxymethyl)-1,3-propanediol see... Tris(hydroxymethyl)nitromethane
4'-Nitro-2,4-dichlorodiphenyl ether see...Nitrofen
4-Nitro-2,6-dichloroaniline see...Dichloran
4-Nitroaniline, 2,6-dichloro- see...Dichloran
Nitrocalcite, see...Calcium Nitrate
Nitrochlor see...Nitrofen
Nitrochloroform see...Chloropicrin
NITROFAN® see...Dinitro-o-cresol (DNOC)
Nitrofene (French) see...Nitrofen
O-p-Nitrofenilfosfato de O,O-dietilo (Spanish) see. Paraoxon
Nitrogen lime see...Calcium Cyanamide
Nitroisobutylglycerol see. Tris(hydroxymethyl)nitromethane
Nitrolime see...Calcium Cyanamide
Nitropentachlorobenzene see...Quintozene
Nitrophen see...Nitrofen
Nitrophene see...Nitrofen
p-Nitrophenyl dieyhylphosphate see...Paraoxon
1-(4-Nitrophenyl)-3-(3-pyridinylmethyl)urea see. Pyriminil
N-(4-Nitrophenyl)-N'-(3-pyridinylmethyl)urea see. Pyriminil
p-Nitrophenyldime thylthionophosphate see...Methyl Parathion
NITROPHOS® see...Fenitrothion
NITROPONE C® see...Dinoseb
Nitrostigmin (German) see...Parathion
NITROSTIGMINE® see...Parathion
Nitrotrichloromethane see...Chloropicrin
Nitrotris(hydroxymethyl)methane see. Tris(hydroxymethyl)nitromethane
Nitrous Acid see...Potassium Salt see...Potassium Nitrite
NITROX® 80 see...Methyl Parathion
NITROX® see...Methyl Parathion
NITROZYME® see...Kinetin (Cytokinin)
NIUIF 100® see...Parathion
NIVRAL® see...Thiodicarb
NIX-SCALD® see...Ethoxyquin
NIX® see...Prometon
NK 136® see...Dithiazanine Iodide
NK 711® see...Leptophos

N-LARGE® see...Gibberellic Acid
N. locustae (Canning) see...Nosema Locustae
NMC® 50 see...Carbaryl
NNF-136® see...Flutolanil
NNI®-850 see...Fenpyroximate
NO BUNT® see...Hexachlorobenzene
NO CRAB® see...Butralin
NO-PEST® see...Dichlorvos
NOA see...Naphthoxyacetic Acid
NOAH GOLD® see...Osmose (USA) see...Copper Carbonate see...Basic
NOAH GOLD® see...Cyproconazole
NOBACTER® see...Bithionol
NOBACTER® see...Triclocarban
NOBECUTAN® see...Thiram
NOCBIN® see...Disulfiram
NOCCELER 22® see...Ethylene Thiourea
NOGOS® see...Dichlorvos
NOLO-BAIT® see...Nosema Locustae
NOLOC®. see...Nosema Locustae
NOLO® BB CONCENTRATE see...Nosema Locustae
NOLTRAN® see...Chlorpyrifos-methyl
NOMERSAN® see...Thiram
1-Nonanecarboxylic acid see...Decanoic Acid
Nonanoic acid see...Pelargonic Acid
6-Nonenamide see...N-[(4-hydroxy-3-methoxyphenyl)methyl]-8-methyl-, (E)- (9CI) see. Capsaicin
6-Nonenamide see...8-methyl-N-vanillyl-, (E)- (8CI) see...Capsaicin
n-Nonoic acid see...Pelargonic Acid
n-Nonylic acid see...Pelargonic Acid
NOR-AM® EP 332 see...Formetanate Hydrochloride
5-Norbornene-2,3-dimethanol see...1,4,5,6,7,7-hexachloro- see...cyclic sulfite see...endo- see. Endosulfan
NOREX® see...Chloroxuron
Norflurazon pyridazine herbicide see...Norflurazon
NORFORMS® see...Phenylmercury Acetate
NORMERSAN® see...Thiram
NOROSAC® see...Dichlobenil
NORTHSTAR® see...Sodium Dicamba
NORTHSTAR® see...Dicamba
NORTHSTAR® see...Primisulfuron-Methyl
NORTRON® see...Ethofumesate
Norway Saltpeter see...Ammonium Nitrate
Norwegian saltpeter, see...Calcium Nitrate
Nosema Locustae Canning see...Nosema Locustae
Nosema Locustae Spores see...Nosema Locustae
NOTT ZINC PHOSPHIDE 93® see...Zinc Phosphide
NOURITHION® see...Parathion
NOVAFOS-M® see...Parathion
NOVAGIB® see...Gibberellic Acid
NOVATHION® see...Fenitrothion
NOVA® see...Myclobutanil
NOVEL® see...Propiconazole
NOVIGAM® see...Lindane
NOVOSIR® N see...Zineb

NOVOTOX® see...Dichlorvos
NOVOZIN® N 50 see...Zineb
NOVOZIR® see...Zineb
NOVOZIR® N see...Zineb
NOVOZIR® N 50 see...Zineb
2-NOXA see...Naphthoxyacetic Acid
β-NOXA see...Naphthoxyacetic Acid
NOXA see...Naphthoxyacetic Acid
NOXALL® see...Prometon
NOXAL® see...Disulfiram
NOXFIRE® see...Rotenone
NOXFISH® see...Rotenone
NP-55® see...Sethoxydim
NPA see...Naptalam
NPD see...Aspon®
NPH-1091® see...Phosalone
NRDC 161® see...Deltamethrin
NRDC 166® see...Cypermethrin
NRDC 107® see...Resmethrin
NRDC 149® see...Cypermethrin
NRDC 119® see...Resmethrin
NRDC 160® see...*alpha*-Cypermethrin
NRDC 160® see...Cypermethrin
NSC 4911 see...Hexachlorophene
NSC see...Quinalphos
NSC 185 see...Cycloheximide
NSC 6738 see...Dichlorvos
NSC 750 see...Busulfan
NSC 423 see...2,4-D
NSC 122023 see...Validamycin
NSC 113243 see...Fentin Hydroxide
NSC 113243 see...Triphenyltin Compounds
NSC 150014 see...Hydrazine Sulfate
NSC 77690 see...Sodium Fluoroacetate
NSC 72005 see...Triclocarban
NSC 163049 see...Prometryn
NSC 167822 see...Carbofuran
NSC 85566 see...Sodium Dimethyldithiocarbamate
NSC 60380 see...Chlorpyrifos-methyl
NSC 1771 see...Thiram
NSC 190935 see...Chlordimeform
NSC 190939 see...Tetramethrin
NSC 190978® see...Phenthoate
NSC 46470 see...Mevinphos
NSC 43675 see...Triphenyltin Compounds
NSC 40486 see...EPTC
NSC 190987 see...Methamidophos
NSC 195022 see...Resmethrin
NSC 39624 see...Dichlorprop
NSC 195106 see...Fenamiphos
NSC 195164 see...Chlorthiophos
NSC 163046 see...Atrazine
NSC 38642 see...Dichlorophene
NSC 202753 see...Dinoseb
NSC 31312 see...Propanil
NSC 7657 see...Acenaphthene
NSC 25999 see...Simazine
NSC 233899 see...Picloram

NSC 263492 see...Carboxin
NSC 303303 see...Triadimefon
NSC 22364 see...Dipropyl Isocinchomeronate
NSC 21914 see...Trichlorobenzoic Acid
NSC 14083 see...Streptomycin
NSC 324552 see...Amitraz
NSC 370785 see...Demeton-methyl
NSC 8819 see...Acrolein
NSC 379698 see...Triallate
NSC 521749 see...2,4-D, isopropyl ester
NSC 525040® see...Thiabendazole
NSC 170810® see...Thiophanate-Methyl
NSC 60282 see...Mecoprop
N-SERVE® see...Nitrapyrin
N-SERVE NITROGEN STABILIZER® see...Nitrapyrin
N-TAC DESSICANT® see...Monocarbamide Dihydrogen Sulfate
NTN 801® see...Mefenacet
NTN 19701® see...Pencycuron
NTN 33893® see...Imidacloprid
NU-BAIT II® see...Methomyl
NUCIDOL® see...Diazinon
NU-COP® see...Copper Hydroxide
NUDRIN® see...Methomyl
NUFLOUR® see...Sodium Fluoride
NU-FLOW® see...Myclobutanil
NUGOR® see...Dimethoate
NU-LAWN WEEDER® see...Bromoxynil
NUOCIDE® see...Chlorothalonil
NUODEX PMA® see...Phenylmercury Acetate
NUP® see...Metsulfuron-methyl
NURATRON® see...Methamidophos
NURELLE see...Cypermethrin
NU-RENE 5 DUST® see...Anilazine
NU REXFORM® see...Lead Arsenate
NUTGRASS KILLER see...Sodium Methanearsonate (MSMA)
NU-TONE® see...1-Naphthaleneacetic Acid
NU-TRAZINE 900 DF® see...Atrazine
NUTRIFOS® STP see...Trisodium Phosphate
NUTRO® see...Anilazine
NUVACRON® see...Monocrotophos
NUVA® see...Dichlorvos
NUVAN® see...Dichlorvos
NUVAND® see...Fenitrothion
NUVANOL® see...Fenitrothion
NUX VOMICA® see...Strychnine
NU-ZINOLE AA® see...Atrazine
NUZONE® see...Imazalil
NW 200® see...Osmose (USA) see...Copper Carbonate see...Basic
NYLAR® see...Pyriproxyfen
NYMERATE® see...Phenylmercury Acetate
NYTEK® see...Copper(II)-8-hydroxyquinoline

- O -

OAKITE® see...Trisodium Phosphate

OAMA see...Octylammonium Methanearsonate
OBLIQUE® see...Resmethrin
OC-11588® see...Benfuracarb
2-(1,3-Octa-4-yl)benzimidazole see...Thiabendazole
1,2,4,5,6,7,8,8-Octachloor-3a,4,7,7a-tetrahydro-4,7-*endo*-methano-indaan (Dutch) see...Chlordane
Octachlor see...Chlordane
OCTACHLOR® see...Chlordane
Octachlorocamphene see...Toxaphene
Octachlorodihydrodicyclopentadiene see...Chlordane
1,2,4,5,6,7,8,8-Octachloro-2,3,3a,4,7,7a-hexahydro-4,7-methanoindene see...Chlordane
1,2,4,5,6,7,8,8-Octachloro-2,3,3a,4,7,7a-hexahydro-4,7-methano-1H-indene see...Chlordane
1,2,4,5,6,7,8,8-Octachloro-3a,4,7,7a-hexahydro-4,7-methylene indane see...Chlordane
1,2,4,5,6,7,8,8-Octachloro-3a,4,7,7a-tetrahydro-4,7-methanoindan see...Chlordane
1,2,4,5,6,7,8,8-Octachloro-3a,4,7,7a-tetrahydro-4,7-methanoindane see...Chlordane
1,2,4,5,6,7,8,8-Octachlor-3a,4,7,7a-tetrahydro-4,7-*endo*-methano-indan (German) see...Chlordane
Octachloro-4,7-methanohydroindane see...Chlordane
1,2,4,5,6,7,8,8-Octachloro-4,7-methano-3a,4,7,7a-tetrahydroindane see...Chlordane
Octachloro-4,7-methanotetrahydroindane see..Chlordane
1,2,4,5,6,7,10,10-Octachloro-4,7,8,9-tetrahydro-4,7-methyleneindane see...Chlordane
2-Octae-4-ylbenzimidazole see...Thiabendazole
3-Octaidinecarboxamide see...5-(4-chlorophenyl)-*N*-cyclohexyl-4-methyl-2-*oxo*-, *trans*- see. Hexythiazox
Octalene see...Aldrin
OCTALENE® see...Aldrin
OCTALOX® see...Dieldrin
Octamethyl-diforzuur-tetramide (Dutch) see..Octamethyl Diphosphoramide
Octamethyl-diphosphorsaeure-tetramid (German) see. Octamethyl Diphosphoramide
Octamethyl pyrophosphortetramide see...Octamethyl Diphosphoramide
Octamethyl tetramido pyrophosphate see...Octamethyl Diphosphoramide
Octametilpirofosforamida (Spanish) see...Octamethyl Diphosphoramide
1-Octanecarboxylic acid see...Pelargonic Acid
1-Octanesulfonamide see...*N*-ethyl-1,1,2,2,3,3,4,4,5,5,6,6,7,7,8,8-heptadecafluoro- see. Sulfluamid
Octan fenylrtutnaty (Czech) see...Phenylmercury Acetate
Octanoic acid see...copper salt see...Copper Octanoate
2-(4-Octayl)benzimidazole see...Thiabendazole
2-(4'-Octayl)benzimidazole see...Thiabendazole
2-(4-Octayl)-1*H*-benzimidazole see...Thiabendazole
OCTHILINONE® see...Octhilinone
Octochlorohexahydromethanoisobenzofuran see. Isobenzan

1,3,4,5,6,8,8-Octochloro-1,3,3a, 4,7,7a-hexahydro-4,7-methanoisobenzofuran see...Isobenzan
1,3,4,5,6,7,10,10-Octochloro-4,7-endo-methylene-4,7,8,9-tetrahydrophthalan see...Isobenzan
1,3,4,5,6,7,8,8-Octochloro-2-oxa-3a,4,7,7a-tetrahydro-4,7-methanoindene see...Isobenzan
OCTOCURE ZDM-50 see...Ziram
Octyl-*O*-(6-chloro-3-phenylpyridazin-4-yl)carbothioate (BSI) see...Pyridate
*n*-Octylisosafrole sulfoxide see...Sulfoxide
2-Octyl-3(2*H*)-isothiazolone see...Octhilinone
2-Octyl-4-isothiazolin-3-one see...Octhilinone
*n*-Octyl sulfoxide isosafrole see...Sulfoxide
ODC-45® see...Diflubenzuron
Odido de etileno (Spanish) see...Ethylene Oxide
ODIX (component of this product) see...Glutaraldehyde
ODM® see...Demeton-methyl
ODYSSEY® see...Imazethapyr
OFHC Cu see...Copper and Copper Compounds
OFIRMOTOX® see...Pyrethrins or Pyrethrum
OFTANOL® see...Isofenphos
OG-25® see...Chloropicrin
Oil of mustard see...artificial see...Allyl Isothiocyanate
OK 174® see...Benfuracarb
OK 622® see...Paraquat
OKO® see...Dichlorvos
OKSISYKLIN® see...Oxytetracycline Calcium
OKTANEX® see...Endrin
OKTATERR® see...Chlordane
OKULTIN® see...MCPA
OLE® see...Chlorothalonil
Olefiant gas see...Ethylene
Oleoakarithion see...Carbophanothion
Oleocuivre (French) see...Cuprous Oxide
OLEODIAZINON® see...Diazinon
OLEOFAC® see...Prothoate
OLEOFOS 20® see...Parathion
OLEOGESAPRIM® see...Atrazine
OLEO NORDOX® see...Cuprous Oxide
OLEOPARATHENE® see...Parathion
OLEOPARATHION® see...Parathion
Oleophosphothion see...Malathion
OLEOSUMIFENE® see...Fenitrothion
Oleovitamin D3 see...Cholecalciferol
OLEOVOFOTOX® see...Methyl Parathion
Oleum sinapis volatile see...Allyl Isothiocyanate
OLIN MATHIESON® 2424 see...Etridiazole
Olipsan see...Quintozene
OLITREF® see...Trifluralin
Olpisan see...Quintozene
OLTITOX® see...Carbaryl
OM® 2424 see...Etridiazole
OMAIT® see...Propargite
OMAL® see...Trichlorophenols
OMEGA see...Fluazinam®
Omethoat see...Omethoate
OMEXAN® see...Bromophos
OMITE® see...Propargite

OMNIPASSIN® see...Dithiazanine Iodide
OMNITOX® see...Lindane
OMNIZOLE® see...Thiabendazole
OMPA see...Octamethyl Diphosphoramide
OMPACIDE® see...Octamethyl Diphosphoramide
OMPATOX® see...Octamethyl Diphosphoramide
OMPAX® see...Octamethyl Diphosphoramide
OMS 14 see...Dichlorvos
OMS 15 see...Phenol, 3-(1-methyethyl)-, methylcarbamate
OMS 16 see...DDT
OMS 19 see...Parathion
OMS 29 see...Carbaryl
OMS 33 see...Propoxur
OMS 43 see...Fenitrothion
OMS 47 see...Mexacarbate
OMS 75 see...Naled
OMS 93 see...Methiocarb
OMS 194 see...Aldrin
OMS 468 see...Allethrins
OMS 570 see...Endosulfan
OMS 595 see...Tetrachlorvinphos
OMS 597 see...Trimethacarb
OMS 629 see...Carbaryl
OMS 658 see...Bromophos
OMS 659 see...Bromophos-ethyl
OMS 716 see...Promecarb
OMS 771 see...Aldicarb
OMS 1075 see...Phenthoate
OMS 1155 see...Chlorpyrifos-methyl
OMS 1191 see...Phosphamidon
OMS 1325 see...Phosphamidon
OMS 1328 see...Chlorfenvinphos
OMS 1342 see...Chlorthiophos
OMS 1394 see...Bendiocarb
OMS 1437 see...Chlordane
OMS 1696 see...Hydroprene
OMS 1804 see...Diflubenzuron
OMS 1809 see...D-Phenothrin
OMS 1810 see...D-Phenothrin
OMS 1820 see...Amitraz
OMS 2007 see...Flucythrinate
OMS 3023 see...Esfenvalerate
OMTAN see...Isobenzan
ONAGER® see...Hexythiazox
ONCOL® see...Benfuracarb
ONCOL 5G® see...Benfuracarb
ONE SHOT® see...Diclofop-methyl
ONESIDE® see...Fluazifop-butyl
ONESIDE EC® see...Fluazifop-butyl
ONMEX® see...Penconazole
ONSLAUGHT® see...Linuron
ONSLAUGHT® see...Trifluralin
ONTRACK® see...Prometon
ONTRACK 8E® see...Metolachlor
ONTRACK WE HERBICIDE® see...Pentachlorophenol
ONYX BTC see...Zilkonium Chloride
OP-THAL-ZIN® see...Zinc Sulfate Heptahydrate

Oplossingen (Dutch) see...Formaldehyde
OPP see...o-Phenylphenol
OPTILL® see...Dimethenamid
OPTIMA® see...Thiram
OPTIMUM® see...Acrylamide
OPTION® see...Fenoxaprop-ethyl
OPTISED® see...Zinc Sulfate Heptahydrate
OPTRAEX® see...Zinc Sulfate Heptahydrate
ORAGULANT® see...Diphacione
ORAZINC® see...Zinc Sulfate Heptahydrate
ORBIT® see...Fentin Hydroxide
ORBIT® see...Propiconazole
ORBIT® see...Triphenyltin Compounds
ORCED® see...Dichloroisocyanuric Acid
ORCEPHATE® see...Acephate
ORCHARD® BRAND ZIRAM see...Ziram
ORDAM® see...Molinate
ORDRAM® see...Molinate
ORGA-414® see...Amitrole
ORIMON® see...Phenothiazine
ORNALIN® see...Vinclozolin
ORNAMEC® see...Fluazifop-butyl
ORNAMENTAL WEEDER® see...Chloramben
ORNAZIN® see...Azadirachtin
ORNITROL® see...Azacosterol Dihydrochloride
ORTHENE® see...Acephate
ORTHENE 755® see...Acephate
ORTHO 4355® see...Naled
ORTHO 5353® see...Brufencarb
ORTHO 5865® see...Captafol
ORTHO 9006® see...Methamidophos
ORTHO 12420® see...Acephate
ORTHOarsenic acid (o-) see...Arsenic Acid
ORTHO C-1 DEFOLIANT & WEED KILLER® see. Sodium Chlorate
OrthoCIDE® see...Captan
OrthoDIBROM® see...Naled
OrthoDIBROMO® see...Naled
ORTHO DIQUAT® see...Diquat
ORTHO GRASS KILLER® see...Propham
Orthohydroxybenzoic acid see...Salicylic Acid
Orthohydroxydiphenyl see...o-Phenylphenol
ORTHO-KLOR® see...Chlordane
ORTHO L-10 DUST® see...Lead Arsenate
ORTHO L-40 DUST® see...Lead Arsenate
ORTHO MALATHION® see...Malathion
ORTHO MC® see...Magnesium Chlorate
ORTHO-MITE® see...Aramite
ORTHO N-4 DUST® see...Nicotine
ORTHO PARAQUAT CL® see...Paraquat
ORTHO P-G BAIT® see...Copper Acetoarsenite
ORTHO P-G BAIT® see...Paris Green
Orthophenylphenol see...o-Phenylphenol
OrthoPHOS® see...Parathion
ORTHO® PHOSPHATE DEFOLIANT see...Tribufos
Orthophosphoric acid see...Phosphoric Acid
ORTHO TRIOX® see...Pentachlorophenol
Orthoxenol see...o-Phenylphenol

ORTRAN® see...Acephate
ORTRIL® see...Acephate
Orvinylcarbinol see...Allyl Alcohol
ORYZA® see...Oryzalin
ORYZALIN® see...Oryzalin
OS 1897® see...Dibromochloropropane
O,S-Dimethyl acetic phosphoramidothioate, N-[Methoxy (methylthio)phosphinoyl] acetamide see...Acephate
O,S-Dimethyl acetylphos-phoramidothioate see.Acephate
O,S-Dimethyl ester of amide of amidothioate see. Methamidophos
OSBAC® see...BPMC
OSBON AC® see...Peracetic Acid
OSDARAN® see...Fenbutatin Oxide
OSMOSALTS® see...Sodium Dichromate
OSMOSE WPC® see...Pentachlorophenol
OSOCIDE® see...Captan
OSP 3506-35® see...Cuprous Thiocyanate
OSSALIN® see...Sodium Fluoride
OSSIN® see...Sodium Fluoride
OSVAN see...Zilkonium Chloride
Othroboric acid see...Borax and Boric acid
OTTAFACT® see...p-Chloro-m-Cresol
OTTASEPT® see...4-Chloro-3,5-xylenol
OTTASEPT® EXTRA see...4-Chloro-3,5-xylenol
1,2,4,5,6,7,8,8-Ottochloro-3a,4,7,7a-tetraidro-4,7-endo-metano-indano (Italian) see...Chlordane
Ottometil-pirofosforammide (Italian) see...Octamethyl Diphosphoramide
OUSTAR® see...Hexazinone
OUSTAR® see...Sulfometuron-Methyl
OUST® see...Sulfometuron-Methyl
OVADEX BW® see...Triallate
OVADOFOS® see...Fenitrothion
OVADZIAK® see...Lindane
OVASYN® see...Amitraz
OVATION® see...Clofentezine
OVERTOP® see...Imazethapyr
OVIDIP® see...Propetamphos
OVIDREX® see...Amitraz
OVITROL® see...Methoprene
OWADZIAK® see...Lindane
7-Oxabicyclo(2.2.1)heptane-2,3-dicarboxylic acid see. Endothall
Oxacyclopropane see...Ethylene Oxide
δ(sup2)2-1,3,4-Oxadiazolin-5-one, 2-tert-butyl-4-(2,4-dichloro-5-isopropoxyphenyl)- see...Oxadiazon
δ(sup2)2-1,3,4-Oxadiazolin-5-one, 2-tert-butyl-4-(2,4-dichloro-5-isopropyloxyphenyl)- see...Oxadiazon
1,3,4-Oxadiazol-2(3H)-one, 3-(2,4-dichloro-5-(1-methylethoxy)phenyl)-5-(1,1-dimethylethyl)- see. Oxadiazon
Oxadiazon see...Oxadiazon
Oxadiazone see...Oxadiazon
OXALIN® see...Carboxin
OXALIS® see...Glyphosate
OXAMYL CARBAMATE INSECTICIDE® see.Oxamyl
Oxane see...Ethylene Oxide

1,4-Oxathiin-3-carboxamide,5,6-dihydro-2-methyl-N-phenyl see...Carboxin
1,4-Oxathiin-3-carboxamide see...5,6-dihydro-2-methyl-N-phenyl-, 4,4-dioxide see...Oxycarboxin
1,4-Oxathiin-3-carboxanilide,5,6-dihydro-2-methyl see. Carboxin
1,4-Oxathiin-3-carboxanilide,5,6-dihydro-2-methyl- see. Carboxin
1,4-Oxathiin-3-carboxanilide, .5,6-dihydro-2-methyl-, 4,4-dioxide see...Oxycarboxin
1,4-Oxathiin-2,3-dihydro-5-carboxanilido-6-methyl- see. Carboxin
1,3,4-Oxazol-2(3H)-one, 3-[2,4-dichloro-5-(1-methylethoxy)phenyl]-5-(1,1-dimethylethyl)- see. Oxadiazon
2,4-Oxazolidinedione, 3-(3,5-dichlorophenyl)-5-ethenyl-5-methyl- see...Vinclozolin
2,4-Oxazolidinedione, 3-(3,5-dichlorophenyl)-5-methyl-5-vinyl- see...Vinclozolin
OXICOB® see...Copper Oxychloride
Oxidimethiin see...Dimethipin
Oxido mercurico amarillo (Spanish) see...Mercuric Oxide
Oxido mercurico rojo (Spanish) see...Mercuric Oxide
Oxidoethane see...Ethylene Oxide
A,β-Oxidoethane see...Ethylene Oxide
Oxime-copper see...Copper(II)-8-hydroxyquinoline
Oxine-copper see...Copper(II)-8-hydroxyquinoline
Oxine-Cu see...Copper(II)-8-hydroxyquinoline
Oxine cuivre (French) see...Copper(II)-8-hydroxyquinoline
Oxiraan (Dutch) see...Ethylene Oxide
Oxirane see...Ethylene Oxide
Oxirene see...Dihydro- see...Ethylene Oxide
Oxitetracyclin see...Oxytetracycline Calcium
OXIVOR® see...Copper Oxychloride
Oxomethane see...Formaldehyde
OXON AMETRYN TECHNICAL® see...Ametryn
OXON ITALIA SIM-TROL® see...Simazine
OXRALOX® see...Dieldrin
OXYATETS® see...Oxytetracycline Calcium
OXY BCP® see...Dibromochloropropane
1,1'-Oxybis(2-chloro)ethane see...Dichloroethyl Ether
Oxycarboxine see...Oxycarboxin
Oxychlorue de cuivre (French) see...Copper Oxychloride
OXYCIL® see...Sodium Chlorate
OXYCLOR® see..Copper Oxychloride
OXYCUR® see..Copper Oxychloride
Oxyde de chlorethyle (French) see...Dichloroethyl Ether
Oxyde de mercure (French) see...Mercuric Oxide
Oxydemeton-methyl see...Demeton-methyl
Oxydemetonmethyl see...Demeton-methyl
Oxydemeton methyl [S-(2-(Ethylsulfinyl)ethyl]O,O-dimethyl ester phosphorothioic acid see...Demeton-methyl
Oxydiazol see...Methazole
Oxydimethiin see...Dimethipin

Oxyfluorfene see...Oxyfluorfen
Oxyfluorofen see...Oxyfluorfen
OXYFUME® see...Ethylene Oxide
OXYFUME 12® see...Ethylene Oxide
OXYMASTER® see...Peracetic Acid
Oxymethylene see...Formaldehyde
OXYMYCIN® see...Oxytetracycline Calcium
OXYMYKOIN® see...Oxytetracycline Calcium
Oxyparathion see...Paraoxon
Oxyquinolinoleate de cuivre (French) see...Copper (II)-8-hydroxyquinoline
OXYTERRACIN® see...Oxytetracycline Calcium
OXYTERRACINE® see...Oxytetracycline Calcium
OXYTERRACYNE® see...Oxytetracycline Calcium
OXYTETRACYCLINE® see...Oxytetracycline Calcium
OXYTETRACYCLINE AMPHOTERIC® see.Oxytetracycline Calcium
OXYTRIL M® see...Bromoxynil

- P -

2,4-PA (in Japan) see...2,4-D
PAARLAN® see...Isopropalin
PAC® see...Parathion
PACE® see...Acephate
PACE® see...Mancozeb
PACE® see...Metalaxyl
PACOL® see...Parathion
PADAN® see...Carboxin
PADAN® see...Cartap Hydrochloride
PADOPHENE® see...Phenothiazine
Painters naphtha see...Naphthas
PAKHTARAN® see...Fluometuron
Pallethrine (France) see...Allethrins
PAMISAN® see...Phenylmercury Acetate
PAMOSOL® 2 FORTE PARZATE® see...Zineb
PANAM® see...Carbaryl
PANAPLATE® see...Dichlorvos
PANCIL® see...Octhilinone
PANDRINOX® see...Methylmercuric Dicyanamide
PANO-DRENCH®-4 see...Methylmercuric Dicyanamide
PANODRIN® A-13 see...Methylmercuric Dicyanamide
PANOGEN® see...Methoxyethylmercuric Acetate
PANOGEN® M see...Methoxyethylmercuric Acetate
PANOGEN® see...Methylmercuric Dicyanamide
PANOGEN® METOX see...Methoxyethylmercuric Acetate
PANOGEN-PX® see...Methylmercuric Dicyanamide
PANOGEN TURF FUNGICIDE® see...Methylmercuric Dicyanamide
PANOGEN TURF SPRAY see...Methylmercuric Dicyanamide
PANORAM® see...Dieldrin
PANORAM 75® see...Thiram
PANORAM D-31® see...Dieldrin
PANOSPRAY® 30 see...Methylmercuric Dicyanamide
PANSOIL® see...Etridiazole

PANTHER® see...Diflufenican
PANTHION® see...Parathion
PANTOX® see...Propanil
PANTOZOL-1® see...Crotoxyphos
PANWARFIN® see...Warfarin
PAP® see...Phenthoate
Papermaker's alum see...Aluminum Sulfate
PAPTHION® see...Phenthoate
PAQEANT® see...Chlorpyrifos
PARA-COL® see...Paraquat
PARA CRYSTALS® see...*para*-Dichlorobenzene
PARA-KILL® see...Parathion
PARA-TOX® see...Parathion
Parachlorocidum see...DDT
Parachlorometacresol see...*p*-Chloro-*m*-Cresol
Parachlorometaxylenol see...4-Chloro-3,5-xylenol
PARACIDE® see...*para*-Dichlorobenzene
PARADERIL® see...Rotenone
PARADI® see...*para*-Dichlorobenzene
Paradichlorobenzene see...*para*-Dichlorobenzene
PARADIGM® see...Clopyralid
PARADIGM® see...Fluroxypyr 1-Methylheptyl Ester
PARADOW® see...*para*-Dichlorobenzene
PARADUST® see...Parathion
PARAMAR® see...Parathion
PARAMINE® see...Paraquat
PARAMOTH® see...*para*-Dichlorobenzene
PARAMOUNT® see...Quinclorac
Paranaphthalene see...Anthracene
PARANUGGETS® see...*para*-Dichlorobenzene
PARAOXONE® see...Paraoxon
PARAPEST M-50® see...Methyl Parathion
PARAPHOS® see...Parathion
Paraqiat-I see...Paraquat Methosulfate
Paraquat bis(methyl sulfate) see...Paraquat Methosulfate
Paraquat chloride see...Paraquat
PARAQUAT CL® see...Paraquat
Paraquat dichloride see...Paraquat
PARAQUAT DICHLORIDE BIPYRIDYLNIUM HERBICIDE® see...Paraquat
Paraquat dimethosulfate see...Paraquat Methosulfate
Paraquat dimethyl sulfate see...Paraquat Methosulfate
Paraquat dimethyl sulphate see...Paraquat Methosulfate
Paraquat methsulfate bipyridylnium herbicide see.Paraquat Methosulfate
PARASOL® see...Copper Hydroxide
Parathene see...Parathion
Parathion-ethyl see...Parathion
Parathion-methyl see...Methyl Parathion
Parathion Metile (Spanish) see...Methyl Parathion
Parathion thiophos see...Parathion
Parationa (Spanish) see...Parathion
PARAWET® see...Parathion
PARAZENE® see...*para*-Dichlorobenzene
Parcipitated chalk see...Calcium Carbonate
PARDNER® see...Bromoxynil
PAR EX® see...Oxadiazon
PAR EX® see...Prodiamine

PAREXAN® see...Pyrethrins or Pyrethrum
PARFLO® see...Quintozene
Paris green see...Copper Acetoarsenite
PARLAY® see...Paclobutrazole
PARMETOL® see...*p*-Chloro-*m*-Cresol
PARMONE® see...1-Naphthaleneacetic Acid
Parrot green see...Copper Acetoarsenite
Parrot green see...Paris Green
PAROXAN® see...Paraoxon
PARRYCOP® see...Copper Oxychloride
PARSEC® see...Amitraz
PARSOL® see...*p*-Chloro-*m*-Cresol
PARSONS® 2,4-D WEED KILLER ISOPROPYL ESTER see...2,4-D, isopropyl ester
PARTEL® see...Dithiazanine Iodide
PARTI-SAN® see...Metsulfuron-methyl
PARTNER® see...Alachlor
PARTOX® see...Chlorophacinone
Partron M see...Methyl Parathion
PARZATE® see...Nabam
PARZATE® ZINEB see...Zineb
PASSPORT® see...Imazethapyr
PASTUREGARD® see...Fluroxypyr 1-methylheptyl Ester
PASTURE® MD see...Metsulfuron-methyl
PATAP® see...Cartap Hydrochloride
Patent green see...Copper Acetoarsenite
Patent green see...Paris Green
PATHCLEAR® see...Diquat
PATHCLEAR® see...Paraquat
PATHFINDER® see...Triclopry
PATHWAY® see...Picloram
PATORAN® see...Metobromuron
PATRIOT® see...Atrazine
PATRIOT® see...Imazethapyr
PATROLE® see...Methamidophos
PATRON® see...Cyromazine
PATTONEX® see...Metobromuron
PAX FUNGUS CONTROL® see...Anilazine
PAXILON® see...Methazole
PAYLOAD® see...Acephate
PAYOFF® see...Flucythrinate
PAY-OFF® see...Pendimethalin
PAYZE® see...Cyanazine
PB see...Piperonyl Butoxide
PBI CROP SAVER® see...Malathion
PBI SLUG GARD® see...Methiocarb
PCA® see...Pyrazon
PCC see...Toxaphene
PCHK see...Toxaphene
PCMC see...*p*-Chloro-*m*-Cresol
PCMX® see...4-Chloro-3,5-xylenol
PCNB see...Quintozene
PCP see...Pentachlorophenol
PCP-sodium see...Sodium Pentachlorophenate
PCP sodium salt see...Sodium Pentachlorophenate
P.C.Q.® see...Diphacione
PD 5® see...Mevinphos

PDCB see...*para*-Dichlorobenzene
PDD 60401® see...Diflubenzuron
PEACH-THIN® see...Naptalam
PEB1 see...DDT
PEBC see...Pebulate
PEDIAFLOR® see...Sodium Fluoride
PEDIDENT® see...Sodium Fluoride
PEDINEX® see...Dinex
PEDRACZAK® see...Lindane
Pelargic acid see...Pelargonic Acid
PELARGON® see...Pelargonic Acid
PELT® see...Thiophanate-Methyl
PEL-TECH® see...Benefin
Penatrol see...Atrazine
PENCAL® see...Calcium Arsenate
Penchlorol see...Pentachlorophenol
PENCYCURONE® see...Pencycuron
Pendimethaline see...Pendimethalin
PENITE® see...Sodium Arsenite
PENNAC CRA® see...Ethylene Thiourea
PENNAMINE® see...2,4-D
PENNANT® see...Metolachlor
PENNCAP E® see...Parathion
PENNCAP M® see...Methyl Parathion
PENNCAP MLS® see...Methyl Parathion
PENNCOZEB® see...Mancozeb
PENNWALT C-4852® see...Fenitrothion
PENNWHITE® see...Sodium Fluoride
PENOXALIN® see...Pendimethalin
PENOXALINE® see...Pendimethalin
Penphene see...Toxaphene
Penta see...Pentachlorophenol
PENTA-KIL® see...Pentachlorophenol
PENTA READY® see...Pentachlorophenol
Pentachloorfenol (Dutch) see...Pentachlorophenol
Pentachlorin see...DDT
Pentachlornirtobenzol (German) see...Quintozene
Pentachlorofenol see...Pentachlorophenol
Pentachloronitrobenzene see...Quintozene
Pentachlorophenate see...Pentachlorophenol
Pentachlorophenate sodium see...Sodium Pentachlorophenate
2,3,4,5,6-Pentachlorophenol see...Pentachlorophenol
Pentachlorophenol, sodium salt see...Sodium Pentachlorophenate
Pentachlorophenol, technical see...Pentachlorophenol
Pentachlorophenoxy sodium see...Sodium Pentachlorophenate
Pentachlorphenol (German) see...Pentachlorophenol
Pentaclorofenato sodico (Spanish) see...Sodium Pentachlorophenate
Pentaclorofenol (Spanish) see...Pentachlorophenol
Pentaclorofenolo (Italian) see...Pentachlorophenol
Pentachlorophenyl chloride see...Hexachlorobenzene
PENTAC® see...Dienochlor
PENTAC® WP see...Dienochlor
PENTACON® see...Pentachlorophenol

1,4-Pentadien-3-one-1,5-bis(α,α,α-trifluoro-*p*-tolyl)-tetrahydro-5,5-dimethyl-2(1*H*)-pyrimidinylidene)hydrazone  see. . .Hydramethylnon
Pentafluorobenzyl  see. . .Fenfluthrin
(+)-(Pentafluorophenyl)methyl (1R-*trans*)-3-(2,2-dichlorovinyl)-2,2-dimethylcyclopropanecarboxylate  see. Fenfluthrin
PENTAGEN®  see. . .Quintozene
Pentamethyleneimine  see. . .Piperidine
Pentanedial  see. . .Glutaraldehyde
1,5-Pentanedial  see. . .Glutaraldehyde
Pentanedinitrile, 2-bromo-2-(bromomethyl)-  see. . .1,2-Dibromo-2,4-dicyanobutane
1,5-Pentanedione  see. . .Glutaraldehyde
Pentaphenate  see. . .Sodium Pentachlorophenate
PENTASOL®  see. . .Pentachlorophenol
PENTECH®  see. . .DDT
PENTHAZINE®  see. . .Phenothiazine
Pentine acid 5431  see. . .Dodecylbenzenesulfonic Acid
*N*-(3-Pentyl)-3,4-dimethyl-2,6-dinitroaniline  see. Pendimethalin
PENWAR®  see. . .Pentachlorophenol
Peperidin (German)  see. . .Piperidine
PEPROSAN®  see. . .Copper Oxychloride
PERATOX®  see. . .Pentachlorophenol
Perchloride of mercury  see. . .Mercuric Chloride
Perchlormethylmerkaptan (Czech)  see. .Perchloromethyl Mercaptan
Perchlorobenzene  see. . .Hexachlorobenzene
Perchlorodihomocubane  see. . .Mirex
Perchloroethane  see. . .Hexachloroethane
Perchloromethane  see. . .Carbon Tetrachloride
Perchloromethanethiol  see. . .Perchloromethyl Mercaptan
Perchloropentacyclodecane  see. . .Mirex
PERCOLATE®  see. . .Phosmet
PERECOT®  see. . .Cuprous Oxide
PERENOX®  see. . .Cuprous Oxide
PERFEKTION®  see. . .Dimethoate
PERFLAN®  see. . .Tebuthiuron
PERGANTENE®  see. . .Sodium Fluoride
PERKACIT ZDMC®  see. . .Ziram
PERLAN®  see. . .6-Benzaldenine
PERMACIDE®  see. . .Pentachlorophenol
PERMAGARD®  see. . .Pentachlorophenol
PERMASAN®  see. . .Pentachlorophenol
PERMASECT C®  see. . .Cypermethrin
PERMATOX DP-2®  see. . .Pentachlorophenol
PERMATOX PENTA®  see. . .Pentachlorophenol
Permethrin  see. . .Aldicarb
PERMIT®  see. . .Halosulfuron-methyl
PERMITE®  see. . .Pentachlorophenol
PEROSIN®  see. . .Zineb
PEROSIN® 75B  see. . .Zineb
Peroxido de Arsenico (Spanish)  see. . .Arsenic Pentoxide
Peroxyacetic acid  see. . .Peracetic Acid
PEROZIN®  see. . .Zineb
PEROZINE®  see. . .Zineb
PERSEVTOX®  see. . .Dinoseb

Persian insect powder  see. . .Pyrethrins or Pyrethrum
PERSIA-PERAZOL®  see. . .*para*-Dichlorobenzene
PERSYST®  see. . .Demeton-methyl
PERTHANE®  see. . .Ethylan
PESTANAL®  see. . .Metobromuron
PESTMASTER®  see. . .Ethylene Dibromide
PESTMASTER® FUMIGANT 1  see. . .Chloropicrin
PESTOX®  see. . .Octamethyl Diphosphoramide
PESTOX 3®  see. . .Octamethyl Diphosphoramide
PESTOX 14®  see. . .Dimefox
PESTOX 101®  see. . .Paraoxon
PESTOX III®  see. . .Octamethyl Diphosphoramide
PESTOX IV®  see. . .Dimefox
PESTOX XIV®  see. . .Dimefox
PESTOX PLUS®  see. . .Parathion
PESTROY®  see. . .Fenitrothion
PETHION®  see. . .Parathion
Petroleum distillates  see. . .Naphthas
Petroleum ether  see. . .Naphthas
Petroleum naphtha  see. . .Naphthas
Petroleum solvent  see. . .Stoddard Solvent
Petroleum spirit  see. . .Naphthas
PF-3  see. . .Isofluorphate
PFLANZOL®  see. . .Lindane
PGR-IV®  see. . .Gibberellic Acid
PGR-IV®  see. . .Indole-3-Butyric Acid
PH 60-40®  see. . .Diflubenzuron
PHANTOM®  see. . .Chlorfenapyr
PHARORID®  see. . .Methoprene
PHASER®  see. . .Endosulfan
PHASOLON®  see. . .Phosalone
PHC  see. . .Propoxur
Phenachlor  see. . .Trichlorophenols
PHENACIDE®  see. . .Toxaphene
PHENADOR-X®  see. . .Biphenyl
Phenamiphos  see. . .Fenamiphos
PHENATOX®  see. . .Toxaphene
PHENDAL®  see. . .Phenthoate
Pheneene germicidal solution and tincture  see.Zilkonium Chloride
PHENEGIC®  see. . .Phenothiazine
Phenitrothion  see. . .Fenitrothion
PHENMAD®  see. . .Phenylmercury Acetate
Phenmediphame  see. . .Phenmedipham
PHENOHEP®  see. . .Hexachloroethane
Phenol, *o-(tert*-butyl)-  see. . .Butylphenols
Phenol,4-*t*-butyl-2-chloro-, ester with methyl methylphosphoramidate  see. . .Crufomate
Phenol, 2-*sec*-butyl-4,6-dinitro-  see. . .Dinoseb
Phenol-2-*tert*-butyl-4,6-dinitro-  see. . .Dinoterb
Phenol, 5-chloro-2-(2,4-dichlorophenoxy)-  see.Triclosan
Phenol, 2-cyclohexyl-4,6-dinitro-  see. . .Dinex
Phenol, 4-(dimethylamino)-3,5-dimethyl-methylcarbamate (ester)  see. . .Mexacarbate
Phenol, 2-(1,1-dimethylethyl)4,6-dinitro-  see. . .Dinoterb
Phenol, 3,5-dimethyl-4-(methylthio)-, methylcarbamate  see. . .Methiocarb

Phenol, 2-methyl-4,6-dinitro- see...Dinitro-o-cresol (DNOC)
Phenol, 2,2'-methylenebis(4-chloro- see...Dichlorophene
Phenol, 2,2'-methylenebis(3,4,6-trichloro)- see.Hexachlorophene
Phenol, 2,2'-methylenebis(3,5,6-trichloro- see.Hexachlorophene
Phenol, 2-(1-methylethoxy)-, methylcarbamate see.Propoxur
Phenol, 2-(1-methylheptyl)-4,6-dinitro-, crotonate (ester) see...Dinocap
Phenol, 3-methyl-5-(1-methylethyl)-, methylcarbamate see...Promecarb
Phenol, 2-(1-methylpropyl)-4,6-dinitro- see...Dinoseb
Phenol, 2-(1-methylpropyl)-, methylcarbamate see. BPMC
Phenol, p-nitro-, O-ester with O,O-dimethyl phosphorothioate see...Methyl Parathion
Phenol, pentachloro- see...Pentachlorophenol
Phenol, pentachloro-, sodium salt see...Sodium Pentachlorophenate
Phenol, pentachloro-, sodium salt, monohydrate see. Sodium Pentachlorophenate
Phenol,4,4'-thiodi-, O,O-diester with O,O-dimethyl phosphorothioate see...Temephos
Phenol, trichloro- see...Trichlorophenols
Phenol, 2,3,4-trichloro- see...Trichlorophenols
Phenol, 2,3,5-trichloro- see...Trichlorophenols
Phenol, 2,3,6-trichloro- see...Trichlorophenols
Phenol, 2,4,6-trichloro- see...Trichlorophenols
Phenomercury acetate see...Phenylmercury Acetate
PHENOSAN® see...Phenothiazine
PHENOSANE® see...Pyrazon
Phenostat-C see...Triphenyltin Compounds
Phenostat-H see...Triphenyltin Compounds
PHENOSTAT®-H see...Fentin Hydroxide
PHENOTAN® see...Dinoseb
Phenothrin see...D-Phenothrin
(+)-cis,trans-Phenothrin see...D-Phenothrin
PHENOVERM® see...Phenothiazine
PHENOVIS® see...Phenothiazine
PHENOX® see...2,4-D
PHENOXUR® see...Phenothiazine
Phenoxyacetic acid, 2,4,5-trichloro- see...2,4,5-Trichlorophenoxyacetic Acid, Esters
3-Phenoxybenzyl cis,trans-chrysanthemate see...D-Phenothrin
3-Phenoxybenzyl (±)-cis,trans-chrysanthemate see...D-Phenothrin
3-Phenoxybenzyl (1RS)-cis,trans-chrysanthemate see.D-Phenothrin
3-Phenoxybenzyl D-cis,trans-chrysanthemate see...D-Phenothrin
3-Phenoxybenzyl d-Z/E chrysanthemate see...D-Phenothrin
3-Phenoxybenzyld-Z/E chrysanthemate see...D-Phenothrin

3-Phenoxybenzyl 2-dimethyl-3-(methylpropenyl)cyclopropanecarboxylate see...D-Phenothrin
m-Phenoxybenzyl 2,2-dimethyl-3-(2-methylpropenyl)cyclopropanecarboxylate see...D-Phenothrin
3-Phenoxybenzyl (1RS,3RS;1RS,3SR)-2,2-dimethyl-3-(2-methylprop-1-enyl)cyclopropanecarboxylate see...D-Phenothrin
3-Phenoxybenzyl(1RS)-(Z),(E)-2,2-dimethyl-3-(2-methylprop-1-enyl)cyclopropanecarboxylate see...D-Phenothrin
3-Phenoxybenzyl(1RS)-cis,trans-2,2-dimethyl-3-(2-methylprop-1-enyl)cyclopropanecarboxylate see...D-Phenothrin
PHENOXYLENE 50® see...MCPA
PHENOXYLENE PLUS® see...MCPA
PHENOXYLENE SUPER® see...MCPA
N-[2-(p-Phenoxyphenoxy)ethyl]carbamic acid see. Fenoxycarb
2-(4-Phenoxyphenoxy)ethylcarbamic acid ethyl ester see. Fenoxycarb
[2-(4-Phenoxyphenoxy)ethyl]carbamic acid ethyl ester see...Fenoxycarb
[2-(4-Phenoxy-phenoxy)-ethyl]carbamic acid ethyl ester see...Fenoxycarb
4-Phenoxyphenyl (RS)-2-(2-pyridyloxy)propyl ether see. Pyriproxyfen
PHENOXYTHRIN® see...D-Phenothrin
Phenthiazine see...Phenothiazine
Phentin acetate see...Triphenyltin Compounds
Phentinoacetate see...Triphenyltin Compounds
Phenvalerate see...Fenvalerate
3-(1'-Phenyl-2'-acetylethyl)-4-hydroxycoumarin see. Warfarin
3-(α-Phenyl-β-acetylethyl)-4-hydroxycoumarin see. Warfarin
(Phenyl-1-acetyl-2-ethyl)-3-hydroxy-4 coumarine (French) see...Warfarin
Phenylacrolein see...Cinnamaldehyde
3-Phenylacrolein see...Cinnamaldehyde
β-Phenylacrolein see...Cinnamaldehyde
1-Phenyl-4-amino-5-chloropyridazon-(6) (German) see. Pyrazon
1-Phenyl-4-amino-5-chloropyridazone-6 see...Pyrazon
1-Phenyl-4-amino-5-chloro-6-pyridazone see...Pyrazon
1-Phenyl-4-amino-5-chlorpyridaz-6-one see...Pyrazon
Phenylbenzene see...Biphenyl
N-Phenylcarbamate d'isopropyle (French) see...Propham
Phenyl carbamic acid-1-methylethyl ester see...Propham
5-–Phenylcarbamoylamino-1,2,3-thiadiazole see. Thidiazuron
2-Phenyl-carbamoyloxy-N-aethyl-propionamid (German) see...Carbetamide
(Phenylcarbamoyloxy)-2-N-ethylpropionamide see. Carbetamide
Phenylcarbamoyloxyphenylcarbamate see.Desmedipham
Phenylcarbinol see...Benzyl Alcohol

Phenyl carboxylic acid  see. . .Benzoic Acid
2-[2-Phenyl-2-(4-chlorophenyl)acetyl]-1,3-indandione  see. . .Chlorophacinone
Phenyl-$N'$-(2-chloro-4-pyridyl)urea  see. .Forchlorfenuron
1-Phenyl-3-($O,O$-diethyl-thionophophoryl)-1,2,4-triazole  see. . .Triazophos
4,4'-$O$-Phenylenebis(ethyl 3-thioallophanate)  see.Thiophanate-Methyl
Phenylenebis(iminocarbonothioyl)]biscarbamic acid diethyl ester  see. . .Thiophanate-Methyl
(1,2-Phenylenebis[iminocarbonothioyl])]biscarbamic acid diethyl ester  see. . .Thiophanate-Methyl
4,4'-$O$-Phenylenebis(3-thioallophanic acid)dimethyl ester  see. . .Thiophanate-Methyl
cis-2-(1-Phenylethoxy)carbonyl-1-methylvinyl dimethylphosphate  see. . .Crotoxyphos
$N$-Phenyl-1-(ethylcarbamoyl-1)-ethylcarbamate, d- isomer  see. . .Carbetamide
Phenyl fluoride  see. . .Fluorobenzene
Phenylformic acid, Benzoic acid  see. . .Benzoic Acid
Phenylglyoxylonitrile oxime $O,O$-diethyl phosphorothioate  see. . .Phoxim
$N$-Phenyl isopropyl carbamate  see. . .Propham
Phenylmercuric acetate  see. . .Phenylmercury Acetate
Phenylmethanol  see. . .Benzyl Alcohol
Phenylmethyl alcohol  see. . .Benzyl Alcohol
Phenyl methyl ether  see. . .Anisole
[5-(Phenylmethyl)-3-furanyl]methyl 2,2-dimethyl-3-furylmethyl-2,2-dimethyl-3-(2-methylpropenyl) cyclopropanecarboxylate  see. . .Resmethrin
5-(Phenylmethyl)-3-furanyl)methyl 2,2-dimethyl-3-(2-methyl-1-propenyl)cyclopropanecarboxylate  see. Resmethrin
[5-(Phenylmethyl)-3-furanyl]methyl 2,2-dimethyl-3-(2-methyl-1-propenyl)cyclopropanecarboxylate)  see. Resmethrin
[5-(Phenylmethyl)-3-furanyl]methyl ester  see.Resmethrin
$N$-(Phenylmethyl)-1$H$-purin-6-amine  see. . .6-Benzaldenine
Phenylmurcuriacetate  see. . .Phenylmercury Acetate
2-Phenylphenol  see. . .$o$-Phenylphenol
Phenylphosphonothioic acid $O$-(4-bromo-2,5-bromo-2,5-dichlorophenyl)$O$-methyl ester  see. . .Leptophos
3-Phenylpropenal  see. . .Cinnamaldehyde
($E$)-3-Phenylpropenal  see. . .Cinnamaldehyde
($E$)-3-Phenyl-2-propenal  see. . .Cinnamaldehyde
3-Phenyl-2-propenal  see. . .Cinnamaldehyde
Phenylquecksilberacetat (German)  see. . .Phenylmercury Acetate
$N$-(4-Phenylsulfonyl-$o$-tolyl)-1,1,1-trifluoromethanesulfonamide  see. . .Perfluidone
1-Phenyl-3-(1,2,3-thiadiazol-5-yl)urea  see. . .Thidiazuron
$N$-Phenyl-$N'$-1,2,3-thiadiazol-5-yl-urea  see. .Thidiazuron
$N$-Phenyl-$N'$-(1,2,3-thiadiazyl)urea  see. . .Thidiazuron
Phenylthiocarbamide  see. . .Phenylthiourea
Phenyl-2-thiourea  see. . .Phenylthiourea
1-Phenylthiourea  see. . .Phenylthiourea
$\alpha$-Phenylthiourea  see. . .Phenylthiourea
$N$-Phenylthiourea  see. . .Phenylthiourea
1-Phenyl-1,2,4-triazolyl-3-($O,O$-diethylthionophosphate)  see. . .Triazophos
PHILIPS-DUPHAR® PH 60-40  see. . .Diflubenzuron
PHIX  see. . .Phenylmercury Acetate
PHOMASAN®  see. . .Quintozene
Phorat (German)  see. . .Phorate
PHORATE-10G®  see. . .Phorate
PHORIL®  see. . .Phorate
PHORTOX®  see. . .2,4,5-T
PHORTOX®  see. . .2,4,5-Trichlorophenoxyacetic Acid, Esters
PHOSALON®  see. . .Phosalone
PHOSALONE®  see. . .Phosalone
Phosazetim  see. . .Phosacetim
Phoschlor  see. . .Trichlorfon
PHOSCHLOR R50®  see. . .Trichlorfon
cis-3-Phosdrin  see. . .Mevinphos
Phosethoprop  see. . .Ethoprop
PHOSETHOPROP®  see. . .Ethoprop
Phoscthyl-Al  see. . .Fosetyl-Al
Phosethyl aluminum  see. . .Fosetyl-Al
Phosfene  see. . .Mevinphos
PHOS-FLUR®  see. . .Sodium Fluoride
PHOSKIL®  see. . .Parathion
PHOSKILL®  see. . .Monocrotophos
Phosphacol  see. . .Paraoxon
Phosphamide  see. . .Dimethoate
PHOSPHATE 10®  see. . .Endothion
Phosphate de $O,O$-diethyle et de $O$-2-chloro-1-(2,4-dichlorophenyl)vinyle (French)  see. . .Chlorfenvinphos
Phosphate de dimethyle et de(2-chloro-2-diethylcarbamoyl-1-methyl-vinyle) (French)  see. Phosphamidon
Phosphate de $O,O$-dimethyle et de $O$-(1,2-dibromo-2,2-dichlorethyle) (French)  see. . .Naled
Phosphate de dimethyle et de 2,2-dichlorovinyle (French)  see. . .Dichlorvos
Phosphate de dimethyle et de 2-dimethylcarbamoyl 1-methyl vinyle (French)  see. . .Dicrotophos
Phosphate de dimethyle et de 2-methylcarbamoyl 1-methyl vinyle (French)  see. . .Monocrotophos
Phosphene  see. . .Mevinphos
Phosphine gas  see. . .Phosphine
Phosphonic acid, (aminocarbonyl)-, monoethyl ester, monoammonium salt  see. . .Fosamine Ammonium
Phosphonic acid, (2-chloroethyl)-  see. . .Ethephon
Phosphonodithioic acid, ethyl-$O$-ethyl  see. . .$S$-phenyl ester  see. . .Fonofos
Phosphonic acid, monoethyl ester  see. . .aluminum salt (3:1)  see. . .Fosetyl-Al
Phosphonic acid, (2,2,2-trichloro-1-hydroxyethyl)-, dimethyl ester  see. . .Trichlorfon
Phosphonodithioimidocarbonic acid, acetimidoyl-, $O,O$-bis($p$-chlorophenyl) ester  see. . .Phosacetim
Phosphonodithioimidocarbonic acid, (1-iminoethyl)-$O,O$-bis($p$-chlorophenyl) ester  see. . .Phosacetim
$N$-(Phosphonomethyl)-glycine  see. . .Glyphosate

Phosphonomethyliminoacetic acid see...Glyphosate
Phosphonothioic acid, ethyl-, *O*-ethyl *O*-(2,4,5-trichlorophenyl) ester see...Trichloronate
Phosphonothioic acid, phenyl-,*O*-(4-bromo-2,5-dichlorophenyl) O-methyl ester see...Leptophos
PHOSPHOPYRON® see...Endothion
PHOSPHOPYRONE® see...Endothion
Phosphoramidothoic acid, *N*-acetyl-,*O,S*,-dimethyl ester see...Acephate
Phosphoramidic acid, 4-*tert*-butyl-2-chlorophenylphosphor amidate see...Crufomate
Phosphoramidic acid, isopropyl-, ethyl 4-(methylthio)-*m*-tolyl ethyl ester see...Fenamiphos
Phosphoramidic acid, methyl-,4-*tert*-butyl-2-chlorophenyl see...Crufomate
Phosphoramidic acid, methyl-,2-chloro-4-(1,1-dimethylethyl)phenyl methyl ester see...Crufomate
Phosphoramidic acid, (1-methylethyl)-, ethyl(3-methyl-4-(methylthio)phenyl)ester see...Fenamiphos
Phosphoramidic acid, (1-methylethyl)-, ethyl 3-methyl-4-(methylthio)phenyl ester see...Fenamiphos
Phosphoramidothioic acid, *N*-ethyl-, *(E)-O*-(2-isopropoxycarbonyl-1-methylvinyl) O-methyl ester see. Propetamphos
Phosphoramidothioic acid, isopropyl-, O-ethyl ester, O-ester with isopropyl salicylate see...Isofenphos
Phosphoramidothioic acid, isopropyl-, *O*-ethyl O-(2-isopropoxycarbonylphenyl) ester see...Isofenphos
Phosphorated hydrogen see...Phosphine
Phosphore blanc (French) see...Phosphorus
*o*-Phosphoric acid see...Phosphoric Acid
Phosphoric acid, 2-chloro-1-(2,4-dichlorophenyl)ethenyldiethyl ester see.Chlorfenvinphos
Phosphoric acid, 2-chloro-3-(diethylamino)-1-methyl-3-oxo-1-propenyl dimethyl ester see...Phosphamidon
Phosphoric acid, 2-chloro-1-(2,3,5-trichlorophenyl)ethenyl dimethyl ester see. Tetrachlorvinphos
Phosphoric acid, 2-chloro-1-(2,4,5-trichlorophenyl)ethenyl, dimethyl ester see. Tetrachlorvinphos
Phosphoric acid, 2-chloro-1-(2,4,5-trichlorophenyl)vinyl, dimethyl ester see...Tetrachlorvinphos
Phosphoric acid, 1,2-dibromo-2,2-dichloroethyl dimethyl ester see...Naled
Phosphoric acid, 2-dichloroethenyl dimethyl ester see. Dichlorvos
Phosphoric acid, 2,2-dichloroethenyl dimethyl ester see. Dichlorvos
Phosphoric acid, 2,2-dichlorovinyl dimethyl ester see. Dichlorvos
Phosphoric acid, *O,O*-diethyl O-6-methyl-2-(1-methylethyl)-4-pyrimidinyl ester see...Diazinon
Phosphoric acid, diethyl 4-nitrophenyl ester see.Paraoxon
Phosphoric acid, diethyl *p*-nitrophenyl ester see.Paraoxon
Phosphoric acid, 3-(dimethylamino)-1-methyl-3-oxo-1-propenyl dimethyl ester, *(E)-* see...Dicrotophos
Phosphoric acid, dimethyl ester with 2-chloro-*N,N*-diethyl-3-hydroxycrotonamide see...Phosphamidon
Phosphoric acid, dimethyl ester with *cis*-3-hydroxy-*N,N*-dimethylcrotonamide see...Dicrotophos
Phosphoric acid, dimethyl ester with *(E)*-3-hydroxy-*N,N*-dimethylcrotonamide see...Dicrotophos
Phosphoric acid, dimethyl ester with *cis*-3-hydroxy-*n*-methylcrotonamide see...Monocrotophos
Phosphoric acid, dimethyl ester with methyl 3-hydroxycrotonate see...Mevinphos
Phosphoric acid, (1-methoxycarboxypropen-2-yl) dimethyl ester see...Mevinphos
Phosphoric acid, tetraethyl ester see...TEPP
Phosphoric acid, trisodium salt see...Trisodium Phosphate
Phosphoroamidic acid, 1,3-dithiolan-2-ylidene-, diethyl ester see...Phosfolan
Phosphorodithioic acid-*O,O*-bis(1-methylethyl)-*S*-[2-((phenylsulfonyl)amino)ethyl]ester see...Bensulide
Phosphorothioic acid, *O*-(4-bromo-2-chlorophenyl)-*O*-ethyl-*S*-propyl ester see...Profenfos
Phosphorothioic acid, *O*-(4-bromo-2,5-dichlorophenyl) *O,O*-diethyl ester see...Bromophos-ethyl
Phosphorothioic acid, *O*-(4-bromo-2,5-dichlorophenyl) *O,O*-dimethyl ester see...Bromophos
Phosphorodithioic acid *S*-[(*tert*-butylthio)methyl]-*O,O*-diethylester see...Terbufos
Phosphorodithioic acid *S*-[2-chloro-1-(1,3-dihydro-1,3-dioxo-2*H*-isoindol-2-yl)ethyl]*O,O*-diethyl ester see. Dialifor
Phosphorodithioic acid, *S*-[(6-chloro-3-(mercaptomethyl)-2-benzoxazolinone] O,O-diethyl ester see...Phosalone
Phosphorodithioic acid, *S*-(chloromethyl) *O,O*-diethyl ester see...Chlormephos
Phosphorodithioic acid-*S*-(2-chloro-1-phthalimidoethyl)-*O,O*-diethyl ester see...Dialifor
Phosphorothioic acid, *O*-(4-cyanophenyl)-*O,O*-dimethyl ester see...Cyanofos
Phosphorodithioic acid, *O,O*-diethyl ester, *S,S*-diester with *p*-dioxane-2,3-dithiol see...Dioxathion
Phosphorodithioic acid, *O,O*-diethyl ester, *S,S*-diester with methanedithiol see...Ethion
Phosphorodithioic acid *O,O*-diethylesters-ester with *N*-isopropyl-2-mercaptoacetamide see...Prothoate
Phosphorodithioic acid, *O,O*-diethyl S-[2-((1-methylethyl)amino]-2-oxoethyl)]ester see...Prothoate
Phosphorodithioic acid, *S*-[(1,3-dihydro-1,3-dioxo-isoindol-2-yl)methyl] *O,O*-dimethyl ester see...Phosmet
Phosphorodithioic acid, *O,O*-dimethyl ester, *S*-ester with ethylmercaptophenylacetate see...Phenthoate
Phosphorodithioic acid, *O,O*-dimethyl ester, *S*-ester with *N*-(mercaptomethyl)phthalimide see...Phosmet
Phosphorodithioic acid *S*-[((1,1-dimethylethyl)thio)methyl]-*O,O*-diethyl ester see.Terbufos
Phosphorodithioic acid, *O,O*-dimethyl S-[2-(methylamino)-2-oxoethyl] ester see...Dimethoate

Phosphorodithioic acid, S,S'-1,4-dioxane-2,3-diyl-O,O,O',O'-tetraethyl ester see...Dioxathion
Phosphorodithioic acid-S,S'-1,4-dioxane-2,3-diyl, O,O,O',O'-tetraethyl ester see...Dioxathion
Phosphorodithioic acid-S,S'-para-dioxane-2,3-diyl, O,O,O',O'-tetraethyl ester see...Dioxathion
Phosphorodithioic acid, O-ethyl S,S-bis(1-methylpropyl) ester see...Cadusafos
Phosphorodithioic acid, O-ethyl S,S-dipropyl ester see. Ethoprop
Phosphorodithioic acid, O-ethyl O-(4-(methylthio)phenyl) S-propyl ester see...Sulprofos
Phosphorodithionic acid, O,O-diethyl S-2-(ethylthio)ethyl ester see...Disulfoton
Phosphorofluoridic acid, diisopropyl ester see. Isofluorphate
Phosphorothioate see...Endothion
Phosphorothioate, O,O-diethyl O-6-(2-isopropyl-4-methylpyrimidyl see...Diazinon
Phosphorothioic acid, O-(3-chloro-4-methyl-2-oxo-2H-1-benzopyran-7-yl) O,O-diethyl ester see...Coumaphos
Phosphorothioic acid, O-(4-cyanophenyl)-9,9-dimethyl ester see...Cyanofos
Phosphorothioic acid, S-[2-(diethylamino)ethyl] O,O-diethyl ester, ethanedioate (1:1) see...Amiton Oxalate
Phosphorothioic acid, S-[(2-diethylamino)ethyl] O,O-diethyl ester, oxalate (1:1) see...Amiton Oxalate
Phosphorothioic acid, O-[2-(diethylamino)-6-methyl-4-pyrimidinyl] O,O-diethyl ester see...Pirimiphos-Ethyl
Phosphorothioic acid, O-[2-(diethylamino)-6-methyl-4-pyrimidinyl] O,O-dimethyl ester see...Pirimiphos-Methyl
Phosphorothioic acid, O,O-diethyl ester, O-ester with 3-chloro-7-hydroxy-4-methylcoumarin see. Coumaphos
Phosphorothioic acid, O,O-diethyl O-(2-ethylthio)ethyl ester see...Demeton
Phosphorothioic acid, O,O-diethyl S-(2-ethylthio)ethyl ester see...Demeton
Phosphorothioic acid, O,O-diethyl O-2-(ethylthio)ethyl ester mixed with O,O-diethyl S-2-(ethylthio)ethyl phosphorothioate see...Demeton
Phosphorothioic acid, O,O-diethyl O-(isopropylmethylpyrimidyl) ester see...Diazinon
Phosphorothioic acid, O,O-diethyl O-(2-isopropyl-6-methyl-4-pyrimidinl) ester see...Diazinon
Phosphorothioic acid, O,O-diethyl O-(6-methyl-2-(1-methylethyl)-4-pyrimidinyl) ester see...Diazinon
Phosphorothioic acid, O,O-diethyl-O-(4-nitrophenyl) ester see...Parathion
Phosphorothioic acid, O,O-diethyl O-(p-nitrophenyl) ester see...Parathion
Phosphorothioic acid, O,O-diethyl O-(1-phenyl-1H-1,2,4-triazol-3-yl) ester see...Triazophos
Phosphorothioic acid-O,O-diethyl-O-2-pyrazinyl ester see...Thionazin
Phosphorothioic acid, O,O-diethyl O-(1,2,2,2-tetrachloroethyl)ester see...Chlorethoxyfos

Phosphorothioic acid, O,O-diethyl O-(3,5,6-trichloro-2-pyridinyl)ester see...Chlorpyrifos
Phosphorothioic acid, O-[4-((dimethylamino)sulfonyl)phenyl] O,O-dimethyl ester see...Famphur
Phosphorothioic acid, O,O'-dimethyl ester, O,O-diester with 4,4'-thiodiphenol see...Temephos
Phosphorothioic acid, O,O-dimethyl ester, O-ester with p-hydroxybenzonitrile see...Cyanofos
Phosphorothioic acid, O,O-dimethyl ester, O-ester with p-hydroxy-N,N-dimethylbenzenesulfonamide see..Famphur
Phosphorothioic acid, O,O-dimethyl ester, S-ester with 2-mercapto-N-methylacetamide see...Omethoate
Phosphorothioic acid, O,O-dimethyl S-[2-(ethylsulfinyl)ethyl] ester see...Demeton-methyl
Phosphorothioic acid, O,O-dimethyl S-[2-(methylamino)-2-oxoethyl]ester see...Omethoate
Phosphorothioic acid, O,O-dimethyl O-(4-nitrophenyl) ester see...Methyl Parathion
Phosphorothioic acid, O,O-dimethyl O-(p-nitrophenyl) ester see...Methyl Parathion
Phosphorothioic acid, O,O-dimethyl O-(4-nitro-m-tolyl)ester see...Fenitrothion
Phosphorothioic acid, O,O-dimethyl O-(3-methyl-4-nitrophenyl)ester see...Fenitrothion
Phosphorothioic acid, O,O-dimethyl O-(2,4,5-trichlorophenyl) ester see...Ronnel
Phosphorothioic acid, O,O-dimethyl O-(3,5,6-trichloro-2-pyridyl)ester see...Chlorpyrifos-methyl
Phosphorothioic acid, O-ethyl O-(4-(methylthio)phenyl) S-propyl ester see...Sulprofos
Phosphorothioic acid, S-[2-(ethylsulfinyl)ethyl] O,O-dimethyl ester see...Demeton-methyl
Phosphorothioic acid, O-(2-(ethylthio)ethyl) O,O-dimethyl ester see...Demeton-methyl
Phosphorothioic acid, O-2-(ethylthio)ethyl O,O-dimethyl ester mixed with S-2-(ethylthio)ethyl O,O-dimethyl phosphorothioate see...Demeton-methyl
Phosphorothioic acid, O,O'-(thiodi-4,1-phenylene)O,O,O',O'-tetramethyl ester see...Temephos
Phosphorothioic acid, O,O'-(thiodi-p-phenylene)O,O,O',O'-tetramethyl ester see...Temephos
Phosphorotrithioic acid, S,S,S-tributyl ester see..Tribufos
Phosphorous hydride see...Phosphine
Phosphorous trihydride see...Phosphine
Phosphorous yellow see...Phosphorus
Phosphorsaeureloesungen (German) see...Phosphoric Acid
Phosphorus elemental white see...Phosphorus
Phosphorus-31 see...Phosphorus
Phosphorwasserstoff (German) see...Phosphine
Phosphostigmine see...Parathion
Phosphothion see...Malathion
Phosphure de calcium (French) see...Calcium Phosphide
Phosphure de zinc (French) see...Zinc Phosphide
Phosphures d'aluminum (French) see...Aluminum Phosphide
Phostoxin see...Phosphine

PHOSTOXIN® see...Aluminum Phosphide
PHOSVEL® see...Leptophos
Phosvin see...Zinc Phosphide
PHOSVIT® see...Dichlorvos
PHOTON® see...Fentin Hydroxide
PHOTON FUNGICIDE® see...Triphenyltin Compounds
Photophor see...Calcium Phosphide
Phoxime see...Phoxim
PHOZALON® see...Phosalone
PHPH see...Biphenyl
Phthalamic acid, N-1-naphthyl- see...Naptalam
Phthalimide, N-(mercaptomethyl)-, S-ester with O,O-dimethylphosphorodithioate see...Phosmet
Phthalimido O,O-dimethyl phosphorodithioate see. Phosmet
Phthalimidomethyl O,O-dimethyl phosphorodithioate see...Phosmet
Phthalophos (USSR) see...Phosmet
d-Phthalthrin see...Tetramethrin
(±)-cis/trans-Phthalthrin see...Tetramethrin
PHTHALTHRIN® see...Tetramethrin
PHYBAN® see...Sodium Methanearsonate (MSMA)
PHYBAN H.C.® see...Sodium Methanearsonate (MSMA)
PHYGON® see...Dichlone
PHYGON® PASTE see...Dichlone
PHYGON® SEED PROTECTANT see...Dichlone
PHYGON® XL see...Dichlone
PHYMONE® see...1-Naphthaleneacetic Acid
PHYOMONE® see...Naphthoxyacetic Acid
Physostol see...Physostigmine
PHYTAR® see...Cacodylic Acid
PHYTAR 138® see...Cacodylic Acid
PHYTAR 560® (with Cacodylic acid) see...Sodium Cacodylate
PHYTAR 560® (with Sodium cacodylate) see..Cacodylic Acid
PHYTAR 600® see...Cacodylic Acid
Phyto-bordeaux see...Copper Sulfate
PHYTOMYCIN® see...Streptomycin Sulfate
PHYTOSOL® see...Trichloronate
PIC-CHLOR® 16 see...Chloropicrin
PICCOLO® see...Paclobutrazole
PICFUME® see...Chloropicrin
Picolinic acid, 4-amino-3,5,6-trichloro- see...Picloram
Picolinic acid, 3,6-dichloro- see...Clopyralid
Picrato amonico (Spanish) see...Ammonium Hexafluorosilicate
PICRIDE® see...Chloropicrin
PICTYL® see...Fenoxycarb
PID® see...Diphacione
PIED PIPER MOUSE SEED® see...Brucine
PIED PIPER MOUSE SEED® see...Strychnine
PIELIK® see...2,4-D
PILARTHENE® see...Acephate
PILLARDIN® see...Monocrotophos
PILLARFURAN® see...Carbofuran
PILLARGON® see...Propoxur

PILLARICH® see...Chlorothalonil
PILLARMATE® see...Methomyl
PILLARON® see...Methamidophos
PILLARQUAT® see...Paraquat
PILLARQUAT® see...Paraquat Methosulfate
PILLARSET® see...Butachlor
PILLARSTIN® see...Carbendazim
PILLARTAN® see...Captafol
PILLARXONE® see...Paraquat
PILLARZO® see...Alachlor
PILOT® see...Chlorpyrifos
PILOT® see...Quizalofop-Ethyl
PIMACOL-SOL® see...1-Naphthaleneacetic Acid
PIN® see...EPN
PINNACLE® see...Thifensulfuron Methyl
PINPOINT® see...Acephate
2,6-Piperidinedione, 4-(2-3,5-dimethyl-2-oxocyclohexyl)-2-hydroxyethyl-, (IS)-[1α(S*),3α,5β]- see.Cycloheximide
Piperonyl butoxyde see...Piperonyl Butoxide
Piperonyl sulfoxide see...Sulfoxide
PIRATE® see...Chlorfenapyr
Piretrina (Spanish) see...Pyrethrins or Pyrethrum
PIRIMIKARB see...Pirimicarb
PIRIMOR® see...Pirimicarb
Pirofos see...Sulfotepp
PITEZIN® see...Atrazine
PIVAL® see...Pindone
Pivaldione (French) see...Pindone
Pivalyl see...Pindone
2-Pivalyl-1,3-indandione see...Pindone
PIVALYL VALONE® see...Pindone
PIVOT® see...Imazethapyr
PIX® see...Mepiquat Chloride
PKhFN see...Sodium Pentachlorophenate
PKhNB® see...Quintozene
PLANOFIX® see...1-Naphthaleneacetic Acid
PLANOTOX® see...2,4-D
PLANT DITHIO AEROSOL® see...Sulfotepp
PLANTDRIN® see...Monocrotophos
PLANTFUME 103 SMOKE GENERATOR® see.Sulfotepp
PLANTGARD® see...2,4-D
PLANTIFOG 160M® see...Maneb
PLANTOMYCIN® see...Streptomycin
PLANTOMYCIN® see...Streptomycin Sulfate
PLANTONIT® see...Terbutryn
PLANT PROTECTION PP511® see...Pirimiphos-Methyl
PLANTULIN® see...Propazine
PLANTVAX® see...Oxycarboxin
PLANT WAX® see...Oxycarboxin
PLATH-LYSE® see...Dichlorophene
PLEDGE® see...Bentazon
PLENUM® see...Pymetrozine
PLEOPARAPHENE® see...Parathion
PLICTRAN® see...Cyhexatin
PLUCKER® see...1-Naphthaleneacetic Acid
Plumbous arsenate see...Lead Arsenate

PLUSBAIT® see...Warfarin
PLYDOX® see...Terbufos
PMA see...Phenylmercury Acetate
PMAC see...Phenylmercury Acetate
Pmacetate see...Phenylmercury Acetate
PMAL see...Phenylmercury Acetate
PMAS see...Phenylmercury Acetate
PMC® see...Phosmet
PMM see...Perchloromethyl Mercaptan
PMP (Japan) see...Phosmet
PMP-787® see...Pyriminil
POAST® see...Sethoxydim
PO-Dimethoate see...Omethoate
POE isooctadecanol see...Arosurf® MSF
POINT TWO® see...Sodium Fluoride
POLACARITOX® see...Tetradifon
Polfoschlor see...Trichlorfon
POLICAR® see...Mancozeb
POLISIN® see...Prometryn
POLIVAL® see...Thiabendazole
POL NU® see...Pentachlorophenol
POLY B RAND DESICCANT® see...Arsenic Acid
Poly(oxy-1,2-ethanediyl), α-isooctadecyl-ω-hydroxy- see...Arosurf® MSF
POLYBOR® see...Sodium Tetraborate
Polychlorcamphene see...Toxaphene
Polychlorinated camphene see...Toxaphene
Polychlorocamphene see...Toxaphene
POLYCLENE® see...Dichlorprop
POLYCRON® see...Profenfos
POLYGRAM® Z see...Zineb
POLYMONE® see...Dichlorprop
Polyoxymethylene glycols see...Formaldehyde
POLYPHASE® see...Carbendazim
POLYRAM M® see...Maneb
POLYRAM ULTRA® see...Thiram
POLYRAM® Z see...Zineb
POLYTANOL® see...Calcium Phosphide
POLYTOX® see...Dichlorprop
POLYTRIN® see...Cypermethrin
POMARSOL® see...Thiram
POMARSOL FORTE® see...Thiram
POMARSOL® Z FORTE see...Ziram
POMASOL® see...Thiram
POMEX® see...Carbaryl
PONDMASTER® see...Glyphosate
PONNAX® see...Mepiquat Chloride
POPCORN® see...Sulfur
Portland stone see...Calcium Carbonate
PO-SAN® see...Maleic Hydrazide
POSMIL® see...Atrazine
POSSE® see...Carbosulfan
POST-KITE® see...Mecoprop
PO-SYSTOX® see...Demeton
Potassium acid arsenate see...Potassium Arsenate
Potassium antimonyl tartrate see...Antimony Potassium Tartrate

Potassium antimonyl-d-tartrate see...Antimony Potassium Tartrate
Potassium arsenate, monobasic see...Potassium Arsenate
Potassium dihydrogen arsenate see...Potassium Arsenate
Potassium fluorure (French) see...Potassium Fluoride
Potassium gibberellate see...Gibberellic Acid
Potassium hydrogen arsenate see...Potassium Arsenate
POTATO SEED PIECE PROTECTANT® see...Captan
Potentiated acid glutaraldehyde see...Glutaraldehyde
POWDER AND ROOT® see...Rotenone
Powder green see...Paris Green
POWER CHLOROTHALONIL® 50 see..Chlorothalonil
POWERDRIVE® see...Vinclozolin
POWERTWIN® see...Phenmedipham
POWERTWIN® see...Ethofumesate
POWER-X® see...Acephate
PP 009® see...Fluazifop-butyl
PP 021® see...Fomesafen
PP 062® see...Pirimicarb
PP 100® see...Diquat Dibromide
PP 145® see...Chlorfluazuron
PP 148® see...Paraquat
PP 192® see...Fluazinam®
PP 211® see...Pirimiphos-Ethyl
PP 321® see...*lamda*-Cyhalothrin
PP 333® see...Paclobutrazole
PP 383® see...Cypermethrin
PP 511® see...Pirimiphos-Methyl
PP 523® see...Hexaconazole
PP 604® see...Tralkoxydim
PP 910® see...Paraquat Methosulfate
PP 993® see...Tefluthrin
PPG-844® see...Lactofen
PPTC see...Vernolate
PRACTIS® see...Propiconazole
PRAMITOL® see...Sodium Metaborate
PRC-1237® see...4-Aminopyridine
PREBANE® see...Terbutryn
Precipitated amorphous silica see...Diatomaceous Earth
PRECISE ACEPHATE® see...Acephate
PRECISION® see...Fenoxycarb
PRECOR® see...Methoprene
PREEGLONE® see...Diquat
PREEGLONE® see...Diquat Dibromide
PRE-EMPT® see...Linuron
PREFAR® see...Bensulide
PREFAR-E® see...Bensulide
PREFIX D® see...Dichlobenil
PREFRIN-Z® see...Zinc Sulfate Heptahydrate
PREGARD® see...Profluralin
PRELUDE® see...Metolachlor
PRELUDE® see...Paraquat
PREMALIN® see...Linuron
PREMALOX® see...Propham
PREMAZINE® see...Simazine
PREMERGE® see...Dinoseb
PREMERLIN 600 CE® see...Trifluralin
PREMIER® see...Imidacloprid

PREMISE® see...Imidacloprid
PRENTOX® see...Aldrin
PRENTOX® see...Dichlorvos
PRENTOX® see...Dieldrin
PRENTOX® see...Malathion
PRENTOX® see...Methoxychlor
PRENTOX® see...Pyrethrins or Pyrethrum
PRENTOX® see...Rotenone
PRENTOX CARBAMATE® see...Propoxur
PREP® see...Ethephon
PREPARATION 125® see...Nitrofen
PRE-SAN® see...Bensulide
PRESCRIBE™ see...Imidacloprid
PRESPERSION, 75 UREA® see...Urea
PREVAIL® see...Cypermethrin
PREVAIL® see...Metalaxyl
PREVENOL® see...Bithionol
PREVENOL 1® see...Trichlorophenols
PREVENTAL® see...Dichlorophene
PREVENTOL® see...p-Chloro-m-Cresol
PREVENTOL® see...Dichlorophene
PREVENTOL® see...Tebuconazole
PREVENTOL GD® see...Dichlorophene
PREVENTOL GDC® see...Dichlorophene
PREVENTOL-O Extra® see...o-Phenylphenol
PREVENTOL P® see...Pentachlorophenol
PREVEX® see...Propamocarb Hydrochloride
PREVICUR® see...Propamocarb Hydrochloride
PREVIEW® see...Chlorimuron-ethyl
PREVIEW® see...Metribuzin
PREZERVIT® see...Dazomet
PRIDE® see...Fluridone
PRILTOX® see...Pentachlorophenol
PRIMACOL® see...1-Naphthaleneacetic Acid
PRIMAGRAM® see...Metolachlor
PRIMAPIN® see...Prometryn
PRIMATEL S® see...Simazine
PRIMATOL® see...Amitrole
PRIMATOL® see...Atrazine
PRIMATOL® see...Prometon
PRIMATOL P® see...Propazine
PRIMATOL-Q® see...Prometryn
PRIMATOL S® see...Simazine
PRIMATOP® see...Atrazine
PRIMAZE® see...Atrazine
PRIMAZE® see...Prometryn
PRIMEXTRA® see...Metolachlor
PRIMICID® see...Pirimiphos-Ethyl
Primifosethyl see...Pirimiphos-Ethyl
PRIMIN® see...Isolan®
PRIMISIL® see...Diatomaceous Earth
PRIMOLE® see...Atrazine
PRIMOTEC® see...Pirimiphos-Ethyl
PRIMO® see...Cimectacarb
PRIMO® WSB see...Cimectacarb
PRINCEP® see...Simazine
PRINCEP® 80W see...Simazine
PRINICID® see...Pirimiphos-Ethyl

PRIODERM® see...Malathion
PRISM® see...Clethodim
PRO 330® CLEAR THIN SPREAD see...Ammonia
PROBE® see...Methazole
PROCLAIM® see...Emamectin Benzoate
Proconazole see...Propiconazole
PROCURE® see...Triflumizole
PROCUTENE® see...Triclocarban
PRODALUMNOL DOUBLE® see...Sodium Arsenite
PRODARAM® see...Ziram
PRODIGY® see...Methoxyfenozide
PROFALON® see...Linuron
PROFAM® see...Propham
PROFILE® see...Paclobutrazole
PROFOS® see...Propham
PROFUME A® see...Chloropicrin
PRO-GIBB® see...Gibberellic Acid
PROGRESS® see...Ethofumesate
PROGRESS® see...Phenmedipham
PROGRESS® see...Desmedipham
PRO-GRO® see...Thiram
PRO GROW® see...Oxadiazon
PROKARBOL® see...Dinitro-o-cresol (DNOC)
PROKIL® see...Sodium Aluminum fluoride
PROKIL® see...Dinoseb
PROKIL® see...Malathion
PROKIL® see...Naled
PROKIL® see...Prometryn
PROKIL® see...Terbutryn
PROKIL® see...Trifluralin
PROKIL® see...Ziram
PROKIL AMETRYNE 80W® see...Ametryn
PROKIL ATRAZINE 80W® see...Atrazine
PROKIL CRYOLITE® see...Sodium Aluminum Fluoride
PRO-KILL NEMATOCIDE® see...1,3-Dichloropropene
PROLATE® see...Phosmet
PROMALIN® see...6-Benzaldenine
Promalin, component of (with Gibberellin D) see...6-Benzaldenine
PROMAR® see...Diphacione
PRO-MATE BARRICADE® see...Prodiamine
PROMET® see...Prometryn
PROMETONE® see...Prometon
Prometrene see...Prometryn
PROMETREX® see...Prometryn
Prometrin see...Prometryn
Prometryne (USDA) see...Prometryn
PROMIDIONE® see...Iprodione
PRONONE® see...Hexazinone
PRO-NOX FISH® see...Rotenone
PROPACET® see...Quinclorac
Propachlore see...Propachlor
Propacloro (Spanish) see...Propachlor
PRO-PACK® see...Bensulfuron-methyl
PROPAL® see...Mecoprop
Propanal, 2-methyl-2-(methylsulfonyl)-, O-[(methylamino)carbonyl]oxime see...Aldoxycarb

Propanal,2-methyl-2-(methythio)-,O-[(methylamino)carbonyl]oxime see...Aldicarb
Propanamide, N-(3,4-Dichlorophenyl)- see...Propanil
Propane, 1-chloro-1-nitro- see...1-Chloro-1-Nitropropane
Propane, 1,2-dibromo-3-chloro- see. Dibromochloropropane
Propane, dibromochloropropane see. Dibromochloropropane
Propane, 1,2-dichloro- see...1,2-Dichloropropane
Propanenitrile, 2-[(4-Chloro-6-(ethylamino)-1,3,5-triazin-2-yl)amino]-2-methyl- see...Cyanazine
Propanenitrile, 2-[(4-Chloro-6-(ethylamino)-s-triazin-2-yl)amino]-2-methyl- see...Cyanazine
PROPANEX® see...Propanil
Propanide see...Propanil
PROPANIL MILENIA® see...Propanil
Propanoic acid, 2-(4-chloro-2-methylphenoxy)- see. Mecoprop
Propanoic acid, 2-[4-((6-chloro-2-quinoxalinyl)oxy)phenoxy]-, ethyl ester see. .Quizalofop-Ethyl
Propanoic acid, 2-[4-((3-chloro-5-(trifluoromethyl)-2-pyridinyl)oxy)phenoxy]-, methyl ester see...Haloxyfop-methyl
Propanoic acid, 2,2-dichloro- see...Dalapon
Propanoic acid, 2-(2,4-dichlorophenoxy)- see. Dichlorprop
Propanoic acid, 2-[4-(2,4-dichlorophenoxy)phenoxy]-, methyl ester see...Diclofop-methyl
Propanoic acid, 2-[4-((5-(trifluoromethyl)-2-pyridinyl)oxy)phenoxy]-,butyl ester see...Fluazifop-butyl
Propanol, oxybis- see...Isofenphos
Propargil see...Propargite
Propargita (Spanish) see...Propargite
Propasin see...Propazine
PROPAZIN® see...Propazine
Propenal see...Acrolein
2-Propenal see...Acrolein
Prop-2-en-1-al see...Acrolein
2-Propenal, 3-phenyl-, (E)- see...Cinnamaldehyde
Propenamide see...Acrylamide
2-Propenamide propenamide see...Acrylamide
1-Propene, 3-bromo- see...Allyl Bromide
Propene, 1,3-dichloro- see...1,3-Dichloropropene
1-Propene, 1,3-dichloro- see...1,3-Dichloropropene
2-Propene-1,1-dioldiacetate see...Acrolein diacetate
1-Propene, 3-isothiocyanato- see...Allyl Isothiocyanate
Propenenitrile see...Acrylonitrile
2-Propenenitrile see...Acrylonitrile
Propenol see...Allyl Alcohol
2-Propenol see...Allyl Alcohol
2-Propen-1-ol see...Allyl Alcohol
Propen-1-ol-3 see...Allyl Alcohol
1-Propen-3-ol see...Allyl Alcohol
2-Propen-1-one see...Acrolein

2-Propene-1-thiol, 2,3-dichloro-,diisopropylcarbamate see...Diallate
2-Propene-1-thiol, 2,3,3-trichloro-, diisopropylcarbamate see...Triallate
Propenyl Alcohol see...Allyl Alcohol
2-Propenyl Alcohol see...Allyl Alcohol
2-Propenyl isothiocyanate see...Allyl Isothiocyanate
PROPIMAX® see...Propiconazole
PROPINEBE® see...Propineb
PROPINEX® see...Propazine
Propionaldehyde, 2-methyl-2-(methylthio)-,O-(methylcarbamoyl)oxime see...Aldicarb
Propionanilide, 3',4'-Dichloro- see...Propanil
Propionic acid, 2-[4-((6-chloro-2-benzoxazolyl)oxy)phenoxy],ethylester, (±)- see. Fenoxaprop-ethyl
Propionic acid, 2-(4-chloro-2-methylphenoxy) see. Mecoprop
Propionic acid, 2-[(4-chloro-o-tolyl)oxy]- see. .Mecoprop
Propionic acid, 3,4-dichloroanilide see...Propanil
Propionic acid, 2-(2,4-dichlorophenoxy)- see. Dichlorprop
Propionic acid, 2-(2-methyl-4-chlorophenoxy)- see. Mecoprop
Propionic acid, 2-[p-((5-(trifluoromethyl)-2-pyridyl)oxy)phenoxy]-, butylester see...Fluazifop-butyl
PROP-JOB® see...Propanil
PROPOGON® see...Propoxur
PROPON® see...Silvex
PROPONEX-PLUS® see...Mecoprop
PRO-PORTION® see...Sodium Fluoride
PROPOTOX® see...Propoxur
PROPOXYLOR® see...Propoxur
PROPROP® see...Dalapon
S-Propyl-N-aethyl-N-butyl-thiocarbamat (German) see. Pebulate
S-Propyl butylethylthiocarbamate see...Pebulate
S-Propyl-N-butyl-N-ethylthiolcarbamate see...Pebulate
N-Propyl-N-(2-chloroethyl)-2,6-dinitro-4-trifluoromethylaniline see...Fluchloralin
N-Propyl-N-(2-chloroethyl)-α,α,α-trifluoro-2,6-dinitro-p-toluidine see...Fluchloralin
Propyl [3-(dimethylamino)propyl]carbamate monohydrochloride see...Propamocarb Hydrochloride
S-Propyl dipropylthiocarbamate see...Vernolate
S-Propyl dipropyl (thiocarbamate) see...Vernolate
N-Propyl-di-N-propylthiolcarbamate see...Vernolate
Propyl-N,N-dipropylthiolcarbamate see...Vernolate
Propylene aldehyde see...Acrolein
Propylenebis(dithiocarbamato)zinc see...Propineb
Propylene chloride see...1,2-Dichloropropane
Propylene dichloride see...1,2-Dichloropropane
α,β-Propylene dichloride see...1,2-Dichloropropane
Propylethylbutylthiocarbamate see...Pebulate
Propyl-ethylbutylthiocarbamate see...Pebulate
Propyl-ethyl-N-butylthiocarbamate see...Pebulate
Propyl-N-ethyl-N-butylthiocarbamate see...Pebulate
N-Propyl-N-ethyl-N-(N-butyl)thiocarbamate see.Pebulate

S-(N-Propyl)-N-ethyl-N-N-butyl)thiocarbamate see. Pebulate
6-(Propylpiperonyl)-butyl carbityl ether see...Piperonyl Butoxide
6-Propylpiperonyl butyl diethylene glycol ether see. Piperonyl Butoxide
Propyl thiopyrophosphate see...Aspon®
5-Propyl-4-(2,5,8-trioxa-dodecyl)-1,3-benzodioxol (German) see...Piperonyl Butoxide
2-Propynyl (E,E)-3,7,11-trimethyl-2,4-dodecadienoate see...Kinoprene
Prop-2-ynyl-3,7,11-trimethyl-2,4-dodecadienoate see. Kinoprene
PROPYON® see...Propoxur
Propyzamide see...Pronamide
PROSEED® see...Hexaconazole
PROSEVOR® 85 see...Carbaryl
PROSPECT® see...Thifensulfuron Methyl
PROSTAR-50 WP® see...Flutolanil
PROTARS® see...Calcium Arsenate
PRO-TECK® see...Oryzalin
PROTEGE® see...Azaxystrobin
PROTEGE ALLEGIANCE BAYTAN® see.Triadimenol
PROTEGE-ALLEGIANCE® WP see...Azaxystrobin
PROTEGE-FL SEED® APPLIED FUNGICIDE see. Azaxystrobin
PRO-TEX® see...Fentin Hydroxide
PRO-TEK® see...Triadimefon
PRO-TEX® see...Triphenyltin Compounds
Prothromadin see...Warfarin
Prothrombin see...Warfarin
PROTREAT® see...Imidacloprid
PROTUGAN® see...Isoproturon
PROTURF® see...Bensulide
PROTURF® see...Iprodione
PRO-TURF® see...Thiophanate-Methyl
PROVADA® see...Methiocarb
PROVADO® see...Imidacloprid
PROWL® see...Pendimethalin
PROXITANE® see...Peracetic Acid
PROXOL® see...Trichlorfon
PROZINEX® see...Propazine
Prussiate of soda see...Sodium Cyanide
Prussic acid see...Hydrogen Cyanide
PRYFON 6® see...Isofenphos
PS® see...Chloropicrin
Pseudourea see...Urea
PSL® see...Leptophos
m-Psopropylphenyl methylcarbamate see...Phenol, 3-(1-methylethyl)-, methylcarbamate
PS-SYSTOX® see...Demeton
PT-515® see...D-Phenothrin
PTC see...Phenylthiourea
PTU, U 6324 see...Phenylthiourea
PTZ® see...Phenothiazine
PUMA® see...Fenoxaprop-ethyl
PUNKASO® see...Ethofenprox
PURALIN® see...Thiram

PURASAN-SC-10® see...Phenylmercury Acetate
PURATURF 10® see...Phenylmercury Acetate
PUREGRO® see...Oxythioquinox
1-H-Purin-6-amine, N-(2-furanylmethyl)- see...Kinetin (Cytokinin)
1H-Purin-6-amine, N-(phenylmethyl)- see...6-Benzaldenine
PURIVEL® see...Metoxuron
PURSUIT® see...Dimethenamid
PURSUIT,® (ammonium salt of) see...Imazethapyr
PURSUIT DG® Herbicide see...Imazethapyr
PYDRIN® see...Fenvalerate
PYLON® see...Chlorfenapyr
PYNAMIN® see...Allethrin
PYRAMIN® see...Pyrazon
PYRAMINE® see...Pyrazon
PYNAMIN-FORTE® see...Allethrin
PYNOSECT® see...Resmethrin
PYRAMITE® see...Pyridaben
Pyrazinol-O-ester with O,O-diethyl phosphorothioate pyrazinol O ester see...Thionazin
1H-Pyrazole-3-carbonitrile, 5-amino-1-(2,6-dichloro-4-(trifluoromethyl)phenyl)-4-[(trifluoromethyl)sulfinyl]- see...Fipronil
1H-Pyrazole-4-carboxylic acid see...Halosulfuron-methyl
PYRAZOLIUM® see...Difenzoquat
1H-Pyrazolium, 1,2-dimethyl-3,5-diphenyl-, methyl sulfate see...Difenzoquat
PYRAZON® see...Pyrazon
PYRAZONE® see...Pyrazon
PYRAZONL® see...Pyrazon
PYRESIN® see...Allethrin
PYRESOTE® see...Sodium Dichromate
PYRESYN® see...Allethrin
PYRETHERM® see...Resmethrin
Pyrethrin I see...Pyrethrins or Pyrethrum
Pyrethrin II see...Pyrethrins or Pyrethrum
Pyrethrum flowers see...Pyrethrins or Pyrethrum
PYRETHRUM INSECTICIDE® see...Pyrethrins or Pyrethrum
PYRETRINER® see...Pyrethrins or Pyrethrum
PYREXCEL® see...Allethrin
3(2H)-Pyridazinone, 4-chloro-2-(1,1-dimethylethyl)-5-[((4-(1,1-dimethylethyl)phenyl)methyl)thio]- see..Pyridaben
3(2H)-Pyridazinone, 4-chloro-5-(methylamino)-2-[3-(trifluoromethyl)phenyl]- see...Norflurazon
3(2H)-Pyridazinone, 4-chloro-5-(methylamino)-2-(α,α,α-trifluoro-m-tolyl)- see...Norflurazon
Pyridimine phosphate see...Pirimiphos-Methyl
4-Pyridinamine see...4-Aminopyridine
Pyridinamine, 3-chloro-N-[3-chloro-2,6-dinitro-4-(trifluoromethyl)phenyl]-5-(trifluoromethyl)- see...Fluazinam®
Pyridin-2,5-dicarbonsaeure-di-N-propylester (German) see...Dipropyl Isocinchomeronate
Pyridine, 4-amino- see...4-Aminopyridine

3-Pyridinecarboxamide, *N*-(2,4-difluorophenyl)-2-[3-(trifluoromethyl)phenoxy]- see...Diflufenican
3-Pyridinecarboxamide, 2-[((((4,6-dimethoxy-2-pyrimidinyl)amino)carbonyl)amino)sulfonyl]-*N,N*-dimethyl- see...Nicosulfuron
2-Pyridine carboxylic acid, 4-amino-3,5,6-trichloro- see. Picloram
2-Pyridinecarboxylic acid, 3,6-dichloro- see. .Clopyralid
3-Pyridinecarboxylic acid, 2-(difluoromethyl)-5-(4,5-dihydro-2-thiazolyl)-4-(2-methylpropyl)-6-(trifluoromethyl)-, methyl ester see...Thiazopry
Pyridine, 2-chloro-6-(trichloromethyl)- see...Nitrapyrin
3,5-Pyridinedicarbothioic acid, 2-(difluoromethyl)-4-(2-methylpropyl)-6-(trifluoromethyl)-, *S,S*-dimethyl ester see...Dithiopyr
2,5-Pyridinedicarboxylic acid, dipropyl ester see. Dipropyl Isocinchomeronate
Pyridine, 2-[1-methyl-2-(4-phenoxyphenoxy)ethoxy]- see...Pyriproxyfen
Pyridine, 3-(1-methyl-2-pyrrolidinyl)- see...Nicotine
Pyridine, (s)-3-(1-methyl-2-pyrrolidinyl)-, and salts see. Nicotine
Pyridine, 3-(1-methyl-2-pyrrolidinyl)-, (*S*)-, sulfate (2:1) see...Nicotine Sulfate
Pyridine, 3-(tetrahydro-1-methylpyrrol-2-yl) see.Nicotine
2-Pyridinol, 3,5,6-trichloro-,*O*-ester with *O,O*-diethylphosphorothioate see...Chlorpyrifos
4(1*H*)-Pyridinone, 1-methyl-3-phenyl-5-[3-(trifluoromethyl)phenyl]- see...Fluridone
4-Pyridylamine see...4-Aminopyridine
Pyridylmethyl-*N'*-*para*-nitrophenyl urea see...Pyriminil
*N*-3-Pyridylmethyl-*N'*-*p*-nitrophenylurea see...Pyriminil
1-(3-Pyridylmethyl)-3-(4-nitrophenyl)urea see. .Pyriminil
β-Pyridyl-α-*N*-methylpyrrolidine see...Nicotine
2-Pyrimidinamine, 4-cyclopropyl-6-methyl-*N*-phenyl- see...Cyprodinil
Pyrimidine, 2-chloro-4-(dimethylamino)-6-methyl- see. Crimidine
2,4(1*H*,3*H*)-Pyrimidinedione see...Bromacil
2,4(1*H*,3*H*)-Pyrimidinedione, 5-chloro-3-(1,1-dimethylethy)-6-methyl- see...Terbacil
5-Pyrimidinemethanol, α-(2-chlorophenyl)-α-(4-chlorophenyl)- see...Fenarimol
5-Pyrimidinemethanol, α-cyclopropyl-α-(4-methoxyphenyl) see...Ancymidol
5-Pyrimidinemethanol, α-(1-methylethyl)-α-[4-(trifluoromethoxy)phenyl]- see...Flurprimidol
4-Pyrimidinol, 2-(diethylamino)-6-methyl-, *O*-ester with *O,O*-dimethyl phosphorothioate see...Pirimiphos-Methyl
4-Pyrimidinol, 2-isopropyl-6-methyl-, *O*-ester with *O,O*-diethylphosphorothioate see...Diazinon
Pyrimidinone see...Hydramethylnon
2(1*H*)-Pyrimidinone, tetrahydro-5,5-dimethyl-, [3-(4-(trifluoromethyl)phenyl]-1-[2-(4-(trifluoromethyl)phenyl)ethenyl]-2-propenylidene]hydrazone see...Hydramethylnon
Pyrimifos see...Pirimiphos-Methyl
PYRIMOR® see...Pirimicarb

PYRINEX® see...Chlorpyrifos
PYRINURON® see...Pyriminil
PYROCIDE® see...Allethrin
Pyrophosphate de tetraethyle (French) see...TEPP
Pyrophosphoric acid, octamethylteraamide see. Octamethyl Diphosphoramide
Pyrophosphorodithioic acid, tetraethyl ester see. Sulfotepp
Pyrophosphorodithioic acid,*O,O,O,O*-tetraethyl ester see. Sulfotepp
Pyrophosphorytetrakisdimethylamide see...Octamethyl Diphosphoramide
1*H*-Pyrrole-3-carbonitrile, 4-bromo-2-(4-chlorophenyl)-1-(ethoxymethyl)-5-(trifluoromethyl)- see...Chlorfenapyr
Pyrrolidine, 1-methyl-2-(3-pyridyl)-, sulfate see. Nicotine Sulfate
PYTHON® see...Flumetsulam
PYTHON® see...Metribuzin
PZEIDAN® see...DDT

- Q -

Q-137® see...Ethylan
QUADMEC® see...Sodium Methanearsonate (MSMA)
QUADRIS OPTI® see...Azaxystrobin
QUARTZ® see...Diflufenican
Quecksilber chlorid (German) see...Mercuric Chloride
QUEL® see...Ancymidol
QUELLADA® see...Lindane
QUESTURAN® see...Dodine
QUICK® see...Chlorophacinone
QUICKPHOS® see...Aluminum Phosphide
QUICKSAN® see...Phenylmercury Acetate
QUICK TOX® see...Aluminum Phosphide
QUILAN® see...Benefin
QUILT® see...Azaxystrobin
QUINALTAF® see...Quinalphos
Quinclorac tech see...Quinclorac
Quinofop-ethyl see...Quizalofop-Ethyl
8-Quinolinecarboxylic acid(8-), 3,7-dichloro- see.Quinclorac
Quinomethionate see...Oxythioquinox
QUINONDO® see...Copper (II)-8-hydroxyquinoline
QUINOPHOS® see...Methyl Parathion
Quinosan see...Quintozene
2,3-Quinoxalinedithiol, 6-methyl-, cyclic carbonate see. Oxythioquinox
2,3-Quinoxalinedithiol,6-methyl-, cyclic dithiocarbonate (ester) see...Oxythioquinox
QUINTAR® see...Dichlone
QUINTAR® 540F see...Dichlone
Quintocene see...Quintozene
Quintoceno (Spanish) see...Quintozene
QUINTOX® see...Cholecalciferol
QUINTOX® see...Dieldrin

## - R -

R 8® see...Methylmercuric Dicyanamide
R 8 FUNGICIDE® see...Methylmercuric Dicyanamide
R 10 see...Carbon Tetrachloride
R 40B1® see...Methyl Bromide
R 50® see...DDT
R 326® see...Dipropyl Isocinchomeronate
R 717® see...Ammonia
R 1303® see...Carbophanothion
R 1504® see...Phosmet
R 1513® see...Azinphos-ethyl
R 1582® see...Azinphos-methyl
R 1607® see...Vernolate
R 1608® see...EPTC
R 1910® see...Butylate
R 2061® see...Pebulate
R 2063® see...Cycloate
R 2170® see...Demeton-methyl
R 4461® see...Bensulide
R 4572® see...Molinate
R 5,158 see...Amiton
R 6700® see...Isobenzan
R 7165® see...Napropamide
R 23979® see...Imazalil
R 42211® see...Pirimiphos-Ethyl
R 151993® see...Tefluthrin
RABON® see...Tetrachlorvinphos
RABOND® see...Tetrachlorvinphos
RACET® see...Acephate
RACUMIN® see...Coumatetralyl
RADAPON® see...Dalapon
RADAPON® see...2,2-Dichloropropionic Acid
RADAR® see...Propiconazole
RADAZIN® see...Atrazine
RAD-E-CATE® see...Sodium Cacodylate
RAD-E-CATE 25® see...Cacodylic Acid
RADIZINE® see...Atrazine
RADOSAN® see...Methoxyethylmercuric Acetate
RADOX® see...Allidochlor
RADOXONE® TL see...Amitrole
RAFEX® see...Dinitro-o-cresol (DNOC)
RAFEX 35® see...Dinitro-o-cresol (DNOC)
RAFLUOR® see...Sodium Fluoride
RAID® see...Resmethrin
RAID® MAX STERILIZER DISCS see...Hydroprene
RAIMONT® see...Phosphoric Acid
RALLY® see...Myclobutanil
RALO 10® see...Cypermethrin
RAMIK® see...Diphacione
RAMIZOL® see...Amitrole
RAMPAGE® see...Cholecalciferol
RAMPART® see...Carbofuran
RAMPART® see...Phorate
RAMROD® see...Propachlor
RAMROD® 65 see...Propachlor
RAMUCIDE® see...Chlorophacinone
RANAC® see...Chlorophacinone
RANCHO® see...Mefenacet
RANDOX® see...Allidochlor
Raney copper see...Copper and Copper Compounds
RANGER® see...Glyphosate
Range oil see...Kerosene
RANKOTEX® see...Mecoprop
RANTOX T® see...Allidochlor
RAPHATOX® see...Dinitro-o-cresol (DNOC)
RAPID® see...Pirimicarb
RAPID KILL 1® see...Diquat Dibromide
RASAYANCHLOR® see...Butachlor
Rasayansulfan see...Endosulfan
RASCHIT® see...p-Chloro-m-Cresol
RASCHIT K® see...p-Chloro-m-Cresol
RASEN-ANICON® see...p-Chloro-m-Cresol
RASIKAL® see...Sodium Chlorate
RASORITE 65® see...Sodium Tetraborate
RASSAPRON® see...Amitrole
RASTOP® see...Difenacoum
RAT & MICE BAIT® see...Warfarin
RATAFIN® see...Coumafuryl
RATAK® see...Difenacoum
RAT ARREST® see...Bromadiolone
RAT-A-WAY® see...Coumafuryl
RAT-A-WAY® see...Warfarin
RATBANE 1080® see...Fluoroacetic Acid
RATBANE 1080® see...Sodium Fluoroacetate
RAT-B-GON® see...Warfarin
RATE® see...lamda-Cyhalothrin
RAT FREE® see...Bromadiolone
RAT-GARD® see...Warfarin
RATIMUS® see...Bromadiolone
Ratindan (Russia) see...Diphacione
RAT KILLER® see...Diphacione
RAT-O-CIDE® see...Warfarin
RATOL® see...Zinc Phosphide
RATOMET® see...Chlorophacinone
RATOX® see...Thallium Sulfate
RATRICK® see...Difenacoum
RATRON® see...Warfarin
RATS-NO-MORE® see...Warfarin
RATSUL SOLUBLE® see...Warfarin
Rattengiftkonserv (German) see...Thallium Sulfate
RATTRACK® see...ANTU
RAT-TU® see...ANTU
RATTUNAL® see...Warfarin
RAUCUMIN® 57 see...Coumatetralyl
RAVE® see...Sodium Dicamba
RAVE® see...Triasulfuron
RAVIAC® see...Chlorophacinone
RAVYON® see...Carbaryl
RAX® see...Warfarin
RAXIL® see...Imazalil
RAXIL® see...Metalaxyl
RAXIL® see...Tebuconazole
RAXIL® see...Thiram
RAZOL DOCK KILLER® see...MCPA
RAZOR® see...Glyphosate

RAZOROOTER II® see...Diquat Dibromide
RB see...Parathion
RBA 777® see...4-Chloro-3,5-xylenol
RCR SQUIRREL KILLER® see...Warfarin
RD 406® see...Dichlorprop
RD 4593® see...Mecoprop
RD 6584® see...Dichloran
RE 12420® see...Acephate
RE 45601® see...Clethodim
READY MASTER® see...Atrazine
READY MASTER® see...Glyphosate
Realgar see...Arsenic and Inorganic Arsenic Compounds
REAL-KILL® see...Diquat Dibromide
REBELATE® see...Dimethoate
RECLAIM® see...Clopyralid
RECLAIM see...Tebuthiuron
RECONOX® see...Phenothiazine
RECOP® see...Copper Oxychloride
RECRUIT® see...Hexaflumuron
Red copper oxide see...Cuprous Oxide
REDDON® see...2,4,5-T
REDDON® see...2,4,5-Trichlorophenoxyacetic Acid, Esters
REDDOX® see...2,4,5-T
REDDOX® see...2,4,5-Trichlorophenoxyacetic Acid, Esters
REDEEM® see...Clopyralid
REDEEM® see...Triclopry
REDEEM® R & P see...Triclopyr, Triethylammonium Salt
Red mercuric oxide see...Mercuric Oxide
Red oxide of mercury see...Mercuric Oxide
RED PANTHER see...Sodium Methanearsonate (MSMA)
Red phosphorus see...Phosphorus
Red precipitate see...Mercuric Oxide
RED-TOP® see...Propargite
REDSKIN® see...Allyl Isothiocyanate
REDUCYMOL® see...Ancymidol
REED® LV 2,4-D see...2,4-D, isooctyl ester
REED® LV 400 2,4-D see...2,4-D, isooctyl ester
REED® LV 600 2,4-D see...2,4-D, isooctyl ester
Refined solvent naphtha see...Naphthas
REFLEX® see...Fomesafen
REFLEX® 2LC see...Fomesafen
Refusal (Netherlands) see...Disulfiram
REGALKADE® see...Prodiamine
REGAL O-O® see...Oxadiazon
REGALSTAR® see...Oxadiazon
REGENT® see...Fipronil
REGLON® see...Diquat
REGLON® see...Diquat Dibromide
REGLONE® see...Diquat
REGLONE® see...Diquat Dibromide
REGLOX® see...Diquat
REGLOX® see...Diquat Dibromide
REGULEX® see...Gibberellic Acid

REGULOX® see...Maleic Hydrazide
REGULOX 50W® see...Maleic Hydrazide
REGULOX W® see...Maleic Hydrazide
RELAX® see...Gibberellic Acid
RELAY® see...Acetochlor
RELDAN® see...Chlorpyrifos-methyl
RELEASE® see...Gibberellic Acid
RELIANCE® see...Chlorimuron-ethyl
RELIANCE® see...Thifensulfuron Methyl
RELY® see...Triclopry
REMASAN CHLOROBLE M® see...Maneb
REMEDY® see...Triclopry
REMELT® see...Ronnel
REMOL TRF® see...o-Phenylphenol
RENEGADE® see...alpha-Cypermethrin
RENOUNCE® see...Cyfluthrin
RENOVATE® see...Triclopyr, Triethylammonium Salt
RENTOKIL® see...Warfarin
RENTOKIL BIOTROL® see...Warfarin
RENTOKIL DEADLINE® see...Bromadiolone
RENTOKIL GASTION® see...Aluminum Phosphide
RENTOKIL FRAM FLY BAIT® see...Methomyl
RENTOKILL® see...Methomyl
REPAR® STREPTOMYCIN see...Streptomycin
REPELLENT-333® see...Dipropyl Isocinchomeronate
REPPER-333® see...Dipropyl Isocinchomeronate
REPULSE® see...Chlorothalonil
RESCUE® see...Naptalam
RESCUE SQUAD® see...Sodium Fluoride
RESIDOX® see...Atrazine
RESISAN® see...Dichloran
RESITOX® see...Coumaphos
Resmethrin, (±)- see...Resmethrin
Resmethrin, (+)-(E),(Z)- see...Resmethrin
Resmethrin, (+)-trans,cis- see...Resmethrin
RESOLVE® see...Imazethapyr
RESOLVE® see...Sodium Dicamba
RESPOND® see...Resmethrin
RESPONSAR® see...Cyfluthrin
RES-Q® see...Hexachlorobenzene
RESTORE® see...Propiconazole
RETACEL® see...Chlormequat Chloride
RETACIL® see...Chlormequat Chloride
RETAIN® see...Aminoethoxyvinylglycine Hydrochloride
RETARD® see...Maleic Hydrazide
RETARDER BA® see...Benzoic Acid
RETARDER W® see...Salicylic Acid
RETARDEX® see...Benzoic Acid
REVENGE® see...Dalapon
REVENGE® see...Potassium Nitrate
REWARD® see...Diquat
REWARD® see...Diquat Dibromide
REWARD® see...Vernolate
REZIFILM® see...Thiram
RH-315 RAPIER® see...Pronamide
RH-787® see...Pyriminil
RH-893® see...Octhilinone

RH-915® see...Oxyfluorfen
RH-2915® see...Oxyfluorfen
RH-3866® see...Myclobutanil
RH-5992® see...Tebufenozide
RH-6201® see...Acifluorfen
RH-7592® see...Fenbuconazole
RH-7592-2F® see...Fenbuconazole
RH-7988® see...Triazamate
RH-7988-25W® see...Triazamate
Rhenogran ETU see...Ethylene Thiourea
RHIZOPON® see...Indole-3-Butyric Acid
RHIZOPON® B ROOTING POWDER see...1-Naphthaleneacetic Acid
RHODACAL ABSA® see...Dodecylbenzenesulfonic Acid
Rhodanin S-62 (Czech) see...Ethylene Thiourea
RHODEN® see...Propoxur
RHODIA® see...2,4-D
RHODIA-6200® see...Amiton
RHODIACHLOR® see...Heptachlor
RHODIACID® see...Ziram
RHODIACUIVRE® see...Copper Oxychloride
RHODIANEHE® see...Maneb
RHODIA-RP-11974® see...Phosalone
RHODIASOL® see...Parathion
RHODIATOX® see...Parathion
RHODIATROX® see...Parathion
RHOMENE® see...MCPA
RHONOX® see...MCPA
RICECO see...Molinate
RICECO TOUCHE® see...Propanil
RICHLORONATE® see...Trichloronate
Richonic acid see...Dodecylbenzenesulfonic Acid
RICIFON® see...Trichlorfon
RICKETON® see...Cholecalciferol
RID-A-MITE® see...Chlorobenzilate
RIDECT® see...Methomyl
RIDEON® see...Diphenamid
RIDOMIL® see...Metalaxyl
RIDOMIL 2E® see...Metalaxyl
RIDOMIL GOLD/BRAVO® see...Chlorothalonil
RIDOMIL GOLD/BRAVO® see...Metalaxyl
RIFLE® see...Primisulfuron-Methyl
RIMIDIN® see...Fenarimol
RIMOSIDEN® see...Streptomycin
RIOGEN® see...Phenylmercury Acetate
RIOMITSIN® see...Oxytetracycline Calcium
RIPCORD® see...Cypermethrin
RIPENTHOL® see...Endothall
RISELECT® see...Propanil
RITSIFON® see...Trichlorfon
RIVAL® see...Thiabendazole
RIVERDALE® see...Clopyralid
RIVERDALE® see...Metsulfuron-methyl
RIVERDALE® see...Sodium Methanearsonate (MSMA)
RIVERDALE® see...Sulfometuron-Methyl
RIVERDALE CORSAIR® see...Chlorsulfuron

RIVERDALE DTDA® SELECTIVE HERBICIDE see. Triclopyr, Triethylammonium Salt
RIVERDALE HORSEPOWER® see...Triclopyr, Triethylammonium Salt
RIVERDALE TAHOE® see...Triclopry
RIVERSIDE® see...Sodium Methanearsonate (MSMA)
RO 13-5223® see...Fenoxycarb
ROACH SALT® see...Sodium Fluoride
ROBAN II AG® see...Zinc Phosphide
ROCKY® see...Endosulfan
ROCYPER® see...Cypermethrin
RODALON® see...Zilkonium Chloride
RODAZIM® see...Carbendazim
RODENT CAKE® see...Diphacione
RODENTEX® see...Warfarin
RODENTIN® see...Coumatetralyl
RODEN-TROL® see...Warfarin
RODEO® see...Glyphosate
RODESCO INSECT POWDER® see...Lindane
RO-DETH® see...Warfarin
RODEX® see...Fluoroacetamide
RO-DEX® see...Strychnine
RODEX® see...Warfarin
RODEX BLOX® see...Warfarin
RODY® see...Fenpropathrin
ROFON® see...Triadimefon
ROGODAN® see...Dimethoate
ROGODIAL® see...Dimethoate
ROGODIAL® see...Phenthoate
ROGOR® see...Dimethoate
ROGUE® see...Propanil
RO-KO® see...Rotenone
ROLL-FRUCT® see...Ethephon
Roman vitriol see...Copper Sulfate
RONALINE-FL® see...Vinclozolin
RONAMID® see...Pronamide
RO-NEET® see...Cycloate
RO-NEET®-6E see...Cycloate
RO-NEET® 10G see...Cycloate
RONILAN® see...Vinclozolin
RONILAN-DF® see...Vinclozolin
RONIT® see...Cycloate
RONONE® see...Rotenone
RONSTAR® see...Oxadiazon
ROOT GUARD see...Diazinon
ROOTGRO® see...Indole-3-Butyric Acid
ROOTONE® see...Indole-3-Butyric Acid
ROOTONE® see...1-Naphthaleneacetamide
ROOTONE® see...1-Naphthaleneacetic Acid
ROOTONE® see...Thiram
ROP 500 F® see...Iprodione
RO-PE® L see...Thymol
ROPHOSATE® see...Glyphosate
ROQUAT® see...Mepiquat Chloride
ROSANIL® see...Propanil
ROSETONE® see...1-Naphthaleneacetamide
RO-SULFIRAM® see...Disulfiram
ROSULFURON® see...Metsulfuron-methyl

ROTACIDE® see...Rotenone
ROTATE® see...Bendiocarb
ROTEFIVE® see...Rotenone
ROTEFOUR® see...Rotenone
Rotenon see...Rotenone
Rotenone, hydrogenated see...Rotenone
Rotenona (Spanish) see...Rotenone
ROTESSENOL® see...Rotenone
ROTILIN® see...Linuron
ROTOX® see...Methyl Bromide
ROTRAZ® see...Amitraz
ROUGH & READY MOUSE MIX® see...Warfarin
ROUNDUP® see...Glyphosate
ROUT® see...Bromacil
ROUT® see...Oryzalin
ROUT® see...Oxyfluorfen
ROVAN® see...Ronnel
ROVRAL® see...Iprodione
ROVOKIL® see...Ethoprop
ROXION® see...Dimethoate
ROYAL BRAND® see...Aldrin
ROYAL BRAND® see...Dieldrin
ROYAL BRAND BEAN TOX 82® see...Toxaphene
ROYAL MH 30® see...Maleic Hydrazide
ROYAL SLO-GRO® see...Maleic Hydrazide
ROYAL TMTD® see...Thiram
ROYSTER® see...Aldrin
ROZOL® see...Chlorophacinone
ROZTOCZOL® see...Tetradifon
ROZTOZOL® see...Tetradifon
RP 11974® see...Phosalone
RP 17623® see...Oxadiazon
RP 26019® see...Iprodione
RPA 90946® see...Cyclanilide
RPA 201772® see...Isoxaflutole
RPH® see...Thiabendazole
RS 141® see...Chlordimeform
(R)-S 3308® see...Diniconazole
RTU 1010® see...Quintozene
RTU-BAYTAN-THIRAM® see...Thiram
RTU FLOWABLE SOYBEAN FUNGICIDE® see. Thiram
RTU-VITAVAX EXTRA® see...Imazalil
RTU-VITAVAX-EXTRA® see...Thiabendazole
RU-11484® see...Resmethrin
RU 22974® see...Deltamethrin
RU-25472® see...Tralomethrin
RU-25474® see...Tralomethrin
Rubber solvent see...Naphthas
RUBIGAN® see...Fenarimol
RUBITOX® see...Phosalone
Ruby arsenic see...Arsenic and Inorganic Arsenic Compounds
RUELENE® see...Crufomate
RUGBY® see...Cadusafos
RUKSEAM® see...DDT
RUMETAN® see...Zinc Phosphide
RUNCATEX® see...Mecoprop

Rutralin see...Butralin
RYCELAN® see...Oryzalin
RYCOPEL® see...Cypermethrin
RYDEX® see...Prodiamine
RYOMYCIN® see...Oxytetracycline Calcium
RYZELAN® see...Oryzalin
RYZUP® see...Gibberellic Acid

- S -

S 95® see...Dalapon
S 112A® see...Fenitrothion
S 276® see...Disulfoton
S 847® see...Barban
S 1065® see...Metolcarb
S 1315® see...Dalapon
S 1844® see...Esfenvalerate
S 1942® see...Bromophos
S 2225® see...Bromophos-ethyl
S 2539® see...D-Phenothrin
S 2703® see...Cyphenothrin
S 2703 FORTE® see...Cyphenothrin
S 2940® see...Phenthoate
S 3206® see...Fenpropathrin
S 3308 L® see...Diniconazole
S 4075® see...Phenmedipham
S 4084® see...Cyanofos
S 5602® see...Fenvalerate
S 5602 ALPHA® see...Esfenvalerate
S 5660® see...Fenitrothion
S 6900® see...Formothion
S 7131® see...Procymidone
S 9318® see...Pyriproxyfen
S 10165® see...Propanil
S 15733® see...Mefluidide
S 31183® see...Pyriproxyfen
SA® see...Salicylic Acid
Saatbenizfungizid (German) see...Hexachlorobenzene
SABER® see...*lamda*-Cyhalothrin
SABET® see...Cycloate
SABRE® see...Bromoxynil
SACEMID® see...Acetochlor
SADH® see...Daminozide
SADOFOS® see...Malathion
SADOPHOS® see...Malathion
SADOPLON® see...Thiram
SAFARI® see...Triflusulfuron-Methyl
SAFAST® see...Omethoate
SAFIDON® see...Phosmet
SAFROTIN® see...Propetamphos
SAKARAT® see...Warfarin
SAKKIMOL® see...Molinate
SALANNIN® see...Azadirachtin
Salicylic acid, isopropyl ester, *O*-ester with *O*-ethyl isopropylphosphoramidothioate see...Isofenphos
SALITHIEX® see...Procymidone
Salt arsenate of lead see...Lead Arsenate
Saltpeter see...Potassium Nitrate

Saltpeter see...Potassium Nitrite
Saltpetre see...Potassium Nitrate
SALVO® see...Cacodylic Acid
SALVO® see...2,4-D
SALVO LIQUID® see...Benzoic Acid
SALVO POWDER® see...Benzoic Acid
SAMURAI® see...*lamda*-Cyhalothrin
SAN-582H® see...Dimethenamid
SAN-619F® see...Cyproconazole
SAN-845H® see...Sodium Dicamba
SAN 244 I® see...Formothion
SAN 6538 I® see...Quinalphos
SAN 6626 I® see...Quinalphos
SAN 6913 I® see...Formothion
SAN 9789 H® see...Norflurazon
SAN-52139® see...Propetamphos
SAN 71071® see...Formothion
SAN 97895® see...Norflurazon
SANACHLOR® see...Alachlor
SANASEED® see...Strychnine
SANCAP® see...Dipropetryn
SANCOPAX® see...Ametryn
SANDEA® see...Halosulfuron-methyl
SANDOLIN® see...Dinitro-o-cresol (DNOC)
SANDOLIN A® see...Dinitro-o-cresol (DNOC)
SANDOZ® 6538 see...Quinalphos
SANDOZ® 52139 see...Propetamphos
SANG GAMMA® see...Lindane
SANHYUUM® see...Ethylene Dibromide
SANICLOR 30® see...Quintozene
SANITIZED® BRAND see...Triclosan
SANITIZED SPG® see...Phenylmercury Acetate
SANMARTON® see...Fenvalerate
SANMITE® see...Pyridaben
SANOCID® see...Hexachlorobenzene
SANOCIDE® see...Hexachlorobenzene
SANOS® see...Glyphosate
SANQUINON® see...Dichlone
SANSEAL® see...Captafol
SANSON® see...Nicosulfuron
SANSPOR® see...Captafol
SANTAR® see...Mercuric Oxide
SANTAR-SM® see...Captafol
SANTOBANE® see...DDT
SANTOBRITE® see...Pentachlorophenol
SANTOBRITE® see...Sodium Pentachlorophenate
SANTOBRITE D® see...Sodium Pentachlorophenate
SANTOCHLOR® see...*para*-Dichlorobenzene
SANTOFLEX A® see...Ethoxyquin
SANTOFLEX AW® see...Ethoxyquin
SANTOPHEN® see...Pentachlorophenol
SANTOQUIN® see...Ethoxyquin
SANTOQUINE® see...Ethoxyquin
SANTOX® see...EPN
SANVEX® see...Carboxin
SANVEX® see...Cartap Hydrochloride
SAOLAN® see...Isolan®
SAP® see...Bensulide

SAPECRON® see...Chlorfenvinphos
SAPECRON® 10FGEC see...Chlorfenvinphos
SAPECRON® 240 see...Chlorfenvinphos
SAPHATE® see...Acephate
SARCLEX® see...Linuron
SAROLEX® see...Diazinon
SATECID® see...Propachlor
SATHON® see...Dienochlor
SATOX 20WSC® see...Trichlorfon
SATURN® see...Thiobencarb
SATURNO® see...Thiobencarb
SAVAGE® see...2,4-D
SAVEY® see...Hexythiazox
SAVIT® see...Carbaryl
SAX® see...Salicylic Acid
SB 1528® see...Perfluidone
SBP 1382/BIOALLETHRIN CONCENTRATE® see. Allethrin
SBP® 1382 see...Resmethrin
*d-trans*-SBP® 1382 see...Resmethrin
SBP®-1390 see...Resmethrin
S.B. PENICK 1382® see...Resmethrin
SC-110® see...Phenylmercury Acetate
SC-0224® see...Sulfosate
SC-12937® see...Azacosterol Dihydrochloride
SCALA® see...Pyrimethanil
SCARCLEX® see...Linuron
SCATHE PEANUT HERBICIDE® see...2,4-DES-Sodium
SCEPTER O.T. HERBICIDE® see...Acifluorfen
Scheele's green see...Copper Arsenite
Scheele's mineral see...Copper Arsenite
SCHERING 4072® see...Phenmedipham
SCHERING® 34615 see...Promecarb
SCHERING® 36056 see...Formetanate Hydrochloride
SCHERING® 36268 see...Chlordimeform
SCHERING® 38107 see...Desmedipham
Schradan see...Octamethyl Diphosphoramide
Schradane (French) see...Octamethyl Diphosphoramide
Schultenite see...Lead Arsenate
Schwefel (German) see...Sulfur
Schwefelkohlenstoff (German) see...Carbon Disulfide
Schweinfurtergruen (German) see...Paris Green
Schweinfurt grun (German) see...Paris Green
Schweinfurt green see...Copper Acetoarsenite
SCIMITAR® see...*lamda*-Cyhalothrin
SCORCH® see...Arsenic Acid
SCORE® see...Difenoconazole
SCORPION® see...Flumetsulam
SCORPION® see...Clopyralid
SCOTLENE® see...Mecoprop
SCOURGE® see...Resmethrin
SCOUT® see...Chlorpyrifos
SCOUT® see...Tralomethrin
SCOUT® X-TRA see...Tralomethrin
SCUTL® see...Phenylmercury Acetate
SCYTHE® see...Pelargonic Acid
SD 345® see...Acrolein diacetate

SD 440® see...Isobenzan
SD 1750® see...Dichlorvos
SD 2794® see...Aldrin
SD 3417® see...Dieldrin
SD 3562® see...Dicrotophos
SD 4072® see...Chlorfenvinphos
SD 4294® see...Crotoxyphos
SD 4901 see...6-Benzaldenine
SD 5532® see...Chlordane
SD 7859® see...Chlorfenvinphos
SD 8447® see...Tetrachlorvinphos
SD 8530® see...Trimethacarb
SD 9129 see...Monocrotophos
SD 9228® see...Methiocarb
SD 14114® see...Fenbutatin Oxide
SD 14999® see...Methomyl
SD 15418® see...Cyanazine
SD 41706® see...Fenpropathrin
SD 43775® see...Fenvalerate
SD 208304® see...Chlorethoxyfos
SDDC see...Sodium Dimethyldithiocarbamate
SEBACIL® see...Phoxim
9,10-Secocholesta-5,7,10(19)-trien-3-β-ol see. Cholecalciferol
9,10-Secocholesta-5,7,10(19)-trien-3-ol, (3.β,5Z,7E)- see...Cholecalciferol
Secondary ammonium phosphate see...Ammonium Phosphate
SECTAGON® see...Metham-Sodium
SECTOR® see...Butralin
SECURITY® see...Calcium Arsenate
SECURITY® see...Lead Arsenate
SECURITY TOX-SOL-6® see...Toxaphene
SEDAGRI TRIFLURALIN® 480 see...Trifluralin
Sedoneural see...Sodium Bromide
SEEDOX® see...Bendiocarb
SEEDTOX® see...Phenylmercury Acetate
SEFFEIN® see...Carbaryl
SEIS-TRES 6-3® see...Methyl Parathion
SEIS-TRES 6-3® see...Parathion
SELECRON® see...Profenfos
SELECT® see...Clethodim
SELECTIN® see...Prometryn
SELECTIN-50® see...Prometryn
SELECT-KIL® see...Sodium Methanearsonate (MSMA)
SELEKTIN® see...Prometryn
SELEPHOS® see...Parathion
SELF POLISHING COPOLYMER® see...Cuprous Thiocyanate
SELINON® see...Dinitro-o-cresol (DNOC)
SELOXONE® see...Mecoprop
SEL-OXONE® see...Mecoprop
SEMASPORE® BAIT see...Nosema Locustae
SEMEVIN® see...Thiodicarb
Semicarbazide, thio- see...Thiosemicarbazide
Semicarbazide, 3-Thio- see...Thiosemicarbazide
SEMPRA CA® see...Halosulfuron-methyl
SENATE® see...Terbutryn

SENCOR® see...Metribuzin
SENCORAL® see...Metribuzin
SENCORER® see...Metribuzin
SENCOREX® see...Metribuzin
SENDER® see...Amitraz
SENDRAN® see...Propoxur
Senf oel (German) see...Allyl Isothiocyanate
SENTINEL® see...*lamda*-Cyhalothrin
SENTINEL® see...Cyproconazole
SENTINEL® see...Zilkonium Chloride
SENTRY SODIUM ACETATE® see...Sodium Diacetate
SEPPIC MMD® see...MCPA
SEPTENE® see...Carbaryl
SEQUEL® see...Fenpyroximate
SERADIX® see...Indole-3-Butyric Acid
SERAPHOS® see...Propetamphos
SERFUME® see...Carbon Disulfide
SERITOX 50® see...Dichlorprop
SES see...2,4-DES-Sodium
SESAGARD® see...Prometryn
Sesone see...2,4-DES-Sodium
Sethoxydim cyclohexanone herbicide see...Sethoxydim
SETRA PROWL® see...Propanil
SETRE FLUOMETURON 80 WP® see...Fluometuron
SEVIGOR® see...Dimethoate
SEVIMOL® see...Carbaryl
SEVIN® see...Carbaryl
SEWARIN® see...Warfarin
SEWIN® see...Carbaryl
SF® 60 see...Malathion
SH-66752® see...Propamocarb Hydrochloride
SHA-105501® see...Tebuthiuron
SHA 486300® see...Hydroprene
SHAMOX® see...MCPA
SHARSTOP 204® see...Sodium Dimethyldithiocarbamate
SHED-A-LEAF® see...Sodium Chlorate
SHELL 345® see...Acrolein diacetate
SHELL 4402® see...Isobenzan
SHELL 4072® see...Chlorfenvinphos
SHELL® ATRAZINE 80W HERBICIDE see...Atrazine
SHELL SD 345® see...Acrolein diacetate
SHELL SD 3562® see...Dicrotophos
SHELL SD 4294® see...Crotoxyphos
SHELL SD 5532® see...Chlordane
SHELL SD 9129® see...Monocrotophos
SHELL SD 14114® see...Fenbutatin Oxide
SHELL UNDRAUTTED A® see...Allyl Alcohol
SHELL WL 1650® see...Isobenzan
SHERMAN® see...Chlormephos
SHERPA® see...Cypermethrin
SHIMMEREX® see...Phenylmercury Acetate
SHIRLAN® see...Fluazinam®
SHOLAY® see...Alachlor
SHORTSTOP® see...EPTC
SHORTSTOP® see...Terbutryn
SHORT-STOP® see...Terbutryn

SIACARB® see...Thiobencarb
SIALEX® see...Procymidone
SIBUTOL® see...Bitertanol
SIBUTOL® see...Fuberidazole
SIBUTROL® see...Fuberidazole
SICLOR® see...Chlorothalonil
SIEGE® see...Hydramethylnon
Silica, amorphous diatomaceous earth see.Diatomaceous Earth
Silicate(2-), hexafluoro-, diammonium see...Ammonium Hexafluorosilicate
Silicofluoruro amonico (Spanish) see...Ammonium Hexafluorosilicate
Silicon dioxide (amorphous) see...Diatomaceous Earth
SILOSAN® see...Pirimiphos-Methyl
SILO® see...Difenacoum
SILVACUR® see...Tebuconazole
SILVANO® see...Hexachlorocyclohexanes
SILVANO® see...Lindane
SILVAPROP® 1 see...2,4-D, butoxyethyl ester
TURFLON® see...2,4-D, butoxyethyl ester
SILVEX HERBICIDE® see...Silvex
SILVICIDE® see...Ammonium Sulfamate
SILVI-RHAP® see...Silvex
SILVISAR® see...Sodium Cacodylate
SILVISAR® see...Sodium Methanearsonate (MSMA)
SILVISAR 510® see...Cacodylic Acid
SILVISAR-550® see...Sodium Methanearsonate (MSMA)
SIMADEX® see...Simazine
SIMANEX® see...Simazine
SIMAZAT® see...Atrazine
SIMAZAT® see...Simazine
Simazina (Spanish) see...Simazine
SIMAZINE® 80W see...Simazine
SIMAZOL® see...Amitrole
SIMFLOW PLUS® see...Amitrole
SIM-TROL® see...Simazine
SINAFID M-48® see...Methyl Parathion
SINBAR® see...Terbacil
SINFLOWAN® see...Trifluralin
SINID® see...Rotenone
SINITUHO® see...Pentachlorophenol
SINOX® see...Dinitro-o-cresol (DNOC)
SINOX GENERAL® see...Dinoseb
SINURON® see...Linuron
SIPAXOL® see...Pendimethalin
SIPCAM® see...Prodiamine
SIPCAM® UK ROVER 5000 see...Chlorothalonil
SIPCAVIT® see...Thiophanate-Methyl
SIPERIN® see...Cypermethrin
SIPTOX I® see...Malathion
Siran hydrazinu (Czech) see...Hydrazine Sulfate
SISTAN® see...Metham-Sodium
SIXTY-THREE SPECIAL E.C. INSECTICIDE® see.Methyl Parathion
SKANE M8® see...Octhilinone
SKATER® see...Metamiton

SKIRMISH® see...Chlorimuron-ethyl
SKW 20010 see...Forchlorfenuron
SKW 83010® see...Cyanamide
SL-236® see...Fluazifop-butyl
Slaked lime see...Calcium Hydroxide
Slaker rejects see...Calcium Carbonate
SLAYMOR® see...Bromadiolone
Slimicide see...Acrolein
SLIMICIDE see...Octhilinone
SLO-GRO® see...Maleic Hydrazide
SLOW-FE® see...Ferrous Sulfate
SLUG-GETA® see...Methiocarb
SLUG-TOX® see...Metaldehyde
SMDC see...Metham-Sodium
SMDC (dihydrate)® see...Metham-Sodium
SMEESANA® see...ANTU
SMFA see...Sodium Fluoroacetate
SMIDAN® see...Phosmet
SMITE® see...Sodium Azide
SMOKE'EM® see...Potassium Nitrate
SMT® see...Fenitrothion
SMUT-GO® see...Hexachlorobenzene
SN 46® see...Dinex
SN 316® see...Promecarb
SN-475® see...Desmedipham
SN 34615® see...Promecarb
SN 36056® see...Formetanate Hydrochloride
SN 36268® see...Chlordimeform
SN-38107® see...Desmedipham
SN 38584® see...Phenmedipham
SN 49537® see...Thidiazuron
SN 100309® see...Pyrimethanil
SNAPSHOT® see...Isoxaben
SNAPSHOT® see...Oryzalin
SNIECIOTOX® see...Hexachlorobenzene
SNIP® see...Dimetilan
SNIP FLY® see...Dimetilan
SNIPPER® see...Indole-3-Butyric Acid
SNP see...Parathion
Sobenate see...Sodium Benzoate
SOBIN® AMMONIUM SULFAMATE see...Ammonium Sulfamate
Soda chlorate see...Sodium Chlorate
SODANIT® see...Sodium Arsenite
Sodio (clorato di) (Italian) see...Sodium Chlorate
Sodio, fluoracetato di (Italian) see...Sodium Fluoroacetate
Sodium acid acetate see...Sodium Diacetate
Sodium acid pyrophosphate see...Trisodium Phosphate
Sodium aluminofluoride see...Sodium Aluminum Fluoride
Sodium aluminum fluoride see...Sodium Aluminum fluoride
Sodium benzoic acid see...Sodium Benzoate
Sodium borate see...Borax and Boric Acid
Sodium borate anhydrous see...Sodium Tetraborate
Sodium borate decahydrate see...Sodium Tetraborate
Sodium borate see...Sodium Metaborate

Sodium cacodylate trihydrate  see. . .Sodium Cacodylate
Sodium (chlorate de) (French)  see. . .Sodium Chlorate
Sodium 2-chloro-6-(4,6-dimethoxypyrimidin-2-ylthio) benzoate  see. . .Pyrithiobac-Sodium
Sodium coumadin  see. . .Warfarin
Sodium cyanide,solid  see. . .Sodium Cyanide
Sodium cyanide, solution  see. . .Sodium Cyanide
Sodium dicambate  see. . .Sodium Dicamba
Sodium 3,6-dichloro-o-anisate  see. . .Sodium Dicamba
Sodium 3,6-dichloro-2-methoxybenzoate  see. . .Sodium Dicamba
Sodium-2-(2,4-dichlorophenoxy)ethyl sulfate  see. . .2,4-DES-Sodium
Sodium-2,4-dichlorophenoxyethyl sulphate  see. . .2,4-DES-Sodium
Sodium-2,4-dichlorophenyl cellosolve sulfate  see. . .2,4-DES-Sodium
Sodium dihydrogen phosphate  see. . .Trisodium Phosphate
Sodium dimethylaminecarbodithioate  see. . .Sodium Dimethyldithiocarbamate
Sodium-4-(dimethylamino)benzenediazosulfonate  see. Fenaminosulf
Sodium-4-(dimethylamino)benzenediazosulphonate  see. Fenaminosulf
Sodium-p-(dimethylamino)benzenediazosulphonate  see. Fenaminosulf
Sodium-p-(dimethylamino)benzenediazosulfonate sodium  see. . .Fenaminosulf
Sodium dimethylaminocarbodithioate  see. . .Sodium Dimethyldithiocarbamate
Sodium-[4-(dimethylamino)phenyl]diazenesulfonate  see. Fenaminosulf
Sodium dimethylarsinate  see. . .Sodium Cacodylate
Sodium dimethyl arsonate  see. . .Sodium Cacodylate
Sodium dimethylcarbamodithioate  see. . .Sodium Dimethyldithiocarbamate
Sodium dimethyldithiocarbamate  see. . .Sodium Dimethyldithiocarbamate
Sodium N,N-dimethyldithiocarbamate  see. . .Sodium Dimethyldithiocarbamate
Sodium dioxoarsenate  see. . .Sodium Arsenite
Sodium fluoacetate  see. . .Sodium Fluoroacetate
Sodium fluoacetic acid  see. . .Sodium Fluoroacetate
Sodium fluoaluminate  see. . .Sodium Aluminum Fluoride
Sodium fluoracetate  see. . .Sodium Fluoroacetate
Sodium fluoracetate de (French)  see. . .Sodium Fluoroacetate
Sodium hexafluoroaluminate  see. . .Sodium Aluminum Fluoride
Sodium hydrofluoride  see. . .Sodium Fluoride
Sodium metaarsenite  see. . .Sodium Arsenite
Sodium metaborate (NaBO$_2$)  see. . .Sodium Metaborate
Sodium metam  see. . .Metham-Sodium
Sodium metham  see. . .Metham-Sodium
Sodium 2-methoxy-3,6-dichlorobenzoate  see. . .Sodium Dicamba
Sodium N-methylaminodithioformate  see. . .Metham-Sodium
Sodium N-methylaminomethanethionothiolate  see. Metham-Sodium
Sodium methylcarbamodithioate  see. . .Metham-Sodium
Sodium methyldithiocarbamate  see. . .Metham-Sodium
Sodium N-methyldithiocarbamate  see. . .Metham-Sodium
Sodium monofluoride  see. . .Sodium Fluoride
Sodium monofluoroacetate  see. . .Sodium Fluoroacetate
Sodium monomethyldithiocarbamate  see. . .Metham-Sodium
Sodium PCP  see. . .Sodium Pentachlorophenate
Sodium pentachlorophenol  see. . .Sodium Pentachlorophenate
Sodium pentachlorophenolate  see. . .Sodium Pentachlorophenate
Sodium pentachlorophenoxide  see. . .Sodium Pentachlorophenate
Sodium, (pentachlorophenoxy)-  see. . .Sodium Pentachlorophenate
Sodium pentachlorphenate  see. . .Sodium Pentachlorophenate
Sodium phosphate  see. . .Trisodium Phosphate
Sodium phosphate, anhydrous  see. . .Trisodium Phosphate
Sodium phosphate, dibasic  see. . .Trisodium Phosphate
Sodium phosphate, monobasic  see. . .Trisodium Phosphate
Sodium phosphate, tribasic  see. . .Trisodium Phosphate
Sodium salt of cacodylic acid  see. . .Sodium Cacodylate
Sodium salt of 3,6-dichloro-o-anisic acid  see. . .Sodium Dicamba
Sodium tellurate(IV)  see. . .Sodium Tellurite
Sodium tetraborate  see. . .Borax and Boric Acid
Sodium tetraborate decahydrate  see. . .Sodium Tetraborate
Sodium warfarin  see. . .Warfarin
SO-FLO®  see. . .Sodium Fluoride
Sofril  see. . .Sulfur
SOGATOX DUST® 22  see. . .Metolcarb
Sohnhofen stone  see. . .Calcium Carbonate
SOILBROM®  see. . .Ethylene Dibromide
SOILFUME®  see. . .Ethylene Dibromide
SOIL FUNGICIDE®-1823  see. . .Chloroneb
SOK®  see. . .Carbaryl
SOLACOL®  see. . .Validamycin
SOLASAN 500®  see. . .Metham-Sodium
SOLDEP®  see. . .Trichlorfon
SOLESAN 500®  see. . .Metham-Sodium
SOLFAC®  see. . .Cyfluthrin
SOLFARIN®  see. . .Warfarin
SOLGARD®  see. . .Pirimiphos-Ethyl
SOLICAM®  see. . .Norflurazon
SOLO®  see. . .Naptalam
SOLO®  see. . .Trifluralin
SOLUBACTER®  see. . .Triclocarban
SOLUBOR®  see. . .Sodium Tetraborate
SOLUCRYL®  see. . .Thiram

SOLUGLACIT® see...Paraoxon
SOLUTION CNCENTREE T271® see...Amitrole
SOLVAN® see...Diphacione
SOLVEZINC® see...Zinc Sulfate Heptahydrate
SOLVEZINK® see...Zinc Sulfate Heptahydrate
SOLVIREX® see...Disulfoton
SOMETAM® see...Metham-Sodium
SOMILAN® see...Ethalfluralin
SOMONIC® see...Methidathion
SOMONIL® see...Methidathion
SONAC® see...Phosphoric Acid
SONACIDE® see...Glutaraldehyde
SONALAN® see...Ethalfluralin
SONALEN® see...Ethalfluralin
SONAR® see...Fluridone
SONET see...Hexaflumuron
SOPRABEL® see...Lead Arsenate
SOPRANEBE® see...Maneb
SOPRATHION® see...Parathion
SOREXA® see...Difenacoum
SOREXA PLUS® see...Warfarin
SOREX CR1® see...Warfarin
SOREX GOLDEN FLY BAIT® see...Methomyl
SOTIPOX® see...Trichlorfon
SOUFRAMINE® see...Phenothiazine
Soufre (French) see...Sulfur
Soufre micronise (French) see...Sulfur
Soufre sublime (French) see...Sulfur
SOWBUG & CUTWORM BAIT® see...Paris Green
SOXINAL-PZ® see...Ziram
SOXINOL 22® see...Ethylene Thiourea
SOXINOL-PZ® see...Ziram
SOYGARD WITH PROTEGE® see...Azaxystrobin
SP-1103 see...Tetramethrin
SPANNIT® see...Chlorpyrifos
SPANON® see...Chlordimeform
SPANONE® see...Chlordimeform
SPARIC® see...Dinoseb
SPARTAN® see...Sulfentrazone
SPEARHEAD® see...Diflufenican
SPECTRACIDE® see...Diazinon
SPECTRO® see...Thiophanate-Methyl
SPENCER S-6538® see...Quinalphos
SPENCER S-6900® see...Formothion
SPERLOX-S® see...Sulfur
SPERLOX®-Z see...Zineb
Spersul thiovit see...Sulfur
SPIKE® see...Tebuthiuron
SPIN-AID® see...Phenmedipham
SPIN-OUT® see...Copper Hydroxide
SPIN OUT 400® see...Copper Carbonate, Basic
SPINNAKER® see...Triadimenol
Spinrite arsenic see...Arsenous Oxide
SPIRE® see...Propiconazole
SPIRIT® see...Primisulfuron-Methyl
Spirit of Hartshorn see...Ammonia
SPLENDOR® see...Tralkoxydim
SPONTOX® see...2,4,5-T

SPONTOX® see...2,4,5-Trichlorophenoxyacetic Acid, Esters
SPOR-KIL® see...Phenylmercury Acetate
SPOTLESS (XE-779L)W® see...Diniconazole
SPOTRETE® see...Thiram
SPOTRETE-F® see...Thiram
Spotting solvent see...Stoddard Solvent
SPRA-CAL® see...Calcium Arsenate
SPRAKIL® see...Tebuthiuron
SPRAY CONCENTRATE® see...Malathion
SPRAY-TROL BRANCH® see...Warfarin
SPRING-BAK® see...Nabam
SPRING-K® see...Ammonium Nitrate
SPRITZ-HORMIN/2,4-D® see...2,4-D
SPRITZ-HORMIT/2,4-D® see...2,4-D
SPRITZ-RAPIDIN® see...Lindane
SPROUT-STOP® see...Maleic Hydrazide
SPRUEHPFLANZOL® see...Lindane
SPUR® see...Fluvalinate
SPURGE® see...Dinoseb
SQ 1489® see...Thiram
SQ 4609® see...6-Benzaldenine
SQUADRON® see...Pendimethalin
SQUADRON AND QUADRANGLE MANEX® see. Maneb
SR 73® see...Clonitralid
SR 406® see...Captan
SRA 5172® see...Methamidophos
SRA-7312® see...Quinalphos
SRA 12869® see...Isofenphos
SRA 128691® see...Isofenphos
SRANAN-SF-X® see...Thiram
SROLEX® see...Diazinon
SSI see...Tebuthiuron
SS-PYDRIN® see...Esfenvalerate
ST-9551® see...Pyridate
STABILAN® see...Chlormequat Chloride
STAFAST® see...1-Naphthaleneacetic Acid
STA-FAST® see...Silvex
STA-FRESH® see...Thiabendazole
STA-FRESH 615® see...Sodium Dimethyldithiocarbamate
STAM® see...Propanil
STAM®F-34 see...Propanil
STAM® LV 10 see...Propanil
STAMPEDE® 3E see...Propanil
STAMPRO® see...Sulfometuron-Methyl
STAM SUPERNOX® see...Propanil
STANDAK® see...Aldoxycarb
Standard lead arsenate see...Lead Arsenate
STANDOUT® see...Glyphosate
STANDOUT® see...Imazethapyr
Stannane, acetoxytriphenyl- (U.S. EPA) see.Triphenyltin Compounds
Stannane, chlorotriphenyl- see...Triphenyltin Compounds
Stannane, hydroxytriphenyl- see...Fentin Hydroxide

Stannane, hydroxytriphenyl- see...Triphenyltin Compounds
Stannane, tributylchloro- see...Tributyltin Chloride
Stannane, tributylfluoro- see...Tributyltin Fluoride
Stannol, triphenyl- see...Triphenyltin Compounds
Stannol, triphenyl- see...Fentin Hydroxide
STAPLE® see...Pyrithiobac-Sodium
STARANE® see...Fluroxypyr 1-methylheptyl Ester
STARBAR CATTLE DUST® see...Phosmet
STARFIRE® see...Paraquat
STATHION® see...Parathion
STATURE® see...Dimethomorph
STATUS® see...Acifluorfen
STAUFFER-2790® see...Fonofos
STAUFFER ASP-51® see...Aspon®
STAUFFER CAPTAN® see...Captan
STAUFFER FERBAM® see...Ferbam
STAUFFER N 521® see...Dazomet
STAUFFER N 3049® see...Trichloronate
STAUFFER R-1303® see...Carbophanothion
STAUFFER R 1504® see...Phosmet
STAUFFER R 1608® see...EPTC
STAUFFER R-1910® see...Butylate
STAUFFER R-2061® see...Pebulate
STAUFFER R 4,572® see...Molinate
STAY-FLO® see...Sodium Fluoride
STAY KLEEN® see...Linuron
STCC 4904210 see...Ammonia
STCC 4921565 see...Ethion
STEADFAST® see...Nicosulfuron
Steenkoolteerrolie-distillaat (Dutch) see...Naphthas
STELADONE® see...Chlorfenvinphos
STELLER® see...Lactofen
STEMPOR® see...Carbendazim
STERIFORM® see...Formaldehyde
STERILITE HOP DEFOLIANT® see...Anthracene
STERILIZING GAS ETHYLENE OXIDE 100%® see. Ethylene Oxide
STERISEAL LIQUID-40® see...Sodium Dimethyldithiocarbamate
STERLING® see...Pymetrozine
Stibnite see...Stibine
STIFLE® see...Butralin
STIK® see...1-Naphthaleneacetic Acid
STIMULATE® see...Gibberellic Acid
STINGER® see...Clopyralid
STIPEND see...Chlorpyrifos
STIROFOS® see...Tetrachlorvinphos
STOCKADE® see...Cypermethrin
STOCK GUARD® see...Flucythrinate
STOMP® see...Pendimethalin
STOPAETHYL® see...Disulfiram
STOP-DROP® see...1-Naphthaleneacetic Acid
STOPETYL® see...Disulfiram
STOPETHYL® see...Disulfiram
STOP-SCALD® see...Ethoxyquin
STORCIDE® see...Chlorpyrifos-methyl
STORM® see...Acifluorfen

STORM® see...Bentazon
STORM® see...Difenacoum
Straight run kerosene see...Kerosene
STRATEGO® see...Propiconazole
STRATEGY® see...Clomazone
STRAZINE® TRIAZINE A 1294 see...Atrazine
STREL® see...Propanil
STREPCEN® see...Streptomycin
STREPCIN® see...Streptomycin Sulfate
STREP-GRAN® see...Streptomycin Sulfate
STREPSULFAT® see...Streptomycin Sulfate
Streptomycin A see...Streptomycin
Streptomycin sesquisulfate see...Streptomycin Sulfate
STREPTOMYCIN SULFATE® see...Streptomycin Sulfate
STREPTOMYCIN SULPHATE B.P.® see.Streptomycin Sulfate
Streptomycine see...Streptomycin
Streptomycinum see...Streptomycin
Streptomyzin (German) see...Streptomycin
STREPTOREX® see...Streptomycin Sulfate
STREPVET® see...Streptomycin Sulfate
STREUNEX® see...Lindane
Stricnina (Italian) see...Strychnine
STRIKER® see...Diuron
STRIKER® IE see...Deltamethrin
STROBANE T® see...Toxaphene
STROBANE T 90® see...Toxaphene
Strychnidin-10-One see...Strychnine
Strychnidin-10-one, 2,3-dimethoxy-(9CI) see...Brucine
Strychnin (German) see...Strychnine
Strychnine, 2,3-dimethoxy- see...Brucine
Strychnos see...Strychnine
STUDAFLOUR® see...Sodium Fluoride
STUNTMAN® see...Maleic Hydrazide
STYPTYSAT® see...Capsaicin
SU SEGURO CARPIDOR® see...Trifluralin
SUBDUE® see...Metalaxyl
SUBITEX® see...Dinoseb
Sublimat (Czech) see...Mercuric Chloride
Succinic acid, dimethyl hydrazide see...Daminozide
Succinic acid, mercapto-, Diethyl ester, S-ester with O,O-dimethyl phosphorodithioate see...Malathion
SUCHLOR® see...Dichlorvos
SUCKER-STUFF® see...Maleic Hydrazide
SUFOS® see...Monocrotophos
SUL-CIDE® see...Sulfur
Sulema (Russian) see...Mercuric Chloride
SULFAMATE® see...Ammonium Sulfamate
Sulfamato amonico (Spanish) see...Ammonium Sulfamate
Sulfamic acid, monoammonium salt see...Ammonium Sulfamate
Sulfaminsaure (German) see...Ammonium Sulfamate
Sulfanilamide, 3,5-dinitro-$N^1,N^1$-dipropyl- see..Oryzalin
Sulfanilylcarbamic acid, methyl ester see...Asulam
Sulfate de cuivre (French) see...Copper Sulfate
Sulfate de nicotine (French) see...Nicotine Sulfate

Sulfate de zinc (French) see...Zinc Sulfate Heptahydrate
Sulfate of copper see...Copper Sulfate
Sulfatep see...Sulfotepp
Sulfato aluminico (Spanish) see...Aluminum Sulfate
Sulfato de cobre (Spanish) see...Copper Sulfate
Sulfato de nicotina (Spanish) see...Nicotine Sulfate
Sulfato de talio (Spanish) see...Thallium Sulfate
Sulfato ferrico (Spanish) see...Ferric Sulfate
Sulfato ferroso (Spanish) see...Ferrous Sulfate
SULFENTRAZONE® see...Sulfentrazone
SULFENTRAZONE® (F6285) 4F see...Sulfentrazone
Sulferrous see...Ferrous Sulfate
SULFIDAL® see...Sulfur
Sulfito amonico (Spanish) see...Ammonium Sulfite
SULFLOX® see...Sulfur
Sulfmethmeton-methyl see...Tribenuron-Methyl
SULFOCARB® see...Aldoxycarb
Sulfometuron methyl see...Sulfometuron-Methyl
Sulfone aldoxycarb see...Aldicarb
Sulfone, p-chlorophenyl 2,4,5-trichlorophenyl see. Tetradifon
Sulfone-2,4,4',5-tetrachlorodiphenyl sulfone see. Tetradifon
SULFONIMIDE® see...Captafol
Sulfonium, trimethyl-, salt with N-(phosphonomethyl)glycine (1:1) see...Sulfosate
Sulfonyl fluoride see...Sulfur Fluoride
SULFORON® see...Sulfur
SULFOX-CIDE® see...Sulfoxide
Sulfoxyl see...Sulfoxide
Sulframin acid 1298 see...Dodecylbenzenesulfonic Acid
Sulfur difluoride dioxide see...Sulfur Fluoride
Sulfuric acid, aluminum salt see...Aluminum Sulfate
Sulfuric acid, cadmium(2+) salt see...Cadmium Sulfate
Sulfuric acid, cadmium(II) salt see...Cadmium Sulfate
Sulfuric acid, copper(2+) Salt (1:1) see...Copper Sulfate
Sulfuric acid, dithallium (+1) salt see...Thallium Sulfate
Sulfuric acid, dithallium (I) salt(8CI,9CI) see...Thallium Sulfate
Sulfuric acid, iron salt (1:1) see...Ferrous Sulfate
Sulfuric acid, iron(2+) salt (1:1) see...Ferrous Sulfate
Sulfuric acid, iron(3+) salt (3:2) see...Ferric Sulfate
Sulfuric acid, iron(II) salt (1:1) see...Ferrous Sulfate
Sulfuric acid, iron(III) salt (3:2) see...Ferric Sulfate
Sulfuric acid, monourea adduct see...Monocarbamide Dihydrogen Sulfate
Sulfuric acid, thallium salt see...Thallium Sulfate
Sulfuric acid, thallium(1+) salt(1:2) see...Thallium Sulfate
Sulfuric acid, thallium(I) salt(1:2) see...Thallium Sulfate
Sulfuric acid, zinc salt see...Zinc Sulfate Heptahydrate
Sulfuric oxyfluoride see...Sulfur Fluoride
Sulfurous acid, 2-(p-tert-butylphenoxy)cyclohexyl-2-propynyl ester see...Propargite
Sulfurous acid 2-(p-tert-butylphenoxy)-1-methylethyl-2-chloroethyl ester see...Aramite
Sulfurous acid cyclic ester with 1,4,5,6,7,7-hexachloro-5-norborene-2,3-dimethanol see...Endosulfan

Sulfurous acid, diammonium salt see...Ammonium Sulfite
Sulfurous acid, 2-[4-(1,1-dimethylethyl)phenoxy] cyclohexyl 2-propynyl ester see. Propargite
Sulfurous acid, monoammonium salt see...Ammonium Sulfite
SULGEN® see...Dodine
SULKOL® see...Sulfur
Sulphabutin see...Busulfan
SULPHEIMIDE® see...Captafol
Sulphocarbonic anhydride see...Carbon Disulfide
SULPHOS® see...Parathion
Sulphoxide see...Sulfoxide
Sulphuric acid, cadmium salt see...Cadmium Sulfate
Sulphuryl difluoride see...Sulfur Fluoride
Sulphuryl fluoride see...Sulfur Fluoride
SULSOL® see...Sulfur
SULTRACOB® see...Copper Sulfate
SUMETHRIN® see...D-Phenothrin
SUMI-ALFA® see...Esfenvalerate
SUMI-ALPHA® see...Esfenvalerate
SUMIBOTO® see...Procymidone
SUMICIDE® see...Fenvalerate
SUMICIDIN® see...Fenvalerate
SUMICIDIN A-ALPHA® see...Esfenvalerate
SUMICIDINE® see...Fenvalerate
SUMICROS® see...Procymidone
SUMI-EIGHT® 12.5 WP see...Diniconazole
SUMIFLEECE® see...Fenvalerate
SUMIFLY® see...Fenvalerate
SUMILARV® see...Chlorfenapyr
SUMILARV® see...Pyriproxyfen
SUMILEX® see...Procymidone
SUMIPOWER® see...Fenvalerate
SUMISCLEX® see...Procymidone
SUMITHION® see...Fenitrothion
SUMITHRIN® see...D-Phenothrin
SUMITICK® see...Fenvalerate
SUMITOMO S® 4084 see...Cyanofos
SUMITOMO® SP-1103 see...Tetramethrin
SUMITOX® see...Malathion
SUMMIT® see...Dicamba
SUMMIT® see...Triadimenol
SUN-BUGGER® see...Resmethrin
SUNCIDE® see...Propoxur
SUNTOL® see...Coumaphos
SUP'ORATS® see...Bromadiolone
SUP'R FLO® see...Diuron
SUP'R FLO® see...Maneb
SUP'R-FLO FERBAM FLOWABLE® see...Ferbam
Superarsonate see...Sodium Methanearsonate (MSMA)
SUPER-CAID® see...Bromadiolone
SUPERCARB,TRITICOL® see...Carbendazim
SUPERCEL 3000® see...Urea
SUPER CRAB-E-RAD-CALAR® see...Calcium Methanearsonate
SUPER DAL-E-RAD® see...Calcium Methanearsonate
SUPER-DENT® see...Sodium Fluoride

SUPER DE-SPROUT® see...Maleic Hydrazide
SUPER D WEEDONE® see...2,4-D
SUPER D WEEDONE® see...2,4,5-T
SUPER D WEEDONE® see...2,4,5-Trichlorophenoxyacetic Acid, Esters
SUPER GREEN AND WEED® see...Mecoprop
SUPER K-GRO® see...Diquat Dibromide
SUPERLYSOFORM® see...Formaldehyde
SUPERMAN MANEB F® see...Maneb
SUPER MOSSTOX® see...Dichlorophene
SUPERNEEM® see...Azadirachtin
SUPERNOX® see...Propanil
SUPERORMONE CONCENTRE® see...2,4-D
SUPERQUIK® see...Monocarbamide Dihydrogen Sulfate
SUPER RODIATOX® see...Parathion
SUPER-ROZOL® see...Bromadiolone
SUPER SPROUT STOP® see...Maleic Hydrazide
SUPERSECT® see...Cypermethrin
SUPER TIN 4L® see...Fentin Hydroxide
SUPER TIN 4L® see...Triphenyltin Compounds
SUPERYACHT® see...Cuprous Thiocyanate
SUPOERTOX® see...Mecoprop
SUPONA® see...Chlorfenvinphos
SUPONE® see...Chlorfenvinphos
SUPRACIDE® see...Methamidophos
SUPRATHION® see...Methidathion
SUPREND® see...Prometryn
SUQUIN® see...Quinalphos
SURCOPUR® see...Propanil
SUREFIRE® see...Paraquat
SURFLAN® see...Oryzalin
SURPASS® see...Acetochlor
SURPASS® see...Vernolate
SURPASS®-E see...Vernolate
SURPRACIDE® see...Methidathion
SURPUR® see...Propanil
SUSVIN® see...Monocrotophos
SUTAN® see...Butylate
SUTATHION® see...Triazophos
SUTAZINE® see...Butylate
SUTOX® see...Copper Oxychloride
SUZI® see...Triphenyltin Compounds
SUZON® see...Diazinon
SUZU H® see...Fentin Hydroxide
SUZU H® see...Triphenyltin Compounds
SVOVEL® see...Sulfur
SVOVL® see...Sulfur
SWAT INSECTICIDE–MITICIDE® see..Phosphamidon
Swebate see...Temephos
Swedish green see...Copper Acetoarsenite
Swedish green see...Copper Arsenite
Swedish green see...Paris Green
SWEEP® see...Chlorothalonil
SWEEP® see...Glyphosate
SWEEP® see...Paraquat
SWIFT'S® GOLD BEAR 44 ESTER see...2,4-D, isopropyl ester

SWIPE® see...Methamidophos
SWIPE 560 EC® see...Mecoprop
SWITCH® see...Cyprodinil
SYBOL® see...Pirimiphos-Methyl
SYLLIT® see...Dodine
SYLLIT® see...Oxythioquinox
SYLLIT® 65 see...Dodine
SYLVICOR® see...Cacodylic Acid
SYNBETAN-P® see...Phenmedipham
SYNCHEMICALS COUCH AND GRASS KILLER® see...Dalapon
SYNCHEMICALS® TOTAL WEED KILLER see.Amitrole
SYNCHRONCY® see...Chlorimuron-ethyl
SYNCHRONY® see...Chlorimuron-ethyl
SYNCHRONY® see...Thifensulfuron Methyl
SYNERGIZED H-10® see...Arsenic Acid
Synfloran see...Trifluralin
SYNKLOR® see...Chlordane
SYNPRAN N® see...Propanil
SYNTHETIC 3956® see...Toxaphene
Synthetic mustard oil see...Allyl Isothiocyanate
Synthetic pyrethrins see...Allethrins
SYNTHRIN® see...Resmethrin
SYNTOX® see...Resmethrin
SYNTOX® TOTAL WEED KILLER see...Amitrole
SYSTAM® see...Octamethyl Diphosphoramide
SYSTEC® see...Thiophanate-Methyl
SYSTEMIC® FUNGICIDE see...Thiophanate-Methyl
SYSTEMOX® see...Demeton
SYSTHANE® see...Myclobutanil
SYSTOPHOS® see...Octamethyl Diphosphoramide
SYSTOX® see...Demeton
SYSTOX THIOL® see...Demeton
SYTAM® see...Octamethyl Diphosphoramide
SZKLARNIAK® see...Dichlorvos

- T -

T-2® see...Trichlorobenzoic Acid
T-5 BRUSH KIL® see...2,4,5-T
T-47® see...Parathion
T-1703® see...Isofluorphate
T-2002® see...Dimefox
TACKLE® see...Acifluorfen
TAC-PLUS® see...Amitraz
TAENIATOL® see...Dichlorophene
TAFABAN® see...Chlorpyrifos
TAFAZINE® see...Simazine
TAFAZINE® 50-W see...Simazine
TAG® see...Diquat
TAG® see...Diquat Dibromide
TAG® see...Phenylmercury Acetate
TAG-HL-33® see...Phenylmercury Acetate
TAHMABON® see...Methamidophos
TAK® see...Malathion
TALBOT® see...Lead Arsenate
TALON® see...Chlorpyrifos

TALSTAR® see...Bifenthrin
TALSTAR LAWN & TREE® see...Bifenthrin
TAMARIZ® see...Thiobencarb
TAMARON® see...Methamidophos
TAME® see...Fenpropathrin
TAMEX® see...Butralin
TAMRAGHOL® see...Copper Oxychloride
TANDEM® see...Ethofumesate
TANDEX® see...Karbutilate
TANONE® see...Phenthoate
TANOS® see...Cymoxanil
TANZENE® see...Karbutilate
TANZINE® see...Simazine
TAOMYCIN® see...Oxytetracycline Calcium
TAOMYXIN® see...Oxytetracycline Calcium
TAP 9VP® see...Dichlorvos
TAP 85® see...Lindane
TAPHAZINE® see...Simazine
TAREDAN® see...Cadusafos
TARGA® see...Quizalofop-Ethyl
TARGET® see...Dicamba
TARGET® see...Mecoprop
TARGET MSMA® see...Sodium Methanearsonate (MSMA)
TARSAN® see...Benomyl
Tartar emetic see...Antimony Potassium Tartrate
Tartaric acid, antimony potassium salt see...Antimony Potassium Tartrate
Tartarized antimony see...Antimony Potassium Tartrate
Tartrated antimony see...Antimony Potassium Tartrate
Tartrato de antimonio y potasio (Spanish) see..Antimony Potassium Tartrate
TASK® see...Dichlorvos
TASPA® see...Propiconazole
TASTOX® see...Antimony Potassium Tartrate
TAT® see...Chlordane
TATA MONO® see...Monocrotophos
TATA PANIDA® see...Pendimethalin
TAT CHLOR® 4 see...Chlordane
TATD® see...Disulfiram
TATTOO® see...Bendiocarb
TATTOO® see...Propamocarb Hydrochloride
TATUZINHO® see...Aldrin
TAUFLUALINATE® see...Fluvalinate
2,3,6-TBA see...Trichlorobenzoic Acid
TBDZ see...Thiabendazole
TBP see...Bithionol
TBTP see...Tribufos
TBZ see...Thiabendazole
TBZ 6® see...Thiabendazole
2,4,5-TC see...Silvex
2,3,6-TCB see...Trichlorobenzoic Acid
2,3,6-TCBA see...Trichlorobenzoic Acid
TCC see...Triclosan
TCC see...Triclocarban
TCNB see...Tecnazene
TCPA® see...Fenac
2,4,5-TCPPA see...Silvex

TCTH see...Cyhexatin
TD 1771® see...Thiophanate-Methyl
TDZ see...Thidiazuron
TEAM see...Benefin
TEAM® see...Trifluralin
TEBUJECT® see...Tebuconazole
TEBULAN® see...Dodine
TEBULAN® see...Tebuthiuron
TECHNETIUM TC 99M® see...Sulfur
Technical chlorinated camphene, 67-69% chlorine see. Toxaphene
TECHNICAL CGA-169374® see...Difenoconazole
Tecnazen (German) see...Tecnazene
Tecnazene see...Tecnazene
TECTO® see...Thiabendazole
TECTO 10P® see...Thiabendazole
TECTO 40F® see...Thiabendazole
TECTO RPH® see...Thiabendazole
TEDION® see...Tetradifon
TEDION V-18® see...Tetradifon
TEDP see...Sulfotepp
TEDTP see...Sulfotepp
Tefluthrine see...Tefluthrin
TEKKAM® see...1-Naphthaleneacetic Acid
TEKTAMER® see...1,2-Dibromo-2,4-dicyanobutane
TEKWAISA® see...Methyl Parathion
TELEFOS® see...Prothoate
TELL® see...Primisulfuron-Methyl
Telluric acid, disodium salt see...Sodium Tellurite
Tellurous acid, disodium salt see...Sodium Tellurite
TELMICID® see...Dithiazanine Iodide
TELMID® see...Dithiazanine Iodide
TELMIDE® see...Dithiazanine Iodide
TELODRIN® see...Isobenzan
TELOK® see...Norflurazon
TELONE® see...Chloropicrin
TELONE® see...D-D mixture
TELONE® see...1,3-Dichloropropene
TELONE II® see...1,3-Dichloropropene
TELONE® C see...Chloropicrin
Telurito sodico (Spanish) see...Sodium Tellurite
TELVAR® see...Diuron
TELVAR® see...Monuron
TELVAR® MONURON see...Monuron
TELVAR®-W MONURON see...Monuron
Temefos (Spanish) see...Temephos
TEMIK® see...Aldicarb
TEMIK 10 G® see...Aldicarb
TEMIK SULFONE® see...Aldoxycarb
Temophos see...Temephos
TEMPO® see...Cyfluthrin
TEMPO® see...Linuron
TEMPO® 20WP see...Cyfluthrin
TEMPO® H see...Cyfluthrin
TEMUS® see...Bromadiolone
TENAC® see...Dichlorvos
TENDEX® see...Propoxur
Tendimethalin see...Pendimethalin

TEN-EIGHTY® see...Sodium Fluoroacetate
TENIATHANE® see...Dichlorophene
TENIATOL® see...Dichlorophene
TENN-PLAS® see...Benzoic Acid
TENORAN® see...Chloroxuron
TENURID® see...Disulfiram
TENUTEX® see...Disulfiram
TEP see...TEPP
TERABOL® see...Methyl Bromide
Teramethylthiuram disulfide see...Thiram
TERBUROX® see...Terbufos
Terbutrazole see...Tebuconazole
TERBUTREX® see...Terbutryn
Terbutryne see...Terbutryn
TERCYL® see...Carbaryl
TERIAL® see...Chlorpyrifos
TERMAFUME® see...Sulfur Fluoride
TERMEX® see...Chlordane
TERMIDE® see...Heptachlor
TERMIDOR® see...Fipronil
TERM-I-TROL® see...Pentachlorophenol
TER-MIL® see...Chlorothalonil
TERNIC® see...Aldicarb
TERPAL® see...Ethephon
TERPAL® see...Mepiquat Chloride
Terpinene see...D-Limonene
TERRA-COAT® see...Quintozene
TERRACHLOR® see...Quintozene
TERRACLOR® see...Phorate
TERRACLOR® see...Quintozene
TERRACLOR 30 G® see...Quintozene
TERRACLOR SUPER X® see...Etridiazole
TERRACOAT® see...Etridiazole
TERRAFLO® see...Etridiazole
TERRAFUN® see...Quintozene
TERRAFUNGINE® see...Oxytetracycline Calcium
TERRAGUARD® see...Triflumizole
TERRAKLENE® see...Paraquat
TERRAMASTER® see...Etridiazole
TERRAMITSIN® see...Oxytetracycline Calcium
TERRAMYCIN® see...Oxytetracycline Calcium
TERRANEB® SP see...Chloroneb
TERRA-SYSTAM® see...Dimefox
TERRA-SYTAM® see...Dimefox
TERRASYTUM® see...Dimefox
TERRATHION GRANULES® see...Phorate
TERRAZAN® see...Quintozene
TERRAZOLE® see...Etridiazole
TERR-O-CIDE® see...Ethylene Dibromide
TERR-O-CIDE® 15 see...Chloropicrin
TERR-O-GAS® see...Chloropicrin
TERR-O-GAS® see...Methyl Bromide
TERSAN 75® see...Thiram
TERSAN-LSR® see...Maneb
TERSAN® SP see...Chloroneb
TERSANTETRAMETHYL DIURANE SULFIDE® see...Thiram
TERSET® see...Mecoprop

$O$-(4-Tertbutyl-2-chloor-fenyl)-$O$-methyl-fosforzuur-$N$-methyl-amide (Dutch) see...Crufomate
Tertiary sodium phosphate see...Trisodium Phosphate
TESTO® see...Thiabendazole
TESULOID® see...Sulfur
TETD see...Disulfiram
TETIDIS® see...Disulfiram
$O,O,O',O'$-Tetraaethyl-bis(dithiophosphat) (German) see. Ethion
$O,O,O,O$-Tetraaethyl-diphosphat see...Bis($O,O$-diaethylphosphorsaeure-anhydrid (German) see...TEPP
(1R,3S)3[(1'RS)(1',2',2',2'-Tetrabromoethyl)]-2,2-dimethylcyclopropanecarboxylic acid, (S)-$\alpha$-cyano-3-phenoxybenzyl ester see...Tralomethrin
Tetrachloorkoolstof (Dutch) see...Carbon Tetrachloride
Tetrachloormetan see...Carbon Tetrachloride
$N$-(1,1,2,2-Tetrachloraethylthio)-cyclohex-4-en-1,4-diacarboximid (German) see...Captafol
2,4,4',5-Tetrachlor-diphenyl-sulfon (German) see. Tetradifon
Tetrachlorkohlenstoff, tetra (German) see...Carbon Tetrachloride
Tetrachlormethan (German) see...Carbon Tetrachloride
2,3,5,6-Tetrachlor-3-nitrobenzol (German) see. Tecnazene
2,4,5,6-Tetrachloro-1,3-benzenedicarbonitrile see .Chlorothalonil
2,3,5,6-Tetrachloro-1,4-benzenedicarboxylic acid, dimethyl ester see...DCPA
Tetrachlorocarbon see...Carbon Tetrachloride
2,4,5,6-Tetrachloro-1,3-dicyanobenzene see. Chlorothalonil
2,4,4',5-Tetrachlorodiphenyl sulfone see...Tetradifon
2,4,5,4'-Tetrachlorodiphenylsulphone see...Tetradifon
$N$-1,1,2,2-Tetrachloroethylmercapto-4-cyclohexene-1,2-carboximide see...Captafol
$N$-[(1,1,2,2-Tetrachloroethyl)sulfenyl]-$cis$-4-cyclohexene-1,2-dicarboximide see...Captafol
$N$-[(1,1,2,2-Tetrachloroethyl)thio]-4-cyclohexene-1,2-dicarboximide see...Captafol
$N$-(1,1,2,2-Tetrachloroethylthio)-4-cyclohexene-1,2-dicarboximide see...Captafol
Tetrachloroisophthalonitrile see...Chlorothalonil
$meta$-Tetrachloroisophthalonitrile see...Chlorothalonil
Tetrachloromethane see...Carbon Tetrachloride
1,2,4,5-Tetrachloro-3-nitrobenzene see...Tecnazene
2,3,5,6-Tetrachloronitrobenzene see...Tecnazene
Tetrachlorophthalodinitrile, $meta$- see...Chlorothalonil
Tetrachloroterephthalic acid, dimethyl ester see...DCPA
2,3,5,6-Tetrachlorphthalsaure-dimethylester (German) see...DCPA
Tetrachlorure de carbone (French) see...Carbon Tetrachloride
Tetraclorometano (Italian) see...Carbon Tetrachloride
Tetracloruro de carbono (Spanish) see...Carbon Tetrachloride
Tetracloruro di carbonio (Italian) see...Carbon Tetrachloride

Tetraclorvinfos (Spanish) see...Tetrachlorvinphos
Tetracycline, 5-hydroxy see...Oxytetracycline Calcium
Tetradichlone see...Tetradifon
TETRADICHLONE® see...Tetradifon
TETRADIN® see...Disulfiram
TETRADINE® see...Disulfiram
Tetradiphon see...Tetradifon
O,O,O,O-Tetraethyl-difosfaat (Dutch) see...TEPP
O,O,O,O-Tetraethyl-dithio-difosfaat (Dutch) see. Sulfotepp
Tetraethyl dithio pyrophosphate see...Sulfotepp
Tetraethyldithiopyrophosphate see...Sulfotepp
O,O,O,O-Tetraethyldithiopyrophosphate see...Sulfotepp
O,O,O',O'-Tetraethyl S,S'-methylenebis(dithiophosphate) see...Ethion
Tetraethyl S,S'-methylene bis(phosphorothiolothionate) see...Ethion
O,O,O',O'-Tetraethyl S,S'-methylenebisphordithioate see...Ethion
O,O,O',O'-Tetraethyl S,S'-methylene di(phosphorodithioate) see...Ethion
Tetraethyl pyrofosfaat (Belgian) see...TEPP
Tetraethyl pyrophosphate see...TEPP
Tetraethyl pyrophosphate, liquid see...TEPP
Tetraethylthioperoxydicarbonic diamide see..Disulfiram
Tetraethylthiram disulphide see...Disulfiram
Tetraethylthiuram see...Disulfiram
Tetraethylthiuram disulfide see...Disulfiram
Tetraethylthiuram disulphide see...Disulfiram
N,N,N',N'-Tetraethylthiuram disulphide see...Disulfiram
Tetraetil (Spanish) see...Disulfiram
O,O,O,O-Tetraetil-di tio-pirofosfato (Italian) see. Sulfotepp
O,O,O,O-Tetraetil-pirofosfato (Italian) see...TEPP
Tetrafenphos see...Temephos
TETRAFINOL® see...Carbon Tetrachloride
(2,3,5,6-Tetrafluoro-4-methylphenyl)methyl cis-3-(2-chloro-3,3,3-trifluoro-1-propenyl)-2,2-dimethylcyclopropanecarboxylate see...Tefluthrin
TETRAFORM® see...Carbon Tetrachloride
Tetrafosfato de hexaetilo (Spanish) see...Hexaethyl Tetraphosphate
Tetrafosfor (Dutch) see...Phosphorus
[2r-(2a,6aa,12aa)]-1,2,12,12a-Tetrahydro-8,9-dimethoxy-2-(1-methylethenyl)[1]-benzopyrano[3,4-b]furo[2,3-H][1]-benzopyran-6(6aH)one see...Rotenone
Tetrahydro-5,5-dimethyl-2(1H)-pyrimidinone[3-(4-(trifluoromethyl)phenyl)-1-[2-(4-(trifluoromethyl)phenyl)ethenyl]-2-propenylidene]hydrazone see...Hydramethylnon
Tetrahydro-5,5-dimethyl-2(1H)-pyrimidinone[1,5-bis(α,α,α-trifluoro-p-tolyl)-1,4-pentadien-3-one]hydrazone see...Hydramethylnon
Tetrahydro-2H-3,5-dimethyl-1,3,5-thiadiazine-2-thione see...Dazomet
Tetrahydro-3,5-dimethyl-1,3,5-thiadiazine-2-thione see. Dazomet
Tetrahydro-3,5-dimethyl-2H-1,3,5-thiadiazine-2-thione see...Dazomet
1,2,3-Tetrahydro-3,6-dioxopyridazine see...Maleic Hydrazide
Tetrahydro-2H-imidazole-2-thione see...Ethylene Thiourea
N-(3,4,5,6-Tetrahydrophthalimido)-methyl dl-cis,trans-chrysanthemate see...Tetramethrin
N-(3,4,5,6-Tetrahydrophthalimido)-methyl dl-(Z),(E)-chrysanthemate see...Tetramethrin
Ttetramethrin, (±)- see...Tetramethrin
2,3,4,5-Tetrahydrophthalimidomethylchrysanthemate see. Tetramethrin
3,4,5,6-Tetrahydrophthalimidomethyl (±)-cis-trans-chrysanthemate see...Tetramethrin
3,4,5,6-Tetrahydrophthalimidomethyl cis and trans dl-chrysanthemummonocarboxylic acid see...Tetramethrin
3,4,5,6-Tetrahydrophthalimidomethyl (±)-(Z)-(E)-chrysanthemate see...Tetramethrin
Tetrakisdimethylaminophosphoric anhydride see. Octamethyl Diphosphoramide
3-(α-Tetral)-4-oxycoumarin see...Coumatetralyl
TETRALEN-PLUS® see...Mecoprop
TETRAM® see...Amiton
TETRAM® 75 see...Amiton Oxalate
TETRAM® ACID OXALATE see...Amiton Oxalate
Tetramethrine see...Tetramethrin
2,2,3,3-Tetramethylcyclopropane carboxylic acid, cyano(3-phenoxyphenyl)methyl ester see..Fenpropathrin
N,N,N',N'-Tetramethyl-diamido-fosforzuur-fluoride (Dutch) see...Dimefox
Tetramethyldiamidophosphoric fluoride see...Dimefox
N,N,N',N'-Tetramethyl-diamido-phosphorsaeure-fluorid (German) see...Dimefox
Tetramethyldiaminodiphenylacetimine see...Auramine
3,7,9,13-Tetramethyl-5,11-dioxa-2,8,14-trithia-4,7,9,12-tetra-azapentadeca-3,12-diene-6,10-dione see.Thiodicarb
Tetramethyldiurane sulphite see...Thiram
Tetramethylene bis(methanesulfonate) see...Busulfan
Tetramethylene dimethane sulfonate see...Busulfan
Tetramethylene thiuram disulfide see...Thiram
Tetramethylene thiuram disulphide see...Thiram
Tetramethylphosphorodiamidic fluoride see...Dimefox
N,N,N,N-Tetramethylphosphorodiamidic fluoride see. Dimefox
Tetramethylthiocarbamoyldisulphide see...Thiram
Tetramethylthioperoxydicarbonic diamide see...Thiram
Tetramethyl-O,O'-thiodi-p-phenylene phosphorothioate see...Temephos
O,O,O',O'-Tetramethyl O,O'-thiodi-p-phenylenebis (phosphorothioate) see...Temephos
O,O,O',O'-Tetramethyl O,O'-thiodi-p-phenylene phosphorothioate see...Temephos
Tetramethylthiuram see...Thiram
Tetramethylthiuram bisulfide see...Thiram
Tetramethylthiuram bisulphide see...Thiram
Tetramethylthiuram disulfide see...Thiram
N,N-Tetramethylthiuram disulfide see...Thiram

*N,N,N',N'*-Tetramethylthiuram disulfide  see...Thiram
Tetramethylthiuram disulphide  see...Thiram
Tetramethylthiuran disulphide  see...Thiram
Tetramethyl thiurane disulfide  see...Thiram
Tetramethyl thiurane disulphide  see...Thiram
Tetramethylthiurum disulfide  see...Thiram
Tetramethylthiurum disulphide  see...Thiram
*N,N,N',N'*-Tetrametil-fosforodiammido-fluoruro (Italian)  see...Dimefox
TETRAM® MONOOXALATE S-  see...Amiton Oxalate
TETRAN®  see...Oxytetracycline Calcium
TETRANOL V18®  see...Tetradifon
TETRA OLIVE N2G®  see...Anthracene
Tetraoxymethylene  see...Formaldehyde
Tetraphosphate hexaethylique (French)  see...Hexaethyl Tetraphosphate
Tetraphosphor (German)  see...Phosphorus
Tetraphosphoric acid, hexaethyl ester  see...Hexaethyl Tetraphosphate
TETRAPOM®  see...Thiram
Tetra-propyl dithiopyrophosphate  see...Aspon®
*O,O,O,O*-Tetrapropyl dithiopyrophosphate  see...Aspon®
Tetra-*n*-propyl dithionopyrophosphate  see...Aspon®
TETRASIPTON®  see...Thiram
TETRASOL®  see...Carbon Tetrachloride
Tetrastigmine  see...TEPP
TETRATHIIN®  see...Dimethipin
Tetrathiuram disulfide  see...Thiram
Tetrathiuram disulphide  see...Thiram
TETRAVOS®  see...Dichlorvos
3-(α-Tetrayl)-4-hydroxycoumarin, Coumatetralyl
Tetramethrin, racemic  see...Tetramethrin
3-(D-Tetrayl)-4-hydroxycoumarin  see...Coumatetralyl
1,2,4,5-Tetrazine, 3,6-bis(2-chlorophenyl)-  see. Clofentezine
TETRON®  see...TEPP
TETRON-100®  see...TEPP
TETROSIN OE®  see...*o*-Phenylphenol
TETROSIN OE-N®  see...*o*-Phenylphenol
1,3,5,7-Tetroxocane, 2,4,6,8-tetramethyl-1,3,5,7-tetraoxacyclooctane  see...Metaldehyde
TETURAM®  see...Disulfiram
TETURAMIN®  see...Disulfiram
T- EXTRA®  see...Ethephon
TF 1169®  see...Fluazifop-butyl
T- FLUORIDE®  see...Sodium Fluoride
T- GAS®  see...Ethylene Oxide
TH 60-40®  see...Diflubenzuron
TH 346-1®  see...Phenthoate
TH 367-1®  see...Fluenetil
TH 3671®  see...Fluenetil
Thallium sulfate  see...Thallium Sulfate
Thallium(1+) sulfate (2:1)  see...Thallium Sulfate
Thallium(I) sulfate (2:1)  see...Thallium Sulfate
Thallium sulphate  see...Thallium Sulfate
Thallous Sulfate  see...Thallium Sulfate
Thalonil  see...Chlorothalonil
THEMET®  see...Phorate

THERA-FLUR-N®  see...Sodium Fluoride
THIABEN®  see...Thiabendazole
THIABENDAZOLUM®  see...Thiabendazole
THIABENZAZOLE®  see...Thiabendazole
THIABENZOLE®  see...Thiabendazole
3-Thiabutan-2-one,*O*-(methylcarbamoyl)oxime  see. Methomyl
4*H*-1,3,5-Thiadiazin-4-one, 2-[(1,1-dimethylethyl)imino]tetrahydro-3-(1-methylethyl)-5-phenyl-  see...Buprofezin
THIADIAZIN®  see...Dazomet
2*H*-1,3,5-Thiadiazine-2-thione, tetrahydro-3,5-dimethyl-  see...Dazomet
1,2,4-Thiadiazole, 5-ethoxy-3-(trichloromethyl)-  see. Etridiazole
(N-1,2,3-Thiadiazolyl-5)-*N*'-phenylurea  see..Thidiazuron
Thiameturon-methyl  see...Thifensulfuron Methyl
THIANOSAN®  see...Thiram
THIBENZOL®  see...Thiabendazole
THIBENZOLE®  see...Thiabendazole
THIBENZOLE 200®  see...Thiabendazole
THIBENZOLE ATT®  see...Thiabendazole
THIDAN®  see...Endosulfan
THIFOR®  see...Endosulfan
THIHEX®  see...Hexachlorobenzene
THILLATE®  see...Thiram
THIMAR®  see...Thiram
THIMENOX®  see...Phorate
THIMER®  see...Thiram
THIMET®  see...Phorate
THIMUL®  see...Endosulfan
THINSEC®  see...Carbaryl
2-Thio-3,5-dimethyltetrahydro-1,3,5-thiadiazine  see. Dazomet
Thioallate  see...Sulfallate
THIOBEL®  see...Carboxin
THIOBEL®  see...Cartap Hydrochloride
Thiobencarbe  see...Thiobencarb
2,2'-Thiobis(4,6-dichlorophenol)  see...Bithionol  see. Bithionol
Thiocarbamic acid, *N*-diisopropyl-, *S*-2,3,3-Trichloroallyl ester  see...Triallate
Thiocarbamic acid-*S*,*S*-[2-(dimethylamino)trimethylene]ester hydrochloride  see. Cartap Hydrochloride
Thiocarbamoylhydrazine  see...Thiosemicarbazide
Thiocarbamylhydrazine  see...Thiosemicarbazide
Thiocyanatoethane  see...Ethylthiocyanate
Thiocyanic acid, ethyl ester  see...Ethylthiocyanate
Thiocyanic acid, methyl ester  see...Methyl Thiocyanate
α-THIODAN®  see...Endosulfan
β-THIODAN®  see...Endosulfan
THIODEMETON®  see...Demeton
THIODEMETON®  see...Disulfoton
THIODEMETRON®  see...Disulfoton
*O,O*'-(Thiodi-4,1-phenylene)bis(*O,O*,dimethyl phosphorothioate)  see...Temephos

O,O'-(Thiodi-4,1-phenylene)phosphorothioic acid O,O,O',O'-tetramethyl ester  see...Temephos
O,O'-(Thiodi-p-phenylene)O,O,O',O'-tetramethylbis(phosphorothioate)  see...Temephos
Thiodifenylamine (Dutch)  see...Phenothiazine
Thiodiphenylamin (German)  see...Phenothiazine
THIODOW®  see...Zineb
Thiofanate  see...Thiophanate-Methyl
Thiofanocarb (South Africa)  see...Thiofanox
THIOFOR®  see...Endosulfan
2-Thioimidazolidine  see...Ethylene Thiourea
THIOKILL®  see...Endosulfan
THIOKNOCK®  see...Thiram
2-Thiol-dihydroglyoxaline  see...Ethylene Thiourea
THIOLMECAPTOPHOS®  see...Demeton
Thiolodemeton  see...Demeton
THIOL-SYSTOX®  see...Demeton
THIOLUX®  see...Sulfur
THIOMEX®  see...Parathion
THIONEX®  see...Endosulfan
α-THIONEX®  see...Endosulfan
β-THIONEX®  see...Endosulfan
Thionobenzenephosphonic acid Ethyl-p-nitrophenyl ester  see...EPN
Thionodemeton  see...Demeton
THIONODEMETON®  see...Demeton
2-Thionoimidazolidine  see...Ethylene Thiourea
THION® 80  see...Sulfur
THION® 95  see...Sulfur
Thioperoxydicarbo NIC diamide  see...Tetramethyl- see. Thiram
Thioperoxydicarbonic acid [[(HO)C(S)]$_2$S$_2$] see.dimethyl ester  see...Tributyltin Chloride
Thiophanate  see...Thiophanate-Methyl
Thiophanox  see...Thiofanox
2-Thiophenecarboxylic acid, 3-[((((4-methoxy-6-methyl-1,3,5-triazin-2-yl)amino)carbonyl)amino)sulfonyl]-, methyl ester  see...Thifensulfuron Methyl
THIOPHENIT®  see...Methyl Parathion
Thiophenite  see...Thiophanate-Methyl
Thiophosphate de O,O-diethyle et de O-(3-chloro-4-methyl-7-coumarinyle) (French)  see...Coumaphos
Thiophosphate de O,O-diethyle et de O-(2,5-dichloro-4-bromo) phenyle (French)  see...Bromophos-ethyl
Thiophosphate de O,O-diethyle et de S-(2-ethylthioethyle)  see...Demeton
Thiophosphate de O,O-dimethyle et de O-4-bromo-2,5-dichlorophenyle (French)  see...Bromophos
Thiophosphate de O,O-dimethyle et de S-2-ethylthioethyle (French)  see...Demeton-methyl
Thiophosphate de O,O-diethyle et de O-2-isopropyl-4-methyl 6-pyrimidyle (French)  see...Diazinon
Thiophosphate de O,O-dimethyle et de S-(N-methylcarbamoyl)methyle (French)  see...Omethoate
Thiophosphate de O,O-dimethyle et de S-[(5-methoxy-4-pyronyl)-methyle] (French)  see...Endothion
Thiophosphate de O,O-dimethyle et de O-(3-methyl-4-nitrophenyle) (French)  see...Fenitrothion
Thiophosphate de O,O-dimethyle et de O-(2,4,5-trichlorophenyle) (French)  see...Ronnel
Thiophosphoric acid 2-isopropyl-4-methyl-6-pyrimidyl diethyl ester  see...Diazinon
Thiophosphorsaeure-O,S-dimethylesteramid (German)  see...Methamidophos
THIOPHOS®  see...Parathion
THIOPHOS® 3422  see...Parathion
Thiopyrophosphoric acid, tetrapropyl ester  see...Aspon®
THIORIT®  see...Sulfur
THIOSAN®  see...Disulfiram
THIOSAN®  see...Thiram
THIOSCABIN®  see...Disulfiram
THIOSCABIN®  see...Thiram
2-Thiosemicarbazide  see...Thiosemicarbazide
3-Thiosemicarbazide  see...Thiosemicarbazide
THIOSTOP-N®  see...Sodium Dimethyldithiocarbamate
THIOSULFAN®  see...Endosulfan
THIOSULFAN THIONEL®  see...Endosulfan
Thiosulfuric acid, diammonium salt  see...Ammonium Thiosulfate
Thiotepp  see...Sulfotepp
THIOTEX®  see...Thiram
THIOTOX®  see...Thiram
Thiourea, N,N'-(1,2-ethanediyl)-  see...Ethylene Thiourea
Thiourea, 1-naphthalenyl-  see...ANTU
Thiovit sulphur  see...Sulfur
THIOXAMYL®  see...Oxamyl
THIRAM 75®  see...Thiram
THIRAM 80®  see...Thiram
THIRAMAD®  see...Thiram
THIRAM B®  see...Thiram
Thirame (French)  see...Thiram
THIRAMPA®  see...Thiram
THIRASAN®  see...Thiram
THIRERANIDE®  see...Disulfiram
THIULIN®  see...Thiram
THIULIX®  see...Thiram
THIURAD®  see...Thiram
Thiuram (Japan)  see...Thiram
THIURAM E®  see...Disulfiram
THIURAMIN®  see...Thiram
THIURAMYL®  see...Thiram
THIURANIDE®  see...Disulfiram
THOMPSON-HAYWARD® 6040  see...Diflubenzuron
THOMPSON'S WOOD FIX®  see...Pentachlorophenol
THREE ELEPHANT V-BOR®  see...Sodium Tetraborate
THYLATE®  see...Thiram
THYLPAR M-50®  see...Methyl Parathion
Thyme camphor  see...Thymol
Thymic acid  see...Thymol
THYMO-CIDE®  see...Thymol
m-Thymol  see...Thymol
TI-1258®  see...Cartap Hydrochloride
Tiabendazol (Spanish)  see...Thiabendazole
TIABENDAZOLE®  see...Thiabendazole
TIAZON®  see...Dazomet

TIEZENE® see...Zineb
TIGREX® see...Diuron
TIKTOK® see...Dicofol
TILCAREX® see...Quintozene
TILLAM® see...Pebulate
TILLER® see...MCPA
TILLRAM® see...Disulfiram
TILT® see...Propiconazole
TIMBERFUME II® see...Chloropicrin
Timet (former USSR) see...Phorate
Tin triphenyl acetate see...Triphenyltin Compounds
Tin, hydroxytriphenyl- see...Fentin Hydroxide
Tin, hydroxytriphenyl- see...Triphenyltin Compounds
TINESTAN® see...Triphenyltin Compounds
TINESTAN 60 WP® see...Triphenyltin Compounds
TINMATE® see...Triphenyltin Compounds
TINTORANE® see...Warfarin
Tiodifenilamina (Italian) see...Phenothiazine
TIOFOS® see...Parathion
TIOLENE® see...Sulfur
Tiosemicarbazida (Spanish) see...Thiosemicarbazide
Tiosulfato amonico (Spanish) see...Ammonium Thiosulfate
TIOVEL® see...Endosulfan
TIPOFF® see...1-Naphthaleneacetic Acid
TIPPON® see...2,4,5-T
TIPPON® see...2,4,5-Trichlorophenoxyacetic Acid, Esters
TIPULA® see...Aldrin
TIRADE® see...Fenvalerate
Tiram (Spanish) see...Thiram
TIRAMPA® see...Thiram
TITAN® see...Hexaconazole
TITAN FL® see...Thiram
TITOFTOROL® see...Zineb
TIURAM® see...Disulfiram
Tiuram (Polish) see...Thiram
Tiuramyl see...Thiram
TIUROLAN® see...Tebuthiuron
TIXIT® see...Propham
TL 314® see...Acrylonitrile
TL 466® see...Isofluorphate
TL 741® see...Ethylene Fluorohydrin
TL 792® see...Dimefox
TL 869® see...Sodium Fluoroacetate
TL 898® see...Mercuric Chloride
TM-4049® see...Malathion
TMTD (former USSR) see...Thiram
TMTDS see...Thiram
TNCS® 53 see...Copper Sulfate
TNIV see...Fentin Hydroxide
TN IV see...Triphenyltin Compounds
T-NOX® see...2,4,5-T
TOBAZ® see...Auramine
TOBAZ® see...Thiabendazole
TOK® see...Nitrofen
TOK-2® see...Nitrofen
TOK E® see...Nitrofen
TOK E 25® see...Nitrofen
TOK E 40® see...Nitrofen
TOKKOM® see...Nitrofen
TOKKORN® see...Nitrofen
TOK WP-50® see...Nitrofen
TOLBAN® see...Profluralin
TOLL® see...Methyl Parathion
Toluene, α-[2-(2-butoxyethoxy)ethoxy]-4,5-(methylenedioxy)-2-propyl- see...Piperonyl Butoxide
Toluene, o-chloro- see...o-Chlorotoluene
α-Toluenethiol, p-chloro-, diethylcarbamate see...Thiobencarb
α-Toluenol see...Benzyl Alcohol
m-(or p-)Toluic acid, 2-(4,5-dihydro-4-methyl-4-isopropyl-5-oxo-1H-imidazol-2-yl)-, methyl ester see...Imazethabenz
m-(or p-)Toluic acid, 6-(4-isopropyl-4-methyl-5-oxo-2-imidazolin-2-yl)-, methyl ester see...Imazethabenz
p-Toluidine, N-butyl-N-ethyl-α,α,α-trifluoro-2,6-dinitro- see...Benefin
p-Toluidine, α,α,α-trifluoro-2,6-dinitro-N,N-dipropyl- see...Trifluralin
m-Tolyester kyseliny methyl karbaminove see...Metolcarb
o-Tolylchloride see...o-Chlorotoluene
3-Tolyl-N-methylcarbamate see...Metolcarb
m-Tolyl-N-methylcarbamate see...Metolcarb
TOMAHAWK® see...Butylate
TOMAHAWK® see...Flucythrinate
TOMAHAWK® see...Fluroxypyr 1-methylheptyl Ester
TOMAHAWK® see...Pirimiphos-Methyl
TOMARIN® see...Coumafuryl
TOMATHREL® see...Ethephon
TOMCAT® see...Diphacione
TOPAS® see...Penconazole
TOPAS-C® see...Penconazole
TOPAS-MZ® see...Penconazole
TOPAZ® see...Penconazole
TOPAZE® see...Penconazole
TOPAZE-C® see...Penconazole
TOPCLIP-PARASOL® see...Cypermethrin
TOP FORM WORMER® see...Thiabendazole
TOPHAND® see...Acetochlor
TOPICHLOR® 20 see...Chlordane
TOPICLOR® see...Chlordane
TOPITOX® see...Chlorophacinone
TOPNOTCH® see...Acetochlor
TOPPS® see...Thymol
TOPSIN® see...Thiophanate-Methyl
TOPSIN-WP METHYL® see...Thiophanate-Methyl
TORAK® see...Dialifor
TORAPRON® see...Amitrole
TORBIN® see...EPTC
TORCH® see...Bromoxynil
TORDON® see...Picloram
TORERO® see...Ethofumesate
TORERO® see...Metamiton
TORMONA® see...2,4,5-T

TORMONA® see...2,4,5-Trichlorophenoxyacetic Acid, Esters
TORMONA® (butyl ester) see...2,4,5-Trichlorophenoxyacetic Acid, Esters
TORNADO® see...Carbaryl
TORNADO® see...Fluazifop-butyl
TORNADO® see...Fomesafen
TORPEDO® see...Diquat
TORPEDO® see...Diquat Dibromide
TORQUE® see...Fenbutatin Oxide
TORSITE® see...o-Phenylphenol
TORUS® see...Fenoxycarb
TOTACOL® see...Paraquat
TOTAMOTT® see...para-Dichlorobenzene
TOTAZINE® see...Simazine
TOUCHDOWN® see...Diquat Dibromide
TOUCHDOWN® see...Glyphosate
TOUCHDOWN® see...Sulfosate
TOUCHE® see...Vinclozolin
TOUGH® see...Pyridate
TOX 47® see...Parathion
TOX-R® see...Rotenone
TOXADRIN® see...Aldrin
TOXADUST® see...Toxaphene
Toxafeen (Dutch) see...Toxaphene
Toxafeno (Spanish) see...Toxaphene
TOXAKIL® see...Toxaphene
Toxaphen (German) see...Toxaphene
TOXASPRAY® see...Toxaphene
TOXER TOTAL® see...Paraquat
TOXICHLOR® see...Chlordane
TOXO® see...Phosphamidon
TOXOL® see...Parathion
TOXON 63® see...Toxaphene
TOXYPHEN® see...Toxaphene
2,4,5-TP see...Silvex
TPN® see...Chlorothalonil
TPN (PESTICIDE)® see...Chlorothalonil
TPTA see...Triphenyltin Compounds
TPTC see...Triphenyltin Compounds
TPTH see...Fentin Hydroxide
TPTH see...Triphenyltin Compounds
TPTH technical see...Fentin Hydroxide
TPTH technical see...Triphenyltin Compounds
TPTOH see...Triphenyltin Compounds
TPZA see...Triphenyltin Compounds
TRACKER® see...Dicamba
TRAILS END® see...Metaldehyde
Tralkoxydime see...Tralkoxydim
Tralomethrine see...Tralomethrin
TRAMETAN® see...Thiram
TRANSAMINE® see...2,4-D
TRANSAMINE® see...2,4,5-T
TRANSAMINE® see...2,4,5-Trichlorophenoxyacetic Acid, Esters
d-TRANS™ INTERMEDIATE 1828® see...Allethrin
TRANSLINE® see...Clopyralid
TRANSPLANTONE® see...1-Naphthaleneacetic Acid
TRANSPLANTONE® see...1-Naphthaleneacetamide
TRANS-VERT® see...Sodium Methanearsonate (MSMA)
TRAPEX® see...Metham-Sodium
TRAVEX® see...Sodium Chlorate
2,3,6-TrCB acid see...Trichlorobenzoic Acid
TRE-HOLD® see...1-Naphthaleneacetic Acid
TRE-HOLD® see...Naphthoxyacetic Acid
TREBON® see...Ethofenprox
TREE TECH® see...Oxytetracycline Calcium
TREFANOCIDE® see...Trifluralin
TREFICON® see...Trifluralin
TREFLAN® see...Trifluralin
TREFLANOCIDE® see...Trifluralin
TREFMID 50W® see...Trifluralin
TREVISSIMO® see...Diuron
TREVISSIMO® see...Glyphosate
TREVI® see...Hexythiazox
TREY® see...Siduron
TRI-6® see...Lindane
TRIACON-10® see...Triacontanol
1-Triacontanol see...Triacontanol
Triadimefon triazole fungicide see...Triadimefon
Triadimefone see...Triadimefon
Triadimeform see...Triadimefon
TRIADIMENOL® see...Triadimenol
TRIAFOL® see...Triadimenol
Tri-allate see...Triallate
TRIANGLE® see...Copper Sulfate
TRIANGLE® see...Kerosene
TRIAPHOL® see...Triadimenol
TRIASYM® see...Anilazine
TRIATIX® see...Amitraz
TRIATOX® see...Amitraz
TRIAZIN® see...Anilazine
1,2,4-Triazin-5-(4H)-one, 4-amino-6-(1,1-dimethylethyl)-3-(methylthio)- see...Metribuzin
1,2,4-Triazin-3(2H)-one, 4,5-dihydro-6-methyl-4-[(3-pyridinylmethylene)amino]-, (E)- (9CI) see..Pymetrozine
TRIAZINE® see...Anilazine
TRIAZINE A 384® see...Simazine
1,3,5-Triazine-2-amine, 4,6-dichloro-N-(2-chlorophenyl)- see...Anilazine
S-Triazine, 2,4-bis(isopropylamino)-6-methoxy- see. Prometon
S-Triazine,4,6-bis(isopropylamino)-2-(methylmercapto)- see...Prometryn
S-Triazine, 2,4-bis(isopropylamino)-6-(methylthio)- see. Prometryn
S-Triazine, 2-(tert-butylamino)-4-(ethylamino)-6-(methylthio)- see...Terbutryn
S-Triazine, 2-chloro-4,6-bis(ethylamino)- see...Simazine
S-Triazine, 2-chloro-4-ethylamino-6-(1-cyano-1-methyl)ethylamino- see...Cyanazine
S-Triazine, 2-chloro-4-(ethylamino)-6-(isopropylamino)- see...Atrazine
1,3,5-Triazine-2,4-diamine, N,N'-bis(1-methylethyl)-6-(methylthio)- see...Prometryn

1,3,5-Triazine-2,4-diamine, 6-chloro-*N*-ethyl-*N*-(1-methylethyl)- see...Atrazine
1,3,5-Triazine-2,4-diamine, 6-chloro-*N,N*'-diethyl- see...Simazine
1,3,5-Triazine-2,4-diamine, *N*-(1,1-dimethylethyl)-*N*'-ethyl-6-(methylthio)- see...Terbutryn
*S*-Triazine, 2,4-dichloro-6-(*o*-chloroanilino)- see...Anilazine
*S*-Triazine-2,4(1*H*,3*H*)-dione, 3-cyclohexyl-6-(dimethylamino)-1-methyl- see...Hexazinone
1,3,5-Triazine-2,4(1*H*,3*H*)-dione, 3-cyclohexyl-6-(dimethylamino)-1-methyl- see...Hexazinone
2-Triazine, 2-ethylamino-4-isopropylamino-6-methylthio- see...Ametryn
1,3,5-Triazine-2,4,6-triamine, *N*-cyclopropyl- see...Cyromazine
Triazolamine see...Amitrole
1,2,4-Triazol-3-amine see...Amitrole
1H-1,2,4-Triazol-3-amine see...Amitrole
1H-1,2,4-Triazol-3-ylamine see...Amitrole
*S*-Triazole, 3-amino- see...Amitrole
1*H*-1,2,4-Triazole, 1-[(*tert*-butylcarbonyl-4-chlorophenoxy)methyl]- see...Triadimefon
1*H*-1,2,4-Triazole, 1-[(2-(2-chloro-4-(4-chlorophenoxy)phenyl)-4-methyl-1,3-dioxolan-2-yl)methyl]- (9CI) see...Difenoconazole
1*H*-1,2,4-Triazole, 1-[2-(2,4-dichlorophenyl)pentyl]- see...Propiconazole
1*H*-1,2,4-Triazole, 1-[(2-(2,4-dichlorophenyl)-4-propyl-1,3-dioxolan-2-yl)methyl]- see...Propiconazole
1*H*-1,2,4-Triazole-1-ethanol, β-[(1,1'-biphenyl)-4-yloxy]-α-(1,1-dimethylethyl)- see...Bitertanol
1*H*-1,2,4-Triazole-1-ethanol, α-butyl-α-(2,4-dichlorophenyl)-, (±)- see...Hexaconazole
1*H*-1,2,4-Triazole-1-ethanol, β-(4-chlorophenoxy)-α-(1,1-dimethylethyl)- see...Triadimenol
1*H*-1,2,4-Triazole-1-ethanol, α-(4-chlorophenyl)-α-(1-cyclopropylethyl)- see...Cyproconazole
1*H*-1,2,4-Triazole-1-ethanol, α-[2-(4-chlorophenyl)ethyl]-α-(1,1-dimethylethyl)-, (±)- see...Tebuconazole
1*H*-1,2,4-Triazole-1-ethanol, β-[(2,4-dichlorophenyl)methylene]-α-(1,1,-dimethylethyl)-, [R-(E)]- see...Diniconazole
1*H*-1,2,4-Triazole-1-propanenitrile, α-butyl-α-(4-chlorophenyl) see...Myclobutanil
1*H*-1,2,4-Triazole-1-propanenitrile, α-[2-(4-chlorophenyl)ethyl]-α-phenyl- see...Fenbuconazole
δ-2-1,2,2,4-Triazoline, 5-imino- see...Amitrole
(1,2,4)Triazolo(1,5-α)pyrimidine-2-sulfonamide, *N*-(2,6-difluorophenyl)-5-methyl- see...Flumetsulam
Triazotion (Russian) see...Azinphos-ethyl
1-(1,2,4-Triazoyl-1)-1-(4-chloro-phenoxy)3,3-dimethylbutanone see...Triadimefon
TRIBAC® see...Trichlorobenzoic Acid
TRI-BAN® see...Pindone
Tri-basic copper sulfate see...Copper Sulfate
Tribasic sodium phosphate see...Trisodium Phosphate
Tribufos see...Tribufos
TRIBUTON® see...2,4-D
TRIBUTON® see...2,4,5-T
TRIBUTON® see...2,4,5-Trichlorophenoxyacetic Acid, Esters
Tributylchlorostannane see...Tributyltin Chloride
Tributylflourostannane see...Tributyltin Fluoride
*S,S,S*-Tributyl phosphorotrithioate see...Tribufos
*S,S,S*-Tributyl trithiophosphate see...Tribufos
Tri-*n*-butyltin chloride see...Tributyltin Chloride
Tri-*n*-butyltin flouride see...Tributyltin Fluoride
Tri-*n*-butylzinn-chlorid (German) see...Tributyltin Chloride
Tricalciumarsenat (German) see...Calcium Arsenate
Tricalcium arsenate see...Calcium Arsenate
Tricalcium diphosphide see...Calcium Phosphide
Tricalcium orthoarsenate see...Calcium Arsenate
TRICARBAMIX® see...Ferbam
TRICARBAMIX Z® see...Ziram
TRICAR® see...Carbaryl
1,1,1-Trichloor-2,2-bis(4-chloorfenyl)-ethaan (Dutch) see...DDT
(2,4,5-Trichloor-fenoxy)-azijnzuur (Dutch) see...2,4,5-T
2(2,4,5-Trichloor-fenoxy)-propionzuur (Dutch) see...Silvex
*O*-(2,4,5-Trichloor-fenyl)-*O,O*-dimethyl-monothiofosfaat (Dutch) see...Ronnel
Trichlor see...Chloropicrin
2,3,6-Trichlorbenzoesaeure (German) see...Trichlorobenzoic Acid
1,1,1-Trichlor-2,2-bis(4-chlor-phenyl)-aethan (German) see...DDT
2,4,6-Trichlorfenol (Czech, Spanish) see...Trichlorophenols
Trichlormethylfos see...Chlorpyrifos-methyl
(2,4,5-Trichlor-phenoxy)-essigsaeure (German) see...2,4,5-T
2-(2,4,5-Trichlor-phenoxy)propionsaeure (German) see...Silvex
*O*-(2,4,5-Trichlor-phenyl)-*O,O*-dimethyl-monothiophosphat (German) see...Ronnel
3,5,6-Trichloro-4-aminopicolinic acid see...Picloram
2,3,3-Trichloroallyl diisopropylthiocarbamate see...Triallate
2,3,3-Trichloroallyl *N,N*-diisopropylthiocarbamate see...Triallate
*S*-(2,3,3-Trichloroallyl) diisopropylthiocarbamate see...Triallate
*S*-2,3,3-Trichloroallyl *N,N*-diisopropylthiocarbamate see...Triallate
2,3,6-Trichlorobenzeneacetic acid see...Fenac
2,3,6-Trichlorobenzoic acid see...Trichlorobenzoic Acid
1,1,1-Trichloro-2,2-bis(*p*-anisyl)ethane see...Methoxychlor
Trichlorobis(4-chlorophenyl)ethane see...DDT
1,1,1-Trichloro-2,2-bis(*p*-chlorophenyl)ethane see...DDT
2,2,2-Trichloro-1,1-bis(4-chlorophenyl)ethanol see...Dicofol
2,2,2-Trichloro-1,1-bis(*p*-chlorophenyl)ethanol see...Dicofol

1,1,1-Trichloro-2,2-bis(4-methoxy-phenyl)aethane (German) see...Methoxychlor
1,1,1-Trichloro-2,2-bis(*p*-methoxyphenyl)ethane see. Methoxychlor
1,1,1-Trichloro-2,2-bis(*p*-methoxyphenol)ethanol see. Methoxychlor
Trichlorocarbanilide see...Triclocarban
3,4,4'-Trichlorocarbanilide see...Triclocarban
2,4,5-Trichloro-α-(chloromethylene)benzyl phosphate see...Tetrachlorvinphos
1,2,4-Trichloro-5-[(4-chlorophenyl)sulfonyl]-benzene see...Tetradifon
2,2,2-Trichloro-1,1-di(4-chlorophenyl)ethanol see. Dicofol
1,1,1-Trichloro-2,2-di(4-chlorophenyl)-ethane see..DDT
1,1,1-Trichloro-2,2-di(4-methoxyphenyl)ethane see. Methoxychlor
1,1,1-Trichloro-2,2-di(*p*-methoxyphenyl)ethane see. Methoxychlor
3,4,4'-Trichlorodiphenylurea see...Triclocarban
1,1-(2,2,2-Trichloroethylidene)bis(4-methoxybenzene) see...Methoxychlor
Trichlorofenol (Czech, Spanish) see...Trichlorophenols
2,3,4-Trichlorofenol (Czech, Spanish) see. Trichlorophenols
2,3,5-Trichlorofenol (Czech, Spanish) see. Trichlorophenols
2,3,6-Trichlorofenol (Czech, Spanish) see. Trichlorophenols
3,4,5-Trichlorofenol (Czech, Spanish) see. Trichlorophenols
Trichlorofon (Dutch) see...Trichlorfon
1,3,5-Trichloro-2-hydroxybenzene see..Trichlorophenols
2,2,2-Trichloro-1-hydroxyethyl-phosphonate, dimethyl ester see...Trichlorfon
(2,2,2-Trichloro-1-hydroxyethyl)phosphonic acid dimethyl ester see...Trichlorfon
2,4,4'-Trichloro-2'-hydroxydiphenyl ether see..Triclosan
Trichlorometafos see...Ronnel
Trichloromethane sulfenyl chloride see..Perchloromethyl Mercaptan
3-(Trichloromethyl)-5-ethoxy-1,2,4-thiadiazole see. Etridiazole
Trichloromethyl sulfur chloride see...Perchloromethyl Mercaptan
*N*-Trichloromethylmercapto-4-cyclohexene-1,2-dicarboximide see...Captan
*N*-(Trichloromethylmercapto)-$\delta^4$-tetrahydrophthalimide see...Captan
Trichloromethylsulfenyl chloride see...Perchloromethyl Mercaptan
Trichloromethylsulphenyl chloride PCV see. .Perchloromethyl Mercaptan
*N*-Trichloromethylthiocyclohex-4-ene-1,2-dicarboximide see...Captan
*N*-Trichloromethylthio-*cis*-$\delta^4$-cyclohexene-1,2-dicarboximide see...Captan

*N*-Trichloromethylthio-3a,4,7,7a-tetrahydrophthalimide see...Captan
Trichloronat see...Trichloronate
Trichloronitromethane see...Chloropicrin
Trichlorophene see...Trichlorfon
Trichlorophenol, 2,3,4- see...Trichlorophenols
Trichlorophenol, 2,3,5- see...Trichlorophenols
Trichlorophenol, 2,3,6- see...Trichlorophenols
Trichlorophenol, 2,4,6- see...Trichlorophenols
Trichlorophenol, 3,4,5- see...Trichlorophenols
2,3,4-Trichlorophenol see...Trichlorophenols
2,3,6-Trichlorophenol see...Trichlorophenols
2,4,6-Trichlorophenol see...Trichlorophenols
3,4,5-Trichlorophenol see...Trichlorophenols
2,4,5-Trichlorophenol, *O*-ester with *O*,*O*-dimethyl phosphorothioate see...Ronnel
2,4,5-Trichlorophenol-*O*-ester with *O*-ethyl ethylphosphonothioate see...Trichloronate
2,4,5-Trichlorophenoxy-α- see...Silvex
2,4,5-Trichlorophenoxyacetic acid see...2,4,5-T
2,4,5-Trichlorophenoxyacetic acid see...2,4,5-Trichlorophenoxyacetic Acid, Esters
2,4,5-Trichlorophenoxyacetic acid, butoxyethanol ester see...2,4,5-Trichlorophenoxyacetic Acid, Esters
2,4,5-Trichlorophenoxyacetic acid, butyl ester see..2,4,5-Trichlorophenoxyacetic Acid, Esters
2,4,5-Trichlorophenoxyacetic acid, 2-ethylhexyl ester see...2,4,5-Trichlorophenoxyacetic Acid, Esters
2,4,5-Trichlorophenoxyacetic acid, isooctyl ester see. 2,4,5-Trichlorophenoxyacetic Acid, Esters
2,4,5-Trichlorophenoxyacetic acid, isopropyl ester see. 2,4,5-Trichlorophenoxyacetic Acid, Esters
2-(2,4,5-Trichlorophenoxy)propanoic acid see...Silvex
α-(2,4,5-Trichlorophenoxy)propanoic acid see...Silvex
2,3,6-Trichlorophenylacetic acid see...Fenac
*S*-(2,3,3-Trichloro-2-propenyl)bis(1-methylethyl)carbamothioate see...Triallate
(3,5,6-Trichloro-2-pyridinyl)oxyacetic acid, triethylamine salt see...Triclopyr, Triethylammonium Salt
[(3,5,6-Trichloro-2-pyridyl)oxy]acetic acid, compound with triethylamine (1:1) see...Triclopyr, Triethylammonium Salt
3,5,6-Trichloro-2-pyridinyloxyacetic acid see..Triclopry
3,5,6-Trichloro-2-pyridinyloxyacetic acid, TEA salt see. Triclopyr, Triethylammonium Salt
[(3,5,6-Trichloro-2-pyridinyl)oxy]acetic acid compounded with *N*,*N*-diethylethanamine (1:1) see. .Triclopyr, Triethylammonium Salt
*O*-3,5,6-Trichloro-2-pyridylphosphorothioate see. .Chlorpyrifos
2,3,6-Trichlorphenylessigsaeure (German) see...Fenac
Trichlorphon see...Trichlorfon
TRICHLORPHON FN® see...Trichlorfon
TRICLOPRY-EZ-JECT® see...Triclopyr, Triethylammonium Salt
Triclopyr, triethylamine see...Triclopyr, Triethylammonium Salt
Triclopyr, triethylamine salt see...Triclopry

1,1,1-Tricloro-2,2-bis(4-cloro-fenil)-etano (Italian) see. DDT
O-(2,4,5-Tricloro-fenil)-O,O-dimetil-monotiofosfato (Italian) see...Ronnel
Triclorofenol see...Trichlorophenols
TRI-CLOR® see...Chloropicrin
TRICLOSAN® see...Triclosan
TRI-CON® see...Chloropicrin
TRICOP 50® see...Copper Oxychloride
Tricyclohexylhydroxystannane and ENT 27395-X see. Cyhexatin
Tricyclohexyltin hydroxide see...Cyhexatin
[1S(1α,4α,5β,6α)]-1,5,6-Trideoxy-4-O-β-D-glucopyranosyl-5-(hydromethyl)-1-[(4,5,6-trihydroxy-3-(hydroxymethyl)-2-cyclohexen-1-yl)amino]-D-chiroinositol see...Validamycin
TRIDIPAM® see...Thiram
TRI-ENDOTHAL® see...Endothall
Trieste flowers see...Pyrethrins or Pyrethrum
TRIESTER II® see...Mecoprop
Triethanolamine DBS see...Triethanolamine Dodecylbenzene Sulfonate
Triethanolamine dodecylbenzenesulfonate see. Norflurazon
Triethanolamine dodecylbenzene sulfonate see. Triethanolamine Dodecylbenzene Sulfonate
Triethylamine triclopyr see...Triclopyr, Triethylammonium Salt
Triethylammonium triclopyr see...Triclopyr, Triethylammonium Salt
TRIFARMON® see...Trifluralin
TRIFARMON FL® see...Linuron
TRI-FEN® see...Fenac
TRIFENE® see...Fenac
Trifenyltinacetaat (Dutch) see...Triphenyltin Compounds
Trifenyl-tin-hydroxyde (Dutch) see...Triphenyltin Compounds
3-(5-Trifluormethylphenyl)-, dimethylharnstoff (German) see...Fluometuron
α,α,α-Trifluoro-2,6-dinitro-N,N-dipropyl-4-toluidine see. Trifluralin
α,α,α-Trifluoro-2,6-dinitro-N,N-dipropyl-p-toluidine see. Trifluralin
α,α,α-Trifluoro-2,6-dinitro-N,N-ethylbutyl-p-toluidine see...Benefin
Trifluoro-3'-isopropoxy-o-toluanalide see...Flutolanil
3-(3-Trifluoromethylphenyl)-1,1-dimethylurea see. Fluometuron
3-(m-Trifluoromethylphenyl)-1,1-dimethylurea see. Fluometuron
N-(m-Trifluoromethylphenyl)-N',N'-dimethylurea see. Fluometuron
N-(3-Trifluoromethylphenyl)-N',N'-dimethylurea see. Fluometuron
1,1,1-Trifluoro-N-[2-methyl-4-(phenylsulfonyl)phenyl]methanesulfonamide see. Perfluidone

2-[4-((5-(Trifluoromethyl)-2-pyridinyl)oxy]-phenoxy]propanoic acid, butyl ester see...Fluazifop-butyl
(RS)-2-[4-(5-Trifluoromethyl-2-pyridyloxy)-phenoxy]propanoic acid, butyl ester see...Fluazifop-butyl
Trifluralina (Spanish) see...Trifluralin
TRIFLURALINA® 600 see...Trifluralin
Trifluraline see...Trifluralin
TRIFLUREX® see...Trifluralin
TRIFLURON® see...Linuron
TRIFMINE® see...Triflumizole
TRI-FORM® see...Chloropicrin
TRIFUME® see...Chloropicrin
TRIFUNGOL® see...Ferbam
TRIFUREX® see...Trifluralin
TRIGARD® see...Cyromazine
TRIGARD® see...Trifluralin
TRIGOSAN® see...Phenylmercury Acetate
TRIHERBIDE® see...Propham
TRIHERBIDE-IPC® see...Propham
2,4a,7-Trihydroxy-1-methyl-8-methylenegibb-3-ene-1,10-dicarboxylic acid, 1,4a-lactone see...Gibberellic Acid
Trihydroxymethylnitromethane see. Tris(hydroxymethyl)nitromethane
TRIKEPIN® see...Trifluralin
TRI-KILL® see...Sodium Metaborate
TRILIN® see...Linuron
TRILIN® see...Trifluralin
TRILIN® 10G see...Trifluralin
TRI-LUX® see...Cuprous Thiocyanate
TRIMANGOL® see...Maneb
TRIMANOC® see...Maneb
TRIMANZONE® see...Zineb
TRIMARAN® see...Trifluralin
TRIMATON (dihydrate)® see...Metham-Sodium
TRIMATRON® see...Metham-Sodium
TRIMAX® see...Imidacloprid
TRIMEC® see...MCPA
TRIMEC 1144 40% SP® see...Mecoprop
TRIMEGOL® see...Captan
Trimethacarb see...Trimethacarb
Trimethoate see...Prothoate
Trimethyl-β-chlorethylammoniumchlorid see. Chlormequat Chloride
Trimethyl-β-chloroethyl ammonium chloride see. Chlormequat Chloride
3,7,11-Trimethyl-2,4-dodecadienoic acid 2-propynyl ester see...Kinoprene
2,2,4-Trimethyl-6-ethoxy-1,2-dihydroquinoline see. Ethoxyquin
Trimethylolnitromethane see..Tris(hydroxymethyl)nitromethane
3,4,5-Trimethylphenyl methylcarbamate and 2,3,5-trimethylphenyl methylcarbamate see...Trimethacarb
Trimethylsulfonium carboxymethylaminomethyl phosphonate see...Sulfosate

Trimethylsulfonium N-phosphonomethylglycine see. Sulfosate
Trimethylsulfonium salt of N-(phosphonomethyl)glycine see...Sulfosate
2-(Trimetil-acetil)-indan-1,3-dione (Italian) see..Pindone
TRIM® see...Trifluralin
TRINATOX-D® see...Ametryn
Trinatriumhexafluoroaluminat (German) see...Sodium Aluminum Fluoride
Trinatriumphosphat (German) see...Trisodium Phosphate
TRINEX® see...Trichlorfon
Trinexapac-ethyl see...Cimectacarb
TRINOXOL® see...2,4-D
TRINOXOL® see...2,4,5-T
TRINOXOL® see...2,4,5-Trichlorophenoxyacetic Acid, Esters
TRIOXAL® see...2,4,5-T
Trioxane see...Formaldehyde
Trioxido de arsenico (Spanish) see...Arsenous Oxide
TRIOX® see...Oxyfluorfen
TRIOXON® see...2,4,5-Trichlorophenoxyacetic Acid, Esters
TRIOXONE® see...2,4,5-T
TRIOXONE® (butyl ester) see...2,4,5-Trichlorophenoxyacetic Acid, Esters
TRIOXONE® see...2,4,5-Trichlorophenoxyacetic Acid, Esters
TRIOXON® see...2,4,5-T
TRIPART® see...Zineb
TRIPART® ATRAZINE 50 SC see...Atrazine
TRIPART® BLUE see...Zineb
TRIPART FABER® see...Chlorothalonil
TRIPART TRIFLURALIN 48 EC see...Trifluralin
TRIPART ULTRAFABER® see...Chlorothalonil
TRIPCNB® see...Quintozene
Triphenylaceto stannane see...Triphenyltin Compounds
Triphenylchlorostannane see...Triphenyltin Compounds
Triphenylchlorotin see...Triphenyltin Compounds
Triphenylstannanol see...Fentin Hydroxide
Triphenylstannanol see...Triphenyltin Compounds
Triphenylstannium hydroxide see...Fentin Hydroxide
Triphenylstannium hydroxide see...Triphenyltin Compounds
Triphenyltin acetate see...Triphenyltin Compounds
Triphenyltin chloride (U.S. EPA) see...Triphenyltin Compounds
Triphenyltin hydroxide organotin fungicide see...Fentin Hydroxide
Triphenyltin hydroxide organotin fungicide see. Triphenyltin Compounds
Triphenyltin(IV) hydroxide see...Triphenyltin Compounds
Triphenyltin(IV) hydroxide see...Fentin Hydroxide
Triphenyltin hydroxide (U.S. EPA) see...Triphenyltin Compounds
Triphenyltin oxide see...Triphenyltin Compounds
Triphenyltin oxide see...Fentin Hydroxide

Triphenyl-zinnacetat (German) see...Triphenyltin Compounds
Triphenyl-zinnhydroxid (German) see...Fentin Hydroxide
Triphenyl-zinnhydroxid (German) see...Triphenyltin Compounds
TRIPLE KILL T® see...EPN
TRIPLE-TIN® see...Fentin Hydroxide
TRIPLE-TIN® see...Triphenyltin Compounds
TRIPLE TIN 4L® see...Fentin Hydroxide
TRIPLE TIN 4L® see...Triphenyltin Compounds
TRIPLET® see...Mecoprop
TRIPOMOL® see...Thiram
TRIQUINTAM® see...Quintozene
Tris(dimethylcarbamodithioato-S,S')iron see...Ferbam
Tris(dimethyldithiocarbamato)iron see...Ferbam
Tris(N,N-dimethyldithiocarbamato)iron(III) see..Ferbam
Tris(hydroxymethyl)nitromethane see. Tris(hydroxymethyl)nitromethane
TRIS NITRO® see...Tris(hydroxymethyl)nitromethane
Trisodium orthophosphate see...Trisodium Phosphate
Trisodium trifluoride see...Sodium Fluoride
TRISTAR® see...Trifluralin
TRITEX-EXTRA® see...Sethoxydim
TRITHAC® see...Maneb
TRITHAC® see...Zineb
TRITHION® MITICIDE see...Carbophanothion
TRITICOL® see...Carbendazim
TRITISAN® see...Quintozene
TRITOFTOROL® see...Zineb
TRITON® K-60 see...Zilkonium Chloride
TRIVEX® see...Dichlorophene
TRIVITAN® see...Cholecalciferol
TRIZILIN® see...Nitrofen
TRIZIMAN® see...Mancozeb
TRIZIMAN-D® see...Mancozeb
TROCLOSENE® see...Dichloroisocyanuric Acid
TROLEN® see...Ronnel
TROLENE® see...Ronnel
TROMETE® see...Trisodium Phosphate
TROOPER® see...Dicamba
TROPAEOLIN D® see...Fenaminosulf
TROP® see...Glyphosate
TROPHY® see...Acetochlor
TROYSAN® 142 see...Dazomet
TROYSAN® COPPER 8% see...Copper Naphthenate
TROYSAN® COPPER 11.5% see...Copper Naphthenate
TRUBAN® see...Etridiazole
TRUENO® see...Hexaflumuron
TRUMPET® see...Naled
TRUST® see...Trifluralin
TRYBEN® see...Trichlorobenzoic Acid
TRYSBEN 200® see...Trichlorobenzoic Acid
TS 219® see...Paraoxon
TS-7236® see...Fluazifop-butyl
TSAR® see...Propetamphos
TSC see...Thiosemicarbazide
TSIDIAL® see...Phenthoate

TSIMAT® see...Ziram
TSINEB® see...Zineb
TSIRAM® (Russia) see...Ziram
TSITREX® see...Dodine
TSITREX® see...Oxythioquinox
TSP see...Trisodium Phosphate
TSUMACIDE® see...Metolcarb
TSUMAUNKA® see...Metolcarb
TSZ see...Thiosemicarbazide
TTD see...Disulfiram
TTD see...Thiram
TTEA see...Triclopyr, Triethylammonium Salt
TTS see...Disulfiram
TUADS® see...Thiram
TUBATOXIN® see...Rotenone
TUBERGRAN® see...Quintozene
TUBERIT® see...Propham
TUBERITE® see...Propham
TUBOTHANE® see...Maneb
TUBOTIN® see...Fentin Hydroxide
TUBOTIN® see...Triphenyltin Compounds
TUDY® see...Amitraz
TUEX® see...Thiram
TUFFCIDE® see...Chlorothalonil
TUGEN® see...Propoxur
TUGON® see...Trichlorfon
TUGON FLY BAIT® see...Trichlorfon
TUGON STABLE SPRAY® see...Trichlorfon
TULISAN® see...Thiram
TUMBLEAF® see...Sodium Chlorate
TUMESCAL OPE® see...o-Phenylphenol
TUNIC® see...Methazole
TUPERSAN® see...Siduron
TUR® see...Chlormequat Chloride
TURBAIR® DICAMATE Z-78® see...Zineb
TURBAIR GRAIN STORAGE INSECTICIDE® see. Fenitrothion
TURBAIR® ZINEB see...Zineb
TURBO® see...Metolachlor
TURBSVIL® see...Terbacil
TURCAM® see...Bendiocarb
TURF BUILDER WITH WEED CONTROL® see. Quinclorac
TURF-CAL® see...Calcium Arsenate
TURFCIDE® see...Quintozene
TURF! EZ® see...Propanil
TURF FERTILIZER® see...Oryzalin
TURFIC® see...Oxadiazon
TURFLAN® see...Trifluralin
TURFLON® see...Triclopry
TURFLON® AMINE see...Triclopyr, Triethylammonium Salt
TURF MANAGER® see...Paclobutrazole
TURNOUT® see...Ammonium Nitrate
TURPLEX® see...Azadirachtin
TWAWPIT® see...Carbon Tetrachloride
TWIN® see...Phenmedipham
TWIN LIGHT RAT AWAY® see...Warfarin

TWINSPAN® see...Chlorpyrifos
TWISTE® see...Fomesafen
TYPHOON® see...Fomesafen

- U -

U-32.104® see...Carbendazim
U 46® see...2,4-D
U46® see...Dichlorprop
U 46® see...MCPA
U 46® see...Mecoprop
U 46® see...2,4,5-Trichlorophenoxyacetic Acid, Esters
U 46 KV-ESTER® see...Mecoprop
U 46 KW® (butyl ester) see...2,4,5-Trichlorophenoxyacetic Acid, Esters
U 46T® (isooctyl ester) see...2,4,5-Trichlorophenoxyacetic Acid, Esters
U 1363® see...Diphacione
U 2069® see...Dichloran
U 3886® see...Sodium Azide
U 4513® see...Diphenamid
U 4527® see...Cycloheximide
U 5043® see...2,4-D
UBI-N 252® see...Dimethipin
UC 9880® see...Promecarb
UC 10854® see...Phenol, 3-(1-methyethyl)-, methylcarbamate
UC 20299® see...Fluenetil
UC 21865® see...Aldoxycarb
UC 27867® see...Trimethacarb
UC 51762® see...Thiodicarb
UC 51769® see...Thiodicarb
UC 62644® see...Chlorfluazuron
UC 80502® see...Thiodicarb
UCAR butylphenol 4-*tert* see...Butylphenols
UCC 974® see...Dazomet
UCECOAT® see...Thiram
UCECRYL® see...Thiram
UL® see...Demeton
ULTIMA® see...Sethoxydim
ULTRA BLAZER® see...Acifluorfen
ULTRACIDE® see...Methidathion
ULTRA FRESH® see...Triclosan
ULTRA ZINC DMC® see...Ziram
ULTRIA® see...Triacontanol
ULVAIR® see...Monocrotophos
UMBETHION® see...Coumaphos
UMET® see...Phorate
UN 1040 see...Ethylene Oxide
UN 1846 see...Carbon Tetrachloride
UNDEN® see...Propoxur
UNDENE® see...Propoxur
UNICROP DNBP® see...Dinoseb
UNICROP MANEB® see...Maneb
UNIDRON® see...Diuron
UNIFOS® see...Dichlorvos
UNIFUME® see...Ethylene Dibromide
UNION CARBIDE 7,744® see...Carbaryl

UNION CARBIDE UC 9880® see...Promecarb
UNION CARBIDE UC 10,854® see...Phenol, 3-(1-methyethyl)-, methylcarbamate
UNION CARBIDE UC 21149® see...Aldicarb
UNIPON® see...Dalapon
UNIQUAT® see...Paraquat
UNIROYAL® 604 see...Dichlone
UNIROYAL® D-014 see...Propargite
UNITED CHEMICAL DEFOLIANT NO. 1® see.Sodium Chlorate
UNITOX® see...Chlorfenvinphos
UNITOX® see...Dichlorvos
UNIVERM see...Carbon Tetrachloride
UNIX® see...Cyprodinil
UPBEET® see...Triflusulfuron-Methyl
UPJOHN U-36059® see...Amitraz
Uracil see...Bromacil
Uracil, 3-*tert*-Butyl-5-chloro-6-methyl- see...Terbacil
URAGAN® see...Bromacil
URAGON® see...Bromacil
URANUS® see...Trifluralin
URANUS® see...Linuron
Urea, *N*-(4-bromophenyl)-*N*-methoxy-*N*-methyl- see..Metobromuron
Urea, 1-(5-*tert*-butyl-1,3,4-thiadiazol-2-yl)-1,3-dimethyl- see...Tebuthiuron
Urea, 2-(5-*tert*-butyl-1,3,4-thiadiazol-2-yl)-1,3-dimethyl- see...Tebuthiuron
Urea, *N*-[2-(2-chloroethoxy)phenylsulfonyl]-*N*'-(6-methoxy-4-methyl-1,3,5-triazinyl-2-yl)- see.Triasulfuron
Urea, *N*'-(3-chloro-4-methylphenyl)-*N*,*N*-dimethyl- see. Metoxuron
Urea, *N*'-[4-(4-chlorophenoxy)phenyl]-*N*,*N*-dimethyl- see. Chloroxuron
Urea, 3-[*p*-(*p*-chlorophenoxy)phenyl]-1,1-dimethyl- see. Chloroxuron
Urea, *N*-(4-chlorophenyl)-*N*'-(3,4-dichlorophenyl)-(9CI) see...Triclocarban
Urea, 1-(*p*-chlorophenyl)-3-(2,6-difluorobenzoyl)- see. Diflubenzuron
Urea, *N*'-(4-chlorophenyl)-*N*,*N*-dimethyl- see...Monuron
Urea, 3-(*p*-chlorophenyl)-1,1-dimethyl- see...Monuron
Urea, *N*-[(4-chlorophenyl)methyl]-*N*-cyclopentyl-*N*'-phenyl- see...Pencycuron
Urea, 1-[(Ochlorophenyl)sulfonyl]-3-(4-methoxy-6-methyl-*S*-triazin-2-yl)- see...Chlorsulfuron
Urea, *N*-(2-chloro-4-pyridinyl)-*N*'-phenyl- see. Forchlorfenuron
Urea, 3-(3,4-dichlorophenyl)-1,1-dimethyl- see..Linuron
Urea, *N*'-(3,4-dichlorophenyl)-*N*,*N*-dimethyl- see.Linuron
Urea, 3-(3,4-dichlorophenyl)-1-methoxy-1-methyl- see. Linuron
Urea, *N*'-(3,4-dichlorophenyl)-*N*-methoxy-*N*-methyl- see. Linuron

Urea dihydrogen sulfate see...Monocarbamide Dihydrogen Sulfate
Urea, *N*,*N*-dimethyl-*N*'-[4-(1-methylethyl)phenyl]- see. Isoproturon
Urea, *N*-(5-(1,1-dimethylethyl)-1,3,4-thiadiazol-2-yl)-*N*,*N*-dimethyl- see...Tebuthiuron
Urea, 1,1-dimethyl-3-(α,α,α-trifluoro-m-tolyl)- see...Fluometuron
Urea, *N*,*N*-dimethyl-*N*'-[3-(trifluoromethyl)phenyl]- see. Fluometuron
Urea, –(2-methylcyclohexyl)-*N*'-phenyl- see...Siduron
Urea,1-(1-naphthyl)-2-thio- see...ANTU
Urea, *N*-(4-nitrophenyl)-*N*'-(3-pyridinylmethyl)- see. Pyriminil
Urea, –phenyl-*N*'-1,2,3-thiadiazol-5-yl (9CI) see. Thidiazuron
Urea, sulfate (1:1) (9CI) see...Monocarbamide Dihydrogen Sulfate
Urea sulfuric acid monoadduct see...Monocarbamide Dihydrogen Sulfate
UREABOR® see...Sodium Metaborate
UREABOR® see...Trichlorobenzoic Acid
UREAPHIL® see...Urea
UREKA® see...Triacontanol
UREOPHIL® see...Urea
UREVERT® see...Urea
UROX D® see...Diuron
UROX® see...Bromacil
UROX B WATER SOLUBLE CONCENTRATE WEED KILLER® see...Bromacil
UROX HX GRANULAR WEED KILLER® see. Bromacil
USAF B-22 see...Bithionol
USAF Cy-2 see...Calcium Cyanamide
USB-3153® see...Prodiamine
USR® 604 see...Dichlone
U.S. RUBBER® 604 see...Dichlone
U.S. RUBBER D-014® see...Propargite
USTAAD® see...Cypermethrin
UVECRYL® see...Thiram
UVON® see...Prometryn
UZGN® see...Benomyl

- V -

V 4X® see...Carboxin
V-18® see...Tetradifon
V-10086® see...Lactofen
V-C 9-104® see...Ethoprop
V.C.S® see...Leptophos
VACATE® see...MCPA
VACOR® see...Pyriminil
VADEN® see...Metolcarb
VAL-DROP® see...Sodium Chlorate
VALENT ORTHENE TECHNICAL® see...Acephate

VALEXONE® see...Phoxim
VALIDACIN® see...Validamycin
Validamycin A see...Validamycin
VALIMON® see...Validamycin
VALINOMICIN® see...Validamycin
VALOR® see...Imazethapyr
VALOR® see...Pendimethalin
VAMPIRINIP® see...Warfarin
VANCIDE® see...Maneb
VANCIDE 51Z® see...Ziram
VANCIDE BL® see...Bithionol
VANCIDE FE95® see...Ferbam
VANCIDE KS® see...Fentin Hydroxide
VANCIDE KS® see...Triphenyltin Compounds
VANCIDE MANEB 80® see...Maneb
VANCIDE MZ-96® see...Ziram
VANCIDA TM-95® see...Thiram
VANCIDE TM® see...Thiram
VANDODINE® see...Dodine
VANGARD® K see...Captan
VANGARD® see...Phenmedipham
VANGUARD® see...Cyprodinil
VANGUARD® K see...Captan
VANICIDE® see...Captan
VANICIDE® 89 see...Captan
VANICIDE® 89RE see...Captan
VANICIDE® P-75 see...Captan
VANQUISH® see...Dicamba
VANTAGE® see...Sethoxydim
VAPAM® see...Metham-Sodium
VAPAM (dihydrate)® see...Metham-Sodium
VAPCOR® see...Metribuzin
VAPCOTHION® see...Dicofol
VAPCOTHION® see...Tetradifon
VAPCOZIN TAKTIC® see...Amitraz
VAPONA® see...Kerosene
VAPONA® see...Dichlorvos
VAPONITE® see...Dichlorvos
VAPOPHOS® see...Parathion
VAPORIN® DAIRY SPRAY see...Crotoxyphos
VAPOROOTER (dihydrate)® see...Metham-Sodium
VAPORPH$_3$OS® see...Phosphine
VAPOTONE® see...TEPP
VARBEX® see...Famphur
VARDHAK® see...1-Naphthaleneacetic Acid
VARFINE® see...Warfarin
VARIKILL® see...Fenoxycarb
Varioform I see...Ammonium Nitrate
VARIOFORM II® see...Urea
Varnish makers' & painters' naphtha see...Naphthas
Varnoline see...Stoddard Solvent
VASSGRO MANEX® see...Maneb
Vaterite (mineral) see...Calcium Carbonate
VCN see...Acrylonitrile
VCS 438® see...Methazole

VCS-506® see...Leptophos
VDM® see...Metham-Sodium
VECTAL® see...Atrazine
VECTRIN® see...Resmethrin
VEGA-RAND® see...Allidochlor
VEGA-RAND® see...Sulfallate
VEGABEN® see...Chloramben
VEGADEX® see...Sulfallate
VEGADEX SUPER® see...Sulfallate
VEGATROLE® see...Diquat
VEGETOX® see...Carboxin
VEGETOX® see...Cartap Hydrochloride
VEGETROLE® see...Diquat Dibromide
Vegfru (Indian) see...Malathion
VEGFRU TARGET® see...Acephate
VEGFRU® see...Phorate
VEGIBEN® see...Chloramben
VEL 3973® see...Mefluidide
VEL 4283® see...Propetamphos
VEL 4284® see...Formothion
VELPAR® see...Hexazinone
VELPAR WEED KILLER® see...Hexazinone
VELSICOL 53 CS 17® see...Heptachlor Epoxide
VELSICOL 58-CS-11® see...Dicamba
VELSICOL 104® see...Heptachlor
VELSICOL 506® see...Leptophos
VELSICOL® 1068 see...Chlordane
VELSICOL COMPOUND R® see...Dicamba
VENDEX® see...Fenbutatin Oxide
VENTOX® see...Carbon Tetrachloride
VENTOX® see...Acrylonitrile
VENTUROL® see...Dodine
VENTUROL VONODINE® see...Oxythioquinox
VEON® see...2,4,5-T
VEON 245® see...2,4,5-T
VEON 245® see...2,4,5-Trichlorophenoxyacetic Acid, Esters
VERAZINC® see...Zinc Sulfate Heptahydrate
VERCIDON® see...Dithiazanine Iodide
Verdan senescence inhibitor see...6-Benzaldenine
VERDICAN® see...Dichlorvos
VERDICT® see...Haloxyfop-methyl
VERDIPOR® see...Dichlorvos
VERDISOL® see...Dichlorvos
VERDONE® see...Mecoprop
VERGEMASTER® see...2,4-D
VERGFRU FORATOX® see...Phorate
VERISAN® see...Iprodione
VERMICIDE BAYER 2349® see...Trichlorfon
VERMITHANA® see...Dichlorophene
VERMITIN® see...Phenothiazine
VERMOESTRICID® see...Carbon Tetrachloride
VERNAM® see...Vernolate
VERNAM®-E see...Vernolate
VERNAM®-G see...Vernolate

VERSAR® see...Sodium Methanearsonate (MSMA)
Vertac Chemical Corp. (USA) see...2,4-D
VERTAC DINITRO WEED KILLER® see...Dinoseb
VERTAC® see...Propanil
VERTAC 90%® see...Toxaphene
VERTAC GENERAL WEED KILLER® see...Dinoseb
VERTAC METHYL PARATHION TECHNISCH 80%® see...Methyl Parathion
VERTAC SELECTIVE WEED KILLER® see...Dinoseb
VERTAC TOXAPHENE 90® see...Toxaphene
VERTAGREEN® see...Oxadiazon
VERTHION® see...Fenitrothion
VERTIMEC® see...Abamectin
VERTON® see...2,4-D
VERTON 2T® see...2,4,5-T
VERTON 2T® see...2,4,5-Trichlorophenoxyacetic Acid, Esters
VESAKONTUHO® see...MCPA
VET-KEM® see...Phosmet
VETIOL® see...Malathion
VETO® see...EPN
VETRAZIN® see...Cyromazine
VETSTREP® see...Streptomycin Sulfate
VI-CAD® see...Cadmium Chloride
VI-PAR® see...Mecoprop
VI-PEX® see...Mecoprop
Vicknite see...Potassium Nitrate
VIDDEN-D® see...D-D mixture
VIDDEN D® see...1,2-Dichloropropane
VIDDEN D® see...1,3-Dichloropropene
VIDON 638® see...2,4-D
Vienna green see...Copper Acetoarsenite
Vienna green see...Paris Green
VIGILANTE® see...Diflubenzuron
VIGOR® see...Gibberellic Acid
VIGORSAN® see...Cholecalciferol
VIKANE® see...Sulfur Fluoride
VIKANE FUMIGANT® see...Sulfur Fluoride
VIKOL® see...Triclosan
VIKROL® RQ see...Zilkonium Chloride
VILLIAUMITE® see...Sodium Aluminum fluoride
VILLIAUMITE® see...Sodium Fluoride
VINCHLOZOLINE® see...Vinclozolin
VINCLOZOLINE® see...Vinclozolin
Vinyl amide see...Acrylamide
Vinyl carbinol see...Allyl Alcohol
Vinyl carbinol,2-propenol see...Allyl Alcohol
Vinyl cyanide see...Acrylonitrile
Vinyl cyanide, propenenitrile see...Acrylonitrile
VINYLOFOS® see...Dichlorvos
VINYLOPHOS® see...Dichlorvos
VINYLPHATE® see...Chlorfenvinphos
VINYZENE® see...Triclosan
Viologen, Methyl- see...Paraquat
VIOXAN® see...Carbaryl

VIOZENE® see...Ronnel
VIPEX® see...Mecoprop
VIRGINIA-CAROLINA VC 9-104® see...Ethoprop
VIRICUIVRE® see...Copper Oxychloride
VIROSIN® see...Antimycin A
VISINE-AC® see...Zinc Sulfate Heptahydrate
VISION® see...Cimectacarb
VISKO-RHAP® see...2,4-D
VISKO-RHAP® see...Dichlorprop
VISKO RHAP® LOW VOLATILE ESTER see...2,4,5-T
VISKO RHAP® LOW VOLATILE ESTER see...2,4,5-Trichlorophenoxyacetic Acid, Esters
VISKO® see...2,4-D
VISOR® see...Thiazopry
VISTAR® see...Mefluidide
VISTA® see...Fluroxypyr 1-methylheptyl Ester
VITAFLO® see...Carboxin
VITAFLO 280® see...Thiram
Vitamin D$_3$ see...Cholecalciferol
VITARON® see...Methamidophos
VITAVAX® see...Carboxin
VITAVAX® see...Maneb
VITAVAX® see...Quintozene
VITAVAX® see...Thiabendazole
VITAVAX® see...Thiram
VITAVAX EXTRA® see...Imazalil
VITAVAX-T® see...Thiram
VITAVEX® see...Oxycarboxin
VITAX MICRO GRAN® see...Ferric Sulfate
VITAX TURF TONIC® see...Ferric Sulfate
VITIGRAN® see...Copper Oxychloride
VITIGRAN BLUE® see...Copper Oxychloride
VITINC DAN-DEE-3® see...Cholecalciferol
VITO SPOT FUNGICIDE® see...Fentin Hydroxide
VITO SPOT FUNGICIDE® see...Triphenyltin Compounds
VITON® see...Lindane
VITREX® see...Parathion
VIVIFUL® see...Prohexadione Calcium
VMI 10-3® see...4-Aminopyridine
Volatile oil of mustard see...Allyl Isothiocyanate
VOLATON® see...Phoxim
VOLCANO® see...Sulfluamid
VOLFARTOL® see...Trichlorfon
VOLFAZOL® see...Crotoxyphos
VOLUNTEERED® see...Dalapon
VOMZLATE® see...Vernolate
VONDALDHYDE® see...Maleic Hydrazide
VONDCAPTAN® see...Captan
VONDODINE® see...Dodine
VONDOZEB PLUS® see...Mancozeb
VONDRAX® see...Maleic Hydrazide
VONDURON® see...Diuron
VORLAND® see...Vinclozolin

VORLEX® see...1,3-Dichloropropene
VORONITE® see...Fuberidazole
VORONIT® see...Fuberidazole
VOROX® see...Amitrole
VOTEXIT® see...Trichlorfon
VPM (dihydrate)® see...Metham-Sodium
VPM® Fungicide see...Metham-Sodium
VPN® see...Metham-Sodium
VUAGT-1-4® see...Thiram
VULCAFOR TMTD® see...Thiram
VULKACIT MTIC® see...Thiram
VULKACIT NPV/C2® see...Ethylene Thiourea
VULKACIT THIURAM® see...Thiram
VULKACIT THIURAM/C® see...Thiram
VULNOPOL® NM see...Sodium Dimethyldithiocarbamate
VYDATE® see...Oxamyl
VYDATE 10G® see...Oxamyl
VYDATE INSECTICIDE/NEMATICIDE® see..Oxamyl
VYDATE L® see...Oxamyl
VYDATE OXAMYL INSECTICIDE/NEMATOCIDE® see...Oxamyl

- W -

W VII/117® see...Fuberidazole
W 491® see...Crimidine
W 6658® see...Simazine
WACKER 14/10® see...Dimefox
WALLOP® see...Propachlor
WARAN® see...Warfarin
WARBEX® see...Famphur
WARCOUMIN® see...Warfarin
WARECURE C® see...Ethylene Thiourea
WARF COMPOUND® see...Warfarin
WARFARAT® see...Warfarin
Warfarine (French) see...Warfarin
WARFILONE® see...Warfarin
WARRIOR® see...*lamda*-Cyhalothrin
WARRIOR® see...Linuron
WATERSHED WP® see...Pentachlorophenol
WAXOLINE YELLOW O® see...Auramine
WAYLAY® see...Napropamide
WAY-UP® see...Pendimethalin
WC-REINIGER® see...Phosphoric Acid
WEC 50® see...Trichlorfon
WEED-108® see...Sodium Methanearsonate (MSMA)
WEED-AG-BAR® see...2,4-D
WEEDAR MCPA CONCENTRATE® see...MCPA
WEEDAR® see...2,4,5-T
WEEDAR® see...2,4,5-Trichlorophenoxyacetic Acid, Esters
WEEDAR® see...MCPA
WEEDAR® see...Amitrole
WEEDAR® see...2,4-D

WEEDAZIN® see...Amitrole
WEEDAZOL® see...Amitrole
WEEDBEADS® see...Sodium Pentachlorophenate
WEED-B-GON® see...Calcium Methanearsonate
WEED-B-GON® see...2,4-D
WEED-BROOM® see...Bromacil
WEED-BROOM® see...Sodium Methanearsonate (MSMA)
WEED-E-RAD® see...Sodium Methanearsonate (MSMA)
WEEDEX® see...Amitrole
WEEDEX® see...Atrazine
WEEDEX® see...Simazine
WEEDEX® see...Triclopyr, Triethylammonium Salt
WEEDEZ WONDER BAR® see...2,4-D
WEED DRENCH® see...Allyl Alcohol
WEED-GO® see...Prometon
WEED HOE® see...Sodium Methanearsonate (MSMA)
WEEDKILLER CONC. D® see...Diquat Dibromide
WEEDMASTER® see...Dicamba
WEEDOCLOR® see...Amitrole
WEEDOL (ICI)® see...Diquat
WEEDOL® see...Diquat Dibromide
WEEDOL® see...Paraquat
WEEDONE® see...2,4-D
WEEDONE® see...2,4,5-T
WEEDONE® see...Pentachlorophenol
WEEDONE® see...2,4,5-Trichlorophenoxyacetic Acid, Esters
WEEDONE® 100 EMULSIFIABLE see...2,4-D, butoxyethyl ester
WEEDONE® 128 see...2,4-D, isopropyl ester
WEEDONE 170® see...Dichlorprop
WEEDONE® 638 see...2,4-D, butoxyethyl ester
WEEDONE DP® see...Dichlorprop
WEEDONE® GARDEN WEEDER see...Chloramben
WEEDONE® LV 4 see...2,4-D, butoxyethyl ester
WEEDONE® LV-6 see...2,4-D, butoxyethyl ester
WEEDONE MCPA ESTER® see...MCPA
WEED-RHAP® see...2,4-D
WEED RHAP® see...MCPA
WEED-RHAP® LV see...2,4-D, butoxyethyl ester
WEED-S-RAD® see...Sodium Methanearsonate (MSMA)
WEED TOX® see...2,4-D
WEEDTRINE®-II see...2,4-D, isooctyl ester
WEEDTRINE-D® see...Diquat
WEEDTRINE-D® see...Diquat Dibromide
WEEDTROL® see...2,4-D
WEEVILTOX® see...Carbon Disulfide
Weiss phosphor (German) see...Phosphorus
WELLCIDE® see...D-Phenothrin
WESLEY® see...Fentin Hydroxide
WESLEY TRIPLE TIN 4L® see...Triphenyltin Compounds

WESPURIL® see...Dichlorophene
WEST AGRO ACID SANITIZER® see...Pelargonic Acid
WHAM! EZ® see...Propanil
WHIP® see...Fenoxaprop-ethyl
White arsenic see...Arsenous Oxide
White-copperas see...Zinc Sulfate Heptahydrate
White phosphoric acid see...Phosphoric Acid
White phosphorus see...Phosphorus
White spirits see...Stoddard Solvent
White-vitriol see...Zinc Sulfate Heptahydrate
Whiting see...Calcium Carbonate
WHITMIRE® PT-110 see...Resmethrin
WHITMIRE PT 527 WITH ALLETHRIN® see.Allethrin
WIDEMATCH® see...Clopyralid
WIDEMATCH® see...Fluroxypyr 1-methylheptyl Ester
WIE OBEN® see...Quinalphos
WILBRO® see...Oxadiazon
WILTHIN® see...Monocarbamide Dihydrogen Sulfate
WILTZ®-65 see...Copper Naphthenate
WING STOP- B® see...Sodium Dimethyldithiocarbamate
WINNER® see...Acetochlor
WINTERWASH® see...Dinitro-o-cresol (DNOC)
WINYLOPHOS® see...Dichlorvos
WIPEOUT® see...Hydramethylnon
WIRKSTOFF® 37289 see...Trichloronate
WITCONATE®-60L see...Triethanolamine Dodecylbenzene Sulfonate
WITCONATE®-60T see...Triethanolamine Dodecylbenzene Sulfonate
WITCONATE®-79S see...Triethanolamine Dodecylbenzene Sulfonate
WITCONATE®-5725 see...Triethanolamine Dodecylbenzene Sulfonate
WITCONATE® S-1280 see...Triethanolamine Dodecylbenzene Sulfonate
WITCONATE® TAB see...Triethanolamine Dodecylbenzene Sulfonate
WITTOX® C see...Copper Naphthenate
WL 1650® see...Isobenzan
WL 18236® see...Methomyl
WL 19805® see...Cyanazine
WL 41706® see...Fenpropathrin
WL 43423® see...Perfluidone
WL 43467® see...Cypermethrin
WL 43775® see...Fenvalerate
WL 85871® see...*alpha*-Cypermethrin
WOCOSIN® see...Propiconazole
WOFATOX 50 EC® see...Methyl Parathion
WOLMAN SALTS® see...Sodium Dichromate
WOLMAN® see...Tebuconazole
WONUK® see...Atrazine
WOODFUME VAPAM® see...Metham-Sodium
WOODLIFE® see...Tebuconazole

WOODTREAT A® see...Pentachlorophenol
WOTEXIT® see...Trichlorfon
WR 62® see...Chlormequat Chloride
WRDC149® see...Cypermethrin
WSB 1 see...Cimectacarb
WURM-THIONAL® see...Phenothiazine

- X -

X 149® see...Busulfan
X-11085® see...Aminoethoxyvinylglycine Hydrochloride
X-ALL® LIQUID see...Amitrole
XDE-565® see...Cloransulam-methyl
XE-938® see...Fenpropathrin
Xenene see...Biphenyl
o-Xenol see...o-Phenylphenol
XL 2G® see...Benefin
XL 2G® see...Oryzalin
XL 7® see...Bithionol
XL-50® see...Phenothiazine
XL ALL INSECTICIDE® see...Nicotine
XR-29® see...Ammonium Hexafluorosilicate
XRD 473® see...Hexaflumuron
XRD-498® see...Flumetsulam
XRM-3972® see...Clopyralid
XRM-5019® see...Flumetsulam
XRM-5084® see...Fluroxypyr 1-methylheptyl Ester
XRM-5202® see...Triclopyr, Triethylammonium Salt
XRM-5313® see...Flumetsulam
3,5-Xylenol, 4-(dimethylamino)-, methylcarbamate see.Mexacarbate
3,4-Xylidine, 2,6-dinitro-*N*-(1-ethylpropyl)- see.Pendimethalin
2,4-Xylidine,*N,N'*-(methyliminodimethylidyne)bis- see.Amitraz
Xylofop-ethyl see...Quizalofop-Ethyl

- Y -

YALAN® see...Molinate
YALTOX® see...Carbofuran
YANOCK® see...Fluoroacetamide
YARDER® see...Fluvalinate
YASOKNOCK® see...Sodium Fluoroacetate
Yellow cuprocide see...Cuprous Oxide
Yellow mercuric oxide see...Mercuric Oxide
Yellow oxide of mercury see...Mercuric Oxide
Yellow phosphorus see...Phosphorus
Yellow precipitate see...Mercuric Oxide
Yellow pyoctanine see...Auramine
YUKON® see...Dicamba
YUKON® see...Sodium Dicamba
YULAN® see...Molinate

## - Z -

Zaclon Discoids  see...Hydrogen Cyanide
ZACTRAN®  see...Mexacarbate
Zaprawa nasienna plynna (Polish)  see...Methylmercuric Dicyanamide
ZAPRAWA NASIENNA SNECIOTOX®  see...Hexachlorobenzene
ZARUR®  see...Diphenamid
ZEAPOS®  see...Atrazine
ZEAPUR®  see...Simazine
ZEAZIN®  see...Atrazine
ZEAZINE®  see...Atrazine
ZEBENIDE®  see...Zineb
ZEBTOX®  see...Zineb
ZECTANE®  see...Mexacarbate
ZECTRAN®  see...Mexacarbate
ZEIDANE®  see...DDT
ZELAN®  see...MCPA
ZELDOX®  see...Hexythiazox
ZELIO®  see...Thallium Sulfate
ZELLEK®  see...Haloxyfop-methyl
ZENDIUM®  see...Sodium Fluoride
ZEOCON®  see...Fluvalinate
ZEOCON®  see...Phosmet
ZEPHEYR®  see...Abamectin
Zephiran chloride  see...Zilkonium Chloride
ZERDANE®  see...DDT
ZERLATE®  see...Ziram
ZERTELL®  see...Chlorpyrifos-methyl
ZEXTRAN®  see...Mexacarbate
ZIARNIK®  see...Phenylmercury Acetate
ZIDAN®  see...Zineb
ZIMANAT®  see...Mancozeb
ZIMANEB®  see...Mancozeb
ZIMATE®  see...Zineb
ZIMATE®  see...Ziram
ZIMMAN-DITHANE®  see...Mancozeb
ZIMTALDEHYDE®  see...Cinnamaldehyde
ZIMTALDEHYDE® LIGHT  see...Cinnamaldehyde
Zinc, bis(dimethylcarbamodithioato-s,s')-, (T-4)-  see...Ziram
Zinc bis(dimethyldithiocarbamoyl)disulphide  see...Ziram
Zinc bis(dimethyldithiocarbamate)  see...Ziram
Zinc $N,N$-dimethyldithiocarbamate  see...Ziram
Zinc dimethyldithiocarbamate  see...Ziram
Zinc ethylene-1,2-bisdithiocarbamate  see...Zineb
Zinc ethylenebisdithiocarbamate  see...Zineb
Zinc ethylenebis (dithiocarbamate)  see...Zineb
Zinc, [ethylenebis(dithiocarbamato)]-  see...Zineb
Zinc fosfid  see...Zinc Phosphide
Zinc ion and manganese ethylenebisdithiocarbamate 80%  see...Mancozeb
Zinc, [((1-methyl-1,2-ethanediyl)bis (carbamodithioato))(2-)]-  see...Propineb
Zinc(phosphure de) (French)  see...Zinc Phosphide
Zinc propylenebis(dithiocarbamate)  see...Propineb
Zinc-1,2-propylene-bisdithiocarbamate  see...Propineb
Zinc [$N,N$'-propylene-1,2-bis(dithiocarbamate)]  see...Propineb
Zinc sulphate heptahydrate  see...Zinc Sulfate Heptahydrate
Zinc sulphate monohydrate  see...Zinc Sulfate Heptahydrate
ZINC-200®  see...Zinc Sulfate Heptahydrate
ZINC-TOX®  see...Zinc Phosphide
ZINC-VITRIOL®  see...Zinc Sulfate Heptahydrate
ZINCATE®  see...Zinc Sulfate Heptahydrate
ZINCFRIN®  see...Zinc Sulfate Heptahydrate
ZINCI-SULFAS®  see...Zinc Sulfate Heptahydrate
ZINCMATE®  see...Ziram
Zinco(fosfuro di) (Italian)  see...Zinc Phosphide
ZINCOMED®  see...Zinc Sulfate Heptahydrate
Zincum-sulfuricum  see...Zinc Sulfate Heptahydrate
Zink-[$N,N$'-aethylen-bis(dithiocarbamat)] (German)  see...Zineb
Zink-bis($N,N$-dimethyl-dithiocarbamat) (German)  see...Ziram
Zinkcarbamate  see...Ziram
Zinkfosfide (Dutch)  see...Zinc Phosphide
ZINK-GROB®  see...Zinc Sulfate Heptahydrate
ZINKLET®  see...Zinc Sulfate Heptahydrate
ZINKOSITE®  see...Zinc Sulfate Heptahydrate
Zinkphosphid (German)  see...Zinc Phosphide
Zink-[$N,N$-propylen-1,2-bis(dithiocarbamat)] (German)  see...Propineb
ZINOCHLOR®  see...Anilazine
ZINOPHOS®  see...Thionazin
ZINOSAN®  see...Zineb
ZIPAK®  see...Bifenthrin
ZIPROMAT®  see...Propineb
ZIRAMVIS®  see...Ziram
ZIRASAN®  see...Ziram
ZIRBERK®  see...Ziram
ZIREX 90®  see...Ziram
ZIRIDE®  see...Ziram
ZIRTHANE®  see...Ziram
ZITHIOL®  see...Malathion
ZITOX®  see...Ziram
ZnDMDC  see...Ziram
Z-O-2-iso-Propoxycarbonyl-1-methylvinyl O-methyl ethyl phosphoramidothioate  see...Propetamphos
ZOBAR®  see...Terbacil
ZOBAR®  see...Trichlorobenzoic Acid
ZOECON®  see...Dienochlor
ZOECON®  see...Hydroprene
ZOECON®  see...Propetamphos
ZOECON® RF-316  see...Ethofenprox
ZOLON®  see...Phosalone
ZOLVIS®  see...Sulfur

Zoocoumarin (Dutch and Russian) see...Warfarin
Zoocoumaring (Russian) see...Warfarin
ZOOLON® see...Phosalone
ZORIAL® see...Norflurazon
ZOTOX® see...Arsenic Acid
ZPP 1560 AS HERBICIDE® see...Glyphosate
ZP® see...Zinc Phosphide
ZR 512® see...Hydroprene
ZR-515® see...Methoprene
ZR-777® see...Kinoprene
ZR 2006® see...Hydroprene
ZR 3210® see...Fluvalinate
Zwavel see...Sulfur
ZYBAN® see...Thiophanate-Methyl
ZYMAFLUOR® see...Sodium Fluoride
ZYTOX® see...Methyl Bromide

# Index 2: Index of EPA Product Codes

000601 see...Carbon Tetrachloride
000601 see...Acrylonitrile
000701 see...Acrolein
004001 see...Allethrins
004002 see...Allethrins
004003 see...Allethrins
004401 see...Amitrole
004501 see...ANTU
004901 see...Allyl Bromide
004901 see...Allyl Isothiocyanate
005101 see...Physostigmine
005101 see...Picloram
005301 see...Ammonia
005302 see...Ammonia
005501 see...Ammonium Sulfamate
006000 see...Copper Napthenate
006101 see...Anthracene
006201 see...Antimony Potassium Tartrate
006202 see...Stibine
006300 see...Copper Napthenate
006304 see...Oxytetracycline Calcium
006306 see...Streptomycin
006310 see...Streptomycin Sulfate
006314 see...Antimycin A
006321 see...Oxytetracycline Calcium
006801 see...Arsenic Acid
006802 see...Arsenic Pentoxide
007001 see...Arsenic and Inorganic Arsenic Compounds
007001 see...Arsenous Oxide
008001 see...Copper Oxychloride
008706 see...Bromophos
008901 see...Hexachlorocyclohexanes
009001 see...Lindane
009101 see...Benzoic Acid
009103 see...Sodium Benzoate
009502 see...Benzyl Alcohol
009801 see...Bensulide
010501 see...Dicofol
011001 see...Borax and Boric Acid
011102 see...Borax and Boric Acid
011102 see...Sodium Tetraborate
011104 see...Sodium Metaborate
011112 see...Sodium Tetraborate
011139 see...Cyanamide
011301 see...Dibromochloropropane (DBCP)
012101 see...Crufomate
012201 see...Binapacryl
012301 see...Bromacil
012501 see...Cacodylic Acid
012502 see...Sodium Cacodylate

012701 see...Terbacil
012902 see...Cadmium Chloride
012905 see...Cadmium Sulfate
013501 see...Calcium Arsenate
013503 see...Lead Arsenate
013602 see...Calcium Arsenite
013603 see...Sodium Arsenite
013803 see...Sodium Methanearsonate (MSMA)
013804 see...Octylammonium Methanearsonate
013806 see...Calcium Methanearsonate
013906 see...Aluminum Sulfate
013907 see...Sodium Bromide
014001 see...Calcium Cyanamide
014002 see...Cyanamide
014503 see...Nabam
014504 see...Mancozeb
014505 see...Maneb
014506 see...Zineb
014601 see...Metiram
015801 see...Mevinphos
016401 see...Carbon Disulfide
016501 see...Carbon Tetrachloride
017002 see...Biphenyl
017302 see...Trichlorobenzoic Acid
017601 see...Barban
018101 see...Chlormequat Chloride
018201 see...Phosphamidon
018301 see...Chlorpropham
018501 see...Phosacetim
019101 see...Propachlor
019301 see...Allidochlor
021007 see...Ampelomyces Quisqualis isolate M10
121701 see...Azadirachtin
022401 see...Copper Arsenite
022501 see...Copper and Copper Compounds
022601 see...Paris Green
022601 see...Copper Acetoarsenite
022703 see...Copper Ammonium Carbonate
022901 see...Copper Carbonate, Basic
023102 see...Copper Napthenate
023306 see...Copper Octanoate
023401 see...Copper Hydroxide
024002 see...Copper(II)-8-hydroxyquinoline
024401 see...Copper Sulfate
024408 see...Copper Sulfate
025501 see...Chloroxuron
025601 see...Cuprous Oxide

025602 see...Cuprous Thiocyanate
025801 see...Cyanogen Chloride
026201 see...Cyclanilide
027301 see...Chloroneb
027401 see...Dichlobenil
027501 see...Dienochlor
027701 see...Chlordecone (Kepone®)
027901 see...Triclocarban
028201 see...Propanil
028801 see...Chlorobenzilate
028901 see...Dalapon
029001 see...1,3-Dichloropropene
029002 see...1,2-Dichloropropane
029003 see...D-D mixture
029201 see...DDT
029501 see...Dichloroethyl Ether
029601 see...Dichlone
029801 see...Dicamba
029806 see...Sodium Dicamba
029901 see...Chloramben
030001 see...2,4-D
030053 see...2,4-D, butoxyethyl ester
030061 see...2,4-D, butoxyethyl ester
030064 see...2,4-D, isooctyl ester
030066 see...2,4-D, isopropyl ester
030501 see...MCPA
030602 see...2,4-DES-Sodium
030702 see...Naptalam
030801 see...2,4-DB
031301 see...Dichloran
031401 see...Dichlorprop
031501 see...Mecoprop
032101 see...Ethylan
032201 see...Diquat
032201 see...Diquat Dibromide
032401 see...Thionazin
032501 see...Disulfoton
034001 see...Methoxychlor
034201 see...Fenaminosulf
034401 see...Naled
034801 see...Ferbam
034804 see...Sodium Dimethyldithiocarbamate
034805 see...Ziram
034902 see...Ferric Sulfate
035001 see...Dimethoate
035002 see...Omethoate
035101 see...Daminozide
035201 see...Dicrotophos
035301 see...Bromoxynil
035302 see...Bromoxynil
035501 see...Monuron
035503 see...Fluometuron

035505  see...Diuron
035506  see...Linuron
035509  see...Siduron
035602  see...Dazomet
035901  see...Metobromuron
036001  see...Dinocap
036101  see...Trifluralin
036501  see...Coumaphos
036601  see...Diphenamid
037501  see...Dinex
037505  see...Dinoseb
037507  see...Dinitro-*o*-cresol (DNOC)
037801  see...Dioxathion
038201  see...Nitrofen
038901  see...Endothall
039001  see...Sulfallate
039003  see...Metham-Sodium
039201  see...Mirex
039501  see...Auramine
040506  see...Cinnamaldehyde
040516  see...Cinnamaldehyde
041101  see...Ethoprop
041301  see...Cycloate
041401  see...EPTC
041402  see...Molinate
041403  see...Pebulate
041404  see...Validamycin
041404  see...Vernolate
041405  see...Butylate
041415  see...Cartap Hydrochloride
041508  see...Methoxyethylmercuric Acetate
041601  see...Endrin
041701  see...Fonofos
041801  see...EPN
041901  see...Ethylene
041901  see...Ethyl Mercuric Chloride
042002  see...Ethylene Dibromide
042003  see...Ethylene Dichloride
042301  see...Ethylene Oxide
043001  see...Formaldehyde
043102  see...Ethyl Formate
043401  see...Cycloheximide
043601  see...Glyodin
043801  see...Gibberellic Acid
043802  see...Gibberellic Acid
043901  see...Glutaraldehyde
044008  see...Sodium Diacetate
044201  see...Mexacarbate
044301  see...Dodine
044801  see...Heptachlor
044801  see...Heptachlor Epoxide
044901  see...Hexachlorophene
045001  see...Dieldrin
045101  see...Aldrin
045201  see...Hexachloroethane

045601  see...Hydrogen Fluoride
045801  see...Hydrogen Cyanide
046701  see...Indole-3-Butyric Acid
047201  see...Dipropyl Isocinchomeronate
047601  see...Propham
047801  see...Phenol, 3-(1-methylethyl)-, methylcarbamate
047802  see...Propoxur
051501  see...Maleic Hydrazide
051909  see...Methylmercuric Dicyanamide
052001  see...Mercuric Chloride
052102  see...Mercuric Oxide
053001  see...Metaldehyde
053201  see...Methyl Bromide
053501  see...Methyl Parathion
053701  see...Methyl Formate
054101  see...Oxythioquinox
054901  see...Triclosan
055001  see...Dichlorophene
055201  see...Tecnazene
055501  see...Ethoxyquin
055601  see...Naphthoxyacetic Acid
056001  see...1-Naphthalene Acetamide
056002  see...1-Naphthaleneacetic Acid
056502  see...Quintozene
056702  see...Nicotine
056703  see...Nicotine Sulfate
056801  see...Carbaryl
057101  see...Sulfoxide
057201  see...Phorate
057201  see...Phenylthiourea
057301  see...Amiton Oxalate
057302  see...Amiton
057401  see...Parathion
057501  see...Parathion
057503  see...Parathion
057601  see...Demeton
057603  see...Demeton-methyl
057701  see...Malathion
057801  see...Diazinon
057901  see...Trichlorfon
058001  see...Azinphos-methyl
058002  see...Azinphos-ethyl
058102  see...Carbophenothion
058201  see...Chlordane
058202  see...Chlordane
058301  see...Ronnel
058401  see...Ethion
058501  see...Isobenzan
058601  see...Octamethyl Diphosphoramide
058701  see...Demeton-methyl
058702  see...Demeton-methyl
058703  see...Demeton-methyl

058801  see...Crotoxyphos
058901  see...Monocrotophos
059001  see...Temephos
059101  see...Chlorpyrifos
059102  see...Chlorpyrifos-Methyl
059201  see...Phosmet
059303  see...Bufencarb
059303  see...Brucine
059701  see...Chlordimeform
059901  see...Famphur
060101  see...Thiabendazole
061001  see...Hexachlorobenzene
061501  see...*para*-Dichlorobenzene
061601  see...Paraquat
061602  see...Paraquat Methosulfate
061603  see...Paraquat
062501  see...Aramite®
063001  see...Pentachlorophenol
063003  see...Sodium Penta chlorophenate
063201  see...Peracetic Acid
063501  see...Kerosene
063503  see...Naphthas
063504  see...Stoddard Solvent
063506  see...Naphthas
064103  see...*o*-Phenylphenol
064113  see...Butylphenols
064201  see...Bithionol
064206  see...p-Chloro-m-Cresol
064210  see...Trichlorophenols
064212  see...Trichlorophenols
064501  see...Phenothiazine
066003  see...Phenylmercury Acetate
066500  see...Phosphine
066501  see...Aluminum Phosphide
066502  see...Phosphorus
066503  see...Calcium Phosphide
067501  see...Piperonyl Butoxide
067701  see...Diphacinone
067703  see...Pindone
067707  see...Chlorophacinone
068304  see...Sodium Dichromate
068401  see...Allyl Alcohol
068402  see...Acrolein Diacetate
068701  see...Propargyl Bromide
069000  see...Pyrethrins or Pyrethrum
069001  see...Pyrethrins or Pyrethrum
069002  see...Pyrethrins or Pyrethrum
069003  see...Tetramethrin
069005  see...D-Phenothrin
069106  see...Zilkonium Chloride
069109  see...Zilkonium Chloride
069201  see...4-Aminopyridine
069203  see...Nitrapyrin

069601 see...Pyrazon
070582 see...Calcium Nitrate
070701 see...Capsaicin
071003 see...Rotenone
072605 see...Diatomaceous Earth
073301 see...Sodium Chlorate
073502 see...Calcium Carbonate
074001 see...Calcium Cyanide
074002 see...Sodium Cyanide
074801 see...Tribufos
075001 see...Fluoroacetic Acid
075002 see...Fluoroacetamide
075003 see...Sodium Fluoroacetate
075101 see...Sodium Aluminum Fluoride
075202 see...Sodium Fluoride
075301 see...Ammonium Hexafluorosilicate
075601 see...Calcium Hydroxide
075604 see...Calcium Oxide
076001 see...Phosphoric Acid
076101 see...Ammonium Nitrate
076102 see...Cupric Nitrate
076103 see...Potassium Nitrate
076203 see...Potassium Nitrite
076406 see...Trisodium Phosphate
076602 see...Salicylic Acid
076901 see...Strychnine
077401 see...Clonitralid
077501 see...Sulfur
078003 see...Sulfuryl Fluoride
078701 see...DCPA
078801 see...Diallate
078802 see...Triallate
078905 see...Pyrithiobac-sodium
079020 see...Triethanolamine Dodecylbenzene Sulfonate
079101 see...ASPON®
079202 see...Tetradifon
079401 see...Endosulfan
079402 see...Endosulfan
079403 see...Endosulfan
079501 see...Sulfotepp
079601 see...TEPP
079701 see...D-Limonene
079801 see...Thiram
080001 see...Thallium Sulfate
080103 see...Ammonium Sulfite
080103 see...Ammonium Thiosulfate
080402 see...Thymol
080501 see...Toxaphene
080801 see...Ametryn
080803 see...Atrazine
080804 see...Prometon
080805 see...Prometryn
080807 see...Simazine
080808 see...Propazine

080811 see...Anilazine
080813 see...Terbutryn
081301 see...Captan
081401 see...Dichloroisocyanuric Acid
081501 see...Chloropicrin
081701 see...Captafol
081702 see...Captafol
081901 see...Chlorothalonil
082001 see...2,4,5-Trichlorophenoxyacetic Acid, Esters
082001 see...2,4,5-T
082053 see...2,4,5-Trichlorophenoxyacetic Acid, Esters
082056 see...2,4,5-Trichlorophenoxyacetic Acid, Esters
082063 see...2,4,5-Trichlorophenoxyacetic Acid, Esters
082066 see...2,4,5-Trichlorophenoxyacetic Acid, Esters
082501 see...Silvex
082601 see...Fenac
083107 see...Tributyltin Chloride
083112 see...Tributyltin Fluoride
083601 see...Triphenyltin Compounds
083601 see...Fentin Hydroxide
083602 see...Triphenyltin Compounds
083701 see...Tetrachlorvinphos
083702 see...Tetrachlorvinphos
083902 see...Tris(hydroxymethyl)nitromethane
084001 see...Dichlorvos
084101 see...Chlorfenvinphos
084301 see...Benefin
084701 see...Etridiazole
085702 see...Urea
086001 see...Coumafuryl
086002 see...Warfarin
086003 see...Warfarin
086801 see...4-Chloro-3,5-xylenol
088601 see...Zinc Phosphide
089001 see...Zinc Sulfate Heptahydrate
090101 see...Dimetilan
090201 see...Carboxin
090202 see...Oxycarboxin
090301 see...Methomyl
090501 see...Alachlor
090601 see...Carbofuran
090602 see...Carbosulfan
097301 see...Formetanate Hydrochloride
097401 see...Karbutilate
097601 see...Propargite
097701 see...Phosalone
097801 see...Resmethrin ; (+)-d-trans-Resmethrin; (+)-cis-Resmethrin; (−)-trans-
097802 see...Resmethrin ; (+)-d-trans-Resmethrin; (+)-cis-Resmethrin; (−)-trans-
097804 see...Resmethrin ; (+)-d-trans-Resmethrin; (+)-cis-Resmethrin; (−)-trans-
097805 see...Deltamethrin
098002 see...Dodecylbenzenesulfonic Acid
098101 see...Azacosterol Dihydrochloride
098301 see...Aldicarb
098701 see...Phenmedipham
099101 see...Benomyl
099801 see...Ethephon
099901 see...Octhilinone
100101 see...Cyanazine
100201 see...Isopropalin
100301 see...Methidathion
100501 see...Methiocarb
100601 see...Fenamiphos
101101 see...Metribuzin
101103 see...Pymetrozine
101201 see...Methamidophos
101601 see...Cyhexatin
101701 see...Pronamide
102001 see...Thiophanate-methyl
102400 see...Trimethacarb
102501 see...Dialifor
103001 see...Napropamide
103301 see...Acephate
103401 see...Thiophanate-methyl
103801 see...Oxamyl
104201 see...Oryzalin
104301 see...Bifenox
104401 see...Dipropetryn
104501 see...Pyriminil
104601 see...Fenbutatin Oxide
104801 see...Desmedipham
104901 see...Phenthoate
105001 see...Terbufos
105201 see...Bendiocarb
105401 see...Methoprene
105501 see...Tebuthiuron
105801 see...Norflurazon
105901 see...Fenitrothion
106001 see...Methazole
106101 see...Pirimicarb
106201 see...Amitraz
106401 see...Difenzoquat
106501 see...Butralin
106601 see...Profluralin
106701 see...Fosamine Ammonium
106901 see...Asulam
107201 see...Hexazinone
107501 see...Kinoprene

107502  see...Kinoprene
107701  see...Sodium Azide
108001  see...Perfluidone
108101  see...Pirimiphos-Ethyl
108102  see...Pirimiphos-Methyl
108201  see...Diflubenzuron
108202  see...Chlorfluazuron
108303  see...Cuprous Chloride
108401  see...Thiobencarb
108501  see...Pendimethalin
108601  see...Ancymidol
108701  see...Fluchloralin
108801  see...Metolachlor
109001  see...Oxadiazon
109101  see...Mepiquat Chloride
109201  see...Thiofanox
109301  see...Fenvalerate
109302  see...Fluvalinate
109303  see...Esfenvalerate
109401  see...Isofenphos
109702  see...Cypermethrin
109702  see...alpha-Cypermethrin
109705  see...Fenfluthrin
109801  see...Iprodione
109901  see...Triadimefon
110201  see...Prodiamine
110401  see...Dodemorph Acetate
110601  see...Ethofumesate
110801  see...Aldoxycarb
110902  see...Diclofop-methyl
111001  see...1,2-Dibromo-2,4-dicyanobutane
111401  see...Profenofos
111501  see...Sulprofos
111601  see...Oxyfluorfen
111811  see...Chlorthiophos
111901  see...Imazalil
112001  see...Bromadiolone
112301  see...Butachlor
112600  see...Prohexadione Calcium
112602  see...C. I. Basic Green 1
112602  see...Cimectacarb
112900  see...Fluridone
113101  see...Ethalfluralin
113201  see...Vinclozolin
113301  see...Fosthietan
113501  see...Metalaxyl
113501  see...Mercury Alkyl Compounds
113601  see...Propetamphos
114001  see...Mefluidide
114401  see...Acifluorfen
114402  see...Acifluorfen
114501  see...Thiodicarb
116001  see...Triclopry
116002  see...Triclopyr, Triethylammonium Salt

116201  see...Triacontanol
116801  see...Kinetin (Cytokinin)
116802  see...Kinetin (Cytokinin)
116901  see...6-Benzaldenine
117001  see...Nosema Locustae
117403  see...Clopyralid
117801  see...Bitertanol
118202  see...Hexaflumuron
118301  see...Flucythrinate
118401  see...Hydramethylnon
118601  see...Chlorsulfuron
119302  see...Propamocarb Hydrochloride
119901  see...Difenacoum
120301  see...Thidiazuron
121000  see...Tralkoxydim
121001  see...Sethoxydim
121011  see...Clethodim
121027  see...Methoxyfenozide
121301  see...Cyromazine
121501  see...Tralomethrin
121601  see...Acetochlor
121702  see...Dihydroazadirachtin
121903  see...Thiafluamide (Flufenacet)
122001  see...Sulfometuron-methyl
122002  see...Sulfometuron-methyl
122010  see...Metsulfuron-methyl
122101  see...Propiconazole
122701  see...Anthraquinone
122804  see...Abamectin
122805  see...Fluazifop-butyl
122806  see...Emamectin Benzoate
122809  see...Fluazifop-butyl
123000  see...Isoxaflutole
123201  see...Benfuracarb
123301  see...Fosetyl-Al
123802  see...Fomesafen
124601  see...Arosurf® MSF
125201  see...Haloxyfop-methyl
125301  see...Fenoxycarb
125401  see...Clomazone
125501  see...Clofentezine
125601  see...Paclobutrazol
125701  see...Flurprimidol
125851  see...Isoxaben
126701  see...Oxadixyl
126801  see...Benazolin Ethyl
127201  see...Triadimenol
127885  see...Anagrapha Falcifera
127901  see...Fenpropathrin
128100  see...Triazamate
128201  see...Quizalofop-ethyl
128501  see...Sulfosate
128701  see...Fenoxaprop-ethyl
128711  see...Quizalofop-ethyl
128721  see...Halosulfuron-methyl
128801  see...Fenoxycarb

128810  see...Azoxystrobin
128819  see...Forchlorfenuron
128820  see...Bensulfuron-methyl
128822  see...Tralomethrin
128823  see...Pencycuron
128825  see...Bifenthrin
128831  see...Cyfluthrin
128834  see...Pyridate
128840  see...Imazaquin
128842  see...Imazethabenz
128845  see...Thifensulfuron Methyl
128847  see...Difenoconazole
128848  see...Imazaquin
128849  see...Hexythiazox
128850  see...Glufosinate-Ammonium
128857  see...Myclobutanil
128864  see...Cadusafos
128872  see...Carbendazim
128876  see...Methanearsonic Acid
128879  see...Triflumizole
128880  see...Bensulfuron-methyl
128887  see...Tribenuron-methyl
128888  see...Lactofen
128897  see...lambda-Cyhalothrin
128901  see...Chlorimuron-ethyl
128912  see...Tefluthrin
128922  see...Imazethapyr
128923  see...Imazethapyr
128925  see...Hexaconazole
128932  see...Diniconazole
128955  see...Decanoic Acid
128961  see...Monocarbamide Dihydrogen Sulfate
128965  see...Ethofenprox
128966  see...Hydroprene
128968  see...Fluroxypyr 1-methylheptyl Ester
128969  see...Triasulfuron
128973  see...Primisulfuron-methyl
128974  see...Quinclorac
128975  see...Flutolanil
128985  see...Triasulfuron
128992  see...Sulfluramid
128993  see...Cyproconazole
128994  see...Dithiopyr
128997  see...Tebuconazole
128999  see...Penconazole
129002  see...Triflusulfuron-methyl
129006  see...Chlorethoxyfos
129008  see...Nicosulfuron
129009  see...Rimsulfuron
129011  see...Fenbuconazole
129013  see...Cyphenothrin
129016  see...Flumetsulam
129024  see...Rimsulfuron
129026  see...Tebufenozide
129031  see...Prosulfuron

129032 see...Pyriproxyfen
129044 see...Procymidone
129051 see...Dimethenamid
129059 see...Imidacloprid
129064 see...alpha-Cypermethrin
129081 see...Sulfentrazone
129091 see...Ethametsulfuron-methyl
129093 see...Chlorfenapyr
129098 see...Fluazinam®
129099 see...Imidacloprid
129104 see...Aminoethoxyvinylglycine hydrochloride
129105 see...Pyridaben
129106 see...Cymoxanil
129116 see...Cloransulam-methyl
129121 see...Fipronil
129131 see...Fenpyroximate
202901 see...Cholecalciferol
206600 see...Fenarimol
207700 see...Nosema Locustae
208100 see...Thidiazuron
208500 see...Bromadiolone
208700 see...Cholecalciferol
208800 see...Thidiazuron
209600 see...alpha-Cypermethrin
209800 see...Acifluorfen
209900 see...Iprodione
210700 see...Profenofos
213600 see...Dodemorph Acetate
214500 see...Bromophos-ethyl
214600 see...Bromadiolone
214900 see...Formic Acid
215900 see...Fluridone
216300 see...o-Chlorotoluene
216800 see...Propetamphos
217500 see...Pelargonic Acid
217700 see...1,2-Dibromo-2,4-dicyanobutane
218500 see...1,2-Dibromo-2,4-dicyanobutane
228400 see...Dinoterb
230000 see...Polyoxin D, Zinc Salt
245100 see...Sodium Azide
251200 see...Ancymidol
259200 see...Carbetamide
268200 see...Cyanophos
268300 see...Phosfolan
268310 see...Mephosfolan
268601 see...Dodemorph Acetate
268800 see...Dimethomorph
271400 see...Promecarb
275100 see...Buprofezin
275200 see...Bentazon
279400 see...Chloroacetic Acid
279500 see...Diethatyl-ethyl
283500 see...Thiobencarb
285200 see...Bifenox
288200 see...Crimidine
288201 see...Pyrimethanil
288202 see...Cyprodinil
288600 see...Oxyfluorfen
288700 see...Metolachlor
294500 see...Phosalone
294600 see...Metolcarb
295300 see...Chlormephos
295700 see...Fenvalerate
296700 see...Fenvalerate
304300 see...Profluralin
323300 see...Methazole
334300 see...Pirimiphos-Methyl
340200 see...Oxydisulfoton
344300 see...Prothoate
344600 see...Triazophos
356100 see...Isofluorphate
362200 see...Methyl Phenkapton
366400 see...Formothion
374350 see...Glutamic Acid
381400 see...Quinalphos
387100 see...Mefluidide
391200 see...Prodiamine
417300 see...Glyphosate
422100 see...Endothion
442200 see...Acenaphthene
443100 see...Dimefox
454300 see...Pendimethalin
456200 see...Trichloronate
460200 see...Fluchloralin
462200 see...Fluenetil
466200 see...Fuberidazole
471300 see...Glyphosate
477300 see...Perchloromethyl Mercaptan
486300 see...Hydroprene
496100 see...Coumatetralyl
496500 see...Triphenyltin Compounds
496700 see...Triphenyltin Compounds
496900 see...Triphenyltin Compounds
511500 see...Isolan®
512200 see...Isoproturon
512400 see...Isofenphos
517200 see...Kinoprene
522200 see...Propineb
525300 see...Leptophos
527200 see...Zinc Sulfate Heptahydrate
530200 see...Magnesium Chlorate
549200 see...Methazole
549500 see...Methoprene
596300 see...Fenbutatin Oxide
597900 see...Oxadiazon
598800 see...Phoxim
599600 see...Potassium Cyanide
600008 see...Acrylamide
600016 see...Ethylene Thiourea
600020 see...Dichlorvos
600021 see...Pentachlorophenol
600023 see...Dinitro-o-cresol (DNOC)
600027 see...Hexachlorophene
600030 see...1,2-Dichloropropane
653502 see...Paraoxon

# Index 3: CAS Number-Cross Index

50-00-0 see...Formaldehyde
50-29-3 see...DDT
50-31-7 see...Trichlorobenzoic Acid
51-03-6 see...Piperonyl Butoxide
52-68-6 see...Trichlorfon
52-85-7 see...Famphur
53-96-3 see...Acetylaminofluorene
54-11-5 see...Nicotine
55-91-4 see...Isofluorphate
55-98-1 see...Busulfan
56-23-5 see...Carbon Tetrachloride
56-38-2 see...Parathion
56-72-4 see...Coumaphos
56-86-0 see...Glutamic Acid
57-06-7 see...Allyl Isothiocyanate
57-13-6 see...Urea
57-24-9 see...Strychnine
57-47-6 see...Physostigmine
57-74-9 see...Chlordane
58-89-9 see...Lindane
59-50-7 see...*p*-Chloro-*m*-Cresol
60-41-3 see...Strychnine
60-51-5 see...Dimethoate
60-57-1 see...Dieldrin
61-82-5 see...Amitrole
62-38-4 see...Phenylmercury Acetate
62-73-7 see...Dichlorvos
62-74-8 see...Sodium Fluoroacetate
63-25-2 see...Carbaryl
64-00-6 see...Phenol, 3-(1-methylethyl)-, methylcarbamate
64-18-6 see...Formic Acid
65-30-5 see...Nicotine Sulfate
65-85-0 see...Benzoic Acid
66-81-9 see...Cycloheximide
67-72-1 see...Hexachloroethane
67-97-0 see...Cholecalciferol
69-72-7 see...Salicylic Acid
70-30-4 see...Hexachlorophene
72-20-8 see...Endrin
72-43-5 see...Methoxychlor
72-56-0 see...Ethylan
74-83-9 see...Methyl Bromide
74-85-1 see...Ethylene
74-90-8 see...Hydrogen Cyanide
75-15-0 see...Carbon Disulfide
75-21-8 see...Ethylene Oxide
75-60-5 see...Cacodylic Acid
75-69-4 see...Trichlorofluoromethane
75-99-0 see...Dalapon
76-06-2 see...Chloropicrin
76-44-8 see...Heptachlor
76-87-9 see...Fentin Hydroxide

76-87-9 see...Triphenyltin Compounds
77-06-5 see...Gibberellic Acid
78-34-2 see...Dioxathion
78-48-8 see...Tribufos
78-53-5 see...Amiton
78-87-5 see...1,2-Dichloropropane
79-06-1 see...Acrylamide
79-11-8 see...Chloroacetic Acid
79-19-6 see...Thiosemicarbazide
79-21-0 see...Peracetic Acid
79-43-6 see...Dichloroacetic Acid
79-57-2 see...Oxytetracycline Calcium
81-81-2 see...Warfarin
82-66-6 see...Diphacinone
82-68-8 see...Quintozene
83-26-1 see...Pindone
83-32-9 see...Acenaphthene
83-79-4 see...Rotenone
84-65-1 see...Anthraquinone
85-00-7 see...Diquat
85-00-7 see...Diquat Dibromide
85-34-7 see...Fenac
86-50-0 see...Azinphos-methyl
86-86-2 see...1-Naphthaleneacetamide
86-87-3 see...1-Naphthaleneacetic Acid
86-88-4 see...ANTU
87-86-5 see...Pentachlorophenol
88-04-0 see...4-Chloro-3,5-xylenol
88-06-2 see...Trichlorophenols
88-18-6 see...Butylphenols
88-85-7 see...Dinoseb
89-72-5 see...Butylphenols
89-83-8 see...Thymol
90-43-7 see...*o*-Phenylphenol
92-52-4 see...Biphenyl
92-84-2 see...Phenothiazine
93-65-2 see...Mecoprop
93-71-0 see...Allidochlor
93-72-1 see...Silvex
93-76-5 see...2,4,5-Trichlorophenoxyacetic Acid, Esters
93-76-5 see...2,4,5-T
93-78-7 see...2,4,5-Trichlorophenoxyacetic Acid, Esters
93-79-8 see...2,4,5-Trichlorophenoxyacetic Acid, Esters
94-11-1 see...2,4-D, Isopropyl Ester
94-74-6 see...MCPA
94-75-7 see...2,4-D
94-82-6 see...2,4-DB
95-06-7 see...Sulfallate

95-49-8 see...*o*-Chlorotoluene
95-95-4 see...Trichlorophenols
96-12-8 see...Dibromochloropropane (DBCP)
96-45-7 see...Ethylene Thiourea
97-18-7 see...Bithionol
97-23-4 see...Dichlorophene
97-77-8 see...Disulfiram
98-54-4 see...Butylphenols
99-30-9 see...Dichloran
99-71-8 see...Butylphenols
100-51-6 see...Benzyl Alcohol
100-66-3 see...Anisole
101-05-3 see...Anilazine
101-20-2 see...Triclocarban
101-21-3 see...Chlorpropham
101-27-9 see...Barban
103-85-5 see...Phenylthiourea
104-55-2 see...Cinnamaldehyde
106-46-7 see...*para*-Dichlorobenzene
106-93-4 see...Ethylene Dibromide
106-95-6 see...Allyl Bromide
106-96-7 see...Propargyl Bromide
107-02-8 see...Acrolein
107-06-2 see...Ethylene Dichloride
107-13-1 see...Acrylonitrile
107-18-6 see...Allyl Alcohol
107-20-0 see...Chloroacetaldehyde
107-27-7 see...Ethyl Mercuric Chloride
107-31-3 see...Methyl Formate
107-49-3 see...TEPP
108-62-3 see...Metaldehyde
109-94-4 see...Ethyl Formate
110-89-4 see...Piperidine
111-44-4 see...Dichloroethyl Ether
112-05-0 see...Pelargonic Acid
114-26-1 see...Propoxur
115-09-3 see...Mercury Alkyl Compounds
115-26-4 see...Dimefox
115-29-7 see...Endosulfan
115-32-2 see...Dicofol
116-06-3 see...Aldicarb
116-29-0 see...Tetradifon
117-18-0 see...Tecnazene
117-52-2 see...Coumafuryl
117-80-6 see...Dichlone
118-74-1 see...Hexachlorobenzene
119-38-0 see...Isolan®
120-12-7 see...Anthracene

120-23-0 see...Naphthoxyacetic Acid
120-36-5 see...Dichlorprop

120-62-7 see...Sulfoxide
121-21-1 see...Pyrethrins or Pyrethrum
121-29-9 see...Pyrethrins or Pyrethrum
121-75-5 see...Malathion
122-14-5 see...Fenitrothion
122-34-9 see...Simazine
122-42-9 see...Propham
123-33-1 see...Maleic Hydrazide
124-58-3 see...Methanearsonic Acid
124-65-2 see...Sodium Cacodylate
125-67-7 see...Gibberellic Acid
126-11-4 see...Tris(hydroxymethyl)nitromethane
126-75-0 see...Demeton
126-96-5 see...Sodium Diacetate
128-04-1 see...Sodium Dimethyldithiocarbamate
129-06-6 see...Warfarin
129-67-9 see...Endothall
131-52-2 see...Sodium Pentachlorophenate
131-89-5 see...Dinex
132-64-9 see...Dibenzofuran
132-66-1 see...Naptalam
133-06-2 see...Captan
133-32-4 see...Indole-3-Butyric Acid
133-90-4 see...Chloramben
136-45-8 see...Dipropyl Isocinchomeronate
136-78-7 see...2,4-DES-Sodium
137-26-8 see...Thiram
137-30-4 see...Ziram
137-42-8 see...Metham-Sodium
139-40-2 see...Propazine
139-89-9 see...N-[2(Biscarboxymethylamino)ethyl]-n-(2-hydroxyethyl)glycine Trisodium Salt
140-56-7 see...Fenaminosulf
140-57-8 see...Aramite®
141-66-2 see...Dicrotophos
142-59-6 see...Nabam
142-71-2 see...Cupric Acetate
143-33-9 see...Sodium Cyanide
143-50-0 see...Chlordecone (Kepone®)
144-49-0 see...Fluoroacetic Acid
145-73-3 see...Endothall
148-79-8 see...Thiabendazole
150-68-5 see...Monuron
151-38-2 see...Methoxyethylmercuric Acetate
151-50-8 see...Potassium Cyanide
152-16-9 see...Octamethyl Diphosphoramide
156-62-7 see...Calcium Cyanamide
297-78-9 see...Isobenzan
297-97-2 see...Thionazin
298-00-0 see...Methyl Parathion
298-02-2 see...Phorate
298-03-3 see...Demeton
298-04-4 see...Disulfoton
299-84-3 see...Ronnel
299-86-5 see...Crufomate
300-76-5 see...Naled
301-12-2 see...Demeton-methyl
309-00-2 see...Aldrin
311-45-5 see...Paraoxon
314-40-9 see...Bromacil
315-18-4 see...Mexacarbate
319-84-6 see...Hexachlorocyclohexanes
319-85-7 see...Hexachlorocyclohexanes
319-86-8 see...Hexachlorocyclohexanes
327-98-0 see...Trichloronate
330-54-1 see...Diuron
330-55-2 see...Linuron
333-41-5 see...Diazinon
334-48-5 see...Decanoic Acid
357-57-3 see...Brucine
371-62-0 see...Ethylene Fluorohydrin
379-52-2 see...Triphenyltin Compounds
404-86-4 see...Capsaicin
420-04-2 see...Cyanamide
462-06-6 see...Fluorobenzene
462-08-8 see...4-Aminopyridine
465-73-6 see...Isodrin
470-90-6 see...Chlorfenvinphos
471-34-1 see...Calcium Carbonate
485-31-4 see...Binapacryl
492-80-8 see...Auramine
497-92-7 see...Allethrins
502-39-6 see...Methylmercuric Dicyanamide
504-24-5 see...4-Aminopyridine
504-29-0 see...4-Aminopyridine
506-68-3 see...Cyanogen Bromide
506-77-4 see...Cyanogen Chloride
510-15-6 see...Chlorobenzilate
514-73-8 see...Dithiazanine Iodide
525-79-1 see...Kinetin (Cytokinin)
532-32-1 see...Sodium Benzoate
534-52-1 see...Dinitro-o-cresol
535-89-7 see...Crimidine
542-75-6 see...1,3-Dichloropropene
542-90-5 see...Ethylthiocyanate
544-92-3 see...Copper Cyanide
556-22-9 see...Glyodin
556-64-9 see...Methyl Thiocyanate
563-12-2 see...Ethion
584-79-2 see...Allethrins
592-01-8 see...Sodium Cyanide
592-01-8 see...Calcium Cyanide
593-50-0 see...Triacontanol
593-74-8 see...Mercury Alkyl Compounds
594-42-3 see...Perchloromethyl Mercaptan
600-25-9 see...1-Chloro-1-Nitropropane
608-73-1 see...Hexachlorocyclohexanes
609-19-8 see...Trichlorophenols
617-65-2 see...Glutamic Acid
633-03-4 see...C. I. Basic Green 1
639-58-7 see...Triphenyltin Compounds
640-19-7 see...Fluoroacetamide
644-64-4 see...Dimetilan
709-98-8 see...Propanil
732-11-6 see...Phosmet
741-58-2 see...Bensulide
757-58-4 see...Hexaethyl Tetraphosphate
759-94-4 see...EPTC
786-19-6 see...Carbophenothion
789-02-6 see...DDT
834-12-8 see...Ametryn
867-27-6 see...Demeton-methyl
869-29-4 see...Acrolein Diacetate
886-50-0 see...Terbutryn
892-20-6 see...Triphenyltin Compounds
900-95-8 see...Triphenyltin Compounds
906-80-5 see...Anthracene
919-86-8 see...Demeton-methyl
933-75-5 see...Trichlorophenols
933-78-8 see...Trichlorophenols
944-22-9 see...Fonofos
947-02-4 see...Phosfolan
950-10-7 see...Mephosfolan
950-37-8 see...Methidathion
957-51-7 see...Diphenamid
959-98-8 see...Endosulfan
961-11-5 see...Tetrachlorvinphos
999-81-5 see...Chlormequat Chloride
1024-57-3 see...Heptachlor Epoxide
1071-83-6 see...Glyphosate
1111-67-7 see...Cuprous Thiocyanate
1111-78-0 see...Ammonium Carbamate
1113-02-6 see...Omethoate

1114-71-2  see...Pebulate
1129-41-5  see...Metolcarb
1134-23-2  see...Cycloate
1194-65-6  see...Dichlobenil
1214-39-7  see...6-Benzaldenine
1249-84-9  see...Azacosterol Dihydrochloride
1303-28-2  see...Arsenic Pentoxide
1303-96-4  see...Borax and Boric Acid
1303-96-4  see...Sodium Tetraborate
1305-62-0  see...Calcium Hydroxide
1305-78-8  see...Calcium Oxide
1305-99-3  see...Calcium Phosphide
1314-84-7  see...Zinc Phosphide
1317-38-0  see...Copper and Copper Compounds
1317-39-1  see...Cuprous Oxide
1317-65-3  see...Calcium Carbonate
1327-52-2  see...Arsenic Acid
1327-53-3  see...Arsenous Oxide
1330-43-4  see...Sodium Tetraborate
1332-40-7  see...Copper Oxychloride
1332-65-6  see...Copper Oxychloride
1338-02-9  see...Copper Napthenate
1420-04-8  see...Clonitralid
1420-04-8  see...Clonitralid
1420-07-1  see...Dinoterb
1461-22-9  see...Tributyltin Chloride
1563-66-2  see...Carbofuran
1582-09-8  see...Trifluralin
1596-84-5  see...Daminozide
1610-18-0  see...Prometon
1638-22-8  see...Butylphenols
1646-88-4  see...Aldoxycarb
1689-84-5  see...Bromoxynil
1689-99-2  see...Bromoxynil
1698-60-8  see...Pyrazon
1702-17-6  see...Clopyralid
1836-75-5  see...Nitrofen
1861-32-1  see...DCPA
1861-40-1  see...Benefin
1897-45-6  see...Chlorothalonil
1910-42-5  see...Paraquat
1912-24-9  see...Atrazine
1918-00-9  see...Dicamba
1918-02-1  see...Picloram
1918-16-7  see...Propachlor
1928-47-8  see...2,4,5-Trichlorophenoxyacetic Acid, Esters
1929-73-3  see...2,4-D, Butoxyethyl Ester
1929-73-3  see...2,4-D, Butoxyethyl Ester
1929-77-7  see...Vernolate
1929-82-4  see...Nitrapyrin
1982-47-4  see...Chloroxuron
1982-49-6  see...Siduron
1982-69-0  see...Sodium Dicamba
1983-10-4  see...Tributyltin Fluoride
2008-41-5  see...Butylate
2032-65-7  see...Methiocarb
2074-50-2  see...Paraquat Methosulfate
2104-64-5  see...EPN
2104-96-3  see...Bromophos
2155-70-6  see...Triphenyltin Compounds
2163-80-6  see...Sodium Methanearsonate (MSMA)
2164-17-2  see...Fluometuron
2212-67-1  see...Molinate
2227-17-0  see...Dienochlor
2275-18-5  see...Prothoate
2303-16-4  see...Diallate
2303-17-5  see...Triallate
2310-17-0  see...Phosalone
2312-35-8  see...Propargite
2385-85-5  see...Mirex
2425-06-1  see...Captafol
2439-01-2  see...Oxythioquinox
2439-10-3  see...Dodine
2465-27-2  see...Auramine
2497-07-6  see...Oxydisulfoton
2540-82-1  see...Formothion
2545-59-7  see...2,4,5-Trichlorophenoxyacetic Acid, Esters
2593-15-9  see...Etridiazole
2597-03-7  see...Phenthoate
2631-37-0  see...Promecarb
2636-26-2  see...Cyanophos
2642-71-9  see...Azinphos-ethyl
2675-77-6  see...Chloroneb
2699-79-8  see...Sulfuryl Fluoride
2778-04-3  see...Endothion
2782-57-2  see...Dichloroisocyanuric Acid
2921-88-2  see...Chlorpyrifos
2939-80-2  see...Captafol
3060-89-7  see...Metobromuron
3180-09-4  see...Butylphenols
3244-90-4  see...ASPON®
3251-23-8  see...Cupric Nitrate
3337-71-1  see...Asulam
3380-34-5  see...Triclosan
3383-96-8  see...Temephos
3687-31-8  see...Lead Arsenate
3689-24-5  see...Sulfotepp
3691-35-8  see...Chlorophacinone
3734-97-2  see...Amiton Oxalate
3735-23-7  see...Methyl Phenkapton
3737-22-2  see...Dipropyl Isocinchomeronate
3766-81-2  see...BPMC
3810-74-0  see...Streptomycin Sulfate
3878-19-1  see...Fuberidazole
4074-43-5  see...Butylphenols
4104-14-7  see...Phosacetim
4147-51-7  see...Dipropetryn
4151-50-2  see...Sulfluramid
4301-50-2  see...Fluenetil
4685-14-7  see...Paraquat
4824-78-6  see...Bromophos-ethyl
4849-32-5  see...Karbutilate
5234-68-4  see...Carboxin
5259-88-1  see...Oxycarboxin
5523-68-6  see...Ethalfluralin
5598-13-0  see...Chlorpyrifos-methyl
5836-29-3  see...Coumatetralyl
5902-51-2  see...Terbacil
5902-95-4  see...Calcium Methanearsenate
6164-98-3  see...Chlordimeform
6379-37-9  see...Octylammonium Methanearsonate
6484-52-2  see...Ammonium Nitrate
6734-80-1  see...Metham-Sodium
6893-26-1  see...Glutamic Acid
6923-22-4  see...Monocrotophos
7085-19-0  see...Mecoprop
7173-51-5  see...Didecyldimethylammonium Chloride
7179-50-2  see...Oxytetracycline Calcium
7287-19-6  see...Prometryn
7440-38-2  see...Arsenic and Inorganic Arsenic Compounds
7440-39-3  see...Barium and Barium Compounds
7440-43-9  see...Cadmium
7440-50-8  see...Copper and Copper Compounds
7446-18-6  see...Thallium Sulfate
7446-19-7  see...Zinc Sulfate heptahydrate
7487-94-7  see...Mercuric Chloride
7601-54-9  see...Trisodium Phosphate
7631-86-9  see...Diatomaceous Earth
7645-25-2  see...Lead Arsenate
7647-15-6  see...Sodium Bromide
7664-38-2  see...Phosphoric Acid
7664-39-3  see...Hydrogen Fluoride
7664-41-7  see...Ammonia
7681-49-4  see...Sodium Fluoride

7696-12-0 see...Tetramethrin
7700-17-6 see...Crotoxyphos
7704-34-9 see...Sulfur
7720-78-7 see...Ferrous Sulfate
7723-14-0 see...Phosphorus
7733-02-0 see...Sulfate heptahydrate
7757-79-1 see...Potassium Nitrate
7758-09-0 see...Potassium Nitrite
7758-89-6 see...Cuprous Chloride
7758-98-7 see...Copper Sulfate
7758-99-8 see...Copper Sulfate
7773-06-0 see...Ammonium Sulfamate
7775-09-9 see...Sodium Chlorate
7775-19-1 see...Sodium Metaborate
7778-39-4 see...Arsenic Acid
7778-44-1 see...Calcium Arsenate
7783-18-8 see...Ammonium Thiosulfate
7783-28-0 see...Ammonium Phosphate
7784-40-9 see...Lead Arsenate
7784-41-0 see...Potassium Arsenate
7784-46-5 see...Sodium Arsenite
7786-34-7 see...Mevinphos
7788-98-9 see...Ammonium Chromate
7789-12-0 see...Sodium Dichromate
7789-23-3 see...Potassium Fluoride
7803-52-3 see...Stibine
8001-35-2 see...Toxaphene
8001-54-5 see...Zilkonium Chloride
8002-05-9 see...Naphthas
8003-19-8 see...D-D mixture
8003-19-8 see...D-D mixture
8003-34-7 see...Pyrethrins or Pyrethrum
8008-20-6 see...Kerosene
8018-01-7 see...Mancozeb
8022-00-2 see...Demeton-methyl
8030-30-6 see...Naphthas
8030-31-7 see...Naphthas
8032-32-4 see...Naphthas
8045-22-5 see...Zilkonium Chloride
8052-41-3 see...Stoddard Solvent
8065-36-9 see...Bufencarb
8065-48-3 see...Demeton
9006-42-2 see...Metiram
10028-22-5 see...Ferric Sulfate
10031-59-1 see...Thallium Sulfate
10034-93-2 see...Hydrazine Sulfate
10043-01-3 see...Aluminum Sulfate
10043-35-3 see...Borax and Boric Acid
10061-01-5 see...1,3-Dichloro propene
10061-02-6 see...1,3-Dichloro propene
10102-20-2 see...Sodium Tellurite
10102-53-1 see...Arsenic Acid
10108-64-2 see...Cadmium Chloride
10124-36-4 see...Cadmium Sulfate
10124-37-5 see...Calcium Nitrate
10192-30-0 see...Ammonium Sulfite
10196-04-0 see...Ammonium Sulfite
10265-92-6 see...Methamidophos
10290-12-7 see...Copper Arsenite
10311-84-9 see...Dialifor
10326-21-3 see...Magnesium Chlorate
10380-28-6 see...Copper(II)-8-hydroxyquinoline
10453-86-8 see...Resmethrin
10588-01-9 see...Sodium Dichromate
10605-21-7 see...Carbendazim
11141-17-6 see...Azadirachtin
12002-03-8 see...Copper Acetoarsenite
12002-03-8 see...Paris Green
12069-69-1 see...Copper Carbonate, Basic
12071-83-9 see...Propineb
12122-67-7 see...Zineb
12407-86-2 see...Trimethacarb
12427-38-2 see...Maneb
12771-68-5 see...Ancymidol
12789-03-6 see...Chlordane
13071-79-9 see...Terbufos
13121-70-5 see...Cyhexatin
13171-21-6 see...Phosphamidon
13194-48-4 see...Ethoprop
13356-08-6 see...Fenbutatin Oxide
13593-03-8 see...Quinalphos
13684-56-5 see...Desmedipham
13684-63-4 see...Phenmedipham
14371-10-9 see...Cinnamaldehyde
14484-64-1 see...Ferbam
14763-77-0 see...Copper Cyanide
14816-18-3 see...Phoxim
15096-52-3 see...Sodium Aluminum Fluoride
15263-52-2 see...Cartap Hydrochloride
15299-99-7 see...Napropamide
15662-33-6 see...Ryanodine
15950-66-0 see...Trichlorophenols
15972-60-8 see...Alachlor
16118-49-3 see...Carbetamide
16672-87-0 see...Ethephon
16752-77-5 see...Methomyl
16919-19-0 see...Ammonium Hexafluorosilicate
17804-35-2 see...Benomyl
19044-88-3 see...Oryzalin
19666-30-9 see...Oxadiazon
19937-59-8 see...Metoxuron
20354-26-1 see...Methazole
20427-59-2 see...Copper Hydroxide
20543-04-8 see...Copper Octanoate
20859-73-8 see...Aluminum Phosphide
21087-64-9 see...Metribuzin
21351-39-3 see...Monocarbamide Dihydrogen Sulfate
21548-32-3 see...Fosthietan
21609-90-5 see...Leptophos
21725-46-2 see...Cyanazine
21908-53-2 see...Mercuric Oxide
21923-23-9 see...Chlorthiophos
22224-92-6 see...Fenamiphos
22248-79-9 see...Tetrachlorvinphos
22781-23-3 see...Bendiocarb
22967-92-6 see...Mercury Alkyl Compounds
23031-36-9 see...Allethrins
23103-98-2 see...Pirimicarb
23135-22-0 see...Oxamyl
23184-66-9 see...Butachlor
23422-53-9 see...Formetanate Hydrochloride
23505-41-1 see...Pirimifos-ethyl
23564-05-8 see...Thiophanate-methyl
23564-06-9 see...Thiophanate-methyl
23950-58-5 see...Pronamide
24017-47-8 see...Triazophos
24307-26-4 see...Mepiquat Chloride
24934-91-6 see...Chlormephos
25057-89-0 see...Bentazon
25059-80-7 see...Benazolin Ethyl
25167-82-2 see...Trichlorophenols
25168-15-4 see...2,4,5-Trichlorophenoxyacetic Acid, Esters
25168-26-7 see...2,4-D, Isooctyl Ester
25311-71-1 see...Isofenphos
25606-41-1 see...Propamocarb Hydrochloride
25954-13-6 see...Fosamine Ammonium
26002-80-2 see...d-Phenothrin
26225-79-6 see...Ethofumesate
26399-36-0 see...Profluralin

26530-20-1 see...Octhilinone
26628-22-8 see...Sodium Azide
27176-87-0 see..Dodecylbenzenesulfonic Acid
27314-13-2 see...Norflurazon
27323-41-7 see...Triethanolamine Dodecylbenzene Sulfonate
28057-48-9 see...Allethrins
28249-77-6 see...Thiobencarb
28300-74-5 see...Antimony Potassium Tartrate
28434-00-6 see...Allethrins
28434-01-7 see...Resmethrin
28772-56-7 see...Bromadiolone
28805-86-9 see...Butylphenols
29091-21-2 see...Prodiamine
29232-93-7 see...Pirimiphos-methyl
30560-19-1 see...Acephate
31218-83-4 see...Propetamphos
31717-87-0 see...Dodemorph Acetate
32809-16-8 see...Procymidone
33089-61-1 see...Amitraz
33113-08-5 see...Copper Ammonium Carbonate
33213-65-9 see...Endosulfan
33245-39-5 see...Fluchloralin
33629-47-9 see...Butralin
33820-53-0 see...Isopropalin
33911-28-3 see...Resmethrin
34014-18-1 see...Tebuthiuron
34123-59-6 see...Isoproturon
34256-82-1 see...Acetochlor
35367-38-5 see...Diflubenzuron
35400-43-2 see...Sulprofos
35554-44-0 see...Imazalil
35691-65-7 see...1,2-Dibromo-2,4-dicyanobutane
35764-59-1 see...Resmethrin
36734-19-7 see...Iprodione
37248-47-8 see...Validamycin
37882-31-8 see...Kinoprene
37924-13-3 see...Perfluidone
38727-55-8 see...Diethatyl-ethyl
39148-24-8 see...Fosetyl-Al
39196-18-4 see...Thiofanox
39300-45-3 see...Dinocap
39515-40-7 see...Cyphenothrin
39515-41-8 see...Fenpropathrin
40487-42-1 see...Pendimethalin
40596-69-8 see...Methoprene
41096-46-2 see...Hydroprene
41198-08-7 see...Profenofos
41394-05-2 see...Metamiton
42576-02-3 see...Bifenox
42588-37-4 see...Kinoprene
42874-03-3 see...Oxyfluorfen

43121-43-3 see...Triadimefon
43222-48-6 see...Difenzoquat
50471-44-8 see...Vinclozolin
50594-66-6 see...Acifluorfen
51218-45-2 see...Metolachlor
51235-04-2 see...Hexazinone
51338-27-3 see...Diclofop-methyl
51630-58-1 see...Fenvalerate
51707-55-2 see...Thidiazuron
52292-17-8 see...Arosurf® MSF
52315-07-8 see...Cypermethrin
52740-16-6 see...Calcium Arsenite
52918-63-5 see...Deltamethrin
53112-28-0 see...Pyrimethanil
53558-25-1 see...Pyriminil
53780-34-0 see...Mefluidide
54593-83-8 see...Chlorethoxyfos
55179-31-2 see...Bitertanol
55219-65-3 see...Triadimenol
55285-14-8 see...Carbosulfan
55290-64-7 see...Dimethipin
55335-06-3 see...Triclopry
55512-33-9 see...Pyridate
55720-26-8 see..Aminoethoxyvinylglycine Hydrochloride
56073-07-5 see...Difenacoum
56425-91-3 see...Flurprimidol
57213-69-1 see...Triclopyr, Triethylammonium Salt
57837-19-1 see...Metalaxyl
57966-95-7 see...Cymoxanil
59669-26-0 see...Thiodicarb
59756-60-4 see...Fluridone
60168-88-9 see...Fenarimol
60207-90-1 see...Propiconazole
60238-56-4 see...Chlorthiophos
61790-53-2 see...Diatomaceous Earth
61792-07-2 see...2,4,5-Trichlorophenoxyacetic Acid, Esters
61792-07-2 see...2,4,5-Trichlorophenoxyacetic Acid, Esters
62476-59-9 see...Acifluorfen
64742-48-9 see...Naphthas
64902-72-3 see...Chlorsulfuron
65195-55-3 see...Avermectin
65195-56-4 see...Avermectin
65733-18-8 see...Hydroprene
65733-20-2 see...Kinoprene
66063-05-6 see...Pencycuron
66215-27-8 see...Cyromazine
66230-04-4 see...Esfenvalerate
66246-88-6 see...Penconazole
66332-96-5 see...Flutolanil
66441-23-4 see...Fenoxaprop-ethyl
66841-24-5 see...Cypermethrin
66841-25-6 see...Tralomethrin

67375-30-8 see...*alpha*-Cypermethrin
67485-29-4 see...Hydramethylnon
68157-60-8 see...Forchlorfenuron
68359-37-5 see...Cyfluthrin
68694-11-1 see...Triflumizole
69327-76-0 see...Buprofezin
69409-94-5 see...Fluvalinate
69806-40-2 see...Haloxyfop-methyl
69806-50-4 see...Fluazifop-butyl
69865-47-0 see...Cypermethrin
70124-77-5 see...Flucythrinate
71422-67-8 see...Chlorfluazuron
71751-41-2 see...Avermectin
72178-02-0 see...Fomesafen
73250-68-7 see...Mefenacet
74051-80-2 see...Sethoxydim
74115-24-5 see...Clofentezine
74222-97-2 see...Sulfometuron-methyl
74223-64-6 see...Metsulfuron-methyl
75867-00-4 see...Fenfluthrin
76578-14-8 see...Quizalofop-ethyl
76738-62-0 see...Paclobutrazol
77182-82-2 see...Glufosinate-Ammonium
77501-63-4 see...Lactofen
77732-09-3 see...Oxadixyl
78587-05-0 see...Hexythiazox
79127-80-3 see...Fenoxycarb
79241-46-6 see...Fluazifop-butyl
79277-27-3 see...Thifensulfuron Methyl
79538-32-2 see...Tefluthrin
79622-59-6 see...Fluazinam®
79983-71-4 see...Hexaconazole
80844-07-1 see...Ethofenprox
81335-37-7 see...Imazaquin
81335-47-9 see...Imazaquin
81335-77-5 see...Imazethapyr
81362-49-4 see...Lactofen
81405-85-8 see...Imazethabenz
81406-37-3 see...Fluroxypyr 1-methylheptyl Ester
81591-81-3 see...Sulfosate
81777-89-1 see...Clomazone
82097-50-5 see...Triasulfuron
82558-50-7 see...Isoxaben
82560-54-1 see...Benfuracarb
82657-04-3 see...Bifenthrin
83055-99-6 see...Bensulfuron-methyl
83164-33-4 see...Diflufenican
83657-18-5 see...Diniconazole
84087-01-4 see...Quinclorac
86209-51-0 see...Primisulfuron-methyl

86479-06-3  see...Hexaflumuron
86752-99-0  see...Cypermethrin
86753-92-6  see...Cypermethrin
87674-68-8  see...Dimethenamid
87674-68-8  see...Dimethenamid
87820-88-0  see...Tralkoxydim
88161-75-5  see...Cypermethrin
88671-89-0  see...Myclobutanil
90982-32-4  see...Chlorimuron-ethyl
91465-08-6  see...*lambda*-Cyhalothrin
94125-34-5  see...Prosulfuron
94361-06-5  see...Cyproconazole
95266-40-3  see...Cimectacarb
95465-99-9  see...Cadusafos
95737-68-1  see...Pyriproxyfen
96489-71-3  see...Pyridaben
97780-06-8  see...Ethametsulfuron-methyl
97886-45-8  see...Dithiopyr
97955-44-7  see...Cypermethrin
98967-40-9  see...Flumetsulam
99129-21-2  see...Clethodim
100784-20-1  see...Halosulfuron-methyl
101200-48-0  see...Tribenuron-methyl
101917-66-2  see...Imazethapyr
102851-06-9  see...Fluvalinate
105827-78-9  see...Imidacloprid
107534-96-3  see...Tebuconazole
108189-58-8  see...Dihydroazadirachtin
108731-70-0  see...Fomesafen
110488-70-5  see...Dimethomorph
111991-09-4  see...Nicosulfuron
112143-82-5  see...Triazamate
112410-23-8  see...Tebufenozide
113136-77-9  see...Cyclanilide
114369-43-6  see...Fenbuconazole
117718-60-2  see...Thiazopyr
119446-68-3  see...Difenoconazole
120068-37-3  see...Fipronil
121552-61-2  see...Cyprodinil
122453-73-0  see...Chlorfenapyr
122836-35-5  see...Sulfentrazone
122931-48-0  see...Rimsulfuron
123312-89-0  see...Pymetrozine
123343-16-8  see...Pyrithiobac-sodium
126535-15-7  see...Triflusulfuron-methyl
127277-53-6  see...Prohexadione Calcium
131860-33-8  see...Azoxystrobin
134098-61-6  see...Fenpyroximate
137512-74-4  see...Emamectin Benzoate
138261-41-3  see...Imidacloprid
141112-29-0  see...Isoxaflutole
142459-58-3  see...Thiafluamide (Flufenacet)
146659-78-1  see...Polyoxin D, Zinc Salt
147150-35-4  see...Cloransulam-methyl
155569-91-8  see...Emamectin Benzoate
161050-58-4  see..Methoxyfenozide
None found  see...Anagrapha Falcifera
None found  see...Antimycin A
None found  see...Ampelomyces Quisqualis isolate M10
None found  see...Nosema Locustae

86479-06-3  see...Hexaflumuron
86752-99-0  see...Cypermethrin
86753-92-6  see...Cypermethrin
87674-68-8  see...Dimethenamid
87674-68-8  see...Dimethenamid
87820-88-0  see...Tralkoxydim
88161-75-5  see...Cypermethrin
88671-89-0  see...Myclobutanil
90982-32-4  see...Chlorimuron-ethyl
91465-08-6  see...*lambda*-Cyhalothrin
94125-34-5  see...Prosulfuron
94361-06-5  see...Cyproconazole
95266-40-3  see...Cimectacarb
95465-99-9  see...Cadusafos
95737-68-1  see...Pyriproxyfen
96489-71-3  see...Pyridaben
97780-06-8  see...Ethametsulfuron-methyl
97886-45-8  see...Dithiopyr
97955-44-7  see...Cypermethrin
98967-40-9  see...Flumetsulam
99129-21-2  see...Clethodim
100784-20-1  see...Halosulfuron-methyl

101200-48-0  see...Tribenuron-methyl
101917-66-2  see...Imazethapyr
102851-06-9  see...Fluvalinate
105827-78-9  see...Imidacloprid
107534-96-3  see...Tebuconazole
108189-58-8  see...Dihydroazadirachtin
108731-70-0  see...Fomesafen
110488-70-5  see...Dimethomorph
111991-09-4  see...Nicosulfuron
112143-82-5  see...Triazamate
112410-23-8  see...Tebufenozide
113136-77-9  see...Cyclanilide
114369-43-6  see...Fenbuconazole
117718-60-2  see...Thiazopyr
119446-68-3  see...Difenoconazole
120068-37-3  see...Fipronil
121552-61-2  see...Cyprodinil
122453-73-0  see...Chlorfenapyr
122836-35-5  see...Sulfentrazone
122931-48-0  see...Rimsulfuron
123312-89-0  see...Pymetrozine
123343-16-8  see...Pyrithiobac-sodium

126535-15-7  see...Triflusulfuron-methyl
127277-53-6  see...Prohexadione Calcium
131860-33-8  see...Azoxystrobin
134098-61-6  see...Fenpyroximate
137512-74-4  see...Emamectin Benzoate
138261-41-3  see...Imidacloprid
141112-29-0  see...Isoxaflutole
142459-58-3  see...Thiafluamide (Flufenacet)
146659-78-1  see...Polyoxin D, Zinc Salt
147150-35-4  see...Cloransulam-methyl
155569-91-8  see...Emamectin Benzoate
161050-58-4  see..Methoxyfenozide
None found  see...Anagrapha Falcifera
None found  see...Antimycin A
None found  see...Ampelomyces Quisqualis isolate M10
None found  see...Nosema Locustae

00060327
Ref.
S
633
.S6265

DATE DUE

| | | | |
|---|---|---|---|
| | | | |
| | | | |
| | | | |
| | | | |
| | | | |
| | | | |
| | | | |
| | | | |
| | | | |

SOUTH UNIVERSITY
709 MALL BLVD.
SAVANNAH, GA 31406